ALGEBRA

Exponents and Radicals

$a^m \cdot a^n = a^{m+n}$

$\frac{a^m}{a^n} = a^{m-n} \quad (a \neq 0)$

$(a^m)^n = a^{mn}$

$(ab)^n = a^n b^n$

$\left(\frac{a}{b}\right)^n = \frac{a^n}{b^n} \quad (b \neq 0)$

$a^0 = 1 \quad (a \neq 0)$

$a^{-n} = \frac{1}{a^n} \quad (a \neq 0)$

$a^{m/n} = \sqrt[n]{a^m} = (\sqrt[n]{a})^m$

$\sqrt{ab} = \sqrt{a}\sqrt{b}$

Special Products

$a(x + y) = ax + ay$

$(x + y)(x - y) = x^2 - y^2$

$(x + y)^2 = x^2 + 2xy + y^2$

$(x - y)^2 = x^2 - 2xy + y^2$

Quadratic Equation and Formula

$ax^2 + bx + c = 0$

$x = \frac{-b \pm \sqrt{b^2 - 4ac}}{2a}$

Properties of Logarithms

$\log_b x + \log_b y = \log_b xy$

$\log_b x - \log_b y = \log_b\left(\frac{x}{y}\right)$

$n \log_b x = \log_b(x^n)$

Complex Numbers

$\sqrt{-a} = j\sqrt{a} \quad (a > 0)$

$x + yj = r(\cos\theta + j\sin\theta)$

$= re^{j\theta} = r\angle\theta$

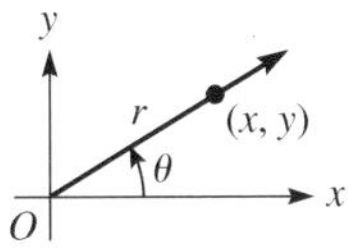

Variation

Direct variation: $y = kx$

Inverse variation: $y = k/x$

TRIGONOMETRY

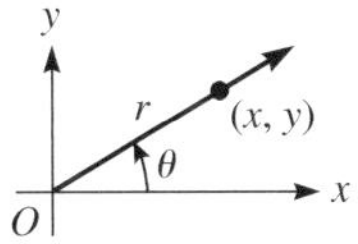

$\sin\theta = \frac{y}{r} \quad \cos\theta = \frac{x}{r} \quad \tan\theta = \frac{y}{x} \quad \cot\theta = \frac{x}{y} \quad \sec\theta = \frac{r}{x} \quad \csc\theta = \frac{r}{y}$

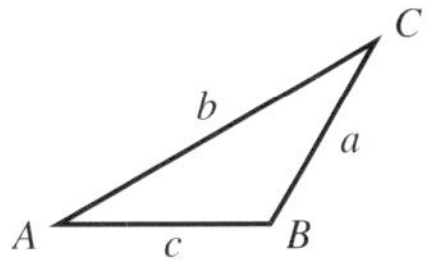

Law of Sines: $\frac{a}{\sin A} = \frac{b}{\sin B} = \frac{c}{\sin C}$

Law of Cosines: $a^2 = b^2 + c^2 - 2bc\cos A$

π rad $= 180°$

Reference Angle

$\theta = \theta_{ref}$ (first quadrant)

$\theta = 180° - \theta_{ref}$ (second quadrant)

$\theta = 180° + \theta_{ref}$ (third quadrant)

$\theta = 360° - \theta_{ref}$ (fourth quadrant)

Basic Identities

$\sin\theta = \frac{1}{\csc\theta} \quad \cos\theta = \frac{1}{\sec\theta} \quad \tan\theta = \frac{1}{\cot\theta} \quad \tan\theta = \frac{\sin\theta}{\cos\theta} \quad \cot\theta = \frac{\cos\theta}{\sin\theta}$

$\sin^2\theta + \cos^2\theta = 1 \quad 1 + \tan^2\theta = \sec^2\theta \quad 1 + \cot^2\theta = \csc^2\theta$

$\sin 2\alpha = 2\sin\alpha\cos\alpha \quad \cos 2\alpha = \cos^2\alpha - \sin^2\alpha = 2\cos^2\alpha - 1 = 1 - 2\sin^2\alpha$

$\sin\frac{\alpha}{2} = \pm\sqrt{\frac{1 - \cos\alpha}{2}} \quad \cos\frac{\alpha}{2} = \pm\sqrt{\frac{1 + \cos\alpha}{2}}$

$\sin(\alpha + \beta) = \sin\alpha\cos\beta + \cos\alpha\sin\beta$

$\cos(\alpha + \beta) = \cos\alpha\cos\beta - \sin\alpha\sin\beta$

$\sin(\alpha - \beta) = \sin\alpha\cos\beta - \cos\alpha\sin\beta$

$\cos(\alpha - \beta) = \cos\alpha\cos\beta + \sin\alpha\sin\beta$

$\tan(\alpha \pm \beta) = \frac{\tan\alpha \pm \tan\beta}{1 \mp \tan\alpha\tan\beta}$

Quantities and Their Associated Units

Quantity	*Quantity Symbol*	*U.S. Customary* Name	*U.S. Customary* Symbol	*Metric (SI)* Name	*Metric (SI)* Symbol	*In Terms of Other SI Units*
Length	*s*	foot	ft	**meter**	m	
Mass	*m*	slug		**kilogram**	kg	
Force	*F*	pound	lb	newton	N	$m \cdot kg/s^2$
Time	*t*	second	s	**second**	s	
Area	*A*		ft^2		m^2	
Volume	*V*		ft^3		m^3	
Capacity	*V*	gallon	gal	liter	L	($1 L = 1 dm^3$)
Velocity	*v*		ft/s		m/s	
Acceleration	*a*		ft/s^2		m/s^2	
Density	*d*, *ρ*		lb/ft^3		kg/m^3	
Pressure	*p*		lb/ft^2	pascal	Pa	N/m^2
Energy, work	*E*, *W*		ft · lb	joule	J	N · m
Power	*P*	horsepower	hp	watt	W	J/s
Period	*T*		s		s	
Frequency	*f*		1/s	hertz	Hz	1/s
Angle	*θ*	radian	rad	radian	rad	
Electric current	*I*, *i*	ampere	A	**ampere**	A	
Electric charge	*q*	coulomb	C	coulomb	C	A · s
Electric potential	*V E*	volt	V	volt	V	J/(A · s)
Capacitance	*C*	farad	F	farad	F	s/Ω
Inductance	*L*	henry	H	henry	H	Ω · s
Resistance	*R*	ohm	Ω	ohm	Ω	V/A
Thermodynamic temperature	*T*			**kelvin**	K	(temp. interval
Temperature	*T*	degrees Fahrenheit	°F	degrees Celsius	°C	1°C = 1 K)
Quantity of heat	*Q*	British thermal unit	Btu	joule	J	
Amount of substance	*n*			**mole**	mol	
Luminous intensity	*I*	candlepower	cp	**candela**	cd	

Special Notes:

1. The SI base units are shown in boldface type.
2. Other units of time, along with their symbols, that are used in this book are the minute (min), hour (h), and day (d).
3. This table includes most of the units used in this book. Occasionally, other units will be noted and used.
4. In addition to the radian, the units degree (°) and revolution (r) are commonly used to measure angles or rotations.
5. Other common U.S. units that are used in this text are the inch (in.), yard (yd), mile (mi), ounce (oz), quart (qt), ton, and acre.
6. Unit symbols are case sensitive. They should be written exactly as shown in the table.
7. The dot symbol is used to show units are multiplied.

BASIC TECHNICAL MATHEMATICS WITH CALCULUS

SI VERSION

BASIC TECHNICAL MATHEMATICS WITH CALCULUS SI VERSION

ELEVENTH EDITION

ALLYN J. WASHINGTON
Dutchess Community College

RICHARD S. EVANS
Corning Community College

MICHELLE BOUÉ
Trent University

ELIZABETH FABBRONI MARTIN
Mohawk College of Applied Arts & Technology

Pearson

Pearson Canada Inc., 26 Prince Andrew Place, North York, Ontario M3C 2H4.

ISBN 978-0-13-428991-5

7 2023

Library and Archives Canada Cataloguing in Publication

Washington, Allyn J., author

Basic technical mathematics with calculus : SI version / Allyn J. Washington, Dutchess Community College, Michelle Boué, Toronto. — Eleventh edition.

Includes indexs.

ISBN 978-0-13-428991-5 (hardcover)

1. Mathematics—Textbooks. 2. Textbooks. I. Boué, Michelle, author II. Title.

QA37.3.W37 2020 510 C2017-903080-9

Pearson

To Douglas, Julia, and Andrea ~Michelle Boué

Moj Piotr, mon coeur ~Elizabeth Fabbroni Martin

In memory of my loving wife, Millie ~Allyn J. Washington

Brief Contents

Contents

Preface

Scope of the Book

Basic Technical Mathematics with Calculus, SI Version, Eleventh edition, is intended primarily for students in technical and pre-engineering technology programs or other programs for which coverage of basic mathematics is required. Chapters 1 through 20 provide the necessary background for further study, with an integrated treatment of algebra and trigonometry. Chapter 21 covers the basic topics of analytic geometry, and Chapter 22 gives an introduction to statistics. Fundamental topics of calculus are covered in Chapters 23 through 31. In the examples and exercises, numerous applications from many fields of technology are included, primarily to indicate where and how mathematical techniques are used. However, it is not necessary that the student have a specific knowledge of the technical area from which any given problem is taken.

Most students using this text will have a background that includes some algebra and geometry. However, the material is presented in adequate detail for those who may need more study in these areas. The material presented here is sufficient for three to four semesters.

One of the principal reasons for the arrangement of topics in this text is to present material in an order that allows a student to take courses concurrently in allied technical areas, such as physics and electricity. These allied courses normally require a student to know certain mathematical topics by certain definite times; yet the traditional order of topics in mathematics courses makes it difficult to attain this coverage without loss of continuity. However, the material in this book can be rearranged to fit any appropriate sequence of topics. Another feature of this text is that certain topics traditionally included for mathematical completeness have been covered only briefly or have been omitted. The approach used in this text is not unduly rigorous mathematically, although all appropriate terms and concepts are introduced as needed and given an intuitive or algebraic foundation. The aim is to help the student develop an understanding of mathematical methods without simply providing a collection of formulas. The text material has been developed with the recognition that it is essential for the student to have a sound background in algebra and trigonometry in order to understand and succeed in any subsequent work in mathematics.

New to This Edition

In this eleventh edition of *Basic Technical Mathematics with Calculus, SI Version*, we have retained all the basic features of successful previous editions and have also introduced a number of improvements, described here.

DESIGN CHANGES

- **Illustrative examples** have been set up side by side with selected definitions and procedural boxes so students can make a clearer connection between the steps of a procedure and the example. Since these examples are placed next to the procedure they illustrate, they do not have descriptive titles. They follow the numbering sequence of all examples, and are indicated with a 🔍 beside the example.
- Content has been reformatted so that figures in definitions appear immediately next to the text that describes them, enabling clearer comparisons between different definitions.
- A step-wise solving framework has been introduced for solving word problems, especially those in the earlier chapters.
- The end of examples has been made more obvious by enclosing examples within lines.

NAMED RULES

Throughout the text, individual rules pertaining to a set have been given names so that they can be more easily referred to. For instance, the rules of exponents include *the product rule, quotient rule, power of a power rule*, and *negative exponent rule*. Similarly, the rules for differentiation include the *constant rule, general power rule, product rule,* and *quotient rule*.

INCREASED BREADTH OF APPLICATIONS

The text features many new applications in the examples and exercises. Here is a sampling of the contexts for these new applications:

- Indigenous culture (Sections 2.4, 2.6, and 21.3)
- Height of One World Trade Center (Section 4.4)
- Multimedia messaging by Canadians (Section 5.1)
- Hydronium ion concentration of a solution (Section 7.3)
- Circadian rhythm in plants (Section 10.3)
- Phase angle for current/voltage lead and lag (Section 10.3)
- Bezier curve roof design (Section 15.3)
- Social networks usage (Section 22.1)
- Video game system market share (Section 22.1)
- Glacier Skywalk shape and area (Sections 21.4 and 26.2)

GRAPHING UTILITIES

The use of technology in this edition follows a flexible approach. Instead of focusing attention on the use of a specific graphing calculator, the text encourages students to solve problems using any graphing utility (such as graphing calculators, computer algebra systems, spreadsheet programs, or statistical software).

EXERCISES

More than 950 exercises are new or have been updated for this edition.

MYLAB MATH

Hundreds of new assignable algorithmic exercises are available to help instructors address the homework needs of students.

ROUNDING

This edition returns to the round half up rule for rounding measured values.

SELECT CONTENT UPDATES

Revisions to content in the eleventh edition were informed by the extensive reviews of the tenth edition. Examples of these include:

- In Section 1.2, visual illustrations of the fundamental operations of algebra were introduced with side-by-side illustrations of the literal rule, a numerical example, and an algebraic example.
- In Section 1.3, explanations of units and unit systems were completely redone, putting greater focus on practical reasons for use.
- In response to reviewer feedback, the section on rectangular coordinates has been moved to the beginning of the chapter, before the introduction to functions, which is now Section 3.2.
- The beginning of Chapter 5 has been reorganized so that systems of equations has a strong introduction in Section 5.2. The prerequisite material needed for systems of equations (linear equations and graphs of linear functions) has been consolidated into Section 5.1.
- In Section 6.3, factoring of trinomials is done more methodically using tables of factors and sums and then grouping rather than by trial and error.
- In Chapter 7, the square root property is explicitly stated and illustrated.
- In Chapter 8, the unit circle definition of the trigonometric functions has been added.

- In Chapter 9, more emphasis had been given to solving equilibrium problems, including those that have more than one unknown.
- In Chapter 10, an example was added to show how the phase angle can be interpreted, and how it is different from the phase shift.
- In Chapter 16, the terminology *row echelon form* is used. Also, the approach for solving systems of equations is to focus on operations on matrices rather than on operations with equations.
- In Chapter 23, the terminology *direct substitution* has been introduced in the context of limits.
- In Chapter 30, the proof of the Fourier coefficients has been moved online.

FUNCTIONAL USE OF COLOUR

The text continues to use colour effectively for didactical purposes. New to this edition is the use of colour within equations to emphasize an important step in a procedure. For example, students can identify visually how one expression is replaced by another equivalent one when both expressions are written in the same colour.

Continuing Features

EXAMPLE DESCRIPTIONS

A brief descriptive title is given with each example that is not an illustrative example. This gives an easy reference for the example, particularly when reviewing the contents of the section. Examples involving applications now describe the area of application explicitly.

PRACTICE EXERCISES

Throughout the text, there are *practice exercises* in the margin. They are included so that a student is more actively involved in the learning process and can check his or her understanding of the material to that point in the section. They can also be used for classroom exercises. The answers to these exercises are given at the end of the exercise set for the section.

CHAPTER INTRODUCTIONS

Each chapter introduction illustrates specific examples of how the development of technology has been related to the development of mathematics. These introductions show that past discoveries in technology led to some of the methods in mathematics, whereas in other cases mathematical topics already known were later very useful in bringing about advances in technology.

SPECIAL EXPLANATORY COMMENTS

Throughout the book, special explanatory comments in colour have been used in the examples to emphasize and clarify certain important points. Arrows are often used to indicate clearly the part of the example to which reference is made.

IMPORTANT FORMULAS

Throughout the book, important formulas are set off and displayed so that they can be easily referenced.

EXERCISES DIRECTLY REFERENCED TO TEXT EXAMPLES

The first few exercises in most of the text sections are referenced directly to a specific example of the section. These exercises are worded so that it is necessary for the student to refer to the example in order to complete the required solution. In this way, the student should be able to review and understand the text material better before attempting to solve the exercises that follow.

WRITING EXERCISES

There are more than 400 writing exercises throughout the book (at least 8 in each chapter) that require at least a sentence or two of explanation as part of the answer. These are noted by a pencil icon next to the exercise number.

KEY FORMULAS, REVIEW EXERCISES, AND PRACTICE TESTS

At the end of each chapter, all important formulas and equations are listed together for easy reference. Each chapter is also followed by a set of review exercises that covers all the material in the chapter. Following the chapter equations and review exercises is a chapter practice test that students can use to check their understanding of the material. Solutions to all practice test problems are provided in the back of the book.

APPLICATIONS

Examples and exercises illustrate the application of mathematics in all fields of technology. Many relate to modern technology such as computer design, electronics, solar energy, lasers, fibre optics, the environment, and space technology. Others relate to technologies such as aeronautics, architecture, automotive, business, chemical, civil, construction, energy, environmental, fire science, machine, medical, meteorology, navigation, police, refrigeration, seismology, and wastewater. A special "Index of Applications" is included near the end of the book.

MARGIN NOTES

Throughout the text, some margin notes briefly point out relevant historical events in mathematics and technology. Other margin notes are used to make specific comments related to the text material. Also, where appropriate, equations from earlier material are shown for reference in the margin.

ANSWERS TO EXERCISES

The answers to all odd-numbered exercises (except the end-of-chapter writing exercises) are given at the back of the book.

FLEXIBILITY OF MATERIAL COVERAGE

The order of material coverage can be changed in many places, and certain sections may be omitted without loss of continuity of coverage. Users of earlier editions have indicated the successful use of numerous variations in coverage. Any changes will depend on the type of course and completeness required.

Technology and Supplements

MYLAB MATH

Personalize learning with MyLab Math®

MyMathLab is an online homework, tutorial, and assessment program designed to work with this text to engage students and improve results. Within its structured environment, students practice what they learn, test their understanding, and pursue a personalized study plan that helps them absorb course material and understand difficult concepts. The MyLab course features hundreds of new algorithmic exercises, tutorial videos, and PowerPoint slides. MyLab can be successfully implemented in any classroom environment—lab-based, hybrid, fully online, or traditional. **By addressing instructor and student needs, MyLab improves student learning.**

Student Engagement and Achievement

Students are motivated to succeed when they're engaged in the learning experience and understand the relevance and power of mathematics. MyLab Math's online homework offers students immediate feedback and tutorial assistance that motivates them to do more, which means they retain more knowledge and improve their test scores.

- **Exercises with immediate feedback**—are based on the textbook exercises, and regenerate algorithmically to give students unlimited opportunity for practice and mastery.

MyLab provides helpful feedback when students enter incorrect answers and includes optional learning aids including Help Me Solve This, View an Example, videos, and the eText.

- **Learning Catalytics™** is a student response tool that uses students' smartphones, tablets, or laptops to engage them in more interactive tasks and thinking. Learning Catalytics fosters student engagement and peer-to-peer learning with real-time analytics.

LEARNING TOOLS FOR STUDENTS

- **Instructional videos**—More than 400 videos in the MyLab Math course provide help for students outside of the classroom. These videos are also available as learning aids within the homework exercises, for students to refer to at point-of-use.
- **The Pearson eText**, available through MyLab Math, gives students **access to their textbook anytime, anywhere**. In addition to note taking, highlighting, and bookmarking, the Pearson eText offers interactive and sharing features. Instructors can share their comments or highlights, and students can add their own, creating a tight community of learners in your class. The eText also includes links to videos.
- The **Student's Solutions Manual** contains revised solutions for every other odd-numbered section exercise. These step-by-step solutions have been expanded for even greater accuracy, clarity, and consistency to improve student problem-solving skills. The Student's Solutions Manual is downloadable from within MyLab Math.
- **Skills for Success Modules** help foster strong study skills in collegiate courses and prepare students for future professions. Topics include "Time Management" and "Stress Management".
- **Accessibility** and achievement go hand in hand. MyLab is compatible with the JAWS screen reader, and enables multiple-choice and free-response problem types to be read and interacted with via keyboard controls and math notation input. MyLab also works with screen enlargers, including ZoomText, MAGic, and SuperNova. And, all MyLab videos have closed-captioning. More information is available at https://www.pearsonmylabandmastering.com/northamerica/mymathlab/accessibility/.

SUPPORT FOR INSTRUCTORS

- **PowerPoint®** files feature animations that are designed to help you better teach key concepts.
- A **comprehensive gradebook** with enhanced reporting functionality allows you to efficiently manage your course.
- The **Reporting Dashboard** provides insight to view, analyze, and report learning outcomes. Student performance data is presented at the class, section, and program levels in an accessible, visual manner so you'll have the information you need to keep your students on track.
- **Item Analysis** tracks class-wide understanding of particular exercises so you can refine your class lectures or adjust the course/department syllabus. Just-in-time teaching has never been easier!

INSTRUCTOR'S SOLUTIONS MANUAL

The Instructor's Solution Manual contains detailed solutions to every section exercise, including review exercises. The manual is available to qualified instructors for download from Pearson Education's online catalogue at www.pearsoned.ca.

ANIMATED POWERPOINT PRESENTATIONS

More than 150 animated slides are available for download from a protected location on Pearson Education's online catalogue, at www.pearsoned.ca. Each slide offers a step-by-step mini lesson on an individual section, or key concept, formula, or equation from the first 28 chapters of the book.

These animated slides offer bite-sized chunks of key information for students to review and process prior to going to the homework questions for practice. Please note that not every section in every chapter is accompanied by an animated slide as some topics lend themselves to this approach more than others.

TESTGEN WITH ALGORITHMICALLY GENERATED QUESTIONS

TestGen enables instructors to build, edit, print, and administer tests using a bank of questions developed to cover all objectives in the text. TestGen is algorithmically based, allowing you to create multiple but equivalent versions of the same question or test. Instructors can also modify test bank questions or add new questions. The TestGen software and accompanying test bank are available to qualified instructors for download from Pearson Education's catalogue, at www.pearsoned.ca.

Acknowledgments

We gratefully acknowledge the unwavering cooperation and support of our team at Pearson Canada—Mark Grzeskowiak, Suzanne Simpson Millar, Emily Dill, Anne Williams, and the many others who have worked tirelessly on this project over the past several years. We also want to thank Allyn J. Washington and Richard S. Evans, authors of the U.S. edition of this text, for their continued support in this project, and their willingness to lend their expertise throughout its development.

Also of great assistance during the production of this edition were Sarah Gallagher, Jessica Hellen, Laurel Sparrow (copyeditor), Cat Haggert (proofreader), and Robert Brooker (accuracy and technical checker).

The authors gratefully acknowledge the contributions of the following reviewers, whose detailed comments and many suggestions were of great assistance in preparing this text:

Cornelia Bica
Northern Alberta Institute of Technology

Peter E. F. Bovell
Red Deer College

Brian Davis
Seneca College

Michael Delgaty
Algonquin College

Mehrdad Hajivandi
Conestoga College

Carol Lai
Seneca College

Niloofar Mani
Seneca College

Paul Matson
Seneca College

Bruce Miller
Georgian College

Wendy Morrison
Sheridan College

Isaac Mulolani
Saskatchewan Polytechnic

Rosalind Osborne
College of the North Atlantic

Punam Parashar
Humber College and Sheridan College

Mark Schneider
Northern Alberta Institute of Technology

Ashley Sembay
Seneca College

Carlo Sgro
Conestoga College

Soobia Siddiqui
Sir Sandford Fleming College

Don Van der Klok
Lambton College

Brea Williams
Red River College

About the Canadian Authors

MICHELLE BOUÉ

Michelle Boué holds M.Sc. and Ph.D. degrees in Applied Mathematics from Brown University. She has more than 25 years of experience teaching mathematics and statistics at different colleges and universities in Canada, the United States, and Mexico, including Trent University, St. Francis Xavier University, the University of Massachusetts (Amherst), Newbury College, and the ITAM (Instituto Tecnológico Autónomo de México). In recent years, she has turned her attention to the promotion of science and mathematics education in elementary and secondary schools. She is currently a presenter for the Scientists in School organization in Ontario.

ELIZABETH FABBRONI MARTIN

Elizabeth Fabbroni Martin is a professor and academic coordinator of the Mechanical Engineering Technology program at Mohawk College in Hamilton, Ontario. Elizabeth has instructed and developed materials for a wide range of foundational and applied quality, mathematics, science, statistics, and research methods courses. In a prior role, Ms. Martin served for three years as the academic coordinator for mathematics for all Engineering Technology Programs at Mohawk, and also served as a representative to the Ontario College Math Committee, and is serving as the secretary of the Ontario College Math Association.

Ms. Martin also has five years of industry experience as a process engineer, automation integrator, and engineering business (project) manager with Intel Corporation, in its Portland Technology Development and Sort Test Technology Development groups in Hillsboro, Oregon. She also briefly served as Director of Operations with the Centre for Surgical Invention & Innovation in Hamilton, Ontario. Ms. Martin's present areas of research interest include translational pedagogies and the effect of applied learning on college level students. Ms. Martin has degrees from Boston College (B.S. Chemistry), the McCormick School of Engineering at Northwestern University (Ph.D. Materials Science & Engineering), and the Richard Ivey School of Business at the University of Western Ontario (MBA).

In her free time, Ms. Martin sits on volunteer boards for adult literacy skills, and research ethics. She also enjoys hiking, camping, and spending time with her family and pets.

1. Basic Algebraic Operations

▲ From the Great Pyramid of Giza, built in Egypt 4500 years ago, to the modern technology of today, mathematics has played a key role in the advancement of civilization. Along the way, important discoveries have been made in areas such as architecture, navigation, transportation, electronics, communication, and astronomy. Mathematics will continue to pave the way for new discoveries.

Interest in things such as the land on which they lived, the structures they built, and the motion of the planets led people in early civilizations to keep records and to create methods of counting and measuring. In turn, some of the early ideas of arithmetic, geometry, and trigonometry were developed. From such beginnings, mathematics has played a key role in the great advances in science and technology.

In today's world, we still benefit from many of the important mathematical models of our past, as well as from many new models. For this reason, the study of technology needs to begin with an understanding of common mathematical models and rules that can be used for problem solving.

Chapter 1 begins with a review of the basic rules of algebra. If mathematics is the foundation of all technology, then algebra is the cement. It helps form the basic rules and principles on which all other areas of mathematics are based. It would be a challenge to find a branch of engineering that does not rely heavily on algebra.

To illustrate how these foundations can be useful in all technical fields, we have included applied problems throughout the chapter (and throughout the text). To solve these applied problems will require a knowledge of the mathematics presented but will *not* require prior knowledge of the field of application.

Chapter 1 also includes a review of numbers and symbols that arise often in engineering and science disciplines. These basic systems will be of use when we apply algebra to the study of geometry, trigonometry, and calculus in the chapters that follow.

LEARNING OUTCOMES

After completion of this chapter, the student should be able to:

- Identify real, imaginary, rational, and irrational numbers
- Perform mathematical operations on integers, decimals, fractions, and radicals
- Use the fundamental laws of algebra in numeric and algebraic equations
- Employ mathematical order of operations
- Understand technical measurement and approximation, as well as the use of significant digits and rounding
- Use scientific and engineering notations
- Convert units of measurement
- Rearrange and solve basic algebraic expressions
- Interpret word problems using algebraic symbols

1.1 Numbers

Mathematics begins with a basic understanding of numbers. In technology, we deal with different kinds of numbers. In some circumstances, only certain number types are allowed to be used for specific purposes. We begin by studying the simplest number types and then build our understanding to include all number types.

Counting numbers are the values we use when counting with our fingers and toes. These numbers, beginning with 1, 2, 3, and so on, are also called **natural numbers** or **positive integers**.

Negative integers represent how much has been taken away, and include -1, -2, -3, and so on. Together, the positive integers, the negative integers, and **zero** make up the set of all **integers**. This means that the integers are the numbers . . . , $-3, -2, -1, 0, 1, 2, 3,$ . . . and so on.

■ Irrational numbers were discussed by the Greek mathematician Pythagoras in about 540 B.C.E.

A **rational number** *is a number that can be expressed as the division of one integer a by another nonzero integer b*, and can be represented by the **fraction** a/b. Here a is the **numerator** and b is the **denominator**. (Note that we have used algebra by letting letters represent numbers.) A number that cannot be written in the form of a fraction that is the division of one integer by another nonzero integer is an **irrational number**.

EXAMPLE 1 Identifying rational numbers and irrational numbers

The numbers 5 and -19 are integers. They are also rational numbers since they can be written as $\frac{5}{1}$ and $\frac{-19}{1}$, respectively. Normally, we do not write the 1's in the denominators.

The numbers $\frac{5}{8}$ and $\frac{-11}{3}$ are rational numbers because the numerator and the denominator of each are integers.

■ For reference, $\pi = 3.14159265\ldots$

The numbers $\sqrt{2}$ and π are irrational numbers. It is not possible to find two integers, one divided by the other, to represent either of these numbers. It can be shown that square roots (and other roots) that cannot be expressed exactly in decimal form are irrational. Also, $\frac{22}{7}$ is sometimes used as an *approximation* for π, but it is not equal *exactly* to π. We must remember that $\frac{22}{7}$ is rational and π is irrational.

The decimal number 1.5 is rational since it can be written as $\frac{3}{2}$. Any such *terminating decimal* is rational. The number 0.6666 . . . , where the 6's continue on indefinitely, is rational since we may write it as $\frac{2}{3}$. In fact, any *repeating decimal* (in decimal form, a specific sequence of digits is repeated indefinitely) is rational. The decimal number 0.673 273 273 2 . . . is a repeating decimal where the sequence of digits 732 is repeated indefinitely ($0.673\,273\,273\,2\ldots = \frac{1121}{1665}$).

LEARNING TIP

A notation that is often used for repeating decimals is to place a bar over the digits that repeat. Using this notation we can write

$$\frac{1121}{1665} = 0.6\overline{732} \text{ and } \frac{2}{3} = 0.\overline{6}$$

Any value with a repeated decimal is rational.

The integers, the rational numbers, and the irrational numbers, including all such numbers that are positive, negative, or zero, make up the **real number system** (see Fig. 1.1). All real numbers can be represented along a single number line.

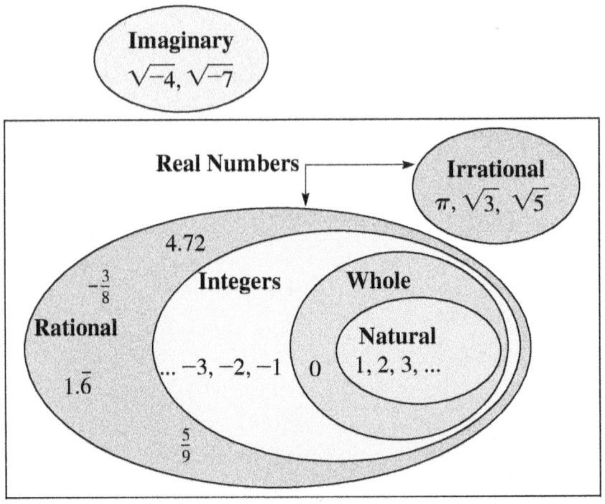

Fig. 1.1

There are times we will encounter an **imaginary number**, the name given to the square root of a negative number. When we study quadratic equations in Chapter 7, we will see how solutions sometimes involve imaginary numbers. Otherwise, the study of imaginary numbers will be limited until complex numbers are introduced in Chapter 12.

EXAMPLE 2 Identifying real numbers and imaginary numbers

(a) The number 7 is an integer. It is also rational since $7 = \frac{7}{1}$, and it is a real number since the real numbers include all the rational numbers.

(b) The number 3π is irrational, and it is real since the real numbers include all the irrational numbers.

(c) The numbers $\sqrt{-10}$ and $-\sqrt{-7}$ are imaginary numbers.

(d) The number $\frac{-3}{7}$ is rational and real. The number $-\sqrt{7}$ is irrational and real.

(e) The number $\frac{\pi}{6}$ is irrational and real. The number $\frac{\sqrt{-3}}{2}$ is imaginary.

■ Real numbers and imaginary numbers are both included in the *complex number system.* See Exercise 39.

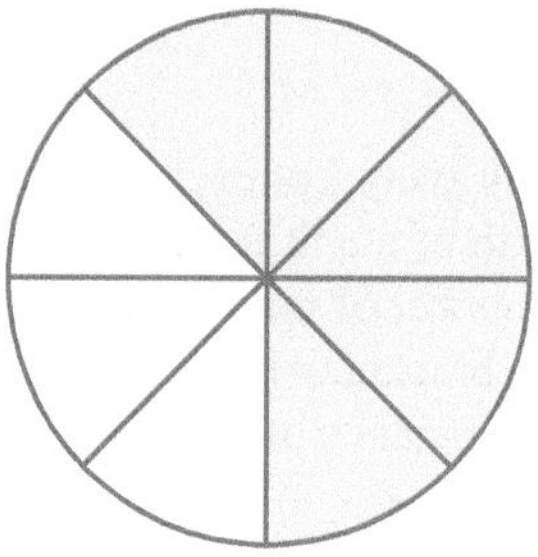

Fig. 1.2

FRACTIONS

We briefly discussed fractions earlier when defining rational numbers. A fraction indicates the division of a numerator by a denominator. However, it is important to note that a fraction can be written with integer, rational, irrational, or imaginary numerator and denominator values.

One useful way to visualize a fraction is as a portion of a whole. For example, if we have a pie that is cut into eight equally sized pieces, and we take three, then we have five portions of the whole, or $\frac{5}{8}$, remaining. This fraction is read "five-eighths." See Fig. 1.2.

EXAMPLE 3 Fractions

(a) The numbers $\frac{2}{7}$ and $\frac{-3}{2}$ are fractions, and they are rational.

(b) The numbers $\frac{\sqrt{2}}{9}$ and $\frac{6}{\pi}$ are fractions, but they are not rational numbers. It is not possible to express either as one integer divided by another integer.

(c) The number $\frac{\sqrt{-5}}{6}$ is a fraction, and it is an imaginary number.

THE NUMBER LINE AND ABSOLUTE VALUE

Real numbers can be represented by points along a line. For many engineering and scientific problems, it is useful to keep track of how far we have moved from a point of interest. This point of interest is often designated by an "O," for **origin**. The integer *zero* is placed at this point (see Fig. 1.3).

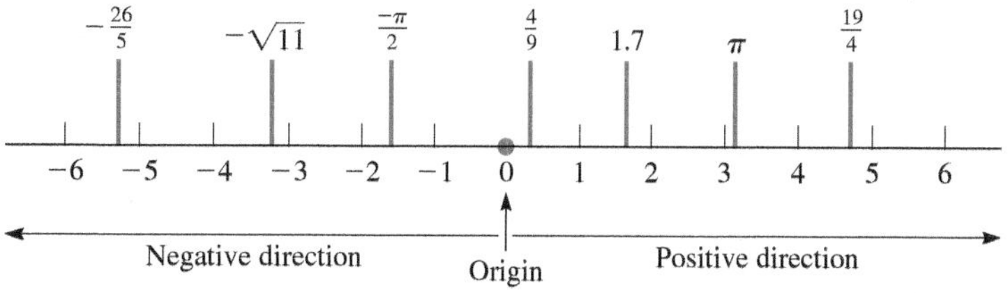

Fig. 1.3

The number line demonstrates the units of measurement away from the origin. Numbers to the right of the origin are positive values, while numbers to the left are negative values. We can specify any integer or rational number exactly on a number line, and we can estimate where an irrational number falls. For this reason, we cannot tell whether a given point on a number line represents a rational number or an irrational one unless it is specifically marked to indicate its value.

The **absolute value** of a number is the numerical value (magnitude) of the number without regard to its sign. The absolute value of a positive number is the number itself, and the absolute value of a negative number is just the number, without the negative sign. On the number line, we may interpret the absolute value of a number as the distance (which is always positive) between the origin and the number. Absolute value is denoted by writing the number between vertical lines, as shown in the following example.

LEARNING TIP

Note that absolute value works for any number in the real number system. If a number is irrational or a decimal, the absolute value is simply telling us how far away the measurement is from the zero point on the number line.

EXAMPLE 4 Absolute value

(a) The absolute value of 6 is 6 because 6 is six units to the right of the origin at zero. We write this as $|6| = 6$. The absolute value of -7 is 7 because -7 is seven units to the left of zero. We write this as $|-7| = 7$. Both examples are illustrated in Fig. 1.4.

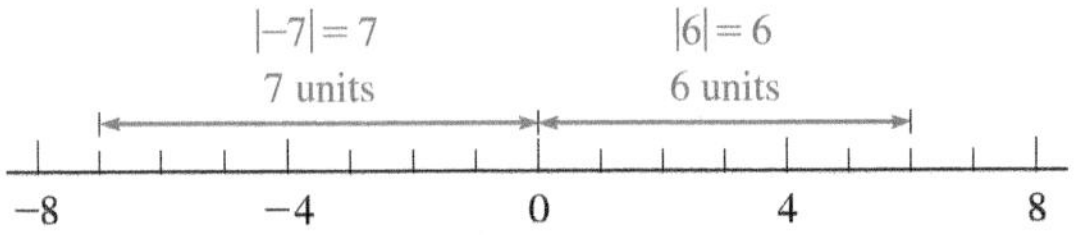

Fig. 1.4

(b) The value of $|-\pi|$ is π because minus pi is pi units away from zero. The value of $-|-\frac{1}{6}|$ is $-\frac{1}{6}$ because minus one-sixth is one-sixth of a unit from zero. When we take the negative of this result, we get a final answer of minus one-sixth. In this sense, an absolute value is treated much like a bracket, where the number inside is evaluated first before working through the balance of the problem.

Practice Exercises

1. $|-4.2| = ?$ **2.** $-\left|-\frac{3}{4}\right| = ?$

EQUALITIES AND INEQUALITIES

The number line provides a visual way of understanding which values are larger than, smaller than, or of the same size as other values. Let us review the symbols used to describe these comparisons. See Fig. 1.5.

Symbol	Meaning	EXAMPLE 5
$=$	The two values are equal	**(a)** $3 = 6/2$ should be read: "Three is equal to six divided by two."
$>$	The value on the left is greater than the value on the right	**(b)** $2 > -4$ should be read: "2 is greater than -4" (2 is to the right of -4 on the number line).
$<$	The value on the left is less than the value on the right	**(c)** $3 < 6$ should be read: "3 is less than 6" (3 is to the left of 6 on the number line).

■ The symbols $=$, $<$, and $>$ were introduced by English mathematicians in the late 1500s.

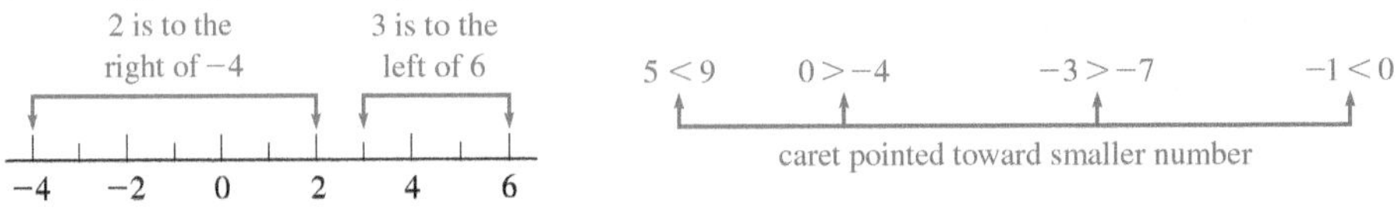

Fig. 1.5

Practice Exercises

Place the correct sign of inequality ($<$ or $>$) between the given numbers.

3. $-5 \quad 4$ **4.** $0 \quad -3$

We can also use the symbol "$\geq$" when the value on the left is *greater than or equal to* the value on the right, and the symbol "$\leq$" when the value on the left is *less than or equal to* the value on the right.

RECIPROCAL VALUES

Every number, except zero, has a **reciprocal**. *The reciprocal of a number is 1 divided by the number.* Also, when the number is multiplied by its reciprocal, the result will equal 1.

EXAMPLE 6 Reciprocals

(a) The reciprocal of 7 is $\frac{1}{7}$ and $7 \times \frac{1}{7} = 1$.

(b) The reciprocal of π is $\frac{1}{\pi}$ and $\pi \times \frac{1}{\pi} = 1$.

(c) The reciprocal of -5 is $\frac{1}{-5}$. Note that the negative sign is retained in the reciprocal of a negative number.

(d) The reciprocal of $\frac{2}{3}$ is $\frac{3}{2}$. Note that the reciprocal of a fraction is found by inverting the fraction and multiplying: $\frac{1}{2/3} = 1 \times \frac{3}{2} = \frac{3}{2}$.

Because the "×" symbol for multiplication can easily be confused with the letter "x" when using algebraic terms, it is useful to indicate multiplication in other ways. The multiplication $7 \times \frac{1}{7}$ can also be written using the dot form $7 \cdot \frac{1}{7}$ or the bracket form $7(\frac{1}{7})$.

DENOMINATE NUMBERS

In applications, *numbers that represent a measurement and are written with units of measurement are called* **denominate numbers**. Denominate numbers play an important role in science and engineering because units add meaning to the numerical value.

EXAMPLE 7 Denominate numbers

■ For reference, see Section 1.3 for units of measurement and the symbols used for them.

(a) To show that a certain HDTV set has mass of 28 kilograms, we write the mass as 28 kg.

(b) To report that a giant redwood tree is 110 units high when our measuring tape is in units of metres, we write the height as 110 m.

(c) To show that the speed of a rocket is 1500 metres per second, we write the speed as 1500 m/s. (Note how m/s represents a ratio, where we keep track of the units in the numerator and the units in the denominator. Also, note the use of s for second. We use s rather than sec.)

(d) To show that the area of a computer chip is 0.75 square centimetres, we write the area as 0.75 cm^2. (The units are raised to the power 2 instead of writing sq. cm.)

(e) To show that the volume of water in a glass tube is 25 cubic centimetres, we write the volume as 25 cm^3. (The units are raised to the power 3 instead of writing cu cm or cc.)

VARIABLES AND THEIR ROLE

A **variable** is a number whose value is not yet known, and which we keep track of with a designated letter, or literal number. Often, we use mathematical equations to describe how one value changes as a result of changes in other values. In this case, we call any value we control an **independent variable**, whereas we call the value that responds to the independent variable(s) a **response variable**.

EXAMPLE 8 Independent and response variables

(a) The resistance of an electric resistor is R. By Ohm's law, the current I in the resistor equals the voltage V divided by R, written as $I = V/R$. For the fixed value of the resistance R, we can adjust the voltage V. Therefore, R is a **constant** (its value does not change), V is the independent variable and I is the response variable.

(b) It costs b dollars per day to run a coffee shop, and each cup of coffee produced costs a dollars. The total daily cost C if we make n cups of coffee is

$$C = an + b$$

Here, a and b are constants (they do not change on a daily basis). The variable n is an independent variable which will vary depending on how many cups of coffee are made in one day. As a result, it will have an impact on the response variable cost (C).

EXERCISES 1.1

In Exercises 1–4, make the given changes in the indicated examples of this section, and then answer the given questions.

1. In the first line of Example 1, change the 5 to -3 and the -19 to 14. What other changes must then be made in the first paragraph?
2. In the first line of Example 4(a), change the 6 to -6. What other changes must then be made in the first sentence?
3. In Example 5(b), change the 2 to -6. What other changes must then be made?
4. In Example 6(d), change the $\frac{2}{3}$ to $\frac{3}{2}$. What other changes must then be made?

In Exercises 5–8, designate each of the given numbers as being an integer, rational, irrational, real, or imaginary. (More than one designation may be correct.)

5. $3, \sqrt{-4}$
6. $\frac{\sqrt{7}}{3}, -6$
7. $-\frac{\pi}{6}, \frac{1}{8}$
8. $-\sqrt{-6}, -2.33$

In Exercises 9 and 10, find the absolute value of each number.

9. $3, \quad -4, \quad -\frac{\pi}{2}$
10. $-0.857, \quad \sqrt{2}, \quad -\frac{19}{4}$

In Exercises 11–18, insert the correct sign of inequality (> or <) between the given numbers.

11. $6 \quad 8$
12. $7 \quad 5$
13. $\pi \quad -3.2$
14. $-4 \quad 0$
15. $-4 \quad -|-3|$
16. $-\sqrt{2} \quad -1.42$
17. $-\frac{1}{3} \quad -\frac{1}{2}$
18. $-0.6 \quad 0.2$

In Exercises 19 and 20, find the reciprocal of each number.

19. $3, \quad -\frac{4}{\sqrt{3}}, \quad \frac{y}{b}$
20. $-\frac{1}{3}, \quad -0.25, \quad x$

In Exercises 21 and 22, locate each number on a number line, as in Fig. 1.3.

21. $2.5, \quad -\frac{12}{5}, \quad \sqrt{3}$
22. $-\frac{\sqrt{2}}{2}, \quad 2\pi, \quad \frac{123}{19}$

In Exercises 23–47, solve the given problems. Refer to Fig. 1.8 for units of measurement and their symbols.

23. Is an absolute value always positive? Explain.
24. Is 2.17 rational? Explain.
25. What is the reciprocal of the reciprocal of any positive or negative number?
26. Find a rational number between -0.9 and -1.0 that can be written with a denominator of 11 and an integer in the numerator.
27. Find a rational number between 0.13 and 0.14 that can be written with a numerator of 3 and an integer in the denominator.
28. If $b > a$ and $a > 0$, is $|b - a| < |b| - |a|$?
29. List the following numbers in numerical order, starting with the smallest: $-1, 9, \pi, \sqrt{5}, |-8|, -|-3|, -3.5$.
30. List the following numbers in numerical order, starting with the smallest: $\frac{-1}{5}, -\sqrt{10}, -|-6|, -4, 0.25, |-\pi|$.
31. If a and b are positive integers and $b > a$, what type of number is represented by the following?
 (a) $b - a$ (b) $a - b$ (c) $\frac{b - a}{b + a}$
32. If a and b represent positive integers, what kind of number is represented by (a) $a + b$, (b) a/b, and (c) $a \times b$?
33. For any positive or negative integer: (a) Is its absolute value always an integer? (b) Is its reciprocal always a rational number?
34. For any positive or negative rational number: (a) Is its absolute value always a rational number? (b) Is its reciprocal always a rational number?
35. Describe the location of a number x on the number line when (a) $x > 0$ and (b) $x < -4$.
36. Describe the location of a number x on the number line when (a) $|x| < 1$ and (b) $|x| > 2$.
37. For a number $x > 1$, describe the location on the number line of the reciprocal of x.
38. For a number $x < 0$, describe the location on the number line of the number with a value of $|x|$.
39. A *complex number* is defined as $a + bj$, where a and b are real numbers and $j = \sqrt{-1}$. For what values of a and b is the complex number $a + bj$ a real number? (All real numbers and all imaginary numbers are also complex numbers.)
40. A sensitive gauge measures the total weight w of a container and the water that forms in it as vapour condenses. It is found that $w = c\sqrt{0.1t + 1}$, where c is the weight of the container and t is the time of condensation. Identify the variables and constants.
41. In an electric circuit, the reciprocal of the total capacitance of two capacitors in series is the sum of the reciprocals of the capacitances. Find the total capacitance of two capacitances of 0.0040 F and 0.0010 F connected in series.
42. Alternating-current (AC) voltages change rapidly between positive and negative values. If a voltage of 100 V changes to -200 V, which is greater in absolute value?
43. The memory of a certain computer has a bits in each byte. Express the number N of bits in n kilobytes in an equation. (A *bit* is a single digit, and bits are grouped in *bytes* in order to represent special characters. Generally, there are eight bits per byte. If necessary, see Fig. 1.9 for the meaning of *kilo*.)
44. The computer design of the base of a truss is x m long. Later it is redesigned and shortened by y cm. Give an equation for the length L, in centimetres, of the base in the second design.
45. In a laboratory report, a student wrote "$-20°C > -30°C$." Is this statement correct? Explain.
46. After 5 s, the pressure on a valve is less than 600 kPa. Using t to represent time and p to represent pressure, this statement can be written "for $t > 5$ s, $p < 600$ kPa." In this way, write the statement "when the current I in a circuit is less than 4 A, the voltage V is greater than 12 V."
47. In Glendon, Ontario, the world's largest perogy is a statue celebrating Ukrainian settlers that stands 8 m tall and has a mass of 2700 kg. If the mass of a Tesla electric car is approximately 2200 kg, write a mathematical statement that describes how these two masses relate.

Answers to Practice Exercises

1. 4.2 2. $-\frac{3}{4}$ 3. < 4. >

1.2 Fundamental Operations of Algebra

■ While the focus of this section is to apply the operations of algebra numerically, algebraic examples have been included for illustrative purposes. These algebraic examples will be discussed further later on in Chapter 1.

The **fundamental laws of algebra** are rules that we must follow when performing operations with numbers or variables. These rules are given below, along with numerical and algebraic illustrations of how they are applied.

The Commutative Law of Addition states that *the sum of two numbers is the same, regardless of the order in which they are added.* Note that subtraction can be thought of as the addition of a negative number, so rules for addition apply equally to subtraction.

The Commutative Law of Addition	
Literal Form of the Rule	$a + b = b + a$ **(1.1)**
Numerical Example	$5 + 3 = 3 + 5$
Algebraic Example	$2x + 4x^2 = 4x^2 + 2x$

The Associative Law of Addition states that *the sum of three or more numbers is the same, regardless of the way in which they are grouped for addition.*

The Associative Law of Addition	
Literal Form of the Rule	$a + (b + c) = (a + b) + c$ **(1.2)**
Numerical Example	$3 + (5 + 6) = (3 + 5) + 6$
Algebraic Example	$-1 + (2x + 4x^2) = (-1 + 2x) + 4x^2$

The Commutative Law of Multiplication states that *the product of two numbers is the same, regardless of the order in which they are multiplied.*

The Commutative Law of Multiplication	
Literal Form of the Rule	$ab = ba$ **(1.3)**
Numerical Examples	$3(-2) = (-2)(3)$ $3 \cdot \frac{1}{5} = \frac{1}{5} \cdot 3$
Algebraic Example	$(2x^3) \cdot y = y \cdot (2x^3)$

The Associative Law of Multiplication states that *the product of three or more numbers is the same, regardless of the way in which they are grouped for multiplication.*

The Associative Law of Multiplication	
Literal Form of the Rule	$a(bc) = (ab)c$ **(1.4)**
Numerical Example	$-2 \cdot (3 \cdot 5) = (-2 \cdot 3) \cdot 5$
Algebraic Example	$3x \cdot (2y \cdot 4z) = (3x \cdot 2y) \cdot 4z$

The Distributive Law states that *the product of one number and the sum of two or more other numbers is equal to the sum of the products of the first number and each of the other numbers of the sum.*

The Distributive Law	
Literal Form of the Rule	$a \cdot (b + c) = ab + ac$ **(1.5)**
Numerical Example	$5(4 + 2) = 5 \cdot 4 + 5 \cdot 2$ Both expressions give the same result of 30.
Algebraic Example	$3x(2x + 4) = 3x \cdot 2x + 3x \cdot 4$

■ Note carefully the difference:
associative law: $5 \times (4 \times 2)$
distributive law: $5 \times (4 + 2)$

Eqs. (1.1)–(1.5) are all **identities**, in that the expression to the left of the = sign equals the expression to the right for any values of a, b, and c.

OPERATIONS ON POSITIVE AND NEGATIVE NUMBERS

When using the basic operations (addition, subtraction, multiplication, division) on positive and negative numbers, we determine the result to be either positive or negative according to the following rules.

■ From Section 1.1, we recall that a positive number is preceded by no sign. Therefore, in using these rules, we show the "sign" of a positive number by simply writing the number itself.

Addition of two numbers of the same sign *Add their absolute values and assign the sum their common sign.*

EXAMPLE 1 Adding numbers of the same sign

(a) $2 + 6 = 8$ — the sum of two positive numbers is positive

(b) $-2 + (-6) = -1(2 + 6) = -8$ — the sum of two negative numbers is negative

The negative number -6 is placed in parentheses since it is also preceded by a plus sign showing addition. It is not necessary to place the -2 in parentheses.

LEARNING TIP
When adding two negative numbers, we make use of the distributive law: adding two negative numbers is the same as grouping out a -1 from each term and adding the magnitudes.

Addition of two numbers of different signs *Determine the magnitude of the difference between the two values and assign to the result the sign of the value of larger magnitude.*

EXAMPLE 2 Adding numbers of different signs

(a) $2 + (-6) = -(6 - 2) = -4$

(b) $-6 + 2 = -(6 - 2) = -4$

(a), (b): the negative 6 has the larger magnitude

(c) $6 + (-2) = 6 - 2 = 4$

(d) $-2 + 6 = 6 - 2 = 4$

(c), (d): the positive 6 has the larger magnitude

($6 - 2$): the subtraction of magnitudes

Subtraction of one number from another *Change the sign of the number being subtracted and change the subtraction to addition. Perform the addition.* Alternatively, we can visualize subtraction using a number line. *Subtracting a positive value* requires us to "take units away" from a value, moving to the left on a number line. This implies that a positive number will become either a smaller positive number, or a negative number, depending on the value subtracted. *Subtracting a negative value* requires us to move to the right on a number line (opposite to the negative value), so the result is the same as if you added the absolute value of the subtracted term.

EXAMPLE 3 Subtracting positive and negative numbers

(a) $12 - 5 = 12 + (-5) = 7$

Note that when 5 units are removed from 12, 7 remain.

(b) $2 - 6 = 2 + (-6) = -(6 - 2) = -4$

Note that after changing the subtraction to addition, and changing the sign of 6 to make it -6, we have precisely the same illustration as Example 2(a).

(c) $12 - (-4) = 12 + 4 = 16$

We see that subtracting a negative number is equivalent to adding a positive number of the same absolute value.

(d) The change in temperature from -12°C to -26°C is

$-26°\text{C} - (-12°\text{C}) = -26°\text{C} + 12°\text{C} = -14°\text{C}$

Multiplication and division of two numbers *The product (or quotient) of two numbers of the same sign is positive. The product (or quotient) of two numbers of different signs is negative.* Note that when values hold opposite signs for multiplication and division, the result will be the same regardless of which value has the negative sign.

EXAMPLE 4 Multiplying and dividing positive and negative numbers

(a) $3(12) = 3 \times 12 = 36$ $\qquad \dfrac{12}{3} = 4$ — result is positive if both numbers are positive

(b) $-3(-12) = 3 \times 12 = 36$ $\qquad \dfrac{-12}{-3} = 4$ — result is positive if both numbers are negative

(c) $3(-12) = -(3 \times 12) = -36$ $\qquad \dfrac{-12}{3} = -\dfrac{12}{3} = -4$ — result is negative if one number is positive and the other is negative

(d) $-3(12) = -(3 \times 12) = -36$ $\qquad \dfrac{12}{-3} = -\dfrac{12}{3} = -4$

ORDER OF OPERATIONS

When mathematical operation symbols separate a series of numbers in an expression, it is important to perform operations in the following unambiguous *order* of operations. In order of priority:

To simplify

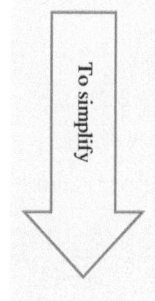

B **1. Bracketed Values:** Perform operations with specific groupings first inside of parentheses ( ), brackets [ ], or absolute values | |.
- If bracketed values occur inside one another, begin with the innermost bracketed value when simplifying.
- Otherwise, evaluate from left to right.

O **2. Evaluate Operators:** Operators can include exponents and roots/radicals, which will be discussed in Section 1.4 and Section 1.6, respectively.

D/M **3. Perform Division and Multiplication Steps** (from left to right).

A/S **4. Perform Addition and Subtraction Steps** (from left to right).

It is very important to remember that this order of operations must take priority over simplifying from left to right.

EXAMPLE 5 Order of operations

■ Note that $20 \div (2 + 3) = \frac{20}{2 + 3}$, whereas $20 \div 2 + 3 = \frac{20}{2} + 3$.

(a) $20 \div (2 + 3)$ is evaluated by simplifying the addition inside the parentheses first, then dividing. Therefore,

B: $20 \div (2 + 3) = 20 \div 5$
D: $20 \div 5 = 4$

(b) $20 \div 2 + 3$ is evaluated by first dividing 20 by 2 and then adding. Division takes priority over addition. Therefore,

D: $20 \div 2 + 3 = 10 + 3$
A: $10 + 3 = 13$

(c) $16 - 2 \times 3$ is evaluated by first multiplying and then subtracting. Multiplication takes priority over subtraction. Therefore,

M: $16 - 2 \times 3 = 16 - 6$
S: $16 - 6 = 10$

(d) $16 \div 2 \times 4$ is evaluated from left to right since multiplication holds the same priority as division. Therefore,

D: $16 \div 2 \times 4 = 8 \times 4$
M: $8 \times 4 = 32$

(e) $|3 - 5| - |-3 - 6|$ is evaluated by first performing the subtractions within the absolute value bars, then evaluating the absolute values, and then subtracting. Therefore,

B: $|3 - 5| - |-3 - 6| = |-2| - |-9| = 2 - 9$
S: $2 - 9 = -7$

Practice Exercises

Evaluate: **1.** $12 - 6 \div 2$
2. $16 \div (2 \times 4)$

EVALUATING EXPRESSIONS WITH SIGNED VALUES

When evaluating expressions, it is generally more convenient to change the operations and numbers so that the result is found by the addition and subtraction of positive numbers. When this is done, we must remember that

$$a + (-b) = a - b \qquad \textbf{(1.6)}$$
$$a - (-b) = a + b \qquad \textbf{(1.7)}$$

LEARNING TIP

Note that when you have more than one term in the numerator or the denominator of a fraction, it is best to surround them with brackets to ensure the correct order of operations (see Example 6d). This is imperative when solving fractions using a calculator.

EXAMPLE 6 Evaluating numerical expressions

(a) $7 + (-3) - 6 = 7 - 3 - 6 = 4 - 6 = -2$ using Eq. (1.6)

(b) $\dfrac{18}{-6} + 5 - (-2)(3) = -3 + 5 - (-6) = 2 + 6 = 8$ using Eq. (1.7)

(c) $\dfrac{|3 - 15|}{-2} - \dfrac{8}{4 - 6} = \dfrac{12}{-2} - \dfrac{8}{-2} = -6 - (-4) = -6 + 4 = -2$

(d) $\dfrac{-12}{2 - 8} + \dfrac{5 - 1}{2(-1)} = \dfrac{-12}{(2 - 8)} + \dfrac{(5 - 1)}{2(-1)} = \dfrac{-12}{-6} + \dfrac{4}{-2} = 2 + (-2) = 2 - 2 = 0$

Simplifying with Eqs. (1.1) and (1.2) is most logically considered immediately before performing addition and subtraction steps in the order of operations. In the examples above, (a) demonstrates simplifying right away because there were no other operations, whereas (b), (c), and (d) show simplification later in the problem, immediately before the A/S step.

Practice Exercises

Evaluate:

3. $2(-3) - \dfrac{4 - 8}{2}$ **4.** $\dfrac{|5 - 15|}{2} - \dfrac{-9}{3}$

EXAMPLE 7 Using a calculator for signed operations

Simplify $2 + (-7)$ using a calculator.

To simplify using addition, we make use of the negative sign key (−). This key indicates that the number that follows it is negative. Therefore we would enter 2 + (−) 7.

To simplify using subtraction (as in Eq. 1.7), we make use of the subtraction key −. This key indicates that the number that follows is being subtracted. Therefore we would enter 2 − 7. Both methods give the same result.

Note that all equations that are entered into a calculator must follow appropriate order of operations. Also, different calculators will have slightly different ways of labelling functions. For example, depending on the calculator model, the =, EXE, or ENTER keys all perform the same task.

EXAMPLE 8 Evaluating—velocity after collision

A 1500-kg van going at 40 km/h ran head-on into a 1000-kg car going at 20 km/h. An insurance investigator determined the velocity of the vehicles immediately after the collision from the following calculation. See Fig. 1.6.

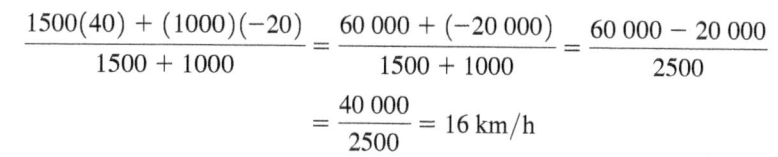

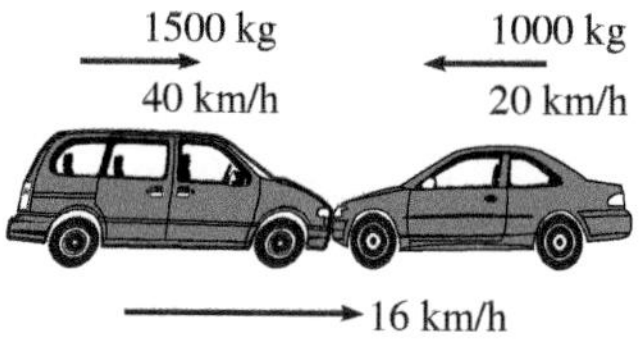

Fig. 1.6

$$\frac{1500(40) + (1000)(-20)}{1500 + 1000} = \frac{60\,000 + (-20\,000)}{1500 + 1000} = \frac{60\,000 - 20\,000}{2500}$$

$$= \frac{40\,000}{2500} = 16 \text{ km/h}$$

The numerator and the denominator must be evaluated before the division is performed. The multiplications in the numerator are performed first, followed by the addition in the denominator and the subtraction in the numerator.

OPERATIONS WITH ZERO

Operations containing zero occur frequently and have unique properties when compared to operations with other integers. For any real number a, the following hold true.

	Adding	Subtracting	Multiplying	Dividing
General Rule(s)	$a + 0 = a$	$a - 0 = a$	$a \times 0 = 0$	$0 \div a = 0$ (if $a \neq 0$) $a \div 0 =$ undefined $0 \div 0 =$ indeterminate
Example(s)	$5 + 0 = 5$	$-6 - 0 = -6$	$7 \times 0 = 0$ $-4 \times 0 = 0$	$0 \div (-3) = 0$ $8 \div 0 =$ undefined $\frac{0}{0} =$ indeterminate

The Indeterminate Case for Division

When a number is divided by zero, the value is undefined because no real value can be associated with that division. When a zero is divided by zero, it is a special case called ***the indeterminate case***, where no specific value for the division can be determined, but many values are possible.

For example, if $c = \frac{0}{0}$, then if each side is multiplied by zero, the equality becomes $0 \times c = 0$, which is true for any value of c. Thus, all real numbers are possible solutions for the division.

While this may be easy to identify when working with numbers, it is important to keep division by zero in mind when working with unknown values. For example, to solve the equation $x \cdot x = 3 \cdot x$, the intuitive approach is to divide each side by x and reduce.

$$\frac{\cancel{x} \cdot x}{\cancel{x}} = \frac{3 \cdot \cancel{x}}{\cancel{x}}, \text{ so } x = 3$$

The issue is that in doing so, we cannot be sure whether we have divided by zero unless we check. In this case, $0 \times 0 = 3 \times 0$ is a true statement, so the valid solution of zero would be missed if simply dividing by x.

To address the issue, whenever you divide by an unknown value, you must first consider whether or not 0 is a valid solution. Dividing by an unknown whose value can be zero is called the ***division by zero error***.

EXERCISES 1.2

In Exercises 1–4, make the given changes in the indicated examples of this section and then solve the resulting problems.

1. In Example 5(c), change 3 to (-3) and then evaluate.

2. In Example 6(b), change 18 to -18 and then evaluate.

3. In Example 6(d), interchange the 2 and 8 in the first denominator and then evaluate.

4. In Example 8, how would the answer change if the smaller car weighed 1400 kg instead of 1000 kg?

In Exercises 5–38, evaluate each of the given expressions by performing the indicated operations.

5. $8 + (-4)$ **6.** $-4 + (-7)$ **7.** $-3 + 9$

8. $18 - 21$ **9.** $-19 - (-16)$ **10.** $8 - (-4)$

11. $8(-3)$ **12.** $-9(3)$ **13.** $-7(-5)$

14. $\dfrac{-9}{3}$ **15.** $\dfrac{-6(20 - 10)}{-3}$ **16.** $\dfrac{28}{-7(6 - 5)}$

17. $-2(4)(-5)$ **18.** $-3(-4)(-6)$ **19.** $2(2 - 7) \div 10$

20. $\dfrac{-64}{-2|4 - 8|}$ **21.** $16 \div 2(-4)$ **22.** $-20 \div 5(-4)$

23. $-9 - |2 - 10|$ **24.** $(7 - 7) \div (5 - 7)$ **25.** $\dfrac{17 - 7}{7 - 7}$

26. $\dfrac{7 - 7}{7 - 7}$ **27.** $8 - 3(-4)$ **28.** $-20 + 8 \div 4$

29. $-2(-6) + \left|\dfrac{8}{-2}\right|$ **30.** $-10 - (-6)(-8)$

31. $10(-8)(-3) \div (10 - 50)$ **32.** $\dfrac{7 - |-5|}{-1(-2)}$

33. $\dfrac{24}{3 + (-5)} - 4(-9)$ **34.** $\dfrac{-18}{3} - \dfrac{4 - |-6|}{-1}$

35. $-7 - \dfrac{|-14|}{2(2 - 3)} - 3|6 - 8|$ **36.** $-7(-3) + \dfrac{6}{-3} - (-9)$

37. $\dfrac{3(-9) - 2(-3)}{3 - 10}$ **38.** $\dfrac{20(-12) - 40(-15)}{98 - |-98|}$

In Exercises 39–46, determine which of the fundamental laws of algebra is demonstrated.

39. $6(7) = 7(6)$ **40.** $6 + 8 = 8 + 6$

41. $6(3 + 1) = 6(3) + 6(1)$ **42.** $4(5 \times \pi) = (4 \times 5)(\pi)$

43. $3 + (5 + 9) = (3 + 5) + 9$ **44.** $8(3 - 2) = 8(3) - 8(2)$

45. $(\sqrt{5} \times 3) \times 9 = \sqrt{5} \times (3 \times 9)$

46. $(3 \times 6) \times 7 = 7 \times (3 \times 6)$

In Exercises 47–50, for numbers a and b, determine which of the following expressions equals the given expression.
(a) $a + b$ (b) $a - b$ (c) $b - a$ (d) $-a - b$

47. $-a + (-b)$ **48.** $b - (-a)$

49. $-b - (-a)$ **50.** $-a - (-b)$

In Exercises 51–70, answer the given questions. Refer to Fig. 1.8 for units of measurement and their symbols.

51. Insert the proper sign $(=, >, <)$ to make the following true: $|5 - (-2)|$ $|-5 - |-2||$

52. Insert the proper sign $(=, >, <)$ to make the following true: $|-3 - |-7||$ $||-3| - 7|$

53. (a) What is the sign of the product of an even number of negative numbers? (b) What is the sign of the product of an odd number of negative numbers?

54. Is subtraction commutative? Explain.

55. Explain why the following definition of the absolute value of a real number x is either correct or incorrect (the symbol $\geq$ means "is equal to or greater than"): If $x \geq 0$, then $|x| = x$; if $x < 0$, then $|x| = -x$.

56. Explain what the error is if the expression $24 - 6 \div 2 \cdot 3$ is evaluated as 27. What is the correct value?

57. Describe the values of x and y for which (a) $-xy = 1$ and (b) $\dfrac{x - y}{x - y} = 1$.

58. Describe the values of x and y for which (a) $|x + y| = |x| + |y|$ and (b) $|x - y| = |x| + |y|$.

59. The changes in the price of a stock (in dollars) for a given week were -0.68, $+0.42$, $+0.06$, -0.11, and $+0.02$. What was the total change in the stock's price that week?

60. Using subtraction of signed numbers, find the difference in the altitude of the bottom of the Dead Sea, 1396 m below sea level, and the bottom of Death Valley, 86 m below sea level.

61. Some solar energy systems are used to supplement the power supplied to a home such that the meter runs backward if the solar energy being generated is greater than the energy being used. With such a system, if the solar power averages 1.5 kW for a 3.0-h period and only 2.1 kW · h is used during this period, what will be the change in the meter reading for this period?

62. A baseball player's batting average (total number of hits divided by total number of at-bats) is expressed in decimal form from .000 (no hits for all at-bats) to 1.000 (one hit for each at-bat). A player's batting average is often shown as .000 before the first at-bat of the season. Is this a correct batting average? Explain.

63. The daily high temperatures (in °C) for Edmonton, Alberta, in the first week in March were recorded as $-7, -3, 2, 3, 1, -4$, and -6. What was the average daily temperature for the week? (Divide the algebraic sum of readings by the number of readings.)

64. A flare is shot up from the top of a tower. Distances above the flare gun are positive and those below it are negative. After 5 s the vertical distance (in metres) of the flare from the flare gun is found by evaluating $(20)(5) + (-5)(25)$. Find this distance.

65. Find the sum of the voltages of the batteries shown in Fig. 1.7. Note the directions in which they are connected.

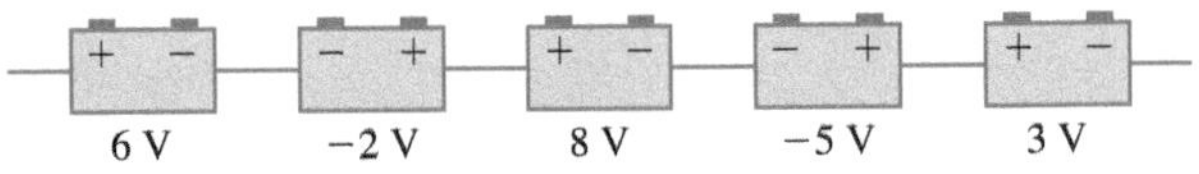

Fig. 1.7

66. The electric current was measured in a given AC circuit at equal intervals as 0.7 mA, −0.2 mA, −0.9 mA, and −0.6 mA. What was the change in the current between (a) the first two readings, (b) the middle two readings, and (c) the last two readings?

67. One oil-well drilling rig drills 100 m deep the first day and 200 m deeper the second day. A second rig drills 200 m deep the first day and 100 m deeper the second day. In showing that the total depth drilled by each rig was the same, state what fundamental law of algebra is illustrated.

68. A water tank leaks 12 L each hour for 7 h, and a second tank leaks 7 L each hour for 12 h. In showing that the total amount leaked is the same for the two tanks, what fundamental law of algebra is illustrated?

69. On each of the 7 days of the week, a person spends 25 min on Facebook and 15 min on Twitter. Set up the expression for the total time spent on these two sites that week. What fundamental law of algebra is illustrated?

70. A jet travels 600 km/h relative to the air. The wind is blowing at 50 km/h. If the jet travels with the wind for 3 h, set up the expression for the distance travelled. What fundamental law of algebra is illustrated?

Answers to Practice Exercises

1. 9 **2.** 2 **3.** −4 **4.** 8

1.3 Measurement, Calculation, and Approximate Numbers

UNITS OF MEASUREMENT

For science and technology, numbers can take on different meanings depending on whether their values are a count, a measurement, or a conversion factor (discussed later in this section).

A **count** represents an exact number of observations, and is not subject to any error or uncertainty. Counts are considered exact values.

In contrast, a **measurement** is an estimate of a physical quantity, and typically contains both a *number* and the *units* with which the measurement was made. Stating the number alone is insufficient because it gives no indication of the unit system used to make the measurement. For example, we would write a length as 12.5 cm to make it clear that a measurement was made in centimetres rather than in metres, feet, or inches.

Measurements are estimates because they are subject to some level of **uncertainty**, or **measurement error**, in the measured value. It is important to measure accurately to avoid large measurement errors that can call results into question. The level of accuracy (trueness) of the value will be discussed later in this chapter.

THE SI UNIT SYSTEM

There are several internationally recognized unit systems. The international standard is called the **International System of Units**, or **Système International** (SI). SI units were developed based on the metric system, or units of base 10. This system, above all others, is the preferred international standard for scientific and engineering communications and should be used unless otherwise specified in a given field. In this text, most calculations will be made using this international standard.

In the SI unit system, there are **seven base units** from which all other units are constructed, a set of **derived units** formed by multiplication or division of one or more base units, and a set of **supplementary units** used for measuring plane and solid angles. Fig. 1.8 presents a summary table of SI units.

It is worth mentioning that although the international standard unit for time is the second, other larger units such as minute (min), hour (h), day (d) and year (y or yr) are also acceptable. For angles, divisions such as the degree, minute of an arc, and second of an arc are also common.

OTHER UNIT SYSTEMS

There are two other systems of units that are noteworthy. Also based on the metric system, the **CGS (centimetre-gram-seconds) system** remains popular for measurements that are small, particularly in microbiology and chemistry. In mechanics, the CGS and the SI system differ only in scale. For other measurements, the derived units may require a unit conversion when converting between CGS and SI units. CGS unit conversions are typically included in most chemistry texts for reference.

The third noteworthy unit system is the **US Customary System** (based on the Imperial unit system). The Imperial unit system was based on units of measure that were historically available, and so they do not all share the same base. This unit system remains important in some fields because materials, plans, and technical papers can originate internationally. Some common examples of base units in the US Customary System are feet and miles (for length), and pounds (for mass). For reference, commonly used US customary units as well as common conversion equivalents between the SI and US customary units have been included inside the back cover of this text.

Fig. 1.8 Physical Quantities and Their SI Units

	Quantity	Quantity Symbol	Unit Name	Unit Symbol	In Terms of Other SI Units
Base Units	Length	s	metre	m	
	Mass	m	kilogram	kg	
	Time	t	second	s	
	Electric current	I, i	ampere	A	
	Thermodynamic temperature	T	kelvin	K	
	Amount of substance	n	mole	mol	
	Luminous intensity	I	candela	cd	
Supplementary Units	Plane angle	q	radian	rad	
	Solid angle	q	steradian	sr	
Derived Units	Area	A		m^2	
	Volume	V		m^3	
	Volume	V	litre	L	$(1000\ L = 1\ m^3)$
	Velocity	v		m/s	
	Acceleration	a		m/s^2	
	Force	F	newton	N	$kg \cdot m/s^2$
	Density	r		kg/m^3	
	Pressure	p	pascal	Pa	$N/m^2 = kg/(m \cdot s^2)$
	Energy, work	E, W	joule	J	$N \cdot m = kg \cdot m^2/s^2$
	Power	P	watt	W	$J/s = kg \cdot m^2/s^3$
	Frequency	f	hertz	Hz	1/s
	Electric charge	q	coulomb	C	$A \cdot s$
	Electric potential	V, E	volt	V	$J/(A \cdot s) = kg \cdot m^2/(A \cdot s^3)$
	Capacitance	C	farad	F	$s/\Omega = s \cdot A/V$
	Inductance	L	henry	H	$\Omega \cdot s = V \cdot s/A$
	Resistance	R	ohm	Ω	V/A
	Heat	Q	joule	J	$N \cdot m = kg \cdot m^2/s^2$
	Temperature	T	degrees Celsius	°C	A change of 1°C = 1 K

CONVENTIONS FOR WRITING UNITS

The following conventions must be followed when expressing units and unit symbols:

Conventions for Writing Units	EXAMPLE 1
1. Do not capitalize **named units** (except for degrees Celsius).	10 square metres, 10 000 pascals, 22 degrees Celsius
2. **Unit symbols** are capitalized when they were named after a person (see Fig. 1.8 for the full list).	14 **N** for 14 newtons, 1000 **Pa** for pascals
3. The **unit symbol** for litres is capitalized to avoid confusion between small "l" and 1.	10 **L** for 10 litres
4. **Include a • symbol** between units that are multiplied to avoid confusion between units and SI prefixes.	$m \cdot m$ or m^2 for metres squared (not mm, which can be confused with millimetres
5. Each unit should be expressed using a **maximum of one division sign**.	m/s^2 for metres over seconds squared (not $m/s/s$)
6. Unit symbols are **upright** (non-italicized) while *physical quantity variable symbols are* ***italicized***.	10 **C** is 10 coulombs, while capacitance is given in an equation as $\boldsymbol{C} = 20$ F.
7. **Unit symbols are not pluralized.** Numerical values will indicate how many without pluralizing.	300 **kg** (not 300 kgs)
8. **Spaces may be used to separate thousands** rather than commas (to avoid confusion over interpretation of commas, used as a period in some countries).	10 000 for ten thousand (not 10,000)

SI PREFIXES

In science, it is common to deal with measurements that are very large or very small in size. This leads to values that contain many zeros when expressed as a decimal value. To manage the leading or trailing zeros, a standard set of **SI prefixes** has been developed. (See Fig. 1.9.) These prefixes substitute for a given multiple of 10, and are written before the unit name. There can never be more than one prefix for a single unit. Typically, prefixes are chosen to reflect the scale of the calculation being performed to make the quantity easy to understand or useful to apply.

Fig. 1.9

Prefix	Factor	Symbol
exa	10^{18}	E
peta	10^{15}	P
tera	10^{12}	T
giga	10^{9}	G
mega	10^{6}	M
kilo	10^{3}	k
hecto	10^{2}	h
deca	10^{1}	da
deci	10^{-1}	d
centi	10^{-2}	c
milli	10^{-3}	m
micro	10^{-6}	μ
nano	10^{-9}	n
pico	10^{-12}	p
femto	10^{-15}	f
atto	10^{-18}	a

EXAMPLE 2 Using SI prefixes

(a) Using SI prefixes, we can rewrite the following measurements:

123 000 000 s	$= 123 \times 10^6$ s	= 123 Ms, or 0.123 Gs
0.000 005 0 m	$= 5.0 \times 10^{-6}$ m	$= 5.0\,\mu$m
85 300 Ω	$= 85.3 \times 10^3\ \Omega$	= 85.3 kΩ

(b) We also use the definitions of the SI prefixes to give the name and meaning of the units corresponding to the following symbols:

ks	= kiloseconds	= 1000 s
mC	= millicoulombs	= 0.001 C
GHz	= gigahertz	= 1 000 000 000 Hz

UNIT CONVERSION

Unit conversion can be accomplished by first identifying statements of equivalence that relate the systems of units, or prefixes of units to their SI unit. Then each statement of equivalence can be used to form a fraction that is equal to 1, where the unit we desire is introduced in the numerator, and the unit we wish to eliminate is placed in the denominator. Finally, all values in the numerator are multiplied and all values in the denominator are multiplied. We then simplify to get the final result.

EXAMPLE 3 Single-step unit conversions

Convert the following units.

(a) 1350 m to km. Statement of equivalence: 1000 m = 1 km.

$$1350\ \cancel{\text{m}} \times \left(\frac{1\ \text{km}}{1000\ \cancel{\text{m}}}\right) = 1.35\ \text{km}$$

(b) 25.2 kg to g. Statement of equivalence: 1000 g = 1 kg.

$$25.2\ \cancel{\text{kg}} \times \left(\frac{1000\ \text{g}}{1\ \cancel{\text{kg}}}\right) = 25\ 200\ \text{g}$$

Often, unit conversion will not be direct, and may require several statements of equivalence to chart a path from the units in the starting value to the equivalent units desired. In these cases, several conversion multiplications are done, where in each step the new units introduced (those that will get us closer to the desired result) are in the numerator, and those that we wish to eliminate are in the denominator.

EXAMPLE 4 Multi-step unit conversions

Convert the following units.

(a) 72.0 km/h into m/s. Statements of equivalence: 1000 m = 1 km, 60 min = 1 h, 1 min = 60 s.

$$\frac{72.0\ \cancel{\text{km}}}{1\ \cancel{\text{h}}} \times \left(\frac{1000\ \text{m}}{1\ \cancel{\text{km}}}\right) \times \left(\frac{1\ \cancel{\text{h}}}{60\ \cancel{\text{min}}}\right) \times \left(\frac{1\ \cancel{\text{min}}}{60\ \text{s}}\right) = 20.0\ \text{m/s}$$

(b) 32 500 ft to km. Statements of equivalence: 1 ft = 0.3048 m, 1000 m = 1 km.

$$32\ 500\ \cancel{\text{ft}} \times \left(\frac{0.3048\ \cancel{\text{m}}}{1\ \cancel{\text{ft}}}\right) \times \left(\frac{1\ \text{km}}{1000\ \cancel{\text{m}}}\right) = 9.91\ \text{km}$$

When working with squared or cubed units, it is best to square or cube the statement of equivalence before substituting into the calculation to avoid confusion as to which terms should be squared or cubed.

EXAMPLE 5 Unit conversions including squared or cubed units

Convert the following units.

(a) 8.75 g/cm^3 to kg/m^3. Statements of equivalence: 1000 g = 1 kg, 1 m = 100 cm, so 1 m^3 = 10^6 cm^3.

$$\frac{8.75\ \cancel{\text{g}}}{\cancel{\text{cm}^3}} \times \left(\frac{1\ \text{kg}}{1000\ \cancel{\text{g}}}\right) \times \left(\frac{10^6\ \cancel{\text{cm}^3}}{1\ \text{m}^3}\right) = 8750\ \text{kg/m}^3$$

(b) 62.8 kPa to N/cm^2. Statements of equivalence: 1 kPa = 1000 N/m^2, 1 m = 100 cm, so 1 m^2 = 10^4 cm^2.

$$62.8\ \cancel{\text{kPa}} \times \left(\frac{1000\ \text{N}/\cancel{\text{m}^2}}{1\ \cancel{\text{kPa}}}\right) \times \left(\frac{1\ \cancel{\text{m}^2}}{10^4\ \text{cm}^2}\right) = 6.28\ \text{N/cm}^2$$

APPROXIMATE NUMBERS AND SIGNIFICANT DIGITS

Most numbers in technical and scientific work are **approximate numbers**, having been determined by some *measurement*. Certain other numbers are **exact numbers**, having been determined by a *definition* or *counting process*.

EXAMPLE 6 Approximate numbers and exact numbers

If a voltage on a voltmeter is read as 116 V, the 116 is approximate. Another voltmeter might show the voltage as 115.7 V. However, the voltage cannot be determined *exactly*.

If a computer prints out the number of names on a list of 97, this 97 is exact. We know it is not 96 or 98. Since 97 was found from precise counting, it is exact.

By definition, 60 s = 1 min, and the 60 and the 1 are exact.

Significant digits are digits in a measurement or result that you can confidently estimate. That is to say, those digits that are not swamped by the error or uncertainty in the measurement are significant. The **accuracy** (or **trueness**) of a measurement refers to the number of significant digits it has.

The measurements 5.00 m and 5.000 m may not seem to be very different, but to a scientist, an engineer, or a technologist, they are *not* the same thing. The first measurement has been measured to the nearest centimetre and the second measurement to the nearest millimetre. The **precision** of a measurement is defined as the last decimal place to which the measurement is expressed, or the decimal place corresponding to the last measured (significant) digit in the measurement. For instance, 5.00 m has precision 0.01 m = 1 cm, and 5.000 m has precision 0.001 m = 1 mm. Therefore, the second measurement is *more precise* (it has a smaller precision). The concept of precision is important when finding the proper significant digits in a calculated result.

To determine the number of significant digits in a single value, the following rules should be applied:

- Start counting at the first nonzero digit, and stop counting once the precision of the measurement is reached.
- All nonzero digits *are* significant.
- Zeros to the left of the first nonzero digit are *not* significant.
- Zeros between nonzero digits *are* significant.
- Trailing zeros (to the right of the last nonzero digit) are generally *not* significant if they fall to the left of the decimal. A cedilla ($\tilde{0}$) must be used if any of the trailing zeros hold significance.
- Trailing zeros *are* significant if they fall to the right of the decimal.

EXAMPLE 7 Significant digits in a single measurement

Measurement	# Significant Digits	Level of Precision
23.0 m	3	The nearest tenth
55.001 cm	5	The nearest thousandth
0.0034 m	2	The nearest ten thousandth
120 cm	2	The nearest ten
0.0120 s	3	The nearest ten thousandth
125 $\tilde{0}$00 s	4	The nearest hundred

EXAMPLE 8 Accuracy and precision

(a) Suppose that an electric current is measured to be 0.31 A on one ammeter and 0.312 A on another ammeter. The measurement 0.312 A is measured to the nearest thousandth ampere, so it is more precise than 0.31 A, which is measured to the nearest hundredth ampere. 0.312 A is also more accurate, since it contains three significant digits, whereas 0.31 A contains only two.

(b) If a concrete driveway is measured to be 135 m long and 0.1 m thick, the measurement 0.1 m (measured to the nearest tenth metre) is more precise than the measurement 135 m (measured to the nearest metre). On the other hand, 135 m is more accurate, since it contains three significant digits, whereas 0.1 m contains only one.

EXAMPLE 9 Significant digits for measured quantities

For the following approximate numbers, specify the number of significant digits.

Number	34.7	0.0039	706.1	5.90	140$\tilde{0}$	960 000	0.0709	700.00	1.070
# Significant Digits	3	2	4	3	4	2	3	5	4

ROUNDING OFF MEASURED VALUES

The last significant digit of an approximate number is not exact. It has usually been determined by estimating or *rounding off*. However, it is not off by more than one-half of a unit in its place value.

There are several methods of rounding that are commonly used. In applied problems, the method of rounding is often related to the intention of the value itself. "Rounding up" or "rounding down" may be necessary depending on how the number will be used. For this reason, it is important to keep in mind that different industries will necessarily use different rounding rules.

■ Using graphing utilities such as calculators, computer algebra systems, or spreadsheet programs, it is possible to set the number of decimal places (to the right of the decimal point) to which results will be rounded off. Note that these devices round half up.

Throughout this text we will use the method of rounding commonly referred to as the **round half up** rule. In this rule, the decimal position immediately to the right of the last significant digit is observed. Then,

- If the *digit is 4 or less*, **round down** (leave the last digit as is, and discard the rest).
- If the *digit is 5 or more*, **round up** (increase the last significant digit by one unit and discard the rest).

For example, to round 168.2437 to the nearest hundredth, we observe that 3 is 4 or less, so we discard the 3 and 7 and round as 168.24. On the other hand, to round 179.8 to the nearest unit, we observe that 8 is greater than 5, so we increase the 179 to 180 and discard the 8, reporting the rounded value as 18$\tilde{0}$.

A method that avoids the slight statistical bias toward rounding up of the round half up rule is **unbiased rounding**, a method in which a value is rounded to the nearest *even* number whenever the last significant digit is a 5 followed by no digits other than 0. The variation is very minor compared with the round half up rule, and the techniques match most of the time.

EXAMPLE 10 Applying the round half up rule

Round the following values off to the stated number of significant digits:

Measured number	70 360	70 430	187.45	187.349	35.003
Significant digits	3	3	4	4	4
Rounded value	70 360 Round up to 70 400	70 430 Round down to 70 400	187.45 Round up* to 187.5	187.349 Round down to 187.3	35.003 Round down to 35.00

* In this case, unbiased rounding would round down to 187.4.

Practice Exercises

Round off each number to three significant digits.

1. 2015 **2.** 0.3004

OPERATIONS WITH APPROXIMATE NUMBERS

When performing operations with approximate numbers, we must express the result to an accuracy or precision that is valid, remembering that the last significant digit may be the result of rounding. Consider the following examples.

EXAMPLE 11 Adding or subtracting measured values

measured length	smallest possible length	largest possible length
16.3 m	16.25 m	16.34 m
0.927 m	0.9265 m	0.9274 m
17.227 m	17.1765 m	17.2674 m

A pipe is made in two sections. One is measured as 16.3 m long and the other as 0.927 m long. What is the total length of the two sections together?

It may appear that we simply add the numbers as shown at the left. However, both numbers are approximate, and adding the smallest possible values and the largest possible values, the result differs by 0.1 (17.2 and 17.3) when rounded off to tenths. Rounded off to hundredths (17.18 and 17.27), they do not agree at all since the tenths digits are different. Thus, we get a good approximation for the total length if it is rounded off to *tenths*, the precision of the least precise length, and it is written as 17.2 m.

EXAMPLE 12 Operations including multiplication and division

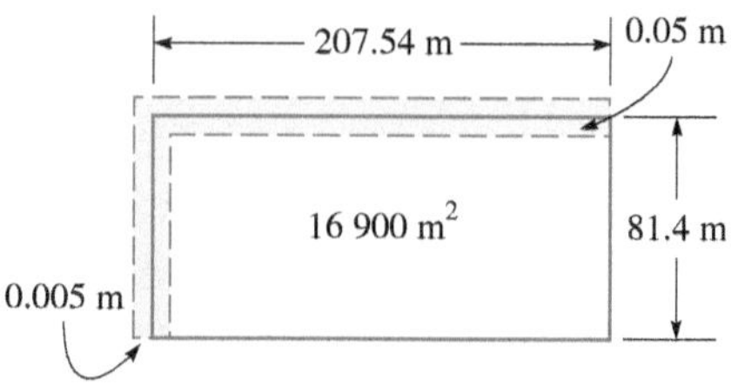

Fig. 1.10

We find the area of the rectangular piece of land in Fig. 1.10 by multiplying the length, 207.54 m, by the width, 81.4 m. Using a calculator, we find that $(207.54)(81.4) = 16\,893.756$. This apparently means the area is $16\,893.756\ \text{m}^2$.

However, the area should not be expressed with this accuracy. Since the length and width are both approximate, we have

$$(207.535\ \text{m})(81.35\ \text{m}) = 16\,882.972\,25\ \text{m}^2 \quad \text{least possible area}$$
$$(207.544\ \text{m})(81.44\ \text{m}) = 16\,902.383\,36\ \text{m}^2 \quad \text{greatest possible area}$$

These values agree when rounded off to three significant digits ($16\,900\ \text{m}^2$) but do not agree when rounded off to a greater accuracy. Thus, we conclude that the result is accurate only to *three* significant digits, the accuracy of the least accurate measurement, and that the area is written as $16\,900\ \text{m}^2$.

Following are the rules used in expressing the result when we perform basic operations on approximate numbers. They are based on reasoning similar to that shown in Examples 11 and 12.

LEARNING TIP

Always express the result of a calculation with the proper accuracy or precision. When using a calculator, if additional digits are displayed, round off the **final** ***result*** (do not round off in any of the intermediate steps).

Operations with Approximate Numbers

1. *When approximate numbers are added or subtracted, the result is expressed with the precision of the least precise number.*
2. *When approximate numbers are multiplied or divided, the result is expressed with the accuracy of the least accurate number.*
3. *When the root of an approximate number is found, the result is expressed with the accuracy of the number.*
4. *When approximate numbers and exact numbers are involved, the accuracy of the result is limited only by the approximate numbers.*

EXAMPLE 13 Adding approximate numbers

■ When rounding off a number, it may seem difficult to discard the extra digits. However, if you keep those digits, you show a number with too great an accuracy, and it is incorrect to do so.

Find the sum of the approximate numbers 73.2, 8.0627, and 93.57.

Showing the addition in the standard way and using a calculator, we have

$$\begin{array}{r} 73.2 \\ 8.0627 \\ 93.57 \\ \hline 174.8327 \end{array}$$

73.2 ← least precise number (expressed to tenths)

174.8327 ← final display must be rounded to tenths

Therefore, the sum of these approximate numbers is 174.8.

EXAMPLE 14 Combined operations

In finding the product of the approximate numbers 2.4832 and 30.5 on a calculator, the final display shows 75.7376. However, since 30.5 has only three significant digits, the product is 75.7.

Similarly, consider the calculation $38.3 - 12.9(-3.58) = 84.482$. If these numbers are approximate, we must round off the result to tenths, which means the sum is 84.5. We see that when there is a combination of operations, we must examine the individual steps of the calculation and determine how many significant digits can carry through to the final result.

Practice Exercise

Evaluate using a calculator.

3. $40.5 + \dfrac{3275}{-60.041}$ (Numbers are approximate.)

EXAMPLE 15 Operations with exact numbers and approximate numbers

Using the exact number 600 and the approximate number 2.7, we express the result to tenths if the numbers are added or subtracted. If they are multiplied or divided, we express the result to two significant digits. Since 600 is exact, the accuracy of the result depends only on the approximate number 2.7.

$$600 + 2.7 = 602.7 \qquad 600 - 2.7 = 597.3$$
$$600 \times 2.7 = 1600 \qquad 600 \div 2.7 = 220$$

You should *make a rough estimate* of the result when using a calculator. An estimation may prevent accepting an incorrect result after using an incorrect calculator sequence, particularly if the calculator result is far from the estimated value.

LEARNING TIP

A note regarding the equal sign (=) is in order. We will use it for its defined meaning of "equals exactly" and when the result is an approximate number that has been properly rounded off. Although $\sqrt{27.8} \approx 5.27$, where ≈ means "equals approximately," we write $\sqrt{27.8} = 5.27$, since 5.27 has been properly rounded off.

EXAMPLE 16 Using estimation as a tool for detecting errors

In Example 14, we found that

$$38.3 - 12.9(-3.58) = 84.482 \quad \text{using exact numbers}$$

When using the calculator, if we forgot to make 3.58 negative, the display would be −7.882, or if we incorrectly entered 38.3 as 83.3, the display would be 129.482.

However, if we estimate the result as

$$40 - 10(-4) = 80$$

we know that a result of −7.882 or 129.482 cannot be correct.

When estimating, we can often use one-significant-digit approximations. If the calculator result is far from the estimate, we should do the calculation again.

EXERCISES 1.3

In Exercises 1–4, make the given changes in the indicated examples of this section, and then solve the given problems.

1. In Example 9, change 0.0039 (the second number discussed) to 0.3900. Is there any change in the conclusion?
2. In the last column of Example 10, change 35.003 to 35.303 and then find the result.
3. In the first paragraph of Example 14, change 2.4832 to 2.483 and then find the result.
4. In Example 16, change 12.9 to 21.9 and then find the estimated value.

In Exercises 5–8, give the symbol and the meaning for the given unit.

5. megahertz 6. kilowatt 7. millimetre 8. picosecond

In Exercises 9–12, give the name and the meaning for the units whose symbols are given.

9. kV 10. GΩ 11. mA 12. pF

In Exercises 13–48, make the indicated changes in units.

13. 1 km to centimetres.
14. 1 kg to milligrams.
15. 20 s to megaseconds.
16. 800 Pa to kilopascals.
17. 250 mm^2 to square metres.
18. 1.75 m^2 to square centimetres.
19. 80.0 m^3 to litres.
20. 0.125 L to millilitres.
21. 45.0 m/s to centimetres per second.
22. 1.32 km/h to metres per second.
23. 9.80 m/s^2 to centimetres per minute squared.
24. 5.10 g/cm^3 to kilograms per cubic metre.
25. 25 h to milliseconds.
26. 5.25 mV to watts per ampere.
27. 15.0 μF to millicoulombs per volt.
28. 42 pounds to kilograms.
29. 27 kilometres to miles.
30. 10 square feet to metres squared.
31. 200 kg/m^2 to lb/ft^2.
32. Determine how many metres light travels in one year.
33. Determine the speed (in km/h) of the earth moving around the sun. Assume it is a circular path of radius 150 000 000 km.
34. At sea level, atmospheric pressure is about 101 300 Pa. How many kilopascals is this?
35. A car's gasoline tank holds 56 L. What is this capacity in cubic centimetres?
36. A hockey puck has a mass of about 0.160 kg. What is its mass in milligrams?
37. The velocity of some seismic waves is 6800 m/s. What is this velocity in kilometres per hour?
38. The memory of a 1985 computer was 64 kB (B is the symbol for byte), and the memory of a 2012 computer is 1.50 TB. How many times greater is the memory of the 2012 computer?
39. The recorded surface area of a DVD is 112 cm^2. What is this area in square metres?
40. A solar panel can generate 0.024 MW · h each day. Convert this to joules.
41. The density of water is 1000 kg/m^3. Change this to grams per litre.
42. Water flows from a kitchen faucet at the rate of 8500 mL/min. What is this rate in cubic metres per second?
43. The speed of sound is about 332 m/s. Change this speed to kilometres per hour.
44. Fifteen grams of a medication are to be dissolved in 0.060 L of water. Express this concentration in milligrams per decilitre.
45. The earth's surface receives energy from the sun at the rate of 1.35 kW/m^2. Reduce this to joules per second per square centimetre.
46. The moon travels about 2 400 000 km in about 28 d in one rotation about the earth. Express its velocity in metres per second.
47. A typical electric current density in a wire is 1.2×10^6 A/m^2. Express this in milliamperes per square centimetre.
48. A certain car travels 24 km on 2.0 L of gas. Express the fuel consumption in litres per 100 kilometres.

In Exercises 49–52, determine whether the given numbers are approximate or exact.

49. A car with eight cylinders travels at 55 km/h.

50. A computer chip 0.002 mm thick is priced at \$7.50.

51. In 24 h there are 1440 min.

52. A calculator has 50 keys, and its battery lasted for 50 h of use.

In Exercises 53–58, determine the number of significant digits in each of the given approximate numbers.

53. 107; 3004; 1040

54. 3600; 730; 2055

55. 6.80; 6.08; 0.068

56. 0.8735; 0.0075; 0.0305

57. 3000; 3000.1; 3000.10

58. 1.00; 0.01; 0.0100

In Exercises 59–64, determine which of the pair of approximate numbers is (a) more precise and (b) more accurate.

59. 30.8; 0.01

60. 0.041; 7.673

61. 0.1; 78.0

62. 7040; 0.004

63. 7000; 0.004

64. 50.060; 8.914

In Exercises 65–72, round off the given approximate numbers (a) to three significant digits and (b) to two significant digits.

65. 4.936

66. 80.53

67. −50.893

68. 7.005

69. 9549

70. 30.96

71. 0.9449

72. 0.9999

In Exercises 73–80, assume that all numbers are approximate. (a) Estimate the result and (b) perform the indicated operations on a calculator and compare with the estimate.

73. $12.78 + 1.0495 - 1.633$

74. $3.64(17.06)$

75. $8.75 + (1.2)(3.84)$

76. $28 - \dfrac{20.955}{2.2}$

77. $\dfrac{8.75(15.32)}{8.75 + 15.32}$

78. $\dfrac{0.693\,78 + 0.049\,97}{257.4 \times 3.216}$

79. $4.52 - \dfrac{2.056(309.6)}{395.2}$

80. $\dfrac{1.00}{0.5926} + \dfrac{3.6957}{2.935 - 1.054}$

In Exercises 81–84, perform the indicated operations. The first number is approximate, and the second number is exact.

81. $0.9788 + 14.9$

82. $17.311 - 22.98$

83. $-3.142(65)$

84. $8.62 \div 1728$

In Exercises 85–88, answer the given questions.

85. The manual for a heart monitor lists the frequency of the ultrasound wave as 2.75 MHz. What are the least possible and the greatest possible frequencies?

86. A car manufacturer states that the engine displacement for a certain model is 2400 cm^3. What should be the least possible and greatest possible displacements?

87. A flash of lightning struck a tower 5.23 km from a person. The thunder was heard 15 s later. The person calculated the speed of sound and reported it as 348.7 m/s. What is wrong with this conclusion?

88. A technician records 4.4 s as the time for a robot arm to swing from the extreme left to the extreme right, 2.72 s as the time for the return swing, and 1.68 s as the difference in these times. What is wrong with this conclusion?

In Exercises 89–104, perform the calculations on a calculator.

89. Evaluate: (a) $2.2 + 3.8 \times 4.5$ (b) $(2.2 + 3.8) \times 4.5$

90. Evaluate: (a) $6.03 \div 2.25 + 1.77$ (b) $6.03 \div (2.25 + 1.77)$

91. Evaluate: (a) $2 + 0$ (b) $2 - 0$ (c) $0 - 2$ (d) 2×0 (e) $2 \div 0$ Compare with operations with zero in Section 1.2.

92. Evaluate: (a) $2 \div 0.0001$ and $2 \div 0$ (b) $0.0001 \div 0.0001$ and $0 \div 0$ (c) Explain how the errors differ.

93. Enter a positive integer x (five or six digits is suggested) and then rearrange the same digits to form another integer y. Evaluate $(x - y) \div 9$. What type of number is the result?

94. Enter the digits in the order 9, 8, 7, 6, 5, 4, 3, 2, 1, 0, using between them any of the operations $(+, -, \times, \div)$ that will lead to a result of 100.

95. Show that π is not equal exactly to (a) 3.1416, or (b) 22/7.

96. At some point in the decimal equivalent of a rational number, some sequence of digits will start repeating endlessly. An irrational number never has an endlessly repeating sequence of digits. Find the decimal equivalents of (a) 8/33 and (b) π. Note the repetition for 8/33 and that no such repetition occurs for π.

97. Following Exercise 96, show that the decimal equivalents of the following fractions indicate they are rational: (a) 1/3 (b) 5/11 (c) 2/5. What is the repeating part of the decimal in (c)?

98. Following Exercise 96, show that the decimal equivalent of the fraction 124/990 indicates that it is rational. Why is the last digit different?

99. In three successive days, a home solar system produced 32.4 MJ, 26.704 MJ, and 36.23 MJ of energy. What was the total energy produced in these three days?

100. Two jets flew at 938 km/h and 1450 km/h, respectively. How much faster was the second jet?

101. If 1 K of computer memory has 1024 bytes, how many bytes are there in 256 K of memory? (All numbers are exact.)

102. Find the voltage in a certain electric circuit by multiplying the sum of the resistances 15.2 Ω, 5.64 Ω, and 101.23 Ω by the current 3.55 A.

103. The percent of alcohol in a certain car engine coolant is found by performing the calculation $\dfrac{100(40.63 + 52.96)}{105.30 + 52.96}$. Find this percent of alcohol. The number 100 is exact.

104. The tension (in N) in a pulley cable lifting a certain crate was found by calculating the value of $\dfrac{50.45(9.80)}{1 + 100.9 \div 23}$, where the 1 is exact. Calculate the tension.

Answers to Practice Exercises

1. 2020 **2.** 0.300 **3.** −14.0

1.4 Exponents

In mathematics and its applications, we often have a number multiplied by itself several times. To show this type of product, we use the notation a^n, where a is the number and n is the number of times it appears. *In the expression a^n, the number a is called the* **base**, *and n is called the* **exponent**; in words, a^n *is read as "the* ***n*th power of *a*.**"

EXAMPLE 1 Meaning of exponents

(a) $4 \times 4 \times 4 \times 4 \times 4 = 4^5$ — the fifth power of 4

(b) $(-2)(-2)(-2)(-2) = (-2)^4$ — the fourth power of −2

(c) $a \times a = a^2$ — the second power of a, called "a squared"

(d) $\left(\frac{1}{5}\right)\left(\frac{1}{5}\right)\left(\frac{1}{5}\right) = \left(\frac{1}{5}\right)^3$ — the third power of $\frac{1}{5}$, called "$\frac{1}{5}$ cubed"

We now state the rules for evaluating expressions with exponents. Here m and n are positive integers.

Rules of Exponents			EXAMPLE 2
Product Rule:	$a^m \times a^n = a^{m+n}$	**(1.8)**	**(a)** $2^4 \times 2^3 = 2^{4+3} = 2^7 = 128$ **(b)** $(r^4)(r^3) = r^{4+3} = r^7$
Quotient Rule:	$\frac{a^m}{a^n} = a^{m-n} \quad (a \neq 0)$	**(1.9)**	**(c)** $\frac{3^6}{3^4} = 3^{6-4} = 3^2 = 9$ **(d)** $\frac{m^6}{m^4} = m^{6-4} = m^2$
Power of a Power Rule:	$(a^m)^n = a^{mn}$	**(1.10)**	**(e)** $(10^4)^3 = 10^{4 \cdot 3} = 10^{12}$ **(f)** $(k^3)^2 = k^{3 \cdot 2} = k^6$
Power of a Product Rule:	$(ab)^n = a^n b^n$	**(1.11)**	**(g)** $(2k)^3 = 2^3k^3 = 8k^3$
Power of a Quotient Rule:	$\left(\frac{a}{b}\right)^n = \frac{a^n}{b^n} \quad (\text{if } b \neq 0)$	**(1.12)**	**(h)** $\left(\frac{x}{5}\right)^3 = \frac{x^3}{5^3} = \frac{x^3}{125}$
Zero Exponent	$a^0 = 1 \quad (a \neq 0)$	**(1.13)**	**(i)** $7^0 = 1$ **(j)** $(-2)^0 = 1$
Negative Exponent	$a^{-n} = \frac{1}{a^n} \quad (a \neq 0)$	**(1.14)**	**(k)** $2^{-4} = \frac{1}{2^4} = \frac{1}{16}$ **(l)** $a^{-9} = \frac{1}{a^9}$

EXAMPLE 3 Illustrating the power and quotient rule

■ In a^3, which equals $a \times a \times a$, each a is called a factor. A more general definition of factor is given in Section 1.7.

(a) Using Eq. (1.8):

add exponents

$$a^3 \times a^5 = a^{3+5} = a^8$$

(b) Using the meaning of exponents:

(3 factors of a)(5 factors of a) → 8 factors of a

$$a^3 \times a^5 = (a \times a \times a)(a \times a \times a \times a \times a) = a^8$$

■ Here we are using the fact that a (not zero) divided by itself equals 1, or $a/a = 1$.

(c) Using Eq. (1.9):

$$\frac{a^5}{a^3} = a^{5-3} = a^2$$

(d) Using the meaning of exponents:

$$\frac{a^5}{a^3} = \frac{\overset{1}{\cancel{a}} \times \overset{1}{\cancel{a}} \times \overset{1}{\cancel{a}} \times a \times a}{\underset{1}{\cancel{a}} \times \underset{1}{\cancel{a}} \times \underset{1}{\cancel{a}}} = a^2$$

EXAMPLE 4 Illustrating the power rules

(a) Using Eq. (1.10):

multiply exponents

$$(a^5)^3 = a^{5(3)} = a^{15}$$

(b) Using the meaning of exponents:

$$(a^5)^3 = (a^5)(a^5)(a^5) = a^{5+5+5} = a^{15}$$

(c) Using Eq. (1.11):

$$(ab)^3 = a^3b^3$$

(d) Using the meaning of exponents:

$$(ab)^3 = (ab)(ab)(ab) = a^3b^3$$

(e) Using Eq. (1.12):

$$\left(\frac{a}{b}\right)^3 = \frac{a^3}{b^3}$$

(f) Using the meaning of exponents:

$$\left(\frac{a}{b}\right)^3 = \left(\frac{a}{b}\right)\left(\frac{a}{b}\right)\left(\frac{a}{b}\right) = \frac{a^3}{b^3}$$

LEARNING TIP

When an expression involves a product or a quotient of different bases, ***only exponents of the same base may be combined.***

EXAMPLE 5 Zero as an exponent

(a) $5^0 = 1$ **(b)** $(-3)^0 = 1$ **(c)** $-(-3)^0 = -1$ **(d)** $(2x)^0 = 1$

(e) $(ax + b)^0 = 1$ **(f)** $(a^2b^0c)^2 = a^4c^2$ ($b^0 = 1$) **(g)** $2t^0 = 2(1) = 2$

We note in illustration (g) that *only t is raised to the zero power.* If the quantity $2t$ were raised to the zero power, it would be written as $(2t)^0$.

LEARNING TIP

Unless otherwise stated, *final answers should be written with a positive exponent.* Nevertheless, the negative exponent form can be very useful in some operations that we will use later.

EXAMPLE 6 Negative exponents

(a) $3^{-1} = \frac{1}{3}$ **(b)** $4^{-2} = \frac{1}{4^2} = \frac{1}{16}$ **(c)** $\frac{1}{a^{-3}} = a^3$ (change signs of exponents)

(d) $\left(\frac{a^3t}{b^2x}\right)^{-2} = \frac{(a^3t)^{-2}}{(b^2x)^{-2}} = \frac{(b^2x)^2}{(a^3t)^2} = \frac{b^4x^2}{a^6t^2}$ **(e)** $3x^{-1} = 3\left(\frac{1}{x}\right) = \frac{3}{x}$

Practice Exercises

Simplify:

1. $\frac{-7^0}{c^{-3}}$ 2. $\frac{(3x)^{-1}}{2a^{-2}}$

COMMON ERROR

In the statement $3x^{-1} = \frac{3}{x}$, the exponent only applies to x, not to 3. Do not confuse this with the statement $(3x)^{-1} = 3^{-1}x^{-1} = \frac{1}{3x}$.

EXAMPLE 7 Other illustrations of exponents

(a) $(-x^2)^3 = [(-1)x^2]^3 = (-1)^3(x^2)^3 = -x^6$

(b) $ax^2(ax)^3 = ax^2(a^3x^3) = a^4x^5$ (exponent of 1; add exponents of a; add exponents of x)

(c) $-3^2 = (-1)(3^2) = -9$

Note that the negative sign represents a multiple of -1 and is not affected by the exponent.

(d) $\dfrac{(3 \times 2)^4}{(3 \times 5)^3} = \dfrac{3^4 2^4}{3^3 5^3} = \dfrac{3 \times 2^4}{5^3}$

(e) $\dfrac{(ry^3)^2}{r(y^2)^4} = \dfrac{r^2y^6}{ry^8} = \dfrac{r}{y^2}$

Practice Exercises

Use Eqs. (1.8)–(1.12) to simplify the given expressions.

3. $ax^3(-ax)^2$ **4.** $\dfrac{(2c)^5}{(3cd)^2}$

COMMON ERROR Note from Example 7(b) that ax^2 *means a times the square of x and does not mean* a^2x^2, whereas $(ax)^3$ *does* mean a^3x^3.

EXAMPLE 8 Exponents—deflection of a beam

In the analysis of the deflection of a beam (the amount it bends), the expression that follows is simplified as shown.

$$\frac{1}{2}\left(\frac{PL}{4EI}\right)\left(\frac{2}{3}\right)\left(\frac{L}{2}\right)^2 = \frac{1}{2}\left(\frac{PL}{4EI}\right)\left(\frac{2}{3}\right)\left(\frac{L^2}{2^2}\right)$$

$$= \frac{\overset{1}{\cancel{2}}PL(L^2)}{\underset{1}{\cancel{2}}(3)(4)(4)EI} = \frac{PL^3}{48EI}$$

L is the length of the beam, and P is the force applied to it. E and I are constants related to the beam. In *simplifying* this expression, we combined exponents of L and divided out the 2 that was in the numerator and in the denominator.

ORDER OF OPERATIONS

Often, exponent rules will be used along with other numerical operators. Simplifying correctly relies heavily on proper use of the order of operations. As we saw in Section 1.2, exponents are evaluated after brackets and before division and multiplication steps. For convenience, we repeat the order of operations here.

To simplify

B **1. Bracketed Values:** Perform operations with specific groupings first inside of parentheses ( ), brackets [ ], or absolute values | |.
- If bracketed values occur inside one another, begin with the innermost bracketed value when simplifying.
- Otherwise, evaluate from left to right.

O **2. Evaluate Operators:** Operators can include exponents (discussed in this section) and roots/radicals (discussed in Section 1.6).

D/M **3. Perform Division and Multiplication Steps** (from left to right).

A/S **4. Perform Addition and Subtraction Steps** (from left to right).

■ The use of exponents is taken up in more detail in Chapter 11.

EXAMPLE 9 Order of operations

(a) $3\left(\dfrac{a^3b^6}{bx}\right)^{-2} = 3\left(\dfrac{a^3b^{6-1}}{x}\right)^{-2}$

$= 3 \cdot \dfrac{a^{3(-2)}b^{(5)(-2)}}{x^{(1)(-2)}}$

$= 3 \cdot \dfrac{a^{-6}b^{-10}}{x^{-2}}$

$= 3 \cdot \dfrac{1}{a^6} \cdot \dfrac{1}{b^{10}}x^2$

$= \dfrac{3x^2}{a^6b^{10}}$

Here we simplified inside the bracket first (**B**), using the quotient rule for exponents. We then used the power rules to raise the expression to the power of -2 and change to positive exponents (**O**). Finally, the multiplication was performed (**M**).

(b) $8 - (-1)^2 - 2(-3)^2 = 8 - 1 - 2(9)$

$= 8 - 1 - 18$

$= -11$

Since there were no groupings (**B**), we first squared -1 and -3 (note that the exponent (**O**) takes priority over multiplication, so *we did not change the sign of* -1 *before squaring*). Next we found the product in the last term (**M**), and finally subtracted (**S**).

EXAMPLE 10 Even and odd powers

Using the meaning of a power of a number, we have

$$(-2)^2 = (-2)(-2) = 4 \qquad (-2)^3 = (-2)(-2)(-2) = -8$$
$$(-2)^4 = 16 \qquad (-2)^5 = -32 \qquad (-2)^6 = 64 \qquad (-2)^7 = -128$$

EVALUATING ALGEBRAIC EXPRESSIONS

An algebraic expression is **evaluated** *by* **substituting** *given numerical values for the letters in the expression and calculating the result.* On a calculator, the [x^2] key is used to square numbers, and the [∧] or [x^y] key is used for other powers.

When entering information on a calculator, pay careful attention to key in the information in a manner that will prompt the correct order of operations.

EXAMPLE 11 Evaluating expressions using a calculator

(a) The expression $\frac{12}{3+9}$ is evaluated with the sequence 12 [÷] [( ] 3 [+] 9 [ )]. Using brackets ensures that the values in the denominator will be added before the division step occurs.

(b) To calculate the value of $20 \times 6 + 200/5 - 3^4$, we use the key sequence

20 [×] 6 [+] 200 [÷] 5 [−] 3 [∧] 4

with a result of 79.

```
4.9*4.2²
              86.436
```

Fig. 1.11

■ When less than half of a calculator screen is needed, the figure for that calculator screen will show only a partial screen.

EXAMPLE 12 Evaluating an expression—free-fall distance

The distance (in m) that an object falls in 4.2 s is found by substituting 4.2 for t in the expression $4.90t^2$ as shown below:

$$4.90(4.2)^2 = 86 \text{ m}$$

The calculator result from Fig. 1.11 has been rounded off to two significant digits, the accuracy of 4.2.

EXAMPLE 13 Evaluating an expression—length of a wire

A wire made of a special alloy has length L (in m) given by $L = a + 0.0115T^3$, where T (in °C) is the temperature (between −4°C and 4°C). To find the wire length L for $a = 8.38$ m and $T = -2.87$°C, we substitute these values to get

$$L = 8.38 + 0.0115(-2.87)^3$$
$$= 8.11 \text{ m}$$

EXERCISES 1.4

In Exercises 1–4, make the given changes in the indicated examples of this section, and then simplify the resulting expression.

1. In Example 2(g), change $(2k)^3$ to $(3k)^2$.
2. In Example 5(d), change $(2x)^0$ to $2x^0$.
3. In Example 6(d), interchange the a^3 and b^2.
4. In Example 9(b), change $(-1)^2$ to $(-1)^3$.

In Exercises 5–48, simplify the given expressions. Express results with positive exponents only.

5. x^3x^4
6. y^2y^7
7. $2b^4b^2$
8. $3k^5k$
9. $\dfrac{m^5}{m^3}$
10. $\dfrac{2x^6}{x}$
11. $\dfrac{-n^5}{7n^9}$
12. $\dfrac{3s}{s^4}$
13. $(P^2)^4$
14. $(x^8)^3$
15. $(2\pi)^3$
16. $(ax)^5$
17. $(aT^2)^{30}$
18. $(3r^2)^3$
19. $\left(\dfrac{2}{b}\right)^3$
20. $\left(\dfrac{F}{t}\right)^{20}$
21. $\left(\dfrac{x^2}{2}\right)^4$
22. $\left(\dfrac{3}{n^3}\right)^3$
23. $(8a)^0$
24. $-v^0$
25. $-3x^0$
26. $-(-2)^0$
27. 6^{-1}
28. $-w^{-5}$
29. $\dfrac{1}{R^{-2}}$
30. $\dfrac{1}{-t^{-48}}$
31. $(-t^2)^7$
32. $(-y^3)^5$
33. $-\dfrac{L^{-3}}{L^{-5}}$
34. $2i^{40}i^{-70}$
35. $\dfrac{2v^4}{(2v)^4}$
36. $\dfrac{x^2x^3}{(x^2)^3}$
37. $\dfrac{(n^2)^4}{(n^4)^2}$
38. $\dfrac{(3t)^{-1}}{3t^{-1}}$
39. $(\pi^0x^2a^{-1})^{-1}$
40. $(3m^{-2}n^4)^{-2}$
41. $(-8g^{-1}s^3)^2$
42. $ax^{-2}(-a^2x)^3$
43. $\left(\dfrac{4x^{-1}}{a^{-1}}\right)^{-3}$
44. $\left(\dfrac{2b^2}{y^5}\right)^{-2}$
45. $\dfrac{15n^2T^5}{3n^{-1}T^6}$
46. $\dfrac{(nRT^{-2})^{32}}{R^{-2}T^{32}}$
47. $\left(\dfrac{16x^2y^{-2}}{xy^{-1}}\right)^{-4}$
48. $\left(\dfrac{7x^{-3}y^5}{x^2y^{-2}}\right)^{-2}$

In Exercises 49–56, evaluate the given expressions. In Exercises 51–56, all numbers are approximate.

49. $7(-4) - (-5)^2$
50. $6 + (-2)^5 - (-2)(8)$
51. $-(-26.5)^2 - (-9.85)^3$
52. $-0.711^2 - (-0.809)^6$
53. $\dfrac{3.07(-1.86)}{(-1.86)^4 + 1.596}$
54. $\dfrac{15.66^2 - (-4.017)^4}{1.044(-3.68)}$
55. $2.38(-10.7)^2 - \dfrac{254}{1.17^3}$
56. $0.513(-2.778) - (-3.67)^3 + 0.889^4/(1.89 - 1.09^2)$

In Exercises 57–68, perform the indicated operations.

57. Does $\left(\dfrac{1}{x^{-1}}\right)^{-1}$ represent the reciprocal of x?
58. Does $\left(\dfrac{0.2 - 5^{-1}}{10^{-2}}\right)^0$ equal 1? Explain.
59. If $a^3 = 5$, then what does a^{12} equal?
60. Is $a^{-2} < a^{-1}$ for any negative value of a? Explain.
61. If a is a positive integer, simplify $(x^a \cdot x^{-a})^5$.
62. If a and b are positive integers, simplify $(y^{a-b} \cdot y^{a+b})^2$.
63. In developing the "big bang" theory of the origin of the universe, the expression $(kT/(hc))^3(GkThc)^2c$ arises. Simplify this expression.
64. In studying planetary motion, the expression $(GmM)(mr)^{-1}(r^{-2})$ arises. Simplify this expression.
65. In designing a cam for a pump, the expression $\pi\left(\dfrac{r}{2}\right)^3\left(\dfrac{4}{3\pi r^2}\right)$ is used. Simplify this expression.
66. For a certain integrated electric circuit, it is necessary to simplify the expression $\dfrac{gM}{2\pi fC(2\pi fM)^2}$. Perform this simplification.
67. If \$2500 is invested at 4.2% interest, compounded quarterly, the amount in the account after six years is $2500(1 + 0.042/4)^{24}$. Calculate this amount. (The values 1, 4, and 24 are exact.)

68. In designing a building, it was determined that the forces acting on an I-beam would deflect the beam an amount (in cm) given by $\frac{x(1000 - 20x^2 + x^3)}{1850}$, where x is the distance (in m) from one end of the beam. Find the deflection for $x = 6.85$ m. (The 1000 and 20 are exact.)

Answers to Practice Exercises

1. $-c^3$ **2.** $\frac{a^2}{6x}$ **3.** a^3x^5 **4.** $\frac{2^5c^3}{3^2d^2} = \frac{32c^3}{9d^2}$

1.5 Scientific Notation

In technical and scientific work, we often encounter numbers that are either very large or very small. Examples of such numbers are:

(a) WiFi standards allow for transfer speeds of 600 000 000 bits per second;
(b) The mass of the earth is about 6 000 000 000 000 000 000 000 000 kg;
(c) One molecule of water has a mass of 0.000 000 000 000 000 000 000 002 99 grams.

Writing numbers such as these is inconvenient in ordinary notation. Instead, alternative notations called **scientific notation** and **engineering notation** are used to record values in ways that are simpler to write and interpret.

Numbers in both scientific and engineering notations are expressed in the following form:

$$P \times 10^k$$

In scientific notation, the coefficient P satisfies $1 \leq P < 10$, and k can be any integer. In this case, P helps specify the number of *significant digits* in the measurement. In engineering notation, the coefficient P is such that $1 \leq P < 1000$, and k must be a multiple of three.

Scientific notation provides a practical way to write very large or very small numbers in a compact form, and to handle calculations with them easily using the laws of exponents. With *engineering notation* we accomplish the same goals, while at the same time setting up the numbers so that they can be easily switched between common measurements in unit systems.

To write a number in scientific notation, move the decimal point so that there is only one nonzero digit before the decimal point. Multiply by a power of 10 (k) equal to the number of places the decimal point has been moved (k is positive if the decimal point is moved to the left and negative if moved to the right). To express a number in scientific notation in ordinary notation, this procedure is reversed.

EXAMPLE 1 Expressing ordinary numbers in scientific notation

Rewrite the following values in scientific notation: **(a)** 34 000; **(b)** 6.82; **(c)** 0.005 03.

(a) $34\,000 = 3.4 \times 10^4$ (two significant digits; the decimal moved four places to the left)

(b) $6.82 = 6.82 \times 10^0$ (three significant digits; the decimal moved zero places)

(c) $0.005\,03 = 5.03 \times 10^{-3}$ (three significant digits; the decimal moved three places to the right)

EXAMPLE 2 Changing scientific notation to ordinary notation

Rewrite the following values in ordinary notation: **(a)** 5.83×10^6; **(b)** 8.06×10^{-3}.

(a) $5.83 \times 10^6 = 5\,830\,000$ (the decimal moves six places to the right, adding trailing zeros to place it)

(b) $8.06 \times 10^{-3} = 0.008\,06$ (the decimal moves three places to the left, adding leading zeros to place it)

Practice Exercises

1. Change 2.35×10^{-3} to ordinary notation.

2. Change 235 to scientific notation.

To write a number in engineering notation, the decimal point is moved *three spaces at a time* until you have a number from 1 to 999. As before, the number of positions moved is reflected in the power of 10.

EXAMPLE 3 Expressing ordinary numbers in engineering notation

Rewrite the following values in engineering notation: **(a)** 19 680 000 000; **(b)** 0.000 45.

(a) $19\,680\,000\,000 = 19.68 \times 10^9$ (the decimal moved nine places to the left, in three groups of three)

(b) $0.000\,45 = 450 \times 10^{-6}$ (the decimal moved six places to the right, in two groups of three, adding a trailing zero to place it)

Practice Exercises

3. Write 0.000 053 5 in engineering notation.

4. Write 9.1×10^{-8} in engineering notation.

EXAMPLE 4 Scientific notation in calculations—processing rate

The processing rate of a computer processing 803 000 bytes of data in 0.000 005 25 s is

$$\frac{803\,000}{0.000\,005\,25} = \frac{8.03 \times 10^5}{5.25 \times 10^{-6}} = \left(\frac{8.03}{5.25}\right) \times 10^{11} = 1.53 \times 10^{11} \text{ bytes/s}$$

$5 - (-6) = 11$

As shown, it is proper to leave the result *(rounded off)* in scientific notation. This method is useful when using a calculator and then estimating the result. In this case, the estimate is $(8 \times 10^5) \div (5 \times 10^{-6}) = 1.6 \times 10^{11}$.

EXAMPLE 5 Scientific notation and significant digits—gravity

In determining the gravitational force between two stars $75\tilde{0}$ 000 000 000 km apart, it is necessary to evaluate $75\tilde{0}\,000\,000\,000^2$. We write

$$75\tilde{0}\,000\,000\,000^2 = (7.50 \times 10^{11})^2 = 7.50^2 \times 10^{2\times 11} = 56.3 \times 10^{22}$$

Since 56.3 is not between 1 and 10, we can write this result in scientific notation with the correct number of significant digits as

$$56.3 \times 10^{22} = (5.63 \times 10)(10^{22}) = 5.63 \times 10^{23}$$

■ The number 7.50×10^{11} cannot be entered on a scientific calculator in the form 750 000 000 000. It could be entered in this form on a graphing utility.

COMMON ERROR When using scientific notation for SI measurements, powers of 10 might end up being mixed with SI prefixes. Generally, try to avoid mixing SI prefixes with scientific notation.

EXAMPLE 6 Scientific notation with SI prefixes

A measurement of a small distance is quoted in units of km as 0.000 045 km. Convert this value to scientific notation and with a single SI prefix.

We begin by writing the value in scientific notation, then eliminating the prefix "kilo" by converting to metres. From the exponent of the result we can identify the correct final prefix.

$$\begin{aligned} 0.000\,045 \text{ km} &= 4.5 \times 10^{-5} \text{ km} \\ &= 4.5 \times 10^{-5} \cancel{\text{km}} \times \frac{10^3 \text{ m}}{1 \cancel{\text{km}}} \\ &= 4.5 \times 10^{-2} \text{ m} \\ &= 4.5 \text{ cm} \end{aligned}$$

Engineering notation, when combined with SI prefixes, allows us to replace the power of 10 notation altogether, as can be seen in the following example.

EXAMPLE 7 Engineering notation and SI prefixes

Write each of the following quantities in engineering notation and replace the power of 10 with the corresponding SI prefix.

(a) $0.000\,009\text{ F} = 9 \times 10^{-6}\text{ F} = 9\ \mu\text{F}$

(b) $62\,900\text{ W} = 62.9 \times 10^{3}\text{ W} = 62.9\text{ kW}$

We can enter numbers in scientific notation on a calculator, as well as have the calculator give results automatically in scientific notation. See the next example.

EXAMPLE 8 Scientific notation on a calculator—laser wavelength

The wavelength λ (in m) of the light in a red laser beam can be found from the following calculation. Note the significant digits in the numerator.

$$\lambda = \frac{3\,000\,000}{4\,740\,000\,000\,000} = \frac{3.00 \times 10^{6}}{4.74 \times 10^{12}} = 6.33 \times 10^{-7}\text{ m}$$

The key sequence is 3 EE 6 ÷ 4.74 EE 12 ENTER. See Fig. 1.12.

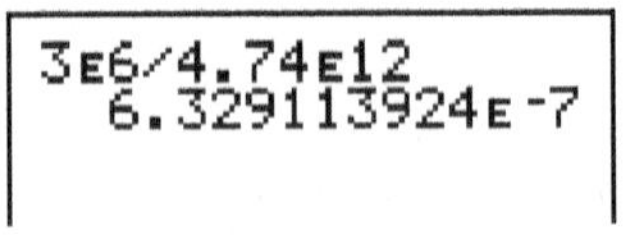

Fig. 1.12

EXERCISES 1.5

In Exercises 1 and 2, make the given changes in the indicated examples of this section and then rewrite the number as directed.

1. In Example 2(b), change the exponent -3 to 3 and then write the number in ordinary notation.
2. In Example 5, change the exponent 2 to -1 and then write the result in scientific notation.

In Exercises 3–10, change the numbers from scientific notation to ordinary notation.

3. 4.5×10^{4} **4.** 6.8×10^{7} **5.** 2.01×10^{-3} **6.** 9.61×10^{-5}

7. 3.23×10^{0} **8.** 8×10^{0} **9.** 1.86×10 **10.** 1×10^{-1}

In Exercises 11–20, change the numbers from ordinary notation to scientific notation.

11. 4000 **12.** 56 000 **13.** 0.0087 **14.** 0.7

15. 609 000 000 **16.** 100 **17.** 0.063 **18.** 0.000 090 8

19. 1 **20.** 10

In Exercises 21–24, perform the indicated calculations using a calculator and by first expressing all numbers in scientific notation.

21. 28 000(2 000 000 000) **22.** 50 000(0.006)

23. $\dfrac{88\,000}{0.0004}$ **24.** $\dfrac{0.000\,03}{6\,000\,000}$

In Exercises 25–28, perform the indicated calculations and then check the result using a calculator. Assume that all numbers are exact.

25. $2 \times 10^{-35} + 3 \times 10^{-34}$ **26.** $5.3 \times 10^{12} - 3.7 \times 10^{10}$

27. $(1.2 \times 10^{29})^{3}$ **28.** $(2 \times 10^{-16})^{-5}$

In Exercises 29–36, perform the indicated calculations using a calculator. All numbers are approximate.

29. 1280(865 000)(43.8) **30.** 0.000 056 9(3 190 000)

31. $\dfrac{0.0732(6710)}{0.001\,34(0.0231)}$ **32.** $\dfrac{0.004\,52}{2430(97\,100)}$

33. $(3.642 \times 10^{-8})(2.736 \times 10^{5})$ **34.** $\dfrac{(7.309 \times 10^{-1})^{2}}{5.9843(2.5036 \times 10^{-20})}$

35. $\dfrac{(3.69 \times 10^{-7})(4.61 \times 10^{21})}{0.0504}$

36. $\dfrac{(9.907 \times 10^{7})(1.08 \times 10^{12})^{2}}{(3.603 \times 10^{-5})(2054)}$

In Exercises 37–46, change numbers in ordinary notation to scientific notation or change numbers in scientific notation to ordinary notation.

37. The Sir Adam Beck Generating Stations in Niagara Falls produce 2 000 000 kW of power.
38. A certain laptop computer has 17 200 000 000 bytes of memory.
39. A fibre-optic system requires 0.000 003 W of power.
40. A red blood cell measures 0.0075 mm across.
41. The frequency of a certain cell phone signal is 1 200 000 000 Hz.
42. The PlayStation 4 game console has a graphic processing unit that can perform 1.84×10^{12} floating point operations per second (1.84 teraflops). (*Source*: playstation.com.)
43. The Gulf of Mexico oil spill in 2010 covered more than 12 000 000 000 m² of ocean surface.

44. A *parsec,* a unit used in astronomy, is about 3.086×10^{16} m.

45. The power of the signal of a laser beam probe is 1.6×10^{-12} W.

46. The electrical force between two electrons is about 2.4×10^{-43} times the gravitational force between them.

In Exercises 47–51, write the numbers from Exercises 37–41 in engineering notation and replace the power of 10 with the corresponding SI prefix.

In Exercises 52–55, solve the given problems.

52. Write the following numbers in engineering notation. (a) 2300 (b) 0.23 (c) 23

53. Write the following numbers in engineering notation. (a) 8 090 000 (b) 809 000 (c) 0.0809

54. A *googol* is defined as 1 followed by 100 zeros. (a) Write this number in scientific notation. (b) A *googolplex* is defined as 10 to the googol power. Write this number using powers of 10, and not the word *googol.* (Note the name of the internet company.)

55. The number of electrons in the universe has been estimated at 10^{79}. How many times greater is a googol than the estimated number of electrons in the universe? (See Exercise 54.)

56. The diameter of the sun, 1.4×10^9 m, is about 109 times the diameter of Earth. Express the diameter of Earth in scientific notation.

57. GB means gigabyte where giga means billion, or 10^9. Actually, 1 GB = 2^{30} bytes. Use a calculator to show that the use of giga is a reasonable choice of terminology.

In Exercises 58–61, perform the indicated calculations.

58. A computer can do an addition in 7.5×10^{-15} s. How long does it take to perform 5.6×10^6 additions?

59. The Intergovernmental Panel on Climate Change predicts that the average temperature of the earth changed by approximately +0.6°C over a 50-year period. At this rate, how much is the earth's temperature changing each day? (*Hint:* Estimate assuming each year has 365 days.)

60. If it takes 0.078 s for a GPS signal travelling at 2.998×10^8 m/s to reach the receiver in a car, find the distance from the receiver to the satellite.

61. (a) Determine the number of seconds in a day in scientific notation. (b) Using the result of part (a), determine the number of seconds in a century (assume 365.25 days/year).

In Exercises 62–65, perform the indicated calculations by first expressing all numbers in scientific notation.

62. One *atomic mass unit* (amu) is 1.66×10^{-27} kg. If one oxygen atom has 16 amu (an exact number), what is the mass of 125 000 000 oxygen atoms?

63. The rate of energy radiation (in W) from an object is found by evaluating the expression kT^4, where T is the thermodynamic temperature. Find this value for the human body, for which $k = 0.000\,000\,057$ W/K^4 and $T = 303$ K.

64. In a microwave receiver circuit, the resistance R of a wire 1 m long is given by $R = k/d^2$, where d is the diameter of the wire. Find R if $k = 0.000\,000\,021\,96\ \Omega \cdot \text{m}^2$ and $d = 0.000\,079\,98$ m.

65. The average distance between the sun and Earth is about 149 600 000 km, and this distance is called an *astronomical unit* (AU). It takes light about 499.0 s to travel 1 AU. What is the speed of light? Compare this with the speed of the GPS signal in Exercise 60.

Answers to Practice Exercises

1. 0.002 35 2. 2.35×10^2
3. 53.5×10^{-6} 4. 91×10^{-9}

1.6 Roots and Radicals

At times, we have to find the *square root* of a number, or maybe some other root of a number, such as a *cube root.* This means we must find a number that when squared, or cubed, and so on equals some given number. For example, to find the square root of 9, we must find a number that when squared equals 9. In this case, either 3 or −3 is an answer. Therefore, *either 3 or −3 is a square root of 9 since* $3^2 = 9$ *and* $(-3)^2 = 9$.

To have a general notation for the square root and have it represent *one* number, *we define the* **principal square root** *of a to be positive if a is positive and represent it by* $\sqrt{a}$. This means $\sqrt{9} = 3$ and not −3.

■ Unless we state otherwise, when we refer to the root of a number, it is the principal root.

The general notation for the **principal *n*th root** of a is $\sqrt[n]{a}$. (When $n = 2$, do not write the 2 for n.) *The* $\sqrt{\ }$ *sign is called a* **radical sign.**

EXAMPLE 1 Roots of numbers

(a) $\sqrt{2}$ (the square root of 2) **(b)** $\sqrt[3]{2}$ (the cube root of 2)
(c) $\sqrt[4]{2}$ (the fourth root of 2) **(d)** $\sqrt[7]{6}$ (the seventh root of 6)

Sign of a	n	Principal nth Root of a
Positive	Even	Positive
Positive	Odd	Positive
Negative	Odd	Negative
Negative	Even	No real root (Imaginary roots)

To have a single defined value for all roots (not just square roots) and to consider only real-number roots, *we define the* **principal *n*th root** *of a to be positive if a is positive and to be negative if a is negative and n is odd.* (If a is negative and n is even, the roots are not real.) See the summary table to the left.

EXAMPLE 2 Principal *n*th root

(a) $\sqrt{169} = \sqrt{13 \cdot 13} = 13$ (b) $-\sqrt{64} = -\sqrt{8 \cdot 8} = -8$

(c) $\sqrt[3]{64} = \sqrt[3]{4 \cdot 4 \cdot 4} = 4$ (d) $\sqrt{0.04} = \sqrt{0.2 \cdot 0.2} = 0.2$

(e) $-\sqrt[4]{81} = -\sqrt[4]{3 \cdot 3 \cdot 3 \cdot 3} = -3$ (f) $\sqrt[3]{-8} = \sqrt[3]{(-2)(-2)(-2)} = -2$

(g) $-\sqrt[3]{8} = -\sqrt[3]{2 \cdot 2 \cdot 2}$

Another property of square roots is developed by noting illustrations such as $\sqrt{36} = \sqrt{4 \times 9} = \sqrt{4} \times \sqrt{9} = 2 \times 3 = 6$. In general, this property states that *the square root of a product of positive numbers is the product of their square roots.*

$$\sqrt{ab} = \sqrt{a}\sqrt{b} \quad (a \text{ and } b \text{ positive real numbers}) \tag{1.15}$$

This property is most useful when a or b is a perfect square, and allows us to separate the perfect square from other factors. Please note that while separating factors in this manner is valid, *it is not valid to separate added or subtracted terms under a radical.* For addition and subtraction under a radical, see Example 5 below.

EXAMPLE 3 Simplifying square roots by factoring

(a) $\sqrt{8} = \sqrt{4 \cdot 2} = \sqrt{4}\sqrt{2} = 2\sqrt{2}$

(b) $\sqrt{75} = \sqrt{(25)(3)} = \sqrt{25}\sqrt{3} = 5\sqrt{3}$

(c) $\sqrt{4 \times 10^2} = \sqrt{(4)(10 \cdot 10)} = \sqrt{4}\sqrt{10 \cdot 10} = 2 \cdot 10 = 20$

In order to represent the square root of a number *exactly*, we use Eq. (1.15) to write it in simplest form. However, a decimal *approximation* is often acceptable, and we use the $\boxed{\sqrt{x}}$ key on a calculator.

EXAMPLE 4 Approximating a square root—rocket descent

After reaching its greatest height, the time (in s) for a rocket to fall h m is found by evaluating $0.45\sqrt{h}$. Find the time for the rocket to fall 360 m.

We calculate $0.45\sqrt{360} = 8.5$. Therefore, the rocket takes 8.5 s to fall 360 m. The result is rounded off to two significant digits, the accuracy of 0.45 (an approximate number).

In simplifying a radical, *all operations under a radical sign must be done before finding the root.* Inside the radical, operations are completed from left to right, following the correct order of operations.

EXAMPLE 5 More on simplifying square roots

(a) $\sqrt{16 + 9} = \sqrt{25}$ first perform the addition 16 + 9
$= 5$

COMMON ERROR However, $\sqrt{16 + 9}$ is *not* $\sqrt{16} + \sqrt{9} = 4 + 3 = 7$.

(b) $\sqrt{2^2 + 6^2} = \sqrt{4 + 36} = \sqrt{40} = \sqrt{4}\sqrt{10} = 2\sqrt{10}$, but
$\sqrt{2^2 + 6^2}$ is ***not*** $\sqrt{2^2} + \sqrt{6^2} = 2 + 6 = 8$.

In defining the principal square root, we did not define the square root of a negative number. However, in Section 1.1, we defined the square root of a negative number to be an **imaginary number**. As stated in the table on page 32, *the even root of a negative number is an imaginary number, and the odd root of a negative number is a negative real number.*

Practice Exercises

Simplify:
1. $\sqrt{12}$ 2. $\sqrt{36 + 144}$

EXAMPLE 6 Imaginary roots and real roots

(a) $\sqrt{-64}$ is imaginary ($n = 2$ is even) **(b)** $\sqrt[3]{-216}$ is real ($n = 3$ is odd)
(c) $\sqrt[5]{-32}$ is real ($n = 5$ is odd) **(d)** $\sqrt[4]{-243}$ is imaginary ($n = 4$ is even)

A much more detailed coverage of imaginary numbers, roots, and radicals is taken up in Chapters 11 and 12.

EXERCISES 1.6

In Exercises 1–4, make the given changes in the indicated examples of this section and then solve the given problems.

1. In Example 2(b), change the square root to a cube root and then evaluate.
2. In Example 3(b), change $\sqrt{(25)(3)}$ to $\sqrt{(15)(5)}$ and explain whether or not this would be a better expression to use.
3. In Example 5(a), change the $+$ to $\times$ and then evaluate.
4. In Example 6(a), place a $-$ sign before the radical. Is there any other change in the statement?

In Exercises 5–36, simplify the given expressions. In each of 5–9 and 12–21, the result is an integer.

5. $\sqrt{49}$	6. $\sqrt{343}$	7. $-\sqrt{45}$	8. $-\sqrt{72}$
9. $-\sqrt{64}$	10. $\sqrt{0.25}$	11. $\sqrt{0.09}$	12. $-\sqrt{900}$
13. $\sqrt[3]{125}$	14. $\sqrt[4]{81}$	15. $\sqrt[3]{-216}$	16. $\sqrt[5]{-32}$
17. $(\sqrt{5})^2$	18. $(\sqrt[3]{31})^3$	19. $(-\sqrt[3]{-47})^3$	20. $(\sqrt[5]{-23})^5$
21. $(-\sqrt[4]{53})^4$	22. $-\sqrt{32}$	23. $\sqrt{18}$	24. $\sqrt{1200}$
25. $2\sqrt{84}$	26. $\dfrac{\sqrt{108}}{2}$	27. $\sqrt{\dfrac{80}{7-3}}$	28. $\sqrt{81 \times 10^2}$
29. $\sqrt[3]{8^2}$	30. $\sqrt[4]{9^2}$	31. $\dfrac{5^3\sqrt{8}}{3^4\sqrt[3]{125}}$	32. $\dfrac{2^4\sqrt[4]{512}}{7\sqrt{121}}$

33. $\sqrt{36 + 64}$ 34. $\sqrt{25 + 144}$
35. $\sqrt{3^2 + 9^2}$ 36. $\sqrt{8^2 - 4^2}$

In Exercises 37–44, find the value of each square root by use of a calculator. Each number is approximate.

37. $\sqrt{85.4}$ 38. $\sqrt{3762}$ 39. $\sqrt{0.8152}$ 40. $\sqrt{0.0627}$
41. (a) $\sqrt{1296 + 2304}$ (b) $\sqrt{1296} + \sqrt{2304}$
42. (a) $\sqrt{10.6276 + 2.1609}$ (b) $\sqrt{10.6276} + \sqrt{2.1609}$
43. (a) $\sqrt{0.0429^2 - 0.0183^2}$ (b) $\sqrt{0.0429^2} - \sqrt{0.0183^2}$
44. (a) $\sqrt{3.625^2 + 0.614^2}$ (b) $\sqrt{3.625^2} + \sqrt{0.614^2}$

In Exercises 45–60, solve the given problems.

45. The speed (in km/h) of a car that skids to a stop on dry pavement is often estimated by $\sqrt{207s}$, where s is the length (in m) of the skid marks. Estimate the speed if $s = 46$ m.
46. The resistance in an amplifier circuit is found by evaluating $\sqrt{Z^2 - X^2}$. Find the resistance for $Z = 5.362\ \Omega$ and $X = 2.875\ \Omega$.
47. The speed (in m/s) of sound in seawater is found by evaluating $\sqrt{B/d}$ for $B = 2.18 \times 10^9$ Pa and $d = 1.03 \times 10^3$ kg/m^3. Find this speed, which is important in locating underwater objects using sonar.
48. The terminal speed (in m/s) of a skydiver can be approximated by $\sqrt{40m}$, where m is the mass (in kg) of the skydiver. Calculate the terminal speed (after reaching this speed, the skydiver's speed remains fairly constant before opening the parachute) of a 75-kg skydiver.
49. An HDTV screen is 93.0 cm wide and 52.1 cm high. The length of a diagonal (the dimension used to describe it—from one corner to the opposite corner) is found by evaluating $\sqrt{w^2 + h^2}$, where w is the width and h is the height. Find the diagonal.
50. A car costs \$22 000 new and is worth \$15 000 two years later. The annual rate of depreciation is found by evaluating $100\left(1 - \sqrt{V/C}\right)$, where C is the cost and V is the value after two years. At what rate did the car depreciate? (100 and 1 are exact.)
51. A *tsunami* is a very high ocean tidal wave (or series of waves) often caused by an earthquake. An equation that approximates the speed v (in m/s) of a tsunami is $v = \sqrt{2gd}$, where $g = 9.8$ m/s^2 and d is the average depth (in m) of the ocean floor. Find v (in km/h) for $d = 3500$ m (valid for many parts of the Indian Ocean and Pacific Ocean).
52. The greatest distance (in km) a person can see from a height h (in m) above the ground is $\sqrt{1.27 \times 10^4 h + h^2}$. What is this distance for the pilot of a plane 9500 m above the ground?
53. Is it always true that $\sqrt{a^2} = a$? Explain.
54. For what values of x is (a) $x > \sqrt{x}$, (b) $x = \sqrt{x}$, and (c) $x < \sqrt{x}$?
55. Using a calculator, find $\sqrt[3]{2140}$ and $\sqrt[3]{-0.214}$.
56. Using a calculator, find $\sqrt[5]{0.382}$ and $\sqrt[7]{-382}$.
57. Determine if the following numbers are real or imaginary:
(a) $\sqrt[4]{-81}$ (b) $\sqrt[7]{-128}$
58. Determine if the following numbers are real or imaginary:
(a) $\sqrt[5]{-32}$ (b) $\sqrt[4]{-64}$

59. The resonance frequency f (in Hz) in an electronic circuit containing inductance L (in H) and capacitance C (in F) is given by $f = \frac{1}{2\pi\sqrt{LC}}$. Find the resonance frequency if $L = 0.250$ H and $C = 40.52 \times 10^{-6}$ F.

60. In statistics, the standard deviation is the square root of the variance. Find the standard deviation if the variance is 80.5 kg^2.

Answers to Practice Exercises

1. $2\sqrt{3}$ **2.** $6\sqrt{5}$

1.7 Addition and Subtraction of Algebraic Expressions

FACTORS AND TERMS IN ALGEBRAIC EXPRESSIONS

Since we use letters to represent numbers, we can see that all operations that can be used on numbers can also be used on literal numbers. In this section, we discuss the methods for adding and subtracting literal numbers.

Addition, subtraction, multiplication, division, and taking of roots are known as **algebraic operations**. *Any combination of numbers and literal symbols that results from algebraic operations is known as an* **algebraic expression**.

Within an algebraic expression, there can be terms and factors. A **term** is any collection of multiples that is added or subtracted in an algebraic expression. A **factor** is made up of values that are multiplied together within a given term. There can be **primary factors**, which cannot be further factored, or **compound factors**, which are the product of two or more primary factors.

The distinction between terms and factors is very important because some operations are only valid for factors, while others are only valid for terms. Many common errors in handling algebraic expressions originate from a poor understanding of these definitions, and the inability to determine when rules are valid.

EXAMPLE 1 Identifying terms and factors

(a) $7a - 2b$ is an expression with two terms: $7a$ and $-7b$. The first term has primary factors 7 and a, and compound factor $7a$. The second term has primary factors -2 and b, and compound factor $-2b$.

(b) $gt^2 - 2vt + 2s$ is an expression with three terms: gt^2, $-2vt$, and $2s$. The first term, gt^2, has a primary factor g and two primary factors t. Any product of these factors (gt, t^2, and gt^2 itself) is a compound factor of gt^2.

(c) $7a(x^2 + 2y) - 6x(5 + x - 3y)$ is an expression with terms $7(x^2 + 2y)$ and $-6x(5 + x - 3y)$. The term $7a(x^2 + 2y)$ has primary factors of 7, a, and $(x^2 + 2y)$, and products of these factors are its compound factors. The factor $x^2 + 2y$ has two terms, x^2 and $2y$.

The term $-6x(5 + x - 3y)$ has factors 2, 3, x, and $(5 + x - 3y)$. The negative sign in front can be treated as a factor of -1. The factor $5 + x - 3y$ has three terms, 5, x, and $-3y$.

POLYNOMIALS

A **polynomial** *is an algebraic expression with only nonnegative integer exponents on one or more variables, and has no variable in a denominator. The* **degree of a term** *is the sum of the exponents of the variables of that term, and the* **degree of the polynomial** *is the degree of the term of highest degree.*

A **multinomial** *is any algebraic expression of more than one term.* Terms like $1/x$ and $\sqrt{x}$ can be included in a multinomial, but not in a polynomial. (Since $1/x = x^{-1}$, the exponent is negative.)

EXAMPLE 2 Degree of a polynomial

Some examples of polynomials are as follows:

(a) $4x^2 - 5x + 3$ (degree 2) **(b)** $2x^6 - x$ (degree 6) **(c)** $3x$ (degree 1)

(d) $xy^3 + 7x - 3$ (degree 4) (add exponents of x and y)

(e) -6 (degree 0)$(-6 = -6x^0)$

From (c), note that a single term can be a polynomial, and from (e), note that a constant can be a polynomial. The expressions in (a), (b), and (d) are also multinomials.

The expression $x^2 + \sqrt{y + 2} - 8$ is a multinomial, but not a polynomial because of the square root term.

A polynomial with one term is called a **monomial**. *A polynomial with two terms is called a* **binomial**, *and one with three terms is called a* **trinomial**. *The numerical factor is called the* **numerical coefficient** (or simply **coefficient**) *of the term. All terms that differ at most in their numerical coefficients are known as* **similar** *or* **like** *terms.* That is, similar terms have the same variables with the same exponents.

EXAMPLE 3 Naming polynomials

(a) $7x^4$ is a monomial. The numerical coefficient is 7. The expressions $-5m$ and $3x^2y^3z$ are also monomials.

(b) $3ab - 6a$ is a binomial. The numerical coefficient of the first term is 3, and the numerical coefficient of the second term is -6. Note that the sign is attached to the coefficient. The expressions $3 + x$ and $x^1 - 1$ are also binomials.

(c) $8cx^3 - x + 2$ is a trinomial. The coefficients of the first two terms are 8 and -1.

(d) $x^2y^2 - 2x + 3y - 9$ is a polynomial with four terms (no special name).

EXAMPLE 4 Similar terms

(a) $8b - 6ab + 81b$ is a trinomial. The first and third terms are similar since they differ only in their numerical coefficients. The middle term is not similar to the others since it has a factor of a.

(b) $4x^2 - 3x$ is a binomial. The terms are not similar since the first term has two factors of x, and the second term has only one factor of x.

(c) $3x^2y^3 - 5y^3x^2 + x^2 - 2y^3$ is a polynomial. The commutative law tells us that $x^2y^3 = y^3x^2$, which means the first two terms are similar.

LEARNING TIP

In adding and subtracting algebraic expressions, we combine similar terms into a single term. The ***simplified*** expression will contain only terms that are not similar.

EXAMPLE 5 Simplifying polynomial expressions

For each of the following polynomials, identify like terms and simplify the polynomial expression where possible.

(a) $3x + 2x - 5y$ We identify similar terms and add their coefficients:

$3x + 2x - 5y = (3 + 2)x - 5y = 5x - 5y$

(b) $6a^2 - 7a + 8ax$ This expression cannot be simplified since there are no like terms.

(c) $6a + 5c + 2a - c$ We group like terms using the commutative law and then simplify: $6a + 5c + 2a - c = 6a + 2a + 5c - 1c = 8a + 4c$

■ Computer algebra systems (CAS) and calculators with CAS features can display algebraic expressions and perform algebraic operations.

SIMPLIFICATION OF EXPRESSIONS WITH GROUPING SYMBOLS

To group terms in an algebraic expression, we use **symbols of grouping**. In this text, we use **parentheses**, ( ), **brackets**, [ ], and **braces**, { }. All operations that occur in the numerator or denominator of a fraction are implied to be inside grouping symbols, as well as all operations under a radical symbol.

In simplifying an expression using the distributive law, remove the symbols of grouping as follows: if a MINUS sign precedes the grouping, change the sign of EVERY term in the grouping; if a plus sign precedes the grouping, each term retains its original sign.

COMMON ERROR

Do not forget to multiply using the distributive law if there is a minus sign in front of a factor with multiple terms.

$$2b + 6a - (3a - 4b) = 2b + 6a - 3a + 4b = 3a + 6b$$

A common error is to forget to change the signs to all the terms in the factor.

$$2b + 6a - (3a - 4b) \neq 2b + 6a - 3a - 4b = 3a - 2b$$

EXAMPLE 6 Simplifying symbols of grouping

(a) $2(a + 2x) = 2a + 2(2x)$ use distributive law

$= 2a + 4x$

(b) $-(+a - 3c) = (-1)(+a - 3c)$ treat − sign as −1

$= (-1)(+a) + (-1)(-3c)$

$= -a + 3c$ note change of signs

Normally, $+a$ would be written simply as a.

Practice Exercises

Use the distributive law.

1. $3(2a + y)$ **2.** $-3(-2r + s)$

In Example 6(a), we see that $2(a + 2x) = 2a + 4x$. This is true for any values of a and x, and is therefore an *identity* (see Section 1.2). In fact, an expression and any proper form to which it may be changed form an identity when they are shown to be equal.

EXAMPLE 7 Simplifying—signs before parentheses

+ sign before parentheses

(a) $3c + (2b - c) = 3c + 2b - c = 2b + 2c$ use distributive law

$2b = +2b$ signs retained

− sign before parentheses

(b) $3c - (2b - c) = 3c - 2b + c = -2b + 4c$ use distributive law

$2b = +2b$ signs changed

(c) $3c - (-2b + c) = 3c + 2b - c = 2b + 2c$ use distributive law

signs changed

■ Note in each case that the parentheses are removed and the sign before the parentheses is also removed.

Practice Exercise

3. Simplify: $2x - 3(4y - x)$

EXAMPLE 8 Simplifying—machine part design

In designing a certain machine part, it is necessary to perform the following simplification.

$$\begin{aligned}16(8-x)-2(8x-x^2)-(64-16x+x^2) &= 128-16x-16x+2x^2-64+16x-x^2\\ &= 64-16x+x^2\end{aligned}$$

LEARNING TIP

At times, we have expressions in which more than one symbol of grouping is to be removed in the simplification. Normally, *when several symbols of grouping are to be removed, it is more convenient to* ***remove the innermost symbols first.***

EXAMPLE 9 Several symbols of grouping

(a) $$\begin{aligned}3ax-[ax-(5s-2ax)] &= 3ax-[ax-5s+2ax] \quad \leftarrow \text{remove parentheses}\\ &= 3ax-ax+5s-2ax \quad \leftarrow \text{remove brackets}\\ &= 5s\end{aligned}$$

(b) $$\begin{aligned}3a^2b-\{[a-(2a^2b-a)]+2b\} &= 3a^2b-\{[a-2a^2b+a]+2b\}\\ &= 3a^2b-\{a-2a^2b+a+2b\}\\ &= 3a^2b-a+2a^2b-a-2b\\ &= 5a^2b-2a-2b\end{aligned}$$

Calculators use only parentheses for grouping symbols, and we often need to use one set of parentheses within another set. These are called **nested parentheses**. In the next example, note that the innermost parentheses are removed first.

EXAMPLE 10 Nested parentheses

$$\begin{aligned}2-(3x-2(5-(7-x))) &= 2-(3x-2(5-7+x))\\ &= 2-(3x-10+14-2x)\\ &= 2-3x+10-14+2x\\ &= -x-2\end{aligned}$$

EXERCISES 1.7

In Exercises 1–4, make the given changes in the indicated examples of this section, and then solve the resulting problems.

1. In Example 5(a), change $2x$ to $2y$.
2. In Example 7(a), change the sign before $(2b-c)$ from + to −.
3. In Example 9(a), change $[ax-(5s-2ax)]$ to $[(ax-5s)-2ax]$.
4. In Example 9(b), change $\{[a-(2a^2b-a)]+2b\}$ to $\{a-[2a^2b-(a+2b)]\}$.

In Exercises 5–46, simplify the given algebraic expressions.

5. $5x+7x-4x$
6. $6t-3t-4t$
7. $2y-y+4x$
8. $4C+L-6C$
9. $2F-2T-2+3F-T$
10. $x-2y+3x-y+z$
11. $a^2b-a^2b^2-2a^2b$
12. $xy^2-3x^2y^2+2xy^2$
13. $s+(3s-4-s)$
14. $5+(3-4n+p)$
15. $v-(4-5x+2v)$
16. $2a-(b-a)$
17. $2-3-(4-5a)$
18. $\sqrt{A}+(h-2\sqrt{A})-3\sqrt{A}$
19. $(a-3)+(5-6a)$
20. $(4x-y)-(-2x-4y)$
21. $-(t-2u)+(3u-t)$
22. $2(x-2y)+(5x-y)$
23. $3(2r+s)-(-5s-r)$
24. $3(a-b)-2(a-2b)$
25. $-7(6-3j)-2(j+4)$
26. $-(5t+a^2)-2(3a^2-2st)$
27. $-[(4-6n)-(n-3)]$
28. $-[(h-k)-(k-h)]$
29. $2[4-(t^2-5)]$
30. $3[-3-(a-4)]$
31. $-2[-x-2a-(a-x)]$
32. $-2[-3(x-2y)+4y]$
33. $aZ-[3-(aZ+4)]$
34. $9v-[6-(v-4)+4v]$
35. $8c-\{5-[2-(3+4c)]\}$
36. $7y-\{y-[2y-(x-y)]\}$
37. $5p-(q-2p)-[3q-(p-q)]$

38. $-(4 - \sqrt{LC}) - [(5\sqrt{LC} - 7) - (6\sqrt{LC} + 2)]$
39. $-2\{-(4 - x^2) - [3 + (4 - x^2)]\}$
40. $-\{-[-(x - 2a) - b] - (a - x)\}$
41. $5V^2 - (6 - (2V^2 + 3))$
42. $-2F + 2((2F - 1) - 5)$
43. $-(3t - (7 + 2t - (5t - 6)))$
44. $a^2 - 2(x - 5 - (7 - 2(a^2 - 2x) - 3x))$
45. $-4[4R - 2.5(Z - 2R) - 1.5(2R - Z)]$
46. $3\{2.1e - 1.3[f - 2(e - 5f)]\}$

In Exercises 47–56, answer the given questions.

47. In determining the size of a V belt to be used with an engine, the expression $3D - (D - d)$ is used. Simplify this expression.
48. When finding the current in a transistor circuit, the expression $i_1 - (2 - 3i_2) + i_2$ is used. Simplify this expression. (The numbers below the i's are *subscripts*. Different subscripts denote different variables.)
49. Research on a plastic building material leads to $[(B + \frac{4}{3}\alpha) + 2(B - \frac{2}{3}\alpha)] - [(B + \frac{4}{3}\alpha) - (B - \frac{2}{3}\alpha)]$. Simplify this expression.
50. One car goes 30 km/h for $t - 1$ hours, and a second car goes 40 km/h for $t + 2$ hours. Find the expression for the sum of the distances travelled by the two cars.
51. A shipment contains x hard drives with 4 terabytes of memory and $x + 25$ hard drives with 8 terabytes. Express the total number of terabytes of memory in the shipment as a variable expression and simplify.
52. Each of two suppliers has $2n + 1$ bundles of shingles costing \$30 each and $n - 2$ bundles costing \$20 each. How much more is the total value of the \$30 bundles than the \$20 bundles?
53. For the expressions $2x^2 - y + 2a$ and $3y - x^2 - b$, find (a) the sum, and (b) the difference if the second is subtracted from the first.
54. For the following expressions, subtract the third from the sum of the first two: $3a^2 + b - c^3$, $2c^3 - 2b - a^2$, $4c^3 - 4b + 3$.
55. For any real numbers a and b, is it true that $|a - b| = |b - a|$? Explain.
56. Is subtraction associative? That is, in general, does $(a - b) - c$ equal $a - (b - c)$? Explain.

Answers to Practice Exercises

1. $6a + 3y$ **2.** $6r - 3s$ **3.** $5x - 12y$

1.8 Multiplication of Algebraic Expressions

MULTIPLYING MONOMIALS

To find the product of two or more monomials, we use the laws of exponents as given in Section 1.4 and the laws for multiplying signed numbers as stated in Section 1.2. We first multiply the numerical coefficients to find the numerical coefficient of the product. Then we multiply the literal numbers, remembering that *the exponents may be combined only if the base is the same.*

EXAMPLE 1 Multiplying monomials

(a) $3c^5(-4c^2) = -12c^7$ multiply numerical coefficients and add exponents of c

(b) $(-2b^2y^3)(-9aby^5) = 18ab^3y^8$ add exponents of same base

(c) $2xy(-6cx^2)(3xcy^2) = -36c^2x^4y^3$

If a product contains a monomial that is raised to a power, *we must first raise it to the indicated power* before proceeding with the multiplication.

EXAMPLE 2 Product containing a power of a monomial

(a) $3(2a^2x)^3(-ax) = 3(8a^6x^3)(-ax) = -24a^7x^4$

(b) $2s^3(-st^4)^2(4s^2t) = 2s^3(s^2t^8)(4s^2t) = 8s^7t^9$

MULTIPLYING POLYNOMIALS

We find the product of a monomial and a polynomial by using the distributive law, which states that we *multiply each term of the polynomial by the monomial.* In doing so, we must be careful to give the correct sign to each term of the product.

EXAMPLE 3 Product of a monomial and a polynomial

(a) $2ax(3ax^2 - 4yz) = 2ax(3ax^2) + (2ax)(-4yz) = 6a^2x^3 - 8axyz$

(b) $5cy^2(-7cx - ac) = (5cy^2)(-7cx) + (5cy^2)(-ac) = -35c^2xy^2 - 5ac^2y^2$

Practice Exercises

Perform the indicated multiplications.

1. $2a^3b(-6ab^2)$ **2.** $-5x^2y^3(2xy - y^4)$

It is generally not necessary to write out the middle step as it appears in the preceding example. We write the answer directly. For instance, Example 3(a) would appear as $2ax(3ax^2 - 4yz) = 6a^2x^3 - 8axyz$.

We find the product of two polynomials by using the distributive law. The result is that *we multiply each term of one polynomial by each term of the other and add the results.* Again we must be careful to give each term of the product its correct sign.

EXAMPLE 4 Product of binomials

$$\begin{aligned}(x - 2)(x + 3) &= x(x) + x(3) + (-2)(x) + (-2)(3)\\ &= x^2 + 3x - 2x - 6\\ &= x^2 + x - 6\end{aligned}$$

■ Note that, using the distributive law, $(x - 2)(x + 3) = (x - 2)(x) + (x - 2)(3)$ leads to the same result.

EXAMPLE 5 Product of polynomials

$$\begin{aligned}(x^2 - 2x + 3)(x + 1) &= x^2(x + 1) - 2x(x + 1) + 3(x + 1)\\ &= x^3 + x^2 - 2x^2 - 2x + 3x + 3\\ &= x^3 - x^2 + x + 3\end{aligned}$$

POWERS OF POLYNOMIALS

Finding the power of a polynomial is equivalent to using the polynomial as a factor the number of times indicated by the exponent. It is sometimes convenient to write the power of a polynomial in this form before multiplying.

EXAMPLE 6 Square of a binomial

two factors

$$\begin{aligned}(x + 5)^2 &= (x + 5)(x + 5) = x^2 + 5x + 5x + 25\\ &= x^2 + 10x + 25\end{aligned}$$

Practice Exercises

Perform the indicated multiplications.

3. $(2s - 5t)(s + 4t)$ **4.** $(3u + 2v)^2$

COMMON ERROR

$$(x + 5)^2 \neq x^2 + 25$$

A common error is to forget to include the middle term of $10x$. By following proper procedure, you must expand the relation as

$$\begin{aligned}(x + 5)^2 &= (x + 5)(x + 5)\\ &= x \cdot x + x \cdot 5 + 5 \cdot x + 5 \cdot 5\\ &= x^2 + 10x + 25\end{aligned}$$

Thus, you cannot simply square each of the terms within the parentheses.

EXAMPLE 7 Power of a polynomial

$$(2a - b)^3 = (2a - b)(2a - b)(2a - b) \quad \text{the exponent 3 indicates three factors}$$
$$= (2a - b)(4a^2 - 2ab - 2ab + b^2)$$
$$= (2a - b)(4a^2 - 4ab + b^2)$$
$$= 8a^3 - 8a^2b + 2ab^2 - 4a^2b + 4ab^2 - b^3$$
$$= 8a^3 - 12a^2b + 6ab^2 - b^3$$

EXAMPLE 8 Simplifying products—telescope lens

Simplify the following expression, used when finding the settings needed to optimize the focus of a telescope lens.

$$a(a + b)^2 + a^3 - (a + b)(2a^2 - s^2)$$
$$= a(a + b)(a + b) + a^3 - (2a^3 - as^2 + 2a^2b - bs^2)$$
$$= a(a^2 + ab + ab + b^2) + a^3 - 2a^3 + as^2 - 2a^2b + bs^2$$
$$= a^3 + a^2b + a^2b + ab^2 - a^3 + as^2 - 2a^2b + bs^2$$
$$= ab^2 + as^2 + bs^2$$

EXERCISES 1.8

In Exercises 1–4, make the given changes in the indicated examples of this section, and then solve the resulting problems.

1. In Example 2(b), change the factor $(-st^4)^2$ to $(-st^4)^3$.
2. In Example 3(a), change the factor $2ax$ to $-2ax$.
3. In Example 4, change the factor $(x + 3)$ to $(x - 3)$.
4. In Example 7, change the exponent 3 to 2.

In Exercises 5–52, perform the indicated multiplications.

5. $(a^2)(ax)$ 6. $(2xy)(x^2y^3)$ 7. $-a^2c^2(a^2cx^3)$
8. $-2cs^2(-4cs)^2$ 9. $(2ax^2)^2(-2ax)$ 10. $6pq^3(3pq^2)^2$
11. $i^2(R + 2r)$ 12. $2x(-p - q)$
13. $-5s(3s^2 - 2t)$ 14. $-3b(2b^2 - b)$
15. $5m(m^2n + 3mn)$ 16. $a^2bc(2ac - 3b^2c)$
17. $3M(-M - N + 2)$ 18. $-4c^2(-9gc - 2c + g^2)$
19. $ax(cx^2)(x + y^3)$ 20. $-2(-3st^3)(3s - 4t)$
21. $(x - 3)(x + 5)$ 22. $(a + 7)(a + 1)$
23. $(x + 5)(2x - 1)$ 24. $(4t_1 + t_2)(2t_1 - 3t_2)$
25. $(y + 8)(y - 8)$ 26. $(z - 4)(z + 4)$
27. $(2a - b)(3a - 2b)$ 28. $(-3 + 4w^2)(3w^2 - 1)$
29. $(2s + 7t)(3s - 5t)$ 30. $(5p - 2q)(p + 8q)$
31. $(x^2 - 1)(2x + 5)$ 32. $(3y^2 + 2)(2y - 9)$
33. $(x - 2y - 4)(x - 2y + 4)$
34. $(2a + 3b + 1)(2a + 3b - 1)$
35. $2(a + 1)(a - 9)$ 36. $-5(y - 3)(y + 6)$
37. $-3(3 - 2T)(3T + 2)$ 38. $2n(5 - N)(6n + 5)$
39. $2L(L + 1)(4 - L)$ 40. $ax(x + 4)(7 - x^2)$
41. $(2x - 5)^2$ 42. $(x - 3y)^2$ 43. $(x_1 + 3x_2)^2$ 44. $(7m + 1)^2$
45. $(xyz - 2)^2$ 46. $(b - 6x^2)^2$ 47. $2(x + 8)^2$ 48. $3(3R + 4)^2$
49. $(2 + x)(3 - x)(x - 1)$ 50. $(3x - c^2)^3$
51. $3T(T + 2)(2T - 1)$ 52. $[(x - 2)^2(x + 2)]^2$

In Exercises 53–66, answer the given questions.

53. Let $x = 3$ and $y = 4$ to show that (a) $(x + y)^2 \neq x^2 + y^2$ and (b) $(x - y)^2 \neq x^2 - y^2$. ($\neq$ means "does not equal.")
54. Evaluate the product $(98)(102)$ by expressing it as $(100 - 2)(100 + 2)$.
55. Square an integer between 1 and 9 and subtract 1 from the result. Explain why the result is the product of the integer before and the integer after the one you chose.
56. Explain how, by appropriate grouping, the product $(x - 2)(x + 3)(x + 2)(x - 3)$ is easier to find. Find the product.
57. By multiplication, show that $(x + y)^3$ is not equal to $x^3 + y^3$.
58. By multiplication, show that $(x + y)(x^2 - xy + y^2) = x^3 + y^3$.
59. In finding the value of a certain savings account, the expression $P(1 + 0.01r)^2$ is used. Multiply out this expression.
60. The force due to the liquid in a parabolic container leads to the expression $w(1 - x)(4 - x^2)$. Multiply out this expression.
61. In using aircraft radar, the expression $(2R - X)^2 - (R^2 + X^2)$ arises. Simplify this expression.
62. In calculating the temperature variation of an industrial area, the expression $(2T^3 + 3)(T^2 - T - 3)$ arises. Perform the indicated multiplication.
63. In a particular computer design containing n circuit elements, n^2 switches are needed. Find the expression for the number of switches needed for $n + 100$ circuit elements.
64. Simplify the expression $(T^2 - 100)(T - 10)(T + 10)$, which arises when analysing the energy radiation from an object.

65. In finding the maximum power in part of a microwave transmitter circuit, the expression $(R_1 + R_2)^2 - 2R_2(R_1 + R_2)$ is used. Multiply and simplify.

66. In determining the deflection of a certain steel beam, the expression $27x^2 - 24(x - 6)^2 - (x - 12)^3$ is used. Multiply and simplify.

Answers to Practice Exercises

1. $-12a^4b^3$ **2.** $-10x^3y^4 + 5x^2y^7$
3. $2s^2 + 3st - 20t^2$ **4.** $9u^2 + 12uv + 4v^2$

1.9 Division of Algebraic Expressions

DIVISION BY MONOMIAL DIVISORS

To find the quotient of one monomial divided by another, we use the laws of exponents and the laws for dividing signed numbers. Again, the *exponents may be combined only if the base is the same.*

EXAMPLE 1 Dividing monomials

(a) $\dfrac{3c^7}{c^2} = 3c^{7-2} = 3c^5$ **(b)** $\dfrac{16x^3y^5}{4xy^2} = \dfrac{16}{4}(x^{3-1})(y^{5-2}) = 4x^2y^3$

divide coefficients — subtract exponents

(c) $\dfrac{-6a^2xy^2}{2axy^4} = \left(\dfrac{-6}{2}\right)a^{2-1}x^{1-1}y^{2-4} = -3a^1y^{-2} = -3a\left(\dfrac{1}{y^2}\right) = \dfrac{-3a}{y^2}$

As shown in illustration (c), we use only positive exponents in the final result unless there are specific instructions otherwise.

Generalizing from arithmetic, we can show how a multinomial is to be divided by a monomial. Recall that to add fractions with like denominators, the numerators are added while the denominator is kept the same, as in $\frac{2}{7} + \frac{3}{7} = \frac{2+3}{7}$. In general, we have $\frac{a}{c} + \frac{b}{c} = \frac{a+b}{c}$ for any $c \neq 0$. Writing this identity from right to left, we see that *the quotient of a multinomial divided by a monomial is found by dividing each term of the multinomial by the monomial and adding the results.* This can be shown as

$$\frac{a+b}{c} = \frac{a}{c} + \frac{b}{c}$$

■ This is an identity and is valid for all values of a and b, and all values of c except zero (which would make it undefined).

COMMON ERROR

Be careful: Although $\dfrac{a+b}{c} = \dfrac{a}{c} + \dfrac{b}{c}$, we must note that $\dfrac{c}{a+b}$ *is not* $\dfrac{c}{a} + \dfrac{c}{b}$.

■ Until you are familiar with the method, it is recommended that you do write out the middle steps.

EXAMPLE 2 Dividing by a monomial

(a) $\dfrac{4a^2 + 8a}{2a} = \dfrac{4a^2}{2a} + \dfrac{8a}{2a} = \left(\dfrac{4}{2}\right)a^{2-1} + \left(\dfrac{8}{2}\right)a^{1-1} = 2a + 4$

each term of numerator divided by denominator

(b) $\dfrac{4x^3y - 8x^3y^2 + 2x^2y}{2x^2y} = \dfrac{4x^3y}{2x^2y} - \dfrac{8x^3y^2}{2x^2y} + \dfrac{2x^2y}{2x^2y}$

$$= \left(\frac{4}{2}\right)x^{3-2}y^{1-1} - \left(\frac{8}{2}\right)x^{3-2}y^{2-1} + \left(\frac{2}{2}\right)x^{2-2}y^{1-1}$$

$$= 2x - 4xy + 1$$

Practice Exercise

Divide: **1.** $\dfrac{4ax^2 - 6a^2x}{2ax}$

We usually do not write out the middle steps as shown in these illustrations. The divisions of the terms of the numerator by the denominator are usually done by inspection (mentally), and the result is shown as it appears in the next example.

EXAMPLE 3 Dividing by a monomial—irrigation pump

The expression $\frac{2p + v^2d + 2ydg}{2dg}$ is used when analysing the operation of an irrigation pump. Performing the indicated division, we have

$$\frac{2p + v^2d + 2ydg}{2dg} = \frac{p}{dg} + \frac{v^2}{2g} + y$$

DIVISION BY POLYNOMIAL DIVISORS

To divide one polynomial by another, use the following steps.

■ This is similar to long division of numbers.

Algebraic Division	EXAMPLE 4
	$(6x^2 + x - 2) \div (2x - 1)$ or $\frac{(6x^2 + x - 2)}{(2x - 1)}$
1. Reorder the polynomials in descending powers of the variable.	Powers are already in the correct order.
2. Set up the division: the divisor to the left and the dividend under the division sign.	$2x - 1\overline{)6x^2 + x - 2}$
3. Divide the first term of the dividend by the first term of the divisor. The result is written above the dividend.	$3x$ ← $\frac{6x^2}{2x}$ $2x - 1\overline{)6x^2 + x - 2}$
4. Multiply each term in the divisor by the value obtained in Step 3. Place the result directly underneath the numerator value.	$3x$ $2x - 1\overline{)6x^2 + x - 2}$ $6x^2 - 3x$
5. Subtract the result from Step 4 from the dividend. It is useful to subtract column by column to avoid sign errors.	$3x$ $2x - 1\overline{)6x^2 + x - 2}$ $-(6x^2 - 3x)$ $0 + 4x$
6. Complete the remainder with any terms from the original dividend that were not involved in the subtraction of Step 5.	$3x$ $2x - 1\overline{)6x^2 + x - 2}$ $-(6x^2 - 3x)$ $0 + 4x - 2$
7. Repeat Steps 3–6 until the remainder is zero or until a term of lower degree than the divisor is found.	$3x + 2$ ← $\frac{4x}{2x}$ $2x - 1\overline{)6x^2 + x - 2}$ $-(6x^2 - 3x)$ $0 + 4x - 2$ $4x - 2$ 0
8. Write the solution as the quotient polynomial (above the division bar) **plus** any remainder **divided by the divisor**.	$\frac{(6x^2 + x - 2)}{(2x - 1)} = 3x + 2 + \frac{0}{2x - 1} = 3x + 2$ Check: $(3x + 2)(2x - 1) = 6x^2 + x - 2$.

Practice Exercise

Divide: **2.** $(6x^2 + 7x - 3) \div (3x - 1)$

EXAMPLE 5 Quotient with a remainder

Perform the division $\frac{8x^3 + 4x^2 + 3}{4x^2 - 1}$. Since there is no x-term in the dividend, we should leave space for any x-terms that might arise (which we will show as $0x$).

$$\begin{array}{rl} & 2x + 1 \\ \text{divisor} \quad 4x^2 - 1 \overline{)\,8x^3 + 4x^2 + 0x + 3} & \text{dividend} \qquad \frac{8x^3}{4x^2} = 2x \\ 8x^3 \qquad\quad - 2x & \text{subtract} \\ \hline 4x^2 + 2x + 3 & \\ 0x - (-2x) = 2x & \qquad\qquad\quad \frac{4x^2}{4x^2} = 1 \\ 4x^2 \qquad - 1 & \text{subtract} \\ \hline 2x + 4 & \text{remainder} \end{array}$$

■ The answer to Example 5 can be checked by showing

$(4x^2 - 1)(2x + 1) + (2x + 4) = 8x^3 + 4x^2 + 3$

Since the degree of the remainder $2x + 4$ is less than that of the divisor, we now show the result in this case as $\frac{8x^3 + 4x^2 + 3}{4x^2 - 1} = 2x + 1 + \frac{2x + 4}{4x^2 - 1}$.

EXERCISES 1.9

In Exercises 1–4, make the given changes in the indicated examples of this section and then perform the indicated divisions.

1. In Example 1(c), change the denominator to $-2a^2xy^5$.
2. In Example 2(b), change the denominator to $2xy^2$.
3. In Example 4, change the dividend to $6x^2 - 7x + 2$.
4. In Example 5, change the sign of the middle term of the numerator from + to −.

In Exercises 5–28, perform the indicated divisions.

5. $\frac{8x^3y^2}{-2xy}$
6. $\frac{-18b^7c^3}{bc^2}$
7. $\frac{-16r^3t^5}{-4r^5t}$
8. $\frac{51mn^5}{17m^2n^2}$
9. $\frac{-5x^{-2}y^6}{10x^3y^2}$
10. $\frac{-20b^5c^{-5}}{4b^{-3}c^{-2}}$
11. $\frac{-200m^3n^5}{-10m^{10}n^{-10}}$
12. $\frac{-81r^7k^{-3}p^{10}}{135r^{-2}k^{-3}p^{-5}}$
13. $\frac{(15x^2)(4bx)(2y)}{30bxy}$
14. $\frac{(5sT)(8s^2T^3)}{10s^3T^2}$
15. $\frac{6(ax)^2}{-ax^2}$
16. $\frac{12a^2b}{(3ab^2)^2}$
17. $\frac{3a^2x + 6xy}{3x}$
18. $\frac{2m^2n - 6mn}{2m}$
19. $\frac{3rst - 6r^2st^2}{3rs}$
20. $\frac{-5a^2n - 10an^2}{5an}$
21. $\frac{4pq^3 + 8p^2q^2 - 16pq^5}{4pq^2}$
22. $\frac{a^2x_1x_2^2 + ax_1^3 - ax_1}{ax_1}$
23. $\frac{2\pi fL - \pi fR^2}{\pi fR}$
24. $\frac{9(aB)^4 - 6aB^4}{3aB^3}$
25. $\frac{-7a^2b + 14ab^2 - 21a^3}{14a^2b^2}$
26. $\frac{2x^{n+2} + 4ax^n}{2x^n}$
27. $\frac{6y^{2n} - 4ay^{n+1}}{2y^n}$
28. $\frac{3a(F + T)b^2 - (F + T)}{a(F + T)}$

In Exercises 29–48, perform the indicated divisions. Express the answer as shown in Example 5 when applicable.

29. $(x^2 + 9x + 20) \div (x + 4)$
30. $(x^2 + 7x - 18) \div (x - 2)$
31. $(2x^2 + 7x + 3) \div (x + 3)$
32. $(3t^2 - 7t + 4) \div (t - 1)$
33. $(x^2 - 3x + 2) \div (x - 2)$
34. $(2x^2 - 5x - 7) \div (x + 1)$
35. $(x - 14x^2 + 8x^3) \div (2x - 3)$
36. $(6 + 7y + 6y^2) \div (2y + 1)$
37. $(4Z^2 - 5Z - 7) \div (4Z + 3)$
38. $(6x^2 - 5x - 9) \div (3x - 4)$
39. $\frac{x^3 + 3x^2 - 4x - 12}{x + 2}$
40. $\frac{3x^3 + 19x^2 + 13x - 20}{3x - 2}$
41. $\frac{2a^4 + 4a^2 - 16}{a^2 - 2}$
42. $\frac{6T^3 + T^2 + 2}{3T^2 - T + 2}$
43. $\frac{x^3 + 8}{x + 2}$
44. $\frac{D^3 - 1}{D - 1}$
45. $\frac{x^2 - 2xy + y^2}{x - y}$
46. $\frac{3r^2 - 5rR + 2R^2}{r - 3R}$
47. $\frac{t^3 - 8}{t^2 + 2t + 4}$
48. $\frac{a^4 + b^4}{a^2 - 2ab + 2b^2}$

In Exercises 49–58, perform the indicated divisions.

49. When $2x^2 - 9x - 5$ is divided by $x + c$, the quotient is $2x + 1$. Find c.
50. When $6x^2 - x + k$ is divided by $3x + 4$, the remainder is zero. Find k.
51. By division show that $\frac{x^4 + 1}{x + 1}$ is not equal to x^3.
52. By division show that $\frac{x^3 + y^3}{x + y}$ is not equal to $x^2 + y^2$.

53. In the optical theory dealing with lasers, the following expression arises: $\dfrac{8A^5 + 4A^3\mu^2E^2 - A\mu^4E^4}{8A^4}$. ($\mu$ is the Greek letter mu.) Perform the division.

54. In finding the total resistance of the resistors shown in Fig. 1.13, the following expression is used.

$$\frac{6R_1 + 6R_2 + R_1R_2}{6R_1R_2}$$

Perform the division.

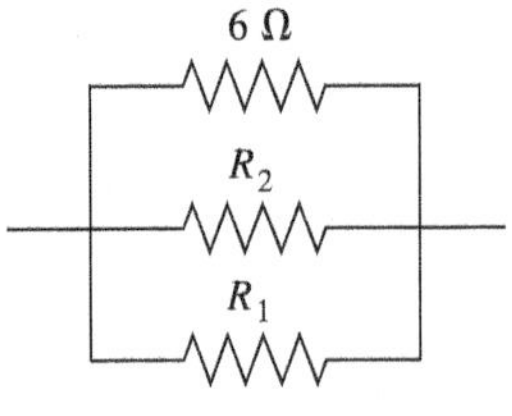

Fig. 1.13

55. When analysing the potential energy associated with gravitational forces, the expression $\dfrac{GMm[(R + r) - (R - r)]}{2rR}$ arises. Perform the indicated division.

56. A computer model shows that the temperature change T in a certain freezing unit is found by using the expression $\dfrac{3T^3 - 8T^2 + 8}{T - 2}$. Perform the indicated division.

57. In analysing the displacement of a certain valve, the expression $\dfrac{s^2 - 2s - 2}{s^4 + 4}$ is used. Find the reciprocal of this expression and then perform the indicated division.

58. In analysing a rectangular computer image, the area and width of the image vary with time such that the length is given by the expression $\dfrac{2t^3 + 94t^2 - 290t + 500}{2t + 100}$. By performing the indicated division, find the expression for the length.

Answers to Practice Exercises

1. $2x - 3a$ **2.** $2x + 3$

1.10 Solving Equations

An **equation** *is an algebraic statement that two algebraic expressions are equal.* Any value of the literal number representing the **unknown** that produces equality when **substituted** in the equation is said to **satisfy** the equation and represents a *valid solution* for the equation.

EXAMPLE 1 Valid values for equations

Determine if $x = 3$ and $x = 2$ are valid solutions of $3x - 5 = x + 1$.

Substituting 3 for x in the equation, we have $3(3) - 5 = 3 + 1$, or $4 = 4$. Since both sides are equal, 3 satisfies the equation and is a valid solution.

Substituting 2 for x in the equation, we have $3(2) - 5 = 2 + 1$, or $1 = 3$. Since both sides are not equal, 2 is not a valid solution.

Terminology for Equations	EXAMPLE 2
An **identity** is an equation that is valid for all values of the unknown(s).	**(a)** $x^2 - 4 = (x - 2)(x + 2)$ is true for all values of x.
A **contradiction** is an equation that is not valid for any value of the unknown(s).	**(b)** $x + 5 = x + 1$ is not valid for any value of x since the left side is always greater than the right.
A **conditional equation** is an equation that is valid only for certain values of the unknown(s).	**(c)** $3x - 5 = x + 1$ is only valid when $x = 3$ (see Example 1). **(d)** $2x + y = 5$ is valid only for certain combinations of x and y. For example, if $x = 2$, then y must be 1 in order to maintain the equality. Similarly, if $x = 1$, then y must be 3.

■ Equations can be solved on many graphing utilities. An estimate (or guess) of the answer may be required to find the solution. See Exercises 47 and 48.

To **solve** *an equation, we find the values of the unknown that satisfy it.* There is one basic rule to follow when solving an equation:

Perform the same operation on both sides of the equation.

We do this to isolate the unknown and thus to find its value.

By performing the same operation on both sides of an equation, the two sides remain equal. Thus,

■ Although we may multiply both sides of an equation by zero, this produces $0 = 0$, which is not useful in finding the solution.

we may add the same number to both sides, subtract the same number from both sides, multiply both sides by the same number, or divide both sides by the same number (not zero).

EXAMPLE 3 Basic operations used in solving

■ The word *algebra* comes from Arabic and means "a restoration." It refers to the fact that when a number has been added to one side of an equation, the same number must be added to the other side to restore equality.

In each of the following equations, we may isolate x, and thereby solve the equation, by performing the indicated operation.

$x - 3 = 12$	$x + 3 = 12$	$\frac{x}{3} = 12$	$3x = 12$
add 3 to both sides	subtract 3 from both sides	multiply both sides by 3	divide both sides by 3
$x - 3 + 3 = 12 + 3$	$x + 3 - 3 = 12 - 3$	$3\left(\frac{x}{3}\right) = 3(12)$	$\frac{3x}{3} = \frac{12}{3}$
$x = 15$	$x = 9$	$x = 36$	$x = 4$

Procedure for Solving Equations

1. *Remove grouping symbols (distributive law).*
2. *Combine any like terms on each side.*
3. *Perform the same operations on both sides to simplify, with the goal of isolating x.*
4. *Repeat Steps 1–3 until x = result is obtained.*
5. *Check the solution in the original equation.*

LEARNING TIP
The solution to any equation can be checked by substitution into the original equation.

EXAMPLE 4 Operations used for solution—checking

Solve the equation $2t - 7 = 9$.

We are to perform basic operations to both sides of the equation to finally isolate t on one side. The steps to be followed are suggested by the form of the equation.

$$2t - 7 = 9 \quad \text{original equation}$$
$$2t - 7 + 7 = 9 + 7 \quad \text{add 7 to both sides}$$
$$2t = 16 \quad \text{combine like terms}$$
$$\frac{2t}{2} = \frac{16}{2} \quad \text{divide both sides by 2}$$
$$t = 8 \quad \text{simplify}$$

■ Note that the solution generally requires a combination of basic operations.

Therefore, we conclude that $t = 8$. Checking *in the original equation,* we have

$$2t - 7 = 9$$
$$2(8) - 7 = 9$$
$$16 - 7 = 9$$
$$9 = 9$$

The solution checks.

■ With simpler numbers, many basic steps are done by inspection and not actually written down.

Practice Exercises

Solve for x.

1. $3x + 4 = x - 6$

2. $2(5 - x) = x - 8$

EXAMPLE 5 First remove parentheses

Solve the equation $x - 7 = 3x - (6x - 8)$.

$x - 7 = 3x - 6x + 8$	parentheses removed	
$x - 7 = -3x + 8$	x-terms combined on right	
$4x - 7 = 8$	$3x$ added to both sides	—by inspection
$4x = 15$	7 added to both sides	—by inspection
$x = \frac{15}{4}$	both sides divided by 4	—by inspection

Checking in the original equation, we obtain (after simplifying) $-\frac{13}{4} = -\frac{13}{4}$.

Note that we ***always check in the original equation.*** This is done since errors may have been made in finding the later equations.

If an equation contains numbers not easily combined by inspection, the best procedure is to *first solve for the unknown algebraically and then perform the calculation.* Rearranging first will limit the number of individual calculations you need to make, and the potential for error due to rounding (which should only be done after the last calculation).

■ Many other types of equations require more advanced methods for solving. These are considered in later chapters.

EXAMPLE 6 First solve for unknown—circuit current

When finding the current i (in A) in a certain radio circuit, the following equation and solution are used.

$$0.0595 - 0.525i - 8.85(i + 0.003\,16) = 0$$

$$0.0595 - 0.525i - 8.85i - 8.85(0.003\,16) = 0$$

$$(-0.525 - 8.85)i = 8.85(0.003\,16) - 0.0595$$

note how the above procedure is followed

$$i = \frac{8.85(0.003\,16) - 0.0595}{-0.525 - 8.85}$$

evaluate

$$= 0.003\,36 \text{ A}$$

When doing the calculation indicated above in the solution for i, be careful to group the numbers in the denominator for the division. Also, be sure to round off the result as shown above, but do not round off values before the final calculation.

RATIO AND PROPORTION

The quotient a/b is also called the **ratio** *of a to b.* For example, if a person earns \$40 in one hour, the ratio could be written as \$40/(1 h). *An equation stating that two ratios are equal is called a* **proportion**. For example, \$40/(1 h) = \$200/(5 h). Since a proportion is an equation, if one of the numbers is unknown, we can solve for its value as with any equation.

■ In general, the proportion $\frac{a}{b} = \frac{c}{d}$ is equivalent to the equation $ad = bc$. Therefore, $\frac{x}{8} = \frac{3}{4}$ can be rewritten as $4x = 24$ in order to remove the fractions.

EXAMPLE 7 Solving a proportion

If the ratio of x to 8 equals the ratio of 3 to 4, we have the proportion

$$\frac{x}{8} = \frac{3}{4}$$

We can solve this equation by multiplying both sides by 8. This gives

$$8\left(\frac{x}{8}\right) = 8\left(\frac{3}{4}\right)$$

$$x = \frac{24}{4} = 6$$

Substituting $x = 6$ into the original proportion gives the proportion $\frac{6}{8} = \frac{3}{4}$. Since these ratios are equal, the solution checks.

Practice Exercise

3. If the ratio of 2 to 5 equals the ratio of x to 30, find x.

EXAMPLE 8 Proportion—roof truss

The supports for a roof are triangular trusses for which the longest side is $8/5$ as long as the shortest side for all of the trusses. If the shortest side of one of the trusses is 3.80 m, what is the length of the longest side of that truss?

If we label the longest side L, since the ratio of sides is $8/5$, we have

$$\frac{L}{3.80} = \frac{8}{5}$$

$$3.80\left(\frac{L}{3.80}\right) = 3.80\left(\frac{8}{5}\right)$$

$$L = 6.08\,\text{m}$$

Checking, we note that $6.08/3.80 = 1.6$ and $8/5 = 1.6$. The solution checks.

■ As in Example 8, units of measurement will not be shown in intermediate steps. The proper units will be shown with the data and final result.

We will use the terms *ratio* and *proportion* (particularly ratio) when studying trigonometry in Chapter 4. A detailed discussion of ratio and proportion is found in Chapter 18. A general method of solving equations involving fractions is given in Chapter 6.

EXERCISES 1.10

In Exercises 1–4, make the given changes in the indicated examples of this section and then solve the resulting problems.

1. In Example 3, change 12 to -12 in each of the four illustrations and then solve.

2. In Example 4, change $2t - 7$ to $7 - 2t$ and then solve.

3. In Example 5, change $(6x - 8)$ to $(8 - 6x)$ and then solve.

4. In Example 8, change $8/5$ to $7/4$ and then solve.

In Exercises 5–34, solve the given equations.

5. $x - 2 = 7$ **6.** $x - 4 = -1$ **7.** $x + 5 = 4$ **8.** $-s - 6 = 14$

9. $-\frac{t}{2} = 5$ **10.** $\frac{x}{-4} = 2$ **11.** $\frac{y - 8}{3} = 4$

12. $\frac{7 - r}{6} = 3$ **13.** $4E = -20$ **14.** $2x = 12$

15. $3t + 5 = -4$ **16.** $5D - 2 = 13$ **17.** $5 - 2y = -3$

18. $8 - 5t = 18$ **19.** $3x + 7 = x$ **20.** $6 + 4L = 5 - 3L$

21. $2(3q + 4) = 5q$ **22.** $3(4 - n) = -n$

23. $6 - (r - 4) = 2r$ **24.** $5 - (x + 2) = 5x$

25. $8(y - 5) = -2y$ **26.** $6(8 - 3F) = -10$

27. $0.1x - 0.5(x - 2) = 2$ **28.** $1.5x - 0.3(x - 4) = 6$

29. $7 - 3(1 - 2p) = 4 + 2p$ **30.** $3 - 6(2 - 3t) = t - 5$

31. $\frac{4x - 2(x - 4)}{3} = 8$ **32.** $2x = \frac{-5(7 - 3x) + 2}{4}$

33. $|x| - 9 = 2$ **34.** $2 - |x| = 4$

In Exercises 35–44, solve the given equations assuming that all numbers are approximate.

35. $5.8 - 0.3(x - 6.0) = 0.5x$

36. $1.9t = 0.5(4.0 - t) - 0.8$

37. $-0.24(C - 0.50) = 0.63$

38. $27.5(5.17 - 1.44x) = 73.4$

39. $\frac{x}{2.0} = \frac{17}{6.0}$ **40.** $\frac{3.0}{7.0} = \frac{R}{42}$

41. $\frac{165}{223} = \frac{13V}{15}$ **42.** $\frac{276x}{17.0} = \frac{1360}{46.4}$

43. $\frac{7.4}{x} = \frac{19.1}{42.7}$ **44.** $\frac{10}{12.5} = \frac{-2.5}{R}$

In Exercises 45–57, solve the given problems.

45. Identify each of the following equations as a conditional equation, an identity, or a contradiction.

(a) $2x + 3 = 3 + 2x$

(b) $2x - 3 = 3 - 2x$

46. For what values of a is the equation $2x + a = 2x$ a conditional equation? Explain.

47. Solve the equation of Example 5 by using the Equation Solver of a graphing utility.

48. Solve the equation of Example 6 by using the Equation Solver of a graphing utility.

49. In finding the speed v (in km/h) at which a certain person was travelling, the equation $2.0v + 40 = 2.5(v + 5.0)$ is to be solved. Find the speed.

50. In finding the rate v (in km/h) at which a polluted stream is flowing, the equation $15(5.5 + v) = 24(5.5 - v)$ is used. Find v.

51. In finding the maximum operating temperature T (in °C) for a computer integrated circuit, the equation $1.1 = (T - 76)/40$ is used. Find the temperature.

52. To find the voltage V in a circuit in a TV remote-control unit, the equation $1.12V - 0.67(10.5 - V) = 0$ is used. Find V.

53. In blending two gasolines of different octanes, in order to find the number n of litres of one octane needed, the equation $0.14n + 0.06(2000 - n) = 0.09(2000)$ is used. Find n, given that 0.06 and 0.09 are exact and the first zero of 2000 is significant.

54. In order to find the distance x such that the weights are balanced on the lever shown in Fig. 1.14, the equation $215(3x) = 55.3x + 38.5(8.25 - 3x)$ must be solved. Find x. (3 is exact.)

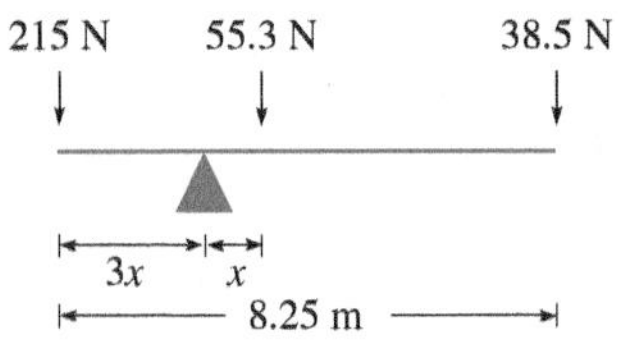

Fig. 1.14

55. The manufacturer of a certain car powered partly by gas and partly by batteries (a hybrid car) claims it can go 2050 km on one full tank of 55 L. How far can the car go on 22 L?

56. A person 1.8 m tall is photographed with a 35-mm camera, and the film image is 20 mm. Under the same conditions, how tall is a person whose film image is 16 mm?

57. An athlete who was jogging and wearing a Fitbit found that she burned 250 calories in 20 minutes. At that rate, how long will it take her to burn 400 calories? Assume all numbers are exact.

Answers to Practice Exercises

1. −5 **2.** 6 **3.** 12

1.11 Formulas and Literal Equations

A **formula** *is an equation that expresses the relationship between two or more related quantities.* For example, Einstein's famous formula $E = mc^2$ shows the equivalence of energy E to the mass m of an object times the speed of light squared, c^2.

LEARNING TIP

We can solve a formula for a particular symbol just as we solve any equation. That is, *we isolate the required symbol by using algebraic operations on literal numbers.*

EXAMPLE 1 Solving for a literal number in a formula—Einstein

In Einstein's formula $E = mc^2$, solve for m.

$$\frac{E}{c^2} = m \quad \text{divide both sides by } c^2$$

$$m = \frac{E}{c^2} \quad \text{switch sides to place } m \text{ at left}$$

By convention, the literal number we are solving for is placed on the left, as shown.

■ The subscript $_0$ makes v_0 a different literal symbol from v. (We have used subscripts in a few of the earlier exercises.)

EXAMPLE 2 Literal number with subscript in a formula—velocity

A formula relating acceleration a, velocity v, initial velocity v_0, and time t is $v = v_0 + at$. Solve for t.

$$v - v_0 = at \quad v_0 \text{ subtracted from both sides}$$

$$t = \frac{v - v_0}{a} \quad \text{both sides divided by } a \text{ and then sides switched}$$

■ A formula can contain a letter in both upper case and lower case, indicating two different unknown quantities. The literal numbers w and W represent different values.

EXAMPLE 3 Literal number in upper and lower case—forces on a beam

In the study of the forces on a certain beam, the equation $W = \frac{L(wL + 2P)}{8}$ is used. Solve for P.

$$8W = \frac{8L(wL + 2P)}{8} \quad \text{multiply both sides by 8}$$

$$8W = L(wL + 2P) \quad \text{simplify right side}$$

$$8W = wL^2 + 2LP \quad \text{remove parentheses}$$

$$8W - wL^2 = 2LP \quad \text{subtract } wL^2 \text{ from both sides}$$

$$P = \frac{8W - wL^2}{2L} \quad \text{divide both sides by } 2L \text{ and switch sides}$$

EXAMPLE 4 Formula with groupings—temperature and volume

The effect of temperature upon measurements is important when measurements must be made with great accuracy. The volume V of a special precision container at temperature T in terms of the volume V_0 at temperature T_0 is given by $V = V_0[1 + b(T - T_0)]$, where b depends on the material of which the container is made. Solve for T.

Since we are to solve for T, we must isolate the term containing T. This can be done by first removing the grouping symbols.

$$V = V_0[1 + b(T - T_0)] \quad \text{original equation}$$
$$V = V_0[1 + bT - bT_0] \quad \text{remove parentheses}$$
$$V = V_0 + bTV_0 - bT_0V_0 \quad \text{remove brackets}$$
$$V - V_0 + bT_0V_0 = bTV_0 \quad \text{subtract } V_0 \text{ and add } bT_0V_0 \text{ to both sides}$$
$$T = \frac{V - V_0 + bT_0V_0}{bV_0} \quad \text{divide both sides by } bV_0 \text{ and switch sides}$$

Practice Exercises

Solve for the indicated letter. Each comes from the indicated area of study.

1. $\theta = kA + \lambda$, for λ (robotics)
2. $P = n(p - c)$, for p (economics)

LEARNING TIP

To determine the values of any literal number in an expression for which we know values of the other literal numbers, we isolate the unknown literal value before substituting the known values. This is good practice in engineering and other technical work for two reasons:
(1) it generates a general formula, and
(2) algebra is easier to troubleshoot than numbers.

EXAMPLE 5 Solve for a symbol, then substitute—temperature and resistance

The electric resistance R (in Ω) of a resistor changes with the temperature T (in °C) according to $R = R_0 + R_0\alpha T$, where R_0 is the resistance at 0°C. For a given resistor, $R_0 = 712\ \Omega$ and $\alpha = 0.004\,55/°\text{C}$. Find the value of T for $R = 825\ \Omega$.

We first solve for T and then substitute the given values.

$$R = R_0 + R_0\alpha T$$
$$R - R_0 = R_0\alpha T$$
$$T = \frac{R - R_0}{\alpha R_0}$$

Now substituting, we have

$$T = \frac{825 - 712}{(0.004\,55)(712)}$$
$$= 34.9°\text{C}$$

EXERCISES 1.11

In Exercises 1–4, solve for the given letter from the indicated example of this section.

1. For the formula in Example 2, solve for a.
2. For the formula in Example 3, solve for w.
3. For the formula in Example 4, solve for T_0.
4. For the formula in Example 5, solve for α. (Do not evaluate.)

In Exercises 5–42, each of the given formulas arises in the technical or scientific area of study shown. Solve for the indicated letter.

5. $E = IR$, for R (electricity)
6. $pV = nRT$, for T (chemistry)
7. $rL = g_2 - g_1$, for g_1 (surveying)
8. $W = S_dT - Q$, for Q (air conditioning)
9. $B = \frac{nTWL}{12}$, for n (construction management)
10. $P = 2\pi Tf$, for T (mechanics)
11. $p = p_a + dgh$, for h (hydrodynamics)
12. $2Q = 2I + A + S$, for I (nuclear physics)
13. $F_c = \frac{mv^2}{r}$, for r (centripetal force)
14. $P = \frac{4F}{\pi D^2}$, for F (automotive trades)
15. $S_T = \frac{A}{5T} + 0.05d$, for A (welding)
16. $A = LH + \frac{1}{2}BH + \frac{\pi L^2}{8}$, for H (architecture)

17. $ct^2 = 0.3t - ac$, for a (medical technology)

18. $2p + dv^2 = 2d(C - W)$, for C (fluid flow)

19. $T = \frac{c + d}{v}$, for d (traffic flow)

20. $B = \frac{\mu_0 I}{2\pi R}$, for R (magnetic field)

21. $\frac{K_1}{K_2} = \frac{m_1 + m_2}{m_1}$, for m_2 (kinetic energy)

22. $f = \frac{F}{d - F}$, for d (photography)

23. $a = \frac{2mg}{M + 2m}$, for M (pulleys)

24. $v = \frac{V(m + M)}{m}$, for M (ballistics)

25. $C_0^2 = C_1^2(1 + 2V)$, for V (electronics)

26. $A_1 = A(M + 1)$, for M (photography)

27. $N = r(A - s)$, for s (engineering)

28. $T = 3(T_2 - T_1)$, for T_1 (oil drilling)

29. $T_2 = T_1 - \frac{h}{100}$, for h (air temperature)

30. $p_2 = p_1 + rp_1(1 - p_1)$, for r (population growth)

31. $Q_1 = P(Q_2 - Q_1)$, for Q_2 (refrigeration)

32. $p - p_a = dg(y_2 - y_1)$, for y_2 (pressure gauges)

33. $N = N_1 T - N_2(1 - T)$, for N_1 (machine design)

34. $t_a = t_c + (1 - h)t_m$, for h (computer access time)

35. $L = \pi(r_1 + r_2) + 2x_1 + x_2$, for r_1 (pulleys)

36. $I = \frac{VR_2 + VR_1(1 + \mu)}{R_1 R_2}$, for μ (electronics)

37. $P = \frac{V_1(V_2 - V_1)}{gJ}$, for V_2 (jet engine power)

38. $W = T(S_1 - S_2) - Q$, for S_2 (refrigeration)

39. $C = \frac{2EAk_1k_2}{d(k_1 + k_2)}$, for E (electronics)

40. $d = \frac{3LPx^2 - Px^3}{6EI}$, for L (beam deflection)

41. $V = C\left(1 - \frac{n}{N}\right)$, for n (property depreciation)

42. $\frac{p}{P} = \frac{AI}{B + AI}$, for B (atomic theory)

In Exercises 43–48, rearrange the equation for the indicated values and then solve.

43. The efficiency η (in %) of a certain engine is given by the formula $\eta = T_2/(T_1 + T_2)$. Find T_1 if $\eta = 45.0$ and $T_2 = 875$ K.

44. A formula used in determining the total transmitted power P_t in an AM radio signal is $P_t = P_c(1 + 0.500\,m^2)$. Find P_c if $P_t = 685$ W and $m = 0.925$.

45. A formula relating the Fahrenheit temperature F and the Celsius temperature C is $F = \frac{9}{5}C + 32$. Find the Celsius temperature that corresponds to 90.2°F. (32 and $\frac{9}{5}$ are exact.)

46. In forestry, a formula used to determine the volume V of a log is $V = \frac{1}{2}L(B + b)$, where L is the length of the log and B and b are the areas of the ends. Find b (in m²) if $V = 1.09$ m³, $L = 4.91$ m, and $B = 0.244$ m². See Fig. 1.15.

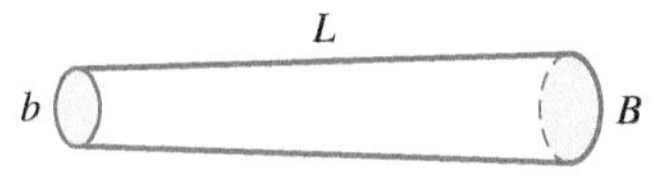

Fig. 1.15

47. The voltage V_1 across resistance R_1 is $V_1 = \frac{VR_1}{R_1 + R_2}$, where V is the voltage across resistances R_1 and R_2. See Fig. 1.16. Find R_2 (in Ω) if $R_1 = 3.56$ Ω, $V_1 = 6.30$ V, and $V = 12.0$ V.

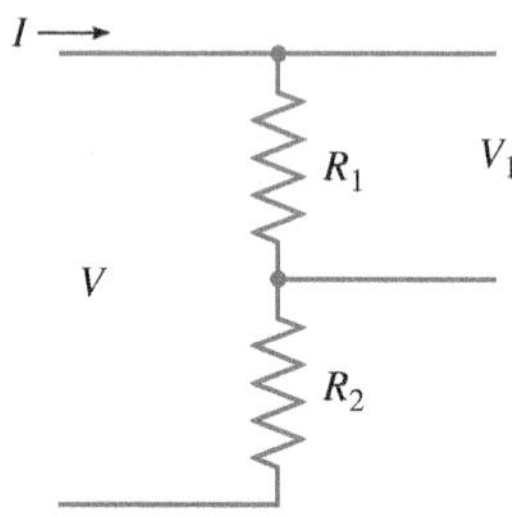

Fig. 1.16

48. The efficiency η of a computer multiprocessor compilation is given by $\eta = \frac{1}{q + p(1 - q)}$, where p is the number of processors and q is the fraction of the compilation that can be performed by the available parallel processors. Find p for $\eta = 0.66$ and $q = 0.83$.

In Exercises 49 and 50, set up the required formula and solve for the indicated letter.

49. One missile travels at a speed of v_2 km/h for 4 h, and another missile travels at a speed of v_1 for $t + 2$ hours. If they travel a total of d km, solve the resulting formula for t.

50. A microwave transmitter can handle x telephone communications, and 15 separate cables can handle y connections each. If the combined system can handle C connections, solve for y.

Answers to Practice Exercises

1. $\theta - kA$ **2.** $\frac{P + nc}{n}$

1.12 Applied Word Problems

Many applied problems are at first word problems, and we must put them into mathematical terms for solution. Usually, the most difficult part in solving a word problem is identifying the information needed for setting up the equation that leads to the solution. To do this, you must read the problem carefully to be sure that you understand all of the terms and expressions used. Following is an approach you should use.

■ See Appendix A, page A-1, for a variation to the method outlined in these steps. You might find it helpful.

Procedure for Solving Word Problems

1. **Read the problem twice.** Read it once quickly to get a general overview. Then, reread slowly and more carefully, paying attention to the information given.
2. **Identify the known quantities.** Write a list of known values and assign an appropriate letter to represent each one.
3. **Identify the unknown quantities.** As in Step 2, assign an appropriate symbol to represent each one.
4. **If possible, draw a picture or graph.** Visualizing can often help in identifying the relationship between known and unknown quantities.
5. **Analyse the problem statement, and select an appropriate equation.** This is often the most difficult step because some information may be implied and not explicitly stated. Again, a very careful reading of the statement is necessary.
6. **Solve the equation algebraically,** clearly stating the solution.
7. **Check the solution** with the original problem statement to ensure the solution is valid.

EXAMPLE 1 Sum of forces on a beam

■ Be sure to carefully identify your choice for the unknown. In most problems, there is really a choice. Using the word *let* clearly shows that a specific choice has been made.

■ The statement after "let x (or some other appropriate letter) =" should be clear. It should completely define the chosen unknown.

A 78-N beam is supported at each end. The supporting force at one end is 10 N more than at the other end. If the beam is not moving, find the values for both forces.

Identify known quantities: We know the downward beam force is 78 N. Let $F_{down} = 78$ N

Identify unknown quantities: We know there are two upward forces, one being 10 N more than the other.

Let $F =$ smaller force (in N)

$F + 10 =$ larger force (in N)

Sketch: See Fig. 1.17.

Fig. 1.17

Select an equation: If the beam is not moving, $F_{up} = F_{down}$. Therefore,

$$F + (F + 10) = 78$$

Solve the equation:

$$F + (F + 10) = 78$$
$$2F + 10 = 78$$
$$2F = 68$$
$$F = 34$$

The smaller force is 34 N, and the larger force is 44 N.

Check: The solution checks: the two forces add to 78 N, with one being 10 N more than the other.

■ "Let $x = 25$ W lights" is incomplete. We want to find out how many there are.

■ See Appendix A, page A-1, for a "sketch" that might be used with this example.

EXAMPLE 2 Office complex energy–efficient lighting

In designing an office complex, an architect planned to install 34 energy-efficient ceiling lights using a total of 1000 W. Two different types of lights—one using 25 W and the other using 40 W—were to be installed. How many of each were planned?

Identify known quantities: The total wattage is $W_{total} = 1000$ W.

Identify unknown quantities: We know there are two types of lights, for a total of 34 lights.

Let x = number of 25 W lights
$34 - x$ = number of 40 W lights

Sketch, if possible: Not possible

Select an equation: Each 25 W light contributes 25 W, and each 40 W light contributes 40 W to the total wattage. Therefore,

$$W_{total} = 25x + 40(34 - x)$$

Solve the equation: Substituting known values and solving,

$$1000 = 25x + 40(34 - x)$$
$$1000 = 25x + 1360 - 40x$$
$$15x = 360$$
$$x = 24$$

There are 24 lights that use 25 W, and 10 lights that use 40 W.

Check: The total wattage is $24(25) + 10(40) = 600 + 400 = 1000$, which checks with the statement of the problem.

EXAMPLE 3 Number of medical slides

A medical researcher finds that a given sample of an experimental drug can be divided into 4 more slides with 5 mg each than with 6 mg each. How many slides with 5 mg each can be made up?

Identify known quantities: None

Identify unknown quantities: Let x = number of slides with 5 mg
$x - 4$ = number of slides with 6 mg

Sketch, if possible: Not possible

Select an equation: The mass of the drug is the same regardless of how it is divided, so

$$5x = 6(x - 4)$$

Solve the equation:

$$5x = 6(x - 4)$$
$$5x = 6x - 24$$
$$-x = -24$$
$$x = 24$$

The sample can be divided into 24 slides with 5 mg each, or 20 slides with 6 mg each.

Check: The total mass, 120 mg, is the same for each set of slides, so the solution checks with the statement of the problem.

Practice Exercise

1. Solve the problem in Example 3 by letting y = number of slides with 6 mg.

EXAMPLE 4 Distance travelled—space travel

A space shuttle maneuvers so that it may "capture" an already orbiting satellite that is 6000 km ahead. If the satellite is moving at 27 000 km/h and the shuttle is moving at 29 500 km/h, how long will it take the shuttle to reach the satellite? (All digits shown are significant.)

Identify known quantities: Velocity: $v_{shuttle} = 29\ 500\ \text{km/h}$

$v_{satellite} = 27\ 000\ \text{km/h}$.

Distance: $d_{between} = 6000\ \text{km}$

Identify unknown quantities: Let t = time (in h) for the shuttle to reach the satellite

Sketch, if possible: See Fig. 1.18.

Select an equation: The distance travelled by the shuttle must be equal to the distance travelled by the satellite plus the distance between them:

$$d_{shuttle} = d_{satellite} + d_{between}$$

We can calculate each distance with the formula $d = vt(\textit{distance} = \textit{velocity} \times \textit{time})$. Therefore,

$$v_{shuttle}t = v_{satellite}t + 6000\ \text{km}$$

Solve the equation: Substituting known values and solving,

$$29\ 500t = 27\ 000t + 6000$$
$$2500t = 6000$$
$$t = 2.400$$

It will take the shuttle 2.400 h to reach the satellite.

Check: In 2.400 h, the shuttle will travel 70 800 km, and the satellite will travel 64 800 km. This checks with the statement of the problem.

■ A maneuver similar to the one in this example was used to repair the Hubble space telescope in 1994 and 2001.

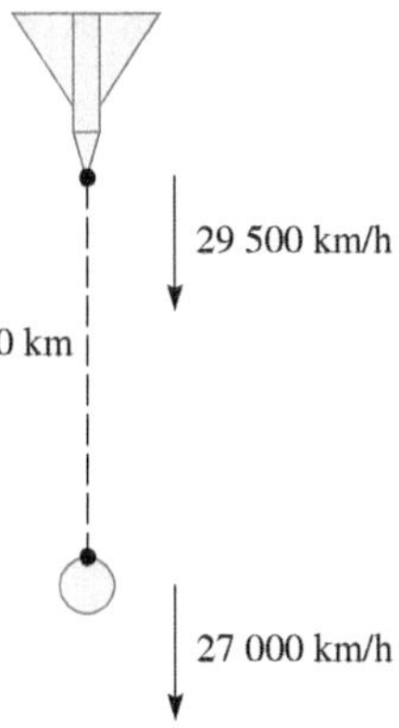

Fig. 1.18

EXAMPLE 5 Mixture—gasoline and methanol

A refinery has 7250 L of a gasoline–methanol blend that is 6.00% methanol. How much pure methanol must be added so that the resulting blend is 10.0% methanol?

Identify known quantities: Initial volume: $V_1 = 7250\ \text{L}$

Concentrations: $c_1 = 0.0600$ and $c_2 = 0.100$

Identify unknown quantities: Let x = litres of methanol added

Sketch, if possible: See Fig. 1.19.

Select an equation: For mixtures, volume or mass of solute must be maintained. The volume of methanol in the first solution plus the added methanol must be the same as the volume of methanol in the final solution. Therefore, $c_1V_1 + x = c_2(V_1 + x)$.

Solve the equation: Substituting known values and solving,

$$0.0600(7250) + x = 0.100(7250 + x)$$
$$435 + x = 725 + 0.100x$$
$$0.900x = 290$$
$$x = 322$$

322 L of pure methanol must be added.

Check: There would be 757 L of methanol in the total of 7570 L of blend (to three significant digits).

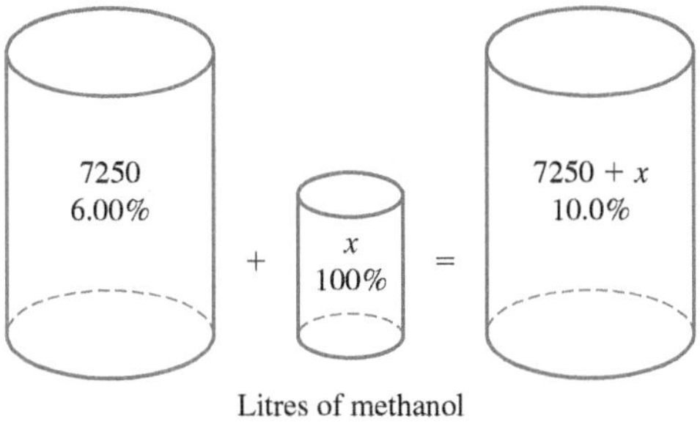

Fig. 1.19

EXERCISES 1.12

In Exercises 1–4, make the given changes in the indicated examples of this section and then solve the resulting problems.

1. In Example 2, in the first line, change 34 to 31.
2. In Example 3, in the second line, change "4 more slides" to "3 more slides."
3. In Example 4, in the second line, change 27 000 km/h to 27 500 km/h.
4. In Example 5, change "pure methanol" to "of a blend with 50.0% methanol."

In Exercises 5–33, set up an appropriate equation and solve. Data are accurate to two significant digits unless greater accuracy is given.

5. A certain new car costs $5000 more today than the same model new car cost six years ago. Together, new models today and six years ago cost $70 000. What was the cost of each?
6. The flow of one stream into a lake is 45 m^3/s more than the flow of a second stream. In 1 h, 4.14×10^5 m^3 flow into the lake from the two streams. What is the flow rate of each?
7. Approximately 4.5 million wrecked cars are recycled in two consecutive years. There were 700 000 more recycled the second year than the first year. How many are recycled each year?
8. A business website was accessed as often on the first day of a promotion as on the next two days combined. The first day hits were four times those on the third. The second day hits were 4000 more than the third. How many were there each day?
9. Petroleum rights to 140 hectares of land are leased for $37 000. Part of the land leases for $200 per hectare, and the remainder for $300 per hectare. How much is leased at each price?
10. A vial contains 2000 mg of a drug, which is to be used for two dosages. One patient is to be administered 660 mg more than another. How much should be administered to each?
11. After installing a pollution control device, a car's exhaust contained the same amount of pollutant after 5.0 h as it had in 3.0 h. Before the installation the exhaust contained 150 ppm/h (parts per million per hour) of the pollutant. By how much did the device reduce the emission?
12. Three meshed spur gears have a total of 107 teeth. If the second gear has 13 more teeth than the first and the third has 15 more teeth than the second, how many teeth does each have?
13. In the design of a bridge, an engineer determines that four fewer 18-m girders are needed for the span length than 15-m girders. How many 18-m girders are needed?
14. A fuel oil storage depot had an eight-week supply on hand. However, cold weather caused the supply to be used in six weeks when 20 kL extra were used each week. How many litres were in the original supply?
15. The sum of three electric currents that come together at a point in a circuit is zero. If the second current is twice the first and the third current is 9.2 μA more than the first, what are the currents? (The sign indicates the direction of flow.)
16. A delivery firm uses one fleet of trucks on daily routes of 8 h. A second fleet, with five more trucks than the first, is used on daily routes of 6 h. Budget allotments allow for 198 h of daily delivery time. How many trucks are in each fleet?
17. A natural gas pipeline feeds into three smaller pipelines, each of which is 2.6 km longer than the main pipeline. The total length of the four pipelines is 35.4 km. How long is each section?
18. At 100% efficiency two generators would produce 750 MW of power. If one generator has 65% efficiency and the second has 75% efficiency, they only produce 530 MW. Use this to determine how much power is generated by each at 100% efficiency.
19. A person installed a solar panel roof with 54 panels. The entire installation cost $8760, with flat panels that cost $150 each and custom panels that cost $180 each. How many of each panel type were installed?
20. A person won a lottery prize of $20 000, from which 25% was deducted for taxes. The remainder was invested, partly for a 40% gain, and the rest for a 10% loss. How much was each investment if there was a $2000 net investment gain?
21. A ski lift takes a skier up a slope at 50 m/min. The skier then skis down the slope at 150 m/min. If a round trip takes 24 min, how long is the slope?
22. Before being put out of service, the supersonic jet Concorde made a trip averaging 100 km/h less than the speed of sound for 1.00 h, and 400 km/h more than the speed of sound for 3.00 h. If the trip covered 5740 km, what is the speed of sound?
23. Trains at each end of the 50.0-km-long Eurotunnel under the English Channel start at the same time into the tunnel. Find their speeds if the train from France travels 8.0 km/h faster than the train from England and they pass in 17.0 min. See Fig. 1.20.

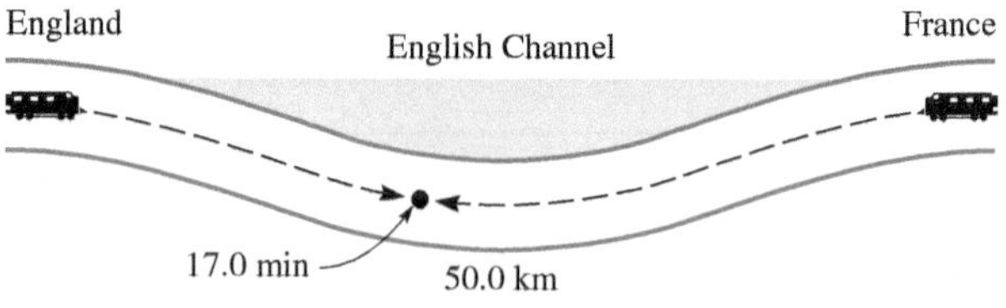

Fig. 1.20

24. An executive would arrive 10.0 min early for an appointment if travelling at 60 km/h, or 5.0 min early if travelling at 45.0 km/h. How much time is there until the appointment?
25. One lap at the Canadian Grand Prix in Montreal is 4.36 km. In a race, a car stalls and then starts 30.0 s after a second car. The first car travels at 79.0 m/s, and the second car travels at 73.0 m/s. How long does it take the first car to overtake the second, and which car will be ahead after eight laps?
26. A computer chip manufacturer produces two types of chips. In testing a total of 6100 chips of both types, 0.50% of one type and 0.80% of the other type were defective. If a total of 38 defective chips were found, how many of each type were tested?
27. Two gasoline distributors, A and B, are 367 km apart on a highway. A charges $1.80/L and B charges $1.68/L. Each charges 0.16¢/L per kilometre for delivery. Where on this highway is the cost to the customer the same?
28. An outboard engine uses a gasoline–oil fuel mixture in the ratio of 15 to 1. How much gasoline must be mixed with a gasoline–oil mixture, which is 75% gasoline, to make 8.0 L of the mixture for the outboard engine?

29. A car's radiator contains 12.0 L of antifreeze at a 25% concentration. How many litres must be drained and then replaced by pure antifreeze to bring the concentration to 50% (the manufacturer's "safe" level)?

30. How much sand must be added to 250 kg of a cement mixture that is 22% sand to have a mixture that is 25% sand?

31. To pass a 20-m long semitrailer travelling at 70 km/h in 10 s, how fast must a 5.0-m long car go?

32. An earthquake emits primary waves moving at 8.0 km/s and secondary waves moving at 5.0 km/s. How far from the epicentre of the earthquake is the seismic station if the two waves arrive at the station 2.0 min apart?

33. If 0.1 L of a 50% concentrated acid is added to an existing 4 L of 10% concentrated acid in a jug, what is the concentration of the resulting mixture in the jug?

Answer to Practice Exercise

1. 24 with 5 mg, 20 with 6 mg

CHAPTER 1 KEY FORMULAS AND EQUATIONS

Commutative Law of Addition: $a + b = b + a$ **(1.1)**

Associative Law of Addition: $a + (b + c) = (a + b) + c$ **(1.2)**

Commutative Law of Multiplication: $ab = ba$ **(1.3)**

Associative Law of Multiplication: $a(bc) = (ab)c$ **(1.4)**

Distributive Law: $a(b + c) = ab + ac$ **(1.5)**

$a + (-b) = a - b$ **(1.6)**

$a - (-b) = a + b$ **(1.7)**

Rules of Exponents

Product Rule: $a^m \times a^n = a^{m+n}$ **(1.8)**

Quotient Rule: $\dfrac{a^m}{a^n} = a^{m-n} \quad (a \neq 0)$ **(1.9)**

Power of a Power Rule: $(a^m)^n = a^{mn}$ **(1.10)**

Power of a Product Rule: $(ab)^n = a^n b^n$ **(1.11)**

Power of a Quotient Rule: $\left(\dfrac{a}{b}\right)^n = \dfrac{a^n}{b^n} \quad (\text{if } b \neq 0)$ **(1.12)**

Zero Exponent $a^0 = 1 \quad (a \neq 0)$ **(1.13)**

Negative Exponent $a^{-n} = \dfrac{1}{a^n} \quad (a \neq 0)$ **(1.14)**

$\sqrt{ab} = \sqrt{a}\sqrt{b}$ (a and b positive real numbers) **(1.15)**

CHAPTER 1 REVIEW EXERCISES

In Exercises 1–12, evaluate the given expressions.

1. $(-2) + (-5) - 3$

2. $6 - 8 - (-4)$

3. $\dfrac{(-5)(6)(-4)}{(-2)(3)}$

4. $\dfrac{(-9)(-12)(-4)}{24}$

5. $-5 - |2(-6)| + \dfrac{-15}{3}$

6. $3 - 5|-3 - 2| - \dfrac{12}{-4}$

7. $\dfrac{18}{3 - 5} - (-4)^2$

8. $-(-3)^2 - \dfrac{-8}{(-2) - |-4|}$

9. $\sqrt{16} - \sqrt{64}$

10. $-\sqrt{81 + 144}$

11. $(\sqrt{7})^2 - \sqrt[3]{8}$

12. $-\sqrt[4]{16} + (\sqrt{6})^2$

In Exercises 13–20, simplify the given expressions. Where appropriate, express results with positive exponents only.

13. $(-2rt^2)^2$

14. $(3a^0b^{-2})^3$

15. $-3mn^{-5}t(8m^{-3}n^4)$

16. $\dfrac{15p^4q^2r}{5pq^5r}$

17. $\dfrac{-16N^{-2}(NT^2)}{-2N^0T^{-1}}$

18. $\dfrac{-35x^{-1}y(x^2y)}{5xy^{-1}}$

19. $\sqrt{45}$

20. $\sqrt{9+36}$

In Exercises 21–24, for each number, (a) determine the number of significant digits and (b) round off each to two significant digits.

21. 8840 **22.** 21 450 **23.** 9.040 **24.** 0.700

In Exercises 25–28, evaluate the given expressions. All numbers are approximate.

25. $37.3 - 16.92(1.067)^2$

26. $\dfrac{8.896 \times 10^{-12}}{3.5954 + 6.0449}$

27. $\dfrac{\sqrt{0.1958 + 2.844}}{3.142(65)^2}$

28. $\dfrac{1}{0.035\,68} + \dfrac{37{,}466}{29.63^2}$

In Exercises 29–60, perform the indicated operations.

29. $a - 3ab - 2a + ab$

30. $xy - y - 5y - 4xy$

31. $6LC - (3 - LC)$

32. $-(2x - b) - 3(-x - 5b)$

33. $(2x - 1)(x + 5)$

34. $(C - 4D)(2C - D)$

35. $(x + 8)^2$

36. $(2r - 9s)^2$

37. $\dfrac{2h^3k^2 - 6h^4k^5}{2h^2k}$

38. $\dfrac{4a^2x^3 - 8ax^4}{-2ax^2}$

39. $4R - [2r - (3R - 4r)]$

40. $3b - [3a - (a - 3b)] + 4a$

41. $2xy - \{3z - [5xy - (7z - 6xy)]\}$

42. $x^2 + 3b + [(b - y) - 3(2b - y + z)]$

43. $(2x + 1)(x^2 - x - 3)$

44. $(x - 3)(2x^2 - 3x + 1)$

45. $-3y(x - 4y)^2$

46. $-s(4s - 3t)^2$

47. $3p[(q - p) - 2p(1 - 3q)]$

48. $3x[2y - r - 4(s - 2r)]$

49. $\dfrac{12p^3q^2 - 4p^4q + 6pq^5}{2p^4q}$

50. $\dfrac{27s^3t^2 - 18s^4t + 9s^2t}{9s^2t}$

51. $(2x^2 + 7x - 30) \div (x + 6)$

52. $(4x^2 - 41) \div (2x + 7)$

53. $\dfrac{3x^3 - 7x^2 + 11x - 3}{3x - 1}$

54. $\dfrac{w^3 - 4w^2 + 7w - 12}{w - 3}$

55. $\dfrac{4x^4 + 10x^3 + 18x - 1}{x + 3}$

56. $\dfrac{8x^3 - 14x + 3}{2x + 3}$

57. $-3\{(r + s - t) - 2[(3r - 2s) - (t - 2s)]\}$

58. $(1 - 2x)(x - 3) - (x + 4)(4 - 3x)$

59. $\dfrac{2y^3 + 9y^2 - 7y + 5}{2y - 1}$

60. $\dfrac{6x^2 + 5xy - 4y^2}{2x - y}$

In Exercises 61–72, solve the given equations.

61. $3x + 1 = x - 8$

62. $4y - 3 = 5y + 7$

63. $\dfrac{5x}{7} = \dfrac{3}{2}$

64. $\dfrac{2(N - 4)}{3} = \dfrac{5}{4}$

65. $6x - 5 = 3(x - 4)$

66. $-2(-4 - y) = 3y$

67. $2s + 4(3 - s) = 6$

68. $2|x| - 1 = 3$

69. $3t - 2(7 - t) = 5(2t + 1)$

70. $-(8 - x) = x - 2(2 - x)$

71. $2.7 + 2.0(2.1x - 3.4) = 0.1$

72. $0.250(6.721 - 2.44x) = 2.08$

In Exercises 73–82, (a) change numbers in ordinary notation to scientific notation or change numbers in scientific notation to ordinary notation. (b) Write the numbers in engineering notation and replace the power of 10 with the corresponding SI prefix.

73. A certain computer has 60 000 000 000 bytes of memory.

74. The escape velocity (the velocity required to leave the earth's gravitational field) is about 40 000 km/h.

75. In 2015, the most distant known object in the solar system, a dwarf planet named V774104, was discovered. It was 15 400 000 000 km from the sun.

76. Police radar has a frequency of 1.02×10^9 Hz.

77. Among the stars nearest the earth, Centaurus A is about 4.05×10^{13} km away.

78. Before its destruction in 2001, the World Trade Center had nearly $10^6\ \text{m}^2$ of office space.

79. The faintest sound that can be heard has an intensity of about $10^{-12}\ \text{W/m}^2$.

80. An optical coating on glass to reduce reflections is about 0.000 000 15 m thick.

81. The maximum safe level of radiation in the air of a home due to radon gas is 1.5×10^{-1} Bq/L. (Bq is the symbol for bequerel, the metric unit of radioactivity, where 1 Bq = 1 decay/s.)

82. A certain virus was measured to have a diameter of about 0.000 000 18 m.

In Exercises 83–96, solve for the indicated letter. Where noted, the given formula arises in the technical or scientific area of study.

83. $V = \pi r^2 L$, for L (oil pipeline volume)

84. $R = \dfrac{2GM}{c^2}$, for G (astronomy: black holes)

85. $P = \dfrac{\pi^2 EI}{L^2}$, for E (mechanics)

86. $f = p(c - 1) - c(p - 1)$, for p (thermodynamics)

87. $Pp + Qq = Rr$, for q (moments of forces)

88. $V = IR + Ir$, for R (electricity)

89. $d = (n - 1)A$, for n (optics)

90. $mu = (m + M)v$, for M (physics: momentum)

91. $N_1 = T(N_2 - N_3) + N_3$, for N_2 (mechanics: gears)

92. $q = \dfrac{KA(B - C)}{L}$, for B (solar heating)

93. $R = \dfrac{A(T_2 - T_1)}{H}$, for T_2 (thermal resistance)

94. $Z^2\left(1 - \dfrac{\lambda}{2a}\right) = k$, for λ (radar design)

95. $d = kx^2[3(a + b) - x]$, for a (mechanics: beams)

96. $V = V_0[1 + 3a(T_2 - T_1)]$, for T_2 (thermal expansion)

In Exercises 97–102, perform the indicated calculations.

97. A computer's memory is 5.25×10^{10} bytes, and that of a model 20 years older is 6.4×10^4 bytes. What is the ratio of the newer computer's memory to the older computer's memory?

98. The time (in s) for an object to fall h metres is given by the expression $0.45\sqrt{h}$. How long does it take a person to fall 22 m from a sixth-floor window into a net while escaping a fire?

99. The CN Tower in Toronto is 0.553 km high. The Willis Tower (formerly the Sears Tower) in Chicago is 442 m high. How much taller is the CN Tower than the Sears Tower?

100. The time (in s) it takes a computer to check n memory cells is found by evaluating $(n/2650)^2$. Find the time to check 48 cells.

101. The combined electric resistance of two parallel resistors is found by evaluating the expression $\frac{R_1R_2}{R_1 + R_2}$. Evaluate this for $R_1 = 0.0275\ \Omega$ and $R_2 = 0.0590\ \Omega$.

102. The distance (in m) from the earth for which the gravitational force of the earth on a spacecraft equals the gravitational force of the sun on it is found by evaluating $1.5 \times 10^{11}\sqrt{m/M}$, where m and M are the masses of the earth and sun, respectively. Find this distance for $m = 5.98 \times 10^{24}$ kg and $M = 1.99 \times 10^{30}$ kg.

In Exercises 103–106, simplify the given expressions.

103. One transmitter antenna is $(x - 2a)$ cm long, and another is $(x + 2a)$ m long. What is the sum, in centimetres, of their lengths?

104. In finding the value of an annuity, the expression $(Ai - R)(1 + i)^2$ is used. Multiply out this expression.

105. A computer analysis of the velocity of a link in an industrial robot leads to the expression $4(t + h) - 2(t + h)^2$. Simplify this expression.

106. When analysing the motion of a communications satellite, the expression $\frac{k^2r - 2h^2k + h^2rv^2}{k^2r}$ is used. Perform the indicated division.

In Exercises 107–114, solve the given problems.

107. Does the value of $3 \times 18 \div (9 - 6)$ change if the parentheses are removed?

108. Does the value of $(3 \times 18) \div 9 - 6$ change if the parentheses are removed?

109. In solving the equation $x - (3 - x) = 2x - 3$, what conclusion can be made?

110. In solving the equation $7 - (2 - x) = x + 2$, what conclusion can be made?

111. Show that $(x - y)^3 = -(y - x)^3$.

112. Is division associative? That is, is it true (if $b \neq 0, c \neq 0$) that $(a \div b) \div c = a \div (b \div c)$?

113. What is the ratio of 8×10^{-3} to 2×10^4?

114. What is the ratio of $\sqrt{4 + 36}$ to $\sqrt{4}$?

In Exercises 115–129, solve the given problems. All data are accurate to two significant digits unless greater accuracy is given.

115. Two computer software programs cost \$190 together. If one costs \$72 more than the other, what is the cost of each?

116. A sponsor pays a total of \$9500 to run a commercial on two different TV stations. One station charges \$1100 more than the other. What does each charge to run the commercial?

117. Three chemical reactions each produce oxygen. If the first produces twice that of the second, the third produces twice that of the first, and the combined total is 560 cm^3, what volume is produced by each?

118. In testing the rate at which a polluted stream flows, a boat that travels at 5.5 km/h in still water took 5.0 h to go downstream between two points, and it took 8.0 h to go upstream between the same two points. What is the rate of flow of the stream?

119. The voltage across a resistor equals the current times the resistance. In a microprocessor series circuit, one resistor is 1200 Ω greater than another. The sum of the voltages across them is 12.0 mV. Find the resistances if the current is 2.4 μA in each.

120. An air sample contains 4.0 ppm (parts per million) of two pollutants. The concentration of one is four times the other. What are the concentrations?

121. One road crew constructs 450 m of road bed in 12 h. How long will it take them to construct another 250 m of road bed?

122. The fuel for a two-cycle motorboat engine is a mixture of gasoline and oil in the ratio of 15 to 1. How many litres of each are in 6.6 L of mixture?

123. A ship enters Lake Superior from Sault Ste. Marie, moving toward Duluth at 17.4 km/h. Two hours later, a second ship leaves Duluth moving toward Sault Ste. Marie at 21.8 km/h. When will the ships pass, given that Sault Ste. Marie is 634 km from Duluth?

124. A helicopter used in fighting a forest fire travels at 175 km/h from the fire to a pond and 115 km/h with water from the pond to the fire. If a round trip takes 30 min, how long does it take from the pond to the fire? See Fig. 1.21.

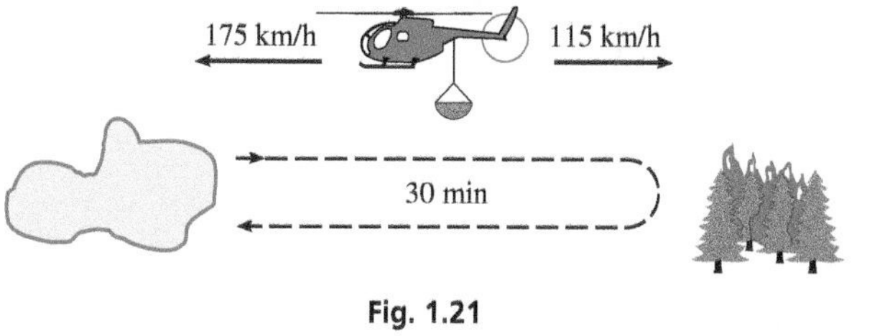

Fig. 1.21

125. One grade of oil has 0.50% of an additive, and a higher grade has 0.75% of the additive. How many litres of each must be used to have 1000 L of a mixture with 0.65% of the additive?

126. How much water must be added to 20.0 mL of a 60.0% saline solution in order to make a 45.0% saline solution?

127. An architect plans to have 25% of the floor area of a house in ceramic tile. In all but the kitchen and entry, there are 205 m^2 of floor area, 15% of which is tile. What area can be planned for the kitchen and entry if each has an all-tile floor?

128. A *karat* equals 1/24 part of gold in an alloy (for example, 9-karat gold is 9/24 gold). How many grams of 9-karat gold must be mixed with 18-karat gold to get 200 g of 14-karat gold?

129. In calculating the simple interest earned by an investment, the equation $P = P_0 + P_0rt$ is used, where P is the value after an initial principal P_0 is invested for t years at interest rate r. Solve for r, and then evaluate r for $P = \$7625$, $P_0 = \$6250$, and $t = 4.000$ years. Write a paragraph or two explaining (a) your method for solving for r, and (b) the calculator steps used to evaluate r, noting the use of parentheses.

CHAPTER 1 PRACTICE TEST

In Problems 1–5, evaluate the given expressions. In Problems 3 and 5, the numbers are approximate.

1. $\sqrt{9 + 36}$

2. $\dfrac{(5)(-31)(6)}{(0)(-4)}$

3. $\dfrac{5.279 \times 10^7}{4.393 \times 10^{-6}}$

4. $\dfrac{(-4)(3) - (-4)(-2)}{|-3 - 1|}$

5. $\dfrac{392.4 - 57.9}{486.2} - \dfrac{0.675^3}{(2.75)(0.113)}$

In Problems 6–12, perform the indicated operations and simplify. When exponents are used, use only positive exponents in the result.

6. $(5a^4b^0c^{-5})^{-2}$

7. $(5x - 2)^2$

8. $4k^3(bk - 5k^2)$

9. $\dfrac{8a^3x^2 - 4a^2x^4}{-2ax^2}$

10. $\dfrac{6x^2 - 13x + 7}{2x - 1}$

11. $(2x - 3)(x + 7)$

12. $[3x - (4x - 3)] - 2x$

In Exercises 13–22, solve the given problems.

13. Solve for y: $3y - 2(3y - 4) = 10$

14. Solve for x: $10(2x - 3) = 2x - (5 - 3d)$

15. Express 0.000 41 in scientific notation.

16. List the numbers -3, $|-4|$, $-\pi$, $\sqrt{2}$, and 0.3 in numerical order.

17. What fundamental law is illustrated by $3(5 + 8) = 3(5) + 3(8)$?

18. **(a)** How many significant digits are in the number 3.0450?
    (b) Round it off to two significant digits.

19. If P dollars is deposited in a bank that compounds interest n times a year, the value of the account after t years is found by evaluating $P(1 + i/n)^{nt}$, where i is the annual interest rate. Find the value of an account for which $P = \$2000$, $i = 2\%$, $n = 12$, and $t = 2$ years (values are exact).

20. In finding the illuminance from a light source, the expression $8(100 - x)^2 + x^2$ is used. Simplify this expression.

21. The equation $L = L_0[1 + \alpha(t_2 - t_1)]$ is used when studying thermal expansion. Solve for t_2.

22. An alloy weighing 20 N is 30% copper. How many newtons of another alloy, which is 80% copper, must be added for the final alloy to be 60% copper?

2. Geometry

Roroto12p/Shutterstock

▲ Repeating geometric patterns can be used to form solid surfaces.

Dmitry Kalinovsky/123RF

▲ Land surveying equipment uses principles of geometry to determine distances, areas, and gradients.

Atomazul/Shutterstock

▲ Geometric features play a crucial role in bridge design (and other feature design), as can be seen in this photograph of the Humber Bay Arch Bridge.

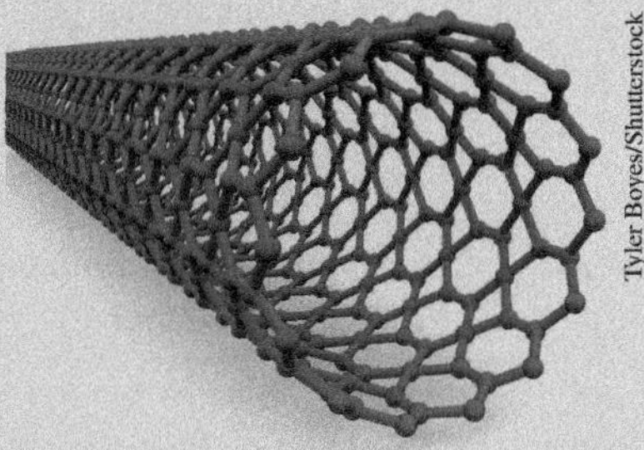

Tyler Boyes/Shutterstock

▲ In chemistry and biology, specific geometries, like those of a repeating nanotube sensor, allow for selective protein bonding.

LEARNING OUTCOMES

After completion of this chapter, the student should be able to:

- Identify perpendicular and parallel lines
- Identify supplementary, complementary, vertical, and corresponding angles
- Determine interior angles and sides in various triangles
- Use the Pythagorean theorem
- Identify and analyse different types of quadrilaterals and polygons
- Identify and analyse circles, arcs, and interior angles of a circle
- Calculate area and perimeter of a geometric shape
- Use an approximation method to estimate an irregular area
- Identify and analyse three-dimensional geometric figures, including volume, angles, and dimensions

When building the pyramids in Egypt nearly 5000 years ago, and today in the deposition of 3D printer geometries, or the identification of molecular geometries in the human genome, the size and shape of an object are measured. Because geometry deals with size and shape, the topics and methods of geometry are important in many of the applications of technology.

The study of geometry dates back to 300 B.C.E., when the Greek mathematician Euclid developed what is known as Euclidean geometry, or the study of properties and measurements of angles, lines, surfaces, and the figures that they form. Today, thousands of years later, we still regularly make use of Euclidean geometry to define space and solve important technical problems. Areas of application include architecture, construction, instrumentation, surveying and civil engineering, mechanical design, product design of various types, and other areas of engineering.

In this chapter, we will review the foundational principles of geometry. In subsequent chapters, these foundational skills will prove important for the study of graphing (in Chapter 3) and trigonometry (in Chapter 4).

2.1 Lines and Angles

A few essential definitions are needed for geometry. This section will focus on dimensionless and one-dimensional (1D) features, and how they sit in 2D space. Consider the following definitions.

■ The use of 360 comes from the ancient Babylonians and their number system based on 60 rather than 10, as we use today. However, the specific reason for the choice of 360 is not known. (One theory is that 360 is divisible by many smaller numbers and is also close to the number of days in a year.)

No dimensions:	A **point** is a location in space. It possesses no width, length, or height. We often make note of a reference point, or point of origin, so that we can identify how far we have moved in space. A "dot" is used to represent a point.
One dimension:	A **line** extends indefinitely along a single dimension in both directions, having a length without limit. A **line segment** is the portion of a line between two specific points in space; it is often named for the two endpoints. A **ray** is a half line, where one end of the line has been fixed but the other end extends without limit.
Two dimensions:	A **plane** is a flat surface that extends infinitely in two dimensions. An infinite number of points and lines can be drawn on the planar surface. The concept of an infinite plane can be useful when considering geometries that repeat indefinitely. A **planar geometry** is a plane that is constrained by sides that have a finite length. In engineering and scientific applications, we will often make use of planar geometries.
Rotations:	An **angle** is the amount of rotation of a ray about its endpoint. The fixed point is the **vertex** of the angle.

Angles are often described depending on the amount of rotation from an initial position to a final position. By convention, positive angles are those where the rotation is counterclockwise, while negative angles are those where the rotation is clockwise. One complete rotation of a ray is an angle with a measure of 360 **degrees**, written as 360°. Some special types of angles are illustrated in the following example.

EXAMPLE 1 Basic angles

Classify the angles in Fig. 2.1 according to their measures as acute, right, obtuse, straight, or reflex.

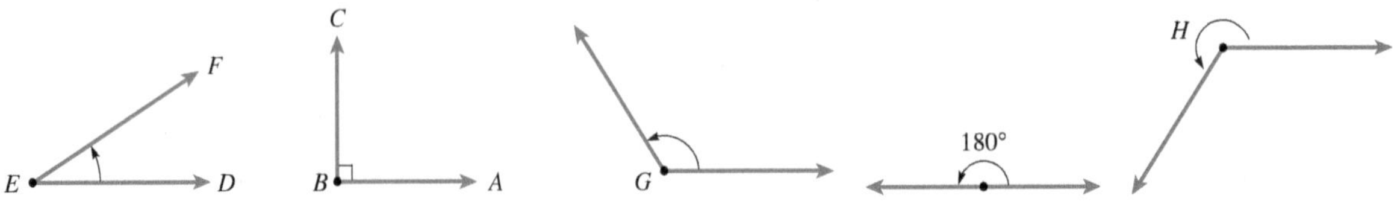

Fig. 2.1(a) Acute angle: between 0° and 90°

Fig. 2.1(b) Right angle: exactly 90° (note the □)

Fig. 2.1(c) Obtuse angle: between 90° and 180°

Fig. 2.1(d) Straight angle: exactly 180°

Fig. 2.1(e) Reflex angle: between 180° and 360°

By convention, *when naming an angle, the vertex position must be specified as the middle letter*. For example, the angle in Fig. 2.1(a) can be written as $\angle FED$. It can also be written as $\angle DEF$, since this expression refers to the same rotation with the same vertex and therefore refers to the same angle.

Parallel and Perpendicular Lines	EXAMPLE 2	
Perpendicular lines cross at one point, forming four right angles. Line segments or rays within a given line are perpendicular to any line the entire line is perpendicular to.	In Fig. 2.2(a), line AC is perpendicular to line DE, written as $AC \perp DE$. Since BA and BC are rays on line AC, $BA \perp DE$ and $BC \perp DE$. Since BD and BE are rays on line DE, $AC \perp BD$ and $AC \perp BE$.	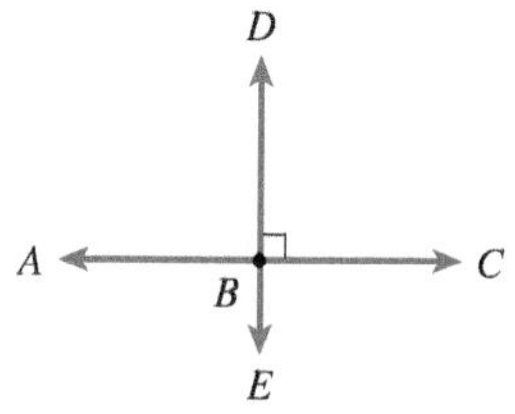 Fig. 2.2(a)
Parallel lines never cross, and remain equidistant from one another.	In Fig. 2.2(b), line AB is parallel to line CD, written as $AB \parallel CD$.	 Fig. 2.2(b)

EXAMPLE 3 Parallel lines and perpendicular lines

In Fig. 2.3, identify which line segments are perpendicular and which are parallel to one another.

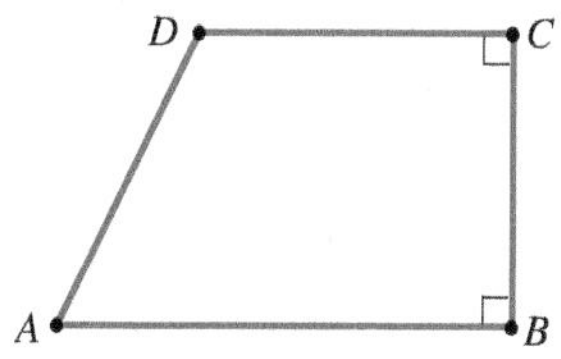

Fig. 2.3

Perpendicular lines: $CB \perp CD$ and $BA \perp CB$
Note that for a line segment, the order of letters does not matter. In Fig 2.3, CB represents the same line as BC.

Parallel lines: $CD \parallel BA$
Note how CD and BA are both perpendicular to the same line CB. In general, lines that are perpendicular to the same line are always parallel.

■ Recognizing complementary angles is important in trigonometry.

Often, we will need to understand how several angles relate to one another. The following definitions provide common language for this purpose.

Complementary and Supplementary Angles	EXAMPLE 4	
Supplementary angles are two angles that sum to 180°. Each angle is the **supplement** of the other.	**(a)** In Fig. 2.4(a), $\angle BAC$ and $\angle DAC$ are supplementary because $\angle BAC + \angle DAC = 180°$ $\angle BAC$ is the supplement of $\angle DAC$. $\angle DAC$ is the supplement of $\angle BAC$.	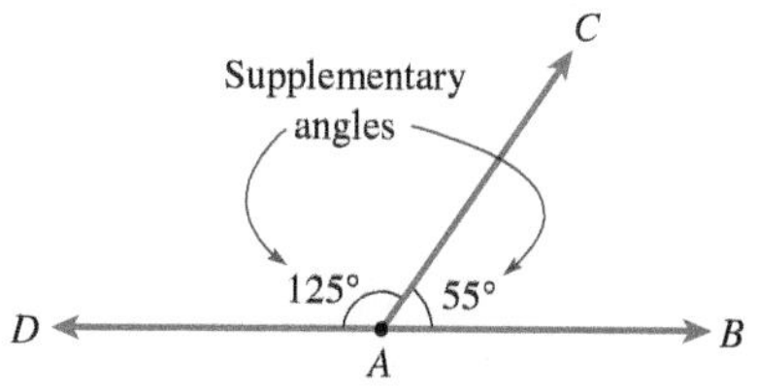 Fig. 2.4(a)
Complementary angles are two angles that sum to 90°. Each angle is the **complement** of the other.	**(b)** In Fig. 2.4(b), $\angle POR$ and $\angle ROQ$ are complementary because $\angle POR + \angle ROQ = 90°$ $\angle POR$ is the complement of $\angle ROQ$. $\angle ROQ$ is the complement of $\angle POR$.	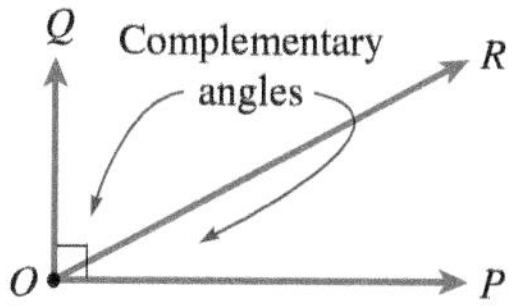 Fig. 2.4(b)

EXAMPLE 5 Supplementary and complementary angles

Find the measures of the supplement and the complement of $\angle BAC = 25°$.

Supplementary Angle:

$\angle BAC + \text{supplement} = 180°$

$\text{supplement} = 180° - \angle BAC$

$= 180° - 25° = 155°$

The supplement of $\angle BAC$ measures 155°.

Complementary Angle:

$\angle BAC + \text{complement} = 90°$

$\text{complement} = 90° - \angle BAC$

$= 90° - 25° = 65°$

The complement of $\angle BAC$ measures 65°.

Practice Exercise

1. What is the measure of the complement of $\angle BAC = 55°$?

In addition to knowing which angles are supplementary or complementary, understanding how angles relate to each other in terms of their placement within the geometry is important when describing complex figures.

Adjacent and Vertical Angles	EXAMPLE 6	
Adjacent angles are two angles that share a common vertex and a common side.	In Fig. 2.5(a), $\angle BAC$ and $\angle CAD$ have a common vertex at A and a common ray AC. This means that $\angle BAC$ and $\angle CAD$ are adjacent.	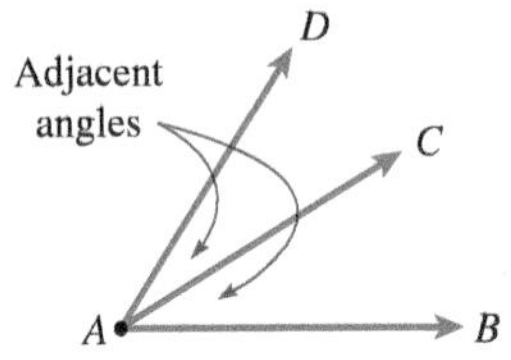 Fig. 2.5(a)
When two lines cross, the angles formed on opposite sides of the point of intersection are called **vertical angles**. Vertical angles are equal.	In Fig. 2.5(b), lines AB and CD intersect at point O. Therefore, $\angle AOC$ and $\angle BOD$ are vertical angles and they are equal. Also, $\angle BOC$ and $\angle AOD$ are vertical angles and are equal.	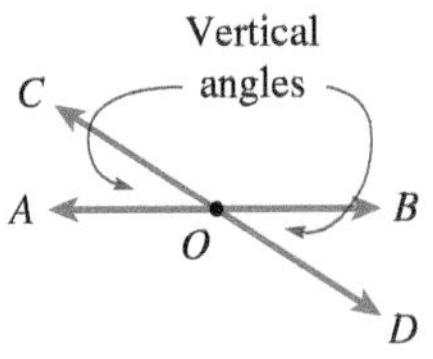 Fig. 2.5(b)

We should also be able to identify *the sides or rays of an angle that are adjacent to the angle*. In Fig. 2.5(a), rays AB and AC are adjacent to $\angle BAC$, and in Fig. 2.5(b), rays OB and OD are adjacent to $\angle BOD$. Identifying sides adjacent and opposite an angle in a triangle is important in trigonometry.

In a plane, *if a line crosses two or more parallel or nonparallel lines, it is called a* **transversal**. If the lines crossed by a transversal are parallel, unique geometric properties result for the angles involved.

Parallel Lines and Transversals	EXAMPLE 7	
Corresponding angles are angles in the same position relative to the parallel lines intersected by a transversal. Corresponding angles are equal.	In Fig. 2.6(a), $\angle 1$ and $\angle 5$ are corresponding angles. Hence $\angle 1 = \angle 5$. Similarly, $\angle 2 = \angle 6$, $\angle 3 = \angle 7$, $\angle 4 = \angle 8$.	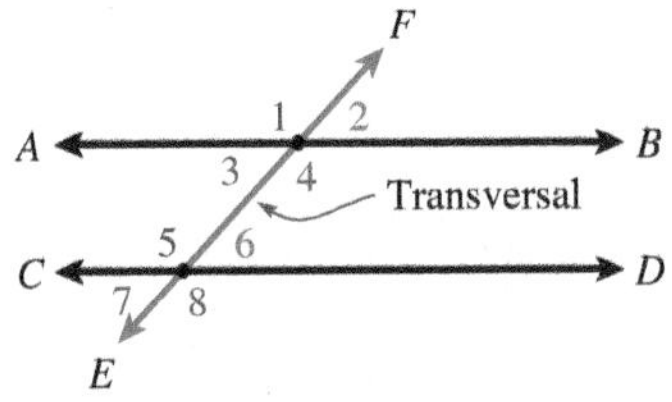 Fig. 2.6(a)

<table>
<tr>
<td>Alternate interior angles are situated inside the two parallel lines on opposite sides of the transversal.
Since alternate interior angles are complementary to equal angles, they are equal.</td>
<td>In Fig. 2.6(b), $\angle 3$ and $\angle 6$ are alternate interior angles. Hence $\angle 3 = \angle 6$.
Similarly, $\angle 4 = \angle 5$.</td>
<td>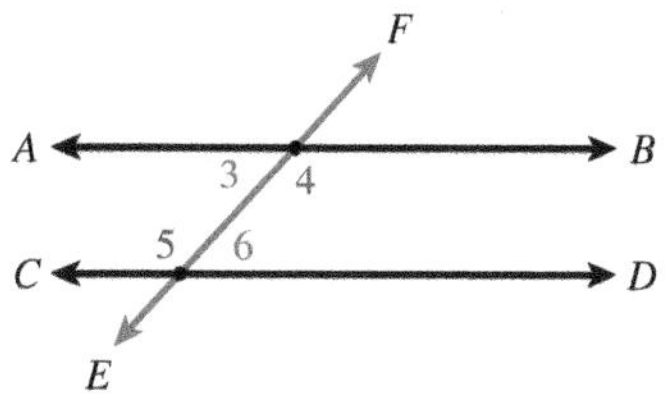

Fig. 2.6(b)</td>
</tr>
<tr>
<td>Alternate exterior angles are situated outside the two parallel lines on opposite sides of the transversal.
Alternate exterior angles are equal.</td>
<td>In Fig. 2.6(c), $\angle 1$ and $\angle 8$ are alternate exterior angles. Hence $\angle 1 = \angle 8$.
Similarly, $\angle 7 = \angle 2$.</td>
<td>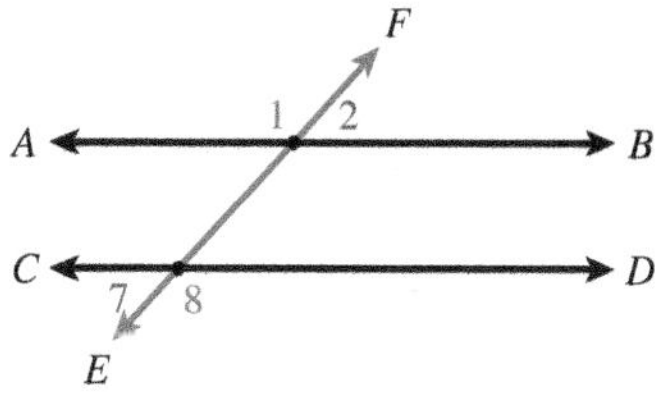

Fig. 2.6(c)</td>
</tr>
</table>

Practice Exercise

2. In Fig. 2.6(a), if $\angle 2 = 42°$, then $\angle 5 = ?$

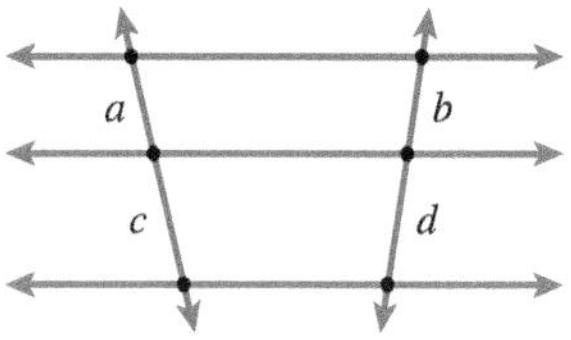

Fig. 2.7

When more than two parallel lines are crossed by *two* transversals, such as is shown in Fig. 2.7, *the segments of the transversals between the same two parallel lines are called* **corresponding segments**. A useful theorem is that *the ratios of corresponding segments of the transversals are equal.* In Fig. 2.7, this means that

$$\frac{a}{b} = \frac{c}{d} \tag{2.1}$$

EXAMPLE 8 Segments of transversals

In Fig. 2.8, part of the beam structure within a building is shown. The vertical beams are parallel, perpendicular to the floor and intersecting the room beam. Some centre-to-centre distances from one beam to the next are shown. Find the value of x.

Using Eq. (2.1), we have

$$\frac{655}{565} = \frac{x}{775}$$

$$x = \frac{655(775)}{565}$$

$$x = 898 \text{ mm}$$

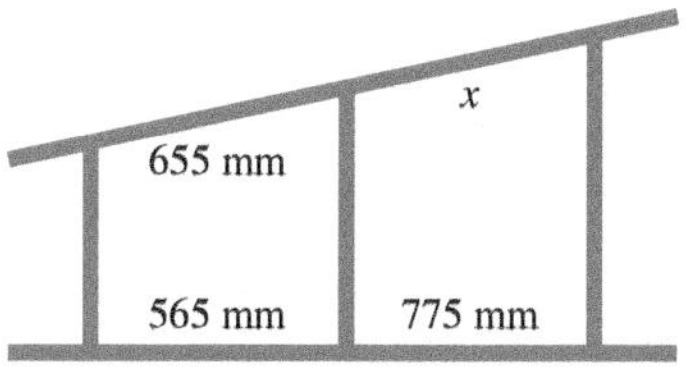

Fig. 2.8

EXERCISES 2.1

In Exercises 1–4, answer the given questions about the indicated examples of this section.

1. In Example 2, what is the measure of $\angle ABE$ in Fig. 2.2(a)?
2. In Example 4(b), if $\angle POR = 32°$ in Fig. 2.4(b), what is the measure of $\angle QOR$?
3. In Example 6, how many different pairs of adjacent angles are there in Fig. 2.5(b)?
4. In Example 8, if the segments of 655 mm and 775 mm are interchanged, (a) what is the answer, and (b) is the beam along which x is measured more nearly vertical or more nearly horizontal?

In Exercises 5–12, identify the indicated angles and sides in Fig. 2.9. In Exercises 9 and 10, also find the measures of the indicated angles.

5. Two acute angles
6. Two right angles

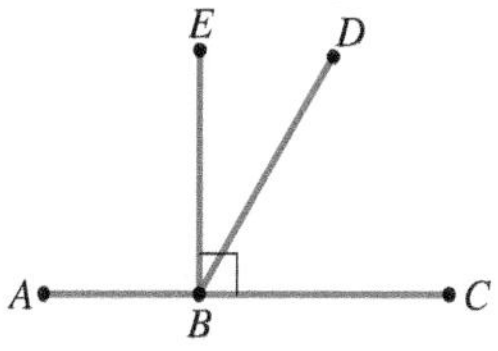

Fig. 2.9

7. The straight angle
8. The obtuse angle
9. If $\angle CBD = 65°$, find its complement.

10. If $\angle CBD = 65°$, find its supplement.
11. The sides adjacent to $\angle DBC$
12. The acute angle adjacent to $\angle DBC$

In Exercises 13–15, use Fig. 2.10. In Exercises 16–18, use Fig. 2.11. Find the measures of the indicated angles.

13. $\angle AOB$ **14.** $\angle AOC$ **15.** $\angle BOD$
16. $\angle 3$ **17.** $\angle 4$ **18.** $\angle 5$

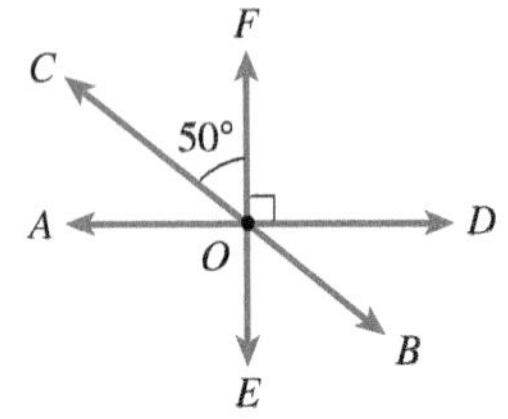

Fig. 2.10

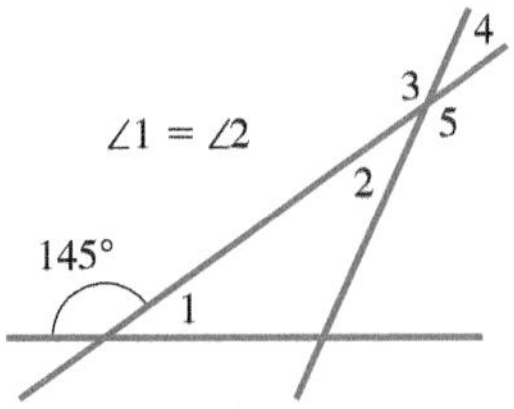

Fig. 2.11

In Exercises 19–24, find the measures of the angles in Fig. 2.12.

19. $\angle 1$ **20.** $\angle 2$ **21.** $\angle 3$ **22.** $\angle 4$ **23.** $\angle 5$ **24.** $\angle 6$

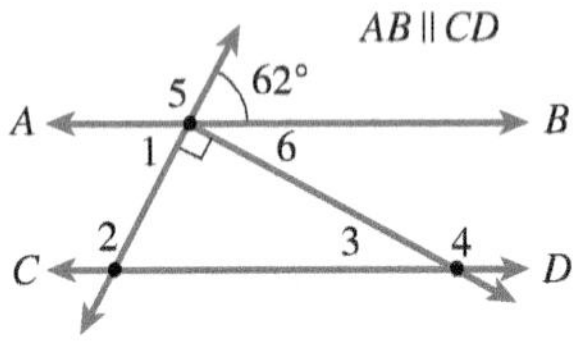

Fig. 2.12

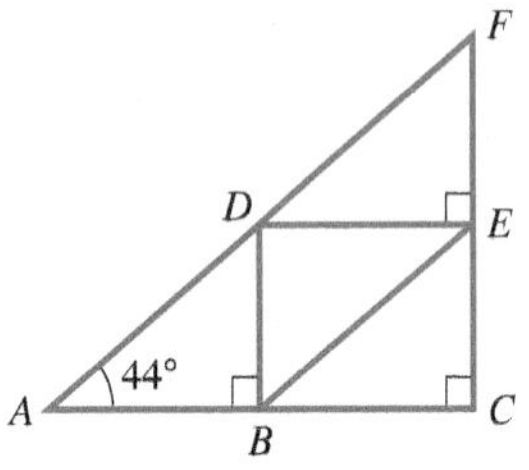

Fig. 2.13

In Exercises 25–30, find the measures of the angles in the truss shown in Fig. 2.13. A truss is a rigid support structure that is used in the construction of buildings and bridges.

25. $\angle BDF$ **26.** $\angle ABE$ **27.** $\angle DEB$
28. $\angle DBE$ **29.** $\angle DFE$ **30.** $\angle ADE$

In Exercises 31–34, find the indicated distances between the straight irrigation ditches shown in Fig. 2.14. The vertical ditches are parallel.

31. a **32.** b **33.** c **34.** d

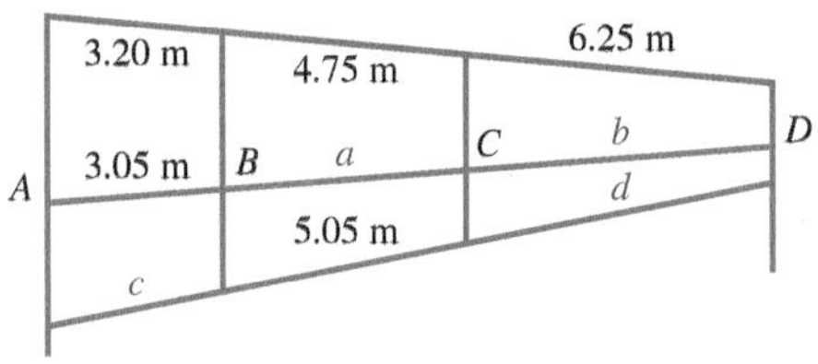

Fig. 2.14

In Exercises 35–40, find all angles of the given measures for the beam support structure shown in Fig. 2.15.

35. 25° **36.** 45° **37.** 65°
38. 70° **39.** 110° **40.** 115°

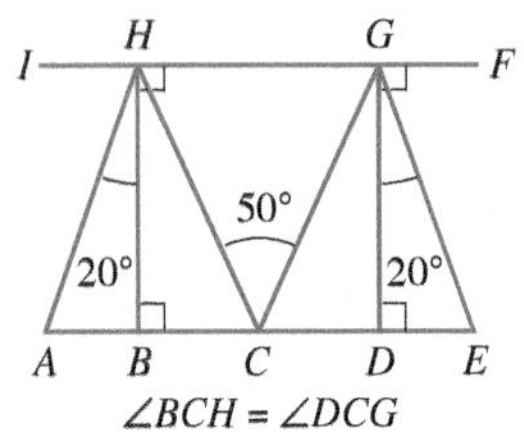

Fig. 2.15

In Exercises 41–44, solve the given problems.

41. A steam pipe is connected in sections AB, BC, and CD as shown in Fig. 2.16. What is the angle between BC and CD if $AB \parallel CD$?

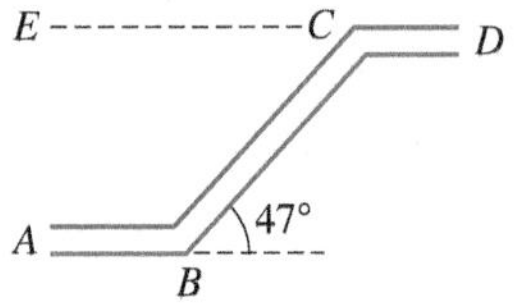

Fig. 2.16

42. Part of a laser beam striking a surface is reflected at the same angle as the original beam, and the remainder is refracted (passes into the new material) at a new angle (see Fig. 2.17). Find the total angle between the reflected beam and the refracted beam and specify the type of angle it is (acute, right, obtuse, straight, or reflex).

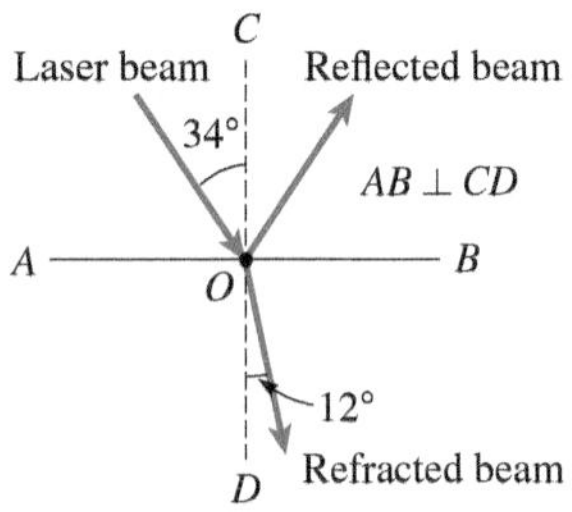

Fig. 2.17

43. Find the distance on Dundas St. W between Dufferin St. and Ossington Ave. in Toronto, as shown in Fig. 2.18. The north–south streets are parallel.

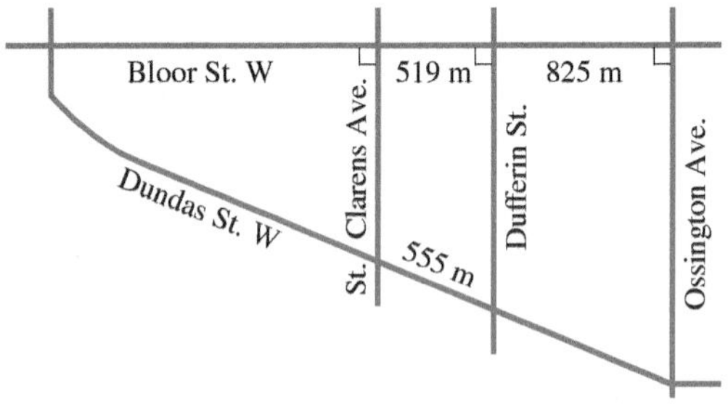

Fig. 2.18

44. An electric circuit board has equally spaced parallel wires with connections at points A, B, and C, as shown in Fig. 2.19. How far is A from C, if $BC = 2.15$ cm?

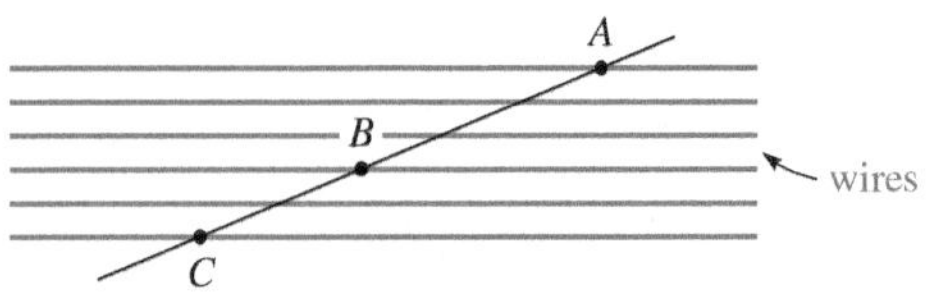

Fig. 2.19

In Exercises 45–48, solve the given problems related to Fig. 2.20.

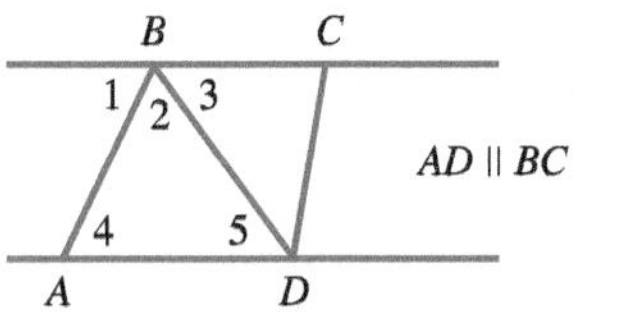

Fig. 2.20

45. $\angle 1 + \angle 2 + \angle 3 = ?$ **46.** $\angle 4 + \angle 2 + \angle 5 = ?$

47. Based on Exercise 46, what conclusion can be drawn about a closed geometric figure like the one with vertices at A, B, and D?

48. Based on Exercises 46 and 47, what conclusion can be drawn about a closed geometric figure like the one with vertices A, B, C, and D?

Answers to Practice Exercises

1. 35° **2.** 138°

2.2 Triangles

When part of a plane is bounded and closed by straight-line segments, it is called a **polygon**, and it is named according to the number of sides it has. A three-sided polygon is called a **triangle**, the subject of this section. Other polygons will be discussed elsewhere in this text.

TYPES AND PROPERTIES OF TRIANGLES

■ The properties of triangles are important in the study of trigonometry, which we start in Chapter 4.

In a **scalene triangle**, no two sides are equal in length. In an **isosceles triangle**, two of the sides are equal in length, and the two *base angles* (the angles opposite the equal sides) are equal. A line drawn through the middle of the angle that is not alike is a line of symmetry of the isosceles triangle, dividing the triangle into two mirror-image halves. In an **equilateral triangle**, the three sides are equal in length, and each of the three angles is 60°.

EXAMPLE 1 Types of triangles

Classify the triangles in Fig. 2.21 according to their sides as scalene, isosceles, or equilateral.

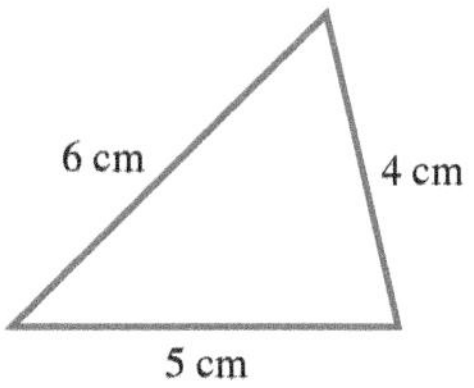

Fig. 2.21(a)

Scalene triangle: three unequal sides; three unequal angles

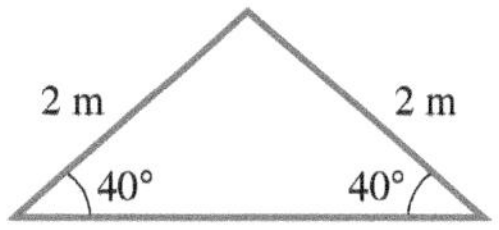

Fig. 2.21(b)

Isosceles triangle: two equal sides; two equal base angles

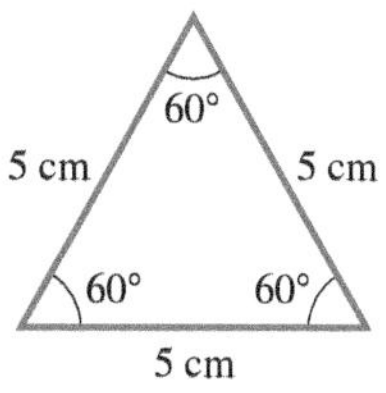

Fig. 2.21(c)

Equilateral triangle: three equal sides; three equal angles

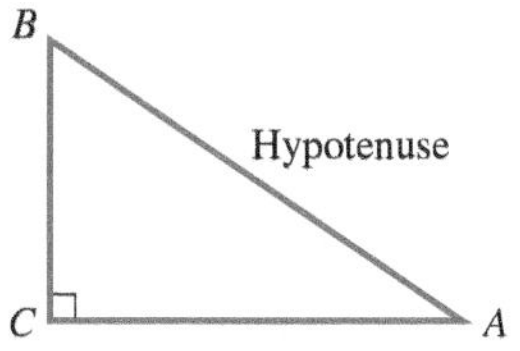

Fig. 2.22

The most important triangle in technical applications is the **right triangle**. *In a right triangle, one of the angles is a right* (90°) *angle. The side opposite the right angle is the* **hypotenuse**, *and the other two sides are called* **legs**. In Fig. 2.22, AC and BC are the legs and AB is the hypotenuse.

Angle Summation Rules for Triangles

Straight Angle Rule: The sum of the measures of angles that together form a straight line is 180°.

Sum of Angle Rule for Triangles: The sum of the measures of the three interior angles of a triangle is 180°.

EXAMPLE 2 Sum of angles of a triangle

Apply the angle summation rules to Fig. 2.23.

Applying the straight angle rule: $\angle 1 + \angle 2 + \angle 3 = 180°$

Applying the sum of angles rule: $\angle 4 + \angle 2 + \angle 5 = 180°$

In either case, if two of the angles are known, the third may be found by subtracting the sum of the first two from 180°.

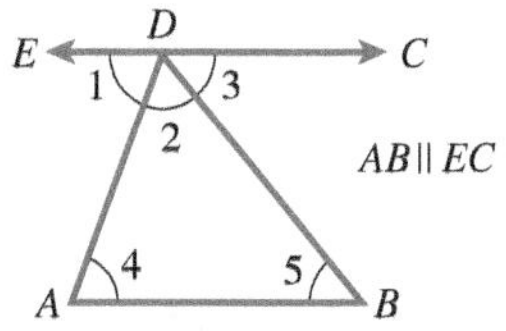

Fig. 2.23

EXAMPLE 3 Sum of angles—airplane flight

An airplane is flying north and then makes a 90° turn to the west at point C. Later, it makes another left turn of 150° at point A. What is the angle of a third left turn that will cause the plane to again fly north? See Fig. 2.24.

Identify known quantities: $\angle ACB = 90°$

$\angle BAC = 180° - 150° = 30°$ (by straight angle rule)

Identify unknown quantities: $\angle ABC$, $\angle CBD$ (required angle)

Select equations: Sum of angles rule: $\angle BAC + \angle ACB + \angle ABC = 180°$

$30° + 90° + \angle ABC = 180°$

Straight angle rule: $\angle ABC + \angle CBD = 180°$

Solve: $\angle ABC = 180° - (30° + 90°) = 60°$

$\angle CBD = 180° - 60° = 120°$

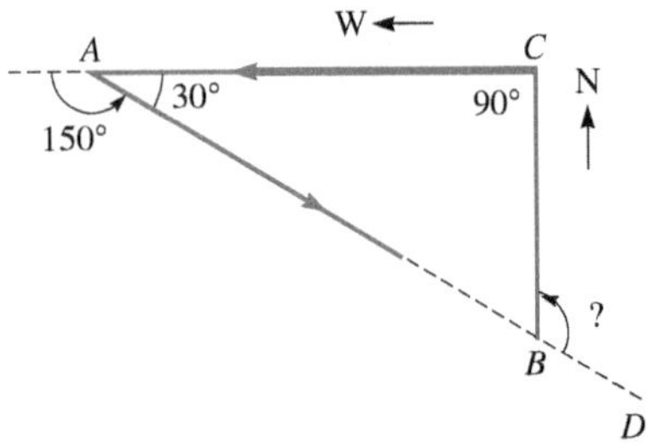

Fig. 2.24

Practice Exercise

1. In a certain isosceles triangle, if the vertex angle (the unequal angle) is 30°, what are the base angles?

A line segment drawn from a vertex of a triangle to the midpoint of the opposite side is called a **median** of the triangle. Each median divides the triangle into two smaller triangles each having the same area. A basic property of a triangle is that the three medians meet at a single point, called the **centroid** (or centre of mass) of the triangle. See Fig. 2.25. Centroids are calculated as an application of calculus in Chapter 26.

An **angle bisector** is defined as a line that splits the angle into two angles of equal size. The three angle bisectors of a triangle meet at a common point (not the centroid). See Fig. 2.26.

An **altitude** (or **height**) of a triangle is the line segment drawn from a vertex perpendicular to the opposite side (or its extension), which is called the **base** of the triangle. The three altitudes of a triangle also meet at a common point (again, not the centroid). See Fig. 2.27.

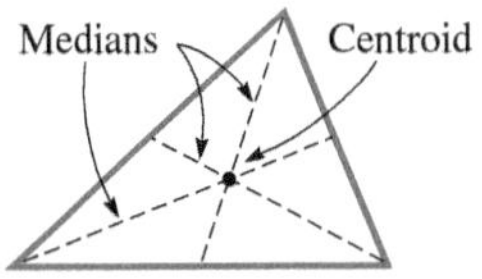

Fig. 2.25

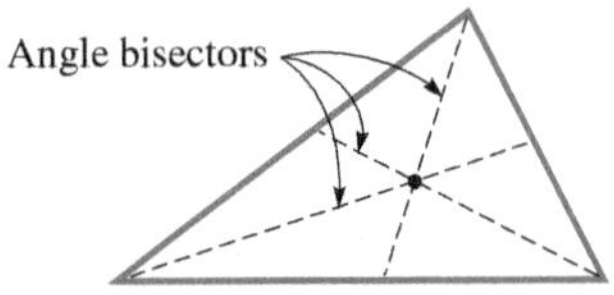

Fig. 2.26

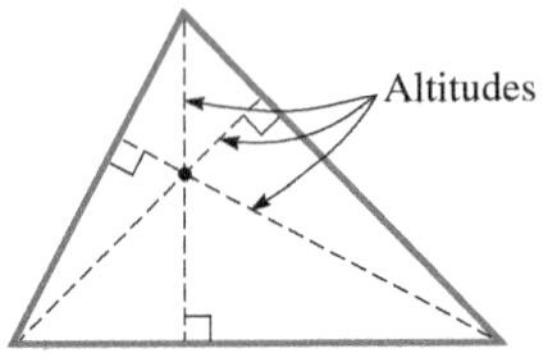

Fig. 2.27

PERIMETER AND AREA OF A TRIANGLE

For plane geometric figures, a common one-dimensional measurement is the **perimeter**, which is the total length of the boundary around the figure. A common two-dimensional measurement is the **area**, which gives the size of the surface of a figure. For triangles, perimeter and area are calculated as follows.

Area and Perimeter of a Triangle	EXAMPLE 4	
Perimeter: $p = a + b + c,$ where a, b and c are the three sides	**(a)** In Fig. 2.28(a), the perimeter p of the triangle is $p = 2.56 + 3.22 + 4.89 = 10.67$ m	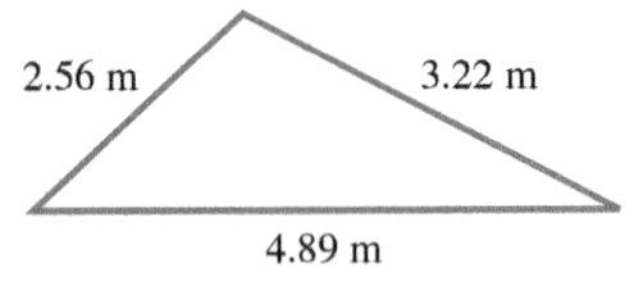 Fig. 2.28(a)

Area (base b and altitude h known):

$$A = \tfrac{1}{2}bh \qquad (2.2)$$

(b) In Fig. 2.28(b) we see that one side (the base) and its altitude are given, so the area is

$$\begin{aligned} A &= \tfrac{1}{2}bh \\ &= \tfrac{1}{2}(16.2)(5.75) \\ &= 46.6 \text{ cm}^2 \end{aligned}$$

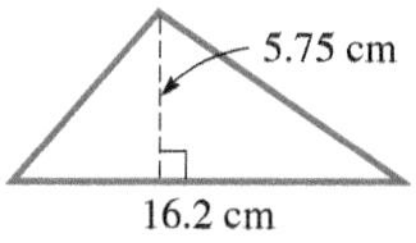

Fig. 2.28(b)

Area by Hero's formula (sides a, b, and c known):

$$A = \sqrt{s(s-a)(s-b)(s-c)}, \qquad (2.3)$$

where s is the half-perimeter

$$s = \tfrac{1}{2}(a+b+c) \qquad (2.4)$$

(c) For the triangle in Fig. 2.28(a) above where the three sides $a = 2.56$ m, $b = 3.22$ m, $c = 4.89$ m are known, we first calculate the half-perimeter s as

$$s = \tfrac{1}{2}(a+b+c) = \tfrac{1}{2}(2.56 + 3.22 + 4.89) = 5.335 \text{ m}$$

Then the area is

$$\begin{aligned} A &= \sqrt{s(s-a)(s-b)(s-c)} \\ &= \sqrt{5.335(5.335-2.56)(5.335-3.22)(5.335-4.89)} \\ &= 3.732\ 793\ 913 \text{ m}^2 \\ &= 3.73 \text{ m}^2 \quad \text{(rounded off)} \end{aligned}$$

■ Hero's formula is named for Hero (or Heron), a first-century Greek mathematician.

EXAMPLE 5 Area of a triangle

Find the area of the triangle in Fig. 2.29.

The base b of the triangle is 16.2 cm, and the altitude h is 5.75 cm. Therefore, even though the shape of the triangle is not the same as in Fig. 2.28(b), the area of the triangle is also

$$A = \tfrac{1}{2}bh = \tfrac{1}{2}(16.2)(5.75) = 46.6 \text{ cm}^2$$

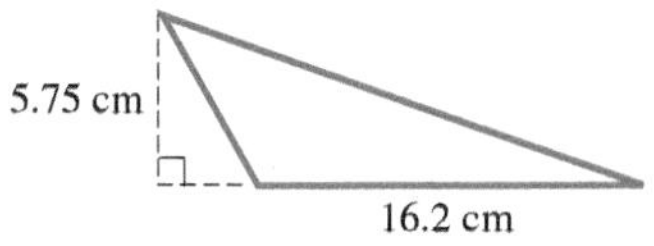

Fig. 2.29

EXAMPLE 6 Hero's formula—area of land parcel

A surveyor measures the sides of a triangular parcel of land between two intersecting straight roads to be 206 m, 293 m, and 187 m. Find the area of this parcel (see Fig. 2.30).

Since three sides are known, we find the area with Hero's formula (Eqs. 2.3 and 2.4). We first find s:

$$\begin{aligned} s &= \tfrac{1}{2}(206 + 293 + 187) \\ &= \tfrac{1}{2}(686) = 343 \text{ m} \end{aligned}$$

Now, substituting in Eq. (2.3), we have

$$\begin{aligned} A &= \sqrt{343(343-206)(343-293)(343-187)} \\ &= 19\ 144.968\ 01 \text{ m}^2 \\ &= 19\ 100 \text{ m}^2 \quad \text{(rounded to three significant digits)} \end{aligned}$$

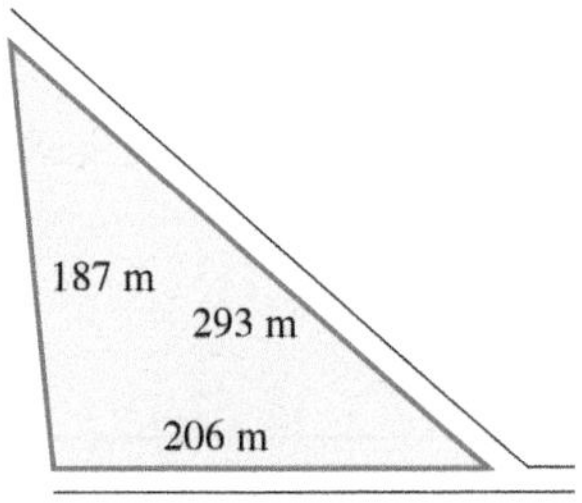

Fig. 2.30

Practice Exercise

2. What is the perimeter of the triangle in Fig. 2.30?

■ The Pythagorean theorem is named for the Greek mathematician Pythagoras (sixth century B.C.E.)

THE PYTHAGOREAN THEOREM

The Pythagorean theorem provides a general rule to relate the lengths of the three sides of a right triangle. As in Fig. 2.31, for a right triangle labelled so that the length of the hypotenuse is c and the lengths of the legs are a and b, the Pythagorean theorem states that

$$c^2 = a^2 + b^2 \qquad (2.5)$$

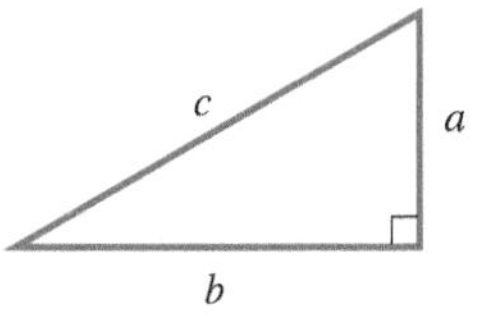

Fig. 2.31

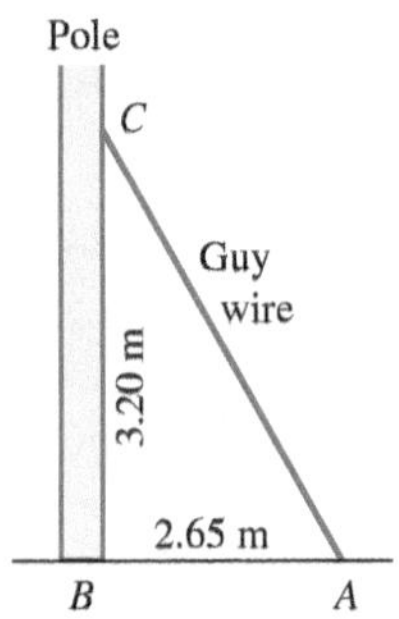

Fig. 2.32

EXAMPLE 7 Pythagorean theorem—length of a guy wire

A pole is perpendicular to the level ground around it. A guy wire is attached 3.20 m up the pole and at a point on the ground 2.65 m from the pole. How long is the guy wire?

We sketch the pole and guy wire as shown in Fig. 2.32. We see that the guy wire is the hypotenuse, of length AC. Using the Pythagorean theorem we have

$$(AC)^2 = (AB)^2 + (BC)^2$$
$$(AC)^2 = (2.65\text{ m})^2 + (3.20\text{ m})^2$$
$$AC = \sqrt{(2.65\text{ m})^2 + (3.20\text{ m})^2} = 4.15\text{ m}$$

Note that the principal root is positive. The guy wire is 4.15 m long.

SIMILAR TRIANGLES

Two triangles are **similar** *if they have the same shape (but not necessarily the same size).* There are two very important properties of similar triangles.

Properties of Similar Triangles

1. The corresponding angles of similar triangles are equal.
2. The corresponding sides of similar triangles are proportional. **Corresponding sides** are the sides, one in each triangle, that are between the same pair of equal corresponding angles.

EXAMPLE 8 Similar triangles

In Fig. 2.33, two similar triangles are shown. (We write $\triangle ABC \sim \triangle A'B'C'$, where $\triangle$ means "triangle" and $\sim$ means "is similar to.") Identify the pairs of corresponding angles and corresponding sides. Use the corresponding sides to create statements of proportionality for the two triangles.

The pairs of corresponding angles are A and A', B and B', and C and C'. This means $A = A'$, $B = B'$, and $C = C'$.

The pairs of corresponding sides are AB and $A'B'$, BC and $B'C'$, and AC and $A'C'$. Statements of proportionality for the two triangles are

$$\frac{AB}{A'B'} = \frac{BC}{B'C'} = \frac{AC}{A'C'} \quad \begin{matrix}\leftarrow \text{sides of } \triangle ABC \\ \leftarrow \text{sides of } \triangle A'B'C'\end{matrix}$$

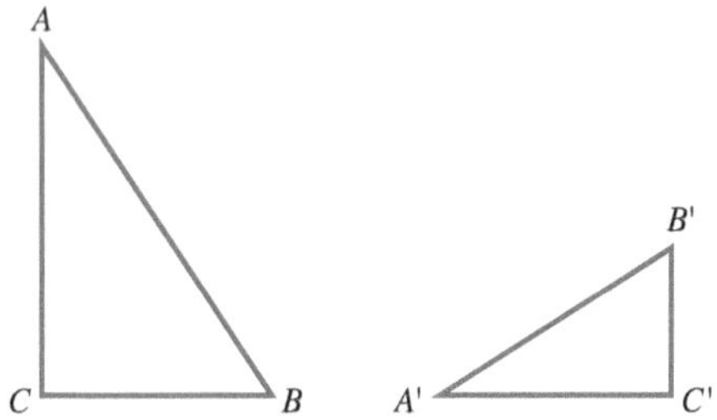

Fig. 2.33

LEARNING TIP

Similar triangles don't have to be right triangles (as in Fig. 2.33). Sides will relate proportionally as long as the triangles have identical angle measures.

If we know that two triangles are similar, we can use the two basic properties of similar triangles to find the unknown parts of one triangle from the known parts of the other triangle. The next example illustrates this in a practical application.

EXAMPLE 9 Similar triangles—finding the height of a silo

On level ground, a silo casts a shadow 24.0 m long. At the same time, a nearby vertical pole 4.00 m high casts a shadow 3.00 m long. How tall is the silo? See Fig. 2.34 (not drawn to scale).

The rays of the sun are essentially parallel. The two triangles in Fig. 2.34 are similar since *each has a right angle and the angles at the tops are equal.* The other angles must be equal since the sum of the angles is 180°. The height h of the silo

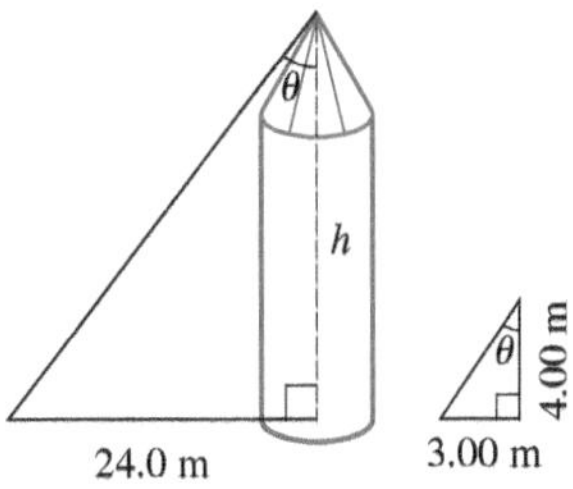

Fig. 2.34

Practice Exercise

3. In Fig. 2.34, knowing the value of h, what is the distance between the top of the silo and the end of its shadow?

corresponds to the height 4.00 of the pole; the shadow of 24.0 m for the silo corresponds to the shadow of 3.00 m for the pole. Therefore,

$$\frac{h}{4.00} = \frac{24.0}{3.00}$$

Multiplying each side by 4.00 and cancelling, we conclude that the silo is 32.0 m high.

SCALE DRAWINGS

In scale drawing, all distances are drawn at a certain ratio of the distances that they represent, and all angles are drawn equal to the angles they represent. Scale drawings such as maps, charts, and architectural blueprints make use of proportionality and similar triangles.

EXAMPLE 10 Scale drawing

The scale on a map is given as 1 cm on the map representing 200 km in real distance. On the map, distances between Chicago and Toronto, and between Toronto and Philadelphia are measured, as shown in Fig. 2.35. Find the actual distances between these cities.

Google and the Google logo are registered trademarks of Google LLC, used with permission.

Fig. 2.35

LEARNING TIP

These calculations can also be done exactly like a unit conversion (see Section 1.3). We are converting between map units and real units, using the scale of the drawing.

$$x = 3.5 \text{ cm} \cdot \left(\frac{200 \text{ km}}{1 \text{ cm}}\right) = 700 \text{ km}$$

scale conversion

Distance from Chicago to Toronto:

$$\frac{200 \text{ km}}{1 \text{ cm on map}} = \frac{x}{3.5 \text{ cm on map}}$$

$$x = (3.5 \text{ cm on map})\left(\frac{200 \text{ km}}{1 \text{ cm on map}}\right) = 700 \text{ km}$$

Distance from Toronto to Philadelphia:

$$\frac{200 \text{ km}}{1 \text{ cm on map}} = \frac{y}{2.7 \text{ cm on map}}$$

$$y = (2.7 \text{ cm on map})\left(\frac{200 \text{ km}}{1 \text{ cm on map}}\right) = 540 \text{ km}$$

Congruent Triangles

Congruent triangles are those that have equal corresponding angles and equal corresponding sides. The areas and perimeters of congruent triangles are also equal.

In other words, we can say that similar triangles have the same shape, whereas congruent triangles have the same shape and the same size.

EXAMPLE 11 Similar and congruent triangles

In Fig. 2.36, determine which triangles are congruent and which are similar.

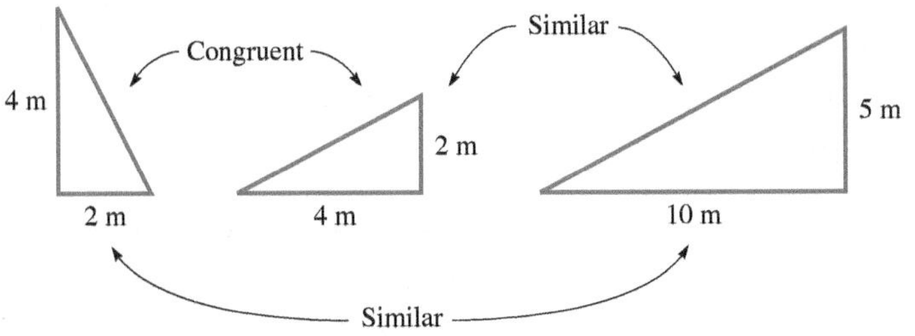

Fig. 2.36

The two triangles to the left have equal sides and angles, and are therefore congruent, even though they are not oriented the same way. The rightmost triangle is similar to the other two since, although it has the same angles, the lengths of its sides are not equal to the lengths of the sides of the others, but only proportional (in the ratio of 1 to 2).

EXERCISES 2.2

In Exercises 1–4, answer the given questions about the indicated examples of this section.

1. In Example 2, if $\angle 1 = 70°$ and $\angle 5 = 45°$ in Fig. 2.23, what is the measure of $\angle 2$?
2. In Example 5, change 16.2 cm to 61.2 cm. What is the answer?
3. In Example 7, change 2.65 m to 6.25 m. What is the answer?
4. In Example 9, interchange 4.00 m and 3.00 m. What is the answer?

In Exercises 5–8, determine $\angle A$ in the indicated figures.

5. Fig. 2.37(a)
6. Fig. 2.37(b)
7. Fig. 2.37(c)
8. Fig. 2.37(d)

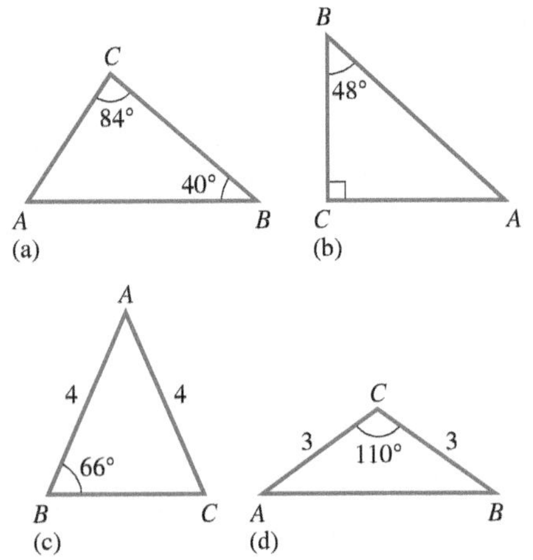

Fig. 2.37

In Exercises 9–16, find the area of each triangle.

9. Fig. 2.38(a)
10. Fig. 2.38(b)
11. Fig. 2.38(c)
12. Fig. 2.38(d)

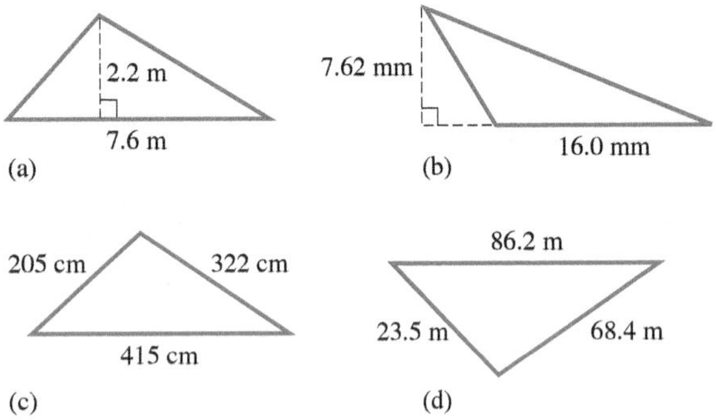

Fig. 2.38

13. Right triangle with legs 3.46 cm and 2.55 cm
14. Right triangle with legs 234 mm and 343 mm
15. Isosceles triangle, equal sides of 0.986 m, third side of 0.884 m
16. Equilateral triangle of sides 322 dm

In Exercises 17–20, find the perimeter of each triangle.

17. Fig. 2.38(c)
18. Fig. 2.38(d)
19. An equilateral triangle of sides 21.5 cm
20. Isosceles triangle, equal sides of 2.45 mm, third side of 3.22 mm

In Exercises 21–24, find the third side of the right triangle shown in Fig. 2.39 for the given values.

21. $a = 13.8$ mm, $b = 22.7$ mm
22. $a = 2.48$ m, $b = 1.45$ m
23. $a = 175$ cm, $c = 551$ cm
24. $b = 0.474$ km, $c = 0.836$ km

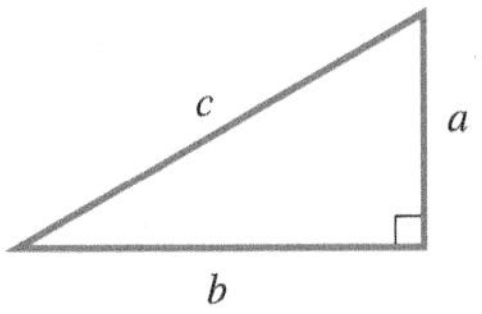

Fig. 2.39

In Exercises 25–28, use the right triangle in Fig. 2.40.

25. Find $\angle B$.

26. Find side c.

27. Find the perimeter.

28. Find the area.

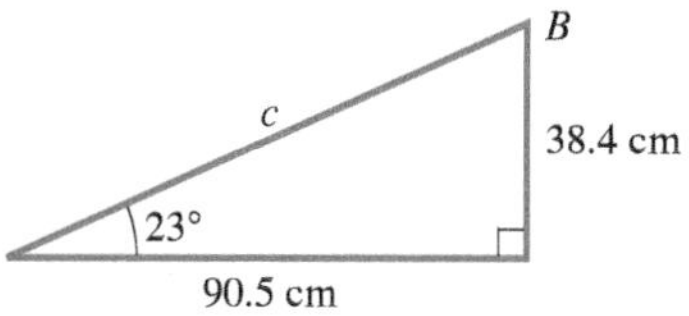

Fig. 2.40

In Exercises 29–54, solve the given problems.

29. What is the angle between the bisectors of the acute angles of a right triangle?

30. If the midpoints of the sides of an isosceles triangle are joined, another triangle is formed. What do you conclude about this inner triangle?

31. For what type of triangle is the centroid the same as the intersection of altitudes and the intersection of angle bisectors?

32. Is it possible that the altitudes of a triangle meet, when extended, outside the triangle? Explain.

33. The altitude to the hypotenuse of a right triangle divides the triangle into two smaller triangles. What do you conclude about these two triangles?

34. The altitude to the hypotenuse of a right triangle divides the triangle into two smaller triangles. What do you conclude about the original triangle and the two new triangles? Explain.

35. In Fig. 2.41, show that $\triangle MKL \sim \triangle MNO$.

36. In Fig. 2.42, show that $\triangle ACB \sim \triangle ADC$.

37. In Fig. 2.41, if $KN = 15$, $MN = 9$, and $MO = 12$, find LM.

38. In Fig. 2.42, if $AD = 9$ and $AC = 12$, find AB.

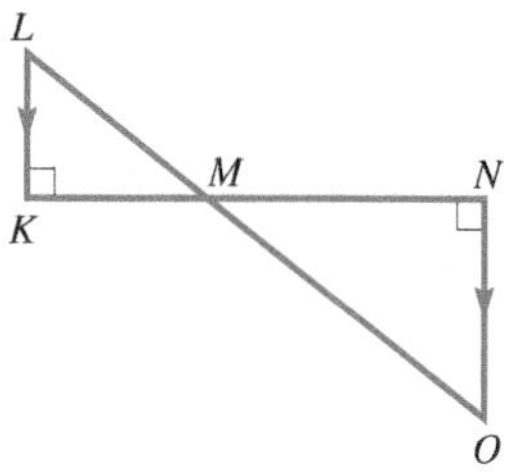

Fig. 2.41

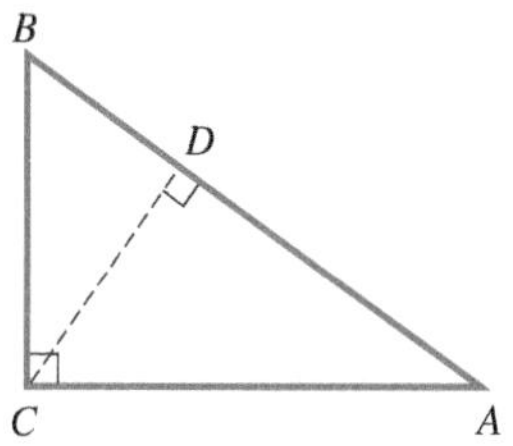

Fig. 2.42

39. The angle between the roof sections of an A-frame house is 50°. What is the angle between either roof section and a horizontal rafter?

40. A transmitting tower is supported by a wire that makes an angle of 52° with the level ground. What is the angle between the tower and the wire?

41. A wall pennant is in the shape of an isosceles triangle. If each equal side is 76.6 cm long and the third side is 30.6 cm, what is the area of the pennant?

42. The Bermuda Triangle is sometimes defined as an equilateral triangle 1600 km on a side, with vertices in Bermuda, Puerto Rico, and the Florida coast. Assuming it is flat, what is its approximate area?

43. The sail of a sailboat is in the shape of a right triangle with sides of 3.2 m, 6.0 m, and 6.8 m. What is the area of the sail?

44. An observer is 550 m horizontally from the launch pad of a rocket. After the rocket has ascended 750 m, how far is it from the observer?

45. The base of a 6.0-m ladder is 1.8 m from a wall. How far up on the wall does the ladder reach?

46. The beach shade shown in Fig. 2.43 is made of 30°-60°-90° triangular sections. Find x. (In a 30°-60°-90° triangle, the side opposite the 30° angle is one-half the hypotenuse.)

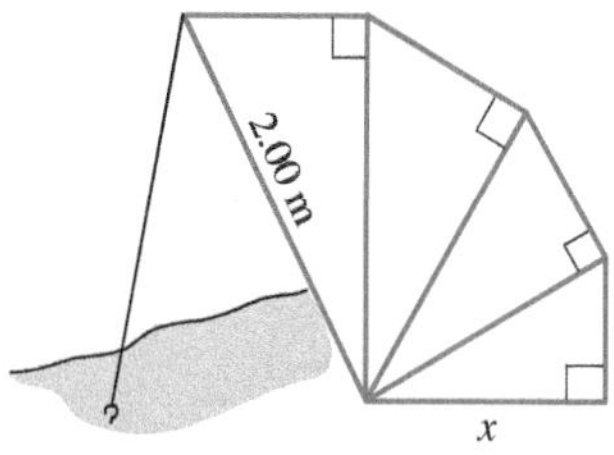

Fig. 2.43

47. A rectangular room is 18.0 m long, 12.0 m wide, and 8.00 m high. What is the length of the longest diagonal from one corner to another corner of the room?

48. On a blueprint, a hallway is 45.6 cm long. The scale is 1.20 cm = 1.00 m. How long is the hallway?

49. Two parallel guy wires are attached to a vertical pole 4.5 m and 5.4 m above the ground. They are secured on the level ground at points 1.2 m apart. How long are the guy wires?

50. The two sections of a folding door, hinged in the middle, are each 0.85 m wide. What width of the doorway is taken up when the sections are at right angles to each other?

51. A 1.5-m wall stands 0.75 m from a building. The ends of a straight pole touch the building and the ground 2.0 m from the wall. A point on the pole touches the top of the wall. How long is the pole? See Fig. 2.44.

52. To find the width ED of a river, a surveyor places markers at A, B, C, and D, as shown in Fig. 2.45. The markers are placed such that $AB \parallel ED$, $BC = 50.0$ m, $DC = 312$ m, and $AB = 80.0$ m. How wide is the river?

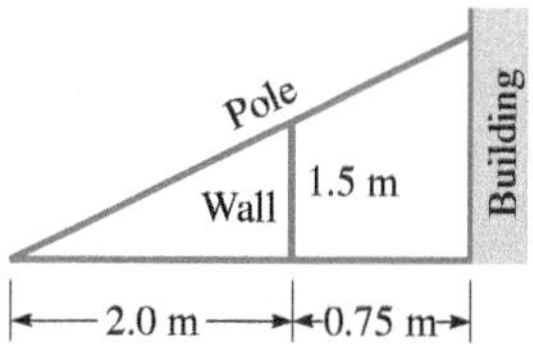

Fig. 2.44

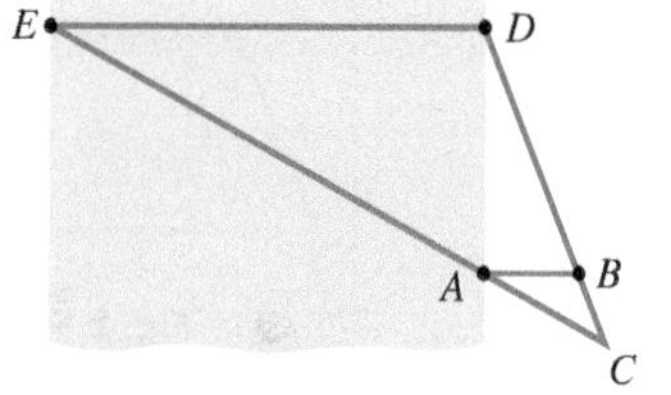

Fig. 2.45

53. A water pumping station is to be built on a river at point P in order to deliver water to points A and B. See Fig. 2.46. The design requires that $\angle APD = \angle BPC$ so that the total length of piping that will be needed is a minimum. Find this minimum length of pipe.

54. The cross-section of a drainage trough has the shape of an isosceles triangle whose depth is 12 cm less than its width. If the depth is increased by 16 cm and the width remains the same, the area of the cross-section is increased by 160 cm^2. Find the original depth and width. See Fig. 2.47.

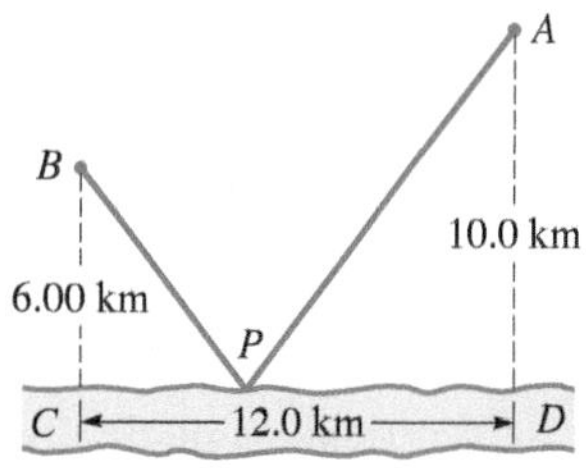

Fig. 2.46

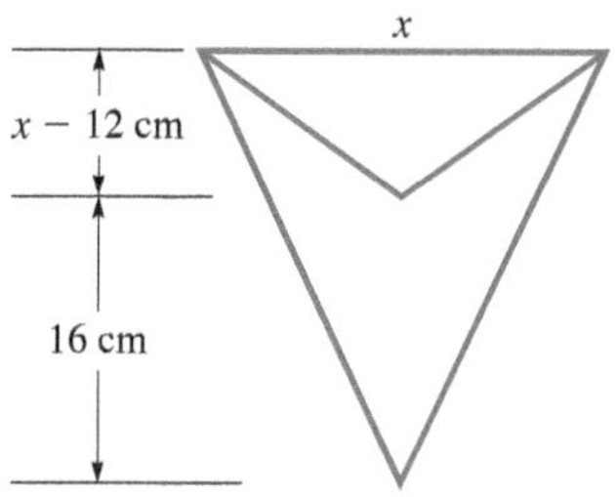

Fig. 2.47

Answers to Practice Exercises

1. 75° **2.** 686 m **3.** 40.0 m

2.3 Quadrilaterals

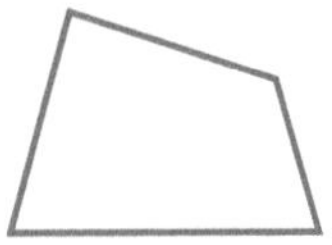

Fig. 2.48

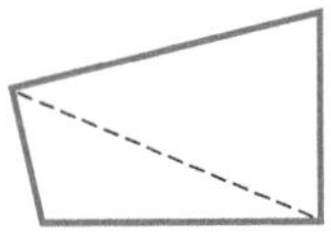

Fig. 2.49

A **quadrilateral** *is a closed plane figure with four sides,* and these four sides form four interior angles. Figs. 2.48 and 2.49 both illustrate general quadrilaterals.

A **diagonal** *of a polygon is a straight line segment joining any two nonadjacent vertices.* The dashed line is one of two diagonals of the quadrilateral in Fig. 2.49.

Quadrilaterals and Their Properties

	Parallelogram	Rhombus	Square	Rectangle	Trapezoid
Sides	Opposite sides are parallel and equal.	All sides are equal.	All sides are equal.	Opposite sides are equal.	One set of opposite sides (called bases) are parallel.
Angles	Opposite angles are equal.	Opposite angles are equal.	Four right angles	Four right angles	No requirement
Perimeter	$p = 2(a + b)$ **(2.6)**	$p = 4s$ **(2.8)**	$p = 4s$ **(2.10)**	$p = 2(l + w)$ **(2.12)**	$p = a + b_1 + b_2 + c$ **(2.14)**
Area	$A = bh$ **(2.7)**	$A = sh$ **(2.9)**	$A = s^2$ **(2.11)**	$A = lw$ **(2.13)**	$A = \frac{1}{2}h(b_1 + b_2)$ **(2.15)**
	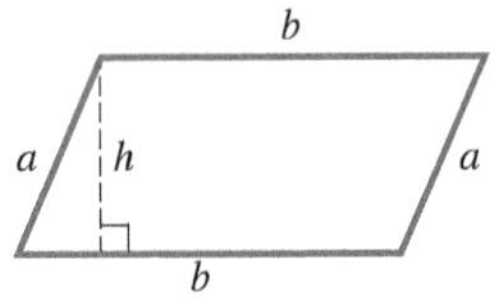 Fig. 2.50(a)	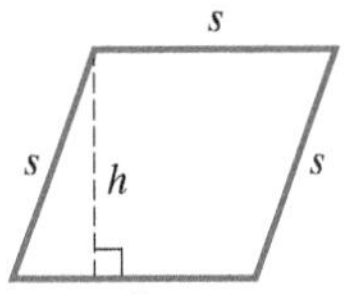 Fig. 2.50(b)	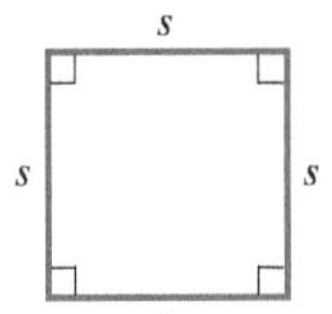 Fig. 2.50(c)	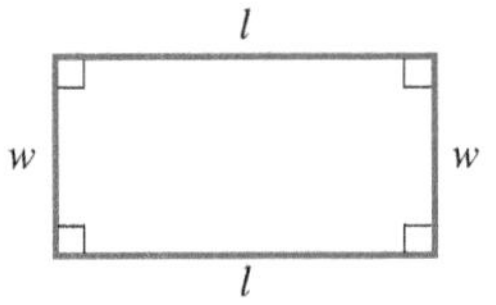 Fig. 2.50(d)	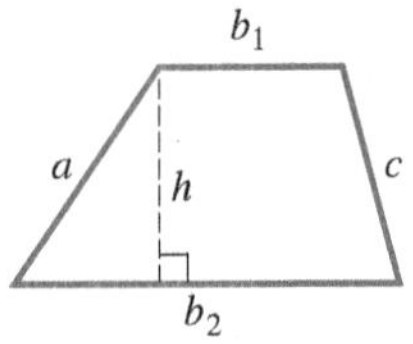 Fig. 2.50(e)

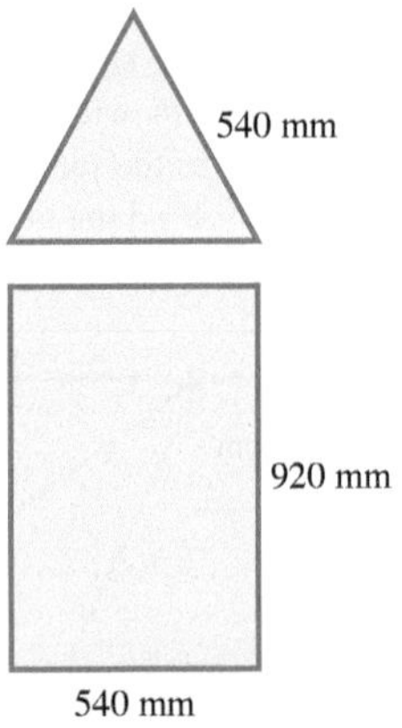

Fig. 2.51

EXAMPLE 1 Perimeter—window moulding

An architect designs a room with a rectangular window 920 mm high and 540 mm wide, with another window above in the shape of an equilateral triangle, 540 mm on a side. See Fig. 2.51. If moulding must be placed around the perimeter of each window, how much moulding is required?

The length of moulding is the sum of the perimeters of the windows. For the rectangular window, the opposite sides are equal, which means the perimeter is twice the sum of the length l plus the width w. For the equilateral triangle, the perimeter is three times the side s. Therefore, we have the following:

Perimeter of equilateral triangle:	$p_1 = 3s = 3(540) = 1620$ mm
Perimeter of rectangle:	$p_2 = 2(l + w) = 2(920 + 540) = 2920$ mm
Total length of moulding required:	$L = p_1 + p_2 = 1620 + 2920 = 4540$ mm

Practice Exercise

1. What is the perimeter of a rhombus of side 13.5 cm?

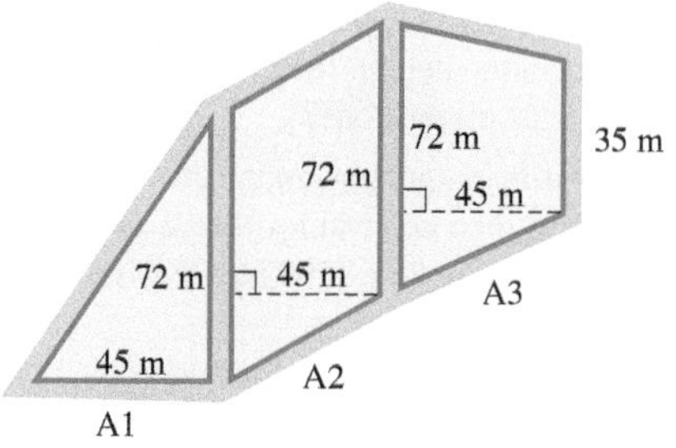

Fig. 2.52

EXAMPLE 2 Area—park design

A city park is designed with lawn areas in the shape of a right triangle, a parallelogram, and a trapezoid, as shown in Fig. 2.52, with walkways between them. Find the area of each section of lawn and the total lawn area.

Area of triangle: $A_1 = \frac{1}{2}bh = \frac{1}{2}(72)(45) = 1600 \text{ m}^2$

Area of parallelogram: $A_2 = bh = (72)(45) = 3200 \text{ m}^2$

Area of trapezoid: $A_3 = \frac{1}{2}h(b_1 + b_2) = \frac{1}{2}(45)(72 + 35) = 2400 \text{ m}^2$

Adding all three areas, the total lawn area is about 7200 m^2.

■ The computer microprocessor chip was first commercially available in 1971.

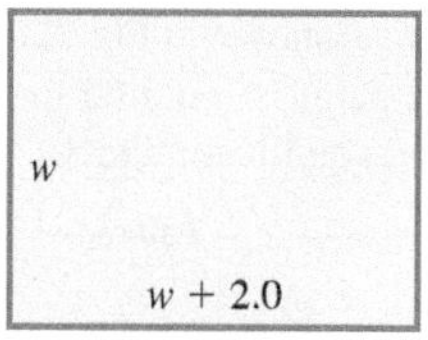

Fig. 2.53

EXAMPLE 3 Perimeter—computer chip dimensions

The length of a rectangular computer chip is 2.0 mm longer than its width. Find the dimensions of the chip if its perimeter is 26.4 mm.

Identify known quantities: Perimeter $p = 26.4$ mm

Identify unknown quantities: Let w = width of chip (in mm)

l = length of chip = $w + 2.0$

Sketch: See Fig. 2.53.

Select equation: $2l + 2w = p$

Solve (substituting l and p):

$$2(w + 2.0) + 2w = 26.4$$
$$2w + 4.0 + 2w = 26.4$$
$$4w = 22.4$$
$$w = 5.6 \text{ mm}$$
$$l = w + 2.0 = 7.6 \text{ mm}$$

Check: A width of 5.6 mm and a length of 7.6 mm do give a perimeter of 26.4 mm, so the solution checks.

Practice Exercise

2. What is the area of the chip in Fig. 2.53?

EXERCISES 2.3

In Exercises 1–2, make the given changes in the indicated examples of this section and then solve the given problems.

1. In Example 2, change the dimension of 45 m to 55 m in each figure and then find the area.
2. In Example 3, change 2.0 mm to 3.0 mm, and then solve.

In Exercises 3–10, find the perimeter of each figure.

3. Square: side of 65 m
4. Rhombus: side of 2.46 km
5. Rectangle: $l = 0.920$ mm, $w = 0.742$ mm
6. Rectangle: $l = 142$ cm, $w = 126$ cm
7. Parallelogram in Fig. 2.54
8. Parallelogram in Fig. 2.55
9. Trapezoid in Fig. 2.56
10. Trapezoid in Fig. 2.57

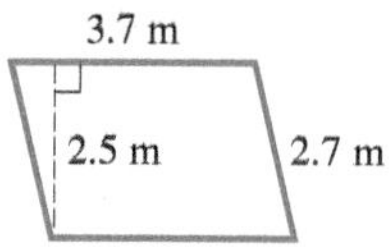

Fig. 2.54

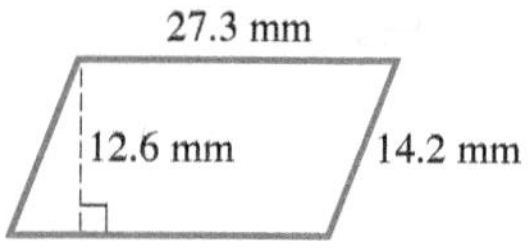

Fig. 2.55

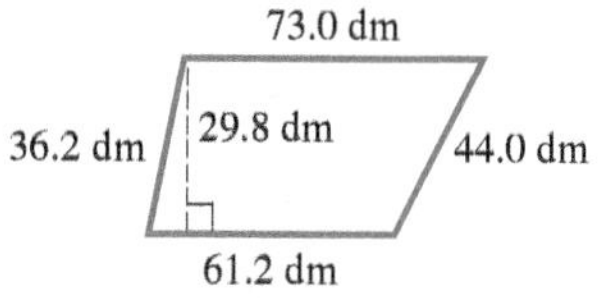

Fig. 2.56

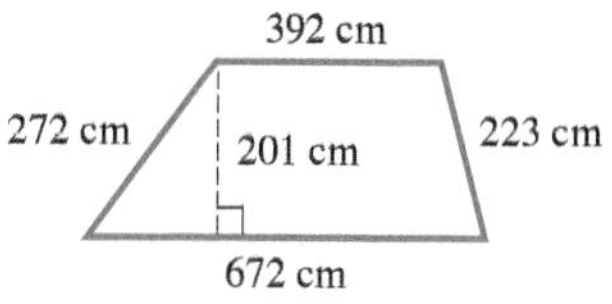

Fig. 2.57

In Exercises 11–20, find the area of each figure.

11. Square: $s = 2.7$ mm
12. Square: $s = 15.6$ m

13. Rectangle: $l = 20$ ft, $w = 7$ ft

14. Rectangle: $l = 27$ in., $w = 12$ in.

15. Rectangle: $l = 0.920$ km, $w = 0.742$ km

16. Rectangle: $l = 142$ cm, $w = 126$ cm

17. Parallelogram in Fig. 2.54

18. Parallelogram in Fig. 2.55

19. Trapezoid in Fig. 2.56

20. Trapezoid in Fig. 2.57

In Exercises 21–24, set up a formula for the indicated perimeter or area. (Do not include dashed lines.)

21. The perimeter of the figure in Fig. 2.58 (a parallelogram and a square attached)

22. The perimeter of the figure in Fig. 2.59 (two trapezoids attached)

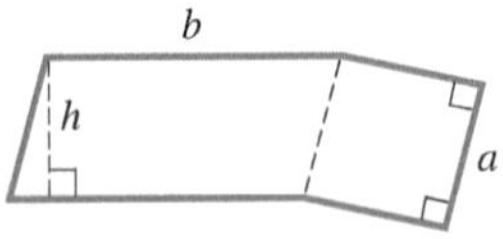

Fig. 2.58

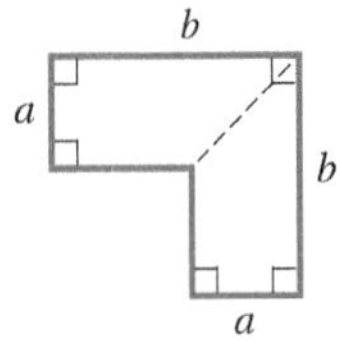

Fig. 2.59

23. Area of figure in Fig. 2.61

24. Area of figure in Fig. 2.62

In Exercises 25–41, solve the given problems.

25. If the angle between adjacent sides of a parallelogram is 90°, what conclusion can you make about the parallelogram?

26. What conclusion can you make about the two triangles formed by the sides and diagonal of a parallelogram? Explain.

27. Find the area of a square whose diagonal is 24.0 cm.

28. In a trapezoid, find the angle between the bisectors of the two angles formed by the bases and one nonparallel side.

29. Noting how a diagonal of a rhombus divides an interior angle, explain why the automobile jack in Fig. 2.60 is in the shape of a rhombus.

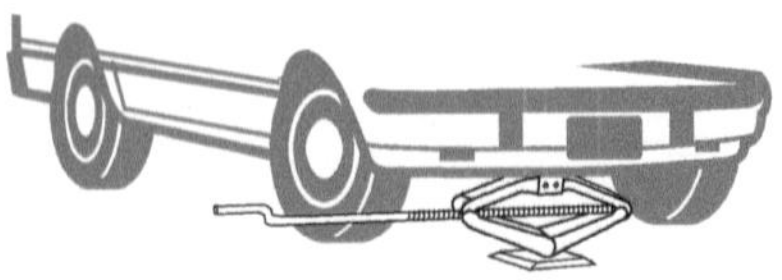

Fig. 2.60

30. Part of an electric circuit is wired in the configuration of a rhombus and one of its altitudes, as shown in Fig. 2.61. What is the length of wire in this part of the circuit?

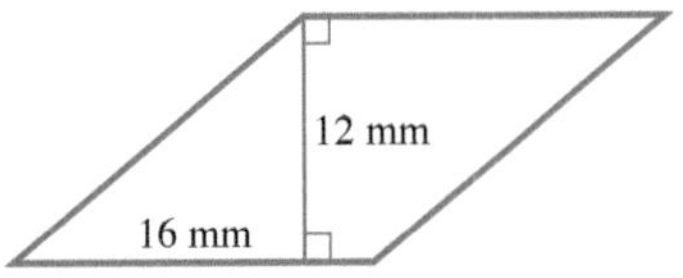

Fig. 2.61

31. A walkway 3.00 m wide is constructed along the outside edge of a square courtyard. If the perimeter of the courtyard is 324 m, what is the perimeter of the square formed by the outer edge of the walkway?

32. An architect designs a rectangular window such that the width of the window is 450 mm less than the height. If the perimeter of the window is 4500 mm, what are its dimensions?

33. A designer plans the top of a rectangular workbench to be four times as long as it is wide and then determines that if the width were 1500 mm greater and the length were 4500 mm less, it would be a square. What are its dimensions?

34. A beam support in a building is in the shape of a parallelogram, as shown in Fig. 2.62. Find the area of the side of the beam shown.

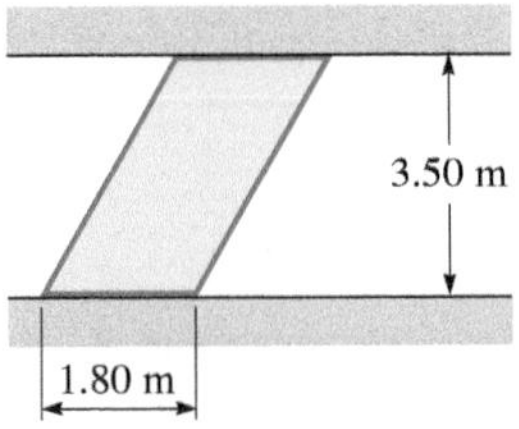

Fig. 2.62

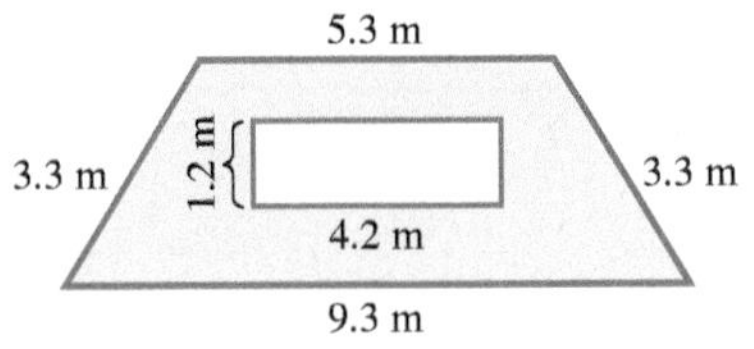

Fig. 2.63

35. Each of two walls (with rectangular windows) of an A-frame house has the shape of a trapezoid, as shown in Fig. 2.63. If a litre of paint covers 12 m^2, how much paint is required to paint these walls? (All data are accurate to two significant digits.)

36. Six equal trapezoidal sections form a conference table in the shape of a hexagon, with a hexagonal opening in the middle. See Fig. 2.64. From the dimensions shown, find the area of the table top.

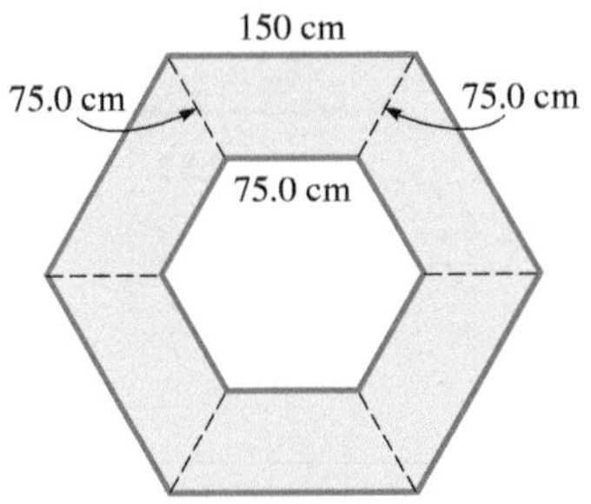

Fig. 2.64

37. A fenced section of a ranch is in the shape of a quadrilateral whose sides are 1.74 km, 1.46 km, 2.27 km, and 1.86 km, the last two sides being perpendicular to each other. Find the area of the section.

38. A rectangular security area is enclosed on one side by a wall, and the other sides are fenced. The length of the wall is twice the width of the area. The total cost of building the wall and fence is \$13 200. If the wall costs \$50.00/m and the fence costs \$5.00/m, find the dimensions of the area.

39. What is the sum of the measures of the interior angles of a quadrilateral? Explain.

40. Find a formula for the area of a rhombus in terms of its diagonals d_1 and d_2. (See Exercise 29.)

41. A house has a trapezoidal front-facing side, with dimensions shown in Fig. 2.65 (not drawn to scale). Excluding the area of the door, how much siding (in m^2) is required to cover the side if the length between the walls is 12 m, and the walls are perpendicular to the ground?

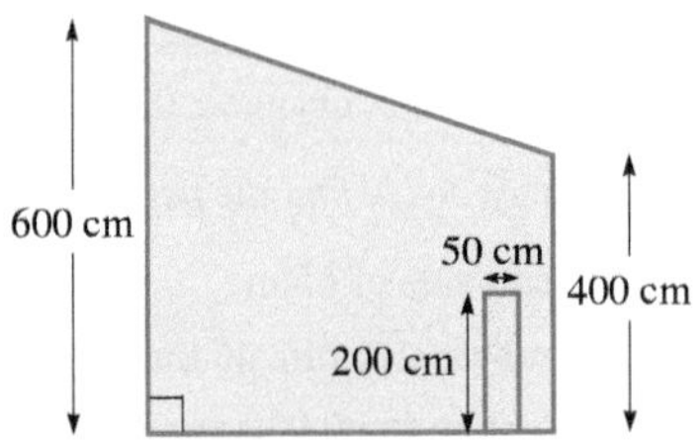

Fig. 2.65

Answers to Practice Exercises

1. 54 cm **2.** 43 mm^2

2.4 Circles

A **circle** is a plane curve in which all points are at the same distance from a fixed point, the centre of the circle.

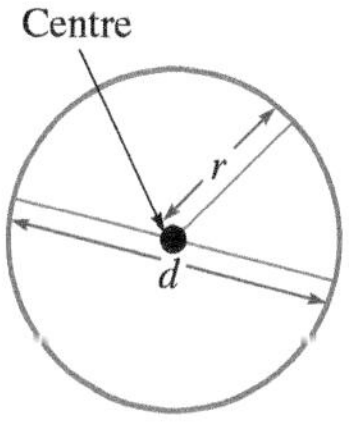

Fig. 2.66

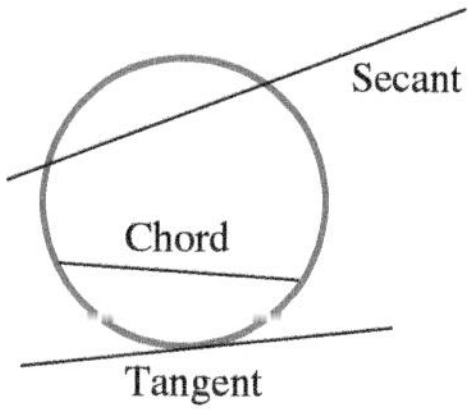

Fig. 2.67

Parts of a Circle

The **radius** ***r*** is the distance from the centre to any point along the edge of the circle.

The **diameter** ***d*** measures the farthest distance across the circle. Since it passes through the centre, *the diameter is twice the radius*, or $d = 2r$. See Fig. 2.66.

A **chord** is a line segment having its endpoints on the circle.

A **secant** is a line that passes through *two points on the circle* and continues in both directions.

A **tangent** is a line that only touches (does not pass through) the circle at *one point.* See Fig. 2.67.

An important property of a tangent is that *a tangent is perpendicular to the radius drawn to the point of contact.* This is illustrated in the following example.

EXAMPLE 1 Tangent line perpendicular to the radius

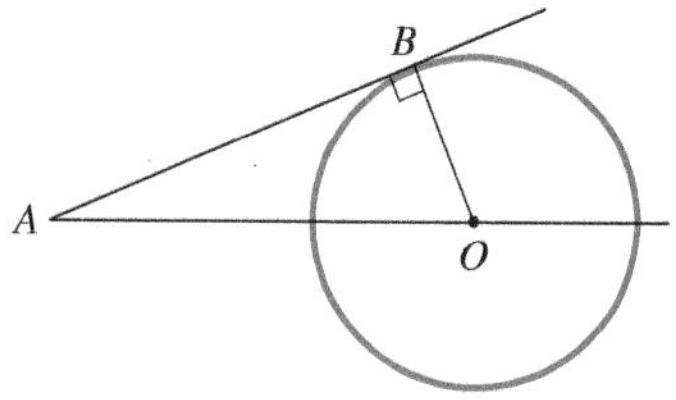

Fig. 2.68

In Fig. 2.68, O is the centre of the circle, and AB is tangent at B. If $\angle OAB = 25°$, find $\angle AOB$.

Since the centre is O, OB is a radius of the circle. A tangent is perpendicular to a radius at the point of tangency, which means $\angle ABO = 90°$.

Since the sum of the angles of a triangle is 180°, we have

$$\angle OAB + \angle ABO + \angle AOB = 180°$$
$$\angle AOB = 180° - 90° - 25°$$
$$\angle AOB = 65°$$

CIRCUMFERENCE AND AREA OF A CIRCLE

The perimeter of a circle is called the **circumference.** The formulas for the circumference and area of a circle, given in terms of both the radius r and the diameter d, are:

■ The symbol π (the Greek letter pi), which we use as a number, was first used in this way in the 1700s.

Circumference of a Circle:	$c = 2\pi r$	**(2.16)**	or	$c = \pi d$	**(2.17)**
Area of a Circle:	$A = \pi r^2$	**(2.18)**	or	$A = \dfrac{\pi d^2}{4}$	**(2.19)**

Here, π is an irrational number that is displayed in most calculators rounded as 3.141 592 654. In using a calculator or a graphing utility, π can be entered to a much greater accuracy by using the $\boxed{\pi}$ key or a π function.

EXAMPLE 2 Area and circumference of a circle—oil spill

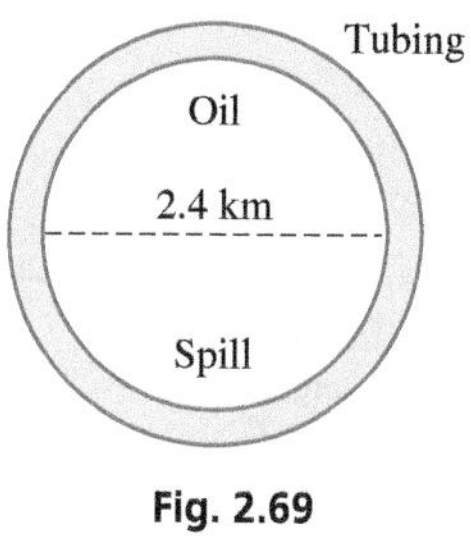

Fig. 2.69

A circular oil spill has a diameter of 2.4 km. It is to be enclosed within special flexible tubing. What is the area of the spill, and how long must the tubing be? See Fig. 2.69.

Since we know the diameter, we use Eq. (2.18). The area is

$$A = \frac{\pi d^2}{4} = \frac{\pi(2.4)^2}{4}$$
$$A = 4.5\ \text{km}^2$$

The length of tubing needed is the circumference of the circle. Eq. (2.17) gives

$$c = \pi d = \pi(2.4)$$
$$c = 7.5\ \text{km} \qquad \text{rounded off}$$

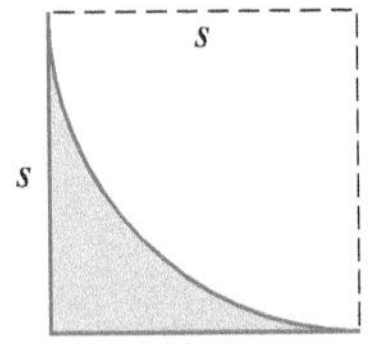

Fig. 2.70

EXAMPLE 3 Perimeter and area—machine part

Using design software, an engineer designs a machine part by taking a flat, square piece of metal and then removing a quarter-circle from the square. The resulting component is shown in Fig. 2.70. If the length of one side of the square is 3.25 cm, find the perimeter and the area of one side of the part.

Identify known quantities: Side of square $s = 3.25$ cm

Radius of circle $r = s = 3.25$ cm

Identify unknowns: Let p and A be the perimeter and the area of the part.

Select equations: The perimeter is the length of the boundary of the part: two sides of the square *plus* one-fourth of the circumference: $p = 2s + \frac{1}{4}(2\pi r)$

The area is the area of the square *minus* one-fourth of the area of the circle: $A = s^2 - \frac{1}{4}(\pi r^2)$

Solve:

$$p = 2(3.25) + \frac{2\pi(3.25)}{4} = 11.6 \text{ cm}$$

$$A = 3.25^2 - \frac{\pi(3.25^2)}{4} = 2.17 \text{ cm}^2$$

Practice Exercises

1. Find the circumference of a circle with a radius of 20.0 cm.
2. Find the area of the circle in Exercise 1.

Circular Arcs and Angles

EXAMPLE 4

An **arc** is a part of a circle. It is typically denoted by its two endpoints. If a third point is indicated in the notation, it makes it clear that the larger arc is the arc of interest.

An angle formed at the centre by two radii is a **central angle**. The **measure of an arc** is the same as the central angle between the radii that define the arc.

A **sector** of a circle is the region bounded by two radii and the arc they intercept.

(a) In Fig. 2.71, the magenta arc is $\overset{\frown}{AB}$ or $\overset{\frown}{BA}$. The blue arc is $\overset{\frown}{ACB}$ or $\overset{\frown}{BCA}$.

(b) In Fig. 2.71, $\angle AOB$ is a central angle that measures 70°. It is formed by the radii OA and OB, whose ends A and B define the arc $\overset{\frown}{AB}$. Therefore, the measure of the arc $\overset{\frown}{AB}$ is 70°.

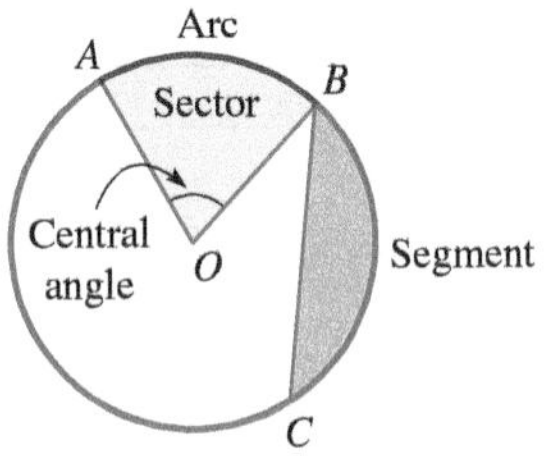

Fig. 2.71

A **segment** of a circle is the region bounded by a chord and its arc.

Every chord creates two segments. The smaller region is called the **minor segment** and the larger area is called the **major segment**.

(c) In Fig. 2.72, the region bounded by the chord DE and the magenta arc is a minor segment; the region bounded by the chord DE and the blue arc is a major segment.

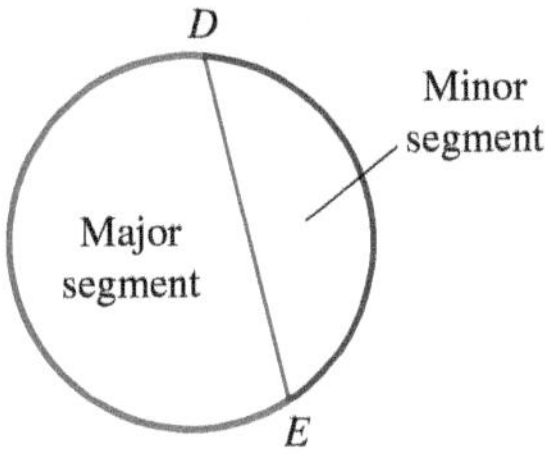

Fig. 2.72

An **inscribed angle** of an arc is formed by two chords that have a common endpoint on the circle, the vertex of the angle.

An important property of a circle is that *the measure of an inscribed angle is one-half the measure of its intercepted arc.*

(d) In Fig. 2.73, $\angle ABC$ is inscribed in arc $\overset{\frown}{ABC}$ The endpoints of the angle other than the vertex define the intercepted arc $\overset{\frown}{AC}$. If the measure of $\overset{\frown}{AC}$ is 60°, then $\angle ABC = 30°$.

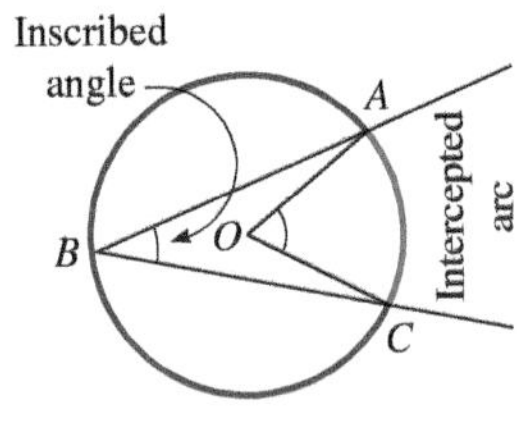

Fig. 2.73

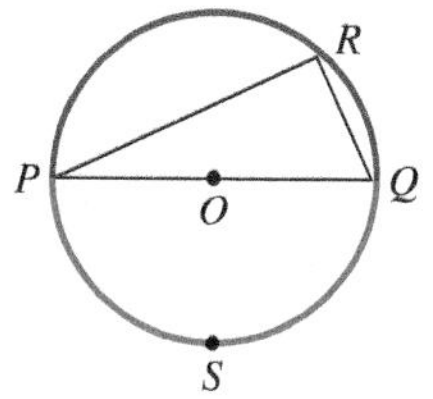

Fig. 2.74

EXAMPLE 5 Inscribed angle

In the circle shown in Fig. 2.74, PQ is a diameter, and $\angle PRQ$ is inscribed in the semicircular $\widehat{PRQ}$. Since the measure of $\widehat{PSQ} = 180°$, $\angle PRQ = 90°$. From this we conclude that *an angle inscribed in a semicircle is a right angle.*

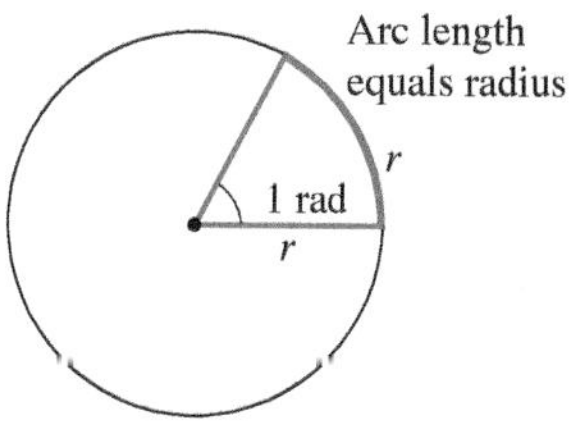

Fig. 2.75

RADIAN MEASURE OF AN ANGLE

In addition to degrees, another common unit of angular measurement is the radian. **One radian** is defined as the *measure of a central angle that intercepts an arc one radius in length.* See Fig. 2.75. The radius can be marked off along the circumference 2π times (about 6.283 times). Thus, 2π rad = 360° (where rad is the symbol for radian), and the basic relationship between radians and degrees is

$$\pi \text{ rad} = 180° \quad \textbf{(2.20)}$$

This equivalence can be used as a unit conversion factor, as discussed in Section 1.3. Note that by convention, *if no unit is given for an angle, the angle is assumed to be reported in radians.*

EXAMPLE 6 Radian measure of an angle

(a) Convert 1 radian to degrees. If we divide each side of Eq. (2.20) by π, we get

$$1 \text{ rad} = 57.3°$$

where the result has been rounded off.

(b) Convert 118.2° to radians. We have

$$118.2° = 118.2°\left(\frac{\pi \text{ rad}}{180°}\right) = 2.063 \text{ rad}$$

Multiplying 118.2° by π rad/180°, the unit that remains is rad since degrees "cancel."

Practice Exercise

3. Express the angle 85.0° in radian measure.

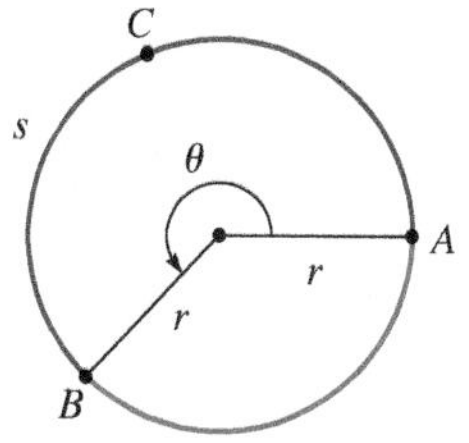

Fig. 2.76

Radian measure is especially useful when determining the length of an arc or the area of a sector formed by a central angle in a circle. Since the arc length for 1 radian of central angle is equivalent to 1 radius, any multiple of radian measure for the central angle will result in the same multiple for the length of the resulting arc (see Fig. 2.76). In general, the arc length s (the length of $\widehat{ACB}$) can be determined from the product of the central angle θ (if expressed in radians) and the radius r.

$$s = \theta r, \quad \text{where } \theta \text{ is given in radians} \quad \textbf{(2.21a)}$$

Moreover, the area of a sector is

$$A = \frac{1}{2}r^2\theta, \quad \text{where } \theta \text{ is given in radians} \quad \textbf{(2.21b)}$$

EXAMPLE 7 Arc length with radian measure

If a circle of radius 2.75 m has a central angle of 3.76 radians, determine the length of the arc formed by the central angle.

$$s = \theta r = (3.76)(2.75 \text{ m}) = 10.3 \text{ m}$$

EXERCISES 2.4

In Exercises 1–4, answer the given questions about the indicated examples of this section.

1. In Example 1, if $\angle AOB = 72°$ in Fig. 2.68, then what is the measure of $\angle OAB$?

2. In the first line of Example 2, if "diameter" is changed to "radius," what are the results?

3. In Example 3, if the machine part is the unshaded part (rather than the shaded part) of Fig. 2.70, what are the results?

4. In Example 4(d), if $\angle ABC = 25°$ in Fig. 2.73, then what is the measure of $\overset{\frown}{AC}$?

In Exercises 5–8, refer to the circle with centre at O in Fig. 2.77. Identify the following.

5. (a) A secant line **(b)** A tangent line

6. (a) Two chords
(b) An inscribed angle

7. (a) Two perpendicular line segments
(b) An isosceles triangle

8. (a) A segment
(b) A sector with an acute central angle

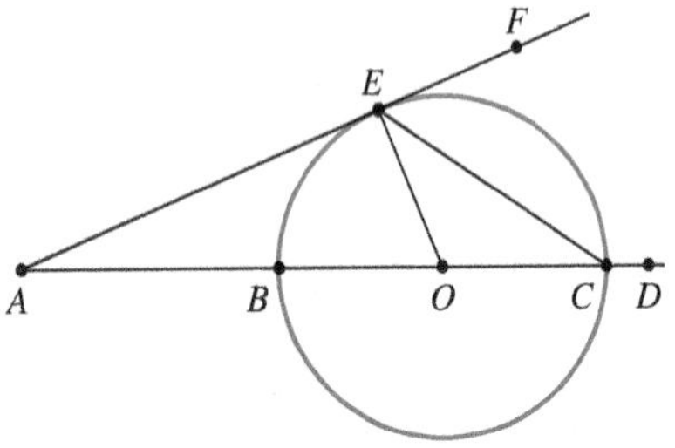

Fig. 2.77

In Exercises 9–12, find the circumference of the circle with the given radius or diameter.

9. $r = 275$ cm **10.** $r = 0.563$ m

11. $d = 23.1$ mm **12.** $d = 8.2$ dm

In Exercises 13–18, find the area of the circle with the given radius, diameter, or circumference.

13. $r = 0.0952$ km **14.** $r = 45.8$ cm **15.** $d = 2.33$ m

16. $d = 1256$ mm **17.** $c = 40.1$ cm **18.** $c = 147$ m

In Exercises 19–22, refer to Fig. 2.78, where AB is a diameter, TB is a tangent line at B, and $\angle ABC = 65°$. Determine the indicated angles.

19. $\angle CBT$

20. $\angle BCT$

21. $\angle CAB$

22. $\angle BTC$

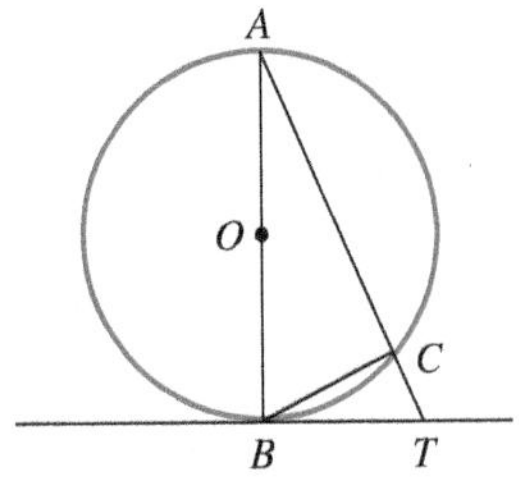

Fig. 2.78

In Exercises 23–26, refer to Fig. 2.79. Determine the indicated arcs and angles.

23. $\overset{\frown}{BC}$ (angle measure)

24. $\overset{\frown}{AB}$ (angle measure)

25. $\angle ABC$

26. $\angle ACB$

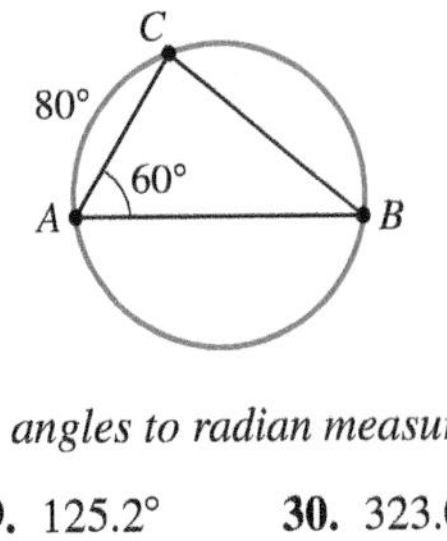

Fig. 2.79

In Exercises 27–30, change the given angles to radian measure.

27. 22.5° **28.** 60.0° **29.** 125.2° **30.** 323.0°

In Exercises 31–34, find a formula for the indicated perimeter or area.

31. The perimeter of the quarter-circle in Fig. 2.80

32. The perimeter of the figure in Fig. 2.81. A quarter-circle is attached to a triangle.

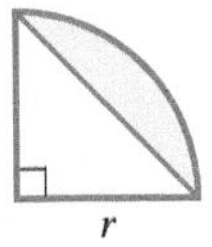

Fig. 2.80

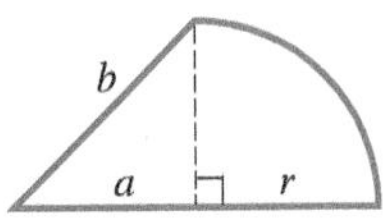

Fig. 2.81

33. The area of the segment of the quarter-circle in Fig. 2.80

34. The area of the figure in Fig. 2.81

In Exercises 35–53, solve the given problems.

35. Find the area of a sector for an interior angle of $\pi/5$ if the diameter is 1.00 m.

36. Find the area of a sector for an interior angle of 42° if the diameter is 5.00 in.

37. Describe the location of the midpoints of a set of parallel chords of a circle.

38. The measure of $\overset{\frown}{AB}$ on a circle of radius r is 45°. What is the length of the arc in terms of r and π?

39. In a circle, a chord connects the ends of two perpendicular radii of 6.00 cm. What is the area of the minor segment?

40. In Fig. 2.82, chords AB and DE are parallel. What is the relationship between $\triangle ABC$ and $\triangle CDE$? Explain.

41. Many Indigenous North Americans often describe drums as the heartbeat of Mother Earth. In the hand drum in Fig 2.83, 15 equally spaced laces connect at the centre of the circle. Find the arc length and area of the sector formed between two of the equally spaced laces if the radius of the drum is 10.0 cm.

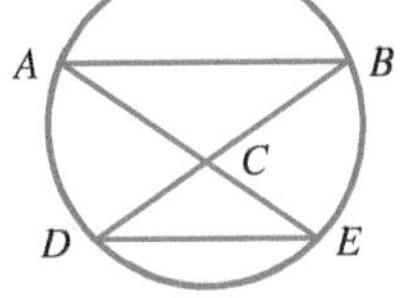

Fig. 2.82

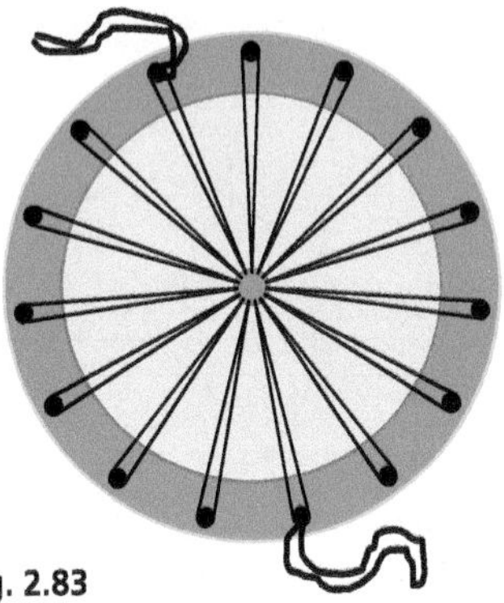
Fig. 2.83

42. A person is in a plane 11.5 km above the shore of the Pacific Ocean. How far from the plane can the person see out on the Pacific? (The radius of Earth is 6378 km.)

43. The radius of the earth's equator is 6375 km. What is the circumference?

44. As a ball bearing rolls along a straight track, it makes 11.0 revolutions while travelling a distance of 109 mm. Find its radius.

45. Using a tape measure, the circumference of a tree is found to be 112 cm. What is the diameter of the tree (assuming a circular cross-section)?

46. With no change in the speed of flow, by what factor should the diameter of a pipe be increased in order to double the amount of water that flows through the pipe?

47. Suppose that a 23 400-N force is applied to a hollow steel cylindrical beam that has the cross-section shown in Fig. 2.84. The stress on the beam is found by dividing the force by the cross-sectional area. Find the stress.

48. The cross-section of a large circular conduit has seven smaller equal circular conduits within it. The conduits are tangent to each other, as shown in Fig. 2.85. What fraction of the large conduit is occupied by the seven smaller conduits?

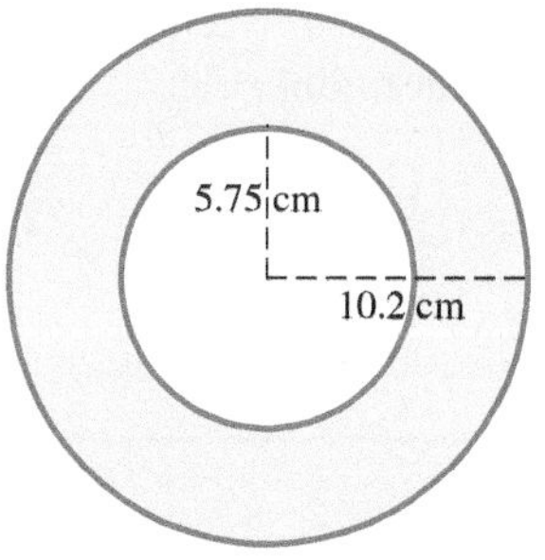

Fig. 2.84

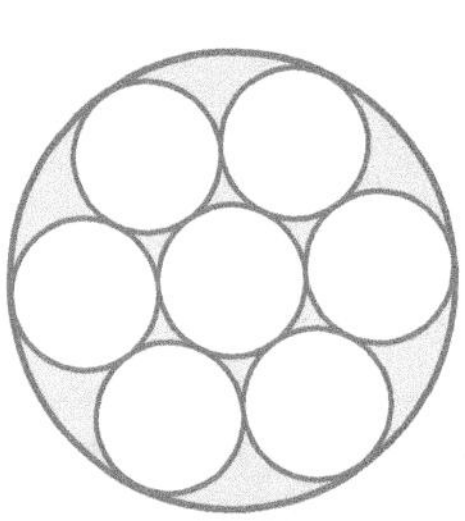

Fig. 2.85

49. Find the area of the room in the plan shown in Fig. 2.86.

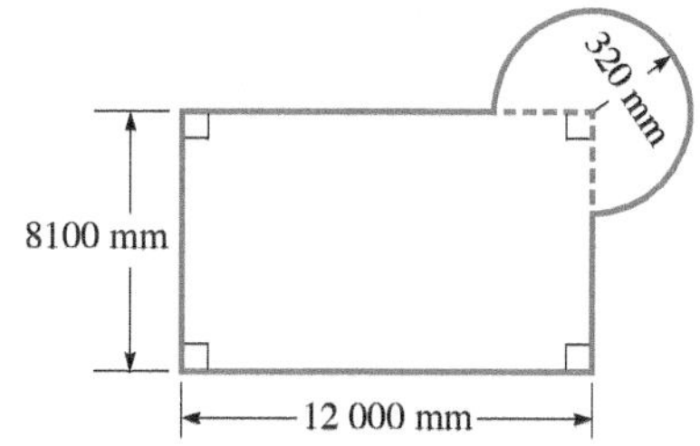

Fig. 2.86

50. Find the length of the pulley belt shown in Fig. 2.87 if the belt crosses at right angles. The radius of each pulley wheel is 5.50 cm.

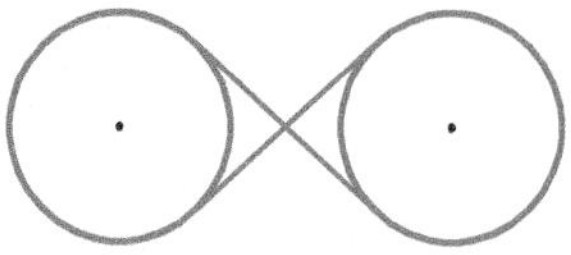

Fig. 2.87

51. The velocity of an object moving in a circular path is directed tangent to the circle in which it is moving. A stone on a string moves in a vertical circle, and the string breaks after 5.5 revolutions. If the string was initially in a vertical position, in what direction does the stone move after the string breaks? Explain.

52. Part of a circular gear with 24 teeth is shown in Fig. 2.88. Find the indicated angle.

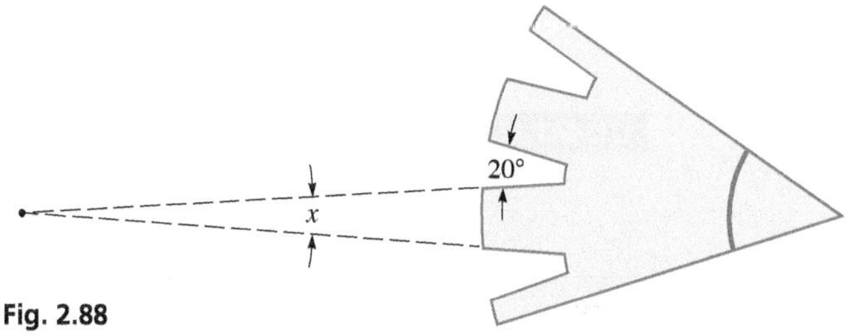

Fig. 2.88

53. The Natapoka Arc is a geological feature located on the eastern shore of Hudson Bay. It is a nearly perfect semicircular arc, covering about 2.8 radians of a circle 450 km in diameter. Determine the length of the arc.

Answers to Practice Exercises

1. 126 cm **2.** 1260 cm^2 **3.** 1.48 rad

2.5 Measurement of Irregular Areas

The figures for which we have found areas are well defined, and the areas can be found by direct use of a specific formula. In practice, however, it may be necessary to find the area of a figure with an irregular perimeter. In this section, we show two methods of finding a very good *approximation* of such an area. These methods are particularly useful in technical areas such as surveying, architecture, and mechanical design.

THE TRAPEZOIDAL RULE

As shown in Fig. 2.89(a), we measure the distance between the edges of a figure and divide the area with parallel lines into strips of equal length h (the smaller the h, the better the approximation will be). We then join the points where the parallel lines intersect the figure to form adjacent trapezoids. The sum of the areas of these trapezoids is a good approximation to the area. Now, in Fig. 2.89(b), labelling the lengths of the parallel lines

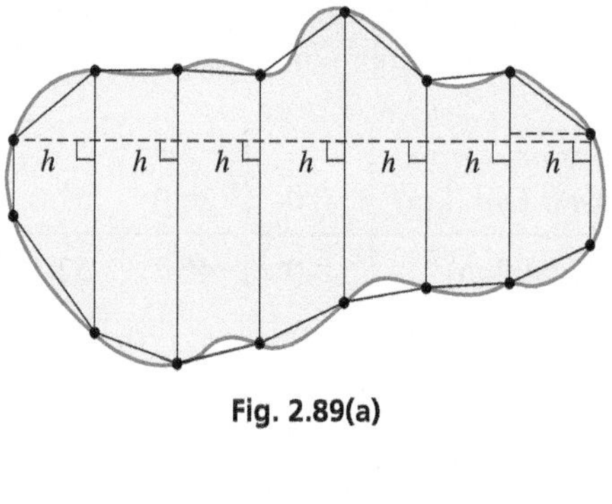

Fig. 2.89(a)

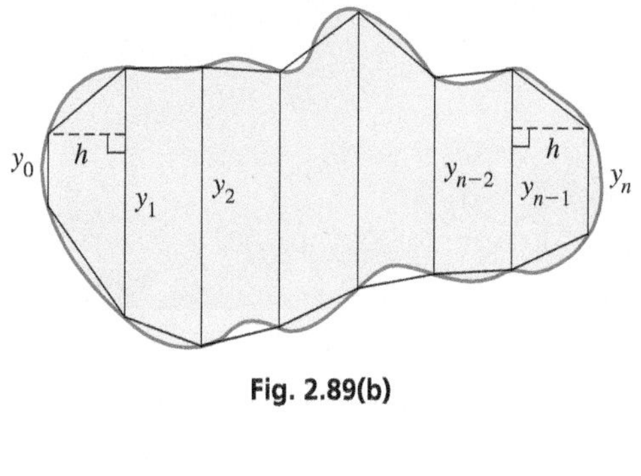

Fig. 2.89(b)

$y_0, y_1, y_2, \ldots, y_n$ and applying Eq. (2.15) to find the area of each trapezoid, the total area of all trapezoids is

$$\underbrace{A = \frac{h}{2}(y_0 + y_1)}_{\text{first trapezoid}} + \underbrace{\frac{h}{2}(y_1 + y_2)}_{\text{second trapezoid}} + \underbrace{\frac{h}{2}(y_2 + y_3)}_{\text{third trapezoid}} + \cdots + \underbrace{\frac{h}{2}(y_{n-2} + y_{n-1})}_{\text{next-to-last trapezoid}} + \underbrace{\frac{h}{2}(y_{n-1} + y_n)}_{\text{last trapezoid}}$$

$$A = \frac{h}{2}(y_0 + y_1 + y_1 + y_2 + y_2 + y_3 + \cdots + y_{n-2} + y_{n-1} + y_{n-1} + y_n)$$

Combining like terms, we obtain what is known as the **trapezoidal rule**:

$$A = \frac{h}{2}(y_0 + 2y_1 + 2y_2 + \cdots + 2y_{n-1} + y_n) \qquad \textbf{(2.22)}$$

COMMON ERROR Note carefully that the values of y_0 and y_n are not multiplied by 2.

EXAMPLE 1 Trapezoidal rule—area of cam

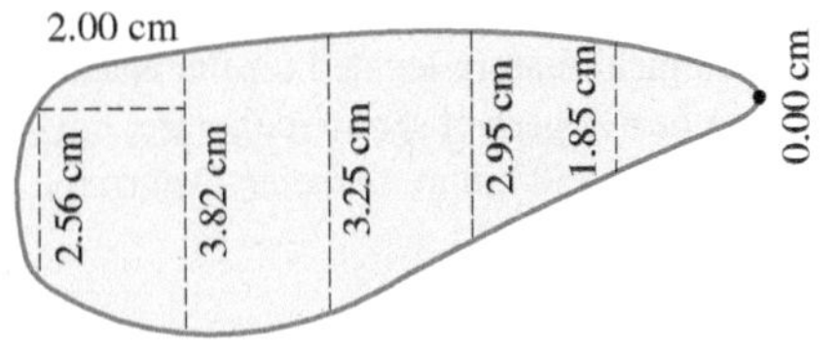

Fig. 2.90

A plate cam for opening and closing a valve is shown in Fig. 2.90. Widths of the face of the cam are shown at 2.00-cm intervals from one end of the cam to the other. Approximate the area of the face of the cam.

We can label the lengths from left to right as

$$y_0 = 2.56 \text{ cm} \qquad y_1 = 3.82 \text{ cm} \qquad y_2 = 3.25 \text{ cm}$$
$$y_3 = 2.95 \text{ cm} \qquad y_4 = 1.85 \text{ cm} \qquad y_5 = 0.00 \text{ cm}$$

We identify $h = 2.00$ cm. Therefore, using the trapezoidal rule, Eq. (2.22), we have

$$A = \frac{2.00}{2}[2.56 + 2(3.82) + 2(3.25) + 2(2.95) + 2(1.85) + 0.00]$$
$$A = 26.3 \text{ cm}^2$$

■ As in Example 1, note that all measurements in the trapezoidal rule must be in the same units.

The area of the face of the cam is approximately 26.3 cm^2.

When approximating the area with trapezoids, we omit small parts of the area for some trapezoids and include small extra areas for other trapezoids. When more strips are used, there are fewer errors of this kind, making the approximation better.

EXAMPLE 2 Trapezoidal rule—approximating the area of Lake Ontario

Fig. 2.91

Lake Ontario is one of the Great Lakes between the United States and Canada. Measurements of the width of the lake were made along its length, starting at the west end, at 26.0-km intervals. The widths are shown in Fig. 2.91 and are given in the table below. Approximate the area of Lake Ontario.

Distance from West End (km)	0.0	26.0	52.0	78.0	104	130	156
Width (km)	0.0	46.7	52.1	59.2	60.4	65.7	73.9

Distance from West End (km)	182	208	234	260	286	312
Width (km)	87.0	75.5	66.4	86.1	77.0	0.0

We label the widths as $y_0 = 0.0$ km, $y_1 = 46.7$ km, $y_2 = 52.1$ km, . . . , and $y_n = 0.0$ km and identify $h = 26$ km. Therefore, using the trapezoidal rule, the approximate area of Lake Ontario is

$$A = \frac{26.0}{2}\left[0.0 + 2(46.7) + 2(52.1) + 2(59.2) + 2(60.4) + 2(65.7)\right.$$
$$\left. +2(73.9) + 2(87.0) + 2(75.5) + 2(66.4) + 2(86.1) + 2(77.0) + 0.0\right]$$
$$A = 19\,500 \text{ km}^2$$

The area of Lake Ontario is actually 19 477 km^2.

Practice Exercise

1. In Example 2, use only the distances from west end of (in km) 0.0, 52.0, 104, 156, 208, 260, and 312. Calculate the area and compare with the answer in Example 2.

SIMPSON'S RULE

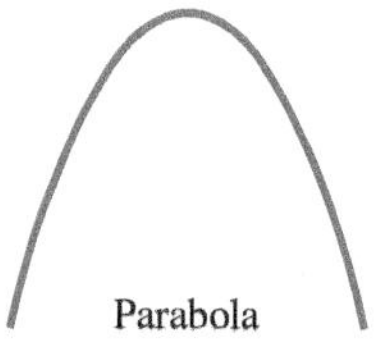

Fig. 2.92

When the edges of an irregular figure are not flat, Simpson's rule typically provides a more accurate estimate of the area than the trapezoidal rule because it estimates the edges of the figure as arcs of parabolas, which are curved as seen in Fig. 2.92. (The path of a ball that has been thrown, going up and then coming down to the ground, and the arc created on the inside of a satellite dish are both examples of parabolas.) Parabolic features will be discussed in Chapter 21.

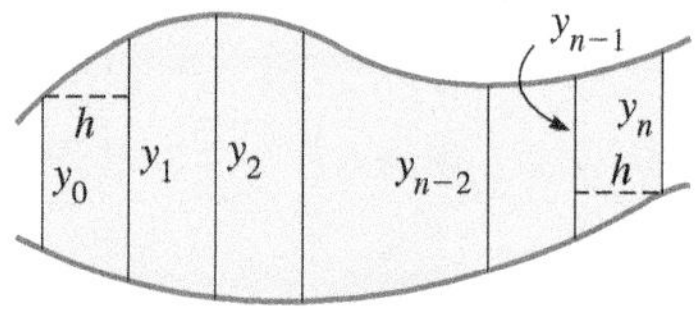

Fig. 2.93

The approximate area of a geometric figure, as illustrated in Fig. 2.93, is calculated by **Simpson's rule** as

$$A = \frac{h}{3}(y_0 + 4y_1 + 2y_2 + 4y_3 + \cdots + 2y_{n-2} + 4y_{n-1} + y_n) \qquad \textbf{(2.23)}$$

In using Eq. (2.23), the number n of intervals of width h must be even. Note the pattern in the coefficients: 1 for the first and last terms; 2 for even-numbered terms; and 4 for odd-numbered terms.

■ Simpson's rule is named for the English mathematician Thomas Simpson (1710–1761).

EXAMPLE 3 Simpson's rule—parking lot area

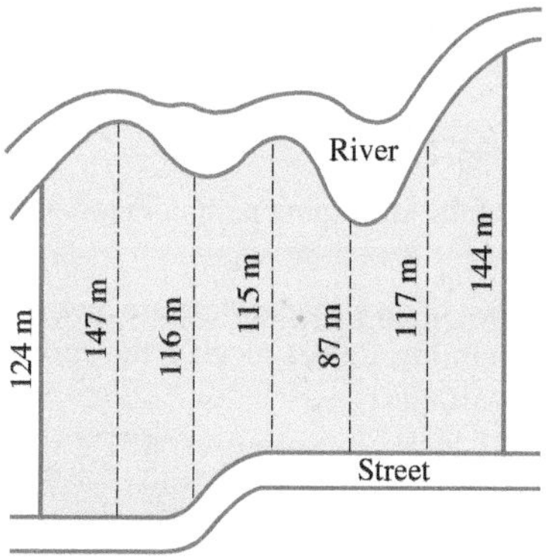

Fig. 2.94

A parking lot is proposed for a riverfront area in a town. The town engineer measured the widths of the area at 30.0-m intervals, as shown in Fig. 2.94. Find the area available for parking.

First, we see that there are six intervals, which means Eq. (2.23) may be used. With $y_0 = 124$ m, $y_1 = 147$ m, . . . , $y_6 = 144$ m, and $h = 30.0$ m, we have

$$A = \frac{30.0}{3}\left[124 + 4(147) + 2(116) + 4(115) + 2(87) + 4(117) + 144\right]$$
$$A = 21\,900 \text{ m}^2$$

For most areas, Simpson's rule gives a somewhat better approximation than the trapezoidal rule. The accuracy of Simpson's rule is also usually improved by using smaller intervals.

EXAMPLE 4 Simpson's rule—Easter Island area

From an aerial photograph, a cartographer determines the widths of Easter Island (in the Pacific Ocean) at 1.50-km intervals as shown in Fig. 2.95. The widths found are as follows:

Distance from South End (km)	0	1.50	3.00	4.50	6.00	7.50	9.00	10.5	12.0	13.5	15.0
Width (km)	0	4.8	5.7	10.5	15.2	18.5	18.8	17.9	11.3	8.8	3.1

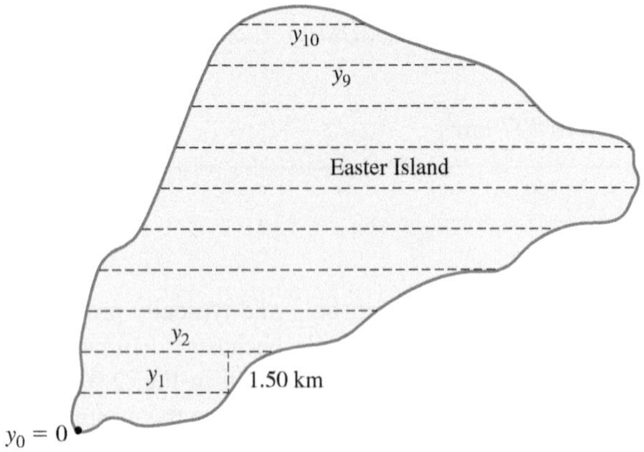

Fig. 2.95

Since there are 10 intervals (an even number), Simpson's rule may be used. From the table, we have the following values: $y_0 = 0$, $y_1 = 4.8$, $y_2 = 5.7, \ldots,$ $y_9 = 8.8$, $y_{10} = 3.1$, and $h = 1.5$. Using Simpson's rule, the cartographer would approximate the area of Easter Island as follows:

$$A = \frac{1.50}{3}(0 + 4(4.8) + 2(5.7) + 4(10.5) + 2(15.2) + 4(18.5)$$
$$+2(18.8) + 4(17.9) + 2(11.3) + 4(8.8) + 3.1)$$
$$A = 174 \text{ km}^2$$

As compared with a more accurate approximation of 165 km^2, Simpson's rule provides a reasonable approximation. However, if a high level of precision is required, computer aided calculations and Simpson's rule with a large number of strips are recommended.

EXERCISES 2.5

In Exercises 1 and 2, answer the given questions related to the indicated examples of this section.

1. In Example 1, if widths of the face of the same cam were given at 1.00-cm intervals (five more widths would be included), from the methods of this section, what is probably the most accurate way of finding the area? Explain.

2. In Example 4, if you use only the data from the south end of (in km) 0, 3.00, 6.00, 9.00, 12.0, and 15.0, would you choose the trapezoidal rule or Simpson's rule to calculate the area? Explain. Do not calculate the area for these data.

In Exercises 3 and 4, answer the given questions related to Fig. 2.96.

3. Which should be more accurate for finding the area, the trapezoidal rule or Simpson's rule? Explain.

4. If the trapezoidal rule is used to find the area, will the result probably be too high, about right, or too little? Explain.

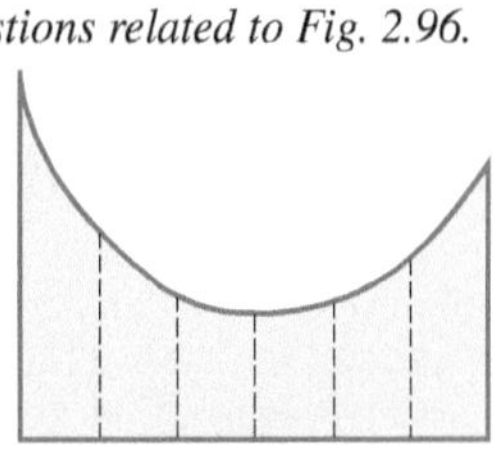
Fig. 2.96

In Exercises 5–16, calculate the indicated areas. All data are accurate to at least two significant digits.

5. The widths of a kidney-shaped swimming pool were measured at 2.0-m intervals, as shown in Fig. 2.97. Calculate the surface area of the pool using the trapezoidal rule.

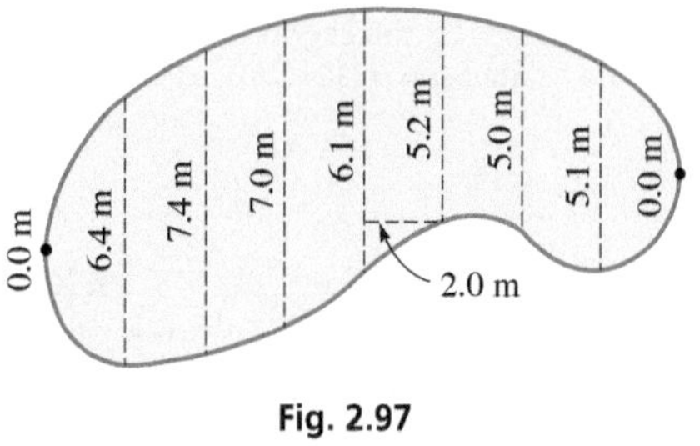

Fig. 2.97

6. Calculate the surface area of the swimming pool in Fig. 2.97 using Simpson's rule.

7. The widths of a cross-section of an airplane wing are measured at 0.30-m intervals, as shown in Fig. 2.98. Calculate the area of the cross-section using Simpson's rule.

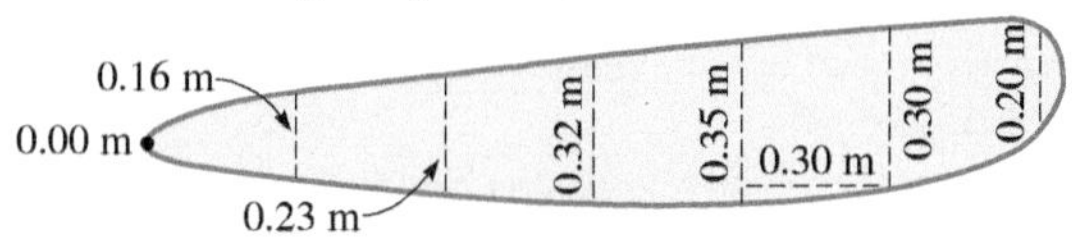

Fig. 2.98

8. Calculate the area of the cross-section of the airplane wing in Fig. 2.98 using the trapezoidal rule.

9. Using aerial photography, the widths of an area burned by a forest fire were measured at 0.5-km intervals, as shown in the following table:

Distance (km)	0.0	0.5	1.0	1.5	2.0	2.5	3.0	3.5	4.0
Width (km)	0.6	2.2	4.7	3.1	3.6	1.6	2.2	1.5	0.8

Determine the area burned by the fire by using the trapezoidal rule.

10. Find the area burned by the forest fire of Exercise 9 using Simpson's rule.

11. A cartographer measured the width of Bruce Peninsula in Ontario at 10-mm intervals on a map (scale 10 mm to 23 km) as shown in Fig. 2.99. The widths are shown in the following list. What is the area of Bruce Peninsula?

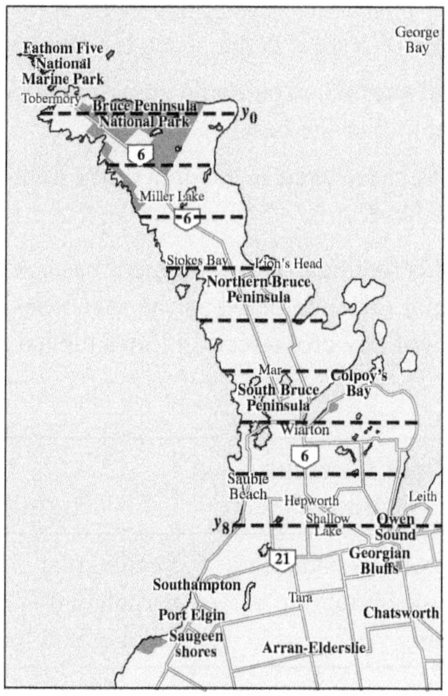

Fig. 2.99

$y_0 = 38\,\text{mm}$ $y_1 = 24\,\text{mm}$ $y_2 = 25\,\text{mm}$

$y_3 = 17\,\text{mm}$ $y_4 = 34\,\text{mm}$ $y_5 = 29\,\text{mm}$

$y_6 = 36\,\text{mm}$ $y_7 = 34\,\text{mm}$ $y_8 = 30\,\text{mm}$

12. A land surveyor measures the elevation at several points along a hill that she wishes to have removed. When a backhoe cuts into the mound, it reveals the profile shown in Figure 2.100. Using the trapezoidal rule, estimate the cross-sectional area of the mound.

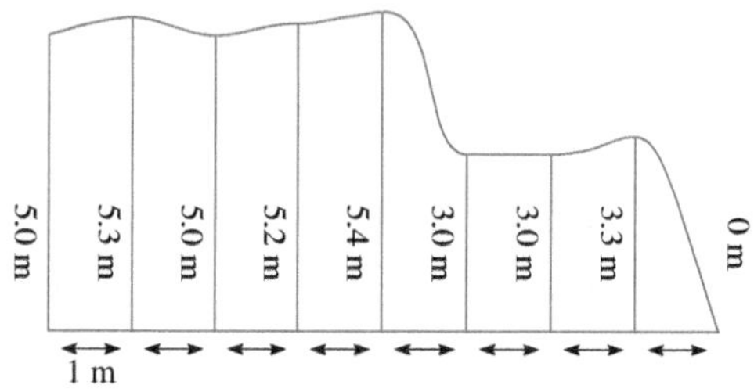

Fig. 2.100

13. The widths of Kejimkujik National Park (and National Historic Site) in Nova Scotia, measured at 2.0-km intervals, are shown in Fig. 2.101. Find the area of the park using the trapezoidal rule.

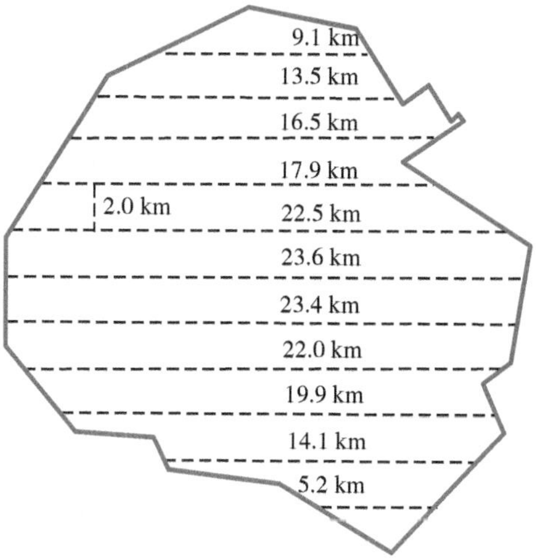

Fig. 2.101

14. Find the area of Kejimkujik National Park (see Exercise 13) using Simpson's rule.

15. Soundings taken across a river channel give the following depths, with the corresponding distances from one shore.

Distance (m)	0	50	100	150	200	250	300	350	400	450	500
Depth (m)	5	12	17	21	22	25	26	16	10	8	0

Find the area of the cross-section of the channel using Simpson's rule.

16. The widths of a bell crank are measured at 2.0-cm intervals, as shown in Fig. 2.102. Find the area of the bell crank if the two connector holes are each 2.50 cm in diameter.

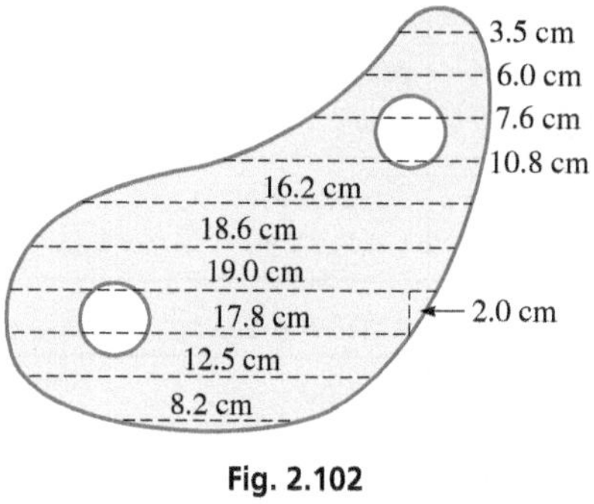

Fig. 2.102

In Exercises 17–20, calculate the area of the circle by the indicated method.

The lengths of parallel chords of a circle that are 0.250 cm apart are given in the following table. The diameter of the circle is 2.000 cm. The distance shown is the distance from one end of a diameter.

Distance (cm)	0.000	0.250	0.500	0.750	1.000	1.250	1.500	1.750	2.000
Length (cm)	0.000	1.323	1.732	1.936	2.000	1.936	1.732	1.323	0.000

Using the formula $A = \pi r^2$, the area of the circle is 3.14 cm^2.

17. Find the area of the circle using the trapezoidal rule and only the values of distance of 0.000 cm, 0.500 cm, 1.000 cm, 1.500 cm, and 2.000 cm with the corresponding values of the chord lengths. Explain why the value found is less than 3.14 cm^2.

18. Find the area of the circle using the trapezoidal rule and all values in the table. Explain why the value found is closer to 3.14 cm^2 than the value found in Exercise 17.

19. Find the area of the circle using Simpson's rule and the same table values as in Exercise 17. Explain why the value found is closer to 3.14 cm^2 than the value found in Exercise 17.

20. Find the area of the circle using Simpson's rule and all values in the table. Explain why the value found is closer to 3.14 cm^2 than the value found in Exercise 19.

Answer to Practice Exercise

1. 18 100 km^2

2.6 Solid Geometric Figures

A **solid geometric figure** fills space in three dimensions (length, width, and height). In this section we discuss basic solid geometric figures. For the most part, we will discuss **right solids**, which are those whose axis or sides are perpendicular to their base (the bottom side of the solid). We are interested in the following properties:

Volume, V	The **volume** of a solid is the space it occupies.
Lateral surface area, S	The **lateral surface area** is the area of all sides of the solid without including the base(s).
Total surface area, A	The **total surface area** is the area of the solid including the area of the base(s).

We begin with formulas for right solids that have two congruent bases. In this case, the height of the figure is the perpendicular distance between the two bases. Note how the volume of these solids equals the area of any cross-section times the height.

	Rectangular Solid	Cube	Right Circular Cylinder	Right Prism
Description	6 rectangular faces; perpendicular intersecting sides	Rectangular solid with 6 square faces	Generated by rotating a rectangle about one of its sides	Rectangular sides perpendicular to 2 polygonal bases
Bases	2 rectangles	2 squares	2 circles	2 equal n-sided polygons of perimeter p
Side(s)	4 rectangles	4 squares	Single curved side	n rectangles
Area of One Base	$B = lw$ **(2.24)**	$B = s^2$ **(2.28)**	$B = \pi r^2$ **(2.32)**	B depends on the type of polygon the bases are.
Lateral Surface Area	$S = 2wh + 2lh$ **(2.25)**	$S = 4s^2$ **(2.29)**	$S = 2\pi rh$ **(2.33)**	$S = ph$ **(2.36)**
Total Surface Area ($A = 2B + S$)	$A = 2lw + 2wh + 2lh$ **(2.26)**	$A = 6s^2$ **(2.30)**	$A = 2\pi r^2 + 2\pi rh$ **(2.34)**	$A = 2B + ph$ **(2.37)**
Volume	$V = lwh$ **(2.27)**	$V = s^3$ **(2.31)**	$V = \pi r^2 h$ **(2.35)**	$V = Bh$ **(2.38)**
	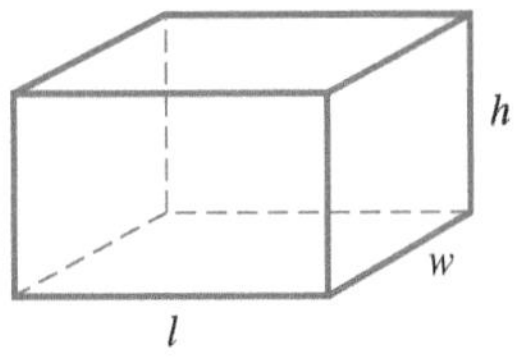 Fig. 2.103(a)	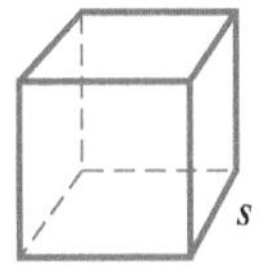 Fig. 2.103(b)	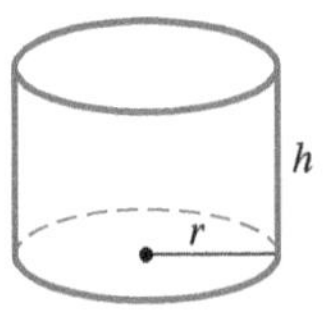 Fig. 2.103(c)	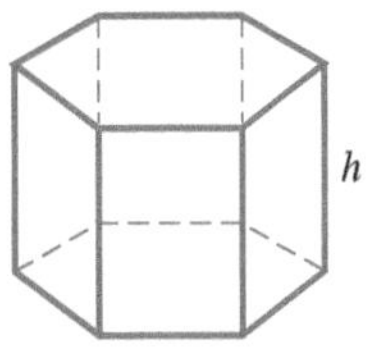 Fig. 2.103(d)

We now consider cones and pyramids, which are right solids that have only one base. The height of these figures is the perpendicular distance between the centre of the base and the vertex, the point at the top of the solid lying directly above the centre of the base. The **slant height** is the distance between the vertex and the base measured along the outside of the figure. The slant height and the height always meet at the vertex, forming a right triangle with the base of the figure. In the table that follows, we also include formulas for the volume and surface area of a sphere and of a frustum of a right circular cone.

	Right Circular Cone	**Regular Pyramid**	**Sphere**	**Frustum of Right Circular Cone**
Description	Generated by rotating a right triangle about one of its legs	Polygonal base and congruent triangular sides	Generated by rotating a circle about a diameter	Figure that remains when the top is cut off by a plane parallel to the base
Base	Circle	Polygon of perimeter p	None	Circle
Side(s)	Single surface	Congruent triangles	None	Single surface
Area of Base(s)	$B = \pi r^2$ **(2.39)**	B depends on the type of polygon the base is.	None	$B = \pi R^2 + \pi r^2$
Lateral Surface Area	$S = \pi rs$ **(2.40)**	$S = \frac{1}{2}ps$ **(2.43)**	None	$S = \pi(R + r)s$ **(2.48)**
Total Surface Area $(A = B + S)$	$A = \pi r^2 + \pi rs$ **(2.41)**	$A = B + \frac{1}{2}ps$ **(2.44)**	$A = 4\pi r^2$ **(2.46)**	$A = \pi(R^2 + r^2) + \pi(R + r)s$ **(2.49)**
Volume	$V = \frac{1}{3}\pi r^2 h$ **(2.42)**	$V = \frac{1}{3}Bh$ **(2.45)**	$V = \frac{4}{3}\pi r^3$ **(2.47)**	$V = \frac{1}{3}\pi h(R^2 + Rr + r^2)$ **(2.50)**
	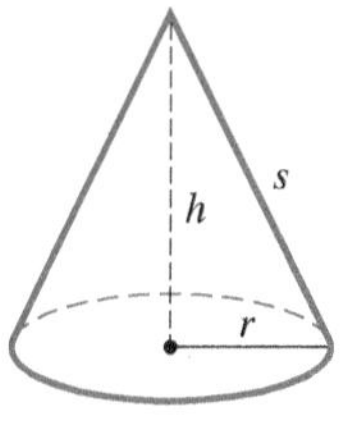 Fig. 2.104(a)	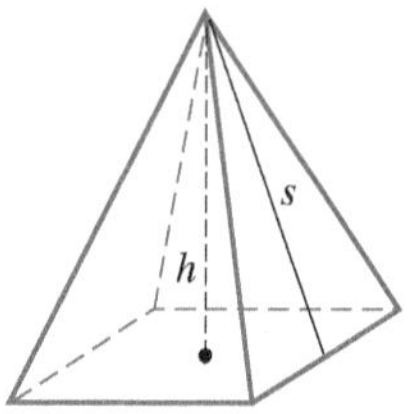 Fig. 2.104(b)	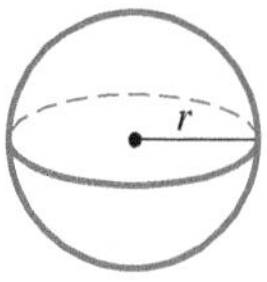 Fig. 2.104(c)	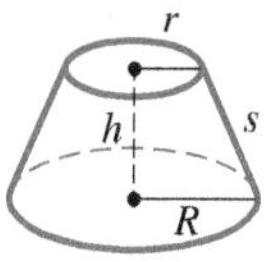 Fig. 2.104(d)

EXAMPLE 1 Volume of a rectangular solid

What volume of concrete is needed for a driveway 25.0 m long, 2.75 m wide, and 0.100 m thick?

The driveway is a rectangular solid for which $l = 25.0$ m, $w = 2.75$ m, and $h = 0.100$ m. Using Eq. (2.27), we have

$$V = (25.0)(2.75)(0.100)$$
$$V = 6.88 \text{ m}^3$$

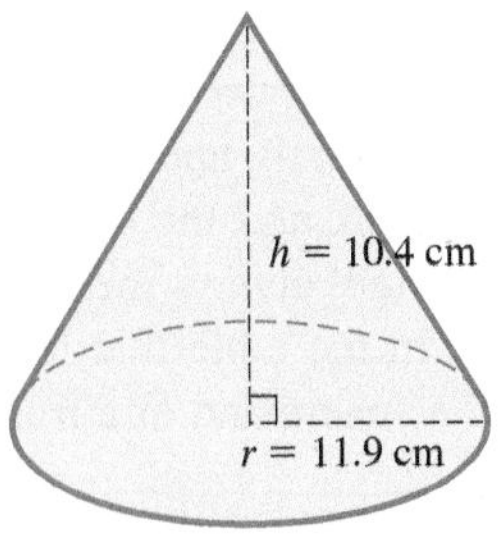

Fig. 2.105

EXAMPLE 2 Total surface area of a right circular cone—protective cover

How many square centimetres of sheet metal are required to make a protective cone-shaped cover if the radius is $r = 11.9$ cm and the height is $h = 10.4$ cm? See Fig. 2.105.

To find the total surface area using Eq. (2.41), we need the radius and the slant height s of the cone. Therefore, we must first find s. The radius and height are legs of a right triangle, and the slant height is the hypotenuse. To find s, we *use the Pythagorean theorem:*

$$s^2 = r^2 + h^2 \quad \text{Pythagorean theorem}$$
$$s = \sqrt{r^2 + h^2} \quad \text{solve for } s$$
$$s = \sqrt{11.9^2 + 10.4^2}$$
$$s = 15.8 \text{ cm}$$

Now, calculating the total surface area, we have

$$A = \pi r^2 + \pi rs \quad \text{Eq. (2.41)}$$
$$A = \pi(11.9)^2 + \pi(11.9)(15.8) \quad \text{substituting}$$
$$A = 1040 \text{ cm}^2$$

Thus, 1040 cm^2 of sheet metal are required.

Practice Exercise

1. Find the volume within the conical cover of Example 2.

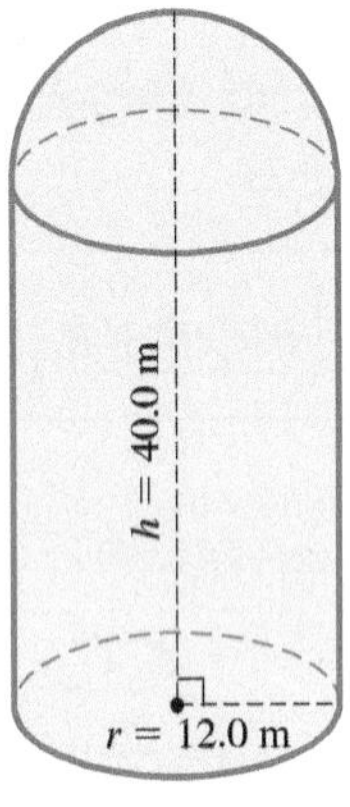

Fig. 2.106

EXAMPLE 3 Volume of a combination of solids

A grain storage building is in the shape of a cylinder surmounted by a hemisphere (*half a sphere*). See Fig. 2.106. Assuming that the width of the building walls is negligible, find the volume of grain that can be stored if the height of the cylinder is 40.0 m and its radius is 12.0 m.

Identify known quantities: Height of cylinder: $h = 40.0$ m
Radius of cylinder and of hemisphere: $r = 12.0$ m

Identify unknowns: Let V_{hem} and V_{cylinder} be the volume of the hemisphere and of the cylinder, respectively. Then $V = V_{\text{hem}} + V_{\text{cylinder}}$

Select equations and solve:

$$V_{\text{hem}} = \frac{1}{2}\left(\frac{4}{3}\pi r^3\right) = \frac{2}{3}\pi(12.0)^3$$
$$V_{\text{cylinder}} = \pi r^2 h = \pi(12.0)^2(40.0)$$
$$V = \frac{2}{3}\pi(12.0)^3 + \pi(12.0)^2(40.0)$$
$$V = 21\,700 \text{ m}^3$$

Practice Exercise

2. Find the surface area (not including the base) of the storage building in Example 3.

EXERCISES 2.6

In Exercises 1–4, answer the given questions about the indicated examples of this section.

1. In Example 1, if the length is doubled and the thickness is doubled, by what factor is the volume changed?
2. In Example 2, if the value of the slant height $s = 17.5$ cm is given instead of the height, what is the height?
3. In Example 2, if the radius is halved and the height is doubled, what is the volume?
4. In Example 3, if h is halved, what is the volume?

In Exercises 5–22, find the volume or area of each solid figure for the given values.

5. Volume of cube: $s = 7.15$ cm
6. Volume of right circular cylinder: $r = 23.5$ cm, $h = 48.4$ cm
7. Total surface area of right circular cylinder: $r = 689$ m, $h = 233$ m
8. Lateral area of right circular cylinder: diameter = 250 mm, $h = 347$ mm
9. Lateral area of regular prism: equilateral triangle base of side 1.092 m, $h = 1.025$ m

10. Volume of right prism: square base of side 29.0 cm, $h = 11.2$ cm
11. Lateral area of right circular cone: $r = 78.0$ cm, $s = 83.8$ cm
12. Volume of right circular cone: $r = 25.1$ mm, $h = 5.66$ mm
13. Total surface area of right circular cone: $r = 3.39$ cm, $h = 0.274$ cm
14. Lateral area of regular pyramid: $p = 345$ m, $s = 272$ m
15. Volume of regular pyramid: square base of side 76 cm, $h = 130$ cm
16. Volume of regular pyramid: square base of side 22.4 m, $s = 14.2$ m
17. Total surface area of pyramid: All faces and base are equilateral triangles of side 3.67 dm.
18. Area of sphere: $r = 0.067$ mm
19. Volume of sphere: $r = 0.877$ m
20. Volume of hemisphere: diameter = 0.83 cm
21. Volume of frustum of right circular cone: $R = 37.3\,\text{mm}, r = 28.2\,\text{mm}, h = 45.1\,\text{mm}$
22. Lateral area of frustum of right circular cone: $R = 3.42\,\text{m}, r = 2.69\,\text{m}, s = 3.25\,\text{m}$

In Exercises 23–45, solve the given problems.

23. A rectangular box is to be used to store radioactive materials. The inside of the box is 12.0 cm long, 9.50 cm wide, and 8.75 cm deep. What is the area of sheet lead that must be used to line the inside of the box?
24. A parfleche is a container made of rawhide that was used for storage by a variety of Plains Indigenous people. A parfleche trunk is in the shape of a rectangular solid that is 53 cm long, 36 cm wide, and 24 cm deep. What is the volume of the parfleche?
25. The Athabasca oil pipeline in Alberta is 540 km long and has a diameter of 0.76 m. What is the maximum volume of the pipeline?
26. A swimming pool is 15.0 m wide, 24.0 m long, 1.00 m deep at one end, and 2.60 m deep at the other end. How many cubic metres of water can it hold? (The slope on the bottom is constant.) See Fig. 2.107.

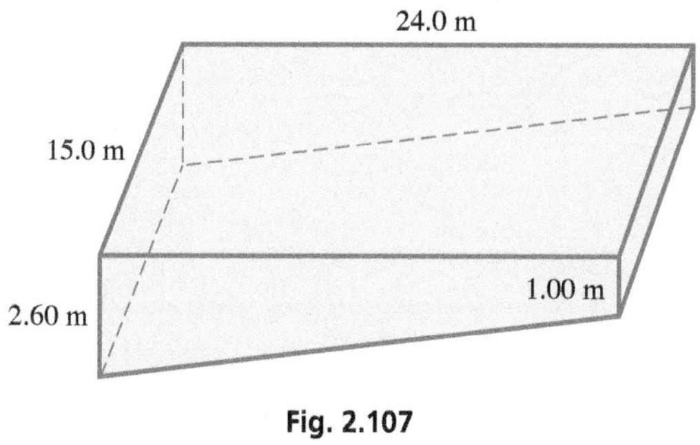

Fig. 2.107

27. A glass prism used in the study of optics has a right triangular base. The legs of the triangle are 3.00 cm and 4.00 cm. The prism is 8.50 cm high. What is the total surface area of the prism? See Fig. 2.108.

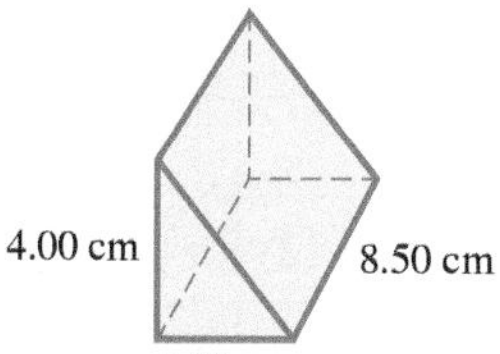

Fig. 2.108

28. The base area of a cone is one-fourth of the total area. Find the ratio of the radius to the slant height.
29. A paper cup is in the shape of a cone, as shown in Fig. 2.109. What is the surface area of the cup?

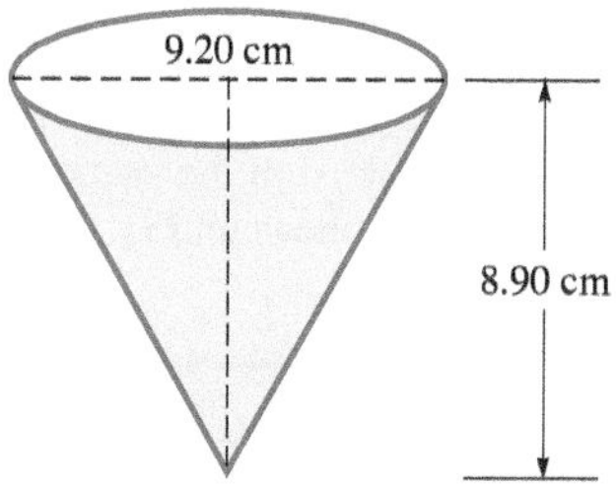

Fig. 2.109

30. A special wedge in the shape of a regular pyramid has a square base 16.0 mm on a side. The height of the wedge is 40.0 mm. What is the total surface area of the wedge?
31. The Great Pyramid of Egypt has a square base approximately 230 m on a side. The height of the pyramid is about 150 m. What is its volume? See Fig. 2.110.

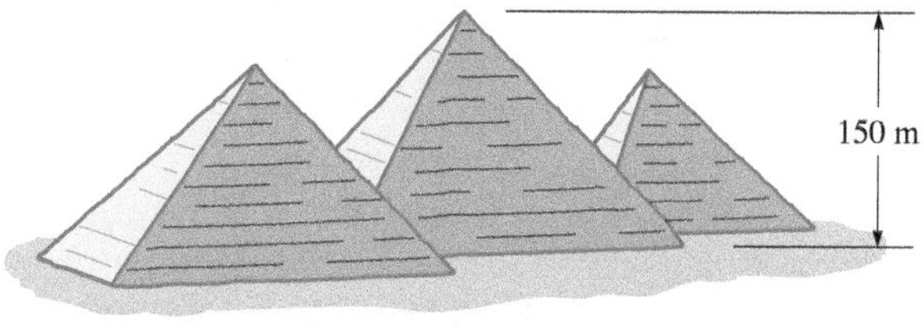

Fig. 2.110

32. Eq. (2.47) expresses the volume V of a sphere in terms of the radius r. Express V in terms of the diameter d.
33. *Spaceship Earth* (shown in Fig. 2.111) at Epcot Center in Florida is a sphere 50.3 m in diameter. What is the volume of *Spaceship Earth*?

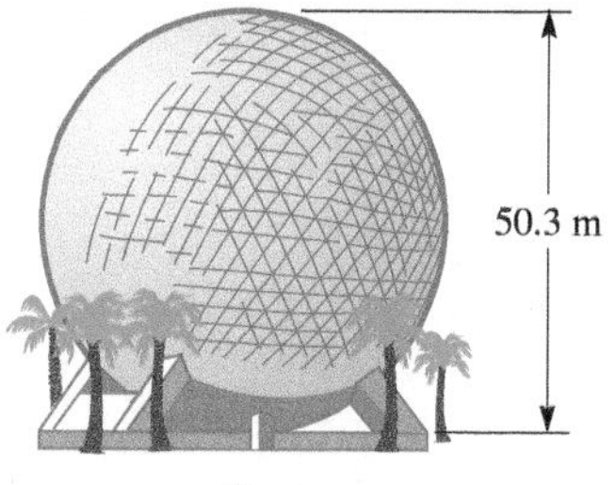

Fig. 2.111

34. In designing a weather balloon, it is decided to double the diameter of the balloon so that it can carry a heavier instrument load. What is the ratio of the final surface area to the original surface area?
35. Derive a formula for the total surface area A of a hemispherical volume of radius r (curved surface and flat surface).
36. A pole supporting a wind turbine is constructed of solid steel and is in the shape of a frustum of a cone. It measures 62.5 m high, and the diameter of the pole at the bottom and top are 3.88 m and 1.90 m, respectively. What is the volume of the pole?
37. During a rainfall of 3.00 cm, what weight of water falls on an area of 1.00 km^2? Each cubic metre of water weighs 9800 N.

38. A lawn roller is a cylinder 0.96 m long and 0.60 m in diameter. How many revolutions of the roller are needed to roll 76 m^2 of lawn?

39. The radius of a cylinder is twice as long as the radius of a cone, and the height of the cylinder is half as long as the height of the cone. What is the ratio of the volume of the cylinder to that of the cone?

40. A propane tank is constructed in the shape of a cylinder with a hemisphere at each end, as shown in Fig. 2.112. Find the volume of the tank.

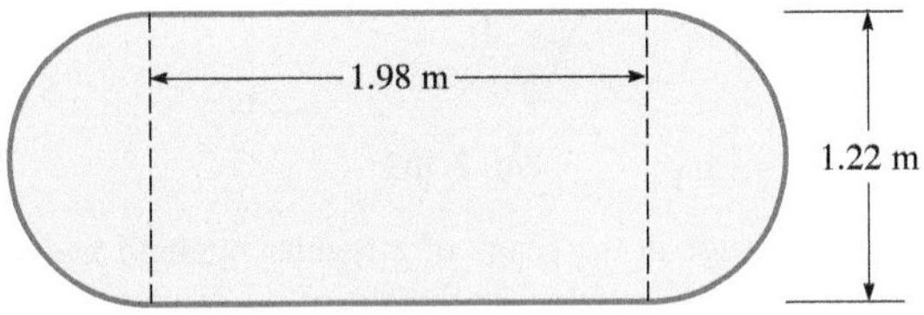

Fig. 2.112

41. The side view of a rivet is shown in Fig. 2.113. It is a conical part on a cylindrical part. Find the volume of the rivet.

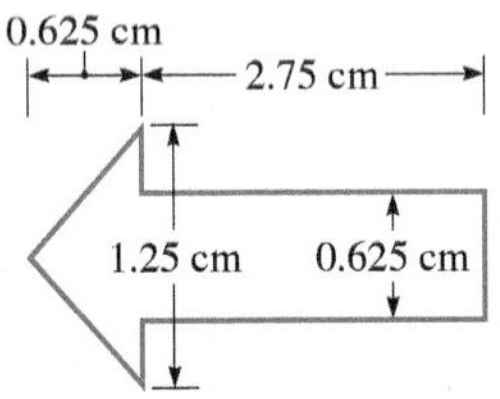

Fig. 2.113

42. What is the area of a paper label that is to cover the lateral surface of a cylindrical can 8.50 cm in diameter and 11.5 cm high? The ends of the label will overlap 0.50 cm when the label is placed on the can.

43. An ice cream cone has a height of 10 cm from bottom to top with a radius of 2.5 cm where it opens up, as shown in Fig. 2.114. If the ice cream is added to fill the cone and form a hemispherical scoop on top, what is the volume of ice cream?

44. A dipstick is made to measure the volume remaining in the conical container shown in Fig. 2.115. How far below the full mark (at the top of the container) on the stick should the mark for half-full be placed?

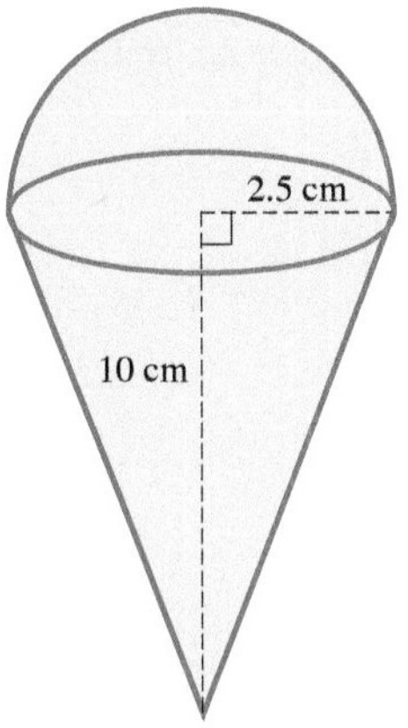

Fig. 2.114

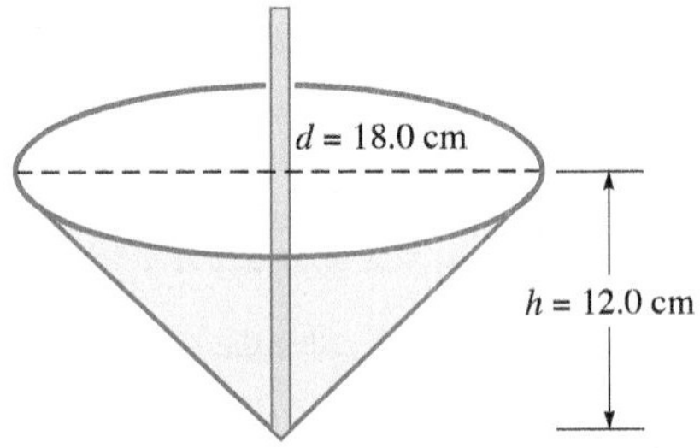

Fig. 2.115

45. The circumference of a basketball is about 75.7 cm. What is its volume?

Answers to Practice Exercises

1. 1540 cm^3 **2.** 3920 m^2

CHAPTER 2 KEY FORMULAS AND EQUATIONS

Line segments Fig. 2.7 $\frac{a}{b} = \frac{c}{d}$ (2.1)

Area of a triangle Fig. 2.28(b)

Area formula $A = \frac{1}{2}bh$ (2.2)

Hero's formula $A = \sqrt{s(s-a)(s-b)(s-c)},$ (2.3)

where $s = \frac{1}{2}(a+b+c)$ (2.4)

Pythagorean theorem Fig. 2.31 $c^2 = a^2 + b^2$ (2.5)

Fig. 2.50	Parallelogram	Rhombus	Square	Rectangle	Trapezoid
Perimeter	$p = 2(a+b)$ (2.6)	$p = 4s$ (2.8)	$p = 4s$ (2.10)	$p = 2(l+w)$ (2.12)	$p = a + b_1 + b_2 + c$ (2.14)
Area	$A = bh$ (2.7)	$A = sh$ (2.9)	$A = s^2$ (2.11)	$A = lw$ (2.13)	$A = \frac{1}{2}h(b_1 + b_2)$ (2.15)

Circle

Circumference		$c = 2\pi r$	**(2.16)**	or	$c = \pi d$	**(2.17)**
Area		$A = \pi r^2$	**(2.18)**	or	$A = \frac{1}{4}\pi d^2$	**(2.19)**
Radian to degree conversion	Fig. 2.75	π rad $= 180°$				**(2.20)**
Arc length	Fig. 2.76	$s = \theta r$	(θ in radians)			**(2.21a)**
Area of sector		$A = \frac{1}{2}\theta r^2$	(θ in radians)			**(2.21b)**

Trapezoidal rule Fig. 2.89

$$A = \frac{h}{2}(y_0 + 2y_1 + 2y_2 + \cdots + 2y_{n-1} + y_n) \quad \textbf{(2.22)}$$

Simpson's rule Fig. 2.93

$$A = \frac{h}{3}(y_0 + 4y_1 + 2y_2 + 4y_3 + \cdots + 2y_{n-2} + 4y_{n-1} + y_n) \quad \textbf{(2.23)}$$

Fig. 2.103	**Rectangular Solid**	**Cube**	**Right Circular Cylinder**	**Right Prism**
Area of One Base	$B = lw$ **(2.24)**	$B = s^2$ **(2.28)**	$B = \pi r^2$ **(2.32)**	B depends on the type of polygon the bases are.
Lateral Surface Area	$S = 2wh + 2lh$ **(2.25)**	$S = 4s^2$ **(2.29)**	$S = 2\pi rh$ **(2.33)**	$S = ph$ **(2.36)**
Total Surface Area ($A = 2B + S$)	$A = 2lw + 2wh + 2lh$ **(2.26)**	$A = 6s^2$ **(2.30)**	$A = 2\pi r^2 + 2\pi rh$ **(2.34)**	$A = 2B + ph$ **(2.37)**
Volume	$V = lwh$ **(2.27)**	$V = s^3$ **(2.31)**	$V = \pi r^2 h$ **(2.35)**	$V = Bh$ **(2.38)**

Fig. 2.104	**Right Circular Cone**	**Regular Pyramid**	**Sphere**	**Frustum of Right Circular Cone**
Area of Base(s)	$B = \pi r^2$ **(2.39)**	B depends on the type of polygon the base is.	None	$B = \pi R^2 + \pi r^2$
Lateral Surface Area	$S = \pi rs$ **(2.40)**	$S = \frac{1}{2}ps$ **(2.43)**	None	$S = \pi(R + r)s$ **(2.48)**
Total Surface Area ($A = B + S$)	$A = \pi r^2 + \pi rs$ **(2.41)**	$A = B + \frac{1}{2}ps$ **(2.44)**	$A = 4\pi r^2$ **(2.46)**	$A = \pi(R^2 + r^2) + \pi(R + r)s$ **(2.49)**
Volume	$V = \frac{1}{3}\pi r^2 h$ **(2.42)**	$V = \frac{1}{3}Bh$ **(2.45)**	$V = \frac{4}{3}\pi r^3$ **(2.47)**	$V = \frac{1}{3}\pi h(R^2 + Rr + r^2)$ **(2.50)**

CHAPTER 2 REVIEW EXERCISES

In Exercises 1–4, use Fig. 2.116. Determine the indicated angles.

1. $\angle CGE$
2. $\angle EGF$
3. $\angle DGH$
4. $\angle EGI$

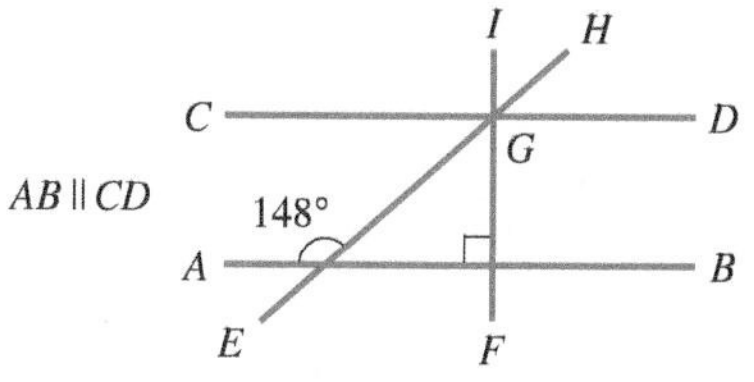

Fig. 2.116

In Exercises 5–12, find the indicated sides of the right triangle shown in Fig. 2.117.

5. $a = 9, b = 40, c = ?$
6. $a = 14, b = 48, c = ?$
7. $a = 400, b = 580, c = ?$
8. $b = 56, c = 65, a = ?$
9. $a = 6.30, b = 3.80, c = ?$
10. $a = 126, b = 25.1, c = ?$
11. $b = 29.3, c = 36.1, a = ?$
12. $a = 0.782, c = 0.885, b = ?$

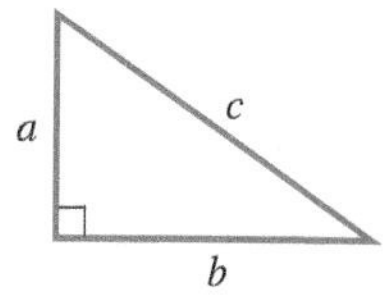

Fig. 2.117

In Exercises 13–20, find the perimeter or area of the indicated figure.

13. Perimeter: equilateral triangle of side 8.5 mm

14. Perimeter: rhombus of side 15.2 cm

15. Area: triangle, $b = 3.25$ m, $h = 1.88$ m

16. Area: triangle of sides 175 cm, 138 cm, 119 cm

17. Circumference of circle: $d = 98.4$ mm

18. Perimeter: rectangle, $l = 2.98$ dm, $w = 1.86$ dm

19. Area: trapezoid, $b_1 = 67.2$ cm, $b_2 = 126.7$ cm, $h = 34.2$ cm

20. Area: circle, $d = 32.8$ m

In Exercises 21–24, find the volume of the indicated solid geometric figure.

21. Prism: base is right triangle with legs 26.0 cm and 34.0 cm, height is 14.0 cm

22. Cylinder: base radius 36.0 cm, height 2.40 cm

23. Pyramid: base area 3850 m^2, height 125 m

24. Sphere: diameter 22.1 mm

In Exercises 25–28, find the surface area of the indicated solid geometric figure.

25. Total area of cube of edge 0.520 m

26. Total area of cylinder: base diameter 1.20 cm, height 5.80 cm

27. Lateral area of cone: base radius 1.82 mm, height 11.5 mm

28. Total area of sphere: $d = 12\,760$ km

In Exercises 29–32, use Fig. 2.118. Line CT is tangent to the circle with centre at O. Find the indicated angles.

29. $\angle BTA$ **30.** $\angle TAB$

31. $\angle BTC$ **32.** $\angle ABT$

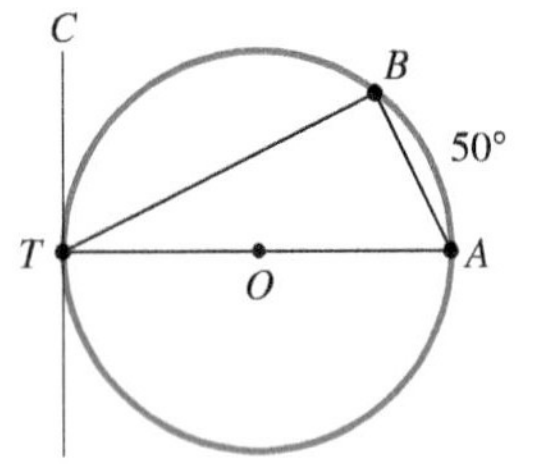

Fig. 2.118

In Exercises 33–36, use Fig. 2.119. Given that $AB = 4$, $BC = 4$, $CD = 6$, and $\angle ADC = 53°$, find the indicated angle and lengths.

33. $\angle ABE$ **34.** AD

35. BE **36.** AE

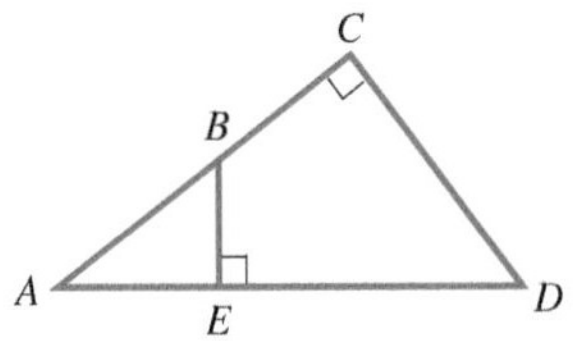

Fig. 2.119

In Exercises 37–40, find the formulas for the indicated perimeters and areas.

37. Perimeter of Fig. 2.120 (a right triangle and semicircle attached)

38. Perimeter of Fig. 2.121 (a square with a quarter-circle at each end)

39. Area of Fig. 2.120 **40.** Area of Fig. 2.121

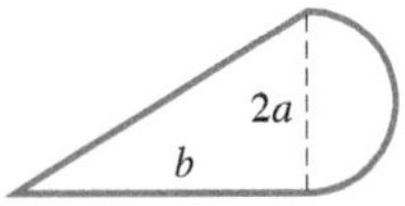

Fig. 2.120

Fig. 2.121

In Exercises 41–46, answer the given questions.

41. Is a square also a rectangle, a parallelogram, and a rhombus?

42. If the measures of two angles of one triangle equal the measures of two angles of a second triangle, are the two triangles similar?

43. If the dimensions of a plane geometric figure are each multiplied by n, by how much is the area multiplied? Explain, using a circle to illustrate.

44. If the dimensions of a solid geometric figure are each multiplied by n, by how much is the volume multiplied? Explain, using a cube to illustrate.

45. What is an equation relating chord segments a, b, c, and d shown in Fig. 2.122? The dashed chords are an aid in the solution.

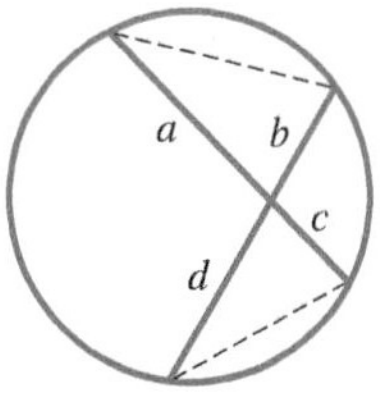

Fig. 2.122

46. From a common point, two line segments are tangent to the same circle. If the angle between the lines segments is 36°, what is the angle between the two radii of the circle drawn from the points of tangency?

In Exercises 47–73, solve the given problems.

47. A tooth on a saw is in the shape of an isosceles triangle. If the angle at the point is 38°, find the two base angles.

48. A lead sphere 1.50 cm in diameter is flattened into a circular sheet 14.0 cm in diameter. How thick is the sheet?

49. A ramp for the disabled is designed so that it rises 1.20 m over a horizontal distance of 7.80 m. How long is the ramp?

50. An airplane is 640 m directly above one end of a 3200-m runway. How far is the plane from the glide-slope indicator on the ground at the other end of the runway?

51. A machine part is in the shape of a square with equilateral triangles attached to two sides (see Fig. 2.123). Find the perimeter of the machine part.

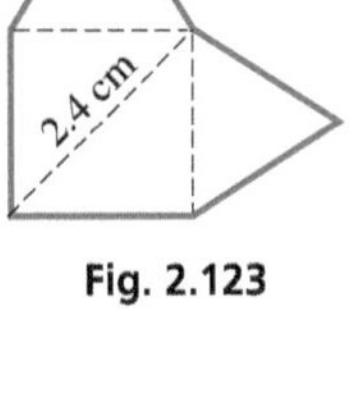

Fig. 2.123

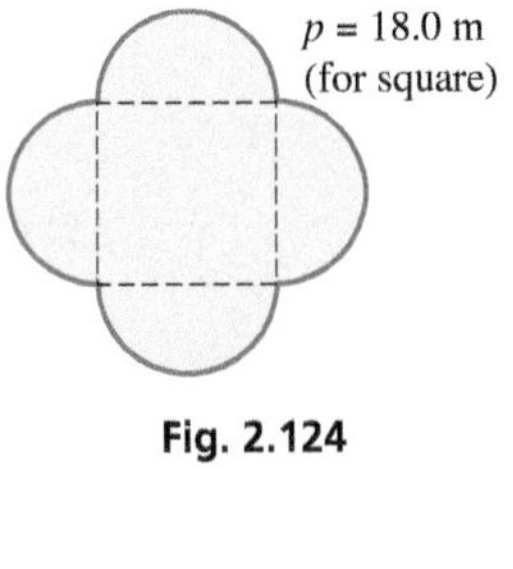

Fig. 2.124

52. A patio is designed with semicircular areas attached to a square, as shown in Fig. 2.124. Find the area of the patio.

53. A radio transmitting tower is supported by guy wires. The tower and three parallel guy wires are shown in Fig. 2.125. Find the distance AB along the tower.

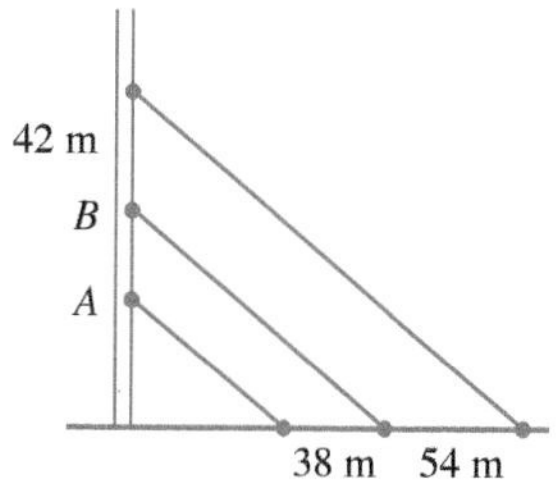

Fig. 2.125

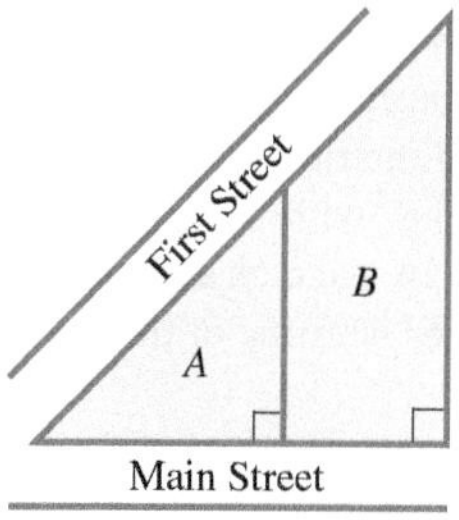

Fig. 2.126

54. Find the areas of lots A and B in Fig. 2.126. A has a frontage on Main St. of 145 m, and B has a frontage on Main St. of 84.0 m. The boundary between lots is 125 m.

55. To find the height of a flagpole, a person places a mirror at M, as shown in Fig. 2.127. The person's eyes at E are 160 cm above the ground at A. From physics, it is known that $\angle AME = \angle BMF$. If $AM = 120$ cm and $MB = 4.5$ m, find the height BF of the flagpole.

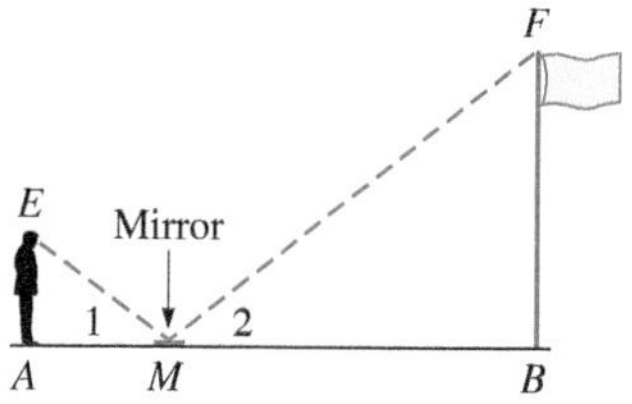

Fig. 2.127

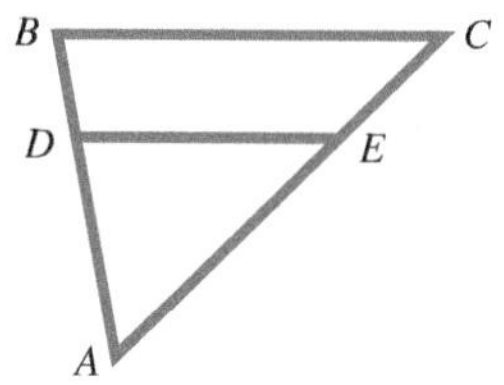

Fig. 2.128

56. The metal support in the form of $\triangle ABC$ shown in Fig. 2.128 is strengthened by brace DE, which is parallel to BC. How long is the brace if $AB = 24.0$ cm, $AD = 16.0$ cm, and $BC = 33.0$ cm?

57. A typical scale for an aerial photograph is 1/18 450. In a 20.0-by-25.0-cm photograph with this scale, what is the longest distance (in km) between two locations in the photograph?

58. For a hydraulic press, the mechanical advantage is the ratio of the large piston area to the small piston area. Find the mechanical advantage if the pistons have diameters of 3.10 cm and 2.25 cm.

59. The diameter of the earth is 12 700 km, and a satellite is in orbit at an altitude of 345 km. How far does the satellite travel in one rotation about the earth?

60. BC Place Stadium in Vancouver, British Columbia, is the world's largest cable-supported retractable-roof stadium. Its opening is rectangular, having length 15.0 m longer than its width, and it has a diagonal of 131 m. Determine the area of the roof opening. *Hint:* You do not need to solve a quadratic equation; the area of the opening appears in the diagonal equation.

61. A rectangular piece of wallboard with two holes cut out for heating ducts is shown in Fig. 2.129. What is the area of the remaining piece?

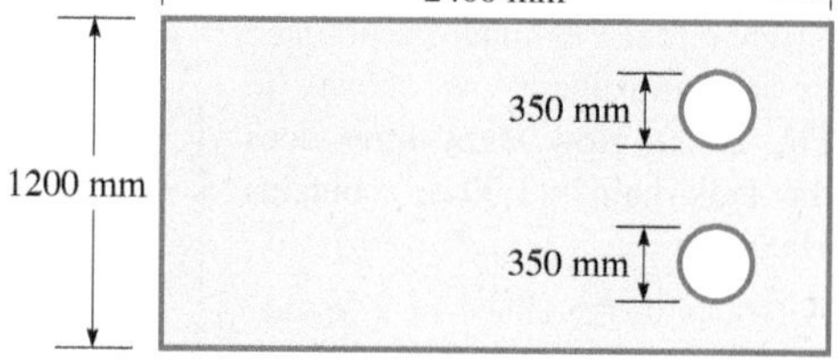

Fig. 2.129

62. The diameter of the sun is 1.38×10^6 km, the diameter of the earth is 1.27×10^4 km, and the distance from the earth to the sun (centre to centre) is 1.50×10^8 km. What is the distance from the centre of the earth to the end of the shadow due to the rays from the sun?

63. Using aerial photography, the width of an oil spill is measured at 250-m intervals, as shown in Fig. 2.130. Using Simpson's rule, find the area of the oil spill.

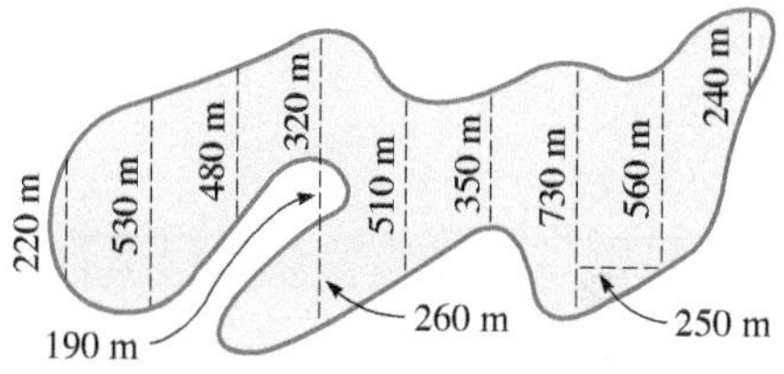

Fig. 2.130

64. To build a highway, it is necessary to cut through a hill. A surveyor measured the cross-sectional areas at 250-m intervals through the cut, as shown in the following table. Using the trapezoidal rule, determine the volume of soil to be removed.

Dist. (m)	0	250	500	750	1000	1250	1500	1750
Area (m^2)	560	1780	4650	6730	5600	6280	2260	230

65. The Hubble space telescope is within a cylinder 4.32 m in diameter and 13.2 m long. What is the volume within this cylinder?

66. A horizontal cross-section of a concrete bridge pier is a regular hexagon (six sides, all equal in length, and all internal angles are equal), each side of which is 2.50 m long. If the height of the pier is 6.75 m, what is the volume of concrete in the pier?

67. A railroad track 1000.00 m long expands 0.20 m (20 cm) during the afternoon (due to an increase in temperature of about 17°C). Assuming that the track cannot move at either end and that the increase in length causes a bend straight up in the middle of the track, how high is the top of the bend?

68. Two persons are talking to each other on cellular phones. If the angle between their signals at the tower is 90°, as shown in Fig. 2.131, how far apart are they?

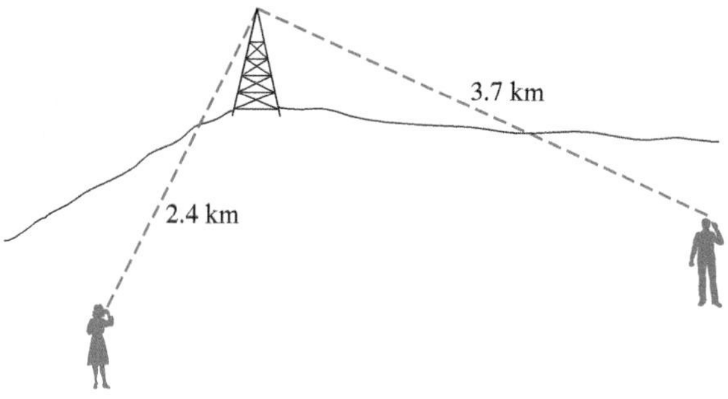

Fig. 2.131

69. A hot-water tank is in the shape of a right circular cylinder surmounted by a hemisphere, as shown in Fig. 2.132. How many litres does the tank hold? (1.00 m^3 contains 1000 L.)

70. A tent is in the shape of a regular pyramid surmounted on a cube. If the edge of the cube is 2.50 m and the total height of the tent is 3.25 m, find the area of the material used in making the tent (not including any floor area).

71. The diagonal of an HDTV screen is 107 cm. If the ratio of the width of the screen to the height of the screen is 16 to 9, what are the width and height?

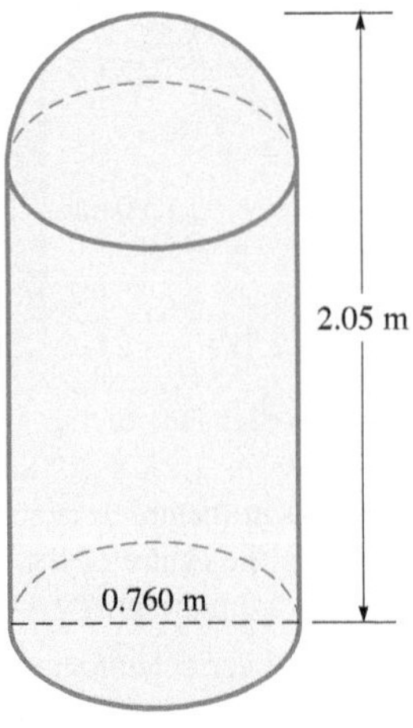

Fig. 2.132

72. A satellite is 1590 km directly above the centre of the eye of a circular (approximately) hurricane that has formed in the Atlantic Ocean. The distance from the satellite to the edge of the hurricane is 1620 km. What area does the hurricane cover? Neglect the curvature of Earth and any possible depth of the hurricane.

73. The U.S. Department of Defense is located inside a large building shaped like a regular pentagon (five sides, all equal in length, and all interior angles are equal). Each side of the building is 281 m long, with a diagonal of length 454 m. Using this information draw a sketch and write one or two paragraphs to explain what the area is inside of the building's floorplan.

CHAPTER 2 PRACTICE TEST

1. In Fig. 2.133, determine $\angle 1$.
2. In Fig. 2.133, determine $\angle 2$.

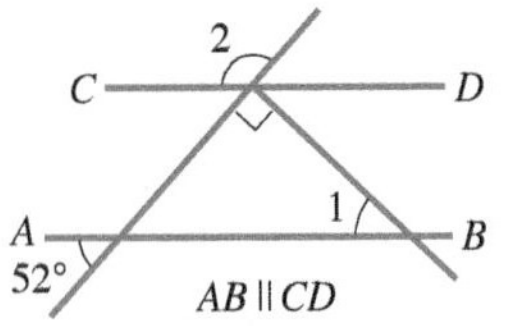

Fig. 2.133

3. A tree is 2.4 m high and casts a shadow 3.0 m long. At the same time, a telephone pole casts a shadow 7.6 m long. How tall is the pole?
4. Find the area of a triangle with sides of 2.46 cm, 3.65 cm, and 4.07 cm.
5. What is the diagonal distance between corners of a rectangular room 3810 mm wide and 5180 mm long?
6. What is the area of the trapezoid shown in Fig. 2.134?

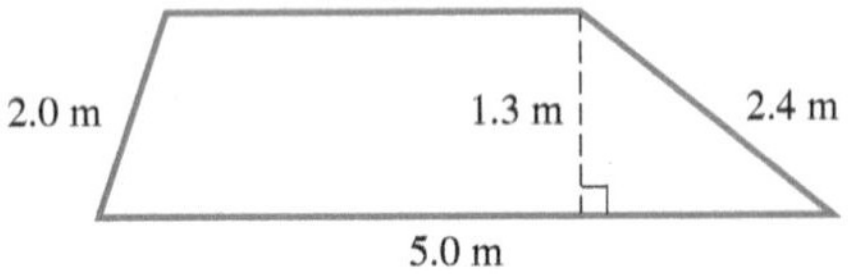

Fig. 2.134

7. The edge of a cube is 4.50 cm. What are (a) the surface area and (b) the volume of the cube?
8. Find the surface area of a tennis ball whose circumference is 21.0 cm.
9. Find the volume of a right circular cone of radius 2.08 m and height 1.78 m.
10. In Fig. 2.135, find $\angle 1$.
11. In Fig. 2.135, find $\angle 2$.

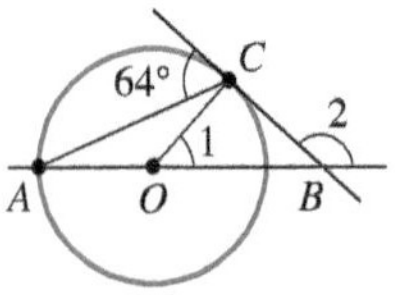

Fig. 2.135

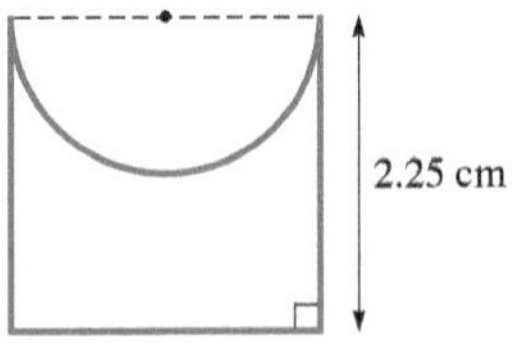

Fig. 2.136

12. In Fig. 2.136, find the perimeter of the figure shown. It is a square with a semicircle removed.
13. In Fig. 2.136, find the area of the figure shown.
14. The width of a marshy area is measured at 50-m intervals, with the results shown in the following table. Using the trapezoidal rule, find the area of the marsh. (All data accurate to two or more significant digits.)

Distance (m)	0	50	100	150	200	250	300
Width (m)	0	90	145	260	205	110	20

3. Functions and Graphs

▲ The electric power P produced by a wind turbine depends on the velocity v of the wind. In Section 3.4 we draw the graph of $P = 0.5C_P\rho Av^3 = 1.304v^3$ for a specific 3.00-m diameter turbine to see this type of relationship.

LEARNING OUTCOMES

After completion of this chapter, the student should be able to:

- Define a function and distinguish between dependent and independent variables
- Use mathematical functional notation
- Interpret a function as a process
- Determine domain and range for a function
- Graph a function using the rectangular coordinate system
- Use a graphing utility to solve an equation graphically
- Graph and interpret sets of data values

Controlled experiments and observations of natural phenomena form part of the *scientific method.* The Italian scientist Galileo was one of the first scientists to realize that claims about how the universe works can be tested mathematically, using mathematical formulas to describe the phenomena. He realized that such formulas provide a way of showing a compact and precise relationship between the observed variables.

We will begin this chapter by graphing the relationship between two variables. We will then discuss how collections of related measurements can be used to generate a **function**, a *general mathematical statement about how the different variables relate*. Examples of such relations in technology and science, as well as in everyday life, are numerous. Plant growth depends on sunlight and rainfall; traffic flow depends on roadway design and number of cars on the road; the sales tax on an item depends on the cost of an item; the time to access the internet depends on how fast a computer processes data and on the data transfer rate of the internet service provider; distance travelled depends on time and speed of travel; electric voltage depends on the current and resistance. These are but a few of the innumerable possibilities.

Observational data can often be described by a mathematical function that shows a compact and precise relationship between the observed variables. In this chapter, we will discuss common graphing techniques, and begin to explore the powerful use of functions to generalize scientific observation into meaningful rules.

3.1 Rectangular Coordinates

■ Rectangular (Cartesian) coordinates were developed by the French mathematician René Descartes (1596–1650).

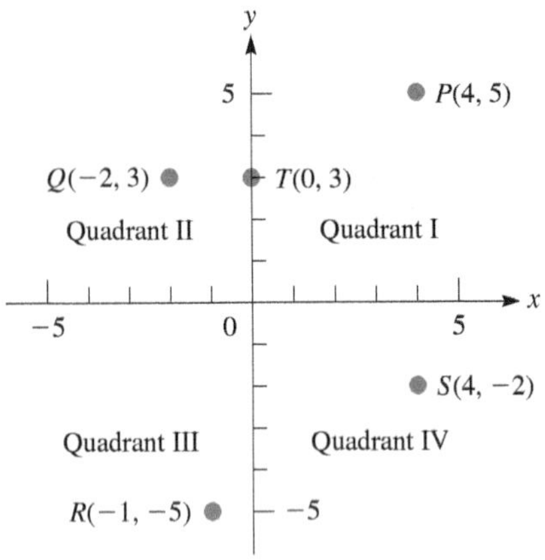

Fig. 3.1

By using graphs, we can obtain a "picture" of how two variables are related, where the value of one variable depends on the other. Recall from Chapter 1 that numbers can be represented by points on a line. When working with two variables, it is necessary to use two different lines to represent the values from each of them. We do this by placing the lines perpendicular to each other.

Place one line horizontally and label it the **x-axis**. The values of an independent variable are normally placed on this axis. *The other line is placed vertically and labelled the* **y-axis**. Normally, the y-axis is used for values of a response variable. *The point of intersection is called the* **origin**. This is the **rectangular coordinate system**.

On the x-axis, positive values are to the right of the origin, and negative values are to the left of the origin. On the y-axis, positive values are above the origin, and negative values are below it. The positive direction is also indicated by an arrow on the axis line. *The four parts into which the plane is divided are called* **quadrants**, which are numbered as in Fig. 3.1.

A point P in the plane is designated by the pair of numbers (x, y), where x is the value of the independent variable and y is the value of the response variable. *The x-value is the perpendicular distance of P from the y-axis, and the y-value is the perpendicular distance of P from the x-axis.* The values of x and y, written as (x, y), are the **coordinates** of the point P.

LEARNING TIP

Note very carefully that the ***x-coordinate is always written first,*** *and the y-coordinate is always written second.* It is very important to keep the proper order when writing the coordinates of a point. This is why the coordinates of a point *P (x, y)* are called an **ordered pair.** (The *x*-coordinate is also known as the *abscissa,* and the *y*-coordinate is also known as the *ordinate.*)

EXAMPLE 1 Locating points

(a) Locate the points $A(2, 1)$ and $B(-4, -3)$ on the rectangular coordinate system.

$A(2, 1)$: the point is 2 units to the *right* of the y-axis and 1 unit *above* the x-axis, as shown in Fig. 3.2.

$B(-4, -3)$: the point is 4 units to the *left* of the y-axis and 3 units *below* the x-axis, as shown.

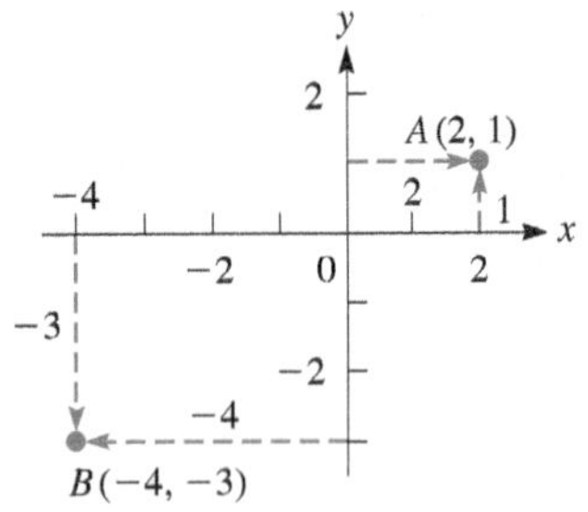

Fig. 3.2

(b) The positions of points $P(4, 5)$, $Q(-2, 3)$, $R(-1, -5)$, $S(4, -2)$, and $T(0, 3)$ are shown in Fig. 3.1. We see that this representation allows for *one point for any pair of values* (x, y). Also note that the point $T(0, 3)$ is on the y-axis. Any point where at least one variable is equal to zero will be on an axis, and not *in* any quadrant.

EXAMPLE 2 Coordinates of the vertices of a rectangle

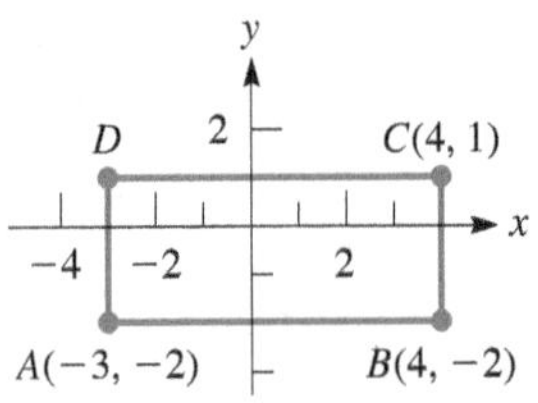

Fig. 3.3

Three vertices of the rectangle in Fig. 3.3 are $A(-3, -2)$, $B(4, -2)$, and $C(4, 1)$. Find the fourth vertex.

We use the fact that opposite sides of a rectangle are equal and parallel to find the solution. Since both vertices of the base AB of the rectangle have a y-coordinate of -2, the base is parallel to the x-axis. Therefore, the top of the rectangle must also be parallel to the x-axis. Thus, the fourth vertex must be on the horizontal line that passes through point C (with y-coordinate of 1). In the same way, the fourth vertex must be on the vertical line that passes through point A (with x-coordinate of -3). Therefore, the fourth vertex is $D(-3, 1)$.

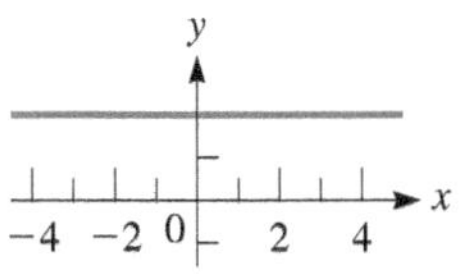

Fig. 3.4

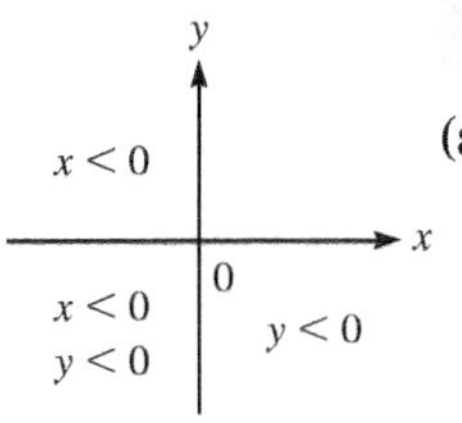

Fig. 3.5

EXAMPLE 3 Locating sets of points

(a) Identify all the points whose y-coordinates equal 2 on a rectangular coordinate system.

Plotting some points where x varies and y equals 2, we see that all such points are two units above the x-axis. We can represent those points with a line 2 units above the x-axis. See Fig. 3.4.

(b) Identify all points (x, y) for which $x < 0$ and $y < 0$.

Recall that $x < 0$ means "x is less than zero," and that $y < 0$ means "y is less than zero." Points that satisfy both conditions are those where both x and y are negative. See Fig. 3.5.

Practice Exercise

1. Where are all points for which $x = 0$ and $y > 0$?

Note from Example 3 that, on the rectangular coordinate system, equalities are represented by lines or other curves, and inequalities are represented by regions on the plane.

EXERCISES 3.1

In Exercises 1 and 2, make the given changes in the indicated examples of this section and then solve the resulting problems.

1. In Example 2, change $A(-3, -2)$ to $A(-1, -2)$ and then find the fourth vertex.

2. In Example 3(b), change $y < 0$ to $y > 0$ and then find the location of the points (x, y).

In Exercises 3 and 4, determine (at least approximately) the coordinates of the points shown in Fig. 3.6.

3. A, B, C **4.** D, E, F

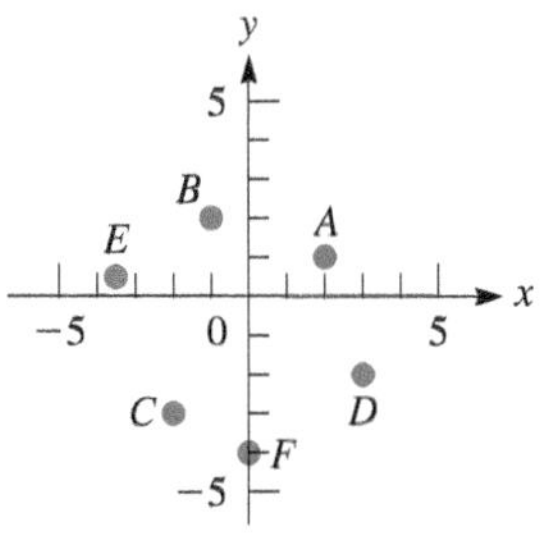

Fig. 3.6

In Exercises 5 and 6, plot the given points.

5. $A(2, 7), B(-1, -2), C(-4, 2)$

6. $A\left(3, \frac{1}{2}\right), B(-6, 0), C\left(-\frac{5}{2}, -5\right)$

In Exercises 7–10, plot the given points and then join these points, in the order given, by straight-line segments. Name the geometric figure formed.

7. $A(-1, 4), B(3, 4), C(1, -2)$

8. $A(0, 3), B(0, -1), C(4, -1)$

9. $A(-2, -1), B(3, -1), C(3, 5), D(-2, 5)$

10. $A(-5, -2), B(4, -2), C(6, 3), D(-3, 3)$

In Exercises 11–14, find the indicated coordinates.

11. Three vertices of a rectangle are $(5, 2)$, $(-1, 2)$, and $(-1, 4)$. What are the coordinates of the fourth vertex?

12. Two vertices of an equilateral triangle are $(7, 1)$ and $(2, 1)$. What is the x-coordinate of the third vertex?

13. P is the point $(3, 2)$. Locate point Q such that the x-axis is the perpendicular bisector of the line segment joining P and Q.

14. P is the point $(-4, 1)$. Locate point Q such that the line segment joining P and Q is bisected by the origin.

In Exercises 15–18, determine the quadrant in which the point (x, y) lies.

15. $x < 0$ and $y > 0$ **16.** $x > 0$ and $y > 0$

17. $x < 0$ and $y < 0$ **18.** $x > 0$ and $y > 0$

In Exercises 19–38, answer the given questions.

19. Where are all points whose x-coordinates are 1?

20. Where are all points whose y-coordinates are -3?

21. Where are all points such that $y = 3$?

22. Where are all points such that $|x| = 2$?

23. Where are all points whose x-coordinates equal their y-coordinates?

24. Where are all points whose x-coordinates equal the negative of their y-coordinates?

25. What is the x-coordinate of all points on the y-axis?

26. What is the y-coordinate of all points on the x-axis?

27. Where are all points for which $x > 0$?

28. Where are all points for which $y < 0$?

29. Where are all points for which $x < -1$?

30. Where are all points for which $y > 4$?

31. Where are all points for which $xy > 0$?

32. Where are all points for which $y/x < 0$?

33. Where are all points for which $xy = 0$?

34. Where are all points for which $x < y$?

35. If the point (a, b) is in the second quadrant, in which quadrant is $(a, -b)$?

36. The points $(3, -1)$, $(3, 0)$, and (x, y) are on the same straight line. Describe this line.

37. On a circular machine part, holes are to be drilled at the points $(2, 2)$, $(-2, 2)$, $(-2, -2)$, and $(2, -2)$, where $(0, 0)$ represents the centre. Plot these points and find the distance between the points in quadrants I and III.

38. Join the points $(-1, -2)$, $(-4, -2)$, $(7, 2)$, $(2, 2)$ and $(-1, -2)$ in order with straight-line segments. Find the distance between successive points and then identify the geometric figure formed.

Answer to Practice Exercise

1. On the positive y-axis

3.2 Introduction to Functions

In the previous section we discussed how points graphed on a rectangular coordinate system can be used to represent the relationship between two variables. When a regular pattern in the relationship between the variables exists, it is often possible to describe the nature of that relationship using a mathematical formula. These formulas define functions, in the sense that the value of the dependent variable is determined by (is a function of) the value of the independent variable.

Definition of a Function

A **function** is a relationship between two variables such that for every input value of the first, there is only one corresponding output value of the second.

The independent variable is sometimes called the **manipulated variable**, since it represents the physical quantity that is varied through choosing its values. The dependent variable is sometimes called the **responding variable**, since it represents the value of the measured quantity that reacts in response to changes in the value of the manipulated variable.

It is important to remember that the independent and dependent variables in a function represent real numbers. Therefore, there may be restrictions on their possible values. This is discussed in Section 3.3.

EXAMPLE 1 Identifying independent and dependent variables

Identify the independent and the dependent variables in each of the following functions.

(a) $y = 2x$

Here y is a function of x, because for each value of x there is only one value of y. For example, if $x = 3$, then $y = 6$ and no other value. Therefore, x is the independent variable, and y is the dependent variable.

(b) $P = 4I^2$, where P is the power developed in a certain resistor by a current I.

P is a function of I: as the current is adjusted, there is a single value of power that results. The dependent variable is P and the independent variable is I.

(c) $V = s^3$, where V is the volume of a cube of edge s (see Fig. 3.7).

V is a function of s, since for each value of s, there is only one value of V. The dependent variable is V and the independent variable is s.

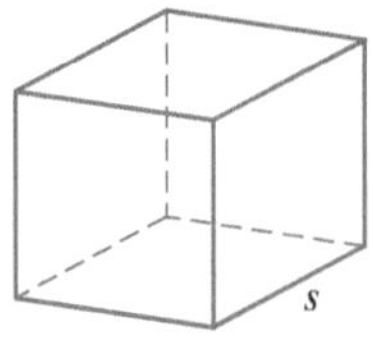

Fig. 3.7

EXAMPLE 2 Manipulated and responding variables

In an experiment, the distance s (in metres) that is travelled by a falling object is suspected to depend on the time elapsed t (in seconds) during the fall. During the experiment, measurements will be taken every 0.1 s, and the distance s will be measured at those times. Identify the manipulated and responding variables.

Time t is the manipulated variable, since the values 0.0 s, 0.1 s, 0.2 s, 0.3 s, . . . are selected by the experimenter. The displacement s is the responding variable, as its value will be measured during the fall in response to the changes in time.

FUNCTIONAL NOTATION

If a variable y depends upon the independent variable x, we will often express this as $y = f(x)$, which is read as "y is a function of x," or alternatively, "y is a function that depends on the values of an input variable x." Here, f denotes *dependence* and does not represent a quantity or a variable.

COMMON ERROR $f(x)$ does not mean f multiplied by x. $f(x)$ describes a variable dependence, in that the function value returned will be determined by the value of variable x.

EXAMPLE 3 Functional notation $f(x)$

Suppose we have a relationship between y and x that is given by the equality $y = 6x^3 - 5x$. Since the value of y depends on the value of x that is substituted into the equation, it is appropriate to state that y is a function of x. Therefore, it would be equally correct to write the function as $f(x) = 6x^3 - 5x$.

One of the most important uses of functional notation is to designate the value of the function for a particular value of the independent variable. That is,

the value of the function $f(x)$ when $x = a$ is written as $f(a)$.

■ A graphing utility (a graphing calculator, a computer algebra system, a spreadsheet, or any graphing software) can be used to evaluate a function in several ways. One is to directly substitute the value into the function. A second is to enter the function as a function and subsequently recall the function and assign values to the manipulated variable. A third way, which is very useful when many values of the manipulated variable will be used, is to enter the function in tabular form so that a series of entries with different values of the manipulated variable can be sequentially created as rows.

EXAMPLE 4 Evaluating a function at a specific value of x

Given the function $f(x) = 3x - 7$, find the value of $f(x)$ for **(a)** $x = 2$, and **(b)** $x = -1.4$.

(a) The value of $f(x)$ for $x = 2$ is written as $f(2)$ and is found by replacing all occurrences of x by 2, so

$$f(2) = 3(2) - 7 = -1$$

Therefore, when x is set equal to 2, the value of the function is -1.

(b) The value of $f(x)$ for $x = -1.4$ is

$$f(-1.4) = 3(-1.4) - 7 = -11.2$$

Therefore, when x is set equal to -1.4, the value of the function is -11.2.

LEARNING TIP

Whatever number a may represent, in order to evaluate $f(a)$, we substitute the value of a for x in the function $f(x)$. This is true even if a is a literal number or is itself a function composed of another variable.

EXAMPLE 5 Evaluating a function at variable values

(a) Given the function $g(t) = \frac{t^2}{2t + 1}$, find $g(r^3)$.

Replacing all occurrences of t by r^3 we get

$$g(r^3) = \frac{(r^3)^2}{2(r^3) + 1} = \frac{r^6}{2r^3 + 1}$$

(b) Given the function $F(y) = 5 - 4y^2$, find $F(a + 1)$ and $F(a) + 1$.

To find $F(a + 1)$, we substitute $a + 1$ for y to get

$$\begin{aligned} F(a + 1) &= 5 - 4(a + 1)^2 \\ &= 5 - 4(a^2 + 2a + 1) \\ &= -4a^2 - 8a + 1 \end{aligned}$$

This is very different from finding $F(a) + 1$, where we *first* substitute a for y and *then* add 1, to get

$$F(a) + 1 = 5 - 4a^2 + 1 = 6 - 4a^2$$

Practice Exercise

1. For the function in Example 5(a), find $g(-1)$.

EXAMPLE 6 Functional notation—electric resistance

The electric resistance R of a certain resistor as a function of the temperature T (in °C) is given by $R = 10.0 + 0.01T + 0.001T^2$. If a given temperature T is increased by 10°C, what is the value of R that results from this change?

We are to determine R for a temperature of $T + 10$. Since

$$R(T) = 10.0 + 0.10T + 0.001T^2$$

then

$$R(T + 10) = 10.0 + 0.10(T + 10) + 0.001(T + 10)^2 \quad \text{substitute } T + 10 \text{ for } T$$

$$R(T + 10) = 10.0 + 0.10T + 1.0 + 0.001T^2 + 0.02T + 0.1$$

$$R(T + 10) = 11.1 + 0.12T + 0.001T^2$$

At times, we need to define more than one function. We then use different symbols, such as $f(x)$ and $g(x)$, to denote these functions.

EXAMPLE 7 Multiple functions notation

Consider the functions $f(x) = 5x - 3$ and $g(x) = ax^2 + x$, where a is a constant. If evaluating both functions at $x = -4$, the naming convention gives us a clear way of telling the two calculations apart.

$$f(-4) = 5(-4) - 3 = -23 \quad \text{substitute } -4 \text{ for } x \text{ in } f(x)$$

$$g(-4) = a(-4)^2 + (-4) = 16a - 4 \quad \text{substitute } -4 \text{ for } x \text{ in } g(x)$$

Practice Exercise

2. For the function in Example 7, find $f(-a^2x)$ and $g(-a^2x)$.

A function may be viewed as a set of instructions that will transform a given input into an output. This is illustrated in the next example.

EXAMPLE 8 Function as a set of instructions

LEARNING TIP
In functional form, this same sequence of steps can easily be coded and used again to explain how the dependent value of $f(x)$ results when the independent variable x is used as an input value.

The function $f(x) = x^2 - 3x$ tells us to "square the value of the independent variable, multiply the value of the independent variable by 3, and subtract the second result from the first." An analogy would be a computer that was programmed so that when a number was entered into the program, it would square the number, then multiply the number by 3, and finally subtract the second result from the first. This is diagrammed in Fig. 3.8.

■ The literal symbol chosen to hold the position of the input variable is arbitrary. For example, in our process outlined in Fig. 3.8, we could call the input variable n, and express the same relationship as $f(n) = n^2 - 3n$.

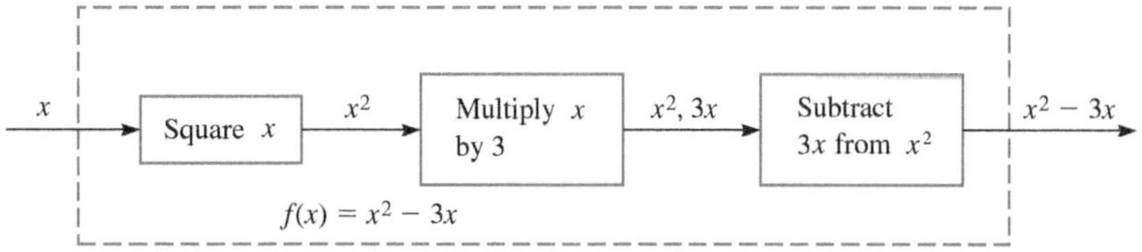

Fig. 3.8

EXERCISES 3.2

In Exercises 1–4, solve the given problems related to the indicated examples of this section.

1. In Example 4, find the value of $f(-2)$.
2. In Example 5(a), evaluate $g(-a^2)$.
3. In Example 6, change "increased" to "decreased" in the second line and then evaluate the function to find the proper expression.
4. In Example 8, change $x^2 - 3x$ to $x^3 + 4x$ and then determine the statements that should be placed in the three boxes in Fig. 3.8.

In Exercises 5–12, find the indicated functions.

5. Express the area A of a circle as a function of (a) its radius r and (b) its diameter d.
6. Express the circumference c of a circle as a function of (a) its radius r and (b) its diameter d.
7. Express the diameter d of a sphere as a function of its volume V.
8. Express the edge s of a cube as a function of its surface area A.
9. Express the area A of a square as a function of its side s; express the side s of a square as a function of its area A.
10. Express the perimeter p of a square as a function of its side s; express the side s of a square as a function of its perimeter p.
11. A circle is inscribed in a square (the circle is tangent to each side of the square). Express the area A of the four corner regions bounded by the circle and the square as a function of the radius r of the circle.
12. Express the area A of an equilateral triangle as a function of its side s.

In Exercises 13–24, evaluate the given functions.

13. $f(x) = 2x + 1$; find $f(1)$ and $f(-1)$.
14. $f(x) = 5x - 9$; find $f(2)$ and $f(-2)$.
15. $f(x) = 5$; find $f(-2)$ and $f(0.4)$.
16. $f(T) = 7.2 - 2.5|T|$; find $f(2.6)$ and $f(-4)$.
17. $\phi(x) = \dfrac{6 - x^2}{2x}$; find $\phi(2\pi)$ and $\phi(-2)$.
18. $H(q) = \dfrac{8}{q} + 2\sqrt{q}$; find $H(4)$ and $H(0.16)$.
19. $g(t) = at^2 - a^2t$; find $g\left(-\frac{1}{2}\right)$ and $g(a)$.
20. $s(y) = 6\sqrt{y + 11} - 3$; find $s(-2)$ and $s(a^2)$.
21. $K(s) = 3s^2 - s + 6$; find $K(-s)$ and $K(2s)$.
22. $T(t) = 5t + 7$; find $T(-2t)$ and $T(t + 1)$.
23. $f(x) = 2x + 4$; find $f(3x) - 3f(x)$.
24. $f(x) = 2x^2 + 1$; find $f(x + 2) - [f(x) + 2]$.

In Exercises 25–28, evaluate the given functions. The values of the independent variable are approximate.

25. Given $f(x) = 5x^2 - 3x$, find $f(3.86)$ and $f(-6.92)$.
26. Given $g(t) = \sqrt{t + 1.0604} - 6t^3$, find $g(0.9261)$.
27. Given $F(H) = \dfrac{2H^2}{H + 36.85}$, find $F(-84.466)$.
28. Given $f(x) = \dfrac{x^4 - 2.0965}{6x}$, find $f(1.9654)$.

In Exercises 29–36, state the instructions of the function in words as in Example 8.

29. $f(x) = x^2 + 2$
30. $f(x) = 2x - 6$
31. $g(y) = 6y - y^3$
32. $\phi(s) = 8 - 5s + s^2$
33. $R(r) = 3(2r + 5) - 1$
34. $f(z) = \dfrac{4z}{5 - z}$
35. $p(t) = \dfrac{2t - 3}{t + 2}$
36. $Y(y) = 2 + \dfrac{5y}{2(y - 3)}$

In Exercises 37–40, write the equation as given by the statement. Then write the indicated function using functional notation.

37. The surface area A of a cubical open-top aquarium equals five times the square of an edge s of the aquarium.

38. A helicopter is at an altitude of 1000 m and is x m horizontally from a fire. Its distance d from the fire is the square root of the sum of 1000 squared and x squared.

39. The area A of the Athabasca Glacier in Jasper National Park, Canada, given that its present area is 6.00 km^2 and that it is melting at the rate of $0.25t$ km^2 where t is the time in centuries.

40. The electrical resistance R of a certain ammeter, in which the resistance of the coil R_c is the product of 10 and R_c divided by the sum of 10 and R_c.

In Exercises 41–48, solve the given problems and round all results to three significant digits.

41. A demolition ball is used to tear down a building. Its distance s (in m) above the ground as a function of time t (in s) after it is dropped is $s = 17.5 - 4.9t^2$. Since $s = f(t)$, find $f(1.2)$.

42. The length L (in m) of a cable hanging between equal supports 100 m apart is $L = 100(1 + 0.0003s^2)$, where s is the sag (in m) in the middle of the cable. Since $L = f(s)$, find $f(15)$.

43. The stopping distance d (in m) of a car going v km/h is given by $d = 0.2v + 0.008v^2$. Since $d = f(v)$, find $f(30)$, $f(2v)$, and $f(60)$, using both $f(v)$ and $f(2v)$.

44. The electric power P (in W) dissipated in a resistor of resistance R (in Ω) is given by the function $P = \frac{200R}{(100 + R)^2}$. Since $P = f(R)$, find $f(R + 10)$.

45. A motorist travels at 55 km/h for t hours. Express the distance d travelled as a function of t.

46. The sales tax in a city is 7%. The price tag on an item shows D dollars. Express the total cost C of the item as a function of D.

47. (a) Explain the meaning of $f[f(x)]$.
(b) Find $f[f(x)]$ for $f(x) = 2x^2$.

48. If $f(x) = x$ and $g(x) = x^2$, find (a) $f[g(x)]$ and (b) $g[f(x)]$.

Answers to Practice Exercises

1. $f(-1) = -1$ **2.** $f(-a^2x) = -5a^2x - 3$; $g(-a^2x) = a^5x^2 - a^2x$

3.3 More about Functions

For a given function, *the complete set of possible values of the independent variable is called the* **domain** *of the function*, and the *complete set of all possible resulting values of the dependent variable is called the* **range** *of the function.*

As noted earlier, *we will be using only* ***real numbers*** when using functions. This means there may be restrictions as to the values that may be used since *values that lead to* ***division by zero*** *or to* ***imaginary numbers may not be included*** *in the domain or the range.*

EXAMPLE 1 Domain and range

Find the domain and range of each of the following functions.

(a) $f(x) = x^2 + 2$

Domain: The function is defined for all real values of x. Therefore, the domain is all real numbers.

Range: Since x^2 is either positive or zero, $x^2 + 2$ is always at least 2. We then say that the range is the set of *all real numbers that are greater than or equal to 2*, or $f(x) \geq 2$.

■ Note that $f(2) = 6$ and $f(-2) = 6$. This is all right. Two different values of x may give the same value of y, but there may not be two different values of y for one value of x.

■ Later in this chapter, we will see that a graphing utility is useful in finding the domain and range of a function.

(b) $g(t) = \frac{1}{t + 2}$

Domain: We find the domain by considering the values of the independent variable t that are not valid, and excluding them from the domain. Here $t = -2$ would lead to division by zero. Therefore, the domain is all real numbers except -2.

Range: Since the numerator of the fraction is never zero, $g(t)$ will never exactly equal zero. Therefore, the range is *all real numbers except 0.*

EXAMPLE 2 Domain and range of radical functions

Find the domain and range of the function $g(s) = \sqrt{3 - s}$.

Domain: We examine the conditions under which the radical will produce a real number. If $s > 3$, the radicand will be negative, resulting in a number that is not real. Therefore, the domain is *all real numbers $s \le 3$*.

Practice Exercise

1. Find the domain and range of the function $f(x) = \sqrt{x + 4}$.

Range: When evaluating the function at $s = 3$, the result is 0. For values of s less than 3, the radicand is positive and greater than zero, leading to a positive principal square root (see Section 1.6). Therefore, the range is *all real numbers $g(s) \ge 0$*.

INTERVAL NOTATION

Interval notation is a shorthand method that can be used to write the domain and range of a function. This notation lists two values within brackets. A square bracket, [or], means the *value at that end is included in the interval*. Round brackets, (or), indicate that *the values go up to but do not include the specified value*. Note that for real numbers that extend indefinitely in either the negative or the positive direction, the infinity symbol ∞ indicates "increase without limit" in that direction.

EXAMPLE 3 Interpreting interval notation

Interpret the interval notation describing each of the following sets.

(a) $[-2, 5]$ is the set of real numbers that are greater than or equal to -2 and less than or equal to 5.

(b) $(-\infty, 3]$ is the set of real numbers that are less than or equal to 3.

(c) $(0, \infty)$ is the set of real numbers that are greater than zero (zero is not included).

(d) $(-\infty, \infty)$ is the set of all real numbers.

EXAMPLE 4 Domain with more than one constraint

Find the domain of the function $f(x) = 16\sqrt{x} + \dfrac{1}{x}$.

From the term $16\sqrt{x}$, we see that x must be greater than or equal to zero in order to have real values. The term $\frac{1}{x}$ indicates that x cannot be zero, because x is in the denominator and it would lead to division by zero. We therefore exclude zero from the domain and state that the domain is *all real numbers $x > 0$*, or $(0, \infty)$.

We have seen that the domain may be restricted since we do not use imaginary numbers or divide by zero. The domain may also be restricted by the definition of the function, or by practical considerations in an application.

EXAMPLE 5 Identifying restricted domains

Discuss the restrictions that exist for the domain of each of the following functions.

(a) $f(x) = x^2 + 4$ (for $x > 2$).

The domain is restricted to real numbers greater than 2 *by definition*. Therefore the domain is $(2, \infty)$. Note that the restriction on the domain results in a restriction on the range. If x can only be greater than 2, then $x^2 + 4$ can only be greater than 8, so the range would be $(8, \infty)$.

(b) $h = 20t - 4.9t^2$, where h is the height (in m) of a projectile given as a function of time t (in s).

The domain is restricted to real numbers greater than or equal to zero by practical considerations, since the independent variable is time and negative values of time have no meaning. Therefore the domain is $[0, \infty)$. In this case, the range would be restricted by the physical system (not by the domain). There would be negative values of h if the projectile were being launched off a cliff, so that heights that occur below the initial height would be valid. In contrast, if the projectile were being launched on a flat surface, it would not be possible for the height to be less than zero.

The following example illustrates a function that is defined differently for different intervals of the domain.

EXAMPLE 6 Function defined for intervals of the domain

In a certain electric circuit, the current i (in mA) is a function of the time t (in s), which means $i = f(t)$. The function is

$$f(t) = \begin{cases} 8 - 2t & \text{(for } 0 \le t \le 4 \text{ s)} \\ 0 & \text{(for } t > 4 \text{ s)} \end{cases}$$

Since negative values of t are not usually meaningful, $f(t)$ is not defined for $t < 0$. Find the current for $t = 3$ s, $t = 6$ s, and $t = -1$ s.

We are to find $f(3), f(6)$, and $f(-1)$, and we see that values of this function are determined differently depending on the value of t. Since 3 is between 0 and 4,

$$f(3) = 8 - 2(3) = 2 \quad \text{or} \quad i = 2 \text{ mA}$$

Since 6 is greater than 4, $f(6) = 0$, or $i = 0$ mA. We see that $i = 0$ mA for all values of t that are 4 or greater.

Since $f(t)$ is not defined for $t < 0, f(-1)$ is not defined.

FUNCTIONS FROM VERBAL STATEMENTS

To find a mathematical function from a verbal statement, we use methods like those for setting up equations in Chapter 1. This is shown in the following examples.

EXAMPLE 7 Function from a verbal statement with cost

The fixed cost for a company to operate a certain plant is $3000 per day. It also costs $4 for each unit produced in the plant. Express the daily cost C of operating the plant as a function of the number n of units produced.

The daily total cost C equals the fixed cost of $3000 plus the cost of producing n units. Since the cost of producing one unit is $4, the cost of producing n units is $4n$. Thus, the total cost C, where $C = f(n)$, is

$$C = 3000 + 4n$$

Here, we know that the domain is all values of $n \ge 0$, with some upper limit on n based on the production capacity of the plant.

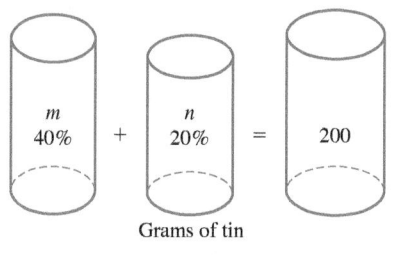

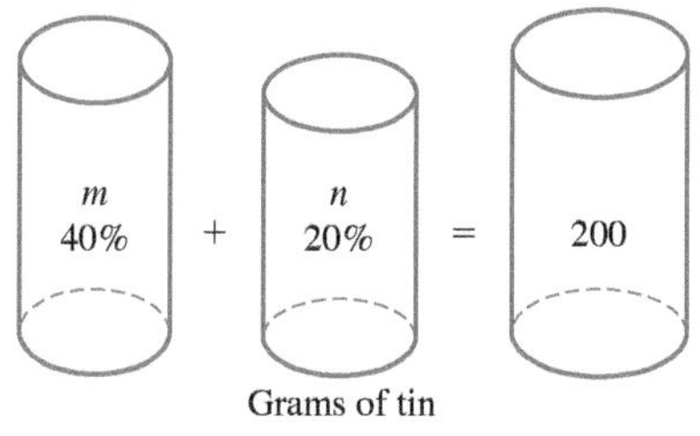

Fig. 3.9

	1st Solder	2nd Solder	Final
% of tin	40%	20%	
g of solder	m	n	
g of tin	$0.40m$	$0.20n$	200

EXAMPLE 8 Function from a verbal statement for a mixture

A metallurgist melts and mixes m grams (g) of solder that is 40% tin with n grams of another solder that is 20% tin to get a final solder mixture that contains 200 g of tin. Express n as a function of m. See Fig. 3.9.

For all mixtures, mass is preserved between the starting and the final mixtures. The contribution to the mass of tin from each solder can be seen in the table to the left. These values lead to the following equation:

$$0.40m \quad + \quad 0.20n \quad = \quad 200$$

Since we want $n = f(m)$, we solve for n to obtain the required function:

$$0.20n = 200 - 0.40m$$
$$n = 1000 - 2m$$

Since mass is never negative, m must be 0 g or greater. Since n cannot be negative, m cannot be larger than 500 g. Therefore, the domain is $[0\text{ g}, 500\text{ g}]$, or $0\text{ g} \le m \le 500\text{ g}$. This means that n can never exceed 1000 g, so the range is $[0\text{ g}, 1000\text{ g}]$, or $0\text{ g} \le n \le 1000\text{ g}$.

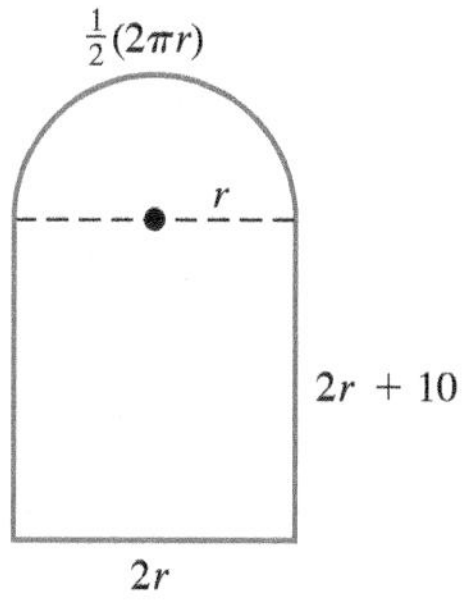

Fig. 3.10

EXAMPLE 9 Function from a verbal statement with perimeter

An architect designs a window in the shape of a rectangle with a semicircle on top, as shown in Fig. 3.10. The base of the window is 10 cm less than the height of the rectangular part. Express the perimeter p of the window as a function of the radius r of the circular part.

The perimeter is the distance around the window. Since the top part is a semicircle, the length of this top circular part is $\frac{1}{2}(2\pi r)$, and the base of the window is $2r$ because it is equal in length to the dashed line (the diameter of the circle). Finally, the base being 10 cm less than the height of the rectangular part tells us that each vertical side of the rectangle is $2r + 10$. Therefore, the perimeter p, where $p = p(r)$, is

$$p = \tfrac{1}{2}(2\pi r) + 2r + 2(2r + 10)$$
$$p = \pi r + 2r + 4r + 20$$
$$p = \pi r + 6r + 20$$

We see that the required function is $p = \pi r + 6r + 20$. Since the radius cannot be negative and there would be no window if $r = 0$, the domain of the function is all values $0 < r \le R$, where R is a maximum possible value of r determined by design considerations, or $(0, R]$. The range can be expressed in terms of R as $(0, R(6 + \pi) + 20]$, or $0 < p \le R(6 + \pi) + 20$.

Practice Exercise

2. In Example 9, find p as a function of r if there is to be a semicircular section of the window at the bottom, as well as at the top.

From the definition of a function, we know that any value of the independent variable must yield only one value of the dependent variable. *If a value of the independent variable yields one or more values of the dependent variable, the relationship is called a* **relation**. A function is a relation for which each value of the independent variable yields only one value of the dependent variable. Therefore, a function is a special type of relation. There are also many relations that are not functions.

EXAMPLE 10 Relation

For $y^2 = 4x^2$, if $x = 2$, then y can be either 4 or -4. Since a value of x yields more than one value for y, we see that $y^2 = 4x^2$ is a relation, but not a function.

EXERCISES 3.3

In Exercises 1–4, solve the given problems related to the indicated examples of this section.

1. In Example 1(a) change x^2 to $-x^2$. What changes result in the domain and range?
2. In Example 4, in the first line, change $\frac{1}{x}$ to $\frac{1}{x-1}$. What changes result in the domain?
3. In Example 6, find $f(2)$ and $f(5)$.
4. In Example 8, interchange 40% and 20% and then find the function.

In Exercises 5–14, find the domain and range of the given functions. In Exercises 11 and 12, explain your answers.

5. $f(x) = x + 5$
6. $g(u) = 3 - u^2$
7. $G(R) = \dfrac{3.2}{R}$
8. $F(r) = \sqrt{r+4}$
9. $f(s) = \dfrac{2}{s^2}$
10. $T(t) = 2t^4 + t^2 - 1$
11. $H(h) = 2h + \sqrt{h} + 1$
12. $f(x) = \dfrac{-6}{\sqrt{2-x}}$
13. $y = |x - 3|$
14. $y = x + |x|$

In Exercises 15–18, find the domain of the given functions.

15. $Y(y) = \dfrac{y+1}{\sqrt{y-2}}$
16. $f(n) = \dfrac{n}{6-2n}$
17. $f(D) = \dfrac{D}{D-2} + \dfrac{4}{D+4}$
18. $g(x) = \dfrac{\sqrt{x-2}}{x-3}$

In Exercises 19–22, evaluate the indicated functions.

$F(t) = 3t - t^2$ (for $t \leq 2$)

$$h(s) = \begin{cases} 2s & (\text{for } s < -1) \\ s+1 & (\text{for } s \geq -1) \end{cases}$$

$$f(x) = \begin{cases} x+1 & (\text{for } x < 1) \\ \sqrt{x+3} & (\text{for } x \geq 1) \end{cases}$$

$$g(x) = \begin{cases} \dfrac{1}{x} & (\text{for } x \neq 0) \\ 0 & (\text{for } x = 0) \end{cases}$$

19. Find $F(2)$ and $F(3)$.
20. Find $h(-8)$ and $h(-0.5)$.
21. Find $f(1)$ and $f(-0.25)$.
22. Find $g(0.2)$ and $g(0)$.

In Exercises 23–36, determine the appropriate functions.

23. A motorist travels at 60 km/h for 2 h and then at 80 km/h for t h. Express the distance d travelled as a function of t.
24. Express the cost C of insulating a cylindrical water tank of height 2 m as a function of its radius r, if the cost of insulation is \$3 per square metre.
25. A rocket burns up at the rate of 2 Mg/min after falling out of orbit into the atmosphere. If the rocket weighed 5500 Mg before reentry, express its weight w as a function of the time t, in minutes, of reentry.
26. A computer part costs \$3 to produce and distribute. Express the profit p made by selling 100 of these parts as a function of the price of c dollars each.
27. Upon ascending, a weather balloon ices up at the rate of 0.5 kg/m after reaching an altitude of 1000 m. If the mass of the balloon below 1000 m is 110 kg, express its mass m as a function of its altitude h if $h > 1000$ m.
28. A chemist adds x L of a solution that is 50% alcohol to 100 L of a solution that is 70% alcohol. Express the number n of litres of alcohol in the final solution as a function of x.
29. A company installs underground cable at a cost of \$500 for the first 50 m (or up to 50 m) and \$5 for each metre thereafter. Express the cost C as a function of the length l of underground cable if $l > 50$ m.
30. The *mechanical advantage* of an inclined plane is the ratio of the length of the plane to its height. Express the mechanical advantage MA of a plane of length 8 m as a function of its height h.
31. The capacities (in L) of two oil-storage tanks are x and y. The tanks are initially full; 1200 L is removed from them by taking 10% of the contents of the first tank and 40% of the contents of the second tank. (a) Express y as a function of x. (b) Find $y(400)$.
32. A province does not tax the first \$30 000 of a resident's income but taxes any amount over \$30 000 at 5%. (a) Find the tax T as a function of a resident's income I and (b) find $T(25\,000)$ and $T(45\,000)$.
33. A company finds that it earns a profit of \$15 on each cell phone and a profit of \$25 on each DVD player that it sells. If x cell phones and y DVD players are sold, the profit is \$2750. Express y as a function of x.
34. A satellite telephone is leased at a cost C of \$200 plus \$10 per minute t the phone is used. Solve for t as a function of C and find $t(500)$.
35. For studying the electric current that is induced in wire rotating through a magnetic field, a piece of wire 60 cm long is cut into two pieces. One of these is bent into a circle and the other into a square. Express the total area A of the two figures as a function of the perimeter p of the square.
36. The cross-section of an air-conditioning duct is in the shape of a square with semicircles on opposite sides. See Fig. 3.11. Express the area A of this cross-section as a function of the diameter d (in cm) of the circular part.

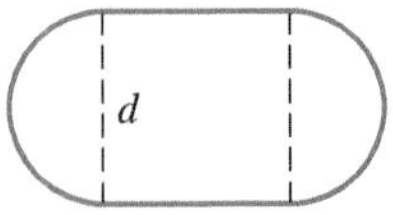

Fig. 3.11

In Exercises 37–50, solve the given problems.

37. A helicopter 120 m from a person takes off vertically. Express the distance d from the person to the helicopter as a function of the height h of the helicopter. What are the domain and the range of $d(h)$? See Fig. 3.12.

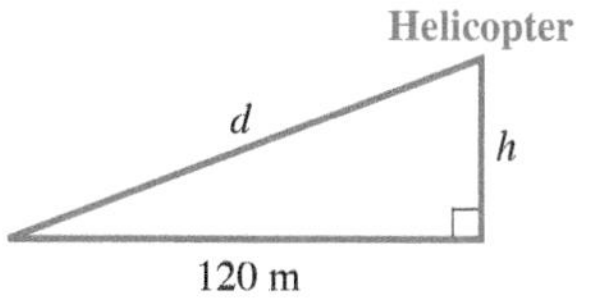

Fig. 3.12

38. A computer program displays a circular image of radius 6 cm. If the radius is decreased by x cm, express the area of the image as a function of x. What are the domain and range of $A(x)$?

39. A truck travels 300 km in t h. Express the average speed v of the truck as a function of t. What are the domain and range of $v(t)$?

40. A rectangular grazing range with an area of 8 km^2 is to be fenced. Express the length l of the field as a function of its width w. What are the domain and range of $l(w)$?

41. The resonant frequency f of a certain electric circuit as a function of the capacitance is $f = \frac{1}{2\pi\sqrt{C}}$. Describe the domain.

42. A jet is travelling directly between Calgary, Alberta, and Halifax, Nova Scotia, which are 3760 km apart. If the jet is x km from Calgary and y km from Halifax, find the domain of $y = f(x)$.

43. Using a piecewise-defined function, express the mass m of the weather balloon in Exercise 27 as a function of any height h.

44. Using a piecewise-defined function, express the cost C of installing any length l of the underground cable in Exercise 29.

45. A rectangular piece of cardboard twice as long as it is wide is to be made into an open box by cutting 5-cm squares from each corner and bending up the sides. (a) Express the volume V of the box as a function of the width w of the piece of cardboard. (b) Find the domain of the function.

46. A spherical buoy 36.0 cm in diameter is floating in a lake and is more than half above the water. (a) Express the circumference c of the circle of intersection of the buoy and water surface as a function of the depth d to which the buoy sinks. (b) Find the domain and range of the function.

47. For $f(x-1) = |x|$, find $f(0)$.

48. For $f(x) = x^2$, find $\frac{f(x+h) - f(x)}{h}$.

49. What is the range of the function $f(x) = |x| + |x-2|$?

50. For $f(x) = \sqrt{x-1}$ and $g(x) = x^2$, find the domain of $g[f(x)]$. Explain.

Answers to Practice Exercises

1. Domain: all real numbers $x \geq -4$; Range: all real numbers $f(x) \geq 0$ **2.** $p = 2\pi r + 4r + 20$

3.4 The Graph of a Function

In scientific and engineering disciplines, we often use the graph of a function to illustrate the way in which variables relate to each other. The graph provides a visual representation of how the independent variable x relates to the dependent response $f(x)$.

The graph of a function is the set of all points whose coordinates (x, y) satisfy the functional relationship $y = f(x)$. Since $y = f(x)$, we can write the coordinates of the points on the graph as $(x, f(x))$. Writing the coordinates in this manner tells us exactly how to find them. *We assume a certain value for x and then find the value of the function of x. These two numbers are the coordinates of the point.*

The following steps are used to plot the graph of a function.

■ As to just what values of x to choose and how many to choose, with a little experience you will usually be able to tell if you have enough points to plot an accurate graph.

Procedure for Plotting the Graph of a Function

1. **Find (x, y) values of the function.** Depending on the region of interest, select several x values, and use the function to calculate the corresponding values of y.
2. **Create a table of values.** Organize the pairs from Step 1 so that values of x are increasing.
3. **Construct a rectangular coordinate system.** Draw the axes, using appropriate labels and scales. Plot each pair (x, y) as a point on this system.
4. **Join the points.** Starting with the point with the smallest x value, join the points from left to right using a **smooth curve** (not short straight-line segments).

Draw the axes with an appropriate scale, plot points, and join from left to right with smooth curve

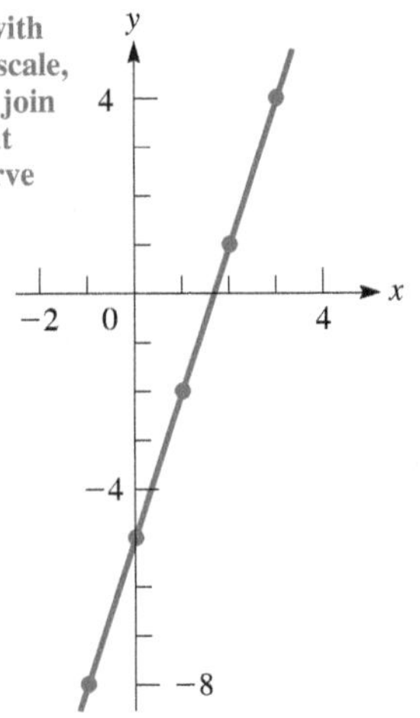

Fig. 3.13

EXAMPLE 1 Graphing a function by plotting points

Graph the function $f(x) = 3x - 5$.

Steps 1 and 2: We select several x values. For efficiency, we order them from lowest to highest in a table ahead of calculating the corresponding y values. We next calculate $y = f(x)$ for each value of x. For example, $f(-1) = 3(-1) - 5 = -8$. We then obtain the following table of values.

x	-1	0	1	2	3
$f(x)$	$f(-1) = -8$	$f(0) = -5$	$f(1) = -2$	$f(2) = 1$	$f(x) = 4$
(x, y)	$(-1, -8)$	$(0, -5)$	$(1, -2)$	$(2, 1)$	$(3, 4)$

Steps 3 and 4: See Fig. 3.13. As illustrated, we begin by connecting the two points farthest to the left, and continue the smooth connecting line through the other points.

■ The table of values for a graph can be seen on a graphing utility using table features.

Draw axes using an appropriate scale, plot points, and join from left to right with smooth curve

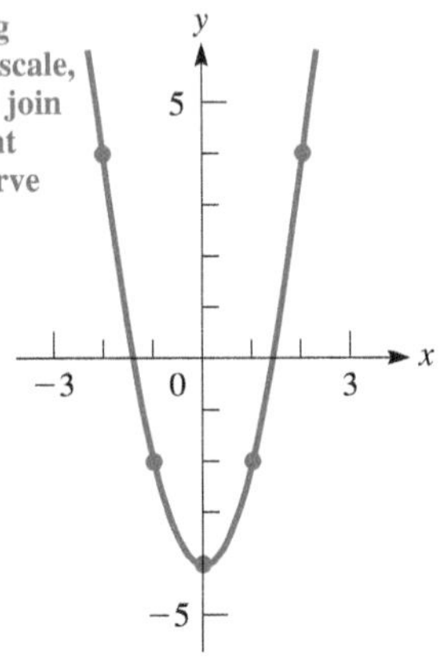

Fig. 3.14

EXAMPLE 2 Plotting with negative numbers

Graph the function $f(x) = 2x^2 - 4$.

Steps 1 and 2: We select several x values and calculate $y = f(x)$ for each value of x. We must pay careful attention to the laws for signed numbers.

x	$f(x)$	(x, y)
-2	$f(-2) = 2(-2)^2 - 4 = 8 - 4 = 4$	$(-2, 4)$
-1	$f(-1) = 2(-1)^2 - 4 = 2 - 4 = -2$	$(-1, -2)$
0	$f(0) = 2(0)^2 - 4 = 0 - 4 = -4$	$(0, -4)$
1	$f(1) = 2(1)^2 - 4 = 2 - 4 = -2$	$(1, -2)$
2	$f(2) = 2(2)^2 - 4 = 8 - 4 = 4$	$(2, 4)$

Steps 3 and 4: See Fig. 3.14. We connect the points, being careful to draw a smooth curve through the points.

COMMON ERROR

Mistakes are common in evaluating $f(x)$ for negative values of x. Ensure that you are careful! As in Example 2, if $f(x) = 2x^2 - 4$, then

$$f(-1) = 2(-1)^2 - 4 = 2(1) - 4 = -2$$

But if $f(x) = -5x^3 + 7x$, then

$$f(-2) = -5(-2)^3 + 7(-2) = -5(-8) - 14 = 40 - 14 = 26$$

There are several important considerations that you should keep in mind when graphing.

Special Notes on Graphing

1. **The graphs of most common functions are smooth.** Any place where the graph changes in a way that is not expected should be checked with care. Discontinuities are often due to miscalculation.
2. **Some values of *x* may not be valid.** Remember to check the domain of the function, and select only x values that are within the domain. Remember that division by zero is not defined, and that only real values of the variables may be used.
3. **Use only values of the variables that have meaning.** In particular, negative values of some variables, such as time, can have no meaning. By convention, we plot only the sections of a function that have meaning.

x	y
−2	−6
−1	−2
0	0
1	0
2	−2
3	−6

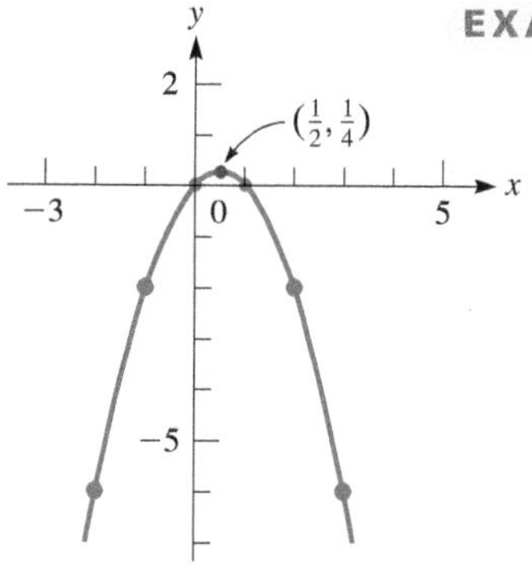

Fig. 3.15

EXAMPLE 3 Finding the value of a specific point on a curve

Graph the function $y = x - x^2$ tabulating points for integer values of x. Review the graph and add additional points if necessary.

First, we determine the values in the table, as shown with Fig. 3.15. Again, we must be careful when dealing with negative values of x. For the value $x = -1$, we have $y = (-1) - (-1)^2 = -1 - 1 = -2$.

Once all values in the table have been plotted, we find that both $x = 0$ and $x = 1$ result in a value of $y = 0$. The plotted points suggest that the turning point of the curve can be identified by using the value $x = \frac{1}{2}$ (exactly between $x = 0$ and $x = 1$). We find $f\left(\frac{1}{2}\right) = \frac{1}{2} - \left(\frac{1}{2}\right)^2 = \frac{1}{4}$, and add the point $\left(\frac{1}{2}, \frac{1}{4}\right)$ to the graph together with its coordinate label to illustrate its importance. The smooth curve can now be completed. See Fig. 3.15.

Note that in plotting these graphs, we do not stop the graph with the points we found, but indicate that the curve continues by drawing the graph past these points. However, this is not true for all graphs. As we will see in later examples, the graph should not include points for values of x not in the domain, and therefore may not continue past a particular point.

EXAMPLE 4 Avoiding division by zero

x	y
−4	3/4
−3	2/3
−2	1/2
−1	0
−1/2	−1
−1/3	−2

x	y
1/3	4
1/2	3
1	2
2	3/2
3	4/3
4	5/4

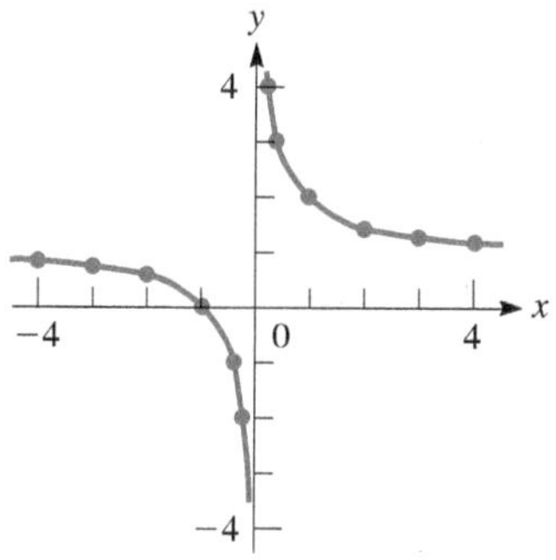

Fig. 3.16

Graph the function $y = 1 + \dfrac{1}{x}$.

In finding the points on this graph, as shown in Fig. 3.16, note that y is not defined for $x = 0$, due to division by zero. Thus, $x = 0$ is not in the domain, and no point on the graph will have an x-coordinate of zero. This means ***the curve will not cross the y-axis***. Although we cannot let $x = 0$, we can choose other values for x between −1 and 1 that are close to zero on either side. In doing so, we find that as x gets closer to zero, the points get closer to the y-axis, although they do not reach or touch it. In this case, the y-axis is called an **asymptote** of the curve.

EXAMPLE 5 Imaginary values cannot be graphed on an *xy* coordinate plane

Graph the function $y = \sqrt{x + 1}$.

When finding the points for the graph, we may not let x take on any value less than −1, for ***all such values would lead to imaginary values for y*** and are not in the domain. Also, since we have the positive square root indicated, the range consists of all values of y that are positive or zero ($y \geq 0$). See Fig. 3.17. Note that the graph starts at $(-1, 0)$.

x	y
−1	0
0	1
1	1.4
2	1.7
3	2

x	y
4	2.2
5	2.4
6	2.6
7	2.8
8	3

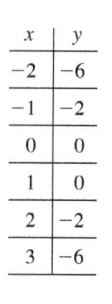

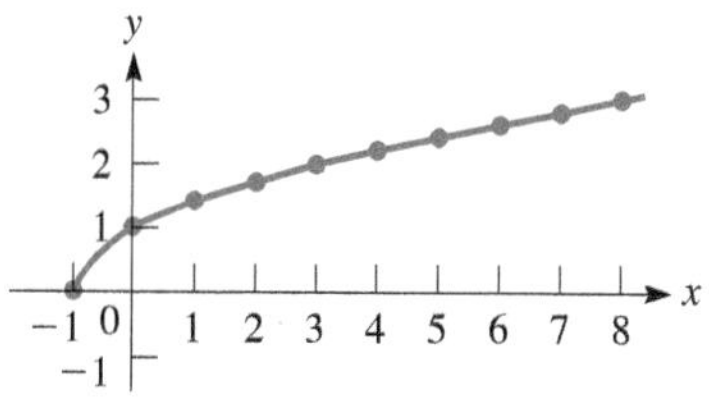

Fig. 3.17

Practice Exercises

Determine for what value(s) of x there are no points on the graph.

1. $y = \dfrac{2x}{x+5}$ **2.** $y = \sqrt{x-3}$

■ See the chapter introduction.

■ According to Betz's law, no turbine can capture more than 59.3% of the power in the wind. This is the maximum value of the power coefficient C_P.

EXAMPLE 6 Graph of wind turbine power

The *extractable power* P (in W) in the wind hitting a wind turbine depends on the velocity v (in m/s) of the wind, the power coefficient C_P, which is related to the efficiency of the turbine, the air density ρ (in kg/m^3), and the sweep area A (in m^2) of the turbine according to $P = 0.5C_P\rho Av^3$. For a specific 3.00-m diameter turbine of $C_P = 30\%$ and $A = 7.069\ \text{m}^2$, with air density $\rho = 1.23\ \text{kg/m}^3$, the extractable power is $P = 1.304v^3$. Plot P as a function of v.

Since negative values of velocity are not relevant, we only plot values of $v > 0$. Fig. 3.18 shows the table and the smooth curve obtained. Note on the graph how the scale on the P-axis differs from that of the v-axis. Different scales are normally used when the variables differ in magnitudes and ranges. The extractable power from the wind increases rapidly as wind velocity increases.

v (m/s)	P (W)
0.00	0.0
2.00	10.4
4.00	83.5
6.00	281.7
8.00	667.6
10.00	1304.0

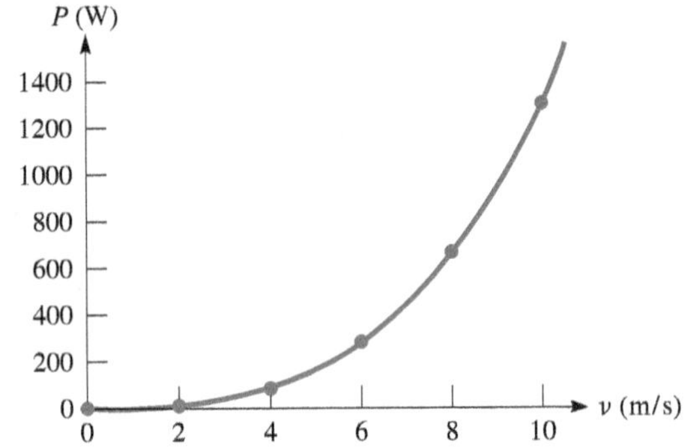

Fig. 3.18

■ The unit of power, the watt (W), is named for James Watt (1736–1819), a British engineer.

EXAMPLE 7 Graph of electric power

The electric power P (in W) delivered by a certain fuel cell as a function of the resistance R (in Ω) in the circuit is given by $P = \dfrac{100R}{(0.50 + R)^2}$. Plot P as a function of R.

Since negative values for the resistance have no physical significance, we should not plot any values of P for negative values of R. The following table is obtained:

$R(\Omega)$	0	0.25	0.50	1.0	2.0	3.0	4.0	5.0	10.0
$P(\text{W})$	0.0	44.4	50.0	44.4	32.0	24.5	19.8	16.5	9.1

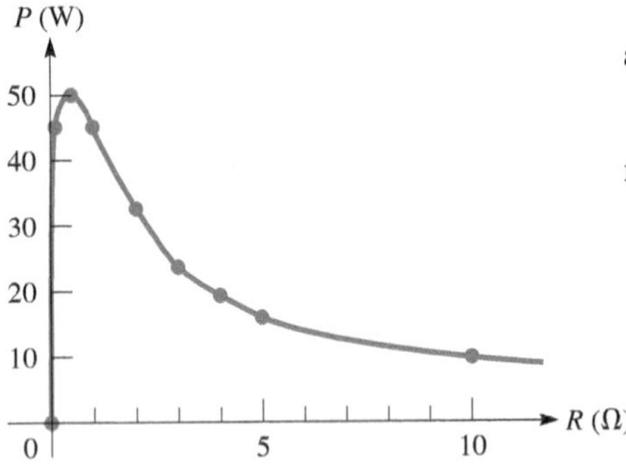

Fig. 3.19

The R values of 0.25 and 0.50 are used since P is less for $R = 2\ \Omega$ than for $R = 1\ \Omega$. The sharp change in direction should be checked by using these additional points to get a smoother curve near $R = 1\ \Omega$. See Fig. 3.19.

EXAMPLE 8 Piecewise-defined function

Graph the function $f(x) = \begin{cases} 2x + 1 & (\text{for } x \le 1) \\ 6 - x^2 & (\text{for } x > 1) \end{cases}$.

First, let $y = f(x)$ and then tabulate the necessary values. In evaluating $f(x)$, we must be careful to use the proper part of the definition. To see where to start the curve for $x > 1$, we evaluate $6 - x^2$ for $x = 1$, but we must realize that ***the curve does not include this point*** $(1, 5)$ and starts immediately to its right. To show that it is not part of the curve, draw it as an open circle. See Fig. 3.20.

A function such as this one, with a "break" in it, is called *discontinuous*.

$f(-2) = 2(-2) + 1 = -3$
$f(-1) = 2(-1) + 1 = -1$
$f(0) = 2(0) + 1 = 1$
$f(1) = 2(1) + 1 = 3$
$f(2) = 6 - 2^2 = 2$
$f(3) = 6 - 3^2 = -3$

x	y
−2	−3
−1	−1
0	1
1	3
2	2
3	−3

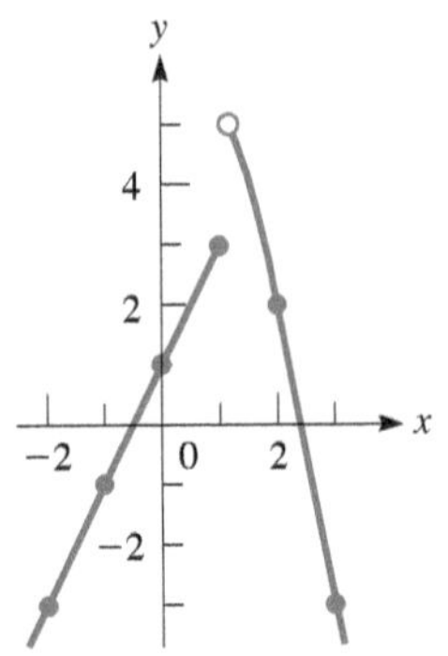

Fig. 3.20

Recall that a function assigns only one value of the dependent variable to each value of the independent variable, but a *relation* may assign more than one. The following test shows how to use a graph to determine whether or not a relation is also a function.

Vertical Line Test

The graph of a relation also represents a function if and only if **every vertical line crosses the graph at most once.**

EXAMPLE 9 Using the vertical line test

Use the vertical line test to determine whether the relation $y^2 = 4x^2$ represents a function.

Making the table as shown at the left, we can draw the graph in Fig. 3.21. We note that there are two values of y for every value of x, except $x = 0$. Since a vertical line crosses this graph twice, this relation is not a function (as we noted in Example 10 of Section 3.3).

x	y
−2	±4
−1	±2
0	0
1	±2
2	±4

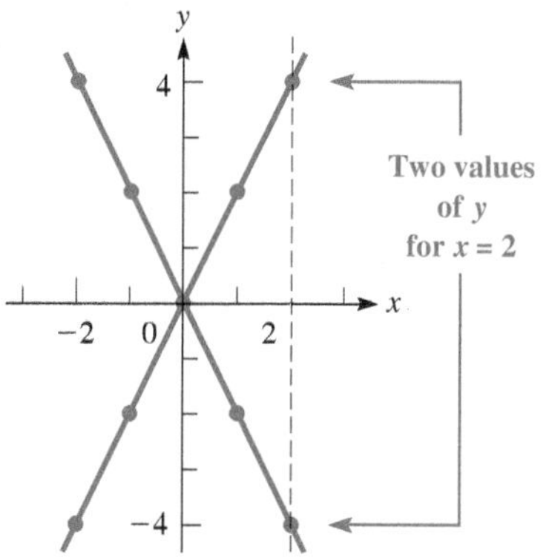

Fig. 3.21

Some functions have regular shapes that occur often enough that they are given a specific name. The four examples in Fig. 3.22 have appeared in the examples of this section:

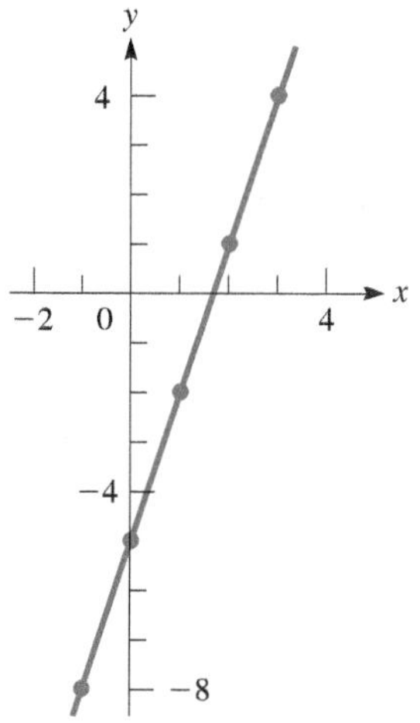

Fig. 3.22(a)

Straight lines (linear functions) will be studied in Chapter 5.

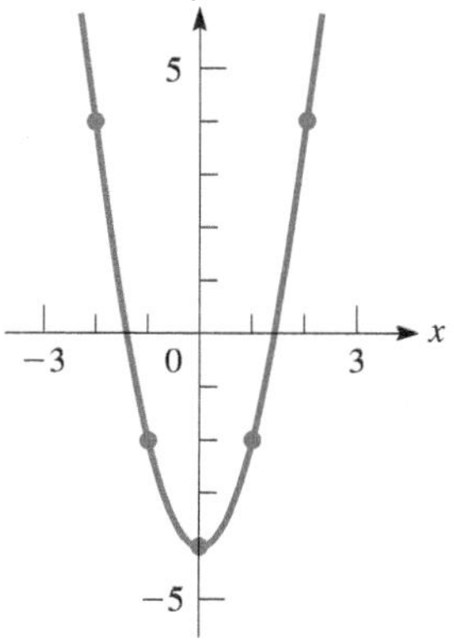

Fig. 3.22(b)

Parabolas (quadratic functions) will be studied in Chapter 7.

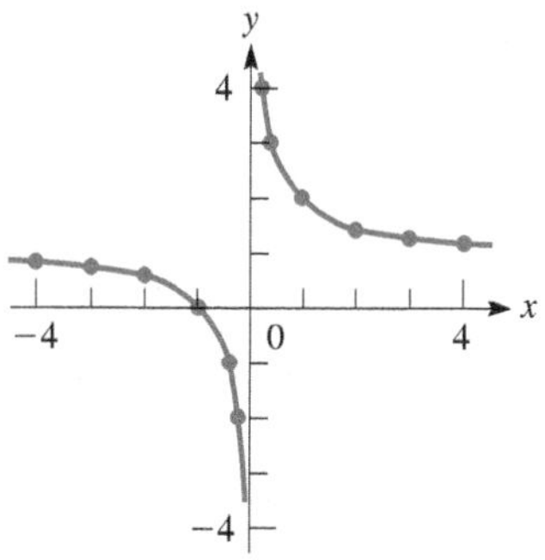

Fig. 3.22(c)

Hyperbolas will be studied in Chapter 21.

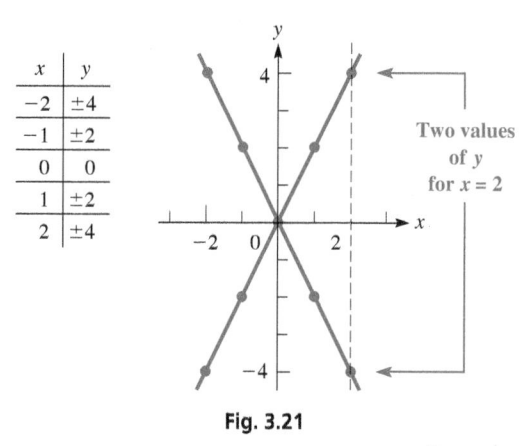

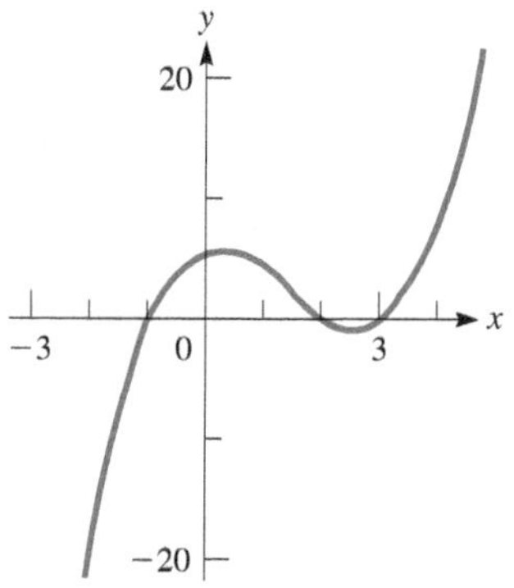

Fig. 3.22(d)

Cubic functions will be graphed in Chapter 24.

Other types of graphs are found in many of the later chapters.

EXERCISES 3.4

In Exercises 1–4, make the given changes in the indicated examples of this section and then plot the graphs.

1. In Example 1, change the − sign to +.
2. In Example 2, change $2x^2 - 4$ to $4 - 2x^2$.
3. In Example 4, change the x in the denominator to $x - 1$.
4. In Example 5, change the + sign to −.

In Exercises 5–36, graph the given functions.

5. $y = 3x$
6. $y = -2x$
7. $y = 2x - 4$
8. $y = 3x + 5$
9. $s = 7 - 2t$
10. $y = -3$
11. $y = \frac{1}{2}x - 2$
12. $A = 6 - \frac{1}{3}r$
13. $y = x^2$
14. $y = -2x^2$
15. $y = 3 - x^2$
16. $y = x^2 - 3$
17. $y = \frac{1}{2}x^2 + 2$
18. $y = 2x^2 + 1$
19. $y = x^2 + 2x$
20. $h = 20t - 5t^2$
21. $y = x^2 - 3x + 1$
22. $y = 2 + 3x + x^2$
23. $V = s^3$
24. $y = -2x^3$
25. $y = x^3 - x^2$
26. $L = 3e - e^3$
27. $D = v^4 - 4v^2$
28. $y = x^3 - x^4$
29. $P = \dfrac{1}{V} + 1$
30. $y = \dfrac{2}{x + 2}$
31. $y = \dfrac{4}{x^2}$
32. $p = \dfrac{1}{n^2 + 0.5}$
33. $y = \sqrt{x}$
34. $y = \sqrt{4 - x}$
35. $v = \sqrt{16 - h^2}$
36. $y = \sqrt{x^2 - 16}$

In Exercises 37–64, graph the indicated functions.

37. In blending gasoline, the number of litres n of 85-octane gas to be blended with m litres of 92-octane gas is given by the equation $n = 0.40m$. Plot n as a function of m.
38. The consumption of fuel c (in L/h) of a certain engine is determined as a function of the number r of revolutions per minute (r/min) of the engine, to be $c = 0.011r + 40$. This formula is valid for 500 r/min to 3000 r/min. Plot c as a function of r. (r is the symbol for revolution.)
39. For a certain model of truck, its resale value V (in dollars) as a function of the distance m it has been driven is $V = 50\,000 - 0.2m$. Plot V as a function of m for $m \leq 100\,000$ km.
40. The resistance R (in Ω) of a resistor as a function of the temperature T (in °C) is given by $R = 250(1 + 0.0032T)$. Plot R as a function of T.
41. The rate H (in W) at which heat is developed in the filament of an electric light bulb as a function of the electric current I (in A) is $H = 240I^2$. Plot H as a function of I.
42. The total annual fraction f of energy supplied by solar energy to a home as a function of the area A (in m²) of the solar collector is $f = 0.065\sqrt{A}$. Plot f as a function of A.
43. The maximum speed v (in km/h) at which a car can safely travel around a circular turn of radius r (in m) is given by $r = 0.55v^2$. Plot r as a function of v.
44. The height h (in m) of a rocket as a function of the time t (in s) is given by the function $h = 1500t - 4.9t^2$. Plot h as a function of t, assuming level terrain.
45. The power P (in W) that a certain windmill generates is given by $P = 0.004v^3$, where v is the wind speed (in km/h). Plot the graph of P versus v.
46. An astronaut weighs 750 N at sea level. The astronaut's weight at an altitude of x km above sea level is given by $w = 750\left(\dfrac{6400}{6400 + x}\right)^2$. Plot w as a function of x for $x = 0$ to $x = 8000$ km.
47. A formula used to determine the amount of lumber V (in dm³) that can be cut from a 2-m section of a log of diameter d (in mm) is $V = 0.0013d^2 - 0.043d$. Plot V as a function of d for values of d from 300 mm to 1000 mm.
48. A copper electrode with a mass of 25.0 g is placed in a solution of copper sulphate. An electric current is passed through the solution, and 1.6 g of copper is deposited on the electrode each hour. Express the total mass m on the electrode as a function of the time t and plot the graph.
49. A land developer is considering several options of dividing a large tract into rectangular building lots, many of which would have perimeters of 200 m. For these, the minimum width would be 30 m and the maximum width would be 70 m. Express the areas A of these lots as a function of their widths w and plot the graph.
50. The distance p (in m) from a camera with a 50-mm lens to the object being photographed is a function of the magnification m of the camera, given by $p = \dfrac{0.05(1 + m)}{m}$. Plot the graph for positive values of m up to 0.50.
51. A measure of the light beam that can be passed through an optic fibre is its numerical aperture N. For a particular optic fibre, N is a function of the index of refraction n of the glass in the fibre, given by $N = \sqrt{n^2 - 1.69}$. Plot the graph for $n \leq 2.00$.
52. $F = x^4 - 12x^3 + 46x^2 - 60x + 25$ is the force (in N) exerted by a cam on the arm of a robot, as shown in Fig. 3.23. Noting that x varies from 1 cm to 5 cm, plot the graph.

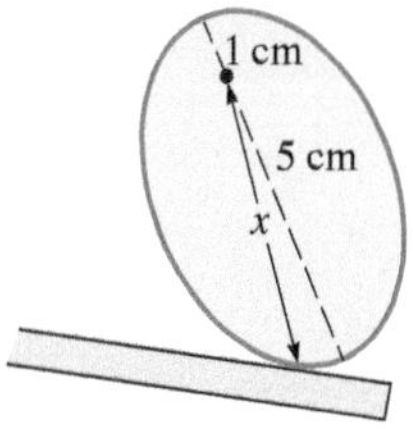

Fig. 3.23

53. The number of times S that a certain computer can perform a computation faster with a multiprocessor than with a uniprocessor is given by $S = \dfrac{5n}{4 + n}$, where n is the number of processors. Plot S as a function of n.

54. The voltage V (in volts) across a capacitor in a certain electric circuit for a 2-s interval is $V = 2t$ during the first second and $V = 4 - 2t$ during the second second. Here, t is the time (in s). Plot V as a function of t.

55. Given that the point (1, 2) is on the graph of $y = f(x)$, must it be true that $f(2) = 1$? Explain.

56. In Fig. 3.24, part of the graph of $y = 2x^2 + 0.5$ is shown. What is the area of the rectangle, if its vertex is on the curve, as shown?

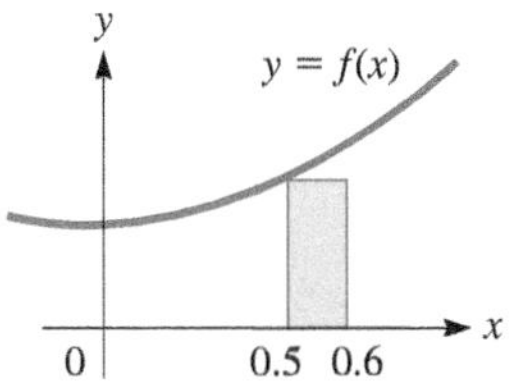

Fig. 3.24

57. On a taxable income of x dollars, a certain province's income tax T is defined as $T = 0.02x$ if $0 < x \le 20\,000$, and $T = 400 + 0.03(x - 20\,000)$ if $x > 20\,000$. Graph $T = f(x)$ for $0 \le x < 100\,000$.

58. The temperature T (in °C) recorded on a day during which a cold front passed through a city was $T = 2 + h$ for $6 < h < 14$, $T = 16 - 0.5h$ for $14 \le h < 20$, where h is the number of hours past midnight. Graph T as a function of h for $6 < h < 20$.

59. Plot the graphs of $y = x$ and $y = |x|$ on the same coordinate system. Explain why the graphs differ.

60. Plot the graphs of $y = 2 - x$ and $y = |2 - x|$ on the same coordinate system. Explain why the graphs differ.

61. Plot the graph of $f(x) = \begin{cases} 3 - x & (\text{for } x < 1) \\ x^2 + 1 & (\text{for } x \ge 1) \end{cases}$.

62. Plot the graph of $f(x) = \begin{cases} \dfrac{1}{x-1} & (\text{for } x < 0) \\ \sqrt{x+1} & (\text{for } x \ge 0) \end{cases}$.

63. Plot the graphs of (a) $y = x + 2$ and (b) $y = \dfrac{x^2 - 4}{x - 2}$. Explain the difference between the graphs.

64. Plot the graphs of (a) $y = x^2 - x + 1$ and (b) $y = \dfrac{x^3 + 1}{x + 1}$. Explain the difference between the graphs.

In Exercises 65–68, determine whether or not the indicated graph is that of a function.

65. Fig. 3.25(a)

66. Fig. 3.25(b)

67. Fig. 3.25(c)

68. Fig. 3.25(d)

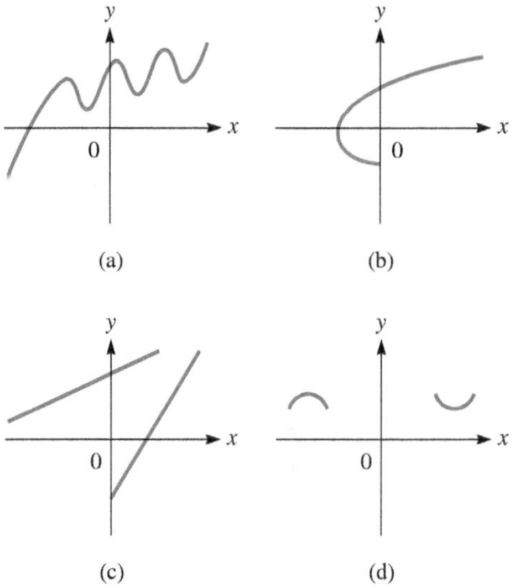

Fig. 3.25

Answers to Practice Exercises

1. $x = -5$ **2.** $x < 3$

3.5 More about Graphs

GRAPHS USING A GRAPHING UTILITY

A graphing utility (a graphing calculator, a computer algebra system, or any graphing software) can display a graph quickly and easily. To use a graphing utility effectively, *you must know the sequence of keys or instructions* for any operation you intend to use. The manual for any particular utility should be used for detailed coverage of its features.

When using a graphing utility to view a graph, the x- and y-display settings selected will dictate what portion of the graph is observed. Unless the settings are appropriate, you may not get a good view of the graph. With some settings, it is even possible that very little or no part of the graph can be seen. Therefore, it is necessary to be careful when choosing the settings. This includes the scale settings for axes, which should be chosen so that several (not too many or too few) axis markers are used.

SOLVING EQUATIONS GRAPHICALLY

An equation can be solved by using a graph. Most of the time, a graphical solution will be approximate. The use of technology can make it possible to get approximations with good accuracy. To do so, follow the procedure at the top of the next page.

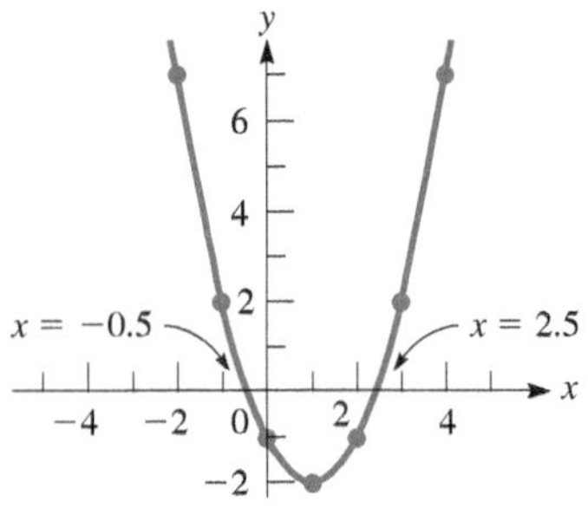

Fig. 3.26

<table>
<tr><th>Procedure for Solving an Equation Graphically</th><th>EXAMPLE 1</th></tr>
<tr><td></td><td>Graphically solve $x^2 - 2x = 1$.</td></tr>
<tr><td>1. Collect all terms on one side of the equal sign. This gives us $f(x) = 0$.</td><td>We rewrite as
$$x^2 - 2x - 1 = 0$$</td></tr>
<tr><td>2. Graph $y = f(x)$.</td><td>Setting $y = x^2 - 2x - 1$, we obtain the following table of values:
<table>
<tr><td>x</td><td>−2</td><td>−1</td><td>0</td><td>1</td><td>2</td><td>3</td><td>4</td></tr>
<tr><td>y</td><td>7</td><td>2</td><td>−1</td><td>−2</td><td>−1</td><td>2</td><td>7</td></tr>
</table>
The plot is shown in Fig. 3.26.</td></tr>
<tr><td>3. Find the points where the graph crosses the x-axis (where $y = 0$). These points are the x-intercepts of the graph.</td><td>As illustrated in Fig. 3.26, the graph crosses the x-axis twice. We estimate the x-values as
$$x = -0.5 \quad \text{and} \quad x = 2.5$$</td></tr>
<tr><td>4. The values of x for which $y = 0$ are the solutions to the equation. These are the zeros of the function $f(x)$.</td><td>The two solutions of the equation are approximately $x = -0.5$ and $x = 2.5$.</td></tr>
</table>

A graphing utility can be used to find the x-intercepts with a high level of accuracy. For instance, in the next example we use the features of a graphing calculator to solve the same equation as in Example 1.

■ Most calculators have an *intersect* feature. The equation in Example 2 can be solved using this feature and finding points where the curves $y = x^2 - 2x$ and $y = 1$ cross.

EXAMPLE 2 Graphing calculator solution of an equation

Using a graphing calculator, solve the equation $x^2 - 2x = 1$.

Following the above procedure, we set $y = x^2 - 2x - 1$. We then display this function on the calculator, as shown in Fig. 3.27(a). We can then use the *zero* feature to find the two values of x for which $y = 0$, which appear to be approximately $x \approx -0.5$ and $x \approx 2.5$.

The *zero* feature looks for the point in a given region that satisfies the equation algebraically, making the estimate as precise as what the calculator can handle. One must move the cursor to choose a left bound (to the left of the x-intercept), a right bound (to the right of the x-intercept), and a guess (near the x-intercept), pressing Enter after each entry. The calculator will then display the x-intercept. This process must be repeated twice, once for each x-intercept. The solutions, which are shown in Fig. 3.27(b) and (c), are $x = -0.414$ and $x = 2.414$ (rounded to the nearest thousandth). Note that these solutions are close to our original estimates.

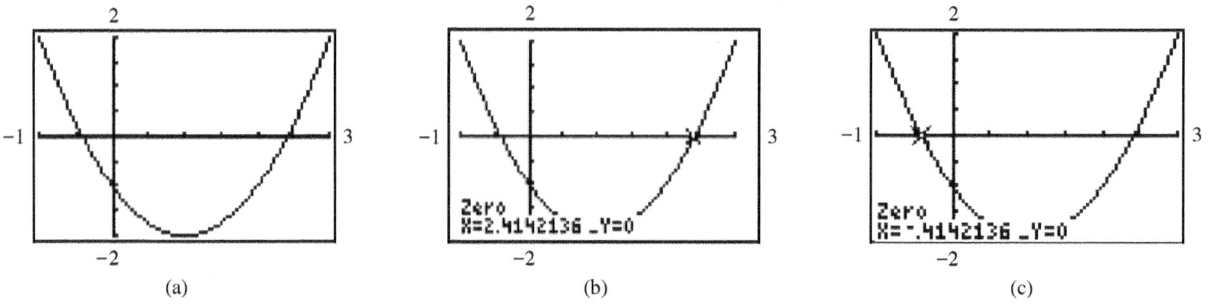

Fig. 3.27

EXAMPLE 3 Solving graphically—container dimensions

A rectangular box whose volume is 30.0 cm^3 is made with a square base and a height that is 2.00 cm less than the length of a side of the base. Find the dimensions of the box by first setting up the necessary equation and then solving it graphically.

Identify known quantities: The volume of the box is $V = 30.0$ cm^3.

Identify unknown quantities: The base is a square. Therefore, we let

$$x = \text{length of the base (in cm)}$$
$$x = \text{width of the base (in cm)}$$
$$x - 2 = \text{height of the box (in cm)}$$

Sketch, if possible: See Fig. 3.28(a)

Select an equation: We express volume as a function of x:

$$V = (x)(x)(x - 2) = x^3 - 2x^2$$

Substituting the known volume

$$x^3 - 2x^2 = 30.0$$

Solve: To solve graphically, first rewrite as $x^3 - 2x^2 - 30.0 = 0$ and then set $y = x^3 - 2x^2 - 30.0$. The graph of this function is shown in Fig. 3.28(b), where we use only positive values of x, because negative values have no meaning. The approximate solution obtained using a graphing utility is $x = 3.94$ cm. The dimensions of the box are 3.94 cm, 3.94 cm, and 1.94 cm.

Check: Checking, these dimensions give a volume of 30.1 cm^3. The difference from 30.0 cm^3 is due to rounding. A better check can be made by using the obtained value before rounding off. However, the final dimensions should be given only to three significant digits.

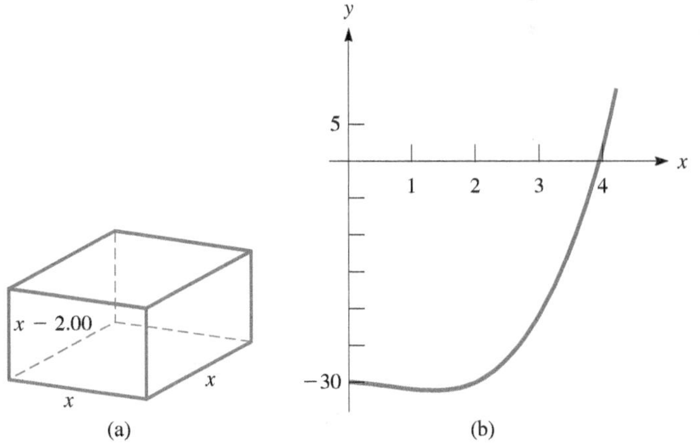

Fig. 3.28

FINDING THE DOMAIN AND RANGE OF A FUNCTION GRAPHICALLY

The graph of a function can be used to find its domain and range. The domain is the set of all possible input values on the x-axis (the horizontal variation in the graph), while the range consists of all possible output values on the y-axis (the vertical variation). If the graph of the function is not given, we can first use a graphing utility to obtain the graph, and then use the graph to find the domain and range.

EXAMPLE 4 Finding the domain and range graphically

For the functions graphed in blue in Fig. 3.29, the domain is shown in green and the range is shown in magenta. A solid dot indicates that the point is included, a hollow dot indicates that the point is not included, and an arrow shows that the graph continues indefinitely in that direction.

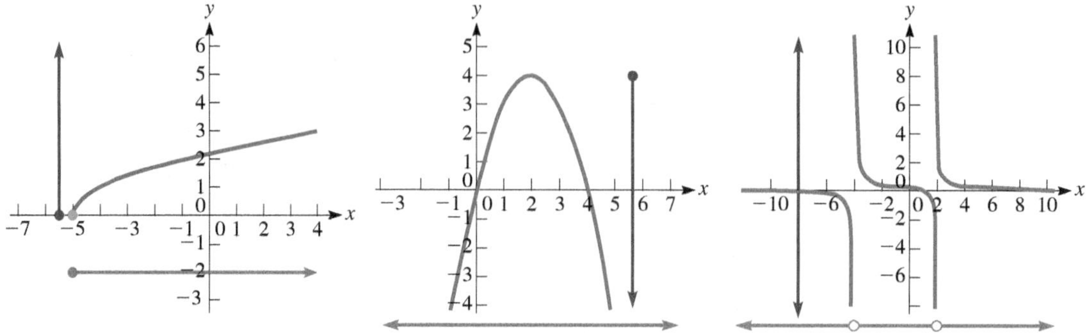

Domain: All values $x \geq -5$
Range: All values $y \geq 0$

Domain: All real numbers
Range: All values $y \leq 4$

Domain: All values x except -4 and 2
Range: All real numbers

Fig. 3.29

SHIFTING A GRAPH

By adding a constant to the right side of the function $y = f(x)$, the graph of the function is *shifted vertically*. If the added constant is positive, the graph is shifted up; if the added constant is negative, the graph is shifted down.

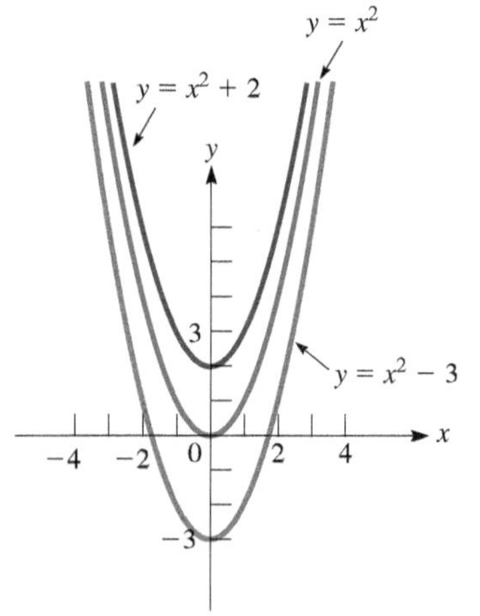

Fig. 3.30

EXAMPLE 5 Shifting a graph vertically (shift in *y*)

Fig. 3.30 shows the graphs of the functions $y = x^2$, $y = x^2 + 2$, and $y = x^2 - 3$. We see that adding 2 to the function $y = x^2$ shifts the graph up 2 units, whereas adding -3 shifts the graph down 3 units.

By adding a constant to x in the function $y = f(x)$, the graph of the function is shifted to the right or to the left. *Adding a positive constant to x shifts the graph to the* ***left****, and adding a negative constant to x shifts the graph to the* ***right****.*

COMMON ERROR

Because an added ***positive*** constant to the ***function*** shifts the graph of the function ***upward***, it is a common error to think that adding a positive constant to the variable of the function shifts the graph to the right (in the positive direction). However, adding a ***positive*** constant to the ***variable*** results in a shift to the ***left*** (to the negative direction), and adding a ***negative*** constant to the ***variable*** results in a shift to the ***right*** (to the positive direction).

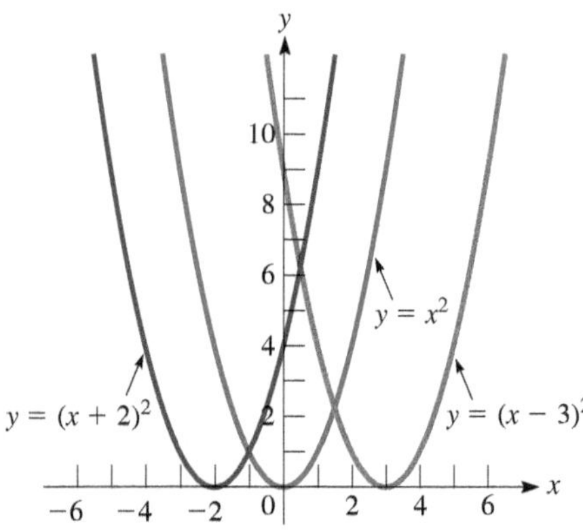

Fig. 3.31

EXAMPLE 6 Shifting a graph horizontally (shift in *x*)

For the function $y = x^2$, if we add 2 to x, we get $y = (x + 2)^2$, or if we add -3 to x, we get $y = (x - 3)^2$. The graphs of these three functions are shown in Fig. 3.31.

We see that the graph of $y = (x + 2)^2$ is 2 units to the *left* of $y = x^2$ and that the graph of $y = (x - 3)^2$ is 3 units to the *right* of $y = x^2$.

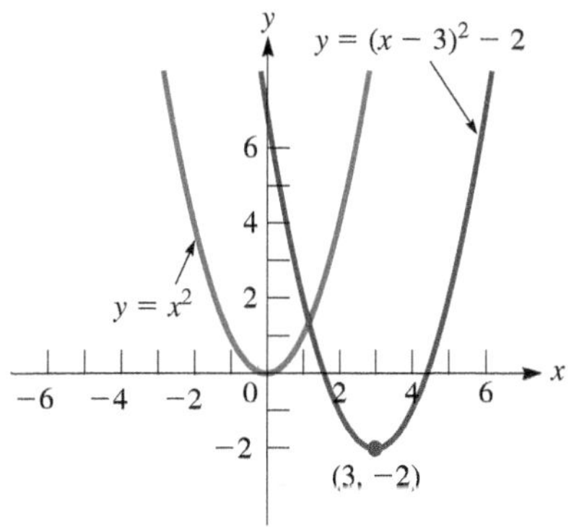

Fig. 3.32

Be very careful when shifting graphs horizontally. To check the direction and magnitude of a horizontal shift, find the value of x that makes the expression in parentheses equal to 0. For example, in Example 6, $y = (x - 3)^2$ is 3 units to the *right* of $y = x^2$ because $x = +3$ makes $x - 3$ equal to zero. The point (3, 0) on $y = (x - 3)^2$ is equivalent to the point (0, 0) on $y = x^2$. In the same way $x + 2 = 0$ for $x = -2$, and the graph of $y = (x + 2)^2$ is shifted 2 units to the *left* of $y = x^2$.

Summarizing how a graph is shifted, we have the following:

Shifting a Graph

Vertical shifts: $y = f(x) + k$ shifts the graph of $y = f(x)$ up k units if $k > 0$ and down k units if $k < 0$.

Horizontal shifts: $y = f(x + k)$ shifts the graph of $y = f(x)$ ***left*** k units if $k > 0$ and ***right*** k units if $k < 0$.

EXAMPLE 7 Shifting both vertically and horizontally

The graph of a function can be shifted both vertically and horizontally. To shift the graph of $y = x^2$ to the right three units and down two units, add -3 to x and add -2 to the resulting function. In this way, we get $y = (x - 3)^2 - 2$. The graph of this function and the graph of $y = x^2$ are shown in Fig. 3.32.

Practice Exercise

1. Describe the graph of $y = (x + 2)^2 + 3$ relative to the graph of $y = x^2$.

EXERCISES 3.5

In Exercises 1–4, make the indicated changes in the given examples of this section and then solve.

1. In Example 1, change the sign on the left side of the equation from − to +.
2. In Example 2, change the sign in front of $2x$ from − to +.
3. In Example 3, change 2.00 cm to 3.00 cm.
4. In Example 7, in the second line, change "to the right three units and down two units" to "to the left two units and down three units."

In Exercises 5–18, display the graphs of the given functions on a graphing utility. Use appropriate display settings.

5. $y = 3x - 1$
6. $y = 4 - 0.5x$
7. $y = x^2 - 4x$
8. $y = 8 - 2x^2$
9. $y = 6 - x^3$
10. $y = x^4 - 6x^2$
11. $y = x^4 - 2x^3 - 5$
12. $y = x^2 - 3x^5 + 3$
13. $y = \dfrac{2x}{x - 2}$
14. $y = \dfrac{3}{x^2 - 4}$
15. $y = x + \sqrt{x + 3}$
16. $y = \sqrt[3]{2x + 1}$
17. $y = 3 + \dfrac{2}{x}$
18. $y = \dfrac{x^2}{\sqrt{3 - x}}$

In Exercises 19–28, solve the given equations graphically to the nearest 0.001.

19. $x^2 + x - 5 = 0$
20. $v^2 - 2v - 4 = 0$
21. $x^3 - 3 = 3x$
22. $x^4 - 2x = 0$
23. $s^3 - 4s^2 = 6$
24. $3x^2 - x^4 = 2 + x$
25. $\sqrt{5R + 2} = 3$
26. $\sqrt{x} + 3x = 7$
27. $\dfrac{1}{x^2 + 1} = 0$
28. $T - 2 = \dfrac{1}{T}$

In Exercises 29–38, find the range of the given functions graphically. (The functions of Exercises 35–38 are the same as the functions of Exercises 15–18 of Section 3.3.)

29. $y = -4x^2 + 8x + 3$
30. $y = 3x^2 + 12x - 4$
31. $y = \dfrac{4}{x^2 - 4}$
32. $y = \dfrac{x + 1}{x^2}$
33. $y = \dfrac{x^2}{x + 1}$
34. $y = \dfrac{x}{x^2 - 4}$
35. $Y(y) = \dfrac{y + 1}{\sqrt{y - 2}}$
36. $f(n) = \dfrac{n}{6 - 2n}$
37. $f(D) = \dfrac{D}{D - 2} + \dfrac{4}{D + 4}$
38. $g(x) = \dfrac{\sqrt{x - 2}}{x - 3}$

In Exercises 39–46, a function and how it is to be shifted is given. Find the shifted function and then graph both functions on the same coordinate system.

39. $y = 3x$, up 1
40. $y = x^3$, down 2
41. $y = \sqrt{x}$, right 3
42. $y = \dfrac{2}{x}$, left 4
43. $y = -2x^2$, down 3, left 2
44. $y = -4x$, up 4, right 3
45. $y = \sqrt{2x + 1}$, up 1, left 1
46. $y = \sqrt{x^2 + 4}$, down 2, right 2

In Exercises 47–50, use the function for which the graph is shown in Fig. 3.33 to sketch graphs of the indicated functions.

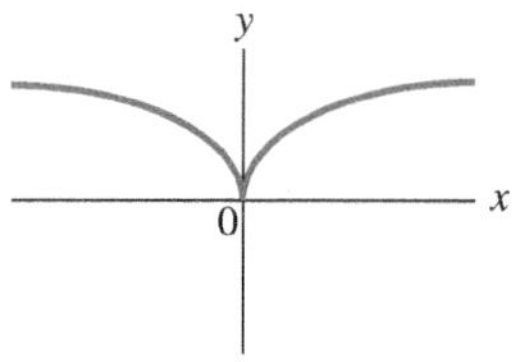

Fig. 3.33

47. $f(x + 2)$
48. $f(x - 2)$
49. $f(x - 2) + 2$
50. $f(x + 2) + 2$

In Exercises 51–58, solve the indicated equations graphically. Assume all data are accurate to three significant digits unless greater accuracy is given.

51. In an electric circuit, the current i (in A) as a function of voltage v is given by $i = 0.01v - 0.06$. Find v for $i = 0$.

52. For tax purposes, a corporation assumes that one of its computers depreciates according to the equation $V = 90\,000 - 12\,000t$, where V is the value (in dollars) of the computer after t years. According to this formula, when will the computer be fully depreciated (have no value)?

53. Two cubical coolers together hold 40.0 L (40 000 cm^3). If the inside edge of one is 5.00 cm greater than the inside edge of the other, what is the inside edge of each?

54. The height h (in m) of a rocket as a function of time t (in s) of flight is given by $h = 15 + 86t - 4.9t^2$. Determine when the rocket is at ground level.

55. The length of a rectangular solar panel is 12 cm more than its width. If its area is 520 cm^2, find its dimensions.

56. A computer model shows that the cost (in dollars) to remove x percent of a pollutant from a lake is $C = \dfrac{8000x}{100 - x}$. What percent can be removed for \$25 000?

57. In finding the illumination at a point x m from one of two light sources that are 100 m apart, it is necessary to solve the equation $9x^3 - 2400x^2 + 240\,000x - 8\,000\,000 = 0$. Find x.

58. A rectangular storage bin is to be made from a rectangular piece of sheet metal 12.0 cm by 10.0 cm by cutting out equal corners of side x and bending up the sides. See Fig. 3.34. Find x if the storage bin is to hold 90.0 cm^3.

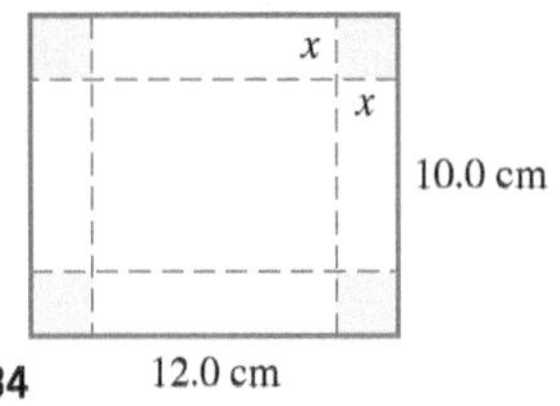

Fig. 3.34

In Exercises 59–66, solve the given problems.

59. In Example 3, assume the data are good to five significant digits and then find the dimensions of the box.

60. Explain how to show the graph of the *relation* $y^2 = 4x^2$ on a graphing utility, and then display it. See Example 9 in Section 3.4.

61. The cutting speed s (in cm/s) of a saw in cutting a particular type of metal piece is given by $s = \sqrt{t - 4t^2}$, where t is the time in seconds. What is the maximum cutting speed in this operation (to three significant digits)? (*Hint*: Find the range.)

62. Referring to Exercise 58, explain how to determine the maximum possible capacity for a storage bin constructed in this way. What is the maximum possible capacity (to three significant digits)?

63. A balloon is being blown up at a constant rate. (a) Sketch a reasonable graph of the radius of the balloon as a function of time. (b) Compare to a typical situation that can be described by $r = \sqrt[3]{3t}$, where r is the radius (in cm) and t is the time (in s).

64. A hot-water faucet is turned on. (a) Sketch a reasonable graph of the water temperature as a function of time. (b) Compare to a typical situation described by $T = \dfrac{t^3 + 80}{0.015t^3 + 4}$, where T is the water temperature (in °C) and t is the time (in s).

65. Display the graph of $y = cx^3$ with $c = -2$ and with $c = 2$. Describe the effect of the value of c.

66. Display the graph of $y = cx^4$, with $c = 4$ and with $c = 1/4$. Describe the effect of the value of c.

Answer to Practice Exercise

1. shifted two units to the left, three units up

3.6 Graphs of Functions Defined by Tables of Data

As we noted in Section 3.1, there are ways other than formulas to show functions. One important way to show the relationship between variables is by means of a table of values found by observation or from an experiment.

Statistical data often give values that are taken for certain intervals or are averaged over various intervals, and there is no meaning to the intervals *between* the points. Such points should be connected by straight-line segments only to make them stand out better and make the graph easier to read. See Example 1.

EXAMPLE 1 Graph using straight-line segments

The electric energy usage (in kW·h) for a certain all-electric house for each month of a year is shown in the following table. Plot these data.

Month	Jan	Feb	Mar	Apr	May	Jun
Energy Usage	10 504	12 363	10 168	7500	4825	3568
Month	July	Aug	Sep	Oct	Nov	Dec
Energy Usage	2548	2887	3301	5748	7302	9706

We know that there is no meaning to the intervals between the month categories, since we have the total number of kilowatt-hours for each month. Therefore, we use straight-line segments, but only to highlight changes in the function. See Fig. 3.35.

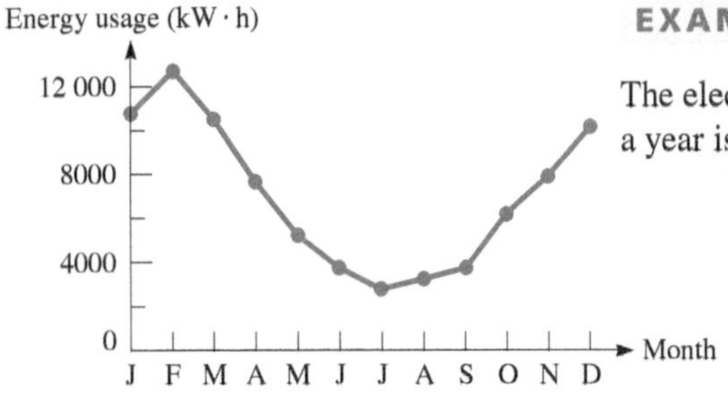

Fig. 3.35

■ Graphs such as the one shown in Fig. 3.35 can be obtained with a spreadsheet, or by using the statistical plotting features of a graphing utility.

Data from experiments in science and technology often indicate that the variables could have a formula relating them, although the formula may not be known. In this case, when plotting the graph, the points should be connected by a smooth curve.

EXAMPLE 2 Graph using a smooth curve

■ The Celsius degree is named for the Swedish astronomer Andres Celsius (1701–1744). He designated 100° as the freezing point of water and 0° as the boiling point. These were later reversed.

Steam in a boiler was heated to 150°C and then allowed to cool. Its temperature T (in °C) was recorded each minute, as shown in the following table. Plot the graph.

Time (min)	0.0	1.0	2.0	3.0	4.0	5.0
Temperature (°C)	150.0	142.8	138.5	135.2	132.7	130.8

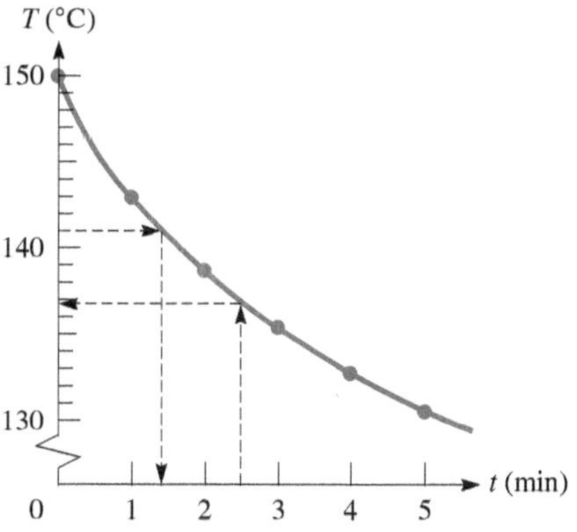

Fig. 3.36

Since the temperature changes in a continuous way, there is meaning to the values in the intervals between points. Therefore, these points are joined by a smooth curve, as in Fig. 3.36. Also, note that most of the data is far from the zero point on the Celsius scale, presented with the indicated break in the scale between 0 and 130.

The dotted arrows in the graph show how we can estimate the value of one variable for a given value of the other. This is called **reading a graph**. For example, after 2.5 min, the temperature is approximately 136.8°C (follow the vertical arrow at $t = 2.5$ to the graph and then horizontally to the T-axis). Also, if the temperature is known to be 141°C, then the time is approximately 1.4 min (follow the horizontal arrow at $T = 141$ to the graph and then vertically to the t-axis).

Note that we cannot read values between points on the graph in Example 1, since there is no real meaning to values between those listed.

■ In Chapter 22, we see how to find an equation that approximates the function relating the variables in a table of values, a process called curve-fitting.

LINEAR INTERPOLATION

In Example 2, we saw that we can estimate values from a graph. However, unless a very accurate graph is drawn with expanded scales for both variables, only very approximate values can be found. There is a method, called *linear interpolation*, that uses the table itself to get more accurate results.

■ Before the use of electronic calculators, interpolation was used extensively in mathematics textbooks for finding values from mathematics tables. It is still of use when using scientific, technical, and statistical tables.

Linear interpolation *assumes that the curve segment connecting two points can be approximated by the straight line connecting the two points.* This provides a reasonable approximation if the values in the table are sufficiently close together.

EXAMPLE 3 Linear interpolation of values

$t_1 = 1.0$ min	$t = 1.4$ min	$t_2 = 2.0$ min
$T_1 = 142.8$ °C	$T = ?$	$T_2 = 138.5$ °C

For the cooling steam in Example 2, we can use interpolation to find its temperature after 1.4 min. The value of $t = 1.4$ min is between $t_1 = 1.0$ min and $t_2 = 2.0$ min, so we will estimate T on the line segment connecting $(1.0, 142.8)$ and $(2.0, 138.5)$. See the table in Fig. 3.37.

The value of $t = 1.4$ is 0.4 of the way between t_1 and t_2, because

$$\frac{t - t_1}{t_2 - t_1} = \frac{1.4 - 1.0}{2.0 - 1.0} = 0.4$$

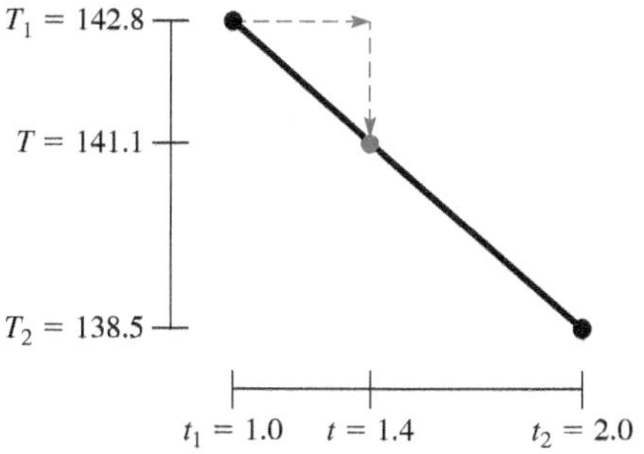

Fig. 3.37

We can then assume that the desired temperature T is 0.4 of the way between T_1 and T_2, so

$$\frac{T - T_1}{T_2 - T_1} = \frac{T - 142.8}{138.5 - 142.8} = 0.4$$

Solving for T,

$$T - 142.8 = 0.4(138.5 - 142.8)$$
$$T = -1.72 + 142.8 = 141.08$$

Rounding off, we would estimate that the required value of T is about 141.1 °C. See Fig. 3.37.

Another method of indicating the interpolation is shown in Fig. 3.38. From the figure, we have the proportion

$t_1 = 1.0$ min	$t = 1.4$ min	$t_2 = 2.0$ min
$T_1 = 142.8$ °C	$T = ?$	$T_2 = 138.5$ °C

$$\frac{0.4}{1.0} = \frac{x}{-4.3}$$

$$x = -1.7 \quad \text{rounded off}$$

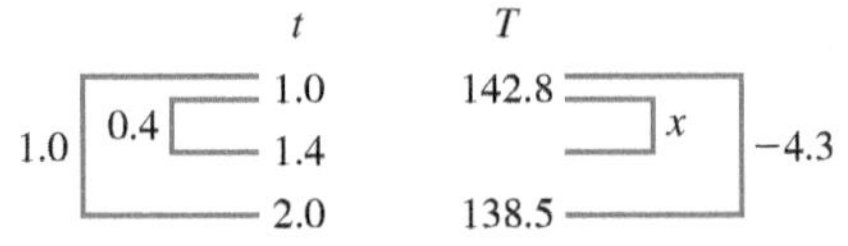

Fig. 3.38

Therefore,

$$142.8 + (-1.7) = 141.1°\text{C}$$

is the required value of T. If the values of T had been increasing, we would have added 1.7 to the value of T for 1.0 min.

Practice Exercise

1. As in Example 3, use the data from Example 2 and interpolate to find T for $t = 4.2$ min.

EXERCISES 3.6

In Exercises 1–8, represent the data graphically.

1. The diesel fuel production (in 1000s of litres) at a certain refinery during an 8-wk period was as follows:

Week	1	2	3	4	5	6	7	8
Production	765	780	840	850	880	840	760	820

2. The *annual average exchange rate* for the number of Canadian dollars equal to one U.S. dollar for 2009–2016 is as follows:

Year	2009	2010	2011	2012	2013	2014	2015	2016
Can. dol.	1.14	1.03	0.99	1.00	1.03	1.10	1.27	1.33

3. The amount of material necessary to make a cylindrical litre container depends on the diameter, as shown in this table:

Diameter (cm)	6.0	8.0	10.0	12.0	14.0	16.0
Material (cm²)	723	601	557	560	594	652

4. An oil burner propels air that has been heated to 90°C. The temperature then drops as the distance from the burner increases, as shown in the following table:

Distance (m)	0.0	1.0	2.0	3.0	4.0	5.0	6.0
Temperature (°C)	90	84	76	66	54	46	41

5. A changing electric current in a coil of wire will induce a voltage in a nearby coil. Important in the design of transformers, the effect is called *mutual inductance*. For two coils, the mutual inductance (in H) as a function of the distance between them is given in the following table:

Distance (cm)	0.0	2.0	4.0	6.0	8.0	10.0	12.0
M. ind. (H)	0.77	0.75	0.61	0.49	0.38	0.25	0.17

6. The temperatures felt by the body as a result of the *wind-chill factor* for an outside temperature of −5°C (as determined by Environment Canada) are given in the following table:

Wind speed (km/h)	10	20	30	40	50	60	70
Temp. felt (°C)	−9	−12	−13	−14	−15	−16	−16

7. The time required for a sum of money to double in value, when compounded annually, is given as a function of the interest rate in the following table:

Rate (%)	4	5	6	7	8	9	10
Time (years)	17.7	14.2	11.9	10.2	9.0	8.0	7.3

8. The torque T of an engine, as a function of the frequency f of rotation, was measured as follows:

f (r/min)	500	1000	1500	2000	2500	3000	3500
T (N · m)	175	90	62	45	34	31	27

In Exercises 9–12, use the graph in Fig. 3.36, which relates the temperature of cooling steam and the time. Find the indicated values by reading the graph.

9. For $t = 4.3$ min, find T.
10. For $t = 1.8$ min, find T.
11. For $T = 145.0$°C, find t.
12. For $T = 133.5$°C, find t.

In Exercises 13 and 14, use the following table, which gives the valve lift L (in mm) of a certain cam as a function of the angle θ (in degrees) through which the cam is turned. Plot the values. Find the indicated values by reading the graph.

θ (°)	0	20	40	60	80	100	120	140
L (mm)	0	1.2	2.3	3.3	3.8	3.0	1.6	0

13. (a) For $\theta = 25°$, find L. (b) For $\theta = 96°$, find L.
14. For $L = 2.0$ mm, find θ.

In Exercises 15–18, find the indicated values by means of linear interpolation.

15. In Exercise 5, find the inductance for $d = 9.2$ cm.
16. In Exercise 6, find the temperature for $s = 12$ km/h.
17. In Exercise 7, find the rate for $t = 10.0$ years.
18. In Exercise 8, find the torque for $t = 2300$ r/min.

In Exercises 19–22, use the following table that gives the rate R of discharge from a tank of water as a function of the height H of water in the tank. For Exercises 19 and 20, plot the graph and find the values from the graph. For Exercises 21 and 22, find the indicated values by linear interpolation.

Height (cm)	0	50	100	200	300	400	600
Rate (m^3/s)	0	1.0	1.5	2.2	2.7	3.1	3.5

19. (a) for $R = 2.0\ m^3/s$, find H. (b) For $H = 240$ cm, find R.

20. (a) For $R = 3.4\ m^3/s$, find H. (b) For $H = 320$ cm, find R.

21. Find R for $H = 80$ cm.

22. Find H for $R = 2.5\ m^3/s$.

In Exercises 23–26, use the following table, which gives the fraction (as a decimal) of the total heating load of a certain system that will be supplied by a solar collector of area A (in m²). Find the indicated values by linear interpolation.

f	0.22	0.30	0.37	0.44	0.50	0.56	0.61
A (m^2)	20	30	40	50	60	70	80

23. For $A = 36\ m^2$, find f.

24. For $A = 52\ m^2$, find f.

25. For $f = 0.59$, find A.

26. For $f = 0.27$, find A.

In Exercises 27–30, a method of finding values beyond those given is considered. By using a straight-line segment to extend a graph beyond the last known point, we can estimate values from the extension of the graph. The method is known as **linear extrapolation**. *Use this method to estimate the required values from the given graphs.*

27. Using Fig. 3.36, estimate T for $t = 5.3$ min.

28. Using the graph for Exercise 7, estimate t for $R = 10.4\%$.

29. Using the graph for Exercises 19–22, estimate R for $H = 700$ cm.

30. Using the graph for Exercises 19–22, estimate R for $H = 800$ cm.

Answer to Practice Exercise

1. $T = 132.3°C$

CHAPTER 3 REVIEW EXERCISES

In Exercises 1–4, determine the appropriate function.

1. The radius of a circular water wave increases at the rate of 2 m/s. Express the area of the circle as a function of the time t (in s).

2. A conical sheet-metal hood is to cover an area 6 m in diameter. Find the total surface area A of the hood as a function of its height h.

3. A person invests x dollars at 5% APR (annual percentage rate) and y dollars at 4% APR. If the total annual income is $2000, solve for y as a function of x.

4. Fencing around a rectangular storage depot area costs twice as much along the front as along the other three sides. The back costs $10 per metre. Express the cost C of the fencing as a function of the width w if the length (along the front) is 20 m longer than the width.

In Exercises 5–12, evaluate the given functions.

5. $f(x) = 7x - 5$; find $f(3)$ and $f(-6)$.

6. $g(I) = 8 - 3I$; find $g\left(\frac{1}{6}\right)$ and $g(-4)$.

7. $H(h) = \sqrt{1 - 2h}$; find $H(-4)$ and $H(2h)$.

8. $\phi(v) = \dfrac{3v - 2}{|v + 1|}$; find $\phi(-2)$ and $\phi(v + 1)$.

9. $F(x) = x^3 + 2x^2 - 3x$; find $F(3 + h) - F(3)$.

10. $f(x) = 3x^2 - 2x + 4$; find $\dfrac{f(x + h) - f(x)}{h}$.

11. $f(x) = 3 - 2x$; find $f(2x) - 2f(x)$.

12. $f(x) = 1 - x^2$; find $[f(x)]^2 - f(x^2)$.

In Exercises 13–16, evaluate the given functions. Values of the independent variable are approximate.

13. $f(x) = 8.07 - 2x$; find $f(5.87)$ and $f(-4.29)$.

14. $g(x) = 7x - x^2$; find $g(45.81)$ and $g(-21.85)$.

15. $G(S) = \dfrac{S - 0.087\,629}{3.0125S}$; find $G(0.174\,27)$ and $G(0.053\,206)$.

16. $h(t) = \dfrac{t^2 - 4t}{t^3 + 564}$; find $h(8.91)$ and $h(-4.91)$.

In Exercises 17–22, determine the domain and the range of the given functions.

17. $f(x) = x^4 + 1$

18. $G(z) = \dfrac{4}{z^3}$

19. $g(t) = \dfrac{2}{\sqrt{t + 4}}$

20. $F(y) = 1 - 2\sqrt{y}$

21. $f(n) = 1 + \dfrac{2}{(n - 5)^2}$

22. $F(x) = 3 - |x|$

In Exercises 23–32, plot the graphs of the given functions.

23. $y = 4x + 2$

24. $y = 5x - 10$

25. $s = 4t - t^2$

26. $y = x^2 - 8x - 5$

27. $y = 3 - x - 2x^2$

28. $V = 3 - 0.5s^3$

29. $A = 2 - s^4$

30. $y = x^4 - 4x$

31. $y = \dfrac{x}{x + 1}$

32. $Z = \sqrt{25 - 2R^2}$

In Exercises 33–40, solve the given equations graphically to the nearest 0.01.

33. $7x - 3 = 0$

34. $3x + 11 = 0$

35. $x^2 + 1 = 6x$

36. $3t - 2 = t^2$

37. $x^3 - x^2 = 2 - x$

38. $5 - x^3 = 2x^2$

39. $\dfrac{1}{v} = 2v$

40. $\sqrt{x} = 2x - 1$

In Exercises 41–44, find the range of the given function graphically.

41. $y = x^4 - 5x^2$

42. $y = x\sqrt{4 - x^2}$

43. $A = w + \dfrac{2}{w}$

44. $y = 2x + \dfrac{3}{\sqrt{x}}$

In Exercises 45–62, solve the given problems.

45. Explain how $A(a, b)$ and $B(b, a)$ may be in different quadrants.

46. Determine the distance from the origin to the point (a, b).

47. Two vertices of an equilateral triangle are $(0, 0)$ and $(2, 0)$. What is the third vertex?

48. The points (1, 2) and $(1, -3)$ are two adjacent vertices of a square. Find the other vertices.

49. Where are all points for which $|y/x| > 0$?

50. Describe the values of x and y for which $(1, -2)$, $(-1, -2)$, and (x, y) are on the same straight line.

51. Sketch the graph of a function for which the domain is $0 \le x \le 4$ and the range is $1 \le y \le 3$.

52. Sketch the graph of a function for which the domain is all values of x and the range is $-2 < y < 2$.

53. If the function $y = \sqrt{x - 1}$ is shifted left 2 and up 1, what is the resulting function?

54. If the function $y = 3 - 2x$ is shifted right 1 and down 3, what is the resulting function?

55. For $f(x) = 2x^3 - 3$, display the graphs of $f(x)$ and $f(-x)$ on a graphing utility. Describe the graphs in relation to the y-axis.

56. For $f(x) = 2x^3 - 3$, display the graphs of $f(x)$ and $-f(x)$ on a graphing utility. Describe the graphs in relation to the x-axis.

57. Is it possible that the points (2, 3), $(5, -1)$, and $(-1, 3)$ are all on the graph of the same function? Explain.

58. For the functions $f(x) = |x|$ and $g(x) = \sqrt{x^2}$, use a graphing utility to determine whether or not $f(x) = g(x)$ for all real x.

59. An equation used in electronics with a transformer antenna is $I = 12.5\sqrt{1 + 0.5m^2}$. For $I = f(m)$, find $f(0.550)$.

60. The percent p of wood lost in cutting it into boards 38.0 mm thick due to the thickness t (in mm) of the saw blade is $p = \dfrac{100t}{t + 38}$. Find p if $t = 5.00$ mm. That is, since $p = f(t)$, find $f(5.00)$.

61. The angle A (in degrees) of a robot arm with the horizontal as a function of time t (for 0.0 s to 6.0 s) is given by $A = 8.0 + 12t^2 - 2.0t^3$. What is the greatest value of A to the nearest 0.1°? See Fig. 3.39. (*Hint*: Find the range.)

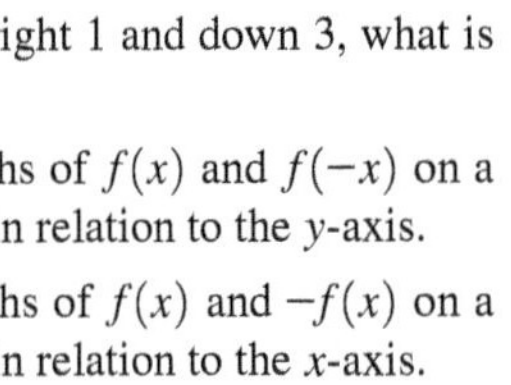

Fig. 3.39

62. The electric power P (in W) produced by a certain battery is $P = \dfrac{24R}{R^2 + 1.40R + 0.49}$, where R is the resistance (in Ω) in the circuit. What is the maximum power produced? (*Hint*: Find the range.) See Example 7 in Section 3.4.

In Exercises 63–78, plot the graphs of the indicated functions.

63. When El Niño, a Pacific Ocean current, moves east and warms the water off South America, weather patterns in many parts of the world change significantly. Special buoys along the equator in the Pacific Ocean send data via satellite to monitoring stations. If the temperature T (in °C) at one of these buoys is $T = 28.0 + 0.15t$, where t is the time in weeks between Jan. 1 and Aug. 1 (30 weeks), plot the graph of $T = f(t)$.

64. The change C (in cm) in the length of a 100-m steel bridge girder from its length at 10°C as a function of the temperature T is given by $C = 0.12(T - 10)$. Plot the graph for $T = -20$°C to $T = 40$°C.

65. The profit P (in dollars) a retailer makes in selling 50 cell phones is given by $P = 50(p - 50)$, where p is the selling price. Plot P as a function of p for $p = \$30$ to $p = \$150$.

66. There are 500 L of oil in a tank that has the capacity of 100 000 L. It is filled at the rate of 7000 L/h. Determine the function relating the number of litres N and the time t while the tank is being filled. Plot N as a function of t.

67. The length L (in cm) of a pulley belt is 12 cm longer than the circumference of one of the pulley wheels. Express L as a function of the radius r of the wheel and plot L as a function of r.

68. The pressure loss P (in kPa per 100 m) in a fire hose is given by $P = 0.000\,12Q^2 + 0.0055Q$, where Q is the rate of flow (in L/min). Plot the graph of P as a function of Q.

69. A *thermograph* measures the infrared radiation from each small area of a person's skin. Since the skin over a tumour radiates more than skin from nearby areas, a thermograph can help detect cancer cells. The emissivity ε (in %) of radiation as a function of skin temperature T (in K) is $\varepsilon = f(T) = 100(T^4 - 307^4)/307^4$, if nearby skin is at 34°C (307K). Find ε for $T = 309$K.

70. If an amount A is due on a certain credit card, the minimum payment p is A, if A is \$20 or less, or \$20 plus 10% of any amount over \$20 that is owed. Find $p = f(A)$, and graph this function.

71. For a certain laser device, the laser output power P (in mW) is negligible if the drive current i is less than 80 mA. From 80 mA to 140 mA, $P = 1.5 \times 10^{-6}i^3 - 0.77$. Plot the graph of $P = f(i)$.

72. It is determined that a good approximation for the cost C (in cents/km) of operating a certain car at a constant speed v (in km/h) is given by $C = 0.025v^2 - 1.4v + 35$. Plot C as a function of v for $v = 10$ km/h to $v = 60$ km/h.

73. A medical researcher exposed a virus culture to an experimental vaccine. It was observed that the number of live cells N in the culture as a function of the time t (in h) after exposure was given by $N = \dfrac{1000}{\sqrt{t + 1}}$. Plot the graph of $N = f(t)$.

74. The electric field E (in V/m) from a certain electric charge is given by $E = 25/r^2$, where r is the distance (in m) from the charge. Plot the graph of $E = f(r)$ for values of r up to 10 cm.

75. To draw the approximate shape of an irregular shoreline, a surveyor measured the distances d from a straight wall to the shoreline at 20-m intervals along the wall, as shown in the following table. Plot the graph of distance d as a function of the distance D along the wall.

D (m)	0	20	40	60	80	100	120	140	160
d (m)	15	32	56	33	29	47	68	31	52

76. The percent p of a computer network that is in use during a particular loading cycle as a function of the time t (in s) is given in the following table. Plot the graph of $p = f(t)$.

t (s)	0.0	0.2	0.4	0.6	0.8	1.0	1.2	1.4	1.6
P (%)	0	45	85	90	85	85	60	10	0

77. The vertical sag s (in m) at the middle of an 800-m power line as a function of the temperature T (in °C) is given in the following table. See Fig. 3.40. For the function $s = f(T)$, find $f(14)$ by linear interpolation.

T (°C)	−10	0	10	20
s(m)	3.11	3.23	3.38	3.57

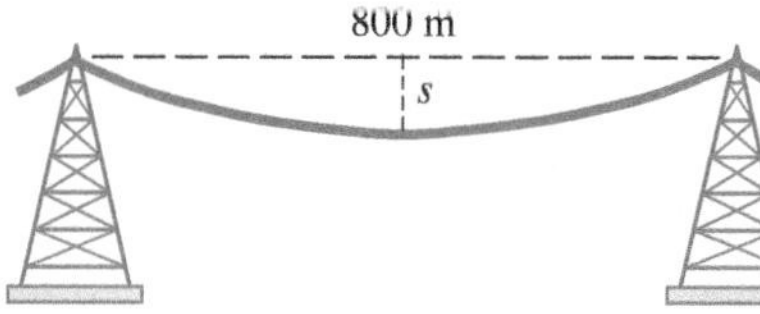

Fig. 3.40

78. In an experiment measuring the pressure p (in kPa) at a given depth d (in m) of seawater, the results in the following table were found. Plot the graph of $p = f(d)$ and from the graph determine $f(10)$.

d (m)	0.0	3.0	6.0	9.0	12	15
p (kPa)	101	131	161	193	225	256

In Exercises 79–87, solve the indicated equations graphically.

79. A person 350 km from home starts toward home and travels at 90 km/h for the first 2.0 h and then slows down to 60 km/h for the rest of the trip. How long does it take the person to be 80 km from home?

80. One industrial cleaner contains 30% of a certain solvent, and another contains 10% of the solvent. To get a mixture containing 50 L of the solvent, 120 L of the first cleaner is used. How much of the second must be used?

81. The solubility s (in kg/m^3 of water) of a certain type of fertilizer is given by $s = 135 + 4.9T + 0.19T^2$, where T is the temperature (in °C). Find T for $s = 500$ kg/m^3.

82. A 2.00-L (2000-cm^3) metal container is to be made in the shape of a right circular cylinder. Express the total area A of metal necessary as a function of the radius r of the base. Then find A for $r = 6.00$ cm, 7.00 cm, and 8.00 cm.

83. In an oil pipeline, the velocity v (in m/s) of the oil as a function of the distance x (in m) from the wall of the pipe is given by $v = 9.6x - 7.5x^2$. Find x for $v = 2.6$ m/s. The diameter of the pipe is 1.20 m.

84. One ball bearing is 1.00 mm more in radius and has twice the volume of another ball bearing. What is the radius of each?

85. A computer, using data from a refrigeration plant, estimates that in the event of a power failure, the temperature (in °C) in the freezers would be given by $T = \frac{4t^2}{t+2} - 20$, where t is the number of hours after the power failure. How long would it take for the temperature to reach 0°C?

86. Two electrical resistors in parallel (see Fig. 3.41) have a combined resistance R_T given by $R_T = \frac{R_1 R_2}{R_1 + R_2}$. If $R_2 = R_1 + 2.00$, express R_T as a function of R_1 and find R_1 if $R_T = 6.00\ \Omega$.

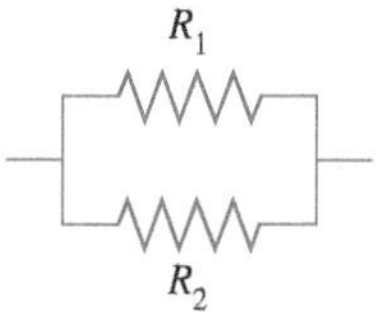

Fig. 3.41

87. Find the inner surface area A of a cylindrical 250.0-cm^3 cup as a function of the radius r of the base. Then if $A = 175.0$ cm^2, solve for r. Write one or two paragraphs explaining your method for setting up the function, and how you used a graphing utility to solve for r with the given value of A.

CHAPTER 3 PRACTICE TEST

1. Given $f(x) = 2x - x^2 + \frac{8}{x}$, find $f(-4)$ and $f(x - 4)$.

2. A rocket has a mass of 2000 Mg at liftoff. If the first-stage engines burn fuel at the rate of 10 Mg/s, find the mass m of the rocket as a function of the time t (in s) while the first-stage engines operate.

3. Plot the graph of the function $f(x) = 4 - 2x$.

4. Solve the equation $2x^2 - 3 = 3x$ graphically to the nearest 0.1.

5. Plot the graph of the function $y = \sqrt{4 + 2x}$.

6. Locate all points (x, y) for which $x < 0$ and $y = 0$.

7. Find the domain and the range of the function $f(x) = \sqrt{6 - x}$.

8. If the function $y = 2x^2 - 3$ is shifted right 1 and up 3, what is the resulting function?

9. Find the range of the function $y = \frac{x^2 + 2}{x + 2}$ graphically.

10. A window has the shape of a semicircle over a square, as shown in Fig. 3.42. Express the area of the window as a function of the radius of the circular part.

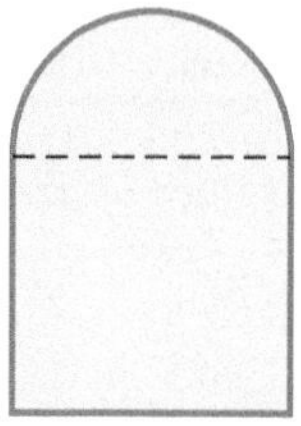

Fig. 3.42

11. The voltage V and current i (in mA) for a certain electrical experiment were measured as shown in the following table. Plot the graph of $i = f(V)$ and from the graph find $f(45.0)$.

Voltage (V)	10.0	20.0	30.0	40.0	50.0	60.0
Current (mA)	145	188	220	255	285	315

12. From the table in Problem 11, find the voltage for $i = 200$ mA.

4. The Trigonometric Functions

LEARNING OUTCOMES

After completion of this chapter, the student should be able to:

- Define positive angle and negative angle
- Express an angle in degrees or radians and convert between the two measurements
- Determine a standard position angle
- Define, calculate, and use the six trigonometric functions
- Use the Pythagorean theorem and trigonometric functions to solve a right triangle
- Employ the inverse trigonometric functions to solve for a missing angle
- Solve application problems involving right triangles

Scoast/iStock/Getty Images

Thinkstock Images/Stockbyte/Getty Images

▲ Using trigonometry, it is often possible to calculate distances that may not be directly measured. In Section 4.5, we show how we can estimate the height of Horseshoe Falls on the Canadian side of Niagara Falls. Also in Section 4.5, using a method similar to one used by Eratosthenes, the circumference and radius of Earth will be estimated.

One of the earliest uses of measurements with triangles was performed by the Greek mathematician and astronomer Eratosthenes (276–194 B.C.E.) while trying to determine the circumference and radius of Earth. Using the length of a cast shadow and the approximate distance between two cities, Eratosthenes was able to estimate the circumference and radius of a presumably spherical Earth. A few decades later, the Greek astronomer and geographer Hipparchus (190–120 B.C.E.) attempted to determine the distance between Earth and the moon. To complete this task, he expanded upon earlier mathematical works by collecting and developing various mathematical techniques that involve triangles into a coherent field of study now known as **trigonometry**.

Trigonometry involves measuring the sides and angles of triangles and determining and using the relationships between those quantities. These relationships are used in geometric problems and applications involving triangles directly, such as navigation, surveying, structural design, and astronomy. Further developments in mathematics have shown how trigonometric relationships are valuable in applications in which an actual triangle is not involved, including electric circuits, analysis of vibrations, acoustics, optics, seismology, and mechanical engineering, to name but a few. Because trigonometry has a great number of applications in so many areas, it is considered one of the most practical branches of mathematics.

In this chapter, we introduce the basic trigonometric functions and study applications that make use of *right triangles*. These techniques will provide the foundational skills required for the study of other types of triangles in later chapters.

4.1 Angles

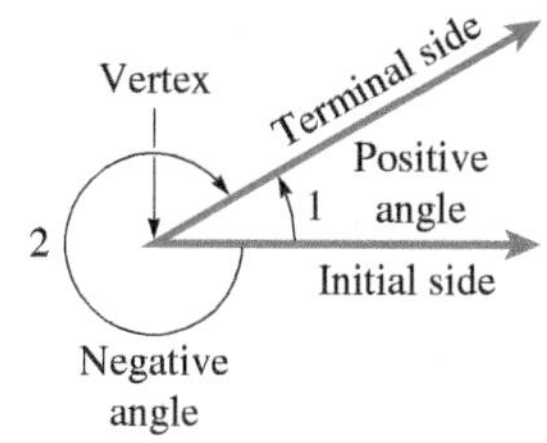

Fig. 4.1

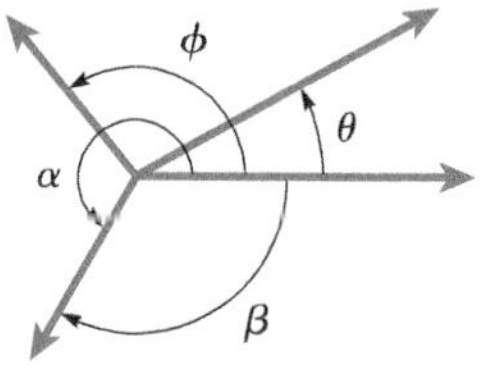

Fig. 4.2

■ The division of a degree (1/360 of a rotation) into 60 minutes of arc and each minute into 60 seconds of arc comes from the ancient Sumerians, who employed a sexagesimal (base-60) counting system. The Babylonians extended this counting system to divide the hours of a day into 60 minutes of time, with each minute containing 60 seconds of time.

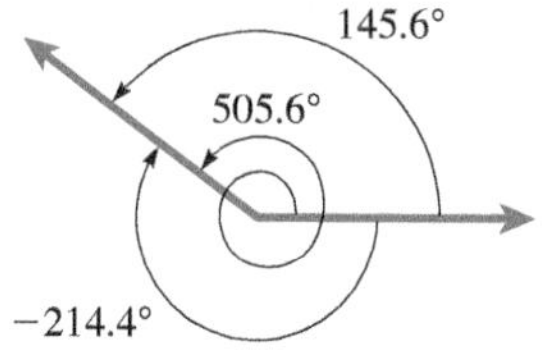

Fig. 4.3

In Chapter 2, we gave a basic definition of an *angle*. In this section, we extend this definition and also give some other important definitions related to angles.

An **angle** *is generated by rotating a ray about its fixed endpoint from an* ***initial position*** *to a* ***terminal position****. The initial position is called the* **initial side** *of the angle, the terminal position is called the* **terminal side**, *and the fixed endpoint is the* **vertex**. The angle itself is *the amount of rotation* from the initial side to the terminal side.

If the rotation of the terminal side from the initial side is ***counterclockwise****, the angle is defined as* ***positive***. If the rotation is ***clockwise***, the angle is ***negative***. In Fig. 4.1, $\angle 1$ is positive and $\angle 2$ is negative.

Many symbols are used to designate angles. Among the most widely used are certain Greek letters such as θ (theta), ϕ (phi), α (alpha), and β (beta). Capital letters representing the vertex (e.g., $\angle A$ or simply A) and other literal symbols, such as x and y, are also used commonly.

In Fig. 4.2, we note that angles α and β have the same initial and terminal sides. *Such angles are called* **coterminal angles**. An understanding of coterminal angles will be important in the use of trigonometric concepts in later chapters.

EXAMPLE 1 Coterminal angles

Determine the measures of four angles that are coterminal with an angle of 145.6°.

Since there are 360° in one complete rotation, coterminal angles are found by adding or subtracting a multiple of 360° to the given angle. In symbols, the angle $\theta + n(360°)$ is coterminal to the angle θ for any integer n.

$$\mathbf{n = 1:}\quad 145.6° + 360° = 505.6° \qquad \mathbf{n = 2:}\quad 145.6° + 2(360°) = 865.6°$$
$$\mathbf{n = -1:}\quad 145.6° - 360° = -214.4° \qquad \mathbf{n = -2:}\quad 145.6° - 2(360°) = -574.4°$$

Therefore, four angles that are coterminal to 145.6° are 505.6°, 865.6°, −214.4°, and −574.4°. Two of these are shown in Fig. 4.3. We could continue to add or subtract 360° to obtain additional coterminal angles.

ANGLE CONVERSIONS

We will primarily use degrees as a measure of angles in this chapter. However, there are several unit systems that can be used, including revolutions, degrees, and radians, which were introduced in Chapter 2. We review some definitions here.

One revolution: One full rotation of the terminal side from its starting location, through the full circle, and back to the same position.

One degree: 1/360 of a revolution. One degree is written as 1°.

One minute: 1/60 of a degree. One minute is written as 1′.

One second: 1/60 of a minute, or 1/3600 of a degree. One second is written as 1″.

One radian: The measure of a central angle subtending an arc equal in length to the radius. When angles are measured in radians, no units are written. π radians equal 180°.

For reasons of precision, minutes and seconds are often used to report degrees, using what is called degrees-minutes-seconds form (or dms units). In this form, the decimal portion of the degree measurement is converted into minutes, and then the decimal portion of the minutes is converted into seconds. Symbolically, 3 degrees, 14 minutes and 25 seconds is written as 3°14′25″, equivalent to 3.240278°.

With so many options for unit systems, it is crucial that you pay careful attention to the setting on your calculator. Most calculators indicate the angular measurement in use visibly on the screen. Also, there are often keys that allow the user to switch quickly between degrees and radians, and in some cases degrees and dms. The examples below illustrate angle conversion directly from the definitions.

COMMON ERROR When computing the values of trigonometric functions, or when solving for angles using inverse trigonometric functions (see Sections 4.2 and 4.3), ***you must know what mode your calculator is in*** to ensure that you are using or obtaining the angle with the correct unit. It is a common error to use the values given by a calculator without ensuring the proper mode is set (radians or degrees).

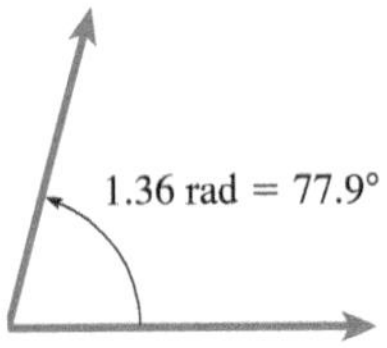

Fig. 4.4

■ In some graphing calculators, functions that can convert radians to degrees (while in degree mode) and degrees to radians (while in radian mode) can be found under the *Angle* menu.

EXAMPLE 2 Convert radians to degrees

Express 1.36 rad in degrees.

We know that π rad = 180°, which becomes our unit conversion fraction (see Section 1.3). Therefore,

$$1.36 \text{ rad} = 1.36 \text{ rad}\left(\frac{180^\circ}{\pi \text{ rad}}\right) = 77.9^\circ \qquad \text{to nearest } 0.1^\circ$$

This angle is shown in Fig. 4.4. We again note that degrees and radians are simply two different ways of measuring an angle.

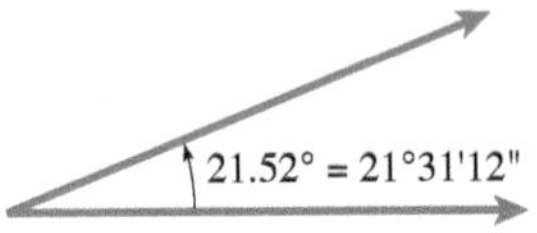

Fig. 4.5

Practice Exercises

1. Change 17°24′ to decimal form, to the nearest 0.01°.
2. Change 38.25° to dms form.

EXAMPLE 3 Converting angles from decimal form to dms form

Convert 21.52° into dms form (Fig. 4.5).

We can use statements of equivalence ($1^\circ = 60'$ and $1' = 60''$) to convert the decimal portion of the measurement to minutes, and then the decimal portion of the minutes to seconds, as shown below.

$$0.52^\circ \times \frac{60'}{1^\circ} = 31.2', \quad \text{and} \quad 0.2' \times \frac{60''}{1'} = 12''$$

Therefore, $21.52^\circ = 21^\circ 31' 12''$.

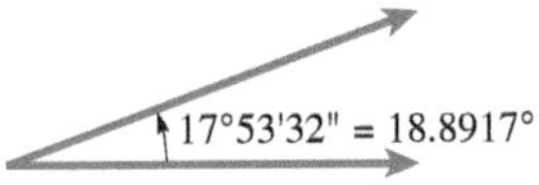

Fig. 4.6

LEARNING TIP

When determining how many decimal places to retain, keep in mind that one second is equal to 1/3600 of a degree. Therefore, for high-precision work where resolution to the second is needed, it is best to report degrees to four decimal places. This preserves the integrity of the information from the minutes and seconds of the calculation. If no seconds are used, decimal conversion of minutes to degrees is 1/60, so two decimal places should be retained.

EXAMPLE 4 Converting angles between dms and decimal forms

Change 17°53′30″ into decimal form (Fig. 4.6).

We use statements of equivalence ($1^\circ = 60'$ and $1' = 60''$) to convert seconds into minutes and minutes into degrees, as shown below.

$$\begin{aligned} 17^\circ 53' 30'' &= 17^\circ + 53'\left(\frac{1^\circ}{60'}\right) + 30''\left(\frac{1'}{60''}\right)\left(\frac{1^\circ}{60'}\right) \\ &= 17^\circ + 0.8833333^\circ + 0.0083333^\circ \\ &= 17.8916666^\circ \end{aligned}$$

Therefore, $17^\circ 53' 30'' = 17.8917^\circ$.

STANDARD POSITION OF AN ANGLE

The rectangular coordinate system gives us a standard way of positioning angles in space. *If the initial side of the angle is the positive x-axis and the vertex is the origin, the angle is said to be in* **standard position**. The angle is then determined by the position of the terminal side. If the terminal side of an angle in standard position coincides with the x- or y-axis, the angle is called a **quadrantal** angle. In Fig 4.7(a), the quadrantal angle of 90 degrees is shown. If the terminal side of an angle in standard position falls inside a quadrant, it is identified by the name of that quadrant. For example, as seen in Fig. 4.7(b), a 60° standard angle has its terminal side in the first quadrant, so it is a **first-quadrant angle**.

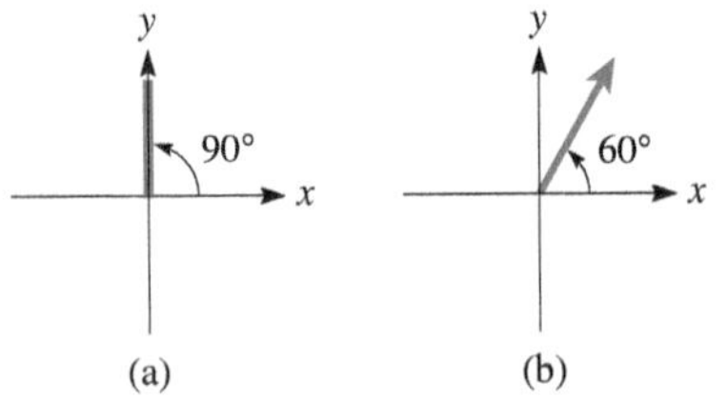

Fig. 4.7

EXAMPLE 5 Angles in standard position

Designate each of the following angles in standard position by the quadrant in which the terminal side lies, or as a quadrantal angle: 130°, 225°, −120°, 180°. See Fig. 4.8.

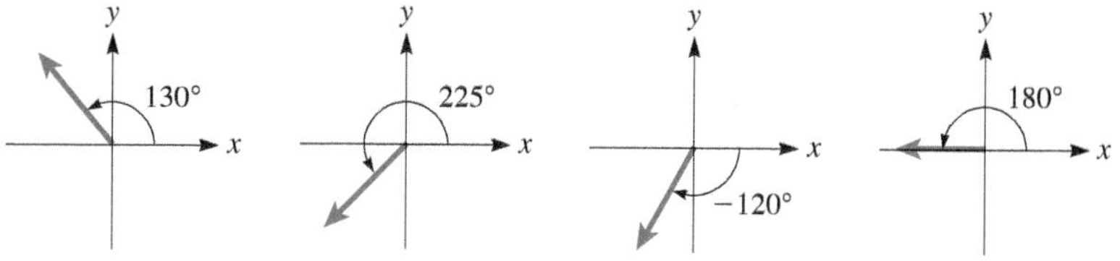

Fig. 4.8(a) 130° is a second-quadrant angle.

Fig. 4.8(b) 225° is a third-quadrant angle.

Fig. 4.8(c) −120° is a third-quadrant angle.

Fig. 4.8(d) 180° is a quadrantal angle.

Practice Exercise

3. In Fig. 4.8, which terminal side is that of a standard position angle of 240°?

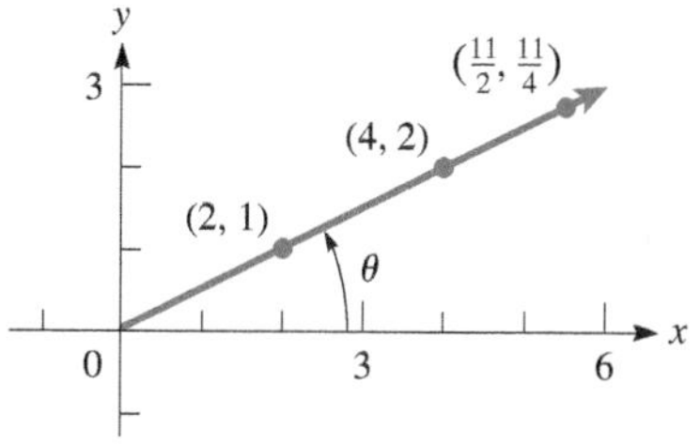

Fig. 4.9

EXAMPLE 6 Standard position—terminal side

In Fig. 4.9, θ is in standard position, and the terminal side is uniquely determined by knowing that it passes through the point $(2, 1)$. The same terminal side passes through the points $(4, 2)$ and $\left(\frac{11}{2}, \frac{11}{4}\right)$, among an unlimited number of other points. Knowing that the terminal side passes through any one of these points makes it possible to determine the terminal side of the angle. The method to calculate this angle θ is shown in Section 4.3.

EXERCISES 4.1

In Exercises 1–4, find the indicated angles in the given examples of this section.

1. In Example 1, find another angle that is coterminal with the given angle.
2. In Example 4, change 53′ to 35′ and then find the decimal form.
3. In Example 5, find another standard-position angle that has the same terminal side as the angle in Fig. 4.8(b).
4. In Example 5, find another standard-position angle that has the same terminal side as the angle in Fig. 4.8(c).

In Exercises 5–8, draw the given angles.

5. 60°, 120°, −90° **6.** 330°, −150°, 450°
7. 50°, −360°, −30° **8.** 45°, 245°, −250°

In Exercises 9–16, determine one positive and one negative coterminal angle for each angle given.

9. 135° **10.** 37°
11. −150° **12.** 462°
13. 30°7′ **14.** 147°53′
15. 278.1° **16.** −197.6°

In Exercises 17–20, by means of the definition of a radian, change the given angles in radians to equal angles expressed in degrees to the nearest 0.01°.

17. 0.652 rad **18.** 0.383 rad
19. 1.447 rad **20.** −3.642 rad

In Exercises 21–24, use a calculator conversion sequence to change the given angles in radians to equal angles expressed in degrees to the nearest 0.01°.

21. 0.239 rad **22.** 8.029 rad
23. −4.110 rad **24.** 6.705 rad

In Exercises 25–28, change the given angles to equal angles expressed in radians to three significant digits.

25. 71.4° **26.** 173.4° **27.** 384.8° **28.** −17.5°

In Exercises 29–32, change the given angles to equal angles expressed to the nearest minute.

29. 54.70° **30.** 517.83° **31.** −5.62° **32.** 142.87°

In Exercises 33–36, change the given angles to equal angles expressed in decimal form to the nearest 0.01°.

33. 15°12′ **34.** 517°39′ **35.** 301°16′ **36.** −94°47′

In Exercises 37–44, draw angles in standard position such that the terminal side passes through the given point.

37. (4, 2) **38.** (−3, 8) **39.** (−3, −5) **40.** (6, −1)
41. (−7, 5) **42.** (−4, −2) **43.** (−2, 0) **44.** (0, 6)

In Exercises 45–52, the given angles are in standard position. Designate each angle by the quadrant in which the terminal side lies, or as a quadrantal angle.

45. 31°, 310° **46.** 180°, 92°
47. 435°, −270° **48.** −5°, 265°

49. 1 rad, 2 rad **50.** 3 rad, π rad

51. 4 rad, $\pi/3$ rad **52.** 12 rad, -2 rad

In Exercises 53 and 54, change the given angles to equal angles expressed in decimal form to the nearest 0.001°. In Exercises 55 and 56, change the given angles to equal angles expressed to the nearest second.

53. 21°42′36″ **54.** 7°16′23″

55. 86.274° **56.** 257.019°

In Exercises 57 and 58, solve the given problems.

57. A circular gear rotates clockwise by exactly 3.5 revolutions. By how many degrees does it rotate?

58. A windmill rotates 15.6 revolutions in a counterclockwise direction. By how many radians does it rotate?

Answers to Practice Exercises

1. 17.40° **2.** 38°15′ **3.** (c)

4.2 Defining the Trigonometric Functions

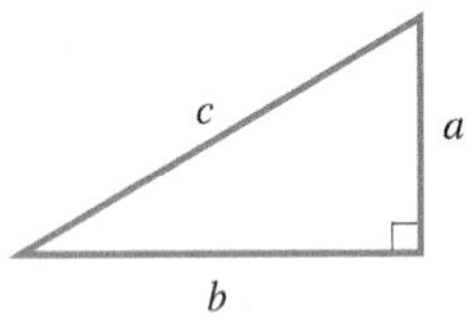

Fig. 4.10

We know that a triangle has three sides and three angles. If one side and any other two of these six parts are known, we can **solve the triangle**, that is, we can find the other three parts. When the triangle is a right triangle, we can solve it by using the following relationships:

1. The sum of angles rule, which states that the sum of the interior angles of a triangle is 180° (to relate the three angles).
2. The Pythagorean theorem, which states that for a right triangle with hypotenuse c and legs a and b (as in Fig. 4.10), $c^2 = a^2 + b^2$ (to relate the three sides).
3. The trigonometric functions (to relate any two sides with an angle).

The sum of angles rule and the Pythagorean theorem have already been discussed in Chapter 2. In this section, we define the trigonometric functions by using properties of similar triangles. We begin by reviewing these properties.

As stated in Section 2.2, *two triangles are* **similar** *if they have the same shape (but not necessarily the same size).* Similar triangles have the following important properties.

Properties of Similar Triangles

1. Corresponding angles are equal.
2. Corresponding sides are proportional.

The **corresponding sides** *are the sides, one in each triangle, that are between the same pair of equal* **corresponding angles**.

EXAMPLE 1 Similar triangles

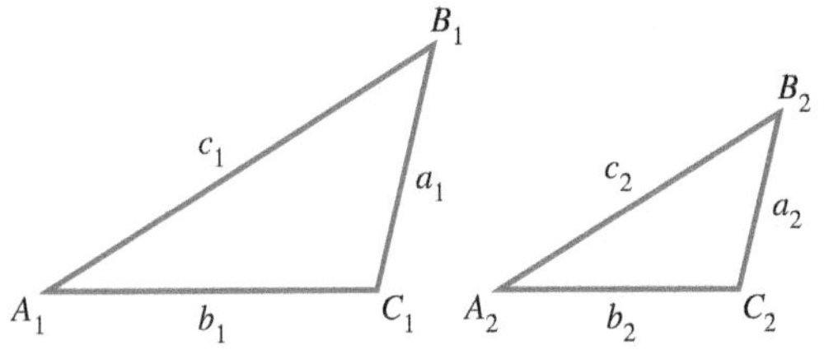

Fig. 4.11

Determine the corresponding angles and corresponding sides of the similar triangles in Fig. 4.11.

Angles A_1 and A_2, angles B_1 and B_2, and angles C_1 and C_2 are pairs of corresponding angles. The corresponding angles are equal, or

$$\angle A_1 = \angle A_2 \qquad \angle B_1 = \angle B_2 \qquad \angle C_1 = \angle C_2$$

The pairs of corresponding sides are a_1 and a_2, b_1 and b_2, and c_1 and c_2. The corresponding sides are proportional, or

$$\frac{a_1}{a_2} = \frac{b_1}{b_2} \qquad \frac{a_1}{a_2} = \frac{c_1}{c_2} \qquad \frac{b_1}{b_2} = \frac{c_1}{c_2}$$

In Example 1, if we multiply both sides of $a_1/a_2 = b_1/b_2$ by a_2/b_1, we get

$$\frac{a_1}{a_2}\left(\frac{a_2}{b_1}\right) = \frac{b_1}{b_2}\left(\frac{a_2}{b_1}\right), \text{ or } \frac{a_1}{b_1} = \frac{a_2}{b_2}$$

Using this, we now proceed to the definitions of the trigonometric functions.

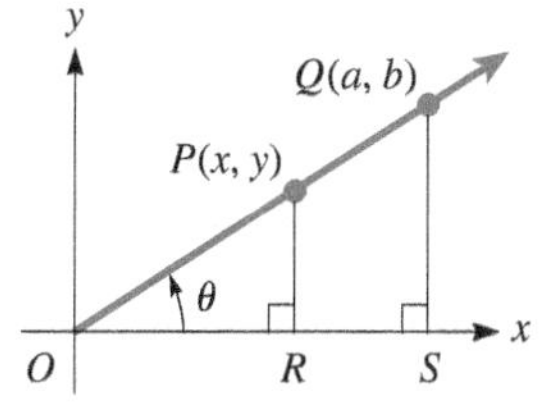

Fig. 4.12

> **LEARNING TIP**
> *When two triangles are similar, the ratio of one side to another side in one triangle is the same as the ratio of the corresponding sides in the other triangle.*

As shown in Fig. 4.12, an angle θ is placed in standard position and perpendiculars from points on the terminal side are dropped to the x-axis (lines PR and QS). The x-axis and the terminal side of θ can be seen as transversals cutting the parallel lines PR and QS. As a result, triangles ORP and OSQ are similar. This means that ratios of the lengths of corresponding sides are equal (see the Learning Tip to the left). For instance, since triangle OPR has a right leg length of y and a base length of x, and triangle OQS has a right leg length of b and a base length of a, we can write

$$\frac{y}{x} = \frac{b}{a}$$

This equality remains true for any position of Q on the terminal side of θ (except at the origin). In other words, the ratio of the ordinate b to the abscissa a is the same for any point $Q(a, b)$ on the terminal side of θ.

Proceeding in the same manner, for any angle θ in standard position it is possible to set up six different ratios of the values of x, y, and r, as shown in Fig. 4.13. Because of the properties of similar triangles, for a given angle θ, each of these ratios is the same for any point on the terminal side. This means that the values of the ratios depend solely on the angle θ, and for each value of θ, there will be *only one* value of each ratio. In other words, *the ratios are **functions** of the angle θ, and are called* the **trigonometric functions** (or trigonometric identities).

DEFINITION OF THE TRIGONOMETRIC FUNCTIONS

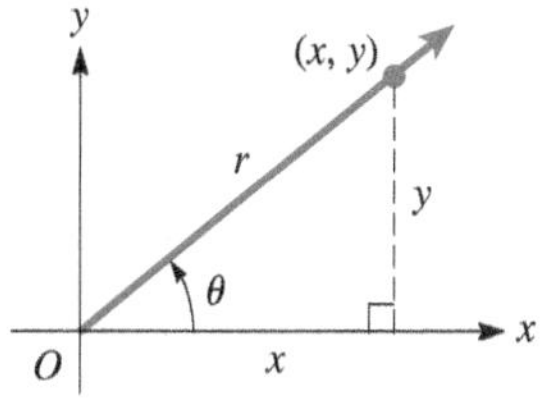

Fig. 4.13

Let (x, y) be any point (other than the origin) on the terminal side of the angle θ in standard position, and let r be the distance between the origin and the given point (see Fig. 4.13). The following names and abbreviations are used to define the six trigonometric functions. Notice that three of the trigonometric functions (cosecant, secant, and cotangent) are the reciprocals of the three primary trigonometric functions (sine, cosine, and tangent, respectively).

Primary Trigonometric Functions			Reciprocal Trigonometric Functions		
sine of θ:	$\sin\theta = \dfrac{y}{r}$	**(4.1a)**	**cosecant of θ:**	$\csc\theta = \dfrac{1}{\sin\theta} = \dfrac{r}{y}$	**(4.1d)**
cosine of θ:	$\cos\theta = \dfrac{x}{r}$	**(4.1b)**	**secant of θ:**	$\sec\theta = \dfrac{1}{\cos\theta} = \dfrac{r}{x}$	**(4.1e)**
tangent of θ:	$\tan\theta = \dfrac{y}{x}$	**(4.1c)**	**cotangent of θ:**	$\cot\theta = \dfrac{1}{\tan\theta} = \dfrac{x}{y}$	**(4.1f)**

The definitions given above are completely general, which means they apply to *all angles*, no matter which of the four quadrants they terminate in. However, in this chapter, we will concentrate on angles between 0° and 90°. In Chapters 8 and 20, we will expand our scope and apply this definition to angles terminating in all four quadrants. In all cases, *the distance r from the origin to the point is always a positive number*, and it is called the ***radius vector***.

It should also be noted that because a denominator cannot be zero, there are points at which some of the trigonometric functions are not defined. For example, the tangent is undefined if $x = 0$, while the cotangent is undefined if $y = 0$. In Chapter 10, we will discuss the domains and graphs of these functions.

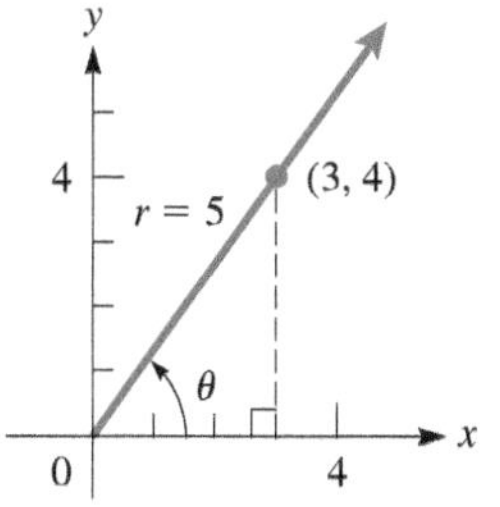

Fig. 4.14

EXAMPLE 2 Values of trigonometric functions as fractions or decimals

Find the values of the trigonometric functions of the standard-position angle θ with its terminal side passing through the point $(3, 4)$.

By placing the angle in standard position, as shown in Fig. 4.14, and drawing the terminal side through $(3, 4)$, we find by use of the Pythagorean theorem

$$r = \sqrt{3^2 + 4^2} = \sqrt{25} = 5$$

Using the values $x = 3$, $y = 4$, and $r = 5$, we find that

$$\sin\theta = \frac{y}{r} = \frac{4}{5} \qquad \cos\theta = \frac{x}{r} = \frac{3}{5} \qquad \tan\theta = \frac{y}{x} = \frac{4}{3}$$

$$\csc\theta = \frac{1}{\sin\theta} = \frac{5}{4} \qquad \sec\theta = \frac{1}{\cos\theta} = \frac{5}{3} \qquad \cot\theta = \frac{1}{\tan\theta} = \frac{3}{4}$$

We have left each of these results in the form of a fraction, which is considered to be an *exact form* in that there has been no approximation made. In writing decimal values, some fractions divide exactly, while others require an approximation. Rounding to two decimal places *when required*, we find

$$\sin\theta = 0.8 \qquad \cos\theta = 0.6 \qquad \tan\theta = 1.33$$

$$\csc\theta = 1.25 \qquad \sec\theta = 1.67 \qquad \cot\theta = 0.75$$

Practice Exercise

1. In Example 2, change $(3, 4)$ to $(4, 3)$ and then find $\tan\theta$ and $\sec\theta$.

EXAMPLE 3 Values of trigonometric functions with decimals

Find the values of the trigonometric functions of the standard-position angle whose terminal side passes through $(7.27, 4.49)$. The coordinates are approximate.

We show the angle and the given point in Fig. 4.15. From the Pythagorean theorem, we have

$$r = \sqrt{7.27^2 + 4.49^2} = 8.544\,764\,479$$

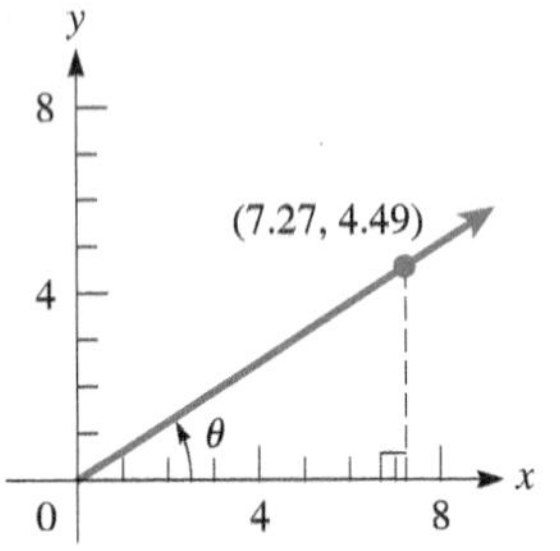

Fig. 4.15

We use r without rounding in all of our calculations (stored in the memory of the calculator). For purposes of illustration, we show r rounded as $r = 8.545$. Therefore, we have the following values:

$$\sin\theta = \frac{y}{r} = \frac{4.49}{8.545} = 0.525 \qquad \cos\theta = \frac{x}{r} = \frac{7.27}{8.545} = 0.851$$

$$\tan\theta = \frac{y}{x} = \frac{4.49}{7.27} = 0.618 \qquad \csc\theta = \frac{1}{\sin\theta} = \frac{8.545}{4.49} = 1.90$$

$$\sec\theta = \frac{1}{\cos\theta} = \frac{8.545}{7.27} = 1.18 \qquad \cot\theta = \frac{1}{\tan\theta} = \frac{7.27}{4.49} = 1.62$$

Since the coordinates are approximate, the results are rounded off to three significant digits.

COMMON ERROR

When one expresses the result of a trigonometric function in form $\sin\theta = 0.525$, it is a common error to omit the angle and report the value as $\sin = 0.525$. This is a meaningless expression because ***the angle for which the trigonometric expression is evaluated must be given.*** The trigonometric function itself indicates which pair of sides in the right triangle to divide, but without an angle for reference, the ratio has no meaning.

EXAMPLE 4 Using one trigonometric function to find the others

For a first-quadrant angle θ, $\sin\theta = 3/7$. Use this information to find the values of the other trigonometric functions in both fractional and decimal forms.

Since $\sin\theta = y/r$, we can use the values provided to set up a reference triangle where $y = 3$ and $r = 7$. The missing side, x, can be found by using the Pythagorean theorem:

$$x^2 + 3^2 = 7^2$$

$$x = \sqrt{7^2 - 3^2} = \sqrt{49 - 9} = \sqrt{40} = 2\sqrt{10}$$

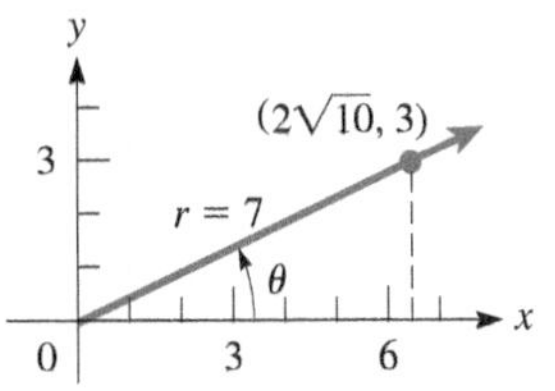

Fig. 4.16

Therefore, the point $(2\sqrt{10}, 3)$ is on the terminal side, as shown in Fig. 4.16.

Using the values $x = 2\sqrt{10}$, $y = 3$, and $r = 7$, we have the other trigonometric functions of θ. They are

$$\cos\theta = \frac{2\sqrt{10}}{7} \quad \tan\theta = \frac{3}{2\sqrt{10}} \quad \cot\theta = \frac{2\sqrt{10}}{3} \quad \sec\theta = \frac{7}{2\sqrt{10}} \quad \csc\theta = \frac{7}{3}$$

These values are *exact*. *Approximate* decimal values found on a calculator are

$$\cos\theta = 0.904 \quad \tan\theta = 0.474 \quad \cot\theta = 2.11$$
$$\sec\theta = 1.11 \quad \csc\theta = 2.33$$

Practice Exercise

2. In Example 4, change $\sin\theta = 3/7$ to $\cos\theta = 3/7$, and then find approximate values of $\sin\theta$ and $\cot\theta$.

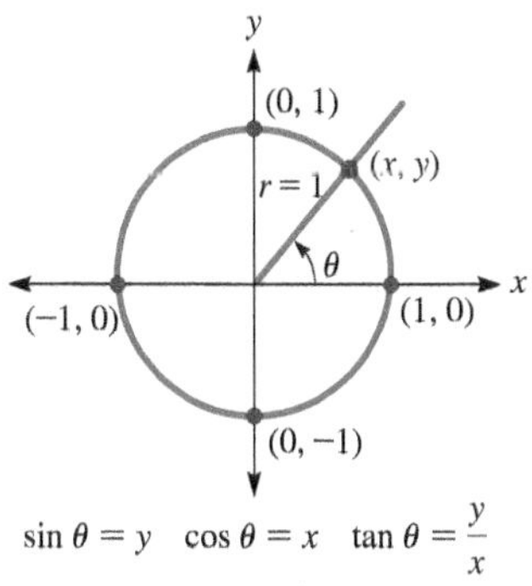

Fig. 4.17

THE UNIT CIRCLE AND THE TRIGONOMETRIC FUNCTIONS

The circle that is centred at the origin and has a radius of 1 is called the ***unit circle***. Suppose the terminal side of θ in standard position intersects the unit circle at the point (x, y) as shown is Fig. 4.17.

Using the definition of the trigonometric functions and the fact that $r = 1$, we have $\sin\theta = \frac{y}{r} = \frac{y}{1} = y$, and $\cos\theta = \frac{x}{r} = \frac{x}{1} = x$. Thus, **the y-coordinate of the point on the unit circle equals** $\sin\theta$, **and the x-coordinate equals** $\cos\theta$. This fact can be very useful in many situations.

In terms of the unit circle, $\tan\theta = \frac{y}{x}$, as in the original definition. Also, the three reciprocal functions are still found by taking the reciprocal of the sine, cosine, or tangent.

EXAMPLE 5 The unit circle

(a) Use the unit circle to find $\cos 90°$ and $\sin 90°$.

A 90° angle, when placed in standard position, intersects the unit circle at the point $(0, 1)$. Therefore, $\cos 90° = x = 0$ and $\sin 90° = y = 1$.

(b) Find $\cos\theta$ and $\csc\theta$ if the terminal side of θ (in standard position) intersects the unit circle at the point $(0.6, 0.8)$.

We know that $\cos\theta = x = 0.6$. Since $\sin\theta = y = 0.8$, $\csc\theta$ is the reciprocal of 0.8.

Thus, $\csc\theta = \frac{1}{0.8} = \frac{10}{8} = \frac{5}{4}$.

EXERCISES 4.2

In Exercises 1 and 2, answer the given questions about the indicated examples of this section.

1. In Example 2, if the point $(4, 3)$ replaces the point $(3, 4)$, what are the values of the trigonometric functions in exact form?

2. In Example 4, if $4/7$ replaces $3/7$, what are the values of the trigonometric functions in exact form?

In Exercises 3–16, find values of the trigonometric functions of the angle (in standard position) whose terminal side passes through the given points. For Exercises 3–14, give answers in exact form. For Exercises 15 and 16, the coordinates are approximate.

3. $(4, 7)$ **4.** $(5, 11)$ **5.** $(15, 8)$

6. $(240, 70)$ **7.** $(0.09, 0.40)$ **8.** $(3.2, 6.0)$

9. $(1, \sqrt{15})$ **10.** $(\sqrt{3}, 2)$ **11.** $(7, 7)$

12. $(840, 130)$ **13.** $(50, 20)$ **14.** $(1, \frac{1}{2})$

15. $(0.687, 0.943)$ **16.** $(37.65, 21.87)$

In Exercises 17–24, find the values of the indicated functions. In Exercises 17–20, give answers in exact form. In Exercises 21–24, the values are approximate. Assume all angles are acute.

17. Given $\cos\theta = 5/9$, find $\sin\theta$ and $\cot\theta$.

18. Given $\sin\theta = 1/2$, find $\cos\theta$ and $\csc\theta$.

19. Given $\tan\theta = 2$, find $\sin\theta$ and $\sec\theta$.

20. Given $\sec\theta = \sqrt{5}/2$, find $\tan\theta$ and $\cos\theta$.

21. Given $\sin\theta = 0.750$, find $\cot\theta$ and $\csc\theta$.

22. Given $\cos\theta = 0.0326$, find $\sin\theta$ and $\tan\theta$.

23. Given $\cot\theta = 0.254$, find $\cos\theta$ and $\tan\theta$.

24. Given $\csc\theta = 1.20$, find $\sec\theta$ and $\cos\theta$.

In Exercises 25–28, each point listed is on the terminal side of an angle. Show that each of the indicated functions is the same for each of the points.

25. $(3, 4)$, $(6, 8)$, $(4.5, 6)$, $\sin\theta$ and $\tan\theta$

26. $(5, 12)$, $(15, 36)$, $(7.5, 18)$, $\cos\theta$ and $\cot\theta$

27. $(0.3, 0.1), (9, 3), (33, 11)$, $\tan\theta$ and $\sec\theta$

28. $(40, 30), (8, 6), (36, 27)$, $\csc\theta$ and $\cos\theta$

Use the unit circle to complete Exercises 29–32 (see Example 5).

29. Find $\cos 0°$.

30. Find $\sin 0°$.

31. If the angle θ intersects the unit circle at exactly $(0.28, 0.96)$, find $\sin\theta$ and $\csc\theta$.

32. For the angle in Exercise 31, find $\cos\theta$ and $\sec\theta$.

In Exercises 33–40, answer the given questions.

33. If $\tan\theta = 3/4$, what is the value of $\sin^2\theta + \cos^2\theta$? $[\sin^2\theta = (\sin\theta)^2]$

34. If $\sin\theta = 2/3$, what is the value of $\sec^2\theta - \tan^2\theta$?

35. If $y = \sin\theta$, what is $\cos\theta$ in terms of y?

36. If $x = \cos\theta$, what is $\tan\theta$ in terms of x?

37. What is x if $(x + 1, 4)$ and $(-2, 6)$ are on the same terminal side of a standard-position angle?

38. What is x if $(2, 5)$ and $(7, x)$ are on the same terminal side of a standard-position angle?

39. From the definitions of the trigonometric functions, it can be seen that $\csc\theta$ is the reciprocal of $\sin\theta$. What function is the reciprocal of $\cos\theta$?

40. Refer to the definitions of the trigonometric functions in Eqs. (4.1a)–(4.1f). Is the quotient of one of the functions divided by $\cos\theta$ equal to $\tan\theta$? Explain.

Answers to Practice Exercises

1. $\tan\theta = 3/4$, $\sec\theta = 5/4$
2. $\sin\theta = 0.904$, $\cot\theta = 0.474$

4.3 Values of the Trigonometric Functions

The trigonometric functions play an important role in the understanding and application of trigonometry. We begin by considering the values of trigonometric functions for the common angles of 30°, 60°, and 90°. For these angles, we can use geometry to find the values of the trigonometric functions.

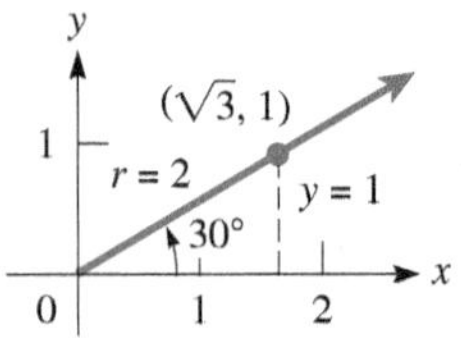

Fig. 4.18(a)

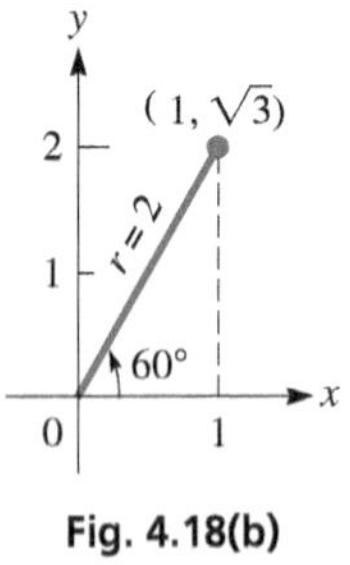

Fig. 4.18(b)

EXAMPLE 1 Function values of 30° and 60°

From geometry, we find that the side opposite a 30° angle in a right triangle is one-half of the hypotenuse. Therefore, letting $y = 1$ and $r = 2$ (see Fig. 4.18a), and using the Pythagorean theorem, we have $x = \sqrt{2^2 - 1^2} = \sqrt{3}$. Now, with $x = \sqrt{3}$, $y = 1$, and $r = 2$,

$$\sin 30° = \frac{1}{2} \qquad \cos 30° = \frac{\sqrt{3}}{2} \qquad \tan 30° = \frac{1}{\sqrt{3}}$$

Using this same method, as shown in Fig. 4.18(b), the side adjacent a 60° angle is one-half of the hypotenuse, so that

$$\sin 60° = \frac{\sqrt{3}}{2} \qquad \cos 60° = \frac{1}{2} \qquad \tan 60° = \sqrt{3}$$

Note that because 30° and 60° are complementary angles, their opposite and adjacent sides are related. This will be discussed further in Section 4.4.

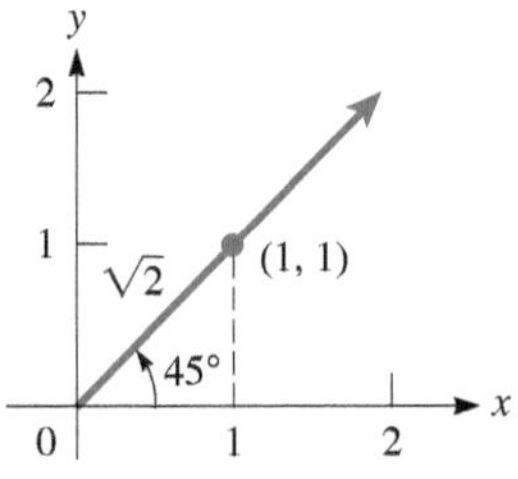

Fig. 4.19

EXAMPLE 2 Trigonometric function values of 45°

Find $\sin 45°$, $\cos 45°$, and $\tan 45°$.

If we place an isosceles right triangle with one of its 45° angles in standard position and hypotenuse along the radius vector (see Fig. 4.19), the terminal side passes through $(1, 1)$, since the legs of the triangle are equal. Using this point, $x = 1$, $y = 1$, and $r = \sqrt{2}$. Thus,

$$\sin 45° = \frac{1}{\sqrt{2}} \qquad \cos 45° = \frac{1}{\sqrt{2}} \qquad \tan 45° = 1$$

In Examples 1 and 2, we have given *exact* values. Decimal approximations are also given in the following table that summarizes the results for 30°, 45°, and 60°.

Trigonometric Functions of 30°, 45°, and 60°

	(exact values)			*(decimal approximations)*		
θ	30°	45°	60°	30°	45°	60°
$\sin\theta$	$\frac{1}{2}$	$\frac{1}{\sqrt{2}}$	$\frac{\sqrt{3}}{2}$	0.500	0.707	0.866
$\cos\theta$	$\frac{\sqrt{3}}{2}$	$\frac{1}{\sqrt{2}}$	$\frac{1}{2}$	0.866	0.707	0.500
$\tan\theta$	$\frac{1}{\sqrt{3}}$	1	$\sqrt{3}$	0.577	1.000	1.732

■ It is helpful to be familiar with these values, as they are used in later sections.

Unfortunately, the geometric methods used above only work for a limited number of angles. Another way to find values of the functions is to use a scale drawing. Measure the angle with a protractor, then measure directly the values of x, y, and r for some point on the terminal side, and finally use the proper ratios to evaluate the functions. However, this method is only approximate. As it turns out, it is possible to find these values to any required accuracy through more advanced methods (using calculus and what are known as *power series*).

Most calculators have $\sin\theta$, $\cos\theta$, and $\tan\theta$ programmed for ease of use. For the remainder of this chapter, *be sure that your calculator is set for* ***degrees*** (not radians).

```
tan(67.36)
        2.397626383
```

Fig. 4.20

EXAMPLE 3 Trigonometric values from a calculator

Using a calculator to find the value of tan 67.36°, first enter the function and then the angle, just as it is written. The resulting display is shown in Fig. 4.20.

Therefore, tan 67.36° = 2.397 626 383. This also means the ratio y/x is approximately 2.398 for this angle.

Not only are we able to find values of the trigonometric functions if we know the angle, but we can also find the angle if we know that value of a function. In doing this, we are actually using another important type of mathematical function, an **inverse function.** *The inverse of a function, in general, undoes the process that the function performs.*

■ Another notation that is used for $\sin^{-1} x$ is $\arcsin x$.

Let us illustrate the meaning of the inverse trigonometric functions through the *inverse sine function*:

$\sin\theta = \frac{y}{r}$ The sine function operates on an angle and returns the value of the ratio of two sides of a triangle, in this case y/r. If the angle θ and one of the sides y or r are known, this function can be used to find the length of the other side.

$\theta = \sin^{-1}(y/r)$ The inverse sine function operates on the ratio of sides (a number) and returns an angle (the angle whose sine is y/r). If the lengths of the sides y and r are known, this function can be used to find the unknown angle θ.

$\theta = \sin^{-1}(\sin\theta)$ The inverse function applied to the function undoes what the function does, so it returns the value of the angle itself. This procedure can be used to solve equations.

Equivalent meanings are given to $\cos^{-1} x$ (the angle whose cosine is x) and $\tan^{-1} x$ (the angle whose tangent is x). These functions are all discussed in detail in Chapter 20. For the purpose of using a calculator at this point, it is sufficient to recognize and understand the notation that is used.

COMMON ERROR

With trigonometric functions, a raised −1 shows that the ***inverse of the function is intended***—it is ***not a negative exponent.***

$\sin^{-1} x$ represents an angle whose sine value is x. It is a common error to confuse this with cosecant, the reciprocal of sine.

$$\sin^{-1}x = \text{inverse sine of ratio } x$$

$$\sin^{-1}x \neq \frac{1}{\sin x} = \csc x$$

EXAMPLE 4 Inverse trigonometric function value from a calculator

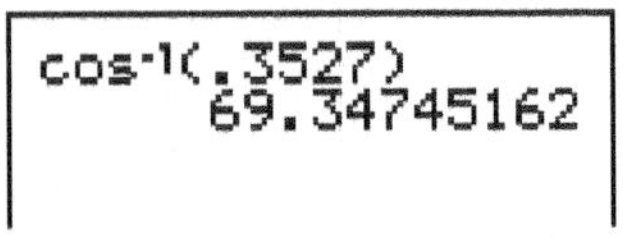

Fig. 4.21

Given the equation $\cos\theta = 0.3527$, solve for θ.

If $\cos\theta = 0.3527$, the definition of the inverse cosine implies that $\theta = \cos^{-1} 0.3527$ (θ is the angle whose cosine is 0.3527). We can then use a graphing calculator to find θ. The display is shown in Fig. 4.21.

Instead of interpreting the definition, we could also solve the equation by taking the inverse cosine of both sides and simplifying: $\cos^{-1}(\cos\theta) = \cos^{-1}(0.3527)$, or $\theta = \cos^{-1}(0.3527)$ as before. Rounding off to four significant digits, $\theta = 69.35°$.

Practice Exercises

1. Find the value of $\sin 12.5°$.
2. Find θ if $\tan\theta = 1.039$.

When using the trigonometric functions, the angle is often *approximate*. Angles of 2.3°, 92.3°, and 182.3° are angles with equal accuracy, which shows that *the accuracy of an angle does not depend on the number of digits shown.* The measurement of an angle and the accuracy of its trigonometric functions are shown in Table 4.1:

Table 4.1 **Angles and Accuracy of Trigonometric Functions**

Measurements of Angle to Nearest	*Accuracy of Trigonometric Function*
1°	2 significant digits
0.1° or 10′	3 significant digits
0.01° or 1′	4 significant digits

We rounded off the result in Example 4 according to this table.

Although we can usually set up the solution of a problem in terms of the sine, cosine, or tangent, there are times when a value of the cotangent, secant, or cosecant is used. Before a calculator can be used to solve, we must first express these functions in terms of the primary trigonometric functions.

EXAMPLE 5 Reciprocal trigonometric function value from a calculator

Find the value of $\sec 27.82°$.

We use the fact that

$$\sec 27.82° = \frac{1}{\cos 27.82°} \quad \text{or} \quad \sec 27.82° = (\cos 27.82°)^{-1}$$

The right-hand side of either of these two expressions can be entered into a calculator to find that $\sec 27.82° = 1.131$ (rounded off).

Note that we calculated the reciprocal $(\cos 27.82°)^{-1}$, and not the angle that would be denoted by using the $\cos^{-1}$ notation.

Practice Exercises

3. Find the value of $\cot 56.4°$.
4. Find θ if $\csc\theta = 1.904$.

EXAMPLE 6 Given a reciprocal trigonometric function, solve for θ

Find the value of θ if $\cot\theta = 0.354$.

We use the fact that

$$\tan\theta = \frac{1}{\cot\theta} = \frac{1}{0.354}$$

This means that $\theta = \tan^{-1}(0.354^{-1})$. Therefore, $\theta = 70.5°$ (rounded off).

EXAMPLE 7 Given one trigonometric function, find another

Find $\sin\theta$ if $\sec\theta = 2.504$.

We first restate the statement as $\frac{1}{\cos\theta} = 2.504$, which is rearranged algebraically as $\cos\theta = 1/2.504$.

This in turn tells us that $\theta = \cos^{-1}(1/2.504)$. Since we are to find the value of $\sin\theta$, we can see that

$$\sin\theta = \sin(\cos^{-1}(1/2.504))$$

Therefore, $\sin\theta = 0.9168$ (rounded off).

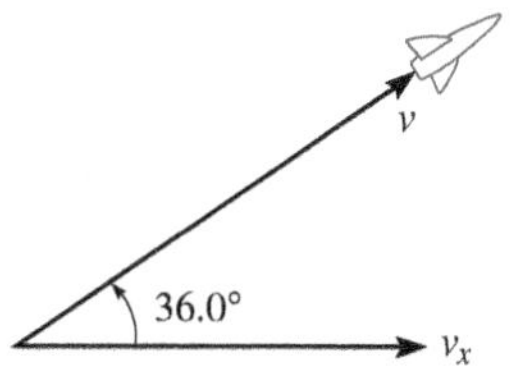

Fig. 4.22

EXAMPLE 8 Evaluating cosine—rocket velocity

When a rocket is launched, its horizontal velocity v_x is related to the velocity v with which it is fired by the equation $v_x = v\cos\theta$ (which means $v(\cos\theta)$). Here, θ is the angle between the horizontal and the direction in which it is fired (see Fig. 4.22). Find v_x if $v = 1250$ m/s and $\theta = 36.0°$.

Substituting the given values of v and θ in $v_x = v\cos\theta$, we have

$$v_x = 1250\cos 36.0°$$
$$v_x = 1010 \text{ m/s}$$

Therefore, the horizontal velocity is 1010 m/s.

ORDER OF OPERATIONS AND NOTATION WITH TRIGONOMETRIC FUNCTIONS

Some standard conventions have been adopted for trigonometric functions regarding order of operations and how the independent variable is written. If in doubt, include the brackets to avoid any confusion.

Notation Conventions for Trigonometric Functions	EXAMPLE 9
1. *Brackets may be omitted* when the argument of a trigonometric function is a **single term.**	**(a)** $\sin(x) = \sin x$ **(b)** $\sin(2x) = \sin 2x$ **(c)** $\sin(x^2) = \sin x^2$
2. *Brackets should never be omitted* if the argument of a trigonometric function has **more than one term.**	**(d)** $\sin(x+5) \neq \sin x + 5$ **(e)** $\sin(2x^3 - 4x) \neq \sin 2x^3 - 4x$
3. **Positive exponents** of a trigonometric function may be placed before the argument of the function, omitting brackets.	**(f)** $(\sin x)^2 = \sin^2 x$ **(g)** $(\sin x)^{-1} = \frac{1}{\sin x}$ and CANNOT be written as $\sin^{-1} x$.

Pay careful attention when applying these conventions. For example, $\sin^2 x$ is not the same as $\sin x^2$. In the first case, the function $\sin x$ is squared; in the second, we first square x, then take the sine.

EXERCISES 4.3

In Exercises 1–4, make the given changes in the indicated examples of this section and then find the indicated values.

1. In Example 4, change $\cos\theta$ to $\sin\theta$ and then find the angle.

2. In Example 5, change sec 27.82° to csc 27.82° and then find the value.

3. In Example 6, change 0.354 to 0.345 and then find the angle.

4. In Example 7, change $\sin\theta$ to $\tan\theta$ and then find the value.

In Exercises 5–8, use a protractor to draw the given angle. Measure off 10 units (centimetres are convenient) along the radius vector. Then measure the corresponding values of x and y. From these values, determine the trigonometric functions of the angle.

5. 40° **6.** 75° **7.** 15° **8.** 53°

In Exercises 9–24, find the values of the trigonometric functions.

9. sin 24.2° **10.** cos 57.2°

11. tan 57.6° **12.** sin 36.0°

13. cos 15.71° **14.** tan 8.653°

15. sin 89.0° **16.** cos 0.700°

17. cot 67.78° **18.** csc 22.81°

19. sec 40.5° **20.** cot 81.4°

21. csc 0.4900° **22.** sec 7.80°

23. cot 85.96° **24.** csc 76.30°

In Exercises 25–40, find one value of θ for each of the given trigonometric functions. Finding all solutions will be discussed in a later section of the text.

25. $\cos\theta = 0.3261$ **26.** $\tan\theta = 2.470$

27. $\sin\theta = 0.9114$ **28.** $\cos\theta = 0.0427$

29. $\tan\theta = 0.207$ **30.** $\sin\theta = 1.09$

31. $\cos\theta = 0.650\,07$ **32.** $\tan\theta = 5.7706$

33. $\csc\theta = 1.245$ **34.** $\sec\theta = 2.045$

35. $\cot\theta = 0.060\,60$ **36.** $\csc\theta = 1.002$

37. $\sec\theta = 0.305$ **38.** $\cot\theta = 14.4$

39. $\csc\theta = 8.26$ **40.** $\cot\theta = 0.1519$

In Exercises 41–44, use a calculator to verify the given relationships or statements. $[\sin^2\theta = (\sin\theta)^2]$

41. $\dfrac{\sin 43.7°}{\cos 43.7°} = \tan 43.7°$ **42.** $\sin^2 77.5° + \cos^2 77.5° = 1$

43. $\tan 70° = \dfrac{\tan 30° + \tan 40°}{1 - (\tan 30°)(\tan 40°)}$

44. $\sin 78.4° = 2(\sin 39.2°)(\cos 39.2°)$

In Exercises 45–48, explain why the given statements are true for an acute angle θ.

45. $\sin\theta$ is always between 0 and 1.

46. $\tan\theta$ can equal any positive real number.

47. $\cos\theta$ decreases in value from 0° to 90°.

48. The value of $\sec\theta$ is never less than 1.

In Exercises 49–52, find the values of the indicated trigonometric functions.

49. Find $\sin\theta$, given $\tan\theta = 1.936$.

50. Find $\cos\theta$, given $\sin\theta = 0.6725$.

51. Find $\tan\theta$, given $\sec\theta = 1.3698$.

52. Find $\csc\theta$, given $\cos\theta = 0.1063$.

In Exercises 53–58, solve the given problems.

53. According to Snell's law, if a ray of light passes from air into water with an angle of incidence of 45.0°, then the angle of refraction θ_r is given by the equation $\sin 45.0° = 1.33 \sin\theta_r$. Find θ_r.

54. If the backup camera on a car is mounted at a height h above the road and is angled downward (from the horizontal) at an angle θ, then the distance x along the road between the car and the point at which the camera is directed is given by $x = \frac{h}{\tan\theta}$. Find x if $h = 61.0$ cm and $\theta = 15.0°$.

55. The sound produced by a jet engine was measured at a distance of 100 m in all directions. The loudness L of the sound (in decibels) was found to be $L = 70.0 + 30.0\cos\theta$, where the 0° line was directed in front of the engine. Calculate L for $\theta = 54.5°$.

56. A brace is used in the structure shown in Fig. 4.23. Its length is $l = a(\sec\theta + \csc\theta)$. Find l if $a = 28.0$ cm and $\theta = 34.5°$.

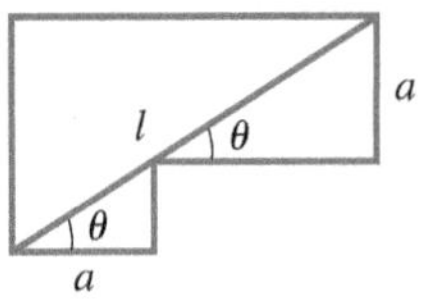

Fig. 4.23

57. The signal from an AM radio station with two antennas d metres apart has a wavelength λ (in m). The intensity of the signal depends on the angle θ as shown in Fig. 4.24. An angle of minimum intensity is given by $\sin\theta = 1.50\,\lambda/d$. Find θ if $\lambda = 200$ m and $d = 400$ m.

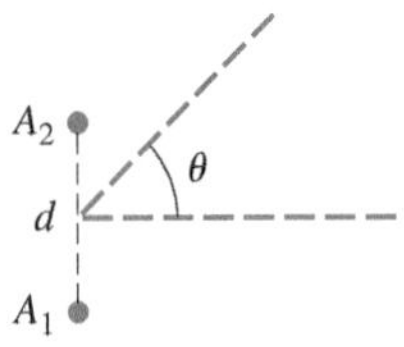

Fig. 4.24

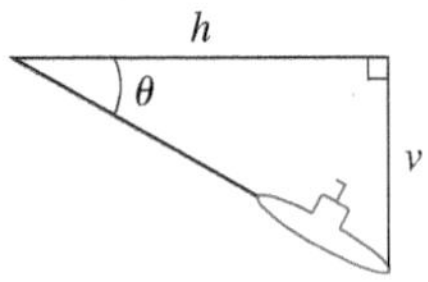

Fig. 4.25

58. A submarine dives such that the horizontal distance h it moves and the vertical distance v it dives are related by $v = h\tan\theta$. Here, θ is the angle of the dive, as shown in Fig. 4.25. Find θ if $h = 2.35$ km and $v = 1.52$ km.

Answers to Practice Exercises

1. 0.216 **2.** 46.10° **3.** 0.664 **4.** 31.68°

4.4 The Right Triangle

As stated earlier, if we know a minimum of three parts of a triangle (including at least one side), it is possible to *solve the triangle* and find the other missing parts. The example below illustrates why one of the known parts must be a side.

EXAMPLE 1 Parts of a triangle

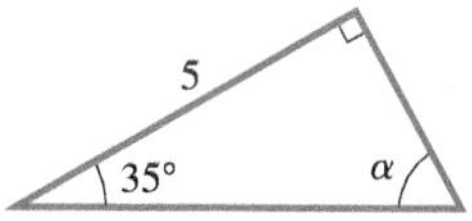

Fig. 4.26

Assume that one side and two angles are known, such as the side of 5 and the angles of 35° and 90° in the triangle in Fig. 4.26. Then we may determine the third angle α by the fact that the sum of the angles of a triangle is always 180°. While there are many triangles having the three angles of 35°, 90°, and 55° that are similar, we have only one with the particular side of 5 between angles of 35° and 90°. This unique combination of parts will allow us to make use of trigonometric functions to solve for the missing lengths exactly.

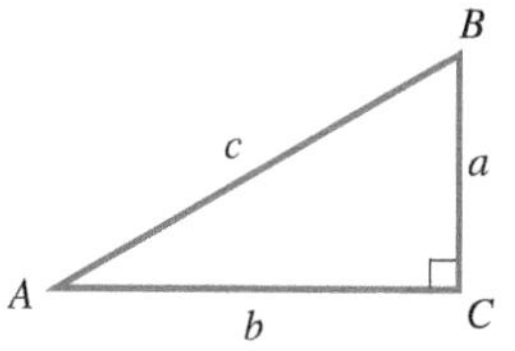

Fig. 4.27

In this section, we are going to demonstrate the method of solving a right triangle. *Since one angle of the triangle is 90°, it is necessary to know one side and one other part.* Also, since the sum of the three angles is 180°, we know that *the sum of the other two angles is 90°, and they are acute angles.* It also means they are **complementary angles**, following the definition in Section 2.1.

For consistency, when we are labelling the parts of the right triangle, *we will use the letters A and B to denote the acute angles and C to denote the right angle. The letters a, b, and c will denote the sides opposite these angles, respectively. Thus, side c is the hypotenuse of the right triangle.* See Fig. 4.27.

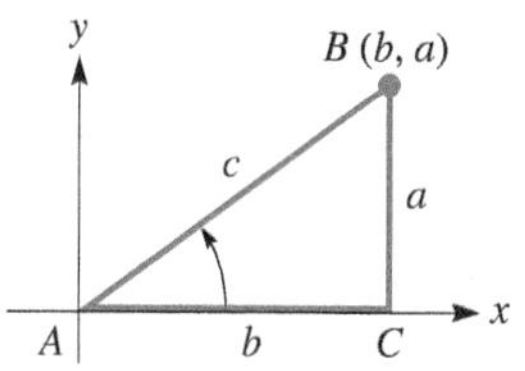

Fig. 4.28

In solving right triangles, we will find it convenient to express the trigonometric functions of the acute angles in terms of the sides. By placing the vertex of angle A at the origin and the vertex of right angle C on the positive x-axis, as shown in Fig. 4.28, we have the following ratios for angle A in terms of the sides of the triangle.

$$\sin A = \frac{a}{c} \qquad \cos A = \frac{b}{c} \qquad \tan A = \frac{a}{b}$$
$$\cot A = \frac{b}{a} \qquad \sec A = \frac{c}{b} \qquad \csc A = \frac{c}{a} \tag{4.2}$$

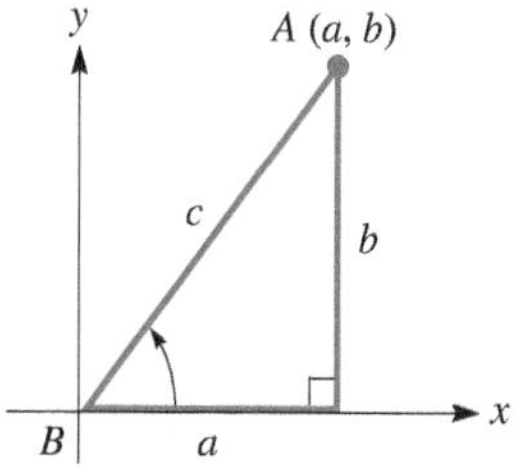

Fig. 4.29

If we should place the vertex of B at the origin, instead of the vertex of angle A, we would obtain the following ratios for the functions of angle B (see Fig. 4.29):

$$\sin B = \frac{b}{c} \qquad \cos B = \frac{a}{c} \qquad \tan B = \frac{b}{a}$$
$$\cot B = \frac{a}{b} \qquad \sec B = \frac{c}{a} \qquad \csc B = \frac{c}{b} \tag{4.3}$$

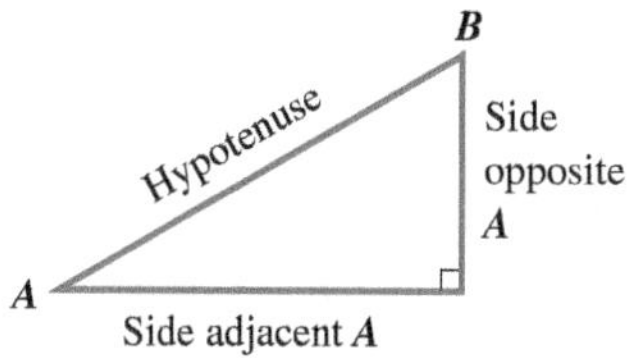

Fig. 4.30

Eqs. (4.2) and (4.3) show that we may generalize our definitions of the trigonometric functions of an acute angle of a right triangle (we have chosen $\angle A$ in Fig. 4.30) to be as follows:

$$\sin A = \frac{\text{side opposite } A}{\text{hypotenuse}} \qquad \csc A = \frac{\text{hypotenuse}}{\text{side opposite } A}$$
$$\cos A = \frac{\text{side adjacent } A}{\text{hypotenuse}} \qquad \sec A = \frac{\text{hypotenuse}}{\text{side adjacent } A} \tag{4.4}$$
$$\tan A = \frac{\text{side opposite } A}{\text{side adjacent } A} \qquad \cot A = \frac{\text{side adjacent } A}{\text{side opposite } A}$$

LEARNING TIP

Which side is adjacent or opposite depends on the angle being considered. In Fig. 4.30, the side opposite *A* is adjacent to *B*, and the side adjacent to *A* is opposite *B*.

Using the definitions in this form, we can solve right triangles without placing the angle in standard position. The angle need only be a part of any right triangle.

In general, regardless of its orientation, any triangle having a right angle and a specified angle of interest θ can have its sides and angles related using the trigonometric function definitions. In Fig. 4.31, if we let the side adjacent to angle θ be a, the side opposite the angle θ be o, and the hypotenuse of the right triangle be h, then we have the following general trigonometric functions:

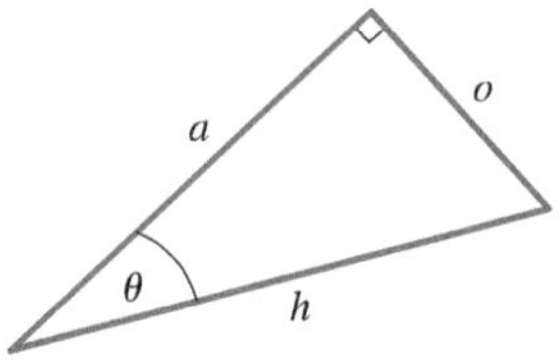

Fig. 4.31

$$\begin{aligned} \sin\theta &= o/h \qquad & \csc\theta &= h/o \\ \cos\theta &= a/h \qquad & \sec\theta &= h/a \\ \tan\theta &= o/a \qquad & \cot\theta &= a/o \end{aligned} \tag{4.5}$$

LEARNING TIP

The first three basic trigonometric relations on the left-hand side of Eq. (4.5) are often remembered with a useful mnemonic, ***soh cah toa***, where each leading letter in the triplet corresponds to the trigonometric function name (sine, cosine, and tangent, respectively), and the other two letters in each triplet correspond to the names of the sides in the ratio that defines each function, i.e., ***soh*** means $\sin\theta = o/h$.

EXAMPLE 2 Trigonometric functions in terms of sides

Given $a = 4$, $b = 7$, and $c = \sqrt{65}$ (see Fig. 4.32), find $\sin A$, $\cos A$, and $\tan A$ in exact form and in approximate decimal form (to three significant digits).

We identify a as the side opposite A, b as the side adjacent A, and c as the hypotenuse. Hence,

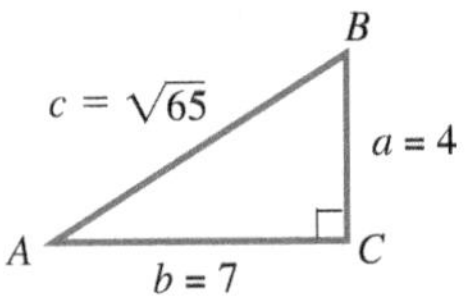

Fig. 4.32

$$\sin A = \frac{4}{\sqrt{65}} = 0.496 \qquad \cos A = \frac{7}{\sqrt{65}} = 0.868 \qquad \tan A = \frac{4}{7} = 0.571$$

Referring again to Fig. 4.30, we note that the side opposite angle A is adjacent angle B, and the side adjacent to angle A is opposite angle B. As a result, the trigonometric functions of the complementary angles A and B are related as follows:

$$\sin A = \frac{a}{c} = \cos B \qquad \cos A = \frac{b}{c} = \sin B \qquad \tan A = \frac{a}{b} = \cot B$$

$$\csc A = \frac{c}{a} = \sec B \qquad \sec A = \frac{c}{b} = \csc B \qquad \cot A = \frac{b}{a} = \tan B$$

We call the related functions **cofunctions**. Sine and cosine are cofunctions, tangent and cotangent are cofunctions, and secant and cosecant are cofunctions. *Cofunctions of complementary angles are equal.*

EXAMPLE 3 Cofunctions of complementary angles

In Fig. 4.32, find $\sin B$, $\cos B$, and $\cot B$ in exact form using the results of Example 2.

Angle A and angle B are complementary. Since cofunctions of complementary angles are equal,

$$\cos B = \sin A = \frac{4}{\sqrt{65}} \qquad \sin B = \cos A = \frac{7}{\sqrt{65}} \qquad \cot B = \tan A = \frac{4}{7}$$

We are now ready to solve right triangles. See how the following procedure is used in the examples that follow.

Procedure for Solving a Right Triangle

1. **Sketch and label** the known and unknown sides of the right triangle.
2. **Solve for the unknown parts** by using the sum of angles rule, the Pythagorean theorem, and the primary trigonometric identities to relate unknown values to known parts.
3. **Check the results** with the sum of angles rule and the Pythagorean theorem. Also, perform a visual inspection to ensure that *the largest side is opposite the largest angle, and the shortest side is opposite the smallest angle.*

EXAMPLE 4 Given angle and side—find other parts

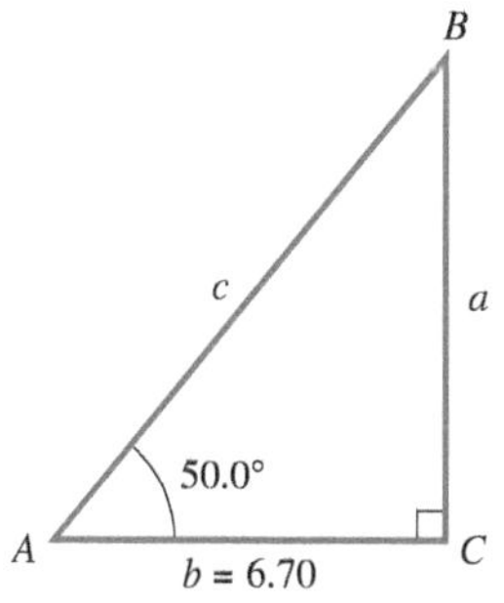

Fig. 4.33

Solve the right triangle with $A = 50.0°$ and $b = 6.70$.

We first sketch the right triangle shown in Fig. 4.33. Identifying what is known and unknown, we have

$$A = 50.0° \qquad B = \text{unknown} \qquad C = 90.0°$$
$$a = \text{unknown} \qquad b = 6.70 \qquad c = \text{unknown}$$

Using the sum of angles rule, we can solve for angle B. We know $A + B + C = 180°$. Therefore,

$$B = 180° - A - C = 180° - 50.0° - 90.0° = 40.0°$$

To find side a, we use the equation $\tan A = \dfrac{\text{side opposite } A}{\text{side adjacent } A}$.

$$\tan 50.0° = \frac{a}{6.70}$$
$$a = 6.70 \tan 50.0° \qquad \text{multiply both sides by 6.70}$$
$$= 7.984\,749\,07$$

To find side c, we use the equation $\cos A = \dfrac{\text{side adjacent } A}{\text{hypotenuse}}$.

$$\cos 50.0° = \frac{6.70}{c}$$
$$c(\cos 50.0°) = 6.70 \qquad \text{multiply both sides by } c$$
$$c = \frac{6.70}{\cos 50.0°} \qquad \text{divide both sides by } \cos 50.0°$$
$$= 10.423\,349\,64$$

Rounding off, $a = 7.98$, $c = 10.4$, and $B = 40.0°$.

Checking the angles: $A + B + C = 50.0° + 40.0° + 90° = 180°$

Checking the sides (without rounding): $c^2 = a^2 + b^2$

$$108.646\,217\,7 = 108.646\,217\,7$$

Also, c is the longest side (opposite the 90° angle) and b is the shortest side (opposite the 40° angle). The solution checks.

Practice Exercise

1. In a right triangle, find a if $B = 20.0°$ and $c = 8.50$.

LEARNING TIP

In finding unknown parts, we express them in terms of known parts as long as it is possible. We do this because *it is best to use given values in calculations.* If we use one computed value to find another computed value, any error in the first would be carried to the value of the second. For instance, in Example 4, if we were to find the value of c by using a rounded value of a, any error in a would cause c to be in error as well.

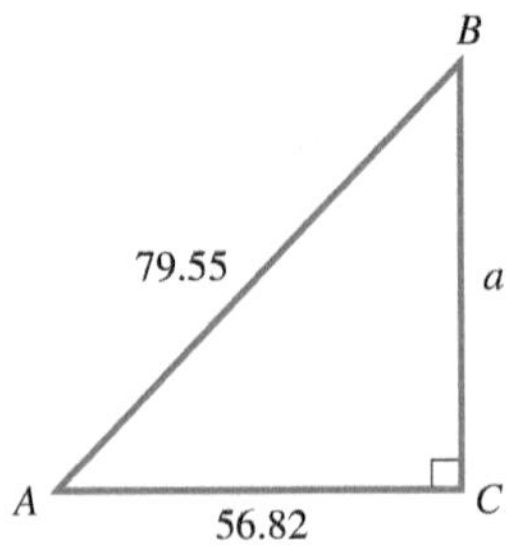

Fig. 4.34

EXAMPLE 5 Given two sides, find other parts

Solve the right triangle with $b = 56.82$ and $c = 79.55$.

Identifying what is known and unknown, we have

$$A = \text{unknown} \qquad B = \text{unknown} \qquad C = 90.0^\circ$$
$$a = \text{unknown} \qquad b = 56.82 \qquad c = 79.55$$

We sketch the right triangle as shown in Fig. 4.34. Since two sides are given, we will use the Pythagorean theorem to find the third side a. Also, we will use the cosine to find $\angle A$.

Using the Pythagorean theorem, we can solve for side a. Since $c^2 = a^2 + b^2$, $a^2 = c^2 - b^2$. Therefore,

$$a = \sqrt{c^2 - b^2} \quad a = \sqrt{79.55^2 - 56.82^2} \quad a = 55.67486$$

Since $\cos A = \dfrac{\text{side adjacent } A}{\text{hypotenuse}}$, we have

$$\cos A = \frac{56.82}{79.55} \quad A = \cos^{-1}\left(\frac{56.82}{79.55}\right) \quad A = 44.42^\circ$$

We can now find B from the fact that $A + B + C = 180^\circ$, or $B = 180^\circ - A - C = 180^\circ - 44.42^\circ - 90^\circ = 45.58^\circ$ (we store and use values without rounding in the calculator).

We have now found that

$$a = 55.67 \qquad A = 44.42^\circ \qquad B = 45.58^\circ$$

Checking the sides and angles, we first note that side a is the shortest side and is opposite the smallest angle, $\angle A$. Also, the hypotenuse is the longest side. Next, using the sine function and values without rounding to check the sides, we have

$$\sin 44.42^\circ = \frac{55.67}{79.55}, \text{ or} \qquad \sin 45.58^\circ = \frac{56.82}{79.55}, \text{ or}$$
$$0.6999 = 0.6999 \qquad 0.7142 = 0.7142$$

This shows that the values check. Note that checking results using rounded values would give only approximate equality of both sides.

Practice Exercise

2. In a right triangle, find B if $a = 20.0$ and $b = 28.0$.

EXAMPLE 6 Unknown parts in terms of known parts

If A and a are known, express the unknown parts of a right triangle in terms of A and a.

We sketch a right triangle as in Fig. 4.35, and then set up the required expressions.

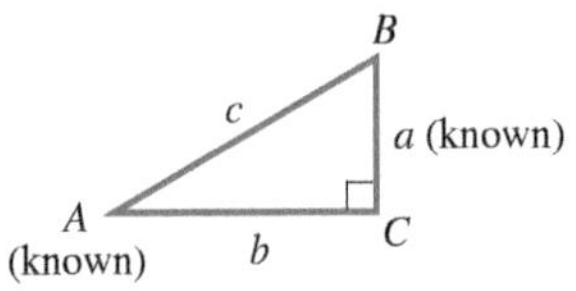

Fig. 4.35

Since $\dfrac{a}{b} = \tan A$, we have $a = b \tan A$, or $b = \dfrac{a}{\tan A}$.

Since $\dfrac{a}{c} = \sin A$, we have $a = c \sin A$, or $c = \dfrac{a}{\sin A}$.

Since $A + B + C = 180^\circ$, $B = 90^\circ - A$.

EXERCISES 4.4

In Exercises 1 and 2, make the given changes in the indicated examples of this section and then find the indicated values.

1. In Example 2, interchange the values for a and b and then find the values.
2. In Example 5, change 56.82 to 65.82 and then solve the triangle.

In Exercises 3–6, draw appropriate figures and verify through observation that only one triangle may contain the given parts (that is, any others which may be drawn will be congruent).

3. A 60° angle included between sides of 3 cm and 6 cm
4. A side of 4 cm included between angles of 40° and 50°

5. A right triangle with a hypotenuse of 5 cm and a leg of 3 cm

6. A right triangle with a 70° angle between the hypotenuse and a leg of 5 cm

In Exercises 7–30, solve the right triangles with the given parts. Round off results. Refer to Fig. 4.36.

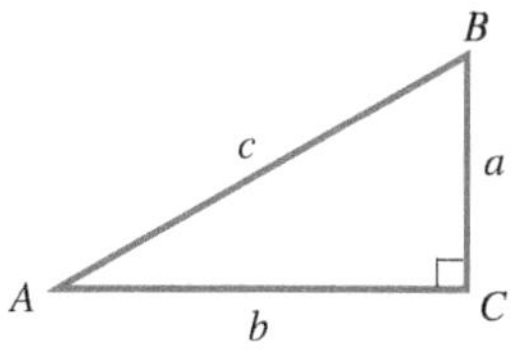

Fig. 4.36

7. $A = 78.7°, a = 7600$

8. $A = 22.5°, c = 0.0782$

9. $a = 123, c = 435$

10. $a = 932, c = 1240$

11. $B = 32.1°, c = 23.8$

12. $B = 64.3°, b = 0.652$

13. $b = 82, c = 88$ **14.** $a = 5920, b = 4110$

15. $A = 32.10°, c = 56.85$ **16.** $B = 12.60°, c = 18.42$

17. $a = 56.73, b = 44.09$ **18.** $a = 9.908, c = 12.63$

19. $B = 37.5°, a = 0.862$ **20.** $A = 87.25°, b = 8.450$

21. $B = 74.18°, b = 1.849$ **22.** $A = 51.36°, a = 3692$

23. $a = 591.87, b = 264.93$ **24.** $b = 2.9507, c = 5.0864$

25. $A = 2.975°, b = 14.592$ **26.** $B = 84.942°, a = 7413.5$

27. $B = 9.56°, c = 0.0973$ **28.** $a = 1.28, b = 16.3$

29. $a = 35.0, C = 90.0°$ **30.** $A = 25.7°, B = 64.3°$

In Exercises 31–34, find the part of the triangle labelled either x or A in the indicated figure.

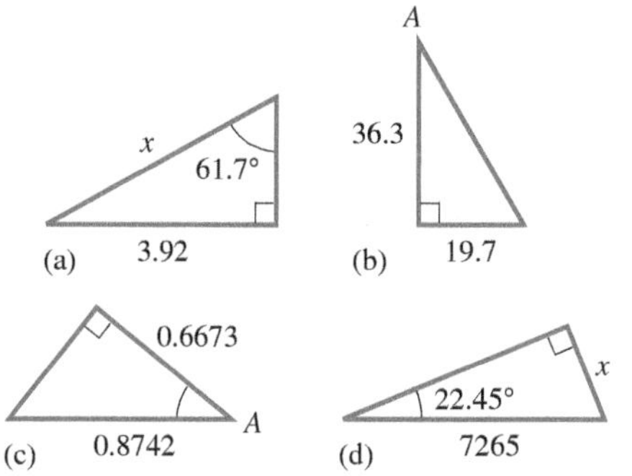

Fig. 4.37

31. Fig. 4.37(a) **32.** Fig. 4.37(b)

33. Fig. 4.37(c) **34.** Fig. 4.37(d)

In Exercises 35–38, find the indicated part of the right triangle that has the given parts.

35. One leg is 25.6, and the hypotenuse is 37.5. Find the smaller acute angle.

36. One leg is 8.50, and the angle opposite this leg is 52.3°. Find the other leg.

37. The hypotenuse is 827, and one angle is 17.6°. Find the longer leg.

38. The legs are 0.596 and 0.842. Find the larger acute angle.

In Exercises 39–42, solve the given problems.

39. Find the exact area of a circle inscribed in a regular hexagon (the circle is tangent to each of the six sides) of perimeter 72.

40. Find the exact perimeter of an equilateral triangle that is inscribed in a circle (each vertex is on the circle) of circumference 20π.

41. **One World Trade Center** refers to the main building of the rebuilt World Trade Center in New York City, which was opened in 2014. A person standing on level ground 106.7 m from the base of the building must look upward (from the ground) at an angle of 78.85° to see the tip of the spire on top of the building. Use a right triangle to find the height of the building to four significant digits. *(This problem is included in memory of those who suffered and died as a result of the terrorist attack of September 11, 2001.)*

42. The screen on a certain Samsung Galaxy tablet has 2048 pixels along its length and 1536 pixels along its width. Using a right triangle, find the angle between the longer side and the diagonal of the screen. If the diagonal measures 245.8 mm, find the length and width of the screen. (*Source*: www.samsung.com.)

Answers to Practice Exercises

1. $a = 7.99$ **2.** 54.5°

4.5 Applications of Right Triangles

Many problems in science, technology, and everyday life can be solved by finding the missing parts of a right triangle. In this section, we illustrate a number of these in the examples and exercises. The first example is a classic application that dates back to the origins of trigonometry: the triangulation method.

■ See the chapter introduction.

EXAMPLE 1 Determining the circumference and radius of Earth

Todos Santos, Mexico, lies on the Tropic of Cancer (latitude 23°26′N). During the summer solstice at noon, the sun in Todos Santos shines directly downward, and vertical objects cast no shadow. At the same time, since the sun's rays are parallel to each other, the sun casts a shadow in Empress, Alberta, situated 3060 km north of Todos Santos (see Fig. 4.38a). A person conducted an experiment by placing a 1.00 m pole vertically in Empress during the summer solstice at noon. The pole cast a shadow that was 52.3 cm long. Using this information, approximate the radius and circumference of Earth.

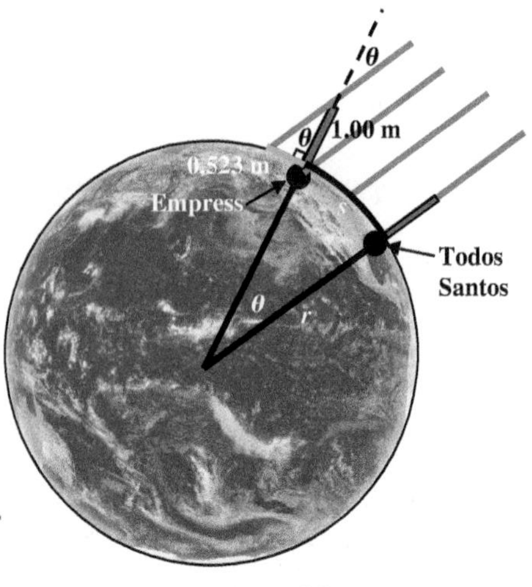

Fig. 4.38(a)

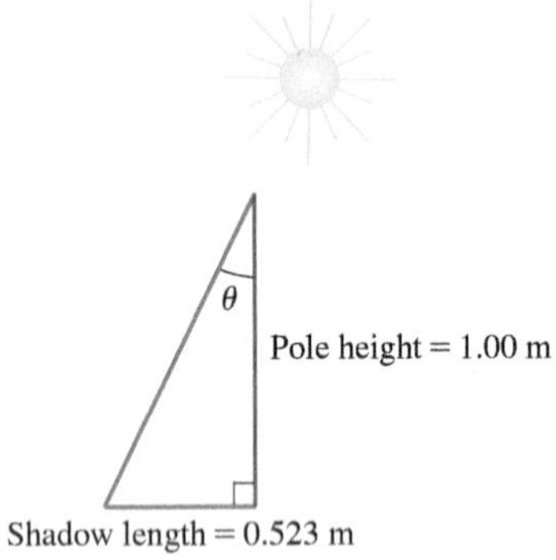

Fig. 4.38(b)

We first sketch the right triangle shown in Fig. 4.38(b) to describe how, at Empress, a vertical pole makes an angle θ with respect to the sun's rays. To make units match, we have converted the length of the shadow into metres, writing 52.3 cm = 0.523 m. We find the angle θ (in radians) from

$$\tan\theta = \frac{\text{side opposite }\theta}{\text{side adjacent }\theta} = \frac{0.523\text{ m}}{1.00\text{ m}}$$

$$\tan^{-1}(\tan\theta) = \tan^{-1}\left(\frac{0.523\text{ m}}{1.00\text{ m}}\right)$$

$$\theta = 0.481\,877\,852\text{ rad}$$

We can now use this angle to approximate the earth's radius. If the sun's rays are parallel to one another, then the angle θ is the same as the central angle subtended by the arc s connecting the two towns. We know that $s = \theta r$ (Eq. 2.12). Rearranging we get

$$r = \frac{s}{\theta} = \frac{3060\text{ km}}{0.481\,877\,852\text{ rad}} = 6350.156\,969\text{ km}$$

Rounding to four significant digits, the earth's radius is approximately 6350 km.

With the radius known, we can approximate the circumference:

$$\begin{aligned} c &= 2\pi r \\ &= 2\pi(6350.156\,569) \\ &= 39\,899.212\,94 \end{aligned}$$

The circumference of Earth is approximately 39 900 km. For comparison, the average radius of Earth is 6370 km, and its circumference is 40 000 km.

In observational science and mechanical and building construction, a convention is often used when we specify the direction of an angle relative to the horizontal line of sight. The **angle of elevation** indicates *how far we have to rotate our view above (or up from) the horizontal line of sight to view a feature of interest.* In contrast, the **angle of depression** indicates *how far we have to rotate our view below (or down from) the horizontal line of sight to view the feature of interest.* The two examples that follow illustrate these naming conventions.

■ See the chapter introduction.

EXAMPLE 2 Angle of elevation

Horseshoe Falls on the Canadian side of Niagara Falls can be seen from a small boat 760 m downstream. The angle of elevation from the observer to the top of Horseshoe Falls is 4.0°. How high are the falls?

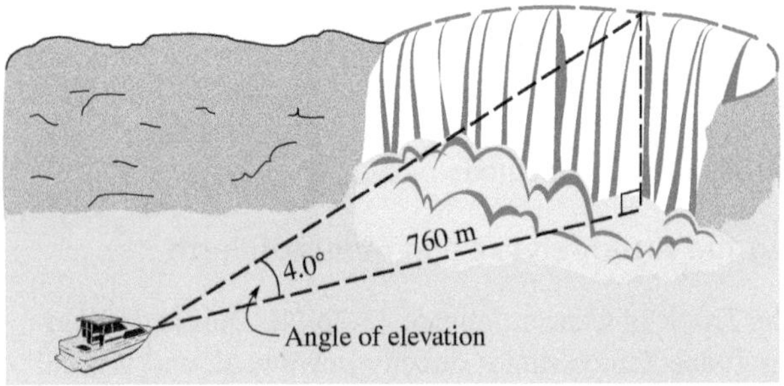

Fig. 4.39

Identify known quantities: Angle of elevation = 4.0°
Distance from boat to falls = 760 m

Identify unknown quantities: Let h = height of the falls (in m)

Sketch: See Fig. 4.39.

Select an equation:

$$\tan 4.0° = \frac{\text{required opposite side}}{\text{given adjacent side}} = \frac{h}{760}$$

Solve: $h = 760\tan 4.0° = 53.144\,377\,08$ m

The approximate height of the falls is 53 m. Note that with as much as 170 000 tonnes of water flowing over the falls each minute, the height of the falls changes over time.

EXAMPLE 3 Angle of depression

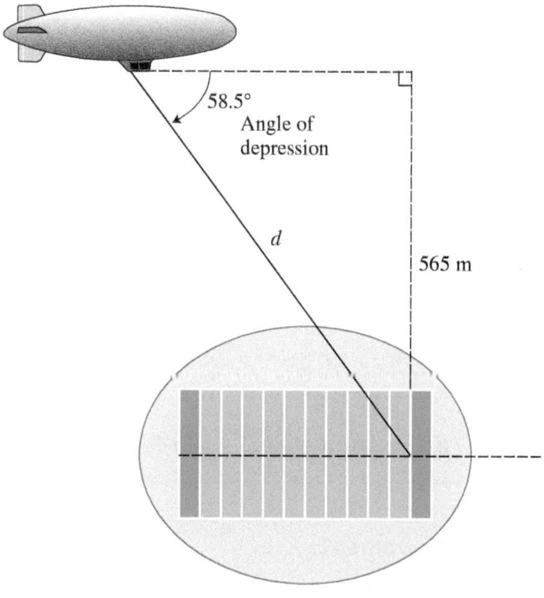

Fig. 4.40

A blimp is 565 m above the ground and south of Olympic Stadium in Montreal during a Grey Cup game. The angle of depression of the north goal line from the blimp is 58.5°. How far is the observer in the blimp from the goal line?

Identify known quantities: Angle of depression = 58.5°
Height of the blimp = 565 m

Identify unknown quantities: Let d = distance from blimp to goal line (in m)

Sketch: See Fig. 4.40.

Select an equation:

$$\sin 58.5° = \frac{\text{known opposite side}}{\text{required hypotenuse}} = \frac{565}{d}$$

Solve:

$$d = \frac{565}{\sin 58.5°} = 662.647\,648\,6 \text{ m}$$

The blimp is approximately 663 m away from the goal line.

In this case, we have rounded off the result to three significant digits, which is the accuracy of the given information.

EXAMPLE 4 Height of a missile

A missile is launched at an angle of 26.55° with respect to the horizontal. If it travels in a straight line over level terrain for 2.000 min and its average speed is 6355 km/h, what is its altitude at this time?

Identify known quantities: Angle with respect to horizontal = 26.55°
Average speed = 6355 km/h
Time = 2.000 min
Distance flown = speed × time
$= (6355 \text{ km}/60 \text{ min}) \times 2.000 \text{ min}$
$= 211.8\overline{3} \text{ km}$

Identify unknown quantities: Let h = altitude of the missile (in km)

Sketch: See Fig. 4.41.

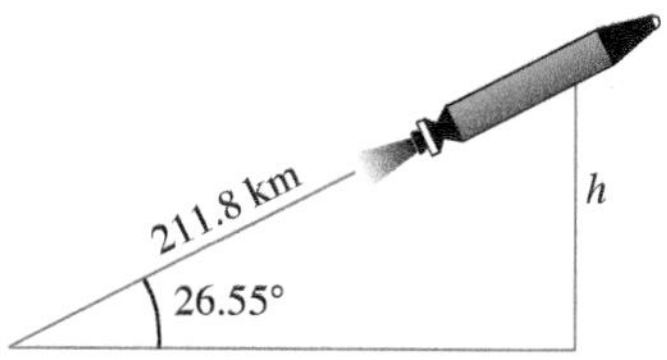

Fig. 4.41

Select an equation:

$$\sin 26.55° = \frac{\text{required opposite side}}{\text{known hypotenuse}} = \frac{h}{211.8\overline{3}}$$

Solve:

$$h = 211.8\overline{3}\,(\sin 26.55°)$$
$$h = 94.684\,971\,15 \text{ km}$$

After 2.000 min, the height of the missile is 94.68 km.

EXAMPLE 5 Measurement of an angle

A driver approaching an intersection sees the word STOP on the pavement, as shown in Fig. 4.42. From the geometry provided, find the angle θ representing the difference in viewing angle from the bottom of the "T" to the top of the "T."

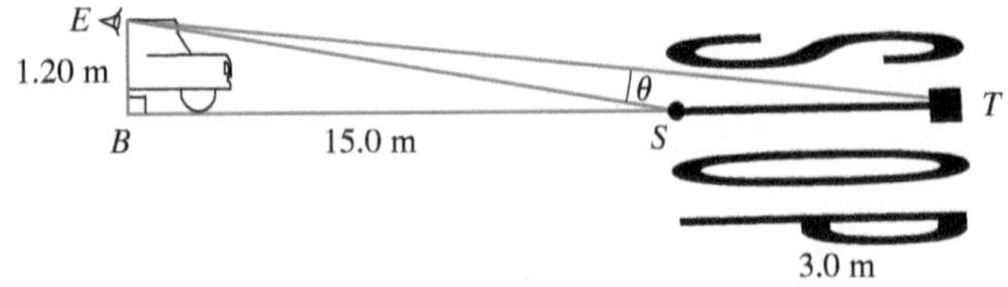

Fig. 4.42

Identify known quantities: $BE = 1.20$ m, $BS = 15.0$ m, $ST = 3.0$ m
$\angle EBS = 90°$, $BT = BS + ST = 18.0$ m

Identify unknown quantities: $\angle TEB$, $\angle SEB$, $\theta = \angle TEB - \angle SEB$

Select equations:

$$\tan \angle TEB = \frac{\text{opposite}}{\text{adjacent}}$$
$$= \frac{18.0}{1.20} = 15$$
$$\tan \angle SEB = \frac{\text{opposite}}{\text{adjacent}}$$
$$= \frac{15.0}{1.20} = 12.5$$

Solve:

$$\angle TEB = \tan^{-1}(15) = 86.2°$$
$$\angle SEB = \tan^{-1}(12.5) = 85.4°$$
$$\theta = \angle TEB - \angle SEB = 86.2° - 85.4° = 0.8°$$

Practice Exercise

1. Find θ if the letters in the road are 2.0 m long, rather than 3.0 m long.

EXAMPLE 6 Surveyor—indirect measurement

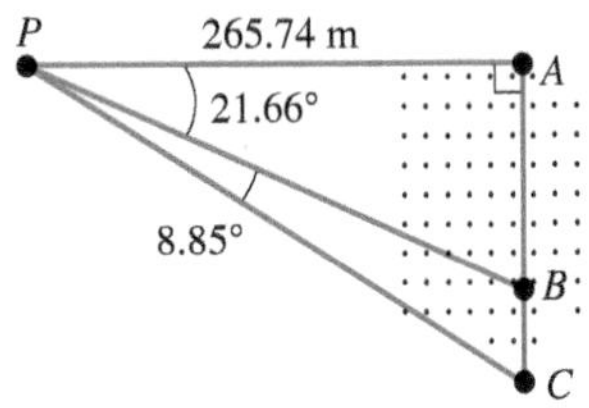

Fig. 4.43

At times it is not practical to take measurements for physical reasons. For example, when land surveyors need to locate a position that falls within a body of water, they make use of lasers and indirect measurements. Fig. 4.43 illustrates such a measurement, where point A is along the edge of a marsh, and points B and C cannot be accessed. Find the distance between B and C.

■ Lasers were first produced in the late 1950s.

Identify known quantities: $PA = 265.74$ m, $\angle PAC = 90°$
$\angle APB = 21.66°$, $\angle BPC = 8.85°$
$\angle APC = \angle APB + \angle BPC = 30.51°$

Identify unknown quantities: AB, AC, $BC = AC - AB$

Select equations:

$$\tan 21.66° = \frac{\text{opposite}}{\text{adjacent}} = \frac{AB}{265.74}$$
$$\tan 30.51° = \frac{\text{opposite}}{\text{adjacent}} = \frac{AC}{265.74}$$

Solve:

$$AC = 265.74 \tan(21.66°) = 105.535\,948\,2 \text{ m}$$
$$AB = 265.74 \tan(30.51°) = 156.595\,302\,3 \text{ m}$$
$$BC = AC - AB = 51.059\,354\,03$$

The distance between B and C is 51.06 m.

EXERCISES 4.5

In Exercises 1 and 2, make the given changes in the indicated examples of this section and then find the indicated values.

1. In Example 3, change the known angle 58.5° to 62.1° and then find the distance.
2. In Example 4, change the known time 2.000 min to 3.000 min and then find the altitude.

In Exercises 3–42, solve the given problems. Sketch an appropriate figure, unless the figure is given.

3. A straight 122-m culvert is built down a hillside that makes an angle of 54.0° with the horizontal. Find the height of the hill.
4. In 2000, about 70 tonnes of soil were removed from under the Leaning Tower of Pisa, and the angle the tower made with the ground was increased by about 0.5°. Before that, a point near the top of the tower was 50.5 m from a point at the base (measured along the tower), and this top point was directly above a point on the ground 4.25 m from the same base point. See Fig. 4.44. How much did the point on the ground move toward the base point?

Fig. 4.44

5. A totem pole has a shadow 2.28 m long when the angle of elevation of the sun is 62.6°. How tall is the totem pole?

6. The straight arm of a robot is 1.25 m long and makes an angle of 13.0° above a horizontal conveyor belt. How high above the belt is the end of the arm? See Fig. 4.45.

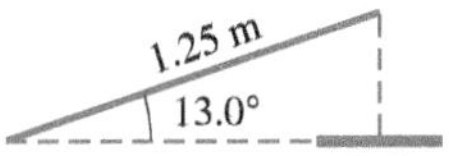

Fig. 4.45

7. The headlights of an automobile are set such that the beam drops 5.10 cm for each 7.50 m in front of the car. What is the angle between the beam and the road?

8. A bullet was fired such that it just grazed the top of a table. It entered a wall, which is 3.84 m from the graze point in the table, at a point 1.41 m above the tabletop. At what angle was the bullet fired above the horizontal? See Fig. 4.46.

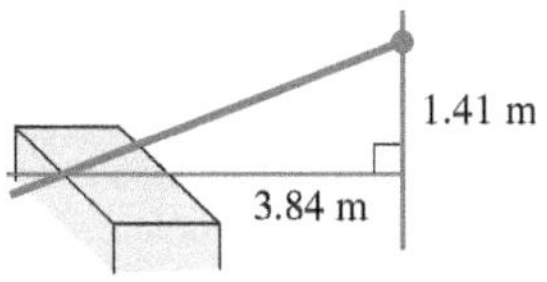

Fig. 4.46

9. A robot is on the surface of Mars. The angle of depression from a camera in the robot to a rock on the surface of Mars is 13.33°. The camera is 196.0 cm above the surface. How far from the camera is the rock?

10. The CN Tower in Toronto can be seen from a point on the ground known to be 1600 m away from the base of the tower. The angle of elevation from the observer to the top of the tower is 19°. How high is the CN Tower? See Fig. 4.47.

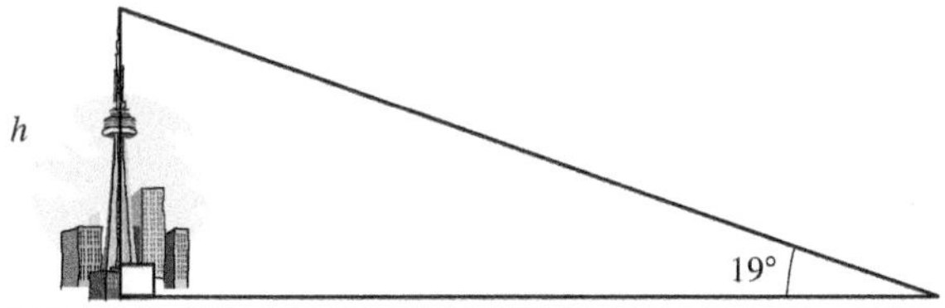

Fig. 4.47

11. In the design of a new building, a doorway is 795 mm above the ground. An entrance ramp, at an angle of 6.00° with the ground, is to be built to the doorway. How long will the ramp be?

12. On a test flight, during the landing of the space shuttle, the ship was 105 m above the end of the landing strip. It then came in on a constant angle of 7.50° with the landing strip. How far from the end of the landing strip did it first touch ground?

13. From the southernmost point in Sleeping Giant Provincial Park in Thunder Bay, Ontario, it is found that the angle of elevation of the highest point on the Sleeping Giant is 3.9°. If the horizontal distance between the points is 5.5 km, how much higher is the point on the summit?

14. What is the steepest angle between the surface of a board 3.50 cm thick and a nail 5.00 cm long if the nail is hammered into the board such that it does not go through?

15. A rectangular piece of plywood 1200 mm by 2400 mm is cut from one corner to an opposite corner. What are the angles between edges of the resulting pieces?

16. A guardrail is to be constructed around the top of a circular observation tower. The diameter of the observation area is 12.3 m. If the railing is constructed with 30 equal straight sections, what should be the length of each section?

17. The angle of inclination of a road is often expressed as *percent grade*, which is the vertical rise divided by the horizontal run (expressed as a percent). See Fig. 4.48. A 6.0% grade corresponds to a road that rises 6.0 m for every 100 m along the horizontal. Find the angle of inclination that corresponds to a 6.0% grade.

Fig. 4.48

18. A tabletop is in the shape of a regular octagon (eight sides). What is the greatest distance across the table if one side of the octagon is 0.750 m?

19. To get a good view of a person in front of a teller's window, it is determined that a surveillance camera at a bank should be directed at a point 5.17 m to the right and 2.25 m below the camera. See Fig. 4.49. At what angle of depression should the camera be directed?

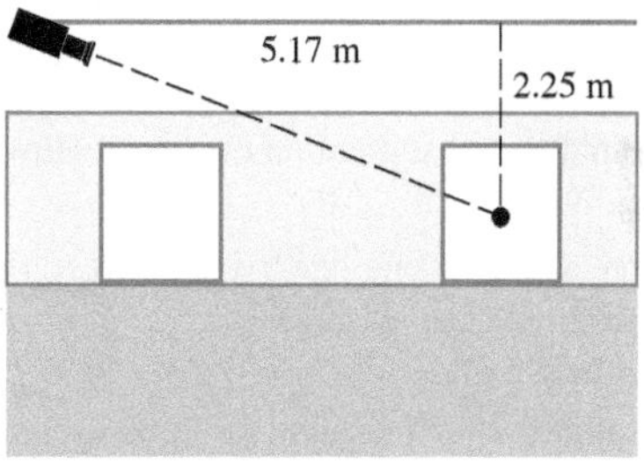

Fig. 4.49

20. A street light is designed as shown in Fig. 4.50. How high above the street is the light?

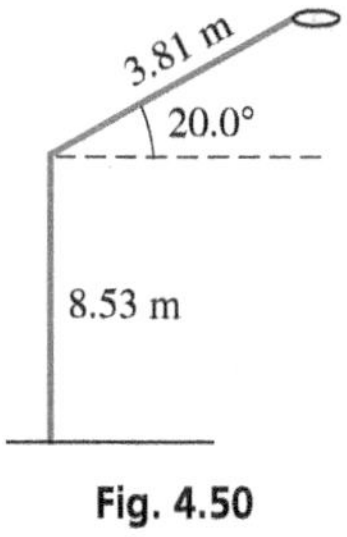

Fig. 4.50

21. A straight driveway is 85.0 m long, and the top is 12.0 m above the bottom. What angle does it make with the horizontal?

22. Part of the Tower Bridge in London is a drawbridge. This part of the bridge is 76.0 m long. When each half is raised, the distance between them is 8.0 m. What angle does each half make with the horizontal? See Fig. 4.51.

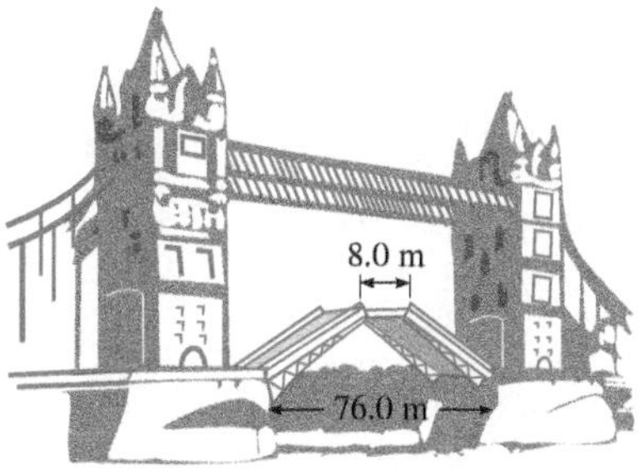

Fig. 4.51

23. A square wire loop is rotating in the magnetic field between two poles of a magnet in order to induce an electric current. The axis of rotation passes through the centre of the loop and is midway between the poles, as shown in the side view in Fig. 4.52. How far is the edge of the loop from either pole if the side of the square is 7.30 cm and the poles are 7.66 cm apart when the angle between the loop and the vertical is 78.0°?

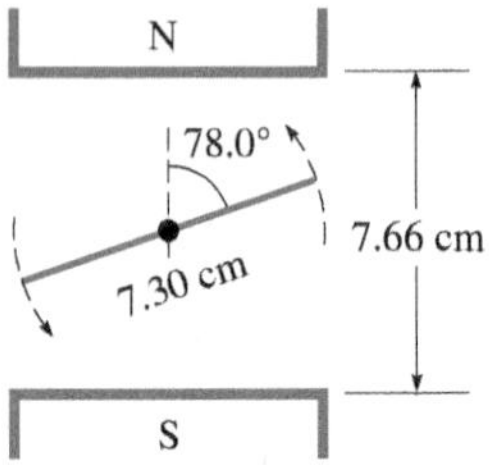

Fig. 4.52

24. From a space probe circling Io, one of Jupiter's moons, at an altitude of 552 km, it was observed that the angle of depression of the horizon was 39.7°. What is the radius of Io?

25. A manufacturing plant is designed to be in the shape of a regular pentagon with 92.5 m on each side. A security fence surrounds the building to form a circle, and each corner of the building is to be 25.0 m from the closest point on the fence. How much fencing is required?

26. A traveller on a boat along the Saguenay River in Quebec wishes to estimate the height of the cliff at Cape Trinity. Fig. 4.53 shows the angle measurements she made at points 250 m apart. How high is the cliff? (In the figure, the triangle containing the height h is vertical and perpendicular to the path followed by the boat. The height of the boat can be ignored.)

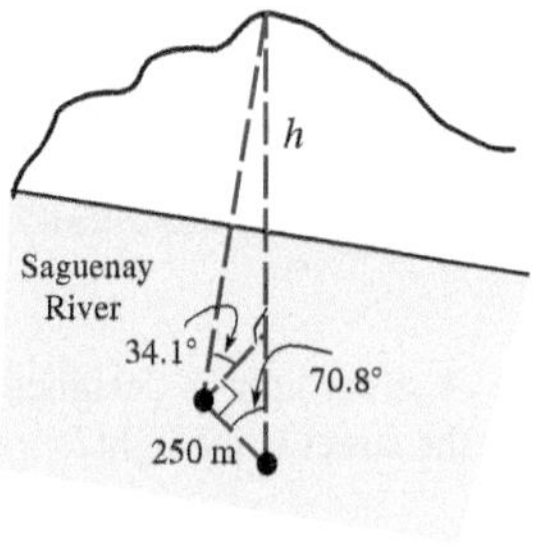

Fig. 4.53

27. Find the angle θ in the taper shown in Fig. 4.54. (The front face is an isosceles trapezoid.)

28. What is the circumference of the Arctic Circle (latitude 66°32′ N)? The radius of the earth is 6370 km.

29. A draftsman sets the legs of a pair of dividers so that the angle between them is 32.0°. If each leg is 11.4 cm long, what is the distance between their ends where they touch the paper?

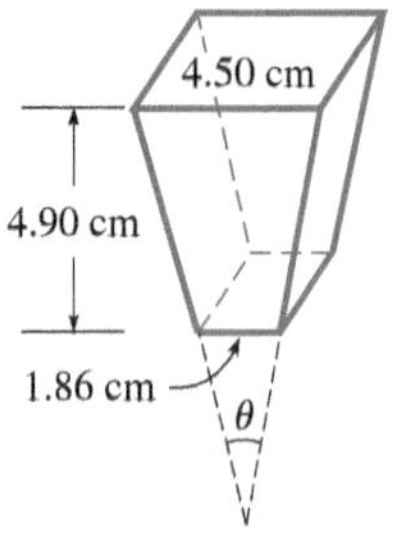

Fig. 4.54

30. The ratio of the width to the height of an HDTV screen is 16 to 9. What is the angle between the width and a diagonal of the screen to the nearest 0.1°?

31. A stairway 1.0 m wide goes from the bottom of a cylindrical storage tank to the top at a point halfway around the tank. The handrail on the outside of the stairway makes an angle of 31.8° with the horizontal, and the radius of the tank is 11.8 m. Find the length of the handrail. See Fig. 4.55.

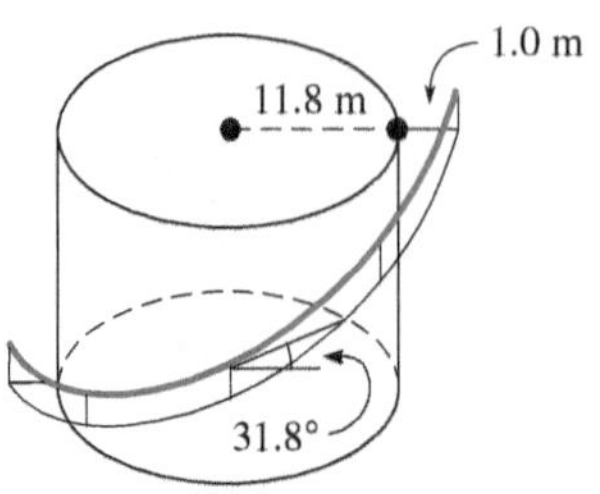

Fig. 4.55

32. An antenna was on the top of the World Trade Center (before it was destroyed in 2001). From a point on the river 2400 m from the Center, the angles of elevation of the top and bottom of the antenna were 12.1° and 9.9°, respectively. How tall was the antenna? (Disregard the small part of the antenna near the base that could not be seen.) The former World Trade Center is shown in Fig. 4.56.

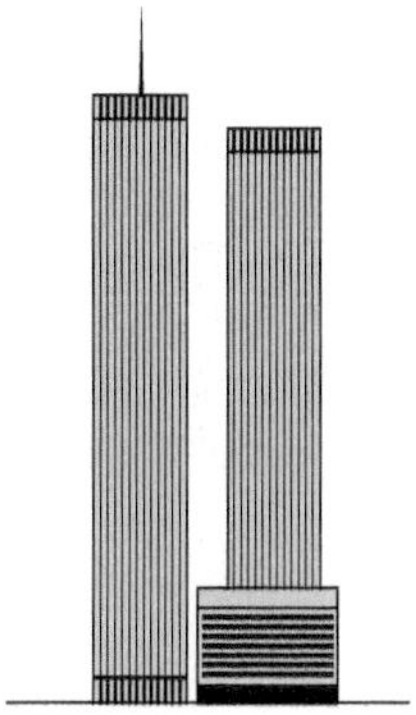
Fig. 4.56

33. Some of the streets of Montreal, Quebec, are shown in Fig. 4.57. (a) How far is it between intersections B and C? (b) How far is it between intersections C and D?

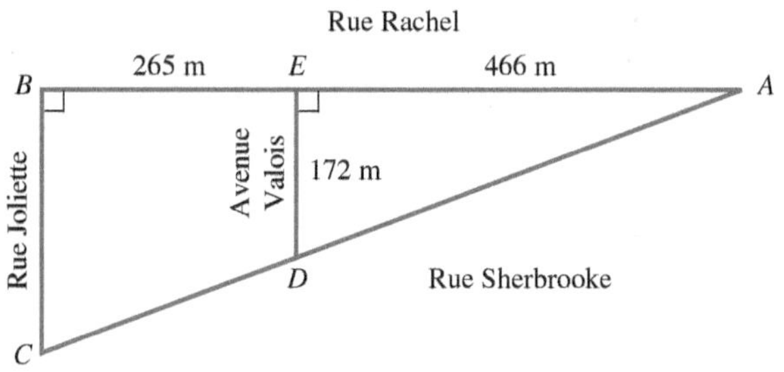

Fig. 4.57

34. A supporting girder structure is shown in Fig. 4.58. Find the length x.

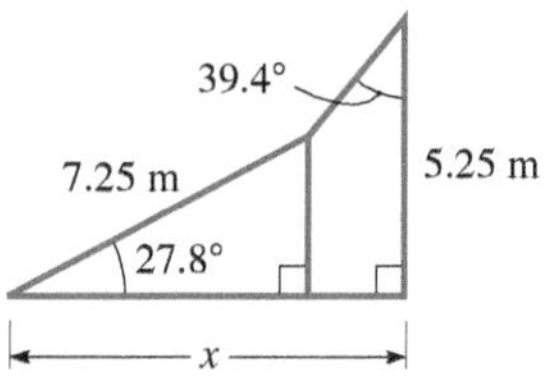

Fig. 4.58

35. The diameter d of a pipe can be determined by noting the distance x on the V-gauge shown in Fig. 4.59. Points A and B indicate where the pipe touches the gauge, and x equals either AV or VB. Find a formula for d in terms of x and θ.

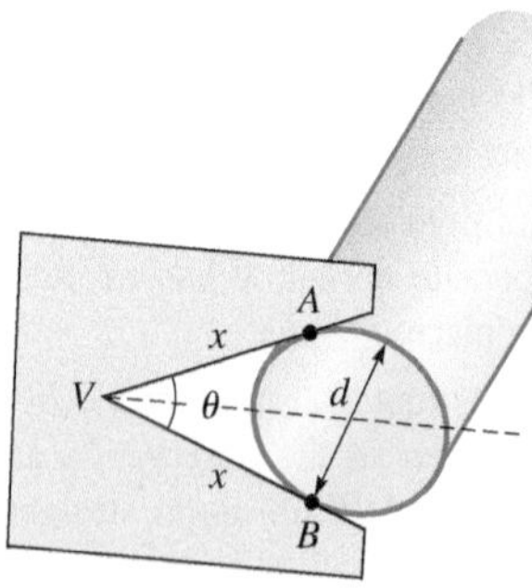

Fig. 4.59

36. The political banner shown in Fig. 4.60 is in the shape of a parallelogram. Find its area.

Fig. 4.60

37. Find a formula for the area of the trapezoidal aqueduct cross-section shown in Fig. 4.61.

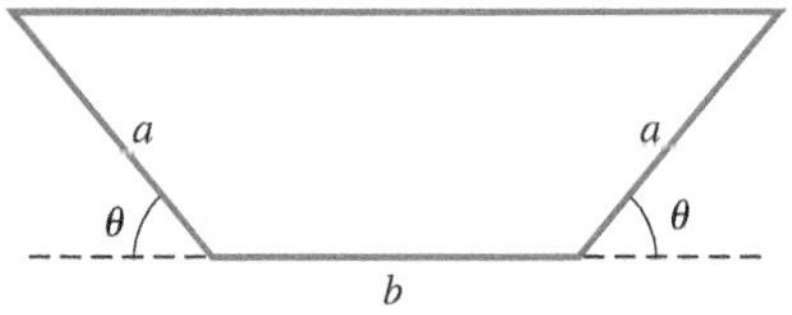

Fig. 4.61

38. A communications satellite is in orbit 35 300 km directly above the earth's equator. What is the greatest latitude ϕ from which a signal can travel from the earth's surface to the satellite in a straight line? The radius of the earth is 6370 km.

39. What is the angle between the base of a cubical glass paperweight and a diagonal of the cube (from one corner to the opposite corner) to three significant digits?

40. Find the angle of view θ of the camera lens (see Fig. 4.62), given the measurements shown in the figure.

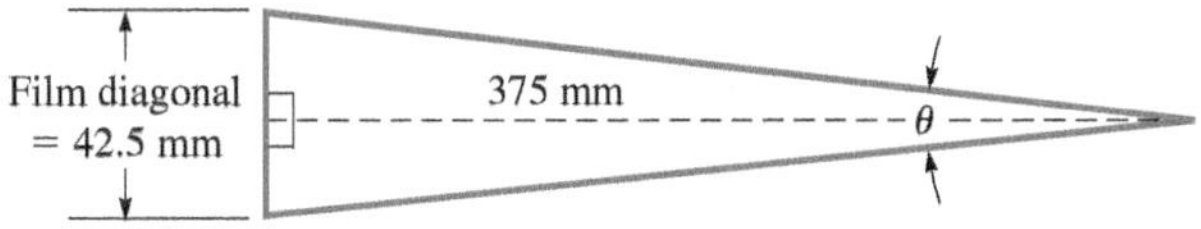

Fig. 4.62

41. Supports on a simple truss bridge are separated 5.00 m from each other, as shown in Fig 4.63. If the centre pole to which all supports are affixed is 8.00 m tall, find the angle θ and the length r of the cross-bracing supports.

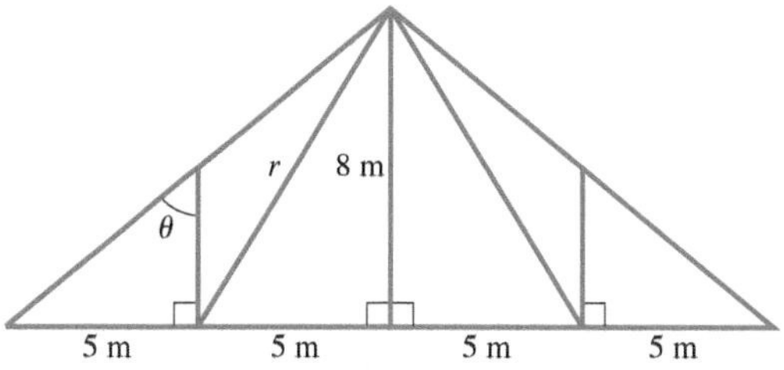

Fig. 4.63

42. A dam of length $L = 10.0$ m is designed to have an inclining angle of 60.0°, as shown in Fig 4.64. If the dam is filled with $h = 6.00$ m of water, calculate the length of the dam wall L_e that remains exposed above the water line.

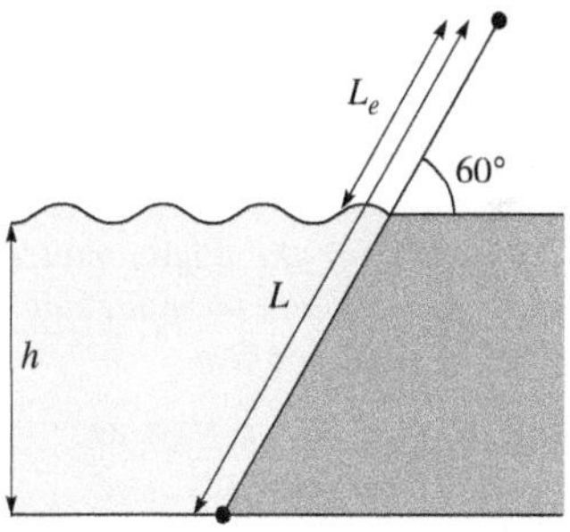

Fig. 4.64

Answer to Practice Exercise

1. $\theta = 0.5°$

CHAPTER 4 KEY FORMULAS AND EQUATIONS

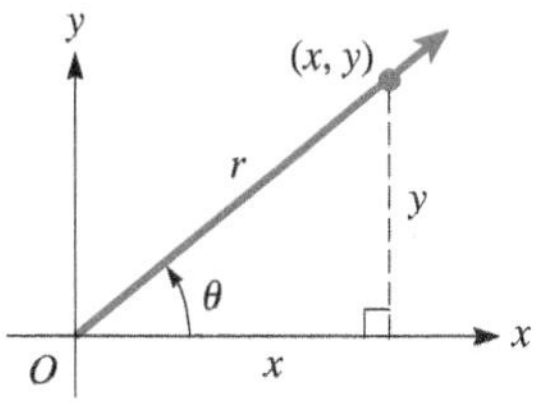

Primary Trigonometric Functions

sine of θ:

$$\sin \theta = \frac{y}{r} \quad \textbf{(4.1a)}$$

cosine of θ:

$$\cos \theta = \frac{x}{r} \quad \textbf{(4.1b)}$$

tangent of θ:

$$\tan \theta = \frac{y}{x} \quad \textbf{(4.1c)}$$

Reciprocal Trigonometric Functions

cosecant of θ:

$$\csc \theta = \frac{1}{\sin \theta} = \frac{r}{y} \quad \textbf{(4.1d)}$$

secant of θ:

$$\sec \theta = \frac{1}{\cos \theta} = \frac{r}{x} \quad \textbf{(4.1e)}$$

cotangent of θ:

$$\cot \theta = \frac{1}{\tan \theta} = \frac{x}{y} \quad \textbf{(4.1f)}$$

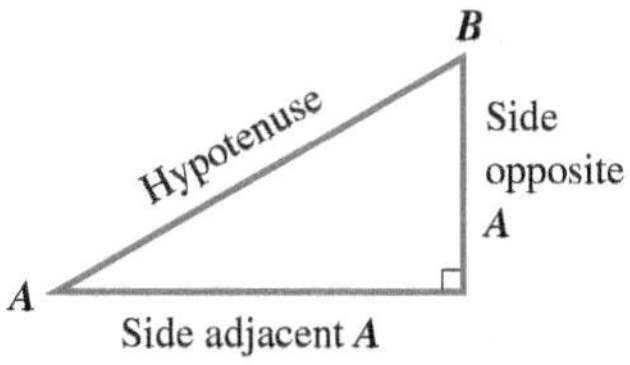

$$\sin A = \frac{\text{side opposite } A}{\text{hypotenuse}} \qquad \csc A = \frac{\text{hypotenuse}}{\text{side opposite } A}$$

$$\cos A = \frac{\text{side adjacent } A}{\text{hypotenuse}} \qquad \sec A = \frac{\text{hypotenuse}}{\text{side adjacent } A} \quad \textbf{(4.4)}$$

$$\tan A = \frac{\text{side opposite } A}{\text{side adjacent } A} \qquad \cot A = \frac{\text{side adjacent } A}{\text{side opposite } A}$$

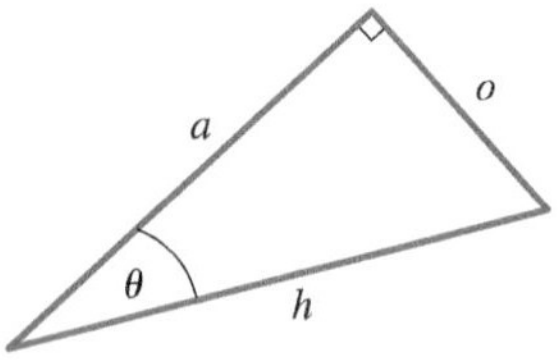

$$\sin\theta = o/h \qquad \csc\theta = h/o$$
$$\cos\theta = a/h \qquad \sec\theta = h/a \qquad (4.5)$$
$$\tan\theta = o/a \qquad \cot\theta = a/o$$

CHAPTER 4 **REVIEW EXERCISES**

In Exercises 1–4, find the smallest positive angle and the smallest negative angle (numerically) coterminal with but not equal to the given angle.

1. $17.0°$ **2.** $248.3°$ **3.** $-217.5°$ **4.** $-7.6°$

In Exercises 5–8, express the given angles in decimal form (to the nearest 0.01°).

5. $31°54'$ **6.** $574°45'$ **7.** $-38°6'$ **8.** $321°27'$

In Exercises 9–12, express the given angles to the nearest minute.

9. $17.5°$ **10.** $-65.4°$ **11.** $749.7°$ **12.** $126.25°$

In Exercises 13–16, determine the trigonometric functions of the angles (in standard position) whose terminal side passes through the given points. Give answers in exact form.

13. $(24, 7)$ **14.** $(5, 4)$ **15.** $(48, 48)$ **16.** $(1.2, 0.5)$

In Exercises 17–20, find the indicated trigonometric functions. Give answers in decimal form, rounded off to three significant digits.

17. Given $\sin\theta = \frac{5}{13}$, find $\cos\theta$ and $\cot\theta$.

18. Given $\cos\theta = \frac{3}{8}$, find $\sin\theta$ and $\tan\theta$.

19. Given $\tan\theta = 2$, find $\cos\theta$ and $\csc\theta$.

20. Given $\cot\theta = 40$, find $\sin\theta$ and $\sec\theta$.

In Exercises 21–28, find the values of the trigonometric functions. Round off results.

21. $\sin 72.1°$ **22.** $\cos 40.3°$ **23.** $\tan 88.64°$

24. $\sin 0.91°$ **25.** $\sec 18.4°$ **26.** $\csc 82.4°$

27. $(\cot 7.06°)(\sin 7.06°) - \cos 7.06°$

28. $(\sec 79.36°)(\sin 79.36°) - \tan 79.36°$

In Exercises 29–40, find θ for each of the given trigonometric functions. Round off results.

29. $\cos\theta = 0.950$ **30.** $\sin\theta = 0.630\,52$ **31.** $\tan\theta = 1.574$

32. $\cos\theta = 0.0135$ **33.** $\csc\theta = 4.713$ **34.** $\cot\theta = 0.7561$

35. $\sec\theta = 25.4$ **36.** $\csc\theta = 1.92$ **37.** $\cot\theta = 7.117$

38. $\sec\theta = 1.006$ **39.** $\sin\theta = 0.9998$ **40.** $\cos\theta = 1.402$

In Exercises 41–52, solve the right triangles with the given parts. Refer to Fig. 4.65.

41. $A = 17.0°$, $b = 6.00$

42. $B = 68.1°$, $a = 1080$

43. $a = 81.0$, $b = 64.5$

44. $a = 106$, $c = 382$

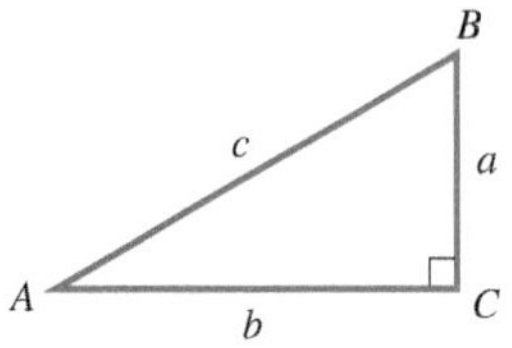

Fig. 4.65

45. $A = 37.5°$, $a = 12.0$ **46.** $B = 15.7°$, $c = 126$

47. $b = 6.508$, $c = 7.642$ **48.** $a = 0.721$, $b = 0.144$

49. $A = 49.67°$, $c = 0.8253$ **50.** $B = 4.38°$, $b = 5682$

51. $a = 11.652$, $c = 15.483$ **52.** $a = 724.39$, $b = 852.44$

In Exercises 53–95, solve the given problems.

53. Find the value of x for the triangle shown in Fig. 4.66.

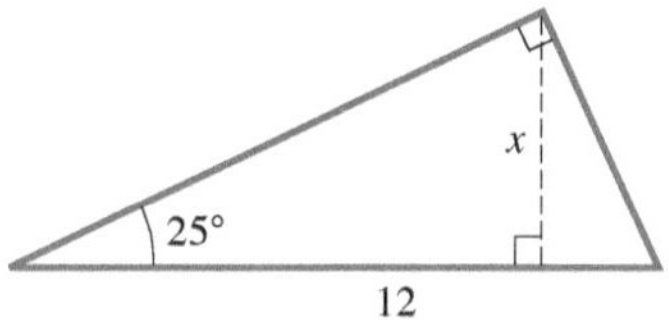

Fig. 4.66

54. Explain three ways in which the value of x can be found for the triangle shown in Fig. 4.67. Which of these methods is the easiest?

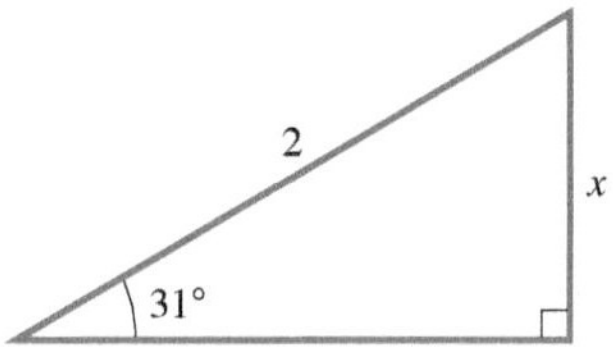

Fig. 4.67

55. Find the perimeter of a regular octagon (eight equal sides with equal interior angles) that is inscribed in a circle (all vertices of the octagon touch the circle) of radius 10.

56. Explain why values of $\sin\theta$ increase as θ increases from 0° to 90°.

57. What is x if $(3, 2)$ and $(x, 7)$ are on the same terminal side of an acute angle?

58. Two legs of a right triangle are 2.607 and 4.517. What is the smaller acute angle?

59. Show that the side c of any triangle ABC is related to the perpendicular h from C to side AB by the equation

$$c = h\cot A + h\cot B.$$

60. For the isosceles triangle shown in Fig. 4.68, show that $c = 2a\sin\frac{A}{2}$.

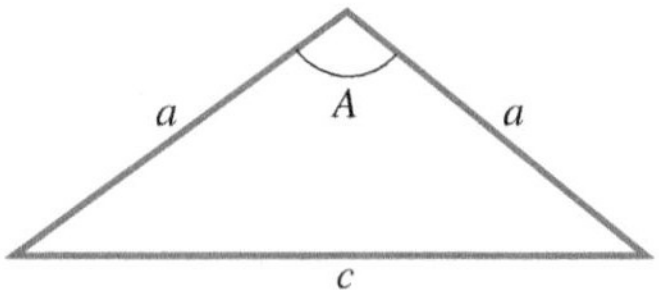

Fig. 4.68

61. If $x = \tan\theta$, what is $\csc\theta$ in terms of x?

62. Find the angle between the line passing through the origin and $(3, 2)$, and the line passing through the origin and $(2, 3)$.

63. A sloped cathedral ceiling is between walls that are 2.50 m high and 4.00 m high. If the walls are 5.00 m apart, at what angle does the ceiling rise?

64. A pendulum 1.25 m long swings through an angle of 5.60°. What is the distance between the extreme positions of the pendulum?

65. The voltage E at any instant in a coil of wire that is turning in a magnetic field is given by $E = E_M \cos \alpha$, where E_M is the maximum voltage and α is the angle the coil makes with the field. Find the acute angle α if $E = 56.9$ V and $E_M = 339$ V.

66. A formula for the area of a quadrilateral is $A = \frac{1}{2}d_1 d_2 \sin \theta$, where d_1 and d_2 are the lengths of the diagonals and θ is the angle between them. Find the area of a four-sided carpet remnant with diagonals 1050 mm and 1330 mm and $\theta = 72.0°$.

67. For a car rounding a curve, the road should be banked at an angle θ according to the equation $\tan \theta = \frac{v^2}{gr}$. Here, v is the speed of the car and r is the radius of the curve in the road. See Fig. 4.69. Find θ for $v = 24.2$ m/s, $g = 9.80$ m/s^2, and $r = 282$ m.

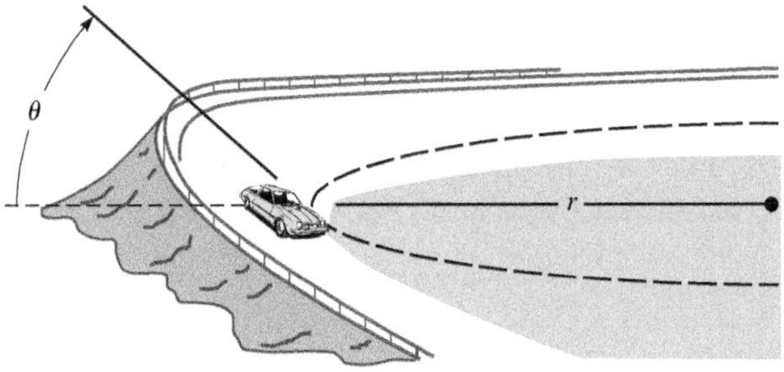

Fig. 4.69

68. The *apparent power* S in an electric circuit in which the power is P and the impedance phase angle is θ is given by $S = P \sec \theta$. Given $P = 12.0$ V · A and $\theta = 29.4°$, find S.

69. A surveyor measures two sides and the included angle of a triangular tract of land to be $a = 31.96$ m, $b = 47.25$ m, and $C = 64.09°$. (a) Show that a formula for the area A of the tract is $A = \frac{1}{2}ab \sin C$. (b) Find the area of the tract.

70. A water channel has the cross-section of an isosceles trapezoid. See Fig. 4.70. (a) Show that a formula for the area of the cross-section is $A = bh + h^2 \cot \theta$. (b) Find A if $b = 12.6$ m, $h = 4.75$ m, and $\theta = 37.2°$.

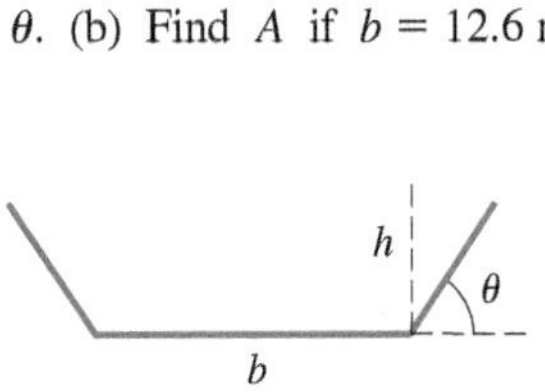

Fig. 4.70

71. In tracking an airplane on radar, it is found that the plane is 27.5 km on a direct line from the control tower, with an angle of elevation of 10.3°. What is the altitude of the plane?

72. A straight emergency chute for an airplane is 5.5 m long. In being tested, the top of the chute is 2.9 m above the ground. What angle does the chute make with the ground?

73. The windshield on an automobile is inclined 42.5° with respect to the horizontal. Assuming that the windshield is flat and rectangular, what is its area if it is 1.50 m wide and the bottom is 0.480 m in front of the top?

74. A water slide at an amusement park is 25.0 m long and is inclined at an angle of 52.0° with the horizontal. How high is the top of the slide above the water level?

75. Find the area of the patio shown in Fig. 4.71.

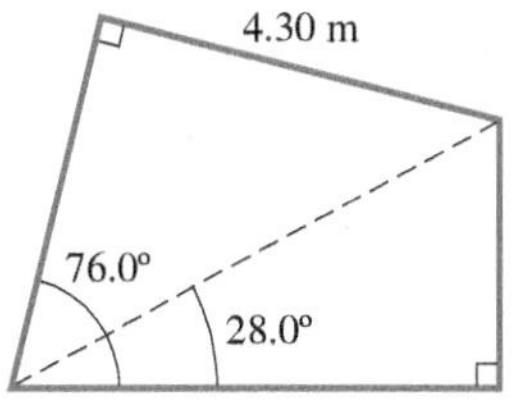

Fig. 4.71

76. The cross-section (a regular trapezoid) of a levee to be built along a river is shown in Fig. 4.72. What is the volume of rock and soil that will be needed for a 1-km length of the levee?

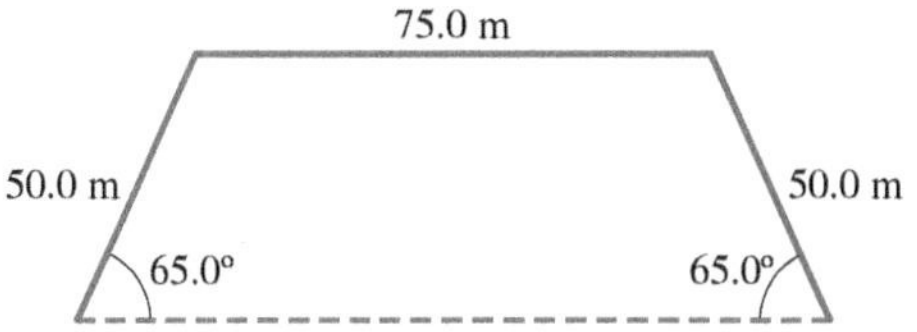

Fig. 4.72

77. The vertical cross-section of an attic room in a house is shown in Fig. 4.73. Find the distance d across the floor.

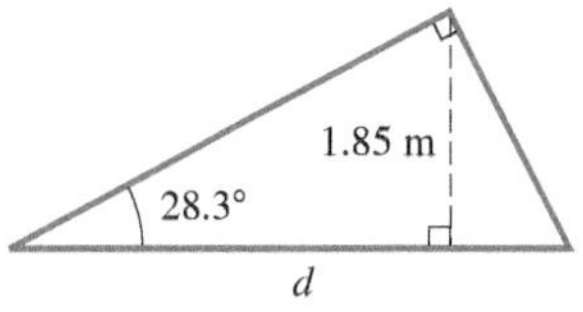

Fig. 4.73

78. The impedance Z and resistance R in an AC circuit may be represented by letting the impedance be the hypotenuse of a right triangle and the resistance be the side adjacent to the phase angle θ. If $R = 1750\ \Omega$ and $\theta = 17.38°$, find Z.

79. A typical aqueduct built by the Romans dropped on average at an angle of about 0.030° to allow gravity to move the water from the source to the city. For such an aqueduct of 65 km in length, how much higher was the source than the city?

80. The distance from the ground level to the underside of a cloud is called the *ceiling*. See Fig. 4.74. A ground observer 950 m from a searchlight aimed vertically notes that the angle of elevation of the spot of light on a cloud is 76°. What is the ceiling?

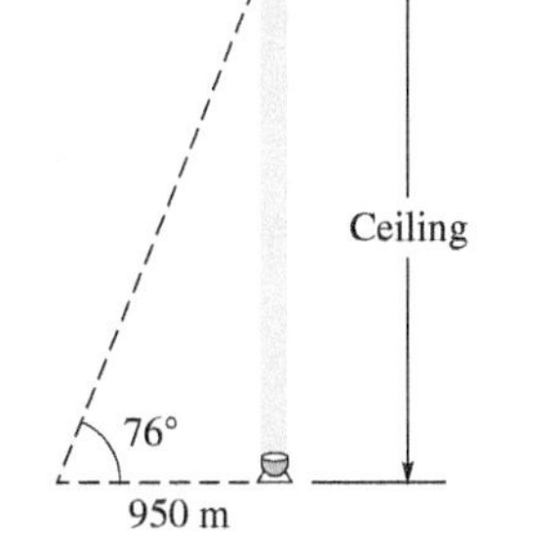

Fig. 4.74

81. The window of a house is shaded as shown in Fig. 4.75. What percent of the window is shaded when the angle of elevation θ of the sun is 65°?

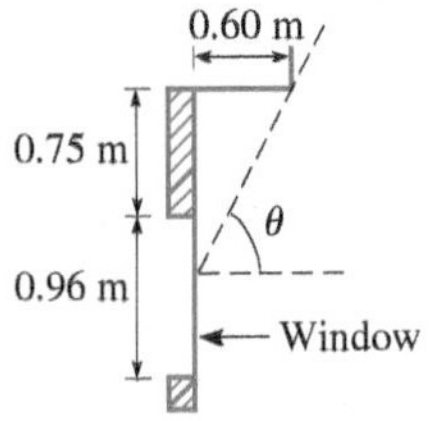

Fig. 4.75

82. A person standing on a level plain hears the sound of a plane and looks in the direction of the sound, but the plane is not there (familiar?). When the sound was heard, it was coming from a point at an angle of elevation of 25°, and the plane was travelling at 720 km/h (201 m/s) at a constant altitude of 850 m along a straight line. If the plane later passes directly over the person, at what angle of elevation should the person have looked directly to see the plane when the sound was heard? (The speed of sound is 340 m/s.) See Fig. 4.76.

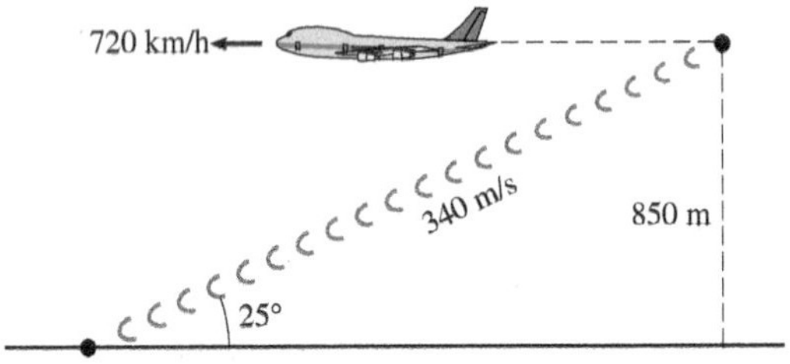

Fig. 4.76

83. For the structural support shown in Fig. 4.77, find x.

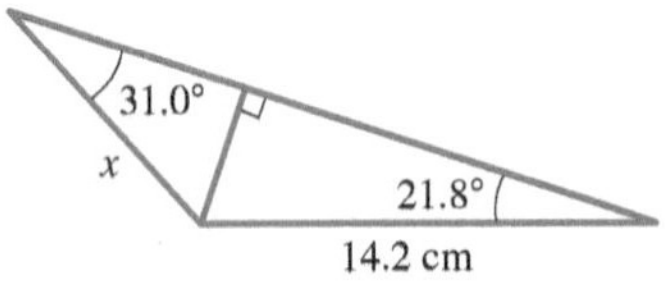

Fig. 4.77

84. A Coast Guard boat 2.75 km from a straight beach can travel at 37.5 km/h. By travelling along a line that is at 69.0° with the beach, how long will it take to reach the beach? See Fig. 4.78.

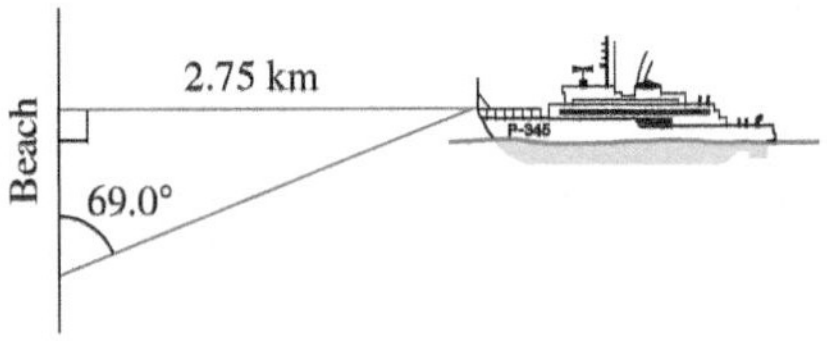

Fig. 4.78

85. One span of the multi-span Thousand Islands Bridge system connecting Canada and the United States over the St. Lawrence River is 230 m long (see Fig. 4.79). The angle *subtended* by the span at the eye of an observer in a helicopter is 2.2°. Show that the distance calculated from the helicopter to the span is about the same if the line of sight is perpendicular to the end or to the middle of the span.

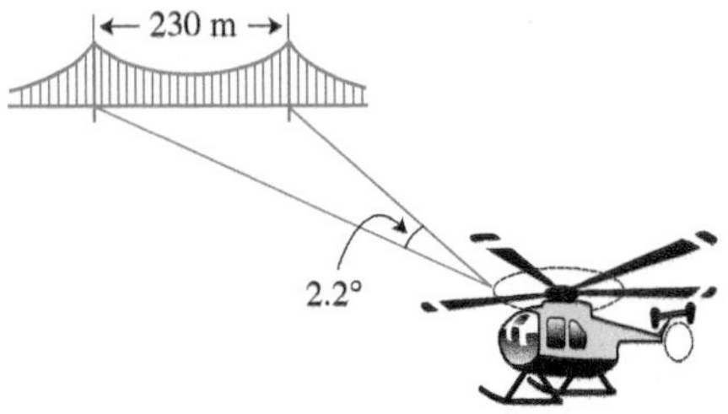

Fig. 4.79

86. Each side piece of the trellis shown in Fig. 4.80 makes an angle of 80.0° with the ground. Find the length of each side piece and the area covered by the trellis.

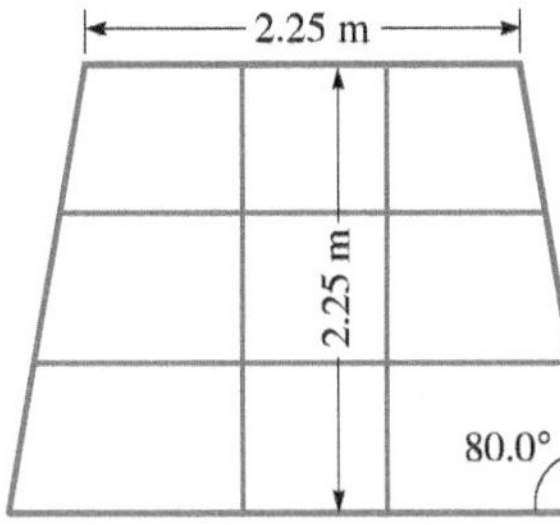

Fig. 4.80

87. A laser beam is transmitted with a "width" of 0.002 00°. What is the diameter of a spot of the beam on an object 52 500 km distant? See Fig. 4.81.

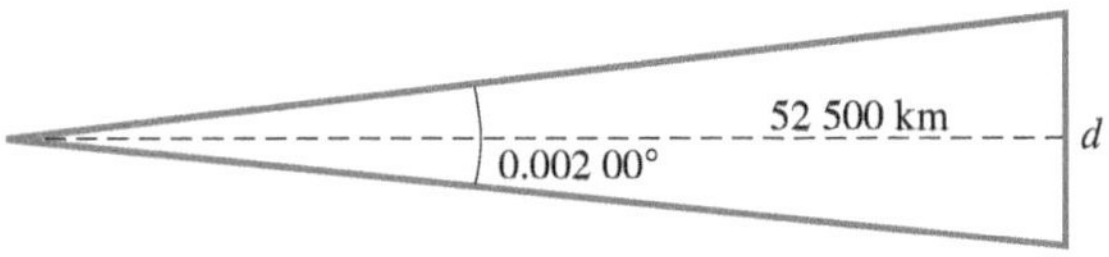

Fig. 4.81

88. Find the gear angle θ in Fig. 4.82 if $t = 0.180$ cm.

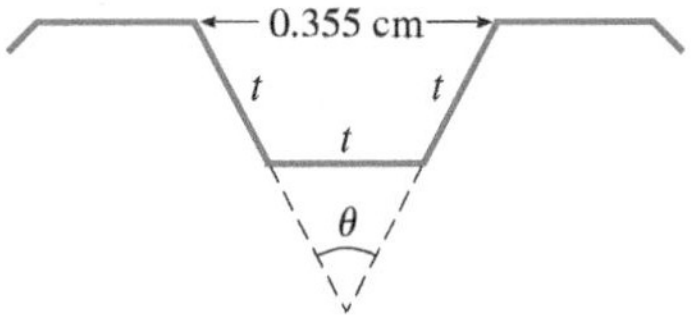

Fig. 4.82

89. The surface of a soccer ball consists of 20 regular hexagons (six sides) interlocked around 12 regular pentagons (five sides). See Fig. 4.83. (a) If the side of each hexagon and pentagon is 45.0 mm, what is the surface area of the soccer ball? (b) Find the surface area, given that the diameter of the ball is 222 mm. (c) Assuming that the given values are accurate, account for the difference in the values found in parts (a) and (b).

Fig. 4.83

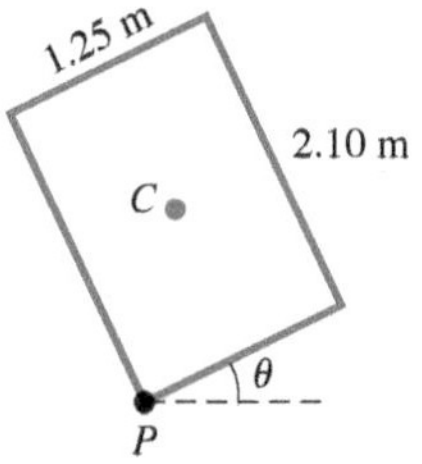

Fig. 4.84

90. Through what angle θ must the crate shown in Fig. 4.84 be tipped in order that its centre of gravity C is directly above the pivot point P?

91. A hang glider is directly above the shore of a lake. An observer on a hill is 375 m along a straight line from the shore. From the observer, the angle of elevation of the hang glider is 42.0°, and the angle of depression of the shore is 25.0°. How far above the shore is the hang glider?

92. A ground observer sights a weather balloon to the east at an angle of elevation of 15.0°. A second observer 2.35 km to the east of the first also sights the balloon to the east at an angle of elevation of 24.0°. How high is the balloon? See Fig. 4.85.

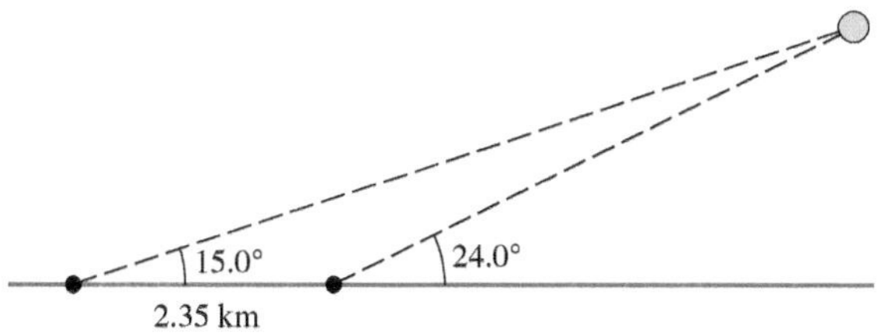

Fig. 4.85

93. A uniform strip of wood 5.0 cm wide frames a trapezoidal window, as shown in Fig. 4.86. Find the left dimension l of the outside of the frame.

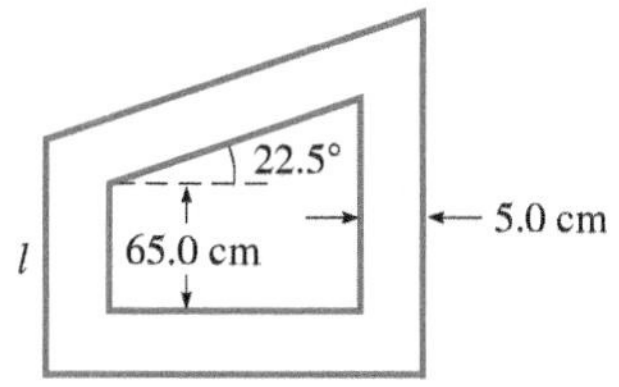

Fig. 4.86

94. A crop-dusting plane flies over a level field at a height of 8.0 m. If the dust leaves the plane through a 30° angle and hits the ground after the plane travels 25 m, how wide a strip is dusted? See Fig. 4.87.

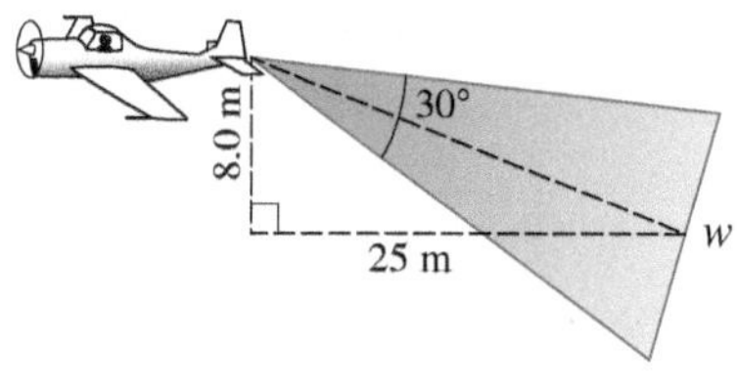

Fig. 4.87

95. A patio is designed in the shape of an isosceles trapezoid with bases 5.0 m and 7.0 m. The other sides are 6.0 m each. Write one or two paragraphs explaining how to use (a) the sine and (b) the cosine to find the internal angles of the patio, and (c) the tangent in finding the area of the patio.

CHAPTER 4 PRACTICE TEST

1. Express 137°29′33″ in decimal form.
2. Find the value of θ if $\cot \theta = 6.2$.
3. Find θ if $\sin \theta = 0.3726$.
4. A ship's captain, desiring to travel due south, discovers that due to an improperly functioning instrument the ship has gone 22.62 km in a direction 4.05° east of south. How far from its course (to the east) is the ship?
5. Find $\tan \theta$ in fractional form if $\sin \theta = \dfrac{2}{3}$.
6. Find $\sec \theta$ if $\tan \theta = 1.294$.
7. Solve the right triangle in Fig. 4.88 if $B = 37.4°$ and $a = 52.8$.
8. Solve the right triangle in Fig. 4.88 if $a = 2.49$ and $c = 3.88$.

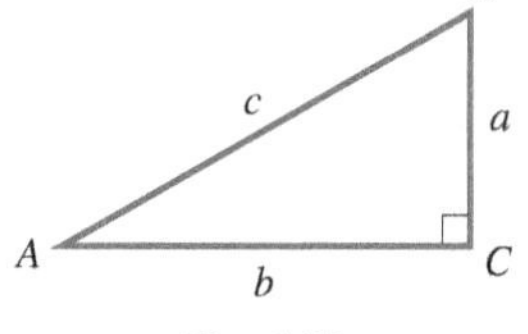

Fig. 4.88

9. The equal sides of an isosceles triangle are each 12.0, and each base angle is 42.0°. What is the length of the third side?
10. If $\tan \theta = 9/40$, find values of $\sin \theta$ and $\cos \theta$. Then evaluate $\sin \theta / \cos \theta$.
11. In finding the wavelength λ (the Greek letter lambda) of light, the equation $\lambda = d \sin \theta$ is used. Find λ if $d = 30.05\ \mu\text{m}$ and $\theta = 1.167°$. (μ is the prefix for 10^{-6}.)
12. Determine the trigonometric functions of an angle in standard position if its terminal side passes through $(5, 2)$. Give answers in exact and decimal forms.
13. The loading ramp at the back of a truck is 3.2 m long. What angle does it make with the ground if the top of the ramp is 1.0 m above the ground?
14. A surveyor sights two points directly ahead. Both are at an elevation 18.525 m lower than the observation point. How far apart are the points if the angles of depression are 13.500° and 21.375°, respectively? See Fig. 4.89.

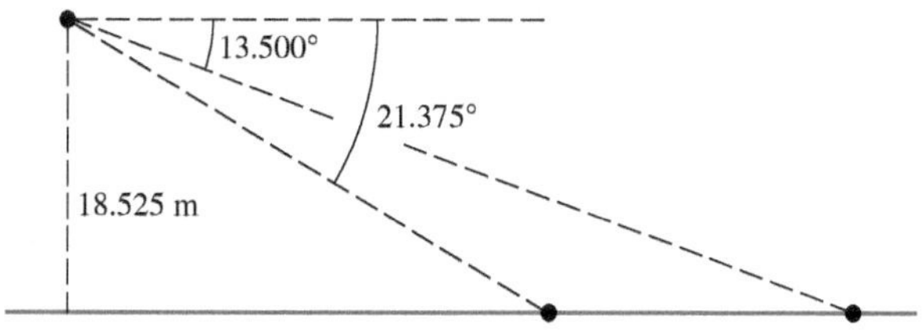

Fig. 4.89

5. Systems of Linear Equations; Determinants

LEARNING OUTCOMES

After completion of this chapter, the student should be able to:

- Identify linear equations
- Calculate the slope of a linear function
- Graph a linear function
- Use the slope–intercept form of a line
- Determine x- and y-intercepts
- Test for a solution of a system of equations
- Solve a system of linear equations graphically
- Identify an inconsistent system of equations, and a dependent system of equations
- Solve a system of two or three linear equations algebraically
- Solve a system of two or three linear equations using determinants (Cramer's rule)
- Solve application problems involving systems of linear equations

▲ Kirchhoff's circuit laws deal with the concepts of conservation of electric charge and energy in electric circuits. The systems of equations generated from these laws for specific circuits will be analysed in this chapter.

When attempting to identify unknown values, we often find that there are several different equations that include a common set of unknown values. For example, when using *Kirchhoff's current law* and *Kirchhoff's voltage law* for electrical circuits (named after the German physicist Gustav Kirchhoff, who formulated these laws in 1848), more than one equation is usually set up. To find the needed information about the circuit, it is necessary to find solutions that satisfy all equations at the same time. The solution of some circuits using Kirchhoff's laws will be shown in this chapter.

Mathematical methods of solving such *systems* of equations were well known to Kirchhoff. In fact, 100 years earlier a book by the English mathematician Colin Maclaurin was published (two years after his death) in which many well organized topics in algebra were covered, including a general method of solving systems of equations. This method is now called *Cramer's rule* (named for the Swiss mathematician Gabriel Cramer, who popularized it in a book he wrote in 1750). We will explain some of the methods of solution, including Cramer's rule, later in the chapter.

Two or more equations that relate variables are found in many fields of science and technology. These include aeronautics, business, transportation, the analysis of forces on a structure, medical doses, and robotics, as well as electric circuits. These applications often require solutions that satisfy all equations simultaneously.

In this chapter, our attention is restricted to *linear* equations (variables occur only to the first power). We will consider systems of two equations with two unknowns and systems of three equations with three unknowns. Systems with other kinds of equations and systems with more unknowns are covered in later chapters.

5.1 Linear Equations and Graphs of Linear Functions

In this chapter, we will discuss various techniques for solving *systems of equations*, which are groups of equations that contain more than one unknown. The methods we study assume that the equations are of a specific type, called *linear equations*. In this first section, we will define a linear equation and describe ways of graphing linear equations in two variables.

LINEAR EQUATIONS

An equation is called **linear** in a given set of variables if:

1. each term contains one variable or is a constant; and
2. each variable is raised to the first power.

EXAMPLE 1 Identifying linear equations

(a) $5x + 6 = 0$ **is** a linear equation. — Each term has one variable or is constant; x is raised to the first power.

(b) $5x^2 - t + 6 = 0$ **is not** a linear equation. — The variable x is squared.

(c) $4x + y = 8$ **is** a linear equation. — Each term has one variable or is constant; both x and y are to the first power.

(d) $4xy = 8$ **is not** a linear equation. — The first term has two variables.

(e) $x - 6y + z = 7$ **is** a linear equation. — Each term has one variable or is constant; x, y, and z are all to the first power.

(f) $x - \dfrac{6}{y} + z - 4w = 7$ **is not** a linear equation. — The variable y appears in the denominator.

Linear equations can have any number of unknowns (or variables). In Example 1, the equation in (a) is a linear equation in one unknown, the equation in (c) is a linear equation in two unknowns, and the equation in (e) is a linear equation in three unknowns. We discussed the solution of linear equations in one unknown in Section 1.10. In this chapter we deal with equations in two and three unknowns.

■ See the chapter introduction.

EXAMPLE 2 Linear equations—Kirchhoff's current and voltage laws

Kirchhoff's current law states that the current going into any junction of an electric circuit equals the current going out. At the top junction in Fig. 5.1, this leads to

$$I_1 = I_2 + I_3$$

Kirchhoff's voltage law states that the sum of the voltages around any closed loop is zero. Going around the loop on the left in Fig. 5.1 (shown in blue) leads to

$$12 - 4I_1 - 3I_2 = 0$$

Both equations in this example are linear equations since the variables I_1, I_2, and I_3 occur only to the first power. Note that a third equation could be found by applying the voltage law to the closed loop on the right in Fig. 5.1.

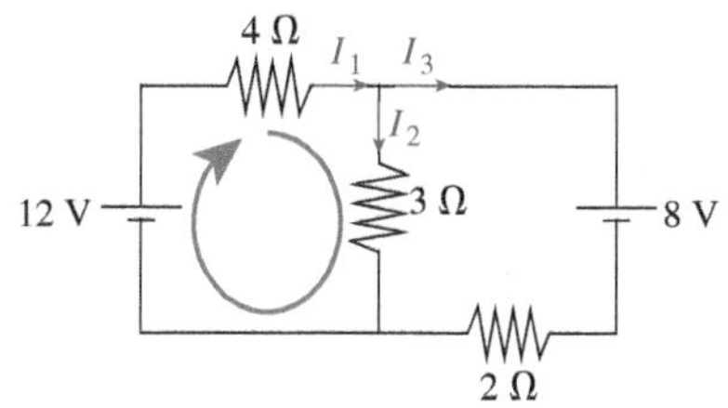

Fig. 5.1

The first method of solving a system of equations that we will discuss relies on graphing linear equations in two unknowns. Before taking this method up in Section 5.2, we will first develop various ways of graphing these equations.

GRAPHING A LINEAR EQUATION IN TWO UNKNOWNS

A **linear equation in two unknowns** is an equation that can be written in the following form:

$$ax + by = c \tag{5.1}$$

LEARNING TIP
A **solution** is any set of numbers, one for each variable, that satisfies the equation.

For each value of x there is a corresponding value of y (found easily by substitution if we write y in terms of x). Each of these ordered pairs of numbers is a *solution* to the equation that can be plotted as a point in the rectangular coordinate system. There are an infinite number of solutions, which together form the **graph** of the equation. As can be seen in the next example, the graph is a straight line.

EXAMPLE 3 Graphing a linear equation in two unknowns

Graph the equation $2x - y - 4 = 0$. Use the points on the line where $x = -1$ and $x = 3$.

Rearrange the equation and write y in terms of x (so y is a **linear function** of x).	We can rewrite the equation as $y = 2x - 4$. Written in this form, we can substitute any number in for x and then find the corresponding value of y that makes the equation true.	
Identify two values of x that lie inside the region of interest.	Here we are given two values: $x = -1$ and $x = 3$. Any other values of x chosen would result in the same line.	
By substitution, find the values of y corresponding to the values of x selected.	Substituting $x = -1$: $y = 2(-1) - 4 = -6$ $(-1, -6)$ is a solution.	Substituting $x = 3$: $y = 2(3) - 4 = 2$ $(3, 2)$ is a solution.
Graph the equation in the rectangular coordinate system.	Plot the two points and connect them with a straight line. See Fig. 5.2.	
Check your work with a different point on the line.	From Fig. 5.2, the point $(1, -2)$ that is on the line should be a solution to the equation. Substituting $x = 1$ and $y = -2$ into the original equation we get $0 = 0$, so it is a valid solution and the graph is correct.	

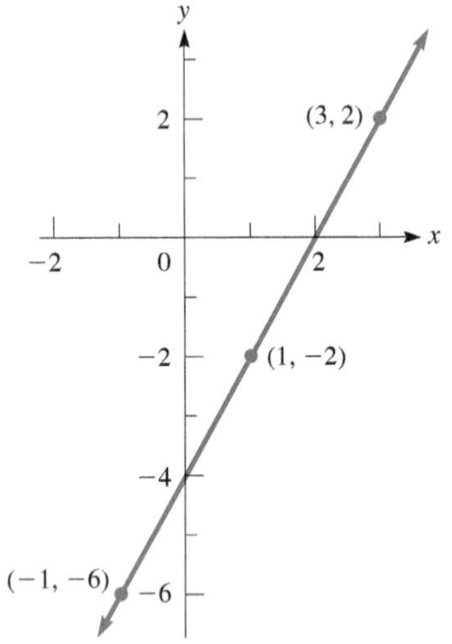

Fig. 5.2

Practice Exercise

1. Is $x = -3$, $y = 2$ a solution to the equation of Example 3?

LINEAR FUNCTIONS AND SLOPE

The **slope of a line** is a constant value that describes how much of a change we anticipate in the dependent variable y as a result of a change in the independent variable x. If we pick two points on the line in Fig. 5.3, point A with coordinates (x_1, y_1), and point B with coordinates (x_2, y_2), the slope is found by taking the difference in the y values divided by the difference in the x values:

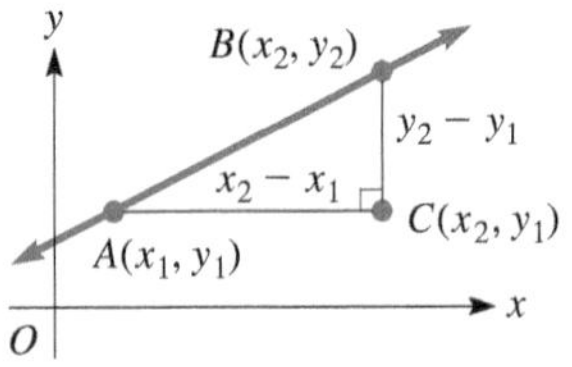

Fig. 5.3

$$m = \frac{y_2 - y_1}{x_2 - x_1} \tag{5.2}$$

An alternative form of the slope equation is $m = \dfrac{\Delta y}{\Delta x}$, where $\Delta y = y_2 - y_1$ is the difference in the y-coordinates, and $\Delta x = x_2 - x_1$ is the difference in the x-coordinates.

The slope is often referred to as the *rise* (vertical change) over the *run* (horizontal change). Note that the slope of a vertical line, for which $x_2 = x_1$, is undefined (for $x_2 = x_1$, the denominator of Eq. 5.2 is zero). Also, the slope of a horizontal line, for which $y_1 = y_2$, is zero.

EXAMPLE 4 Positive slope of a line through two points

The line drawn in Fig 5.4 includes the points $(2, -3)$ and $(5, 3)$. Determine the slope and direction of the line.

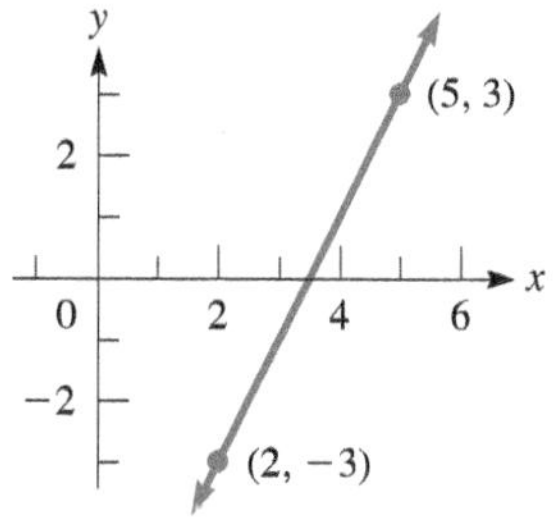

Fig. 5.4

Assign subscripts to points. — Let $(x_1, y_1) = (2, -3)$ and $(x_2, y_2) = (5, 3)$. The assignment of subscripts is arbitrary, but once the assignments are made, the order must be maintained to avoid sign errors.

Use Eq. (5.2) to find the slope. — $$m = \frac{y_2 - y_1}{x_2 - x_1} = \frac{3 - (-3)}{5 - 2} = \frac{6}{3} = 2$$

Interpret the slope. — As the value of x increases, for each change of one unit in x, the value of y will rise (increase) by two units. The **positive slope** indicates that the line moves upward when going from left to right.

EXAMPLE 5 Negative slope of a line through two points

The line drawn in Fig 5.5 includes the points $(-1, 2)$ and $(3, -1)$. Determine the slope and direction of the line.

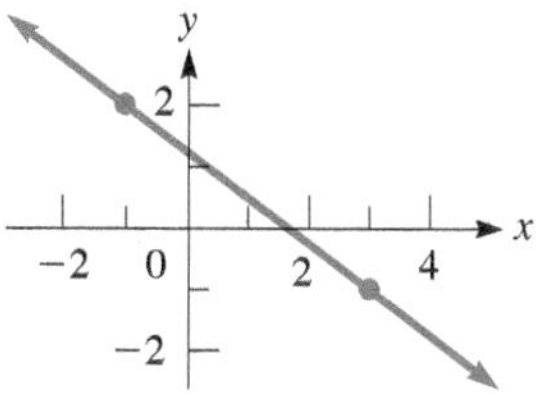

Fig. 5.5

Assign subscripts to points. — Let $(x_1, y_1) = (-1, 2)$ and $(x_2, y_2) = (3, -1)$.

Use Eq. (5.2) to find the slope. — $$m = \frac{y_2 - y_1}{x_2 - x_1} = \frac{-1 - 2}{3 - (-1)} = -\frac{3}{4}$$

Interpret the slope. — As the value of x increases, for each change of one unit in x, the value of y will fall (decrease) by ¾ of a unit. Alternatively, for an increase of four units in x, there is a decrease of three units in y. The **negative slope** indicates that the line moves downward when going from left to right.

Practice Exercise

2. Find the slope of the line through $(-2, 4)$ and $(-5, -6)$.

LEARNING TIP
The larger the absolute value of the slope, the steeper is the line. The closer the slope is to zero, the more the line approaches a horizontal line.

EXAMPLE 6 Slope and steepness of a line

For each of the lines in Fig. 5.6, calculate the slope and describe the direction and steepness of the line.

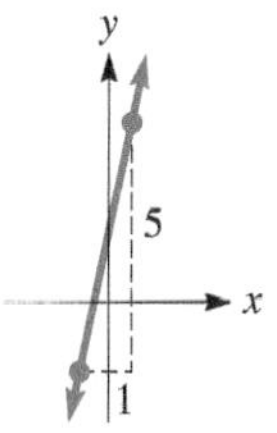

Fig. 5.6(a)

$$m = \frac{\Delta y}{\Delta x} = \frac{5}{1} = 5$$
The line rises sharply.

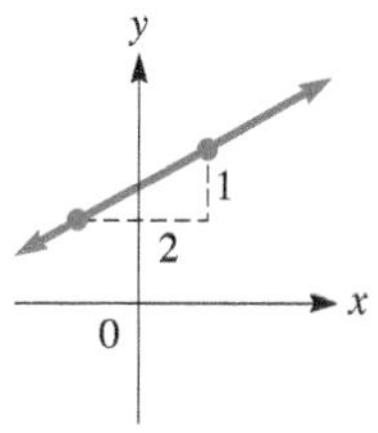

Fig. 5.6(b)

$$m = \frac{\Delta y}{\Delta x} = \frac{1}{2} = 0.5$$
The line rises slowly.

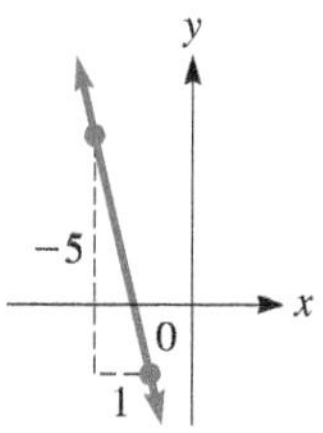

Fig. 5.6(c)

$$m = \frac{\Delta y}{\Delta x} = \frac{-5}{1} = -5$$
The line falls sharply.

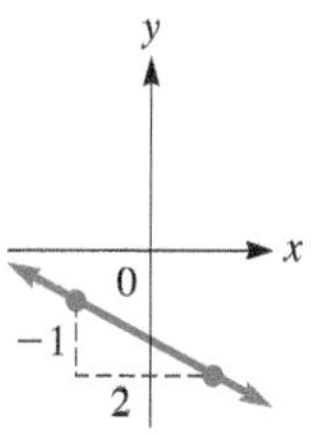

Fig. 5.6(d)

$$m = \frac{\Delta y}{\Delta x} = \frac{-1}{2} = -0.5$$
The line falls slowly.

SLOPE–INTERCEPT FORM OF THE EQUATION OF A STRAIGHT LINE

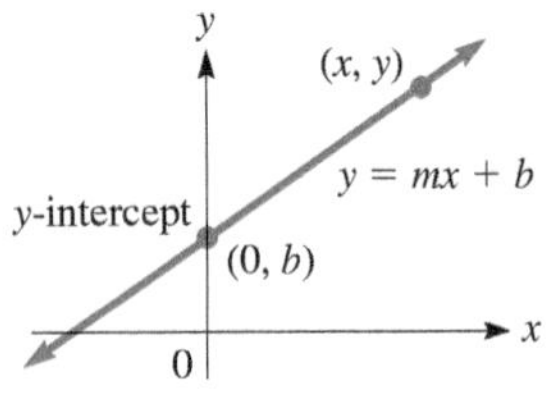

Fig. 5.7

In Fig. 5.7, if we have two points, a general point (x, y) and the point $(0, b)$ where the line crosses the y-axis, the slope is

$$m = \frac{y - b}{x - 0}$$

Simplifying and rearranging algebraically, we obtain the **slope–intercept** form of a line:

$$y = mx + b \tag{5.3}$$

In Eq. (5.3), m is the slope, and b is the y-coordinate of the point where it crosses the y-axis. *This point is the* ***y-intercept*** *of the line,* and its coordinates are $(0, b)$. As illustrated in the following example, the line can be graphed using the y-intercept and another point found using the interpretation of the slope.

LEARNING TIP

$y = mx + b$ is the **slope–intercept** form of the equation of a straight line. The coefficient of x is the slope, and the added constant is the ordinate of the y-intercept. The point $(0, b)$ and simply b are both referred to as the y-intercept.

EXAMPLE 7 Graphing from the slope–intercept form

Graph the line $y = \frac{3}{2}x - 3$ using the y-intercept and the slope.

Identify the slope and the y-intercept.	The equation is in slope–intercept form. The slope is the coefficient of x, so $m = 3/2$. The constant term is -3, so the y-intercept is the point $(0, -3)$.
Interpret the slope.	Since $m = 3/2$, for every two units of increase in x, there will be an increase of three units in y.
Find another point on the line.	Starting at the y-intercept $(0, -3)$, increase x by two units, and increase y by three units, giving the new point $(2, 0)$.
Graph the line.	The resulting line is graphed in Fig 5.8.

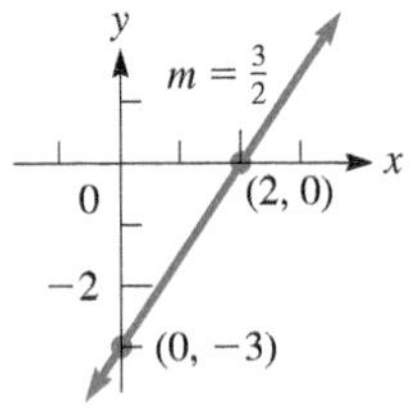

Fig. 5.8

EXAMPLE 8 Graphing from the slope–intercept form of an equation

Graph the line $2x + 3y = 4$ using the y-intercept and the slope.

Identify the slope and the y-intercept. — We first rearrange the equation algebraically into slope–intercept form. Solving for y, we have

$$3y = -2x + 4$$
$$y = -\frac{2}{3}x + \frac{4}{3}$$

The slope is the coefficient of x, so $m = -2/3$. The constant term is $4/3$, so the y-intercept is the point $(0, 4/3)$.

Interpret the slope.	Since $m = -2/3$, for every three units of increase in x, there will be a decrease of two units in y.
Find another point on the line.	Starting at the y-intercept $(0, 4/3)$, increase x by three units, and decrease y by two units, giving the new point $(3, -2/3)$.
Graph the line.	The resulting line is graphed in Fig 5.9.

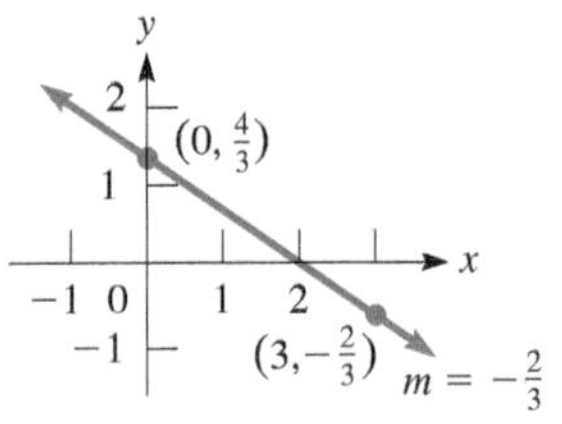

Fig. 5.9

Practice Exercise

3. Write the equation $3x - 5y - 15 = 0$ in slope–intercept form.

EXAMPLE 9 Slope–intercept form—Fahrenheit and Celsius temperatures

The equation $F = \frac{9}{5}C + 32$ relates degrees Fahrenheit (F) and degrees Celsius (C). Graph this equation using the slope–intercept form.

Identify the dependent and independent variables.	This equation is in slope–intercept form, expressing F as a function of C. Therefore F is the dependent variable (graphed on the y-axis), and C is the independent variable (graphed on the x-axis).
Identify the slope and the y-intercept.	The slope is the coefficient of C, so $m = 9/5$. The constant term is 32, so the y-intercept is the point $(0, 32)$. It represents the freezing point of water $(0°C = 32°F)$.
Interpret the slope.	Since $m = 9/5$, for every five degrees of increase in C, there will be an increase of nine degrees in F.
Find another point on the line.	Starting at the y-intercept $(0, 32)$, increase C by five units, and increase F by nine units, giving the new point $(5, 41)$.
Graph the line.	The resulting line is graphed in Fig. 5.10.

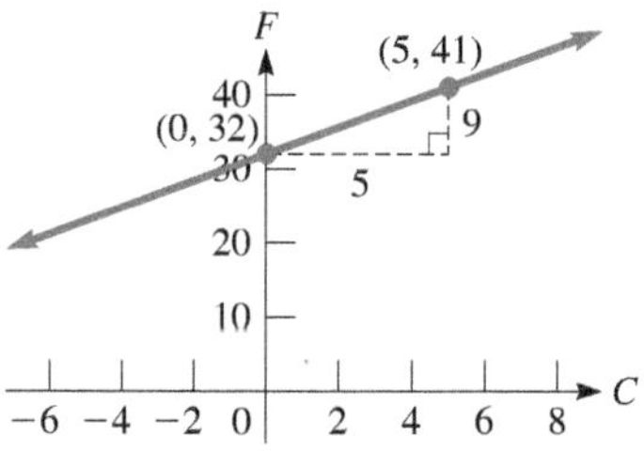

Fig. 5.10

■ Further details and discussion of slope and the graphs of linear equations are found in Chapter 21.

SKETCHING LINES BY INTERCEPTS

Another way of sketching the graph of a straight line is to find the points where the line crosses the y-axis and the x-axis and then draw the line through these points. We already know that the point where it crosses the y-axis is the y-intercept. In the same way, *the point where it crosses the x-axis is called the* ***x*-intercept**.

x-intercept: It is the point where the graph crosses the x-axis. Set $y = 0$ and solve for x.

y-intercept: It is the point where the graph crosses the y-axis. Set $x = 0$ and solve for y.

EXAMPLE 10 Sketching a line by intercepts

Sketch the graph of the line $2x - 3y = 6$ by finding its intercepts and one check point.

x-intercept: Set $y = 0$ and solve for x.

$$2x - 3(0) = 6$$
$$x = 3$$

The x-intercept is $(3, 0)$.

y-intercept: Set $x = 0$ and solve for y.

$$2(0) - 3y = 6$$
$$y = -2$$

The y-intercept is $(0, -2)$.

The intercepts are enough to sketch the line as shown in Fig. 5.11. To find a check point, we can use any value for x other than 3 or any value of y other than -2. Choosing $x = 1$, we find that $y = -\frac{4}{3}$. This means that the point $\left(1, -\frac{4}{3}\right)$ should be on the line. In Fig. 5.11, we can see that it is on the line.

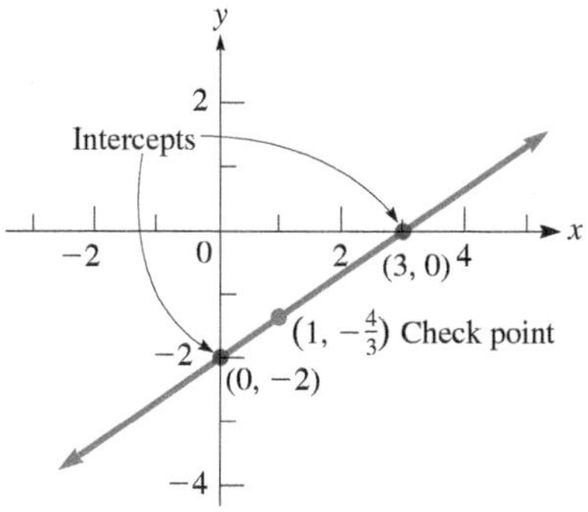

Fig. 5.11

Practice Exercise

4. Find the intercepts of the line $3x - 5y - 15 = 0$.

EXERCISES 5.1

In Exercises 1–4, answer the given questions about the indicated examples of this section.

1. In Example 3, what is the solution if $x = 4$?

2. In Example 5, if the first y-coordinate is changed to -2, what changes result in the example?

3. In the first line of Example 8, if $+$ is changed to $-$, what changes result in the example?

4. In the first line of Example 10, if $-$ is changed to $+$, what changes occur in the graph?

In Exercises 5–8, determine whether or not the given equation is linear.

5. $8x - 3y = 12$ **6.** $2v + 3t^2 = 60$

7. $2l + 3w = 4lw$ **8.** $I_1 + I_3 = I_2$

In Exercises 9–12, determine whether or not the given pair of values is a solution to the given linear equation in two unknowns.

9. $2x + 3y = 9$; $(3, 1), (5, \frac{1}{3})$

10. $5x + 2y = 1$; $(0.2, -1), (1, -2)$

11. $-3x + 5y = 13$; $(-1, 2), (4, 5)$

12. $4y - x = -10$; $(2, -2), (2, 2)$

In Exercises 13–16, for each given value of x, determine the value of y that gives a solution to the given linear equation in two unknowns.

13. $3x - 2y = 12$; $x = 2, x = -3$

14. $6y - 5x = 60$; $x = -10, x = 8$

15. $x - 4y = 2$; $x = 3, x = -0.4$

16. $3x - 2y = 9$; $x = \frac{2}{3}, x = -3$

In Exercises 17–26, find the slope of the line that passes through the given points.

17. $(1, 0), (3, 8)$ **18.** $(3, 1), (2, -7)$

19. $(-1, 2), (-4, 17)$ **20.** $(-1, -2), (2, 10)$

21. $(5, -3), (-2, -5)$ **22.** $(-3, 4), (-7, -4)$

23. $(0.4, 0.5), (-0.2, 0.2)$ **24.** $(-2.8, 3.4), (1.2, 4.2)$

25. $(3.2, -4.1), (-1.5, -10.2)$ **26.** $(-2.5, -8.9), (-6.4, -3.1)$

In Exercises 27–36, sketch the line with the given slope and y-intercept.

27. $m = 2, (0, -1)$ **28.** $m = 3, (0, 1)$ **29.** $m = 0, (0, 2)$

30. $m = -4, (0, -2)$ **31.** $m = \frac{1}{2}, (0, 0)$ **32.** $m = \frac{2}{3}, (0, -1)$

33. $m = -9, (0, 20)$ **34.** $m = -0.3, (0, -1.4)$

35. $m = -\frac{4}{5}, (0, 7)$ **36.** $m = 4.2, (0, -5)$

In Exercises 37–44, find the slope and the y-intercept of the line with the given equation and sketch the graph using the slope and the y-intercept.

37. $y = -2x + 1$ **38.** $y = -4x$ **39.** $y = x - 4$

40. $y = \frac{4}{5}x + 2$ **41.** $5x - 2y = 40$ **42.** $-2y = 7$

43. $24x + 40y = 15$ **44.** $1.5x - 2.4y = 3.0$

In Exercises 45–52, find the x-intercept and the y-intercept of the line with the given equation. Sketch the line using the intercepts.

45. $x + 2y = 4$ **46.** $3x + y = 3$

47. $4x - 3y = 12$ **48.** $5y - x = 5$

49. $y = 3x + 6$ **50.** $y = -2x - 4$

51. $y = -12x + 30$ **52.** $y = 0.25x + 4.5$

In Exercises 53–56, solve the given problems.

53. Find the intercepts of the line $\frac{x}{a} + \frac{y}{b} = 1$.

54. Find the intercepts of the line $y = mx + b$.

55. Do the points $(1, -2), (3, -3), (5, -4), (7, -6)$ and $(11, -7)$ lie on the same straight line?

56. The points $(-2, 5)$ and $(3, 7)$ are on the same straight line.

(a) Find another point on the line for which $x = -12$.
(b) Find another point on the line for which $y = -3$.

In Exercises 57–61, sketch the indicated lines.

57. The diameter of the large end, d (in cm), of a certain type of machine tool can be found from the equation $d = 0.2l + 1.2$, where l is the length of the tool. Sketch d as a function of l, for values of l to 10 cm. See Fig. 5.12.

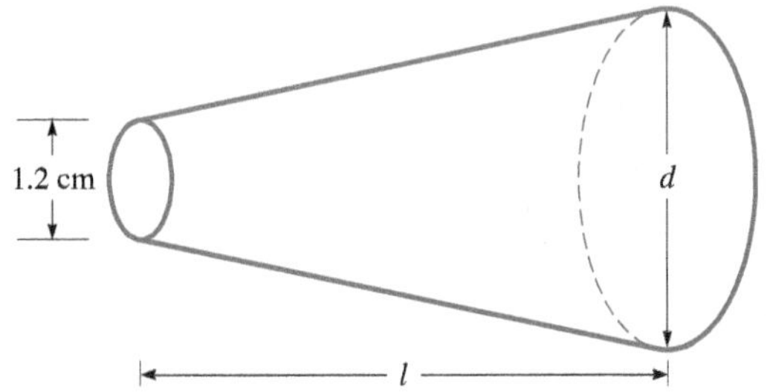

Fig. 5.12

58. In mixing 85-octane gasoline and 93-octane gasoline to produce 91-octane gasoline, the equation $0.85x + 0.93y = 910$ is used. Sketch the graph.

59. Two electric currents, I_1 and I_2 (in mA), in part of a circuit in a computer are related by the equation $4I_1 - 5I_2 = 2$. Sketch I_2 as a function of I_1. These currents can be negative.

60. The total number of messages sent and received by Canadians via multimedia messaging services (MMS) and short message services (SMS) in 2015 was 195 billion messages. In 2016, the total had reached 200 billion. Assuming the growth is linear, set up a function for the total number N (in billions) of MMS and SMS messages at time t (in years), with $t = 0$ being 2015. Sketch the graph for $t = 0$ to $t = 5$.

61. Chemists often use an Arrhenius plot, which linearly relates $\ln k$ to $1/T$ with the equation $\ln k = -\frac{E_a}{RT} + \ln A$, where $\ln k$ is taken as the dependent variable, and $\frac{1}{T}$ as the independent one. Using the equation provided, what are the slope and the y-intercept of the Arrhenius equation?

Answers to Practice Exercises

1. No **2.** 10/3 **3.** $y = \frac{3}{5}x - 3$ **4.** $(5, 0)$ and $(0, -3)$

5.2 Systems of Equations and Graphical Solutions

In many applications, there is *more than one unknown* that we wish to solve for. To accomplish this, we must also have more than one equation that involves these unknowns. In fact, we need as many equations as there are unknowns. A group of two or more equations involving two or more unknowns is called a **system of equations**. In the remaining sections of this chapter, we will discuss various techniques for solving systems of linear equations. We begin by studying systems of two equations with two unknowns.

Two linear equations, each containing the same two unknowns,

$$\begin{aligned} a_1x + b_1y &= c_1 \\ a_2x + b_2y &= c_2 \end{aligned} \qquad \textbf{(5.4)}$$

are said to form a **system of linear equations in two unknowns**. *A* **solution of the system** *is a pair of numbers (x, y) that make* ***both equations true***. Except in special circumstances, there will only be one pair of numbers that make *both* equations true. The following two examples illustrate the meaning of a solution of a system.

EXAMPLE 1 Testing for solution of a system

Is $(2, 8)$ a solution to the system $x + 2y = 18$ and $-5x + 2y = 6$?

We substitute the values into the two equations to ensure they produce valid statements. Both statements must produce a valid check; otherwise, the proposed solution is not valid.

$x + 2y = 18$	$-5x + 2y = 6$
$2 + 2(8) = 18$	$-5(2) + 2(8) = 6$
$18 = 18$	$6 = 6$
Valid check	Valid check

Since both statements are valid, $x = 2$, $y = 8$ is a solution of the system.

EXAMPLE 2 Testing for solution of a system—finding forces

Are the forces in the following system given by $F_1 = 18$ N and $F_2 = 41$ N?

$$\begin{aligned} 2F_1 + 4F_2 &= 200 \\ F_2 &= 2F_1 \end{aligned}$$

We enter the values into the two equations to ensure they produce valid statements.

$2F_1 + 4F_2 = 200$	$F_2 = 2F_1$
$2(18) + 4(41) = 200$	$41 = 2(18)$
$200 = 200$	$41 = 36$
Valid check	Invalid statement

Since one of these statements is not true, $F_1 = 18$ N, $F_2 = 41$ N is *not* a solution of the system.

SOLUTION BY GRAPHING

In Section 5.1, we saw that the graph of a linear equation in two unknowns is a straight line. A system of two equations with two shared unknowns can thus be graphed as two straight lines. When the lines cross, there will be one solution at the pair (x, y) that satisfies both equations (the **point of intersection**). When the lines are parallel and never cross, there will be no valid solution to the system. When both equations represent the same line, there will be an infinite number of solutions.

It is important to note that the graphical approach will often provide only an estimate of the point of intersection. Technology can be used to find the point more accurately. Algebraic approaches, which will be discussed in later sections, should be applied when an exact solution is required.

EXAMPLE 3 Determine the point of intersection

Find the (x, y) pair that solves the system $y = x - 3$ and $y = -2x + 1$.

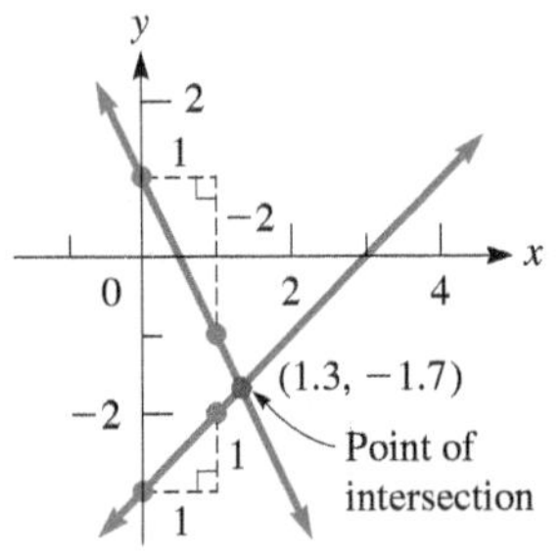

Fig. 5.13

Graph: From the slope–intercept form of the first equation we note that $m = 1$ and the y-intercept is $(0, -3)$. From $(0, -3)$, we increase one unit in x and one unit in y to find a second point, then plot the line by connecting the two points.
For the second equation, $m = -2$ and the y-intercept is $(0, 1)$. From $(0, 1)$ we increase one unit in x and decrease two units in y to find a second point, then plot the line by connecting the two points. See Fig. 5.13.

Find the point of intersection: From the figure, it can be seen that the two lines cross at approximately $(x, y) = (1.3, -1.7)$. (The exact algebraic solution is $x = \frac{4}{3}, y = -\frac{5}{3}$.)

Check: When $x = 1.3$ and $y = -1.7$ are substituted into both equations,
$y = x - 3$ gives $-1.7 = -1.7$ (valid check), and
$y = -2x + 1$ gives $-1.7 = -1.6$ (not a check, but close).
Note that the lack of an exact match is a reminder that while our estimate is close, it is not the exact solution.

COMMON ERROR **When checking the solution, ensure that you substitute the values into *both* equations.** It is a common error to substitute the values into only one equation, which, even if it checks, may not illustrate a valid solution for both equations simultaneously.

EXAMPLE 4 Determine the point of intersection

Find the (x, y) pair that solves the system $2x + 5y = 10$ and $3x - y = 6$.

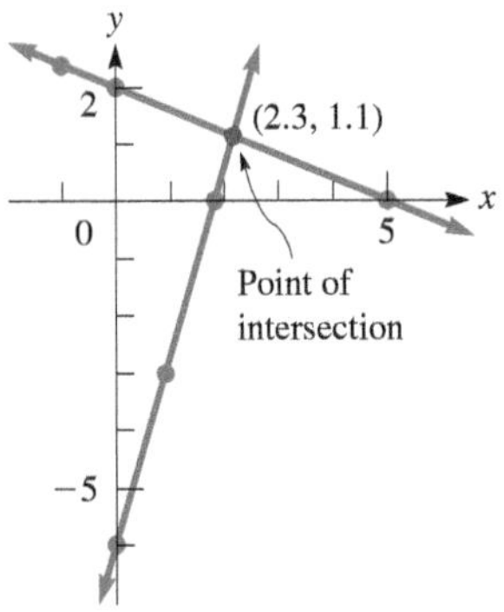

Fig. 5.14

Graph: For each equation, we set $y = 0$ to find the x-intercept, and $x = 0$ to find the y-intercept.
The intercepts of $2x + 5y = 10$ are $(5, 0)$ and $(0, 2)$.
The intercepts of $3x - y = 6$ are $(2, 0)$ and $(0, -6)$.
Straight lines are drawn connecting each pair of intercepts. See Fig. 5.14.

Find the point of intersection: From the figure, it can be seen that the two lines cross at approximately $(x, y) = (2.3, 1.1)$. (The exact algebraic solution is $x = \frac{40}{17}, y = \frac{18}{17}$.)

Check: We substitute both values into both equations:

$$\begin{aligned} 2x + 5y &= 10 \qquad & 3x - y &= 6 \\ 2(2.3) + 5(1.1) &= 10 & 3(2.3) - 1.1 &= 6 \\ 10.1 &= 10 & 5.8 &= 6 \end{aligned}$$

This shows that the solution is correct to the accuracy we can get from the graph.

Practice Exercise

1. In Example 4, change the 6 to 3, and then solve. (This can be done using Fig. 5.14.)

SOLVING SYSTEMS OF EQUATIONS USING TECHNOLOGY

A graphing utility can be used to find the point of intersection with much greater accuracy than is possible by hand-sketching the lines. Once we locate the point of intersection, the features of the utility allow us to get the accuracy we need.

EXAMPLE 5 Using a graphing utility to solve a system of equations

Use a graphing utility to determine the solution to the system of equations in Example 3 to four significant digits.

Let $y_1 = x - 3$ and $y_2 = -2x + 1$. After displaying the graphs in the proper display settings, identify the point of intersection. Several features such as *snap to intersection*, or *trace* and *zoom* can be used to improve the precision of the point you read off where the lines cross. Using these features, we find the improved approximation to be $x = 1.333$, $y = -1.667$.

APPLICATIONS INVOLVING TWO LINEAR EQUATIONS

Linear equations in two unknowns are often useful in solving applied problems. Just as in Section 1.12, *we must read the statement carefully in order to identify the unknowns and the information for setting up the equations.*

EXAMPLE 6 Solving a system

A driver travelled for 1.5 h at a constant speed along a highway. Then, through a construction zone, the driver reduced the car's speed by 40 km/h for 30 min. If 200 km were covered in the 2.0 h, determine the speed along the highway, and the speed through the construction zone.

Identify known quantities: Highway time: $t_h = 1.5$ h; construction zone time: $t_c = 0.5$ h; total distance: $d = 200$ km

Identify unknown quantities: Let v_h = highway speed, and v_c = speed in the construction zone (in km/h)

Select two equations:

1. The two velocities are related directly by

$$v_c = v_h - 40$$

2. The total distance can be found by multiplying the travel time by the speed in each portion of the trip and then adding:

$$d = v_h t_h + v_c t_c$$
$$200 = 1.5v_h + 0.5v_c$$

Identify dependent and independent variables and sketch the graph: We arbitrarily choose v_h as the independent variable and v_c as the dependent variable. The intercepts are $(40, 0)$ and $(0, -40)$ for the first equation, and $(400/3, 0)$ and $(0, 400)$ for the second equation. The sketch is shown in Fig. 5.15.

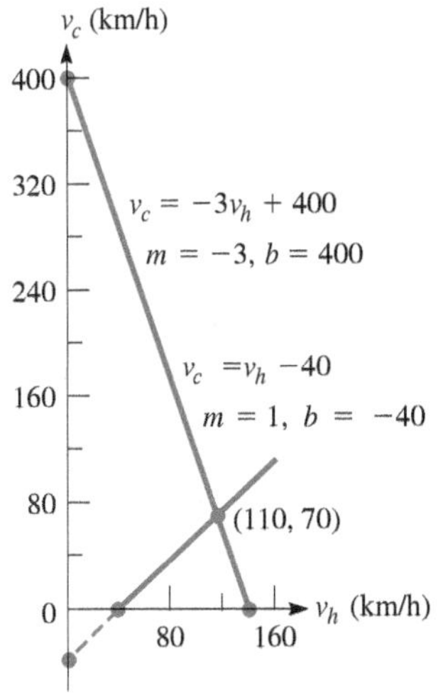

Fig. 5.15

Solve the two equations: From the figure we see that the point of intersection is $(110, 70)$, which means that the highway speed was $v_h = 110$ km/h and the speed in the construction zone was $v_c = 70$ km/h.

Check: Checking in the statement of the problem, we have

$(1.5\text{ h})(110\text{ km/h}) + (0.5\text{ h})(70\text{ km/h}) = 200$ km and
$70\text{ km/h} = 110\text{ km/h} - 40\text{ km/h}$

Both the quantities and the units match.

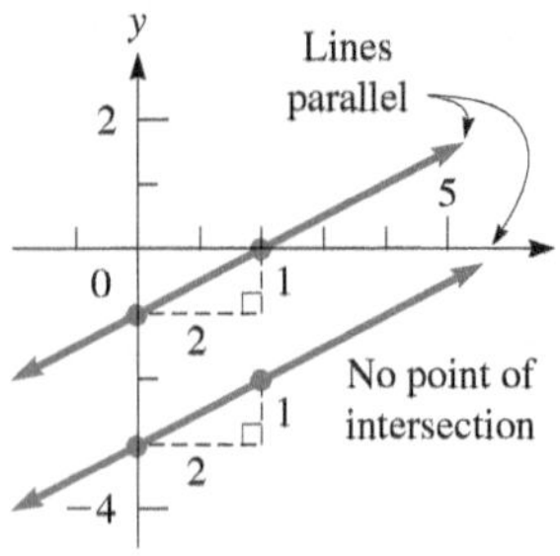

Fig. 5.16

INCONSISTENT AND DEPENDENT SYSTEMS

When the two lines of a system of equations cross at a single point, the system is called **consistent** and **independent**. However, *not all systems have just one such solution.* The following examples illustrate the two other possible scenarios: a system that has *no solution* and a system that has an *infinite number of solutions.*

EXAMPLE 7 Inconsistent system

Solve the system of equations $x = 2y + 6$ and $6y = 3x - 6$.
Writing each of these equations in slope–intercept form, we have

$$y = \frac{1}{2}x - 3 \text{ and } y = \frac{1}{2}x - 1$$

From these, we see that both lines have a slope of $\frac{1}{2}$. As a result, *the lines are parallel and do not intersect,* as shown in Fig. 5.16. This means that ***there are no solutions*** for this system of equations. *Such a system is called* **inconsistent**.

Practice Exercise

2. Is the system in Example 7 inconsistent if the −6 in the second equation is changed to −12?

EXAMPLE 8 Dependent system

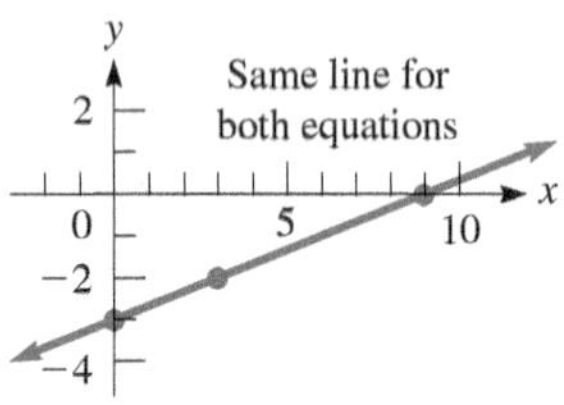

Fig. 5.17

Solve the system of equations $x - 3y = 9$ and $-2x + 6y = -18$.

We write each equation in slope–intercept form. This gives us the equation $y = \frac{1}{3}x - 3$ for both lines. This means that the two equations do not provide two different conditions relating the two variables. They describe only one, and are therefore the same line. See Fig. 5.17.

Since the lines are the same, *the coordinates of any point on this common line constitute a solution of the system—that is to say, there are an infinite number of solutions.* Since ***no unique solution can be determined,*** *the system is called* **dependent**.

Later in the chapter, we will show algebraic ways of finding out whether a given system is consistent, inconsistent, or dependent.

EXERCISES 5.2

In Exercises 1–4, answer the given questions about the indicated examples of this section.

1. In Example 2, is $F_1 = 20$ N, $F_2 = 40$ N a solution?
2. In the second equation of Example 4, if − is replaced with +, what is the solution?
3. In Example 7, what changes occur if the 6 in the first equation is changed to a 2?
4. In Example 8, by changing what one number in the first equation does the system become (a) inconsistent? (b) consistent?

In Exercises 5–12, determine whether or not the given pair of values is a solution of the given system of simultaneous linear equations.

5. $x - y = 5$, $2x + y = 7$ $x = 4, y = -1$
6. $2x + y = 8$, $3x - y = -13$ $x = -1, y = 10$
7. $A + 5B = -7$, $3A - 4B = -4$ $A = -2, B = 1$
8. $-30x + 5y = 1$, $6x - 3y = -4$ $x = \frac{1}{3}, y = 2$
9. $2x - 5y = 0$, $4x + 10y = 4$ $x = \frac{1}{2}, y = -\frac{1}{5}$
10. $6i_1 + i_2 = 5$, $3i_1 - 4i_2 = -1$ $i_1 = 1, i_2 = -1$
11. $3x - 2y = 2.2$, $5x + y = 2.8$ $x = 0.6, y = -0.2$
12. $s - 7t = -3.2$, $2s + t = 2.5$ $s = -1.1, t = 0.3$

In Exercises 13–30, solve each system of equations by sketching the graphs. Use the slope and the y-intercept or both intercepts. Estimate each result to the nearest 0.1 if necessary.

13. $y = -x + 4$, $y = x - 2$
14. $y = \frac{1}{2}x - 1$, $y = -x + 8$
15. $y = 2x - 6$, $y = -\frac{1}{3}x + 1$
16. $y = \frac{1}{2}x - 4$, $y = 2x + 2$
17. $3x + 2y = 6$, $x - 3y = 3$
18. $4R - 3V = -8$, $6R + V = 6$

19. $2x - 5y = 10$
$3x + 4y = -12$

20. $-5x + 3y = 15$
$2x + 7y = 14$

21. $s - 4t = 8$
$2s = t + 4$

22. $y = 4x - 6$
$y = 2x + 4$

23. $y = -x + 3$
$y = -2x + 3$

24. $p - 6 = 6v$
$v = 3 - 3p$

25. $x - 4y = 6$
$2y = x + 4$

26. $x + y = 3$
$3x - 2y = 14$

27. $-2r_1 + 2r_2 = 7$
$4r_1 - 2r_2 = 1$

28. $2x - 3y = -5$
$3x + 2y = 12$

29. $15y = 20x - 48$
$32x + 21y = 60$

30. $7A = 8B + 17$
$5B = 12A - 22$

In Exercises 31–42, solve each system of equations graphically to the nearest 0.001 for each variable.

31. $x = 4y + 2$
$3y = 2x + 3$

32. $1.2x - 2.4y = 4.8$
$3.0x = -2.0y + 7.2$

33. $4.0x - 3.5y = 1.5$
$0.7y + 0.1x = 0.7$

34. $5F - 2T = 7$
$3F + 4T = 8$

35. $x - 5y = 10$
$2x - 10y = 20$

36. $18x - 3y = 7$
$2y = 1 + 12x$

37. $1.9v = 3.2t$
$1.2t - 2.6v = 6$

38. $3y = 14x - 9$
$12x + 23y = 0$

39. $5x = y + 3$
$4x = 2y - 3$

40. $0.75u + 0.67v = 5.9$
$2.1u - 3.9v = 4.8$

41. $7R = 18V + 13$
$-1.4R + 3.6V = 2.6$

42. $y = 6x + 2$
$12x - 2y = -4$

In Exercises 43–46, solve the given problems.

43. In planning a search pattern from an aircraft carrier, a pilot plans to fly at p km/h relative to a wind that is blowing at w km/h. Travelling with the wind, the ground speed would be 300 km/h, and against the wind the ground speed would be 220 km/h. This leads to two equations:

$$p + w = 300$$
$$p - w = 220$$

Are the speeds 260 km/h and 40 km/h?

44. The electric resistance R of a certain resistor is a function of the temperature T given by the equation $R = aT + b$, where a and b are constants. If $R = 1200\ \Omega$ when $T = 10.0°C$ and $R = 1280\ \Omega$ when $T = 50.0°C$, we can find the constants a and b by substituting and obtaining the equations

$$1200 = 10.0a + b$$
$$1280 = 50.0a + b$$

Are the constants $a = 4.00\ \Omega/°C$ and $b = 1160\ \Omega$?

45. The forces acting on part of a structure are shown in Fig. 5.18. An analysis of the forces leads to the equations

$$0.80F_1 + 0.50F_2 = 50$$
$$0.60F_1 - 0.87F_2 = 12$$

Are the forces 45 N and 28 N?

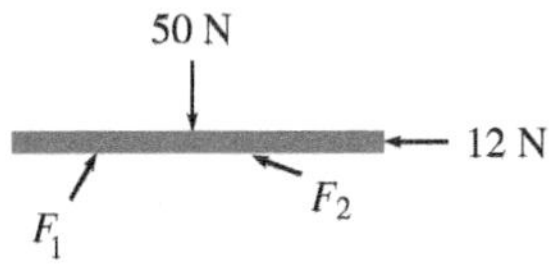

Fig. 5.18

46. A student earned \$4000 during the summer and decided to put half into a registered retirement savings plan (RRSP). If the RRSP was invested in two accounts earning 4.0% and 5.0%, the total income for the first year is \$92. The equations to determine the amounts of x and y are

$$x + y = 2000$$
$$0.040x + 0.050y = 92$$

Are the amounts $x = \$1200$ and $y = \$800$?

In Exercises 47–52, graphically solve the given problems.

47. Chains support a crate, as shown in Fig. 5.19. The equations relating tensions T_1 and T_2 are given below. Determine the tensions to the nearest 1 N from the graph.

$$0.8T_1 - 0.6T_2 = 12$$
$$0.6T_1 + 0.8T_2 = 68$$

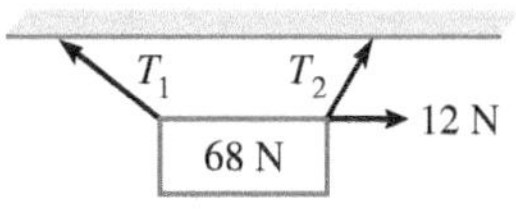

Fig. 5.19

48. The equations relating the currents i_1 and i_2 shown in Fig. 5.20 are given below. Find the currents to the nearest 0.1 A.

$$2i_1 + 6(i_1 + i_2) = 12$$
$$4i_2 + 6(i_1 + i_2) = 12$$

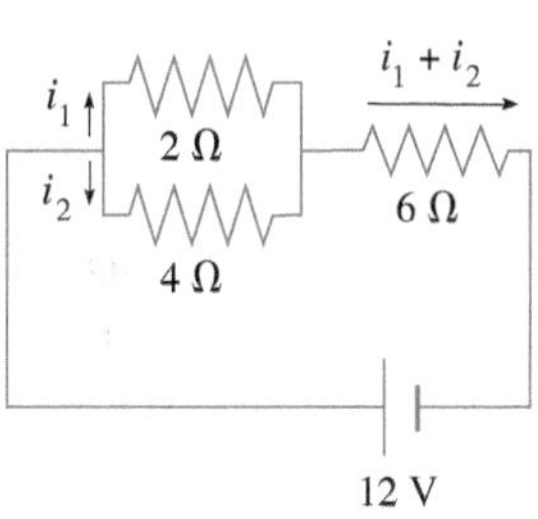

Fig. 5.20

49. An architect designing a parking lot has a row 62.0 m wide to divide into spaces for compact cars and full-size cars. The architect determines that 16 compact car spaces and 6 full-size car spaces use the width, or that 12 compact car spaces and 9 full-size car spaces use all but 0.2 m of the width. What are the widths (to the nearest 0.1 m) of the spaces being planned? See Fig. 5.21.

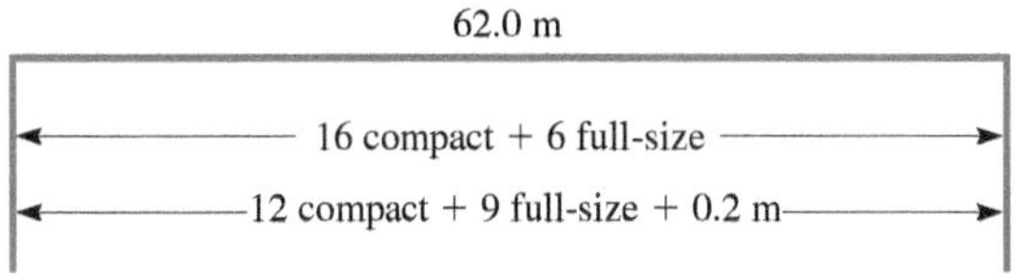

Fig. 5.21

50. An airplane flies into a headwind with an effective ground speed of 140 km/h. On the return trip it flies with the tailwind and has an effective ground speed of 240 km/h. Find the speed p of the plane in still air, and the speed w of the wind (to the nearest 1 km/h).

51. A construction company placed two orders, at the same prices, with a lumber retailer. The first order was for 8 sheets of plywood and 40 framing studs for a total cost of \$304. The second order was for 25 sheets of plywood and 12 framing studs for a total cost of \$498. Find the price of one sheet of plywood and one framing stud.

52. A certain car uses 7.2 L/100 km in city driving and 5.4 L/100 km in highway driving. If 36 L of gas are used in travelling 598 km, how many kilometres were driven in the city, and how many were driven on the highway (assuming that only the given rates of usage were actually used, to the nearest km)?

Answers to Practice Exercises

1. $x = 1.5, y = 1.4$ **2.** yes

5.3 Solving Systems of Two Linear Equations in Two Unknowns Algebraically

The graphical method of solving two linear equations in two unknowns is good for getting a "picture" of the solution. One problem is that graphical methods usually give *approximate* results, although the accuracy obtained when using a graphing utility is excellent. If *exact* solutions are required, we turn to other methods. In this section, we present two algebraic methods of solution.

SOLUTION OF A SYSTEM OF EQUATIONS BY SUBSTITUTION

The first method involves the **substitution of one equation into the other**. Stepwise, the procedure is shown below together with an example.

Solution of Two Linear Equations by Substitution	EXAMPLE 1
	Solve the system $x - 3y = 6$ and $2x + 3y = 3$.
1. Solve one of the equations for one of the unknowns.	We solve the first equation for x: $x - 3y = 6 \rightarrow x = 3y + 6$
2. Substitute this solution into the other equation and simplify.	Substituting $x = 3y + 6$ into $2x + 3y = 3$ and simplifying, we get $2(3y + 6) + 3y = 3$ $6y + 12 + 3y = 3$ $9y + 12 = 3$
3. Solve the resulting equation for the value of the unknown it contains.	$9y + 12 = 3$ $9y = -9$ $y = -1$
4. Substitute this value into the equation from Step 1 to solve for the other unknown.	Substituting $y = -1$ into $x = 3y + 6$: $x = 3(-1) + 6 = 3$ The solution is $x = 3$, $y = -1$.
5. Check the values in **both original equations.**	We substitute $x = 3$ and $y = -1$ into both equations: $x - 3y = 6$ $2x + 3y = 3$ $3 - 3(-1) = 6$ $2(3) + 3(-1) = 3$ $6 = 6$ $3 = 3$ Valid check Valid check

LEARNING TIP

In following the substitution technique, you must first choose an unknown for which to solve. Often it makes little difference, but if it is easier to solve a particular equation for one of the unknowns, that is the one to use.

EXAMPLE 2 Solution by substitution

Solve the following system by substitution: $-5x + 2y = -4$ and $10x + 6y = 3$

It makes little difference which equation or which unknown is chosen for the first step.

Solve one equation for one unknown: We solve the first equation for y:

$$-5x + 2y = -4$$
$$2y = 5x - 4$$
$$y = \frac{5}{2}x - 2$$

Substitute into the other equation, simplify and solve:	Substituting into $10x + 6y = 3$ we get $$10x + 6\left(\frac{5}{2}x - 2\right) = 3$$ $$10x + 15x - 12 = 3$$ $$25x = 15$$ $$x = \frac{15}{25} = \frac{3}{5}$$
Substitute into the equation from Step 1 and solve:	Substituting $x = 3/5$ into $y = \frac{5}{2}x - 2$, $$y = \frac{5}{2}\left(\frac{3}{5}\right) - 2 = \frac{3}{2} - 2 = -\frac{1}{2}$$ The solution is $x = \frac{3}{5}$ and $y = -\frac{1}{2}$.
Check in both original equations:	$-5x + 2y = -4$ $10x + 6y = 3$ $-5\left(\frac{3}{5}\right) + 2\left(-\frac{1}{2}\right) = -4$ $10\left(\frac{3}{5}\right) + 6\left(-\frac{1}{2}\right) = 3$ $-4 = -4$ $3 = 3$

Practice Exercise

1. Solve the system in Example 2 by first solving for x and then substituting.

SOLUTION BY ELIMINATION

The method of substitution is useful if one of the equations can easily be solved for one of the unknowns. However, often a fraction results (such as in Example 2), and this leads to additional algebraic steps to find the solution. Even worse, if the coefficients are themselves decimals or fractions, the algebraic steps are even more involved.

Therefore, we now present another algebraic method of solving a system of equations, **elimination** *of a variable by addition or subtraction* (also called **the addition or subtraction method**). Following is the basic procedure to be used.

Solution of Two Linear Equations by the Elimination Method	**EXAMPLE 3**
	Solve the system $x - 3y = 6$ and $2x + 3y = 3$.
1. Write the equations in the general form $ax + by = c$.	No rearrangement is needed since both equations are in the correct form.
2. If necessary, multiply all terms of each equation by a constant chosen so that **the coefficients of one unknown are numerically the same but differ in sign** in both equations.	$x - 3y = 6$ $2x + 3y = 3$ The coefficients of y (-3 and 3) are already numerically the same and differ in sign.
3. Add the terms on each side of the resulting equations. One variable will be eliminated.	$(x - 3y) + (2x + 3y) = 6 + 3$ $3x = 9$
4. Solve the resulting linear equation for the other unknown.	$3x = 9$ $x = 3$
5. Substitute the resulting value into one of the **original** equations to solve for the other unknown.	Substituting $x = 3$ into $x = 3y + 6$ we get $3 = 3y + 6$ $-3 = 3y$ $y = -1$
6. Check by substituting both values into both original equations.	We check the solution $x = 3$, $y = -1$: $3 = 3(-1) + 6$ $2(3) + 3(-1) = 3$ $3 = 3$ $3 = 3$

■ In the elimination method, it is also possible to have the coefficients of one unknown be the same in both equations and then use subtraction of terms to eliminate that variable. The method presented here always uses addition to avoid possible errors that may be caused when subtracting.

■ Some calculators and online solvers have a specific feature for solving simultaneous linear equations. It is necessary only to enter the coefficients and constants to get the solution.

EXAMPLE 4 Solution by elimination

Solve the following system by elimination: $3x - 2y = 4$ and $3y = -x + 2$.

Rewrite as $ax + by = c$: $3x - 2y = 4$ is in the correct form. $3y = -x + 2$ is rearranged as $x + 3y = 2$.

Multiply **both sides** of equation(s) as necessary: We choose to make the coefficients of x be numerically the same but different in sign. For that, we multiply the second equation by -3:

$$\left.\begin{matrix} 3x - 2y = 4 \\ -3(x + 3y) = -3(2) \end{matrix}\right\} \longrightarrow \begin{matrix} 3x - 2y = 4 \\ -3x - 9y = -6 \end{matrix}$$

Add terms on each side:

$$3x - 2y + (-3x - 9y) = 4 + (-6)$$
$$-11y = -2$$

Solve:

$$y = \frac{2}{11}$$

Substitute into one of the original equations: Substituting $y = \frac{2}{11}$ into $3y = -x + 2$:

$$3\left(\frac{2}{11}\right) = -x + 2$$
$$x = 2 - \frac{6}{11}$$
$$x = \frac{16}{11}$$

The solution is $x = \frac{16}{11}$, $y = \frac{2}{11}$. See Fig. 5.22.

Check:

$$3x - 2y = 4 \qquad 3y = -x + 2$$
$$3\left(\tfrac{16}{11}\right) - 2\left(\tfrac{2}{11}\right) = 4 \qquad 3\left(\tfrac{2}{11}\right) = -\left(\tfrac{16}{11}\right) + 2$$
$$\frac{54}{11} = 4 \qquad \frac{6}{11} = \frac{6}{11}$$
$$4 = 4$$

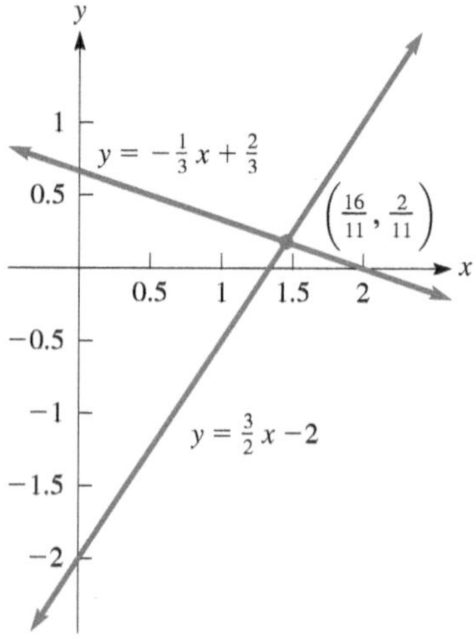

Fig. 5.22

Practice Exercise

2. Solve the system in Example 4 by multiplying the terms of the first equation by 3 and those of the second equation (in general form) by 2, and then adding.

COMMON ERROR When multiplying any equation by a constant, **ensure that you multiply the constant with the terms on *both* sides of the equation**. It is a common error to forget to multiply the value on the right.

EXAMPLE 5 Solution by elimination—mixing metal alloys

By weight, one alloy is 70% copper and 30% zinc. Another alloy is 40% copper and 60% zinc. How many grams of each are required to make 300 g of an alloy that is 60% copper and 40% zinc?

Identify known quantities: Total mass of final alloy = 300 g
Grams of copper in final alloy = 0.60(300) = 180 g
Grams of zinc in final alloy = 0.40(300) = 120 g

Identify unknown quantities: Let A = required number of grams of first alloy
B = required number of grams of second alloy

Select two equations:

1. Total mass must be conserved, so

$$A + B = 300$$

2. Total mass of copper must be conserved, so

$$0.70A + 0.40B = 180$$

■ The second equation is based on the amount of copper. We could have used the amount of zinc, which would have led to the equation $0.30A + 0.60B = 120$. Since we need only two equations, we may use any two of these three equations to find the solution.

Solve the two equations: To make the coefficients of B be numerically the same but different in sign, we multiply the first equation by -0.40:

$$\left.\begin{aligned} -0.40(A + B) &= -0.40(300) \\ 0.70A + 0.40B &= 180 \end{aligned}\right\} \rightarrow \begin{aligned} -0.40A - 0.40B &= -120 \\ 0.70A + 0.40B &= 180 \end{aligned}$$

Adding terms on each side and solving, we get

$$-0.40A - 0.40B + 0.70A + 0.40B = -120 + 180$$
$$0.30A = 60$$
$$A = 200$$

We substitute $A = 200$ into $A + B = 300$ to find B:

$$200 + B = 300$$
$$B = 100$$

The solution is $A = 200$ g and $B = 100$ g.

Check: Checking in the statement of the problem, using the percentages of zinc, we have

$$0.30(200)\text{ g} + 0.60(100)\text{ g} = 120\text{ g}$$
$$120\text{ g} = 120\text{ g}$$

Both the quantities and the units match.

Practice Exercise

3. Solve the problem in Example 5 by using the equation shown in the note above.

EXAMPLE 6 Inconsistent system with elimination

Solve the following system by elimination: $4x = 2y + 3$ and $-y + 2x - 2 = 0$.

Rewrite as $ax + by = c$: $4x = 2y + 3$ is rearranged as $4x - 2y = 3$.
$-y + 2x - 2 = 0$ is rearranged as $2x - y = 2$.

Multiply **both sides** of equation(s) as necessary: We choose to make the coefficients of x be numerically the same but different in sign. For that, we multiply the second equation by -2:

$$\left.\begin{aligned} 4x - 2y &= 3 \\ -2(2x - y) &= -2(2) \end{aligned}\right\} \rightarrow \begin{aligned} 4x - 2y &= 3 \\ -4x + 2y &= -4 \end{aligned}$$

Add terms on each side:

$$4x - 2y + (-4x + 2y) = 3 + (-4)$$
$$0 = -1$$

LEARNING TIP

- When the result is $0 = a$ $(a \neq 0)$, the system is *inconsistent*. From Section 5.2, we know this means the lines representing the equations are parallel.
- If the result of solving a system is $0 = 0$, the system is *dependent*. As shown in Section 5.2, this means there is an unlimited number of solutions, and the lines that represent the equations are really the same line.

Since 0 does not equal -1, we conclude that there is no solution. See Fig. 5.23, which shows that the lines representing the equations are parallel and do not intersect.

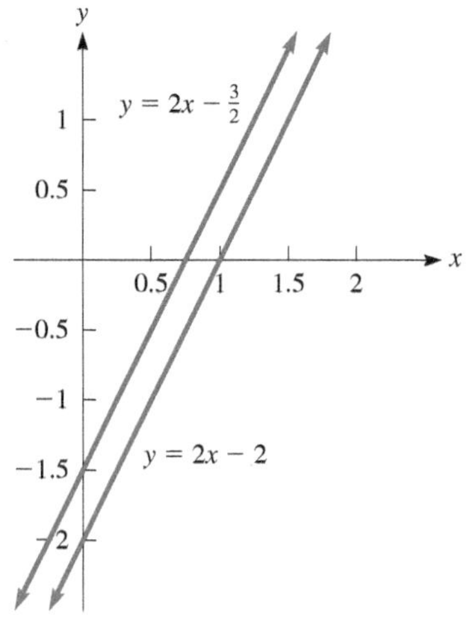

Fig. 5.23

EXERCISES 5.3

In Exercises 1–4, make the given changes in the indicated examples of this section and then solve the resulting problems.

1. In Example 1, change the + to − in the second equation and then solve the system of equations.
2. In Example 3, change the + to − in the second equation and then solve the system of equations.
3. In Example 4, change $-x$ to $-2x$ in the second equation and then solve the system of equations.
4. In Example 6, change 3 to 4 in the first equation and then find if there is any change in the conclusion that is drawn.

In Exercises 5–14, solve the given systems of equations by the method of substitution.

5. $x = y + 3$, $x - 2y = 5$
6. $x = 2y + 1$, $2x - 3y = 4$
7. $p = V - 4$, $V + p = 10$
8. $y = 2x + 10$, $2x + y = -2$
9. $x + y = -5$, $2x - y = 2$
10. $3x + y = 1$, $3x - 2y = 16$
11. $2x + 3y = 7$, $6x - y = 1$
12. $2s + 2t = 1$, $4s - 2t = 17$
13. $33x + 2y = 34$, $40y = 9x + 11$
14. $3A + 3B = -1$, $5A = -6B - 1$

In Exercises 15–24, solve the given systems of equations by the method of elimination.

15. $x + 2y = 5$, $x - 2y = 1$
16. $x + 3y = 7$, $2x + 3y = 5$
17. $2x - 3y = 4$, $2x + y = -4$
18. $R - 4r = 17$, $3R + 4r = 3$
19. $12t + 9y = 14$, $6t = 7y - 16$
20. $3x - y = 3$, $4x = 3y + 14$
21. $v + 2t = 7$, $2v + 4t = 9$
22. $3x - y = 5$, $-9x + 3y = -15$
23. $2x - 3y - 4 = 0$, $3x + 2 = 2y$
24. $3i_1 + 5 = -4i_2$, $3i_2 = 5i_1 - 2$

In Exercises 25–36, solve the given systems of equations by either method of this section.

25. $2x - y = 5$, $6x + 2y = -5$
26. $3x + 2y = 4$, $6x - 6y = 13$
27. $6x + 3y + 4 = 0$, $5y = -9x - 6$
28. $1 + 6q = 5p$, $3p - 4q = 7$
29. $15x + 10y = 11$, $20x - 25y = 7$
30. $2x + 6y = -3$, $-6x - 18y = 5$
31. $12V + 108 = -84C$, $36C + 48V + 132 = 0$
32. $66x + 66y = -77$, $33x - 132y = 143$
33. $44A = 1 - 15B$, $5B = 22 + 7A$
34. $3.0x - 2.0y = 4.0$, $2.9x - 2.0y = 8.0$
35. $2b = 6a - 16$, $33a = 4b + 39$
36. $30P = 55 - Q$, $19P + 14Q + 32 = 0$

In Exercises 37–42, in order to make the coefficients easier to work with, first multiply each term of the equation or divide each term of the equation by a number selected by inspection. Then proceed with the solution of the system by an appropriate algebraic method.

37. $0.3x - 0.7y = 0.4$, $0.2x + 0.5y = 0.7$
38. $250R + 225Z = 400$, $375R - 675Z = 325$
39. $40s - 30t = 60$, $20s - 40t = -50$
40. $0.060x + 0.048y = -0.084$, $0.013x - 0.065y = -0.078$

41. $\frac{x}{3} + \frac{2y}{3} = 2$

$\frac{x}{2} - 2y = \frac{5}{2}$

42. $\frac{2x}{5} - \frac{y}{5} = 1$

$\frac{3x}{4} - y = \frac{5}{2}$

In Exercises 43–46, solve the given systems of equations by an appropriate algebraic method.

43. Find the voltages V_1 and V_2 of the batteries shown in Fig. 5.24. The terminals are aligned in the same direction in Fig. 5.24(a) and in the opposite directions in Fig. 5.24(b).

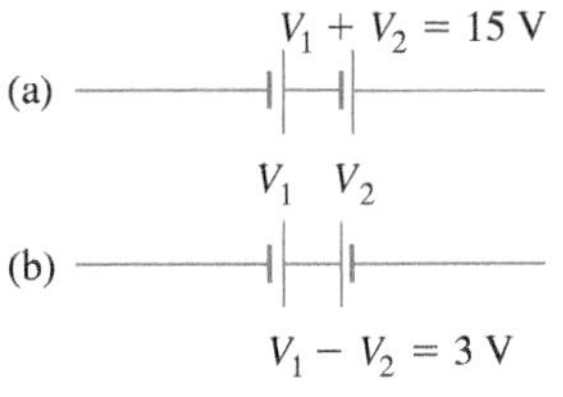

Fig. 5.24

44. A spring of length L is stretched x cm for each newton of weight hung from it. Weights of 3 N and then 5 N are hung from the spring, leading to the equations

$$L + 3x = 18$$
$$L + 5x = 22$$

Solve for L and x.

45. Two grades of gasoline are mixed to make a blend with 1.50% of a special additive. Combining x litres of a grade with 1.80% of the additive to y litres of a grade with 1.00% of the additive gives 10 000 L of the blend. The equations relating x and y are

$$x + y = 10\,000$$
$$0.0180x + 0.0100y = 0.0150(10\,000)$$

Find x and y (to three significant digits).

46. A 6.0% solution and a 15.0% solution of a drug are added to 200 mL of a 20.0% solution to make 1200 mL of a 12.0% solution for a proper dosage. The equations relating the number of millilitres of the added solutions are

$$x + y + 200 = 1200$$
$$0.060x + 0.150y + 0.200(200) = 0.120(1200)$$

Find x and y (to three significant digits).

In Exercises 47–58, set up appropriate systems of two linear equations and solve the systems algebraically. All data are accurate to at least two significant digits.

47. A person's email for a week contained a total of 79 messages. The number of spam messages was four more than twice the number of other messages. How many were spam?

48. A 150-m cable is cut into two pieces such that one piece is four times as long as the other. How long is each piece?

49. The weight W_f supported by the front wheels of a certain car and the weight W_r supported by the rear wheels together equal the weight of the car, 17 700 N. See Fig. 5.25. Also, the ratio of W_r to W_f is 0.847. What are the weights supported by each set of wheels (to three significant digits)?

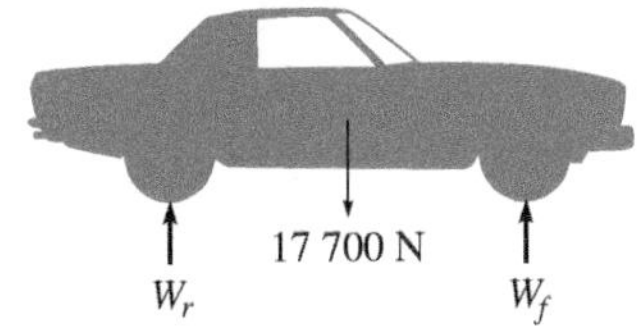

Fig. 5.25

50. A sprinkler system is used to water two areas. If the total water flow is 980 L/h and the flow through one sprinkler is 65% as much as the other, what is the flow in each?

51. In a test of a heat-seeking rocket, a first rocket is launched at 600 m/s, and the heat-seeking rocket is launched along the same flight path 12 s later at a speed of 960 m/s. Find the times t_1 and t_2 of flight of the rockets until the heat-seeking rocket destroys the first rocket.

52. The *moment* of a force (also called *torque*) is the product of the force and the perpendicular distance from a specified axis of rotation. If a lever is supported at only one point (called the *fulcrum*) and is in balance, the sum of the torques (about the support) of forces acting on one side of the support must equal the sum of the torques of the forces acting on the other side. Find the forces F_1 and F_2 that are in the positions shown in Fig. 5.26(a) and then move to the positions in Fig. 5.26(b). The lever weighs 20 N and is in balance in each case.

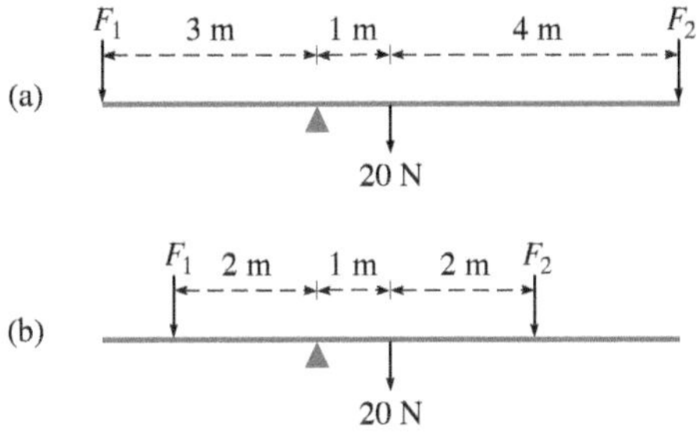

Fig. 5.26

53. There are two types of offices in an office building, and a total of 54 offices. One type rents for $900/month and the other type rents for $1250/month. If all offices are occupied and the total rental income is $55 600/month, how many of each type are there?

54. An underwater (but near the surface) explosion is detected by sonar on a ship 30 s before it is heard on the deck. If sound travels at 1500 m/s in water and 330 m/s in air, how far is the ship from the explosion (to three significant digits)? See Fig. 5.27.

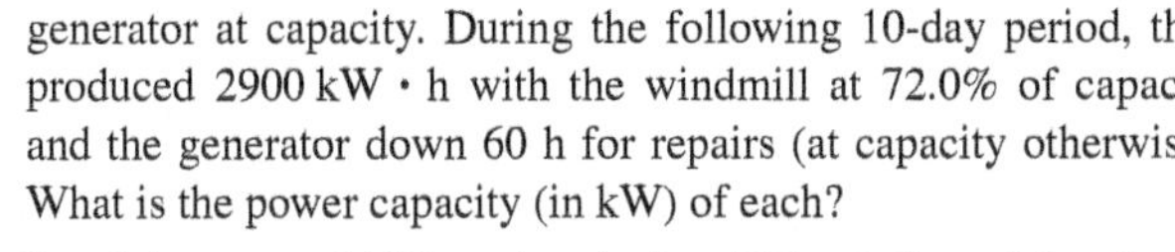

Fig. 5.27

55. A small isolated farm uses a windmill and a gas generator for power. During a 10-day period, they produced 3010 kW · h of energy with the windmill operating at 45.0% of capacity and the generator at capacity. During the following 10-day period, they produced 2900 kW · h with the windmill at 72.0% of capacity and the generator down 60 h for repairs (at capacity otherwise). What is the power capacity (in kW) of each?

56. In mixing a weed-killing chemical, a 40% solution of the chemical is mixed with an 85% solution to get 20 L of a 60% solution. How much of each solution is needed?

57. What conclusion can you draw from a sales report that states that "sales this month were $8000 more than last month, which means that total sales for both months are $4000 more than twice the sales last month"?

58. For an electric circuit, a report stated that current i_1 is twice current i_2 and that twice the sum of the two currents less 6 times i_2 is 6 mA. Explain your conclusion about the values of the currents found from this report.

In Exercises 59–62, answer the given questions.

59. What condition(s) must be placed on the constants of the system of equations

$$ax + y = c$$
$$bx + y = d$$

such that there is a unique solution for x and y?

60. What conditions must be placed on the constants of the system of equations in Exercise 59 such that the system is (a) inconsistent? (b) dependent?

61. For the dependent system of Example 8 of Section 5.2, both equations can be written as $y = \frac{1}{3}x - 3$. The solution for the system can then be shown in terms of a general point on the graph as $(x, \frac{1}{3}x - 3)$. This is referred to as the solution with arbitrary x. Find the solutions in this form for this system for $x = -3$, and $x = 9$.

62. For the dependent system of Example 8 of Section 5.2, write the form for the solution with arbitrary y. See Exercise 61.

Answers to Practice Exercises

1. $x = 3/5, y = -1/2$ **2.** $x = 16/11, y = 2/11$
3. $A = 200$ g, $B = 100$ g

5.4 Solving Systems of Two Linear Equations in Two Unknowns by Determinants

■ Determinants were invented by the German mathematician Gottfried Wilhelm Leibniz (1646–1716).

In this section we discuss another way to solve systems of linear equations with two unknowns. It involves a quantity called a **determinant**.

A determinant is a square array of numbers (written within a pair of vertical lines) which represents a certain sum of products. A **determinant of the second order** has two rows and two columns and is denoted by

$$\begin{vmatrix} a_1 & b_1 \\ a_2 & b_2 \end{vmatrix}$$

The numbers a_1 and b_1 are called the **elements** *of the first* **row** *of the determinant. The numbers a_1 and a_2 are the elements of the first* **column** *of the determinant.* In the same manner, the numbers a_2 and b_2 are the elements of the second row, and the numbers b_1 and b_2 are the elements of the second column. *The numbers a_1 and b_2 are the elements of the* **principal diagonal**, *and the numbers a_2 and b_1 are the elements of the* **secondary diagonal**.

We can find the value of a second order determinant as follows:

$$\begin{vmatrix} a_1 & b_1 \\ a_2 & b_2 \end{vmatrix} = a_1b_2 - a_2b_1 \tag{5.5}$$

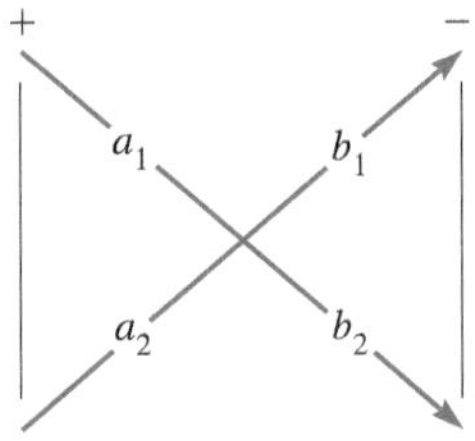

Fig. 5.28

In terms of the diagonals defined earlier, the value of a second order determinant is found by multiplying the elements of the principal diagonal (those marked with a "+" sign in Fig. 5.28) and *subtracting* the product of the elements of the secondary diagonal (those marked with a "−" sign).

■ Determinants can be evaluated using a graphing utility. The use of a graphing calculator for this purpose is described in Section 5.6.

EXAMPLE 1 Evaluating a second-order determinant

Find the value of the determinant $\begin{vmatrix} -5 & 8 \\ 3 & 7 \end{vmatrix}$.

Using Eq. (5.5), we have

$$\begin{vmatrix} -5 & 8 \\ 3 & 7 \end{vmatrix} = (-5 \times 7) - (3 \times 8) = -35 - 24 = -59$$

COMMON ERROR Pay careful attention to signs when multiplying. Using brackets can help keep track of the signs multiplied together, as demonstrated in Example 1. Also, if a diagonal contains a zero, the product is zero (see Example 2b.)

EXAMPLE 2 Evaluating second-order determinants

(a) $\begin{vmatrix} 4 & 6 \\ 3 & 17 \end{vmatrix} = (4 \times 17) - (3 \times 6) = 68 - 18 = 50$

(b) $\begin{vmatrix} 4 & 0 \\ -3 & 17 \end{vmatrix} = (4 \times 17) - (-3 \times 0) = 68 - 0 = 68$

(c) $\begin{vmatrix} 3.6 & 6.1 \\ -3.2 & -17.2 \end{vmatrix} = 3.6(-17.2) - (-3.2 \times 6.1) = -42.4$

Practice Exercise

1. Evaluate the determinant: $\begin{vmatrix} 2 & -4 \\ 3 & 5 \end{vmatrix}$

SOLVING TWO EQUATIONS BY DETERMINANTS

Consider two linear equations in two unknowns, as given in Eq. (5.4):

$$\begin{aligned} a_1x + b_1y &= c_1 \\ a_2x + b_2y &= c_2 \end{aligned} \quad \textbf{(5.4)}$$

If we multiply the first of these equations by b_2 and the second by $(-b_1)$, we obtain

$$\begin{aligned} a_1b_2x + b_1b_2\,y &= c_1b_2 \\ -a_2b_1x - b_2b_1\,y &= -c_2b_1 \end{aligned} \quad \textbf{(5.6)}$$

We see that the coefficients of y are the same but differ in sign. Thus, adding both equations, we can solve for x. The solution can be shown to be

$$x = \frac{c_1b_2 - c_2b_1}{a_1b_2 - a_2b_1} \quad \textbf{(5.7)}$$

In the same manner, we may show that

$$y = \frac{a_1c_2 - a_2c_1}{a_1b_2 - a_2b_1} \quad \textbf{(5.8)}$$

We note that the numerators and denominators of Eqs. (5.7) and (5.8) may be written as determinants. The numerators of the equations are

$$\begin{vmatrix} c_1 & b_1 \\ c_2 & b_2 \end{vmatrix} \quad \text{and} \quad \begin{vmatrix} a_1 & c_1 \\ a_2 & c_2 \end{vmatrix}$$

Therefore, the solutions for x and y can be written directly in terms of determinants, which is called **Cramer's rule**.

■ Cramer's rule is named for the Swiss mathematician Gabriel Cramer (1704–1752).

Cramer's Rule

For a system of equations of the form of Eq. (5.4), the solutions for x and y are given by

$$x = \frac{\begin{vmatrix} c_1 & b_1 \\ c_2 & b_2 \end{vmatrix}}{\begin{vmatrix} a_1 & b_1 \\ a_2 & b_2 \end{vmatrix}} \quad \text{and} \quad y = \frac{\begin{vmatrix} a_1 & c_1 \\ a_2 & c_2 \end{vmatrix}}{\begin{vmatrix} a_1 & b_1 \\ a_2 & b_2 \end{vmatrix}} \quad \textbf{(5.9)}$$

provided the determinant in the denominator is not equal to zero.

LEARNING TIP

Since the same determinant appears in each denominator, *it needs to be evaluated only once.* This means that three determinants are to be evaluated in order to solve the system.

The following steps must be followed to solve a system of two equations in two unknowns by determinants.

Solution of Two Linear Equations by Determinants	**EXAMPLE 3**
	Solve the system $2x + y = 1$ and $5x - 2y = -11$.
1. Write the equations in the general form $ax + by = c$.	No rearrangement is needed: $2x + 1y = 1$ $5x - 2y = -11$
2. Find the value of the determinant for the denominator, D, which consists of the coefficients of x and y in the system.	The columns of coefficients for x and y are read directly from the general form. We have $D = \begin{vmatrix} a_1 & b_1 \\ a_2 & b_2 \end{vmatrix} = \begin{vmatrix} 2 & 1 \\ 5 & -2 \end{vmatrix}$ $= 2(-2) - (5 \cdot 1) = -4 - 5 = -9$
3. Find the numerator for the solution for x by replacing the first column in D (the coefficients of x) by the constants on the right sides of the equations.	$N_x = \begin{vmatrix} c_1 & b_1 \\ c_2 & b_2 \end{vmatrix} = \begin{vmatrix} 1 & 1 \\ -11 & -2 \end{vmatrix}$ $= 1(-2) - (-11 \cdot 1) = -2 + 11 = 9$
4. Find the numerator for the solution for y by replacing the second column in D (the coefficients of y) by the constants.	$N_y = \begin{vmatrix} a_1 & c_1 \\ a_2 & c_2 \end{vmatrix} = \begin{vmatrix} 2 & 1 \\ 5 & -11 \end{vmatrix}$ $= 2(-11) - (5 \cdot 1) = -22 - 5 = -27$
5. Solve for x and y.	$x = \frac{N_x}{D} = \frac{9}{-9} = -1$ and $y = \frac{N_y}{D} = \frac{-27}{-9} = 3$
6. Check by substituting both values into both original equations.	We check the solution $x = -1$, $y = 3$: $2(-1) + 3 = 1$ $5(-1) - 2(3) = -11$ $1 = 1$ $-11 = -11$

LEARNING TIP

The equations must be in the form of Eq. (5.3) before the determinants in Steps 2–4 are set up. The specific positions of the values in the determinants are based on that form of writing the system. If either unknown is missing from an equation, a *zero* must be placed in the proper position.

Practice Exercise

2. Solve the following system by determinants.

$x + 2y = 4$

$3x - y = -9$

EXAMPLE 4 Solving a system using Cramer's rule

Solve the following system of equations by determinants. Numbers are approximate.

$$5.3x = 4.5 - 7.2y$$
$$-6.9y + 3.2x = 5.7$$

Rewrite as $ax + by = c$: Rewrite the system as:

$$5.3x + 7.2y = 4.5$$
$$3.2x - 6.9y = 5.7$$

Find D:

$$D = \begin{vmatrix} a_1 & b_1 \\ a_1 & b_2 \end{vmatrix} = \begin{vmatrix} 5.3 & 7.2 \\ 3.2 & -6.9 \end{vmatrix}$$
$$= 5.3(-6.9) - (3.2 \cdot 7.2) = -59.61$$

Find N_x:

$$N_x = \begin{vmatrix} c_1 & b_1 \\ c_2 & b_2 \end{vmatrix} = \begin{vmatrix} 4.5 & 7.2 \\ 5.7 & -6.9 \end{vmatrix}$$
$$= 4.5(-6.9) - (5.7 \cdot 7.2) = -72.09$$

Find N_y:

$$N_y = \begin{vmatrix} a_1 & c_1 \\ a_2 & c_2 \end{vmatrix} = \begin{vmatrix} 5.3 & 4.5 \\ 3.2 & 5.7 \end{vmatrix}$$

$$= (5.3 \cdot 5.5) - (3.2 \cdot 4.5) = 15.81$$

Solve for x and y:

$$x = \frac{N_x}{D} = \frac{-72.09}{-59.61} = 1.2 \text{ and}$$

$$y = \frac{N_y}{D} = \frac{15.81}{-59.61} = -0.27$$

Note that we round only the final answers.

Check: When substituted into the original equations, the solutions check reasonably well. The small differences are due to rounding.

LEARNING TIP

Determinants are often easier to use than other algebraic methods when the coefficients for the variables in the equations are decimals.

EXAMPLE 5 Solving a system using Cramer's rule—investment income

Two investments totalling \$18 000 yield an annual income of \$700. If the first investment has an interest rate of 5.5% and the second a rate of 3.0%, what is the value of each?

Identify known quantities: Total investment = \$18 000
Annual income = \$700

Identify unknown quantities: Let x = amount invested at 5.5%
y = amount invested at 3.0%

Select two equations:

1. Total amount invested is fixed, so

$$x + y = 18\,000$$

2. Total annual income is fixed, so

$$0.055x + 0.030y = 700$$

Solve the two equations:

$$D = \begin{vmatrix} 1 & 1 \\ 0.055 & 0.03 \end{vmatrix} = 0.03 - 0.055 = -0.025$$

$$N_x = \begin{vmatrix} 18\,000 & 1 \\ 700 & 0.03 \end{vmatrix} = 540 - 700 = -160$$

$$N_y = \begin{vmatrix} 1 & 18\,000 \\ 0.055 & 700 \end{vmatrix} = 700 - 990 = -290$$

$$x = \frac{-160}{-0.025} = 6400 \text{ and } y = \frac{-290}{-0.025} = 11\,600$$

Check: The total investment is \$6400 + \$11 600 = \$18 000, and the total income is \$6400(0.055) + \$11 600(0.030) = \$700, which agree with the statement of the problem.

Practice Exercise

3. In Example 5, change 5.5% to 5.0% and solve for the investments.

INCONSISTENT OR DEPENDENT SOLUTIONS WITH DETERMINANTS

If the determinant of the denominator is zero, we do not have a unique solution since this would require division by zero. If the determinant of the denominator is zero and that of the numerator is not zero, the system is *inconsistent*. If the determinants of both numerator and denominator are zero, the system is *dependent*.

EXAMPLE 6 Inconsistent and dependent systems using determinants

(a) Solve the following system of equations using determinants.

$$x - 4y = 6$$
$$-3x + 12y = -6$$

$$x = \frac{\begin{vmatrix} 6 & -4 \\ -6 & 12 \end{vmatrix}}{\begin{vmatrix} 1 & -4 \\ -3 & 12 \end{vmatrix}} = \frac{72 - 24}{12 - 12} = \frac{48}{0}$$

$$y = \frac{\begin{vmatrix} 1 & 6 \\ -3 & -6 \end{vmatrix}}{\begin{vmatrix} 1 & -4 \\ -3 & 12 \end{vmatrix}} = \frac{-6 - (-18)}{12 - 12} = \frac{12}{0}$$

Both solutions yield a division by zero error, so the system is inconsistent. (There are no solutions.)

(b) Solve the following system of equations using determinants.

$$5x + 2y = -3$$
$$15x + 6y = -9$$

$$x = \frac{\begin{vmatrix} -3 & 2 \\ -9 & 6 \end{vmatrix}}{\begin{vmatrix} 5 & 2 \\ 15 & 6 \end{vmatrix}} = \frac{-18 - (-18)}{30 - 30} = \frac{0}{0}$$

$$y = \frac{\begin{vmatrix} 5 & -3 \\ 15 & -9 \end{vmatrix}}{\begin{vmatrix} 1 & -4 \\ -3 & 12 \end{vmatrix}} = \frac{-45 - (-45)}{30 - 30} = \frac{0}{0}$$

Both solutions yield an indeterminate 0/0 division, so the system is dependent. (There are an infinite number of intersections since the two lines are the same line.)

EXERCISES 5.4

In Exercises 1–4, make the given changes in the indicated examples of this section and then solve the resulting problems.

1. In Example 2(a), change the 6 to -6 and then evaluate.
2. In Example 2(a), change the 4 to -4 and the 6 to -6 and then evaluate.
3. In Example 3, change the + to − in the first equation and then solve the system of equations.
4. In Example 5, change \$700 to \$830 and then solve for the values of the investments.

In Exercises 5–16, evaluate the given determinants.

5. $\begin{vmatrix} 2 & 4 \\ 3 & 1 \end{vmatrix}$ 6. $\begin{vmatrix} -1 & 3 \\ 2 & 6 \end{vmatrix}$ 7. $\begin{vmatrix} 3 & -5 \\ 7 & -2 \end{vmatrix}$

8. $\begin{vmatrix} -4 & 7 \\ 1 & -3 \end{vmatrix}$ 9. $\begin{vmatrix} 8 & -10 \\ 0 & 4 \end{vmatrix}$ 10. $\begin{vmatrix} -4 & -3 \\ -8 & -6 \end{vmatrix}$

11. $\begin{vmatrix} -20 & 110 \\ -70 & -80 \end{vmatrix}$ 12. $\begin{vmatrix} -6.5 & 12.2 \\ -15.5 & 34.6 \end{vmatrix}$ 13. $\begin{vmatrix} 0.75 & -1.32 \\ 0.15 & 1.18 \end{vmatrix}$

14. $\begin{vmatrix} 0.20 & -0.05 \\ 0.28 & 0.09 \end{vmatrix}$ 15. $\begin{vmatrix} 16 & -8 \\ 42 & -15 \end{vmatrix}$ 16. $\begin{vmatrix} 43 & -7 \\ -81 & 16 \end{vmatrix}$

In Exercises 17–26, solve the given systems of equations by determinants. (These are the same as those for Exercises 15–24 of Section 5.3.)

17. $x + 2y = 5$
 $x - 2y = 1$

18. $x + 3y = 7$
 $2x + 3y = 5$

19. $2x - 3y = 4$
 $2x + y = -4$

20. $R - 4r = 17$
 $3R + 4r = 3$

21. $12t + 9y = 14$
 $6t = 7y - 16$

22. $3x - y = 3$
 $4x = 3y + 14$

23. $v + 2t = 7$
$2v + 4t = 9$

24. $3x - y = 5$
$-9x + 3y = -15$

25. $2x - 3y - 4 = 0$
$3x + 2 = 2y$

26. $3i_1 + 5 = -4i_2$
$3i_2 = 5i_1 - 2$

In Exercises 27–34, solve the given systems of equations by determinants. All numbers are approximate. (Exercises 27–30 are the same as Exercises 37–40 of Section 5.3.)

27. $0.3x - 0.7y = 0.4$
$0.2x + 0.5y = 0.7$

28. $250R + 225Z = 400$
$375R - 675Z = 325$

29. $40s - 30t = 60$
$20s - 40t = -50$

30. $0.060x + 0.048y = -0.084$
$0.013x - 0.065y = -0.078$

31. $301x - 529y = 1520$
$385x - 741y = 2540$

32. $0.25d + 0.63n = -0.37$
$-0.61d - 1.80n = 0.55$

33. $1.2y + 10.8 = -8.4x$
$3.5x + 4.8y + 12.9 = 0$

34. $6541x + 4397y = -7732$
$3309x - 8755y = 7622$

In Exercises 35–38, answer the given questions about the determinant to the right.

$$\begin{vmatrix} a & b \\ c & d \end{vmatrix}$$

35. What is the value of the determinant if $c = d = 0$?

36. What change in value occurs if rows and columns are interchanged (b and c are interchanged)?

37. What is the value of the determinant if $a = kb$ and $c = kd$?

38. How does the value change if a and c are doubled?

In Exercises 39–42, solve the given systems of equations by determinants. All numbers are accurate to at least two significant digits.

39. The forces acting on a link of an industrial robot are shown in Fig. 5.29. The equations for finding forces F_1 and F_2 are

$$F_1 + F_2 = 21$$
$$2F_1 = 5F_2$$

Find F_1 and F_2.

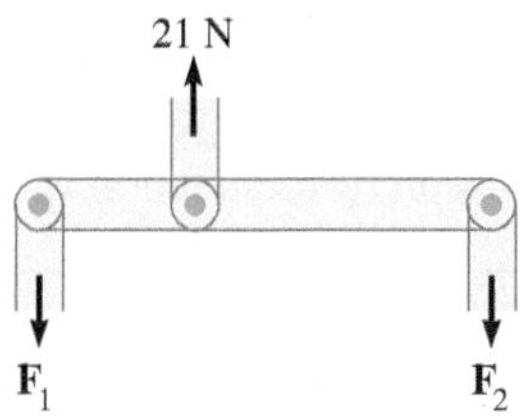

Fig. 5.29

40. The area of a quadrilateral is

$$A = \frac{1}{2}\left(\begin{vmatrix} x_0 & x_1 \\ y_0 & y_1 \end{vmatrix} + \begin{vmatrix} x_1 & x_2 \\ y_1 & y_2 \end{vmatrix} + \begin{vmatrix} x_2 & x_3 \\ y_2 & y_3 \end{vmatrix} + \begin{vmatrix} x_3 & x_0 \\ y_3 & y_0 \end{vmatrix}\right)$$

where (x_0, y_0), (x_1, y_1), (x_2, y_2), and (x_3, y_3) are the rectangular coordinates of the vertices of the quadrilateral, listed counterclockwise. (This *surveyor's formula* can be generalized to find the area of any polygon.)

A surveyor records the locations of the vertices of a quadrilateral building lot on a rectangular coordinate system as (12.79, 0.00), (67.21, 12.30), (53.05, 47.12), and (10.09, 53.11), where distances are in metres. Find the area of the lot.

41. An airplane begins a flight with a total of 144 L of fuel stored in two separate wing tanks. During the flight, 25.0% of the fuel in one tank is used, and in the other tank 37.5% of the fuel is used. If the total fuel used is 44.8 L, the amounts x and y used from each tank can be found by solving the system of equations

$$x + y = 144$$
$$0.250x + 0.375y = 44.8$$

Find x and y.

42. In applying Kirchhoff's laws (see the chapter introduction; for the equations, see, for example, Beiser, *Modern Technical Physics*, 6th ed., p. 550) to the electric circuit shown in Fig. 5.30, the following equations are found. Find the indicated currents I_1 and I_2 (in A).

$$52I_1 - 27I_2 = -420$$
$$-27I_1 + 76I_2 = 210$$

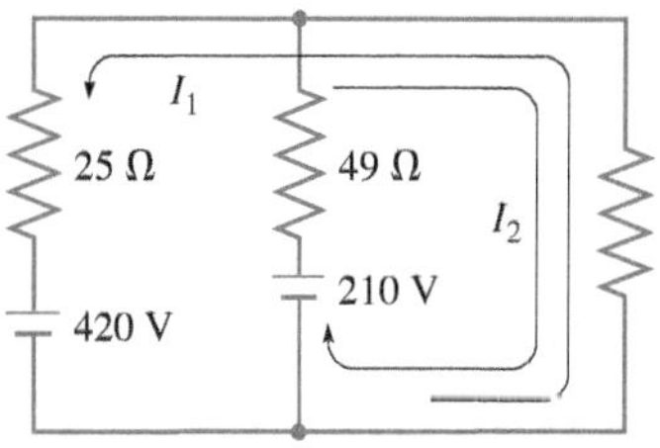

Fig. 5.30

In Exercises 43–52, set up appropriate systems of two linear equations in two unknowns and then solve the systems by determinants. All numbers are accurate to at least two significant digits.

43. A new development has 3-bedroom homes and 4-bedroom homes. The developer's profit was \$25 000 from each 3-br home, and \$35 000 from each 4-br home, totalling \$6 800 000. Total annual property taxes are \$560 000, with \$2000 from each 3-br home and \$3000 from each 4-br home. How many of each were built?

44. Two joggers are 2.0 km apart. If they jog toward each other, they will meet in 12 min. If they jog in the same direction, the faster one will overtake the slower one in 2.0 h. At what rate does each jog?

45. A shipment of 320 cell phones and radar detectors was destroyed due to a truck accident. On the insurance claim, the shipper stated that each phone was worth \$110, each detector was worth \$160, and their total value was \$40 700. How many of each were in the shipment?

46. Two types of electromechanical carburetors are being assembled and tested. Each of the first type requires 15 min of assembly time and 2 min of testing time. Each of the second type requires 12 min of assembly time and 3 min of testing time. If 222 min of assembly time and 45 min of testing time are available, how many of each type can be assembled and tested, if all the time is used?

47. A machinery sales representative receives a fixed salary plus a sales commission each month. If \$6200 is earned on sales of \$70 000 in one month and \$4700 is earned on sales of \$45 000 in the following month, what are the fixed salary and the commission percent?

48. A moving walkway at an airport is 65.0 m long. A child running at a constant speed takes 20.0 s to run along the walkway in the direction it is moving, and then 52.0 s to run all the way back. What are the speed of the walkway and the speed of the child?

49. A boat carrying illegal drugs leaves a port and travels at 63 km/h. A Coast Guard cutter leaves the port 24 min later and travels at 75 km/h in pursuit of the boat. Find the times each has travelled when the cutter overtakes the boat with drugs. See Fig. 5.31.

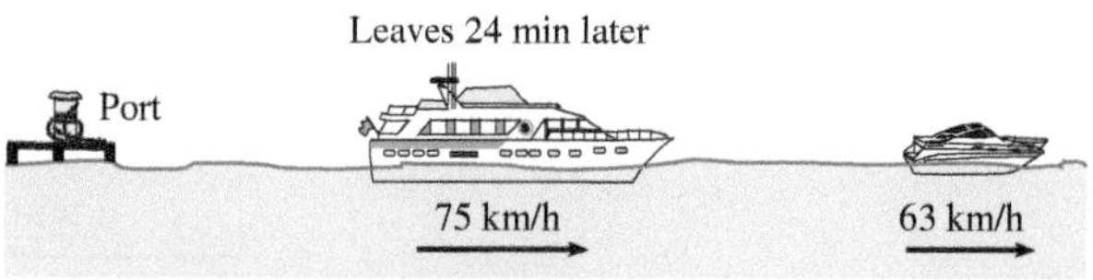

Fig. 5.31

50. Sterling silver is 92.5% silver and 7.5% copper. One silver-copper alloy is 94.0% silver, and a second silver-copper alloy is 85.0% silver. How much of each should be used in order to make 100 g of sterling silver?

51. In an experiment, a variable voltage V is in a circuit with a fixed voltage V_0 and a resistance R. The voltage V is related to the current i in the circuit by $V = Ri - V_0$. If $V = 5.8$ V for $i = 2.0$ A and $V = 24.7$ V for $i = 6.2$ A, find V as a function of i.

52. The velocity of sound in steel is 4850 m/s faster than the velocity of sound in air. One end of a long steel bar is struck, and an instrument at the other end measures the time it takes for the sound to reach it. The sound in the bar takes 0.0120 s, and the sound in the air takes 0.180 s. What are the velocities of sound in air and in steel?

Answers to Practice Exercises

1. 22 **2.** $x = -2, y = 3$ **3.** $x = \$8000, y = \$10\,000$

5.5 Solving Systems of Three Linear Equations in Three Unknowns Algebraically

Many technical problems involve systems of linear equations with more than two unknowns. In this section, we solve systems with three unknowns, and later we will show how systems with even more unknowns are solved.

Solving such systems is very similar to solving systems in two unknowns. In this section, we will show the algebraic method, and in the next section we will show how determinants are used. Graphical solutions are not typically used since a linear equation in three unknowns represents a plane in space, and drawing the intersection of three such planes is difficult. Nevertheless, graphical interpretations of systems of three linear equations will be briefly shown at the end of this section.

A system of three linear equations in three unknowns written in the form

$$\begin{aligned} a_1x + b_1y + c_1z &= d_1 \\ a_2x + b_2y + c_2z &= d_2 \\ a_3x + b_3y + c_3z &= d_3 \end{aligned} \qquad (5.10)$$

has as its solution the set of values x, y, and z that satisfy all three equations simultaneously. The method of solution by elimination is similar to the one used for two equations.

Solution of Three Linear Equations Algebraically	EXAMPLE 1
	Solve the system
1. Write the equations in the general form $ax + by + cz = d$.	(1) $4x + y + 3z = 1$ (2) $2x - 2y + 6z = 11$ (3) $-6x + 3y + 12z = -4$
2. (a) Eliminate one of the unknowns from one pair of equations.	We can choose any one of the three unknowns to eliminate first. We choose y. Multiplying (1) by 2 and adding to (2) we get $8x + 2y + 6z = 2$ $2x - 2y + 6z = 11$ (4) $10x + 12z = 13$
(b) Eliminate the *same unknown* from a *different pair* of the original equations.	Multiplying (1) by -3 and adding to (3) we get $-12x - 3y - 9z = -3$ $-6x + 3y + 12z = -4$ (5) $-18x + 3z = -7$

Practice Exercise

1. Solve the system in Example 1 by first eliminating x.

3. Step 2 generates two linear equations in two unknowns. Solve them by any of the methods previously discussed.	We solve (4) and (5) by elimination. Adding (4) to (5) multiplied by -4 we have $10x + 12z = 13$ $\underline{72x - 12z = 28}$ $82x \quad = 41$ (6) $x = \frac{1}{2}$ Substituting (6) into (5), we solve for z: $-18\left(\frac{1}{2}\right) + 3z = -7$ $3z = 2$ (7) $z = \frac{2}{3}$
4. Substitute the two solved variables into one of the original equations to solve for the third unknown.	Substituting (6) and (7) into (1): $4\left(\frac{1}{2}\right) + y + 3\left(\frac{2}{3}\right) = 1$ $2 + y + 2 = 1$ $y = -3$
5. Check by substituting all values into the three original equations.	We check the solution $x = \frac{1}{2}$. $y = -3$, $z = \frac{2}{3}$:
$4\left(\frac{1}{2}\right) + (-3) + 3\left(\frac{2}{3}\right) = 1$ $1 = 1$ $-6\left(\frac{1}{2}\right) + 3(-3) + 12\left(\frac{2}{3}\right) = -4$ $-4 = -4$	$2\left(\frac{1}{2}\right) - 2(-3) + 6\left(\frac{2}{3}\right) = 11$ $11 = 11$

■ Some graphing utilities and online solvers have a specific feature for solving simultaneous linear equations. It is necessary only to enter the coefficients and constants to get the solution.

EXAMPLE 2 Solving a system—robotic forces

An industrial robot and the forces acting on its main link are shown in Fig. 5.32. An analysis of the forces leads to the following equations relating the forces. Determine the forces.

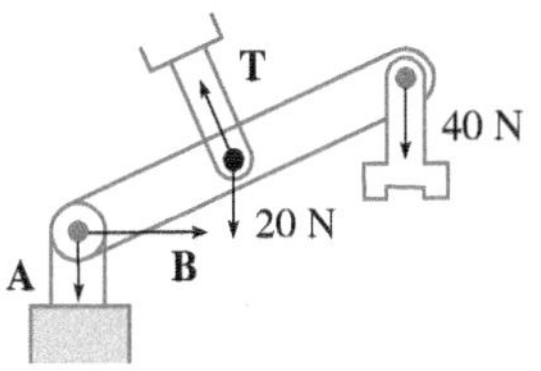

Fig. 5.32

$$(1) \qquad A + 60 = 0.8T$$
$$(2) \qquad B = 0.6T$$
$$(3) \qquad 8A + 6B + 80 = 5T$$

Write in general form:

As we rewrite, we also multiply (1) and (2) by 5 to avoid working with decimals:

$$(4) \quad 5A - 4T = -300$$
$$(5) \quad 5B - 3T = 0$$
$$(6) \quad 8A + 6B - 5T = -80$$

Eliminate T from (4) and (5):

We choose to eliminate T (see the note to the left). Multiplying (4) by 3, (5) by -4, and adding, we get

$$15A - 12T = -900$$
$$\underline{-20B + 12T = 0}$$
$$(7) \quad 15A - 20B = -900$$

■ We have chosen to eliminate T to show all the steps of the procedure (it is the only variable that appears in all three equations). Eliminating A or B first leads to a simpler solution because each variable already does not appear in one of the equations.

Eliminate T from (4) and (6):

Multiplying (4) by 5, (6) by -4, and adding, we get

$$25A - 20T = -1500$$
$$\underline{-32A - 24B + 20T = 320}$$
$$(8) \quad -7A - 24B = -1180$$

Solve (7) and (8) by elimination: Multiplying (7) by 6 and (8) by -5 and adding we have

$$\begin{aligned} 90A - 120B &= -5400 \\ 35A + 120B &= 5900 \\ \hline 125A \qquad &= 11\,300 \\ (9) \qquad A &= 4 \end{aligned}$$

Substituting (9) into (7), we solve for B:

$$\begin{aligned} 15(4) - 20B &= -900 \\ 60 - 20B &= -900 \\ (10) \qquad B &= 48 \end{aligned}$$

Substitute and solve for T: Substituting (9) into (4):

$$\begin{aligned} 5(4) - 4T &= -300 \\ 20 - 4T &= -300 \\ T &= 80 \end{aligned}$$

Therefore, the forces are $A = 4$ N, $B = 48$ N, and $T = 80$ N. The solution checks when substituted into the original equations.

Practice Exercise

2. Solve the system in Example 2 by first eliminating A. See the note above.

EXAMPLE 3 Setting up and solving a system—voltage

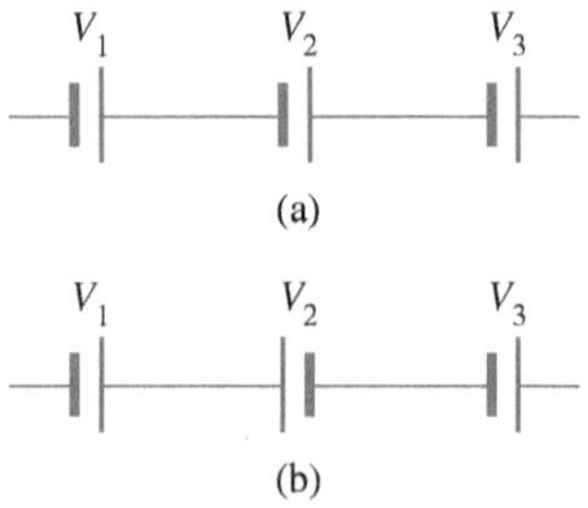

Fig. 5.33

Three voltages, V_1, V_2, and V_3, where V_3 is three times V_1, are in series with the same polarity (see Fig. 5.33a) and have a total voltage of 85 mV. If V_2 is reversed in polarity (see Fig. 5.33b), the voltage is 35 mV. Find the voltages.

Since the voltages are in series with the same polarity, $V_1 + V_2 + V_3 = 85$. Then, since V_3 is three times V_1, we have $V_3 = 3V_1$. Then, with the reversed polarity of V_2, we have $V_1 - V_2 + V_3 = 35$. Writing these equations in standard form, we have the following solution:

(1)	$V_1 + V_2 + V_3 = 85$	
(2)	$3V_1 \qquad - V_3 = 0$	rewriting second equation
(3)	$V_1 - V_2 + V_3 = 35$	
(4)	$2V_1 \qquad + 2V_3 = 120$	adding (1) and (3)
(5)	$6V_1 \qquad - 2V_3 = 0$	(2) multiplied by 2
(6)	$8V_1 \qquad = 120$	adding
(7)	$V_1 = 15$ mV	
(8)	$3(15) - V_3 = 0$	substituting (7) in (2)
(9)	$V_3 = 45$ mV	
(10)	$15 + V_2 + 45 = 85$	substituting (7) and (9) in (1)
(11)	$V_2 = 25$ mV	

Therefore, the three voltages are 15 mV, 25 mV, and 45 mV.

Checking the solution, the sum of the three voltages is 85 mV, V_3 is three times V_1, and the sum of V_1 and V_3 less V_2 is 35 mV.

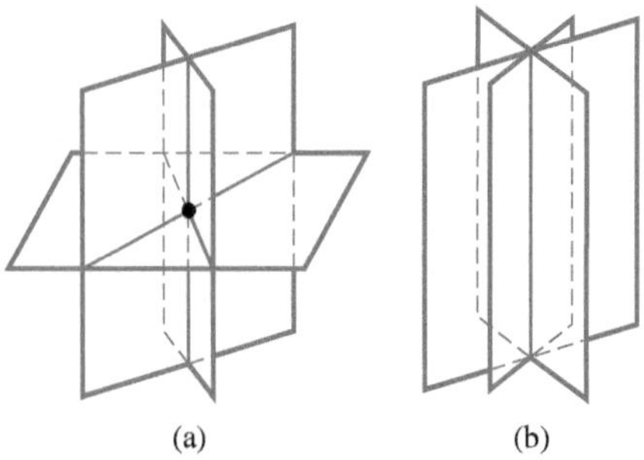

Fig. 5.34

Linear systems with more than two unknowns may have an unlimited number of solutions or be inconsistent. After eliminating unknowns, if we have $0 = 0$, there is an unlimited number of solutions. If we have $0 = a$ $(a \neq 0)$, the system is inconsistent, and there is no solution. (See Exercises 33–36.)

As noted earlier, a linear equation in three unknowns represents a plane in space. For three linear equations in three unknowns, if the planes intersect at a point, there is a unique solution (Fig. 5.34a); if they intersect in a line, there is an unlimited number of solutions (Fig. 5.34b). If the planes do not have a common intersection, the system is inconsistent.

The planes can be parallel (Fig. 5.35a), two can be parallel (Fig. 5.35b), or they can intersect in three parallel lines (Fig. 5.35c). If one plane is coincident with another plane, the system has an unlimited number of solutions if they intersect with the third plane, or is inconsistent otherwise.

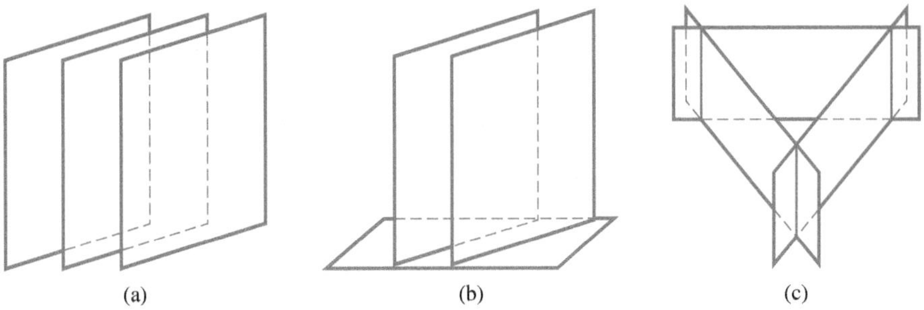

Fig. 5.35

EXTENDING TO FOUR OR MORE UNKNOWNS

Systems with four or more unknowns are solved in a manner similar to that used for three unknowns. With four unknowns, one is eliminated between three different pairs of equations, and the resulting three equations are then solved. The three solved variables are substituted back into one original equation to solve for the fourth one. Technology is often used for these problems. (See Exercises 17 and 18.)

EXERCISES 5.5

In Exercises 1 and 2, make the given changes in Illustrative Example 1 of this section and then solve the resulting system of equations.

1. In the second equation, change the constant to the right of the = sign from 11 to 12, and in the third equation, change the constant to the right of the = sign from −4 to −14.
2. Change the second equation to $8x + 9z = 10$ (no y-term).

In Exercises 3–18, solve the given systems of equations.

3. $x + y + z = 2$
   $x - z = 1$
   $x + y = 1$

4. $x + y - z = -3$
   $x + z = 2$
   $2x - y + 2z = 3$

5. $2x + 3y + z = 2$
   $-x + 2y + 3z = -1$
   $-3x - 3y + z = 0$

6. $2x + y - z = 4$
   $4x - 3y - 2z = -2$
   $8x - 2y - 3z = 3$

7. $5l + 6w - 3h = 6$
   $4l - 7w - 2h = -3$
   $3l + w - 7h = 1$

8. $3r + s - t = 2$
   $r - 2s + t = 0$
   $4r - s + t = 3$

9. $2x - 2y + 3z = 5$
   $2x + y - 2z = -1$
   $4x - y - 3z = 0$

10. $2u + 2v + 3w = 0$
    $3u + v + 4w = 21$
    $-u - 3v + 7w = 15$

11. $3x - 7y + 3z = 6$
    $3x + 3y + 6z = 1$
    $5x - 5y + 2z = 5$

12. $18x + 24y + 4z = 46$
    $63x + 6y - 15z = -75$
    $-90x + 30y - 20z = -55$

13. $10x + 15y - 25z = 35$
    $40x - 30y - 20z = 10$
    $16x - 2y + 8z = 6$

14. $2i_1 - 4i_2 - 4i_3 = 3$
    $3i_1 + 8i_2 + 2i_3 = -11$
    $4i_1 + 6i_2 - i_3 = -8$

15. $2x + 4y + 5z = 12.9$
    $3x + 3y + 3z = 14.1$
    $5x - 5y + z = -12.7$

16. $3x - 2y + 4z = 53\pi/6$
    $-2x - 2y + 7z = 37\pi/3$
    $2x + 3y - 5z = -8\pi$

17. $r - s - 3t - u = 1$
    $2r + 4s - 2u = 2$
    $3r + 4s - 2t = 0$
    $r + 2t - 3u = 3$

18. $3x + 2y - 4z + 2t = 3$
    $5x - 3y - 5z + 6t = 8$
    $2x - y + 3z - 2t = 1$
    $-2x + 3y + 2z - 3t = -2$

In Exercises 19 and 20, find the indicated functions.

19. Find the function $f(x) = ax^2 + bx + c$, if $f(1) = 3$, $f(-2) = 15$, and $f(3) = 5$.
20. Find the function $f(x) = ax^2 + bx + c$, if $f(1) = -3$, $f(-3) = -35$, and $f(3) = -11$.

In Exercises 21–32, solve the systems of equations. In Exercises 25–32, it is necessary to set up the appropriate equations. All numbers are accurate to at least three significant digits.

21. A medical supply company has 1150 worker-hours for production, maintenance, and inspection. Using this and other factors, the number of hours used for each operation, P, M, and I, respectively, is found by solving the following system of equations:

$$P + M + I = 1150$$
$$P = 4I - 100$$
$$P = 6M + 50$$

22. Three oil pumps fill three different tanks. The pumping rates of the pumps (in L/h) are r_1, r_2, and r_3, respectively. Because of malfunctions, they do not operate at capacity each time. Their rates can be found by solving the following system of equations:

$$\begin{aligned} r_1 + r_2 + r_3 &= 14\,000 \\ r_1 + 2r_2 \qquad &= 13\,000 \\ 3r_1 + 3r_2 + 2r_3 &= 36\,000 \end{aligned}$$

23. The forces acting on a certain girder, as shown in Fig. 5.36, can be found by solving the following system of equations:

$$\begin{aligned} 0.707F_1 - 0.800F_2 \qquad\quad &= 0 \\ 0.707F_1 + 0.600F_2 - F_3 &= 10.0 \\ 3.00\ F_2 - 3.00\ F_3 &= 20.0 \end{aligned}$$

Find the forces, in newtons.

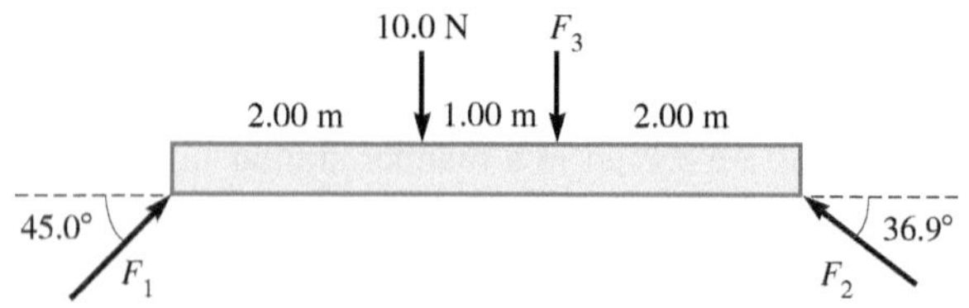

Fig. 5.36

24. Using Kirchhoff's laws (see the chapter introduction; for the equations, see, for example, Beiser, *Modern Technical Physics,* 6th ed., p. 550) with the electric circuit shown in Fig. 5.37, the following equations are found. Find the indicated currents (in A) I_1, I_2, and I_3.

$$\begin{aligned} 1.0I_1 + 3.0(I_1 - I_3) &= 12 \\ 2.0I_2 + 4.0(I_2 + I_3) &= 12 \\ 1.0I_1 - 2.0I_2 + 3.0I_3 &= 0 \end{aligned}$$

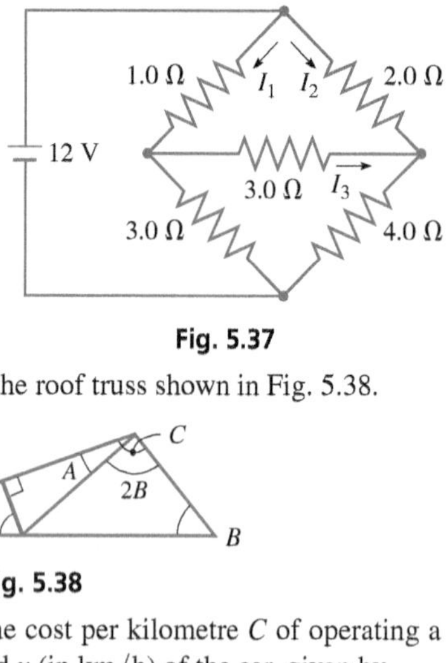

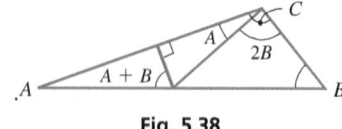

Fig. 5.37

25. Find angles A, B, and C in the roof truss shown in Fig. 5.38.

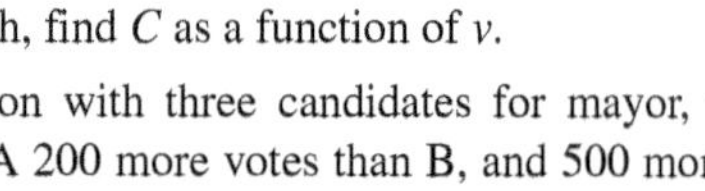

Fig. 5.38

26. Under certain conditions, the cost per kilometre C of operating a car is a function of the speed v (in km/h) of the car, given by $C = av^2 + bv + c$. If $C = 28$ ¢/km for $v = 10$ km/h, $C = 22$ ¢/km for $v = 50$ km/h, and $C = 24$ ¢/km for $v = 80$ km/h, find C as a function of v.

27. In an election with three candidates for mayor, the initial vote count gave A 200 more votes than B, and 500 more votes than C. An error was found, and 1.0% of A's initial votes went to B, and 2.0% of A's initial votes went to C, such that B had 100 more votes than C. How many votes did each have in the final tabulation?

28. A person invests $22 500, partly at 5.00%, partly at 6.00%, and the remainder at 6.50%, with total annual interest of $1308. If the interest received at 5.00% equals the interest received at 6.00%, how much is invested at each rate?

29. The angle θ (in degrees) between two links of a robot arm is given by $\theta = at^3 + bt^2 + ct$, where t is the time during an 11.8-s cycle. If $\theta = 19.0°$ for $t = 1.00$ s, $\theta = 30.9°$ for $t = 3.00$ s, and $\theta = 19.8°$ for $t = 5.00$ s, find the equation $\theta = f(t)$. See Fig. 5.39.

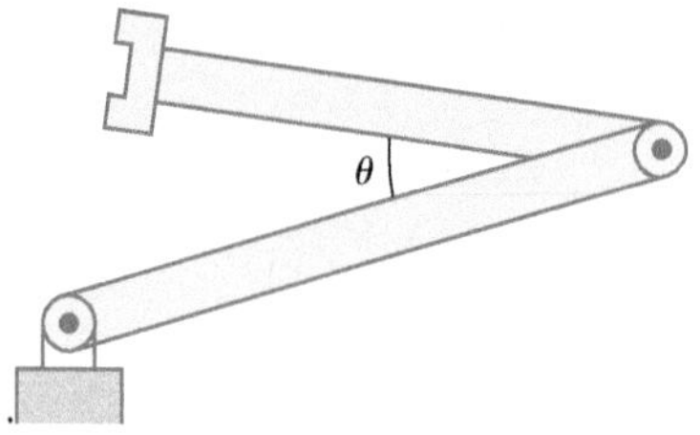

Fig. 5.39

30. The computer systems at three weather bureaus have a combined hard-disk memory capacity of 8.0 TB (terabytes). Systems A and C have 0.2 TB more memory capacity than twice that of system B, and twice the sum of the memory capacities of systems A and B is three times that of system C. What are the memory capacities of each of these computer systems?

31. By weight, one fertilizer is 20% potassium, 30% nitrogen, and 50% phosphorus. A second fertilizer has percents of 10, 20, and 70, respectively, and a third fertilizer has percents of 0, 30, and 70, respectively. How much of each must be mixed to get 200 kg of fertilizer with percents of 12, 25, and 63, respectively?

32. The average traffic flow (number of vehicles) from noon until 1 P.M. in a certain section of one-way streets in a city is shown in Fig. 5.40. Explain why an analysis of the flow through these intersections is not sufficient to obtain unique values for x, y, and z.

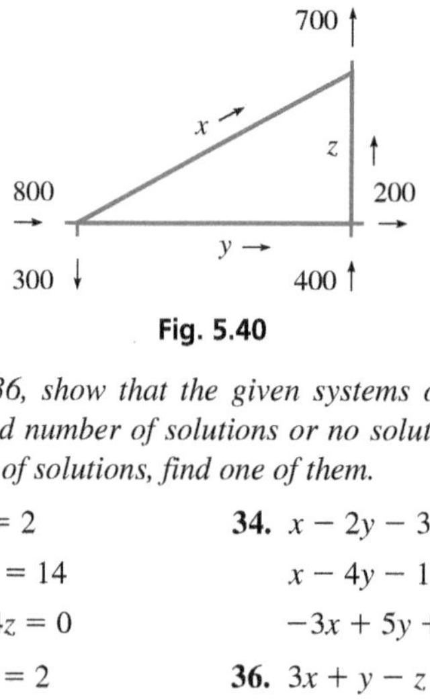

Fig. 5.40

In Exercises 33–36, show that the given systems of equations have either an unlimited number of solutions or no solution. If there is an unlimited number of solutions, find one of them.

33. $x - 2y - 3z = 2$
$x - 4y - 13z = 14$
$-3x + 5y + 4z = 0$

34. $x - 2y - 3z = 2$
$x - 4y - 13z = 14$
$-3x + 5y + 4z = 2$

35. $3x + 3y - 2z = 2$
$2x - y + z = 1$
$x - 5y + 4z = -3$

36. $3x + y - z = -3$
$x + y - 3z = -5$
$-5x - 2y + 3z = -7$

Answers to Practice Exercises

1. $x = 1/2, y = -3, z = 2/3$ **2.** $A = 4$ N, $B = 48$ N, $T = 80$ N

5.6 Solving Systems of Three Linear Equations in Three Unknowns by Determinants

A **determinant of the third order** is a determinant with three rows and three columns. We can find its value as follows:

$$\begin{vmatrix} a_1 & b_1 & c_1 \\ a_2 & b_2 & c_2 \\ a_3 & b_3 & c_3 \end{vmatrix} = a_1b_2c_3 + a_3b_1c_2 + a_2b_3c_1 - a_3b_2c_1 - a_1b_3c_2 - a_2b_1c_3 \qquad \textbf{(5.11)}$$

An easy method of finding this value is as follows:

■ This rewriting method is used only for third-order determinants. It does not work for determinants of order higher than three. A method of finding the value of a determinant of third or higher order is covered in Section 16.6.

Finding the Value of a Third-Order Determinant

1. Rewrite the first two columns to the right of the determinant.	$\begin{vmatrix} a_1 & b_1 & c_1 \\ a_2 & b_2 & c_2 \\ a_3 & b_3 & c_3 \end{vmatrix} \begin{matrix} a_1 & b_1 \\ a_2 & b_2 \\ a_3 & b_3 \end{matrix}$
2. Add the products of the elements of the downward diagonals, beginning with each element in the first row of the determinant. See Fig. 5.41(a).	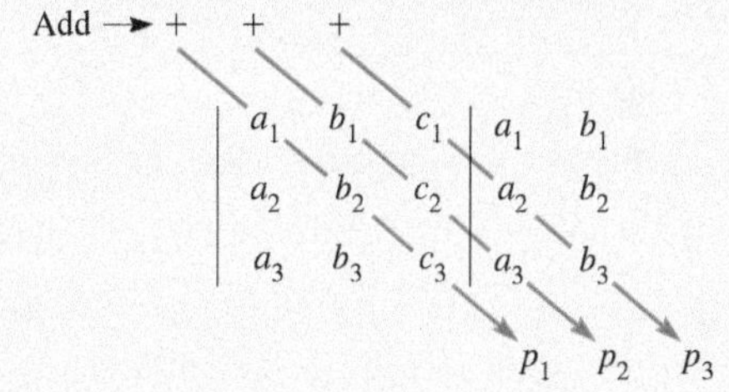 (a)
3. Subtract the products of the upward diagonals, beginning with each element in the third row of the determinant. See Fig. 5.41(b).	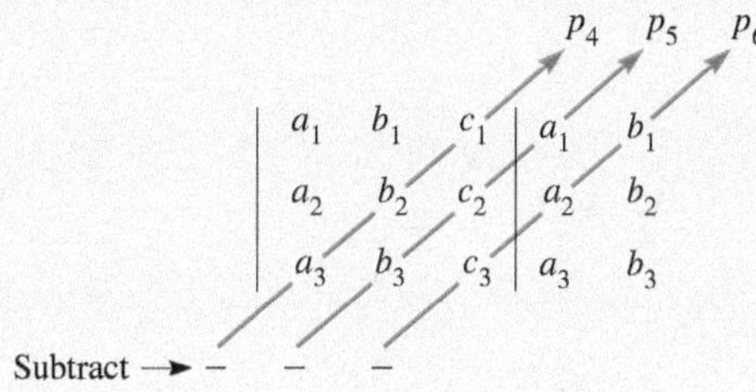 (b) Fig. 5.41
4. The algebraic sum of these six products gives the value of the determinant.	$\begin{vmatrix} a_1 & b_1 & c_1 \\ a_2 & b_2 & c_2 \\ a_3 & b_3 & c_3 \end{vmatrix} = p_1 + p_2 + p_3 - p_4 - p_5 - p_6$

EXAMPLE 1 Evaluating a third-order determinant

(a) $$\begin{vmatrix} 1 & 5 & 4 \\ -2 & 3 & -1 \\ 2 & -1 & 5 \end{vmatrix} = \begin{vmatrix} 1 & 5 & 4 \\ -2 & 3 & -1 \\ 2 & -1 & 5 \end{vmatrix} \begin{matrix} 1 & 5 \\ -2 & 3 \\ 2 & -1 \end{matrix}$$

$$= p_1 + p_2 + p_3 - p_4 - p_5 - p_6$$

$$= 15 + (-10) + (8) - (24) - (1) - (-50) = 38$$

(b) $$\begin{vmatrix} 3 & -2 & 8 \\ -5 & 5 & 0 \\ 4 & 9 & -6 \end{vmatrix} = \begin{vmatrix} 3 & -2 & 8 \\ -5 & 5 & 0 \\ 4 & 9 & -6 \end{vmatrix} \begin{matrix} 3 & -2 \\ -5 & 5 \\ 4 & 9 \end{matrix}$$

$$= -90 + 0 + (-360) - (160) - 0 - (-60) = -550$$

Practice Exercise

1. Evaluate the determinant:

$$\begin{vmatrix} 2 & -2 & 0 \\ -1 & 5 & 1 \\ 3 & 4 & 5 \end{vmatrix}$$

■ Determinants can be evaluated on a calculator using the *matrix* feature. The numbers are entered into a *matrix* (an array of numbers; see Section 16.1), and the determinant is calculated using the *det* instruction.

SOLVING A SYSTEM OF THREE LINEAR EQUATIONS BY DETERMINANTS

Just as systems of two linear equations in two unknowns can be solved by determinants, so can systems of three linear equations in three unknowns. The system

$$
\begin{aligned}
a_1 x + b_1 y + c_1 z &= d_1 \\
a_2 x + b_2 y + c_2 z &= d_2 \\
a_3 x + b_3 y + c_3 z &= d_3
\end{aligned}
\tag{5.10}
$$

can be solved in general terms by the method of elimination. This leads to the following solutions for x, y, and z.

$$
\begin{aligned}
x &= \frac{d_1 b_2 c_2 + d_3 b_1 c_2 + d_2 b_3 c_1 - d_3 b_2 c_1 - d_1 b_3 c_2 - d_2 b_1 c_3}{a_1 b_2 c_3 + a_3 b_1 c_2 + a_2 b_3 c_1 - a_3 b_2 c_1 - a_1 b_3 c_2 - a_2 b_1 c_3} \\
y &= \frac{a_1 d_2 c_3 + a_3 d_1 c_2 + a_2 d_3 c_1 - a_3 d_2 c_1 - a_1 d_3 c_2 - a_2 d_1 c_3}{a_1 b_2 c_3 + a_3 b_1 c_2 + a_2 b_3 c_1 - a_3 b_2 c_1 - a_1 b_3 c_2 - a_2 b_1 c_3} \\
z &= \frac{a_1 b_2 d_3 + a_3 b_1 d_2 + a_2 b_3 d_1 - a_3 b_2 d_1 - a_1 b_3 d_2 - a_2 b_1 d_3}{a_1 b_2 c_3 + a_3 b_1 c_2 + a_2 b_3 c_1 - a_3 b_2 c_1 - a_1 b_3 c_2 - a_2 b_1 c_3}
\end{aligned}
\tag{5.12}
$$

Inspection of Eq. (5.12) reveals that the solutions may be written in terms of determinants. Thus, we may write the general solution to a system of three equations in three unknowns using determinants, which again is Cramer's rule.

Cramer's Rule: Three Unknowns

For a system of equations in three unknowns of the form of Eq. (5.10), the solutions are given by

$$
x = \frac{\begin{vmatrix} d_1 & b_1 & c_1 \\ d_2 & b_2 & c_2 \\ d_3 & b_3 & c_3 \end{vmatrix}}{\begin{vmatrix} a_1 & b_1 & c_1 \\ a_2 & b_2 & c_2 \\ a_3 & b_3 & c_3 \end{vmatrix}} \quad
y = \frac{\begin{vmatrix} a_1 & d_1 & c_1 \\ a_2 & d_2 & c_2 \\ a_3 & d_3 & c_3 \end{vmatrix}}{\begin{vmatrix} a_1 & b_1 & c_1 \\ a_2 & b_2 & c_2 \\ a_3 & b_3 & c_3 \end{vmatrix}} \quad
z = \frac{\begin{vmatrix} a_1 & b_1 & d_1 \\ a_2 & b_2 & d_2 \\ a_3 & b_3 & d_3 \end{vmatrix}}{\begin{vmatrix} a_1 & b_1 & c_1 \\ a_2 & b_2 & c_2 \\ a_3 & b_3 & c_3 \end{vmatrix}}
\tag{5.13}
$$

provided the determinant in the denominator is not equal to zero.

An analysis of Eq. (5.13) shows that the situation is precisely the same as it was when we were using determinants to solve systems of two linear equations. That is, the determinant in the denominator of each is made up of the coefficients of x, y, and z, and the determinants in the numerators are the same as that in the denominator, except that the column of d's replaces the column of coefficients of the unknown for which we are solving. We can define the determinants

$$
D = \begin{vmatrix} a_1 & b_1 & c_1 \\ a_2 & b_2 & c_2 \\ a_3 & b_3 & c_3 \end{vmatrix} \quad
N_x = \begin{vmatrix} d_1 & b_1 & c_1 \\ d_2 & b_2 & c_2 \\ d_3 & b_3 & c_3 \end{vmatrix} \quad
N_y = \begin{vmatrix} a_1 & d_1 & c_1 \\ a_2 & d_2 & c_2 \\ a_3 & d_3 & c_3 \end{vmatrix} \quad
N_z = \begin{vmatrix} a_1 & b_1 & d_1 \\ a_2 & b_2 & d_2 \\ a_3 & b_3 & d_3 \end{vmatrix}
$$

Then the solution to the system in Eq. (5.10) is given by

$$
x = \frac{N_x}{D} \quad y = \frac{N_y}{D} \quad z = \frac{N_z}{D}
$$

As with systems of two equations, if all four determinants are zero, there is an *unlimited number of solutions*. If the D determinant is zero and any of the determinants of the numerators is not zero, the system is *inconsistent*, and there is *no solution*.

EXAMPLE 2 Solving a system using Cramer's rule

Solve the following system by determinants:

$$\begin{aligned} 3x + 2y - 5z &= -1 \\ 2x - 3y - z &= 11 \\ 5x - 2y + 7z &= 9 \end{aligned}$$

The equations are already in the correct form. We can thus read the coefficients for the determinant in the numerator.

$$D = \left|\begin{array}{rrr|rr} 3 & 2 & -5 & 3 & 2 \\ 2 & -3 & -1 & 2 & -3 \\ 5 & -2 & 7 & 5 & -2 \end{array}\right. = (-63) + (-10) + (20) - (75) - (6) - (28) = -162$$

Since D is not zero, there is a unique solution and we can set up the determinants for the numerators.

$$N_x = \left|\begin{array}{rrr|rr} -1 & 2 & -5 & -1 & 2 \\ 11 & -3 & -1 & 11 & -3 \\ 9 & -2 & 7 & 9 & -2 \end{array}\right. \qquad N_y = \left|\begin{array}{rrr|rr} 3 & -1 & -5 & 3 & -1 \\ 2 & 11 & -1 & 2 & 11 \\ 5 & 9 & 7 & 5 & 9 \end{array}\right.$$

$$N_z = \left|\begin{array}{rrr|rr} 3 & 2 & -1 & 3 & 2 \\ 2 & -3 & 11 & 2 & -3 \\ 5 & -2 & 9 & 5 & -2 \end{array}\right.$$

Evaluating, we have

$$\begin{aligned} N_x &= (21) + (-18) + (110) - (135) - (-2) - (154) = -174 \\ N_y &= (231) + (5) + (-90) - (-275) - (-27) - (-14) = 462 \\ N_z &= (-81) + (110) + (4) - (15) - (-66) - (36) = 48 \end{aligned}$$

Solving for the three unknowns:

$$x = \frac{-174}{-162} = \frac{29}{27} \quad y = \frac{462}{-162} = -\frac{77}{27} \quad z = \frac{48}{-162} = -\frac{8}{27}$$

Substituting in each of the original equations shows that the solution checks:

$$3\left(\frac{29}{27}\right) + 2\left(-\frac{77}{27}\right) - 5\left(-\frac{8}{27}\right) = \frac{87 - 154 + 40}{27} = \frac{-27}{27} = -1$$

$$2\left(\frac{29}{27}\right) - 3\left(-\frac{77}{27}\right) - \left(-\frac{8}{27}\right) = \frac{58 + 231 + 8}{27} = \frac{297}{27} = 11$$

$$5\left(\frac{29}{27}\right) - 2\left(-\frac{77}{27}\right) + 7\left(-\frac{8}{27}\right) = \frac{145 + 154 - 56}{27} = \frac{243}{27} = 9$$

After the values of x and y were determined, we could also have evaluated z by substituting the values of x and y into one of the original equations.

EXAMPLE 3 Setting up and solving a system—mixing acid solutions

An 8.0% solution, an 11% solution, and an 18% solution of nitric acid are to be mixed to get 150 mL of a 12% solution. If the volume of acid from the 8.0% solution equals half the volume of acid from the other two solutions, how much of each is needed?

Identify known and unknown quantities:

Total volume of 12% solution = 150 mL
Let x = volume of 8.0% solution needed
y = volume of 11% solution needed
z = volume of 18% solution needed

Select three equations:

1. Total volume is $x + y + z = 150$
2. Total volume of nitric acid is

$$0.080x + 0.11y + 0.18z = 0.12(150)$$

3. Volume of acid in the 8.0% solution is half that of the other two combined: $0.080x = 0.5(0.11y + 0.18z)$

Rewrite in the general form and solve using Cramer's rule:

$$\begin{aligned} x + \quad y + \quad z &= 150 \\ 0.080x + 0.11y + 0.18z &= 18 \\ 0.080x - 0.055y - 0.090z &= 0 \end{aligned}$$

$$D = \begin{vmatrix} 1 & 1 & 1 \\ 0.080 & 0.11 & 0.18 \\ 0.080 & -0.055 & -0.090 \end{vmatrix} \begin{matrix} 1 & 1 \\ 0.080 & 0.11 \\ 0.080 & -0.055 \end{matrix}$$

$$N_x = \begin{vmatrix} 150 & 1 & 1 \\ 18 & 0.11 & 0.18 \\ 0 & -0.055 & -0.090 \end{vmatrix} \begin{matrix} 150 & 1 \\ 18 & 0.11 \\ 0 & -0.055 \end{matrix}$$

$$N_y = \begin{vmatrix} 1 & 150 & 1 \\ 0.080 & 18 & 0.18 \\ 0.080 & 0 & -0.090 \end{vmatrix} \begin{matrix} 1 & 150 \\ 0.080 & 18 \\ 0.080 & 0 \end{matrix}$$

$$N_z = \begin{vmatrix} 1 & 1 & 150 \\ 0.080 & 0.11 & 18 \\ 0.080 & -0.055 & 0 \end{vmatrix} \begin{matrix} 1 & 1 \\ 0.080 & 0.11 \\ 0.080 & -0.055 \end{matrix}$$

Evaluating, we get:

$$D = -0.0099 + 0.0144 - 0.0044 - 0.0088 + 0.0099 + 0.0072 = 0.0084$$

$$N_x = -1.485 + 0 - 0.990 - 0 + 1.485 + 1.620 = 0.630$$

$$N_y = -1.620 + 2.160 + 0 - 1.440 - 0 + 1.080 = 0.180$$

$$N_z = 0 + 1.440 - 0.660 - 1.320 + 0.990 - 0 = 0.450$$

Solving for the three unknowns:

$$x = \frac{0.630}{0.0084} = 75, \quad y = \frac{0.180}{0.0084} = 21, \quad z = \frac{0.450}{0.0084} = 54$$

Therefore, 75 mL of the 8.0% solution, 21 mL of the 11% solution, and 54 mL of the 18% solution are required to make the 12% solution. Results have been rounded off to two significant digits, the accuracy of the data. Checking with the statement of the problem, we see that these volumes total 150 mL.

EXERCISES 5.6

In Exercises 1 and 2, make the given change in the indicated examples of this section and then solve the resulting problems.

1. In Example 1, interchange the first and second rows of the determinant and then evaluate it.
2. In Example 2, change the constant to the right of the = sign in the first equation from −1 to −3, change the constant to the right of the = sign in the third equation from 9 to 11, and then solve the resulting system of equations.

In Exercises 3–16, evaluate the given third-order determinants.

3. $\begin{vmatrix} 5 & 4 & -1 \\ -2 & -6 & 8 \\ 7 & 1 & 1 \end{vmatrix}$

4. $\begin{vmatrix} -7 & 0 & 0 \\ 2 & 4 & 5 \\ 1 & 4 & 2 \end{vmatrix}$

5. $\begin{vmatrix} 8 & 9 & -6 \\ -3 & 7 & 2 \\ 4 & -2 & 5 \end{vmatrix}$

6. $\begin{vmatrix} -2 & 4 & -1 \\ 5 & -10 & 4 \\ 4 & -8 & 2 \end{vmatrix}$

7. $\begin{vmatrix} -3 & -4 & -8 \\ 5 & -1 & 0 \\ 2 & 10 & -1 \end{vmatrix}$

8. $\begin{vmatrix} 10 & 2 & -7 \\ -2 & -3 & 6 \\ 6 & 5 & -2 \end{vmatrix}$

9. $\begin{vmatrix} 4 & -3 & -11 \\ -9 & 2 & -2 \\ 0 & 1 & -5 \end{vmatrix}$

10. $\begin{vmatrix} 9 & -2 & 0 \\ -1 & 3 & -6 \\ -4 & -6 & -2 \end{vmatrix}$

11. $\begin{vmatrix} 25 & 18 & -50 \\ -30 & 40 & -12 \\ -20 & 55 & -22 \end{vmatrix}$

12. $\begin{vmatrix} 20 & 0 & -15 \\ -4 & 30 & 1 \\ 6 & -1 & 40 \end{vmatrix}$

13. $\begin{vmatrix} 0.1 & -0.2 & 0 \\ -0.5 & 1 & 0.4 \\ -2 & 0.8 & 2 \end{vmatrix}$

14. $\begin{vmatrix} 0.25 & -0.54 & -0.42 \\ 1.20 & 0.35 & 0.28 \\ -0.50 & 0.12 & -0.44 \end{vmatrix}$

15. $\begin{vmatrix} \pi & -\frac{\pi}{2} & \frac{\pi}{3} \\ \frac{\pi}{4} & 2\pi & -\frac{\pi}{5} \\ -\frac{\pi}{2} & \frac{3\pi}{4} & -\frac{2\pi}{5} \end{vmatrix}$

16. $\begin{vmatrix} x^2y & 3y & 4x \\ 7x & -5y & 2x \\ y & 2y & x \end{vmatrix}$

In Exercises 17–32, solve the given systems of equations by use of determinants. (Exercises 19–28 are the same as Exercises 3–12 of Section 5.5.) Assume all numbers are accurate to three significant digits.

17. $2x + 3y + z = 4$
$3x - z = -3$
$x - 2y + 2z = -5$

18. $4x + y + z = 2$
$2x - y - z = 4$
$3y + z = 2$

19. $x + y + z = 2$
$x - z = 1$
$x + y = 1$

20. $x + y - z = -3$
$x + z = 2$
$2x - y + 2z = 3$

21. $2x + 3y + z = 2$
$-x + 2y + 3z = -1$
$-3x - 3y + z = 0$

22. $2x + y - z = 4$
$4x - 3y - 2z = -2$
$8x - 2y - 3z = 3$

23. $5l + 6w - 3h = 6$
$4l - 7w - 2h = -3$
$3l + w - 7h = 1$

24. $3r + s - t = 2$
$r - 2s + t = 0$
$4r - s + t = 3$

25. $2x - 2y + 3z = 5$
$2x + y - 2z = -1$
$4x - y - 3z = 0$

26. $2u + 2v + 3w = 0$
$3u + v + 4w = 21$
$-u - 3v + 7w = 15$

27. $3x - 7y + 3z = 6$
$3x + 3y + 6z = 1$
$5x - 5y + 2z = 5$

28. $18x + 24y + 4z = 46$
$63x + 6y - 15z = -75$
$-90x + 30y - 20z = -55$

29. $p + 2q + 2r = 0$
$2p + 6q - 3r = -1$
$4p - 3q + 6r = -8$

30. $9x + 12y + 2z = 23$
$21x + 2y - 5z = -25$
$-18x + 6y - 4z = -11$

31. $3.0x + 4.5y - 7.5z = 10.5$
$4.8x - 3.6y - 2.4z = 1.2$
$4.0x - 0.5y + 2.0z = 1.5$

32. $26L - 52M - 52N = 39$
$45L + 96M + 40N = -80$
$55L + 62M - 11N = -48$

The three points $(x_1, y_1), (x_2, y_2)$, and (x_3, y_3) *are collinear (they lie on a straight line) if* $\begin{vmatrix} x_1 & y_1 & 1 \\ x_2 & y_2 & 1 \\ x_3 & y_3 & 1 \end{vmatrix} = 0$. *In Exercises 33 and 34, use this fact to decide if the given points are collinear.*

33. $(-2, 1), (2, 3), (6, 5)$

34. $(-8, 20), (-3, 6), (1, -8)$

In Exercises 35–38, use the determinant at the right. Answer the questions about the determinant for the changes given in each exercise.

$$\begin{vmatrix} 4 & 2 & 1 \\ 3 & 6 & 5 \\ 7 & 9 & 8 \end{vmatrix} = 19$$

35. How does the value change if the first two rows are interchanged?

36. What is the value if the second row is replaced with the first row (the first row remains unchanged—the first and second rows are the same)?

37. How does the value change if the elements of the first row are added to the corresponding elements of the second row (the first row remains unchanged)?

38. How does the value change if each element of the first row is multiplied by 2?

In Exercises 39–49, solve the given problems by determinants. In Exercises 42–49, set up appropriate systems of equations. All numbers are accurate to at least two significant digits.

39. In analysing the forces on the bell-crank mechanism shown in Fig. 5.42, the following equations are found. Find the forces.

$A - 0.60F = 80$
$B - 0.80F = 0$
$6.0A - 10F = 0$

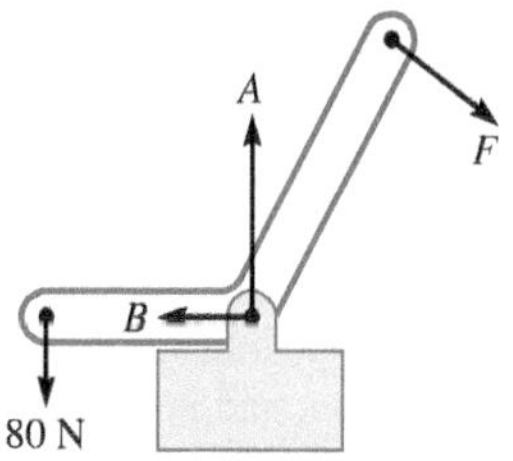

Fig. 5.42

40. Using Kirchhoff's laws (see the chapter introduction; the equations can be found in most physics textbooks) with the electric circuit shown in Fig. 5.43, the following equations are found. Find the indicated currents (in A) I_1, I_2, and I_3.

$19I_1 - 12I_2 = 60$
$12I_1 - 18I_2 + 6.0I_3 = 0$
$6.0I_2 - 18I_3 = 0$

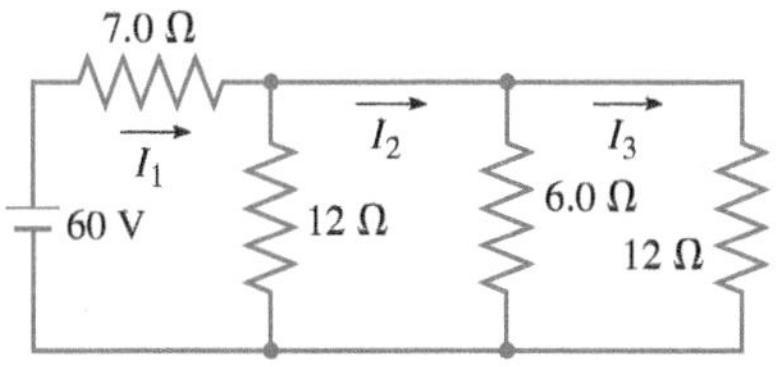

Fig. 5.43

41. In a laboratory experiment to measure the acceleration of an object, the distances travelled by the object were recorded for three different time intervals. These data led to the following equations:

$s_0 + 2v_0 + 2a = 20$
$s_0 + 4v_0 + 8a = 54$
$s_0 + 6v_0 + 18a = 104$

Here, s_0 is the initial displacement (in m), v_0 is the initial velocity (in m/s), and a is the acceleration (in m/s^2). Find s_0, v_0, and a.

42. A certain 18-hole golf course has par-3, par-4, and par-5 holes, and there are twice as many par-4 holes as par-5 holes. How many holes of each type are there if a golfer has par on every hole for a score of 70?

43. The voltage V across part of an electric circuit is a function of the temperature T (in °C) given by $V = a + bT + cT^2$, where a, b, and c are constants. By evaluating a, b, and c, find $V = f(T)$ if $V = 6.4$ V for $T = 2.0$°C, $V = 8.6$ V for $T = 4.0$°C, and $V = 11.6$ V for $T = 6.0$°C.

44. A mail-order company charges \$4 for shipping orders of less than \$50, \$6 for orders from \$50 to \$200, and \$8 for orders over \$200. One day the total shipping charges were \$2160 for 384 orders. Find the number of orders shipped at each rate if the number of orders under \$50 was 12 more than twice the number of orders over \$200.

45. An alloy used in electrical transformers contains nickel (Ni), iron (Fe), and molybdenum (Mo). The percent of Ni is 1% less than five times the percent of Fe. The percent of Fe is 1% more than three times the percent of Mo. Find the percent of each in the alloy.

46. A person invests \$20 000, partly at 5.00%, partly at 6.00%, and the remainder at 6.50%. The total annual interest is \$1170. Three times the amount invested at 6.00% equals the amount invested at 5.00% and 6.50% combined. How much is invested at each rate?

47. A person spent 1.10 h in a car going to an airport, 1.95 h flying in a jet, and 0.520 h in a taxi to reach the final destination. The jet's speed averaged 12.0 times that of the car, which averaged 15.0 km/h more than the taxi. What was the average speed of each if the trip covered 1140 km?

48. An intravenous aqueous solution is made from three mixtures to get 500 mL with 6.0% of one medication, 8.0% of a second medication, and 86% water. The percents in the mixtures are, respectively, 5.0, 20, 75 (first); 0, 5.0, 95 (second); and 10, 5.0, 85 (third). How much of each is used?

49. Three unknown forces are exerted at the pin joint shown in a static structure (see Fig. 5.44). It is known that the sum of the magnitudes of the three forces is 125 N. Analysing the force components in the horizontal and vertical directions yields two more equations:

$$-0.174F_1 - 0.985F_2 + 0.927F_3 = 0$$
$$0.985F_1 - 0.174F_2 - 0.375F_3 = 0$$

Solve for the forces.

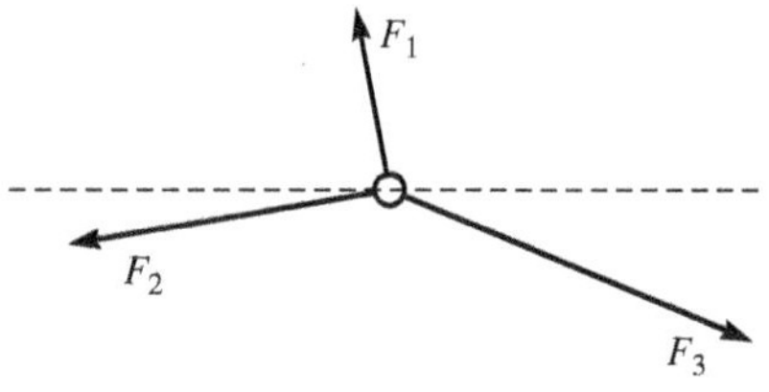

Fig. 5.44

Answer to Practice Exercise

1. 26

CHAPTER 5 KEY FORMULAS AND EQUATIONS

Linear equation in two unknowns $\quad ax + by = c \quad$ (5.1)

Definition of slope $\quad m = \dfrac{y_2 - y_1}{x_2 - x_1} \quad$ (5.2)

Slope–intercept form $\quad y = mx + b \quad$ (5.3)

System of two linear equations (5.4)

$$a_1x + b_1y = c_1$$
$$a_2x + b_2y = c_2$$

Second-order determinant (5.5)

$$\begin{vmatrix} a_1 & b_1 \\ a_2 & b_2 \end{vmatrix} = a_1b_2 - a_2b_1$$

Cramer's rule (5.9)

$$x = \frac{\begin{vmatrix} c_1 & b_1 \\ c_2 & b_2 \end{vmatrix}}{\begin{vmatrix} a_1 & b_1 \\ a_2 & b_2 \end{vmatrix}} \quad \text{and} \quad y = \frac{\begin{vmatrix} a_1 & c_1 \\ a_2 & c_2 \end{vmatrix}}{\begin{vmatrix} a_1 & b_1 \\ a_2 & b_2 \end{vmatrix}}$$

System of three linear equations (5.10)

$$a_1x + b_1y + c_1z = d_1$$
$$a_2x + b_2y + c_2z = d_2$$
$$a_3x + b_3y + c_3z = d_3$$

Third-order determinant (5.11)

$$\begin{vmatrix} a_1 & b_1 & c_1 \\ a_2 & b_2 & c_2 \\ a_3 & b_3 & c_3 \end{vmatrix} = a_1b_2c_3 + a_3b_1c_2 + a_2b_3c_1 - a_3b_2c_1 - a_1b_3c_2 - a_2b_1c_3$$

Cramer's rule (5.13)

$$x = \frac{\begin{vmatrix} d_1 & b_1 & c_1 \\ d_2 & b_2 & c_2 \\ d_3 & b_3 & c_3 \end{vmatrix}}{\begin{vmatrix} a_1 & b_1 & c_1 \\ a_2 & b_2 & c_2 \\ a_3 & b_3 & c_3 \end{vmatrix}} \quad y = \frac{\begin{vmatrix} a_1 & d_1 & c_1 \\ a_2 & d_2 & c_2 \\ a_3 & d_3 & c_3 \end{vmatrix}}{\begin{vmatrix} a_1 & b_1 & c_1 \\ a_2 & b_2 & c_2 \\ a_3 & b_3 & c_3 \end{vmatrix}} \quad z = \frac{\begin{vmatrix} a_1 & b_1 & d_1 \\ a_2 & b_2 & d_2 \\ a_3 & b_3 & d_3 \end{vmatrix}}{\begin{vmatrix} a_1 & b_1 & c_1 \\ a_2 & b_2 & c_2 \\ a_3 & b_3 & c_3 \end{vmatrix}}$$

CHAPTER 5 REVIEW EXERCISES

In Exercises 1–4, evaluate the given determinants.

1. $\begin{vmatrix} -2 & 5 \\ 3 & 1 \end{vmatrix}$

2. $\begin{vmatrix} 40 & 10 \\ -20 & -60 \end{vmatrix}$

3. $\begin{vmatrix} -18 & -33 \\ -21 & 44 \end{vmatrix}$

4. $\begin{vmatrix} 0.91 & -1.2 \\ 0.73 & -5.0 \end{vmatrix}$

In Exercises 5–8, find the slopes of the lines that pass through the given points.

5. $(2, 0), (4, -8)$

6. $(-1, -5), (-4, 4)$

7. $(40, -20), (-30, -40)$

8. $\left(-6, \frac{1}{2}\right), \left(1, -\frac{7}{2}\right)$

In Exercises 9–12, find the slopes and the y-intercepts of the lines with the given equations. Sketch the graphs.

9. $y = -2x + 4$

10. $2y = \frac{2}{3}x - 3$

11. $8x - 2y = 5$

12. $3x = 8 + 3y$

In Exercises 13–20, solve the given systems of equations graphically.

13. $y = 2x - 4$
$y = -\frac{3}{2}x + 3$

14. $y = -3x + 3$
$y = 2x - 6$

15. $4A - B = 6$
$3A + 2B = 12$

16. $2x - 5y = 10$
$3x + y = 6$

17. $7x = 2y + 14$
$y = -4x + 4$

18. $5v = 15 - 3t$
$t = 6v - 12$

19. $3M + 4N = 6$
$2M - 3N = 2$

20. $5x + 2y = 5$
$2x - 4y = 3$

In Exercises 21–30, solve the given systems of equations by using an appropriate algebraic method.

21. $x + 2y = 5$
$x + 3y = 7$

22. $2x - y = 7$
$x + y = 2$

23. $4x + 3y = -4$
$y = 2x - 3$

24. $r = -3s - 2$
$-2r - 9s = 2$

25. $10i - 27v = 29$
$40i + 33v = 69$

26. $3x - 6y = 5$
$7x + 2y = 4$

27. $7x = 2y - 6$
$7y = 12 - 4x$

28. $3R = 8 - 5I$
$6I = 8R + 11$

29. $90x - 110y = 40$
$30x - 15y = 25$

30. $0.42x - 0.56y = 1.26$
$0.98x - 1.40y = -0.28$

In Exercises 31–40, solve the given systems of equations by determinants. (These are the same as for Exercises 21–30.)

31. $x + 2y = 5$
$x + 3y = 7$

32. $2x - y = 7$
$x + y = 2$

33. $4x + 3y = -4$
$y = 2x - 3$

34. $r = -3s - 2$
$-2r - 9s = 2$

35. $10i - 27v = 29$
$40i + 33v = 69$

36. $3x - 6y = 5$
$7x + 2y = 4$

37. $7x = 2y - 6$
$7y = 12 - 4x$

38. $3R = 8 - 5I$
$6I = 8R + 11$

39. $90x - 110y = 40$
$30x - 15y = 25$

40. $0.42x - 0.56y = 1.26$
$0.98x - 1.40y = -0.28$

In Exercises 41–43, choose an exercise from among Exercises 31–40 that you think is most easily solved by the indicated method. In Exercises 41–44, explain your answers.

41. Substitution

42. Addition or subtraction

43. Determinants

44. Considering the slopes m and the y-intercepts b, for a system of two equation with two unknowns, explain how to determine if there is (a) a unique solution, (b) an inconsistent solution, or (c) a dependent solution.

In Exercises 45–48, evaluate the given determinants.

45. $\begin{vmatrix} 4 & -1 & 8 \\ -1 & 6 & -2 \\ 2 & 1 & -1 \end{vmatrix}$

46. $\begin{vmatrix} -500 & 0 & -500 \\ 250 & 300 & -100 \\ -300 & 200 & 200 \end{vmatrix}$

47. $\begin{vmatrix} -2.2 & -4.1 & 7.0 \\ 1.2 & 6.4 & -3.5 \\ -7.2 & 2.4 & -1.0 \end{vmatrix}$

48. $\begin{vmatrix} 30 & 22 & -12 \\ 0 & -34 & 44 \\ 35 & -41 & -27 \end{vmatrix}$

In Exercises 49–54, solve the given systems of equations algebraically. In Exercises 53 and 54, the numbers are approximate, and are accurate to three significant digits.

49. $2x + y + z = 4$
$x - 2y - z = 3$
$3x + 3y - 2z = 1$

50. $x + 2y + z = 2$
$3x - 6y + 2z = 2$
$2x - z = 8$

51. $2r + s + 2t = 8$
$3r - 2s - 4t = 5$
$-2r + 3s + 4t = -3$

52. $4u + 4v - 2w = -4$
$20u - 15v + 10w = -10$
$24u - 12v - 9w = 39$

53. $3.6x + 5.2y - z = -2.2$
$3.2x - 4.8y + 3.9z = 8.1$
$6.4x + 4.1y + 2.3z = 5.1$

54. $32t + 24u + 63v = 32$
$42t - 31u + 19v = 132$
$48t + 12u + 11v = 0$

In Exercises 55–60, solve the given systems of equations by determinants. In Exercises 59 and 60, the numbers are approximate. (These systems are the same as for Exercises 49–54.)

55. $2x + y + z = 4$
$x - 2y - z = 3$
$3x + 3y - 2z = 1$

56. $x + 2y + z = 2$
$3x - 6y + 2z = 2$
$2x - z = 8$

57. $2r + s + 2t = 8$
$3r - 2s - 4t = 5$
$-2r + 3s + 4t = -3$

58. $4u + 4v - 2w = -4$
$20u - 15v + 10w = -10$
$24u - 12v - 9w = 39$

59. $3.6x + 5.2y - z = -2.2$
$3.2x - 4.8y + 3.9z = 8.1$
$6.4x + 4.1y + 2.3z = 5.1$

60. $32t + 24u + 63v = 32$
$42t - 31u + 19v = 132$
$48t + 12u + 11v = 0$

In Exercises 61–64, solve for x. Here, we see that we can solve an equation in which the unknown is an element of a determinant.

61. $\begin{vmatrix} 2 & 5 \\ 1 & x \end{vmatrix} = 3$ **62.** $\begin{vmatrix} -1 & x \\ 3 & 4 \end{vmatrix} = 7$

63. $\begin{vmatrix} x & 1 & 2 \\ 0 & -1 & 3 \\ -2 & 2 & 1 \end{vmatrix} = 5$ **64.** $\begin{vmatrix} 1 & 2 & -1 \\ -2 & 3 & x \\ -1 & 2 & -2 \end{vmatrix} = -3$

In Exercises 65–68, let $1/x = u$ and $1/y = v$. Solve for u and v and then solve for x and y. In this way we see how to solve systems of equations involving reciprocals.

65. $\frac{1}{x} - \frac{1}{y} = \frac{1}{2}$
$\frac{1}{x} + \frac{1}{y} = \frac{1}{4}$

66. $\frac{1}{x} + \frac{1}{y} = 3$
$\frac{2}{x} + \frac{1}{y} = 1$

67. $\frac{2}{x} + \frac{3}{y} = 3$
$\frac{5}{x} - \frac{6}{y} = 3$

68. $\frac{3}{x} - \frac{2}{y} = 4$
$\frac{2}{x} + \frac{4}{y} = 1$

In Exercises 69 and 70, determine the value of k that makes the system dependent. In Exercises 71 and 72, determine the value of k that makes the system inconsistent.

69. $3x - ky = 6$
$x + 2y = 2$

70. $5x + 20y = 15$
$2x + ky = 6$

71. $kx - 2y = 5$
$4x + 6y = 1$

72. $2x - 5y = 7$
$kx + 10y = 2$

In Exercises 73 and 74, solve the given systems of equations by any appropriate method. All numbers in Exercise 73 are accurate to two significant digits, and in Exercise 74 they are accurate to three significant digits.

73. A 20-m crane arm with a supporting cable and with a 9000-N box suspended from its end has forces acting on it, as shown in Fig. 5.45. Find the forces (in N) from the following equations.

$$F_1 + 2.0F_2 = 26\,000$$
$$0.87F_1 - F_3 = 0$$
$$3.0F_1 - 4.0F_2 = 54\,000$$

Fig. 5.45

74. In applying Kirchhoff's laws (see Exercise 24 of Section 5.5) to the electric circuit shown in Fig. 5.46, the following equations result. Find the indicated currents (in A).

$$i_1 + i_2 + i_3 = 0$$
$$5.20i_1 - 3.25i_2 = 8.33 - 6.45$$
$$3.25i_2 - 2.62i_3 = 6.45 - 9.80$$

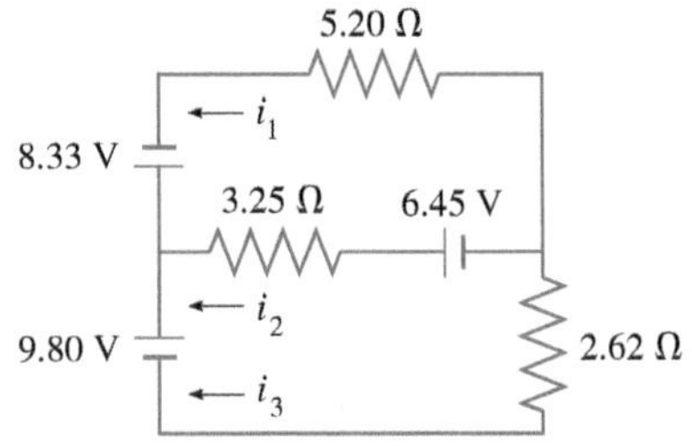

Fig. 5.46

In Exercises 75–94, set up systems of equations and solve by any appropriate method. All numbers are accurate to at least three significant digits.

75. A study found that the fuel consumption for transportation contributes a percent p_1 that is 16% less than the percent p_2 of all other sources combined. Find p_1 and p_2.

76. A certain amount of a fuel contains 150 MJ of potential heat. Part is burned at 80% efficiency, and the rest is burned at 70% efficiency, such that the total amount of heat actually delivered is 114 MJ. Find the amounts burned at each efficiency.

77. The sales representatives of a company have a choice of being paid 10% of their sales in a month, or \$2400 plus 4% of their sales in the month. For what monthly sales is the income the same, and what is that income?

78. A person invested a total of \$20 900 into two bonds, one with an annual interest rate of 6.00% and the other with an annual interest rate of 5.00% per year. If the total annual interest from the bonds is \$1170, how much is invested in each bond?

79. A total of 42 tonnes of two types of ore is to be loaded into a smelter. The first type contains 6.0% copper, and the second contains 2.4% copper. Find the necessary amounts of each ore to produce two tonnes of copper.

80. As a 40-m pulley belt makes one revolution, one of the two pulleys makes one more revolution than the other. Another pulley of half the radius replaces the smaller pulley and makes six more revolutions than the larger pulley for one revolution of the belt. Find the circumferences of the pulleys. See Fig. 5.47.

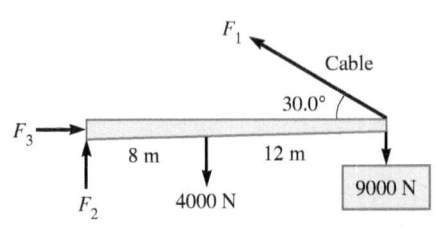

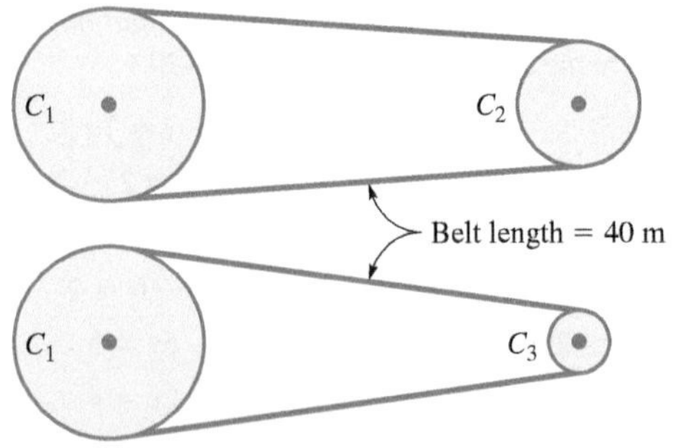

Fig. 5.47

81. A computer analysis showed that the temperature T of the ocean water within 1000 m of a nuclear-plant discharge pipe was given by $T = \frac{a}{x + 100} + b$, where x is the distance from the pipe and a and b are constants. If $T = 14°C$ for $x = 0$ and $T = 10°C$ for $x = 900$ m, find a and b.

82. In measuring the angles of a triangular parcel, a surveyor noted that one of the angles equalled the sum of the other two angles. What conclusion can be drawn?

83. A satellite is to be launched from a space shuttle. It is calculated that the satellite's speed will be 24 200 km/h if launched directly ahead of the shuttle or 21 400 km/h if launched directly to the rear of the shuttle. What is the speed of the shuttle and the launching speed of the satellite relative to the shuttle? See Fig. 5.48.

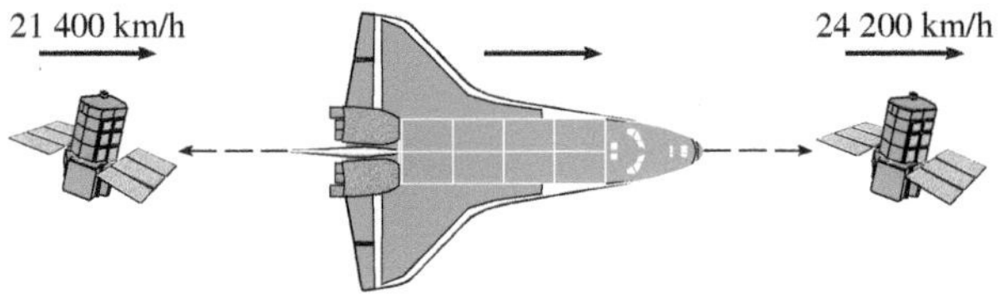

Fig. 5.48

84. The velocity v of sound is a function of the temperature T according to the function $v = aT + b$, where a and b are constants. If $v = 337.5$ m/s for $T = 10.0°C$ and $v = 346.6$ m/s for $T = 25.0°C$, find v as a function of T.

85. The power P (in W) dissipated in an electric resistance R (in Ω) equals the resistance times the square of the current I (in A). If 1.0 A flows through resistance R_1 and 3.0 A flows through resistance R_2, the total power dissipated is 14.0 W. If 3.0 A flows through R_1 and 1.0 A flows through R_2, the total power dissipated is 6.0 W. Find R_1 and R_2.

86. Twelve equal rectangular ceiling panels are placed as shown in Fig. 5.49. If each panel is 150 mm longer than it is wide and a total of 40.0 m of edge and middle strips is used, what are the dimensions of the room?

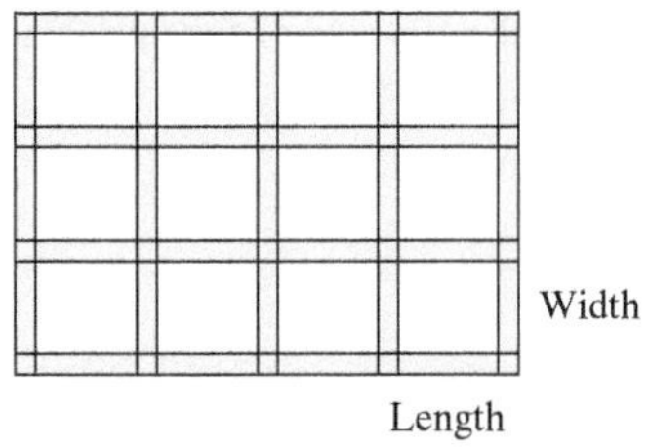

Fig. 5.49

87. The weight of a lever may be considered to be at its centre. A 10-m lever of weight w is balanced on a fulcrum 4 m from one end by a load L at that end. If a load of $4L$ is placed at that end, it requires a 20-N weight at the other end to balance the lever. What are the initial load L and the weight w of the lever? (See Exercise 52 of Section 5.3.) See Fig. 5.50.

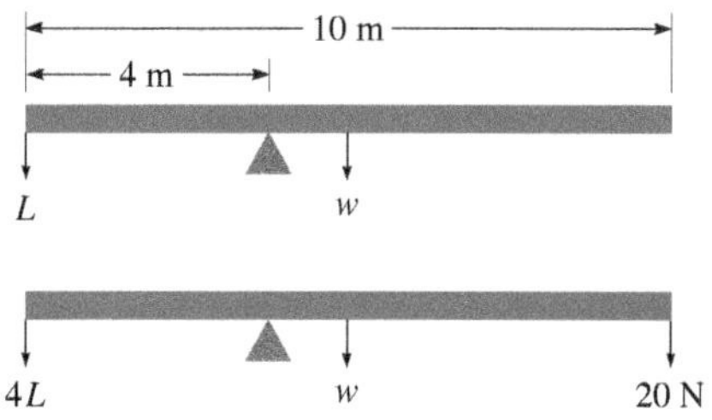

Fig. 5.50

88. Two fuel mixtures, one of 2.0% oil and 98.0% gasoline and another of 8.0% oil and 92.0% gasoline, are to be used to make 10.0 L of a fuel that is 4.0% oil and 96.0% gasoline for use in a chain saw. How much of each mixture is needed?

89. For the triangular truss shown in Fig. 5.51, $\angle A$ is 55° less than twice $\angle B$, and $\angle C$ is 25° less than $\angle B$. Find the measures of the three angles.

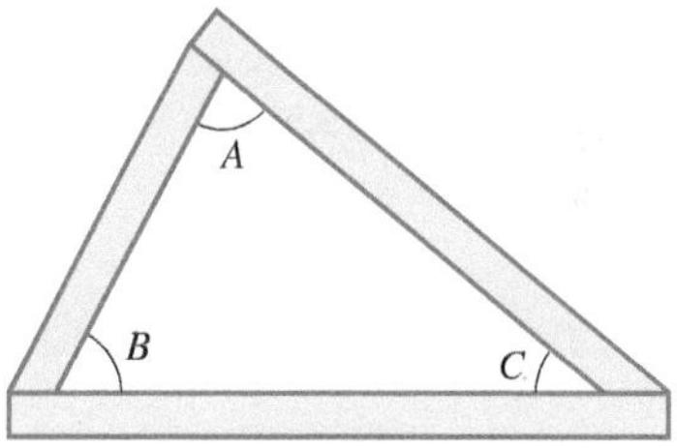

Fig. 5.51

90. A manufacturer produces three models of DVD players in a year. Four times as many of model A are produced as model C, and 7000 more of model B than model C. If the total production for the year is 97 000 units, how many of each are produced?

91. Three computer programs A, B, and C, require a total of 140 MB (megabytes) of hard-disk memory. If three other programs, two requiring the same memory as B and one the same as C, are added to a disk with A, B, and C, a total of 236 MB are required. If three other programs, one requiring the same memory as A and two the same memory as C, are added to a disk with A, B, and C, a total of 304 MB are required. How much memory is required for each of A, B, and C?

92. One ampere of electric current is passed through a solution of sulphuric acid, silver nitrate, and cupric sulphate, releasing hydrogen gas, silver, and copper. A total mass of 1.750 g is released. The mass of silver deposited is 3.40 times the mass of copper deposited, and the mass of copper and 70.0 times the mass of hydrogen combined equals the mass of silver deposited less 0.037 g. How much of each is released?

93. Gold loses about 5.3%, and silver about 10%, of its weight when immersed in water. If a gold-silver alloy weighs 6.0 N in air and 5.6 N in water, find the weight in air of the gold and the silver in the alloy.

94. Three unknown forces are exerted at the pin joint shown in a static structure (see Fig. 5.52). It is known that the sum of the magnitudes of forces F_2 and F_3 is 71.0 kN. Analysing the force components in the horizontal and vertical directions yields two more equations:

$$0.766F_1 - 0.985F_2 - 0.309F_3 = 0$$
$$0.643F_1 - 0.174F_2 - 0.951F_3 = 0$$

Solve for the forces.

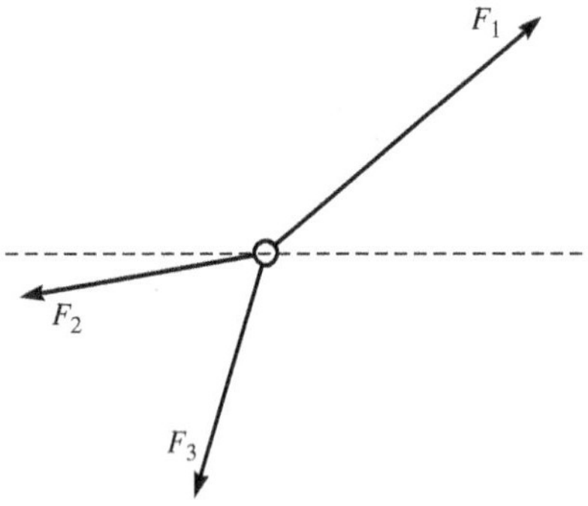

Fig. 5.52

95. Write one or two paragraphs giving reasons for choosing a particular method of solving the following problem. If a first pump is used for 2.2 h and a second pump is used for 2.7 h, 32 m^3 can be removed from a wastewater-holding tank. If the first pump is used for 1.4 h and the second for 2.5 h, 24 m^3 can be removed. How much can each pump remove in 1.0 h? (What is the result to three significant digits?)

CHAPTER 5 PRACTICE TEST

1. Find the slope of the line through $(2, -5)$ and $(-1, 4)$.

2. Is $x = -2, y = 3$ a solution to the system of equations

$$2x + 5y = 11$$
$$y - 5x = 12?$$

Explain.

3. Evaluate:

$$\begin{vmatrix} -1 & 3 & -2 \\ 4 & -3 & 0 \\ 5 & -4 & 2 \end{vmatrix}$$

4. Solve by substitution:

$$x + 2y = 5$$
$$4y = 3 - 2x$$

5. Solve by determinants:

$$3x - 2y = 4$$
$$2x + 5y = -1$$

6. By finding the slope and y-intercept, sketch the graph of $2x + y = 4$.

7. The perimeter of a rectangular ranch is 24 km, and the length is 6.0 km more than the width. Set up equations relating the length l and the width w and then solve for l and w.

8. Solve by addition or subtraction:

$$6N - 2P = 13$$
$$4N + 3P = -13$$

9. In testing an anticholesterol drug, it was found that each milligram of drug administered reduced a person's blood cholesterol level by 2 units. Set up the function relating the cholesterol level C as a function of the dosage d for a person whose cholesterol level is 310 before taking the drug. Sketch the graph.

10. Solve the following system of equations graphically. Determine the values of x and y to the nearest 0.1.

$$2x - 3y = 6$$
$$4x + y = 4$$

11. Solve algebraically:

$$x + y + z = 4$$
$$x + y - z = 6$$
$$2x + y + 2z = 5$$

12. By volume, one alloy is 60% copper, 30% zinc, and 10% nickel. A second alloy has percents of 50, 30, and 20, respectively. A third alloy is 30% copper and 70% nickel. How much of each alloy is needed to make 100 cm^3 of a resulting alloy with percents of 40, 15, and 45, respectively?

13. Solve for y by determinants:

$$3x + 2y - z = 4$$
$$2x - y + 3z = -2$$
$$x + 4z = 5$$

6. Factoring and Fractions

NASA/JPL

▲ After travelling through space for five years, NASA's unmanned *Juno* spacecraft entered Jupiter's orbit in July 2016. In Section 6.8 we illustrate the use of algebraic operations in the study of planetary motion.

LEARNING OUTCOMES

After completion of this chapter, the student should be able to:

- Identify and perform various algebraic special products
- Factor algebraic expressions using common factors, difference of squares, sum and difference of cubes, and other techniques
- Simplify expressions involving algebraic fractions
- Solve equations involving algebraic fractions

In this chapter, we further develop operations with products, quotients, and fractions and show how they are used in solving equations. These additional algebraic methods are needed in later chapters, and they will be useful in solving many application problems.

The development of the symbols now used in algebra in itself led to advances in mathematics and science. Until about 1500, most problems and their solutions were stated in words, which made them very lengthy and often difficult to follow. For example, to solve the equation $3x + 7 = 8(2x - 5)$, the problem would be stated something like "find a number such that seven added to three times the number is equal to the product of eight and the quantity of five subtracted from twice the number." Then the solution would be written out, word by word.

As time went on, writers abbreviated some of the words in a problem, but they still essentially wrote out the solution in words. Some symbols did start to come into use. For example, the $+$ and $-$ signs first appeared in a published book in the late 1400s, the square root symbol $\sqrt{\ }$ was first used in 1525, and the $=$ sign was introduced in 1557. Then, in the late 1500s, the French lawyer François Viète wrote several articles in which he used symbols, including letters to represent numbers. He so improved the symbolism of algebra that he is often called "the father of algebra." By the mid-1600s, the notation being used was reasonably similar to what we use today.

The ability to describe physical phenomena using mathematical statements has allowed significant advancements in all technical fields. For example, as the *Juno* spacecraft orbits around Jupiter (pictured above), its mission depends on many complex mathematical models, including models related to electronics, mechanical design, thermodynamics, fluid mechanics, and computer algorithms, all of which start with mathematical statements and symbolism. Almost every significant feat of modern engineering relies heavily on mathematical problem-solving and models.

6.1 Special Products

Certain types of algebraic products are used so often that we should be very familiar with them. These are shown below, starting with the very important distributive law and continuing with six other special products that are obtained by using that law.

SPECIAL ALGEBRAIC PRODUCTS

Distributive law:	$a(x + y) = ax + ay$	**(6.1)**
Square of a binomial:	$(x + y)^2 = x^2 + 2xy + y^2$	**(6.2)**
	$(x - y)^2 = x^2 - 2xy + y^2$	**(6.3)**
Product of two binomials:	$(x + a)(x + b) = x^2 + (b + a)x + ab$	**(6.4)**
	$(ax + b)(cx + d) = acx^2 + (ad + bc)x + bd$	**(6.5)**
Product of sum and difference:	$(x + y)(x - y) = x^2 - y^2$	**(6.6)**
	$(ax + by)(ax - by) = (ax)^2 - (by)^2$	**(6.7)**

Each special product can be used in two ways:

1. *When expanding statements to simplify algebraic expressions*, these rules can be identified as they appear on the left and expanded to the statement on the right.
2. *When factoring algebraic statements*, each special product, when used in reverse, gives us a method for factoring.

In this section we see how to apply the special products when expanding algebraic statements. The next three sections are devoted to discussing how these products give us different techniques for factoring expressions.

EXAMPLE 1 Using special products (distributive law and product of sum and difference)

Find the product.

(a) $6(3r + 2s)$

We use the distributive law Eq. (6.1) to distribute the factor 6 to both terms by multiplication.

$$6(3r + 2s) = 6(3r) + 6(2s) = 18r + 12s$$

(b) $(3r + 2s)(3r - 2s)$

We recognize the product of the sum and difference of identical terms. Eq. (6.7) gives

$$(3r + 2s)(3r - 2s) = (3r)^2 - (2s)^2 = 9r^2 - 4s^2$$

LEARNING TIP

When expanding the square of a binomial, three terms result: the square of the first term, the square of the last term, and a middle term of the products of the two terms multiplied by 2:

$$(3x - 2y)^2 = (3x)^2 + 2(3x)(-2y) + (-2y)^2$$

First term squared — Twice product — Last term squared

$$(3x - 2y)^2 = 9x^2 - 12xy + 4y^2$$

EXAMPLE 2 Using special products involving the square of a binomial

Find the product.

(a) $(5a + 2)^2$

We recognize the square of a binomial. Eq. (6.2) gives

$$(5a + 2)^2 = (5a)^2 + 2(5a)(2) + 2^2 = 25a^2 + 20a + 4$$

(b) $(5a - 2)^2$

We recognize the square of a binomial with a negative second term. Eq. (6.3) gives

$$(5a - 2)^2 = (5a)^2 - 2(5a)(2) + 2^2 = 25a^2 - 20a + 4$$

Note how the sign of the second term of the binomial dictates the sign of the middle term of the resulting trinomial.

COMMON ERROR

It should be emphasized that $(5a + 2)^2$ is **not** $(5a)^2 + 2^2$, or $25a^2 + 4$. Ensure that you use the forms of Eqs. (6.2) and (6.3) very carefully by **including the middle term, 20*a*.**

EXAMPLE 3 Finding the product of two binomials

Find the following products using the distributive law and adding like terms.

(a) $(x + 5)(x - 3)$

$$(x + 5)(x - 3) = x(x - 3) + 5(x - 3)$$
$$= x^2 - 3x + 5x - 15 = x^2 + 2x - 15$$

Note that by recognizing $a = 5$ and $b = -3$ in Eq. (6.4), we could have obtained the product quickly without having to distribute each term. We would find the coefficient of x in the result as $a + b = 5 + (-3) = 2$.

(b) $(4x + 5)(2x - 3)$

$$(4x + 5)(2x - 3) = 4x(2x - 3) + 5(2x - 3)$$
$$= 8x^2 - 12x + 10x - 15 = 8x^2 - 2x - 15$$

Here, recognizing $a = 4$, $b = 5$, $c = 2$, and $d = -3$ in Eq. (6.5) could give us the result directly without having to write down intermediate steps.

As you become more familiar with these special products, it will become easier to find the middle term mentally and write down the result directly, as shown in the next example.

EXAMPLE 4 Finding the middle term mentally

For each of the following expressions, identify the special product and find the middle term mentally.

(a) $(y - 5)(y + 5)$

This is a product of the sum and difference of identical terms, Eq. (6.6). There is no middle term.

(b) $(3x - 2)^2$

This is the square of a binomial, Eq. (6.3). The middle term is twice the product of the first and second terms, or $2(3x)(-2) = -12x$.

(c) $(x^2 - 4)(x^2 + 7)$

This is the product of two binomials, Eq. (6.4), but with variable x^2 instead of x. The middle term is the algebraic sum of the two constants times the variable, or $(-4 + 7)x^2 = 3x^2$.

Practice Exercises

Find the indicated products

1. $(3x + 4)(3x - 4)$
2. $(3x + 4)^2$
3. $(3x + 4)(x - 2)$

At times, it is necessary to use more than one of the special products to simplify an algebraic expression. When this happens, it may be necessary to write down an intermediate step. This is illustrated in the following examples.

EXAMPLE 5 Applications of special products

(a) When analysing the forces on a certain type of beam, the expression $Fa(L - a)(L + a)$ occurs. In expanding this expression, we first multiply $L - a$ by $L + a$ by use of Eq. (6.6). The expansion is completed by using Eq. (6.1), the distributive law.

$$Fa(L - a)(L + a) = Fa(L^2 - a^2) \quad \text{Eq. (6.6)}$$
$$= FaL^2 - Fa^3 \quad \text{Eq. (6.1)}$$

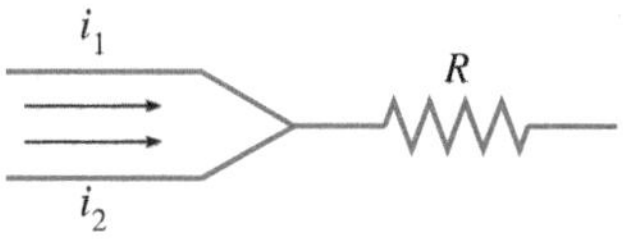

Fig. 6.1

(b) The electrical power delivered to the resistor R in Fig. 6.1 is $R(i_1 + i_2)^2$. Here, i_1 and i_2 are electric currents. To expand this expression, we first perform the square by use of Eq. (6.2) and then complete the expansion by use of Eq. (6.1).

$$R(i_1 + i_2)^2 = R(i_1^2 + 2i_1i_2 + i_2^2) \quad \text{Eq. (6.2)}$$
$$= Ri_1^2 + 2Ri_1i_2 + Ri_2^2 \quad \text{Eq. (6.1)}$$

EXAMPLE 6 Group terms, then use special products

Find the product $(x + y - 3)^2$.

We group the quantity $(x + y)$ in an intermediate step. This leads to

$$(x + y - 3)^2 = [(x + y) - 3]^2$$
$$= (x + y)^2 - 2(x + y)(3) + 3^2 \quad \text{Eq. (6.3)}$$
$$= x^2 + 2xy + y^2 - 6x - 6y + 9 \quad \text{Eqs. (6.2) and (6.1)}$$

SPECIAL PRODUCTS INVOLVING CUBES

Four other special products occur less frequently. However, they are sufficiently important that patterns of the resulting polynomials should be readily recognized. They are shown in Eqs. (6.8)–(6.11).

Cube of a binomial:	$(x + y)^3 = x^3 + 3x^2y + 3xy^2 + y^3$	**(6.8)**
	$(x - y)^3 = x^3 - 3x^2y + 3xy^2 - y^3$	**(6.9)**
Result is a sum of cubes:	$(x + y)(x^2 - xy + y^2) = x^3 + y^3$	**(6.10)**
Result is a difference of cubes:	$(x - y)(x^2 + xy + y^2) = x^3 - y^3$	**(6.11)**

EXAMPLE 7 Special products involving cubes

Find the product.

(a) $(x + 4)^3 = x^3 + 3(x^2)(4) + 3(x)(4^2) + 4^3$ Eq. (6.8)

$= x^3 + 12x^2 + 48x + 64$

(b) $(2x - 5)^3 = (2x)^3 - 3(2x)^2(5) + 3(2x)(5^2) - 5^3$ Eq. (6.9)

$= 8x^3 - 60x^2 + 150x - 125$

(c) $(x + 3)(x^2 - 3x + 9) = x^3 + 3^3$ Eq. (6.10)

$= x^3 + 27$

(d) $(x - 2)(x^2 + 2x + 4) = x^3 - 2^3$ Eq. (6.11)

$= x^3 - 8$

EXERCISES 6.1

In Exercises 1–4, make the given changes in the indicated examples of this section and then solve the resulting problems.

1. In Example 1(b), interchange the + and − signs and then find the product.
2. In Example 3(b), interchange the + and − signs and then find the product.
3. In Example 6, group $y - 3$ instead of $x + y$ and then find the product.
4. In Example 7(a), change + to − in the original expression and then find the product.

In Exercises 5–36, find the indicated products directly by inspection. It should not be necessary to write down intermediate steps [except possibly when using Eq. (6.5)].

5. $30(x - y)$
6. $4x(2a - 5)$
7. $7x^3(x - 5)$
8. $6a^2(3a + 2)$
9. $(T + 4)(T - 4)$
10. $(s + 3t)(s - 3t)$
11. $(4v - 3)(4v + 3)$
12. $(ab - c)(ab + c)$
13. $(5x - 4y)(5x + 4y)$
14. $(7s + 2t)(7s - 2t)$
15. $(12 + 5ab)(12 - 5ab)$
16. $(2xy - 11)(2xy + 11)$
17. $(4f + 5)^2$
18. $(2x - 7)^2$
19. $(2x + 13)^2$
20. $(8a + 9b)^2$
21. $(L^2 - 1)^2$
22. $(b^2 - 6)^2$
23. $(4a + 7xy)^2$
24. $(3A + 10z)^2$
25. $(0.5s - 0.1t)^2$
26. $(0.4p - 0.3q)^2$
27. $(x + 1)(x + 5)$
28. $(y - 8)(y + 5)$
29. $(3k + 6C^2)(6k + C^2)$
30. $(8 - e^3)(3 - e^3)$
31. $(4x - 5)(5x + 1)$
32. $(2y - 1)(3y - 1)$
33. $(10v - 3)(4v + 15)$
34. $(7s + 12)(8s + 5)$
35. $(30x + 7y)(20x - 9y)$
36. $(24x - y)(15x + 4y)$

Use the special products of this section to determine the products of Exercises 37–62. You may need to write down one or two intermediate steps.

37. $2(x - 2)(x + 2)$
38. $25(V - 5)(V + 5)$
39. $2a(2a - 1)(2a + 1)$
40. $4c(2c - 3)(2c + 3)$
41. $6a(x + 2b)^2$
42. $7r(5r + 2b)^2$
43. $5n^2(2n + 5)^2$
44. $8T(T - 7)^2$
45. $[(2R + 3r)(2R - 3r)]^2$
46. $[(6t - y)(6t + y)]^2$
47. $(x + y + 1)^2$
48. $(x + 2 + 3y)^2$
49. $(3 - x - y)^2$
50. $2(x - y + 1)^2$
51. $(5 - t)^3$
52. $(2s + 3)^3$
53. $(3L + 7R)^3$
54. $(2A - 5B)^3$
55. $(w + h - 1)(w + h + 1)$
56. $(2a - c + 2)(2a - c - 2)$
57. $(x + 2)(x^2 - 2x + 4)$
58. $(s - 3)(s^2 + 3s + 9)$
59. $(4 - 3x)(16 + 12x + 9x^2)$
60. $(2x + 3a)(4x^2 - 6ax + 9a^2)$
61. $(x + y)^2(x - y)^2$
62. $(x - y)(x + y)(x^2 + y^2)$

In Exercises 63–72, use the special products of this section to determine the products. Each comes from the technical area indicated.

63. $P_1(P_0c + G)$ (computers)
64. $akT(t_2 - t_1)$ (heat conduction)
65. $4(p + DA)^2$ (photography)
66. $(2J + 3)(2J - 1)$ (lasers)
67. $\frac{1}{2}\pi(R + r)(R - r)$ (architecture)
68. $k(T - T_0)(T + T_0)(T^2 + T_0^2)$ (radiation)
69. $\frac{L}{6}(x - a)^3$ (mechanics: beams)
70. $(1 - z)^2(1 + z)$ (motion: gyroscope)
71. $L_0[1 + a(T - T_0)]$ (thermal expansion)
72. $(s + 1 + j)(s + 1 - j)$ (electricity)

In Exercises 73–82, solve the given problems.

73. Multiply 49 by 51 by writing $(49)(51) = (50 - 1)(50 + 1)$ and using one of the special products.
74. Multiply 82 by 78 by writing $(82)(78) = (80 + 2)(80 - 2)$ and using one of the special products.
75. The length of a piece of rectangular floor tile is 3 cm more than twice the side x of a second square piece of tile. The width of the rectangular piece is 3 cm less than twice the side of the square piece. Find the area of the rectangular piece in terms of x (in expanded form).
76. The radius of a circular oil spill is r. It then increases in radius by 40 m before being contained. Find the area of the oil spill at the time it is contained in terms of r (in expanded form). See Fig. 6.2.

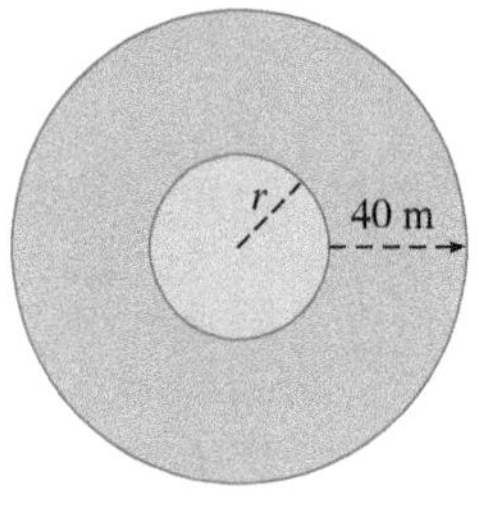

Fig. 6.2

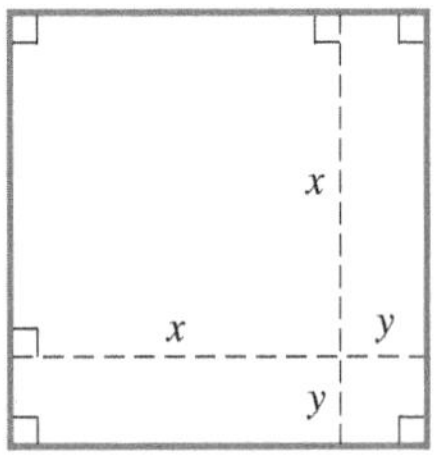

Fig. 6.3

77. Referring to Fig. 6.3, find (a) an expression for the area of the large square (the entire figure) as the power of a binomial; (b) the sum of the areas of the four figures within the large square as a trinomial. Noting that the results of (a) and (b) should be equal, what special product is shown geometrically?
78. If $(x + k)(x - 1) = x^2 + k(x + 2) - (x - 3)$, find the value of k.
79. Find the product $(2x - y - 2)(2x + y + 2)$ by first grouping the last two terms of each factor. (See Example 6.)

80. A contractor is installing tiles on a wall. The manufacturer changes the size of the tiles to be 2 cm wider and 3 cm higher. Write an algebraic expression to determine the area of the new tiles in terms of the old width w and the old height h.

81. A rectangular cement deck of length l, width w, and thickness 10 mm is being installed. Write an algebraic expression to describe the volume of cement required if the deck surrounds a circular hot tub of diameter 2 m and a square baby pool of side 3 m.

82. Verify Eqs. (6.9) and (6.10) by multiplication. Then explain why we may refer to Eqs. (6.1)–(6.11) as *identities*.

Answers to Practice Exercises

1. $9x^2 - 16$ **2.** $9x^2 + 24x + 16$ **3.** $3x^2 - 2x - 8$

6.2 Factoring: Common Factor and Difference of Squares

When an algebraic expression is the product of two or more quantities, each of these quantities is called a ***factor*** of the expression. Finding these factors, which is essentially reversing the process of multiplication, is called **factoring**. Factoring is needed in many technical applications that require us to solve higher-degree equations, perform operations on algebraic fractions, or solve certain literal equations.

In our work on factoring, we will consider only the factoring of polynomials (see Section 1.7) that have integers as coefficients for all terms. Also, all factors will have integral coefficients. *A polynomial or a factor is called* **prime** *if it contains no factors other than* $+1$ *or* -1 *and plus or minus itself.* Also, we say that an *expression is* **factored completely** *if it is expressed as a product of its prime factors.*

To factor expressions easily, we must know how to do algebraic multiplication and really know the special products of the previous section. The special products also give us a way of checking answers and deciding whether a given factor is prime.

LEARNING TIP
The ability to factor algebraic expressions depends heavily on the proper recognition of the special products.

FACTORING OUT THE GREATEST COMMON FACTOR (GCF)

We will be discussing several factoring techniques in this chapter. The first technique we discuss, which should always be the first step in factoring, is **factoring out the greatest common factor** (or GCF, for short). This technique is a result of the distributive law of Eq. (6.1), written in reverse as

$$ax + ay = a(x + y) \tag{6.12}$$

In Eq. (6.12), a is the GCF and it has been *factored out* of the expression.

To factor out the GCF, we first find the greatest monomial that is common to each term of the expression. This is the GCF. We then use Eq. (6.12) to write the expression as the product of the GCF and the remaining expression. The following examples illustrate this method.

EXAMPLE 1 Factoring out the GCF

Factor: $6x - 2y$.

We factor each term into its prime factors and notice that each term contains a factor of 2, and that there are no other common factors. Therefore, 2 is the GCF and we can factor as

$$6x - 2y = 2 \cdot 3 \cdot x - 2 \cdot y = 2(3x - y)$$

We check the result by multiplication. We have

$$2(3x - y) = 6x - 2y$$

Since the result of the multiplication gives the original expression, the factored form is correct.

LEARNING TIP
You can check your factoring result by multiplication.

LEARNING TIP

Usually, the factorization of common monomial factors is done by inspection. However, once the common factor is found, the other factor can also be determined by dividing the original expression by the common factor. For example, $2abc + 4ab$ has a common factor of $2ab$ in each term, so

$$2abc + 4ab = 2ab\left(\frac{2abc}{2ab} + \frac{4ab}{2ab}\right)$$

$$2abc + 4ab = 2ab(c + 2)$$

Practice Exercises

Factor: **1.** $3cx^3 - 9cx$ **2.** $9cx^3 - 3cx$

The next example illustrates the case where the common factor is the same as one of the terms. In this situation, we can rewrite any quantity a as the product $a(1)$. This simple substitution will help us complete the factoring correctly.

EXAMPLE 2 Greatest common factor same as term

Factor: $4ax^2 + 2ax$.

The numerical factor 2 and the literal factors a and x are common to each term. Therefore, the greatest common factor of $4ax^2 + 2ax$ is $2ax$. Before factoring, we rewrite the second term to include a factor of 1. We have

$$4ax^2 + 2ax = 2ax(2x) + 2ax(1) = 2ax(2x + 1)$$

Note that had we not factored a 1 explicitly in the second term, it could appear that the second term vanished after factoring, which would be incorrect.

COMMON ERROR

When the GCF is the same as one of the terms in the expression, remember that the term divides to 1, so do not omit the 1 and cancel the term out to nothing!

$$5rt + 5r = 5r(t + 1)$$

Not

$$5rt + 5r \neq 5r(t + 0)$$

EXAMPLE 3 Greatest common factor by inspection

Factor: $6a^5x^2 - 9a^3x^3 + 3a^3x^2$.

After inspecting each term, we determine that each contains a factor of 3, a^3, and x^2. Thus, the common monomial factor is $3a^3x^2$. This means that

$$6a^5x^2 - 9a^3x^3 + 3a^3x^2 = 3a^3x^2(2a^2 - 3x + 1)$$

When factoring, it is important to factor out the *greatest* common factor, not just any common factor. Factoring out something other than the GCF will lead to an expression that is not factored completely. The following example illustrates this point.

EXAMPLE 4 Factoring completely

Factor: $12x + 6x^2$.

We note that each term contains a factor of 2, so we factor as

$$12x + 6x^2 = 2(6x + 3x^2)$$

However, this is not factored completely. The factor $6x + 3x^2$ is not prime because it may be further factored as $6x + 3x^2 = 3x(2 + x)$.

If we instead factor out the *greatest* common factor $6x$, we get $12x + 6x^2 = 6x(2 + x)$. This is now factored completely since the factor $2 + x$ is prime.

In these examples, note that factoring an expression does not actually change the expression, although it does change the *form* of the expression. In equating the expression to its factored form, we write an *identity.*

It is often necessary to use factoring when solving an equation. This is illustrated in the following example.

EXAMPLE 5 Using factoring in solving an equation—FM reception

An equation used in the analysis of FM reception is $R_F = \alpha(2R_A + R_F)$. Solve for R_F.

The steps in the solution are as follows:

$$R_F = \alpha(2R_A + R_F) \quad \text{original equation}$$
$$R_F = 2\alpha R_A + \alpha R_F \quad \text{use distributive law}$$
$$R_F - \alpha R_F = 2\alpha R_A \quad \text{subtract } \alpha R_F \text{ from both sides}$$
$$R_F(1 - \alpha) = 2\alpha R_A \quad \text{factor out } R_F \text{ on left}$$
$$R_F = \frac{2\alpha R_A}{1 - \alpha} \quad \text{divide both sides by } 1 - \alpha$$

We see that we collected both terms containing R_F on the left so that we could factor and thereby solve for R_F.

FACTORING THE DIFFERENCE OF TWO SQUARES

After factoring out the GCF, the next simplest factoring method relies upon identifying the pattern that results from the product of a sum and difference rule. When written in reverse, this rule gives us a method of factoring called **factoring the difference of two squares.**

$$u^2 - v^2 = (u + v)(u - v) \quad \textbf{(6.13)}$$

Eq. (6.13) can be used to factor the difference between two perfect squares as shown in the following examples.

LEARNING TIP

Usually, in factoring an expression of this type, where it is very clear what numbers are squared, we do not actually write out the middle step as shown. However, if in doubt, write it out.

EXAMPLE 6 Factoring the difference of two squares

In factoring $x^2 - 16$, note that x^2 is the square of x and 16 is the square of 4. Therefore,

$$x^2 - 16 = x^2 - 4^2 = (x + 4)(x - 4)$$

(squares: x^2 and 4^2; difference; sum; difference)

When the factors in a difference of squares are composite, what expressions are being squared is not always obvious. Factoring each term into prime factors can help in those situations.

EXAMPLE 7 Factoring the difference of squares for composite factors

Factor each of the following expressions.

(a) $4x^2 - 9 = 2 \cdot 2 \cdot x \cdot x - 3 \cdot 3 = (2x)^2 - 3^2 = (2x + 3)(2x - 3)$

(b) $(y - 3)^2 - 16x^4 = (y - 3)^2 - 2 \cdot 2 \cdot 2 \cdot 2 \cdot x \cdot x \cdot x \cdot x = (y - 3)^2 - (4x^2)^2$
$= (y - 3 + 4x^2)(y - 3 - 4x^2)$

COMPLETE FACTORING

As noted before, ***the greatest common factor should be factored out first***. However, we must be careful to *see if the other factor can itself be factored.* It is possible, for example, that the other factor is the difference of squares. This means that **complete factoring** often requires more than one step. Be sure to include only prime factors in the final result.

COMMON ERROR

Always factor your equations completely. It is a common error to stop before the expression involves only prime factors.

For example, $4a^2 - a^2x^2$ can be factored to $a^2(4 - x^2)$, and it is a common error to think you are now finished. However, the second factor is a difference of squares. The complete factorization is

$$4a^2 - a^2x^2 = a^2(4 - x^2)$$
$$= a^2(2 + x)(2 - x)$$

EXAMPLE 8 Factoring completely

Factor completely.

(a) $20x^2 - 45$

Factor out a common factor of 5: $20x^2 - 45 = 5(4x^2 - 9)$

Factor the difference of squares: $= 5(2x + 3)(2x - 3)$

(b) $x^4 - y^4$

Factor the difference of squares: $x^4 - y^4 = (x^2 + y^2)(x^2 - y^2)$

Factor the other difference of squares: $= (x^2 + y^2)(x + y)(x - y)$

Practice Exercises

Factor: **3.** $9c^2 - 64$ **4.** $18c^2 - 128$

COMMON ERROR

The factor $x^2 + y^2$ is prime. It is **not** equal to $(x + y)^2$. Remember that the square of a binomial includes a middle term.

FACTORING BY GROUPING

Terms in an expression can sometimes be grouped and then factored by the methods of this section. A pattern will sometimes emerge when grouping into smaller sets of terms. The following example illustrates this method of *factoring by grouping.* In the next section, we discuss another type of expression that can be factored by grouping.

EXAMPLE 9 Factoring by grouping

Factor: $2x - 2y + ax - ay$.

We see that there is no common factor to all four terms, but that each of the first two terms contains a factor of 2, and each of the third and fourth terms contains a factor of a. Grouping terms this way and then factoring each group, we have

$$2x - 2y + ax - ay = 2(x - y) + a(x - y)$$

Now note the common factor of $x - y$, which can be factored out, grouping the other terms within parentheses:

$$2(x - y) + a(x - y) = (x - y)(2 + a)$$

EXERCISES 6.2

In Exercises 1–4, make the given changes in the indicated examples of this section and then solve the indicated problems.

1. In Example 2, change the + sign to − and then factor.

2. In Example 2, set the given expression equal to B and then solve for a.

3. In Example 8(a), change the coefficient of the first term from 20 to 5 and then factor.

4. In Example 9, change both − signs to + and then factor.

In Exercises 5–46, factor the given expressions completely.

5. $7x - 7y$ **6.** $5b + 5a$ **7.** $6b - 24$

8. $3x^2 - 6$ **9.** $15x^2 - 3x$ **10.** $5s^2 + 25$

11. $7w^2h - 28w$ **12.** $5r^2 - 20rh$ **13.** $288n^2 + 24n$

14. $90p^3 - 15p^2$ **15.** $2x + 4y - 8z$ **16.** $23a - 46b + 69c$

17. $3ab^2 - 6ab + 12ab^3$ **18.** $4pq - 14q^2 - 16pq^2$

19. $6pq^2 - 5pq - 14pq^3$ **20.** $27a^2b - 24ab - 9a$

21. $2a^2 - 2b^2 + 4c^2 - 6d^2$ **22.** $5a + 10ax - 5ay - 20az$

23. $x^2 - 9$ **24.** $r^2 - 25$ **25.** $100 - 9A^2$

26. $49 - Z^4$ **27.** $36a^4 + 1$ **28.** $324z^2 - 4$

29. $162s^2 - 50t^2$ **30.** $36s^2 - 121t^2$ **31.** $144n^2 - 169p^4$

32. $36a^2b^2 + 169c^2$ **33.** $(x + y)^2 - 9$ **34.** $(a - b)^2 - 1$

35. $32 - 18x^2$ **36.** $5a^2 - 125$ **37.** $300x^2 - 2700z^2$

38. $28x^2 - 700y^2$ **39.** $2(I - 3)^2 - 8$ **40.** $a(x + 2)^2 - ay^2$

41. $x^4 - 16$ **42.** $y^4 - 81$ **43.** $x^8 - 1$

44. $2x^4 - 8y^4$ **45.** $v^2(i_1 - 4)^2 - 25(i_2)^2$

46. $rm + nm - rn + n^2$

In Exercises 47–52, solve for the indicated letter.

47. $2a - b = ab + 3$, for a **48.** $n(x + 1) = 5 - x$, for x

49. $3 - 2s = 2(3 - st)$, for s **50.** $k(2 - y) = y(2k - 1)$, for y

51. $(x + 2k)(x - 2) = x^2 + 3x - 4k$, for k

52. $(2x - 3k)(x + 1) = 2x^2 - x - 3$, for k

In Exercises 53–60, factor the given expressions by grouping as illustrated in Example 9.

53. $3x - 3y + bx - by$ **54.** $am + an + cn + cm$

55. $a^2 + ax - ab - bx$ **56.** $2y - y^2 - 6y^4 + 12y^3$

57. $x^3 + 3x^2 - 4x - 12$ **58.** $S^3 - 5S^2 - S + 5$

59. $x^2 - y^2 + x - y$ **60.** $4p^2 - q^2 + 2p + q$

In Exercises 61 and 62, evaluate the given expressions by using factoring. The results may be checked with a calculator.

61. $\dfrac{8^9 - 8^8}{7}$ **62.** $\dfrac{5^9 - 5^7}{7^2 - 5^2}$

In Exercises 63 and 64, give the required explanations.

63. Factor $n^2 + n$, and then explain why it represents a positive even integer if n is a positive integer.

64. Factor $n^3 - n$, and then explain why it represents a multiple of 6 if n is an integer greater than 1.

In Exercises 65–74, factor the expressions completely. In Exercises 71 and 72, it is necessary to set up the proper expression. Each expression comes from the technical area indicated.

65. $2Q^2 + 2$ (fire science)

66. $4d^2D^2 - 4d^3D - d^4$ (machine design)

67. $Rv + Rv^2 + Rv^3$ (business)

68. $PbL^2 - Pb^3$ (architecture)

69. $rR^2 - r^3$ (pipeline flow)

70. $p_1R^2 - p_1r^2 - p_2R^2 + p_2r^2$ (fluid flow)

71. A square as large as possible is cut from a circular metal plate of radius r. Express in factored form the area of the metal pieces that are left.

72. A pipe of outside diameter d is inserted into a pipe of inside radius r. Express in factored form the cross-sectional area within the larger pipe that is outside the smaller pipe.

73. A spherical float has a volume of air within it of radius r_1, and the outer radius of the float is r_2. Express in factored form the difference in areas of the outer surface and the inner surface.

74. The kinetic energy of an object of mass m travelling at velocity v is given by $\frac{1}{2}mv^2$. Suppose a car of mass m_0 equipped with a crash-avoidance system automatically applies the brakes to avoid a collision and slows from a velocity of v_1 to a velocity of v_2. Find an expression, in factored form, for the difference between the original and final kinetic energy.

In Exercises 75–80, solve for the indicated letter. Each equation comes from the technical area indicated.

75. $i_1R_1 = (i_2 - i_1)R_2$, for i_1 (electricity: ammeter)

76. $nV + n_1v = n_1V$, for n_1 (acoustics)

77. $3BY + 5Y = 9BS$, for B (physics: elasticity)

78. $Sq + Sp = Spq + p$, for q (computer design)

79. $ER = AtT_0 - AtT_1$, for t (energy conservation)

80. $R = kT_2^4 - kT_1^4$, for k (factor resulting denominator) (radiation)

Answers to Practice Exercises

1. $3cx(x^2 - 3)$ **2.** $3cx(3x^2 - 1)$

3. $(3c - 8)(3c + 8)$ **4.** $2(3c - 8)(3c + 8)$

6.3 Factoring Trinomials

In the previous section, we discussed factoring expressions by reversing two special products: the distributive law and the product of a sum and difference. In this section we discuss factoring **trinomials** (expressions with three terms) by reversing the product of two binomials. In other words, we will find two binomials that, when multiplied together, will equal the given trinomial. We begin with the simplest case, which is factoring trinomials in which the coefficient of x^2, called the **leading coefficient**, is 1.

FACTORING TRINOMIALS WITH A LEADING COEFFICIENT OF 1

■ For reference, Eq. (6.4) is $(x + a)(x + b) = x^2 + (a + b)x + ab$.

Starting with Eq. (6.4), we notice that

1. the coefficient of x in the trinomial is the *sum of the constants in the two binomial factors*; and
2. the constant term in the trinomial is the *product of the constants in the two binomial factors*.

Reversing this process will allow us to factor a trinomial when the coefficient of x^2 is 1 as shown in the diagram below:

sum

$$\text{coefficient} = 1 \longrightarrow x^2 + (a + b)x + ab = (x + a)(x + b)$$

product

Essentially, *we need to find two integers a and b that have a product equal to the last term of a given trinomial and a sum equal to the coefficient of x.*

EXAMPLE 1 Factoring a trinomial with leading coefficient of 1

Factor the following trinomials.

(a) $x^2 + 3x + 2$

Table 6.1

Factors of $+2$	Sum
$+2, +1$	$+3$
$-2, -1$	-3

All pairs of numbers whose product is $+2$ (the constant term) are given in Table 6.1. The numbers whose sum is $+3$ (the coefficient of x) are $+2$ and $+1$. Therefore,

$$x^2 + 3x + 2 = (x + 2)(x + 1)$$

(b) $x^2 - 3x + 2$

The constant term is again $+2$, but now we require numbers whose sum is -3. From Table 6.1 we see that they are -2 and -1. Therefore,

$$x^2 - 3x + 2 = (x - 2)(x - 1)$$

We can make the following observations regarding the signs of the terms of the trinomial and the signs of the numbers a and b:

1. If the sign of the constant term is positive, a and b have the same sign (both are positive or both are negative).
2. If the sign of the constant term is negative, a and b differ in sign (one is positive and one is negative).
3. The sign of the number with larger magnitude is the same as the sign of the middle term.

LEARNING TIP

Not all trinomials with a leading coefficient of 1 can be factored using this procedure. For example, the expression $x^2 + 4x + 2$ is not factorable because there are no two numbers whose product is $+2$ that add to $+4$.

EXAMPLE 2 Factoring trinomials with x^2 coefficient of 1

Factor each of the following expressions.

(a) $x^2 + 7x - 8$

Table 6.2

Factors of -8	Sum (Need $+7$)
$-8, +1$	-7
$+8, -1$	$+7$
$-4, +2$	-2
$+4, -2$	$+2$

All pairs of numbers whose product is -8 (the constant term) are given in Table 6.2. The numbers whose sum is $+7$ (the coefficient of x) are $+8$ and -1. Therefore,

$$x^2 + 7x - 8 = (x + 8)(x - 1)$$

Note that the sign of the constant term is negative, so the two required numbers differ in sign. The middle term is positive so the larger number is positive.

Table 6.3

Factors of −12	Sum (Need −1)
−12, +1	−11
+12, −1	+11
−6, +2	−4
+6, −2	+4
−4, +3	−1
+4, −3	+1

(b) $x^2 - x - 12$

The sign of the constant term is negative, so the two required numbers differ in sign. The middle term is negative, so the larger value should be negative. See Table 6.3. We get

$$x^2 - x - 12 = (x - 4)(x + 3)$$

(c) $x^2 - 5xy + 6y^2$

In this situation, we need to find binomials with second terms whose product is $6y^2$ (the last term) and whose sum is $-5y$ (the coefficient of x). This means that each second term must have a factor of y. From Table 6.4, we get

$$x^2 - 5xy + 6y^2 = (x - 3y)(x - 2y)$$

Table 6.4

Factors of $6y^2$	Sum (Need $-5y$)
$-6y, -y$	$-7y$
$+6y, +y$	$+7y$
$-3y, -2y$	$-5y$
$+3y, +2y$	$+5y$

COMMON ERROR

Ensure that you have the **proper signs** on the two numbers chosen to make the product and the sum leading to the trinomial coefficients. To factor $x^2 - x - 6$, we require two numbers that multiply to −6. The pairs +2 and −3 and −2 and +3 both work. However, only one of those pairs adds to the coefficient of the middle term (+2 and −3). Therefore,

$$x^2 - x - 6 = (x + 2)(x - 3)$$
$$x^2 - x - 6 \neq (x - 2)(x + 3)$$

A **perfect square trinomial** is the result of the product of two identical binomial factors. The following example illustrates how the same process of trinomial factoring can be used to factor trinomials of this type.

EXAMPLE 3 Factoring perfect square trinomials

Factor the following trinomials.

(a) $x^2 + 10x + 25$

The sign of the constant term is positive, so the two required numbers have the same sign. The middle term is positive, so we consider only pairs of positive numbers. From Table 6.5, we have

$$x^2 + 10x + 25 = (x + 5)(x + 5) = (x + 5)^2$$

Note that the first and third terms of the trinomial are perfect squares.

Table 6.5

Positive Factors of +25	Sum (Need +10)
+25, +1	+26
+5, +5	+10

(b) $A^4 - 20A^2 + 100$

Although the required pair of numbers could be found from Table 6.6, we note that the first and third terms of the trinomial are perfect squares, with $A^4 = (A^2)^2$ and $100 = 10^2$. The middle term is negative, so the required numbers must be negative too. Since the sum of −10 and −10 is −20, we can write

$$A^4 - 20A^2 + 100 = (A^2 - 10)^2$$

The factorization is complete if using only integers. Note that it *can* be factored further as a difference of squares if factors can include any real numbers. In that case we would write

$$A^4 - 20A^2 + 100 = [(A + \sqrt{10})(A - \sqrt{10})]^2$$

Table 6.6

Negative Factors of +100	Sum
−100, −1	−101
−50, −2	−52
−25, −4	−29
−20, −5	−25
−10, −10	−20

(c) $x^2 - 29x + 100$

Although the first and third terms of the trinomial are perfect squares, the sum of -10 and -10 is not -29. From Table 6.6 in the previous page we see that the numbers -25 and -4 do add to -29, so the correct factorization is

$$x^2 - 29x + 100 = (x - 25)(x - 4)$$

This shows that just because the first and last terms of a trinomial are perfect squares, we can't *assume* the expression is a perfect square trinomial. We must check the middle term.

Practice Exercises

Factor: **1.** $x^2 - x - 2$ **2.** $x^2 - 4x + 4$

FACTORING GENERAL TRINOMIALS

■ For reference, Eq. (6.5) is $(ax + b)(cx + d) = acx^2 + (ad + bc)x + bd$.

We will now turn our attention to factoring *general trinomials*, where *the coefficient of the squared term can be any integer* (not just 1 as in the previous examples). Rewriting Eq. (6.5) with sides reversed, we have

product ← coefficients

$$acx^2 + (ad + bc)x + bd = (ax + b)(cx + d)$$

product ← integers

Note that the coefficient of x is the sum of two products. Therefore, we need to find two integer products ad and bc whose sum is equal to the coefficient of x, and whose product is equal to the product of the leading coefficient and the constant term of the trinomial.

EXAMPLE 4 Factoring a general trinomial

Factor: $2x^2 + 11x + 5$.

All coefficients and constants are positive, so only positive values need to be considered. The product of the leading coefficient and the constant term of the trinomial is $2 \cdot 5 = 10$. Therefore, we need to find two values whose product is 10 and whose sum is 11 (the coefficient of x). From Table 6.7, we see that the correct values are $+10$ and $+1$.

Table 6.7

Positive Factors of +10	Sum (Need +11)
+10, +1	+11
+5, +2	+7

Instead of factoring these quantities by trial and error to identify the values of a, b, c, and d, we rewrite the middle term of the trinomial as the sum of two x-terms having these quantities as coefficients, and then factor by grouping. We have

$$\begin{aligned} 2x^2 + 11x + 5 &= 2x^2 + 10x + 1x + 5 \\ &= 2x(x + 5) + 1(x + 5) = (x + 5)(2x + 1) \end{aligned}$$

Note that writing the middle term in a different order gives the same result, with factors appearing in a different order:

$$\begin{aligned} 2x^2 + 11x + 5 &= 2x^2 + 1x + 10x + 5 \\ &= x(2x + 1) + 5(2x + 1) = (2x + 1)(x + 5) \end{aligned}$$

EXAMPLE 5 Factoring a general trinomial

Factor the following trinomials.

(a) $4x^2 + 4x - 3$.

The product of the leading coefficient and the constant term of the trinomial is $4(-3) = -12$. Therefore, we need to find two values whose product is -12 (they will have different signs) and whose sum is $+4$ (the larger value will be positive). From Table 6.8, we see that the correct values are $+6$ and -2. Rewriting the middle term of the trinomial as a sum and then factoring by grouping, we get

Table 6.8

Factors of −12	Sum (Need +4)
+12, −1	+11
+6, −2	+4
+4, −3	+1

$$\begin{aligned} 4x^2 + 4x - 3 &= 4x^2 + 6x - 2x - 3 \\ &= 2x(2x + 3) - 1(2x + 3) = (2x - 1)(2x + 3) \end{aligned}$$

Table 6.9

Factors of -120	Sum (Need $+7$)
$+120, -1$	$+119$
$+60, -2$	$+58$
$+40, -3$	$+37$
$+30, -4$	$+26$
$+24, -5$	$+19$
$+20, -6$	$+14$
$+15, -8$	$+7$
$+12, -10$	$+2$

(b) $6x^2 + 7x - 20$.

We need to find two values whose product is $6(-20) = -120$ (they will have different signs) and whose sum is $+7$ (the larger value will be positive). From Table 6.9, we see that the correct values are $+15$ and -8. Rewriting the middle term of the trinomial as a sum and then factoring by grouping, we get

$$\begin{aligned} 6x^2 + 7x - 20 &= 6x^2 + 15x - 8x - 20 \\ &= 3x(2x + 5) - 4(2x + 5) = (2x + 5)(3x - 4) \end{aligned}$$

Practice Exercise

3. Factor: $4x^2 + x - 5$

Table 6.10

Negative Factors of 252	Sum (Need -32)
$-252, -1$	-253
$-126, -2$	-128
$-84, -3$	-87
$-63, -4$	-67
$-42, -6$	-48
$-36, -7$	-43
$-28, -9$	-37
$-21, -12$	-33
$-18, -14$	-32

EXAMPLE 6 Factoring trinomials with two variables—beam deflection

An expression that arises when analysing the deflection of beams is $9x^2 - 32Lx + 28L^2$. Factor this expression.

In this case we need to find two values whose product is $9(28) = 252$ (they will have equal signs) and whose sum is -32 (they will both be negative). Table 6.10 shows that there are numerous possible combinations for a product of 252, but the only pair that gives the correct middle term is -18 and -14. Rewriting the middle term of the trinomial as a sum and then factoring by grouping, we get

$$\begin{aligned} 9x^2 - 32xL + 28L^2 &= 9x^2 - 18xL - 14xL + 28L^2 \\ &= 9x(x - 2L) - 14L(x - 2L) = (x - 2L)(9x - 14L) \end{aligned}$$

EXAMPLE 7 Factoring general perfect square trinomials

(a) Factor: $9x^2 - 6x + 1$. We note that $9x^2$ is the square of $3x$ and 1 is the square of 1. To see if this is a perfect square trinomial, we check the middle term. Since $-2(3x)(1) = -6x$, the middle term checks. Therefore,

$$9x^2 - 6x + 1 = (3x - 1)^2 \qquad \text{check middle term: } -2(3x)(-1) = -6x$$

(b) Factor: $36x^2 + 84xy + 49y^2$. In the same way, we have

$$36x^2 + 84xy + 49y^2 = (6x + 7y)^2 \qquad \text{check middle term: } 2(6x)(7y) = 84xy$$

since $2(6x)(7y) = 84xy$. *We must be careful to include the variable y in the second terms of the factors.*

COMMON ERROR

When the last term in a trinomial is not just a constant but also has a variable dependence, it is common to forget this variable when stating the factors:

$$2x^2 + 5xy - 3y^2 = (2x - y)(x + 3y)$$
$$2x^2 + 5xy - 3y^2 \neq (2x - 1)(x + 3)$$

As noted before, we must be careful to factor an expression completely. *We first look for common monomial factors* and *then check each resulting factor to see if it can be factored* when we complete each step.

EXAMPLE 8 General trinomial with common monomial factor

Factor $2x^2 + 6x - 8$ completely.

First note the common monomial factor of 2. This leads to

$$2x^2 + 6x - 8 = 2(x^2 + 3x - 4)$$

Now, notice that $x^2 + 3x - 4$ is also factorable. The two numbers whose product is -4 and whose sum is $+3$ are $+4$ and -1. Hence

$$2x^2 + 6x - 8 = 2(x + 4)(x - 1)$$

Note that if you attempt this example without factoring the common factor first, the solution is more difficult and takes more steps. **It is always best to factor out the greatest common factor first.**

Practice Exercise

4. Factor: $6x^2 + 9x - 6$

Table 6.11

Factors of -100	Sum (Need $+15$)
$+100, -1$	$+99$
$+50, -2$	$+48$
$+25, -4$	$+37$
$+20, -5$	$+15$
$+10, -10$	0

■ Liquid-fuel rockets were designed in the United States in the 1920s but were developed by German engineers. They were first used in the 1940s during World War II.

EXAMPLE 9 Factoring completely—rocket flight

A study of the path of a certain rocket leads to the expression $16t^2 + 240t - 1600$, where t is the time of flight. Factor this expression.

An inspection shows that there is a common factor of 16. (This might be found by noting successive factors of 2 or 4.) Factoring out 16 leads to

$$\begin{aligned} 16t^2 + 240t - 1600 &= 16(t^2 + 15t - 100) \\ &= 16(t + 20)(t - 5) \end{aligned}$$

Here, the values of $+20$ and -5 were found in Table 6.11.

Note that by first factoring out the common factor of 16, the resulting trinomial is *much easier* to factor than the original trinomial.

EXAMPLE 10 Factoring expressions with more terms by grouping

Factor: $x^2 - 4xy + 4y^2 - 9$.

This is not a trinomial. However, we can focus on the first three terms and factor them as a trinomial first.

We note that the first and third terms of $x^2 - 4xy + 4y^2$ are perfect squares. Checking the middle term, $-2(x)(2y) = -4xy$, so the middle term checks for a perfect square trinomial. Therefore,

$$x^2 - 4xy + 4y^2 - 9 = (x^2 - 4xy + 4y^2) - 9 = (x - 2y)^2 - 9$$

This expression can now be factored as a difference of squares. The complete factorization is

$$x^2 - 4xy + 4y^2 - 9 = (x - 2y - 3)(x - 2y + 3)$$

Not all groupings work. Noting the factorable combination $4y^2 - 9$, and then grouping the first two terms and the last two terms, would not have led to the factorization.

EXERCISES 6.3

In Exercises 1–4, make the given changes in the indicated examples of this section and then factor.

1. In Example 1(a), change the 3 to 4 and the 2 to 3.

2. In Example 2(a), change the + before $7x$ to −.

3. In Example 4, change the + before $11x$ to −.

4. In Example 8, change the 8 to 36.

In Exercises 5–54, factor the given expressions completely.

5. $x^2 + 6x + 5$
6. $x^2 - 6x - 7$
7. $s^2 - s - 56$
8. $x^2 + 12x - 45$
9. $t^2 + 3t - 28$
10. $r^3 - 11r^2 + 18r$
11. $x^2 + 2x + 1$
12. $D^2 + 8D + 16$
13. $L^2 - 4LK + 4K^2$
14. $b^2 - 12bc + 36c^2$
15. $3x^2 - 5x - 2$
16. $2n^2 - 13n - 7$
17. $12y^2 - 32y - 12$
18. $25x^2 + 45x - 10$
19. $2s^2 + 13s + 11$
20. $7y^2 - 12y + 5$
21. $3f^4 - 16f^2 + 5$
22. $5R^4 - 3R^2 - 2$
23. $2t^2 + 7t - 15$
24. $3n^2 - 20n + 20$
25. $3t^2 - 7tu + 4u^2$
26. $3x^2 + xy - 14y^2$
27. $4x^2 - 3x - 7$
28. $2z^2 + 13z - 5$
29. $9x^2 + 7xy - 2y^2$
30. $4r^2 + rs - 14s^2$
31. $4m^2 + 20m + 25$
32. $16q^2 + 24q + 9$
33. $8x^2 - 24x + 18$
34. $3a^2c^2 - 6ac + 3$
35. $9t^2 - 15t + 4$
36. $6t^4 + t^2 - 12$
37. $8b^6 + 31b^3 - 4$
38. $12n^4 + 8n^2 - 15$
39. $4p^2 - 25pq + 6q^2$
40. $12x^2 + 4xy - 5y^2$
41. $12x^2 + 47xy - 4y^2$
42. $8r^2 - 14rs - 9s^2$
43. $12 - 14x + 2x^2$
44. $6y^2 - 33y - 18$
45. $4x^5 + 14x^3 - 8x$
46. $12B^2 + 22BH - 4H^2$
47. $ax^3 + 4a^2x^2 - 12a^3x$
48. $6x^4 - 13x^3 + 5x^2$
49. $a^2 + 2ab + b^2 - 4$
50. $x^2 - 6xy + 9y^2 - 4z^2$
51. $25a^2 - 25x^2 - 10xy - y^2$
52. $r^2 - s^2 + 2st - t^2$
53. $4x^{2n} + 13x^n - 12$
54. $12B^{2n} + 19B^nH - 10H^2$

In Exercises 55–66, factor the given expressions completely using integers. Each is from the technical area indicated.

55. $16t^2 - 80t + 64$ (projectile motion)
56. $9x^2 - 33Lx + 30L^2$ (civil engineering)
57. $d^4 - 10d^2 + 16$ (magnetic field)
58. $3\eta^2 + 18\eta - 1560$ (fuel efficiency)
59. $200n^2 - 2100n - 3600$ (biology)
60. $bT^2 - 40bT + 400b$ (thermodynamics)
61. $V^2 - 2nBV + n^2B^2$ (chemistry)
62. $a^4 + 8a^2\pi^2f^2 + 16\pi^4f^4$ (periodic motion: energy)
63. $wx^4 - 5wLx^3 + 6wL^2x^2$ (beam design)
64. $1 - 2r^2 + r^4$ (lasers)
65. $3Adu^2 - 4Aduv + Adv^2$ (water power)
66. $k^2A^2 + 2k\lambda A + \lambda^2 - \alpha^2$ (robotics)

In Exercises 67–71, solve the given problems.

67. Find the integral value of k that makes $4x^2 + 4x - k$ a perfect square trinomial, and express the result in factored form.
68. Find the integral value of k that makes $49x^2 - 70x + k$ a perfect square trinomial, and express the result in factored form.
69. Often, the sum of squares cannot be factored, but $x^4 + 4$ can be factored. Add and subtract $4x^2$ and then use factoring by grouping.
70. Using the method of Exercise 69, factor $x^4 + x^2 + 1$. (First add and subtract x^2.)
71. A yard was planned to be a square of width w. After property lines and sidewalks were added, the final area of usable space can be represented by the equation $w^2 - 6w + 8$ (in m^2). By how much were the width and the length of the yard altered?

Answers to Practice Exercises

1. $(x - 2)(x + 1)$ 2. $(x - 2)^2$
3. $(4x + 5)(x - 1)$ 4. $3(2x - 1)(x + 2)$

6.4 The Sum and Difference of Cubes

We have seen that the difference of squares can be factored, but that the sum of squares often cannot be factored. We now turn our attention to the sum and difference of cubes, both of which can be factored. Reversing the order of Eqs. (6.10) and (6.11), we have

■ Note carefully the positions of the + and − signs.

$$x^3 + y^3 = (x + y)(x^2 - xy + y^2) \quad (6.14)$$

$$x^3 - y^3 = (x - y)(x^2 + xy + y^2) \quad (6.15)$$

In these equations, the second factors are prime, assuming x and y are themselves prime factors.

Table of Cubes

$1^3 =$	1
$2^3 =$	8
$3^3 =$	27
$4^3 =$	64
$5^3 =$	125
$6^3 =$	216

EXAMPLE 1 Factoring the sum and difference of cubes

Factor the following expressions.

(a) $x^3 + 8 = x^3 + 2^3 = (x + 2)(x^2 - 2x + 2^2) = (x + 2)(x^2 - 2x + 4)$

(b) $x^3 - 1 = x^3 - 1^3 = (x - 1)(x^2 + (x)(1) + 1^2) = (x - 1)(x^2 + x + 1)$

(c) $8 - 27x^3 = 2^3 - (3x)^3 = (2 - 3x)(2^2 + 2(3x) + (3x)^2)$

$= (2 - 3x)(4 + 6x + 9x^2)$

EXAMPLE 2 Factoring a difference of cubes with an initial common factor

Factor: $ax^5 - ax^2$.

First, as always, we factor out the common factor from both terms to get

$$ax^5 - ax^2 = ax^2(x^3 - 1)$$

The second factor is a difference of cubes, already solved in Example 1(b). Therefore,

$$ax^5 - ax^2 = ax^2(x^3 - 1) = ax^2(x - 1)(x^2 + x + 1)$$

Practice Exercises

Factor: **1.** $x^3 + 216$ **2.** $x^3 - 216$

EXAMPLE 3 Factoring a difference of cubes—volume of steel bearing

The volume of material used to make a steel bearing with a hollow core is given by $\frac{4}{3}\pi R^3 - \frac{4}{3}\pi r^3$. Factor this expression.

$$\frac{4}{3}\pi R^3 - \frac{4}{3}\pi r^3 = \frac{4}{3}\pi(R^3 - r^3) \quad \text{common factor of } \frac{4}{3}\pi$$

$$= \frac{4}{3}\pi(R - r)(R^2 + Rr + r^2) \quad \text{using Eq. (6.10)}$$

We have now discussed several methods of factoring. It is important to know which methods to use and the order in which to use them. The diagram below shows a general strategy that can be used to factor most expressions.

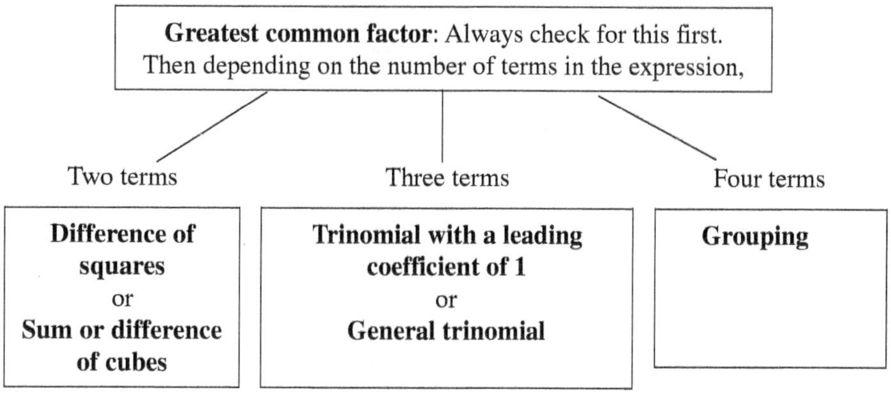

Always be sure the expression is factored **completely**.

EXERCISES 6.4

In Exercises 1 and 2, make the given changes in the indicated examples of this section and then factor.

1. In Example 1(a), change the + before the 8 to −.
2. In Example 2, change $-ax^2$ to $+a^4x^2$.

In Exercises 3–28, factor the given expressions completely.

3. $x^3 + 1$
4. $R^3 + 27$
5. $8 - t^3$
6. $y^3 - 125$
7. $27x^3 - 8y^3$
8. $64x^4 + 125x$
9. $4x^3 + 32$
10. $3y^3 - 81$
11. $7n^5 - 7n^2$
12. $8s^9 - 64$
13. $162x^3y - 6x^3y^4$
14. $12a^3 + 96a^3b^3$
15. $x^6y^3 + x^3y^6$
16. $16r^3 - 432$
17. $3a^6 - 3a^2$
18. $x^6 - 27y^3$
19. $0.001R^3 - 0.064r^3$
20. $0.027x^3 + 0.125$
21. $27L^6 + 216L^3$
22. $a^3s^5 - 8000a^3s^2$
23. $(a + b)^3 + 64$
24. $125 + (2x + y)^3$
25. $64 - x^6$
26. $a^6 - 27b^6$
27. $125h^2w^2 - 8h^2w^5$
28. $\frac{\pi^2r^3}{8} - \frac{\pi^5R^6}{27}$

In Exercises 29–34, factor the given expressions completely. Each is from the technical area indicated.

29. $32x - 4x^4$ (rocket trajectory)
30. $kT^3 - kT_0^3$ (thermodynamics)
31. $D^4 - d^3D$ (machine design)
32. $(h + 2t)^3 - h^3$ (container design)
33. $QH^4 + Q^4H$ (thermodynamics)
34. $\left(\frac{s}{r}\right)^{12} - \left(\frac{s}{r}\right)^6$ (molecular interaction)

In Exercises 35 and 36, perform the indicated operations.

35. Perform the division $(x^5 - y^5) \div (x - y)$. Noting the result, determine the quotient $(x^7 - y^7) \div (x - y)$, without dividing. From these results, factor $x^5 - y^5$ and $x^7 - y^7$.
36. Perform the division $(x^5 + y^5) \div (x + y)$. Noting the result, determine the quotient $(x^7 + y^7) \div (x + y)$, without dividing. From these results, factor $x^5 + y^5$ and $x^7 + y^7$.

In Exercises 37–40, solve the given problems.

37. Factor $x^6 - y^6$ as the difference of cubes.
38. Factor $x^6 - y^6$ as the difference of squares. Then explain how the result of Exercise 37 can be shown to be the same as the result in this exercise.
39. By factoring $(n^3 + 1)$, explain why this expression represents a number that is not prime if n is an integer greater than one.
40. A certain wind turbine produces a power of $1200v^3$ (in watts), where v is the wind speed. Find an expression, in factored form, for the difference in power for the wind speeds v_1 and v_2.

Answers to Practice Exercises

1. $(x + 6)(x^2 - 6x + 36)$ 2. $(x - 6)(x^2 + 6x + 36)$

6.5 Equivalent Fractions

When we deal with algebraic expressions, we must be able to work effectively with fractions. Since algebraic expressions are representations of numbers, the basic operations on fractions from arithmetic form the basis of our algebraic operations. In this section, we demonstrate a very important property of fractions, and in the following two sections, we establish the basic algebraic operations with fractions.

The following property of fractions, often referred to as the **fundamental principle of fractions**, is very important when working with fractions.

Fundamental Principle of Fractions

The value of a fraction does not change if both the numerator and the denominator are multiplied or divided by the same nonzero number.

Two fractions are said to be **equivalent** if one can be obtained from the other by use of the fundamental principle.

EXAMPLE 1 Equivalent arithmetic fractions

Find two fractions equivalent to $\frac{6}{8}$.

Multiplying the numerator and the denominator by 2, we obtain the equivalent fraction

$$\frac{6}{8} = \frac{6(2)}{8(2)} = \frac{12}{16}$$

Dividing the numerator and the denominator by 2 we obtain the equivalent fraction

$$\frac{6}{8} = \frac{\frac{6}{2}}{\frac{8}{2}} = \frac{3}{4}$$

Obviously, there is an unlimited number of other fractions that are equivalent to these fractions.

EXAMPLE 2 Equivalent algebraic fractions

Find two fractions equivalent to $\frac{a^2x}{2a}$.

Dividing the numerator and the denominator by a, we obtain the equivalent fraction

$$\frac{a^2x}{2a} = \frac{a^2\,x/a}{2a/a} = \frac{ax}{2}$$

We are not restricted to monomials for obtaining equivalent fractions. Multiplying the numerator and the denominator by $3a - 1$, we get

$$\frac{a^2x}{2a} = \frac{(a^2x)(3a-1)}{2a(3a-1)} = \frac{3a^3x - a^2x}{6a^2 - 2a}$$

SIMPLEST FORM, OR LOWEST TERMS, OF A FRACTION

One of the most important operations with a fraction is that of ***reducing*** it to its **simplest form**, or **lowest terms**.

Simplest Form of a Fraction

A fraction is said to be in its simplest form if the numerator and the denominator have no common factors other than $+1$ or -1.

In reducing a fraction to its simplest form, we use the fundamental principle of fractions and *divide* both the numerator and the denominator by all ***factors*** that are common to each.

Note that we will always assume that no unknown factor occurring in the denominator is equal to zero so as to avoid division by zero. This can also be indicated by stating explicitly which factors cannot equal zero.

EXAMPLE 3 Reducing a fraction to its lowest terms

Reduce the fraction $\dfrac{16ab^3c^2}{24ab^2c^5}$.

We can reduce the fraction by first writing numerator and denominator in prime factors using exponents, and then using the laws of exponents. We have

$$\frac{16ab^3c^2}{24ab^2c^5} = \frac{2 \cdot 2 \cdot 2 \cdot 2ab^3c^2}{2 \cdot 2 \cdot 2 \cdot 3ab^2c^5} = \frac{2^4 3^0 a^1 b^3 c^2}{2^3 3^1 a^1 b^2 c^5}$$
$$= 2^{4-3}3^{0-1}a^{1-1}b^{3-2}c^{2-5}$$
$$= 2^1 3^{-1} a^0 b^1 c^{-3} = \frac{2b}{3c^3}$$

This fraction is in its lowest terms, because there are no common factors in the numerator and the denominator other than $+1$ or -1.

We can also reduce the fraction by dividing both the numerator and the denominator by their common factor, in this case $8ab^2c^2$. We have

$$\frac{16ab^3c^2}{24ab^2c^5} = \frac{2b(8ab^2c^2)}{3c^3(8ab^2c^2)} = \frac{2b}{3c^3}$$

For both methods we have assumed that a, b, and c are not equal to zero to avoid division by zero.

Practice Exercise

1. Reduce to lowest terms: $\dfrac{9xy^5}{15x^3y^2}$

As shown in the previous example, we can reduce a fraction to its simplest form by dividing both its numerator and denominator by the common factor. This process, which produces an equivalent fraction, is often referred to as **cancellation**.

COMMON ERROR It is very important that you cancel only **factors** that appear in the numerator and denominator. This means that the numerator and denominator must both be in **factored form** before making any cancellations. It is a common error to cancel expressions without first factoring the numerator and denominator. In particular, **terms** (expressions separated by + or − signs) **cannot be cancelled.**

EXAMPLE 4 Cancel factors only

Simplify the expression $\dfrac{x^2(x-2)}{x^2-4}$.

We note that the quantity x^2 appears in both the numerator and the denominator. However, while the x^2 in the numerator appears as a factor, the x^2 in the denominator appears only in the first term and cannot be factored out. Therefore, the quantity x^2 cannot be cancelled.

In order to simplify the above fraction properly, we must first factor the denominator. Note that $x^2 - 4$ is a difference of squares that can be factored as $(x+2)(x-2)$. Substituting into the denominator of the fraction, we get

$$\frac{x^2(x-2)}{(x+2)(x-2)}$$

Now $(x-2)$ is a common ***factor*** and can be cancelled:

$$\frac{x^2\overset{1}{\cancel{(x-2)}}}{\underset{1}{\cancel{(x-2)}}(x+2)} = \frac{x^2}{x+2}, \text{ where } x \neq 2$$

Practice Exercise

2. Reduce to lowest terms: $\dfrac{4a}{4a-2x}$

EXAMPLE 5 Distinguishing between term and factor

In the following expressions, determine if $2a$ is a term or a factor, and simplify if possible.

(a) $\dfrac{2a}{2a + x}$. In the denominator, the quantity $2a$ is a term, not a factor shared with the second term. The fraction cannot be simplified because there are no common factors in the numerator and the denominator.

(b) $\dfrac{2a}{2ax}$. Here $2a$ is a common factor of the numerator and the denominator and can be divided out. Therefore, $\dfrac{2a}{2ax} = \dfrac{1}{x}$, if $a \neq 0$.

EXAMPLE 6 Remaining factor of 1 in the denominator

Simplify the expression $\dfrac{2x^2 + 8x}{x + 4}$.

Factoring the numerator and cancelling common factors from the numerator and denominator, we get

$$\frac{2x^2 + 8x}{x + 4} = \frac{2x\cancel{(x + 4)}}{1\cancel{(x + 4)}} = \frac{2x}{1} = 2x$$

Writing the factor of 1 in the denominator ensures that the place is preserved after cancellation. Generally, a 1 in the denominator is not written in the final result.

EXAMPLE 7 Cancelling factors after factoring

Reduce the fraction $\dfrac{x^2 - 4x + 4}{x^2 - 4}$ to lowest terms.

We begin by factoring the numerator as a perfect square trinomial and the denominator as a difference of squares. We can then cancel common factors. For $x \neq 2$,

$$\frac{x^2 - 4x + 4}{x^2 - 4} = \frac{(x - 2)\cancel{(x - 2)}}{(x + 2)\cancel{(x - 2)}} = \frac{x - 2}{x + 2}$$

In the final form, neither the x's nor the 2's may be cancelled, since they are not common *factors*.

Practice Exercise

3. Reduce to lowest terms: $\dfrac{x^2 - x - 2}{x^2 + 3x + 2}$

EXAMPLE 8 Reducing a fraction—mechanical vibration

Simplify $\dfrac{8s + 12}{4s^2 + 26t + 30}$, an expression that arises in the mathematical analysis of the vibrations in a certain mechanical system.

We begin by factoring out GCFs and then factoring the trinomial in the denominator. We can then cancel out the common factors.

$$\frac{8s + 12}{4s^2 + 26s + 30} = \frac{4(2s + 3)}{2(2s^2 + 13s + 15)} = \frac{\cancel{2} \cdot 2 \cdot \cancel{(2s + 3)}}{\cancel{2}\cancel{(2s + 3)}(s + 5)}$$

$$= \frac{2}{s + 5}, \quad s \neq -\frac{3}{2}$$

FACTORS THAT DIFFER ONLY IN SIGN

In simplifying fractions, we must be able to distinguish between factors that differ only in ***sign***. Since $-(y - x) = -y + x = x - y$, we have

$$x - y = -(y - x) \qquad \textbf{(6.16)}$$

Here, the ***factors*** $x - y$ ***and*** $y - x$ ***differ only in sign***. The following examples illustrate the simplification of fractions where a change of signs is necessary.

EXAMPLE 9 Factors that differ only in sign

Simplify $\dfrac{x^2 - 1}{1 - x}$.

We factor the numerator as a difference of squares, and factor -1 out of $1 - x$ and rearrange the terms in order of descending powers of x. Then

$$\frac{x^2 - 1}{1 - x} = \frac{(x + 1)(x - 1)}{-1(-1 + x)} = \frac{(x + 1)\cancel{(x - 1)}}{-1\cancel{(x - 1)}} = -(x + 1), \quad x \neq 1$$

It should be clear that the factors $1 - x$ and $x - 1$ differ only in sign.

Practice Exercise

4. Reduce to lowest terms: $\dfrac{y - 2}{4 - y^2}$

COMMON ERROR When simplifying fractions, it is important to look for factors that differ only in sign. They can provide a cancellation that often goes unnoticed, yielding a fraction which is **not in its lowest terms.**

EXAMPLE 10 Factors that differ only in sign

Simplify $\dfrac{2x^4 - 128x}{20 + 7x - 3x^2}$.

We begin by reordering the denominator terms as $-3x^2 + 7x + 20$ (in order of descending powers of x). To factor completely, we factor out GCFs, and then factor the numerator as a difference of cubes and the denominator as a general trinomial. For $x \neq 4$, we have

$$\frac{2x^4 - 128x}{-3x^2 + 7x + 20} = \frac{2x(x^3 - 64)}{-1(3x^2 - 7x - 20)} = \frac{2x\cancel{(x - 4)}(x^2 + 4x + 16)}{-1\cancel{(x - 4)}(3x + 5)}$$

$$= -\frac{2x(x^2 + 4x + 16)}{3x + 5}$$

EXERCISES 6.5

In Exercises 1 and 2, make the given changes in the indicated examples of this section and then solve the resulting problems.

1. In Example 7, change the $-$ sign in the numerator to $+$ and then simplify.
2. In Example 10, change the numerator to $2x^4 - 32x^2$ and then simplify.

In Exercises 3–10, multiply the numerator and the denominator of each fraction by the given factor and obtain an equivalent fraction.

3. $\dfrac{2}{3}$ (by 7) **4.** $\dfrac{7}{5}$ (by 9)

5. $\dfrac{ax}{y}$ (by $2x$) **6.** $\dfrac{2x^2y}{3n}$ (by $2xn^2$)

7. $\frac{2}{x+3}$ (by $x-2$)
8. $\frac{7}{a-1}$ (by $a+2$)
9. $\frac{a(x-y)}{x-2y}$ (by $x+y$)
10. $\frac{B-1}{B+1}$ (by $B-1$)

In Exercises 11–18, divide the numerator and the denominator of each fraction by the given factor and obtain an equivalent fraction.

11. $\frac{28}{44}$ (by 4)
12. $\frac{25}{65}$ (by 5)
13. $\frac{4x^2y}{8xy^2}$ (by $2x$)
14. $\frac{6a^3b^2}{9a^5b^4}$ (by $3a^2b^2$)
15. $\frac{2(R-1)}{(R-1)(R+1)}$ (by $R-1$)
16. $\frac{(x+5)(x-3)}{3(x+5)}$ (by $x+5$)
17. $\frac{s^2-3s-10}{2s^2+3s-2}$ (by $s+2$)
18. $\frac{6x^2+13x-5}{6x^3-2x^2}$ (by $3x-1$)

In Exercises 19–26, replace the A with the proper expression such that the fractions are equivalent.

19. $\frac{3x}{2y}=\frac{A}{6y^2}$
20. $\frac{2R}{R+T}=\frac{2R^2T}{A}$
21. $\frac{7}{a+5}=\frac{7a-35}{A}$
22. $\frac{a+1}{5a^2c}=\frac{A}{5a^3c-5a^2c}$
23. $\frac{2x^3+2x}{x^4-1}=\frac{A}{x^2-1}$
24. $\frac{n^2-1}{n^3+1}=\frac{A}{n^2-n+1}$
25. $\frac{x^2+3bx-4b^2}{x-b}=\frac{x+4b}{A}$
26. $\frac{4y^2-1}{4y^2+6y-4}=\frac{A}{2y+4}$

In Exercises 27–62, reduce each fraction to simplest form.

27. $\frac{2a}{8a}$
28. $\frac{6x}{15x}$
29. $\frac{18x^2y}{24xy}$
30. $\frac{2a^2xy}{6axyz^2}$
31. $\frac{a+b}{5a^2+5ab}$
32. $\frac{t-a}{t^2-a^2}$
33. $\frac{4a-4b}{4a-2b}$
34. $\frac{20s-5r}{10r-5s}$
35. $\frac{4x^2+1}{4x^2-1}$
36. $\frac{x^2-y^2}{x^2+y^2}$
37. $\frac{3x^2-6x}{x-2}$
38. $\frac{10T^2+15T}{2T+3}$
39. $\frac{3+2y}{4y^3+6y^2}$
40. $\frac{6-3t}{4t^3-8t^2}$
41. $\frac{x^2-10x+25}{x^2-25}$
42. $\frac{4a^2+12ab+9b^2}{4a^2+6ab}$
43. $\frac{2w^4+5w^2-3}{w^4+11w^2+24}$
44. $\frac{3y^3+7y^2+4y}{y^2+5y+4}$
45. $\frac{5x^2-6x-8}{x^3+x^2-6x}$
46. $\frac{5s^2+8rs-4s^2}{6r^2-17rs+5s^2}$
47. $\frac{N^4-16}{8N-16}$
48. $\frac{3+x(4+x)}{3+x}$
49. $\frac{t+4}{(2t+9)t+4}$
50. $\frac{2A^3+8A^4+8A^5}{4A+2}$
51. $\frac{(x-1)(3+x)}{(3-x)(1-x)}$
52. $\frac{(2x-1)(x+6)}{(x-3)(1-2x)}$
53. $\frac{y^2-x^2}{2x-2y}$
54. $\frac{x^2-y^2-4x+4y}{x^2-y^2+4x-4y}$
55. $\frac{x^3+x^2-x-1}{x^3-x^2-x+1}$
56. $\frac{3a^2-13a-10}{5+4a-a^2}$
57. $\frac{(x+5)(x-2)(x+2)(3-x)}{(2-x)(5-x)(3+x)(2+x)}$
58. $\frac{(2x-3)(3-x)(x-7)(3x+1)}{(3x+2)(3-2x)(x-3)(7+x)}$
59. $\frac{x^3+y^3}{2x+2y}$
60. $\frac{w^3-8}{w^2+2w+4}$
61. $\frac{6x^2+2x}{27x^3+1}$
62. $\frac{24-3a^3}{a^2-4a+4}$

In Exercises 63–66, after finding the simplest form of each fraction, explain why it cannot be simplified more.

63. (a) $\frac{x^2(x+2)}{x^2+4}$ (b) $\frac{x^4+4x^2}{x^4-16}$
64. (a) $\frac{2x+3}{2x+6}$ (b) $\frac{2(x+6)}{2x+6}$
65. (a) $\frac{x^2-x-2}{x^2-x}$ (b) $\frac{x^2-x-2}{x^2+x}$
66. (a) $\frac{x^3-x}{1-x}$ (b) $\frac{2x^2+4x}{2x^2+4}$

In Exercises 67–72, reduce each fraction to simplest form. Each is from the indicated area of application.

67. $\frac{mu^2-mv^2}{mu-mv}$ (nuclear energy)
68. $\frac{16(t^2-2tt_0+t_0^2)(t-t_0-3)}{3t-3t_0}$ (rocket motion)
69. $\frac{E^2R^2-E^2r^2}{(R^2+2Rr+r^2)^2}$ (electricity)
70. $\frac{r_0^3-r_i^3}{r_0^2-r_i^2}$ (machine design)
71. $\frac{v^2-v_0^2}{vt-v_0t}$ (rectilinear motion)
72. $\frac{a^3-b^3}{a^3-ab^2}$ (beam design)

Answers to Practice Exercises

1. $\frac{3y^3}{5x^2}$ 2. $\frac{2a}{2a-x}$ 3. $\frac{x-2}{x+2}, x\neq -1$ 4. $-\frac{1}{y+2}, y\neq 2$

6.6 Multiplication and Division of Fractions

From arithmetic, recall that *the product of two fractions is a fraction whose numerator is the product of the numerators and whose denominator is the product of the denominators of the given fractions.* Also, *the quotient of two fractions is found by inverting the denominator and proceeding as in multiplication.* Symbolically, these operations are shown as

Multiplication and Division of Fractions

$$\frac{a}{b} \times \frac{c}{d} = \frac{ac}{bd}$$

$$\frac{a}{b} \div \frac{c}{d} = \frac{\frac{a}{b}}{\frac{c}{d}} = \frac{a}{b} \times \frac{d}{c} = \frac{ad}{bc}$$

The rule for division may be verified by multiplying the numerator and the denominator of the fraction by the reciprocal of the denominator to obtain an equivalent fraction with a denominator of 1, effectively converting the division problem into a multiplication problem. We have

$$\frac{\frac{a}{b}}{\frac{c}{d}} = \frac{\frac{a}{b} \times \frac{d}{c}}{\frac{c}{d} \times \frac{d}{c}} = \frac{\frac{ad}{bc}}{1} = \frac{ad}{bc}$$

The following three examples illustrate the multiplication of fractions.

EXAMPLE 1 Multiplying basic fractions

(a) $\frac{3}{5} \times \frac{2}{7} = \frac{(3)(2)}{(5)(7)} = \frac{6}{35}$ ← multiply numerators / ← multiply denominators

(b) $\frac{3a}{5b} \times \frac{15b^2}{a} = \frac{3 \cdot a}{5 \cdot b} \times \frac{3 \cdot 5 \cdot b \cdot b}{a} = \frac{3 \cdot 3 \cdot \cancel{5} \cdot \cancel{a} \cdot \cancel{b} \cdot b}{1 \cdot \cancel{5} \cdot \cancel{a} \cdot \cancel{b}} = 9b$ where $a, b \neq 0$

(c) $(6x)\left(\frac{2y}{3x^2}\right) = \left(\frac{6x}{1}\right)\left(\frac{2y}{3x^2}\right) = \left(\frac{2 \cdot 3 \cdot x}{1}\right)\left(\frac{2 \cdot y}{3 \cdot x \cdot x}\right) = \frac{2 \cdot 2 \cdot \cancel{3} \cdot \cancel{x} \cdot y}{1 \cdot \cancel{3} \cdot \cancel{x} \cdot x} = \frac{4y}{x}$

where $x \neq 0$

LEARNING TIP

When multiplying fractions, since all the factors in the numerators and denominators are to be multiplied, it is best to ***factor*** the numerator and denominator and ***cancel any common factors before multiplying.***

This technique will simplify the fraction before the multiplication expansion, and make it easier to see how to factor and simplify the result.

When multiplying fractions, we usually want to express the final result in simplest form. This means we will have to express the numerator and denominator of the final answer in factored form and cancel any common factors. For this reason, when multiplying fractions, it is best to only *indicate* the multiplications between the numerators and denominators rather than actually perform them. By doing this, the numerator and denominator of the answer will already be partially factored. We can then factor them completely and cancel any common factors to simplify the answers. If we actually perform the multiplications, the answer will be very difficult to simplify. The following example illustrates this point.

EXAMPLE 2 Factor and simplify first

Perform the multiplication $\frac{3(x-y)}{(x-y)^2} \times \frac{(x^2-y^2)}{6x+9y}$.

We first indicate the multiplications without actually performing them. Then we factor the resulting numerator and denominator, and cancel any common factors. Doing this, we have

$$\frac{3(x-y)}{(x-y)^2} \times \frac{(x^2-y^2)}{6x+9y} = \frac{3\cancel{(x-y)}}{(x-y)\cancel{(x-y)}} \times \frac{(x+y)(x-y)}{3(2x+3y)} \quad \text{factor and cancel within each fraction}$$

$$= \frac{3(x+y)(x-y)}{3(x-y)(2x+3y)} \quad \text{indicate multiplication}$$

$$= \frac{\cancel{3}(x+y)\cancel{(x-y)}}{\cancel{3}\cancel{(x-y)}(2x+3y)} = \frac{x+y}{2x+3y} \quad \text{cancel common factors}$$

Note that if we had multiplied out the numerators and the denominators before performing any factoring, we would have had to simplify the fraction $\frac{3x^3-3x^2y-3xy^2+3y^3}{6x^3-3x^2y-12xy^2+9y^3}$.

Practice Exercise

1. Multiply: $\frac{2x}{4x+8} \times \frac{x+2}{xy-x}$

EXAMPLE 3 Multiplying algebraic fractions

Simplify $\left(\frac{2x-4}{4x+12}\right)\left(\frac{2x^2+x-15}{3x-1}\right)$.

$$\left(\frac{2x-4}{4x+12}\right)\left(\frac{2x^2+x-15}{3x-1}\right) = \frac{2(x-2)(2x-5)(x+3)}{2 \cdot 2(x+3)(3x-1)} \quad \text{indicate multiplication and factor}$$

$$= \frac{\cancel{2}(x-2)(2x-5)\cancel{(x+3)}}{2 \cdot \cancel{2}\cancel{(x+3)}(3x-1)} \quad \text{cancel common factors}$$

$$= \frac{(x-2)(2x-5)}{2(3x-1)}$$

LEARNING TIP

It is permissible to multiply out the final form of the numerator and denominator, but it is preferable to leave the numerator and denominator in factored form.

The factored form makes it easier to identify important features of a function, such as zeros or vertical asymptotes. These ideas will become important in later chapters.

The following examples illustrate the division of fractions.

EXAMPLE 4 Dividing basic algebraic fractions

(a) $\dfrac{\frac{6x}{7}}{\frac{5}{3}} = \frac{6x}{7} \times \frac{3}{5} = \frac{18x}{35}$ (multiply; invert)

(b) $\dfrac{\frac{3a^2}{5c}}{\frac{2c^2}{a}} = \frac{3a^2}{5c} \times \frac{a}{2c^2} = \frac{3a^3}{10c^3}$ (multiply; invert)

EXAMPLE 5 Division by a fraction—centre of mass

When finding the centre of mass (CM) of a uniform flat semicircular metal plate as shown in Fig. 6.4, the equation $X = \frac{4\pi r^3}{3} \div \left(\frac{\pi r^2}{2} \times 2\pi\right)$ is derived. Simplify the right side of this equation to find X as a function of r in simplest form.

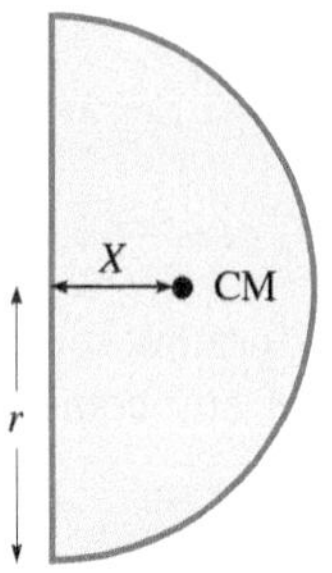

Fig. 6.4

The parentheses indicate that we should perform the multiplication first:

$$X = \frac{4\pi r^3}{3} \div \left(\frac{\pi r^2}{2} \times 2\pi\right) = \frac{4\pi r^3}{3} \div \left(\frac{2\pi^2 r^2}{2}\right) \qquad 2\pi = \frac{2\pi}{1}$$

$$X = \frac{4\pi r^3}{3} \div (\pi^2 r^2) = \frac{4\pi r^3}{3} \times \frac{1}{\pi^2 r^2} \qquad \text{invert to multiply}$$

$$X = \frac{4\pi r^3}{3\pi^2 r^2} = \frac{4r}{3\pi} \qquad \text{divide out the common factor of } \pi r^2$$

This is the exact solution. Approximately, $X = 0.424r$.

EXAMPLE 6 Dividing algebraic fractions

Simplify $(x + y) \div \frac{2x + 2y}{6x + 15y}$.

We begin by writing the first expression as an equivalent fraction, placing it over a denominator of one. Then we can indicate the multiplication, factor and simplify. We get

$$(x + y) \div \frac{2x + 2y}{6x + 15y} = \frac{(x + y)}{1} \div \frac{2(x + y)}{3(2x + 5y)} = \frac{(x + y)}{1} \times \frac{3(2x + 5y)}{2(x + y)}$$

divide by 1 — invert

$$= \frac{3(2x + 5y)\cancel{(x + y)}}{2\cancel{(x + y)}} = \frac{3(2x + 5y)}{2}, \quad x \neq -y$$

Practice Exercise

2. Divide: $\frac{3x}{a + 1} \div \frac{x^2 + 2x}{a^2 + a}$

EXAMPLE 7 Dividing compound algebraic fractions

$$\frac{\dfrac{4 - x^2}{x^2 - 3x + 2}}{\dfrac{x + 2}{x^2 - 9}} = \frac{\dfrac{-1(x + 2)\cancel{(x - 2)}}{\cancel{(x - 2)}(x - 1)}}{\dfrac{(x + 2)}{(x + 3)(x - 3)}} \qquad \text{factor and cancel within each fraction}$$

$$= \frac{-1(x + 2)}{(x - 1)} \times \frac{(x + 3)(x - 3)}{(x + 2)} \qquad \text{invert}$$

$$= \frac{-1\cancel{(x + 2)}(x + 3)(x - 3)}{(x - 1)\cancel{(x + 2)}} = \frac{-1(x + 3)(x - 3)}{(x - 1)} \qquad \text{indicate multiplication; cancel common factors}$$

EXERCISES 6.6

In Exercises 1 and 2, make the given changes in the indicated examples of this section and then solve the resulting problems.

1. In Example 3, change the first denominator to $4x - 10$ and do the multiplication.
2. In Example 7, change the numerator $x + 2$ of the divisor to $x + 3$ and then simplify.

In Exercises 3–42, simplify the given expressions involving the indicated multiplications and divisions.

3. $\frac{3}{8} \times \frac{2}{7}$
4. $11 \times \frac{13}{33}$
5. $\frac{4x}{3y} \times \frac{9y^2}{2}$
6. $\frac{-6x}{5} \cdot \frac{2}{x^8}$
7. $\frac{2}{9} \div \frac{4}{7}$
8. $\frac{5}{16} \div \frac{25}{-13}$

9. $\dfrac{xy}{az} \div \dfrac{bz}{ay}$

10. $\dfrac{sr^2}{2t} \div \dfrac{st}{4}$

11. $\dfrac{4x + 12}{5} \times \dfrac{15t}{3x + 9}$

12. $\dfrac{2y^2 + 6y}{6y} \times \dfrac{y^3}{y^2 - 9}$

13. $\dfrac{u^2 - v^2}{u + 2v}(3u + 6v)$

14. $(x - y)\dfrac{x + 2y}{x^2 - y^2}$

15. $\dfrac{2a + 8}{15} \div \dfrac{a^2 + 8a + 16}{125}$

16. $\dfrac{a^2 - a}{3a + 9} \div \dfrac{a^2 - 2a + 1}{18 - 2a^2}$

17. $\dfrac{x^4 - 9}{x^2} \div (x^2 + 3)^2$

18. $\dfrac{9B^2 - 16}{B + 1} \div (4 - 3B)$

19. $\dfrac{3ax^2 - 9ax}{10x^2 + 5x} \times \dfrac{2x^2 + x}{a^2x - 3a^2}$

20. $\dfrac{4R^2 - 36}{R^3 - 25R} \times \dfrac{7R - 35}{3R^2 + 9R}$

21. $\left(\dfrac{x^4 - 1}{8x + 16}\right)\left(\dfrac{2x^2 - 8x}{x^3 + x}\right)$

22. $\left(\dfrac{2x^2 - 4x - 6}{x^2 - 3x}\right)\left(\dfrac{x^3 - 4x^2}{4x^2 - 4x - 8}\right)$

23. $\dfrac{\dfrac{x^2 + ax}{2b - cx}}{\dfrac{a^2 + 2ax + x^2}{2bx - cx^2}}$

24. $\dfrac{\dfrac{x^4 - 11x^2 + 28}{2x^2 + 6}}{\dfrac{4 - x^2}{2x^2 + 3}}$

25. $\dfrac{35a + 25}{12a + 33} \div \dfrac{28a + 20}{36a + 99}$

26. $\dfrac{2a^3 + a^2}{2b^3 + b^2} \div \dfrac{2ab + a}{2ab + b}$

27. $\dfrac{x^2 - 6x + 5}{4x^2 - 17x - 15} \times \dfrac{6x + 21}{2x^2 + 5x - 7}$

28. $\dfrac{n^2 + 5n}{3n^2 + 8n + 4} \times \dfrac{2n^2 - 8}{n^3 + 3n^2 - 10n}$

29. $\dfrac{\dfrac{6T^2 - NT - N^2}{2V^2 - 9V - 35}}{\dfrac{8T^2 - 2NT - N^2}{20V^2 + 26V - 60}}$

30. $\dfrac{\dfrac{4L^3 - 9L}{8L^2 + 10L - 3}}{2L^3 - 3L^2}$

31. $\dfrac{7x^2}{3a} \div \left(\dfrac{a}{x} \times \dfrac{a^2x}{x^2}\right)$

32. $\left(\dfrac{3u}{8v^2} \div \dfrac{9u^2}{2w^2}\right) \times \dfrac{2u^4}{15vw}$

33. $\left(\dfrac{4t^2 - 1}{t - 5} \div \dfrac{2t + 1}{2t}\right) \times \dfrac{2t^2 - 50}{4t^2 + 4t + 1}$

34. $\dfrac{2x^2 - 5x - 3}{x - 4} \div \left(\dfrac{x - 3}{x^2 - 16} \times \dfrac{1}{3 - x}\right)$

35. $\dfrac{x^3 - y^3}{2x^2 - 2y^2} \times \dfrac{y^2 + 2xy + x^2}{x^2 + xy + y^2}$

36. $\dfrac{2M^2 + 4M + 2}{6M - 6} \div \dfrac{5M + 5}{M^2 - 1}$

37. $\left(\dfrac{ax + bx + ay + by}{p - q}\right)\left(\dfrac{3p^2 + 4pq - 7q^2}{a + b}\right)$

38. $\dfrac{x^4 + x^5 - 1 - x}{x - 1} \div \dfrac{x - 1}{x}$

39. $\dfrac{x}{2x + 4} \times \dfrac{x^2 - 4}{3x^2}$

40. $\dfrac{4x^2 - 25}{4x^2} \div \dfrac{4x + 10}{8}$

41. $\dfrac{2x^2 + 3x - 2}{2 + 3x - 2x^2} \div \dfrac{5x + 10}{4x + 2}$

42. $\dfrac{16x^2 - 8x + 1}{9x} \times \dfrac{12x + 3}{1 - 16x^2}$

In Exercises 43–46, simplify the given expressions. The technical application of each is indicated.

43. $\dfrac{c\lambda^2 - c\lambda_0^2}{\lambda_0^2} \div \dfrac{\lambda^2 + \lambda_0^2}{\lambda_0^2}$ (cosmology)

44. $\left(\dfrac{8\pi n^2 eu}{mv^2 - mvu^2}\right)\left(\dfrac{mv^2}{2mv^2 - 2\pi ne^2}\right)$ (electromagnetism)

45. $\dfrac{2\pi}{\lambda}\left(\dfrac{a + b}{2ab}\right)\left(\dfrac{ab\lambda}{2a + 2b}\right)$ (optics)

46. $(p_1 - p_2) \div \left(\dfrac{\pi a^4 p_1 - \pi a^4 p_2}{81u}\right)$ (hydrodynamics)

In Exercises 47–48, answer the given problems.

47. The speed v of a satellite can be found from the equation $v^2 = \dfrac{GmM}{r^2} \div \dfrac{m}{r}$. Simplify the right side of the equation and then solve for v.

48. Simplify the following expression involving units (the number 2 is exact):

$$\frac{\left(28\dfrac{\text{m}}{\text{s}}\right)^2}{2(0.90)\left(9.8\dfrac{\text{m}}{\text{s}^2}\right)}$$

Answers to Practice Exercises

1. $\dfrac{1}{2(y - 1)}, x \neq 0, -2$ **2.** $\dfrac{3a}{x + 2}, x \neq 0, a \neq -1$

6.7 Addition and Subtraction of Fractions

From arithmetic, recall that *the sum of a set of fractions that all have the same denominator is the sum of the numerators divided by the common denominator.*

Addition and Subtraction of Fractions

$$\frac{a}{c} + \frac{b}{c} = \frac{a + b}{c}$$

$$\frac{a}{c} - \frac{b}{c} = \frac{a - b}{c}$$

Since algebraic expressions represent numbers, this fact is also true in algebra. Addition and subtraction of such fractions are illustrated in the following example.

EXAMPLE 1 Combining basic fractions

(a) $\frac{5}{9}+\frac{2}{9}-\frac{4}{9}=\frac{5+2-4}{9}=\frac{3}{9}=\frac{1\cdot \cancel{3}}{3\cdot \cancel{3}}=\frac{1}{3}$

(b) $\frac{b}{ax}+\frac{1}{ax}-\frac{2b-1}{ax}=\frac{b+(1)-(2b-1)}{ax}=\frac{b+1-2b+1}{ax}=\frac{-b+2}{ax}$

$=\frac{-1(b-2)}{ax}$

COMMON ERROR When subtracting a fraction with multiple terms in the numerator, ensure that the negative sign is applied to ***all*** terms. The signs of all the terms must be changed before they can be combined with other terms. It is a common error to forget to apply the subtraction to all terms. It is useful to place parentheses around each numerator while grouping to ensure that the correct sign convention is used.

LOWEST COMMON DENOMINATOR

If the fractions to be combined do not all have the same denominator, we must first change each to an equivalent fraction so that the resulting fractions do have the same denominator. We call the lowest multiple of prime factors that produce a shared denominator the **lowest common denominator** (abbreviated as **LCD**).

LEARNING TIP
The LCD is the product of all the prime factors that appear in the denominators, with each factor raised to the ***highest power*** *to which it appears* ***in any one*** *of the denominators.* This means that the lowest common denominator is the *simplest* algebraic expression into which all given denominators will divide exactly.

Procedure for Finding the Lowest Common Denominator

1. Factor each denominator into its prime factors.
2. For each different prime factor that appears, note the highest power to which it is raised in any one of the denominators.
3. Multiply all the different prime factors, each raised to the power found in Step 2. This product is the lowest common denominator.

The examples that follow illustrate the method of finding the LCD.

EXAMPLE 2 Lowest common denominator (LCD) with numerical fractions

Find the LCD of the fractions. $\frac{1}{8}\quad\frac{3}{4}\quad\frac{7}{20}$

1. Factor each denominator:

$$8=2\cdot 2\cdot 2=2^3\qquad 4=2\cdot 2=2^2\qquad 20=2\cdot 2\cdot 5=2^2\cdot 5^1$$

2. The highest power of 2 that appears is 3, and the highest power of 5 that appears is 1.
3. LCD $=2^3\cdot 5^1=40$

Equivalent fractions with the LCD as denominator are

$$\frac{1(5)}{8(5)}=\frac{5}{40}\qquad\frac{3(10)}{4(10)}=\frac{30}{40}\qquad\frac{7(2)}{20(2)}=\frac{14}{40}$$

EXAMPLE 3 Lowest common denominator (LCD) with algebraic fractions

Find the LCD of the fractions $\frac{3}{4a^2b}$ $\frac{5}{6ab^3}$ $\frac{1}{4ab^2}$

1. Factor each denominator:

$$4a^2b = 2^2 \cdot a^2 \cdot b^1 \qquad 6ab^3 = 2 \cdot 3 \cdot a^1 \cdot b^3 \qquad 4ab^2 = 2^2a^1b^2$$

2. Prime factors and their highest power: 2^2 3^1 a^2 b^3
3. LCD $= 2^2 \cdot 3^1 \cdot a^2 \cdot b^3 = 12a^2b^3$

Equivalent fractions with the LCD as denominator are

$$\frac{3(3b^2)}{4a^2b^1(3b^2)} = \frac{9b^2}{12a^2b^3} \qquad \frac{5(2a)}{6a^1b^3(2a)} = \frac{10a}{12a^2b^3} \qquad \frac{1(3ab)}{4a^1b^2(3ab)} = \frac{3ab}{12a^2b^3}$$

Finding the LCD of a set of fractions can be a source of difficulty. Remember, it is necessary to find *all of the prime factors* that may be in any of the denominators, and then find *the highest power to which each is raised in any of the denominators.* The following example illustrates this when factoring the denominators is the first step.

EXAMPLE 4 Lowest common denominator with factored denominators

Find the LCD of the fractions: $\frac{x-4}{x^2-2x+1}$ $\frac{1}{x^2-1}$ $\frac{x+3}{x^2-x}$

1. Factor each denominator:

$$x^2 - 2x + 1 = (x-1)^2 \qquad x^2 - 1 = (x+1)(x-1) \qquad x^2 - x = x(x-1)$$

2. Prime factors and their highest power: x $(x+1)$ $(x-1)^2$
3. LCD $= x(x+1)(x-1)^2$

Equivalent fractions with the LCD as denominator are

$$\frac{x-4}{x^2-2x+1} = \frac{(x-4)[x(x+1)]}{(x-1)^2[x(x+1)]} = \frac{x(x-4)(x+1)}{x(x+1)(x-1)^2},$$

$$\frac{1}{x^2-1} = \frac{1[x(x-1)]}{(x+1)(x-1)[x(x-1)]} = \frac{x(x-1)}{x(x+1)(x-1)^2},$$

$$\frac{x+3}{x^2-x} = \frac{(x+3)[(x+1)(x-1)]}{x(x-1)[(x+1)(x-1)]} = \frac{(x+3)(x+1)(x-1)}{x(x+1)(x-1)^2}$$

Practice Exercise

1. Find the LCD of the following fractions: $\frac{x}{2x^2+2x}$, $\frac{1}{x^2+2x+1}$, $\frac{2x+4}{4x^2}$

ADDITION AND SUBTRACTION OF FRACTIONS

Once we have found the LCD for the fractions and expressed them as equivalent fractions with the LCD as denominator, it is necessary only to add the numerators, place this result over the common denominator, and simplify.

EXAMPLE 5 Adding fractions using LCD

Combine: $\frac{2}{3r^2} + \frac{4}{rs^3} - \frac{5}{3s}$.

By looking at the denominators, notice that the factors necessary in the LCD are 3, r, and s. The 3 appears only to the first power, the largest exponent of r is 2, and the largest exponent of s is 3. Therefore, the LCD is $3r^2s^3$.

We now write equivalent fractions with the LCD as denominator and then combine the numerators. We get

$$\frac{2}{3r^2} + \frac{4}{rs^3} - \frac{5}{3s} = \frac{2(s^3)}{(3r^2)(s^3)} + \frac{4(3r)}{(rs^3)(3r)} - \frac{5(r^2s^2)}{(3s)(r^2s^2)}$$ change to equivalent fractions with LCD

$$= \frac{2s^3}{3r^2s^3} + \frac{12r}{3r^2s^3} - \frac{5r^2s^2}{3r^2s^3} = \frac{2s^3 + 12r - 5r^2s^2}{3r^2s^3}$$ combine numerators over LCD

Practice Exercise

2. Combine: $\frac{5}{2ab} - \frac{3}{4a^2}$

EXAMPLE 6 Adding fractions using LCD, binomial factors

Add: $\frac{a}{x-1} + \frac{a}{x+1}$.

In this case each denominator is a prime factor so the LCD is $(x-1)(x+1)$. Writing equivalent fractions with the LCD as denominator and combining numerators we get

$$\frac{a}{x-1} + \frac{a}{x+1} = \frac{a(x+1)}{(x-1)(x+1)} + \frac{a(x-1)}{(x+1)(x-1)}$$ change to equivalent fractions with LCD

$$= \frac{ax + a + ax - a}{(x+1)(x-1)}$$ combine numerators over LCD

$$= \frac{2ax}{(x+1)(x-1)}$$ simplify

When we multiply each fraction by the quantity required to obtain the proper denominator, we do not actually have to write the common denominator under each numerator. Placing all the products that appear in the numerators over the common denominator is sufficient. To illustrate, this example would appear as

$$\frac{a}{x-1} + \frac{a}{x+1} = \frac{a(x+1) + a(x-1)}{(x-1)(x+1)} = \frac{ax + a + ax - a}{(x-1)(x+1)} = \frac{2ax}{(x-1)(x+1)}$$

Practice Exercise

3. Combine: $\frac{5}{x-3} - \frac{2}{x+2}$

EXAMPLE 7 Combining fractions

$$\frac{3x}{2x^2 - 2x - 24} - \frac{x-1}{x^2 - 8x + 16} = \frac{3x}{2(x-4)(x+3)} - \frac{x-1}{(x-4)^2}$$ factor denominators

$$= \frac{3x(x-4) - (x-1)(2)(x+3)}{2(x-4)^2(x+3)}$$ change to equivalent fractions with LCD

$$= \frac{(3x^2 - 12x) - (2x^2 + 4x - 6)}{2(x-4)^2(x+3)}$$ expand in numerator

$$= \frac{3x^2 - 12x - 2x^2 - 4x + 6}{2(x-4)^2(x+3)} = \frac{x^2 - 16x + 6}{2(x-4)^2(x+3)}$$ simplify

LEARNING TIP

As in Example 7, remember that if a minus sign precedes an expression inside parentheses, the signs of *all* terms must be changed before they can be combined with other terms.

EXAMPLE 8 Combining fractions—missile firing

When analysing the dynamics of a missile firing, the following expression is found: $\frac{1}{s} - \frac{1}{s+4} + \frac{8}{s^2+8s+16}$. Simplify by addition and subtraction.

$$\frac{1}{s} - \frac{1}{s+4} + \frac{8}{s^2+8s+16} = \frac{1}{s} - \frac{1}{s+4} + \frac{8}{(s+4)^2}$$ factor third denominator

$$= \frac{1(s+4)^2 - 1(s)(s+4) + 8s}{s(s+4)^2}$$ LCD has one factor of s and two factors of $(s+4)$

$$= \frac{s^2+8s+16-s^2-4s+8s}{s(s+4)^2}$$ expand terms of the numerator

$$= \frac{12s+16}{s(s+4)^2} = \frac{4(3s+4)}{s(s+4)^2}$$ simplify/factor

Practice Exercise

4. Combine: $\frac{3}{2x+2} - \frac{2}{x^2+2x+1}$

COMPLEX FRACTIONS

A **complex fraction** *is one in which the numerator, the denominator, or both the numerator and the denominator contain fractions* (it is a "fraction within a fraction"). When a numerator (or denominator) includes added or subtracted terms where one or more of the values is a fraction, all terms in that numerator (or denominator) must be combined over a common denominator before simplifying using the reciprocal.

EXAMPLE 9 Complex numerical fraction

Simplify $\dfrac{\frac{1}{4}+\frac{1}{6}}{\frac{3}{7}+2}$.

Equivalent fractions for the numerator and the denominator using an LCD are

$$\frac{1}{4}+\frac{1}{6} = \frac{1(3)+1(2)}{12} = \frac{5}{12} \qquad \frac{3}{7}+2 = \frac{3}{7}+\frac{2}{1} = \frac{3(1)+2(7)}{7} = \frac{17}{7}$$

Substituting and inverting the denominator leads to

$$\frac{\frac{1}{4}+\frac{1}{6}}{\frac{3}{7}+2} = \frac{\frac{5}{12}}{\frac{17}{7}} = \frac{5}{12} \times \frac{7}{17} = \frac{35}{204}$$

EXAMPLE 10 Complex fractions

Simplify: **(a)** $\dfrac{\dfrac{2}{x}}{1-\dfrac{4}{x}}$ **(b)** $\dfrac{1-\dfrac{2}{x}}{\dfrac{1}{x}+\dfrac{2}{x^2+4x}}$

■ The original complex fraction can be written as a division as follows:

$$\frac{2}{x}\div\left(1-\frac{4}{x}\right)$$

(a) We rewrite the denominator as $\dfrac{1}{1}-\dfrac{4}{x}=\dfrac{1(x)-4(1)}{x}=\dfrac{x-4}{x}$. Inverting the divisor and multiplying, we get

$$\frac{\dfrac{2}{x}}{\dfrac{x-4}{x}}=\frac{2}{x}\times\frac{x}{x-4}=\frac{2\cancel{x}}{\cancel{x}(x-4)}=\frac{2}{x-4},\ x\neq 0$$

(b) We rewrite the numerator as $\dfrac{1}{1}-\dfrac{2}{x}=\dfrac{1(x)-2(1)}{x}=\dfrac{x-2}{x}$. We also simplify the denominator as

$$\frac{1}{x}+\frac{2}{x^2+4x}=\frac{1}{x}+\frac{2}{x(x+4)}=\frac{1(x+4)+2}{x(x+4)}=\frac{x+6}{x(x+4)}$$

Substituting and inverting the denominator leads to

$$\frac{\dfrac{(x-2)}{x}}{\dfrac{(x+6)}{x(x+4)}}=\frac{(x-2)}{\cancel{x}}\times\frac{\cancel{x}(x+4)}{(x+6)}=\frac{(x-2)(x+4)}{(x+6)},\ x\neq 0$$

EXERCISES 6.7

In Exercises 1–4, make the given changes in the indicated examples of this section and then solve the resulting problems.

1. In Example 3, change the denominator of the third fraction to $4a^2b^2$ and then find the LCD.
2. In Example 6, add the fraction $\dfrac{2}{x^2-1}$ to those being added and then find the result.
3. In Example 8, change the denominator of the third fraction to s^2+4s and then find the result.
4. In Example 10(a), change the fraction in the numerator to $2/x^2$ and then simplify.

In Exercises 5–44, perform the indicated operations and simplify.

5. $\dfrac{5}{6}+\dfrac{7}{6}$
6. $\dfrac{2}{13}+\dfrac{6}{13}$
7. $\dfrac{1}{x}+\dfrac{7}{x}$
8. $\dfrac{2}{a}+\dfrac{3}{a}$
9. $\dfrac{1}{2}+\dfrac{3}{4}$
10. $\dfrac{5}{9}-\dfrac{1}{3}$
11. $\dfrac{3}{4x}+\dfrac{7a}{4}+2$
12. $\dfrac{t-3}{a}-\dfrac{t}{2a}$
13. $\dfrac{a}{x}-\dfrac{b}{x^2}$
14. $\dfrac{3}{2s^2}+\dfrac{5}{4s}$
15. $\dfrac{6}{5x^3}+\dfrac{a}{25x}$
16. $\dfrac{a}{6y}-\dfrac{2b}{3y^4}$
17. $\dfrac{2}{5a}+\dfrac{1}{a}-\dfrac{a}{10}$
18. $\dfrac{1}{2A}-\dfrac{6}{B}-\dfrac{9}{4C}$
19. $\dfrac{x+1}{2x}-\dfrac{y-3}{4y}-\dfrac{2-x}{xy}$
20. $\dfrac{1-x}{6y}-\dfrac{3+x}{4y}$
21. $\dfrac{x}{2x-2}+\dfrac{4}{3x-3}$
22. $\dfrac{5}{6y+3}-\dfrac{a}{8y+4}$
23. $\dfrac{4}{x(x+1)}-\dfrac{3}{2x}$
24. $\dfrac{3}{ax+ay}-\dfrac{1}{a^2}$
25. $\dfrac{s}{2s-6}+\dfrac{1}{4}-\dfrac{3s}{4s-12}$
26. $\dfrac{2}{x+2}-\dfrac{3-x}{x^2+2x}+\dfrac{1}{x}$
27. $\dfrac{3R}{R^2-9}-\dfrac{2}{3R+9}$
28. $\dfrac{2}{n^2+4n+4}-\dfrac{3}{4+2n}$
29. $\dfrac{3}{x^2-8x+16}-\dfrac{2}{4-x}$
30. $\dfrac{2a-b}{c-3d}-\dfrac{b-2a}{3d-c}$
31. $\dfrac{v+4}{v^2+5v+4}-\dfrac{v-2}{v^2-5v+6}$
32. $\dfrac{N-1}{2N^3-4N^2}-\dfrac{5}{2-N}$
33. $\dfrac{x-1}{3x^2-13x+4}-\dfrac{3x+1}{4-x}$

34. $\dfrac{x}{4x^2 - 12x + 5} + \dfrac{2x - 1}{4x^2 - 4x - 15}$

35. $\dfrac{t}{t^2 - t - 6} - \dfrac{2t}{t^2 + 6t + 9} + \dfrac{t}{9 - t^2}$

36. $\dfrac{5}{2x^3 - 3x^2 + x} - \dfrac{x}{x^4 - x^2} + \dfrac{2 - x}{2x^2 + x - 1}$

37. $\dfrac{1}{w^3 + 1} + \dfrac{1}{w + 1} - 2$

38. $\dfrac{2}{8 - x^3} + \dfrac{1}{x^2 - x - 2}$

39. $\dfrac{\frac{1}{x}}{1 - \frac{1}{x}}$

40. $\dfrac{x - \frac{1}{x}}{1 - \frac{1}{x}}$

41. $\dfrac{\frac{x}{y} - \frac{y}{x}}{1 + \frac{y}{x}}$

42. $\dfrac{\frac{V^2 - 9}{V}}{\frac{1}{V} - \frac{1}{3}}$

43. $\dfrac{\frac{3}{x} + \frac{1}{x^2 + x}}{\frac{1}{x + 1} - \frac{1}{x - 1}}$

44. $\dfrac{\frac{1}{u - v} + \frac{1}{2u + 2v}}{\frac{2u}{2u^2 - 3uv + v^2} + \frac{2}{2u - v}}$

The expression $f(x + h) - f(x)$ is frequently used in the study of calculus. (If necessary, refer to Section 3.2 for a review of functional notation.) In Exercises 45–48, determine and then simplify this expression for the given functions.

45. $f(x) = \dfrac{x}{x + 1}$

46. $f(x) = \dfrac{3}{2x - 1}$

47. $f(x) = \dfrac{1}{x^2}$

48. $f(x) = \dfrac{2}{x^2 + 4}$

In Exercises 49–57, simplify the given expressions. In Exercise 58, answer the given question.

49. Using the definitions of the trigonometric functions given in Section 4.2, find an expression that is equivalent to $(\tan\theta)(\cot\theta) + (\sin\theta)^2 - \cos\theta$, in terms of x, y, and r.

50. Using the definitions of the trigonometric functions given in Section 4.2, find an expression that is equivalent to $\sec\theta - (\cot\theta)^2 + \csc\theta$, in terms of x, y, and r.

51. If $f(x) = 2x - x^2$, find $f\left(\dfrac{1}{a}\right)$.

52. If $f(x) = x^2 + x$, find $f\left(a + \dfrac{1}{a}\right)$.

53. If $f(x) = x - \dfrac{2}{x}$, find $f(a + 1)$.

54. If $f(x) = 2x - 3$, find $f[1/f(x)]$.

55. The sum of two numbers a and b is divided by the sum of their reciprocals. Simplify the expression for this quotient.

56. Find A and B if $\dfrac{2x - 9}{x^2 - x - 6} = \dfrac{A}{x - 3} + \dfrac{B}{x + 2}$.

57. If $x = \dfrac{mn}{m + n}$ and $y = \dfrac{mn}{m - n}$, show that $\dfrac{y^2 - x^2}{y^2 + x^2} = \dfrac{2mn}{m^2 + n^2}$.

58. When adding fractions, explain why it is better to find the lowest common denominator rather than any denominator that is common to the fractions.

In Exercises 59–68, perform the indicated operations. Each expression occurs in the indicated area of application.

59. $\dfrac{3}{4\pi} - \dfrac{3H_0}{4\pi H}$ (transistor theory)

60. $1 + \dfrac{9}{128T} - \dfrac{27p}{64T^3}$ (thermodynamics)

61. $\dfrac{2n^2 - n - 4}{2n^2 + 2n - 4} + \dfrac{1}{n - 1}$ (optics)

62. $\dfrac{b}{x^2 + y^2} - \dfrac{2bx^2}{x^4 + 2x^2y^2 + y^4}$ (magnetic field)

63. $\left(\dfrac{3Px}{2L^2}\right)^2 + \left(\dfrac{P}{2L}\right)^2$ (force of a weld)

64. $\dfrac{a}{b^2h} + \dfrac{c}{bh^2} - \dfrac{1}{6bh}$ (strength of materials)

65. $\dfrac{\frac{L}{C} + \frac{R}{sC}}{sL + R + \frac{1}{sC}}$ (electronics)

66. $\dfrac{\frac{m}{c}}{1 - \frac{p^2}{c^2}}$ (airfoil design)

67. $\dfrac{1}{R^2} + \left(\omega C - \dfrac{1}{\omega L}\right)^2$ (electricity)

68. $\dfrac{\frac{x}{h_1} + \frac{x - L}{h_2}}{1 + \frac{x(L - x)}{h_1h_2}}$ (optics)

In Exercises 69 and 70, find the required expressions.

69. A boat that travels at v km/h in still water travels 5 km upstream and then 5 km downstream in a stream that flows at w km/h. Write an expression for the total time the boat travelled as a single fraction.

70. A person travels a distance d at an average speed v_1, and then returns over the same route at an average speed v_2. Write an expression in simplest form for the average speed of the round trip.

Answers to Practice Exercises

1. $4x^2(x + 1)^2$
2. $\dfrac{10a - 3b}{4a^2b}$
3. $\dfrac{3x + 16}{(x - 3)(x + 2)}$
4. $\dfrac{3x - 1}{2(x + 1)^2}$

6.8 Equations Involving Fractions

Many important equations in science, engineering, and technology contain fractions. In order to solve these types of equations, it is beneficial to first rewrite them in a form that contains no fractions. This can be done by using the following procedure:

Solving Equations Involving Fractions

1. Multiply each term of the equation by the LCD of all the denominators. This should eliminate all fractions from the equation.
2. Solve the resulting equation using previous methods.
3. Check the solution in the original equation.

The following examples illustrate this procedure.

LEARNING TIP

Note very carefully that the method of multiplying each term of the equation by the LCD to eliminate the denominators may be used only when there is an *equation*. If there is no equation, do not multiply by the LCD to eliminate denominators.

■ Note carefully that we are multiplying both sides of the equation by the LCD and not combining terms over a common denominator as when adding fractions.

EXAMPLE 1 Multiply each term by the LCD

Solve for x: $\frac{x}{12} - \frac{1}{8} = \frac{x+2}{6}$.

First, note that the LCD of the denominators $12 = 2^2 \cdot 3$, $8 = 2^3$, and $6 = 2 \cdot 3$ is $2^3 \cdot 3$. Therefore, multiply each term by 24. This gives

$$\frac{24(x)}{12} - \frac{24(1)}{8} = \frac{24(x+2)}{6} \quad \text{each term multiplied by LCD}$$

Reduce each term to its lowest terms and solve the resulting equation:

$$\begin{aligned} 2x - 3 &= 4(x+2) && \text{each term reduced} \\ 2x - 3 &= 4x + 8 && \text{expand} \\ -2x &= 11 && \text{subtract } 2x + 8 \text{ from each side and simplify} \\ x &= -\frac{11}{2} && \text{divide each side by 2} \end{aligned}$$

When we check this solution in the original equation, we obtain $-7/12$ on each side of the equal sign. Therefore, the solution is correct.

EXAMPLE 2 Solving a literal equation

Solve for x: $\frac{x}{2} - \frac{1}{b^2} = \frac{x}{2b}$.

First, determine that the LCD of the terms of the equation is $2b^2$. Then multiply each term by $2b^2$ and continue with the solution:

$$\begin{aligned} \frac{2b^2(x)}{2} - \frac{2b^2(1)}{b^2} &= \frac{2b^2(x)}{2b} && \text{each term multiplied by LCD} \\ b^2x - 2 &= bx && \text{each term reduced} \\ b^2x - bx &= 2 && \text{subtract } bx - 2 \text{ from each side and simplify} \\ x(b^2 - b) &= 2 && \text{factor} \\ x &= \frac{2}{b^2 - b} = \frac{2}{b(b-1)} \end{aligned}$$

Checking shows that each side of the original equation equals $\frac{1}{b^2(b-1)}$.

EXAMPLE 3 Solving an equation—focal length of a lens

An equation relating the focal length f of a lens with the object distance p and the image distance q is:

$$f = \frac{pq}{p+q} \quad \text{given equation}$$

(see Fig. 6.5). Solve for q.

Since the only denominator is $p + q$, the LCD is also $p + q$. By first multiplying each term by $p + q$, the solution is completed as follows:

$$\begin{aligned}
f(p+q) &= \frac{pq(p+q)}{p+q} && \text{each term multiplied by LCD}\\
fp + fq &= pq && \text{reduce term on right}\\
fq - pq &= -fp && \text{subtract } fp + pq \text{ from each side}\\
q(f-p) &= -fp && \text{factor}\\
q &= -\frac{fp}{f-p} && \text{divide by } f-p\\
q &= \frac{fp}{p-f} && \text{use } -(f-p) = p-f
\end{aligned}$$

■ The first telescope was invented by Lippershay, a Dutch lens maker, in about 1608.

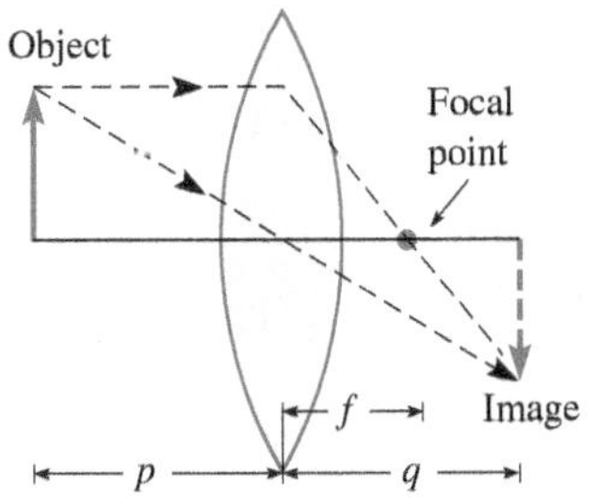

Fig. 6.5

EXAMPLE 4 Solving an equation—planetary motion

In astronomy, when developing the equations that describe the motion of the planets, the equation $\frac{1}{2}v^2 - \frac{GM}{r} = -\frac{GM}{2a}$ is found. Solve for M.

First, determine that the LCD of the terms of the equation is $2ar$. Multiplying each term by $2ar$ and proceeding, we have

$$\begin{aligned}
\frac{2ar(v^2)}{2} - \frac{2ar(GM)}{r} &= -\frac{2ar(GM)}{2a} && \text{each term multiplied by LCD}\\
arv^2 - 2aGM &= -rGM && \text{each term reduced}\\
rGM - 2aGM &= -arv^2 && \text{add } rGM - arv^2 \text{ to each side}\\
M(rG - 2aG) &= -arv^2 && \text{factor}\\
M &= -\frac{arv^2}{rG - 2aG} && \text{divide by } rG - 2aG\\
M &= \frac{arv^2}{2aG - rG} && \text{use } -(rG - 2aG) = 2aG - rG
\end{aligned}$$

■ See the chapter introduction.

■ In 1609, the Italian scientist Galileo (1564–1642) learned of the invention of the telescope and developed it for astronomical observations. Among his first discoveries were the four largest moons of the planet Jupiter.

Practice Exercise

1. Solve for x: $\frac{x-1}{3a} = \frac{5}{6} + \frac{x}{a^2}$

EXAMPLE 5 Extraneous solution

Solve for x: $\frac{2}{x+1} - \frac{1}{x} = -\frac{2}{x^2+x}$.

Note first that $x = 0$ and $x = -1$ make a denominator equal to zero, so these are not possible values of the solution. Now, the LCD of $x + 1$, x, and $x^2 + x = x(x + 1)$ is $x(x + 1)$. Therefore,

$$\begin{aligned}
\frac{2(x)(x+1)}{x+1} - \frac{x(x+1)}{x} &= -\frac{2x(x+1)}{x(x+1)} && \text{multiply each side by the LCD}\\
2x - (x+1) &= -2 && \text{simplify each fraction}\\
2x - x - 1 &= -2 && \text{complete the solution}\\
x &= -1
\end{aligned}$$

As we said earlier, $x = -1$ cannot be a solution. Therefore, *there is no solution to this equation.*

LEARNING TIP

Whenever we multiply each term by a common denominator that ***contains the unknown****, it is possible to obtain a value that is not a solution of the original equation. Such a value is termed an* **extraneous solution**. Only certain equations will lead to extraneous solutions, but we must be careful to identify them when they occur.

COMMON ERROR It is a common error to give a value that makes the denominator zero as a solution to an equation that involves fractions. Such an error can be avoided by stating explicitly from the beginning the values that are not valid solutions (because they make a denominator equal to zero), as we did in Example 5. Checking your solution in the original equation also helps identify values that cannot be solutions.

A number of stated problems give rise to equations involving fractions. The following examples illustrate the solution of such problems.

EXAMPLE 6 Setting up an equation—computer processing

An industrial firm uses a computer system that processes and prints out its data for an average day in 20 min. To process the data more rapidly and to handle increased future computer needs, the firm plans to add new components to the system. One set of new components can process the data in 12 min, without the present system. How long would it take the new system, a combination of the present system and the new components, to process the data?

■ The first large-scale electronic computer was the ENIAC. It was constructed at the University of Pennsylvania in the mid-1940s and used until 1955. It had 18 000 vacuum tubes and occupied 1400 m^2 of floor space.

Identify known quantities: Time for present system to process data = 20 min
Time for new components to process data = 12 min

Identify the unknown: Let x = time (in min) for combined system to process data

Select an equation: The present system processes $\frac{1}{20}$ of the data in 1 min, or $\frac{1}{20}x$ of the data in x min. Similarly, the new components process $\frac{1}{12}x$ of the data in x min. After x min, the sum of these two amounts is 1 (a complete set of data will have been processed). Therefore,

■ A military programmable computer called *Colossus* was used to break the German codes in World War II. It was developed, in secret, before ENIAC.

$$\frac{x}{20} + \frac{x}{12} = 1$$

Solve the equation: The LCD of $20 = 2^2 \cdot 5$ and $12 = 2^2 \cdot 3$ is $2^2 \cdot 3 \cdot 5 = 60$. Multiplying each term by 60,

$$\frac{60x}{20} + \frac{60x}{12} = 60(1)$$

$$3x + 5x = 60$$

$$8x = 60$$

$$x = \frac{60}{8} = 7.5 \text{ min}$$

Therefore, the new system should take about 7.5 min to process the data.

Practice Exercise

2. What is the result in Example 6, if the new system can process the data in 6.0 min?

EXAMPLE 7 Setting up an equation—rate of travel

A bus averaging 80.0 km/h takes 1.50 h longer to travel from Montreal, QC, to London, ON, than a train that averages 96.0 km/h. How far apart are Montreal and London?

Identify known quantities: $v_{bus} = 80.0\text{ km/h}$, $v_{train} = 96.0\text{ km/h}$

Identify the unknown: Let d = distance (in km) between Montreal and London

Select an equation: Since *distance* = *rate* × *time*, the time the train takes is $d/96.0$ h, and the time the bus takes is $d/80.0$ h. This means

$$t_{bus} - t_{train} = 1.50\text{ h}, \quad \text{or} \quad \frac{d}{80.0} - \frac{d}{96.0} = 1.50$$

Solve the equation: The LCD of $80 = 2^4 \cdot 5$ and $96 = 2^5 \cdot 3$ is $2^5 \cdot 3 \cdot 5 = 480$. Therefore,

$$\frac{480d}{80.0} - \frac{480d}{96.0} = 480(1.50)$$

$$6.00d - 5.00d = 720$$

$$d = 720\text{ km}$$

EXERCISES 6.8

In Exercises 1–4, make the given changes in the indicated examples of this section and then solve for the indicated variable.

1. In Example 2, change the second denominator from b^2 to b and then solve for x.
2. In Example 3, solve for p.
3. In Example 4, solve for G.
4. In Example 5, change the numerator on the right to 1 and then solve for x.

In Exercises 5–32, solve the given equations and check the results.

5. $\frac{x}{2} + 6 = 2x$
6. $\frac{x}{5} + 2 = \frac{15 + x}{10}$
7. $\frac{x}{6} - \frac{1}{2} = \frac{x}{3}$
8. $\frac{3N}{8} - \frac{3}{4} = \frac{N - 4}{2}$
9. $1 - \frac{t - 5}{6} = \frac{3}{4}$
10. $\frac{2x - 7}{3} + 5 = \frac{1}{5}$
11. $\frac{3x}{7} - \frac{5}{21} = \frac{2 - x}{14}$
12. $\frac{F - 3}{12} - \frac{2}{3} = \frac{1 - 3F}{2}$
13. $\frac{3}{T} + 2 = \frac{5}{3}$
14. $\frac{1}{2y} - \frac{1}{2} = 4$
15. $3 - \frac{x - 2}{5x} = \frac{1}{5}$
16. $\frac{1}{2R} - \frac{1}{3} = \frac{2}{3R}$
17. $\frac{2y}{y - 1} = 5$
18. $\frac{x}{2x - 3} = 4$
19. $\frac{2}{s} = \frac{3}{s - 1}$
20. $\frac{5}{n + 2} = \frac{3}{2n}$
21. $\frac{5}{2x + 4} + \frac{3}{6x + 12} = 2$
22. $\frac{3}{4x - 6} + \frac{1}{4} = \frac{5}{3 - 2x}$
23. $\frac{2}{Z - 5} - \frac{3}{10 - 2Z} = 3$
24. $\frac{4}{4 - x} + 2 - \frac{2}{12 - 3x} = \frac{1}{3}$
25. $\frac{1}{4x} + \frac{3}{2x} = \frac{2}{x + 1}$
26. $\frac{3}{t + 3} - \frac{1}{t} = \frac{5}{6 + 2t}$
27. $\frac{7}{y} = \frac{3}{y - 4} + \frac{7}{2y^2 - 8y}$
28. $\frac{1}{2x + 3} = \frac{5}{2x} - \frac{4}{2x^2 + 3x}$
29. $\frac{1}{x^2 - x} - \frac{1}{x} = \frac{1}{x - 1}$
30. $\frac{2}{x^2 - 1} - \frac{2}{x + 1} = \frac{1}{x - 1}$
31. $\frac{2}{B^2 - 4} - \frac{1}{B - 2} = \frac{1}{2B + 4}$
32. $\frac{2}{2x^2 + 5x - 3} - \frac{1}{4x - 2} + \frac{3}{2x + 6} = 0$

In Exercises 33–48, solve for the indicated letter. In Exercises 37–48, each of the given formulas arises in the technical or scientific area of study listed.

33. $2 - \frac{1}{b} + \frac{3}{c} = 0$, for c
34. $\frac{2}{3} - \frac{h}{x} = \frac{1}{6x}$, for x
35. $\frac{t - 3}{b} - \frac{t}{2b - 1} = \frac{1}{2}$, for t

36. $\frac{1}{a^2 + 2a} - \frac{y}{2a} = \frac{2y}{a + 2}$, for y

37. $\frac{s - s_0}{t} = \frac{v + v_0}{2}$, for v (velocity of object)

38. $S = \frac{P}{A} + \frac{Mc}{I}$, for A (machine design)

39. $V = 1.2\left(5.0 + \frac{8.0R}{8.0 + R}\right)$, for R (electric resistance in Fig. 6.6)

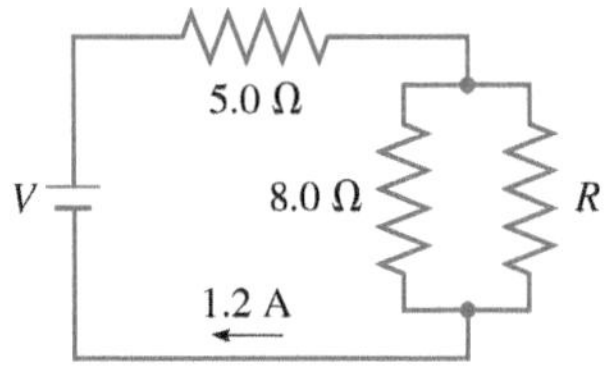

Fig. 6.6

40. $C = \frac{7p}{100 - p}$, for p (environmental pollution)

41. $z = \frac{1}{g_m} - \frac{jX}{g_m R}$, for R (FM transmission)

42. $A = \frac{1}{2}wp - \frac{1}{2}w^2 - \frac{\pi}{8}w^2$, for p (architecture)

43. $p = \frac{RT}{V - b} - \frac{a}{V^2}$, for T (thermodynamics)

44. $\frac{1}{x} + \frac{1}{nx} = \frac{1}{f}$, for f (photography)

45. $\frac{1}{C} = \frac{1}{C_2} + \frac{1}{C_1 + C_3}$, for C_1 (electricity: capacitors)

46. $D = \frac{wx^4}{24EI} - \frac{wLx^3}{6EI} + \frac{wL^2x^2}{4EI}$, for w (beam design)

47. $f(n - 1) = \frac{1}{\frac{1}{R_1} + \frac{1}{R_2}}$, for R_1 (optics)

48. $P = \frac{\frac{1}{1 + i}}{1 - \frac{1}{1 + i}}$, for i (business)

In Exercises 49–58, set up appropriate equations and solve the given stated problems. All numbers are accurate to at least three significant digits.

49. One pump can empty an oil tanker in 5.00 h, and a second pump can empty the tanker in 8.00 h. How long would it take the two pumps working together to empty the tanker?

50. One company determines that it will take its crew 450 h to clean up a chemical dump site, and a second company determines that it will take its crew 600 h to clean up the site. How long will it take the two crews working together?

51. One automatic packaging machine can package 100 boxes of machine parts in 12.0 min, and a second machine can do it in 10.0 min. A newer model machine can do it in 8.00 min. How long will it take the three machines working together?

52. A painting crew can paint a structure in 12.0 h, or the crew can paint it in 7.20 h when working with a second crew. How long would it take the second crew to do the job if working alone?

53. An elevator travelled from the first floor to the top floor of a building at an average speed of 2.00 m/s and returned to the first floor at 2.20 m/s. If it was on the top floor for 90.0 s and the total elapsed time was 5.00 min, how far above the first floor is the top floor?

54. A commuter rapid transit train travels 24.0 km farther between stops A and B than between stops B and C. If it averages 60.0 km/h from A to B and 30.0 km/h between B and C, and an express averages 50.0 km/h between A and C (not stopping at B), how far apart are stops A and C?

55. A jet takes the same time to travel 2580 km with the wind as it does to travel 1800 km against the wind. If its speed relative to the air is 450 km/h, what is the speed of the wind?

56. An engineer travels from St. John's, NL, to the Hibernia oil field in the North Atlantic Ocean on a ship that averages 28.0 km/h. After spending 6.00 h at the field, the engineer returns to St. John's in a helicopter that averages 140 km/h. If the total trip takes 19.5 h, how far is the Hibernia oil field from St. John's?

57. The current through each of the resistances R_1 and R_2 in Fig. 6.7 equals the voltage V divided by the resistance. The sum of the currents equals the current i in the rest of the circuit. Find the voltage if $i = 1.20$ A, $R_1 = 2.70\ \Omega$, and $R_2 = 6.00\ \Omega$.

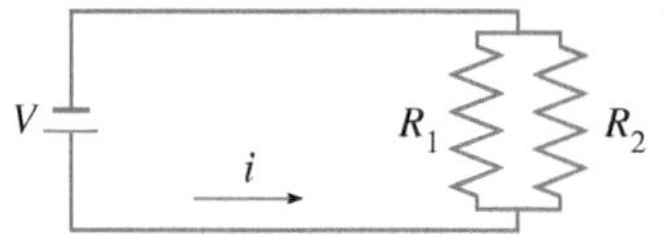

Fig. 6.7

58. A fox, pursued by a greyhound, has a start of 60 leaps. He makes 9 leaps while the greyhound makes but 6; but, 3 leaps of the greyhound are equivalent to 7 of the fox. How many leaps must the greyhound make to overcome the fox? (Copied from Davies, Charles, *Elementary Algebra,* New York: A. S. Barnes & Burr, 1852.) (*Hint*: Let the unit of distance be one fox leap.)

In Exercises 59 and 60, find constants A and B such that the equation is true.

59. $\frac{x - 12}{x^2 + x - 6} = \frac{A}{x + 3} + \frac{B}{x - 2}$

60. $\frac{23 - x}{2x^2 + 7x - 4} = \frac{A}{2x - 1} - \frac{B}{x + 4}$

Answers to Practice Exercises

1. $x = \frac{5a^2 + 2a}{2a - 6}$ 2. 4.6 min

CHAPTER 6 KEY FORMULAS AND EQUATIONS

Distributive law:	$a(x + y) = ax + ay$	**(6.1)**
Square of a binomial:	$(x + y)^2 = x^2 + 2xy + y^2$	**(6.2)**
	$(x - y)^2 = x^2 - 2xy + y^2$	**(6.3)**
Product of two binomials:	$(x + a)(x + b) = x^2 + (b + a)x + ab$	**(6.4)**
	$(ax + b)(cx + d) = acx^2 + (ad + bc)x + bd$	**(6.5)**
Product of sum and difference:	$(x + y)(x - y) = x^2 - y^2$	**(6.6)**
	$(ax + by)(ax - by) = (ax)^2 - (by)^2$	**(6.7)**
Cube of a binomial:	$(x + y)^3 = x^3 + 3x^2y + 3xy^2 + y^3$	**(6.8)**
	$(x - y)^3 = x^3 - 3x^2y + 3xy^2 - y^3$	**(6.9)**
Result is a sum of cubes:	$(x + y)(x^2 - xy + y^2) = x^3 + y^3$	**(6.10)**
Result is a difference of cubes:	$(x - y)(x^2 + xy + y^2) = x^3 - y^3$	**(6.11)**
Factor a difference of squares:	$u^2 - v^2 = (u + v)(u - v)$	**(6.13)**
Factors that differ only in sign:	$x - y = -(y - x)$	**(6.16)**

CHAPTER 6 REVIEW EXERCISES

In Exercises 1–12, find the products by inspection. No intermediate steps should be necessary.

1. $3a(4x + 5a)$
2. $-7xy(4x^2 - 7y)$
3. $(2a + 7b)(2a - 7b)$
4. $(x - 4z)(x + 4z)$
5. $(2a + 1)^2$
6. $(5C - 2D)^2$
7. $(b - 4)(b + 7)$
8. $(y - 5)(y - 7)$
9. $(2x + 5)(x - 9)$
10. $(4ax - 3)(5ax + 7)$
11. $(2c^a + d)(2c^a - d)$
12. $(3s^n - 2t)(3s^n + 2t)$

In Exercises 13–44, factor the given expressions completely.

13. $3s + 9t$
14. $7x - 28y$
15. $a^2x^2 + a^2$
16. $3ax - 6ax^4 - 9a$
17. $W^2b^{x+2} - 144b^x$
18. $900n^a - n^{a+4}$
19. $16(x + 2)^2 - t^4$
20. $25s^4 - 36t^2$
21. $36t^2 - 24t + 4$
22. $4x^2 - 12x + 9$
23. $25t^2 + 10t + 1$
24. $4c^2 + 36cd + 81d^2$
25. $x^2 + x - 56$
26. $x^2 - 4x - 45$
27. $t^4 - 5t^2 - 36$
28. $3N^4 - 33N^2 + 30$
29. $2k^2 - k - 36$
30. $5x^2 + 2x - 3$
31. $8x^2 - 8x - 70$
32. $27F^3 + 21F^2 - 48F$
33. $10b^2 + 23b - 5$
34. $12x^2 - 7xy - 12y^2$
35. $4x^2 - 64y^2$
36. $4a^2x^2 + 26a^2x + 36a^2$
37. $250 - 16y^6$
38. $8a^4 + 64a$
39. $8x^3 + 27$
40. $R^3 - 125r^3$
41. $ab^2 - 3b^2 + a - 3$
42. $axy - ay + ax - a$
43. $nx + 5n - x^2 + 25$
44. $ty - 4t + y^2 - 16$

In Exercises 45–68, perform the indicated operations and express results in simplest form.

45. $\dfrac{48ax^3y^6}{9a^3xy^6}$
46. $\dfrac{-39r^2s^4t^8}{52rs^5t}$
47. $\dfrac{6x^2 - 7x - 3}{4x^2 - 8x + 3}$
48. $\dfrac{p^4 - 4p^2 - 4}{p^4 - p^2 - 12}$
49. $\dfrac{4x + 4y}{35x^2} \times \dfrac{28x}{x^2 - y^2}$
50. $\left(\dfrac{6x - 3}{x^2}\right)\left(\dfrac{4x^2 - 12x}{6 - 12x}\right)$
51. $\dfrac{18 - 6L}{L^2 - 6L + 9} \div \dfrac{L^2 - 2L - 15}{L^2 - 9}$
52. $\dfrac{6x^2 - xy - y^2}{2x^2 + xy - y^2} \div \dfrac{16y^2 - 4x^2}{2x^2 + 6xy + 4y^2}$
53. $\dfrac{\dfrac{3x}{7x^2 + 13x - 2}}{\dfrac{6x^2}{x^2 + 4x + 4}}$
54. $\dfrac{\dfrac{3x - 3y}{2x^2 + 3xy - 2y^2}}{\dfrac{3x^2 - 3y^2}{x^2 + 4xy + 4y^2}}$
55. $\dfrac{x + \dfrac{1}{x} + 1}{x^2 - \dfrac{1}{x}}$
56. $\dfrac{\dfrac{4}{y} - 4y}{2 - \dfrac{2}{y}}$

57. $\frac{4}{9x} - \frac{5}{12x^2}$

58. $\frac{3}{10a^2} + \frac{1}{4a^3}$

59. $\frac{6}{x} - \frac{7}{2x} + \frac{3}{xy}$

60. $\frac{T}{T^2 + 2} - \frac{1}{2T + T^3}$

61. $\frac{a+1}{a+2} - \frac{a+3}{a}$

62. $\frac{y}{y+2} - \frac{1}{y^2 + 2y}$

63. $\frac{2x}{x^2 + 2x - 3} - \frac{1}{6x + 2x^2}$

64. $\frac{x}{4x^2 + 4x - 3} - \frac{3}{4x^2 - 9}$

65. $\frac{3x}{2x^2 - 2} - \frac{2}{4x^2 - 5x + 1}$

66. $\frac{2n - 1}{4 - n} + \frac{n + 2}{5n - 20}$

67. $\frac{3x}{x^2 + 2x - 3} - \frac{2}{x^2 + 3x} + \frac{x}{x - 1}$

68. $\frac{3}{y^4 - 2y^3 - 8y^2} + \frac{y - 1}{y^2 + 2y} - \frac{y - 3}{y^2 - 4y}$

In Exercises 69–76, factor the given expressions. In Exercises 69–72, disregard the restriction that coefficients and exponents must be integers. The expressions in Exercises 73–76 can be factored if they are first rewritten in a different form.

69. $x^2 - 5$

70. $x - y$

71. $x - 1$

72. $7 - x^2$

73. $x + y$ (one of two factors is to be x)

74. $x + y$ (one of two factors is to be a)

75. $x - 2y$ (one of two factors is to be 2)

76. $x - y$ (one of two factors is to be $-y$)

In Exercises 77–84, solve the given equations.

77. $\frac{x}{2} - 3 = \frac{x - 10}{4}$

78. $\frac{2x}{c} - \frac{1}{2c} = \frac{3}{c} - x$, for x

79. $\frac{2}{t} - \frac{1}{at} = 2 + \frac{a}{t}$, for t

80. $\frac{3}{a^2y} - \frac{1}{ay} = \frac{9}{a}$, for y

81. $\frac{2x}{2x^2 - 5x} - \frac{3}{x} = \frac{1}{4x - 10}$

82. $\frac{3}{x^2 + 3x} - \frac{1}{x} = \frac{1}{3 + x}$

83. Given $f(x) = \frac{1}{x + 2}$, solve for x if $f(x + 2) = 2f(x)$.

84. Given $f(x) = \frac{x}{x + 1}$, solve for x if $3f(x) + f\left(\frac{1}{x}\right) = 2$.

In Exercises 85–102, solve the given problems. Where indicated, the expression is found in the stated technical area.

85. In an algebraic fraction, what effect on the result is made if the sign is changed of (a) an odd number of factors? (b) an even number of factors?

86. Algebraically show that the reciprocal of the reciprocal of the number x is x.

87. Show that $xy = \frac{1}{4}[(x + y)^2 - (x - y)^2]$.

88. Show that $x^2 + y^2 = \frac{1}{2}[(x + y)^2 + (x - y)^2]$.

89. Multiply: $2zS(S + 1)$ (solid-state physics)

90. Expand: $kr(R - r)$ (blood flow)

91. Expand: $[2b + (n - 1)\lambda]^2$ (optics)

92. Factor: $\pi r_1^2 l - \pi r_2^2 l$ (jet plane fuel supply)

93. Factor: $cT_2 - cT_1 + RT_2 - RT_1$ (pipeline flow)

94. Factor: $9600t + 8400t^2 - 1200t^3$ (solar energy)

95. Simplify and express in factored form: $(2R - r)^2 - (r^2 + R^2)$ (aircraft radar)

96. Express in factored form: $2R(R + r) - (R + r)^2$ (electricity: power)

97. Expand and simplify: $(n + 1)^3(2n + 1)^3$ (fluid flow in pipes)

98. Expand and simplify: $2(e_1 - e_2)^2 + 2(e_2 - e_3)^2$ (mechanical design)

99. Expand and simplify: $10a(T - t) + a(T - t)^2$ (instrumentation)

100. Expand the third term and then factor by grouping: $pa^2 + (1 - p)b^2 - [pa + (1 - p)b]^2$ (nuclear physics)

101. A metal cube of edge x is heated and each edge increases by 4 mm. Express the increase in volume in factored form.

102. Express the difference in volumes of two ball bearings of radii r mm and 3 mm in factored form. ($r > 3$ mm)

In Exercises 103–114, perform the indicated operations and simplify the given expressions. Each expression is from the indicated technical area of application.

103. $\left(\frac{2wtv^2}{Dg}\right)\left(\frac{b\pi^2D^2}{n^2}\right)\left(\frac{6}{bt^2}\right)$ (machine design)

104. $\frac{m}{c} \div \left[1 - \left(\frac{p}{c}\right)^2\right]$ (airfoil design)

105. $\frac{\frac{\pi ka}{2}(R^4 - r^4)}{\pi ka(R^2 - r^2)}$ (flywheel rotation)

106. $\frac{V}{kp} - \frac{RT}{k^2p^2}$ (electric motors)

107. $1 - \frac{d^2}{2} + \frac{d^4}{24} - \frac{d^6}{120}$ (aircraft emergency locator transmitter)

108. $\frac{wx^2}{2T_0} + \frac{kx^4}{12T_0}$ (bridge design)

109. $\frac{4k - 1}{4k - 4} + \frac{1}{2k}$ (spring stress)

110. $\frac{Am}{k} - \frac{g}{2}\left(\frac{m}{k}\right)^2 + \frac{AML}{k}$ (rocket fuel)

111. $1 - \frac{3a}{4r} - \frac{a^3}{4r^3}$ (hydrodynamics)

112. $\frac{1}{F} + \frac{1}{f} - \frac{d}{fF}$ (optics)

113. $\frac{\frac{u^2}{2g} - x}{\frac{1}{2gc^2} - \frac{u^2}{2g} + x}$ (mechanism design)

114. $\frac{V}{\frac{1}{2R} + \frac{1}{2R + 2}}$ (electricity)

In Exercises 115–124, solve for the indicated letter. Each equation is from the indicated technical area of application.

115. $W = mgh_2 - mgh_1$, for m (work done on object)

116. $R_1(V_2 - V_1) + R_2V_2 = 3R_1R_2$, for V_2 (electricity)

117. $R = \dfrac{wL}{H(w + L)}$, for L (architecture)

118. $\dfrac{J}{T} = \dfrac{t}{\omega_1 - \omega_2}$, for ω_2 (rotational torque)

119. $E = V_0 + \dfrac{(m + M)V^2}{2} + \dfrac{p^2}{2I}$, for I (nuclear physics)

120. $\dfrac{q_2 - q_1}{d} = \dfrac{f + q_1}{D}$, for q_1 (photography)

121. $s^2 + \dfrac{cs}{m} + \dfrac{kL^2}{mb^2} = 0$, for c (mechanical vibrations)

122. $I = \dfrac{A}{x^2} + \dfrac{B}{(10 - x)^2}$, for A (optics)

123. $V = \dfrac{i}{sC} + \dfrac{V_0}{s}$, for C (electricity)

124. $C = \dfrac{k}{n(1 - k)}$, for k (reinforced concrete design)

In Exercises 125–132, set up appropriate equations and solve the given stated problems. All numbers are accurate to at least three significant digits. In Exercise 133, answer the given question.

125. If a certain car's lights are left on, the battery will be dead in 4.00 h. If only the radio is left on, the battery will be dead in 24.0 h. How long will the battery last if both the lights and the radio are left on?

126. Two pumps are being used to fight a fire. One pumps 5000 L in 20 min, and the other pumps 5000 L in 25.0 min. How long will it take the two pumps together to pump 5000 L?

127. A number m is the *harmonic mean* of number x and y if $1/m$ equals the average of $1/x$ and $1/y$ (divide $1/x + 1/y$ by 2). Find the harmonic mean of the musical notes that have frequencies of 400 Hz and 1200 Hz.

128. An auto mechanic can do a certain motor job in 3.00 h, and with an assistant the mechanic can do it in 2.10 h. How long would it take the assistant to do the job alone?

129. The *relative density,* also known as specific gravity sg, of an object may be defined as its weight in air w_a, divided by the difference of its weight in air and its weight when submerged in water, w_w. For a lead weight, $w_a = 1.097\, w_w$. Find the specific gravity of lead.

130. A car travels halfway to its destination at 80.0 km/h and the remainder of the distance at 60.0 km/h. What is the average speed of the car for the trip?

131. For electric resistors in parallel, the reciprocal of the combined resistance equals the sum of the reciprocals of the individual resistances. For three resistors of 12.0 Ω, R ohms, and $2R$ ohms, in parallel, the combined resistance is 6.00 Ω. Find R.

132. An ambulance averaged 36.0 km/h going to an accident and 48.0 km/h on its return to the hospital. If the total time for the round-trip was 40.0 min, including 5.00 min at the accident scene, how far from the hospital was the accident?

133. In analysing a certain electric circuit, the expression $\dfrac{\left(1 + \dfrac{1}{s}\right)\left(1 + \dfrac{1}{s/2}\right)}{3 + \dfrac{1}{s} + \dfrac{1}{s/2}}$ is used. Simplify this expression, and write a paragraph describing your procedure. When you "cancel," explain what basic operation is being performed.

CHAPTER 6 PRACTICE TEST

1. Find the product: $2x(2x - 3)^2$.

2. The following equation is used in electricity.

Solve for R_1: $\dfrac{1}{R} = \dfrac{1}{R_1 + r} + \dfrac{1}{R_2}$

3. Reduce to simplest form: $\dfrac{2x^2 + 5x - 3}{2x^2 + 12x + 18}$

4. Factor: $4x^2 - 16y^2$

5. Factor: $pb^3 + 8a^3p$ (business application)

6. Factor: $2a - 4T - ba + 2bT$

7. Factor: $36x^2 + 14x - 16$

In Problems 8–10, perform the indicated operations and simplify.

8. $\dfrac{3}{4x^2} - \dfrac{2}{x^2 - x} - \dfrac{x}{2x - 2}$

9. $\dfrac{x^2 + x}{2 - x} \div \dfrac{x^2}{x^2 - 4x + 4}$

10. $\dfrac{1 - \dfrac{3}{2x + 2}}{\dfrac{x}{5} - \dfrac{1}{2}}$

11. If one riveter can do a job in 12 days, and a second riveter can do it in 16 days, how long would it take for them to do it together?

12. Solve for x: $\dfrac{3}{2x^2 - 3x} + \dfrac{1}{x} = \dfrac{3}{2x - 3}$

7. Quadratic Equations

After completion of this chapter, the student should be able to:

- Identify quadratic functions
- Solve quadratic functions by factoring, by completing the square, and by using the quadratic formula
- Solve engineering and technology problems involving quadratic equations
- Graph a quadratic function and identify important points on the curve

Panther Media GmbH/Alamy Stock Photo

▲ The path followed by a projectile (for example, a jumping snowboarder) is described by a quadratic function. In Section 7.4 we see that the graph of a quadratic function is a parabola.

In this chapter, we develop algebraic and graphical methods for solving an important type of equation called a *quadratic equation*, where variables occur squared. We will also see that the graph of a *quadratic function* is a *parabola*, which has many modern applications in science and technology.

One important application of a parabola is a television satellite dish. Because of its parabolic design, incoming signals are reflected off the surface of the dish and directed to a common point called the *focus*, where the receiver is placed.

The parabolic shape also occurs in the supporting cables of suspension bridges. It can be shown that when a hanging cable is subjected to a uniformly distributed load, such as the deck of a suspension bridge, the cable takes the form of a parabola.

Another important application of quadratic functions is that of projectile motion. A tossed baseball, a fired artillery shell, a jumping snowboarder, and a rocket in motion are all examples of projectiles. When Galileo showed that the time an object falls does not depend on its mass, he discovered that the distance fallen depends on the square of the time of fall (so that distance is a quadratic function of time). Consequently, the path of a projectile is parabolic (not considering air resistance), and finding the location of the projectile at any given time requires the solution of a quadratic equation.

The Babylonians developed methods of solving quadratic equations nearly 4000 years ago. However, their use in applied situations was limited, as they had little real-world relevance. Today, in addition to those mentioned above, quadratic equations have applications in architecture, electric circuits, mechanical systems, forces on structures, and product design.

7.1 Quadratic Equations; Solution by Factoring

Given that a, b, and c are constants $(a \neq 0)$, *the equation*

$$ax^2 + bx + c = 0 \qquad \textbf{(7.1)}$$

is called the **general quadratic equation in x**. Any equation that can be simplified and then written in the form of Eq. (7.1) is a quadratic equation in one unknown.

Observe that the left side of Eq. (7.1) is a polynomial function of degree 2. If we use it to define a function then we obtain what is known as a **quadratic function**. For example, in finding the time t of flight of a projectile, we have the equation $s = s_0 + v_0t + \frac{1}{2}at^2$; in analysing the electric current i in a circuit, the function $f(i) = Ei - Ri^2$ is found; and in determining the forces at a distance x along a beam, the function $f(x) = ax^2 + bLx + cL^2$ is used.

Since it is the term with x^2 in Eq. (7.1) that distinguishes the quadratic equation from other types of equations, the equation is not quadratic if $a = 0$. However, either b or c (or both) may be zero, and the equation is still quadratic. No power of x higher than two may be present in a quadratic equation. Also, we should be able to identify a quadratic equation properly, even when it does not initially appear in the form of Eq. (7.1).

EXAMPLE 1 Examples of quadratic equations

■ If either (or both) the first-power term or the constant is missing, the quadratic is called *incomplete*.

Determine the coefficients of the following quadratic equations.

(a) $x^2 - 4x - 5 = 0$ Rewriting as $1x^2 + (-4)x + (-5) = 0$, we see that $a = 1, b = -4$, and $c = -5$.

(b) $3x^2 - 6 = 0$ Since there is no term with x, $a = 3, b = 0$, and $c = -6$.

(c) $2x^2 + 7x = 0$ Since there is no constant term, $a = 2, b = 7$, and $c = 0$.

(d) $(m - 3)x^2 - mx + 7 = 0$ These coefficients include literal expressions: $a = m - 3, b = -m$, and $c = 7$.

(e) $4x^2 - 2x = x^2$ Collecting terms on the left side, we get $3x^2 - 2x = 0$. Hence $a = 3, b = -2$, and $c = 0$.

(f) $(x + 1)^2 = 4$ We first expand and collect terms to get $x^2 + 2x - 3 = 0$. Therefore $a = 1, b = 2$, and $c = -3$.

EXAMPLE 2 Examples of equations that are not quadratic

Explain why the following are not quadratic equations.

(a) $bx - 6 = 0$ There is no term with x^2.

(b) $x^3 - x^2 - 5 = 0$ There should be no term of degree higher than 2. Thus, there can be no term with x^3 in a quadratic equation.

(c) $x^2 + x - 7 = x^2$ When terms are collected, there will be no term with x^2.

SOLUTIONS OF A QUADRATIC EQUATION

Recall that *the* **solution** *of an equation consists of all numbers* (**roots**) *which, when substituted in the equation, give equality.* There are *two* roots for a quadratic equation. At times, these roots are equal (see Example 3b), so only one number is actually a solution. Also, the roots can contain imaginary numbers. If this happens, all we wish to do at this point is to recognize that there are no real roots.

■ Quadratic equations can be solved directly using a graphing utility (for example, with the *Solver* feature of a graphing calculator).

EXAMPLE 3 Solutions (roots) of a quadratic equation

(a) The quadratic equation $3x^2 - 7x + 2 = 0$ has roots $x = 1/3$ and $x = 2$. This is seen by substituting these numbers in the equation.

$$3\left(\tfrac{1}{3}\right)^2 - 7\left(\tfrac{1}{3}\right) + 2 = 3\left(\tfrac{1}{9}\right) - \tfrac{7}{3} + 2 = \tfrac{1}{3} - \tfrac{7}{3} + 2 = \tfrac{0}{3} = 0$$
$$3(2)^2 - 7(2) + 2 = 3(4) - 14 + 2 = 14 - 14 = 0$$

LEARNING TIP
The **multiplicity** of a root refers to how many times a particular root forms a solution of an equation. For double roots of quadratics, they are said to have a **multiplicity of two.**

(b) The quadratic equation $4x^2 - 4x + 1 = 0$ has a **double root** (*both roots are the same*) of $x = 1/2$. Showing that this number is a solution, we have

$$4\left(\tfrac{1}{2}\right)^2 - 4\left(\tfrac{1}{2}\right) + 1 = 4\left(\tfrac{1}{4}\right) - 2 + 1 = 1 - 2 + 1 = 0$$

(c) The quadratic equation $x^2 + 9 = 0$ has roots $x = 3\sqrt{-1}$ and $x = -3\sqrt{-1}$. These are **imaginary roots** because they involve taking the square root of a negative number. There are **no real roots.**

SOLVING A QUADRATIC EQUATION BY FACTORING

We now describe a method for solving quadratic equations whose quadratic expression is factorable. We use the **zero product rule**, which states that a product is zero if and only if any of its factors is zero.

Solving a Quadratic Equation by Factoring	EXAMPLE 4
	Solve: $x^2 - x - 12 = 0$
1. Collect all terms on the left and **rearrange** into the general form.	$x^2 - x - 12 = 0$
2. Factor the quadratic expression.	$(x - 4)(x + 3) = 0$
3. Set each factor equal **to zero.**	$x - 4 = 0$ or $x + 3 = 0$
4. Solve the two resulting linear equations. The solutions are the roots of the quadratic equation.	$x = 4$ or $x = -3$
5. Check the solutions in the original equation.	$(4)^2 - (4) - 12 = 0 \rightarrow 0 = 0$ $(-3)^2 - (-3) - 12 = 0 \rightarrow 0 = 0$ Both solutions satisfy the original equation.

Practice Exercise

1. Solve for x: $x^2 + 4x - 21 = 0$

LEARNING TIP
It is usually easier to have integer coefficients and $a > 0$ in a quadratic equation. This can be done by multiplying each term by the LCD if there are fractions, and to change the sign of all terms if $a < 0$.

EXAMPLE 5 Solving a quadratic equation by factoring with $a \neq 1$

Solve: $-2N^2 - 7N + 4 = 0$.

Rearrange into general form (change the sign of all terms to get $a > 0$): $2N^2 + 7N - 4 = 0$

Factor the expression: $(2N - 1)(N + 4) = 0$

Set each factor to zero: $2N - 1 = 0 \quad N + 4 = 0$

Solve the two equations: $N = \dfrac{1}{2} \quad N = -4$

Check:
$$-2\left(\tfrac{1}{2}\right)^2 - 7\left(\tfrac{1}{2}\right) + 4 = 0 \rightarrow 0 = 0$$
$$-2(-4)^2 - 7(-4) + 4 = 0 \rightarrow 0 = 0$$

LEARNING TIP

When solving a quadratic equation, it is important to write the equation in the general form as a first step, setting the quadratic equation to be equal to zero. If the product is equal to zero, then setting each factor equal to zero will lead to the roots. If the product is not zero, then it is likely that neither factor will give a correct root.

Practice Exercise

2. Solve for x: $9x^2 + 1 = 6x$

EXAMPLE 6 Rearranging an equation into general quadratic form

Solve: $x^2 + 4 = 4x$.

Rearrange into general form (subtract $4x$ from both sides):	$x^2 - 4x + 4 = 0$
Factor the expression:	$(x - 2)^2 = 0$
Set to zero and solve:	$x - 2 = 0, \quad x = 2$ ($x = 2$ is a double root)
Check:	$2^2 + 4 = 4(2) \rightarrow 8 = 8$

EXAMPLE 7 Quadratic equation with $c = 0$

Solve: $3x^2 - 12x = 0$.

We note a common factor of 3, and because 3 is a constant, we can divide each term by 3. We also note the common factor x, but because we are solving for x, we *cannot* divide each term by x. If we were to divide by x, we would lose one of the two roots. We solve as follows.

Rearrange into general form (divide the equation by the common factor 3):	$x^2 - 4x = 0$
Factor the expression:	$x(x - 4) = 0$
Set each factor to zero and solve:	$x = 0 \quad x - 4 = 0$ $x = 4$
Check:	$3(0)^2 - 12(0) = 0$ and $3(4)^2 - 12(4) = 0 \rightarrow 0 = 0$

A number of equations involving fractions lead to quadratic equations after the fractions are eliminated. The following two examples illustrate the process of solving such equations with fractions.

EXAMPLE 8 Fractional equation solved as a quadratic

Solve for x:

$$\frac{1}{x} + 3 = \frac{2}{x + 2}$$

Note first that $x \neq 0, -2$. To eliminate the fractions, we multiply each term of the original equation by the LCD.

$$\frac{x(x+2)}{x} + 3x(x+2) = \frac{2x(x+2)}{x+2} \quad \text{multiply each term by the LCD, } x(x+2)$$

$$x + 2 + 3x^2 + 6x = 2x \quad \text{reduce each term}$$

$$3x^2 + 5x + 2 = 0 \quad \text{collect terms on left}$$

$$(3x + 2)(x + 1) = 0 \quad \text{factor}$$

$$3x + 2 = 0, \quad x = -\frac{2}{3} \quad \text{set each factor equal to zero and solve}$$

$$x + 1 = 0, \quad x = -1$$

Both of these solutions check when substituted into the original equation.

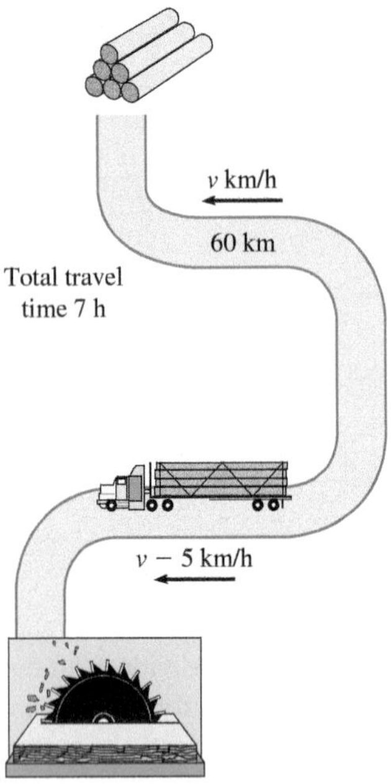

Fig. 7.1

EXAMPLE 9 Fractional equation—finding average speed

A lumber truck travels 60.0 km from a sawmill to a lumber camp and then back in 7.00 h travel time. If the truck averages 5.00 km/h less on the return trip than on the trip to the camp, find its average speed to the camp. See Fig. 7.1.

Identify known quantities: Distance travelled each way: $d = 60.0$ km
Total time of travel: $t_{\text{total}} = 7.00$ h

Identify unknown quantities: Let $v =$ the average speed (in km/h) of the truck going to the camp. Then the average speed during the return trip is $(v - 5.00)$ km/h.

Select an equation: We can relate distance, speed, and time by $d = vt$ (distance equals speed multiplied by time), so that $t = d/v$. Hence

time from camp $= 60.0/v$

time to camp $= 60.0/(v - 5.00)$

total time $= 60.0/v + 60.0/(v - 5.00) = 7.00$

Solve by factoring:

$$60(v - 5) + 60v = 7v(v - 5)$$ multiply each term by $v(v - 5)$

$$7v^2 - 155v + 300 = 0$$ collect terms on the left

$$(7v - 15)(v - 20) = 0$$ factor

$$7v - 15 = 0, \quad v = \frac{15}{7}$$ set each factor equal to zero and solve

$$v - 20 = 0, \quad v = 20$$

The value $v = 15/7$ km/h cannot be a solution since the return speed of 5.00 km/h less would be negative. Thus, the solution is $v = 20.0$ km/h, which means the return speed was 15.0 km/h. The trip to the camp took 3.00 h, and the return took 4.00 h, which shows that the solution checks.

EXERCISES 7.1

In Exercises 1 and 2, make the given changes in the indicated examples of this section and then solve the resulting quadratic equations.

1. In Example 5, change the − sign before $7N$ to + and then solve.
2. In Example 8, change the numerator of the first term to 2 and the numerator of the term on the right to 1 and then solve.

In Exercises 3–8, determine whether or not the given equations are quadratic. If the resulting form is quadratic, identify a, b, and c, with $a > 0$. Otherwise, explain why the resulting form is not quadratic.

3. $x(x - 2) = 4$
4. $(3x - 2)^2 = 2$
5. $x^2 = (x + 2)^2$
6. $x(2x^2 + 5) = 7 + 2x^2$
7. $n(n^2 + n - 1) = n^3$
8. $(T - 7)^2 = (2T + 3)^2$

In Exercises 9–38, solve the given quadratic equations by factoring.

9. $x^2 - 25 = 0$
10. $B^2 - 400 = 0$
11. $4x^2 = 9$
12. $2x^2 = 0.32$
13. $x^2 - 5x - 14 = 0$
14. $x^2 + x - 6 = 0$
15. $R^2 + 12 = 7R$
16. $x^2 + 30 = 11x$
17. $40x - 16x^2 = 0$
18. $15L = 20L^2$
19. $27m^2 = 3$
20. $a^2x^2 = 9$
21. $3x^2 - 13x + 4 = 0$
22. $A^2 + 8A + 16 = 0$
23. $7x^2 + 3x = 4$
24. $4x^2 + 25 = 20x$
25. $6x^2 = 13x - 6$
26. $6z^2 = 6 + 5z$
27. $4x(x + 1) = 3$
28. $9t^2 = 9 - t(43 + t)$
29. $6y^2 + by = 2b^2$
30. $2x^2 - 7ax + 4a^2 = a^2$
31. $8s^2 + 16s = 90$
32. $18t^2 = 48t - 32$
33. $(x + 2)^3 = x^3 + 8$
34. $V(V^2 - 4) = V^2(V - 1)$
35. $(x + a)^2 - b^2 = 0$
36. $2x^2 = 2b^2 - 3xb$
37. $x^2 + 2ax = b^2 - a^2$
38. $x^2(a^2 + 2ab + b^2) = x(a + b)$

In Exercises 39–44, solve the given problems.

39. In Eq. (7.1), for $a = 2$, $b = -7$, and $c = 3$, show that the sum of the roots is $-b/a$.

40. For the equation of Exercise 39, show that the product of the roots is c/a.

41. The voltage V (in V) across a semiconductor in a computer is given by $V = \alpha I + \beta I^2$, where I is the current (in A). If a 6.00-V battery is conducted across the semiconductor, find the current if $\alpha = 2.00\ \Omega$ and $\beta = 0.500\ \Omega/\text{A}$.

42. The mass m (in Mg) of the fuel supply in the first-stage booster of a rocket is $m = 135 - 6t - t^2$, where t is the time (in s) after launch. When does the booster run out of fuel? (Round to three significant digits.)

43. The power P (in MW) produced between midnight and noon by a nuclear power plant is $P = 4h^2 - 48h + 744$, where h is the hour of the day. At what time is the power 664 MW?

44. In determining the speed v (in km/h) of a car while studying its fuel economy, the equation $v^2 - 16v = 3072$ is used. Find v to the nearest km/h.

In Exercises 45 and 46, although the equations are not quadratic, factoring will lead to one quadratic factor and the solution can be completed by factoring as with a quadratic equation. Find the three roots of each equation.

45. $x^3 - x = 0$

46. $x^3 - 4x^2 - x + 4 = 0$

In Exercises 47–50, solve the given equations involving fractions.

47. $\dfrac{1}{x-3} + \dfrac{4}{x} = 2$

48. $2 - \dfrac{1}{x} = \dfrac{3}{x+2}$

49. $\dfrac{x+4}{x-3} = \dfrac{-2x+27}{x^2-3x}$

50. $\dfrac{x}{2} + \dfrac{1}{x-3} = 3$

In Exercises 51–54, set up the appropriate quadratic equations and solve.

51. The spring constant k is the force F divided by the amount x the spring stretches ($k = F/x$). See Fig. 7.2(a). For two springs in series (see Fig. 7.2b), the reciprocal of the spring constant k_c for the combination equals the sum of the reciprocals of the individual spring constants. Find the spring constants for each of two springs in series if $k_c = 2$ N/cm and one spring constant is 3 N/cm more than the other.

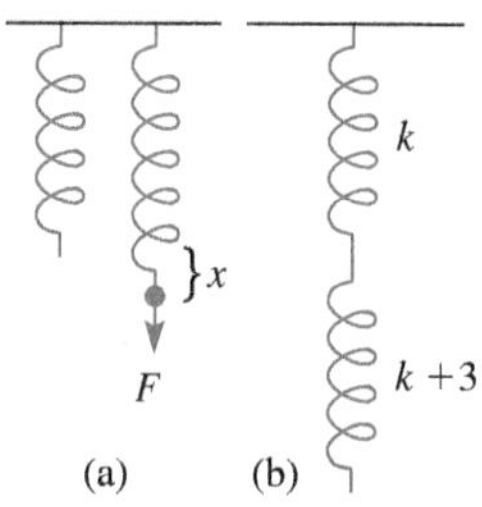

Fig. 7.2

52. The reciprocal of the combined resistance R of two resistances R_1 and R_2 connected in parallel (see Fig. 7.3a) is equal to the sum of the reciprocals of the individual resistances. If the two resistances are connected in series (see Fig. 7.3b), their combined resistance is the sum of their individual resistances. If two resistances connected in parallel have a combined resistance of 3.0 Ω and the same two resistances have a combined resistance of 16 Ω when connected in series, what are the resistances?

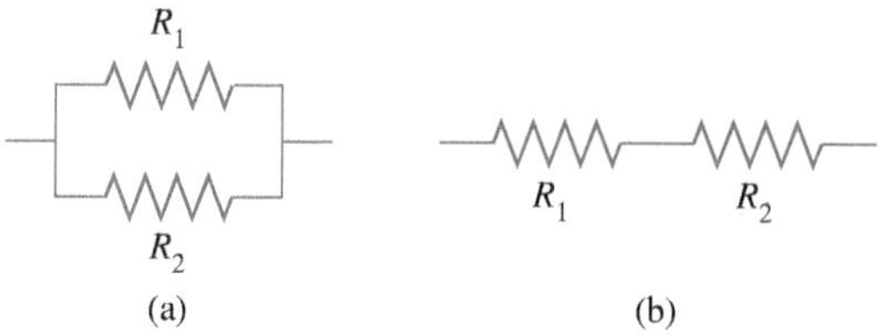

Fig. 7.3

53. A hydrofoil made the round-trip of 120 km between two islands in 3.5 h of travel time. If the average speed going was 10 km/h less than the average speed returning, find these speeds.

54. A rectangular solar panel is 20 cm by 30 cm. By adding the same amount to each dimension, the area is doubled. How much is added?

Answers to Practice Exercises

1. $x = -7, x = 3$ **2.** $x = 1/3, x = 1/3$

7.2 Completing the Square

Most quadratic equations that arise in applications cannot be solved by factoring. Therefore, we now develop a method, called **completing the square**, that can be used to solve any quadratic equation. In the next section, this method is used to develop a general formula that can be used to solve any quadratic equation. Solving an equation by completing the square relies on the following property:

Square Root Property
If $x^2 = k$, then $x = \pm\sqrt{k}$.

This property allows us to solve an equation by taking square roots on both sides and equating the square root of the left side to the ***positive and negative square roots of the right side***, giving us the two roots $x = \sqrt{k}$ and $x = -\sqrt{k}$. The following example illustrates how this property is used.

EXAMPLE 1 Solving a quadratic by taking square roots

Solve the following equations by taking square roots on both sides.

(a) $x^2 = 16$	$x = \pm 4$	Solutions: $x = 4$ and $x = -4$
(b) $(x + 3)^2 = 16$	$x + 3 = \pm 4$	Solutions: $x = 1$ and $x = -7$
(c) $(x - 3)^2 = 17$	$x - 3 = \pm\sqrt{17}$	Solutions: $x = 3 + \sqrt{17}$ and $x = 3 - \sqrt{17}$

In each of the equations of Example 1, the left side was a perfect square of the form $(x + d)^2$ or $(x - d)^2$. To apply the same method of solution to the general equation $ax^2 + bx + c = 0$, we can rewrite the equation so that the left side becomes a perfect square (hence the name *completing the square*). We will use the special products in Eqs. (6.3) and (6.4), rewritten here for convenience:

$$(x + d)^2 = x^2 + 2dx + d^2 \quad (7.2)$$

$$(x - d)^2 = x^2 - 2dx + d^2 \quad (7.3)$$

Note that the coefficient of x^2 in the perfect squares is 1, so the **first step for completing the square is to divide both sides of the equation by a**, the coefficient of x^2. Also, the coefficient of x in the perfect squares is numerically $2d$, and the constant term is d^2. **Thus, if we take half the coefficient of the x-term (this is now $b/2a$) and square it, we have the number that completes the square ($b^2/4a^2$).** As long as we add $b^2/4a^2$ to *both sides* of the equation, we will now be able to write the left-hand side as a perfect square. We summarize these steps below.

LEARNING TIP

When solving the general equation $ax^2 + bx + c = 0$, the number that completes the square is the square of one-half of the coefficient of x *after dividing each side by a*. In symbols, you must find $b/2a$ and then square the result. This means that you must add $(b/2a)^2$ to both sides of the equation.

Steps for Solving by Completing the Square	EXAMPLE 2
	Solve: $x^2 - 6x - 8 = 0$
1. Divide each side by a (the coefficient of x^2).	$x^2 - 6x - 8 = 0 \quad (a = 1)$
2. Rewrite the equation with the constant on the right side.	$x^2 - 6x = 8$
3. Complete the square: add the square of one-half of the coefficient of x to both sides.	$x^2 - 6x + 9 = 8 + 9$ $\left(\frac{-6}{2} = -3, \text{ and } (-3)^2 = 9\right)$
4. Factor the left side as a square and simplify the right side.	$(x - 3)^2 = 17$
5. Take square roots, writing the positive and negative square roots of the right side.	$x - 3 = \pm\sqrt{17}$
6. Solve the two resulting linear equations.	$x = 3 \pm \sqrt{17}$ The solutions are $x = 3 + \sqrt{17}$ and $x = 3 - \sqrt{17}$.

Practice Exercise

1. Solve by completing the square: $x^2 + 4x - 12 = 0$

EXAMPLE 3 Method of completing the square

Solve $2x^2 + 16x - 9 = 0$ by completing the square.

■ The method of completing the square is used later in Section 7.4 and also in Chapters 21, 28, and 31.

Divide each side by a: $x^2 + 8x - \frac{9}{2} = 0 \quad (a = 2)$

Rewrite with constant on the right: $x^2 + 8x = \frac{9}{2}$

Add $(b/2a)^2$ to both sides: $\left(\frac{b}{2a}\right)^2 = \left(\frac{16}{2(2)}\right)^2 = 16$:

$$x^2 + 8x + 16 = \frac{9}{2} + 16$$

Factor and simplify: $(x + 4)^2 = \frac{41}{2}$

Take square roots (write positive and negative roots): $x + 4 = \pm\sqrt{\frac{41}{2}}$

Solve: $x = -4 \pm \sqrt{\frac{41}{2}}$

The solutions are

$$x = -4 + \sqrt{\frac{41}{2}} \text{ and } x = -4 - \sqrt{\frac{41}{2}}.$$

The method of completing the square is useful in situations other than the solutions of quadratic equations. For example, writing an expression as the sum of two squares is an important algebraic step when computing some integrals (Section 28.6) and when finding inverse Laplace transforms (Section 31.11). Example 4 illustrates how the square is completed for such cases.

EXAMPLE 4 Writing an expression as the sum of two squares

Write the expression $x^2 + 4x + 13$ as the sum of two squares.

Here $(b/2a)^2 = 4$, so 4 is the number that completes the square. We rewrite 13 as the sum $13 = 4 + 9$, so that

$$\begin{aligned} x^2 + 4x + 13 &= x^2 + 4x + 4 + 9 \\ &= (x + 2)^2 + 9 \\ &= (x + 2)^2 + 3^2 \end{aligned}$$

EXERCISES 7.2

In Exercises 1 and 2, make the given changes in the indicated examples of this section and then solve the resulting quadratic equations by completing the square.

1. In Example 2, change the − sign before $6x$ to +.
2. In Example 3, change the coefficient of the second term from 16 to 12.

In Exercises 3–10, solve the given quadratic equations by finding appropriate square roots as in Example 1.

3. $x^2 = 25$
4. $x^2 = 100$
5. $x^2 = 7$
6. $x^2 = 15$
7. $(x - 2)^2 = 25$
8. $(x + 2)^2 = 10$
9. $(y + 3)^2 = 7$
10. $\left(x - \frac{5}{2}\right)^2 = 100$

In Exercises 11–30, solve the given quadratic equations by completing the square. Exercises 11–14 and 17–20 may be checked by factoring.

11. $x^2 + 2x - 15 = 0$
12. $x^2 - 8x - 20 = 0$
13. $D^2 + 3D + 2 = 0$
14. $t^2 + 5t - 6 = 0$
15. $n^2 = 4n - 2$
16. $(R + 9)(R + 1) = 13$
17. $v(v + 2) = 15$
18. $Z^2 + 12 = 8Z$
19. $2s^2 + 5s = 3$
20. $4x^2 + x = 3$

21. $3y^2 = 3y + 2$ **22.** $3x^2 = 3 - 4x$

23. $2y^2 - y - 2 = 0$ **24.** $9v^2 - 6v - 2 = 0$

25. $10T - 5T^2 = 4$ **26.** $4V^2 + 9 = 12V$

27. $9x^2 + 6x + 1 = 0$ **28.** $2x^2 - 3x + 2a = 0$

29. $x^2 + 2bx + c = 0$ **30.** $px^2 + qx + r = 0$

In Exercises 31 and 32, use completing the square to write the given expression as the sum of two squares.

31. $x^2 + 6x + 13$ **32.** $x^2 - 8x + 17$

In Exercises 33–36, use completing the square to solve the given problems.

33. The voltage V (in V) across a certain electronic device is related to the temperature T (in °C) by $V = 4.00T - 0.200T^2$. For what temperature(s) is $V = 15.0$ V?

34. A flare is shot vertically into the air such that its distance s (in m) above the ground is given by $s = 20.0t - 5.00t^2$, where t is the time (in s) after it was fired. Find t for $s = 15.0$ m.

35. A surveillance camera is 12.0 m on a direct line from an ATM. The camera is 5.00 m more to the right of the ATM than it is above the ATM. How far above the ATM is the camera?

36. A rectangular storage area is 8.00 m longer than it is wide. If the area is 28.0 m^2, what are its dimensions?

Answer to Practice Exercise

1. $x = -6, x = 2$

7.3 The Quadratic Formula

We now use the method of completing the square to derive a general formula that may be used for the solution of any quadratic equation.

Consider Eq. (7.1), the general quadratic equation:

$$ax^2 + bx + c = 0 \quad (a \neq 0)$$

When we divide through by a, we obtain

$$x^2 + \frac{b}{a}x + \frac{c}{a} = 0$$

Subtracting c/a from each side, we have

$$x^2 + \frac{b}{a}x = -\frac{c}{a}$$

Half of b/a is $b/2a$, which squared is $b^2/4a^2$. Adding $b^2/4a^2$ to each side gives us

$$x^2 + \frac{b}{a}x + \frac{b^2}{4a^2} = -\frac{c}{a} + \frac{b^2}{4a^2}$$

Writing the left side as a perfect square and combining fractions on the right side, we have

$$\left(x + \frac{b}{2a}\right)^2 = \frac{b^2 - 4ac}{4a^2}$$

Equating $x + \dfrac{b}{2a}$ to the principal square root of the right side and its negative,

$$x + \frac{b}{2a} = \frac{\pm\sqrt{b^2 - 4ac}}{2a}$$

■ The quadratic formula, with the $\pm$ sign, means that the solutions to the quadratic equation

$$ax^2 + bx + c = 0$$

are

$$x = \frac{-b + \sqrt{b^2 - 4ac}}{2a}$$

and

$$x = \frac{-b - \sqrt{b^2 - 4ac}}{2a}$$

When we subtract $b/2a$ from each side and simplify the resulting expression, we obtain the **quadratic formula**:

$$x = \frac{-b \pm \sqrt{b^2 - 4ac}}{2a} \qquad (7.4)$$

The quadratic formula gives us a quick general way of solving any quadratic equation. We need only write the equation in the standard form of Eq. (7.1), substitute these numbers into the formula, and simplify.

EXAMPLE 1 Quadratic formula—rational roots

$$\text{Solve: } \underset{a=1}{x^2} - \underset{b=-5}{5x} + \underset{c=6}{6} = 0.$$

Here, using the indicated values of a, b, and c in the quadratic formula, we have

$$x = \frac{-(-5) \pm \sqrt{(-5)^2 - 4(1)(6)}}{2(1)} = \frac{5 \pm \sqrt{25 - 24}}{2} = \frac{5 \pm 1}{2}$$

$$x = \frac{5+1}{2} = 3 \quad \text{or} \quad x = \frac{5-1}{2} = 2$$

■ The fact that the square root in the quadratic formula simplified to a whole number tells us that it is also possible to solve this equation by factoring.

The roots $x = 3$ and $x = 2$ check when substituted in the original equation.

COMMON ERROR It must be emphasized that, in using the quadratic formula, the entire expression $-b \pm \sqrt{b^2 - 4ac}$ is divided by $2a$. *It is a relatively common error to divide only the radical* $\sqrt{b^2 - 4ac}$ by $2a$.

EXAMPLE 2 Quadratic formula—irrational roots

$$\text{Solve: } \underset{a=2}{2x^2} - \underset{b=-7}{7x} - \underset{c=-5}{5} = 0.$$

Substituting the values for a, b, and c in the quadratic formula, we have

$$x = \frac{-(-7) \pm \sqrt{(-7)^2 - 4(2)(-5)}}{2(2)} = \frac{7 \pm \sqrt{49 + 40}}{4} = \frac{7 \pm \sqrt{89}}{4}$$

$$x = \frac{7 + \sqrt{89}}{4} = 4.108 \quad \text{or} \quad x = \frac{7 - \sqrt{89}}{4} = -0.6085$$

The exact roots are $x = \dfrac{7 \pm \sqrt{89}}{4}$ (this form is often used when the roots are irrational). Approximate decimal values are $x = 4.11$ and $x = -0.608$.

Practice Exercise

1. Solve using the quadratic formula: $3x^2 + x - 5 = 0$

EXAMPLE 3 Quadratic formula—double root

Solve: $9x^2 + 24x + 16 = 0$.

In this example, $a = 9$, $b = 24$, and $c = 16$. Thus,

$$x = \frac{-24 \pm \sqrt{24^2 - 4(9)(16)}}{2(9)} = \frac{-24 \pm \sqrt{576 - 576}}{18} = \frac{-24 \pm 0}{18} = -\frac{4}{3}$$

Here, both roots are $-\frac{4}{3}$, and we write the result as $x = -\frac{4}{3}$ and $x = -\frac{4}{3}$. We will get a double root when $b^2 = 4ac$, as in this case.

EXAMPLE 4 Quadratic formula—no real roots

Solve: $3x^2 - 5x + 4 = 0$.

In this example, $a = 3$, $b = -5$, and $c = 4$. Therefore,

$$x = \frac{-(-5) \pm \sqrt{(-5)^2 - 4(3)(4)}}{2(3)} = \frac{5 \pm \sqrt{25 - 48}}{6} = \frac{5 \pm \sqrt{-23}}{6}$$

These roots contain imaginary numbers. This happens if $b^2 < 4ac$.

Examples 1–4 illustrate the *character of the roots* of a quadratic equation. If a, b, and c are rational numbers (see Section 1.1), by noting the value of $b^2 - 4ac$ (called the **discriminant**), we have the following:

Character of the Roots of a Quadratic Equation		EXAMPLE 5
Discriminant $b^2 - 4ac$	**Character of the Roots**	
Positive and a perfect square	Real, rational, and unequal	**(a)** $x^2 - 5x + 6 = 0$ $(b^2 - 4ac = 1$, as in Example 1)
Positive but not a perfect square	Real, irrational, and unequal	**(b)** $2x^2 - 7x - 5 = 0$ $(b^2 - 4ac = 89$, as in Example 2)
Zero	Real, rational, and equal	**(c)** $9x^2 + 24x + 16 = 0$ $(b^2 - 4ac = 0$, as in Example 3)
Negative	Complex (containing imaginary numbers) and unequal. No real roots.	**(d)** $3x^2 - 5x + 4 = 0$ $(b^2 - 4ac = -23$, as in Example 4)

LEARNING TIP

We can use the value of $b^2 - 4ac$ to help in checking the roots or in finding the character of the roots without having to solve the equation completely.

EXAMPLE 6 Quadratic formula—literal numbers

The equation $s = s_0 + v_0t - \frac{1}{2}gt^2$ is used in the analysis of projectile motion (see Fig. 7.4). Solve for t.

$$gt^2 - 2v_0t - 2(s_0 - s) = 0 \quad \text{multiply by } -2\text{, put in general form}$$

In this form, we see that $a = g$, $b = -2v_0$, and $c = -2(s_0 - s)$:

$$t = \frac{-(-2v_0) \pm \sqrt{(-2v_0)^2 - 4g(-2)(s_0 - s)}}{2g} = \frac{2v_0 \pm \sqrt{4(v_0^2 + 2gs_0 - 2gs)}}{2g}$$

$$t = \frac{2v_0 \pm 2\sqrt{v_0^2 + 2gs_0 - 2gs}}{2g} = \frac{v_0 \pm \sqrt{v_0^2 + 2gs_0 - 2gs}}{g}$$

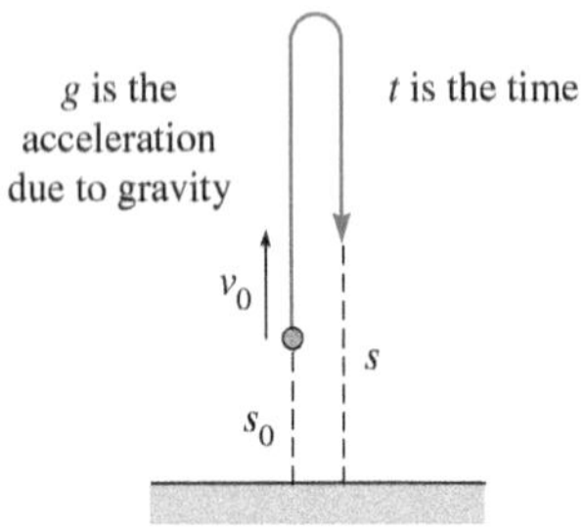

Fig. 7.4

EXAMPLE 7 Quadratic formula—designing a swimming pool

A rectangular area 17.0 m long and 12.0 m wide is to be used for a patio with a rectangular pool. One end and one side of the patio area around the pool are to be the same width. The other end with the diving board is to be twice as wide, and the other side is to be three times as wide as the narrow side. The pool area is to be 96.5 m^2. What are the widths of the patio ends and sides, and the dimensions of the pool?

Identify known quantities: Length = 17.0 m; width = 12.0 m
Total area of the pool $A = 96.5$ m^2

Identify unknown quantities: Let x = width of the narrow end and side of the patio. Then $2x$ is the width of the diving board end, and $3x$ is the width of the other side, so that the length and width of the pool are $17.0 - 3x$, and $12.0 - 2x$, respectively.

Sketch: See Fig. 7.5.

Select an equation: The area of the pool is A = length × width, so
$(17.0 - 3x)(12.0 - 4x) = 96.5$

Solve with the quadratic formula:

$$204 - 68.0x - 36.0x + 12x^2 = 96.5$$

$$12x^2 - 104.0x + 107.5 = 0$$

$$x = \frac{-(-104.0) \pm \sqrt{(-104.0)^2 - 4(12)(107.5)}}{2(12)} = \frac{104.0 \pm \sqrt{5656}}{24}$$

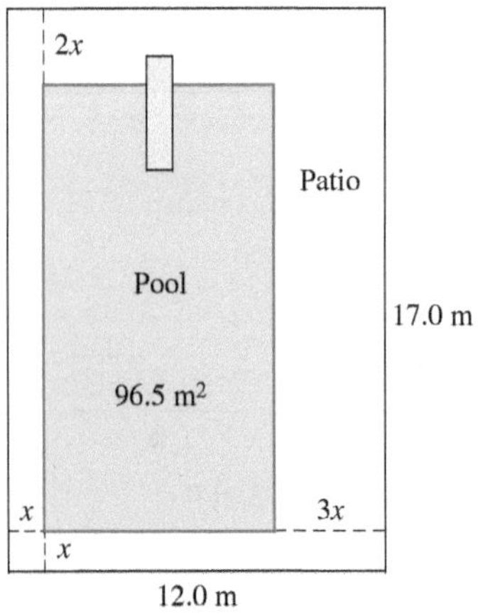

Fig. 7.5

Evaluating, we get $x = 7.47$ m and $x = 1.20$ m. The value 7.47 m cannot be the required result since the width of the patio would be greater than the width of the entire area. For $x = 1.20$ m, the pool would have a length 13.4 m and width 7.20 m. These give an area of 96.5 m^2, which checks. The widths of the patio area are then 1.20 m, 1.20 m, 2.40 m, and 3.60 m.

EXERCISES 7.3

In Exercises 1–2, make the given changes in the indicated examples of this section and then solve the resulting equations by the quadratic formula.

1. In Example 1, change the − sign before $5x$ to +.

2. In Example 2, change the coefficient of x^2 from 2 to 3.

In Exercises 3–34, solve the given quadratic equations, using the quadratic formula. Exercises 3–6 are the same as Exercises 11–14 of Section 7.2.

3. $x^2 + 2x - 15 = 0$ **4.** $x^2 - 8x - 20 = 0$

5. $D^2 + 3D + 2 = 0$ **6.** $t^2 + 5t - 6 = 0$

7. $x^2 - 4x + 2 = 0$ **8.** $x^2 + 10x - 4 = 0$

9. $v^2 = 15 - 2v$ **10.** $16V - 24 = 2V^2$

11. $2s^2 + 5s = 3$ **12.** $4x^2 + x = 3$ **13.** $3y^2 = 3y + 2$

14. $3x^2 = 3 - 4x$ **15.** $z + 2 = 2z^2$ **16.** $2 + 6v = 9v^2$

17. $30y^2 + 23y - 40 = 0$ **18.** $40x^2 - 62x - 63 = 0$

19. $5t^2 + 3 = 7t$ **20.** $2d(d - 2) = -7$

21. $s^2 = 9 + s(1 - 2s)$ **22.** $20r^2 = 20r + 1$

23. $25y^2 = 121$ **24.** $37T = T^2$

25. $15 + 4z = 32z^2$ **26.** $4x^2 - 12x = 7$

27. $x^2 - 0.200x - 0.400 = 0$ **28.** $3.20x^2 = 2.50x + 7.60$

29. $0.290Z^2 - 0.180 = 0.630Z$ **30.** $12.5x^2 + 13.2x = 15.5$

31. $x^2 + 2cx - 1 = 0$ **32.** $x^2 - 7x + (6 + a) = 0$

33. $b^2x^2 + 1 - a = (b + 1)x$ **34.** $c^2x^2 - x - 1 = x^2$

In Exercises 35–38, without solving the given equations, determine the character of the roots.

35. $2x^2 - 7x = -8$ **36.** $3x^2 + 19x = 14$

37. $3.6t^2 + 2.1 = 7.7t$ **38.** $0.45s^2 + 0.33 = 0.12s$

In Exercises 39–58, solve the given problems. All numbers are accurate to at least three significant digits.

39. Find k if the equation $x^2 + 4x + k = 0$ has a real double root.

40. Find the smallest positive integral value of k if the equation $x^2 + 3x + k = 0$ has roots with imaginary numbers.

41. Solve the equation $x^4 - 5x^2 + 4 = 0$ for x. (*Hint:* The equation can be written as $(x^2)^2 - 5(x^2) + 4 = 0$. First solve for x^2.)

42. Without drawing the graph or completely solving the equation, explain how to find the number of x-intercepts of a quadratic function.

43. In machine design, in finding the outside diameter D_0 of a hollow shaft, the equation $D_0^2 - DD_0 - 0.250D^2 = 0$ is used. Solve for D_0 if $D = 3.625$ cm.

44. A missile is fired vertically into the air. The distance s (in m) above the ground as a function of time t (in s) is given by $s = 100 + 500t - 4.9t^2$. (a) When will the missile hit the ground? (b) When will the missile be 1000 m above the ground?

45. In analysing the deflection of a certain beam, the equation $8x^2 - 15Lx + 6L^2 = 0$ is used. Solve for x, if $x < L$.

46. A homeowner wants to build a rectangular patio with an area of 20.0 m², such that the length is 2.0 m more than the width. What should the dimensions be?

47. For a rectangle, if the ratio of the length to the width equals the ratio of the length plus the width to the length, the ratio is called the *golden ratio*. Find the value of the golden ratio, which the ancient Greeks thought had the most pleasing properties to look at.

48. When focusing a camera, the distance r the lens must move from the infinity setting is given by $r = f^2/(p - f)$, where p is the distance from the object to the lens, and f is the focal length of the lens. Solve for f.

49. In calculating the current in an electric circuit with an inductance L, a resistance R, and a capacitance C, it is necessary to solve the equation $Lm^2 + Rm + 1/C = 0$. Solve for m in terms of L, R, and C. See Fig. 7.6.

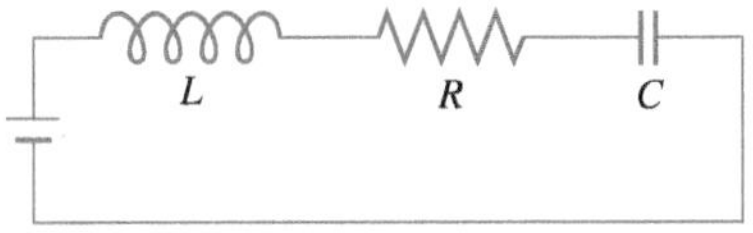

Fig. 7.6

50. In finding the radius r of a circular arch of height h and span b, an architect used the following formula. Solve for h.

$$r = \frac{b^2 + 4h^2}{8h}$$

51. A computer monitor has a viewing screen that is 37.0 cm wide and 31.0 cm high, with a uniform edge around it. If the edge covers 20.0% of the monitor front, what is the width of the edge?

52. An investment of \$2000 is deposited at a certain annual interest rate. One year later, \$3000 is deposited in another account at the same rate. At the end of the second year, the accounts have a total value of \$5319.05. What is the interest rate?

53. The length of a tennis court is 12.8 m more than its width. If the area of the tennis court is 262 m², what are its dimensions? See Fig. 7.7.

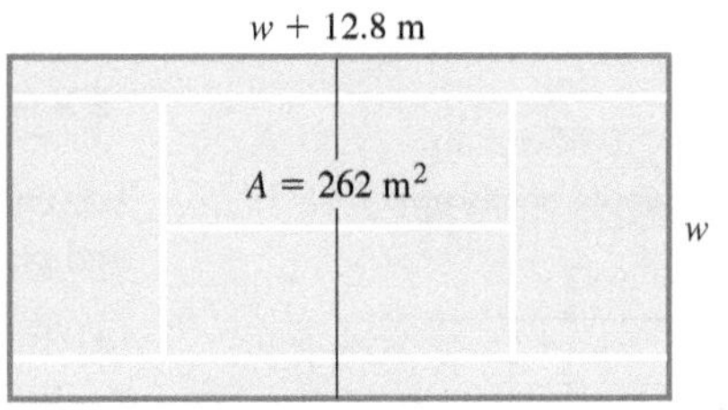

Fig. 7.7

54. Two circular oil spills are tangent to each other. If the distance between centres is 800 m and they cover a combined area of 1.02×10^6 m², what is the radius of each? See Fig. 7.8.

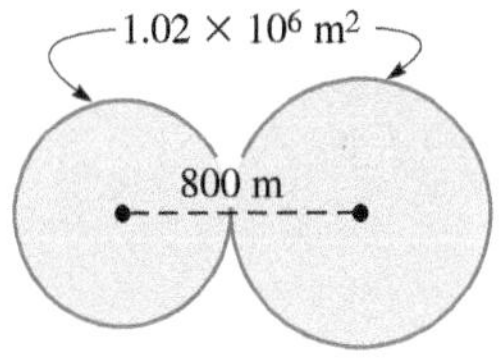

Fig. 7.8

55. In remodelling a house, an architect finds that by adding the same amount to each dimension of a 3.80-m by 5.00-m rectangular room, the area would be increased by 11.0 m². How much must be added to each dimension?

56. Two pipes together drain a wastewater-holding tank in 6.00 h. If used alone to empty the tank, one takes 2.00 h longer than the other. How long does each take to empty the tank if used alone?

57. In order to have the proper strength, the angle iron shown in Fig. 7.9 must have a cross-sectional area of $53.5\,\text{cm}^2$. Find the required thickness x.

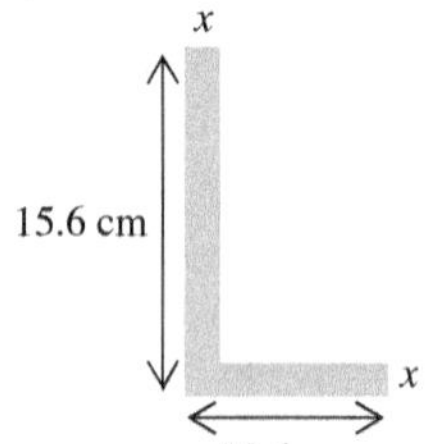

Fig. 7.9

58. The hydronium ion concentration x (in mol/L) in a vinegar solution of concentration C_i mol/L satisfies the equation $1.8 \times 10^{-5} = x^2/(C_i - x)$. Find x when $C_i = 0.30$ mol/L.

Answer to Practice Exercise

1. $x = \dfrac{-1 \pm \sqrt{61}}{6}$

7.4 The Graph of the Quadratic Function

In this section, we discuss the graph of the quadratic function $ax^2 + bx + c$ and show the graphical solution of a quadratic equation. By letting $y = ax^2 + bx + c$, we can graph this function, as in Chapter 3.

EXAMPLE 1 Graphing a quadratic function

Graph the function $f(x) = x^2 + 2x - 3$.

First, let $y = x^2 + 2x - 3$. We can now set up a table of values and graph the function as shown in Fig. 7.10.

x	y
−4	5
−3	0
−2	−3
−1	−4
0	−3
1	0
2	5

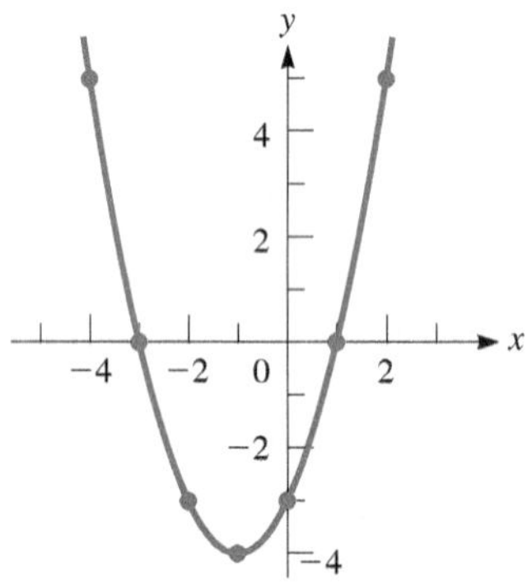

Fig. 7.10

The shape of the graph in Fig. 7.10 is called a **parabola**, and *the graph of any quadratic function $y = ax^2 + bx + c$ will have the same basic shape.* A parabola can open upward (as in Example 1) or downward. The location of the parabola and how it opens depends on the values of a, b, and c.

In Example 1, the parabola has a minimum point at $(-1, -4)$, and the curve opens upward. *All parabolas have an* **extreme point** *of this type known as* the **vertex**.

For $y = ax^2 + bx + c = 0$, the sign of a determines whether the vertex is a minimum or a maximum, as shown in Fig. 7.11. Note also that a parabola is *symmetric* with respect to a vertical line through the vertex.

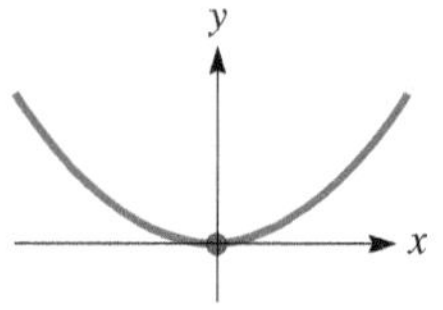

(a)
$a > 0$ Minimum
(parabola opens up)

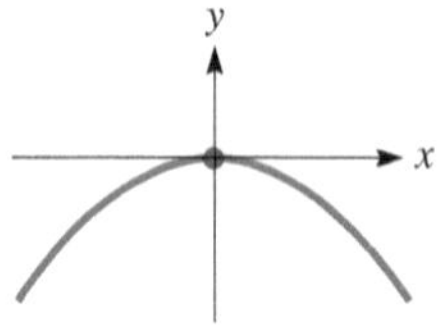

(b)
$a < 0$ Maximum
(parabola opens down)

Fig. 7.11

EXAMPLE 2 Parabola—extreme points

For each of the following quadratic functions, consider the sign of the quadratic term and classify the vertex as a minimum or a maximum.

(a) $y = 2x^2 - 8x + 6$ **(b)** $y = -2x^2 + 8x - 6$

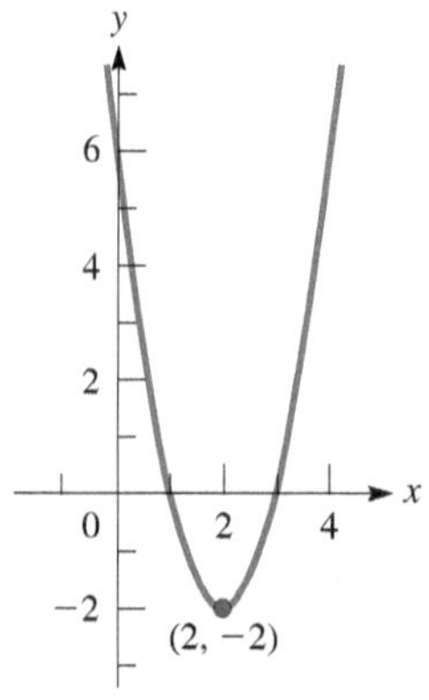

Fig. 7.12(a)
$a = 2$, so $a > 0$, the vertex is a minimum, and the parabola opens up.

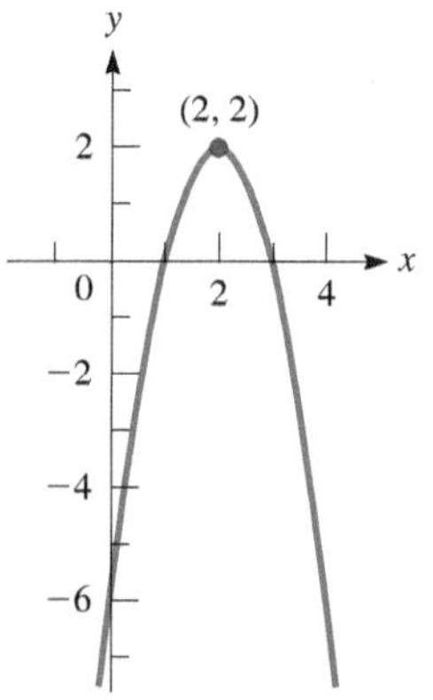

Fig. 7.12(b)
$a = -2$, so $a < 0$, the vertex is a maximum, and the parabola opens down.

We can sketch a graph of a parabola by using its basic shape and symmetry, and knowing two or three points, including the vertex.

In order to find the coordinates of the extreme point, start with the quadratic function

$$y = ax^2 + bx + c$$

Then factor a from the two terms containing x, obtaining

$$y = a\left(x^2 + \frac{b}{a}x\right) + c$$

Now, completing the square of the terms within parentheses, we have

$$y = a\left(x^2 + \frac{b}{a}x + \frac{b^2}{4a^2}\right) + c - \frac{b^2}{4a}$$

$$y = a\left(x + \frac{b}{2a}\right)^2 + c - \frac{b^2}{4a}$$

This form of the function shows a *horizontal shift of* $-b/2a$, which means that the x-coordinate of the vertex is $-b/2a$. *The y-coordinate is found by substituting this x-value into the original function.*

Another easily found point is the y-intercept. By substituting $x = 0$ into $y = ax^2 + bx + c$, we get $y = c$. *This means that the point* $(0, c)$ *is the y-intercept.* In summary:

Vertex and y-Intercept of the Quadratic Function $y = f(x) = ax^2 + bx + c$

Vertex: $\left(-\frac{b}{2a}, f\left(-\frac{b}{2a}\right)\right)$ **(7.5)**

y-Intercept: $(0, c)$

EXAMPLE 3 Graphing a parabola—vertex and y-intercept

For the graph of the function $y = 2x^2 - 8x + 6$, find the vertex and y-intercept and sketch the graph. (This function is the same as Example 2(a).)

We identify $a = 2$, $b = -8$, and $c = 6$. Therefore,

x-coordinate of the vertex: $-\frac{b}{2a} = \frac{-(-8)}{2(2)} = \frac{8}{4} = 2$

y-coordinate of the vertex: $y = f(2) = 2(2^2) - 8(2) + 6 = -2$

Type of extreme point: $a > 0$, so $(2, -2)$ is a minimum.

y-intercept: $(0, c) = (0, 6)$

Plotting the minimum point $(2, -2)$ and the y-intercept $(0, 6)$, and using the fact that the graph is a parabola, symmetric to a vertical line through the vertex, we sketch the graph in Fig. 7.13. It is the same graph as that shown in Fig. 7.12(a).

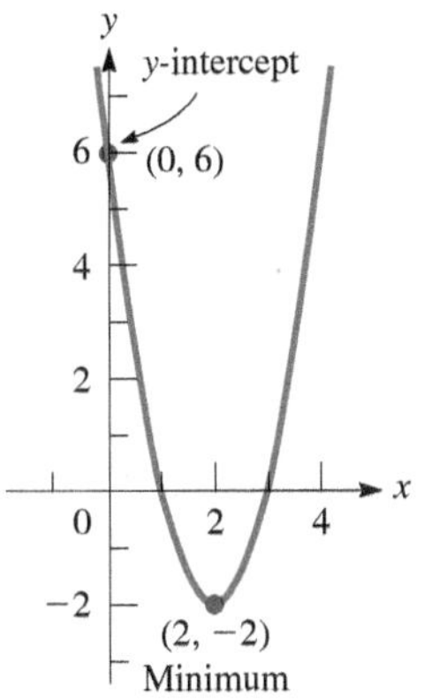

Fig. 7.13

If we are not using a graphing utility, we may need one or two additional points to get a good sketch of a parabola. This would be true if the y-intercept is close to the vertex. Two points we can find are the x-intercepts, if the parabola crosses the x-axis (one point if the vertex is on the x-axis). They are found by setting $y = 0$ and solving the quadratic equation $ax^2 + bx + c = 0$. Also, we may simply find one or two points other than the vertex and the y-intercept. Sketching a parabola in this way is shown in the following two examples.

EXAMPLE 4 Graphing a parabola—using the vertex and intercepts

Sketch the graph of $y = -x^2 + x + 6$.

We first note that $a = -1$ and $b = 1$. Therefore, the x-coordinate of the maximum point $(a < 0)$ is $-\frac{1}{2(-1)} = \frac{1}{2}$. The y-coordinate is $-\left(\frac{1}{2}\right)^2 + \frac{1}{2} + 6 = \frac{25}{4}$. This means that the maximum point is $\left(\frac{1}{2}, \frac{25}{4}\right)$. Also, $c = 6$, so the y-intercept is $(0, 6)$.

We can see from Fig. 7.14 that the vertex and y-intercept are too close together for us to get a good idea of the shape of the parabola, so additional points are needed. We will find the x-intercepts.

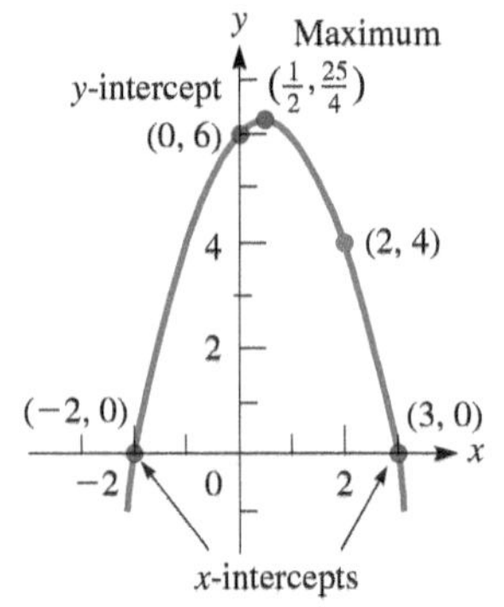

Fig. 7.14

$$-x^2 + x + 6 = 0 \qquad \text{set } y = 0$$
$$x^2 - x - 6 = 0 \qquad \text{multiply both sides by } -1$$
$$(x - 3)(x + 2) = 0 \qquad \text{factor}$$
$$x - 3 = 0 \qquad x + 2 = 0 \qquad \text{set each factor equal to 0}$$
$$x = 3 \qquad x = -2 \qquad \text{solve}$$

This means that the x-intercepts are $(3, 0)$ and $(-2, 0)$, as shown in Fig. 7.14.

Also, rather than finding the x-intercepts, we can select an arbitrary point $(x, f(x))$. In Fig. 7.14 we have plotted the point $(2, 4)$.

Note that the y-coordinate of the vertex can be used to find the range of a quadratic function. In this case, the range is all values $y \leq 25/4$. The domain is all real numbers.

Practice Exercise

1. **(a)** Find the vertex and y-intercept for the graph of the function $y = 3x^2 + 12x - 4$.
   (b) Is the vertex a minimum or a maximum point?

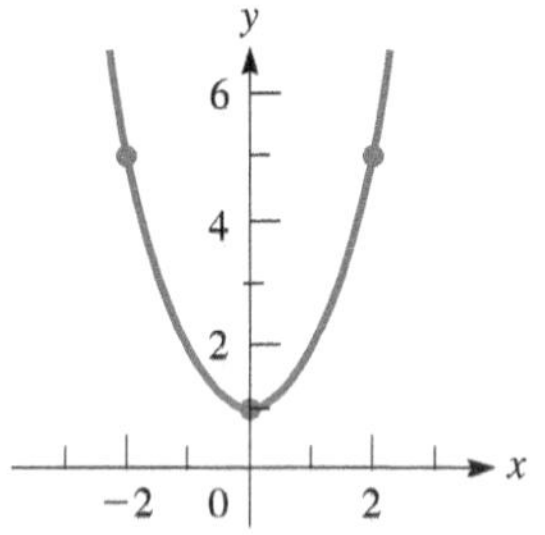

Fig. 7.15

EXAMPLE 5 Graphing a parabola—no x-term

Sketch the graph of $y = x^2 + 1$.

Since there is no x-term, $b = 0$. This means that the x-coordinate of the minimum point $(a > 0)$ is 0 and that the minimum point and the y-intercept are both $(0, 1)$. We know that the graph opens upward, since $a > 0$, which in turn means that it does not cross the x-axis. Now, letting $x = 2$ and $x = -2$, we find the points $(2, 5)$ and $(-2, 5)$ on the graph, which is shown in Fig. 7.15.

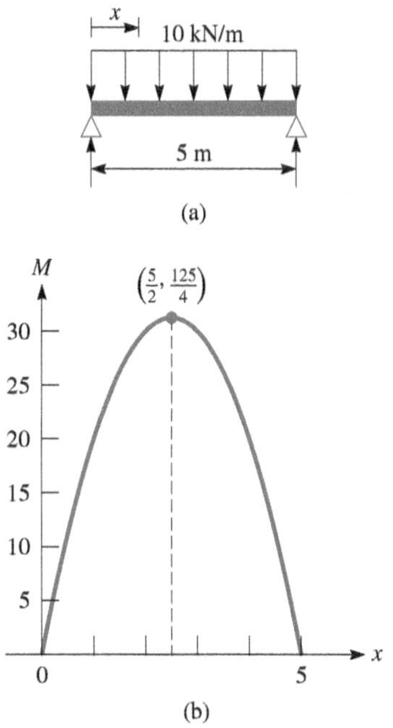

Fig. 7.16

EXAMPLE 6 Graphing a parabola—bending moment along a beam

The bending moment M (in kN · m) at any point along a beam is the algebraic sum of the moments of all the forces acting on the portion of the beam to the right or to the left of that point. Consider a simply supported 5 m beam that supports a uniformly distributed load of 10 kN/m, as shown in Fig. 7.16(a). For this beam, the bending moment x metres from one end of the beam is given by $M = 25x - 5x^2$, where $0 \leq x \leq 5$. Graph this function and determine where the bending moment is zero and where the maximum bending moment occurs.

Since $c = 0$, the M-intercept is at the origin. We also have that $a = -5$ and $b = 25$. Therefore, the parabola opens downward and the x-coordinate of the maximum point is $-\dfrac{b}{2a} = -\dfrac{25}{2(-5)} = \dfrac{5}{2}$. Substituting into the function, the M-coordinate of the maximum is $25\left(\dfrac{5}{2}\right) - 5\left(\dfrac{5}{2}\right)^2 = \dfrac{125}{4}$.

Because of symmetry, this is enough information to obtain the graph, as shown in Fig. 7.16(b). From this graph we see that the bending moment is zero for $x = 0$ and $x = 5$ (at both ends of the beam), and that the maximum bending moment occurs at the centre of the beam.

SOLVING QUADRATIC EQUATIONS GRAPHICALLY

In Chapter 3, we showed that to solve an equation graphically, we (1) collect all terms on the left side of the equation, with zero on the right side, (2) graph the function on the left side, and (3) determine the x-coordinate(s) of the x-intercepts, which will be the roots of the equation. This procedure works for *all* equations, including quadratic equations.

EXAMPLE 7 Solving an equation graphically—projectile motion

A projectile is fired vertically upward from the ground with a velocity of 38 m/s. Its distance above the ground is given by $s = -4.9t^2 + 38t$, where s is the distance (in m) and t is the time (in s). Graph the function and from the graph determine (1) when the projectile hits the ground and (2) how long it takes to reach 45 m above the ground.

Since $c = 0$, the s-intercept is at the origin. Using Eq. (7.5), we find that the vertex is at (3.9, 74), which will help us select display settings if using a graphing utility. The graph can be seen in Fig. 7.17.

From the graph we see that: (1) the projectile hits the ground at about 7.8 s (when $s = 0$ m at the right t-intercept); and (2) there are two times when the projectile reaches 45 m above the ground (or $s = 45$ m): $t = 1.5$ s and $t = 6.3$ s.

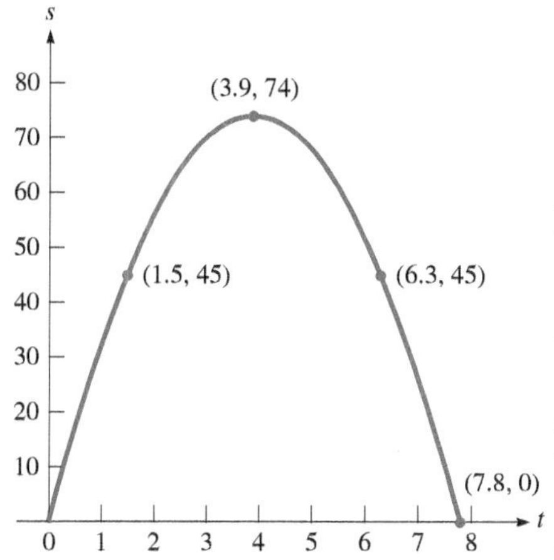

Fig. 7.17

EXERCISES 7.4

In Exercises 1 and 2, make the given changes in the indicated examples of this section and then solve the resulting problems.

1. In Example 3, change the − sign before $8x$ to + and then sketch the graph.
2. In Example 5, change the sign of both terms on the right to − and then sketch the graph.

In Exercises 3–8, sketch the graph of each parabola by using only the vertex and the y-intercept.

3. $y = x^2 - 6x + 5$
4. $y = -x^2 - 4x - 3$
5. $y = -3x^2 + 10x - 4$
6. $s = 2t^2 + 8t - 5$
7. $R = v^2 - 4v$
8. $y = -2x^2 - 5x$

In Exercises 9–12, sketch the graph of each parabola by using the vertex, the y-intercept, and the x-intercepts.

9. $y = x^2 - 4$
10. $y = x^2 + 3x$
11. $y = -2x^2 - 6x + 8$
12. $u = -3v^2 + 12v - 9$

In Exercises 13–16, sketch the graph of each parabola by using the vertex, the y-intercept, and two other points, not including the x-intercepts.

13. $y = 2x^2 + 3$
14. $s = t^2 + 2t + 2$
15. $y = -2x^2 - 2x - 6$
16. $y = -3x^2 - x$

In Exercises 17–24, solve the given equations graphically. If there are no real roots, state this as the answer.

17. $2x^2 - 3 = 0$
18. $5 - x^2 = 0$
19. $-3x^2 + 11x - 5 = 0$
20. $2t^2 = 7t + 4$
21. $x(2x - 1) = -3$
22. $2w - 5 = w^2$
23. $6R^2 = 18 - 7R$
24. $3x^2 - 25 = 20x$

In Exercises 25–30, graph all three parabolas on the same coordinate system. In Exercises 25–28, describe (a) the shifts of $y = x^2$ that occur and (b) how each parabola opens. In Exercises 29 and 30, describe (a) the shifts and (b) how the shape of each parabola changes.

25. **(a)** $y = x^2$ **(b)** $y = x^2 + 3$ **(c)** $y = x^2 - 3$
26. **(a)** $y = x^2$ **(b)** $y = (x - 3)^2$ **(c)** $y = (x + 3)^2$
27. **(a)** $y = x^2$ **(b)** $y = (x - 2)^2 + 3$ **(c)** $y = (x + 2)^2 - 3$
28. **(a)** $y = x^2$ **(b)** $y = -x^2$ **(c)** $y = -(x - 2)^2$
29. **(a)** $y = x^2$ **(b)** $y = 3x^2$ **(c)** $y = \frac{1}{3}x^2$
30. **(a)** $y = x^2$ **(b)** $y = -3(x - 2)^2$ **(c)** $y = \frac{1}{3}(x + 2)^2$

In Exercises 31–46, solve the given applied problem.

31. A quadratic equation $f(x) = 0$ has a solution $x = -1$. Its graph has its vertex at (3, 4). What is the other solution?
32. Find c such that $y = x^2 - 12x + c$ has exactly one real root.
33. Find the smallest integral value of c such that $y = 2x^2 - 4x - c$ has at least one real root.
34. Find the smallest integral value of c such that $y = 3x^2 - 12x + c$ has no real roots.
35. The vertical distance d (in cm) of the end of a robot arm above a conveyor belt in its 8-s cycle is given by $d = 2t^2 - 16t + 47$. Sketch the graph of $d = f(t)$.
36. When mineral deposits form a uniform coating 1 mm thick on the inside of a pipe of radius r (in mm), the cross-sectional area A through which water can flow is $A = \pi(r^2 - 2r + 1)$. Sketch $A = f(r)$.
37. An equipment company determines that the area A (in m^2) covered by a rectangular tarpaulin is given by $A = w(8 - w)$, where w is the width of the tarpaulin and its perimeter is 16 m. Sketch the graph of A as a function of w.
38. Under specified conditions, the pressure loss L (in kPa per 100 m), in the flow of water through a water line in which the flow is Q L/min, is given by $L = 0.0001Q^2 + 0.005Q$. Sketch the graph of L as a function of Q, for $Q < 400$ L/min.

39. A computer analysis of the power P (in W) used by a pressing machine shows that $P = 50i - 3i^2$, where i is the current (in A). Sketch the graph of $P = f(i)$.

40. Tests show that the power P (in kW) of an automobile engine as a function of r (in r/min) is given by $P = -5.0 \times 10^{-6}r^2 + 0.050r - 45(1500 < r < 6000\ \text{r/min})$. Sketch the graph of P vs. r and find the maximum power that is produced.

41. The height h (in m) of a fireworks shell shot vertically upward as a function of time t (in s) is $h = -4.9t^2 + 68t + 2$. How long should the fuse last so that the shell explodes at the top of its trajectory?

42. In a certain electric circuit, the resistance R (in Ω) that gives resonance is found by solving the equation $25R = 3(R^2 + 4)$. Solve this equation graphically (to 0.1 Ω).

43. If the radius of a circular solar cell is increased by 1.00 cm, its area is 96.0 cm^2. What was the original radius?

44. The diagonal of a rectangular floor is 1.00 m less than twice the length of one of the sides. If the other side is 5.00 m long, what is the area of the floor?

45. A security fence is to be built around a rectangular parking area of 2000 m^2. If the front side of the fence costs \$60/m and the other three sides cost \$30/m, solve graphically for the dimensions (to 0.1 m) of the parking area if the fence is to cost \$7500. See Fig. 7.18.

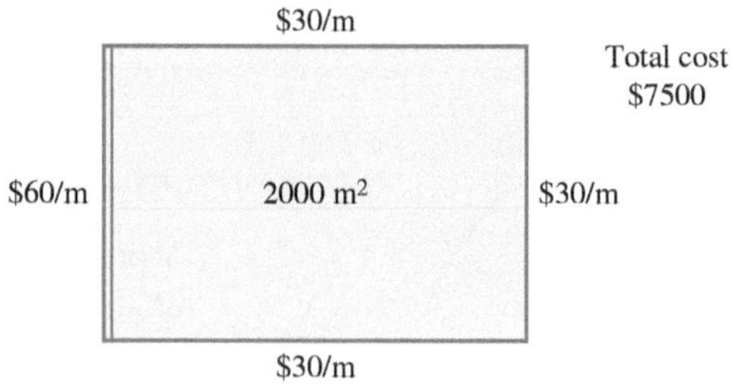

Fig. 7.18

46. An airplane pilot could decrease the time t (in h) needed to travel the 5400 km from Ottawa to London by 60 min if the plane's speed v is increased by 40 km/h. Set up the appropriate equation and solve graphically for v (to two significant digits).

Answers to Practice Exercise

1. **(a)** $(-2, -16)$, $(0, -4)$ **(b)** minimum

CHAPTER 7 KEY FORMULAS AND EQUATIONS

Quadratic equation	$ax^2 + bx + c = 0$	**(7.1)**
Quadratic formula	$x = \dfrac{-b \pm \sqrt{b^2 - 4ac}}{2a}$	**(7.4)**
Vertex coordinates	$\left(-\dfrac{b}{2a}, f\left(-\dfrac{b}{2a}\right)\right)$	**(7.5)**

CHAPTER 7 REVIEW EXERCISES

In Exercises 1–12, solve the given quadratic equations by factoring.

1. $x^2 + 3x - 4 = 0$ **2.** $x^2 + 3x - 10 = 0$

3. $x^2 - 10x + 21 = 0$ **4.** $P^2 - 27 = 6P$

5. $3x^2 + 11x = 4$ **6.** $11y = 6y^2 + 3$

7. $6t^2 = 13t - 5$ **8.** $3x^2 + 5x + 2 = 0$

9. $4s^2 = 18s$ **10.** $23n + 35 = 6n^2$

11. $4B^2 = 8B + 21$ **12.** $6\pi^2x^2 = 8 - 47\pi x$

In Exercises 13–24, solve the given quadratic equations by using the quadratic formula.

13. $x^2 - x - 110 = 0$ **14.** $x^2 + 3x - 18 = 0$

15. $m^2 + 2m = 6$ **16.** $1 + 7D = D^2$

17. $2x^2 - x = 36$ **18.** $3x^2 = 14 - x$

19. $18s + 12 = 24s^2$ **20.** $2 - 7x = 5x^2$

21. $2.1x^2 + 2.3x + 5.5 = 0$ **22.** $0.30R^2 - 0.42R = 0.15$

23. $6x^2 = 9 - 4x$ **24.** $25t = 24t^2 - 20$

In Exercises 25–36, solve the given quadratic equations by any appropriate algebraic method.

25. $x^2 + 4x - 4 = 0$ **26.** $x^2 + 3x + 1 = 0$

27. $3x^2 + 8x + 2 = 0$ **28.** $3p^2 = 28 - 5p$

29. $4v^2 = v + 5$ **30.** $n - 2 = 6n^2$

31. $7 + 3C = -2C^2$ **32.** $4y^2 - 5y = 8$

33. $a^2x^2 + 2ax + 2 = 0$ **34.** $16r^2 = 8r - 1$

35. $ay^2 = a - 3y$ **36.** $2bx = x^2 - 3b$

In Exercises 37–40, solve the given quadratic equations by completing the square.

37. $x^2 - x - 30 = 0$ **38.** $x^2 = 2x + 5$

39. $2t^2 = t + 4$ **40.** $4x^2 - 8x = 3$

In Exercises 41–44, solve the given equations.

41. $\frac{x-4}{x-1} = \frac{2}{x}$

42. $\frac{V-1}{3} = \frac{5}{V} + 1$

43. $\frac{x^2-3x}{x-3} = \frac{x^2}{x+2}$

44. $\frac{x-2}{x-5} = \frac{15}{x^2-5x}$

In Exercises 45–48, sketch the graphs of the given functions by using the vertex, the y-intercept, and one or two other points.

45. $y = 2x^2 - x - 1$

46. $y = -4x^2 - 1$

47. $y = x - 3x^2$

48. $y = 2x^2 + 8x - 10$

In Exercises 49–52, solve the given equations graphically. If there are no real roots, state this as the answer.

49. $2x^2 + x - 4 = 0$

50. $-4x^2 - x - 1 = 0$

51. $3x^2 = -x - 2$

52. $x(15x - 12) = 8$

In Exercises 53 and 54, solve the given problems.

53. A quadratic equation $f(x) = 0$ has a solution $x = 2$. Its graph has its vertex at $(-1, 6)$. What is the other solution?

54. Find c such that $y = 2x^2 + 16x + c$ has exactly one real root.

In Exercises 55–68, solve the given quadratic equations by any appropriate method. All numbers are accurate to at least three significant digits.

55. The bending moment M of a simply supported beam of length L with a uniform load of w kg/m at a distance x from one end is $M = 0.5wLx - 0.5wx^2$. For what values of x is $M = 0$?

56. A nuclear power plant supplies a fixed power level at a constant voltage. The current I (in A) is found by solving the equation $I^2 - 17I - 12 = 0$. Solve for $I > 0$.

57. A computer analysis shows that the cost C (in dollars) for a company to make x units of a certain product is given by $C = 0.1x^2 + 0.8x + 7$. How many units can be made for \$50?

58. For laminar flow of fluids, the coefficient K used to calculate energy loss due to sudden enlargements (where pipe size suddenly increases) is given by $K = 1.00 - 2.67R + R^2$, where R is the ratio of cross-sectional areas. If $K = 0.500$, what is the value of R?

59. At an altitude h (in m) above sea level, the boiling point of water is lower by T°C $(T > 0)$ than the boiling point at sea level, which is 100.0°C. The difference can be approximated by solving the equation $T^2 + 244T - h = 0$. What is the boiling point in Calgary, Alberta (altitude 1050 m)?

60. In a natural gas pipeline, the velocity v (in m/s) of the gas as a function of the distance x (in cm) from the wall of the pipe is given by $v = 5.20x - x^2$. Determine x for $v = 4.80$ m/s.

61. The height h of an object shot at an angle θ moving with velocity v is given by $h = vt\sin\theta - 4.9t^2$, where t is the time of flight. Find t (to the nearest 0.01s) if $v = 15.0$ m/s, $\theta = 65.0°$, and $h = 6.00$ m.

62. In studying the emission of light, in order to determine the angle at which the intensity is a given value, the equation $\sin^2 A - 4\sin A + 1 = 0$ must be solved. Find angle A (to 0.1°) $(\sin^2 A = (\sin A)^2)$.

63. A computer analysis shows that the number n of electronic components a company should produce for supply to equal demand is found by solving

$$\frac{n^2}{500\,000} = 144 - \frac{n}{500}. \text{ Find } n.$$

64. To determine the resistances of two resistors that are to be in parallel in an electric circuit, it is necessary to solve the equation $\frac{20}{R} + \frac{20}{R+10} = \frac{1}{5}$. Find R (to nearest 1 Ω).

65. In designing a cylindrical container, the formula $A = 2\pi r^2 + 2\pi rh$ (Eq. 2.34) is used. Solve for r.

66. In determining the number of bytes b that can be stored on a hard disk, the equation $b = kr(R - r)$ is used. Solve for r.

67. In the study of population growth, the equation $p_2 = p_1 + rp_1(1 - p_1)$ occurs. Solve for p_1.

68. In the study of the velocities of deep-water waves, the equation $v^2 = k^2\left(\frac{L}{C} + \frac{C}{L}\right)$ occurs. Solve for L.

In Exercises 69–83, set up the necessary equations where appropriate and solve the given problems. All numbers are accurate to at least two significant digits.

69. In testing the effects of a drug, the percentage of the drug in the blood was given by $p = 0.090t - 0.015t^2$, where t is the time (in h) after the drug was administered. Sketch the graph of $p = f(t)$.

70. In an electric circuit, the voltage V (in V) as a function of the time t (in min) is given by $V = 9.8 - 9.2t + 2.3t^2$. Sketch the graph of $V = f(t)$, for $t \leq 5$ min.

71. By adding the same amount to its length and its width, a developer increased the area of a rectangular lot by 3000 m² to make it 80 m by 100 m. What were the original dimensions of the lot? See Fig. 7.19.

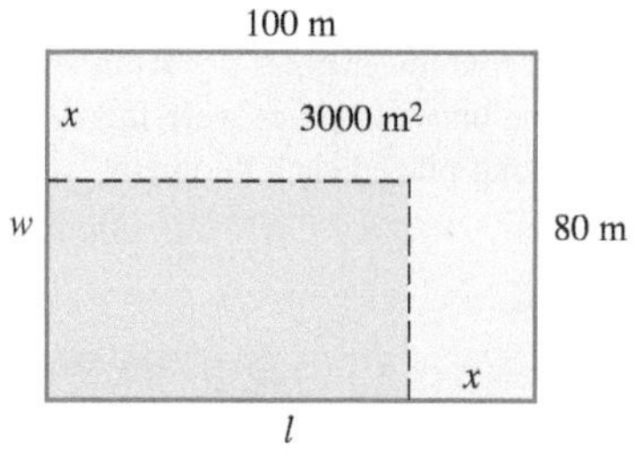

Fig. 7.19

72. A thick-walled, cylindrical pipe of mass m, outer radius r, and thickness t is to be rotated along its axis of symmetry. The pipe possesses a resistance to a change in its rotation called moment of inertia I, given by $I = mr^2\left(1 - \frac{t}{r} + \frac{t^2}{2r^2}\right)$. A particular pipe is measured to have a mass of 8.00 kg, an outer radius of 0.250 m, and a moment of inertia 0.410 kg · m². Determine the thickness of the wall of the pipe.

73. Concrete contracts as it dries. If the volume of a cubical concrete block is 29.0 cm³ less and each edge is 0.100 cm less after drying, what was the original length of an edge of the block?

74. A jet flew 1250 km with a tailwind of 50.0 km/h. The tailwind then changed to 20.0 km/h for the remaining 575 km of the flight. If the total time of the flight was 3.00 h, find the speed of the jet relative to the air.

75. A food company increased the profit on a package of frozen vegetables by decreasing the volume, and keeping the price the same. The thickness of the rectangular package remained at 4.00 cm, but the length and width were reduced by equal amounts from 16.0 cm and 12.0 cm such that the volume was reduced by 10.0%. What are the dimensions of the new container?

76. A military jet flies directly over and at right angles to the straight course of a commercial jet. The military jet is flying at 200 km/h faster than four times the speed of the commercial jet. How fast is each going if they are 2050 km apart (on a direct line) after 1.00 h?

77. The width of a rectangular TV screen is 14.5 cm more than the height. If the diagonal is 68.6 cm, find the dimensions of the screen. See Fig. 7.20.

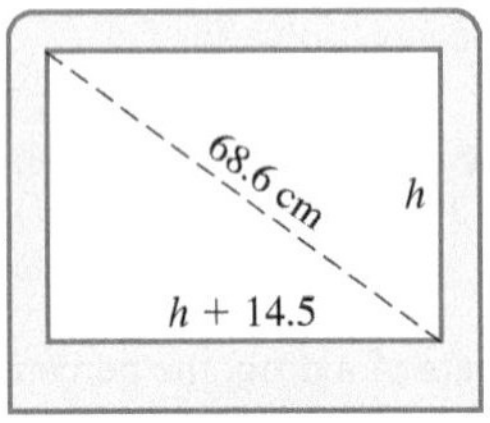

Fig. 7.20

78. Find the exact value of x that is defined in terms of the *continuing fraction* at the right (the pattern continues endlessly). Explain your method of solution.

$$x = 2 + \cfrac{1}{2 + \cfrac{1}{2 + \cfrac{1}{2 + \cdots}}}$$

79. An electric utility company is placing utility poles along a road. It is determined that five fewer poles per kilometre would be necessary if the distance between poles were increased by 10 m. How many poles are being placed each kilometre?

80. An architect is designing a Norman window (a semicircular part over a rectangular part) as shown in Fig. 7.21. If the area of the window is to be 1.75 m^2 and the height of the rectangular part is 1.35 m, find the radius of the circular part.

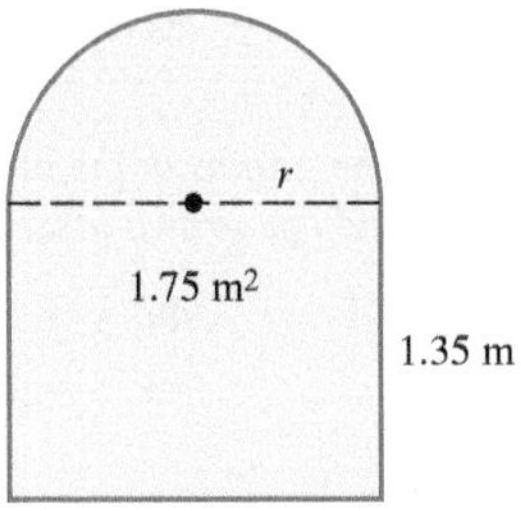

Fig. 7.21

81. A testing station found p parts per million (ppm) of sulfur dioxide in the air as a function of the hour h of the day to be $p = 0.001\,74(10 + 24h - h^2)$. Sketch the graph of $p = f(h)$ and, from the graph, find the time when $p = 0.205$ ppm.

82. A compact disc (CD) is made such that it is 53.0 mm from the edge of the centre hole to the edge of the disc. Find the radius of the hole if 1.36% of the disc is removed in making the hole. See Fig. 7.22.

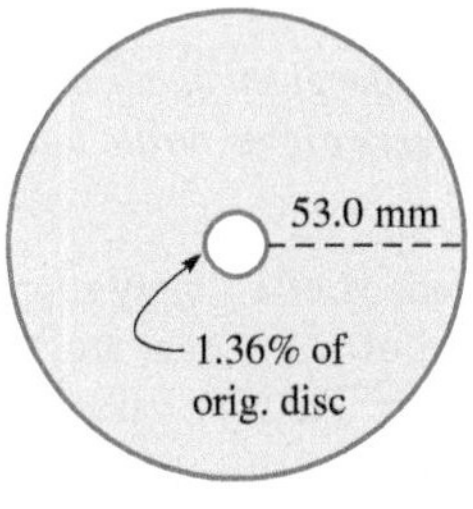

Fig. 7.22

83. An electronics student is asked to solve the equation $\frac{1}{2} = \frac{1}{R} + \frac{1}{R + 1}$ for R. Write one or two paragraphs explaining your procedure for the solution, including a discussion of what methods of the chapter may be used in completing the solution, assuming $R > 0$.

CHAPTER 7 PRACTICE TEST

1. Solve by factoring: $2x^2 + 5x = 12$.
2. Solve by using the quadratic formula: $x^2 = 3x + 5$.
3. Solve graphically using a graphing utility: $4x^2 - 5x - 3 = 0$.
4. Solve algebraically: $2x^2 - x = 6 - 2x(3 - x)$.
5. Solve algebraically: $\frac{3}{x} - \frac{2}{x + 2} = 1$.
6. Sketch the graph of $y = 2x^2 + 8x + 5$ using the extreme point and the y-intercept.
7. In electricity, the formula $P = EI - RI^2$ is used. Solve for I in terms of E, P, and R.
8. Solve by completing the square: $x^2 - 6x - 9 = 0$.
9. The perimeter of a rectangular window is 8.4 m, and its area is 3.8 m^2. Find its dimensions.
10. If $y = x^2 - 8x + 8$, sketch the graph using the vertex and any other useful points and find graphically the values of x for which $y = 0$.

8. Trigonometric Functions of Any Angle

Jallfree/iStock/Getty Images

▲ Gears and pulleys are common elements in the design of mechanical systems. In Section 8.4, we will see how radian measure of an angle is used to analyse the rotational properties of gears and pulleys.

LEARNING OUTCOMES

After completion of this chapter, the student should be able to:

- Determine the magnitude and sign of any trigonometric function of any angle
- Identify reference angles and use them to solve for trigonometric functions of angles in any quadrant
- Express an angle in degrees or radians and convert between the two measurements
- Given the value of a trigonometric function of an angle, find the angle
- Use radian measure of an angle to find arc length
- Solve problems involving area of a sector of a circle
- Solve problems involving angular velocity
- Solve application problems involving trigonometric functions of any angle

When we introduced the trigonometric functions in Chapter 4, we defined them in general but used them only with acute angles. In this chapter, we show how trigonometric functions are used with angles of any size and with angles measured in radians.

By the mid-1700s, the trigonometric functions had been used for many years as ratios, as in Chapter 4. It was also known that they were useful in describing the behaviour of periodic functions (functions for which values repeat at specific intervals) without reference to actual triangles. In about 1750, this led Swiss mathematician Leonhard Euler to include, for the first time in a textbook, the trigonometric functions of numbers (not angles). As will be shown in this chapter, this is equivalent to using these functions with arguments of angles of any size measured in radians.

Euler wrote over 70 volumes in mathematics and applied subjects such as astronomy, mechanics, and music. (Many of these volumes were dictated, as he was blind for the last 17 years of his life.) Noted as one of the greatest mathematicians of all time, Euler's interest in applied subjects often led him to study and develop topics in mathematics that were used in these applications.

Today, trigonometric functions of numbers are important to many applications, such as electric circuits, mechanical vibrations, and rotational motion. Although electronics were unknown in the 1700s, the trigonometric functions of numbers developed at that time played an important role in leading us into the electronic age of today.

8.1 Signs of the Trigonometric Functions

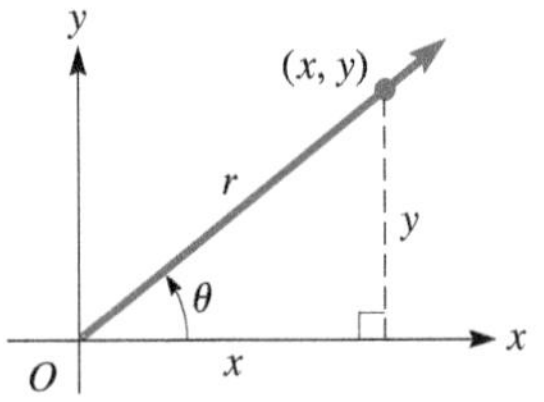

Fig. 8.1

Recall the definitions of the trigonometric functions that were given in Section 4.2. *Here, the point (x, y) is a point on the terminal side of angle θ, and r is the radius vector.* See Fig. 8.1.

Trigonometric Functions

$$\sin\theta = \frac{y}{r} \qquad \csc\theta = \frac{1}{\sin\theta} = \frac{r}{y}$$

$$\cos\theta = \frac{x}{r} \qquad \sec\theta = \frac{1}{\cos\theta} = \frac{r}{x} \qquad \textbf{(8.1)}$$

$$\tan\theta = \frac{y}{x} \qquad \cot\theta = \frac{1}{\tan\theta} = \frac{x}{y}$$

These definitions are valid for a standard-position angle of any size. In this section, we determine the *sign* of the trigonometric functions in each of the four quadrants.

We can find the values of the functions if we know one pair of coordinates (x, y) on the terminal side and the magnitude of the radius vector r. *Since r is always taken to be positive, the functions will vary in sign, depending on the values of x and y.* If either x or y is zero in the denominator, the function is undefined. We will consider this in the next section.

Since $\sin\theta = y/r$, *the sign of* $\sin\theta$ *depends on the sign of* y. Similarly, since $\cos\theta = x/r$, the sign of $\cos\theta$ *depends on the sign of* x. For example, in quadrant II, where x is negative and y is positive, $\cos\theta$ is negative and $\sin\theta$ is positive. Signs for all four quadrants are summarized in the grid to the left, and illustrated in Fig. 8.2.

Quadrant	I	II	III	IV
$\sin\theta$	+	+	−	−
$\cos\theta$	+	−	−	+

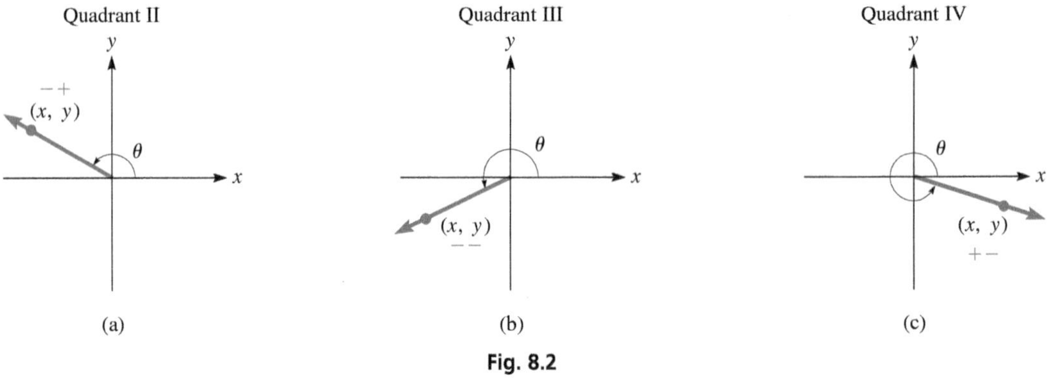

Fig. 8.2

EXAMPLE 1 Sign of sine and cosine in each quadrant

For each of the following angles, determine the sign of $\sin\theta$ and $\cos\theta$ depending on the quadrant where the terminal side falls.

(a) $\theta = 20°$	Quadrant I	$x > 0, y > 0$	$\sin 20° > 0, \cos 20° > 0$
(b) $\theta = 160°$	Quadrant II	$x < 0, y > 0$	$\sin 160° > 0, \cos 160° < 0$
(c) $\theta = 200°$	Quadrant III	$x < 0, y < 0$	$\sin 200° < 0, \cos 200° < 0$
(d) $\theta = 340°$	Quadrant IV	$x > 0, y < 0$	$\sin 340° < 0, \cos 340° > 0$

Since $\tan\theta = y/x$, *the sign of* $\tan\theta$ *depends on the ratio of* y *to* x. Therefore, when the signs of x and y are the same, $\tan\theta$ is *positive*, but when x and y differ in sign, $\tan\theta$ is *negative*. In other words, the tangent is positive in quadrants I and III, and it is negative in quadrants II and IV. See Fig. 8.2.

Quadrant	I	II	III	IV
$\tan\theta$	+	−	+	−

EXAMPLE 2 Sign of tan θ in each quadrant

For each of the following angles, determine the sign of tan θ depending on the quadrant where the terminal side falls.

(a) $\theta = 20°$	Quadrant I	$x > 0, y > 0$	$\tan 20° > 0$
(b) $\theta = 160°$	Quadrant II	$x < 0, y > 0$	$\tan 160° < 0$
(c) $\theta = 200°$	Quadrant III	$x < 0, y < 0$	$\tan 200° > 0$
(d) $\theta = 340°$	Quadrant IV	$x > 0, y < 0$	$\tan 340° < 0$

Since csc θ is defined in terms of y and r, as in sin θ, csc θ has the same sign as sin θ. For similar reasons, cot θ has the same sign as tan θ, and sec θ has the same sign as cos θ. Therefore, as shown in Fig. 8.3(a):

All functions of first-quadrant angles are positive. Sin θ and csc θ are positive for second-quadrant angles. Tan θ and cot θ are positive for third-quadrant angles. Cos θ and sec θ are positive for fourth-quadrant angles. All others are negative.

A helpful memory aid for the signs of the trigonometric functions is the word CAST, spelled counterclockwise from Quadrant IV (see Fig. 8.3b). Each letter represents the basic trigonometric function that is positive in that quadrant. You could also start in the first quadrant and remember the phrase "All Students Take Calculus."

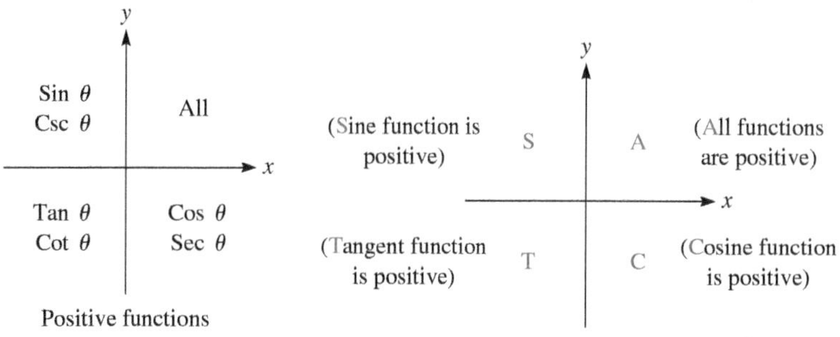

Fig. 8.3 (a) (b)

This discussion does not include the *quadrantal angles*, those angles with terminal sides on one of the axes. They will be discussed in the next section.

EXAMPLE 3 Positive and negative functions

Determine the sign of the following functions.

(a) sin 170°
Sine in QII: $\sin 170° > 0$

(b) cos 190°
Cosine in QIII: $\cos 190° < 0$

(c) tan 350°
Tangent in QIV: $\tan 350° < 0$

(d) cot 260°
Cotangent in QIII: $\cot 260° > 0$

(e) sec 280°
Secant in QIV: $\sec 280° > 0$

(f) csc 100°
Cosecant in QII: $\csc 100° > 0$

Practice Exercise

1. Determine the sign of the given functions: **(a)** sin 140° **(b)** tan 255° **(c)** sec 175°

EXAMPLE 4 Determining the quadrant of θ

For the given conditions, determine the quadrant(s) in which the terminal side of the angle lies.

(a) $\sin \theta > 0$ — Sine is positive in QI and QII.

(b) $\sec \theta < 0$ — Secant is negative in QII and QIII.

(c) $\cos \theta > 0, \tan \theta < 0$ — Cosine is positive in QI and QIV; tangent is negative in QII and QIV. Both conditions are true in QIV.

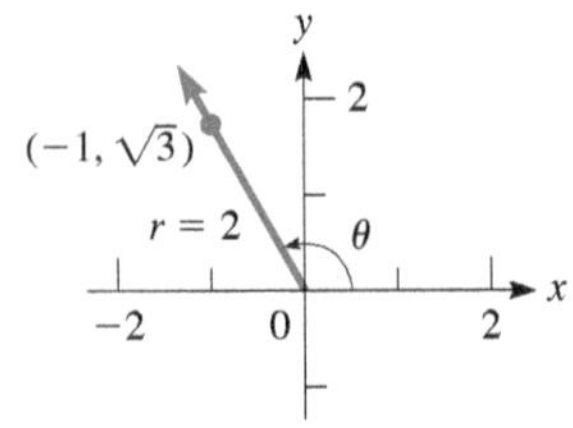

Fig. 8.4

EXAMPLE 5 Evaluating functions

Determine the trigonometric functions of θ to three significant digits if the terminal side of θ passes through $(-1, \sqrt{3})$. See Fig. 8.4.

We know that $x = -1$, $y = \sqrt{3}$, and from the Pythagorean theorem, we find that $r = 2$. Therefore, the trigonometric functions of θ are

$$\sin\theta = \frac{\sqrt{3}}{2} = 0.866 \qquad \cos\theta = -\frac{1}{2} = -0.500 \qquad \tan\theta = -\sqrt{3} = -1.73$$

$$\cot\theta = -\frac{1}{\sqrt{3}} = -0.577 \qquad \sec\theta = -2 = -2.00 \qquad \csc\theta = \frac{2}{\sqrt{3}} = 1.16$$

The point $(-1, \sqrt{3})$ is in the second quadrant, and the signs of the functions of θ are those of a second-quadrant angle.

Practice Exercise

2. Determine the value of $\cos\theta$ if the terminal side of θ passes through $(-1, 4)$.

When dealing with negative angles, or angles greater than 360°, the values of the trigonometric functions (and therefore their signs) are the same as those of a coterminal angle between 0° and 360°. For example, to determine the sign of $\sin(-120°)$, we note that $-120°$ is coterminal with $-120° + 360° = 240°$, a third-quadrant angle. Therefore, $\sin(-120°) < 0$. Also, for the same reason, $\sin 600° < 0$ (also coterminal to 240°).

EXERCISES 8.1

In Exercises 1 and 2, answer the given questions about the indicated examples of this section.

1. In Example 3, if 90° is added to each angle, what is the sign of each resulting function?
2. In Example 5, if the point $(-1, \sqrt{3})$ is replaced with the point $(1, -\sqrt{3})$, what are the resulting values?

In Exercises 3–16, determine the sign of the given functions.

3. tan 135°, sec 50°
4. sin 240°, cos300°
5. sin 290°, cos 200°
6. tan 320°, sec 185°
7. csc 98°, cot 82°
8. cos 260°, csc 290°
9. sec 150°, tan 220°
10. sin 335°, cot 265°
11. cos 348°, csc 238°
12. cot 110°, sec 309°
13. tan 460°, sin(−185°)
14. csc(−200°), cos 550°
15. cot(−95°), cos 710°
16. sin 539°, tan(−480°)

In Exercises 17–24, find the trigonometric functions of θ if the terminal side of θ passes through the given point. All coordinates are exact.

17. (2, 1)
18. (−5, 5)
19. (−2, −3)
20. (16, −12)
21. (−0.5, 1.2)
22. (−39, −80)
23. (20, −8)
24. (0.9, 4)

In Exercises 25–30, for the given values, determine the quadrant(s) in which the terminal side of the angle lies.

25. $\sin\theta = 0.500$
26. $\cos\theta = 0.866$
27. $\tan\theta = -2.50$
28. $\sin\theta = -0.866$
29. $\cos\theta = -0.500$
30. $\tan\theta = 0.427$

In Exercises 31–40, determine the quadrant in which the terminal side of θ lies, subject to both given conditions.

31. $\sin\theta > 0$, $\cos\theta < 0$
32. $\tan\theta > 0$, $\cos\theta < 0$
33. $\sec\theta < 0$, $\cot\theta < 0$
34. $\cos\theta > 0$, $\csc\theta < 0$
35. $\csc\theta < 0$, $\tan\theta < 0$
36. $\tan\theta < 0$, $\cos\theta > 0$
37. $\sin\theta > 0$, $\tan\theta > 0$
38. $\sec\theta > 0$, $\csc\theta < 0$
39. $\sin\theta > 0$, $\cot\theta < 0$
40. $\tan\theta > 0$, $\csc\theta < 0$

In Exercises 41–44, with (x, y) in the given quadrant, determine whether the given ratio is positive or negative.

41. III, $\frac{x}{r}$
42. II, $\frac{y}{r}$
43. IV, $\frac{y}{x}$
44. III, $\frac{x}{y}$

Answers to Practice Exercises

1. (a) + (b) + (c) − 2. $-1/\sqrt{17}$

8.2 Trigonometric Functions of Any Angle

As stated earlier, the trigonometric functions of negative angles or angles greater than 360° are the same as those of their coterminal angle whose measure is between 0° and 360°. Therefore, we can evaluate the functions of an angle of any size if we can find the trigonometric functions of angles between 0° and 360°. In this section we will focus our attention on these angles and establish important connections between certain angles in different quadrants.

Let us begin by taking the sine of the angles 30°, 150°, 210°, and 330°. We have

$$\sin 30° = \frac{1}{2} \qquad \sin 150° = \frac{1}{2} \qquad \sin 210° = -\frac{1}{2} \qquad \sin 330° = -\frac{1}{2}$$

A very important thing to notice is that **the numerical values (disregarding the signs) are the same for all four angles**. The reason for this is that these angles have something in common, called a **reference angle**, which is defined below:

> The **reference angle**, labelled θ_{ref}, of a given angle θ is the **positive, acute angle formed between the terminal side of θ and the x-axis.**

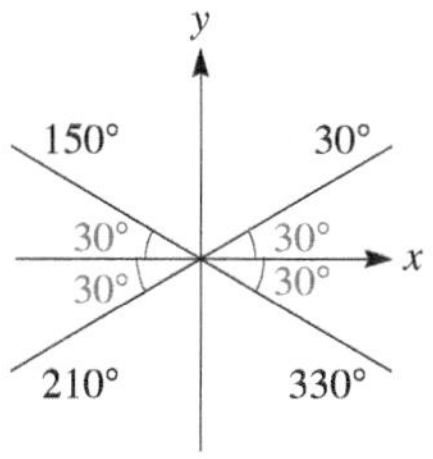

Fig. 8.5

■ In diagrams, we may choose to draw the arcs for reference angles without arrows since we know reference angles are always positive.

The four angles given have a reference angle of 30°. Fig. 8.5 shows the terminal sides of these angles drawn in standard position along with their reference angles. Because the values of x, y, and r in the definition of the trigonometric functions will be the same for all four angles (except for the signs of x and y, which depend on the quadrant), they will all have the same trigonometric function values, neglecting the sign.

For this reason, reference angles are very important in trigonometry. To find the reference angle of a given angle, we sketch the angle in standard position and then determine the acute angle between its terminal side and the x-axis. The calculations for each quadrant are illustrated in the following example.

EXAMPLE 1 Finding reference angles

Find the reference angle for each of the following angles. Sketch the angle and its reference angle.

(a) 60°
Quadrant: I
Measure of θ_{ref}: θ
$\theta_{\text{ref}} = 60°$

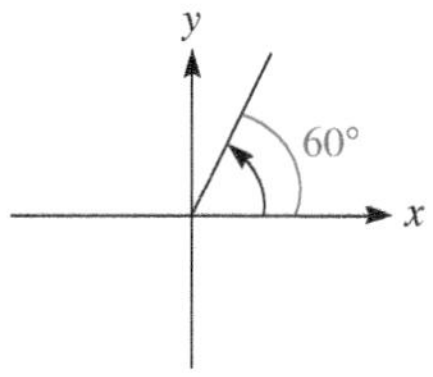

Fig. 8.6(a)

(b) 150°
Quadrant: II
Measure of θ_{ref}: $180° - \theta$
$\theta_{\text{ref}} = 180° - 150° = 30°$

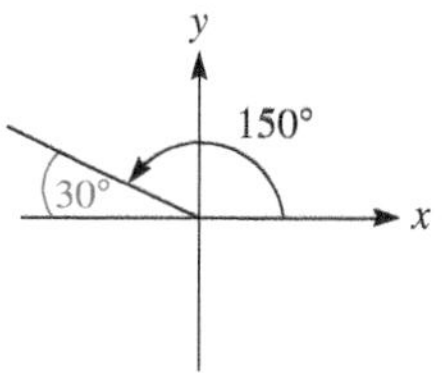

Fig. 8.6(b)

(c) 230°
Quadrant: III
Measure of θ_{ref}: $\theta - 180°$
$\theta_{\text{ref}} = 230° - 180° = 50°$

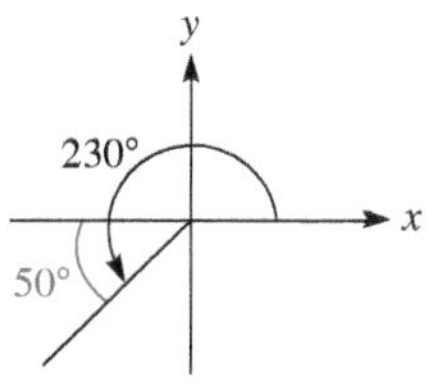

Fig. 8.6(c)

(d) 285°
Quadrant: IV
Measure of θ_{ref}: $360° - \theta$
$\theta_{\text{ref}} = 360° - 285° = 75°$

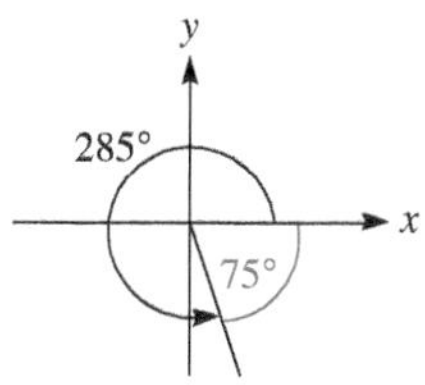

Fig. 8.6(d)

We have observed that the trigonometric functions of different angles with the same reference angle will be the same, except for the sign, possibly. Since the reference angle is an acute angle, it will always have a positive function value. The signs in the other quadrants will follow the rules stated in Section 8.1. This is summarized below:

> Let θ be any angle in standard position, and let θ_{ref} be its reference angle. If we let "trig" stand for any of the six trigonometric functions, then
>
> $$\text{trig } \theta = \pm \text{trig } \theta_{ref} \qquad (8.2)$$
>
> where the sign is determined by the quadrant of θ.

According to the statement above, $\sin\theta = \pm\sin\theta_{ref}$, $\cos\theta = \pm\cos\theta_{ref}$, $\tan\theta = \pm\tan\theta_{ref}$, etc. This means that a trigonometric function of any angle can be found by first finding the trigonometric function of its reference angle and then attaching the correct sign, depending on the quadrant. The following examples illustrate this process.

EXAMPLE 2 Trigonometric functions in terms of reference angles

Express each of the functions of 315° in terms of its reference angle and evaluate (in exact form and to three significant digits).

As shown in Fig. 8.7, $\theta_{ref} = 360° - 315° = 45°$. In quadrant IV, $\cos\theta$ and $\sec\theta$ are (+), but $\sin\theta$, $\csc\theta$, $\tan\theta$, and $\cot\theta$ are (−). Therefore

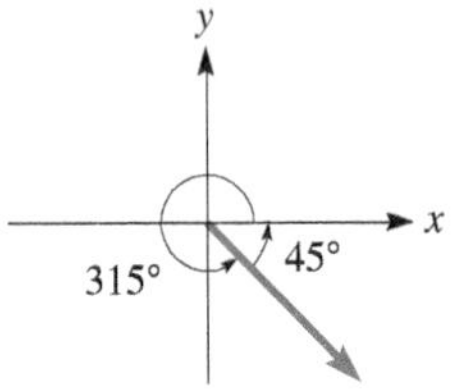

Fig. 8.7

$$\sin 315° = -\sin 45° = -\frac{1}{\sqrt{2}} = -0.707 \qquad \text{sine is } (-) \text{ in QIV}$$

$$\cos 315° = +\cos 45° = -\frac{1}{\sqrt{2}} = +0.707 \qquad \text{cosine is } (+) \text{ in QIV}$$

$$\tan 315° = -\tan 45° = -1 = -1.00 \qquad \text{tangent is } (-) \text{ in QIV}$$

$$\csc 315° = -\csc 45° = -\sqrt{2} = -1.41 \qquad \text{cosecant is } (-) \text{ in QIV}$$

$$\sec 315° = +\sec 45° = +\sqrt{2} = +1.41 \qquad \text{secant is } (+) \text{ in QIV}$$

$$\cot 315° = -\cot 45° = -1 = -1.00 \qquad \text{cotangent is } (-) \text{ in QIV}$$

Practice Exercise

1. Express each of the functions of 225° in terms of the reference angle.

EXAMPLE 3 Evaluating trigonometric functions using reference angles

> **LEARNING TIP**
> In most examples, we will round off trigonometric function values to three significant digits. However, if the angle is approximate, use the guidelines in Table 4.1 of Section 4.3 for rounding off values.

Express each of the following functions in terms of the reference angle and evaluate. Sketch the angle and its reference angle.

(a) sin 160°

$\theta_{ref} = 180° - 160° = 20°$
Sine is (+) in QII.

$\sin 160° = +\sin 20°$
$= +0.342$

(b) tan 110°

$\theta_{ref} = 180° - 110° = 70°$
Tangent is (−) in QII.

$\tan 110° = -\tan 110°$
$= -2.75$

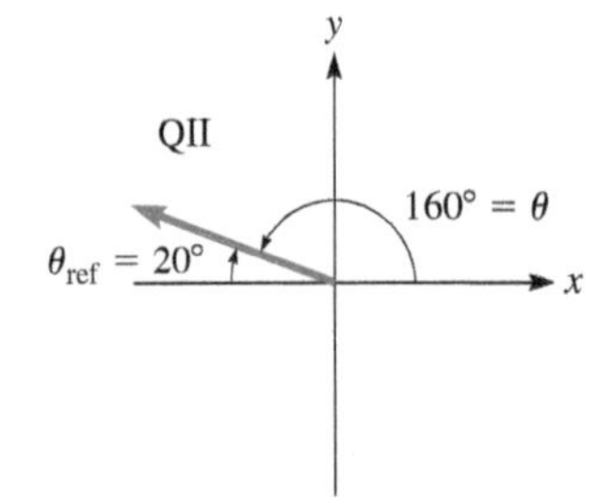

Fig. 8.8(a)

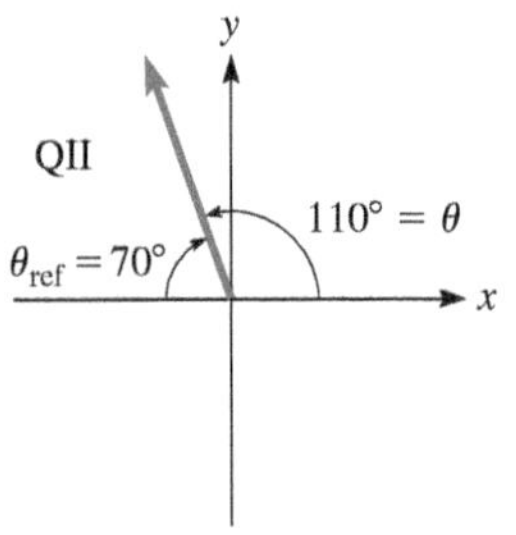

Fig. 8.8(b)

(c) cos 225°

$\theta_{ref} = 225° - 180° = 45°$
Cosine is (−) in QIII.

$\cos 225° = -\cos 45°$
$= -0.707$

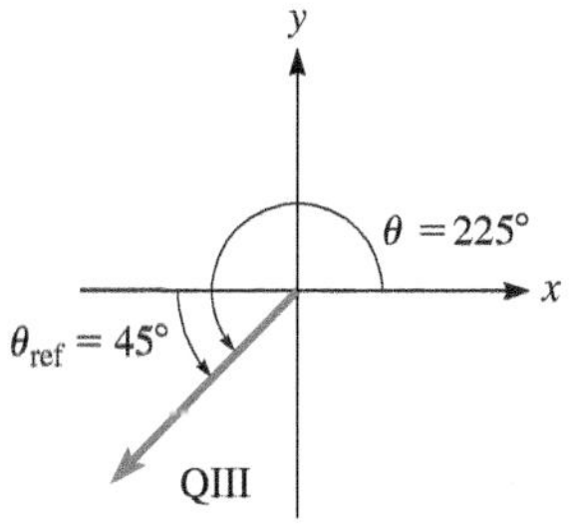

Fig. 8.8(c)

(d) cot 260°

$\theta_{ref} = 260° - 180° = 80°$
Cotangent is (+) in QIII.

$\cot 260° = +\cot 80°$
$= +0.176$

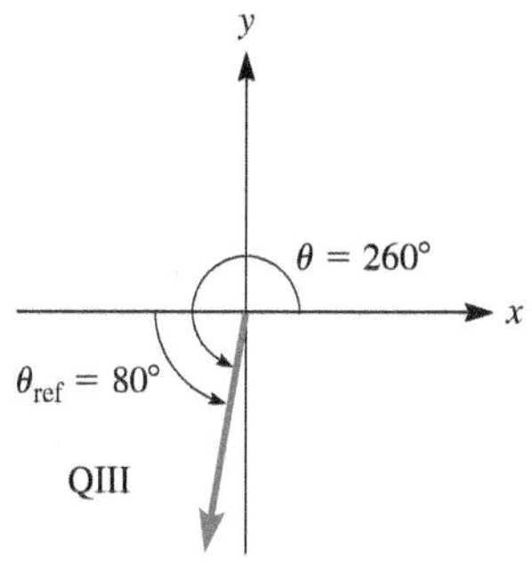

Fig. 8.8(d)

(e) sec 304°

$\theta_{ref} = 360° - 304° = 56°$
Secant is (+) in QIV.

$\sec 304° = +\sec 56°$
$= +1.79$

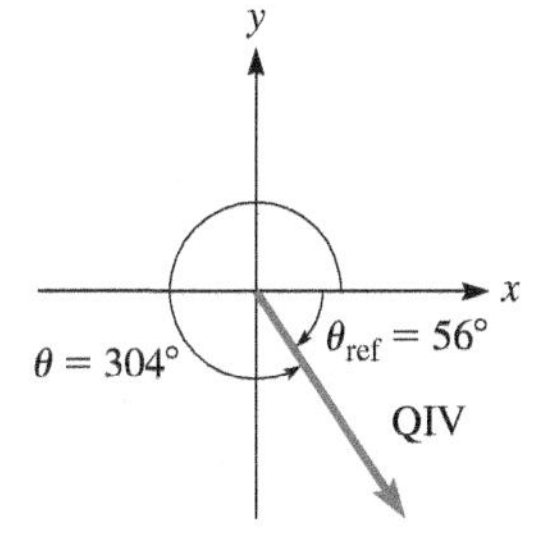

Fig. 8.8(e)

(f) sin 357°

$\theta_{ref} = 360° - 357° = 3°$
Sine is (−) in QIV.

$\sin 357° = -\sin 3°$
$= -0.0523$

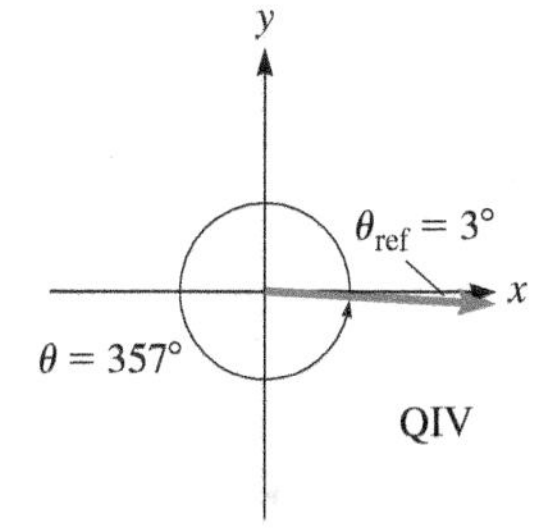

Fig. 8.8(f)

A calculator will give values, with the proper signs, of functions like those in Examples 2 and 3. To find cot θ, sec θ, and csc θ, we must take the reciprocal of the corresponding function. For example, cot 304° = 1/cos 304°.

EXAMPLE 4 Quadrant II angle—land area

A formula for finding the area of a triangle, knowing sides a and b and the included $\angle C$, is $A = \frac{1}{2}ab \sin C$. A surveyor uses this formula to find the area of a triangular tract of land for which $a = 173.2$ m, $b = 156.3$ m, and $C = 112.51°$. See Fig. 8.9.

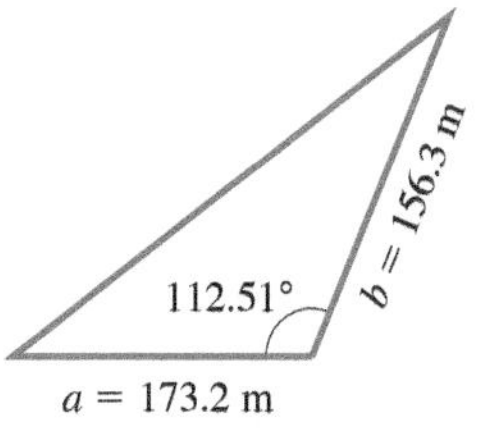

Fig. 8.9

$$A = \tfrac{1}{2}(173.2)(156.3) \sin 112.51°$$
$$= 12\,500 \text{ m}^2 \quad \text{rounded to four significant digits}$$

The calculator automatically uses a positive value for sin 112.51°.

■ The reason that the calculator displays these particular angles is shown in Chapter 20, when the inverse trigonometric functions are discussed in detail.

Reference angles are especially important when we wish to find an angle that has a given trigonometric function value. This is because the inverse trigonometric functions on a calculator are programmed to give angles within the specific intervals shown below:

$\sin^{-1} x$ always returns an angle between −90° and 90°
$\cos^{-1} x$ always returns an angle between 0° and 180°
$\tan^{-1} x$ always returns an angle between −90° and 90°

COMMON ERROR

When we wish to find an angle that is *not* within the intervals given above, we must use reference angles. For example, if we need to find a second-quadrant angle for which $\sin\theta = 0.853$, then taking $\sin^{-1}(0.8532)$ **will not return the correct angle.** In this type of situation, we must use the reference angle to find the correct angle.

EXAMPLE 5 Finding angles given sin θ

Solve $\sin\theta = 0.225$ for θ between 0° and 360°.

Identify the quadrants where the sign of the function is correct: Sine is positive in QI and QII.

Use the inverse function to find a solution: $\theta = \sin^{-1}(0.225) = 13.0°$
θ is in QI, so it is the reference angle for the solution in QII.

Use the reference angle to find the other solution: In QII, $\theta = 180° - \theta_{ref}$. Therefore,

$$\theta = 180° - 13.0° = 167.0°$$

is the other solution. See Fig. 8.10.

Check: $\sin 13.0° = 0.225$ and $\sin 127.0° = 0.225$

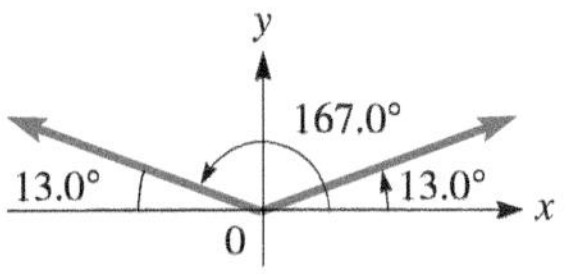

Fig. 8.10

EXAMPLE 6 Finding angles given sec θ

Solve $\sec\theta = -2.722$ for θ between 0° and 360°.

Identify the quadrants where the sign of the function is correct: Secant is negative in QII and QIII.

Rewrite using the reciprocal:

$$\frac{1}{\cos\theta} = -2.722, \quad \text{so} \quad \cos\theta = \left(\frac{1}{-2.722}\right)$$

Use the inverse function to find a solution:

$$\theta = \cos^{-1}\left(\frac{1}{-2.722}\right) = 111.55°$$

θ is in QII, so it cannot be the reference angle for the solution in QIII. The common reference angle is

$$\theta_{ref} = 180° - 111.55° = 68.45°$$

Use the reference angle to find the other solution: In QIII, $\theta = 180° + \theta_{ref}$. Therefore,

$$\theta = 180° + 68.45° = 248.45°$$

is the other solution. See Fig. 8.11.

Check: $\sec 111.55° = -2.722$ and $\sec 68.45° = -2.722$

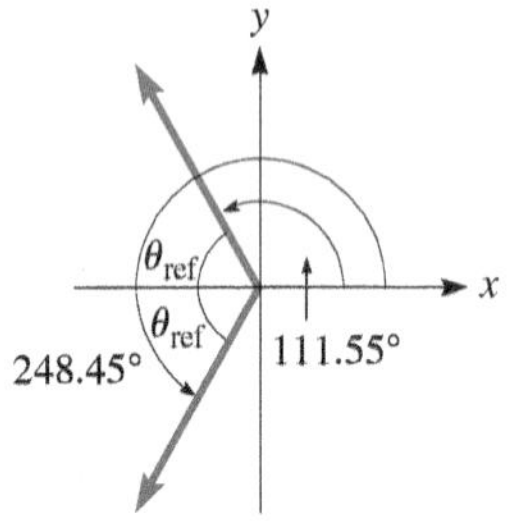

Fig. 8.11

EXAMPLE 7 Finding angles given tan θ

Solve $\tan\theta = 2.05$ for θ between 0° and 360° if $\cos\theta < 0$.

Identify the quadrants where the signs of the functions are correct: Tangent is positive in QI and QIII. Cosine is negative in QII and QIII. Both will be correct only in QIII.

Use the inverse function to find a solution: $\theta = \tan^{-1}(2.05) = 64.0°$
θ is in QI, so it is the reference angle for the correct solution in QIII.

Use the reference angle to find the correct solution: In QIII, $\theta = 180° + \theta_{ref}$. Therefore,

$$\theta = 180° + 64.0° = 244.0°$$

is the correct solution. See Fig. 8.12.

Check: $\tan 244.0° = 2.05$ and $\cos 244.0° = -0.44 < 0$

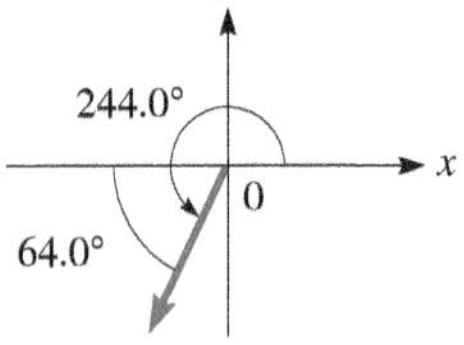

Fig. 8.12

Practice Exercise

2. If $\cos\theta = 0.5736$, find θ for $0° \le \theta < 360°$.

> **LEARNING TIP**
> The calculator gives the reference angle (disregarding any minus signs) in all cases except when $\cos\theta$ is negative. To avoid confusion from the angle displayed by the calculator, *a good procedure is to find the reference angle first.* Then it can be used to determine the angle required by the problem.

We can *find the reference angle by entering the absolute value of the function. The displayed angle will be the reference angle.* The required angle θ is found by using the reference angle as described earlier and as shown in Eq. (8.3).

$$\begin{aligned} \theta &= \theta_{ref} && \text{(first quadrant)} \\ \theta &= 180^\circ - \theta_{ref} && \text{(second quadrant)} \\ \theta &= 180^\circ + \theta_{ref} && \text{(third quadrant)} \\ \theta &= 360^\circ - \theta_{ref} && \text{(fourth quadrant)} \end{aligned} \qquad \textbf{(8.3)}$$

EXAMPLE 8 Using the reference angle

Solve $\cos\theta = -0.1298$ for θ between 0° and 360°.

Identify the quadrants where the sign of the function is correct: Cosine is negative in QII and QIII.

Use the inverse function of *the absolute value* to find θ_{ref}: $\theta_{ref} = \cos^{-1}(0.1298) = 82.54^\circ$

Use the reference angle to find the two solutions:
In QII, $\theta = 180^\circ - 82.54^\circ = 97.46^\circ$
In QIII, $\theta = 180^\circ + 82.54^\circ = 262.54^\circ$

Check: $\cos 97.46^\circ = -0.1298$ and $\cos 262.54^\circ = -0.1298$

In Section 4.2, we showed how the unit circle could be used to evaluate the trigonometric functions. We now restate this as a definition, which is a special case of the general definition in Eq. (8.1) with the restriction that $r = 1$.

Unit Circle Definition of the Trigonometric Functions

Suppose the terminal side of θ in standard position intersects the unit circle at the point (x, y) as shown is Fig. 8.13. Then the six trigonometric functions are defined as follows:

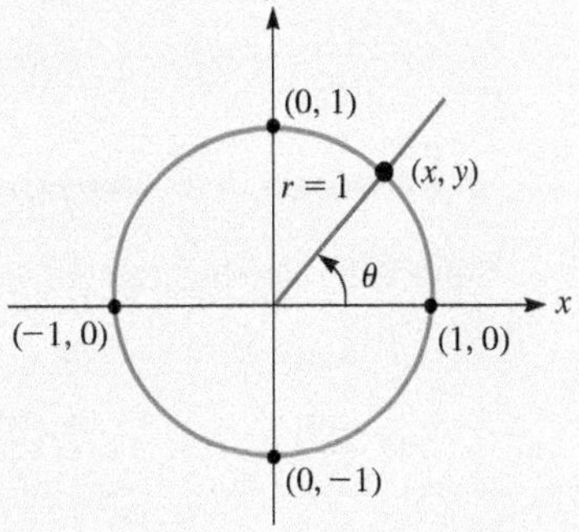

Fig. 8.13

$$\begin{aligned} \sin\theta &= y & \csc\theta &= \frac{1}{y} \\ \cos\theta &= x & \sec\theta &= \frac{1}{x} \\ \tan\theta &= \frac{y}{x} & \cot\theta &= \frac{x}{y} \end{aligned} \qquad \textbf{(8.4)}$$

There are many situations when the unit circle definition is useful, including evaluating trigonometric functions of **quadrantal angles**, which have their terminal side along an axis (see Fig. 8.14). The sine of θ is simply the y-coordinate on the unit circle, and the cosine of θ is the x-coordinate. The other functions are found by taking ratios, and if the denominator is zero, the function is undefined. The table that follows provides the values of the trigonometric functions at quadrantal angles.

θ	$\sin\theta$	$\cos\theta$	$\tan\theta$	$\cot\theta$	$\sec\theta$	$\csc\theta$
0°	0.000	1.000	0.000	undef.	1.000	undef.
90°	1.000	0.000	undef.	0.000	undef.	1.000
180°	0.000	−1.000	0.000	undef.	−1.000	undef.
270°	−1.000	0.000	undef.	0.000	undef.	−1.000
360°	Same as the functions of 0° (same terminal side)					

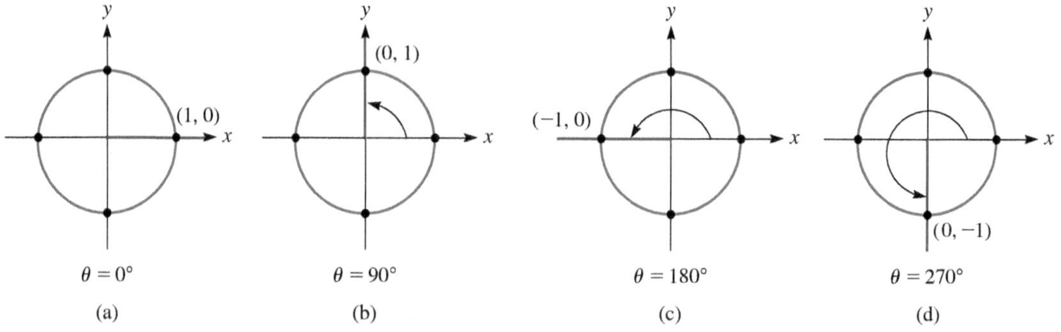

Fig. 8.14

EXAMPLE 9 Quadrantal angles and the unit circle

Find the indicated values at quadrantal angles using Fig. 8.14, or state that they are undefined.

(a) $\sin 0°$. As shown in Fig. 8.14(a), $\theta = 0°$ intersects the unit circle at $(1, 0)$. Hence $\sin 0° = y = 0$.

(b) $\tan 90°$. As shown in Fig. 8.14(b), $\theta = 90°$ intersects the unit circle at $(0, 1)$. Hence $\tan 90° = \frac{y}{x} = \frac{1}{0}$ and $\tan 90°$ is undefined.

(c) $\cos 180°$. As shown in Fig. 8.14(c), $\theta = 180°$ intersects the unit circle at $(-1, 0)$. Hence $\cos 180° = x = -1$.

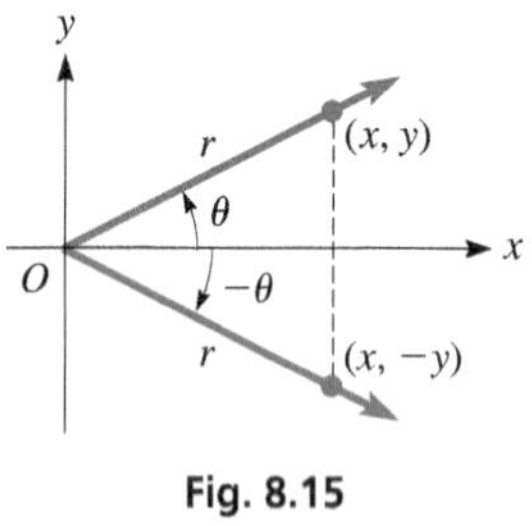

Fig. 8.15

To evaluate functions of negative angles, we can use functions of corresponding positive angles, if we use the correct ***sign***. In Fig. 8.15, note that $\sin\theta = y/r$, and $\sin(-\theta) = -y/r$, which means $\sin(-\theta) = -\sin\theta$. In the same way, we can get all of the relations between functions of $-\theta$ and the functions of θ. Therefore,

$$\sin(-\theta) = -\sin\theta \qquad \cos(-\theta) = \cos\theta \qquad \tan(-\theta) = -\tan\theta$$
$$\csc(-\theta) = -\csc\theta \qquad \sec(-\theta) = \sec\theta \qquad \cot(-\theta) = -\cot\theta \tag{8.5}$$

EXAMPLE 10 Negative angles

Solve the following trigonometric functions of negative angles, to three significant digits:

$$\sin(-60°) = -\sin 60° = -0.866 \qquad \cos(-60°) = \cos 60° = 0.500$$
$$\tan(-60°) = -\tan 60° = -1.73 \qquad \cot(-60°) = -\cot 60° = -0.577$$
$$\sec(-60°) = \sec 60° = 2.00 \qquad \csc(-60°) = -\csc 60° = -1.16$$

EXERCISES 8.2

In Exercises 1 and 2, make the given changes in the indicated examples of this section and then solve the resulting problems.

1. In Example 3, add 40° to each angle, express in terms of the same function of a positive acute angle, and then evaluate.
2. In Example 8, change the − to + and then find θ.

In Exercises 3–8, express the given trigonometric function in terms of the same function of a positive acute angle.

3. $\sin 155°$, $\cos 220°$
4. $\tan 91°$, $\sec 345°$
5. $\tan 105°$, $\csc 328°$
6. $\cos 190°$, $\tan 290°$
7. $\sec 425°$, $\sin(-520°)$
8. $\tan 920°$, $\csc(-550°)$

In Exercises 9–42, the given angles are approximate. In Exercises 9–16, find the values of the given trigonometric functions by finding the reference angle and attaching the proper sign.

9. $\sin 195°$ **10.** $\tan 311°$ **11.** $\cos 106.3°$

12. $\sin 93.4°$ **13.** $\sec 328.33°$ **14.** $\cot 516.53°$

15. $\tan(-109.1°)$ **16.** $\csc(-108.4°)$

In Exercises 17–24, find the values of the given trigonometric functions directly from a calculator.

17. $\cos(-62.7°)$ **18.** $\cos 141.4°$

19. $\sin 310.36°$ **20.** $\tan 242.68°$

21. $\csc 194.82°$ **22.** $\sec 441.08°$

23. $\tan 148.25°$ **24.** $\sin(-215.5°)$

In Exercises 25–38, find θ for $0° \le \theta < 360°$.

25. $\sin\theta = -0.8480$ **26.** $\tan\theta = -1.830$

27. $\cos\theta = 0.4003$ **28.** $\sec\theta = -1.637$

29. $\cot\theta = -0.0122$ **30.** $\csc\theta = -8.09$

31. $\sin\theta = 0.870$, $\cos\theta < 0$ **32.** $\tan\theta = 0.932$, $\sin\theta < 0$

33. $\cos\theta = -0.12$, $\tan\theta > 0$ **34.** $\sin\theta = -0.192$, $\tan\theta < 0$

35. $\csc\theta = -1.366$, $\cos\theta > 0$ **36.** $\cos\theta = 0.0726$, $\sin\theta < 0$

37. $\sec\theta = 2.047$, $\cot\theta < 0$ **38.** $\cot\theta = -0.3256$, $\csc\theta > 0$

In Exercises 39–42, determine the function that satisfies the given conditions.

39. Find $\tan\theta$ when $\sin\theta = -0.5736$ and $\cos\theta > 0$.

40. Find $\sin\theta$ when $\cos\theta = 0.422$ and $\tan\theta < 0$.

41. Find $\cos\theta$ when $\tan\theta = -0.809$ and $\csc\theta > 0$.

42. Find $\cot\theta$ when $\sec\theta = 6.122$ and $\sin\theta < 0$.

In Exercises 43–46, insert the proper sign, > or < or =, between the given expressions. Explain your answers.

43. $\sin 90°\quad 2\sin 45°$ **44.** $\cos 360°\quad 2\cos 180°$

45. $\tan 180°\quad \tan 0°$ **46.** $\sin 270°\quad 3\sin 90°$

In Exercises 47–50, solve the given problems. In Exercises 49 and 50, assume $0° < \theta < 90°$. *(Hint: Review cofunctions.)*

47. Using the fact that $\sin 75° = 0.9659$, evaluate $\cos 195°$.

48. Using the fact that $\cot 20° = 2.747$, evaluate $\tan 290°$.

49. Express $\tan(270° - \theta)$ in terms of $\cot\theta$.

50. Express $\cos(90° + \theta)$ in terms of $\sin\theta$.

In Exercises 51–54, evaluate the given expressions.

51. The current i in an alternating-current circuit is given by $i = i_m \sin\theta$, where i_m is the maximum current in the circuit. Find i if $i_m = 0.0259$ A and $\theta = 495.2°$.

52. The force F that a rope exerts on a crate is related to force F_x directed along the x-axis by $F = F_x \sec\theta$, where θ is the standard-position angle for F. See Fig. 8.16. Find F if $F_x = -365$ N and $\theta = 127.0°$.

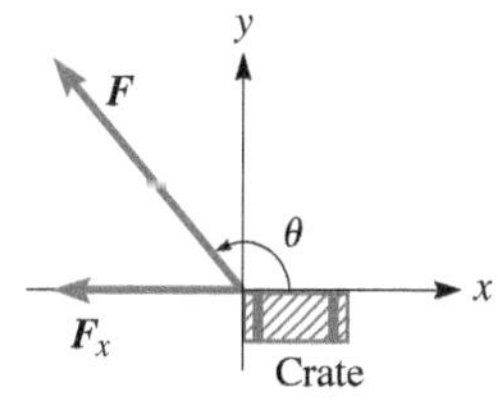

Fig. 8.16

53. For the slider mechanism shown in Fig. 8.17, $y\sin\alpha = x\sin\beta$. Find y if $x = 6.78$ cm, $\alpha = 31.3°$, and $\beta = 104.7°$.

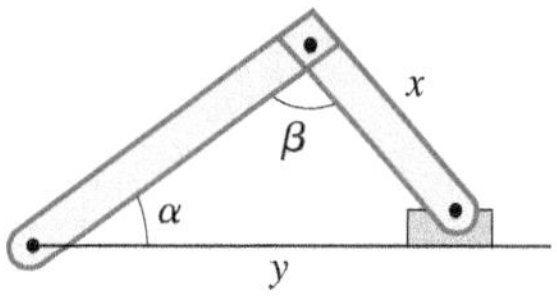

Fig. 8.17

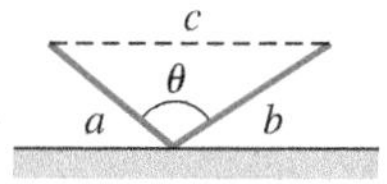

Fig. 8.18

54. A laser follows the path shown in Fig. 8.18. The angle θ is related to the distances a, b, and c by $2ab\cos\theta = a^2 + b^2 - c^2$. Find θ if $a = 12.9$ cm, $b = 15.3$ cm, and $c = 24.6$ cm.

Answers to Practice Exercises

1. $\sin 225° = -\sin 45°$ $\cos 225° = -\cos 45°$
$\tan 225° = \tan 45°$ $\cot 225° = \cot 45°$
$\sec 225° = -\sec 45°$ $\csc 225° = -\csc 45°$

2. 55.00°, 305.00°

8.3 Radians

■ The use of 360 comes from the ancient Babylonians and their number system based on 60 rather than 10, as we use today. However, the specific reason for the choice of 360 is not known. (One theory is that 360 is divisible by many smaller numbers and is also close to the number of days in a year.)

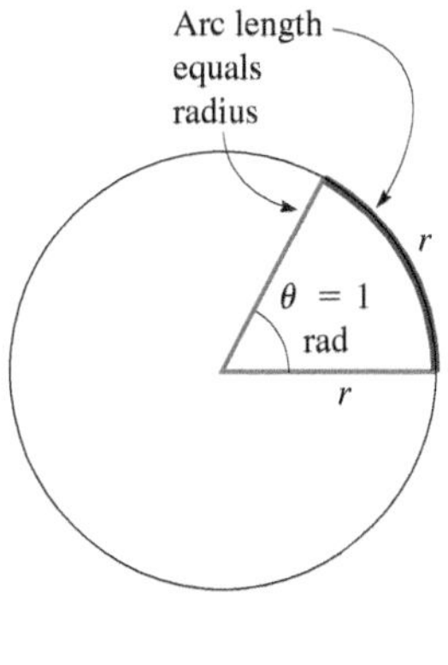

Fig. 8.19

For many problems in which trigonometric functions are used, particularly those involving the solution of triangles, degree measurements of angles are convenient and quite sufficient. However, division of a circle into 360 equal parts is by definition, and it is arbitrary and artificial.

In numerous other types of applications and in more theoretical discussions, the *radian* is a more meaningful measure of an angle. We defined the radian in Chapter 2 and reviewed it briefly in Chapter 4. In this section, we discuss the radian in detail and start by reviewing its definition.

A **radian** *is the measure of an angle with its vertex at the centre of a circle and with an intercepted arc on the circle equal in length to the radius of the circle.* See Fig. 8.19.

Since the circumference of any circle in terms of its radius is given by $c = 2\pi r$, the ratio of the circumference to the radius is $c/r = 2\pi$. This means that the radius may be laid out 2π (about 6.28) times along the circumference, regardless of the length of the radius. Therefore, note that radian measure is independent of the radius of the circle. The definition of a radian is based on an important property of a circle and is therefore

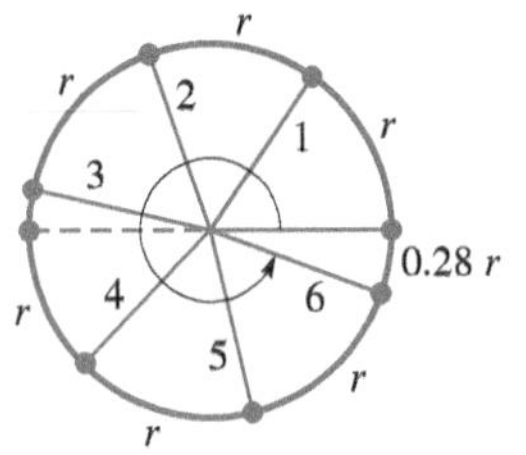

Fig. 8.20

a more natural measure of an angle. In Fig. 8.20, the numbers on each of the radii indicate the number of radians in the angle measured in standard position. The circular arrow shows an angle of 6 radians.

Since the radius may be laid out 2π times along the circumference, it follows that there are 2π radians in one complete rotation. Also, there are 360° in one complete rotation. Therefore, 360° is *equivalent* to 2π radians. It then follows that the relation between degrees and radians is 2π rad = 360°, or

Converting Angles

$$\pi \text{ rad} = 180° \qquad (8.6)$$

Degrees to Radians

$$1° = \frac{\pi}{180} \text{ rad} = 0.017\,45 \text{ rad} \qquad (8.7)$$

Radians to Degrees

$$1 \text{ rad} = \frac{180°}{\pi} = 57.30° \qquad (8.8)$$

Note that Eq. (8.6) gives us two statements of equivalence which we can use to convert angle measurements from degrees to radians or radians to degrees using a procedure identical to that of all unit conversions (see Section 1.3). We use the statement of equivalence to multiply by one. After cancellation, the unit of measurement is different, but the angle is the same.

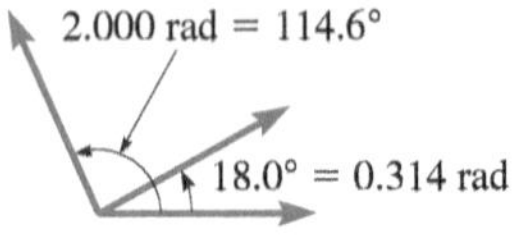

Fig. 8.21

Procedure for Converting Angle Measurements	EXAMPLE 1
To convert an angle measured in degrees to the same angle measured in radians, multiply the number of degrees by $\left(\frac{\pi \text{ rad}}{180°}\right)$.	**(a)** Convert 18.0° to radians. See Fig. 8.21. $18.0°\left(\frac{\pi \text{ rad}}{180°}\right) = \frac{\pi}{10} \text{rad} = 0.314 \text{ rad}$
To convert an angle measured in radians to the same angle measured in degrees, multiply the number of radians by $\left(\frac{180°}{\pi \text{ rad}}\right)$.	**(b)** Convert 2.000 rad to degrees. See Fig. 8.21. $2.000 \text{ rad}\left(\frac{180°}{\pi \text{ rad}}\right) = \frac{360.0°}{\pi} = 114.6°$

Because of the definition of the radian, it is common to express radians in terms of π, particularly for angles whose degree measure is a fraction of 180°.

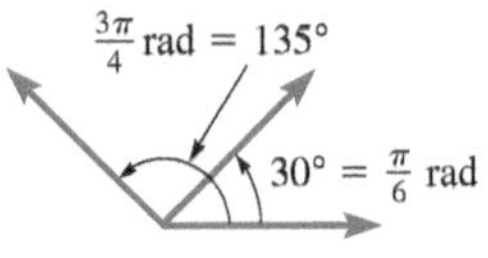

Fig. 8.22

EXAMPLE 2 Radians in terms of π

(a) Convert 30° to radians.

$$30° = 30°\left(\frac{\pi \text{ rad}}{180°}\right) = \frac{\pi}{6} \text{rad} \quad \text{(See Fig. 8.22.)}$$

(b) Convert $3\pi/4$ rad to degrees.

$$\frac{3\pi}{4} \text{ rad} = \frac{3\pi}{4} \text{ rad}\left(\frac{180°}{\pi \text{ rad}}\right) = 135° \quad \text{(See Fig. 8.22.)}$$

Practice Exercises

1. Convert 36° to radians in terms of π.
2. Convert $7\pi/9$ to degrees.

LEARNING TIP
When the angle is measured in radians, it is customary that no units are shown. The radian is understood to be the unit of measurement. If the symbol rad is used, it is only to emphasize that the angle is in radians.

We wish now to make a very important point. Since π is a number (a little greater than 3) that is the ratio of the circumference of a circle to its diameter, it is the ratio of one length to another. This means *radians have no units*, and *radian measure amounts to measuring angles in terms of real numbers*. It is this property that makes radians useful in many applications.

EXAMPLE 3 No angle units indicates radians

(**a**) $60.0° = 60.0°\left(\frac{\pi}{180°}\right) = \frac{\pi}{3.00} = 1.05$ 1.05 = 1.05 rad

no units indicates radian measure

(**b**) $3.80 = 3.80\left(\frac{180°}{\pi}\right) = 218°$

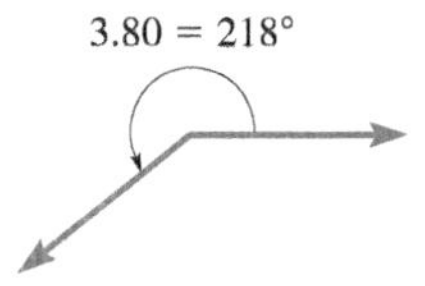

Fig. 8.23

Because no units are shown for 1.05 and 3.80, they are known to be measured in radians. See Fig. 8.23.

■ Most calculators have a function that switches between angular measurements in degrees and angular measurement in radians. Consult your calculator manual for details.

We can use a calculator to find the value of a function of an angle in radians. If the calculator is in radian mode, it then uses values in radians directly and will *consider any angle entered to be in radians*. The mode can be changed as needed, but *always be careful to have your calculator in the proper mode*.

COMMON ERROR
Check the setting in the *mode* feature on the calculator. If you are working in degrees, use the degree mode, but if you are working in radians, use the radian mode. It is a common error to use the values given by a calculator without ensuring that the proper mode is set (radians or degrees).

EXAMPLE 4 Calculator evaluations

Find the values of sin 0.7538, tan 0.9977, and cos 2.074.

The angles are provided with no units, so they are measured in radians. Setting the calculator in radian mode, we evaluate

$$\sin 0.7538 = 0.6844 \qquad \tan 0.9977 = 1.550 \qquad \cos 2.074 = -0.4822$$

Practice Exercise

3. Find the value of sin 3.56.

In the following application, the resulting angle is a unitless number, and it is therefore in radian measure.

EXAMPLE 5 Angles in radians—harmonic motion

The velocity v of an object undergoing simple harmonic motion at the end of a spring (Fig. 8.24) is given by

$$v = A\sqrt{\frac{k}{m}}\cos\left(\sqrt{\frac{k}{m}}(t)\right) \qquad \text{the angle is } \left(\sqrt{\frac{k}{m}}(t)\right)$$

Here, m is the mass of the object (in g), k is a constant depending on the spring, A is the maximum distance the object moves, and t is the time (in s). Find the velocity (in cm/s) after 0.100 s of a 36.0-g object at the end of a spring for which $k = 400\ \text{g/s}^2$, if $A = 5.00$ cm.

Substituting, we have

$$v = (5.00\ \text{cm})\sqrt{\frac{400\ \text{g/s}^2}{36.0\ \text{g}}}\cos\left(\sqrt{\frac{400\ \text{g/s}^2}{36.0\ \text{g}}}(0.100\ \text{s})\right)$$

Using calculator memory for $\sqrt{\frac{400}{36.0}}$, and with the calculator in radian mode, we have

$$v = 15.7\ \text{cm/s}$$

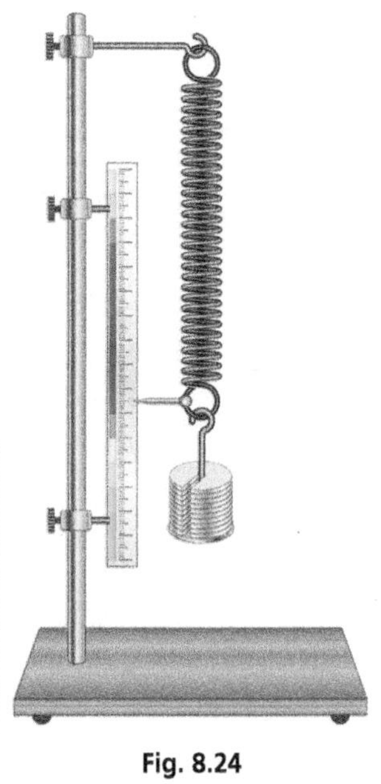
Source: Fouad A. Saad/Shutterstock
Fig. 8.24

If we need a reference angle in radians, recall that $\pi/2 = 90°$, $\pi = 180°$, $3\pi/2 = 270°$, and $2\pi = 360°$ (see Fig. 8.25). These and their decimal approximations are shown in Table 8.1.

Table 8.1 Quadrantal Angles

Degrees	Radians	Radians (decimal)
90°	$\frac{1}{2}\pi$	1.571
180°	π	3.142
270°	$\frac{3}{2}\pi$	4.712
360°	2π	6.283

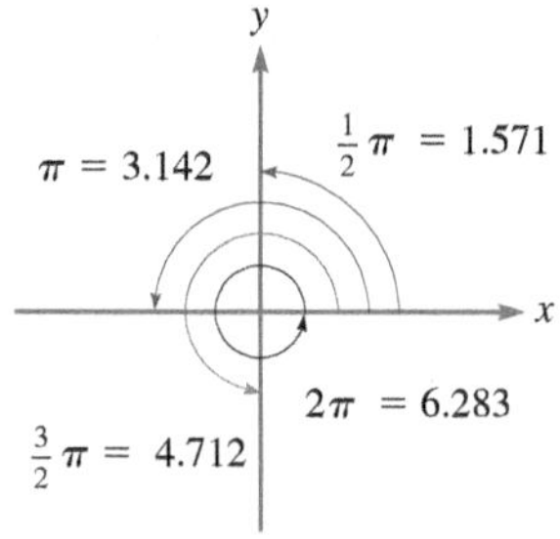

Fig. 8.25

EXAMPLE 6 Angles of any size in radians and reference angles in radians

Find the reference angle for the following angles: **(a)** 3.402 **(b)** 5.210

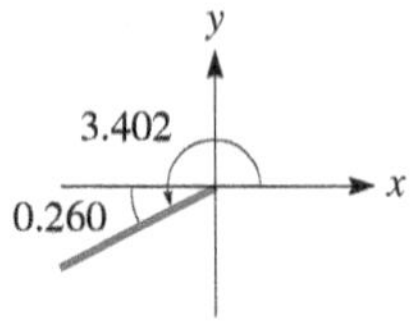

Fig. 8.26

(a) An angle of 3.402 is between π and $3\pi/2$. Thus, it is a third-quadrant angle, and the reference angle is $3.402 - \pi = 0.260$. The calculator π key can be used. See Fig. 8.26.

(b) An angle of 5.210 is between 4.712 and 6.283. Therefore, it is in the fourth quadrant, and the reference angle is $2\pi - 5.210 = 1.073$.

EXAMPLE 7 Find an angle in radians for a given function

Express θ in radians, such that $\cos\theta = 0.8829$ and $0 \le \theta < 2\pi$.

Since $\cos\theta$ is positive and θ is between 0 and 2π, we want a first-quadrant angle and a fourth-quadrant angle. With the calculator in radian mode,

$\cos^{-1} 0.8829 = 0.4888$ first-quadrant angle

$2\pi - 0.4888 = 5.794$ fourth-quadrant angle

$\theta = 0.4888$ or $\theta = 5.794$ see Fig. 8.27

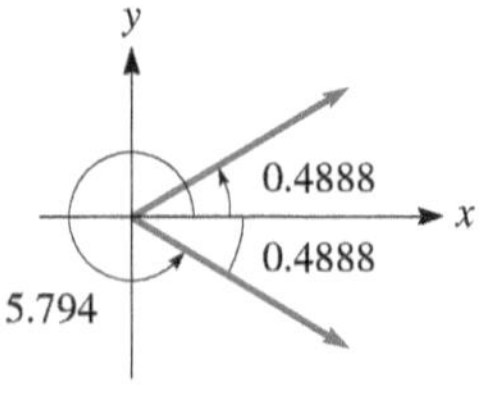

Fig. 8.27

COMMON ERROR When one first encounters radian measure, *expressions such as* sin 1 *and* $\sin\theta = 1$ *are often confused.* The first is equivalent to sin 57.30° (since 1 radian = 57.30°). The second means θ is the angle for which the sine is 1. Since sin 90° = 1, we can say that $\theta = 90°$ or $\theta = \pi/2$.

EXAMPLE 8 Angle in radians and value of a trigonometric function relationship

(a) $\sin\pi/3 = \sqrt{3}/2$

(b) $\cos\theta = 0.5$, $\theta = 60° = \pi/3$ (smallest positive θ)

(c) $\tan 2 = \tan 114.6°$

(d) $\tan\theta = 2$, $\theta = 1.107$ (smallest positive θ)

Practice Exercise

4. If $\sin\theta = 0.4235$, find the smallest positive θ (in radians).

EXERCISES 8.3

In Exercises 1 amd 2, make the given changes in the indicated examples of this section and then solve the resulting problems.

1. In Example 3(b), change 3.80 to 2.80.

2. In Example 7, change cos to sin.

In Exercises 3–10, express the given angle measurements in radian measure in terms of π.

3. 15°, 120° **4.** 12°, 225° **5.** 75°, 330° **6.** 36°, 315°

7. 210°, 99° **8.** 5°, 300° **9.** 720°, −9° **10.** −66°, 540°

In Exercises 11–18, the given numbers express angle measure. Express the measure of each angle in terms of degrees.

11. $\frac{3\pi}{5}, \frac{3\pi}{2}$ **12.** $\frac{3\pi}{10}, \frac{11\pi}{6}$ **13.** $\frac{5\pi}{9}, \frac{7\pi}{4}$ **14.** $\frac{8\pi}{15}, \frac{4\pi}{3}$

15. $\frac{7\pi}{18}, \frac{5\pi}{6}$ **16.** $\frac{\pi}{40}, \frac{5\pi}{4}$ **17.** $-\frac{\pi}{15}, \frac{3\pi}{20}$ **18.** $\frac{9\pi}{2}, \frac{-4\pi}{15}$

In Exercises 19–26, express the given angles in radian measure. Round off results to the number of significant digits in the given angle.

19. 84.0° **20.** 54.3° **21.** 252° **22.** 104°

23. −333.5° **24.** 268.7° **25.** 478.5° **26.** −86.1°

In Exercises 27–34, the given numbers express the angle measure. Express the measure of each angle in terms of degrees, with the same accuracy as the given value.

27. 0.750 **28.** 0.240 **29.** 3.407 **30.** 1.703

31. 12.4 **32.** −34.4 **33.** −16.42 **34.** 100.00

In Exercises 35–42, evaluate the given trigonometric functions by first changing the radian measure to degree measure. Round off results to four significant digits.

35. $\sin\frac{\pi}{4}$ **36.** $\cos\frac{\pi}{6}$ **37.** $\tan\frac{5\pi}{12}$ **38.** $\sin\frac{71\pi}{36}$

39. $\cos\frac{5\pi}{6}$ **40.** $\tan\left(-\frac{7\pi}{3}\right)$ **41.** sec 4.5920 **42.** cot 3.2732

In Exercises 43–50, evaluate the given trigonometric functions directly, without first changing to degree measure.

43. tan 0.7359 **44.** cos 1.4308 **45.** sin 4.24 **46.** tan 3.47

47. sec 2.07 **48.** sin(−1.34) **49.** cot(−4.86) **50.** csc 6.19

In Exercises 51–58, find θ to four significant digits for $0 \leq \theta < 2\pi$.

51. $\sin\theta = 0.3090$ **52.** $\cos\theta = -0.9135$

53. $\tan\theta = -0.2126$ **54.** $\sin\theta = -0.0436$

55. $\cos\theta = 0.6742$ **56.** $\cot\theta = 1.860$

57. $\sec\theta = -1.307$ **58.** $\csc\theta = 3.940$

In Exercises 59–64, solve the given problems. (Hint: Review cofunctions.)

59. Find the radian measure of an angle at the centre of a circle of radius 12 cm that intercepts an arc of 15 cm on the circle.

60. Find the length of arc of a circle of radius 10 cm that is intercepted from the centre of the circle by an angle of 3 radians.

61. Using the fact that $\sin\frac{\pi}{8} = 0.3827$, find the value of $\cos\frac{5\pi}{8}$. (A calculator should be used only to check the result.)

62. Using the fact that $\tan\frac{\pi}{6} = 0.5774$, find the value of $\cot\frac{5\pi}{3}$. (A calculator should be used only to check the result.)

63. Express $\tan\left(\frac{\pi}{2} + \theta\right)$ in terms of $\cot\theta$. $\left(0 < \theta < \frac{\pi}{2}\right)$

64. Express $\cos\left(\frac{3\pi}{2} + \theta\right)$ in terms of $\sin\theta$. $\left(0 < \theta < \frac{\pi}{2}\right)$

In Exercises 65–72, evaluate the given problems.

65. A unit of angle measurement used in artillery is the *mil*, which is defined as a central angle of a circle that intercepts an arc equal in length to 1/6400 of the circumference. How many mils are in a central angle of 34.4°?

66. Through how many radians does the minute hand of a clock move in 25 min?

67. After the brake was applied, a bicycle wheel went through 1.60 rotations. Through how many radians did a spoke rotate?

68. La Grande roue de Montréal is a Ferris wheel inaugurated in 2017. It is 60 m high and has 42 air-conditioned/heated passenger gondolas, each able to hold eight people. Through how many radians does it move after loading gondola 1, until gondola 29 is reached?

69. A flat plate of weight W oscillates as shown in Fig. 8.28. Its potential energy V is given by $V = \frac{1}{2}Wb\theta^2$, where θ is measured in radians. Find V if $W = 8.75$ N, $b = 0.75$ m, and $\theta = 5.5°$.

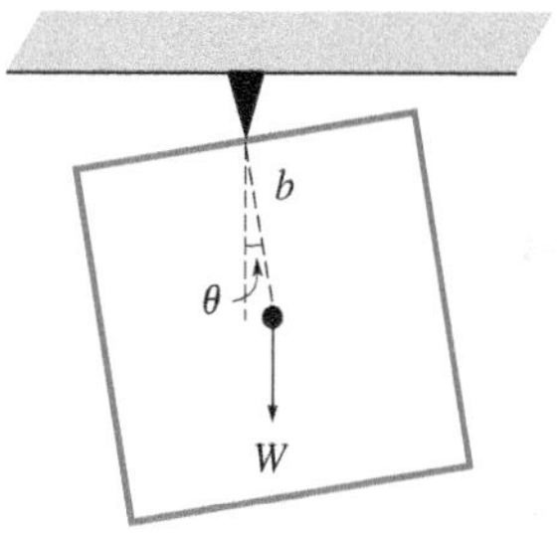

Fig. 8.28

70. The charge q (in C) on a capacitor as a function of time is $q = A\sin\omega t$. If t is measured in seconds, in what units is ω measured? Explain.

71. The height h of a rocket launched 1200 m from an observer is found to be $h = 1200\tan\frac{5t}{3t + 10}$ for $t < 10$ s, where t is the time after launch. Find h for $t = 8.0$ s.

72. The electric intensity I (in W/m^2) from the two radio antennas in Fig. 8.29 is a function of θ given by $I = 0.023\cos^2(\pi\sin\theta)$. Find I for $\theta = 40.0°$. ($\cos^2\alpha = (\cos\alpha)^2$.)

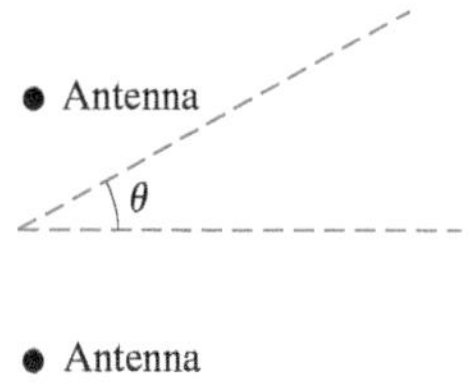

Fig. 8.29

Answers to Practice Exercises

1. $\pi/5$ **2.** 140° **3.** −0.406 **4.** 0.4373 rad

8.4 Applications of Radian Measure

Radian measure has numerous applications in mathematics and technology. In this section, we discuss applications involving circular arc length, the area of a sector, and the relationship between linear and angular velocity.

ARC LENGTH

From geometry, we know that *the length of an arc on a circle is proportional to the central angle* formed by the radii that intercept the arc. The length of arc of a complete circle is the circumference. Letting s represent the length of arc, we may state that $s = 2\pi r$ for a complete circle. Since 2π is the central angle (in radians) of the complete circle, the length s of any circular arc with central angle θ (in radians) is given by

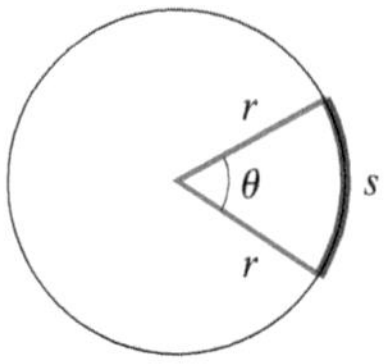

$$s = \theta r \quad (\theta \text{ in radians}) \qquad \textbf{(8.9)}$$

Therefore, if we know the central angle θ ***in radians*** and the radius of a circle, we can find the length of a circular arc directly by using Eq. (8.9). See Fig. 8.30.

Fig. 8.30

EXAMPLE 1 Arc length

Find the length of arc on a circle of radius $r = 3.00$ cm, for which the central angle $\theta = \pi/6$. See Fig. 8.31.

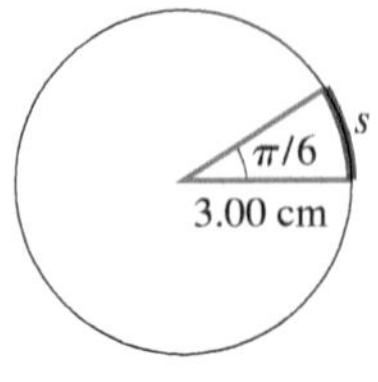

Fig. 8.31

θ in radians

$$s = \theta r$$

$$s = \left(\frac{\pi}{6}\right)(3.00 \text{ cm}) = \frac{\pi}{2.00} \text{ cm}$$

$$s = 1.57 \text{ cm}$$

Therefore, the length of arc s is 1.57 cm.

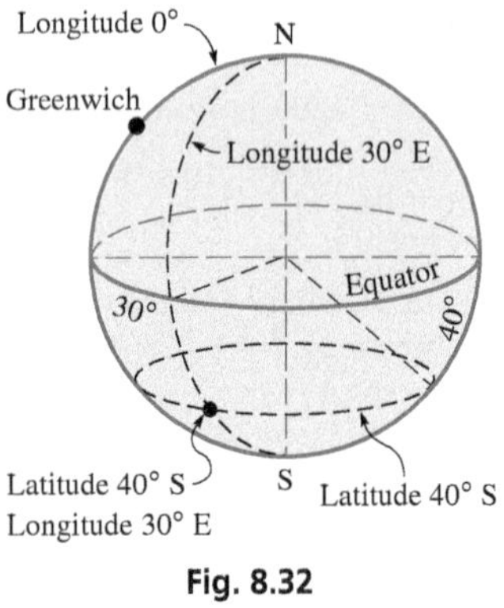

Fig. 8.32

Among the important applications of arc length are distances on the earth's surface. For most purposes, the earth may be regarded as a sphere (the diameter at the equator is slightly greater than the distance between the poles). A *great circle* of the earth (or any other sphere) is the circle of intersection of the surface of the sphere and a plane that passes through the centre.

The equator is a great circle and is designated as 0° *latitude*. Other *parallels of latitude* are parallel to the equator with diameters decreasing to zero at the poles, which are 90° N and 90° S. See Fig. 8.32.

Meridians of longitude are half great circles between the poles. The *prime meridian* through Greenwich, England, is designated as 0°, with meridians to 180° measured east and west from Greenwich. Positions on the surface of the earth are designated by longitude and latitude coordinates.

EXAMPLE 2 Arc length—nautical mile

The traditional definition of a *nautical mile* is the length of arc along a great circle of the earth for a central angle of 1′. The modern international definition is a distance of 1852 m. What measurement of the earth's radius does this definition use?

Here, $\theta = 1' = (1/60)°$, and $s = 1852$ m. Solving for r, we have

$$r = \frac{s}{\theta} = \frac{1852 \text{ m}}{\left(\frac{1}{60}\right)^{\circ}\left(\frac{\pi}{180^{\circ}}\right)} = 6.367 \times 10^6 \text{ m}$$

$$r = 6367 \text{ km}$$

Historically, the fact that the earth is not a perfect sphere has led to many variations in the distance used for a nautical mile.

Practice Exercise

1. Find θ in degrees if $s = 2.50$ m and $r = 1.75$ m.

AREA OF A SECTOR OF A CIRCLE

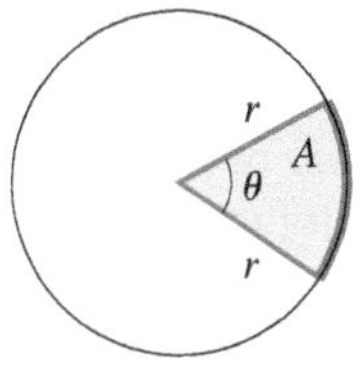

Fig. 8.33

Another application of radians is finding the area of a sector of a circle (see Fig. 8.33). Recall from geometry that areas of sectors of circles are proportional to their central angles. The area of a circle is $A = \pi r^2$, which can be written as $A = \frac{1}{2}(2\pi)r^2$. Since the angle for a complete circle is 2π, *the area of any sector of a circle in terms of the radius and central angle (in radians) is*

$$A = \frac{1}{2}\theta r^2 \qquad (\theta \text{ in radians}) \tag{8.10}$$

EXAMPLE 3 Area of a sector of a circle—text messages

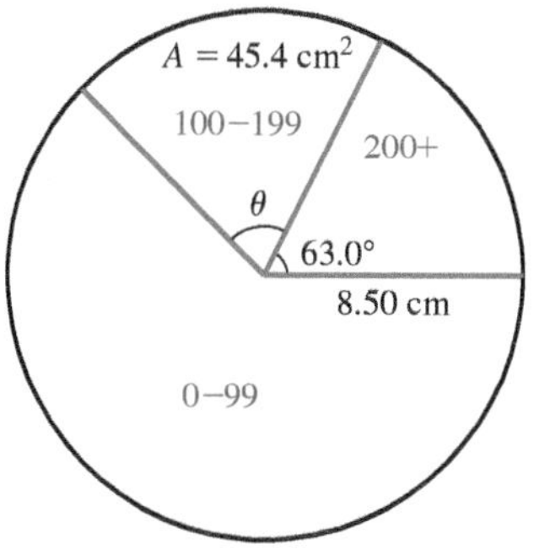

Fig. 8.34

(a) A pie chart of radius 8.50 cm shows that 17.5% of high school students send at least 200 text messages per day. Find the area of the sector (see Fig. 8.34).

The central angle of the sector is 63.0° (17.5% of 360°). Therefore, its area is

$$A = \frac{1}{2}(63.0°)\left(\frac{\pi}{180°}\right)(8.50\,\text{cm})^2 = 39.7\ \text{cm}^2$$

(θ in radians)

(b) The area of the pie chart in Fig. 8.34 that shows the percentage of students who send 100–199 text messages per day is 45.4 cm². Find the central angle of this sector.

We solve for θ in Eq. (8.10). This gives

$$\theta = \frac{2A}{r^2} = \frac{2(45.4\ \text{cm}^2)}{(8.50\ \text{cm})^2} = 1.26$$

no units indicates radian measure

This means that the central angle is 1.26 rad, or 72.2°.

Practice Exercise

2. Find A if $r = 17.5$ cm and $\theta = 125°$.

LINEAR AND ANGULAR VELOCITY

The average velocity of a moving object is defined by $v = s/t$, where v is the average velocity, s is the distance travelled, and t is the elapsed time. For an object moving in a circular path with constant speed, the distance travelled is the length of arc through which it moves. Therefore, if we divide both sides of Eq. (8.9) by t, we obtain

$$\frac{s}{t} = \frac{\theta r}{t} = \frac{\theta}{t}r$$

where we define

$$\omega = \theta/t \tag{8.11}$$

as the angular velocity. Therefore,

$$v = \omega r \tag{8.12}$$

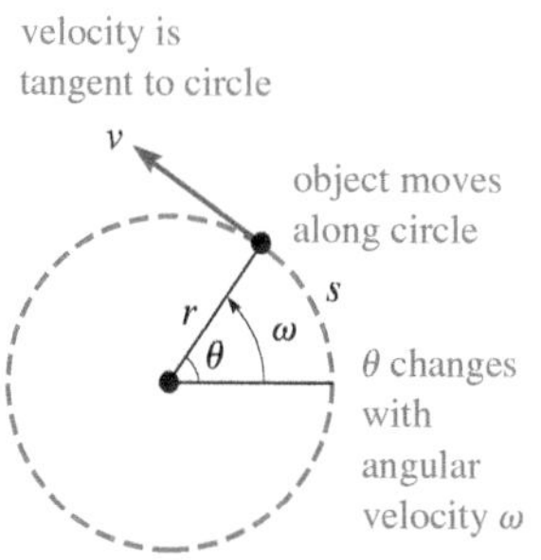

Fig. 8.35

Eq. (8.12) expresses the relationship between the **linear velocity** v *and the* **angular velocity** ω *of an object moving around a circle of radius r.* See Fig. 8.35. In the figure, v is shown directed tangent to the circle, for that is its direction for the position shown. The direction of v changes constantly.

The standard SI units for ω are radians per second (rad/s). In this way, the formula can be used directly. However, in practice, ω is often given in revolutions per minute or in some similar unit. In these cases, it is necessary to convert the units of ω to radians per unit of time before substituting in Eq. (8.12).

■ Some of the typical units used for angular velocity are

rad/s rad/min rad/h
°/s °/min
r/s r/min

(r represents revolutions.)

(r/min is the same as rpm. This text does not use rpm.)

EXAMPLE 4 Angular velocity—hang glider motion

A person on a hang glider is moving in a horizontal circular arc of radius 90.0 m with an angular velocity of 0.125 rad/s. The person's linear velocity is

$$v = (0.125\ \text{rad/s})(90.0\ \text{m}) = 11.3\ \text{m/s}$$

(Remember that radians are numbers and are not included in the final set of units.) This means that the person is moving along the circumference of the arc at 11.3 m/s (40.7 km/h).

EXAMPLE 5 Angular velocity—geosynchronous satellites

A communications satellite remains at an altitude of 35 920 km above a point on the equator. If the radius of the earth is 6370 km, what is the velocity of the satellite?

In order for the satellite to remain over a point on the equator, it must rotate exactly once each day around the centre of the earth (and it must remain at an altitude of 35 920 km). Since there are 2π radians in each revolution, the angular velocity is

$$\omega = \frac{1\ \text{r}}{1\ \text{day}}\left(\frac{1\ \text{day}}{24\ \text{h}}\right)\left(\frac{2\pi\ \text{rad}}{1\ \text{r}}\right) = 0.2618\ \text{rad/h}$$

The radius of the circle through which the satellite moves is its altitude plus the radius of the earth, or 35 920 + 6370 = 42 290 km. Thus, the velocity is

$$v = \omega r$$
$$v = (0.2618\ \text{rad/h})(42\ 290\ \text{km})$$
$$v = 11\ 070\ \text{km/h}$$
$$v = 11\ 070\ \text{km/h}\left(\frac{1000\ \text{m}}{1\ \text{km}}\right)\left(\frac{1\ \text{h}}{3600\ \text{s}}\right)$$
$$v = 3075\ \text{m/s}$$

Practice Exercise

3. Find r if $v = 25.0$ m/s and the angular velocity is 6.0°/min.

Mechanical elements that are connected on their edges (like pulleys with a belt, or meshed gears) have the same linear velocity v on their edges. Mechanical elements that are mounted on the same shaft have the same rotational properties (same amount of rotation, same rate of rotation).

EXAMPLE 6 Angular velocity—gears and pulleys

As illustrated in Fig. 8.36, a pulley belt is 2.00 m long and takes 0.800 s to make one complete revolution around pulleys A and B. Pulley B and gear C are mounted on the same shaft, and the teeth of gear D mesh with those of gear C. The radius of pulley A is 0.150 m, the radius of pulley B is 0.100 m, the radius of gear C is 0.300 m, and the radius of gear D is 0.200 m. What is the resulting angular velocity (in rad/s and in r/min) of gear D?

Since the linear velocity of a point on the edge of pulleys A and B is the same as the linear velocity of the belt,

$$v = s/t$$
$$v = 2.00\ \text{m}/0.800\ \text{s}$$
$$v = 2.50\ \text{m/s}$$

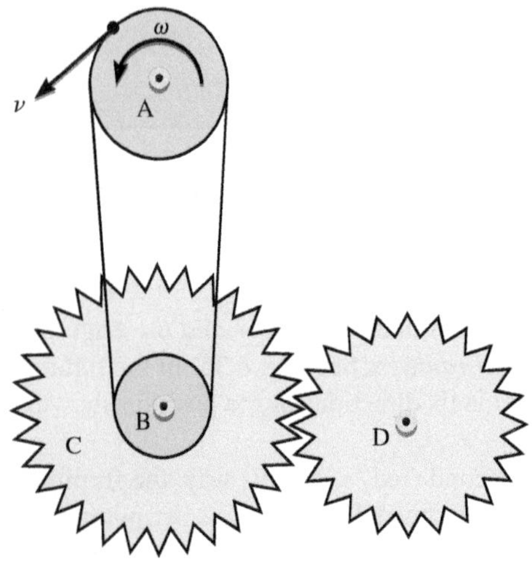

Fig. 8.36

The angular velocity ω for pulley B is then

$$\omega_B = v_B/r_B$$
$$\omega_B = 2.50\ \text{m/s}/0.100\ \text{m}$$
$$\omega_B = 25.0\ \text{rad/s}$$

Since gear C and pulley B are on the same shaft, they have the same angular velocity, $\omega_C = \omega_B = 25.0 \text{ rad/s}$, so the linear velocity of the edge of gear C can be found.

$$v_C = \omega_C r_C$$
$$v_C = 25.0 \text{ rad/s } (0.300 \text{ m})$$
$$v_C = 7.50 \text{ m/s}$$

■ See the chapter introduction.

Since gear C and gear D are meshed, the linear velocities of their edges (their teeth) must be the same, so $v_D = v_C = 7.50$ m/s. The angular velocity of gear D can then be found.

$$\omega_D = v_D/r_D$$
$$\omega_D = 7.50 \text{ m/s}/0.200 \text{ m}$$
$$\omega_D = 37.5 \text{ rad/s}$$
$$\omega_D = 37.5 \text{ rad/s}\left(\frac{1 \text{ r}}{2\pi \text{ rad}}\right)\left(\frac{60 \text{ s}}{1 \text{ min}}\right)$$
$$\omega_D = 358 \text{ r/min}$$

Since A, B, and C all spin counterclockwise, at the teeth connection point between gears C and D, the teeth will be moving vertically upward. Thus, the rotation of gear D must be clockwise.

ANGULAR ACCELERATION

The average acceleration of a moving object is defined by $a = \frac{v_f - v_i}{t}$, where a is the average acceleration, v_f and v_i are the final and initial velocities respectively, and t is the elapsed time. For an object moving in a circular path with a changing angular velocity, if we divide the acceleration equation by the radius r, we obtain

$$\frac{a}{r} = \frac{\frac{v_f}{r} - \frac{v_i}{r}}{t}$$

However, from Eq. (8.12) we see that the ratio of linear velocity to radius is angular velocity ω. We thus define

$$\alpha = \frac{\omega_f - \omega_i}{t} \qquad \textbf{(8.13)}$$

where α is the angular acceleration. Comparing the above two equations gives

$$a = \alpha r \qquad \textbf{(8.14)}$$

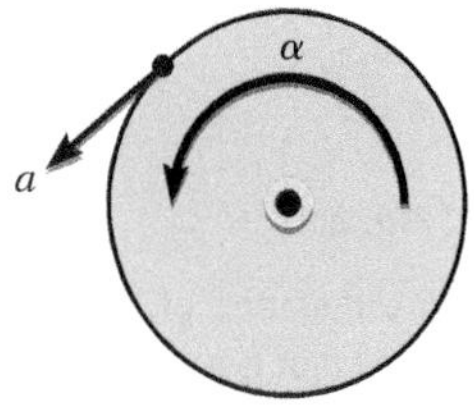

Fig. 8.37

Eq. (8.14) represents the relationship between the linear acceleration a and the angular acceleration α of an object moving in a circle with a changing magnitude for its velocity. See Fig. 8.37. The standard SI units for α are rad/s^2.

EXAMPLE 7 Angular acceleration—motion of a flywheel

A flywheel has a radius 40.0 cm, and its edge is accelerating at the rate of 4.25 m/s^2. Determine its angular acceleration α.

The wheel radius expressed in metres is $r = 40.0 \text{ cm} = 0.400 \text{ m}$. Eq. (8.14) gives

$$a = \alpha r$$
$$\alpha = \frac{a}{r}$$
$$\alpha = \frac{4.25 \text{ m/s}^2}{0.400 \text{ m}}$$
$$\alpha = 10.6 \text{ rad/s}^2$$

COMMON ERROR When dealing with arc length, area of a sector of a circle, angular velocity, or angular acceleration, the equations require that the angle θ be expressed in radians. A common error is to use θ in degrees.

EXAMPLE 8 Application to electric current

The current at any time in a certain alternating-current electric circuit is given by $i = I\sin 120\pi t$, where I is the maximum current and t is the time in seconds. Given that $I = 0.0685$ A, find i for $t = 0.005\,00$ s.

Substituting, with the calculator in radian mode, we get

$$i = (0.0685\text{ A})\sin[(120\pi)(0.005\,00)]$$
$$i = 0.0651\text{ A}$$

$(120\pi)(0.005\,00)$ is a pure number, and therefore is an angle in radians.

EXERCISES 8.4

In Exercises 1–4, make the given changes in the indicated examples of this section, and then solve the resulting problems.

1. In Example 1, change $\pi/6$ to $\pi/4$.
2. In Example 3(a), change 17.5% to 18.5%
3. In Example 4, change 90.0 m to 115 m.
4. In Example 6, change 0.800 s to 1.20 s.

In Exercises 5–16, for an arc length s, area of sector A, and central angle θ of a circle of radius r, find the indicated quantity for the given values.

5. $r = 5.70$ cm, $\theta = \pi/3$, $s = ?$
6. $r = 21.2$ cm, $\theta = 2.65$, $s = ?$
7. $s = 915$ mm, $\theta = 136.0°$, $r = ?$
8. $s = 0.3456$ m, $\theta = 73.61°$, $A = ?$
9. $s = 0.3913$ km, $r = 0.9449$ km, $A = ?$
10. $s = 3.19$ m, $r = 2.29$ m, $\theta = ?$
11. $r = 3.8$ cm, $\theta = 6.7$, $A = ?$
12. $r = 46.3$ dm, $\theta = 2\pi/5$, $A = ?$
13. $A = 0.0119$ m^2, $\theta = 326.0°$, $r = ?$
14. $A = 1200$ mm^2, $\theta = 17°$, $s = ?$
15. $A = 165$ m^2, $r = 40.2$ m, $s = ?$
16. $A = 67.8$ km^2, $r = 67.8$ km, $\theta = ?$

In Exercises 17–56, solve the given problems.

17. In travelling three-fourths of the way around a traffic circle, a car travels 0.203 km. What is the radius of the traffic circle? See Fig. 8.38.

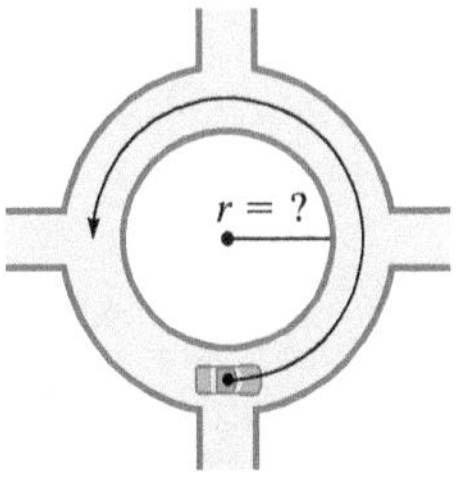

Fig. 8.38

18. The latitude of Sudbury, Ontario, is 46° N, and the latitude of Orlando, Florida, is 28° N. Both are at a longitude of 81° W. What is the distance between Sudbury and Orlando? Explain how the angle used in the solution is found. The radius of the earth is 6370 km.
19. Of the estimated natural gas reserves in North America, 8.73×10^{12} m^3 are in the United States, 1.89×10^{12} m^3 are in Canada, and 4.84×10^{12} m^3 are in Mexico. In making a *pie chart* with a radius of 4.00 cm for these data (see Example 3), what are the central angle and area of the sector that represents Canada's reserves?
20. A pizza is cut into eight equal pieces, the area of each being 88 cm^2. What was the diameter of the original pizza?
21. When between 12:00 noon and 1:00 P.M. are the minute and hour hands of a clock 180° apart?
22. A cam is in the shape of a circular sector, as shown in Fig. 8.39. What is the perimeter of the cam?

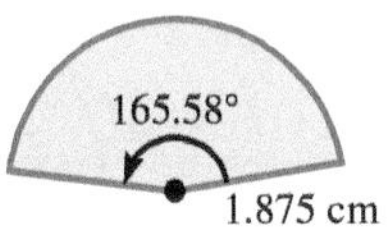

Fig. 8.39

23. A lawn sprinkler can water up to a distance of 25.0 m. It turns through an angle of 115.0°. What area can it water?
24. A spotlight beam sweeps through a horizontal angle of 75.0°. If the range of the spotlight is 115 m, what area can it cover?
25. If a car makes a U-turn in 6.0 s, what is its average angular velocity in the turn?
26. A ceiling fan has blades 61.0 cm long. What is the linear velocity of the tip of a blade when the fan is rotating at 8.50 r/s?
27. What is the floor area of the hallway shown in Fig. 8.40? The outside and inside of the hallway are circular arcs.

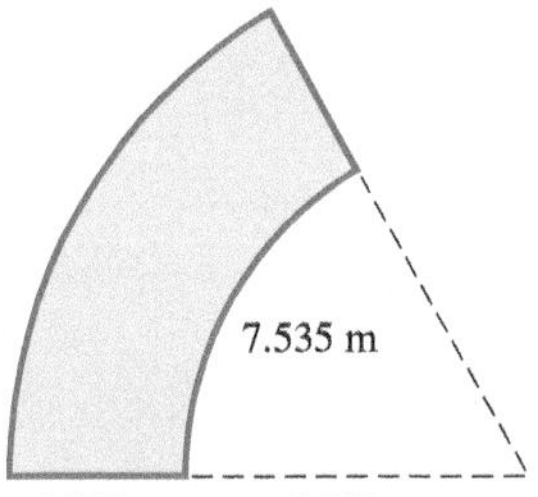

Fig. 8.40

28. The arm of a car windshield wiper is 32.4 cm long and is attached at the middle of a 38.1-cm blade. (Assume that the arm and blade are in line.) What area of the windshield is cleaned by the wiper if it swings through 110.0° arcs?

29. Part of a railroad track follows a circular arc with a central angle of 28.0°. If the radius of the arc of the inner rail is 28.55 m and the rails are 1.44 m apart, how much longer is the outer rail than the inner rail?

30. A wrecking ball is dropped as shown in Fig. 8.41. Its velocity at the bottom of its swing is $v = \sqrt{2gh}$, where g is the acceleration due to gravity. What is its angular velocity at the bottom if $g = 9.80\ \text{m/s}^2$ and $h = 4.80$ m?

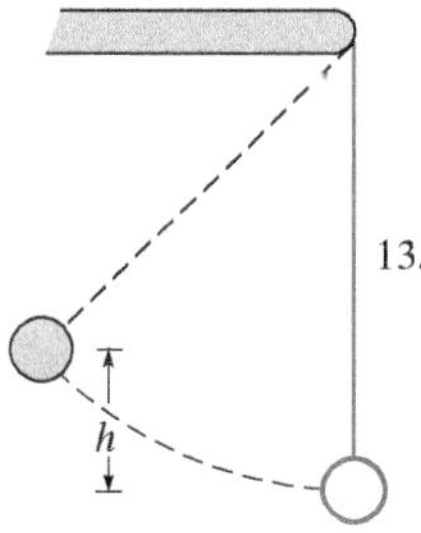

Fig. 8.41

31. Part of a security fence is built 2.50 m from a cylindrical storage tank 11.2 m in diameter. What is the area between the tank and this part of the fence if the central angle of the fence is 75.5°? See Fig. 8.42.

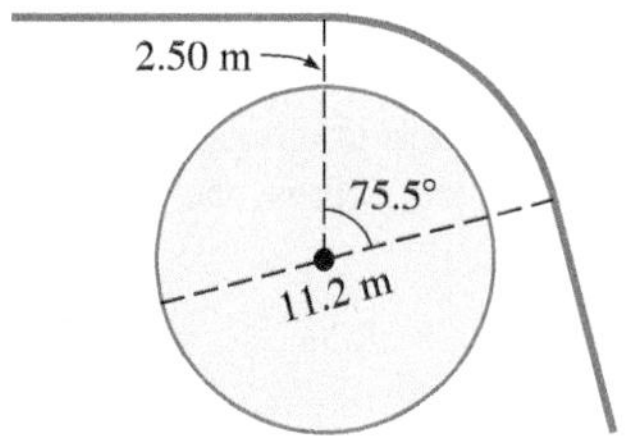

Fig. 8.42

32. Through what angle does the drum in Fig. 8.43 turn in order to lower the crate 10.3 m? If the crate accelerates downward at $3.75\ \text{m/s}^2$, what is the angular acceleration of the drum?

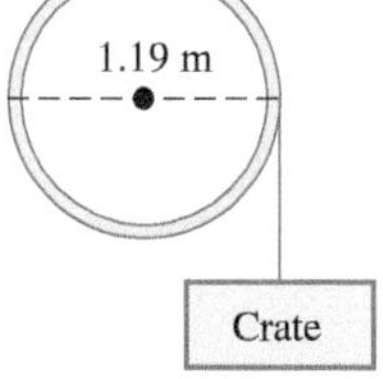

Fig. 8.43

33. A section of road follows a circular arc with a central angle of 15.6°. The radius of the inside of the curve is 285.0 m, and the road is 15.2 m wide. What is the volume of the concrete in the road if it is 0.305 m thick?

34. The propeller of the motor on a motorboat is rotating at 133 rad/s. What is the linear velocity of a point on the tip of a blade if it is 22.5 cm long?

35. A storm causes a pilot to follow a circular-arc route, with a central angle of 12.8°, from city A to city B rather than the straight-line route of 185.0 km. How much farther does the plane fly due to the storm?

36. A freeway interchange exit is a circular arc 330 m long with a central angle of 79.4°. What is the radius of curvature of the exit?

37. The paddles of a riverboat have a radius of 2.59 m and revolve at 20.0 r/min. What is the speed of a tip of one of the paddles (in m/s)?

38. The sweep second hand of a watch is 15.0 mm long. What is the linear velocity of the tip?

39. A DVD has a diameter of 12.1 cm and rotates at 360.0 r/min. What is the linear velocity of a point on the outer edge? At startup the DVD accelerates from rest to 360.0 r/min in 0.200 s. Find the linear acceleration of a point on its edge.

40. The *Singapore Flyer* is a Ferris wheel that has 28 air-conditioned capsules, each able to hold 28 people. It is 165 m high, with a 150-m-diameter wheel, and makes one revolution in 37 min. Find the speed (in cm/s) of a capsule.

41. GPS satellites orbit Earth twice per day with an orbital radius of 26 600 km. What is the linear velocity (in km/s) of these satellites?

42. Assume that Earth rotates around the sun in a circular orbit of radius 150 000 000 km (which is approximately correct). What is Earth's linear velocity (in km/h)? (Use 1 yr = 365.25 d.)

43. The sprocket assembly for a 70.0-cm bike is shown in Fig. 8.44. How fast (in r/min) does the rider have to pedal in order to go 25.0 km/h on level ground?

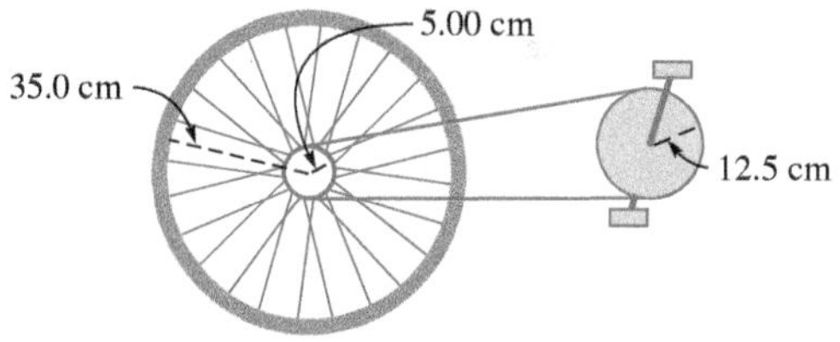

Fig. 8.44

44. The flywheel of a car engine is 0.36 m in diameter. If it is revolving at 750 r/min, through what distance does a point on the rim move in 2.00 s?

45. Two streets meet at an angle of 82.0°. What is the length of the piece of curved curbing at the intersection if it is constructed along the arc of a circle 5.50 m in radius? See Fig. 8.45.

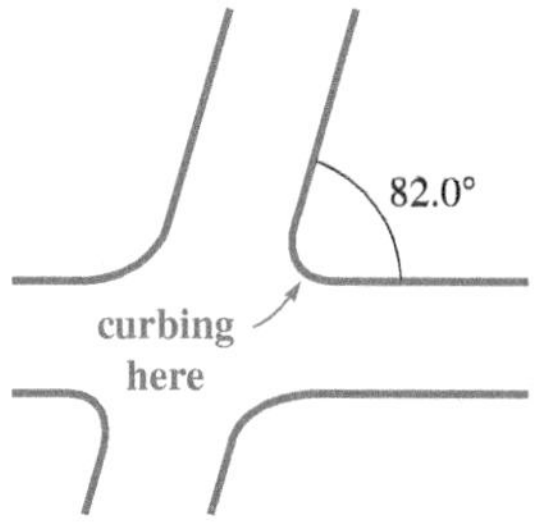

Fig. 8.45

46. An ammeter needle is deflected 52.00° by a current of 0.2500 A. The needle is 3.750 cm long, and a circular scale is used. How long is the scale for a maximum current of 1.500 A?

47. A drill bit 9.53 mm in diameter rotates at 1260 r/min. What is the linear velocity of a point on its circumference (in mm/s)?

48. A helicopter blade is 2.75 m long and from rest accelerates for 5.00 s until it is rotating at 420 r/min. What is the average linear acceleration of the tip of the blade (in m/s^2)?

49. Most DVD players use *constant linear velocity*. This means that the angular velocity of the disk continually changes, but the linear velocity of the point being read by the laser remains the same. If a DVD rotates at 1590 r/min for a point 2.25 cm from the centre, find the number of revolutions per minute for a point 5.51 cm from the centre.

50. A jet is travelling westward with the sun directly overhead (the jet is on a line between the sun and the centre of the earth). How fast must the jet fly in order to keep the sun directly overhead? (Assume that the earth's radius is 6370 km, the altitude of the jet is 11 km, and the earth rotates about its axis once in 24.0 h.)

51. What is the linear velocity of a point in Winnipeg, Manitoba, which is at a latitude of 49°53′ N? The radius of the earth is 6370 km.

52. Through what total angle does the drive shaft of a car rotate in 1.0 s when the tachometer reads 2400 r/min? If it starts from rest, what is the drive shaft's average angular acceleration if it takes 3.25 s to reach 2400 r/min?

53. A baseball field is designed such that the outfield fence is along the arc of a circle with its centre at second base. If the radius of the circle is 85.0 m, what is the playing area of the field? See Fig. 8.46.

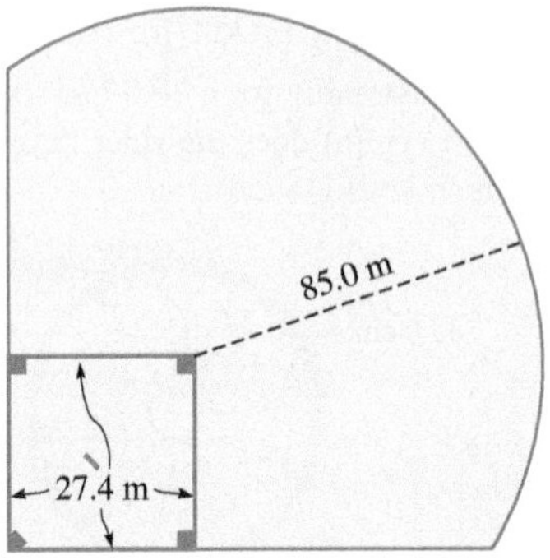

Fig. 8.46

54. A patio is in the shape of a circular sector with a central angle of 160.0°. It is enclosed by a railing of which the circular part is 11.6 m long. What is the area of the patio?

55. An oil storage tank 4.25 m long has a flat bottom as shown in Fig. 8.47. The radius of the circular part is 1.10 m. What volume of oil does the tank hold?

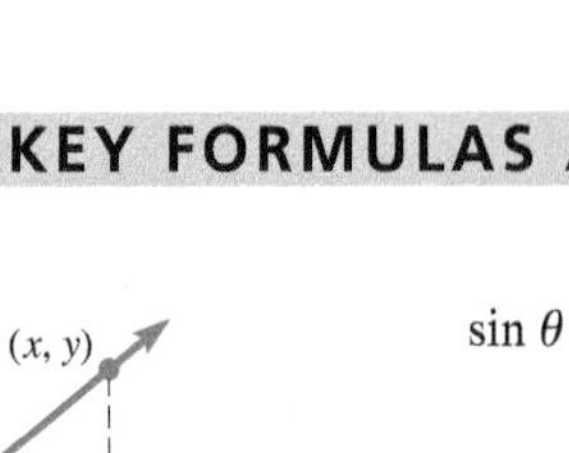

Fig. 8.47

56. Two equal beams of light illuminate the area shown in Fig. 8.48. What area is lit by both beams?

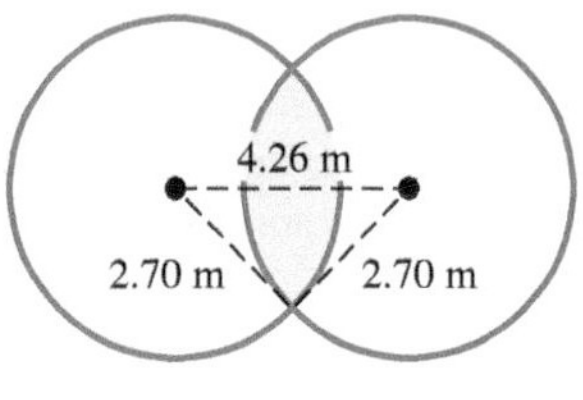

Fig. 8.48

In Exercises 57–60, another use of radians is illustrated.

57. Use a calculator (in radian mode) to evaluate the ratios $(\sin\theta)/\theta$ and $(\tan\theta)/\theta$ for $\theta = 0.1, 0.01, 0.001$, and 0.0001. From these values, explain why it is possible to say that

$$\sin\theta = \tan\theta = \theta \qquad \textbf{(8.15)}$$

approximately for very small angles.

58. Using Eq. (8.15), evaluate tan 0.001°. Compare with a calculator value.

59. An astronomer observes that a star 12.5 light-years away moves through an angle of 0.2″ in 1 year. Assuming it moved in a straight line perpendicular to the initial line of observation, how many kilometres did the star move? (1 light-year $= 9.46 \times 10^{12}$ km.) Use Eq. (8.15) and express the answer to three significant digits.

60. In calculating a back line of a lot, a surveyor discovers an error of 0.05° in an angle measurement. If the lot is 136.0 m deep, by how much is the back-line calculation in error? See Fig. 8.49. Use Eq. (8.15) and express the answer to three significant digits.

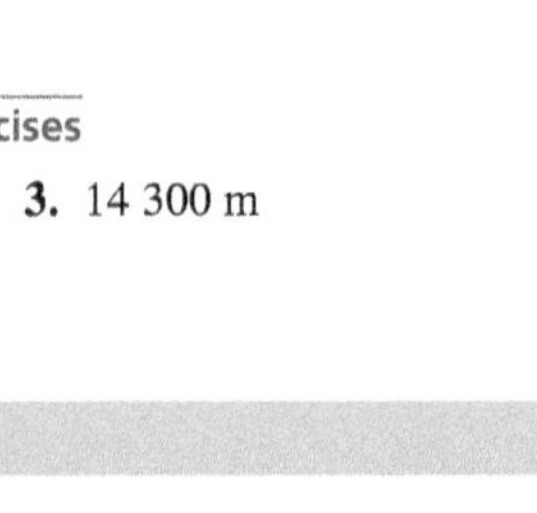

Fig. 8.49

Answers to Practice Exercises

1. 81.9° **2.** 334 cm^2 **3.** 14 300 m

CHAPTER 8 KEY FORMULAS AND EQUATIONS

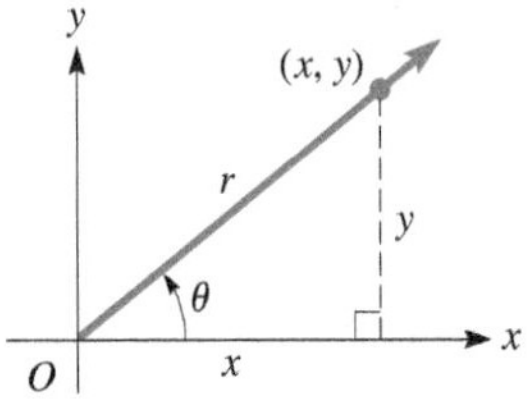

$$\sin\theta = \frac{y}{r} \qquad \csc\theta = \frac{1}{\sin\theta} = \frac{r}{y}$$

$$\cos\theta = \frac{x}{r} \qquad \sec\theta = \frac{1}{\cos\theta} = \frac{r}{x} \qquad \textbf{(8.1)}$$

$$\tan\theta = \frac{y}{x} \qquad \cot\theta = \frac{1}{\tan\theta} = \frac{x}{y}$$

$\text{trig}\,\theta = \pm\text{trig}\,\theta_{\text{ref}}$ ("trig" is any trigonometric function) **(8.2)**

Using reference angles	$\theta = \theta_{ref}$ (first quadrant) $\theta = 180° - \theta_{ref}$ (second quadrant) $\theta = 180° + \theta_{ref}$ (third quadrant) $\theta = 360° - \theta_{ref}$ (fourth quadrant)		**(8.3)**

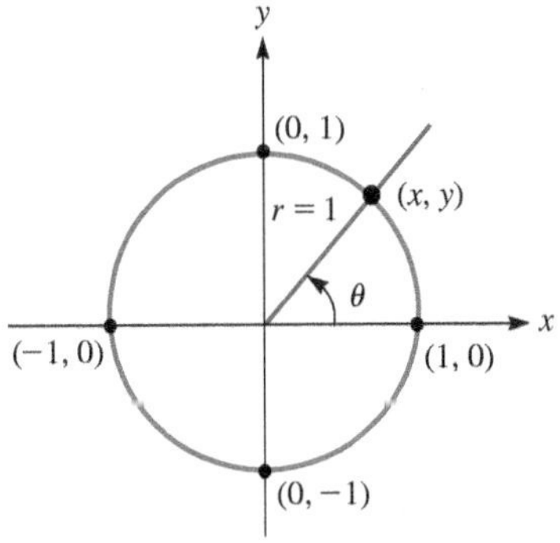

$$\sin\theta = y \qquad \csc\theta = \frac{1}{y}$$

$$\cos\theta = x \qquad \sec\theta = \frac{1}{x} \qquad \textbf{(8.4)}$$

$$\tan\theta = \frac{y}{x} \qquad \cot\theta = \frac{x}{y}$$

Negative angles

$\sin(-\theta) = -\sin\theta$ $\quad \cos(-\theta) = \cos\theta$ $\quad \tan(-\theta) = -\tan\theta$

$\csc(-\theta) = -\csc\theta$ $\quad \sec(-\theta) = \sec\theta$ $\quad \cot(-\theta) = -\cot\theta$ **(8.5)**

Radian degree conversions $\quad \pi \text{ rad} = 180°$ **(8.6)**

$$1° = \frac{\pi}{180} \text{ rad} = 0.017\,45 \text{ rad} \qquad \textbf{(8.7)}$$

$$1 \text{ rad} = \frac{180°}{\pi} = 57.30° \qquad \textbf{(8.8)}$$

Circular arc length $\quad s = \theta r \quad (\theta \text{ in radians})$ **(8.9)**

Circular sector area $\quad A = \frac{1}{2}\theta r^2 \quad (\theta \text{ in radians})$ **(8.10)**

Angular velocity $\quad \omega = \frac{\theta}{t}$ **(8.11)**

Linear and angular velocity $\quad v = \omega r$ **(8.12)**

Angular acceleration $\quad \alpha = \frac{\omega_f - \omega_i}{t}$ **(8.13)**

Linear and angular acceleration $\quad a = \alpha r$ **(8.14)**

Small angle approximation $\quad \sin\theta = \tan\theta = \theta \quad$ (for small θ in radians) **(8.15)**

CHAPTER 8 **REVIEW EXERCISES**

In Exercises 1–4, find the trigonometric functions of θ. The terminal side of θ passes through the given point.

1. (6, 8) **2.** (−12, 5) **3.** (42, −12) **4.** (−0.2, −0.3)

In Exercises 5–8, express the given trigonometric functions in terms of the same function of a positive acute angle.

5. cos 132°, tan 194° **6.** sin 243°, cot 318°

7. sin 289°, sec(−15°) **8.** cos 463°, csc(−100°)

In Exercises 9–12, express the given angle measurements in terms of π.

9. 40°, 153° **10.** 22.5°, 324° **11.** 408°, 202.5° **12.** 12°, −162°

In Exercises 13–20, the given numbers represent angle measure. Express the measure of each angle in degrees.

13. $\frac{7\pi}{5}, \frac{13\pi}{18}$ **14.** $\frac{3\pi}{8}, \frac{7\pi}{20}$ **15.** $\frac{\pi}{15}, \frac{11\pi}{6}$ **16.** $\frac{27\pi}{10}, \frac{5\pi}{4}$

17. 0.560 **18.** −1.354 **19.** −36.07 **20.** 14.5

In Exercises 21–28, express the given angles in radians (not in terms of π).

21. 102° **22.** 305° **23.** 20.25° **24.** 148.38°

25. 262.05° **26.** −18.72° **27.** −636.2° **28.** 385.4°

In Exercises 29–48, determine the values of the given trigonometric functions directly on a calculator. The angles are approximate. Express answers to Exercises 41–44 to four significant digits.

29. $\cos 237.4°$ **30.** $\sin 141.3°$ **31.** $\cot 295°$

32. $\tan 184°$ **33.** $\csc 247.82°$ **34.** $\sec 96.17°$

35. $\sin 542.8°$ **36.** $\cos 326.72°$ **37.** $\tan 301.4°$

38. $\sin 703.9°$ **39.** $\tan 436.42°$ **40.** $\cos(-162.32°)$

41. $\sin \frac{9\pi}{5}$ **42.** $\sec \frac{5\pi}{8}$ **43.** $\cos -\frac{7\pi}{6}$

44. $\tan \frac{23\pi}{12}$ **45.** $\sin 0.5906$ **46.** $\tan 0.8035$

47. $\csc 2.153$ **48.** $\cos(-7.190)$

In Exercises 49–52, find θ in degrees for $0° \leq \theta < 360°$.

49. $\tan \theta = 0.1817$ **50.** $\sin \theta = -0.9323$

51. $\cos \theta = -0.4730$ **52.** $\cot \theta = 1.196$

In Exercises 53–56, find θ in radians for $0 \leq \theta < 2\pi$.

53. $\cos \theta = 0.8387$ **54.** $\csc \theta = 9.569$

55. $\sin \theta = -0.8650$ **56.** $\tan \theta = 8.480$

In Exercises 57–60, find θ in degrees for $0° \leq \theta < 360°$.

57. $\cos \theta = -0.672$, $\sin \theta < 0$ **58.** $\tan \theta = -1.683$, $\cos \theta < 0$

59. $\cot \theta = 0.4291$, $\cos \theta < 0$ **60.** $\sin \theta = 0.2626$, $\tan \theta < 0$

In Exercises 61–68, for an arc of length s, area of sector A, and central angle θ of circle of radius r, find the indicated quantity for the given values.

61. $s = 20.3$ cm, $\theta = 107.5°$, $r = ?$

62. $s = 5840$ m, $r = 1060$ m, $\theta = ?$

63. $A = 265$ mm^2, $r = 12.8$ mm, $\theta = ?$

64. $A = 0.908$ km^2, $\theta = 234.5°$, $r = ?$

65. $r = 4.62$ m, $A = 32.8$ m^2, $s = ?$

66. $\theta = 98.5°$, $A = 0.493$ dm^2, $s = ?$

67. $\theta = 0.85°$, $s = 7.94$ cm, $A = ?$

68. $r = 254$ cm, $s = 76.1$ cm, $A = ?$

In Exercises 69–93, solve the given problems.

69. In Fig. 8.50, show that the area of a design label (a segment of a circle) intercepted by angle θ is $A = \frac{1}{2}r^2(\theta - \sin \theta)$. Find the area if $r = 4.00$ cm and $\theta = 1.45$.

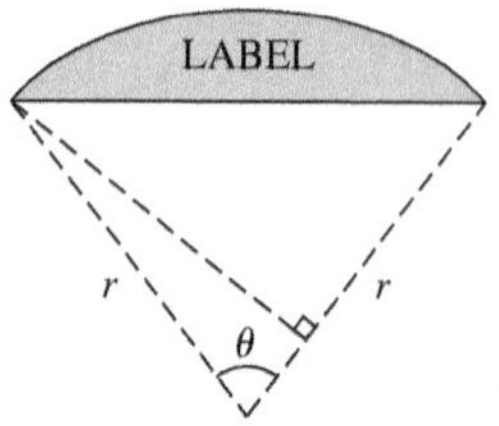

Fig. 8.50

70. The cross section of a tunnel is the major segment of a circle of radius 12.0 m wide. The base of the tunnel is 20.0 m wide. What is the area of the cross section? See Exercise 69.

71. Find the area of the parcel of land shown in Fig. 8.51. It is a right triangle attached to a circle sector.

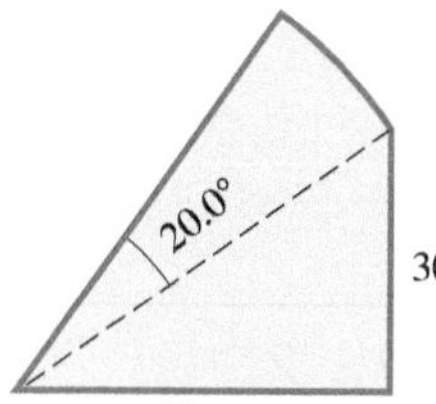

Fig. 8.51

72. The speedometer of a car is designed to be accurate with tires that are 35.6 cm in radius. If the tires are changed to 38.1 cm in radius, and the speedometer shows 88.0 km/h, how fast is the car actually going?

73. The instantaneous power P (in W) input to a resistor in an alternating-current circuit is $P = P_m \sin^2 377t$, where P_m is the maximum power input and t is the time (in s). Find P for $P_m = 0.120$ W and $t = 2.00$ ms. ($\sin^2\theta = (\sin \theta)^2$.)

74. The horizontal distance x through which a pendulum moves is given by $x = a(\theta + \sin \theta)$, where a is a constant and θ is the angle between the vertical and the pendulum. Find x for $a = 45.0$ cm and $\theta = 0.175$.

75. A sector gear with a pitch radius of 8.25 cm and a 6.60-cm arc of contact is shown in Fig. 8.52. What is the sector angle θ (in °)?

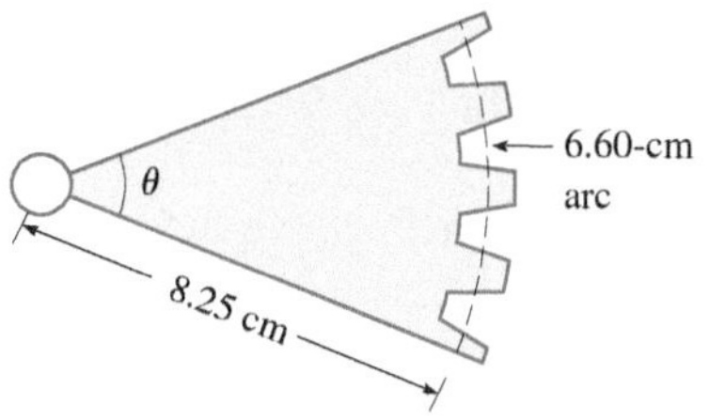

Fig. 8.52

76. Two pulleys have radii of 10.0 cm and 6.00 cm, and their centres are 40.0 cm apart. If the pulley belt is uncrossed, what must be the length of the belt?

77. A special vehicle for travelling on glacial ice in Jasper National Park, Alberta, has tires that are 1.36 m in diameter. If the vehicle travels at 5.60 km/h, what is the angular velocity (in r/min) of the tire?

78. A rotating circular restaurant at the top of a hotel normally completes one revolution in 24.0 min. What is the angular acceleration required to accelerate the restaurant from rest to normal speed in 5.00 min?

79. Find the velocity (in km/h) of the moon as it revolves about the earth. Assume it takes 28 days for one revolution at a distance of 390 000 km from the earth.

80. The stopboard of a shot-put circle is a circular arc 1.22 m in length. The radius of the circle is 1.06 m. What is the central angle (in degrees)?

81. The longitude of Whitehorse, Yukon, is 135° W, and the longitude of St. Petersburg, Russia, is 30° E. Both cities are at a latitude of 60° N. (a) Find the great circle distance (see Section 8.4) from Whitehorse to St. Petersburg over the north pole. (b) Find the distance between them along the 60° N latitude arc. The radius of the earth is 6370 km. What do the results show?

82. A piece of circular filter paper 15.0 cm in diameter is folded such that its effective filtering area is the same as that of a sector with central angle of 220°. What is the filtering area?

83. To produce an electric current, a circular loop of wire of diameter 25.0 cm is rotating about its diameter at 60.0 r/s in a magnetic field. What is the greatest linear velocity of any point on the loop (in m/s)?

84. Find the area of the decorative glass panel shown in Fig. 8.53. The panel is made of two equal circular sectors and an isosceles triangle.

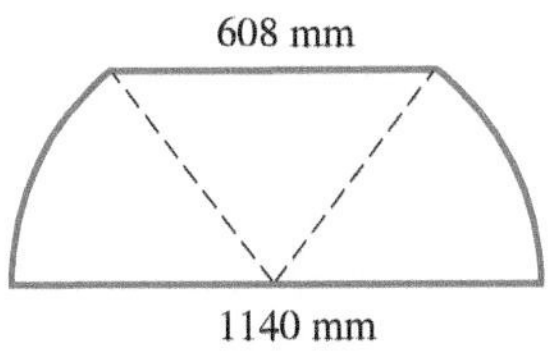

Fig. 8.53

85. A circular hood is to be used over a piece of machinery. It is to be made from a circular piece of sheet metal 1.08 m in radius. A hole 0.25 m in radius and a sector of central angle 80.0° are to be removed to make the hood. What is the area of the top of the hood?

86. The chain on a chain saw is driven by a sprocket 7.50 cm in diameter. If the chain is 108 cm long and makes one revolution in 0.250 s, what is the angular velocity (in r/s) of the sprocket?

87. An *ultracentrifuge*, used to observe the sedimentation of particles such as proteins, may rotate as fast as 80 000 r/min. If it rotates at this rate and is 7.20 cm in diameter, what is the linear velocity of a particle at the outer edge (in m/s to three significant digits)?

88. A computer is programmed to shade in a sector of a pie chart 2.44 cm in radius. If the perimeter of the shaded sector is 7.32 cm, what is the central angle (in degrees) of the sector? See Fig. 8.54.

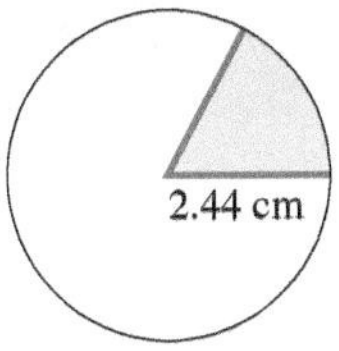

Fig. 8.54

89. A Gothic arch, commonly used in medieval European structures, is formed by two circular arcs. In one type, each arc is one-sixth of a circle, with the centre of each at the base on the end of the other arc. See Fig. 8.55. Therefore, the width of the arch equals the radius of each arc. For such an arch, find the area of the opening if the width is 15.0 m.

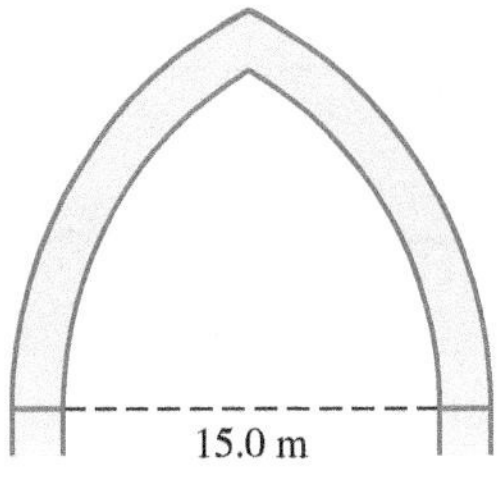

Fig. 8.55

90. The Trans-Alaska Pipeline was assembled in sections 12.2 m long and 1.22 m in diameter. If the depth of the oil in one horizontal section is 0.305 m, what is the volume of oil in this section?

91. A laser beam is transmitted with a "width" of 0.0008° and makes a circular spot of radius 2.50 km on a distant object. How far is the object from the source of the laser beam? Use Eq. (8.15).

92. The planet Venus subtends an angle of 15″ to an observer on earth. If the distance between Venus and Earth is 167 Gm, what is the diameter of Venus? Use Eq. (8.15).

93. Write a paragraph explaining how you determine the units for the result of the following problem: An astronaut in a spacecraft circles the moon once each 1.95 h. If the altitude of the spacecraft is constant at 113 km, what is its velocity? The radius of the moon is 1740 km. (What is the answer?)

CHAPTER 8 **PRACTICE TEST**

1. Change 150° to radians in terms of π.

2. Determine the sign of (a) sec 285°, (b) sin 245°, (c) cot 265°.

3. Express sin 205° in terms of the sine of a positive acute angle. Do not evaluate.

4. Find $\sin\theta$ and $\sec\theta$ if θ is in standard position and the terminal side passes through $(-9, 12)$.

5. An airplane propeller blade is 1.40 m long and rotates at 2200 r/min. What is the linear velocity of a point on the tip of the blade?

6. Given that 3.572 is the measure of an angle, express the angle in degrees.

7. If $\tan\theta = 0.2396$, find θ, in degrees, for $0° \le \theta < 360°$.

8. If $\cos\theta = -0.8244$ and $\csc\theta < 0$, find θ in radians for $0 \le \theta < 2\pi$.

9. The floor of a sunroom is in the shape of a circular sector of arc length 16.0 m and radius 4.25 m. What is the area of the floor?

10. A circular sector has an area of 38.5 cm^2 and a diameter of 12.2 cm. What is the arc length of the sector?

9. Vectors and Oblique Triangles

Mbolina/Fotolia

After completion of this chapter, the student should be able to:

- Distinguish between a scalar and a vector
- Add vectors by the polygon (or head-to-tail) method
- Add vectors by the parallelogram (or tail-to-tail) method
- Resolve a vector into its *x*- and *y*-components
- Add vectors by components
- Solve application problems involving vectors
- Solve oblique triangles using the law of sines and/or the law of cosines
- Solve application problems involving oblique triangles

▲ Vectors can be used to find certain forces when a system is in equilibrium. In Section 9.4 we see how to analyse the forces acting on a mountain climber.

In many applications, we often deal with quantities such as forces and velocities which cannot be described by a single number, and which must have a *direction* associated with them. In general, a quantity for which we must specify both *magnitude* ("how big" or "how much") and a *direction in space* is called a *vector*.

In this text, we deal only with vectors on the plane. These are usually represented by an arrow whose length shows the vector's magnitude, and whose direction shows the vector's direction. This type of representation has been common since the 1800s, when mathematicians defined vectors more generally, opening up new areas of study in advanced mathematics. Vectors are an excellent example of a mathematical concept that was derived from a basic physics concept.

In this chapter we also study the *law of cosines* and the *law of sines* to solve *oblique* triangles (triangles that are not right triangles). Although not in their modern forms, the law of cosines was known to the Greek astronomer Ptolemy (90–170 C.E.), while the law of sines was known to Islamic mathematicians of the 1100s.

Vectors and oblique triangles are of great importance in many fields of science, engineering, and technology. These include physics, structural design, surveying, and navigation.

9.1 Introduction to Vectors

We deal with many quantities that may be described by a number that shows only the magnitude. These include lengths, areas, time intervals, monetary amounts, and temperatures. *Quantities such as these, described only by the* ***magnitude****, are known as* **scalars**.

Many other quantities, called **vectors**, *are fully described only when both the* ***magnitude*** *and* ***direction*** *are specified.* The following example shows the difference between scalars and vectors.

EXAMPLE 1 Scalars and vectors—speed vs. velocity

Determine whether each of the following statements refers to a scalar or to a vector.

(a) "A jet is travelling at 600 km/h." From this statement alone, we know only the *speed* of the jet. *Speed is a scalar quantity*, and it tells us only the *magnitude* of the rate. Knowing only the speed of the jet, we know the rate at which it is moving, but we do not know where it is headed.

(b) "A jet is travelling at 600 km/h in a direction 10° south of west." Here, we specify the direction of travel as well as the speed. We then know the *velocity* of the jet; that is, the *direction* of travel as well as the rate at which it is moving. ***Velocity*** *is a* ***vector*** *quantity.*

EXAMPLE 2 Action of two vectors—riverboat velocity

Consider a boat moving in a river that is 0.4 km wide. We will assume that the boat is driven by a motor that can move it at 8 km/h in still water and that the river flows at 6 km/h downstream. Determine the velocity of the boat if it is headed (a) downstream; (b) upstream; and (c) across the river.

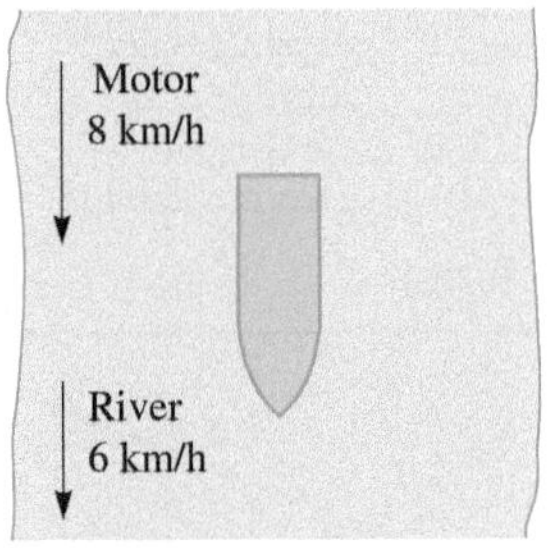

Fig. 9.1(a): Downstream

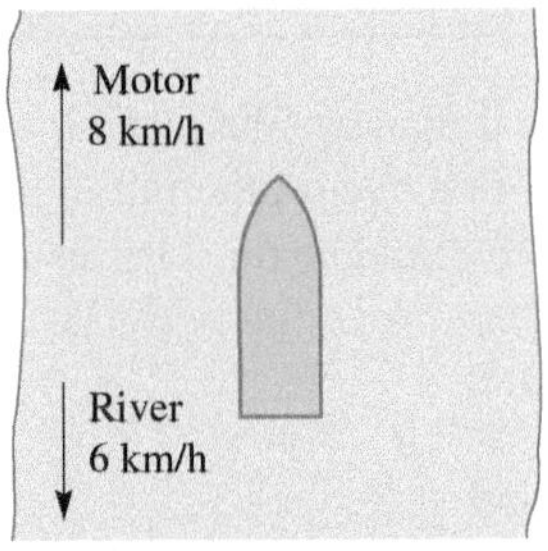

Fig. 9.1(b): Upstream

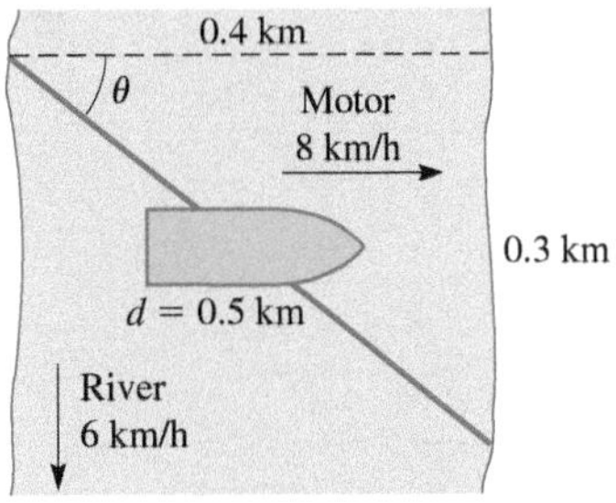

Fig. 9.1(c): Across

(a) If the boat heads downstream, as in Fig. 9.1(a), then the water is moving at 6 km/h and the motor is moving the boat at 8 km/h in the same direction as the water, so the boat is moving at 14 km/h.

(b) If the boat heads upstream as in Fig. 9.1(b), it moves at only 2 km/h, since the river is acting directly against the motor.

(c) If the boat heads directly across the river, as in Fig. 9.1(c), then the river is moving the boat downstream *at the same time* as the boat moves across the river. Therefore, the point the boat reaches on the other side is not directly opposite the point from which it started. Since it takes 0.05 h (0.4 km ÷ 8 km/h = 0.05 h) to cross, in that time the river will carry the boat 0.3 km (0.05 h × 6 km/h = 0.3 km) downstream. From the Pythagorean theorem, we see that it went 0.5 km from its starting point to its finishing point:

$$d^2 = 0.4^2 + 0.3^2 = 0.25$$
$$d = 0.5 \text{ km}$$

Since the 0.5 km was travelled in 0.05 h, the magnitude of the velocity (the *speed*) of the boat was actually

$$v = \frac{d}{t} = \frac{0.5 \text{ km}}{0.05 \text{ h}} = 10 \text{ km/h}$$

Also, note that the direction of this velocity can be represented along a line that makes an angle θ with the line directed directly across the river, as shown in Fig. 9.1(c). We can find this angle by noting that

$$\tan\theta = \frac{0.3 \text{ km}}{0.4 \text{ km}} = 0.75$$
$$\theta = \tan^{-1}0.75 = 37°$$

Therefore, when headed directly across the river, the boat's velocity is 10 km/h directed at an angle of 37° downstream from a line directly across the river.

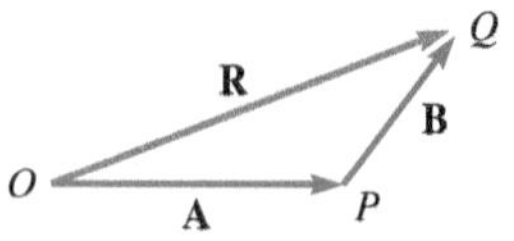

R = A + B

Fig. 9.2

> **LEARNING TIP**
> Since a vector is determined by its magnitude and direction, two vectors in different locations are considered the same if they have the same magnitude and direction.

ADDITION OF VECTORS

We have just seen two velocity vectors being *added.* Note that these vectors are not added the way numbers are added. We must take into account their directions as well as their magnitudes. Reasoning along these lines, let us now define the sum of two vectors.

We will represent a vector quantity by a letter printed in **boldface** type. The same letter in *italic* (lightface) type represents the magnitude only. Thus, **A** is a vector of magnitude *A*. In handwriting, one usually places an arrow over the letter to represent a vector, such as $\vec{A}$.

Let **A** and **B** represent vectors directed from *O* to *P* and *P* to *Q*, respectively (see Fig. 9.2). *The vector sum* **A** + **B** *is the vector* **R**, *from the* **initial point** *O to the* **terminal point** *Q. Here, vector* **R** *is called the* **resultant**. *In general, a resultant is a single vector that is the vector sum of any number of other vectors.*

There are two common methods of adding vectors by means of a diagram: the polygon method and the parallelogram method. For both methods, vectors must be drawn with reasonable accuracy and moved to specific locations without changing their magnitude and direction.

The polygon method is illustrated in Fig. 9.3. It can be described as follows.

Polygon (or Head-to-Tail) Method of Adding Two Vectors	EXAMPLE 3
To add two vectors **A** and **B**:	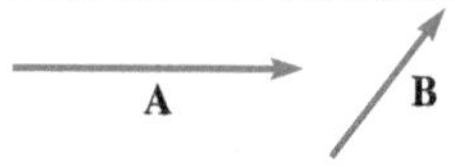 Fig. 9.3(a)
1. Move **B** so that where the head of **A** ends, the tail of **B** begins (hence the name *head-to-tail*).	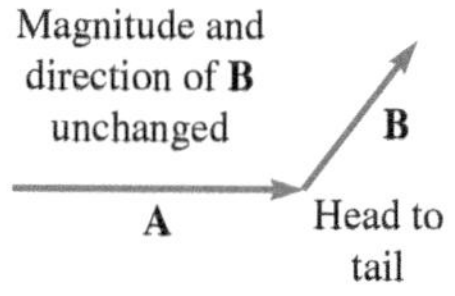 Fig. 9.3(b)
2. The vector sum **A** + **B** is the resultant vector **R**, which is drawn from the tail of **A** to the head of **B**. The order in which the vectors are added does not matter.	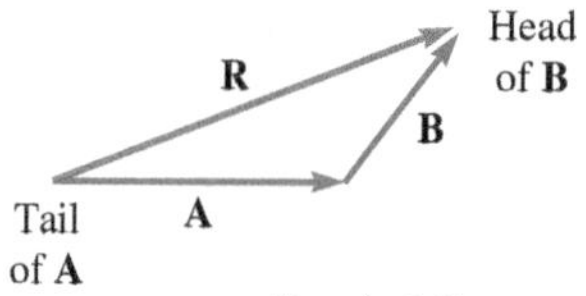 Fig. 9.3(c) R = A + B

Three or more vectors are added in the same general manner. We place the initial point of the second vector at the terminal point of the first vector, the initial point of the third vector at the terminal point of the second vector, and so on. The resultant is the vector from the initial point of the first vector to the terminal point of the last vector. Again, the order in which they are added does not matter.

EXAMPLE 4 Adding vectors—polygon method

The addition of vectors **A**, **B**, and **C** is shown in Fig. 9.4 as **A** + **B** + **C**, and as **A** + **C** + **B**. Other orderings are also possible, all with the same resultant **R**.

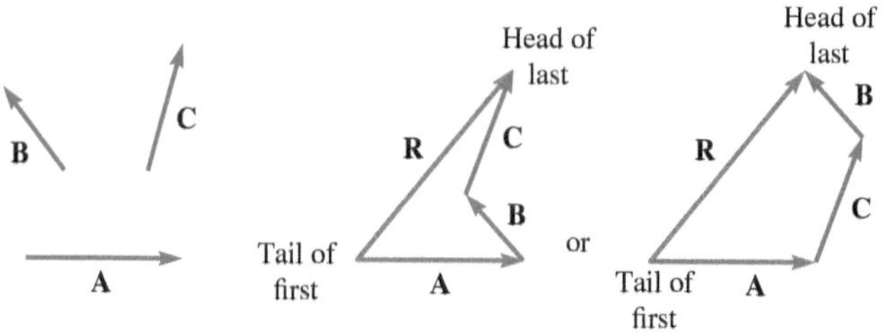

R = A + B + C   R = A + C + B

Fig. 9.4

Practice Exercise

1. For the vectors in Example 4, show that **R** = **B** + **C** + **A**.

The parallelogram method is illustrated in Fig. 9.5. It can be described as follows.

Parallelogram (or Head-to-Tail Method of Adding Two Vectors)	**EXAMPLE 5**
To add two vectors **A** and **B**:	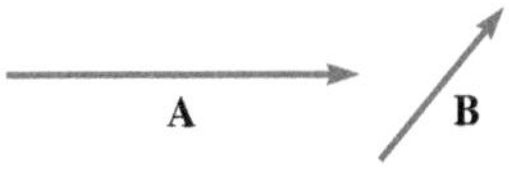 Fig. 9.5(a)
1. Move both vectors so that they have a common initial point (hence the name *tail-to-tail*).	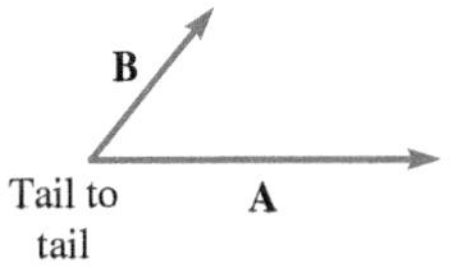 Fig. 9.5(b)
2. Let the vectors be the sides of a parallelogram.	 Fig. 9.5(c)
3. The vector sum **A** + **B** is the resultant vector **R**, which corresponds to the *diagonal* of the parallelogram.	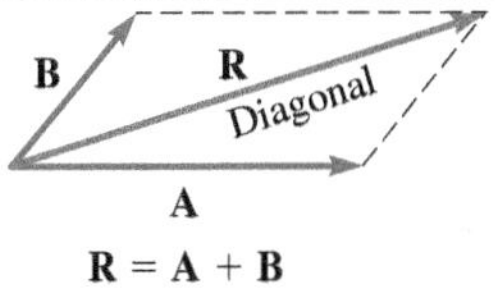 Fig. 9.5(d)

If vector **C** is in the same direction as vector **A** and **C** has a magnitude n times that of **A**, then **C** = n**A**, where *the vector n***A** *is called the* **scalar multiple** *of vector* **A**. This means that 2**A** is a vector that is twice as long as **A** but is *in the same direction*. Note carefully that only the magnitudes of **A** and 2**A** are different, and their directions are the same. The addition of scalar multiples of vectors is illustrated in the following example.

EXAMPLE 6 Scalar multiple of a vector

For vectors **A** and **B** in Fig. 9.6, find vector 3**A** + 2**B**.

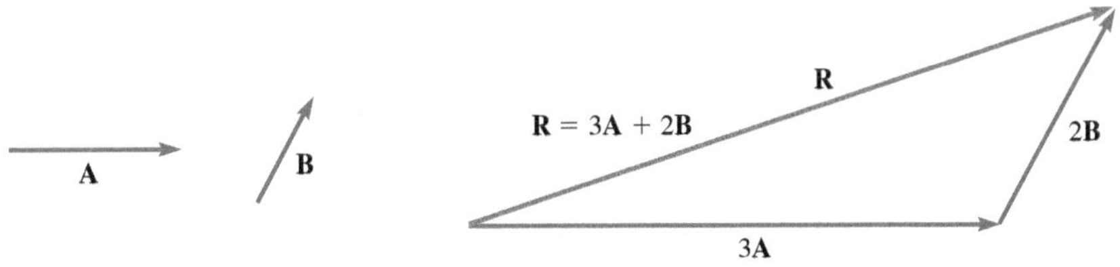

Fig. 9.6

Vector **B** is subtracted from vector **A** by reversing the direction of **B** and proceeding as in vector addition. Thus, **A** − **B** = **A** + (−**B**), where *the minus sign indicates that vector* −**B** *has the opposite direction of vector* **B**. Vector subtraction is illustrated in the following example.

EXAMPLE 7 Subtracting vectors

For vectors **A** and **B** in Fig. 9.7, find vector 2**A** − **B**.

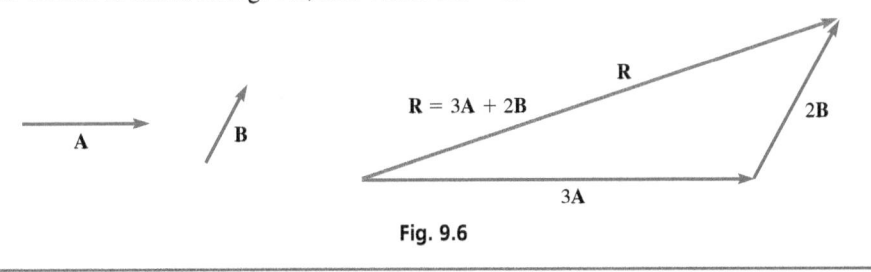

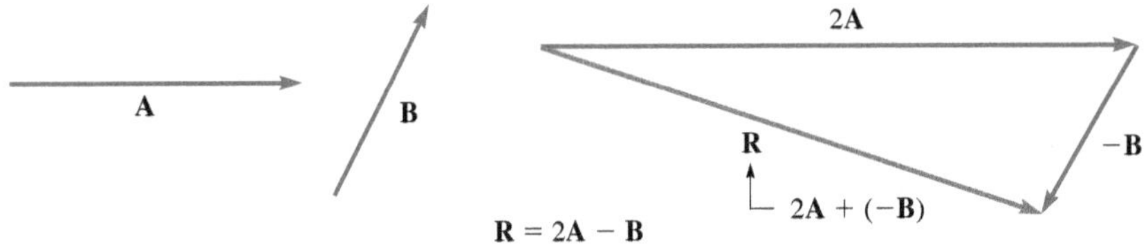

Fig. 9.7

Practice Exercise

2. For the vectors in Example 7, find vector **B** − 3**A**.

Among the most important applications of vectors is that of the forces acting on a structure or on an object. The next example shows the addition of forces by using the parallelogram method.

EXAMPLE 8 Adding vectors—pulling a car

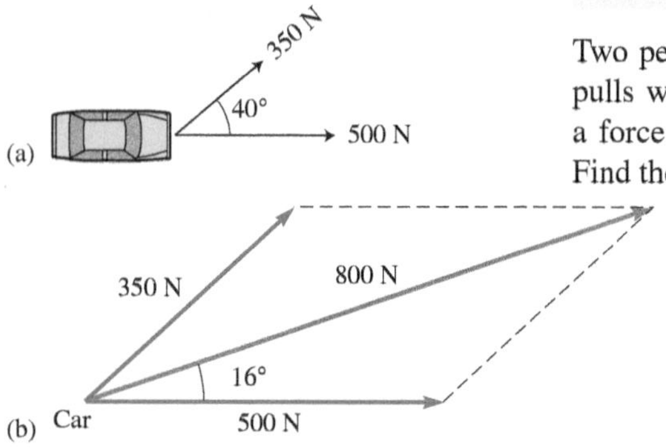

Fig. 9.8

Two people pull horizontally on ropes attached to a car mired in mud. One person pulls with a force of 500 N directly to the right, while the other person pulls with a force of 350 N at an angle of 40° from the first force, as shown in Fig. 9.8(a). Find the resultant force on the car.

We make a scale drawing of the forces as shown in Fig. 9.8(b), measuring the magnitudes of the forces with a ruler and the angles with a protractor. (The scale drawing of the forces is made larger and with a different scale than that in Fig. 9.8(a) in order to get better accuracy.) We then complete the parallelogram and draw in the diagonal that represents the resultant force. Finally, we find that the resultant force is about 800 N and that it acts at an angle of about 16° from the first force.

Two other important vector quantities are *velocity* and *displacement*. Velocity as a vector is illustrated in Examples 1 and 2. *The* **displacement** *of an object is the change in its position. Displacement is given by the distance from a reference point and the angle from a reference direction.* The following example illustrates the difference between *distance* and *displacement*.

EXAMPLE 9 Adding vectors—distance vs. displacement

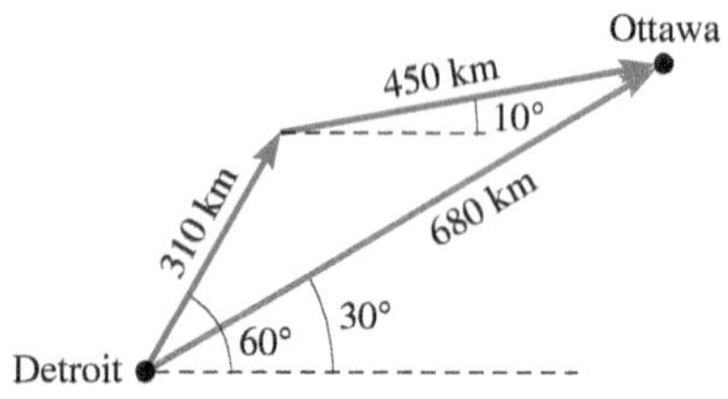

Fig. 9.9

To avoid a storm, a jet travels at 60° north of east from Detroit for 310 km and then turns to a direction of 10° north of east for 450 km to Ottawa. Find the displacement of Ottawa from Detroit.

We make a scale drawing in Fig. 9.9 to show the route taken by the jet. Measuring distances with a ruler and angles with a protractor, we find that Ottawa is about 680 km from Detroit, at an angle of about 30° north of east. By giving both the magnitude and the ***direction***, we have given the displacement.

If the jet returned directly from Ottawa to Detroit, its *displacement* from Detroit would be *zero*, although it travelled a *distance* of about 1440 km.

EXERCISES 9.1

In Exercises 1–4, find the resultant vectors if the given changes are made in the indicated examples of this section.

1. In Example 4, what is the resultant of the three vectors if the direction of vector **A** is reversed?

2. In Example 6, for vectors **A** and **B**, what is vector 2**A** + 3**B**?

3. In Example 7, for vectors **A** and **B**, what is vector 2**B** − **A**?

4. In Example 8, if 20° replaces 40°, what is the resultant force?

In Exercises 5–8, determine whether a scalar or a vector is described in (a) and (b). Explain your answers.

5. **(a)** A soccer player runs 15 m from the centre of the field.
(b) A soccer player runs 15 m from the centre of the field toward the opponents' goal.

6. **(a)** A small-craft warning reports 35 km/h winds.
(b) A small-craft warning reports 35 km/h winds from the north.

7. **(a)** An arm of an industrial robot pushes with a 10-N force downward on a part.
(b) A part is being pushed with a 10-N force by an arm of an industrial robot.

8. **(a)** A ballistics test shows that a bullet hit a wall at a speed of 100 m/s.
(b) A ballistics test shows that a bullet hit a wall at a velocity of 100 m/s perpendicular to the wall.

In Exercises 9–14, add the given vectors by drawing the appropriate resultant.

9.

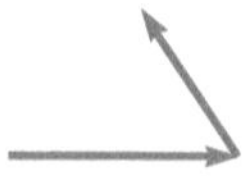

10.

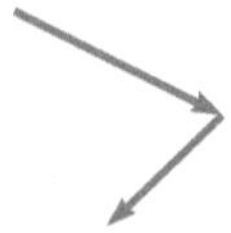

11.

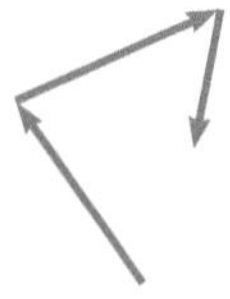

12. 

13. 

14. 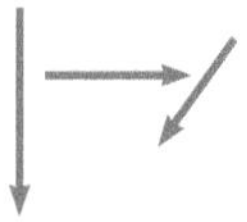

In Exercises 15–18, draw the given vectors and find their sum graphically. The magnitude is shown first, followed by the direction as an angle in standard position.

15. 3.6 cm, 0°; 4.3 cm, 90°
16. 2.3 cm, 45°; 5.2 cm, 120°
17. 6.0 cm, 150°; 1.8 cm, 315°
18. 7.5 cm, 240°; 2.3 cm, 30°

In Exercises 19–40, find the indicated vector sums and differences with the given vectors by means of diagrams. (You might find graph paper to be helpful.)

A B C D E

19. A + B
20. B + C
21. C + D
22. D + E
23. A + C + E
24. B + D + A
25. 2A + D + E
26. B + E + A
27. B + 3E
28. A + 2C
29. 3C + E
30. 2D + A
31. A − B
32. C − D
33. E − B
34. C − 2A
35. 3B + $\frac{1}{2}$A
36. 2B − $\frac{3}{2}$E
37. B + 2C − E
38. A + 2D − 3B
39. C − B − $\frac{3}{4}$A
40. D − 2C − $\frac{1}{2}$E

In Exercises 41–48, solve the given problems. Use a ruler and protractor as in Examples 8 and 9.

41. Two forces that act on an airplane wing are called the *lift* and the *drag*. Find the resultant of these forces acting on the airplane wing in Fig. 9.10.

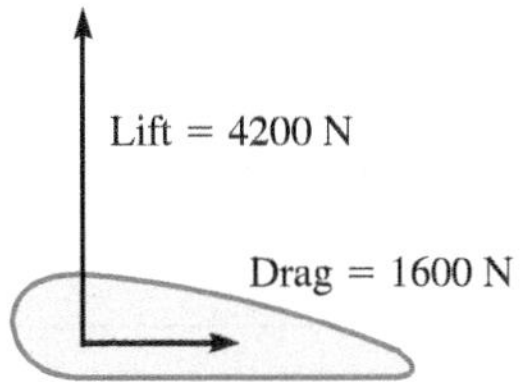

Fig. 9.10

42. Two electric charges create an electric field intensity, a vector quantity, at a given point. The field intensity is 30 kN/C to the right and 60 kN/C at an angle of 45° above the horizontal to the right. Find the resultant electric field intensity at this point.

43. A ski tow is moving skiers vertically upward at 24 m/min and horizontally at 44 m/min. What is the velocity of the tow?

44. A small plane travels at 180 km/h in still air. It is headed due south in a wind of 50 km/h from the northeast. What is the resultant velocity of the plane?

45. A driver takes the wrong road at an intersection and travels 4 km north, then 6 km east, and finally 10 km to the southeast to reach the home of a friend. What is the displacement of the friend's home from the intersection?

46. A ship travels 20 km in a direction of 30° south of east and then turns due south for another 40 km. What is the ship's displacement from its initial position?

47. A rope holds a helium-filled balloon in place with a tension of 510 N at an angle of 80° with the ground due to a wind. The weight (a vertical force) of the balloon and contents is 400 N, and the upward buoyant force is 900 N. The wind creates a horizontal force of 90 N on the balloon. What is the resultant force on the balloon?

48. While unloading a crate weighing 610 N, the chain from a crane supports it with a force of 650 N at an angle of 20° from the vertical. What force must a horizontal rope exert on the crate so that the total force (including its weight) on the crate is zero?

Answers to Practice Exercises

1. Same as **R** in Fig. 9.4 **2.** (Half scale)

9.2 Components of Vectors

In this section, we show how to write a vector as the sum of two other vectors. In the next section we show how this allows us to add vectors with any required degree of accuracy.

Two vectors that, when added together, have a resultant equal to the original vector are called **components** *of the original vector.* In the illustration of the boat in Fig. 9.1(c), the velocities of 8 km/h across the river and 6 km/h downstream are components of the 10 km/h vector directed at the angle θ.

Certain components of a vector are of particular importance. If the initial point of a vector is placed at the origin of a rectangular coordinate system and its direction is given by an angle in standard position, we may find its **x- and y-components**. *These components are vectors directed along the axes that, when added together, equal the given vector.* The initial points of these components are at the origin, and the terminal points are at the points where perpendicular lines from the terminal point of the given vector cross the axes. *Finding these component vectors is called* **resolving the vector into its components**.

The steps used in finding the x- and y-components of a vector, together with an illustrative example, are summarized here.

Resolving a Vector into x- and y-Components	EXAMPLE 1	
1. Place the vector **A** at the origin with its direction angle θ in standard position.	**A** is a vector of magnitude 7.25 units directed at an angle of 62.0°. Place **A** at the origin as in Fig. 9.11(a).	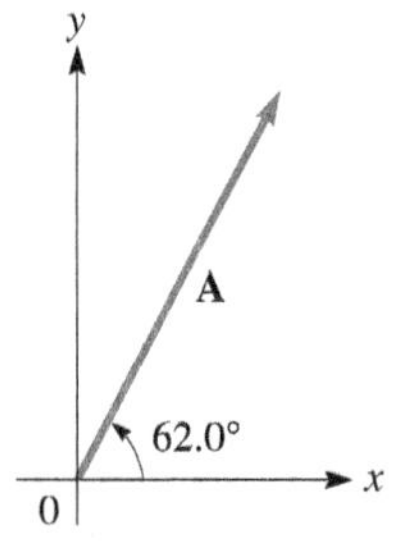 **Fig. 9.11(a)**
2. The x-component $\mathbf{A}_x$ is directed along the x-axis and has magnitude $A_x = A \cos \theta$ The y-component $\mathbf{A}_y$ is directed along the y-axis and has magnitude $A_y = A \sin \theta$ (You may use the reference angle if you note the direction of the component.)	$\mathbf{A}_x$ is directed along the x-axis to the right and has magnitude $A_x = 7.25 \cos 62.0° = 3.40$ $\mathbf{A}_y$ is directed along the y-axis upward and has magnitude $A_y = 7.25 \sin 62.0° = 6.40$	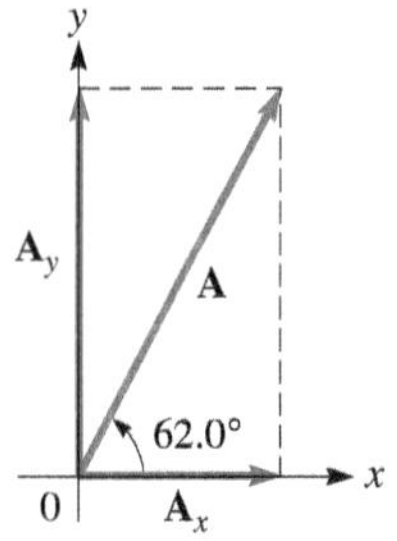 **Fig. 9.11(b)**
3. Check graphically with the original vector that the direction and magnitude of the components are correct.	The two component vectors $\mathbf{A}_x$ and $\mathbf{A}_y$ can replace vector **A**, since the effect they have is the same as **A**. See Fig. 9.11(b).	

Practice Exercise

1. Find the x- and y-components of a vector of magnitude 7.25 units directed at an angle of 25.0°.

EXAMPLE 2 Components of a second-quadrant vector—land survey

A surveyor's marker is located 14.4 m from the set pin at a 126.0° standard-position angle as shown in Fig. 9.12. Find the x- and y-components of the displacement of the marker from the pin.

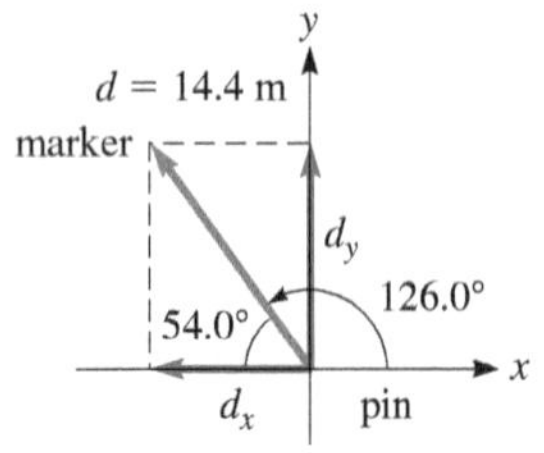

Fig. 9.12

Let the pin be at the origin. Place the displacement vector at the origin with its direction angle in standard position, as in Fig. 9.12. The x-component $\mathbf{d}_x$ is directed to the left along the x-axis and has magnitude

$$d_x = d \cos 126.0° = -8.46$$

The y-component $\mathbf{d}_y$ is directed upward along the y-axis and has magnitude

$$d_y = d \sin 126.0° = 11.6$$

Checking with the original vector in Fig. 9.12, the components do have the correct magnitude and direction.

Observe that we can obtain the same magnitudes using the reference angle, as long as we note the direction of the component:

$$d_x = -14.4 \cos 54.0° = -8.46 \qquad d_y = +14.4 \sin 54.0° = 11.6$$

(−: directed along negative x-axis; +: directed along positive y-axis)

The minus sign shows that the x-component is directed to the left.

Practice Exercise

2. For the vector in Example 2, change the angle to 306.0° and find the x- and y-components.

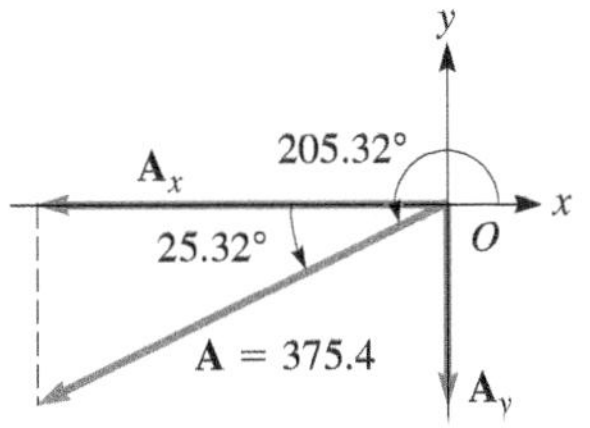

Fig. 9.13

EXAMPLE 3 Components of a third-quadrant vector

Resolve vector **A**, of magnitude 375.4 and direction $\theta = 205.32°$, into its x- and y-components. See Fig. 9.13.

By placing **A** such that θ is in standard position, we see that

$$A_x = A\cos 205.32° = 375.4\cos 205.32° = -339.3$$

and

directed along negative axis

$$A_y = A\sin 205.32° = 375.4\sin 205.32° = -160.5$$

The angle 205.32° places the vector in the third quadrant, and each of the components is directed along the negative axis. This must be the case for a third-quadrant angle. Also, the reference angle is 25.32°, and we see that the magnitude of A_x is greater than the magnitude of A_y, which must be true for a reference angle that is less than 45°.

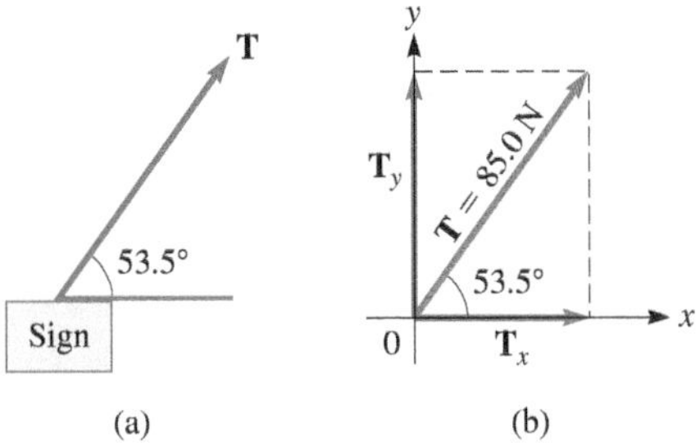

Fig. 9.14

EXAMPLE 4 Vector components—tension

The tension **T** in a cable supporting the sign shown in Fig. 9.14(a) is 85.0 N. If the cable makes an angle of 53.5° with the horizontal, find the horizontal and vertical components of the tension.

The tension is the force the cable exerts on the sign. Showing the tension in Fig. 9.14(b), we see that

$$T_y = T\sin 53.5° = 85.0\sin 53.5°$$
$$= 68.3\text{ N}$$
$$T_x = T\cos 53.5° = 85.0\cos 53.5°$$
$$= 50.6\text{ N}$$

Note that $T_y > T_x$, which should be the case for an acute angle greater than 45°.

Practice Exercise

3. Find the components of the tension in Example 4, if the cable makes an angle of 143.5° with the horizontal.

EXERCISES 9.2

In Exercises 1–4, find the component vectors if the given changes are made in the indicated examples of this section.

1. In Example 2, change 126.0° to 216.0°.

2. In Example 3, change 205.32° to 295.32°.

3. In Example 3, change 205.32° to 270.00°.

4. In Example 4, change 53.5° to 45.0°.

In Exercises 5–10, find the horizontal and vertical components of the vectors shown in the given figures. In each, the magnitude of the vector is 750.

5.

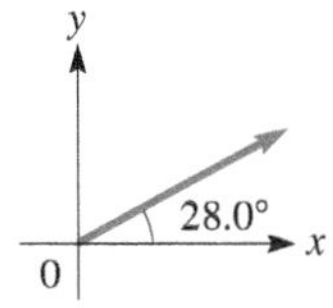

6.

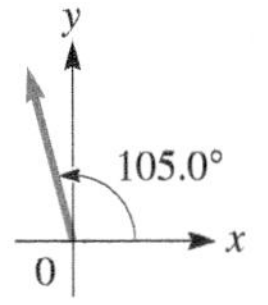

7.

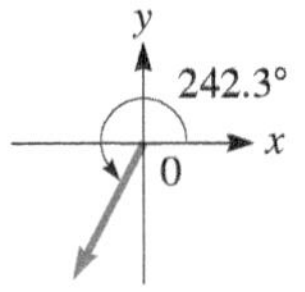

8.

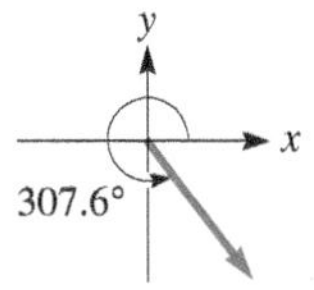

9.

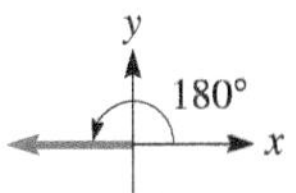

10.

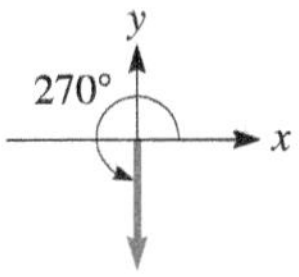

In Exercises 11–20, find the x- and y-components of the given vectors by use of the trigonometric functions. The magnitude is shown first, followed by the direction as an angle in standard position.

11. 8.60 N, $\theta = 68.0°$

12. 9750 N, $\theta = 243.0°$

13. 76.8 m/s, $\theta = 145.0°$

14. 0.0998 dm/s, $\theta = 296.0°$

15. 9040 mm/s^2, $\theta = 270.0°$

16. 16.4 cm/s^2, $\theta = 156.5°$

17. 2.65 mN, $\theta = 197.3°$

18. 678 N, $\theta = 22.5°$

19. 0.8734 dm, $\theta = 157.83°$

20. 509.4 m, $\theta = 360.0°$

In Exercises 21–34, find the required horizontal and vertical components of the given vectors.

21. A nuclear submarine approaches the surface of the ocean at 25.0 km/h and at an angle of 17.3° with the surface. What are the components of its velocity? See Fig. 9.15.

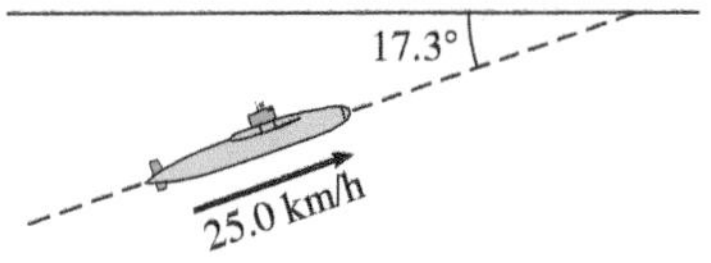

Fig. 9.15

22. A wind-blown fire with a speed of 5.5 m/s is moving toward a highway at an angle of 66.4° with the highway. What is the component of the velocity toward the highway?

23. A car is being unloaded from a ship. It is supported by a cable from a crane and guided into position by a horizontal rope. If the tension in the cable is 12 400 N and the cable makes an angle of 3.50° with the vertical, what are the weight W of the car and the tension T in the rope? (The weight of the cable is negligible to that of the car.) See Fig. 9.16.

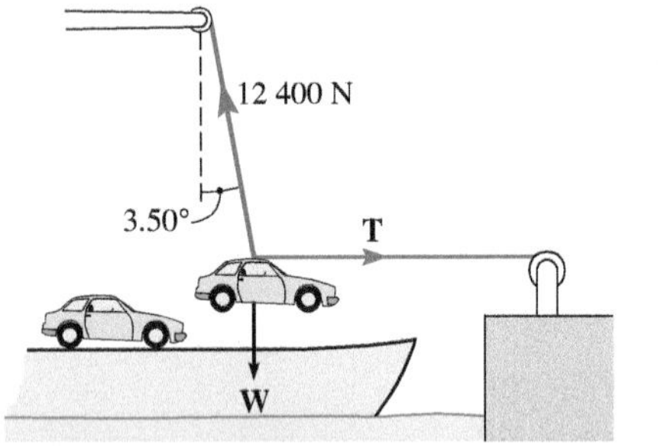

Fig. 9.16

24. A car travelling at 25 km/h hits a stone wall at an angle of 65° with the wall. What is the component of the car's velocity that is perpendicular to the wall?

25. With the sun directly overhead, a plane is taking off at 125 km/h at an angle of 22.0° above the horizontal. How fast is the plane's shadow moving along the runway?

26. A jet is at a position 145 km and 37.5° north of east of Vancouver. What are the components of the jet's displacement from Vancouver?

27. The end of a robot arm is 1.20 m on a line 78.6° above the horizontal from the point where it does a weld. What are the components of the displacement from the end of the robot arm to the welding point?

28. The tension in a rope attached to a boat is 55.0 N. The rope is attached to the boat 4.00 m below the level at which it is being drawn in. At the point where there is 12.0 m of rope out, what force tends to bring the boat toward the wharf, and what force tends to raise the boat? See Fig. 9.17.

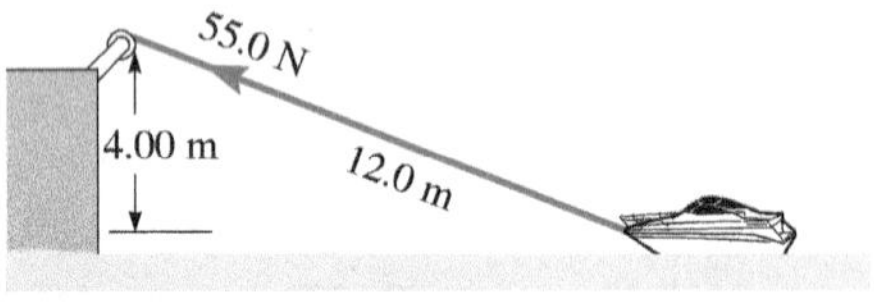

Fig. 9.17

29. A person applies a force of 210 N perpendicular to a jack handle that is at an angle of 25° above the horizontal. What are the horizontal and vertical components of the force?

30. A water skier is pulled up the ramp shown in Fig. 9.18 at 8.5 m/s. How fast is the skier rising when leaving the ramp?

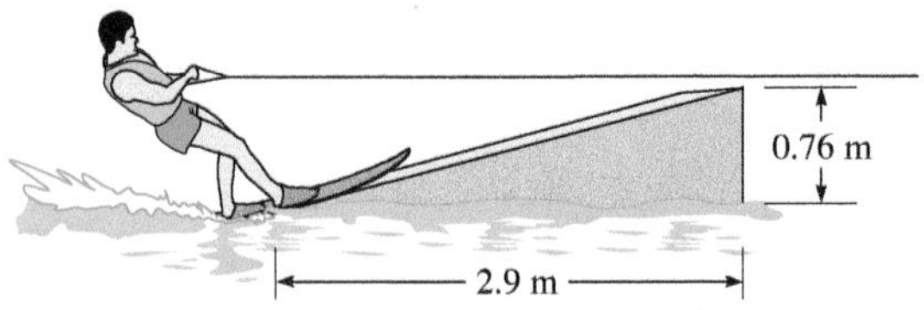

Fig. 9.18

31. A wheelbarrow is being pushed with a force of 310 N directed along the handles, which are at an angle of 20° above the horizontal. What is the effective force for moving the wheelbarrow forward?

32. Two upward forces are acting on a bolt. One force of 60.5 N acts at an angle of 82.4° above the horizontal, and the other force of 37.2 N acts at an angle of 50.5° below the first force. What is the total upward force on the bolt? See Fig. 9.19.

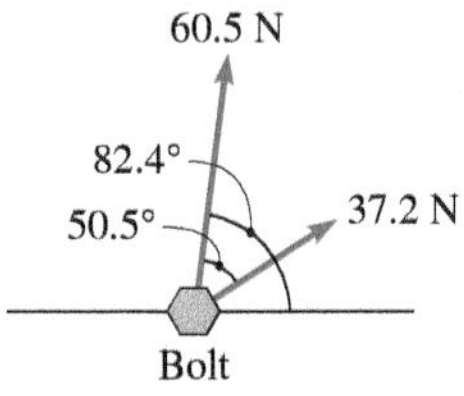

Fig. 9.19

33. Vertical wind shear in the lowest 100 m above the ground is of great importance to aircraft when taking off or landing. It is defined as the rate at which the wind velocity changes per metre above ground. If the vertical wind shear at 50 m above the ground is 0.75 (km/h)/m directed at an angle of 40° above the ground, what are its vertical and horizontal components?

34. At one point, the *Juno* space probe was entering the gravitational field of Jupiter at an angle of 2.550° below the horizontal with a velocity of 29 870 km/h. What were the components of its velocity?

Answers to Practice Exercises

1. $A_x = 6.57$, $A_y = 3.06$
2. $V_x = 8.46$, $V_y = -11.6$
3. $T_x = -68.3$ N, $T_y = 50.6$ N

9.3 Vector Addition by Components

Now that we have developed the meaning of the components of a vector, we are able to add vectors to any degree of required accuracy. To do this, we use the components of the vector, the Pythagorean theorem, and the tangent of the standard-position angle of the resultant. In the following example, two vectors at right angles are added.

EXAMPLE 1 Adding vectors at right angles

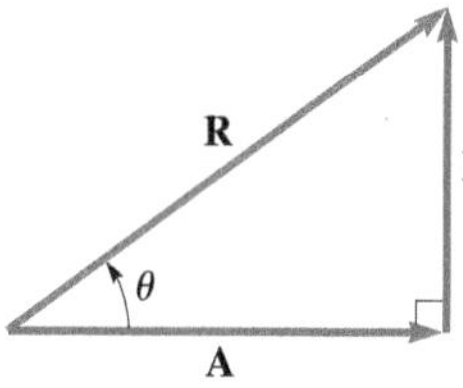

Fig. 9.20

Add vectors **A** and **B**, with $A = 14.5$ and $B = 9.10$. The vectors are at right angles, as shown in Fig. 9.20.

We can find the magnitude R of the resultant vector **R** by use of the Pythagorean theorem. This leads to

$$R = \sqrt{A^2 + B^2} = \sqrt{(14.5)^2 + (9.10)^2}$$
$$= 17.1$$

We now determine the direction of the resultant vector **R** by specifying its direction as the angle θ in Fig. 9.20, that is, the angle that **R** makes with vector **A**. Therefore,

$$\tan\theta = \frac{B}{A} = \frac{9.10}{14.5}$$
$$\theta = 32.1°$$

Therefore, we see that **R** is a vector of magnitude $R = 17.1$ and in a direction 32.1° from vector **A**.

Note that Fig. 9.20 shows vectors **A** and **B** as horizontal and vertical, respectively. This means they are the horizontal and vertical components of the resultant vector **R**. However, we would find the resultant of any two vectors *at right angles* in the same way.

COMMON ERROR *Remember, a vector is not completely specified unless both its magnitude and its* ***direction*** *are specified.* A common error is to determine the magnitude, but not to find the angle θ that is used to define its direction.

If the vectors being added are not at right angles, the resultant is found by first finding the components of the given vectors and then combining components.

The general procedure is summarized below, together with an illustrative example.

Adding Vectors by Components	EXAMPLE 2	
1. Place the vectors on a coordinate system with the tail of each at the origin.	Add **A** and **B** such that $A = 1220,\ \theta_A = 270.0°$ $B = 1750,\ \theta_B = 115.0°$ The correct placement of the vectors is as in Fig. 9.21(a).	Fig. 9.21 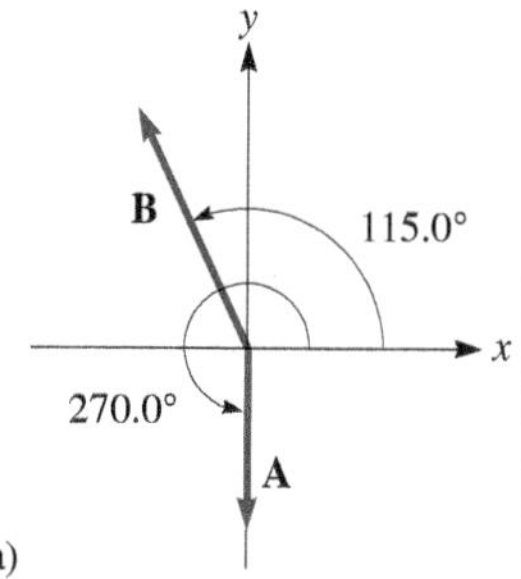 (a)
2. Resolve each vector into its x- and y-components.	$A_x = A\cos\theta_A = 1220\cos 270.0°$ $= 0$ $A_y = A\sin\theta_A = 1220\sin 270.0°$ $= -1220$ (Note: **A** is vertical, so $A_x = 0$.) $B_x = B\cos\theta_B = 1750\cos 115.0°$ $= -739.6$ $B_y = B\sin\theta_B = 1750\sin 115.0°$ $= 1586$	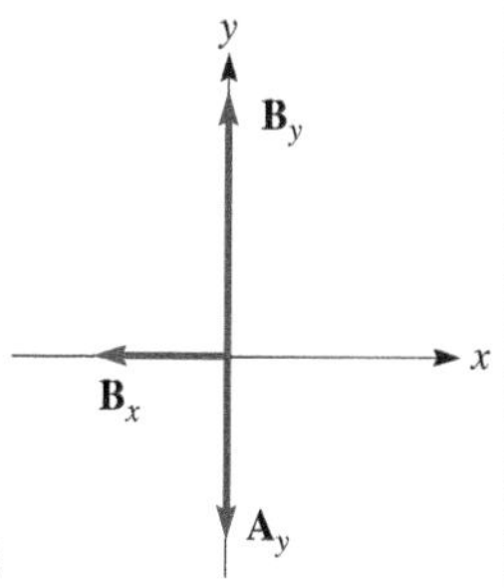 (b)

3. Add the magnitudes of the x-components to obtain R_x, the x-component of the resultant.	$R_x = A_x + B_x$ $= 0 - 739.6 = -739.6$
4. Add the magnitudes of the y-components to obtain R_y, the y-component of the resultant.	$R_y = A_y + B_y$ $= -1220 + 1556 = 366$
5. Find the magnitude of the resultant **R** as $R = \sqrt{R_x^2 + R_y^2}$	$R = \sqrt{R_x^2 + R_y^2}$ $= \sqrt{(-739.6)^2 + 366^2}$ $= 825$
6. Find the standard-position angle θ. If the angle is not in the first quadrant, find the reference angle $\theta_{ref} = \tan^{-1}\left\lvert\frac{R_y}{R_x}\right\rvert$ Then determine θ considering the quadrant of **R**.	$\theta_{ref} = \tan^{-1}\left\lvert\frac{R_y}{R_x}\right\rvert$ $= \tan^{-1}\left\lvert\frac{366}{-739.6}\right\rvert$ $= 26.3°$ **R** is in the second quadrant, so $\theta = 180° - 26.3° = 153.7°$

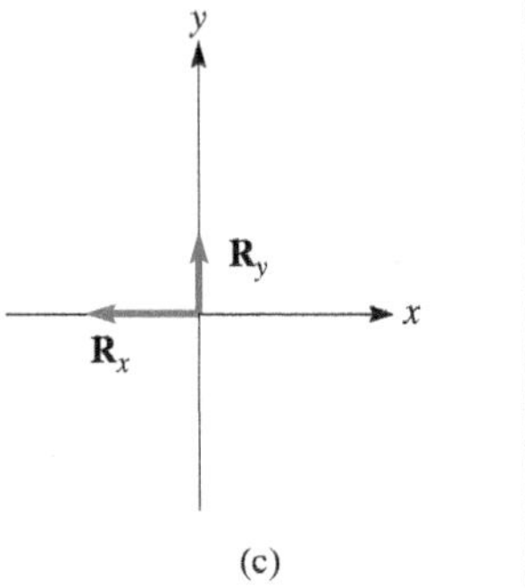

(c)

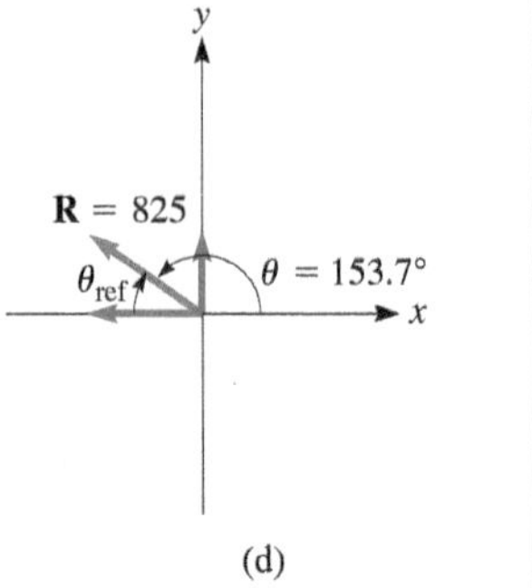

(d)

Practice Exercise

1. In Example 2, change 270° to 90° and then add the vectors.

LEARNING TIP

To reduce the accumulation of rounding errors when using a calculator, calculate each of the components R_x and R_y in one step and store their values without rounding for the calculation of R and θ. (We follow this practice throughout the chapter, even though values of R_x and R_y are shown rounded off.)

EXAMPLE 3 Adding by first combining components

Find the resultant **R** of the two vectors shown in Fig. 9.22(a), **A** of magnitude 8.075 and standard-position angle of 237.26° and **B** of magnitude 5.437 and standard-position angle of 142.15°.

In Fig. 9.22(b), we show the components of vectors **A** and **B**, and then in Fig. 9.22(c), we show the resultant and its components.

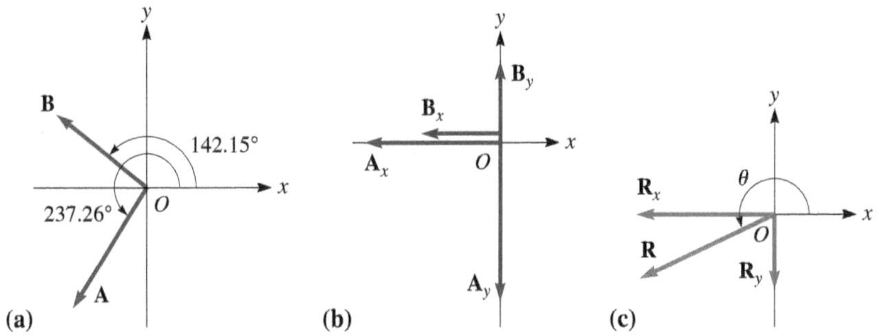

Fig. 9.22

Vector	*Magnitude*	*Angle*	*Magnitude of x-Component*	*Magnitude of y-Component*
A	8.075	237.26°	$A_x = 8.075 \cos 237.26° = -4.367$	$A_y = 8.075 \sin 237.26° = -6.792$
B	5.437	142.15°	$B_x = 5.437 \cos 142.15° = -4.293$	$B_y = 5.437 \sin 142.15° = 3.336$
R			$R_x = A_x + B_x = -8.660$	$R_y = A_y + B_y = -3.456$

$$R = \sqrt{R_x^2 + R_y^2} = \sqrt{(-8.660)^2 + (-3.456)^2} = 9.324$$

$$\theta_{ref} = \tan^{-1}\left|\frac{R_y}{R_x}\right| = \tan^{-1}\left|\frac{-3.456}{-8.660}\right| = 21.76°$$

R is in the third quadrant (R_x and R_y are both negative), so

$$\theta = 180° + \theta_{ref} = 180° + 21.76° = 201.76°$$

The resultant vector is 9.324 units long and is directed at a standard-position angle of 201.76°, as shown in Fig. 9.22(c).

For convenience, we summarize the general formulas for addition of vectors here. For a given vector **A**, directed at an angle θ, of magnitude A,

$$A_x = A\cos\theta \quad A_y = A\sin\theta \tag{9.1}$$

If R_x and R_y are the components of the resultant vector, then

$$R = \sqrt{R_x^2 + R_y^2} \tag{9.2}$$

$$\theta_{ref} = \tan^{-1}\frac{|R_y|}{|R_x|} \tag{9.3}$$

The standard-position angle θ is found by using the reference angle from Eq. (9.3) and the quadrant in which the resultant lies.

EXAMPLE 4 Addition of three vectors

Find the resultant of the three given vectors in Fig. 9.23. The magnitudes of these vectors are $T = 422$, $U = 405$, and $V = 211$.

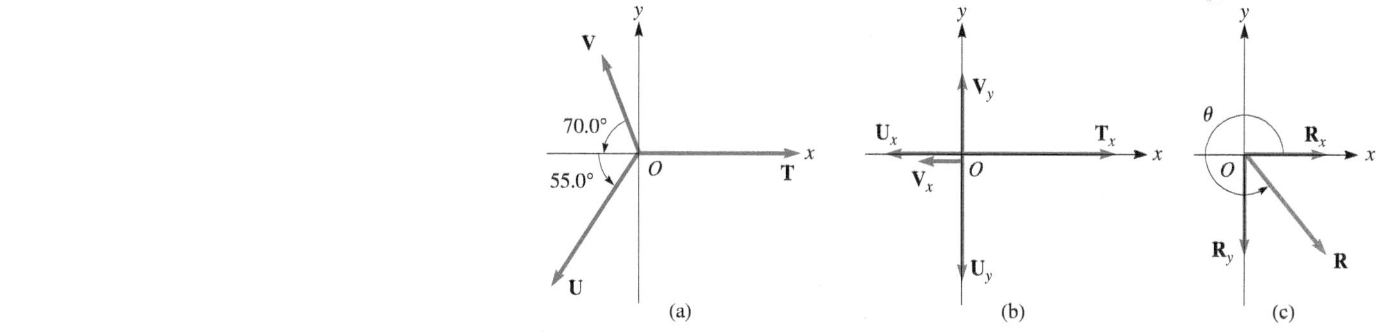

Fig. 9.23

We could change the given angles to standard-position angles. However, we will use the given angles, *being careful to give the proper sign to each component.* In the following table, we show the x- and y-components of the given vectors, and the sums of these components give us the components of **R**.

Vector	*Magnitude*	*Ref. Angle*	*Magnitude of x-Component*	*Magnitude of y-Component*
T	422	0°	422 cos 0° = 422.0	422 sin 0° = 0.0
U	405	55.0°	−405 cos 55.0° = −232.3	−405 sin 55.0° = −331.8
V	211	70.0°	−211 cos 70.0° = −72.2	+211 sin 70.0° = 198.3
R			117.5	−133.5

From these components, we find R and θ_{ref}:

$$\mathbf{R} = \sqrt{R_x^2 + R_y^2} = \sqrt{(117.5)^2 + (-133.5)^2} = 178$$

$$\theta_{\text{ref}} = \tan^{-1}\left(\frac{|-133.5|}{|117.5|}\right) = 48.6°$$

Since R_x is positive and R_y is negative, we know that θ is a fourth-quadrant angle. Therefore, to find θ, we subtract θ_{ref} from 360°. This means

$$\theta = 360° - 48.6° = 311.4°$$

Practice Exercise

2. In Example 4, find the resultant of vectors **U** and **V**. (Do not include vector **T**.)

EXERCISES 9.3

In Exercises 1 and 2, find the resultant vectors if the given changes are made in the indicated examples of this section.

1. In Example 2, find the resultant if θ_A is changed to 0°.

2. In Example 3, find the resultant if θ_B is changed to 232.15°.

In Exercises 3–6, vectors **A** *and* **B** *are at right angles. Find the magnitude and direction (the angle from vector* **A***) of the resultant.*

3. $A = 14.7$, $B = 19.2$

4. $A = 592$, $B = 195$

5. $A = 3.086$, $B = 7.143$

6. $A = 1734$, $B = 3297$

In Exercises 7–14, with the given sets of components, find R and θ.

7. $R_x = 5.18$, $R_y = 8.56$

8. $R_x = 89.6$, $R_y = -52.0$

9. $R_x = -0.982$, $R_y = 2.56$

10. $R_x = -729$, $R_y = -209$

11. $R_x = -646$, $R_y = 2030$

12. $R_x = -31.2$, $R_y = -41.2$

13. $R_x = 6941$, $R_y = -1246$

14. $R_x = 7.627$, $R_y = -6.353$

In Exercises 15–30, add the given vectors by components.

15. $A = 18.0$, $\theta_A = 0.00°$; $B = 12.0$, $\theta_B = 27.0°$

16. $F = 154$, $\theta_F = 90.0°$; $T = 128$, $\theta_T = 43.0°$

17. $C = 5650$, $\theta_C = 76.0°$; $D = 1280$, $\theta_D = 160.0°$

18. $A = 6.89$, $\theta_A = 123.0°$; $B = 29.0$, $\theta_B = 260.0°$

19. $A = 9.821$, $\theta_A = 34.27°$; $B = 17.45$, $\theta_B = 752.50°$

20. $E = 1653$, $\theta_E = 36.37°$; $F = 9807$, $\theta_F = 253.06°$

21. $A = 21.9$, $\theta_A = 236.2°$; $B = 96.7$, $\theta_B = 11.5°$; $C = 62.9$, $\theta_C = 143.4°$

22. $R = 630$, $\theta_R = 189.6°$; $F = 176$, $\theta_F = 320.1°$; $T = 324$, $\theta_T = 75.4°$

23. $U = 0.364$, $\theta_U = 175.7°$; $V = 0.596$, $\theta_V = 319.5°$; $W = 0.129$, $\theta_W = 100.6°$

24. $A = 6.4$, $\theta_A = 126°$; $B = 5.9$, $\theta_B = 238°$; $C = 3.2$, $\theta_C = 72°$

25. The displacement vectors in Fig. 9.24.

26. The velocity vectors in Fig. 9.25.

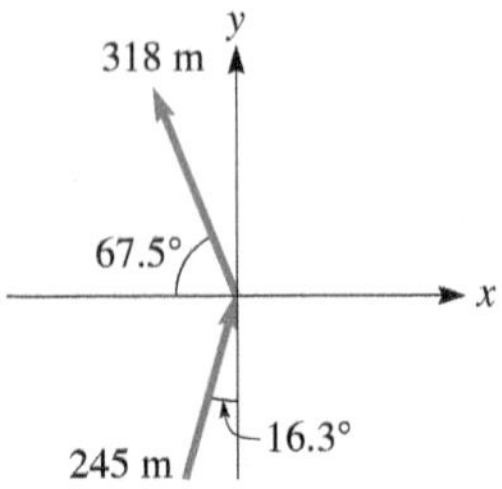

Fig. 9.24

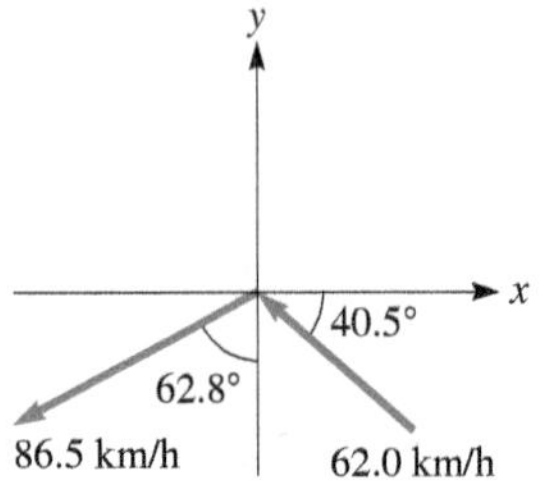

Fig. 9.25

27. The force vectors in Fig. 9.26.

28. The displacement vectors in Fig. 9.27.

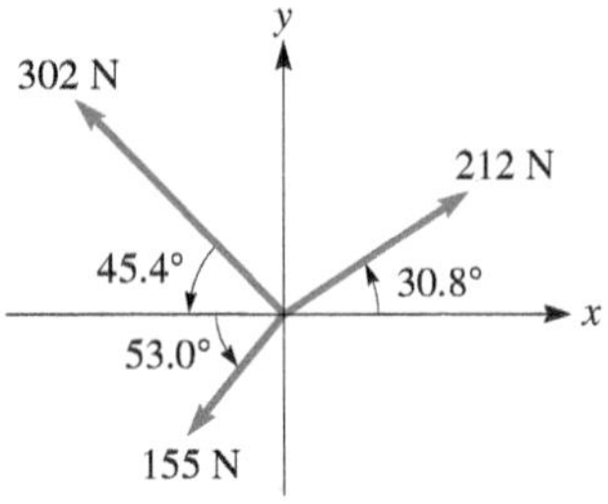

Fig. 9.26

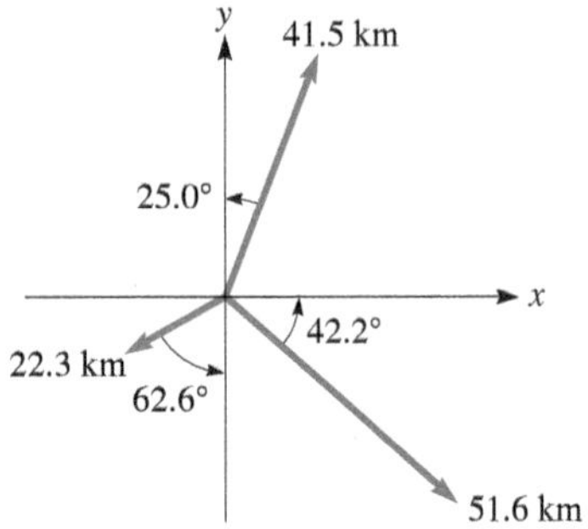

Fig. 9.27

29. In order to move an ocean liner into the channel, three tugboats exert the forces shown in Fig. 9.28. What is the resultant of these forces?

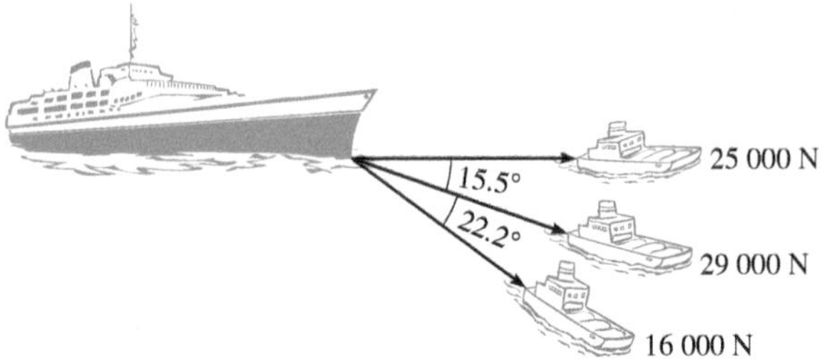

Fig. 9.28

30. A naval cruiser on maneuvers travels 54.0 km at 18.7° west of north, then turns and travels 64.5 km at 15.6° south of east, and finally turns to travel 72.4 km at 38.1° east of south. Find its displacement from its original position. See Fig. 9.29.

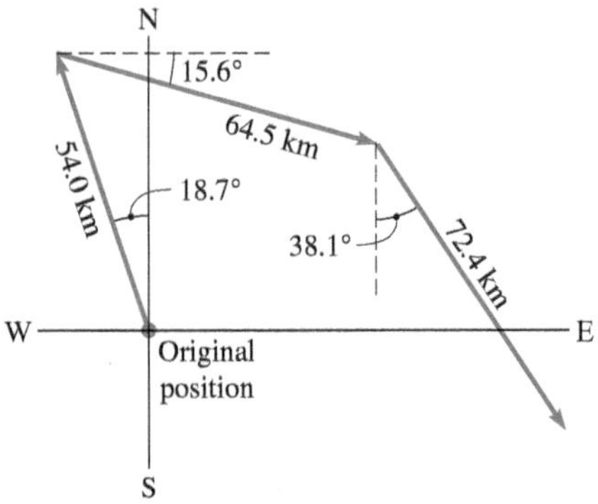

Fig. 9.29

In Exercises 31–34, solve the given problems.

31. The angle between two forces of 17 N and 25 N is 37° when placed tail to tail. What is the magnitude of the resultant force?

32. What is the magnitude of a velocity vector that makes an angle of 18.0° with its horizontal component of 420 km/h?

33. The angle between two equal-momentum vectors of 15.0 kg · m/s in magnitude is 72.0°. What is the magnitude of the resultant?

34. The resultant **R** of displacements **A** and **B** is perpendicular to **A**. If $A = 25$ m, B = 35 m, and $R = 25$ m, find the angle between **A** and **B**.

35. Four forces of 110 N, 180 N, 120 N and 140 N act on a bracket as shown in Fig. 9.30. Find the resultant force.

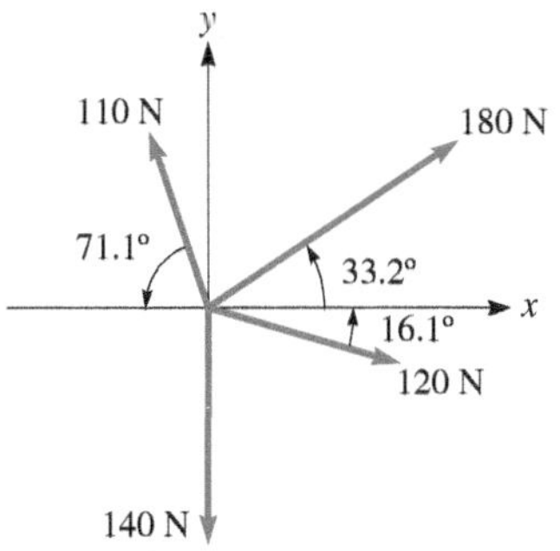

Fig. 9.30

36. Find the magnitude and direction of the force **F** if the resultant of the three forces in Fig. 9.31 has magnitude 1360 N and is directed upward along the y-axis.

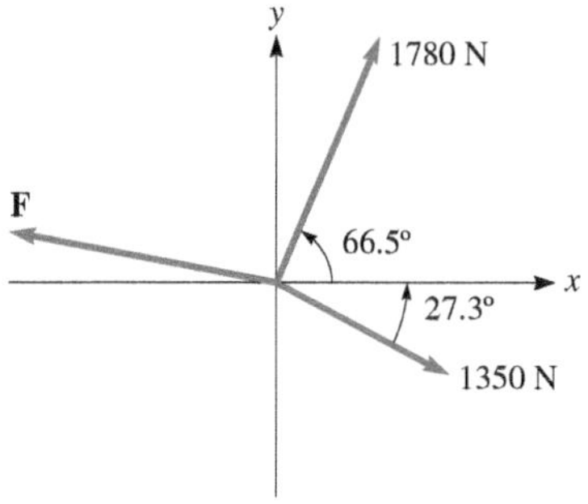

Fig. 9.31

Answers to Practice Exercises

1. $R = 2900$, $\theta = 104.8°$
2. $R = 332$, $\theta = 203.7°$

9.4 Applications of Vectors

In Section 9.1, we introduced the important vector quantities of force, velocity, and displacement, and we found vector sums by use of diagrams. Now, we can use the method of Section 9.3 to find sums of these kinds of vectors and others and to use them in various types of applications.

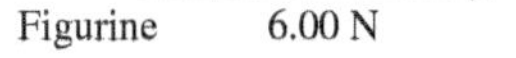

Fig. 9.32

■ The metric unit of force, the newton (N), is named for the great English mathematician and physicist Sir Isaac Newton (1642–1727). His name will appear on other pages of this text, as some of his many accomplishments are noted.

EXAMPLE 1 Forces at right angles

In centring a figurine on a table, two persons apply forces on it. These forces are at right angles and have magnitudes of 6.00 N and 8.00 N. The angle between their lines of action is 90.0°. What is the resultant of these forces on the figurine?

Identify known quantities: It is convenient to choose the x- and y-axes to fit the problem. Since the angle between the forces is 90°, we choose to have the x-axis in the direction of the 6.00-N force and the y-axis in the direction of the 8.00-N force. Therefore,

$$F_x = 6.00 \text{ N, and } F_y = 8.00 \text{ N.}$$

Sketch: See Fig. 9.32. Note that the two given forces are the x- and y-components of the resultant.

Solve:

$$F = \sqrt{(6.00)^2 + (8.00)^2} = 10.0 \text{ N}$$

$$\theta = \tan^{-1}\frac{F_y}{F_x} = \tan^{-1}\left(\frac{8.00}{6.00}\right)$$

$$= 53.1°$$

The resultant has a magnitude of 10.0 N and acts at an angle of 53.1° from the 6.00-N force.

EXAMPLE 2 Ship displacement

A ship sails 32.50 km due east and then turns 41.25° north of east. After sailing another 16.28 km, where is it with reference to the starting point?

Identify known quantities: First displacement: vector **A**, of magnitude 32.50 km and direction 0° (east corresponds to the positive x-direction).
Second displacement: vector **B**, of magnitude 16.28 km and direction 41.25° (north of east).

Identify unknown quantities: Let **R** be the resultant displacement of the ship from the two given displacements.

Sketch: See Fig. 9.33. Note that the x-component of the resultant is $\mathbf{A} + \mathbf{B}_x$ and the y-component of the resultant is $\mathbf{B}_y$.

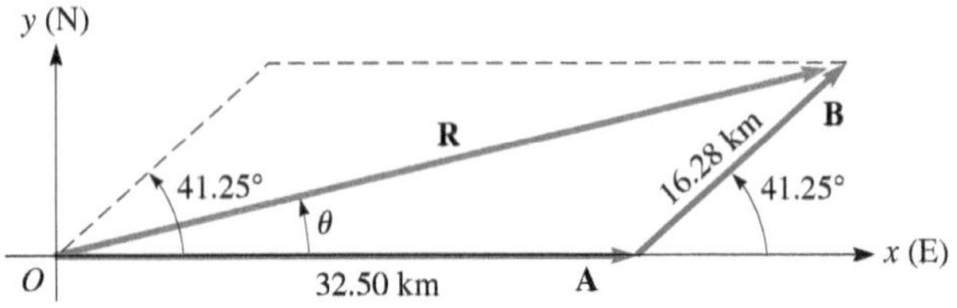

Fig. 9.33

Solve:

$$R_x = A + B_x = 32.50 + 16.28\cos 41.25$$
$$= 32.50 + 12.24$$
$$= 44.74\text{ km}$$
$$R_y = 16.28\sin 41.25° = 10.73\text{ km}$$
$$R = \sqrt{(44.74)^2 + (10.73)^2} = 46.01\text{ km}$$
$$\theta = \tan^{-1}\left(\frac{10.73}{44.74}\right)$$
$$= 13.49°$$

Therefore, the ship is 46.01 km from the starting point, in a direction 13.49° north of east.

EXAMPLE 3 Airplane velocity

■ Recall that a bar over a digit indicates it is significant.

An airplane headed due east is in a wind blowing from the northeast. What is the resultant velocity of the plane with respect to the surface of the earth if the velocity of the plane with respect to the air is $60\bar{0}$ km/h and that of the wind is $10\bar{0}$ km/h? See Fig. 9.34.

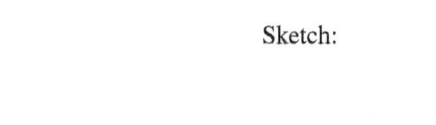

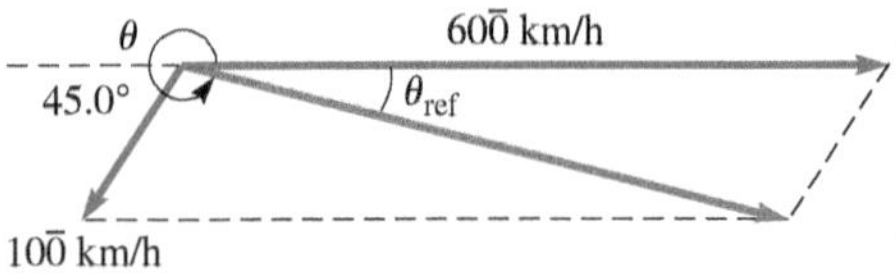

Fig. 9.34

If **A** is the plane's air vector (air speed and directional heading) and **W** is the wind vector (wind speed and wind direction), the calculations below can be used to find the magnitude and direction of the resultant ground vector **R** (ground speed and ground course). Note that we have written each given angle as a standard-position angle (abbreviated SP angle).

Vector	*Magnitude*	*SP Angle*	*x-Component*	*y-Component*
A	600 km/h	0°	$600\cos 0° = 600$ km/h	$600\sin 0° = 0.00$ km/h
W	100 km/h	225.0°	$100\cos 225.0° = -70.7$ km/h	$100\sin 225.0° = -70.7$ km/h
R			529 km/h	−70.7 km/h

$$R = \sqrt{(529)^2 + (-70.7)^2} = 534$$
$$\theta_{ref} = \tan^{-1}\left|\frac{R_y}{R_x}\right| = \tan^{-1}\left|\frac{-70.7}{529}\right| = 7.61°$$

The plane is travelling at 534 km/h and is flying in a direction 7.61° south of east. From this, we observe that a plane does not necessarily head in the direction of its destination.

Practice Exercise

1. In Example 3, find the plane's resultant velocity if the wind is from the southeast.

As we have seen, an important vector quantity is the force acting on an object. One of the most important applications of vectors involves forces that are in **equilibrium**. *For an object to be in equilibrium, the net force acting on it in any direction must be zero.* This condition is satisfied if the sum of the x-components of the force is zero and the sum of the y-components of the force is also zero. The following two examples illustrate forces in equilibrium.

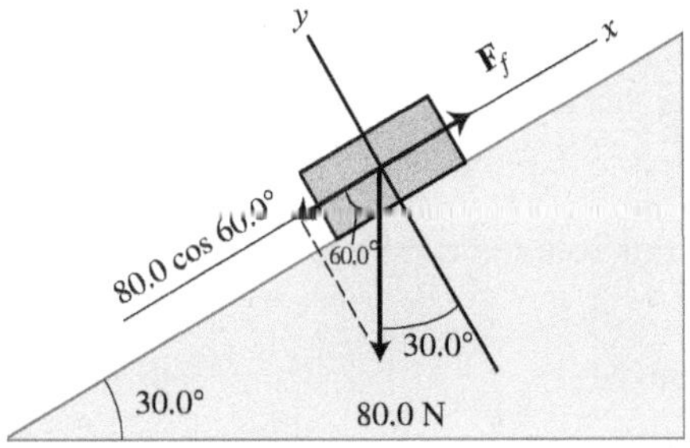

Fig. 9.35

■ Newton's *third law of motion* states that when an object exerts a force on another object, the second object exerts on the first object a force of the same magnitude but in the opposite direction. The force exerted by the plank on the block is an illustration of this law.

EXAMPLE 4 Forces on a block on an inclined plane

A cement block is resting on a straight inclined plank that makes an angle of 30.0° with the horizontal. If the block weighs 80.0 N, what is the force of friction between the block and the plank?

Identify known quantities: The weight of the cement block is the force exerted on the block due to gravity. Therefore, the weight has magnitude 80.0 N and is directed vertically downward.

Identify unknown quantities: The frictional force $\mathbf{F}_f$ tends to oppose the motion of the block and is directed upward along the plank. For the block to be at rest (not moving), it must be sufficient to counterbalance that component of the weight of the block that is directed down the plank.

Sketch: A convenient set of coordinates is one with the origin at the centre of the block and with the x-axis directed up the plank and the y-axis perpendicular to the plank. See Fig. 9.35.

Solve: The magnitude of the frictional force is

$$F_f = 80.0 \cos 60.0^\circ$$
$$= 40.0 \text{ N}$$

We have used the 60.0° angle since it is the reference angle. We could have expressed the frictional force as $\mathbf{F}_f = 80.0 \sin 30.0^\circ$.

Here, it is assumed that the block is small enough that we may calculate all forces as though they act at the centre of the block (although we know that the frictional force acts at the surface between the block and the plank).

■ See the chapter introduction.

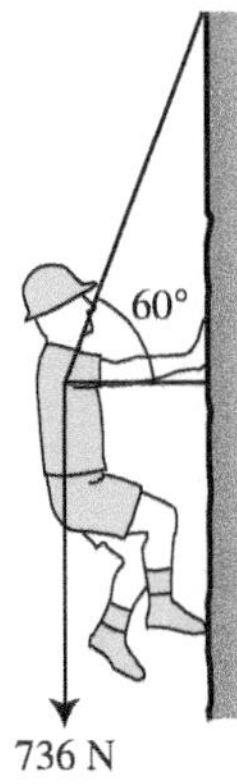

Fig. 9.36

EXAMPLE 5 Equilibrium—forces on a climber

A 736-N mountain climber is suspended by a rope that is angled 60.0° from the horizontal. See Fig. 9.36. If T is the tension in the rope and F is the horizontal force with which the climber pushes on the side of a cliff, find the magnitudes of the two forces, T and F.

We will use the ***free-body diagram*** in Fig. 9.37 to analyse all the forces acting on the climber's body. *A free-body diagram shows a body (an object) removed from its environment along with all the external forces acting directly on the body.* We have:

- a large dot representing the object of interest (in this case, the climber's body);
- the climber's weight $\mathbf{W}$, with magnitude 736 N and direction 270° (vertically downward);
- the tension in the rope $\mathbf{T}$, with unknown magnitude T and direction 60°;
- the reaction force $\mathbf{F}$ of the cliff, with unknown magnitude F and direction 180° (exerted by the cliff against the climber's body);
- equilibrium of forces, so the sum of both the x- and y-components of the three vectors must be zero.

Free-body diagram

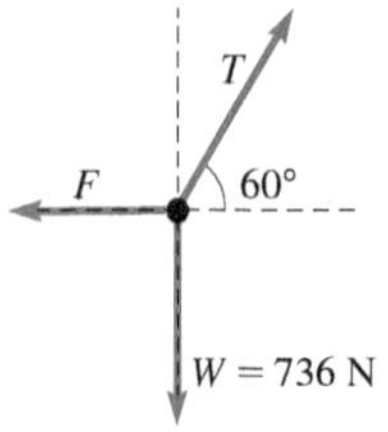

Fig. 9.37

The analysis of these vectors by components is shown below:

Vector	*Magnitude*	θ	*x-Component* (N)	*y-Component* (N)
W	736	270°	$736 \cos 270° = 0$	$736 \sin 270° = -736$
T	T	60°	$T \cos 60° = 0.5T$	$T \sin 60° = \sqrt{3}T/2$
F	F	180°	$F \cos 180° = -F$	$F \sin 180° = 0$
R			0	0

The equilibrium conditions give us the following two equations:

$$0.5T - F = 0$$
$$-736 + \frac{\sqrt{3}}{2}T = 0$$

The second equation can be solved for T, and this solution can be substituted into the first equation to find F:

$$T = \frac{736(2)}{\sqrt{3}} = 849.9 \text{ N}$$
$$0.5(849.9) - F = 0$$
$$F = 425 \text{ N}$$

Therefore, the tension in the rope is $T = 850$ N and the force with which the climber pushes against the cliff is $F = 425$ N.

EXAMPLE 6 Equilibrium involving a system—cable tension

A 3850-N crate of building supplies is hanging from two cables as shown in Fig. 9.38. Find the tensions T_1 and T_2 assuming that the forces are in equilibrium.

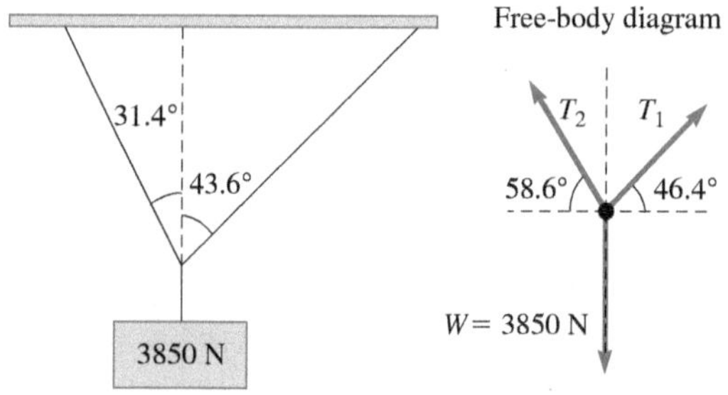

Fig. 9.38

The table below shows the standard-position angles and the x- and y-components of the three vectors:

Vector	*Magnitude*	*SP Angle*	*x-Component* (N)	*y-Component* (N)
W	3850	270°	$3850 \cos 270° = 0$	$3850 \sin 270° = -3850$
$\mathbf{T_1}$	T_1	46.4°	$T_1 \cos 46.4° = 0.6896T_1$	$T_1 \sin 46.4° = 0.7242T_1$
$\mathbf{T_2}$	T_2	121.4°	$T_2 \cos 121.4° = -0.5210T_2$	$T_2 \sin 121.4° = 0.8536T_2$
R			0	0

Setting the sums of the x- and y-components equal to zero gives us the following system of equations:

$$0.6896T_1 - 0.5210T_2 = 0$$
$$-3850 + 0.7242T_1 + 0.8536T_2 = 0$$

We solve the first equation for T_1 and substitute the result into the second equation to find T_2:

$$T_1 = \frac{0.5210}{0.6896}T_2 = 0.7555T_2$$

$$-3850 + 0.7242(0.7555T_2) + 0.8536T_2 = 0$$

$$1.401T_2 = 3850$$

$$T_2 = \frac{3850}{1.401} = 2749 \text{ N}$$

The solution for T_2 can be used to find T_1:

$$T_1 = 0.7555(4949) = 2077 \text{ N}$$

Therefore, the tensions in the two cables are $T_1 = 2750$ N and $T_2 = 2080$ N.

EXERCISES 9.4

In Exercises 1 and 2, find the necessary quantities if the given changes are made in the indicated examples of this section.

1. In Example 2, find the resultant location if the 41.25° angle is changed to 31.25°.
2. In Example 3, find the resultant velocity if the wind is from the southwest, rather than the northeast.

In Exercises 3–36, solve the given problems.

3. Two hockey players strike the puck at the same time, hitting it with horizontal forces of 34.5 N and 19.5 N that are perpendicular to each other. Find the resultant of these forces.
4. To straighten a small tree, two horizontal ropes perpendicular to each other are attached to the tree. If the tensions in the ropes are 82.3 N and 102 N, what is the resultant force on the tree?
5. In lifting a heavy piece of equipment from the mud, a cable from a crane exerts a vertical force of 6500 N, and a cable from a truck exerts a force of 8300 N at 10.0° above the horizontal. Find the resultant of these forces.
6. At a point in the plane, two electric charges create an electric field (a vector quantity) of 25.9 kN/C at 10.8° above the horizontal to the right and 12.6 kN/C at 83.4° below the horizontal to the right. Find the resultant electric field.
7. A motorboat leaves a dock and travels 1580 m due west, then turns 35.0° to the south and travels another 1640 m to a second dock. What is the displacement of the second dock from the first dock?
8. Regina is 530 km at 6.9° north of west from Winnipeg. Saskatoon is 230 km at 53° north of west from Regina. What is the displacement of Saskatoon from Winnipeg?
9. From a fixed point, a surveyor locates a pole at 215.6 m due east and a building corner at 358.3 m at 37.72° north of east. What is the displacement of the building from the pole?
10. A rocket is launched with a vertical component of velocity of 2840 km/h and a horizontal component of velocity of 1520 km/h. What is its resultant velocity?
11. In testing the behaviour of a tire on ice, a force of 2080 N is exerted to the side, and a force of 3120 N is exerted to the front. What is the resultant force on the tire?
12. To raise a crate, two ropes are attached to its top. If the force in one rope is 960 N at 25° from the vertical, what must be the force in the second rope at 35° from the vertical in order to lift the crate straight upward?
13. A storm front is moving east at 22.0 km/h and south at 12.5 km/h. Find the resultant velocity of the front.
14. To move forward, a helicopter pilot tilts the helicopter forward. If the rotor generates a force of 14 000 N, with a horizontal component (thrust) of 1900 N, what is the vertical component (lift)?
15. The acceleration (a vector quantity) of gravity on a sky diver is 9.8 m/s^2. If the force of the wind also causes an acceleration of 1.2 m/s^2 at an angle of 14° above the horizontal, what is the resultant acceleration of the sky diver?
16. In an accident, a truck with momentum (a vector quantity) of 22 100 kg · m/s strikes a car with momentum of 17 800 kg · m/s from the rear. The angle between their directions of motion is 25.0°. What is the resultant momentum?
17. In an automobile safety test, a shoulder and lap seat belt exerts a force of 425 N directly backward and a force of 368 N backward at an angle of 20.0° below the horizontal on a dummy. If the belt holds the dummy from moving farther forward, what force did the dummy exert on the belt? See Fig. 9.39.

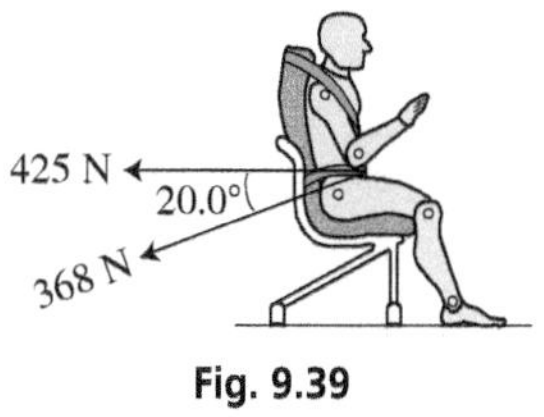

Fig. 9.39

18. Two perpendicular forces act on a ring at the end of a chain that passes over a pulley and holds an automobile engine. If the forces have the values shown in Fig. 9.40, what is the weight of the engine?

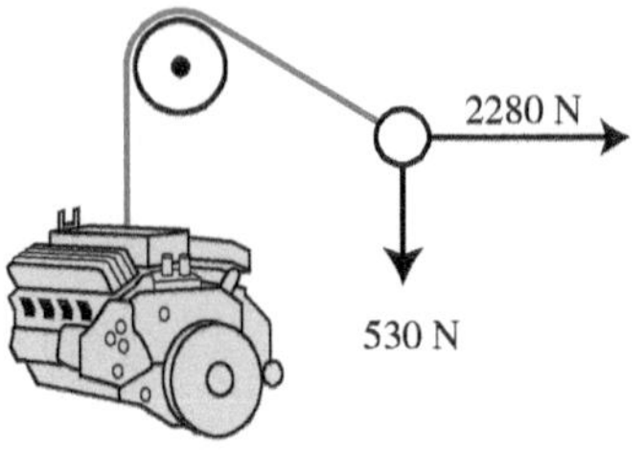

Fig. 9.40

19. A plane flies at 550 km/h into a headwind of 60 km/h at 78° with the direction of the plane. Find the resultant velocity of the plane with respect to the ground. See Fig. 9.41.

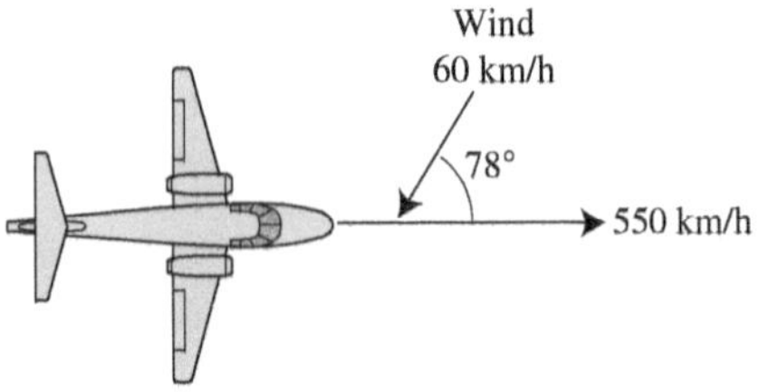

Fig. 9.41

20. A ship's navigator determines that the ship is moving through the water at 17.5 km/h with a heading of 26.3° north of east, but the ship is actually moving at 19.3 km/h in a direction of 33.7° north of east. What is the velocity of the current?

21. A launch vehicle is moving in orbit at 29 370 km/h. A satellite is launched to the rear at 190 km/h at an angle of 5.20° from the direction of the launcher. Find the velocity of the satellite.

22. A block of ice slides down a (frictionless) ramp with an acceleration of 5.3 m/s^2. If the ramp makes an angle of 32.7° with the horizontal, find g, the acceleration due to gravity. See Fig. 9.42.

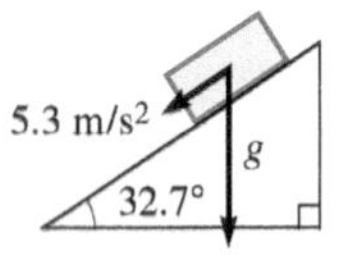

Fig. 9.42

23. In navigation, one method of expressing *bearing* is to give a single angle measured clockwise from due north. A radar station notes the bearing of two planes, one 15.5 km away on a bearing of 22.0°, and the other 12.0 km away on a bearing of 180.0°. Find the displacement of the second plane from the first. See Fig. 9.43.

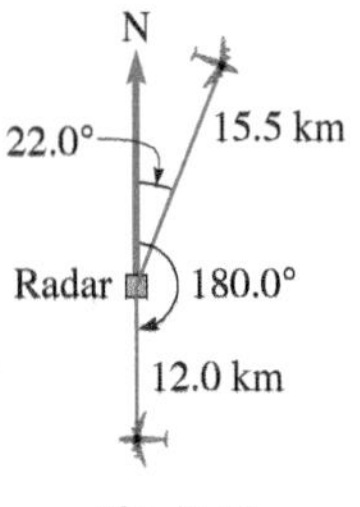

Fig. 9.43

24. As a satellite travels around Earth at an altitude of 35 940 km, the magnitude of its linear velocity (see Section 8.4) remains unchanged, but its *direction* changes continually, which means there is a change in velocity. Using a diagram based on Fig. 9.44, draw the difference in velocities (subtract the first velocity from the second) for the satellite as it travels (a) one-sixth, and (b) one-quarter of its orbit. Draw the vectors from the point at which their lines of action cross. What conclusion can be drawn about the direction of the change in velocity?

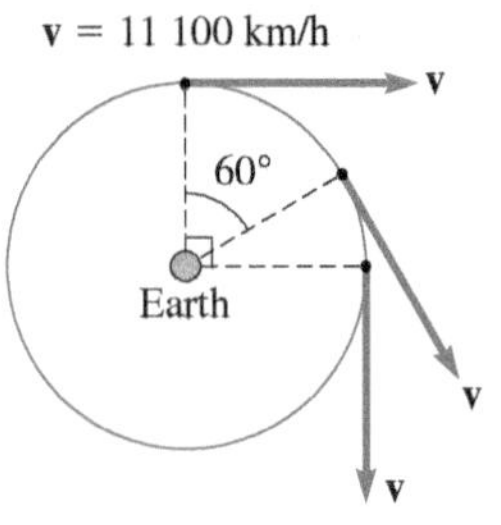

Fig. 9.44

25. A passenger on a cruise ship travelling due east at a speed of 32 km/h notes that the smoke from the ship's funnels makes an angle of 15° with the ship's wake. If the wind is from the southwest, find the speed of the wind.

26. A mine shaft goes due west 75 m (measured horizontally) from the opening at an angle of 25° below the horizontal surface. It then becomes horizontal and turns 30° north of west and continues for another 45 m. What is the displacement of the end of the tunnel from the opening?

27. A crowbar 1.5 m long is supported at a fulcrum 1.2 m from the end, with the other end under a boulder. If the crowbar is at an angle of 18° with the horizontal and a person pushes down, perpendicular to the crowbar, with a force of 240 N, what vertical force is applied to the boulder? See Fig. 9.45. (The forces are related by $1.2\mathbf{F}_1 = 0.3\mathbf{F}_2$.)

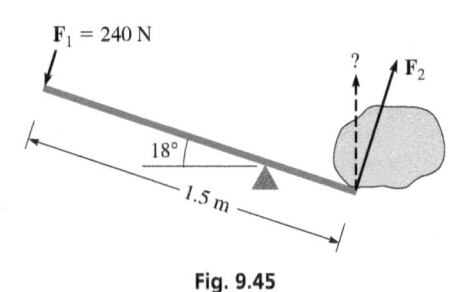

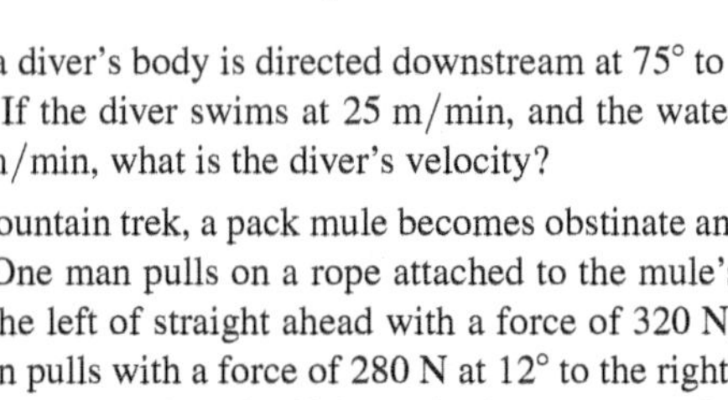

Fig. 9.45

28. A scuba diver's body is directed downstream at 75° to the bank of a river. If the diver swims at 25 m/min, and the water is moving at 5.0 m/min, what is the diver's velocity?

29. On a mountain trek, a pack mule becomes obstinate and refuses to move. One man pulls on a rope attached to the mule's harness at 15° to the left of straight ahead with a force of 320 N, and a second man pulls with a force of 280 N at 12° to the right. With what force does the mule resist if the mule does not move?

30. A boat travels across a river, reaching the other bank at a point directly opposite that from which it left. If the boat travels 6.00 km/h in still water and the river flows at 3.00 km/h, what was the velocity of the boat in the water?

31. In searching for a boat lost at sea, a Coast Guard cutter leaves St. John's and travels 75.0 km due east. It then turns 65° north of east and travels another 75.0 km, and finally turns another 65.0° toward the west and travels another 75.0 km. What is its displacement from the port?

32. A car is held stationary on a ramp by two forces. One is the force of 2140 N by the brakes, which hold it from rolling down the ramp. The other is a reaction force by the ramp of 9850 N, perpendicular to the ramp. This force keeps the car from going through the ramp. See Fig. 9.46. What is the weight of the car, and at what angle with the horizontal is the ramp inclined?

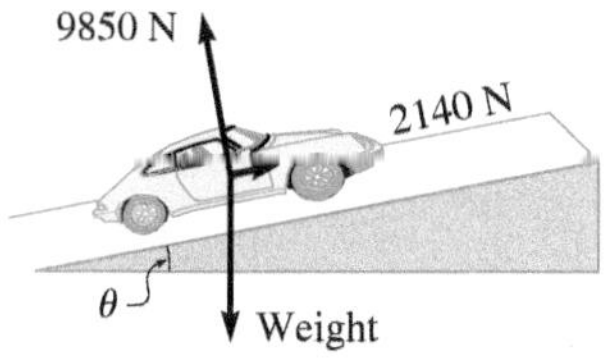

Fig. 9.46

33. A plane is moving at 75.0 m/s, and a package with weather instruments is ejected horizontally from the plane at 15.0 m/s, perpendicular to the direction of the plane. If the vertical velocity v_v (in m/s), as a function of time t (in s) of fall, is given by $v_v = 9.80t$, what is the velocity of the package after 2.00 s (before its parachute opens)?

34. A flat rectangular barge, 48.0 m long and 20.0 m wide, is headed directly across a stream at 4.5 km/h. The stream flows at 3.8 km/h. What is the velocity, relative to the riverbed, of a person walking diagonally across the barge at 5.0 km/h while facing the opposite upstream bank?

35. In Fig. 9.47, a long, straight conductor perpendicular to the plane of the paper carries an electric current i. A bar magnet having poles of strength m lies in the plane of the paper. The vectors $\mathbf{H}_i$, $\mathbf{H}_N$, and $\mathbf{H}_S$ represent the components of the magnetic intensity $\mathbf{H}$ due to the current and to the N and S poles of the magnet, respectively. The magnitudes of the components of $\mathbf{H}$ are given by

$$H_i = \frac{1}{2\pi}\frac{i}{a} \qquad H_N = \frac{1}{4\pi}\frac{m}{b^2} \qquad H_S = \frac{1}{4\pi}\frac{m}{c^2}$$

Given that $a = 0.300$ m, $b = 0.400$ m, $c = 0.300$ m, the length of the magnet is 0.500 m, $i = 4.00$ A, and $m = 2.00$ A · m, calculate the resultant magnetic intensity $\mathbf{H}$. The component $\mathbf{H}_i$ is parallel to the magnet.

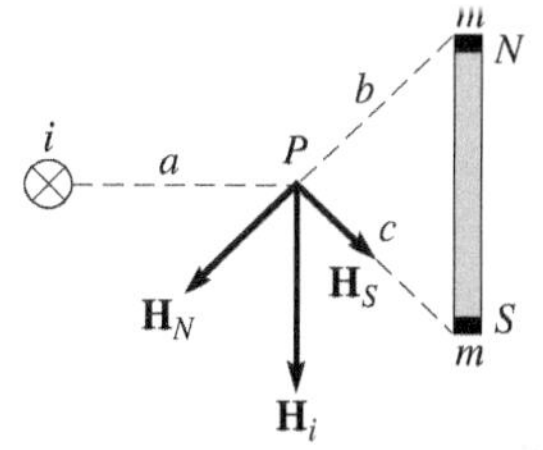

Fig. 9.47

36. Solve the problem of Exercise 35 if $\mathbf{H}_i$ is directed away from the magnet, making an angle of 10.0° with the direction of the magnet.

Answer to Practice Exercise

1. 534 km/h, 7.6° north of east

9.5 Oblique Triangles, the Law of Sines

To this point, we have limited our study of triangle solution to right triangles. However, *many triangles that require solution do not contain a right angle. Such triangles are called* **oblique triangles**. We now discuss solutions of oblique triangles.

In Section 4.4, we stated that we need to know three parts, at least one of them a side, to solve a triangle. There are four possible such combinations of parts, and these combinations are as follows:

Case 1. *Two angles and one side (AAS)*

Case 2. *Two sides and the angle opposite one of them (SSA)*

Case 3. *Two sides and the included angle (SAS)*

Case 4. *Three sides (SSS)*

There are several ways in which oblique triangles may be solved, but we restrict our attention to the two most useful methods: the **law of sines** and the **law of cosines**. In this section, we derive the law of sines and show how it may be used to solve Case 1 (AAS) and Case 2 (SSA).

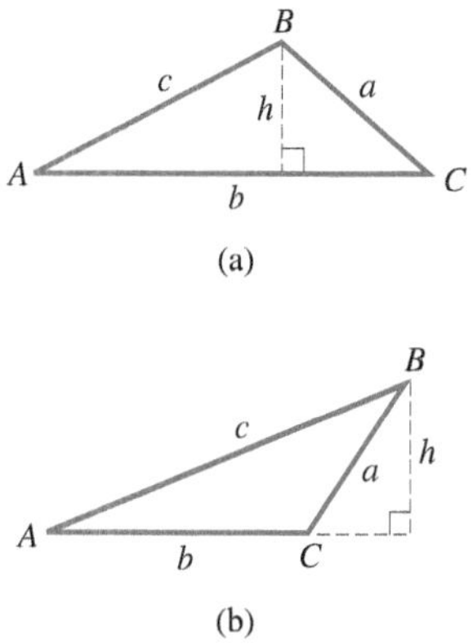

Fig. 9.48

Let ABC be an oblique triangle with sides a, b, and c opposite angles A, B, and C, respectively. By drawing a perpendicular h from B to side b, as shown in Fig. 9.48(a), note that $h/c = \sin A$ and $h/a = \sin C$, and therefore

$$h = c \sin A \quad \text{or} \quad h = a \sin C \tag{9.4}$$

This relationship also holds if one of the angles is obtuse. By drawing a perpendicular h to the extension of side b, as shown in Fig. 9.48(b), we see that $h/c = \sin A$.

Also, in Fig. 9.48(b), $h/a = \sin(180° - C)$, where $180° - C$ is also the reference angle for the obtuse angle C. For a second-quadrant reference angle $180° - C$, we have

$\sin[180° - (180° - C)] = \sin C$. Putting this all together, $\sin(180° - C) = \sin C$. This means that

$$h = c \sin A \quad \text{or} \quad h = a \sin(180° - C) = a \sin C \tag{9.5}$$

The results are precisely the same in Eqs. (9.4) and (9.5). Setting the results for h equal to each other, we have

$$c \sin A = a \sin C$$

Dividing each side by $\sin A$ multiplied by $\sin C$ and reversing sides gives us

$$\frac{a}{\sin A} = \frac{c}{\sin C} \tag{9.6}$$

By dropping a perpendicular from A to a, we also derive the result

$$c \sin B = b \sin C$$

which, when each side is divided by $\sin B$ multiplied by $\sin C$, and reversing sides, gives us

$$\frac{b}{\sin B} = \frac{c}{\sin C} \tag{9.7}$$

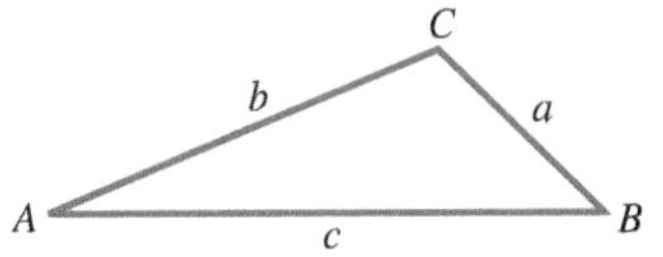

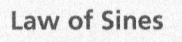

Fig. 9.49

If we had dropped a perpendicular to either side a or side c, equations similar to Eqs. (9.6) and (9.7) would have resulted.

Combining Eqs. (9.6) and (9.7), *for any triangle with sides a, b, and c, opposite angles A, B, and C, respectively,* such as the one shown in Fig. 9.49, *we have the* **law of sines**:

Law of Sines

$$\frac{a}{\sin A} = \frac{b}{\sin B} = \frac{c}{\sin C} \tag{9.8}$$

> **LEARNING TIP**
> Note that *there are actually three equations combined in Eq. (9.8).* Of these, we use the one with three known parts of the triangle, and we find the fourth part. In finding the complete solution of a triangle, it may be necessary to use two of the three equations.

Another form of the law of sines can be obtained by equating the reciprocals of each of the fractions in Eq. (9.8). The law of sines is a statement of proportionality between the sides of a triangle and the sines of the angles opposite them.

CASE 1: TWO ANGLES AND ONE SIDE (AAS)

Now, we see how the law of sines is used in the solution of a triangle in which two angles and one side are known. If two angles are known, the third may be found from the fact that the sum of the angles in a triangle is 180°. At this point, we must be able to find the ratio between the given side and the sine of the angle opposite it. Then, by use of the law of sines, we may find the other two sides.

EXAMPLE 1 Case 1: Two angles and one side (AAS)

Given $c = 6.00$, $A = 60.0°$, and $B = 40.0°$, find a, b, and C.

First, we can see that

$$C = 180.0° - (60.0° + 40.0°) = 80.0°$$

We now know side c and angle C, which allows us to use Eq. (9.8). Therefore, using the equation relating a, A, c, and C, we have

$$\frac{a}{\sin 60.0°} = \frac{6.00}{\sin 80.0°} \quad \text{or} \quad a = \frac{6.00 \sin 60.0°}{\sin 80.0°} = 5.28$$

(annotations: A → $\sin 60.0°$; C → 6.00 and $\sin 80.0°$)

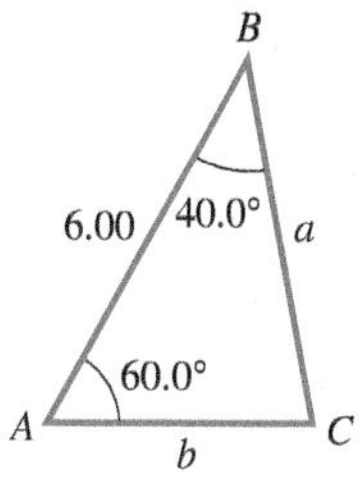

Fig. 9.50

Now, using the equation relating b, B, c, and C, we have

$$\frac{b}{\sin 40.0°} = \frac{6.00}{\sin 80.0°} \quad \text{or} \quad b = \frac{6.00 \sin 40.0°}{\sin 80.0°} = 3.92$$

Thus, $a = 5.28$, $b = 3.92$, and $C = 80.0°$. See Fig. 9.50. We could also have used the form of Eq. (9.8) relating a, A, b, and B in order to find b, but any error in calculating a would make b in error as well.

Practice Exercise

1. In Example 1, change the value of B to $65.0°$, and then find a.

EXAMPLE 2 Case 1: Two angles and one side (AAS)

Solve the triangle with the following given parts: $a = 63.71$, $A = 56.29°$, and $B = 97.06°$. See Fig. 9.51.

From the figure, we see that we are to find angle C and sides b and c. We first determine angle C:

$$C = 180° - (A + B) = 180° - (56.29° + 97.06°)$$
$$= 26.65°$$

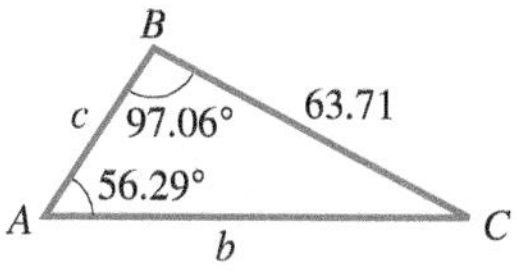

Fig. 9.51

Now using the ratio $a/\sin A$ of Eq. (9.8) (the law of sines) to find sides b and c, we have

$$\frac{b}{\sin 97.06°} = \frac{63.71}{\sin 56.29°} \quad \text{or} \quad b = \frac{63.71 \sin 97.06°}{\sin 56.29°} = 76.01$$

$$\frac{c}{\sin 26.65°} = \frac{63.71}{\sin 56.29°} \quad \text{or} \quad c = \frac{63.71 \sin 26.65°}{\sin 56.29°} = 34.35$$

Thus, $b = 76.01$, $c = 34.35$, and $C = 26.65°$. Note that the resulting values match what we should expect from the given information: c is the shortest side (opposite the smallest angle C), and b is the longest side (opposite the largest angle B).

LEARNING TIP

The solution of a triangle can be checked approximately by noting that *the smallest angle is opposite the shortest side, and the largest angle is opposite the longest side.* Note that this is the case in Example 2, where c (shortest side) is opposite C (smallest angle), and b (longest side) is opposite B (largest angle).

If the given information is appropriate, the law of sines may be used to solve applied problems. The following example illustrates the use of the law of sines in such a problem.

EXAMPLE 3 Case 1 (AAS)—distance to a helicopter

Two observers A and B sight a helicopter due east. The observers are 1540 m apart, and the angles of elevation they each measure to the helicopter are 32.0° and 44.0°, respectively. How far is observer A from the helicopter? See Fig. 9.52.

■ The first successful helicopter was made in the United States by Igor Sikorsky in 1939.

Letting H represent the position of the helicopter, we see that angle B within the triangle ABH is $180° - 44.0° = 136.0°$. This means that the angle at H within the triangle is

$$H = 180° - (32.0° + 136.0°) = 12.0°$$

Now, using the law of sines to find side b, we have

$$\frac{b}{\sin 136.0°} = \frac{1540}{\sin 12.0°}$$

(required side opposite known angle → b; known side opposite known angle ← 1540)

or

$$b = \frac{1540 \sin 136.0°}{\sin 12.0°} = 5150 \text{ m}$$

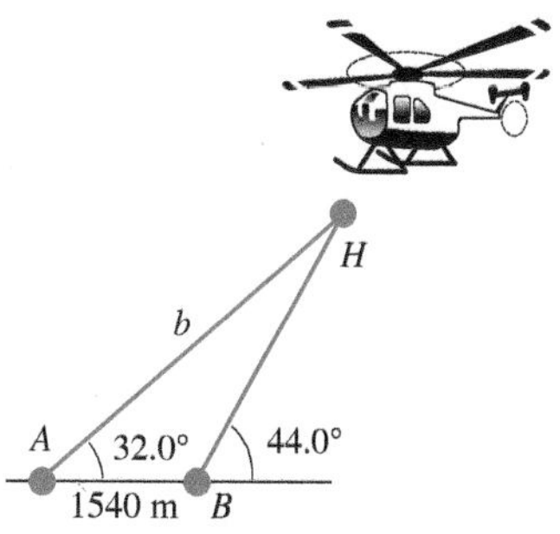

Fig. 9.52

Thus, observer A is about 5150 m from the helicopter.

CASE 2: TWO SIDES AND THE ANGLE OPPOSITE ONE OF THEM (SSA)

For a triangle in which we know two sides and the angle opposite one of the given sides, the solution will be either *one triangle*, or *two triangles*, or even possibly *no triangle*. The following examples illustrate how each of these results is possible.

EXAMPLE 4 Case 2: Two sides and angle opposite (SSA)

Solve the triangle with the following given parts: $a = 40.0$, $b = 60.0$, and $A = 30.0°$.

First, make a good scale drawing (Fig. 9.53a) by drawing angle A and measuring off 60 for b. This will more clearly show that side $a = 40.0$ will intersect side c at either position B or B'. This means there are two triangles that satisfy the given values. Using the law of sines, we solve the case for which B is an acute angle:

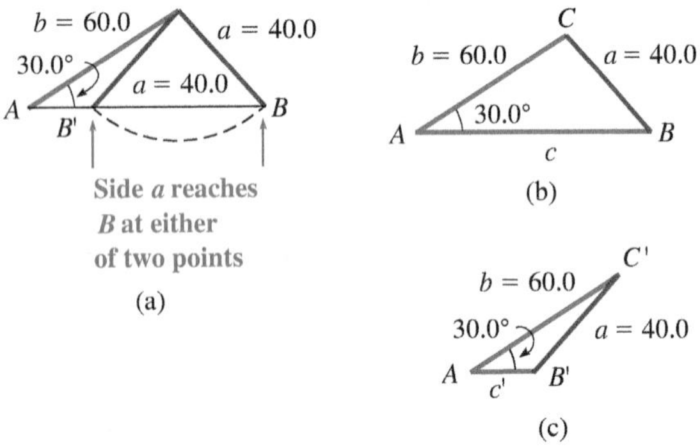

Fig. 9.53

$$\frac{60.0}{\sin B} = \frac{40.0}{\sin 30.0°} \quad \text{or} \quad \sin B = \frac{60.0 \sin 30.0°}{40.0}$$

$$B = \sin^{-1}\left(\frac{60.0 \sin 30.0°}{40.0}\right) = 48.6°$$

$$C = 180° - (30.0° + 48.6°) = 101.4°$$

Therefore, $B = 48.6°$ and $C = 101.4°$. Using the law of sines again to find c, we have

$$\frac{c}{\sin 101.4°} = \frac{40.0}{\sin 30.0°}$$

$$c = \frac{40.0 \sin 101.4°}{\sin 30.0°}$$

$$= 78.4$$

Thus, $B = 48.6°$, $C = 101.4°$, and $c = 78.4$. See Fig. 9.53(b).

The other solution is the case in which B', opposite side b, is an obtuse angle. Therefore,

$$B' = 180° - B = 180° - 48.6°$$
$$= 131.4°$$
$$C' = 180° - (30.0° + 131.4°)$$
$$= 18.6°$$

Using the law of sines to find c', we have

$$\frac{c'}{\sin 18.6°} = \frac{40.0}{\sin 30.0°}$$

$$c' = \frac{40.0 \sin 18.6°}{\sin 30.0°}$$

$$= 25.5$$

This means that the second solution is $B' = 131.4°$, $C' = 18.6°$, and $c' = 25.5$. See Fig. 9.53(c).

EXAMPLE 5 Case 2 (SSA)—possible solutions

In Example 4, if $a > 60.0$, only one solution would result. In this case, side a would intercept side c at B. It also intercepts the extension of side c, but this would require that angle A not be included in the triangle (see Fig. 9.54). Thus, only one solution may result if $a > b$.

In Example 4, there would be *no solution* if side a were not at least 30.0. If this were the case, side a would not be long enough to even touch side c. It can be seen that a must at least equal $b \sin A$. If it is just equal to $b \sin A$, there is *one solution*, a right triangle. See Fig. 9.55.

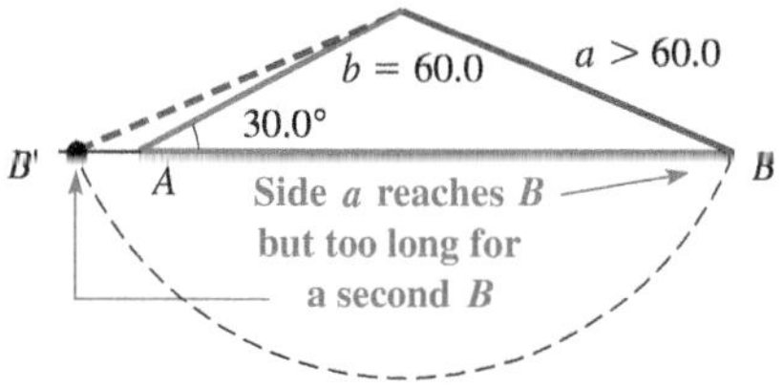

Fig. 9.54

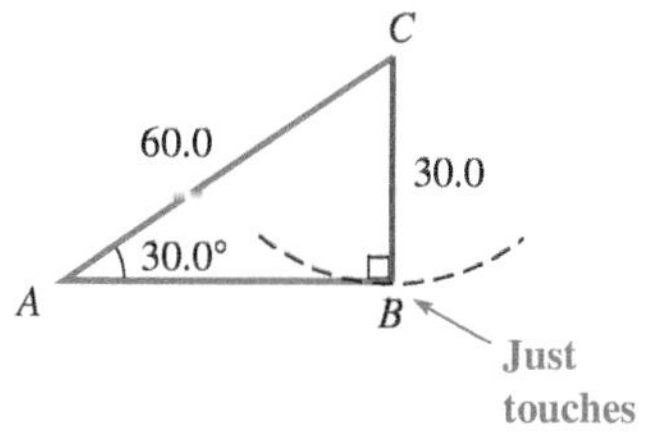

Fig. 9.55

Practice Exercise

2. Determine which of the four possible solution types occurs if $a = 28$, $b = 48$, and $A = 30°$.

Summarizing the results for Case 2 (SSA) as illustrated in Examples 4 and 5, we make the following conclusions for when sides a and b and angle A [assuming here that a and A ($A < 90°$) are corresponding parts] are known.

Summary of Solutions:
Two Sides and the Angle Opposite One of Them (SSA)

Conditions	Number of Possible Solutions	Fig. 9.56
$a < b \sin A$	0	(a) 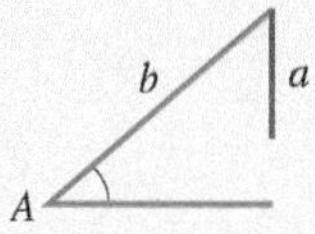
$a = b \sin A$	1	(b) 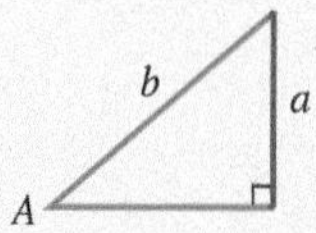
$b \sin A < a < b$	2	(c) 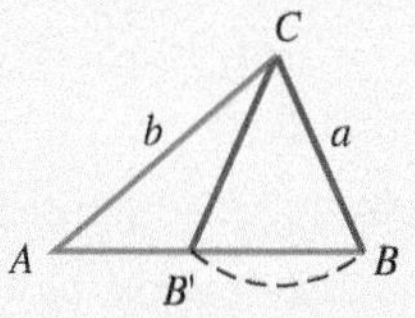
$a > b$	1	(d) 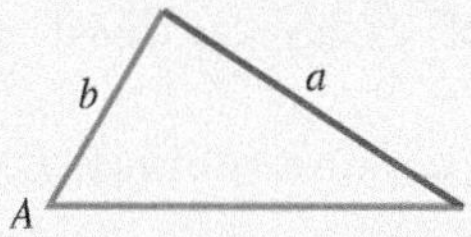

LEARNING TIP

Note that *in order to have two solutions, we must know two sides and the angle opposite one of the sides, and the shorter side must be opposite the known angle.*

If there is *no solution*, the calculator will indicate an *error*. If the solution is a *right triangle*, the calculator will show an angle of *exactly* 90° (no extra decimal digits will be displayed).

For the reason that two solutions may result from it, Case 2 is called the **ambiguous case**. The following example illustrates Case 2 in an applied problem.

EXAMPLE 6 Case 2 (SSA)—heading of a plane

Edmonton is 35.2° north of east of Vancouver. What should be the heading of a plane from Vancouver to Edmonton if the wind is from the west at 40.0 km/h and the plane's speed with respect to the air is 300 km/h?

The heading should be set so that the resultant of the plane's velocity with respect to the air $\mathbf{v}_{pa}$ and the velocity of the wind $\mathbf{v}_w$ will be in the direction from Vancouver to Edmonton. This means that the resultant velocity $\mathbf{v}_{pg}$ of the plane with respect to the ground must be at an angle of 35.2° north of east of Vancouver.

Using the given information, we draw the vector triangle shown in Fig. 9.57. In the triangle, we know that the angle at Edmonton is 35.2° by noting the alternate-interior angles (see Section 2.1). By finding θ, the required heading can be found. There can be only one solution, since $v_{pa} > v_w$ (see Fig. 9.55d). Using the law of sines, we have

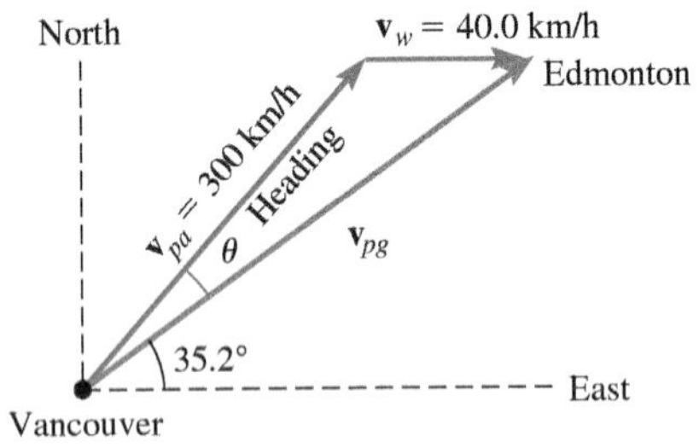

Fig. 9.57

$$\frac{40.0}{\sin\theta} = \frac{300}{\sin 35.2^\circ}$$

(known side opposite required angle → 40.0, $\sin\theta$; known side opposite known angle → 300, $\sin 35.2^\circ$)

$$\sin\theta = \frac{40.0 \sin 35.2^\circ}{300}, \qquad \theta = 4.4^\circ$$

Therefore, the heading should be 35.2° + 4.4° = 39.6° north of east.

If we try to use the law of sines for Case 3 or Case 4, we find that we do not have enough information to complete any of the ratios. These cases can, however, be solved by the law of cosines as shown in the next section.

EXAMPLE 7 Cases 3 (SAS) and 4 (SSS) not solvable by the law of sines

Given (Case 3) two sides and the included angle of a triangle $a = 2$, $b = 3$, $C = 45^\circ$, and (Case 4) the three sides $a = 5$, $b = 6$, $c = 7$, we set up the ratios

$$\text{(Case 3)}\ \frac{2}{\sin A} = \frac{3}{\sin B} = \frac{c}{\sin 45^\circ} \quad \text{and} \quad \text{(Case 4)}\ \frac{5}{\sin A} = \frac{6}{\sin B} = \frac{7}{\sin C}$$

The solution cannot be found since each of the three possible equations in either Case 3 or Case 4 contains two unknowns.

EXERCISES 9.5

In Exercises 1 and 2, solve the resulting triangles if the given changes are made in the indicated examples of this section.

1. In Example 2, solve the triangle if the value of B is changed to 82.94°.
2. In Example 4, solve the triangle if the value of b is changed to 70.0.

In Exercises 3–22, solve the triangles with the given parts.

3. $a = 45.7$, $A = 65.0^\circ$, $B = 49.0^\circ$
4. $b = 3.07$, $A = 26.0^\circ$, $C = 120.0^\circ$
5. $c = 4380$, $A = 37.4^\circ$, $B = 34.6^\circ$
6. $a = 932$, $B = 0.9^\circ$, $C = 82.6^\circ$
7. $a = 4.601$, $b = 3.107$, $A = 18.23^\circ$
8. $b = 362.2$, $c = 294.6$, $B = 69.37^\circ$
9. $b = 7751$, $c = 3642$, $B = 20.73^\circ$
10. $a = 150.4$, $c = 250.9$, $C = 76.43^\circ$
11. $b = 0.0742$, $B = 51.0^\circ$, $C = 3.40^\circ$
12. $c = 729$, $B = 121.0^\circ$, $C = 44.2^\circ$
13. $a = 63.8$, $B = 58.4^\circ$, $C = 22.2^\circ$
14. $a = 0.130$, $A = 55.2^\circ$, $B = 117.5^\circ$
15. $b = 4384$, $B = 47.43^\circ$, $C = 64.56^\circ$
16. $b = 283.2$, $B = 13.79^\circ$, $C = 76.38^\circ$
17. $a = 5.240$, $b = 4.446$, $B = 48.13^\circ$
18. $a = 89.45$, $c = 37.36$, $C = 15.62^\circ$
19. $b = 2880$, $c = 3650$, $B = 31.4^\circ$
20. $a = 0.841$, $b = 0.965$, $A = 57.1^\circ$
21. $a = 450$, $b = 1260$, $A = 64.8^\circ$
22. $a = 20$, $c = 10$, $C = 30^\circ$

In Exercises 23–41, use the law of sines to solve the given problems.

23. A small island is approximately a triangle in shape. If the longest side of the island is 520 m, and two of the angles are 45° and 55°, what is the length of the shortest side?

24. A boat followed a triangular route going from dock A, to dock B, to dock C, and back to dock A. The angles turned were 135° at B and 125° at C. If B is 875 m from A, how far is it from B to C?

25. The loading ramp at a package delivery service is 3.2 m long and makes a 22.5° angle with the level ground. If this ramp is replaced with one that is 7.5 m long, what angle does the new ramp make with the ground?

26. The longest side of a triangular parcel of land is 172 m long, and the shortest side is 105 m long. If the largest angle is 82.0°, what is the length of the third side?

27. The floor of the Bastion in Nanaimo, British Columbia, is a regular octagon, 2.3 m on each side. Find the greatest distance across the floor (that is, find the length of the longest diagonal).

28. Two ropes hold a 175-N crate as shown in Fig. 9.58. Find the tensions $\mathbf{T}_1$ and $\mathbf{T}_2$ in the ropes. (*Hint:* Move the vectors so that they are tail to head to form a triangle. The vector sum $\mathbf{T}_1 + \mathbf{T}_2$ must equal 175 N for equilibrium.)

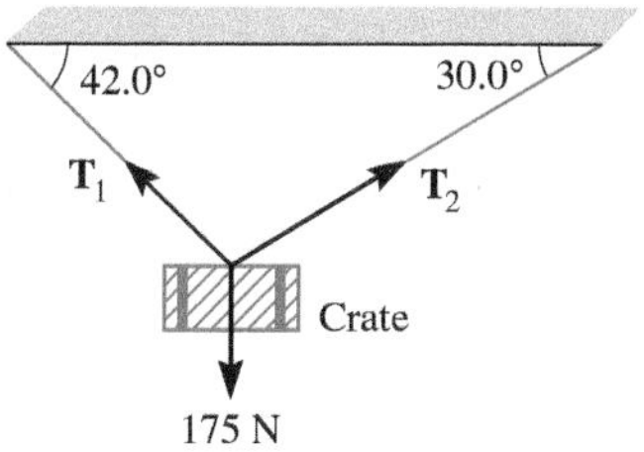

Fig. 9.58

29. Find the tension **T** in the left guy wire attached to the top of the tower shown in Fig. 9.59. (*Hint:* The horizontal components of the tensions must be equal and opposite for equilibrium. Thus, move the tension vectors tail to head to form a triangle with a vertical resultant. This resultant equals the upward force at the top of the tower for equilibrium. This last force is not shown and does not have to be calculated.)

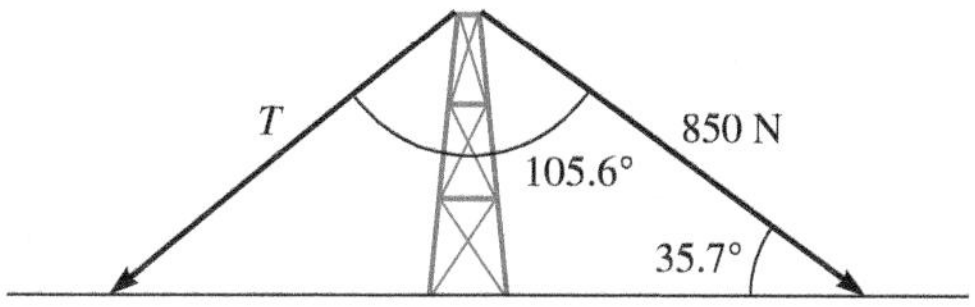

Fig. 9.59

30. Find the distance from Golden to Radium, from Fig. 9.60. The route that joins these landmarks across three national parks is known as the Golden Triangle.

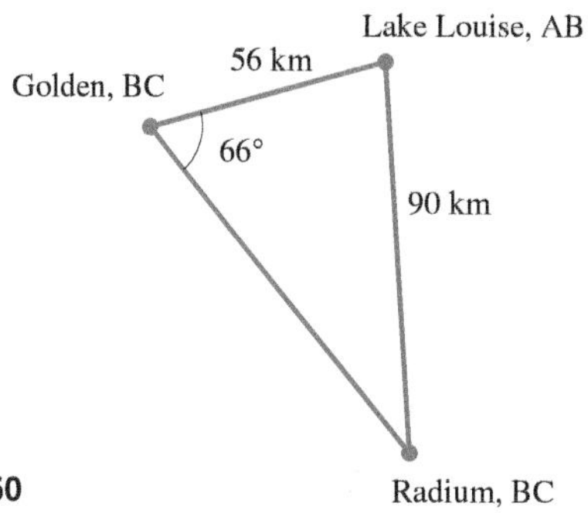

Fig. 9.60

31. An Amazon robot begins at point O and travels in a straight line across a warehouse floor to point A where it picks up some merchandise. It then turns a 35° corner and travels 32.5 m to point B where it drops off the merchandise. See Fig. 9.61. If the robot must now turn a 29.0° corner to return to its original position, how far must it travel to get there?

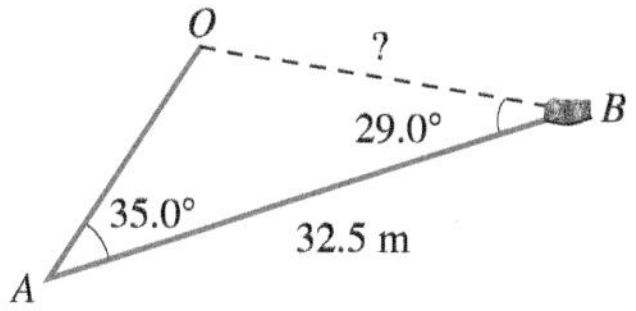

Fig. 9.61

32. When an airplane is landing on a 2510-m runway, the angles of depression to the ends of the runway are 10.0° and 13.5°. How far is the plane from the near end of the runway?

33. Find the total length of the path of the laser beam that is shown in Fig. 9.62.

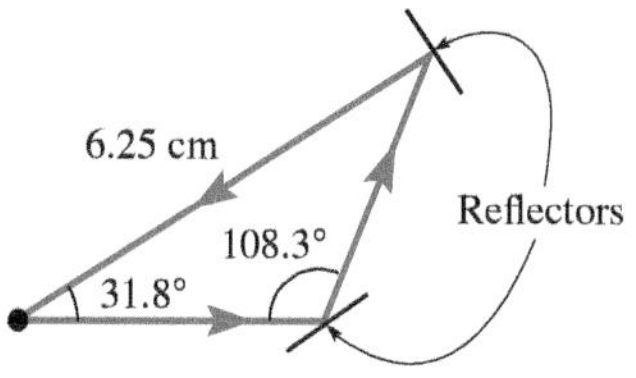

Fig. 9.62

34. In widening a highway, it is necessary for a construction crew to cut into the bank along the highway. The present angle of elevation of the straight slope of the bank is 23.0°, and the new angle is to be 38.5°, leaving the top of the slope at its present position. If the slope of the present bank is 66.0 m long, how far horizontally into the bank at its base must they dig?

35. A communications satellite is directly above the extension of a line between receiving towers A and B. It is determined from radio signals that the angle of elevation of the satellite from tower A is 89.2°, and the angle of elevation from tower B is 86.5°. See Fig. 9.63. If A and B are 1290 km apart, how far is the satellite from A? (Ignore the curvature of the earth.)

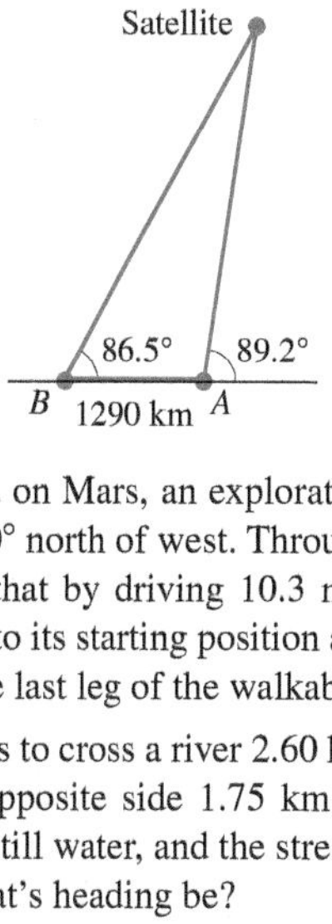

Fig. 9.63

36. During a walkabout on Mars, an exploration rover drives 15.4 m in the direction 22.0° north of west. Through what angle must the rover then turn so that by driving 10.3 m farther, the rover can turn again to return to its starting position along an east–west line? How long would the last leg of the walkabout be?

37. A boat owner wishes to cross a river 2.60 km wide and go directly to a point on the opposite side 1.75 km downstream. The boat goes 8.00 km/h in still water, and the stream flows at 3.50 km/h. What should the boat's heading be?

38. A motorist travelling along a level highway at 75 km/h directly toward a mountain notes that the angle of elevation of the mountain top changes from about 20° to about 30° in a 20-min period. How much closer on a direct line did the mountain top become?

39. A hillside is inclined at 23° with the horizontal. From a given point on the slope, it has been found that a vein of gold is 55 m directly below. At what point downhill and at what angle below the hillside slope must a straight 65-m shaft be dug to reach the vein?

40. Point P on the mechanism shown in Fig. 9.64 is driven back and forth horizontally. If the minimum value of angle θ is 32.0°, what is the distance between extreme positions of P? What is the maximum possible value of angle θ?

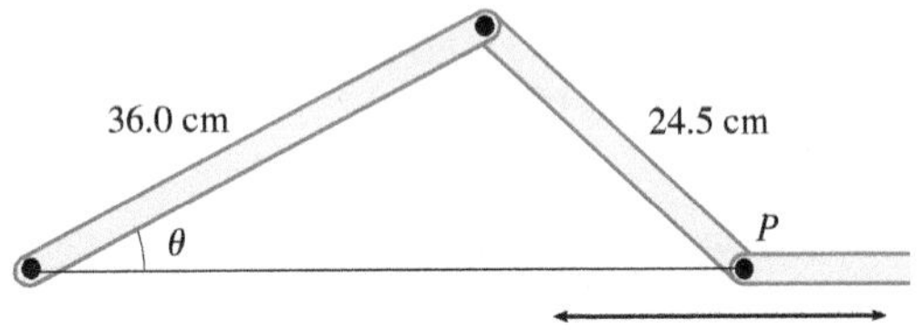

Fig. 9.64

41. The floor of the Winnipeg Art Gallery is in the shape of a triangle. Find the length of the side facing Memorial Boulevard, from Fig. 9.65.

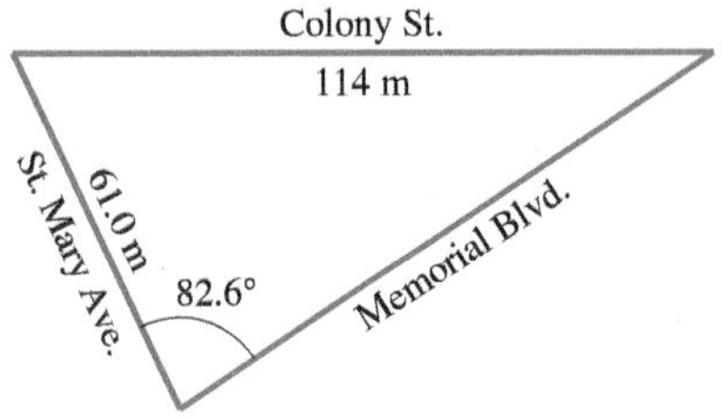

Fig. 9.65

Answers to Practice Exercises

1. $a = 6.34$ **2.** Two solutions

9.6 The Law of Cosines

As noted in the last section, the law of sines cannot be used for Case 3 (two sides and the included angle—SAS) and Case 4 (three sides—SSS). In this section, we develop the *law of cosines*, which can be used for Cases 3 and 4. After finding another part of the triangle using the law of cosines, we will often find it easier to complete the solution using the law of sines.

Consider any oblique triangle—for example, either triangle shown in Fig. 9.66. For each triangle, $h/b = \sin A$, or $h = b \sin A$. Also, using the Pythagorean theorem, we obtain $a^2 = h^2 + x^2$ for each triangle. Therefore [with $(\sin A)^2 = \sin^2 A$],

$$a^2 = b^2 \sin^2 A + x^2 \tag{9.9}$$

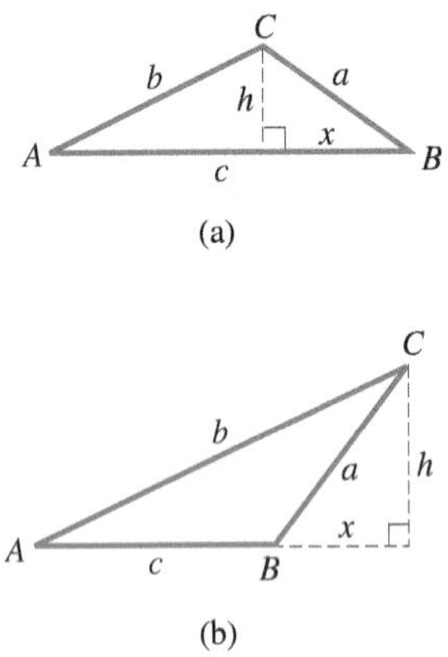

Fig. 9.66

In Fig. 9.66(a), note that $(c - x)/b = \cos A$, or $c - x = b \cos A$. Solving for x, we have $x = c - b \cos A$. In Fig. 9.66(b), $c + x = b \cos A$, and solving for x, we have $x = b \cos A - c$. Substituting these relations into Eq. (9.9), we obtain

$$a^2 = b^2 \sin^2 A + (c - b \cos A)^2$$

and

$$a^2 = b^2 \sin^2 A + (b \cos A - c)^2 \tag{9.10}$$

respectively. When expanded, these both give

$$\begin{aligned} a^2 &= b^2 \sin^2 A + b^2 \cos^2 A + c^2 - 2bc \cos A \\ &= b^2(\sin^2 A + \cos^2 A) + c^2 - 2bc \cos A \end{aligned} \tag{9.11}$$

Recalling the definitions of the trigonometric functions, we know that $\sin \theta = y/r$ and $\cos \theta = x/r$. Thus, $\sin^2\theta + \cos^2\theta = (y^2 + x^2)/r^2$. However, $x^2 + y^2 = r^2$, which means

$$\sin^2\theta + \cos^2\theta = 1 \tag{9.12}$$

This equation is valid for any angle θ, since we have made no assumptions as to the properties of θ. Thus, by substituting Eq. (9.12) into Eq. (9.11), we arrive at the **law of cosines**:

Law of Cosines

$$a^2 = b^2 + c^2 - 2bc \cos A \tag{9.13}$$

The law of cosines can also be written in the following forms:

$$b^2 = a^2 + c^2 - 2ac \cos B$$
$$c^2 = a^2 + b^2 - 2ab \cos C$$

CASE 3: TWO SIDES AND THE INCLUDED ANGLE (SAS)

If two sides and the included angle of a triangle are known, the forms of the law of cosines show that we may directly solve for the side opposite the given angle. Then, as noted earlier, the solution may be completed using the law of sines.

EXAMPLE 1 Case 3: Two sides and included angle (SAS)

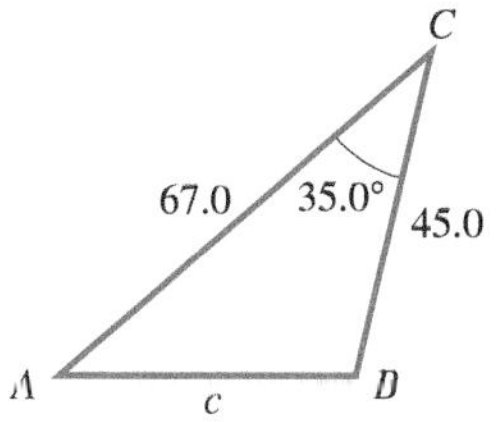

Fig. 9.67

Solve the triangle with $a = 45.0$, $b = 67.0$, and $C = 35.0°$. See Fig. 9.67.

Since angle C is known, first solve for side c, using the law of cosines in the form $c^2 = a^2 + b^2 - 2ab\cos C$. Substituting, we have

$$c^2 = 45.0^2 + 67.0^2 - 2(45.0)(67.0)\cos 35.0°$$

(unknown side opposite known angle: c and $35.0°$; known sides: 45.0, 67.0, 45.0, 67.0)

$$c = \sqrt{45.0^2 + 67.0^2 - 2(45.0)(67.0)\cos 35.0°} = 39.7$$

From the law of sines, we now have

$$\frac{45.0}{\sin A} = \frac{67.0}{\sin B} = \frac{39.7}{\sin 35.0°} \quad \leftarrow \text{sides opposite angles}$$

$$\sin A = \frac{45.0 \sin 35.0°}{39.7}, \qquad A = 40.6°$$

Finally, rather than use the law of sines again, we obtain angle B by subtraction. We find that $B = 180° - (35.0° + 40.6°) = 104.4°$. Therefore, $c = 39.7$, $A = 40.6°$, and $B = 104.4°$.

Practice Exercise

1. In Example 1, change the value of a to 95.0, and find A and B. (*Hint*: Solve for the smaller angle first.)

COMMON ERROR

If after finding side c we had solved for angle B rather than angle A, the calculator would have shown 75.6°. This would have given us the value of the reference angle for angle B, and not the correct value for the obtuse angle B ($B = 180° - 75.6° = 104.4°$).

Since only the largest angle of a triangle can be greater than 90°, we can avoid this possible source of error if we ***first solve for the smaller unknown angle*** (the angle opposite the shorter known side). The larger unknown angle can then be found by subtraction.

EXAMPLE 2 Solving Case 3—forces on a bolt

Two forces are acting on a bolt. One is a 78.0-N force acting horizontally to the right, and the other is a force of 45.0 N acting upward to the right, 15.0° from the vertical. Find the resultant force **F**. See Fig. 9.68.

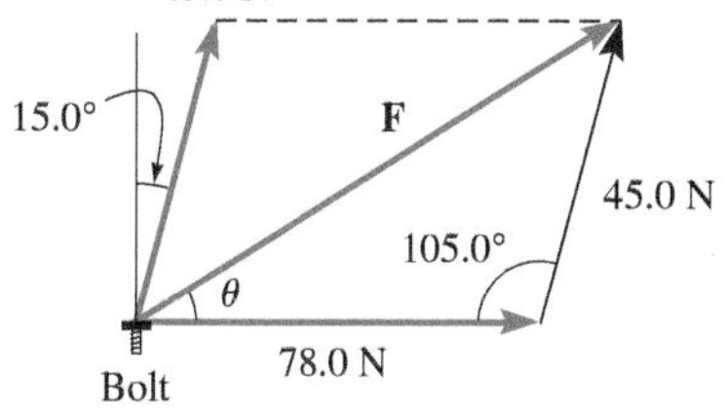

Fig. 9.68

Moving the 45.0-N vector to the right and using the lower triangle with the 105.0° angle, the magnitude of **F** is

$$F = \sqrt{78.0^2 + 45.0^2 - 2(78.0)(45.0)\cos 105.0°}$$
$$= 99.6 \text{ N}$$

To find θ, use the law of sines:

$$\frac{45.0}{\sin\theta} = \frac{99.6}{\sin 105.0°}, \qquad \sin\theta = \frac{45.0 \sin 105.0°}{99.6}$$

This gives us $\theta = 25.9°$.

We can also solve this problem using vector components.

CASE 4: THREE SIDES (SSS)

Given the three sides of a triangle, we may solve for the angle opposite any side using the law of cosines.

LEARNING TIP

The best procedure for Case 4 (SSS) is to find the largest angle first. This avoids the ambiguous case if we switch to the law of sines and there is an obtuse angle. *The largest angle is opposite the longest side.* Another procedure is to use the law of cosines to find two angles.

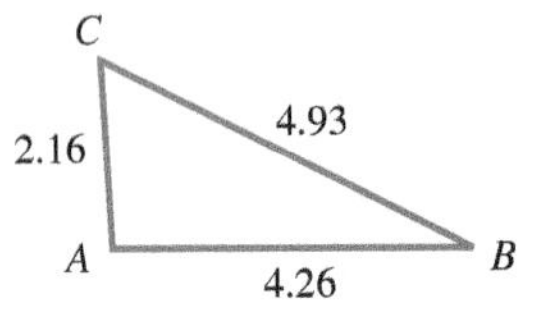

Fig. 9.69

EXAMPLE 3 Case 4 (SSS): Three sides—roof angles

For a triangular roof truss with sides 4.93 m, 2.16 m, and 4.26 m, find the angles between the sides. See Fig. 9.69.

First, we let the sides be a, b, and c, with opposite angles of A, B, and C. Then, because the longest side is $a = 4.93$, we first solve for angle A:

$$a^2 = b^2 + c^2 - 2bc \cos A$$

$$\cos A = \frac{b^2 + c^2 - a^2}{2bc} = \frac{2.16^2 + 4.26^2 - 4.93^2}{2(2.16)(4.26)}$$

$$A = 94.6°$$

From the law of sines, we now have

$$\frac{4.93}{\sin A} = \frac{2.16}{\sin B} = \frac{4.26}{\sin C}$$

Therefore,

$$B = \sin^{-1}\left(\frac{2.16 \sin 94.65°}{4.93}\right) = 25.9°$$

and $C = 180° - (94.6° + 25.9°) = 59.5°$.

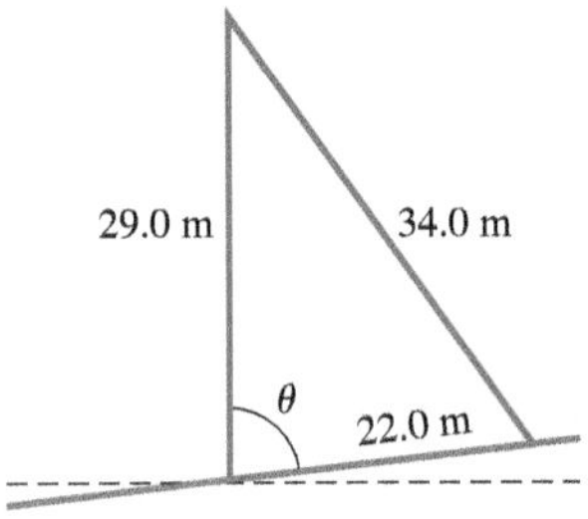

Fig. 9.70

EXAMPLE 4 Solving Case 4—securing an antenna

A vertical radio antenna is to be built on a hillside with a constant slope. A guy wire is to be attached at a point 29.0 m up the antenna, and at a point 22.0 m from the base of the antenna up the hillside. If the guy wire is 34.0 m long, what angle does the antenna make with the hillside?

From Fig. 9.70, we can set up the equation necessary for the solution.

$$34.0^2 = 22.0^2 + 29.0^2 - 2(22.0)(29.0)\cos\theta$$

$$\theta = \cos^{-1}\left(\frac{22.0^2 + 29.0^2 - 34.0^2}{2(22.0)(29.0)}\right) = 82.4°$$

Practice Exercise

2. Using Fig. 9.70, and only the data given in Example 4, find the angle between the guy wire and the hillside.

Solving Oblique Triangles

Given	Common Steps for Solving
Case 1 (AAS): Two angles and one side	**1.** Find the unknown angle by subtracting the sum of the known angles from 180°. **2.** Use the **law of sines** to find the unknown sides.
Case 2 (SSA): Two sides and the angle opposite one of them	**1.** Use the **law of sines** to find the unknown angle opposite the known side. **2.** Find the third angle using the fact that the sum of the angles is 180°. **3.** Use the **law of sines** to find the third side. **CAUTION:** This is the *ambiguous* case; there can be zero, one, or two solutions.
Case 3 (SAS): Two sides and the included angle	**1.** Use the **law of cosines** to find the third side. **2.** Use the **law of sines** to find the *smaller* unknown angle (opposite the shorter side). **3.** Find the third angle using the fact that the sum of the angles is 180°.
Case 4 (SSS): three sides	**1.** Use the **law of cosines** to find the *largest* angle (opposite the longest side). **2.** Use the **law of sines** to find a second angle. **3.** Find the third angle using the fact that the sum of the angles is 180°.

Other variations in finding the solutions can be used. For example, after finding the third side in Case 3 or finding the largest angle in Case 4, the solution can be completed by using the law of sines. All angles in Case 4 can be found by using the law of cosines. The methods shown above are those most commonly used.

EXERCISES 9.6

In Exercises 1 and 2, solve the resulting triangles if the given changes are made in the indicated examples of this section.

1. In Example 1, solve the triangle if the value of C is changed to 145°.

2. In Example 3, solve the triangle if 4.93 is changed to 2.93.

In Exercises 3–22, solve the triangles with the given parts.

3. $a = 6.00$, $b = 7.56$, $C = 54.0°$

4. $b = 87.3$, $c = 34.0$, $A = 130.0°$

5. $a = 4530$, $b = 924$, $C = 98.0°$

6. $a = 0.0845$, $c = 0.116$, $B = 85.0°$

7. $a = 395.3$, $b = 452.2$, $c = 671.5$

8. $a = 2.331$, $b = 2.726$, $c = 2.917$

9. $a = 385.4$, $b = 467.7$, $c = 800.9$

10. $a = 0.2433$, $b = 0.2635$, $c = 0.1538$

11. $a = 319$, $b = 848$, $C = 158.0°$

12. $b = 18.3$, $c = 27.1$, $A = 8.70°$

13. $a = 2140$, $c = 428$, $B = 86.3°$

14. $a = 1.13$, $b = 0.510$, $C = 77.6°$

15. $b = 103.7$, $c = 159.1$, $C = 104.67°$

16. $a = 49.32$, $b = 54.55$, $B = 114.36°$

17. $a = 0.4937$, $b = 0.5956$, $c = 0.6398$

18. $a = 69.72$, $b = 49.30$, $c = 22.29$

19. $a = 723$, $b = 598$, $c = 158$

20. $a = 1.78$, $b = 6.04$, $c = 4.80$

21. $a = 1500$, $A = 15°$, $B = 140°$

22. $a = 17$, $b = 24$, $c = 42$. Explain your answer.

In Exercises 23–42, use the law of cosines to solve the given problems.

23. An iRobot Roomba vacuum cleaner travels 2.5 m across the floor where it then hits a wall and turns on a 47° corner. If it travels 1.5 m in this new direction, how far is it from its starting point?

24. Set up equations (do not solve) to solve the triangle in Fig. 9.71 by the law of cosines. Why is the law of sines easier to use?

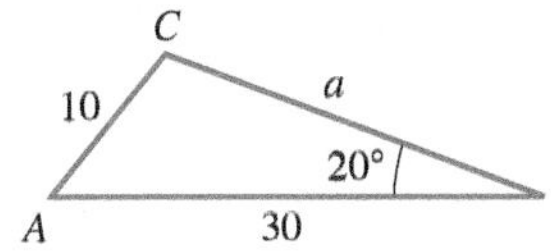

Fig. 9.71

25. Write the form of the law of cosines, given sides a and b, and angle $C = 90°$.

26. Three circles of radii 24 cm, 32 cm, and 42 cm are externally tangent to each other (each is tangent to the other two). Find the largest angle of the triangle formed by joining their centres.

27. A nuclear submarine leaves its base and travels at 23.5 km/h. For 2.00 h, it travels along a course of 32.1° north of west. It then turns an additional 21.5° north of west and travels for another 1.00 h. How far from its base is it?

28. The robot arm shown in Fig. 9.72 places packages on a conveyor belt. What is the distance x?

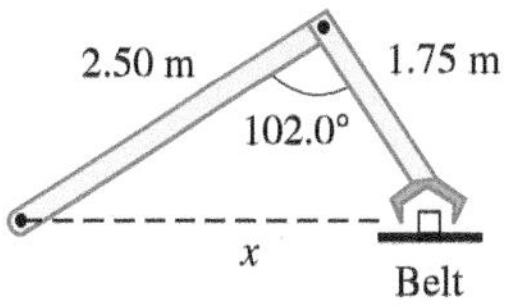

Fig. 9.72

29. Find the angle between the front legs and the back legs of the folding chair shown in Fig. 9.73.

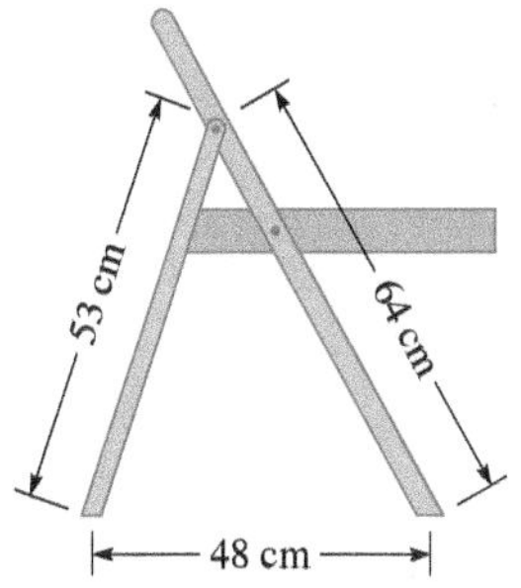

Fig. 9.73

30. In a baseball field, the four bases are at the vertices of a square 90.0 ft on a side. The pitching rubber is 60.5 ft from home plate. See Fig. 9.74. How far is it from the pitching rubber to first base? (1 ft = 0.3048 m)

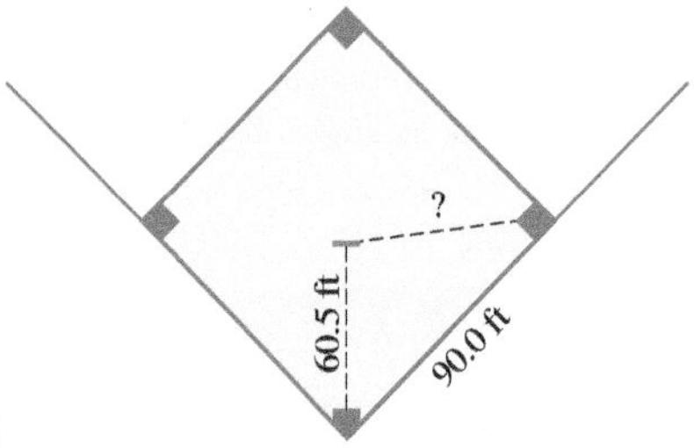

Fig. 9.74

31. A plane leaves an airport and travels 624 km due east. It then turns toward the north and travels another 326 km. It then turns again less than 180° and travels another 846 km directly back to the airport. Through what angles did it turn?

32. The apparent depth of an object submerged in water is less than its actual depth. A coin is actually 5.0 cm from an observer's eye just above the surface, but it appears to be only 4.3 cm. The real light ray from the coin makes an angle with the surface that is 8.1° greater than the angle the apparent ray makes. How much deeper is the coin than it appears to be? See Fig. 9.75.

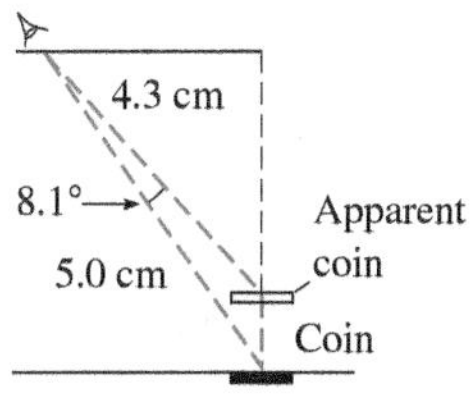

Fig. 9.75

33. A nut is in the shape of a regular hexagon (six sides). If each side is 9.53 mm, what opening on a wrench is necessary to tighten the nut? See Fig. 9.76.

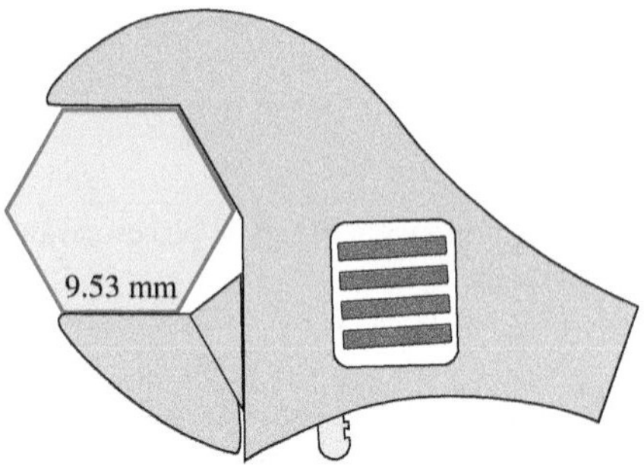

Fig. 9.76

34. The GPS of a plane over Lake Erie indicates that the plane's position is 190 km from Detroit and 110 km from London, Ontario, which are known to be 160 km apart. What is the angle between the plane's directions to Detroit and London?

35. A ferryboat travels at 11.5 km/h with respect to the water. Because of the river current, it is travelling at 12.7 km/h with respect to the land in the direction of its destination. If the ferryboat's heading is 23.6° from the direction of its destination, what is the velocity of the current?

36. Find the distance across Georgian Bay from Nottawasaga Island to Christian Island, from Fig. 9.77.

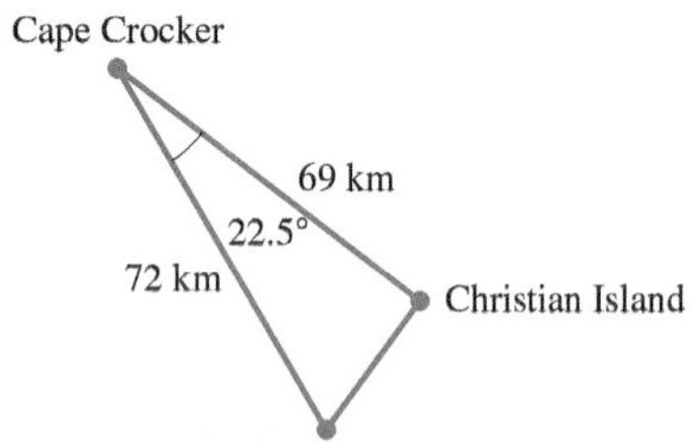

Fig. 9.77

37. An air traffic controller sights two planes that are due east from the control tower and headed toward each other. One is 15.8 km from the tower at an angle of elevation of 26.4°, and the other is 32.7 km from the tower at an angle of elevation of 12.4°. How far apart are the planes?

38. A ship's captain notes that a second ship is 14.5 km away at a bearing measured clockwise from true north of 46.3°, and that a third ship is at a distance of 21.7 km at a bearing of 201.0°. How far apart are the second and third ships?

39. Google's self-driving car uses a laser and detects a pedestrian a distance of 15.8 m away. A split second later (assume the car has not moved), the laser rotates by 56.3° and detects a street light pole 12.1 m away. How far is the pedestrian from the street light pole?

40. A triangular machine part has sides of 5 cm and 8 cm. Explain why the law of sines, or the law of cosines, is used to start the solution of the triangle if the third known part is (a) the third side, (b) the angle opposite the 8-cm side, or (c) the angle between the 5-cm and 8-cm sides.

41. Find the length of the chord intercepted by a central angle of 54.2° in a circle or radius 18.0 cm.

42. Two people are talking to each other on cell phones. The angle between their signals at the tower is 132°, as shown in Fig. 9.78. How far apart are they?

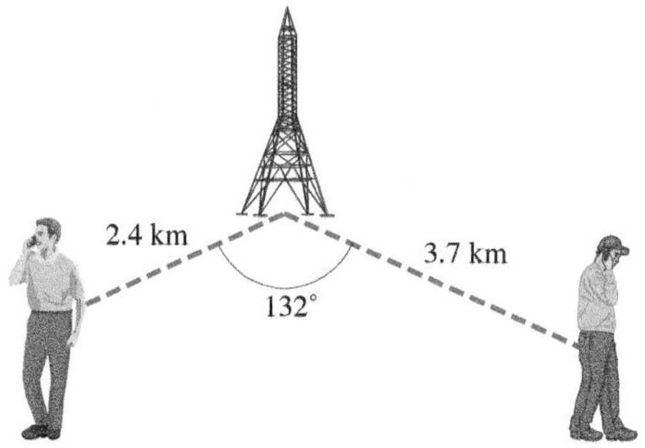

Fig. 9.78

Answers to Practice Exercises

1. $B = 43.8°$, $A = 101.2°$ **2.** 57.7°

CHAPTER 9 KEY FORMULAS AND EQUATIONS

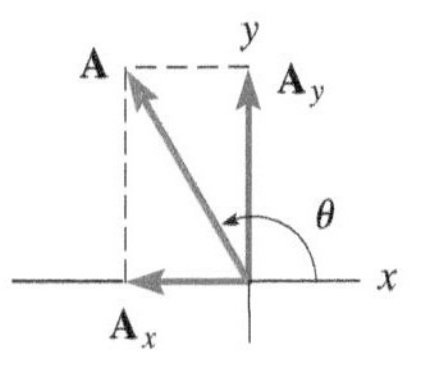

Vector components $A_x = A\cos\theta \qquad A_y = A\sin\theta$ **(9.1)**

$$R = \sqrt{R_x^2 + R_y^2} \qquad (9.2)$$

$$\theta_{\text{ref}} = \tan^{-1}\frac{|R_y|}{|R_x|} \qquad (9.3)$$

Law of sines

$$\frac{a}{\sin A} = \frac{b}{\sin B} = \frac{c}{\sin C} \qquad (9.8)$$

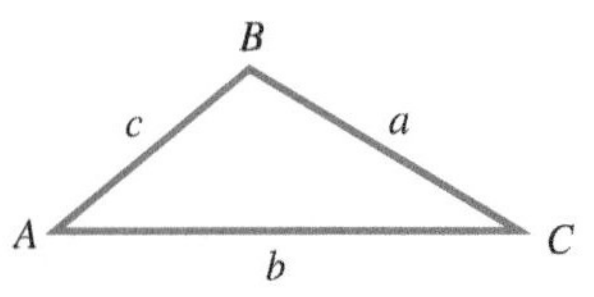

Law of cosines

$$a^2 = b^2 + c^2 - 2bc\cos A \qquad (9.13)$$

$$b^2 = a^2 + c^2 - 2ac\cos B$$

$$c^2 = a^2 + b^2 - 2ab\cos C$$

CHAPTER 9 **REVIEW EXERCISES**

In Exercises 1–4, find the x- and y-components of the given vectors by use of the trigonometric functions.

1. $A = 65.0, \theta_A = 28.0°$

2. $A = 8050, \theta_A = 149.0°$

3. $A = 0.9204, \theta_A = 215.59°$

4. $A = 657.1, \theta_A = 343.74°$

In Exercises 5–8, vectors **A** *and* **B** *are at right angles. Find the magnitude and direction of the resultant.*

5. $A = 327$
$B = 505$

6. $A = 6.8$
$B = 2.9$

7. $A = 4964$
$B = 3298$

8. $A = 26.52$
$B = 89.86$

In Exercises 9–16, add the given vectors by components.

9. $A = 778, \theta_A = 28.0°$
$B = 346, \theta_B = 320.0°$

10. $J = 0.0120, \theta_J = 370.5°$
$K = 0.007\,81, \theta_K = 260.0°$

11. $A = 22.51, \theta_A = 130.16°$
$B = 7.604, \theta_B = 200.09°$

12. $A = 18\,760, \theta_A = 110.43°$
$B = 4835, \theta_B = 350.20°$

13. $Y = 51.33, \theta_Y = 12.25°$
$Z = 42.61, \theta_Z = 291.77°$

14. $A = 703.1, \theta_A = 122.54°$
$B = 302.9, \theta_B = 214.82°$

15. $A = 0.750, \theta_A = 15.0°$
$B = 0.265, \theta_B = 192.4°$
$C = 0.548, \theta_C = 344.7°$

16. $S = 8120, \theta_S = 141.9°$
$T = 1540, \theta_T = 165.2°$
$U = 3470, \theta_U = 296.0°$

In Exercises 17–36, solve the triangles with the given parts.

17. $A = 48.0°, B = 68.0°, a = 145$

18. $A = 132.0°, b = 7.50, C = 32.0°$

19. $a = 22.8, B = 33.5°, C = 125.3°$

20. $A = 71.0°, B = 48.5°, c = 8.42$

21. $A = 17.85°, B = 154.16°, c = 7863$

22. $a = 1.985, b = 4.189, c = 3.652$

23. $b = 7607, c = 4053, B = 110.09°$

24. $A = 77.06°, a = 12.07, c = 5.104$

25. $b = 14.5, c = 13.0, C = 56.6°$

26. $B = 40.6°, b = 7.00, c = 18.0$

27. $a = 186, B = 130.0°, c = 106$

28. $b = 750, c = 1100, A = 56°$

29. $a = 7.86, b = 2.45, C = 2.50°$

30. $a = 0.208, c = 0.697, B = 165.4°$

31. $A = 67.16°, B = 96.84°, c = 532.9$

32. $A = 43.12°, a = 7.893, b = 4.113$

33. $a = 17, b = 12, c = 25$

34. $a = 9064, b = 9953, c = 1106$

35. $a = 0.530, b = 0.875, c = 1.25$

36. $a = 47.4, b = 40.0, c = 45.5$

In Exercises 37–72, solve the given problems.

37. For any triangle ABC show that

$$\frac{a^2 + b^2 + c^2}{2abc} = \frac{\cos A}{a} + \frac{\cos B}{b} + \frac{\cos C}{c}$$

38. In solving a triangle for Case 3 (two sides and the included angle—SAS), explain what type of solution is obtained if the included angle is a right angle.

39. Three straight streets intersect each other such that two of the angles of intersection are 22° and 112°, and the shortest distance between any two intersections is 540 m. What is the longest distance between any two intersections?

40. A triangular brace is designed such that sides meet at angles of 42.0° and 59.5°, with the longest side being 5.00 cm longer than the shortest side. What is the perimeter of the brace?

41. An architect determines the two acute angles and one of the legs of a right triangular wall panel. Show that the area A_t is

$$A_t = \frac{a^2 \sin B}{2 \sin A}$$

42. A surveyor determines the three angles and one side of a triangular tract of land. (a) Show that the area A_t can be found from $A_t = \dfrac{a^2 \sin B \sin C}{2 \sin A}$. (b) For a right triangle, show that this agrees with the formula in Exercise 41.

43. Find the horizontal and vertical components of the force shown in Fig. 9.79.

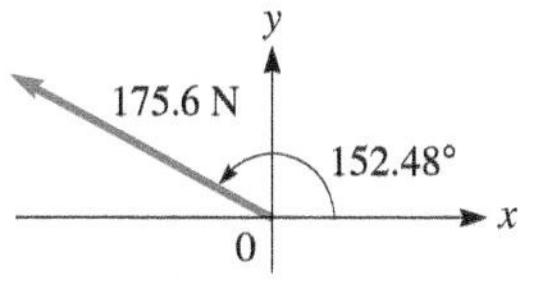

Fig. 9.79

44. Find the horizontal and vertical components of the velocity shown in Fig. 9.80.

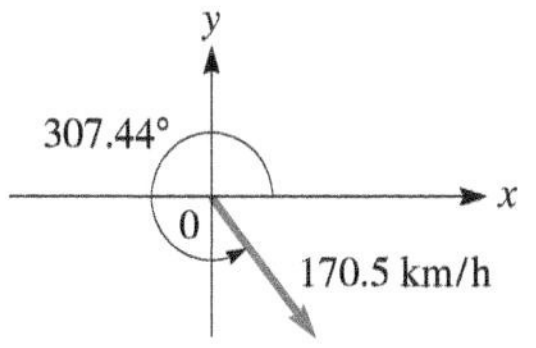

Fig. 9.80

45. In a ballistics test, a bullet was fired into a block of wood with a velocity of 670 m/s and at an angle of 71.3° with the surface of the block. What was the component of the velocity perpendicular to the surface?

46. A storm cloud is moving at 15 km/h from the northwest. A television tower is 60° south of east of the cloud. What is the component of the cloud's velocity toward the tower?

47. During a 3.00-min period after taking off, the Concorde supersonic jet travelled at 480 km/h at an angle of 24.0° above the horizontal. What was its gain in altitude during this period?

48. A rocket is launched at an angle of 42.0° with the horizontal and with a speed of 760 m/s. What are its horizontal and vertical components of velocity?

49. Three forces of 3200 N, 1300 N, and 2100 N act on a bolt as shown in Fig. 9.81. Find the resultant force.

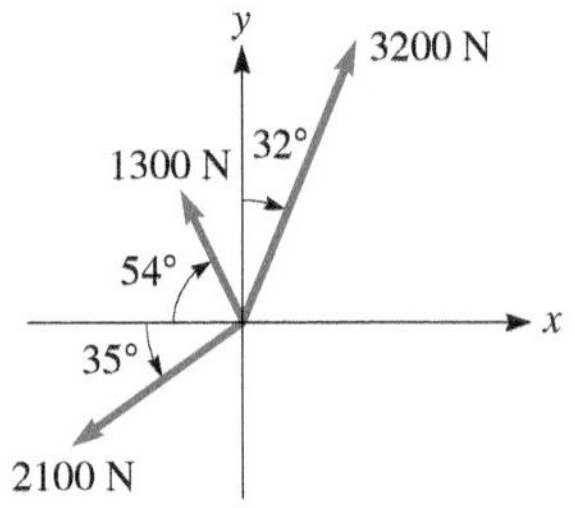

Fig. 9.81

50. In Fig. 9.82, force **F** represents the total surface tension force around the circumference on the liquid in the capillary tube. The vertical component of **F** holds up the liquid in the tube above the liquid surface outside the tube. What is the vertical component of **F**?

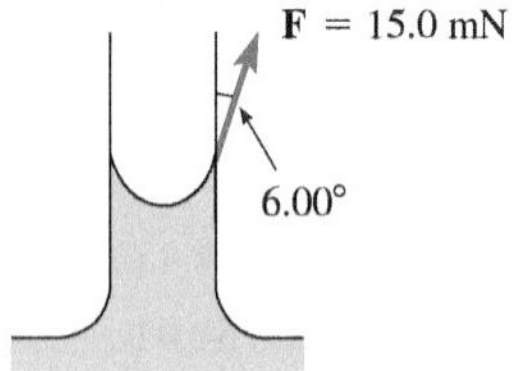

Fig. 9.82

51. A helium-filled balloon rises vertically at 3.5 m/s as the wind carries it horizontally at 5.0 m/s. What is the resultant velocity of the balloon?

52. A shearing pin is designed to break and disengage gears before damage is done to a machine. In a test, a vertically upward force of 8250 N and a force of 7520 N at 35.0° below the horizontal are applied to a shearing pin. What is the resultant force?

53. A magnetic force of 0.15 N is applied at an angle of 22.5° above the horizontal on an iron bar. A second magnetic force of 0.20 N is applied from the opposite side at an angle of 15.0° above the horizontal. What is the upward force on the bar?

54. At a certain point, the angle of elevation of the cliff at Head-Smashed-In Buffalo Jump, Alberta (a UNESCO World Heritage Site), is measured as 55°. Measured a distance 11.0 m farther away from the cliff, the angle of elevation is 29°. Find the height of the cliff. See Fig. 9.83.

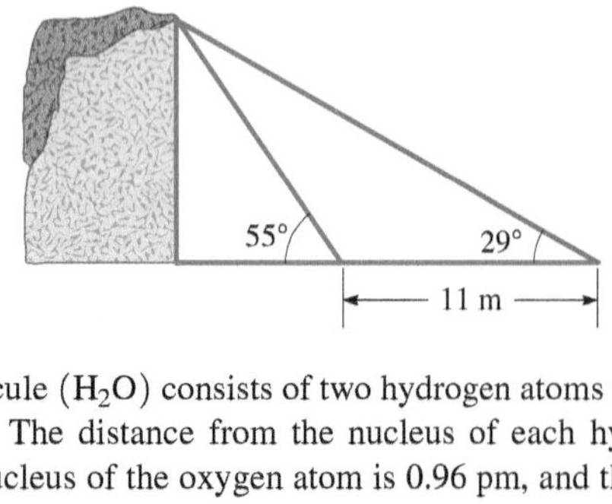

Fig. 9.83

55. A water molecule (H_2O) consists of two hydrogen atoms and one oxygen atom. The distance from the nucleus of each hydrogen atom to the nucleus of the oxygen atom is 0.96 pm, and the bond angle (see Fig. 9.84) is 105°. How far is one hydrogen nucleus from the other?

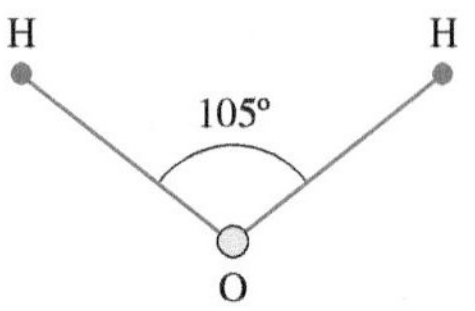

Fig. 9.84

56. A crate weighing 2500 N hangs from two ropes that are angled at 41.5° and 37.2° from the vertical as shown in Fig. 9.85. Make a free-body diagram, and then find the tensions T_L and T_R in the left and right ropes, respectively.

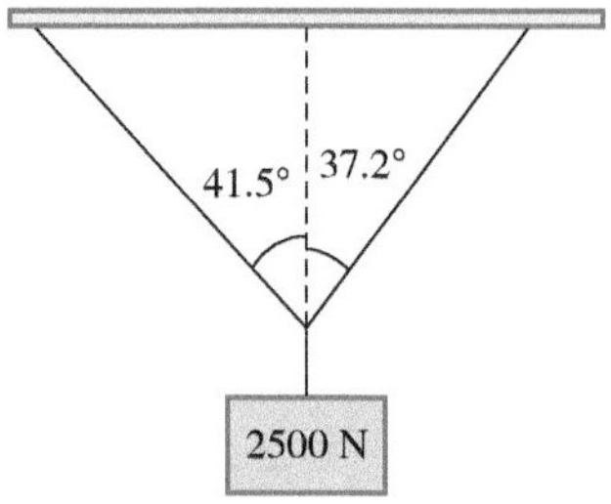

Fig. 9.85

57. In Fig. 9.86, a damper mechanism in an air-conditioning system is shown. If $\theta = 27.5°$ when the spring is at its shortest and longest lengths, what are these lengths?

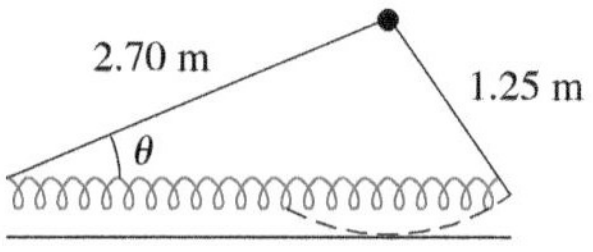

Fig. 9.86

58. A bullet is fired from the ground of a level field at an angle of 39.0° above the horizontal. It travels in a straight line at 670 m/s for 0.20 s when it strikes a target. The sound of the strike is recorded 0.32 s later on the ground. If sound travels at 350 m/s, where is the recording device located?

59. In order to get around an obstruction, an oil pipeline is constructed in two straight sections, one 3.756 km long and the other 4.675 km long, with an angle of 168.85° between the sections where they are joined. How much more pipeline was necessary due to the obstruction?

60. Three pipes of radii 2.50 cm, 3.25 cm, and 4.25 cm are welded together lengthwise. See Fig. 9.87. Find the angles between the centre-to-centre lines.

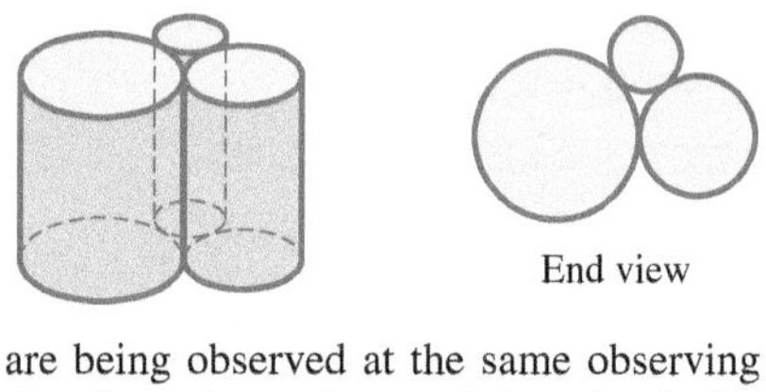

Fig. 9.87

61. Two satellites are being observed at the same observing station. One is 36 200 km from the station, and the other is 30 100 km away. The angle between their lines of observation is 105.4°. How far apart are the satellites?

62. Find the side x in the truss in Fig. 9.88.

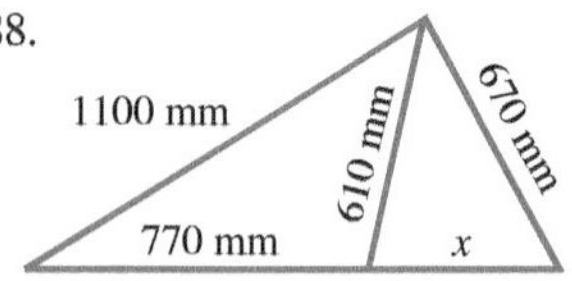

Fig. 9.88

63. The angle of depression of a fire noticed west of a fire tower is 6.2°. The angle of depression of a pond, also west of the tower, is 13.5°. If the fire and pond are at the same altitude, and the tower is 2.25 km from the pond on a direct line, how far is the fire from the pond?

64. A surveyor wishes to find the distance between two points between which there is a security-restricted area. The surveyor measures the distances from each of these points to a third point and finds them to be 226.73 m and 185.12 m. If the angle between the lines of sight from the third point to the other points is 126.724°, how far apart are the two points?

65. In Australia, Adelaide is 805 km and 69.0° south of east of Alice Springs. The pilot of an airplane due north of Adelaide radios Alice Springs and finds the plane is on a line 10.5° south of east from Alice Springs. How far is the plane from Alice Springs?

66. In going around a storm, a plane flies 125 km south, then 140 km at 30.0° south of west, and finally 225 km at 15.0° north of west. What is the displacement of the plane from its original position?

67. One end of a 725-m bridge is sighted from a distance of 1630 m. The angle between the lines of sight of the ends of the bridge is 25.2°. From these data, how far is the observer from the other end of the bridge?

68. A plane is travelling horizontally at 400 m/s. A missile is fired horizontally from it 30.0° from the direction in which the plane is travelling. If the missile leaves the plane at 650 m/s, what is its velocity 10.0 s later if the vertical component is given by $v_V = -9.80t$ (in m/s)?

69. A sailboat is headed due north, and its sail is set perpendicular to the wind, which is from the south of west. The component of the force of the wind in the direction of the heading is 480 N, and the component perpendicular to the heading (the *drift* component) is 650 N. What is the force exerted by the wind, and what is the direction of the wind? See Fig. 9.89.

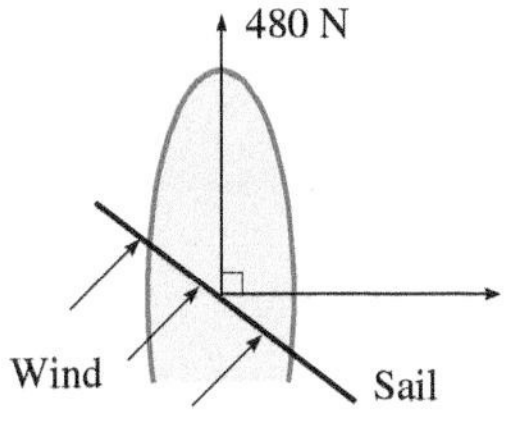

Fig. 9.89

70. Boston is 650 km and 21.0° south of west from Halifax, Nova Scotia. Radio signals locate a ship 10.5° east of south from Halifax and 5.6° north of east from Boston. How far is the ship from each city?

71. A bridge 90 m long spans a valley, as in Fig. 9.90. From one end, the angle of depression of a certain point at the bottom of the valley is 33°. From the other end, the angle of depression of the same point is 67°. Find the height of the bridge.

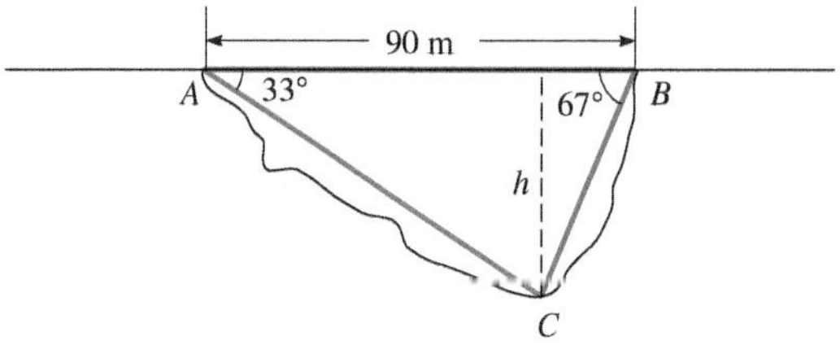

Fig. 9.90

72. The resultant of three horizontal forces—45 N, 35 N, and 25 N—that act on a bolt is zero. Write a paragraph or two explaining how to find the angles between the forces.

CHAPTER 9 **PRACTICE TEST**

In all triangle solutions, sides a, b, and c are opposite angles A, B, and C, respectively.

1. By use of a diagram, find the vector sum $2\mathbf{A} + \mathbf{B}$ for the given vectors.

2. For the triangle in which $a = 22.5$, $B = 78.6°$, and $c = 30.9$, find b.

3. A surveyor locates a tree 36.50 m to the northeast of a set position. The tree is 21.38 m north of a utility pole. What is the displacement of the utility pole from the set position?

4. For the triangle in which $A = 18.9°$, $B = 104.2°$, and $a = 426$, find c.

5. Solve the triangle in which $a = 9.84$, $b = 3.29$, and $c = 8.44$.

6. For vector **R**, find R and standard-position θ if $R_x = -235$ and $R_y = 152$.

7. Find the horizontal and vertical components of a vector of magnitude 871 that is directed at a standard-position angle of 284.3°.

8. A ship leaves a port and travels due west. At a certain point it turns 31.5° north of west and travels an additional 42.0 km to a point 63.0 km on a direct line from the port. How far from the port is the point where the ship turned?

9. Find the sum of the vectors for which $A = 449$, $\theta_A = 74.2°$, $B = 285$, and $\theta_B = 208.9°$ by components.

10. Solve the triangle for which $a = 22.3$, $b = 29.6$, and $A = 36.5°$.

11. Assuming the vectors in Fig. 9.91 are in equilibrium, find T_1 and T_2.

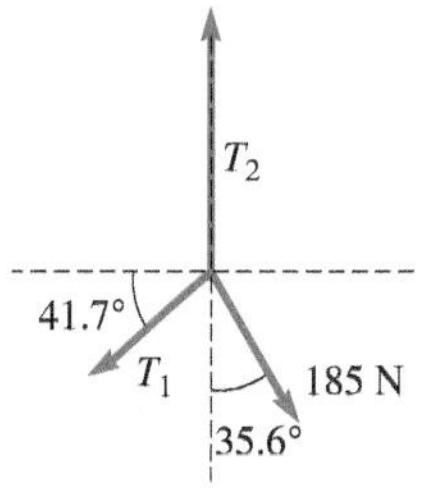

Fig. 9.91

10. Graphs of the Trigonometric Functions

LEARNING OUTCOMES

After completion of this chapter, the student should be able to:

- Graph the functions $y = \sin x$ and $y = \cos x$ and identify their important values for sketching
- Define amplitude, period, and displacement
- Graph the functions $y = a \sin x$ and $y = a \cos x$
- Graph the functions $y = a \sin bx$ and $y = a \cos bx$
- Graph the functions $y = a \sin (bx + c)$ and $y = a \cos (bx + c)$
- Graph the functions $y = \tan x$, $y = \cot x$, $y = \sec x$, and $y = \csc x$
- Solve application problems involving graphs of trigonometric functions
- Graph composite trigonometric curves by addition of coordinates
- Graph Lissajous figures

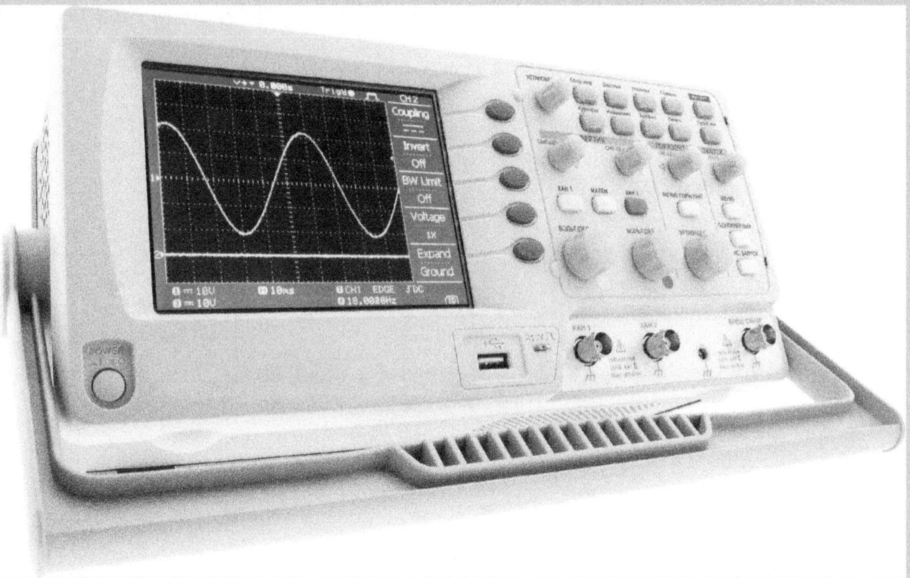

ALEXEY FILATOV/Shutterstock

▲ The graphs of electrical signals (including sinusoidal signals) can be displayed on an oscilloscope, and properties of the signal—such as period, frequency, phase, and amplitude—can be measured. In Section 10.6, we show the patterns that result when two signals are combined and displayed.

The electronics era is thought by many to have started in the 1880s, with the discovery of the vacuum tube by the American inventor Thomas Edison and the discovery of radio waves by the German physicist Heinrich Hertz. Then, in the 1890s, the cathode-ray tube was developed and, as an oscilloscope, has been used since that time to analyse various types of wave forms, such as sound waves and radio waves. Since the mid-1900s, devices similar to a cathode-ray tube have been used in TV picture tubes and computer displays.

What is seen on the screen of an oscilloscope are electric signals that are represented by graphs of trigonometric functions. As noted earlier, the basic method of graphing was developed in the mid-1600s, and using trigonometric functions of numbers has been common since the mid-1700s. Therefore, the graphs of the trigonometric functions were well known in the late 1800s and became very useful in the development of electronics.

The graphs of the trigonometric functions are useful in many areas of application, particularly those that involve wave motion and periodic values. Filtering electronic signals in communications, mixing musical sounds in a recording studio, studying the seasonal temperatures of an area, and analysing ocean waves and tides are some of the many applications of periodic motion.

Aside from their use in applications, the graphs of the trigonometric functions give us one of the clearest ways of showing the properties of the various functions. Therefore, in this chapter, we show graphs of these functions, with emphasis on the sine and cosine functions.

10.1 Graphs of $y = a \sin x$ and $y = a \cos x$

> **LEARNING TIP**
> Throughout this chapter, make sure that your calculator is in *radian* mode.

When angles are expressed in radian measure, both the independent variable and the dependent variable of trigonometric functions are real numbers. For this reason, *we only consider angles in radians* when plotting and sketching the graphs of trigonometric functions. If necessary, review Section 8.3 on radian measure.

In this section, we discuss the graphs of the sine and cosine functions. We begin by making a table of values of x and y for the function $y = \sin x$, where x and y are used in the standard way as the *independent variable* and *dependent variable.* We plot the points to obtain the graph in Fig. 10.1.

x	0	$\frac{\pi}{6}$	$\frac{\pi}{3}$	$\frac{\pi}{2}$	$\frac{2\pi}{3}$	$\frac{5\pi}{6}$	π
y	0	0.5	0.87	1	0.87	0.5	0

x	$\frac{7\pi}{6}$	$\frac{4\pi}{3}$	$\frac{3\pi}{2}$	$\frac{5\pi}{3}$	$\frac{11\pi}{6}$	2π
y	−0.5	−0.87	−1	−0.87	−0.5	0

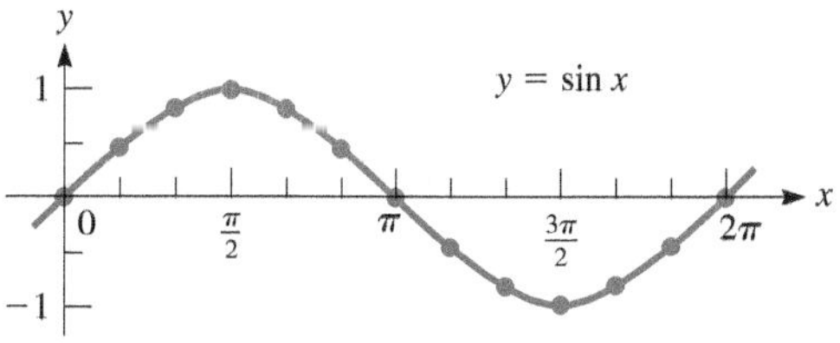

Fig. 10.1

The graph of $y = \cos x$ may be drawn in the same way. The next table gives values for plotting the graph of $y = \cos x$, and the graph is shown in Fig. 10.2.

x	0	$\frac{\pi}{6}$	$\frac{\pi}{3}$	$\frac{\pi}{2}$	$\frac{2\pi}{3}$	$\frac{5\pi}{6}$	π
y	1	0.87	0.5	0	−0.5	−0.87	−1

x	$\frac{7\pi}{6}$	$\frac{4\pi}{3}$	$\frac{3\pi}{2}$	$\frac{5\pi}{3}$	$\frac{11\pi}{6}$	2π
y	−0.87	−0.5	0	0.5	0.87	1

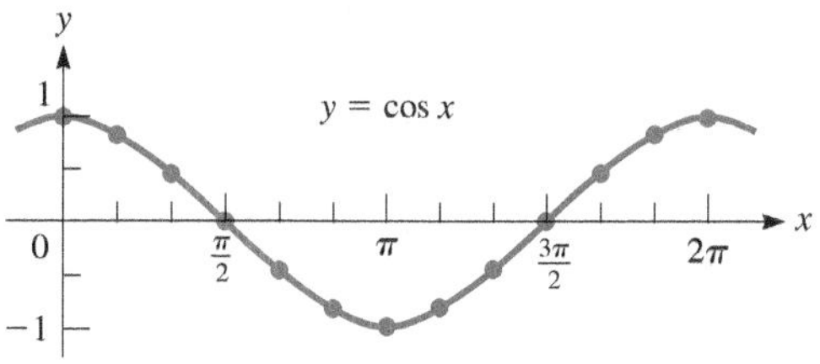

Fig. 10.2

The graphs are continued beyond the values shown in the tables to indicate that *they continue indefinitely in each direction.* To show this more clearly, in Figs. 10.3 and 10.4, note the graphs of $y = \sin x$ and $y = \cos x$ from $x = -10$ to $x = 10$.

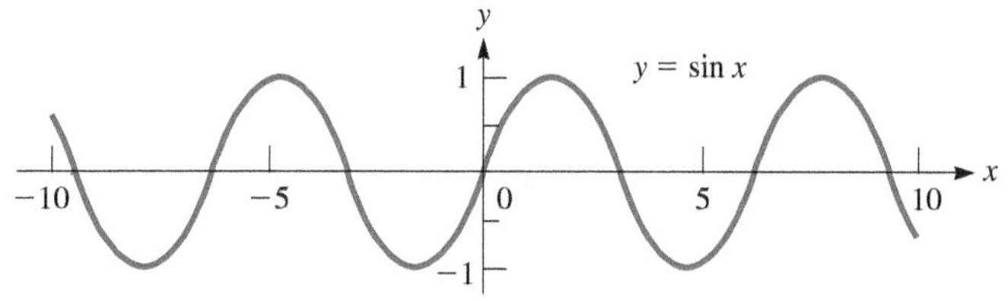

Fig. 10.3

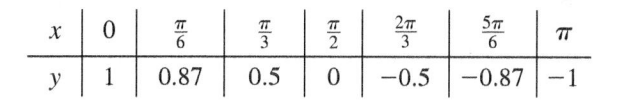

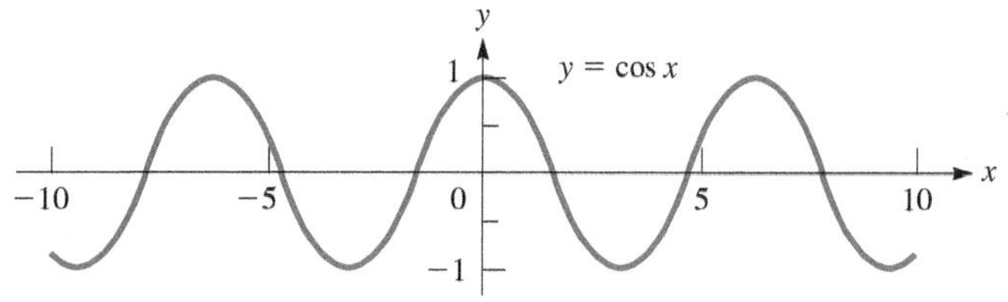

Fig. 10.4

From these tables and graphs, we can see that the graphs of $y = \sin x$ and $y = \cos x$ have the basic features listed in Table 10.1. These features (illustrated in Fig. 10.5) will be especially valuable in *sketching* similar curves. Since some of these features remain the same, it will not be necessary to plot numerous points every time we wish to sketch such a curve.

Table 10.1

Key Values	$y = \sin x$	$y = \cos x$
0	0	1
$\frac{\pi}{2}$	1	0
π	0	-1
$\frac{3\pi}{2}$	-1	0
2π	0	1

Basic Features of the Graphs of $y = \sin x$ and $y = \cos x$

1. The domain is all values of x.
2. The range is $-1 \leq y \leq 1$, or $[-1, 1]$.
3. The **amplitude** (half the distance between the maximum value and the minimum value) is 1.
4. Both graphs are exactly the same shape (called **sinusoidal**).
5. The graph of the cosine curve is shifted $\pi/2$ units to the left of the sine curve.
6. For both graphs, the values of y repeat every 2π units of x. We therefore say that the functions are **periodic** with **period** 2π.
7. The functions have zeros, maximum values, and minimum values when x is a multiple of $\pi/2$ (that is, when x is a quadrantal angle). The behaviour from 0 to 2π at these key values is summarized in Table 10.1.

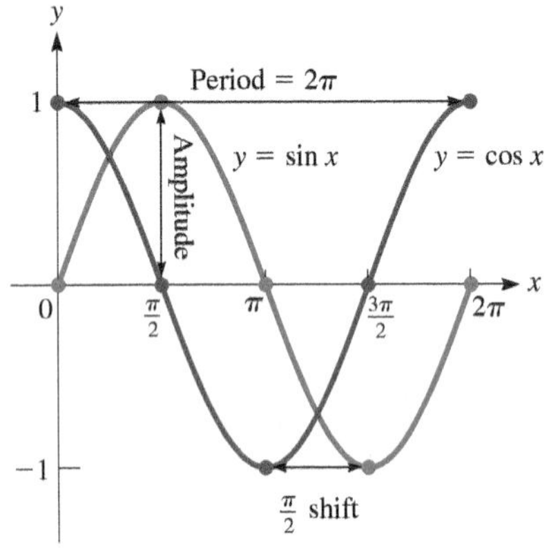

Fig. 10.5

To obtain the graph of $y = a \sin x$, note that all the y-values obtained for the graph of $y = \sin x$ are to be multiplied by the number a. In this case, the greatest value of the sine function is $|a|$, and the curve will have no value less than $-|a|$. Therefore, the range is $[-|a|, |a|]$, and the amplitude of the curve is $|a|$. This is also true for $y = a \cos x$.

EXAMPLE 1 Plotting the graph of $y = a \sin x$

Plot the graph of $y = 2 \sin x$.

Since $a = 2$, the amplitude of this curve is $|2| = 2$. This means that the maximum value of y is 2 and the minimum value is $y = -2$. The table of values follows, and the curve is shown in Fig. 10.6.

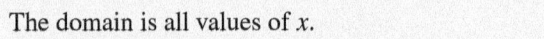

x	0	$\frac{\pi}{6}$	$\frac{\pi}{3}$	$\frac{\pi}{2}$	$\frac{2\pi}{3}$	$\frac{5\pi}{6}$	π
y	0	1	1.73	2	1.73	1	0

x	$\frac{7\pi}{6}$	$\frac{4\pi}{3}$	$\frac{3\pi}{2}$	$\frac{5\pi}{3}$	$\frac{11\pi}{6}$	2π
y	-1	-1.73	-2	-1.73	-1	0

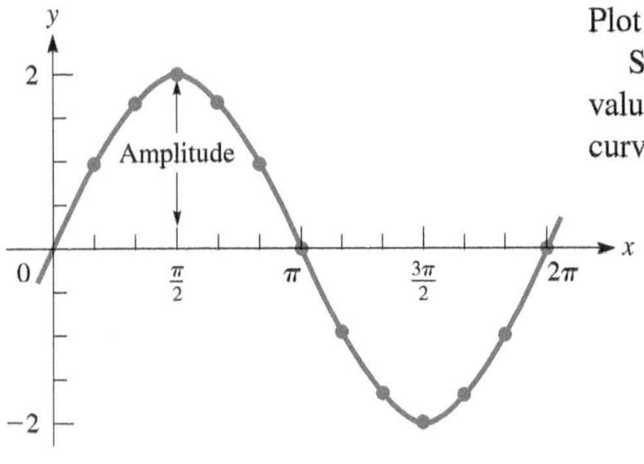

Fig. 10.6

EXAMPLE 2 Plotting the graph of $y = a \cos x$

Plot the graph of $y = -3 \cos x$.

In this case, $a = -3$, and this means that the amplitude is $|-3| = 3$. Therefore, the maximum value of y is 3, and the minimum value of y is -3. The table of values follows, and the curve is shown in Fig. 10.7.

x	0	$\frac{\pi}{6}$	$\frac{\pi}{3}$	$\frac{\pi}{2}$	$\frac{2\pi}{3}$	$\frac{5\pi}{6}$	π
y	-3	-2.6	-1.5	0	1.5	2.6	3

x	$\frac{7\pi}{6}$	$\frac{4\pi}{3}$	$\frac{3\pi}{2}$	$\frac{5\pi}{3}$	$\frac{11\pi}{6}$	2π
y	2.6	1.5	0	-1.5	-2.6	-3

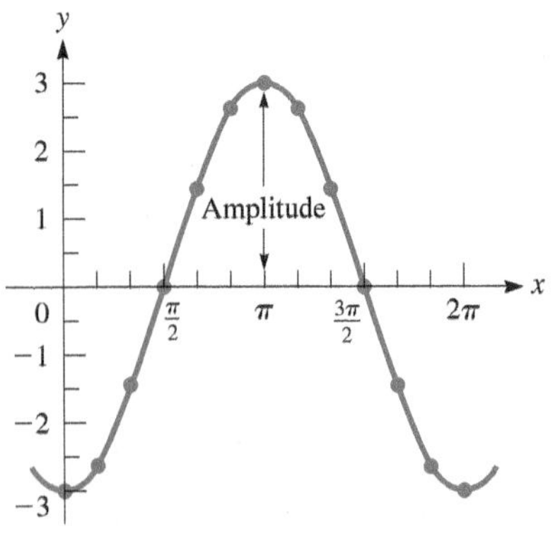

Fig. 10.7

Note that *the effect of the negative sign with the number a is to* ***invert*** *the curve about the x-axis.*

Table 10.2

Key Values	$y = a \sin x$	$y = a \cos x$
0	0	a
$\frac{\pi}{2}$	a	0
π	0	$-a$
$\frac{3\pi}{2}$	$-a$	0
2π	0	a

Apart from the fact that the range of the functions $y = a \sin x$ and $y = a \cos x$ is $[-|a|, |a|]$, we can see from the previous examples that the number a has no other effect on the basic features of these functions as compared to those of the functions $y = \sin x$ and $y = \cos x$. In particular, they have the same sinusoidal shape and the same period, and their zeros, maximum points, and minimum points are located at the same key values at the quadrantal angles as before (see Table 10.2). It follows that by knowing the basic features of the sine and cosine functions, we can *sketch* the graphs of functions of the form $y = a \sin x$ and $y = a \cos x$ quickly by simply using the appropriate amplitude and inverting the curve when necessary.

EXAMPLE 3 Using key values to sketch a graph

Sketch the graph of $y = 40 \cos x$.

First, we set up a table of values for the points where the curve has its zeros, maximum points, and minimum points:

x	0	$\frac{\pi}{2}$	π	$\frac{3\pi}{2}$	2π
y	40 max.	0	-40 min.	0	40 max.

Now, we plot these points and join them, knowing the basic sinusoidal shape of the curve. See Fig. 10.8.

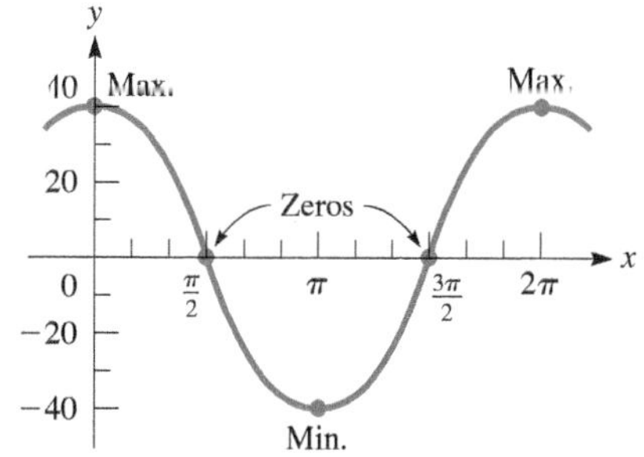

Fig. 10.8

EXAMPLE 4 Using key values to sketch a graph

Sketch the graph of $y = -2 \sin x$.

The key values between 0 and 2π are the following:

x	0	$\frac{\pi}{2}$	π	$\frac{3\pi}{2}$	2π
y	0	-2 min.	0	2 max.	0

The graph from $x = -\frac{5\pi}{2}$ to $x = \frac{5\pi}{2}$ is shown in Fig. 10.9, plotted in the same set of axes as the function $y = \sin x$. The effect of the constant $a = -2$ in terms of the change in amplitude and the inversion of the curve can be readily seen.

Practice Exercise

1. For the graph of $y = -6 \sin x$, set up a table of key values for $0 \leq x \leq 2\pi$.

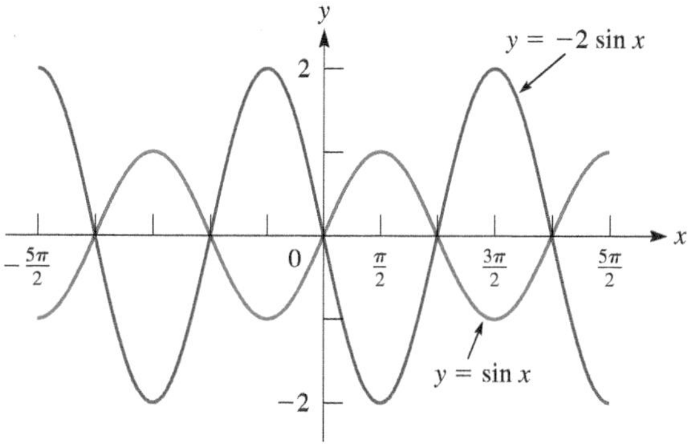

Fig. 10.9

EXERCISES 10.1

In Exercises 1 and 2, graph the function if the given changes are made in the indicated examples of this section.

1. In Example 2, if the sign of the coefficient of $\cos x$ is changed, plot the graph of the resulting function.
2. In Example 4, if the sign of the coefficient of $\sin x$ is changed, display the graph of the resulting function.

In Exercises 3–6, complete the following table for the given functions and then plot the resulting graphs.

x	$-\pi$	$-\frac{3\pi}{4}$	$-\frac{\pi}{2}$	$-\frac{\pi}{4}$	0	$\frac{\pi}{4}$	$\frac{\pi}{2}$	$\frac{3\pi}{4}$	π
y									

x	$\frac{5\pi}{4}$	$\frac{3\pi}{2}$	$\frac{7\pi}{4}$	2π	$\frac{9\pi}{4}$	$\frac{5\pi}{2}$	$\frac{11\pi}{4}$	3π
y								

3. $y = \sin x$ **4.** $y = \cos x$

5. $y = 3\cos x$ **6.** $y = -4\sin x$

In Exercises 7–22, sketch the graphs of the given functions.

7. $y = 3\sin x$ **8.** $y = 5\sin x$

9. $y = \frac{5}{2}\sin x$ **10.** $y = 35\sin x$

11. $y = 200\cos x$ **12.** $y = 0.25\cos x$

13. $y = 0.8\cos x$ **14.** $y = \frac{3}{2}\cos x$

15. $y = -\sin x$ **16.** $y = -300\sin x$

17. $y = -1500\sin x$ **18.** $y = -0.2\sin x$

19. $y = -\cos x$ **20.** $y = -8\cos x$

21. $y = -50\cos x$ **22.** $y = -0.4\cos x$

Although units of π are convenient, we must remember that π is only a number. Numbers that are not multiples of π may be used. In Exercises 23–26, plot the indicated graphs by finding the values of y that correspond to values of x of 0, 1, 2, 3, 4, 5, 6, and 7 on a calculator. (Remember, the numbers 0, 1, 2, and so on represent radian measure.)

23. $y = \sin x$ **24.** $y = -30\sin x$

25. $y = 12\cos x$ **26.** $y = 2\cos x$

In Exercises 27–32, solve the given problems.

27. Find the function and graph it for a function of the form $y = a\sin x$ that passes through $(\pi/2, -2)$.

28. Find the function and graph it for a function of the form $y = a\sin x$ that passes through $(3\pi/2, -2)$.

29. Find the function and graph it for a function of the form $y = a\cos x$ that passes through $(\pi, 2)$.

30. Find the function and graph it for a function of the form $y = a\cos x$ that passes through $(2\pi, -2)$.

31. The graph displayed on an oscilloscope can be represented by $y = -0.05\sin x$. Display this curve with a graphing utility.

32. The displacement y (in cm) of the end of a robot arm for welding is $y = 4.75\cos t$, where t is the time (in s). Display this curve with a graphing utility.

In Exercises 33–36, the graph of a function of the form $y = a\sin x$ or $y = a\cos x$ is shown. Determine the specific function of each.

33.

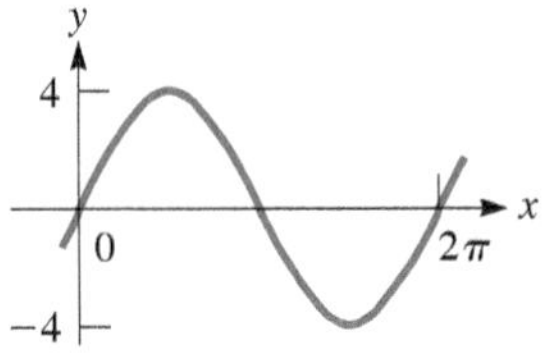

34.

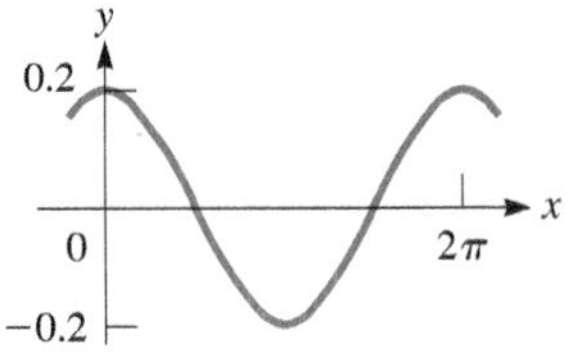

35.

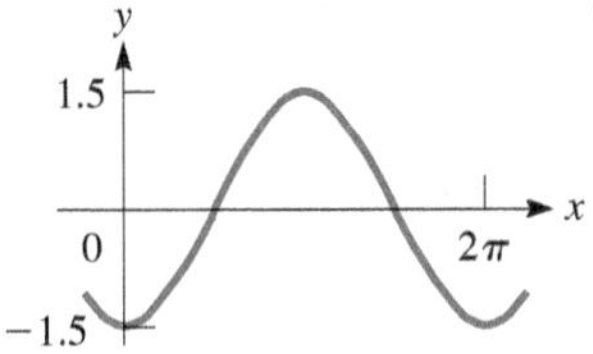

36.

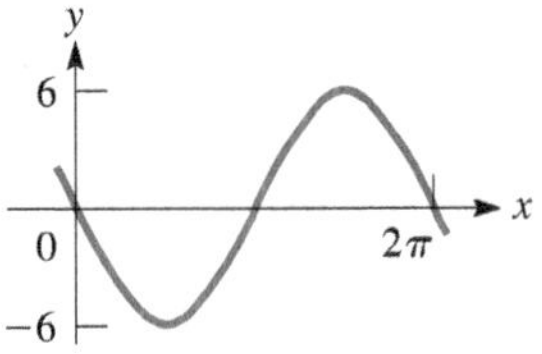

In Exercises 37–40, find the value of a for either $y = a\sin x$ or $y = a\cos x$, whichever is correct, such that the given point is on the graph. Determine the function knowing that the amplitude of each function is 2.50. (All points are located such that the x value is between $-\pi$ and π.)

37. $(0.67, -1.55)$ **38.** $(-1.20, 0.91)$

39. $(2.07, 1.20)$ **40.** $(-2.47, -1.56)$

Answer to Practice Exercise

1.

x	0	$\frac{\pi}{2}$	π	$\frac{3\pi}{2}$	2π
y	0	-6 min.	0	6 max.	0

10.2 Graphs of $y = a \sin bx$ and $y = a \cos bx$

In graphing the functions $y = a\sin x$ and $y = a\cos x$, we see that the values repeat every 2π units of x, making them periodic with period 2π. More generally, we say that a function F has period P if $F(x) = F(x + P)$ for all x in the domain of F, and P is the smallest such number. In other words, the period is the x-distance between a point and the next corresponding point after which the values of y repeat.

Let us now plot the curve $y = \sin 2x$. This means that we choose a value of x, multiply this value by 2, and find the sine of the result. This leads to the following table of values for this function:

x	0	$\frac{\pi}{8}$	$\frac{\pi}{4}$	$\frac{3\pi}{8}$	$\frac{\pi}{2}$	$\frac{5\pi}{8}$	$\frac{3\pi}{4}$	$\frac{7\pi}{8}$	π	$\frac{9\pi}{8}$	$\frac{5\pi}{4}$
$2x$	0	$\frac{\pi}{4}$	$\frac{\pi}{2}$	$\frac{3\pi}{4}$	π	$\frac{5\pi}{4}$	$\frac{3\pi}{2}$	$\frac{7\pi}{4}$	2π	$\frac{9\pi}{4}$	$\frac{5\pi}{2}$
y	0	0.7	1	0.7	0	−0.7	−1	−0.7	0	0.7	1

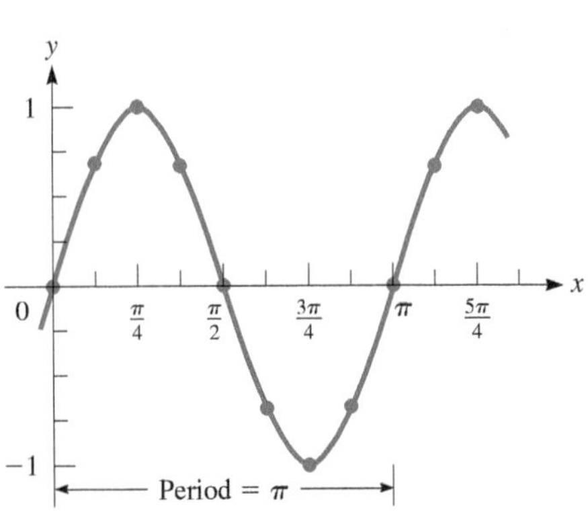

Fig. 10.10

Plotting these points, we have the curve shown in Fig. 10.10.

From the table and Fig. 10.10, note that $y = \sin 2x$ repeats after π units of x. The effect of the 2 is that the period of $y = \sin 2x$ is half the period of the curve of $y = \sin x$.

LEARNING TIP
If the period of a function $F(x)$ is P, then the period of $F(bx)$ is P/b. Since each of the functions $\sin x$ and $\cos x$ has a period of 2π, *each of the functions* $\sin bx$ *and* $\cos bx$ *has a period of* $2\pi/b$.

EXAMPLE 1 Finding the period of a function

Determine the period of each of the following functions.

(a) $\cos 4x$. The period is $\frac{2\pi}{4} = \frac{\pi}{2}$. **(b)** $\sin 3\pi x$. The period is $\frac{2\pi}{3\pi} = \frac{2}{3}$.

(c) $\sin \frac{1}{2}x$. The period is $\dfrac{2\pi}{\frac{1}{2}} = 4\pi$. **(d)** $\cos \frac{\pi}{4}x$. The period is $\dfrac{2\pi}{\frac{\pi}{4}} = 8$.

In (a), the period tells us that the curve of $y = \cos 4x$ will repeat every $\pi/2$ (approximately 1.57) units of x. In (b), we see that the curve of $y = \sin 3\pi x$ will repeat every 2/3 of a unit. In (c) and (d), the periods are longer than those of $y = \sin x$ and $y = \cos x$.

Practice Exercises

Find the period of each function.

1. $y = \sin \pi x$ **2.** $y = \cos \frac{1}{3}x$

Combining the value of the period with the value of the amplitude from Section 10.1, we conclude that *the functions* $y = a \sin bx$ *and* $y = a \cos bx$ *have an amplitude of* $|a|$ *and a period of* $2\pi/b$. These properties are very useful in sketching these functions.

EXAMPLE 2 Sketching the graph of $y = a \sin bx$

Sketch the graph of $y = 3 \sin 4x$ for $0 \leq x \leq \pi$.

Since $a = 3$, the amplitude is 3. The $4x$ tells us that the period is $2\pi/4 = \pi/2$. This means that $y = 0$ for $x = 0$ and for $y = \pi/2$. Since this sine function is zero halfway between $x = 0$ and $x = \pi/2$, we find that $y = 0$ for $x = \pi/4$. Also, the fact that the graph of the sine function reaches its maximum and minimum values halfway between zeros means that $y = 3$ for $x = \pi/8$, and $y = -3$ for $x = 3\pi/8$. Note that the values of x in the following table are those for which $4x = 0$, $\pi/2$, π, $3\pi/2$, 2π, and so on, which correspond to the key values listed in Tables 10.1 and 10.2.

x	0	$\frac{\pi}{8}$	$\frac{\pi}{4}$	$\frac{3\pi}{8}$	$\frac{\pi}{2}$	$\frac{5\pi}{8}$	$\frac{3\pi}{4}$	$\frac{7\pi}{8}$	π
y	0	3	0	−3	0	3	0	−3	0

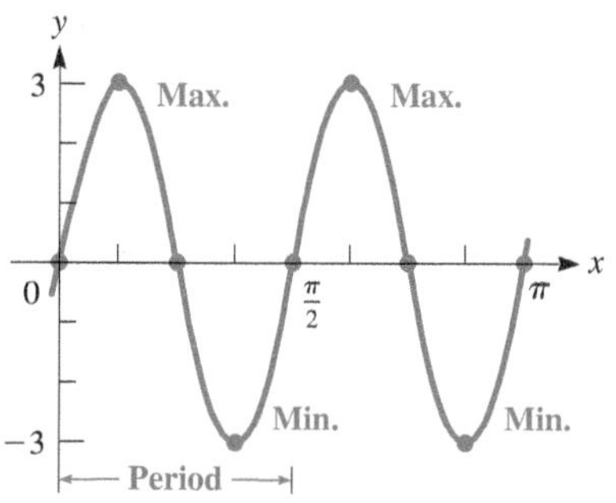

Fig. 10.11

Using the values from the table and the fact that the curve is sinusoidal in form, we sketch the graph of this function in Fig. 10.11. We see that the key values where the function has zeros, maxima, and minima occur when x is a multiple of $\pi/8$, which is exactly one-fourth of the period.

Practice Exercise

3. Find the amplitude and period of the function $y = -8 \sin \dfrac{\pi x}{4}$.

Note from Example 2 that *an important distance in sketching a sine curve or a cosine curve is one-fourth of the period.* For $y = a \sin bx$, it is one-fourth of the period from the origin to the first value of x where y is at its maximum (or minimum) value. Then we proceed another one-fourth period to a zero, another one-fourth period to the next minimum (or maximum) value, another to the next zero (this is where the period is completed), and so on.

Similarly, one-fourth of the period is useful in sketching the graph of $y = \cos bx$. For this function, the maximum (or minimum) value occurs for $x = 0$. At the following one-fourth-period values, there is a zero, a minimum (or maximum), a zero, and a maximum (or minimum) at the start of the next period.

LEARNING TIP
By finding one-fourth of the period, we can easily find the important values for sketching the curve.

We now summarize the important values for sketching the graphs of $y = a \sin bx$ and $y = a \cos bx$.

Important Values for Sketching $y = a \sin bx$ and $y = a \cos bx$

1. The amplitude: $|a|$
2. The period: $2\pi/b$
3. Values of the function for each one-fourth period

EXAMPLE 3 Using important values to sketch a graph

Sketch the graph of $y = -2 \cos 3x$ for $0 \le x \le 2\pi$.

Amplitude: $|a| = 2$

Period: $\frac{2\pi}{b} = \frac{2\pi}{3}$ $\left[\frac{1}{4}(\text{period}) = \frac{1}{4}\left(\frac{2\pi}{3}\right) = \frac{\pi}{6}\right]$

Key values for cosine: Minimum at $x = 0$ (because $a < 0$)

Zero at $x = \frac{\pi}{6}$

Maximum at $x = \frac{\pi}{3}$

Continuing the key values at multiples of $\frac{\pi}{6}$, we have the following table:

x	0	$\frac{\pi}{6}$	$\frac{\pi}{3}$	$\frac{\pi}{2}$	$\frac{2\pi}{3}$	$\frac{5\pi}{6}$	π	$\frac{7\pi}{6}$	$\frac{4\pi}{3}$	$\frac{3\pi}{2}$	$\frac{5\pi}{3}$	$\frac{11\pi}{6}$	2π
y	-2	0	2	0	-2	0	2	0	-2	0	2	0	-2

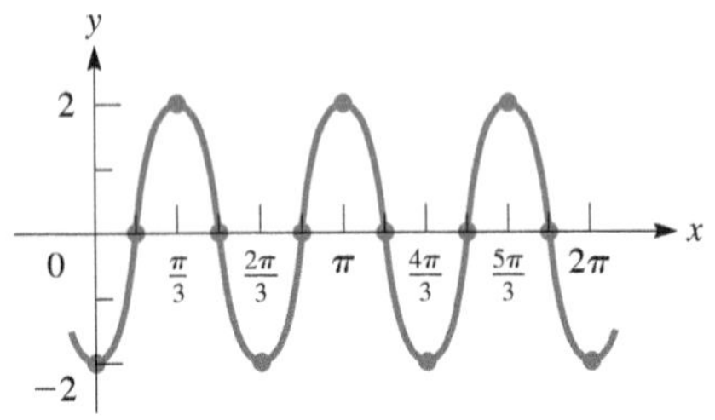

Fig. 10.12

Using this table and the sinusoidal shape of the cosine curve, we sketch the function in Fig. 10.12.

For a periodic function, a **cycle** is *any section of the graph that includes exactly one period.* Fig. 10.11 shows two cycles, whereas Fig. 10.12 shows three.

EXAMPLE 4 Graph of $y = a \cos bx$—generator voltage

A generator produces a voltage $V = 200 \cos 50\pi t$, where t is the time in seconds (50π is angular velocity, so it has units of rad/s; thus, $50\pi t$ is an angle in radians). Graph V as a function of t for $0 \le t \le 0.06$ s.

Amplitude: $|a| = 200$

Period: $\frac{2\pi}{b} = \frac{2\pi}{50\pi} = 0.04$ $\left[\frac{1}{4}(\text{period}) = \frac{1}{4}(0.04) = 0.01\right]$

Key values for cosine: Maximum $x = 0$

Zero at $x = 0.01$

Minimum at $x = 0.02$

Continuing the key values at multiples of 0.01, we have

t(s)	0	0.01	0.02	0.03	0.04	0.05	0.06
V(V)	200	0	−200	0	200	0	−200

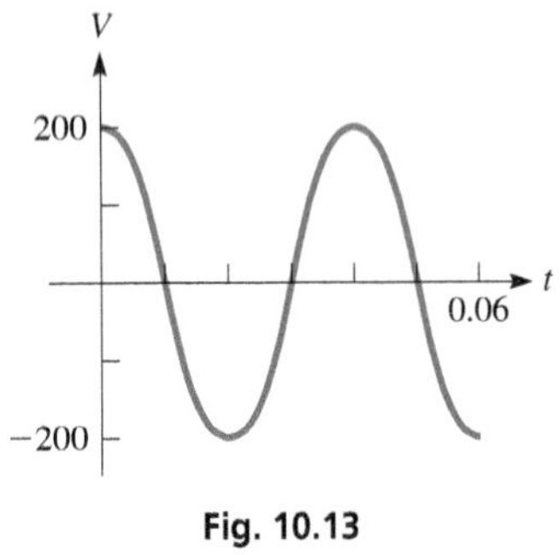

Fig. 10.13

The graph is shown in Fig. 10.13. Note that between 0 and 0.06 s, the function completes 1.5 cycles. We do not consider negative values of t, for they have no real meaning in this problem.

EXERCISES 10.2

In Exercises 1 and 2, graph the function if the given changes are made in the indicated examples of this section.

1. In Example 2, if the coefficient of x is changed from 4 to 6, sketch the graph of the resulting function.
2. In Example 3, if the coefficient of x is changed from 3 to 4, sketch the graph of the resulting function.

In Exercises 3–22, find the amplitude and period of each function and then sketch its graph.

3. $y = 2 \sin 6x$
4. $y = 4 \sin 2x$
5. $y = 3 \cos 8x$
6. $y = 28 \cos 10x$
7. $y = -2 \sin 12x$
8. $y = -\frac{1}{5} \sin 5x$
9. $y = -\cos 16x$
10. $y = -4 \cos 2x$
11. $y = 520 \sin 2\pi x$
12. $y = 2 \sin 3\pi x$
13. $y = 3 \cos 4\pi x$
14. $y = 4 \cos 10\pi x$
15. $y = 15 \sin \frac{1}{3}x$
16. $y = -25 \sin \frac{2}{5}x$
17. $y = -\frac{1}{2} \cos \frac{2}{3}x$
18. $y = \frac{1}{3} \cos \frac{1}{4}x$
19. $y = 0.4 \sin \dfrac{2\pi x}{3}$
20. $y = 1.5 \cos \dfrac{\pi x}{10}$
21. $y = 3.3 \cos \pi^2 x$
22. $y = -12.5 \sin \frac{2x}{\pi}$

In Exercises 23–26, the period is given for a function of the form $y = \sin bx$. Write the function corresponding to the given period.

23. $\dfrac{\pi}{3}$
24. $\dfrac{2\pi}{5}$
25. $\dfrac{1}{3}$
26. 6

In Exercises 27–30, graph the given functions. In Exercises 27 and 28, use the equations for negative angles in Section 8.2 to first rewrite the function with a positive angle, and then graph the resulting function.

27. $y = 3 \sin(-2x)$
28. $y = -5 \cos(-4\pi x)$
29. $y = 8\left|\cos\left(\frac{\pi}{2}x\right)\right|$
30. $y = 0.4 \left|\sin 6x\right|$

In Exercises 31–40, solve the given problems.

31. By noting the periods of $\sin 2x$ and $\sin 3x$, find the period of the function $y = \sin 2x + \sin 3x$ by finding the least common multiple of the individual periods.
32. By noting the period of $\cos \frac{1}{2}x$ and $\cos \frac{1}{3}x$, find the period of the function $y = \cos \frac{1}{2}x + \cos \frac{1}{3}x$ finding the least common multiple of the individual periods.
33. Find the function and graph it for a function of the form $y = -2 \sin bx$ that passes through $(\pi/4, -2)$ and for which b has the smallest possible positive value.
34. Find the function and graph it for a function of the form $y = 2 \sin bx$ that passes through $(\pi/6, 2)$ and for which b has the smallest possible positive value.
35. Find the function and graph it for a function of the form $y = 2 \cos bx$ that passes through $(\pi, 0)$ and for which b has the smallest possible positive value.
36. Find the function and graph it for a function of the form $y = -2 \cos bx$ that passes through $(\pi/2, 2)$ and for which b has the smallest possible positive value.
37. The standard electric voltage in a 60-Hz alternating-current circuit is given by $V = 170 \sin 120\pi t$, where t is the time in seconds. Sketch the graph of V as a function of t for $0 \le t \le 0.05$ s.
38. To tune the instruments of an orchestra before a concert, an A note is struck on a piano. The piano wire vibrates with a displacement y (in mm) given by $y = 3.2 \cos 880\pi t$, where t is in seconds. Sketch the graph of y vs. t for $0 \le t \le 0.01$ s.
39. The velocity v (in cm/s) of a piston is $v = 450 \cos 3600t$, where t is in seconds. Sketch the graph of v vs. t for $0 \le t \le 0.006$ s.
40. On a certain day in Saint John, New Brunswick, the difference between high tide and low tide was 6.4 m. The period was about 12.4 h. Find a cosine function that describes these tides if high tide was at midnight.

In Exercises 41 and 42, use the fact that the frequency, in cycles/s (or Hz), is the reciprocal of the period (in s). (Frequency is discussed in more detail in Section 10.5.)

41. The carrier signal transmitted by a certain FM radio station is given by $y = \sin(6.60 \times 10^8 t)$. Find the frequency and express it in MHz($1\text{ MHz} = 10^6\text{ Hz}$).
42. The current in a certain alternating-current circuit is given by $i = 2.5 \sin(120\pi t)$. Find the period and the frequency.

In Exercises 43–46, the graph of a function of the form $y = a \sin bx$ or $y = a \cos bx$ is shown. Determine the specific function of each.

43.
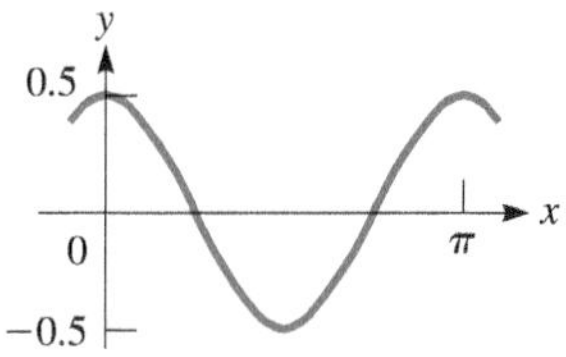

44.
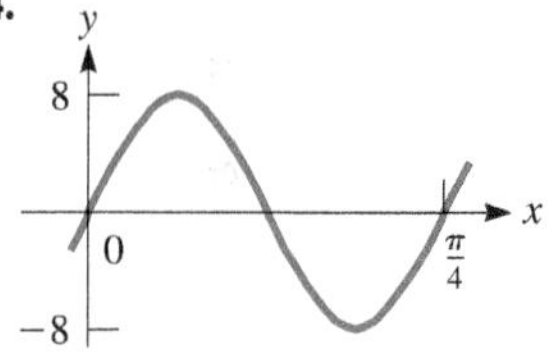

45.
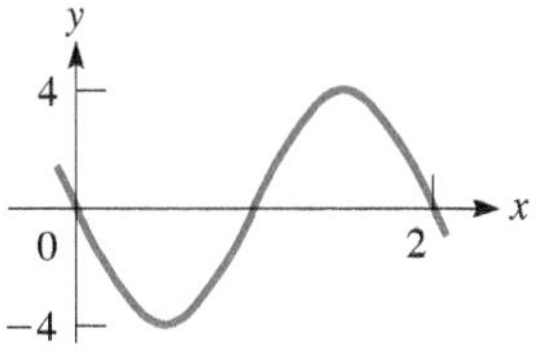

46.
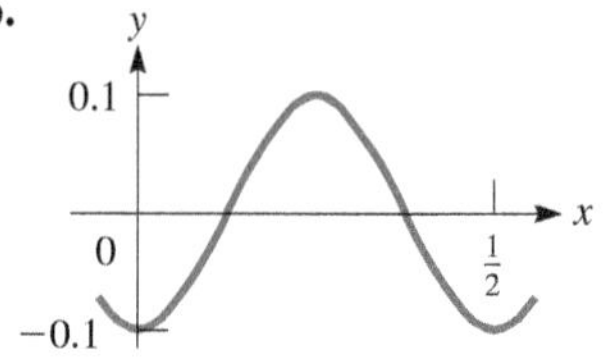

Answers to Practice Exercises

1. 2 2. 6π 3. amp.: 8; per.: 8

10.3 Graphs of $y = a \sin(bx + c)$ and $y = a \cos(bx + c)$

In the function $y = a \sin(bx + c)$, c represents the **phase angle**. It is another very important quantity in graphing the sine and cosine functions. Its meaning is illustrated in the following example.

EXAMPLE 1 Sketch of a function with phase angle

Sketch the graph of $y = \sin\left(2x + \frac{\pi}{4}\right)$.

Here, $c = \pi/4$. Therefore, in order to obtain values for the table, we assume a value for x, multiply it by 2, add $\pi/4$ to this value, and then find the sine of the result. The values shown are those for which $2x + \pi/4 = 0, \pi/4, \pi, 2, 3\pi/4, \pi$, and so on, which are the important values for $y = \sin 2x$.

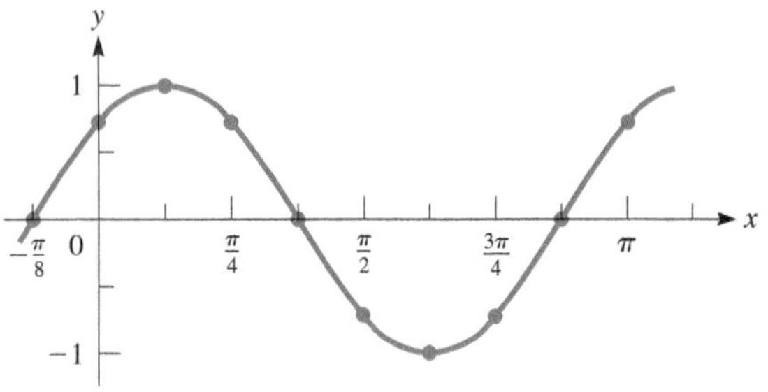

Fig. 10.14

x	$-\frac{\pi}{8}$	0	$\frac{\pi}{8}$	$\frac{\pi}{4}$	$\frac{3\pi}{8}$	$\frac{\pi}{2}$	$\frac{5\pi}{8}$	$\frac{3\pi}{4}$	$\frac{7\pi}{8}$	π
y	0	0.7	1	0.7	0	−0.7	−1	−0.7	0	0.7

Solving $2x + \pi/4 = 0$, we get $x = -\pi/8$, which gives $y = \sin 0 = 0$. The other values for y are found in the same way. See Fig. 10.14.

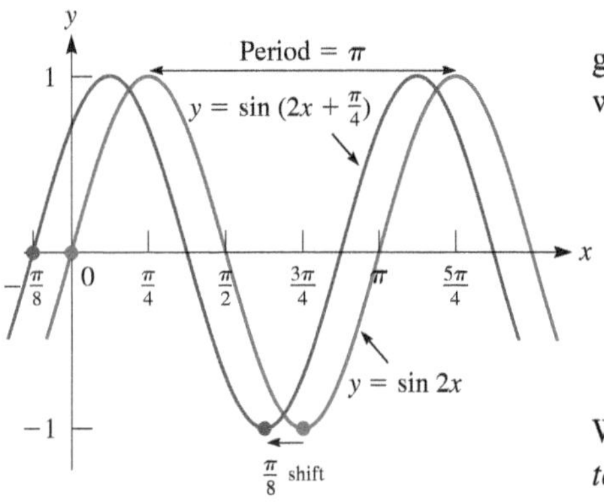

Fig. 10.15

Fig. 10.15 shows the graph of $y = \sin(2x + \frac{\pi}{4})$ from Example 1 together with the graph of $y = \sin 2x$. We see that the two graphs have exactly the same shape. Moreover, we can compare some key points on the two graphs:

	$y = \sin 2x$	$y = \sin(2x + \frac{\pi}{4})$
Key point (zero)	$(0, 0)$	$\left(-\frac{\pi}{8}, 0\right)$
Key point (minimum)	$\left(\frac{3\pi}{4}, -1\right)$	$\left(\frac{5\pi}{8}, -1\right)$

We see that the graph of $y = \sin\left(2x + \frac{\pi}{4}\right)$ is the graph of $y = \sin 2x$ ***shifted*** *$\pi/8$ units to the left.*

More generally, the function $y = a \sin(bx + c)$ corresponds to the function $y = a \sin bx$ evaluated at $x + \frac{c}{b}$. As we learned in Section 3.5, this means that the graph of $y = a \sin(bx + c)$ is the graph of $y = a \sin bx$ *shifted horizontally* $-c/b$ units (to the left if $c > 0$, and to the right if $c < 0$). The quantity $-c/b$ is called the **displacement** (or **phase shift**). It can be obtained by solving for x in the equation $bx + c = 0$.

We can use the displacement combined with the amplitude and the period along with the other information from the previous sections to sketch curves of the functions $y = a \sin(bx + c)$ and $y = a \cos(bx + c)$, where $b > 0$.

LEARNING TIP

Note that the constant c and the displacement $-c/b$ differ in sign:

- If $c > 0$, the graph is shifted to the **left**, and displacement is **negative**.
- If $c < 0$, the graph is shifted to the **right**, and displacement is **positive**.

Important Quantities for Sketching Graphs of $y = a \sin(bx + c)$ and $y = a \cos(bx + c)$

1. Amplitude $= |a|$
2. Period $= \dfrac{2\pi}{b}$ **(10.1)**
3. Displacement $= -\dfrac{c}{b}$

These quantities allow us to evaluate the function at key values each one-fourth period. Table 10.3 summarizes these key values for the cycle that starts at $x = -c/b$ and is completed at $x = -c/b + \text{period}$. The graphs of $y = a \sin(bx + c)$ for the two cases $c > 0$ and $c < 0$ are shown in Fig. 10.16.

LEARNING TIP

Carefully note the difference between $y = \sin(bx + c)$ and $y = \sin bx + c$. Writing $\sin(bx + c)$ means to find the sine of the quantity $bx + c$, whereas $\sin bx + c$ means to find the sine of bx and then add the value c.

Table 10.3

Key Values (one cycle)	$y = a \sin(bx + c)$	$y = a \cos(bx + c)$
$-\frac{c}{b}$	0	a
$-\frac{c}{b} + \frac{\text{period}}{4}$	a	0
$-\frac{c}{b} + \frac{\text{period}}{2}$	0	$-a$
$-\frac{c}{b} + \frac{3 \cdot \text{period}}{4}$	$-a$	0
$-\frac{c}{b} + \text{period}$	0	a

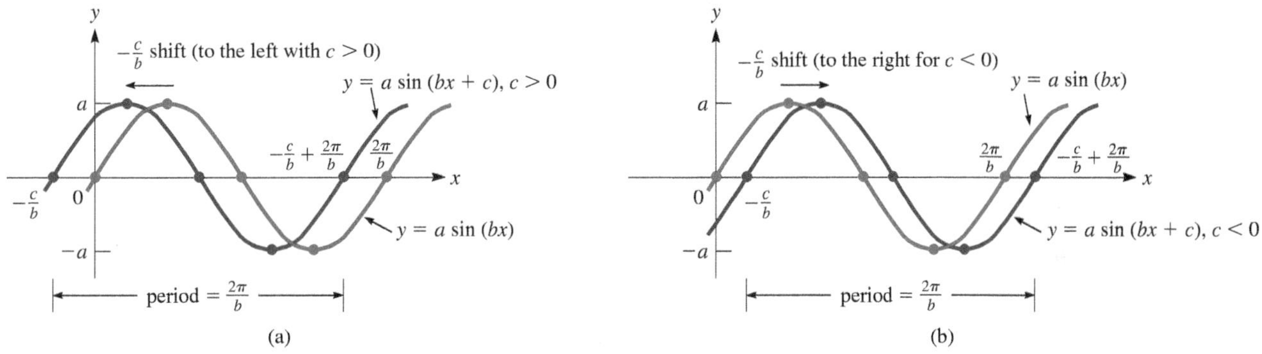

Fig. 10.16

EXAMPLE 2 Sketching the graph of $y = a \sin(bx + c)$

Sketch the graph of $y = 2 \sin(3x - \pi)$.

Amplitude: $|a| = 2$

Period: $\frac{2\pi}{b} = \frac{2\pi}{3}$ $\quad \left[\frac{1}{4}(\text{period}) = \frac{1}{4}\left(\frac{2\pi}{3}\right) = \frac{\pi}{6}\right]$

Displacement: $-\frac{c}{b} = -\frac{(-\pi)}{3} = \frac{\pi}{3}$

Therefore, key values for one full cycle start at $\pi/3$, end at π, and are found every $\pi/6$ units in between. This cycle is represented by the magenta portion of the graph in Fig. 10.17, obtained from the following table of important values.

x	0	$\frac{\pi}{6}$	$\frac{\pi}{3}$	$\frac{\pi}{2}$	$\frac{2\pi}{3}$	$\frac{5\pi}{6}$	π
y	0	-2	0	2	0	-2	0

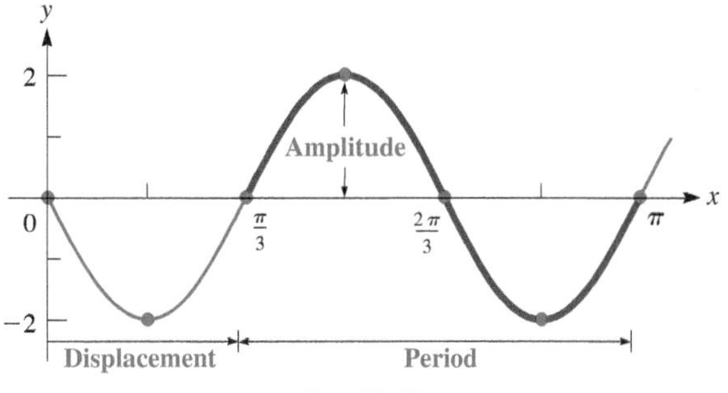

Fig. 10.17

EXAMPLE 3 Sketching the graph of $y = a\cos(bx + c)$

Sketch the graph of the function $y = -\cos\left(2x + \frac{\pi}{6}\right)$.

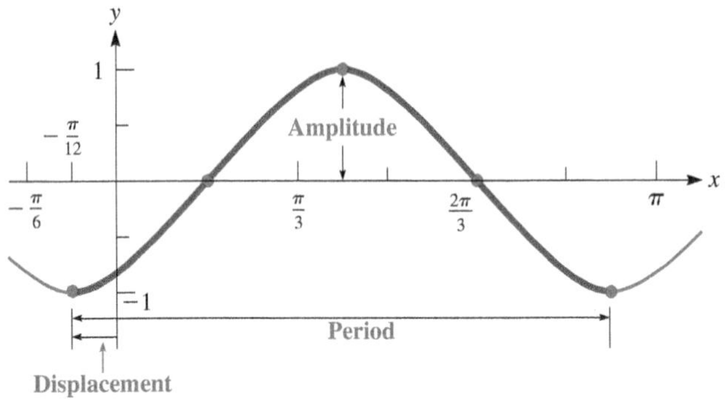

Fig. 10.18

Amplitude:	$\lvert a\rvert = 1$	
Period:	$\frac{2\pi}{2} = \pi$	$\left[\frac{1}{4}(\text{period}) = \frac{1}{4}\left(\frac{2\pi}{3}\right) = \frac{\pi}{6}\right]$
Displacement:	$-c \div b = -\frac{\pi}{6} \div 2 = -\frac{\pi}{12}$	

Therefore, key values for one cycle are found every $\pi/4$ units beginning at $-\frac{\pi}{12}$ and ending at $\frac{11\pi}{12}$. This cycle is represented by the magenta portion of the graph in Fig. 10.18, sketched from the following table:

x	$-\frac{\pi}{12}$	$\frac{\pi}{6}$	$\frac{5\pi}{12}$	$\frac{2\pi}{3}$	$\frac{11\pi}{12}$
y	-1	0	1	0	-1

Practice Exercise

1. For the graph of $y = 8\sin(2x - \pi/3)$, determine amplitude, period, and displacement.

EXAMPLE 4 Graph with $b < 0$

Sketch the graph of the function $y = \sin\left(-2x + \frac{\pi}{3}\right)$.

We first rewrite the function with a positive angle. We use the fact that $\sin(-x) = -\sin x$ (see Section 8.2) to write

$$y = \sin\left(-2x + \frac{\pi}{3}\right) = \sin\left(-\left(2x - \frac{\pi}{3}\right)\right) = -\sin\left(2x - \frac{\pi}{3}\right)$$

Therefore, we have $a = -1$, $b = 2$, $c = -\frac{\pi}{3}$. We determine that the amplitude is 1, the period is $2\pi \div 2 = \pi$, and the displacement is $-\left(-\frac{\pi}{3}\right) \div 2 = \frac{\pi}{6}$. Also, one-fourth of the period is $\frac{\pi}{4}$, so key values are $\frac{\pi}{4}$ units apart, starting at $\frac{\pi}{6}$ and ending at $\frac{7\pi}{6}$. We now make a table of important values and sketch the graph in Fig. 10.19.

x	$\frac{\pi}{6}$	$\frac{5\pi}{12}$	$\frac{2\pi}{3}$	$\frac{11\pi}{12}$	$\frac{7\pi}{6}$
y	0	-1	0	1	0

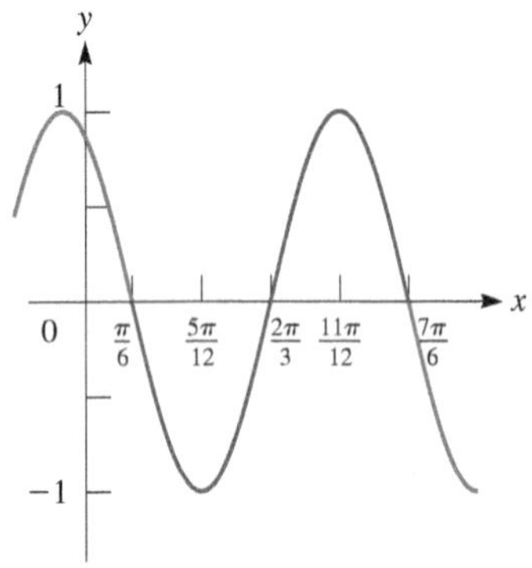

Fig. 10.19

When the range of a sine or a cosine function is not $[-a, a]$, the graph can be shifted vertically by adding a constant to the function (see Section 3.5). Consider the following example.

EXAMPLE 5 Trigonometric function with vertical shift—daylight hours

At 48° N latitude, the number of daylight hours during the longest day of the year (June 21) is 16.1 hours. During the shortest day of the year (December 21), the number of daylight hours is 8.31 hours. Approximate the number of hours h of daylight each day during a year with a function of the form $h = a\sin(bt + c) + d$, where time is measured in days. Cities at 48° N latitude include Paris, Vienna, and Victoria, British Columbia.

We determine the values of a, b, c, and d as follows:

1. The amplitude is half the distance between the maximum and the minimum daylight hours, so $a = \dfrac{16.1 - 8.31}{2} = 3.90$.

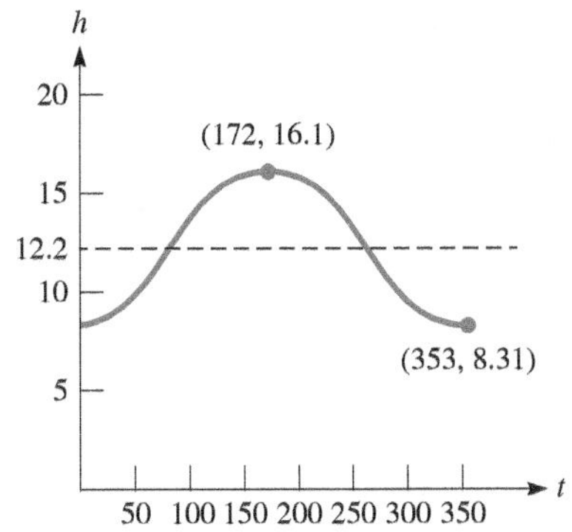

Fig. 10.20

2. The period is 365 days. Therefore, $\frac{2\pi}{b} = 365$ and $b = \frac{2\pi}{365}$.
3. The maximum of the sine function occurs at one-fourth of the period, which is $\frac{365}{4} = 91.25$. Since the maximum for the data at hand occurs at $t = 172$ (June 21), we must shift the sine function to the right $172 - 91.25 = 80.75$ days. This gives $-\frac{c}{b} = 80.75$, so that $c = -80.75 \times \frac{2\pi}{365} = -1.39$.
4. Finally, the range is $[8.31, 16.1]$ instead of $[-3.90, 3.90]$, so we must shift the sine function $16.1 - 3.90 = 12.2$ hours vertically by setting $d = 12.2$.

We can now state our sine model and sketch the graph in Fig. 10.20.

$$h = 3.90\sin\left(\frac{2\pi}{365}t - 1.39\right) + 12.2$$

EXAMPLE 6 Phase angle—AC voltage/current lead and lag

In alternating-current (AC) circuits, voltage (v) and current (i) can both be represented by sine waves with the same period. Depending on the existence of resistors, capacitors, or inductors in the circuit, these waves can be *in phase* (maxima and minima occur at the same time) or one can *lead or lag* the other. See Fig. 10.21.

■ In Chapter 12, we show how the phase angle depends on the resistance, capacitance, and inductance in a circuit.

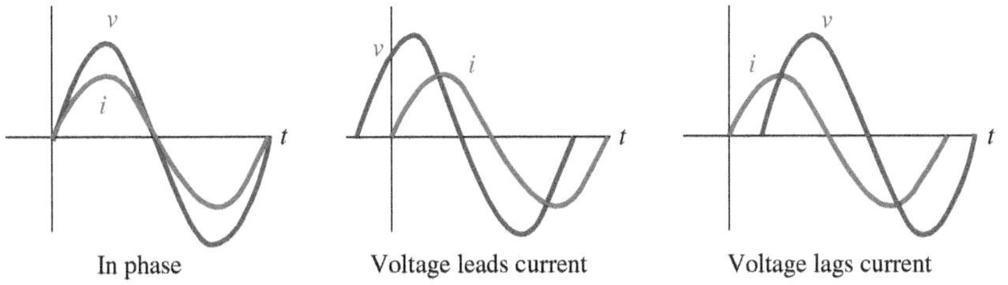

Fig. 10.21

If we assume that the sine wave for current passes through the origin, it can be represented as $i = I_m \sin(2\pi ft)$, where f is the frequency (in cycles/s) and t is time (in s). The voltage can then be represented by $v = V_m \sin(2\pi ft + \phi)$, where ϕ is the phase angle. The **phase angle**, often written in degrees, tells us how many degrees (assuming 360° is one complete cycle) the voltage leads (if ϕ is positive) or lags (if ϕ is negative) the current by.

For example, if $i = 15\sin(120\pi t)$ and $v = 170\sin(120\pi t + 90°)$, then voltage leads current by 90°, or one-quarter of the period. See Fig. 10.22. It is somewhat unusual that the voltage function combines radians ($2\pi ft$) and degrees (90°), but this is commonly done in this particular application. If we convert the phase angle to $\frac{\pi}{2}$ radians, we can then find the displacement, which gives the time shift in seconds between the two waves:

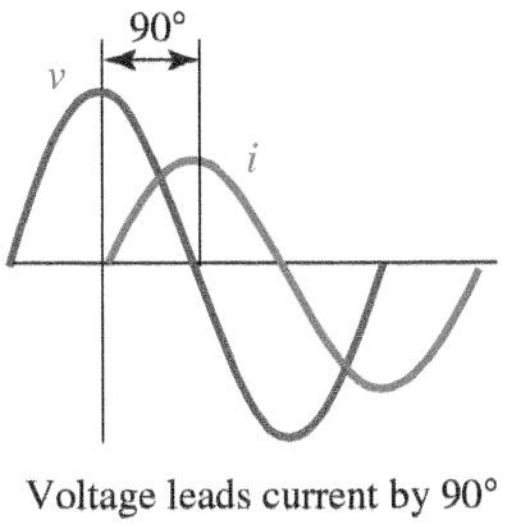

Voltage leads current by 90°

Fig. 10.22

$$\text{Displacement} = -\frac{c}{b} = -\frac{\pi/2}{120\pi} = -\frac{1}{240}\text{ s}$$

Therefore, voltage leads current by $\frac{1}{240}$ s. This is one-quarter of the period, with the period equal to $\frac{2\pi}{120\pi} = \frac{1}{60}$ s. Note that in order to graph the voltage function on a graphing utility, the phase angle must be expressed in radians.

Practice Exercise

2. In Example 6, find the function for voltage if voltage lags current by 45° and all other information remains the same.

EXERCISES 10.3

In Exercises 1 and 2, graph the function if the given changes are made in the indicated examples of this section.

1. In Example 3, if the sign before $\pi/6$ is changed, sketch the graph of the resulting function.

2. In Example 4, if the sign before $\pi/3$ is changed, sketch the graph of the resulting function.

In Exercises 3–26, determine the amplitude, period, and displacement for each function. Then sketch the graphs of the functions.

3. $y = \sin\left(x - \frac{\pi}{6}\right)$

4. $y = 3\sin\left(x + \frac{\pi}{4}\right)$

5. $y = \cos\left(x + \frac{\pi}{6}\right)$

6. $y = 2\cos\left(x - \frac{\pi}{8}\right)$

7. $y = 0.2\sin\left(2x + \frac{\pi}{2}\right)$

8. $y = -\sin\left(3x - \frac{\pi}{2}\right)$

9. $y = -\cos(2x - \pi)$

10. $y = 0.4\cos\left(3x + \frac{\pi}{3}\right)$

11. $y = \frac{1}{2}\sin\left(\frac{1}{2}x - \frac{\pi}{4}\right)$

12. $y = 2\sin\left(\frac{1}{4}x + \frac{\pi}{2}\right)$

13. $y = 30\cos\left(\frac{1}{3}x + \frac{\pi}{3}\right)$

14. $y = \frac{1}{3}\cos\left(\frac{1}{2}x - \frac{\pi}{8}\right)$

15. $y = \sin\left(\pi x + \frac{\pi}{8}\right)$

16. $y = -2\sin(2\pi x - \pi)$

17. $y = 0.08\cos\left(4\pi x - \frac{\pi}{5}\right)$

18. $y = 25\cos\left(3\pi x + \frac{\pi}{2}\right)$

19. $y = -0.6\sin(2\pi x - 1)$

20. $y = 1.8\sin\left(\pi x + \frac{1}{3}\right)$

21. $y = 40\cos(3\pi x + 2)$

22. $y = 360\cos(6\pi x - 1)$

23. $y = \sin(\pi^2 x - \pi)$

24. $y = -\frac{1}{2}\sin\left(2x - \frac{1}{\pi}\right)$

25. $y = -\frac{3}{2}\cos\left(\pi x + \frac{\pi^2}{6}\right)$

26. $y = \pi\cos\left(\frac{1}{\pi}x + \frac{1}{3}\right)$

In Exercises 27–30, write the equation for the given function with the given amplitude, period, and displacement, respectively ($a > 0$).

27. sine, 4, 3π, $-\pi/4$

28. cosine, 8, $2\pi/3$, $\pi/3$

29. cosine, 12, $1/2$, $1/8$

30. sine, 18, 4, -1

In Exercises 31 and 32, show that the given equations are identities. The method using a graphing utility is indicated in Exercise 31.

31. By viewing the graphs of $y_1 = \sin x$ and $y_2 = \cos(x - \pi/2)$, show that $\cos(x - \pi/2) = \sin x$.

32. Show that $\cos(2x - 3\pi/8) = \cos(3\pi/8 - 2x)$.

In Exercises 33–40, solve the given problems.

33. The current in an alternating-current circuit is given by $i = 12\sin(120\pi t)$. Find a function for the voltage v if the amplitude is 18V and voltage lags current by 60°. Then find the displacement. See Example 6.

34. The current in an alternating-current circuit is given by $i = 8.50\sin(120\pi t)$. Find a function for the voltage v if the amplitude is 28V and voltage leads current by 45°. Then find the displacement. See Example 6.

35. A wave travelling in a string may be represented by the equation $y = A\sin 2\pi\left(\frac{t}{T} - \frac{x}{\lambda}\right)$. Here, A is the amplitude, t is the time the wave has travelled, x is the distance from the origin, and T is the time required for the wave to travel one *wavelength* λ (the Greek letter lambda). Sketch three cycles of the wave for which $A = 2.00$ cm, $T = 0.100$ s, $\lambda = 20.0$ cm, and $x = 5.00$ cm.

36. The electric current i (in μA) in a certain circuit is given by $i = 3.8\cos 2\pi(t + 0.20)$, where t is the time in seconds. Sketch three cycles of this function.

37. A certain satellite circles the earth such that its distance y, in kilometres north or south (altitude is not considered) from the equator, is $y = 7200\cos(0.025t - 0.25)$, where t is the time (in min) after launch. Use a graphing utility to graph two cycles of the curve.

38. In performing a test on a patient, a medical technician used an ultrasonic signal given by the equation $I = A\sin(\omega t + \theta)$. Use a graphing utility to view two cycles of the graph of I vs. t if $A = 5\ \text{nW/m}^2$, $\omega = 2 \times 10^5$ rad/s, and $\theta = 0.4$.

39. In Sydney, Australia, the number of daylight hours during the shortest day of the year (June 21) is 9.90 hours. During the longest day of the year (December 21), the number of daylight hours is 14.4 hours. Approximate the number of hours h of daylight each day during a year with a function of the form $h = a\sin(bt + c) + d$, where time is measured in days.

40. Carbon assimilation in plants has a *circadian rhythm* (it oscillates with a period of approximately 24 hours). For a certain species, carbon assimilation oscillates between a minimum of 5.8 mmol $CO_2\ m^{-2}\ s^{-1}$ at midnight and a maximum of 8.2 mmol $CO_2\ m^{-2}\ s^{-1}$ at noon. Approximate the carbon assimilation y with a function of the form $y = a\sin(bt + c) + d$, where t is measured in hours after midnight.

In Exercises 41–44, give the specific form of the equation by evaluating a, b, and c through an inspection of the given curve. Explain how a, b, and c are found.

41. $y = a\sin(bx + c)$ Fig. 10.23

42. $y = a\cos(bx + c)$ Fig. 10.23

43. $y = a\cos(bx + c)$ Fig. 10.24

44. $y = a\sin(bx + c)$ Fig. 10.24

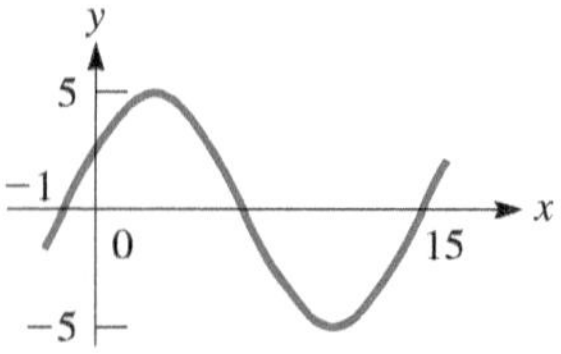

Fig. 10.23

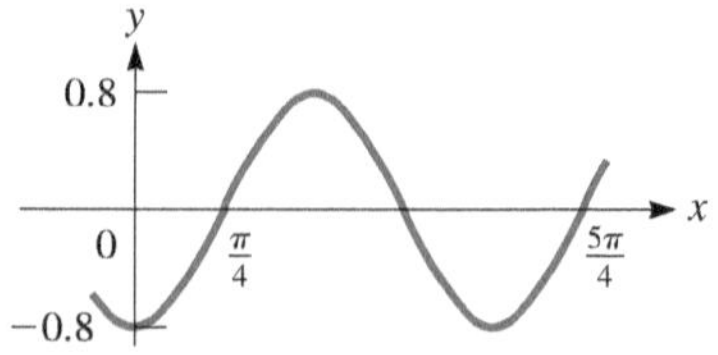

Fig. 10.24

Answers to Practice Exercises

1. amp. = 8, per. = π, disp. = $\pi/6$ **2.** $v = 170\sin(120\pi t - 45°)$

10.4 Graphs of $y = \tan x$, $y = \cot x$, $y = \sec x$, $y = \csc x$

The graph of $y = \tan x$ is shown in Fig. 10.25. Because from Section 4.2 we know that $\csc x = 1/\sin x$, $\sec x = 1/\cos x$, and $\cot x = 1/\tan x$, we are able to find the values of $y = \csc x$, $y = \sec x$, and $y = \cot x$ and graph these functions, as shown in Figs. 10.25–10.28.

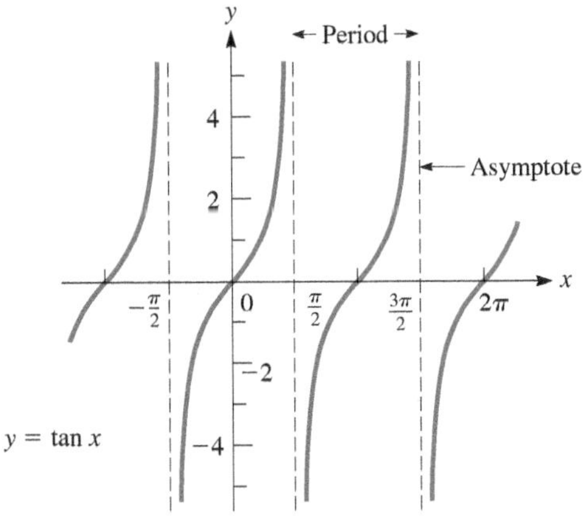

Fig. 10.25

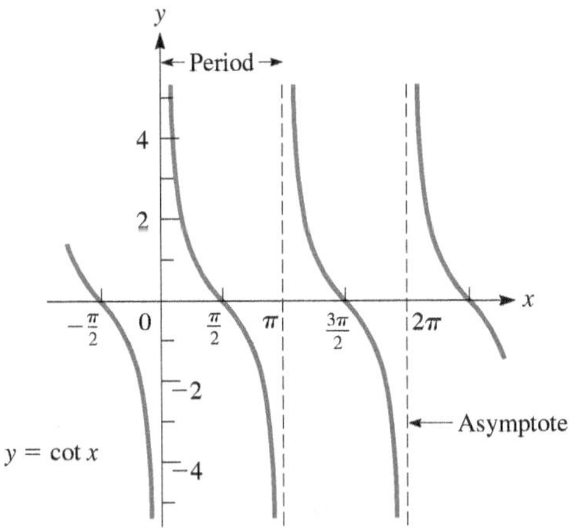

Fig. 10.26

■ Repeating from above the reciprocal relationships that are the basis for graphing $y = \cot x$, $y = \sec x$, and $y = \csc x$, we have:

$$\csc x = \frac{1}{\sin x} \qquad \sec x = \frac{1}{\cos x}$$

$$\cot x = \frac{1}{\tan x} \tag{10.2}$$

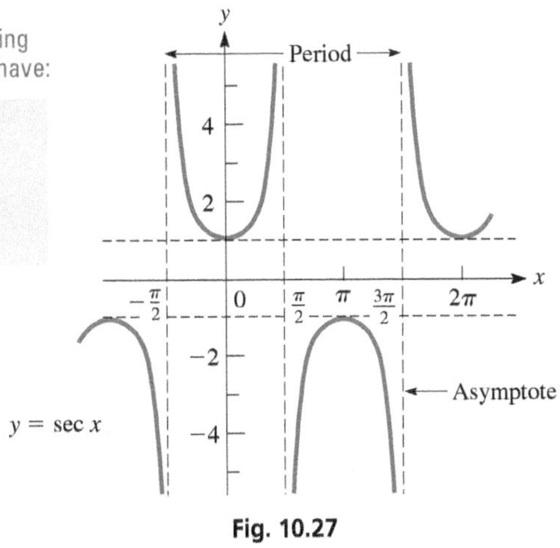

Fig. 10.27

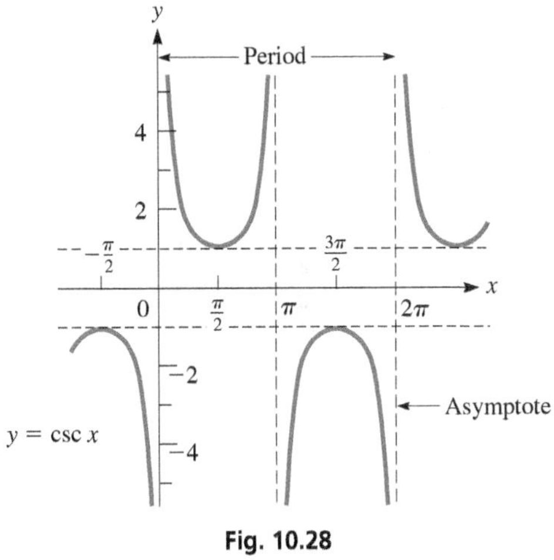

Fig. 10.28

From these graphs, note that $y = \tan x$ and $y = \cot x$ have period π and have all real numbers as their range. The functions $y = \sec x$ and $y = \csc x$ have period 2π, but their ranges do not include the real numbers between -1 and 1.

The vertical dashed lines on the graphs are **vertical asymptotes** (see Section 3.4). The curves *approach* these lines but never actually reach them. The values of x for which the curve has an asymptote are not included in the domain of the function.

As we can see from the graphs, the functions have asymptotes, zeros, and maximum or minimum values when x is a multiple of $\pi/2$, that is to say, at the same key values which allowed us to sketch the graphs of the sine and cosine functions easily. The characteristics of the trigonometric functions at the key values from 0 to 2π are compared in Table 10.4. For example, note that when the sine is zero, its reciprocal (the cosecant) has an asymptote.

Table 10.4

Key Values	$y = \sin x$	$y = \cos x$	$y = \tan x$
0	0	1	0
$\frac{\pi}{2}$	1	0	asymptote
π	0	-1	0
$\frac{3\pi}{2}$	-1	0	asymptote
2π	0	1	0
Key Values	$y = \csc x$	$y = \sec x$	$y = \cot x$
0	asymptote	1	asymptote
$\frac{\pi}{2}$	1	asymptote	0
π	asymptote	-1	asymptote
$\frac{3\pi}{2}$	-1	asymptote	0
2π	asymptote	1	asymptote

To sketch functions such as $y = a \sec x$, first sketch $y = \sec x$ and then multiply the y-values by a. *Here, a is not an amplitude,* since the ranges of these functions are not limited in the same way they are for the sine and cosine functions.

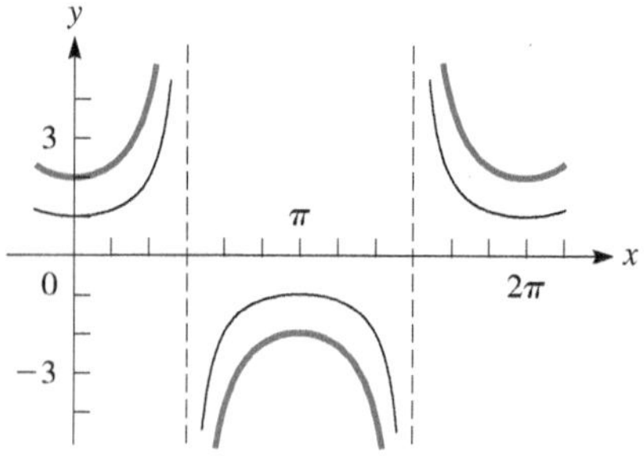

Fig. 10.29

EXAMPLE 1 Sketching the graph of $y = a \sec x$

Sketch the graph of $y = 2 \sec x$.

First, we sketch in $y = \sec x$, shown as the light curve in Fig. 10.29. Then we multiply the y-values of this secant function by 2. Although we can only estimate these values and do this approximately, a reasonable graph can be sketched this way. The desired curve is shown in Fig. 10.29.

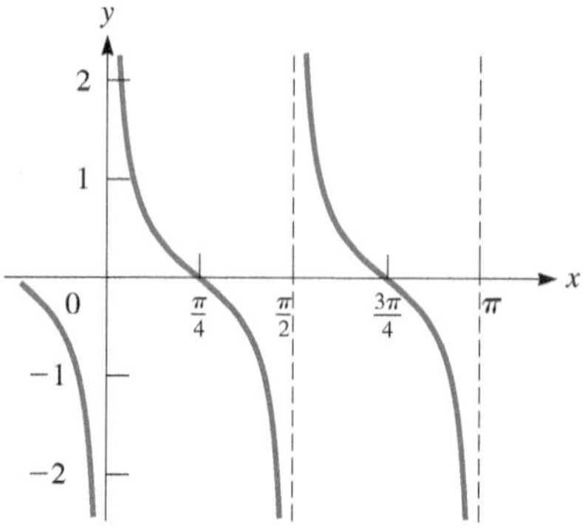

Fig. 10.30

EXAMPLE 2 Graph of $y = a \cot bx$

Sketch the graph of $y = 0.5 \cot 2x$.

Since the period of $y = \cot x$ is π, the period of $y = \cot 2x$ is $\pi/2$. Therefore, we have the following table of key values:

x	0	$\frac{\pi}{4}$	$\frac{\pi}{2}$	$\frac{3\pi}{4}$	π
y	asymptote	0	asymptote	0	asymptote

Since $a = 0.5$, the function increases more slowly than $y = \cot x$. We sketch the graph as shown in Fig. 10.30.

Using a graphing utility (a graphing calculator in *radian* mode, a computer algebra system, or any graphing software), we can display the graphs of these functions more easily and more accurately than by sketching them. By knowing the general shape and period of the function, the values for the display settings can be determined.

EXAMPLE 3 Calculator graph of $y = a \csc(bx + c)$

View at least two periods of the graph of $y = 2\csc(2x + \pi/4)$ on a graphing calculator.

Since the period of $\csc x$ is 2π, the period of $\csc(2x + \pi/4)$ is $2\pi/2 = \pi$. Recalling that $\csc x = (\sin x)^{-1}$, the curve will have the same displacement as $y = \sin(2x + \pi/4)$. Therefore, displacement is $-\frac{\pi/4}{2} = -\frac{\pi}{8}$. There is some flexibility in choosing the *window* settings, and as an example, we choose the following settings:

Xmin $= -0.5$ (the displacement is $-\pi/8 = -0.4$)

Xmax $= 6$ (displacement $= -\pi/8$; period $= \pi$, $-\pi/8 + 2\pi = 15\pi/8 = 5.9$)

Ymin $= -6$, Ymax $= 6$ (there is no curve between $y = -2$ and $y = 2$)

With $y_1 = 2(\sin(2x + \pi/4))^{-1}$, and with the calculator in radian mode, Fig. 10.31 shows the calculator view.

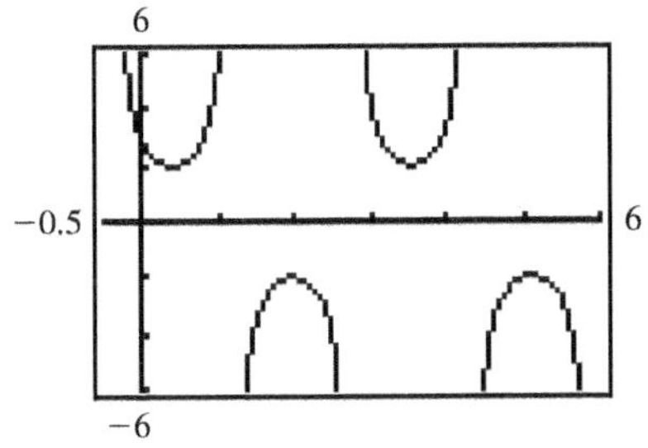

Fig. 10.31

EXERCISES 10.4

In Exercises 1 and 2, view the graphs on a graphing utility if the given changes are made in the indicated examples of this section.

1. In Example 2, change 0.5 to 5.
2. In Example 3, change the sign before $\pi/4$.

In Exercises 3–6, fill in the following table for each function and plot the graph from these points.

x	$-\frac{\pi}{2}$	$-\frac{\pi}{3}$	$-\frac{\pi}{4}$	$-\frac{\pi}{6}$	0	$\frac{\pi}{6}$	$\frac{\pi}{4}$	$\frac{\pi}{3}$	$\frac{\pi}{2}$	$\frac{2\pi}{3}$	$\frac{3\pi}{4}$	$\frac{5\pi}{6}$	π
y													

3. $y = \tan x$
4. $y = \cot x$
5. $y = \sec x$
6. $y = \csc x$

In Exercises 7–14, sketch the graphs of the given functions by use of the basic curve forms (Figs. 10.25, 10.26, 10.27, and 10.28). See Example 1.

7. $y = 2\tan x$
8. $y = 3\cot x$
9. $y = \frac{1}{2}\sec x$
10. $y = \frac{3}{2}\csc x$
11. $y = -8\cot x$
12. $y = -0.1\tan x$
13. $y = -3\csc x$
14. $y = -60\sec x$

In Exercises 15–24, view at least two cycles of the graphs of the given functions on a graphing utility.

15. $y = \tan 2x$
16. $y = 2\cot 3x$
17. $y = \frac{1}{2}\sec 3x$
18. $y = 0.4\csc 2x$
19. $y = 2\cot\left(2x + \dfrac{\pi}{6}\right)$
20. $y = \tan\left(3x - \dfrac{\pi}{2}\right)$
21. $y = 18\csc\left(3x - \dfrac{\pi}{3}\right)$
22. $y = 12\sec\left(2x + \dfrac{\pi}{4}\right)$
23. $y = 75\tan\left(0.5x - \dfrac{\pi}{8}\right)$
24. $y = 0.5\sec\left(0.2x + \dfrac{\pi}{25}\right)$

In Exercises 25 and 26, solve the given problems. In Exercises 27–30, sketch the appropriate graphs.

25. Write the equation of a secant function with zero displacement and a period of 4π, and that passes through $(0, -3)$.
26. Use a graphing utility to show that $\sin x < \tan x$ for $0 < x < \pi/2$, although $\sin x$ and $\tan x$ are nearly equal for the values near zero.
27. Near Antarctica, an iceberg with a vertical face 200 m high is seen from a small boat. At a distance x from the iceberg, the angle of elevation θ of the top of the iceberg can be found from the equation $x = 200\cot\theta$. Sketch x as a function of θ.
28. In a laser experiment, two mirrors move horizontally equal and opposite distances from point A. The laser path from and to point B is shown in Fig. 10.32. From the figure, we see that $x = a\tan\theta$. Sketch the graph of $x = f(\theta)$ for $a = 5.00$ cm.

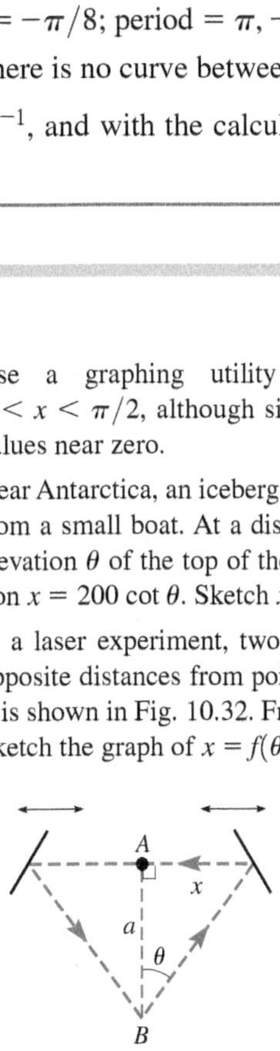

Fig. 10.32

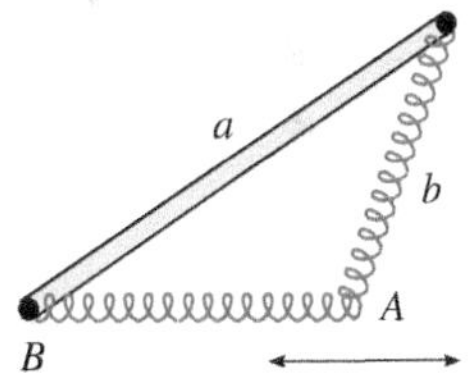

Fig. 10.33

29. A mechanism with two springs is shown in Fig. 10.33, where point A is restricted to move horizontally. From the law of sines, we see that $b = (a\sin B)\csc A$. Sketch the graph of b as a function of A for $a = 4.00$ cm and $B = \pi/4$.
30. A cantilever column of length L will buckle if too large a downward force P is applied d units off centre. The horizontal deflection x (see Fig. 10.34) is $x = d(\sec(kL) - 1)$, where k is a constant depending on P, and $0 < kL < \pi/2$. For a constant d, sketch the graph of x as a function of kL.

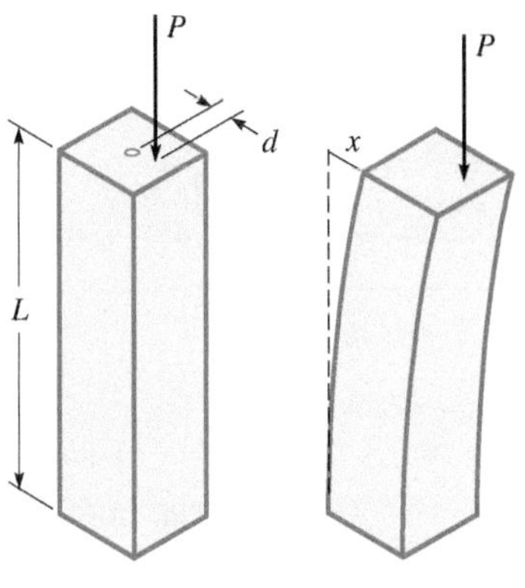

Fig. 10.34

10.5 Applications of the Trigonometric Graphs

When an object moves in a circular path with constant velocity (see Section 8.4), its *projection* on a diameter moves with **simple harmonic motion**. For example, the shadow of a ball at the end of a string and moving at a constant rate moves with simple harmonic motion. We now consider this physical concept and some of its applications.

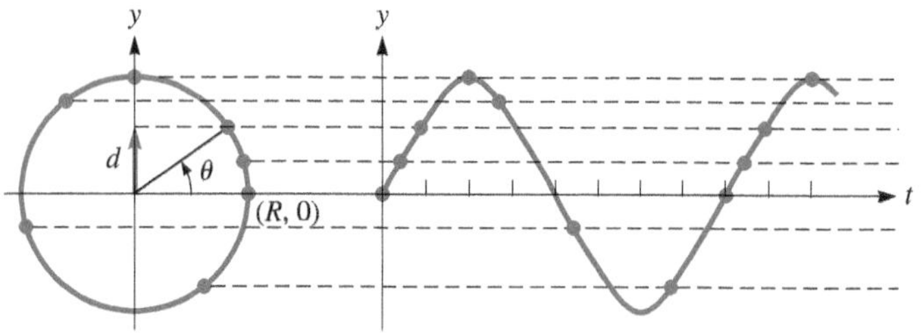

Fig. 10.35

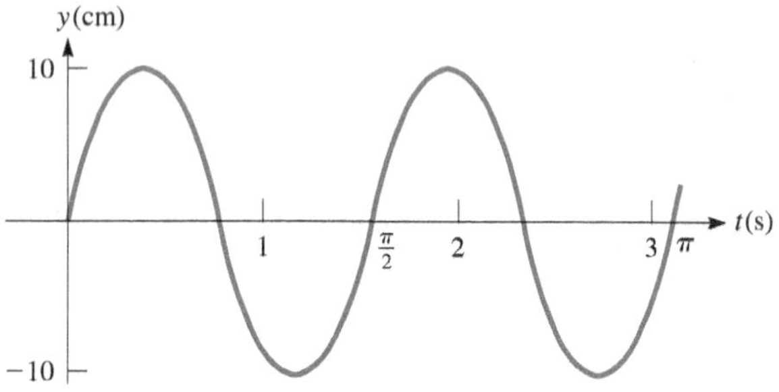

Fig. 10.36

EXAMPLE 1 Simple harmonic motion

In Fig. 10.35, assume that a particle starts at the end of the radius at $(R, 0)$ and moves counterclockwise around the circle with constant angular velocity ω. *The displacement of the projection on the y-axis is d* and is given by $d = R \sin \theta$. The displacement is shown for a few different positions of the end of the radius.

Since $\theta/t = \omega$, or $\theta = \omega t$, we have

$$d = R \sin \omega t \qquad \textbf{(10.3)}$$

as the equation for the displacement of this projection, with time t as the independent variable.

For the case where $R = 10.0$ cm and $\omega = 4.00$ rad/s, we have

$$d = 10.0 \sin 4.00t$$

By sketching or viewing the graph of this function, we can find the displacement d of the projection for a given time t. The graph is shown in Fig. 10.36.

In Example 1, note that *time is the independent variable.* This is motion for which the object (the end of the projection) remains at the same horizontal position ($x = 0$) and moves only vertically according to a sinusoidal function. In the previous sections, we dealt with functions in which y is a sinusoidal function of the horizontal displacement x. Think of a water wave. At *one point* of the wave, the motion is only vertical and sinusoidal with time. At *one given time,* a picture would indicate a sinusoidal movement from one horizontal position to the next.

EXAMPLE 2 Simple harmonic motion of a rotating blade

A windmill is used to pump water. The radius of the blade is 2.5 m, and it is moving with constant angular velocity. If the vertical displacement of the end of the blade is timed from the point it is at an angle of 45° ($\pi/4$ rad) from the horizontal (see Fig. 10.37a), the displacement d is given by

$$d = 2.5 \sin\left(\omega t + \frac{\pi}{4}\right)$$

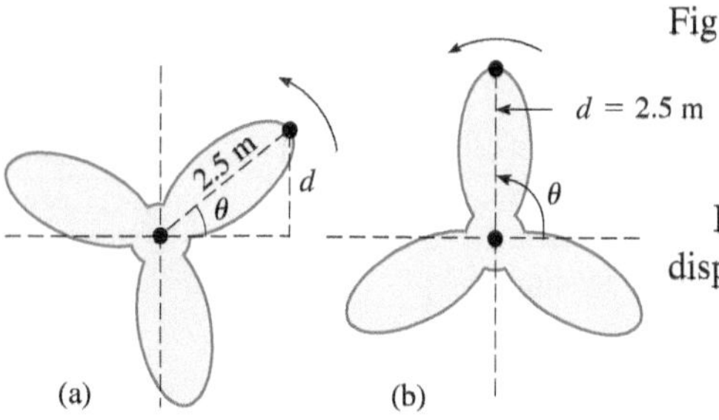

Fig. 10.37

If the blade makes an angle of 90° ($\pi/2$ rad) when $t = 0$ (see Fig. 10.37b), the displacement d is given by

$$d = 2.5 \sin\left(\omega t + \frac{\pi}{2}\right) \quad \text{or} \quad d = 2.5 \cos \omega t$$

If timing started at the first maximum for the displacement, the resulting curve for the displacement would be that of the cosine function.

Practice Exercise

1. If the windmill blade in Example 2 starts at an angle of 135°, what equation gives its displacement in terms of the cosine function?

Other examples of simple harmonic motion are (1) the movement of a pendulum bob through its arc (a very close approximation to simple harmonic motion), (2) the motion of an object "bobbing" in water, (3) the movement of the end of a vibrating rod (which we hear as sound), and (4) the displacement of a weight moving up and down on a spring. Other phenomena that give rise to equations like those for simple harmonic motion are found in the fields of optics, sound, and electricity. The equations for such phenomena have the same mathematical form because they result from vibratory movement or motion in a circle.

EXAMPLE 3 Alternating current

As shown in Section 10.3, a very important use of the trigonometric curves arises in the study of alternating current, which is caused by the motion of a wire passing through a magnetic field. If the wire is moving in a circular path, with angular velocity ω, the current i in the wire at time t is given by an equation of the form

$$i = I_m \sin(\omega t + \alpha)$$

where I_m is the maximum current attainable and α is the phase angle.

The current may be represented by a sinusoidal wave. Given that $I_m = 6.00$ A, $\omega = 120\pi$ rad/s, and $\alpha = \pi/6$, we have the equation

$$i = 6.00 \sin\left(120\pi t + \frac{\pi}{6}\right)$$

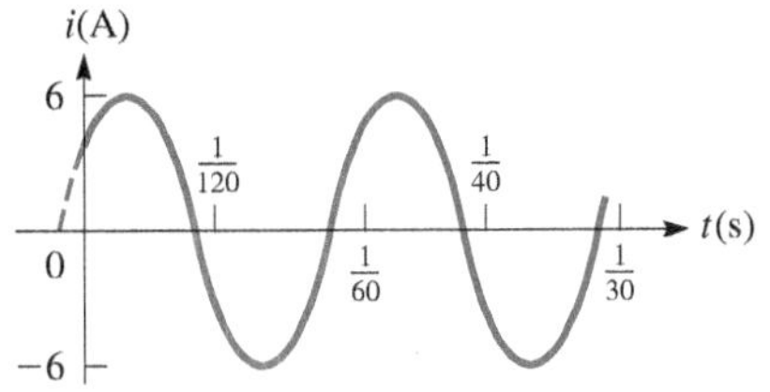

Fig. 10.38

From this equation, note that the amplitude is 6.00 A, the period is $\frac{1}{60}$ s, and the displacement is $-\frac{1}{720}$ s. From these values, we draw the graph as shown in Fig. 10.38. Since the current takes on both positive and negative values, we conclude that it moves alternately in one direction and then the other.

It is a common practice to express the rate of rotation in terms of *the* **frequency** f, the *number of cycles per second,* rather than directly in terms of the angular velocity ω, the number of radians per second. *The unit for frequency is the* **hertz** (Hz), *and* 1 Hz = 1 cycle/s. Since there are 2π rad in one cycle, we have

■ The unit *Hertz* is named for the German physicist Heinrich Hertz (1857–1894).

$$\omega = 2\pi f \qquad \textbf{(10.4)}$$

It is the frequency f that is referred to in electric current, on radio stations, for musical tones, and so on.

It is important to note that **the frequency and the period are reciprocals of each other**. The following example illustrates this point.

EXAMPLE 4 Frequency—hertz

For the electric current in Example 3, $\omega = 120\pi$ rad/s. The corresponding frequency f is

$$f = \frac{120\pi}{2\pi} = 60 \text{ Hz}$$

This means that 120π rad/s corresponds to 60 cycles/s. This is the standard frequency used for alternating current in North America. We have already shown that the period is 1/60 s. We can see here that the frequency and the period are reciprocals.

Practice Exercise

2. In Example 1, what is the frequency?

EXERCISES 10.5

In Exercises 1 and 2, answer the given questions about the indicated examples of this section.

1. In Example 1, what is the equation relating d and t if the end of the radius starts at $(0, R)$?
2. In Example 2, if the blade starts at an angle of $-45°$, what is the equation relating d and t as (a) a sine function? (b) a cosine function?

A graphing utility may be used in the following exercises.

In Exercises 3 and 4, sketch two cycles of the curve of the projection of Example 1 as a function of time for the given values.

3. $R = 2.40$ cm, $\omega = 2.00$ rad/s
4. $R = 1.80$ m, $f = 0.250$ Hz

In Exercises 5 and 6, a point on a cam is 8.30 cm from the centre of rotation. The cam is rotating with a constant angular velocity, and the vertical displacement $d = 8.30$ cm for $t = 0$ s. See Fig. 10.39. Sketch two cycles of d as a function of t for the given values.

5. $f = 3.20$ Hz
6. $\omega = 3.20$ rad/s

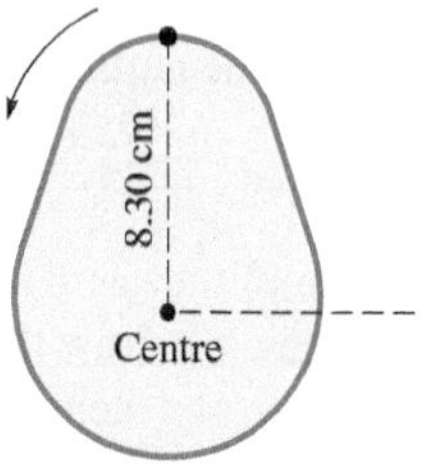

Fig. 10.39

In Exercises 7 and 8, a satellite is orbiting the earth such that its displacement D north of the equator (or south if $D < 0$) is given by $D = A\sin(\omega t + \alpha)$. Sketch two cycles of D as a function of t for the given values.

7. $A = 500$ km, $\omega = 3.60$ rad/h, $\alpha = 0$
8. $A = 850$ km, $f = 1.6 \times 10^{-4}$ Hz, $\alpha = \pi/3$

In Exercises 9 and 10, for an alternating-current circuit in which the voltage V is given by $V = E\cos(\omega t + \alpha)$, sketch two cycles of the voltage as a function of time for the given values.

9. $E = 170$ V, $f = 60.0$ Hz, $\alpha = -\pi/3$
10. $E = 80$ V, $\omega = 377$ rad/s, $\alpha = \pi/2$

In Exercises 11 and 12, refer to the wave in the string described in Exercise 35 of Section 10.3. For a point on the string, the displacement y is given by $y = A\sin 2\pi\left(\frac{t}{T} - \frac{x}{\lambda}\right)$. We see that each point on the string moves with simple harmonic motion. Sketch two cycles of y as a function of t for the given values.

11. $A = 3.20$ cm, $T = 0.050$ s, $\lambda = 40.0$ cm, $x = 5.00$ cm
12. $A = 0.750$ cm, $T = 0.250$ s, $\lambda = 24.0$ cm, $x = 20.0$ cm

In Exercises 13 and 14, the air pressure within a plastic container changes above and below the external atmospheric pressure by $p = p_0 \sin 2\pi ft$. Sketch two cycles of p as a function of t for the given values.

13. $p_0 = 280$ kPa, $f = 2.30$ Hz
14. $p_0 = 45.0$ kPa, $f = 0.450$ Hz

In Exercises 15–22, sketch the required curves.

15. The angular displacement θ of a certain pendulum bob in terms of its initial displacement θ_0 is $\theta = \theta_0 \cos \omega t$. If $\omega = 2.00$ rad/s, and $\theta_0 = \pi/30$ rad, draw two cycles for the resulting equation.
16. A study found that, when breathing normally, the increase in volume V (in L) of air in a person's lungs as a function of the time t (in s) is $V = 0.30 \sin 0.50\pi t$. Sketch two cycles.
17. Sketch two cycles of the radio signal $V = 0.014\cos(2\pi ft + \pi/4)$ (V in volts, f in hertz, and t in seconds) for a station broadcasting with $f = 950$ kHz ("95" on the AM radio dial).
18. Sketch two cycles of the acoustical intensity I of the sound wave for which $I = A\cos(2\pi ft - \alpha)$, given that t is in seconds, $A = 0.027$ W/cm^2, $f = 240$ Hz, and $\alpha = 0.80$.
19. The rotating beacon of a parked police car is 12 m from a straight wall. (a) Sketch the graph of the length L of the light beam, where $L = 12 \sec \pi t$, for $0 \le t \le 2.0$ s. (b) Which part(s) of the graph show meaningful values? Explain.
20. The motion of a piston of a car engine approximates simple harmonic motion. Given that the stroke (twice the amplitude) is 0.100 m, the engine runs at 2800 r/min, and the piston starts at the middle of its stroke, find the equation for the displacement d as a function of t. Sketch two cycles.
21. The paddle wheel of the *S. S. Beaver*, Canada's first steamship, had a radius of 1.98 m and rotated at 30 r/min when moving at top speed. Find the equation of motion of the vertical displacement (from the centre of the wheel) y of the end of a paddle as a function of the time t (in s) if the paddle was initially horizontal. Sketch two cycles.
22. The sinusoidal electromagnetic wave emitted by an antenna in a cellular phone system has a frequency of 7.5×10^9 Hz and an amplitude of 0.045 V/m. Find the equation representing the wave if it starts at the origin. Sketch two cycles.

Answers to Practice Exercises

1. $d = 2.5\cos(\omega t + \pi/4)$
2. $f = 2/\pi$ Hz

10.6 Composite Trigonometric Curves

Many applications involve functions that in themselves are a combination of two or more simpler functions. In this section, we discuss methods by which the curve of such a function can be found.

ADDITION OF ORDINATES

One way to sketch the resulting graph is to *first sketch the two simpler curves and then add the* y-values *graphically. This method is called* **addition of ordinates** and is illustrated in the following example.

EXAMPLE 1 Addition of ordinates

Sketch the graph of $y = 2\cos x + \sin 2x$.

On the same set of coordinate axes, we sketch the curves $y = 2\cos x$ and $y = \sin 2x$. These are shown as dashed and solid light curves in Fig. 10.40. For various values of x, we determine the distance above or below the x-axis of each curve and add these distances, noting that those above the axis are positive and those below the axis are negative. We thereby *graphically* ***add*** *the* y-values of these curves to get points on the resulting curve, shown in the thicker line in Fig. 10.40.

At A, add the two lengths (shown side-by-side for clarity) to get the length for y. At B, both lengths are negative, and the value for y is the sum of these negative values. At C, one is positive and the other negative, and we must subtract the lower length from the upper one to get the length for y.

We combine these lengths for enough x-values to get a good curve. Some points are easily found. Where one curve crosses the x-axis, its value is zero, and the resulting curve has its point on the other curve. Here, where $\sin 2x$ is zero, the points for the resulting curve lie on the curve of $2\cos x$.

We should also add values where each curve is at its maximum or its minimum. *Extra care should be taken for those values of x for which one curve is* ***positive*** *and the other is* ***negative***.

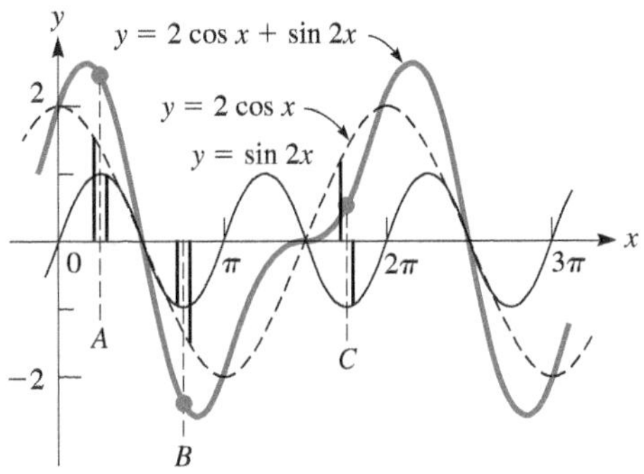

Fig. 10.40

We have seen how a fairly complex curve can be sketched graphically. Graphs that are difficult to sketch can be constructed much more easily and with greater accuracy by using a graphing utility.

EXAMPLE 2 Graphing utility addition of ordinates

Use a graphing utility to display the graph of $y = \frac{x}{2} - \cos x$.

Here, we note that the curve is a combination of the straight line $y = x/2$ and the trigonometric curve $y = \cos x$. To see a little more than one period of $\cos x$, we make the following choices for the display settings:

Xmin $= -1$ (to start to left of y-axis)

Xmax $= 7$ (period of $\cos x$ is $2\pi = 6.3$)

Ymin $= -2$ (line passes through $(0, 0)$; amplitude of $y = \cos x$ is 1)

Ymax $= 4$ (slope of line is $1/2$)

Fig. 10.41(a) shows the display on a graphing calculator. The graphs of $y = x/2 - \cos x$, $y = x/2$, and $y = -\cos x$ are shown in Fig. 10.41(b). We show $y = -\cos x$ rather than $y = \cos x$ because for *addition of ordinates, it is easier to add graphic values than to subtract them.*

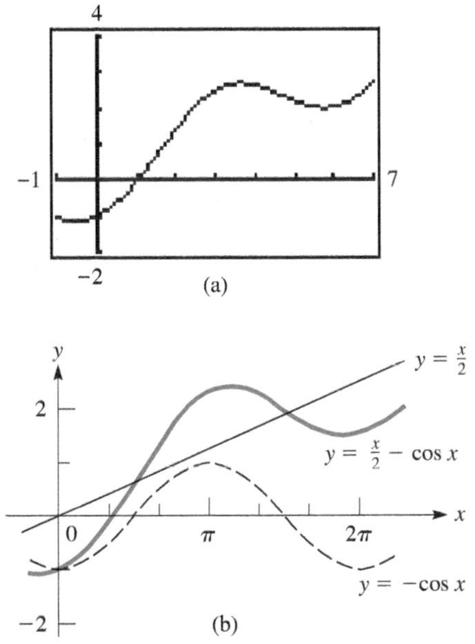

Fig. 10.41

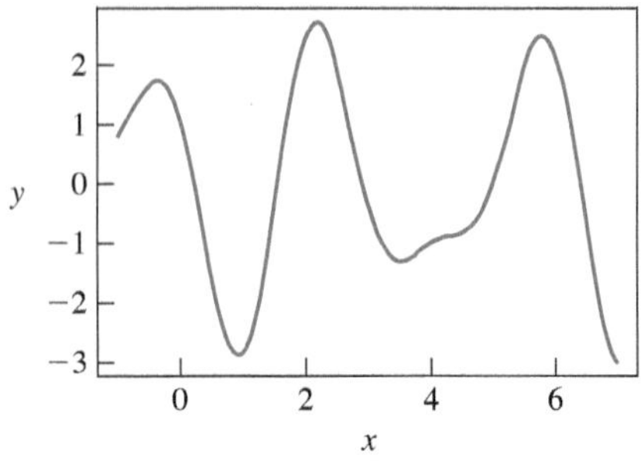

Fig. 10.42

EXAMPLE 3 Graphing utility graph of a composite curve

View the graph of $y = \cos \pi x - 2 \sin 2x$ with a graphing utility.

To combine $y = \cos \pi x$ and $y = 2 \sin 2x$, we make the following choices for the display settings:

Xmin $= -1$ (to start to the left of the y-axis)

Xmax $= 7$ (the periods are 2 and π; this shows at least two periods of each)

Ymin $= -3$, Ymax $= 3$ (the sum of the amplitudes is 3)

The graph obtained using the software R (www.r-project.org) is shown in Fig. 10.42. When using the same display settings, any graphing utility will produce a similar graph. Depending on how much of the curve is to be viewed, many possible choices for Xmin and Xmax are possible. However, the graph will always be contained between $y = -3$ and $y = 3$.

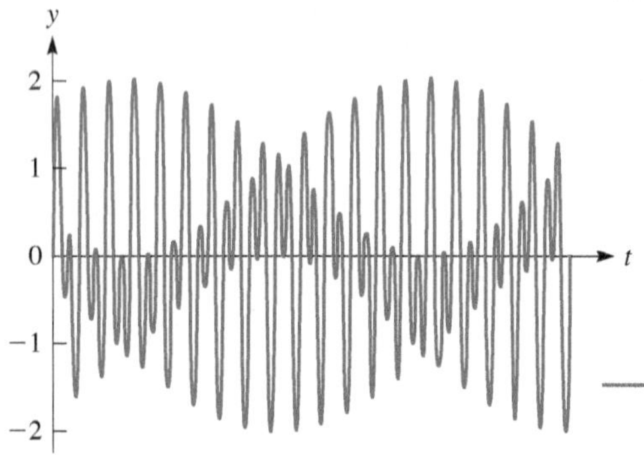

Fig. 10.43

EXAMPLE 4 Composite trigonometric curves—touch tone telephones

When a key is pressed, a touch tone phone produces a dual-tone multi-frequency signal (DTMF) that uniquely identifies the key. This signal is the combination of two sinusoidal signals at different frequencies (hence the name). For $0 \leq t \leq 0.03$ s, view the graph of $y = \sin(2\pi(1336t)) + \sin(2\pi(697t))$, the signal produced by the "2" key.

The graph is difficult to construct by addition of ordinates, so it is best to use a graphing utility. The sum of the amplitudes is 2, so we have used Ymin $= -2$, Ymax $= 2$ for the display settings. The graph is shown is Fig. 10.43.

LISSAJOUS FIGURES

An important application of trigonometric curves is made when they are added at *right angles.* The methods for doing this are shown in the following examples.

EXAMPLE 5 Graphing parametric equations

Plot the graph for which the values of x and y are given by the equations $y = \sin 2\pi t$ and $x = 2 \cos \pi t$. *Equations given in this form—x and y in terms of a third variable—are called* **parametric equations**.

Since both x and y are in terms of t, by assuming values of t, we find corresponding values of x and y and use these values to plot the graph. Since the periods of $\sin 2\pi t$ and $2 \cos \pi t$ are $t = 1$ and $t = 2$, respectively, we will use values of $t = 0, 1/4, 1/2, 3/4, 1$, and so on. These give us convenient values of $0, \pi/4, \pi/2, 3\pi/4, \pi$, and so on to use in the table. We plot the points in Fig. 10.44.

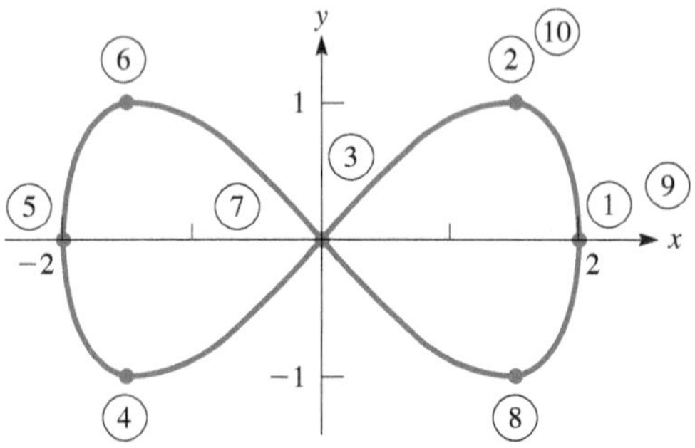

Fig. 10.44

t	0	$\frac{1}{4}$	$\frac{1}{2}$	$\frac{3}{4}$	1	$\frac{5}{4}$	$\frac{3}{2}$	$\frac{7}{4}$	2	$\frac{9}{4}$
x	2	1.4	0	−1.4	−2	−1.4	0	1.4	2	1.4
y	0	1	0	−1	0	1	0	−1	0	1
Point number	1	2	3	4	5	6	7	8	9	10

Since x and y are trigonometric functions of a third variable t, and since the x-axis is at right angles to the y-axis, values of x and y obtained in this way result in a combination of two trigonometric curves at right angles. *Figures obtained in this way are called*

■ Lissajous figures are named for the French physicist Jules Lissajous (1822–1880).

Lissajous figures. Note that the Lissajous figure in Fig. 10.44 *is not a function* since there are *two* values of y for each value of x (except $x = -2, 0, 2$) in the domain.

■ See the chapter introduction.

In practice, Lissajous figures can be displayed on an *oscilloscope* by applying an electric signal between a pair of horizontal plates and another signal between a pair of vertical plates. These signals are then seen on the screen of the oscilloscope. This type of screen (a cathode-ray tube) is similar to that used on most older TV sets.

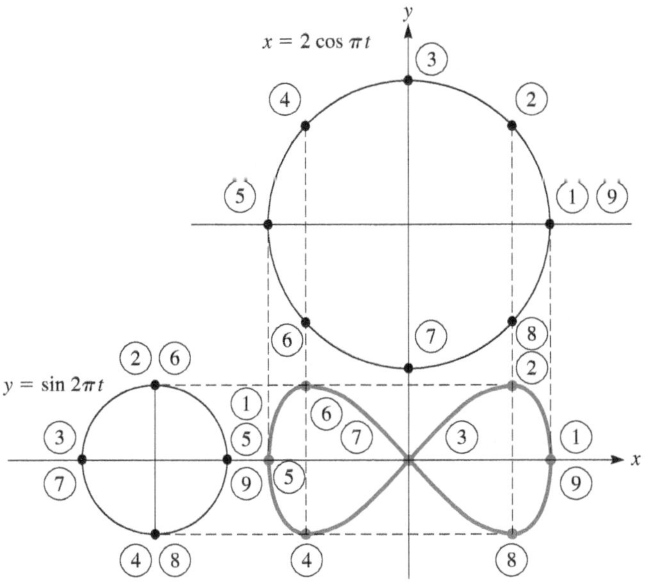

Fig. 10.45

EXAMPLE 6 Graphing Lissajous figures

If a circle is placed on the x-axis and another on the y-axis, we may represent the coordinates (x, y) for the curve of Example 5 by the lengths of the projections (see Example 1 of Section 10.5) of a point moving around each circle. A careful study of Fig. 10.45 will clarify this. We note that the radius of the circle giving the x-values is 2 and that the radius of the circle giving the y-values is 1. This is due to the way in which x and y are defined. Also, due to these definitions, the point revolves around the y-circle twice as fast as the corresponding point around the x-circle.

■ On an oscilloscope, the curve would result when two electric signals are used, with the first having twice the amplitude and half the frequency of the second.

Graphing utilities can be used to display a curve defined by parametric equations. In a graphing calculator, use the *mode* feature and select *parametric equations* as described in the calculator's manual.

EXAMPLE 7 Graphing calculator Lissajous figures

Use a graphing calculator to display the graph defined by the parametric equations $x = 3\sin(3\pi t)$ and $y = 2\sin(2\pi t + 1)$.

First, select the parametric equation option from the *mode* feature and enter the parametric equations $x_{1T} = 3\sin(3\pi t)$ and $y_{1T} = 2\sin(2\pi t + 1)$. Then make the following *window* settings:

Tmin = 0 (standard default settings, and the usual choice)
Tmax = 2 (the periods are 2/3 and 1; the smallest common multiple is 2)
Tstep = 0.01 (so as to obtain a smooth curve)
Xmin = −3, Xmax = 3 (smallest and largest possible values of x), Xscl = 1
Ymin = −2, Ymax = 2 (smallest and largest possible values of y), Yscl = 1

The calculator graph is shown in Fig. 10.46.

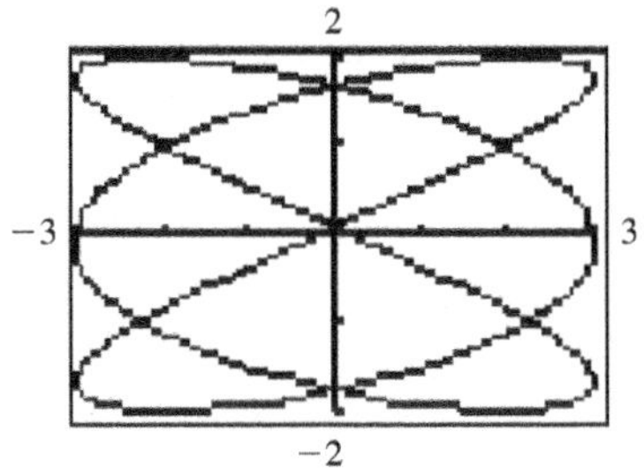

Fig. 10.46

EXERCISES 10.6

In Exercises 1–8, sketch the curves of the given functions by addition of ordinates.

1. $y = 1 + \sin x$
2. $y = 3 - 2\cos x$
3. $y = \frac{1}{3}x + \sin 2x$
4. $y = x - \sin x$
5. $y = \frac{1}{10}x^2 - \sin \pi x$
6. $y = \frac{1}{4}x^2 + \cos 3x$
7. $y = \sin x + \cos x$
8. $y = \sin x + \sin 2x$

In Exercises 9–20, display the graphs of the given functions with a graphing utility.

9. $y = x^3 + 10\sin 2x$
10. $y = \dfrac{1}{x^2 + 1} - \cos \pi x$
11. $y = \sin x - 1.5\sin 2x$
12. $y = \cos 3x - 3\sin x$

13. $y = 20\cos 2x + 30\sin x$

14. $y = \frac{1}{2}\sin 4x + \cos 2x$

15. $y = 2\sin x - \cos 1.5x$

16. $y = 8\sin 0.5x - 12\sin x$

17. $y = \sin \pi x - \cos 2x$

18. $y = 2\cos 4x - \cos\left(x - \frac{\pi}{4}\right)$

19. $y = 2\sin\left(2x - \frac{\pi}{6}\right) + \cos\left(2x + \frac{\pi}{3}\right)$

20. $y = 3\cos 2\pi x + \sin \frac{\pi}{2}x$

In Exercises 21–24, plot the Lissajous figures.

21. $x = 3\sin t,\ y = 2\sin t$

22. $x = 2\cos t,\ y = \cos(t + 4)$

23. $x = 2\cos 2\pi t,\ y = \sin \pi t$

24. $x = \cos\left(t + \frac{\pi}{4}\right),\ y = \sin 2t$

In Exercises 25–32, use a graphing utility to display the Lissajous figures.

25. $x = \cos \pi\left(t + \frac{1}{6}\right),\ y = 2\sin \pi t$

26. $x = \sin^2 \pi t,\ y = \cos \pi t$

27. $x = 2\cos 3t,\ y = \cos 2t$

28. $x = 2\sin \pi t,\ y = 3\sin 3\pi t$

29. $x = \sin(t + 1),\ y = \sin 5t$

30. $x = 5\cos t,\ y = 3\sin 5t$

31. $x = 2\cos \pi t,\ y = 3\sin\left(2\pi t - \frac{\pi}{4}\right)$

32. $x = 1.5\cos 3\pi t,\ y = 0.5\cos 5\pi t$

In Exercises 33–44, sketch the appropriate curves. Any graphing utility may be used.

33. An object oscillating on a spring has a displacement (in m) given by $y = 0.4\sin 4t + 0.3\cos 4t$, where t is the time (in s). Sketch the graph.

34. The voltage V in a certain electric circuit is given by $V = 50\sin 50\pi t + 80\sin 60\pi t$, where t is the time (in s). Sketch the graph.

35. An analysis of the temperature records for Montreal indicates that the average daily temperature T (in °C) during the year is approximately $T = 6 - 15\cos\left[\frac{\pi}{6}(x - 0.5)\right]$, where x is measured in months ($x = 0.5$ is Jan. 15, etc.). Sketch the graph of T vs. x for one year.

36. An analysis of data shows that the mean density d (in mg/cm^3) of a calcium compound in the bones of women is given by $d = 139.3 + 48.6\sin(0.0674x - 0.210)$, where x represents the ages of women ($20 \le x \le 80$ years). (A woman is considered to be osteoporotic if $d < 115$ mg/cm^3.) Sketch the graph.

37. A normal person with a pulse rate of 60 beats/min has a blood pressure of "120 over 80." This means the pressure is oscillating between a high (systolic) of 120 mm of mercury (shown as mmHg) and a low (diastolic) of 80 mmHg. Assuming a sinusoidal type of function, find the pressure p as a function of the time t if the initial pressure is 120 mmHg. Sketch the graph for the first 5 s.

38. The world's highest tides occur in the Bay of Fundy on the Atlantic coast of Canada. One day in August 2012, the first low tide at Minas Basin, Nova Scotia, occurred at 6:21 A.M. The water level at low tide was 1.8 m; later, at high tide, it was 14.9 m. The next low tide occurred at 6:45 P.M. Assuming a sinusoidal type of function, find the height of the water h (in m) as a function of time t (in hours since midnight). Sketch the graph for one day.

39. The electric current i (in mA) in a certain circuit is given by $i = 0.32 + 0.50\sin t - 0.20\cos 2t$, where t is in milliseconds. Sketch two cycles of i as a function of t.

40. The available solar energy depends on the amount of sunlight, and the available time in a day for sunlight depends on the time of the year. An approximate correction factor (in min) to standard time is $C = 10\sin\frac{1}{29}(n - 80) - 7.5\cos\frac{1}{58}(n - 80)$, where n is the number of the day of the year. Sketch C as a function of n.

41. Two signals are seen on an oscilloscope as being at right angles. The equations for the displacements of these signals are $x = 4\cos \pi t$ and $y = 2\sin 3\pi t$. Sketch the figure that appears on the oscilloscope.

42. In the study of optics, light is said to be *elliptically polarized* if certain optic vibrations are out of phase. These may be represented by Lissajous figures. Determine the Lissajous figure for two light waves given by $w_1 = \sin \omega t$ and $w_2 = \sin\left(\omega t + \frac{\pi}{4}\right)$.

43. In checking electric circuit elements, a *square wave* such as that shown in Fig. 10.47 may be displayed on an oscilloscope. Display the graph of $y = 1 + \frac{4}{\pi}\sin\left(\frac{\pi x}{4}\right) + \frac{4}{3\pi}\sin\left(\frac{3\pi x}{4}\right)$ with a graphing utility, and compare it with Fig. 10.47. This equation gives the first three terms of a *Fourier series*. As more terms of the series are added, the approximation to a square wave is better.

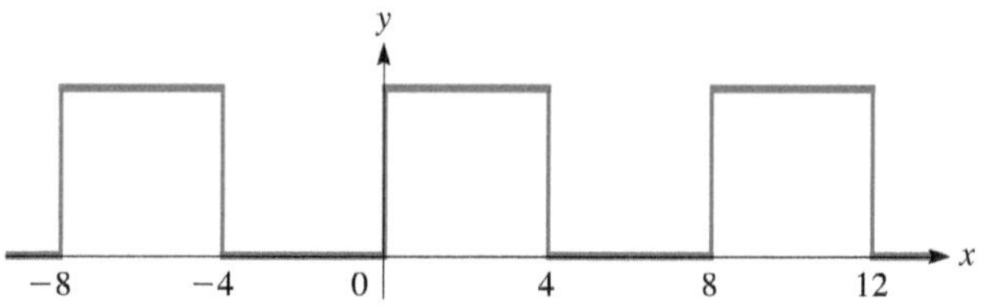

Fig. 10.47

44. Another type of display on an oscilloscope may be a *sawtooth wave* such as that shown in Fig. 10.48. Display the graph of $y = 1 - \frac{8}{\pi^2}\left(\cos\frac{\pi x}{2} + \frac{1}{9}\cos\frac{3\pi x}{2}\right)$ with a graphing utility and compare it with Fig. 10.48. See Exercise 43.

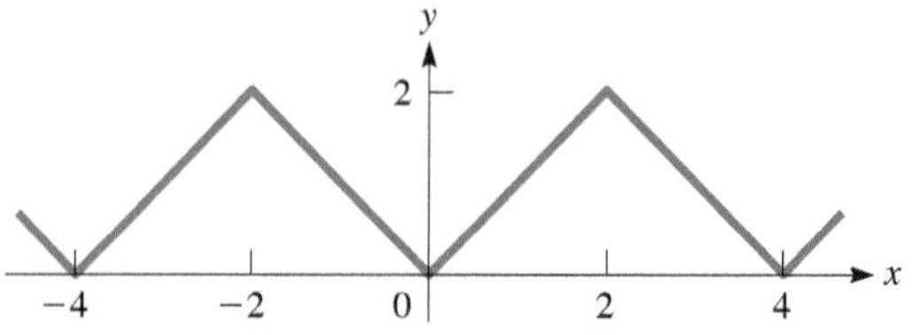

Fig. 10.48

CHAPTER 10 KEY FORMULAS AND EQUATIONS

For the graphs of $y = a\sin(bx + c)$ and $y = a\cos(bx + c)$ Amplitude $= |a|$ Period $= \dfrac{2\pi}{b}$ Displacement $= -\dfrac{c}{b}$ **(10.1)**

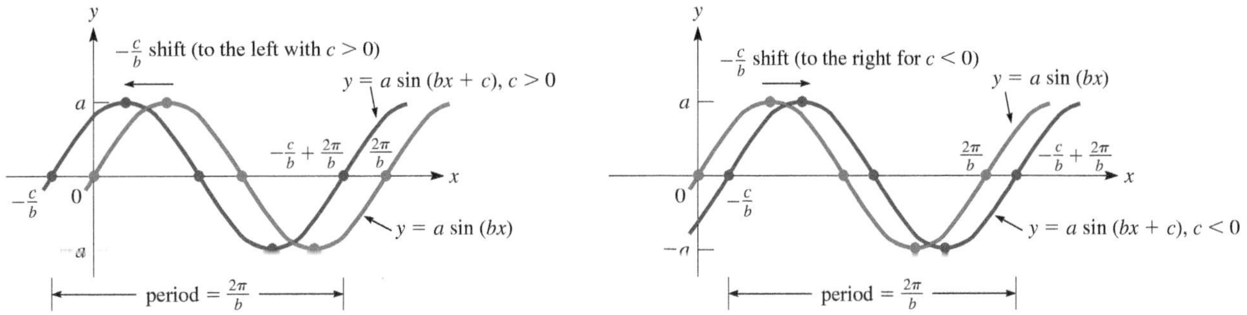

(a) (b)

Reciprocal relationships $\csc x = \dfrac{1}{\sin x}$ $\sec x = \dfrac{1}{\cos x}$ $\cot x = \dfrac{1}{\tan x}$ **(10.2)**

Simple harmonic motion $d = R\sin\omega t$ **(10.3)**

Angular velocity and frequency $\omega = 2\pi f$ **(10.4)**

CHAPTER 10 REVIEW EXERCISES

In Exercises 1–28, sketch the curves of the given trigonometric functions.

1. $y = \frac{2}{3}\sin x$ **2.** $y = -4\sin x$

3. $y = -2\cos x$ **4.** $y = 2.3\cos(-x)$

5. $y = 2\sin 3x$ **6.** $y = 4.5\sin 12x$

7. $y = 0.4\cos 4x$ **8.** $y = 24\cos 6x$

9. $y = 3\cos\frac{1}{3}x$ **10.** $y = 3\sin(-0.5x)$

11. $y = \sin\pi x$ **12.** $y = 36\sin 4\pi x$

13. $y = 5\cos\left(\frac{\pi x}{2}\right)$ **14.** $y = -\cos 6\pi x$

15. $y = -0.5\sin\left(-\frac{\pi x}{6}\right)$ **16.** $y = 8\sin\frac{\pi}{4}x$

17. $y = 2\sin\left(3x - \dfrac{\pi}{2}\right)$ **18.** $y = 3\sin\left(\dfrac{x}{2} + \dfrac{\pi}{2}\right)$

19. $y = -2\cos(4x + \pi)$ **20.** $y = 0.8\cos\left(\dfrac{x}{6} - \dfrac{\pi}{2}\right)$

21. $y = -\sin\left(\pi x + \dfrac{\pi}{6}\right)$ **22.** $y = 250\sin(3\pi x - \pi)$

23. $y = 8\cos\left(4\pi x - \dfrac{\pi}{2}\right)$ **24.** $y = 3\cos(2\pi x + \pi)$

25. $y = 0.3\tan 0.5x$ **26.** $y = \frac{1}{4}\sec x$

27. $y = -\frac{1}{3}\csc x$ **28.** $y = -5\cot\pi x$

In Exercises 29–32, sketch the curves of the given functions by addition of ordinates.

29. $y = 2 + \frac{1}{2}\sin 2x$ **30.** $y = \frac{1}{2}x - \cos\frac{1}{3}x$

31. $y = \sin 2x + 3\cos x$ **32.** $y = \sin 3x + 2\cos 2x$

In Exercises 33–40, display the curves of the given functions with a graphing utility.

33. $y = 2\sin x - \cos 2x$ **34.** $y = 10\sin 3x - 20\cos x$

35. $y = \cos\left(x + \dfrac{\pi}{4}\right) - 0.4\sin 2x$

36. $y = 2\cos\pi x + \cos(2\pi x - \pi)$

37. $y = \dfrac{\sin x}{x}$ **38.** $y = \sqrt{x}\,\sin 0.5x$

39. $y = \sin^2 x + \cos^2 x$ $(\sin^2 x = (\sin x)^2)$

What conclusion can be drawn from the graph?

40. $y = \sin\left(x + \dfrac{\pi}{4}\right) - \cos\left(x - \dfrac{\pi}{4}\right) + 1$

What conclusion can be drawn from the graph?

In Exercises 41–44, give the specific form of the indicated equation by evaluating a, b, and c through an inspection of the given curve.

41. $y = a\sin(bx + c)$ (Fig. 10.49) **42.** $y = a\cos(bx + c)$ (Fig. 10.49)

43. $y = a\cos(bx + c)$ (Fig. 10.50) **44.** $y = a\sin(bx + c)$ (Fig. 10.50)

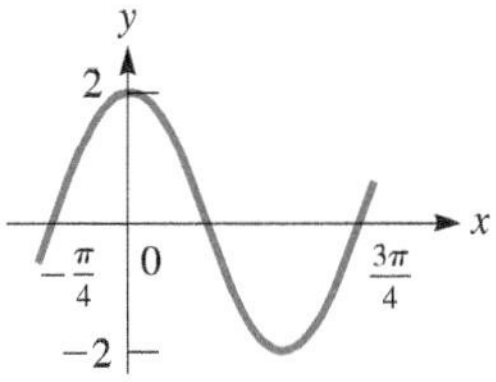

Fig. 10.49

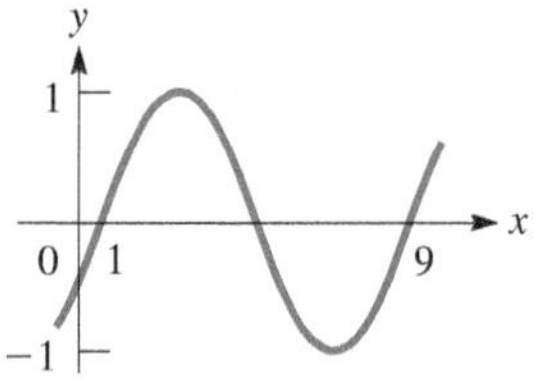

Fig. 10.50

In Exercises 45–48, display the Lissajous figures with a graphing utility.

45. $x = -\cos 2\pi t,\ y = 2\sin \pi t$

46. $x = \sin\left(t + \frac{\pi}{6}\right),\ y = \sin t$

47. $x = 2\cos\left(2\pi t + \frac{\pi}{4}\right),\ y = \cos \pi t$

48. $x = \cos\left(t - \frac{\pi}{6}\right),\ y = \cos\left(2t + \frac{\pi}{3}\right)$

In Exercises 49–60, solve the given problems.

49. Display the function $y = 2|\sin 0.2\pi x| - |\cos 0.4\pi x|$ with a graphing utility.

50. Display the function $y = 0.2|\tan 2x|$ with a graphing utility.

51. Show that $\cos\left(x + \frac{\pi}{4}\right) = \sin\left(\frac{\pi}{4} - x\right)$ with a graphing utility.

52. Show that $\tan\left(x - \frac{\pi}{3}\right) = -\tan\left(\frac{\pi}{3} - x\right)$ with a graphing utility.

53. What is the period of the function $y = 2\cos 0.5x + \sin 3x$?

54. What is the period of the function $y = \sin \pi x + 3\sin 0.25\pi x$?

55. Find the function and graph it, if it is of the form $y = a\sin x$ and passes through $(5\pi/2, 3)$.

56. Find the function and graph it, if it is of the form $y = a\cos x$ and passes through $(4\pi, -3)$.

57. Find the function and graph it, if it is of the form $y = 3\cos bx$ and passes through $(\pi/3, -3)$ and b has the smallest possible positive value.

58. Find the function and graph it, if it is of the form $y = 3\sin bx$ and passes through $(\pi/3, 0)$ and b has the smallest possible positive value.

59. Find the function and graph it for a function of the form $y = 3\sin(\pi x + c)$ that passes through $(-0.25, 0)$ and for which c has the smallest possible positive value.

60. Write the equation of the cosecant function with zero displacement and a period of 2, and that passes through $(0.5, 4)$.

In Exercises 61–82, sketch the appropriate curves. Any graphing utility may be used. In Exercise 83, answer the given question.

61. The range R of a rocket is given by $R = \dfrac{v_0^2 \sin 2\theta}{g}$. Sketch R as a function of θ for $v_0 = 1000$ m/s and $g = 9.8$ m/s^2. See Fig. 10.51.

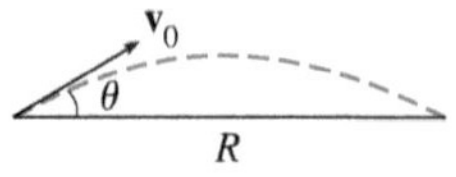

Fig. 10.51

62. The blade of a sabre saw moves vertically up and down at 18 strokes per second. The vertical displacement y (in cm) is given by $y = 1.2\sin 36\pi t$, where t is in seconds. Sketch at least two cycles of the graph of y vs. t.

63. The velocity v (in cm/s) of a piston in a certain engine is given by $v = \omega D\cos \omega t$, where ω is the angular velocity of the crankshaft in radians per second and t is the time in seconds. Sketch the graph of v vs. t if the engine is at 3000 r/min and $D = 3.6$ cm.

64. A light wave for the colour yellow can be represented by the equation $y = A\sin 3.4 \times 10^{15}\, t$. With A as a constant, sketch two cycles of y as a function of t (in s).

65. The electric current i (in A) in a circuit in which there is a *full-wave rectifier* is $i = 10\,|\sin 120\pi t|$. Sketch the graph of $i = f(t)$ for $0 \le t \le 0.05$ s. What is the period of the current?

66. A circular disc suspended by a thin wire attached to the centre of one of its flat faces is twisted through an angle θ. Torsion in the wire tends to turn the disc back in the opposite direction (thus, the name *torsion pendulum* is given to this device). The angular displacement θ (in rad) as a function of time t (in s) is $\theta = \theta_0\cos(\omega t + \alpha)$, where θ_0 is the maximum angular displacement, ω is a constant that depends on the properties of the disc and wire, and α is the phase angle. Sketch the graph of θ vs. t if $\theta_0 = 0.100$ rad, $\omega = 2.50$ rad/s, and $\alpha = \pi/4$. See Fig. 10.52.

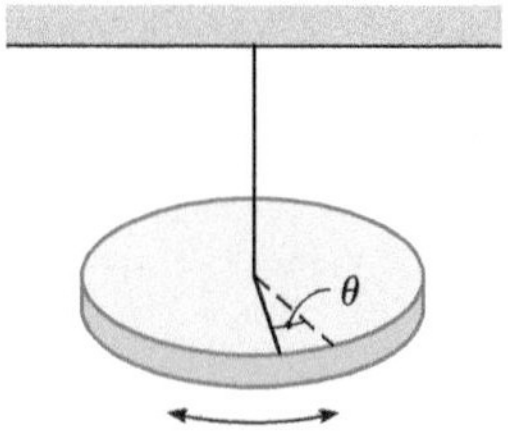

Fig. 10.52

67. The vertical displacement y of a point at the end of a propeller blade of a small boat is $y = 14.0\sin 40.0\pi t$. Sketch two cycles of y (in cm) as a function of t (in s).

68. In optics, two waves are said to interfere destructively if, when they both pass through a medium, the amplitude of the resulting wave is zero. Sketch the graph of $y = \sin x + \cos(x + \pi/2)$ and explain whether or not it would represent destructive interference of two waves.

69. The vertical displacement y (in dm) of a buoy floating in water is given by $y = 3.0\cos 0.2t + 1.0\sin 0.4t$, where t is in seconds. Sketch the graph of y as a function of t for the first 40 s.

70. The vertical motion of a rubber raft on a lake approximates simple harmonic motion due to the waves. If the amplitude of the motion is 0.250 m and the period is 3.00 s, find an equation for the vertical displacement y as a function of the time t. Sketch two cycles.

71. A drafting student draws a circle through the three vertices of a right triangle. The hypotenuse of the triangle is the diameter d of the circle, and from Fig. 10.53, we see that $d = a\sec\theta$. Sketch the graph of d as a function of θ for $a = 3.00$ cm.

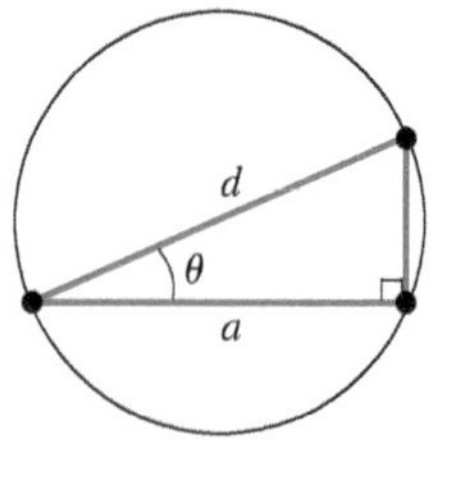

Fig. 10.53

72. The height h (in m) of a certain rocket ascending vertically is given by $h = 800\tan\theta$, where θ is the angle of elevation from an observer 800 m from the launch pad. Sketch h as a function of θ.

73. The number of hours h of daylight each day during the year in the city of Toronto, Ontario, can be approximated by $h = 12.2 + 3.3\sin\left(\frac{2\pi}{365}t - \frac{160\pi}{365}\right)$, where t is measured in days ($t = 15$ is January 15, etc.). Sketch the graph of h vs. t for one year.

74. The equation in Exercise 73 can be used to approximate the number of hours of daylight in Christchurch, New Zealand—at approximately the same latitude as Toronto, but in the southern hemisphere. Explain what change is necessary and determine the proper equation. Sketch the graph for one year.

75. If the upper end of a spring is not fixed and is being moved with a sinusoidal motion, the motion of the bob at the end of the spring is affected. Sketch the curve if the motion of the upper end of a spring is being moved by an external force and the bob moves according to the equation $y = 4 \sin 2t - 2 \cos 2t$.

76. The loudness L (in decibels) of a fire siren as a function of the time t (in s) is approximately $L = 40 - 35 \cos 2t + 60 \sin t$. Sketch this function for $0 \le t \le 10$ s.

77. The path of a roller mechanism used in an assembly-line process is given by $x = \theta - \sin \theta$ and $y = 1 - \cos \theta$. Sketch the path for $0 \le \theta \le 2\pi$.

78. The equations for two microwave signals that give a resulting curve on an oscilloscope are $x = 6 \sin \pi t$ and $y = 4 \cos 4\pi t$. Sketch the graph of the curve displayed on the oscilloscope.

79. The impedance Z (in Ω) and resistance R (in Ω) for an alternating-current circuit are related by $Z = R \sec \theta$, where θ is called the phase angle. Sketch the graph for Z as a function of θ for $-\pi/2 < \theta < \pi/2$.

80. For an object sliding down an inclined plane at constant speed, the coefficient of friction μ between the object and the plane is given by $\mu = \tan \theta$, where θ is the angle between the plane and the horizontal. Sketch the graph of μ vs. θ.

81. The charge q (in C) on a certain capacitor as a function of the time t (in s) is given by $q = 0.0003(3 - 2 \sin 100t \cos 100t)$. Sketch two cycles of q vs. t.

82. The instantaneous power P (in W) in an electric circuit is defined as the product of the instantaneous voltage V and the instantaneous current i (in A). If we have $V = 100 \cos 200t$ and $i = 2 \cos\left(200t + \frac{\pi}{4}\right)$, plot the graph V vs. t and the graph of i vs. t on the same coordinate system. Then sketch the graph of P vs. t by multiplying appropriate values of V and i.

83. A wave passing through a string can be described at any instant by the equation $y = a \sin(bx + c)$. Write one or two paragraphs explaining the change in the wave (a) if a is doubled, (b) if b is doubled, and (c) if c is doubled.

CHAPTER 10 **PRACTICE TEST**

1. Determine the amplitude, period, and displacement of the function $y = -3 \sin(4\pi x - \pi/3)$.

In Problems 2–5, sketch the graphs of the given functions.

2. $y = 0.5 \cos \frac{\pi}{2} x$

3. $y = 2 + 3 \sin x$

4. $y = 3 \sec x$

5. $y = 2 \sin\left(2x - \frac{\pi}{3}\right)$

6. A wave is travelling in a string. The displacement y (in cm) as a function of the time t (in s) from its equilibrium position is given by $y = A \cos(2\pi/T)t$. T is the period (in s) of the motion. If $A = 0.200$ cm and $T = 0.100$ s, sketch two cycles of y vs t.

7. Sketch the graph of $y = 2 \sin x + \cos 2x$ by addition of ordinates.

8. Use a graphing utility to display the Lissajous figure for which $x = \sin \pi t$ and $y = 2 \cos 2\pi t$.

9. Sketch two cycles of the curve of a projection of the end of a radius on the y-axis. The radius is of length R and it is rotating counterclockwise about the origin at 2.00 rad/s. It starts at an angle of $\pi/6$ with the positive x-axis.

10. Find the function of the form $y = 2 \sin bx$ if its graph passes through $(\pi/3, 2)$ and b is the smallest possible positive value. Then graph the function.

11. In Iqaluit, Nunavut, the number of daylight hours during the longest day of the year (June 21) is 20.82 hours. During the shortest day of the year (December 21), the number of daylight hours is 4.33 hours. Approximate the number of hours h of daylight each day during a year with a function of the form $h = a \sin(bt + c) + d$, where time is measured in days.

11. Exponents and Radicals

LEARNING OUTCOMES

After completion of this chapter, the student should be able to:

- Use the laws of exponents to simplify expressions
- Convert radicals to fractional exponents and vice versa
- Perform mathematical operations with exponents and radicals
- Simplify expressions containing radicals
- Rationalize a denominator containing radicals

▲ In finding the rate at which solar radiation changes at a solar-energy collector, the following expression is found:

$$\frac{(t^4 + 100)^{1/2} - 2t^3(t + 6)(t^4 + 100)^{-1/2}}{[(t^4 + 100)^{1/2}]^2}$$

In Section 11.2, we show that this can be written in a much simpler form.

Exponents and radicals are useful in many technical applications. To set the foundation to study these applications, further development of the connection between fractional exponents and radicals must be undertaken. For example, the concept $\sqrt{x} = x^{1/2}$ will be illustrated. The use of fractional exponents can be more convenient than the use of radicals in advanced mathematical operations.

As we noted in Chapter 6, the use of symbols led to advances in mathematics and science. As letters began to be used as algebraic symbols for numbers in the 1600s, it was common to write, for example, x^3 as xxx. For larger powers, this is obviously inconvenient, and the modern use of exponents came into use. The first to use exponents consistently was the French mathematician René Descartes in the 1630s.

The meaning of negative and fractional exponents was first defined by the English mathematician Wallis in the 1650s, although he did not write them as we do today. In the 1670s, it was the great English mathematician and physicist Isaac Newton who first used all exponents (positive, negative, and fractional) with modern notation. This improvement in notation made the development of many areas of mathematics, particularly calculus, easier. In this way, it also facilitated many advances in the applications of mathematics.

As we develop the various operations with exponents and radicals, we will show their uses in some technical areas of application. They are used in a number of formulas in areas such as electronics, hydrodynamics, optics, solar energy, and machine design.

11.1 Simplifying Expressions with Integral Exponents

The laws of exponents were given in Section 1.4. For reference, they are

Product Rule:	$a^m \times a^n = a^{m+n}$	**(11.1)**
Quotient Rule:	$\frac{a^m}{a^n} = a^{m-n} \quad (a \neq 0)$	**(11.2)**
Power of a Power Rule:	$(a^m)^n = a^{mn}$	**(11.3)**
Power of a Product Rule:	$(ab)^n = a^n b^n$	**(11.4)**
Power of a Quotient Rule:	$\left(\frac{a}{b}\right)^n = \frac{a^n}{b^n} \quad (\text{if } b \neq 0)$	**(11.5)**
Zero Exponent	$a^0 = 1 \quad (a \neq 0)$	**(11.6)**
Negative Exponent	$a^{-n} = \frac{1}{a^n} \quad (a \neq 0)$	**(11.7)**

These equations can be used in many different combinations, and are valid for all integral exponents. Later in this chapter, we will show how these rules apply to fractional exponents.

EXAMPLE 1 Simplifying basic expressions

(a) Simplify $a^3 \times a^{-5}$.

$$a^3 \times a^{-5} = a^{3-5} = a^{-2} \quad \text{Eq. (11.1)}$$
$$= \frac{1}{a^2} \quad \text{Eq. (11.7)}$$

(b) Simplify $(10^3 \times 10^{-4})^2$.

$$(10^3 \times 10^{-4})^2 = (10^{3-4})^2 = (10^{-1})^2 \quad \text{Eq. (11.1)}$$
$$= 10^{(-1)(2)} = 10^{-2} \quad \text{Eq. (11.3)}$$
$$= \frac{1}{10^2} = \frac{1}{100} \quad \text{Eq. (11.7)}$$

Note that negative exponents are generally not used in the final solution, unless specified otherwise. Also, the result is in proper form in either of the two formats in blue. When the exponent is large, it is common to leave the exponent in the answer.

LEARNING TIP

Although negative exponents are not often used in the final result, if ***scientific notation is being used, then the form with the negative exponent is indicated.*** For example,

$$0.00625 \text{ m} = 6.25 \times \frac{1}{1000} \text{ m} = 6.25 \times \frac{1}{10^3} \text{ m} = 6.25 \times 10^{-3} \text{ m}$$

EXAMPLE 2 Simplifying basic expressions with multiple variables

(a) Simplify $\frac{a^2b^3c^0}{ab^7}$.

$$\frac{a^2b^3c^0}{ab^7} = a^{2-1}b^{3-7}c^0 \quad \text{Eq. (11.2)}$$
$$= a^1b^{-4} \quad \text{Eq. (11.6)}$$
$$= \frac{a}{b^4} \quad \text{Eq. (11.7)}$$

(b) Simplify $(x^{-2}y)^3$.

$$(x^{-2}y)^3 = (x^{-2})^3(y)^3 \quad \text{Eq. (11.4)}$$
$$= x^{-6}y^3 \quad \text{Eq. (11.3)}$$
$$= \frac{y^3}{x^6} \quad \text{Eq. (11.7)}$$

Often, *several different combinations of Eqs. (11.1)–(11.7) can be used to simplify an expression.* This is illustrated in the next example.

EXAMPLE 3 Simplification done in different ways

Simplify $(x^2y)^2\left(\frac{2}{x}\right)^{-2}$ by two different methods.

(a) $(x^2y)^2\left(\frac{2}{x}\right)^{-2} = \frac{(x^2y)^2}{\left(\frac{2}{x}\right)^2}$ Eq. (11.7)

$= \frac{x^{2\cdot 2}y^{1\cdot 2}}{\frac{2^2}{x^2}}$ Eqs. (11.4), (11.5)

$= \frac{x^4y^2}{1} \times \frac{x^2}{4}$ invert

$= \frac{x^6y^2}{4}$ Eq. (11.1)

(b) $(x^2y)^2\left(\frac{2}{x}\right)^{-2} = (x^{2\cdot 2}y^{1\cdot 2})\left(\frac{2^{1(-2)}}{x^{1(-2)}}\right)$ Eqs. (11.4), (11.5)

$= (x^4y^2)\left(\frac{x^2}{2^2}\right)$ Eq. (11.7)

$= \frac{x^{4+2}y^2}{4}$ Eq. (11.1)

$= \frac{x^6y^2}{4}$

Practice Exercise

1. Simplify: $x\left(\frac{3}{x}\right)^{-3}$

When writing a denominate number, if units of measurement appear in the denominator they can be written using negative exponents.

■ The unit *pascal* is named for the French mathematician and scientist Blaise Pascal (1623–1662).

■ The unit *joule* is named for the English physicist James Prescott Joule (1818–1899).

EXAMPLE 4 Exponents and units of measurement

(a) The metric unit for pressure is the *pascal*, where $1\text{ Pa} = 1\text{ N/m}^2$. This can be written as

$$1\text{ Pa} = 1\text{ N/m}^2 = 1\text{ N}\cdot\text{m}^{-2}$$

where $1/\text{m}^2 = \text{m}^{-2}$.

(b) The metric unit for energy is the *joule*, where $1\text{ J} = 1\text{ kg}\cdot(\text{m}\cdot\text{s}^{-1})^2$, or

$$1\text{ J} = 1\text{ kg}\cdot\text{m}^2/\text{s}^2 = 1\text{ kg}\cdot\text{m}^2\cdot\text{s}^{-2}$$

LEARNING TIP

The special case 0^0 requires its own discussion. For many purposes, 0^0 can be *defined* to have a definite value. For example, some series notations and the binomial theorem (discussed in Chapter 19) require that $0^0 = 1$. For the power rule of differential calculus (discussed in Chapter 23), it is required that $0^0 = 0$. But in evaluating some limits (limits are also discussed in Chapter 23), 0^0 can be treated as indeterminate (it has no definite value). It depends on the mathematical context.

Care must be taken to apply the laws of exponents properly. Certain common problems are pointed out in the following examples.

EXAMPLE 5 Order of operations with exponents

Simplify the following pairs of frequently confused statements.

(a) $(-5x)^0 = 1$ **(b)** $-5x^0 = -5\cdot 1 = -5$

(c) $(-2)^2 = (-2)(-2) = 4$ **(d)** $-2^2 = -1(2^2) = -1(4) = -4$

In parts (a) and (c), the brackets indicate that the whole expression is raised to the power outside the bracket. In parts (b) and (d), the power affects only the value immediately before it; it does not affect the coefficients of that value.

Practice Exercise

2. Simplify: $3/x^0$

COMMON ERROR

Order of operations requires exponents to be performed before multiplication. A common error is to apply the exponent to all multiplied factors instead of just the one to which the exponent is applied.

$$2x^{-1} = \frac{2}{x} \quad \textbf{not} \quad 2x^{-1} \neq \frac{1}{2x}$$

All factors must be raised to the exponent for it to be applied to all:

$$(2x)^{-1} = \frac{1}{2x}$$

EXAMPLE 6 Simplification done in different ways

Simplify $(2a + b^{-1})^{-2}$ to an expression without negative exponents by two different methods.

(a)
$$(2a + b^{-1})^{-2} = \frac{1}{(2a + b^{-1})^2} \quad \text{Eq. (11.7)}$$
$$= \frac{1}{\left(2a + \frac{1}{b}\right)^2} \quad \text{Eq. (11.7)}$$
$$= \frac{1}{\left(\frac{2ab + 1}{b}\right)^2} \quad \text{use LCD}$$
$$= \frac{1}{\frac{(2ab + 1)^2}{b^2}} \quad \text{Eq. (11.5)}$$
$$= \frac{b^2}{(2ab + 1)^2} \quad \text{invert}$$

(b)
$$(2a + b^{-1})^{-2} = \left(2a + \frac{1}{b}\right)^{-2} \quad \text{Eq. (11.7)}$$
$$= \left(\frac{2ab + 1}{b}\right)^{-2} \quad \text{use LCD}$$
$$= \frac{(2ab + 1)^{-2}}{b^{-2}} \quad \text{Eq. (11.5)}$$
$$= \frac{b^2}{(2ab + 1)^2} \quad \text{Eq. (11.7)}$$

COMMON ERROR

A common error occurs when simplifying an expression in which multiple terms are raised to an exponent. Remember that

$(2a + b^{-1})^{-2}$ **is *not* equal to** $(2a)^{-2} + (b^{-1})^{-2}$ or $\frac{1}{4a^2} + b^2$

As in Section 6.1, *when raising any multinomial to a power, we* **cannot** *simply raise each term to the power applied to obtain the result.*

Remember, when raising a product of factors to a power, we use the equation $(ab)^n = a^n b^n$. Thus,

$$(2ab^{-1})^{-2} = (2a)^{-2}(b^{-1})^{-2} = \frac{b^2}{(2a)^2} = \frac{b^2}{4a^2}$$

We see that we must be careful to distinguish between the power of a sum of terms and the power of a product of factors.

LEARNING TIP

When a factor is moved from the denominator to the numerator of a fraction (or vice versa), the **sign** *of the exponent is changed.* We should carefully note the word *factor;* this rule does not apply to moving *terms* in the numerator or the denominator.

EXAMPLE 7 Factor moved from numerator to denominator

(a) $$3L^{-1} - (2L)^{-2} = \frac{3}{L} - \frac{1}{(2L)^2} = \frac{3}{L} - \frac{1}{4L^2}$$

$$= \frac{12L - 1}{4L^2}$$

(b) $$3^{-1}\left(\frac{4^{-2}}{3 - 3^{-1}}\right) = \frac{1}{3}\left(\frac{1}{4^2}\right)\left(\frac{1}{3 - \frac{1}{3}}\right) = \frac{1}{3 \times 4^2}\left(\frac{1}{\frac{9-1}{3}}\right)$$

$$= \frac{1}{3 \times 4^2}\left(\frac{3}{8}\right) = \frac{1}{128}$$

Practice Exercise

3. Simplify: $(3a)^{-1} - 3a^{-2}$

EXAMPLE 8 Simplifying complex expressions

terms

$$\frac{1}{x^{-1}}\left(\frac{x^{-1} - y^{-1}}{x^2 - y^2}\right) = \frac{x}{1}\left(\frac{\frac{1}{x} - \frac{1}{y}}{x^2 - y^2}\right) = x\left(\frac{\frac{y-x}{xy}}{x^2 - y^2}\right)$$

$$= \frac{\frac{x(y-x)}{xy}}{(x-y)(x+y)}$$

$$= \frac{x(y-x)}{xy} \times \frac{1}{(x-y)(x+y)}$$

$$= \frac{x(y-x)}{xy(x-y)(x+y)} = \frac{-(x-y)}{y(x-y)(x+y)}$$

$$= -\frac{1}{y(x+y)}$$

Note that in this example, ***the x^{-1} and y^{-1} in the numerator could not be moved directly to the denominator with positive exponents*** because they are only ***terms*** of the original numerator.

EXAMPLE 9 Simplifying an expression from calculus

In finding out the rate at which a quantity is changing, it may be necessary to simplify an expression found by using the advanced mathematics of calculus. Simplify the following expression, which is derived using calculus.

$$3(x+4)^2(x-3)^{-2} - 2(x-3)^{-3}(x+4)^3$$

$$= \frac{3(x+4)^2}{(x-3)^2} - \frac{2(x+4)^3}{(x-3)^3}$$

$$= \frac{3(x-3)(x+4)^2 - 2(x+4)^3}{(x-3)^3}$$

$$= \frac{(x+4)^2[3(x-3) - 2(x+4)]}{(x-3)^3}$$

$$= \frac{(x+4)^2(x-17)}{(x-3)^3}$$

EXERCISES 11.1

In Exercises 1–4, solve the resulting problems if the given changes are made in the indicated examples of this section.

1. In Example 3, change the factor x^2 to x^{-2} and then find the result.
2. In Example 6, change the term $2a$ to $2a^{-1}$ and then find the result.
3. In Example 7(b), change the 3^{-1} in the denominator to 3^{-2} and then find the result.
4. In Example 8, change the sign in the numerator from − to + and then find the result.

In Exercises 5–58, express each of the given expressions in simplest form with only positive exponents.

5. x^9x^{-5}
6. y^0y^{-4}
7. $2a^2a^{-6}$
8. $5ss^{-5}$
9. $\frac{c^7}{c^{-2}}$
10. $\frac{t^{-8}}{t^{-3}}$
11. $\frac{x^5}{y^{-2}}$
12. $\frac{n^{-6}}{m^{-4}}$
13. $5^0 \times 5^{-3}$
14. $(3^2 \times 4^{-3})^3$
15. $(3\pi x^{-3})^2$
16. $(4x^2y^{-5})^4$
17. $2(5an^{-2})^{-1}$
18. $4(6s^2t^{-1})^{-2}$
19. $(-4)^0$
20. -4^0
21. $-7x^0$
22. $(-7x)^0$
23. $6x^{-4}$
24. $(2x^3)^{-4}$
25. $(7a^{-1}x)^{-3}$
26. $7a^{-1}x^{-3}$
27. $\left(\frac{2a}{b^{-3}}\right)^{-4}$
28. $\left(\frac{10n^{-6}}{D^{-3}}\right)^{-3}$
29. $\left(\frac{2}{n^3}\right)^{-3}$
30. $\left(\frac{3}{x^3}\right)^{-2}$
31. $3\left(\frac{a}{b^{-2}}\right)^{-3}$
32. $5\left(\frac{2n^{-2}}{D^{-1}}\right)^{-2}$
33. $(a+b)^{-1}$
34. $a^{-1}+b^{-1}$
35. $3x^{-2}+2y^{-2}$
36. $(3x+2y)^{-2}$
37. $(2a^{-n})^2\left(\frac{3}{2a^n}\right)^{-1}$
38. $(7 \times 3^{-a})\left(\frac{3^a}{7}\right)^2$
39. $\left(\frac{3a^2}{4b}\right)^{-3}\left(\frac{4}{a}\right)^{-5}$
40. $(2np^{-2})^{-2}(4^{-1}p^2)^{-1}$
41. $\left(\frac{V^{-1}}{2t}\right)^{-2}\left(\frac{t^2}{V^{-2}}\right)^{-3}$
42. $ab\left(\frac{a^{-2}}{b^2}\right)^{-3}\left(\frac{a^{-3}}{b^5}\right)^2$
43. $2a^{-2}+(2a^{-2})^4$
44. $3(a^{-1}z^2)^{-3}+c^{-2}z^{-1}$
45. $2 \times 3^{-1} + 4 \times 3^{-2}$
46. $5 \times 2^{-2} - 3^{-1} \times 2^3$
47. $(R_1^{-1}+R_2^{-1})^{-1}$
48. $2(2a-b^{-2})^{-1}$
49. $(n^{-2}-2n^{-1})^2$
50. $2(2^{-3}-4^{-1})^{-2}$
51. $\frac{6^{-1}}{4^{-2}+2}$
52. $\frac{x-y^{-1}}{x^{-1}-y}$
53. $\frac{x^{-2}-y^{-2}}{x^{-1}-y^{-1}}$
54. $\frac{ax^{-2}+a^{-2}x}{a^{-1}+x^{-1}}$
55. $2t^{-2}+t^{-1}(t+1)$
56. $3x^{-1}-x^{-3}(y+2)$
57. $(D-1)^{-1}+(D+1)^{-1}$
58. $4(2x-1)(x+2)^{-1}-(2x-1)^2(x+2)^{-2}$

In Exercises 59–79, solve the given problems.

59. If $x < 0$, is it ever true that $x^{-2} < x^{-1}$?
60. Is it true that $(a+b)^0 = 1$ for all values of a and b?
61. Express $4^2 \times 64$ (a) as a power of 4 and (b) as a power of 2.
62. Express $1/81$ (a) as a power of 9 and (b) as a power of 3.
63. (a) By use of Eqs. (11.4), (11.5), and (11.7), show that
$$\left(\frac{a}{b}\right)^{-n} = \left(\frac{b}{a}\right)^n$$
(b) Verify the equation in part (a) by evaluating each side with $a = 3.576$, $b = 8.091$, and $n = 7$.
64. Evaluate $(8^{19})^{12}/(8^{16})^{14}$. What happens when you try to evaluate this on a calculator?
65. Is it true that $[-2^0 - (-1)^0]^0 = 1$?
66. Is it true that, if $x \neq 0$, $[(-x^{-2})^{-2}]^{-2} = 1/x^2$?
67. If $f(x) = 4^x$, find $f(a+2)$.
68. If $f(x) = 2(9^x)$, find $f(2-a)$.
69. Solve for x: $2^{5x} = 2^7(2^{2x})^2$.
70. In analysing the tuning of an electronic circuit, the expression $[\omega\omega_0^{-1} - \omega_0\omega^{-1}]^2$ is used. Expand and simplify this expression.
71. The metric unit of energy, the *joule* (J), can be expressed as $kg \cdot s^{-2} \cdot m^2$. Simplify these units and include *newtons* (see Section 1.3) and only positive exponents in the final result.
72. The units for the electric quantity called *permittivity* are $C^2 \cdot N^{-1} \cdot m^{-2}$. Given that $1\ F = 1\ C^2 \cdot J^{-1}$, show that the units of permittivity are F/m. See Section 1.3.
73. When studying a solar energy system, the units encountered are $kg \cdot s^{-1}(m \cdot s^{-2})^2$. Simplify these units and include *joules* (see Example 4) and only positive exponents in the final result.
74. The metric units for the velocity v of an object are $m \cdot s^{-1}$, and the units for the acceleration a of the object are $m \cdot s^{-2}$. What are the units for v/a?
75. Given that $v = a^pt^r$, where v is the velocity of an object, a is its acceleration, and t is the time, use the metric units given in Exercise 74 to show that $p = r = 1$.
76. An idealized model of the thermodynamic process in a gasoline engine is the *Otto cycle*. The efficiency η of the process is
$$\eta = \frac{\frac{T_1r^\gamma}{r} - \frac{T_2r^\gamma}{r} - T_1 + T_2}{\frac{T_1r^\gamma}{r} - \frac{T_2r^\gamma}{r}}.\quad \text{Show that } \eta = 1 - \frac{1}{r^{\gamma-1}}.$$
77. An expression encountered in finance is
$$\frac{p(1+i)^{-1}[(1+i)^{-n}-1]}{(1+i)^{-1}-1}$$
where n is an integer. Simplify this expression.
78. In optics, the combined focal length F of two lenses is given by $F = [f_1^{-1} + f_2^{-1} + d(f_1f_2)^{-1}]^{-1}$, where f_1 and f_2 are the focal lengths of the lenses and d is the distance between them. Simplify the right side of this equation.
79. Solve for x: $3^{4x} = 3^{10}(3^{2x})^3$

Answers to Practice Exercises

1. $\frac{x^4}{27}$ 2. 3 3. $\frac{a-9}{3a^2}$

11.2 Fractional Exponents

In Section 11.1, we reviewed the use of integral exponents, including exponents that are negative integers and zero. We now show how rational numbers may be used as exponents. With the appropriate definitions, all the laws of exponents are valid for all rational numbers as exponents.

Eq. (11.3) states that $(a^m)^n = a^{mn}$. If we were to let $m = \frac{1}{2}$ and $n = 2$, we would have $(a^{1/2})^2 = a^1$. However, we already have a way of writing a quantity that, when squared, equals a. This is written as $\sqrt{a}$. To be consistent with previous definitions and to allow the laws of exponents to hold, we define

$$a^{1/n} = \sqrt[n]{a} \qquad \textbf{(11.8)}$$

In order that Eqs. (11.3) and (11.8) may hold at the same time, we define

$$a^{m/n} = \sqrt[n]{a^m} = \left(\sqrt[n]{a}\right)^m \qquad \textbf{(11.9)}$$

These definitions are valid for all the laws of exponents. We must note that Eqs. (11.8) and (11.9) are valid as long as $\sqrt[n]{a}$ does not involve the even root of a negative number. Such numbers are imaginary and are considered in Chapter 12.

COMMON ERROR

The radical $\sqrt[n]{a}$ is the **nth root** of the number a. Do not confuse it with $n\sqrt{a}$, which is **n multiplied by the root** of the number a.

■ For reference, Eq. (11.1) is $a^m \times a^n = a^{m+n}$.

■ Eq. (11.3) is $(a^m)^n = a^{mn}$.

LEARNING TIP

We may interpret $a^{m/n}$ as the mth power of the nth root of a, as well as the nth root of the mth power of a, or

$$a^{m/n} = \left(a^{1/n}\right)^m = \left(\sqrt[n]{a}\right)^m$$

or

$$a^{m/n} = (a^m)^{1/n} = \sqrt[n]{a^m}$$

EXAMPLE 1 Meaning of a fractional exponent

Verify that Eqs. (11.1) and (11.3) hold for the above definitions if $n = 4$.

$$a^{1/4}a^{1/4}a^{1/4}a^{1/4} = a^{(1/4)+(1/4)+(1/4)+(1/4)} = a^1$$

Now, $a^{1/4} = \sqrt[4]{a}$ by definition. Also, by definition, $\sqrt[4]{a}\,\sqrt[4]{a}\,\sqrt[4]{a}\,\sqrt[4]{a} = a$. Eq. (11.1) is thereby verified.

Eq. (11.3) is verified by the following:

$$(a^{1/4})(a^{1/4})(a^{1/4})(a^{1/4}) = (a^{1/4})^4 = a^1 = \left(\sqrt[4]{a}\right)^4$$

EXAMPLE 2 Interpretation of a fractional exponent

$$8^{2/3} = \left(\sqrt[3]{8}\right)^2 = (2)^2 = 4 \quad \text{or} \quad 8^{2/3} = \sqrt[3]{8^2} = \sqrt[3]{64} = 4$$

Although both interpretations of Eq. (11.9) are possible, as indicated in Example 2, in evaluating numerical expressions involving fractional exponents without a calculator, it is almost always best to *find the root first, as indicated by the denominator* of the fractional exponent. This allows us to find the power of the smaller number, which is normally easier to find.

EXAMPLE 3 Order of operations for fractional exponents

To evaluate $(64)^{5/2}$, we should proceed as follows:

$$(64)^{5/2} = \left[(64)^{1/2}\right]^5 = 8^5 = 32\,768$$

If we raised 64 to the fifth power first, we would have

$$(64)^{5/2} = (64^5)^{1/2} = (1\ 073\ 741\ 824)^{1/2} = 32\ 768$$

Although both methods give the same result, it is preferable to find the indicated root first in order to avoid working with large values.

EXAMPLE 4 Evaluating fractional exponents

(a) $(16)^{3/4} = (16^{1/4})^3 = 2^3 = 8$

(b) $4^{-1/2} = \frac{1}{4^{1/2}} = \frac{1}{2}$ **(c)** $9^{3/2} = (9^{1/2})^3 = 3^3 = 27$

We note in (b) that Eq. (11.7) must also hold for negative rational exponents. In writing $4^{-1/2}$ as $\frac{1}{4^{1/2}}$, the *sign* of the exponent is changed.

Fractional exponents allow us to find roots of numbers on a calculator. By use of the appropriate key (on most calculators x^y or ∧), we may raise any positive number to any power. For roots, we use the equivalent fractional exponent. Powers that are fractions or decimal in form are entered directly.

Practice Exercises

Evaluate: **1.** $81^{5/4}$ **2.** $27^{-4/3}$

EXAMPLE 5 Fractional exponents—thermodynamic temperature

■ The unit *kelvin* is named for the Scottish physicist Lord Kelvin (1824–1907).

The thermodynamic temperature T (in kelvin (K)) is related to the pressure p (in kPa) of a gas by the equation $T = 80.5p^{2/7}$. Find the value of T for $p = 750$ kPa.

Substituting, we have

$$T = 80.5(750)^{2/7} = 534 \text{ K}$$

When finding powers of negative numbers, some calculators and computer software will show an error. If this is the case, enter the positive value of the number and then enter a negative sign for the result when appropriate. From Section 1.6, we recall that *an even root of a negative number is imaginary and an odd root of a negative number is negative.* If it is an integral power, the basic laws of signs are used.

EXAMPLE 6 Fractional exponents and graphs

Plot the graph of the function $y = 2x^{1/3}$.

In obtaining points for the graph, we use $(1/3)$ as the power when using the calculator. If your calculator does not evaluate powers of negative numbers, enter positive values for the negative values of x and then make the results negative. Since $x^{1/3} = \sqrt[3]{x}$, we know that $x^{1/3}$ is negative for negative values of x because we have an *odd* root of a negative number. We get the following table of values:

x	-3	-2	-1	0	1	2	3
y	-2.9	-2.5	-2.0	0	2.0	2.5	2.9

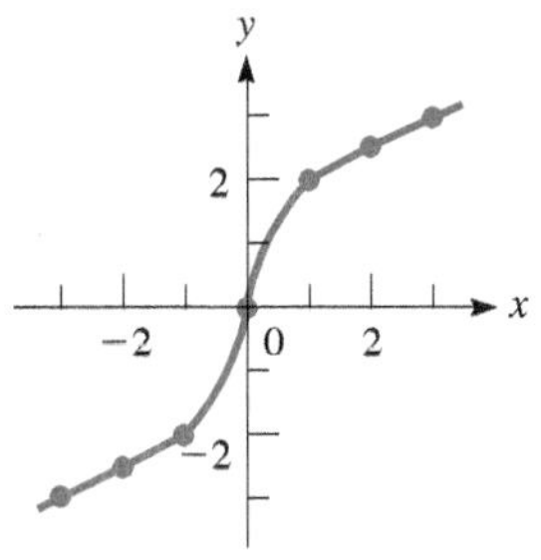

Fig. 11.1

The graph is shown in Fig. 11.1. Of course, this curve can easily be shown on a graphing utility.

Fractional exponents are often easier to use in more complex expressions involving roots. This is true in algebra and in topics from more advanced mathematics. Any expression with radicals can also be expressed with fractional exponents and then simplified.

EXAMPLE 7 Simplifying expressions with fractional exponents

(a) $(8a^2b^4)^{1/3} = [(8^{1/3})(a^2)^{1/3}(b^4)^{1/3}]$ using $(ab)^n = a^nb^n$

$= 2a^{2/3}b^{4/3}$ using $a^{1/n} = \sqrt[n]{a}$ and $(a^m)^n = a^{mn}$

(b) $a^{3/4}a^{4/5} = a^{(3/4)+(4/5)} = a^{31/20}$ using $a^m \times a^n = a^{m+n}$

(c) $\left(\frac{4^{-3/2}x^{2/3}y^{-7/4}}{2^{3/2}x^{-1/3}y^{3/4}}\right)^{2/3} = \left(\frac{x^{2/3}x^{1/3}}{2^{3/2}4^{3/2}y^{3/4}y^{7/4}}\right)^{2/3}$ using $a^{-n} = \frac{1}{a^n}$

$= \left(\frac{x^{(2/3)+(1/3)}}{2^{3/2}4^{3/2}y^{(3/4)+(7/4)}}\right)^{2/3}$ using $a^m \times a^n = a^{m+n}$

$= \frac{x^{(1)(2/3)}}{2^{(3/2)(2/3)}4^{(3/2)(2/3)}y^{(10/4)(2/3)}}$ using $(ab)^n = a^nb^n$

$= \frac{x^{2/3}}{8y^{5/3}}$

(d) $(4x^4)^{-1/2} - 3x^{-3} = \frac{1}{(4x^4)^{1/2}} - \frac{3}{x^3}$ using $a^{-n} = \frac{1}{a^n}$

$= \frac{1}{2x^2} - \frac{3}{x^3} = \frac{x-6}{2x^3}$ using $a^{1/n} = \sqrt[n]{a}$; common denominator

Practice Exercises

Simplify: **3.** $(64a^2c^3)^{2/3}$
4. $2x^{-4} - (8x^3)^{-1/3}$

It is common to write fractional exponents as radicals as the last step in the simplification of expressions. We will see how this is done in the next section.

EXAMPLE 8 Simplifying an expression—solar radiation

■ See the chapter introduction.

■ Solar collectors supply the power for most space satellites.

During a day, the rate R at which the radiation changes at a solar-radiation collector is

$$R = \frac{(t^4+100)^{1/2} - 2t^3(t+6)(t^4+100)^{-1/2}}{[(t^4+100)^{1/2}]^2}$$

Here, R is measured in $kW/(m^2 \cdot h)$, t is the number of hours from noon, and $-6\text{ h} \le t \le 8\text{ h}$. Express the right side of this equation in simpler form and find R for $t = 0$ (noon) and for $t = 4$ h (4 P.M.).

Performing the simplification, we have the following steps:

$$R = \frac{(t^4+100)^{1/2} - \frac{2t^3(t+6)}{(t^4+100)^{1/2}}}{(t^4+100)}$$

← using $a^{-n} = \frac{1}{a^n}$
← using $(a^m)^n = a^{mn}$

$$R = \frac{\frac{(t^4+100)^{1/2}(t^4+100)^{1/2} - 2t^3(t+6)}{(t^4+100)^{1/2}}}{(t^4+100)}$$

← common denominator

$$R = \frac{(t^4+100) - 2t^3(t+6)}{(t^4+100)^{1/2}} \times \frac{1}{t^4+100}$$

← invert divisor and multiply

$$R = \frac{100 - 12t^3 - t^4}{(t^4+100)^{1/2}(t^4+100)} = \frac{100 - 12t^3 - t^4}{(t^4+100)^{3/2}}$$

← using $a^m \times a^n = a^{m+n}$

For $t = 0$: $R = \frac{100 - 12(0^3) - 0^4}{(0^4+100)^{3/2}} = 0.100\ kW/(m^2 \cdot h)$

For $t = 4$ h: $R = \frac{100 - 12(4^3) - 4^4}{(4^4+100)^{3/2}} = -0.138\ kW/(m^2 \cdot h)$

This shows that the radiation is increasing at noon and decreasing at 4 P.M.

EXERCISES 11.2

In Exercises 1–4, solve the resulting problems if the given changes are made in the indicated examples of this section.

1. In Example 2, change the exponent to $4/3$.
2. In Example 4(b), change the exponent to $-3/2$.
3. In Example 6, change the exponent to $-1/3$.
4. In Example 7(d), change the exponent $-1/2$ to $-3/2$.

In Exercises 5–28, evaluate the given expressions.

5. $36^{1/2}$ 6. $216^{1/3}$ 7. $81^{1/4}$ 8. $125^{2/3}$

9. $100^{25/2}$ 10. $-16^{5/4}$ 11. $8^{-1/3}$ 12. $16^{-1/4}$

13. $64^{-2/3}$ 14. $-32^{-4/5}$ 15. $5^{1/6}5^{8/6}$ 16. $(4^5)^{4/3}$

17. $(3^6)^{2/3}$ 18. $\dfrac{121^{-1/2}}{100^{1/2}}$ 19. $\dfrac{1000^{1/3}}{-400^{-1/2}}$ 20. $\dfrac{-7^{-1/2}}{6^{-1}7^{1/2}}$

21. $\dfrac{21^{3/5}}{3^2 21^{-1/5}}$ 22. $\dfrac{(-64)^{1/3}}{5}$ 23. $\dfrac{(-8)^{2/3}}{-2}$ 24. $\dfrac{-4^{-1/2}}{(-64)^{-2/3}}$

25. $125^{-2/3} - 100^{-3/2}$ 26. $32^{0.4} + 25^{-0.5}$

27. $\dfrac{16^{-0.2}}{3} + \dfrac{2^{-0.4}}{2^{0.6}}$ 28. $\dfrac{3^{-1}}{49^{-1/2}} - \dfrac{4^{-1/2}}{4^{1/2}}$

In Exercises 29–32, use a calculator to evaluate each expression.

29. $17.98^{1/4}$ 30. $(-750.81)^{2/3}$

31. $4.0187^{-4/9}$ 32. $0.1863^{-1/6}$

In Exercises 33–56, simplify the given expressions. Express all answers with positive exponents.

33. $B^{4/5}B^{2/3}$ 34. $x^{3/5}x^{-1/4}$ 35. $\dfrac{y^{-1/2}}{-y^{2/5}}$

36. $\dfrac{s^{1/4}s^{2/3}}{s^{-1}}$ 37. $\dfrac{x^{3/10}}{x^{-1/5}x^2}$ 38. $\dfrac{R^{-2/5}R^2}{R^{-3/10}}$

39. $(8a^3b^6)^{1/3}$ 40. $(8b^{-4}c^2)^{2/3}$

41. $(16a^4b^3)^{-3/4}$ 42. $(32C^5D^4)^{-2/5}$

43. $\left(\dfrac{a^{5/7}}{a^{2/3}}\right)^{7/4}$ 44. $\left(\dfrac{4a^{5/6}b^{-1/5}}{a^{2/3}b^2}\right)^{-1/2}$

45. $\dfrac{1}{2}(4x^2+1)^{-1/2}(8x)$ 46. $\dfrac{(x+x^{1/2})(x-x^{1/2})}{x}$

47. $\dfrac{y^{3/8}(y^{5/8}-y^{13/8})}{y^{1/2}(y^{1/2}-y^{-1/2})}$ 48. $\dfrac{3^{-1}a^{1/2}}{4^{-1/2}b} \div \dfrac{9^{1/2}a^{-1/3}}{2b^{-1/4}}$

49. $(T^{-1}+2T^{-2})^{-1/2}$ 50. $(a^{-2}-a^{-4})^{-1/4}$

51. $(a^3)^{-4/3} + a^{-2}$ 52. $(4N^6)^{-1/2} - 2N^{-1}$

53. $[(a^{1/2}-a^{-1/2})^2+4]^{1/2}$ 54. $4x^{1/2} + \dfrac{1}{2}x^{-1/2}(4x+1)$

55. $x^2(2x-1)^{-1/2} + 2x(2x-1)^{1/2}$

56. $(3n-1)^{-2/3}(1-n) - (3n-1)^{1/3}$

In Exercises 57–60, graph the given functions.

57. $f(x) = 3x^{1/2}$ 58. $f(x) = 2x^{2/3}$

59. $f(t) = t^{-4/5}$ 60. $f(V) = 4V^{3/2}$

In Exercises 61–69, perform the indicated operations.

61. Simplify $(x^{n-1} \div x^{n-3})^{1/3}$ and express the result as a radical.

62. (a) Simplify $(x^2-4x+4)^{1/2}$. (b) For what values of x is your answer in part (a) valid? Explain.

63. A factor used in determining the performance of a solar-energy storage system is $(A/S)^{-1/4}$, where A is the actual storage capacity and S is a standard storage capacity. If this factor is 0.5, explain how to find the ratio A/S.

64. A factor used in measuring the loudness sensed by the human ear is $(I/I_0)^{0.3}$, where I is the intensity of the sound and I_0 is a reference intensity. Evaluate this factor for $I = 3.20 \times 10^{-6}\ \text{W/m}^2$ (ordinary conversation) and $I_0 = 10^{-12}\ \text{W/m}^2$.

65. The period T of a satellite circling Earth is given by $T^2 = kR^3\left(1+\dfrac{d}{R}\right)^3$, where R is the radius of Earth, d is the distance of the satellite above Earth, and k is a constant. Solve for R, using fractional exponents in the result.

66. The withdrawal resistance R of a nail of diameter d indicates its holding power. One formula for R is $R = k(sg)^{5/2}dh$, where k is a constant, sg is the specific gravity of the wood, and h is the depth of the nail in the wood. Solve for sg using fractional exponents in the result.

67. The electric current i (in A) in a circuit with a battery of voltage E, a resistance R, and an inductance L is $i = \dfrac{E}{R}(1-e^{-Rt/L})$, where t is the time after the circuit is closed. See Fig. 11.2. Find i for $E = 6.20$ V, $R = 1.20\ \Omega$, $L = 3.24$ H, and $t = 0.001\ 00$ s. (The number e is an irrational constant and can be found from the calculator.)

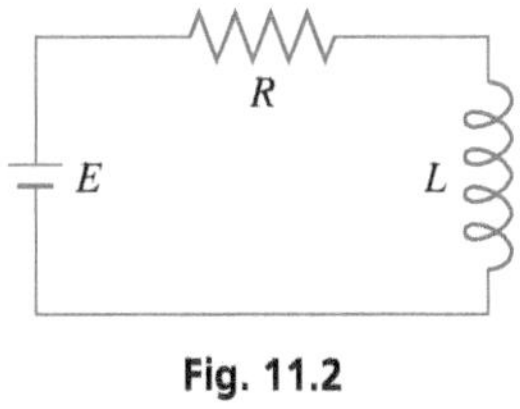

Fig. 11.2

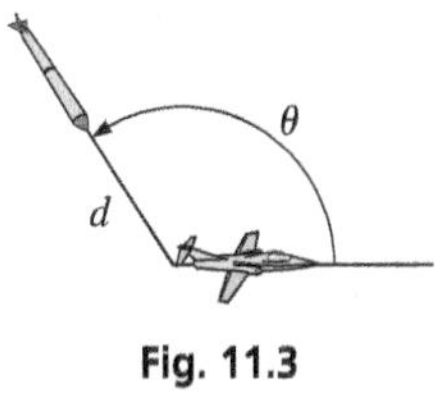

Fig. 11.3

68. For a heat-seeking rocket in pursuit of an aircraft, the distance d (in km) from the rocket to the aircraft is $d = \dfrac{500(\sin\theta)^{1/2}}{(1-\cos\theta)^{3/2}}$, where θ is shown in Fig. 11.3. Find d for $\theta = 125.0°$.

69. Carbon-14 has a half-life of approximately 5730 years. If A_0 is the original amount present, then the amount present after t years is given by $A(t) = A_0 2^{-t/5730}$. If 55.0 mg is present initially, how much is present after 3250 years?

Answers to Practice Exercises

1. 243 2. $\dfrac{1}{81}$ 3. $16a^{4/3}c^2$ 4. $\dfrac{4-x^3}{2x^4}$

11.3 Simplest Radical Form

As we have said, any expression with radicals can also be expressed with fractional exponents. For adding or subtracting radicals, there is little advantage to changing form, but with multiplication or division of radicals, fractional exponents have some advantages. Therefore, we now define operations with radicals so that they are consistent with the laws of exponents. This will let us use the form that is more convenient for the operation being performed.

Operations with Radicals		EXAMPLE 1
$\sqrt[n]{a^n} = (\sqrt[n]{a})^n = a$	**(11.10)**	$\sqrt[5]{4^5} = (4^{1/5})^5 = 4$
$\sqrt[n]{a}\sqrt[n]{b} = \sqrt[n]{ab}$	**(11.11)**	$\sqrt[3]{2}\sqrt[3]{3} = \sqrt[3]{2 \times 3} = \sqrt[3]{6}$
$\sqrt[m]{\sqrt[n]{a}} = \sqrt[mn]{a}$	**(11.12)**	$\sqrt[3]{\sqrt{5}} = \sqrt[3 \cdot 2]{5} = \sqrt[6]{5}$
$\dfrac{\sqrt[n]{a}}{\sqrt[n]{b}} = \sqrt[n]{\dfrac{a}{b}}$ where $(b \neq 0)$	**(11.13)**	$\dfrac{\sqrt{7}}{\sqrt{3}} = \sqrt{\dfrac{7}{3}}$

The number under the radical is called the **radicand**, *and the number indicating the root being taken is called the* **order** *(or* **index***) of the radical.* To avoid difficulties with imaginary numbers (which are considered in the next chapter), *we will assume that all letters represent positive numbers.*

EXAMPLE 2 Root of a sum and root of a product

It is important to distinguish between the root of a sum and the root of a product.

The root of a sum: $\sqrt{16 + 9} = \sqrt{25} = 5$, but $\sqrt{16 + 9} \neq \sqrt{16} + \sqrt{9}$

The root of a product: $\sqrt{16 \times 9} = \sqrt{144} = 12$, and $\sqrt{16 \times 9} = \sqrt{16} \times \sqrt{9}$

COMMON ERROR

Because multiplication of factors inside radicals allows separation into two factors in separate radicals $\sqrt[n]{ab} = \sqrt[n]{a} \cdot \sqrt[n]{b}$, it is a common error to apply this procedure to addition or subtraction terms inside radicals. Remember that $\sqrt[n]{a + b} \neq \sqrt[n]{a} + \sqrt[n]{b}$. ***Be careful to distinguish between the root of a sum of terms and the root of a product of factors.***

LEARNING TIP

Factors of the form r^n (perfect *n*th power factors) should always be identified and factored from the radicand of a radical of order *n*. By writing $(r^n)^{1/n} = r^{n/n} = r$, the factor r^n is factored as the root *r outside the radical.*

EXAMPLE 3 Removing perfect square factors

Simplify $\sqrt{75}$.

We know that $75 = (25)(3)$ and that $\sqrt{25} = 5$ (25 is a perfect square). As in Section 1.6 and now using Eq. (11.11), we write

$$\sqrt{75} = \sqrt{(25)(3)} = \sqrt{25}\sqrt{3} = 5\sqrt{3}$$

This illustrates how factoring can be used in simplifying radicals.

EXAMPLE 4 Removing perfect *n*th-power factors

(a) $\sqrt{72} = \sqrt{(36)(2)} = \sqrt{36}\sqrt{2} = 6\sqrt{2}$

(36 is a perfect square)

(b) $\sqrt{a^3b^2} = \sqrt{(a^2)(a)(b^2)} = \sqrt{a^2}\sqrt{a}\sqrt{b^2} = ab\sqrt{a}$

(a^2 and b^2 are perfect squares)

(c) cube root → $\sqrt[3]{40} = \sqrt[3]{(8)(5)} = \sqrt[3]{8}\sqrt[3]{5} = 2\sqrt[3]{5}$

(8 is a perfect cube)

(d) fifth root → $\sqrt[5]{64x^8y^{12}} = \sqrt[5]{(32)(2)(x^5)(x^3)(y^{10})(y^2)}$

(32, x^5, y^{10} are perfect fifth powers)

$= \sqrt[5]{(32)(x^5)(y^{10})}\sqrt[5]{2x^3y^2}$

$= 2xy^2\sqrt[5]{2x^3y^2}$

Practice Exercises

Simplify: **1.** $\sqrt{8x^5}$ **2.** $\sqrt[3]{16a^7b^2}$

The next example illustrates another procedure used to simplify radicals. It is to *reduce the order of the radical,* when it is possible to do so.

EXAMPLE 5 Reducing the order of a radical using fractional exponents

(a) $\sqrt[6]{8} = (8)^{1/6} = (2^3)^{1/6} = 2^{3/6} = 2^{1/2} = \sqrt{2}$

(b) $\sqrt[8]{16} = (16)^{1/8} = (2^4)^{1/8} = 2^{4/8} = 2^{1/2} = \sqrt{2}$

(c) $\dfrac{\sqrt[4]{9}}{\sqrt{3}} = \dfrac{9^{1/4}}{3^{1/2}} = \dfrac{(3^2)^{1/4}}{3^{1/2}} = \dfrac{3^{1/2}}{3^{1/2}} = 1$

(d) $\dfrac{\sqrt[6]{8}}{\sqrt{7}} = \dfrac{(2^3)^{1/6}}{7^{1/2}} = \dfrac{2^{1/2}}{7^{1/2}} = \sqrt{\dfrac{2}{7}}$

(e) $\sqrt{81a^{10}b^8} = \sqrt{3^4}\sqrt{a^{10}}\sqrt{b^8} = 3^{4/2}a^{10/2}b^{8/2} = 9a^5b^4$

(f) $\sqrt[9]{27x^6y^{12}} = \sqrt[9]{3^3}\sqrt[9]{x^6}\sqrt[9]{y^{12}} = 3^{3/9}x^{6/9}y^{12/9} = 3^{1/3}x^{2/3}y^{4/3}$

$= 3^{1/3}x^{2/3}y^1y^{1/3} = y(3x^2y)^{1/3} = y\sqrt[3]{3x^2y}$

If a radical is to be written in its *simplest form,* the two operations illustrated in the last four examples must be performed. Therefore, we have the following:

Steps to Reduce a Radical to Simplest Form

1. Remove all perfect *n*th-power factors from a radical of order *n*.
2. If possible, reduce the order of the radical using fractional exponents where necessary.

RATIONALIZING THE DENOMINATOR

The simplest form of a fraction with radicals is typically considered to be a fraction that contains no radicals in the denominator. The procedure of writing a fraction in this form is called **rationalizing the denominator**. A fraction with a monomial term in the denominator is rationalized by multiplying both the numerator and the denominator by the term that transforms the radicand in the denominator into a perfect nth power, which can then be simplified without a radical.

EXAMPLE 6 Rationalizing a denominator with a square root

Simplify: **(a)** $\sqrt{\frac{2}{5}}$ **(b)** $\frac{5}{\sqrt{18}}$

(a) The radicand in the denominator will be transformed into a perfect square if we multiply numerator and denominator by $\sqrt{5}$. We get

$$\frac{\sqrt{2}}{\sqrt{5}} = \frac{\sqrt{2}(\sqrt{5})}{\sqrt{5}(\sqrt{5})} = \frac{\sqrt{10}}{\sqrt{25}} = \frac{\sqrt{10}}{5}$$

(b) Before rationalizing, we first take out perfect square factors from the radicand. Therefore,

$$\frac{5}{\sqrt{18}} = \frac{5}{\sqrt{3^2 \cdot 2}} = \frac{5}{3\sqrt{2}} = \frac{5(\sqrt{2})}{3\sqrt{2}(\sqrt{2})} = \frac{5\sqrt{2}}{3\sqrt{2^2}} = \frac{5\sqrt{2}}{3 \cdot 2} = \frac{5\sqrt{2}}{6}$$

EXAMPLE 7 Rationalizing a denominator with a cube root

Simplify the cube roots: **(a)** $\sqrt[3]{\frac{2}{3}}$ **(b)** $\sqrt[3]{\frac{4000}{1029}}$

(a) The radicand in the denominator will be transformed into a perfect *cube* if we multiply numerator and denominator by $\sqrt[3]{3^2}$. We get

$$\sqrt[3]{\frac{2}{3}} = \frac{\sqrt[3]{2}(\sqrt[3]{3^2})}{\sqrt[3]{3}(\sqrt[3]{3^2})} = \frac{\sqrt[3]{2 \cdot 3^2}}{\sqrt[3]{3^3}} = \frac{\sqrt[3]{18}}{3}$$

(b) Before rationalizing, we first take out perfect *cube* factors from the radicand. Therefore,

$$\sqrt[3]{\frac{4000}{1029}} = \frac{\sqrt[3]{4 \cdot 10^3}}{\sqrt[3]{3 \cdot 7^3}} = \frac{10\sqrt[3]{4}}{7\sqrt[3]{3}} = \frac{10\sqrt[3]{4}(\sqrt[3]{3^2})}{7\sqrt[3]{3}(\sqrt[3]{3^2})} = \frac{10\sqrt[3]{4 \cdot 3^2}}{7\sqrt[3]{3^3}} = \frac{10\sqrt[3]{36}}{21}$$

Practice Exercise

Rationalize the denominator: **3.** $\sqrt{\frac{a}{3b}}$

EXAMPLE 8 Rationalizing an expression—pendulum period

The period T (in s) for one cycle of a simple pendulum is given by $T = 2\pi\sqrt{L/g}$, where L is the length of the pendulum and g is the acceleration due to gravity. Rationalize the denominator on the right side of this equation.

$$2\pi\sqrt{\frac{L}{g}} = \frac{2\pi\sqrt{L}\sqrt{g}}{\sqrt{g}\sqrt{g}} = \frac{2\pi\sqrt{Lg}}{g}$$

Therefore, the equation is rewritten with a rationalized denominator as $T = \frac{2\pi\sqrt{Lg}}{g}$.

EXAMPLE 9 Simplifying an expression with multiple terms in a radical

Simplify $\sqrt{\dfrac{1}{2a^2} + 2b^{-2}}$ and rationalize the denominator.

$$\sqrt{\frac{1}{2a^2} + 2b^{-2}} = \sqrt{\frac{1}{2a^2} + \frac{2}{b^2}} \qquad \text{positive exponent, Eq. (11.7)}$$

$$= \sqrt{\frac{b^2 + 4a^2}{2a^2b^2}} \qquad \text{lowest common denominator}$$

$$= \frac{\sqrt{b^2 + 4a^2}}{ab\sqrt{2}} \qquad \text{split the radical (Eq. 11.13) and take out perfect squares}$$

$$= \frac{\sqrt{b^2 + 4a^2}\sqrt{2}}{ab\sqrt{2}\sqrt{2}} = \frac{\sqrt{2(b^2 + 4a^2)}}{2ab} \qquad \text{rationalize the denominator}$$

Since the sum of squares cannot be factored, we cannot simplify any further.

EXERCISES 11.3

In Exercises 1–4, simplify the resulting expressions if the given changes are made in the indicated examples of this section.

1. In Example 4(b), change the exponent of b to 4 and then find the resulting expression.
2. In Example 5(d), change the 8 to 27 and then find the resulting expression.
3. In Example 7(a), change the root to a fourth root and then find the resulting expression.
4. In Example 9, replace b^{-2} with b^{-4} and then find the resulting expression.

In Exercises 5–66, write each expression in simplest radical form. If a radical appears in the denominator, rationalize the denominator.

5. $\sqrt{28}$
6. $\sqrt{216}$
7. $\sqrt{72}$
8. $\sqrt{162}$
9. $\sqrt{x^4y^7}$
10. $\sqrt{pq^5r^3}$
11. $\sqrt{x^4y^2z^5}$
12. $\sqrt{15a^3b^8}$
13. $\sqrt{54R^3T^6V^3}$
14. $\sqrt{63M^2N^9}$
15. $\sqrt[3]{16}$
16. $\sqrt[4]{48}$
17. $\sqrt[5]{96}$
18. $\sqrt[3]{-512}$
19. $\sqrt[3]{8a^2}$
20. $\sqrt[3]{5a^4b^2}$
21. $\sqrt[4]{64r^3s^4t^5}$
22. $\sqrt[5]{16x^5y^3z^{11}}$
23. $\sqrt[5]{8}\sqrt[5]{-4}$
24. $\sqrt[7]{4}\sqrt[7]{64}$
25. $\sqrt[3]{P}\sqrt[3]{P^2V}$
26. $\sqrt[6]{3m^5n^8}\sqrt[6]{9mn}$
27. $\dfrac{2}{\sqrt{3}}$
28. $\dfrac{1}{\sqrt{2}}$
29. $\sqrt{\dfrac{3}{2}}$
30. $\sqrt{\dfrac{11}{12}}$
31. $\sqrt[3]{\dfrac{9}{12}}$
32. $\sqrt[4]{\dfrac{2}{25}}$
33. $\sqrt[5]{\dfrac{1}{9}}$
34. $\sqrt[6]{\dfrac{5}{4}}$
35. $\sqrt[4]{400}$
36. $\sqrt[8]{81}$
37. $\sqrt[6]{64}$
38. $\sqrt[9]{27}$
39. $\sqrt{4 \times 10^4}$
40. $\sqrt{4 \times 10^5}$
41. $\sqrt{4 \times 10^6}$
42. $\sqrt[3]{16 \times 10^5}$
43. $\sqrt[4]{4a^2}$
44. $\sqrt[6]{b^2c^4}$
45. $\sqrt[4]{\dfrac{1}{4}}$
46. $\dfrac{\sqrt[4]{320}}{\sqrt[4]{5}}$
47. $\sqrt[4]{\sqrt[3]{16}}$
48. $\sqrt[5]{\sqrt[4]{9}}$
49. $\sqrt{\sqrt{\sqrt{n}}}$
50. $\sqrt{b^4\sqrt{a}}$
51. $\sqrt{28u^3v^{-5}}$
52. $\sqrt{98x^6y^{-7}}$
53. $\sqrt{64 + 144}$
54. $\sqrt{9 + 81}$
55. $\sqrt{\dfrac{2x}{3c^4}}$
56. $\sqrt{\dfrac{n}{m^3}}$
57. $\sqrt{\dfrac{5}{4} - \dfrac{1}{8}}$
58. $\sqrt{\dfrac{1}{a^2} + \dfrac{1}{b}}$
59. $\sqrt{xy^{-1} + x^{-1}y}$
60. $\sqrt{x^2 + 4^{-1}}$
61. $\sqrt{\dfrac{C-2}{C+2}}$
62. $\sqrt{a^2 + 2ab + b^2}$
63. $\sqrt{a^2 + b^2}$
64. $\sqrt{4x^2 - 1}$
65. $\sqrt{9x^2 - 6x + 1}$
66. $\sqrt{\dfrac{1}{2} + 2r + 2r^2}$

In Exercises 67–76, perform the required operation.

67. Change to radicals of the same order: $\sqrt{a}$; $\sqrt[3]{b}$; $\sqrt[6]{c}$.
68. Change to radicals of the same order: $3x^{2/3}$; $2y^{1/2}$; $(5z)^{1/4}$.
69. Display the graphs of $y_1 = \sqrt{x+2}$ and $y_2 = \sqrt{x} + \sqrt{2}$ on a graphing utility to show that $\sqrt{x+2}$ is not equal to $\sqrt{x} + \sqrt{2}$.
70. The *escape velocity* required to overcome the gravitational pull of a planet or star is $v_{esc} = \sqrt{\dfrac{2GM}{R}}$. Write this equation with a rationalized denominator.
71. In designing musical instruments, the equation $f = \sqrt{\dfrac{T}{4\mu L^2}}$ arises for the frequency of vibration of strings. Write this equation with a rationalized denominator.
72. An approximate equation for the efficiency η (in percent) of an engine is $\eta = 100\left(1 - 1/\sqrt[5]{R^2}\right)$, where R is the compression ratio. Explain how this equation can be written with fractional exponents and then find η for $R = 7.35$.
73. When analysing the velocity of an object falling through a great distance, the expression $a\sqrt{2g/a}$ arises. Show by rationalizing the denominator that this expression takes on a simpler form.

74. According to the Doppler effect, the observed frequency of light f_o when the observer is moving away with a velocity v from the light source emitting a signal of frequency f is given by $f_o = \frac{fc}{c-v}\sqrt{1-\left(\frac{v}{c}\right)^2}$, where c is the speed of light. Simplify the right side of the equation, leaving a radical in the denominator.

75. In analysing an electronic filter circuit, the expression $\frac{8A}{\pi^2\sqrt{1+(f_0/f)^2}}$ is used. Rationalize the denominator, expressing the answer without the fraction f_0/f.

76. The expression $\frac{1}{2L}\sqrt{R^2-\frac{4L}{C}}$ occurs in the study of electric circuits. Simplify this expression by combining terms under the radical and rationalizing the denominator.

Answers to Practice Exercises

1. $2x^2\sqrt{2x}$ **2.** $2a^2\sqrt[3]{2ab^2}$ **3.** $\frac{\sqrt{3ab}}{3b}$

11.4 Addition and Subtraction of Radicals

When we add or subtract algebraic expressions, we combine similar terms, those that differ only in numerical coefficients. This is true when adding or subtracting radicals. *The radicals must be similar to perform the addition*, rather than simply indicate the addition. *Radicals are* **similar** *if they differ only in their numerical coefficients.* This means they must have the same order and have the same radicand.

EXAMPLE 1 Adding similar radicals

For each of the following expressions, use the distributive law to group similar terms and then simplify.

(a) $2\sqrt{7}-5\sqrt{7}+\sqrt{7}=(2-5+1)\sqrt{7}=-2\sqrt{7}$

(b) $\sqrt[5]{6}+4\sqrt[5]{6}-2\sqrt[5]{6}=(1+4-2)\sqrt[5]{6}=3\sqrt[5]{6}$

(c) $\sqrt{5}+2\sqrt{3}-5\sqrt{5}=(1-5)\sqrt{5}+2\sqrt{3}=-4\sqrt{5}+2\sqrt{3}$ — not all terms are similar

EXAMPLE 2 Adding radicals that must first be factored

Simplify the radicals and then perform the following additions.

(a) $\sqrt{2}+\sqrt{8}=\sqrt{2}+\sqrt{2^2\cdot 2}=\sqrt{2}+2\sqrt{2}=(1+2)\sqrt{2}=3\sqrt{2}$

(b) $\sqrt[3]{24}+\sqrt[3]{81}=\sqrt[3]{2^3\cdot 3}+\sqrt[3]{3^3\cdot 3}=2\sqrt[3]{3}+3\sqrt[3]{3}=(2+3)\sqrt[3]{3}=5\sqrt[3]{3}$

Note that $\sqrt{2}+\sqrt{8}$ **is not equal to** $\sqrt{2+8}$. By simplifying the radicals before adding, this common error can be avoided. Moreover, after simplification we are able to recognize similar radicals that do not initially appear to be similar.

EXAMPLE 3 Simplifying each and combining similar radicals

Simplify the radicals, rationalize the denominator if necessary, and then perform the following operations by combining similar radicals.

(a) $3\sqrt{125}-\sqrt{20}+\sqrt{27}=3\sqrt{5^2\cdot 5}-\sqrt{2^2\cdot 5}+\sqrt{3^2\cdot 3}$

$=3\cdot 5\sqrt{5}-2\sqrt{5}+3\sqrt{3}$ — not similar to others

$=(15-2)\sqrt{5}+3\sqrt{3}=13\sqrt{5}+3\sqrt{3}$

(b) $\sqrt{24}+\sqrt{\frac{3}{2}}=\sqrt{2^2\cdot 6}+\frac{\sqrt{3}\sqrt{2}}{\sqrt{2}\sqrt{2}}$ — rationalize denominator

$=2\sqrt{6}+\frac{\sqrt{6}}{2}=\left(2+\frac{1}{2}\right)\sqrt{6}=\frac{5}{2}\sqrt{6}$ — combine similar radicals

Practice Exercise

Combine: **1.** $5\sqrt{44}-\sqrt{99}$

EXAMPLE 4 Radicals with literal numbers

$$\sqrt{\frac{2}{3a}} - 2\sqrt{\frac{3}{2a}} = \frac{\sqrt{2}\sqrt{3a}}{\sqrt{3a}\sqrt{3a}} - \frac{2\sqrt{3}\sqrt{2a}}{\sqrt{2a}\sqrt{2a}} \quad \text{rationalize denominator}$$

$$= \frac{\sqrt{6a}}{3a} - \frac{2\sqrt{6a}}{2a} \quad \text{cancel common factors}$$

$$= \left(\frac{1}{3a} - \frac{1}{a}\right)\sqrt{6a} \quad \text{combine similar radicals}$$

$$= \left(\frac{1-3}{3a}\right)\sqrt{6a} = \frac{-2\sqrt{6a}}{3a}$$

Practice Exercise

Combine: **2.** $\sqrt{\frac{8x}{5y}} - \sqrt{10xy}$

EXAMPLE 5 Adding radicals—roof truss

A roof truss for a house has been designed as shown in Fig. 11.4. Find an expression for the exact number of linear metres of wood needed to construct the truss.

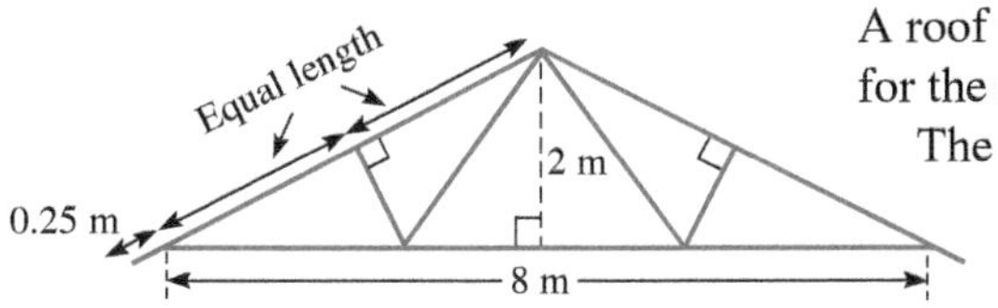

Fig. 11.4

The left side of the truss is shown in Fig. 11.5. By the Pythagorean theorem,

$$x^2 = 4^2 + 2^2$$
$$x = \sqrt{20} = \sqrt{2^2 \cdot 5} = 2\sqrt{5}$$

Therefore, $a = \frac{2\sqrt{5}}{2} = \sqrt{5}$.

Since triangles ADE and ACB are similar triangles, $\frac{b}{2} = \frac{\sqrt{5}}{4}$, or $b = \frac{\sqrt{5}}{2}$. Side c can then be found using the Pythagorean theorem:

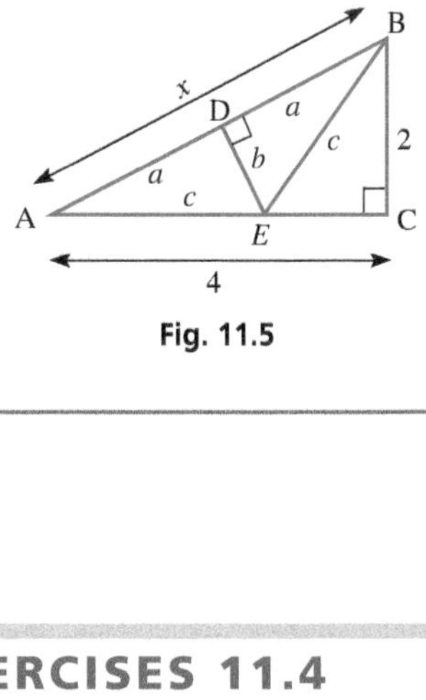

Fig. 11.5

$$c^2 = a^2 + b^2 = \left(\sqrt{5}\right)^2 + \left(\frac{\sqrt{5}}{2}\right)^2 = 5 + \frac{5}{4} = \frac{25}{4}$$

Thus, $c = \sqrt{\frac{25}{4}} = \frac{5}{2}$. The total number of linear metres needed for the truss is

$$8 + 0.25 + 0.25 + \frac{5}{2} + \frac{5}{2} + \sqrt{5} + \sqrt{5} + \frac{\sqrt{5}}{2} + \frac{\sqrt{5}}{2}$$
$$= 13.5 + 3\sqrt{5} \text{ (approximately 20 m)}$$

EXERCISES 11.4

In Exercises 1 and 2, simplify the resulting expressions if the given changes are made in the indicated examples of this section.

1. In Example 3(a), change $\sqrt{27}$ to $\sqrt{45}$ and then find the resulting simplified expression.

2. In Example 4, change $\sqrt{\frac{2}{3a}}$ to $\sqrt{\frac{1}{6a}}$ and then find the resulting simplified expression.

In Exercises 3–42, express each radical in simplest form, rationalize denominators, and perform the indicated operations.

3. $4\sqrt{11} + 3\sqrt{11}$

4. $9\sqrt{13} - 4\sqrt{13}$

5. $\sqrt{32} + \sqrt{5} - 3\sqrt{8}$

6. $8\sqrt{8} - \sqrt{24} - 5\sqrt{8}$

7. $\sqrt{5} + \sqrt{16 + 4}$

8. $\sqrt{7} + \sqrt{36 + 27}$

9. $2\sqrt{3t^2} - 3\sqrt{12t^2}$

10. $4\sqrt{2n^2} - \sqrt{50n^2}$

11. $\sqrt{8a} - \sqrt{32a}$

12. $\sqrt{27x} + 2\sqrt{18x}$

13. $2\sqrt{28} + 3\sqrt{175}$

14. $\sqrt{100 + 25} - 7\sqrt{80}$

15. $2\sqrt{200} - \sqrt{1250} - \sqrt{450}$

16. $2\sqrt{44} - \sqrt{99} + \sqrt{2}\sqrt{88}$

17. $3\sqrt{75R} + 2\sqrt{48R} - 2\sqrt{18R}$

18. $2\sqrt{28} - \sqrt{108} - 2\sqrt{175}$

19. $\sqrt{60} + \sqrt{\frac{5}{3}}$

20. $3\sqrt{84} - \sqrt{\frac{3}{7}}$

21. $\sqrt{\frac{1}{2}} + \sqrt{\frac{25}{2}} - 4\sqrt{18}$

22. $\sqrt{6} - \sqrt{\frac{2}{3}} - \sqrt{18}$

23. $\sqrt[3]{81} + \sqrt[3]{3000}$

24. $\sqrt[3]{-16} + \sqrt[3]{54}$

25. $\sqrt[4]{32} - \sqrt[8]{4}$

26. $\sqrt[6]{\sqrt{2}} - \sqrt[12]{2^{13}}$

27. $5\sqrt{a^3b} - \sqrt{4ab^5}$

28. $\sqrt{2R^2I} + \sqrt{8}\sqrt{I^3}$

29. $\sqrt{6}\sqrt{5}\sqrt{3} - \sqrt{40a^2}$

30. $3\sqrt{60b^2n} - b\sqrt{135n}$

31. $\sqrt[3]{24a^2b^4} - \sqrt[3]{3a^5b}$

32. $\sqrt[5]{32a^6b^4} + 3a\sqrt[5]{243ab^9}$

33. $\sqrt{\frac{R}{H^5}} - \sqrt{\frac{H}{R^3}}$

34. $\sqrt{\frac{2x}{3y}} + \sqrt{\frac{27y}{8x}}$

35. $\sqrt[3]{ab^{-1}} - \sqrt[3]{8a^{-2}b^2}$

36. $\sqrt{\frac{2xy^{-1}}{3}} + \sqrt{\frac{27x^{-1}y}{8}}$

37. $\sqrt{\frac{T-V}{T+V}} - \sqrt{\frac{T+V}{T-V}}$

38. $\sqrt{\frac{16}{x} + 8 + x} - \sqrt{1 - \frac{1}{x}}$

39. $\sqrt{\frac{\sqrt{4RT}}{12R^{4/3}}}$

40. $\sqrt{\sqrt{\frac{6x^4y^4}{24x^2y^6}} - 1}$

41. $\sqrt{4x + 8} + 2\sqrt{9x + 18}$

42. $3\sqrt{50y - 75} - \sqrt{8y - 12}$

In Exercises 43–48, express each radical in simplest form, rationalize denominators, and perform the indicated operations. Then use a calculator to verify the result.

43. $3\sqrt{45} + 3\sqrt{75} - 2\sqrt{500}$

44. $2\sqrt{40} + 3\sqrt{90} - 5\sqrt{250}$

45. $2\sqrt{\frac{2}{3}} + \sqrt{24} - 5\sqrt{\frac{3}{2}}$

46. $\sqrt{\frac{2}{7}} - 2\sqrt{\frac{7}{2}} + 5\sqrt{56}$

47. $2\sqrt[3]{16} - \sqrt[3]{\frac{1}{4}}$

48. $5\sqrt[4]{810} - \sqrt[4]{\frac{5}{8}}$

In Exercises 49–56, solve the given problems.

49. Find the exact sum of the positive roots of $x^2 - 2x - 2 = 0$ and $x^2 + 2x - 11 = 0$.

50. For the quadratic equation $ax^2 + bx + c = 0$, if a, b, and c are integers, the sum of the roots is a rational number. Explain.

51. Without calculating the actual value, determine whether $10\sqrt{11} - \sqrt{1000}$ is positive or negative. Explain.

52. The adjacent sides of a parallelogram are $\sqrt{12}$ and $\sqrt{27}$ units long. What is the perimeter of the parallelogram?

53. The two legs of a right triangle are $2\sqrt{2}$ and $2\sqrt{6}$ units long. What is the perimeter of the triangle?

54. The current I (in A) passing through a resistor R (in Ω) in which P watts of power are dissipated is $I = \sqrt{P/R}$. If the power dissipated in the resistors shown in Fig. 11.6 is W watts, what is the sum of the currents in radical form?

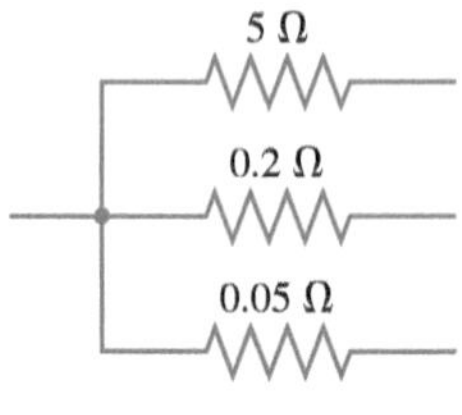

Fig. 11.6

55. A rectangular piece of plywood 1200 mm by 2400 mm has corners cut from it, as shown in Fig. 11.7. Find the perimeter of the remaining piece in exact form and in decimal form (to three significant digits).

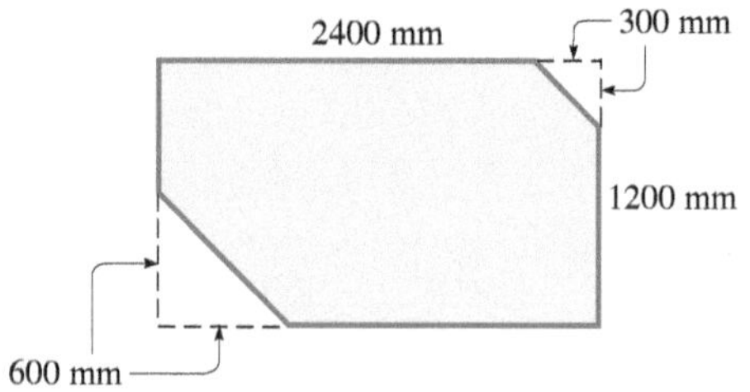

Fig. 11.7

56. Three squares with areas of 150 cm^2, 54 cm^2, and 24 cm^2 are displayed on a computer monitor. What is the sum (in radical form) of the perimeters of these squares?

Answers to Practice Exercises

1. $7\sqrt{11}$ **2.** $\frac{(2 - 5y)\sqrt{10xy}}{5y}$

11.5 Multiplication and Division of Radicals

When multiplying expressions containing radicals, we use $\sqrt[n]{a}\sqrt[n]{b} = \sqrt[n]{ab}$, along with the normal procedures of algebraic multiplication. *Note that the product of two radicals can be combined only if the orders of the radicals are the same.* The following examples illustrate the method.

EXAMPLE 1 Multiplying monomial radicals

(a) $\sqrt{5}\sqrt{2} = \sqrt{5 \cdot 2} = \sqrt{10}$

(b) $\sqrt{33}\sqrt{3} = \sqrt{33 \cdot 3} = \sqrt{11 \cdot 3 \cdot 3} = 3\sqrt{11}$

(c) $\sqrt[3]{6}\sqrt[3]{4} = \sqrt[3]{6 \cdot 4} = \sqrt[3]{3 \cdot 2^3} = 2\sqrt[3]{3}$

(d) $\sqrt[5]{8a^3b^4}\sqrt[5]{8a^2b^3} = \sqrt[5]{8a^3b^4 \cdot 8a^2b^3} = \sqrt[5]{2^3 \cdot 2^3 \cdot a^{3+2}b^{4+3}}$
$= \sqrt[5]{2 \cdot 2^5 \cdot a^5 \cdot b^5 \cdot b^2}$
$= 2ab\sqrt[5]{2b^2}$

EXAMPLE 2 Multiplying binomial radicals

(a) $\sqrt{2}(3\sqrt{5} - 4\sqrt{2}) = 3\sqrt{2}\sqrt{5} - 4\sqrt{2}\sqrt{2} = 3\sqrt{10} - 4\sqrt{4}$
$= 3\sqrt{10} - 4(2) = 3\sqrt{10} - 8$

(b) $(5\sqrt{7} - 2\sqrt{3})(4\sqrt{7} + 3\sqrt{3}) = (5\sqrt{7})(4\sqrt{7}) + (5\sqrt{7})(3\sqrt{3})$
$+ (-2\sqrt{3})(4\sqrt{7}) + (-2\sqrt{3})(3\sqrt{3})$
$= 20\sqrt{7^2} + 15\sqrt{21} - 8\sqrt{21} - 6\sqrt{3^2}$
$= 20 \cdot 7 + (15 - 8)\sqrt{21} - 6 \cdot 3$
$= 140 + 7\sqrt{21} - 18 = 122 + 7\sqrt{21}$

When raising a single-term radical expression to a power, the power can be applied to each factor of the term separately. However, when raising a binomial to a power, the entire binomial must be multiplied by itself the indicated number of times. This is illustrated in the following example.

EXAMPLE 3 Power of a radical

(a) $(2\sqrt{7})^2 = (2\sqrt{7})(2\sqrt{7}) = 2 \cdot 2 \cdot \sqrt{7 \cdot 7} = 2 \cdot 2 \cdot 7 = 28$

(b) $(2\sqrt{7})^3 = (2\sqrt{7})(2\sqrt{7})(2\sqrt{7}) = 2^3\sqrt{7 \cdot 7 \cdot 7} = 2^3 \cdot 7\sqrt{7} = 56\sqrt{7}$

(c) $(3 + \sqrt{5})^2 = (3 + \sqrt{5})(3 + \sqrt{5}) = 3 \cdot 3 + 3\sqrt{5} + 3\sqrt{5} + \sqrt{5 \cdot 5}$
$= 9 + 2(3\sqrt{5}) + 5 = 14 + 6\sqrt{5}$

(d) $(\sqrt{a} - \sqrt{b})^2 = (\sqrt{a} - \sqrt{b})(\sqrt{a} - \sqrt{b})$
$= \sqrt{a \cdot a} - 2\sqrt{a \cdot b} + \sqrt{b \cdot b} = a - 2\sqrt{ab} + b$

Practice Exercises

Multiply: **1.** $2\sqrt{5}(\sqrt{15} - 3\sqrt{3})$
2. $(2\sqrt{6} - \sqrt{3})(3\sqrt{6} + 2\sqrt{3})$

EXAMPLE 4 Use of fractional exponents

(a) $\sqrt[3]{2}\sqrt{5} = 2^{1/3}5^{1/2} = 2^{2/6}5^{3/6} = (2^2 5^3)^{1/6} = \sqrt[6]{500}$

(b) $\sqrt[3]{4a^2b}\sqrt[4]{8a^3b^2} = (2^2a^2b)^{1/3}(2^3a^3b^2)^{1/4} = (2^2a^2b)^{4/12}(2^3a^3b^2)^{3/12}$
$= (2^8a^8b^4)^{1/12}(2^9a^9b^6)^{1/12} = (2^{17}a^{17}b^{10})^{1/12}$
$= 2a(2^5a^5b^{10})^{1/12}$
$= 2a\sqrt[12]{32a^5b^{10}}$

LEARNING TIP

Remember, *to multiply radicals and combine them under one radical sign, **it is necessary that the order of the radicals be the same.*** As in Example 4, when the orders of the radicals are not the same, fractional exponents can be used to make them the same.

DIVISION OF RADICALS

If a fraction involving a radical is to be changed in form, *rationalizing the denominator* or *rationalizing the numerator* is the principal step. We now consider rationalizing when the denominator (or numerator) to be rationalized has more than one term.

LEARNING TIP

If the denominator of a fraction is a binomial where one or both terms are radicals, the denominator is rationalized by multiplying both the numerator and denominator by the conjugate of the denominator, obtained by changing the middle sign of the binomial to its opposite. This forces the denominator to become a difference of squares, thus eliminating the square root(s). In rationalizing a binomial numerator, the same process works using the conjugate of the numerator.

EXAMPLE 5 Rationalizing the denominator

Rationalize the denominator of $\dfrac{1}{\sqrt{3} - \sqrt{2}}$.

We change the middle sign of $\sqrt{3} - \sqrt{2}$ to get the conjugate $\sqrt{3} + \sqrt{2}$. Multiplying numerator and denominator by the conjugate leads to

$$\frac{1(\sqrt{3} + \sqrt{2})}{(\sqrt{3} - \sqrt{2})(\sqrt{3} + \sqrt{2})} = \frac{\sqrt{3} + \sqrt{2}}{\sqrt{3^2} + \sqrt{6} - \sqrt{6} - \sqrt{2^2}} = \frac{\sqrt{3} + \sqrt{2}}{3 - 2} = \sqrt{3} + \sqrt{2}$$

We note that, after rationalizing the denominator, the result can have a simpler form than the original expression.

EXAMPLE 6 Rationalizing a denominator with literal numbers

Rationalize the denominator of $\dfrac{4\sqrt{x}}{\sqrt{7} + \sqrt{x}}$ and simplify the result.

$$\frac{4\sqrt{x}(\sqrt{7} - \sqrt{x})}{(\sqrt{7} + \sqrt{x})(\sqrt{7} - \sqrt{x})} = \frac{4\sqrt{7x} - 4x}{7 - \sqrt{7x} + \sqrt{7x} - x}$$
$$= \frac{4(\sqrt{7x} - x)}{7 - x}$$

change sign

Practice Exercise

Rationalize the denominator and simplify:

3. $\dfrac{3 - 2\sqrt{3}}{2 + 7\sqrt{3}}$

As we noted earlier, in certain types of algebraic operations, it may be necessary to rationalize the numerator of an expression. This procedure is illustrated in the following example.

EXAMPLE 7 Rationalizing—semiconductor expression

In studying the properties of a semiconductor, the expression $\dfrac{C_1 + C_2\sqrt{1 + 2V}}{\sqrt{1 + 2V}}$ is used. Here, C_1 and C_2 are constants and V is the voltage across a junction of the semiconductor. Rationalize the numerator of this expression.

Multiplying numerator and denominator by the conjugate of the numerator, $C_1 - C_2\sqrt{1 + 2V}$, we have

$$\frac{C_1 + C_2\sqrt{1 + 2V}}{\sqrt{1 + 2V}} = \frac{(C_1 + C_2\sqrt{1 + 2V})(C_1 - C_2\sqrt{1 + 2V})}{\sqrt{1 + 2V}(C_1 - C_2\sqrt{1 + 2V})}$$
$$= \frac{C_1^2 - C_2^2(\sqrt{1 + 2V})^2}{C_1\sqrt{1 + 2V} - C_2\sqrt{1 + 2V}\sqrt{1 + 2V}}$$
$$= \frac{C_1^2 - C_2^2(1 + 2V)}{C_1\sqrt{1 + 2V} - C_2(1 + 2V)}$$

EXERCISES 11.5

In Exercises 1–4, perform the indicated operations on the resulting expressions if the given changes are made in the indicated examples of this section.

1. In Example 2(a), change $4\sqrt{2}$ to $4\sqrt{8}$ and then perform the multiplication.
2. In Example 4(a), change $\sqrt{5}$ to $\sqrt[6]{5}$ and then perform the multiplication.
3. In Example 5, change the sign in the denominator from $-$ to $+$ and then rationalize the denominator.
4. In Example 6, rationalize the numerator of the given expression.

In Exercises 5–48, perform the indicated operations, expressing answers in simplest form with rationalized denominators.

5. $\sqrt{3}\sqrt{7}$ **6.** $\sqrt{2}\sqrt{17}$ **7.** $\sqrt{10}\sqrt{2}$ **8.** $\sqrt{7}\sqrt{21}$

9. $\sqrt[3]{4}\sqrt[3]{2}$ **10.** $\sqrt[5]{4}\sqrt[5]{16}$ **11.** $(5\sqrt{2})^2$ **12.** $2(3\sqrt{5})^3$

13. $\sqrt{8}\sqrt{\frac{5}{2}}$ **14.** $\sqrt{\frac{6}{7}}\sqrt{\frac{2}{3}}$

15. $\sqrt{3}(\sqrt{2}-\sqrt{5})$ **16.** $3\sqrt{5}(\sqrt{15}-2\sqrt{5})$

17. $(2-\sqrt{5})(2+\sqrt{5})$ **18.** $4(2-\sqrt{7})^2$

19. $(3\sqrt{30}-2\sqrt{3})(6\sqrt{30}+7\sqrt{3})$

20. $(3\sqrt{7a}-\sqrt{8})(\sqrt{7a}+\sqrt{2})$

21. $(3\sqrt{11}-\sqrt{x})(2\sqrt{11}+5\sqrt{x})$

22. $(2\sqrt{10}+3\sqrt{15})(\sqrt{10}-7\sqrt{15})$

23. $(2\sqrt{5}-3\sqrt{x})(2\sqrt{5}+3\sqrt{x})$

24. $\left(\sqrt{\frac{3}{x}}+\sqrt{\frac{4}{y}}\right)\left(\sqrt{\frac{3}{x}}-\sqrt{\frac{4}{y}}\right)$

25. $\sqrt{a}(\sqrt{ab}+\sqrt{c^3})$ **26.** $\sqrt{3x}(\sqrt[3]{3x}-\sqrt{xy})$

27. $\dfrac{\sqrt{6}-3}{\sqrt{6}}$ **28.** $\dfrac{5-\sqrt{10}}{\sqrt[4]{10}}$

29. $(\sqrt{2a}-\sqrt{b})(\sqrt{2a}+3\sqrt{b})$ **30.** $(2\sqrt{mn}+3\sqrt{n})^2$

31. $\sqrt{2}\sqrt[3]{3}$ **32.** $\sqrt[5]{16}\sqrt[3]{8}$

33. $\dfrac{1}{\sqrt{7}+\sqrt{3}}$ **34.** $\dfrac{6\sqrt[4]{25}}{5-2\sqrt{5}}$

35. $\dfrac{\sqrt{2}-1}{\sqrt{7}-3\sqrt{2}}$ **36.** $\dfrac{2\sqrt{15}-3}{\sqrt{15}+4}$

37. $\dfrac{2\sqrt{3}-5\sqrt{5}}{\sqrt{3}+2\sqrt{5}}$ **38.** $\dfrac{\sqrt{15}-3\sqrt{5}}{2\sqrt{15}-\sqrt{5}}$

39. $\dfrac{2\sqrt{x}}{\sqrt{x}-\sqrt{5}}$ **40.** $\dfrac{\sqrt{2c}+3d}{\sqrt{2c}-d}$

41. $\left(\sqrt[5]{\sqrt{6}-\sqrt{5}}\right)\left(\sqrt[5]{\sqrt{6}+\sqrt{5}}\right)$

42. $\sqrt[3]{5-\sqrt{17}}\sqrt[3]{5+\sqrt{17}}$

43. $\left(\sqrt{\frac{2}{R}}+\sqrt{\frac{R}{2}}\right)\left(\sqrt{\frac{2}{R}}-2\sqrt{\frac{R}{2}}\right)$

44. $(3+\sqrt{6-2a})(2-\sqrt{6-2a})$

45. $\dfrac{\sqrt{x+y}}{\sqrt{x-y}-\sqrt{x}}$ **46.** $\dfrac{\sqrt{1+a}}{a-\sqrt{1-a}}$

47. $\dfrac{\sqrt{a}+\sqrt{a-2}}{\sqrt{a}-\sqrt{a-2}}$ **48.** $\dfrac{\sqrt{T^4-V^4}}{\sqrt{V^{-2}-T^{-2}}}$

In Exercises 49–52, perform the indicated operations, expressing answers in simplest form with rationalized denominators. Then verify the result with a calculator.

49. $(\sqrt{11}+\sqrt{6})(\sqrt{11}-2\sqrt{6})$

50. $(2\sqrt{5}-\sqrt{7})(3\sqrt{5}+\sqrt{7})$

51. $\dfrac{2\sqrt{6}-\sqrt{5}}{3\sqrt{6}-4\sqrt{5}}$ **52.** $\dfrac{\sqrt{7}-4\sqrt{2}}{5\sqrt{7}-4\sqrt{2}}$

In Exercises 53–56, combine the terms into a single fraction, but do not rationalize the denominators.

53. $2\sqrt{x}+\dfrac{1}{\sqrt{x}}$ **54.** $\dfrac{3}{2\sqrt{3x-4}}-\sqrt{3x-4}$

55. $\dfrac{x^2}{\sqrt{2x+1}}+2x\sqrt{2x+1}$ **56.** $4\sqrt{x^2+1}-\dfrac{4x}{\sqrt{x^2+1}}$

In Exercises 57–60, rationalize the numerator of each fraction.

57. $\dfrac{\sqrt{5}+2\sqrt{2}}{3\sqrt{10}}$ **58.** $\dfrac{\sqrt{19}-3}{5}$

59. $\dfrac{\sqrt{x+h}-\sqrt{x}}{h}$ **60.** $\dfrac{\sqrt{3x+4}+\sqrt{3x}}{8}$

In Exercises 61–74, solve the given problems.

61. By substitution, show that $x = 1-\sqrt{2}$ is a solution of the equation $x^2-2x-1=0$.

62. For the quadratic equation $ax^2+bx+c=0$, if a, b, and c are integers, the product of the roots is a rational number. Explain.

63. Determine the relationship between a and c in $ax^2+bx+c=0$ if the roots of the equation are reciprocals.

64. Evaluate r^2-s^2 if $r=-b+\sqrt{b^2-4ac}$ and $s=-b-\sqrt{b^2-4ac}$.

65. Rationalize the denominator of $\dfrac{1}{\sqrt[3]{x^2}+\sqrt[3]{x}+1}$. (*Hint:* See Eq. 6.10.)

66. One leg of a right triangle is $2\sqrt{7}$, and the hypotenuse is 6. What is the area of the triangle?

67. For an object oscillating at the end of a spring and on which there is a force that retards the motion, the equation $m^2+bm+k^2=0$ must be solved. Here, b is a constant related to the retarding force, and k is the spring constant. By substitution, show that $m=\frac{1}{2}(\sqrt{b^2-4k^2}-b)$ is a solution.

68. Among the products of a specialty furniture company are tables with tops in the shape of a regular octagon (eight sides). Express the area A of a table top as a function of the side s of the octagon. (*Hint:* Draw a square around the octagon.)

69. An expression used in determining the characteristics of a spur gear is $\dfrac{50}{50+\sqrt{V}}$. Rationalize the denominator.

70. When studying the orbits of Earth satellites, the expression $\left(\dfrac{GM}{4\pi^2}\right)^{1/3}T^{2/3}$ arises. Express it in simplest rationalized radical form.

71. In analysing a tuned amplifier circuit, the expression $\dfrac{2Q}{\sqrt{\sqrt{2}-1}}$ is used. Rationalize the denominator.

72. When analysing the ratio of resultant forces when forces with magnitudes F and T act on a structure, the expression $\sqrt{F^2+T^2}/\sqrt{F^{-2}+T^{-2}}$ arises. Simplify this expression.

73. The resonant frequency ω of a capacitance C in parallel with a resistance R and inductance L (see Fig. 11.8) is $\omega = \dfrac{1}{\sqrt{LC}}\sqrt{1-\dfrac{R^2C}{L}}$. Combine terms under the radical, rationalize the denominator, and simplify.

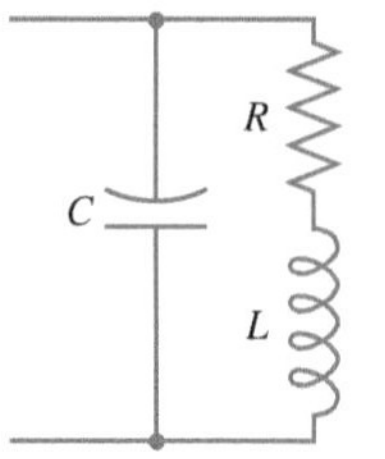

Fig. 11.8

74. In fluid dynamics, the expression $\dfrac{a}{1+m/\sqrt{r}}$ arises. Combine terms in the denominator, rationalize the denominator, and simplify.

Answers to Practice Exercises

1. $2(5\sqrt{3}-3\sqrt{15})$ **2.** $3(10+\sqrt{2})$ **3.** $\dfrac{-48+25\sqrt{3}}{143}$

CHAPTER 11 KEY FORMULAS AND EQUATIONS

Product Rule:	$a^m \times a^n = a^{m+n}$	**(11.1)**
Quotient Rule:	$\dfrac{a^m}{a^n} = a^{m-n} \quad (a \neq 0)$	**(11.2)**
Power of Power:	$(a^m)^n = a^{mn}$	**(11.3)**
Power of Product:	$(ab)^n = a^n b^n$	**(11.4)**
Power of Quotient:	$\left(\dfrac{a}{b}\right)^n = \dfrac{a^n}{b^n} \quad (\text{if } b \neq 0)$	**(11.5)**
Zero Exponent:	$a^0 = 1 \quad (a \neq 0)$	**(11.6)**
Negative Exponent:	$a^{-n} = \dfrac{1}{a^n} \quad (a \neq 0)$	**(11.7)**
Fractional Exponents:	$a^{1/n} = \sqrt[n]{a}$	**(11.8)**
	$a^{m/n} = \sqrt[n]{a^m} = \left(\sqrt[n]{a}\right)^m$	**(11.9)**
Radicals:	$\sqrt[n]{a^n} = \left(\sqrt[n]{a}\right)^n = a$	**(11.10)**
	$\sqrt[n]{a}\sqrt[n]{b} = \sqrt[n]{ab}$	**(11.11)**
	$\sqrt[m]{\sqrt[n]{a}} = \sqrt[mn]{a}$	**(11.12)**
	$\dfrac{\sqrt[n]{a}}{\sqrt[n]{b}} = \sqrt[n]{\dfrac{a}{b}} \quad (b \neq 0)$	**(11.13)**

CHAPTER 11 REVIEW EXERCISES

In Exercises 1–28, express each expression in simplest form with only positive exponents.

1. $2a^{-2}b^0$ **2.** $(2c)^{-1}z^{-2}$ **3.** $\dfrac{9c^{-1}}{d^{-3}}$ **4.** $\dfrac{-5x^0}{3y^{-1}}$

5. $3(25)^{3/2}$ **6.** $32^{2/5}$ **7.** $400^{-3/2}$ **8.** $7(1000)^{-2/3}$

9. $\left(\dfrac{3}{t^2}\right)^{-2}$ **10.** $\left(\dfrac{2x^3}{3}\right)^{-3}$ **11.** $\dfrac{-8^{2/3}}{49^{-1/2}}$ **12.** $\dfrac{4^{-45}}{2^{-90}}$

13. $(2a^{1/3}b^{5/6})^6$ **14.** $(ax^{-1/2}y^{1/4})^8$ **15.** $(-32m^{15}n^{10})^{3/5}$

16. $(27x^{-6}y^9)^{2/3}$ **17.** $2L^{-2}-4C^{-1}$ **18.** $C^4C^{-1}+(C^4)^{-1}$

19. $\dfrac{2x^{-1}}{2x^{-1}+y^{-1}}$ **20.** $\dfrac{9a}{(2a)^{-1}-a}$ **21.** $(a-3b^{-1})^{-1}$

22. $(2s^{-2}+t)^{-2}$ **23.** $(x^3-y^{-3})^{1/3}$

24. $(8a^3)^{2/3}(a^{-2}+1)^{1/2}$ **25.** $(W^2+2WH+H^2)^{-1/2}$

26. $\left[\dfrac{(9a)^0(4x^2)^{1/3}(3b^{1/2})}{(2b^0)^2}\right]^{-6}$

27. $2x(x-1)^{-2}-2(x^2+1)(x-1)^{-3}$

28. $4(1-T^2)^{1/2}-(1-T^2)^{-1/2}$

In Exercises 29–78, perform the indicated operations and express the result in simplest radical form with rationalized denominators.

29. $\sqrt{68}$ **30.** $4\sqrt{96}$ **31.** $\sqrt{ab^5c^2}$ **32.** $\sqrt{x^3y^4z^6}$

33. $\sqrt{9a^3b^4}$ **34.** $\sqrt{8x^5y^2}$ **35.** $\sqrt{84st^3u^{-2}}$ **36.** $\sqrt{52L^2C^{-5}}$

37. $\dfrac{5}{\sqrt{2s}}$ **38.** $\dfrac{3a}{\sqrt{5x}}$ **39.** $\sqrt{\dfrac{11}{27}}$ **40.** $\sqrt{\dfrac{7}{8V}}$

41. $\sqrt[4]{8m^6n^9}$ **42.** $\sqrt[3]{9a^7b^{-3}}$ **43.** $\sqrt[4]{\sqrt[3]{64}}$

44. $\sqrt{a^{-3}\sqrt[5]{b^{12}}}$ **45.** $\sqrt{36+4}-2\sqrt{10}$

46. $2\sqrt{68x}-\sqrt{153x}$ **47.** $\sqrt{63}-2\sqrt{112}-\sqrt{28}$

48. $7\sqrt{20}-\sqrt{80}-2\sqrt{125}$ **49.** $a\sqrt{2x^3}+\sqrt{8a^2x^3}$

50. $2\sqrt{m^2n^3}-\sqrt{n^5}$ **51.** $\sqrt[3]{8a^4}+b\sqrt[3]{a}$

52. $\sqrt[4]{2xy^5}-\sqrt[4]{32xy}$ **53.** $5\sqrt{5}(6\sqrt{5}-\sqrt{35})$

54. $2\sqrt{8}(5\sqrt{2}-\sqrt{6})$ **55.** $2\sqrt{2}(\sqrt{6}-\sqrt{10})$

56. $3\sqrt{5}(\sqrt{15m^2} + 2\sqrt{35m^2})$ **57.** $(2 - 3\sqrt{17B})(3 + \sqrt{17B})$

58. $(5\sqrt{6} - 4)(3\sqrt{6} + 5)$

59. $(2\sqrt{7} - 3\sqrt{a})(3\sqrt{7} + \sqrt{a})$

60. $(3\sqrt{2} - \sqrt{13})(5\sqrt{2} + 3\sqrt{13})$

61. $\dfrac{\sqrt{3x}}{2\sqrt{3x} - \sqrt{y}}$ **62.** $\dfrac{5\sqrt{a}}{2\sqrt{a} - c}$ **63.** $\dfrac{\sqrt{2}}{\sqrt{3} - 4\sqrt{2}}$

64. $\dfrac{4}{3 - 2\sqrt{7}}$ **65.** $\dfrac{\sqrt{7} - \sqrt{5}}{\sqrt{5} + 3\sqrt{7}}$ **66.** $\dfrac{7 - 2\sqrt{6}}{3 + 2\sqrt{6}}$

67. $\dfrac{2\sqrt{x} - a}{3\sqrt{x} + 5a}$ **68.** $\dfrac{2\sqrt{3y} - \sqrt{2}}{9\sqrt{3y} - 5\sqrt{2}}$ **69.** $\sqrt{4b^2 + 1}$

70. $x^{3/8}(x^{5/8} - 8x^{13/8})$ **71.** $(3x^{1/2} - 2x)(3x^{1/2} + 2x)$

72. $(x^{3n-1}/x^{n-1})^{1/n}$

73. $(1 + 6^{1/2})(3^{1/2} + 2^{1/2})(3^{1/2} - 2^{1/2})$

74. $\sqrt{a^{-2} + \dfrac{1}{b^2}}$

75. $\left(\dfrac{2 - \sqrt{15}}{2}\right)^2 - \left(\dfrac{2 - \sqrt{15}}{2}\right)$

76. $\sqrt{2 + a^{-1}b + ab^{-1}} + \sqrt{a^4b^2 + 2a^3b^2 + a^2b^2}$

77. $\sqrt{3 + n}(\sqrt{3 + n} - \sqrt{n})^{-1}$

78. $\sqrt{1 + \sqrt{2}}(\sqrt{1 + \sqrt{2}} + \sqrt{2})^{-1}$

In Exercises 79 and 80, perform the indicated operations and express the result in simplified radical form with rationalized denominators. In Exercises 81 and 82, without a calculator, show that the given equations are true. Then verify each result with a calculator.

79. $(\sqrt{7} - 2\sqrt{15})(3\sqrt{7} - \sqrt{15})$ **80.** $\dfrac{2\sqrt{3} - 7\sqrt{14}}{3\sqrt{3} + 2\sqrt{14}}$

81. $\sqrt{\sqrt{2} - 1}(\sqrt{2} + 1) = \sqrt{\sqrt{2} + 1}$

82. $\sqrt{\sqrt{3} + 1}(\sqrt{3} - 1) = \sqrt{2(\sqrt{3} - 1)}$

In Exercises 83–100, perform the indicated operations. In Exercise 101, answer the given question.

83. The legs of a right triangle are $\sqrt{18}$ and $\sqrt{32}$. Find the perimeter of the triangle.

84. One of the legs of a right triangle is $3\sqrt{5}$ and the hypotenuse is $5\sqrt{3}$. Find the area of the triangle.

85. Evaluate $3x^2 - 2x + 5$ for $x = \frac{1}{2}(2 - \sqrt{3})$.

86. Evaluate x if $x^{-1/3} = 0.200$.

87. The average annual increase i (in %) of the cost of living over n years is given by $i = 100[(C_2/C_1)^{1/n} - 1]$, where C_1 is the cost of living index for the first year and C_2 is the cost of living index for the last year of the period. Evaluate i if $C_1 = 130.7$ and $C_2 = 172.0$ are the values for 1990 and 2000, respectively.

88. Kepler's third law of planetary motion may be given as $T = kr^{3/2}$, where T is the time for one revolution of a planet around the sun, r is its mean radius from the sun, and $k = 5.46 \times 10^{-13}$ year/km$^{3/2}$. Find the time for one revolution of Venus about the sun if $r = 1.08 \times 10^8$ km.

89. The speed v of a ship of weight w whose engines produce power P is $v = k\sqrt[3]{P/w}$. Express this equation (a) with a fractional exponent and (b) as a radical with the denominator rationalized.

90. In analysing the orbit of an Earth satellite, the expression $\sqrt{1 + \dfrac{2E}{m}\left(\dfrac{h}{GM}\right)^2}$ is used. Combine terms under the radical and express the result with the denominator rationalized.

91. A square plastic sheet of side x is stretched by an amount equal to $\sqrt{x}$ horizontally and vertically. Find the expression for the percent increase in the area of the sheet.

92. The value P of an object originally worth P_0 and that appreciates at an annual rate of r for t years is given by $P = P_0(1 + r)^t$. Solve for r.

93. The expression $(1 - v^2/c^2)^{-1/2}$ arises in the theory of relativity. Simplify, using only positive exponents in the result.

94. The compression ratio r of a certain gasoline engine is related to the efficiency η (in %) of the engine by $r = \left(\dfrac{100 - \eta}{100}\right)^{-2.5}$. What compression ratio is necessary to have an efficiency of 55.0%?

95. A square is decreasing in size on a computer screen. To find how fast the side of the square changes, we must rationalize the numerator of $\dfrac{\sqrt{A + h} - \sqrt{A}}{h}$. Perform this operation.

96. A plane takes off 200 km east of its destination, and a 50 km/h wind is from the south. If the plane's velocity is 250 km/h and its heading is always toward its destination, its path can be described as $y = 100[(0.005x)^{0.8} - (0.005x)^{1.2}]$, $0 \le x \le 200$ km. Sketch the path and check using a graphing calculator.

97. In an experiment, a laser beam follows the path shown in Fig. 11.9. Express the length of the path in simplest radical form.

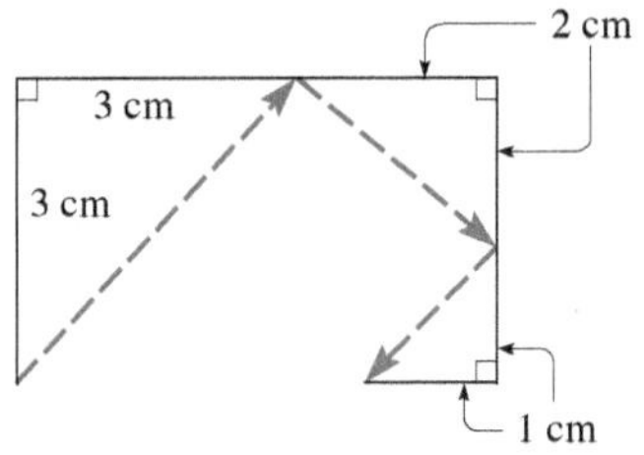

Fig. 11.9

98. The flow rate v (in m/s) through a storm drain pipe is found from $v = 33.5(0.550)^{2/3}(0.00180)^{1/2}$. Find the value of v.

99. The frequency of a certain electric circuit is given by $f = \dfrac{1}{2\pi\sqrt{\dfrac{LC_1C_2}{C_1 + C_2}}}$. Express this in simplest rationalized radical form.

100. A computer analysis of an experiment showed that the fraction f of viruses surviving X-ray dosages was given by $f = \dfrac{20}{d + \sqrt{3d + 400}}$, where d is the dosage. Express this with the denominator rationalized.

101. In calculating the forces on a tower by the wind, it is necessary to evaluate $0.018^{0.13}$. Write a paragraph explaining why this form is preferable to the equivalent radical form.

CHAPTER 11 PRACTICE TEST

In Problems 1–14, simplify the given expressions. For those with exponents, express each result with only positive exponents. For the radicals, rationalize the denominator where applicable.

1. $-5y^0$ **2.** $(3\pi x^{-4})^{-2}$ **3.** $\dfrac{100^{3/2}}{8^{-2/3}}$ **4.** $(as^{-1/3}t^{3/4})^{12}$

5. $2\sqrt{20} - \sqrt{125}$ **6.** $(2x^{-1} + y^{-2})^{-1}$ **7.** $(\sqrt{2x} - 3\sqrt{y})^2$

8. $\sqrt[3]{\sqrt[4]{4}}$ **9.** $\dfrac{3 - 2\sqrt{2}}{2\sqrt{x}}$ **10.** $\sqrt{27a^4b^3}$

11. $2\sqrt{2}(3\sqrt{10} - \sqrt{6})$

12. $(2x + 3)^{1/2} + (x + 1)(2x + 3)^{-1/2}$

13. $\left(\dfrac{4a^{-1/2}b^{3/4}}{b^{-2}}\right)\left(\dfrac{b^{-1}}{2a}\right)$ **14.** $\dfrac{2\sqrt{15} + \sqrt{3}}{\sqrt{15} - 2\sqrt{3}}$

15. Express $\dfrac{3^{-1/2}}{2}$ in simplest radical form with a rationalized denominator.

16. In the study of fluid flow in pipes, the expression $0.220N^{-1/6}$ is found. Evaluate this expression for $N = 64 \times 10^6$.

12. Complex Numbers

Africa Studio/Fotolia

▲ In Section 12.7, we see that tuning in a radio station involves a basic application of complex numbers to electricity.

LEARNING OUTCOMES

After completion of this chapter, the student should be able to:

- Express complex numbers in rectangular form, polar form, exponential form, and graphically, and convert from one form to another
- Perform mathematical operations on complex numbers in rectangular, polar, and exponential forms
- Add complex numbers graphically
- Solve algebraic equations involving complex numbers
- Find *n*th roots of a complex number using DeMoivre's theorem
- Solve application problems involving complex numbers, including the analysis of alternating-current circuits

Electromagnetic waves were predicted mathematically by James Clerk Maxwell in 1865, and were first produced and observed in the laboratory in the form of radio waves by Heinrich Hertz in 1887. Today, extensive use of electromagnetic waves is made in a great variety of situations, including radar, microwave ovens, X-ray machines, and the transmission of signals for radio, television, and cell phones.

Important in the study of electromagnetic waves and in many areas of electricity and electronics are *complex numbers*, which we study in this chapter. Complex numbers include imaginary numbers as well as real numbers. Other than defining complex numbers in Chapter 1, and describing how they arise when solving some quadratic equations in Chapter 7, we have purposely avoided discussing them until now.

Despite their names, complex numbers and imaginary numbers have very real and useful applications. Besides electricity and electronics, they are used in the study of mechanical vibrations, optics, acoustics, and fluid mechanics, as well as to generate beautiful geometric images called fractals.

12.1 Basic Definitions

Since the square of a positive number or a negative number is positive, it is not possible to square any real number and have a negative result. Therefore, we must define a number system to include square roots of negative numbers. We will find that such numbers can be used to great advantage in certain applications.

If the radicand in a square root is negative, we can express the indicated root as the product of $\sqrt{-1}$ and the square root of a positive number. *The symbol* $\sqrt{-1}$ *is defined as the* **imaginary unit** *and is denoted by the symbol j.* Therefore, we have

$$j = \sqrt{-1} \quad \text{and} \quad j^2 = -1 \tag{12.1}$$

LEARNING TIP

For any positive real number a we have $\sqrt{-a} = \sqrt{(-1)(a)} = j\sqrt{a}$, or simply

$$\sqrt{-a} = j\sqrt{a} \; (a > 0) \tag{12.2}$$

Note that we write the j before the radical. This clearly shows that the j is not *under* the radical.

Generally, mathematicians use the symbol i for $\sqrt{-1}$, and therefore most nontechnical textbooks use i. However, one of the major technical applications of complex numbers is in electronics, where i represents electric current. Therefore, we will use j for $\sqrt{-1}$, which is also the standard symbol in electronics textbooks.

Practice Exercises

Simplify:

1. $\sqrt{-5}\sqrt{-20}$ **2.** $-\sqrt{-48}$

EXAMPLE 1 Square roots in terms of j

Express the following numbers in terms of j.

(a) $\sqrt{-9} = \sqrt{(9)(-1)} = \sqrt{9}\sqrt{-1} = 3j$

(b) $\sqrt{-0.25} = \sqrt{0.25}\sqrt{-1} = 0.5j$

(c) $\sqrt{-5} = \sqrt{5}\sqrt{-1} = \sqrt{5}j = j\sqrt{5}$ the form $j\sqrt{5}$ is better than $\sqrt{5}j$

(d) $-\sqrt{-75} = -\sqrt{(5^2)(3)(-1)} = -5j\sqrt{3}$

COMMON ERROR

Note that $-\sqrt{-75}$ is not $\sqrt{75}$.

We can avoid this common error if we keep in mind the rules for order of operations. The square root is performed *before* multiplying by -1, so writing the expression in terms of j is done before changing sign.

When performing operations involving square roots of negative numbers, these must always be expressed in terms of j first. Consider the following example.

EXAMPLE 2 Be careful when simplifying

Evaluate $(\sqrt{-4})^2$.

We write $(\sqrt{-4})^2$ as $(\sqrt{-4})(\sqrt{-4})$. However, we **cannot now apply the product rule** $\sqrt{ab} = \sqrt{a}\sqrt{b}$, since that rule is valid only if a and b are not negative. Following correct order of operations, we first express the square roots in terms of j and then perform the product. We get

$$(\sqrt{-4})^2 = (\sqrt{-4})(\sqrt{-4}) = (j\sqrt{4})(j\sqrt{4}) = 4j^2 = -4$$

COMMON ERROR

We cannot overemphasize that

$(\sqrt{-4})^2$ **is not equal to** $\sqrt{(-4)(-4)}$

The square root is the first operation applied on -4, so we immediately write $\sqrt{-4}$ as $j\sqrt{4}$. After that, the product (or the square) can be performed, as in Example 2.

EXAMPLE 3 Order of operations is key

Simplify $\sqrt{-3}\sqrt{-12}$ and $\sqrt{(-3)(-12)}$, paying close attention to the operation that must be performed first in each case.

Expression	Operation performed first	Simplification
$\sqrt{-3}\sqrt{-12}$	Square root (product of square roots)	$\sqrt{-3}\sqrt{-12} = j\sqrt{3}j\sqrt{12} = j^2\sqrt{36}$ $= (-1)(6) = -6$
$\sqrt{(-3)(-12)}$	Product (square root of a product)	$\sqrt{(-3)(-12)} = \sqrt{36} = 6$

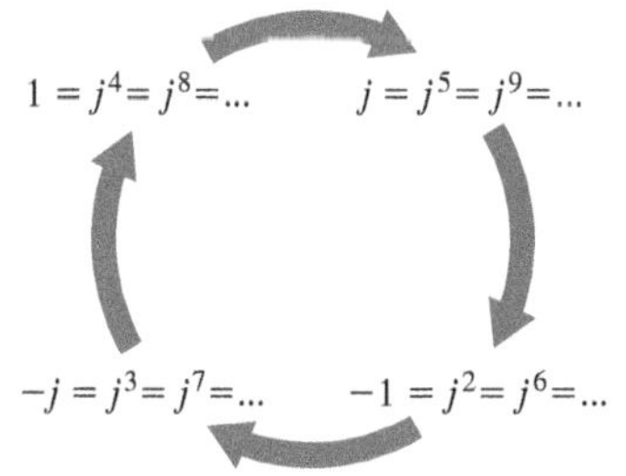

Fig. 12.1

At times, we need to raise imaginary numbers to some power. Using the definitions of exponents and of j, we have the following results:

$$\begin{aligned} j &= j & j^5 &= j^4 j = j \\ j^2 &= -1 & j^6 &= j^4 j^2 = (1)(-1) = -1 \\ j^3 &= j^2 j = -j & j^7 &= j^4 j^3 = (1)(-j) = -j \\ j^4 &= j^2 j^2 = (-1)(-1) = 1 & j^8 &= j^4 j^4 = (1)(1) = 1 \end{aligned}$$

The powers of j go through the cycle, $j, -1, -j, 1, j, -1, -j, 1$, and so forth. See Fig. 12.1. Noting this—and the fact that j raised to a power that is a multiple of 4 equals 1—allows us to raise j to any integral power easily.

EXAMPLE 4 Powers of *j*

(a) $j^{10} = j^8 j^2 = (1)(-1) = -1$

(b) $j^{45} = j^{44} j = (1)(j) = j$

(c) $j^{531} = j^{528} j^3 = (1)(-j) = -j$

exponents 8, 44, 528 are multiples of 4

Practice Exercise

Simplify:

3. $-j^{25}$

RECTANGULAR FORM OF A COMPLEX NUMBER

■ Even negative numbers were not widely accepted by mathematicians until late in the sixteenth century.

■ Imaginary numbers were so named because the French mathematician René Descartes (1596–1650) referred to them as "imaginaires." Most of their mathematical development occurred in the eighteenth century.

■ Complex numbers were named by the German mathematician Karl Friedrich Gauss (1777–1855).

Using real numbers and the imaginary unit j, we define a new kind of number. *A* **complex number** *is any number that can be written in the form* $a + bj$, *where* a *and* b *are real numbers. If* $a = 0$ *and* $b \neq 0$, *we have a number of the form* bj, *which is a* **pure imaginary number**. *If* $b = 0$, *then* $a + bj$ *is a real number. The form* $a + bj$ *is known as the* **rectangular form** *of a complex number, where* a *is known as the* **real part** *and* b *is known as the* **imaginary part**. We see that complex numbers include all real numbers and all pure imaginary numbers.

A comment here about the words *imaginary* and *complex* is in order. The choice of the names of these numbers is historical in nature, and unfortunately it leads to some misconceptions about the numbers. The use of *imaginary* does not imply that the numbers do not exist. Imaginary numbers do in fact exist, as they are defined above. In the same way, the use of *complex* does not imply that the numbers are complicated and therefore difficult to understand. With the proper definitions and operations, we can work with complex numbers just as with any other type of number.

The real part of a complex number is positive or negative (or zero), and the same is true of the imaginary part. However, the complex number itself is not positive or negative in the usual sense. Also, two complex numbers are equal if and only if the real parts are equal and the imaginary parts are equal, or $a + bj = x + yj$ if $a = x$ and $b = y$.

EXAMPLE 5 Equality of complex numbers

(a) $a + bj = 3 + 4j$ if $a = 3$ and $b = 4$

real part (3) — imaginary part (4)

(b) $x + 3j = yj - 5$

$x - yj = -5 - 3j$ if $x = -5$ and $y = 3$

(c) What values of x and y satisfy the equation $x + 3(xj + y) = 5 - j - yj$?

Rewriting each side in the form $a + bj$, and equating real parts and equating imaginary parts, we have

$$(x + 3y) + 3xj = 5 + (-1 - y)j$$

For equality, $x + 3y = 5$ and $3x = -1 - y$. The solution of this system of equations is $x = -1$ and $y = 2$.

Practice Exercise

Evaluate x and y:

4. $4 - 6j - x = j + yj$

The **conjugate** *of the complex number $a + bj$ is the complex number $a - bj$.* We see that the sign of the imaginary part is changed to obtain the conjugate.

EXAMPLE 6 Conjugates

Find the conjugate of each complex number.

(a) $3 - 2j$ Conjugate: $3 + 2j$

(b) $3 + 2j$ Conjugate: $3 - 2j$

(c) $-2 - 5j$ Conjugate: $-2 + 5j$

(d) $6j$ Conjugate: $-6j$

(e) 3 Conjugate: 3 (imaginary part is 0)

EXERCISES 12.1

In Exercises 1–4, perform the indicated operations on the resulting expressions if the given changes are made in the indicated examples of this section.

1. In Example 1(c), put a j in front of the radical and then simplify.

2. In Example 3, put a − sign before the first radical of the first illustration and then simplify.

3. In Example 4(a), add 40 to the exponent and then evaluate.

4. In Example 5(c), change the 5 on the right side of the equation to −5 and then solve.

In Exercises 5–16, express each number in terms of j.

5. $\sqrt{-81}$ **6.** $\sqrt{-121}$ **7.** $-\sqrt{-4}$ **8.** $-\sqrt{-49}$

9. $\sqrt{-0.36}$ **10.** $-\sqrt{-0.01}$ **11.** $4\sqrt{-8}$ **12.** $3\sqrt{-48}$

13. $\sqrt{-\frac{7}{4}}$ **14.** $-\sqrt{-\frac{5}{9}}$ **15.** $-\sqrt{-4e^2}$ **16.** $\sqrt{-\pi^4}$

In Exercises 17–32, simplify each of the given expressions.

17. (a) $(\sqrt{-7})^2$ **(b)** $\sqrt{(-7)^2}$

18. (a) $\sqrt{(-15)^2}$ **(b)** $(\sqrt{-15})^2$

19. (a) $\sqrt{(-2)(-8)}$ **(b)** $\sqrt{-2}\sqrt{-8}$

20. (a) $\sqrt{-9}\sqrt{-16}$ **(b)** $\sqrt{(-9)(-16)}$

21. $\sqrt{-\frac{1}{15}}\sqrt{-\frac{27}{5}}$ **22.** $-\sqrt{\left(-\frac{4}{7}\right)\left(-\frac{49}{16}\right)}$

23. $-\sqrt{-5}\sqrt{-2}\sqrt{-10}$ **24.** $-\sqrt{(-3)(-7)}\sqrt{-21}$

25. (a) $-j^6$ **(b)** $(-j)^6$ **26. (a)** $-j^{21}$ **(b)** $(-j)^{21}$

27. $j^2 - j^6$ **28.** $2j^5 - 1/j^{-2}$ **29.** $j^{15} - j^{13}$

30. $3j^{48} + j^{200}$ **31.** $-\sqrt{(-j)^2}$ **32.** $-\sqrt{-j^2}$

In Exercises 33–44, perform the indicated operations and simplify each complex number to its rectangular form.

33. $2 + \sqrt{-9}$ **34.** $-26 + \sqrt{-64}$ **35.** $3j - \sqrt{-100}$

36. $-\sqrt{1} - \sqrt{-400}$ **37.** $\sqrt{-4j^2} + \sqrt{-4}$ **38.** $5 - 2\sqrt{25j^2}$

39. $2j^2 + 3j$ **40.** $j^3 - 6$ **41.** $\sqrt{18} - \sqrt{-8}$

42. $\sqrt{-27} + \sqrt{12}$ **43.** $(\sqrt{-2})^2 + j^4$ **44.** $(2\sqrt{2})^2 - j^6$

In Exercises 45–48, find the conjugate of each complex number.

45. (a) $6 - 7j$ **(b)** $8 + j$ **46. (a)** $-3 + 2j$ **(b)** $-9 - j$

47. (a) $2j$ **(b)** -4 **48. (a)** 6 **(b)** $-5j$

In Exercises 49–54, find the values of x and y that satisfy the given equations.

49. $7x - 2yj = 14 + 4j$ **50.** $2x + 3yj = -6 + 12j$

51. $6j - 7 = 3 - x - yj$ **52.** $9 - j = xj + 1 - y$

53. $x - 2j^2 + 7j = yj + 2xj^3$

54. $2x - 6xj^3 - 3j^2 = yj - y + 7j^5$

In Exercises 55–58, solve the given equations for x. Express the answer in simplified form in terms of j.

55. $x^2 + 32 = 0$

56. $x^2 - 2x + 2 = 0$

57. $x^2 + 2x + 7 = 0$

58. $3x^2 - 6x + 4 = 0$

In Exercises 59–68, answer the given questions.

59. How is a number changed if it is multiplied by (a) j^4; (b) j^2?

60. Evaluate: (a) j^{-8}; (b) j^{-6}.

61. Are $8j$ and $-8j$ the solutions to the equation $x^2 + 64 = 0$?

62. Are $2j$ and $-2j$ solutions to the equation $x^4 + 16 = 0$?

63. Evaluate $j + j^2 + j^3 + j^4 + j^5 + j^6 + j^7 + j^8$.

64. What is the smallest positive value of n for which $j^{-1} = j^n$?

65. Is it possible that a given complex number and its conjugate are equal? Explain.

66. Is it possible that a given complex number and the negative of its conjugate are equal? Explain.

67. Show that the real part of $x + yj$ equals the imaginary part of $j(x + yj)$.

68. Explain why a real number is a complex number, but a complex number may not be a real number.

Answers to Practice Exercises

1. -10 **2.** $-4j\sqrt{3}$ **3.** $-j$ **4.** $x = 4, y = -7$

12.2 Basic Operations with Complex Numbers

The basic operations of addition, subtraction, multiplication, and division of complex numbers are based on the operations for binomials with real coefficients. In performing these operations, we treat j as we would any other literal number, although we must properly handle any powers of j that might occur. However, ***we must be careful to express all complex numbers in terms of j before performing these operations.*** We have the following definitions.

Basic Operations on Complex Numbers

Addition:

$$(a + bj) + (c + dj) = (a + c) + (b + d)j \quad \textbf{(12.3)}$$

Subtraction:

$$(a + bj) - (c + dj) = (a - c) + (b - d)j \quad \textbf{(12.4)}$$

Multiplication:

$$(a + bj)(c + dj) = (ac - bd) + (ad + bc)j \quad \textbf{(12.5)}$$

Division:

$$\frac{a + bj}{c + dj} = \frac{(a + bj)(c - dj)}{(c + dj)(c - dj)} = \frac{(ac + bd) + (bc - ad)j}{c^2 + d^2} \quad \textbf{(12.6)}$$

Eqs. (12.3) and (12.4) show that we *add and subtract complex numbers by combining the real parts and combining the imaginary parts.*

EXAMPLE 1 Adding and subtracting complex numbers

(a) $(3 - 2j) + (-5 + 7j) = (3 - 5) + (-2 + 7)j$
$= -2 + 5j$

(b) $(7 + 9j) - (6 - 4j) = (7 - 6) + (9 - (-4))j$
$= 1 + 13j$

When complex numbers are multiplied, Eq. (12.5) indicates that *we express numbers in terms of j, proceed as with any algebraic multiplication, and note that* $j^2 = -1$.

EXAMPLE 2 Multiplying complex numbers

(a) $(6 - \sqrt{-4})(\sqrt{-9}) = (6 - 2j)(3j)$ — write in terms of j

$= 18j - 6j^2 = 18j - 6(-1)$

$= 6 + 18j$

(b) $(-9.4 - 6.2j)(2.5 + 1.5j) = (-9.4)(2.5) + (-9.4)(1.5j)$

$+ (-6.2j)(2.5) + (-6.2j)(1.5j)$

$= -23.5 - 14.1j - 15.5j - 9.3j^2$

$= -23.5 - 29.6j - 9.3(-1)$

$= -14.2 - 29.6j$

Practice Exercise

Multiply: **1.** $(3 - 7j)(9 + 2j)$

LEARNING TIP

To divide by a complex number, multiply the numerator and the denominator by the conjugate of the denominator.

The procedure shown in Eq. (12.6) for dividing by a complex number is the same as that used for rationalizing the denominator of a fraction (that is, multiplying the numerator and denominator by the conjugate of the denominator). The result is in the proper form of a complex number.

EXAMPLE 3 Dividing complex numbers

■ Many calculators are programmed for complex numbers. They use i instead of j.

(a) $\frac{7 - 2j}{3 + 4j} = \frac{(7 - 2j)(3 - 4j)}{(3 + 4j)(3 - 4j)}$ — multiply by conjugate of denominator

$= \frac{21 - 28j - 6j + 8j^2}{9 - 16j^2} = \frac{21 - 34j + 8(-1)}{9 - 16(-1)}$

$= \frac{13 - 34j}{25}$

This could be written in the form $a + bj$ as $\frac{13}{25} - \frac{34}{25}j$, but this type of result is generally left as a single fraction. In decimal form, the result would be expressed as $0.52 - 1.36j$.

(b) $\frac{1}{j} + \frac{2}{3 + j} = \frac{(3 + j) + 2j}{j(3 + j)} = \frac{3 + 3j}{3j - 1} = \frac{3 + 3j}{-1 + 3j} \cdot \frac{-1 - 3j}{-1 - 3j}$

$= \frac{-3 - 12j - 9j^2}{1 - 9j^2} = \frac{6 - 12j}{10}$

$= \frac{3 - 6j}{5}$

Practice Exercise

Divide: **2.** $\frac{5 - 3j}{2 + 7j}$

EXAMPLE 4 Dividing complex numbers—alternating-current circuits

In an alternating-current circuit, the voltage V is given by $V = IZ$, where I is the current (in A) and Z is the impedance (in Ω). Each of these can be represented by complex numbers. Find the complex number representation for I if $V = 4.20 - 3.00j$ volts and $Z = 5.30 + 2.65j$ ohms. (This type of circuit is discussed in Section 12.7.)

Since $I = V/Z$, we have

$$I = \frac{4.20 - 3.00j}{5.30 + 2.65j} = \frac{(4.20 - 3.00j)(5.30 - 2.65j)}{(5.30 + 2.65j)(5.30 - 2.65j)}$$ — multiply by conjugate of denominator

$$= \frac{22.26 - 11.13j - 15.90j + 7.95j^2}{5.30^2 - 2.65^2j^2} = \frac{22.26 - 7.95 - 27.03j}{5.30^2 + 2.65^2}$$

$$= \frac{14.31 - 27.03j}{35.11} = 0.408 - 0.770j \text{ amperes}$$

EXERCISES 12.2

In Exercises 1–4, perform the indicated operations on the resulting expressions if the given changes are made in the indicated examples of this section.

1. In Example 1(b), change the sign in the first parentheses from + to − and then perform the subtraction.
2. In Example 2(b), change the sign before $6.2j$ from − to + and then perform the multiplication.
3. In Example 3(a), change the sign in the denominator from + to − and then simplify.
4. In Example 3(b), change the sign in the second denominator from + to − and then simplify.

In Exercises 5–42, perform the indicated operations, expressing all answers in the form $a + bj$.

5. $(3 - 7j) + (2 - j)$
6. $(-4 - j) + (-7 - 4j)$
7. $(7j - 6) - (19 - 3j)$
8. $(5.4 - 3.4j) - (2.9j + 5.5)$
9. $0.23 - (0.46 - 0.19j) + 0.67j$
10. $(7 - j) - (4 - 4j) + (6 - j)$
11. $(12j - 21) - (15 - 18j) - 9j$
12. $(0.062j - 0.073) - 0.030j - (0.121 - 0.051j)$
13. $(7 - j)(7j)$
14. $(-2.2j)(1.5j - 4.0)$
15. $(4 - j)(5 + 2j)$
16. $(8j - 5)(7 + 4j)$
17. $(\sqrt{-18}\sqrt{-4})(3j)$
18. $\sqrt{-6}\sqrt{-12}\sqrt{30}$
19. $7j^3 - 7\sqrt{-9}$
20. $6j - 5j^2\sqrt{-63}$
21. $j\sqrt{-7} - j^6\sqrt{112} + 3j$
22. $j^2\sqrt{-7} - \sqrt{-28} + 8$
23. $(3 - 7j)^2$
24. $(8j + 20)^2$
25. $(4 - j)(5 + 2j)$
26. $(8 + 3j)(8 - 3j)$
27. $\dfrac{6j}{2 - 5j}$
28. $\dfrac{0.25}{3.0 - \sqrt{-1.0}}$
29. $\dfrac{1 - j}{3j}$
30. $\dfrac{12 + 10j}{6 - 8j}$
31. $\dfrac{j\sqrt{2} - 5}{j\sqrt{2} + 3}$
32. $\dfrac{j^5 - j^3}{3 + j}$
33. $\dfrac{j^2 - j}{2j - j^8}$
34. $\dfrac{3}{2j} + \dfrac{5}{6 - j}$
35. $\dfrac{4j}{1 - j} - \dfrac{8 + j}{2 + 3j}$
36. $\dfrac{(6j + 5)(2 - 4j)}{(5 - j)(4j + 1)}$
37. $(4j^5 - 5j^4 + 2j^3 - 3j^2)^2$
38. $(2j^2 - 3j^3 + 2j^4 - 2j^5)^6$
39. $(3j^2 - 5j^4)(4j^3 - 6j)$
40. $(5j - 4j^2 + 3j^7)(2j^{12} - j^{13})$
41. $\dfrac{(2 - j^3)^4}{(j^8 - j^6)^3} + j$
42. $(1 + j)^{-3}(2 - j)^{-2}$

In Exercises 43–54, answer the given problems.

43. Show that $-1 - j$ is a solution to the equation $x^2 + 2x + 2 = 0$.
44. Show that $1 - j\sqrt{3}$ is a solution to the equation $x^2 + 4 = 2x$.
45. Multiply $-3 + j$ by its conjugate.
46. Divide $2 - 3j$ by its conjugate.
47. Write the reciprocal of $3 - j$ in rectangular form.
48. Write the reciprocal of $2 + 5j$ in rectangular form.
49. Write $j^{-2} + j^{-3}$ in rectangular form.
50. Solve for x: $(x + 2j)^2 = 5 + 12j$
51. Solve for x: $(x + 3j)^2 = 7 - 24j$
52. For $\dfrac{3}{5} + \dfrac{4}{5}j$, find: (a) the conjugate; (b) the reciprocal.
53. If $f(x) = x + \dfrac{1}{x}$, find $f(1 + 3j)$.
54. When finding the current in a certain electric circuit, the expression $(s + 1 + 4j)(s + 1 - 4j)$ occurs. Simplify this expression.

In Exercises 55–58, solve the given problems. Refer to Example 4.

55. If $I = 0.835 - 0.427j$ amperes and $Z = 250 + 170j$ ohms, find the complex-number representation for V.
56. If $V = 5.70 - 3.65j$ volts and $I = 0.360 - 0.525j$ amperes, find the complex-number representation for Z.
57. If $V = 85 + 74j$ volts and $Z = 2500 - 1200j$ ohms, find the complex-number representation for I.
58. In an alternating-current circuit, two impedances Z_1 and Z_2 have a total impedance Z_T of $Z_T = \dfrac{Z_1Z_2}{Z_1 + Z_2}$. Find Z_T for $Z_1 = 3200 + 4800j$ ohms and $Z_2 = 4800 - 6400j$ ohms.

In Exercises 59–62, answer or explain as indicated.

59. What type of number is the result of (a) adding a complex number to its conjugate and (b) subtracting a complex number from its conjugate?
60. If the reciprocal of $a + bj$ equals $a - bj$, what condition must a and b satisfy?
61. Explain why the product of a complex number and its conjugate is real and nonnegative.
62. Explain how to show that the reciprocal of the imaginary unit is the negative of the imaginary unit.

Answers to Practice Exercises

1. $41 - 57j$ 2. $\dfrac{-11 - 41j}{53}$

12.3 Graphical Representation of Complex Numbers

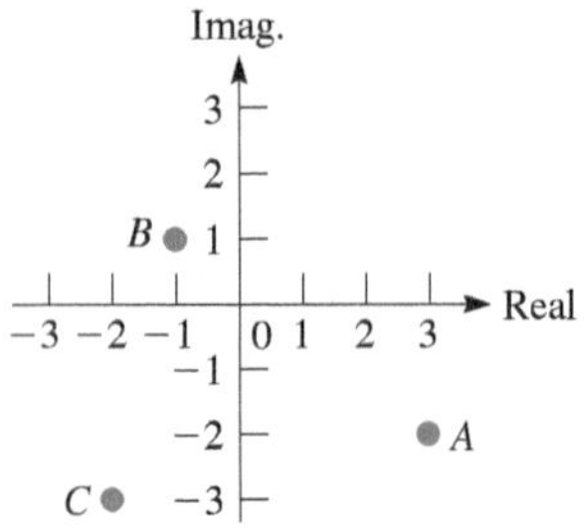

Fig. 12.2

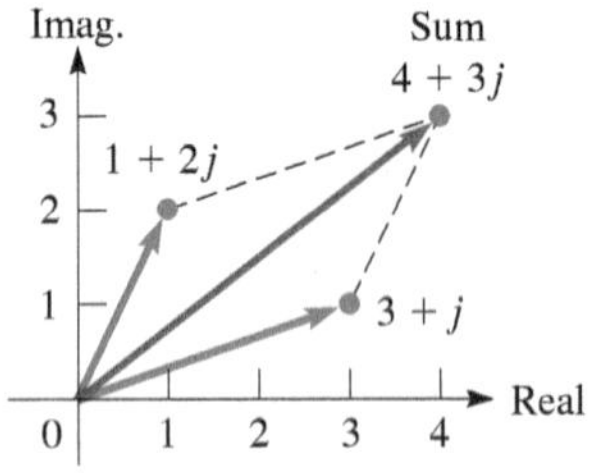

Fig. 12.3

Graphically, we represent a complex number as a ***point***, designated as $a + bj$, in the rectangular coordinate system. *The **real part** is the **x-value*** of the point, and the ***imaginary part** is the **y-value*** of the point. Used in this way, *the coordinate system is called the* **complex plane**, *the horizontal axis is the* **real axis**, and *the vertical axis is the* **imaginary axis**.

EXAMPLE 1 Complex number in the complex plane

In Fig. 12.2, A represents the complex number $3 - 2j$, B represents $-1 + j$, and C represents $-2 - 3j$. These complex numbers are represented by the points $(3, -2)$, $(-1, 1)$, and $(-2, -3)$, respectively, of the standard rectangular coordinate system.

In the complex plane, consider two complex numbers—for example, $1 + 2j$ and $3 + j$ —and their sum $4 + 3j$. Drawing lines from the origin to these points (see Fig. 12.3), note that if we think of the complex numbers as being vectors, their sum is the vector sum. Because complex numbers can be used to represent vectors, the numbers are particularly important. *Any complex number can be thought of as representing a vector from the origin to its point in the complex plane.* This leads to the method used to add complex numbers graphically.

LEARNING TIP

Keep in mind that the meaning given to the points representing complex numbers in the complex plane is different from the meaning given to the points in the standard rectangular coordinate system. *A point in the complex plane represents a* single complex number, whereas *a point in the rectangular coordinate system represents a* pair of real numbers.

Steps to Add Complex Numbers Graphically

1. Find the point corresponding to one of the numbers and draw a line from the origin to this point.
2. Repeat Step 1 for the second number.
3. Complete a parallelogram with the lines drawn as adjacent sides. The resulting fourth vertex is the point representing the sum.

Note that *this is equivalent to adding vectors by graphical means.*

EXAMPLE 2 Adding complex numbers graphically

(a) Add the complex numbers $5 - 2j$ and $-2 - j$ graphically.

The solution is shown in Fig. 12.4. We see that the fourth vertex of the parallelogram is at $3 - 3j$. This is, of course, the algebraic sum of these two complex numbers.

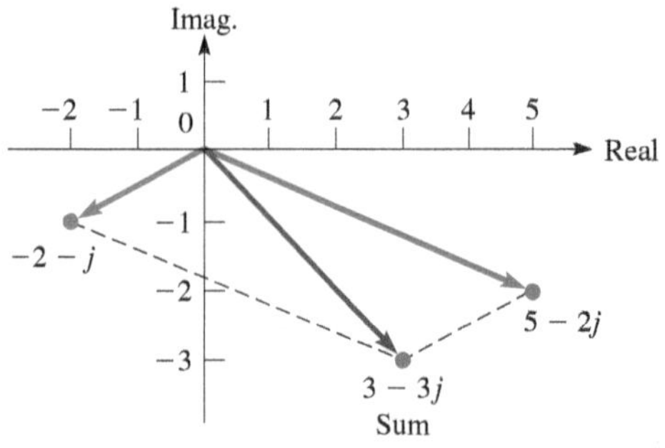

Fig. 12.4

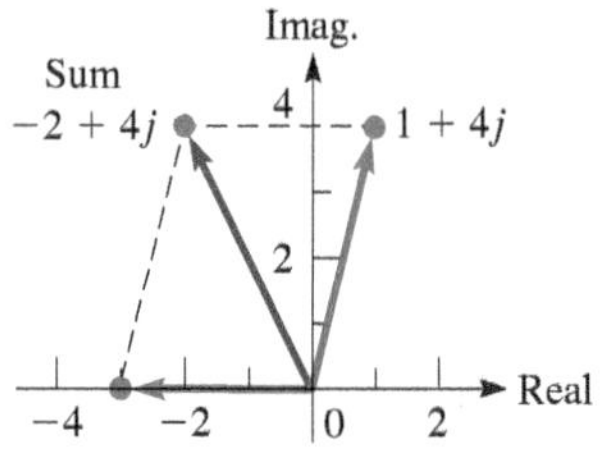

Fig. 12.5

(b) Add the complex numbers -3 and $1 + 4j$.

First, note that $-3 = -3 + 0j$, which means that the point representing -3 is on the negative real axis. In Fig. 12.5, we show the numbers -3 and $1 + 4j$ on the graph and complete the parallelogram. From the graph, we see that the sum is $-2 + 4j$.

EXAMPLE 3 Subtracting complex numbers graphically

Subtract $4 - 2j$ from $2 - 3j$ graphically.

Subtracting $4 - 2j$ is equivalent to adding $-4 + 2j$. Therefore, we complete the solution by adding $-4 + 2j$ and $2 - 3j$, as shown in Fig. 12.6. The result is $-2 - j$.

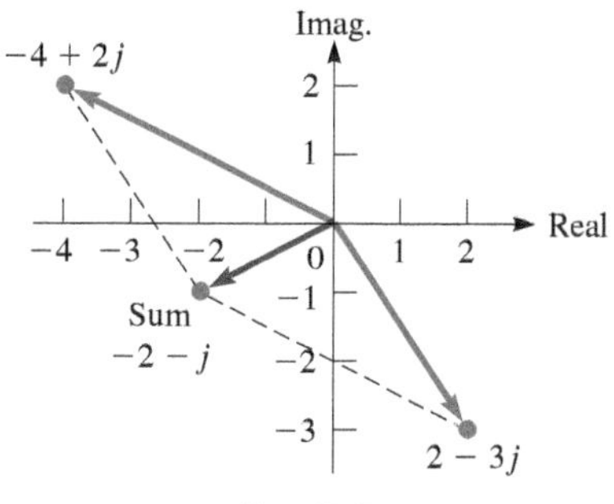

Fig. 12.6

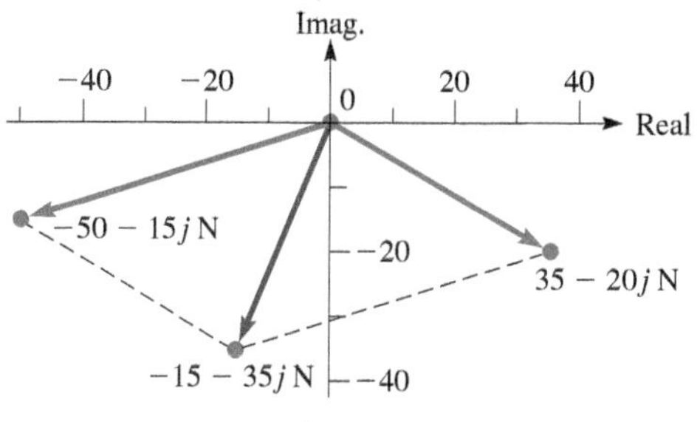

Fig. 12.7

EXAMPLE 4 Adding complex numbers—forces on a bolt

Two forces acting on an overhead bolt can be represented by $35 - 20j$ N and $-50 - 15j$ N. Find the resultant force graphically.

The forces are shown in Fig. 12.7. From the graph, we can see that the sum of the forces, which is the resultant force, is $-15 - 35j$ N.

EXERCISES 12.3

In Exercises 1 and 2, perform the indicated operations for the resulting complex numbers if the given changes are made in the indicated examples of this section.

1. In Example 2(a), change the sign of the imaginary part of the second complex number and then add the numbers graphically.
2. In Example 3, change the sign of the imaginary part of the second complex number and do the subtraction graphically.

In Exercises 3–8, locate the given numbers in the complex plane.

3. $2 + 6j$
4. $-5 + j$
5. $-4 - 3j$
6. 10
7. $-3j$
8. $3 - 4j$

In Exercises 9–28, perform the indicated operations graphically. Check them algebraically.

9. $2 + (3 + 4j)$
10. $2j + (-2 + 3j)$
11. $(5 - j) + (3 + 2j)$
12. $(3 - 2j) + (-1 - j)$
13. $5j - (1 - 4j)$
14. $(0.2 - 0.1j) - 0.1$
15. $(2 - 4j) + (-2 + j)$
16. $(-1 - 2j) + (6 - j)$
17. $(3 - 2j) - (4 - 6j)$
18. $(-25 - 40j) - (20 - 55j)$
19. $(80 + 300j) - (260 + 150j)$
20. $(-j - 2) - (-1 - 3j)$
21. $(1.5 - 0.5j) + (3.0 + 2.5j)$
22. $(3.5 + 2.0j) - (1.5j - 4.0)$
23. $(3 - 6j) - (-1 - 8j)$
24. $(-6 - 3j) + (2 - 7j)$
25. $(2j + 1) - 3j - (j + 1)$
26. $(6 - j) - 9 - (2j - 3)$
27. $(j - 6) - j + (j - 7)$
28. $j - (1 - j) + (3 + 2j)$

In Exercises 29–32, show the given number, its negative, and its conjugate on the same coordinate system.

29. $3 + 2j$
30. $-2 + 4j$
31. $-3 - 5j$
32. $5 - j$

In Exercises 33 and 34, show the numbers $a + bj$, $3(a + bj)$, and $-3(a + bj)$ on the same coordinate system. The multiplication of a complex number by a real number is called **scalar multiplication** *of the complex number.*

33. $3 - j$
34. $-10 - 30j$

In Exercises 35–38, perform the indicated vector operations graphically on the complex number $2 + 4j$.

35. Graph the complex number and its conjugate. Describe the relative positions.
36. Add the number and its conjugate. Describe the result.
37. Subtract the conjugate from the number. Describe the result.
38. Graph the number, the number multiplied by j, the number multiplied by j^2, and the number multiplied by j^3 on the same graph. Describe the result of multiplying a complex number by j.

In Exercises 39 and 40, perform the indicated vector additions graphically. Check them algebraically.

39. Two ropes hold a boat at a dock. The tensions in the ropes can be represented by $40 + 10j$ N and $50 - 25j$ N. Find the resultant force.
40. Relative to the air, a plane heads north of west with a velocity that can be represented by $-480 + 210j$ km/h. The wind is blowing from south of west with a velocity that can be represented by $60 + 210j$ km/h. Find the resultant velocity of the plane.

12.4 Polar Form of a Complex Number

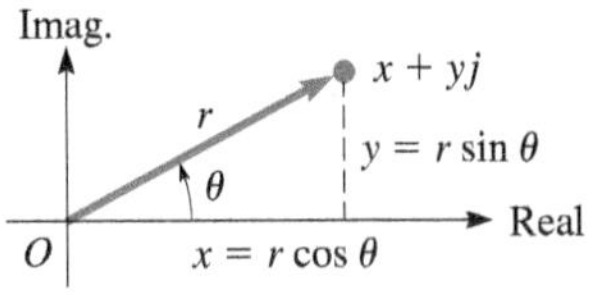

Fig. 12.8

In this section, we use the fact that a complex number can be represented by a vector to write complex numbers in another form. This form has advantages when multiplying and dividing complex numbers, and we will discuss these operations later in this chapter.

In the complex plane, by drawing a vector from the origin to the point that represents the number $x + yj$, an angle θ in standard position is formed. The point $x + yj$ is r units from the origin. In fact, *we can find any point in the complex plane by knowing the angle θ and the value of r.* The equations relating x, y, r, and θ are similar to those developed for vectors in Chapter 9. Referring to Fig. 12.8, we see that

$$x = r \cos \theta \qquad y = r \sin \theta \tag{12.7}$$

$$r^2 = x^2 + y^2 \qquad \tan \theta = \frac{y}{x} \tag{12.8}$$

■ Recall that θ and θ_{ref} are related in the four quadrants as follows:

$\theta = \theta_{ref}$ (first quadrant)
$\theta = 180° - \theta_{ref}$ (second quadrant)
$\theta = 180° + \theta_{ref}$ (third quadrant)
$\theta = 360° - \theta_{ref}$ (fourth quadrant)

Note that, except for the case of applications to circuits, we will find angles that are not in the first quadrant by disregarding the signs of x and y and using the tangent to find a reference angle. Depending on the quadrant, the required angle θ is then found as we learned in Chapter 8.

Substituting Eq. (12.7) into the rectangular form $x + yj$ of a complex number, we have $x + yj = r \cos \theta + j(r \sin \theta)$, or

$$x + yj = r(\cos \theta + j \sin \theta) \tag{12.9}$$

The right side of Eq. (12.9) is called the **polar form** (sometimes the **trigonometric form**). *The length r is the* **absolute value**, or *the* **modulus**, and *θ is the* **argument** *of the complex number.* Eq. (12.9), along with Eq. (12.8), defines the polar form of a complex number.

EXAMPLE 1 Representing a number in polar form

Represent the complex number $3 + 4j$ graphically and give its polar form.

From the rectangular form $3 + 4j$, we see that $x = 3$ and $y = 4$. Using Eq. (12.8), we have

$$r = \sqrt{3^2 + 4^2} = 5 \qquad \theta = \tan^{-1}\frac{4}{3} = 53.1°$$

Thus, the polar form is $5(\cos 53.1° + j \sin 53.1°)$. See Fig. 12.9.

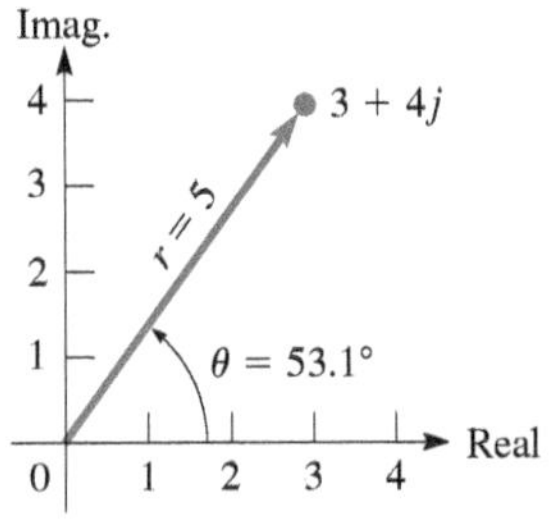

Fig. 12.9

In Example 1, in expressing the polar form as $5(\cos 53.1° + j \sin 53.1°)$, we rounded the angle to the nearest 0.1°, as it is not possible to express the result *exactly* in degrees. In dealing with inexact numbers, we will express the angle to the nearest 0.1°. Other results that cannot be written exactly will be expressed to three significant digits, unless a different accuracy is given in the problem. Of course, in applied situations, most numbers are approximate, as they are derived through measurement.

Another convenient and widely used notation for the polar form is $r\underline{/\theta}$. We must remember in using this form that it represents a complex number and is simply a shorthand way of writing $r(\cos \theta + j \sin \theta)$. Therefore,

$$r\underline{/\theta} = r(\cos \theta + j \sin \theta) \tag{12.10}$$

Practice Exercise

Write in polar form: **1.** $15 - 8j$

EXAMPLE 2 Polar form $r\underline{/\theta}$

(a) $3(\cos 40° + j \sin 40°) = 3\underline{/40°}$

(b) $6.26(\cos 217.3° + j \sin 217.3°) = 6.26\underline{/217.3°}$

(c) $5\underline{/120°} = 5(\cos 120° + j \sin 120°)$

(d) $14.5\underline{/306.2°} = 14.5(\cos 306.2° + j \sin 306.2°)$

EXAMPLE 3 Rectangular form to polar form

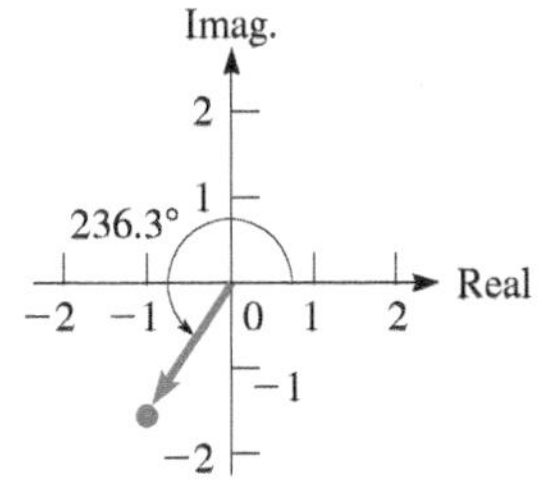

Fig. 12.10

Represent the complex number $-1.04 - 1.56j$ graphically and give its polar form.

The graphical representation is shown in Fig. 12.10. From Eq. (12.8), we have

$$r = \sqrt{(-1.04)^2 + (-1.56)^2} = 1.87$$

$$\theta_{ref} = \tan^{-1}\frac{1.56}{1.04} = 56.3^\circ \qquad \theta = 180^\circ + 56.3^\circ = 236.3^\circ$$

Since both the real and imaginary parts are negative, θ is a third-quadrant angle. Therefore, we found the reference angle before finding θ. This means the polar forms are

$$1.87(\cos 236.3^\circ + j\sin 236.3^\circ) = 1.87\angle 236.3^\circ$$

EXAMPLE 4 Polar form to rectangular form—impedance

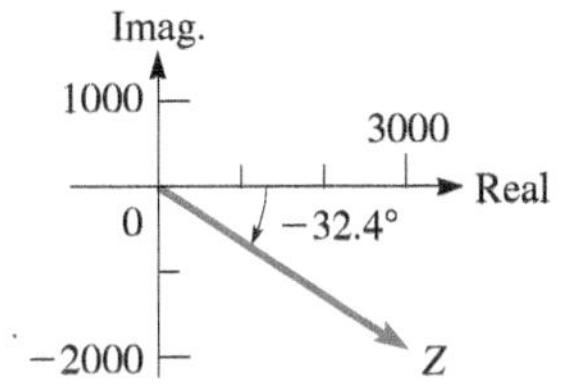

Fig. 12.11

The impedance Z (in Ω) in an alternating-current circuit is given by $Z = 3560\angle -32.4^\circ$. Express this in rectangular form.

From the polar form, we have $r = 3560\ \Omega$ and $\theta = -32.4^\circ$ (it is common to use negative angles in this type of application). This means that we can also write

$$Z = 3560(\cos(-32.4^\circ) + j\sin(-32.4^\circ))$$

See Fig. 12.11. This means that

$$x = 3560\cos(-32.4^\circ) = 3010$$

$$y = 3560\sin(-32.4^\circ) = -1910$$

Therefore, the rectangular form is $Z = 3010 - 1910j\ \Omega$.

The following table summarizes the polar forms of real and imaginary numbers. Note how the polar form angle is always a multiple of 90°.

Complex number	Position on complex plane	Argument	Polar form
$a,\ a > 0$	positive real axis	0°	$a(\cos 0^\circ + j\sin 0^\circ) = a\ \angle 0^\circ$
$a,\ a < 0$	negative real axis	180°	$\lvert a\rvert(\cos 180^\circ + j\sin 180^\circ) = a\ \angle 180^\circ$
$bj,\ b > 0$	positive imaginary axis	90°	$b(\cos 90^\circ + j\sin 90^\circ) = b\ \angle 90^\circ$
$bj,\ b < 0$	negative imaginary axis	270°	$\lvert b\rvert(\cos 270^\circ + j\sin 270^\circ) = a\ \angle 270^\circ$

Practice Exercise

Write in rectangular form:

2. $2.50\angle 120^\circ$

EXAMPLE 5 Polar form—real and imaginary numbers

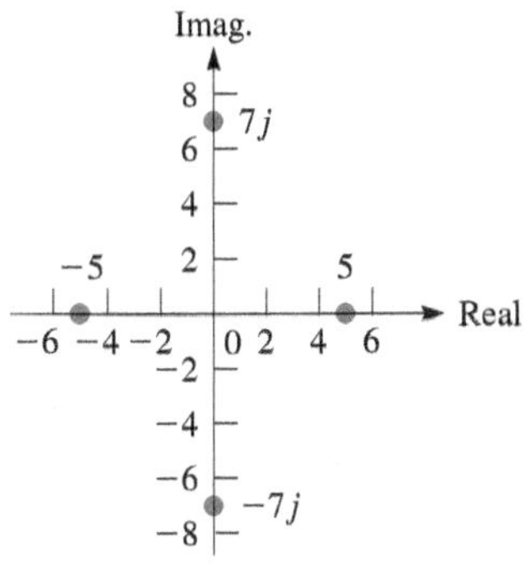

Fig. 12.12

Represent the numbers 5, −5, 7*j*, and −7*j* in polar form. See Fig. 12.12.

$$5 = 5(\cos 0^\circ + j\sin 0^\circ) = 5\angle 0^\circ$$

$$-5 = 5(\cos 180^\circ + j\sin 180^\circ) = 5\angle 180^\circ$$

$$7j = 7(\cos 90^\circ + j\sin 90^\circ) = 7\angle 90^\circ$$

$$-7j = 7(\cos 270^\circ + j\sin 270^\circ) = 7\angle 270^\circ$$

EXERCISES 12.4

In Exercises 1 and 2, change the sign of the real part of the complex number in the indicated example of this section and then perform the indicated operations for the resulting complex number.

1. Example 1 **2.** Example 3

In Exercises 3–18, represent each complex number graphically and give the polar form of each.

3. $8 + 6j$ **4.** $-8 - 15j$

5. $30 - 40j$ **6.** $-5 + 12j$

7. $-2.00 + 3.00j$ **8.** $7.00 - 5.00j$

9. $-0.55 - 0.24j$ **10.** $460 - 460j$

11. $1 + j\sqrt{3}$ **12.** $\sqrt{2} - j\sqrt{2}$

13. $3.514 - 7.256j$ **14.** $62.31 + 95.27j$

15. -3 **16.** 60 **17.** $9j$ **18.** $-2j$

In Exercises 19–36, represent each complex number graphically and give the rectangular form of each.

19. $5.00(\cos 54.0° + j \sin 54.0°)$

20. $6(\cos 180° + j \sin 180°)$

21. $160(\cos 150.0° + j \sin 150.0°)$

22. $2.50(\cos 315.0° + j \sin 315.0°)$

23. $3.00(\cos 232.0° + j \sin 232.0°)$

24. $220.8(\cos 155.13° + j \sin 155.13°)$

25. $0.08(\cos 360° + j \sin 360°)$

26. $15(\cos 0° + j \sin 0°)$

27. $120(\cos 270° + j \sin 270°)$

28. $\cos 600.0° + j \sin 600.0°$

29. $4.75\underline{/172.8°}$ **30.** $1.50\underline{/62.3°}$

31. $0.9326\underline{/229.54°}$ **32.** $277.8\underline{/-342.63°}$

33. $7.32\underline{/-270°}$ **34.** $18.3\underline{/540.0°}$

35. $86.42\underline{/94.62°}$ **36.** $4629\underline{/182.44°}$

In Exercises 37–44, solve the given problems.

37. What is the argument for any negative real number?

38. For $x + yj$, what is the argument if $x = y < 0$?

39. Show that the conjugate of $r\underline{/\theta}$ is $r\underline{/-\theta}$.

40. Find r and $\tan(\theta_{ref})$ for the complex number $a^b(a + bj)$. $(a > 0)$

41. The voltage of a certain generator is represented by $2.84 - 1.06j$ kV. Write this voltage in polar form.

42. Find the magnitude and direction of a force on a bolt that is represented by $40.5 + 24.5j$ newtons.

43. The electric field intensity of a light wave can be described by $12.4\underline{/78.3°}$ V/m. Write this in rectangular form.

44. The current in a certain microprocessor circuit is represented by $3.75\underline{/15.0°}$ μA. Write this in rectangular form.

Answers to Practice Exercises

1. $17(\cos 331.9° + j \sin 331.9°)$
2. $-1.25 + 2.17j$

12.5 Exponential Form of a Complex Number

Another important form of a complex number is the *exponential form.* It is commonly used in electronics, engineering, and physics applications. As we will see in the next section, it is also convenient for multiplication and division of complex numbers, as the rectangular form is for addition and subtraction.

■ The number *e* is named for the Swiss mathematician Leonhard Euler (1707–1783). His works in mathematics and other fields filled about 70 volumes.

The **exponential form** *of a complex number is written as* $re^{j\theta}$, where r and θ have the same meanings as given in the previous section, although θ is expressed in radians. *The number e is a special irrational number and has an approximate value:*

$$e = 2.718\,281\,828\,459\,045\,2$$

This number e is very important in mathematics, and we will see it again in the next chapter. For now, it is necessary to accept the value for e, although in calculus its meaning is shown along with the reason it has the above value. We can find its value on a calculator by using the e^x key, with $x = 1$, or by using the e key.

In advanced mathematics, it is shown that

$$re^{j\theta} = r(\cos \theta + j \sin \theta) \tag{12.11}$$

LEARNING TIP
We will always express θ in radians when using the exponential form.

By expressing θ in radians, the expression $j\theta$ is an exponent that can be shown to obey all the laws of exponents as discussed in Chapter 11.

EXAMPLE 1 Polar form to exponential form

Express the number $8.50\angle 136.3^\circ$ in exponential form.

Since this complex number is in polar form, we note that $r = 8.50$ and that we must express 136.3° in radians. Changing 136.3° to radians, we have

$$136.3^\circ = 136.3^\circ\left(\frac{\pi \text{ rad}}{180^\circ}\right) = 2.38 \text{ rad}$$

Therefore, the required exponential form is $8.50e^{2.38j}$. This means that

$$8.50\angle 136.3^\circ = 8.50e^{2.38j}$$

degrees to radians

value of r

LEARNING TIP

The principal step in changing from polar form to exponential form is to change the argument θ from degrees to radians.

Practice Exercise

1. Express $25.0(\cos 127.0^\circ + j \sin 127.0^\circ)$ in exponential form.

EXAMPLE 2 Rectangular form to exponential form

Express the number $3.07 - 7.43j$ in exponential form.

From the rectangular form, we have $x = 3.07$ and $y = -7.43$. Therefore, using Eq. (12.8), we get

$$r = \sqrt{(3.07)^2 + (-7.43)^2} = 8.04$$

$$\theta_{ref} = \tan^{-1}\frac{7.43}{3.07} = 67.6^\circ \qquad \theta = 360^\circ - 67.6^\circ = 292.4^\circ$$ θ is in the fourth quadrant

Since $292.4^\circ = 5.10$ rad, the exponential form is $8.04e^{5.10j}$. This means that

$$3.07 - 7.43j = 8.04e^{5.10j}$$

Note that we could have used radian mode directly to obtain θ_{ref} in radians and then find θ by subtracting from 2π—that is, $\theta = 2\pi - \tan^{-1}(7.43/3.07) = 5.10$.

Practice Exercise

2. Express $-20.5 - 16.8j$ in exponential form.

EXAMPLE 3 Exponential form to other forms

Express the complex number $2.00e^{4.80j}$ in polar and rectangular forms.

We first express 4.80 rad as 275.0°. From the exponential form, we know that $r = 2.00$. Thus, the polar form is

$$2.00(\cos 275.0^\circ + j \sin 275.0^\circ)$$

Using the distributive law, we rewrite the polar form and then evaluate. Thus,

$$\begin{aligned} 2.00e^{4.80j} &= 2.00(\cos 275.0^\circ + j \sin 275.0^\circ) \\ &= 2.00 \cos 275.0^\circ + (2.00 \sin 275.0^\circ)j \\ &= 0.174 - 1.99j \end{aligned}$$

■ Calculators that operate with complex numbers treat the exponential and polar forms as being the same, since each uses r and θ.

EXAMPLE 4 Power of a number in exponential form

Express $(1.846e^{1.229j})^2$ in polar and rectangular forms.

First, we note that $(1.846e^{1.229j})^2 = 1.846^2e^{2.458j}$. Since 2.458 rad $= 140.83^\circ$, the polar form is $1.846^2\angle 140.83^\circ = 3.408\angle 140.83^\circ$.

For the rectangular form, using radian mode, we have

$$\begin{aligned} (1.846e^{1.229j})^2 &= 1.846^2e^{2.458j} = 1.846^2(\cos 2.458 + j \sin 2.458) \\ &= -2.642 + 2.152j \end{aligned}$$

As we noted earlier, an important application of the use of complex numbers is in alternating-current analysis. When an alternating current flows through a given circuit, usually the current and voltage have different phases. That is, they do not reach their peak values at the same time. Therefore, one way of accounting for the magnitude as well as the phase of an electric current or voltage is to write it as a complex number. Here, the modulus is the actual magnitude of the current or voltage, and the argument θ is a measure of the phase.

EXAMPLE 5 Exponential form—alternating-current circuits

A current of $2.00 - 4.00j$ amperes flows in a given circuit. Express this current in exponential form and find the magnitude of the current.

From the rectangular form, we have $x = 2.00$ and $y = -4.00$. Therefore,

$$r = \sqrt{(2.00)^2 + (-4.00)^2} = 4.47 \text{ A}$$

$$\theta = \tan^{-1}\frac{-4.00}{2.00} = -63.4^\circ$$

It is normal to express the phase in terms of negative angles when the imaginary part is negative. Therefore, $63.4^\circ = 1.11$ rad, which means the exponential form is $4.47e^{-1.11j}$. The modulus of 4.47 means the magnitude of the current is 4.47 A.

We now summarize the three important forms of a complex number. See Fig. 12.13.

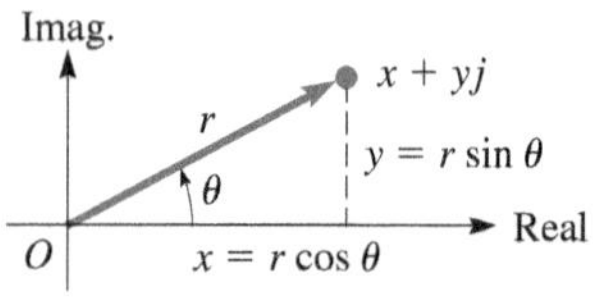

Fig. 12.13

Rectangular:	$x + yj$
Polar:	$r(\cos\theta + j\sin\theta) = r\angle\theta$
Exponential:	$re^{j\theta}$

It follows that

$$x + yj = r(\cos\theta + j\sin\theta) = r\angle\theta = re^{j\theta} \tag{12.12}$$

where

$$r^2 = x^2 + y^2 \qquad \tan\theta = \frac{y}{x} \tag{12.8}$$

In Eq. (12.12), the argument θ is the same for exponential and polar forms. It is usually expressed in radians in exponential form and in degrees in polar form.

EXERCISES 12.5

In Exercises 1–2, perform the indicated operations for the resulting complex numbers if the given changes are made in the indicated examples of this section.

1. In Example 1, change 136.3° to 226.3° and then find the exponential form.

2. In Example 3, change the exponent to $3.80j$ and then find the polar and rectangular forms.

In Exercises 3–22, express the given numbers in exponential form.

3. $3.00(\cos 60.0^\circ + j\sin 60.0^\circ)$ **4.** $575(\cos 135^\circ + j\sin 135^\circ)$

5. $0.450(\cos 282.3^\circ + j\sin 282.3^\circ)$

6. $2.10(\cos 588.7^\circ + j\sin 588.7^\circ)$

7. $375.5[\cos(-95.46^\circ) + j\sin(-95.46^\circ)]$

8. $16.7[\cos(-7.14^\circ) + j\sin(-7.14^\circ)]$

9. $0.5150\angle 198.3^\circ$ **10.** $4650\angle 326.5^\circ$ **11.** $4.06\angle -61.4^\circ$

12. $0.0192\angle 76.7^\circ$ **13.** $9245\angle 296.3^\circ$ **14.** $82.76\angle 470.1^\circ$

15. $3 - 4j$ **16.** $-1 - 5j$ **17.** $-30 + 20j$

18. $600 + 100j$ **19.** $5.90 + 2.40j$ **20.** $47.3 - 10.9j$

21. $-634.6 - 528.2j$ **22.** $-8573 + 5477j$

In Exercises 23–30, express the given complex numbers in polar and rectangular forms.

23. $3.00e^{0.500j}$ **24.** $20.0e^{1.00j}$ **25.** $464e^{1.85j}$

26. $2.50e^{3.84j}$ **27.** $3.20e^{-5.41j}$ **28.** $0.800e^{3.00j}$

29. $0.1724e^{2.391j}$ **30.** $820.7e^{-3.492j}$

In Exercises 31–34, perform the indicated operations and express results in rectangular and polar forms.

31. $(4.55e^{1.32j})^2$ **32.** $(0.926e^{0.253j})^3$

33. $(6.25e^{3.46j})(4.40e^{1.22j})$ **34.** $(18.0e^{5.13j})(25.5e^{0.770j})$

In Exercises 35–38, perform the indicated operations.

35. The impedance in an antenna circuit is $375 + 111j$ ohms. Write this in exponential form and find the magnitude of the impedance.

36. The intensity of a radar microwave signal is $37.0[\cos(-65.3°) + j\sin(-65.3°)]$ V/m. Write this in exponential form.

37. In an electric circuit, the *admittance* is the reciprocal of the impedance. If the impedance is $2800 - 1450j$ ohms in a certain circuit, find the exponential form of the admittance.

38. In dealing with the superposition of waves, the expression $E_1E_2(e^{j(\alpha-\beta)} + e^{-j(\alpha-\beta)})$ occurs. Show that it can be written as $2E_1E_2\cos(\alpha - \beta)$.

Answers to Practice Exercises

1. $25.0e^{2.22j}$ **2.** $26.5e^{3.83j}$

12.6 Products, Quotients, Powers, and Roots of Complex Numbers

The operations of multiplication and division can be performed with complex numbers in polar and exponential forms, as well as rectangular form. Doing them in these forms is convenient, but also leads to finding powers and roots of complex numbers.

Using the exponential form and the laws of exponents, we multiply two complex numbers as

$$[r_1e^{j\theta_1}] \times [r_2e^{j\theta_2}] = r_1r_2e^{j\theta_1+j\theta_2} = r_1r_2e^{j(\theta_1+\theta_2)}$$

We use this equation to express the product of two complex numbers in polar form:

$$[r_1e^{j\theta_1}] \times [r_2e^{j\theta_2}] = [r_1(\cos\theta_1 + j\sin\theta_1)] \times [r_2(\cos\theta_2 + j\sin\theta_2)]$$

and

$$r_1r_2e^{j(\theta_1+\theta_2)} = r_1r_2[\cos(\theta_1+\theta_2) + j\sin(\theta_1+\theta_2)]$$

The polar expressions are equal, which means that *the product of two complex numbers is*

$$\begin{aligned} &r_1(\cos\theta_1 + j\sin\theta_1)r_2(\cos\theta_2 + j\sin\theta_2) \\ &= r_1r_2[\cos(\theta_1+\theta_2) + j\sin(\theta_1+\theta_2)] \\ &(r_1\underline{/\theta_1})(r_2\underline{/\theta_2}) = r_1r_2\underline{/\theta_1+\theta_2} \end{aligned} \quad \textbf{(12.13)}$$

That is, the magnitudes are multiplied, and the angles are added.

EXAMPLE 1 Multiplication in polar form

Multiply the complex numbers $3.61\angle 36.3°$ and $1.41\angle 315.0°$

Multiplying the magnitudes and adding the angles, we get

Multiply

$$(3.61\underline{/56.3°})(1.41\underline{/315.0°}) = (3.61)(1.41)\underline{/56.3° + 315.0°}$$

add

$$= 5.09\underline{/371.3°}$$
$$= 5.09\underline{/11.3°}$$

Note that the angle in the final result is between 0° and 360°. This is usually the case, although in some applications, it is expressed as a negative angle.

Practice Exercise

1. Find the polar form product: $(3\underline{/50°})(5\underline{/65°})$

If we wish to *divide* one complex number in exponential form by another, we arrive at the following result:

$$r_1e^{j\theta_1} \div r_2e^{j\theta_2} = \frac{r_1}{r_2}e^{j(\theta_1-\theta_2)} \tag{12.14}$$

Therefore, *the result of dividing one complex number in polar form by another is given by*

$$\frac{r_1(\cos\theta_1 + j\sin\theta_1)}{r_2(\cos\theta_2 + j\sin\theta_2)} = \frac{r_1}{r_2}[\cos(\theta_1 - \theta_2) + j\sin(\theta_1 - \theta_2)]$$
$$\frac{r_1\angle\theta_1}{r_2\angle\theta_2} = \frac{r_1}{r_2}\angle\theta_1 - \theta_2 \tag{12.15}$$

That is, the magnitudes are divided, and the angles are subtracted.

EXAMPLE 2 Division in polar form

Divide the first complex number of Example 1 by the second.

Using the polar forms, we have

divide

$$\frac{3.61(\cos 56.3^\circ + j\sin 56.3^\circ)}{1.41(\cos 315.0^\circ + j\sin 315.0^\circ)} = \frac{3.61}{1.41}[\cos(56.3^\circ - 315.0^\circ) + j\sin(56.3^\circ - 315.0^\circ)]$$

subtract

$$= 2.56[\cos(-258.7^\circ) + j\sin(-258.7^\circ)]$$
$$= 2.56(\cos 101.3^\circ + j\sin 101.3^\circ)$$

Practice Exercise

2. Find the polar form quotient:

$$\frac{3(\cos 50^\circ + j\sin 50^\circ)}{5(\cos 65^\circ + j\sin 65^\circ)}$$

LEARNING TIP

To add or subtract numbers in polar form, we must change each number to rectangular form and then add or subtract.

EXAMPLE 3 Addition in polar form

Perform the addition $1.563\angle 37.56^\circ + 3.827\angle 146.23^\circ$.

In order to do this addition, we must change each number to rectangular form:

$$\begin{aligned}
&1.563\angle 37.56^\circ + 3.827\angle 146.23^\circ\\
&= 1.563(\cos 37.56^\circ + j\sin 37.56^\circ) + 3.827(\cos 146.23^\circ + j\sin 146.23^\circ)\\
&= 1.2390 + 0.9528j - 3.1813 + 2.1273j\\
&= -1.9423 + 3.0801j
\end{aligned}$$

Now, we change this to polar form:

$$r = \sqrt{(-1.9423)^2 + (3.0801)^2} = 3.641$$

$$\tan\theta = \frac{3.0801}{-1.9423} \qquad \theta = 122.24^\circ$$

Therefore,

$$1.563\angle 37.56^\circ + 3.827\angle 146.23^\circ = 3.641\angle 122.24^\circ$$

DEMOIVRE'S THEOREM

To raise a complex number to a power, we may use the exponential form of the number along with the properties of exponents $(a^m)^n = a^{mn}$. This leads to

$$(re^{j\theta})^n = r^ne^{jn\theta} \tag{12.16}$$

Extending this to polar form, we have

$$[r(\cos\theta + j\sin\theta)]^n = r^n(\cos n\theta + j\sin n\theta)$$
$$(r\angle\theta)^n = r^n\angle n\theta \qquad \textbf{(12.17)}$$

That is, the magnitude is raised to the power, and the angle is multiplied by the power.

■ DeMoivre's theorem is named for the mathematician Abraham DeMoivre (1667–1754).

Eq. (12.17) is known as **DeMoivre's theorem** *and is valid for all real values of n. It is also used for finding roots of complex numbers if n is a fractional exponent.*

EXAMPLE 4 Power by DeMoivre's theorem

Using DeMoivre's theorem, find $(2 + 3j)^3$.

Converting $2 + 3j$ to polar form: $r = \sqrt{2^2 + 3^2} = 3.61$, $\tan\theta = \frac{3}{2}$, $\theta = 56.3°$. Therefore,

$$\begin{aligned}[3.61(\cos 56.3° + j\sin 56.3°)]^3 &= (3.61)^3[\cos(3\times 56.3°) + j\sin(3\times 56.3°)]\\ &= 47.0(\cos 168.9° + j\sin 168.9°)\\ &= 47.0\angle 168.9°\end{aligned}$$

Expressing θ in radians, we have $\theta = 56.3° = 0.983$ rad. Therefore,

$$\begin{aligned}(3.61e^{0.983j})^3 = (3.61)^3e^{3\times 0.983j} &= 47.0e^{2.95j}\\ (2+3j)^3 &= 47.0(\cos 168.9° + j\sin 168.9°)\\ &= 47.0\angle 168.9°\\ &= 47.0e^{2.95j} = -46 + 9j\end{aligned}$$

Practice Exercise

3. Find the polar form power: $(3\cos 50°)^8$

EXAMPLE 5 Cube roots by DeMoivre's theorem

Find the cube root of -1.

Since we know that -1 is a real number, we can find its cube root by means of the definition. That is, $(-1)^3 = -1$. We check this by DeMoivre's theorem. Writing -1 in polar form, we have

$$-1 = 1(\cos 180° + j\sin 180°)$$

Applying DeMoivre's theorem, with $n = \frac{1}{3}$, we obtain

$$\begin{aligned}(-1)^{1/3} = 1^{1/3}(\cos\tfrac{1}{3}180° + j\sin\tfrac{1}{3}180°) &= \cos 60° + j\sin 60°\\ &= \frac{1}{2} + j\frac{\sqrt{3}}{2} \quad \text{exact answer}\\ &= 0.5000 + 0.8660j \quad \text{decimal approximation}\end{aligned}$$

Observe that we did not obtain -1 as the answer. If we check the answer, in the form $\frac{1}{2} + j\frac{\sqrt{3}}{2}$, by actually cubing it, we obtain -1. Therefore, it is a correct answer.

We should note that it is possible to take $\frac{1}{3}$ of any angle up to 1080° and still have an angle less than 360°. Since 180° and 540° have the same terminal side, let us try writing -1 as $1(\cos 540° + j\sin 540°)$. Using DeMoivre's theorem, we have

$$(-1)^{1/3} = 1^{1/3}(\cos\tfrac{1}{3}540° + j\sin\tfrac{1}{3}540°) = \cos 180° + j\sin 180° = -1$$

We have found the answer we originally anticipated.

Angles of 180° and 900° also have the same terminal side, so we try

$$\begin{aligned}(-1)^{1/3} = 1^{1/3}(\cos\tfrac{1}{3}900° + j\sin\tfrac{1}{3}900°) &= \cos 300° + j\sin 300°\\ &= \frac{1}{2} - j\frac{\sqrt{3}}{2} \quad \text{exact answer}\\ &= 0.5000 - 0.8660j \quad \text{decimal approximation}\end{aligned}$$

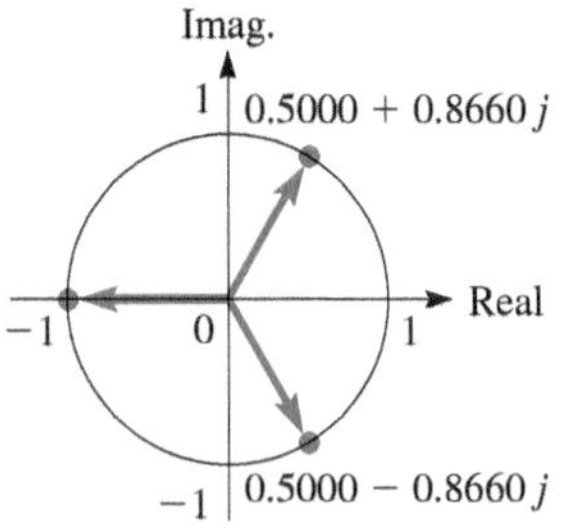

Fig. 12.14

Checking this, we find that it is also a correct root. We may try 1260°, but $\frac{1}{3}(1260°) = 420°$, which has the same functional values as 60°, and would give us the answer $0.5000 + 0.8660j$ again.

We have found, therefore, *three cube roots* of -1. They are

$$-1, \quad \frac{1}{2} + j\frac{\sqrt{3}}{2}, \quad \frac{1}{2} - j\frac{\sqrt{3}}{2}$$

These roots are graphed in Fig. 12.14. Note that they are equally spaced on the circumference of a circle of radius 1.

When the results of Example 5 are generalized, it can be proven that *there are n nth roots of a complex number.* When graphed, these roots are on a circle of radius $r^{1/n}$ and are equally spaced $360°/n$ apart. Following is the method for finding these n roots.

Using DeMoivre's Theorem to Find the *n* *n*th Roots of a Complex Number

1. Express the number in polar form.
2. Express the root as a fractional exponent.
3. Use Eq. (12.17) with θ to find one root.
4. Use Eq. (12.17) and add 360° to θ, $n - 1$ times, to find the other roots.

EXAMPLE 6 Square roots by DeMoivre's theorem

Find the two square roots of $2j$.

First, we write $2j$ in polar form as $2j = 2(\cos 90° + j \sin 90°)$. To find square roots, we use the exponent $1/2$. The first square root is

$$(2j)^{1/2} = 2^{1/2}\left(\cos\frac{90°}{2} + j\sin\frac{90°}{2}\right) = \sqrt{2}(\cos 45° + j\sin 45°) = 1 + j$$

To find the other square root, we add 360° to 90°. This gives us

$$(2j)^{1/2} = 2^{1/2}\left(\cos\frac{450°}{2} + j\sin\frac{450°}{2}\right) = \sqrt{2}(\cos 225° + j\sin 225°) = -1 - j$$

Therefore, the two square roots of $2j$ are $1 + j$ and $-1 - j$. We see in Fig. 12.15 that they are on a circle of radius $\sqrt{2}$ and 180° apart.

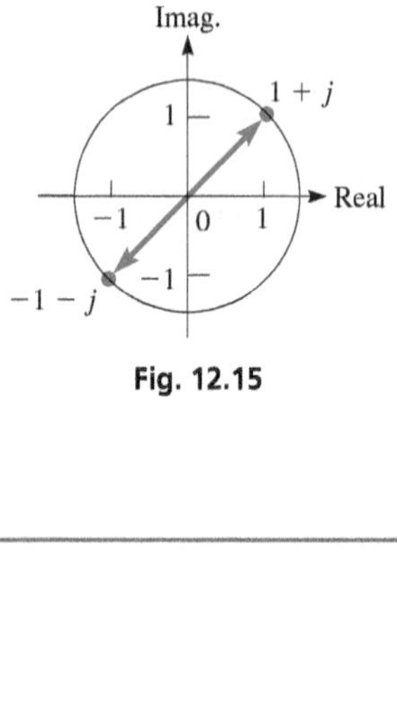

Fig. 12.15

EXAMPLE 7 Sixth roots by DeMoivre's theorem

Find all the roots of the equation $x^6 - 64 = 0$.

Solving for x, we have $x^6 = 64$, or $x = \sqrt[6]{64}$. Therefore, we have to find the six sixth roots of 64. Writing 64 in polar form, we have $64 = 64(\cos 0° + j \sin 0°)$. Using the exponent $1/6$ for the sixth root, we have the following solutions:

First root: $$64^{1/6} = 64^{1/6}\left(\cos\frac{0°}{6} + j\sin\frac{0°}{6}\right) = 2(\cos 0° + j\sin 0°) = 2$$

add 360°

Second root: $$64^{1/6} = 64^{1/6}\left(\cos\frac{0° + 360°}{6} + j\sin\frac{0° + 360°}{6}\right)$$

$$= 2(\cos 60° + j\sin 60°) = 1 + j\sqrt{3}$$

Third root: $64^{1/6} = 64^{1/6}\left(\cos\dfrac{0° + 720°}{6} + j\sin\dfrac{0° + 720°}{6}\right)$ (add 2 × 360°)

$= 2(\cos 120° + j\sin 120°) = -1 + j\sqrt{3}$

Fourth root: $64^{1/6} = 64^{1/6}\left(\cos\dfrac{0° + 1080°}{6} + j\sin\dfrac{0° + 1080°}{6}\right)$ (add 3 × 360°)

$= 2(\cos 180° + j\sin 180°) = -2$

Fifth root: $64^{1/6} = 64^{1/6}\left(\cos\dfrac{0° + 1440°}{6} + j\sin\dfrac{0° + 1440°}{6}\right)$ (add 4 × 360°)

$= 2(\cos 240° + j\sin 240°) = -1 - j\sqrt{3}$

Sixth root: $64^{1/6} = 64^{1/6}\left(\cos\dfrac{0° + 1800°}{6} + j\sin\dfrac{0° + 1800°}{6}\right)$ (add 5 × 360°)

$= 2(\cos 300° + j\sin 300°) = 1 - j\sqrt{3}$

These roots are graphed in Fig. 12.16. Note that they are equally spaced 60° apart on the circumference of a circle of radius 2.

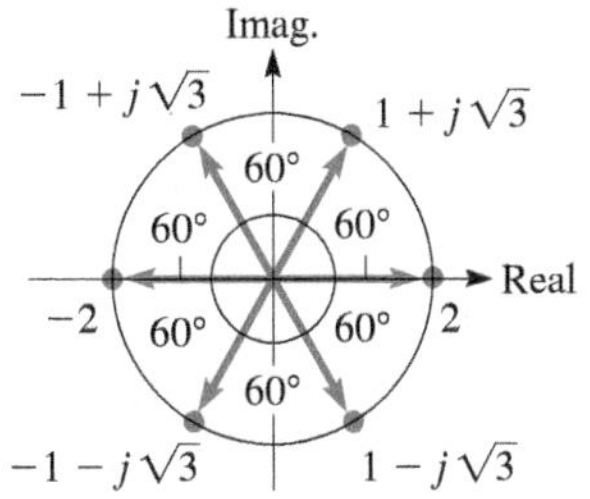

Fig. 12.16

From the text and examples of this and previous sections, we are able see the uses and advantages of the different forms of complex numbers. These can be summarized as follows:

Rectangular form: Used for all operations; best for addition and subtraction.
Polar form: Used for multiplication, division, powers, roots.
Exponential form: Used for multiplication, division, powers, and theoretical purposes (e.g., deriving DeMoivre's theorem)

EXERCISES 12.6

In Exercises 1–4, perform the indicated operations for the resulting complex numbers if the given changes are made in the indicated examples of this section.

1. In Example 1, change the sign of the complex part of the second complex number and then perform the multiplication.
2. In Example 2, change the sign of the angle in the second complex number and then divide.
3. In Example 4, change the exponent to 5 and then find the result.
4. In Example 6, replace $2j$ with $-2j$ and then find the roots.

In Exercises 5–20, perform the indicated operations. Leave the result in polar form.

5. $[4(\cos 60° + j\sin 60°)][2(\cos 20° + j\sin 20°)]$
6. $[3(\cos 120° + j\sin 120°)][5(\cos 45° + j\sin 45°)]$
7. $(0.5\angle 140°)(6\angle 110°)$
8. $(0.4\angle 320°)(5.5\angle -150°)$
9. $\dfrac{8(\cos 100° + j\sin 100°)}{4(\cos 65° + j\sin 65°)}$
10. $\dfrac{9(\cos 230° + j\sin 230°)}{45(\cos 80° + j\sin 80°)}$
11. $\dfrac{12\angle 320°}{5\angle -210°}$
12. $\dfrac{2\angle 90°}{4\angle 75°}$
13. $[0.2(\cos 35° + j\sin 35°)]^3$
14. $[3(\cos 120° + j\sin 120°)]^4$
15. $(2\angle 135°)^8$
16. $(1\angle 142°)^{10}$
17. $\dfrac{(50\angle 236°)(2\angle 84°)}{125\angle 47°}$
18. $\dfrac{36\angle 274°}{(2\angle 141°)(6\angle 195°)}$
19. $\dfrac{(4\angle 24°)(10\angle 326°)}{(1\angle 186°)(8\angle 77°)}$
20. $\dfrac{(25\angle 194°)(6\angle 239°)}{(30\angle 17°)(10\angle 29°)}$

In Exercises 21–24, perform the indicated operations. Express results in polar form. See Example 5.

21. $2.78\angle 56.8° + 1.37\angle 207.3°$
22. $15.9\angle 142.6° - 18.5\angle 71.4°$
23. $7085\angle 115.62° - 4667\angle 296.34°$
24. $307.5\angle 326.54° + 726.3\angle 96.41°$

In Exercises 25–36, change each number to polar form and then perform the indicated operations. Express the result in rectangular and polar forms. Check by performing the same operation in rectangular form.

25. $(3 + 4j)(5 - 12j)$ **26.** $(-2 + 5j)(-1 - j)$

27. $(7 - 3j)(8 + j)$ **28.** $(1 + 5j)(4 + 2j)$

29. $\frac{7}{1 - 3j}$ **30.** $\frac{40j}{7 + 2j}$ **31.** $\frac{30 + 40j}{5 - 12j}$ **32.** $\frac{-2 + 5j}{-1 - j}$

33. $(3 + 4j)^4$ **34.** $(-1 - j)^8$ **35.** $(2 + 3j)^5$ **36.** $(1 - 2j)^6$

In Exercises 37–42, use DeMoivre's theorem to find all the indicated roots. Be sure to find all roots.

37. The two square roots of $4(\cos 60° + j \sin 60°)$

38. The three cube roots of $27(\cos 120° + j \sin 120°)$

39. The three cube roots of $3 - 4j$

40. The two square roots of $-5 + 12j$

41. The square roots of $1 + j$

42. The cube roots of $\sqrt{3} + j$

In Exercises 43–48, find all of the roots of the given equations.

43. $x^4 - 1 = 0$ **44.** $x^3 - 8 = 0$ **45.** $x^3 + 27j = 0$

46. $x^4 - j = 0$ **47.** $x^5 + 32 = 0$ **48.** $x^6 + 8 = 0$

In Exercises 49–58, perform the indicated operations.

49. Using the results of Example 5, find the cube roots of -125.

50. Using the results of Example 6, find the square roots of $32j$.

51. In Example 5, we showed that one cube root of -1 is $\frac{1}{2} - \frac{1}{2}j\sqrt{3}$. Cube this number in rectangular form and show that the result is -1.

52. Explain why the two square roots of a complex number are negatives of each other.

53. The cube roots of -1 can be found by solving the equation $x^3 + 1 = 0$. Find these roots by factoring $x^3 + 1$ as the sum of cubes and compare with Example 5.

54. The cube roots of 8 can be found by solving the equation $x^3 - 8 = 0$. Find these roots by factoring $x^3 - 8$ as the difference of cubes and compare with Exercise 44.

55. The electric power P (in W) supplied to an element in a circuit is the product of the voltage V (in V) and the current I (in A). Find the expression for the power supplied if $V = 6.80\angle 56.3°$ volts and $I = 0.0705\angle -15.8°$ amperes.

56. The displacement d (in cm) of a weight suspended on a system of two springs is $d = 6.03\angle 22.5° + 3.26\angle 76.0°$ cm. Perform the addition and express the answer in polar form.

57. The voltage across a certain inductor is $V = (8.66\angle 90.0°)(50.0\angle 135.0°)/(10.0\angle 60.0°)$ volts. Simplify this expression and find the magnitude of the voltage.

58. In a microprocessor circuit, the current is $I = 3.75\angle 15.0°$ μA and the impedance is $Z = 2500\angle -35.0°$ ohms. Find the voltage V in rectangular form. (See Example 4 of Section 12.2.)

Answers to Practice Exercises

1. $15\angle 115°$ **2.** $0.6\angle 345°$ **3.** $6561\angle 40°$

12.7 An Application to Alternating-Current (AC) Circuits

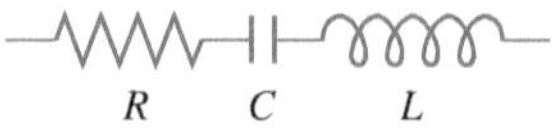

Fig. 12.17

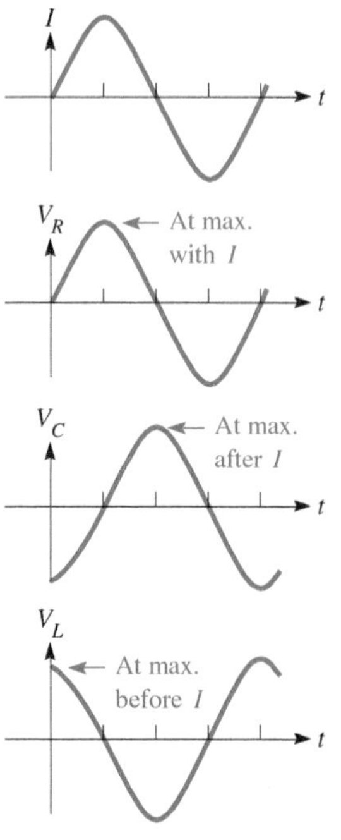

Fig. 12.18

We now show an application of complex numbers in a basic type of alternating-current circuit. We show how the voltage is measured between any two points in a circuit containing a resistance, a capacitance, and an inductance. This circuit is similar to one noted in earlier examples and exercises in this chapter.

A *resistance* is any part of a circuit that tends to obstruct the flow of electric current through the circuit. It is denoted by R (units in ohms, Ω) and in diagrams by , as shown in Fig. 12.17. A *capacitance* is two nonconnected plates in a circuit; no current actually flows across the gap between them. In an AC circuit, an electric charge is continually going to and from each plate and, therefore, the current in the circuit is not effectively stopped. It is denoted by C (units in farads, F) and in diagrams by (see Fig. 12.17). An *inductance* is basically a coil of wire in which current is induced because the current is continually changing in the circuit. It is denoted by L (units in henrys, H) and in diagrams by (see Fig. 12.17). All these elements affect the voltage in an alternating-current circuit. We state here the relation each has to the voltage and current in the circuit.

In Chapter 10, we noted that the current and voltage in an AC circuit could be represented by a sine or a cosine curve. Therefore, each reaches peak values periodically. *If they reach their respective peak values at the same time, they are* ***in phase****. If the voltage reaches its peak before the current, the voltage* ***leads*** *the current. If the voltage reaches its peak after the current, the voltage* ***lags*** *the current.*

In the study of electricity, it is shown that the voltage across a resistance is in phase with the current. The voltage across a capacitor lags the current by 90°, and the voltage across an inductance leads the current by 90°. This is shown in Fig. 12.18, where, in a given circuit, I represents the current, V_R is the voltage across a resistor, V_C is the voltage across a capacitor, V_L is the voltage across an inductor, and t represents time.

Each element in an AC circuit tends to offer a type of resistance to the flow of current. *The effective resistance of any part of the circuit is called the* **reactance**, and it is

■ Eq. (12.18) is based on Ohm's law, which states that the current is proportional to the voltage for a constant resistance. It is named for the German physicist Georg Ohm (1787–1854). The ohm (Ω) is also named for him.

denoted by X. The voltage across any part of the circuit whose reactance is X is given by $V = IX$, where I is the current (in amperes) and V is the voltage (in volts). Therefore,

the voltage V_R across a resistor with resistance R,
the voltage V_C across a capacitor with reactance X_C, and
the voltage V_L across an inductor with reactance X_L

are, respectively,

$$V_R = IR \qquad V_C = IX_C \qquad V_L = IX_L \tag{12.18}$$

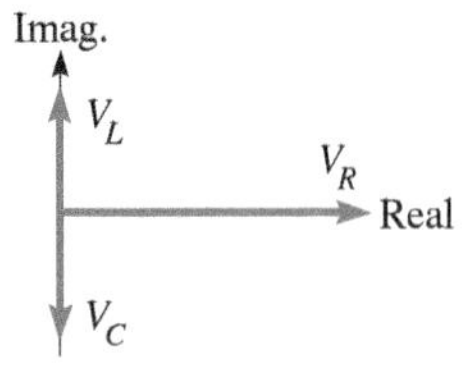

Fig. 12.19

To determine the voltage across a combination of these elements of a circuit, we must account for the reactance, as well as the phase of the voltage across the individual elements. Since the voltage across a resistor is in phase with the current, we represent V_R along the positive real axis as a real number. Since the voltage across an inductance leads the current by 90°, we represent this voltage as a positive, pure imaginary number. In the same way, by representing the voltage across a capacitor as a negative, pure imaginary number, we show that the voltage *lags* the current by 90°. These representations are meaningful since the positive imaginary axis is +90° from the positive real axis, and the negative imaginary axis is −90° from the positive real axis. See Fig. 12.19.

The circuit elements shown in Fig. 12.17 are in *series,* and all circuits we consider (except Section 12.7 Exercises 22 and 23) are series circuits. The total voltage across a series of all three elements is given by $V_R + V_L + V_C$, which we represent by V_{RLC}. Therefore,

$$V_{RLC} = IR + IX_Lj - IX_Cj = I[R + j(X_L - X_C)]$$

■ In the 1880s, it was decided that alternating current (favoured by George Westinghouse) would be used to distribute electric power. Thomas Edison had argued for the use of direct current.

This expression is also written as

$$V_{RLC} = IZ \tag{12.19}$$

where the symbol Z is called the **impedance** *of the circuit. It is the total effective resistance to the flow of current by a combination of the elements in the circuit,* taking into account the phase of the voltage in each element. From its definition, we see that Z is a complex number.

$$Z = R + j(X_L - X_C) \tag{12.20}$$

with magnitude

$$|Z| = \sqrt{R^2 + (X_L - X_C)^2} \tag{12.21}$$

Also, as a complex number, it makes an angle θ with the x-axis, given by

$$\theta = \tan^{-1}\frac{X_L - X_C}{R} \tag{12.22}$$

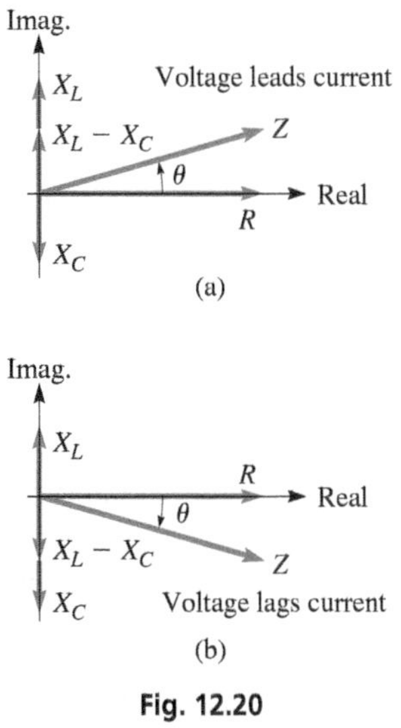

Fig. 12.20

All these equations are based on phase relations of voltages with respect to the current. Therefore, *the angle θ represents the phase angle between the current and the voltage.* The standard way of expressing θ is to *use a positive angle if the voltage leads the current* and *use a negative angle if the voltage lags the current.* Using Eq. (12.19), a calculator will give the correct angle even when $\tan\theta < 0$.

If the voltage leads the current, then $X_L > X_C$ as shown in Fig. 12.20(a). If the voltage lags the current, then $X_L < X_C$ as shown in Fig. 12.20(b).

In the examples and exercises of this section, the commonly used units and symbols for them are used. For a summary of these units and symbols, including prefixes, see Section 1.3.

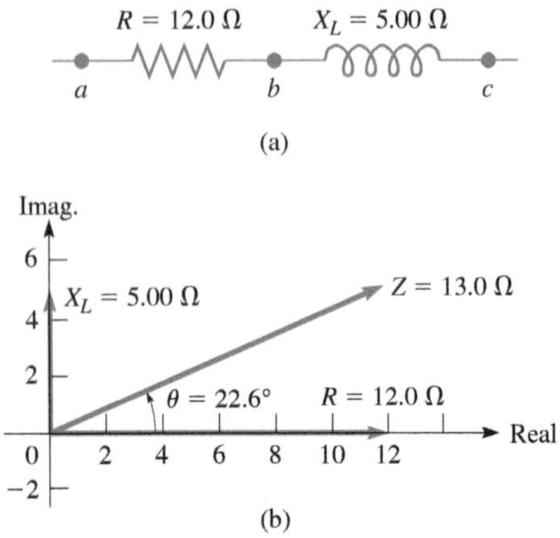

Fig. 12.21

EXAMPLE 1 Finding the impedance and the voltage

In the series circuit shown in Fig. 12.21(a), $R = 12.0\ \Omega$ and $X_L = 5.00\ \Omega$. A current of 2.00 A is in the circuit. Find the voltage across each element, the impedance, the voltage across the combination, and the phase angle between the current and the voltage.

The voltage across the resistor (between points a and b) is the product of the current and the resistance ($V = IR$). This means $V_R = (2.00)(12.0) = 24.0$ V. The voltage across the inductor (between points b and c) is the product of the current and the reactance, or $V_L = (2.00)(5.00) = 10.0$ V.

To find the voltage across the combination, between points a and c, we must first find the magnitude of the impedance. Note that ***the voltage is not the arithmetic sum of V_R and V_L***, as we must account for the phase.

By Eq. (12.20), the impedance is (there is no capacitor)

$$Z = 12.0 + 5.00j$$

with magnitude

$$|Z| = \sqrt{R^2 + X_L^2} = \sqrt{(12.0)^2 + (5.00)^2} = 13.0\ \Omega$$

Thus, the magnitude of the voltage across the combination of the resistor and the inductance is

$$|V_{RL}| = (2.00)(13.0) = 26.0\text{ V}$$

The phase angle between the voltage and the current is found by Eq. (12.22). This gives

$$\theta = \tan^{-1}\frac{5.00}{12.0} = 22.6°$$

The voltage *leads* the current by 22.6°, and this is shown in Fig. 12.21(b).

EXAMPLE 2 Finding the impedance and the phase angle

For a circuit in which $R = 8.00\ \Omega$, $X_L = 7.00\ \Omega$, and $X_C = 13.0\ \Omega$, find the impedance and the phase angle between the current and the voltage.

By the definition of impedance, Eq. (12.20), we have

$$Z = 8.00 + (7.00 - 13.0)j = 8.00 - 6.00j$$

where the magnitude of the impedance is

$$|Z| = \sqrt{(8.00)^2 + (-6.00)^2} = 10.0\ \Omega$$

The phase angle is found by

$$\theta = \tan^{-1}\frac{-6.00}{8.00} = -36.9°$$

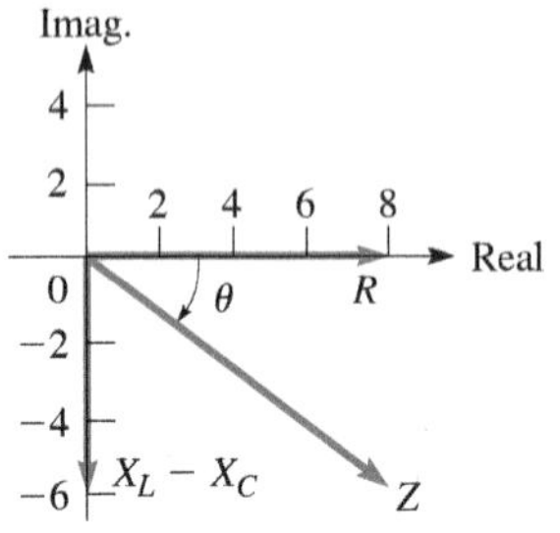

Fig. 12.22

The angle $\theta = -36.9°$ is given directly by the calculator, and it is the angle we want. As we noted after Eq. (12.22), we express θ as a negative angle if the voltage lags the current, as it does in this example. See Fig. 12.22.

From the values above, we write the impedance in polar form as $Z = 10.0\angle{-36.9°}$ ohms.

Note that the resistance is represented in the same way as a vector along the positive x-axis. Actually, resistance is not a vector quantity but is represented in this manner in order to assign an angle as the phase of the current. The important concept in this analysis is that *the phase **difference** between the current and voltage is constant,* and therefore any direction may be chosen arbitrarily for one of them. Once this choice is made, other phase angles are measured with respect to this direction. A common choice, as above, is to make the phase angle of the current zero. If an arbitrary angle is chosen, it is necessary to treat the current, voltage, and impedance as complex numbers.

EXAMPLE 3 Finding the voltage

In a particular circuit, the current (I) is $2.00 - 3.00j$ A and the impedance (Z) is $6.00 + 2.00j$ ohms. The voltage across this part of the circuit is

$$V = IZ$$
$$\begin{aligned} V &= (2.00 - 3.00j)(6.00 + 2.00j) \\ &= 12.0 - 14.0j - 6.00j^2 \\ &= 12.0 - 14.0j + 6.00 \\ &= 18.0 - 14.0j \text{ volts} \end{aligned}$$

The magnitude of the voltage is

$$|V| = \sqrt{(18.0)^2 + (-14.0)^2} = 22.8 \text{ V}$$

■ The ampere (A) is named for the French physicist André Ampère (1775–1836).

■ The volt (V) is named for the Italian physicist Alessandro Volta (1745–1827).

■ The farad (F) is named for the British physicist Michael Faraday (1791–1867).

■ The henry (H) is named for the U.S. physicist Joseph Henry (1797–1878).

■ The coulomb (C) is named for the French physicist Charles Coulomb (1736–1806).

Since the voltage across a resistor is in phase with the current, this voltage can be represented as having a phase difference of zero with respect to the current. Therefore, the resistance is indicated as an arrow in the positive real direction, denoting the fact that the current and the voltage are in phase. *Such a representation is called a* **phasor**. The arrow denoted by R, as in Fig. 12.21, is actually the phasor representing the voltage across the resistor. Remember, the positive real axis is arbitrarily chosen as the direction of the phase of the current.

To show properly that the voltage across an inductance leads the current by 90°, its reactance (effective resistance) is multiplied by j. We know that there is a positive 90° angle between a positive real number and a positive imaginary number. In the same way, by multiplying the capacitive reactance by $-j$, we show the 90° difference in phase between the voltage and the current in a capacitor, with the current leading. Therefore, jX_L represents the phasor for the voltage across an inductor and $-jX_C$ is the phasor for the voltage across the capacitor. The phasor for the voltage across the combination of the resistance, inductance, and capacitance is Z, where the phase difference between the voltage and the current for the combination is the angle θ.

From this, we see that *multiplying a phasor by j means to perform the operation of rotating it through* 90°. For this reason, j is also called the *j-operator.*

EXAMPLE 4 Multiplication by *j*

Multiplying a positive real number A by j, we have $A \times j = Aj$, which is a positive imaginary number. In the complex plane, Aj is 90° from A, which means that by multiplying A by j we rotated A by 90°. Similarly, we see that $Aj \times j = Aj^2 = -A$, which is a negative real number, rotated 90° from Aj. Therefore, successive multiplications of A by j give us

$A \times j = Aj$	positive imaginary number
$Aj \times j = Aj^2 = -A$	negative real number
$-A \times j = -Aj$	negative imaginary number
$-Aj \times j = -Aj^2 = A$	positive real number

See Fig. 12.23. (See Exercise 38 in Section 10.3.)

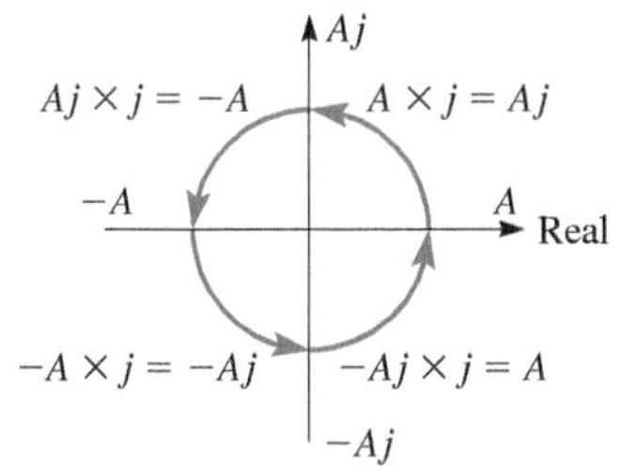

Fig. 12.23

An alternating current is produced by a coil of wire rotating through a magnetic field. If the angular velocity of the wire is ω, the capacitive and inductive reactances are given by

$$X_C = \frac{1}{\omega C} \quad \text{and} \quad X_L = \omega L \tag{12.23}$$

Therefore, if ω, C, and L are known, the reactance of the circuit can be found.

EXAMPLE 5 Finding the voltage–current phase difference

If $R = 12.0\ \Omega$, $L = 0.300$ H, $C = 250\ \mu$F, and $\omega = 80.0$ rad/s, find the impedance and the phase difference between the current and the voltage.

$$X_C = \frac{1}{(80.0)(250 \times 10^{-6})} = 50.0\ \Omega$$

$$X_L = (0.300)(80.0) = 24.0\ \Omega$$

$$Z = 12.0 + (24.0 - 50.0)j = 12.0 - 26.0j$$

$$|Z| = \sqrt{(12.0)^2 + (-26.0)^2} = 28.6\ \Omega$$

$$\theta = \tan^{-1}\frac{-26.0}{12.0} = -65.2^\circ$$

$$Z = 28.6\angle{-65.2^\circ}\ \Omega$$

The voltage *lags* the current (see Fig. 12.24).

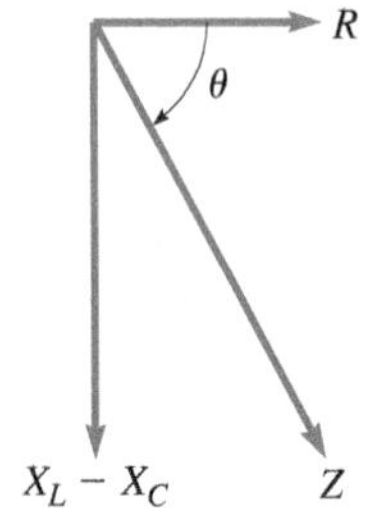

Fig. 12.24

Recall from Section 10.5 that the angular velocity ω is related to the frequency f by the relation $\omega = 2\pi f$. It is very common to use frequency when discussing alternating current.

An important concept in the application of this theory is **resonance**. *For resonance, the impedance of any circuit is a minimum, or the total impedance is R.* Thus, $X_L - X_C = 0$. Also, it can be seen that the current and the voltage are in phase under these conditions. Resonance is required for the tuning of radio and television receivers.

■ See the chapter introduction.

EXAMPLE 6 Resonance

In the antenna circuit of a radio, the inductance is 4.20 mH and the capacitance is variable. What range of values of capacitance is necessary for the radio to receive the AM band of radio stations, with frequencies from 530 kHz to 1600 kHz?

For proper tuning, the circuit should be in resonance, or $X_L = X_C$. This means that

$$2\pi fL = \frac{1}{2\pi fC} \quad \text{or} \quad C = \frac{1}{(2\pi f)^2 L}$$

■ From Section 1.3, the following prefixes are defined as follows:

Prefix	*Factor*	*Symbol*
pico	10^{-12}	p
milli	10^{-3}	m
kilo	10^{3}	k

For $f_1 = 530$ kHz $= 5.30 \times 10^5$ Hz and $L = 4.20$ mH $= 4.20 \times 10^{-3}$ H,

$$C_1 = \frac{1}{(2\pi)^2(5.30 \times 10^5)^2(4.20 \times 10^{-3})} = 2.15 \times 10^{-11}\ \text{F} = 21.5\ \text{pF}$$

and for $f_2 = 1600$ kHz $= 1.60 \times 10^6$ Hz and $L = 4.20 \times 10^{-3}$ H, we have

$$C_2 = \frac{1}{(2\pi)^2(1.60 \times 10^6)^2(4.20 \times 10^{-3})} = 2.36 \times 10^{-12}\ \text{F} = 2.36\ \text{pF}$$

The capacitance should be capable of varying from 2.36 pF to 21.6 pF.

EXERCISES 12.7

In Exercises 1 and 2, perform the indicated operations if the given changes are made in the indicated examples of this section.

1. In Example 1, change the value of X_L to 16.0 Ω and then solve the given problem.
2. In Example 5, double the values of L and C and then solve the given problem.

In Exercises 3–6, use the circuit shown in Fig. 12.25. The current in the circuit is 5.75 mA. Determine the indicated quantities.

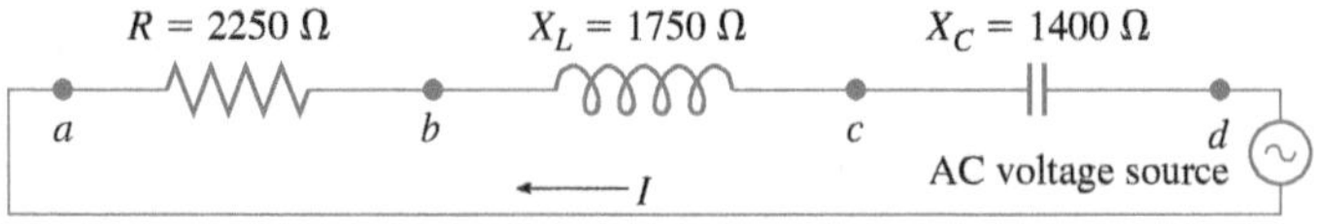

Fig. 12.25

3. The voltage across the resistor (between points a and b).

4. The voltage across the inductor (between points b and c).

5. **(a)** The magnitude of the impedance across the resistor and the inductor (between points a and c).
(b) The phase angle between the current and the voltage for this combination.
(c) The voltage across this combination.

6. **(a)** The magnitude of the impedance across the resistor, inductor, and capacitor (between points a and d).
(b) The phase angle between the current and the voltage for this combination.
(c) The voltage across this combination.

In Exercises 7–10, an AC circuit contains the given combination of circuit elements from among a resistor $(R = 45.0\ \Omega)$, *a capacitor* $(C = 86.2\ \mu\text{F})$, *and an inductor* $(L = 42.9\ \text{mH})$. *If the frequency in the circuit is* $f = 60.0$ Hz, *find (a) the magnitude of the impedance and (b) the phase angle between the current and the voltage.*

7. The circuit has the inductor and the capacitor (an LC circuit).

8. The circuit has the resistor and the capacitor (an RC circuit).

9. The circuit has the resistor and the inductor (an RL circuit).

10. The circuit has the resistor, the inductor, and the capacitor (an RLC circuit).

In Exercises 11–24, solve the given problems.

11. Given that the current in a given circuit is $3.90 - 6.04j$ mA and the impedance is $5.16 + 1.14j\ \text{k}\Omega$, find the magnitude of the voltage.

12. Given that the voltage in a given circuit is $8.375 - 3.140j$ V and the impedance is $2.146 - 1.114j\ \Omega$, find the magnitude of the current.

13. A resistance $(R = 25.3\ \Omega)$ and a capacitance $(C = 2.75\ \text{nF})$ are in an AM radio circuit. If $f = 1200$ kHz, find the impedance across the resistor and the capacitor.

14. A resistance $(R = 64.5\ \Omega)$ and an inductance $(L = 1.08\ \text{mH})$ are in a telephone circuit. If $f = 8.53$ kHz, find the impedance across the resistor and inductor.

15. The reactance of an inductor is $1200\ \Omega$ for $f = 280$ Hz. What is the inductance?

16. A resistor, an inductor, and a capacitor are connected in series across an AC voltage source. A voltmeter measures 12.0 V, 15.5 V, and 10.5 V, respectively, when placed across each element separately. What is the voltage of the source?

17. An inductance of 12.5 μH and a capacitance of 47.0 nF are in series in an amplifier circuit. Find the frequency for resonance.

18. A capacitance $(C = 95.2\ \text{nF})$ and an inductance are in series in the circuit of a receiver for navigation signals. Find the inductance if the frequency for resonance is 50.0 kHz.

19. In Example 6, what should be the capacitance in order to receive a 680-kHz radio signal?

20. A 220-V source with $f = 60.0$ Hz is connected in series to an inductance $(L = 2.05\ \text{H})$ and a resistance R in an electric-motor circuit. Find R if the current is 0.250 A.

21. The power P (in W) supplied to a series combination of elements in an AC circuit is $P = VI \cos\theta$, where V is the effective voltage, I is the effective current, and θ is the phase angle between the current and voltage. If $V = 225$ mV across the resistor, capacitor, and inductor combination in Exercise 10, determine the power supplied to these elements.

22. For two impedances Z_1 and Z_2 in parallel, the reciprocal of the combined impedance Z_C is the sum of the reciprocals of Z_1 and Z_2. Find the combined impedance for the parallel circuit elements in Fig. 12.26 if the current in the circuit has a frequency of 60.0 Hz.

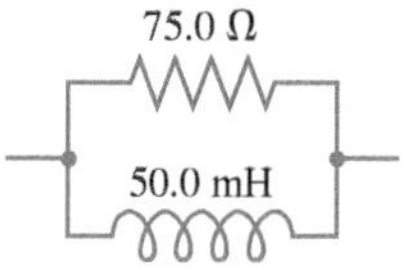

Fig. 12.26

23. Find the combined impedance of the circuit elements in Fig. 12.27. The frequency of the current in the circuit is 60.0 Hz. See Exercise 22.

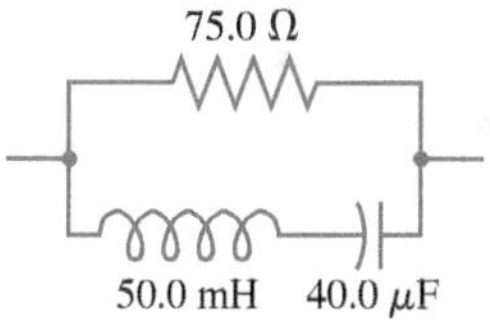

Fig. 12.27

24. (a) If the complex number j, in polar form, is multiplied by itself, what is the resulting number in polar and rectangular forms?
(b) In the complex plane, where is the resulting complex number in relation to j?

CHAPTER 12 KEY FORMULAS AND EQUATIONS

Chapter Equations for Complex Numbers

Imaginary unit

$$j = \sqrt{-1} \quad \text{and} \quad j^2 = -1 \tag{12.1}$$

$$\sqrt{-a} = j\sqrt{a} \quad (a > 0) \tag{12.2}$$

Basic operations

$$(a + bj) + (c + dj) = (a + c) + (b + d)j \tag{12.3}$$

$$(a + bj) - (c + dj) = (a - c) + (b - d)j \tag{12.4}$$

$$(a + bj)(c + dj) = (ac - bd) + (ad + bc)j \tag{12.5}$$

$$\frac{a + bj}{c + dj} = \frac{(a + bj)(c - dj)}{(c + dj)(c - dj)} = \frac{(ac + bd) + (bc - ad)j}{c^2 + d^2} \tag{12.6}$$

Complex number forms

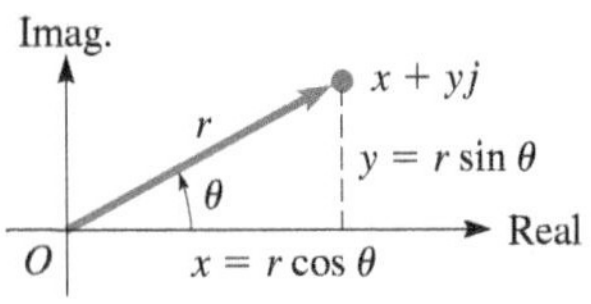

Rectangular: $x + yj$

Polar: $r(\cos\theta + j\sin\theta) = r\angle\theta$

Exponential: $re^{j\theta}$

$$x = r\cos\theta \qquad y = r\sin\theta \tag{12.7}$$

$$r^2 = x^2 + y^2 \qquad \tan\theta = \frac{y}{x} \tag{12.8}$$

$$x + yj = r(\cos\theta + j\sin\theta) = r\angle\theta = re^{j\theta} \tag{12.12}$$

Product in polar form

$$r_1(\cos\theta_1 + j\sin\theta_1)r_2(\cos\theta_2 + j\sin\theta_2) = r_1r_2[\cos(\theta_1 + \theta_2) + j\sin(\theta_1 + \theta_2)] \tag{12.13}$$

$$(r_1\angle\theta_1)(r_2\angle\theta_2) = r_1r_2\angle\theta_1 + \theta_2$$

Quotient in polar form

$$\frac{r_1(\cos\theta_1 + j\sin\theta_1)}{r_2(\cos\theta_2 + j\sin\theta_2)} = \frac{r_1}{r_2}[\cos(\theta_1 - \theta_2) + j\sin(\theta_1 - \theta_2)]$$

$$\frac{r_1\angle\theta_1}{r_2\angle\theta_2} = \frac{r_1}{r_2}\angle\theta_1 - \theta_2 \tag{12.15}$$

DeMoivre's theorem

$$[r(\cos\theta + j\sin\theta)]^n = r^n(\cos n\theta + j\sin n\theta) \tag{12.17}$$

$$(r\angle\theta)^n = r^n\angle n\theta$$

Chapter Equations for Alternating-Current Circuits

Voltage, current, reactance

$$V_R = IR \qquad V_C = IX_C \qquad V_L = IX_L \tag{12.18}$$

Impedance

$$V_{RLC} = IZ \tag{12.19}$$

$$Z = R + j(X_L - X_C) \tag{12.20}$$

$$|Z| = \sqrt{R^2 + (X_L - X_C)^2} \tag{12.21}$$

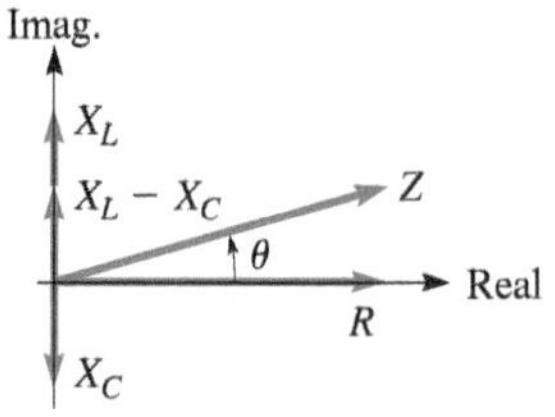

Phase angle

$$\theta = \tan^{-1}\frac{X_L - X_C}{R} \tag{12.22}$$

Capacitive reactance and inductive reactance

$$X_C = \frac{1}{\omega C} \quad \text{and} \quad X_L = \omega L \tag{12.23}$$

CHAPTER 12 REVIEW EXERCISES

In Exercises 1–16, perform the indicated operations, expressing all answers in simplest rectangular form.

1. $(6 - 2j) + (4 + j)$
2. $(12 + 7j) + (-8 + 6j)$
3. $(18 - 3j) - (12 - 5j)$
4. $(-4 - 2j) - \sqrt{-49}$
5. $(2 + j)(4 - j)$
6. $(-5 + 3j)(8 - 4j)$
7. $(2j^5)(6 - 3j)(4 + 3j)$
8. $j(3 - 2j) - (j^3)(5 + j)$
9. $\dfrac{3}{7 - 6j}$
10. $\dfrac{24j}{2 + 9j}$
11. $\dfrac{6 - \sqrt{-16}}{\sqrt{-4}}$
12. $\dfrac{3 + \sqrt{-4}}{4 - j}$
13. $\dfrac{5j - (3 - j)}{4 - 2j}$
14. $\dfrac{2 + (j - 6)}{1 - 2j}$
15. $\dfrac{j(7 - 3j)}{2 - j^7}$
16. $\dfrac{(2 - j)(3 + 2j)}{4 + 3j^3}$

In Exercises 17–20, find the values of x and y for which the equations are valid.

17. $3x - 2j = yj - 9$
18. $2xj - 2y = (y + 3)j - 3$
19. $(3 - 2j)(x + yj) = 4 + j^9$
20. $(x + yj)(7j - 4) = j(x - 5)$

In Exercises 21–24, perform the indicated operations graphically. Check them algebraically.

21. $(-1 + 5j) + (4 + 6j)$
22. $(7 - 2j) + (-5 + 4j)$
23. $(9 + 2j) - (5 - 6j)$
24. $(8 + 4j) - (11 - 3j)$

In Exercises 25–32, give the polar and exponential forms of each of the complex numbers.

25. $1 - j$
26. $4 + 3j$
27. $-22 - 77j$
28. $60 - 20j$
29. $1.07 + 4.55j$
30. $-327 + 158j$
31. 5000
32. $-4j^5$

In Exercises 33–44, give the rectangular form of each number.

33. $2(\cos 225° + j\sin 225°)$
34. $48(\cos 60° + j\sin 60°)$
35. $5.011(\cos 123.82° + j\sin 123.82°)$
36. $2.417(\cos 656.26° + j\sin 656.26°)$
37. $0.62\angle{-72°}$
38. $20\angle 160°$
39. $27.08\angle 346.27°$
40. $1.689\angle 194.36°$
41. $2.00e^{0.25j}$
42. $e^{-3.62j}$
43. $(35.37e^{1.096j})^2$
44. $(13.6e^{2.158j})(3.27e^{3.888j})$

In Exercises 45–60, perform the indicated operations. Leave the result in polar form.

45. $[3(\cos 32^\circ + j\sin 32^\circ)][5(\cos 52^\circ + j\sin 52^\circ)]$

46. $[2.5(\cos 162^\circ + j\sin 162^\circ)][8(\cos 115^\circ + j\sin 115^\circ)]$

47. $(40\angle 18^\circ)(0.5\angle 245^\circ)$

48. $(0.1254\angle 172.38^\circ)(27.17\angle 204.34^\circ)$

49. $\dfrac{24(\cos 165^\circ + j\sin 165^\circ)}{3(\cos 106^\circ + j\sin 106^\circ)}$

50. $\dfrac{18(\cos 403^\circ + j\sin 403^\circ)}{4(\cos 192^\circ + j\sin 192^\circ)}$

51. $\dfrac{245.6\angle 326.44^\circ}{17.19\angle 192.83^\circ}$

52. $\dfrac{4\angle 206^\circ}{100\angle -320^\circ}$

53. $0.983\angle 47.2^\circ + 0.366\angle 95.1^\circ$

54. $17.8\angle 110.4^\circ - 14.9\angle 226.3^\circ$

55. $7644\angle 294.36^\circ - 6871\angle 17.86^\circ$

56. $4.944\angle 327.49^\circ + 8.009\angle 7.37^\circ$

57. $[2(\cos 16^\circ + j\sin 16^\circ)]^{10}$

58. $[3(\cos 36^\circ + j\sin 36^\circ)]^6$

59. $(7\angle 110.5^\circ)^3$

60. $(536\angle 220.3^\circ)^4$

In Exercises 61–64, change each number to polar form and then perform the indicated operations. Express the final result in rectangular and polar forms. Check by performing the same operation in rectangular form.

61. $(1 - j)^{10}$

62. $(\sqrt{3} + j)^8(1 + j)^5$

63. $\dfrac{(5 + 5j)^4}{(-1 - j)^6}$

64. $(\sqrt{3} - j)^{-8}$

In Exercises 65–68, find all the roots of the given equations.

65. $x^3 + 8 = 0$

66. $x^3 - 1 = 0$

67. $x^4 + j = 0$

68. $x^5 - 32j = 0$

In Exercises 69–72, determine the rectangular form and the polar form of the complex number for which the graphical representation is shown in the given figure.

69.
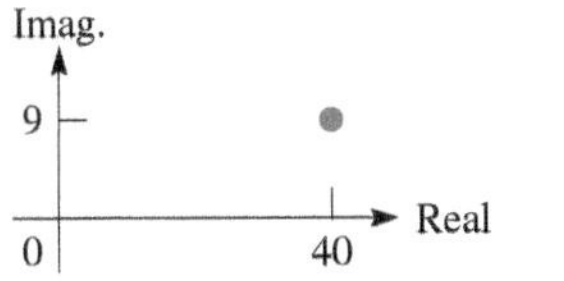

70.
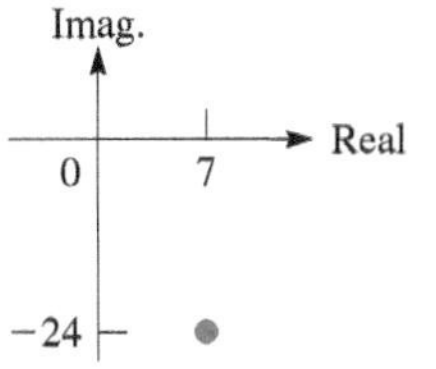

71.
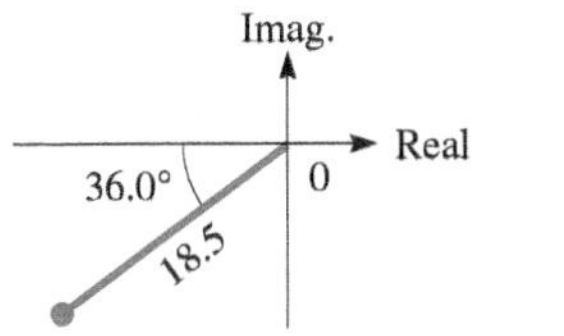

72.
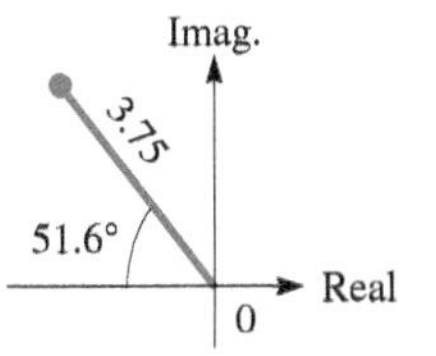

In Exercises 73–84, solve the given problems.

73. Evaluate $x^2 - 2x + 4$ for $x = 5 - 2j$.

74. Evaluate $2x^2 + 5x - 7$ for $x = -8 + 7j$.

75. Using the quadratic formula, solve for x: $x^2 + 3jx - 2 = 0$.

76. Using the quadratic formula, solve for x: $jx^2 - 2x - 3j = 0$.

77. Find a quadratic equation with roots $2 + j$ and $2 - j$.

78. Find a quadratic equation with roots $-3 + 4j$ and $-3 - 4j$.

79. Are $1 - j$ and $-1 - j$ solutions to the equation $x^2 - 2x + 2 = 0$?

80. Show that $\frac{1}{2}(1 + j\sqrt{3})$ is the reciprocal of its conjugate.

81. Solve for x: $(1 + jx)^2 = 1 + j - x^2$.

82. What is the argument for any negative imaginary number?

83. If $f(x) = 2x - (x - 1)^{-1}$, find $f(1 + 2j)$.

84. If $f(x) = x^{-2} + 3x^{-1}$, find $f(4 + j)$.

In Exercises 85–96, find the required quantities. In Exercises 97, answer the given question.

85. A 60-V AC voltage source is connected in series across a resistor, an inductor, and a capacitor. The voltage across the inductor is 60 V, and the voltage across the capacitor is 60 V. What is the voltage across the resistor?

86. In a series AC circuit with a resistor, an inductor, and a capacitor, $R = 6.50\ \Omega$, $X_C = 3.74\ \Omega$, and $Z = 7.50\ \Omega$. Find X_L.

87. In a series AC circuit with a resistor, an inductor, and a capacitor, $R = 6250\ \Omega$, $Z = 6720\ \Omega$, and $X_L = 1320\ \Omega$. Find the phase angle θ.

88. A coil of wire rotates at 120.0 r/s. If the coil generates a current in a circuit containing a resistance of 12.07 Ω, an inductance of 0.1405 H, and an impedance of 22.35 Ω, what must be the value of a capacitor (in F) in the circuit?

89. What is the frequency f for resonance in a circuit for which $L = 2.65$ H and $C = 18.3\ \mu$F?

90. The displacement of an electromagnetic wave is given by $d = A(\cos \omega t + j\sin \omega t) + B(\cos \omega t - j\sin \omega t)$. Find the expressions for the magnitude and phase angle of d.

91. Two cables lift a crate. The tensions in the cables can be represented by $2100 - 1200j$ N and $1200 + 5600j$ N. Express the resultant tension in polar form.

92. A boat is headed across a river with a velocity (relative to the water) that can be represented as $6.5 + 1.7j$ km/h. The velocity of the river current can be represented as $-1.1 - 4.3j$ km/h. Express the resultant velocity of the boat in polar form.

93. In the study of shearing effects in the spinal column, the expression $\dfrac{1}{\mu + j\omega n}$ is found. Express this in rectangular form.

94. In the theory of light reflection on metals, the expression $\dfrac{\mu(1 - kj) - 1}{\mu(1 - kj) + 1}$ is encountered. Simplify this expression.

95. Show that $e^{j\pi} = -1$.

96. Show that $(e^{j\pi})^{1/2} = j$.

97. A computer programmer is writing a program to determine the n nth roots of a real number. Part of the program is to show the number of real roots and the number of pure imaginary roots. Write one or two paragraphs explaining how these numbers of roots can be determined without actually finding the roots.

CHAPTER 12 PRACTICE TEST

1. Add, expressing the resultant in rectangular form: $(3 - \sqrt{-4}) + (5\sqrt{-9} - 1)$.
2. Multiply, expressing the resultant in polar form: $(2\underline{/130°})(3\underline{/45°})$.
3. Express $2 - 7j$ in polar form.
4. Express in terms of j: (a) $-\sqrt{-64}$ (b) $-j^{15}$.
5. Add graphically: $(4 - 3j) + (-1 + 4j)$.
6. Simplify, expressing the result in rectangular form: $\frac{2 - 4j}{5 + 3j}$.
7. Express $2.56(\cos 125.2° + j \sin 125.2°)$ in exponential form.
8. For an AC circuit in which $R = 3.50\ \Omega$, $X_L = 6.20\ \Omega$, and $X_C = 7.35\ \Omega$, find the impedance and the phase angle between the current and the voltage.
9. Express $3.47 - 2.81j$ in exponential form.
10. Find the values of x and y: $x + 2j - y = yj - 3xj$.
11. What is the capacitance of the circuit in a radio that has an inductance of 8.75 mH if it is to receive a station with frequency 600 kHz?
12. Find the cube roots of j.

13. Exponential and Logarithmic Functions

Rawpixel.com/Shutterstock

▲ The growth of population can often be measured using an exponential function. This is illustrated in Section 13.6.

LEARNING OUTCOMES

After completion of this chapter, the student should be able to:

- Evaluate and graph an exponential function
- Recognize the connection between properties of logarithms and laws of exponents
- Change equations from exponential form to logarithmic form and vice versa
- Evaluate and graph a logarithmic function
- Identify the exponential and logarithmic functions as inverse functions
- Solve logarithmic and exponential equations
- Change a logarithm in one base to a logarithm in another base
- Solve application problems involving logarithmic and exponential functions
- Graph functions on logarithmic or semilogarithmic paper

By the early 1600s, astronomy had progressed to the point of finding accurate information about the motion of the heavenly bodies. Also, navigation had led to a more systematic exploration of Earth. In making the accurate measurements needed in astronomy and navigation, many lengthy calculations involving large numbers had to be performed, and all such calculations had to be done by hand.

Noting that astronomers' calculations usually involved sines of angles, John Napier (1550–1617), a Scottish mathematician, constructed a table of values that allowed multiplication of these sines by addition of values from the table. These tables of *logarithms* first appeared in 1614. Therefore, logarithms were essentially invented to turn the more difficult operation of multiplication into the easier process of addition.

Napier's logarithms did not really make use of a number base. It was the English mathematician Henry Briggs (1561–1630) who, recognizing the usefulness of logarithms, suggested to Napier that logarithms should use the base 10 in order to make the calculations even easier.

In 1624, Briggs published the first extensive table of base 10 logarithms, which was enthusiastically received by the scientific community as a long-needed tool for lengthy calculations. The great French mathematician Pierre Laplace (1749–1827) was hardly exaggerating when he said that logarithms "by shortening the labors, doubled the life of the astronomer." Logarithms were commonly used for calculations until the 1970s, when the scientific calculator came into use.

In this chapter, we study the *logarithmic function* and the *exponential function.* Although logarithms are no longer used directly for calculations, they are of great importance in many scientific and technical applications and in advanced mathematics. For example, they are used to measure the intensity of sound, the intensity of earthquakes, the power gains and losses in electrical transmission lines, and to distinguish between a base and an acid. Exponential functions are used in electronics, mechanical systems, thermodynamics, and nuclear physics, in biology to study population growth, and in business to calculate compound interest.

13.1 Exponential Functions

For any numbers a and b such that $a \neq 0$, and $b > 0$, $b \neq 1$, an **exponential function** is a function of the form

$$y = b^x \quad \textbf{(13.1a)} \qquad \text{or} \qquad y = ab^x \quad \textbf{(13.1b)}$$

Here b is called the **base** and x is any real number. The base b is restricted to be positive so that the function is a real number for all real values of x. Also, the coefficient a cannot be zero, and b cannot be zero or one; those three values would result in a constant function rather than an exponential function. Note that Eq. (13.1a) is a special case of Eq. (13.1b), with $a = 1$.

EXAMPLE 1 Exponential functions

Determine whether each of the following expressions defines an exponential function.

(a) $y = 3^x$ This is an exponential function with $b = 3$.

(b) $y = (-3)^x$ This is not an exponential function because $b = -3 < 0$.

(c) $y = -3^x$ This is an exponential function with $a = -1$ and $b = 3$.

(d) $y = 3^{-2x}$ To identify the base, we rewrite as $y = 3^{-2x} = (3^{-2})^x = \left(\frac{1}{9}\right)^x$. This is therefore an exponential function with $b = \frac{1}{9}$. It is common to write exponential functions with negative exponents rather than fractional bases.

EXAMPLE 2 Evaluating an exponential function

Evaluate the function $y = -2(4^x)$ for the given values of x.

(a) If $x = 2$, $y = -2(4^2) = -2(16) = -32$.

(b) If $x = -2$, $y = -2(4^{-2}) = -2/16 = -1/8$.

(c) If $x = 3/2$, $y = -2(4^{3/2}) = -2(8) = -16$.

(d) If $x = \sqrt{2}$, $y = -2\left(4^{\sqrt{2}}\right) = -14.206$.

Practice Exercises

Evaluate $y = 16^x$ for:

1. $x = 3/2$

2. $x = -0.5$

GRAPHING EXPONENTIAL FUNCTIONS

We now show some representative graphs of the exponential function.

EXAMPLE 3 Graphing an exponential function

Plot the graph of $y = 3^{-2x}$.

For this function, we have the values in the following table:

x	0	$\frac{1}{2}$	1	$\frac{3}{2}$	2
$2x$	0	1	2	3	4
y	1	$3^{-1} = \frac{1}{3}$	$3^{-2} = \frac{1}{9}$	$3^{-3} = \frac{1}{27}$	$3^{-4} = \frac{1}{81}$

The curve is shown in Fig. 13.1. The values from the table can be seen as enlarged points on the plot, all corresponding to integer exponents. Using all the real numbers for exponents, including the irrational numbers, results in the smooth curve shown in Fig. 13.1.

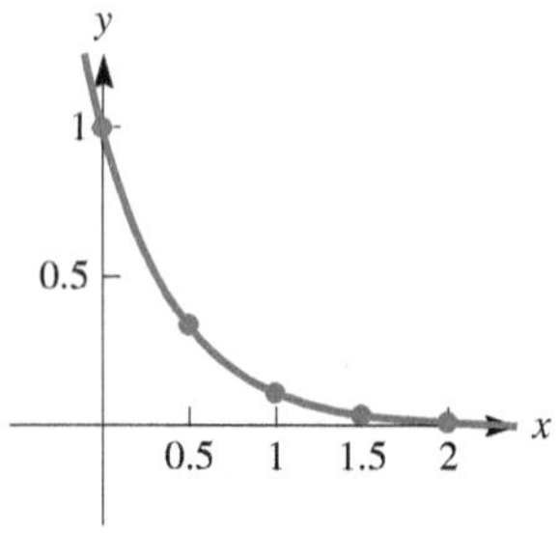

Fig. 13.1

We see that the x-axis is an *asymptote* of the curve. As x increases, the values of y decay and the points on the graph get closer and closer to the x-axis, although they never touch it.

We note that setting the exponents to be integers and then choosing the x values that result in those exponents is a common way of graphing exponential functions. The values chosen here ensure that the section of the graph where the function transitions from a steep slope to a small slope is displayed.

EXAMPLE 4 Graphing an exponential function

Plot the graph of $i = 30e^{4t}$, where t is time in seconds, and i is in milliamperes.

Choosing t values that result in integer exponents, we have the values in the following table:

t	0	$\frac{1}{4}$	$\frac{1}{2}$	$\frac{3}{4}$	1
$4t$	0	1	2	3	4
i	30	$30e = 81.6$	$30e^2 = 221.7$	$30e^3 = 602.6$	$30e^4 = 1637.9$

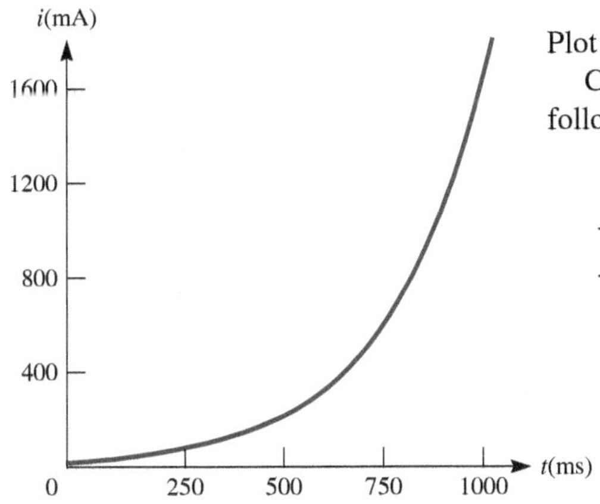

Fig. 13.2

Since all values of t that are of interest are less than 1 s, we use a smaller unit (milliseconds) for the plot. The resulting graph is shown in Fig. 13.2.

In this case, the t-axis is an asymptote for negative values of t, which are meaningless in the context of the problem. As t increases, the function grows without bound.

Any exponential decay curve will be similar in shape to that shown in Fig. 13.1. For the same base, the rate at which the graph decays depends on the exponent. Functions with larger negative coefficients in the exponent decay to zero faster, as illustrated in Fig. 13.3(a).

Similarly, any exponential growth curve will be similar in shape to that shown in Fig. 13.2. For the same base, the rate at which the graph grows also depends on the exponent. Functions with larger coefficients in the exponent grow at a faster rate, as illustrated in Fig. 13.3(b).

From this discussion and the examples above, we can draw the following general conclusions about exponential functions.

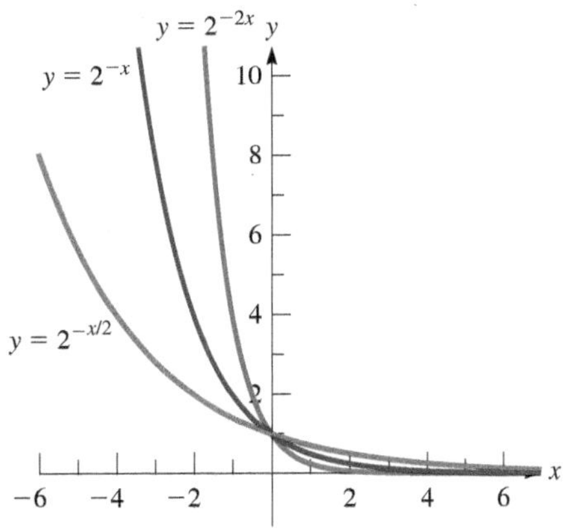

Fig. 13.3(a)

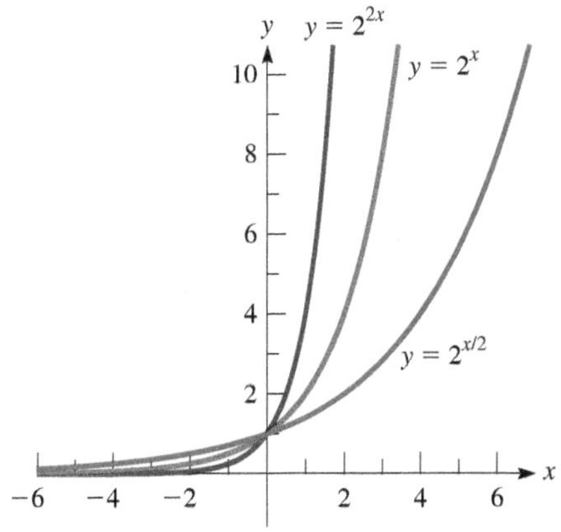

Fig. 13.3(b)

Basic Features of Exponential Functions of the Form $y = b^x$

1. The domain is all values of x; the range is $y > 0$.
2. The x-axis is an asymptote of the graph.
3. As x increases, the function increases if $b > 1$ and decreases if $b < 1$.

Exponential functions are important in many applications. We now illustrate an application in the next example, and others are shown in the exercises.

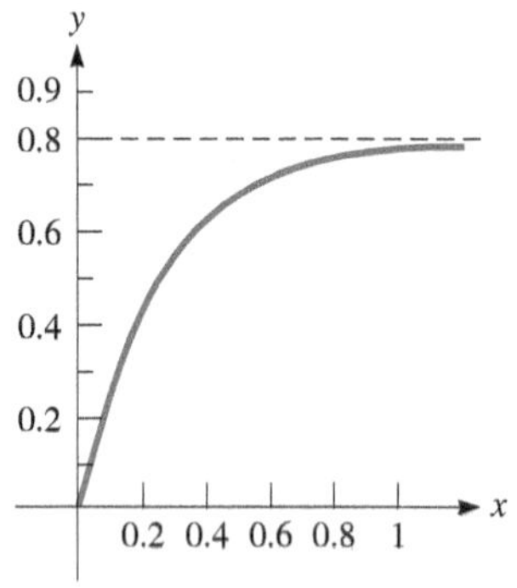

Fig. 13.4

EXAMPLE 5 Exponential function—electric current

In an electric circuit in which there is a battery, an inductor, and a resistor, the current i (in A) as a function of the time t (in s) is $i = 0.8(1 - e^{-4t})$. Graph this function. Here, e is equal to approximately 2.718.

Here, $e^{-4t} = 1$ for $t = 0$, and this means $i = 0$ for $t = 0$. Also, e^{-4t} becomes very small in a very short time, and this means i cannot be greater than 0.8 A, and that $i = 0.8$ A is an asymptote. The curve is shown in Fig. 13.4.

EXERCISES 13.1

In Exercises 1 and 2, perform the indicated operations if the given changes are made in the indicated examples of this section.

1. In Example 2(c), change the sign of x and then evaluate.
2. In Example 3, change the sign of the exponent and then plot the graph.

In Exercises 3–6, use a calculator to evaluate (to three significant digits) the given numbers.

3. $3^{\sqrt{5}}$ 4. 1.5^{π} 5. $(2\pi)^{-e}$ 6. $(2e)^{-\sqrt{2}}$

In Exercises 7–10, determine if the given functions are exponential functions.

7. (a) $y = 5^x$ (b) $y = 5^{-x}$
8. (a) $y = -7^x$ (b) $y = (-7)^{-x}$
9. (a) $y = -7(-5)^{-x}$ (b) $y = -7(5^{-x})$
10. (a) $y = (\sqrt{5})^{-x}$ (b) $y = -(\sqrt{-5})^x$

In Exercises 11–16, evaluate the exponential function $y = 7^x$ for the given values of x.

11. $x = 0.5$ 12. $x = 4$ 13. $x = -2$
14. $x = -0.5$ 15. $x = -3/2$ 16. $x = 5/2$

In Exercises 17–22, plot the graphs of the given functions.

17. $y = 4^x$ 18. $y = 0.25^x$ 19. $y = 0.2(10^{-x})$
20. $y = -5(1.6^{-x})$ 21. $y = 0.5\pi^x$ 22. $y = 2e^x$

In Exercises 23–28, display the graphs of the given functions on a graphing utility.

23. $y = 0.3(2.55)^x$ 24. $y = -1.5(4.15)^x$
25. $y = 0.1(0.25^{2x})$ 26. $y = 0.4(0.95)^x$
27. $i = 1.2(2 + 6^{-t})$ 28. $y = 0.5e^{-x}$

In Exercises 29–42, solve the given problems. Where necessary, round results to three significant digits.

29. Find the base b of the function $y = b^x$ if its graph passes through the point $(3, 64)$.
30. Find the base b of the function $y = b^x$ if its graph passes through the point $(-2, 64)$.
31. For the function $f(x) = b^x$, show that $f(c + d) = f(c) \cdot f(d)$.
32. For the function $f(x) = b^x$, show that $f(c - d) = f(c)/f(d)$.
33. Use a graphing utility to graph the function $y = 2^{|x|}$.
34. To show the *damping effect* of an exponential function, use a graphing utility to display the graph of $y = (x^3)(2^{-x})$. Be sure to use appropriate display settings.
35. Use a graphing utility to find the value(s) for which $x^2 = 2^x$.
36. Use a graphing utility to find the integral values of x for which $x^3 > 3^x$.
37. The value V of a bank account in which \$250 is invested at 5.00% interest, compounded annually, is $V = 250(1.0500)^t$, where t is the time in years. Find the value of the account after 4 years.

38. The intensity I of an earthquake is given by $I = I_0(10)^R$, where I_0 is a minimum intensity reference value, and R is the Richter scale magnitude of the earthquake. Evaluate I in terms of I_0 if $R = 5.5$.

39. The electric current i (in mA) in the circuit shown in Fig. 13.5 is $i = 2.5(1 - e^{-0.10t})$, where t is the time (in s). Evaluate i for $t = 5.0$ ms.

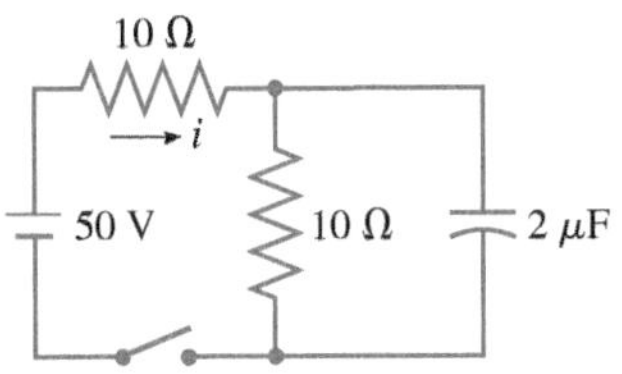

Fig. 13.5

40. The strength I of a certain cable signal is given by $I = I_0 e^{-0.0015x}$, where I_0 is the signal strength at the source and x is the distance (in km) from the source. What percent of the signal strength is lost 15 km from the source?

41. The flash unit on a camera operates by releasing the stored charge on a capacitor. For a particular unit, the charge q (in μC) as a function of the time t (in s) is $q = 100e^{-10t}$. Display the graph on a graphing utility.

42. The height y (in m) of the Gateway Arch in St. Louis (see Fig. 13.6) is given by $y = 230.9 - 19.5(e^{x/38.9} + e^{-x/38.9})$, where x is the distance (in m) from the point on the ground level directly below the top. Display the graph on a graphing utility.

Fig. 13.6

Answers to Practice Exercises

1. 64 **2.** 1/4

13.2 Logarithmic Functions

Any time we wish to solve for the value of an unknown exponent, it is necessary to express equations in logarithmic form. One useful way to think about a logarithm is as the exponent in an exponential function. In fact, every exponential function can be expressed as an equivalent logarithmic equation as follows.

$$\text{If } y = b^x, \text{ then } x = \log_b y. \quad \textbf{(13.2)}$$

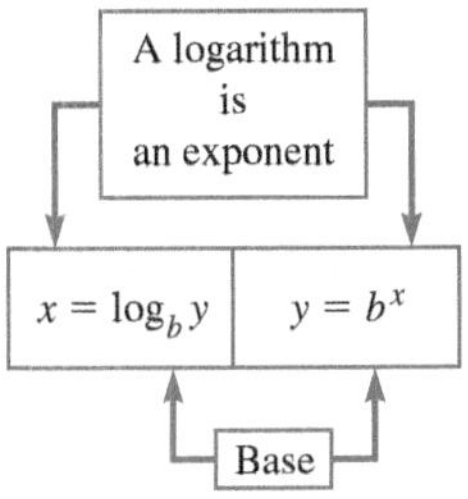

Fig. 13.7

This means that x is the power to which the base b must be raised in order to equal the number y. As with the exponential function, for the equation $x = \log_b y$, x may be any real number, b is a positive number other than 1, and y is a positive real number. In Eq. (13.2),

$y = b^x$ is the **exponential form**, and $x = \log_b y$ is the **logarithmic form**.

See Fig. 13.7. The exponential form is useful if we know x and need to obtain y; the logarithmic form is useful if we know y and need to obtain x.

EXAMPLE 1 Exponential form and logarithmic form

For each of the following, determine the change of form that is needed to solve the problem.

(a) $y = 2^x$; solve for x.

This exponential form must be written in logarithmic form to solve for x because the unknown is an exponent. The equivalent logarithmic form is $x = \log_2 y$.

(b) $\log_4 y = x$; solve for y.

This logarithmic form must be written in exponential form to find the value of y. The equivalent exponential form is $y = 4^x$.

EXAMPLE 2 Changing between forms

In each of the following expressions, identify the base and the exponent. Write each exponential form in logarithmic form or each logarithmic form in exponential form.

(a) $2 = \log_3 9$. The base is 3, the exponent is 2, so the exponential form is $3^2 = 9$.

(b) $4^{-1} = 1/4$. The base is 4, the exponent is -1, so the logarithmic form is $-1 = \log_4\left(\frac{1}{4}\right)$.

(c) $(64)^{1/3} = 4$. The base is 64, the exponent is $1/3$, so the logarithmic form is $\frac{1}{3} = \log_{64} 4$.

(d) $\log_2 32 = 5$. The base is 2, the exponent is 5, so the exponential form is $2^5 = 32$.

(e) $(32)^{3/5} = 8$. The base is 32, the exponent is $3/5$, so the logarithmic form is $\frac{3}{5} = \log_{32} 8$.

(f) $\log_6\left(\frac{1}{36}\right) = -2$. The base is 6, the exponent is -2, so the exponential form is $\frac{1}{36} = 6^{-2}$.

LEARNING TIP

Remember, ***the exponent may be negative.*** The *base* must be positive.

EXAMPLE 3 Solving for unknowns by changing form

(a) Find b, given that $-4 = \log_b\left(\frac{1}{81}\right)$.

Writing this in exponential form, we have $\frac{1}{81} = b^{-4}$. Thus, $\frac{1}{81} = \frac{1}{b^4}$ or $\frac{1}{3^4} = \frac{1}{b^4}$. Therefore, $b = 3$.

(b) Find y given that $\log_4 y = 1/2$. In exponential form this becomes $y = 4^{1/2}$, or $y = 2$.

Practice Exercises

1. Change $125^{2/3} = 25$ to logarithmic form.

2. Change $\log_3(1/3) = -1$ to exponential form.

The variable that we wish to solve for will dictate whether the exponential form or the logarithmic form is to be used. For this reason, it is important that you learn to transform readily from one form to the other.

EXAMPLE 4 Solving for unknowns—satellite power

The power supply P (in W) of a certain satellite is given by $P = 75e^{-0.005t}$, where t is the time (in days) after launch. By writing this equation in logarithmic form, solve for t.

In order to have the equation in the exponential form of Eq. (13.1a), we must have only $e^{-0.005t}$ on the right. Therefore, by dividing by 75, we have

$$\frac{P}{75} = e^{-0.005t}$$

Writing this in logarithmic form, we have $\log_e(P/75) = -0.005t$, or

$$t = \frac{\log_e\left(\dfrac{P}{75}\right)}{-0.005} = -200 \log_e\left(\frac{P}{75}\right) \qquad \frac{1}{-0.005} = -200$$

We recall from Section 12.5 that e is the special irrational number equal to about 2.718. It is important as a base of logarithms, and this will be discussed in Section 13.5.

LEARNING TIP

In order to change a function of the form $y = ab^x$ into logarithmic form, we must first write it as $y/a = b^x$. ***The coefficient of b^x must be equal to 1,*** which is the form of Eq. (13.1a). In the same way, the coefficient of $\log_b y$ must be 1 in order to change it into exponential form.

THE LOGARITHMIC FUNCTION

When we are working with functions, we must keep in mind that a function is defined by the operation being performed on the independent variable, and not by the letter chosen to represent it. However, for consistency, it is standard practice to let y represent

LEARNING TIP

Eqs. (13.2) and (13.3) do not represent different *functions,* due to the difference in location of the variables, since they represent the *same operation* on the independent variable that appears in each. However, Eq. (13.3) expresses the function with the standard dependent and independent variables.

the dependent variable and x represent the independent variable. Therefore, *the* **logarithmic function** *is*

$$y = \log_b x \tag{13.3}$$

As with the exponential function, $b > 0$ and $b \neq 1$.

EXAMPLE 5 Evaluating a logarithmic function

Evaluate the logarithmic function $y = \log_2 x$ for the given value of x.

(a) $x = 16$. Here $y = \log_2 16$, which means that $y = 4$, since $2^4 = 16$.
(b) $x = \frac{1}{16}$. Here $y = \log_2\left(\frac{1}{16}\right)$, which means that $y = -4$, since $2^{-4} = \frac{1}{16}$.

Practice Exercise

3. For the function in Example 5, evaluate y for $x = 8$.

GRAPHING LOGARITHMIC FUNCTIONS

We now show some representative graphs for the logarithmic function. From these graphs, we can see the basic properties of the logarithmic function.

EXAMPLE 6 Graphing a logarithmic function

Plot the graph of $y = \log_2 x$.

We can find the points for this graph more easily if we first put the equation in exponential form: $x = 2^y$. By assuming values for y, we can find the corresponding values for x.

$2^{-2} = \frac{1}{4}$ $\qquad$ $2^2 = 4$

x	$\frac{1}{4}$	$\frac{1}{2}$	1	2	4	8
y	−2	−1	0	1	2	3

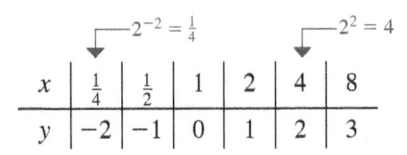

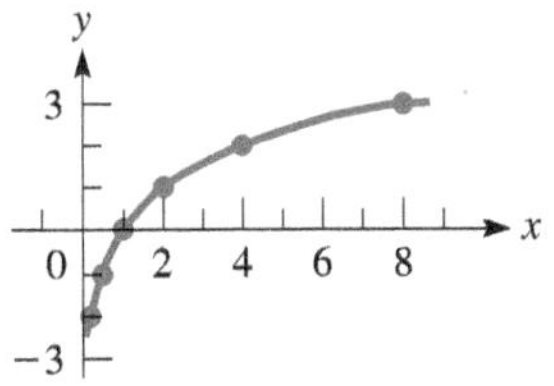

Fig. 13.8

Using these values, we construct the graph seen in Fig. 13.8.

From the graph in Example 6, we can see that logarithmic functions have the following features. We consider only the features for $b > 1$, for these are the bases of greatest importance.

Basic Features of Logarithmic Functions of the Form $y = \log_b x$ $(b > 1)$

1. The domain is $x > 0$; the range is all values of y.
2. The negative y-axis is an asymptote of the graph.
3. If $0 < x < 1$, $\log_b x < 0$; if $x = 1$, $\log_b x = 0$; if $x > 1$, $\log_b x > 0$.
4. If $x > 1$, $\log_b x$ increases more slowly than x. If $0 < x < 1$, $\log_b x$ increases more rapidly than x.

We just noted that if $b > 1$ and $x > 1$, $\log_b x$ increases more slowly than x. It is also true that b^x increases more rapidly than x. Actually, as x becomes larger, $\log_b x$ increases very slowly, and b^x increases very rapidly. Using $\log_2 x$ and 2^x and a calculator, we have the table of values, Table 13.1. This shows that we must choose values of x carefully when graphing these functions.

Table 13.1

x	1	4	16	64
$\log_2 x$	0	2	4	6
2^x	2	16	65 536	1.8×10^{19}

INVERSE FUNCTIONS

The exponential and the logarithmic functions are **inverses** of each other. Recall that, in general, the inverse of a function undoes the process that the function performs. In this case, the exponential function operates on a base to raise it to a certain exponent and return a value, while the logarithmic function operates on a value and returns the exponent to which the base has to be raised to produce that number.

Procedure for Finding the Inverse of a Function

1. Solve for the independent variable as a function of the dependent variable.
2. Interchange the variables.

LEARNING TIP

Note that the x and y coordinates of inverse functions are interchanged. As a result, the graphs of inverse functions are mirror images of each other across the line $y = x$.

For a function, there is exactly one value of y in the range for each value of x in the domain. This must also hold for the inverse function. Thus, for a function to have an inverse, there must be only one x for each y. This is true for $y = b^x$ and $y = \log_b x$, as we have seen earlier in this section.

In the following example we use the procedure for finding an inverse to verify that an exponential function and a logarithmic function are inverses of each other. Other examples are found in the exercises.

EXAMPLE 7 Inverse functions

The functions $y = 2^x$ and $y = \log_2 x$ are inverse functions. Show this by solving $y = 2^x$ for x and then interchanging x and y.

Writing $y = 2^x$ in logarithmic form gives us $x = \log_2 y$. Then interchanging x and y, we have $y = \log_2 x$, which is the inverse function.

Making a table of values for each function, we have

$y = 2^x$:

x	-3	-2	-1	0	1	2	3
y	$\frac{1}{8}$	$\frac{1}{4}$	$\frac{1}{2}$	1	2	4	8

$y = \log_2 x$:

x	$\frac{1}{8}$	$\frac{1}{4}$	$\frac{1}{2}$	1	2	4	8
y	-3	-2	-1	0	1	2	3

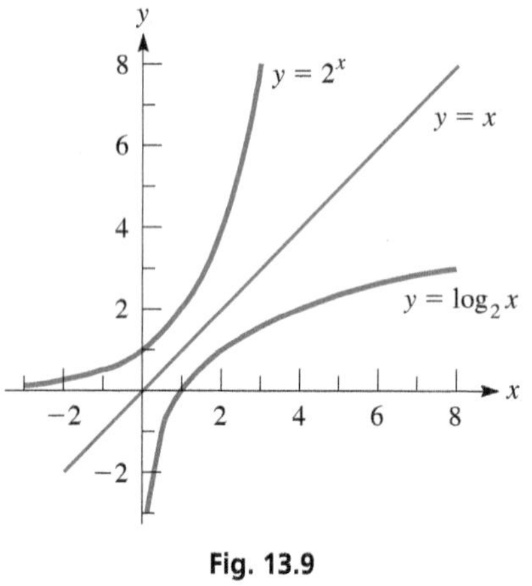

Fig. 13.9

We see that the coordinates are interchanged. In Fig. 13.9, note that the graphs of these two functions reflect each other across the line $y = x$.

EXERCISES 13.2

In Exercises 1–4, perform the indicated operations if the given changes are made in the indicated examples of this section.

1. In Example 2(e), change the exponent to $4/5$ and then make any other necessary changes.
2. In Example 3(b), change the $1/2$ to $5/2$ and then make any other necessary changes.
3. In Example 5, change the logarithm base to 4 and then make any other necessary changes.
4. In Example 6, change the logarithm base to 4 and then plot the graph.

In Exercises 5–16, express the given equations in logarithmic form.

5. $3^4 = 81$ **6.** $5^4 = 625$

7. $4^5 = 1024$ **8.** $2^9 = 512$

9. $7^{-2} = \frac{1}{49}$ **10.** $3^{-2} = \frac{1}{9}$

11. $2^{-6} = \frac{1}{64}$ **12.** $(12)^0 = 1$

13. $8^{1/3} = 2$ **14.** $(81)^{3/4} = 27$

15. $\left(\frac{1}{4}\right)^2 = \frac{1}{16}$ **16.** $\left(\frac{1}{2}\right)^{-2} = 4$

In Exercises 17–28, express the given equations in exponential form.

17. $\log_3 81 = 4$ **18.** $\log_{11} 121 = 2$

19. $\log_9 9 = 1$ **20.** $\log_{15} 1 = 0$

21. $\log_{25} 5 = \frac{1}{2}$ **22.** $\log_8 16 = \frac{4}{3}$

23. $5 \log_{243} 3 = 1$ **24.** $\log_{32}\left(\frac{1}{8}\right) = -0.6$

25. $\log_{10} 0.1 = -1$ **26.** $\log_7\left(\frac{1}{49}\right) = -2$

27. $\log_{0.5} 16 = -4$ **28.** $\log_{1/3} 3 = -1$

In Exercises 29–44, determine the value of the unknown.

29. $\log_4 16 = x$ **30.** $\log_5 125 = x$

31. $\log_{10} 0.01 = x$ **32.** $\log_{16}\left(\frac{1}{4}\right) = x$

33. $\log_7 y = 3$ **34.** $\log_8(N + 1) = 3$

35. $3 \log_8(A - 2) = -2$ **36.** $\log_7 y = -2$

37. $\log_b 5 = 2$ **38.** $\log_b 625 = 4$

39. $\log_b 4 = -\frac{1}{3}$ **40.** $\log_b 4 = \frac{2}{3}$

41. $\log_{10} 10^{0.2} = x$ **42.** $\log_5 5^{2.3} = R + 1$

43. $\log_3 27^{-1} = x + 1$ **44.** $\log_b\left(\frac{1}{4}\right) = -0.5$

In Exercises 45–50, plot the graphs of the given functions.

45. $y = \log_3 x$ **46.** $y = \log_4 x$

47. $y = \log_{0.5} x$ **48.** $y = 3 \log_2 x$

49. $N = 0.2 \log_4 v$ **50.** $A = 2.4 \log_{10}(2r)$

In Exercises 51–54, display the graphs of the given functions with a graphing utility.

51. $y = 3 \log_e x$ **52.** $y = 5 \log_{10}|x|$

53. $y = -\log_{10}(-x)$ **54.** $y = -\log_e(-2x)$

In Exercises 55–72, perform the indicated operations.

55. Evaluate $\log_4 x$ for (a) $x = 1/64$ and (b) $x = -1/2$.

56. Evaluate: (a) $\log_b b$ (b) $\log_b 1$

57. If $f(x) = \log_5 x$, find: (a) $f(\sqrt{5})$ (b) $f(0)$

58. If $f(x) = \log_b x$ and $f(3) = 2$, find $f(9)$.

59. With a graphing utility, display the graphs of: (a) $y_1 = \log_{10} x$ (b) $y_2 = \log_{10}(x + 2)$ (c) $y_3 = \log_{10}(x - 2)$.

60. With a graphing utility, display the graphs of $y_1 = 2 \log_{10} x$ and $y_2 = \log_{10} x^2$. Describe any similarities or differences.

61. Using a graphing utility, solve the equation $\log_{10} x = x - 2$.

62. Use a graphing utility to display the graphs (on the same plot) of $y = \log_e(x^2 + c)$, with $c = -4, 0, 4$. Describe the results.

63. The magnitudes (visual brightnesses), m_1 and m_2, of two stars are related to their (actual) brightnesses, b_1 and b_2, by the equation $m_1 - m_2 = 2.5 \log_{10}(b_2/b_1)$. Solve for b_2.

64. The velocity v of a rocket when its fuel is completely burned is given by $v = u \log_e(w_0/w)$, where u is the exhaust velocity, w_0 is the liftoff weight, and w is the burnout weight. Solve for w.

65. An equation relating the number N of atoms of radium at any time t in terms of the number N_0 of atoms at $t = 0$ is $\log_e(N/N_0) = -kt$, where k is a constant. Solve for N.

66. The capacitance C of a cylindrical capacitor is given by $C = k/\log_e(R_2/R_1)$, where R_1 and R_2 are the inner and outer radii. Solve for R_1.

67. The work W done by a sample of nitrogen gas during an isothermal (constant temperature) change from volume V_1 to volume V_2 is given by $W = k \log_e(V_2/V_1)$. Solve for V_1.

68. An equation used in measuring the flow of water in a channel is $C = -a \log_{10}(b/R)$. Solve for R.

69. The time t (in ps) required for N calculations by a certain computer design is $t = N + \log_2 N$. Sketch the graph of this function.

70. When a tractor-trailer turns a right-angle corner, its rear wheels follow a curve called a *tractrix*, the equation for which is $y = \log_e\left(\dfrac{1 + \sqrt{1 - x^2}}{x}\right) - \sqrt{1 - x^2}$. Display the curve on a graphing utility.

71. Graph the functions $y = (1/3)^x$ and $y = 3^{-x}$ and then compare them. Explain what your comparison shows.

72. In Exercise 47, the graph of $y = \log_{0.5} x$ is plotted. By inspecting the graph and noting the properties of $\log_{0.5} x$, describe some of the differences between logarithms to a base less than 1 and those to a base greater than 1.

In Exercises 73–76, show that the given functions are inverse functions of each other. Then display the graphs of each function and the line $y = x$ with a graphing utility and verify that each is the mirror image of the other across $y = x$.

73. $y = 10^{x/2}$ and $y = 2 \log_{10} x$

74. $y = e^x$ and $y = \log_e x$

75. $y = 3x$ and $y = x/3$

76. $y = 2x + 4$ and $y = 0.5x - 2$

Answers to Practice Exercises

1. $2/3 = \log_{125} 25$ **2.** $1/3 = 3^{-1}$ **3.** 3

13.3 Properties of Logarithms

Since a logarithm is an exponent, it must follow the laws of exponents. The laws used in this section to derive the very useful properties of logarithms are listed here for reference.

■ A logarithm is an exponent. However, a historical curiosity is that logarithms were developed before exponents were used.

Product Rule for Exponents: $b^u b^v = b^{u+v}$ **(13.4)**

Quotient Rule for Exponents: $\dfrac{b^u}{b^v} = b^{u-v}$ **(13.5)**

Power Rule for Exponents: $(b^u)^n = b^{nu}$ **(13.6)**

We now use these rules of exponents to derive equivalent rules for logarithms. For all three derivations, we let $x = b^u$ and $y = b^v$, which written in logarithmic form give $u = \log_b x$ and $v = \log_b y$.

Deriving the Product Rule for Logarithms

Form the product of x and y and simplify: $xy = b^u b^v$ which simplifies as $xy = b^{u+v}$

Write the expression in logarithmic form: $\log_b xy = u + v$

Replace u and v with their logarithmic forms: $\log_b xy = \log_b x + \log_b y$

In words, the logarithm of the product of two numbers is equal to the sum of the logarithms of the numbers.

Deriving the Quotient Rule for Logarithms

Form the quotient of x and y and simplify: $\frac{x}{y} = \frac{b^u}{b^v}$ which simplifies as $\frac{x}{y} = b^{u-v}$

Write the expression in logarithmic form: $\log_b\left(\frac{x}{y}\right) = u - v$

Replace u and v with their logarithmic forms: $\log_b\left(\frac{x}{y}\right) = \log_b x - \log_b y$

In words, the logarithm of the quotient of two numbers is equal to the logarithm of the numerator minus the logarithm of the denominator.

Deriving the Power Rule for Logarithms

Form the nth power of x and simplify: $x^n = (b^u)^n$, which simplifies as $x^n = b^{nu}$

Write the expression in logarithmic form: $\log_b x^n = nu$

Replace u with its logarithmic form: $\log_b x^n = n \log_b x$

In words, the logarithm of the nth power of a number is equal to n times the logarithm of the number. The exponent n may be any real number, which, of course, includes all rational and irrational numbers.

In summary,

Product Rule for Logarithms: $\log_b xy = \log_b x + \log_b y$ **(13.7)**

Quotient Rule for Logarithms: $\log_b\left(\frac{x}{y}\right) = \log_b x - \log_b y$ **(13.8)**

Power Rule for Logarithms: $\log_b x^n = n\log_b x$ **(13.9)**

LEARNING TIP

In Section 13.2, we showed that the base b of logarithms must be a positive number. Since $x = b^u$ and $y = b^v$, this means that x and y are also positive numbers. Therefore, the *properties of logarithms that have just been derived are valid only for positive values of x and y.*

COMMON ERROR

The logarithm of a product is the sum of the logarithms. However, this property **does not** give us a way to rewrite the logarithm of a sum. For instance,

$\log_2(3x)$ can be rewritten as $\log_2(3x) = \log_2 3 + \log_2 x$, but
$\log_2(3 + x)$ **cannot** be rewritten in terms of $\log_2 3$ and $\log_2 x$.

Similarly, the logarithm of a quotient is the difference of the logarithms. However, this property **does not** give us a way to rewrite the logarithm of a difference. For instance,

$\log_3\left(\frac{5}{x}\right)$ can be rewritten as $\log_3\left(\frac{5}{x}\right) = \log_3 5 - \log_3 x$, but
$\log_3(5 - x)$ **cannot** be rewritten in terms of $\log_3 5$ and $\log_3 x$.

The properties of logarithms can be used to expand logarithmic expressions, rewriting the logarithm of a single argument as the combination of logarithms of simpler arguments.

EXAMPLE 1 Using the properties of logarithms

Expand each of the following logarithms.

(a) $\log_4 15$. We factor 15 as $3 \cdot 5$ and then use the product rule:

$$\log_4 15 = \log_4(3 \cdot 5) = \log_4 3 + \log_4 5$$

(b) $\log_4\left(\frac{5}{3}\right)$. Using the quotient rule:

$$\log_4\left(\frac{5}{3}\right) = \log_4 5 - \log_4 3$$

(c) $\log_4 t^2$. Using the power rule with $n = 2$:

$$\log_4 t^2 = 2 \log_4 t$$

(d) $\log_4\left(\frac{xy}{z}\right)$. Using the quotient rule and then the product rule:

$$\log_4\left(\frac{xy}{z}\right) = \log_4(xy) - \log_4 z$$
$$= \log_4 x + \log_4 y - \log_4 z$$

(e) $\log_3\left(\frac{\sqrt{5x^2y}}{xz}\right)$. There are several ways to expand this logarithm. We begin by simplifying the fraction and writing the radical as a fractional exponent. Then we use the quotient rule, the power rule, and the product rule in turn. We get

$$\log_3\left(\frac{\sqrt{5x^2y}}{xz}\right) = \log_3\left(\frac{(5y)^{1/2}}{z}\right) = \log_3((5y)^{1/2}) - \log_3 z$$
$$= \frac{1}{2}\log_3(5y) - \log_3 z = \frac{1}{2}\log_3 5 + \frac{1}{2}\log_3 y - \log_3 z$$

Practice Exercises

Express as a sum or difference of logarithms.

1. $\log_3 10$ **2.** $\log_3(2a/5)$

EXAMPLE 2 Sum of logarithms as a single quantity

Write each of the following expressions as the logarithm of a single quantity.

(a) $\log_4 3 + \log_4 x = \log_4(3 \times x) = \log_4 3x$ — product rule

(b) $\log_4 3 - \log_4 x = \log_4\left(\frac{3}{x}\right)$ — quotient rule

(c) $\log_4 3 + 2 \log_4 x = \log_4 3 + \log_4(x^2) = \log_4 3x^2$ — power rule then product rule

(d) $\log_4 3 + 2 \log_4 x - \log_4 y = \log_4\left(\frac{3x^2}{y}\right)$ — power rule, then product and quotient rules

Note that when condensing expressions into the logarithm of a single quantity, we use the power rule first so as to set up in the correct form for using the product and quotient rules. Also, it is easier to avoid sign errors if the product rule is used to group terms before the quotient rule is applied.

Because any value raised to the power of zero equals one, we have that $\log_b 1 = 0$. Also, since $b = b^1$ in logarithmic form is $\log_b b = 1$, we have $\log_b(b^x) = x \log_b b = x(1) = x$. Also, the logarithmic form of $\log_b x = \log_b x$ is $b^{\log_b x} = x$.

Summarizing these properties, we have

$$\log_b 1 = 0 \qquad \log_b b = 1 \tag{13.10}$$

$$\log_b(b^x) = x \tag{13.11}$$

$$b^{\log_b x} = x \tag{13.12}$$

These equations can be used to simplify certain expressions.

EXAMPLE 3 Exact values for certain logarithms

Find the exact value of each of the following logarithms.

(a) $\log_3 9$. We write 9 as a power of 3 and use Eq. (13.11):

$$\log_3 9 = \log_3 3^2 = 2$$

(b) $\log_3 3^{0.4}$. Without evaluating $3^{0.4}$, we can evaluate this expression using Eq. (13.11). We get

$$\log_3 3^{0.4} = 0.4$$

Practice Exercise

3. Find the exact value of $2 \log_2 8$.

EXAMPLE 4 Using the properties of logarithms

(a) $\log_2 6 = \log_2(2 \times 3) = \log_2 2 + \log_2 3 = 1 + \log_2 3$
(b) $\log_5 \frac{1}{5} = \log_5 1 - \log_5 5 = 0 - 1 = -1$
(c) $\log_7 \sqrt{7} = \log_7(7^{1/2}) = \frac{1}{2}\log_7 7 = \frac{1}{2}$

EXAMPLE 5 Evaluation of logarithms in two ways

The following illustration shows the evaluation of a logarithm in two different ways. Either method is appropriate.

(a) $\log_5\left(\frac{1}{25}\right) = \log_5 1 - \log_5 25 = 0 - \log_5(5^2) = -2$
(b) $\log_5\left(\frac{1}{25}\right) = \log_5(5^{-2}) = -2$

When solving equations containing logarithms, two general approaches can be applied:

1. Rewrite the equation so that logarithms of the same base of two expressions are equal to each other. The two expressions must then be equal.
2. Combine all terms involving logarithms as the isolated logarithm of a single quantity containing the unknown. Rewrite in exponential form.

The following two examples illustrate these approaches.

EXAMPLE 6 Solving an equation with logarithms

Use the basic properties of logarithms to solve the following equation for y in terms of x: $\log_b y = 2 \log_b x + \log_b a$.

Using the power rule and then the product rule, we have

$$\log_b y = \log_b(x^2) + \log_b a = \log_b(ax^2)$$

Since we have the logarithm to the base b of different expressions on each side of the resulting equation, the expressions must be equal. Therefore,

$$y = ax^2$$

EXAMPLE 7 Solving an equation—radioactive decay

An equation encountered in the study of radioactive elements is $\log_e N - \log_e N_0 = kt$. Here, N is the amount of element present at any time t, and N_0 is the original amount. Solve for N as a function of t.

Using the quotient rule, we rewrite the left side of this equation, obtaining

$$\log_e\left(\frac{N}{N_0}\right) = kt$$

Rewriting this in exponential form, we have

$$\frac{N}{N_0} = e^{kt} \quad \text{or} \quad N = N_0\, e^{kt}$$

EXERCISES 13.3

In Exercises 1–8, perform the indicated operations on the resulting expressions if the given changes are made in original expressions of the indicated examples of this section.

1. In Example 1(a), change the 15 to 21.
2. In Example 1(d), change the x to 2 and the z to 3.
3. In Example 2(b), change the 3 to 5.
4. In Example 2(c), change the 2 to 3.
5. In Example 3(a), change the 9 to 27.
6. In Example 4(a), change the 6 to 10.
7. In Example 4(c), change the 7's to 5's.
8. In Example 6, change the 2 to 3.

In Exercises 9–20, expand each as a sum, difference, or multiple of logarithms. See Example 1.

9. $\log_5 34$ **10.** $\log_3 21$

11. $\log_7 \frac{11}{9}$ **12.** $\log_3 \frac{3}{5}$

13. $\log_7 a^4$ **14.** $2 \log_6 n^7$

15. $\log_5(2abc)$ **16.** $\log_4 \dfrac{x^2y^4}{z^5}$

17. $8 \log_7 \sqrt[4]{y^3}$ **18.** $\log_5 \sqrt[9]{x}$

19. $\log_3 \dfrac{\sqrt{x}}{a^5}$ **20.** $\log_2 \dfrac{\sqrt{y}}{9}$

In Exercises 21–28, express each as the logarithm of a single quantity. See Example 2.

21. $\log_b H + \log_b M$ **22.** $\log_2 4 + \log_2 R$

23. $\log_5 10 - \log_5 5$ **24.** $-\log_7 R + \log_7 V$

25. $-\log_b \sqrt{x} + \log_b x^2$ **26.** $\log_4 3^3 + \log_4 9$

27. $2 \log_e 2 + 3 \log_e \pi - \log_e 3$

28. $\frac{1}{2}\log_b x + 2 \log_b 5 - 3 \log_b x$

In Exercises 29–36, determine the exact value of each of the given logarithms.

29. $\log_2\left(\frac{1}{32}\right)$ **30.** $\log_3\left(\frac{1}{81}\right)$

31. $\log_2(2^{2.5})$ **32.** $\log_5(5^{0.1})$

33. $6 \log_7 \sqrt{7}$ **34.** $\pi \log_6 \sqrt[3]{6}$

35. $4^{\log_4 8}$ **36.** $10^{2 \log_{10} 3}$

In Exercises 37–44, express each as a sum, difference, or multiple of logarithms. In each case, part of the logarithm may be determined exactly.

37. $\log_3 18$ **38.** $\log_5 75$

39. $\log_2\left(\frac{1}{6}\right)$ **40.** $\log_{10}(0.05)$

41. $\log_3 \sqrt{6}$ **42.** $\log_2 \sqrt[3]{24}$

43. $\log_{10} 3000$ **44.** $\log_{10}(40^2)$

In Exercises 45–56, solve for y in terms of x.

45. $\log_b y = \log_b 3 + \log_b x$

46. $\log_b y = \log_b 8 - \log_b 3x$

47. $\log_4 y = \log_4 x - \log_4 7 + \log_4 2$

48. $\log_3 y = \log_3(x - 4)^{-5} + \log_3 9$

49. $\log_{10} y = 2 \log_{10} 7 - 3 \log_{10} x$

50. $\log_b y = 3 \log_b \sqrt{x} + 2 \log_b 10$

51. $5 \log_2 y - \log_2 x = 3 \log_2 4 + \log_2 a$

52. $4 \log_2 x - 3 \log_2 y = \log_2 27$

53. $\log_2 x + \log_2 y = 1$ **54.** $\pi \log_4 x + \log_4 y = 1$

55. $\dfrac{2 \log_5 x}{\log_5 3} - \log_5 y = 2$ **56.** $\log_8 x = 2 \log_8 y + 4$

In Exercises 57–68, solve the given problems.

57. Explain why $\log_{10}(x + 3)$ is not equal to $\log_{10} x + \log_{10} 3$.

58. Evaluate $2 \log_2(2x) - \log_2 x^2$. For what values of x is the value of this expression valid? Explain.

59. Display the graphs of $y = \log_e(e^2x)$ and $y = 2 + \log_e x$ with a graphing utility and explain why they are the same.

60. If $x = \log_b 2$ and $y = \log_b 3$, express $\log_b 12$ in terms of x and y.

61. If $\log_b x = 2$ and $\log_b y = 3$, find $\log_b \sqrt{x^2y^4}$.

62. Is it true that $\log_b (ab)^x = x \log_b a + x$?

63. Using a graphing utility, display the graphs of

$y_1 = \log_{10} x - \log_{10}(x^2 + 1)$ and $y_2 = \log_{10} \dfrac{x}{x^2 + 1}$ on the same plot. What conclusion can be drawn from the display?

64. The use of the insecticide DDT was banned in Canada in 1970. A computer analysis shows that an expression relating the amount A still present in an area, the original amount A_0, and the time t (in years) since 1970 is $\log_{10} A = \log_{10} A_0 + 0.1t \log_{10} 0.8$. Solve for A as a function of t.

65. A study of urban density shows that the population density D (in persons/km^2) is related to the distance r (in km) from the city centre by $\log_e D = \log_e a - br + cr^2$, where a, b, and c are positive constants. Solve for D as a function of r.

66. When a person ingests a medication capsule, it is found that the rate R (in mg/min) at which it enters the bloodstream at time t (in min) is given by $\log_{10} R - \log_{10} 5 = t \log_{10} 0.95$. Solve for R as a function of t.

67. A container of water is heated to 90°C and then placed in a room at 0°C. The temperature T of the water is related to the time t (in min) by $\log_e T = \log_e 90.0 - 0.23\,t$. Find T as a function of t.

68. In analysing the power gain in an electric circuit, the equation $N = 10(2 \log_{10} I_1 - 2 \log_{10} I_2 + \log_{10} R_1 - \log_{10} R_2)$ is used. Express this with a single logarithm on the right side.

Answers to Practice Exercises

1. $\log_3 2 + \log_3 5$ **2.** $\log_3 2 + \log_3 a - \log_3 5$ **3.** 6

13.4 Logarithms to the Base 10

In Section 13.2, we stated that a base of logarithms must be a positive number, not equal to 1. In the examples and exercises of the previous sections, we used a number of different bases. However, only two bases are generally used. They are 10 and e, where e is the irrational number approximately equal to 2.718 that we introduced in Section 12.5 and have used in the previous sections of this chapter. We discuss logarithms to the base 10 in this section and those to the base e in the next section.

Logarithms to the base 10 are called **common logarithms**. They may be found directly by use of a calculator, and the [log] key is used for this purpose.

LEARNING TIP

1. When no base is shown, the logarithm is assumed to be to the base 10.
2. The *decimal* part of a logarithm is expressed with the same number of significant digits as there are in the number whose logarithm is being found.

EXAMPLE 1 Base 10 logarithms

Find the following logarithms using a calculator. Write the equivalent exponential form.

(a) $\log 426 = 2.649$. Here we have rounded the decimal part of the answer to three significant digits because 426 has three significant digits. In exponential form, this is written as

$$10^{2.629} = 426$$

Note that 426 falls between $10^2 = 100$ and $10^3 = 1000$.

(b) $\log 2 = 0.3$. In exponential form, we write

$$10^{0.3} = 2$$

(c) $\log 0.036\,54 = -1.4372$. In exponential form, we have

$$10^{-1.4372} = 0.036\,54$$

Note that the logarithm here is negative. Raising 10 to a negative power gives us a number between 0 and 1; in this case, it is between $10^{-2} = 0.01$ and $10^{-1} = 0.1$.

We may also use a calculator to find a number N if we know $\log N$. *In this case, we refer to N as the* **antilogarithm** *of log N*. On the calculator, we use the [10^x] key. We note that it shows the basic definition of a logarithm. (On many scientific calculators, the key sequence [inv] [log] is used. Note that this sequence makes it clear that the exponential and logarithmic functions are inverse functions.)

EXAMPLE 2 Antilogarithm—inverse logarithm

Given the following logarithms, find the antilogarithm.

(a) $\log N = 1.1854$

Since $1 < \log N < 2$, we know that N must be a number between $10^1 = 10$ and $10^2 = 100$. We find the exact number as an antilogarithm. We have

$$N = 10^{1.1854} = 15.32$$

where the result has been rounded off to four significant digits because the decimal part of the logarithm had four significant digits.

(b) $\log x = -2.231$

The antilogarithm is $x = 10^{-2.231} = 0.005\ 87$, rounded to three significant digits.

The following example illustrates an application in which a measurement requires the direct use of the value of a logarithm.

■ The unit of sound intensity level (used for power gain), the bel (B), is named for the inventor Alexander Graham Bell (1847–1922). The decibel is the commonly used unit.

Practice Exercises

Evaluate x using a calculator.

1. $x = \log 0.5392$

2. $\log x = 2.2901$

EXAMPLE 3 Base 10 logarithm—power gain

The power gain G (in decibels (dB)) of an electronic device is given by $G = 10 \log(P_0/P_i)$, where P_0 is the output power (in W) and P_i is the input power. Determine the power gain for an amplifier for which $P_0 = 15.8$ W and $P_i = 0.625$ W.

Substituting the given values, we have

$$G = 10 \log \frac{15.8}{0.625} = 14.0 \text{ dB}$$

As noted in the chapter introduction, logarithms were first used to make complicated calculations simpler and make large numbers more manageable. Performing calculations in this way provides an opportunity to understand better the meaning and properties of logarithms. Also, certain calculations cannot be done directly on a calculator but can be done by logarithms.

EXAMPLE 4 Calculation using logarithms

A certain computer design has 64 different sequences of 10 binary digits so that the total number of possible states is $(2^{10})^{64} = 1024^{64}$. Evaluate 1024^{64} using logarithms and write it in scientific notation. (It is very possible that your calculator cannot do this calculation directly.)

Define the quantity as N:	$N = 1024^{64}$
Take logarithms on both sides:	$\log N = \log 1024^{64}$
Use the power rule for logarithms:	$\log N = 64 \log 1024$
Evaluate using a calculator:	$\log N = 192.659\ 197\ 2$
Express as an antilogarithm:	$N = 10^{192.659\ 197\ 2}$
Use the product rule for exponents:	$N = 10^{192} \times 10^{0.659\ 197\ 2}$
Find the antilogarithm of the decimal exponent:	$N = 10^{192} \times 4.5624$
Write in scientific notation:	$N = 4.5624 \times 10^{192}$

Note that although we have used a calculator to find a logarithm and an antilogarithm, the only actual calculation we have done is a multiplication by 64. The use of logarithms has allowed us to change the calculation of a power into the simpler process of multiplication.

EXERCISES 13.4

In Exercises 1 and 2, find the indicated values if the given changes are made in the indicated examples of this section.

1. In Example 1, change 0.036 54 to 0.3654 and then find the required value.

2. In Example 2, change 1.1854 to 2.1854 and then find the required value.

In Exercises 3–10, find the common logarithm of each of the given numbers by using a calculator.

3. 675 **4.** 0.0460

5. 4.29×10^5 **6.** 4.19^4

7. 1.741^{-3} **8.** 2.034×10^{-6}

9. $\sqrt{447}$ **10.** $\log_3 81$

In Exercises 11–18, find the antilogarithm of each of the given logarithms by using a calculator.

11. 7.742 **12.** 0.695 **13.** −1. 4503 **14.** −4.215

15. 3.301 12 **16.** 8.824 36 **17.** −2.237 46 **18.** −10.336

In Exercises 19–22, use logarithms to evaluate the given expressions. All exponents are exact.

19. $(5.98)(14.3)$ **20.** $\dfrac{895}{73.4^{86}}$

21. $(\sqrt[10]{7.32})(2470)^{30}$ **22.** $\dfrac{126\,000^{20}}{2.632^{2.5}}$

In Exercises 23–26, use a calculator to verify the given values.

23. $\log 14 + \log 0.5 = \log 7$ **24.** $\log 500 - \log 20 = \log 25$

25. $\log 81 = 4 \log 3$ **26.** $\log 6 = 0.5 \log 36$

In Exercises 27–31, find the logarithms of the given numbers.

27. The signal used by some cell phones has a frequency of 9.00×10^8 Hz.

28. A large sunspot may be 3.5×10^4 km in diameter.

29. About $1.3 \times 10^{-14}\%$ of carbon atoms are carbon-14, the radioactive isotope used in determining the age of samples from ancient sites.

30. In an air sample taken in an urban area, $5/10^6$ of the air was carbon monoxide.

31. The mass of an electron is 9.11×10^{-31} kg.

In Exercises 32–35, find the indicated values.

32. Find T (in K) if $\log T = 8$, where T is the temperature sufficient for nuclear fission.

33. Find v (in m/s) if $\log v = 7.423$, where v is the speed of an electron in a TV picture tube.

34. Find η if $\log \eta = -0.35$, where η is the efficiency of a certain gasoline engine.

35. Find E (in J) if $\log E = -18.49$, where E is the energy of a photon of visible light.

In Exercises 36–42, solve the given problems by evaluating the appropriate logarithms or antilogarithms.

36. Simplify: $\dfrac{\log_b x^2}{\log 100}$.

37. Evaluate: $\log(\log 10^{100})$.

38. Evaluate: $2(10^{\log 0.1}) + 3(10^{\log 0.01})$.

39. A stereo amplifier has an input power of 0.750 W and an output power of 25.0 W. What is the power gain? (See Example 3.)

40. The percent transmittance ($\%T$) of a substance is related to its absorbance (A) by $A = -\log(\%T/100)$. Find $\%T$ if $A = 0.27$.

41. Measured on the Richter scale, the magnitude of an earthquake of intensity I is defined as $R = \log(I/I_0)$, where I_0 is a minimum level for comparison. What is the Richter scale reading for the 2007 Peruvian earthquake for which $I = 79\,000\,000 I_0$?

42. The moment magnitude scale M_W was devised as a successor to the Richter scale to measure the size of very large earthquakes adequately. In terms of the seismic moment of the earthquake M_0 (in N·m), it is given by $M_W = (\log M_0 + 7)/1.5 - 10.7$. What is the moment magnitude scale reading for the 2010 Haiti earthquake for which $M_0 = 3.54 \times 10^{19}$ N·m?

In Exercises 43 and 44, use logarithms to perform the indicated calculations.

43. A certain type of optical switch in a fibre-optic system allows a light signal to continue in either of two fibres. How many possible paths could a light signal follow if it passes through 400 such switches? Express your answer in scientific notation.

44. The peak current I_m (in A) in an alternating-current circuit is given by $I_m = \sqrt{\dfrac{2P}{Z \cos \theta}}$, where P is the power developed, Z is the magnitude of the impedance, and θ is the phase angle between the current and voltage. Evaluate I_m for $P = 5.25$ W, $Z = 320\ \Omega$, and $\theta = 35.4°$.

Answers to Practice Exercises

1. −0.2683 **2.** 195.0

13.5 Natural Logarithms

As we have noted, another number important as a base of logarithms is the number e. *Logarithms to the base e are called* **natural logarithms**. Recall that e is an irrational number. Its value can be approximated as 2.718 282 828 with a nine-digit calculator. While it may seem unusual to choose e as a base for logarithms, base e calculations are used regularly to express continuous growth and decay, for instance. In calculus, the reason for its choice and the fact that it is a very natural number for a base of logarithms are shown.

Just as $\log x$ refers to logarithms to the base 10, the notation $\ln x$ is used to denote logarithms to the base e. Due to the extensive use of natural logarithms, the notation $\ln x$ is more convenient than $\boldsymbol{\log_e x}$, although they mean the same thing.

Since more than one base is important, at times it is useful to change a logarithm from one base to another. We now develop a formula for changing the base of a logarithm. Therefore, even when values are provided which are not expressed as common or natural logarithms, we can change the expressions to include the base of our choosing. For this derivation, we let $u = \log_b x$, which written in exponential form gives $b^u = x$.

Deriving the Change-of-Base Formula for Logarithms

Starting with the exponential form, take logarithms to the base a on both sides: $\log_a b^u = \log_a x$

Use the power rule for logarithms: $u \log_a b = \log_a x$

Isolate u: $u = \dfrac{\log_a x}{\log_a b}$

Replacing u by its logarithmic form, we obtain

$$\log_b x = \frac{\log_a x}{\log_a b} \tag{13.13}$$

Since natural logarithms are used extensively, it is often convenient to have the change-of-base formula written specifically for use with logarithms to the base 10 and natural logarithms. First, using $a = 10$ and $b = e$, we have

$$\ln x = \frac{\log x}{\log e} \tag{13.14}$$

Then with $a = e$ and $b = 10$, we have

$$\log x = \frac{\ln x}{\ln 10} \tag{13.15}$$

EXAMPLE 1 Change of base to find natural log

Find ln 20 using only logarithms to the base 10.

$$\ln 20 = \log_e 20 = \frac{\log 20}{\log e} = \frac{1.301}{0.434} = 2.996$$

This means that $e^{2.996} = 20$.

COMMON ERROR

A common error is to confuse the quotient rule for logarithms with the change-of-base formula. Make sure that you understand the difference between these expressions:

$\log_3 x = \frac{\log x}{\log 3}$ evaluates a logarithm to the base 3 using logarithms to the base 10.

$\log\left(\frac{x}{3}\right) = \log x - \log 3$ expresses the logarithm of a quotient as the difference of logarithms.

EXAMPLE 2 Change of base to find log to base 5

Find $\log_5 560$.

In Eq. (13.13), if we let $a = 10$, $b = 5$, and $x = 560$, we have

$$\log_5 560 = \frac{\log 560}{\log 5} = 3.932$$

From the definition of a logarithm, this means that

$$5^{3.932} = 560$$

Practice Exercise

1. Find $\log_3 23$.

Values of natural logarithms can be found directly on a calculator. The [ln] key is used for this purpose. In order to find the antilogarithm of a natural logarithm, we use the [e^x] key. The following example illustrates finding a natural logarithm and an antilogarithm on a calculator.

EXAMPLE 3 Evaluating natural logarithms

(a) We evaluate ln 236.5 directly on the calculator. We find that

$$\ln 236.5 = 5.4659$$

which means that $e^{5.4659} = 236.5$.

(b) Given that $\ln N = -0.8729$, we determine N by finding $e^{-0.8729}$ on the calculator. This gives us

$$N = 0.4177$$

Practice Exercise

2. Use a calculator to evaluate ln 57.2.

Using the change-of-base formulas we can display the graph of a logarithmic function to any base with a graphing utility, as we show in the next example.

EXAMPLE 4 Graphing a logarithm to the base 2

Graph the function $y = 3 \log_2 x$.

The function $y = \log_2 x$ cannot be graphed directly, so we must use the fact that

$$\log_2 x = \frac{\log x}{\log 2} \quad \text{or} \quad \log x = \frac{\ln x}{\ln 2}$$

Therefore, we graph the function

$$y = \frac{3 \log x}{\log 2} \quad \left(\text{or } y = \frac{3 \ln x}{\ln 2}\right)$$

The curve is shown in Fig. 13.10.

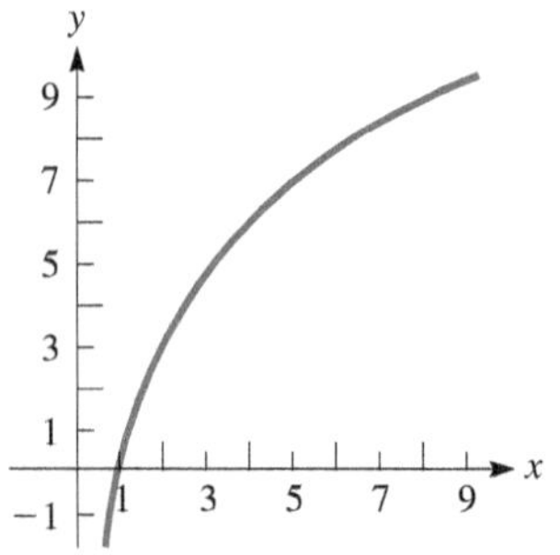

Fig. 13.10

In Section 13.3, we introduced the properties $\log_b(b^x) = x$ and $b^{\log_b x} = x$. If we let $b = e$, we get the following special cases of these properties, which can be used to simplify certain expressions:

$$\ln(e^x) = x \qquad (13.16)$$

$$e^{\ln x} = x \qquad (13.17)$$

EXAMPLE 5 Simplifying with base e

Using Eq. (13.16), $\ln(e^2) = 2$ and $\ln(e^{0.03t}) = 0.03t$
Using Eq. (13.17), $e^{\ln 7} = 7$ and $e^{\ln 2y} = 2y$

Applications of natural logarithms are found in many fields of technology. One such application is shown in the next example, and others are found in the exercises.

EXAMPLE 6 Natural log—electric current

The electric current i in a circuit containing a resistance and an inductance (see Fig. 13.11) is given by $\ln(i/I) = -Rt/L$, where I is the current at $t = 0$, R is the resistance, t is the time, and L is the inductance. Calculate how long (in s) it takes i to reach 0.430 A, if $I = 0.750$ A, $R = 7.50\ \Omega$, and $L = 1.25$ H.

Solving for t and then evaluating, we have

$$t = -\frac{L\ln(i/I)}{R} = -\frac{L(\ln i - \ln I)}{R} \qquad \text{either form can be used}$$

$$= -\frac{1.25(\ln 0.430 - \ln 0.750)}{7.50} = 0.0927 \text{ s} \qquad \text{evaluating}$$

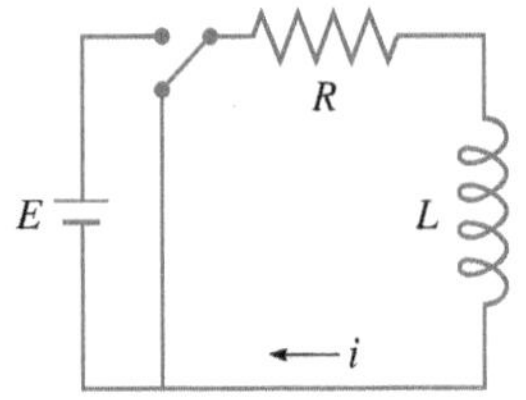

Fig. 13.11

Therefore, the current changes from 0.750 A to 0.430 A in 0.0927 s.

EXERCISES 13.5

In Exercises 1 and 2, find the indicated values if the given changes are made in the indicated examples of this section.

1. In Example 1, change 20 to 200 and then evaluate.

2. In Example 2, change the base to 4 and then evaluate.

In Exercises 3–8, use logarithms to the base 10 to find the natural logarithms of the given numbers.

3. 26.0 **4.** 631 **5.** 1.562

6. 0.5017 **7.** 0.007 326 7 **8.** 0.000 443 48

In Exercises 9–14, use logarithms to the base 10 to find the indicated logarithms.

9. $\log_7 42$ **10.** $\log_2 86$ **11.** $\log_\pi 245$

12. $\log_{12} 122$ **13.** $\log_{40} 750$ **14.** $\log_{100} 3720$

In Exercises 15–22, find the natural logarithms of the given numbers.

15. 45.1 **16.** 392 **17.** 3.914

18. 5662 **19.** 0. 7719 **20.** 0.006 802

21. $(0.019\ 273)^3$ **22.** $\sqrt{0.000\ 080\ 606}$

In Exercises 23–26, use Eq. (13.15) to find the common logarithms of the given numbers.

23. 45.17 **24.** 8765

25. 0.685 28 **26.** 0.001 429 8

In Exercises 27–34, find the natural antilogarithms of the given logarithms.

27. 2.190 **28.** 5.420 **29.** 0.008 421 0

30. 0.632 **31.** −0.7429 **32.** −2.942 18

33. −23.504 **34.** −0.008 04

In Exercises 35–38, use a graphing utility to display the indicated graphs.

35. The graph of $y = \log_5 x$

36. The graph of $y = 2 \log_8 x$

37. Verify that $y = 2^x$ and $y = \log_2 x$ are inverse functions by displaying the graphs of each function and the line $y = x$ on a graphing utility. (See Example 7 of Section 13.2.)

38. Explain what is meant by the expression $\ln(\ln x)$. Display the graph of $y = \ln(\ln x)$ with a graphing utility.

In Exercises 39–42, use a calculator to verify the given values.

39. $\ln 5 + \ln 8 = \ln 40$

40. $2 \ln 6 - \ln 3 = \ln 12$

41. $4 \ln 3 = \ln 81$

42. $\ln 5 - 0.5 \ln 25 = \ln 1$

In Exercises 43–56, solve the given problems.

43. Evaluate: $\sqrt{\ln e^9}$.

44. Solve for y in terms of x: $\ln y + 2 \ln x = 1 + \ln 5$.

45. Solve for x: $\ln(\log x) = 0$.

46. If $\ln x = 3$ and $\ln y = 4$, find $\sqrt{x^2 y}$.

47. If $x = \ln 4$ and $y = \ln 5$, express $\ln 80$ in terms of x and y.

48. Simplify: $2 \ln(e^{3x+1}) - e^{\ln(2x-3)}$.

49. Find f (in Hz) if $\ln f = 21.619$, where f is the frequency of the microwaves in a microwave oven.

50. Find k (in 1/Pa) if $\ln k = -21.504$, where k is the compressibility of water.

51. If interest is compounded continuously (daily compounded interest closely approximates this), with an interest rate i, a bank account will double in t years according to $i = (\ln 2)/t$. Find i if the account is to double in 8.5 years. (2 is exact.)

52. World population is currently growing by 1.1% annually. If it continues at this rate, the time (in years) for the population to double in size is given by $t = \dfrac{\ln 2}{\ln 1.011}$. How many years is this (to the nearest year)?

53. For the electric circuit of Example 6, find how long it takes the current to reach 10% of the initial value of 0.750 A.

54. The velocity v (in m/s) of a rocket increases as fuel is consumed and ejected. Considering fuel as part of the mass m of a rocket in flight and m_0 as its original mass, the velocity is given by $v = 2500(\ln m_0 - \ln m)$. If 0.75 of the original mass is fuel, find v when all the fuel is used.

55. The distance x travelled by a motorboat t seconds after the engine is cut off is given by $x = k^{-1} \ln(kv_0 t + 1)$, where v_0 is the velocity of the boat at the time the engine is cut and k is a constant. Find how long it takes a boat to go 150 m if $v_0 = 12.0$ m/s and $k = 6.80 \times 10^{-3}$/m.

56. The electric current i (in A) in a circuit that has a 1-H inductor, a 10-Ω resistor, and a 6-V battery, and the time t (in s) are related by the equation $10t = -\ln(1 - i/0.6)$. Solve for i.

Answers to Practice Exercises

1. 2.854 **2.** 4.047

13.6 Exponential and Logarithmic Equations

EXPONENTIAL EQUATIONS

An equation in which the variable occurs in an exponent is called an **exponential equation**. Although some may be solved by changing to logarithmic form, they are generally solved by *taking the logarithm of each side* and then using the properties of logarithms.

EXAMPLE 1 Two ways to solve an exponential equation

Solve the equation $2^x = 8$.

(a) Changing to logarithmic form

Write the equation in logarithmic form: $x = \log_2 8$

Evaluate the logarithm to find x: $x = 3$

(b) Taking logarithms of each side

Take the logarithm to any proper base of both sides (here we use common logarithms): $\log 2^x = \log 8$

Use the power rule for logarithms: $x \log 2 = \log 8$

Isolate x: $x = \dfrac{\log 8}{\log 2}$

Evaluate using a calculator to find x: $x = 3$

Note that we would have arrived at the same answer if we had used natural logarithms instead of common logarithms.

EXAMPLE 2 Solving an exponential equation

Solve the equation $3^{x-2} = 5$.

(a) Changing to logarithmic form

Write the equation in logarithmic form:	$\log_3 5 = x - 2$
Isolate x:	$x = \log_3 5 + 2$
Evaluate x with the change of base formula:	$x = \dfrac{\log 5}{\log 3} + 2 = 3.465$
Check using a calculator:	$3^{3.465-2} = 3^{1.465} = 5$

(b) Taking logarithms of each side

Take common logarithms of both sides:	$\log(3^{x-2}) = \log 5$
Use the power rule for logarithms:	$(x - 2)\log 3 = \log 5$
Isolate x:	$x = \dfrac{\log 5}{\log 3} + 2 = 3.465$

Practice Exercise

1. Solve for x: $2^{x+1} = 7$

EXAMPLE 3 Solving an exponential equation

Solve the equation $2(4^{x-1}) = 17^x$.

Take common logarithms of both sides:	$\log[2(4^{x-1})] = \log 17^x$
Use the product rule for logarithms:	$\log 2 + \log 4^{x-1} = \log 17^x$
Use the power rule for logarithms:	$\log 2 + (x - 1)\log 4 = x \log 17$
Use the distributive law:	$\log 2 + x \log 4 - \log 4 = x \log 17$
Isolate x:	$x \log 4 - x \log 17 = \log 4 - \log 2$
	$x(\log 4 - \log 17) = \log 4 - \log 2$
	$x = \dfrac{\log 4 - \log 2}{\log 4 - \log 17}$
Evaluate using a calculator:	$x = -0.479$

EXAMPLE 4 Exponential equation—application to atmospheric pressure

At constant temperature, the atmospheric pressure p (in Pa) at an altitude h (in m) is given by $p = p_0e^{kh}$, where p_0 is the pressure where $h = 0$ (usually taken as sea level). Given that $p_0 = 101.3$ kPa (atmospheric pressure at sea level) and $p = 68.9$ kPa for $h = 3050$ m, find the value of k.

Take natural logarithms of both sides:	$\ln p = \ln (p_0e^{kh})$
Use the product rule for logarithms:	$\ln p = \ln p_0 + \ln (e^{kh})$
Use the power rule for logarithms:	$\ln p = \ln p_0 + kh \ln e$
Isolate k (recall $\ln e = 1$):	$kh = \ln p - \ln p_0$
	$k = \dfrac{1}{h}(\ln p - \ln p_0)$
Substitute values:	$k = \dfrac{1}{3050}(\ln 68.9 - \ln 101.3)$
	$= -0.000\,126/\text{m}$

LOGARITHMIC EQUATIONS

Some of the important measurements in scientific and technical work are defined in terms of logarithms. Using these formulas can lead to solving a **logarithmic equation**, *which is an equation with the logarithm of an expression involving the variable.* In solving logarithmic equations, we use the basic properties of logarithms to help change them into a usable form. There is, however, no general algebraic method for solving such equations, and we consider only some special cases.

EXAMPLE 5 Logarithmic equation—application to sound intensity

The human ear responds to sound over a broad range of sound intensities. Therefore, a logarithmic intensity scale is usually used, where the intensity of sound is compared to a reference value. In particular, the sound intensity level SL (measured in dB) is defined by the equation $SL = 10\log\left(\frac{I}{I_0}\right)$, where I is the intensity of the sound and I_0 is the minimum detectable intensity.

A jackhammer has a sound level of 100 dB, and a busy street has a sound level of 70 dB. To find how many times greater the intensity I_j of the sound of a jackhammer is than the intensity I_c of the sound of the city street, we begin by substituting the given sound levels and their corresponding intensities separately into the above equation. This gives

$$70 = 10\log\left(\frac{I_c}{I_0}\right) \quad \text{and} \quad 100 = 10\log\left(\frac{I_j}{I_0}\right)$$

To solve for I_c and I_j, we divide each side by 10 and then use exponential form:

$$\begin{aligned} 7.0 &= \log\left(\frac{I_c}{I_0}\right) & \quad \text{and} \quad 10 &= \log\left(\frac{I_j}{I_0}\right) \\ \frac{I_c}{I_0} &= 10^{7.0} & \frac{I_j}{I_0} &= 10^{10} \\ I_c &= I_0(10^{7.0}) & I_j &= I_0(10^{10}) \end{aligned}$$

Since we want the number of times I_j is greater than I_c, we divide I_j by I_c:

■ This demonstrates that sound intensities change much more than sound levels. (See Exercise 50.)

$$\frac{I_j}{I_c} = \frac{I_0(10^{10})}{I_0(10^{7.0})} = \frac{10^{10}}{10^{7.0}} = 10^{3.0} \quad \text{or} \quad I_j = 10^{3.0}I_c = 1000I_c$$

Thus, the sound of a jackhammer is 1000 times as intense as the sound of the city street.

■ See the chapter introduction.

EXAMPLE 6 Logarithmic equation—application to population growth

Under a medium-growth projection scenario, an analysis of the population of Canada from 1981 to 2008 led to the equation $\log_2 P = \log_2 33.7 + 0.378$, where P is the projected population (in millions) in 2036. Determine this population.

We solve the logarithmic equation as follows:

$$\begin{aligned} \log_2 P &= \log_2 33.7 + 0.378 & \\ \log_2(P/33.7) &= 0.378 & \text{using the quotient rule for logarithms} \\ P/33.7 &= 2^{0.378} & \text{changing to exponential form} \\ P &= 33.7(2^{0.378}) & \\ P &= 43.8 \text{ million} & \text{projected 2036 population} \end{aligned}$$

EXAMPLE 7 Solving a logarithmic equation

Solve the logarithmic equation $2 \ln 2 + \ln x = \ln 3$.

Using the properties of logarithms, we have the following solution:

$$2 \ln 2 + \ln x = \ln 3$$

$$\ln 2^2 + \ln x - \ln 3 = 0 \quad \text{using the power rule}$$

$$\ln \frac{4x}{3} = 0 \quad \text{using the product and the quotient rules}$$

$$\frac{4x}{3} = e^0 = 1 \quad \text{changing to exponential form}$$

$$4x = 3$$

$$x = 3/4 = 0.75$$

Since $\ln(3/4) = \ln 3 - \ln 4$, this solution checks in the *original equation.*

LEARNING TIP

Many exponential and logarithmic equations cannot be solved algebraically, as we have done in the examples in this section. However, they can be solved graphically using a graphing utility as described in Section 3.5.

For instance, to solve the equation $2 \ln 2 + \ln x = \ln 3$ from Example 7 graphically, we can find the *zero* of the function $y_1 = 2 \ln 2 + \ln x - \ln 3$, or the *intersection* of the functions $y_1 = 2 \ln 2 + \ln x$ and $y_2 = \ln 3$.

Solutions can also be approximated numerically.

EXAMPLE 8 Solving a logarithmic equation

Solve the logarithmic equation $2 \log x - 1 = \log(1 - 2x)$.

$$\log x^2 - \log(1 - 2x) = 1 \quad \text{using the power rule}$$

$$\log \frac{x^2}{1 - 2x} = 1 \quad \text{using the quotient rule}$$

$$\frac{x^2}{1 - 2x} = 10^1 \quad \text{changing to exponential form}$$

$$x^2 = 10 - 20x$$

$$x^2 + 20x - 10 = 0$$

$$x = \frac{-20 \pm \sqrt{400 + 40}}{2} = -10 \pm \sqrt{110} \quad \text{using the quadratic formula from Chapter 7}$$

Since logarithms of negative numbers are not defined and $-10 - \sqrt{110}$ is negative and cannot be used in the first term of the original equation, we have that the unique solution is

$$x = -10 + \sqrt{110} = 0.488$$

Practice Exercise

2. Solve for x: $3 \log 2 - \log(x + 1) = 1$

EXERCISES 13.6

In Exercises 1 and 2, find the indicated values if the given changes are made in the indicated examples of this section.

1. In Example 2, change the sign in the exponent from − to + and then solve the equation.
2. In Example 7, change ln 3 to ln 6 and then solve the equation.

In Exercises 3–30, solve the given equations. When necessary, round expressions involving logarithms of integers to three significant digits in their decimal part.

3. $2^x = 16$
4. $3^x = 1/81$
5. $5^x = 0.3$
6. $\pi^{2x} = 20$
7. $5^{-3x} = 0.525$
8. $e^{-3x} = 100$
9. $5^{x-4} = 78$
10. $6^{x+3} = 0.005$
11. $4(12^x) = 300$
12. $0.6^x = 0.5$
13. $0.3^x = 4^{x^2}$
14. $12.2^{x+3} = 18^x$
15. $5 \log_4 x = -4$
16. $6 \log_{30} x = -3$
17. $\log x^2 = (\log x)^2$
18. $x^{\log x} = 1000x^2$
19. $\log_2 x + \log_2 7 = \log_2 21$
20. $2 \log_2 3 - \log_2 x = \log_2 45$
21. $2 \log(3 - x) = 1$
22. $3 \log(2x - 1) = 1$
23. $\log 12x^2 - \log 3x = 3$
24. $\ln x - \ln(1/3) = 1$
25. $3 \ln 2 + \ln(x - 1) = \ln 24$
26. $\log_2 x + \log_2(x + 2) = 3$
27. $\frac{1}{3}\log(x - 2) + \log 7 = 2$
28. $2 \log_x 3 + \log_3 x = 3$

29. $\log(3x - 2) + \log(x - 3) = 3$

30. $\ln(4x - 1) - 3 \ln 9 = 2 \ln 3$

In Exercises 31–38, use a graphing utility to solve the given equations. Round your answers to three significant digits.

31. $15^{-x} = 1.326$

32. $e^{2x} = 3.625$

33. $4(3^x) = 5$

34. $5^{x+2} = e^{2x}$

35. $3 \ln 2x = 2$

36. $\log 4x + \log x = 2$

37. $2 \ln 2 - \ln x = -1$

38. $\log(x - 3) + \log x = \log 4$

In Exercises 39–56, find the indicated quantities.

39. Solve for x: $e^x + e^{-x} = 3$. (*Hint:* Multiply each term by e^x and then it can be treated as a quadratic equation in e^x.)

40. Solve for x: $3^x + 3^{-x} = 4$. See Exercise 39.

41. If $y = 1.5e^{-0.90x}$, find y when $x = 7.1$.

42. What values of x cannot be solutions of the equation $y = \log(2x - 5) + \log(x^2 + 1)$?

43. Use logarithms to find the x-intercept of the graph of $y = 3 - 4^{x+2}$.

44. Use a graphing utility to find the point of intersection of the curves of $3x + 5y = 6$ and $y = 1.5 \ln(x + 3)$.

45. In computer design, the number N of bits of memory is often expressed as a power of 2, or $N = 2^x$. Find x if $N = 2.68 \times 10^8$ bits.

46. Referring to Exercise 64 in Section 13.3, in what year will the amount of DDT be 25% of the original amount?

47. Forensic scientists determine the temperature T (in °C) of a body t hours after death from the equation $T = T_0 + (37 - T_0)0.97^t$, where T_0 is the air temperature. If a body is discovered at midnight with a body temperature of 27°C in a room at 22°C, at what time did death occur?

48. When a camera flash goes off, the batteries recharge the flash's capacitor to a charge Q according to $Q = Q_0(1 - e^{-kt})$, where Q_0 is the maximum charge and t is in seconds. How long does it take to recharge the capacitor to 90% of capacity if $k = 0.5$?

49. In chemistry, the pH value of a solution is a measure of its acidity. The pH value is defined by $\text{pH} = -\log(\text{H}^+)$, where H^+ is the hydrogen-ion concentration (in mol/L). If the pH of a sample of rainwater is 4.764, find the hydrogen-ion concentration. (If $\text{pH} < 7$, the solution is acid. If $\text{pH} > 7$, the solution is basic.) Acid rain has a pH between 4 and 5, and normal rain is slightly acidic with a pH of about 5.6.

50. Referring to Example 5, show that if the difference in sound level of two sounds is d decibels, the louder sound is $10^{d/10}$ more intense than the quieter sound.

51. The moment magnitude scale M_w was devised as a successor to the Richter scale to measure the size of very large earthquakes adequately. In terms of the seismic moment of the earthquake M_0 (in N·m), it is given by $M_w = (\log M_0 + 7)/1.5 - 10.7$. What was the seismic moment of the 2011 earthquake off the coast of Japan whose moment magnitude was 9.0?

52. How many times stronger (in terms of seismic moment) was the 2004 Indian Ocean earthquake off the coast of Sumatra ($M_W = 9.1$) than the 2012 Haida Gwaii earthquake off the coast of British Columbia ($M_w = 7.7$)? (See Exercise 51.)

53. Studies have shown that the concentration c (in mg/cm³ of blood) of aspirin in a typical person is related to the time t (in h) after the aspirin reaches maximum concentration by the equation $\ln c = \ln 15 - 0.20t$. Solve for c as a function of t.

54. In an electric circuit containing a resistor and a capacitor with an initial charge q_0, the charge q on the capacitor at any time t after closing the switch can be found by solving the equation $\ln q = -\dfrac{t}{RC} + \ln q_0$. Here, R is the resistance and C is the capacitance. Solve for q as a function of t.

55. An Earth satellite loses 0.1% of its remaining power each week. An equation relating the power P, the initial power P_0, and the time t (in weeks) is $\ln P = t \ln 0.999 + \ln P_0$. Solve for P as a function of t.

56. In finding the path of a certain plane, the equation $\ln r = \ln a - \ln \cos \theta - (v/w) \ln(\sec \theta + \tan \theta)$ is used. Solve for r. (The conditions of the plane's flight are similar to those described in Chapter 11 Review Exercise 96.)

In Exercises 57–60, solve the given equations graphically. Round your answers to three significant digits.

57. $2^x + 3^x = 50$

58. $4^x + x^2 = 25$

59. The curve in which a uniform wire or rope hangs under its own weight is called a *catenary*. An example of a catenary that we see every day is a wire strung between utility poles, as shown in Fig. 13.12. For a particular wire, the equation of the catenary it forms is $y = 2(e^{x/4} + e^{-x/4})$, where (x, y) is a point on the curve. Find x for $y = 5.8$ m.

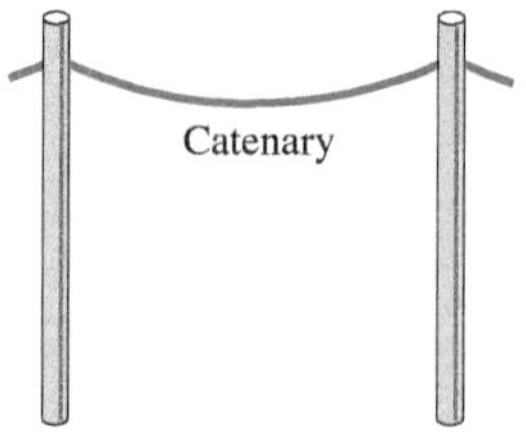

Fig. 13.12

60. In finding the current i in a certain electric circuit, the equation $i = 2 \ln(t + 2) - t$ relates the current and the time t (in s). Find t for $i = 0.05$ A.

Answers to Practice Exercises

1. 1.8 **2.** $-1/5$

13.7 Logarithmic and Semilogarithmic Graphs

When constructing the graphs of some functions, one of the variables changes much more rapidly than the other. We saw this in graphing the exponential and logarithmic functions in Sections 13.1 and 13.2. The following example illustrates this point.

EXAMPLE 1 Graph of an exponential function

Plot the graph of $y = 4(3^x)$.

Constructing the following table of values,

x	-1	0	1	2	3	4	5
y	1.3	4	12	36	108	324	972

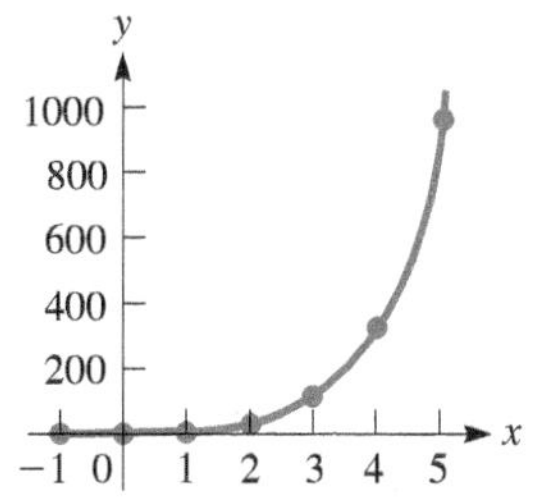

Fig. 13.13

we then plot these values as shown in Fig. 13.13.

We see that as x changes from -1 to 5, y changes much more rapidly, from about 1 to nearly 1000. Also, because of the scale that must be used, we see that it is not possible to show accurately the differences in the y-values on the graph.

It is possible to graph a function with a large change in values, for one or both variables, more accurately than can be done on the standard rectangular coordinate system. This is done by using a scale marked off in distances proportional to the logarithms of the values being represented. *Such a scale is called a* **logarithmic scale**. For example, to plot the interval between 1 and 10 on a logarithmic scale, we have

n	1	2	3	4	5	6	7	8	9	10
$\log n$	0	0.301	0.477	0.602	0.699	0.778	0.845	0.903	0.954	1

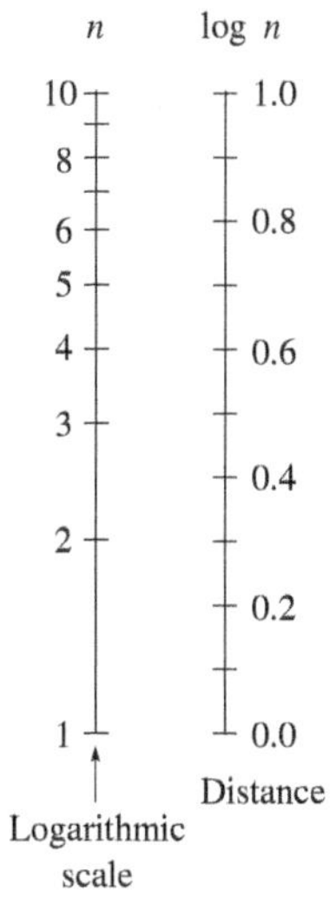

Fig. 13.14

Thus, on a logarithmic scale, the 2 is placed 0.301 unit of distance (or 30.1% of the way) from the 1 to the 10, the 3 is placed at 0.477 unit, and so on. Fig. 13.14 shows the corresponding logarithmic scale with the numbers represented and the distance used for each.

On a logarithmic scale, the distances between the integers are not equal, but this scale does allow for a much greater range of values and much greater accuracy for many of the values. There is another advantage to using logarithmic scales. Many equations that would have more complex curves when graphed on the standard rectangular coordinate system will have simpler curves, often straight lines, when graphed using logarithmic scales. In many cases, this makes the analysis of the curve much easier.

Zero and negative numbers do not appear on the logarithmic scale because we cannot take logarithms of those numbers. Thus, the logarithmic scale must start at some number greater than zero. This number must be a power of 10 and can be very small, say, $10^{-6} = 0.000\,001$, but still positive.

If we wish to use a large range of values for only one of the variables, we use what is known as a **semilogarithmic**, or **semilog**, graph. On this type of graph, only one axis (usually the y-axis) uses a logarithmic scale. If we wish to use a large range of values for both variables, we use a **logarithmic**, or **log-log**, graph. Both axes are marked with logarithmic scales.

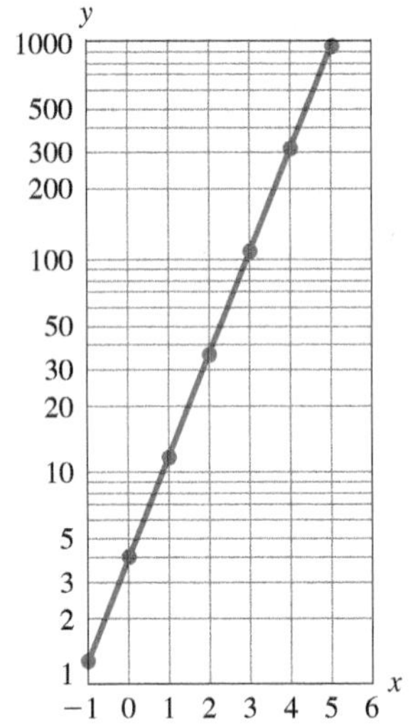

Fig. 13.15

EXAMPLE 2 Graphing a semilogarithmic graph

Construct a semilogarithmic graph of $y = 4(3^x)$.

This is the same function as in Example 1, and we repeat the table of values, together with the logarithms of y:

x	−1	0	1	2	3	4	5
y	1.3	4	12	36	108	324	972
$\log y$	0.114	0.602	1.079	1.556	2.033	2.511	2.988

Plotting the (x, y) points on the graph, we obtain Fig. 13.15. Although the y-axis represents actual values of y, the spacing between them is obtained from $\log y$. For example, $y = 1.3$ is plotted 11.4% of the way between 1 and 10, and $y = 324$ is plotted 51.1% of the way between 100 and 1000.

To see why the resulting graph is a straight line, we take logarithms on each side of the equation. We have

$$\log y = \log[4(3^x)] = \log 4 + \log 3^x \qquad \text{using } \log_b xy = \log_b x + \log_b y$$
$$= \log 4 + x \log 3 \qquad \text{using } \log_b(x^n) = n \log_b x$$

Letting $u = \log y$, the graph represents

$$u = \log 4 + x \log 3$$

Because $\log 3$ and $\log 4$ are constants, this equation is of the form $u = mx + b$, which is a straight line (see Section 5.1).

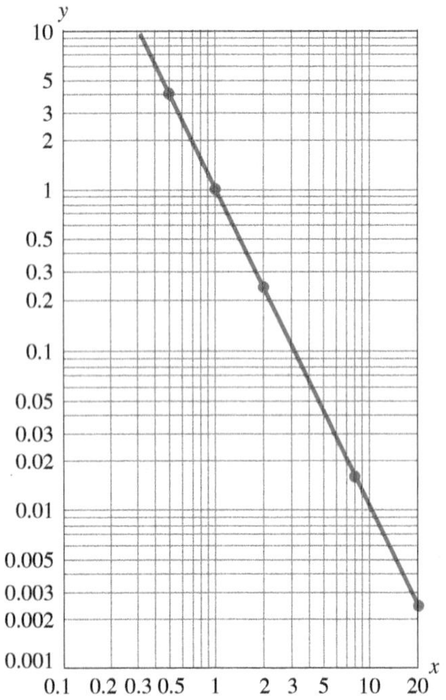

Fig. 13.16

EXAMPLE 3 Graphing a logarithmic graph

Construct the logarithmic graph (sometimes called the log-log plot) of $x^4y^2 = 1$.

First, we solve for y and make a table of values. Considering positive values of x and y, we have

$$y = \sqrt{\frac{1}{x^4}} = \frac{1}{x^2}$$

x	0.5	1	2	8	20
y	4	1	0.25	0.0156	0.0025

We plot these values on axes where both scales are logarithmic, as shown in Fig. 13.16. We again see that we have a straight line. To see why, we take logarithms of both sides of the equation. We have

$$\log(x^4y^2) = \log 1$$
$$\log x^4 + \log y^2 = 0 \qquad \text{using } \log_b xy = \log_b x + \log_b y$$
$$4 \log x + 2 \log y = 0 \qquad \text{using } \log_b (x^n) = n \log_b x$$

If we let $u = \log y$ and $v = \log x$, we then have

$$4v + 2u = 0 \quad \text{or} \quad u = -2v$$

which is the equation of a straight line, as shown in Fig. 13.16. Note, however, that not all graphs that use a logarithmic scale are straight lines.

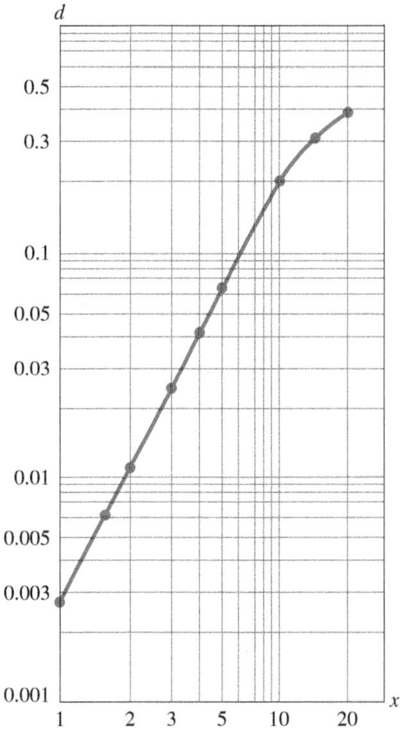

Fig. 13.17

EXAMPLE 4 Deflection of a beam as a log-log plot

The deflection (in m) of a certain cantilever beam as a function of the distance x (in m) from one end is

$$d = 0.0001(30x^2 - x^3)$$

If the beam is 20.0 m long, plot a graph of d as a function of x as a log-log plot.

Constructing a table of values, we have

$x(m)$	1.00	1.50	2.00	3.00	4.00
$d(m)$	0.002 90	0.006 41	0.0112	0.0243	0.0416

$x(m)$	5.00	10.0	15.0	20.0
$d(m)$	0.0625	0.200	0.338	0.400

Since the beam is 20.0 m long, there is no meaning to values of x greater than 20.0 m. The graph is shown in Fig. 13.17.

Logarithmic and semilogarithmic graphs may be useful for plotting data derived from experimentation. Often, the data cover too large a range of values to be plotted on on a rectangular coordinate system. The next example illustrates the use of a semilogarithmic graph to plot data.

EXAMPLE 5 Vapour pressure on a semilog graph

The vapour pressure of water depends on the temperature. The following table gives the vapour pressure (in kPa) for the corresponding values of temperature (in °C):

T(°C)	10	20	40	60	80	100	120	140	160
p (kPa)	1.19	2.33	7.34	19.9	47.3	101	199	361	617

These data are then plotted on a semilogarithmic graph, as shown in Fig. 13.18. Intermediate values of temperature and pressure can then be read directly from the graph.

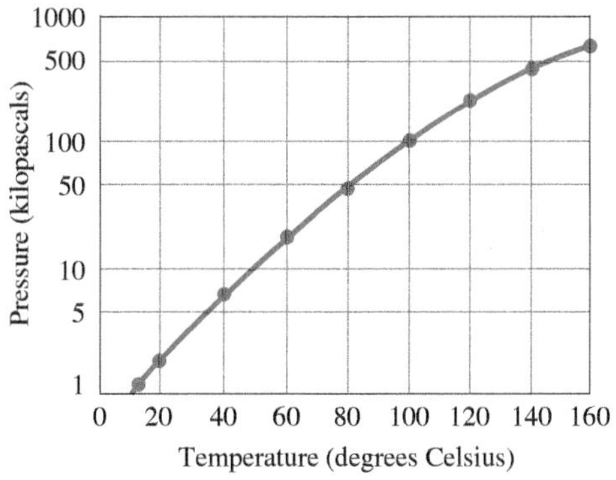

Fig. 13.18

EXERCISES 13.7

In Exercises 1 and 2, make the given changes in the indicated examples, and then draw the graphs.

1. In Example 2, change the 4 to 2 and then make the graph.

2. In Example 3, change the 1 to 4 and then make the graph.

In Exercises 3–10, plot the graphs of the given functions using a semilogarithmic scale.

3. $y = 2^x$ **4.** $y = 5^x$ **5.** $y = 5(4^{-x})$

6. $y = 6^{-x}$ **7.** $y = x^3$ **8.** $y = 2x^4$

9. $y = 2x^3 + 6x$ **10.** $y = 4x^3 + 2x^2$

In Exercises 11–18, plot the graphs of the given functions using a log-log scale.

11. $y = 0.01x^4$ **12.** $y = \sqrt{x}$ **13.** $y = x^{2/3}$

14. $y = 8x^{0.25}$ **15.** $xy = 40$ **16.** $x^2y^3 = 16$

17. $x^2y^2 = 25$ **18.** $x^3y = 8$

In Exercises 19–26, determine the type of scale on which the graph of the given function is a straight line. Using the appropriate scale, sketch the graph.

19. $y = 3^{-x}$ **20.** $y = 0.2x^3$ **21.** $y = 3x^6$

22. $y = 5(10^{-x})$ **23.** $y = 4^{x/2}$ **24.** $xy^3 = 10$

25. $x\sqrt{y} = 4$ **26.** $y(2^x) = 3$

In Exercises 27–38, plot the indicated graphs.

27. On the moon, the distance s (in m) a rock will fall due to gravity is $s = 0.81t^2$, where t is the time (in s) of fall. Plot the graph of s as a function of t for $0 \leq t \leq 10$ s on (a) a regular rectangular coordinate system and (b) a logarithmic coordinate system.

28. By pumping, the air pressure in a tank is reduced by 18% each second. Thus, the pressure p (in kPa) in the tank is given by $p = 101(0.82)^t$, where t is the time (in s). Plot the graph of p as a function of t for $0 \leq t \leq 30$ s on (a) a regular rectangular coordinate system and (b) a semilogarithmic coordinate system.

29. Strontium-90 decays according to the equation $N = N_0e^{-0.028t}$, where N is the amount present after t years and N_0 is the original amount. Plot N as a function of t on a semilog scale if $N_0 = 1000$ g.

30. The electric power P (in W) in a certain battery as a function of the resistance R (in Ω) in the circuit is given by $P = \dfrac{100R}{(0.50 + R)^2}$. Plot P as a function of R on a semilog scale, using the logarithmic scale for R and values of R from 0.01 Ω to 10 Ω. Compare the graph with that in Fig. 3.19.

31. The acceleration g (in m/s^2) produced by the gravitational force of the earth on a spacecraft is given by $g = 3.99 \times 10^{14}/r^2$, where r is the distance from the centre of the earth to the spacecraft. On a log-log scale, graph g as a function of r from $r = 6.37 \times 10^6$ m (the earth's surface) to $r = 3.91 \times 10^8$ m (the distance to the moon).

32. In undergoing an adiabatic (no *heat* gained or lost) expansion of a gas, the relation between the pressure p (in kPa) and the volume v (in m^3) is $p^2v^3 = 850$. Using a log-log scale, graph p as a function of v from $v = 0.10$ m^3 to $v = 10$ m^3.

33. The number of cell phone subscribers in Canada from 1985 to 2011 is shown in the following table. Plot N as a function of the year on a semilog scale.

Year	1985	1990	1995	2000	2005	2010	2011
$N\ (\times 10^6)$	0.006	0.584	2.59	8.73	17.0	25.8	27.4

34. The period T (in years) and mean distance d (given as a ratio of that of Earth) from the sun to the planets (Mercury, Venus, Earth, Mars, Jupiter, Saturn, Uranus, Neptune) are given below. Plot T as a function of d using a log-log scale.

Planet	M	V	E	M	J	S	U	N
d	0.39	0.72	1.00	1.52	5.20	9.54	19.2	30.1
T	0.24	0.62	1.00	1.88	11.9	29.5	84.0	165

35. The intensity level B (in dB) and the frequency f (in Hz) for a sound of constant loudness were measured as shown in the table that follows. Plot the data for B as a function of f as a semilog graph, using the log scale for f.

f (Hz)	100	200	500	1000	2000	5000	10 000
B (dB)	40	30	22	20	18	24	30

36. The atmospheric pressure p (in kPa) at a given altitude h (in km) is given in the following table. Plot p as a function of h as a semilog graph.

h (km)	0	10	20	30	40
p (kPa)	101	25	6.3	2.0	0.53

37. One end of a very hot steel bar is sprayed with a stream of cool water. The rate of cooling R (in °C/s) as a function of the distance d (in cm) from one end of the bar is then measured, with the results shown in the following table. Using a log-log scale, plot R as a function of d. Such experiments are made to determine the hardness of steel.

d (cm)	0.063	0.13	0.19	0.25
R (°C/s)	600	190	100	72

d (cm)	0.38	0.50	0.75	1.0	1.5
R (°C/s)	46	29	17	10	6.0

38. The magnetic intensity H (in A/m) and flux density B (in teslas) of annealed iron are given in the following table. Plot H as a function of B as a log-log graph.

B (T)	0.0042	0.043	0.67	1.01
H (A/m)	10	50	100	150

B (T)	1.18	1.44	1.58	1.72
H (A/m)	200	500	1000	10 000

In Exercises 39 and 40, plot the indicated semilogarithmic graphs for the following application.

In a particular electric circuit, called a low-pass filter, the input voltage V_i is across a resistor and a capacitor, and the output voltage V_0 is across the capacitor (see Fig. 13.19). The voltage gain G (in dB) is given by

$$G = 20 \log \frac{1}{\sqrt{1 + (\omega T)^2}}$$

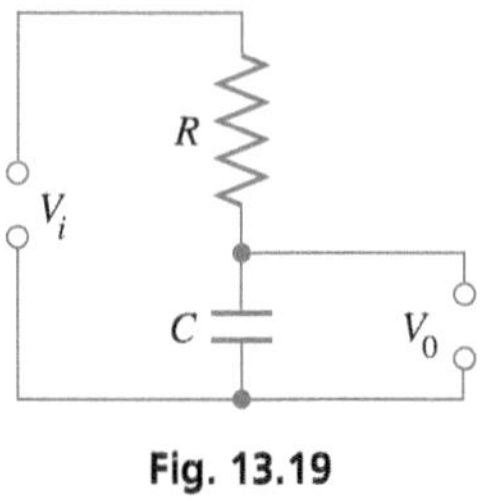

Fig. 13.19

where $\tan \phi = -\omega T$.

Here, ϕ is the phase angle of V_0/V_i. For values of ωT of 0.01, 0.10, 0.30, 1.0, 3.0, 10.0, 30.0, *and* 100, *plot the indicated graphs. These graphs are called a Bode diagram for the circuit.*

39. Calculate values of G for the given values of ωT and plot a semi-logarithmic graph of G vs. ωT.

40. Calculate values of ϕ (as negative angles) for the given values of ωT and plot a semilogarithmic graph of ϕ vs. ωT.

CHAPTER 13 KEY FORMULAS AND EQUATIONS

Exponential function	$y = b^x$	**(13.1a)**
	$y = ab^x$	**(13.1b)**
Logarithmic form	$x = \log_b y$	**(13.2)**
Logarithmic function	$y = \log_b x$	**(13.3)**
Product rule for exponents	$b^u b^v = b^{u+v}$	**(13.4)**
Quotient rule for exponents	$\frac{b^u}{b^v} = b^{u-v}$	**(13.5)**
Power rule for exponents	$(b^u)^n = b^{nu}$	**(13.6)**
Product rule for logarithms	$\log_b xy = \log_b x + \log_b y$	**(13.7)**
Quotient rule for logarithms	$\log_b\left(\frac{x}{y}\right) = \log_b x - \log_b y$	**(13.8)**
Power rule for logarithms	$\log_b(x^n) = n \log_b x$	**(13.9)**
Identity rules for logarithms	$\log_b 1 = 0 \quad \log_b b = 1$	**(13.10)**
	$\log_b(b^x) = x$	**(13.11)**
	$b^{\log_b x} = x$	**(13.12)**
Change-of-base formulas	$\log_b x = \frac{\log_a x}{\log_a b}$	**(13.13)**
	$\ln x = \frac{\log x}{\log e}$	**(13.14)**
	$\log x = \frac{\ln x}{\ln 10}$	**(13.15)**
Base *e* properties	$\ln e^x = x$	**(13.16)**
	$e^{\ln x} = x$	**(13.17)**

CHAPTER 13 REVIEW EXERCISES

In Exercises 1–12, determine the value of x.

1. $\log_{10} x = 4$
2. $\log_9 x = 3$
3. $\log_5 x = -1$
4. $\log_4(\sin x) = -0.5$
5. $2 \log_2 8 = x$
6. $\log_{12} 144 = x - 3$
7. $\log_8 32 = x$
8. $\log_9 27 = x$
9. $\log_x 36 = 2$
10. $\log_x(1/243) = 5$
11. $\log_x 10 = \frac{1}{2}$
12. $\log_x 8 = 0$

In Exercises 13–24, express each as a sum, difference, or multiple of logarithms. Wherever possible, evaluate logarithms of the result.

13. $\log_3 2x$
14. $\log_5\left(\frac{7}{a}\right)$
15. $\log_3(t^2)$
16. $\log_6 \sqrt{5}$
17. $\log_2 28$
18. $\log_7 98$
19. $\log_4 \sqrt{48}$
20. $\log_2 \sqrt[4]{32y}$

21. $\log_3\left(\frac{9}{x}\right)$ **22.** $\log_6\left(\frac{5}{36}\right)$

23. $\log_{10}(1000x^4)$ **24.** $\log_3(9^2 \times 6^3)$

In Exercises 25–36, solve for y in terms of x.

25. $\log_6 y = \log_6 4 - \log_6 x$ **26.** $2 \ln y = \ln e^2 - 3 \ln x$

27. $3 \ln y = 2 + 3 \ln x$ **28.** $2(\log_9 y + 2 \log_9 x) = 1$

29. $\log_3 y = \frac{1}{2} \log_3 7 + \frac{1}{2} \log_3 x$

30. $(\log_2 3)(\log_2 y - \log_2 x) = 3$

31. $\log_5 x + \log_5 y = \log_5 3 + 1$

32. $\log_7 y = 2 \log_7 5 + \log_7 x + 2$

33. $2(\log_4 y - 3 \log_4 x) = 3$ **34.** $\frac{\log_7 x}{\log_7 4} - \log_7 y = 1$

35. $2^y = e^x$ **36.** $10^y = 3^{x+1}$

In Exercises 37–44, display the graphs of the given functions on a graphing utility.

37. $y = 0.5(5^x)$ **38.** $y = 3(2^{-x})$

39. $R = 0.2 \log_4 r$ **40.** $y = 10 \log_{16} x$

41. $y = \log_{3.15} x$ **42.** $y = 0.1 \log_{4.05} x$

43. $y = 1 - e^{-|x|}$ **44.** $s = 2(1 - e^{-0.2t})$

In Exercises 45–48, use logarithms to the base 10 to find the natural logarithms of the given numbers.

45. 8.86 **46.** 0.303 **47.** sin 2.07 **48.** $\sqrt{0.542}$

In Exercises 49–52, use natural logarithms to find logarithms to the base 10 of the given numbers.

49. 65.89 **50.** 0.0781 **51.** 0.1197^3 **52.** log 1000

In Exercises 53–60, solve the given equations. Round all results to three significant digits.

53. $e^{2x} = 5$ **54.** $2(5^x) = 15$

55. $3^{x+2} = 5^x$ **56.** $2^x/3^{1-x} = 12^{x+1}$

57. $\log_4 z + \log_4 6 = \log_4 12$ **58.** $\log_8(x + 2) = 2 - \log_8 2$

59. $2 \log_3 2 - \log_3(x + 1) = \log_3 5$

60. $\log(n + 2) + \log n = 0.4771$

In Exercises 61 and 62, plot the graphs of the given functions using a semilogarithmic scale. In Exercises 63 and 64, plot the graphs of the given functions on a log-log scale.

61. $y = 8^x$ **62.** $y = 5x^3$ **63.** $y = \sqrt[3]{x}$ **64.** $xy^4 = 16$

In Exercises 65–68, evaluate the given expressions using the property $b^{\log_b x} = x$.

65. $10^{\log 4}$ **66.** $2e^{\ln 7.5}$ **67.** $3e^{2\ln 2}$ **68.** $5(10^{2\log 3})$

In Exercises 69–104, solve the given problems. When necessary, round results to three significant digits. In Exercises 105, answer the given question.

69. Use a calculator to verify that $2 \log 3 - \log 6 = \log 1.5$.

70. Use a calculator to verify that $3 \ln 2 + 0.5 \ln 64 = 3 \ln 4$.

71. Evaluate $\sqrt[3]{\ln e^8} - \sqrt{\log 10^4}$.

72. Solve for x: $2^x + 32(2^{-x}) = 12$.

73. If $f(x) = 2 \log_b x$ and $f(8) = 3$, find $f(2)$.

74. Evaluate: $\log(2 \log 1000)$.

75. Evaluate: $7(10^{\log 0.1}) + 6000(100^{\log 0.001})$.

76. If $x = \log_b 7$ and $y = \log_b 2$, express $\log_b 42$ in terms of x and y.

77. Solve the equation $\log_5 x = 2x - 7$ graphically.

78. Use logarithms to find the x-intercept of the graph of $y = 2 - 5^{2x-3}$.

79. If an amount of P dollars is invested at an annual interest rate r (expressed as a decimal), the value V of the investment after t years is $V = P(1 + r/n)^{nt}$, if interest is compounded n times a year. If \$1000 is invested at an annual interest rate of 6%, compounded semiannually, express V as a function of t and solve for t.

80. The current i (in A) in a certain electric circuit is given by $i = 16(1 - e^{-250t})$, where t is the time (in s). Solve for t.

81. The formula $\ln(I/I_0) = -\beta h$ is used in estimating the thickness of the ozone layer. Here, I_0 is the intensity of a wavelength of sunlight before reaching the earth's atmosphere, I is the intensity of the light after passing through h cm of the ozone layer, and β is a constant. Solve for I.

82. The amount A of cesium-137 (a dangerous radioactive element) remaining after t years is given by $A = A_0(0.5^{t/30.3})$, where A_0 is the initial amount. In what year will the cesium-137 be 10% of the amount released at the Fukushima Daiichi nuclear power plant in Japan after the March 2011 earthquake and tsunami damaged its cooling systems?

83. The bending moment M (in N · m) of a particular concrete column is given by $\log M = 6.663$. What is the value of M?

84. A lottery pays \$500 for a \$1 ticket if a person picks the correct three-digit number determined by the random draw of three numbered balls. The probability p of winning this lottery at least once in x attempts is given by $p = 1 - 0.999^x$. How many attempts are necessary for a person to have a 50% chance ($p = 0.50$) of winning this lottery at least once?

85. The amount A of alcohol in a person's bloodstream is given by $A = A_0e^{kt}$, where A_0 is an initial amount (mg alcohol/mL blood), t is the time (in h) after drinking the alcohol, and k is a constant depending on the person. If $A_0 = 0.24$ mg/mL, and $A = 0.16$ mg/mL in 2.0 h, how long should the person wait to drive if the legal limit is 0.08 mg/mL?

86. A computer analysis of the luminous efficiency E (in lumens/W) of a tungsten lamp as a function of its input power P (in W) is given by $E = 22.0(1 - 0.65e^{-0.008P})$. Sketch the graph of E as a function of P for $0 \le P \le 1000$ W.

87. An original amount of 100 mg of radium radioactively decomposes such that N mg remain after t years. The function relating t and N is $t = 2350(\ln 100 - \ln N)$. Sketch the graph.

88. The time t (in s) to chemically change 5 kg of a certain substance into another is given by $t = -5 \log\left(\frac{5 - x}{5}\right)$, where x is the number of kilograms that have been changed at any time. Sketch the graph.

89. An equation that may be used for the angular velocity ω of the slider mechanism in Fig. 13.20 is $2 \ln \omega = \ln 3g + \ln \sin \theta - \ln l$, where g is the acceleration due to gravity. Solve for $\sin \theta$.

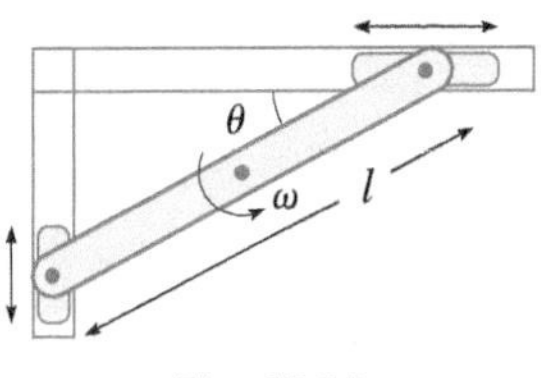

Fig. 13.20

90. Taking into account the weight loss of fuel, the maximum velocity v_m of a rocket is $v_m = u(\ln m_0 - \ln m_s) - gt_f$, where m_0 is the initial mass of the rocket and fuel, m_s is the mass of the rocket shell, t_f is the time during which fuel is expended, u is the velocity of the expelled fuel, and g is the acceleration due to gravity. Solve for m_0.

91. An equation used to calculate the capacity C (in bits/s) of a telephone channel is $C = B \log_2(1 + R)$, where B is the bandwidth of the channel (in Hz) and R is the signal-to-noise ratio. Solve for R.

92. An equation used in studying the action of a protein molecule is $\ln A = \ln \theta - \ln(1 - \theta)$. Solve for θ.

93. The magnitudes (visual brightnesses), m_1 and m_2, of two stars are related to their (actual) brightnesses, b_1 and b_2, by the equation $m_1 - m_2 = 2.5 \log(b_2/b_1)$. As a result of this definition, magnitudes may be negative, and *magnitudes decrease as brightnesses increase.* The magnitude of the brightest star, Sirius, is -1.4, and the magnitudes of the faintest stars observable with the naked eye are about 6.0. How much brighter is Sirius than these faintest stars?

94. The power gain of an electronic device such as an amplifier is defined as $n = 10 \log(P_0/P_i)$, where n is measured in decibels, P_0 (in W) is the power output, and P_i (in W) is the power input. If $P_0 = 10.0$ W and $P_i = 0.125$ W, calculate the power gain. (See Example 4 in Section 13.4.)

95. In studying the frictional effects on a flywheel, the revolutions per minute R that it makes as a function of the time t (in min) are given by $R = 4520(0.750)^{2.50t}$. Find t for $R = 1950$ r/min.

96. The efficiency η of a gasoline engine as a function of its compression ratio r is given by $\eta = 1 - r^{1-\gamma}$, where γ is a constant. Find γ for $\eta = 0.55$ and $r = 7.5$.

97. The intensity I of light decreases from its value I_0 as it passes a distance x through a medium. Given that $x = k(\ln I_0 - \ln I)$, where k is a constant depending on the medium, find x for $I = 0.850I_0$ and $k = 5.00$ cm.

98. According to Benford's law, in lists of numbers from many real-life sources of data (such as tables of physical constants, accounting data, or scientific calculations), the leading digit 1 occurs much more often than the others. In particular, the probability that the first digit of a randomly selected number from such a list is n is given by

$$P(N) = \log\left(\frac{n+1}{n}\right)$$

How many times more likely is a 1 than a 9 as the leading digit of a number in a data set that satisfies Benford's law?

99. Under certain conditions, the temperature T and pressure p are related by the following equation, where T_0 is the temperature at pressure p_0: $\frac{T}{T_0} = \left(\frac{p}{p_0}\right)^{\frac{k-1}{k}}$. Solve for k.

100. The temperature T of an object with an initial temperature T_1 in water at temperature T_0 as a function of the time t is given by $T = T_1 + (T_0 - T_1)e^{-kt}$. Solve for t.

101. Pure water is running into a brine solution, and the same amount of solution is running out. The number n of kilograms of salt in the solution after t min is found by solving the equation $\ln n = -0.04t + \ln 20$. Solve for n as a function of t.

102. For the circuit in Fig. 13.21, the current i (in mA) is given by $i = 1.6\,e^{-100t}$. Plot the graph of i as a function of t for the first 0.05 s using a semilogarithmic scale.

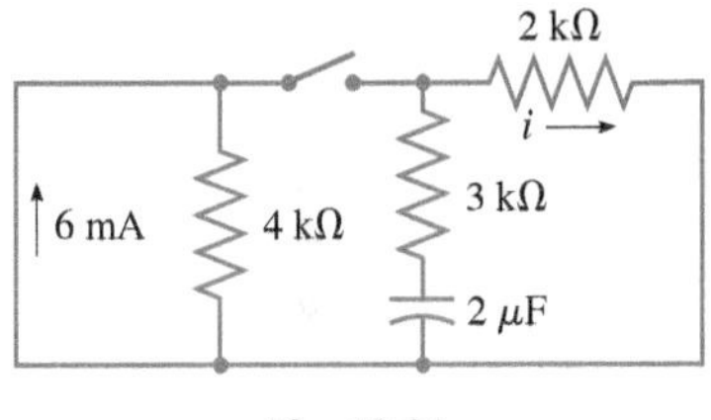

Fig. 13.21

103. For a particular solar-energy system, the collector area A required to supply a fraction F of the total energy is given by $A = 480\,F^{2.2}$. Plot A (in m^2) as a function of F, from $F = 0.1$ to $F = 0.9$ using a semilogarithmic scale.

104. The current I (in μA) and resistance R (in Ω) were measured as follows in a certain microcomputer circuit:

R (Ω)	100	200	500	1000	2000	5000	10 000
I (μA)	81	41	16	8.2	4.0	1.6	0.8

Plot I as a function of R as a log-log plot.

105. While checking logarithmic curves on a calculator, a machine-design student noted that a certain robotic arm was shaped like part of the graph of $y = \ln(2x^{2/3})$. As a check, the student rewrote the equation as $y = (2 \ln x)/3 + \ln 4 - \ln(\ln e^2)$. Write a paragraph explaining (a) whether the second equation is equivalent to the first, and (b) if the graphs of the two equations are identical.

CHAPTER 13 PRACTICE TEST

In Problems 1–4, determine the value of x.

1. $\log_9 x = -\frac{1}{2}$

2. $\log_3 x - \log_3 2 = 2$

3. $\log_x 64 = 3$

4. $3^{3x+1} = 8$

5. Graph the function $y = 2 \log_4 x$.

6. Graph the function $y = 2(3^x)$ on semilog paper.

7. Express $\log_5\left(\frac{4a^3}{7}\right)$ as a combination of a sum, difference, and multiple of logarithms, including $\log_5 2$.

8. Solve for y in terms of x: $3 \log_7 x - \log_7 y = 2$.

9. An equation used for a certain electric circuit is $\ln i - \ln I = -t/RC$. Solve for i.

10. Evaluate: $\frac{2 \ln 0.9523}{\log 6066}$.

11. Evaluate: $\log_5 7.32$.

12. If A_0 dollars are invested at 8%, compounded continuously for t years, the value A of the investment is given by $A = A_0e^{0.08t}$. Determine how long it takes for the investment to double in value.

14. Additional Types of Equations and Systems of Equations

LEARNING OUTCOMES

After completion of this chapter, the student should be able to:

- Graph nonlinear equations using a graphing utility
- Solve systems involving nonlinear equations graphically
- Solve systems involving nonlinear equations algebraically by the method of substitution or by the method of elimination
- Solve application problems involving systems of nonlinear equations
- Solve equations in quadratic form
- Solve equations with radicals
- Solve application problems involving equations in quadratic form or equations with radicals

Stephen Meese/Fotolia

jhvephoto/Fotolia

▲These pictures allow us to visualize the graphical solution of a system of two equations. On the left, the Seal Island Bridge (Cape Breton, Nova Scotia) shows a parabola and a line, a system with two real solutions. On the right, the Huron St. Bridge (Stratford, Ontario) and its reflection show a circle and a line, a system with no real solutions.

In this chapter we discuss methods for solving systems of equations where the equations involved are not all linear. We begin by presenting graphical solutions, a method which has been used extensively since the mid-1600s when René Descartes used the intersection of curves as a way of solving higher order equations. The approach is the same as what we have already used in Chapters 3 and 7, except that we will encounter graphs of equations that do not necessarily represent a function (circles and ellipses, for example). We will then show how to solve these systems algebraically.

We will also consider solutions of two special types of equations: equations in quadratic form and equations involving radicals. One of the methods presented here has been used in the study of optics since 1836, when the French mathematician Cauchy developed the dispersion of light equation that bears his name.

Applications of the types of equations and systems of equations of this chapter are found in many fields of science and technology. These include physics, electricity, business, and structural design.

14.1 Graphical Solution of Systems of Equations

■ The graph of a relation also represents a function if and only if every vertical line crosses the graph at most once; this rule is known as the vertical line test.

In this section we will discuss graphical solutions of systems of equations involving nonlinear equations. In these systems we often encounter second-degree equations that represent curves such as circles and ellipses, which are not the graph of a single function (*two* values of y are assigned to the same value of x). Therefore, we first show how the graphs of such equations can be obtained.

EXAMPLE 1 Plotted graph—circle

Plot the graph of the equation $x^2 + y^2 = 25$.

We first solve this equation for y, and we obtain $y = \sqrt{25 - x^2}$, or $y = -\sqrt{25 - x^2}$, which we write as $y = \pm\sqrt{25 - x^2}$. We now assume values for x and find the corresponding values for y.

x	0	± 1	± 2	± 3	± 4	± 5
y	± 5	± 4.9	± 4.6	± 4	± 3	0

If $x > 5$, values of y are imaginary. We cannot plot these because x and y must both be real. When we show $y = \pm 3$ for $x = \pm 4$, this is a short way of representing four points. These points are $(4, 3)$, $(4, -3)$, $(-4, 3)$, and $(-4, -3)$.

In Fig. 14.1, *the resulting curve is a* **circle**. A circle with its centre at the origin results from an equation of the form $x^2 + y^2 = r^2$, where r is the radius.

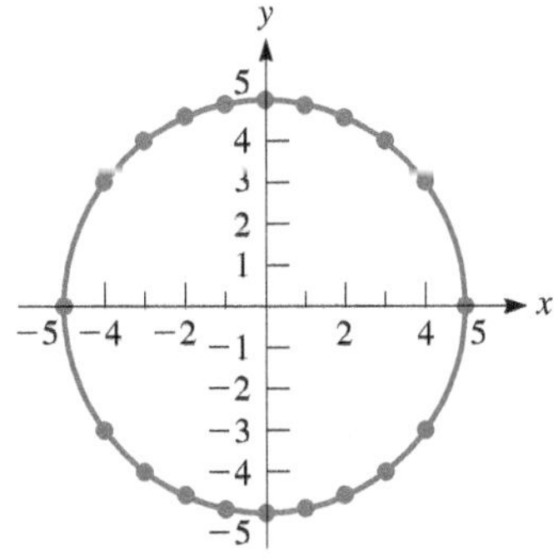

Fig. 14.1

EXAMPLE 2 Graph with a graphing utility—ellipse

Graph the equation $2x^2 + 5y^2 = 10$ using a graphing utility.

First, solving for y, we get $y = \pm\sqrt{\dfrac{10 - 2x^2}{5}}$. To display the graph of this equation using a graphing utility, *we must enter both functions,* one as $y_1 = \sqrt{(10 - 2x^2)/5}$, and the other as $y_2 = -\sqrt{(10 - 2x^2)/5}$.

The display settings are chosen noting that the domain is from $-\sqrt{5}$ to $\sqrt{5}$ and the range is from $-\sqrt{2}$ to $\sqrt{2}$. On a graphing calculator, we obtain the display shown in Fig. 14.2.

The curve is an **ellipse**. An ellipse centred at the origin will be the resulting curve if the equation is of the form $ax^2 + by^2 = c$, where the constants a, b, and c have the same sign, and for which $a \neq b$.

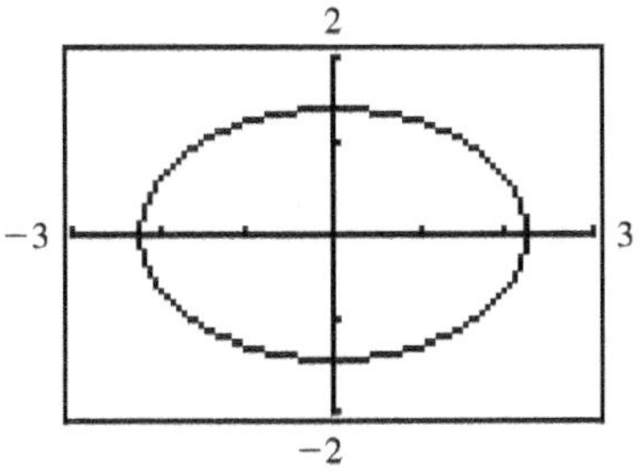

Fig. 14.2

EXAMPLE 3 Graphing a hyperbola

Graph the equation $2x^2 - y^2 = 4$ using a graphing utility.

Solving for y, we get

$$y = \pm\sqrt{2x^2 - 4}$$

As in Example 2, we must enter two functions, one with the plus sign and the other with the minus sign. Each one is displayed in a different colour in Fig. 14.3. The *display* settings are chosen by noting that the values $-\sqrt{2} < x < \sqrt{2}$ are not in the domain of either $y_1 = \sqrt{2x^2 - 4}$ or $y_2 = -\sqrt{2x^2 - 4}$. These values of x would lead to imaginary values of y.

The curve is a **hyperbola** centred at the origin, which results from an equation of the form $ax^2 + by^2 = c$, if a and b have *different* signs.

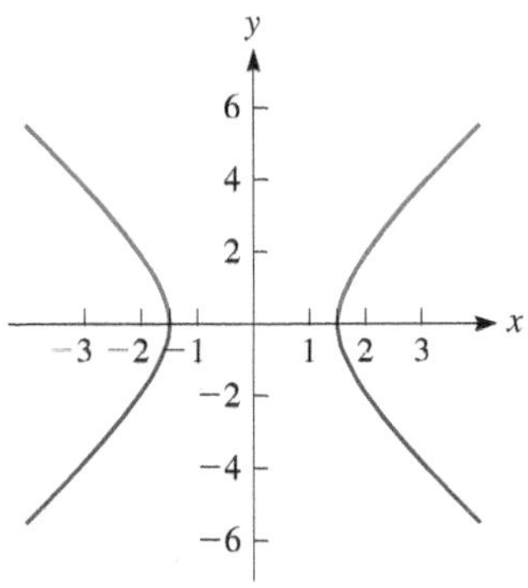

Fig. 14.3

■ Circles, ellipses, hyperbolas and parabolas are conics. They are discussed in detail in Chapter 21.

Note that when these types of curves are displayed on a graphing utility, it is possible that there are small gaps in the curve. The appearance of such gaps depends on the pixel width and the functional values used by the utility.

SOLVING SYSTEMS OF EQUATIONS GRAPHICALLY

As in solving systems of linear equations, we solve any system by finding the values of x and y that satisfy both equations at the same time. To solve a system graphically, we graph the equations and find the coordinates of all points of intersection. If the curves do not intersect, the system has no real solutions. In that case, complex solutions can be found algebraically using the methods of the next section.

EXAMPLE 4 Solving a system graphically

Graphically solve the following system of equations to the nearest 0.01:

$$y = 3^x$$
$$9x^2 + 4y^2 = 36$$

To graph the first equation, note that the function $y = 3^x$ is always positive, so we can restrict the display settings to non-negative values of y.

We graph the second equation by solving for y and considering the two functions $y = \pm\frac{1}{2}\sqrt{36 - 9x^2}$. As in Example 2, this is an ellipse, extending from $x = -2$ to $x = 2$ ($9x^2$ cannot be greater than 36). As shown in Fig. 14.4, only the upper part of the ellipse (using the + sign) appears when displaying non-negative values of y.

The features of a graphing utility can be used to find an approximate solution. For example, using the *intersect* feature (or *trace* and *zoom* features) of a graphing calculator, we find $x = -2.00$, $y = 0.11$ and $x = 0.90$, $y = 2.68$.

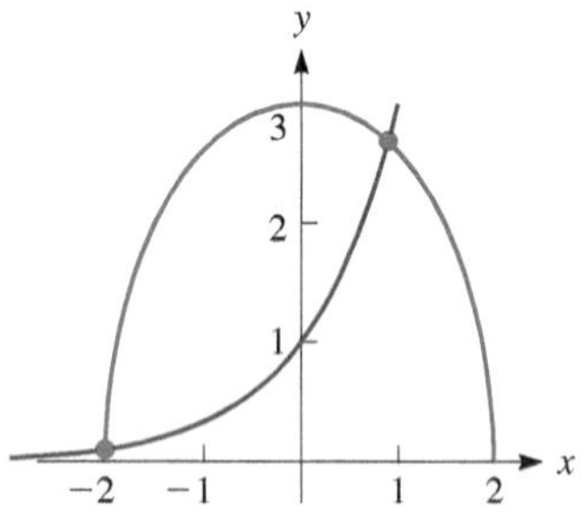

Fig. 14.4

EXAMPLE 5 Solving a system—dimensions of a vent

For proper ventilation, the vent for a hot-air heating system is to have a rectangular cross-sectional area of 2.3 m^2 and is to be made from sheet metal 6.4 m wide. Find the dimensions of this cross-sectional area of the vent.

Identify knowns and unknowns: From Fig. 14.5, let l = length and w = width
First equation: Area is given as 2.3 m^2, so $lw = 2.3$
Second equation: Perimeter is given as 6.4 m (the width of the sheet when unfolded), so $2l + 2w = 6.4$

Graph: Graph $l = 2.3/w$ and $l = 3.2 - w$ (using x for w and y for l). The display settings of Fig. 14.6 are chosen because negative l and w have no meaning, and the line has intercepts $(3.2, 0)$ and $(0, 3.2)$.

Solve: Using the *intersect* feature, we find the solutions $(1.1, 2.1)$ and $(2.1, 1.1)$. Using the length as the longer dimension, we have

$$l = 2.1 \text{ m and } w = 1.1 \text{ m}$$

Check: The solution checks with the statement of the problem.

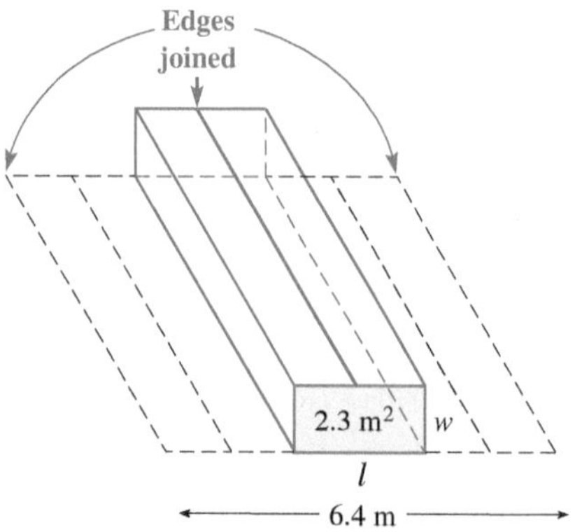

Fig. 14.5

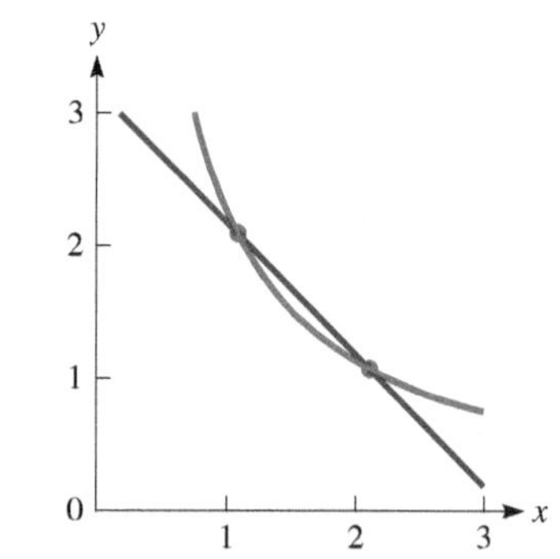

Fig. 14.6

In Example 5, we graphed the equation $xy = 2.3$ (having used x for w and y for l). The graph of this equation is also a *hyperbola*, another form of which is $xy = c$.

EXAMPLE 6 System of equations with no solution—rocket paths

In a computer game, two rockets follow paths described by the equations $x^2 = 2y$ and $3x - y = 5$. Determine if the rocket paths ever cross.

Solving the equation of the path of the first rocket, we get $y = x^2/2$. From Chapter 7, we know that this is a parabola with vertex at the origin, and opening upward ($a > 0$). Any $y < 0$ cannot be a solution.

The straight line rocket path $3x - y = 5$ has intercepts of $(0, -5)$, and $(5/3, 0)$. Any intersection must therefore be in the first quadrant, and we restrict the display settings there. See Fig. 14.7.

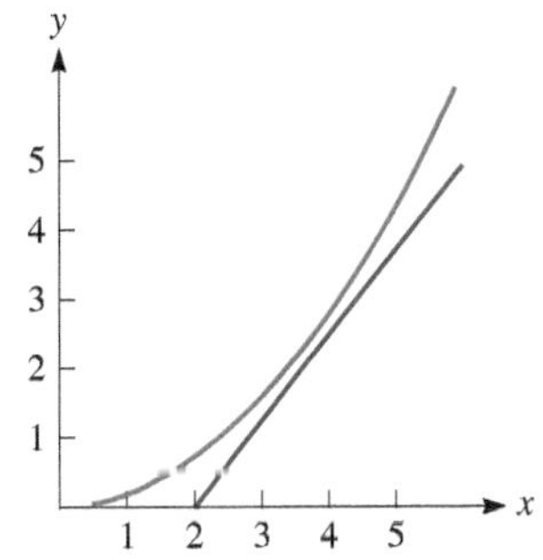

Fig. 14.7

Inspecting the figure, we see that these curves do not intersect. The system has *no real solutions*. The two complex solutions of the system can be found algebraically using the methods of the next section.

EXERCISES 14.1

In Exercises 1–4, make the given changes in the indicated examples of this section, and then perform the indicated operations.

1. In Example 1, change the + sign before y^2 to −.
2. In Example 3, change the − sign before y^2 to +.
3. In Example 4, change the coefficient of x^2 to 25.
4. In Example 6, change the coefficient of y in the first equation to 3.

In Exercises 5–30, solve the given systems of equations graphically by using a graphing utility. Find all values to the nearest 0.01.

5. $y = 2x$, $x^2 + y^2 = 16$
6. $3x - y = 4$, $y = 6 - 2x^2$
7. $x^2 + 2y^2 = 8$, $x - 2y = 4$
8. $y = 3x - 6$, $xy = 6$
9. $y = x^2 - 2$, $4y = 12x - 17$
10. $4x^2 + 25y^2 = 21$, $10y = 31 - 9x$
11. $8y = 11x^2$, $xy = 30$
12. $y = -2x^2$, $y = x^2 - 6$
13. $y = -x^2 + 4$, $x^2 + y^2 = 9$
14. $y = x^3$, $x^2 + 2y^2 = 16$
15. $x^2 - 4y^2 = 16$, $x^2 + y^2 = 1$
16. $y = 2x^2 - 4x$, $x^2y = -4$
17. $2x^2 + 3y^2 = 19$, $x^2 + y^2 = 9$
18. $x^2 - y^2 = 4$, $2x^2 + y^2 = 16$
19. $x^2 + y^2 = 7$, $y(x + 2) = 3$
20. $x^2 + y^2 = 4$, $y^2 = x - 4$
21. $y = x^2$, $y = \sin x$
22. $y = 3 + 2x - x^2$, $y = 2\cos 2x$
23. $y = e^{-x}$, $y = x^{2/3}$
24. $y = 2^{x+1}$, $x^2 + y^2 = 4$
25. $x^2 - y^2 = 7$, $y = 4\log_2 x$
26. $x^2 + 4y^2 = 16$, $y = 2\ln x$
27. $y = \ln(x - 1)$, $y = \sin\frac{1}{2}x$
28. $y = \cos x$, $y = \log_3 x$
29. $10^{x+y} = 150$, $y = x^2$
30. $e^{x^2+y^2} = 20$, $xy = 4$

In Exercise 31, draw the appropriate figures. In Exercises 32–38, set up systems of equations and solve them graphically.

31. By drawing rough sketches, show that a parabola and an ellipse can have zero, one, two, three, or four possible points of intersection.
32. A rectangular security area is to be enclosed by fencing and divided into two equal parts of 1600 m^2 each by a fence parallel to the shorter sides. Find the dimensions of the security area if the total amount of fencing is 280 m.
33. A helicopter is located 5.2 km north of east of a radio tower such that it is three times as far north as it is east from the tower. Find the northern and eastern components of the displacement from the tower.
34. A 4.60-m insulating strip is placed completely around a rectangular solar panel with an area of 1.20 m^2. What are the dimensions of the panel?
35. The power developed in an electric resistor is i^2R, where i is the current. If a first current passes through a 2.0-Ω resistor and a second current passes through a 3.0-Ω resistor, the total power produced is 12 W. If the resistors are reversed, the total power produced is 16 W. Find the currents (in A) if $i > 0$.
36. A circular hot tub is located on the square deck of a home. The side of the deck is 7.3 m more than the radius of the hot tub, and there are 72.5 m^2 of deck around the tub. Find the radius of the hot tub and the length of the side of the deck. Explain your answer.

37. Assume Earth is a sphere, with $x^2 + y^2 = 41$ as the equation of a circumference (distance in thousands of kilometres). If a meteorite approaching Earth has a path described as $y^2 = 20x + 140$, will the meteorite strike Earth? If so, where?

38. Two people, with one walking 1.0 km/h faster than the other, are on straight roads that are perpendicular. After meeting at an intersection, each continues on straight. How fast is each walking if they are 7.0 km apart (on a straight line) 1.0 h after meeting?

14.2 Algebraic Solution of Systems of Equations

Often, using the graphical method is the easiest way to solve a system of equations. With a graphing utility, it is possible to find the result with good accuracy. However, the graphical method does not usually give the *exact* answer. Using algebraic methods to find exact solutions for some systems of equations is either not possible or quite involved. There are systems, however, for which there are relatively simple algebraic solutions. In this section, we consider two useful methods, both of which we discussed before when we were studying systems of linear equations.

SOLUTION BY SUBSTITUTION

The first method is *substitution.* If we can solve one of the equations for one of its variables, we can substitute this solution into the other equation. We then have only one unknown in the resulting equation, and we can then solve this equation by methods discussed in earlier chapters.

Solving a System of Equations by Substitution	EXAMPLE 1
	Solve the system $2x - y = 4$ and $x^2 - y^2 = 4$.
1. Solve one of the equations for one of the unknowns.	Solving the first equation for y: $y = 2x - 4$
2. Substitute into the other equation and simplify.	Substitute $y = 2x - 4$ for y in $x^2 - y^2 = 4$ and then simplify: $x^2 - (2x - 4)^2 = 4$ $x^2 - (4x^2 - 16x + 16) = 4$ $-3x^2 + 16x - 20 = 0$
3. Solve the resulting equation for the value of the unknown it contains.	$x = \dfrac{-16 \pm \sqrt{256 - 4(-3)(-20)}}{-6}$ $= \dfrac{-16 \pm \sqrt{16}}{-6} = \dfrac{-16 \pm 4}{-6}$ $x = 10/3$ or $x = 2$
4. Substitute the value(s) into the equation from Step 1 to solve for the other unknown.	$y = 2x - 4 \rightarrow y = 2(10/3) - 4 = 8/3$ or $\rightarrow y = 2(2) - 4 = 0$ The solutions are $x = 10/3$, $y = 8/3$, and $x = 2$, $y = 0$.
5. Check the values in both original equations.	We can check that these values indeed satisfy both equations. We can also check the solutions graphically in Fig. 14.8.

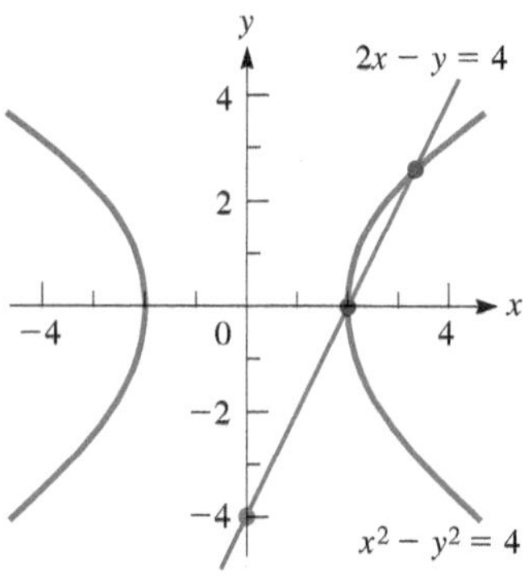

Fig. 14.8

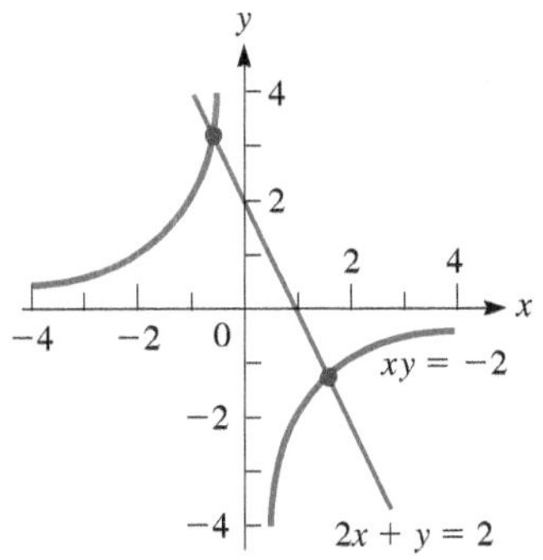

Fig. 14.9

EXAMPLE 2 Solution by substitution

By substitution, solve the system of equations $xy = -2$ and $2x + y = 2$.

Solve one of the equations for one of the unknowns:	Solving the first equation for y: $y = -2/x$
Substitute into the other equation and simplify:	Substitute $y = -2/x$ for y in $2x + y = 2$: $2x + \left(-\frac{2}{x}\right) = 2$ Simplifying: $2x^2 - 2 = 2x$, $x^2 - x - 1 = 0$
Solve the resulting equation for the value of the unknown it contains:	$x = \dfrac{1 \pm \sqrt{1 + 4)}}{2} = \dfrac{1 \pm \sqrt{5}}{2}$
Substitute the value(s) into the equation from Step 1 to solve for the other unknown:	$y = -2/x \rightarrow y = 1 - \sqrt{5}$ or $\rightarrow y = 1 + \sqrt{5}$ The approximate solutions are $x = 1.62$, $y = -1.24$, and $x = -0.618$ and $y = 3.24$.
Check the values in both original equations:	We can check that these values indeed satisfy both equations. We can also check the solutions graphically from Fig. 14.9.

Practice Exercise

1. Solve by substitution:

$$2x^2 + y^2 = 3$$
$$x - y = 2$$

SOLUTION BY ELIMINATION

The other algebraic method is that of elimination by *addition or subtraction.* This method is most useful if both equations have only squared terms and constants.

Solving a System of Equations by Elimination	EXAMPLE 3
1. Multiply both sides of each equation by a constant so that **the coefficients of one unknown are numerically the same but differ in sign** in both equations.	Solve the system $2x^2 + y^2 = 9$ and $x^2 - y^2 = 3$. The terms $+y^2$ and $-y^2$ are numerically the same but differ in sign. No multiplying is required.
2. Add the terms on each side of the resulting equation. Solve the resulting equation for the other unknown.	$2x^2 + y^2 = 9$ $x^2 - y^2 = 3$ $3x^2 \quad = 12$ $x^2 = 4$ $x = \pm 2$
3. Substitute the value(s) into one of the original equations.	For $x = 2$, $2^2 - y^2 = 3 \rightarrow y = \pm 1$ For $x = -2$, $(-2)^2 - y^2 = 3 \rightarrow y = \pm 1$ The four solutions are: $x = 2, y = 1 \quad x = 2, y = -1$ $x = -2, y = 1 \quad x = -2, y = -1$
4. Check in both equations.	Each solution checks in the original equation. We can also check graphically in Fig. 14.10.

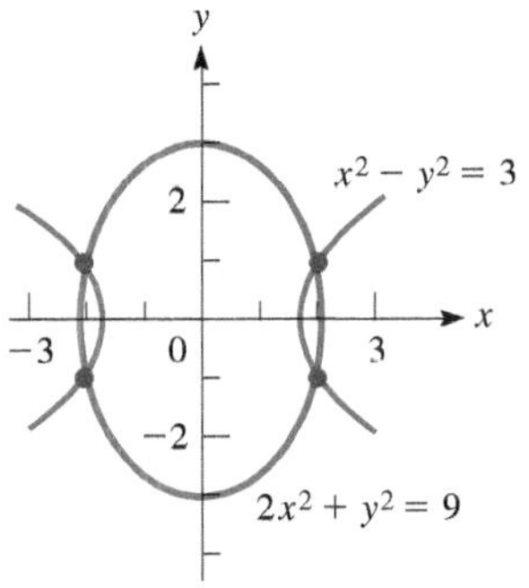

Fig. 14.10

EXAMPLE 4 Solution by elimination

By elimination, solve the system of equations

$$5x^2 + 2y^2 = 17$$
$$x^2 + y^2 = 4$$

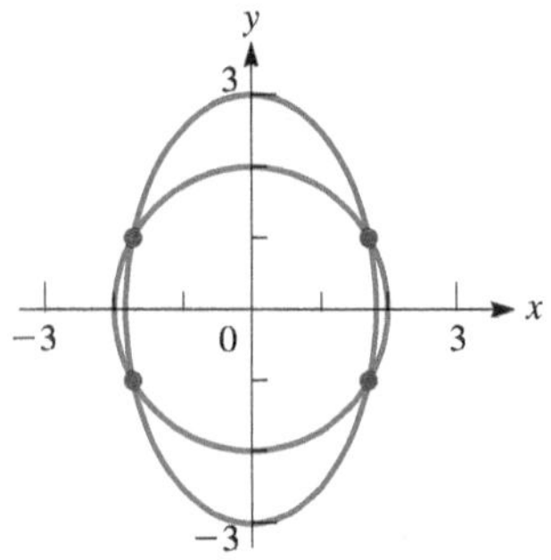

Fig. 14.11

Multiply both sides of each equation as necessary: To allow the y^2 terms to cancel, we multiply the second equation by −2:

$$-2x^2 - 2y^2 = -8$$

Add two lines and solve for one unknown:

$$\begin{aligned} 5x^2 + 2y^2 &= 17 \\ -2x^2 - 2y^2 &= -8 \\ \hline 3x^2 \qquad &= 9 \\ x^2 &= 3 \\ x &= \pm\sqrt{3} \end{aligned}$$

Substitute the value(s) into one of the original equations:

For $x = +\sqrt{3}$, $(\sqrt{3})^2 + y^2 = 4 \rightarrow y = \pm 1$

For $x = -\sqrt{3}$, $(-\sqrt{3})^2 + y^2 = 4 \rightarrow y = \pm 1$

The four solutions are:

$x = \sqrt{3}, y = 1 \qquad x = \sqrt{3}, y = -1$

$x = -\sqrt{3}, y = 1 \qquad x = -\sqrt{3}, y = -1$

Check in both equations: Each solution checks in the original equation. We can also check graphically in Fig. 14.11.

Practice Exercise

2. Solve by addition or subtraction:

$$x^2 + y^2 = 6$$
$$2x^2 - y^2 = 6$$

EXAMPLE 5 Algebraic solution—cost of machine parts

A certain number of machine parts cost \$1000. If they cost \$5 less per part, 10 additional parts could be purchased for the same amount of money. What is the cost of each part?

Let c = the cost per part, and n = the number of parts. From the first statement, we see that $cn = 1000$. From the second statement, $(c - 5)(n + 10) = 1000$. Rewriting these equations, we have

$$n = 1000/c$$
$$cn + 10c - 5n - 50 = 1000$$

$$c\left(\frac{1000}{c}\right) + 10c - 5\left(\frac{1000}{c}\right) - 50 = 1000 \qquad \text{substituting first equation in second equation}$$

$$1000 + 10c - \frac{5000}{c} - 50 = 1000$$

$$10c - \frac{5000}{c} - 50 = 0$$

$$c^2 - 5c - 500 = 0$$

$$(c + 20)(c - 25) = 0$$

$$c = -20, 25$$

Since a negative answer has no significance in this particular situation, we see that the solution is c = \$25 per part. Checking with the original statement of the problem, we see that this is correct.

EXERCISES 14.2

In Exercises 1–4, make the given changes in the indicated examples of this section and then solve the resulting systems of equations.

1. In Example 1, change the sign before y in the first equation from $-$ to $+$ and then solve the system.

2. In Example 2, change the right side of the second equation from 2 to 3 and then solve the system.

3. In Example 3, change the coefficient of x^2 in the first equation from 2 to 1 and then solve the system.

4. In Example 4, change the left side of the first equation from $5x^2 + 2y^2$ to $2x^2 + 5y^2$ and then solve the system.

In Exercises 5–28, solve the given systems of equations algebraically. (Hint for Exercises 27 and 28: First, subtract one equation from the other to get an equation relating x and y. Then substitute this equation in either given equation.)

5. $y = x + 1$; $y = x^2 + 1$

6. $y = 2x - 1$; $y = 2x^2 + 2x - 3$

7. $x + 2y = 3$; $x^2 + y^2 = 26$

8. $p^2 + 4h^2 = 4$; $h = p + 1$

9. $x + y = 1$; $x^2 - y^2 = 1$

10. $x + y = 2$; $2x^2 - y^2 = 1$

11. $2x - y = 2$; $2x^2 + 3y^2 = 4$

12. $6y - x = 6$; $x^2 + 3y^2 = 36$

13. $wh = 1$; $w + h = 2$

14. $xy = 100$; $x + y = 20$

15. $xy = 4$; $4x - 3y = 2$

16. $xy = -4$; $2x + y = -2$

17. $y = x^2$; $y = 3x^2 - 50$

18. $M = L^2 - 1$; $2L^2 - M^2 = 2$

19. $x^2 - y = -1$; $x^2 + y^2 = 5$

20. $s^2 + t^2 = 8$; $3s - t = 2$

21. $D^2 - 1 = R$; $D^2 - 2R^2 = 1$

22. $2y^2 - 4x = 7$; $y^2 + 2x^2 = 3$

23. $x^2 + y^2 = 25$; $x^2 - 2y^2 = 7$

24. $3x^2 - y^2 = 4$; $x^2 + 4y^2 = 10$

25. $x^2 + 3y^2 = 37$; $2x^2 - 9y^2 = 14$

26. $5x^2 - 4y^2 = 15$; $3y^2 + 4x^2 = 12$

27. $x^2 + y^2 + 4x = 1$; $x^2 + y^2 - 2y = 9$

28. $x^2 + y^2 - 4x - 2y + 4 = 0$; $x^2 + y^2 - 2x - 4y + 4 = 0$

In Exercises 29–44, solve the indicated systems of equations algebraically. In Exercises 33–44, it is necessary to set up the systems of equations.

29. Solve for x and y: $x^2 - y^2 = a^2 - b^2$; $x - y = a - b$.

30. For what value of b are the two solutions of the system $x^2 - 2y = 5$; $y = 3x + b$ equal to each other? For this value of b, what is true of the graphs of the two functions?

31. A rocket is fired from behind a ship and follows the path given by $h = 3x - 0.05x^2$, where h is its altitude (in km) and x is the horizontal distance travelled (in km). A missile fired from the ship follows the path given by $h = 0.8x - 15$. For $h > 0$ and $x > 0$, find where the paths of the rocket and missile cross.

32. A 2-kg block collides with an 8-kg block. Using the physical laws of conservation of energy and conservation of momentum, along with given conditions, the following equations involving the velocities are established:

$$v_1^2 + 4v_2^2 = 41$$
$$2v_1 + 8v_2 = 12$$

Find these velocities (in m/s) if $v_2 > 0$.

33. One face of a washer has an area of 37.7 cm^2. The inner radius is 2.00 cm less than the outer radius. What are the radii?

34. A right triangular sail has a perimeter of 9.30 m and a hypotenuse of 4.17 m. Find the lengths of the sides of the sail.

35. The edges of a rectangular piece of plastic sheet are joined together to make a plastic tube. If the area of the plastic sheet is 216 cm^2, and the volume of the resulting tube is 224 cm^3, what are the dimensions of the plastic sheet?

36. On a map, the path of an underground stream can be approximated by the parabola $y = x^2 - 20x + 84$. Wells need to be dug at two locations on the line $y = 2x + 12$. Find the coordinates of the points where the wells should be dug.

37. A roof truss is in the shape of a right triangle. If there are 4.60 m of lumber in the truss and the longest side is 2.20 m long, what are the lengths of the other two sides of the truss?

38. In a certain roller mechanism, the radius of one steel ball is 2.00 cm greater than the radius of a second steel ball. If the difference in their masses is 7100 g, find the radii of the balls. The density of steel is 7.70 g/cm^3.

39. A set of equal electrical resistors in series has a total resistance (the sum of the resistances) of 78.0 Ω. Another set of two fewer equal resistors in series also has a total resistance of 78.0 Ω. If the resistance of each resistor in the second set is 1.3 Ω greater than that of each resistor in the first, how many resistors are in each set?

40. Security fencing encloses a rectangular storage area of 1600 m^2 that is divided into two sections by additional fencing parallel to the shorter sides. Find the dimensions of the storage area if 220 m of fencing are used.

41. An open liner for a box is to be made from a rectangular sheet of cardboard of area 216 cm^2 by cutting equal 2.00-cm squares from each corner and bending up the sides. If the volume within the liner is 224 cm^3, what are the dimensions of the cardboard sheet?

42. Two guy wires, one 140 m long and the other 120 m long, are attached at the same point of a TV tower with the longer one secured in the (level) ground 30 m farther from the base of the tower than the shorter one. How high up on the tower are they attached?

43. A jet travels at 990 km/h relative to the air. It takes the jet 1.4 h longer to travel the 5200 km from London to Montreal against the wind than it takes from Montreal to London with the wind. Find the velocity of the wind.

44. In a marketing survey, a company found that the total gross income for selling t tables at a price of p dollars each was \$35 000. It then increased the price of each table by \$100 and found that the total income was only \$27 000 because 40 fewer tables were sold. Find p and t.

Answers to Practice Exercises

1. $x = 1/3,\ y = -5/3;\ x = 1,\ y = -1$

2. $x = 2,\ y = \sqrt{2};\ x = 2,\ y = -\sqrt{2};\ x = -2,\ y = \sqrt{2};$ $x = -2,\ y = -\sqrt{2}$

14.3 Equations in Quadratic Form

Often, we encounter equations that can be solved by methods applicable to quadratic equations, even though these equations are not actually quadratic. They do have the property, however, that *with a proper substitution* ***they may be written in the form of a quadratic equation***. All that is necessary is that the equation have terms including some variable quantity, its square, and perhaps a constant term. The following example illustrates these types of equations.

EXAMPLE 1 Identifying a quadratic form

Rewrite each of the following as a quadratic equation using the proper substitution.

	Substitution	Quadratic Equation
(a) $x - 2\sqrt{x} - 5 = 0$	$y = \sqrt{x}$ (since $y^2 = x$)	$y^2 - 2y - 5 = 0$
(b) $t^{-4} - 5t^{-2} + 3 = 0$	$y = t^{-2}$ (since $y^2 = t^{-4}$)	$y^2 - 5y + 3 = 0$
(c) $t^3 - 3t^{3/2} - 7 = 0$	$y = t^{3/2}$ (since $y^2 = t^3$)	$y^2 - 3y - 7 = 0$
(d) $(x+1)^4 - (x+1)^2 - 1 = 0$	$y = (x+1)^2$ (since $y^2 = (x+1)^4$)	$y^2 - y - 1 = 0$
(e) $x^{10} - 2x^5 + 1 = 0$	$y = x^5$ (since $y^2 = x^{10}$)	$y^2 - 2y + 1 = 0$

A method for solving equations in quadratic form is summarized as follows.

Procedure for Solving an Equation in Quadratic Form

1. **Identify the equation as quadratic:** Find a quantity involving the unknown variable whose square is included in the equation. (Some algebraic manipulation may be necessary.) Let this quantity be y.
2. **Rewrite as a quadratic:** Substitute y into the original equation to rewrite it in the form $ay^2 + by + c = 0$.
3. **Solve the quadratic equation for y.**
4. **Solve for the original variable:** Substitute each of the solutions for y into the equation relating y and the original variable, and solve for the original variable.
5. **Check all answers in the original equation.** The substitution could have added extraneous roots (that is, roots of a subsequent equation that are not roots of the original equation). Extraneous roots must be identified and discarded.

EXAMPLE 2 Solving an equation in quadratic form

Solve the equation $2x^4 + 7x^2 = 4$.

Identify the equation as quadratic: Let $y = x^2$ (since $y^2 = x^4$)

Rewrite as a quadratic: $2y^2 + 7y - 4 = 0$

Solve the quadratic:

$$2y^2 + 7y - 4 = 0$$
$$(2y - 1)(y + 4) = 0$$
$$y = \tfrac{1}{2} \quad \text{or} \quad y = -4$$

Solve for the original variable: Substitute each value of y into $y = x^2$ and solve for x:

$$\tfrac{1}{2} = x^2 \qquad \text{or} \qquad -4 = x^2$$
$$x = \pm\frac{1}{\sqrt{2}} \qquad\qquad x = \pm 2j$$

Check in the original equation: Each solution checks in the **original** equation.

As shown in Fig. 14.12, we can also solve the equation graphically by graphing $y = 2x^4 + 7x^2 - 4$ and finding the x-intercepts. We can verify the two real solutions $x = \pm 1/\sqrt{2} = \pm 0.707$. However, the imaginary solutions cannot be found graphically.

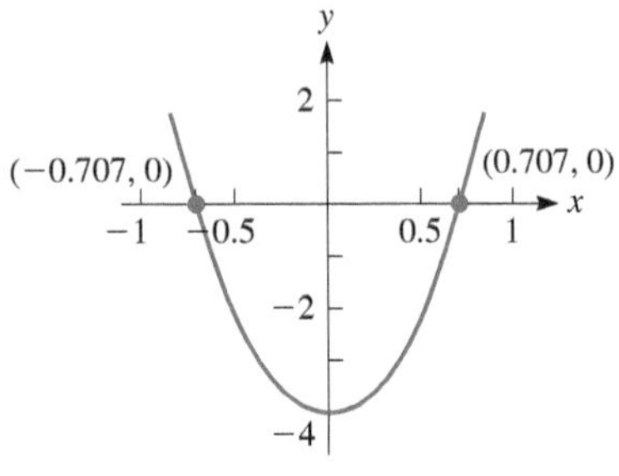

Fig. 14.12

Two of the solutions in Example 2 are complex numbers. We were able to find these solutions directly from the definition of the square root of a negative number. In some cases (see Exercise 22 of this section), it is necessary to use DeMoivre's theorem (see Section 12.6) to find such complex-number solutions.

EXAMPLE 3 Solving an equation containing a square root

Solve the equation $x - \sqrt{x} - 2 = 0$.

By letting $y = \sqrt{x}$, we have

$$y^2 - y - 2 = 0$$
$$(y - 2)(y + 1) = 0$$
$$y = 2 \quad \text{or} \quad y = -1$$

Since $y = \sqrt{x}$, we note that y cannot be negative, and this means $y = -1$ cannot lead to a solution. For $y = 2$, we have $x = 4$. Checking, we find that $x = 4$ satisfies the original equation. Therefore, the only solution is $x = 4$.

The graph of $f(x) = x - \sqrt{x} - 2$ is shown in Fig. 14.13. Note that the only zero is at $x = 4$, confirming that $y = -1$ is an extraneous root and should be discarded.

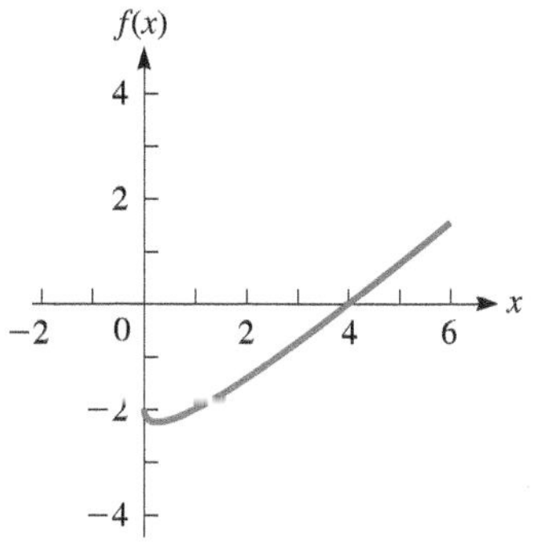

Fig. 14.13

COMMON ERROR It is a common error to give an extraneous root as the solution to an equation. As we said earlier, it is important to check all answers in the original equation so that extraneous roots can be identified and discarded.

EXAMPLE 4 Solving an equation containing negative exponents

Solve the equation $x^{-2} + 3x^{-1} + 1 = 0$.

By substituting $y = x^{-1}$, we have $y^2 + 3y + 1 = 0$. To solve this equation, we use the quadratic formula:

$$y = \frac{-3 \pm \sqrt{9 - 4}}{2} = \frac{-3 \pm \sqrt{5}}{2}$$

Since $x = 1/y$, we have

$$x = \frac{2}{-3 + \sqrt{5}} \quad \text{or} \quad x = \frac{2}{-3 - \sqrt{5}}$$

These answers, to the nearest 0.001, are

$$x \approx -2.618 \quad \text{or} \quad x \approx -0.382$$

These results check when substituted in the original equation.

Practice Exercise

1. Solve for x: $x^{-4} - 8x^{-2} + 16 = 0$

LEARNING TIP

Note that when checking decimal answers in the original equation, it is more accurate to store values in the memory of the calculator and use them without rounding than to use rounded values.

EXAMPLE 5 Solving an equation containing grouped terms

Solve the equation $(x^2 - x)^2 - 8(x^2 - x) + 12 = 0$.

By substituting $y = x^2 - x$, we have

$$y^2 - 8y + 12 = 0$$
$$(y - 2)(y - 6) = 0$$
$$y = 2 \quad \text{or} \quad y = 6$$
$$x^2 - x = 2 \quad \text{or} \quad x^2 - x = 6 \qquad y = x^2 - x$$

Solving each of these equations, we have

$$x^2 - x - 2 = 0 \qquad\qquad x^2 - x - 6 = 0$$
$$(x - 2)(x + 1) = 0 \qquad (x - 3)(x + 2) = 0$$
$$x = 2 \quad \text{or} \quad x = -1 \qquad x = 3 \quad \text{or} \quad x = -2$$

Each value checks when substituted in the original equation.

EXAMPLE 6 Quadratic form—dimensions of a photograph

Fig. 14.14

An old rectangular photograph has an area of 120 cm^2. The diagonal of the photograph is 17 cm. Find its length and width. See Fig. 14.14.

Let l = the length of the photograph and let w = its width. Since the area is 120 cm^2, $lw = 120$. Also, using the Pythagorean theorem and the fact that the diagonal is 17 cm, we have the equation $l^2 + w^2 = 17^2 = 289$. Therefore, we are to solve the system of equations

$$lw = 120 \qquad l^2 + w^2 = 289$$

Solving the first equation for l, we have $l = 120/w$. Substituting this into the second equation, we have

$$\left(\frac{120}{w}\right)^2 + w^2 = 289$$

$$\frac{14\,400}{w^2} + w^2 = 289$$

$$14\,400 + w^4 = 289w^2$$

■ The solution to the problem of finding the length and width of a rectangle given the diagonal and the area was known to Old Babylonian mathematicians around 1770 B.C.E.

Let $y = w^2$.

$$y^2 - 289y + 14\,400 = 0$$

$$y = \frac{-(-289) \pm \sqrt{(-289)^2 - 4(1)(14\,400)}}{2(1)}$$

$$y = 225 \quad \text{or} \quad y = 64$$

Therefore, $w^2 = 225$ or $w^2 = 64$.

■ The first permanent photograph was taken in 1816 by the French inventor Joseph Niépce (1765–1833).

Solving for w, we get $w = \pm 15$ or $w = \pm 8.0$. Only the positive values are meaningful in this problem, which means if $w = 8.0$ cm, then $l = 15$ cm, and if $w = 15$ cm, $l = 8.0$ cm. Therefore, the picture is 15 cm by 8.0 cm, and the solution satisfies the statement of the problem.

EXERCISES 14.3

In Exercises 1 and 2, make the given changes in the indicated examples of this section and then solve the resulting equations.

1. In Example 2, change the + before the $7x^2$ to − and then solve the equation.

2. In Example 3, change the 2 to 6 and then solve the equation.

In Exercises 3–26, solve the given equations algebraically.

3. $x^4 - 10x^2 + 9 = 0$

4. $4R^4 + 15R^2 = 4$

5. $3x^{-2} - 7x^{-1} - 6 = 0$

6. $10x^{-2} + 3x^{-1} - 1 = 0$

7. $x^{-4} + 2x^{-2} = 24$

8. $x^{-1} - x^{-1/2} = 2$

9. $2x - 7\sqrt{x} + 5 = 0$

10. $4x + 3\sqrt{x} = 1$

11. $3\sqrt[3]{x} - 5\sqrt[6]{x} + 2 = 0$

12. $\sqrt{x} + 3\sqrt[4]{x} = 28$

13. $x^{2/3} - 2x^{1/3} - 15 = 0$

14. $x^3 + 2x^{3/2} - 80 = 0$

15. $8n^{1/2} - 20n^{1/4} = 12$

16. $4x^{4/3} + 9 = 13x^{2/3}$

17. $(x - 1) - \sqrt{x - 1} = 20$

18. $(C + 1)^{-2/3} + 5(C + 1)^{-1/3} - 6 = 0$

19. $(x^2 - 2x)^2 - 11(x^2 - 2x) + 24 = 0$

20. $(x^2 - 1)^2 + (x^2 - 1)^{-2} = 2$

21. $x - 3\sqrt{x - 2} = 6$ (Let $y = \sqrt{x - 2}$.)

22. $x^6 + 7x^3 - 8 = 0$

23. $\frac{1}{s^2 + 1} + \frac{2}{s^2 + 3} = 1$

24. $\left(x + \frac{2}{x}\right)^2 - 6x - \frac{12}{x} = -9$

25. $e^{2x} - e^x = 0$

26. $10^{2x} - 2(10^x) = 0$

In Exercises 27–32, solve the given equations algebraically and check the solutions graphically.

27. $x^4 - 20x^2 + 64 = 0$

28. $x^{-2} - x^{-1} - 42 = 0$

29. $x + 2 = 3\sqrt{x}$

30. $x^{2/3} - 4x^{1/3} = 12$

31. $(\log x)^2 - 3\log x + 2 = 0$

32. $2^x + 32(2^{-x}) = 12$

In Exercises 33–40, solve the given problems algebraically.

33. Solve for x: $\log(x^4 + 4) - \log(5x^2) = 0$.

34. If $f(x) = x^2 + 1$, find x if $f(x^2 - 3) = 5$.

35. The equivalent resistance R_T of two resistors R_1 and R_2 in parallel is given by $R_T^{-1} = R_1^{-1} + R_2^{-1}$. If $R_T = 1.00\ \Omega$ and $R_2 = \sqrt{R_1}$, find R_1 and R_2.

36. An equation used in the study of the dispersion of light is $\mu = A + B\lambda^{-2} + C\lambda^{-4}$. Solve for λ.

37. In the theory dealing with optical interferometers, the equation $\sqrt{F} = 2\sqrt{p}/(1 - p)$ is used. Solve for p if $F = 16$.

38. A special washer is made from a circular disc 3.50 cm in radius by removing a rectangular area of 12.0 cm^2 from the centre. If each corner of the rectangular area is 0.50 cm from the outer edge of the washer, what are the dimensions of the area that is removed?

39. A rectangular TV screen has an area of 2240 cm^2 and a diagonal of 68.6 cm. Find the dimensions of the screen.

40. The impedance Z in an alternating-current circuit is 2.00 Ω. If the resistance R is numerically equal to the square of the reactance X, find R and X. See Section 12.7.

Answer to Practice Exercise

1. $1/2, 1/2, -1/2, -1/2$

14.4 Equations with Radicals

Equations with radicals can usually be solved using the following method.

Procedure for Solving an Equation with Radicals

1. Isolate the radical (or one of the radicals if more than one radical is present) by rewriting the equation with the radical on one side and all other terms on the other side.
2. Raise both sides of the equation to the order of the radical (for instance, square both sides when the radical is a square root). **Be careful to raise the complete expression on each side, not just the terms separately.**
3. Repeat Steps 1 and 2 until the equation does not contain radicals.
4. Solve the resulting equation.
5. **Check all solutions in the original equation.** Identify and discard extraneous roots.

EXAMPLE 1 Solve by squaring both sides

Solve the equation $\sqrt{x-4} = 2$.

The radical is already isolated. By squaring both sides of the equation, we have

$$(\sqrt{x-4})^2 = 2^2$$
$$x - 4 = 4$$
$$x = 8$$

This solution checks when put into the original equation.

EXAMPLE 2 Solve by squaring both sides

Solve the equation $2\sqrt{3x-1} = 3x$.

The radical is already isolated. Squaring both sides of the equation gives us

$$(2\sqrt{3x-1})^2 = (3x)^2 \quad \text{don't forget to square the 2}$$
$$4(3x-1) = 9x^2$$
$$12x - 4 = 9x^2$$
$$9x^2 - 12x + 4 = 0$$
$$(3x-2)^2 = 0$$
$$x = \frac{2}{3} \quad \text{(double root)}$$

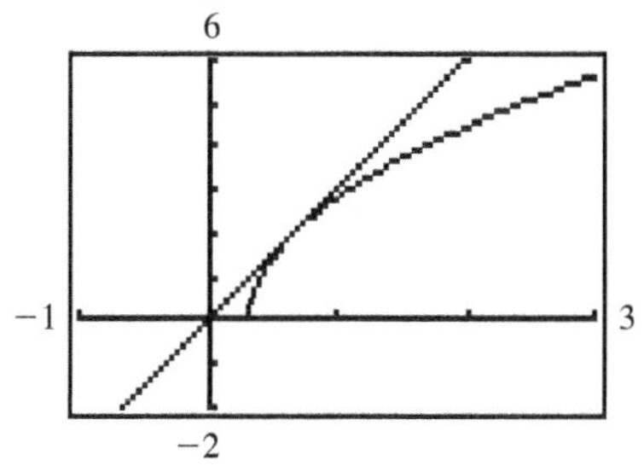

Fig. 14.15

Checking this solution in the original equation, we have

$$2\sqrt{3\left(\tfrac{2}{3}\right) - 1} = 3\left(\tfrac{2}{3}\right), \qquad 2\sqrt{2-1} = 2, \qquad 2 = 2$$

Therefore, the solution $x = \frac{2}{3}$ checks.

We can check this solution graphically by letting $y_1 = 2\sqrt{3x-1}$ and $y_2 = 3x$. The calculator display is shown in Fig. 14.15. The *intersect* feature shows that the only x-value that the curves have in common is $x = 0.6667$, which agrees with the solution of $x = 2/3$. This also means the line y_2 is tangent to the curve of y_1.

Practice Exercise

1. Solve for x: $\sqrt{2x+3} = x$

EXAMPLE 3 Solve by cubing both sides

Solve the equation $\sqrt[3]{x-8} = 2$.

The radical is already isolated. Cubing both sides of the equation, we have

$$x - 8 = 8$$
$$x = 16$$

Checking this solution in the original equation, we get

$$\sqrt[3]{16-8} = 2, \qquad 2 = 2$$

Therefore, the solution checks.

EXAMPLE 4 Solve by isolating the radical

Solve the equation $\sqrt{x-1} + 3 = x$.

We first isolate the radical by subtracting 3 from each side. This gives us

$$\sqrt{x-1} = x - 3$$

We now square both sides and proceed with the solution:

$$(\sqrt{x-1})^2 = (x-3)^2 \qquad \text{square the expression on each side, not just the terms separately}$$
$$x - 1 = x^2 - 6x + 9$$
$$x^2 - 7x + 10 = 0$$
$$(x-5)(x-2) = 0$$
$$x = 5 \quad \text{or} \quad x = 2$$

The solution $x = 5$ checks, but the solution $x = 2$ gives $4 = 2$. Thus, the solution is $x = 5$. The value $x = 2$ is an extraneous root.

Practice Exercise

2. Solve for x: $\sqrt{x+4} + 2 = x$

EXAMPLE 5 Solve by isolating one radical at a time

Solve the equation $\sqrt{x+1} + \sqrt{x-4} = 5$.

We start by isolating the first radical and then squaring both sides of the resulting equation:

$$\sqrt{x+1} = 5 - \sqrt{x-4}$$
$$(\sqrt{x+1})^2 = (5 - \sqrt{x-4})^2$$
$$x + 1 = 25 - 10\sqrt{x-4} + (\sqrt{x-4})^2 \qquad \text{square of binomial}$$
$$= 25 - 10\sqrt{x-4} + x - 4$$

Now, isolating the radical on one side of the equation and squaring again, we have

$$10\sqrt{x-4} = 20$$
$$\sqrt{x-4} = 2 \qquad \text{divide by 10}$$
$$x - 4 = 4 \qquad \text{square both sides}$$
$$x = 8$$

This solution checks.

COMMON ERROR

It should be emphasized that when squaring both sides of an equation, the complete expression is squared, and not just the terms separately.

$(5 - \sqrt{x-4})^2$ is **not** $25 + (x - 4)$. You are calculating the square of a binomial, so you must have three terms, including the middle term $-10\sqrt{x-4}$ (which still contains a radical).

EXAMPLE 6 Nested radical equation

Solve the equation $\sqrt{7 + \sqrt{x}} - 1 = \sqrt{x}$.

We first **isolate the outer radical** and then square both sides. We then isolate the remaining radical and square both sides again. The steps are shown below.

$$\sqrt{7 + \sqrt{x}} = \sqrt{x} + 1 \qquad \text{isolate the outer radical}$$
$$\left(\sqrt{7 + \sqrt{x}}\right)^2 = \left(\sqrt{x} + 1\right)^2 \qquad \text{square the expression on each side}$$
$$7 + \sqrt{x} = x + 2\sqrt{x} + 1$$
$$6 - x = \sqrt{x} \qquad \text{isolate the remaining radical}$$
$$(6 - x)^2 = \left(\sqrt{x}\right)^2 \qquad \text{square the expression on each side}$$
$$36 - 12x + x^2 = x$$
$$x^2 - 13x + 36 = 0$$
$$(x - 9)(x - 4) = 0$$
$$x = 9 \quad \text{or} \quad x = 4$$

When checking, $x = 4$ satisfies the original equation but $x = 9$ does not. Therefore, the only solution is $x = 4$ ($x = 9$ is an extraneous root).

EXAMPLE 7 System with a radical—holograph dimensions

Each cross-section of a holographic image is in the shape of a right triangle. The perimeter of the cross-section is 60 cm, and its area is 120 cm^2. Find the length of each of the three sides.

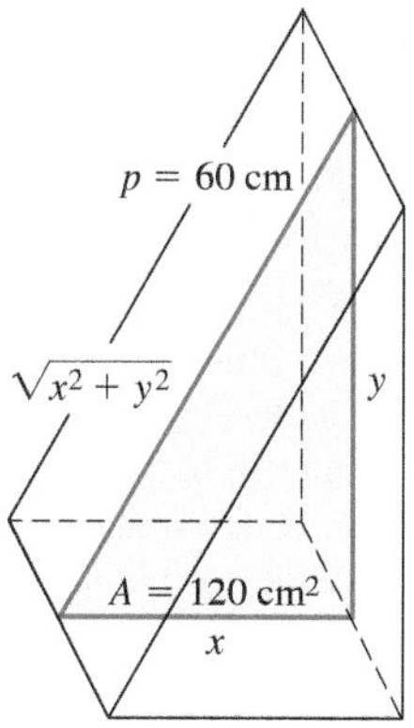

Fig. 14.16

If we let the two legs of the triangle be x and y, as shown in Fig. 14.16, from the formulas for the perimeter p and the area A of a triangle, we have

$$p = x + y + \sqrt{x^2 + y^2} \quad \text{and} \quad A = \tfrac{1}{2}xy$$

where the hypotenuse was found by use of the Pythagorean theorem. Using the information given in the statement of the problem, we arrive at the equations

$$x + y + \sqrt{x^2 + y^2} = 60 \quad \text{and} \quad xy = 240$$

Isolating the radical in the first equation and then squaring both sides, we have

$$\sqrt{x^2 + y^2} = 60 - x - y$$
$$x^2 + y^2 = 3600 - 120x - 120y + x^2 + 2xy + y^2$$
$$0 = 3600 - 120x - 120y + 2xy$$

Solving the second of the original equations for y, we have $y = 240/x$. Substituting, we have

$$0 = 3600 - 120x - 120\left(\frac{240}{x}\right) + 2x\left(\frac{240}{x}\right)$$
$$0 = 3600x - 120x^2 - 120(240) + 480x \qquad \text{multiply by } x$$
$$0 = 30x - x^2 - 240 + 4x \qquad \text{divide by 120}$$
$$x^2 - 34x + 240 = 0 \qquad \text{collect terms on left}$$
$$(x - 10)(x - 24) = 0$$
$$x = 10 \text{ cm} \quad \text{or} \quad x = 24 \text{ cm}$$

If $x = 10$ cm, then $y = 24$ cm, or if $x = 24$ cm, then $y = 10$ cm. Therefore, the legs of the holographic cross-section are 10 cm and 24 cm, and the hypotenuse is 26 cm. For these sides, $p = 60$ cm and $A = 120$ cm^2. We see that these values check with the statement of the problem.

■ Holography is a method of producing a three-dimensional image without the use of a lens. The theory of holography was developed in the late 1940s by the British engineer Dennis Gabor (1900–1979). After the invention of lasers, the first holographs were produced in the early 1960s.

EXERCISES 14.4

In Exercises 1–4, make the given changes in the indicated examples of this section, and then solve the resulting equations.

1. In Example 2, change the $3x$ on the right to 3.
2. In Example 3, change the 8 under the radical to 19.
3. In Example 4, change the 3 on the left to 7.
4. In Example 5, change the 4 under the second radical to 14.

In Exercises 5–34, solve the given equations. In Exercises 18 and 19, explain how the extraneous roots are introduced.

5. $\sqrt{x-8}=2$
6. $\sqrt{x+4}=3$
7. $\sqrt{15-2x}=x$
8. $2\sqrt{2P+5}=P$
9. $\sqrt{3x+2}=3x$
10. $\sqrt{5x-1}+3=x$
11. $2\sqrt{3-x}-x=5$
12. $x-3\sqrt{2x+1}=-5$
13. $\sqrt[3]{y-7}=2$
14. $\sqrt[4]{5-x}=2$
15. $5\sqrt{s}-6=s$
16. $3\sqrt{x}+2=2x$
17. $\sqrt{x^2-11}=5$
18. $t^2=3-\sqrt{2t^2-3}$
19. $\sqrt{x+4}+8=x$
20. $\sqrt{x^2-x-4}=x+2$
21. $\sqrt{5+\sqrt{x}}=\sqrt{x}-1$
22. $\sqrt{13+\sqrt{x}}=\sqrt{x}+1$
23. $3\sqrt{1-2t}+1=2t$
24. $1-2\sqrt{y+4}=y$
25. $2\sqrt{x+2}-\sqrt{3x+4}=1$
26. $\sqrt{x-1}+\sqrt{x+2}=3$
27. $\sqrt{5x+1}-1=3\sqrt{x}$
28. $\sqrt{x-7}=\sqrt{x}-7$
29. $\sqrt{6x-5}-\sqrt{x+4}=2$
30. $\sqrt{5x-4}-\sqrt{x}=2$
31. $\sqrt{x-9}=\dfrac{36}{\sqrt{x-9}}-\sqrt{x}$
32. $\sqrt[4]{x+10}=\sqrt{x-2}$
33. $\sqrt{x-2}=\sqrt[4]{x-2}+12$
34. $\sqrt{3x+\sqrt{3x+4}}=4$

In Exercises 35–38, solve the given equations algebraically and check the solutions graphically.

35. $\sqrt{3x+4}=x$
36. $\sqrt{x-2}+3=x$
37. $\sqrt{2x+1}+3\sqrt{x}=9$
38. $\sqrt{2x+1}-\sqrt{x+4}=1$

In Exercises 39–52, solve the given problems.

39. If $f(x)=\sqrt{x+3}$, find x if $f(x+6)=5$.
40. Solve $\sqrt{x}-\sqrt{2x}=2$ algebraically, and check the solution graphically.
41. Solve $\sqrt{x-1}+x=3$ algebraically. Then compare the solution with that of Example 4. Noting that the algebraic steps after isolating the radical are identical, why is the solution different?
42. The resonant frequency f in an electric circuit with an inductance L and a capacitance C is given by $f=\dfrac{1}{2\pi\sqrt{LC}}$. Solve for L.
43. A formula used in calculating the range r for radio communication is $R=\sqrt{2rh+h^2}$. Solve for h.
44. An equation used in analysing a certain type of concrete beam is $k=\sqrt{2np+(np)^2}-np$. Solve for p.
45. In the study of spur gears in contact, the equation $kC=\sqrt{R_1^2-R_2^2}+\sqrt{r_1^2-r_2^2}-A$ is used. Solve for r_1^2.
46. The speed s (in m/s) at which a tsunami wave moves is related to the depth d (in m) of the ocean according to $s=\sqrt{gd}$, where g is the acceleration of gravity (9.8 m/s^2). If a wave from the 2004 Indian Ocean tsunami was travelling at 195 m/s, estimate the depth of the ocean at that point.
47. If the value of a home increases from v_1 to v_2 over n years, the average annual rate of growth (as a decimal) is given by $r=\sqrt[n]{\dfrac{v_2}{v_1}}-1$. Suppose the value of a home increases on average by 3.6% per year over 10 years. If its value at the end of the 10-year period is \$325 000, find its value at the beginning of the period.
48. The smaller of two cubical boxes is centred on the larger box, and they are taped together with a wide adhesive that just goes around both boxes (see Fig. 14.17). If the edge of the larger box is 1.00 cm greater than that of the smaller box, what are the lengths of the edges of the boxes if 100.0 cm of tape is used?

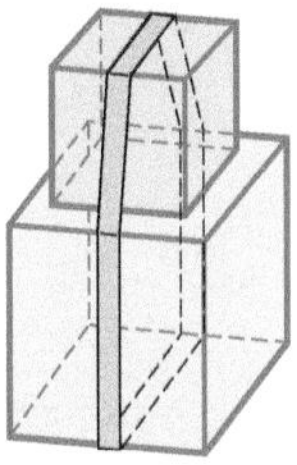

Fig. 14.17

49. A freighter is 5.2 km farther from a Coast Guard station on a straight coast than from the closest point A on the coast. If the station is 8.3 km from A, how far is it from the freighter?
50. The velocity v of an object that falls through a distance h is given by $v=\sqrt{2gh}$, where g is the acceleration due to gravity. Two objects are dropped from heights that differ by 10.0 m such that the sum of their velocities when they strike the ground is 20.0 m/s. Find the heights from which they are dropped if $g=9.80\ m/s^2$.
51. A point T on Thorah Island in Lake Simcoe, Ontario, is 3.8 km from Beaverton (B). A person in a motorboat travels straight from T to a point on Alsop's Beach, x km from B, and then travels x km farther along the beach away from B. (Assume that the coast is straight, which is nearly the case.) Find x if the person travelled a total of 5.4 km. See Fig. 14.18.

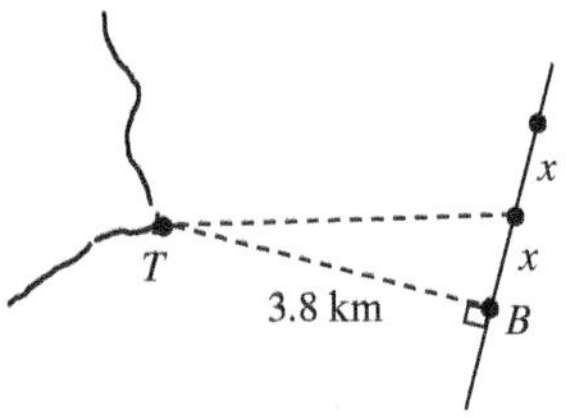

Fig. 14.18

52. The length of the roller belt in Fig. 14.19 is 28.0 m. Find x.

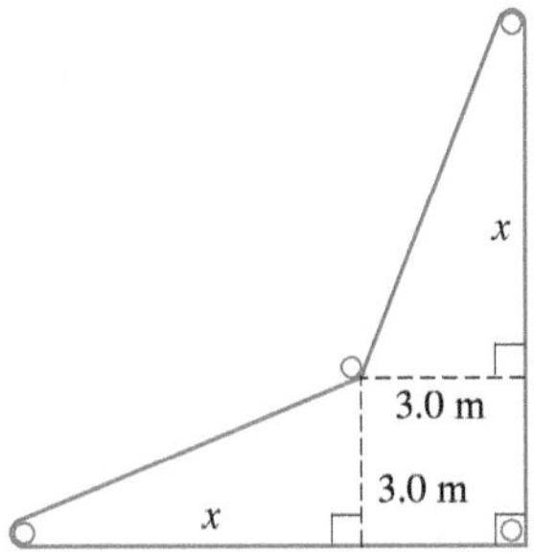

Fig. 14.19

Answers to Practice Exercises

1. 3 **2.** 5

CHAPTER 14 **REVIEW EXERCISES**

In Exercises 1–10, solve the given systems of equations by use of a graphing utility. Find all values to the nearest 0.01.

1. $x + 2y = 6$
$y = 4x^2$

2. $x + y = 3$
$x^2 + y^2 = 25$

3. $3x + 2y = 6$
$x^2 + 4y^2 = 4$

4. $x^2 - 3y = 0$
$3x - 2y = 6$

5. $y = x^2 + 1$
$4x^2 + 16y^2 = 29$

6. $\frac{x^2}{4} + y^2 = 1$
$x^2 - y^2 = 1$

7. $y = 11 - x^2$
$y = 2x^2 - 1$

8. $x^2y = 63$
$y = 25 - 2x^2$

9. $y = x^2 - 2x$
$y = 1 - e^{-x}$

10. $y = \ln x$
$y = \sin x$

In Exercises 11–20, solve each of the given systems of equations algebraically.

11. $y = 4x^2$
$y = 8x$

12. $x + y = 12$
$xy = 20$

13. $2R = L^2$
$R^2 + L^2 = 3$

14. $y = x^2$
$2x^2 - y^2 = 1$

15. $4u^2 + v = 3$
$2u + 3v = 1$

16. $x^2 + 7y^2 = 56$
$2x^2 - 8y^2 = 90$

17. $4x^2 - 7y^2 = 21$
$x^2 + 2y^2 = 99$

18. $s - t = 6$
$\sqrt{s} - \sqrt{t} = 1$

19. $4x^2 + 3xy = 4$
$x + 3y = 4$

20. $\frac{6}{x} + \frac{3}{y} = 4$
$\frac{36}{x^2} + \frac{36}{y^2} = 13$

In Exercises 21–38, solve the given equations.

21. $x^4 - 20x^2 + 64 = 0$

22. $t^6 - 26t^3 - 27 = 0$

23. $x^{3/2} - 9x^{3/4} + 8 = 0$

24. $x^{1/2} + 3x^{1/4} - 28 = 0$

25. $D^{-2} + 4D^{-1} - 21 = 0$

26. $2x - 3\sqrt{x} - 5 = 0$

27. $4(\ln x)^2 - \ln x^2 = 0$

28. $e^x + e^{-x} = 2$

29. $\frac{4}{r^2 + 1} + \frac{7}{2r^2 + 1} = 2$

30. $(x^2 + 5x)^2 - 5(x^2 + 5x) = 6$

31. $3\sqrt{2Z + 4} = 2Z$

32. $\sqrt[3]{x - 2} = 3$

33. $\sqrt{5x + 9} + 1 = x$

34. $2\sqrt{5x - 3} - 1 = 2x$

35. $\sqrt{x + 1} + \sqrt{x} = 2$

36. $\sqrt{3x^2 - 2} - \sqrt{x^2 + 7} = 1$

37. $\sqrt{n + 4} + 2\sqrt{n + 2} = 3$

38. $\sqrt{3x - 2} - \sqrt{x + 7} = 1$

In Exercises 39 and 40, find the value of the constant exactly. Explain your method. (In each, the expression on the right is called a ***continued radical.*** *Also, . . . means that the pattern continues indefinitely.) Hint: Square both sides.*

39. $x = \sqrt{2 + \sqrt{2 + \sqrt{2 + \cdots}}}$

40. $\tau = \sqrt{1 + \sqrt{1 + \sqrt{1 + \cdots}}}$

The constant τ is known as the *golden ratio.* It has many applications in mathematics, architecture, biology, music, and art.

In Exercises 41–46, solve the given equations algebraically and check the solutions graphically.

41. $x^3 - 2x^{3/2} - 48 = 0$

42. $(x + 1)^4 - 54 = 3(x + 1)^2$

43. $2\sqrt{3x + 1} - \sqrt{x - 1} = 6$

44. $3\sqrt{x} + \sqrt{x - 9} = 11$

45. $\sqrt[3]{x^3 - 7} = x - 1$

46. $\sqrt{x^2 + 7} + \sqrt[4]{x^2 + 7} = 6$

In Exercises 47–58, solve the given problems.

47. Solve for x and y: $x^2 - y^2 = 2a + 1$; $x - y = 1$.

48. Solve for x: $\log(\sqrt{x} + 38) - \log x = 1$.

49. Solve $\sqrt{\sqrt{x} - 1} = 2$ for x. Check graphically.

50. If $f(x) = \sqrt{8 - 2x}$ and $f(x + 1) = 2$, find x.

51. Use a graphing utility to solve the following system of three equations: $x^2 + y^2 = 13$, $y = x - 1$, $xy = 6$.

52. Algebraically solve the following system of three equations: $y = -x^2$, $y = x - 1$, $xy = 1$. Explain the results.

53. In the study of atomic structure, the equation $L = \frac{h}{2\pi}\sqrt{l(l + 1)}$ is used. Solve for l $(l > 0)$.

54. The frequency ω of a certain *RLC* circuit is given by

$$\omega = \frac{\sqrt{R^2 + 4(L/C)} + R}{2L}.$$ Solve for C.

55. In the theory dealing with a suspended cable, the equation $y = \sqrt{s^2 - m^2} - m$ is used. Solve for m.

56. The equation $V = e^2cr^{-2} - e^2Zr^{-1}$ is used in spectroscopy. Solve for r.

57. In an experiment, an object is allowed to fall, stops, and then falls for twice the initial time. The total distance the object falls is 392 cm. The equations relating the times t_1 and t_2 (in s) of fall are $490t_1^2 + 490t_2^2 = 392$ and $t_2 = 2t_1$. Find the times of fall.

58. If two objects collide and the kinetic energy remains constant, the collision is termed perfectly elastic. Under these conditions, if an object of mass m_1 and initial velocity u_1 strikes a second object (initially at rest) of mass m_2, such that the velocities after collision are v_1 and v_2, the following equations are found:

$$m_1u_1 = m_1v_1 + m_2v_2$$
$$\tfrac{1}{2}m_1u_1^2 = \tfrac{1}{2}m_1v_1^2 + \tfrac{1}{2}m_2v_2^2$$

Solve these equations for m_2 in terms of u_1, v_1, and m_1.

In Exercises 59–70, set up the appropriate equations and solve them. In Exercise 71, answer the given question.

59. A wrench is dropped by a worker at a construction site. Four seconds later the worker hears it hit the ground below. How high is the worker above the ground? (The velocity of sound is 331 m/s, and the distance the wrench falls as a function of time is $s = 4.9t^2$.)

60. The rectangular screen for a laptop computer has an area of 840 cm^2 and a perimeter of 119 cm. Find the dimensions of the screen.

61. The perimeter of a banner in the shape of a tall isosceles triangle (that is, with height longer than its base) is 72 dm, and its area is 240 dm^2. Graphically find the lengths of the sides of the banner.

62. A rectangular field is enclosed by fencing and a wall along one long side and half of an adjacent side. See Fig. 14.20. If the area of the field is 9000 m^2 and 240 m of fencing are used, what are the dimensions of the field?

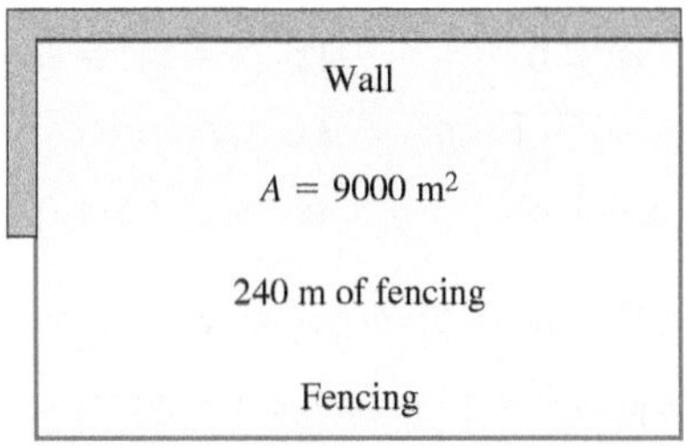

Fig. 14.20

63. A circuit on a computer chip is designed to be within the area shown in Fig. 14.21. If this part of the chip has an area of 9.0 mm^2 and a perimeter of 16 mm, find x and y.

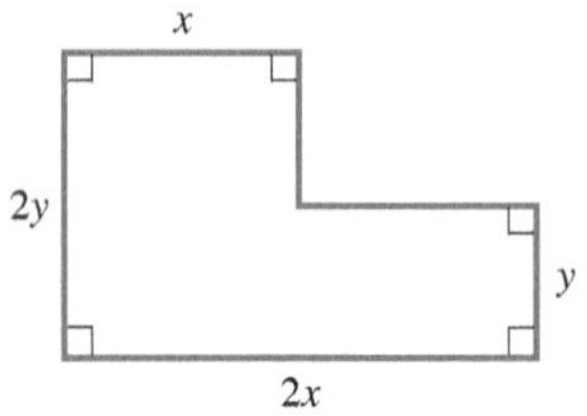

Fig. 14.21

64. For the plywood piece shown in Fig. 14.22, find x and y.

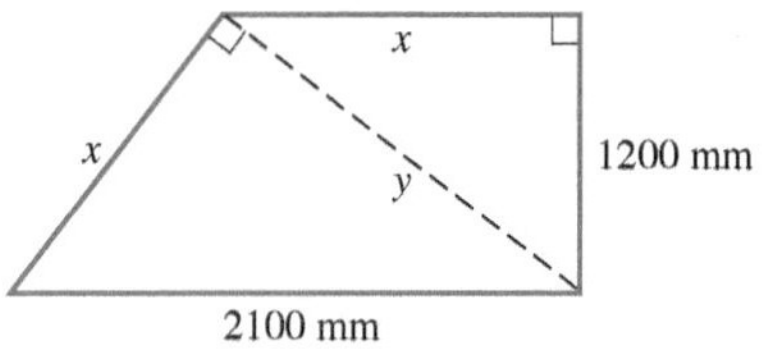

Fig. 14.22

65. The viewing window on a graphing calculator has an area of 1770 mm^2 and a diagonal of 62 mm. What are the length and width of the rectangle?

66. A trough is made from a piece of sheet metal 12.0 cm wide. The cross-section of the trough is shown in Fig. 14.23. Find x.

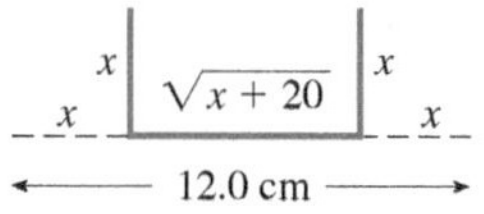

Fig. 14.23

67. The circular solar cell and square solar cell shown in Fig. 14.24 have a combined surface area of 40.0 cm^2. Find the radius of the circular cell and the side of the square cell.

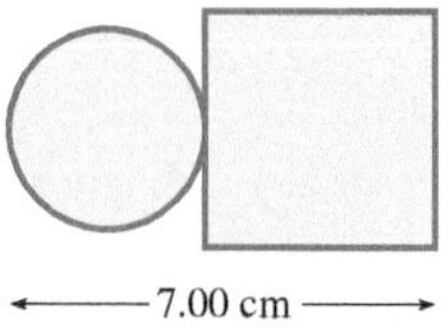

Fig. 14.24

68. A plastic band 19.0 cm long is bent into the shape of a triangle with sides $\sqrt{x-1}$, $\sqrt{5x-1}$, and 9. Find x.

69. A ferry travels from Digby, Nova Scotia, to Saint John, New Brunswick, and later it returns to Digby at an average speed that is 3.2 km/h slower. If Digby is 72 km from Saint John, and the total travel time is 5 hours and 42 minutes, find the average speed of the ferry in each direction.

70. Two trains are approaching the same crossing on tracks that are at right angles to each other. Each is travelling at 60.0 km/h. If one is 6.00 km from the crossing when the other is 3.00 km from it, how much later will they be 4.00 km apart (on a direct line and before reaching the crossing)?

71. Using a computer, an engineer designs a triangular support structure with sides (in m) of x, $\sqrt{x-1}$, and 4.00 m. If the perimeter is to be 9.00 m, the equation to be solved is $x + \sqrt{x-1} + 4.00 = 9.00$. Write one or two paragraphs explaining how to solve this equation by two different methods discussed in this chapter.

CHAPTER 14 PRACTICE TEST

1. Solve for x: $x^{1/2} - 2x^{1/4} = 3$.
2. Solve for x: $3\sqrt{x-2} - \sqrt{x+1} = 1$.
3. Solve for x: $x^4 - 17x^2 + 16 = 0$.
4. Solve for x and y algebraically:

$$x^2 - 2y = 5$$
$$2x + 6y = 1$$

5. Solve for x: $\sqrt[3]{2x+5} = 5$.
6. The velocity v of an object falling under the influence of gravity in terms of its initial velocity v_0, the acceleration due to gravity g, and the height h fallen is given by $v = \sqrt{v_0^2 + 2gh}$. Solve for h.
7. Solve for x and y graphically: $x^2 - y^2 = 4$

$$xy = 2$$

8. A rectangular desktop has a perimeter of 14.0 m and an area of 10.0 m^2. Find the length and the width of the desktop.

15. Equations of Higher Degree

▲ Bezier curves are frequently used in computer graphics, animation, modelling, computer-aided design (CAD), and other related fields. In Section 15.3, we show how to find the height of a Bezier curve roof design by solving a polynomial equation.

After completion of this chapter, the student should be able to:

- Use the remainder theorem to evaluate polynomials and to find remainders
- Use the factor theorem to identify factors and zeros of polynomials
- Perform synthetic division
- Use the fundamental theorem of algebra to determine the number of roots of an equation
- Solve equations given at least one root
- Determine the possible rational roots of an equation
- Determine the maximum possible number of positive and negative roots by using Descartes' rule of signs
- Solve polynomial equations of degree three and higher
- Solve application problems involving polynomial equations

The desire to eliminate errors in the computation of mathematical tables led the British mathematician Charles Babbage (1792–1871) to design a machine for solving polynomial equations. Nevertheless, despite years of government financing, his *difference engine* was not realized in its entirety during his lifetime. (The first working difference engine was built in 1991, faithful to one of Babbage's designs.) Babbage also designed an *analytical engine*, which he hoped would perform many kinds of calculations. Although never built, it did have the important features of a modern computer: input, storage, control unit, and output. Because of this design, Babbage is often considered the father of the computer.

We see that polynomials played an important role in the development of computers. Today, among many other things, computers are used to solve equations, including polynomial equations of higher degree, the topic of this chapter.

Algebraic methods for solving third- and fourth-degree polynomial equations were developed in the 1500s, but the search for an algebraic method for solving equations of degree five continued for a few hundred years. Although known prior to 1650, the *fundamental theorem of algebra* (which we state in this chapter) was proven in 1799 by Karl Friedrich Gauss, considered by many as the greatest mathematician of all time. The theorem guaranteed the existence of solutions to all polynomial equations, although it said nothing about the existence of exact formulas for solving them. In the early nineteenth century, it was finally established that no general algebraic formula exists for solving polynomials of degree higher than four.

In this chapter we study some methods for solving higher-degree polynomial equations. The solutions we find include all possible roots, including complex-number roots. Applications of higher-degree equations arise in a number of technical areas. Among them we find the calculation of resistances in an electric circuit, of the dimensions of a container or structure, and of various business production costs.

15.1 The Remainder and Factor Theorems; Synthetic Division

In this section, we present two theorems and a simplified method for algebraic division. These will help us in solving polynomial equations later in the chapter.

Any function of the form

$$f(x) = a_0x^n + a_1x^{n-1} + \cdots + a_n \quad \textbf{(15.1)}$$

where $a_0 \neq 0$ *and n is a positive integer or zero is called a* **polynomial function**. We will be considering only polynomials in which the coefficients $a_0, a_1, \ldots, a_n$ are real numbers.

If we divide a polynomial by $x - r$, we can rewrite it in the form

$$f(x) = (x - r)q(x) + R \quad \textbf{(15.2)}$$

where $q(x)$ is the quotient and R is the remainder.

$3x + 11$
$x - 2 \,\overline{)\, 3x^2 + 5x - 8}$
$3x^2 - 6x$
$11x - 8$
$11x - 22$
14

EXAMPLE 1 Division with remainder

Divide $f(x) = 3x^2 + 5x - 8$ by $x - 2$.

The division is shown at the left, and it shows that

$$3x^2 + 5x - 8 = (x - 2)(3x + 11) + 14$$

where, for this function $f(x)$ with $r = 2$, we identify $q(x)$ and R as

$$q(x) = 3x + 11 \qquad R = 14$$

If we now set $x = r$ in Eq. (15.2), we have $f(r) = q(r)(r - r) + R = q(r)(0) + R$, which leads us to the following theorem.

The Remainder Theorem

If a polynomial $f(x)$ is divided by $(x - r)$, then the remainder R is a constant given by $f(r)$. That is,

$$f(r) = R \quad \textbf{(15.3)}$$

In other words, without having to perform the division, we can find the remainder by evaluating the function at $x = r$. Similarly, without having to evaluate the function, we can find $f(r)$ by performing the division and noting the remainder.

EXAMPLE 2 Verifying the remainder theorem

In Example 1, $f(x) = 3x^2 + 5x - 8$, $R = 14$, and $r = 2$.

We find that

$$\begin{aligned} f(2) &= 3(2^2) + 5(2) - 8 \\ &= 12 + 10 - 8 \\ &= 14 \end{aligned}$$

Therefore, $f(2) = 14$ verifies that $f(r) = R$ for this example.

EXAMPLE 3 Using the remainder theorem

By using the remainder theorem, determine the remainder when $3x^3 - x^2 - 20x + 5$ is divided by $x + 4$.

We start by identifying the value of r. We write $x + 4 = x - (-4)$, which means that $r = -4$. We therefore evaluate the function $f(x) = 3x^3 - x^2 - 20x + 5$ for $x = -4$, or find $f(-4)$:

$$\begin{aligned} f(-4) &= 3(-4)^3 - (-4)^2 - 20(-4) + 5 \\ &= -192 - 16 + 80 + 5 \\ &= -123 \end{aligned}$$

The remainder is -123 when $3x^3 - x^2 - 20x + 5$ is divided by $x + 4$.

The remainder theorem allows us to establish an important connection between zeros of polynomials and factors of polynomials. We see in Eq. (15.2) that if the remainder $R = 0$, then $f(x) = (x - r)q(x)$, and this shows that $x - r$ is a factor of $f(x)$. Therefore, we have the following theorem.

The Factor Theorem

If $f(x)$ is a polynomial and $f(r) = 0$, then $x - r$ is a factor of $f(x)$.

Consequently, if $f(r) = 0$, then

$x = r$ is a **zero** of $f(x)$

$x - r$ is a **factor** of $f(x)$

$x = r$ is a **root** of the equation $f(x) = 0$

EXAMPLE 4 Using the factor theorem

Determine whether **(a)** $t + 1$ and **(b)** $t + 2$ are factors of $f(t) = t^3 + 2t^2 - 5t - 6$.

(a) We write $t + 1 = t - (-1)$ and identify $r = -1$ to use in the factor theorem. We have

$$f(-1) = (-1)^3 + 2(-1)^2 - 5(-1) - 6 = -1 + 2 + 5 - 6 = 0$$

Because $f(-1) = 0$, $t + 1$ is a factor of $f(t)$.

(b) However, $t + 2$ is not a factor of $f(t)$ because $f(-2)$ is not zero, as we now show:

$$f(-2) = (-2)^3 + 2(-2)^2 - 5(-2) - 6 = -8 + 8 + 10 - 6 = 4$$

Practice Exercise

1. Use the factor theorem to determine whether $x - 1/2$ is a factor of $f(x) = 2x^3 + 9x^2 - 11x + 3$.

SYNTHETIC DIVISION

In the sections that follow, we will find that division of a polynomial by the factor $x - r$ is also useful in solving polynomial equations. Therefore, we now develop a simplified form of long division, known as **synthetic division**. It allows us to easily find the coefficients of the quotient and the remainder. If the degree of the equation is high, it is easier to use synthetic division than to calculate $f(r)$. The method for synthetic division is developed in the following example.

EXAMPLE 5 Developing synthetic division

Divide $x^4 + 4x^3 - x^2 - 16x - 14$ by $x - 2$.

We first perform this division in the usual manner:

$$\begin{array}{r|l}
 & x^3 + 6x^2 + 11x + 6 \\
\hline
x - 2 & x^4 + 4x^3 - x^2 - 16x - 14 \\
 & \underline{x^4 - 2x^3} \\
 & \quad 6x^3 - x^2 \\
 & \quad \underline{6x^3 - 12x^2} \\
 & \qquad 11x^2 - 16x \\
 & \qquad \underline{11x^2 - 22x} \\
 & \qquad\qquad 6x - 14 \\
 & \qquad\qquad \underline{6x - 12} \\
 & \qquad\qquad\qquad -2
\end{array}$$

$$\begin{array}{r|rrrrr}
 & 1 & 6 & 11 & 6 & \\
\hline
-2 & 1 & 4 & -1 & -16 & -14 \\
 & & -2 & & & \\
 & & 6 & & & \\
 & & & -12 & & \\
 & & & 11 & & \\
 & & & & -22 & \\
 & & & & 6 & \\
 & & & & & -12 \\
 & & & & & -2
\end{array}$$

In doing the division, notice that we repeat many terms and that the only important numbers are the coefficients. This means there is no need to write in the powers of x. To the left of the division, we write it without x's and without identical terms.

$$\begin{array}{r|rrrrr}
-2 & 1 & 4 & -1 & -16 & -14 \\
 & & -2 & -12 & -22 & -12 \\
\hline
 & & 6 & 11 & 6 & -2
\end{array}$$

All numbers below the dividend may be written in two lines. Then all coefficients of the quotient, except the first, appear in the bottom line. Therefore, the line above the dividend is omitted, and we have the table at the left.

$$\begin{array}{rrrrr|l}
1 & 4 & -1 & -16 & -14 & \underline{2} \\
 & -2 & -12 & -22 & -12 & \\
\hline
1 & 6 & 11 & 6 & -2 &
\end{array}$$

Now, write the first coefficient (in this case, 1) in the bottom line. Also, change the -2 to 2, which is the actual value of r. Then in the table at the left, write the 2 on the right. *In this table, the* 1, 6, 11, *and* 6 *are the coefficients of the* x^3, x^2, x, *and constant term of the quotient. The* -2 *is the remainder.*

$$\begin{array}{rrrrr|l}
1 & 4 & -1 & -16 & -14 & \underline{2} \\
 & 2 & 12 & 22 & 12 & \\
\hline
1 & 6 & 11 & 6 & -2 &
\end{array}$$

Finally, it is easier to use addition rather than subtraction in the process, so we change the signs of the numbers in the middle row. Remember that originally the bottom line was found by subtraction. Therefore, we have the last table on the left.

In the last table at the left, we have 1 (of the bottom row) $\times$ 2 ($= r$) $= 2$, the first number of the middle row. In the second column, $4 + 2 = 6$, the second number in the bottom row. Then, 6×2 ($= r$) $= 12$, the second number of the second row; $-1 + 12 = 11$; $11 \times 2 = 22$; $22 + (-16) = 6$; $6 \times 2 = 12$; and $12 + (-14) = -2$.

We read the bottom line of the last table, the one we use in *synthetic division*, as

$$1x^3 + 6x^2 + 11x + 6 \text{ with a remainder of } -2$$

The method of synthetic division shown in the last table is outlined below, accompanied by an example.

Procedure for Synthetic Division of the Polynomial $f(x)$ by $x - r$	**EXAMPLE 6**
	Divide $x^5 + 2x^4 - 4x^2 + 3x - 4$ by $x + 3$.
1. Write the coefficients of $f(x)$ in descending order, and with zeros inserted for missing powers. Identify the value of r and place it to the right of the coefficients.	Write $x + 3$ as $x - (-3)$, so $r = -3$. coefficients → 1 2 0 −4 3 −4 $\lfloor\underline{-3}$ ← r ↑ (under 0) x^3 term missing

2. Carry down the left coefficient, multiply it by r, and place this product under the second coefficient of the top line.	$\begin{array}{rrrrrr|l} 1 & 2 & 0 & -4 & 3 & -4 & \underline{-3} \\ & -3 & & & & & \\ \hline 1 & & & & & & \end{array}$ $1 \times -3 = -3$
3. Add the two numbers in the second column and place the result below. Multiply this sum by r and place the product under the third coefficient of the top line.	add $\begin{array}{rrrrrr|l} 1 & 2 & 0 & -4 & 3 & -4 & \underline{-3} \\ & -3 & 3 & & & & \\ \hline 1 & -1 & & & & & \end{array}$ $-1 \times -3 = 3$
4. Continue the process until the bottom row has as many numbers as the top row.	$\begin{array}{rrrrrr|l} 1 & 2 & 0 & -4 & 3 & -4 & \underline{-3} \\ & -3 & 3 & -9 & 39 & -126 & \\ \hline 1 & -1 & 3 & -13 & 42 & -130 & \end{array}$
5. Read the coefficients of $x^{n-1}, \ldots, x$, the constant term and the remainder from the last line of the table.	$1x^4 - 1x^3 + 3x^2 - 13x + 42$ is the quotient; -130 is the remainder.

Practice Exercise

2. Use synthetic division to divide $3x^3 - 5x + 6$ by $x + 2$.

EXAMPLE 7 Checking a factor with synthetic division

By synthetic division, determine whether or not $t + 4$ is a factor of $t^4 + 2t^3 - 15t^2 - 32t - 16$.

$$\begin{array}{rrrrr|l} 1 & 2 & -15 & -32 & -16 & \underline{-4} \\ & -4 & 8 & 28 & 16 & \\ \hline 1 & -2 & -7 & -4 & 0 & \end{array}$$

Since the remainder is zero, $t + 4$ is a factor. We may also conclude that

$$f(t) = (t + 4)(t^3 - 2t^2 - 7t - 4)$$

Practice Exercise

3. Use synthetic division to determine whether $x + 2$ is a factor of $2x^3 + x^2 - 12x - 8$.

EXAMPLE 8 Checking a rational factor

By using synthetic division, determine whether $2x - 3$ is a factor of $2x^3 - 3x^2 + 8x - 12$.

We first note that the coefficient of x in the divisor is not 1. Thus we write $2x - 3 = 2\left(x - \frac{3}{2}\right)$, and identify $r = \frac{3}{2}$ from the factorization. We have

$$\begin{array}{rrrr|l} 2 & -3 & 8 & -12 & \underline{\tfrac{3}{2}} \\ & 3 & 0 & 12 & \\ \hline 2 & 0 & 8 & 0 & \end{array}$$

Since the remainder is zero, $x - \frac{3}{2}$ is a factor. The result $2x^2 + 8$ can now be divided by 2 (the coefficient of x in the divisor) to complete the division. We obtain the quotient $x^2 + 4$, with remainder 0. Thus, $2\left(x - \frac{3}{2}\right) = 2x - 3$ is indeed a factor of the function, and we have that

$$2x^3 - 3x^2 + 8x - 12 = (2x - 3)(x^2 + 4)$$

LEARNING TIP

The procedure for synthetic division requires that the divisor be of the form $x - r$. If the coefficient of x in the divisor is not 1, the coefficient needs to be factored out and the value of r identified from the factorization. The resulting quotient is then divided by the value of the factored coefficient to complete the division.

For example, if the divisor is $3x + 1$, we rewrite it as $3\left(x - \left(-\frac{1}{3}\right)\right)$, and divide by $\left(x - \left(-\frac{1}{3}\right)\right)$. The result is then divided by 3 to complete the division.

EXAMPLE 9 Checking a zero

Determine whether or not -12.5 is a zero of the function $f(x) = 6x^3 + 61x^2 - 171x + 100$.

We use synthetic division to determine whether $x - (-12.5)$ is a factor of $f(x)$. We have

$$\begin{array}{rrrr|l} 6 & 61 & -171 & 100 & \underline{-12.5} \\ & -75 & 175 & -50 & \\ \hline 6 & -14 & 4 & 50 & \end{array}$$

Since the remainder is not zero, $x - (-12.5)$ is not a factor of $f(x)$, and therefore -12.5 is not a zero of $f(x)$.

EXERCISES 15.1

In Exercises 1–4, make the given changes in the indicated examples of this section, and then perform the indicated operations.

1. In Example 3, change the $x + 4$ to $x + 3$ and then find the remainder.
2. In Example 4(a), change the $t + 1$ to $t - 1$ and then determine if $t - 1$ is a factor.
3. In Example 6, change the $x + 3$ to $x + 2$ and then perform the synthetic division.
4. In Example 8, change the $2x - 3$ to $2x + 3$ and then determine whether $2x + 3$ is a factor.

In Exercises 5–10, find the remainder by long division.

5. $(x^3 + 2x + 3) \div (x + 1)$
6. $(x^4 - 4x^3 - x^2 + x - 100) \div (x + 3)$
7. $(2x^5 - x^2 + 8x + 44) \div (x + 2)$
8. $(4s^3 - 9s^2 - 24s - 17) \div (s - 5)$
9. $(2x^4 - 3x^3 - 2x^2 - 15x - 16) \div (2x - 3)$
10. $(2x^4 - 10x^2 + 30x - 60) \div (x + 4)$

In Exercises 11–16, find the remainder using the remainder theorem. Do not use synthetic division.

11. $(R^4 + R^3 - 9R^2 + 3) \div (R + 4)$
12. $(4x^4 - x^2 + 5x - 7) \div (x - 3)$
13. $(2x^4 - 7x^3 - x^2 + 8) \div (x - 3)$
14. $(3n^4 - 13n^2 + 10n - 10) \div (n + 4)$
15. $(x^5 - 3x^3 + 5x^2 - 10x + 6) \div (x - 2)$
16. $(3x^4 - 12x^3 - 60x + 4) \div (x - 0.5)$

In Exercises 17–22, use the factor theorem to determine whether or not the second expression is a factor of the first expression. Do not use synthetic division.

17. $4x^3 + x^2 - 16x - 4,\ x - 2$
18. $3x^3 + 14x^2 + 7x - 4,\ x + 4$
19. $3V^4 - 7V^3 + V + 8,\ V - 2$
20. $x^5 - 2x^4 + 3x^3 - 6x^2 - 4x + 8,\ x - 2$
21. $x^{61} - 1,\ x + 1$
22. $x^7 - 128^{-1},\ x + 2^{-1}$

In Exercises 23–32, perform the indicated divisions by synthetic division.

23. $(x^3 + 2x^2 - x - 2) \div (x - 1)$
24. $(x^3 - 3x^2 - x + 2) \div (x - 2)$
25. $(x^3 + 2x^2 - 3x + 4) \div (x + 1)$
26. $(2x^3 - 4x^2 + x - 1) \div (x + 2)$
27. $(p^6 - 6p^3 - 2p^2 - 6) \div (p - 2)$
28. $(x^5 + 4x^4 - 8) \div (x + 1)$
29. $(x^7 - 128) \div (x - 2)$
30. $(20x^4 + 11x^3 - 89x^2 + 60x - 77) \div (x + 2.75)$
31. $(2x^4 + x^3 + 3x^2 - 1) \div (2x - 1)$
32. $(6t^4 + 5t^3 - 10t + 4) \div (3t - 2)$

In Exercises 33–40, use the factor theorem and synthetic division to determine whether or not the second expression is a factor of the first.

33. $2x^5 - x^3 + 3x^2 - 4;\quad x + 1$
34. $t^5 - 3t^4 - t^2 - 6;\quad t - 3$
35. $4x^3 - 6x^2 + 2x - 2;\quad x - \frac{1}{2}$
36. $3x^3 - 5x^2 + x + 1;\quad x + \frac{1}{3}$
37. $2Z^4 - Z^3 - 4Z^2 + 1;\quad 2Z - 1$
38. $6x^4 + 5x^3 - x^2 + 6x - 2;\quad 3x - 1$
39. $4x^4 + 2x^3 - 8x^2 + 3x + 12;\quad 2x + 3$
40. $3x^4 - 2x^3 + x^2 + 15x + 4;\quad 3x + 4$

In Exercises 41–44, use synthetic division to determine whether or not the given numbers are zeros of the given functions.

41. $x^4 - 5x^3 - 15x^2 + 5x + 14;\quad 7$
42. $r^4 + 5r^3 - 18r - 8;\quad -4$
43. $85x^3 + 348x^2 - 263x + 120;\quad -4.8$
44. $2x^3 + 13x^2 + 10x - 4;\quad \frac{1}{2}$

In Exercises 45–59, solve the given problems.

45. If $f(x) = 2x^3 + 3x^2 - 19x - 4$, and $f(x) = (x + 4)g(x)$, find $g(x)$.
46. Using synthetic division, divide $ax^2 + bx + c$ by $x + 1$.
47. By division, show that $2x - 1$ is a factor of $f(x) = 4x^3 + 8x^2 - x - 2$. May we therefore conclude that $f(1) = 0$? Explain.

48. By division, show that $x^2 + 2$ is a factor of $f(x) = 3x^3 - x^2 + 6x - 2$. May we therefore conclude that $f(-2) = 0$? Explain.

49. For what value of k is $x - 2$ a factor of $f(x) = 2x^3 + kx^2 - x + 14$?

50. For what value of k is $x + 1$ a factor of $f(x) = 3x^4 + 3x^3 + 2x^2 + kx - 4$?

51. Use synthetic division: $(x^3 - 3x^2 + x - 3) \div (x + j)$.

52. Use synthetic division: $(2x^3 - 7x^2 + 10x - 6) \div [x - (1 + j)]$.

53. If $f(x) = -g(x)$, do the functions have the same zeros? Explain.

54. Do the functions $f(x)$ and $f(-x)$ have the same zeros? Explain.

55. In finding the electric current in a certain circuit, it is necessary to factor the denominator of $\frac{2s}{s^3 + 5s^2 + 4s + 20}$. Is (a) $(s - 2)$ or (b) $(s + 5)$ a factor?

56. In the theory of the motion of a sphere moving through a fluid, the function $f(r) = 4r^3 - 3ar^2 - a^3$ is used. Is (a) $r = a$ or (b) $r = 2a$ a zero of $f(r)$?

57. In finding the volume V (in cm^3) of a certain gas in equilibrium with a liquid, it is necessary to solve the equation $V^3 - 6V^2 + 12V = 8$. Use synthetic division to determine if $V = 2$ cm^3.

58. An architect is designing a window in the shape of a segment of a circle. An approximate formula for the area is $A = \frac{h^3}{2w} + \frac{2wh}{3}$, where A is the area, w is the width, and h is the height of the segment. If the width is 1.500 m and the area is 0.5417 m^2, use synthetic division to show that $h = 0.500$ m.

59. The length of a rectangular box is 3 cm longer than its width. If the volume as a function of the width is $f(w) = 2w^3 + 5w^2 - 3w$, find the height of the box.

Answers to Practice Exercises

1. Yes $(R = 0)$ 2. Quotient: $3x^2 - 6x + 7$, $R = -8$
3. No $(R = 4)$

15.2 The Roots of an Equation

In this section, we present certain theorems that are useful in determining the number of roots in the equation $f(x) = 0$ and the nature of some of these roots. In dealing with polynomial equations of higher degree, it is helpful to have as much of this kind of information as we can find before actually solving for all of the roots.

The first of these theorems is so important that it is called the **fundamental theorem of algebra**. As we will see, the other two theorems are a direct consequence of the fundamental theorem.

■ The fundamental theorem of algebra was first proven in 1799 by the German mathematician Karl Gauss (1777–1855) for his doctoral thesis. During his lifetime, Gauss gave three other different proofs of the theorem. See the chapter introduction.

The Fundamental Theorem of Algebra and Related Theorems

1. **(The fundamental theorem of algebra)** Every polynomial equation has at least one real or complex root.
2. A polynomial of degree n can be factored into n linear factors.
3. A polynomial equation of degree n has exactly n roots.

The proof of the fundamental theorem is of an advanced nature, and therefore we accept its validity at this time. However, using the fundamental theorem, we can show the validity of the other two statements.

Let us assume that we have a polynomial equation $f(x) = 0$ and that we are looking for its roots. By the fundamental theorem, we know that it has at least one root. Assuming that we can find this root by some means (the factor theorem, for example), we call this root r_1. Thus,

$$f(x) = (x - r_1)f_1(x)$$

where $f_1(x)$ is the polynomial quotient found by dividing $f(x)$ by $(x - r_1)$. However, since the fundamental theorem states that any polynomial equation has at least one root, this must apply to $f_1(x) = 0$ as well. Let us assume that $f_1(x) = 0$ has the root r_2. Therefore, this means that $f(x) = (x - r_1)(x - r_2)f_2(x)$. Continuing this process until one of the quotients is a constant a, we have

$$f(x) = a(x - r_1)(x - r_2)\cdots(x - r_n)$$

Note that one linear factor appears each time a root is found and that the degree of the quotient is one less each time. Therefore, if $f(x)$ is of degree n, there are n linear factors, with one root associated with each of them.

EXAMPLE 1 Illustrating the fundamental theorem

The equation $f(x) = 2x^4 - 3x^3 - 12x^2 + 7x + 6 = 0$ has roots 3, −2, 1, and $-\frac{1}{2}$ (we will see how to find these roots in the next section). Verify that each of the theorems above hold for this polynomial equation.

1. We are given four roots, so the polynomial has at least one real root.
2. If $x = 3$ is a root, then $x - 3$ is a factor of $f(x)$, and similarly for all the given roots. Dividing in sequence by each of the factors, we have

$$2x^4 - 3x^3 - 12x^2 + 7x + 6 = (x-3)(2x^3 + 3x^2 - 3x - 2)$$
$$2x^3 + 3x^2 - 3x - 2 = (x+2)(2x^2 - x - 1)$$
$$2x^2 - x - 1 = (x-1)(2x+1)$$
$$2x + 1 = 2\left(x + \tfrac{1}{2}\right)$$

Therefore,

$$2x^4 - 3x^3 - 12x^2 + 7x + 6 = 2(x-3)(x+2)(x-1)\left(x+\tfrac{1}{2}\right)$$

The polynomial $f(x)$ of degree four has been factored into four linear factors.

3. We cannot factor $f(x)$ any further, so $f(x)$ has exactly four roots.

LEARNING TIP

If the coefficients of the equation $f(x) = 0$ are real and $a + bj (b \neq 0)$ is a complex root, then its conjugate, $a - bj$, is also a root.

It is not necessary for each root of an equation to be different from the other roots. For example, the equation $(x - 1)^2 = 0$ has two roots, both of which are 1. Such roots are referred to as *multiple* (or *repeated*) *roots.*

When we solve the equation $x^2 + 1 = 0$, the roots are j and $-j$. In fact, for any quadratic equation (with real coefficients) that has a root of the form $a + bj$ $(b \neq 0)$, there is also a root of the form $a - bj$.

The result is also true for polynomial equations of any degree.

EXAMPLE 2 Illustrating complex roots

Find the five roots of the equation $f(x) = (x - 1)^3(x^2 + x + 1) = 0$.

The factor $(x - 1)^3$ shows that there is a triple root of 1, and there is a total of five roots, since the highest-power term would be x^5 if we were to multiply out the function. To find the other two roots, we use the quadratic formula on the *factor* $(x^2 + x + 1)$. This is permissible, since we are finding the values of x for

$$x^2 + x + 1 = 0$$

For this, we have

$$x = \frac{-1 \pm \sqrt{1-4}}{2}$$

Thus,

$$x = \frac{-1 + j\sqrt{3}}{2} \quad \text{and} \quad x = \frac{-1 - j\sqrt{3}}{2}$$

Therefore, the roots of $f(x) = 0$ are 1, 1, 1, $\frac{-1 + j\sqrt{3}}{2}$, and $\frac{-1 - j\sqrt{3}}{2}$. As shown in Fig. 15.1, the graph of $f(x)$ intersects the x-axis at $x = 1$, so that $x = 1$ is verified as a root of $f(x) = 0$. However, it is not possible to tell from the graph that this is a triple root or that there are complex roots.

Fig. 15.1

LEARNING TIP

From Example 2, we can see that *whenever enough roots are known so that the remaining factor is quadratic, it is possible to find the remaining roots from the quadratic formula.* This is true for finding real or complex roots.

3	10	−16	−32	$\lfloor -\frac{4}{3}$
	−4	−8	32	
3	6	−24	0	

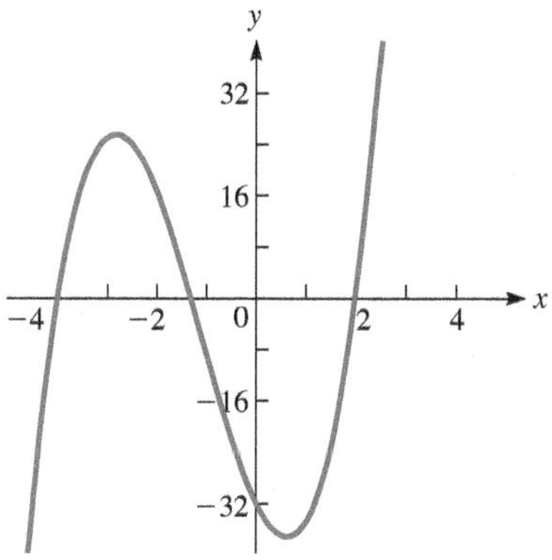

Fig. 15.2

EXAMPLE 3 Solving an equation given one root

Solve the equation $3x^3 + 10x^2 - 16x - 32 = 0$; $-\frac{4}{3}$ is a root.

Using synthetic division and the given root, we have the table shown at the left. From this, we see that

$$3x^3 + 10x^2 - 16x - 32 = \left(x + \tfrac{4}{3}\right)(3x^2 + 6x - 24)$$

The second factor can be factored as

$$3x^2 + 6x - 24 = 3(x^2 + 2x - 8) = 3(x + 4)(x - 2)$$

Therefore, we have

$$3x^3 + 10x^2 - 16x - 32 = 3\left(x + \tfrac{4}{3}\right)(x + 4)(x - 2)$$

This means the roots are $-\frac{4}{3}$, -4, and 2. The three real roots are the three intersections of $f(x)$ with the x-axis, as shown in Fig. 15.2.

EXAMPLE 4 Solving an equation given two roots

1	3	−4	−10	−4	$\lfloor -1$
	−1	−2	6	4	
1	2	−6	−4	0	

1	2	−6	−4	$\lfloor 2$
	2	8	4	
1	4	2	0	

Solve $x^4 + 3x^3 - 4x^2 - 10x - 4 = 0$; -1 and 2 are roots.

Using synthetic division and the root -1, the first table at the left shows that

$$x^4 + 3x^3 - 4x^2 - 10x - 4 = (x + 1)(x^3 + 2x^2 - 6x - 4)$$

We now know that $x - 2$ must be a factor of $x^3 + 2x^2 - 6x - 4$, since it is a factor of the original function. Again, using synthetic division and this time the root 2, we have the second table at the left. Thus,

$$x^4 + 3x^3 - 4x^2 - 10x - 4 = (x + 1)(x - 2)(x^2 + 4x + 2)$$

Since the original equation can now be written as

$$(x + 1)(x - 2)(x^2 + 4x + 2) = 0$$

the remaining two roots are found by solving

$$x^2 + 4x + 2 = 0$$

by the quadratic formula. This gives us

$$x = \frac{-4 \pm \sqrt{16 - 8}}{2} = \frac{-4 \pm 2\sqrt{2}}{2} = -2 \pm \sqrt{2}$$

Therefore, the roots are -1, 2, $-2 + \sqrt{2}$, and $-2 - \sqrt{2}$.

Practice Exercise

1. Solve $x^4 - x^3 - 2x^2 - 4x - 24 = 0$, given that -2 and 3 are roots.

EXAMPLE 5 Solving an equation given a double root

3	−26	63	−36	−20	$\lfloor 2$
	6	−40	46	20	
3	−20	23	10	0	

3	−20	23	10	$\lfloor 2$
	6	−28	−10	
3	−14	−5	0	

Solve the equation $3x^4 - 26x^3 + 63x^2 - 36x - 20 = 0$, given that 2 is a double root.

Using synthetic division, we have the first table at the left. It tells us that

$$3x^4 - 26x^3 + 63x^2 - 36x - 20 = (x - 2)(3x^3 - 20x^2 + 23x + 10)$$

Also, since 2 is a double root, it must be a root of $3x^3 - 20x^2 + 23x + 10 = 0$. Using synthetic division again, we have the second table at the left. This second quotient $3x^2 - 14x - 5$ factors into $(3x + 1)(x - 5)$. The roots are 2, 2, $-\frac{1}{3}$, and 5.

Since the quotient of the first division is the dividend for the second division, both divisions can be done without rewriting the first quotient as follows:

	3	−26	63	−36	−20	$\lfloor 2$
first division →		6	−40	46	20	
	3	−20	23	10	0	$\lfloor 2$
second division →		6	−28	−10		
	3	−14	−5	0		

EXAMPLE 6 Solving an equation given a complex root

Solve the equation $2x^4 - 5x^3 + 11x^2 - 3x - 5 = 0$, given that $1 + 2j$ is a root.

Since $1 + 2j$ is a root, we know that $1 - 2j$ is also a root. Using synthetic division twice, we can then reduce the remaining factor to a quadratic function.

2	−5	11	−3	−5	$\lfloor 1 + 2j$
	$2 + 4j$	$-11 - 2j$	$4 - 2j$	5	
2	$-3 + 4j$	$-2j$	$1 - 2j$	0	$\lfloor 1 - 2j$
	$2 - 4j$	$-1 + 2j$	$-1 + 2j$		
2	−1	−1	0		

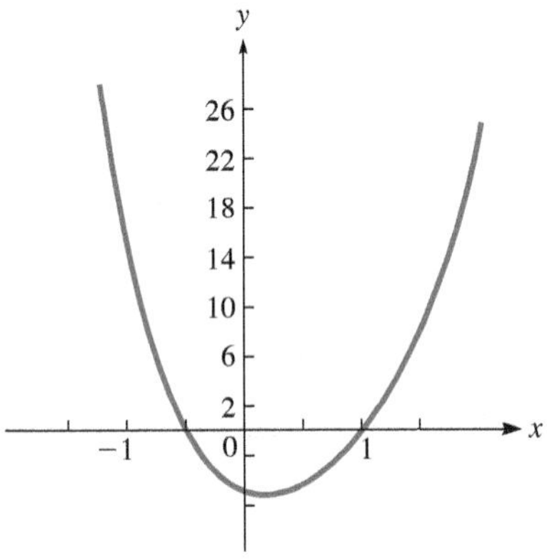

Fig. 15.3

The quadratic factor $2x^2 - x - 1$ factors into $(2x + 1)(x - 1)$. Therefore, the roots of the equation are $1 + 2j$, $1 - 2j$, 1, and $-\frac{1}{2}$. As we can see, the graph in Fig. 15.3 shows only the real roots.

The following table summarizes the relationship between the types of roots of a polynomial function of degree n and the features of its graph.

Types of Roots of a Polynomial of Degree n	Features of the Graph	EXAMPLE 7
All real and different	Crosses the x-axis and changes sign n times	$y = f(x)$ $= x^4 - 5x^3 + 5x^2 + 5x - 6$ $= (x + 1)(x - 1)(x - 2)(x - 3)$ Here $n = 4$, there are four real and different roots, so the graph crosses the x-axis and changes sign four times (at $x = -1$, $x = 1$, $x = 2$, and $x = 3$). 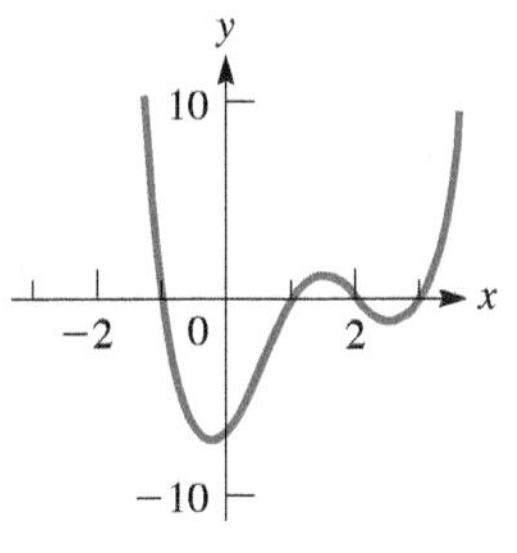 Fig. 15.4(a)
k pairs of non-real complex roots, degree n is odd	Crosses the x-axis and changes sign $n - 2k$ times (at least once since n is odd)	$y = f(x)$ $= x^3 - 2x^2 + 3$ $= (x + 1)(x^2 - 3x + 3)$ Here $n = 3$, there is one pair of complex roots ($k = 1$), so the graph crosses the x-axis and changes sign $n - 2k = 3 - 2(1) = 1$ time (at $x = -1$). 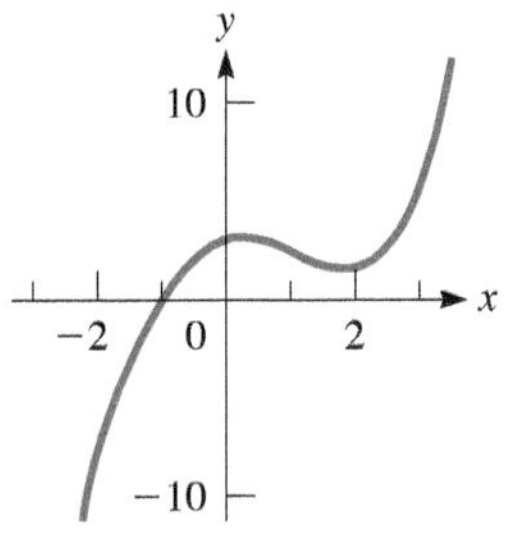 Fig. 15.4(b)

k pairs of non-real complex roots, degree n is even	Crosses the x-axis and changes sign $n - 2k$ times (possibly none since n is even)	$y = f(x)$ $= x^4 + 4$ $= (x^2 - 2x + 2)(x^2 + 2x + 2)$ Here $n = 4$, there are two pairs of complex roots ($k = 2$), so the graph does not cross the x-axis ($n - 2k = 4 - 2(2) = 0$). 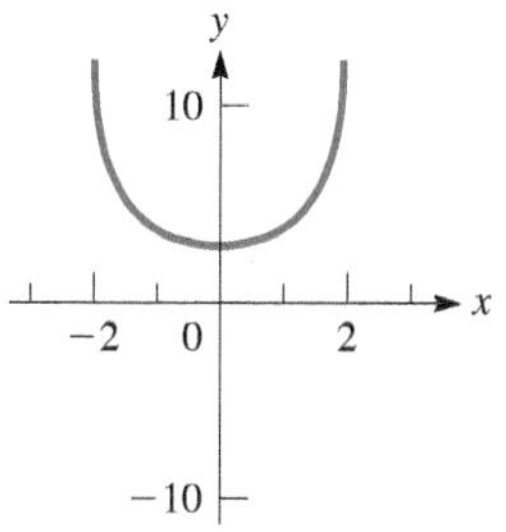 Fig. 15.4(c)
Multiple roots	Crosses the x-axis once if the multiple is odd, or is tangent to the x-axis if the multiple is even	$y = f(x)$ $= x^5 + x^4 - 2x^3 - 2x^2 + x + 1$ $= (x + 1)^3(x - 1)^2$ Here $n = 5$, there is one real root with multiplicity 3 (so the graph crosses the axis and changes sign at $x = -1$), and one real root with multiplicity 2 (so the graph is tangent to the x-axis at $x = 1$). 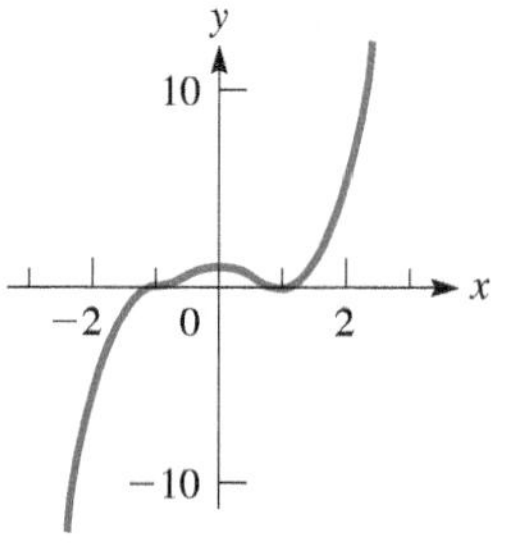 Fig. 15.4(d)

EXERCISES 15.2

In Exercises 1 and 2, make the given changes in the indicated examples of this section and then solve the resulting equation.

1. In Example 2, change the middle term of the second factor to $2x$.

2. In Example 6, change the middle three terms to the left of the = sign to $-3x^3 + 7x^2 + 7x$, given the same root.

In Exercises 3–6, find the roots of the given equations by inspection.

3. $(x + 3)(x^2 - 4) = 0$

4. $x(2x + 5)^2(x^2 - 64) = 0$

5. $(x^2 + 6x + 9)(x^2 + 4) = 0$

6. $(4x^2 + 9)(4x^2 + 4x + 1) = 0$

In Exercises 7–26, solve the given equations using synthetic division, given the roots indicated.

7. $x^3 - 5x^2 + 2x + 8 = 0 \quad (r_1 = 2)$

8. $R^3 + 1 = 0 \quad (r_1 = -1)$

9. $2x^3 + 11x^2 + 20x + 12 = 0 \quad \left(r_1 = -\frac{3}{2}\right)$

10. $4x^3 - 20x^2 - x + 5 = 0 \quad \left(r_1 = \frac{1}{2}\right)$

11. $3x^3 + 2x^2 + 3x + 2 = 0 \quad (r_1 = j)$

12. $x^3 + 5x^2 + 9x + 5 = 0 \quad (r_1 = -2 + j)$

13. $t^4 + t^3 - 2t^2 + 4t - 24 = 0 \quad (r_1 = 2, r_2 = -3)$

14. $x^4 - 2x^3 - 20x^2 - 8x - 96 = 0 \quad (r_1 = 6, r_2 = -4)$

15. $2x^4 - 19x^3 + 39x^2 + 35x - 25 = 0$ (5 is a double root)

16. $4n^4 + 28n^3 + 61n^2 + 42n + 9 = 0$ (-3 is a double root)

17. $6x^4 + 5x^3 - 15x^2 + 4 = 0 \quad \left(r_1 = -\frac{1}{2}, r_2 = \frac{2}{3}\right)$

18. $6x^4 - 5x^3 - 14x^2 + 14x - 3 = 0 \quad \left(r_1 = \frac{1}{3}, r_2 = \frac{3}{2}\right)$

19. $2x^4 - x^3 - 4x^2 + 10x - 4 = 0 \quad (r_1 = 1 + j)$

20. $s^4 - 8s^3 - 72s - 81 = 0 \quad (r_1 = 3j)$

21. $x^5 - 3x^4 + 4x^3 - 4x^2 + 3x - 1 = 0$ (1 is a triple root)

22. $12x^5 - 7x^4 + 41x^3 - 26x^2 - 28x + 8 = 0$
$\left(r_1 = 1, r_2 = \frac{1}{4}, r_3 = -\frac{2}{3}\right)$

23. $P^5 - 3P^4 - P + 3 = 0 \quad (r_1 = 3, r_2 = j)$

24. $4x^5 + x^3 - 4x^2 - 1 = 0 \quad \left(r_1 = 1, r_2 = \frac{1}{2}j\right)$

25. $x^6 + 2x^5 - 4x^4 - 10x^3 - 41x^2 - 72x - 36 = 0$
(-1 is a double root; $2j$ is a root)

26. $x^6 - x^5 - 2x^3 - 3x^2 - x - 2 = 0$ (j is a double root)

In Exercises 27 and 28, answer the given questions.

27. Why cannot a third-degree polynomial function with real coefficients have zeros of 1, 2, and j?

28. How can the graph of a fourth-degree polynomial equation have its only x-intercepts as 0, 1, and 2?

In Exercises 29 and 30, form the indicated equations.

29. Form a polynomial equation of degree 3 and with integral coefficients, having a root of $1 + j$, and for which $f(2) = 4$.

30. Form a polynomial equation of the smallest possible degree and with integral coefficients, having a double root of 3 and a root of j.

31. Given points $P(-1, 1)$ and $Q(0, 2)$ on the elliptic curve $y^2 = x^3 + 2x + 4$, find the coordinates of R, as in Fig. 15.5. This type of operation is used in *elliptic curve cryptography* (ECC). *Hint*: Write the equation of the line and substitute y into the elliptic curve. Solve for x to find the three roots (you already know two).

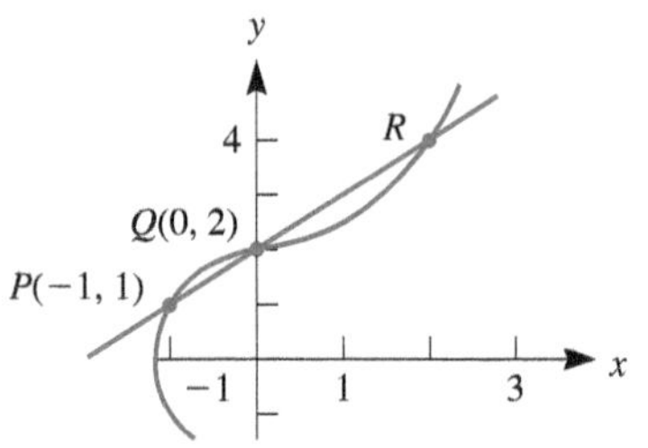

Fig. 15.5

Answer to Practice Exercise

1. $-2,\ 3,\ 2j,\ -2j$

15.3 Rational and Irrational Roots

The product $(x + 2)(x - 4)(x + 3)$ equals $x^3 + x^2 - 14x - 24$. Note how the constant 24 is determined only by the 2, 4, and 3. These numbers represent the roots of the equation if the given function is set equal to zero. In fact, if we find all the integral roots of an equation with integral coefficients and represent the equation in the form

$$f(x) = (x - r_1)(x - r_2)\cdots(x - r_k)f_{k+1}(x) = 0$$

where all the roots indicated are integers, the constant term of $f(x)$ must have factors of $r_1, r_2, \ldots, r_k$. This leads us to the result in the Learning Tip to the left.

LEARNING TIP
In a polynomial equation $f(x) = 0$ with integral coefficients, if the coefficient of the highest power is 1, then any integral roots are factors of the constant term of $f(x)$.

EXAMPLE 1 Possible integral roots if $a_0 = 1$

The equation $x^5 - 4x^4 - 7x^3 + 14x^2 - 44x + 120 = 0$ can be written as

$$(x - 5)(x + 3)(x - 2)(x^2 + 4) = 0$$

We now note that $5(3)(2)(4) = 120$. Thus, the roots 5, -3, and 2 are numerical factors of $|120|$. The theorem states nothing about the signs involved.

If the coefficient a_0 of the highest-power term of $f(x)$ is an integer not equal to 1, the polynomial equation $f(x) = 0$ may have rational roots that are not integers. We can factor a_0 from every term of $f(x)$. Thus, any polynomial equation $f(x) = a_0x^n + a_1x^{n-1} + \cdots + a_n = 0$ with integral coefficients can be written in the form

$$f(x) = a_0\left(x^n + \frac{a_1}{a_0}x^{n-1} + \cdots + \frac{a_n}{a_0}\right) = 0$$

Since a_n and a_0 are integers, a_n/a_0 is a rational number. Using the same reasoning as with integral roots applied to the polynomial within the parentheses, we see that any rational roots are factors of a_n/a_0. This leads to the following theorem:

Rational Roots of a Polynomial Equation

Any rational root r_r of a polynomial equation (with integral coefficients)

$$f(x) = a_0x^n + a_1x^{n-1} + \cdots + a_n = 0$$

is an integral factor of a_n divided by an integral factor of a_0. Therefore,

$$r_r = \frac{\text{integral factor of } a_n}{\text{integral factor of } a_0} \qquad \textbf{(15.4)}$$

EXAMPLE 2 Possible rational roots

Find the possible rational roots of $f(x) = 4x^3 - 3x^2 - 25x - 6$.

Coefficient	**Integral factors**	**Possible rational roots**
$a_n = 6$	1, 2, 3, 6	All possible positive and negative quotients:
$a_0 = 4$	1, 2, 4	$\pm 1, \pm\frac{1}{2}, \pm\frac{1}{4}, \pm 2, \pm 3, \pm\frac{3}{2}, \pm\frac{3}{4}, \pm 6$

There are 16 different possible rational roots in Example 2, but we cannot tell which of these are in fact roots (the roots are actually -2, 3, and $-\frac{1}{4}$). As we now discuss, the changes in sign in the equation can help us find these roots.

If $f(x)$ has all positive terms, there cannot be a positive root. This is because any positive number substituted in $f(x)$ will give a positive (i.e., nonzero) value for $f(x)$. Hence, for a positive number to be a root, there must be at least one negative and one positive term in the function. A similar argument applies for negative roots and the sign changes of $f(-x)$. We thus have the following procedure, using what is known as *Descartes' rule of signs.*

■ Descartes' rule of signs is named for the French mathematician René Descartes (1596–1650).

Maximum Number of Positive and Negative Roots by Descartes' Rule of Signs	EXAMPLE 3
	Find the numbers of positive and negative roots of $f(x) = 3x^3 - x^2 - x + 4$.
1. For a polynomial equation $f(x)$, count the number of changes in sign in $f(x)$ when going from one term to the next.	$f(x)$ has two sign changes: $f(x) = 3x^3 - x^2 - x + 4$ (sign changes marked 1 and 2)
2. The number of *positive roots* of $f(x) = 0$ cannot exceed the number from Step 1. If there is just one sign change, there is one positive root.	There are *no more than two* positive roots of $f(x) = 0$.
3. Obtain $f(-x)$. Count the number of changes in sign in $f(-x)$ when going from one term to the next.	$f(-x)$ has one sign change: $f(-x) = 3(-x)^3 - (-x)^2 - (-x) + 4$ $= -3x^3 - x^2 + x + 4$ (sign change marked 1)
4. The number of *negative roots* of $f(x) = 0$ cannot exceed the number from Step 3. If there is just one sign change in $f(-x)$, there is one negative root.	There is just one sign change, so there is one negative root of $f(x) = 0$.

Practice Exercise

1. Determine the maximum possible numbers of positive and negative roots of $9x^4 - 12x^3 - 5x^2 + 12x - 4 = 0$.

EXAMPLE 4 Using Descartes' rule of signs

For the equation $4x^5 - x^4 - 4x^3 + x^2 - 5x - 6 = 0$, we write

$$f(x) = 4x^5 - x^4 - 4x^3 + x^2 - 5x - 6$$

← three sign changes

$$f(-x) = -4x^5 - x^4 + 4x^3 + x^2 + 5x - 6$$

← two sign changes

Thus, there are no more than three positive and two negative roots.

We now put together all the information that can be used to solve a polynomial equation $f(x) = 0$ of degree n with real coefficients, and state a method for finding all the roots. We include one rule that we have not seen yet, but which will be illustrated in the example. This procedure will allow us to find *all* the roots, including *at most two complex roots*, or the exact values of *at most two irrational roots*. When a polynomial equation has more than two irrational roots, approximate values must be found (see Examples 7 and 8 below).

Roots of a Polynomial Equation of Degree n	**EXAMPLE 5**
	Find the roots of $f(x) = 2x^3 + x^2 + 5x - 3 = 0$.
1. Determine the number of roots (n).	Since $n = 3$, there are three roots.
2. Use Descartes' rule of signs to determine the maximum number of positive roots (number of sign changes in $f(x)$) and the maximum number of negative roots (number of sign changes in $f(-x)$).	$f(x) = 2x^3 + x^2 + 5x - 3$ $f(-x) = -2x^3 + x^2 - 5x - 3$ There is one positive root and at most two negative roots.
3. List all possible rational roots. They must be factors of the constant term divided by factors of the coefficient of the highest-power term.	The possible roots are $\pm 1, \pm\frac{1}{2}, \pm\frac{3}{2}, \pm 3$
4. Use synthetic division to try possible roots. **a.** When trying a positive root, if the bottom row of the synthetic division table contains all positive numbers, then there are no roots larger than the value tried. **b.** When trying a negative root, if the bottom row of the synthetic division table contains alternating signs, then there are no roots less than the value tried.	We try the root 1. The synthetic division gives $\begin{array}{rrrr\|l} 2 & 1 & 5 & -3 & 1 \\ & 2 & 3 & 8 & \\ \hline 2 & 3 & 8 & 5 & \end{array}$ The remainder of 5 tells us 1 is not a root. The bottom row contains all positive numbers, so there are no roots larger than 1 (they would give larger positive values on the table than these). We try the root $+\frac{1}{2}$. The synthetic division gives $\begin{array}{rrrr\|l} 2 & 1 & 5 & -3 & \frac{1}{2} \\ & 1 & 1 & 3 & \\ \hline 2 & 2 & 6 & 0 & \end{array}$ The remainder of 0 tells us $+\frac{1}{2}$ is a root (Fig. 15.6).
5. When a root is found, continue working with the quotient, which is one degree less than the degree of the dividend. If $n - 2$ roots have been found, use factoring or the quadratic formula to find the remaining two roots.	The quotient from the previous step is $2x^2 + 2x + 6 = 0$ We can use the quadratic formula to find the remaining two roots. $x = \dfrac{-2 \pm \sqrt{4 - 48}}{4}$ $= \dfrac{-1 \pm j\sqrt{11}}{2}$ The three roots are $\frac{1}{2}, \frac{-1 \pm j\sqrt{11}}{2}$.

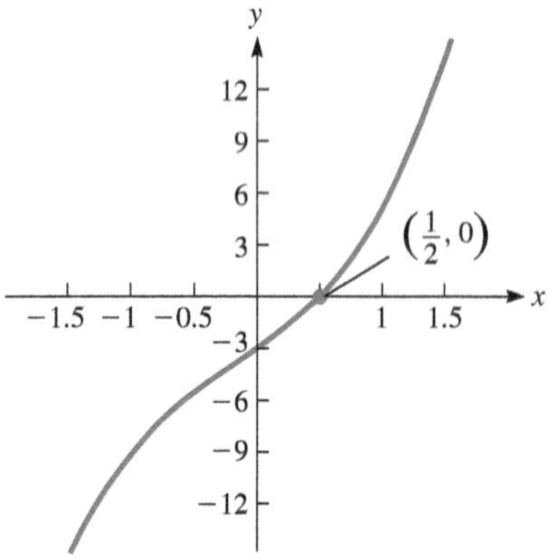

Fig. 15.6

Practice Exercise

2. Find the roots of the equation $9x^4 - 12x^3 - 5x^2 + 12x - 4 = 0$.

EXAMPLE 6 Finding roots—Bezier curve design

■ See the chapter introduction. Bezier curves were made popular by French engineer Pierre Bézier, who used them to design automobile bodies at Renault in the 1960s.

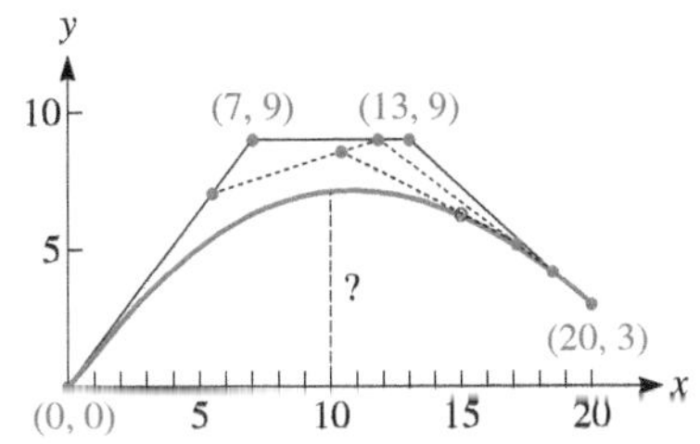

Fig. 15.7

Bezier curves are used frequently in computer graphics to generate smooth curves and surfaces. In particular, cubic Bezier curves are formed by four points: two that define the curve's start and end (at $t = 0$ and $t = 1$), and two that "stretch" the middle part of the graph. The blue curve in Fig. 15.7 shows a Bezier curve design for the roofline of a new art museum. It is defined by the parametric equations shown below, where the units for x and y are in metres.

$$x = 2t^3 - 3t^2 + 21t$$
$$y = 3t^3 - 27t^2 + 27t$$
$$0 \le t \le 1$$

Find the height of the roofline at the point that is 10 m from each of the edges.

To find this height, we will substitute 10 for x in the top parametric equation, solve for t, and then substitute the result into the bottom equation to find y.

$$10 = 2t^3 - 3t^2 + 21t$$
$$2t^3 - 3t^2 + 21t - 10 = 0$$

If the above equation has any rational roots, they must be of the form $\frac{\pm 1, \pm 2, \pm 5, \pm 10}{\pm 1, \pm 2}$. Since we know $0 \le t \le 1$, the only possibilities are $t = \frac{1}{2}$ or $t = 1$. We also know that $t = 1$ represents an endpoint, so it is not the solution we desire. Thus, we will try $t = \frac{1}{2}$ (or 0.5) using synthetic division.

$$\begin{array}{rrrr|l} 2 & -3 & 21 & -10 & \underline{0.5} \\ & 1 & -1 & 10 & \\ \hline 2 & -2 & 20 & 0 & \end{array}$$

Since the remainder is zero, $t = 0.5$ is a root. To find the value of y, we substitute $t = 0.5$ into the second parametric equation:

$$y = 3\left(\tfrac{1}{2}\right)^3 - 27\left(\tfrac{1}{2}\right)^2 + 27\left(\tfrac{1}{2}\right) = 7.125 \text{ m}$$

Therefore, the height of the roofline is 7.125 m at the desired location.

EXAMPLE 7 Finding roots with a graphing utility

Find the roots of the equation $x^3 - 2x^2 - 3x + 2 = 0$, using a graphing utility.

Since the degree of the equation is 3, we know there are three roots, and since the degree is odd, there is at least one real root. Using Descartes' rule of signs, we have

$$f(x) = x^3 - 2x^2 - 3x + 2 \qquad \text{two changes of sign}$$
$$f(-x) = -x^3 - 2x^2 + 3x + 2 \qquad \text{one change of sign}$$

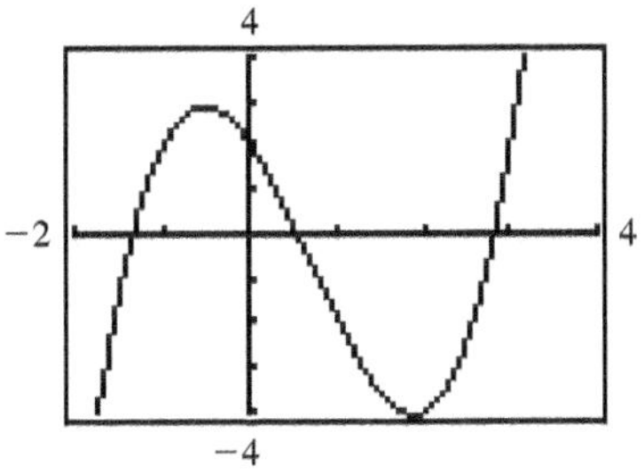

Fig. 15.8

This means there is one negative root and no more than two positive roots, if any. To use a graphing calculator, we set

$$y = x^3 - 2x^2 - 3x + 2$$

and use the *window* settings (after two or three trials) shown in Fig. 15.8. The view shows one negative and two positive roots. Using the *zero* feature, these roots are $-1.34, 0.53, 2.81$.

The roots of a polynomial equation can be approximated by successive evaluation of the function. The method can be summarized as follows.

Approximating Roots by Successive Evaluation

1. Find an interval (a, b) such that $f(a)$ and $f(b)$ have opposite signs.
2. Choose a point c in (a, b) and evaluate $f(c)$.
3. If $f(c) = 0$, then c is a root. Otherwise, if $f(a)$ and $f(c)$ have opposite signs, then the root is in (a, c). If $f(c)$ and $f(b)$ have opposite signs, then the root is in (c, b).
4. Repeat Steps 2 and 3 until the interval is sufficiently small.

EXAMPLE 8 Successive evaluation—box design

The bottom part of a box to hold a jigsaw puzzle is to be made from a rectangular piece of cardboard 37.0 cm by 31.0 cm by cutting out equal squares from the corners, bending up the sides, and taping the corners. See Fig. 15.9. If the volume of the bottom part is to be 2770 cm^3, find the side of the square that is to be cut out.

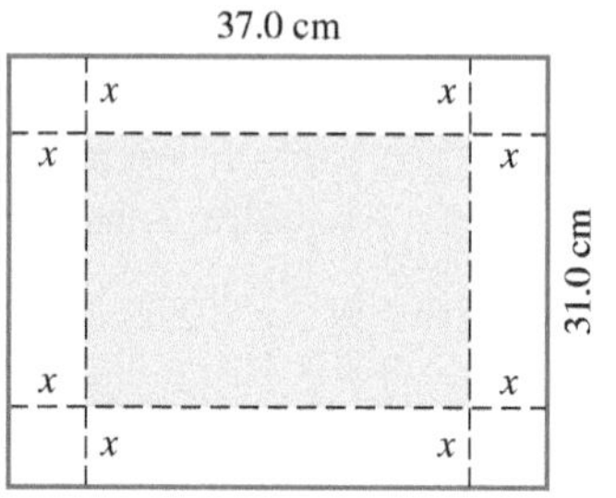

Fig. 15.9

Let $x =$ the side of the square to be cut out. This means

$$2770 = x(37.0 - 2x)(31.0 - 2x) \quad \text{volume} = 2770 \text{ cm}^3$$
$$= 4x^3 - 136x^2 + 1147x$$
$$4x^3 - 136x^2 + 1147x - 2770 = 0 \quad \text{simplify with terms on the left}$$

We evaluate $f(x) = 4x^3 - 136x^2 + 1147x - 2770$ for integral values of x and find that $f(4) = -102$ and $f(5) = 65$. Since $f(4)$ and $f(5)$ have opposite signs, this means that there is a root in $(4, 5)$. We pick $c = 4.5$ and evaluate $f(4.5) = 2$. Now $f(4)$ and $f(4.5)$ have opposite signs, so there is a root in $(4, 4.5)$. As we continue the evaluations, we obtain the following table:

Interval	c	$f(c)$	New interval
$(4, 5)$	4.5	2	$(4, 4.5)$
$(4, 4.5)$	4.4	-15.42	$(4.4, 4.5)$
$(4.4, 4.5)$	4.48	-1.353	$(4.48, 4.5)$
$(4.48, 4.5)$	4.49	0.3318	$(4.48, 4.49)$

From these values, $x = 4.49$ cm, to three significant digits. This value gives a volume

$$V = 4.49(37.0 - 2(4.49))(31.0 - 2(4.49)) = 2770 \text{ cm}^3$$

which means that the solution checks.

Checking values greater than $x = 5$, we find that there is another root between 6 and 7. Using the same procedure as above, we find that $x = 6.79$ cm is also a solution. (A third root $x = 22.7$ cm is obviously too large.) Therefore, there are two possible squares that can be cut from each corner to get a bottom part with a volume of 2770 cm^3.

EXERCISES 15.3

In Exercises 1 and 2, make the given changes in the indicated examples of this section and then perform the indicated operation.

1. In Example 4, change all signs between terms.
2. In Example 5, change the + sign before the $5x$ to $-$.

In Exercises 3–20, solve the given equations.

3. $x^3 + 5x^2 + 2x - 8 = 0$
4. $2x^3 + 5x^2 - x + 6 = 0$
5. $x^3 + 2x^2 - 5x - 6 = 0$
6. $t^3 - 12t - 16 = 0$
7. $3x^3 - x - 2 = 0$
8. $3t^3 + 8t^2 - 1 = 0$
9. $2x^3 - 3x^2 - 3x + 2 = 0$
10. $4x^3 - 5x^2 - 23x + 6 = 0$
11. $x^4 - 11x^2 - 12x + 4 = 0$
12. $8x^4 - 32x^3 - x + 4 = 0$
13. $5n^4 - 2n^3 + 40n - 16 = 0$
14. $4n^4 - 17n^2 + 14n - 3 = 0$
15. $12x^4 + 44x^3 + 21x^2 - 11x = 6$
16. $9x^4 - 3x^3 + 34x^2 - 12x = 8$
17. $D^5 + D^4 - 9D^3 - 5D^2 + 16D + 12 = 0$
18. $x^6 - x^4 - 14x^2 + 24 = 0$
19. $4x^5 - 24x^4 + 49x^3 - 38x^2 + 12x - 8 = 0$
20. $2x^5 + 5x^4 - 4x^3 - 19x^2 - 16x = 4$

In Exercises 21–24, use a graphing utility to solve the given equations to the nearest 0.01.

21. $x^3 - 8x + 3 = 0$
22. $2x^4 - 15x^2 - 7x + 3 = 0$
23. $8x^4 + 36x^3 + 35x^2 - 4x - 4 = 0$
24. $2x^5 - 3x^4 + 8x^3 - 4x^2 - 4x + 2 = 0$

In Exercises 25–28, find the irrational root (to the nearest 0.01) that lies between the given values by successive evaluation. (See Example 8.)

25. $x^3 - 6x^2 + 10x - 4 = 0$ (0 and 1)
26. $r^4 - r^3 - 3r^2 - r - 4 = 0$ (2 and 3)
27. $3x^3 + 13x^2 + 3x - 4 = 0$ (-1 and 0)
28. $3x^4 - 3x^3 - 11x^2 - x - 4 = 0$ (-2 and -1)

In Exercises 29–46, solve the given problems. Use a graphing utility if necessary.

29. Solve the following system algebraically:
$y = x^4 - 11x^2;\quad y = 12x - 4$
30. Find rational values of a such that $(x - a)$ will divide into $x^3 + x^2 - 4x - 4$ with a remainder of zero.
31. Where does the graph of the function $f(x) = 4x^3 + 3x^2 - 20x - 15$ cross the x-axis?
32. Where does the graph of the function $f(s) = 2s^4 - s^3 - 5s^2 + 7s - 6$ cross the s-axis?
33. The angular acceleration α (in rad/s^2) of the wheel of a car is given by $\alpha = -0.2t^3 + t^2$, where t is the time (in s). For what values of t is $\alpha = 2.0\ \text{rad/s}^2$?
34. In finding one of the dimensions d (in cm) of the support columns of a building, the equation $3d^3 + 5d^2 - 400d - 18\,000 = 0$ is found. What is this dimension?
35. The deflection y of a beam at a horizontal distance x from one end is given by $y = k(x^4 - 2Lx^3 - L^3x)$, where L is the length of the beam and k is a constant. For what values of x is the deflection zero?
36. The specific gravity s of a sphere of radius r that sinks to a depth h in water is given by $s = \dfrac{3rh^2 - h^3}{4r^3}$. Find the depth to which a spherical buoy of radius 4.0 cm sinks if $s = 0.50$.
37. Cubic Bezier curves are commonly used to control the timing of animations. A certain "ease in" curve is given by $x = 0.1t^3 - 1.2t^2 + 2.1t$, $y = -2t^3 + 3t^2$ for $0 \le t \le 1$, where x represents the percentage of elapsed time for the animation and y represents the percentage of the progression of the animation (both as decimals). What percentage of the animation will be completed after 50% of the time has elapsed? (See the chapter introduction.)
38. The pressure difference p (in kPa) at a distance x (in km) from one end of an oil pipeline is given by $p = x^5 - 3x^4 - x^2 + 7x$. If the pipeline is 4 km long, where is $p = 0$?
39. A rectangular tray is made from a square piece of sheet metal 10.0 cm on a side by cutting equal squares from each corner, bending up the sides, and then welding them together. How long is the side of the square that must be cut out if the volume of the tray is 70.0 cm^3?
40. The angle θ of a robot arm with the horizontal as a function of time t (in s) is given by $\theta = 15 + 20t^2 - 4t^3$ for $0 \le t \le 5$ s. Find t for $\theta = 40°$.
41. The radii of four different-sized ball bearings differ by 1.0 mm in radius from one size to the next. If the volume of the largest equals the volumes of the other three combined, find the radii.
42. A rectangular safe is to be made of steel of uniform thickness, including the door. The inside dimensions are 1.20 m, 1.20 m, and 2.00 m. If the volume of steel is 1.25 m^3, find its thickness.
43. For electrical resistors connected in parallel, the reciprocal of the combined resistance equals the sum of the reciprocals of the individual resistances. If three resistors are connected in parallel such that the second resistance is 1400 Ω more than the first and the third is 5600 Ω more than the first, find the resistances for a combined resistance of 1400 Ω.
44. Each of three revolving doors has a perimeter of 6.60 m and revolves through a volume of 9.50 m^3 in one revolution about their common vertical side. What are the doors' dimensions?
45. If a, b, and c are positive integers, find the combinations of the possible positive, negative, and complex roots if $f(x) = ax^3 - bx^2 + c = 0$.
46. An equation $f(x) = 0$ involves only odd powers of x with positive coefficients. Explain why this equation has no real root except $x = 0$.

Answers to Practice Exercises

1. 3 positive, 1 negative 2. -1, $2/3$, $2/3$, 1

CHAPTER 15 KEY FORMULAS AND EQUATIONS

Polynomial function	$f(x) = a_0x^n + a_1x^{n-1} + \cdots + a_n$	**(15.1)**
Remainder theorem	$f(x) = (x - r)q(x) + R$	**(15.2)**
	$f(r) = R$	**(15.3)**
Rational roots	$r_r = \dfrac{\text{integral factor of } a_n}{\text{integral factor of } a_0}$	**(15.4)**

CHAPTER 15 REVIEW EXERCISES

In Exercises 1–4, find the remainder of the indicated division by the remainder theorem.

1. $(2x^3 - 4x^2 - x + 4) \div (x - 1)$

2. $(x^3 - 2x^2 + 9) \div (x + 2)$

3. $(4n^3 + n + 4) \div (n + 3)$

4. $(x^4 - 5x^3 + 8x^2 + 15x - 2) \div (x - 3)$

In Exercises 5–8, use the factor theorem to determine whether or not the second expression is a factor of the first.

5. $x^4 + x^3 + x^2 - 2x - 3; \quad x + 1$

6. $2s^3 - 6s - 4; \quad s - 2$

7. $2t^4 - 10t^3 - t^2 - 3t + 10; \quad t + 5$

8. $9v^3 + 6v^2 + 4v + 2; \quad 3v + 1$

In Exercises 9–16, use synthetic division to perform the indicated divisions.

9. $(x^3 + 3x^2 + 6x + 1) \div (x - 1)$

10. $(3x^3 - 2x^2 + 7) \div (x - 3)$

11. $(2x^3 - 3x^2 - 4x + 3) \div (x + 2)$

12. $(3D^3 + 8D^2 - 16) \div (D + 4)$

13. $(x^4 + 3x^3 - 20x^2 - 2x + 56) \div (x + 6)$

14. $(x^4 - 6x^3 + x - 8) \div (x - 3)$

15. $(2m^5 - 46m^3 + m^2 - 9) \div (m - 5)$

16. $(x^6 + 63x^3 + 5x^2 - 9x - 8) \div (x + 4)$

In Exercises 17–20, use synthetic division to determine whether or not the given numbers are zeros of the given functions.

17. $y^3 + 5y^2 - 6; \quad -3$

18. $8y^4 - 32y^3 - y + 4; \quad 4$

19. $2x^4 - x^3 + 2x^2 + x - 1; \quad \frac{1}{2}$

20. $6W^4 - 7W^3 + 2W^2 - 9W - 6; \quad -\frac{2}{3}$

In Exercises 21–32, find all the roots of the given equations, using synthetic division and the given roots.

21. $x^3 - 4x^2 - 7x + 10 = 0 \quad (r_1 = 5)$

22. $3B^3 - B^2 - 24B + 28 = 0 \quad (r_1 = 2)$

23. $x^4 - 10x^3 + 35x^2 - 50x + 24 = 0 \quad (r_1 = 1, r_2 = 2)$

24. $x^4 - x^3 - 5x^2 - x - 6 = 0 \quad (r_1 = 3, r_2 = -2)$

25. $4p^4 - p^2 - 18p + 9 = 0 \quad \left(r_1 = \frac{1}{2}, r_2 = \frac{3}{2}\right)$

26. $15x^4 + 4x^3 + 56x^2 + 16x - 16 = 0$
$\left(r_1 = \frac{2}{5}, r_2 = -\frac{2}{3}\right)$

27. $4x^4 + 4x^3 + x^2 + 4x - 3 = 0 \quad (r_1 = j)$

28. $x^4 + 2x^3 - 4x - 4 = 0 \quad (r_1 = -1 + j)$

29. $s^5 + 3s^4 - s^3 - 11s^2 - 12s - 4 = 0 \quad (-1 \text{ is a triple root})$

30. $24x^5 + 10x^4 + 7x^2 - 6x + 1 = 0$
$\left(r_1 = -1, r_2 = \frac{1}{4}, r_3 = \frac{1}{3}\right)$

31. $V^5 + 4V^4 + 5V^3 - V^2 - 4V - 5 = 0$
$(r_1 = 1, r_2 = -2 + j)$

32. $2x^5 - x^4 + 8x - 4 = 0 \quad \left(r_1 = \frac{1}{2}, r_2 = 1 + j\right)$

In Exercises 33–40, solve the given equations.

33. $x^3 + x^2 - 10x + 8 = 0$

34. $x^3 - 8x^2 + 20x = 16$

35. $2r^3 - 3r^2 + 1 = 0$

36. $2x^3 - 3x^2 - 11x + 6 = 0$

37. $6x^3 - x^2 - 12x = 5$

38. $6y^3 + 19y^2 + 2y = 3$

39. $4t^4 - 17t^2 + 14t - 3 = 0$

40. $2x^4 + 5x^3 - 14x^2 - 23x + 30 = 0$

In Exercises 41–65, solve the given problems. Where appropriate, set up the required equations.

41. What are the possible numbers of real zeros (double roots count as two, etc.) for a polynomial with real coefficients and of degree five?

42. What are the possible combinations of real and complex zeros (double roots count as two, etc.) of a fourth-degree polynomial?

43. If a calculator shows a real root, how many complex roots are possible for a sixth-degree polynomial equation $f(x) = 0$?

44. Find rational values of a such that $(x - a)$ will divide into $x^3 - 13x + 3$ with a remainder of -9.

45. Explain how to find k if $x + 2$ is a factor of $f(x) = 3x^3 + kx^2 - 8x - 8$. What is k?

46. Explain how to find k if $x - 3$ is a factor of $f(x) = kx^4 - 15x^2 - 5x - 12$. What is k?

47. Form a polynomial equation of degree three with integral coefficients and having roots of j and 5.

48. If $f(x) = 3x^4 - 18x^3 - 2x^2 + 13x - 6$, and $f(x) = g(x)(x - 6)$, find $g(x)$.

49. Solve the following system algebraically: $x^2 = y + 3$; $xy = 2$.

50. Where does the graph of the function
$f(x) = 2x^4 - 7x^3 + 11x^2 - 28x + 12$ cross the x-axis?

51. Find the irrational root of the equation $3x^3 - x^2 - 8x - 2 = 0$ that lies between 1 and 2.

52. A propane tank is in the shape of a cylinder with hemispheres on each end. The cylinder and the hemispheres have the same radius. If the total length of the tank is 4.0 m and the volume is 7.0 m^3, what is the radius?

53. The bending moment M (in kN · m) along a beam with a nonuniformly distributed load is given by $M = -\frac{25}{6}x^3 + 600x$, where x is the horizontal distance from one end. For what values of x is the bending moment 850 kN · m?

54. A company determined that the number s (in thousands) of computer chips that it could supply at a price p of less than \$5 is given by $s = 4p^2 - 25$, whereas the demand d (in thousands) for the chips is given by $d = p^3 - 22p + 50$. For what price is the supply equal to the demand?

55. In order to find the diameter d (in cm) of a helical spring subject to given forces, it is necessary to solve the equation $64d^3 - 144d^2 + 108d - 27 = 0$. Solve for d.

56. A cubical tablet for purifying water is wrapped in a sheet of foil 0.500 mm thick. The total volume of tablet and foil is 33.1% greater than the volume of the tablet alone. Find the length of the edge of the tablet.

57. For the mirror shown in Fig. 15.10, the reciprocal of the focal distance f equals the sum of the reciprocals of the object distance p and image distance q (in cm). Find p, if $q = p + 4$ and $f = (p + 1)/p$.

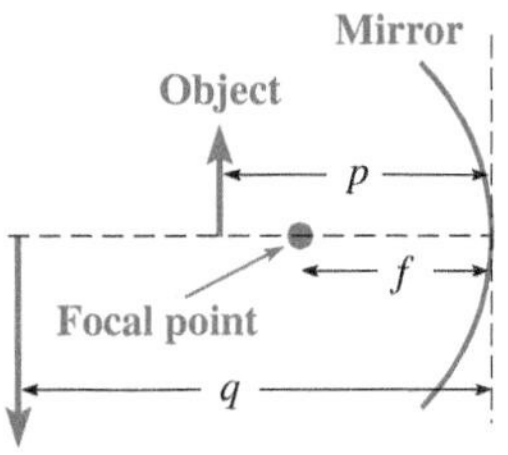

Fig. 15.10 Image

58. Three electric capacitors are connected in series. The capacitance of the second is 1 μF more than that of the first, and the third is 2 μF more than the second. The capacitance of the combination is 1.33 μF. The equation used to determine C, the capacitance of the first capacitor, is

$$\frac{1}{C} + \frac{1}{C+1} + \frac{1}{C+3} = \frac{3}{4}$$

Find the values of the capacitances.

59. The height of a cylindrical oil tank is 3.2 m more than the radius. If the volume of the tank is 680 m^3, what are the radius and the height of the tank?

60. A grain storage bin has a square base, each side of which is 5.5 m longer than the height of the bin. If the bin holds 160 m^3 of grain, find its dimensions.

61. A rectangular door has a diagonal brace that is 0.300 m longer than the height of the door. If the area of the door is 2.70 m^2, find its dimensions.

62. The radius of one ball bearing is 1.0 mm greater than the radius of a second ball bearing. If the sum of their volumes is 100 mm^3, find the radius of each.

63. The edge of a cube is 10 cm greater than the radius of a sphere. If the volumes of the figures are equal, what is this volume?

64. A computer analysis of the number of crimes committed each month in a certain city for the first 10 months of a year showed that $n = x^3 - 9x^2 + 15x + 600$. Here, n is the number of monthly crimes and x is the number of the month (as of the last day). In what month were 580 crimes committed?

65. A computer science student is to write a computer program that will print out the values of n for which $x + r$ is a factor of $x^n + r^n$. Write a paragraph that states the values of n and explains how they are found.

CHAPTER 15 **PRACTICE TEST**

1. Is -3 a zero for the function $2x^3 + 3x^2 + 7x - 6$? Explain.

2. Find the remaining roots of the equation
$x^4 - 2x^3 - 7x^2 + 20x - 12 = 0$; 2 is a double root.

3. Use synthetic division to perform the division
$(x^3 - 5x^2 + 4x - 9) \div (x - 3)$.

4. Use the factor theorem and synthetic division to determine whether or not $2x + 1$ is a factor of $2x^4 + 15x^3 + 23x^2 - 16$.

5. Use the remainder theorem to find the remainder of the division $(x^3 + 4x^2 + 7x - 9) \div (x + 4)$.

6. Solve for x: $2x^4 - x^3 + 5x^2 - 4x - 12 = 0$.

7. The ends of a 10-m beam are supported at different levels. The deflection y of the beam is given by $y = kx^2(x^3 + 436x - 4000)$, where x is the horizontal distance from one end and k is a constant. Find the values of x for which the deflection is zero.

8. A cubical metal block is heated such that its edge increases by 1.0 mm and its volume is doubled. Find the edge of the cube to the nearest tenth of a millimetre.

16. Matrices; Systems of Linear Equations

LEARNING OUTCOMES

After completion of this chapter, the student should be able to:

- Add and subtract matrices
- Multiply a matrix by a scalar
- Solve a matrix equation
- Perform matrix multiplication
- Find the inverse of a matrix
- Solve a system of linear equations by using the inverse of the matrix of coefficients
- Solve a system of linear equations by using Gaussian elimination
- Identify a system of linear equations as having a unique solution, no solution, or an unlimited number of solutions
- Evaluate determinants of any order
- Solve a system of linear equations by determinants

Stocktrek Images/Getty Images

▲ The robotic arm Canadarm2 was used to assemble the International Space Station while in space. It is routinely used to move supplies, equipment, and even astronauts. In Section 16.2 we show how the translational and rotational displacements of robotic arms are obtained by multiplication of matrices.

While working with systems of linear transformations in the 1850s, the English mathematician Arthur Cayley developed the use of a *matrix*. (As we will see, a *matrix* is simply a rectangular array of numbers.) Cayley's interests, as well as those of the many other mathematicians who contributed to the theory of matrices during the nineteenth century, were on the purely mathematical aspects of the theory, with little or no reference to possible applications.

It was not until the 1920s that one of the first major applications of matrices was made, when physicists used them in developing theories about the elementary particles within the atom. Their work is still very important in atomic and nuclear physics.

In the 1930s, many authors started using matrix notation to describe Gaussian elimination, an important method for solving systems of linear equations. (We will discuss Gaussian elimination in Section 16.5.) The matrix interpretation of Gaussian elimination became very important after World War II, with the development of the first electronic computers. The method was used to study the accuracy of calculations made by the machines.

Although Gaussian elimination is named in honour of German mathematician Karl Friedrich Gauss, the method first appeared in a third-century B.C.E. Chinese text. Isaac Newton is responsible for an independent development of the method in seventeenth-century Europe. Thanks to his influence, the method was a standard lesson in algebra textbooks by the turn of the nineteenth century. Around 1800, Gauss developed his systematic version of Gaussian elimination in order to solve astronomical problems by the method of least squares (see Section 22.6).

Since about 1950, matrices have become a very important and useful tool in many other areas of application, such as social science and economics. They are now used extensively in business and industry in making appropriate decisions in research, development, and production.

16.1 Matrices: Definitions and Basic Operations

In this section, we introduce the definitions and some basic operations with matrices. In the three sections that follow, we develop additional operations and show how they are used in solving systems of linear equations. This is only an introduction to the use of matrices, and additional uses and operations with matrices are shown in other sources.

A **matrix** is an ordered rectangular array of numbers (called the **elements** of the matrix). The **size** of a matrix is given by the number of rows (first) and columns (second). If the matrix has n rows and m columns, it is an $n \times m$ matrix. When $n = m$, the matrix is called a **square matrix**. It is convenient to designate a given matrix by a capital letter (for example, matrix A).

EXAMPLE 1 Illustrations of matrices

Determine the size of each of the following matrices.

$$A = \begin{bmatrix} 5 & 0 & -1 \\ 1 & 2 & 6 \\ 0 & 4 & -5 \end{bmatrix} \quad B = \begin{bmatrix} 9 \\ 8 \\ 1 \\ 5 \end{bmatrix} \quad C = \begin{bmatrix} -1 & 6 & 8 & 9 \end{bmatrix} \quad O = \begin{bmatrix} 0 & 0 \\ 0 & 0 \end{bmatrix}$$

(**zero** matrix)

3×3 (square) $\quad 4 \times 1 \quad 1 \times 4 \quad 2 \times 2$ (square)

To be able to refer to specific elements of a matrix and to give a general representation, a double-subscript notation is usually employed. That is,

$$A = \begin{bmatrix} a_{11} & a_{12} & a_{13} \\ a_{21} & a_{22} & a_{23} \\ a_{31} & a_{32} & a_{33} \end{bmatrix}$$

row — column

We see that the first subscript refers to the row in which the element lies and the second subscript refers to the column in which the element lies.

Two matrices are said to be **equal** *if and only if they are identical.* That is, they must have the same number of columns and the same number of rows, and the elements must respectively be equal.

EXAMPLE 2 Equality of matrices

(a) $\begin{bmatrix} a_{11} & a_{12} & a_{13} \\ a_{21} & a_{22} & a_{23} \end{bmatrix} = \begin{bmatrix} 1 & -5 & 0 \\ 4 & 6 & -3 \end{bmatrix}$

if and only if $a_{11} = 1$, $a_{12} = -5$, $a_{13} = 0$, $a_{21} = 4$, $a_{22} = 6$, and $a_{23} = -3$.

(b) The matrices

$$\begin{bmatrix} 1 & 2 & 3 \\ -1 & -2 & -5 \end{bmatrix} \text{ and } \begin{bmatrix} 1 & 2 & -5 \\ -1 & -2 & 3 \end{bmatrix}$$

are not equal, since the elements in the third column are reversed.

(c) The matrices

$$\begin{bmatrix} 2 & 3 \\ -1 & 5 \end{bmatrix} \text{ and } \begin{bmatrix} 2 & 3 & 0 \\ -1 & 5 & 0 \end{bmatrix}$$

are not equal, since the numbers of columns are different. This is true despite the fact that both elements of the third column are zeros.

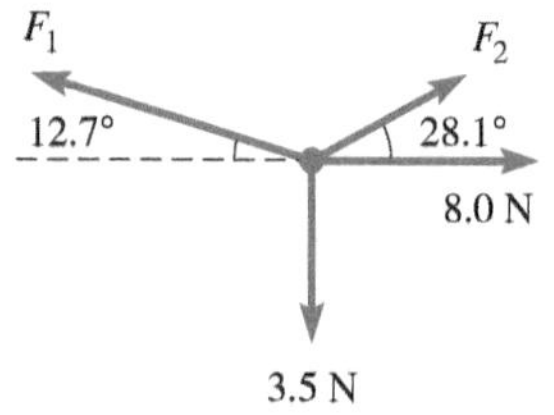

Fig. 16.1

EXAMPLE 3 Matrix equation—equilibrium forces

The forces acting on a bolt are in equilibrium, as shown in Fig. 16.1. Analysing the horizontal and vertical components as in Section 9.4, we find the following matrix equation. Find forces F_1 and F_2.

$$\begin{bmatrix} 0.98F_1 - 0.88F_2 \\ 0.22F_1 + 0.47F_2 \end{bmatrix} = \begin{bmatrix} 8.0 \\ 3.5 \end{bmatrix}$$

From the equality of matrices, we know that $0.98F_1 - 0.88F_2 = 8.0$ and $0.22F_1 + 0.47F_2 = 3.5$. Therefore, to find the forces F_1 and F_2, we must solve the system of equations

$$0.98F_1 - 0.88F_2 = 8.0$$
$$0.22F_1 + 0.47F_2 = 3.5$$

Using determinants, we have

$$F_1 = \frac{\begin{vmatrix} 8.0 & -0.88 \\ 3.5 & 0.47 \end{vmatrix}}{\begin{vmatrix} 0.98 & -0.88 \\ 0.22 & 0.47 \end{vmatrix}} = \frac{8.0(0.47) - 3.5(-0.88)}{0.98(0.47) - 0.22(-0.88)} = 10.5 \text{ N}$$

Using determinants again, or by substituting this value into either equation, we find that $F_2 = 2.6$ N. These values check when substituted into the original matrix equation.

MATRIX ADDITION AND SUBTRACTION

If two matrices have the same number of rows and the same number of columns, their **sum** *is defined as the matrix consisting of the sums of the corresponding elements.* If the numbers of rows or the numbers of columns of the two matrices are not equal, they cannot be added.

EXAMPLE 4 Adding matrices

(a)

$$\begin{bmatrix} 8 & 1 & -5 & 9 \\ 0 & -2 & 3 & 7 \end{bmatrix} + \begin{bmatrix} -3 & 4 & 6 & 0 \\ 6 & -2 & 6 & 5 \end{bmatrix} = \begin{bmatrix} 8+(-3) & 1+4 & -5+6 & 9+0 \\ 0+6 & -2+(-2) & 3+6 & 7+5 \end{bmatrix}$$
$$= \begin{bmatrix} 5 & 5 & 1 & 9 \\ 6 & -4 & 9 & 12 \end{bmatrix}$$

(b) The matrices

$$\begin{bmatrix} 3 & -5 & 8 \\ 2 & 9 & 0 \\ 4 & -2 & 3 \end{bmatrix} \quad \text{and} \quad \begin{bmatrix} 3 & -5 & 8 & 0 \\ 2 & 9 & 0 & 0 \\ 4 & -2 & 3 & 0 \end{bmatrix}$$

cannot be added since the second matrix has one more column than the first matrix. This is true even though the extra column contains only zeros.

Practice Exercise

1. For matrices A and B, find $A + B$.

$$A = \begin{bmatrix} 2 & -4 \\ -7 & 5 \end{bmatrix} \quad B = \begin{bmatrix} -9 & -6 \\ 7 & 3 \end{bmatrix}$$

The product of a number and a matrix (known as **scalar multiplication** *of a matrix) is defined as the matrix whose elements are obtained by multiplying each element of the given matrix by the given number.* Thus, we obtain matrix kA by multiplying the elements of matrix A by k. In this way, $A + A$ and $2A$ are the same matrix.

EXAMPLE 5 Scalar multiplication

For the given matrix A, find $2A$:

$$A = \begin{bmatrix} -5 & 7 \\ 3 & 0 \end{bmatrix} \qquad 2A = \begin{bmatrix} 2(-5) & 2(7) \\ 2(3) & 2(0) \end{bmatrix} = \begin{bmatrix} -10 & 14 \\ 6 & 0 \end{bmatrix}$$

By combining the definitions for the addition of matrices and for the scalar multiplication of a matrix, we can define the subtraction of matrices. That is, the **difference** of matrices A and B is given by $A - B = A + (-B)$. Therefore, we change the sign of each element of B, and proceed as in addition.

EXAMPLE 6 Subtracting matrices

$$\begin{bmatrix} 7 & -4 \\ -9 & 3 \end{bmatrix} - \begin{bmatrix} -2 & 6 \\ -8 & 5 \end{bmatrix} = \begin{bmatrix} 7 & -4 \\ -9 & 3 \end{bmatrix} + \begin{bmatrix} 2 & -6 \\ 8 & -5 \end{bmatrix} = \begin{bmatrix} 9 & -10 \\ -1 & -2 \end{bmatrix}$$

The operations of addition, subtraction, and multiplication of a matrix by a number are like those for real numbers. For these operations, the algebra of matrices is like the algebra of real numbers. We see that the following laws hold for matrices:

Practice Exercise

2. For matrices A and B in Practice Exercise 1, find $A - 2B$.

$$A + B = B + A \quad \text{(commutative law)} \tag{16.1}$$

$$A + (B + C) = (A + B) + C \quad \text{(associative law)} \tag{16.2}$$

$$k(A + B) = kA + kB \tag{16.3}$$

$$A + O = A \tag{16.4}$$

Here, we have let O represent the zero matrix. We will find in the next section that not all laws for matrix operations are like those for real numbers.

EXERCISES 16.1

In Exercises 1 and 2, make the given changes in the indicated examples of this section and then perform the indicated operations.

1. In Example 4(a), interchange the second and third columns of the second matrix and then add the matrices.

2. In Example 5, find the matrix $-2A$.

In Exercises 3–10, determine the value of the literal numbers in each of the given matrix equalities.

3. $\begin{bmatrix} a & b \\ c & d \end{bmatrix} = \begin{bmatrix} 1 & -3 \\ 4 & 7 \end{bmatrix}$

4. $\begin{bmatrix} x \\ x+y \end{bmatrix} = \begin{bmatrix} 2 \\ 5 \end{bmatrix}$

5. $\begin{bmatrix} x & 2y & z \\ r/4 & -s & -5t \end{bmatrix} = \begin{bmatrix} -2 & 10 & -9 \\ 12 & -4 & 5 \end{bmatrix}$

6. $\begin{bmatrix} a+bj & 2c-dj & 3e+fj \end{bmatrix} = \begin{bmatrix} 5j & a+6 & 3b+c \end{bmatrix}$ $(j = \sqrt{-1})$

7. $\begin{bmatrix} C+D \\ 2C-D \\ D-2E \end{bmatrix} = \begin{bmatrix} 5 \\ 4 \\ 6 \end{bmatrix}$

8. $\begin{bmatrix} 2x-3y \\ x+4y \end{bmatrix} = \begin{bmatrix} 13 \\ 1 \end{bmatrix}$

9. $\begin{bmatrix} x-3 & x+y \\ x-z & y+z \\ x+t & y-t \end{bmatrix} = \begin{bmatrix} 5 & 3 \\ 4 & -1 \end{bmatrix}$

10. $\begin{bmatrix} x \\ x+y \end{bmatrix} = \begin{bmatrix} 2 & 0 \\ 4 & -3 \end{bmatrix}$

In Exercises 11–14, find the indicated sums of matrices.

11. $\begin{bmatrix} 2 & 3 \\ -5 & 4 \end{bmatrix} + \begin{bmatrix} -1 & 7 \\ 5 & -2 \end{bmatrix}$

12. $\begin{bmatrix} 1 & 0 & 9 \\ 3 & -5 & -2 \end{bmatrix} + \begin{bmatrix} 4 & -1 & 7 \\ 2 & 0 & -3 \end{bmatrix}$

13. $\begin{bmatrix} 50 & -82 \\ -34 & 57 \\ -15 & 62 \end{bmatrix} + \begin{bmatrix} -55 & 82 \\ 45 & 14 \\ 26 & -67 \end{bmatrix}$

14. $\begin{bmatrix} 4.7 & 2.1 & -9.6 \\ -6.8 & 4.8 & 7.4 \\ -1.9 & 0.7 & 5.9 \end{bmatrix} + \begin{bmatrix} -4.9 & -9.6 & -2.1 \\ 3.4 & 0.7 & 0.0 \\ 5.6 & 10.1 & -1.6 \end{bmatrix}$

In Exercises 15–34, use matrices A, B, C, and D to find the indicated matrices. If the operations cannot be performed, explain why.

$$A = \begin{bmatrix} 6 & -3 \\ 4 & -5 \end{bmatrix} \quad B = \begin{bmatrix} 3 & 12 \\ -9 & -6 \end{bmatrix}$$

$$C = \begin{bmatrix} -1 & 4 & -7 \\ 2 & -6 & 11 \end{bmatrix} \quad D = \begin{bmatrix} 7 & 9 & -6 \\ 4 & -1 & -8 \end{bmatrix}$$

15. $A + B$ **16.** $A - B$

17. $A + C$ **18.** $2A + B$

19. $C - D$ **20.** $2C + D$

21. $-2C + D$ **22.** $2B - D$

23. $D - 4C$ **24.** $-C - 2D$

25. $B - 3A$ **26.** $2A - \frac{1}{3}B$

27. $3A$ **28.** $-2C$

29. $C + 3D$ **30.** $B - 0.5A$

31. $5A - 4B$ **32.** $-4C - 3D$

33. $-6B - 4A$ **34.** $A - B + 2C$

In Exercise 35–38, use matrices A and B to show that the indicated laws hold for these matrices. In Exercise 35, explain the meaning of the result.

$$A = \begin{bmatrix} -1 & 2 & 3 & 7 \\ 0 & -3 & -1 & 4 \\ 9 & -1 & 0 & -2 \end{bmatrix} \quad B = \begin{bmatrix} 4 & -1 & -3 & 0 \\ 5 & 0 & -1 & 1 \\ 1 & 11 & 8 & 2 \end{bmatrix}$$

35. $A + B = B + A$ **36.** $A + O = A$

37. $-(A - B) = B - A$ **38.** $3(A + B) = 3A + 3B$

In Exercises 39 and 40, find the unknown quantities in the given matrix equations.

39. An airplane is flying in a direction 21.0° north of east at 235 km/h but is headed 14.5° north of east. The wind is from the southeast. Find the speed of the wind v_w and the speed of the plane v_p relative to the wind from the given matrix equation. See Fig. 16.2.

$$\begin{bmatrix} v_p \cos 14.5° - v_w \cos 45.0° \\ v_p \sin 14.5° + v_w \sin 45.0° \end{bmatrix} = \begin{bmatrix} 235 \cos 21.0° \\ 235 \sin 21.0° \end{bmatrix}$$

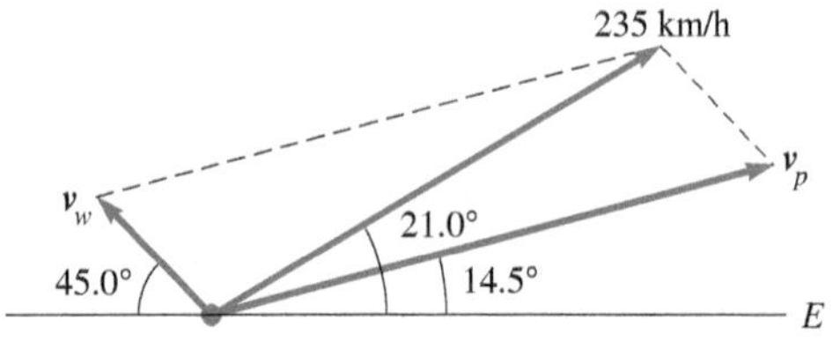

Fig. 16.2

40. Find the electric currents shown in Fig. 16.3 by solving the following matrix equation:

$$\begin{bmatrix} I_1 + I_2 + I_3 \\ -2I_1 + 3I_2 \\ -3I_2 + 6I_3 \end{bmatrix} = \begin{bmatrix} 0 \\ 24 \\ 0 \end{bmatrix}$$

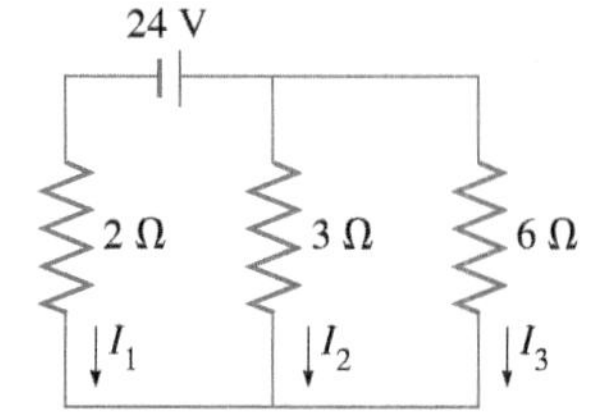

Fig. 16.3

In Exercises 41–44, perform the indicated matrix operations.

41. The contractor of a housing development constructs four different types of houses, with either a carport, a one-car garage, or a two-car garage. The following matrix shows the number of houses of each type and the type of garage.

	Type A	Type B	Type C	Type D
Carport	96	75	0	0
1-car garage	62	44	24	0
2-car garage	0	35	68	78

If the contractor builds two additional identical developments, find the matrix showing the total number of each house–garage type built in the three developments.

42. The inventory of a drug supply company shows that the following numbers of cases of bottles of vitamins C and B_3 (niacin) are in stock: vitamin C—25 cases of 100-mg bottles, 10 cases of 250-mg bottles, and 32 cases of 500-mg bottles; vitamin B_3—30 cases of 100-mg bottles, 18 cases of 250-mg bottles, and 40 cases of 500-mg bottles. This is represented by matrix *A* below. After two shipments are sent out, each of which can be represented by matrix *B* below, find the matrix that represents the remaining inventory.

$$A = \begin{bmatrix} 25 & 10 & 32 \\ 30 & 18 & 40 \end{bmatrix} \quad B = \begin{bmatrix} 10 & 5 & 6 \\ 12 & 4 & 8 \end{bmatrix}$$

43. One serving of brand K of breakfast cereal provides the given percentages of the given vitamins and minerals: vitamin A, 15%; vitamin C, 25%; calcium, 10%; iron, 25%. One serving of brand G provides: vitamin A, 10%; vitamin C, 10%; calcium, 10%; iron, 45%. One serving of tomato juice provides: vitamin A, 15%; vitamin C, 30%; calcium, 3%; iron, 3%. One serving of orange–pineapple juice provides vitamin A, 0%; vitamin C, 100%; calcium, 2%; iron, 2%. Set up a two-row, four-column matrix *B* to represent the data for the cereals and a similar matrix *J* for the juices.

44. Referring to Exercise 43, find the matrix $B + J$ and explain the meaning of its elements.

Answers to Practice Exercises

1. $\begin{bmatrix} -7 & -10 \\ 0 & 8 \end{bmatrix}$ **2.** $\begin{bmatrix} 20 & 8 \\ -21 & -1 \end{bmatrix}$

16.2 Multiplication of Matrices

Unlike matrix addition and scalar multiplication, matrix multiplication is not an intuitive operation. Therefore, we begin this section by motivating the definition through the solution of a system of linear equations.

Consider the system of equations

$$\begin{aligned} 5x + y &= 7 \\ 7x + 3y &= 5 \end{aligned}$$

The solution is $x = 2$, $y = -3$. Checking the solution in each of the equations, we get

$$\begin{aligned} 5(2) + 1(-3) &= 7 \\ 7(2) + 3(-3) &= 5 \end{aligned}$$

Let us represent the coefficients of the equations by the matrix $\begin{bmatrix} 5 & 1 \\ 7 & 3 \end{bmatrix}$, the solutions by the matrix $\begin{bmatrix} 2 \\ -3 \end{bmatrix}$, and the right-side constants by the matrix $\begin{bmatrix} 7 \\ 5 \end{bmatrix}$. The definition of matrix multiplication is such that when we multiply the matrix of coefficients and the matrix of solutions, the result should be the matrix of right-side constants. The last display written in matrix form shows us how to combine the rows and columns of the two matrices in order to achieve that:

$$\begin{bmatrix} 5 & 1 \\ 7 & 3 \end{bmatrix}\begin{bmatrix} 2 \\ -3 \end{bmatrix} = \begin{bmatrix} 5(2) + 1(-3) \\ 7(2) + 3(-3) \end{bmatrix} = \begin{bmatrix} 7 \\ 5 \end{bmatrix}$$

Note that we have multiplied across rows of the first matrix and down the column of the second matrix, element by element, and then added the products. Moreover, we could not have done this if the number of columns in the first matrix were not equal to the number of rows in the second matrix. These remarks lead us to the following definition.

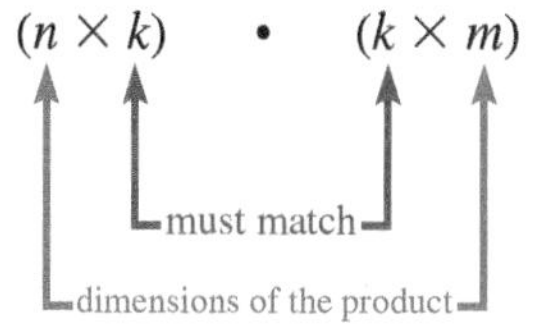

Fig. 16.4

Multiplication of Matrices

- If A is an $n \times k$ matrix, and B is a $k \times m$ matrix, then their product AB is an $n \times m$ matrix. See Fig. 16.4.
- The ij^{th} element of AB is formed by multiplying each element of row i in A by the corresponding element of column j in B and then adding these products.
- The product BA requires that $n = m$, so just because you can form the product AB does not mean that you can form the product BA. Clearly, $AB \neq BA$ in general, so that **matrix multiplication is not commutative.**

EXAMPLE 1 Multiplying matrices

Find the product AB and the product BA (if possible), where

$$A = \begin{bmatrix} 2 & 1 \\ -3 & 0 \\ 1 & 2 \end{bmatrix} \qquad B = \begin{bmatrix} -1 & 6 & 5 & -2 \\ 3 & 0 & 1 & -4 \end{bmatrix}$$

Matrix A is 3×2, and matrix B is 2×4, so the product AB can be formed, resulting in a 3×4 matrix. The calculations are shown below, with the elements used to form the element AB_{11} and the element AB_{32} outlined in colour.

$$\begin{bmatrix} 2 & 1 \\ -3 & 0 \\ 1 & 2 \end{bmatrix}\begin{bmatrix} -1 & 6 & 5 & -2 \\ 3 & 0 & 1 & -4 \end{bmatrix} = \begin{bmatrix} 2(-1)+1(3) & 2(6)+1(0) & 2(5)+1(1) & 2(-2)+1(-4) \\ -3(-1)+0(3) & -3(6)+0(0) & -3(5)+0(1) & -3(-2)+0(-4) \\ 1(-1)+2(3) & 1(6)+2(0) & 1(5)+2(1) & 1(-2)+2(-4) \end{bmatrix}$$

$$= \begin{bmatrix} 1 & 12 & 11 & -8 \\ 3 & -18 & -15 & 6 \\ 5 & 6 & 7 & -10 \end{bmatrix}$$

In trying to form the product BA, we see that B has four columns and A has three rows. Since these numbers are not the same, the product cannot be formed.

Practice Exercise

1. Find the product AB.

$$A = \begin{bmatrix} -1 & 4 \\ 8 & -2 \\ 0 & 12 \end{bmatrix} \quad B = \begin{bmatrix} 5 & -3 \\ -7 & 10 \end{bmatrix}$$

EXAMPLE 2 Multiplying matrices

■ Note that, if the second matrix had been to the left of the first, then the product would have had four rows and four columns.

The product of two matrices below may be formed because the first matrix has four columns and the second matrix has four rows. The matrix is formed as shown.

$$\begin{bmatrix} -1 & 9 & 3 & -2 \\ 2 & 0 & -7 & 1 \end{bmatrix}\begin{bmatrix} 6 & -2 \\ 1 & 0 \\ 3 & -5 \\ 3 & 9 \end{bmatrix} = \begin{bmatrix} -1(6)+9(1)+3(3)+(-2)(3) & -1(-2)+9(0)+3(-5)+(-2)(9) \\ 2(6)+0(1)+(-7)(3)+1(3) & 2(-2)+0(0)+(-7)(-5)+1(9) \end{bmatrix}$$

$$= \begin{bmatrix} -6+9+9-6 & 2+0-15-18 \\ 12+0-21+3 & -4+0+35+9 \end{bmatrix}$$

$$= \begin{bmatrix} 6 & -31 \\ -6 & 40 \end{bmatrix}$$

IDENTITY MATRIX

There are two special matrices of particular importance in the multiplication of square matrices. The first of these is the **identity matrix I**, *which is a square matrix with 1's for elements of the principal diagonal with all other elements zero.* (The principal diagonal starts with the element a_{11}.) It has the property that if it is multiplied by another square matrix with the same number of rows and columns, then the second matrix equals the product matrix.

EXAMPLE 3 Identity matrix

Show that $AI = IA = A$ for the matrix

$$A = \begin{bmatrix} 2 & -3 \\ 4 & 1 \end{bmatrix}$$

Since A is 2×2, we choose I to be 2×2. Therefore, for this case,

$$I = \begin{bmatrix} 1 & 0 \\ 0 & 1 \end{bmatrix}$$ elements of principal diagonal are 1's

Forming the indicated products, we have results as follows:

$$AI = \begin{bmatrix} 2 & -3 \\ 4 & 1 \end{bmatrix}\begin{bmatrix} 1 & 0 \\ 0 & 1 \end{bmatrix}$$

$$= \begin{bmatrix} 2(1) + (-3)(0) & 2(0) + (-3)(1) \\ 4(1) + 1(0) & 4(0) + 1(1) \end{bmatrix} = \begin{bmatrix} 2 & -3 \\ 4 & 1 \end{bmatrix}$$

$$IA = \begin{bmatrix} 1 & 0 \\ 0 & 1 \end{bmatrix}\begin{bmatrix} 2 & -3 \\ 4 & 1 \end{bmatrix}$$

$$= \begin{bmatrix} 1(2) + 0(4) & 1(-3) + 0(1) \\ 0(2) + 1(4) & 0(-3) + 1(1) \end{bmatrix} = \begin{bmatrix} 2 & -3 \\ 4 & 1 \end{bmatrix}$$

Therefore, we see that $AI = IA = A$.

INVERSE OF A MATRIX

For a given square matrix A, its **inverse** A^{-1} (when it exists) is the other important special matrix. *The matrix A and its inverse A^{-1} have the property that*

$$AA^{-1} = A^{-1}A = I \qquad \textbf{(16.5)}$$

If the product of two square matrices equals the identity matrix, the matrices are called inverses of each other. In the next section, we develop the procedure for finding the inverse of a square matrix, and the section that follows shows how the inverse is used in the solution of systems of equations. At this point, we simply show that the product of certain matrices equals the identity matrix and that therefore these matrices are inverses of each other.

EXAMPLE 4 Inverse matrix

For the given matrices A and B, show that $AB = BA = I$, and therefore that $B = A^{-1}$:

$$A = \begin{bmatrix} 1 & -3 \\ -2 & 7 \end{bmatrix} \qquad B = \begin{bmatrix} 7 & 3 \\ 2 & 1 \end{bmatrix}$$

Forming the products AB and BA, we have the following:

$$AB = \begin{bmatrix} 1 & -3 \\ -2 & 7 \end{bmatrix}\begin{bmatrix} 7 & 3 \\ 2 & 1 \end{bmatrix} = \begin{bmatrix} 7-6 & 3-3 \\ -14+14 & -6+7 \end{bmatrix} = \begin{bmatrix} 1 & 0 \\ 0 & 1 \end{bmatrix}$$

$$BA = \begin{bmatrix} 7 & 3 \\ 2 & 1 \end{bmatrix}\begin{bmatrix} 1 & -3 \\ -2 & 7 \end{bmatrix} = \begin{bmatrix} 7-6 & -21+21 \\ 2-2 & -6+7 \end{bmatrix} = \begin{bmatrix} 1 & 0 \\ 0 & 1 \end{bmatrix}$$

Since $AB = I$ and $BA = I$, $B = A^{-1}$ and $A = B^{-1}$.

Practice Exercise

2. Show that $AB = BA = I$.

$$A = \begin{bmatrix} 5 & -7 \\ -2 & 3 \end{bmatrix} \quad B = \begin{bmatrix} 3 & 7 \\ 2 & 5 \end{bmatrix}$$

EXAMPLE 5 Matrix multiplication—manufacturing

A company makes three types of automobile transmissions: 4-gear manual (type X), 4-gear automatic (type Y), and 5-gear automatic (type Z). In one day, it produces 40 of type X, 50 of type Y, and 80 of type Z. Required for production are 4 units of material and 1 worker-hour for type X, 5 units of material and 2 worker-hours for type Y, and 3 units of material and 2 worker-hours for type Z.

Letting matrix A represent the number of each type produced, and matrix B represent the parts and time requirements, we have

$$A = \begin{bmatrix} 40 & 50 & 80 \end{bmatrix} \qquad B = \begin{bmatrix} 4 & 1 \\ 5 & 2 \\ 3 & 2 \end{bmatrix}$$

(In A, the columns are type X, type Y, type Z: number of each type produced. In B, the columns are units of material and worker-hours; the rows are type X, type Y, type Z: material and time required for each.)

The product AB gives the total number of units of material and the total number of worker-hours needed for the day's production in a one-row, two-column matrix:

$$AB = \begin{bmatrix} 40 & 50 & 80 \end{bmatrix} \begin{bmatrix} 4 & 1 \\ 5 & 2 \\ 3 & 2 \end{bmatrix}$$

$$= \begin{bmatrix} 160 + 250 + 240 & 40 + 100 + 160 \end{bmatrix} = \begin{bmatrix} 650 & 300 \end{bmatrix}$$

(total units of material; total worker-hours)

Therefore, 650 units of material and 300 worker-hours are required.

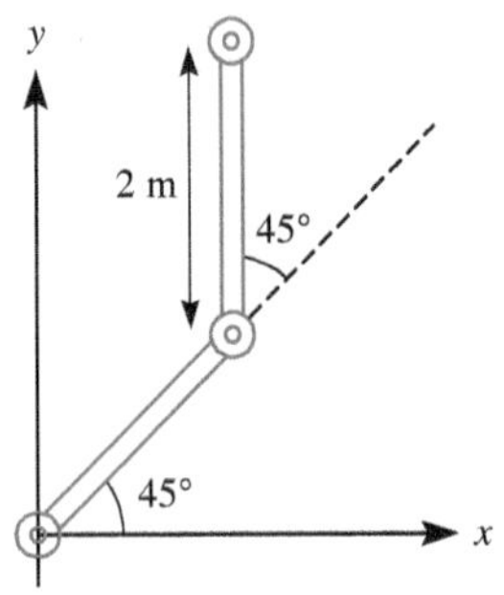

Fig. 16.5

■ See the chapter introduction.

EXAMPLE 6 Multiplication of matrices—robotic arms

Consider the two-dimensional robotic arm in Fig. 16.5, consisting of two links, each one 2.0 m long. The first link has been rotated 45° with respect to the x-axis, and the second link has been rotated 45° with respect to the first link. We can find the coordinates of the end of the arm by applying transformation matrices to the origin of our coordinate system.

We start by representing a point (x, y) as the vector (a 3×1 matrix) $\begin{bmatrix} x \\ y \\ 1 \end{bmatrix}$.

In particular, the origin is represented by the vector $\begin{bmatrix} 0 \\ 0 \\ 1 \end{bmatrix}$. If the coordinate system is rotated by an angle θ and then translated by m units in the x-direction and n units in the y-direction, the new coordinates of the origin are found by the matrix multiplication

$$\begin{bmatrix} \cos\theta & -\sin\theta & m \\ \sin\theta & \cos\theta & n \\ 0 & 0 & 1 \end{bmatrix} \begin{bmatrix} 0 \\ 0 \\ 1 \end{bmatrix}$$

For our example, $\theta = 45°$, $m = 2$ (the length of a link), and $n = 0$, so the transformation matrix becomes

$$\begin{bmatrix} \cos 45° & -\sin 45° & 2 \\ \sin 45° & \cos 45° & 0 \\ 0 & 0 & 1 \end{bmatrix} = \begin{bmatrix} \frac{1}{\sqrt{2}} & -\frac{1}{\sqrt{2}} & 2 \\ \frac{1}{\sqrt{2}} & \frac{1}{\sqrt{2}} & 0 \\ 0 & 0 & 1 \end{bmatrix}$$

Since the process is repeated twice (once for each link), the location of the end of the arm is given by

$$\begin{bmatrix} \frac{1}{\sqrt{2}} & -\frac{1}{\sqrt{2}} & 2 \\ \frac{1}{\sqrt{2}} & \frac{1}{\sqrt{2}} & 0 \\ 0 & 0 & 1 \end{bmatrix} \begin{bmatrix} \frac{1}{\sqrt{2}} & -\frac{1}{\sqrt{2}} & 2 \\ \frac{1}{\sqrt{2}} & \frac{1}{\sqrt{2}} & 0 \\ 0 & 0 & 1 \end{bmatrix} \begin{bmatrix} 0 \\ 0 \\ 1 \end{bmatrix} = \begin{bmatrix} 0 & -1 & 2 + \sqrt{2} \\ 1 & 0 & \sqrt{2} \\ 0 & 0 & 1 \end{bmatrix} \begin{bmatrix} 0 \\ 0 \\ 1 \end{bmatrix} = \begin{bmatrix} 2 + \sqrt{2} \\ \sqrt{2} \\ 1 \end{bmatrix}$$

which represents the point $(2 + \sqrt{2}, \sqrt{2})$.

This method can be generalized to robotic arms in three dimensions by including more variables. Moreover, transformation matrices are also useful for computer graphics and CAD.

We have seen that *matrix multiplication is not commutative* (see Example 1); that is, $AB \neq BA$ in general. This differs from the multiplication of real numbers. Another difference is that it is possible that $AB = O$, even though neither A nor B is O. Moreover, the only number that does not have a multiplicative inverse is 0, yet there are many nonzero matrices that do not have an inverse. There are also some similarities, for example, $AI = A$, where I and the number 1 are analogous. Also, *the distributive property* $A(B + C) = AB + AC$ *holds for matrix multiplication.*

EXERCISES 16.2

In Exercises 1 and 2, make the given changes in the indicated examples of this section and then perform the indicated multiplications.

1. In Example 2, interchange columns 1 and 2 in matrix A and then do the multiplication.
2. In Example 4, in A change -2 to 2 and -3 to 3, in B change 2 to -2 and 3 to -3, and then do the multiplications.

In Exercises 3–14, perform the indicated multiplications.

3. $\begin{bmatrix} 4 & -2 \end{bmatrix}\begin{bmatrix} -1 & 0 \\ 2 & 6 \end{bmatrix}$

4. $\begin{bmatrix} -\frac{1}{2} & 6 & -\frac{2}{3} \end{bmatrix}\begin{bmatrix} 4 & 8 \\ \frac{1}{3} & -\frac{1}{2} \\ 0 & 9 \end{bmatrix}$

5. $\begin{bmatrix} 2 & -3 & 1 \\ 0 & 7 & -3 \end{bmatrix}\begin{bmatrix} 90 \\ -25 \\ 50 \end{bmatrix}$

6. $\begin{bmatrix} 0 & -1 & 2 \\ 4 & 11 & 2 \end{bmatrix}\begin{bmatrix} 3 & -1 \\ 1 & 2 \\ 6 & 1 \end{bmatrix}$

7. $\begin{bmatrix} -8 & \frac{3}{4} \\ \frac{7}{2} & -8 \\ -6 & \frac{4}{5} \end{bmatrix}\begin{bmatrix} \frac{1}{4} & -3 \\ 2 & 5 \end{bmatrix}$

8. $\begin{bmatrix} 12 & -47 \\ 43 & -18 \\ 36 & -22 \end{bmatrix}\begin{bmatrix} 25 \\ 66 \end{bmatrix}$

9. $\begin{bmatrix} -1 & 7 \\ 3 & 5 \\ 10 & -1 \\ -5 & 12 \end{bmatrix}\begin{bmatrix} 2 & 1 \\ 5 & -3 \end{bmatrix}$

10. $\begin{bmatrix} 5 & 4 \end{bmatrix}\begin{bmatrix} 4 & -4 \\ -5 & 5 \end{bmatrix}$

11. $\begin{bmatrix} 2 & -3 \\ 5 & -1 \end{bmatrix}\begin{bmatrix} 3 & 0 & -1 \\ 7 & -5 & 8 \end{bmatrix}$

12. $\begin{bmatrix} -7 & 8 \\ 5 & 0 \end{bmatrix}\begin{bmatrix} -90 & 100 \\ 10 & 40 \end{bmatrix}$

13. $\begin{bmatrix} -9.2 & 2.3 & 0.5 \\ -3.8 & -2.4 & 9.2 \end{bmatrix}\begin{bmatrix} 6.5 & -5.2 \\ 4.9 & 1.7 \\ -1.8 & 6.9 \end{bmatrix}$

14. $\begin{bmatrix} 1 & 2 & -6 & 6 & 1 \\ -2 & 4 & 0 & 1 & 2 \end{bmatrix}\begin{bmatrix} 1 \\ -1 \\ 0 \\ 5 \\ 2 \end{bmatrix}$

In Exercises 15–18, find, if possible, AB and BA. If it is not possible, explain why.

15. $A = \begin{bmatrix} 1 & -3 & 8 \end{bmatrix} \quad B = \begin{bmatrix} -1 \\ 5 \\ 7 \end{bmatrix}$

16. $A = \begin{bmatrix} -3 & 2 & 0 \\ 1 & -4 & 5 \end{bmatrix} \quad B = \begin{bmatrix} -2 & 0 \\ 4 & -6 \\ 5 & 1 \end{bmatrix}$

17. $A = \begin{bmatrix} -10 & 25 & 40 \\ 42 & -5 & 0 \end{bmatrix} \quad B = \begin{bmatrix} 6 \\ -15 \\ 12 \end{bmatrix}$

18. $A = \begin{bmatrix} -2 & 1 & 7 \\ 3 & -1 & 0 \\ 0 & 2 & -1 \end{bmatrix} \quad B = \begin{bmatrix} 4 & -1 & 5 \end{bmatrix}$

In Exercises 19–22, show that $AI = IA = A$.

19. $A = \begin{bmatrix} 1 & 8 \\ -2 & 2 \end{bmatrix}$

20. $A = \begin{bmatrix} -15 & 28 \\ -5 & 64 \end{bmatrix}$

21. $A = \begin{bmatrix} 1 & 3 & -5 \\ 2 & 0 & 1 \\ 1 & -2 & 4 \end{bmatrix}$

22. $A = \begin{bmatrix} -1 & 2 & 0 \\ 4 & -3 & 1 \\ 2 & 1 & 3 \end{bmatrix}$

In Exercises 23–26, determine whether or not $B = A^{-1}$.

23. $A = \begin{bmatrix} 5 & -2 \\ -2 & 1 \end{bmatrix} \quad B = \begin{bmatrix} 1 & 2 \\ 2 & 5 \end{bmatrix}$

24. $A = \begin{bmatrix} 3 & -4 \\ 5 & -7 \end{bmatrix} \quad B = \begin{bmatrix} 7 & -4 \\ 5 & -2 \end{bmatrix}$

25. $A = \begin{bmatrix} 1 & -2 & 3 \\ 2 & -5 & 7 \\ -1 & 3 & -5 \end{bmatrix} \quad B = \begin{bmatrix} 4 & -1 & 1 \\ 3 & -2 & -1 \\ 1 & -1 & -1 \end{bmatrix}$

26. $A = \begin{bmatrix} 1 & -1 & 3 \\ 3 & -4 & 8 \\ -2 & 3 & -4 \end{bmatrix} \quad B = \begin{bmatrix} 8 & -5 & -4 \\ 4 & -2 & -1 \\ -1 & 1 & 1 \end{bmatrix}$

In Exercises 27–30, determine by matrix multiplication whether or not A is the proper matrix of solution values.

27. $3x - 2y = -1$
$4x + y = 6$ $\quad A = \begin{bmatrix} 1 \\ 2 \end{bmatrix}$

28. $4x + y = -5$
$3x + 4y = 6$ $\quad A = \begin{bmatrix} -2 \\ 3 \end{bmatrix}$

29. $3x + y + 2z = 1$
$x - 3y + 4z = -3$
$2x + 2y + z = 1$ $\quad A = \begin{bmatrix} -1 \\ 2 \\ 1 \end{bmatrix}$

30. $2x - y + z = 7$
$x - 3y + 2z = 6$
$3x + y - z = 8$ $\quad A = \begin{bmatrix} 3 \\ -2 \\ -1 \end{bmatrix}$

In Exercises 31–34, perform the indicated matrix multiplications, using the following matrices. For matrix A, $A^2 = A \times A$.

$$A = \begin{bmatrix} 2 & -3 & -5 \\ -1 & 4 & 5 \\ 1 & -3 & -4 \end{bmatrix} \quad B = \begin{bmatrix} 1 & -2 & -6 \\ -3 & 2 & 9 \\ 2 & 0 & -3 \end{bmatrix}$$

$$C = \begin{bmatrix} 1 & -3 & -4 \\ -1 & 3 & 4 \\ 1 & -3 & -4 \end{bmatrix}$$

31. Show that $A^2 = A$.

32. Show that $C^2 = O$.

33. Show that $B^3 = B$.

34. Show that $A^3B^3 = AB$.

In Exercises 35–47, perform the indicated matrix multiplications.

35. For matrices $A = \begin{bmatrix} a & b \\ b & a \end{bmatrix}$ and $B = \begin{bmatrix} c & d \\ d & c \end{bmatrix}$, show that $AB = BA$.

36. For matrix $A = \begin{bmatrix} 1 & 2 & 2 \\ 2 & 1 & 2 \\ 2 & 2 & 1 \end{bmatrix}$, show that $A^2 - 4A - 5I = O$.

37. Using two rows and columns, show that $(-I)^2 = I$.

38. For $J = \begin{bmatrix} j & 0 \\ 0 & j \end{bmatrix}$, where $j = \sqrt{-1}$, show that $J^2 = -I$, $J^3 = -J$, and $J^4 = I$. Explain the similarity with j^2, j^3, and j^4.

39. Show that $A^2 - I = (A + I)(A - I)$ for $A = \begin{bmatrix} 2 & 4 \\ 3 & 5 \end{bmatrix}$.

40. In the study of polarized light, the matrix product $\begin{bmatrix} 1 & 0 \\ 0 & -j \end{bmatrix}\begin{bmatrix} 1 & 0 \\ 1 & -j \end{bmatrix}\begin{bmatrix} 1 \\ 1 \end{bmatrix}$ occurs ($j = \sqrt{-1}$). Find this product.

41. In studying the motion of electrons, one of the Pauli spin matrices used is $\sigma_y = \begin{bmatrix} 0 & -j \\ j & 0 \end{bmatrix}$, where $j = \sqrt{-1}$. Show that $\sigma_y^2 = I$.

42. In analysing the motion of a robotic mechanism, the following matrix multiplication is used. Perform the multiplication and evaluate each element of the result. (See Example 6.)

$$\begin{bmatrix} \cos 60° & -\sin 60° & 0 \\ \sin 60° & \cos 60° & 0 \\ 0 & 0 & 1 \end{bmatrix}\begin{bmatrix} 2 \\ 4 \\ 1 \end{bmatrix}$$

43. In an *ammeter,* nearly all the electric current flows through a *shunt,* and the remaining known fraction of current is measured by the meter. See Fig. 16.6. From the given matrix equation, find voltage V_2 and current i_2 in terms of V_1, i_1, and resistance R, whichever may be applicable.

$$\begin{bmatrix} V_2 \\ i_2 \end{bmatrix} = \begin{bmatrix} 1 & 0 \\ -\frac{1}{R} & 1 \end{bmatrix}\begin{bmatrix} V_1 \\ i_1 \end{bmatrix}$$

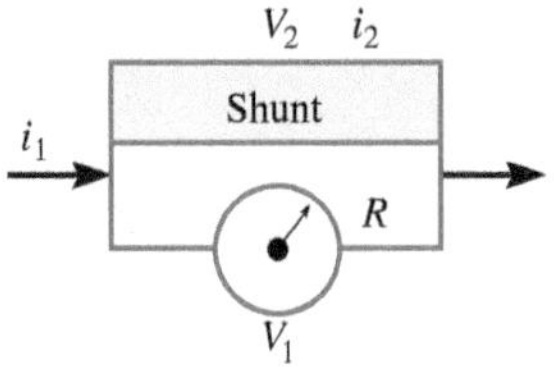

Fig. 16.6

44. In the theory related to the reproduction of colour photography, the equation

$$\begin{bmatrix} X \\ Y \\ Z \end{bmatrix} = \begin{bmatrix} 1.0 & 0.1 & 0 \\ 0.5 & 1.0 & 0.1 \\ 0.3 & 0.4 & 1.0 \end{bmatrix}\begin{bmatrix} x \\ y \\ z \end{bmatrix}$$

is found. The X, Y, and Z represent the red, green, and blue densities of the reproductions, respectively, and the x, y, and z represent the red, green, and blue densities, respectively, of the subject. Give the equations relating X, Y, and Z and x, y, and z.

45. The path of an Earth satellite can be written as

$$\begin{bmatrix} x & y \end{bmatrix}\begin{bmatrix} 7.10 & -1 \\ 1 & 7.23 \end{bmatrix}\begin{bmatrix} x \\ y \end{bmatrix} = \begin{bmatrix} 5.13 \times 10^8 \end{bmatrix}$$

where distances are in kilometres. What type of curve is represented? (See Section 14.1.)

46. Using Kirchhoff's laws on the circuit shown in Fig. 16.7, the following matrix equation is found. By matrix multiplication, find the resulting system of equations.

$$\begin{bmatrix} R_1+R_2 & -R_2 & 0 \\ -R_2 & R_2+R_3+R_4 & -R_4 \\ 0 & -R_4 & R_4+R_5 \end{bmatrix}\begin{bmatrix} I_1 \\ I_2 \\ I_3 \end{bmatrix} = \begin{bmatrix} V_1 \\ 0 \\ -V_2 \end{bmatrix}$$

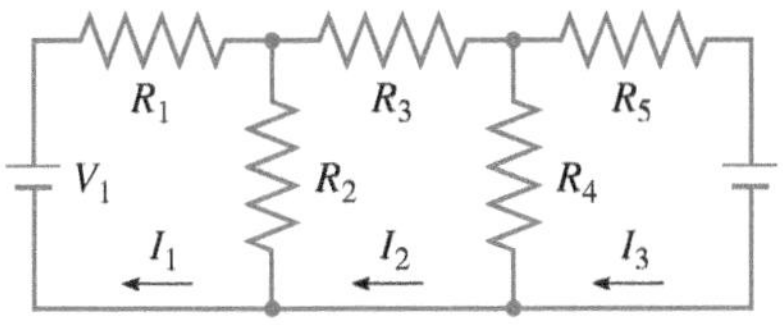

Fig. 16.7

47. The matrix A represents the connectivity between airports in five cities.

$$A = \begin{bmatrix} 0 & 1 & 0 & 1 & 0 \\ 1 & 0 & 0 & 0 & 1 \\ 0 & 0 & 0 & 1 & 1 \\ 1 & 0 & 1 & 0 & 0 \\ 0 & 1 & 1 & 0 & 0 \end{bmatrix}$$

The entry a_{ij} is 1 if there is a direct flight from city i to city j, and 0 otherwise. Calculate A^2, where the element i, j represents the number of one-stop flights between city i and city j. What would the elements of the matrix $A + A^2$ represent?

Answers to Practice Exercises

1. $\begin{bmatrix} -33 & 43 \\ 54 & -44 \\ -84 & 120 \end{bmatrix}$ 2. $AB = BA = \begin{bmatrix} 1 & 0 \\ 0 & 1 \end{bmatrix}$

16.3 Finding the Inverse of a Matrix

In this section we discuss two methods for finding the inverse of a square matrix. The first method is straightforward, but it can only be used for 2 × 2 matrices.

Finding the Inverse of a 2 × 2 Matrix	EXAMPLE 1
	Find the inverse of $A = \begin{bmatrix} 2 & -3 \\ 4 & -7 \end{bmatrix}$.
1. Evaluate the determinant. If the determinant is zero, the inverse does not exist.	$\begin{vmatrix} 2 & -3 \\ 4 & -7 \end{vmatrix} = -14 - (-12) = -2$ Since it is not zero, the inverse exists.
2. Interchange the elements on the principal diagonal.	$\begin{bmatrix} -7 & -3 \\ 4 & 2 \end{bmatrix}$ ← elements interchanged
3. Change the signs of the off-diagonal elements.	$\begin{bmatrix} -7 & 3 \\ -4 & 2 \end{bmatrix}$ ← signs changed
4. Divide each resulting element by the determinant.	$A^{-1} = \begin{bmatrix} \frac{-7}{-2} & \frac{3}{-2} \\ \frac{-4}{-2} & \frac{2}{-2} \end{bmatrix} = \begin{bmatrix} \frac{7}{2} & -\frac{3}{2} \\ 2 & -1 \end{bmatrix}$
5. Check by multiplication that $AA^{-1} = I$.	$AA^{-1} = \begin{bmatrix} 2 & -3 \\ 4 & -7 \end{bmatrix}\begin{bmatrix} \frac{7}{2} & -\frac{3}{2} \\ 2 & -1 \end{bmatrix}$ $= \begin{bmatrix} 7-6 & -3+3 \\ 14-14 & -6+7 \end{bmatrix} = \begin{bmatrix} 1 & 0 \\ 0 & 1 \end{bmatrix} = I$

Practice Exercise

1. Find the inverse:

$A = \begin{bmatrix} 3 & -8 \\ 1 & -2 \end{bmatrix}$.

■ In Exercise 37, you are asked to verify the method used in Example 1.

■ The Gauss–Jordan method is named for the German mathematician Karl Gauss (1777–1855) and the German geodesist Wilhelm Jordan (1842–1899).

The second method, called the *Gauss–Jordan method*, is applicable for square matrices of any size.

The Inverse of a Matrix by the Gauss–Jordan Method

1. Set up the given matrix and the identity matrix of the same size side by side.
2. Transform the given matrix into the identity matrix by performing any of the following allowable **row operations**:
   a. Any two rows may be interchanged.
   b. Every element in any row may be multiplied by any number other than zero.
   c. Any row may be replaced by a row whose elements are the sum of a nonzero multiple of itself and a nonzero multiple of another row.

   Work one column at a time, transforming the columns in order from left to right.
3. At every step, perform the same row operations on the identity matrix. The resulting matrix will be the required inverse.

Note that these are **row operations**, not column operations, and that they are the operations used in solving a system of equations by elimination.

Practice Exercise

2. Find the inverse using the Gauss–Jordan method: $A = \begin{bmatrix} 3 & -8 \\ 1 & -2 \end{bmatrix}$.

LEARNING TIP

As in Example 2, (1) always work one column at a time, from left to right, (2) never undo the work in a previously completed column, and (3) make the element on the principal diagonal for the column 1 first and then make all other elements in the column 0.

EXAMPLE 2 2 × 2 inverse—Gauss–Jordan method

Find the inverse of the matrix.

$$A = \begin{bmatrix} 2 & -3 \\ 4 & -7 \end{bmatrix}$$ this is the same matrix as in Example 1

First, we set up the given matrix with the identity matrix side by side.

$$\left[\begin{array}{cc|cc} 2 & -3 & 1 & 0 \\ 4 & -7 & 0 & 1 \end{array}\right]$$

The vertical line simply shows the separation of the two matrices.

We wish to transform the left matrix into the identity matrix. We work one column at a time, from left to right. The row operations required at every step (summarized in red) are as follows:

Objective	Row Operation	Resulting Setup
Make a_{11} a 1:	Divide the first row by 2 $\left(\frac{1}{2}R_1\right)$	$\left[\begin{array}{cc\|cc} 1 & -\frac{3}{2} & \frac{1}{2} & 0 \\ 4 & -7 & 0 & 1 \end{array}\right]$
Make a_{21} a 0:	Subtract 4 times row 1 from row 2, replacing row 2 $(R_2 - 4R_1)$	$\left[\begin{array}{cc\|cc} 1 & -\frac{3}{2} & \frac{1}{2} & 0 \\ 4-4(1) & -7-4\left(-\frac{3}{2}\right) & 0-4\left(\frac{1}{2}\right) & 1-4(0) \end{array}\right]$ or $\left[\begin{array}{cc\|cc} 1 & -\frac{3}{2} & \frac{1}{2} & 0 \\ 0 & -1 & -2 & 1 \end{array}\right]$
Make a_{22} a 1:	Multiply row 2 by -1 $(-R_2)$	$\left[\begin{array}{cc\|cc} 1 & -\frac{3}{2} & \frac{1}{2} & 0 \\ 0 & 1 & 2 & -1 \end{array}\right]$
Make a_{12} a 0:	Add $\frac{3}{2}$ times row 2 to row 1, replacing row 1 $\left(R_1 + \frac{3}{2}R_2\right)$	$\left[\begin{array}{cc\|cc} 1+\frac{3}{2}(0) & -\frac{3}{2}+\frac{3}{2}(1) & \frac{1}{2}+\frac{3}{2}(2) & 0+\frac{3}{2}(-1) \\ 0 & 1 & 2 & -1 \end{array}\right]$ or $\left[\begin{array}{cc\|cc} 1 & 0 & \frac{7}{2} & -\frac{3}{2} \\ 0 & 1 & 2 & -1 \end{array}\right]$

At this point, we have transformed the given matrix into the identity matrix, and the identity matrix into the inverse. Therefore, the matrix to the right of the vertical bar in the last setup is the required inverse. Thus,

$$A^{-1} = \begin{bmatrix} \frac{7}{2} & -\frac{3}{2} \\ 2 & -1 \end{bmatrix}$$

EXAMPLE 3 2 × 2 inverse—Gauss–Jordan method

Find the inverse of the matrix $\begin{bmatrix} -3 & 6 \\ 4 & 5 \end{bmatrix}$.

original setup; make top left entry 1; make other entry in first column 0

$$\left[\begin{array}{cc|cc} -3 & 6 & 1 & 0 \\ 4 & 5 & 0 & 1 \end{array}\right] \xrightarrow{-\frac{1}{3}R_1} \left[\begin{array}{cc|cc} 1 & -2 & -\frac{1}{3} & 0 \\ 4 & 5 & 0 & 1 \end{array}\right] \xrightarrow{-4R_1 + R_2} \left[\begin{array}{cc|cc} 1 & -2 & -\frac{1}{3} & 0 \\ 0 & 13 & \frac{4}{3} & 0 \end{array}\right]$$

make second entry in bottom row 1; make other entry in second column 0

$$\xrightarrow{-\frac{1}{13}R_1} \left[\begin{array}{cc|cc} 1 & -2 & -\frac{1}{3} & 0 \\ 0 & 1 & \frac{4}{39} & \frac{1}{13} \end{array}\right] \xrightarrow{-2R_1 + R_2} \left[\begin{array}{cc|cc} 1 & 0 & -\frac{5}{39} & \frac{2}{13} \\ 0 & 1 & \frac{4}{39} & \frac{1}{13} \end{array}\right]$$

Therefore, $A^{-1} = \begin{bmatrix} -\frac{5}{39} & \frac{2}{13} \\ \frac{4}{39} & \frac{1}{13} \end{bmatrix}$, which can be checked by multiplication.

EXAMPLE 4 3 × 3 inverse—Gauss–Jordan method

Find the inverse of the matrix $\begin{bmatrix} 1 & 2 & -1 \\ 3 & 5 & -1 \\ -2 & -1 & -2 \end{bmatrix}$.

original setup: top left entry is already 1; make other entries in first column 0; make second entry in middle row 1

$$\left[\begin{array}{ccc|ccc} 1 & 2 & -1 & 1 & 0 & 0 \\ 3 & 5 & -1 & 0 & 1 & 0 \\ -2 & -1 & -2 & 0 & 0 & 1 \end{array}\right] \xrightarrow[2R_1 + R_3]{-3R_1 + R_2} \left[\begin{array}{ccc|ccc} 1 & 2 & -1 & 1 & 0 & 0 \\ 0 & -1 & 2 & -3 & 1 & 0 \\ 0 & 3 & -4 & 2 & 0 & 1 \end{array}\right] \xrightarrow{-1R_3} \left[\begin{array}{ccc|ccc} 1 & 2 & -1 & 1 & 0 & 0 \\ 0 & 1 & -2 & 3 & -1 & 0 \\ 0 & 3 & -4 & 2 & 0 & 1 \end{array}\right]$$

make other entries in second column 0; make third entry in bottom row 1; make other entries in third column 0

$$\xrightarrow[-3R_2 + R_3]{-2R_2 + R_1} \left[\begin{array}{ccc|ccc} 1 & 0 & 3 & -5 & 2 & 0 \\ 0 & 1 & -2 & 3 & -1 & 0 \\ 0 & 0 & 2 & -7 & 3 & 0 \end{array}\right] \xrightarrow{\frac{1}{2}R_3} \left[\begin{array}{ccc|ccc} 1 & 0 & 3 & -5 & 2 & 0 \\ 0 & 1 & -2 & 3 & -1 & 0 \\ 0 & 0 & 1 & -\frac{7}{2} & \frac{3}{2} & \frac{1}{2} \end{array}\right] \xrightarrow[2R_3 + R_2]{-3R_3 + R_1} \left[\begin{array}{ccc|ccc} 1 & 0 & 0 & \frac{11}{2} & -\frac{5}{2} & -\frac{3}{2} \\ 0 & 1 & 0 & -4 & 2 & 1 \\ 0 & 0 & 1 & -\frac{7}{2} & \frac{3}{2} & \frac{1}{2} \end{array}\right]$$

Therefore, the required inverse matrix is

$$\begin{bmatrix} \frac{11}{2} & -\frac{5}{2} & -\frac{3}{2} \\ -4 & 2 & 1 \\ -\frac{7}{2} & \frac{3}{2} & \frac{1}{2} \end{bmatrix}$$

which may be checked by multiplication.

It is possible to find the inverse of a matrix using a calculator, a computer algebra system, or an online tool. It is advised that you consult the manual for details on how to operate the matrix features of the technology of your choice.

EXAMPLE 5 4 × 4 inverse using technology

By using technology, find the inverse of the matrix.

$$A = \begin{bmatrix} 2 & -2 & 3 & 2 \\ 3 & 1 & 5 & 2 \\ -2 & 5 & 2 & -3 \\ 4 & -5 & -1 & 4 \end{bmatrix}$$

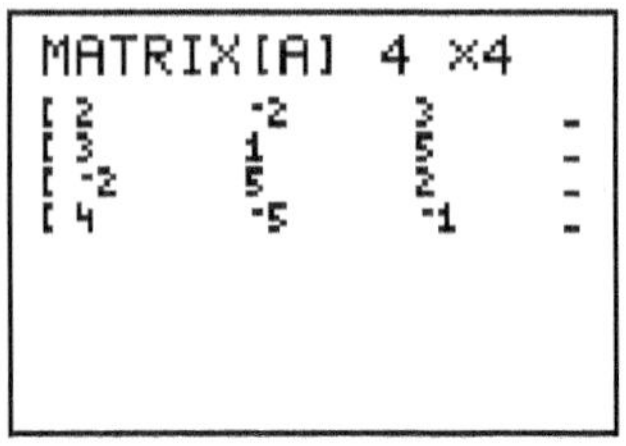

Fig. 16.8

Here we obtain the inverse using a graphing calculator. After selecting the matrix to edit from the *matrix* menu, the elements are entered as shown in Fig. 16.8. Scrolling by use of the arrow key is often necessary to enter all values.

To obtain the inverse, the matrix name is first selected from the *matrix* menu (here [A]). Then the inverse is found with the key x^{-1}. The result, displayed in fractions, is

$$\begin{bmatrix} 10/23 & -15/23 & 38/23 & 31/23 \\ -17/23 & 14/23 & -14/23 & -9/23 \\ 9/23 & -2/23 & 2/23 & -2/23 \\ -29/23 & 32/23 & -55/23 & -37/23 \end{bmatrix}$$

The elements of A^{-1} can be checked directly on the calculator by showing $AA^{-1} = I$ (if rounded decimal values are used, the values of I will be approximate, but should be sufficiently close to 1 in the principal diagonal and 0 everywhere else).

EXERCISES 16.3

In Exercises 1 and 2, make the given changes in the indicated examples of this section and then find the matrix inverses.

1. In Example 1, change the element -7 to -5 and then find the inverse using the same method.
2. In Example 3, change the element -3 to -2 and then find the the inverse using the same method.

In Exercises 3–10, find the inverse of each of the given matrices by the method of Example 1 of this section.

3. $\begin{bmatrix} 2 & -5 \\ -2 & 4 \end{bmatrix}$ **4.** $\begin{bmatrix} -6 & 3 \\ 3 & -2 \end{bmatrix}$ **5.** $\begin{bmatrix} -1 & 5 \\ 4 & 10 \end{bmatrix}$

6. $\begin{bmatrix} 8 & -1 \\ -4 & -5 \end{bmatrix}$ **7.** $\begin{bmatrix} 0 & -4 \\ 2 & 6 \end{bmatrix}$ **8.** $\begin{bmatrix} 7 & -3 \\ -6 & 4 \end{bmatrix}$

9. $\begin{bmatrix} -50 & -45 \\ 26 & 80 \end{bmatrix}$ **10.** $\begin{bmatrix} 7.2 & -3.6 \\ -1.3 & -5.7 \end{bmatrix}$

In Exercises 11–24, find the inverse of each of the given matrices by transforming the identity matrix, as in Examples 2–4.

11. $\begin{bmatrix} 1 & 2 \\ 2 & 3 \end{bmatrix}$ **12.** $\begin{bmatrix} 1 & 5 \\ -1 & -4 \end{bmatrix}$

13. $\begin{bmatrix} 2 & 4 \\ -1 & -1 \end{bmatrix}$ **14.** $\begin{bmatrix} -2 & 6 \\ 3 & -4 \end{bmatrix}$

15. $\begin{bmatrix} 2 & 5 \\ -1 & 2 \end{bmatrix}$ **16.** $\begin{bmatrix} -2 & 3 \\ -3 & 5 \end{bmatrix}$

17. $\begin{bmatrix} 25 & 30 \\ -10 & -14 \end{bmatrix}$ **18.** $\begin{bmatrix} 1 & -3 \\ 7 & -5 \end{bmatrix}$

19. $\begin{bmatrix} 1 & -3 & -2 \\ -2 & 7 & 3 \\ 1 & -1 & -3 \end{bmatrix}$ **20.** $\begin{bmatrix} 1 & 2 & -1 \\ 3 & 7 & -5 \\ -1 & -2 & 0 \end{bmatrix}$

21. $\begin{bmatrix} 1 & 3 & 2 \\ -2 & -5 & -1 \\ 2 & 4 & 0 \end{bmatrix}$ **22.** $\begin{bmatrix} 1 & 3 & 4 \\ -1 & -4 & -2 \\ 4 & 9 & 20 \end{bmatrix}$

23. $\begin{bmatrix} 2 & 4 & 0 \\ 3 & 4 & -2 \\ -1 & 1 & 2 \end{bmatrix}$ **24.** $\begin{bmatrix} -2 & 6 & 1 \\ 0 & 3 & -3 \\ 4 & -7 & 3 \end{bmatrix}$

In Exercises 25–34, find the inverse of each of the given matrices by using technology, as in Example 5. The matrices in Exercises 27–29 are the same as those in Exercises 21–23.

25. $\begin{bmatrix} 2 & 8 \\ -1 & 6 \end{bmatrix}$ **26.** $\begin{bmatrix} 20 & -45 \\ -12 & 24 \end{bmatrix}$

27. $\begin{bmatrix} 1 & 3 & 2 \\ -2 & -5 & -1 \\ 2 & 4 & 0 \end{bmatrix}$ **28.** $\begin{bmatrix} 1 & 3 & 4 \\ -1 & -4 & -2 \\ 4 & 9 & 20 \end{bmatrix}$

29. $\begin{bmatrix} 2 & 4 & 0 \\ 3 & 4 & -2 \\ -1 & 1 & 2 \end{bmatrix}$ **30.** $\begin{bmatrix} 10 & -5 & 30 \\ -2 & 4 & -5 \\ 20 & 5 & 5 \end{bmatrix}$

31. $\begin{bmatrix} 1 & -2 & 1 & 0 \\ 1 & -2 & 2 & -3 \\ 0 & 1 & -1 & 1 \\ -2 & 3 & -2 & 3 \end{bmatrix}$

32. $\begin{bmatrix} 3 & -2 & -1 & 4 \\ 2 & 0 & 5 & 1 \\ -1 & 2 & 1 & -2 \\ 4 & 1 & 3 & 5 \end{bmatrix}$

33. $\begin{bmatrix} 0.2 & 1.2 & -0.8 & -0.5 \\ -0.4 & 3.0 & -1.6 & 0.4 \\ 1.0 & -2.4 & 3.2 & 1.5 \\ -0.1 & 0.4 & 0.0 & 3.0 \end{bmatrix}$

34. $\begin{bmatrix} 12.5 & -2.6 & 1.2 & 7.6 \\ -4.6 & 10.0 & -4.7 & -6.8 \\ 5.7 & -3.7 & 7.3 & 11.0 \\ 8.8 & 6.8 & 14.0 & 4.7 \end{bmatrix}$

In Exercises 35–42, solve the given problems.

35. Show that the matrix $\begin{bmatrix} 1 & 1 \\ 1 & 1 \end{bmatrix}$ has no inverse.

36. Find the determinant of the matrix $\begin{bmatrix} 1 & -2 & 0 \\ -2 & 4 & 8 \\ 3 & -6 & 6 \end{bmatrix}$.

Explain what this tells us about its inverse.

37. For the matrix $A = \begin{bmatrix} a & b \\ c & d \end{bmatrix}$, show that

$$\frac{1}{ad - bc}\begin{bmatrix} a & b \\ c & d \end{bmatrix}\begin{bmatrix} d & -b \\ -c & a \end{bmatrix} = \begin{bmatrix} 1 & 0 \\ 0 & 1 \end{bmatrix}$$

This verifies the method of Illustrative Example 1.

38. Describe the relationship between the elements of the matrix $\begin{bmatrix} a & 0 & 0 \\ 0 & b & 0 \\ 0 & 0 & c \end{bmatrix}$ and the elements of its inverse.

39. The matrix $A = \begin{bmatrix} 2 & -1 & 1 \\ -1 & 4 & -3 \\ 1 & -3 & 2 \end{bmatrix}$ is *symmetric* (note the elements on opposite sides of the main diagonal are equal). Show that A^{-1} is also symmetric.

40. For the *four-terminal network* shown in Fig. 16.9, it can be shown that the voltage matrix V is related to the coefficient matrix A and the current matrix I by $V = A^{-1}I$, where

$$V = \begin{bmatrix} v_1 \\ v_2 \end{bmatrix} \qquad A = \begin{bmatrix} a_{11} & a_{12} \\ a_{21} & a_{22} \end{bmatrix} \qquad I = \begin{bmatrix} i_1 \\ i_2 \end{bmatrix}$$

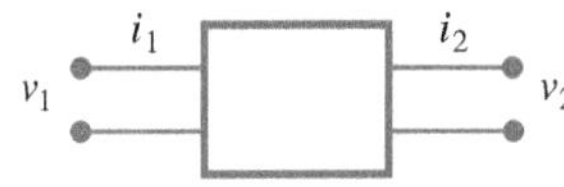

Fig. 16.9

Find the individual equations for v_1 and v_2 that give each in terms of i_1 and i_2.

41. The rotations of a robot arm such as that shown in Fig. 16.10 are often represented by matrices. The values represent trigonometric functions of the angles of rotation. For the following rotation matrix R, find R^{-1}.

$$R = \begin{bmatrix} 0.8 & 0.0 & -0.6 \\ 0.0 & 1.0 & 0.0 \\ 0.6 & 0.0 & 0.8 \end{bmatrix}$$

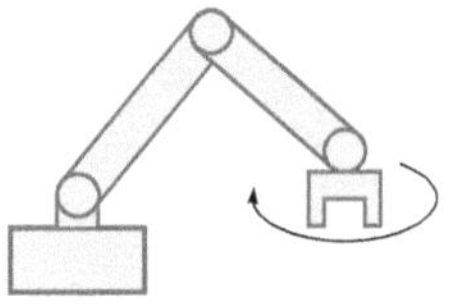

Fig. 16.10

42. In cryptography, one type of code makes use of a matrix (the encoding matrix) to encode a message. The receiver of the message decodes it using the inverse of the matrix (the decoding matrix). Given the following encoding matrix, find the decoding matrix.

$$A = \begin{bmatrix} 6 & 1 & 1 \\ 2 & 1 & 0 \\ 1 & -1 & 1 \end{bmatrix}$$

Answers to Practice Exercises

1. $\begin{bmatrix} -1 & 4 \\ -\frac{1}{2} & \frac{3}{2} \end{bmatrix}$ **2.** $\begin{bmatrix} -1 & 4 \\ -\frac{1}{2} & \frac{3}{2} \end{bmatrix}$

16.4 Matrices and Linear Equations

As we stated at the beginning of Section 16.1, matrices can be used to solve systems of linear equations. In this section we show how this can be done using the inverse of the matrix of coefficients. Since the inverse can be found using a calculator, a computer algebra system, or an online tool, systems with more than three equations can be solved more readily than with earlier methods.

Matrix Solution for an $n \times n$ System of Linear Equations	EXAMPLE 1
	Solve: $2x - y = 7$ $5x - 3y = 18$
1. Write the system as a matrix equation $AX = C$ **(16.6)** A is the $n \times n$ matrix of coefficients, X is the $n \times 1$ matrix of variables, and C is the $n \times 1$ matrix of constants.	$A = \begin{bmatrix} 2 & -1 \\ 5 & -3 \end{bmatrix}$, $X = \begin{bmatrix} x \\ y \end{bmatrix}$, $C = \begin{bmatrix} 7 \\ 18 \end{bmatrix}$

2. Find the inverse A^{-1} by any of the methods of Section 16.3.	$A^{-1} = \begin{bmatrix} 3 & -1 \\ 5 & -2 \end{bmatrix}$
3. The solution is found by multiplying the matrix of constants by A^{-1} **on the left.** That is, $X = A^{-1}C$ **(16.7)**	$A^{-1}C = \begin{bmatrix} 3 & -1 \\ 5 & -2 \end{bmatrix}\begin{bmatrix} 7 \\ 18 \end{bmatrix} = \begin{bmatrix} 3 \\ -1 \end{bmatrix}$ Therefore, $X = \begin{bmatrix} x \\ y \end{bmatrix} = \begin{bmatrix} 3 \\ -1 \end{bmatrix}$
4. Check the solution in the original equations. You can also verify that Eq. (16.6) holds.	$\begin{bmatrix} 2 & -1 \\ 5 & -3 \end{bmatrix}\begin{bmatrix} 3 \\ -1 \end{bmatrix} = \begin{bmatrix} 7 \\ 18 \end{bmatrix}$ The solution checks.

LEARNING TIP

Note that

$X = A^{-1}C$ and ***not*** CA^{-1}

as the order of matrix multiplication must be carefully followed.

Practice Exercise

1. Use matrices to solve the system of equations

$$x - 3y = 6$$
$$2x + y = 5$$

To derive Eq. (16.7), we multiply both sides of Eq. (16.6) by A^{-1} on the left. That leads to $A^{-1}AX = A^{-1}C$. Since $A^{-1}A = I$, and $IX = X$, we get $A^{-1}AX = IX = X = A^{-1}C$, as stated in Eq. (16.7).

EXAMPLE 2 Matrix solution for two equations—electric circuits

For the electric circuit shown in Fig. 16.11, the equations used to find the currents (in amperes) i_1 and i_2 are

$$\begin{aligned} 2.30i_1 + 6.45(i_1 + i_2) &= 15.0 \\ 1.25i_2 + 6.45(i_1 + i_2) &= 12.5 \end{aligned} \quad \text{or} \quad \begin{aligned} 8.75i_1 + 6.45i_2 &= 15.0 \\ 6.45i_1 + 7.70i_2 &= 12.5 \end{aligned}$$

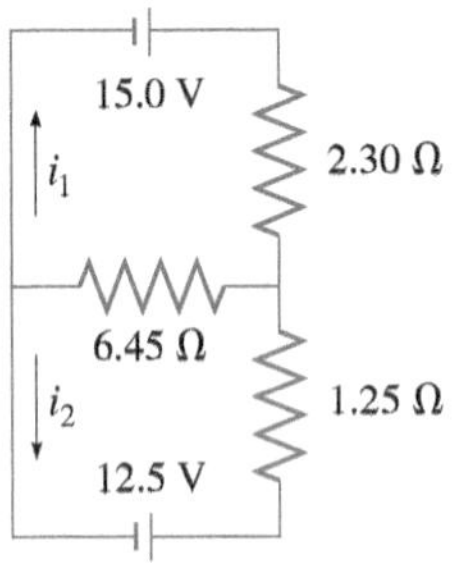

Fig. 16.11

Use matrices to solve this system of equations.

Write the system as a matrix equation $AX = C$:

$$A = \begin{bmatrix} 8.75 & 6.45 \\ 6.45 & 7.70 \end{bmatrix},$$

$$X = \begin{bmatrix} i_1 \\ i_2 \end{bmatrix}, \quad C = \begin{bmatrix} 15.0 \\ 12.5 \end{bmatrix}$$

Find the inverse A^{-1}: By any of the methods of the previous section,

$$A^{-1} = \begin{bmatrix} 0.2988 & -0.2503 \\ -0.2503 & 0.3395 \end{bmatrix}$$

Set $X = A^{-1}C$:

$$A^{-1}C = \begin{bmatrix} 0.2900 & -0.2503 \\ -0.2503 & 0.3395 \end{bmatrix}\begin{bmatrix} 15.0 \\ 12.5 \end{bmatrix} = \begin{bmatrix} 1.35 \\ 0.49 \end{bmatrix}$$

Therefore,

$$X = \begin{bmatrix} i_1 \\ i_2 \end{bmatrix} = \begin{bmatrix} 1.35 \\ 0.49 \end{bmatrix}$$

Check:

$$\begin{bmatrix} 8.75 & 6.45 \\ 6.45 & 7.70 \end{bmatrix}\begin{bmatrix} 1.35 \\ 0.49 \end{bmatrix} = \begin{bmatrix} 15.0 \\ 12.5 \end{bmatrix}$$

The solution checks.

The required currents are $i_1 = 1.35$ A and $i_2 = 0.49$ A.

EXAMPLE 3 Matrix solution—three equations

Use matrices to solve the system of equations

$$\begin{aligned} x + 4y - z &= 4 \\ x + 3y + z &= 8 \\ 2x + 6y + z &= 13 \end{aligned}$$

Write the system as a matrix equation $AX = C$:

$$A = \begin{bmatrix} 1 & 4 & -1 \\ 1 & 3 & 1 \\ 2 & 6 & 1 \end{bmatrix}, \quad X = \begin{bmatrix} x \\ y \\ z \end{bmatrix}, \quad C = \begin{bmatrix} 4 \\ 8 \\ 13 \end{bmatrix}$$

Find the inverse A^{-1} (the steps are shown below):

$$A^{-1} = \begin{bmatrix} -3 & -10 & 7 \\ 1 & 3 & -2 \\ 0 & 2 & -1 \end{bmatrix}$$

Set $X = A^{-1}C$:

$$A^{-1}C = \begin{bmatrix} -3 & -10 & 7 \\ 1 & 3 & -2 \\ 0 & 2 & -1 \end{bmatrix} \begin{bmatrix} 4 \\ 8 \\ 13 \end{bmatrix} = \begin{bmatrix} -1 \\ 2 \\ 3 \end{bmatrix}$$

Therefore,

$$X = \begin{bmatrix} x \\ y \\ z \end{bmatrix} = \begin{bmatrix} -1 \\ 2 \\ 3 \end{bmatrix}$$

Check:

$$\begin{bmatrix} 1 & 4 & -1 \\ 1 & 3 & 1 \\ 2 & 6 & 1 \end{bmatrix} \begin{bmatrix} -1 \\ 2 \\ 3 \end{bmatrix} = \begin{bmatrix} 4 \\ 8 \\ 13 \end{bmatrix}$$

The steps for finding A^{-1} are:

$$\left[\begin{array}{ccc|ccc} 1 & 2 & -1 & 1 & 0 & 0 \\ 3 & 5 & -1 & 0 & 1 & 0 \\ -2 & -1 & -2 & 0 & 0 & 1 \end{array}\right] \xrightarrow[-2R_1 + R_3]{-R_1 + R_2} \left[\begin{array}{ccc|ccc} 1 & 4 & -1 & 1 & 0 & 0 \\ 0 & -1 & 2 & -1 & 1 & 0 \\ 0 & -2 & 3 & -2 & 0 & 1 \end{array}\right] \xrightarrow{-R_2} \left[\begin{array}{ccc|ccc} 1 & 4 & -1 & 1 & 0 & 0 \\ 0 & 1 & -2 & 1 & -1 & 0 \\ 0 & -2 & 3 & -2 & 0 & 1 \end{array}\right]$$

$$\xrightarrow[2R_2 + R_3]{-4R_2 + R_1} \left[\begin{array}{ccc|ccc} 1 & 0 & 7 & -3 & 4 & 0 \\ 0 & 1 & -2 & 1 & -1 & 0 \\ 0 & 0 & 1 & 0 & 2 & -1 \end{array}\right] \xrightarrow[2R_3 + R_2]{-7R_3 + R_1} \left[\begin{array}{ccc|ccc} 1 & 0 & 0 & -3 & -10 & 7 \\ 0 & 1 & 0 & 1 & 3 & -2 \\ 0 & 0 & 1 & 0 & 2 & -1 \end{array}\right]$$

EXAMPLE 4 Matrix solution—four equations

Solve the following system of equations:

$$\begin{aligned} 2r + 4s - t + u &= 5 \\ r - 2s + 3t - u &= -4 \\ 3r + s + 2t - 4u &= 8 \\ 4r + 5s - t + 3u &= -1 \end{aligned}$$

Write the system as a matrix equation $AX = C$:

$$A = \begin{bmatrix} 2 & 4 & -1 & 1 \\ 1 & -2 & 3 & -1 \\ 3 & 1 & 2 & -4 \\ 4 & 5 & -1 & 3 \end{bmatrix}, \quad X = \begin{bmatrix} r \\ s \\ t \\ u \end{bmatrix}, \quad C = \begin{bmatrix} 5 \\ -4 \\ 8 \\ -1 \end{bmatrix}$$

Find the inverse A^{-1} and set $X = A^{-1}C$:

We solve the system using technology, in which case there is no need to record or display A^{-1}. After entering the matrices A and C, the product $A^{-1}C$ is computed directly:

$$X = \begin{bmatrix} r \\ s \\ t \\ u \end{bmatrix} = A^{-1}C = \begin{bmatrix} -2 \\ 3 \\ 0.5 \\ -2.5 \end{bmatrix}$$

Check:

We can check using the same technology that

$$\begin{bmatrix} 2 & 4 & -1 & 1 \\ 1 & -2 & 3 & -1 \\ 3 & 1 & 2 & -4 \\ 4 & 5 & -1 & 3 \end{bmatrix}\begin{bmatrix} -2 \\ 3 \\ 0.5 \\ -2.5 \end{bmatrix} = \begin{bmatrix} 5 \\ -4 \\ 8 \\ -1 \end{bmatrix}$$

EXERCISES 16.4

In Exercises 1 and 2, make the given changes in the indicated examples of this section and then solve the systems of equations.

1. In Example 1, change the 18 to 19 and then solve the system of equations.

2. In Example 3, change the 8 to 7 and the 13 to 12 and then solve the system of equations.

In Exercises 3–10, solve the given systems of equations by using the inverse of the coefficient matrix. The numbers in parentheses refer to exercises from Section 16.3, where the inverses may be checked.

3. $2x - 5y = -14$ (3)
$-2x + 4y = 11$

4. $-x + 5y = 4$ (5)
$4x + 10y = -4$

5. $x + 2y = 7$ (11)
$2x + 3y = 11$

6. $2x + 4y = -9$ (13)
$-x - y = 2$

7. $2x + 5y = -6$ (15)
$-x + 2y = -6$

8. $x - 3y - 2z = -8$ (19)
$-2x + 7y + 3z = 19$
$x - y - 3z = -3$

9. $x + 3y + 2z = 5$ (21)
$-2x - 5y - z = -1$
$2x + 4y = -2$

10. $2x + 4y = -2$ (23)
$3x + 4y - 2z = -6$
$-x + y + 2z = 5$

In Exercises 11–18, solve the given systems of equations by using the inverse of the coefficient matrix.

11. $2x - 3y = 3$
$4x - 5y = 4$

12. $x + 2y = 3$
$3x + 4y = 11$

13. $2.5x + 2.8y = -3.0$
$3.5x - 1.6y = 9.6$

14. $12x - 5y = -400$
$31x + 25y = 180$

15. $x + 2y + 2z = -4$
$4x + 9y + 10z = -18$
$-x + 3y + 7z = -7$

16. $x - 4y - 2z = -7$
$-x + 5y + 5z = 18$
$3x - 7y + 10z = 38$

17. $2x + 4y + z = 5$
$-2x - 2y - z = -6$
$-x + 2y + z = 0$

18. $4x + y = 2$
$-2x - y + 3z = -18$
$2x + y - z = 8$

In Exercises 19–26, solve the given systems of equations by using the inverse of the coefficient matrix. You may use technology to perform the necessary matrix operations. See Example 4.

19. $2x - y - z = 7$
$4x - 3y + 2z = 4$
$3x + 5y + z = -10$

20. $6x + 2y + 9z = 13$
$7x + 6y - 6z = 6$
$5x - 4y + 3z = 15$

21. $u - 3v - 2w = 9$
$3u + 2v + 6w = 20$
$4u - v + 3w = 25$

22. $2x + y - z = 1$
$3x - 2y - 8z = -3$
$x + 3y + z = 10$

23. $x - 5y + 2z - t = -18$
$3x + y - 3z + 2t = 17$
$4x - 2y + z - t = -1$
$-2x + 3y - z + 4t = 11$

24. $2p + q + 5r + s = 5$
$p + q - 3r - 4s = -1$
$3p + 6q - 2r + s = 8$
$2p + 2q + 2r - 3s = 2$

25. $2v + 3w + x - y - 2z = 6$
$6v - 2w - x + 3y - z = 21$
$v + 3w - 4x + 2y + 3z = -9$
$3v - w - x + 7y + 4z = 5$
$v + 6w + 6x - 4y - z = -4$

26. $4x - y + 2z - 2t + u = -15$
$8x + y - z + 4t - 2u = 26$
$2x - 6y - 2z + t - u = 10$
$2x + 5y + z - 3t + 8u = -22$
$4x - 3y + 2z + 4t + 2u = -4$

In Exercises 27–39, solve the indicated systems of equations using the inverse of the coefficient matrix. In Exercises 33–38, it is necessary to set up the appropriate equations.

27. For the following system of equations, solve for x^2 and y using the matrix methods of this section, and then solve for x and y.

$$x^2 + y = 2$$
$$2x^2 - y = 10$$

28. For the following system of equations, solve for x^2 and y^2 using the matrix methods of this section, and then solve for x and y.

$$x^2 - y^2 = 8$$
$$x^2 + y^2 = 10$$

29. Solve any pair of the system of equations $2x - y = 4$, $3x + y = 1$, and $x - 2y = 5$. Show that the solution is valid for any pair chosen. What conclusions can be drawn about the graphs of the three equations?

30. In solving the system of equations $3x - 4y = 5$, $8y - 6x = 7$, what conclusion can be drawn?

31. Forces $\mathbf{F}_1$ and $\mathbf{F}_2$ hold up a beam that weighs 2540 N, as shown in Fig. 16.12. The equations used to find the forces are

$$F_1 \sin 47.2° + F_2 \sin 64.4° = 2540$$
$$F_1 \cos 47.2° - F_2 \cos 64.4° = 0$$

Find the magnitude of each force.

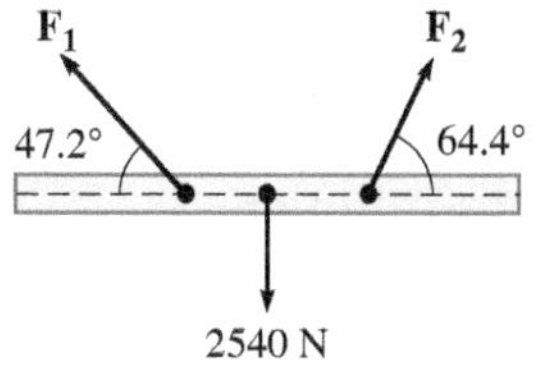

Fig. 16.12

32. In applying Kirchhoff's laws to the circuit shown in Fig. 16.13, the following equations are found. Determine the indicated currents (in A).

$$I_A + I_B + I_C = 0$$
$$2I_A - 5I_B = 6$$
$$5I_B - I_C = -3$$

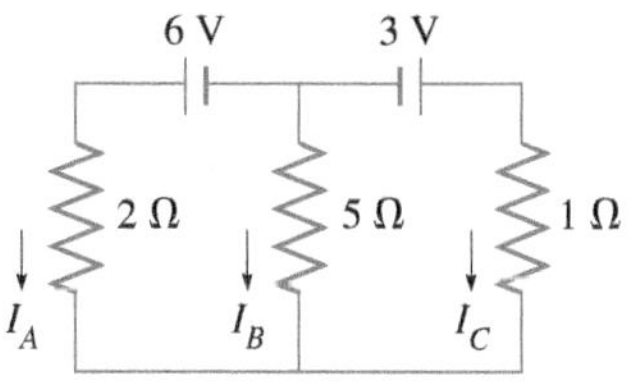

Fig. 16.13

33. Two batteries in an electric circuit have a combined voltage of 18 V, and one battery produces 6 V less than twice the other. What is the voltage of each?

34. An investment of \$5000 is invested partly at 3.00% and the remainder at 3.50%. If the total annual income is \$167, how much is invested at each rate?

35. What volume of each of a 20% acid solution and a 50% acid solution should be combined to form 48 mL of a 25% solution?

36. Two computer programs together require 51.8 megabytes of memory. If one program requires 2.0 megabytes more than twice the other, what are the memory requirements of each?

37. A research chemist wants to make 10.0 L of gasoline containing 2.0% of a new experimental additive. Gasoline without additive and two mixtures of gasoline with additive, one with 5.0% and the other with 6.0%, are to be used. If four times as much gasoline without additive as the 5.0% mixture is to be used, how much of each is needed?

38. A river tour boat takes 5.0 h to cruise downstream and 7.0 h for the return upstream. If the river flows at 4.0 km/h, how fast does the boat travel in still water, and how far downstream does the boat go before starting the return trip?

39. A logging company in British Columbia harvests in four different regions in order to supply its clients with Douglas fir, lodgepole pine, western red cedar, and ponderosa pine. Considering species mix and yield per region, the number of hectares that must be logged in each region for a certain week can be found by solving the system of equations

$$\begin{aligned} 324x + 240y + 224z + 274t &= 3010 \\ 80y + 28z + 20t &= 378 \\ 28z + 98t &= 434 \\ 36x &= 90 \end{aligned}$$

Find x, y, z, and t.

Answer to Practice Exercise

1. $x = 3$, $y = -1$

16.5 Gaussian Elimination

We dedicate this section to **Gaussian elimination**, a general method that can be used to solve a system of linear equations. The procedure is based on the same row operations used when finding the inverse of a matrix in Section 16.3. Starting with an **augmented** matrix (obtained by appending the columns of the matrix of coefficients and those of the matrix of constants), the objective is to transform it into an equivalent matrix that has the following specific form.

Row–Echelon Form of a Matrix

A matrix is in row–echelon form when the following conditions hold:

1. If a row consists entirely of zeros, then it is at the bottom of the matrix.
2. The first nonzero element of any row is a one (called a **leading one**).
3. The leading one of any row is to the right of the leading one of the previous row.

EXAMPLE 1 Row–echelon form

Determine whether each of the following matrices is in row–echelon form. If it is not, specify the condition that fails.

(a)
$$\begin{bmatrix} 1 & -3 & -1 & 1 \\ 0 & 1 & 2 & 5 \\ 0 & 0 & 1 & 2 \end{bmatrix}$$
Row–echelon form

(b)
$$\begin{bmatrix} 1 & 3 & 8 \\ 0 & -2 & 4 \end{bmatrix}$$
No: the leading number is not one.

(c)
$$\begin{bmatrix} 1 & -6 & 2 \\ 0 & 0 & 1 \\ 0 & 0 & 0 \end{bmatrix}$$
Row–echelon form

(d)
$$\begin{bmatrix} 1 & -4 & 3 & 1 \\ 0 & 0 & 1 & -2 \\ 0 & 1 & 3 & 7 \end{bmatrix}$$
No: the leading one is to the left of the previous leading one.

A matrix in row–echelon form represents a system of equations that can be solved by back-substitution. The next example shows you how.

EXAMPLE 2 Back-substitution

From the augmented matrix $\begin{bmatrix} 1 & -3 & -1 & 1 \\ 0 & 1 & 2 & 5 \\ 0 & 0 & 1 & 2 \end{bmatrix}$, use back-substitution to solve the corresponding system of linear equations.

$$\begin{aligned} x - 3y - z &= 1 \\ y + 2z &= 5 \\ z &= 2 \end{aligned}$$

The augmented matrix represents the system to the left (the first three columns are the coefficients of the variables, while the last column contains the constants on the right). We see that the third equation directly gives us the value $z = 2$. Since the second equation contains only y and z, we can now substitute $z = 2$ into the second equation to get $y = 1$. Then we can find x by substituting $y = 1$ and $z = 2$ into the first equation, and get $x = 6$.

We now state the method of Gaussian elimination, together with an example. The allowable row operations for Step 2 are:

1. Any two rows may be interchanged (equivalent to interchanging two equations).
2. Every element in any row may be multiplied by any number other than zero (equivalent to multiplying both sides of an equation by a nonzero constant).
3. Any row may be replaced by a row whose elements are the sum of a nonzero multiple of itself and a nonzero multiple of another row (equivalent to adding a nonzero multiple of one equation to another equation).

$$\begin{aligned} 2x + y &= 4 \\ 3x - 2y &= 3 \\ \hline x + \tfrac{1}{2}y &= 2 \\ 3x - 2y &= 2 \\ \hline x + \tfrac{1}{2}y &= 2 \\ -\tfrac{7}{2}y &= -3 \\ \hline x + \tfrac{1}{2}y &= 2 \\ y &= \tfrac{6}{7} \end{aligned}$$

Solving a Linear System by Gaussian Elimination	EXAMPLE 3
	Solve the system. $\begin{aligned} 2x + y &= 4 \\ 3x - 2y &= 3 \end{aligned}$
1. Write the augmented matrix.	Putting together coefficients and constants gives $\left[\begin{array}{cc|c} 2 & 1 & 4 \\ 3 & -2 & 3 \end{array}\right]$
2. Transform the matrix into row–echelon form by performing any of the allowable row operations. Work from left to right, one column at a time.	$\left[\begin{array}{cc|c} 2 & 1 & 4 \\ 3 & -2 & 3 \end{array}\right] \xrightarrow{\frac{1}{2}R_1} \left[\begin{array}{cc|c} 1 & \frac{1}{2} & 2 \\ 3 & -2 & 3 \end{array}\right] \xrightarrow{-3R_1 + R_2} \left[\begin{array}{cc|c} 1 & \frac{1}{2} & 2 \\ 0 & -\frac{7}{2} & -3 \end{array}\right]$ $\xrightarrow{-\frac{2}{7}R_2} \left[\begin{array}{cc|c} 1 & \frac{1}{2} & 2 \\ 0 & 1 & \frac{6}{7} \end{array}\right]$ This matrix is now in row–echelon form. The row operations on the complete system are shown to the left.
3. Convert the matrix back into a system of linear equations.	The equivalent system of equations is $\begin{aligned} x + \tfrac{1}{2}y &= 2 \\ y &= \tfrac{6}{7} \end{aligned}$

4. Use back-substitution to obtain the solution.	The second equation gives the value of $y = \frac{6}{7}$ directly. Substituting in the first equation, $x + \frac{1}{2}\left(\frac{6}{7}\right) = 2$, so $x = \frac{11}{7}$.
5. Check the solutions in the original equation.	The solution checks: $\begin{aligned} 2\left(\tfrac{11}{7}\right) + \quad \tfrac{6}{7} &= 4 \\ 3\left(\tfrac{11}{7}\right) - 2\left(\tfrac{6}{7}\right) &= 3 \end{aligned}$

EXAMPLE 4 Solving a system of three equations

Solve the following system of equations by Gaussian elimination:

$$\begin{aligned} x + 3y - 2z &= -5 \\ 2x - y + 4z &= 7 \\ -3x + 2y - 3z &= -1 \end{aligned}$$

We form the augmented matrix and use row operations to transform into row–echelon form:

$$\left[\begin{array}{rrr|r} 1 & 3 & -2 & -5 \\ 2 & -1 & 4 & 7 \\ -3 & 2 & -3 & -1 \end{array}\right] \xrightarrow[3R_1 + R_3]{-2R_1 + R_2} \left[\begin{array}{rrr|r} 1 & 3 & -2 & -5 \\ 0 & -7 & 8 & 17 \\ 0 & 11 & -9 & -16 \end{array}\right] \xrightarrow{-\frac{1}{7}R_2} \left[\begin{array}{rrr|r} 1 & 3 & -2 & -5 \\ 0 & 1 & -\frac{8}{7} & -\frac{17}{7} \\ 0 & 11 & -9 & -16 \end{array}\right]$$

$$\xrightarrow{-11R_2 + R_3} \left[\begin{array}{rrr|r} 1 & 3 & -2 & -5 \\ 0 & 1 & -\frac{8}{7} & -\frac{17}{7} \\ 0 & 0 & \frac{25}{7} & \frac{75}{7} \end{array}\right] \xrightarrow{\frac{7}{25}R_3} \left[\begin{array}{rrr|r} 1 & 3 & -2 & -5 \\ 0 & 1 & -\frac{8}{7} & -\frac{17}{7} \\ 0 & 0 & 1 & 1 \end{array}\right]$$

$$\begin{aligned} x + 3y - 2z &= -5 \\ 2x - y + 4z &= 7 \\ -3x + 2y - 3z &= -1 \end{aligned}$$

$$\begin{aligned} x + 3y - 2z &= -5 \\ -7y + 8z &= 17 \\ 11y - 9z &= -16 \end{aligned}$$

$$\begin{aligned} x + 3y - 2z &= -5 \\ y - \tfrac{8}{7}z &= -\tfrac{17}{7} \\ 11y - 9z &= -16 \end{aligned}$$

$$\begin{aligned} x + 3y - 2z &= -5 \\ y - \tfrac{8}{7}z &= -\tfrac{17}{7} \\ \tfrac{25}{7}z &= \tfrac{75}{7} \end{aligned}$$

$$\begin{aligned} x + 3y - 2z &= -5 \\ y - \tfrac{8}{7}z &= -\tfrac{17}{7} \\ z &= 3 \end{aligned}$$

For comparison, the row operations on the complete system are shown to the left. We use back-substitution on the last set of equations to obtain the solution. The last line shows that $z = 3$. This can be substituted into the second equation to find y, and then the values of z and y can be substituted into the first equation to find x.

$$\begin{aligned} y - \tfrac{8}{7}(3) &= -\tfrac{17}{7} \qquad & x + 3(1) - 2(3) &= -5 \\ y &= 1 & x &= -2 \end{aligned}$$

The solution is $x = -2$, $y = 1$, $z = 3$. This solution checks in the original equations.

EXAMPLE 5 Solving a system with an unlimited number of solutions

Solve the following system of equations by Gaussian elimination:

$$\begin{aligned} 4y + z &= 2 \\ 2x + 6y - 2z &= 3 \\ 4x + 8y - 5z &= 4 \end{aligned}$$

We form the augmented matrix and use row operations to write it in row–echelon form. We start by interchanging the first two equations.

$$\left[\begin{array}{rrr|r} 2 & 6 & -2 & 3 \\ 0 & 4 & 1 & 2 \\ 4 & 8 & -5 & 4 \end{array}\right] \xrightarrow{\frac{1}{2}R_1} \left[\begin{array}{rrr|r} 1 & 3 & -1 & \frac{3}{2} \\ 0 & 4 & 1 & 2 \\ 4 & 8 & -5 & 4 \end{array}\right] \xrightarrow{-4R_1 + R_3} \left[\begin{array}{rrr|r} 1 & 3 & -1 & \frac{3}{2} \\ 0 & 4 & 1 & 2 \\ 0 & -4 & -1 & -2 \end{array}\right]$$

$$\xrightarrow{\frac{1}{4}R_2} \left[\begin{array}{rrr|r} 1 & 3 & -1 & \frac{3}{2} \\ 0 & 1 & \frac{1}{4} & \frac{1}{2} \\ 0 & -4 & -1 & -2 \end{array}\right] \xrightarrow{4R_1 + R_3} \left[\begin{array}{rrr|r} 1 & 3 & -1 & \frac{3}{2} \\ 0 & 1 & \frac{1}{4} & \frac{1}{2} \\ 0 & 0 & 0 & 0 \end{array}\right]$$

The last row of zeros indicates that z can be any number. But it is still possible to express both x and y in terms of z. We first solve the equation in the second row for y

in terms of z, which results in $y = \frac{1}{2} - \frac{1}{4}z$. Then this expression can be substituted into the equation in the top row to solve for x in terms of z

$$x + 3\left(\frac{1}{2} - \frac{1}{4}z\right) - z = \frac{3}{2}$$
$$x = \frac{7}{4}z$$

Therefore, the solution is given by $x = \frac{7}{4}z$ and $y = \frac{1}{2} - \frac{1}{4}z$, where z can be any number. This means **there are an unlimited number of solutions**. For example, if $z = 4$, then $x = 7$ and $y = -\frac{1}{2}$.

Practice Exercise

1. Use Gaussian elimination to solve the system of equations.

$2x - y = 7$
$4x + 3y = -1$

Possible Number of Solutions

When we solved systems of linear equations in Chapter 5, we found that not all systems have unique solutions. Systems can also have no solution or an unlimited number of solutions. When using Gaussian elimination, the characteristics of the final matrix in row–echelon form determine the number of solutions.

Number of Solutions from the Row–Echelon Form	EXAMPLE 6
Unique solution The system has at least as many equations as variables. The matrix in row–echelon form has the same number of nonzero rows (leading ones) as variables.	**(a)** $\left[\begin{array}{ccc\|c} 1 & -\frac{1}{3} & \frac{2}{3} & 1 \\ 0 & 1 & \frac{5}{2} & \frac{3}{2} \\ 0 & 0 & 1 & \frac{2}{3} \end{array}\right]$ Solution: $x = \frac{1}{2}, y = -\frac{1}{6}, z = \frac{2}{3}$.
Unlimited number of solutions The matrix in row–echelon form has more variables than nonzero rows. There is usually a row of zeros, although this is not necessarily so.	**(b)** $\left[\begin{array}{ccc\|c} 1 & \frac{2}{3} & -\frac{1}{3} & 1 \\ 0 & 1 & 1 & 0 \\ 0 & 0 & 0 & 0 \end{array}\right]$ Solution: z arbitrary, $x = 1 + z, y = -z$.
No solution The matrix in row–echelon form has a row with zeros on the left and a nonzero right side.	**(c)** $\left[\begin{array}{ccc\|c} 1 & \frac{1}{7} & -\frac{4}{7} & 1 \\ 0 & 1 & 4 & -5 \\ 0 & 0 & 0 & 2 \end{array}\right]$ No solution.

If a system has more equations than unknowns, it is inconsistent unless enough equations become $0 = 0$ such that at least one solution is found. The following example illustrates two systems in which there are more equations than unknowns.

EXAMPLE 7 Consistent and inconsistent systems

Identify the type of solution for each of the following systems given the equivalent row–echelon form obtained after Gaussian elimination.

(a) $x + 2y = 5$
$3x - y = 1$
$4x + y = 6$

Row–echelon form:

$$\left[\begin{array}{cc|c} 1 & 2 & 5 \\ 0 & 1 & 2 \\ 0 & 0 & 0 \end{array}\right]$$

(b) $x + 2y = 5$
$3x - y = 1$
$4x + y = 2$

Row–echelon form:

$$\left[\begin{array}{cc|c} 1 & 2 & 5 \\ 0 & 1 & 2 \\ 0 & 0 & -4 \end{array}\right]$$

Same number of leading ones as variables, so the system has a unique solution: $x = 1$, $y = 2$.

Row of zeros on the left with a nonzero right side. The third equation becomes $0 = -4$, so the system is inconsistent and there is no solution.

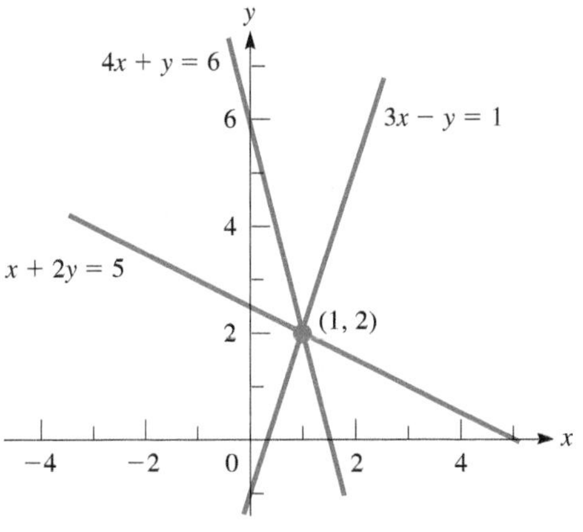

Fig. 16.14(a)

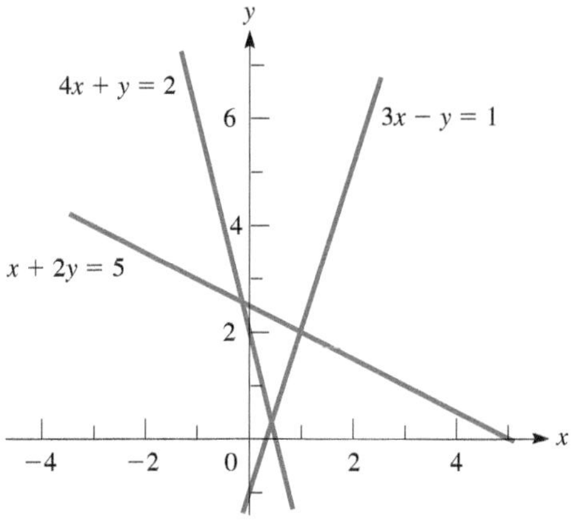

Fig. 16.14(b)

EXERCISES 16.5

In Exercises 1 and 2, make the given changes in the indicated examples of this section, and then solve the resulting systems using Gaussian elimination.

1. In Example 3, change the signs of both coefficients of x to $-$.

2. In Example 7, change the third equation of the second system to $5x + 3y = 11$.

In Exercises 3–28, solve the given systems of equations by Gaussian elimination. If there is an unlimited number of solutions, find two of them.

3. $\begin{aligned} x + 2y &= 4 \\ 3x - y &= 5 \end{aligned}$

4. $\begin{aligned} 2x + y &= 1 \\ 5x + 2y &= 1 \end{aligned}$

5. $\begin{aligned} 5x - 3y &= 2 \\ -2x + 4y &= 3 \end{aligned}$

6. $\begin{aligned} -3x + 2y &= 4 \\ 4x + y &= -5 \end{aligned}$

7. $\begin{aligned} 2x - 3y + z &= 4 \\ 6y - 4x - 2z &= 9 \end{aligned}$

8. $\begin{aligned} 3s + 4t - u &= -5 \\ 2u - 6s - 8t &= 10 \end{aligned}$

9. $\begin{aligned} x + 3y + 3z &= -3 \\ 2x + 2y + z &= -5 \\ -2x - y + 4z &= 6 \end{aligned}$

10. $\begin{aligned} 3x - y + 2z &= 3 \\ 4x - 2y + z &= 3 \\ 6x + 6y + 3z &= 4 \end{aligned}$

11. $\begin{aligned} w + 2x - y + 3z &= 12 \\ 2w - 2y - z &= 3 \\ 3x - y - z &= -1 \\ -w + 2x + y + 2z &= 3 \end{aligned}$

12. $\begin{aligned} 3x - 2y &= -11 \\ 5x + y &= -1 \\ 2x + 3y &= 10 \\ x - 5y &= -21 \end{aligned}$

13. $\begin{aligned} x - 4y + z &= 2 \\ 3x - y + 4z &= -4 \end{aligned}$

14. $\begin{aligned} 4x + z &= 6 \\ 2x - y - 2z &= -2 \end{aligned}$

15. $\begin{aligned} 2x - y + z &= 5 \\ 3x + 2y - 2z &= 4 \\ 5x + 8y - 8z &= 5 \end{aligned}$

16. $\begin{aligned} 3u + 6v + 2w &= -2 \\ u + 3v - 4w &= 2 \\ 2u - 3v - 2w &= -2 \end{aligned}$

17. $\begin{aligned} x + 3y + z &= 4 \\ 2x - 6y - 3z &= 10 \\ 4x - 9y + 3z &= 4 \end{aligned}$

18. $\begin{aligned} 30x + 20y - 10z &= 30 \\ 4x - 2y - 6z &= 4 \\ -5x + 20y - 25z &= -5 \end{aligned}$

19. $\begin{aligned} 2x - 4y &= 7 \\ 3x + 5y &= -6 \\ 9x - 7y &= 15 \end{aligned}$

20. $\begin{aligned} 4x - y &= 5 \\ 2x + 2y &= 3 \\ 6x - 4y &= 7 \\ 2x + y &= 4 \end{aligned}$

21. $\begin{aligned} 6x + 10y &= -4 \\ 24x - 18y &= 13 \\ 15x - 33y &= 19 \\ 6x + 68y &= -33 \end{aligned}$

22. $\begin{aligned} x + 3y - z &= 1 \\ 3x - y + 4z &= 4 \\ -2x + 2y + 3z &= 17 \\ 3x + 7y + 5z &= 23 \end{aligned}$

23. $\begin{aligned} 4x - 8y - 8z &= 12 \\ 10x + 5y + 15z &= 20 \\ -6x - 3y - 3z &= 15 \\ 3x + 3y - 2z &= 2 \end{aligned}$

24. $\begin{aligned} 2x - y - 2z - t &= 4 \\ 4x + 2y + 3z + 2t &= 3 \\ -2x - y + 4z \quad &= -2 \end{aligned}$

25. $\begin{aligned} s + 2t - 3u &= 2 \\ 6t + 3s - 9u &= 6 \\ 7s + 14t - 21u &= 13 \end{aligned}$

26. $\begin{aligned} x + 2y - 3z + 2t &= 3 \\ 4y - 6z + 4t &= 1 \end{aligned}$

27. $\begin{aligned} r - s - 3t - u &= 1 \\ 2r + 4s - 2u &= 2 \\ r + 5s + 3t - u &= 1 \\ 3r + 4s - 2t &= 0 \\ r + 2t - 3u &= 3 \end{aligned}$

28. $\begin{aligned} x + 2y - 3z &= 4 \\ 2x - y - 6z + 2t &= 2 \\ x + 3y + 3z - t &= 1 \end{aligned}$

In Exercises 29 and 30, solve the given problems using Gaussian elimination.

29. Solve the system $a_1x + b_1y = c_1$, $a_2x + b_2y = c_2$ and show that the result is the same as that obtained using determinants as in Section 5.5.

30. Solve the system $x + 2y = 6,\ 2x + ay = 4$ and show that the solution depends on the value of a. What value of a does the solution show may not be used?

In Exercises 31–34, set up systems of equations and solve by Gaussian elimination.

31. Two jets are 2370 km apart and travelling toward each other, one at 720 km/h and the other at 860 km/h. How far does each travel before they pass?

32. The voltage across an electric resistor equals the current (in A) times the resistance (in Ω). If a current of 3.00 A passes through each of two resistors, the sum of the voltages is 10.5 V. If 2.00 A passes through the first resistor and 4.00 A passes through the second resistor, the sum of the voltages is 13.0 V. Find the resistances.

33. Three machines together produce 650 parts each hour. Twice the production of the second machine is 10 parts/h more than the sum of the production of the other two machines. If the first operates for 3.00 h and the others operate for 2.00 h, 1550 parts are produced. Find the production rate of each machine.

34. A total of $12 000 is invested—part at 6.5%, part at 6.0%, and part at 5.5%—yielding total annual interest of $726. The income from the 6.5% part yields $128 less than that for the other two parts combined. How much is invested at each rate?

Answer to Practice Exercise

1. $x = 2,\ y = -3$

16.6 Higher-Order Determinants

In Chapter 5, we limited our discussion of determinants to those of the second and third orders. We now show some methods of evaluating higher-order determinants. We will then be able to use Cramer's rule to solve systems of equations of higher order.

From Section 5.7, we recall the definition of a third-order determinant. Then we rewrite it in a form that can be generalized to determinants of any order. We have

$$\begin{vmatrix} a_1 & b_1 & c_1 \\ a_2 & b_2 & c_2 \\ a_3 & b_3 & c_3 \end{vmatrix} = a_1b_2c_3 + a_3b_1c_2 + a_2b_3c_1 - a_3b_2c_1 - a_1b_3c_2 - a_2b_1c_3$$

$$= a_1(b_2c_3 - b_3c_2) - a_2(b_1c_3 - b_3c_1) + a_3(b_1c_2 - b_2c_1) \tag{16.8}$$

$$= a_1\begin{vmatrix} b_2 & c_2 \\ b_3 & c_3 \end{vmatrix} - a_2\begin{vmatrix} b_1 & c_1 \\ b_3 & c_3 \end{vmatrix} + a_3\begin{vmatrix} b_1 & c_1 \\ b_2 & c_2 \end{vmatrix}$$

In Eq. (16.8), the third-order determinant is expanded as products of the elements of the first column and second-order determinants, known as *minors*. In general, *the* **minor** *of an element of a determinant is the determinant that results by deleting the row and column in which the element lies.*

EXAMPLE 1 Minors

Determine the minor of the circled element in each of the following determinants:

(a) $\begin{vmatrix} \textcircled{1} & 2 & 3 \\ 4 & 5 & 6 \\ 7 & 8 & 9 \end{vmatrix}$. The minor is $\begin{vmatrix} 5 & 6 \\ 8 & 9 \end{vmatrix}$. (b) $\begin{vmatrix} 1 & 2 & 3 \\ 4 & 5 & \textcircled{6} \\ 7 & 8 & 9 \end{vmatrix}$. The minor is $\begin{vmatrix} 1 & 2 \\ 7 & 8 \end{vmatrix}$.

The following procedure generalizes the expansion by minors expressed in Eq. (16.8) in two ways. First, the determinant can in fact be expanded along any column or any row, not just the first column. Moreover, the expansion is valid for determinants of any order.

Expansion of Determinant by Minors	EXAMPLE 2
	Expand by minors. $\begin{vmatrix} 3 & -2 & 8 \\ -5 & 5 & 0 \\ 4 & 9 & -6 \end{vmatrix}$
1. Select any row or column along which to expand. (A row or column containing zeros will require less numerical work.)	We select to expand along the second row (note the zero in the third column).
2. Form the product of each of the n elements in that row or column and its minor.	For the second row, the products are: $(-5)\begin{vmatrix} -2 & 8 \\ 9 & -6 \end{vmatrix}, (5)\begin{vmatrix} 3 & 8 \\ 4 & -6 \end{vmatrix}, (0)\begin{vmatrix} 3 & -2 \\ 4 & 9 \end{vmatrix}$
3. Give a product a plus sign if the sum of its row number and its column number is even, and a minus sign if the sum is odd. The signs for all positions form a "checkerboard" pattern, as in Fig. 16.15.	The products with their corresponding signs are: $-(-5)\begin{vmatrix} -2 & 8 \\ 9 & -6 \end{vmatrix}, +(5)\begin{vmatrix} 3 & 8 \\ 4 & -6 \end{vmatrix}, -(0)\begin{vmatrix} 3 & -2 \\ 4 & 9 \end{vmatrix}$
4. Add the signed products to obtain the value of the determinant.	$-(-5)(12-72)+(5)(-18-32)-(0)=-550$

$$\begin{vmatrix} + & - & + & - & \cdots \\ - & + & - & + & \cdots \\ + & - & + & - & \cdots \\ - & + & - & + & \cdots \\ \vdots & \vdots & \vdots & \vdots & \end{vmatrix}$$

Fig. 16.15

Practice Exercise

1. Use expansion by minors to evaluate the determinant.

$$\begin{vmatrix} -4 & -1 & 0 & 2 \\ 2 & -3 & -2 & 1 \\ 1 & 0 & 0 & -2 \\ 3 & 0 & 5 & 4 \end{vmatrix}$$

EXAMPLE 3 Expansion by minors

In evaluating the following determinant, we expand along the third column because it has two zeros. Therefore,

$$\begin{vmatrix} 3 & -2 & 0 & 2 \\ 1 & 0 & -1 & 4 \\ -3 & 1 & 2 & -2 \\ 2 & -1 & 0 & -1 \end{vmatrix} = +(0)\begin{vmatrix} 1 & 0 & 4 \\ -3 & 1 & -2 \\ 2 & -1 & -1 \end{vmatrix} - (-1)\begin{vmatrix} 3 & -2 & 2 \\ -3 & 1 & -2 \\ 2 & -1 & -1 \end{vmatrix} + (2)\begin{vmatrix} 3 & -2 & 2 \\ 1 & 0 & 4 \\ 2 & -1 & -1 \end{vmatrix} - (0)\begin{vmatrix} 3 & -2 & 2 \\ 1 & 0 & 4 \\ -3 & 1 & -2 \end{vmatrix}$$

$$= \begin{vmatrix} 3 & -2 & 2 \\ -3 & 1 & -2 \\ 2 & -1 & -1 \end{vmatrix} + 2\begin{vmatrix} 3 & -2 & 2 \\ 1 & 0 & 4 \\ 2 & -1 & -1 \end{vmatrix}$$

$$= \left[+3\begin{vmatrix} 1 & -2 \\ -1 & -1 \end{vmatrix} - (-3)\begin{vmatrix} -2 & 2 \\ -1 & -1 \end{vmatrix} + 2\begin{vmatrix} -2 & 2 \\ 1 & -2 \end{vmatrix}\right] + 2\left[-(-2)\begin{vmatrix} 1 & 4 \\ 2 & -1 \end{vmatrix} + 0\begin{vmatrix} 3 & 2 \\ 2 & -1 \end{vmatrix} - (-1)\begin{vmatrix} 3 & 2 \\ 1 & 4 \end{vmatrix}\right]$$

expanding first determinant by first column; expanding second determinant by second column

$$= [3(-1-2)+3(2+2)+2(4-2)] + 2[2(-1-8)+(12-2)]$$

$$= [-9+12+4] + 2[-18+10] = 7 + 2(-8) = -9$$

PROPERTIES OF DETERMINANTS

We can expand any determinant by minors, but even a fourth-order determinant may require many calculations. We now state some properties of determinants with which they can be evaluated, often with less work than with minors, and not requiring a calculator or computer.

Properties of Determinants	EXAMPLE 4
1. If each element above or each element below the principal diagonal is zero, the value of the determinant is the product of the elements in the principal diagonal.	**(a)** $\begin{vmatrix} 2 & 1 & 8 \\ 0 & -5 & 9 \\ 0 & 0 & 6 \end{vmatrix} = 2(-5)(6) = 60$
2. If all the rows of a determinant are changed into columns (or vice versa), the value of the determinant is unchanged.	**(b)** $\begin{vmatrix} 1 & 3 & -1 \\ 2 & 0 & 4 \\ -2 & 5 & -6 \end{vmatrix} = \begin{vmatrix} 1 & 2 & -2 \\ 3 & 0 & 5 \\ -1 & 4 & -6 \end{vmatrix} = -18$
3. If two columns (or rows) of a determinant are identical, the determinant is zero.	**(c)** $\begin{vmatrix} 3 & 5 & 2 \\ -4 & 6 & 9 \\ -4 & 6 & 9 \end{vmatrix} = 0$
4. If two columns (or rows) of a determinant are interchanged, the value of the determinant changes in sign.	**(d)** $\begin{vmatrix} 3 & 0 & 2 \\ 1 & 1 & 5 \\ 2 & 1 & 3 \end{vmatrix} = -8$ and $\begin{vmatrix} 2 & 0 & 3 \\ 5 & 1 & 1 \\ 3 & 1 & 2 \end{vmatrix} = 8$
5. If all elements of a column (or row) are multiplied by the same number k, the value of the determinant is multiplied by k.	**(e)** $\begin{vmatrix} -1 & 0 & 6 \\ 2 & 1 & -2 \\ 0 & 5 & 3 \end{vmatrix} = 47$ $\begin{vmatrix} -1 & 0 & 6 \\ 6 & 3 & -6 \\ 0 & 5 & 3 \end{vmatrix} = 3\begin{vmatrix} -1 & 0 & 6 \\ 2 & 1 & -2 \\ 0 & 5 & 3 \end{vmatrix} = 3(47) = 141$
6. If all elements of any column (or row) are multiplied by the same number k, and the resulting numbers are added to the corresponding elements of another column (or row), the value of the determinant is unchanged.	**(f)** $\begin{vmatrix} 4 & -1 & 3 \\ 2 & 2 & 1 \\ 1 & 0 & -3 \end{vmatrix} \overset{2R_1 + R_2}{=} \begin{vmatrix} 4 & -1 & 3 \\ 10 & 0 & 7 \\ 1 & 0 & -3 \end{vmatrix} = -37$

LEARNING TIP

With the use of these six properties, determinants of higher order can be evaluated much more easily. The technique is to

obtain zeros in a given column (or row) in all positions except one.

We can then expand by this column (or row), thereby reducing the order of the determinant. Property 6 is probably the most valuable for obtaining the zeros.

EXAMPLE 5 Evaluating a determinant using properties 1–6

Using the properties of determinants, evaluate $\begin{vmatrix} 3 & 2 & -1 & 1 \\ -1 & 1 & 2 & 3 \\ 2 & 2 & 1 & 4 \\ 0 & -1 & -2 & 2 \end{vmatrix}$.

We use row operations to obtain zeros in all elements but one of the first column and then expand by minors, repeating the process. In each step we identify the row operation and/or the property used.

$$\begin{vmatrix} 3 & 2 & -1 & 1 \\ -1 & 1 & 2 & 3 \\ 2 & 2 & 1 & 4 \\ 0 & -1 & -2 & 2 \end{vmatrix} \underset{\text{property 6}}{\overset{3R_2 + R_1}{=}} \begin{vmatrix} 0 & 5 & 5 & 10 \\ -1 & 1 & 2 & 3 \\ 2 & 2 & 1 & 4 \\ 0 & -1 & -2 & 2 \end{vmatrix} \underset{\text{property 6}}{\overset{2R_2 + R_3}{=}} \begin{vmatrix} 0 & 5 & 5 & 10 \\ -1 & 1 & 2 & 3 \\ 0 & 4 & 5 & 10 \\ 0 & -1 & -2 & 2 \end{vmatrix}$$

$$\underset{\text{expand by minors}}{=} -(-1)\begin{vmatrix} 5 & 5 & 10 \\ 4 & 5 & 10 \\ -1 & -2 & 2 \end{vmatrix} \underset{\text{property 5}}{=} 5\begin{vmatrix} 1 & 1 & 2 \\ 4 & 5 & 10 \\ -1 & -2 & 2 \end{vmatrix} \underset{\text{property 5}}{=} 2(5)\begin{vmatrix} 1 & 1 & 1 \\ 4 & 5 & 5 \\ -1 & -2 & 1 \end{vmatrix}$$

$$\underset{\text{property 6}}{\overset{-4R_1 + R_2}{=}} 10\begin{vmatrix} 1 & 1 & 1 \\ 0 & 1 & 1 \\ -1 & -2 & 1 \end{vmatrix} \underset{\text{property 6}}{\overset{R_1 + R_3}{=}} 10\begin{vmatrix} 1 & 1 & 1 \\ 0 & 1 & 1 \\ 0 & -1 & 2 \end{vmatrix} \underset{\text{expand by minors}}{=} 10(1)\begin{vmatrix} 1 & 1 \\ -1 & 2 \end{vmatrix} = 10(2 + 1) = 30$$

The methods of this section and that of expansion by minors illustrate the way to compute determinants by hand. However, determinants can be easily calculated with the use of a calculator, a computer algebra system, or an online tool.

SOLVING SYSTEMS OF LINEAR EQUATIONS BY DETERMINANTS

We can use the expansion of determinants by minors when solving systems of linear equations. **Cramer's rule** *for solving systems of linear equations, as stated in Section 5.7, is valid for any system of n equations in n unknowns.*

EXAMPLE 6 Solving a system of four equations

Solve the following system of equations:

$$
\begin{aligned}
x + 2y + z &= 5 \\
2x + z + 2t &= 1 \\
x - y + 3z + 4t &= -6 \\
4x - y - 2t &= 0
\end{aligned}
$$

constants

$$
x = \frac{\begin{vmatrix} 5 & 2 & 1 & 0 \\ 1 & 0 & 1 & 2 \\ -6 & -1 & 3 & 4 \\ 0 & -1 & 0 & -2 \end{vmatrix}}{\begin{vmatrix} 1 & 2 & 1 & 0 \\ 2 & 0 & 1 & 2 \\ 1 & -1 & 3 & 4 \\ 4 & -1 & 0 & -2 \end{vmatrix}} =
$$

expanding by fourth row

$$
\frac{-(0)\begin{vmatrix} 2 & 1 & 0 \\ 0 & 1 & 2 \\ -1 & 3 & 4 \end{vmatrix} + (-1)\begin{vmatrix} 5 & 1 & 0 \\ 1 & 1 & 2 \\ -6 & 3 & 4 \end{vmatrix} - (0)\begin{vmatrix} 5 & 2 & 0 \\ 1 & 0 & 2 \\ -6 & -1 & 4 \end{vmatrix} + (-2)\begin{vmatrix} 5 & 2 & 1 \\ 1 & 0 & 1 \\ -6 & -1 & 3 \end{vmatrix}}{(1)\begin{vmatrix} 0 & 1 & 2 \\ -1 & 3 & 4 \\ -1 & 0 & -2 \end{vmatrix} - 2\begin{vmatrix} 2 & 1 & 2 \\ 1 & 3 & 4 \\ 4 & 0 & -2 \end{vmatrix} + (1)\begin{vmatrix} 2 & 0 & 2 \\ 1 & -1 & 4 \\ 4 & -1 & -2 \end{vmatrix} - (0)\begin{vmatrix} 2 & 0 & 1 \\ 1 & -1 & 3 \\ 4 & -1 & 0 \end{vmatrix}}
$$

expanding by first row

$$
= \frac{-(-26) - 2(-14)}{1(0) - 2(-18) + 1(18)} = \frac{26 + 28}{36 + 18} = \frac{54}{54} = 1
$$

Note that we chose to expand the determinant in the numerator by the fourth row since it contains two zeros. In the denominator, we expanded by the first row since it contains a zero, and no other row or column contains more than one zero.

In solving for y, we again note the two zeros in the fourth row:

$$
y = \frac{\begin{vmatrix} 1 & 5 & 1 & 0 \\ 2 & 1 & 1 & 2 \\ 1 & -6 & 3 & 4 \\ 4 & 0 & 0 & -2 \end{vmatrix}}{54} = \frac{-4\begin{vmatrix} 5 & 1 & 0 \\ 1 & 1 & 2 \\ -6 & 3 & 4 \end{vmatrix} + (-2)\begin{vmatrix} 1 & 5 & 1 \\ 2 & 1 & 1 \\ 1 & -6 & 3 \end{vmatrix}}{54}
$$

expanding by fourth row

$$
= \frac{-4(-26) - 2(-29)}{54} = \frac{104 + 58}{54} = \frac{162}{54} = 3
$$

Substituting $x = 1$ and $y = 3$ in the first equation gives us $z = -2$. Then substituting $x = 1$ and $y = 3$ in the fourth equation gives us $t = 1/2$. Therefore, the required solution is $x = 1$, $y = 3$, $z = -2$, $t = 1/2$. We can check the solution by substituting in the second or third equation (we used the first and fourth to *find* values of z and t).

Practice Exercise

2. Solve the system in Example 6 if the 5 in the first equation is changed to a 2.

EXERCISES 16.6

In Exercises 1 and 2, make the given changes in the indicated examples of this section. Then evaluate the resulting determinants.

1. In Example 4(d), change the third column of the first determinant to 0 0 3.

2. In Example 4(e), change the second row of the second determinant to −10 −5 10.

In Exercises 3–6, evaluate each determinant by inspection. Observation will allow evaluation by using the properties of this section.

3. $\begin{vmatrix} 4 & -5 & 8 \\ 0 & 3 & -8 \\ 0 & 0 & -5 \end{vmatrix}$ **4.** $\begin{vmatrix} 3 & 0 & 0 \\ 0 & 10 & 0 \\ -9 & -1 & -5 \end{vmatrix}$

5. $\begin{vmatrix} 3 & -2 & 4 & 2 \\ 5 & -1 & 2 & -1 \\ 3 & -2 & 4 & 2 \\ 0 & 3 & -6 & 0 \end{vmatrix}$ **6.** $\begin{vmatrix} -12 & -24 & -24 & 15 \\ 12 & 32 & 32 & -35 \\ -22 & 18 & 18 & 18 \\ 44 & 0 & 0 & -26 \end{vmatrix}$

In Exercises 7–10, use the given value of the determinant at the right and the properties of this section to evaluate the following determinants.

$$\begin{vmatrix} 2 & -3 & 1 \\ -4 & 1 & 3 \\ 1 & -3 & -2 \end{vmatrix} = 40$$

7. $\begin{vmatrix} 2 & 1 & -3 \\ -4 & 3 & 1 \\ 1 & -2 & -3 \end{vmatrix}$ **8.** $\begin{vmatrix} 2 & -3 & 1 \\ -4 & 1 & 3 \\ 2 & -6 & -4 \end{vmatrix}$

9. $\begin{vmatrix} 2 & -3 & -1 \\ -4 & 1 & -3 \\ 1 & -3 & 2 \end{vmatrix}$ **10.** $\begin{vmatrix} 2 & -4 & 1 \\ -3 & 1 & -3 \\ 1 & 3 & -2 \end{vmatrix}$

In Exercises 11–20, evaluate the given determinants by expansion by minors.

11. $\begin{vmatrix} 3 & 0 & 0 \\ -2 & 1 & 4 \\ 4 & -2 & 5 \end{vmatrix}$ **12.** $\begin{vmatrix} 10 & 0 & -3 \\ -2 & -4 & 1 \\ 3 & 0 & 2 \end{vmatrix}$

13. $\begin{vmatrix} 3 & 1 & 0 \\ -2 & 3 & -1 \\ 4 & 2 & 5 \end{vmatrix}$ **14.** $\begin{vmatrix} -40 & 30 & -20 \\ -8 & 8 & 16 \\ -15 & 75 & -45 \end{vmatrix}$

15. $\begin{vmatrix} 4 & 3 & 6 & 0 \\ 3 & 0 & 0 & 4 \\ 5 & 0 & 1 & 2 \\ 2 & 1 & 1 & 7 \end{vmatrix}$ **16.** $\begin{vmatrix} 6 & -3 & -6 & 3 \\ -2 & 1 & 2 & -1 \\ 18 & 7 & -1 & 5 \\ 0 & -1 & 10 & 10 \end{vmatrix}$

17. $\begin{vmatrix} 5 & 3 & 0 & 5 \\ 4 & 2 & 1 & 2 \\ 3 & 2 & -2 & 2 \\ 0 & 1 & 2 & -1 \end{vmatrix}$ **18.** $\begin{vmatrix} -2 & 2 & 1 & 3 \\ 1 & 4 & 3 & 1 \\ 4 & 3 & -2 & -2 \\ 3 & -2 & 1 & 5 \end{vmatrix}$

19. $\begin{vmatrix} 1 & 2 & 0 & 1 & 0 \\ 0 & 2 & 1 & 0 & 1 \\ 1 & 0 & -1 & 1 & -1 \\ -2 & 0 & -1 & 2 & 1 \\ 1 & 0 & 2 & -1 & -2 \end{vmatrix}$ **20.** $\begin{vmatrix} -1 & 3 & 5 & 0 & -5 \\ 0 & 1 & 7 & 3 & -2 \\ 5 & -2 & -1 & 0 & 3 \\ -3 & 0 & 2 & -1 & 3 \\ 6 & 2 & 1 & -4 & 2 \end{vmatrix}$

In Exercises 21–30, use the determinants for Exercises 11–20 and evaluate each using the properties of determinants. Do not evaluate directly more than one second-order determinant.

In Exercises 31–34, solve the given systems of equations by determinants. Evaluate by expansion by minors.

31. $x + t = 0$
$3x + y + z = -1$
$2y - z + 3t = 1$
$2z - 3t = 1$

32. $2x + y + z = 4$
$2y - 2z - t = 3$
$3y - 3z + 2t = 1$
$6x - y + t = 0$

33. $x + 2y - z = 6$
$y - 2z - 3t = -5$
$3x - 2y + t = 2$
$2x + y + z - t = 0$

34. $2p + 3r + s = 4$
$p - 2r - 3s + 4t = -1$
$3p + r + s - 5t = 3$
$-p + 2r + s + 3t = 2$

In Exercises 35–38, solve the given systems of equations by determinants. Evaluate by using the properties of determinants.

35. $2x + y + z = 2$
$3y - z + 2t = 4$
$y + 2z + t = 0$
$3x + 2z = 4$

36. $2x + y + z = 0$
$x - y + 2t = 2$
$2y + z + 4t = 2$
$5x + 2z + 2t = 4$

37. $D + E + 2F = 1$
$2D - E + G = -2$
$D - E - F - 2G = 4$
$2D - E + 2F - G = 0$

38. $3x + y + t = 0$
$3z + 2t = 8$
$6x + 2y + 2z + t = 3$
$3x - y - z - t = 0$

In Exercises 39–42, make the indicated changes in the determinant at below, and then solve the indicated problem. Assume the elements are nonzero, unless otherwise specified.

$$\begin{vmatrix} a & b & c \\ d & e & f \\ g & h & i \end{vmatrix}$$

39. Evaluate the determinant if $a = c$, $d = f$, and $g = i$.

40. Evaluate the determinant if $b = c = f = 0$.

41. By what factor is the value of the determinant changed if all elements are doubled?

42. How is the value changed if a is added to g, b added to h, and c added to i?

In Exercises 43–48, solve the given problems by using determinants.

43. In applying Kirchhoff's laws to the circuit shown in Fig. 16.16, the following equations are found. Determine the indicated currents (in A).

$$I_A + I_B + I_C + I_D + I_E = 0$$
$$-2I_A + 3I_B = 0$$
$$3I_B - 3I_C = 6$$
$$-3I_C + I_D = 0$$
$$-I_D + 2I_E = 0$$

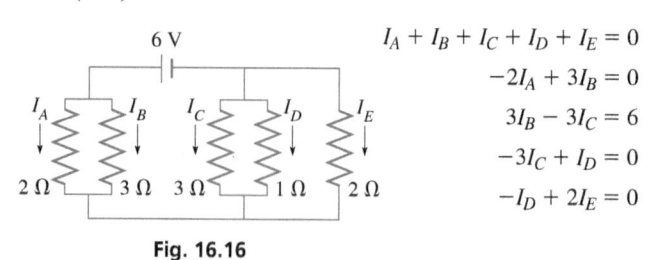

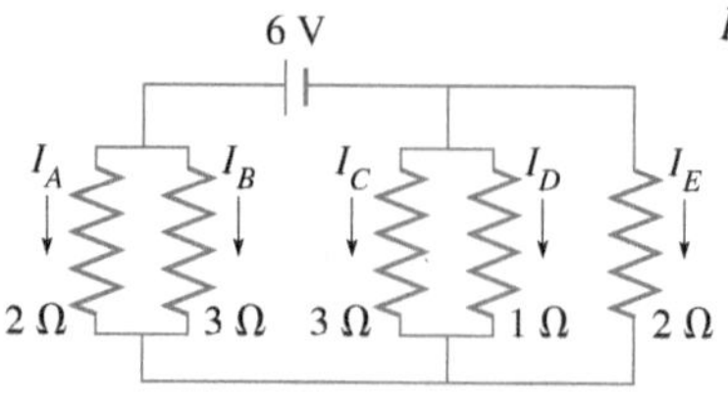

Fig. 16.16

44. In analysing the forces A, B, C, and D shown on the beam in Fig. 16.17, the following equations are used. Find these forces.

$$A + B = 850$$
$$A + B + 400 = 0.8C + 0.6D$$
$$0.6C = 0.8D$$
$$5A - 5B + 4C - 3D = 0$$

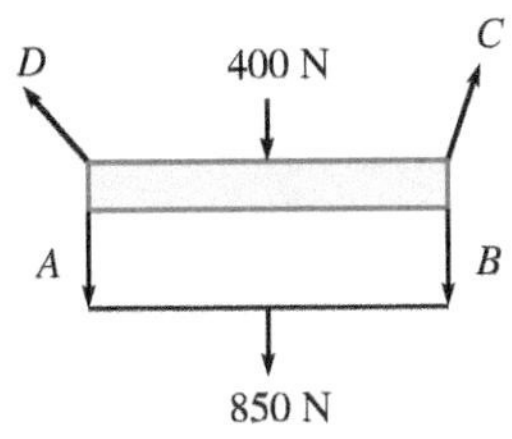

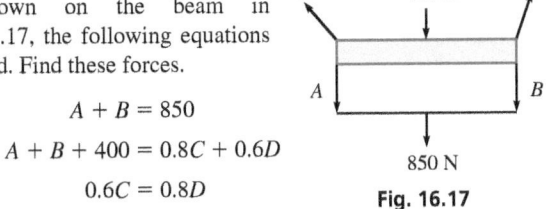

Fig. 16.17

45. The area of a triangle with vertices (x_1, y_1), (x_2, y_2), and (x_3, y_3) is given by $A = \pm\frac{1}{2}\begin{vmatrix} x_1 & y_1 & 1 \\ x_2 & y_2 & 1 \\ x_3 & y_3 & 1 \end{vmatrix}$. (If the answer comes out negative, take the absolute value.) A natural gas company locates the three corners of a triangular-shaped shale natural gas reserve at $(0, 0)$, $(8.45, 3.64)$, and $(1.82, 5.70)$, where all measurements are in kilometres. Find the area of the reserve.

46. An alloy is to be made from four other alloys containing copper (Cu), nickel (Ni), zinc (Zn), and iron (Fe). The first is 80% Cu and 20% Ni. The second is 60% Cu, 20% Ni, and 20% Zn. The third is 30% Cu, 60% Ni, and 10% Fe. The fourth is 20% Ni, 40% Zn, and 40% Fe. How much of each is needed so that the final alloy has 56 g Cu, 28 g Ni, 10 g Zn, and 6 g Fe?

47. In testing for air pollution, a given air sample contained 6.0 parts per million (ppm) of four pollutants, sulfur dioxide (SO_2), nitric oxide (NO), nitrogen dioxide (NO_2), and carbon monoxide (CO). The ppm of CO was 10 times that of SO_2, which in turn equaled those of NO and NO_2. There was a total of 0.8 ppm of SO_2 and NO. How many ppm of each were present in the air sample?

48. A tablet with a 32-GB hard drive starts out with three different apps A, B, and C, which use up 4% of the tablet's memory. Two more apps are added, each using the same amount of memory as app A, and the apps then use a total of 6% of the tablet's memory. Then, in addition to those, three more apps requiring the same memory as app B are added. All eight apps combined use up 13.5% of the tablet's memory. Find the number of megabytes (MB) of memory required for each app A, B, and C. (1 GB = 1000 MB.)

Answers to Practice Exercises

1. 135 2. $x = 1, y = 2, z = -3, t = 1$

CHAPTER 16 KEY FORMULAS AND EQUATIONS

Basic laws for matrices

$A + B = B + A$ (commutative law) **(16.1)**

$A + (B + C) = (A + B) + C$ (associative law) **(16.2)**

$k(A + B) = kA + kB$ **(16.3)**

$A + O = A$ **(16.4)**

Inverse matrix

$AA^{-1} = A^{-1}A = I$ **(16.5)**

Solving systems of equations by matrices

$AX = C$ **(16.6)**

$X = A^{-1}C$ **(16.7)**

A is the $n \times n$ matrix of coefficients, X is the $n \times 1$ matrix of variables, and C is the $n \times 1$ matrix of constants.

Expansion by minors of third-order determinants

$$\begin{vmatrix} a_1 & b_1 & c_1 \\ a_2 & b_2 & c_2 \\ a_3 & b_3 & c_3 \end{vmatrix} = a_1\begin{vmatrix} b_2 & c_2 \\ b_3 & c_3 \end{vmatrix} - a_2\begin{vmatrix} b_1 & c_1 \\ b_3 & c_3 \end{vmatrix} + a_3\begin{vmatrix} b_1 & c_1 \\ b_2 & c_2 \end{vmatrix} \quad \textbf{(16.8)}$$

CHAPTER 16 REVIEW EXERCISES

In Exercises 1–6, determine the values of the literal numbers.

1. $\begin{bmatrix} 2a \\ a - b \end{bmatrix} = \begin{bmatrix} 8 \\ 5 \end{bmatrix}$

2. $\begin{bmatrix} x - y \\ 2x + 2z \\ 4y + z \end{bmatrix} = \begin{bmatrix} 1 \\ 3 \\ -1 \end{bmatrix}$

3. $\begin{bmatrix} 2x & 3y & 2z \\ x + y & 2y + z & z - x \end{bmatrix} = \begin{bmatrix} 4 & -9 & 5 \\ a & b & c \end{bmatrix}$

4. $\begin{bmatrix} a + bj & b \\ aj & b - aj \end{bmatrix} = \begin{bmatrix} 6j & 2d \\ 2cj & ej^2 \end{bmatrix} \quad (j = \sqrt{-1})$

5. $\begin{bmatrix} \cos\pi & \sin\frac{\pi}{6} \\ x + y & x - y \end{bmatrix} = \begin{bmatrix} x & y \\ a & b \end{bmatrix}$

6. $\begin{bmatrix} \ln e & \log 100 \\ a^2 & b^2 \end{bmatrix} = \begin{bmatrix} a + b & a - b \\ x & y \end{bmatrix}$

In Exercises 7–12, use the given matrices and perform the indicated operations.

$$A = \begin{bmatrix} 2 & -3 \\ 4 & 1 \\ -5 & 0 \\ 2 & -3 \end{bmatrix} \quad B = \begin{bmatrix} -1 & 0 \\ 4 & -6 \\ -3 & -2 \\ 1 & -7 \end{bmatrix} \quad C = \begin{bmatrix} 5 & -6 \\ 2 & 8 \\ 0 & -2 \end{bmatrix}$$

7. $A + B$

8. $2C$

9. $B - A$

10. $2C - B$

11. $2A - 3B$

12. $2(A - B)$

In Exercises 13–16, perform the indicated matrix multiplications.

13. $\begin{bmatrix} 2 & -1 \\ -2 & 1 \end{bmatrix}\begin{bmatrix} 1 & -1 \\ 2 & -2 \end{bmatrix}$

14. $\begin{bmatrix} 6 & -4 & 1 & 0 \\ 2 & 0 & -4 & 3 \end{bmatrix}\begin{bmatrix} 7 & -1 & 6 \\ 4 & 0 & 1 \\ 3 & -2 & 5 \\ 9 & 1 & 0 \end{bmatrix}$

15. $\begin{bmatrix} -0.1 & 0.7 \\ 0.2 & 0.0 \\ 0.4 & -0.1 \end{bmatrix}\begin{bmatrix} 0.1 & -0.4 & 0.5 \\ 0.5 & 0.1 & 0.0 \end{bmatrix}$

16. $\begin{bmatrix} 0 & -1 & 6 \\ 8 & 1 & 4 \\ 7 & -2 & -1 \end{bmatrix}\begin{bmatrix} 5 & -1 & 7 & 1 & 5 \\ 0 & 1 & 0 & 4 & 1 \\ 1 & -2 & 3 & 0 & 1 \end{bmatrix}$

In Exercises 17–24, find the inverses of the given matrices.

17. $\begin{bmatrix} 2 & -5 \\ 2 & -4 \end{bmatrix}$

18. $\begin{bmatrix} -1 & -6 \\ 2 & 10 \end{bmatrix}$

19. $\begin{bmatrix} 0.07 & -0.01 \\ 0.04 & 0.08 \end{bmatrix}$

20. $\begin{bmatrix} 50 & -12 \\ 42 & -80 \end{bmatrix}$

21. $\begin{bmatrix} 1 & 1 & -2 \\ -1 & -2 & 1 \\ 0 & 3 & 4 \end{bmatrix}$

22. $\begin{bmatrix} -1 & -1 & 2 \\ 2 & 3 & 0 \\ 1 & 4 & 1 \end{bmatrix}$

23. $\begin{bmatrix} 2 & -4 & 3 \\ 4 & -6 & 5 \\ -2 & 1 & -1 \end{bmatrix}$

24. $\begin{bmatrix} 3 & 1 & -4 \\ -3 & 1 & -2 \\ -6 & 0 & 3 \end{bmatrix}$

In Exercises 25–32, solve the given systems of equations using the inverse of the coefficient matrix.

25. $2x - 3y = -9$
$4x - y = -13$

26. $5D - 7E = 62$
$6D + 5E = -6$

27. $33x + 52y = -450$
$45x - 62y = 1380$

28. $0.24x - 0.26y = -3.1$
$0.40x + 0.34y = -1.3$

29. $2u - 3v + 2w = 7$
$3u + v - 3w = -6$
$u + 4v + w = -13$

30. $2x + 2y - z = 8$
$x + 4y + 2z = 5$
$3x - 2y + z = 17$

31. $x + 2y + 3z = 1$
$3x - 4y - 3z = 2$
$7x - 6y + 6z = 2$

32. $3x + 2y + z = 2$
$2x + 3y - 6z = 3$
$x + 3y + 3z = 1$

In Exercises 33–40, solve the given systems of equations by Gaussian elimination. For Exercises 33–38, use those that are indicated from Exercises 25–32.

33. Exercise 25

34. Exercise 26

35. Exercise 29

36. Exercise 30

37. Exercise 31

38. Exercise 32

39. $2x + 3y - z = 10$
$x - 2y + 6z = -6$
$5x + 4y + 4z = 14$

40. $x - 3y + 4z - 2t = 6$
$2x + y - 2z + 3t = 7$
$3x - 9y + 12z - 6t = 12$

In Exercises 41–44, solve the systems of equations by determinants, using the properties of determinants. As indicated, use the systems from Exercises 29–32.

41. Exercise 29

42. Exercise 30

43. Exercise 31

44. Exercise 32

In Exercises 45–52, solve the given systems of equations by using the coefficient matrix. You may use technology to perform the necessary matrix operations.

45. $3x - 2y + z = 6$
$2x + 3z = 3$
$4x - y + 5z = 6$

46. $7n + p + 2r = 3$
$4n - 2p + 4r = -2$
$2n + 3p - 6r = 3$

47. $2x - 3y + z - t = -8$
$4x + 3z + 2t = -3$
$2y - 3z - t = 12$
$x - y - z + t = 3$

48. $3x + 2y - 2z - 2t = 0$
$5y + 3z + 4t = 3$
$6y - 3z + 4t = 9$
$6x - y + 2z - 2t = -3$

49. $3x - y + 6z - 2t = 8$
$2x + 5y + z + 2t = 7$
$4x - 3y + 8z + 3t = -17$
$3x + 5y - 3z + t = 8$

50. $A + B + 2C - 3D = 15$
$3A + 3B - 8C - 2D = 9$
$6A - 4B + 6C + D = -6$
$2A + 2B - 4C - 2D = 8$

51. $4r - s + 8t - 2u + 4v = -1$
$3r + 2s - 4t + 3u - v = 4$
$3r + 3s + 2t + 5u + 6v = 13$
$6r - s + 2t - 2u + v = 0$
$r - 2s + 4t - 3u + 3v = 1$

52. $v + 3w + 2x - 2y + 5z = 2$
$7v + 8w + 3x + y - 4z = 7$
$v - 2w - 4x - 4y - 8z = -20$
$3v - w + 7x + 5y - 3z = -3$
$4v + 5w + x + 3y - 6z = 14$

In Exercises 53–56, use matrices A and B.

$$A = \begin{bmatrix} 1 & 0 \\ 3 & 4 \end{bmatrix} \quad B = \begin{bmatrix} 0 & 1 & 0 \\ 0 & 0 & 1 \\ 1 & 0 & 0 \end{bmatrix}$$

53. Find A^2, A^3, and A^4.

54. Show that $(A^2)^2 = A^4$.

55. Show that $B^3 = I$.

56. Show that $B^4 = B$.

In Exercises 57–60, evaluate the given determinants by expansion by minors.

57. $\begin{vmatrix} 4 & 2 & 3 \\ 1 & -5 & -2 \\ -3 & 4 & -3 \end{vmatrix}$

58. $\begin{vmatrix} 4 & -5 & -1 \\ 6 & 1 & 6 \\ -2 & 4 & -3 \end{vmatrix}$

59. $\begin{vmatrix} 1 & 4 & 0 & -3 \\ 3 & 1 & 2 & 5 \\ -2 & -2 & -4 & 1 \\ -1 & 6 & 3 & -4 \end{vmatrix}$

60. $\begin{vmatrix} -2 & 6 & 6 & -1 \\ 1 & -2 & -5 & 2 \\ 5 & -4 & 4 & 3 \\ -3 & 1 & -2 & -3 \end{vmatrix}$

In Exercises 61–64, use the determinants for Exercises 57–60 and evaluate each using properties of determinants. Do not evaluate directly more than one second-order determinant.

In Exercises 65 and 66, use the matrix N.

$$N = \begin{bmatrix} 0 & -1 \\ 1 & 0 \end{bmatrix}$$

65. Show that $N^{-1} = -N$.

66. Show that $N^2 = -I$.

In Exercises 67 and 68, solve the given problems.

67. For any real number n, show that $\begin{bmatrix} n & 1+n \\ 1-n & -n \end{bmatrix}^2 = I$.

68. For the matrix $N = \begin{bmatrix} 1 & 1 \\ 1 & 1 \end{bmatrix}$, find (a) N^2, (b) N^3, (c) N^4. What is N^{20}? Explain.

In Exercises 69–72, use matrices A and B.

$$A = \begin{bmatrix} 1 & -2 \\ 0 & 3 \end{bmatrix} \qquad B = \begin{bmatrix} -3 & 1 \\ 2 & -1 \end{bmatrix}$$

69. Show that $(A + B)(A - B) \neq A^2 - B^2$.

70. Show that $(A + B)^2 \neq A^2 + 2AB + B^2$.

71. Show that the inverse of $2A$ is $A^{-1}/2$.

72. Show that the inverse of $B/2$ is $2B^{-1}$.

In Exercises 73–76, solve the given systems of equations by use of matrices as in Section 16.4.

73. Two electric resistors, R_1 and R_2, are tested with currents and voltages such that the following equations are found:

$$2R_1 + 3R_2 = 26$$
$$3R_1 + 2R_2 = 24$$

Find the resistances R_1 and R_2 (in Ω).

74. A company produces two products, each of which is processed in two departments. Considering the worker time available, the numbers x and y of each product produced each week can be found by solving the system of equations

$$4.0x + 2.5y = 1200$$
$$3.2x + 4.0y = 1200$$

Find x and y.

75. A beam is supported as shown in Fig. 16.18. Find the magnitudes of the force $\boldsymbol{F}$ and the tension $\boldsymbol{T}$ by solving the following system of equations:

$$0.500F = 0.866T$$
$$0.866F + 0.500T = 350$$

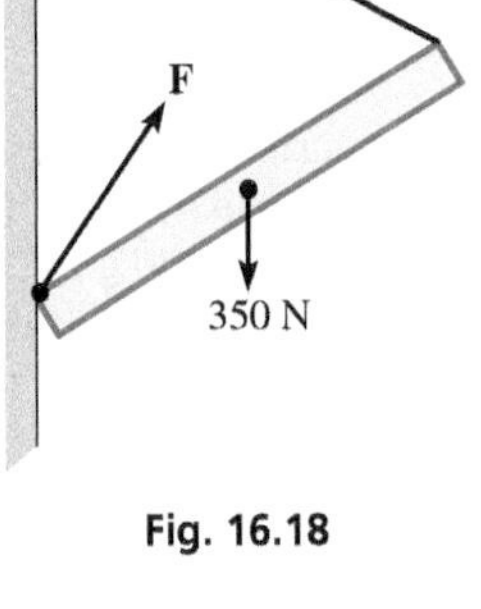

Fig. 16.18

76. To find the electric currents (in A) indicated in Fig. 16.19, it is necessary to solve the following equations.

$$I_A + I_B + I_C = 0$$
$$5I_A - 2I_B = -4$$
$$2I_B - I_C = 0$$

Find I_A, I_B, and I_C.

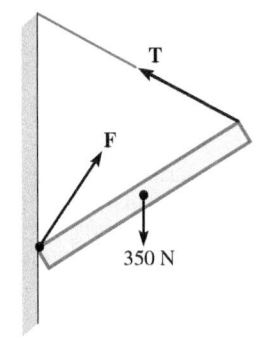

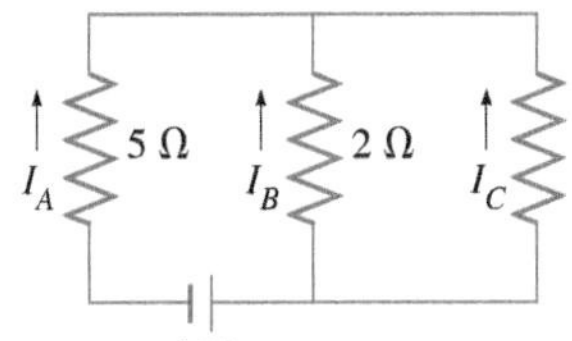

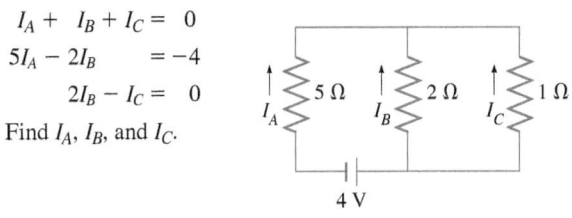

Fig. 16.19

In Exercises 77–80, solve the system of equations in Exercises 73–76 by Gaussian elimination.

In Exercises 81–85, solve the given problems by setting up the necessary equations and solving them by any appropriate method of this chapter.

81. A crime suspect passes an intersection in a car travelling at 180 km/h. The police pass the intersection 3.0 min later in a car travelling at 225 km/h. How long is it before the police overtake the suspect?

82. A contractor needs a backhoe and a generator for two different jobs. Renting the backhoe for 5.0 h and the generator for 6.0 h costs \$425 for one job. On the other job, renting the backhoe for 2.0 h and the generator for 8.0 h costs \$310. What are the hourly charges for the backhoe and the generator?

83. By mass, three alloys have the following percentages of lead, zinc, and copper:

	Lead	*Zinc*	*Copper*
Alloy A	60%	30%	10%
Alloy B	40%	30%	30%
Alloy C	30%	70%	

How many grams of each of alloys A, B, and C must be mixed to get 100 g of an alloy that is 44% lead, 38% zinc, and 18% copper?

84. On a 1014-km trip from Halifax to Ottawa that took a total of 5.0 h, a person took a shuttle to the airport, then a plane, and finally a taxi to reach the final destination. The shuttle took twice as long as the taxi, and the time for connections was as long as all other legs of the trip combined. The shuttle averaged 61 km/h, the plane averaged 640 km/h, and the taxi averaged 40 km/h. How long did each leg of the trip and the connections take?

85. In a Markov chain brand-switching model for three brands of detergent, the long-run market share of each brand (given as a proportion) is found by solving the system of equations

$$-0.3x + 0.5y + 0.32z = 0$$
$$0.2x - 0.8y + 0.3z = 0$$
$$0.1x + 0.3y - 0.62z = 0$$
$$x + y + z = 1$$

Find x, y, and z.

In Exercises 86–90, perform the indicated matrix operations.

86. An automobile maker has two assembly plants at which cars with either 4, 6, or 8 cylinders and with either standard or automatic transmission are assembled. The annual production at the first plant of cars with the number of cylinders–transmission type (standard, automatic) is as follows:

4: 12 000, 15 000; 6: 24 000, 8000; 8: 4000, 30 000

At the second plant the annual production is

4: 15 000, 20 000; 6: 12 000, 3000; 8: 2000, 22 000

Set up matrices for this production and by matrix addition, find the matrix for the total production by the number of cylinders and type of transmission.

87. Set up a matrix representing the information given in Exercise 83. A given shipment contains 500 g of alloy A, 800 g of alloy B, and 700 g of alloy C. Set up a matrix for this information. By multiplying these matrices, obtain a matrix that gives the total weight of lead, zinc, and copper in the shipment.

88. The matrix equation

$$\left[\begin{bmatrix} R_1 & -R_2 \\ -R_2 & R_1 \end{bmatrix} + R_2\begin{bmatrix} 1 & 0 \\ 0 & 1 \end{bmatrix}\right]\begin{bmatrix} i_1 \\ i_2 \end{bmatrix} = \begin{bmatrix} 6 \\ 0 \end{bmatrix}$$

may be used to represent the system of equations relating the currents and resistances of the circuit in Fig. 16.20. Find this system of equations by performing the indicated matrix operations.

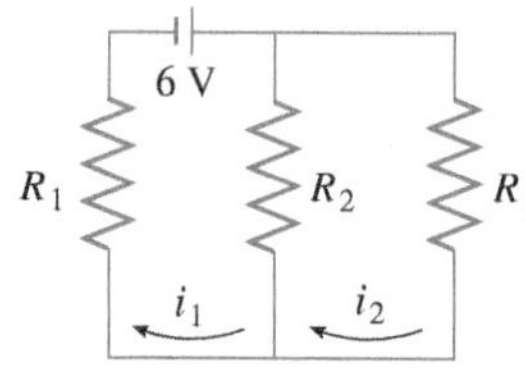

Fig. 16.20

89. A person prepared a meal of the following items, each having the given number of grams of protein, carbohydrates, and fat, respectively. Beef stew: 25, 21, 22; coleslaw: 3, 10, 10; (light) ice cream: 7, 25, 6. If the kilojoule count of each gram of protein, carbohydrate, and fat is 17 kJ/g, 16 kJ/g, and 37 kJ/g, respectively, find the total kilojoule count of each item by matrix multiplication.

90. A hardware company has 60 different retail stores in which 1500 different products are sold. Write a paragraph or two explaining why matrices provide an efficient method of inventory control, and what matrix operations in this chapter would be of use.

CHAPTER 16 PRACTICE TEST

1. For matrices A and B, find $A - 2B$.

$$A = \begin{bmatrix} 3 & -1 & 4 \\ 2 & 0 & -2 \end{bmatrix} \qquad B = \begin{bmatrix} 1 & 4 & 5 \\ -1 & -2 & 3 \end{bmatrix}$$

2. Evaluate the literal symbols.

$$\begin{bmatrix} 2x & x - y & z \\ x + z & 2y & y + z \end{bmatrix} = \begin{bmatrix} 6 & -2 & 4 \\ a & b & c \end{bmatrix}$$

3. For matrices C and D, find CD and DC.

$$C = \begin{bmatrix} 1 & 0 & 4 \\ 2 & -2 & 1 \\ -1 & 3 & 2 \end{bmatrix} \qquad D = \begin{bmatrix} 2 & -2 \\ 4 & -5S \\ 6 & 1 \end{bmatrix}$$

4. Determine whether or not $B = A^{-1}$.

$$A = \begin{bmatrix} 2 & -5 \\ 1 & -2 \end{bmatrix} \qquad B = \begin{bmatrix} -2 & 5 \\ -1 & 2 \end{bmatrix}$$

5. For matrix C of Problem 3, find C^{-1}.

6. Solve by using the inverse of the coefficient matrix.

$$\begin{aligned} 2x - 3y &= 11 \\ x + 2y &= 2 \end{aligned}$$

7. Solve the following system of equations by Gaussian elimination.

$$\begin{aligned} x + y - z + 2t &= 6 \\ 2x - 2y + 3z - t &= 0 \\ -x + 3y - 5z + t &= 1 \\ 5y - z + 4t &= 3 \end{aligned}$$

8. Evaluate the following determinant by expansion by minors.

$$\begin{vmatrix} 2 & -4 & -3 \\ -3 & 6 & 2 \\ 5 & -1 & 5 \end{vmatrix}$$

9. Evaluate the determinant in Problem 8 by using the properties of determinants.

10. Solve the following system of equations by using the inverse of the coefficient matrix. You may use technology to perform the necessary matrix operations.

$$\begin{aligned} 7x - 2y + z &= 6 \\ 2x + 3y - 4z &= 6 \\ 4x - 5y + 2z &= 10 \end{aligned}$$

11. Fifty shares of stock A and 30 shares of stock B cost \$2600. Thirty shares of stock A and 40 shares of stock B cost \$2000. What is the price per share of each stock? Solve by setting up the appropriate equations and then using the inverse of the coefficient matrix.

12. Solve the following system of equations using determinants.

$$\begin{aligned} p + 2r + 2s \qquad &= 6 \\ -p \qquad + 4s + 3t &= 1 \\ r + s - 3t &= 5 \\ 4p \qquad\qquad + 6t &= 2 \end{aligned}$$

13. An alloy is to be made from four other alloys containing copper (Cu), nickel (Ni), zinc (Zn), and iron (Fe). The first is 70% Cu and 30% Ni. The second is 40% Cu, 30% Ni, and 30% Zn. The third is 30% Cu, 60% Ni, and 10% Fe. The fourth is 20% Ni, 40% Zn, and 40% Fe. How much of each is needed so that the final alloy has 50 g Cu, 32 g Ni, 13 g Zn, and 5 g Fe? Solve by setting up the appropriate equations and solving by any method of this chapter (you may use technology).

17. Inequalities

Lindenblade/iStock/Getty Images

▲ In Section 17.6, we use inequalities to show how an airline can minimize operating costs when determining how many planes it needs.

LEARNING OUTCOMES

After completion of this chapter, the student should be able to:

- Distinguish between a conditional and an absolute inequality
- Graph solutions of inequalities on the number line
- Represent solutions of inequalities in interval notation
- Apply the properties of inequalities to solve linear inequalities
- Solve linear inequalities with three members
- Solve nonlinear inequalities algebraically and graphically
- Solve inequalities involving absolute values
- Solve inequalities and systems of inequalities with two variables graphically
- Understand the concepts of constraint, objective function, and feasible point in the context of linear programming
- Solve linear programs with two variables graphically
- Solve application problems involving inequalities

Having devoted a great deal of time to the solution of equations and systems of equations, we now turn our attention to solving inequalities and systems of inequalities. In doing so, we will find it necessary to find *all values* of the variable or variables that satisfy the inequality or system of inequalities.

There are numerous technical applications of inequalities. For example, in electricity it might be necessary to find the values of a current that are *greater than* a specified value. In designing a link in a robotic mechanism, it might be necessary to find the forces that are *less than* a specified value. Computers can be programmed to switch from one part of a program to another, based upon a result that is greater than (or less than) some given value.

We dedicate the last section of this chapter to linear programming, an important method for the solution of systems of linear inequalities. Linear programming was developed in 1947 by George Danzig for solving military logistic problems. Today, linear programming is widely used in business and industry in order to set production levels for maximizing profits and minimizing costs.

17.1 Properties of Inequalities

In Chapter 1, we first introduced the symbols of inequality. Up to this point, we have used them only in a basic way to describe certain intervals associated with a variable (for example, when describing the domain and range of a function). We now encounter them in the context of solving inequalities.

An inequality is a statement that contains one of the symbols: $<$, $>$, $\leq$, or $\geq$ (their meaning is reviewed below). The direction in which the inequality symbol points is known as the **sense** of the inequality. Two inequalities are said to have the same sense if the signs of inequality point in the same direction. They are said to have the opposite sense if the signs of inequality point in opposite directions. *The two sides of the inequality are called* **members** *of the inequality.*

EXAMPLE 1 Sense of an inequality

The inequalities $x + 3 > 2$ and $x + 1 > 0$ have the same sense, as do the inequalities $3x - 1 < 4$ and $x^2 - 1 < 3$.

The inequalities $x - 4 < 0$ and $x > -4$ have the opposite sense, as do the inequalities $2x + 4 > 1$ and $3x^2 - 7 < 1$.

The **solution** *of an inequality in one or more variables consists of all real values of the variables that make the inequality a true statement.* For inequalities in one variable, it is often useful to show the solution on the number line. We now show how this is done.

Symbol	Meaning	Number Line Symbol		EXAMPLE 2
$>$	Greater than		The *open circle* shows that the point is not part of the solution.	(a) $x > 2$ 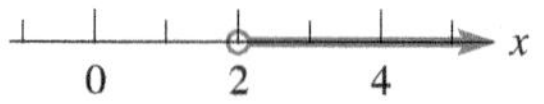 Fig. 17.1(a)
$<$	Less than			
$\geq$	Greater than or equal to		The *solid circle* shows that the point is part of the solution.	(b) $x \leq 1$  Fig. 17.1(b)
$\leq$	Less than or equal to			

LEARNING TIP
When we have an equation like $x - 1 = 0$, there is a unique solution to the equation. The value $x = 1$ is the only one that makes the statement true. In contrast, when we have an inequality like $x + 1 > 0$, there is more than one value that makes the statement true. As we see in Example 3(a), x can be any number in the interval $(-1, \infty)$. Any and all values of x within this interval are solutions of the inequality.

Most inequalities we will work with are **conditional inequalities,** *which are true for some, but not all, values of the variable(s). Inequalities which are true for all values of the variable(s) are* **absolute inequalities.**

EXAMPLE 3 Conditional and absolute inequalities

Determine whether each of the following is a conditional or an absolute inequality.

(a) $x + 1 > 0$ is a conditional inequality, valid for all x greater than -1, or x in $(-1, \infty)$.

(b) $x^2 + 1 > 0$ is an absolute inequality, valid for all real values of x.

PROPERTIES OF INEQUALITIES

We now show the basic operations performed on inequalities. These are the same operations as those performed on equations, but in certain cases the results take on a different form. *The following are the* **properties of inequalities**:

Properties of Inequalities		EXAMPLE 4	
Property	**Symbolically**		
1. The sense of an inequality is not changed when the same number is added or subtracted to both members of the inequality.	If $a > b$, then $a + c > b + c$	**(a)** $9 > 6$ add 4 to each member $9 + 4 > 6 + 4$ $13 > 10$	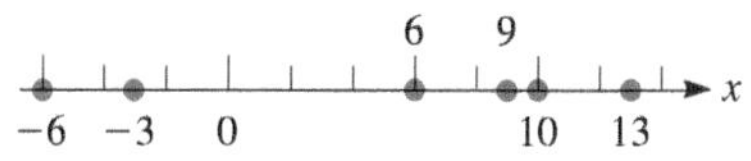 **Fig. 17.2(a)** We see that 9 is to the right of 6, 13 is to the right of 10, and -3 is to the right of -6.
	If $a > b$, then $a - c > b - c$	**(b)** $9 > 6$ subtract 12 from each member $9 - 12 > 6 - 12$ $-3 > -6$	
2. The sense of an inequality is not changed when both members are multiplied or divided by the same **positive** number.	If $a > b$ and $c > 0$, then $ac > bc$	**(c)** $8 < 12$ multiply both members by 2 $2(8) < 2(12)$ $16 < 24$	 **Fig.17.2(b)** We see that 8 is to the left of 12, 16 is to the left of 24, and 4 is to the left of 6.
	If $a > b$ and $c > 0$, then $\frac{a}{c} > \frac{b}{c}$	**(d)** $8 < 12$ divide both members by 2 $\frac{8}{2} < \frac{12}{2}$ $4 < 6$	
3. The sense of an inequality is **reversed** when both members are multiplied or divided by the same **negative** number.	If $a > b$ and $c < 0$, then $ac < bc$	**(e)** $4 > -2$ multiply both members by -3; sense is reversed $-3(4) < -3(-2)$ $-12 < 6$	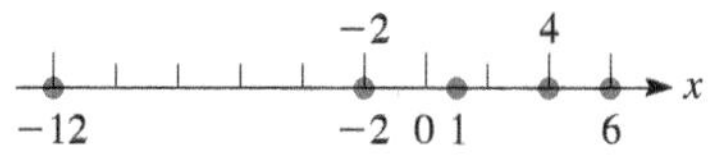 **Fig. 17.2(c)** We see that 4 is to the right of -2 but that -12 *is to the left of* 6, and that -2 *is to the left of* 1.
	If $a > b$ and $c < 0$, then $\frac{a}{c} < \frac{b}{c}$	**(f)** $4 > -2$ divide both members by -2; sense is reversed $\frac{4}{-2} < \frac{-2}{-2}$ $-2 < 1$	
4. If both members of an inequality are positive numbers and n is a positive number, then the inequality formed by taking the nth power or the nth root of both members has the same sense as the given inequality.	If $a > b$, then $a^n > b^n$, provided that $a > 0, b > 0, n > 0$	**(g)** $16 > 9$ square both members $16^2 > 9^2$ $256 > 81$	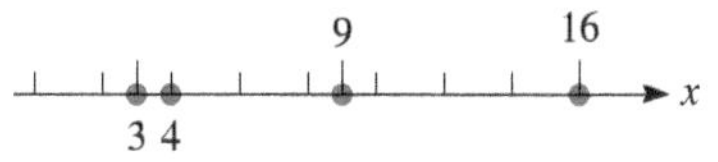 **Fig. 17.2(d)** We see that 16 is to the right of 9, and 4 is to the right of 3. Also, 256 is to the right of 81 (not shown).
	If $a > b$, then $\sqrt[n]{a} > \sqrt[n]{b}$, provided that $a > 0, b > 0, n > 0$	**(h)** $16 > 9$ take square root of both members $\sqrt{16} > \sqrt{9}$ $4 > 3$	

COMMON ERROR

In using Properties 2 and 3, be very careful to note that *the inequality sign remains the same if both members are multiplied or divided by a positive number,* but that *the* ***inequality sign changes if both members are multiplied or divided by a negative number.*** Most of the errors made in dealing with inequalities occur when multiplying or dividing by negative numbers.

Practice Exercises

For the inequality $-6 < 3$, state the inequality that results.

1. Multiply both members by 4.
2. Divide both members by -3.

Many inequalities have more than two members. In fact, inequalities with three members are common, and care must be used in stating these inequalities.

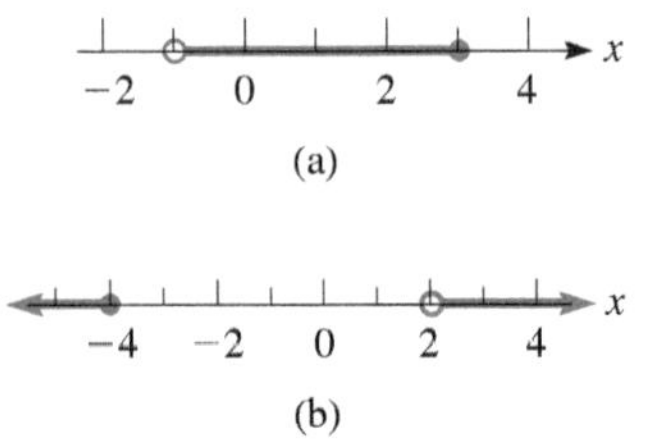

Fig. 17.3

EXAMPLE 5 Inequalities with three members

(a) To state that 5 is less than 6, and also greater than 2, we may write $2 < 5 < 6$, or $6 > 5 > 2$. (Generally, the *less than* form is preferred.)

(b) To state that a number x may be greater than -1 *and* also less than or equal to 3, we write $-1 < x \leq 3$, or as the interval $(-1, 3]$. (It can also be written as $x > -1$ *and* $x \leq 3$.) This is shown in Fig. 17.3(a). Note the use of the open circle and the solid circle.

(c) By writing $x \leq -4$ *or* $x > 2$, we state that x is less than or equal to -4, *or* greater than 2. In interval notation we write $(-\infty, -4]$ or $(2, \infty)$. See Fig. 17.3(b).

COMMON ERROR

If x is less than or equal to -4 *or* greater than 2, **it may not be written as the single expression $2 < x \leq -4$.** This statement would say that x is less than or equal to -4, while at the same time being greater than 2, and *no such numbers exist.* Moreover, it would follow that -4 is greater than 2, which is absurd.

Practice Exercise

3. Graph the inequality $-1 < x \leq 3$ on the number line.

LEARNING TIP

Note carefully that ***and*** *is used when the solution consists of values that make* ***both*** *statements true. The word* ***or*** *is used when the solution consists of values that make* ***either*** *statement true.*

EXAMPLE 6 Meaning of *and/or*

The inequality $x^2 - 3x + 2 > 0$ is satisfied if x is either greater than 2 *or* less than 1. This is written as $x > 2$ *or* $x < 1$, or as $(-\infty, 1)$ *or* $(2, \infty)$. Once again, it is incorrect to write the single expression $1 > x > 2$. However, it is correct to say that the inequality is *not* satisfied for $1 \leq x \leq 2$, or in $[1, 2]$. In other words, the inequality does not hold for values of x greater than or equal to 1 *and* less than or equal to 2.

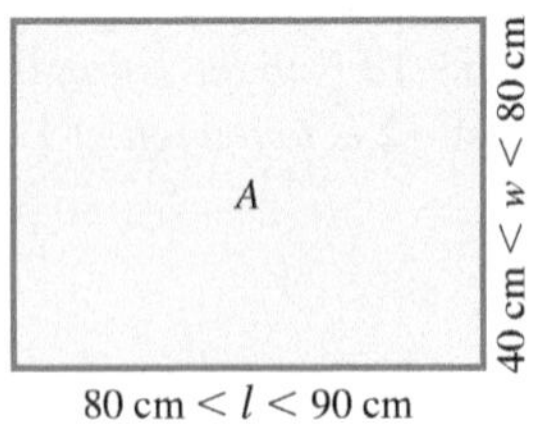

Fig. 17.4

EXAMPLE 7 Setting up an inequality—solar panels

The design of a rectangular solar panel shows that the length l is between 80 cm and 90 cm and the width w is between 40 cm and 80 cm. See Fig. 17.4. Find the values of area the panel may have.

Since l is to be less than 90 cm and w less than 80 cm, the area must be less than $(90\text{ cm})(80\text{ cm}) = 7200\text{ cm}^2$. Also, since l is to be greater than 80 cm and w greater than 40 cm, the area must be greater than $(80\text{ cm})(40\text{ cm}) = 3200\text{ cm}^2$. Therefore, the area A is in the interval $(3200\text{ cm}^2, 7200\text{ cm}^2)$, or

$$3200\text{ cm}^2 < A < 7200\text{ cm}^2$$

This means the area is greater than 3200 cm^2 *and* less than 7200 cm^2.

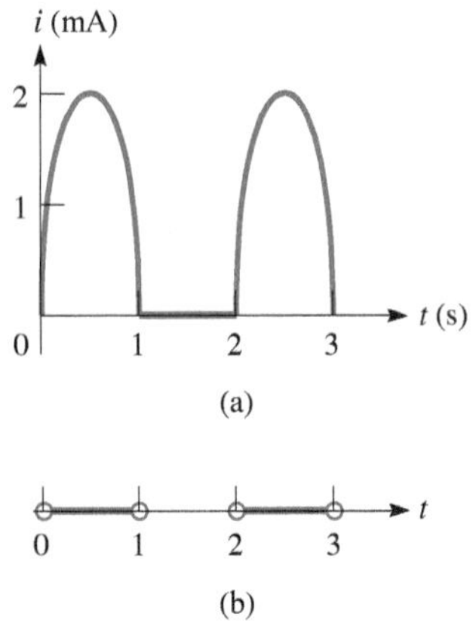

Fig. 17.5

EXAMPLE 8 Setting up an inequality—current through a diode

A semiconductor *diode* has the property that an electric current flows through it in only one direction. If it is an alternating-current circuit, the current in the circuit flows only during the half-cycle when the diode allows it to flow. If a source of current given by $i = 2 \sin \pi t$ (i in mA, t in seconds) is connected in series with a diode, write the inequalities for the current and the time. Assume that the source is on for 3.0 s and a positive current passes through the diode.

We are to find the values of t that correspond to $i > 0$. From the properties of the sine function, we know that $2 \sin \pi t$ has a period of $2\pi/\pi = 2.0$ s. Therefore, the current is zero for $t = 0$, 1.0 s, 2.0 s, and 3.0 s.

The source current is positive for $0 < t < 1.0$ s and for $2.0\ s < t < 3.0$ s.

The source current is negative for $1.0\text{ s} < t < 2.0$ s.

Therefore, in the circuit

$$i > 0 \text{ for } \quad 0 < t < 1.0 \text{ s} \quad \text{and} \quad 2.0 \text{ s} < t < 3.0 \text{ s}$$
$$i = 0 \text{ for } \quad t = 0, 1.0 \text{ s} \leq t \leq 2.0 \text{ s}$$

A graph of the current in the circuit as a function of time is shown in Fig. 17.5(a). In Fig. 17.5(b), the values of t for which $i > 0$ are shown.

EXERCISES 17.1

In Exercises 1–4, make the given changes in the indicated examples of this section and then perform the indicated operations.

1. In Example 2(b), change $\leq$ to $>$ and then graph the resulting inequality.

2. In Example 4(e), change the inequality to $-2 > -4$ and then perform the required operation.

3. In Example 4(h), change the inequality to $4 < 25$ and then perform the required operation.

4. In Example 5(b), change the -1 to -3 and the 3 to 1 and then write the two forms in which an inequality represents the statement.

In Exercises 5–12, for the inequality $4 < 9$, state the inequality that results when the given operations are performed on both members.

5. Add 5.
6. Subtract 16.
7. Multiply by 4.
8. Multiply by -2.
9. Divide by -1.
10. Divide by 0.5.
11. Square both.
12. Take square roots.

In Exercises 13–24, give the inequalities equivalent to the following statements about the number x.

13. Greater than -2
14. Less than 0.7
15. Less than or equal to 45
16. Greater than or equal to -6
17. Greater than 1 and less than 7
18. Greater than or equal to -200 and less than 650
19. Less than -9, or greater than or equal to -4
20. Less than or equal to 8, or greater than or equal to 12
21. Less than 1, or greater than 3 and less than or equal to 5
22. Greater than or equal to 0 and less than or equal to 2, or greater than 5
23. Greater than -2 and less than 2, or greater than or equal to 3 and less than 4
24. Less than -4, or greater than or equal to 0 and less than or equal to 1, or greater than or equal to 5

In Exercises 25–28, give verbal statements equivalent to the given inequalities involving the number x.

25. $0 < x \leq 2$
26. $x < 5$ or $x > 7$
27. $x < -10$ or $10 \leq x < 20$
28. $-1 \leq x < 3$ or $5 < x < 7$

In Exercises 29–44, express the given inequalities in interval notation and graph them on the number line.

29. $x < 3$
30. $x \geq -1$
31. $x \leq -1$ or $x > 0.5$
32. $x < -300$ or $x \geq 0$
33. $0 \leq x < 5$
34. $-4 < y < -2$
35. $x \geq -3$ and $x < 5$
36. $x > 4$ and $x < 3$
37. $x < -1$ or $1 \leq x < 4$
38. $-3 < x < 0$ or $x > 3$
39. $-3 < x < -1$ or $1 < x \leq 3$
40. $1 < x \leq 2$ or $3 \leq x < 4$
41. $t \leq -5$ and $t \geq -5$
42. $x < 1$ or $1 < x \leq 4$
43. $(x \leq 5$ or $x \geq 8)$ and $(3 < x < 10)$
44. $(x < 7$ and $x > 2)$ or $(x > 10$ or $x < 1)$

In Exercises 45–48, answer the given questions about the inequality $0 < a < b$.

45. Is $a^2 < b^2$ a conditional inequality or an absolute inequality?

46. Is $|a - b| < b - a$?

47. If each member of the inequality $2 > 1$ is multiplied by $a - b$, is the result $2(a - b) > (a - b)$?

48. What is wrong with the following sequence of steps?
$$a < b,\ ab < b^2,\ ab - b^2 < 0,\ b(a - b) < 0,\ b < 0$$

In Exercises 49–52, solve the given problems.

49. Write the relationship between $(|x| + |y|)$ and $|x + y|$ if $x > 0$ and $y < 0$.

50. Write the relationship between $|xy|$ and $|x||y|$ if $x > 0$ and $y < 0$.

51. Explain the error in the following "proof" that $3 < 2$:
(1) $1/8 < 1/4$ (2) $0.5^3 < 0.5^2$ (3) $\log 0.5^3 < \log 0.5^2$
(4) $3 \log 0.5 < 2 \log 0.5$ (5) $3 < 2$

52. If $x \neq y$, show that $x^2 + y^2 > 2xy$.

In Exercises 53–60, some applications of inequalities are shown.

53. An electron microscope can magnify an object from 2000 times to 1 000 000 times. Assuming that these values are exact, express these magnifications M as an inequality and graph them.

54. A busy person glances at a digital clock that shows 9:36. Another glance a short time later shows the clock at 9:44. Express the amount of time t (in min) that could have elapsed between glances by use of inequalities. Graph these values of t.

55. An Earth satellite put into orbit near Earth's surface will have an elliptic orbit if its velocity v is between 29 000 km/h and 40 000 km/h. Write this as an inequality and graph these values of v.

56. Fossils found in Jurassic rocks indicate that dinosaurs flourished during the Jurassic geological period, 140 MY (million years ago) to 200 MY. Write this as an inequality, with t representing past time. Graph the values of t.

57. In executing a program, a computer must perform a set of calculations. Any one of the calculations takes no more than 2565 steps. Express the number n of steps required for a given calculation by an inequality. (Note that n is a positive *integer.*)

58. The velocity v of an ultrasound wave in soft human tissue may be represented as 1550 ± 60 m/s, where the ± 60 m/s gives the possible variation in the velocity. Express the possible velocities by an inequality.

59. The electric intensity E within a charged spherical conductor is zero. The intensity on the surface and outside the sphere equals a constant k divided by the square of the distance r from the centre of the sphere. State these relations for a sphere of radius a by using inequalities, and graph E as a function of r.

60. If the current from the source in Example 8 is $i = 5 \cos 4\pi t$ and the diode allows only negative current to flow, write the inequalities and draw the graph for the current in the circuit as a function of time for $0 \leq t \leq 1$ s.

Answers to Practice Exercises

1. $-24 < 12$ **2.** $2 > -1$ **3.** 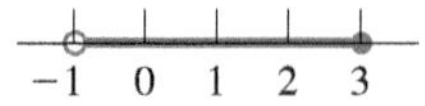

17.2 Solving Linear Inequalities

Using the properties and definitions discussed in Section 17.1, we can now proceed to solve inequalities. In this section, we solve linear inequalities in one variable. Similar to linear functions as defined in Chapter 5, *a* **linear inequality** *is one in which each term contains only one variable and the exponent of each variable is 1.* We will consider linear inequalities in two variables in Section 17.5.

LEARNING TIP

The procedure for solving a linear inequality in one variable is the same as the one used for solving linear equations, with the added restriction that every time we multiply or divide by a negative number, we have to change the sense of the inequality. In other words, the objective is to isolate the variable by performing the same operations on each member of the inequality, keeping in mind the basic properties given in Section 17.1.

EXAMPLE 1 Solutions using the basic operations

Solve each of the following inequalities by performing the necessary operation to isolate x.

$x + 2 < 4$	$\frac{x}{2} > 4$	$2x \leq 4$
Subtract 2 from each member.	Multiply each member by 2.	Divide each member by 2.
$x < 2$	$x > 8$	$x \leq 2$
$(-\infty, 2)$	$(8, \infty)$	$(-\infty, 2]$

Each solution can be checked by substituting any number in the indicated interval into the original inequality. For example, any value less than 2 will satisfy the first inequality, whereas 2 or any number less than 2 will satisfy the third inequality.

EXAMPLE 2 Solving a linear inequality

Solve the following inequality: $3 - 2x \geq 15$.

We have the following solution:

$$3 - 2x \geq 15$$ original inequality

$$-2x \geq 12$$ subtract 3 from each member

inequality reversed →

$$x \leq -6$$ divide each member by −2

$$(-\infty, -6]$$

Again, carefully note that ***the sign of inequality was reversed when each number was divided by* −2**. We check the solution by substituting −7 in the original inequality, obtaining $17 \geq 15$.

EXAMPLE 3 Solving a linear inequality

Solve the inequality $2x \leq 3 - x$.

The solution proceeds as follows:

$$2x \leq 3 - x \quad \text{original inequality}$$
$$3x \leq 3 \quad \text{add } x \text{ to each member}$$
$$x \leq 1 \quad \text{divide each member by 3}$$
$$(-\infty, 1]$$

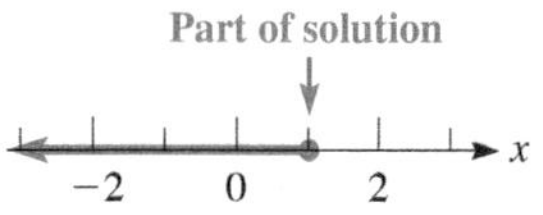

Fig. 17.6

This solution checks and is represented in Fig. 17.6, as we showed in Section 17.1.

This inequality could have been solved by combining x-terms on the right. In doing so, we would obtain $1 \geq x$. Since this might be misread, it is best to combine the variable terms on the left, as we did above.

EXAMPLE 4 Solving a linear inequality

Solve the inequality $\frac{3}{2}(1 - x) > \frac{1}{4} - x$.

$$\frac{3}{2}(1 - x) > \frac{1}{4} - x \quad \text{original inequality}$$
$$6(1 - x) > 1 - 4x \quad \text{multiply each member by 4}$$
$$6 - 6x > 1 - 4x \quad \text{remove parentheses}$$
$$-6x > -5 - 4x \quad \text{subtract 6 from each member}$$
$$-2x > -5 \quad \text{add } 4x \text{ to each member}$$
$$x < \frac{5}{2} \quad \text{divide each member by } -2$$
$$\left(-\infty, \frac{5}{2}\right)$$

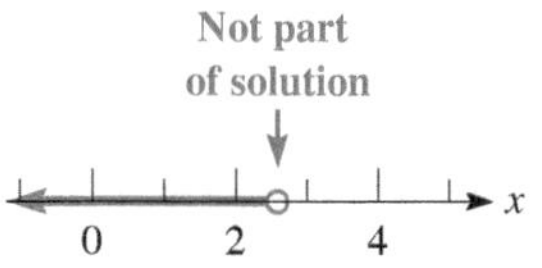

Fig. 17.7

Note that the sense of the inequality was reversed when we divided by -2. This solution is shown in Fig. 17.7. Any value of $x < 5/2$ checks when substituted into the original inequality.

Practice Exercise

1. Solve the inequality $2(3 - x) > 5 + 4x$.

The following example illustrates an application that involves the solution of an inequality.

EXAMPLE 5 Linear inequality—missile velocity

The velocity v (in m/s) of a missile in terms of the time t (in s) is given by $v = 490 - 9.8t$. For how long is the velocity positive? (Since velocity is a vector, this can also be interpreted as asking, "How long is the missile moving upward?")

In terms of inequalities, we are asked to find the values of t for which $v > 0$. This means that we must solve the inequality $490 - 9.8t > 0$. The solution is as follows:

$$490 - 9.8t > 0 \quad \text{original inequality}$$
$$-9.8t > -490 \quad \text{subtract 490 from each member}$$
$$t < 50 \text{ s} \quad \text{divide each member by } -9.8$$

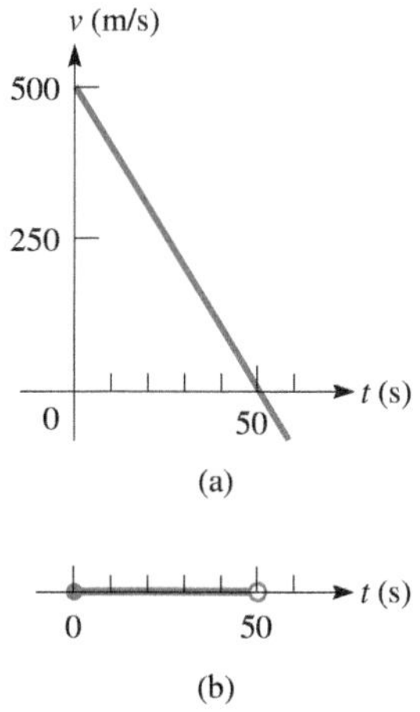

Fig. 17.8

Negative values of t have no meaning in this problem. Checking $t = 0$, we find that $v = 490$ m/s. Therefore, the complete solution is $0 \leq t < 50$ s, or $[0, 50 \text{ s})$.

In Fig. 17.8(a), we show the graph of $v = 490 - 9.8t$, and in Fig. 17.8(b), we show the solution $0 \leq t < 50$ s on the number line (which is really the t-axis in this case). Note that the values of v are above the t-axis for those values of t that are part of the solution. This shows the relationship of the graph of v as a function of t, and the solution as graphed on the number line (the t-axis).

INEQUALITIES WITH THREE MEMBERS

EXAMPLE 6 Solving an inequality with three members

Solve $-1 < 2x + 3 < 6$.

We have the following solution:

$$-1 < 2x + 3 < 6 \quad \text{original inequality}$$

$$-4 < 2x < 3 \quad \text{subtract 3 from each member}$$

$$-2 < x < \frac{3}{2} \quad \text{divide each member by 2}$$

$$\left(-2, \tfrac{3}{2}\right)$$

The solution is shown in Fig. 17.9.

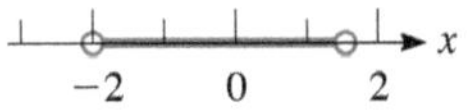

Fig. 17.9

EXAMPLE 7 Solving an inequality with three members

Solve the inequality $2x < x - 4 \le 3x + 8$.

Since we cannot isolate x in the middle member (or in any member), we rewrite the inequality as

$$2x < x - 4 \quad \text{and} \quad x - 4 \le 3x + 8$$

We then solve each of the inequalities, keeping in mind that the solution must satisfy both of them. Therefore, we have

$$2x < x - 4 \quad \text{and} \quad x - 4 \le 3x + 8$$

$$-2x \le 12$$

$$x < -4 \qquad\qquad x \ge -6$$

We see that the solution is $x < -4$ *and* $x \ge -6$, which can also be written as $-6 \le x < -4$, or $[-6, -4)$. Either of these latter two forms is generally preferred since they are more concise and more easily interpreted. The solution checks and is shown in Fig. 17.10.

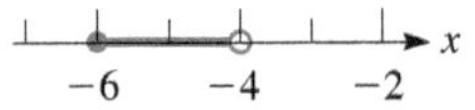

Fig. 17.10

Practice Exercise

2. Solve the inequality $-2 \le 4x - 3 < 5$.

EXAMPLE 8 Inequality with three members—pump rates

In emptying a wastewater tank, one pump can remove no more than 40 L/min. If it operates for 8.0 min and a second pump operates for 5.0 min, what must be the pumping rate of the second pump if 480 L are to be removed?

Let $x =$ the pumping rate of the first pump and $y =$ the pumping rate of the second pump. Since the first operates for 8.0 min and the second for 5.0 min to remove 480 L, we have

first pump, second pump, total ← amounts pumped

$$8.0x + 5.0y = 480$$

Since we know that the first pump can remove no more than 40 L/min, which means that $0 \le x \le 40$ L/min, we solve for x, then substitute in this inequality:

$$x = 60 - 0.625y \quad \text{solve for } x$$

$$0 \le 60 - 0.625y \le 40 \quad \text{substitute in inequality}$$

$$-60 \le -0.625y \le -20 \quad \text{subtract 60 from each member}$$

$$96 \ge y \ge 32 \quad \text{divide each member by } -0.625$$

$$32 \le y \le 96 \text{ L/min} \quad \text{use} \le \text{symbol (optional step)}$$

$$[32 \text{ L/min}, 96 \text{ L/min}]$$

This means that the second pump must be able to pump at least 32 L/min and no more than 96 L/min. See Fig. 17.11.

Although this was a three-member inequality combined with equalities, the solution was done in the same way as with a two-member inequality.

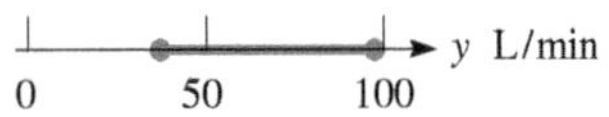

Fig. 17.11

EXERCISES 17.2

In Exercises 1–4, make the given changes in the indicated examples of this section and then perform the indicated operations.

1. In Example 2, change the 3 to 21 and then solve the resulting inequality.

2. In Example 4, change the $1/4$ to $7/4$ and then solve and display the resulting inequality.

3. In Example 6, change the $+$ in the middle member to $-$ and then solve the resulting inequality. Graph the solution.

4. In Example 7, change the middle member to $-x + 6$ and then solve the resulting inequality. Graph the solution.

In Exercises 5–28, solve the given inequalities. Graph each solution.

5. $x - 3 > -4$ **6.** $x + 2 \leq 6$ **7.** $\frac{1}{2}x < 32$

8. $-4t > 12$ **9.** $3x - 5 \leq -11$ **10.** $\frac{1}{3}x + 2 \geq 1$

11. $12 - 2y > 16$ **12.** $32 - 5x < -8$ **13.** $\frac{4x - 5}{2} \leq x$

14. $1.50 - 5.24x > 3.75 + 2.25x$

15. $180 - 6(T + 12) > 14T + 285$

16. $-2[x - (3 - 2x)] > \frac{1 - 5x}{3} + 2$

17. $2.50(1.50 - 3.40x) < 3.84 - 8.45x$

18. $(2x - 7)(x + 1) \leq 4 - x(1 - 2x)$

19. $\frac{1}{3} - \frac{L}{2} < L + \frac{3}{2}$ **20.** $\frac{x}{5} - 2 > \frac{2}{3}(x + 3)$

21. $-1 \leq 2x + 1 \leq 3$ **22.** $2 < 3R + 1 \leq 8$

23. $-4 \leq 1 - x < -1$ **24.** $0 \leq 3 - 2x \leq 6$

25. $2x < x - 1 \leq 3x + 5$ **26.** $x + 19 \leq 25 - x < 2x$

27. $2s - 3 < s - 5 < 3s - 3$

28. $0 < 1 - x \leq 3$ or $-1 < 2x - 3 < 5$

In Exercises 29–46, solve the given problems by setting up and solving appropriate inequalities. Graph each solution. When necessary, round to three significant digits.

29. Determine the values of x that are in the domain of the function $f(x) = \sqrt{2x - 10}$.

30. Determine the values of x that are in the domain of the function $f(x) = 1/\sqrt{3 - 0.5x}$.

31. Determine the values of x that are in the domain of the function $f(x) = \log(2x + 10)$.

32. Determine the values of x that are in the domain of the function $f(x) = \ln(7x + 4)$.

33. For what values of k are the roots of the equation $x^2 - kx + 9 = 0$ imaginary?

34. For what values of k are the roots of the equation $2x^2 + 3x + k = 0$ real and unequal?

35. A contractor is considering two similar jobs, each of which is estimated to take n hours to complete. One pays \$350 plus \$15 per hour, and the other pays \$25 per hour. For what values of n will the contractor make more at the second position?

36. In designing plastic pipe, if the inner radius r is increased by 5.00 cm, and the inner cross-sectional area is increased by between 125 cm^2 and 175 cm^2, what are the possible inner radii of the pipe?

37. The relation between the temperature in degrees Fahrenheit F and degrees Celsius C is $9C = 5(F - 32)$. What temperatures F correspond to temperatures between 10°C and 20°C?

38. The voltage drop V (in volts) across a resistor is the product of the current i (in A) and the resistance R (in Ω). Find the possible voltage drops across a variable resistor R if the minimum and maximum resistances are 1.6 kΩ and 3.6 kΩ, respectively, and the current is constant at 2.5 mA.

39. A rectangular PV (photovoltaic) solar panel is designed to be 1.42 m long and supply 130 W/m^2 of power. What must the width of the panel be in order to supply between 100 W and 150 W?

40. A beam is supported at each end, as shown in Fig. 17.12. Analysing the forces leads to the equation $F_1 = 13 - 3d$. For what values of d is F_1 more than 6 N?

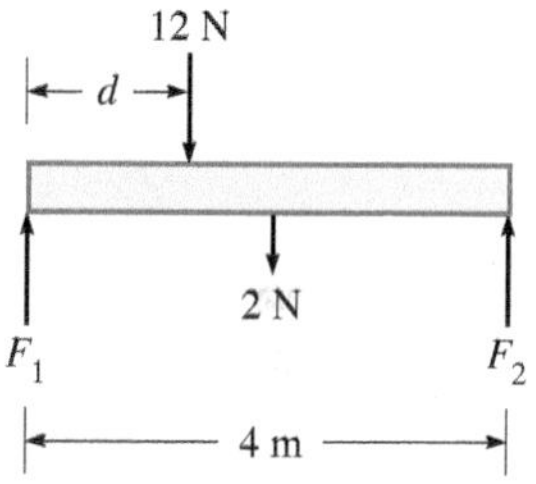

Fig. 17.12

41. The mass m (in g) of silver plate on a dish is increased by electroplating. The mass of silver on the plate is given by $m = 125 + 15.0t$, where t is the time (in h) of electroplating. For what values of t is m between 131 g and 164 g?

42. For a ground temperature of T_0 (in °C), the temperature T (in °C) at a height h (in m) above the ground is given approximately by $T = T_0 - 0.010h$. If the ground temperature is 25°C, for what heights is the temperature above 10°C?

43. During a given rush hour, the numbers of vehicles shown in Fig. 17.13 go in the indicated directions in a one-way-street section of a city. By finding the possible values of x and the equation relating x and y, find the possible values of y.

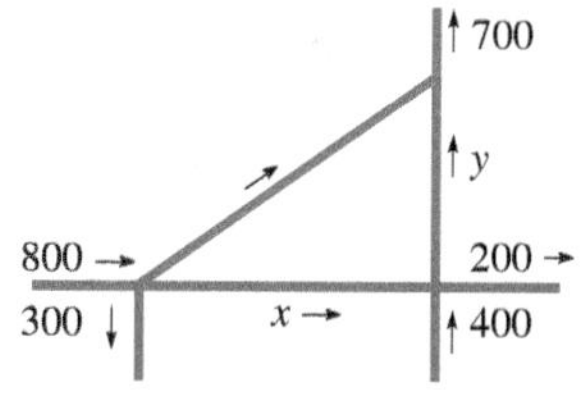

Fig. 17.13

44. The minimum legal speed on a certain highway is 70 km/h, and the maximum legal speed is 110 km/h. What legal distances can a motorist travel in 4 h on this highway without stopping?

45. The route of a rapid transit train is 40 km long, and the train makes five stops of equal length. If the train is actually moving for 1 h and each stop must be at least 2 min, what are the lengths of the stops if the train maintains an average speed of at least 30 km/h, including stop times?

46. An oil company plans to install eight storage tanks, each with a capacity of x litres, and five additional tanks, each with a capacity of y litres, such that the total capacity of all tanks is 440 000 L. If capacity y will be at least 40 000 L, what are the possible values of capacity x?

Answers to Practice Exercises

1. $x < 1/6$ **2.** $1/4 \leq x < 2$

17.3 Solving Nonlinear Inequalities

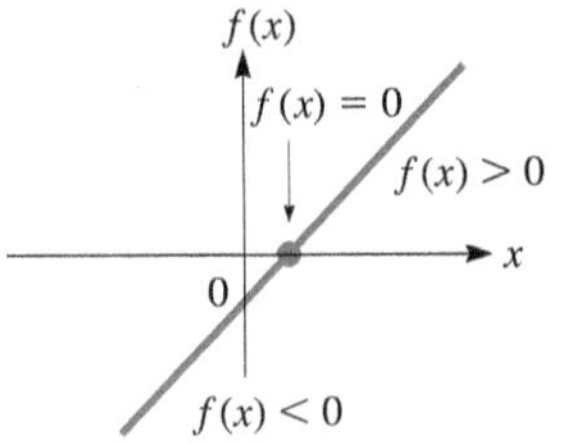

Fig. 17.14

In this section, we develop methods of solving inequalities with polynomials, rational expressions (expressions involving fractions), and nonalgebraic expressions. To develop the basic method for solving these types of inequalities, we now take another look at a linear inequality.

In Fig. 17.14, we see that all values of the linear function $f(x) = ax + b\ (a \neq 0)$ are positive on one side of the point at which $f(x) = 0$, and all values of $f(x)$ are negative on the opposite side of the same point. This means that *we can solve a linear inequality by expressing it with* ***zero*** *on the right and then finding the* ***sign*** *of the resulting function on either side of zero.*

EXAMPLE 1 Sign of a function with zero on the right

Solve the inequality $2x - 5 > 1$.

Finding the equivalent inequality with zero on the right, we have $2x - 6 > 0$. Setting the left member equal to zero, we have

$$2x - 6 = 0 \quad \text{for} \quad x = 3$$

which means $f(x) = 2x - 6$ has one sign for $x < 3$, and the other sign for $x > 3$. Testing values in these intervals, we find, for example, that

$$f(x) = -2 \quad \text{for} \quad x = 2 \qquad \text{and} \qquad f(x) = +2 \quad \text{for} \quad x = 4$$

Therefore, the solution to the original inequality is $x > 3$, or $(3, \infty)$. The solution in Fig. 17.15(b) corresponds to the positive values of $f(x)$ in Fig. 17.15(a).

We could have solved this inequality by methods of the previous section, but the important idea here is to use the ***sign*** *of the function,* with zero on the right.

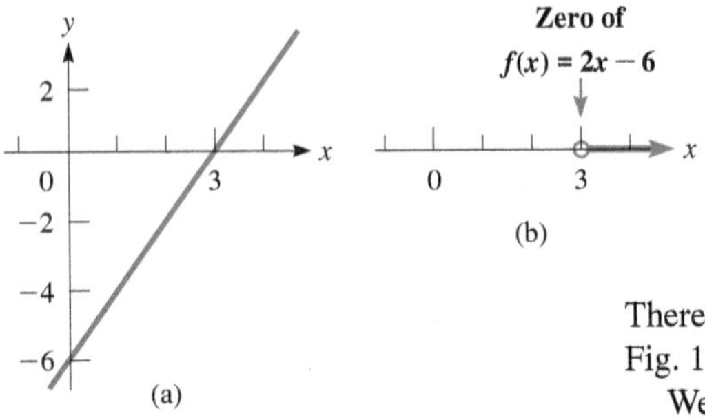

Fig. 17.15

Critical Values

We can extend this method to solving inequalities with polynomials of higher degree. The key point is that a function can change sign only at real values of x for which it is either zero or undefined (we call those values **critical values**). An equality can thus be solved by first finding the equivalent inequality with zero on the right, then factoring to find the critical values, and then examining the sign of the equivalent function for all real values of x. We outline the method below. Several examples follow.

Using Critical Values to Solve an Inequality

1. Determine the equivalent inequality with zero on the right.
2. Find all linear factors of the function.
3. To find the critical values, set each linear factor equal to zero and solve for x.
4. Determine the sign of the function to the left of the leftmost critical value, between critical values, and to the right of the rightmost critical value.
5. Those intervals in which the function has the proper sign satisfy the inequality.

EXAMPLE 2 Solving a quadratic inequality

Solve the inequality $x^2 - 3 > 2x$.

We first find the equivalent inequality with zero on the right. Therefore, we have $x^2 - 2x - 3 > 0$. This means that the solution of the inequality will be the values for which the function $f(x) = x^2 - 2x - 3$ is positive (see Fig. 17.16a).

We factor the left side of the inequality as

$$(x + 1)(x - 3) > 0$$

Each of these factors is set to zero to find the critical values:

$$x + 1 = 0 \qquad \text{or} \qquad x - 3 = 0$$
$$x = -1 \qquad\qquad x = 3$$

The critical values are thus -1 and 3. These values are the only possible places where the function can change in sign.

We now construct a table where we determine the sign of each factor and of the function to the left of -1, between -1 and 3, and to the right of 3. The signs of the factors in each interval are determined by choosing an arbitrary *test value* within the interval and substituting it into the factor to determine its sign.

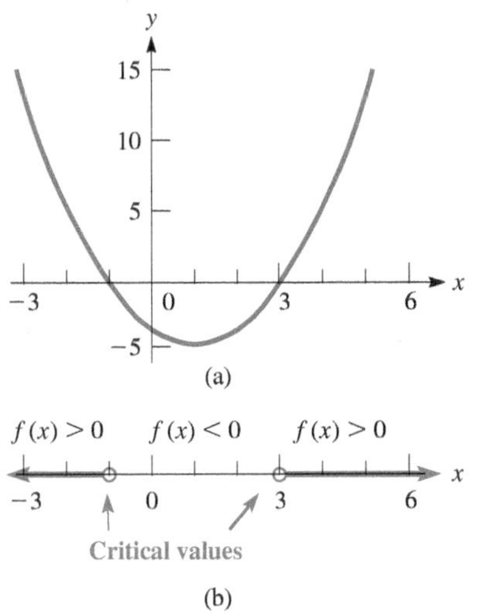

Fig. 17.16

Interval	$(x+1)$	$(x-3)$	*Sign of* $(x+1)(x-3)$
$x < -1$ $(-\infty, -1)$	$-$	$-$	$+$ (both factors negative)
$-1 < x < 3$ $(-1, 3)$	$+$	$-$	$-$ (one factor positive, one negative)
$x > 3$ $(3, \infty)$	$+$	$+$	$+$ (both factors positive)

Because we seek the intervals where the product is positive, the solution to the inequality is $x < -1$ or $x > 3$ or, in interval notation, $(-\infty, -1)$ or $(3, \infty)$. See Fig. 17.16(b).

EXAMPLE 3 Solving a cubic inequality

Solve the inequality $x^3 - 4x^2 + x + 6 < 0$.

By methods developed in Chapter 15, we factor the function on the left and obtain $(x + 1)(x - 2)(x - 3) < 0$. The critical values are $-1, 2, 3$. We wish to determine the sign of the left member for the intervals $x < -1, -1 < x < 2, 2 < x < 3$, and $x > 3$. The following table shows the information.

$$f(x) = (x + 1)(x - 2)(x - 3)$$

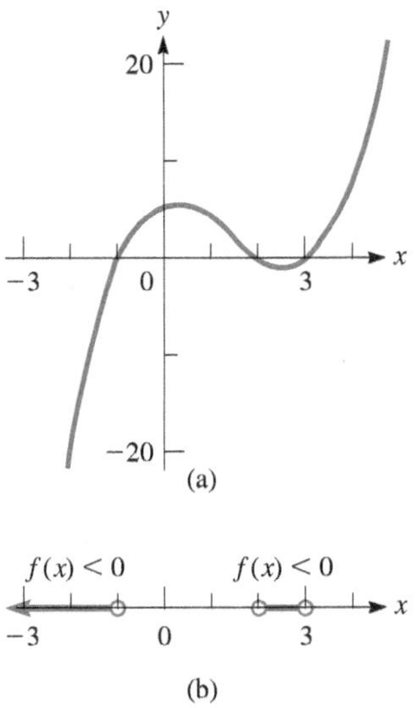

Fig. 17.17

Interval	$(x+1)$	$(x-2)$	$(x-3)$	*Sign of* $f(x)$
$x < -1$ $(-\infty, -1)$	$-$	$-$	$-$	$-$
$-1 < x < 2$ $(-1, 2)$	$+$	$-$	$-$	$+$
$2 < x < 3$ $(2, 3)$	$+$	$+$	$-$	$-$
$x > 3$ $(3, \infty)$	$+$	$+$	$+$	$+$

Since we want values for $f(x) < 0$, the solution is $(-\infty, -1)$ or $(2, 3)$, that is, $x < -1$ or $2 < x < 3$. The solution shown in Fig. 17.17(b) corresponds to the values of $f(x) < 0$ in Fig. 17.17(a).

The following example illustrates an applied situation that involves the solution of an inequality.

EXAMPLE 4 Solving a quadratic inequality—force on a cam

The force F (in N) acting on a cam varies according to the time t (in s), and it is given by the function $F = 2t^2 - 12t + 20$. For what values of t, $0 \leq t \leq 6$ s, is the force at least 4 N?

For a force of at least 4 N, we know that $F \geq 4$ N, or $2t^2 - 12t + 20 \geq 4$. This means we are to solve the inequality $2t^2 - 12t + 16 \geq 0$, and the solution is as follows:

$$2t^2 - 12t + 16 \geq 0$$
$$t^2 - 6t + 8 \geq 0$$
$$(t-2)(t-4) \geq 0$$

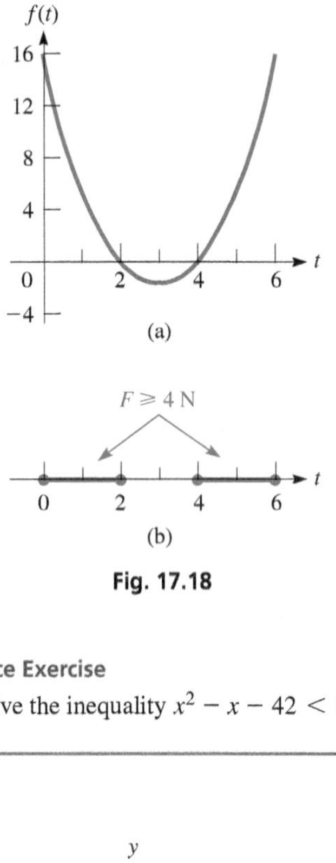

Fig. 17.18

The critical values are $t = 2$ and $t = 4$, which lead to the following table:

Interval	$(t-2)(t-4)$		*Sign of* $(t-2)(t-4)$
$0 \leq t < 2$ $[0, 2)$	$-$	$-$	$+$
$2 < t < 4$ $(2, 4)$	$+$	$-$	$-$
$4 < t \leq 6$ $(4, 6]$	$+$	$+$	$+$

We see that the values of t that satisfy the *greater than* part of the problem are $[0, 2)$ and $(4, 6]$. Since we know that $(t-2)(t-4) = 0$ for $t = 2$ and $t = 4$, the solution is $[0, 2]$ or $[4, 6]$, that is,

$$0 \leq t \leq 2 \text{ s} \quad \text{or} \quad 4 \text{ s} \leq t \leq 6 \text{ s}$$

The graph of $f(t) = 2t^2 - 12t + 16$ is shown in Fig. 17.18(a). The solution, shown in Fig. 17.18(b), corresponds to the values of $f(t)$ that are zero or positive or for which $F \geq 4$ N.

Practice Exercise

1. Solve the inequality $x^2 - x - 42 < 0$.

EXAMPLE 5 Solving a cubic inequality

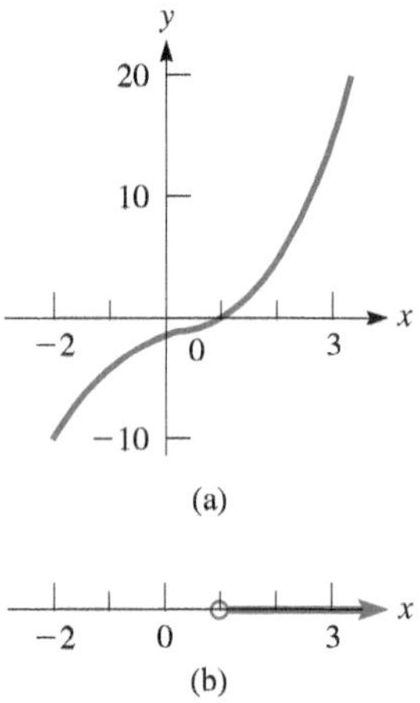

Fig. 17.19

Solve the inequality $x^3 - x^2 + x - 1 > 0$.

Although we can use the methods of Chapter 15, this function is factorable by grouping. We have

$$x^3 - x^2 + x - 1 = x^2(x-1) + 1(x-1) = (x^2+1)(x-1) > 0$$

The factor $x^2 + 1$ is always positive, so the only critical value is found when $x - 1 = 0$, or $x = 1$. This is the only value of x where the function can change sign.

Since $x - 1$ is positive for $x > 1$ and negative for $x < 1$, the solution to the inequality is $x > 1$, or $(1, \infty)$.

The solution is shown in Fig. 17.19(b), and we see that this solution corresponds to the positive values of $f(x) = x^3 - x^2 + x - 1$ shown in Fig. 17.19(a).

EXAMPLE 6 Solving a rational inequality

Find the values of x for which $\sqrt{\dfrac{x-3}{x+4}}$ represents a real number.

For the expression to represent a real number, the fraction under the radical must be greater than or equal to zero. This means we must solve the inequality

$$\frac{x-3}{x+4} \geq 0$$

The critical values are found from the factors that are in the numerator or in the denominator. Thus, the critical values are -4 and 3. Note that if $x = -4$ the fraction is undefined, so x may not equal -4. The table of signs is constructed as follows.

Interval	$\frac{x-3}{x+4}$	*Sign of* $\frac{x-3}{x+4}$
$x < -4$ $(-\infty, -4)$	$\frac{-}{-}$	$+$
$-4 < x < 3$ $(-4, 3)$	$\frac{-}{+}$	$-$
$x > 3$ $(3, \infty)$	$\frac{+}{+}$	$+$

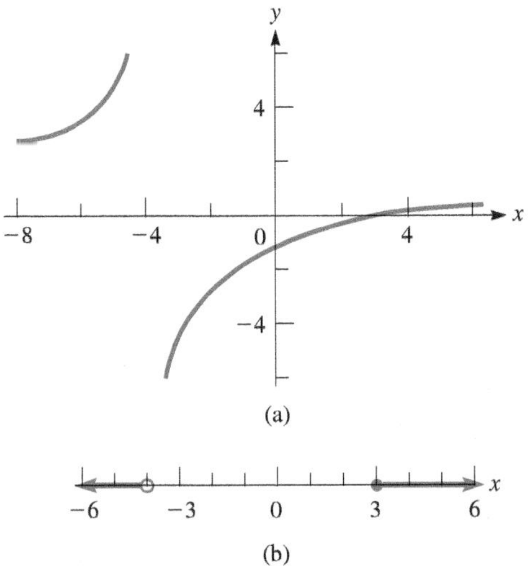

Fig. 17.20

Thus, the values that satisfy the *greater than* part of the problem are $(-\infty, -4)$ or $(3, \infty)$. Moreover, $x = 3$ is also a valid solution for the equality part, for the fraction is zero.

Therefore, the solution of the inequality is $(-\infty, 4)$ or $[3, \infty]$, that is, $x < -4$ or $x \geq 3$. This means these are the values for which the original expression represents a real number. The graph of $f(x) = \dfrac{x-3}{x+4}$ is shown in Fig. 17.20(a), and the graph of the solution is shown in Fig. 17.20(b).

Practice Exercise

2. Solve the inequality $\dfrac{x+3}{2x-5} \geq 0$.

EXAMPLE 7 Solving a rational inequality

Solve the inequality $\dfrac{(x-2)^2(x+3)}{4-x} < 0$.

The critical values are -3, 2, and 4. Thus, we have the following table:

Interval	$\frac{(x-2)^2(x+3)}{4-x}$	*Sign of* $\frac{(x-2)^2(x+3)}{4-x}$
$x < -3$ $(-\infty, -3)$	$\frac{+\ -}{+}$	$-$
$-3 < x < 2$ $(-3, 2)$	$\frac{+\ +}{+}$	$+$
$2 < x < 4$ $(2, 4)$	$\frac{+\ +}{+}$	$+$
$x > 4$ $(4, \infty)$	$\frac{+\ +}{-}$	$-$

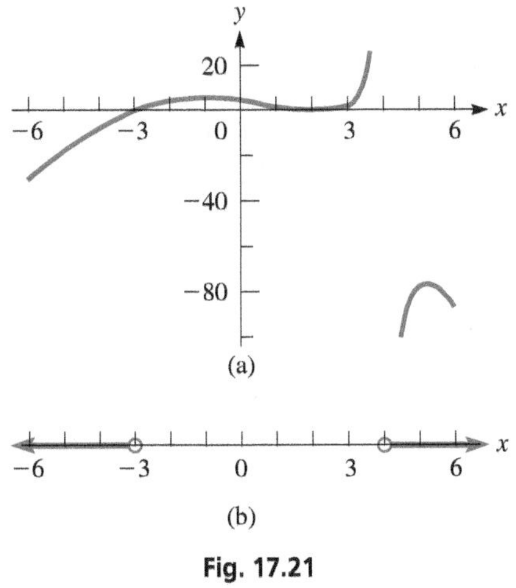

Fig. 17.21

Thus, the solution is $(-\infty, -3)$ or $(4, \infty)$, or $x < -3$ or $x > 4$. The graph of the function is shown in Fig. 17.21(a), and the graph of the solution is shown in Fig. 17.21(b).

EXAMPLE 8 Solving a rational inequality

Solve the inequality $\dfrac{x+3}{x-1} > 2.$

We first subtract 2 from each member and combine terms on the left. We have

$$\frac{x+3}{x-1} - 2 > 0 \qquad \text{subtract 2 from both members}$$

$$\frac{x+3-2(x-1)}{x-1} > 0 \qquad \text{combine over the common denominator}$$

$$\frac{5-x}{x-1} > 0 \qquad \text{this is the form to use}$$

The critical values are 1 and 5, and we have the following table of signs:

Interval	$(5-x)/(x-1)$	*Sign of* $(5-x)/(x-1)$
$x < 1$ $(-\infty, 1)$	$+/-$	$-$
$1 < x < 5$ $(1, 5)$	$+/+$	$+$
$x > 5$ $(5, \infty)$	$-/+$	$-$

Thus, the solution is $(1, 5)$, or $1 < x < 5$. The graph of $f(x) = (5-x)/(x-1)$ is shown in Fig. 17.22(a), and the graph of the solution is shown in Fig. 17.22(b).

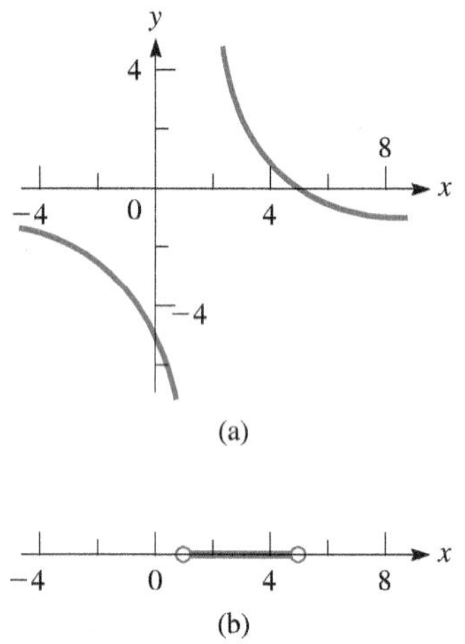

Fig. 17.22

COMMON ERROR We stress that to solve a rational inequality, we ***do not*** start by multiplying by the least common denominator, as we did with equations involving fractions. Since we do not know the value of x, we do not know if the sense of the inequality would change after multiplication or not. It is for this reason that we solve rational inequalities using critical values.

Note also that to avoid division by zero, the solution ***never*** includes the critical value that makes the denominator zero.

LEARNING TIP

When a rational inequality involves more than one nonzero term, the first step is to get a zero on the right and combine terms using the least common denominator so as to obtain a single rational expression. We can then proceed to use critical values as before, without multiplying both sides by the least common denominator. Example 8 illustrates the procedure.

SOLVING INEQUALITIES GRAPHICALLY

We can get an approximate solution of an inequality from the graph of a function, including functions that are not factorable or not algebraic. The method follows several of the earlier examples. First, write the equivalent inequality with zero on the right and then graph this function. Those values of x corresponding to the proper values of y (either above or below the x-axis) are those that satisfy the inequality.

EXAMPLE 9 Solving an inequality graphically

Use a graphing utility to solve the inequality $x^3 > x^2 - 3$.

Finding the equivalent inequality with zero on the right, we have $x^3 - x^2 + 3 > 0$. We then let $y_1 = x^3 - x^2 + 3$ and graph this function. On a calculator, this gives the display shown in Fig. 17.23.

We can see that the curve crosses the x-axis only once, between $x = -2$ and $x = -1$. Using the *zero* feature, we find that this value is approximately $x = -1.17$. Since we want values of x that correspond to positive values of y, we see that the solution is $x > -1.17$, or $(-1.17, \infty)$.

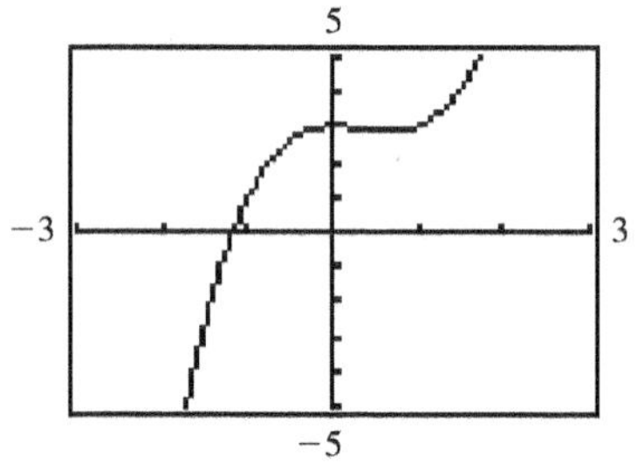

Fig. 17.23

EXERCISES 17.3

In Exercises 1–4, make the given changes in the indicated examples of this section and then solve the resulting inequalities.

1. In Example 2, change the $-$ sign before the 3 to $+$ and change $2x$ to $4x$, then solve the resulting inequality, and graph the solution.
2. In Example 5, change both $-$ signs to $+$, then solve the resulting inequality, and graph the solution.
3. In Example 7, change the exponent on $(x - 2)$ from 2 to 3, then solve the resulting inequality, and graph the solution.
4. In Example 8, on the right, change the 2 to 3, then solve the resulting inequality, and graph the solution.

In Exercises 5–32, solve the given inequalities. Graph each solution.

5. $x^2 - 16 < 0$
6. $x^2 + 3x \geq 0$
7. $2x^2 \leq 4x$
8. $x^2 - 4x > 21$
9. $2x^2 - 12 \leq -5x$
10. $9t^2 + 6t > -1$
11. $x^2 + 4x \leq -4$
12. $6x^2 + 1 < 5x$
13. $R^2 + 4 > 0$
14. $x^4 + 2 < 1$
15. $x^3 + x^2 - 2x < 0$
16. $x^3 - 2x^2 + x \geq 0$
17. $s^3 + 2s^2 - s \geq 2$
18. $n^4 - 2n^3 + 8n + 12 \leq 7n^2$
19. $3x^2 + 5x \geq 2$
20. $12x^2 + x > 1$
21. $\dfrac{2x - 3}{x + 6} \leq 0$
22. $\dfrac{x + 15}{x - 9} > 0$
23. $\dfrac{x^2 - 6x - 7}{x + 5} > 0$
24. $\dfrac{(x - 2)^2(5 - x)}{(4 - x)^3} \leq 0$
25. $\dfrac{x}{x + 1} > 1$
26. $\dfrac{2p}{p - 1} > 3$
27. $\dfrac{T - 8}{3 - T} \leq 0$
28. $\dfrac{3x + 1}{x + 3} \geq 0$
29. $\dfrac{6 - x}{3 - x - 4x^2} \geq 0$
30. $\dfrac{4 - x}{3 + 2x - x^2} > 0$
31. $\dfrac{x^4(9 - x)(x - 5)(2 - x)}{(4 - x)^5} > 0$
32. $\dfrac{2}{x - 3} < 4$

In Exercises 33–36, determine the values of x for which the radicals represent real numbers.

33. $\sqrt{(x - 1)(x + 2)}$
34. $\sqrt{x^2 - 3x}$
35. $\sqrt{-x - x^2}$
36. $\sqrt{\dfrac{x^3 + 6x^2 + 8x}{3 - x}}$

In Exercises 37–44, solve the given inequalities graphically by using a graphing calculator. Approximate the critical values to the nearest 0.01. See Example 9.

37. $x^3 - x > 2$
38. $0.5x^3 < 3 - 2x^2$
39. $x^4 < x^2 - 2x - 1$
40. $3x^4 + x + 1 > 5x^2$
41. $2^x > x + 2$
42. $\log x < 1 - 2x^2$
43. $\sin x < 0.1x^2 - 1$
44. $4 \cos 2x > 2x - 3$

In Exercises 45–52, use inequalities to solve the given problems. Where necessary, approximate the critical values to the nearest 0.01.

45. Is $x^2 > x$ for all x? Explain.
46. Is $x > 1/x$ for all x? Explain.
47. Find an inequality of the form $ax^2 + bx + c < 0$ with $a > 0$ for which the solution is $-1 < x < 4$.
48. Find an inequality of the form $ax^3 + bx < 0$ with $a > 0$ for which the solution is $x < -1$ or $0 < x < 1$.
49. Algebraically find the values of x for which $2^{x+2} > 3^{2x-3}$.
50. Graphically find the values of x for which $2 \log_2 x < \log_3(x + 1)$.
51. For what values of real numbers a and b does the inequality $(x - a)(x - b) < 0$ have real solutions?
52. Algebraically find the intervals for which $f(x) = 2x^4 - 5x^3 + 3x^2$ is positive and those for which it is negative. Using only this information, draw a rough sketch of the graph of the function.

In Exercises 53–64, answer the given questions by solving the appropriate inequalities.

53. The electric power P (in W) delivered to part of a circuit is given by $P = 6i - 4i^2$, where i is the current (in A). For what positive values of i is the power greater than 2 W?
54. The mass m (in Mg) of fuel in a rocket after launch is $m = 2000 - t^2 - 140t$, where t is the time (in min). During what period of time is the mass of fuel greater than 500 Mg?
55. The weekly sales S (in thousands of units) of a certain product t weeks after it is introduced to the market are given by $S = 200t/(t^2 + 64)$. When will sales be at least 10 000 units per week?
56. The object distance p (in cm) and image distance q (in cm) for a camera of focal length 3.00 cm is given by $p = 3.00q/(q - 3.00)$. For what values of q is $p > 12.0$ cm?
57. The total capacitance C of capacitors C_1 and C_2 in series is $C^{-1} = C_1^{-1} + C_2^{-1}$. If $C_2 = 4.00\ \mu F$, find C_1 if $C > 1.00\ \mu F$.
58. A rectangular field is to be enclosed by a fence and divided down the middle by another fence. The middle fence costs \$4/m and the other fence costs \$8/m. If the area of the field is to be 8000 m^2, and the cost of the fence cannot exceed \$4000, what are the possible dimensions of the field?
59. The weight w (in N) of an object h metres above the surface of earth is $w = r^2w_0/(r + h)^2$, where r is the radius of Earth and w_0 is the weight of the object at sea level. Given that $r = 6380$ km, if an object weighs 200 N at sea level, for what altitudes is its weight less than 100 N?
60. One type of machine part costs \$10 each, and a second type costs \$20 each. How many of the second type can be purchased if at least 30 of the first type are purchased and a total of \$1000 is spent?
61. The length of a rectangular microprocessor chip is 2.0 mm more than its width. If its area is less than 35 mm^2, what values are possible for the width if it must be at least 3.0 mm?

62. A laser source is 2.0 cm from the nearest point P on a flat mirror, and the laser beam is directed at a point Q that is on the mirror and is x cm from P. The beam is then reflected to the receiver, which is x cm from Q. What is x if the total length of the beam is greater than 6.5 cm? See Fig. 17.24.

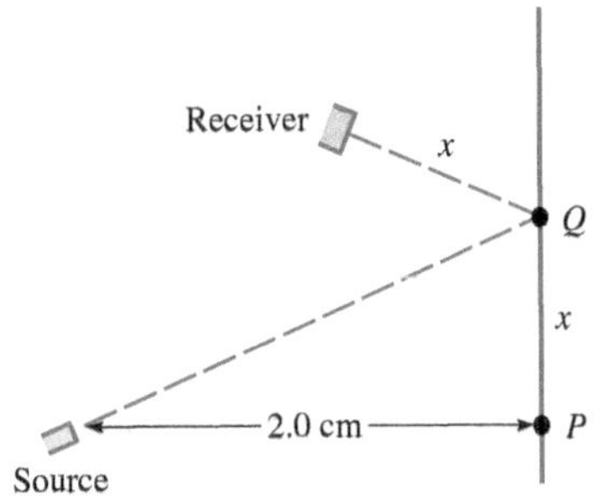

Fig. 17.24

63. A plane takes off from Winnipeg and flies due east at 620 km/h. At the same time, a second plane takes off from the surface of Lake Winnipeg 310 km due north of Winnipeg and flies due north at 560 km/h. For how many hours are the planes less than 1000 km apart?

64. An open box (no top) is formed from a piece of cardboard 8.00 cm square by cutting equal squares from the corners, turning up the resulting sides, and taping the edges together. Find the edges of the squares that are cut out in order that the volume of the box is greater than 32.0 cm^3. See Fig. 17.25.

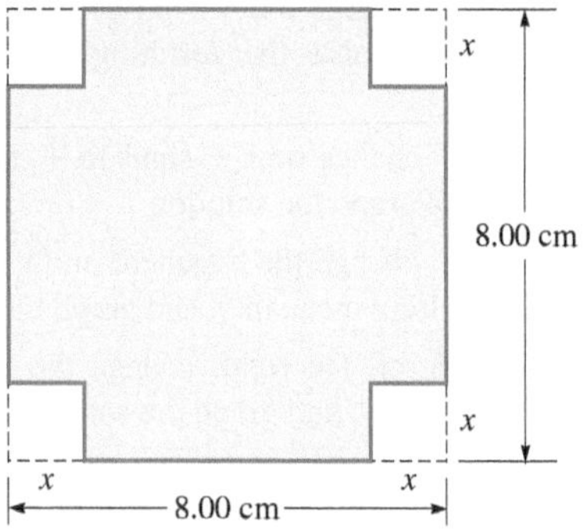

Fig. 17.25

Answers to Practice Exercises

1. $-6 < x < 7$, or $(-6, 7)$

2. $x \le -3$ or $x > 5/2$, or $(-\infty, -3]$ or $\left(\frac{5}{2}, \infty\right)$

17.4 Inequalities Involving Absolute Values

Inequalities involving absolute values are often useful in later topics in mathematics such as calculus and in applications such as the accuracy of measurements. In this section, we show the meaning of such inequalities and how they are solved.

Examining the inequality $|x| > 1$, note that we are considering values of x that are *numerically* larger than 1. Thus, we may write this inequality in the equivalent form $x < -1$ or $x > 1$. This means that *the original inequality, with an absolute-value sign, can be written in terms of two equivalent inequalities, neither involving absolute values.* Similarly, we can write the inequality $|x| < 1$ without absolute-value signs as $-1 < x < 1$, since we are considering values of x numerically less than 1. More generally, the following relations hold for functions $f(x)$ and $n > 0$.

Inequalities Involving Absolute Values			EXAMPLE 1
Inequality with Absolute Value	**Equivalent Inequality without Absolute Value**		
$\lvert f(x)\rvert > n$	$f(x) < -n$ or $f(x) > n$	**(17.1)**	$\lvert x + 3\rvert > 7$ is equivalent to $x + 3 > 7$ or $x + 3 < -7$
$\lvert f(x)\rvert < n$	$-n < f(x) < n$	**(17.2)**	$\lvert 3x - 5\rvert < 1$ is equivalent to $-1 < 3x - 5 < 1$

EXAMPLE 2 Absolute value less than a number

Solve the inequality $|x - 3| < 2$.

Here, we want values of x such that $x - 3$ is numerically smaller than 2, or the values of x within 2 units of $x = 3$. These are given by the inequality $1 < x < 5$. Now, using Eq. (17.2), we have

$$-2 < x - 3 < 2$$

By adding 3 to all three members of this inequality, we have

$$1 < x < 5$$

or $(1, 5)$, which is the proper interval. See Fig. 17.26.

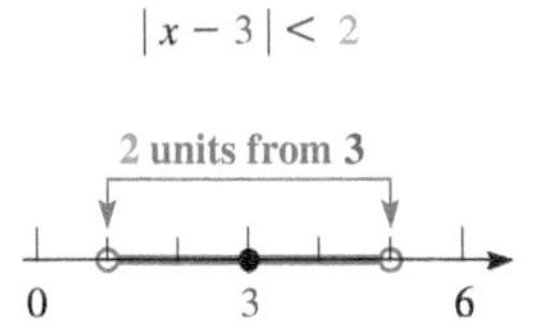

Fig. 17.26

EXAMPLE 3 Absolute value greater than a number

Solve the inequality $|2x - 1| > 5$.

By using Eq. (17.1), we have

$$2x - 1 < -5 \quad \text{or} \quad 2x - 1 > 5$$

Completing the solution, we have

$$2x < -4 \quad \text{or} \quad 2x > 6 \qquad \text{add 1 to each member}$$
$$x < -2 \qquad\qquad x > 3 \qquad \text{divide each member by 2}$$
$$(-\infty, -2) \qquad (3, \infty)$$

This means that the given inequality is satisfied for $x < -2$ or for $x > 3$, or in $(-\infty, -2)$ or $(3, \infty)$. We must be very careful to remember that *we cannot write this as* $3 < x < -2$. The solution is shown in Fig. 17.27.

The meaning of this inequality is that the numerical value of $2x - 1$ is greater than 5. By considering values in these intervals, we can see that this is true for values of x less than -2 or greater than 3.

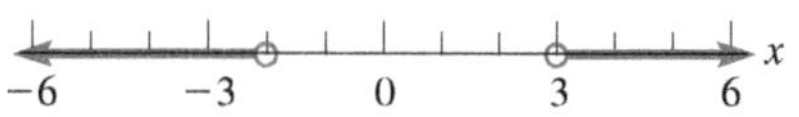

Fig. 17.27

EXAMPLE 4 Absolute value greater than or equal to a number

Solve the inequality $2\left|\dfrac{2x}{3} + 1\right| \ge 4$.

The solution is as follows:

$$2\left|\frac{2x}{3} + 1\right| \ge 4 \qquad \text{original inequality}$$
$$\left|\frac{2x}{3} + 1\right| \ge 2 \qquad \text{divide each member by 2}$$
$$\frac{2x}{3} + 1 \le -2 \quad \text{or} \quad \frac{2x}{3} + 1 \ge 2 \qquad \text{using Eq. (17.1)}$$
$$2x + 3 \le -6 \qquad 2x + 3 \ge 6$$
$$2x \le -9 \qquad 2x \ge 3$$
$$x \le -\frac{9}{2} \qquad x \ge \frac{3}{2} \qquad \text{solution}$$
$$\left(-\infty, -\frac{9}{2}\right] \qquad \left[\frac{3}{2}, \infty\right)$$

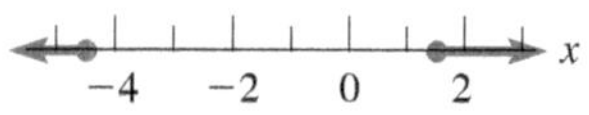

Fig. 17.28

This solution is shown in Fig. 17.28. Note that the sign of equality does not change the method of solution. It simply indicates that $-\frac{9}{2}$ and $\frac{3}{2}$ are included in the solution.

Practice Exercise

1. Solve the inequality $|2x - 9| > 3$.

EXAMPLE 5 Absolute value less than a number

Solve the inequality $|3 - 2x| < 3$.

We have the following solution:

$$|3 - 2x| < 3 \qquad \text{original inequality}$$
$$-3 < 3 - 2x < 3 \qquad \text{using Eq. (17.2)}$$
$$-6 < -2x < 0$$
$$3 > x > 0 \qquad \text{divide by } -2 \text{ and reverse signs of inequality}$$
$$0 < x < 3 \qquad \text{solution}$$
$$(0, 3)$$

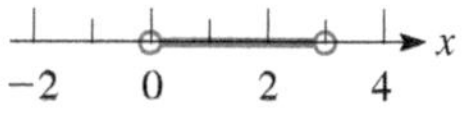

Fig. 17.29

The meaning of the inequality is that the numerical value of $3 - 2x$ is less than 3. This is true for values of x between 0 and 3. The solution is shown in Fig. 17.29.

Practice Exercise

2. Solve the inequality $|4 - x| \le 2$.

EXAMPLE 6 Absolute value—accuracy of measurement

A technician measures an electric current and reports that it is 0.036 A with a possible error of ± 0.002 A. Write this result for the current i, using an inequality with absolute values.

The statement of the problem tells us that the current is no less than 0.034 A and no more than 0.038 A. Another way of stating this is that the numerical difference between the true value of i (unknown exactly) and the measured value, 0.036 A, is less than or equal to 0.002 A. Using an absolute-value inequality, this is written as

$$|i - 0.036| \le 0.002$$

where values are in amperes.

We can see that this inequality is correct by using (Eq. 17.2):

$$-0.002 \le i - 0.036 \le 0.002$$

$$0.034 \le i \le 0.038 \qquad \text{add 0.036 to each member}$$

Fig. 17.30

This verifies that i should not be less than 0.034 A or more than 0.038 A. The solution is shown in Fig. 17.30.

EXERCISES 17.4

In Exercises 1 and 2, make the given changes in the indicated examples of this section and then solve the resulting inequalities.

1. In Example 3, change the $>$ to $<$, solve the resulting inequality, and graph the solution.
2. In Example 5, change the $<$ to $>$, solve the resulting inequality, and graph the solution.

In Exercises 3–24, solve the given inequalities. Graph each solution.

3. $|x - 4| < 1$
4. $|x + 4| < 6$
5. $|5x + 4| > 6$
6. $\left|\frac{1}{2}N - 1\right| > 1$
7. $1 + |6x - 5| \le 5$
8. $|30 - 42x| \le 0$
9. $|3 - 4x| > 3$
10. $3 + |3x + 1| \ge 5$
11. $\left|\frac{t + 1}{5}\right| < 5$
12. $\left|\frac{2x - 9}{4}\right| < 1$
13. $|20x + 85| \le 43$
14. $|2.6x - 9.1| > 10.4$
15. $2|x - 24| > 84$
16. $3|4 - 3x| \le 10$
17. $8 + 3|3 - 2x| < 11$
18. $5 - 4|1 - 7x| > 13$
19. $4|2 - 5x| \ge 6$
20. $2.5|7.1 - 2.0x| \le 6.5$
21. $\left|\frac{3R}{5} + 1\right| < 8$
22. $\left|\frac{4x}{3} - 5\right| \ge 7$
23. $\left|6.5 - \frac{x}{2}\right| \ge 2.3$
24. $\left|\frac{2w + 5}{w + 1}\right| \ge 1$

In Exercises 25–28, solve the given quadratic inequalities.

25. $|x^2 + x - 4| > 2$
(After using Eq. 17.1, you will have two inequalities. The solution includes the values of x that satisfy *either* of the inequalities.)
26. $|x^2 + 3x - 1| > 3$ (See Exercise 25.)
27. $|x^2 + x - 4| < 2$
(Use Eq. 17.2, then treat the resulting inequality as two inequalities of the form $f(x) > -n$ and $f(x) < n$. The solution includes the values of x that satisfy *both* of the inequalities.)
28. $|x^2 + 3x - 1| < 3$ (See Exercise 27.)

In Exercises 29–34, solve the given problems.

29. Solve for x if $|x| < a$ and $a \le 0$. Explain.
30. Solve for x if $|x - 1| < 4$ and $x \ge 0$.
31. Solve for x: $a - |bx| < c$ given that $a - c > 0$.
32. Solve for x: $1 < |x - 2| < 3$.
33. The thickness t (in km) of Earth's crust varies and can be described as $|t - 27| \le 23$. What are the minimum and maximum values of the thickness of Earth's crust?
34. The temperature T (in °C) at which a certain integrated-circuit computer chip is designed to operate is given by the inequality $|T - 110| \le 10$. What are the minimum and maximum temperatures of this range?

In Exercises 35–43, use inequalities involving absolute values to solve the given problems.

35. The production p (in barrels) of oil at a refinery is estimated at 2 000 000 $\pm$ 200 000. Express p using an inequality with absolute values and describe the production in a verbal statement.
36. According to the Waze navigation app, the time required for a driver to reach his destination is 52 min. If this time is accurate to ± 3 min, express the travel time t using an inequality with absolute values.
37. The temperature T (in °C) at which a certain machine can operate properly is 70 $\pm$ 20. Express the temperature T for proper operation using an inequality with absolute values.

38. The Mach number M of a moving object is the ratio of its velocity v to the velocity of sound v_s, and v_s varies with temperature. A jet travelling at 1650 km/h changes its altitude from 500 m to 5500 m. At 500 m (with the temperature at 27°C), $v_s = 1250$ km/h, and at 5500 m (−3°C), $v_s = 1180$ km/h. Express the range of M, using an inequality with absolute values.

39. The diameter d of a certain type of tubing is 3.675 cm with a tolerance of 0.002 cm. Express this as an inequality with absolute values.

40. The moon is at a maximum distance of 4.07×10^5 km from Earth and at a minimum distance of 3.57×10^5 km. Express this as an inequality with absolute values.

41. The Peterborough Lift Lock in Peterborough, Ontario (the highest hydraulic lift lock in the world), raises and lowers boats as the mass m of its two identical caissons changes from 1542 t to 1672.6 t. Express the range of m using an inequality with absolute values.

42. The voltage V in a certain circuit is given by $V = 6.0 - 200i$, where i is the current (in A). For what values of the current is the absolute value of the voltage less than 2.0 V?

43. A rocket is fired from a plane flying horizontally at 3000 m. The height h (in m) of the rocket above the plane is given by $h = 190t - 4.9t^2$, where t is the time (in s) of flight of the rocket. When is the rocket more than 1300 m above or below the plane? Round to three significant digits. See Fig. 17.31.

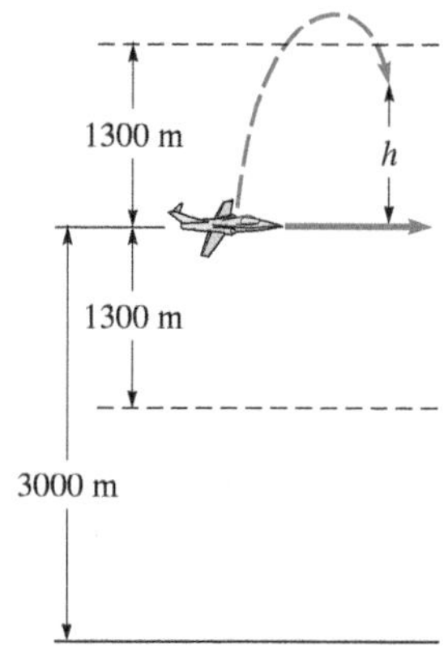

Fig. 17.31

Answers to Practice Exercises

1. $x < 3$ or $x > 6$, or $(-\infty, 3)$ or $(6, \infty)$
2. $2 \le x \le 6$, or $[2, 6]$

17.5 Graphical Solution of Inequalities with Two Variables

To this point, we have considered inequalities with one variable and certain methods of solving them. We may also graphically solve inequalities involving two variables, such as x and y. In this section, we consider the solution of such inequalities.

Let us consider the function $y = f(x)$. We know that the coordinates of points on the graph satisfy the equation $y = f(x)$. However, for points above the graph of the function, we have $y > f(x)$, and for points below the graph of the function, we have $y < f(x)$. Consider the following example.

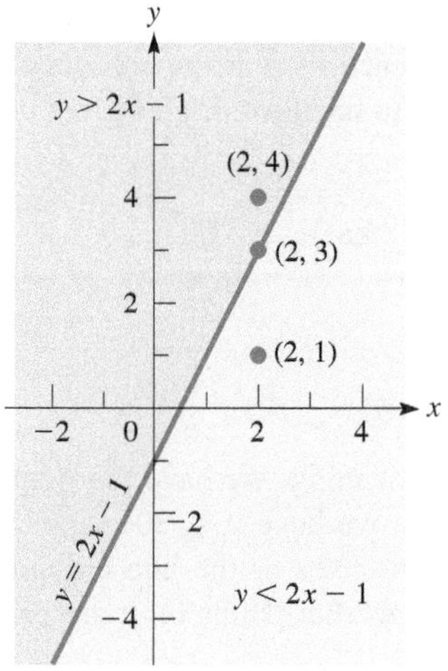

Fig. 17.32

EXAMPLE 1 Checking points above and below a line

Consider the linear function $y = 2x - 1$. Determine whether each of the following points is on the line, contained in the region above the line, or contained in the region below the line.

(a) $(2, 4)$ is contained in the region above the line, since $4 > 2(2) - 1$, or $4 > 3$. Points above the line satisfy the inequality $y > 2x - 1$.

(b) $(2, 3)$ is on the line, since $3 = 2(2) - 1$, or $3 = 3$. Points on the line satisfy the equality $y = 2x - 1$.

(c) $(2, 1)$ is contained in the region below the line, since $1 < 2(2) - 1$, or $1 < 3$. Points below the line satisfy the inequality $y < 2x - 1$.

The line for which $y = 2x - 1$ and the regions for which $y > 2x - 1$ and for which $y < 2x - 1$ are shown in Fig. 17.32.

The illustration in Example 1 leads us to the graphical method of indicating the points that satisfy an inequality with two variables.

Procedure for Solving an Inequality with Two Variables

1. Solve the inequality for y by isolating y.
2. Consider the equation $y = f(x)$ obtained by changing the inequality sign to an equality sign. Determine the graph of $y = f(x)$ and draw it as a solid or dashed curve, as follows:
   a. If the original inequality is strict ($>$ or $<$), draw $f(x)$ as a dashed curve. This indicates that the graph of $f(x)$ **is not** part of the solution.
   b. If the original inequality is not strict ($\geq$ or $\leq$), draw $f(x)$ as a solid curve. This indicates that the graph of $f(x)$ **is** part of the solution.
3. The solution to the inequality consists of all points in an entire region of the plane, which is shaded in as follows:
   a. Shade in the region **above the curve** if the inequality is of the form $y > f(x)$ or $y \geq f(x)$.
   b. Shade in the region **below the curve** if the inequality is of the form $y < f(x)$ or $y \leq f(x)$.

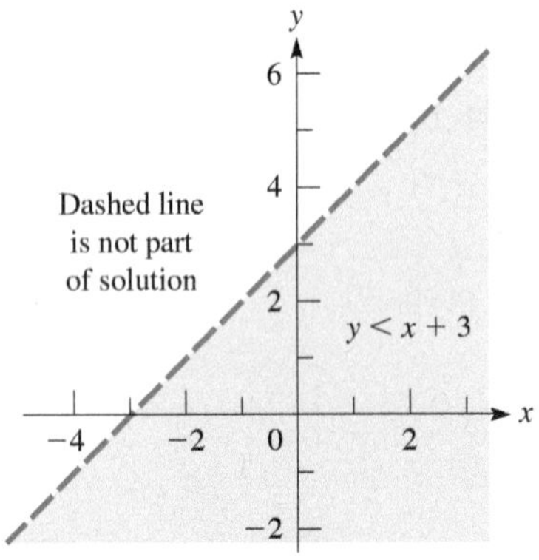

Fig. 17.33

EXAMPLE 2 Sketching an inequality of the form $y < f(x)$

Draw a sketch of the graph of the inequality $y < x + 3$.

The variable y is already isolated, so the first step is to draw the function $y = x + 3$. Since the original inequality is strict, the function is represented by a dashed line, as shown in Fig. 17.33. The solution to the inequality is then shaded in as the region *below* the line.

EXAMPLE 3 Sketching an inequality—possible plowing routes

After a snowstorm, it is estimated that it will take 30 min to plow each kilometre of Route 15 and 45 min to plow each kilometre of Route 80. If no more than 60 plowing-hours are available, what combinations of Route 15 and Route 80 can be plowed?

Let $x =$ kilometres of Route 15 that can be plowed and $y =$ kilometres of Route 80 that can be plowed. The time to plow along each route is the product of the time for each kilometre and the number of kilometres to be plowed. This gives us

$$\underbrace{(0.50\text{ h/km})}_{\text{30 min}}(x\text{ km}) + \underbrace{(0.75\text{ h/km})}_{\text{45 min}}(y\text{ km}) \leq 60\text{ h}$$

(time to plow Rt. 15; time to plow Rt. 80; max. available time)

$$0.50x + 0.75y \leq 60$$
$$y \leq 80 - 0.67x$$

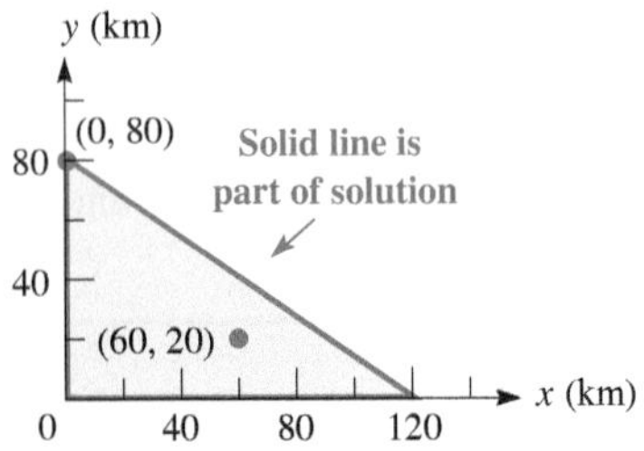

Fig. 17.34

Noting that negative values of x and y do not have meaning, we have the graph in Fig. 17.34, shading in the region below the line since we have $y < 80 - 0.67x$ for that region. Any point in the shaded region, or on the axes or the line around the shaded region, gives a solution. The *solid line* indicates that points on it are part of the solution.

The point (0, 80), for example, is a solution and tells us that 80 km of Route 80 can be plowed if none of Route 15 is plowed. In this case, all 60 h of plowing time are used for Route 80. Another possibility is shown by the point (60, 20), which indicates that 60 km of Route 15 and 20 km of Route 80 can be plowed. In this case, not all of the 60 plowing hours are used.

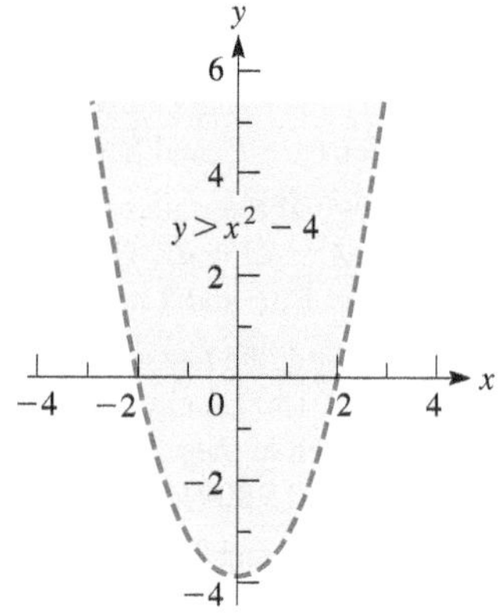

Fig. 17.35

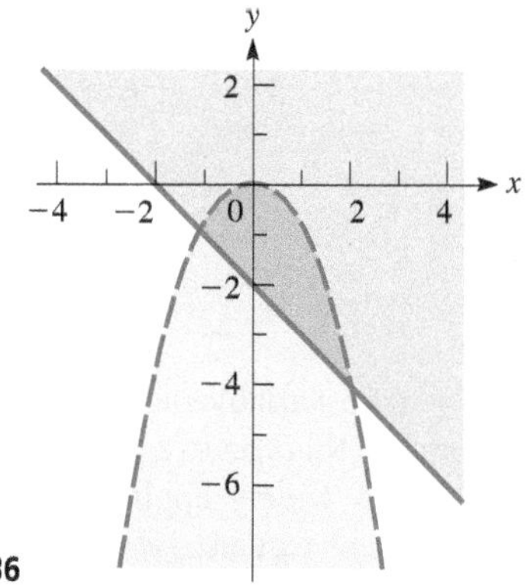

Fig. 17.36

EXAMPLE 4 Sketching a nonlinear inequality with two variables

Draw a sketch of the graph of the inequality $y > x^2 - 4$. Although the graph of $y = x^2 - 4$ is not a straight line, the method of solution is the same. We graph the function $y = x^2 - 4$ as a dashed curve, since it is not part of the solution, as shown in Fig. 17.35. We then shade in the region above the curve to indicate the points that satisfy the inequality.

EXAMPLE 5 Solution of a system of inequalities

Draw a sketch of the region that is defined by the system of inequalities $y \geq -x - 2$ and $y + x^2 < 0$.

Similar to the solution of a system of equations, *the solution of a system of inequalities is any pair of values* (x, y) *that satisfies both inequalities.* This means we want the region common to both inequalities. In Fig. 17.36, we first shade in the region above the line $y = -x - 2$ and then shade in the region below the parabola $y = -x^2$. The region defined by this system is the darkly shaded region below the parabola that is also above and on the line.

Note that the region defined by $y \geq -x - 2$ *or* $y + x^2 < 0$ consists of both shaded regions and all points on the line since that is where either of the inequalities is true.

A graphing utility can be used to display the solution of an inequality or of a system of inequalities involving two variables. Some examples can be found in the exercises.

EXERCISES 17.5

In Exercises 1 and 2, make the given changes in the indicated examples of this section and then draw the graph of the resulting inequality.

1. In Example 2, change $x + 3$ to $3 - x$ and then draw the graph of the resulting inequality.
2. In Example 4, change $x^2 - 4$ to $4 - x^2$ and then draw the graph of the resulting inequality.

In Exercises 3–22, draw a sketch of the graph of the given inequality.

3. $y > x - 1$
4. $y < 3x - 2$
5. $y \geq 2x + 5$
6. $y \leq 15 - 3x$
7. $3x + 2y + 6 > 0$
8. $x + 4y - 8 < 0$
9. $y < x^2$
10. $y \leq 2x^2 - 3$
11. $2x^2 - 4x - y > 0$
12. $y \leq x^3 - 1$
13. $y < 32x - x^4$
14. $y \leq \sqrt{2x + 5}$
15. $y > \dfrac{10}{x^2 + 1}$
16. $y < \ln x$
17. $y > 1 + \sin 2x$
18. $y > |x| - 3$
19. $-4 < y \leq -2$
20. $y > |x + 3|$
21. $|x| < |y|$
22. $|x - 3y| > 2$

In Exercises 23–32, draw a sketch of the graph of the region in which the points satisfy the given system of inequalities.

23. $y > x$, $y > 1 - x$
24. $y \leq 2x$, $y \geq x - 1$
25. $y \leq 2x^2$, $y > x - 2$
26. $y > x^2$, $y < x + 4$
27. $y > \frac{1}{2}x^2$, $y \leq 4x - x^2$
28. $y < 4 - x$, $y < \sqrt{16 - x^2}$
29. $y \geq 0$, $y \leq \sin x$, $0 \leq x \leq 3\pi$
30. $y > 0$, $y > 1 - x$, $y < e^x$
31. $|y + 2| < 5$, $|x - 3| \leq 2$
32. $16x + 3y - 12 > 0$, $y > x^2 - 2x - 3$, $|2x - 3| < 3$

In Exercises 33–42, use a graphing utility to display the solution of the given inequality or system of inequalities.

33. $2x + y < 5$
34. $4x - y > 1$
35. $y \geq 1 - x^2$
36. $y < |4 - 2x|$
37. $y > 2x - 1$, $y < x^4 - 8$
38. $y < 3 - x$, $y > 3x - x^3$
39. $y > x^2 + 2x - 8$, $y < \dfrac{1}{x} - 2$
40. $y > -2x^2$, $y < 1 - e^{-x}$
41. $y \leq |2x - 3|$, $y > 1 - 2x^2$
42. $y \geq |4 - x^2|$, $y < 2 \ln |x|$

In Exercises 43–48, solve the given problems.

43. By an inequality, define the region below the line $4x - 2y + 5 = 0$.
44. By an inequality, define the region that is bounded by or includes the parabola $x^2 - 2y = 0$, and that contains the point $(1, 0.4)$.
45. For $Ax + By > C$, if $B < 0$, would you shade above or below the line?

46. Find a system of inequalities that would describe the region within the triangle with vertices $(0, 0)$, $(0, 4)$, and $(2, 0)$.

47. Draw a graph of the solution of the system $y \geq 2x^2 - 6$ and $y = x - 3$.

48. Draw a graph of the solution of the system $y < |x + 2|$ and $y = x^2$.

In Exercises 49–54, set up the necessary inequalities and sketch the graph of the region in which the points satisfy the indicated system of inequalities.

49. A telephone company is installing two types of fibre-optic cable in an area. It is estimated that no more than 300 m of type A cable, and at least 200 m but no more than 400 m of type B cable, are needed. Graph the possible lengths of cable that are needed.

50. A refinery can produce gasoline and diesel fuel, in amounts of any combination, except that equipment restricts total production to 7500 L/day. Graph the different possible production combinations of the two fuels.

51. The elements of an electric circuit dissipate P watts of power. The power P_R dissipated by a resistor in the circuit is given by $P_R = Ri^2$, where R is the resistance (in Ω) and i is the current (in A). Graph the possible values of P and i for $P > P_R$ and $R = 0.5\ \Omega$.

52. The cross-sectional area A (in m^2) of a certain trapezoid culvert in terms of its depth d (in m) is $A = 2d + d^2$. Graph the possible values of d and A if A is between 1 m^2 and 2 m^2.

53. One pump can remove wastewater at the rate of 250 L/min, and a second pump works at the rate of 150 L/min. Graph the possible values of the time (in min) that each of these pumps operates such that together they pump more than 15 000 L.

54. A rectangular computer chip is being designed such that its perimeter is no more than 30 mm, its width at least 3 mm, and its length at least 8 mm. Graph the possible values of the width w and the length l.

17.6 Linear Programming

An important area in which inequalities with two or more variables are used is in the branch of mathematics known as **linear programming** (in this context, "programming" does not mean computer programming). This subject is widely applied in industry, business, economics, and technology. Many social problems can also be analysed using linear programming.

■ This section is an introduction to linear programming. Methods for applications involving many decision variables can be found in a linear programming or operations research text.

Linear programming is used to analyse problems such as those related to maximizing profits, minimizing costs, or the use of materials with certain constraints of production. We begin by introducing the basic terminology of linear programming through an example.

EXAMPLE 1 Maximum value of F

Find the maximum value of F, where $F = 2x + 3y$ and x and y are subject to the conditions that

$$x \geq 0, y \geq 0$$
$$x + y \leq 6$$
$$x + 2y \leq 8$$

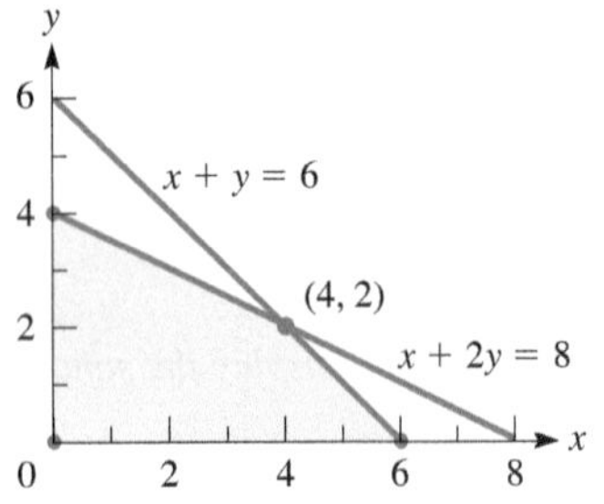

Fig. 17.37

The variables x and y are the **decision variables** of the problem. Their values are restricted by the four given inequalities, known as the **constraints**. The function F to be maximized is known as the **objective function**.

We now graph this set of inequalities, as shown in Fig. 17.37. Each point in the shaded region (including the line segments on the edges) satisfies all the constraints and is known as a **feasible point**.

The maximum value of F must be found at one of the feasible points. Testing for values at the vertices of the region, we have the values in the following table:

Point	(0, 0)	(6, 0)	(4, 2)	(0, 4)
Value of F	0	12	14	12

If we evaluate F at any other feasible point, we will find that $F < 14$. Therefore, the maximum value of F under the given constraints is 14. Since $(4, 2)$ is the feasible point with the largest value of the objective function, we say that $(4, 2)$ is the **optimal solution**.

Practice Exercise

1. In Example 1, what is the maximum value of F if the third constraint is changed to $x + y \leq 5$?

In Example 1, we found the maximum value of a *linear* objective function, subject to *linear* constraints (thus the name *linear programming*). We found the maximum value of F, subject to the given constraints, to be at one of the vertices of the region of feasible points. In fact, in the theory of linear programming the following result is established.

Optimal Solution of a Linear Programming Problem

If a linear programming problem has an optimal solution, then it will occur at a vertex of the region of feasible points.

If the problem has more than one solution, all points along a line segment connecting two vertices will be optimal.

In either case, the value of the objective function is unique.

EXAMPLE 2 Maximum and minimum values of *F*

Find the maximum and minimum values of the objective function $F = 3x + y$, subject to the constraints

$$x \geq 2,\ y \geq 0$$
$$x + y \leq 8$$
$$2y - x \leq 1$$

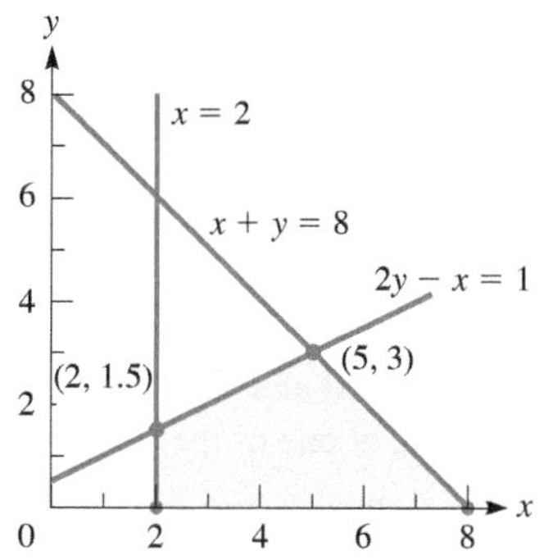

Fig. 17.38

The constraints are graphed as shown in Fig. 17.38. We then locate the vertices and evaluate F at these vertices as follows:

Vertex	(2, 0)	(8, 0)	(5, 3)	(2, 1.5)
Value of F	6	24	18	7.5

Therefore, we see that the maximum value of F is 24, and the minimum value of F is 6. If we check the value of F at any other feasible point, we will find a value between 6 and 24.

The following two examples show the use of linear programming in finding a maximum value and a minimum value in applied situations.

EXAMPLE 3 Linear programming—maximizing profit

■ The loudspeaker was developed in the early 1920s by the U.S. inventor Kellogg Rice.

A company makes two types of stereo speaker systems: their good-quality system and their highest-quality system. The production of these systems requires assembly of the speaker system itself and the production of the cabinets in which they are installed. The good-quality system requires three worker-hours for speaker assembly and two worker-hours for cabinet production for each complete system. The highest-quality system requires four worker-hours for speaker assembly and six worker-hours for cabinet production for each complete system. Available skilled labour allows for a maximum of 480 worker-hours per week for speaker assembly and a maximum of 540 worker-hours per week for cabinet production. It is anticipated that all systems will be sold and that the profit will be \$30 for each good-quality system and \$75 for each highest-quality system. How many of each system should be produced to provide the greatest profit?

First, let x = the number of good-quality systems and y = the number of highest-quality systems made in one week. Thus, the profit P is given by

$$P = 30x + 75y$$

We know that negative numbers are not valid for either x or y, and therefore we have $x \geq 0$ and $y \geq 0$.

The number of available worker-hours per week for each part of the production also restricts the number of systems that can be made. In the speaker-assembly shop, $3x$ worker-hours are needed for each good-quality system and $4y$ worker-hours are needed for each highest-quality system. The constraint that only 480 hours are available in the speaker-assembly shop means that $3x + 4y \leq 480$.

In the cabinet shop, it takes $2x$ worker-hours for a good-quality system and $6y$ worker-hours for a highest-quality system. The constraint that only 540 hours are available in the cabinet shop means that $2x + 6y \leq 540$.

Therefore, we want to maximize the profit $P = 30x + 75y$, which is the objective function, under the constraints

$x \geq 0, y \geq 0$	number of systems produced cannot be negative
$3x + 4y \leq 480$	worker-hours for speaker assembly
$2x + 6y \leq 540$	worker-hours for cabinet production

The constraints are graphed as shown in Fig. 17.39. We locate the vertices and evaluate the profit P at these points as follows:

Vertex	(0, 0)	(160, 0)	(72, 66)	(0, 90)
Profit ($)	0	4800	7110	6750

Therefore, we see that the greatest profit of $7110 is made by producing 72 good-quality systems and 66 highest-quality systems.

A way of showing the number of each system to be made for the greatest profit is to assume values of the profit P and graph these lines. For example, for $P = \$3000$ or $P = \$6000$, we have the lines shown in Fig. 17.39. Both amounts of profit are possible with various combinations of speaker systems being produced. However, we note that the line for $P = \$6000$ passes through feasible points farther from the origin. It is also clear that the lines for the profits are parallel and that *the greatest profit attainable is given by the line passing through A,* where $3x + 4y = 40$ and $2x + 6y = 540$ intersect. This also illustrates why the greatest profit is found at one of the vertices of the region of feasible points.

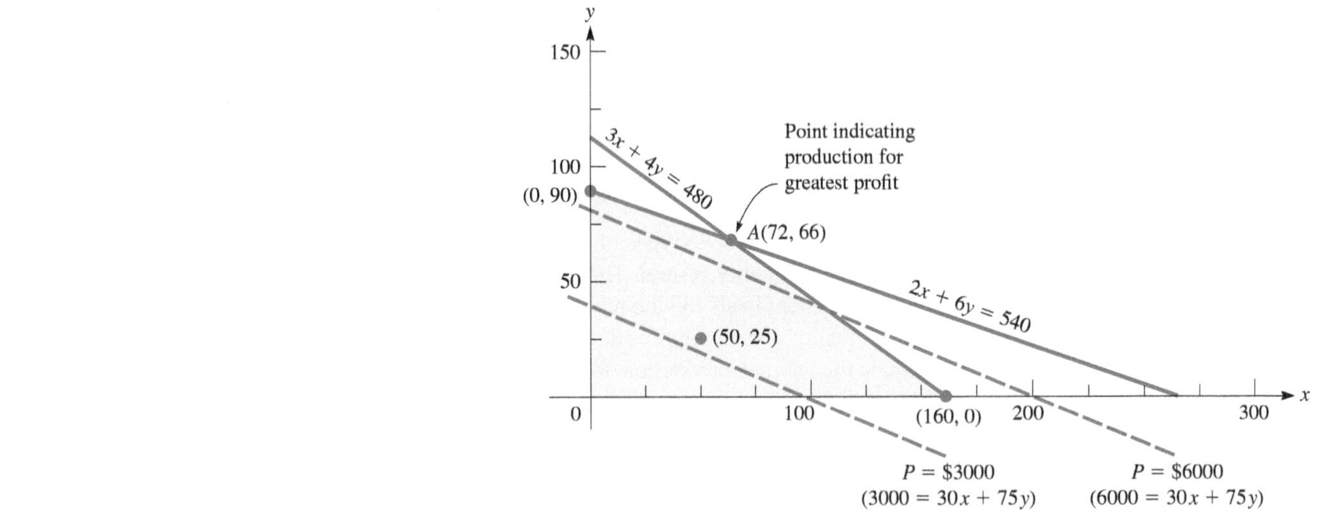

Fig. 17.39

EXAMPLE 4 Linear programming—minimizing cost

■ See the chapter introduction.

An airline plans to open new routes and use two types of planes, A and B, on these routes. It is expected there would be at least 400 first-class passengers and 2000 economy-class passengers on these routes each day. Plane A costs $18 000/day to operate and has seats for 40 first-class and 80 economy-class passengers. Plane B costs $16 000/day to operate and has seats for 20 first-class and 160 economy-class passengers. How many of each type of plane should be used for these routes to minimize operating costs? (Assume that the planes will be used only on these routes.)

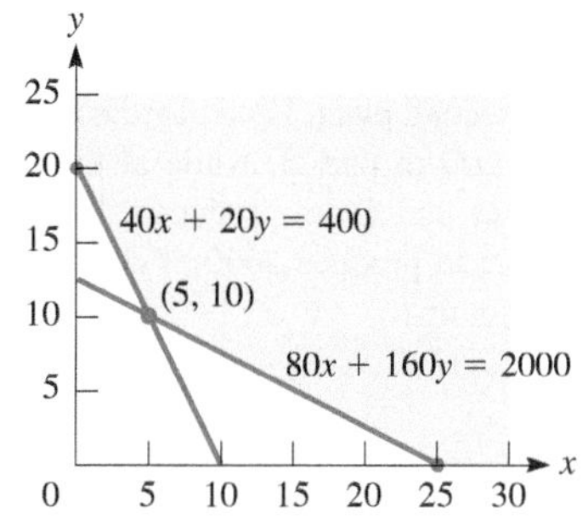

Fig. 17.40

We first let $x =$ the number of A planes and $y =$ the number of B planes to be used. Then the operating cost $C = 18\,000x + 16\,000y$ is the objective function. The constraints are

$x \geq 0, y \geq 0$ — number of planes cannot be negative

$40x + 20y \geq 400$ — at least 400 first-class passengers

$80x + 160y \geq 2000$ — at least 2000 economy-class passengers

The constraints are graphed as shown in Fig. 17.40. We see that the region of feasible points is unlimited (there could be more than 2400 passengers), but we still want to evaluate the operating cost C at the vertices. Evaluating C at these points, we have

Vertex	(25, 0)	(5, 10)	(0, 20)
Cost ($)	450 000	250 000	320 000

Therefore, 5 type-A planes and 10 type-B planes should be used on these routes to keep the operating costs at a minimum of $250 000.

The problems in linear programming that arise in business and industry involve many more variables than in the simplified examples in this text. They are solved by computers using matrices, but the basic idea is the same as that shown in this section.

EXERCISES 17.6

In Exercises 1 and 2, make the given changes in the indicated examples of this section and then find the indicated values.

1. In Example 1, change the last constraint to $2x + y \leq 8$. Then graph the feasible points and find the maximum value of F.

2. In Example 2, change the last constraint to $3y - x \leq 4$. Then graph the feasible points and find the maximum and minimum values of F.

In Exercises 3–16, find the indicated maximum and minimum values by the linear programming method of this section. For Exercises 5–16, the constraints are shown below the objective function.

3. Graphing the constraints of a linear programming problem shows the consecutive vertices of the region of feasible points to be (0, 0), (12, 0), (10, 7), (0, 5), and (0, 0). What are the maximum and minimum values of the objective function $F = 3x + 4y$ in this region?

4. Graphing the constraints of a linear programming problem shows the consecutive vertices of the region of feasible points to be (1, 3), (8, 0), (9, 7), (5, 8), (0, 6), and (1, 3). What are the maximum and minimum values of the objective function $F = 2x + 5y$ in this region?

5. Maximum P:
$P = 3x + 5y$
$x \geq 0, y \geq 0$
$2x + y \leq 6$

6. Maximum P:
$P = 2x + 7y$
$x \geq 1, y \geq 0$
$x + 4y \leq 8$

7. Maximum P:
$P = 5x + 9y$
$x \geq 0, y \geq 0$
$x + 2y \leq 6$

8. Minimum C:
$C = 10x + 20y$
$x \geq 0, y \geq 0$
$3x + 5y \geq 30$

9. Minimum C:
$C = 4x + 6y$
$x \geq 0, y \geq 0$
$x + y \geq 5$
$x + 2y \geq 7$

10. Minimum C:
$C = 6x + 4y$
$x \geq 0, y \geq 0$
$2x + y \geq 6$
$x + y \geq 5$

11. Maximum and minimum F:
$F = x + 3y$
$x \geq 1, y \geq 2$
$y - x \leq 3$
$y + 2x \leq 8$

12. Maximum and minimum F:
$F = 3x - y$
$x \geq 1, y \geq 0$
$x + 4y \leq 8$
$4x + y \leq 8$

13. Maximum P:
$P = 9x + 2y$
$x \geq 0, y \geq 0$
$2x + 5y \leq 10$
$4x + 3y \leq 12$

14. Minimum C:
$C = 3x + 8y$
$x \geq 0, y \geq 0$
$6x + y \geq 6$
$x + 4y \geq 4$

15. Minimum C:
$C = 6x + 4y$
$y \geq 2$
$x + y \leq 12$
$x + 2y \geq 12$
$2x + y \geq 12$

16. Maximum P:
$P = 3x + 4y$
$2x + y \geq 2$
$x + 2y \geq 2$
$x + y \leq 2$

In Exercises 17–22, solve the given linear programming problems.

17. A person wants to invest no more than $9000, part at 6% and part at 5%, with no more than twice as much being invested at 6% as at 5%. How much should be invested at each rate to maximize the income from the investments?

18. An oil refinery refines types A and B of crude oil and can refine as much as 4000 barrels each week. Type A crude has 2 kg of impurities per barrel, type B has 3 kg of impurities per barrel, and the refinery can handle no more than 9000 kg of these impurities each week. How much of each type should be refined in order to maximize profits, if the profit is $4/barrel for type A and $5/barrel for type B?

19. A manufacturer produces a business calculator and a graphing calculator. Each calculator is assembled in two sets of operations, where each operation is in production 8 h during each day. The average time required for a business calculator in the first operation is 3 min, and 6 min is required in the second operation. The graphing calculator averages 6 min in the first operation and 4 min in the second operation. All calculators can be sold; the profit for a business calculator is $8, and the profit for a graphing calculator is $10. How many of each type of calculator should be made each day in order to maximize profit?

20. Using the information given in Example 3, with the one change that the profit on each good-quality speaker system is \$60 (instead of \$30), how many of each system should be made? Explain why this one change in data makes such a change in the solution.

21. Brands A and B of breakfast cereal are both enriched with vitamins P and Q. The necessary information about these cereals is as follows:

	Cereal A	*Cereal B*	*RDA*
Vitamin P	1 unit/serving	2 units/serving	10 units
Vitamin Q	5 units/serving	3 units/serving	30 units
Cost	12¢/serving	18¢/serving	

(RDA is the *recommended dietary allowance.*) Find the amounts of each cereal that together satisfy the RDA of vitamins P and Q at the lowest cost. (There are 30 g of cereal in one serving.)

22. A computer company makes parts A and B in each of two different plants. It costs \$4000 per day to operate the first plant and \$5000 per day to operate the second plant. Each day the first plant produces 100 of part A and 200 of part B, while at the second plant 250 of part A and 100 of part B are produced. How many days should each plant operate to produce 2000 of each part and keep operating costs at a minimum?

Answer to Practice Exercise

1. 13 (at(2, 3))

CHAPTER 17 KEY FORMULAS AND EQUATIONS

If $|f(x)| > n$, then $f(x) < -n$ or $f(x) > n$. **(17.1)**

If $|f(x)| < n$, then $-n < f(x) < n$. **(17.2)**

CHAPTER 17 REVIEW EXERCISES

In Exercises 1–24, solve each of the given inequalities algebraically. Graph each solution.

1. $2x - 12 > 0$

2. $2.4(T - 4.0) \geq 5.5 - 2.4T$

3. $5 - 3x < 0$

4. $6 \leq 4x - 2 < 9$

5. $4 < 2x - 1 < 11$

6. $2x < x + 1 < 4x + 7$

7. $2 \leq \dfrac{4n - 2}{3} < 3$

8. $3x < 2x + 1 < x - 5$

9. $5x^2 + 9x < 2$

10. $x^2 + 2x > 63$

11. $6n^2 - n > 35$

12. $2x^3 + 4 \leq x^2 + 8x$

13. $\dfrac{(2x - 1)(3 - x)}{x + 4} > 0$

14. $x^4 + x^2 \leq 0$

15. $\dfrac{8}{x} < 2$

16. $\dfrac{1}{x - 2} < \dfrac{1}{4}$

17. $\dfrac{8 - R}{2R + 1} \leq 0$

18. $\dfrac{(3 - x)^2}{2x + 7} \leq 0$

19. $|3x + 2| \leq 4$

20. $|4 - 3x| \geq 7$

21. $|3 - 5x| > 7$

22. $\left|2 - \dfrac{y}{2}\right| \leq 0$

23. $|x - 30| > 48$

24. $2|2x - 9| < 8$

In Exercises 25–28, use a graphing utility to solve the given inequalities. Graph the appropriate function and from the graph determine the solution. Approximate roots to the nearest hundredth.

25. $x^3 + x + 1 < 0$

26. $\dfrac{2}{R + 2} > 3$

27. $e^{-t} > 0.5$

28. $\sin 2x < 0.8 \quad (0 < x < 4)$

In Exercises 29–40, draw a sketch of the region in which the points satisfy the given inequality or system of inequalities.

29. $y > 12 - 3x$

30. $y < \dfrac{1}{2}x + 2$

31. $2y - 3x - 4 \leq 0$

32. $3y - x + 6 \geq 0$

33. $y > 6 - 2x^2$

34. $y \leq \dfrac{6}{x^2 - 4}$

35. $y - |x + 1| < 0$

36. $2y + 2x^3 + 6x > 3$

37. $y > x + 1$
$y < 4 - x^2$

38. $y > 2x - x^2$
$y \geq -2$

39. $y \leq \dfrac{4}{x^2 + 1}$
$y < x - 3$

40. $y < \cos \dfrac{1}{2}x$
$y > \dfrac{1}{2}e^x$
$-\pi < x < \pi$

In Exercises 41–48, use a graphing utility to display the region in which the points satisfy the given inequality or system of inequalities.

41. $y < 3x + 5$

42. $y > 2 - \dfrac{1}{4}x$

43. $y > 8 + 7x - x^2$

44. $y < x^3 + 4x^2 - x - 4$

45. $y < 32x - x^4$

46. $y > 2x - 1$
$y < 6 - 3x^2$

47. $y > 1 - x \sin 2x$
$y < 5 - x^2$

48. $y > |x - 1|$
$y < 4 + \ln x$

In Exercises 49–52, determine the values of x for which the given radicals represent real numbers.

49. $\sqrt{3-x}$ **50.** $\sqrt{x+5}$ **51.** $\sqrt{x^2+4x}$ **52.** $\sqrt{\dfrac{x-1}{x+2}}$

In Exercises 53–56, find the indicated maximum and minimum values by the method of linear programming. The constraints are shown below the objective function.

53. Maximum P:
$P = 2x + 9y$
$x \geq 0,\ y \geq 0$
$x + 4y \leq 13$
$3y - x \leq 8$

54. Maximum P:
$P = x + 2y$
$x \geq 1,\ y \geq 0$
$3x + y \leq 6$
$2x + 3y \leq 8$

55. Minimum C:
$C = 3x + 4y$
$x \geq 0,\ y \geq 1$
$2x + 3y \geq 6$
$4x + 2y \geq 5$

56. Minimum C:
$C = 2x + 4y$
$x \geq 0,\ y \geq 0$
$x + 3y \geq 6$
$4x + 7y \geq 18$

In Exercises 57–82, solve the given problems using inequalities. (All data are accurate to at least two significant digits.) In Exercise 83, answer the given question.

57. Under what conditions is $|a + b| < |a| + |b|$?

58. Is $|a - b| < |a| + |b|$ always true? Explain.

59. If $a > 0$, $a \neq 1$, show that $a + a^{-1} > 2$.

60. Solve for x: $a + |bx| < c$, given that $a - c < 0$.

61. Form an inequality of the form $ax^2 + bx + c < 0$ with $a > 0$ for which the solution is $-2 < x < 5$.

62. By means of an inequality, define the region above the line $x - 3y - 6 = 0$.

63. Draw a graph of the system $y < 1 - x^2$ and $y = x^2$.

64. Find a system of inequalities that would describe the region within the quadrilateral with vertices $(0, 0)$, $(4, 4)$, $(0, -3)$, and $(4, -3)$.

65. Find the values for which $f(x) = (x - 2)(x - 3)$ is positive, zero, and negative. Use this information along with $f(0)$ and $f(5)$ to make a rough sketch of the graph of $f(x)$.

66. Follow the same instructions as in Exercise 65 for the function $f(x) = (x - 2)/(x - 3)$.

67. If two adjacent sides of a square design on a sheet of metal are extended 6.0 cm and 10.0 cm, respectively, how long is each side of the square if the perimeter of the rectangle is at least twice as long as the perimeter of the square? See Fig. 17.41.

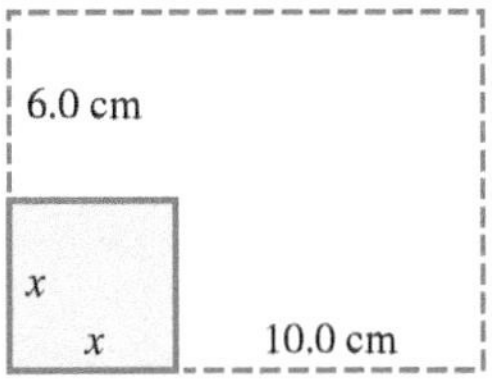

Fig. 17.41

68. The value V (in \$) of each building lot in a development is estimated as $V = 65\,000 + 5000t$, where t is the time in years from now. For how long is the value of each lot no more than \$90 000?

69. The cost C of producing two of one type of calculator and five of a second type is \$50. If the cost of producing each of the second type is between \$5 and \$8, what are the possible costs of producing each of the first type?

70. Ottawa is 450 km from Québec City. One car starts from Ottawa for Québec City one hour before a second car. The first car averages 60 km/h, and the second car averages 80 km/h for the trip. For what times after the first car starts is the second car ahead of the first car?

71. The pressure p (in kPa) at a depth d (in m) in the ocean is given by $p = 101 + 10.1d$. For what values of d is $p > 500$ kPa?

72. After conducting tests, it was determined that the stopping distance x (in m) of a car travelling at 90 km/h was $|x - 95| \leq 10$. Express this inequality without absolute values and find the interval of stopping distances that were found in the tests.

73. A heating unit with 80% efficiency and a second unit with 90% efficiency deliver 360 MJ of heat to an office complex. If the first unit consumes an amount of fuel that contains no more than 261 MJ, what is the MJ content of the fuel consumed by the second unit?

74. A rectangular parking lot is to have a perimeter of 100 m and an area no greater than 600 m^2. What are the possible dimensions of the lot?

75. The electric power P (in W) dissipated in a resistor is given by $P = Ri^2$, where R is the resistance (in Ω) and i is the current (in A). For a given resistor, $R = 12.0$ Ω, and the power varies between 2.50 W and 8.00 W. Find the values of the current.

76. The reciprocal of the total resistance of two electric resistances in parallel equals the sum of the reciprocals of the resistances. If a 2.0-Ω resistance is in parallel with a resistance R, with a total resistance greater than 0.5 Ω, find R.

77. The efficiency E (in %) of a certain gasoline engine is given by $E = 100(1 - r^{-0.4})$, where r is the *compression ratio* for the engine. For what values of r is $E > 50\%$?

78. A rocket is fired such that its height h (in km) is given by $h = 41t - t^2$. For what values of t (in min) is the height greater than 400 km?

79. In developing a new product, a company estimates that it will take no more than 1200 min of computer time for research and no more than 1000 min of computer time for development. Graph the possible combinations of the computer times that are needed.

80. A natural-gas supplier has a maximum of 120 worker-hours per week for delivery and for customer service. Graph the possible combinations of times available for these two services.

81. A company produces two types of cell phones: the regular model and the deluxe model. For each regular model produced, there is a profit of \$8, and for each deluxe model the profit is \$15. The same amount of materials is used to make each model, but the supply is sufficient only for 450 cell phones per day. The deluxe model requires twice the time to produce as the regular model. If only regular models were made, there would be time enough to produce 600 per day. Assuming all cell phones will be sold, how many of each model should be produced if the profit is to be a maximum?

82. A company that manufactures DVD/CD players gets two different parts, A and B, from two different suppliers. Each package of parts from the first supplier costs \$2.00 and contains six of each type of part. Each package of parts from the second supplier costs \$1.50 and contains four of A and eight of B. How many packages should be bought from each supplier to keep the total cost to a minimum, if production requirements are 600 of A and 900 of B?

83. In planning a new city development, an engineer uses a rectangular coordinate system to locate points within the development. A park in the shape of a quadrilateral has corners at $(0, 0)$, $(0, 20)$, $(40, 20)$, and $(20, 40)$ (measurements in metres). Write two or three paragraphs explaining how to describe the park region with inequalities and find these inequalities.

CHAPTER 17 PRACTICE TEST

1. State conditions on x and y in terms of inequalities if the point (x, y) is in the second quadrant.

In Problems 2–7, solve the given inequalities algebraically and graph each solution.

2. $\frac{-x}{2} \geq 3$
3. $3x + 1 < -5$
4. $-1 < 1 - 2x < 5$
5. $\frac{x^2 + x}{x - 2} \leq 0$
6. $|2x + 1| \geq 3$
7. $|2 - 3x| < 8$
8. Sketch the region in which the points satisfy the following system of inequalities:

$$y < x^2$$
$$y \geq x + 1$$

9. Determine the values of x for which $\sqrt{x^2 - x - 6}$ represents a real number.
10. The length of a rectangular lot is 20 m more than its width. If the area is to be at least 4800 m^2, what values may the width be?
11. Type A wire costs \$0.10 per metre, and type B wire costs \$0.20 per metre. Show the possible combinations of lengths of wire that can be purchased for less than \$5.00.
12. The range of the visible spectrum in terms of the wavelength λ of light ranges from about $\lambda = 400$ nm (violet) to about $\lambda = 700$ nm (red). Express these values using an inequality with absolute values.
13. Solve the inequality $x^2 > 12 - x$ graphically.
14. By using linear programming, find the maximum value of the objective function $P = 5x + 3y$, subject to the following constraints: $x \geq 0$, $y \geq 0$, $2x + 3y \leq 12$, $4x + y \leq 8$.

18. Variation

Sergei Domashenko/Shutterstock

▲ Pressure varies inversely as area. In Section 18.2 we see how pressure exerted on snow decreases when wearing snowshoes instead of boots.

LEARNING OUTCOMES

After completion of this chapter, the student should be able to:

- Solve application problems involving ratios and proportions
- Set up relationships between variables in terms of direct variation, inverse variation, or joint variation
- Solve application problems involving the different types of variation

The concepts of *ratio* and *proportion*, which were first introduced in Chapter 1, have many uses in mathematics, science, and technology. We have used them extensively up to this point, whether implicitly or explicitly. For instance, we defined the number π as the ratio of the circumference of a circle to its diameter; we defined the trigonometric functions as ratios of the lengths of sides of a triangle; and we established the law of sines as a statement of proportionality between the sides of a triangle and the sines of the angles opposite them. We have also used ratios implicitly when working with measurements (a measurement is the ratio of a measured magnitude to an accepted unit of measurement), or when working with average velocity (the ratio of distance travelled over time elapsed).

In this chapter we will present a more detailed discussion of ratios and proportions. These concepts will then lead us to the concept of *variation*, the principal topic of this chapter. Using the language of variation, we can describe how one variable changes as other related variables change. Newton's *universal law of gravitation* is an example of a scientific law that can be stated in terms of variation.

Variation allows us to establish different types of relationships between variables. Many of these relationships are based on experimentation and observation, just as in the case of gravitation. Applications of variation are found in all areas of technology. It is used in acoustics, biology, chemistry, computer technology, economics, electronics, environmental technology, hydrodynamics, mechanics, navigation, optics, physics, space technology, and thermodynamics, among others.

18.1 Ratio and Proportion

In order to develop the meaning of *variation,* we now review and expand our discussion of *ratio* and *proportion.* First, from Chapter 1, recall that *the quotient* a/b *is the* **ratio** *of* a *to* b. Therefore, a fraction is a ratio.

Ratios can be of two types. On the one hand, ratios can compare quantities of the same kind. In this case, both the numerator and the denominator have identical units and the result is a dimensionless number. Examples include the definition of π, the trigonometric ratios, or comparisons between two parts of a particular whole.

On the other hand, ratios can express the division of magnitudes of different quantities. In this case the ratio is also called a **rate**. Examples include the definitions of density (weight/volume), power (work/time), and pressure (force/area).

EXAMPLE 1 Units of measurement when calculating ratios

The approximate distance by air from Toronto to Los Angeles is 3500 km, and the approximate distance by air from Toronto to Miami is 2000 km. The ratio of these distances is

$$\frac{3500 \text{ km}}{2000 \text{ km}} = \frac{7}{4}$$

■ The first jet-propelled airplane was flown in Germany in 1928.

Since both units are kilometres, the resulting ratio is a dimensionless number.

If a jet travels from Toronto to Los Angeles in 5 h, its average speed is

$$\frac{3500 \text{ km}}{5 \text{ h}} = 700 \text{ km/h}$$

In this case, we must attach the proper units to the resulting ratio.

EXAMPLE 2 Ratio of measurements of the same kind

The length of a certain room is 8 m, and the width of the room is 6 m. Therefore, the ratio of the length to the width is $\frac{8}{6}$, or $\frac{4}{3}$.

If the width of the room is expressed as 6000 mm, we have the ratio 8 m/6000 mm = 1 m/750 mm. However, this does not clearly show the ratio. It is better and more meaningful first to change the units of one of the measurements to the units of the other measurement. Changing the length from 8 m to 8000 mm, we express the ratio as $\frac{4}{3}$, as we saw above. From this ratio, we can easily see that the length is $\frac{4}{3}$ as great as the width.

From Chapter 1, also recall that *an equation stating that two ratios are equal is called a* **proportion**. By this definition, a proportion is

$$\frac{a}{b} = \frac{c}{d} \tag{18.1}$$

Consider the following example.

EXAMPLE 3 Proportion—scale model

On a certain map, 1 cm represents 10 km. Find the distance represented by 3.5 cm on the map.

The ratio of distances on the map is 1 cm/10 km. We can therefore set up the proportion

$$\frac{3.5\text{ cm}}{x} = \frac{1\text{ cm}}{10\text{ km}}$$

map distances (numerators); land distances (denominators)

$$(10x)\left(\frac{3.5}{x}\right) = 10x\left(\frac{1}{10}\right) \qquad \text{multiply each side by LCD} = 10x$$

$$35 = x \quad \text{or} \quad x = 35\text{ km}$$

The ratio 1 cm/10 km is the *scale* of the map and has a special meaning, relating map distances in centimetres to land distances in kilometres. In a case like this, we should not change either unit to the other, even though they are both units of length.

In Section 1.3, we discussed a method for converting measurements from one set of units to another. As we see in the next example, unit conversion can also be done by setting up a proportion.

EXAMPLE 4 Changing units of measurement by using a proportion

Given that 1 in. = 2.54 cm ("in." is the symbol for the unit of length the *inch*), what is the length in centimetres of the diagonal of a laptop computer screen that is 17.0 in. long? See Fig. 18.1.

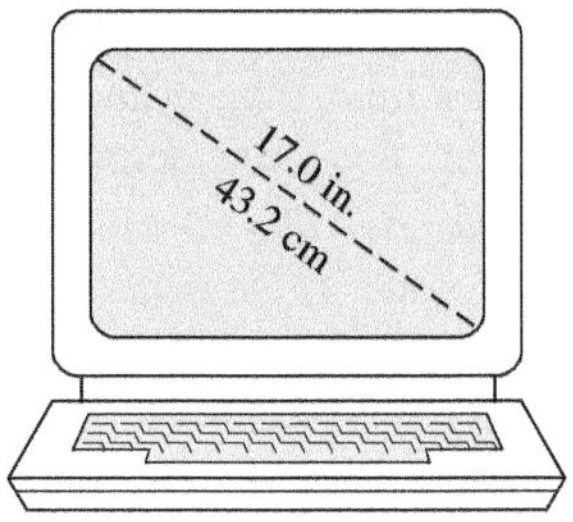

Fig. 18.1

If we equate the ratio of known lengths to the ratio of the given length to the required length, we can find the required length by solving the resulting proportion (which is an equation). This gives us

$$\frac{1\text{ in.}}{2.54\text{ cm}} = \frac{17.0\text{ in.}}{x\text{ cm}}$$

$$x = (17.0)(2.54)$$

$$= 43.2\text{ cm}$$

Therefore, the diagonal of the laptop computer screen is 17.0 in., or 43.2 cm.

EXAMPLE 5 Proportion—magnitude of an electric field

The magnitude of an electric field E is the ratio of the force F on a charge q to the magnitude of q. We can write this as $E = F/q$. If we know the force exerted on a particular charge at some point in the field, we can determine the force that would be exerted on another charge placed at the same point. For example, if we know that a force of 10 nN is exerted on a charge of 4.0 nC, we can then determine the force that would be exerted on a charge of 6.0 nC by the proportion

$$\frac{10 \times 10^{-9}}{4.0 \times 10^{-9}} = \frac{F}{6.0 \times 10^{-9}}$$

forces at point (numerators); charges at point (denominators)

$$F = \frac{(6.0 \times 10^{-9})(10 \times 10^{-9})}{4.0 \times 10^{-9}}$$

$$= 15 \times 10^{-9}$$

$$= 15\text{ nN}$$

Practice Exercise

1. In a certain electric field a force of 21 nN is exerted on a charge of 6.0 nC. At the same point, what is the force on a charge of 16 nC?

EXAMPLE 6 Proportion—alloy composition

An alloy is 5 parts tin and 3 parts lead. How many grams of each are in 40 g of the alloy?

First, let $x =$ the number of grams of tin in 40 g of the alloy. Next, we note that there are 8 total parts of alloy, of which 5 are tin. Thus, 5 is to 8 as x is to 40. Therefore,

parts tin → 5, total parts → 8; x ← grams of tin, 40 ← total grams

$$\frac{5}{8} = \frac{x}{40}$$

$$x = 40\left(\frac{5}{8}\right) = 25 \text{ g}$$

There are 25 g of tin and 15 g of lead. The ratio 25/15 is the same as 5/3.

Practice Exercise

2. An alloy is 7 parts zinc and 5 parts lead. How much lead is there in 48 g of the alloy?

EXERCISES 18.1

In Exercises 1 and 2, make the given changes in the indicated examples of this section and then solve the indicated problem.

1. In Example 3, change 10 km to 16 km.
2. In Example 6, change 5 parts tin to 7 parts tin.

In Exercises 3–10, express the ratios in the simplest form.

3. 18 V to 3 V
4. 63 m^2 to 18 m^2
5. 96 h to 3 days
6. 120 s to 4 min
7. 6500 cL to 2.6 L
8. 25 cm^2 to 75 mm^2
9. 0.14 kg to 3500 mg
10. 2000 μm to 6 mm

In Exercises 11–24, find the required ratios.

11. The *efficiency* of a power amplifier is defined as the ratio of the power output to the power input. Find the efficiency of an amplifier for which the power output is 2.6 W and the power input is 9.6 W.
12. A virus 3.0×10^{-5} cm long appears to be 1.2 cm long through a microscope. What is the *magnification* (ratio of image length to object length) of the microscope?
13. The *coefficient of friction* for two contacting surfaces is the ratio of the frictional force between them to the perpendicular force that presses them together. If it takes 450 N to overcome friction to move a 1.10-kN crate along the floor, what is the coefficient of friction between the crate and the floor? See Fig. 18.2.

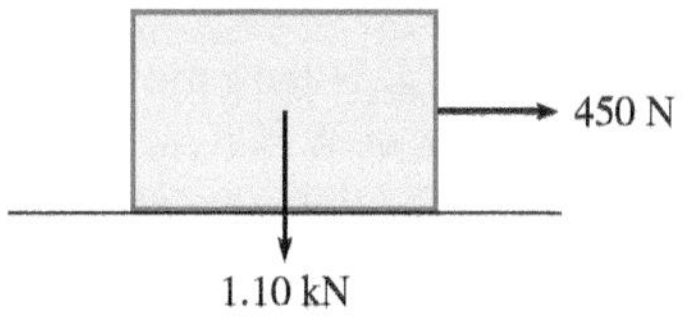

Fig. 18.2

14. The *Mach number* of a moving object is the ratio of its velocity to the velocity of sound (1200 km/h). Find the Mach number of a jet travelling at 1500 km/h.
15. The *capacitance* C of a capacitor is defined as the ratio of its charge q (in C) to the voltage V in volts. Find C (in F) for which $q = 5.00\ \mu$C and $V = 200$ V. (1 F = 1 C/1 V.)
16. The *atomic mass* of an atom of carbon is defined to be 12 u. The ratio of the atomic mass of an atom of nitrogen to that of an atom of carbon is $\frac{7}{6}$. What is the atomic mass of an atom of nitrogen? (The symbol u represents the *unified atomic mass unit*, where $1 \text{ u} = 1.66 \times 10^{-27}$ kg.)
17. An important design feature of an aircraft wing is its *aspect ratio*. It is defined as the ratio of the square of the span of the wing (wingtip to wingtip) to the total area of the wing. If the span of the wing for a certain aircraft is 10.0 m and the area is 18.0 m^2, find the aspect ratio.
18. For an automobile engine, the ratio of the cylinder volume to compressed volume is the *compression ratio*. If the cylinder volume of 820 cm^3 is compressed to 110 cm^3, find the compression ratio.
19. The specific gravity (or relative density) of a substance is the ratio of its density to the density of water. If the density of water is 1.00 g/cm^3, find the specific gravity of a western Canada crude oil that has density of 931 kg/m^3.
20. The *percent grade* of a road is the ratio of vertical rise to the horizontal change in distance (expressed in percent). If a highway rises 75 m for each 1.2 km along the horizontal, what is the percent grade?
21. The *percent error* in a measurement is the ratio of the error in the measurement to the theoretical value that is accepted as correct, expressed as a percent. What is the percent error in a measured value of 8.11 g/cm^3 for the density of iron if the theoretical value is 7.87 g/cm^3?
22. The electric *current* in a given circuit is the ratio of the voltage to the resistance. What is the current ($1\text{ V}/1\ \Omega = 1$ A) for a circuit where the voltage is 24.0 mV and the resistance is 10.0 Ω?
23. The *mass* of an object is the ratio of its weight to the acceleration g due to gravity. If a space probe weighs 8.46 kN on Earth, where $g = 9.80$ m/s^2, find its mass.
24. *Power* is defined as the ratio of work done to the time required to do the work. If an engine performs 2.92 kJ of work in 12.0 s, find the power developed by the engine.

In Exercises 25–28, find the required quantities from the given proportions.

25. According to Boyle's law, the relation $p_1/p_2 = V_1/V_2$ holds for pressures p_1 and p_2 and volumes V_1 and V_2 of a gas at constant temperature. Find V_1 if $p_1 = 36.6$ kPa, $p_2 = 84.4$ kPa, and $V_2 = 0.0447\ m^3$.

26. For two connected gears, the relation

$$\frac{d_1}{d_2} = \frac{N_1}{N_2}$$

holds, where d is the diameter of the gear and N is the number of teeth. Find N_1 if $d_1 = 2.60$ cm, $d_2 = 11.7$ cm, and $N_2 = 45$. The ratio N_2/N_1 is called the *gear ratio*. See Fig. 18.3.

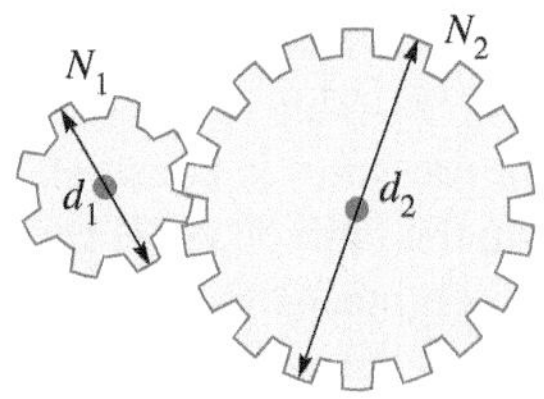

Fig. 18.3

27. In an electric instrument called a "Wheatstone bridge," electric resistances are related by

$$\frac{R_1}{R_2} = \frac{R_3}{R_4}$$

Find R_2 if $R_1 = 6.00\ \Omega$, $R_3 = 62.5\ \Omega$, and $R_4 = 15.0\ \Omega$. See Fig. 18.4.

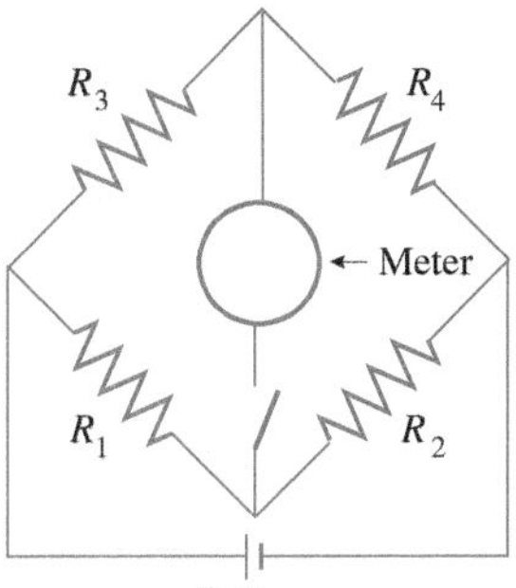

Fig. 18.4

28. In a transformer, an electric current in one coil of wire induces a current in a second coil. For a transformer,

$$\frac{i_1}{i_2} = \frac{t_2}{t_1}$$

where i is the current and t is the number of windings in each coil. In a neon sign amplifier, $i_1 = 1.2$ A and the *turns ratio* $t_2/t_1 = 160$. Find i_2. See Fig. 18.5.

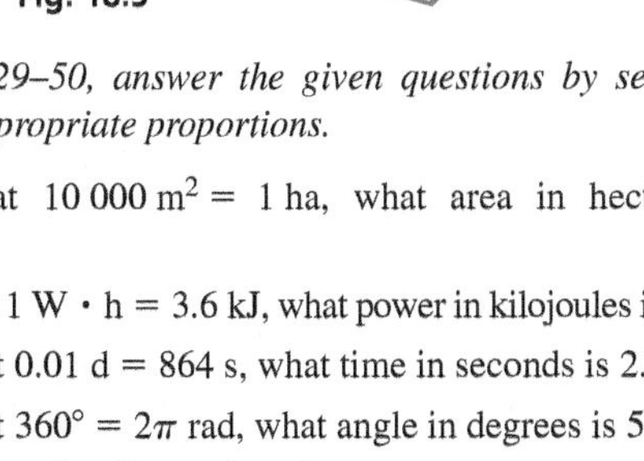

Fig. 18.5

In Exercises 29–50, answer the given questions by setting up and solving the appropriate proportions.

29. Given that $10\,000\ \text{m}^2 = 1$ ha, what area in hectares (ha) is $4500\ \text{m}^2$?

30. Given that 1 W · h = 3.6 kJ, what power in kilojoules is 250 W · h?

31. Given that 0.01 d = 864 s, what time in seconds is 2.75 d?

32. Given that $360° = 2\pi$ rad, what angle in degrees is 5.00 rad?

33. Given that $10^4\ \text{cm}^2 = 10^6\ \text{mm}^2$, what area in square centimetres is $2.50 \times 10^5\ \text{mm}^2$?

34. How many metres per second are equivalent to 45.0 km/h?

35. How many kilolitres per hour are equivalent to 540 mL/min?

36. The water/cement ratio of a concrete mix is the ratio of the mass of water in the mix to the mass of cement in the mix. If the ratio of a mix is specified as 0.40, how much water should be added to a mixture with 25 kg of cement?

37. A particular type of automobile engine produces $62\,500\ \text{cm}^3$ of carbon monoxide in 2.00 min. How much carbon monoxide is produced in 45.0 s?

38. An airplane consumes 140 L of gasoline in flying 680 km. Under similar conditions, how far can it fly on 240 L?

39. Two separate sections of a roof have the same slope. If the rise and run on one section are, respectively, 3.0 m and 6.3 m, what is the run on the other section if its rise is 4.2 m?

40. When a bullet is fired from a loosely held rifle, the ratio of the mass of the bullet to that of the rifle equals the negative of the reciprocal of the ratio of the velocity of the bullet to that of the rifle. If a 3.0 kg rifle fires a 5.0 g bullet and the velocity of the bullet is 300 m/s, what is the recoil velocity of the rifle?

41. A board 7.5 m long is cut into two pieces, the lengths of which are in the ratio 2/3. Find the lengths of the pieces.

42. If c/d is in *inverse ratio* to a/b, then $a/b = d/c$ (see Exercises 27 and 28). The current i (in A) in an electric circuit is in inverse ratio to the resistance R (in Ω). If $i = 0.25$ mA when $R = 2.8\ \Omega$, what is i when $R = 7.2\ \Omega$?

43. By mass, the ratio of chlorine to sodium in table salt is 35.46 to 23.00. How much sodium is contained in 50.00 kg of salt?

44. An industrial cleaner is diluted 3 parts of cleaner to 5 parts of water. What volume (in mL) of cleaner should be used to get 480 mL of diluted solution?

45. In testing for quality control, it was found that 17 of every 500 computer chips produced by a company in a day were defective. If a total of 595 defective parts were found, what was the total number of chips produced during that day?

46. An electric current of 0.772 mA passes into two wires in which it is divided into currents in the ratio of 2.83 to 1.09. What are the currents in the two wires?

47. The ratio of the width to the height of an HDTV screen is 16/9. What is the length of the diagonal of an HDTV screen if it is 98.0 cm wide?

48. Of the earth's water area, the Pacific Ocean covers 46.0%, and the Atlantic Ocean covers 23.9%. Together they cover a total of $2.53 \times 10^8\ \text{km}^2$. What is the area of each?

49. After a race, a runner's GPS watch shows the length of the race course to be 5058 m. If the actual length of the course is 4998 m, find the percent error in the watch's reading. See Exercise 21.

50. The weight of a person on Earth and the weight of the same person on Mars are proportional. If an astronaut weighs 920 N on Earth and 350 N on Mars, what is the weight of another astronaut on Mars if the astronaut weighs 640 N on Earth?

Answers to Practice Exercises

1. 56 nN **2.** 20 g

18.2 Variation

Scientific laws are often stated in terms of ratios and proportions. For example, Charles' law can be stated as "for a perfect gas under constant pressure, the ratio of any two volumes this gas may occupy equals the ratio of the absolute temperatures." Symbolically, this could be stated as $V_1/V_2 = T_1/T_2$. Thus, if the ratio of the volumes and one of the values of the temperature are known, we can easily find the other temperature.

■ Charles' law is named for the French physicist Jacques Charles (1746–1823).

By multiplying both sides of the proportion of Charles' law by V_2/T_1, we can change the form of the proportion to $V_1/T_1 = V_2/T_2$. This statement says that the ratio of the volume to the temperature (for constant pressure) is constant. Thus, if any pair of values of volume and temperature is known, the ratio of V_1/T_1 can be calculated. This ratio of V_1/T_1 can be called a constant k, which means that Charles' law can be written as $V/T = k$. We now have the statement that the ratio of the volume to temperature is always constant; or, as it is normally stated, "The volume is proportional to the temperature." Therefore, we write $V = kT$, the clearest and most informative statement of Charles' law.

Thus, *for any two quantities always in the same proportion, we say that one is* **proportional to** (*or* **varies directly as**) *the second.* This type of relationship is called **direct variation.**

Direct Variation

We say that y is **proportional to x** (or y **varies directly as x**) if

$$y = kx \quad \textbf{(18.2)}$$

where k is called the **constant of proportionality**.

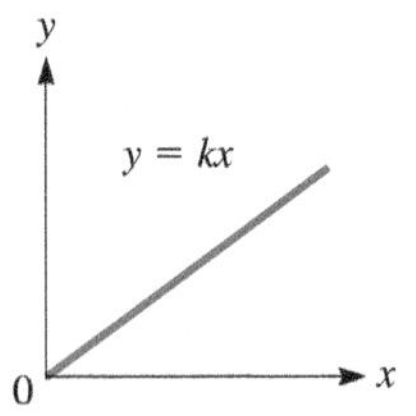

Fig. 18.6

In Fig. 18.6, the graph of the equation for direct variation $y = kx$, where $x \geq 0$, is shown. It is a straight line of slope k $(k > 0)$ and y-intercept of 0. We see that y increases as x increases, or y decreases as x decreases.

EXAMPLE 1 Direct variation—applications

(a) The circumference c of a circle is proportional to (varies directly as) the radius r. We write this as $c \propto r$, or $c = kr$. Since we know that $c = 2\pi r$ for a circle, we know that in this case the constant of proportionality is $k = 2\pi$.

(b) The fact that the resistance R of a wire varies directly as (is proportional to) its length s is written as $R \propto s$, or $R = ks$. As the length of the wire increases (or decreases), this equation tells us that the resistance increases (or decreases) proportionally. In this case, the constant of proportionality is different for different wires, although it remains constant for any given wire.

It is very common that, when two quantities are related, the product of the two quantities remains constant. This type of relationship is called **inverse variation.**

Inverse Variation

We say that **y is inversely proportional to x** (or **y varies inversely as x**) if

$$y = \frac{k}{x} \quad \textbf{(18.3)}$$

where k is the constant of proportionality.

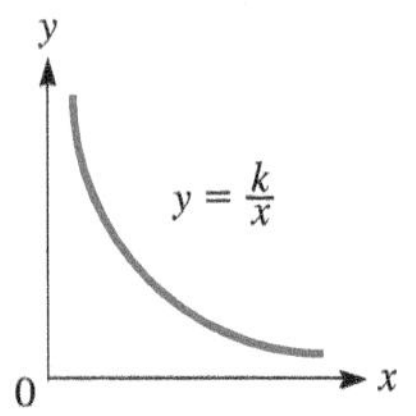

Fig. 18.7

In Fig. 18.7, the graph of the equation for inverse variation $y = k/x$ $(k > 0, x > 0)$ is shown. (It is a *hyperbola*.) As x increases, y decreases, or as x decreases, y increases.

EXAMPLE 2 Inverse variation—Boyle's law

■ Boyle's law is named for the English physicist Robert Boyle (1627–1691).

Boyle's law states that "at a given temperature, the pressure p of an ideal gas varies inversely as the volume V." We write this as $p = k/V$. In this case, as the volume of the gas increases, the pressure decreases, or as the volume decreases, the pressure increases.

For many relationships, one quantity varies as a specific power of another quantity. The terms *varies directly* and *varies inversely as* are used in the following example with a specific power of the independent variable.

EXAMPLE 3 Direct variation with powers—applications

(a) The statement that the volume V of a sphere varies directly as the cube of its radius is written as $V = kr^3$. In this case, we know that $k = 4\pi/3$. We see that as the radius increases, the volume increases much more rapidly. For example, if $r = 2.00$ cm, $V = 33.5\text{ cm}^3$, and if $r = 3.00$ cm, $V = 113\text{ cm}^3$.

(b) A company finds that the number n of units of a product that are sold is inversely proportional to the square of the price p of the product. This is written as $n = k/p^2$. As the price is raised, the number of units that are sold decreases much more rapidly.

One quantity may vary as the product of two or more other quantities. Such variation is called **joint variation**.

Joint Variation

We say that **y varies jointly as x and z** if

$$y = kxz \tag{18.4}$$

where k is the constant of proportionality.

EXAMPLE 4 Joint variation—production cost

The cost C of a piece of sheet metal varies jointly as the area A of the piece and the cost c per unit area. This is written as $C = kAc$. Here, C increases as the product Ac increases, and decreases as the product Ac decreases.

Direct, inverse, and joint variations may be combined. A given relationship may be a combination of two or all three of these types of variation.

EXAMPLE 5 Combined variation—universal law of gravitation

■ Newton's universal law of gravitation was formulated by the great English mathematician and physicist Isaac Newton (1642–1727).

Newton's *universal law of gravitation* can be stated as follows: "The force F of gravitation between two objects varies jointly as the masses m_1 and m_2 of the objects, and inversely as the square of the distance r between their centres." We write this as

$$F = \frac{Gm_1m_2}{r^2}$$

(force varies jointly as masses and inversely as the square of the distance)

where G is the constant of proportionality.

Practice Exercise

1. Express the relationship that y varies directly as the square of x and inversely as z.

COMMON ERROR Note the use of the word *and* in the third line of Example 5. It is used to indicate that F varies in more than one way, but *it is* **not** *interpreted as addition.*

Procedure for Solving Problems Involving Variation	EXAMPLE 6
	Find y for $x = 7.0$ if y varies directly as the square root of x, and $y = 17$ when $x = 12$.
1. Use the given statement to set up a general equation in terms of the variables and the constant of proportionality.	The general equation showing direct variation as $\sqrt{x}$ is $$y = k\sqrt{x}$$
2. Substitute one complete set of known values into the general equation and solve to find the constant of proportionality.	We substitute $x = 12$ and $y = 17$, then solve for k: $$17 = k\sqrt{12}, \quad k = 4.9$$
3. Substitute the constant of proportionality into the general equation to find the specific equation relating the variables.	The specific equating relating y and x is $$y = 4.9\sqrt{x}$$
4. Use the specific equation to find the value of any of the variables given any set of the others.	We can now evaluate y for $x = 7.0$: $$y = 4.9\sqrt{7.0} = 13$$

Practice Exercise

2. If y varies directly as x, and $y = 5$ when $x = 20$, find the value of y when $x = 12$.

LEARNING TIP

In applied situations such as Example 7, the constant of proportionality usually has a set of units associated with it. As long as all values used for the same variable are expressed in the same units, the units for the final variable evaluated will be the appropriate ones.

■ See the chapter introduction.

EXAMPLE 7 Inverse variation of pressure with area

The pressure p exerted by a person standing on snow varies inversely as the total area of the soles of their footwear. A woman exerts a pressure of 1.15 N/cm^2 when wearing boots that have a sole area of 243 cm^2 each. Find the pressure on the snow when she wears snowshoes, each with an area of 1290 cm^2.

$$p = \frac{k}{A} \quad \text{set up general equation}$$

$$1.15 \text{ N/cm}^2 = \frac{k}{2(243 \text{ cm}^2)} \quad \text{substitute given values (for two boots) and evaluate } k$$

$$k = 559 \text{ N} \quad \text{(the constant of proportionality is the weight of the woman)}$$

$$p = \frac{559 \text{ N}}{A} \quad \text{substitute value of } k \text{ to get specific equation}$$

$$p = \frac{559 \text{ N}}{2(1290 \text{ cm}^2)} \quad \text{evaluate } p \text{ for } A = 2 \times 1290 \text{ cm}^2$$

$$p = 0.217 \text{ N/cm}^2$$

EXAMPLE 8 Variation as a square—heat in a resistor

The heat H developed in an electric resistor varies jointly as the time t and the square of the current i in the resistor. If H_0 joules of heat are developed in t_0 seconds with i_0 amperes passing through the resistor, how much heat is developed if both the time and the current are doubled?

$$H = kti^2 \quad \text{set up general equation}$$

$$H_0 \text{ J} = k(t_0 \text{ s})(i_0 \text{ A})^2 \quad \text{substitute given values and evaluate } k$$

$$k = \frac{H_0}{t_0 i_0^2} \text{J/(s} \cdot \text{A}^2)$$

$$H = \frac{H_0 t i^2}{t_0 i_0^2} \quad \text{substitute for } k \text{ to get specific equation}$$

We are asked to determine H when both the time and the current are doubled. This means we are to substitute $t = 2t_0$ and $i = 2i_0$. Making this substitution,

$$H = \frac{H_0(2t_0)(2i_0)^2}{t_0 i_0^2} = \frac{8H_0 t_0 i_0^2}{t_0 i_0^2} = 8H_0$$

Since $H = H_0$ when $t = t_0$ and $i = i_0$, when time and current are doubled, the heat developed is eight times that for the original values of i and t.

EXAMPLE 9 Application of Newton's universal law

■ The first landing on the moon was by the crew of the U. S. spacecraft *Apollo 11* in July 1969.

A spacecraft is travelling from the earth to the moon, which are 390 000 km apart. The mass of the moon is 0.0123 that of the earth. How far from the earth is the gravitational force of the earth on the spacecraft equal to the gravitational force of the moon on the spacecraft?

From Example 5, we have the gravitational force between two objects as

$$F = \frac{Gm_1m_2}{r^2}$$

where the constant of proportionality G is the same for any two objects. Since we want the force between the earth and the spacecraft to equal the force between the moon and the spacecraft, we have

$$\frac{Gm_sm_e}{r^2} = \frac{Gm_sm_m}{(390\,000 - r)^2}$$

where m_s, m_e, and m_m are the masses of the spacecraft, the earth, and the moon, respectively; r is the distance from the earth to the spacecraft; and $390\,000 - r$ is the distance from the moon to the spacecraft. Since $m_m = 0.0123m_e$, we have

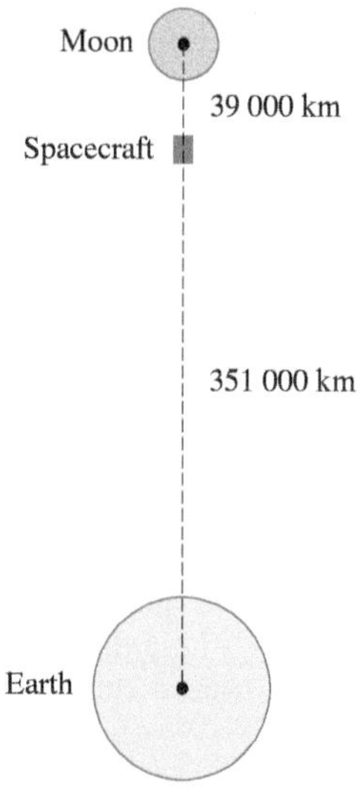

Fig. 18.8

$$\frac{Gm_sm_e}{r^2} = \frac{Gm_s(0.0123m_e)}{(390\,000 - r)^2}$$

$$\frac{1}{r^2} = \frac{0.0123}{(390\,000 - r)^2} \quad \text{divide each side by } Gm_sm_e$$

$$(390\,000 - r)^2 = 0.0123r^2 \quad \text{multiply each side by LCD}$$

$$390\,000^2 - 780\,000r + r^2 = 0.0123r^2$$

$$390\,000^2 - 780\,000r + 0.9877r^2 = 0$$

$$r = \frac{780\,000 \pm \sqrt{780\,000^2 - 4(0.9877)(390\,000^2)}}{2(0.9877)}$$

$$r_1 = 439\,000 \text{ km}$$

$$r_2 = 351\,000 \text{ km}$$

The only valid solution is $r^2 = 351\,000$, so that the spacecraft is 351 000 km from the earth and 39 000 km from the moon when the gravitational forces are equal. See Fig. 18.8.

EXERCISES 18.2

In Exercises 1–4, make the given changes in the indicated examples of this section and solve the indicated problems.

1. In Example 1(a), change radius r to diameter d, write the appropriate equation, and find the value of the constant of proportionality.
2. In Example 6, change "directly as the square root of x" to "inversely as the square of x" and then solve the resulting problem.
3. In Example 7, change 1.15 N to 1.35 N and then solve the resulting problem.
4. In Example 8, change "both the time and current are doubled" to "the time is halved and the current is doubled" and then solve the resulting problem.

In Exercises 5–12, set up the general equations from the given statements.

5. The speed v at which a galaxy is moving away from Earth varies directly as its distance r from Earth.

6. The demand D for a product varies inversely as its price P.

7. The electric resistance R of a wire varies inversely as the square of its diameter d.

8. The volume V of silt carried by a river is proportional to the sixth power of the velocity v of the river.

9. In a tornado, the pressure p that a roof will withstand is inversely proportional to the square root of the area A of the roof.

10. During an adiabatic (no heat loss or gain) expansion of a gas, the pressure p is inversely proportional to the $3/2$ power of the volume V.

11. The stiffness S of a beam varies jointly as its width w and the cube of its depth d.

12. The average electric power P entering a load varies jointly as the resistance R of the load and the square of the effective voltage V, and inversely as the square of the impedance Z.

In Exercises 13–16, express the meaning of the given equation in a verbal statement, using the language of variation. (k and π are constants.)

13. $A = \pi r^2$ **14.** $s = \dfrac{k}{t^{1.2}}$

15. $f = \dfrac{kL}{\sqrt{m}}$ **16.** $V = \pi r^2 h$

In Exercises 17–20, give the specific equation relating the variables after evaluating the constant of proportionality for the given set of values.

17. V varies directly as the square of H, and $V = 2$ when $H = 64$.

18. n is inversely proportional to the cube root of p, and $n = 4$ when $p = 27$.

19. p is proportional to q and inversely proportional to the cube of r, and $p = 8$ when $q = 4$ and $r = 2$.

20. v is proportional to t and the square root of s, and $v = 80$ when $s = 4$ and $t = 5$.

In Exercises 21–28, find the required value by setting up the general equation and then evaluating.

21. Find y when $x = 10$ if y varies directly as x, and $y = 200$ when $x = 16$.

22. Find y when $x = 6$ if y varies directly as the square of x, and $y = 12$ when $x = 3$.

23. Find s when $t = 720$ if s is inversely proportional to t, and $s = 800$ when $t = 0.030$.

24. Find p for $q = 0.8$ if p is inversely proportional to the square of q, and $p = 18$ when $q = 0.2$.

25. Find y for $x = 6$ and $z = 5$ if y varies directly as x and inversely as z, and $y = 60$ when $x = 4$ and $z = 10$.

26. Find r when $n = 16$ if r varies directly as the square root of n and $r = 4$ when $n = 25$.

27. Find f when $p = 2$ and $c = 4$ if f varies jointly as p and the cube of c, and $f = 8$ when $p = 4$ and $c = 0.1$.

28. Find v when $r = 2$, $s = 3$, and $t = 4$ if v varies jointly as r and s and inversely as the square of t, and $v = 8$ when $r = 2$, $s = 6$, and $t = 6$.

In Exercises 29 and 30, A varies directly as x, and B varies directly as x, although not in the same proportion as A. All numbers are positive.

29. Show that $A + B$ varies directly as x.

30. Show that $\sqrt{AB}$ varies directly as x.

In Exercises 31–62, solve the given applied problems involving variation.

31. The time t it takes to download a program onto a computer hard drive is proportional to the memory M required for the program. If it takes 45.0 min to download a program with 27.5 MB (megabytes) of memory required, what is the rate of downloading in kB (kilobytes) per second?

32. The volume V of carbon dioxide (CO_2) that is exhausted from a room in a given time varies directly as the initial volume V_0 that is present. If 75 m^3 of CO_2 are removed in 1 h from a room with an initial volume of 160 m^3, how much is removed in 1 h if the initial volume is 130 m^3?

33. The amount of heat H required to melt ice is proportional to the mass m of ice that is melted. If it takes 2.93×10^5 J to melt 875 g of ice, how much heat is required to melt 625 g?

34. In electroplating, the mass m of the material deposited varies directly as the time t during which the electric current is on. Set up the equation for this relationship if 2.50 g are deposited in 5.25 h.

35. Hooke's law states that the force needed to stretch a spring is proportional to the amount the spring is stretched. If 10.0 N stretches a certain spring 4.00 cm, how much will the spring be stretched by a force of 6.00 N?

36. The rate H of heat removal by an air conditioner is proportional to the electric power input P. The constant of proportionality is the *performance coefficient*. Find the performance coefficient of an air conditioner for which $H = 1.8$ kW and $P = 720$ W.

37. The energy E available daily from a solar collector varies directly as the percent p that the sun shines during the day. If a collector provides 1200 kJ for 75% sunshine, how much does it provide for a day during which there is 35% sunshine?

38. The distance d that can be seen from horizon to horizon from an airplane varies directly as the square root of the altitude h of the airplane. If $d = 213$ km for $h = 3950$ m, find d for $h = 4320$ m.

39. The time t required to empty a wastewater-holding tank is inversely proportional to the cross-sectional area A of the drainage pipe. If it takes 2.0 h to empty a tank with a drainage pipe for which $A = 48\ cm^2$, how long will it take to empty the tank if $A = 68\ cm^2$?

40. The time t required to make a particular trip is inversely proportional to the average speed v. If a jet takes 2.75 h at an average speed of 520 km/h, how long will it take at an average speed of 620 km/h? Explain the meaning of the constant of proportionality.

41. In a physics experiment, a given force was applied to three objects. The mass m and the resulting acceleration a were recorded as follows:

m (g)	2.0	3.0	4.0
a (cm/s^2)	30	20	15

(a) Is the relationship $a = f(m)$ one of direct or inverse variation? Explain. (b) Find $a = f(m)$.

42. The lift L of each of three model airplane wings of width w was measured and recorded as follows:

w (cm)	20	40	60
L (N)	10	40	90

(a) Is the relationship $L = f(w)$ one of direct or inverse variation? Explain. (b) Find $L = f(w)$.

43. The power P required to propel a ship varies directly as the cube of the speed s of the ship. If 3.88 MW will propel a ship at 19.3 km/h, what power is required to propel it at 24.2 km/h?

44. The f-number lens setting of a camera varies directly as the square root of the time t that the sensor (or film) is exposed. If the f-number is 8 (written as $f/8$) for $t = 0.0200$ s, find the f-number for $t = 0.0098$ s.

45. The force F on the blade of a wind generator varies jointly as the blade area A and the square of the wind velocity v. Find the equation relating F, A, and v if $F = 76.5$ N when $A = 0.372$ m^2 and $v = 9.42$ m/s.

46. The escape velocity v a spacecraft needs to leave the gravitational field of a planet varies directly as the square root of the product of the planet's radius R and its acceleration due to gravity g. For Mars and Earth, $R_M = 0.533R_e$ and $g_M = 0.400g_e$. Find v_M for Mars if $v_e = 11.2$ km/s.

47. The force F between two parallel wires carrying electric currents is inversely proportional to the distance d between the wires. If a force of 0.750 N exists between wires that are 1.25 cm apart, what is the force between them if they are separated by 1.75 cm?

48. The frequency f of vibration of a wire varies directly as the square root of the tension T of the wire. If $f = 420$ Hz when $T = 1.14$ N, find f when $T = 3.40$ N.

49. The average speed s of oxygen molecules in the air is directly proportional to the square root of the absolute temperature T. If the speed of the molecules is 460 m/s at 273 K, what is the speed at 300 K?

50. The time t required to test a computer memory unit varies directly as the square of the number n of memory cells in the unit. If a unit with 4800 memory cells can be tested in 15.0 s, how long does it take to test a unit with 8400 memory cells?

51. The electric resistance R of a wire varies directly as its length l and inversely as its cross-sectional area A. Find the relation between resistance, length, and area for a wire that has a resistance of 0.200 Ω for a length of 60.0 m and cross-sectional area of 0.007 80 cm^2.

52. The general gas law states that the pressure p of an ideal gas varies directly as the thermodynamic temperature T and inversely as the volume V. If $p = 610$ kPa for $V = 10.0$ cm^3 and $T = 290$ K, find V for $p = 400$ kPa and $T = 400$ K.

53. The power P in an electric circuit varies jointly as the resistance R and the square of the current I. If the power is 10.0 W when the current is 0.500 A and the resistance is 40.0 Ω, find the power if the current is 2.00 A and the resistance is 20.0 Ω.

54. The difference $m_1 - m_2$ in magnitudes (visual brightnesses) of two stars varies directly as the base 10 logarithm of the ratio b_2/b_1 of their actual brightnesses. For two particular stars, if $b_2 = 100b_1$ for $m_1 = 7$ and $m_2 = 2$, find the equation relating m_1, m_2, b_1, and b_2.

55. The power gain G by a parabolic microwave dish varies directly as the square of the diameter d of the opening and inversely as the square of the wavelength λ of the wave carrier. Find the equation relating G, d, and λ if $G = 5.5 \times 10^4$ for $d = 2.9$ m and $\lambda = 3.0$ cm.

56. The intensity I of sound varies directly as the power P of the source and inversely as the square of the distance r from the source. Two sound sources are separated by a distance d, and one has twice the power output of the other. Where should an observer be located on a line between them such that the intensities of both sounds are the same?

57. The x-component of the acceleration of an object moving around a circle with constant angular velocity ω varies jointly as $\cos \omega t$ and the square of ω. If the x-component of the acceleration is -11.4 cm/s^2 when $t = 1.00$ s for $\omega = 0.524$ rad/s, find the x-component of the acceleration when $t = 2.00$ s.

58. The tangent of the proper banking angle θ of the road for a car making a turn is directly proportional to the square of the car's velocity v and inversely proportional to the radius r of the turn. If 7.75° is the proper banking angle for a car travelling at 20.0 m/s around a turn of radius 300 m, what is the proper banking angle for a car travelling at 30.0 m/s around a turn of radius 250 m? See Fig. 18.9.

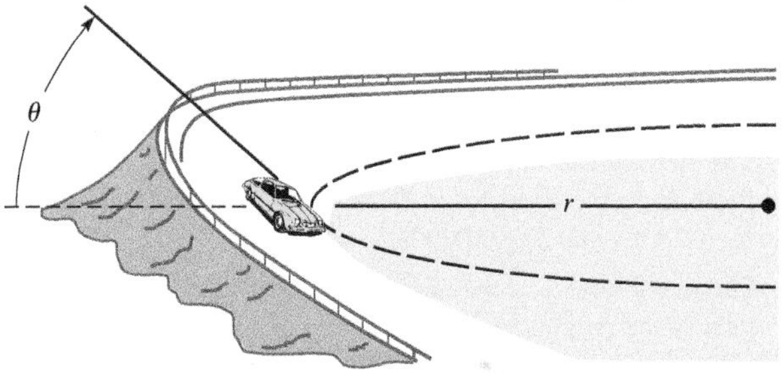

Fig. 18.9

59. The acoustical intensity I of a sound wave is proportional to the square of the pressure amplitude p and inversely proportional to the velocity v of the wave. If $I = 0.474$ W/m^2 for $p = 20.0$ Pa and $v = 346$ m/s, find I if $p = 15.0$ Pa and $v = 320$ m/s.

60. To cook a certain vegetable mix in a microwave oven, the instructions are to cook 120 g for 2.5 min or 240 g for 3.5 min. Assuming the cooking time t is proportional to some power (not necessarily integral) of the weight w, use logarithms to find t as a function of w.

61. Noting the general gas law in Exercise 52, if V remains constant, (a) express p as a function of T. (b) A car tire is filled with air at 10.0°C with a pressure of 225 kPa, and after a while the air temperature in the tire is 60.0°C. Assuming no change in volume, what is the pressure? [*Hint*: You must convert the temperatures to thermodynamic temperatures (in K) using $T_K = T_C + 273.15$.]

62. The weight of an object is the force F on the object due to gravity, where $F = mg$. State how g (the acceleration due to gravity) varies with respect to the mass m_b of the spherical body on which the object lies, and the radius of that body (assume all of the mass is at its centre). (See Example 5.)

Answers to Practice Exercises

1. $y = kx^2/z$ 2. $y = 3$

CHAPTER 18 KEY FORMULAS AND EQUATIONS

Proportion	$\frac{a}{b} = \frac{c}{d}$	**(18.1)**
Direct variation	$y = kx$	**(18.2)**
Inverse variation	$y = \frac{k}{x}$	**(18.3)**
Joint variation	$y = kxz$	**(18.4)**

CHAPTER 18 REVIEW EXERCISES

In Exercises 1–14, find the indicated ratios.

1. 4 Mg to 20 kg
2. 300 nm to 6 μm
3. 375 mL to 25 cL
4. 12 ks to 2 h

5. The number π equals the ratio of the circumference c of a circle to its diameter d. To check the value of π, a technician used computer simulation to measure the circumference and diameter of a metal cylinder and found the values to be $c = 4.2736$ cm and $d = 1.3603$ cm. What value of π did the technician get?

6. The ratio of the diagonal d of a square to the side s of the square is $\sqrt{2}$. The diagonal and side of the face of a glass cube are found to be $d = 35.375$ mm and $s = 25.014$ mm. What is the value of $\sqrt{2}$ found from these measurements?

7. The mechanical advantage of a lever is the ratio of the output force F_o to the input force F_i. Find the mechanical advantage if $F_o = 28$ kN and $F_i = 5000$ N.

8. For an automobile, the ratio of the number n_1 of teeth on the ring gear to the number n_2 of teeth on the pinion gear is the *rear axle ratio* of the car. Find this ratio if $n_1 = 64$ and $n_2 = 20$.

9. The pressure p exerted on a surface is the ratio of the force F on the surface to its area A. Find the pressure on a square patch, 2.25 cm on a side, on a tank if the force on the patch is 37.4 N.

10. The electric resistance R of a resistor is the ratio of the voltage V across the resistor to the current i in the resistor. Find R if $V = 0.632$ V and $i = 2.03$ mA.

11. The *heat of vapourization* of a substance is the amount of heat needed to change a unit amount from liquid to vapour. Experimentation shows 7910 J are needed to change 3.50 g of water to steam. What is the heat of vapourization of water?

12. The ratio of the effective length of a column to its radius of gyration is the *slenderness ratio*. What is the slenderness ratio of a column of effective length 3.15 m and radius of gyration 60.0 mm?

13. The ratio of the commission for selling a home to the selling price of the home is the commission rate. What is this rate if the commission of \$20 900 is charged for selling a home priced at \$380 000?

14. A total cholesterol level of 200 mg/dL is considered too high, but many doctors consider the ratio of the total cholesterol to HDL (high density lipids—the "good" cholesterol) a more important measure of risk to coronary heart disease. What is this ratio if the total cholesterol is 179 mg/dL and the HDL is 39 mg/dL? (Average for women is about 4.5, and for men is about 5.0.)

In Exercises 15 and 16, given that $a/b = c/d$ (b and d not zero), show that the indicated proportions are correct.

15. $\frac{a+b}{b} = \frac{c+d}{d}$
16. $\frac{a-b}{b} = \frac{c-d}{d}$

In Exercises 17–32, answer the given questions by setting up and solving the appropriate proportions.

17. On a map of Australia, 37 mm represents 300 km. If the distance on the map between Melbourne and Hobart, Tasmania, is 78 mm, how far is Hobart from Melbourne?

18. Given that 1.000 kg = 1000 g, what is the mass in kilograms of a 14.0-g computer disk?

19. Given that 1.00 kJ = 10^6 mJ, how much heat in kilojoules is produced by a heating element that produces 2660 mJ?

20. Given that 1.00 L = 1000 cm^3, what capacity in litres has a cubical box that is 3.23 cm along an edge?

21. A laser printer can print 50 pages in 1 min 11 s. How many pages can it print in 9.0 min?

22. A solar heater with a collector area of 58.0 m^2 is required to heat 2560 kg of water. Under the same conditions, how much water can be heated by a rectangular solar collector 9.50 m by 8.75 m?

23. The dosage of a certain medicine is 25 mL for each 10 kg of the patient's weight. What is the dosage for a person weighing 56 kg?

24. A woman invests \$50 000 and a man invests \$20 000 in a partnership. If profits are to be shared in the ratio that each invested in the partnership, how much does each receive from \$10 500 in profits?

25. On a certain blueprint, a measurement of 25.0 m is represented by 2.00 mm. What is the actual distance between two points if they are 7.25 mm apart on the blueprint?

26. The chlorine concentration in a water supply is 0.12 part per million. How much chlorine is there in a cylindrical holding tank 4.22 m in radius and 5.82 m high filled from the water supply?

27. One fibre-optic cable carries 60.0% as many messages as another fibre-optic cable. Together they carry 12 000 messages. How many does each carry?

28. Two types of roadbed material, one 50% rock and the other 100% rock, are used in the ratio of 4 to 1 to form a roadbed. If a total of 150 Mg are used, how much rock is in the roadbed?

29. To neutralize 80.0 kg of sodium hydroxide, 98.0 kg of sulfuric acid are needed. How much sodium hydroxide can be neutralized with 37.0 kg of sulfuric acid?

30. For analysis, 105 mg of DNA sample is divided into two vials in the ratio of 2 to 5. How many milligrams are in each vial?

31. A total of 322 bolts are in two containers. The ratio of the number of bolts in the first container to the number in the second container is 5/9. How many are in each container?

32. A gasoline company sells octane-87 gas and octane-91 gas in the ratio of 9 to 2. How many of each octane are sold of a total of 16.5 million litres?

In Exercises 33–36, give the specific equation relating the variables after evaluating the constant of proportionality for the given set of values.

33. y varies directly as the square of x, and $y = 27$ when $x = 3$.

34. f varies inversely as l, and $f = 5$ when $l = 8$.

35. v is directly proportional to x and inversely proportional to the cube of y, and $v = 10$ when $x = 5$ and $y = 4$.

36. r varies jointly as u, v, and the square of w, and $r = 8$ when $u = 2$, $v = 4$, and $w = 3$.

In Exercises 37–78, solve the given applied problems.

37. For the lever balanced at the fulcrum, the relation

$$\frac{F_1}{F_2} = \frac{L_2}{L_1}$$

holds, where F_1 and F_2 are forces on opposite sides of the fulcrum at distances L_1 and L_2 (see Fig. 18.10). Find L_2 if $F_1 = 4.50$ N, $F_2 = 6.75$ N, and $L_1 = 17.5$ cm.

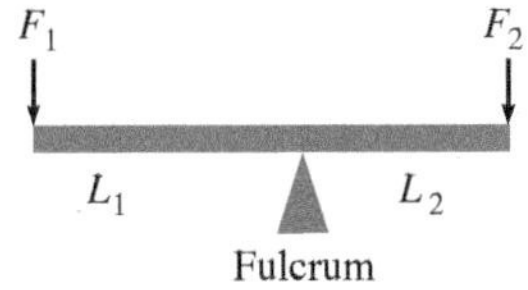

Fig. 18.10

38. A company finds that the volume V of sales of a certain item and the price P of the item are related by

$$\frac{P_1}{P_2} = \frac{V_2}{V_1}$$

Find V_2 if $P_1 = \$8.00$, $P_2 = \$6.00$, and $V_1 = 3000$ items per week.

39. The image height h and object height H for the lens shown in Fig. 18.11 are related to the image distance q and object distance p by

$$\frac{h}{H} = \frac{q}{p}$$

Find q if $h = 24$ cm, $H = 84$ cm, and $p = 36$ cm.

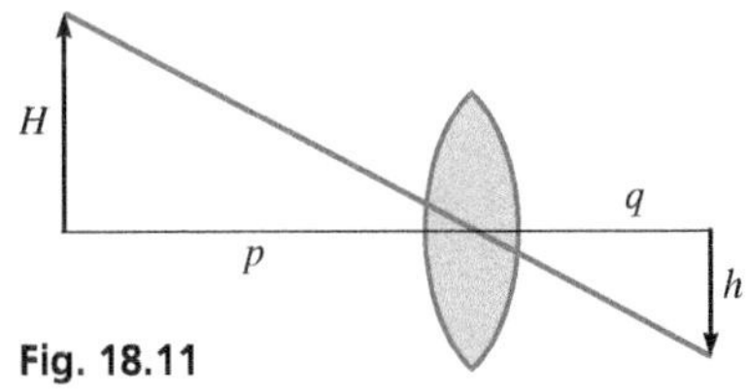

Fig. 18.11

40. For two pulleys connected by a belt, the relation

$$\frac{d_1}{d_2} = \frac{n_2}{n_1}$$

holds, where d is the diameter of the pulley and n is the number of revolutions per unit time it makes. Find n_2 if $d_1 = 4.60$ cm, $d_2 = 8.30$ cm, and $n_1 = 18.0$ r/min.

41. An apartment owner charges rent R proportional to the floor area A of the apartment. Find the equation relating R and A if an apartment of 120 m^2 rents for \$1560/month.

42. The number r of aluminum cans that can be made by recycling n used cans is proportional to n. How many cans can be made from 50 000 used cans if $r = 115$ cans for $n = 125$ cans?

43. Under certain conditions, the rate of increase v of bacteria is proportional to the number N of bacteria present. Find v for $N = 7500$ bacteria, if $v = 800$ bacteria/h for $N = 4000$ bacteria.

44. The index of refraction n of a medium varies inversely as the velocity of light v within it. For quartz, $n = 1.46$ and $v = 2.05 \times 10^8$ m/s. What is n for a diamond, in which $v = 1.24 \times 10^8$ m/s?

45. The force F needed to tighten a bolt varies as the length L of the wrench handle. See Fig. 18.12. If $F = 250$ N for $L = 22$ cm, and $F = 550$ N for $L = 10$ cm, determine the equation relating F and L.

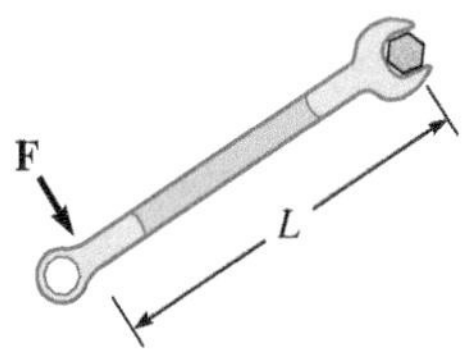

Fig. 18.12

46. In the two-base method of barometric levelling, bases are established at two points of known elevation: one at the lowest point and one at the highest point. An altimeter at an unknown elevation is read at the same time that both the upper and lower base altimeters are read. The relation

$$\frac{H_R - H_L}{H_U - H_L} = \frac{R_R - R_L}{R_U - R_L}$$

holds, where H_L, H_U, and H_R are the elevations at the lower base, upper base, and roving altimeter stations (in mm), and R_L, R_U, and R_R are the altimeter readings at the lower base, upper base, and roving altimeter stations (in m). Find H_R if $H_U = 412.3$ m, $H_L = 107.5$ m, $R_U = 360$ mm, $R_L = 108$ m, and $R_R = 135$ mm. See Fig. 18.13.

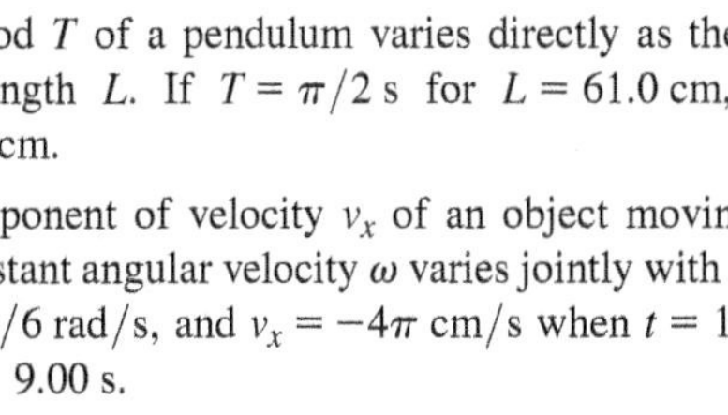

Fig. 18.13

47. The period T of a pendulum varies directly as the square root of its length L. If $T = \pi/2$ s for $L = 61.0$ cm, find T for $L = 122$ cm.

48. The component of velocity v_x of an object moving in a circle with constant angular velocity ω varies jointly with ω and $\sin \omega t$. If $\omega = \pi/6$ rad/s, and $v_x = -4\pi$ cm/s when $t = 1.00$ s, find v_x when $t = 9.00$ s.

49. The charge C on a capacitor varies directly as the voltage V across it. If the charge is 6.3 μC with a voltage of 220 V across a capacitor, what is the charge on it with a voltage of 150 V across it?

50. The amount of natural gas burned is proportional to the amount of oxygen consumed. If 24.0 kg of oxygen is consumed in burning 15.0 kg of natural gas, how much air, which is 23.2% oxygen by weight, is consumed to burn 50.0 kg of natural gas?

51. The power P of a gas engine is proportional to the area A of the piston. If an engine with a piston area of 50.0 cm^2 can develop 22.5 kW, what power is developed by an engine with a piston area of 40.0 cm^2?

52. The decrease in temperature above a region is directly proportional to the altitude above the region. If the temperature T at the base of the Rock of Gibraltar is 22.0°C and a plane 3.50 km above notes that the temperature is 1.0°C, what is the temperature at the top of Gibraltar, the altitude of which is 430 m? (Assume there are no other temperature effects.)

53. The distance d an object falls under the influence of gravity varies directly as the square of the time t of fall. A chunk of ice falls from the top of the CN Tower and travels 19.62 m in 2.00 s. It hits the ground after 10.43 s. How tall is the tower?

54. The kinetic energy E of a moving object varies jointly as the mass m of the object and the square of its velocity v. If a 5.00-kg object, travelling at 10.0 m/s, has a kinetic energy of 250 J, find the kinetic energy of an 8.00-kg object moving at 50.0 m/s.

55. In a particular computer design, N numbers can be sorted in a time proportional to the square of log N. How many times longer does it take to sort 8000 numbers than to sort 2000 numbers?

56. The velocity v of a jet of fluid flowing from an opening in the side of a container is proportional to the square root of the depth d of the opening. If the velocity of the jet from an opening at a depth of 1.22 m is 4.88 m/s, what is the velocity of a jet from an opening at a depth of 7.62 m? See Fig. 18.14.

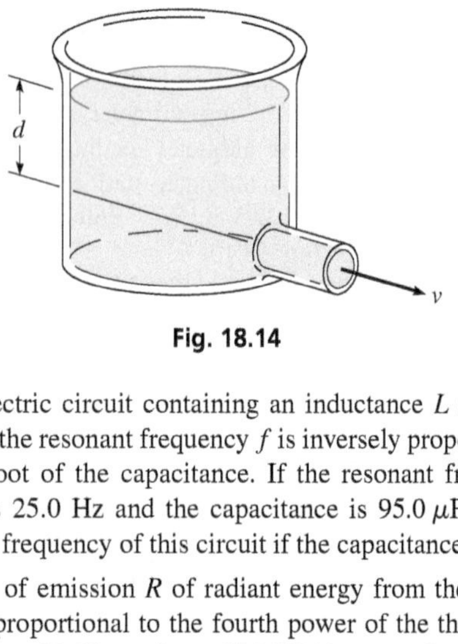

Fig. 18.14

57. In an electric circuit containing an inductance L and a capacitance C, the resonant frequency f is inversely proportional to the square root of the capacitance. If the resonant frequency in a circuit is 25.0 Hz and the capacitance is 95.0 μF, what is the resonant frequency of this circuit if the capacitance is 25.0 μF?

58. The rate of emission R of radiant energy from the surface of a body is proportional to the fourth power of the thermodynamic temperature T. Given that a 25.0-W (the rate of emission) lamp has an operating temperature of 2500 K, what is the operating temperature of a similar 40.0-W lamp?

59. The frequency f of a radio wave is inversely proportional to its wavelength λ. The constant of proportionality is the velocity of the wave, which equals the speed of light. Find this velocity if an FM radio wave has a frequency of 90.9 MHz and a wavelength of 3.29 m.

60. The acceleration of gravity g on a satellite in orbit around the earth varies inversely as the square of its distance r from the centre of the earth. If $g = 8.7 \text{ m/s}^2$ for a satellite at an altitude of 400 km above the surface of the earth, find g if it is 1000 km above the surface. The radius of the earth is 6.4×10^6 m.

61. If the weight w of an airplane varies directly as the cube of its length L, and $w = 15\,400$ N for $L = 15$ m, what would be the weight of a plane that is 550 cm long?

62. The thrust T of a propeller varies jointly as the square of the number n of revolutions per second and the fourth power of its diameter d. What is the effect on T if n is doubled and d is halved?

63. Using *holography* (a method of producing an image without using a lens), an image of concentric circles is formed. The radius r of each circle varies directly as the square root of the wavelength λ of the light used. If $r = 3.56$ cm for $\lambda = 575$ nm, find r if $\lambda = 483$ nm.

64. A metal circular ring has a circular cross-section of radius r. If R is the radius of the ring (measured to the middle of the cross-section), the volume V of metal in the ring varies directly as R and the square of r. If $V = 2550 \text{ mm}^3$ for $r = 2.32$ mm and $R = 24.0$ mm, find V for $r = 3.50$ mm and $R = 32.0$ mm. See Fig. 18.15.

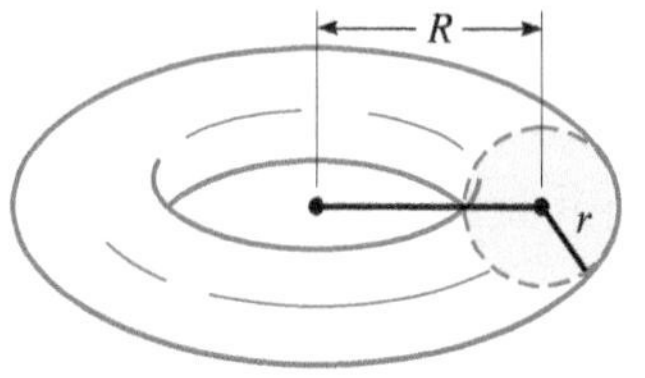

Fig. 18.15

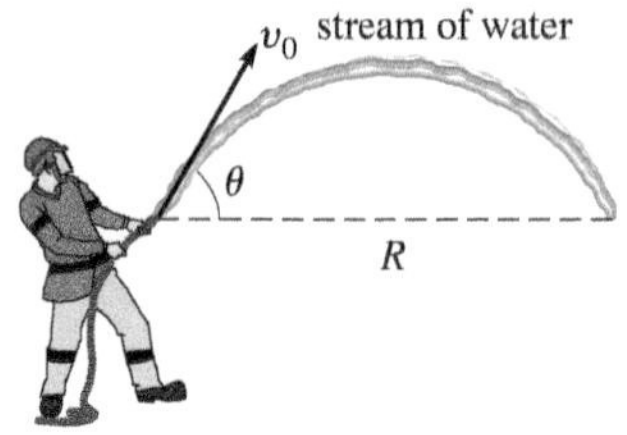

Fig. 18.16

65. The range R of a stream of water from a fire hose varies jointly as the square of its initial velocity v_0 and the sine of twice the angle θ from the horizontal from which it is released. See Fig. 18.16. A stream for which $v_0 = 35$ m/s and $\theta = 22°$ has a range of 85 m. Find the range if $v_0 = 28$ m/s and $\theta = 43°$.

66. Kepler's third law of planetary motion states that the square of the period of any planet is proportional to the cube of the mean radius (about the sun) of that planet, with the constant of proportionality being the same for all planets. Using the fact that the period of the earth is 1 year and its mean radius is 150×10^6 km, calculate the mean radius for Venus, given that its period is 7.38 months.

67. The stopping distance d of a car varies directly as the square of the velocity v of the car when the brakes are applied. A car moving at 48 km/h can stop in 15 m. What is the stopping distance for the car if it is moving at 85 km/h?

68. The load L that a helical spring can support varies directly as the cube of its wire diameter d and inversely as its coil diameter D. A spring for which $d = 0.120$ cm and $D = 0.953$ cm can support 45.0 N. What is the coil diameter of a similar spring that supports 78.5 N and for which $d = 0.156$ cm?

69. The volume rate of flow R of blood through an artery varies directly as the fourth power of the radius r of the artery and inversely as the distance d along the artery. If an operation is successful in effectively increasing the radius of an artery by 25% and decreasing its length by 2%, by how much is the volume rate of flow increased?

70. The safe, uniformly distributed load L on a horizontal beam, supported at both ends, varies jointly as the width w and the square of the depth d and inversely as the distance D between supports. Given that one beam has double the dimensions of another, how many times heavier is the safe load it can support than the first can support?

71. A bank statement exactly 30 years old is discovered. It states, "This 10-year-old account is now worth \$185.03 and pays 4% interest compounded annually." An investment with annual compound interest varies directly as $1 + r$ to the power n, where r is the interest rate expressed as a decimal and n is the number of years of compounding. What was the value of the original investment, and what is it worth now?

72. The distance s that an object falls due to gravity varies jointly as the acceleration g due to gravity and the square of the time t of fall. The acceleration due to gravity on the moon is 0.172 of that on Earth. If a rock falls for t_0 seconds on Earth, how many times farther would the rock fall on the moon in $3t_0$ seconds?

73. The heat loss L through fibreglass insulation varies directly as the time t and inversely as the thickness d of the fibreglass. If the loss through 20.0 cm of fibreglass is 1.20 MJ in 30 min, what is the loss through 15.0 cm in 1 h 30 min?

74. A quantity important in analysing the rotation of an object is its *moment of inertia I*. For a ball bearing, the moment of inertia varies directly as its mass m and the square of its radius r. Find the general expression for I if $I = 39.9 \text{ g} \cdot \text{cm}^2$ for $m = 63.8$ g and $r = 1.25$ cm.

75. In the study of polarized light, the intensity I is proportional to the square of the cosine of the angle θ of transmission. If $I = 0.025 \text{ W/m}^2$ for $\theta = 12.0°$, find I for $\theta = 20.0°$.

76. The force F that acts on a pendulum bob is proportional to the mass m of the bob and the sine of the angle θ the pendulum makes with the vertical. If $F = 0.120$ N for $m = 0.350$ kg and $\theta = 2.00°$, find F for $m = 0.750$ kg and $\theta = 3.50°$.

77. On a map of Canada, 12.5 mm represents 200 km. If the distance on the map between Edmonton and Calgary is 18 mm, how far is Edmonton from Calgary?

78. A fruit-packing company plans to reduce the size of its fruit juice can (a right circular cylinder) by 10% and keep the price of each can the same (effectively raising the price). The radius and the height of the new can are to be equally proportional to those of the old can. Write one or two paragraphs explaining how to determine the percent decrease in the radius and the height of the old can that is required to make the new can.

CHAPTER 18 PRACTICE TEST

1. Express the ratio of 180 s to 4 min in simplest form.

2. A person 1.8 m tall is photographed, and the film image is 20.0 mm high. Under the same conditions, how tall is a person whose film image is 14.5 mm high?

3. The change L in length of a copper rod varies directly as the change T in temperature of the rod. Set up an equation for this relationship if $L = 2.7$ cm for $T = 150°\text{C}$.

4. Given that 1.00 in. = 2.54 cm, what length in inches is 7.24 cm?

5. The perimeter of a rectangular solar panel is 210.0 cm. The ratio of the length to the width is 7 to 3. What are the dimensions of the panel?

6. The difference p in pressure in a fluid between that at the surface and that at a point below varies jointly as the density d of the fluid and the depth h of the point. The density of water is 1000 kg/m^3, and the density of alcohol is 800 kg/m^3. This difference in pressure at a point 0.200 m below the surface of water is 1.96 kPa. What is the difference in pressure at a point 0.300 m below the surface of alcohol? (All data are accurate to three significant digits.)

7. The crushing load L of a pillar varies directly as the fourth power of its radius r and inversely as the square of its length l. If one pillar has twice the radius and three times the length of a second pillar, what is the ratio of the crushing load of the first pillar to that of the second pillar?

19. Sequences and the Binomial Theorem

LEARNING OUTCOMES

After completion of this chapter, the student should be able to:

- Determine if a sequence forms an arithmetic sequence
- Find the terms, common difference, number of terms, or sum of an arithmetic sequence
- Determine if a sequence forms a geometric sequence
- Find the terms, common ratio, number of terms, or sum of a finite geometric sequence
- Find the sum of an infinite geometric series
- Find the fraction equal to a given repeating decimal
- Evaluate expressions involving factorials
- Expand binomials to any power, using Pascal's triangle and the binomial theorem
- Obtain terms of a binomial series
- Solve application problems involving sequences and series

▲ Sequences are basic to many calculations in business. The pattern in the picture illustrates the arithmetic sequence formed by the accumulated interest of an investment that pays simple interest.

If a person invests \$100 at 5% per year simple interest (so that only the principal invested earns interest), the value of the investment is \$105 after the first year, \$110 after the second year, \$115 after the third year, and so on. We can see that the annual values of the investment follow a pattern, with consecutive values having a *common difference* of \$5 (the interest earned in one year). The annual values of the investment form what is called an *arithmetic sequence*.

If a person invests \$100 at 5% per year compound interest (so that the interest is added to the investment each year), the value of the investment is \$105 after the first year, \$110.25 after the second year, \$115.76 after the third year, and so on. Once again the annual values of the investment follow a pattern, but with consecutive values having a common *ratio* of 1.05. In this situation, the annual values of the investment form what is called a *geometric sequence*.

A *sequence* is a set of numbers arranged in some specific way and usually follows a pattern. Sequences have been of interest to people for centuries. Records dating back to at least 1700 B.C.E. show calculations involving sequences. Euclid, in his *Elements*, dealt with sequences in about 300 B.C.E. More advanced forms of sequences were used extensively in the study of advanced mathematics in the 1700s and 1800s. These advances in mathematics have been very important in many areas of science and technology.

We dedicate this chapter to the study of some basic types of sequences, and to the sum of their terms (which we call series). Applications include the study of radioactivity, population growth, and, of course, the value of investments.

19.1 Arithmetic Sequences

A **sequence** is an ordered list of numbers, each of which is said to be a **term** of the sequence. While there are sequences with no relationship between their terms, sequences that are useful follow a pattern. In this chapter we consider only sequences of real numbers or of literal numbers representing real numbers.

An **arithmetic sequence** (*or* **arithmetic progression**) *is a set of numbers in which each number after the first can be obtained from the preceding one by adding to it a fixed number called the* **common difference**. This definition can be expressed in terms of the *recursion formula*

$$a_n = a_{n-1} + d \tag{19.1}$$

where a_n is any term, a_{n-1} is the preceding term, and d is the common difference.

EXAMPLE 1 Illustrations of arithmetic sequences

Find the common difference for each of the following arithmetic sequences. (The three dots after the last term listed indicate that the sequences continue.)

Sequence	Consecutive differences	Common difference
(a) 2, 5, 8, 11, 14, . . .	$5 - 2 = 3$, or $11 - 8 = 3$	$d = 3$
(b) 7, 2, −3, −8, . . .	$2 - 7 = -5$, or $-8 - (-3) = -5$	$d = -5$

If we know the first term of an arithmetic sequence, we can find any other term by adding the common difference enough times to get the desired term. This, however, is very inefficient, and there is a general way of finding a particular term.

If a_1 is the first term and d is the common difference, the second term is $a_1 + d$, the third term is $a_1 + 2d$, and so on. For the nth term, we need to add d to the first term $n - 1$ times. Therefore, *the nth term, a_n, of the arithmetic sequence is given by*

$$a_n = a_1 + (n - 1)d \tag{19.2}$$

Eq. (19.2) can be used to find any given term in any arithmetic sequence. We can refer to a_n as the *last term* of an arithmetic sequence if no terms beyond it are included in the sequence. Such a sequence is called a ***finite*** sequence. If the terms in a sequence continue without end, the sequence is called an ***infinite*** sequence.

EXAMPLE 2 Finding a specified term—simple interest

■ See the chapter introduction.

The annual value of an investment of \$100 at 5% per year simple interest for 10 years follows the arithmetic sequence 100, 105, 110, 115, . . . , a_{11}. Find the last term of the sequence (the value of the investment after the 10 years).

The first term is $a_1 = 100$, and by subtracting any term from the next term, we find the common difference $d = 5$. Therefore, the 11th term, a_{11}, is

$$\begin{aligned} a_{11} &= \underset{\overset{\downarrow}{a_1}}{100} + (\underset{\overset{\downarrow}{n}}{11} - 1)\underset{\overset{\downarrow}{d}}{5} = 100 + (10)(5) \\ &= 150 \end{aligned}$$

The value of the investment after 10 years is \$150.

EXAMPLE 3 Finding the common difference

Find the common difference between successive terms of the arithmetic sequence for which the third term is 5 and the 34th term is -119.

In order to use Eq. (19.2), we could calculate a_1, but we can also treat the third term as a_1 and the 34th term as a_{32}. This gives us the same sequence from 5 to -119. Therefore, using the values $a_1 = 5$, $a_{32} = -119$, and $n = 32$, we can find the value of d. Substituting these values in Eq. (19.2) gives

$$-119 = 5 + (32 - 1)d$$
$$31d = -124$$
$$d = -4$$

There is no information as to whether this is a finite or an infinite sequence. The solution is the same in either case.

Practice Exercise

1. Find d if $a_1 = 4$ and $a_{13} = 34$.

EXAMPLE 4 Finding the number of terms

How many integers between 10 and 1000 are divisible by 6?

We must first find the smallest and the largest numbers in this range that are divisible by 6. These numbers are 12 and 996. Obviously, the common difference between one multiple of 6 and the next is 6. Thus, we can solve this as an arithmetic sequence with $a_1 = 12$, $a_n = 996$, and $d = 6$. Substituting these values in Eq. (19.2), we have

$$996 = 12 + (n - 1)6$$
$$6n = 990$$
$$n = 165$$

Thus, 165 numbers between 10 and 1000 are divisible by 6.

All the positive multiples of 6 are included in the infinite arithmetic sequence 6, 12, 18, . . . , whereas those between 10 and 1000 are included in the finite arithmetic sequence 12, 18, 24, . . . , 996.

EXAMPLE 5 Arithmetic sequence—velocity with constant acceleration

A package delivery company uses a metal (very low friction) ramp to slide packages from the sorting area down to the loading area below. If a given package is pushed such that it starts down the ramp at 25 cm/s, and the package accelerates as it slides down the ramp such that it gains 35 cm/s during each second, after how many seconds is its velocity 305 cm/s? See Fig. 19.1.

Here, we see that the velocity (in cm/s) of the package after each second is

$$60, 95, 130, \ldots, 305, \ldots$$

Therefore, $a_1 = 60$ (the 25 cm/s was at the beginning, that is, after 0 s), $d = 35$, $a_n = 305$, and we are to find n, which in this case represents the number of seconds during which the velocity increases. Therefore,

$$305 = 60 + (n - 1)(35)$$
$$245 = 35n - 35$$
$$35n = 280$$
$$n = 8.0$$

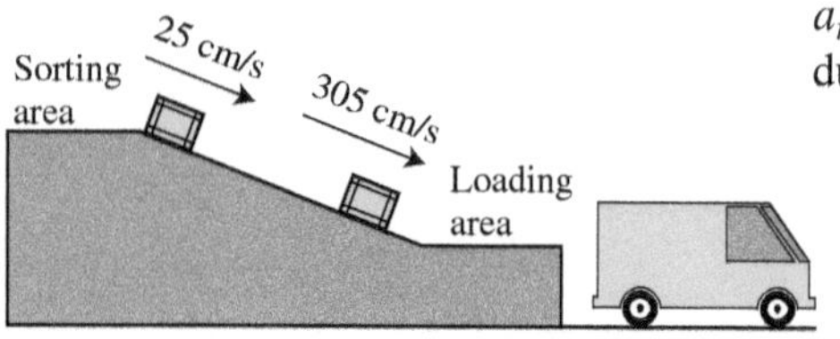

Fig. 19.1

This means that the velocity of a package sliding down the ramp is 305 cm/s after 8.0 s.

SUM OF *n* TERMS

Another important quantity related to an arithmetic sequence is the sum of the first n terms. We can indicate this sum by starting the sum either with the first term or with the last term, as shown by these two equations:

$$S_n = a_1 + (a_1 + d) + (a_1 + 2d) + \cdots + (a_n - d) + a_n$$

or

$$S_n = a_n + (a_n - d) + (a_n - 2d) + \cdots + (a_1 + d) + a_1$$

If we now add the corresponding members of these two equations, we obtain the result

$$2S_n = (a_1 + a_n) + (a_1 + a_n) + (a_1 + a_n) + \cdots + (a_1 + a_n) + (a_1 + a_n)$$

Each term on the right in parentheses has the same expression $(a_1 + a_n)$, and there are n such terms. This tells us that *the sum of the first n terms is given by*

$$S_n = \frac{n}{2}(a_1 + a_n) \tag{19.3}$$

The use of Eq. (19.3) is illustrated in the following examples.

■ Karl Friedrich Gauss (1777–1856) is considered one of the greatest mathematicians of all time. A popular story about him is that at the age of 10 his schoolmaster asked his class to add all the numbers from 1 to 100, assuming it would take some time to do. He had barely finished stating the problem when Gauss handed in the correct answer of 5050. He had done it essentially as we did the solution in Example 6.

EXAMPLE 6 Finding the sum of terms

Find the sum of the first 1000 positive integers.

The first 1000 positive integers form a finite arithmetic sequence for which $a_1 = 1$, $a_{1000} = 1000$, $n = 1000$, and $d = 1$. Substituting these values in Eq. (19.3) (in which we do not use the value of d), we have

$$S_{1000} = \frac{1000}{2}(1 + 1000) = 500(1001)$$
$$= 500\ 500$$

EXAMPLE 7 Finding the sum of terms

Find the sum of the first 10 terms of the arithmetic sequence in which the first term is 4 and the common difference is -5.

We are to find S_n, given that $n = 10$, $a_1 = 4$, and $d = -5$. Since Eq. (19.3) uses the value of a_n but not the value of d, we first find a_{10} by using Eq. (19.2). This gives us

$$a_{10} = 4 + (10 - 1)(-5) = 4 - 45$$
$$= -41$$

Now, we can solve for S_{10} by using Eq. (19.3):

$$S_{10} = \frac{10}{2}(4 - 41) = 5(-37)$$
$$= -185$$

Practice Exercise

2. Find S_9 if $d = -2$ and $a_9 = 1$.

If any three of the five values a_1, a_n, n, d, and S_n are given for a particular arithmetic sequence, the other two may be found from Eqs. (19.2) and (19.3). Consider the following example.

EXAMPLE 8 Finding n and a_n

For an arithmetic sequence, given that $a_1 = 2$, $d = 1.5$, and $S_n = 72$, find n and a_n.

First, we substitute the given values in Eqs. (19.2) and (19.3) in order to identify what is known, and how to proceed. Making these substitutions, we have

$$a_n = 2 + (n - 1)(1.5) \quad \text{substituting in Eq. (19.2)}$$

$$72 = \frac{n}{2}(2 + a_n) \quad \text{substituting in Eq. (19.3)}$$

$$72 = \frac{n}{2}[2 + 2 + (n - 1)(1.5)] \quad \text{substituting first equation into second equation}$$

$$= 2n + 0.75n(n - 1)$$

$$288 = 8n + 3n^2 - 3n \quad \text{multiplying each term by 4 to clear decimals}$$

$$3n^2 + 5n - 288 = 0$$

$$n = \frac{-5 \pm \sqrt{25 - 4(3)(-288)}}{6} = \frac{-5 \pm \sqrt{3481}}{6} = \frac{-5 \pm 59}{6}$$

Since n must be a positive integer, we find that $n = (-5 + 59)/6 = 9$. Therefore, since $n = 9$, $a_9 = 2 + (9 - 1)(1.5) = 14$. The values $n = 9$ and $a_9 = 14$ check with the given values.

EXAMPLE 9 Sum of terms—voltage

The voltage across a resistor increases such that during each second the increase is 0.002 mV less than during the previous second. Given that the increase during the first second is 0.350 mV, what is the total voltage increase during the first 10.0 s?

We are asked to find the sum of the voltage increases 0.350 mV, 0.348 mV, 0.346 mV, . . . , so as to include 10 increases. This means we want the sum of an arithmetic sequence for which $a_1 = 0.350$, $d = -0.002$, and $n = 10$. Since we need a_n to use Eq. (19.3), we first calculate it, using Eq. (19.2):

$$a_{10} = 0.350 + (10 - 1)(-0.002) = 0.332 \text{ mV}$$

Now, we use Eq. (19.3) to find the sum with $a_1 = 0.350$, $a_{10} = 0.332$, and $n = 10$:

$$S_{10} = \frac{10}{2}(0.350 + 0.332) = 3.410 \text{ mV}$$

Thus, the total voltage increase is 3.410 mV.

EXERCISES 19.1

In Exercises 1–2, make the given changes in the indicated examples of this section and then solve the resulting problems.

1. In Example 3, change −119 to −88 and then find the common difference.
2. In Example 6, change 1000 to 500 and then find the sum.

In Exercises 3–6, write the first five terms of the arithmetic sequence with the given values.

3. $a_1 = 4, d = 2$
4. $a_1 = 6, d = -\frac{1}{2}$
5. $a_1 = 2.5, a_5 = -1.5$
6. $a_2 = -2, a_5 = 43$

In Exercises 7–14, find the nth term of the arithmetic sequence with the given values.

7. $1, 4, 7, \ldots; n = 10$
8. $-6, -4, -2, \ldots; n = 8$
9. $\frac{7\pi}{4}, \frac{3\pi}{2}, \frac{5\pi}{4}, \ldots; n = 17$
10. $2, 0.5, -1, \ldots; n = 25$
11. $a_1 = -0.7, d = 0.4, n = 80$
12. $a_1 = \frac{3}{2}, d = \frac{1}{6}, n = 601$
13. $a_1 = b, d = 2b, n = 25$
14. $a_1 = -c, d = 3c, n = 30$

In Exercises 15–18, find the sum of the n terms of the indicated arithmetic sequence.

15. $4 + 8 + 12 + \cdots + 64$

16. $27 + 24 + 21 + \cdots + (-9)$

17. $-2, -\frac{5}{2}, -3, \ldots; n = 10$

18. $3k, \frac{10}{3}k, \frac{11}{3}k, \ldots; n = 40$

In Exercises 19–30, find any of the values of a_1, d, a_n, n, or S_n that are missing for an arithmetic sequence.

19. $5, 13, 21, \ldots, 45$

20. $-2, -1.5, -1, \ldots, 28$

21. $a_1 = \frac{5}{3},\ n = 20,\ S_{20} = \frac{40}{3}$

22. $a_1 = 0.1,\ a_n = -5.9,\ S_n = -8.7$

23. $d = 3,\ n = 40,\ S_{40} = 3100$

24. $d = 9,\ a_n = 86,\ S_n = 455$

25. $a_1 = 7.4,\ d = -0.5,\ a_n = -23.1$

26. $a_1 = -\frac{9}{7}, n = 19,\ a_{19} = -\frac{36}{7}$

27. $a_1 = 5k,\ d = 0.5k,\ S_n = 104k$

28. $d = -2c,\ n = 50,\ S_{50} = 0$

29. $a_1 = -c,\ a_n = \dfrac{b}{2},\ S_n = 2b - 4c$

30. $a_1 = 3b,\ n = 7,\ d = \dfrac{b}{3}$

In Exercises 31–60, find the indicated quantities for the appropriate arithmetic sequence.

31. $a_6 = 560, a_{10} = 720$ (find a_1, d, S_n for $n = 10$)

32. $a_{17} = -91, a_2 = -73$ (find a_1, d, S_n for $n = 40$)

33. Is ln 3, ln 6, ln 12, . . . an arithmetic sequence? Explain. If it is, what is the fifth term?

34. Is sin 2°, sin 4°, sin 6°, . . . an arithmetic sequence? Explain. If it is, what is the fifth term?

35. Find a formula with variable n for the nth term of the arithmetic sequence with $a_1 = 3$ and $a_{n+1} = a_n + 2$ for $n = 1, 2, 3, \ldots$.

36. If a, b, and c are the first three terms of an arithmetic sequence, find their sum in terms of b only.

37. The terms a, m, b form an arithmetic sequence. Express a formula for the common difference d in terms of a and b. (The term m is the *arithmetic mean* of the terms a and b.)

38. The terms a, m, n, b form an arithmetic sequence. Express m and n in terms of a and b. (The terms m and n are *arithmetic means* of the terms a and b.)

39. Write the first five terms of the arithmetic sequence for which the second term is b and the third term is c.

40. If the sum of the first two terms of an arithmetic sequence equals the sum of the first three terms, find the sum of the first five terms.

41. Find the sum of the first 100 positive integers. (See the margin note about Gauss beside Example 6.)

42. Find the number of multiples of 8 between 99 and 999.

43. Find x if $3 - x, -x$, and $\sqrt{9 - 2x}$ are the first three terms of an arithmetic sequence.

44. Is $x, x + 2y, 2x + 3y, \ldots$ an arithmetic sequence? If it is, find the sum of the first 100 terms.

45. A beach now has an area of 9500 m^2 but is eroding such that it loses 100 m^2 more of its area each year than during the previous year. If it lost 400 m^2 during the last year, what would you expect its area to be eight years from now?

46. During a period of heavy rains, on a given day 4500 m^3/s of water was being released from a dam. In order to minimize downstream flooding, engineers then reduced the releases by 500 m^3/s each day thereafter. How much water was released during the first week of these releases?

47. At a logging camp, 15 layers of logs are so piled that there are 20 logs in the bottom layer, and each layer has one log fewer than the layer below it. How many logs are in the pile?

48. In order to prevent an electric current surge in a circuit, the resistance R in the circuit is stepped down by 4.0 Ω after each 0.1 s. If the voltage V is constant at 120 V, do the resulting currents I (in A) form an arithmetic sequence if $V = IR$?

49. There are 12 seats in the first row around a semicircular stage. Each row behind the first has four more seats than the row in front of it. How many rows of seats are there if there is a total of 300 seats?

50. A bank loan of \$8000 is repaid in annual payments of \$1000 plus 10% interest on the unpaid balance. What is the total amount of interest paid?

51. A car depreciates \$1800 during the first year after it is bought. Each year thereafter it depreciates \$150 less than the year before. How many years after it was bought will it be considered to have no value, and what was the original cost?

52. The sequence of ships' bells is as follows: 12:30 A.M. one bell is rung, and each half hour later one more bell is rung than the previous time until eight bells are rung. The sequence is then repeated starting at 4:30 A.M., again until eight bells are rung. This pattern is followed throughout the day. How many bells are rung in one day?

53. If a stone released from a spacecraft near the surface of Mars falls 1.85 m during the first second, 5.55 m during the second second, 9.25 m during the third second, 12.95 m during the fourth second, and so on, how far from the surface is the spacecraft if it takes the stone 10.0 s to reach the surface?

54. In preparing a bid for constructing a new building, a contractor determines that the foundation and basement will cost \$605 000 and the first floor will cost \$360 000. Each floor above the first will cost \$15 000 more than the one below it. How much will the building cost if it is to be 18 floors high?

55. A college graduate is offered two positions. A computer company offers an annual salary of \$42 000 with a guaranteed annual raise of \$1200. A marketing company offers an annual salary of \$44 000 with a guaranteed annual raise of \$600. Which company will pay more for the first six years of employment, and how much more?

56. A person has a \$5000 balance due on a credit card account that charges 1% interest per month on the unpaid balance B_n. Assuming no extra charges, if \$250 is paid each month, find (a) a recursion formula [similar to Eq. (19.1)] for B_n and (b) B_n after two months.

57. Derive a formula for S_n in terms of n, a_1, and d.

58. A *harmonic sequence* is a sequence of numbers whose reciprocals form an arithmetic sequence. Is a harmonic sequence also an arithmetic sequence? Explain.

59. Show that the sum of the first n positive integers is $\frac{1}{2}n(n + 1)$.

60. Show that the sum of the first n positive odd integers is n^2.

Answers to Practice Exercises

1. $d = 2.5$ **2.** $S_9 = 81$

19.2 Geometric Sequences

A second type of important sequence of numbers is the **geometric sequence** (or **geometric progression**). *In a geometric sequence, each number after the first can be obtained from the preceding one by multiplying it by a fixed number, called the* **common ratio.** We can express this definition in terms of the *recursion formula*

$$a_n = ra_{n-1} \qquad \textbf{(19.4)}$$

where a_n is any term, a_{n-1} is the preceding term, and r is the common ratio. One important application of geometric sequences is in computing compound interest on savings accounts. Other applications are found in areas such as biology and physics.

EXAMPLE 1 Illustrations of geometric sequences

Find the common ratio for each of the following geometric sequences.

Sequence	Consecutive quotients	Common ratio
(a) $2, 4, 8, 16, \ldots$	$4/2 = 2$, or $8/4 = 2$, or $16/8 = 2$	$r = 2$
(b) $9, -3, 1, -1/3, \ldots$	$-3/9 = -1/3$, or $-(1/3)/1 = -1/3$	$r = -1/3$

If we know the first term, we can find any other term by multiplying by the common ratio a sufficient number of times. When we do this for a general geometric sequence, we can find the nth term in terms of the first term a_1, the common ratio r, and n. Thus, the second term is a_1r, the third term is a_1r^2, and so forth. In general, the expression for the **nth term of a geometric sequence** is

$$a_n = a_1r^{n-1} \qquad \textbf{(19.5)}$$

EXAMPLE 2 Finding a specified term

Find the eighth term of the geometric sequence 8, 4, 2, ...

By dividing any term by the previous term, we find the common ratio to be $\frac{1}{2}$. From the terms given, we see that $a_1 = 8$. From the statement of the problem, we know that $n = 8$. Thus, we substitute into Eq. (19.5) to find a_8:

$$a_8 = \underset{a_1}{8}\left(\underset{r}{\frac{1}{2}}\right)^{\overset{n}{8}-1} = \frac{8}{2^7} = \frac{1}{16}$$

EXAMPLE 3 Finding the common ratio

Find the common ratio r if $a_1 = \frac{8}{625}$ and $a_{10} = -\frac{3125}{64}$.

Using Eq. (19.5), we have

$$-\frac{3125}{64} = \frac{8}{625}r^{10-1} = \frac{8}{625}r^9 \qquad \text{substituting in Eq. (19.5)}$$

$$r^9 = -\frac{(3125)(625)}{(64)(8)} \qquad \text{solving for } r$$

$$r = \left[-\frac{(3125)(625)}{(64)(8)}\right]^{1/9} = -2.5$$

EXAMPLE 4 Finding a specified term

Find the seventh term of the geometric sequence for which the second term is 3, the fourth term is 9, and $r > 0$.

To get from the second term to the fourth term, we must multiply by the common ratio twice. Thus,

$$9 = 3r^2, \qquad r = \sqrt{3} \qquad (\text{since } r > 0)$$

Now, to get from the fourth term to the seventh term, we must multiply by the common ratio three times. Therefore,

$$a_7 = 9r^3 = 9\left(\sqrt{3}\right)^3 = 27\sqrt{3}$$

Practice Exercise

1. Find the sixth term of the geometric sequence for which $a_1 = 81$ and $r = -1/3$.

EXAMPLE 5 Geometric sequence—chemical change

In an experiment, 22.0% of a substance changes chemically each 10.0 min. If there is originally 120 g of the substance, how much will remain after 45.0 min?

Let $a_n =$ the amount of the substance remaining after $n - 1$ periods of 10.0 minutes. From the statement of the problem, we know that $a_1 = 120\text{g}$. Moreover, since $(100 - 22.0)\% = 78.0\%$ remains after each 10.0 min period, $r = 0.780$. Therefore, $a_n = 120(0.780)^{n-1}$.

The number of 10.0-min periods in 45.0 min is $45.0/10.0 = 4.50$ periods. This means that the amount of substance after 45.0 min is

$$a_{5.50} = 120(0.780)^{4.50} = 39.2 \text{ g}$$

SUM OF *n* TERMS

A general expression for the sum S_n of the first n terms of a geometric sequence may be found by directly forming the sum and multiplying this equation by r:

$$S_n = a_1 + a_1r + a_1r^2 + \cdots + a_1r^{n-1}$$
$$rS_n = a_1r + a_1r^2 + a_1r^3 + \cdots + a_1r^n$$

If we now subtract the second of these equations from the first, we get $S_n - rS_n = a_1 - a_1r^n$. All other terms cancel by subtraction. Now, factoring S_n from the terms on the left and a_1 from the terms on the right, we solve for S_n. Thus, the **sum S_n of the first *n* terms of a geometric sequence** is

$$S_n = \frac{a_1(1 - r^n)}{1 - r} \qquad (r \neq 1) \tag{19.6}$$

EXAMPLE 6 Finding the sum of terms

Find the sum of the first seven terms of the geometric sequence for which the first term is 2 and the common ratio is $1/2$.

We are to find S_n, given that $a_1 = 2$, $r = \frac{1}{2}$, and $n = 7$. Using Eq. (19.6), we have

$$S_7 = \frac{2\left(1 - \left(\frac{1}{2}\right)^7\right)}{1 - \frac{1}{2}} = \frac{2\left(1 - \frac{1}{128}\right)}{\frac{1}{2}} = 4\left(\frac{127}{128}\right) = \frac{127}{32}$$

Practice Exercise

2. Find the sum of the six terms of the geometric sequence in Practice Exercise 1.

EXAMPLE 7 Finding the sum of terms—compound interest

■ See the chapter introduction and Example 2 from Section 19.1.

If \$100 is invested each year at 5% interest compounded annually, what would be the total amount of the investment after 10 years (before the 11th deposit is made)?

After one year, the amount invested will have added to it the interest for the year. Therefore, for the last (10th) \$100 invested, its value will become

$$\$100(1 + 0.05) = \$100(1.05) = \$105$$

The next to last \$100 will have interest added twice. After one year, its value becomes \$100(1.05), and after two years it is $\$100(1.05)(1.05) = \$100(1.05)^2$. In the same way, the value of the first \$100 becomes $\$100(1.05)^{10}$, since it will have interest added 10 times. This means that we are to find the sum of the sequence

$$\underset{\text{1 year in account}}{100(1.05)} + \underset{\text{2 years in account}}{100(1.05)^2} + \underset{\text{3 years in account}}{100(1.05)^3} + \cdots + \underset{\text{10 years in account}}{100(1.05)^{10}}$$

or

$$100\left[1.05 + (1.05)^2 + (1.05)^3 + \cdots + (1.05)^{10}\right]$$

For the sequence in the brackets, we have $a_1 = 1.05$, $r = 1.05$, and $n = 10$. Thus,

$$S_{10} = \frac{1.05\left[1 - (1.05)^{10}\right]}{1 - 1.05} = 13.2068$$

The total value of the \$100 investments is $100(13.2068) = \$1320.68$. We see that \$320.68 in interest has been earned.

EXERCISES 19.2

In Exercises 1 and 2, make the given changes in the indicated examples of this section and then solve the given problems.

1. In Example 4, change "seventh" to "10th."
2. In Example 6, change $\frac{1}{2}$ to $\frac{1}{3}$ and then find the sum.

In Exercises 3–6, write down the first five terms of the geometric sequence with the given values.

3. $a_1 = 6400,\ r = 0.25$
4. $a_1 = 0.09,\ r = -\frac{2}{3}$
5. $a_1 = \frac{1}{6},\ r = 3$
6. $a_1 = -3,\ r = 2$

In Exercises 7–14, find the nth term of the geometric sequence with the given values.

7. $\frac{1}{2}, 1, 2, \ldots; n = 9$
8. $10, 1, 0.1, \ldots; n = 8$
9. $125, -25, 5, \ldots; n = 7$
10. $0.1, 0.3, 0.9, \ldots; n = 5$
11. $a = -2700,\ r = -\frac{1}{3},\ n = 10$
12. $a_1 = 48,\ r = \frac{1}{2},\ n = 12$
13. $10^{100}, -10^{98}, 10^{96}, \ldots; n = 51$
14. $-2, 4k, -8k^2, \ldots; n = 6$

In Exercises 15–20, find the sum of the first n terms of the indicated geometric sequence with the given values.

15. $\frac{1}{8} + \frac{1}{2} + 2 + \cdots + 32$
16. $162 - 54 + 18 - \cdots - \frac{2}{3}$
17. $384, 192, 96, \ldots; n = 7$
18. $a_1 = 9,\ a_n = -243,\ n = 4$
19. $a_1 = 96,\ r = -\frac{k}{2},\ n = 10$
20. $\log 2, \log 4, \log 16, \ldots; n = 6$

In Exercises 21–28, find any of the values of a_1, r, a_n, n, or S_n that are missing.

21. $\frac{1}{16}, \frac{1}{4}, 1, \ldots, 64$
22. $5, 1, 0.2, \ldots, 0.00032$
23. $r = \frac{3}{2},\ n = 5,\ S_5 = 211$
24. $r = -\frac{1}{2},\ a_n = \frac{1}{8},\ n = 7$
25. $a_n = 27,\ n = 4,\ S_4 = 40$
26. $a_1 = 3,\ n = 7,\ a_7 = 192$
27. $a_1 = 75,\ r = \frac{1}{5},\ a_n = \frac{3}{625}$
28. $r = -2,\ n = 6,\ S_6 = 42$

In Exercises 29–56, find the indicated quantities.

29. Is $3, 3^{x+1}, 3^{2x+1}, \ldots$ a geometric sequence? Explain. If it is, find a_{20}.
30. Show that a, $\sqrt{ab}$, b are three successive terms of a geometric sequence ($a > 0, b > 0$).
31. Find x if $2, 6, 3x + 12$ are successive terms of a geometric sequence.
32. Write the first five terms of a geometric sequence of which the first two terms are a and b.
33. Using Eq. (19.6), find the sum of the terms of a geometric sequence if $r = -1$, and n is (a) odd, (b) even.
34. If a positive number c is added to each term of a geometric sequence, is the resulting sequence geometric?
35. Each stroke of a pump removes 8.2% of the remaining air from a container. What percent of the air remains after 40 strokes?
36. In 2009, the population of Canada was 33.7 million. According to a low-growth scenario, it is estimated that the population will increase at an average annual rate of 0.65% until 2036. What is the estimated 2036 population, according to this scenario?

37. An electric current decreases by 12.5% each 1.00 μs. If the initial current is 3.27 mA, what is the current after 8.20 μs?

38. A copying machine is set to reduce the dimensions of material copied by 10%. A drawing 17.0 cm wide is reduced, and then the copies are in turn reduced. What is the width of the drawing on the sixth reduction?

39. How much is an investment of \$250 worth after eight years if it earns annual interest of 1.2% compounded monthly? (1.2% annual interest compounded monthly means that 0.1% (1.2%/12) interest is added each month.)

40. A chemical spill pollutes a stream. A monitoring device finds 620 ppm (parts per million) of the chemical 1.0 km below the spill, and the readings decrease by 12.5% for each kilometre farther downstream. How far downstream is the reading 100 ppm?

41. Measurements show that the temperature of a distant star is presently 9800°C and is decreasing by 10% every 800 years. What will its temperature be in 4000 years?

42. The chlorine in a swimming pool was measured (in ppm, parts per million) to be 1.80, 1.53, 1.30, and 1.10 on four successive days. Noting that these values approximate a geometric sequence, what would be the reading three days after the last reading?

43. The strength of a signal in a fibre-optic cable decreases 12% for every 15 km along the cable. What percent of the signal remains after 100 km?

44. A series of deposits, each of value A and made at equal time intervals, earns an interest rate of i for the time interval. The deposits have a total value of

$$A(1 + i) + A(1 + i)^2 + A(1 + i)^3 + \cdots + A(1 + i)^n$$

after n time intervals (just before the next deposit). Find a formula for this sum.

45. A thermometer is removed from hot water at 100.0°C into a room at 20.0°C. The temperature difference D between the thermometer and the air decreases by 35.0% each minute. What is the temperature reading on the thermometer 10.0 min later?

46. The power on a space satellite is supplied by a radioactive isotope. On a given satellite, the power decreases by 0.2% each day. What percent of the initial power remains after one year?

47. If you decided to save money by putting away 1¢ on a given day, 2¢ one week later, 4¢ a week later, and so on, how much would you have to put away six months (26 weeks) after putting away the 1¢?

48. How many direct ancestors (parents, grandparents, and so on) does a person have in the 10 generations that preceded him or her (assuming that no ancestor appears in more than one line of descent)?

49. A professional hockey player is offered a contract for an annual salary of \$5 000 000 for six years. Also offered is a bonus (based on performance) of either \$400 000 each year, or a 5.00% increase in salary each year. Which bonus option pays more over the term of the contract, and how much more?

50. If 85% of an aspirin remains in the bloodstream after 3.0 h, how long after an 80-mg aspirin is taken does it take for there to be 35 mg in the bloodstream?

51. Derive a formula for S_n in terms of a_1, r, and a_n.

52. Write down several terms of a general geometric sequence. Then take the logarithm of each term. Explain why the resulting sequence is an arithmetic sequence.

53. Do the squares of the terms of a geometric sequence also form a geometric sequence? Explain.

54. If $a_1, a_2, a_3, \ldots$ is an arithmetic sequence, explain why $2^{a_1}, 2^{a_2}, 2^{a_3}, \ldots$ is a geometric sequence.

55. The numbers 8, x, y form an arithmetic sequence, and the numbers x, y, 36 form a geometric sequence. Find all of the possible sequences.

56. For $f(x) = ab^x$, with $b > 0, b \neq 1$, and $a \neq 0$, is $f(1), f(3), f(5)$ an arithmetic sequence or a geometric sequence? What is the common difference (or common ratio)?

Answers to Practice Exercises

1. $a_6 = -1/3$ 2. $S_6 = 182/3$

19.3 Infinite Geometric Series

In the previous sections, we developed formulas for the sum of the first n terms of an arithmetic sequence and of a geometric sequence. *The sum of the terms of a sequence is called a* **series**.

EXAMPLE 1 Illustrations of series

(a) The sum of the terms of the arithmetic sequence 2, 5, 8, 11, 14, . . . is the series $2 + 5 + 8 + 11 + 14 + \cdots$.

(b) The sum of the terms of the geometric sequence $1, \frac{1}{2}, \frac{1}{4}, \frac{1}{8}, \ldots$ is the series $1 + \frac{1}{2} + \frac{1}{4} + \frac{1}{8} + \cdots$.

The series associated with a finite sequence will sum up to a real number. The series associated with an infinite arithmetic sequence will not sum up to a real number, as the terms being added become larger and larger numerically. The sum is unbounded, as we can see in Example 1(a). The series associated with an infinite geometric sequence may or may not sum up to a real number, as we now show.

Let us consider the sum of the first n terms of the infinite geometric sequence $1, \frac{1}{2}, \frac{1}{4}, \ldots$. This is the sum of the n terms of the associated geometric series

$$1 + \frac{1}{2} + \frac{1}{4} + \cdots + \frac{1}{2^{n-1}}$$

Here, $a_1 = 1$ and $r = \frac{1}{2}$, and we find that we get the values of S_n for the given values of n in the following table:

n	2	3	4	5	6	7	8	9	10
S_n	$\frac{3}{2}$	$\frac{7}{4}$	$\frac{15}{8}$	$\frac{31}{16}$	$\frac{63}{32}$	$\frac{127}{64}$	$\frac{255}{128}$	$\frac{511}{256}$	$\frac{1023}{512}$

$1 + \frac{1}{2} + \frac{1}{4} + \frac{1}{8}$ (for $n = 4$)

$1 + \frac{1}{2} + \frac{1}{4} + \frac{1}{8} + \frac{1}{16} + \frac{1}{32} + \frac{1}{64}$ (for $n = 7$)

The series for $n = 4$ and $n = 7$ are shown. We see that as n gets larger, the numerator of each fraction comes closer to being twice the denominator. In fact, we find that if we continue to compute S_n as n becomes larger, S_n can be found as close to the value 2 as desired, although it will never actually reach the value 2. For example, if $n = 100$, $S_{100} = 2 - 1.6 \times 10^{-30}$, which could be written as

$$1.999\,999\,999\,999\,999\,999\,999\,999\,999\,998\,4$$

to 32 significant digits. For the sum of the first n terms of a geometric sequence

$$S_n = a_1\frac{1 - r^n}{1 - r}$$

the term r^n becomes exceedingly small if $|r| < 1$, and if n is sufficiently large, this term is effectively zero. *If this term were exactly zero*, the sum would be

$$S_n = 1\frac{1 - 0}{1 - \frac{1}{2}} = 2$$

The symbol ∞ for infinity was first used by the English mathematician John Wallis (1616–1703).

The only problem is that we cannot find any number large enough for n to make $\left(\frac{1}{2}\right)^n$ zero. There is, however, an accepted notation for this. This notation is

$$\lim_{n \to \infty} r^n = 0 \quad (\text{if } |r| < 1)$$

and it is read as "the limit, as n *approaches* infinity, of r to the nth power is zero."

COMMON ERROR

The symbol ∞ is read as **infinity,** ***but it must not be thought of as a number.*** It is simply a symbol that stands for a *process* of considering numbers that become large without bound. The number called the limit of the sums is simply the number the sums get closer and closer to, as n is considered to approach infinity. This notation and terminology are of particular importance in calculus.

If we consider values of r such that $|r| < 1$ and let the values of n become unbounded, we find that $\lim_{n \to \infty} r^n = 0$. The formula for the **sum of the terms of an infinite geometric series** then becomes

$$S = \frac{a_1}{1 - r} \quad (|r| < 1) \tag{19.7}$$

where a_1 is the first term and r is the common ratio. If $|r| \geq 1$, S is unbounded in value.

EXAMPLE 2 Sum of an infinite geometric series

Find the sum of the infinite geometric series

$$4 - \frac{1}{2} + \frac{1}{16} - \frac{1}{128} + \cdots$$

Here, we see that $a_1 = 4$. We find r by dividing any term by the previous term, and we find that $r = -\frac{1}{8}$. We then find the sum by substituting in Eq. (19.7). This gives us

$$S = \frac{4}{1 - \left(-\frac{1}{8}\right)} = \frac{4}{1 + \frac{1}{8}}$$
$$= \frac{4}{1} \times \frac{8}{9} = \frac{32}{9}$$

Practice Exercise

1. Find the sum of the infinite geometric series $9 + 3 + 1 + 1/3 + \ldots$.

EXAMPLE 3 Writing a repeating decimal as a fraction

Find the fraction that has as its decimal form 0.121 212 . . . .

This decimal form can be considered as being

$$0.12 + 0.0012 + 0.000\,012 + \cdots$$

which means that we have an infinite geometric series in which $a_1 = 0.12$ and $r = 0.01$. Thus,

$$S = \frac{0.12}{1 - 0.01} = \frac{0.12}{0.99}$$
$$= \frac{4}{33}$$

Therefore, the decimal 0.121 212 . . . and the fraction $\frac{4}{33}$ represent the same number.

Practice Exercise

2. Find the fraction that has as its decimal form 0.272727 . . . .

The decimal in Example 3 is called a **repeating decimal** because *a particular sequence of digits in the decimal form repeats endlessly.* This example verifies the theorem that any repeating decimal represents a rational number. However, not all repeating decimals start repeating immediately. If the numbers never do repeat, the decimal represents an irrational number. For example, there are no repeating decimals that represent π, $\sqrt{2}$, or e. The decimal form of the number e does repeat at one point, but the repetition stops. As we noted in Section 12.5, the decimal form of e to 16 decimal places is 2.7 1828 1828 4590 452. We see that the sequence of digits 1828 repeats only once.

EXAMPLE 4 Repeating decimal

■ A calculator can be used to change a repeating decimal to a fraction.

Find the fraction that has as its decimal form the repeating decimal 0.503 453 453 45 . . . .

We first separate the decimal into the beginning, nonrepeating part, and the infinite repeating decimal, which follows. Thus, we have

$$0.503\,453\,453\,45\ldots = 0.50 + 0.003\,453\,453\,45\ldots$$

This means that we are to add $\frac{50}{100}$ to the fraction that represents the sum of the terms of the infinite geometric series $0.003\,45 + 0.000\,003\,45 + \cdots$. For this series, $a_1 = 0.003\,45$ and $r = 0.001$. We find this sum to be

$$S = \frac{0.003\,45}{1 - 0.001} = \frac{0.003\,45}{0.999} = \frac{115}{33\,300} = \frac{23}{6660}$$

Therefore,

$$0.503\,453\,45\ldots = \frac{5}{10} + \frac{23}{6660} = \frac{5(666) + 23}{6660} = \frac{3353}{6660}$$

EXAMPLE 5 Infinite geometric series—pendulum distance

Each swing of a certain pendulum bob is 95% as long as the previous swing. How far does the bob travel in coming to rest if the first swing is 40.0 cm long?

We are to find the sum of the terms of an infinite geometric series for which $a_1 = 40.0$ and $r = 95\% = \frac{19}{20}$. Substituting these values into Eq. (19.7), we obtain

$$S = \frac{40.0}{1 - \frac{19}{20}} = \frac{40.0}{\frac{1}{20}} = (40.0)(20) = 800 \text{ cm}$$

The pendulum bob travels 800 cm in coming to rest.

Practice Exercise

3. In Example 5, how far does the bob travel if each swing is 99% as long as the previous swing?

EXERCISES 19.3

In Exercises 1 and 2, make the given changes in the indicated examples of this section, and then solve the resulting problems.

1. In Example 2, change all − signs to + in the series.
2. In Example 3, change the decimal form to 0.012012012 . . . .

In Exercises 3–6, find the indicated quantity for an infinite geometric series.

3. $a_1 = 4,\ r = \frac{1}{2},\ S = ?$
4. $a_1 = 68,\ r = -\frac{1}{3},\ S = ?$
5. $a_1 = 0.5,\ S = 0.625,\ r = ?$
6. $S = 4 + 2\sqrt{2},\ r = \frac{1}{\sqrt{2}},\ a_1 = ?$

In Exercises 7–14, find the sums of the given infinite geometric series.

7. $20 - 1 + 0.05 - \cdots$
8. $9 + 8.1 + 7.29 + \cdots$
9. $1 + \frac{5}{8} + \frac{25}{64} + \cdots$
10. $5 - 3 + \frac{9}{5} - \cdots$
11. $1 + 10^{-4} + 10^{-8} + \cdots$
12. $1000 - 300 + 90 - \cdots$
13. $(2 + \sqrt{3}) + 1 + (2 - \sqrt{3}) + \cdots$
14. $(1 + \sqrt{2}) - 1 + (\sqrt{2} - 1) - \cdots$

In Exercises 15–28, find the fractions equal to the given decimals.

15. 0.333 33 . . .
16. 0.272 727 . . .
17. 0.499 999 . . .
18. 0.999 999 . . .
19. 0.606 060 . . .
20. 0.080 808 . . .
21. 0.181 818 . . .
22. 0.336 336 336 . . .
23. 0.027 327 327 3 . . .
24. 0.822 22 . . .
25. 0.366 666 . . .
26. 0.664 242 42 . . .
27. 0.100 841 841 841 . . .
28. 0.184 561 845 618 456 . . .

In Exercises 29–39, solve the given problems by use of the sum of an infinite geometric series.

29. There are two infinite geometric series with $a_1 = 50$ and $a_3 = 2$. Find the sum of each.
30. Explain why there is no infinite geometric series with $a_1 = 5$ and $S = 2$.
31. Liquid is continuously collected in a wastewater-holding tank such that during a given hour only 92.0% as much liquid is collected as in the previous hour. If 28.0 L are collected in the first hour, what must be the minimum capacity of the tank?
32. If 70% of all aluminum cans are recycled, find the total number of recycled cans that can be made from the 325 million beverage aluminum cans that are recycled in one year in British Columbia. Assume cans are recycled over and over until all the aluminum is used up, and no aluminum is lost in the recycling process.
33. The amounts of plutonium-237 that decay each day because of radioactivity form a geometric sequence. Given that the amounts that decay during each of the first four days are 5.882 g, 5.782 g, 5.684 g, and 5.587 g, respectively, what total amount will decay?
34. A helium-filled balloon rose 36.0 m in 1.0 min. Each minute after that, it rose 75% as much as in the previous minute. What was its maximum height?
35. A bicyclist travelling at 10 m/s coasts to a stop as the bicycle travels 0.90 as far each second as in the previous second. How far does the bicycle travel in coasting to a stop?
36. A square has sides of 20 cm. Another square is inscribed in the first square by joining the midpoints of the sides. Assuming that such inscribed squares can be formed endlessly, find the sum of the areas of all the squares and explain how the sum is found. See Fig. 19.2.

Fig. 19.2

37. Find x if the sum of the terms of the infinite geometric series $1 + 2x + 4x^2 + \ldots$ is $2/3$.
38. Find the sum of the terms of the infinite series $1 + 2x + 3x^2 + 4x^3 + \ldots$ for $|x| < 1$. (*Hint:* Use $S - xS$.)
39. In a "torture test," a light switch is turned on and off until it fails. For a certain switch, the probability that the switch will fail after it has been turned on or off 1000 times is given by

$$(0.999)^{1000}(0.001) + (0.999)^{1001}(0.001) + (0.999)^{1002}(0.001) + \ldots$$

Find this probability.

Answers to Practice Exercises

1. $S = 27/2$ 2. $3/11$ 3. 4000 cm

19.4 The Binomial Theorem

We end this chapter by studying another useful and important series, the **binomial series**. The binomial series is the infinite series generalization of the **binomial formula**, which we develop first. The binomial formula allows us to expand binomials to any positive integer power without direct multiplication. It is used in technical applications and in many areas of mathematics.

By direct multiplication, we may obtain the following expansions of the binomial $a + b$:

$$(a+b)^0 = 1$$
$$(a+b)^1 = a + b$$
$$(a+b)^2 = a^2 + 2ab + b^2$$
$$(a+b)^3 = a^3 + 3a^2b + 3ab^2 + b^3$$
$$(a+b)^4 = a^4 + 4a^3b + 6a^2b^2 + 4ab^3 + b^4$$
$$(a+b)^5 = a^5 + 5a^4b + 10a^3b^2 + 10a^2b^3 + 5ab^4 + b^5$$

Inspection shows that these expansions have certain properties, and it can be shown that these properties are valid for the expansion of $(a + b)^n$, where n is any positive integer.

Properties of the Binomial $(a + b)^n$	EXAMPLE 1
	Expand $(a + b)^5$.
1. There are $n + 1$ terms.	Here $n = 5$, so there are six terms.
2. The first term is a^n, and the last term is b^n.	The first term is a^5, and the last term is b^5.
3. Progressing one term at a time from the first term to the last, the exponent of a decreases by 1, the exponent of b increases by 1, and the sum of both exponents is n.	Variables in term 2: a^4b Variables in term 3: a^3b^2 Variables in term 4: a^2b^3 Variables in term 5: ab^4
4. The coefficients of terms equidistant from the ends are equal.	We only need to obtain the coefficients for terms 2 and 3 because the coefficients of terms 4 and 5 are the same as those of terms 3 and 2, respectively.
5. If the coefficient of any term is multiplied by the exponent of a in that term and then divided by the number of that term, we obtain the coefficient of the next term.	Term 1: a^5 (coefficient is 1, exponent of a is 5). The coefficient of term 2 is $\frac{1 \times 5}{1} = 5$ (and so is the coefficient of term 4). Term 2: $5a^4b$ (coefficient is 5, exponent of a is 4). The coefficient of term 3 is $\frac{5 \times 4}{2} = 10$ (and so is the coefficient of term 5). Putting it all together, $(a + b)^5 = a^5 + 5a^4b + 10a^3b^2 + 10a^2b^3 + 5ab^4 + b^5$.

It is not necessary to use the above properties directly to expand a given binomial. If they are applied to $(a + b)^n$, a general formula for the expansion of a binomial may be obtained. In developing and stating the general formula, it is convenient to use the **factorial notation $n!$**, where

$$n! = n(n-1)(n-2)\cdots(2)(1) \quad \textbf{(19.8)}$$

We see that $n!$, read "n factorial," represents the product of the first n positive integers. We also define $0! = 1$. (See Exercise 39.)

EXAMPLE 2 Evaluating factorials

(a) $3! = (3)(2)(1) = 6$

(b) $5! = (5)(4)(3)(2)(1) = 120$

(c) $8! = (8)(7)(6)(5)(4)(3)(2)(1) = 40\,320$

(d) $3! + 5! = 6 + 120 = 126$

(e) $\dfrac{4!}{2!} = \dfrac{(4)(3)(2)(1)}{(2)(1)} = 12$

Practice Exercise

1. Evaluate $9!/7!$.

COMMON ERROR In evaluating factorials, we must remember that they represent products of numbers. In parts (d) and (e) of Example 2, we see that

$3! + 5!$ is not $(3 + 5)!$ and $4!/2!$ is not $(4/2)!$.

THE BINOMIAL FORMULA

Based on the binomial properties, *the* **binomial theorem** *states that the following* **binomial formula** *is valid for all positive integer values of n* (the binomial theorem is proven through advanced methods).

$$(a + b)^n = a^n + na^{n-1}b + \frac{n(n-1)}{2!}a^{n-2}b^2 + \frac{n(n-1)(n-2)}{3!}a^{n-3}b^3 + \cdots + b^n \qquad \textbf{(19.9)}$$

EXAMPLE 3 Using the binomial formula

Using the binomial formula, expand $(2x + 3)^6$.

In using the binomial formula for $(2x + 3)^6$, we use $2x$ for a, 3 for b, and 6 for n. Thus,

$$(2x+3)^6 = (2x)^6 + 6(2x)^5(3) + \frac{(6)(5)}{2}(2x)^4(3)^2 + \frac{(6)(5)(4)}{(2)(3)}(2x)^3(3)^3 + \frac{(6)(5)(4)(3)}{(2)(3)(4)}(2x)^2(3)^4 + \frac{(6)(5)(4)(3)(2)}{(2)(3)(4)(5)}(2x)(3)^5 + 3^6$$

(annotations: $n = 6$, $n - 1$, n, $n - 1$, $n - 2$ above; $2!$, $3!$, $4!$, $5!$ below the denominators)

$$= 64x^6 + 576x^5 + 2160x^4 + 4320x^3 + 4860x^2 + 2916x + 729$$

Pascal's Triangle

For the first few integral powers of a binomial $a + b$, the coefficients (called the **binomial** coefficients) can be obtained by setting them up in the following pattern, known as **Pascal's triangle**.

■ Pascal's triangle is named after the French scientist, mathematician, and philosopher Blaise Pascal (1623–1662). However, it was already known in China and in medieval Islam during the eleventh century.

$n = 0$	1
$n = 1$	1 1
$n = 2$	1 2 1
$n = 3$	1 3 3 1
$n = 4$	1 4 6 4 1
$n = 5$	1 5 10 10 5 1
$n = 6$	1 6 15 20 15 6 1

see expansions at the beginning of Section 19.4

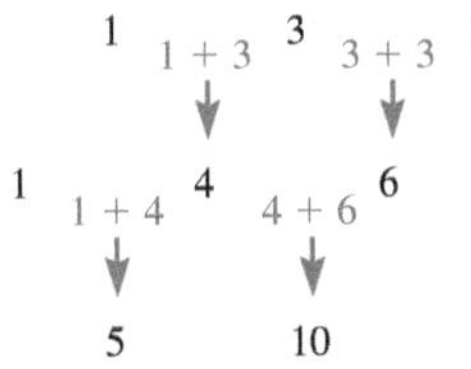

Fig. 19.3

We note that the first and last coefficients shown in each row are 1, and the second and next-to-last coefficients are equal to n. Other coefficients are obtained by adding the two nearest coefficients in the row above, as illustrated in Fig. 19.3 for the indicated section of Pascal's triangle. This pattern may be continued indefinitely, although use of Pascal's triangle is cumbersome for high values of n.

EXAMPLE 4 Using Pascal's triangle

Using Pascal's triangle, expand $(5s - 2t)^4$.

Here, we note that $n = 4$. Thus, the coefficients of the five terms are 1, 4, 6, 4, and 1, respectively. Also, here we use $5s$ for a and $-2t$ for b. We are expanding this expression as $[(5s) + (-2t)]^4$. Therefore, using Pascal's triangle for $n = 4$,

$$\begin{aligned}(5s - 2t)^4 &= 1(5s)^4 + 4(5s)^3(-2t) + 6(5s)^2(-2t)^2 + 4(5s)(-2t)^3 + 1(-2t)^4 \\ &= 625s^4 - 1000s^3t + 600s^2t^2 - 160st^3 + 16t^4\end{aligned}$$

In certain uses of a binomial expansion, it is not necessary to obtain all terms. Only the first few terms are required. The following example illustrates finding the first four terms of an expansion.

EXAMPLE 5 Using the binomial formula

Find the first four terms of the expansion of $(x + 7)^{12}$.

Here, we use x for a, 7 for b, and 12 for n. Thus, from the binomial formula, we have

$$\begin{aligned}(x + 7)^{12} &= x^{12} + 12x^{11}(7) + \frac{(12)(11)}{2}x^{10}(7)^2 + \frac{(12)(11)(10)}{(2)(3)}x^9(7)^3 + \cdots \\ &= x^{12} + 84x^{11} + 3234x^{10} + 75\,460x^9 + \cdots\end{aligned}$$

Practice Exercise

2. Find the first three terms of the expansion of $(x - 4)^9$.

If we let $a = 1$ and $b = x$ in the binomial formula, we obtain the **binomial series**

Binomial Series

$$(1 + x)^n = 1 + nx + \frac{n(n-1)}{2!}x^2 + \frac{n(n-1)(n-2)}{3!}x^3 + \cdots \quad \textbf{(19.10)}$$

which, through advanced methods, can be shown to be valid for any real number n if $|x| < 1$. When n is either negative or a fraction, we obtain an infinite series. In such a case, we calculate as many terms as may be needed, although such a series is not obtainable through direct multiplication. The binomial series may be used to develop important expressions that are used in applications and more advanced mathematics topics.

EXAMPLE 6 Binomial series—beam analysis

In the analysis of forces on beams, the expression $1/(1 + m^2)^{3/2}$ is used. Use the binomial series to find the first four terms of the expansion.

Using negative exponents, we have

$$1/(1 + m^2)^{3/2} = (1 + m^2)^{-3/2}$$

Now, in using Eq. (19.10), we have $n = -3/2$ and $x = m^2$:

$$(1 + m^2)^{-3/2} = 1 + \left(-\frac{3}{2}\right)(m^2) + \frac{\left(-\frac{3}{2}\right)\left(-\frac{3}{2} - 1\right)}{2!}(m^2)^2 + \frac{\left(-\frac{3}{2}\right)\left(-\frac{3}{2} - 1\right)\left(-\frac{3}{2} - 2\right)}{3!}(m^2)^3 + \cdots$$

Therefore,

$$\frac{1}{(1 + m^2)^{3/2}} = 1 - \frac{3}{2}m^2 + \frac{15}{8}m^4 - \frac{35}{16}m^6 + \cdots$$

EXAMPLE 7 Evaluation using the binomial series

Approximate the value of 0.97^7 by use of the binomial series.

We note that $0.97 = 1 - 0.03$, which means $0.97^7 = [1 + (-0.03)]^7$. Using four terms of the binomial series, we have

$$\begin{aligned} 0.97^7 &= [1 + (-0.03)]^7 \\ &= 1 + 7(-0.03) + \frac{7(6)}{2!}(-0.03)^2 + \frac{7(6)(5)}{3!}(-0.03)^3 \\ &= 1 - 0.21 + 0.0189 - 0.000\,945 = 0.807\,955 \end{aligned}$$

From a calculator, we find that $0.97^7 = 0.807\,983$ (to six decimal places), which means these values agree to four significant digits with a value 0.8080. Greater accuracy can be found using the binomial series if more terms are used.

EXERCISES 19.4

In Exercises 1 and 2, make the given changes in the indicated examples of this section and then solve the given problems.

1. In Example 3, change the exponent from 6 to 5.
2. In Example 7, change 0.97 to 0.98.

In Exercises 3–12, expand and simplify the given expressions by use of the binomial formula.

3. $(t + 4)^3$
4. $(x - 2)^3$
5. $(3x - 1)^4$
6. $(x^2 + 7)^4$
7. $(6 + 0.1)^5$
8. $(xy - z)^5$
9. $(n + 2\pi)^5$
10. $(1 - j)^6$ $(j = \sqrt{-1})$
11. $(2a - b^2)^6$
12. $\left(\frac{a}{x} + x\right)^6$

In Exercises 13–16, expand and simplify the given expressions by use of Pascal's triangle.

13. $(5x - 3)^4$
14. $(x - 4)^5$
15. $(2a + 1)^6$
16. $(x - 3)^7$

In Exercises 17–24, find the first four terms of the indicated expansions.

17. $(x + 2)^{10}$
18. $(x - 4)^8$
19. $(2a - 1)^7$
20. $(3b + 2)^9$
21. $(x^{1/2} - 4y)^{12}$
22. $(2a - x^{-1})^{11}$
23. $\left(b^2 + \frac{1}{2b}\right)^{20}$
24. $\left(2x^2 + \frac{y}{3}\right)^{15}$

In Exercises 25–28, approximate the value of the given expression to three decimal places by using three terms of the appropriate binomial series. Check using a calculator.

25. 1.04^6
26. $\sqrt{0.927}$
27. $\sqrt[3]{1.045}$
28. 0.99^{-8}

In Exercises 29–36, find the first four terms of the indicated expansions by use of the binomial series.

29. $(1 + x)^8$
30. $(1 + x)^{-1/3}$
31. $(1 - 3x)^{-2}$
32. $(1 - 2\sqrt{x})^9$
33. $\sqrt{1 + x}$
34. $\frac{1}{\sqrt{1 - x}}$
35. $\frac{1}{\sqrt{9 - 9x}}$
36. $\sqrt{4 + x^2}$

In Exercises 37–40, solve the given problems involving factorials.

37. Using a calculator, evaluate (a) $17! + 4!$, (b) $21!$, (c) $17! \times 4!$, and (d) $68!$.
38. Using a calculator, evaluate (a) $8! - 7!$, (b) $8!/7!$, (c) $8! \times 7!$, and (d) $56!$.
39. Show that $n! = n \times (n - 1)!$ for $n \geq 2$. To use this equation for $n = 1$, explain why it is necessary to define $0! = 1$.
40. Show that $\frac{(n + 1)!}{(n - 2)!} = n^3 - n$ for $n \geq 2$. See Exercise 39.

In Exercises 41–44, find the indicated terms by use of the following information. The $r + 1$ term of the expansion of $(a + b)^n$ is given by

$$\frac{n(n - 1)(n - 2)\cdots(n - r + 1)}{r!}a^{n-r}b^r$$

41. The term involving b^5 in $(a + b)^8$
42. The term involving y^6 in $(x + y)^{10}$
43. The fifth term of $(2x - 3b)^{12}$
44. The sixth term of $(\sqrt{a} - \sqrt{b})^{14}$

In Exercises 45–58, solve the given problems.

45. Explain why $n!$ ends in a zero if $n > 4$.
46. Expand $[(a + b) + c]^3$ using the binomial theorem. Group terms as indicated.
47. In the expansion of $(a - x)^n$, where n is a positive integer, show that the sum of the coefficients is zero.
48. If x is very small, show that $(1 + x)^{-n}$ is approximately $(1 - nx)$.
49. A mechanical system has four redundant engines, each with a 95% probability of functioning properly. When the binomial theorem is used to expand $(0.95 + 0.05)^4$, the resulting terms give the probabilities of exactly four, three, two, one, and zero engines functioning properly. Find the probability of exactly three engines functioning properly (the term containing 0.95^3). Round to the third decimal place.
50. In Exercise 49, find the probability that *at least one* engine functions properly (the sum of the terms containing 0.95^4, 0.95^3, 0.95^2, and 0.95^1). Round to the third decimal place.
51. Approximate $\sqrt{6}$ to hundredths by noting that $\sqrt{6} = \sqrt{4(1.5)} = 2\sqrt{1 + 0.5}$ and using four terms of the appropriate binomial series.
52. Approximate $\sqrt[3]{10}$ by using the method of Exercise 51.

53. A company purchases a piece of equipment for A dollars, and the equipment depreciates at a rate of r each year. Its value V after n years is $V = A(1 - r)^n$. Expand this expression for $n = 5$.

54. In finding the rate of change of emission of energy from the surface of a body at temperature T, the expression $(T + h)^4$ is used. Expand this expression.

55. In the theory associated with the magnetic field due to an electric current, the expression $1 - \dfrac{x}{\sqrt{a^2 + x^2}}$ is found. By expanding $(a^2 + x^2)^{-1/2}$, find the first three nonzero terms that could be used to approximate the given expression.

56. In the theory related to the dispersion of light, the expression $1 + \dfrac{A}{1 - \lambda_0^2/\lambda^2}$ arises. (a) Find the first four terms of the expansion of $(1 - x)^{-1}$. (b) Find the same expansion by using long division. (c) Let $x = \lambda_0^2/\lambda^2$ and write the original expression in expanded form, using the results of (a) and (b).

57. A cubic Bezier curve is defined by four points (P_0 through P_3) from the parametric equations given by $(1 - t)^3P_0 + 3(1 - t)^2tP_1 + 3(1 - t)t^2P_2 + t^3P_3$, where $0 \le t \le 1$. Note that the points P_0 through P_3 are multiplied by terms of the binomial expansion $[(1 - t) + t]^3$. Using a similar process, find the equation of the fourth-degree Bezier curve defined by the five points P_0 through P_4 (see Example 6 in Section 15.3).

58. Find the first four terms of the expansion of $(1 + x)^{-1}$ and then divide $1 + x$ into 1. Compare the results.

Answers to Practice Exercises

1. 72 **2.** $x^9 - 36x^8 + 576x^7 - \cdots$

CHAPTER 19 KEY FORMULAS AND EQUATIONS

Arithmetic sequences

Recursion formula $a_n = a_{n-1} + d$ (19.1)

nth term $a_n = a_1 + (n - 1)d$ (19.2)

Sum of n terms $S_n = \dfrac{n}{2}(a_1 + a_n)$ (19.3)

Geometric sequences

Recursion formula $a_n = ra_{n-1}$ (19.4)

nth term $a_n = a_1r^{n-1}$ (19.5)

Sum of n terms $S_n = \dfrac{a_1(1 - r^n)}{1 - r}$ $(r \ne 1)$ (19.6)

Sum of a geometric series $S = \dfrac{a_1}{1 - r}$ $(|r| < 1)$ (19.7)

Factorial notation $n! = n(n - 1)(n - 2)\cdots(2)(1)$ (19.8)

Binomial formula $(a + b)^n = a^n + na^{n-1}b + \dfrac{n(n - 1)}{2!}a^{n-2}b^2 + \dfrac{n(n - 1)(n - 2)}{3!}a^{n-3}b^3 + \cdots + b^n$ (19.9)

Binomial series $(1 + x)^n = 1 + nx + \dfrac{n(n - 1)}{2!}x^2 + \dfrac{n(n - 1)(n - 2)}{3!}x^3 + \cdots$ (19.10)

CHAPTER 19 REVIEW EXERCISES

In Exercises 1–8, find the indicated term of each sequence.

1. $1, 6, 11, \ldots$ (17th)
2. $1, -3, -7, \ldots$ (21st)
3. $500, 100, 20, \ldots$ (9th)
4. $0.025, 0.01, 0.004, \ldots$ (7th)
5. $-1, 3.5, 8, \ldots$ (25th)
6. $-1, -\frac{5}{3}, -\frac{7}{3}, \ldots$ (16th)
7. $\frac{3}{4}, \frac{1}{2}, \frac{1}{3}, \ldots$ (7th)
8. $5^{-2}, 5^0, 5^2, \ldots$ (7th)

In Exercises 9–12, find the sum of each sequence with the indicated values.

9. $a_1 = -4,\ n = 15,\ a_{15} = 17$ (arith.)
10. $a_1 = 300,\ d = -\frac{20}{3},\ n = 10$
11. $a_1 = 16,\ r = -\frac{1}{2},\ n = 14$
12. $a_1 = 64,\ a_n = 729,\ n = 7$ (geom., $r > 0$)

In Exercises 13–24, find the indicated quantities for the appropriate sequences.

13. $a_1 = 17,\ d = -2,\ n = 9,\ S_9 = ?$

14. $d = \frac{4}{3},\ a_1 = -3,\ a_n = 17,\ n = ?$

15. $a_1 = 4,\ r = \sqrt{2},\ n = 7,\ a_7 = ?$

16. $a_n = \frac{49}{8},\ r = -\frac{2}{7},\ S_n = \frac{1911}{32},\ a_1 = ?$

17. $a_1 = 80,\ a_n = -25,\ S_n = 220,\ d = ?$

18. $a_1 = 2,\ d = 0.2,\ n = 11,\ S_{11} = ?$

19. $n = 5,\ r = -0.25,\ S_5 = 205,\ a_5 = ?$

20. $a_1 = 9,\ r = 0.1,\ S_n = 9.9999,\ n = ?$

21. $a_1 = -1,\ a_n = 32,\ n = 12,\ S_{12} = ?$ (arith.)

22. $a_1 = 100,\ a_n = 6400,\ S_n = 32\,500,\ n = ?$ (arith.)

23. $a_1 = 1,\ n = 7,\ a_7 = 64,\ S_7 = ?$

24. $a_1 = \frac{1}{4},\ n = 6,\ a_6 = 8,\ S_6 = ?$

In Exercises 25–28, find the sums of the given infinite geometric series.

25. $0.9 + 0.6 + 0.4 + \cdots$

26. $1280 - 320 + 80 - \cdots$

27. $1 + 1.02^{-1} + 1.02^{-2} + \cdots$

28. $3 - \sqrt{3} + 1 - \cdots$

In Exercises 29–32, find the fractions equal to the given decimals.

29. 0.030 303 . . .

30. 0.636 636 . . .

31. 0.027 272 7 . . .

32. 0.253 993 993 99 . . .

In Exercises 33–36, expand and simplify the given expression. In Exercises 37–40, find the first four terms of the appropriate expansion.

33. $(x - 2)^4$

34. $(3 + 0.1)^4$

35. $(x^2 + 4)^5$

36. $(3n^{1/2} - a)^6$

37. $(a + 2e)^{10}$

38. $\left(\frac{x}{4} - y\right)^{12}$

39. $\left(p^2 - \frac{q}{6}\right)^9$

40. $\left(2s^2 - \frac{3}{2}t^{-1}\right)^{14}$

In Exercises 41–48, find the first four terms of the indicated expansions by use of the binomial series.

41. $(1 + x)^{12}$

42. $(1 - 2x)^{10}$

43. $\sqrt{1 + x^2}$

44. $\left(4 - 4\sqrt{x}\right)^{-1}$

45. $\sqrt{1 - a^2}$

46. $\sqrt{1 + 2b^4}$

47. $(2 - 4x)^{-3}$

48. $(1 + 4x)^{-1/4}$

In Exercises 49–93, solve the given problems by use of an appropriate sequence or expansion. All numbers are accurate to at least two significant digits.

49. Find the sum of the first 1000 positive even integers.

50. How many integers divisible by 4 lie between 23 and 121?

51. What is the fifth term of the arithmetic sequence in which the first term is a and the second term is b?

52. Find three consecutive numbers in an arithmetic sequence such that their sum is 15 and the sum of their squares is 77.

53. For a geometric sequence, is it possible that $a_3 = 6$, $a_5 = 9$, and $a_7 = 12$?

54. Find three consecutive numbers in a geometric sequence such that their product is 64 and the sum of their squares is 84.

55. Find a formula with variable n of the arithmetic sequence with $a_1 = -5,\ a_{n+1} = a_n - 3$, for $n = 1, 2, 3, \ldots$.

56. Find the ninth term of the sequence $(a + 2b), b, -a, \ldots$.

57. Approximate the value of $(1.06)^{-6}$ by using three terms of the appropriate binomial series. Check using a calculator.

58. Approximate the value of $\sqrt[4]{0.94}$ by using three terms of the appropriate binomial series. Check using a calculator.

59. Approximate the value of $\sqrt{30}$ by noting that $\sqrt{30} = \sqrt{25(1.2)} = 5\sqrt{1 + 0.2}$ and using three terms of the appropriate binomial series.

60. Approximate $\sqrt[3]{29.7}$ by using the *method* of Exercise 59.

61. Each stroke of a pile driver moves a post 2 cm less than the previous stroke. If the first stroke moves the post 24 cm, which stroke moves the post 4 cm?

62. During each hour, an exhaust fan removes 15.0% of the carbon dioxide present in the air in a room at the beginning of the hour. What percent of the carbon dioxide remains after 10.0 h?

63. Each 1.0 mm of a filter through which light passes reduces the intensity of the light by 12%. How thick should the filter be to reduce the intensity of the light to 20%?

64. A pile of dirt and 10 holes are in a straight line. It is 20 m from the dirt pile to the nearest hole, and the holes are 8 m apart. If a backhoe takes two trips to fill each hole, how far must it travel in filling all the holes if it starts and ends at the dirt pile?

65. A roof support with equally spaced vertical pieces is shown in Fig. 19.4. Find the total length of the vertical pieces if the shortest one is 254 mm long.

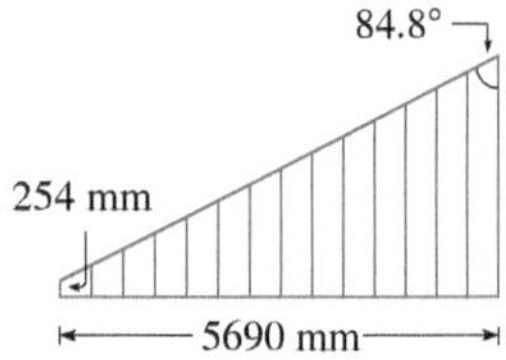

Fig. 19.4

66. During each microsecond, the current in an electric circuit decreases by 9.3%. If the initial current is 2.45 mA, how long does it take to reach 0.50 mA?

67. A machine that costs \$8600 depreciates 1.0% in value each month. What is its value five years after it was purchased?

68. The level of chemical pollution in a lake is 4.50 ppb (parts per billion). If the level increases by 0.20 ppb in the following month and by 5.0% less each month thereafter, what will be the maximum level?

69. A piece of paper 0.015 cm thick is cut in half. These two pieces are then placed one on the other and cut in half. If this is repeated such that the paper is cut in half 40 times, how high will the pile be?

70. After the power is turned off, an object on a nearly frictionless surface slows down such that it travels 99.9% as far during 1 s as during the previous second. If it travels 100 cm during the first second after the power is turned off, how far does it travel while stopping?

71. Under gravity, an object falls 4.9 m during the first second, 14.7 m during the second second, 24.5 m during the third second, and so on. How far will it fall during the 20th second?

72. For the object in Exercise 71, what is the total distance fallen during the first 20 s?

73. An object suspended on a spring is oscillating up and down. If the first oscillation is 10.0 cm and each oscillation thereafter is 9/10 of the preceding one, find the total distance the object travels in coming to rest.

74. During each oscillation, a pendulum swings through 85% of the distance of the previous oscillation. If the pendulum swings through 80.8 cm in the first oscillation, through what total distance does it move in 12 oscillations?

75. A person invests \$1000 each year at the beginning of the year. What is the total value of these investments after 20 years if they earn 7.5% annual interest, compounded semiannually?

76. A well driller charges \$10.00 for drilling the first metre of a well and for each metre thereafter charges 0.20% more than for the preceding metre. How much is charged for drilling a 150-m well?

77. When new, an article cost \$250. It was then sold at successive yard sales at 40% of the previous price. What was the price at the fourth yard sale?

78. Each side of an equilateral triangle is 2 cm in length. The midpoints are joined to form a second equilateral triangle. The midpoints of the second triangle are joined to form a third equilateral triangle. Find the sum of all of the perimeters of the triangles if this process is continued indefinitely. See Fig. 19.5.

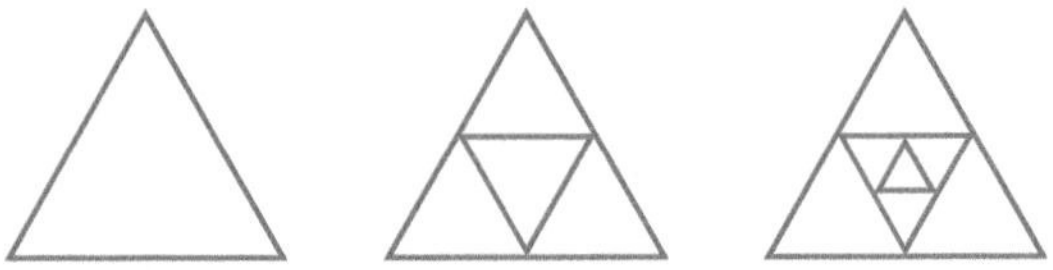

Fig. 19.5

79. In testing a type of insulation, the temperature in a room was made to fall to $2/3$ of the initial temperature after 1.0 h, to $2/5$ of the initial temperature after 2.0 h, to $2/7$ of the initial temperature after 3.0 h, and so on. If the initial temperature was 50.0°C, what was the temperature after 12.0 h? (This is an illustration of a *harmonic sequence.*)

80. Two competing businesses make the same item and sell it initially for \$100. One increases the price by \$8 each year for five years, and the other increases the price by 8% each year for five years. What is the difference in price after five years?

81. In hydrodynamics, while studying compressible fluid flow, the expression $\left(1 + \frac{a-1}{2}m^2\right)^{a/(a-1)}$ arises. Find the first three terms of the expansion of this expression.

82. The maximum tension T of a suspension bridge cable is given by

$$T = \frac{wL}{2}\sqrt{1 + \left(\frac{L}{4s}\right)^2}$$

Find the first three terms of the expansion of this expression.

83. In finding the partial pressure P_F of fluorine gas under certain conditions, the equation

$$P_F = \frac{(1 + 2 \times 10^{-10}) - \sqrt{1 + 4 \times 10^{-10}}}{2} \text{ atm}$$

is found. By using three terms of the expansion for $\sqrt{1+x}$, approximate the value of this expression.

84. On a highway with a steep incline, the runaway truck ramp is constructed so that a vehicle that has lost its brakes can stop. The ramp is designed to slow a truck in succeeding 20-m distances by 10 km/h, 12 km/h, 14 km/h, ... If the ramp is 160 m long, will it stop a truck moving at 120 km/h when it reaches the ramp?

85. Each application of an insecticide destroys 75% of a certain insect population. How many applications are needed to destroy at least 99.9% of the insects?

86. A house valued at \$375 000 increases in value 5.00%/year for the next six years and then decreases 3.00%/year for the following four years. What is its value at the end of the 10 years?

87. Oil pumped from a certain oil field decreases 10% each year. How long will it take for production to be 10% of the first year's production?

88. A wire hung between two poles is parabolic in shape. To find the length of wire between two points on the wire, the expression $\sqrt{1 + 0.08x^2}$ is used. Find the first three terms of the binomial expansion of this expression.

89. Show that the middle term of a finite arithmetic sequence equals S/n, if n is odd.

90. The sum of three terms of a geometric sequence is -3, and the second is 6 more than the first. Find these terms.

91. Do the reciprocals of the terms of a geometric sequence form a geometric sequence? Explain.

92. The terms a, $a + 12$, $a + 24$ form an arithmetic sequence, and the terms a, $a + 24$, $a + 12$ form a geometric sequence. Find these sequences.

93. Derive a formula for the value V after one year of an amount A invested at $r\%$ (as a decimal) annual interest, compounded n times during the year. If $A = \$1000$ and $r = 0.10$ (10%), write two or three paragraphs explaining why the amount of interest increases as n increases and stating your approach to finding the maximum possible amount of interest.

CHAPTER 19 PRACTICE TEST

1. Write the first five terms of the sequence for which (a) $a_1 = 8$ and $d = -1/2$; (b) $a_1 = 8$ and $r = -1/2$.
2. Find the sum of the first seven terms of the sequence $6, -2, \frac{2}{3}, \ldots$.
3. For a given sequence, $a_1 = 6$, $d = 4$, and $S_n = 126$. Find n.
4. Find the fraction equal to the decimal 0.454 545 . . . .
5. Find the first three terms of the expansion of $\sqrt{1 - 4x}$.
6. Expand and simplify the expression $(2x - y)^5$.
7. What is the value after 20 years of an investment of \$2500 if it draws 5% annual interest compounded annually?
8. Find the sum of the first 100 even integers.
9. A ball is dropped from a height of 8.00 m, and on each rebound, it rises to $1/2$ of the height it last fell. If it bounces indefinitely, through what total distance will it move?

20. Additional Topics in Trigonometry

LEARNING OUTCOMES

After completion of this chapter, the student should be able to:

- Recognize the basic trigonometric identities and use them to prove other trigonometric identities and simplify trigonometric expressions
- Recognize and apply the formulas for trigonometric functions of sums and differences of angles
- Recognize and apply the formulas for trigonometric functions of half and double angles
- Solve trigonometric equations
- Evaluate inverse trigonometric functions in their defined range
- Find algebraic expressions for expressions involving inverse trigonometric functions
- Solve application problems involving basic trigonometric identities, other trigonometric formulas, and inverse trigonometric functions

Allyn J. Washington

▲ Tesla electric cars are powered by a three-phase alternating-current engine. In Section 20.2, we see how certain trigonometric identities can be used to show an important property of a three-phase generator.

As the use of electricity became widespread in the 1880s, there was a serious debate over the best way to distribute electric power. The American inventor Thomas Edison favoured the use of direct current because it was safer and did not vary with time. Another American inventor and engineer, George Westinghouse, favoured alternating current because the voltage could be stepped up and down with transformers during transmission.

Also favouring the use of alternating current was Nikola Tesla, an American (born in Croatia) electrical engineer, and he had a strong influence on the fact that alternating current came to be used for transmission. Tesla developed the *polyphase generator* that allowed alternating current to be transmitted with constant instantaneous power. Using this type of generator, power losses are greatly reduced in transmission lines, which allows for smaller conductors, and the power can be generated far from where it is used.

Three-phase systems are used in most commercial electric generators, as well as in some electric cars, including those produced by Tesla Motors, a company named in honour of Nikola Tesla. In Section 20.2 we show how a relationship involving trigonometric functions can be used to show a basic property of the current produced by a three-phase generator. Many such relationships among the trigonometric functions can be found from the definitions and other known relationships.

The trigonometric relationships that we develop in this chapter are important for a number of reasons. In fact, we already made use of some of them in Section 10.4 when we graphed certain trigonometric functions, and in Chapter 9 in deriving the law of cosines. In calculus, certain problems use trigonometric relationships, even including some in which these functions do not appear in the initial problem or final answer. Also, they are useful in a number of technical applications in areas such as electronics, optics, solar energy, and robotics.

Later in the chapter, we see that various trigonometric relationships are used in solving equations with trigonometric functions. Also, we develop the concept of the inverse trigonometric functions that were introduced in Chapter 4.

20.1 Fundamental Trigonometric Identities

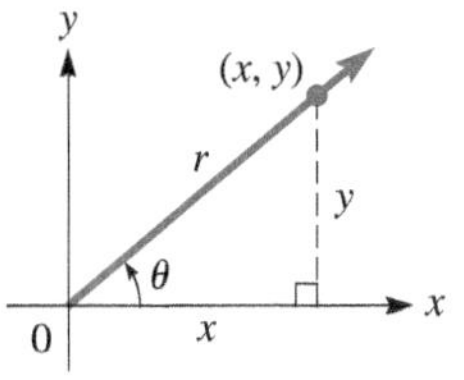

Fig. 20.1

From Chapters 4 and 8, from the definitions, recall that $\sin\theta = y/r$ and $\csc\theta = r/y$ (see Fig. 20.1). Since $y/r = 1/(r/y)$, we see that $\sin\theta = 1/\csc\theta$. These definitions hold for *any* angle, which means that $\sin\theta = 1/\csc\theta$ is true for *any* angle. *This type of relation, which is true for any value of the variable, is called an* **identity**. Of course, values where division by zero would be indicated are excluded.

In this section, we develop several important identities among the trigonometric functions. We also show how the basic identities are used to verify other identities.

From the definitions, we have

$$\sin\theta\csc\theta = \frac{y}{r}\times\frac{r}{y} = 1 \quad \text{or} \quad \sin\theta = \frac{1}{\csc\theta} \quad \text{or} \quad \csc\theta = \frac{1}{\sin\theta}$$

$$\cos\theta\sec\theta = \frac{x}{r}\times\frac{r}{x} = 1 \quad \text{or} \quad \cos\theta = \frac{1}{\sec\theta} \quad \text{or} \quad \sec\theta = \frac{1}{\cos\theta}$$

$$\tan\theta\cot\theta = \frac{y}{x}\times\frac{x}{y} = 1 \quad \text{or} \quad \tan\theta = \frac{1}{\cot\theta} \quad \text{or} \quad \cot\theta = \frac{1}{\tan\theta}$$

$$\frac{\sin\theta}{\cos\theta} = \frac{y/r}{x/r} = \frac{y}{x} = \tan\theta \qquad \frac{\cos\theta}{\sin\theta} = \frac{x/r}{y/r} = \frac{x}{y} = \cot\theta$$

Also, from the definitions and the Pythagorean theorem in the form of $x^2 + y^2 = r^2$, we arrive at the following identities.

By dividing the Pythagorean relation through by r^2, we have

$$\left(\frac{x}{r}\right)^2 + \left(\frac{y}{r}\right)^2 = 1, \quad \text{which leads us to} \quad \cos^2\theta + \sin^2\theta = 1$$

By dividing the Pythagorean relation by x^2, we have

$$1 + \left(\frac{y}{x}\right)^2 = \left(\frac{r}{x}\right)^2, \quad \text{which leads us to} \quad 1 + \tan^2\theta = \sec^2\theta$$

By dividing the Pythagorean relation by y^2, we have

$$\left(\frac{x}{y}\right)^2 + 1 = \left(\frac{r}{y}\right)^2, \quad \text{which leads us to} \quad \cot^2\theta + 1 = \csc^2\theta$$

The term $\cos^2\theta$ is the common way of writing $(\cos\theta)^2$, and it means to square the value of the cosine of the angle. Obviously, the same holds true for the other functions.

Summarizing these results, we have the following important identities:

Reciprocal Identities		Quotient Identities		Pythagorean Identities	
$\sin\theta = \dfrac{1}{\csc\theta}$	**(20.1)**	$\tan\theta = \dfrac{\sin\theta}{\cos\theta}$	**(20.4)**	$\sin^2\theta + \cos^2\theta = 1$	**(20.6)**
$\cos\theta = \dfrac{1}{\sec\theta}$	**(20.2)**	$\cot\theta = \dfrac{\cos\theta}{\sin\theta}$	**(20.5)**	$1 + \tan^2\theta = \sec^2\theta$	**(20.7)**
$\tan\theta = \dfrac{1}{\cot\theta}$	**(20.3)**			$1 + \cot^2\theta = \csc^2\theta$	**(20.8)**

In using the basic identities, θ may stand for any angle or number or expression representing an angle or a number.

EXAMPLE 1 Using the basic identities

(a) $\sin ax = \dfrac{1}{\csc ax}$ using Eq. (20.1) **(b)** $\tan 157^\circ = \dfrac{\sin 157^\circ}{\cos 157^\circ}$ using Eq. (20.4)

(c) $\sin^2 \frac{\pi}{4} + \cos^2 \frac{\pi}{4} = 1$ using Eq. (20.6)

EXAMPLE 2 Checking identities with numerical values

Let us check the last two illustrations of Example 1 for the given values of θ.

(a) Using a calculator, we find that

$$\frac{\sin 157^\circ}{\cos 157^\circ} = \frac{0.390\ 731\ 128\ 5}{-0.920\ 504\ 853\ 5} = -0.424\ 474\ 816\ 2 \text{ and}$$

$$\tan 157^\circ = -0.424\ 474\ 816\ 2$$

We see that $\sin 157^\circ/\cos 157^\circ = \tan 157^\circ$.

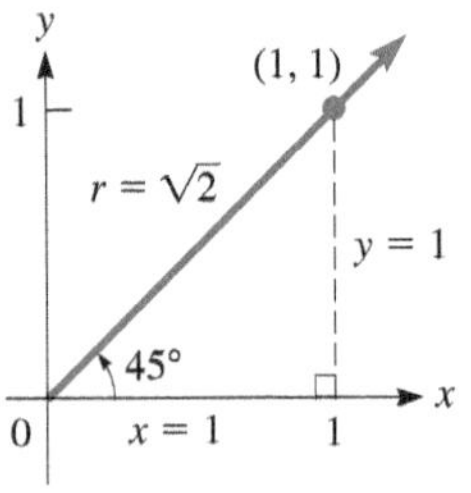

Fig. 20.2

(b) Checking Example 1(c), refer to Fig. 20.2.

$$\sin \tfrac{\pi}{4} = \sin 45^\circ = \tfrac{1}{\sqrt{2}} = \tfrac{\sqrt{2}}{2} \quad \text{and} \quad \cos \tfrac{\pi}{4} = \cos 45^\circ = \tfrac{\sqrt{2}}{2}$$

$$\sin^2 \tfrac{\pi}{4} + \cos^2 \tfrac{\pi}{4} = \left(\tfrac{\sqrt{2}}{2}\right)^2 + \left(\tfrac{\sqrt{2}}{2}\right)^2 = \tfrac{1}{2} + \tfrac{1}{2} = 1$$

We see that this checks with Eq. (20.6) for these values.

EXAMPLE 3 Simplifying basic expressions

(a) Multiply and simplify the expression $\sin\theta \tan\theta(\csc\theta + \cot\theta)$.

$$\begin{aligned}\sin\theta\tan\theta(\csc\theta + \cot\theta) &= \sin\theta\tan\theta\csc\theta + \sin\theta\tan\theta\cot\theta\\ &= \left(\tfrac{1}{\csc\theta}\right)\tan\theta\csc\theta + \sin\theta\left(\tfrac{1}{\cot\theta}\right)\cot\theta \quad \text{using Eqs. (20.1) and (20.4)}\\ &= \tan\theta + \sin\theta\end{aligned}$$

(b) Factor and simplify the expression $\tan^3\alpha + \tan\alpha$.

$$\begin{aligned}\tan^3\alpha + \tan\alpha &= \tan\alpha(\tan^2\alpha + 1)\\ &= \tan\alpha \sec^2\alpha \quad \text{using Eq. (20.7)}\end{aligned}$$

We see that when algebraic operations are performed on trigonometric terms, they are handled in just the same way as with algebraic terms. This is true for the basic operations of addition, subtraction, multiplication, and division, as well as other operations used in simplifying expressions, such as factoring and taking roots.

Practice Exercise

1. Multiply and simplify $\cos x(\sec x + \tan x)$.

LEARNING TIP

The ability to prove trigonometric identities depends to a large extent on being very familiar with the basic identities so that you can *recognize them in somewhat different forms.*

If you do not learn these basic identities well, you will have difficulty in following the examples and doing the exercises. The more readily you recognize these forms, the more easily you will be able to prove such identities.

PROVING TRIGONOMETRIC IDENTITIES

A great many identities exist among the trigonometric functions. We are going to use the reciprocal, quotient, and Pythagorean identities developed in Eqs. (20.1) through (20.8), along with a few additional ones developed in later sections to prove the validity of still other identities.

We begin with examples of how the reciprocal and quotient identities allow us to write simple products and quotients in terms of sines and cosines, some of which then cancel out. We then list some basic strategies for more general identities.

EXAMPLE 4 Using quotient identities

Prove the identity $\sin x = \dfrac{\cos x}{\cot x}$.

We know that $\cot x = \dfrac{\cos x}{\sin x}$. We make the substitution on the right to write the expression in terms of sines and cosines.

$$\sin x = \frac{\cos x}{\cot x} = \frac{\cos x}{\frac{\cos x}{\sin x}} = \frac{\cos x}{1} \times \frac{\sin x}{\cos x} = \frac{\cos x \sin x}{\cos x} = \sin x$$

Eq. (20.5) — invert — cancel $\cos x$ factors

Eq. (20.5); invert; cancel $\cos x$ factors

By showing that the right side may be changed exactly to $\sin x$, the expression on the left side, we have proved the identity.

EXAMPLE 5 Changing to sines and cosines

Prove that $\tan\theta \csc\theta = \sec\theta$.

We know that $\tan\theta = \dfrac{\sin\theta}{\cos\theta}$ and also that $\csc\theta = \dfrac{1}{\sin\theta}$. Changing the left side, we have

$$\tan\theta\csc\theta = \frac{\sin\theta}{\cos\theta} \cdot \frac{1}{\sin\theta} = \frac{\sin\theta}{\cos\theta\sin\theta} = \frac{1}{\cos\theta} = \sec\theta$$

Eq. (20.4) — cancel $\sin\theta$ factors — Eq. (20.2)

Having changed the left side into the form on the right side, we have proven the identity.

Practice Exercise

2. Prove the identity $\dfrac{\cos x \csc x}{\cot^2 x} = \tan x$.

Some Guidelines for Proving Trigonometric Identities

There is no specific procedure for proving trigonometric identities. However, the following points are important:

- The basic identities that may be useful must be readily recognized, keeping in mind equivalent forms of the same identity. For example, $\sin^2\theta = 1 - \cos^2\theta$ is an equivalent form of the identity $\sin^2\theta + \cos^2\theta = 1$.
- Algebraic operations such as substituting, factoring, and simplifying fractions are frequently used, and they must be done carefully and correctly.
- Either side of the identity can be changed to the form on the other side. It is usually easier to change the form of the more complicated side to the same form as the less complicated side. (It is also possible to prove an identity by working on both sides at the same time.)
- Substitutions should be selected looking ahead and keeping in mind the side that is not being changed, which is the goal.
- Multiplying both numerator and denominator of an expression by the same quantity can be useful to obtain a difference of squares. For example, if an expression contains $1 - \sin\theta$, multiplication by $1 + \sin\theta$ gives $1 - \sin^2\theta$, which can be replaced by $\cos^2\theta$.
- A given identity can be proven through a variety of procedures.

EXAMPLE 6 Using a Pythagorean identity

Prove the identity $\tan y = \dfrac{\sec^2 y}{\cot y} - \tan^3 y$.

Here, we simplify the right side. The presence of a square suggests a Pythagorean identity, so we use Eq. (20.7). Also, the reciprocal identity $\cot y = 1/\tan y$ can remove $\cot y$ from the denominator. Therefore,

$$\frac{\sec^2 y}{\cot y} - \tan^3 y = \frac{1 + \tan^2 y}{\frac{1}{\tan y}} - \tan^3 y = \tan y(1 + \tan^2 y) - \tan^3 y$$

Eq. (20.7)

Eq. (20.3)

$$= \tan y + \tan^3 y - \tan^3 y = \tan y$$

EXAMPLE 7 A trigonometric identity—difference of squares

Prove the identity $\dfrac{1 - \sin x}{\sin x \cot x} = \dfrac{\cos x}{1 + \sin x}$.

The combination $1 - \sin x$ suggests $1 - \sin^2 x$, since multiplying $(1 - \sin x)$ by $(1 + \sin x)$ gives $1 - \sin^2 x$, which can then be replaced by $\cos^2 x$. Thus, changing only the left side, we have

$$\frac{1 - \sin x}{\sin x \cot x} = \frac{(1 - \sin x)(1 + \sin x)}{\sin x\left(\frac{\cos x}{\sin x}\right)(1 + \sin x)}$$

multiply numerator and denominator by $1 + \sin x$

Eq. (20.5)

cancel $\sin x$

Eq. (20.6)

$$= \frac{1 - \sin^2 x}{\cos x(1 + \sin x)} = \frac{\cos^2 x}{\cos x(1 + \sin x)}$$

$$= \frac{\cos x}{1 + \sin x}$$

cancel $\cos x$

EXAMPLE 8 A trigonometric identity—radiation rate

In finding the radiation rate of an accelerated electric charge, it is necessary to show that $\sin^3 \theta = \sin \theta - \sin \theta \cos^2 \theta$. Show this by changing the left side.

Since each term on the right has a factor of $\sin \theta$, we see that we can proceed by writing $\sin^3 \theta$ as $\sin \theta(\sin^2 \theta)$. Then the factor $\sin^2 \theta$ and the $\cos^2 \theta$ on the right suggest the use of Eq. (20.6). Thus, we have

$$\sin^3 \theta = \sin \theta(\sin^2 \theta) = \sin \theta(1 - \cos^2 \theta)$$

$$= \sin \theta - \sin \theta \cos^2 \theta$$

multiplying

Since we substituted for $\sin^2 \theta$, we used Eq. (20.6) in the form $\sin^2 \theta = 1 - \cos^2 \theta$.

EXAMPLE 9 Simplifying a trigonometric expression

Simplify the expression $\dfrac{\csc x}{\tan x + \cot x}$.

We proceed with a simplification in a manner similar to proving an identity, although we do not know what the result should be. Following is one procedure for this simplification, writing the expression in terms of sines and cosines:

$$\frac{\csc x}{\tan x + \cot x} = \frac{\csc x}{\dfrac{\sin x}{\cos x} + \dfrac{\cos x}{\sin x}} \quad \text{Eq. (20.4), Eq. (20.1)}$$

$$= \frac{\csc x}{\dfrac{\sin^2 x + \cos^2 x}{\sin x \cos x}} = \frac{\dfrac{1}{\sin x} \cdot \sin x \cos x}{1} \quad \text{Eq. (20.5), Eq. (20.6)}$$

$$= \cos x \quad \text{cancel } \sin x$$

A graphing utility can be used to check an identity or a simplification. This is done by graphing the function on each side of an identity, or the initial expression and the final expression for simplification. If the two graphs are the same, the identity or simplification can be seen to be correct, although it would not be a proof. Examples are left as exercises throughout the chapter.

EXERCISES 20.1

In Exercises 1 and 2, make the given changes in the indicated examples of this section and then prove the resulting identities.

1. In Example 4, change the right side to $\tan x/\sec x$.

2. In Example 6, change the first term on the right to $\sin y/\cos^3 y$.

In Exercises 3–6, use a calculator to check the indicated basic identities for the given angles.

3. Eq. (20.4) for $\theta = 62°$

4. Eq. (20.5) for $\theta = 310°$

5. Eq. (20.6) for $\theta = 4\pi/3$

6. Eq. (20.7) for $\theta = 5\pi/6$

In Exercises 7–12, multiply and simplify. In Exercises 13–18, factor and simplify.

7. $\cos x(\tan x - \sec x)$

8. $\tan y(\cot y + 3\cos y)$

9. $\cos\theta\cot\theta(\sec\theta - 2\tan\theta)$

10. $(\csc x - 1)(\csc x + 1)$

11. $\tan u(\cot u + \tan u)$

12. $\cos^2 t(1 + \tan^2 t)$

13. $\sin x + \sin x\tan^2 x$

14. $\sec\theta + \sec^3\theta$

15. $\sin^3 t\cos t + \sin t\cos^3 t$

16. $\cot^2 u\csc^2 u - \cot^4 u$

17. $\csc^4 y - 1$

18. $\sin x + \sin x\tan^2 x$

In Exercises 19–42, prove the given identities.

19. $\dfrac{\sin x}{\tan x} = \cos x$

20. $\dfrac{\csc\theta}{\sec\theta} = \cot\theta$

21. $\sin x\sec x = \tan x$

22. $\cot\theta\sec\theta = \csc\theta$

23. $\csc^2 x(1 - \cos^2 x) = 1$

24. $\sec\theta(1 - \sin^2\theta) = \cos\theta$

25. $\sin x(1 + \cot^2 x) = \csc x$

26. $\csc x(\csc x - \sin x) = \cot^2 x$

27. $\cos\theta\cot\theta + \sin\theta = \csc\theta$

28. $\csc x\sec x - \tan x = \cot x$

29. $\cot\theta\sec^2\theta - \cot\theta = \tan\theta$

30. $\sin y + \sin y\cot^2 y = \csc y$

31. $\tan x + \cot x = \sec x\csc x$

32. $\tan x + \cot x = \tan x\csc^2 x$

33. $\cos^2 x - \sin^2 x = 1 - 2\sin^2 x$

34. $\dfrac{1 + \cos x}{\sin x} = \dfrac{\sin x}{1 - \cos x}$

35. $\dfrac{\sin\theta}{\csc\theta} + \dfrac{\cos\theta}{\sec\theta} = 1$

36. $\dfrac{\sec\theta}{\cos\theta} - \dfrac{\tan\theta}{\cot\theta} = 1$

37. $2\sin^4 x - 3\sin^2 x + 1 = \cos^2 x(1 - 2\sin^2 x)$

38. $\dfrac{\sin^2\theta + 2\cos\theta - 1}{\sin^2\theta + 3\cos\theta - 3} = \dfrac{1}{1 - \sec\theta}$

39. $\dfrac{1}{2}\sin\pi t\left(\dfrac{\sin\pi t}{1 - \cos\pi t} + \dfrac{1 - \cos\pi t}{\sin\pi t}\right) = 1$

40. $\dfrac{\cot\omega t}{\sec\omega t - \tan\omega t} - \dfrac{\cos\omega t}{\sec\omega t + \tan\omega t} = \sin\omega t + \csc\omega t$

41. $1 + \sin^2 x + \sin^4 x + \cdots = \sec^2 x\left(-\frac{\pi}{2} < x < \frac{\pi}{2}\right)$

42. $1 - \tan^2 x + \tan^4 x - \cdots = \cos^2 x\left(-\frac{\pi}{4} < x < \frac{\pi}{4}\right)$

In Exercises 43–50, simplify the given expressions. The result will be one of $\sin x$, $\cos x$, $\tan x$, $\cot x$, $\sec x$, *or* $\csc x$.

43. $\dfrac{\tan x\csc^2 x}{1 + \tan^2 x}$

44. $\dfrac{\cos x - \cos^3 x}{\sin x - \sin^3 x}$

45. $\cot x(\sec x - \cos x)$

46. $\sin x(\tan x + \cot x)$

47. $\dfrac{\tan x + \cot x}{\csc x}$

48. $\dfrac{1 + \tan x}{\sin x} - \sec x$

49. $\dfrac{\cos x + \sin x}{1 + \tan x}$

50. $\dfrac{\sec x - \cos x}{\tan x}$

In Exercises 51–54, use a graphing utility to verify the given identities by comparing the graphs of both sides.

51. $\sin x(\csc x - \sin x) = \cos^2 x$ **52.** $\cos y(\sec y - \cos y) = \sin^2 y$

53. $\dfrac{\sec x + \csc x}{1 + \tan x} = \csc x$ **54.** $\dfrac{\cot x + 1}{\cot x} = 1 + \tan x$

In Exercises 55–58, use a graphing utility to determine whether the given equations are identities.

55. $\sec\theta \tan\theta \csc\theta = \tan^2\theta + 1$ **56.** $\sin x \cos x \tan x = \cos^2 x - 1$

57. $\dfrac{2\cos^2 x - 1}{\sin x \cos x} = \tan x - \cot x$

58. $\cos^3 x \csc^3 x \tan^3 x = \csc^2 x - \cot^2 x$

In Exercises 59–62, solve the given problems involving trigonometric identities.

59. When designing a solar-energy collector, it is necessary to account for the latitude and longitude of the location, the angle of the sun, and the angle of the collector. In doing this, the equation $\cos\theta = \cos A \cos B \cos C + \sin A \sin B$ is used. If $\theta = 90°$, show that $\cos C = -\tan A \tan B$.

60. The path of a point on the circumference of a circle, such as a point on the rim of a bicycle wheel as it rolls along, traces out a curve called a *cycloid*. To find the distance through which the point moves, it is necessary to simplify the expression $(1 - \cos\theta)^2 + \sin^2\theta$. Perform this simplification.

61. Show that the length l of the straight brace shown in Fig. 20.3 can be found from the equation

$$l = \frac{a(1 + \tan\theta)}{\sin\theta}$$

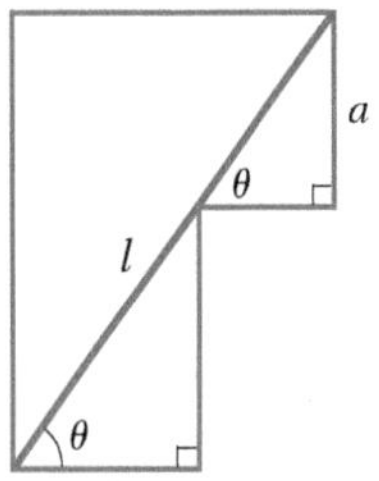

Fig. 20.3

62. In determining the path of least time between two points under certain conditions, it is necessary to show that

$$\sqrt{\frac{1 + \cos\theta}{1 - \cos\theta}}\sin\theta = 1 + \cos\theta$$

Show this by transforming the left-hand side.

In Exercises 63–68, solve the given problems.

63. Explain how to transform $\sin\theta \tan\theta + \cos\theta$ into $\sec\theta$.

64. Explain how to transform $\tan^2\theta \cos^2\theta + \cot^2\theta \sin^2\theta$ into 1.

65. Show that $\sin^2 x(1 - \sec^2 x) + \cos^2 x(1 + \sec^4 x)$ has a constant value.

66. Show that $\cot y \csc y \sec y - \csc y \cos y \cot y$ has a constant value.

67. Prove that $\sec^2\theta + \csc^2\theta = \sec^2\theta \csc^2\theta$ by expressing each function in terms of its x, y, and r definition.

68. Prove that $\dfrac{\csc\theta}{\tan\theta + \cot\theta} = \cos\theta$ by expressing each function in terms of its x, y, and r definition.

In Exercises 69–72, use the given substitutions to show that the given equations are valid. In each, $0 < \theta < \pi/2$.

69. If $x = \cos\theta$, show that $\sqrt{1 - x^2} = \sin\theta$.

70. If $x = 3\sin\theta$, show that $\sqrt{9 - x^2} = 3\cos\theta$.

71. If $x = 8\tan\theta$, show that $\sqrt{64 + x^2} = 8\sec\theta$.

72. If $x = 9\sec\theta$, show that $\sqrt{x^2 - 81} = 9\tan\theta$.

Answers to Practice Exercises

1. $1 + \sin x$ **2.** $\dfrac{\cos x \cdot \frac{1}{\sin x}}{\cot^2 x} = \dfrac{\cot x}{\cot^2 x} = \tan x$

20.2 The Sum and Difference Formulas

There are other important relations among the trigonometric functions. The most important and useful relations are those that involve twice an angle and half an angle. To obtain these relations, in this section we derive the expressions for the sine and cosine of the sum and difference of two angles. These expressions will lead directly to the desired relations of double and half angles that we will derive in the following sections.

Equation (12.13), shown in the margin, gives the polar (or trigonometric) form of the product of two complex numbers. We can use this formula to derive the expressions for the sine and cosine of the sum and difference of two angles.

■ For reference, Eq. (12.13) is
$r_1(\cos\theta_1 + j\sin\theta_1)r_2(\cos\theta_2 + j\sin\theta_2)$
$= r_1 r_2[\cos(\theta_1 + \theta_2) + j\sin(\theta_1 + \theta_2)]$

Using Eq. (12.13) to find the product of the complex numbers $\cos\alpha + j\sin\alpha$ and $\cos\beta + j\sin\beta$, which are represented in Fig. 20.4, we have

$$(\cos\alpha + j\sin\alpha)(\cos\beta + j\sin\beta) = \cos(\alpha + \beta) + j\sin(\alpha + \beta)$$

Expanding the left side, and then switching sides, we have

$$\cos(\alpha + \beta) + j\sin(\alpha + \beta) = (\cos\alpha\cos\beta - \sin\alpha\sin\beta) + j(\sin\alpha\cos\beta + \cos\alpha\sin\beta)$$

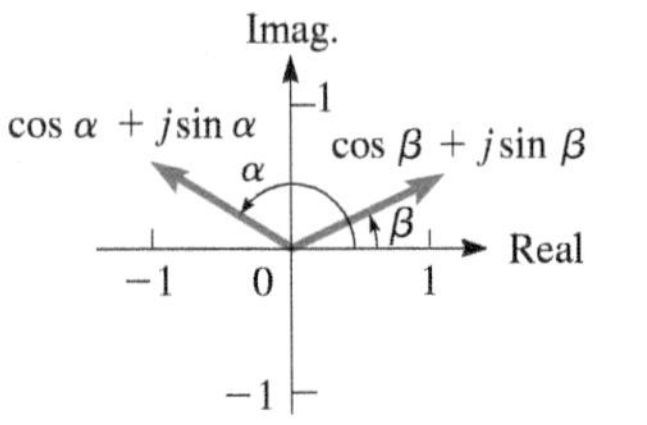

Fig. 20.4

Since two complex numbers are equal if their real parts are equal and their imaginary parts are equal, we have the following formulas:

$$\sin(\alpha + \beta) = \sin\alpha\cos\beta + \cos\alpha\sin\beta \quad (20.9)$$
$$\cos(\alpha + \beta) = \cos\alpha\cos\beta - \sin\alpha\sin\beta \quad (20.10)$$

EXAMPLE 1 Verifying the $\sin(\alpha + \beta)$ formula

Verify that $\sin 90° = 1$, by finding $\sin(60° + 30°)$.

$$\sin 90° = \sin(60° + 30°) = \sin 60° \cos 30° + \cos 60° \sin 30° \quad \text{using Eq. (20.9)}$$

$$= \frac{\sqrt{3}}{2} \cdot \frac{\sqrt{3}}{2} + \frac{1}{2} \cdot \frac{1}{2} \quad \text{for values, see Section 4.3}$$

$$= \frac{3}{4} + \frac{1}{4} = 1$$

COMMON ERROR It should be obvious from this example that $\sin(\alpha + \beta)$ *is* **not** *equal to* $\sin \alpha + \sin \beta$. If we used such a formula, we would get $\sin 90° = \frac{1}{2}\sqrt{3} + \frac{1}{2} = 1.366$ for the combination $(60° + 30°)$. This is not possible since the values of the sine never exceed 1 in value.

EXAMPLE 2 Using $\cos(\alpha + \beta)$ with numerical values

Given that $\sin \alpha = \frac{5}{13}$ (α in the first quadrant) and $\sin \beta = -\frac{3}{5}$ (for β in the third quadrant), find $\cos(\alpha + \beta)$.

Since $\sin \alpha = \frac{5}{13}$ for α in the first quadrant, from Fig. 20.5, we have $\cos \alpha = \frac{12}{13}$. Also, since $\sin \beta = -\frac{3}{5}$ for β in the third quadrant, from Fig. 20.5, we also have $\cos \beta = -\frac{4}{5}$.

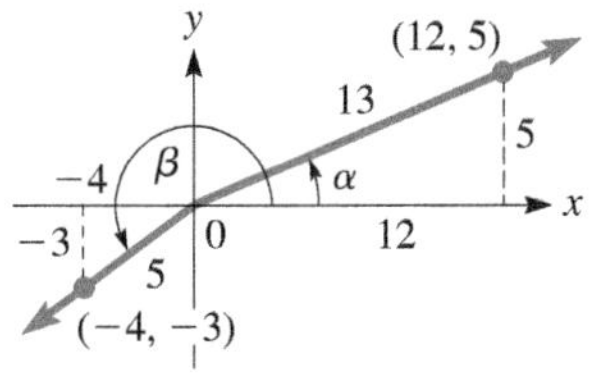

Fig. 20.5

Then, by using Eq. (20.10), we have

$$\cos(\alpha + \beta) = \cos \alpha \cos \beta - \sin \alpha \sin \beta$$

$$= \frac{12}{13}\left(-\frac{4}{5}\right) - \frac{5}{13}\left(-\frac{3}{5}\right)$$

$$= -\frac{48}{65} + \frac{15}{65} = -\frac{33}{65}$$

From Eqs. (20.9) and (20.10), we can easily find expressions for $\sin(\alpha - \beta)$ and $\cos(\alpha - \beta)$. This is done by finding $\sin[\alpha + (-\beta)]$ and $\cos[\alpha + (-\beta)]$. Thus, we have

$$\sin(\alpha - \beta) = \sin[\alpha + (-\beta)] = \sin \alpha \cos(-\beta) + \cos \alpha \sin(-\beta)$$

Since $\cos(-\beta) = \cos \beta$ and $\sin(-\beta) = -\sin \beta$ (see Eq. 8.7), we have

$$\sin(\alpha - \beta) = \sin \alpha \cos \beta - \cos \alpha \sin \beta \quad \textbf{(20.11)}$$

In the same manner, we find that

$$\cos(\alpha - \beta) = \cos \alpha \cos \beta + \sin \alpha \sin \beta \quad \textbf{(20.12)}$$

EXAMPLE 3 Using the $\cos(\alpha - \beta)$ formula

Find $\cos 15°$ from $\cos(45° - 30°)$.

$$\cos 15° = \cos(45° - 30°) = \cos 45° \cos 30° + \sin 45° \sin 30° \quad \text{using Eq. (20.12)}$$

$$= \frac{\sqrt{2}}{2} \cdot \frac{\sqrt{3}}{2} + \frac{\sqrt{2}}{2} \cdot \frac{1}{2} = \frac{\sqrt{6} + \sqrt{2}}{4} \quad \text{(exact)}$$

$$= 0.966$$

EXAMPLE 4 $\sin(\alpha - \beta)$ formula—simple harmonic motion

In analysing the motion of an object oscillating up and down at the end of a spring, the expression $\sin(\omega t + \alpha)\cos\alpha - \cos(\omega t + \alpha)\sin\alpha$ occurs. Simplify this expression.

If we let $x = \omega t + \alpha$, the expression becomes $\sin x \cos\alpha - \cos x \sin\alpha$, which is the form for $\sin(x - \alpha)$. Therefore,

$$\sin(\omega t + \alpha)\cos\alpha - \cos(\omega t + \alpha)\sin\alpha = \sin x\cos\alpha - \cos x\sin\alpha = \sin(x - \alpha)$$
$$= \sin(\omega t + \alpha - \alpha) = \sin \omega t$$

EXAMPLE 5 Using the $\cos(\alpha + \beta)$ formula

Evaluate $\cos 23° \cos 67° - \sin 23° \sin 67°$.

We note that this expression fits the form of the right side of Eq. (20.10), so

$$\cos 23° \cos 67° - \sin 23° \sin 67° = \cos(23° + 67°)$$
$$= \cos 90°$$
$$= 0$$

Again, we are able to evaluate this expression by ***recognizing the form*** of the given expression. Evaluation by a calculator will verify the result.

Practice Exercise

1. Evaluate $\sin 115° \cos 25° - \cos 115° \sin 25°$.

By dividing the right side of Eq. (20.9) by that of Eq. (20.10), we can determine expressions for $\tan(\alpha + \beta)$, and by dividing the right side of Eq. (20.11) by that of Eq. (20.12), we can determine an expression for $\tan(\alpha - \beta)$. The derivation of these formulas is Exercise 33 of this section. These formulas can be written together, as

$$\tan(\alpha \pm \beta) = \frac{\tan\alpha \pm \tan\beta}{1 \mp \tan\alpha\tan\beta} \qquad \textbf{(20.13)}$$

The formula for $\tan(\alpha + \beta)$ uses the upper signs, and the formula for $\tan(\alpha - \beta)$ uses the lower signs.

Certain trigonometric identities can be proven by the formulas derived in this section. The following examples illustrate this use of these formulas.

EXAMPLE 6 Proving a trigonometric identity with $\tan(\alpha \pm \beta)$

Show that $\tan(\alpha + \beta)\tan(\alpha - \beta) = \dfrac{\tan^2\alpha - \tan^2\beta}{1 - \tan^2\alpha\tan^2\beta}$.

Using Eq. (20.13), we have

$$\tan(\alpha + \beta)\tan(\alpha - \beta) = \left(\frac{\tan\alpha + \tan\beta}{1 - \tan\alpha\tan\beta}\right)\left(\frac{\tan\alpha - \tan\beta}{1 + \tan\alpha\tan\beta}\right)$$
$$= \frac{\tan^2\alpha - \tan^2\beta}{1 - \tan^2\alpha\tan^2\beta}$$

EXAMPLE 7 Using the $\sin(\alpha + \beta)$ formula

Prove that $\sin(180° + x) = -\sin x$.

$$\sin(180° + x) = \sin 180° \cos x + \cos 180° \sin x$$ using Eq. (20.9)
$$= (0)\cos x + (-1)\sin x$$ $\sin 180° = 0, \cos 180° = -1$
$$= -\sin x$$

Although x may or may not be an acute angle, this agrees with the results for the sine of a third-quadrant angle, as discussed in Section 8.2. See Fig. 20.6.

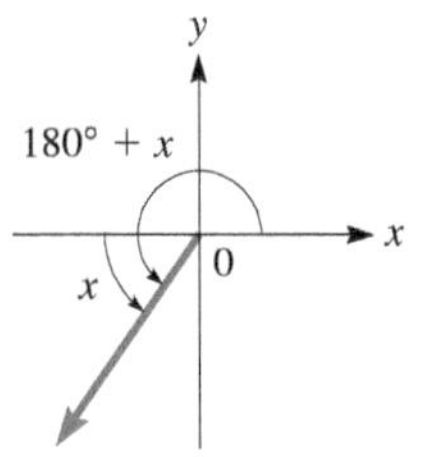

Fig. 20.6

EXAMPLE 8 A trigonometric simplification using $\sin(\alpha - \beta)$

Simplify the expression $\dfrac{\sin(\alpha - \beta)}{\sin\alpha\sin\beta}$.

$$\frac{\sin(\alpha-\beta)}{\sin\alpha\sin\beta} = \frac{\sin\alpha\cos\beta - \cos\alpha\sin\beta}{\sin\alpha\sin\beta} \quad \text{using Eq. (20.11)}$$

$$= \frac{\sin\alpha\cos\beta}{\sin\alpha\sin\beta} - \frac{\cos\alpha\sin\beta}{\sin\alpha\sin\beta}$$

$$= \frac{\cos\beta}{\sin\beta} - \frac{\cos\alpha}{\sin\alpha}$$

$$= \cot\beta - \cot\alpha \quad \text{using Eq. (20.5)}$$

Practice Exercise

2. Simplify $\tan(180° + x)$.

EXAMPLE 9 Using $\sin(\alpha - \beta)$—three-phase generators

Alternating electric current is produced essentially by a coil of wire rotating in a magnetic field, and this is the basis for designing generators of alternating current. A three-phase generator uses three coils of wire and thereby produces three electric currents at the same time. This is the most widely used type of *polyphase generator* as mentioned in the chapter introduction.

■ See the chapter introduction.

The voltages induced in a three-phase generator can be represented as

$$E_1 = E_0 \sin\omega t \qquad E_2 = E_0\sin\left(\omega t - \tfrac{2\pi}{3}\right) \qquad E_3 = E_0 \sin\left(\omega t - \tfrac{4\pi}{3}\right)$$

where E_0 is the maximum voltage and ω is the angular velocity of rotation. Show that the sum of these voltages at any time t is zero.

Setting up the sum $E_1 + E_2 + E_3$ and using Eq. (20.11), we have

$$\begin{aligned} E_1 + E_2 + E_3 &= E_0\left[\sin\omega t + \sin\left(\omega t - \tfrac{2\pi}{3}\right) + \sin\left(\omega t - \tfrac{4\pi}{3}\right)\right] \\ &= E_0\left(\sin\omega t + \sin\omega t\cos\tfrac{2\pi}{3} - \cos\omega t\sin\tfrac{2\pi}{3} + \sin\omega t\cos\tfrac{4\pi}{3} - \cos\omega t\sin\tfrac{4\pi}{3}\right) \\ &= E_0\left[\sin\omega t + (\sin\omega t)\left(-\tfrac{1}{2}\right) - (\cos\omega t)\left(\tfrac{1}{2}\sqrt{3}\right) + (\sin\omega t)\left(-\tfrac{1}{2}\right) - (\cos\omega t)\left(-\tfrac{1}{2}\sqrt{3}\right)\right] \\ &= E_0\left[(\sin\omega t)\left(1 - \tfrac{1}{2} - \tfrac{1}{2}\right) + (\cos\omega t)\left(\tfrac{1}{2}\sqrt{3} - \tfrac{1}{2}\sqrt{3}\right)\right] = 0 \end{aligned}$$

EXERCISES 20.2

In Exercises 1 and 2, make the given changes in the indicated examples of this section and then solve the given problems.

1. In Example 2, change $\frac{5}{13}$ to $\frac{12}{13}$ and then find the value of $\cos(\alpha + \beta)$.
2. In Example 7, change $180° + x$ to $180° - x$ and then determine what other changes result.

In Exercises 3–6, determine the values of the given functions as indicated. If using a calculator, round to three significant digits.

3. Find $\sin 105°$ by using $105° = 60° + 45°$.
4. Find $\tan 75°$ by using $75° = 30° + 45°$.
5. Find $\cos 15°$ by using $15° = 60° - 45°$.
6. Find $\sin 15°$ by using $15° = 45° - 30°$.

In Exercises 7–10, evaluate the given functions with the following information: $\sin\alpha = 4/5$ *(α in first quadrant) and* $\cos\beta = -12/13$ *(β in second quadrant).*

7. $\sin(\alpha + \beta)$
8. $\tan(\beta - \alpha)$
9. $\cos(\alpha + \beta)$
10. $\sin(\alpha - \beta)$

In Exercises 11–20, simplify the given expressions.

11. $\sin x\cos 2x + \sin 2x\cos x$
12. $\sin 3x\cos x - \sin x\cos 3x$
13. $\cos\pi\cos x + \sin\pi\sin x$
14. $\dfrac{\tan(x-y) + \tan x}{1 - \tan(x-y)\tan y}$
15. $\sin(360° - x)$
16. $\cos(2\pi - x)$
17. $\tan(x - \pi)$
18. $\sin(x + \pi/2)$

19. $\sin 3x \cos(3x - \pi) - \cos 3x \sin(3x - \pi)$

20. $\cos(x + \pi)\cos(x - \pi) + \sin(x + \pi)\sin(x - \pi)$

In Exercises 21–24, evaluate each expression by first changing the form. Verify each by use of a calculator.

21. $\sin 122° \cos 32° - \cos 122° \sin 32°$

22. $\cos 250° \cos 70° + \sin 250° \sin 70°$

23. $\cos\frac{\pi}{5}\cos\frac{3\pi}{10} - \sin\frac{\pi}{5}\sin\frac{3\pi}{10}$ **24.** $\dfrac{\tan\frac{\pi}{10} + \tan\frac{3\pi}{20}}{1 - \tan\frac{\pi}{10}\tan\frac{3\pi}{20}}$

In Exercises 25–28, prove the given identities.

25. $\sin(x + y)\sin(x - y) = \sin^2 x - \sin^2 y$

26. $\cos(x + y)\cos(x - y) = \cos^2 x - \sin^2 y$

27. $\cos(\alpha + \beta) + \cos(\alpha - \beta) = 2\cos\alpha\cos\beta$

28. $\tan(90° + x) = -\cot x$ (Explain why Eq. (20.13) cannot be used for this, but Eqs. (20.9) and (20.10) can be used.)

In Exercises 29–32, verify each identity by comparing the graph of the left side with the graph of the right side on a graphing utility.

29. $\cos(30° + x) = \dfrac{\sqrt{3}\cos x - \sin x}{2}$

30. $\sin(120° - x) = \dfrac{\sqrt{3}\cos x + \sin x}{2}$

31. $\tan\left(\frac{\pi}{4} + x\right) = \dfrac{1 + \tan x}{1 - \tan x}$

32. $\cos\left(\frac{\pi}{2} - x\right) = \sin x$

In Exercises 33–36, derive the given equations as indicated. Equations (20.14)–(20.16) are known as the product formulas.

33. By dividing the right side of Eq. (20.9) by that of Eq. (20.10), and dividing the right side of Eq. (20.11) by that of Eq. (20.12), derive Eq. (20.13).

$$\tan(\alpha \pm \beta) = \frac{\tan\alpha \pm \tan\beta}{1 \mp \tan\alpha\tan\beta} \quad (20.13)$$

(*Hint:* Divide numerator and denominator by $\cos\alpha\cos\beta$.)

34. By adding Eqs. (20.9) and (20.11), derive the equation

$$\sin\alpha\cos\beta = \tfrac{1}{2}[\sin(\alpha + \beta) + \sin(\alpha - \beta)] \quad (20.14)$$

35. By adding Eqs. (20.10) and (20.12), derive the equation

$$\cos\alpha\cos\beta = \tfrac{1}{2}[\cos(\alpha + \beta) + \cos(\alpha - \beta)] \quad (20.15)$$

36. By subtracting Eq. (20.10) from Eq. (20.12), derive

$$\sin\alpha\sin\beta = \tfrac{1}{2}[\cos(\alpha - \beta) - \cos(\alpha + \beta)] \quad (20.16)$$

In Exercises 37–40, derive the given equations by letting $\alpha + \beta = x$ and $\alpha - \beta = y$, which leads to $\alpha = \frac{1}{2}(x + y)$ and $\beta = \frac{1}{2}(x - y)$. The resulting equations are known as the factor formulas.

37. Use Eq. (20.14) and the substitutions above to derive the equation

$$\sin x + \sin y = 2\sin\tfrac{1}{2}(x + y)\cos\tfrac{1}{2}(x - y) \quad (20.17)$$

38. Use Eqs. (20.9) and (20.11) and the substitutions above to derive the equation

$$\sin x - \sin y = 2\sin\tfrac{1}{2}(x - y)\cos\tfrac{1}{2}(x + y) \quad (20.18)$$

39. Use Eq. (20.15) and the substitutions above to derive the equation

$$\cos x + \cos y = 2\cos\tfrac{1}{2}(x + y)\cos\tfrac{1}{2}(x - y) \quad (20.19)$$

40. Use Eq. (20.16) and the substitutions above to derive the equation

$$\cos x - \cos y = -2\sin\tfrac{1}{2}(x + y)\sin\tfrac{1}{2}(x - y) \quad (20.20)$$

In Exercises 41–50, solve the given problems.

41. Show that $\frac{\sin 2x}{\sin x} = 2\cos x$. (*Hint:* $\sin 2x = \sin(x + x)$.)

42. Explain how the exact value of $\sin 75°$ can be found using either Eq. (20.9) or Eq. (20.11).

43. Express $\cos(A + B + C)$ in terms of $\sin A$, $\sin B$, $\sin C$, $\cos A$, $\cos B$, and $\cos C$.

44. The design of a certain three-phase alternating-current generator uses the fact that the sum of the currents $I\cos(\theta + 30°)$, $I\cos(\theta + 150°)$, and $I\cos(\theta + 270°)$ is zero. Verify this.

45. For voltages $V_1 = 20\sin 120\pi t$ and $V_2 = 20\cos 120\pi t$, show that $V = V_1 + V_2 = 20\sqrt{2}\sin(120\pi t + \pi/4)$.

46. The displacements y_1 and y_2 of two waves travelling through the same medium are given by $y_1 = A\sin 2\pi(t/T - x/\lambda)$ and $y_2 = A\sin 2\pi(t/T + x/\lambda)$. Find an expression for the displacement $y_1 + y_2$ of the combination of the waves.

47. An alternating electric current i is given by the equation $i = i_0\sin(\omega t + \alpha)$. Show that this can be written as $i = i_1\sin\omega t + i_2\cos\omega t$, where $i_1 = i_0\cos\alpha$ and $i_2 = i_0\sin\alpha$.

48. A weight **w** is held in equilibrium by forces **F** and **T**, as shown in Fig. 20.7. Equations relating w, F, and T are

$$F\cos\theta = T\sin\alpha$$
$$w + F\sin\theta = T\cos\alpha$$

Show that $w = \dfrac{T\cos(\theta + \alpha)}{\cos\theta}$.

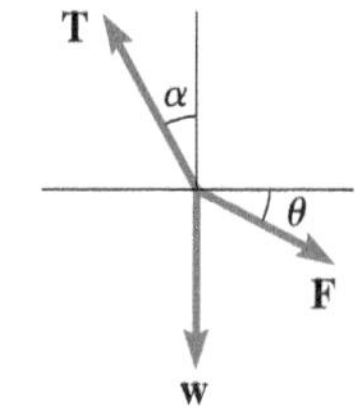

Fig. 20.7

49. For the two bevel gears shown in Fig. 20.8, the equation $\tan\alpha = \dfrac{\sin\beta}{R + \cos\beta}$ is used.
Here, R is the ratio of gear 1 to gear 2. Show that $R = \dfrac{\sin(\beta - \alpha)}{\sin\alpha}$.

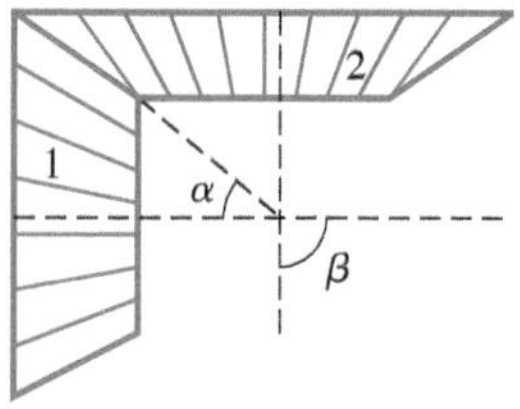

Fig. 20.8

50. In the analysis of the angles of incidence i and reflection r of a light ray subject to certain conditions, the following expression is found:

$$E_2\left(\frac{\tan r}{\tan i} + 1\right) = E_1\left(\frac{\tan r}{\tan i} - 1\right)$$

Show that $E_2 = E_1\dfrac{\sin(r - i)}{\sin(r + i)}$.

Answers to Practice Exercises

1. 1 **2.** $\tan x$

20.3 Double-Angle Formulas

If we let $\beta = \alpha$ in the sum formulas for sine, cosine, and tangent (given in Section 20.2), we can derive the important double-angle formulas:

$$\sin(\alpha + \alpha) = \sin(2\alpha) = \sin\alpha\cos\alpha + \cos\alpha\sin\alpha = 2\sin\alpha\cos\alpha$$

$$\cos(\alpha + \alpha) = \cos(2\alpha) = \cos\alpha\cos\alpha - \sin\alpha\sin\alpha = \cos^2\alpha - \sin^2\alpha$$

$$\tan(\alpha + \alpha) = \tan(2\alpha) = \frac{\tan\alpha + \tan\alpha}{1 - \tan\alpha\tan\alpha} = \frac{2\tan\alpha}{1 - \tan^2\alpha}$$

Then using the basic identity $\sin^2 x + \cos^2 x = 1$, other forms of the equation for $\cos 2\alpha$ may be derived. Summarizing these forms, we have

$$\sin 2\alpha = 2\sin\alpha\cos\alpha \qquad (20.21)$$

$$\cos 2\alpha = \cos^2\alpha - \sin^2\alpha \qquad (20.22)$$

$$= 2\cos^2\alpha - 1 \qquad (20.23)$$

$$= 1 - 2\sin^2\alpha \qquad (20.24)$$

$$\tan 2\alpha = \frac{2\tan\alpha}{1 - \tan^2\alpha} \qquad (20.25)$$

These double-angle formulas are widely used in applications of trigonometry, especially in calculus. They should be recognized quickly in any of the above forms.

EXAMPLE 1 Using double-angle formulas

(a) If $\alpha = 30°$, we have

$$\cos 60° = \cos 2(30°) = \cos^2 30° - \sin^2 30° = \left(\frac{\sqrt{3}}{2}\right)^2 - \left(\frac{1}{2}\right)^2 = \frac{1}{2} \quad \text{using Eq. (20.22)}$$

(b) If $\alpha = 3x$, we have

$$\sin 6x = \sin 2(3x) = 2\sin 3x\cos 3x \quad \text{using Eq. (20.21)}$$

(c) If $2\alpha = x$, we may write $\alpha = x/2$, which means that

$$\sin x = \sin 2\left(\frac{x}{2}\right) = 2\sin\frac{x}{2}\cos\frac{x}{2} \quad \text{using Eq. (20.21)}$$

(d) If $\alpha = \frac{\pi}{6}$, we have

$$\tan\frac{\pi}{3} = \tan 2\left(\frac{\pi}{6}\right) = \frac{2\tan\frac{\pi}{6}}{1 - \tan^2\left(\frac{\pi}{6}\right)} = \frac{2(\sqrt{3}/3)}{1 - (\sqrt{3}/3)^2} = \sqrt{3} \quad \text{using Eq. (20.25)}$$

EXAMPLE 2 Simplification using the cos 2α formula

Simplify the expression $\cos^2 2x - \sin^2 2x$.

Since this is the difference of the square of the cosine of an angle and the square of the sine of the same angle, it fits the right side of Eq. (20.22). Therefore, letting $\alpha = 2x$, we have

$$\cos^2 2x - \sin^2 2x = \cos 2(2x) = \cos 4x$$

EXAMPLE 3 Using the sin 2α formula—area of land

To find the area A of a right triangular tract of land, a surveyor may use the formula $A = \frac{1}{4}c^2\sin 2\theta$, where c is the hypotenuse and θ is *either* of the acute angles. Derive this formula.

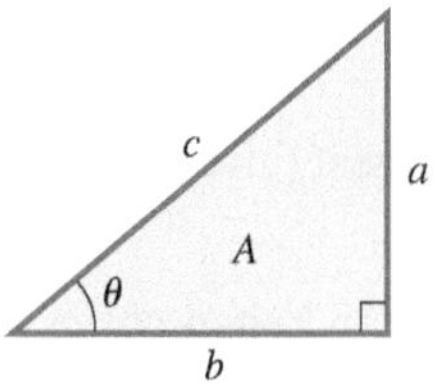

Fig. 20.9

In Fig. 20.9, we see that $\sin\theta = a/c$ and $\cos\theta = b/c$, which gives us

$$a = c\sin\theta \quad \text{and} \quad b = c\cos\theta$$

The area is given by $A = \frac{1}{2}ab$, which leads to the solution

$$\begin{aligned} A &= \frac{1}{2}ab = \frac{1}{2}(c\sin\theta)(c\cos\theta) \\ &= \frac{1}{2}c^2\sin\theta\cos\theta = \frac{1}{2}c^2\left(\frac{1}{2}\sin 2\theta\right) \qquad \text{using Eq. (20.21)} \\ &= \frac{1}{4}c^2\sin 2\theta \end{aligned}$$

In using Eq. (20.21), we divided both sides by 2 to get $\sin\theta\cos\theta = \frac{1}{2}\sin 2\theta$.

If we had labelled the upper acute angle in Fig. 20.9 as θ, we would have $a = c\cos\theta$ and $b = c\sin\theta$. Using these values in the formula for the area gives the same solution.

EXAMPLE 4 Verifying values

(a) Verifying the value of sin 90°, using the functions of 45°, we have

$$\sin 90° = \sin 2(45°) = 2\sin 45°\cos 45° = 2\left(\frac{\sqrt{2}}{2}\right)\left(\frac{\sqrt{2}}{2}\right) = 1 \qquad \text{using Eq. (20.21)}$$

(b) Using Eq. (20.25), $\tan 142° = \dfrac{2\tan 71°}{1 - \tan^2 71°}$. Using a calculator to verify this, we have

$$\tan 142° = -0.781\,285\,626\,5 \quad \text{and} \quad \frac{2\tan 71°}{1 - \tan^2 71°} = -0.781\,285\,626\,5$$

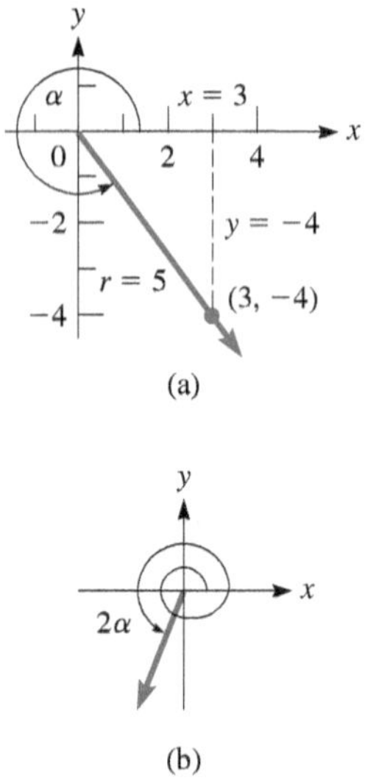

Fig. 20.10

EXAMPLE 5 Evaluation using the sin 2α formula

Knowing that $\cos\alpha = 3/5$ for an angle in the fourth quadrant, we see from Fig. 20.10(a) that $\sin\alpha = -4/5$. Therefore, we have

$$\begin{aligned} \sin 2\alpha &= 2\sin\alpha\cos\alpha \qquad \text{Eq. (20.21)} \\ &= 2\left(-\frac{4}{5}\right)\left(\frac{3}{5}\right) = -\frac{24}{25} \end{aligned}$$

In Fig. 20.10(b), angle 2α is shown. It is a third-quadrant angle, which verifies the sign of the result. (Since $\cos\alpha = 3/5$, $\alpha = 307°$ and $2\alpha = 614°$, which is a third-quadrant angle.)

EXAMPLE 6 Simplification using cos 2α

Simplify the expression $\dfrac{2}{1 + \cos 2x}$.

$$\begin{aligned} \frac{2}{1 + \cos 2x} &= \frac{2}{1 + (2\cos^2 x - 1)} \qquad \text{using Eq. (20.23)} \\ &= \frac{2}{2\cos^2 x} = \sec^2 x \qquad \text{using Eq. (20.2)} \end{aligned}$$

Practice Exercise

1. Simplify $\dfrac{\sin 2x}{\cos^2 x - \sin^2 x}$.

EXERCISES 20.3

In Exercises 1–4, make the given changes in the indicated examples of this section and then solve the resulting problems.

1. In Example 1(d), change $\frac{\pi}{6}$ to $\frac{\pi}{3}$ and then evaluate $\tan \frac{2\pi}{3}$.
2. In Example 2, change $2x$ to $3x$ and then simplify.
3. In Example 5, change $3/5$ to $4/5$ and then evaluate $\sin 2\alpha$.
4. In Example 6, change the $+$ in the denominator to $-$ and then simplify the expression.

In Exercises 5–8, determine the values of the indicated functions in the given manner.

5. Find sin 60° by using the functions of 30°.
6. Find sin 120° by using the functions of 60°.
7. Find tan 120° by using the functions of 60°.
8. Find cos 60° by using the functions of 30°.

In Exercises 9–14, use a calculator to verify the values found by using the double-angle formulas. Round to three significant digits.

9. Find sin 258° directly and by using functions of 129°.
10. Find tan 84° directly and by using functions of 42°.
11. Find cos 96° directly and by using functions of 48°.
12. Find cos 276° directly and by using functions of 138°.
13. Find $\tan \frac{2\pi}{5}$ directly and by using functions of $\frac{\pi}{5}$.
14. Find $\sin(0.2\pi)$ directly and by using functions of 0.1π.

In Exercises 15–18, evaluate the indicated functions with the given information.

15. Find $\sin 2x$ if $\cos x = \frac{4}{5}$ (in first quadrant).
16. Find $\cos 2x$ if $\sin x = -\frac{12}{13}$ (in third quadrant).
17. Find $\tan 2x$ if $\sin x = 0.5$ (in second quadrant).
18. Find $\sin 4x$ if $\sin x = 0.6$ (in first quadrant).

In Exercises 19–30, simplify the given expressions.

19. $6 \sin 5x \cos 5x$
20. $4 \sin^2 x \cos^2 x$
21. $1 - 2 \sin^2 4x$
22. $\dfrac{4 \tan 4\theta}{1 - \tan^2 4\theta}$
23. $2 \cos^2 \frac{1}{2}x - 1$
24. $2 \sin \frac{1}{2}x \cos \frac{1}{2}x$
25. $4 \sin^2 2x - 2$
26. $\cos 3x \sin 3x$
27. $\dfrac{\sin 4\theta}{\sin 2\theta}$
28. $\cos^4 u - \sin^4 u$
29. $\dfrac{\sin 3x}{\sin x} - \dfrac{\cos 3x}{\cos x}$
30. $\dfrac{\cos 3x}{\sin x} + \dfrac{\sin 3x}{\cos x}$

In Exercises 31–40, prove the given identities.

31. $\cos^2 \alpha - \sin^2 \alpha = 2 \cos^2 \alpha - 1$
32. $\cos^2 \alpha - \sin^2 \alpha = 1 - 2 \sin^2 \alpha$
33. $\dfrac{\cos x - \tan x \sin x}{\sec x} = \cos 2x$
34. $2 + \dfrac{\cos 2\theta}{\sin^2 \theta} = \csc^2 \theta$
35. $\dfrac{\sin 2\theta}{1 + \cos 2\theta} = \tan \theta$
36. $\dfrac{2 \tan \alpha}{1 + \tan^2 \alpha} = \sin 2\alpha$
37. $1 - \cos 2\theta = \dfrac{2}{1 + \cot^2 \theta}$
38. $\dfrac{\cos^3 \theta + \sin^3 \theta}{\cos \theta + \sin \theta} = 1 - \dfrac{1}{2} \sin 2\theta$
39. $\ln(1 - \cos 2x) - \ln(1 + \cos 2x) = 2 \ln \tan x$
40. $\log(20 \sin^2 \theta + 10 \cos 2\theta) = 1$

In Exercises 41–44, verify each identity by comparing the graph of the left side with the graph of the right side on a graphing utility.

41. $\tan 2\theta = \dfrac{2}{\cot \theta - \tan \theta}$
42. $\dfrac{1 - \tan^2 x}{\sec^2 x} = \cos 2x$
43. $(\sin x + \cos x)^2 = 1 + \sin 2x$
44. $2 \csc 2x \tan x = \sec^2 x$

In Exercises 45–60, solve the given problems.

45. Express $\sin 3x$ in terms of $\sin x$ only.
46. Express $\cos 3x$ in terms of $\cos x$ only.
47. Express $\cos 4x$ in terms of $\cos x$ only.
48. Express $\sin 4x$ in terms of $\sin x$ and $\cos x$.
49. Find the exact value of $\cos^4 x - \sin^4 x - \cos 2x$.
50. For an acute angle θ, show that $2 \sin \theta > \sin 2\theta$.
51. Without graphing, determine the amplitude and period of the function $y = 4 \sin x \cos x$. Explain.
52. Without graphing, determine the amplitude and period of the function $y = \cos^2 x - \sin^2 x$.
53. The path of a bouncing ball is given by $y = \sqrt{(\sin x + \cos x)^2}$. Show that this path can also be written as $y = \sqrt{1 + \sin 2x}$. Use a graphing utility to demonstrate that this can also be expressed as $y = |\sin x + \cos x|$.
54. The equation for the trajectory of a missile fired into the air at an angle α with velocity v_0 is $y = x \tan \alpha - \dfrac{g}{2v_0^2 \cos^2 \alpha} x^2$. Here, g is the acceleration due to gravity. On the right of the equal sign, combine terms and simplify.
55. The CN Tower in Toronto is 553 m high, and it has an observation deck at the 335-m level. How far from the top of the CN Tower must a helicopter at an altitude of 553 m be in order that the angle subtended at the helicopter by the part of the tower above the deck equals the angle subtended by the part of the tower below the deck?
56. The cross-section of a radio-wave reflector is defined by $x = \cos 2\theta$, $y = \sin \theta$. Find the relation between x and y by eliminating θ.
57. To find the horizontal range R of a projectile, the equation $R = vt \cos \alpha$ is used, where α is the angle between the line of fire and the horizontal, v is the initial velocity of the projectile, and t is the time of flight. It can be shown that $t = (2v \sin \alpha)/g$, where g is the acceleration due to gravity. Show that $R = (v^2 \sin 2\alpha)/g$. See Fig. 20.11.

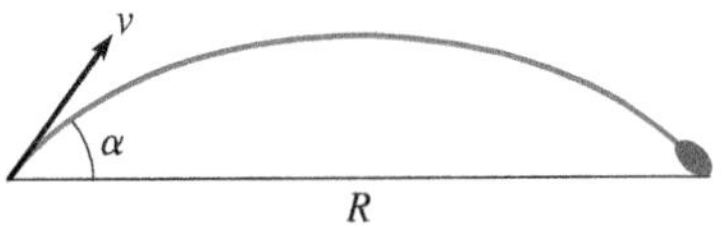

Fig. 20.11

58. In analysing light reflection from a cylinder onto a flat surface, the expression $3\cos\theta - \cos 3\theta$ arises. Show that this equals $2\cos\theta\cos 2\theta + 4\sin\theta\sin 2\theta$.

59. The instantaneous electric power P in an inductor is given by the equation $P = vi\sin\omega t\sin(\omega t - \pi/2)$. Show that this equation can be written as $P = -\frac{1}{2}vi\sin 2\omega t$.

60. In the study of the stress at a point in a bar, the equation $s = a\cos^2\theta + b\sin^2\theta - 2t\sin\theta\cos\theta$ arises. Show that this equation can be written as $s = \frac{1}{2}(a+b) + \frac{1}{2}(a-b)\cos 2\theta - t\sin 2\theta$.

Answer to Practice Exercise

1. $\tan 2x$

20.4 Half-Angle Formulas

If we let $\theta = \alpha/2$ in the identity $\cos 2\theta = 1 - 2\sin^2\theta$ and then solve for $\sin(\alpha/2)$,

$$\sin\frac{\alpha}{2} = \pm\sqrt{\frac{1-\cos\alpha}{2}} \qquad \textbf{(20.26)}$$

Also, with the same substitution in the identity $\cos 2\theta = 2\cos^2\theta - 1$, which is then solved for $\cos(\alpha/2)$, we have

$$\cos\frac{\alpha}{2} = \pm\sqrt{\frac{1+\cos\alpha}{2}} \qquad \textbf{(20.27)}$$

In each of Eqs. (20.26) and (20.27), ***the sign chosen depends on the quadrant in which*** $\frac{\alpha}{2}$ ***lies.***

EXAMPLE 1 Evaluation using the cos(α/2) formula

We can find $\cos 165°$ by using the relation

$$\cos 165° = -\sqrt{\frac{1+\cos 330°}{2}} \qquad \text{using Eq. (20.27)}$$

$$= -\sqrt{\frac{1+0.866}{2}} = -0.966$$

Here, the minus sign is used, since 165° is in the second quadrant, and the cosine of a second-quadrant angle is negative.

EXAMPLE 2 Evaluation using the sin(α/2) formula

Simplify $\sqrt{\frac{1-\cos 114°}{2}}$ by expressing the result in terms of one-half the given angle. Then, using a calculator, show that the values are equal.

We note that the given expression fits the form of the right side of Eq. (20.26), which means that

$$\sqrt{\frac{1-\cos 114°}{2}} = \sin\tfrac{1}{2}(114°) = \sin 57°$$

Using a calculator shows that

$$\sqrt{\frac{1-\cos 114°}{2}} = 0.838\,670\,567\,9 \quad \text{and} \quad \sin 57° = 0.838\,670\,567\,9$$

which verifies the equation for these values.

EXAMPLE 3 Simplification using the cos(α/2) formula

Simplify the expression $\sqrt{\dfrac{9 + 9\cos 6x}{2}}$.

$$\sqrt{\frac{9 + 9\cos 6x}{2}} = \sqrt{\frac{9(1 + \cos 6x)}{2}} = 3\sqrt{\frac{1 + \cos 6x}{2}}$$

$$= 3\cos\tfrac{1}{2}(6x) \quad \text{using Eq. (20.27) with } \alpha = 6x$$

$$= 3\cos 3x$$

Noting the original expression, we see that $\cos 3x$ cannot be negative.

Practice Exercise

1. Simplify: $\sqrt{\dfrac{25 - 25\cos 4x}{2}}$

EXAMPLE 4 Using sin(α/2)—kinetic theory of gases

In the kinetic theory of gases, the expression $\sqrt{(1 - \cos\alpha)^2 + \sin^2\alpha}$ is found. Show that this expression equals $2\sin\frac{1}{2}\alpha$.

$$\sqrt{(1 - \cos\alpha)^2 + \sin^2\alpha} = \sqrt{1 - 2\cos\alpha + \cos^2\alpha + \sin^2\alpha} \quad \text{expanding}$$

$$= \sqrt{1 - 2\cos\alpha + 1} \quad \text{using Eq. (20.26)}$$

$$= \sqrt{2 - 2\cos\alpha}$$

$$= \sqrt{2(1 - \cos\alpha)} \quad \text{factoring}$$

This last expression is very similar to that for $\sin\frac{1}{2}\alpha$, except that no 2 appears in the denominator. Therefore, multiplying the numerator and the denominator under the radical by 2 leads to the solution:

$$\sqrt{2(1 - \cos\alpha)} = \sqrt{\frac{4(1 - \cos\alpha)}{2}} = 2\sqrt{\frac{1 - \cos\alpha}{2}}$$

$$= 2\sin\tfrac{1}{2}\alpha \quad \text{using Eq. (20.26)}$$

Noting the original expression, we see that $\sin\frac{1}{2}\alpha$ cannot be negative.

EXAMPLE 5 Evaluation using the cos(α/2) formula

Given that $\tan\alpha = \frac{8}{15}(180° < \alpha < 270°)$, find $\cos\left(\frac{\alpha}{2}\right)$.

Knowing that $\tan\alpha = \frac{8}{15}$ for a third-quadrant angle, we determine from Fig. 20.12 that $\cos\alpha = -\frac{15}{17}$. This means

$$\cos\frac{\alpha}{2} = -\sqrt{\frac{1 + (-15/17)}{2}} = -\sqrt{\frac{2}{34}} \quad \text{using Eq. (20.27)}$$

$$= -\frac{1}{17}\sqrt{17} = -0.2425$$

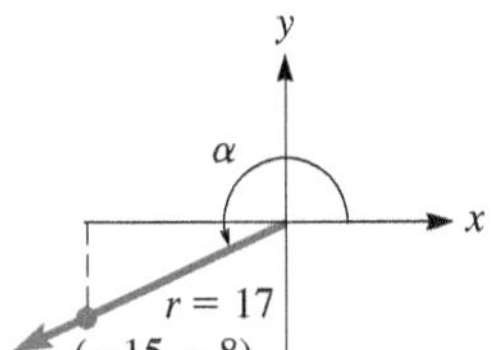

Fig. 20.12

Since $180° < \alpha < 270°$, we know that $90° < \frac{\alpha}{2} < 135°$, and therefore $\frac{\alpha}{2}$ is in the second quadrant. Since the cosine is negative for second-quadrant angles, we use the negative value of the radical.

EXAMPLE 6 Simplification—using the cos(α/2) formula

Show that $2\cos^2\dfrac{x}{2} - \cos x = 1$.

The first step is to substitute for $\cos\frac{x}{2}$, which will result in each term on the left being in terms of x and no $\frac{x}{2}$ terms will exist. This will allow us to combine terms. We have for the left side

$$2\cos^2\frac{x}{2} - \cos x = 2\left(\frac{1 + \cos x}{2}\right) - \cos x \quad \text{using Eq. (20.27) with both sides squared}$$

$$= 1 + \cos x - \cos x = 1$$

Practice Exercise

2. Find the formula for $\csc(x/2)$.

EXAMPLE 7 Formulas for other functions of $\alpha/2$

Find a formula for $\sec\frac{\alpha}{2}$ by first writing the expression in terms of $\sin\left(\frac{\alpha}{2}\right)$ or $\cos\left(\frac{\alpha}{2}\right)$.

$$\sec\frac{\alpha}{2} = \frac{1}{\cos\frac{\alpha}{2}} = \pm\frac{1}{\sqrt{\frac{1+\cos\alpha}{2}}} \quad \text{using Eq. (20.27)}$$

$$= \pm\sqrt{\frac{2}{1+\cos\alpha}}$$

EXERCISES 20.4

In Exercises 1 and 2, make the given changes in the indicated examples of this section and then solve the given problem.

1. In Example 2, change the $-$ sign in the numerator to $+$.

2. In Example 5, change $\frac{8}{15}$ $(180° < \alpha < 270°)$ to $-\frac{8}{15}$ $(270° < \alpha < 360°)$.

In Exercises 3–8, use the half-angle formulas to evaluate the given functions. When using a calculator, round to three significant digits.

3. $\cos 15°$

4. $\sin 22.5°$

5. $\sin 105°$

6. $\cos 112.5°$

7. $\cos\frac{7\pi}{8}$

8. $\sin\frac{11\pi}{12}$

In Exercises 9–12, simplify the given expressions by giving the results in terms of one-half the given angle. Then use a calculator to verify the result.

9. $\sqrt{\frac{1-\cos 192°}{2}}$

10. $\sqrt{\frac{1+\cos 176°}{2}}$

11. $\sqrt{1+\cos 164°}$

12. $\sqrt{2-2\cos 328°}$

In Exercises 13–20, simplify the given expressions.

13. $\sqrt{\frac{1-\cos 6x}{2}}$

14. $\sqrt{\frac{4+4\cos 8\beta}{2}}$

15. $\sqrt{8+8\cos 4x}$

16. $\sqrt{2-2\cos 16x}$

17. $\sqrt{4-4\cos 10\theta}$

18. $\sqrt{18+18\cos 1.4x}$

19. $2\sin^2\frac{x}{2}+\cos x$

20. $2\cos^2\frac{\theta}{2}\sec\theta$

In Exercises 21–24, evaluate the indicated functions.

21. Find the value of $\sin\left(\frac{\alpha}{2}\right)$ if $\cos\alpha = \frac{7}{8}$ $(0° < \alpha < 90°)$.

22. Find the value of $\cos\left(\frac{\alpha}{2}\right)$ if $\sin\alpha = -\frac{15}{17}$ $(180° < \alpha < 270°)$.

23. Find the value of $\cos\left(\frac{\alpha}{2}\right)$ if $\tan\alpha = -0.292$ $(90° < \alpha < 180°)$.

24. Find the value of $\sin\left(\frac{\alpha}{2}\right)$ if $\cos\alpha = 0.471$ $(270° < \alpha < 360°)$.

In Exercises 25–28, derive the required expressions.

25. Derive an expression for $\csc\left(\frac{\alpha}{2}\right)$ in terms of $\sec\alpha$.

26. Derive an expression for $\sec\left(\frac{\alpha}{2}\right)$ in terms of $\sec\alpha$.

27. Derive an expression for $\tan\left(\frac{\alpha}{2}\right)$ in terms of $\sin\alpha$ and $\cos\alpha$.

28. Derive an expression for $\cot\left(\frac{\alpha}{2}\right)$ in terms of $\sin\alpha$ and $\cos\alpha$.

In Exercises 29–32, prove the given identities.

29. $\sin\frac{\alpha}{2} = \frac{1-\cos\alpha}{2\sin\frac{\alpha}{2}}$

30. $\cos\frac{\theta}{2} = \frac{\sin\theta}{2\sin\frac{\theta}{2}}$

31. $2\cos\frac{x}{2} = (1+\cos x)\sec\frac{x}{2}$

32. $\cos^2\frac{x}{2}\left[1+\left(\frac{\sin x}{1+\cos x}\right)^2\right] = 1$

In Exercises 33–36, verify each identity by comparing the graph of the left side with the graph of the right side on a graphing utility.

33. $2\sin^2\frac{\alpha}{2} - \cos^2\frac{\alpha}{2} = \frac{1-3\cos\alpha}{2}$

34. $\cos^2\frac{A}{2} - \sin^2\frac{A}{2} = \frac{\sin 2A}{2\sin A}$

35. $2\sin^2\frac{\theta}{2} = \frac{\sin^2\theta}{1+\cos\theta}$

36. $\tan\frac{\alpha}{2} = \frac{\sin\alpha}{1+\cos\alpha}$

In Exercises 37–44, use the half-angle formulas to solve the given problems.

37. Find $\tan\theta$ if $\sin(\theta/2) = 20/29$.

38. In a right triangle with sides and angles as shown in Fig. 20.13, show that $\sin^2\frac{A}{2} = \frac{c-b}{2c}$.

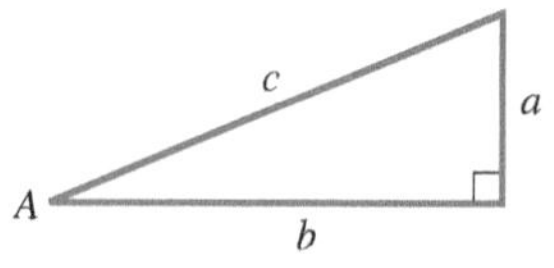

Fig. 20.13

39. In finding the path of a sliding particle, the expression $\sqrt{8-8\cos\theta}$ is used. Simplify this expression.

40. In designing a track for a railway system, the equation $d = 4r\sin^2\frac{A}{2}$ is used. Solve for d in terms of $\cos A$.

41. In electronics, in order to find the *root-mean-square current* in a circuit, it is necessary to express $\sin^2\omega t$ in terms of $\cos 2\omega t$. Show how this is done.

42. In studying interference patterns of radio signals, the expression $2E^2 - 2E^2\cos(\pi-\theta)$ arises. Show that this can be written as $4E^2\cos^2(\theta/2)$.

43. The index of refraction n, the angle A of a prism, and the minimum angle of deflection ϕ are related by $n = \dfrac{\sin\frac{1}{2}(A+\phi)}{\sin\frac{1}{2}A}$. See Fig. 20.14. Show that an equivalent expression is

$$n = \sqrt{\frac{1 - \cos A \cos\phi + \sin A \sin\phi}{1 - \cos A}}.$$

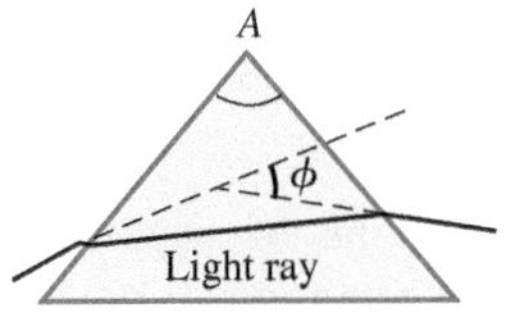

Fig. 20.14

44. For the structure shown in Fig. 20.15, show that $x = 2l\sin^2\frac{1}{2}\theta$.

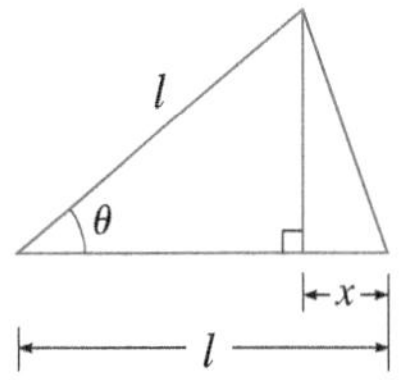

Fig. 20.15

Answers to Practice Exercises

1. $5\sin 2x$ **2.** $\pm\sqrt{\frac{2}{1-\cos x}}$

20.5 Solving Trigonometric Equations

One of the most important uses of the trigonometric identities is in the solution of equations involving trigonometric functions. The solution of this type of equation consists of the angles that satisfy the equation. When solving for the angle, we generally first solve for a value of a function of the angle and then find the angle from this value of the function.

When equations are written in terms of more than one function, the identities provide a way of changing many of them to equations or factors involving only one function of the same angle. Thus, *the solution is found by using algebraic methods and trigonometric identities and values.* From Chapter 8, recall that we must be careful regarding the sign of the value of a trigonometric function in finding the angle. Fig. 20.16 shows again the quadrants in which the functions are positive. Functions not listed are negative.

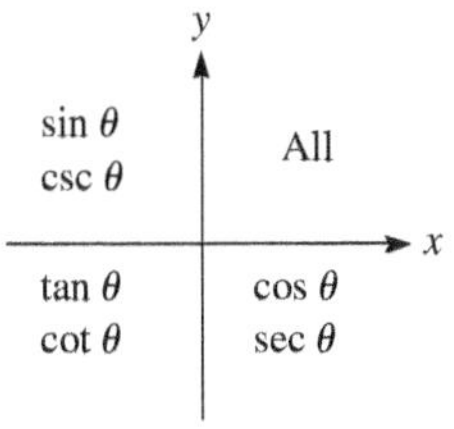

Positive functions

Fig. 20.16

EXAMPLE 1 Solving a linear equation

Solve the equation $2\cos\theta - 1 = 0$ for all values of θ such that $0 \le \theta < 2\pi$.

Solving the equation for $\cos\theta$, we obtain $\cos\theta = \frac{1}{2}$. The problem asks for all values of θ from 0 to 2π that satisfy the equation. We know that the cosines of angles in the first and fourth quadrants are positive. Also, we know that $\cos\frac{\pi}{3} = \frac{1}{2}$, which means that $\frac{\pi}{3}$ is the reference angle. Therefore, the solution proceeds as follows:

$$2\cos\theta - 1 = 0$$
$$2\cos\theta = 1 \qquad \text{solve for } \cos\theta$$
$$\cos\theta = \frac{1}{2}$$
$$\theta = \frac{\pi}{3}, \frac{5\pi}{3} \qquad \theta \text{ in quadrants I and IV}$$

Practice Exercise

1. Solve for x $(0 \le x < 2\pi)$: $2\sin x + 1 = 0$

EXAMPLE 2 Solution using a trigonometric identity and factoring

Solve the equation $2\cos^2 x - \sin x - 1 = 0$ $\quad (0 \le x < 2\pi)$.

By use of the identity $\sin^2 x + \cos^2 x = 1$, this equation may be put in terms of $\sin x$ only. Thus, we have

$$2(1 - \sin^2 x) - \sin x - 1 = 0 \qquad \text{use identity}$$
$$-2\sin^2 x - \sin x + 1 = 0 \qquad \text{multiply and simplify}$$
$$2\sin^2 x + \sin x - 1 = 0$$
$$(2\sin x - 1)(\sin x + 1) = 0 \qquad \text{factor}$$

Setting each factor equal to zero, we find $\sin x = 1/2$ or $\sin x = -1$. For the domain 0 to 2π, $\sin x = 1/2$ gives $x = \pi/6, 5\pi/6$, and $\sin x = -1$ gives $x = 3\pi/2$. Therefore,

$$x = \frac{\pi}{6}, \frac{5\pi}{6}, \frac{3\pi}{2}$$

These values check when substituted in the original equation.

Graphical Solutions

As with algebraic equations, graphical solutions of trigonometric equations are approximate, whereas algebraic solutions are often exact. As before, *we collect all terms on the left of the equal sign, with zero on the right.* We then *graph the function on the left to find its zeros by finding the values of x where the graph crosses (or is tangent to) the x-axis.*

EXAMPLE 3 Solution using a graphing utility

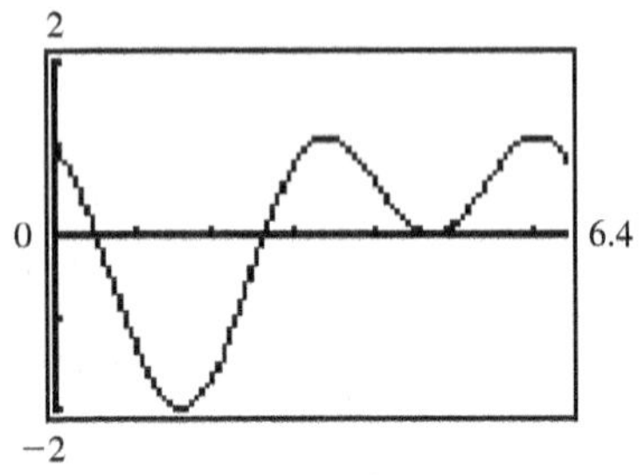

Fig. 20.17

Graphically solve the equation $2\cos^2 x - \sin x - 1 = 0$ $(0 \leq x < 2\pi)$. (This is the same equation as in Example 2.)

Since all the terms of the equation are on the left, with zero on the right, we now set $y = 2\cos^2 x - \sin x - 1$ and graph it. The graph in a graphing calculator is displayed in Fig 20.17. Using the *zero* feature, we find that $y = 0$ for

$$x = 0.52, 2.62, 4.71$$

These values are the same as in Example 2. We note that for $x = 4.71$, the curve *touches* the x-axis but does not cross it. This means it is *tangent* to the x-axis.

EXAMPLE 4 Solution by squaring—extraneous solutions

Solve the equation $\cos(x/2) = 1 + \cos x (0 \leq x < 2\pi)$.

By using the half-angle formula for $\cos(x/2)$ and then squaring both sides of the resulting equation, this equation can be solved:

$$\pm\sqrt{\frac{1 + \cos x}{2}} = 1 + \cos x \quad \text{using identity}$$

$$\frac{1 + \cos x}{2} = 1 + 2\cos x + \cos^2 x \quad \text{squaring both sides}$$

$$2\cos^2 x + 3\cos x + 1 = 0 \quad \text{simplifying}$$

$$(2\cos x + 1)(\cos x + 1) = 0 \quad \text{factoring}$$

$$\cos x = -\frac{1}{2}, -1$$

$$x = \frac{2\pi}{3}, \frac{4\pi}{3}, \pi$$

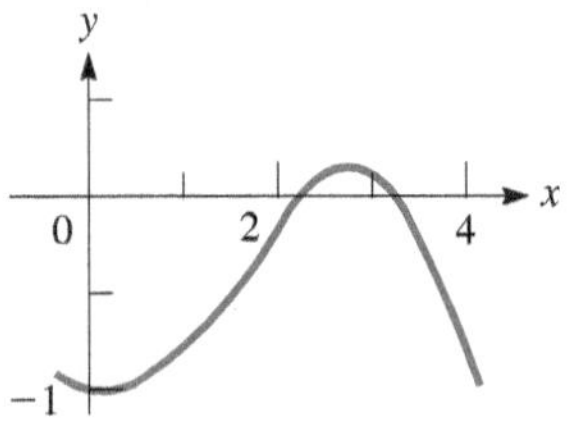

Fig. 20.18

In finding this solution, we squared both sides of the original equation. In doing this, we may have introduced extraneous solutions (see Section 14.3). Thus, we must check each solution in the original equation to see if it is valid. Hence,

$$\cos\frac{\pi}{3} = 1 + \cos\frac{2\pi}{3} \quad \text{or} \quad \frac{1}{2} = 1 + \left(-\frac{1}{2}\right) \quad \text{or} \quad \frac{1}{2} = \frac{1}{2}$$

$$\cos\frac{2\pi}{3} = 1 + \cos\frac{4\pi}{3} \quad \text{or} \quad -\frac{1}{2} = 1 + \left(-\frac{1}{2}\right) \quad \text{or} \quad -\frac{1}{2} = \frac{1}{2} \quad \text{not a valid statement}$$

$$\cos\frac{\pi}{2} = 1 + \cos\pi \quad \text{or} \quad 0 = 1 - 1 \quad \text{or} \quad 0 = 0$$

LEARNING TIP

When checking solutions, we must use the original equation and not one that has been derived after an algebraic step (such as squaring both sides) that could have changed the original function.

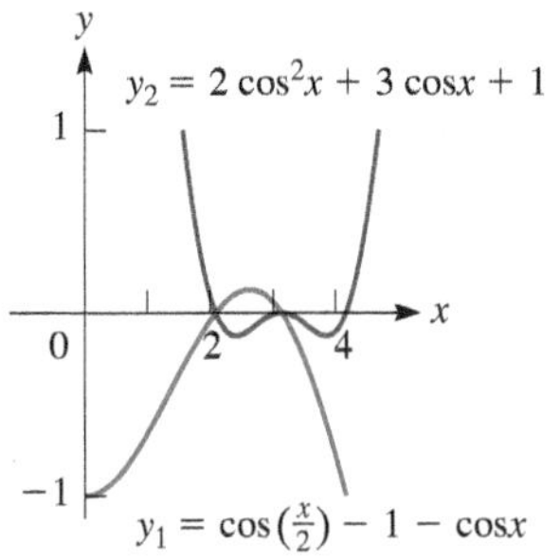

Fig. 20.19

Thus, the apparent solution $x = \frac{4\pi}{3}$ is not a solution of the original equation. The correct solutions are $x = \frac{2\pi}{3}$ and $x = \pi$.

We can see that these values agree with the values for x for which the graph of $y_1 = \cos(x/2) - 1 - \cos x$ crosses the x-axis in Fig. 20.18. In Fig. 20.19 the graph of this function and the graph of $y_2 = 2\cos^2 x + 3\cos x + 1$ are compared. We see $x = 0$ for both curves at $x = 2\pi/3$ and $x = \pi$, but that the curve for y_2 is quite different from that of y_1, and $y_2 = 0$ for $x = 4\pi/3$. When we squared both sides of the equation, we introduced a zero that is not a solution to the original equation.

EXAMPLE 5 A trigonometric equation—spring displacement

The vertical displacement y of an object at the end of a spring, which itself is being moved up and down, is given by $y = 3.50\sin t + 1.20\sin 2t$. Find the first two values of t (in seconds) for which $y = 0$.

Using the double-angle formula for $\sin 2t$ leads to the solution.

$$\begin{aligned} 3.50\sin t + 1.20\sin 2t &= 0 && \text{setting } y = 0 \\ 3.50\sin t + 2.40\sin t\cos t &= 0 && \text{using identities} \\ \sin t(3.50 + 2.40\cos t) &= 0 && \text{factoring} \\ \sin t &= 0 \quad \text{or} \quad \cos t = -1.46 \\ t &= 0.00, 3.14, \ldots \end{aligned}$$

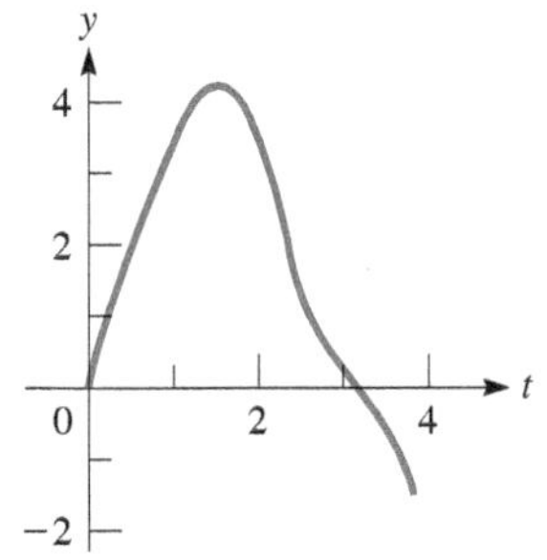

Fig. 20.20

Since $\cos t$ cannot be numerically larger than 1, there are no values of t for which $\cos t = -1.46$. Thus, the required times are $t = 0.00$ s, 3.14 s.

We can see that these values agree with the values of t for which the graph of $y = 3.50\sin t + 1.20\sin 2t$ crosses the t-axis in Fig. 20.20.

EXAMPLE 6 An equation with a double angle

Solve the equation $\tan 2\theta - \cot 2\theta = 0 \quad (0 \le \theta < 2\pi)$.

We first solve for 2θ and then for θ.

$$\begin{aligned} \tan 2\theta - \frac{1}{\tan 2\theta} &= 0 && \text{using } \cot 2\theta = \frac{1}{\tan 2\theta} \\ \tan^2 2\theta &= 1 && \text{multiplying by } \tan 2\theta \text{ and adding 1 to each side} \\ \tan 2\theta &= \pm 1 && \text{taking square roots} \end{aligned}$$

For $0 \le \theta < 2\pi$, we must have values of 2θ such that $0 \le 2\theta < 4\pi$. Therefore,

$$2\theta = \frac{\pi}{4}, \frac{3\pi}{4}, \frac{5\pi}{4}, \frac{7\pi}{4}, \frac{9\pi}{4}, \frac{11\pi}{4}, \frac{13\pi}{4}, \frac{15\pi}{4}$$

This means that the solutions are

$$\theta = \frac{\pi}{8}, \frac{3\pi}{8}, \frac{5\pi}{8}, \frac{7\pi}{8}, \frac{9\pi}{8}, \frac{11\pi}{8}, \frac{13\pi}{8}, \frac{15\pi}{8}$$

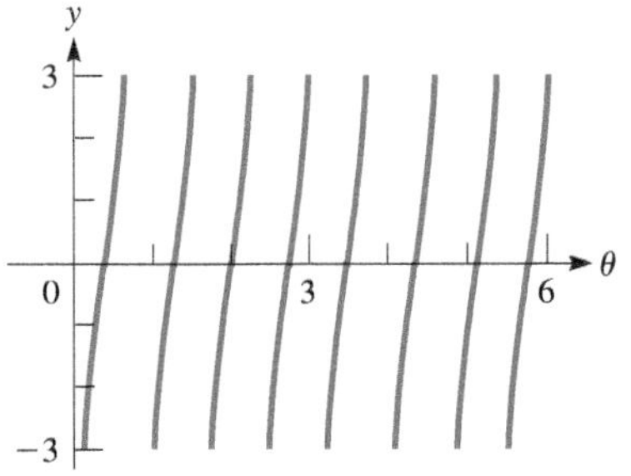

Fig. 20.21

These values satisfy the original equation. Since we multiplied through by $\tan 2\theta$ in the solution, any value of θ that leads to $\tan 2\theta = 0$ would not be valid since this would indicate division by zero in the original equation.

We see that these solutions agree with the values of θ for which the graph of $y = \cos 2\theta - \cot 2\theta$ crosses the θ-axis in Fig. 20.21.

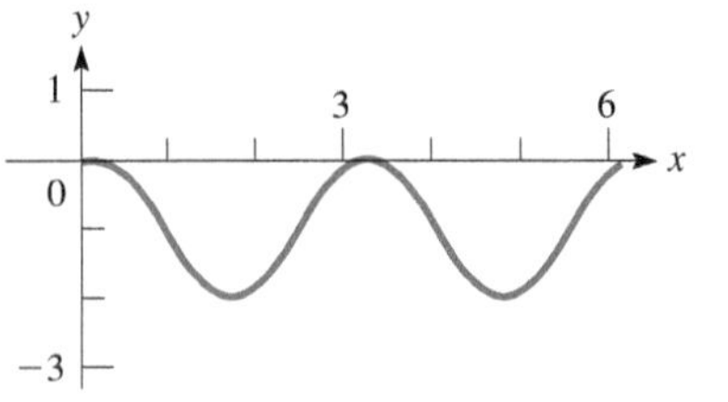

Fig. 20.22

Practice Exercise

2. Solve for $x\,(0 \leq x < 2\pi)$: $\sec^2 x + 2\tan x = 0$

EXAMPLE 7 Solution using a trigonometric identity

Solve the equation $\cos 3x \cos x + \sin 3x \sin x = 1 (0 \leq x < 2\pi)$.

The left side of this equation is of the general form $\cos(A - x)$, where $A = 3x$. Therefore,

$$\cos 3x \cos x + \sin 3x \sin x = \cos(3x - x) = \cos 2x$$

The original equation becomes

$$\cos 2x = 1$$

This equation is satisfied if $2x = 0$ or $2x = 2\pi$. The solutions are $x = 0$ and $x = \pi$. Only through recognition of the proper trigonometric form can we readily solve this equation.

We see that these solutions agree with the two values of x for which the graph of $y = \cos 3x \cos x + \sin 3x \sin x - 1$ touches the x-axis in Fig. 20.22.

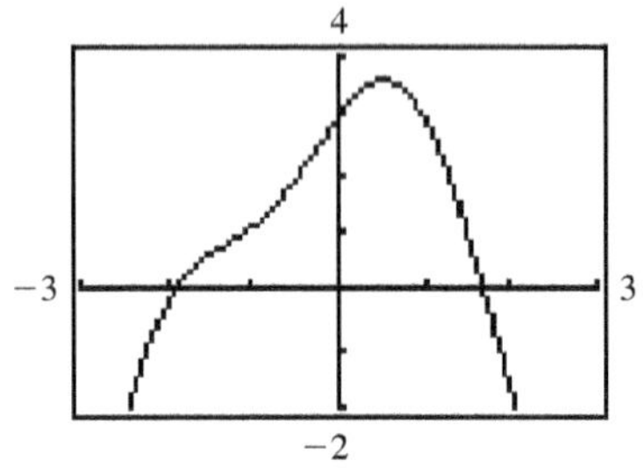

Fig. 20.23

EXAMPLE 8 Solution using a graphing utility

Solve the equation $\sin 2x + 3 = x^2$.

Although we can substitute for $\sin 2x$, we cannot express x^2 in terms of a trigonometric function. However, we can graphically find an approximate solution.

Collecting terms on the left, we have $\sin 2x + 3 - x^2 = 0$. Then we let $y = \sin 2x + 3 - x^2$ and graph it. The graph in a graphing calculator is displayed in Fig 20.23. Using the *zero* feature, we find that $x = -1.90$ and $x = 1.67$ are the only values for which $y = 0$.

EXERCISES 20.5

In Exercises 1–4, make the given changes in the indicated examples of this section and then solve the resulting problems.

1. In Example 1, change $2\cos\theta$ to $\tan\theta$.
2. In Example 2, change $2\cos^2 x$ to $2\sin^2 x$.
3. In Example 4, change $\cos\frac{1}{2}x$ to $\sin\frac{1}{2}x$ and on the right of the equal sign change the + to −.
4. In Example 7, change the + to −.

In Exercises 5–20, solve the given trigonometric equations analytically (using identities when necessary for exact values when possible) for values of x where $0 \leq x < 2\pi$.

5. $\sin x - 1 = 0$
6. $2\cos x + 1 = 0$
7. $4\sin x + 7 = 5$
8. $8\tan x + 2 = 5(1 + \tan x)$
9. $4 - \sec^2 x = 0$
10. $3\tan^2 x - 1 = 0$
11. $2\sin^2 x - \sin x = 0$
12. $\sin 4x - \sin 2x = 0$
13. $\sin 2x \sin x + \cos x = 0$
14. $\sin x - \sin\frac{x}{2} = 0$
15. $2\cos^2 x - 2\cos 2x - 1 = 0$
16. $\tan^2 x + 10 = 7\tan x$
17. $4\tan x - \sec^2 x = 0$
18. $\sin x \sin\frac{1}{2}x = 1 - \cos x$
19. $\sin 2x \cos x - \cos 2x \sin x = 0$
20. $\cos 3x \cos x - \sin 3x \sin x = 0$

In Exercises 21–38, solve the given trigonometric equations analytically. Use values of x for $0 \leq x < 2\pi$.

21. $\tan x + 1 = 0$
22. $2\sin x + 1 = 0$
23. $3 - 4\cos x = 7 - (2 - \cos x)$
24. $7\sin x - 2 = 3(2 - \sin x)$
25. $8\cos^2 x - 2 = 0$
26. $|\sin x| = \frac{1}{2}$
27. $\sin 4x - \cos 2x = 0$
28. $3\cos x - 6\cos^2 x = 0$
29. $2\sin x = \tan x$
30. $\cos 2x + \sin^2 x = 0$
31. $\sin^2 x - 2\sin x = 1$
32. $2\cos^2 2x + 1 = 3\cos 2x$
33. $\tan x + 3\cot x = 4$
34. $\tan^2 x + 4 = 2\sec^2 x$
35. $\sin 2x + \cos 2x = 0$
36. $2\sin 4x + \csc 4x = 3$
37. $2\sin 2x - \cos x \sin^3 x = 0$
38. $\tan^4 x = 1$

In Exercises 39–52, solve the indicated equations analytically.

39. $\sin 3x + \sin x = 0$ (*Hint:* See Eq. 20.17.)
40. $\cos 3x - \cos x = 0$ (*Hint:* See Eq. 20.20.)
41. Is there any positive acute angle θ for which $\sin\theta + \cos\theta + \tan\theta + \cot\theta + \sec\theta + \csc\theta = 1$? Explain.
42. Use a graphing utility to determine the minimum value of the function to the left of the equal sign in Exercise 41 (for a positive acute angle).
43. Solve the system of equations $r = \sin\theta$, $r = \sin 2\theta$, for $0 \leq \theta < 2\pi$.

44. If two musical tones of frequencies 220 Hz and 223 Hz are played together, beats will be heard. This can be represented by $y = \sin 440\pi t + \sin 446\pi t$. Graph this function and estimate t (in s) when $y = 0$ between beats for $0.15 < t < 1.15$ s.

45. The acceleration due to gravity g (in m/s^2) varies with latitude, approximately given by $g = 9.7805(1 + 0.0053 \sin^2 \theta)$, where θ is the latitude in degrees. Find θ for $g = 9.8000$.

46. Under certain conditions, the electric current i (in A) in the circuit shown in Fig. 20.24 is given below. For what value of t (in s) is the current first equal to zero?

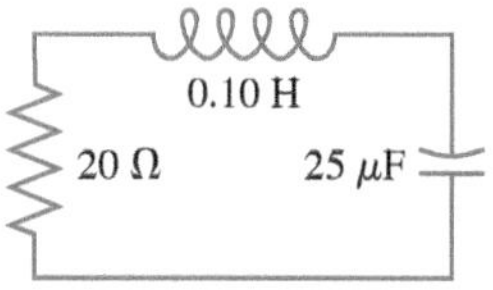

Fig. 20.24

$$i = -e^{-100t}(32.0 \sin 624t + 0.200 \cos 624t)$$

47. The vertical displacement y (in m) of the end of a robot arm is given by $y = 2.30 \cos 0.1t - 1.35 \sin 0.2t$. Find the first four values of t (in s) for which $y = 0$.

48. In finding the maximum illuminance from a point source of light, it is necessary to solve the equation $\cos\theta \sin 2\theta - \sin^3\theta = 0$. Find θ if $0 < \theta < 90°$.

49. To find the acute angle θ subtended by a certain object on a camera film, it is necessary to solve the equation $\dfrac{p^2 \tan\theta}{0.0063 + p\tan\theta} = 1.6$, where p is the distance from the camera to the object. Find θ if $p = 4.8$ m.

50. The velocity of a certain piston is maximum when the acute crank angle θ satisfies the equation $8\cos\theta + \cos 2\theta = 0$. Find this angle.

51. Resolve a force of 500.0 N into two components, perpendicular to each other, for which the sum of their magnitudes is 700.0 N, by using the angle between a component and the resultant.

52. A search and rescue helicopter flew from Gander, Newfoundland and Labrador, and travelled 160 km east. It then turned and flew due south, finally making a final turn to fly directly back to Gander. If the total distance flown was 480 km, how long were the final two legs of the flight? Solve by setting up and solving an appropriate trigonometric equation. (The Pythagorean theorem may be used only as a check.)

In Exercises 53–60, solve the given equations graphically.

53. $3\sin x - x = 0$

54. $4\cos x + 3x = 0$

55. $2\sin 2x = x^2 + 1$

56. $\sqrt{x} - \sin 3x = 1$

57. $2\ln x = 1 - \cos 2x$

58. $e^x = 1 + \sin x$

59. In finding the frequencies of vibration of a vibrating wire, the equation $x\tan x = 2.00$ occurs. Find x if $0 < x < \pi/2$.

60. An equation used in astronomy is $\theta - e\sin\theta = M$. Solve for θ for $e = 0.25$ and $M = 0.75$.

Answers to Practice Exercises

1. $x = 7\pi/6, 11\pi/6$ **2.** $x = 3\pi/4, 7\pi/4$

20.6 The Inverse Trigonometric Functions

LEARNING TIP

Throughout this section, make sure that your calculator is in *radian* mode.

As we discussed in the previous section, at times it is necessary to solve for the value of an angle in a trigonometric function. In those situations, we make use of the **inverse trigonometric functions**. You are already familiar with the notation and definitions for using these functions to solve for a single angle. In this section, we study their properties as inverse functions of the trigonometric functions.

We define the **inverse sine function**

$$y = \sin^{-1} x \quad \left(-\frac{\pi}{2} \le y \le \frac{\pi}{2}\right) \tag{20.28}$$

where y is the angle whose sine is x. This means that x is the value of the sine of the angle y, or $x = \sin y$. (The reason for explicitly showing the range as $-\pi/2 \le y \le \pi/2$ will be explained shortly.)

COMMON ERROR

In Eq. (20.28), the −1 is **not** *an exponent. The −1 in $\sin^{-1} x$ is the notation showing the inverse function.*

The notations Arcsin x, arcsin x, and $\text{Sin}^{-1} x$ are also used to designate the inverse sine. Some calculators use the [inv] key and then the [sin] key to find values of the inverse sine. However, since most calculators use the notation $\sin^{-1}$ as a second function on the [sin] key, we will continue to use $\sin^{-1} x$ for the inverse sine.

Similar definitions are used for the other inverse trigonometric functions. They also have meanings similar to that of Eq. (20.28).

EXAMPLE 1 Meaning of inverse trigonometric functions

For each of the following functions, give the meaning and inverse relationship it satisfies.

	Meaning	Inverse Relationship
(a) $y = \cos^{-1} x$	y is the angle whose cosine is x.	$x = \cos y$
(b) $y = \tan^{-1} 2x$	y is the angle whose tangent is $2x$.	$2x = \tan y$
(c) $y = \csc^{-1}(1 - x)$	y is the angle whose cosecant is $1 - x$.	$1 - x = \csc y$, or $x = 1 - \csc y$

We have seen that $y = \sin^{-1} x$ means that $x = \sin y$. From our previous work with the trigonometric functions, we know that there is an unlimited number of possible values of y for a given value of x in $x = \sin y$. Consider the following example.

EXAMPLE 2 Multiple angles for a single value

(a) When finding y such that $\sin y = \frac{1}{2}$, we know that

$$\sin\frac{\pi}{6} = \frac{1}{2} \quad \text{and} \quad \sin\frac{5\pi}{6} = \frac{1}{2}$$

In fact, $\sin y = \frac{1}{2}$ also for angles y equal to $-\frac{7\pi}{6}, \frac{13\pi}{6}, \frac{17\pi}{6}$, and so on.

(b) When finding y such that $\cos y = 1$, we know that

$$\cos 0 = 1 \quad \text{and} \quad \cos 2\pi = 1$$

In fact, $\cos y = 1$ for y equal to any even multiple of π.

From Chapter 3, we know that *to have a properly defined* ***function****, there must be only one value of the dependent variable for a given value of the independent variable.* As we just saw, there are an infinite number of angles that result in the same value of a trigonometric function. Therefore, in order to have only one value of y for each value of x in the domain of the inverse trigonometric functions, we must restrict the values of y in the range, as in Eq. (20.28). The following table lists the domain and range of all the inverse trigonometric functions.

Function	Domain	Range
$y = \sin^{-1} x$	$-1 \le x \le 1$	$-\frac{\pi}{2} \le y \le \frac{\pi}{2}$
$y = \cos^{-1} x$	$-1 \le x \le 1$	$0 \le y \le \pi$
$y = \tan^{-1} x$	$-\infty < x < \infty$	$-\frac{\pi}{2} < y < \frac{\pi}{2}$
$y = \csc^{-1} x$	$x \le -1$ or $x \ge 1$	$-\frac{\pi}{2} \le y \le \frac{\pi}{2}, y \ne 0$
$y = \sec^{-1} x$	$x \le -1$ or $x \ge 1$	$0 \le y \le \pi, y \ne \frac{\pi}{2}$
$y = \cot^{-1} x$	$-\infty < x < \infty$	$0 < y < \pi$

(20.29)

We must choose a value of y in the range as defined in Eq. (20.29) that corresponds to a given value of x in the domain. We will discuss the reasons for these definitions following the next example.

EXAMPLE 3 Evaluating an inverse trigonometric function

Explain each of the following values.

(a) $\sin^{-1}\left(\frac{1}{2}\right) = \frac{\pi}{6}$ first-quadrant angle

The possible values are $y = -\frac{7\pi}{6}, \frac{\pi}{6}, \frac{5\pi}{6}, \frac{13\pi}{6}, \ldots$ The only value within $\left[-\frac{\pi}{2}, \frac{\pi}{2}\right]$ is $\frac{\pi}{6}$.

(b) $\cos^{-1}\left(-\frac{1}{2}\right) = \frac{2\pi}{3}$ second-quadrant angle

The possible values are $y = -\frac{2\pi}{3}, \frac{2\pi}{3}, \frac{4\pi}{3}, \ldots$ The only value within $[0, \pi]$ is $\frac{2\pi}{3}$.

(c) $\tan^{-1}(-1) = -\frac{\pi}{4}$ fourth-quadrant angle

The possible values are $y = -\frac{5\pi}{4}, -\frac{\pi}{4}, \frac{3\pi}{4}, \frac{7\pi}{4}, \ldots$ The only value within $\left(-\frac{\pi}{2}, \frac{\pi}{2}\right)$ is $-\frac{\pi}{4}$.

The ranges in Eq. (20.29) are chosen so that if x in the domain is positive, the resulting value is an angle in the first quadrant. However, if x in the domain is negative, the resulting value is either a quadrantal angle, or an angle in a quadrant adjacent to the first (that is, either the second or the fourth quadrants). This guarantees a continuous range of values for the function. For example, if x is negative, negative angles in the fourth quadrant give a continuous range of values for $\sin^{-1} x$ (the sine is negative in the third and fourth quadrants), whereas angles in the second quadrant give a continuous range of values for $\cos^{-1}x$ (the cosine is negative in the second and third quadrants).

The graphs of the inverse trigonometric functions can be used to show the domains and ranges. We can obtain the graph of the inverse sine function by first sketching the sine curve $x = \sin y$ *along the y-axis*. We then mark the specific part of this curve for which $-\frac{\pi}{2} \le y \le \frac{\pi}{2}$ as the graph of the inverse sine function. The graphs of the other inverse trigonometric functions are found in the same manner. In Figs. 20.25, 20.26, and 20.27, the graphs $x = \sin y$, $x = \cos y$, and $x = \tan y$, respectively, are shown. The heavier, coloured portions indicate the graphs of the respective inverse trigonometric functions.

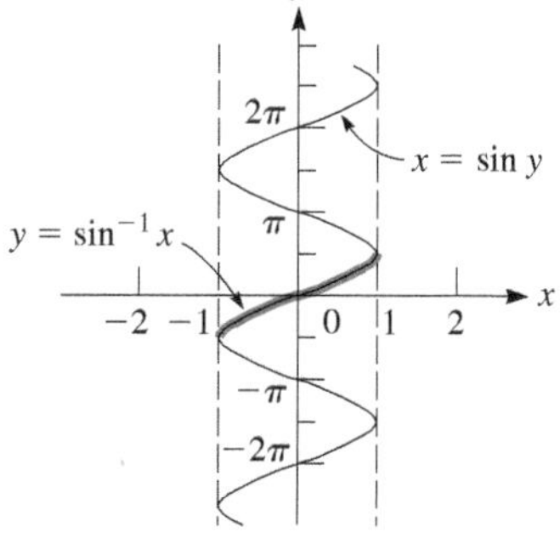

Fig. 20.25

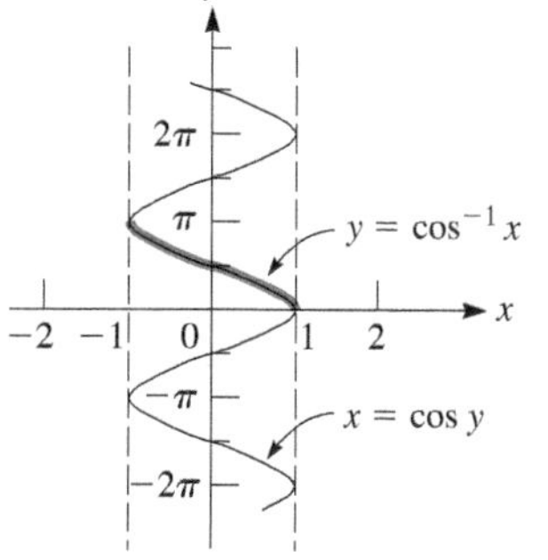

Fig. 20.26

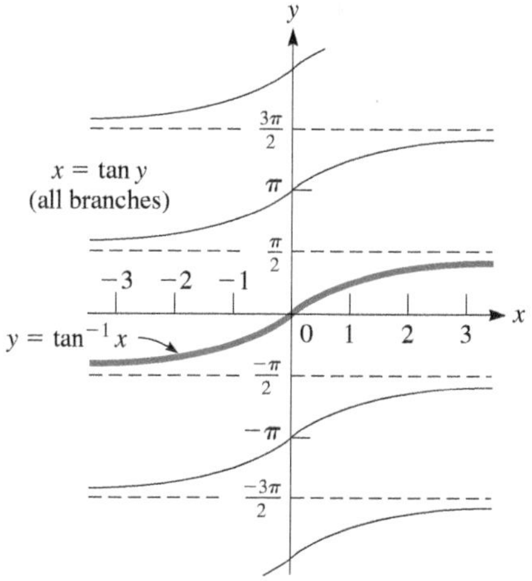

Fig. 20.27

The following examples further illustrate the values and meanings of the inverse trigonometric functions.

EXAMPLE 4 Values of inverse trigonometric functions

(a) $\sin^{-1}\left(-\sqrt{3}/2\right) = -\pi/3$ **(b)** $\cos^{-1}(-1) = \pi$

(c) $\tan^{-1} 0 = 0$ **(d)** $\tan^{-1}\left(\sqrt{3}\right) = \pi/3$

Practice Exercise

1. Evaluate $\tan^{-1}\left(-\sqrt{3}/3\right)$.

Using a calculator in radian mode, we find the following values:

(e) $\sin^{-1} 0.6294 = 0.6808$ **(f)** $\sin^{-1}(-0.1568) = -0.1574$

(g) $\cos^{-1}(-0.8026) = 2.5024$ **(h)** $\tan^{-1}(-1.9268) = -1.0921$

We note that in each case the calculator gives the value of the inverse function in the defined range for the given function.

EXAMPLE 5 Given an inverse function, solve for *x*

Given that $y = \pi - \sec^{-1} 2x$, solve for x.

We first find the expression for $\sec^{-1} 2x$ and then use the meaning of the inverse secant. The solution follows:

$$y = \pi - \sec^{-1} 2x$$

$$\sec^{-1} 2x = \pi - y \qquad \text{isolate } \sec^{-1} 2x$$

$$2x = \sec(\pi - y) \qquad \text{use meaning of inverse secant}$$

$$x = -\frac{1}{2}\sec y \qquad \sec(\pi - y) = -\sec y$$

As $\sec 2x$ and $2 \sec x$ are different functions, $\sec^{-1} 2x$ and $2\sec^{-1} x$ are also different functions. Since the values of $\sec^{-1} 2x$ are restricted, so are the resulting values of y.

EXAMPLE 6 Solve for an angle—power in an inductor

The instantaneous power P in an electric inductor is given by the equation $P = vi \sin \omega t \cos \omega t$. Solve for t.

Noting the product $\sin \omega t \cos \omega t$ suggests using $\sin 2\alpha = 2 \sin \alpha \cos \alpha$. Then, using the meaning of the inverse sine, we can complete the solution:

$$P = vi \sin \omega t \cos \omega t$$

$$= \frac{1}{2} vi \sin 2\omega t \qquad \text{using double-angle formula}$$

$$\sin 2\omega t = \frac{2P}{vi}$$

$$2\omega t = \sin^{-1}\left(\frac{2P}{vi}\right) \qquad \text{using meaning of inverse sine}$$

$$t = \frac{1}{2\omega}\sin^{-1}\left(\frac{2P}{vi}\right)$$

If we know the value of one of the inverse functions, we can find the trigonometric functions of the angle. If general relations are desired, a representative triangle is very useful. The following examples illustrate these methods.

EXAMPLE 7 Angle in terms of inverse functions

(a) Find $\cos(\sin^{-1} 0.5)$.

Knowing that the values of inverse trigonometric functions are *angles*, we see that $\sin^{-1} 0.5$ is a first-quadrant angle. Thus, we find $\sin^{-1} 0.5 = \pi/6$. The problem is now to find $\cos(\pi/6)$. This is, of course, $\sqrt{3}/2$, or 0.8660. Thus,

$$\cos(\sin^{-1} 0.5) = \cos(\pi/6) = 0.8660$$

(b) $\sin(\cot^{-1}1) = \sin(\pi/4)$ first-quadrant angle

$$= \frac{\sqrt{2}}{2} = 0.7071$$

(c) $\tan[\cos^{-1}(-1)] = \tan\pi$ quadrantal angle

$$= 0$$

(d) $\cos[\sin^{-1}(-0.2395)] = 0.9709$ using a calculator

Practice Exercise

2. Evaluate $\sin[\cos^{-1}(-0.5)]$.

EXAMPLE 8 Trigonometric function of an inverse trigonometric function

Find $\sin(\tan^{-1}x)$.

We know that $\tan^{-1}x$ is another way of stating "the angle whose tangent is x." Thus, let us draw a right triangle (as in Fig. 20.28) and label one of the acute angles as θ, the side opposite θ as x, and the side adjacent to θ as 1. In this way, we see that, by definition, $\tan\theta = \frac{x}{1}$, or $\theta = \tan^{-1}x$, which means θ is the desired angle. By the Pythagorean theorem, the hypotenuse of this triangle is $\sqrt{x^2+1}$. Now, we find that $\sin\theta$, which is the same as $\sin(\tan^{-1}x)$, is $x/\sqrt{x^2+1}$. Thus,

$$\sin(\tan^{-1}x) = \frac{x}{\sqrt{x^2+1}}$$

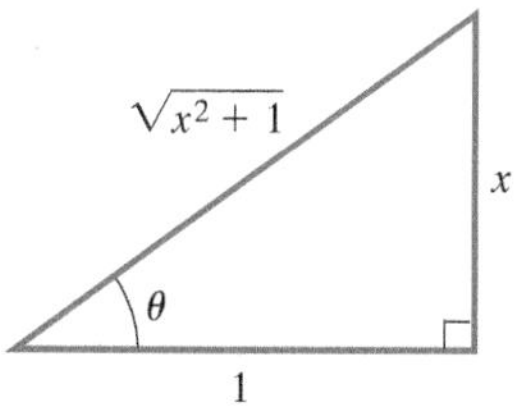

Fig. 20.28

EXAMPLE 9 Trigonometric function of an inverse trigonometric function

Find $\cos(2\sin^{-1}x)$.

From Fig. 20.29, we see that $\theta = \sin^{-1}x$. From the double-angle formulas, we have

$$\cos 2\theta = 1 - 2\sin^2\theta$$

Thus, since $\sin\theta = x$, we have

$$\cos(2\sin^{-1}x) = 1 - 2x^2$$

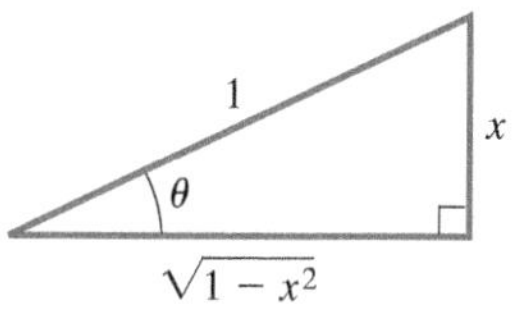

Fig. 20.29

EXAMPLE 10 Inverse trigonometric function—shelf support angles

A triangular brace of sides a, b, and c supports a shelf, as shown in Fig. 20.30. Find the expression for the angle between sides b and c.

The law of cosines leads to the solution:

$$a^2 = b^2 + c^2 - 2bc\cos A \quad \text{law of cosines}$$

$$2bc\cos A = b^2 + c^2 - a^2 \quad \text{solving for } \cos A$$

$$\cos A = \frac{b^2+c^2-a^2}{2bc}$$

$$A = \cos^{-1}\left(\frac{b^2+c^2-a^2}{2bc}\right) \quad \text{using meaning of inverse cosine}$$

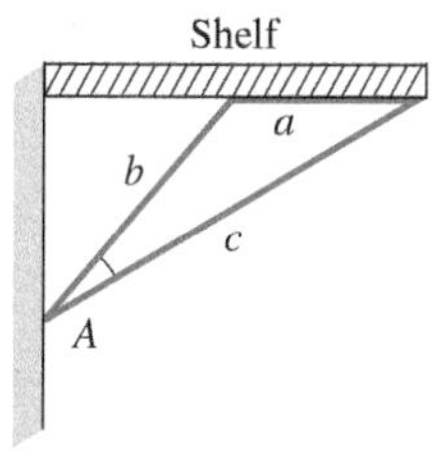

Fig. 20.30

EXERCISES 20.6

In Exercises 1–4, make the given changes in the indicated examples of this section and then solve the resulting problems.

1. In Example 1(b), change $2x$ to $3A$.
2. In Example 3(a), change $1/2$ to -1.
3. In Example 7(a), change 0.5 to 1.
4. In Example 8, change sin to cos.

In Exercises 5–10, write down the meaning of each of the given equations. See Example 1.

5. $y = \cot^{-1} 3x$ **6.** $y = \csc^{-1} 4x$
7. $y = 2 \sin^{-1} x$ **8.** $y = 3 \tan^{-1} x$
9. $y = 5 \cos^{-1}(2x - 1)$ **10.** $y = 4 \sin^{-1}(3x + 2)$

In Exercises 11–28, evaluate the given expressions exactly.

11. $\cos^{-1} 0.5$ **12.** $\sin^{-1} 1$ **13.** $\tan^{-1} 1$
14. $\sin^{-1} 2$ **15.** $\tan^{-1}(-\sqrt{3})$ **16.** $\sin^{-1}(-0.5)$
17. $\csc^{-1}(1/3)$ **18.** $\cot^{-1}\left(\frac{\sqrt{3}}{3}\right)$ **19.** $\sin^{-1}(-\sqrt{2}/2)$
20. $\cos^{-1}(-\sqrt{3}/2)$ **21.** $\cos(\tan^{-1}\sqrt{3})$ **22.** $\tan[\sin^{-1}(2/3)]$
23. $\cos^{-1}[\cos(-\pi/4)]$ **24.** $\tan^{-1}[\tan(2\pi/3)]$
25. $\cos[\tan^{-1}(-5)]$ **26.** $\sec[\cos^{-1}(-0.5)]$
27. $\cos(2 \sin^{-1} 1)$ **28.** $\sin(2 \tan^{-1}(5/12))$

In Exercises 29–32, find the exact value of x.

29. $\tan^{-1} x = \sin^{-1} \frac{2}{5}$ **30.** $\cot^{-1} x = \cos^{-1} \frac{1}{3}$
31. $\sec^{-1} x = -\sin^{-1}\left(-\frac{1}{2}\right)$ **32.** $\sin^{-1} x = -\tan^{-1}(-1)$

In Exercises 33–40, use a calculator to evaluate the given expressions.

33. $\tan^{-1}(-3.1568)$ **34.** $\cos^{-1}(-0.1008)$
35. $\sin^{-1} 0.0219$ **36.** $\tan^{-1} 0.2846$
37. $\tan[\cos^{-1}(-0.6281)]$ **38.** $\cos[\tan^{-1}(-7.2256)]$
39. $\sin[\tan^{-1}(-0.2297)]$ **40.** $\tan[\sin^{-1}(-0.3019)]$

In Exercises 41–46, solve the given equations for x.

41. $y = \sin 5x$ **42.** $y = \cos(2x - \pi)$
43. $y = \tan^{-1}(x/4)$ **44.** $y = 2 \sin^{-1}(x/6)$
45. $1 - y = \cos^{-1}(1 - x)$ **46.** $2y = \cot^{-1} 3x - 9$

In Exercises 47–54, find an algebraic expression for each of the given expressions.

47. $\tan(\sin^{-1} x)$ **48.** $\sin(\cos^{-1} x)$
49. $\sin(\sin^{-1} x + \cos^{-1} y)$ **50.** $\cos\left(\tan^{-1} \frac{x}{3}\right)$
51. $\sec(\csc^{-1} 3x)$ **52.** $\tan(\sin^{-1} 5x)$
53. $\sin(2 \sin^{-1} x)$ **54.** $\cos(2 \tan^{-1} x)$

In Exercises 55–60, solve the given problems with the use of the inverse trigonometric functions.

55. Is $\sin^{-1}(\sin x) = x$ for all x? Explain.

56. Show that the area A of a segment of a circle of radius r, bounded by a chord at a distance d from the centre, is given by $A = r^2\cos^{-1}(d/r) - d\sqrt{r^2 - d^2}$.

57. In the analysis of ocean tides, the equation $y = A \cos 2(\omega t + \phi)$ is used. Solve for t.

58. For an object of weight w on an inclined plane that is at an angle θ to the horizontal, the equation relating w and θ is $\mu w \cos \theta = w \sin \theta$, where μ is the coefficient of friction between the surfaces in contact. Solve for θ.

59. The electric current in a certain circuit is given by $i = I_m[\sin(\omega t + \alpha)\cos \phi + \cos(\omega t + \alpha)\sin \phi]$. Solve for t.

60. The time t as a function of the displacement d of a piston is given by $t = \frac{1}{2\pi f}\cos^{-1}\frac{d}{A}$. Solve for d.

In Exercises 61 and 62, prove that the given expressions are equal. Use the relation for $\sin(\alpha + \beta)$ and show that the sine of the sum of the angles on the left equals the sine of the angle on the right.

61. $\sin^{-1}\frac{3}{5} + \sin^{-1}\frac{5}{13} = \sin^{-1}\frac{56}{65}$ **62.** $\tan^{-1}\frac{1}{3} + \tan^{-1}\frac{1}{2} = \frac{\pi}{4}$

In Exercises 63–66, evaluate the given expressions.

63. $\sin^{-1} 0.5 + \cos^{-1} 0.5$ **64.** $\tan^{-1}\sqrt{3} + \cot^{-1}\sqrt{3}$
65. $\sin^{-1} x + \sin^{-1}(-x)$ **66.** $\sin^{-1} x + \cos^{-1} x$

In Exercises 67–70, solve for the angle A for the given triangles in the given figures in terms of the given sides and angles.

67.

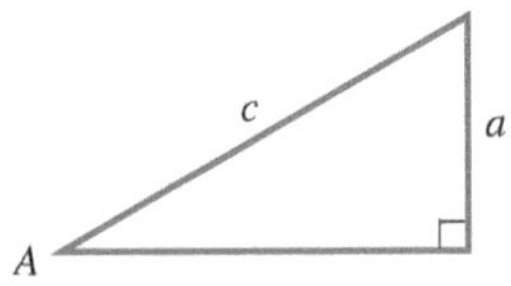

Fig. 20.31

68.

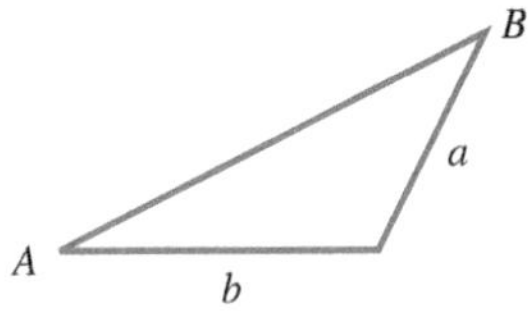

Fig. 20.32

69.

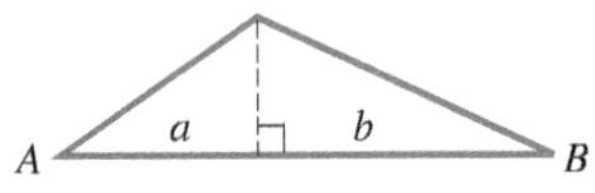

Fig. 20.33

70.

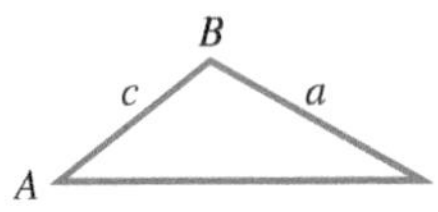

Fig. 20.34

In Exercises 71–74, solve the given problems.

71. The Terry Fox monument in Thunder Bay, Ontario, is a 2.7-m bronze statue that stands on a granite base. From a point at a horizontal distance d from the monument, the angles of elevation of the top of the statue and the top of the base are α and β, respectively. Show that

$$\alpha = \tan^{-1}\left(\frac{2.7}{d} + \tan \beta\right).$$

72. Explain why $\sin^{-1} 2x$ is not equal to $2 \sin^{-1} x$.

73. If a TV camera is x m from a launch pad of a 50-m rocket that is y m above the ground, find an expression for θ, the angle subtended by the rocket at the camera lens.

74. Show that the length L of the pulley belt shown in Fig. 20.35 is $L = 24 + 11\pi + 10 \sin^{-1}\frac{5}{13}$.

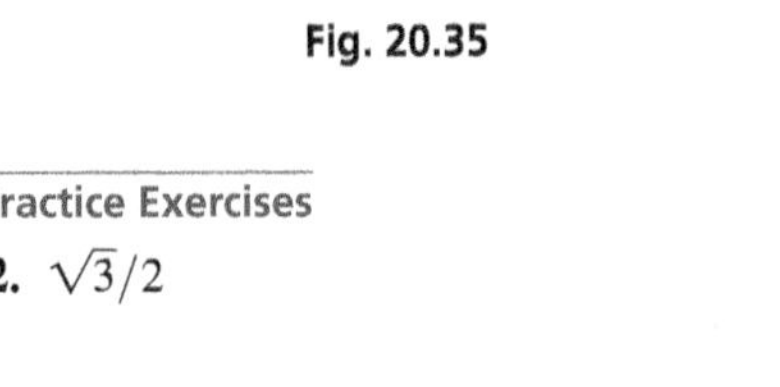

Fig. 20.35

Answers to Practice Exercises

1. $-\pi/6$ **2.** $\sqrt{3}/2$

CHAPTER 20 KEY FORMULAS AND EQUATIONS

Basic trigonometric identities

Reciprocal identities

$$\sin\theta = \frac{1}{\csc\theta} \quad \textbf{(20.1)}$$

$$\cos\theta = \frac{1}{\sec\theta} \quad \textbf{(20.2)}$$

$$\tan\theta = \frac{1}{\cot\theta} \quad \textbf{(20.3)}$$

Quotient identities

$$\tan\theta = \frac{\sin\theta}{\cos\theta} \quad \textbf{(20.4)}$$

$$\cot\theta = \frac{\cos\theta}{\sin\theta} \quad \textbf{(20.5)}$$

Pythagorean identities

$$\sin^2\theta + \cos^2\theta = 1 \quad \textbf{(20.6)}$$

$$1 + \tan^2\theta = \sec^2\theta \quad \textbf{(20.7)}$$

$$1 + \cot^2\theta = \csc^2\theta \quad \textbf{(20.8)}$$

Sum and difference identities

$$\sin(\alpha + \beta) = \sin\alpha\cos\beta + \cos\alpha\sin\beta \quad \textbf{(20.9)}$$

$$\cos(\alpha + \beta) = \cos\alpha\cos\beta - \sin\alpha\sin\beta \quad \textbf{(20.10)}$$

$$\sin(\alpha - \beta) = \sin\alpha\cos\beta - \cos\alpha\sin\beta \quad \textbf{(20.11)}$$

$$\cos(\alpha - \beta) = \cos\alpha\cos\beta + \sin\alpha\sin\beta \quad \textbf{(20.12)}$$

$$\tan(\alpha \pm \beta) = \frac{\tan\alpha \pm \tan\beta}{1 \mp \tan\alpha\tan\beta} \quad \textbf{(20.13)}$$

Double-angle formulas

$$\sin 2\alpha = 2\sin\alpha\cos\alpha \quad \textbf{(20.21)}$$

$$\cos 2\alpha = \cos^2\alpha - \sin^2\alpha \quad \textbf{(20.22)}$$

$$= 2\cos^2\alpha - 1 \quad \textbf{(20.23)}$$

$$= 1 - 2\sin^2\alpha \quad \textbf{(20.24)}$$

$$\tan 2\alpha = \frac{2\tan\alpha}{1 - \tan^2\alpha} \quad \textbf{(20.25)}$$

Half-angle formulas

$$\sin\frac{\alpha}{2} = \pm\sqrt{\frac{1 - \cos\alpha}{2}} \quad \textbf{(20.26)}$$

$$\cos\frac{\alpha}{2} = \pm\sqrt{\frac{1 + \cos\alpha}{2}} \quad \textbf{(20.27)}$$

Inverse trigonometric functions

$$y = \sin^{-1}x \left(-\frac{\pi}{2} \le y \le \frac{\pi}{2}\right) \quad \textbf{(20.28)}$$

(20.29)

Function	Domain	Range
$y = \sin^{-1}x$	$-1 \le x \le 1$	$-\frac{\pi}{2} \le y \le \frac{\pi}{2}$
$y = \cos^{-1}x$	$-1 \le x \le 1$	$0 \le y \le \pi$
$y = \tan^{-1}x$	$-\infty < x < \infty$	$-\frac{\pi}{2} < y < \frac{\pi}{2}$
$y = \csc^{-1}x$	$x \le -1$ or $x \ge 1$	$-\frac{\pi}{2} \le y \le \frac{\pi}{2}, y \ne 0$
$y = \sec^{-1}x$	$x \le -1$ or $x \ge 1$	$0 \le y \le \pi, y \ne \frac{\pi}{2}$
$y = \cot^{-1}x$	$-\infty < x < \infty$	$0 < y < \pi$

CHAPTER 20 REVIEW EXERCISES

In Exercises 1–8, determine the values of the indicated functions in the given manner.

1. Find $\sin 120°$ by using $120° = 90° + 30°$.
2. Find $\cos 30°$ by using $30° = 90° - 60°$.
3. Find $\sin 315°$ by using $315° = 360° - 45°$.
4. Find $\tan\frac{5\pi}{4}$ by using $\frac{5\pi}{4} = \pi + \frac{\pi}{4}$.
5. Find $\cos\pi$ by using $\pi = 2\left(\frac{\pi}{2}\right)$.
6. Find $\sin 180°$ by using $180° = 2(90°)$.
7. Find $\tan 60°$ by using $60° = 2(30°)$.
8. Find $\cos 45°$ by using $45° = \frac{1}{2}(90°)$.

In Exercises 9–16, simplify the given expressions by using one of the basic formulas of the chapter. Then use a calculator to verify the result by finding the value of the original expression and the value of the simplified expression.

9. $\sin 14° \cos 38° + \cos 14° \sin 38°$

10. $\cos^2 148° - \sin^2 148°$

11. $2 \sin \frac{\pi}{12} \cos \frac{\pi}{12}$

12. $1 - 2 \sin^2 \frac{\pi}{8}$

13. $\cos 73° \cos(-142°) + \sin 73° \sin(-142°)$

14. $\cos 3° \cos 215° - \sin 3° \sin 215°$

15. $\dfrac{4 \tan 12°}{1 - \tan^2 12°}$

16. $\sqrt{\dfrac{1 - \cos 166°}{2}}$

In Exercises 17–24, simplify each of the given expressions. Expansion of any term is not necessary; recognition of the proper form leads to the proper result.

17. $\sin 2x \cos 3x + \cos 2x \sin 3x$

18. $\cos 7x \cos 3x + \sin 7x \sin 3x$

19. $8 \sin 6x \cos 6x$

20. $\dfrac{\tan x + \tan 2x}{1 - \tan x \tan 2x}$

21. $2 - 4 \sin^2 6x$

22. $\cos^2 2x - \sin^2 2x$

23. $\sqrt{2 + 2\cos 2x}$

24. $\sqrt{32 - 32 \cos 4x}$

In Exercises 25–32, evaluate the given expressions.

25. $\sin^{-1}(-1)$

26. $\sec^{-1} \sqrt{2}$

27. $\cos^{-1} 0.9659$

28. $\tan^{-1}(-6.249)$

29. $\tan[\sin^{-1}(-0.5)]$

30. $\cos[\tan^{-1}(-\sqrt{3})]$

31. $\sin^{-1}[\sin(7\pi/6)]$

32. $\cos^{-1}[\tan(-\pi/4)]$

In Exercises 33–36, simplify the given expressions.

33. $\dfrac{\sec y}{\cos y} - \dfrac{\tan y}{\cot y}$

34. $\dfrac{\sin 2\theta}{2 \csc \theta} - \cos^3 \theta$

35. $\sin x(\csc x - \sin x)$

36. $\cos y(\sec y - \cos y)$

In Exercises 37–44, prove the given identities.

37. $\dfrac{\sec^4 x - 1}{\tan^2 x} = 2 + \tan^2 x$

38. $\cos^2 y - \sin^2 y = \dfrac{1 - \tan^2 y}{1 + \tan^2 y}$

39. $2 \csc 2x \cot x = 1 + \cot^2 x$

40. $\sin x \cot^2 x = \csc x - \sin x$

41. $\dfrac{1 - \sin^2 \theta}{1 - \cos^2 \theta} = \cot^2 \theta$

42. $\dfrac{\cos 2\theta}{\cos^2 \theta} = 1 - \tan^2 \theta$

43. $\sin \dfrac{\theta}{2} \cos \dfrac{\theta}{2} = \dfrac{\sin \theta}{2}$

44. $\cos(x - y)\cos y - \sin(x - y)\sin y = \cos x$

In Exercises 45–52, simplify the given expressions. The result will be one of $\sin x$, $\cos x$, $\tan x$, $\cot x$, $\sec x$, *or* $\csc x$.

45. $\dfrac{\sec x}{\sin x} - \sec x \sin x$

46. $\cos x \cot x + \sin x$

47. $\sin x \tan x + \cos x$

48. $\dfrac{\tan x \csc x}{\sin x} - \cot x$

49. $\dfrac{\sin x \cot x + \cos x}{2 \cot x}$

50. $\dfrac{1 + \cos 2x}{2 \cos x}$

51. $\dfrac{\sin 2x \sec x}{2}$

52. $(\sec x + \tan x)(1 - \sin x)$

In Exercises 53–60, verify each identity by comparing the graph of the left side with the graph of the right side using a graphing utility.

53. $\dfrac{\cos \theta - \sin \theta}{\cos \theta + \sin \theta} = \dfrac{\cot \theta - 1}{\cot \theta + 1}$

54. $\sin 3y \cos 2y - \cos 3y \sin 2y = \sin y$

55. $\sin 4x(\cos^2 2x - \sin^2 2x) = \dfrac{\sin 8x}{2}$

56. $\csc 2x + \cot 2x = \cot x$

57. $\dfrac{\sin x}{\csc x - \cot x} = 1 + \cos x$

58. $\cos x - \sin \dfrac{x}{2} = \left(1 - 2 \sin \dfrac{x}{2}\right)\left(1 + \sin \dfrac{x}{2}\right)$

59. $\tan \dfrac{\alpha}{2} = \csc \alpha - \cot \alpha$

60. $\sec \dfrac{x}{2} + \csc \dfrac{x}{2} = \dfrac{2\left(\sin \frac{x}{2} + \cos \frac{x}{2}\right)}{\sin x}$

In Exercises 61–64, solve for x.

61. $y = 2 \cos 2x$

62. $y - 2 = 2 \tan\left(x - \dfrac{\pi}{2}\right)$

63. $y = \dfrac{\pi}{4} - 3 \sin^{-1} 5x$

64. $2y = \sec^{-1} 4x - 2$

In Exercises 65–76, solve the given equations for x $(0 \le x < 2\pi)$.

65. $3(\tan x - 2) = 1 + \tan x$

66. $5 \sin x = 3 - (\sin x + 2)$

67. $2(1 - 2 \sin^2 x) = 1$

68. $\sec x = 2 \tan^2 x$

69. $2 \sin^2 \theta + 3 \cos \theta - 3 = 0$

70. $2 \sin 2x + 1 = 0$

71. $\sin x = \sin \frac{x}{2}$

72. $\cos 2x = \sin(-x)$

73. $\sin 2x = \cos 3x$

74. $\cos 3x \cos x + \sin 3x \sin x = 0$

75. $\sin^2\left(\dfrac{x}{2}\right) - \cos x + 1 = 0$

76. $\sin x + \cos x = 1$

In Exercises 77–80, determine whether the equality is an identity or an equation. If it is an identity, prove it. If it is an equation, solve it for $0 \le x < 2\pi$.

77. $\tan x + \cot x = \csc x \sec x$

78. $\tan x - \sin^2 x = \cos^2 x - \sec^2 x$

79. $\sin x \cos x - 1 = \cos x - \sin x$

80. $2 \tan x = \sin 2x \sec^2 x$

In Exercises 81–84, solve the given equations graphically.

81. $x + \ln x - 3 \cos^2 x = 2$

82. $e^{\sin x} - 2 = x \cos^2 x$

83. $2 \tan^{-1} x + x^2 = 3$

84. $3 \sin^{-1} x = 6 \sin x + 1$

In Exercises 85–90, find an algebraic expression for each of the given expressions.

85. $\tan(\cot^{-1} x)$
86. $\cos(\csc^{-1} x)$
87. $\sin(2\cos^{-1} x)$
88. $\cos(\pi - \tan^{-1} x)$
89. $\cos(\sin^{-1} x + \tan^{-1} y)$
90. $\sin(\cos^{-1} x - \tan^{-1} y)$

In Exercises 91–94, use the given substitutions to show that the equations are valid for $0 \le \theta < \pi/2$.

91. If $x = 2\cos\theta$, show that $\sqrt{4 - x^2} = 2\sin\theta$.

92. If $x = 2\sec\theta$, show that $\sqrt{x^2 - 4} = 2\tan\theta$.

93. If $x = \tan\theta$, show that $\dfrac{x}{\sqrt{1 + x^2}} = \sin\theta$.

94. If $x = \cos\theta$, show that $\dfrac{\sqrt{1 - x^2}}{x} = \tan\theta$.

In Exercises 95–115, use the methods and formulas of this chapter to solve the given problems.

95. Prove $(\cos\theta + j\sin\theta)^2 = \cos 2\theta + j\sin 2\theta$ $(j = \sqrt{-1})$.

96. Prove $(\cos\theta + j\sin\theta)^3 = \cos 3\theta + j\sin 3\theta$ $(j = \sqrt{-1})$.

97. Solve the inequality $\sin 2x > 2\cos x$ for $0 \le x < 2\pi$.

98. Show that $(\cos 2\alpha + \sin^2\alpha)\sec^2\alpha$ has a constant value.

99. Show that $y = A\sin 2t + B\cos 2t$ may be written as $y = C\sin(2t + \alpha)$, where $C = \sqrt{A^2 + B^2}$ and $\tan\alpha = B/A$. (*Hint:* Let $A/C = \cos\alpha$ and $B/C = \sin\alpha$.)

100. In a right triangle with sides a and b and hypotenuse c, show that $\sin(A/2) = \sqrt{(c - b)/2c}$, where angle A is opposite a.

101. Forces **A** and **B** act on a bolt such that **A** makes an angle θ with the x-axis and **B** makes an angle θ with the y-axis as shown in Fig. 20.36. The resultant **R** has components $R_x = A\cos\theta - B\sin\theta$ and $R_y = A\sin\theta + B\cos\theta$. Using these components, show that $R = \sqrt{A^2 + B^2}$.

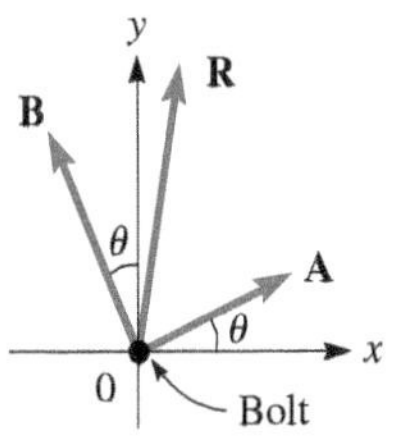

Fig. 20.36

102. For a certain alternating-current generator, the expression $I\cos\theta + I\cos(\theta + 2\pi/3) + I\cos(\theta + 4\pi/3)$ arises. Simplify this expression.

103. Some comets follow a parabolic path that can be described by the equation $r = (k/2)\csc^2(\theta/2)$, where r is the distance to the sun and k is a constant. Show that this equation can be written as $r = k/(1 - \cos\theta)$.

104. In studying the interference of light waves, the identity $\dfrac{\sin\frac{3}{2}x}{\sin\frac{1}{2}x}\sin x = \sin x + \sin 2x$ is used. Prove this identity. $\left(\textit{Hint: } \sin\frac{3}{2}x = \sin\left(x + \frac{1}{2}x\right).\right)$

105. In the study of chemical spectroscopy, the equation $\omega t = \sin^{-1}\dfrac{\theta - \alpha}{R}$ arises. Solve for θ.

106. The power P in a certain electric circuit is given by $P = 2.5[\cos\alpha\sin(\omega t + \phi) - \sin\alpha\cos(\omega t + \phi)]$. Solve for t.

107. In surveying, when determining an azimuth (a measure used for reference purposes), it might be necessary to simplify the expression $(2\cos\alpha\cos\beta)^{-1} - \tan\alpha\tan\beta$. Perform this operation by expressing it in the simplest possible form when $\alpha = \beta$.

108. In analysing the motion of an automobile universal joint, the equation $\sec^2 A - \sin^2 B\tan^2 A = \sec^2 C$ is used. Show that this equation is true if $\tan A\cos B = \tan C$.

109. The instantaneous power P in a certain electric circuit is given by $P = VI\cos\phi\cos^2\omega t - VI\sin\phi\cos\omega t\sin\omega t$. Simplify this expression.

110. In studying waveforms, a sawtooth wave may often be approximated by $y = \sin\pi x + \frac{1}{2}\sin 2\pi x + \frac{1}{3}\sin 3\pi x$. For what values of x, $0 \le x < 2$, is the sum of the first two terms equal to zero?

111. A roof truss is in the shape of an isosceles triangle of height 3.2 m. If the total length of the three members is 25.6 m, what is the length of a rafter? Solve by setting up and solving an appropriate trigonometric equation. (The Pythagorean theorem may be used only as a check.)

112. The angle of elevation of the top of the Peace Tower on Parliament Hill in Ottawa from a point on level ground 77 m from the tower is twice the angle of elevation of the top from a point 120 m farther away. Find the height of the Peace Tower.

113. If a plane surface inclined at angle θ moves horizontally, the angle for which the lifting force of the air is a maximum is found by solving the equation $2\sin\theta\cos^2\theta - \sin^3\theta = 0$, where $0 < \theta < 90°$. Solve for θ.

114. To determine the angle between two sections of a certain robot arm, the equation $1.20\cos\theta + 0.135\cos 2\theta = 0$ is to be solved. Find the required angle θ if $0° < \theta < 180°$.

115. In checking the angles of a section of a bridge support, an engineer finds the expression $\cos(2\sin^{-1} 0.40)$. Write a paragraph explaining how the value of this expression can be found without the use of a calculator.

CHAPTER 20 PRACTICE TEST

1. Prove that $\sec\theta - \dfrac{\tan\theta}{\csc\theta} = \cos\theta$.

2. Solve for $x (0 \le x < 2\pi)$ analytically, using trigonometric relations where necessary: $\sin 2x + \sin x = 0$.

3. Find an algebraic expression for $\cos(\sin^{-1} x)$.

4. The electric current as a function of the time for a particular circuit is given by $i = 8.00e^{-20t}(1.73\cos 10.0t - \sin 10.0t)$. Find the time (in s) when the current is first zero.

5. Prove that $\dfrac{\tan\alpha + \tan\beta}{\tan\alpha - \tan\beta} = \dfrac{\sin(\alpha + \beta)}{\sin(\alpha - \beta)}$.

6. Prove that $\cot^2 x - \cos^2 x = \cot^2 x\cos^2 x$.

7. Find $\cos\frac{1}{2}x$ if $\sin x = -\frac{3}{5}$ and $270° < x < 360°$.

8. The intensity of a certain type of polarized light is given by $I = I_0\sin 2\theta\cos 2\theta$. Solve for θ.

9. Solve graphically: $x - 2\cos x = 5$.

10. Find the exact value of x: $\cos^{-1} x = -\tan^{-1}(-1)$.

21. Plane Analytic Geometry

LEARNING OUTCOMES

After completion of this chapter, the student should be able to:

- Find the distance between two points in the coordinate plane
- Find the slope and the inclination of a line
- Understand the relationship between the slopes of parallel and perpendicular lines
- Determine the equation and properties of a circle, a parabola, an ellipse, and a hyperbola
- Sketch the graph of a line, a circle, a parabola, an ellipse, or a hyperbola
- Find the equation of a conic from the definition of its locus
- Identify the conic section from the general second-degree equation
- Recognize conic sections as intersections of planes and cones
- Obtain a new coordinate system by translation and/or rotation of axes
- Convert from rectangular coordinates to polar coordinates and vice versa
- Graph functions in polar coordinates
- Use algebra to solve geometric problems
- Solve application problems involving lines, conics, and polar coordinates

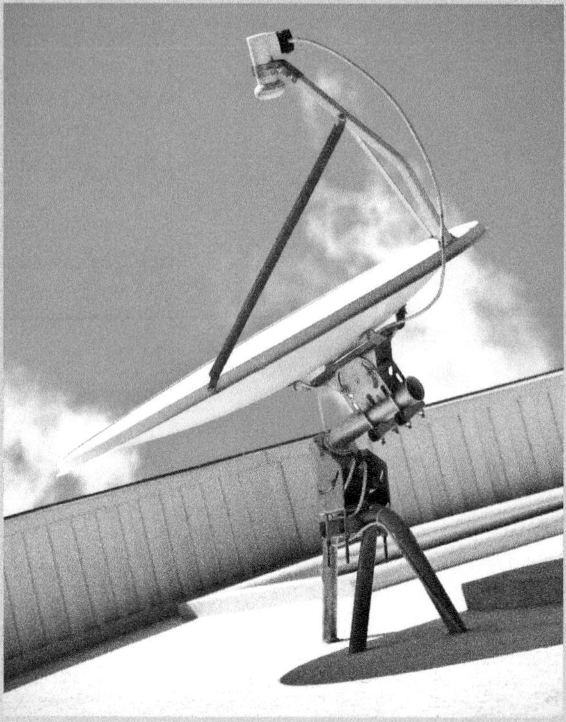

◀ In Section 21.4, we show an important application of analytic geometry in the design of a TV satellite dish.

The development of geometry made little progress from the time of the ancient Greeks until the 1600s. Then in the 1630s, two French mathematicians, Rene Descartes and Pierre de Fermat, independently introduced algebra into the study of geometry. Each used algebraic notation and a coordinate system, and this allowed them to analyse the properties of curves. Fermat briefly indicated his methods in a letter he wrote in 1636. However, in 1637 Descartes included a much more complete work as an appendix to his *Discourse on Method*, and he is therefore now considered the founder of *analytic geometry*.

At the time, it was known, for example, that a projectile follows a parabolic path and that the orbits of the planets are ellipses. There was a renewed interest in curves of various kinds because of their applications. However, Descartes was primarily interested in studying the relationship of algebra to geometry. In doing so, he started the development of analytic geometry, which has many applications and also was very important in the invention of calculus that came soon thereafter. In turn, through these advances in mathematics, many more areas of science and technology were able to be developed.

The underlying principle of analytic geometry is the relationship of an algebraic equation and the geometric properties of the curve that represents the equation. In this chapter, we develop equations for a number of important curves and find their properties through an analysis of their equations. Most important among these curves are the *conic sections*, which we briefly introduced in Chapter 14.

As we have noted, analytic geometry has applications in the study of projectile motion and planetary orbits. Other important applications range from the design of gears, airplane wings, and automobile headlights to the construction of bridges and nuclear towers.

21.1 Basic Definitions

As we have noted, analytic geometry deals with the relationship between an algebraic equation and the geometric curve it represents. In this section, we develop certain basic concepts that will be needed for future use in establishing the proper relationships between an equation and a curve.

THE DISTANCE FORMULA

The first of these concepts involves the distance between any two points in the coordinate plane. If these points lie on a line parallel to the x-axis, the distance from the first point (x_1, y) to the second point (x_2, y) is $|x_2 - x_1|$. The absolute value is used since we are interested only in the magnitude of the distance. Therefore, we could also denote the distance as $|x_1 - x_2|$. Similarly, the distance between two points (x, y_1) and (x, y_2) that lie on a line parallel to the y-axis is $|y_2 - y_1|$ or $|y_1 - y_2|$.

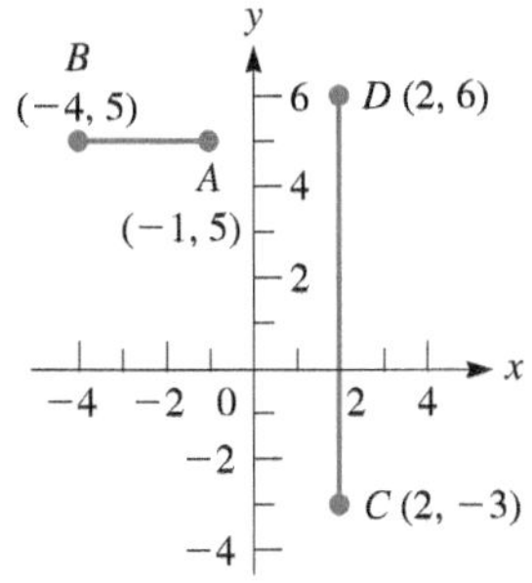

Fig. 21.1

EXAMPLE 1 Distance between points that share one coordinate

(a) The line segment joining $A(-1, 5)$ and $B(-4, 5)$ in Fig. 21.1 is parallel to the x-axis. Find the distance between these points.

$$d = |-4 - (-1)| = 3 \quad \text{or} \quad d = |-1 - (-4)| = 3$$

(b) The line segment joining $C(2, -3)$ and $D(2, 6)$, also in Fig. 21.1, is parallel to the y-axis. Find the distance d between these points.

$$d = |6 - (-3)| = 9 \quad \text{or} \quad d = |-3 - 6| = 9$$

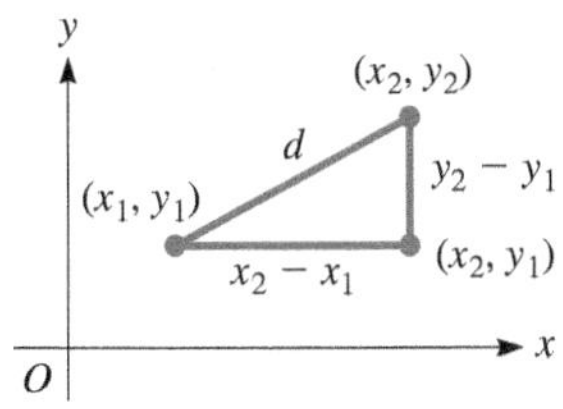

Fig. 21.2

We now wish to find the length of a line segment joining any two points in the plane. If these points are on a line that is not parallel to either axis (see Fig. 21.2), we use the Pythagorean theorem to find the distance between them. By making a right triangle with the line segment joining the points as the hypotenuse and line segments parallel to the axes as legs, we have *the* **distance formula**, *which gives the distance between any two points in the plane.* This formula is

$$d = \sqrt{(x_2 - x_1)^2 + (y_2 - y_1)^2} \tag{21.1}$$

Here, we choose the positive square root since we are concerned only with the magnitude of the length of the line segment.

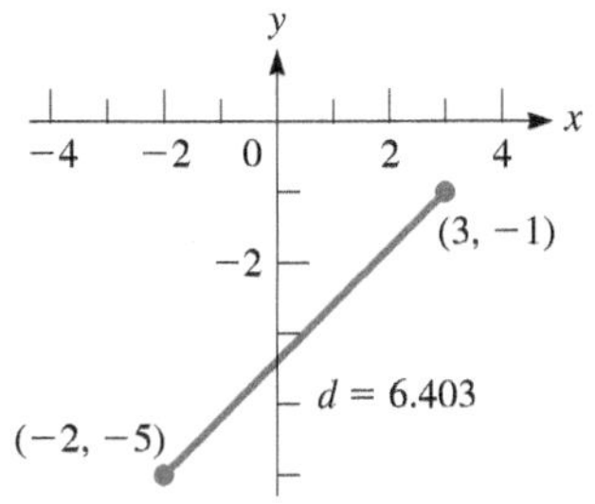

Fig. 21.3

EXAMPLE 2 Distance formula

Find the distance between $(3, -1)$ and $(-2, -5)$. See Fig. 21.3.

$$\begin{aligned} d &= \sqrt{[(-2) - 3]^2 + [(-5) - (-1)]^2} \\ &= \sqrt{(-5)^2 + (-4)^2} = \sqrt{25 + 16} \\ &= \sqrt{41} = 6.40 \end{aligned}$$

It makes no difference which point is chosen as (x_1, y_1) and which is chosen as (x_2, y_2), since the differences in the x-coordinates and the y-coordinates are squared. We obtain the same value for the distance when we calculate it as

$$\begin{aligned} d &= \sqrt{[3 - (-2)]^2 + [(-1) - (-5)]^2} \\ &= \sqrt{5^2 + 4^2} = \sqrt{41} = 6.40 \end{aligned}$$

Practice Exercise

1. Find the distance between $(-2, 6)$ and $(5, -3)$.

THE SLOPE OF A LINE

Another important quantity for a line is its *slope*, which we defined in Chapter 5. Here, we review its definition and develop its meaning in more detail.

The **slope** *of a line through two points is defined as the difference in the y-coordinates (rise) divided by the difference in the x-coordinates (run).* Therefore, the slope, m, which gives a measure of the direction of a line, is defined as

$$m = \frac{y_2 - y_1}{x_2 - x_1} \qquad (21.2)$$

> **LEARNING TIP**
> The larger the numerical value of the slope of a line, the more nearly vertical is the line. Also, ***a line rising to the right has a positive slope, and a line falling to the right has a negative slope.***

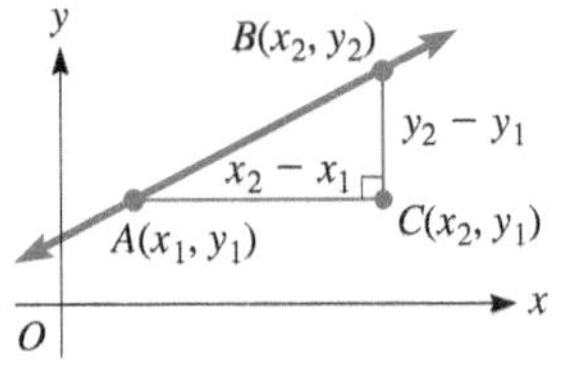

Fig. 21.4

See Fig. 21.4.

Note that we may interpret either of the points as (x_1, y_1) and the other as (x_2, y_2), although *we must be careful to place the $y_2 - y_1$ in the numerator and $x_2 - x_1$ in the denominator.* Moreover, when the line is horizontal, $y_2 - y_1 = 0$ and $m = 0$. When the line is vertical, $x_2 - x_1 = 0$, and the slope is undefined.

EXAMPLE 3 Slope of a line

Determine the slope of the line between each pair of points and interpret its magnitude and sign.

(a) $(3, -5)$ and $(-2, -6)$

We assign values to subscripts: $(x_1, y_1) = (3, -5)$ and $(x_2, y_2) = (-2, -6)$. Then

$$m = \frac{y_2 - y_1}{x_2 - x_1} = \frac{-6 - (-5)}{-2 - 3} = \frac{1}{5}$$

We can also obtain the slope of this same line from

$$m = \frac{-5 - (-6)}{3 - (-2)} = \frac{1}{5}$$

The slope is positive and small in magnitude, so the line rises slowly to the right as seen in Fig. 21.5.

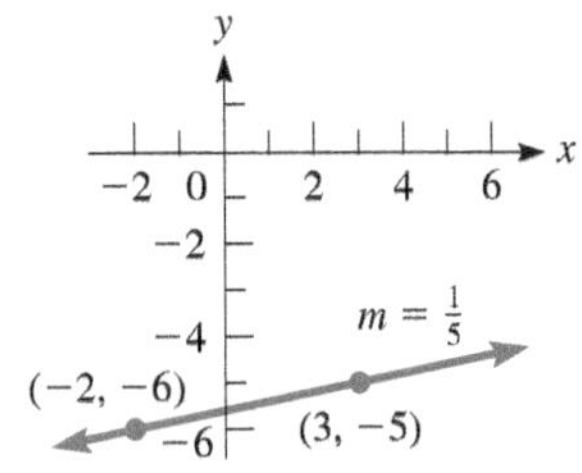

Fig. 21.5

(b) $(3, 4)$ and $(4, -6)$

We have

$$m = \frac{4 - (-6)}{3 - 4} = -10 \quad \text{or} \quad m = \frac{-6 - 4}{4 - 3} = -10$$

The slope is negative and large in magnitude, so the line falls sharply to the right as seen in Fig. 21.6.

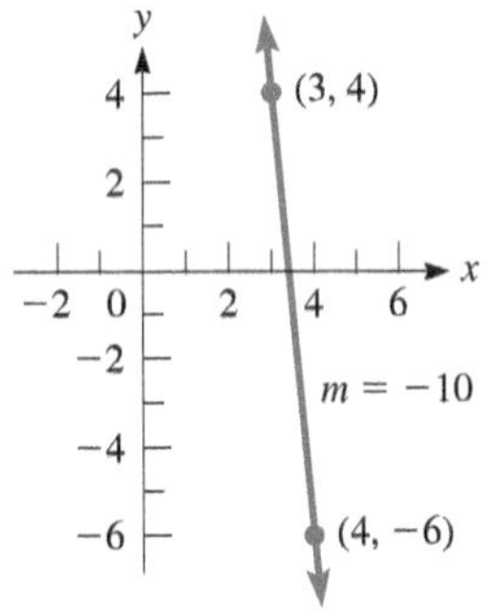

Fig. 21.6

If a given line is extended indefinitely in either direction, it must cross the x-axis at some point unless it is parallel to the x-axis. *The angle measured from the x-axis in a positive direction to the line is called the* **inclination** *of the line* (see Fig. 21.7). The inclination of a line parallel to the x-axis is defined to be zero. *An alternative definition of slope, in terms of the inclination α, is*

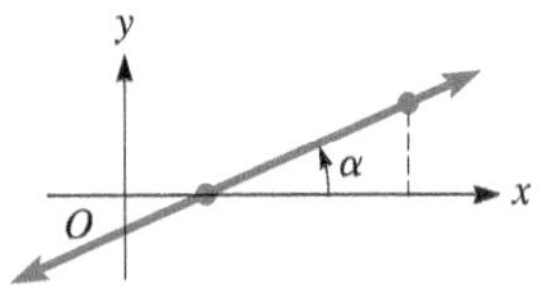

Fig. 21.7

$$m = \tan\alpha \quad (0° \le \alpha < 180°) \qquad (21.3)$$

Since the slope can be defined in terms of any two points on the line, we can choose the x-intercept and any other point. Therefore, from the definition of the tangent of an angle, we see that Eq. (21.3) is in agreement with Eq. (21.2).

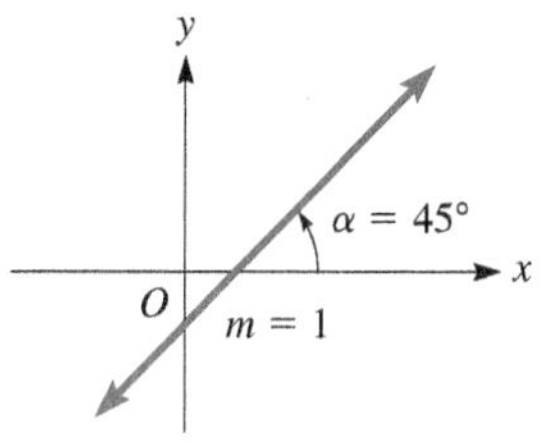

Fig. 21.8

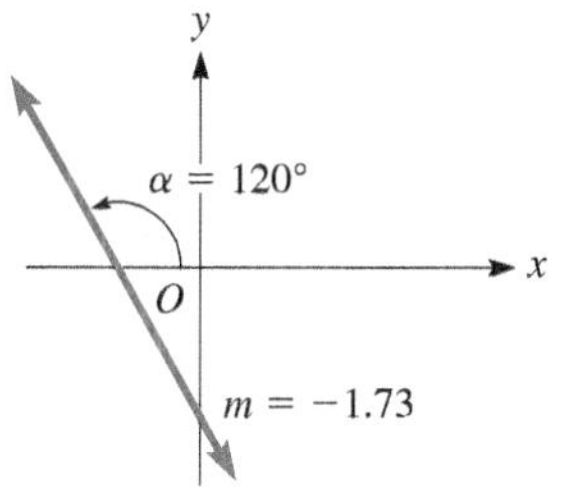

Fig. 21.9

EXAMPLE 4 Inclination of a line

For the following lines, given the slope, find the inclination; or given the inclination, find the slope.

(a) Inclination: $\alpha = 45°$. The slope is

$$m = \tan 45° = 1$$

As seen in Fig. 21.8, when the inclination is an acute angle, the slope is positive and the line rises to the right.

(b) Slope: $m = -1.73$. To find the inclination, we know that $\tan \alpha = -1.73$. Since $\tan \alpha$ is negative, α must be a second-quadrant angle, with reference angle $\tan^{-1} 1.73 = 60°$. Therefore,

$$\alpha = 180° - 60° = 120°$$

As seen in Fig. 21.9, when the inclination is an obtuse angle, the slope is negative and the line falls to the right.

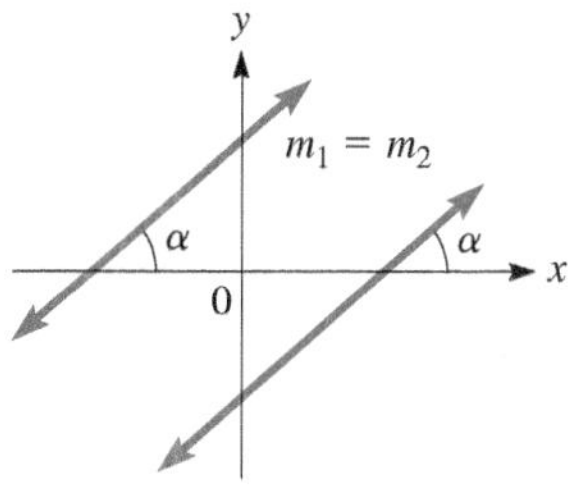

Fig. 21.10

Any two parallel lines crossing the x-axis have the same inclination. Therefore, as shown in Fig. 21.10, the *slopes of parallel lines are equal.* This can be stated as

$$m_1 = m_2 \qquad \text{(for } \parallel \text{ lines)} \qquad \textbf{(21.4)}$$

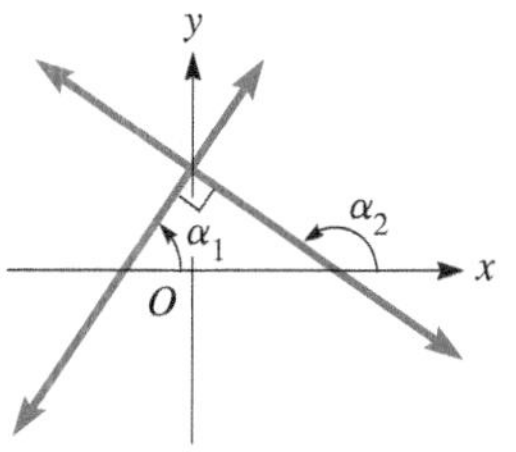

Fig. 21.11

If two lines are perpendicular, this means that there must be 90° between their inclinations (Fig. 21.11). The relation between their inclinations is

$$\alpha_2 = \alpha_1 + 90°$$
$$90° - \alpha_2 = -\alpha_1$$

If neither line is vertical (the slope of a vertical line is undefined) and we take the tangent in this last relation, we have

$$\tan(90° - \alpha_2) = \tan(-\alpha_1)$$

We know $\tan(90° - \alpha) = \cot \alpha$, and $\tan(-\alpha) = -\tan(\alpha)$. Hence

$$\cot \alpha_2 = -\tan \alpha_1$$

But $\cot \alpha = 1/\tan \alpha$, which means $1/\tan \alpha_2 = -\tan \alpha_1$. Using the inclination definition of slope, we have *as the relation between slopes of perpendicular lines,*

$$m_2 = -\frac{1}{m_1} \quad \text{or} \quad m_1 m_2 = -1 \qquad \text{for } \perp \text{ lines} \qquad \textbf{(21.5)}$$

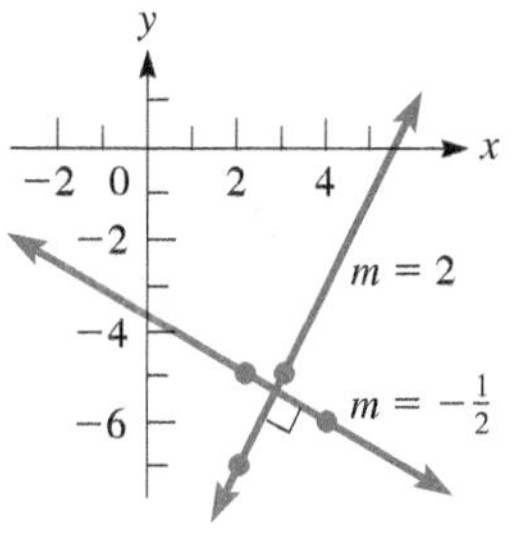

Fig. 21.12

EXAMPLE 5 Perpendicular lines

Verify that the line through $(3, -5)$ and $(2, -7)$ and the line through $(4, -6)$ and $(2, -5)$ are perpendicular.

The slopes of the two lines are

$$m_1 = \frac{-5 + 7}{3 - 2} = 2 \quad \text{and} \quad m_2 = \frac{-6 - (-5)}{4 - 2} = -\frac{1}{2}$$

Since the slopes of the two lines are negative reciprocals (or $m_1 m_2 = -1$), the lines are perpendicular. See Fig. 21.12.

Practice Exercise

2. What is the slope of a line perpendicular to the line through $(-1, 6)$ and $(-3, -3)$?

Using the formulas for distance and slope, we can show certain basic geometric relationships. The following examples illustrate the use of the formulas and thereby show the use of algebra in solving problems that are basically geometric. They illustrate the methods of analytic geometry.

EXAMPLE 6 Use of algebra in geometric problems

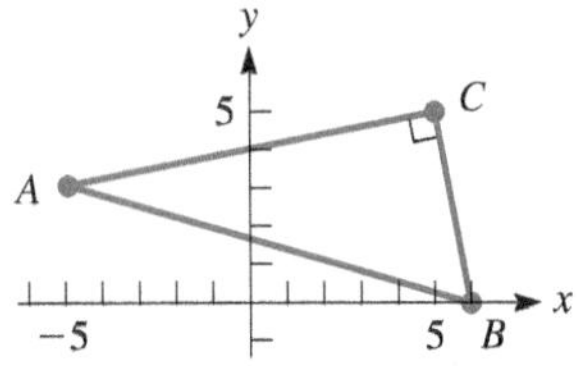

Fig. 21.13

(a) Show that the line segments joining $A(-5, 3)$, $B(6, 0)$, and $C(5, 5)$ form a right triangle. See Fig. 21.13.

If these points are vertices of a right triangle, the slopes of two of the sides must be negative reciprocals. This would show perpendicularity. These slopes are

$$m_{AB} = \frac{3-0}{-5-6} = -\frac{3}{11} \qquad m_{AC} = \frac{3-5}{-5-5} = \frac{1}{5} \qquad m_{BC} = \frac{0-5}{6-5} = -5$$

We see that the slopes of AC and BC are negative reciprocals, which means $AC \perp BC$. From this we conclude that the triangle is a right triangle.

(b) Find the area of the triangle in part (a). See Fig. 21.14.

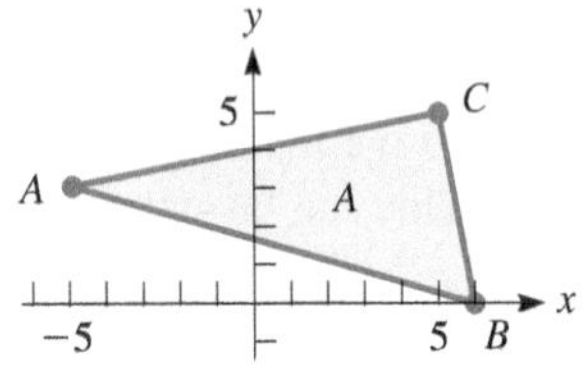

Fig. 21.14

Since the right angle is at C, the legs of the triangle are AC and BC. The area is one-half the product of the lengths of the legs of a right triangle. The lengths of the legs are

$$d_{AC} = \sqrt{(-5-5)^2 + (3-5)^2} = \sqrt{104} = 2\sqrt{26}$$
$$d_{BC} = \sqrt{(6-5)^2 + (0-5)^2} = \sqrt{26}$$

Therefore, the area is $A = \frac{1}{2}(2\sqrt{26})(\sqrt{26}) = 26$.

EXERCISES 21.1

In Exercises 1–4, make the given changes in the indicated examples of this section and then solve the resulting problems.

1. In Example 2, change $(-2, -5)$ to $(-2, 5)$.
2. In Example 3(a), change $(3, -5)$ to $(-3, -5)$.
3. In Example 4(b), change -1.73 to -0.577.
4. In Example 6(a), change $B(6, 0)$ to $B(-4, -2)$.

In Exercises 5–14, find the distance between the given pairs of points.

5. $(3, 8)$ and $(-1, -2)$
6. $(-1, 3)$ and $(-8, -4)$
7. $(4, -5)$ and $(4, -8)$
8. $(-3, 7)$ and $(2, 7)$
9. $(-12, 20)$ and $(32, -13)$
10. $(23, -9)$ and $(-25, 11)$
11. $(\sqrt{32}, -\sqrt{18})$ and $(-\sqrt{50}, \sqrt{8})$
12. $(e, -\pi)$ and $(-2e, -\pi)$
13. $(1.22, -3.45)$ and $(-1.07, -5.16)$
14. (a, h^2) and $(a + h, (a + h)^2)$

In Exercises 15–24, find the slopes of the lines through the points in Exercises 5–14.

In Exercises 25–28, find the slopes of the lines with the given inclinations.

25. 30°
26. 62.5°
27. 163°
28. 93.5°

In Exercises 29–32, find the inclinations of the lines with the given slopes.

29. 0.364
30. 0.824
31. −6.69
32. −0.721

In Exercises 33–36, determine whether the lines through the two pairs of points are parallel or perpendicular.

33. $(6, -1)$ and $(4, 3)$; $(-5, 2)$ and $(-7, 6)$
34. $(-3, 9)$ and $(4, 4)$; $(9, -1)$ and $(4, -8)$
35. $(-1, -4)$ and $(2, 3)$; $(-5, 2)$ and $(-19, 8)$
36. $(-a, -2b)$ and $(3a, 6b)$; $(2a, -6b)$ and $(5a, 0)$

In Exercises 37–40, determine the value of k.

37. The distance between $(-1, 3)$ and $(11, k)$ is 13.
38. The distance between $(k, 0)$ and $(0, 2k)$ is 10.
39. Points $(6, -1)$, $(3, k)$, and $(-3, -7)$ are on the same line.
40. The points in Exercise 39 are the vertices of a right triangle, with the right angle at $(3, k)$.

In Exercises 41–44, show that the given points are vertices of the given geometric figures.

41. $(2, 3)$, $(4, 9)$, and $(-2, 7)$ are the vertices of an isosceles triangle.
42. $(-1, 3)$, $(3, 5)$, and $(5, 1)$ are the vertices of a right triangle.
43. $(-5, -4)$, $(7, 1)$, $(10, 5)$, and $(-2, 0)$ are the vertices of a parallelogram.
44. $(-5, 6)$, $(0, 8)$, $(-3, 1)$, and $(2, 3)$ are the vertices of a square.

In Exercises 45–48, find the indicated areas and perimeters.

45. Find the area of the triangle in Exercise 42.
46. Find the area of the square in Exercise 44.

47. Find the perimeter of the triangle in Exercise 41.

48. Find the perimeter of the parallelogram in Exercise 43.

In Exercises 49–52, use the following definition to find the midpoint between the given points on a straight line.

The *midpoint* between points (x_1, y_1) and (x_2, y_2) on a straight line is the point

$$\left(\frac{x_1 + x_2}{2}, \frac{y_1 + y_2}{2}\right)$$

49. $(-4, 9)$ and $(6, 1)$ **50.** $(-1, 6)$ and $(-13, -8)$

51. $(-12.4, 25.7)$ and $(6.8, -17.3)$ **52.** $(2.6, 5.3)$ and $(-4.2, -2.7)$

In Exercises 53–62, solve the given problems.

53. Find the relation between x and y such that (x, y) is always 3 units from the origin.

54. Find the relation between x and y such that (x, y) is always equidistant from the y-axis and $(2, 0)$.

55. Show that the diagonals of a square are perpendicular to each other. (*Hint:* Use $(0, 0)$, $(a, 0)$, and $(0, a)$ as three of the vertices.)

56. The centre of a circle is $(2, -3)$, and one end of a diameter is $(-1, 2)$. What are the coordinates of the other end of the diameter?

57. A line segment has a slope of 3 and one endpoint at $(-2, 5)$. If the other endpoint is on the x-axis, what are its coordinates?

58. The points $(-1, 3)$, $(5, y)$, and $(2, 4)$ are collinear (on the same line). Find y.

59. Find the coordinates of the point on the y-axis that is equidistant from $(-3, -5)$ and $(2, 4)$.

60. The *grade* of a highway is its slope expressed as a percent (a 5% grade means the slope is $\frac{5}{100}$). If the grade of a certain highway is 6%, find (a) its angle of inclination and (b) the change in elevation (in m) of a car driving for 2.00 km uphill along this highway.

61. On a computer drawing showing the specifications for a mounting bracket, holes are to be drilled at the points $(32.5, 25.5)$ and $(88.0, 62.5)$, where all measurements are in mm. Find the distance between the centres of the two holes.

62. A person is working out on a treadmill inclined at 12% (the slope of the treadmill expressed in percent). What is the angle between the treadmill and the horizontal?

Answers to Practice Exercises

1. $d = \sqrt{130} = 11.4$ **2.** $m = -2/9$

21.2 The Straight Line

In Chapter 5, we discussed the *general form* of the linear equation with two unknowns, and graphed it as a straight line by writing it in the *slope-intercept form* of the equation of a straight line. In this section we will revisit these equations, but we will arrive at them by using the methods of analytic geometry. We will define a straight line as a curve that satisfies certain geometric conditions, and then derive the general type of equation that represents it.

A straight line can be defined as a *curve* with a constant slope. This means that the value for the slope is the same for any two different points on the line that might be chosen. Thus, considering point (x_1, y_1) on a line to be fixed (Fig. 21.15) and another point $P(x, y)$ that *represents* any other point on the line, we have

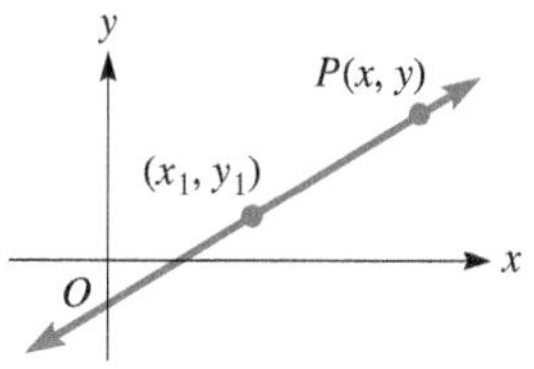

Fig. 21.15

$$m = \frac{y - y_1}{x - x_1}$$

which can be written as

$$y - y_1 = m(x - x_1) \qquad \textbf{(21.6)}$$

Eq. (21.6) is the **point–slope form** *of the equation of a straight line.* It is useful when we know the slope of a line and some point through which the line passes.

Note that when Eq. (21.6) is simplified as much as possible, we end up with a term in x, a term in y, and a constant term. Therefore, we can always rewrite the equation of a line in the form

■ In Eq. (21.7), A and B cannot both be zero.

$$Ax + By + C = 0 \qquad \textbf{(21.7)}$$

Eq. (21.7) is known as the **general form** *of the equation of the straight line.* We saw this form before in Chapter 5. Now we have shown why it represents a straight line.

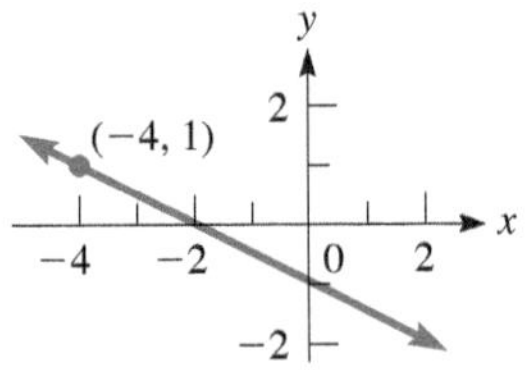

Fig. 21.16

EXAMPLE 1 Point–slope and general forms

Find the equation of the line that passes through $(-4, 1)$ with a slope of $-1/2$. See Fig. 21.16.

Substituting in Eq. (21.6), we find that

$$y - 1 = \left(-\tfrac{1}{2}\right)[x - (-4)]$$

slope

coordinates

Simplifying to obtain the general form, we have

$$2y - 2 = -x - 4$$
$$2y + x + 2 = 0$$

Practice Exercise

1. Write the general form of the equation of the straight line that passes through $(0, -1)$ with a slope of 3.

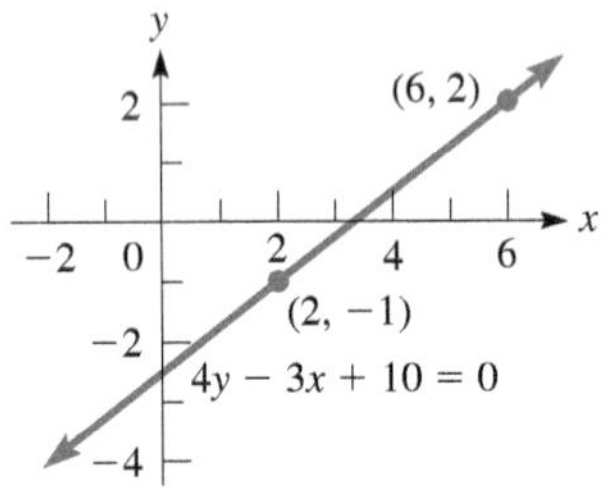

Fig. 21.17

EXAMPLE 2 Equation of a line through two points

Find the equation of the line through $(2, -1)$ and $(6, 2)$.

We first find the slope of the line through these points:

$$m = \frac{2 + 1}{6 - 2} = \frac{3}{4}$$

Then by using either of the two known points and Eq. (21.6), we can find the equation of the line (see Fig. 21.17):

$$y - (-1) = \frac{3}{4}(x - 2)$$
$$4y + 4 = 3x - 6$$
$$4y - 3x + 10 = 0$$

Practice Exercise

2. Find the equation of the straight line that passes through $(-5, 2)$ and $(-1, -6)$.

Eq. (21.6) can be used for any line except for one parallel to the y-axis. For this special case, Eq. (21.7) can still be used, taking a very specific form. The equation of a line parallel to the x-axis also takes a specific form.

Vertical Lines (parallel to y-axis) $x = a$ **(21.8)** All points have the same x-coordinate, so the slope is undefined.	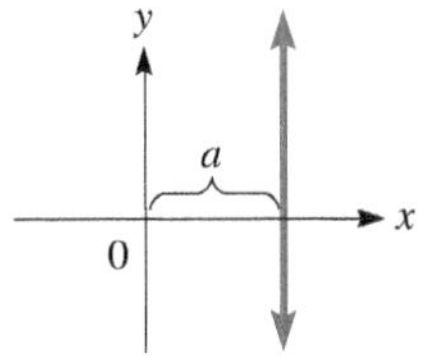 Fig. 21.18(a)
Horizontal Lines (parallel to x-axis) $y = b$ **(21.9)** All points have the same y-coordinate, so the slope is zero.	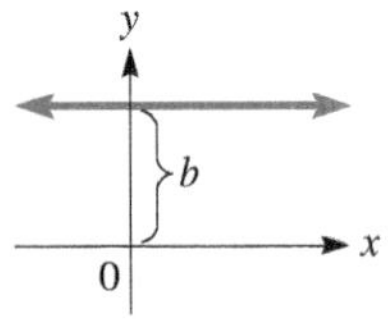 Fig. 21.18(b)

EXAMPLE 3 Vertical line, horizontal line

Find the equations of the **(a)** vertical and **(b)** horizontal lines passing through $(2, -2)$.

(a) Vertical line:

$$x = 2$$

(all points have x-coordinate 2)

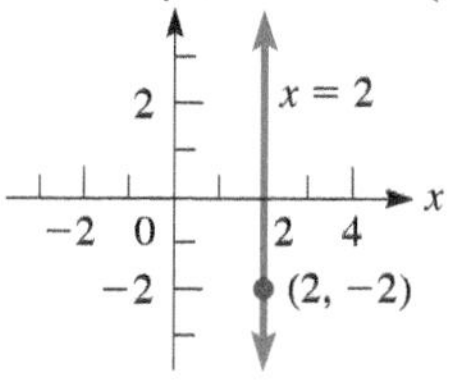

Fig. 21.19(a)

(b) Horizontal line:

$$y = -2$$

(all points have y-coordinate -2)

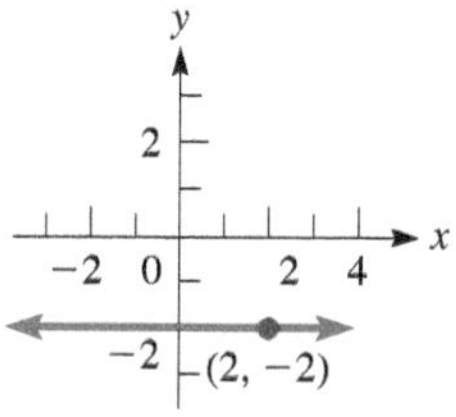

Fig. 21.19(b)

If we choose the special point $(0, b)$, which is the y-intercept of the line, as the point to use in Eq. (21.6), we have $y - b = m(x - 0)$, or

$$y = mx + b \qquad \textbf{(21.10)}$$

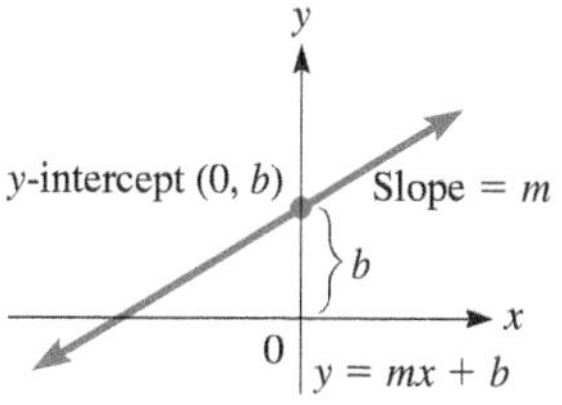

Fig. 21.20

Eq. (21.10) is the familiar **slope–intercept form** of the equation of a straight line which we first discussed in Chapter 5. When the equation of a line is written in this form, we readily know that the slope of the line is the coefficient of the x-term, and that it crosses the y-axis at the coordinate indicated by the constant term. See Fig. 21.20.

EXAMPLE 4 Slope–intercept form

Find the slope and the y-intercept of the straight line for which the equation is $2y + 4x - 5 = 0$.

We write this equation in slope–intercept form:

$$2y = -4x + 5$$

slope ⟶ y-coordinate of intercept

$$y = -2x + \frac{5}{2}$$

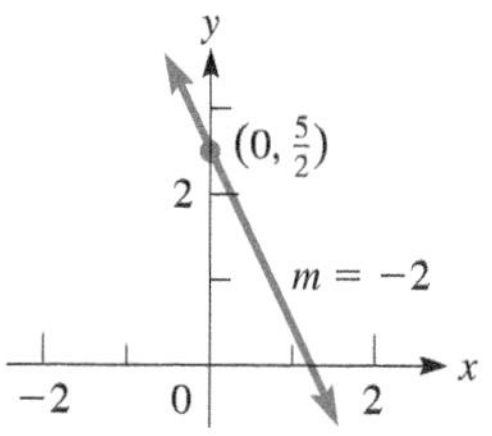

Fig. 21.21

Since the coefficient of x in this form is -2, the slope is -2. The constant on the right is $5/2$, which means that the y-intercept is $(0, 5/2)$. See Fig. 21.21.

EXAMPLE 5 Parallel lines

Find the general form of the equation of the line parallel to the line $3x + 2y - 6 = 0$ that passes through the point $(-1, 2)$.

Since the line whose equation we want is parallel to the line $3x + 2y - 6 = 0$, it has the same slope. Thus, writing $3x + 2y - 6 = 0$ in slope–intercept form,

$$2y = -3x + 6 \qquad \text{solving for } y$$

$$y = -\frac{3}{2}x + 3$$

Since the slope of $3x + 2y - 6 = 0$ is $-3/2$, the slope of the required line is also $-3/2$. Using $m = -3/2$, the point $(-1, 2)$, and the point–slope form, we have

$$y - 2 = -\frac{3}{2}(x + 1)$$

$$2y - 4 = -3(x + 1)$$

$$3x + 2y - 1 = 0$$

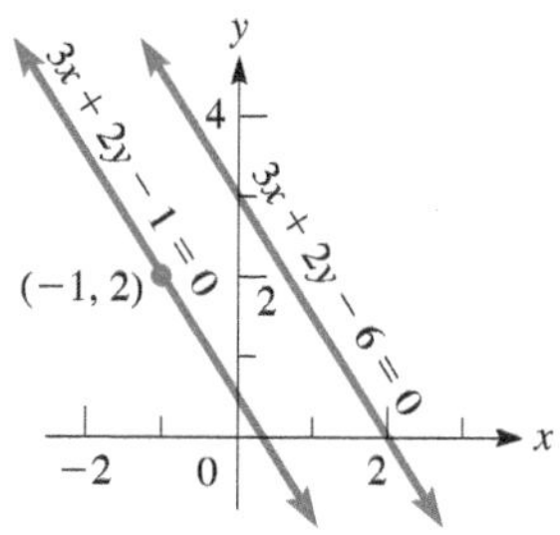

Fig. 21.22

This is the general form of the equation. Both lines are shown in Fig. 21.22.

In many physical situations, a linear relationship exists between variables. A few examples of this are (1) the distance travelled by an object and the elapsed time, when the velocity is constant, (2) the amount a spring stretches and the force applied, (3) the change in electric resistance and the change in temperature, (4) the force applied to an object and the resulting acceleration, and (5) the pressure at a certain point within a liquid and the depth of the point.

EXAMPLE 6 Straight line—variation of pressure with depth

The pressure p_0 at the surface of a body of water (due to the atmosphere) is 101 kPa. The pressure p at a depth of 10.0 m is 199 kPa. In general, the pressure difference $p - p_0$ below the surface varies directly as the depth h. Find the equation of p as a function of h, and sketch its graph.

Following the procedure for problems involving variation (see Section 18.2),

$$p - p_0 = kh \quad \text{set up general equation}$$
$$199 - 101 = k(10.0) \quad \text{substitute given values}$$
$$k = 9.80 \text{ kPa/m} \quad \text{solve for constant of proportionality}$$
$$p - 101 = 9.80h \quad \text{substitute in first equation}$$
$$p = 9.80h + 101$$

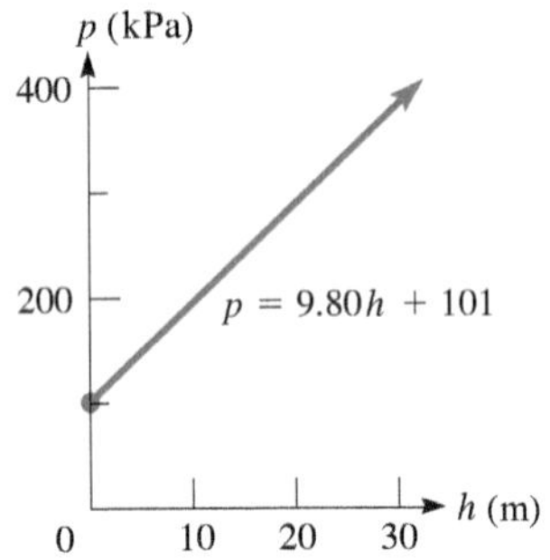

Fig. 21.23

We see that this is the equation of a straight line. The slope is 9.80, and the p-intercept is $(0, 101)$. Negative values do not have any physical meaning. The graph is shown in Fig. 21.23.

EXAMPLE 7 Straight line—slope as acceleration

For a period of 6.0 s, the velocity v of a rocket varies linearly with the elapsed time t. If $v = 40$ m/s when $t = 1.0$ s and $v = 55$ m/s when $t = 4.0$ s, find the equation relating v and t and graph the function. From the graph, find the initial velocity and the velocity after 6.0 s. What is the meaning of the slope of the line?

With v as the dependent variable and t as the independent variable, the slope is

$$m = \frac{v_2 - v_1}{t_2 - t_1}$$

Using the information given in the statement of the problem, we have

$$m = \frac{55 - 40}{4.0 - 1.0} = 5.0$$

Then using the point–slope form of the equation of a straight line, we have

$$v - 40 = 5.0(t - 1.0)$$
$$v = 5.0t + 35$$

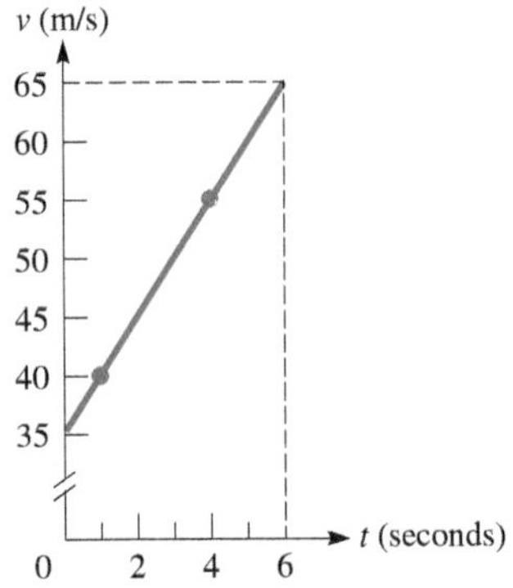

Fig. 21.24

The given values are sufficient to graph the line in Fig. 21.24. There is no need to include negative values of t, since they have no physical meaning. We see that the line crosses the v-axis at 35. This means that the initial velocity (for $t = 0$) is 35 m/s. Also, when $t = 6.0$ s, we see that $v = 65$ m/s.

The slope is the ratio of the change in velocity to the change in time. This is the rocket's *acceleration*. Here, the speed of the rocket increases 5.0 m/s each second. We can express this acceleration as 5.0 (m/s)/s = 5.0 m/s^2.

EXERCISES 21.2

In Exercises 1–4, make the given changes in the indicated examples of this section and then solve the resulting problems.

1. In Example 1, change $(-4, 1)$ to $(4, -1)$.

2. In Example 2, change $(2, -1)$ to $(-2, 1)$.

3. In Example 4, change the $+$ before $4x$ to $-$.

4. In Example 5, change the $+$ before $2y$ to $-$.

In Exercises 5–20, find the equation of each of the lines with the given properties. Sketch the graph of each line.

5. Passes through $(-3, 8)$ with a slope of 4

6. Passes through $(-2, -1)$ with a slope of -2

7. Passes through $(2, -5)$ and $(4, 2)$

8. Has an x-intercept $(4, 0)$ and a y-intercept $(0, -6)$

9. Passes through $(-7, 12)$ with an inclination of 45°

10. Has a y-intercept $(0, -2)$ and an inclination of 120°

11. Passes through $(5.3, -2.7)$ and is parallel to the x-axis

12. Passes through $(-15, 9)$ and is perpendicular to the x-axis

13. Is parallel to the y-axis and is 3 units to the left of it

14. Is parallel to the x-axis and is 4.1 units below it

15. Is perpendicular to line with slope of -3; passes through $(1, -2)$

16. Is parallel to line through $(-1, 7)$ and $(3, 1)$; passes through $(1, 2)$

17. Has equal intercepts and passes through $(5, 2)$

18. Is perpendicular to the line $6.0x - 2.4y - 3.9 = 0$ and passes through $(7.5, -4.7)$

19. Has a slope of -3 and passes through the intersection of the lines $5x - y = 6$ and $x + y = 12$

20. Passes through the point of intersection of $2x + y - 3 = 0$ and $x - y - 3 = 0$ and through the point $(4, -3)$

In Exercises 21–28, reduce the equations to slope–intercept form and find the slope and the y-intercept. Sketch each line.

21. $4x - y = 8$

22. $2x - 3y - 6 = 0$

23. $3x + 5y - 10 = 0$

24. $4y = 6x - 9$

25. $3x - 2y - 1 = 0$

26. $4x + 2y - 5 = 0$

27. $11.2x + 1.6 = 3.2y$

28. $11.5x + 4.60y = 5.98$

In Exercises 29–36, determine whether the given lines are parallel, perpendicular, or neither.

29. $3x - 2y + 5 = 0$ and $4y = 6x - 1$

30. $8x - 4y + 1 = 0$ and $4x + 2y - 3 = 0$

31. $6x - 3y - 2 = 0$ and $x + 2y - 4 = 0$

32. $3y - 2x = 4$ and $6x - 9y = 5$

33. $5x + 2y - 3 = 0$ and $10y = 7 - 4x$

34. $48y - 36x = 71$ and $52x = 17 - 39y$

35. $4.5x - 1.8y = 1.7$ and $2.4x + 6.0y = 0.3$

36. $3.5y = 4.3 - 1.5x$ and $3.6x + 8.4y = 1.7$

In Exercises 37–60, solve the given problems. Exercises 49–60 show some applications of straight lines.

37. Find k if the lines $4x - ky = 6$ and $6x + 3y + 2 = 0$ are parallel.

38. Find k if the lines given in Exercise 37 are perpendicular.

39. Find k if the lines $3x - y = 9$ and $kx + 3y = 5$ are perpendicular. Explain how this value is found.

40. Find k such that the line through $(k, 2)$ and $(3, 1 - k)$ is perpendicular to the line $x - 2y = 5$. Explain your method.

41. Find the slope of the line joining points on the graph of $y = x^2$ that have x-coordinates of $-a$ and b $(a > 0, b > 0)$.

42. Show that the *intercept form* $\frac{x}{a} + \frac{y}{b} = 1$ is the equation of a line with x-intercept $(a, 0)$ and y-intercept $(0, b)$.

43. Find the distance from $(4, 1)$ to the line $4x - 3y + 12 = 0$.

44. Find the acute angle between the lines $x + y = 3$ and $2x - 5y = 4$.

45. Show that the following lines intersect to form a parallelogram. $8x + 10y = 3$; $2x - 3y = 5$; $4x - 6y = -3$; $5y + 4x = 1$.

46. For nonzero values of a, b, and c, find the intercepts of the line $ax + by + c = 0$.

47. For nonzero values of a, b, c, and d, show that (a) lines $ax + by + c = 0$ and $ax + by + d = 0$ are parallel, and (b) lines $ax + by + c = 0$ and $bx - ay + d = 0$ are perpendicular.

48. Find the equation of the line with positive intercepts that passes through $(3, 2)$ and forms with the axes a triangle of area 12.

49. In the 1700s, the French physicist Réaumur established a temperature scale on which the freezing point of water was 0° and the boiling point was 80°. Set up an equation for the Celsius temperature T (freezing point 0°, boiling point 100°) as a function of the Réaumur temperature R.

50. The voltage V across part of an electric circuit is given by $V = E - iR$, where E is a battery voltage, i is the current, and R is the resistance. If $E = 6.00$ V and $V = 4.35$ V for $i = 9.17$ mA, find V as a function of i. Sketch the graph. (i and V may be negative.)

51. The velocity of sound v increases 0.607 m/s for each increase in temperature T of 1.00°C. If $v = 343$ m/s for $T = 20.0$°C, express v as a function of T.

52. An acid solution is made from x litres of a 20% solution and y litres of a 30% solution. If the final solution contains 20 L of acid, find the equation relating x and y.

53. A wall is 15 cm thick. At the outside, the temperature is 3°C, and at the inside, it is 23°C. If the temperature changes at a constant rate through the wall, write an equation of the temperature T in the wall as a function of the distance x from the outside to the inside of the wall. What is the meaning of the slope of the line?

54. An oil-storage tank is emptied at a constant rate. At 10:00 A.M., 1800 barrels remain, and at 2:00 P.M., 600 barrels remain. If pumping started at 8:00 A.M., find the equation relating the number of barrels n at time t (in h) from 8:00 A.M. When will the tank be empty?

55. After taking off, a plane gains altitude at 600 m/min for 5.0 min and then continues to gain altitude at 300 m/min for 15 min. It then continues at a constant altitude. Find the altitude h as a function of time t for the first 20 min, and sketch the graph of $h = f(t)$.

56. The length of a rectangular solar cell is 10 cm more than the width w. Express the perimeter p of the cell as a function of w. What is the meaning of the slope of the line?

57. A light beam is reflected off the edge of an optic fibre at an angle of 0.0032°. The diameter of the fibre is 48 μm. Find the equation of the reflected beam with the x-axis (at the centre of the fibre) and the y-axis as shown in Fig. 21.25.

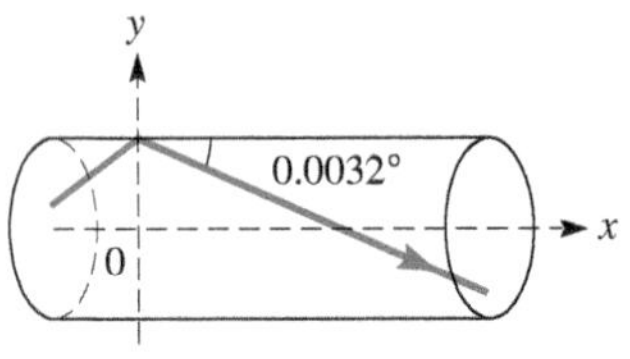

Fig. 21.25

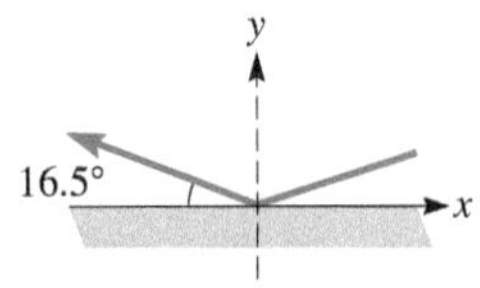

Fig. 21.26

58. A police report stated that a bullet caromed upward off a floor at an angle of 16.5° with the floor, as shown in Fig. 21.26. What is the equation of the bullet's path after impact?

59. A survey of the traffic on a particular highway showed that the number of cars passing a particular point each minute varied linearly from 6:30 A.M. to 8:30 A.M. on workday mornings. The study showed that an average of 45 cars passed the point in 1 min at 7:00 A.M. and that 115 cars passed in 1 min at 8:00 A.M. If n is the number of cars passing the point in 1 min, and t is the number of minutes after 6:30 A.M., find the equation relating n and t and graph the equation. From the graph, determine n at 6:30 A.M. and at 8:30 A.M. What is the meaning of the slope of the line?

60. In a research project on cancer, a tumour was determined to weigh 30 mg when first discovered. While being treated, it grew smaller by 2 mg each month. Find the equation relating the weight w of the tumour as a function of the time t in months. Graph the equation.

In Exercises 61–64, treat the given nonlinear functions as linear functions in order to sketch their graphs. At times, this can be useful in showing certain values of a function. For example, $y = 2 + 3x^2$ can be shown as a straight line by graphing y as a function of x^2. A table of values for this graph is shown along with the corresponding graph in Fig. 21.27.

x	0	1	2	3	4	5
x^2	0	1	4	9	16	25
y	2	5	14	29	50	77

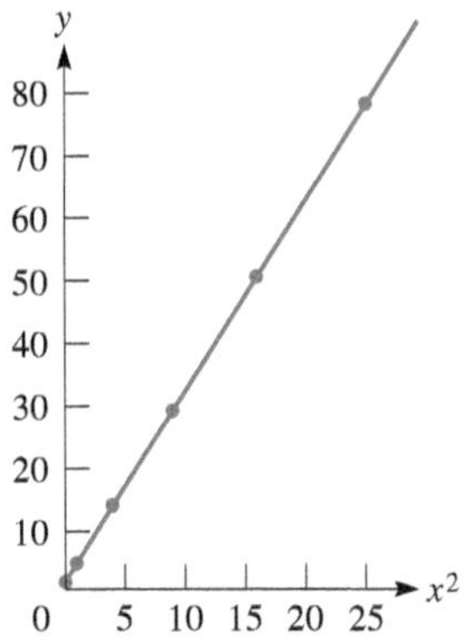

Fig. 21.27

61. The number n of memory cells of a certain computer that can be tested in t seconds is given by $n = 1200\sqrt{t}$. Sketch n as a function of $\sqrt{t}$.

62. The force F (in N) applied to a lever to balance a certain weight on the opposite side of the fulcrum is given by $F = 40/d$, where d is the distance (in m) of the force from the fulcrum. Sketch F as a function of $1/d$.

63. A spacecraft is launched such that its altitude h (in km) is given by $h = 300 + 2t^{3/2}$ for $0 \le t < 100$ s. Sketch this as a linear function.

64. The current i (in A) in a certain electric circuit is given by $i = 6(1 - e^{-t})$. Sketch this as a linear function.

In Exercises 65–68, show that the given nonlinear functions are linear when plotted on a semilogarithmic or a logarithmic scale. In Section 13.7, we noted that graphs on these scales often become straight lines.

65. A function of the form $y = ax^n$ is straight when plotted on a logarithmic scale, since $\log y = \log a + n \log x$ is in the form of a straight line. The variables are $\log y$ and $\log x$; the slope can be found from $(\log y - \log a)/\log x = n$, and the intercept is $\log a$. (To get the slope from the graph, we calculate $(\log y - \log a)/\log x$ for some set values of x and y. The log y-intercept is found where $\log x = 0$, and this occurs when $x = 1$.) Plot $y = 3x^4$ on a logarithmic scale to verify this analysis.

66. A function of the form $y = a(b^x)$ is straight when plotted on a semilogarithmic scale, since $\log y = \log a + x \log b$ is in the form of a straight line. The variables are $\log y$ and x, the slope is $\log b$, and the intercept is a. (To get the slope from the graph, we calculate $(\log y - \log a)/x$ for some set values of x and y. The intercept is read directly off the graph where $x = 0$.) Plot $y = 3(2^x)$ on a semilogarithmic scale to verify this analysis.

67. If experimental data are plotted on a logarithmic scale and the points lie on a straight line, it is possible to determine the function (see Exercise 65). The following data come from an experiment to determine the functional relationship between the pressure p and the volume V of a gas undergoing an adiabatic (no heat loss) change. From the graph on a logarithmic scale, determine p as a function of V.

V (m^3)	0.100	0.500	2.00	5.00	10.0
p (kPa)	20.1	2.11	0.303	0.0840	0.0318

68. If experimental data are plotted on a semilogarithmic scale, and the points lie on a straight line, it is possible to determine the function (see Exercise 66). The following data come from an experiment designed to determine the relationship between the voltage across an inductor and the time after the switch is opened. Determine V as a function of t.

V(V)	40	15	5.6	2.2	0.8
t (ms)	0.0	20	40	60	80

Answers to Practice Exercises

1. $3x - y - 1 = 0$ **2.** $2x + y + 8 = 0$

21.3 The Circle

We can obtain a general equation that represents a straight line by considering a fixed point on the line and then a general point $P(x, y)$ that can represent any other point on the same line. Mathematically, we can state this as "the line is the **locus** of a point $P(x, y)$ that *moves* from a fixed point with constant direction." That is, the point $P(x, y)$ can be considered a variable point that moves along the line.

In this way, we can define a number of important curves. *A* **circle** *is defined as the locus of a point $P(x, y)$ that moves so that it is always equidistant from a fixed point. We call this fixed distance the* **radius**, *and we call the fixed point the* **centre** *of the circle.* Thus, using this definition, calling the fixed point (h, k) and the radius r, we have

$$\sqrt{(x - h)^2 + (y - k)^2} = r$$

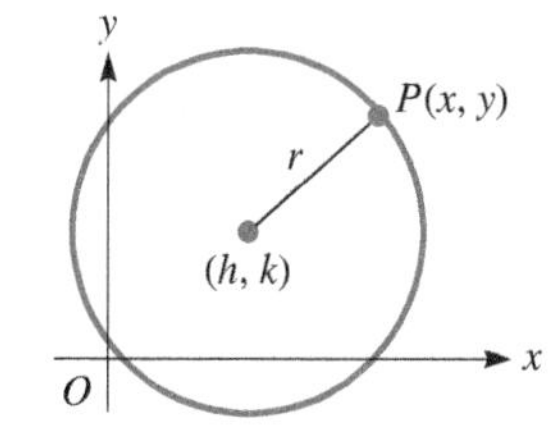

Fig. 21.28

or, by squaring both sides, we have

$$(x - h)^2 + (y - k)^2 = r^2 \qquad \textbf{(21.11)}$$

Eq. (21.11) is called the **standard equation** *of a circle with centre at (h, k) and radius r.* See Fig. 21.28.

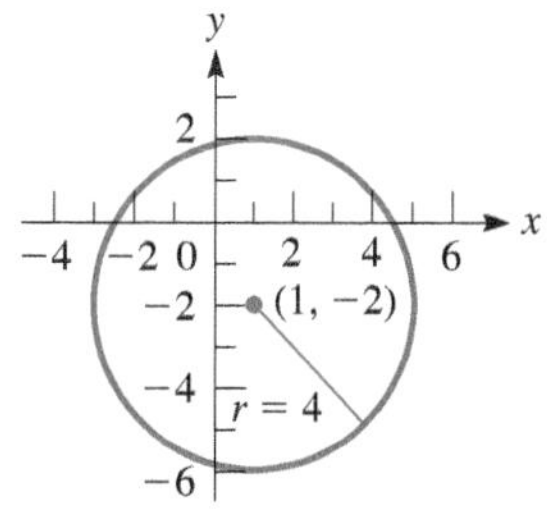

Fig. 21.29

EXAMPLE 1 Standard equation

Find the centre and radius of a circle with equation $(x - 1)^2 + (y + 2)^2 = 16$.

We rewrite the equation in the form of Eq. (21.11) as

$$(x - 1)^2 + [y - (-2)]^2 = 4^2$$

(form requires − signs; coordinates of centre; radius)

Therefore, the centre is at $(1, -2)$ and the radius is 4. This circle is shown in Fig. 21.29.

LEARNING TIP

It is important to pay close attention to signs when finding the coordinates of the centre. *We must have a minus sign before each of the coordinates.* If the expression contains a plus sign, we must rewrite it to have a minus sign (just as we wrote +2 as − (−2) in Example 1).

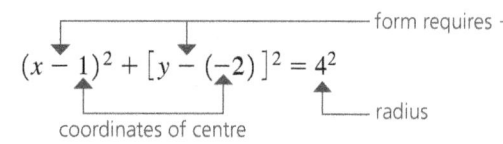

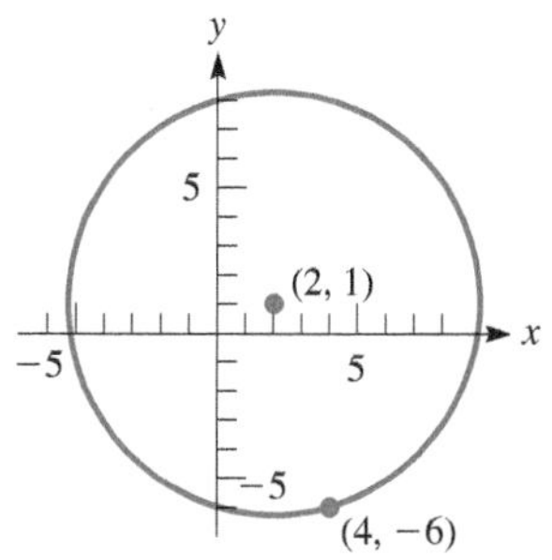

Fig. 21.30

EXAMPLE 2 Find the equation of a circle

Find the equation of the circle with centre at $(2, 1)$ that passes through $(4, -6)$.

Use the coordinate of the centre to set up the form of Eq. (21.11): The centre is at $(2, 1)$, so $h = 2$, $k = 1$, and

$$(x - 2)^2 + (y - 1)^2 = r^2$$

Substitute the given point and solve for r^2: The given point is $(4, -6)$, so

$$(4 - 2)^2 + (-6 - 1)^2 = r^2 \text{ or } r^2 = 53$$

Write the standard equation:

$$(x - 2)^2 + (y - 1)^2 = 53$$

The circle is shown in Fig. 21.30.

Practice Exercise

1. Find the centre and radius of the circle $(x + 7)^2 + (y - 2)^2 = 1$.

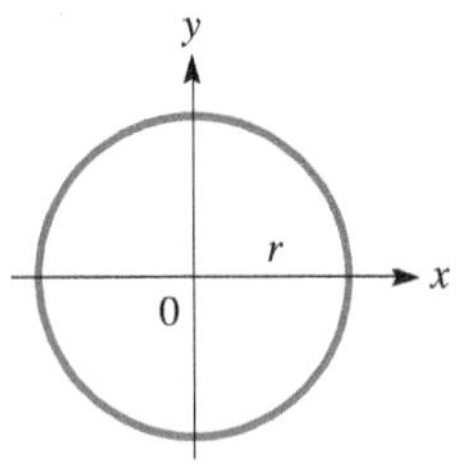

Fig. 21.31

If the centre of the circle is at the origin, which means that the coordinates of the centre are $(0, 0)$, the equation of the circle (see Fig. 21.31) becomes

$$x^2 + y^2 = r^2 \tag{21.12}$$

The following example illustrates an application using this type of circle and one with its centre not at the origin.

EXAMPLE 3 Equations of circles—friction drive sketch

A student is drawing a friction drive in which two circular discs are in contact with each other. They are represented by circles in the drawing. The first has a radius of 10.0 cm, and the second has a radius of 12.0 cm. What is the equation of each circle if the origin is at the centre of the first circle and the positive x-axis passes through the centre of the second circle? See Fig. 21.32.

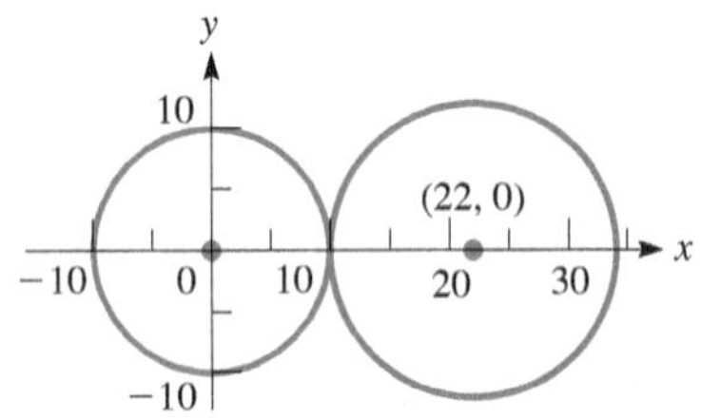

Fig. 21.32

Since the centre of the smaller circle is at the origin, we can use Eq. (21.12). Given that the radius is 10.0 cm, we have as its equation

$$x^2 + y^2 = 100$$

The fact that the two discs are in contact tells us that they meet at the point $(10.0, 0)$. Knowing that the radius of the larger circle is 12.0 cm tells us that its centre is at $(22.0, 0)$. Thus, using Eq. (21.11) with $h = 22.0$, $k = 0$, and $r = 12.0$, we get

$$(x - 22.0)^2 + (y - 0)^2 = 12.0^2$$

or

$$(x - 22.0)^2 + y^2 = 144$$

as the equation of the larger circle.

A circle with its centre at the origin exhibits three types of symmetry that are often of interest when graphing equations. These are: symmetry about the x-axis, symmetry about the y-axis, and symmetry about the origin.

Symmetry	EXAMPLE 4
	Test the symmetry of $x^2 + y^2 = 16$.
Symmetry about the x-axis The part of the curve graphed above the x-axis is a mirror image of the part of the curve graphed below it. **Test:** Substitution of y with $-y$ leads to the same equation.	Substituting y with $-y$, $x^2 + (-y)^2 = 16$ $x^2 + y^2 = 16$ We obtain the same equation, so the graph is symmetric about the x-axis. 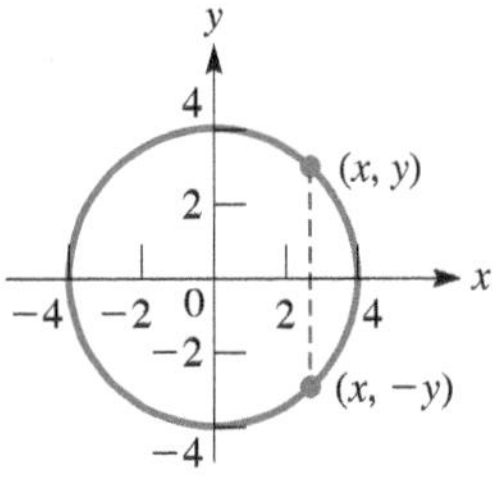 Fig. 21.33(a)
Symmetry about the y-axis The part of the curve graphed to the right of the y-axis is a mirror image of the part of the curve graphed to the left of it. **Test:** Substitution of x with $-x$ leads to the same equation.	Substituting x with $-x$, $(-x)^2 + y^2 = 16$ $x^2 + y^2 = 16$ We obtain the same equation, so the graph is symmetric about the y-axis. 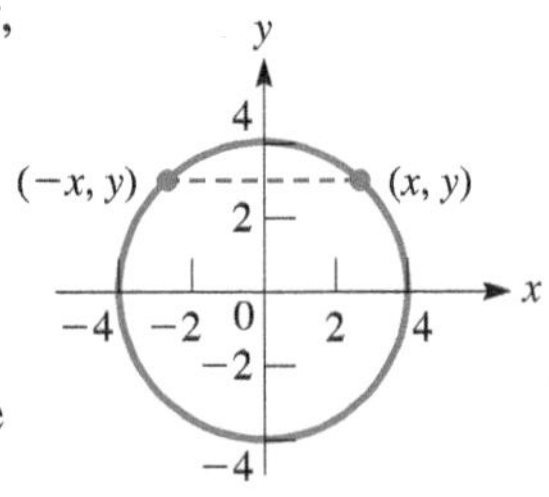 Fig. 21.33(b)

Symmetry about the origin The origin is the midpoint of points (x, y) and $(-x, -y)$ on the curve. **Test:** Substitution of x with $-x$ *and* of y with $-y$ leads to the same equation.	Substituting x with $-x$ and y with $-y$, $(-x)^2 + (-y)^2 = 16$ $x^2 + y^2 = 16$ We obtain the same equation, so the graph is symmetric about the origin. 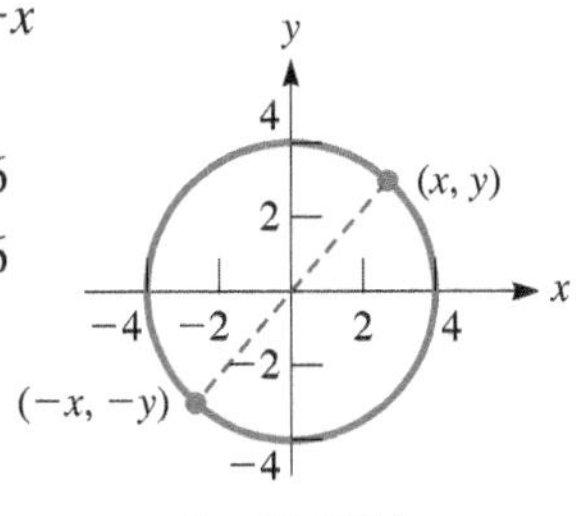 Fig. 21.33(c)

If we multiply out each of the terms in Eq. (21.11), we may combine the resulting terms to obtain

$$x^2 - 2hx + h^2 + y^2 - 2ky + k^2 = r^2$$

$$x^2 + y^2 - 2hx - 2ky + (h^2 + k^2 - r^2) = 0 \qquad \textbf{(21.13)}$$

Since each of h, k, and r is constant for any given circle, the coefficients of x and y and the term within parentheses in Eq. (21.13) are constants. Eq. (21.13) can then be written as

LEARNING TIP

If we know the general equation of a circle, we can find the centre and radius by writing the equation in standard form. To do so, *we must complete the square both in the x-terms and in the y-terms* (see Section 7.2).

$$x^2 + y^2 + Dx + Ey + F = 0 \qquad \textbf{(21.14)}$$

Eq. (21.14) is called the **general equation** *of the circle.* It tells us that any equation that can be written in that form will represent a circle (or possibly a single point or the empty set).

EXAMPLE 5 General equation of a circle

Find the centre and radius of the circle with general equation

$$x^2 + y^2 - 6x + 8y - 24 = 0$$

We write the given equation in standard form by completing the squares. This is done by first writing the equation in the form

$$(x^2 - 6x \quad) + (y^2 + 8y \quad) = 24$$

To complete the square of the x-terms, we take half of -6, which is -3, square it, and add the result, 9, to each side of the equation. In the same way, we complete the square of the y-terms by adding 16 to each side of the equation, which gives

$\left(\frac{-6}{2}\right)^2$ $\quad$ $\left(\frac{8}{2}\right)^2$ $\quad$ add to both sides

$$(x^2 - 6x + 9) + (y^2 + 8y + 16) = 24 + 9 + 16$$

$$(x - 3)^2 + (y + 4)^2 = 49$$

$$(x - 3)^2 + (y - (-4))^2 = 7^2$$

coordinates of centre $\quad$ radius

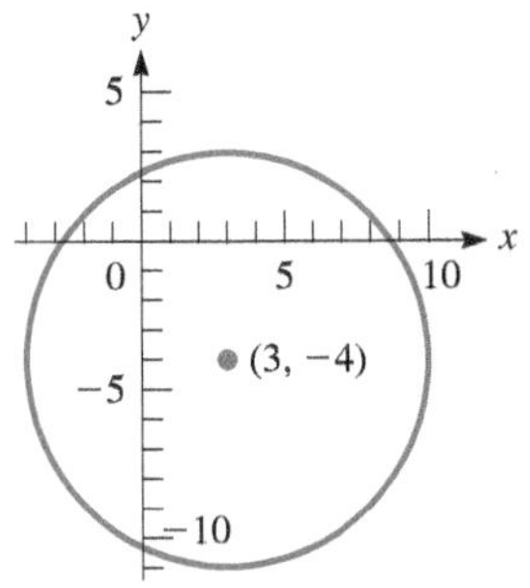

Fig. 21.34

Thus, the centre is $(3, -4)$, and the radius is 7 (see Fig. 21.34).

Practice Exercise

2. Find the centre and radius of the circle $x^2 + y^2 - 8x + 6y + 21 = 0$.

EXAMPLE 6 Circle—pendulum motion

A certain pendulum is found to swing through an arc of the circle $3x^2 + 3y^2 - 9.60y - 2.80 = 0$. What is the length (in m) of the pendulum, and from what point is it swinging?

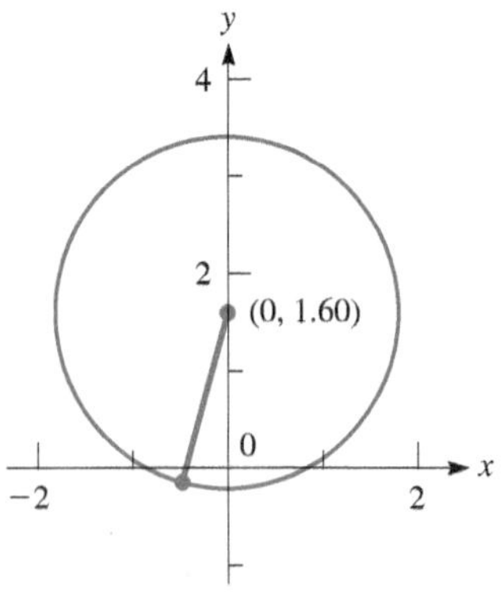

Fig. 21.35

We see that this equation represents a circle by dividing through by 3. This gives us $x^2 + y^2 - 3.20y - 2.80/3 = 0$. The length of the pendulum is the radius of the circle, and the point from which it swings is the centre. These are found as follows:

$$x^2 + (y^2 - 3.20y + 1.60^2) = 1.60^2 + 2.80/3 \quad \text{complete squares in both } x\text{- and } y\text{-terms}$$
$$x^2 + (y - 1.60)^2 = 3.493 \quad \text{standard form}$$

Since $\sqrt{3.493} = 1.87$, the length of the pendulum is 1.87 m. The point from which it is swinging is (0, 1.60). See Fig. 21.35.

Replacing x with $-x$, the equation does not change. Replacing y with $-y$, the equation does change (the $3.20y$ term changes sign). Thus, the circle is symmetric only to the y-axis.

In Section 14.1 we noted that the equation of a circle does not represent a function since there are two values of y for most values of x in the domain. However, the circle is a combination of two functions, each of which has a semicircle as its graph. The top and bottom semicircle functions can be found using the quadratic formula, as illustrated in the following example.

EXAMPLE 7 A circle as a graph of two functions

Determine the two semicircle functions of the circle $3x^2 + 3y^2 + 6y - 20 = 0$.

To fit the form of a quadratic equation in y, we write

$$3y^2 + 6y + (3x^2 - 20) = 0$$

Now, using the quadratic formula to solve for y, we let

$$a = 3 \quad b = 6 \quad c = 3x^2 - 20$$

$$y = \frac{-6 \pm \sqrt{6^2 - 4(3)(3x^2 - 20)}}{2(3)}$$

which means we get the *two functions*

$$y_1 = \frac{-6 + \sqrt{276 - 36x^2}}{6} \quad \text{and} \quad y_2 = \frac{-6 - \sqrt{276 - 36x^2}}{6}$$

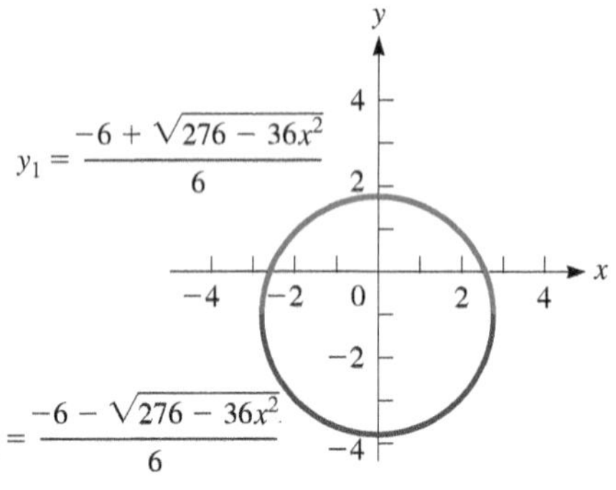

Fig. 21.36

The two functions are graphed together in Fig. 21.36. When using a graphing utility, these two functions would be entered and graphed together in order to display a circle.

EXERCISES 21.3

In Exercises 1–4, make the given changes in the indicated examples of this section and then solve the resulting problems.

1. In Example 1, change $(y + 2)^2$ to $(y + 1)^2$.
2. In Example 2, change $(2, 1)$ to $(-2, 1)$.
3. In Example 5, change the + before $8y$ to −.
4. In Example 7, change the + before $6y$ to −.

In Exercises 5–8, determine the centre and the radius of each circle.

5. $(x - 2)^2 + (y - 1)^2 = 25$
6. $(x - 3)^2 + (y + 4)^2 = 49$
7. $4(x + 1)^2 + 4y^2 = 121$
8. $9x^2 + 9(y - 6)^2 = 64$

In Exercises 9–24, find the standard and general equations of each of the circles from the given information.

9. Centre at (0, 0), radius 3
10. Centre at (0, 0), radius 12
11. Centre at (2, 3), radius 4
12. Centre at $\left(\frac{3}{2}, -2\right)$, radius $\frac{5}{2}$
13. Centre at (12, −15), radius 18
14. Centre at (−3, −5), radius $2\sqrt{3}$
15. The origin and (−6, 8) are ends of a diameter.
16. The points (3, 8) and (−3, 0) are the ends of a diameter.
17. Concentric with the circle $(x - 2)^2 + (y - 1)^2 = 4$ and passes through (4, −1)

18. Concentric with the circle $x^2 + y^2 + 2x - 8y + 8 = 0$ and passes through $(-2, 3)$

19. Centre at $(-3, 5)$ and tangent to the line $y = 10$

20. Centre at $(-7, 1)$ and tangent to the y-axis

21. Tangent to both axes and the lines $y = 4$ and $x = -4$

22. Tangent to the lines $y = 2$ and $y = 8$, centre on the line $y = x$

23. Centre at the origin, tangent to the line $x + y = 2$

24. Centre at $(5, 12)$, tangent to the line $y = 2x - 3$

In Exercises 25–36, determine the centre and radius of each circle. Sketch each circle.

25. $x^2 + (y - 3)^2 = 4$

26. $(x - 2)^2 + (y + 3)^2 = 49$

27. $4(x + 1)^2 + 4(y - 5)^2 = 81$

28. $4(x + 7)^2 + 4(y + 11)^2 = 169$

29. $x^2 + y^2 - 2x - 8 = 0$

30. $x^2 + y^2 - 4x - 6y - 12 = 0$

31. $x^2 + y^2 + 4.20x - 2.60y = 3.51$

32. $2x^2 + 2y^2 + 44x + 28y = 52$

33. $4x^2 + 4y^2 - 16y = 9$

34. $9x^2 + 9y^2 + 18y = 7$

35. $2x^2 + 2y^2 - 4x - 8y - 1 = 0$

36. $3x^2 + 3y^2 - 12x + 4 = 0$

In Exercises 37–40, determine whether the circles with the given equations are symmetric to either axis or to the origin.

37. $x^2 + y^2 = 100$

38. $x^2 + y^2 - 4x - 5 = 0$

39. $3x^2 + 3y^2 + 24y = 8$

40. $5x^2 + 5y^2 - 10x + 20y = 3$

In Exercises 41–65, solve the given problems.

41. A square is inscribed in the circle $x^2 + y^2 = 32$ (all four vertices are on the circle). Find the area of the square.

42. Find the intercepts of the circle $x^2 + y^2 - 6x + 5 = 0$.

43. Determine whether the circle $x^2 - 6x + y^2 - 7 = 0$ crosses the x-axis.

44. Find the points of intersection of the circle $x^2 + y^2 - x - 3y = 0$ and the line $y = x - 1$.

45. Find the locus of a point $P(x, y)$ that moves so that its distance from $(2, 4)$ is twice its distance from $(0, 0)$. Describe the locus.

46. Find the equation of the locus of a point $P(x, y)$ that moves so that the line joining it and $(2, 0)$ is always perpendicular to the line joining it and $(-2, 0)$. Describe the locus.

47. Use a graphing utility to view the circle $x^2 + y^2 + 5y - 4 = 0$.

48. Use a graphing utility to view the circle $2x^2 + 2y^2 + 2y - x - 1 = 0$.

49. What type of graph is represented by each of the following equations?

(a) $y = \sqrt{9 - (x - 2)^2}$ (b) $y = -\sqrt{9 - (x - 2)^2}$

(c) Are the equations in parts (a) and (b) functions? Explain.

50. What type of graph is represented by each of the following equations?

(a) $x^2 + (y - 1)^2 = 0$ (b) $x^2 + (y - 1)^2 = -1$

51. Is the point $(0.1, 3.1)$ inside, outside, or on the circle $x^2 + y^2 - 2x - 4y + 3 = 0$?

52. Find the equation of the line along which the diameter of the circle $x^2 + y^2 - 2x - 4y - 4 = 0$ lies, if the diameter is parallel to the line $3x + 5y = 4$.

53. For the equation $(x - h)^2 + (y - k)^2 = p$, how does the value of p indicate whether the graph is a circle, is a point, or does not exist?

54. Determine whether the graph of $x^2 + y^2 - 2x + 10y + 29 = 0$ is a circle, is a point, or does not exist.

55. The inner and outer circles of the cross-section of a pipe are represented by the equations $2.00x^2 + 2.00y^2 = 5.73$ and $2.80x^2 + 2.80y^2 = 8.91$. How thick (in cm) is the pipe wall?

56. A 4-m pole leaning against a wall slips to the ground. Find the equation that represents the path of the midpoint of the pole. See Fig. 21.37.

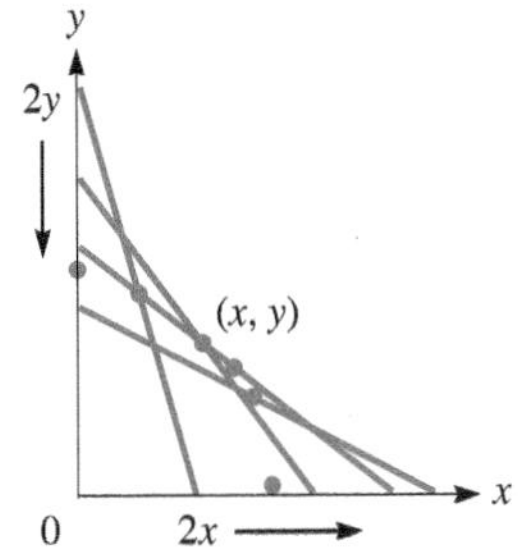

Fig. 21.37

57. In a hoisting device, two of the pulley wheels may be represented by $x^2 + y^2 = 14.5$ and $x^2 + y^2 - 19.6y + 86.0 = 0$. How far apart (in cm) are the wheels?

58. The design of a machine part shows it as a circle represented by the equation $x^2 + y^2 = 42.5$, with a circular hole represented by $x^2 + y^2 + 3.06y - 1.24 = 0$ cut out. What is the least distance (in cm) from the edge of the hole to the edge of the machine part?

59. A wire is rotating in a circular path through a magnetic field to induce an electric current in the wire. The wire is rotating at 60.0 Hz, with a constant velocity of 37.7 m/s. Taking the origin at the centre of the circle of rotation, find the equation of the path of the wire.

60. A communications satellite remains stationary at an altitude of 36 200 km over a point on the earth's equator. It therefore rotates once each day about the earth's centre. Its velocity is constant, but the horizontal and vertical components, v_H and v_V, of the velocity constantly change. Show that the equation relating v_H and v_V (in km/h) is that of a circle. The radius of the earth is 6370 km.

61. Find the equation describing the rim of a circular porthole 0.80 m in diameter if the top is 2.0 m below the surface of the water. Take the origin at the water surface directly above the centre of the porthole.

62. An earthquake occurred 37° north of east of a seismic recording station. If the tremors travel at 4.8 km/s and were recorded 25 s later at the station, find the equation of the circle that represents the tremor recorded at the station. Take the station to be at the centre of the coordinate system.

63. In analysing the strain on a beam, *Mohr's circle* is often used. To form it, normal strain is plotted as the x-coordinate and shear strain is plotted as the y-coordinate. The centre of the circle is midway between the minimum and maximum values of normal strain on the x-axis. Find the equation of Mohr's circle if the minimum normal strain is 100×10^{-6} and the maximum normal strain is 900×10^{-6} (strain is unitless). Sketch the graph.

64. An architect designs a Norman window, which has the form of a semicircle surmounted on a rectangle, as in Fig. 21.38. Find the area (in m^2) of the window if the circular part is on the circle $x^2 + y^2 - 3.00y + 1.25 = 0$.

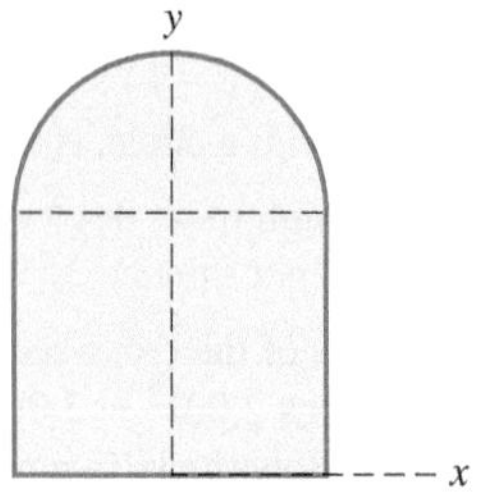

Fig. 21.38

65. The cobblestone circle of the Majorville Cairn and Medicine Wheel in Alberta is represented on a map by the circle $4x^2 + 4y^2 - 64x - 96y + 103 = 0$ (measurements in metres). Find the diameter of the monument, one of the oldest religious monuments in the world (it has been used continuously for the last 4500 years).

Answers to Practice Exercises

1. $C(-7, 2), r = 1$ **2.** $C(4, -3), r = 2$

21.4 The Parabola

In Chapter 7, we showed that the graph of a quadratic function is a *parabola.* We now define the parabola more generally and find the general form of its equation.

A **parabola** *is defined as the locus of a point $P(x, y)$ that moves so that it is always equidistant from a given line (the* **directrix***) and a given point (the* **focus***). The line through the focus that is perpendicular to the directrix is the* **axis** *of the parabola. The point midway between the focus and directrix is the* **vertex**.

Using the definition, we now find the equation of the parabola with the focus at $(p, 0)$ and the directrix $x = -p$. With these choices, we find a general equation of a parabola with its vertex at the origin.

From the definition, the distance from $P(x, y)$ on the parabola to the focus $(p, 0)$ must equal the distance from $P(x, y)$ to the directrix $x = -p$. The distance from P to the focus is found from the distance formula. The distance from P to the directrix is the perpendicular distance and is along a line parallel to the x-axis. These distances are shown in Fig. 21.39. Therefore, we have

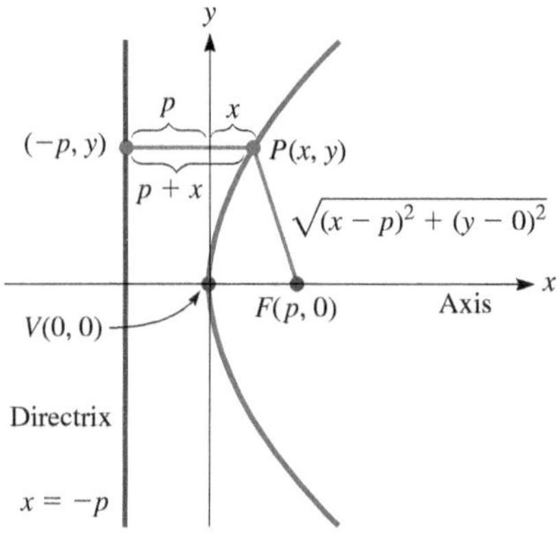

Fig. 21.39

$$\sqrt{(x-p)^2 + (y-0)^2} = x + p$$

$$(x-p)^2 + y^2 = (x+p)^2 \quad \text{squaring both sides}$$

$$x^2 - 2px + p^2 + y^2 = x^2 + 2px + p^2$$

Simplifying, we obtain

$$y^2 = 4px \tag{21.15}$$

Eq. (21.15) is called the **standard form** *of the equation of a parabola with its axis along the x-axis and the vertex at the origin.* It is symmetric about the x-axis since $(-y)^2 = 4px$ is the same as $y^2 = 4px$.

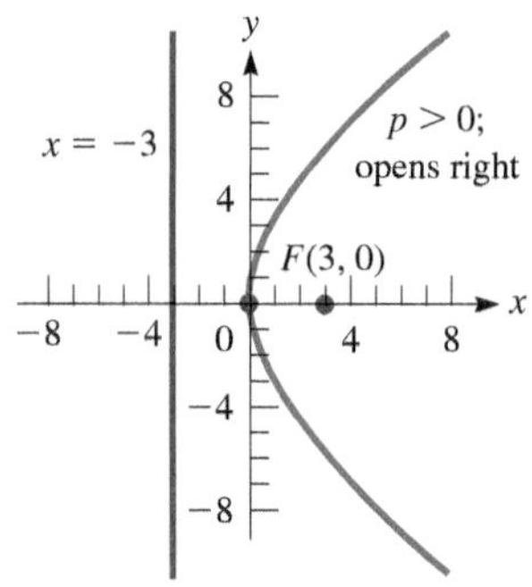

Fig. 21.40

EXAMPLE 1 Focus and directrix of a parabola

Find the coordinates of the focus and the equation of the directrix and sketch the graph of the parabola $y^2 = 12x$.

Since the equation of this parabola fits the form of Eq. (21.15), we know that the vertex is at the origin. The coefficient of 12 tells us that

$$4p = 12, \qquad p = 3$$

The focus is the point $(3, 0)$, and the directrix is the line $x = -3$, as shown in Fig. 21.40.

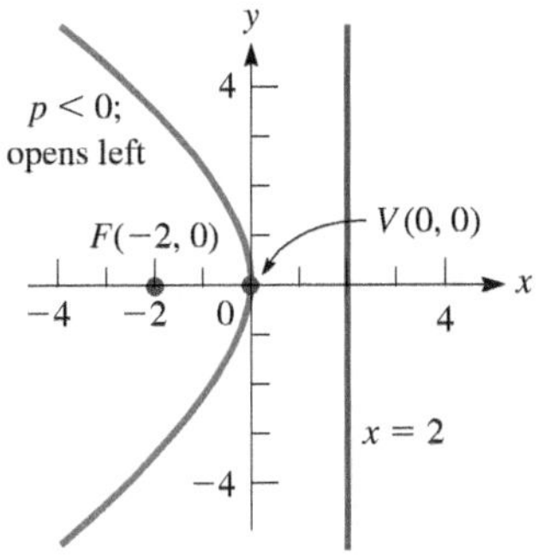

Fig. 21.41

EXAMPLE 2 Find the equation, given the directrix and the vertex

Find the equation of a parabola with vertex at the origin and directrix $x = 2$.

When the vertex is at the origin, a directrix parallel to the y-axis indicates that the parabola's axis is along the x-axis and that Eq. (21.15) applies. The equation of the directrix is of the form $x = p$, so we rewrite $x = 2$ as $x = -(-2)$ and find that $p = -2$. Therefore, $4p = -8$ and Eq. (21.15) becomes

$$y^2 = -8x$$

The negative coefficient of the x-term tells us that the parabola opens to the left, as shown in Fig. 21.41.

If we chose the focus as the point $(0, p)$ and the directrix as the line $y = -p$, we would find that the resulting equation is $x^2 = 4py$, which is symmetric about the y-axis because $(-x^2) = 4py$ is the same as $x^2 = 4py$. We summarize these properties below.

Standard Equations of a Parabola with Vertex at (0, 0)

Equation	$y^2 = 4px$	**(21.15)**	$x^2 = 4py$	**(21.16)**
Axis	x-axis		y-axis	
Focus	$(p, 0)$		$(0, p)$	
Directrix	$x = -p$		$y = -p$	
Symmetry	x-axis		y-axis	
Graph	If $p > 0$, opens right	If $p < 0$, opens left	If $p > 0$, opens up	If $p < 0$, opens down

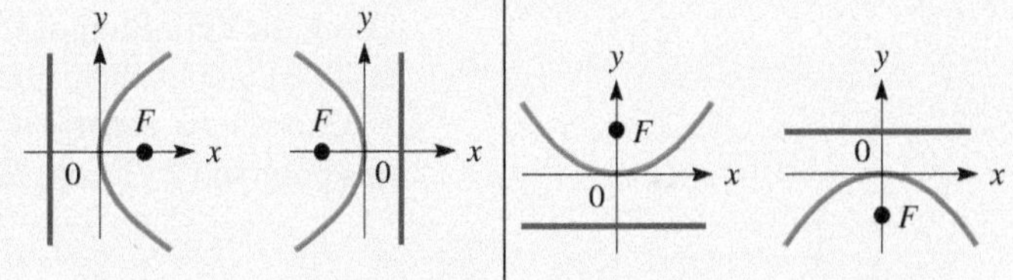

Fig. 21.42(a) Fig. 21.42(b) Fig. 21.42(c) Fig. 21.42(d)

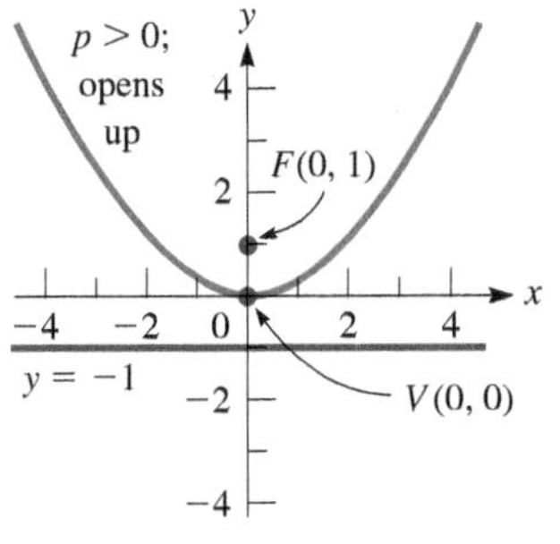

Fig. 21.43

EXAMPLE 3 Standard form—axis along the y-axis

(a) The parabola $x^2 = 4y$ fits the form of Eq. (21.16). Therefore, its axis is along the y-axis, and its vertex is at the origin. From the equation, we find the value of p, which in turn tells us the location of the vertex and the directrix.

$$x^2 = 4y \qquad 4p = 4, \qquad p = 1$$

Focus $(0, p)$ is $(0, 1)$; directrix $y = -p$ is $y = -1$. The parabola is shown in Fig. 21.43, and we see in this case that it opens upward.

(b) The parabola $2x^2 = -9y$ fits the form of Eq. (21.16) if we write it in the form

$$x^2 = -\tfrac{9}{2}y$$

Here, we see that $4p = -9/2$. Therefore, its axis is along the y-axis, and its vertex is at the origin. Since $4p = -9/2$, we have

$$p = -\frac{9}{8} \qquad \text{focus}\left(0, -\frac{9}{8}\right) \qquad \text{directrix } y = \frac{9}{8}$$

The parabola opens downward, as shown in Fig. 21.44.

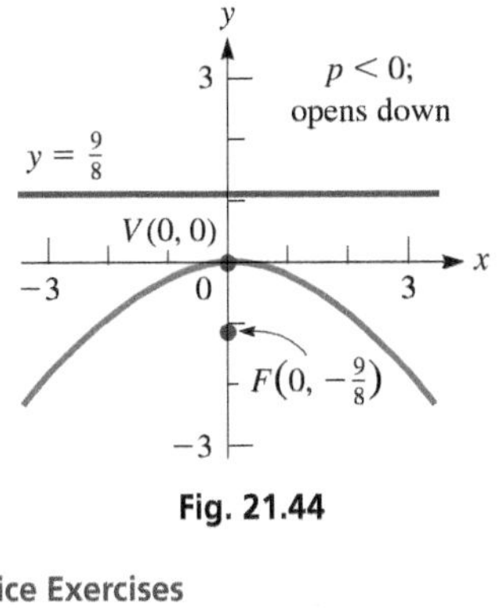

Fig. 21.44

Practice Exercises

Find the coordinates of the focus and the equation of the directrix for each parabola.

1. $y^2 = -24x$ **2.** $x^2 = 40y$

EXAMPLE 4 Parabola—satellite dish

■ See the chapter introduction.

In calculus, it can be shown that an electromagnetic wave (light wave, television signal, etc.) parallel to the axis of a parabolic reflector will pass through the focus of the parabola. Applications of this property of a parabola are many, including a satellite television dish and a spotlight (the light from the focus is reflected as a beam).

A parabolic satellite television dish is 46.0 cm across and 4.50 cm deep, as shown in Fig. 21.45. Find where the receiver should be located to receive all of the waves reflected off the parabolic surface.

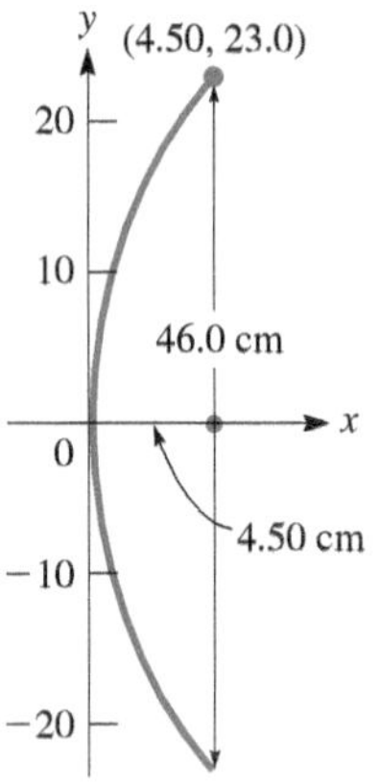

Fig. 21.45

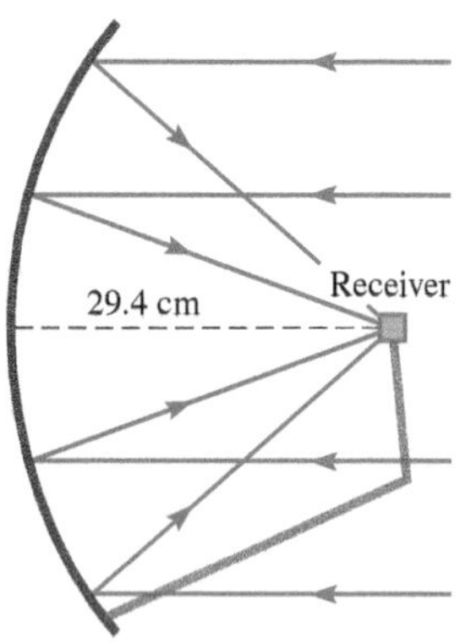

Fig. 21.46

With the vertex at the origin and the axis along the x-axis, we will use the general form $y^2 = 4px$ of the parabola. Since the parabolic opening is 46.0 cm across and 4.50 cm deep, the point $(4.50, 23.0)$ will be on any of the parabolic cross-sections. Therefore, we find p by substituting $(4.50, 23.0)$ in the equation. This means

$$23.0^2 = 4p(4.50), \qquad p = 29.4 \text{ cm}$$

and the receiver should be placed 29.4 cm from the vertex as shown in Fig. 21.46. The equation of the parabolic cross-section is $y^2 = 118x$.

Eqs. (21.15) and (21.16) give the general forms of the equation of a parabola with vertex at the origin and focus on one of the coordinate axes. We now use the definition to find the equation of a parabola that has its vertex at a point other than the origin.

EXAMPLE 5 Find an equation from the definition

Using the definition of the parabola, find the equation of the parabola with its focus at $(2, 3)$ and its directrix the line $y = -1$. See Fig. 21.47.

Choosing a general point $P(x, y)$ on the parabola and equating the distances from this point to $(2, 3)$ and to the line $y = -1$, we have

$$\sqrt{(x-2)^2 + (y-3)^2} = y + 1$$

distance P to F = distance P to $y = -1$

$$(x-2)^2 + (y-3)^2 = (y+1)^2 \qquad \text{squaring both sides}$$

$$x^2 - 4x + 4 + y^2 - 6y + 9 = y^2 + 2y + 1$$

$$8y = 12 - 4x + x^2$$

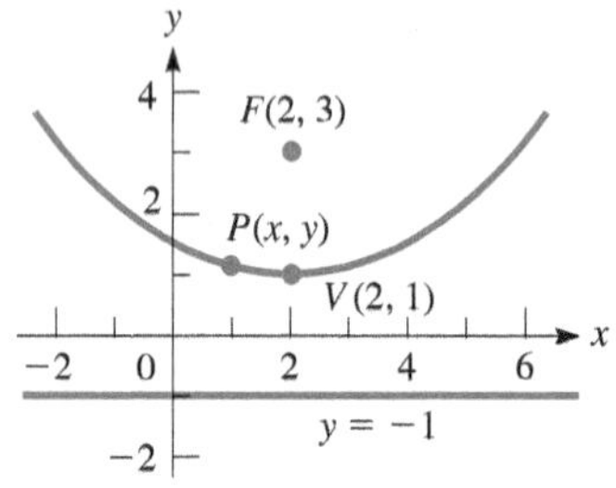

Fig. 21.47

We note that this type of equation has appeared frequently in earlier chapters. The x-term and the constant (12 in this case) are characteristic of a parabola that does not have its vertex at the origin if the directrix is parallel to the x-axis.

A parabola whose axis is along the y-axis or parallel to it, as in Examples 3 and 5, is the graph of the quadratic function obtained when solving for y. In contrast, a parabola whose axis is along the x-axis or parallel to it, as in Examples 1, 2, and 4, is not a function but a combination of two functions. This is similar to Example 7 in Section 21.3. These cases are shown in the next example.

EXAMPLE 6 A parabola as a graph of one or two functions

Graph each of the following parabolas by using functions.

(a) $8y = 12 - 4x + x^2$

Solving for y, we obtain the function $y = (12 - 4x + x^2)/8$. The graph of this function is the same parabola shown in Fig. 21.47.

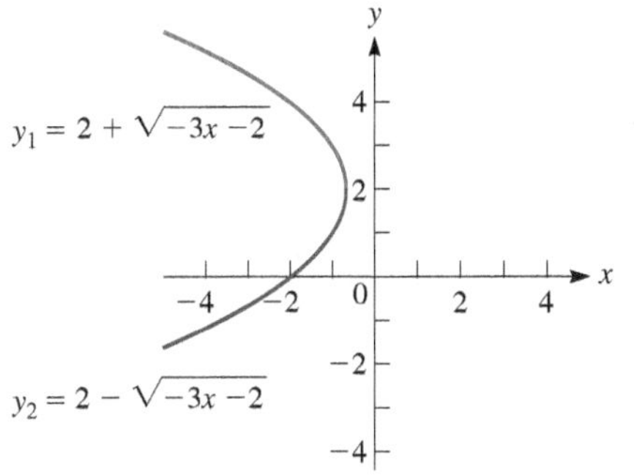

Fig. 21.48

(b) $y^2 + 3x - 4y + 6 = 0$

Solving for y using the quadratic formula, we get

$$a = 1 \quad b = -4 \quad c = 3x + 6$$

$$y = \frac{4 \pm \sqrt{(-4)^2 - 4(1)(3x+6)}}{2}$$

Therefore, we obtain the two functions

$$y_1 = 2 + \sqrt{-3x-2} \quad \text{and} \quad y_2 = 2 - \sqrt{-3x-2}$$

The parabola is thus not the graph of a single function, but the graph of these two functions graphed together, as shown in Fig. 21.48.

We see that *the equation of a parabola is characterized by the square of either x or y (but not both) and a first power term of the other variable.* We will consider the parabola further in Sections 21.7, 21.8, and 21.9.

The parabola has numerous technical applications. The reflection property illustrated in Example 4 has other important applications, such as the design of a radar antenna. Other examples of parabolas include the path of a projectile and the cables of a suspension bridge.

EXERCISES 21.4

In Exercises 1–4, make the given changes in the indicated examples of this section and then solve the resulting problems. In each, find the focus and directrix, and sketch the parabola.

1. In Example 1, change $12x$ to $20x$.

2. In Example 2, change $-8x$ to $-20x$.

3. In Example 3(a), change $4y$ to $-6y$.

4. In Example 3(b), change $-9y$ to $7y$.

In Exercises 5–16, determine the coordinates of the focus and the equation of the directrix of the given parabolas. Sketch each curve.

5. $y^2 = 4x$

6. $y^2 = 16x$

7. $y^2 = -4x$

8. $y^2 = -36x$

9. $x^2 = 72y$

10. $x^2 = y$

11. $x^2 = -4y$

12. $x^2 + 12y = 0$

13. $2y^2 - 5x = 0$

14. $3x^2 = 8y$

15. $y = 0.48x^2$

16. $x = 7.6y^2$

In Exercises 17–30, find the equations of the parabolas satisfying the given conditions. The vertex of each is at the origin.

17. Focus $(3, 0)$

18. Focus $(0, 0.4)$

19. Focus $(0, -0.5)$

20. Focus $(2.5, 0)$

21. Directrix $y = -0.16$

22. Directrix $x = 20$

23. Directrix $x = -84$

24. Directrix $y = 2.3$

25. Axis $x = 0$, passes through $(-1, 8)$

26. Symmetric to x-axis, passes through $(2, -1)$

27. Passes through $(3, 5)$ and $(3, -5)$

28. Passes through $(6, -1)$ and $(-6, -1)$

29. Passes through $(3, 3)$ and $(12, 6)$

30. Passes through $(-5, -5)$ and $(-10, -20)$

In Exercises 31–60, solve the given problems.

31. Sketch the graph of the inequality $x^2 < 8y$.

32. Sketch the graph of the inequality $9y \leq 4x^2$.

33. At what point(s) do the parabolas $y^2 = 2x$ and $x^2 = -16y$ intersect?

34. Using trigonometric identities, show that the parametric equations $x = \sin t$, $y = 2(1 - \cos^2 t)$ are the equations of a parabola.

35. Find the equation of the parabola with focus $(6, 1)$ and directrix $x = 0$, by use of the definition. Sketch the curve.

36. Find the equation of the parabola with focus $(1, 1)$ and directrix $y = 5$, by use of the definition. Sketch the curve.

37. Use a graphing utility to view the parabola $y^2 + 2x + 8y + 13 = 0$.

38. Use a graphing utility to view the parabola $y^2 - 2x - 6y + 19 = 0$.

39. The equation of a parabola with vertex (h, k) and axis parallel to the x-axis is $(y - k)^2 = 4p(x - h)$. (This is shown in Section 21.7.) Sketch the parabola for which (h, k) is $(2, -3)$ and $p = 2$.

40. The equation of a parabola with vertex (h, k) and axis parallel to the y-axis is $(x - h)^2 = 4p(y - k)$. (This is shown in Section 21.7.) Sketch the parabola for which (h, k) is $(-1, 2)$ and $p = -3$.

41. The chord of a parabola that passes through the focus and is parallel to the directrix is called the *latus rectum* of the parabola. Find the length of the latus rectum of the parabola $y^2 = 4px$.

42. Find the equation of the circle that has the focus and the vertex of the parabola $x^2 = 8y$ as the ends of a diameter.

43. For either standard form of the equation of a parabola, describe what happens to the shape of the parabola as $|p|$ increases.

44. To ensure proper drainage, the surface of a lacrosse field with synthetic turf is parabolic. The field is 55.0 m wide and 66.0 cm higher in the centre than at the sides. Find the equation that represents the surface if the origin is at the centre of the field.

45. The Lions' Gate Bridge in Vancouver, British Columbia, is a suspension bridge, and its supporting cables are parabolic. See Fig. 21.49. With the origin at the low point of the cable, what equation represents the cable if the supporting towers are 473 m apart and the maximum sag is 55.0 m?

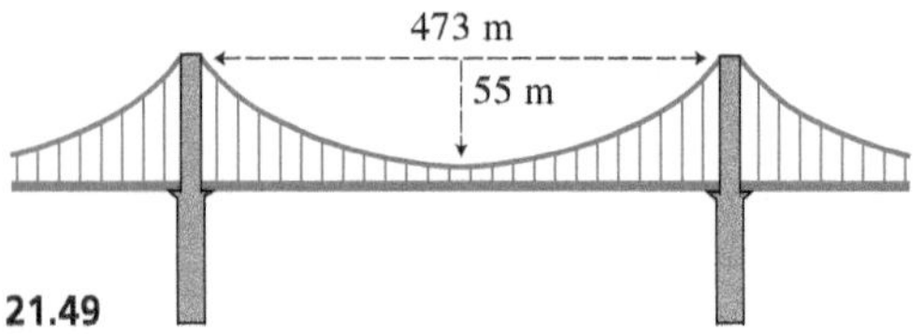

Fig. 21.49

46. Each arch of the Allen Lambert Galleria in Brookfield Place in Toronto is a parabolic arch 26 m high at the centre and 13.7 m wide at the base. What equation represents the arch if the origin is at the top of the arch?

47. A television satellite dish measures 80.0 cm across its opening and is 12.5 cm deep. Find the distance between the vertex and the focus (where the receiver is placed).

48. Part of the steel frame of the Glacier Skywalk over Jasper National Park can be approximated by a parabola with vertex at the origin, axis along the y-axis, and passing through the point (8.55 m, −10.65 m). Find the equation of the parabola.

49. The rate of development of heat H (in W) in a resistor of resistance R (in Ω) of an electric circuit is given by $H = Ri^2$, where i is the current (in A) in the resistor. Sketch the graph of H vs. i, if $R = 6.0\ \Omega$.

50. What is the length of the horizontal bar across the parabolically shaped window shown in Fig. 21.50?

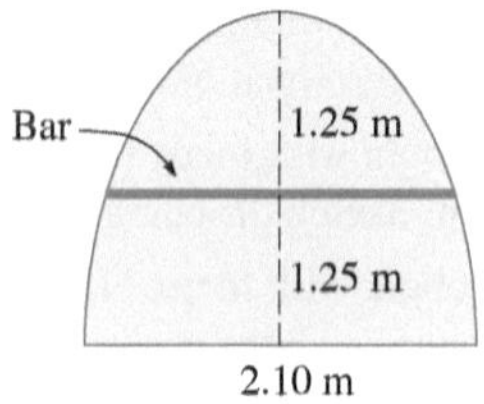

Fig. 21.50

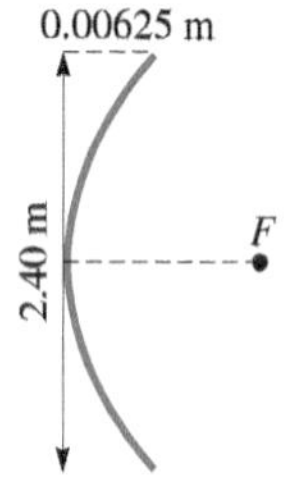

Fig. 21.51

51. The primary mirror in the Hubble space telescope has a parabolic cross-section, which is shown in Fig. 21.51. What is the focal length (vertex to focus) of the mirror?

52. A rocket is fired horizontally from a plane. Its horizontal distance x and vertical distance y from the point at which it was fired are given by $x = v_0 t$ and $y = \frac{1}{2}gt^2$, where v_0 is the initial velocity of the rocket, t is the time, and g is the acceleration due to gravity. Express y as a function of x and show that it is the equation of a parabola.

53. To launch a spacecraft to the moon, it is first put into orbit around the earth and then into a parabolic path toward the moon. Assume the parabolic path is represented by $x^2 = 4py$ and the spacecraft is later observed at (10, 3) (units in thousands of km) after launch. Will this path lead directly to the moon at (110, 340)?

54. Under certain load conditions, a beam fixed at both ends is approximately parabolic in shape. If a beam is 4.0 m long and the deflection in the middle is 2.0 cm, find an equation to represent the shape of the beam.

55. A spotlight with a parabolic reflector is 15.0 cm wide and 6.50 cm deep. See Fig. 21.52 and Example 4. Where should the filament of the bulb be located so as to produce a beam of light?

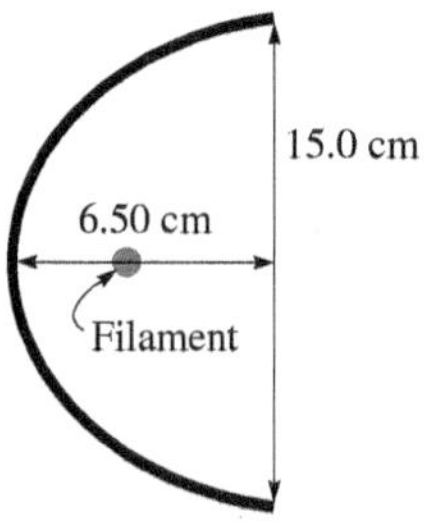

Fig. 21.52

56. A wire is fastened 12.0 m up on each of two telephone poles that are 60.0 m apart. Halfway between the poles, the wire is 10.0 m above the ground. Assuming the wire is parabolic, find the height of the wire 15.0 m from either pole.

57. The total annual fraction f of energy supplied by solar energy to a home is given by $f = 0.065\sqrt{A}$, where A is the area of the solar collector. Sketch the graph of f as a function of A $(0 < A \leq 200\ \text{m}^2)$.

58. The velocity v (in m/s) of a jet of water flowing from an opening in the side of a certain container is given by $v = 4.4\sqrt{h}$, where h is the depth (in m) of the opening. Sketch a graph of v vs. h.

59. A small island is 4 km north of a straight shoreline. A ship channel is equidistant between the island and the shoreline. Write an equation for the channel.

60. Under certain circumstances, the maximum power P (in W) in an electric circuit varies as the square of the voltage of the source E_0 and inversely as the internal resistance R_i (in Ω) of the source. If 10 W is the maximum power for a source of 2.0 V and internal resistance of 0.10 Ω, sketch the graph of P vs. E_0 if R_i remains constant.

Answers to Practice Exercises

1. $F(-6, 0)$; dir. $x = 6$ **2.** $F(0, 10)$; dir. $y = -10$

21.5 The Ellipse

The next important curve is the ellipse. *An* **ellipse** *is defined as the locus of a point $P(x, y)$ that moves so that the sum of its distances from two fixed points is constant. These fixed points are the* **foci** *of the ellipse.* Letting this sum of distances be $2a$ and the foci be the points $(-c, 0)$ and $(c, 0)$, we have

$$\sqrt{(x-c)^2 + y^2} + \sqrt{(x+c)^2 + y^2} = 2a$$

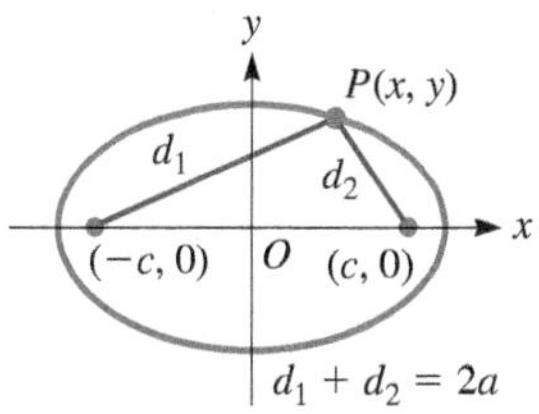

Fig. 21.53

See Fig. 21.53. The ellipse has its centre at the origin so that c is the length of the line segment from the centre to a focus. We will also see that a has a special meaning. Now, from Section 14.4, we see that in order to remove the radicals we should move one radical to the right and then square each side. This leads to the following steps:

$$\sqrt{(x+c)^2+y^2} = 2a - \sqrt{(x-c)^2+y^2}$$
$$(x+c)^2+y^2 = 4a^2 - 4a\sqrt{(x-c)^2+y^2} + \left(\sqrt{(x-c)^2+y^2}\right)^2$$
$$x^2+2cx+c^2+y^2 = 4a^2 - 4a\sqrt{(x-c)^2+y^2} + x^2 - 2cx + c^2 + y^2$$
$$4a\sqrt{(x-c)^2+y^2} = 4a^2 - 4cx$$
$$a\sqrt{(x-c)^2+y^2} = a^2 - cx$$
$$a^2(x^2-2cx+c^2+y^2) = a^4 - 2a^2cx + c^2x^2$$
$$(a^2-c^2)x^2 + a^2y^2 = a^2(a^2-c^2)$$

We now define $a^2 - c^2 = b^2$. (The reason for this will be clear shortly.) Therefore,

$$b^2x^2 + a^2y^2 = a^2b^2$$

Dividing through by a^2b^2, we have

$$\frac{x^2}{a^2} + \frac{y^2}{b^2} = 1 \qquad \textbf{(21.17)}$$

A graphical analysis of this equation is found below.

The x-intercepts are $(-a, 0)$ and $(a, 0)$. This means that $2a$ (the sum of distances used in the derivation) is also the distance between the x-intercepts. *The points $(a, 0)$ and $(-a, 0)$ are the* **vertices** *of the ellipse, and the line between them is the* **major axis** [see Fig. 21.54(a)]. Thus, *a is the length of the* **semimajor axis**.

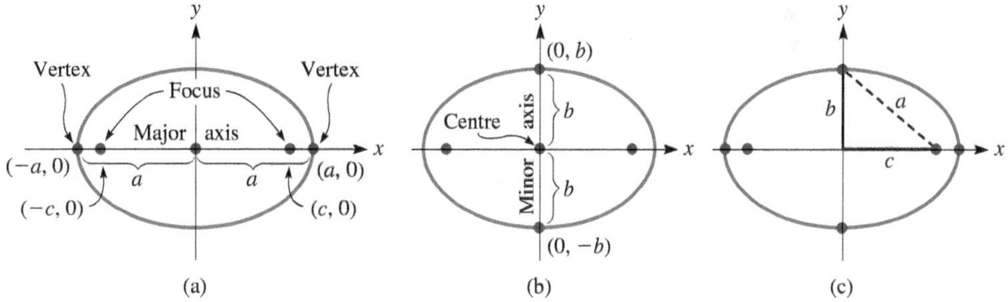

Fig. 21.54

We can now state that *Eq. (21.17) is called the* **standard equation** *of the ellipse with its major axis along the x-axis and its centre at the origin.*

The y-intercepts of this ellipse are $(0, -b)$ and $(0, b)$. *The line joining these intercepts is called the* **minor axis** *of the ellipse* [Fig. 21.54(b)], *which means b is the length of the* **semiminor** axis. The intercept $(0, b)$ is equidistant from $(-c, 0)$ and $(c, 0)$. Since the sum of the distances from these points to $(0, b)$ is $2a$, the distance $(c, 0)$ to $(0, b)$ must be a. Thus, we have a right triangle with line segments of lengths a, b, and c, with a as hypotenuse [Fig. 21.54(c)]. Therefore,

$$a^2 = b^2 + c^2 \qquad \textbf{(21.18)}$$

is the relation between distances a, b, and c. This also shows why b was defined as it was in the derivation of Eq. (21.17).

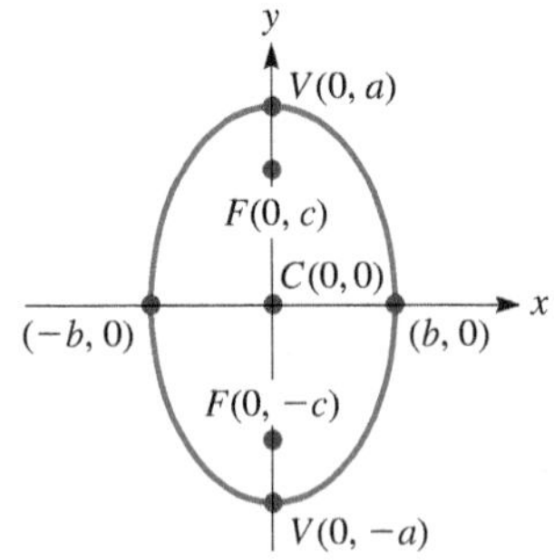

Fig. 21.55

If we choose points on the y-axis as the foci, *the standard equation of the ellipse, with its centre at the origin and its major axis along the y-axis, is*

$$\frac{y^2}{a^2} + \frac{x^2}{b^2} = 1 \tag{21.19}$$

See Fig. 21.55.

We now summarize the properties of ellipses with centre at the origin.

Standard Equation of an Ellipse with Centre at (0, 0)

Equation	$\frac{x^2}{a^2} + \frac{y^2}{b^2} = 1$ **(21.17)** $a > b > 0$	$\frac{y^2}{a^2} + \frac{x^2}{b^2} = 1$ **(21.19)** $a > b > 0$
Geometric relationship	$a^2 = b^2 + c^2$ **(21.18)**	$a^2 = b^2 + c^2$ **(21.18)**
Major axis	x-axis (length $2a$)	y-axis (length $2a$)
Minor axis	y-axis (length $2b$) $(0, -b)$ to $(0, b)$	x-axis (length $2b$) $(-b, 0)$ to $(b, 0)$
Vertices	$(-a, 0)$, $(a, 0)$	$(0, -a)$, $(0, a)$
Foci	$(-c, 0)$, $(c, 0)$	$(0, -c)$, $(0, c)$
Symmetry and graph	x-axis y-axis origin	x-axis y-axis origin

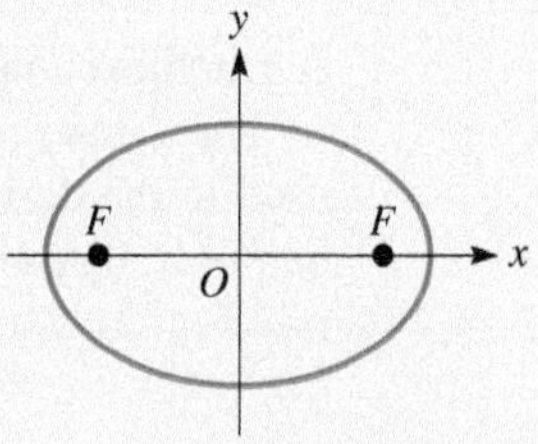

Fig. 21.56(a)

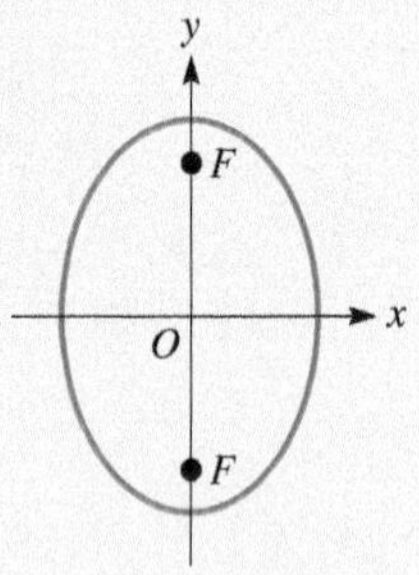

Fig. 21.56(b)

EXAMPLE 1 Standard equation—major axis along the x-axis

Find the coordinates of the vertices and the foci of the ellipse $\frac{x^2}{25} + \frac{y^2}{9} = 1$.

The equation fits the form of Eq. (21.17) because the larger square, 25, appears under x^2. Therefore, $a^2 = 25$ and $b^2 = 9$, or $a = 5$ and $b = 3$. This means that the vertices are $(5, 0)$ and $(-5, 0)$, and the minor axis extends from $(0, -3)$ to $(0, 3)$. See Fig. 21.57. We find c from the relation $c^2 = a^2 - b^2$. This means that $c^2 = 16$ and the foci are $(4, 0)$ and $(-4, 0)$.

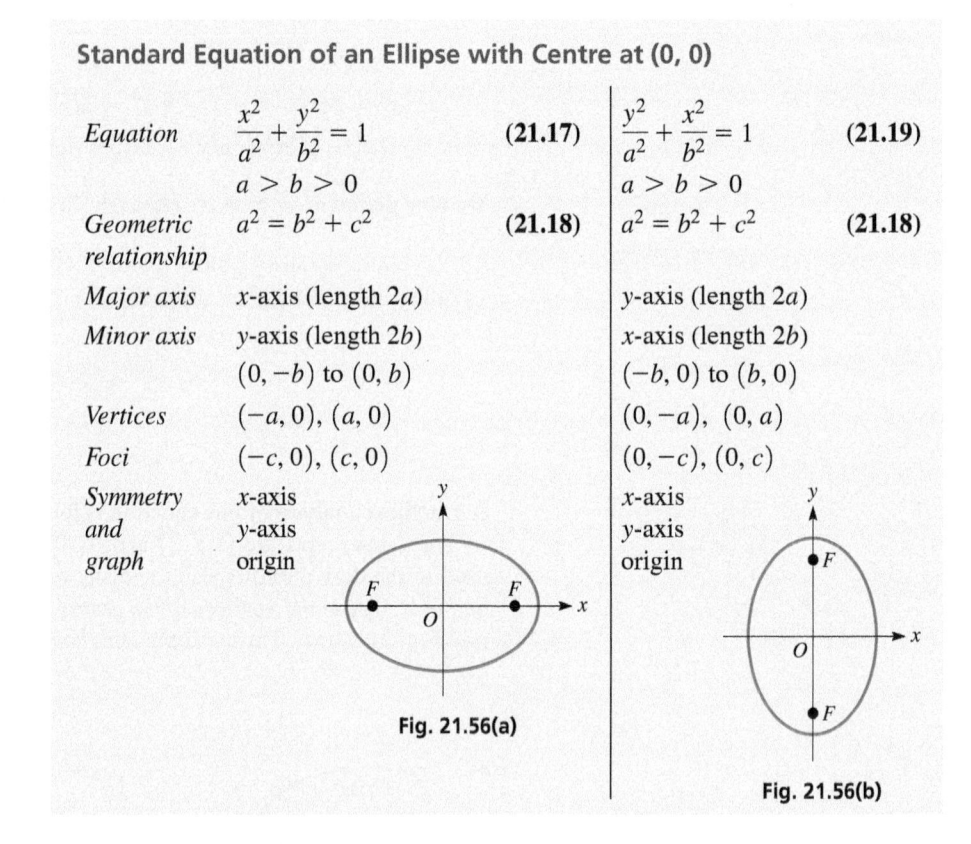

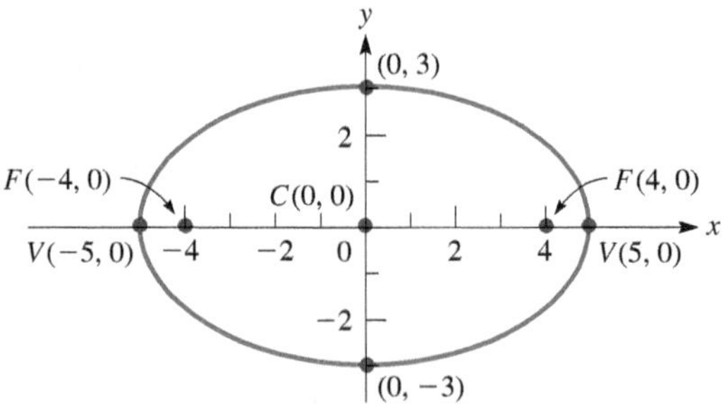

Fig. 21.57

EXAMPLE 2 Standard equation—major axis along the *y*-axis

Find the coordinates of the vertices and the foci of the ellipse $\frac{x^2}{4} + \frac{y^2}{9} = 1$.

The equation has the larger denominator, 9, under the y^2. Therefore, it fits the form of Eq. (21.19) with $a^2 = 9$ and $b^2 = 4$. This means that the vertices are $(0, 3)$ and $(0, -3)$, and the minor axis extends from $(-2, 0)$ to $(2, 0)$. In turn, we know that $c^2 = a^2 - b^2 = 9 - 4 = 5$ and that the foci are $(0, \sqrt{5})$ and $(0, -\sqrt{5})$. This ellipse is shown in Fig. 21.58.

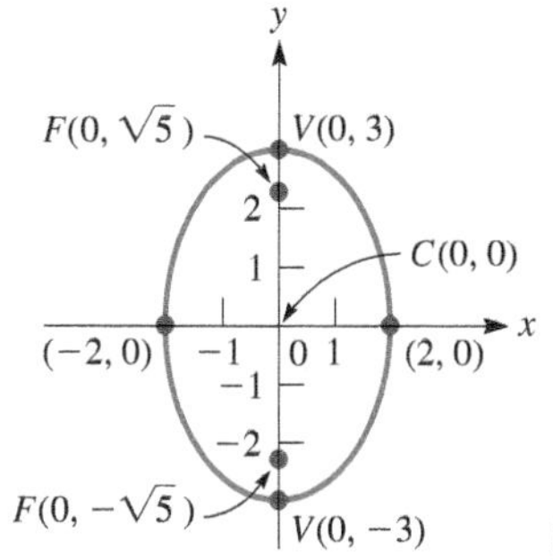

Fig. 21.58

EXAMPLE 3 Find vertices, foci, minor axis

Find the coordinates of the vertices, the ends of the minor axis, and the foci of the ellipse $4x^2 + 16y^2 = 64$.

This equation must be put in standard form first, which we do by dividing through by 64. When this is done, we obtain

$$\frac{x^2}{16} + \frac{y^2}{4} = 1$$

form requires + and 1

We see that $a^2 = 16$ and $b^2 = 4$, which tells us that $a = 4$ and $b = 2$. Then $c = \sqrt{16 - 4} = \sqrt{12} = 2\sqrt{3}$. Since a^2 appears under x^2, the vertices are $(4, 0)$ and $(-4, 0)$. The ends of the minor axis are $(0, 2)$ and $(0, -2)$, and the foci are $(2\sqrt{3}, 0)$ and $(-2\sqrt{3}, 0)$. See Fig. 21.59.

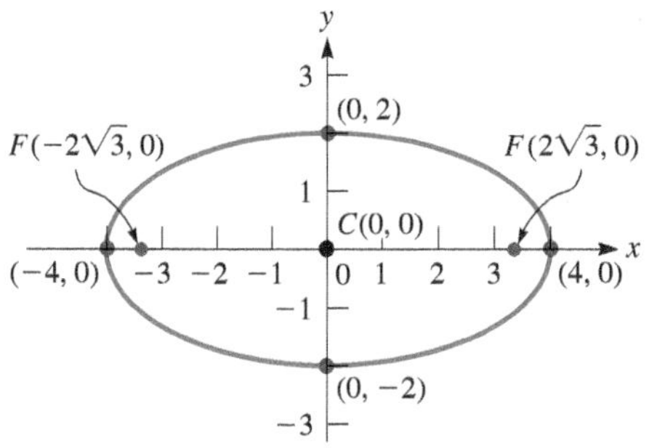

Fig. 21.59

EXAMPLE 4 Ellipse—satellite orbit

A satellite to study the earth's atmosphere has a minimum altitude of 1000 km and a maximum altitude of 3200 km. If the path of the satellite about the earth is an ellipse with the centre of the earth at one focus, what is the equation of its path? Assume that the radius of the earth is 6400 km.

We set up the coordinate system such that the centre of the ellipse is at the origin and the centre of the earth is at the right focus, as shown in Fig. 21.60. We know that the distance between vertices is

$$2a = 3200 + 6400 + 6400 + 1000 = 17\,000 \text{ km}$$
$$a = 8500 \text{ km}$$

The distance from the right focus to the right vertex is 7400 km. This tells us

$$c = a - 7400 = 8500 - 7400 = 1100 \text{ km}$$

We can now calculate b^2 as

$$b^2 = a^2 - c^2 = 8500^2 - 1100^2 = 7.10 \times 10^7 \text{ km}^2$$

Since $a^2 = 8500^2 = 7.23 \times 10^7 \text{ km}^2$, the equation is

$$\frac{x^2}{7.23 \times 10^7} + \frac{y^2}{7.10 \times 10^7} = 1$$

or

$$7.10x^2 + 7.23y^2 = 5.13 \times 10^8$$

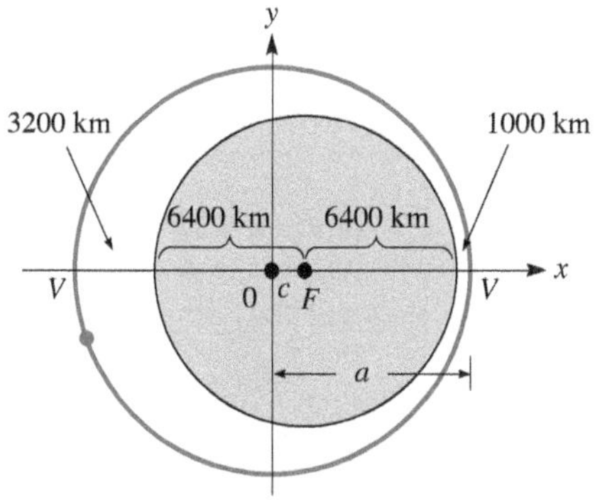

Fig. 21.60

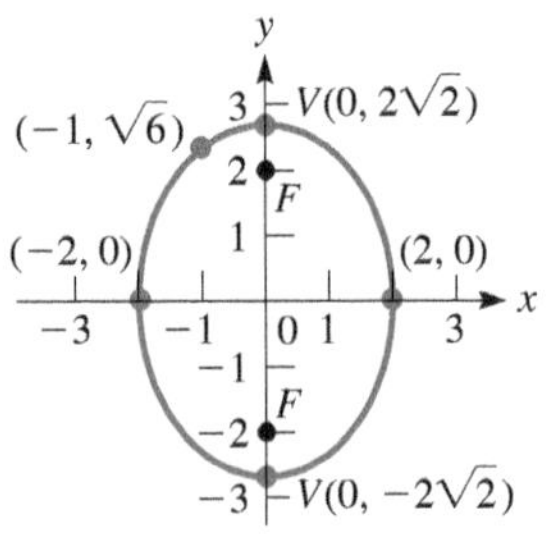

Fig. 21.61

Practice Exercises

Find the vertices and foci of each ellipse.

1. $\frac{x^2}{9} + y^2 = 1$ 2. $25x^2 + 4y^2 = 25$

EXAMPLE 5 Find the equation given points

Find the equation of the ellipse with its centre at the origin and an end of its minor axis at $(2, 0)$ and which passes through $(-1, \sqrt{6})$.

Since the centre is at the origin and an end of the minor axis is at $(2, 0)$, we know that the ellipse is of the form of Eq. (21.19) and that $b = 2$. Thus, we have

$$\frac{y^2}{a^2} + \frac{x^2}{2^2} = 1$$

In order to find a^2, we use the fact that the ellipse passes through $(-1, \sqrt{6})$. This means that these coordinates satisfy the equation of the ellipse. This gives

$$\frac{(\sqrt{6})^2}{a^2} + \frac{(-1)^2}{4} = 1, \qquad \frac{6}{a^2} = \frac{3}{4}, \qquad a^2 = 8$$

Therefore, the equation of the ellipse, shown in Fig. 21.61, is $\frac{y^2}{8} + \frac{x^2}{4} = 1$.

The following example illustrates the use of the definition of the ellipse to find the equation of an ellipse with its centre at a point other than the origin.

EXAMPLE 6 Find the equation from the definition—two functions

Using the definition, find the equation of the ellipse with foci at $(1, 3)$ and $(9, 3)$, with major axis of 10.

Recalling that the sum of distances in the definition equals the length of the major axis, we now use the same method as in the derivation of Eq. (21.17).

$$\sqrt{(x-1)^2 + (y-3)^2} + \sqrt{(x-9)^2 + (y-3)^2} = 10 \quad \text{use definition of ellipse}$$

$$\sqrt{(x-1)^2 + (y-3)^2} = 10 - \sqrt{(x-9)^2 + (y-3)^2} \quad \text{isolate a radical}$$

$$x^2 - 2x + 1 + y^2 - 6y + 9 = 100 - 20\sqrt{(x-9)^2 + (y-3)^2} + x^2 - 18x + 81 + y^2 - 6y + 9 \quad \text{square both sides and simplify}$$

$$20\sqrt{(x-9)^2 + (y-3)^2} = 180 - 16x \quad \text{isolate radical}$$

$$5\sqrt{(x-9)^2 + (y-3)^2} = 45 - 4x \quad \text{divide by 4}$$

$$25(x^2 - 18x + 81 + y^2 - 6y + 9) = 2025 - 360x + 16x^2 \quad \text{square both sides}$$

$$9x^2 - 90x + 25y^2 - 150y + 225 = 0 \quad \text{simplify}$$

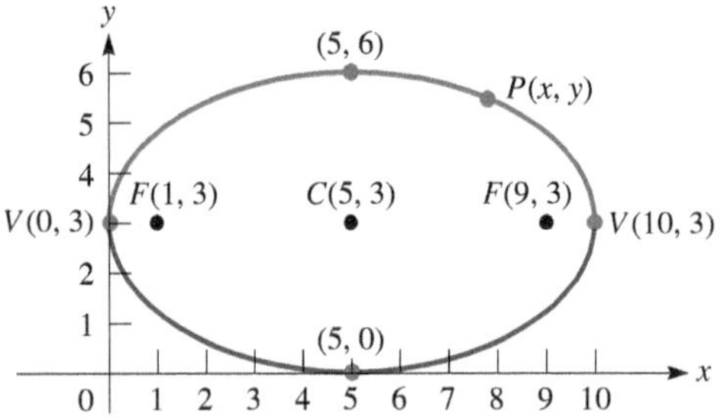

The additional x- and y-terms are characteristic of the equation of an ellipse whose centre is not at the origin (see Fig. 21.62).

The ellipse is shown in Fig. 21.62, with each half drawn in a different colour to represent a different function. Solving the original equation for y, the two functions are $y = \frac{15 \pm 3\sqrt{10x - x^2}}{5}$.

Fig. 21.62

We see that *the equation of an ellipse is characterized by both an x^2-term and a y^2-term, having different coefficients (in value but not in sign).* We note that the coefficients of the squared terms differ, whereas for the circle, they are the same. We will consider the ellipse further in Sections 21.7, 21.8, and 21.9.

The ellipse has many applications. The orbits of the planets about the sun are elliptical. Gears, cams, and springs are often elliptical in shape. Arches are often constructed in the form of a semiellipse. Ellipses are also important for blood pattern analysis in forensic science, for parameter estimation in statistics, and in computer vision.

EXERCISES 21.5

In Exercises 1 and 2, make the given changes in the indicated examples of this section and then solve the given problems.

1. In Example 1, change the 9 to 36; find the vertices, ends of the minor axis, and foci; and sketch the ellipse.
2. In Example 2, interchange the 4 and 9; find the vertices, ends of the minor axis, and foci; and sketch the ellipse.

In Exercises 3–16, find the coordinates of the vertices and foci of the given ellipses. Sketch each curve.

3. $\frac{x^2}{4} + \frac{y^2}{1} = 1$
4. $\frac{x^2}{100} + \frac{y^2}{64} = 1$
5. $\frac{x^2}{25} + \frac{y^2}{144} = 1$
6. $\frac{x^2}{49} + \frac{y^2}{81} = 1$
7. $\frac{4x^2}{25} + \frac{y^2}{4} = 1$
8. $x^2 + \frac{9y^2}{25} = 1$
9. $4x^2 + 9y^2 = 324$
10. $x^2 + 36y^2 = 144$
11. $49x^2 + 4y^2 = 196$
12. $y^2 = 25(1 - x^2)$
13. $y^2 = 8(2 - x^2)$
14. $2x^2 + 3y^2 = 600$
15. $4x^2 + 25y^2 = 0.25$
16. $9x^2 + 4y^2 = 0.09$

In Exercises 17–28, find the equations of the ellipses satisfying the given conditions. The centre of each is at the origin.

17. Vertex $(15, 0)$, focus $(9, 0)$
18. Minor axis 8, vertex $(0, -5)$
19. Focus $(0, 8)$, major axis 34
20. Vertex $(0, 13)$, focus $(0, -5)$
21. End of minor axis $(0, 12)$, focus $(8, 0)$
22. Sum of lengths of major and minor axes 18, focus $(3, 0)$
23. Vertex $(8, 0)$, passes through $(2, 3)$
24. Focus $(0, 2)$, passes through $(-1, \sqrt{3})$
25. Passes through $(2, 2)$ and $(1, 4)$
26. Passes through $(-2, 2)$ and $(1, \sqrt{6})$
27. The sum of distances from (x, y) to $(6, 0)$ and $(-6, 0)$ is 20.
28. The sum of distances from (x, y) to $(0, 2)$ and $(0, -2)$ is 5.

In Exercises 29–56, solve the given problems.

29. Find any point(s) of intersection of the graphs of the ellipse $4x^2 + 9y^2 = 40$ and the parabola $y^2 = 4x$.
30. Find the equation of the circle that has the same centre as the ellipse $4x^2 + 9y^2 = 36$ and is internally tangent to the ellipse.
31. Find the equation of the ellipse with foci $(-2, 1)$ and $(4, 1)$ and a major axis of length 10, by use of the definition. Sketch the curve.
32. Find the equation of the ellipse with foci $(1, 4)$ and $(1, 0)$ that passes through $(4, 4)$, by use of the definition. Sketch the curve.
33. Use a graphing utility to view the ellipse $4x^2 + 3y^2 + 16x - 18y + 31 = 0$.
34. Use a graphing utility to view the ellipse $4x^2 + 8y^2 + 4x - 24y + 1 = 0$.
35. The equation of an ellipse with centre (h, k) and major axis parallel to the x-axis is $\frac{(x-h)^2}{a^2} + \frac{(y-k)^2}{b^2} = 1$. (This is shown in Section 21.7.) Sketch the ellipse that has a major axis of 6, a minor axis of 4, and for which (h, k) is $(2, -1)$.
36. The equation of an ellipse with centre (h, k) and major axis parallel to the y-axis is $\frac{(y-k)^2}{a^2} + \frac{(x-h)^2}{b^2} = 1$. (This is shown in Section 21.7.) Sketch the ellipse that has a major axis of 8, a minor axis of 6, and for which (h, k) is $(1, 3)$.
37. For what values of k does the ellipse $x^2 + k^2 = 1$ have its vertices on the y-axis? Explain how these values are found.
38. For what value of k does the ellipse $x^2 + k^2y^2 = 25$ have a focus at $(3, 0)$? Explain how this value is found.
39. Show that the ellipse $2x^2 + 3y^2 - 8x - 4 = 0$ is symmetric to the x-axis.
40. Show that the ellipse $5x^2 + y^2 - 3y - 7 = 0$ is symmetric to the y-axis.
41. Graph the inequality $100x^2 + 49y^2 \leq 4900$.
42. Graph the inequality $5x^2 + 4y^2 > 20$.
43. Show that the parametric equations $x = 2 \sin t$ and $y = 3 \cos t$ define an ellipse.
44. For what values of k does $y^2 = 1 + kx^2$ represent an ellipse with foci on the y-axis?
45. The electric power P (in W) dissipated in a resistance R (in Ω) is given by $P = Ri^2$, where i is the current (in A) in the resistor. Find the equation for the total power of 64 W dissipated in two resistors, with resistances 2.0 Ω and 8.0 Ω, respectively, and with currents i_1 and i_2, respectively. Sketch the graph, assuming that negative values of current are meaningful.
46. The *eccentricity* e of an ellipse is defined as $e = c/a$. A cam in the shape of an ellipse can be described by the equation $x^2 + 9y^2 = 81$. Find the eccentricity of this elliptical cam.
47. A space object (dubbed 2003 UB313), larger and more distant than Pluto, was discovered in 2003. In its elliptical orbit with the sun at one focus, it is 5.6×10^9 km from the sun at the closest, and 14.4×10^9 km at the farthest. What is the eccentricity of its orbit? See Exercise 46.
48. Halley's Comet has an elliptical orbit with $a = 17.94$ AU (AU is astronomical unit, 1 AU $= 1.5 \times 10^8$ km) and $b = 4.552$ AU, with the sun at one focus. What is the closest that the comet comes to the sun? Explain your method.
49. A draftsman draws a series of triangles with a base from $(-3, 0)$ to $(3, 0)$ and a perimeter of 14 cm (all measurements in centimetres). Find the equation of the curve on which all of the third vertices of the triangles are located.
50. A lithotripter is used to break up a kidney stone by placing the kidney stone at one focus of an ellipsoid end-section and a source of shock waves at the focus of the other end-section. If the vertices of the end-sections are 30.0 cm apart and a minor axis of the ellipsoid is 6.0 cm, how far apart are the foci? In a lithotripter, the shock waves are reflected as the sound waves noted in Exercise 51.
51. An ellipse has a focal property such that a light ray or sound wave emanating from one focus will be reflected through the other focus. Many buildings and structures are built with elliptical ceilings or walls, so that a sound from one focus is easily heard at the other focus. (Some examples include Statuary Hall in the U.S. Capitol, the Echo Wall in the Temple of Heaven in Beijing, the Whispering Wall at the Barossa Reservoir in

South Australia, and the Whispering Wall on the grounds of Parliament Hill in Ottawa.) If a building has a ceiling whose cross-sections are part of an ellipse that can be described by the equation $36x^2 + 225y^2 = 8100$ (measurements in metres), how far apart must two people stand in order to whisper to each other using this focal property?

52. An airplane wing is designed such that a certain cross-section is an ellipse 2.80 m wide and 0.40 m thick. Find the equation that can be used to describe the perimeter of this cross-section.

53. A road passes through a tunnel with a semielliptical cross-section 19.6 m wide and 5.5 m high at the centre. What is the height of the tallest vehicle that can pass through the tunnel at a point 6.7 m from the centre? See Fig. 21.63.

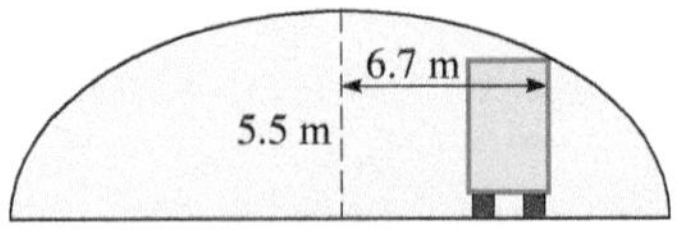

Fig. 21.63

54. An architect designs a window in the shape of an ellipse 1.50 m wide and 1.10 m high. Find the perimeter of the window from the formula $p = \pi(a + b)$. This formula gives a good *approximation* for the perimeter when a and b are nearly equal.

55. The ends of a horizontal tank 20.0 m long are ellipses, which can be described by the equation $9x^2 + 20y^2 = 180$, where x and y are measured in metres. The area of an ellipse is $A = \pi ab$. Find the volume of the tank.

56. A laser beam 6.80 mm in diameter is incident on a plane surface at an angle of 62.0°, as shown in Fig. 21.64. What is the elliptical area that the laser covers on the surface? (See Exercise 55.)

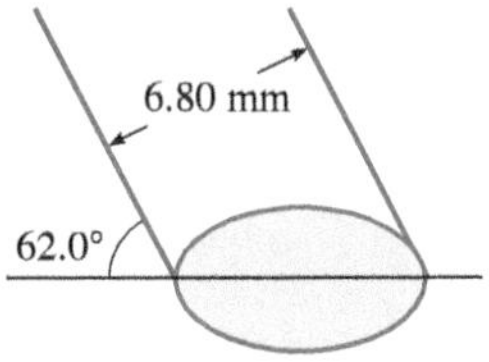

Fig. 21.64

Answers to Practice Exercises

1. $V(3, 0), V(-3, 0); F(2\sqrt{2}, 0), F(-2\sqrt{2}, 0)$
2. $V(0, 5/2), V(0, -5/2); F(0, \sqrt{21}/2), F(0, -\sqrt{21}/2)$

21.6 The Hyperbola

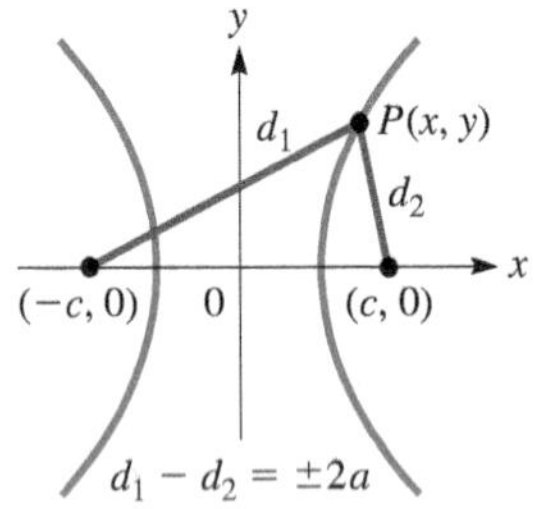

Fig. 21.65

A **hyperbola** *is defined as the locus of a point $P(x, y)$ that moves so that the difference of the distances from two fixed points (the* **foci***) is constant.* We choose the foci to be $(-c, 0)$ and $(c, 0)$ (see Fig. 21.65), and the constant difference to be $2a$. As with the ellipse, c is the length of the line segment from the centre to a focus, and a (as we will see) is the length of the line segment from the centre to a vertex. Therefore,

$$\sqrt{(x + c)^2 + y^2} - \sqrt{(x - c)^2 + y^2} = 2a$$

Following the same procedure as with the ellipse, the **standard equation** of the hyperbola with centre at the origin is

$$\frac{x^2}{a^2} - \frac{y^2}{b^2} = 1 \qquad (21.20)$$

When we derive this equation, ***we have a definition of the relation between a, b, and c that is different from that for the ellipse.*** This relation is

$$c^2 = a^2 + b^2 \qquad (21.21)$$

In Eq. (21.20), if we let $y = 0$, we find that the x-intercepts are $(-a, 0)$ and $(a, 0)$, just as they are for the ellipse. *These are the* **vertices** *of the hyperbola.* For $x = 0$, we find that we have imaginary solutions for y, which means there are no points on the curve that correspond to a value of $x = 0$.

To find the meaning of b, we solve Eq. (21.20) for y in the special form:

$$\frac{y^2}{b^2} = \frac{x^2}{a^2} - 1$$

$$= \frac{x^2}{a^2} - \frac{a^2x^2}{a^2x^2} = \frac{x^2}{a^2}\left(1 - \frac{a^2}{x^2}\right)$$

$$y^2 = \frac{b^2x^2}{a^2}\left(1 - \frac{a^2}{x^2}\right)$$

multiply through by b^2 and take the square root of each side

$$y = \pm\frac{bx}{a}\sqrt{1 - \frac{a^2}{x^2}} \qquad (21.22)$$

We note that if large values of x are assumed in Eq. (21.22), the quantity under the radical becomes approximately 1. In fact, the larger x becomes, the nearer to 1 this expression becomes since the x^2 in the denominator of a^2/x^2 makes this term nearly zero. Thus, for large values of x, Eq. (21.22) is approximately

$$y = \pm\frac{bx}{a} \tag{21.23}$$

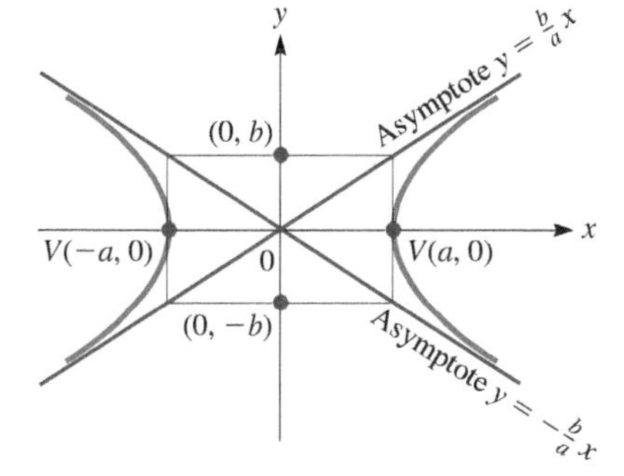

Fig. 21.66

Eq. (21.23) is seen to represent two straight lines, each of which passes through the origin. One has a slope of b/a, and the other has a slope of $-b/a$. ***These lines are called the* asymptotes *of the hyperbola.*** *An* **asymptote** *is a line that the curve approaches as one of the variables approaches some particular value.* The graph of the tangent function also has asymptotes, as we saw in Fig. 10.23. We can designate this limiting procedure with notation introduced in Chapter 19 by

$$y \rightarrow \pm\frac{bx}{a} \text{ as } x \rightarrow \pm\infty$$

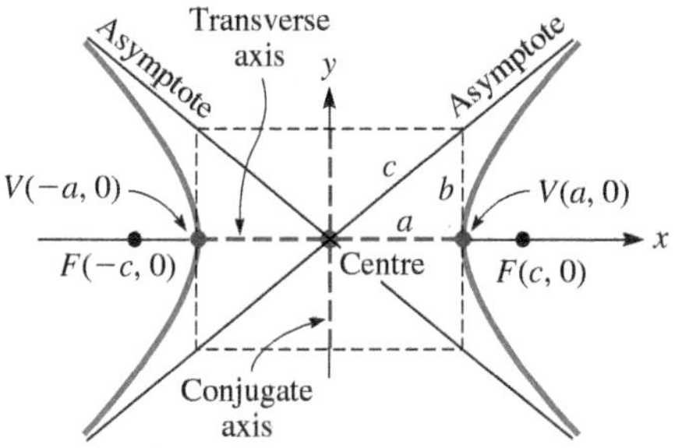

Fig. 21.67

The easiest way to sketch a hyperbola is to draw its asymptotes and then draw the hyperbola out from each vertex so that it comes closer and closer to each asymptote as x becomes numerically larger. To draw the asymptotes, first draw a small rectangle $2a$ by $2b$ with the origin at the centre, as shown in Fig. 21.66. Then straight lines, the asymptotes, are drawn through opposite vertices of the rectangle. This shows us that the significance of the value of b is in the slopes of the asymptotes.

A hyperbola in the form of Eq. (21.20) *has a* **transverse axis** *of length* $2a$ *along the* x*-axis and a* **conjugate axis** *of length* $2b$ *along the* y*-axis.* This means that a represents the length of the semitransverse axis and b represents the length of the semiconjugate axis. See Fig. 21.67. From the definition of c, it is the length of the line segment from the centre to a focus. Also, c is the length of the semidiagonal of the rectangle, as shown in Fig. 21.67. This shows us the geometric meaning of the relationship among a, b, and c given in Eq. (21.21).

If the transverse axis is along the y-axis and the conjugate axis is along the x-axis, the equation of the hyperbola with its centre at the origin (see Fig. 21.68) is

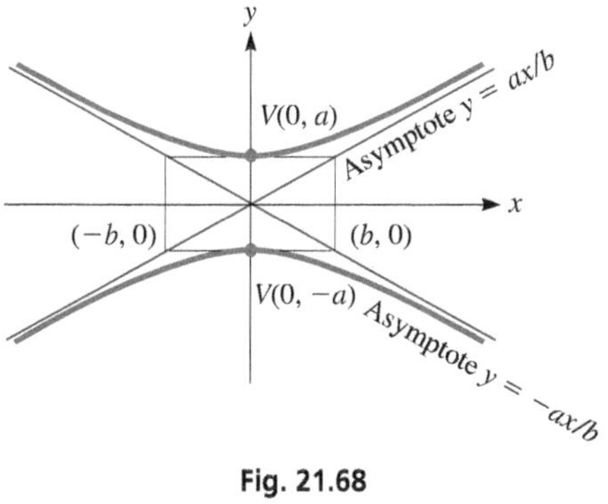

Fig. 21.68

$$\frac{y^2}{a^2} - \frac{x^2}{b^2} = 1 \tag{21.24}$$

In this case, a similar argument as the one leading to Eq. (21.23) shows that the equations of the asymptotes are $y = \pm(a/b)x$.

We now summarize the properties of hyperbolas with centre at the origin.

Standard Equation of a Hyperbola with Centre at (0, 0)

Equation	$\frac{x^2}{a^2} - \frac{y^2}{b^2} = 1$ **(21.20)**	$\frac{y^2}{a^2} - \frac{x^2}{b^2} = 1$ **(21.24)**
Geometric relationship	$c^2 = a^2 + b^2$	$c^2 = a^2 + b^2$
Transverse axis	x-axis (length $2a$)	y-axis (length $2a$)
Conjugate axis	y-axis (length $2b$) $(0, -b)$ to $(0, b)$	x-axis (length $2b$) $(-b, 0)$ to $(b, 0)$
Vertices	$(-a, 0), (a, 0)$	$(0, -a), (0, a)$
Foci	$(-c, 0), (c, 0)$	$(0, -c), (0, c)$

Asymptotes	$y = \pm \dfrac{bx}{a}$	$y = \pm \dfrac{ax}{b}$
Symmetry and graph	x-axis y-axis origin	x-axis y-axis origin

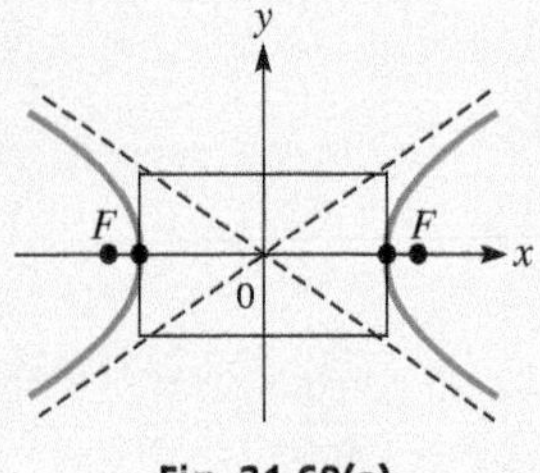

Fig. 21.69(a)

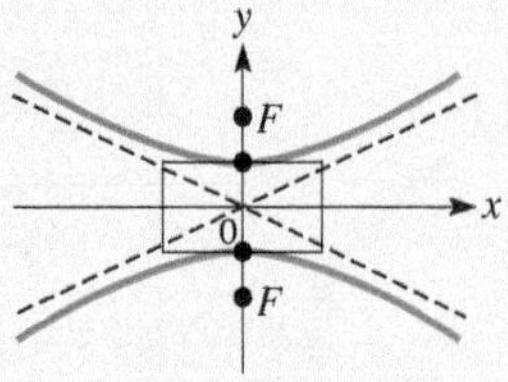

Fig. 21.69(b)

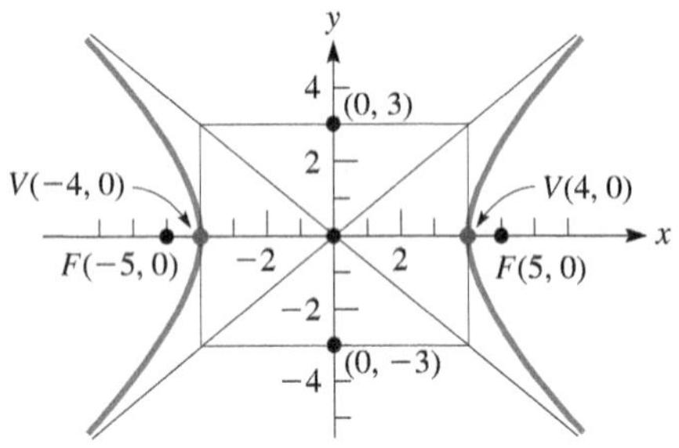

Fig. 21.70

EXAMPLE 1 Standard equation—transverse axis on the x-axis

Find the vertices and foci of the hyperbola $\dfrac{x^2}{16} - \dfrac{y^2}{9} = 1.$

This equation fits the form of Eq. (21.20). We know that it fits Eq. (21.20) and not Eq. (21.24) since the x^2-term is the positive term with 1 on the right. From the equation, we see that $a^2 = 16$ and $b^2 = 9$, or $a = 4$ and $b = 3$. In turn, this means the vertices are $(4, 0)$ and $(-4, 0)$ and the conjugate axis extends from $(0, -3)$ to $(0, 3)$.

Since $c^2 = a^2 + b^2$, we find that $c^2 = 25$, or $c = 5$. The foci are $(-5, 0)$ and $(5, 0)$.

Drawing the rectangle and the asymptotes in Fig. 21.70, we then sketch in the hyperbola from each vertex toward each asymptote.

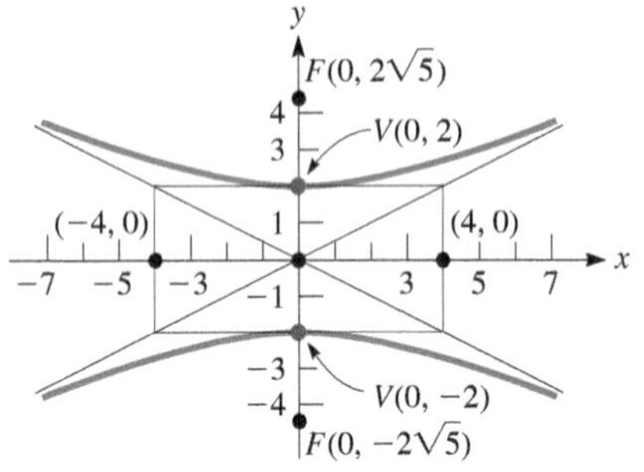

Fig. 21.71

EXAMPLE 2 Standard equation—transverse axis on the y-axis

Find the vertices and the foci of the hyperbola $\dfrac{y^2}{4} - \dfrac{x^2}{16} = 1.$

The hyperbola has vertices at $(0, -2)$ and $(0, 2)$. Its conjugate axis extends from $(-4, 0)$ to $(4, 0)$. The foci are $\left(0, -2\sqrt{5}\right)$ and $\left(0, 2\sqrt{5}\right)$. We find this directly from the equation since the y^2- term is the positive term with 1 on the right. This means the equation fits the form of Eq. (21.24) with $a^2 = 4$ and $b^2 = 16$. Also, $c^2 = 20$, which means that $c = \sqrt{20} = 2\sqrt{5}$. The hyperbola is shown in Fig. 21.71.

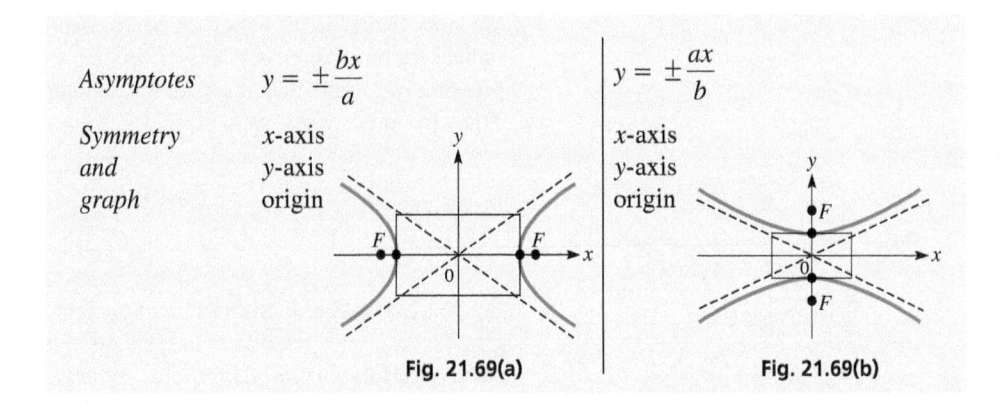

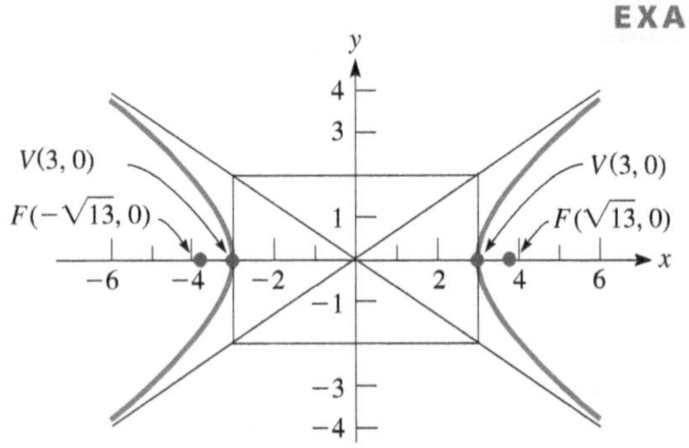

Fig. 21.72

EXAMPLE 3 Find vertices, foci

Determine the coordinates of the vertices and foci of the hyperbola

$$4x^2 - 9y^2 = 36$$

First, by dividing through by 36, we have

$$\frac{x^2}{9} - \frac{y^2}{4} = 1$$

form requires − and 1

From this form, we see that $a^2 = 9$ and $b^2 = 4$. In turn, this tells us that $a = 3$, $b = 2$, and $c = \sqrt{9 + 4} = \sqrt{13}$. Since a^2 appears under x^2, the equation fits the form of Eq. (21.20). Therefore, the vertices are $(-3, 0)$ and $(3, 0)$ and the foci are $\left(-\sqrt{13}, 0\right)$ and $\left(\sqrt{13}, 0\right)$. The hyperbola is shown in Fig. 21.72.

Practice Exercises

Find the vertices and foci of each hyperbola.

1. $\frac{x^2}{36} + \frac{y^2}{13} = 1$ **2.** $4y^2 - x^2 = 4$

EXAMPLE 4 Hyperbola—shape of a water pipe

In physics, it is shown that where the velocity of a fluid is greatest, the pressure is the least. In designing an experiment to study this effect in the flow of water, a pipe is constructed such that its lengthwise cross-section is hyperbolic. The pipe is 1.0 m long, 0.2 m in diameter at the narrowest point in the middle, and 0.4 m in diameter at each end. What is the equation that represents the cross-section of the pipe as shown in Fig. 21.73?

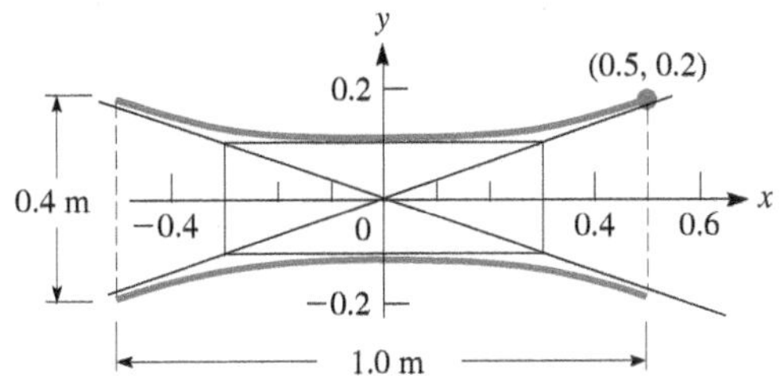

Fig. 21.73

As shown, the hyperbola has its transverse axis along the y-axis and its centre at the origin. This means the general equation is given by Eq. (21.24). Since the radius at the middle of the pipe is 0.1 m, we know that $a = 0.1$ m. Also, since it is 1.0 m long and the radius at the end is 0.2 m, we know the point $(0.5, 0.2)$ is on the hyperbola. This point must satisfy the equation

$$\frac{y^2}{a^2} - \frac{x^2}{b^2} = 1 \quad \text{Eq. (21.24)}$$

point (0.5, 0.2) satisfies equation

$$a = 0.1 \rightarrow \frac{0.2^2}{0.1^2} - \frac{0.5^2}{b^2} = 1$$

$$4 - \frac{0.25}{b^2} = 1, \qquad 3b^2 = 0.25, \qquad b^2 = 0.083$$

$$\frac{y^2}{0.1^2} - \frac{x^2}{0.083} = 1 \quad \text{substituting } a = 0.1,\ b^2 = 0.083 \text{ in Eq. (21.24)}$$

$$100y^2 - 12x^2 = 1 \quad \text{equation of cross-section}$$

Note in Fig. 21.73 that the hyperbola is the graph of two functions, which are found when we solve the last equation for y:

$$y = \pm\sqrt{(12x^2 + 1)/100}$$

One represents the upper half of the hyperbola, and the other represents the lower half. Both functions have to be graphed together to obtain the hyperbola shown in Fig. 21.73.

Equations (21.20) and (21.24) give us the standard forms of the equation of the hyperbola with its centre at the origin and its foci on one of the coordinate axes. There is another important equation form that represents a hyperbola:

$$xy = k \tag{21.25}$$

The asymptotes of this hyperbola are the coordinate axes, and the foci are on the line $y = x$ if k is positive or on the line $y = -x$ if k is negative. We encountered this hyperbola when studying inverse variation in Chapter 18.

The hyperbola represented by Eq. (21.25) is symmetric to the origin, for if $-x$ replaces x and $-y$ replaces y at the same time, we obtain $(-x)(-y) = k$, or $xy = k$. The equation is unchanged. However, if $-x$ replaces x or $-y$ replaces y, but not both, the sign on the left is changed. This means it is not symmetric to either axis. The following two examples deal with hyperbolas of this type.

EXAMPLE 5 $xy = k$ hyperbola

Plot the graph of the hyperbola $xy = 4$.

We find the values in the table below and then plot the appropriate points. Here, it is permissible to use a limited number of points, since we know the equation represents a hyperbola. Therefore, using $y = 4/x$, we obtain the values

x	-8	-4	-1	$-\frac{1}{2}$	$\frac{1}{2}$	1	4	8
y	$-\frac{1}{2}$	-1	-4	-8	8	4	1	$\frac{1}{2}$

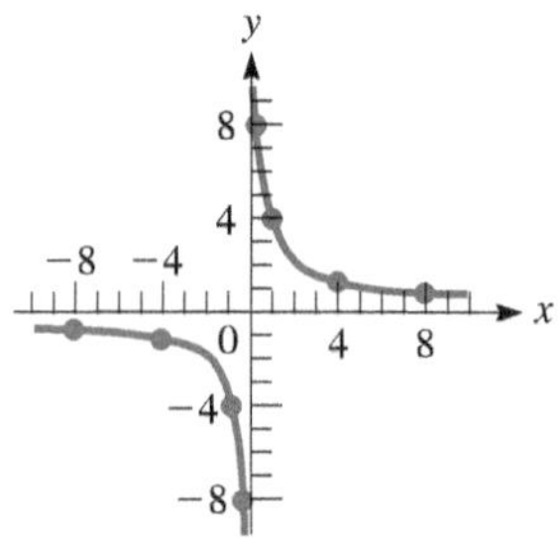

Fig. 21.74

Note that neither x nor y may equal zero. The hyperbola is shown in Fig. 21.74.

If the constant on the right is negative (for example, if $xy = -4$), then the two branches of the hyperbola are in the second and fourth quadrants.

EXAMPLE 6 $xy = k$ hyperbola—wavelength of a light wave

For a light wave, the product of its frequency f of vibration and its wavelength λ is a constant (equal to the speed of light). For green light, for which $f = 600$ THz, $\lambda = 500$ nm. Graph λ as a function of f for any light wave.

From the statement above, we know that $f\lambda = k$, and from the given values, we have

$$(600\text{ THz})(500\text{ nm}) = (6.0 \times 10^{14}\text{ Hz})(5.0 \times 10^{-7}\text{ m}) = 3.0 \times 10^{8}\text{ m/s}$$

which means $k = 3.0 \times 10^8$ m/s. We are to sketch $f\lambda = 3.0 \times 10^8$. Solving for λ as $\lambda = 3.0 \times 10^8/f$, **we have the table at the left** (only positive values have meaning). See Fig. 21.75. (Violet light has wavelengths of about 400 nm, orange light has wavelengths of about 600 nm, and red light has wavelengths of about 700 nm.)

f (THz)	750	600	500	430
λ (nm)	400	500	600	700

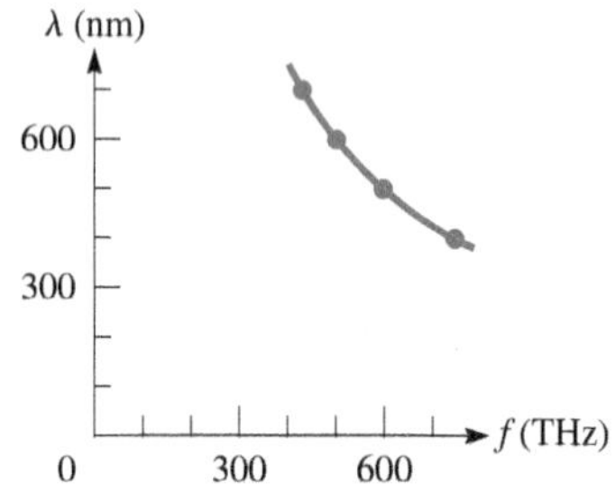

Fig. 21.75

We can conclude that *the equation of a hyperbola is characterized by the presence of both an x^2-term and a y^2-term, having different signs, or by the presence of an xy-term with no squared terms.* We will consider the equation of the hyperbola further in Sections 21.7, 21.8, and 21.9.

The hyperbola has some very useful applications. The LORAN radio navigation system is based on the use of hyperbolic paths. Some reflecting telescopes use hyperbolic mirrors. The paths of comets that escape the solar system are hyperbolic. Some additional applications are illustrated in the exercises.

■ The first reasonable measurement of the speed of light was made by the Danish astronomer Olaf Roemer (1644–1710). He measured the time required for light to come from the moons of Jupiter across the earth's orbit.

EXERCISES 21.6

In Exercises 1 and 2, make the given changes in the indicated examples of this section and then solve the resulting problems.

1. In Example 2, interchange the denominators of 4 and 16; find the vertices, ends of the conjugate axis, and foci. Sketch the curve.
2. In Example 3, change $-9y^2$ to $-y^2$ and then follow the same instructions as in Exercise 1.

In Exercises 3–16, find the coordinates of the vertices and the foci of the given hyperbolas. Sketch each curve.

3. $\frac{x^2}{25} - \frac{y^2}{144} = 1$
4. $\frac{x^2}{16} - \frac{y^2}{4} = 1$
5. $\frac{y^2}{9} - \frac{x^2}{1} = 1$
6. $\frac{y^2}{4} - \frac{x^2}{21} = 1$
7. $\frac{4x^2}{25} - \frac{y^2}{4} = 1$
8. $\frac{9y^2}{25} - x^2 = 1$
9. $4x^2 - y^2 = 4$
10. $x^2 - 9y^2 = 1$
11. $2y^2 - 5x^2 = 10$
12. $9y^2 - 16x^2 = 144$
13. $y^2 = 4(x^2 + 1)$
14. $y^2 = 9(x^2 - 1)$
15. $4x^2 - y^2 = 0.64$
16. $9y^2 - x^2 = 0.36$

In Exercises 17–28, find the equations of the hyperbolas satisfying the given conditions. The centre of each is at the origin.

17. Vertex $(3, 0)$, focus $(5, 0)$
18. Vertex $(0, 1)$, focus $\left(0, \sqrt{3}\right)$
19. Length of conjugate axis = 48, vertex $(0, 10)$
20. Sum of lengths of transverse and conjugate axes 28, focus $(10, 0)$
21. Passes through $(2, 3)$, focus $(2, 0)$

22. Passes through $(8, \sqrt{3})$, vertex $(4, 0)$

23. Passes through $(5, 4)$ and $(3, \frac{4}{5}\sqrt{5})$

24. Passes through $(1, 2)$ and $(2, 2\sqrt{2})$

25. Asymptote $y = 2x$, vertex $(1, 0)$

26. Asymptote $y = -4x$, vertex $(0, 4)$

27. The difference of distances to (x, y) from $(10, 0)$ and $(-10, 0)$ is 12.

28. The difference of distances to (x, y) from $(0, 4)$ and $(0, -4)$ is 6.

In Exercises 29–54, solve the given problems.

29. Sketch the graph of the hyperbola $xy = 2$.

30. Sketch the graph of the hyperbola $xy = -4$.

31. Show that the parametric equations $x = \sec t$, $y = \tan t$ define a hyperbola.

32. Show that all hyperbolas $\dfrac{x^2}{\cos^2\theta} - \dfrac{y^2}{\sin^2\theta} = 1$ have foci at $(\pm 1, 0)$ for all values of θ.

33. Find any points of intersection of the ellipse $2x^2 + y^2 = 17$ and the hyperbola $y^2 - x^2 = 5$.

34. Find any points of intersection of the hyperbolas $x^2 - 3y^2 = 22$ and $xy = 5$.

35. Find the equation of the hyperbola with foci $(1, 2)$ and $(11, 2)$, and a transverse axis of length 8, by use of the definition. Sketch the curve.

36. Find the equation of the hyperbola with vertices $(-2, 4)$ and $(-2, -2)$, and a conjugate axis of length 4, by use of the definition. Sketch the curve.

37. Use a graphing utility to view the hyperbola
$x^2 - 4y^2 + 4x + 32y - 64 = 0$.

38. Use a graphing utility to view the hyperbola
$5y^2 - 4x^2 + 8x + 40y + 56 = 0$.

39. The equation of a hyperbola with centre (h, k) and transverse axis parallel to the x-axis is $\dfrac{(x-h)^2}{a^2} - \dfrac{(y-k)^2}{b^2} = 1$. (This is shown in Section 21.7.) Sketch the hyperbola that has a transverse axis of 4 and a conjugate axis of 6, and for which (h, k) is $(-3, 2)$.

40. The equation of a hyperbola with centre (h, k) and transverse axis parallel to the y-axis is $\dfrac{(y-k)^2}{a^2} - \dfrac{(x-h)^2}{b^2} = 1$. (This is shown in Section 21.7.) Sketch the hyperbola that has a transverse axis of 2 and a conjugate axis of 8, and for which (h, k) is $(5, 0)$.

41. Two concentric (same centre) hyperbolas are called conjugate hyperbolas if the transverse and conjugate axes of one are, respectively, the conjugate and transverse axes of the other. What is the equation of the hyperbola conjugate to the hyperbola in Exercise 18?

42. As with an ellipse, the *eccentricity* e of a hyperbola is defined as $e = c/a$. Find the eccentricity of the hyperbola $2x^2 - 3y^2 = 24$.

43. Graph the inequality $100x^2 - 49y^2 \leq 4900$.

44. Draw a sketch of the graph of the region in which the points satisfy the system of inequalities $xy < 4$, $y > x$.

45. Find the equation of the hyperbola that has the same foci as the ellipse $\frac{x^2}{169} + \frac{y^2}{144} = 1$ and passes through $(4\sqrt{2}, 3)$.

46. Two persons, 1000 m apart, heard an explosion, one hearing it 4.0 s before the other. Explain why the location of the explosion can be on one of the points of a hyperbola.

47. The cross-section of the roof of a storage building shown in Fig. 21.76 is hyperbolic, with the horizontal beam passing through the focus. Find the equation of the hyperbola such that its centre is at the origin.

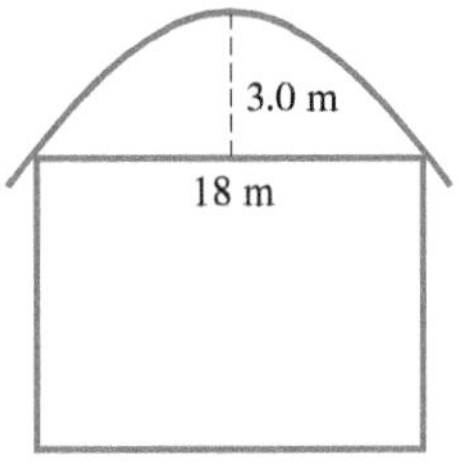

Fig. 21.76

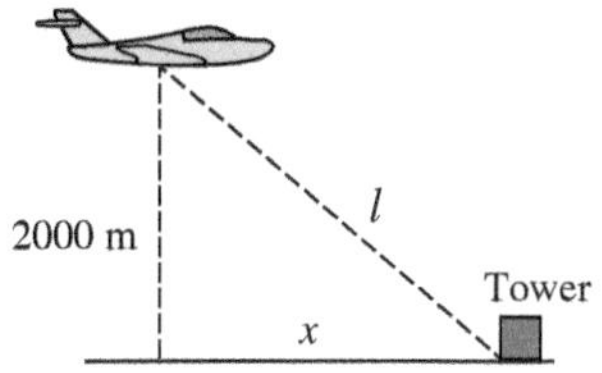

Fig. 21.77

48. A plane is flying at a constant altitude of 2000 m. Show that the equation relating the horizontal distance x and the direct-line distance l from a control tower to the plane is that of a hyperbola. Sketch the graph of l as a function of x. See Fig. 21.77.

49. A jet travels 600 km at a speed of v km/h for t hours. Graph the equation relating v as a function of t.

50. A drain pipe 100 m long has an inside diameter d (in m) and an outside diameter D (in m). If the volume of material of the pipe itself is 0.50 m^3, what is the equation relating d and D? Graph D as a function of d.

51. Ohm's law in electricity states that the product of the current i and the resistance R equals the voltage V across the resistance. If a battery of 6.00 V is placed across a variable resistor R, find the equation relating i and R and sketch the graph of i as a function of R.

52. A ray of light directed at one focus of a hyperbolic mirror is reflected toward the other focus. Find the equation that represents the hyperbolic mirror shown in Fig. 21.78.

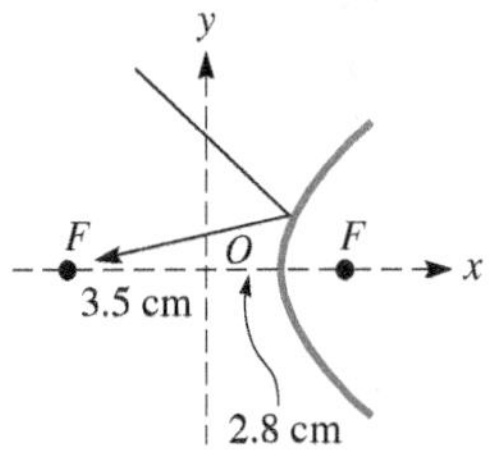

Fig. 21.78

53. A radio signal is sent simultaneously from stations A and B 600 km apart on the Labrador coast. A ship receives the signal from A 1.20 ms before it receives the signal from B. Given that radio signals travel at 300 km/ms, draw a graph showing the possible locations of the ship. This problem illustrates the basis of LORAN (long range navigation).

54. Maximum intensity for monochromatic (single-colour) light from two sources occurs where the difference in distances from the sources is an integral number of wavelengths. Find the equation of the curves of maximum intensity in a thin film between the sources where the difference in paths is two wavelengths and the sources are four wavelengths apart. Let the sources be on the x-axis and the origin midway between them. Use units of one wavelength for both x and y.

Answers to Practice Exercises

1. $V(6, 0)$, $V(-6, 0)$; $F(7, 0)$, $F(-7, 0)$
2. $V(0, 1)$, $V(0, -1)$; $F(0, \sqrt{5})$, $F(0, -\sqrt{5})$

21.7 Translation of Axes

The equations we have considered for the parabola, the ellipse, and the hyperbola are those for which the centre of the ellipse or hyperbola, or the vertex of the parabola, is at the origin. In this section, we consider, without specific use of the definition, the equations of these curves for the cases in which the axis of the curve is parallel to one of the coordinate axes. This is done by **translation of axes**.

In Fig. 21.79, we choose a point (h, k) in the xy-coordinate plane as the origin of another coordinate system, the $x'y'$-coordinate system. The x'-axis is parallel to the x-axis and the y'-axis is parallel to the y-axis. Every point now has two sets of coordinates (x, y) and (x', y'). We see that

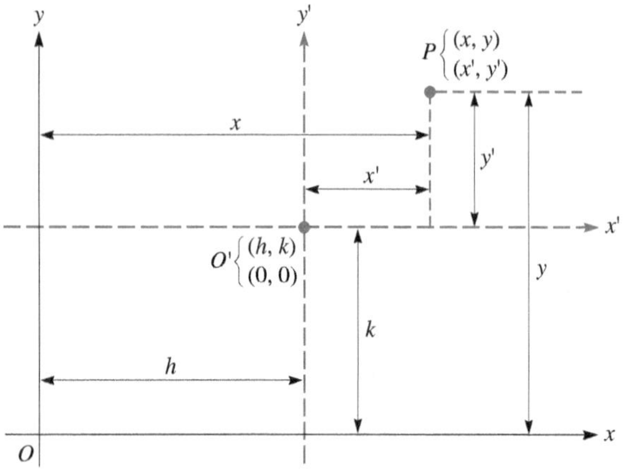

Fig. 21.79

$$x = x' + h \quad \text{and} \quad y = y' + k \tag{21.26}$$

Eq. (21.26) can also be written in the form

$$x' = x - h \quad \text{and} \quad y' = y - k \tag{21.27}$$

EXAMPLE 1 Find the equation—given vertex and focus

Find the equation of the parabola with vertex $(2, 4)$ and focus $(4, 4)$.

We let the origin of the $x'y'$-coordinate system be the point $(2, 4)$, so $h = 2$ and $k = 4$. The new coordinates of the focus are $x' = 4 - h = 4 - 2 = 2$ and $y' = 4 - k = 4 - 4 = 0$. This means $p = 2$ and $4p = 8$. See Fig. 21.80. In the $x'y'$-system, the equation is

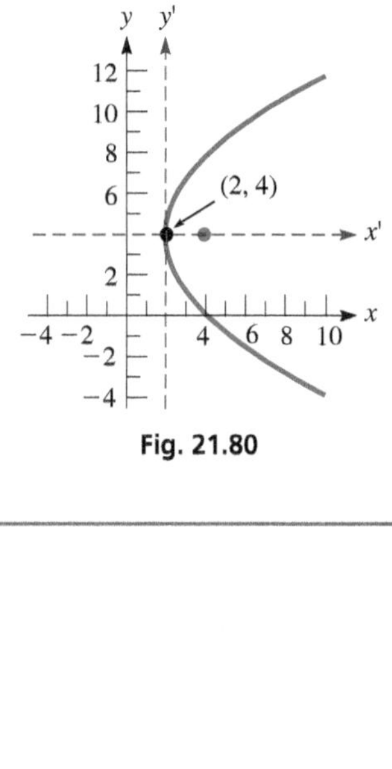

Fig. 21.80

$$(y')^2 = 8(x')$$

Using Eq. (21.27), we have

$$(y - 4)^2 = 8(x - 2)$$

coordinates of vertex $(2, 4)$

as the equation of the parabola in the xy-coordinate system.

Following the method of Example 1, by writing the equation of the curve in the $x'y'$-system and then using Eq. (21.27), we have the following more general forms of the equations of the parabola, ellipse, and hyperbola:

Parabola, vertex (h, k):

$$(y - k)^2 = 4p(x - h) \quad \text{(axis parallel to } x\text{-axis)} \tag{21.28}$$

$$(x - h)^2 = 4p(y - k) \quad \text{(axis parallel to } y\text{-axis)} \tag{21.29}$$

Ellipse, centre (h, k):

$$\frac{(x - h)^2}{a^2} + \frac{(y - k)^2}{b^2} = 1 \quad \text{(major axis parallel to } x\text{-axis)} \tag{21.30}$$

$$\frac{(y - k)^2}{a^2} + \frac{(x - h)^2}{b^2} = 1 \quad \text{(major axis parallel to } y\text{-axis)} \tag{21.31}$$

Hyperbola, centre (h, k):

$$\frac{(x - h)^2}{a^2} - \frac{(y - k)^2}{b^2} = 1 \quad \text{(transverse axis parallel to } x\text{-axis)} \tag{21.32}$$

$$\frac{(y - k)^2}{a^2} - \frac{(x - h)^2}{b^2} = 1 \quad \text{(transverse axis parallel to } y\text{-axis)} \tag{21.33}$$

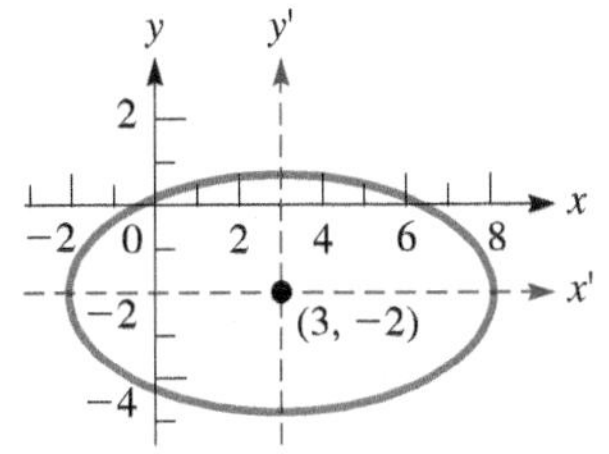

Fig. 21.81

EXAMPLE 2 Describing a curve given its equation

Describe the curve defined by the equation

$$\frac{(x-3)^2}{25}+\frac{(y+2)^2}{9}=1$$

We see that this equation fits the form of Eq. (21.30) with $h = 3$ and $k = -2$. It is the equation of an ellipse with its centre at $(3, -2)$ and its major axis parallel to the x-axis. The semimajor axis has length $a = 5$, and the semiminor axis has length $b = 3$. The ellipse is shown in Fig. 21.81.

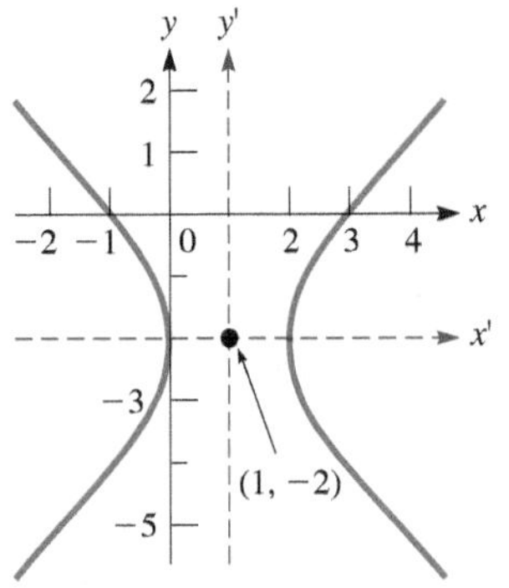

Fig. 21.82

EXAMPLE 3 Finding the centre of a hyperbola

Find the centre of the hyperbola $2x^2 - y^2 - 4x - 4y - 4 = 0$.

To analyse this curve, we first complete the square in the x-terms and in the y-terms. This will allow us to recognize properly the choice of h and k.

$$\begin{aligned} 2x^2 - 4x - y^2 - 4y &= 4 \\ 2(x^2 - 2x \qquad) - (y^2 + 4y \qquad) &= 4 \\ 2(x^2 - 2x + 1) - (y^2 + 4y + 4) &= 4 + 2 - 4 \end{aligned}$$

We note here that when we added 1 to complete the square of the x-terms within the parentheses, ***we were actually adding 2 to the left side.*** Thus, we added 2 to the right side. Similarly, when we added 4 to the y-terms within the parentheses, ***we were actually subtracting 4 from the left side.*** Continuing, we have

$$2(x-1)^2 - (y+2)^2 = 2$$

coordinates of centre $(1, -2)$

$$\frac{(x-1)^2}{1}-\frac{(y+2)^2}{2}=1$$

Therefore, the centre of the hyperbola is $(1, -2)$. See Fig. 21.82.

Practice Exercise

1. Find the centre and foci of the ellipse $2x^2 + 3y^2 - 8x + 18y + 29 = 0$.

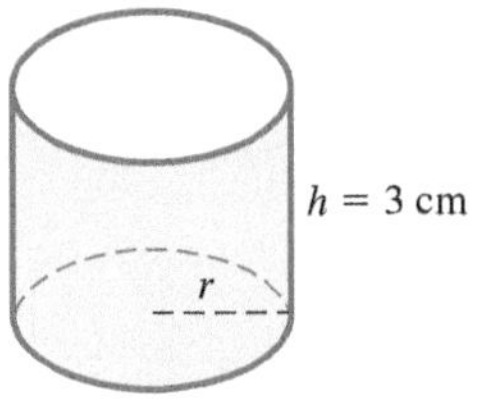

Fig. 21.83

EXAMPLE 4 Translation of axes—surface area

Cylindrical glass beakers are to be made with a height of 3 cm. Express the surface area in terms of the radius of the base and sketch the curve.

The total surface area S of a beaker is the sum of the area of the base and the lateral surface area of the side. In general, S in terms of the radius r of the base and height h of the side is $S = \pi r^2 + 2\pi rh$. Since $h = 3$ cm, we have

$$S = \pi r^2 + 6\pi r$$

which is the desired relationship. See Fig. 21.83.

To analyse the equation relating S and r, we complete the square of the r terms:

$$\begin{aligned} S &= \pi(r^2 + 6r) \\ S + 9\pi &= \pi(r^2 + 6r + 9) \qquad \text{complete the square} \\ S + 9\pi &= \pi(r+3)^2 \end{aligned}$$

vertex $(-3, -9\pi)$

$$(r+3)^2 = \frac{1}{\pi}(S + 9\pi)$$

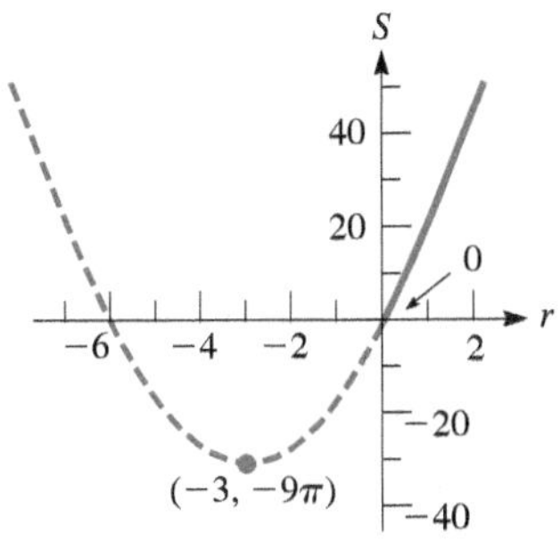

Fig. 21.84

This represents a parabola with vertex $(-3, -9\pi)$. Since $4p = 1/\pi$, then $p = 1/(4\pi)$, and the focus is $\left(-3, \frac{1}{4\pi} - 9\pi\right)$, as shown in Fig. 21.84. The part of the graph for negative r is dashed since only positive values have meaning.

EXERCISES 21.7

In Exercises 1 and 2, make the given changes in the indicated examples of this section and then solve the resulting problem.

1. In Example 2, change $y + 2$ to $y - 2$ and change the sign before the second term from $+$ to $-$. Then describe and sketch the curve.
2. In Example 3, change the fourth and fifth terms from $-4y - 4$ to $+6y - 9$ and then find the centre.

In Exercises 3–10, describe the curve represented by each equation. Identify the type of curve and its centre (or vertex if it is a parabola). Sketch each curve.

3. $(y-2)^2 = 4(x+1)$
4. $\dfrac{(x+4)^2}{4} + \dfrac{(y-1)^2}{1} = 1$
5. $\dfrac{(x-1)^2}{4} - \dfrac{(y-2)^2}{9} = 1$
6. $(y+5)^2 = -8(x-2)$
7. $\dfrac{(x+1)^2}{1} + \dfrac{y^2}{9} = 1$
8. $\dfrac{(y-4)^2}{16} - \dfrac{(x+2)^2}{4} = 1$
9. $(x+3)^2 = -12(y-1)$
10. $\dfrac{x^2}{0.16} + \dfrac{(y+1)^2}{0.25} = 1$

In Exercises 11–22, find the equation of each of the curves described by the given information.

11. Parabola: vertex $(-1, 3)$, focus $(3, 3)$
12. Parabola: focus $(2, -5)$, directrix $y = 3$
13. Parabola: axis, directrix are coordinate axes, focus $(12, 0)$
14. Parabola: vertex $(4, 4)$, vertical directrix, passes through $(0, 1)$
15. Ellipse: centre $(-2, 2)$, focus $(-5, 2)$, vertex $(-7, 2)$
16. Ellipse: centre $(0, 3)$, focus $(12, 3)$, major axis 26 units
17. Ellipse: centre $(-2, 1)$, vertex $(-2, 5)$, passes through $(0, 1)$
18. Ellipse: foci $(1, -2)$ and $(1, 10)$, minor axis 5 units
19. Hyperbola: vertex $(-1, 1)$, focus $(-1, 4)$, centre $(-1, 2)$
20. Hyperbola: foci $(2, 1)$ and $(8, 1)$, conjugate axis 6 units
21. Hyperbola: vertices $(2, 1)$ and $(-4, 1)$, focus $(-6, 1)$
22. Hyperbola: centre $(1, -4)$, focus $(1, 1)$, transverse axis 8 units

In Exercises 23–40, determine the centre (or vertex, if the curve is a parabola) of the given curve. Sketch each curve.

23. $x^2 + 2x - 4y - 3 = 0$
24. $y^2 - 2x - 2y - 9 = 0$
25. $x^2 + 4y = 24$
26. $x^2 + 4y^2 = 32y$
27. $4x^2 + 9y^2 + 24x = 0$
28. $2x^2 + y^2 + 8x = 8y$
29. $9x^2 - y^2 + 8y = 7$
30. $2x^2 - 4x = 9y - 2$
31. $5x^2 - 4y^2 + 20x + 8y = 4$
32. $0.04x^2 + 0.16y^2 = 0.01y$
33. $4x^2 - y^2 + 32x + 10y + 35 = 0$
34. $2x^2 + 2y^2 - 24x + 16y + 95 = 0$
35. $16x^2 + 25y^2 - 32x + 100y - 284 = 0$
36. $9x^2 - 16y^2 - 18x + 96y - 279 = 0$
37. $5x^2 - 3y^2 + 95 = 40x$
38. $5x^2 + 12y + 18 = 2y^2$
39. $9x^2 + 9y^2 + 14 = 6x + 24y$
40. $4y^2 + 29 = 15x + 12y$

In Exercises 41–52, solve the given problems.

41. Find the equation of the hyperbola with asymptotes $x - y = -1$ and $x + y = -3$ and vertex $(3, -1)$.
42. The circle $x^2 + y^2 + 4x - 5 = 0$ passes through the foci and the ends of the minor axis of an ellipse that has its major axis along the x-axis. Find the equation of the ellipse.
43. The vertex and focus of one parabola are, respectively, the focus and vertex of a second parabola. Find the equation of the first parabola, if $y^2 = 4x$ is the equation of the second.
44. Identify the curve represented by $4y^2 - x^2 - 6x - 2y = 14$ and view it on a graphing utility.
45. What is the general form of the equation of a *family* of parabolas if each vertex and focus is on the x-axis?
46. What is the general form of the equation of a *family* of ellipses with foci on the y-axis if each passes through the origin?
47. An electric current (in A) is $i = 2 + \sin\left(2\pi t - \frac{\pi}{3}\right)$. What is the equation for the current if the origin of the (t', i') system is taken as $\left(\frac{1}{6}, 2\right)$ of the (t, i) system?
48. The stopping distance d (in m) of a car travelling at v km/h is represented by $d = 0.005v^2 + 0.2v$. Where is the vertex of the parabola that represents d?
49. The stream from a fire hose follows a parabolic curve and reaches a maximum height of 18 m at a horizontal distance of 28 m from the nozzle. Find the equation that represents the stream, with the origin at the nozzle. Sketch the graph.
50. The vertical cross-section of a culvert under a road is elliptical. The culvert is 18 m wide and 12 m high. Find an equation to represent the perimeter of the culvert with the origin at road level and 2.0 m above the top of the culvert.
51. Two wheels in a friction drive assembly are equal ellipses, as shown in Fig. 21.85. They are always in contact, with the left wheel fixed in position and the right wheel able to move horizontally. Find the equation that can be used to represent the circumference of each wheel in the position shown.

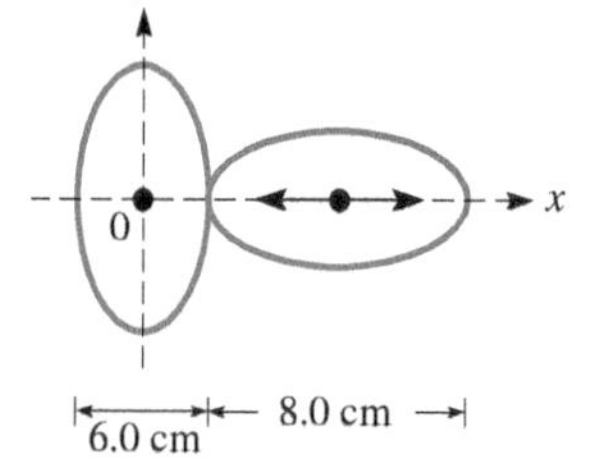

Fig. 21.85

52. An agricultural test station is to be divided into rectangular sections, each with a perimeter of 480 m. Express the area A of each section in terms of its width w and identify the type of curve represented. Sketch the graph of A as a function of w. For what value of w is A the greatest?

Answer to Practice Exercise

1. $C(2, -3)$, $F(1, -3)$, $F(3, -3)$

21.8 The Second-Degree Equation

The equations of the circle, parabola, ellipse, and hyperbola are all special cases of the same general equation. In this section, we discuss this equation and how to identify the particular form it takes when it represents a specific type of curve.

Each of these curves can be represented by a **second-degree equation** *of the form*

$$Ax^2 + Bxy + Cy^2 + Dx + Ey + F = 0 \quad \textbf{(21.34)}$$

The coefficients of the second-degree equation terms determine the type of curve that results. Recalling the discussions of the general forms of the equations of the circle, parabola, ellipse, and hyperbola from the previous sections of this chapter, Eq. (21.34) represents the indicated curve for given conditions of A, B, and C, as follows:

Identifying a Curve from Coefficients		EXAMPLE 1
Characteristics of the Coefficients	**Curve**	
$A = C$, $B = 0$ (two squared terms with identical coefficients)	Circle	$4x^2 + 4y^2 - 9 = 0$
$A \neq C$ (with the same sign), $B = 0$ (two squared terms with the same sign)	Ellipse	$5x^2 + 2y^2 - 3y - 7 = 0$
$A \neq C$ (with different signs), $B = 0$ (two squared terms with opposite signs)	Hyperbola	$-4x^2 + 5y^2 + 8x + 40y + 56 = 0$
$A = 0$ or $C = 0$ (but not both), $B = 0$ (only one squared term)	Parabola	$2x^2 - 4x - 9y + 2 = 0$
$A = 0$, $C = 0$, $B \neq 0$ (a term in xy and no squared term)	Hyperbola	$2xy - x - 4y - 1 = 0$
Special cases such as a single point or no real locus can also result.		$2x^2 + 2y^2 = 0$ is the point $(0, 0)$

We have considered only one special case where $B \neq 0$. The other cases will be dealt with in the next section when we discuss rotation of axes. Also, we have not mentioned the coefficients D and E since they do not have an effect on the type of curve. When at least one of them is nonzero, the centre of the curve (or vertex of a parabola) is not at the origin.

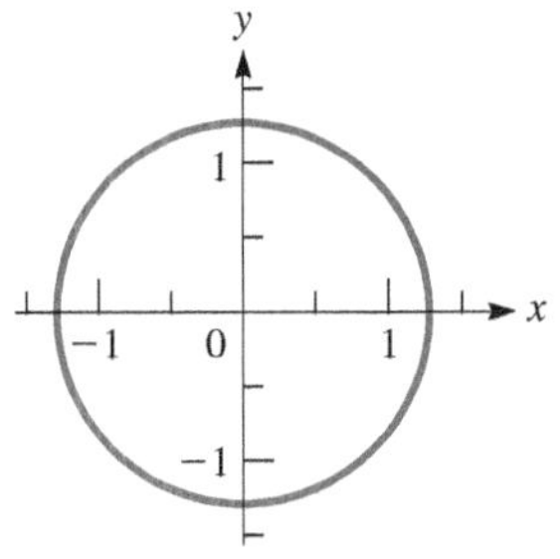

Fig. 21.86

EXAMPLE 2 Identify a circle from an equation

Identify the curve represented by $2x^2 = 3 - 2y^2$.

Writing the equation in the form of Eq. (21.34), we get

$$2x^2 + 2y^2 - 3 = 0$$

We see that $A = C$. Also, since there is no xy-term, we know that $B = 0$. This means that the equation represents a circle. If we write it as $x^2 + y^2 = \frac{3}{2}$, we see that it fits the form of Eq. (21.12). The circle is shown in Fig. 21.86.

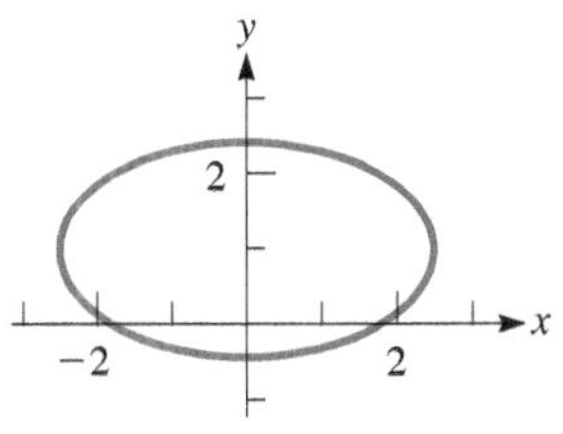

Fig. 21.87

EXAMPLE 3 Identify an ellipse from an equation

Identify the curve represented by $6y^2 = 12y - 2x^2 + 6$.

Writing the equation in the form of Eq. (21.34), we get

$$2x^2 + 6y^2 - 12y - 6 = 0$$

Here, we see that $B = 0$, A and C have the same sign, and $A \neq C$. Therefore, it is an ellipse. The $-12y$ term indicates that the centre of the ellipse is not at the origin. This ellipse is shown in Fig. 21.87.

EXAMPLE 4 Identify a hyperbola from an equation

Identify the curve represented by $2x^2 + 12x = y^2 - 14$. Determine the appropriate quantities for the curve, and sketch the graph.

Writing this equation in the form of Eq. (21.34), we have

$$2x^2 - 1y^2 + 12x + 14 = 0$$

We identify this equation as representing a hyperbola, since A and C have different signs and $B = 0$. We now write it in the standard form of a hyperbola:

$$2x^2 + 12x - y^2 = -14$$

$$2(x^2 + 6x \quad\quad) - y^2 = -14 \quad \text{complete the square}$$

$$2(x^2 + 6x + 9) - y^2 = -14 + 18$$

$$2(x + 3)^2 - y^2 = 4$$

centre is $(-3, 0)$

$$\frac{(x+3)^2}{2} - \frac{y^2}{4} = 1 \qquad \frac{x'^2}{2} - \frac{y'^2}{4} = 1$$

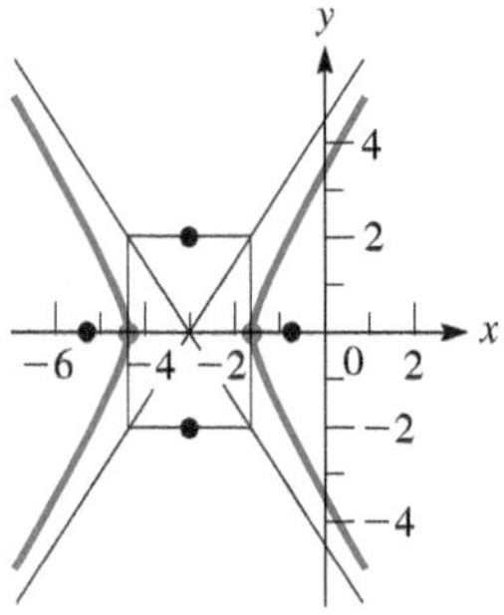

Fig. 21.88

Thus, we see that the centre (h, k) of the hyperbola is the point $(-3, 0)$. Also, $a = \sqrt{2}$ and $b = 2$. This means that the vertices are $(-3 + \sqrt{2}, 0)$ and $(-3 - \sqrt{2}, 0)$, and the conjugate axis extends from $(-3, 2)$ to $(-3, -2)$. Also, $c^2 = 2 + 4 = 6$, which means that $c = \sqrt{6}$. The foci are $(-3 + \sqrt{6}, 0)$ and $(-3 - \sqrt{6}, 0)$. The graph is shown in Fig. 21.88.

Practice Exercises

Identify the type of curve represented by each equation.

1. $2y^2 - 4y + 5 = 4x - x^2$
2. $x^2 + y^2 - 8y + 12x = x^2 - y^2$

EXAMPLE 5 Identify a parabola from an equation

Identify the curve represented by $4y^2 - 23 = 4(4x + 3y)$ and find the appropriate important quantities.

Writing the equation in the form of Eq. (21.34), we have

$$4y^2 - 16x - 12y - 23 = 0$$

Therefore, we recognize the equation as representing a parabola, since $A = 0$ and $B = 0$. Now, writing the equation in the standard form of a parabola, we have

$$4y^2 - 12y = 16x + 23$$

$$4(y^2 - 3y \quad\quad) = 16x + 23 \quad \text{complete the square}$$

$$4\left(y^2 - 3y + \frac{9}{4}\right) = 16x + 23 + 9$$

$$4\left(y - \frac{3}{2}\right)^2 = 16(x + 2)$$

vertex $\left(-2, \frac{3}{2}\right)$

$$\left(y - \frac{3}{2}\right)^2 = 4(x + 2) \quad \text{or} \quad y'^2 = 4x'$$

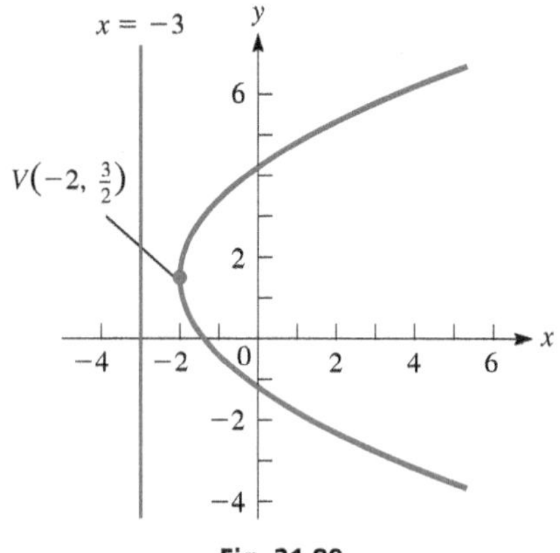

Fig. 21.89

We now note that the vertex is $(-2, 3/2)$ and that $p = 1$. This means that the focus is $(-1, 3/2)$ and the directrix is $x = -3$. The graph is shown in Fig. 21.89.

The circle, the parabola, the ellipse and the hyperbola are all **conic sections**. If a plane is passed through a cone, the intersection of the plane and the cone results in one of these curves; the curve formed depends on the angle of the plane with respect to the axis of the cone. This is shown in Fig. 21.90.

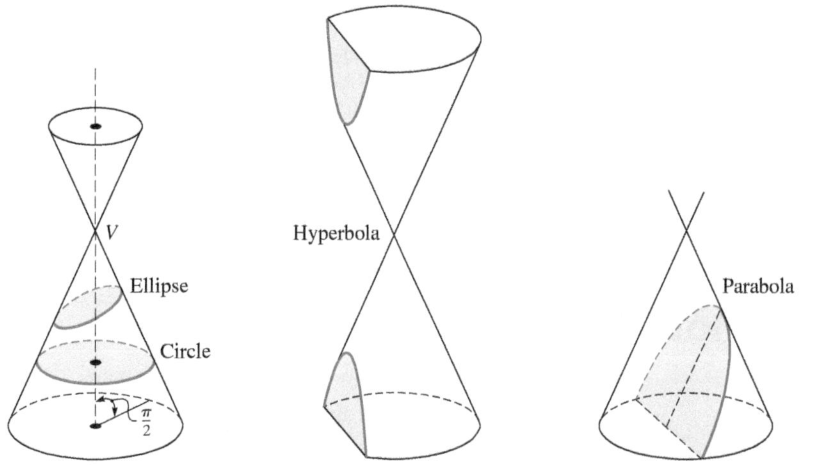

Fig. 21.90

EXERCISES 21.8

In Exercises 1 and 2, make the given changes in the indicated examples of this section and then solve the indicated problem.

1. In Example 2, change the $-$ before the $2y^2$ to $+$ and then determine what type of curve is represented.
2. In Example 4, change $y^2 - 14$ to $14 - y^2$ and then determine what type of curve is represented.

In Exercises 3–24, identify each of the equations as representing either a circle, a parabola, an ellipse, a hyperbola, or none of these.

3. $x^2 + 2y^2 - 2 = 0$
4. $x^2 - y = 0$
5. $2x^2 - y^2 - 1 = 0$
6. $y(y + x^2) = 4$
7. $8x^2 + 2y^2 = 6y(1 - y)$
8. $x(x - 3) = y(1 - 2y^2)$
9. $2.2x^2 - x - y = 1.6$
10. $2x^2 + 4y^2 = y + 2x$
11. $x^2 = (y - 1)(y + 1)$
12. $3.2x^2 = 2.1y(1 - 2y)$
13. $36x^2 = 12y(1 - 3y) + 1$
14. $y = 3(1 - 2x)(1 + 2x)$
15. $y(3 - 2x) = x(5 - 2y)$
16. $x(13 - 5x) = 5y^2$
17. $2xy + x - 3y = 6$
18. $(y + 1)^2 = x^2 + y^2 - 1$
19. $2x(x - y) = y(3 - y - 2x)$
20. $15x^2 = x(x - 12) + 4y(y - 6)$
21. $(x + 1)^2 + (y + 1)^2 = 2(x + y + 1)$
22. $(2x + y)^2 = 4x(y - 2) - 16$
23. $x(y + 3x) = x^2 + xy - y^2 + 1$
24. $4x(x - 1) = 2x^2 - 2y^2 + 3$

In Exercises 25–30, identify the curve represented by each of the given equations. Determine the appropriate important quantities for the curve and sketch the graph.

25. $x^2 = 8(y - x - 2)$
26. $x^2 = 6x - 4y^2 - 1$
27. $y^2 = 2(x^2 - 2x - 2y)$
28. $4x^2 + 4 = 9 - 8x - 4y^2$
29. $y^2 + 42 = 2x(10 - x)$
30. $x^2 - 4y = y^2 + 4(1 - x)$

In Exercises 31–36, identify the type of curve for each equation, and then view it on a graphing utility.

31. $x^2 + 2y^2 - 4x + 12y + 14 = 0$
32. $4y^2 - x^2 + 40y - 4x + 60 = 0$
33. $4(y^2 - 4x - 2) = 5(4y - 5)$
34. $2(2x^2 - y) = 8 - y^2$
35. $4(y^2 + 6y + 1) = x(x - 4) - 24$
36. $8x + 31 - xy = y(y - 2 - x)$

In Exercises 37–42, use the given values to determine the type of curve represented.

37. For the equation $x^2 + ky^2 = a^2$, what type of curve is represented if (a) $k = 1$, (b) $k < 0$, and (c) $k > 0$ $(k \neq 1)$?
38. For the equation $\frac{x^2}{4 - k} - \frac{y^2}{k} = 1$, what type of curve is represented if (a) $k < 0$ and (b) $0 < k < 4$? (For $k > 4$, see Exercise 40.)
39. In Eq. (21.34), if $A > C > 0$ and $B = D = E = F = 0$, describe the locus of the equation.
40. For the equation in Exercise 38, describe the locus of the equation if $k > 4$.
41. In Eq. (21.34), if $A = B = C = 0$, $D \neq 0$, $E \neq 0$, and $F \neq 0$, describe the locus of the equation.
42. In Eq. (21.34), if $A = -C \neq 0$, $B = D = E = 0$, and $F = C$, describe the locus of the equation if $C > 0$.

In Exercises 43–48, determine the type of curve from the given information.

43. The diagonal brace in a rectangular metal frame is 3.0 cm longer than the length of one of the sides. Determine the type of curve represented by the equation relating the lengths of the sides of the frame.

44. One circular solar cell has a radius that is 2.0 cm less than the radius r of a second circular solar cell. Determine the type of curve represented by the equation relating the total area A of both cells and r.

45. A flashlight emits a cone of light onto the floor. What type of curve is the perimeter of the lighted area on the floor, if the flashlight is directed at the floor and perpendicular to it?

46. What type of curve is the perimeter of the lighted area on the floor of the cone of light in Exercise 45, if the upper edge of the cone is parallel to the floor?

47. A supersonic jet creates a conical shock wave behind it. What type of curve is outlined on the surface of a lake by the shock wave if the jet is flying horizontally?

48. In Fig. 21.90, if the plane cutting the cones passes through the intersection of the upper and lower cones, what type of curve is the intersection of the plane and cones?

Answers to Practice Exercises

1. Ellipse **2.** Parabola

21.9 Rotation of Axes

So far in this chapter, we have mostly discussed conics whose axes are parallel to the coordinate axes. For such curves, the xy-term in the second-degree equation

$$Ax^2 + Bxy + Cy^2 + Dx + Ey + F = 0 \qquad \textbf{(21.34)}$$

has coefficient $B = 0$. In this section we study curves whose axes are not parallel to the coordinate axes. Although these curves have $B \neq 0$, we will see how we can rewrite the equation without the xy-term by rotating the coordinate axes. The absence of an xy-term simplifies the analysis of the graph.

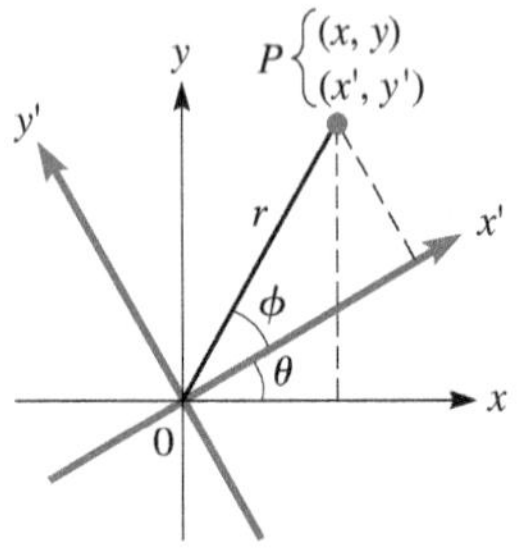

Fig. 21.91

If a set of axes is rotated about the origin through an angle θ, as shown in Fig. 21.91, we say that there has been a **rotation of axes**. In this case, each point P in the plane has two sets of coordinates, (x, y) in the original system and (x', y') in the rotated system.

If we now let r equal the distance from the origin O to point P and let ϕ be the angle between the x'-axis and the line OP, we have

$$x' = r\cos\phi \qquad y' = r\sin\phi \qquad \textbf{(21.35)}$$

$$x = r\cos(\theta + \phi) \qquad y = r\sin(\theta + \phi) \qquad \textbf{(21.36)}$$

Using the cosine and sine of the sum of two angles, we can write Eq. (21.36) as

$$\begin{aligned} x &= r\cos\phi\cos\theta - r\sin\phi\sin\theta \\ y &= r\cos\phi\sin\theta + r\sin\phi\cos\theta \end{aligned} \qquad \textbf{(21.37)}$$

Now, using Eq. (21.35), we have

$$\begin{aligned} x &= x'\cos\theta - y'\sin\theta \\ y &= x'\sin\theta + y'\cos\theta \end{aligned} \qquad \textbf{(21.38)}$$

In our derivation, we have used the special case when θ is acute and P is in the first quadrant of both sets of axes. When simplifying equations of curves using Eq. (21.38), we find that a rotation through a positive acute angle θ is sufficient. It can be shown, however, that Eq. (21.38) holds for any θ and position of P.

EXAMPLE 1 Rotation through 45°

Transform $x^2 - y^2 + 8 = 0$ by rotating the axes through 45°.

When $\theta = 45°$, the rotation Eq. (21.38) becomes

$$x = x'\cos 45° - y'\sin 45° = \frac{x'}{\sqrt{2}} - \frac{y'}{\sqrt{2}}$$

$$y = x'\sin 45° + y'\cos 45° = \frac{x'}{\sqrt{2}} + \frac{y'}{\sqrt{2}}$$

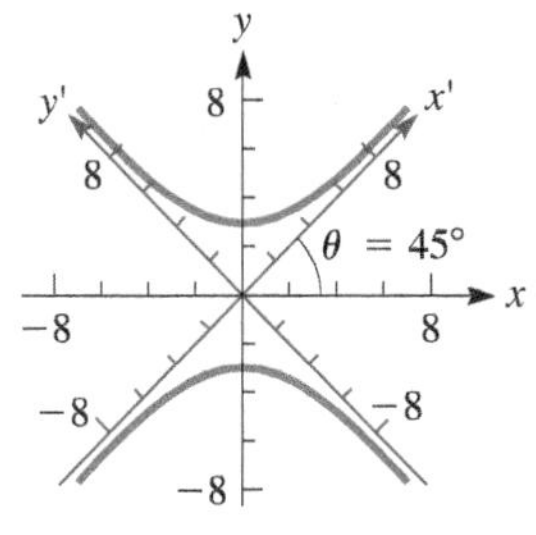

Fig. 21.92

Substituting into the equation $x^2 - y^2 + 8 = 0$ gives

$$\left(\frac{x'}{\sqrt{2}} - \frac{y'}{\sqrt{2}}\right)^2 - \left(\frac{x'}{\sqrt{2}} + \frac{y'}{\sqrt{2}}\right)^2 + 8 = 0$$

$$\frac{1}{2}x'^2 - x'y' + \frac{1}{2}y'^2 - \frac{1}{2}x'^2 - x'y' - \frac{1}{2}y'^2 + 8 = 0$$

$$x'y' = 4$$

The graph and both sets of axes are shown in Fig. 21.92. We see that the $xy = k$ type of hyperbola is obtained by a 45° rotation of a hyperbola with axes parallel to the coordinate axes.

As we said earlier, a rotation of axes can transform Eq. (21.34) with $B \neq 0$ into an equation containing no xy-term. We now show how this is done.

By substituting Eq. (21.38) into Eq. (21.34) and then simplifying, we have

$$(A\cos^2\theta + B\sin\theta\cos\theta + C\sin^2\theta)x'^2 + [B\cos 2\theta - (A - C)\sin 2\theta]x'y'$$
$$+ (A\sin^2\theta - B\sin\theta\cos\theta + C\cos^2\theta)y'^2 + (D\cos\theta + E\sin\theta)x'$$
$$+ (E\cos\theta - D\sin\theta)y' + F = 0 \qquad \textbf{(21.39)}$$

If there is to be no $x'y'$-term, its coefficient must be zero. This means that $B\cos 2\theta - (A - C)\sin 2\theta = 0$. Thus, the following formula can be used to find the angle of rotation.

Angle of Rotation

$$\tan 2\theta = \frac{B}{A - C} \qquad (A \neq C) \qquad \textbf{(21.40)}$$

Eq. (21.40) gives the angle of rotation except when $A = C$. In this case, the coefficient of the $x'y'$-term is $B\cos 2\theta$, which is zero if $2\theta = 90°$. Thus,

$$\theta = 45° \qquad (A = C) \qquad \textbf{(21.41)}$$

Consider the following example.

EXAMPLE 2 Rotating axes—eliminating the *xy*-term

By rotation of axes, transform $8x^2 + 4xy + 5y^2 = 9$ into a form without an xy-term. Identify and sketch the curve.

Here, $A = 8$, $B = 4$, and $C = 5$. Therefore, using Eq. (21.40), we have

$$\tan 2\theta = \frac{4}{8 - 5} = \frac{4}{3}$$

Since $\tan 2\theta$ is positive, we may take 2θ as an acute angle, which means θ is also acute. For the transformation, we need $\sin\theta$ and $\cos\theta$. We can find these values *exactly* if we use trigonometric identities. We first find the value of $\cos 2\theta$:

$$\cos 2\theta = \frac{1}{\sec 2\theta} = \frac{1}{\sqrt{1 + \tan^2 2\theta}} = \frac{1}{\sqrt{1 + (\frac{4}{3})^2}} = \frac{3}{5} \qquad \text{using Eqs. (20.2) and (20.7)}$$

Now, using the half-angle formulas, Eqs. (20.26) and (20.27), we have

$$\sin\theta = \sqrt{\frac{1 - \cos 2\theta}{2}} = \sqrt{\frac{1 - \frac{3}{5}}{2}} = \frac{1}{\sqrt{5}} \qquad \cos\theta = \sqrt{\frac{1 + \cos 2\theta}{2}} = \sqrt{\frac{1 + \frac{3}{5}}{2}} = \frac{2}{\sqrt{5}}$$

■ For reference:

Eq. (20.2) is $\cos\theta = \frac{1}{\sec\theta}$

Eq. (20.7) is $1 + \tan^2\theta = \sec^2\theta$

Eq. (20.26) is $\sin\frac{\alpha}{2} = \pm\sqrt{\frac{1 - \cos\alpha}{2}}$

Eq. (20.27) is $\cos\frac{\alpha}{2} = \pm\sqrt{\frac{1 + \cos\alpha}{2}}$

Here, θ is about 26.6°. Next, substituting these values into Eq. (21.38), we have

$$x = x'\left(\frac{2}{\sqrt{5}}\right) - y'\left(\frac{1}{\sqrt{5}}\right) = \frac{2x' - y'}{\sqrt{5}} \qquad y = x'\left(\frac{1}{\sqrt{5}}\right) + y'\left(\frac{2}{\sqrt{5}}\right) = \frac{x' + 2y'}{\sqrt{5}}$$

Finally, substituting into the equation $8x^2 + 4xy + 5y^2 = 9$ and simplifying, we get

$$8\left(\frac{2x' - y'}{\sqrt{5}}\right)^2 + 4\left(\frac{2x' - y'}{\sqrt{5}}\right)\left(\frac{x' + 2y'}{\sqrt{5}}\right) + 5\left(\frac{x' + 2y'}{\sqrt{5}}\right)^2 = 9$$

$$8(4x'^2 - 4x'y' + y'^2) + 4(2x'^2 + 3x'y' - 2y'^2) + 5(x'^2 + 4x'y' + 4y'^2) = 45$$

$$45x'^2 + 20y'^2 = 45$$

$$\frac{x'^2}{1} + \frac{y'^2}{\frac{9}{4}} = 1$$

This is an ellipse with semimajor axis of length 3/2 and semiminor axis of length 1. See Fig. 21.93.

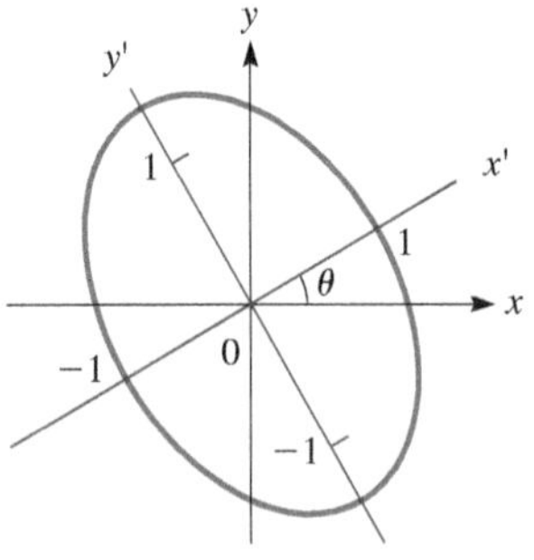

Fig. 21.93

In Example 2, tan 2θ was positive, and we made 2θ and θ positive. If, when using Eq. (21.40), tan 2θ is negative, we then make 2θ obtuse ($90° < 2\theta < 180°$). In this case, cos 2θ will be negative, but θ will be acute ($45° < \theta < 90°$).

In Section 21.8, we identified a conic section by inspecting the values of A and C when $B = 0$. If $B \neq 0$, these curves are identified as follows:

Practice Exercises

Identify the type of curve represented by each equation.

1. $3x^2 - 2xy + 3y^2 = 8$
2. $x^2 - 4xy + y^2 = -5$
3. $4x^2 + 4xy + y^2 - 24x + 38y - 19 = 0$

Identifying a Conic, $B \neq 0$		EXAMPLE 3
Value of $B^2 - 4AC$	**Curve**	
$B^2 - 4AC > 0$	Hyperbola	$xy = 4$, $B^2 - 4AC = 1 > 0$ (see Example 1)
$B^2 - 4AC < 0$	Ellipse	$8x^2 + 4xy + 5y^2 - 9 = 0$, $B^2 - 4AC = -144 < 0$ (see Example 2)
$B^2 - 4AC = 0$	Parabola	$16x^2 - 24xy + 9y^2 + 20x - 140y - 300 = 0$, $B^2 - 4AC = 0$ (see Example 4)

To see why the result above is true, we note that if A', B', and C' represent the coefficients of x'^2, $x'y'$, and y'^2, respectively, in Eq. (21.39), then it can be shown that $B^2 - 4AC = B'^2 - 4A'C'$. This means that if the axes have been rotated so that $B' = 0$, then $B^2 - 4AC = -4'C'$. In other words, the sign of $B^2 - 4AC$ is determined by the values of A' and C', and these are used to identify a conic as in Section 21.8. For example, if the conic is a hyperbola, A' and C' differ in sign, so their product is negative and $-4A'C' > 0$, which gives $B^2 - 4AC > 0$. The other cases are analysed similarly.

It is possible that both a translation of axes and a rotation of axes are needed to write an equation in standard form, as seen in the following example.

EXAMPLE 4 Translation after rotation

For the equation $16x^2 - 24xy + 9y^2 + 20x - 140y - 300 = 0$, identify the curve and simplify it to standard form. Sketch the graph.

With $A = 16$, $B = -24$, and $C = 9$, using Eq. (21.40), we have

$$\tan 2\theta = \frac{-24}{16 - 9} = -\frac{24}{7}$$

In this case, tan 2θ is negative, and we take 2θ to be an obtuse angle. We then find that $\cos 2\theta = -7/25$. In turn, we find that $\sin\theta = 4/5$ and $\cos\theta = 3/5$. Here, θ is about 53.1°. Using these values in Eq. (21.38), we find that

$$x = \frac{3x' - 4y'}{5} \qquad y = \frac{4x' + 3y'}{5}$$

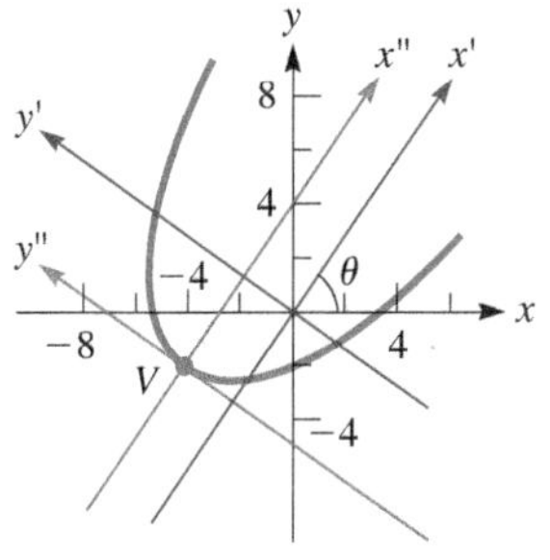

Fig. 21.94

Substituting these into the original equation and simplifying, we get

$$y'^2 - 4x' - 4y' - 12 = 0$$

This equation represents a parabola with its axis parallel to the x'-axis. The vertex is found by completing the square:

$$(y' - 2)^2 = 4(x' + 4)$$

The vertex is the point $(-4, 2)$ in the $x'y'$-rotated system. Therefore,

$$y''^2 = 4x''$$

is the equation in the $x''y''$-rotated and then translated system. The graph and the coordinate systems are shown in Fig. 21.94.

EXERCISES 21.9

In Exercises 1–4, transform the given equations by rotating the axes through the given angle. Identify and sketch each curve.

1. $x^2 - y^2 = 25, \theta = 45°$ **2.** $x^2 + y^2 = 16, \theta = 60°$

3. $8x^2 - 4xy + 5y^2 = 36, \theta = \tan^{-1} 2$

4. $2x^2 + 24xy - 5y^2 = 8, \theta = \tan^{-1} \frac{3}{4}$

In Exercises 5–10, identify the type of curve that each equation represents by evaluating $B^2 - 4AC$.

5. $x^2 + 2xy + x - y - 3 = 0$ **6.** $8x^2 - 4xy + 2y^2 + 7 = 0$

7. $x^2 - 2xy + y^2 + 3y = 0$

8. $4xy + 3y^2 - 8x + 16y + 19 = 0$

9. $13x^2 + 10xy + 13y^2 + 6x - 42y - 27 = 0$

10. $x^2 - 4xy + 4y^2 + 36x + 28y + 24 = 0$

In Exercises 11–16, transform each equation to a form without an xy-term by a rotation of axes. Identify and sketch each curve.

11. $x^2 + 2xy + y^2 - 2x + 2y = 0$ **12.** $5x^2 - 6xy + 5y^2 = 32$

13. $3x^2 + 4xy = 4$

14. $9x^2 - 24xy + 16y^2 - 320x - 240y = 0$

15. $11x^2 - 6xy + 19y^2 = 20$ **16.** $x^2 + 4xy - 2y^2 = 6$

In Exercises 17 and 18, transform each equation to a form without an xy-term by a rotation of axes. Then transform the equation to a standard form by a translation of axes. Identify and sketch each curve.

17. $16x^2 - 24xy + 9y^2 - 60x - 80y + 400 = 0$

18. $73x^2 - 72xy + 52y^2 + 100x - 200y + 100 = 0$

In Exercises 19 and 20, solve the given problems.

19. (a) In Eq. (21.34), if A and C have opposite signs, what type of curve is represented? (b) If $B \neq 0$, and either A or C is zero, what type of curve is represented?

20. An elliptical cam can be represented by the equation $x^2 - 3xy + 5y^2 - 13 = 0$. Through what angle is the cam rotated from its standard position?

Answers to Practice Exercises

1. Ellipse **2.** Hyperbola **3.** Parabola

21.10 Polar Coordinates

LEARNING TIP
Throughout this section, make sure that your calculator is in *radian* mode.

To this point, we have graphed all curves in the rectangular coordinate system. However, for certain types of curves, other coordinate systems are better adapted. We discuss one of these systems here.

Instead of designating a point by its x- and y-coordinates, we can specify its location by its radius vector and the angle the radius vector makes with the x-axis. Thus, the r and θ that are used in the definitions of the trigonometric functions can also be used as the coordinates of points in the plane. The important aspect of choosing coordinates is that, for each set of values, there must be only one point that corresponds to this set. We can see that this condition is satisfied by the use of r and θ as coordinates. *In* **polar coordinates**, *the origin is called the* **pole**, *and the half-line for which the angle is zero (equivalent to the positive x-axis) is called the* **polar axis**. The coordinates of a point are designated as (r, θ). See Fig. 21.95.

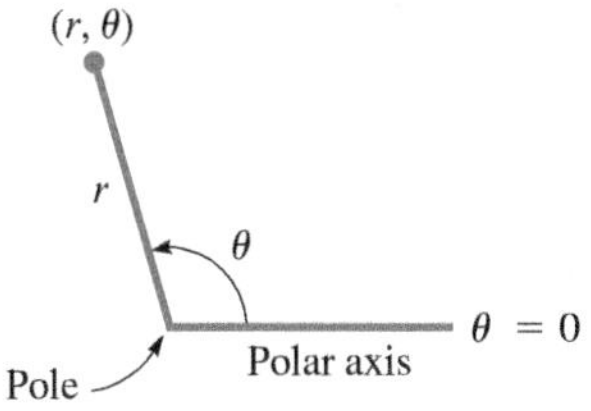

Fig. 21.95

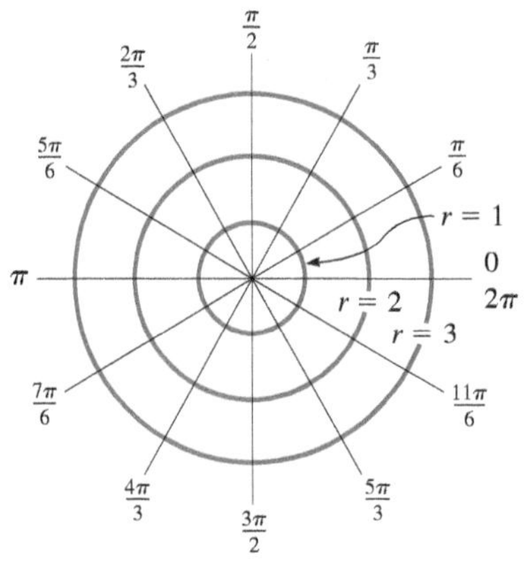

Fig. 21.96

When using polar coordinates, we generally label the lines for some of the values of θ; namely, those for $\theta = 0$ (the polar axis), $\theta = \pi/2$ (equivalent to the positive y-axis), $\theta = \pi$ (equivalent to the negative x-axis), $\theta = 3\pi/2$ (equivalent to the negative y-axis), and possibly others. In Fig. 21.96, these lines and those for multiples of $\pi/6$ are shown. Also, the circles for $r = 1$, $r = 2$, and $r = 3$ are shown in this figure.

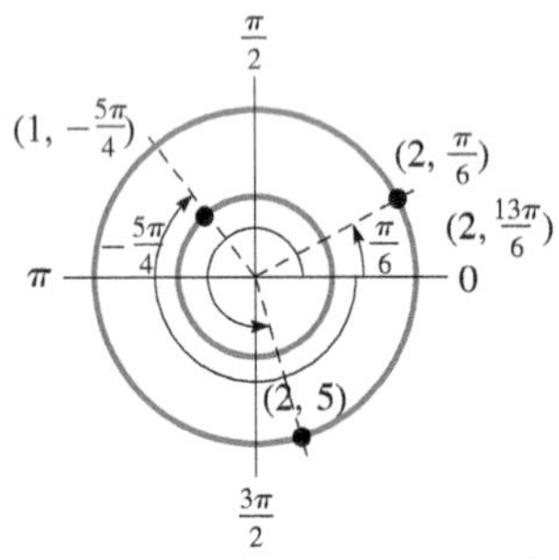

Fig. 21.97

EXAMPLE 1 Locating points in polar coordinates

Locate the following points in the polar coordinate system by first locating the terminal side of θ and then measuring r along this terminal side.

(a) $(2, \frac{\pi}{6})$

Move from the polar axis counterclockwise until you reach an angle of $\frac{\pi}{6}$. Plot the point on the circle of radius 2 as shown in Fig. 21.97.

(b) $(2, 5)$

Move from the polar axis counterclockwise until you reach an angle of 5 *radians*. Measure a radius of 2.

(c) $\left(1, -\frac{5\pi}{4}\right)$

The angle is negative, so move from the polar axis clockwise $\frac{5\pi}{4}$ radians. Plot the point on the circle of radius 1.

(d) $\left(2, \frac{13\pi}{6}\right)$

The angles $\frac{\pi}{6}$ and $\frac{13\pi}{6}$ are coterminal. The point is the same as $\left(2, \frac{\pi}{6}\right)$. There are limitless possibilities for representing a single point.

The restriction that r be positive in the definition of the trigonometric functions does not apply to points in polar coordinates. That is, r is allowed to take positive and negative values.

LEARNING TIP

If r is negative, θ is located as before, but ***the point is found r units from the pole on the opposite side*** from that on which it is positive.

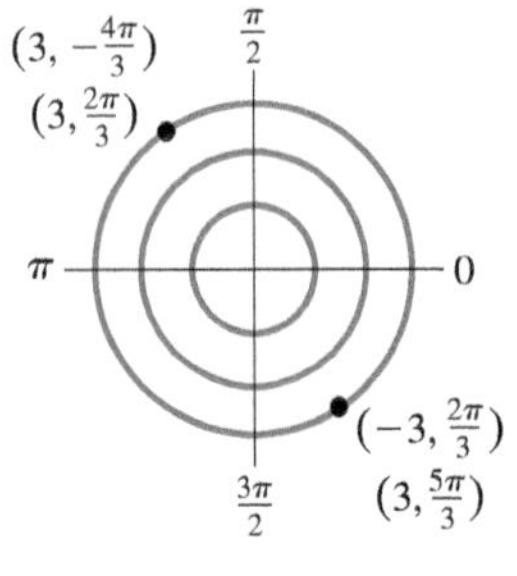

Fig. 21.98

EXAMPLE 2 Polar coordinates—negative *r*

Locate the points $(3, 2\pi/3)$ and $(-3, 2\pi/3)$ in the polar coordinate system.

To locate $(3, 2\pi/3)$ we move from the polar axis counterclockwise to find the terminal side of $2\pi/3$ and then measure $r = 3$. The point $(3, -4\pi/3)$ represents the same point.

To locate $(-3, 2\pi/3)$ we move to the same terminal side of $2\pi/3$, but we measure $r = -3$ on the opposite side of the pole. The point $(3, 5\pi/3)$ represents the same point. These points are shown in Fig. 21.98.

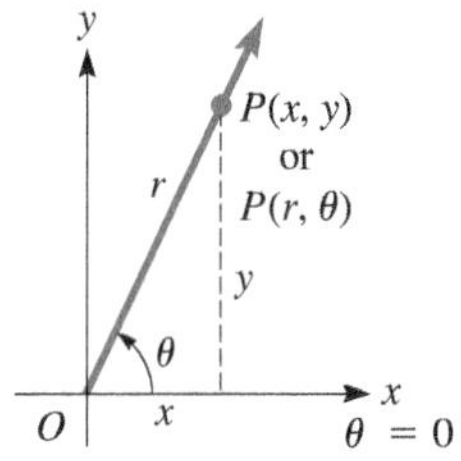

Fig. 21.99

POLAR AND RECTANGULAR COORDINATES

The relationships between the polar coordinates of a point and the rectangular coordinates of the same point come from the definitions of the trigonometric functions. Those most commonly used are (see Fig. 21.99)

$$x = r\cos\theta \quad y = r\sin\theta \tag{21.42}$$

$$\tan\theta = \frac{y}{x} \quad r = \sqrt{x^2 + y^2} \tag{21.43}$$

■ Calculators are programmed to make conversions between rectangular and polar coordinates.

As we have done in previous chapters, when θ is not in the first quadrant, we first find the reference angle and then use it to find θ, depending on the quadrant.

The following examples show the use of Eqs. (21.42) and (21.43) in changing coordinates in one system to coordinates in the other system. Also, these equations are used to transform equations from one system to the other.

EXAMPLE 3 Polar to rectangular coordinates

Transform the polar coordinates $(4, \pi/4)$ into rectangular coordinates.

Using Eq. (21.42) with $r = 4$ and $\theta = \pi/4$, we get

$$x = 4\cos\frac{\pi}{4} = 4\left(\frac{\sqrt{2}}{2}\right) = 2\sqrt{2} \quad \text{and} \quad y = 4\sin\frac{\pi}{4} = 4\left(\frac{\sqrt{2}}{2}\right) = 2\sqrt{2}$$

See Fig. 21.100.

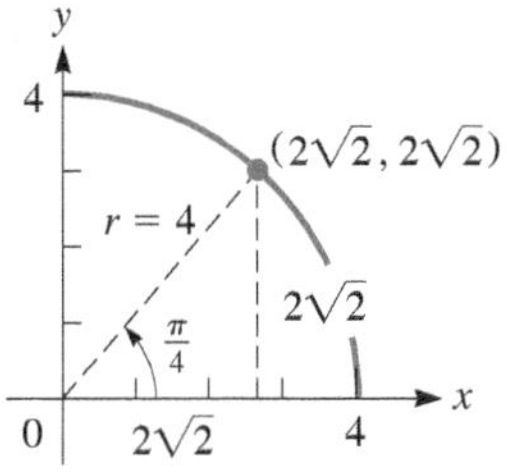

Fig. 21.100

EXAMPLE 4 Rectangular to polar coordinates

Transform the rectangular coordinates $(3, -5)$ into polar coordinates.

Using Eq. (21.43) with $x = 3$ and $y = -5$, we get

$$\tan\theta_{ref} = \frac{5}{3},\ \theta_{ref} = 1.03,\ \theta = 2\pi - 1.03 = 5.25\ (\text{or} -1.03)$$

$$r = \sqrt{3^2 + (-5)^2} = 5.83$$

We know that θ is a fourth-quadrant angle since x is positive and y is negative. Therefore, the point $(3, -5)$ in rectangular coordinates can be expressed as the point $(5.83, 5.25)$ in polar coordinates (see Fig. 21.101). Other polar coordinates for the point are also possible.

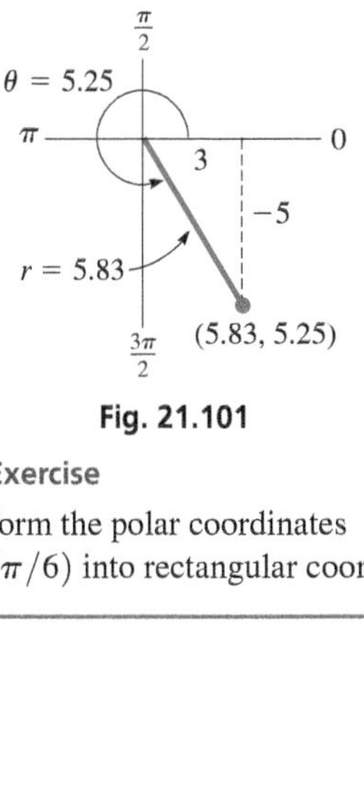

Fig. 21.101

Practice Exercise

1. Transform the polar coordinates $(-4, 5\pi/6)$ into rectangular coordinates.

EXAMPLE 5 Rectangular to polar equation of a curve

■ The cyclotron was invented in 1931 at the University of California. It was the first accelerator to deflect particles into circular paths.

If an electrically charged particle enters a magnetic field at right angles to the field, the particle follows a circular path. This fact is used in the design of nuclear particle accelerators.

A proton (positively charged) enters a magnetic field such that its path may be described by the rectangular equation $x^2 + y^2 = 2x$, where measurements are in metres. Find the polar equation of this circle.

We change this equation expressed in the rectangular coordinates x and y into an equation expressed in the polar coordinates r and θ by using the relations $r^2 = x^2 + y^2$ and $x = r\cos\theta$, as follows:

$$x^2 + y^2 = 2x \quad \text{rectangular equation}$$

$$r^2 = 2r\cos\theta \quad \text{substitute}$$

$$r = 2\cos\theta \quad \text{divided by } r$$

This is the polar equation of the circle, which is shown in Fig. 21.102.

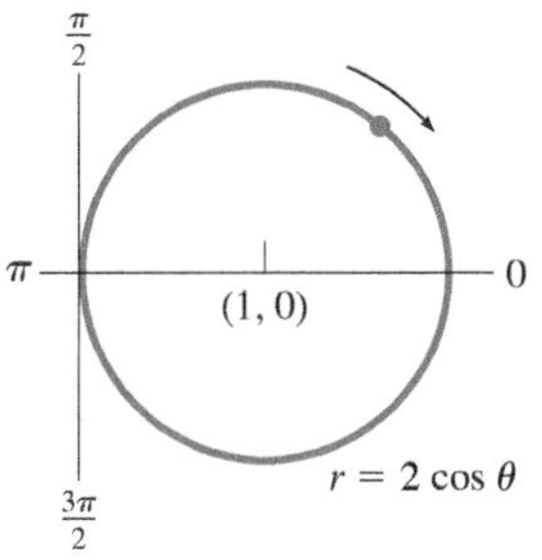

Fig. 21.102

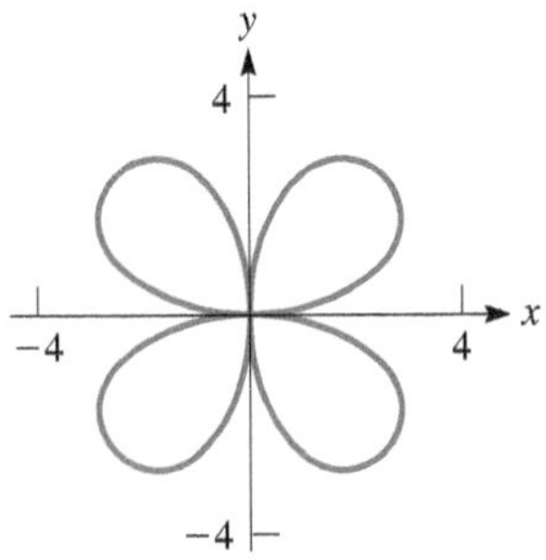

Fig. 21.103

EXAMPLE 6 Polar to rectangular equation of a curve

Find the rectangular equation of the *rose* $r = 4 \sin 2\theta$.

Using the trigonometric identity $\sin 2\theta = 2 \sin \theta \cos \theta$ and Eqs. (21.42) and (21.43) leads to the solution:

$$r = 4 \sin 2\theta \quad \text{polar equation}$$

$$= 4(2 \sin \theta \cos \theta) = 8 \sin \theta \cos \theta \quad \text{using identity}$$

$$\sqrt{x^2 + y^2} = 8\left(\frac{y}{r}\right)\left(\frac{x}{r}\right) = \frac{8xy}{r^2} = \frac{8xy}{x^2 + y^2} \quad \text{using Eqs. (21.42) and (21.43)}$$

$$x^2 + y^2 = \frac{64x^2y^2}{(x^2 + y^2)^2} \quad \text{squaring both sides}$$

$$(x^2 + y^2)^3 = 64x^2y^2 \quad \text{simplifying}$$

Plotting the graph of this equation from the rectangular equation would be complicated. However, as we will see in the next section, plotting this graph in polar coordinates is quite simple. The curve is shown in Fig. 21.103.

Practice Exercise

2. Find the polar equation of the circle $x^2 + y^2 + 2x = 0$.

EXERCISES 21.10

In Exercises 1–4, make the given changes in the indicated examples of this section and then solve the indicated problems.

1. In Example 2, change $2\pi/3$ to $\pi/3$ and then find another set of coordinates for each point, similar to those shown for the points in the example.
2. In Example 3, change $\pi/4$ to $\pi/6$.
3. In Example 4, change 3 to -3 and -5 to 5.
4. In Example 6, change sin to cos.

In Exercises 5–16, plot the given polar coordinate points on polar coordinate paper.

5. $\left(3, \frac{\pi}{6}\right)$ 6. $(2, \pi)$ 7. $\left(\frac{5}{2}, -\frac{2\pi}{5}\right)$

8. $\left(5, -\frac{\pi}{3}\right)$ 9. $\left(-8, \frac{7\pi}{6}\right)$ 10. $\left(-5, \frac{\pi}{4}\right)$

11. $\left(-3, -\frac{5\pi}{4}\right)$ 12. $\left(-4, -\frac{5\pi}{3}\right)$ 13. $(2, 2)$

14. $(-6, -6)$ 15. $(0.5, -8.4)$ 16. $(-2.2, 18.8)$

In Exercises 17–22, find a set of polar coordinates for each of the points for which the rectangular coordinates are given.

17. $(\sqrt{3}, 1)$ 18. $(-3, 3)$

19. $\left(-\frac{\sqrt{3}}{2}, -\frac{1}{2}\right)$ 20. $(-10, 8)$

21. $(0, 4)$ 22. $(-5, 0)$

In Exercises 23–28, find the rectangular coordinates for each of the points for which the polar coordinates are given.

23. $\left(8, \frac{4\pi}{3}\right)$ 24. $(-4, -\pi)$

25. $(3.0, -0.40)$ 26. $(-3.0, 3.0)$

27. $(8.0, -8.0)$ 28. $(\pi, 4.0)$

In Exercises 29–40, find the polar equation of each of the given rectangular equations.

29. $x = 3$ 30. $y = -x$

31. $x + 2y + 3 = 0$ 32. $x^2 + y^2 = 0.81$

33. $x^2 + (y - 2)^2 = 4$ 34. $x^2 - y^2 = 0.01$

35. $x^2 + 4y^2 = 4$ 36. $y^2 = 4x$

37. $x^2 + y^2 = 6y$ 38. $xy = 9$

39. $x^3 + y^3 - 4xy = 0$ 40. $y = \frac{2x}{x^2 + 1}$

In Exercises 41–52, find the rectangular equation of each of the given polar equations. In Exercises 41–48, identify the curve that is represented by the equation.

41. $r = \sin \theta$ 42. $r = 4 \cos \theta$

43. $r \cos \theta = 4$ 44. $r = -2 \csc \theta$

45. $r = \frac{2}{\cos \theta - 3 \sin \theta}$ 46. $r = e^{r \cos \theta} \csc \theta$

47. $r = 4 \cos \theta + 2 \sin \theta$ 48. $r \sin(\theta + \pi/6) = 3$

49. $r = 2(1 + \cos \theta)$ 50. $r = 1 - \sin \theta$

51. $r^2 = \sin 2\theta$ 52. $r^2 = 16 \cos 2\theta$

In Exercises 53–64, solve the given problems. All coordinates given are polar coordinates.

53. Is the point $(2, 3\pi/4)$ on the curve $r = 2 \sin 2\theta$?
54. Is the point $(1/2, 3\pi/2)$ on the curve $r = \sin(\theta/3)$?
55. Show that the polar coordinate equation $r = a \sin \theta + b \cos \theta$ represents a circle by changing it to a rectangular equation.
56. Find the distance between the points $(3, \pi/6)$ and $(4, \pi/2)$ by using the law of cosines.
57. The centre of a regular hexagon is at the pole with one vertex at $(2, \pi)$. What are the coordinates of the other vertices?
58. Find the distance between the points $(4, \pi/6)$ and $(5, 5\pi/3)$.

59. Under certain conditions, the x- and y-components of a magnetic field B are given by the equations

$$B_x = \frac{-ky}{x^2 + y^2} \quad \text{and} \quad B_y = \frac{kx}{x^2 + y^2}$$

Write these equations in terms of polar coordinates.

60. In designing a domed roof for a building, an architect uses the equation $x^2 + \frac{y^2}{k^2} = 1$, where k is a constant. Write this equation in polar form.

61. The shape of a cam can be described by the polar equation $r = 3 - \sin\theta$. Find the rectangular equation for the shape of the cam.

62. The polar equation of the path of a weather satellite of the earth is $r = \frac{7600}{1 + 0.14\cos\theta}$, where r is measured in kilometres. Find the rectangular equation of the path of this satellite. The path is an ellipse, with the earth at one of the foci.

63. The control tower of an airport is taken to be at the pole, and the polar axis is taken as due east in a polar coordinate graph. How far apart (in km) are planes, at the same altitude, if their positions on the graph are $(6.10, 1.25)$ and $(8.45, 3.74)$?

64. The perimeter of a certain type of machine part can be described by the equation $r = a\sin\theta + b\cos\theta$ $(a > 0, b > 0)$. Explain why all such machine parts are circular.

Answers to Practice Exercises

1. $(2\sqrt{3}, -2)$ **2.** $r = -2\cos\theta$

21.11 Curves in Polar Coordinates

The basic method for finding a curve in polar coordinates is the same as in rectangular coordinates. We assume values of the independent variable—in this case, θ—and then find the corresponding values of the dependent variable r. These points are then plotted and joined, thereby forming the curve that represents the relation in polar coordinates.

Before using the basic method, it is useful to point out that certain basic curves can be sketched directly from the equation. This is done by noting the meaning of each of the polar coordinate variables, r and θ. This is illustrated in the following example.

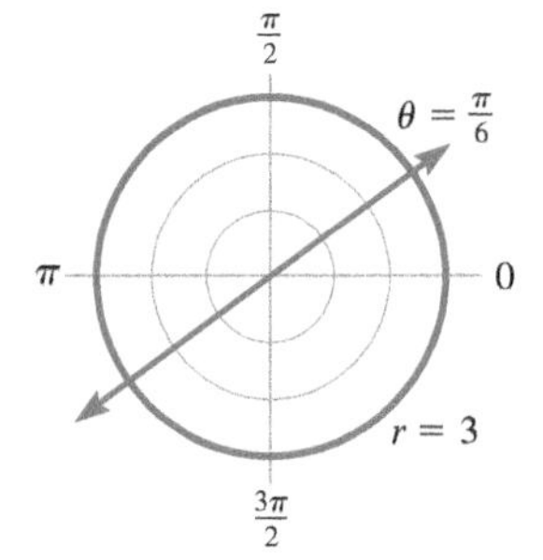

Fig. 21.104

EXAMPLE 1 Curves sketched from equations

(a) The graph of the polar equation $r = 3$ is a circle of radius 3, with centre at the pole. This can be seen to be the case, since $r = 3$ for all possible values of θ. It is not necessary to find specific points for this circle, which is shown in Fig. 21.104.

(b) The graph of $\theta = \pi/6$ is a straight line through the pole. It represents all points for which $\theta = \pi/6$ for all possible values of r. This line is shown in Fig. 21.104.

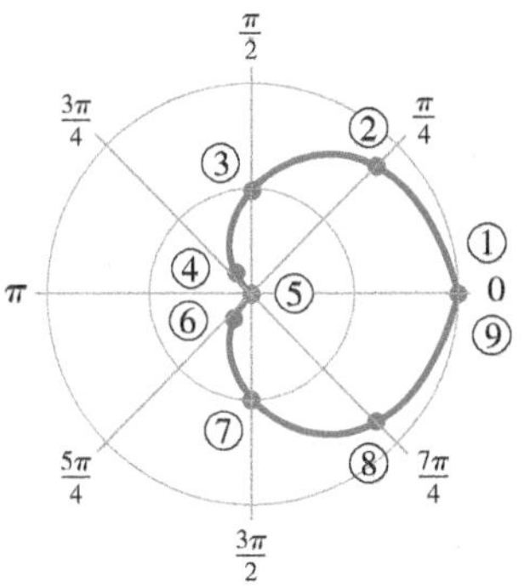

Fig. 21.105

EXAMPLE 2 Plotting a cardioid

Plot the graph of $r = 1 + \cos\theta$.

We find the following values of r corresponding to the chosen values of θ.

θ	0	$\frac{\pi}{4}$	$\frac{\pi}{2}$	$\frac{3\pi}{4}$	π	$\frac{5\pi}{4}$	$\frac{3\pi}{2}$	$\frac{7\pi}{4}$	2π
r	2	1.7	1	0.3	0	0.3	1	1.7	2
Point Number	1	2	3	4	5	6	7	8	9

We now see that the points are repeating, and it is unnecessary to find additional points. This curve is called a **cardioid** and is shown in Fig. 21.105.

EXAMPLE 3 Plotting a limaçon

Plot the graph of $r = 1 - 2\sin\theta$.

Choosing values of θ and then finding the corresponding values of r, we find the following table of values.

θ	0	$\frac{\pi}{4}$	$\frac{\pi}{2}$	$\frac{3\pi}{4}$	π	$\frac{5\pi}{4}$	$\frac{3\pi}{2}$	$\frac{7\pi}{4}$	2π
r	1	-0.4	-1	-0.4	1	2.4	3	2.4	1
Point Number	1	2	3	4	5	6	7	8	9

Particular care should be taken in plotting the points for which r is negative. This curve is known as a **limaçon** and is shown in Fig. 21.106.

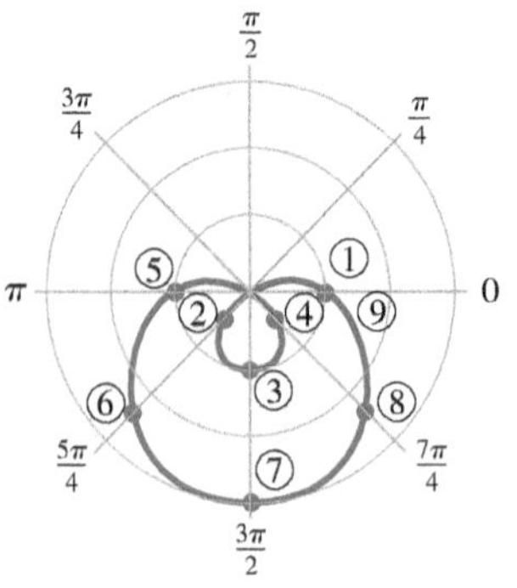

Fig. 21.106

Practice Exercises

Determine the type of curve represented by each polar coordinate equation.

1. $\theta = 2\pi/5$ **2.** $r = -4$

EXAMPLE 4 Polar curve—application

A cam is shaped such that the edge of the upper "half" is represented by the equation $r = 2.0 + \cos\theta$ and the lower "half" by the equation $r = \dfrac{3.0}{2.0 - \cos\theta}$, where measurements are in centimetres. Plot the curve that represents the shape of the cam.

We get the points for the edge of the cam by using values of θ from 0 to π for the upper "half" and from π to 2π for the lower "half." The table of values follows:

$r = 2.0 + \cos\theta$

θ	0	$\frac{\pi}{4}$	$\frac{\pi}{2}$	$\frac{3\pi}{4}$	π
r	3.0	2.7	2.0	1.3	1.0
Point Number	1	2	3	4	5

$r = \dfrac{3.0}{2.0 - \cos\theta}$

θ	π	$\frac{5\pi}{4}$	$\frac{3\pi}{2}$	$\frac{7\pi}{4}$	2π
r	1.0	1.1	1.5	2.3	3.0
Point Number	6	7	8	9	10

The upper "half" is part of a limaçon and the lower "half" is a semiellipse. The cam is shown in Fig. 21.107.

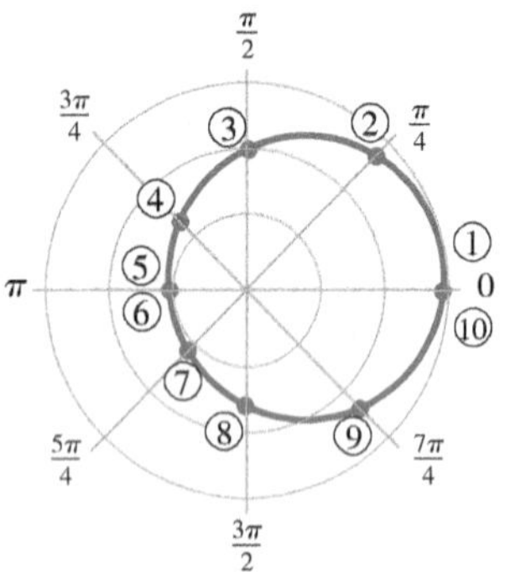

Fig. 21.107

EXAMPLE 5 Plotting a rose

Plot the graph of $r = 2\cos 2\theta$.

In finding values of r, we must be careful first to multiply the values of θ by 2 before finding the cosine of the angle. Also, for this reason, we take values of θ as multiples of $\pi/12$, so as to get enough useful points. The table of values follows:

θ	0	$\frac{\pi}{12}$	$\frac{\pi}{6}$	$\frac{\pi}{4}$	$\frac{\pi}{3}$	$\frac{5\pi}{12}$	$\frac{\pi}{2}$
r	2	1.7	1	0	-1	-1.7	-2

θ	$\frac{7\pi}{12}$	$\frac{2\pi}{3}$	$\frac{3\pi}{4}$	$\frac{5\pi}{6}$	$\frac{11\pi}{12}$	π
r	-1.7	-1	0	1	1.7	2

For values of θ starting with π, the values of r repeat. We have a four-leaf **rose**, as shown in Fig. 21.108.

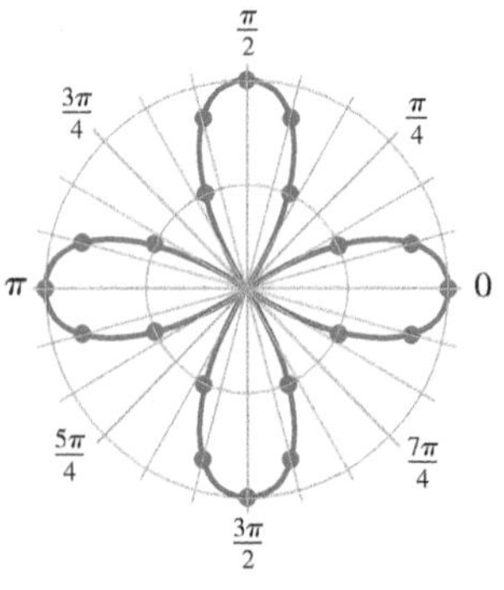

Fig. 21.108

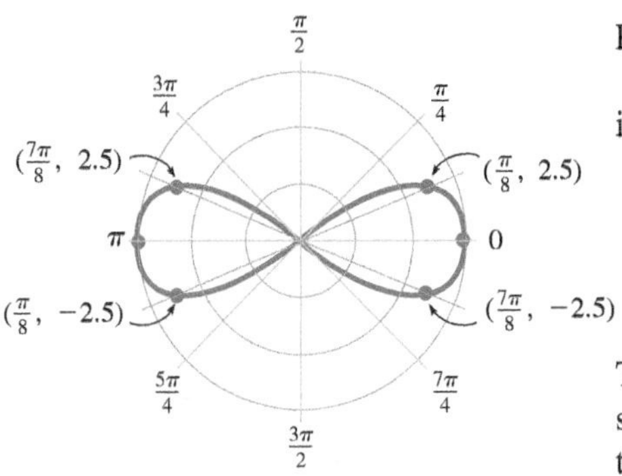

Fig. 21.109

EXAMPLE 6 Plotting a lemniscate

Plot the graph of $r^2 = 9 \cos 2\theta$.

Choosing the indicated values of θ, we get the values of r as shown in the following table:

θ	0	$\frac{\pi}{8}$	$\frac{\pi}{4}$	$\cdots$	$\frac{3\pi}{4}$	$\frac{7\pi}{8}$	π
r	± 3	± 2.5	0		0	± 2.5	± 3

There are no values of r corresponding to values of θ in the range $\pi/4 < \theta < 3\pi/4$, since twice these angles are in the second and third quadrants and the cosine is negative for such angles. The value of r^2 cannot be negative. Also, the values of r repeat for $\theta > \pi$. The figure is called a **lemniscate** and is shown in Fig. 21.109.

EXAMPLE 7 Polar curve using a graphing utility

View the graph of $r = 1 - 2 \cos \theta$ using a graphing utility.

With a graphing calculator, for example, a polar curve can be displayed using the polar graph mode or the parametric graph mode, depending on the calculator. The graph displayed will be the same with either method.

With the polar graph option, the function is entered directly. The values for the viewing window are determined by settings for x, y and the angles θ that will be used. These values are set in a manner similar to those used for parametric equations. (See Section 10.6 for an example of graphing parametric equations.)

With the parametric graph option, to graph $r = f(\theta)$, we note that $x = r \cos \theta$ and $y = r \sin \theta$. This tells us that

$$x = f(\theta) \cos \theta$$
$$y = f(\theta) \sin \theta$$

Thus, for $r = 1 - 2 \cos \theta$, by using

$$x = (1 - 2 \cos \theta) \cos \theta$$
$$y = (1 - 2 \cos \theta) \sin \theta$$

the graph can be displayed, as shown in Fig. 21.110.

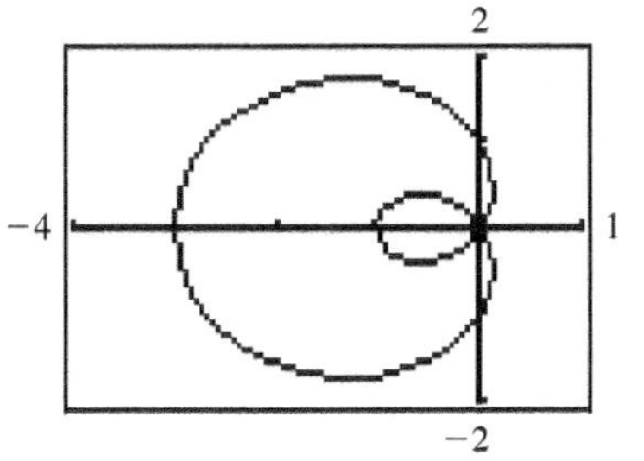

Fig. 21.110

EXERCISES 21.11

In Exercises 1–4, make the given changes in the indicated examples of this section and then make the indicated graphs.

1. In Example 1(b), change $\pi/6$ to $5\pi/6$.
2. In Example 3, change the $-$ before the $2 \sin \theta$ to $+$.
3. In Example 5, change cos to sin.
4. In Example 7, change the $-$ before the $2 \cos \theta$ to $+$.

In Exercises 5–32, plot the curves of the given polar equations in polar coordinates.

5. $r = 5$
6. $r = -2$
7. $\theta = 3\pi/4$
8. $\theta = -1.5$
9. $r = 4 \sec \theta$
10. $r = 4 \csc \theta$
11. $r = 2 \sin \theta$
12. $r = 3 \cos \theta$
13. $1 - r = \cos \theta$ (cardioid)
14. $r + 1 = \sin \theta$ (cardioid)
15. $r = 2 - \cos \theta$ (limaçon)
16. $r = 2 + 3 \sin \theta$ (limaçon)
17. $r = 4 \sin 2\theta$ (rose)
18. $r = 2 \sin 3\theta$ (rose)
19. $r^2 = 4 \sin 2\theta$ (lemniscate)
20. $r^2 = 2 \sin \theta$
21. $r = 2^{\theta}$ (spiral)
22. $r = 1.5^{-\theta}$ (spiral)
23. $r = 4|\sin 3\theta|$
24. $r = 2 \sin \theta \tan \theta$ (cissoid)
25. $r = \dfrac{3}{2 - \cos \theta}$ (ellipse)
26. $r = \dfrac{2}{1 - \cos \theta}$ (parabola)
27. $r - 2r \cos \theta = 6$ (hyperbola)
28. $3r - 2r \sin \theta = 6$ (ellipse)
29. $r = 4 \cos \frac{1}{2}\theta$
30. $r = 2 + \cos 3\theta$
31. $r = 2[1 - \sin(\theta - \pi/4)]$
32. $r = 4 \tan \theta$

In Exercises 33–42, view the curves of the given polar equations on a graphing utility.

33. $r = \theta \quad (-20 \le \theta \le 20)$

34. $r = 0.5^{\sin\theta}$

35. $r = 2\sec\theta + 1$

36. $r = 2\cos(\cos 2\theta)$

37. $r = 3\cos 4\theta$

38. $r = 3\sin 5\theta$

39. $r = 2\cos\theta + 3\sin\theta$

40. $r = 1 + 3\cos\theta - 2\sin\theta$

41. $r + 2 = \cos 2\theta$

42. $2r\cos\theta + r\sin\theta = 2$

In Exercises 43–54, solve the given problems and sketch or display the indicated curves.

43. What is the graph of $\tan\theta = 1$?

44. Find the polar equation of the line through the polar points $(1, 0)$ and $(2, \pi/2)$.

45. Using a graphing utility, show that the curves $r = 2\sin\theta$ and $r = 2\cos\theta$ intersect at right angles. Proper *display* settings are necessary.

46. Using a graphing utility, determine what type of graph is displayed by $r = 3\sec^2(\theta/2)$.

47. An architect designs a patio shaped such that it can be described as the area within the polar curve $r = 4.0 - \sin\theta$, where measurements are in metres. Sketch the curve that represents the perimeter of the patio.

48. The radiation pattern of a certain television transmitting antenna can be represented by $r = 120(1 + \cos\theta)$, where distances (in km) are measured from the antenna. Sketch the radiation pattern.

49. The joint between two links of a robot arm moves in an elliptical path (in cm), given by $r = \frac{25}{10 + 4\cos\theta}$. Sketch the path.

50. A missile is fired at an airplane and is always directed toward the airplane. The missile is travelling at twice the speed of the airplane. An equation that describes the distance r between the missile and the airplane is $r = \dfrac{70\sin\theta}{(1 - \cos\theta)^2}$, where θ is the angle between their directions at all times. See Fig. 21.111. This is a *relative pursuit curve.* Sketch the graph of this equation for $\pi/4 \le \theta \le \pi$.

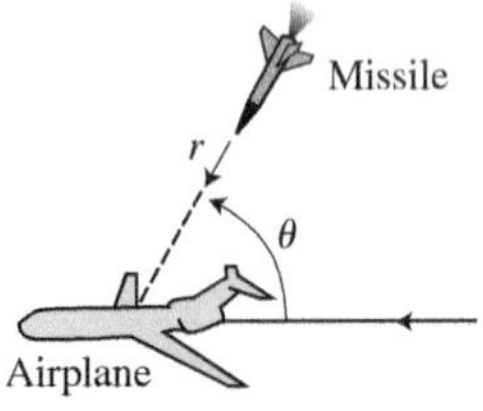

Fig. 21.111

51. In studying the photoelectric effect, an equation used for the rate R at which photoelectrons are ejected at various angles θ is $R = \dfrac{\sin^2\theta}{(1 - 0.5\cos\theta)^2}$. Sketch the graph.

52. In order to display the graph of the rectangular equation $4(x^6 + 3x^4y^2 + 3x^2y^4 + y^6 - x^4 - 2x^2y^2 - y^4) + y^2 = 0$ on a computer screen, it is first transformed into polar coordinates. Transform the equation and then sketch the graph. (*Hint:* The equation can be written as $4(x^2 + y^2)^3 - 4(x^2 + y^2)^2 + y^2 = 0$.)

53. Noting the graphs in Exercises 17, 18, 37, and 38, what conclusion do you draw about the value of n and the graph of $y = a\sin n\theta$ or $y = a\cos n\theta$?

54. (a) Solve the equations $r = 2\sin\theta$ and $r = 1 + \sin\theta$ simultaneously to find a point of intersection. (b) Display the graphs of the equations on a graphing utility and note another common point, the coordinates of which do not satisfy both equations. Explain why this occurs.

Answers to Practice Exercises

1. Straight line **2.** Circle

CHAPTER 21 KEY FORMULAS AND EQUATIONS

Distance formula	Fig. 21.2	$d = \sqrt{(x_2 - x_1)^2 + (y_2 - y_1)^2}$	**(21.1)**
Slope	Fig. 21.4	$m = \dfrac{y_2 - y_1}{x_2 - x_1}$	**(21.2)**
	Fig. 21.7	$m = \tan\alpha \quad (0° \le \alpha < 180°)$	**(21.3)**
	Fig. 21.10	$m_1 = m_2$ (for $\mid\mid$ lines)	**(21.4)**
	Fig. 21.11	$m_2 = -\dfrac{1}{m_1}$ or $m_1m_2 = -1$ (for $\perp$ lines)	**(21.5)**
Straight line	Fig. 21.15	$y - y_1 = m(x - x_1)$	**(21.6)**
		$Ax + By + C = 0$	**(21.7)**
	Fig. 21.18(a)	$x = a$	**(21.8)**
	Fig. 21.18(b)	$y = b$	**(21.9)**
	Fig. 21.20	$y = mx + b$	**(21.10)**
Circle	Fig. 21.28	$(x - h)^2 + (y - k)^2 = r^2$	**(21.11)**
	Fig. 21.31	$x^2 + y^2 = r^2$	**(21.12)**
		$x^2 + y^2 + Dx + Ey + F = 0$	**(21.14)**

Parabola	Fig. 21.39	$y^2 = 4px$	**(21.15)**
	Fig. 21.42(c)	$x^2 = 4py$	**(21.16)**
Ellipse	Fig. 21.54	$\frac{x^2}{a^2} + \frac{y^2}{b^2} = 1$	**(21.17)**
		$a^2 = b^2 + c^2$	**(21.18)**
	Fig. 21.55	$\frac{y^2}{a^2} + \frac{x^2}{b^2} = 1$	**(21.19)**
Hyperbola	Fig. 21.65	$\frac{x^2}{a^2} - \frac{y^2}{b^2} = 1$	**(21.20)**
		$c^2 = a^2 + b^2$	**(21.21)**
	Fig. 21.66	$y = \pm\frac{bx}{a}$ (asymptotes)	**(21.23)**
	Fig. 21.68	$\frac{y^2}{a^2} - \frac{x^2}{b^2} = 1$	**(21.24)**
	Fig. 21.74	$xy = k$	**(21.25)**
Translation of axes	Fig. 21.79	$x = x' + h$ and $y = y' + k$	**(21.26)**
		$x' = x - h$ and $y' = y - k$	**(21.27)**
Parabola, vertex (*h*, *k*)		$(y - k)^2 = 4p(x - h)$ (axis parallel to x-axis)	**(21.28)**
		$(x - h)^2 = 4p(y - k)$ (axis parallel to y-axis)	**(21.29)**
Ellipse, centre (*h*, *k*)		$\frac{(x-h)^2}{a^2} + \frac{(y-k)^2}{b^2} = 1$ (major axis parallel to x-axis)	**(21.30)**
		$\frac{(y-k)^2}{a^2} + \frac{(x-h)^2}{b^2} = 1$ (major axis parallel to y-axis)	**(21.31)**
Hyperbola, centre (*h*, *k*)		$\frac{(x-h)^2}{a^2} - \frac{(y-k)^2}{b^2} = 1$ (transverse axis parallel to x-axis)	**(21.32)**
		$\frac{(y-k)^2}{a^2} - \frac{(x-h)^2}{b^2} = 1$ (transverse axis parallel to y-axis)	**(21.33)**
Second-degree equation		$Ax^2 + Bxy + Cy^2 + Dx + Ey + F = 0$	**(21.34)**
Rotation of axes	Fig. 21.91	$x = x'\cos\theta - y'\sin\theta \quad y = x'\sin\theta + y'\cos\theta$	**(21.38)**
Angle of rotation		$\tan 2\theta = \frac{B}{A - C} \quad (A \neq C)$	**(21.40)**
		$\theta = 45^\circ \quad (A = C)$	**(21.41)**
Polar coordinates	Fig. 21.99	$x = r\cos\theta \quad y = r\sin\theta$	**(21.42)**
		$\tan\theta = \frac{y}{x} \quad r = \sqrt{x^2 + y^2}$	**(21.43)**

CHAPTER 21 REVIEW EXERCISES

In Exercises 1–12, find the equation of the indicated curve, subject to the given conditions. Sketch each curve.

1. Straight line: passes through $(1, -7)$ with a slope of 4
2. Straight line: passes through $(-1, 5)$ and $(-2, -3)$
3. Straight line: perpendicular to $3x - 2y + 8 = 0$ and has a y-intercept of $(0, -1)$
4. Straight line: parallel to $2x - 5y + 1 = 0$ and has an x-intercept of $(2, 0)$
5. Circle: concentric with $x^2 + y^2 = 6x$, passes through $(4, -3)$
6. Circle: tangent to lines $x = 3$ and $x = 9$, centre on line $y = 2x$
7. Parabola: focus $(3, 0)$, vertex $(0, 0)$
8. Parabola: vertex $(0, 0)$, passes through $(1, 1)$ and $(-2, 4)$
9. Ellipse: vertex $(10, 0)$, focus $(8, 0)$, tangent to $x = -10$
10. Ellipse: centre $(0, 0)$, passes through $(0, 3)$ and $(2, 1)$
11. Hyperbola: $V(0, 13)$, $C(0, 0)$, conj. axis of 24
12. Hyperbola: vertex $(0, 8)$, asymptotes $y = 2x$, $y = -2x$

In Exercises 13–26, find the indicated quantities for each of the given equations. Sketch each curve.

13. $x^2 + y^2 + 6x - 7 = 0$, centre and radius

14. $2x^2 + 2y^2 + 4x - 8y - 15 = 0$, centre and radius

15. $x^2 = -20y$, focus and directrix

16. $y^2 = 24x$, focus and directrix

17. $4x^2 + y^2 = 1$, vertices and foci

18. $2y^2 - 9x^2 = 18$, vertices and foci

19. $2x^2 - 5y^2 = 0.25$, vertices and foci

20. $2x^2 + 25y^2 = 800$, vertices and foci

21. $x^2 - 8x - 4y - 16 = 0$, vertex and focus

22. $y^2 - 4x + 4y + 24 = 0$, vertex and directrix

23. $4x^2 + y^2 - 16x + 2y + 13 = 0$, centre

24. $x^2 - 2y^2 + 4x + 4y + 6 = 0$, centre

25. $x^2 - 2xy + y^2 + 4x + 4y = 0$, vertex

26. $3x^2 - 3xy + 7y^2 - 5 = 0$, centre

In Exercises 27–34, plot the given curves in polar coordinates.

27. $r = 4(1 + \sin\theta)$

28. $r = 1 - 3\cos\theta$

29. $r = 4\cos 3\theta$

30. $r = 3\sin\theta - 4\cos\theta$

31. $r = \dfrac{3}{\sin\theta + 2\cos\theta}$

32. $r = \dfrac{1}{2(\sin\theta - 1)}$

33. $r = 2\sin\left(\dfrac{\theta}{2}\right)$

34. $r = 1 - \cos 2\theta$

In Exercises 35–38, find the polar equation of each of the given rectangular equations.

35. $y = 2x$

36. $2xy = 1$

37. $x^2 + xy + y^2 = 2$

38. $x^2 + (y + 3)^2 = 16$

In Exercises 39–42, find the rectangular equation of each of the given polar equations.

39. $r = 2\sin 2\theta$

40. $r^2 = 9\sin\theta$

41. $r = \dfrac{4}{2 - \cos\theta}$

42. $r = 4\tan\theta\sec\theta$

In Exercises 43–48, determine the number of real solutions of the given systems of equations by sketching the indicated curves. (See Section 14.1.)

43. $x^2 + y^2 = 9$
$4x^2 + y^2 = 16$

44. $y = e^x$
$x^2 - y^2 = 1$

45. $x^2 + y^2 - 4y - 5 = 0$
$y^2 - 4x^2 - 4 = 0$

46. $x^2 - 4y^2 + 2x - 3 = 0$
$y^2 - 4x - 4 = 0$

47. $y = 2\sin x$
$y = 2 - x^2$

48. $y = 4\ln x$
$xy = 6$

In Exercises 49–58, view the curves of the given equations using a graphing utility.

49. $x^2 + 3y + 2 - (1 + x)^2 = 0$

50. $y^2 = 4x + 6$

51. $2x^2 + 2y^2 + 4y - 3 = 0$

52. $2x^2 + (y - 3)^2 - 5 = 0$

53. $x^2 - 4y^2 + 4x + 24y - 48 = 0$

54. $x^2 + 2xy + y^2 - 3x + 8y = 0$

55. $r = 3\cos(3\theta/2)$

56. $r = 5 - 2\sin 4\theta$

57. $r = 2 - 3\csc\theta$

58. $r = 2\sin(\cos 3\theta)$

In Exercises 59–62, find the equation of the locus of a point $P(x, y)$ that moves as stated.

59. Always 4 units from $(3, -4)$

60. Passes through $(7, -5)$ with a constant slope of -2

61. The sum of its distances from $(1, -3)$ and $(7, -3)$ is 8.

62. The difference of its distances from $(3, -1)$ and $(3, -7)$ is 4.

In Exercises 63–107, solve the given problems.

63. Considering Eq. (21.30) of an ellipse, describe the graph if $a = b$.

64. Show that the ellipse $x^2 + 9y^2 = 9$ has the same foci as the hyperbola $x^2 - y^2 = 4$.

65. The points $(-2, -5)$, $(3, -3)$, and $(13, x)$ are collinear. Find x.

66. For the polar coordinate point $(-5, \pi/4)$, find another set of polar coordinates such that $r < 0$ and $-2\pi < \theta < 0$.

67. Find the distance between the polar coordinate points $(3, \pi/6)$ and $(6, -\pi/3)$.

68. Show that the parametric equations $y = \cot\theta$ and $x = \csc\theta$ define a hyperbola.

69. In two ways, show that the line segments joining $(-3, 11)$, $(2, -1)$, and $(14, 4)$ form a right triangle.

70. Find the equation of the circle that passes through $(3, -2)$, $(-1, -4)$, and $(2, -5)$.

71. Graph the inequality $y > 4(x + 2)^2$.

72. Graph the inequality $4x^2 + 9(y - 2)^2 < 36$.

73. What type of curve is represented by $(x + jy)^2 + (x - jy)^2 = 2$? ($j = \sqrt{-1}$)

74. For the ellipse in Fig. 21.112, show that the product of the slopes PA and PB is $-b^2/a^2$.

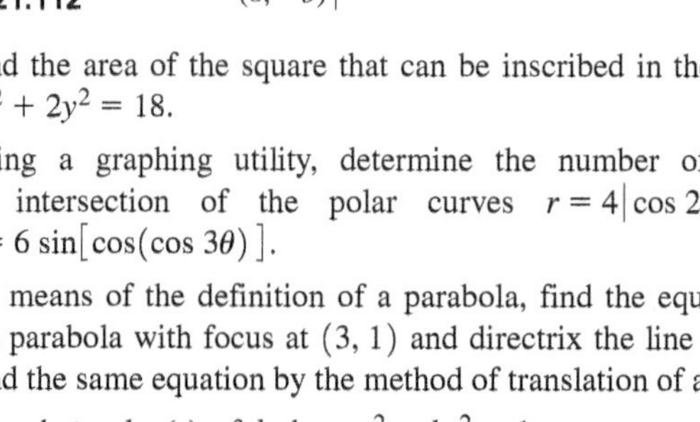

Fig. 21.112

75. Find the area of the square that can be inscribed in the ellipse $7x^2 + 2y^2 = 18$.

76. Using a graphing utility, determine the number of points of intersection of the polar curves $r = 4|\cos 2\theta|$ and $r = 6\sin[\cos(\cos 3\theta)]$.

77. By means of the definition of a parabola, find the equation of the parabola with focus at $(3, 1)$ and directrix the line $y = -3$. Find the same equation by the method of translation of axes.

78. For what value(s) of k does $x^2 - ky^2 = 1$ represent an ellipse with vertices on the y-axis?

79. The total resistance R_T of two resistances in series in an electric circuit is the sum of the resistances. If a variable resistor R is in series with a 2.5-Ω resistor, express R_T as a function of R and sketch the graph.

80. The acceleration of an object is defined as the change in velocity v divided by the corresponding change in time t. Find the equation relating the velocity v and time t for an object for which the acceleration is $6.0\ \text{m/s}^2$ and $v = 5.0$ m/s when $t = 0$ s.

81. The velocity v of a crate sliding down a ramp is given by $v = v_0 + at$, where v_0 is the initial velocity, a is the acceleration, and t is the time. If $v_0 = 1.92$ m/s and $v = 6.20$ m/s when $t = 5.50$ s, find v as a function of t. Sketch the graph.

82. An airplane touches down when landing at 150 km/h. Its velocity v while coming to a stop is given by $v = 150 - 20\,000t$, where t is the time in hours. Sketch the graph of v vs. t.

83. It takes 2.010 kJ of heat to raise the temperature of 1.000 kg of steam by 1.000°C. In a steam generator, a total of y kJ is used to raise the temperature of 50.00 kg of steam from 100°C to T°C. Express y as a function of T and sketch the graph.

84. The temperature in a certain region is 27°C, and at an altitude of 2500 m above the region it is 12°C. If the equation relating the temperature T and the altitude h is linear, find the equation.

85. The radar gun on a police helicopter 170 m above a multilane highway is directed vertically down onto the highway. If the radar gun signal is cone-shaped with a vertex angle of 14°, what area of the highway is covered by the signal?

86. The Niagara Sky Wheel in Niagara Falls, Ontario, has a diameter of 50.5 m. Find the equation representing its circumference. Place the origin of the coordinate system 2.50 m below the bottom of the wheel and the centre of the wheel on the y-axis.

87. The arch of a small bridge across a stream is parabolic. If, at water level, the span of the arch is 80 m and the maximum height above water level is 20 m, what is the equation that represents the arch? Choose the most convenient point for the origin of the coordinate system.

88. A laser source is 2.00 cm from a spherical surface of radius 3.00 cm, and the laser beam is tangent to the surface. By placing the centre of the sphere at the origin, and the source on the positive x-axis, find the equation of the line along which the beam shown in Fig. 21.113 is directed.

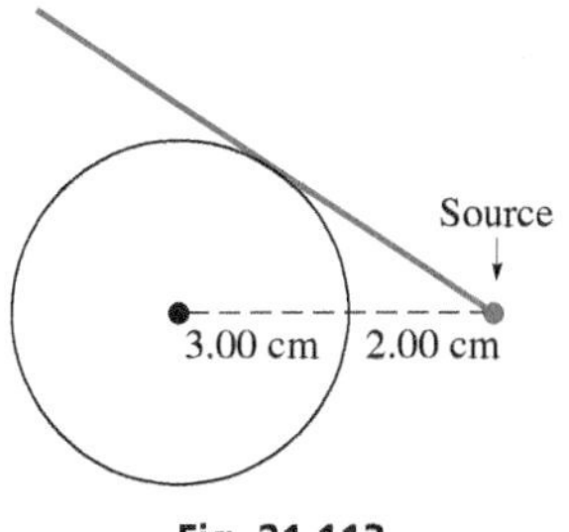

Fig. 21.113

89. The top horizontal cross-section of a dam is parabolic. The open area within this cross-section is 80 m across and 50 m from front to back. Find the equation of the edge of the open area with the vertex at the origin of the coordinate system and the axis along the x-axis.

90. The *quality factor* Q of a series resonant electric circuit with resistance R, inductance L, and capacitance C is given by $Q = \frac{1}{R}\sqrt{\frac{L}{C}}$. Sketch the graph of Q and L for a circuit in which $R = 1000\ \Omega$ and $C = 4.00\ \mu\text{F}$.

91. At very low temperatures, certain metals have an electric resistance of zero. This phenomenon is called *superconductivity*. A magnetic field also affects the superconductivity. A certain level of magnetic field H_T, the threshold field, is related to the thermodynamic temperature T by $H_T/H_0 = 1 - (T/T_0)^2$, where H_0 and T_0 are specifically defined values of magnetic field and temperature. Sketch the graph of H_T/H_0 vs. T/T_0.

92. A rectangular parking lot is to have a perimeter of 600 m. Express the area A in terms of the width w and sketch the graph.

93. The electric power P (in W) supplied by a battery is given by $P = 12.0i - 0.500i^2$, where i is the current (in A). Sketch the graph of P vs. i.

94. The technical ring of the Olympic Stadium in Montreal, Quebec, is in the shape of an ellipse 175 m long and 104 m wide. Find the area of the technical ring. ($A = \pi ab$ for an ellipse.)

95. A specialty electronics company makes an ultrasonic device to repel animals. It emits a 20–25 kHz sound (above those heard by people), which is unpleasant to animals. The sound covers an elliptical area starting at the device, with the longest dimension extending 36 m from the device and the focus of the area 5 m from the device. Find the area covered by the signal. ($A = \pi ab$)

96. A study indicated that the fraction f of cells destroyed by various dosages d of X rays is given by the graph in Fig. 21.114. Assuming that the curve is a quarter-ellipse, find the equation relating f and d for $0 \le f \le 1$ and $0 < d \le 10$ units.

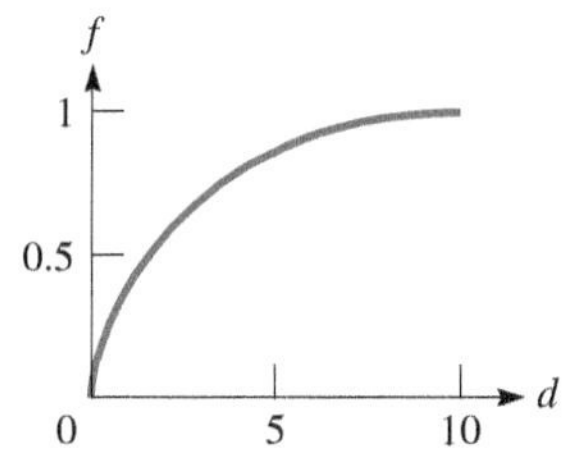

Fig. 21.114

97. A machine-part designer wishes to make a model for an elliptical cam by placing two pins in a design board, putting a loop of string over the pins, and marking off the outline by keeping the string taut. (Note that the definition of the ellipse is being used.) If the cam is to measure 10 cm by 6 cm, how long should the loop of string be and how far apart should the pins be?

98. Soon after reaching the vicinity of the moon, *Apollo 11* (the first spacecraft to land a man on the moon) went into an elliptical lunar orbit. The closest the craft was to the moon in this orbit was 110 km, and the farthest it was from the moon was 310 km. What was the equation of the path if the centre of the moon was at one of the foci of the ellipse? Assume that the major axis is along the x-axis and that the centre of the ellipse is at the origin. The radius of the moon is 1740 km.

99. The vertical cross-section of the cooling tower of a nuclear power plant is hyperbolic, as shown in Fig. 21.115. Find the radius r of the smallest circular horizontal cross-section.

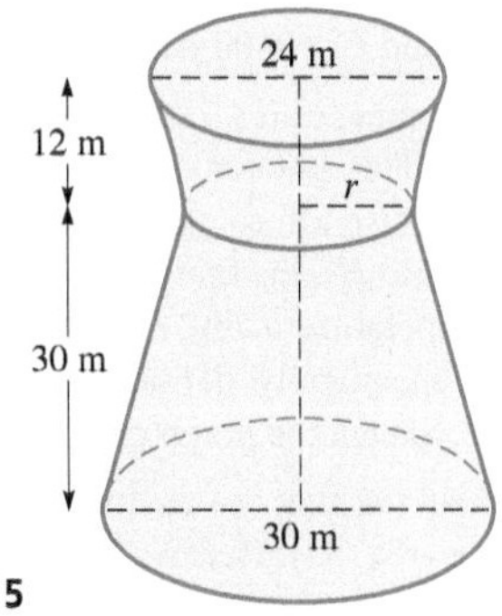

Fig. 21.115

100. Tremors from an earthquake are recorded at the Giralia seismograph station 36 s before they are recorded at the Morawa seismograph station, both in Western Australia. If the stations are 740 km apart and the shock waves from the tremors travel at 5.0 km/s, what is the curve on which lies the point where the earthquake occurred?

101. An electronic instrument located at point P records the sound of a rifle shot and the impact of the bullet striking the target at the same instant. Show that P lies on a branch of a hyperbola. (The bullet travels faster than the speed of sound.)

102. A 20-m rope passes over a pulley 4 m above the ground, and a crate on the ground is attached at one end. The other end of the rope is held at a level of 1 m above the ground and is drawn away from the pulley. Express the height of the crate over the ground in terms of the distance the person is from directly below the crate. Sketch the graph of distance and height. See Fig. 21.116. (Neglect the thickness of the crate.)

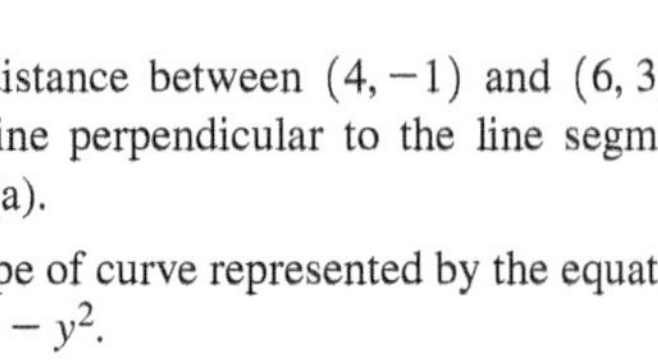

Fig. 21.116

103. A satellite makes one revolution per day around the centre of the earth. The projection on the earth of its path can be approximated by the curve $r^2 = R^2\cos 2(\theta + \frac{\pi}{2})$, where R is the radius of the earth. Sketch the path of the projection.

104. The vertical cross-sections of two pipes as drawn on a drawing board are shown in Fig. 21.117. Find the polar equation of each.

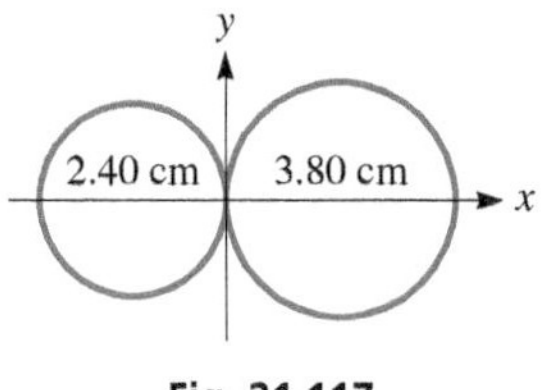

Fig. 21.117

105. The path of a certain plane is $r = 200(\sec\theta + \tan\theta)^{-5}/\cos\theta$, $0 < \theta < \pi/2$. Sketch the path and check it on a graphing utility.

106. The sound produced by a jet engine was measured at a distance of 100 m in all directions. The loudness of the sound d (in decibels) was found to be $d = 115 + 10\cos\theta$, where the 0° line for the angle θ is directed in front of the engine. Sketch the graph of d vs. θ in polar coordinates (use d as r).

107. Under a force that varies inversely as the square of the distance from an attracting object (such as the sun exerts on the earth), it can be shown that the equation of the path an object follows is given in general by

$$\frac{1}{r} = a + b\cos\theta$$

where a and b are constants for a particular path. First, transform this equation into rectangular coordinates. Then write one or two paragraphs explaining why this equation represents one of the conic sections, depending on the values of a and b. It is through this kind of analysis that we know the paths of the planets and comets are conic sections.

CHAPTER 21 PRACTICE TEST

1. (a) Find the distance between $(4, -1)$ and $(6, 3)$. (b) Find the slope of the line perpendicular to the line segment joining the points in part (a).

2. Identify the type of curve represented by the equation $2(x^2 + x) = 1 - y^2$.

3. Sketch the graph of the straight line $4x - 2y + 5 = 0$ by finding its slope and y-intercept.

4. Find the polar equation of the curve whose rectangular equation is $x^2 = 2x - y^2$.

5. Find the vertex and the focus of the parabola $x^2 = -12y$. Sketch the graph.

6. Find the equation of the circle with centre at $(-1, 2)$ that passes through $(2, 3)$.

7. Find the equation of the straight line that passes through $(-4, 1)$ and $(2, -2)$.

8. Where is the focus of a parabolic reflector that is 12.0 cm across and 4.00 cm deep?

9. A hallway 6.0 m wide has a ceiling whose cross-section is a semi-ellipse. The ceiling is 3.0 m high at the walls and 4.0 m high at the centre. Find the height of the ceiling 1.0 m from each wall.

10. Plot the polar curve $r = 3 + \cos\theta$.

11. Find the centre and vertices of the conic section $4y^2 - x^2 - 4x - 8y - 4 = 0$. Show completely the sketch of the curve.

12. (a) What type of curve is represented by $8x^2 - 4xy + 5y^2 = 36$? (b) Through what angle must the curve in part (a) be rotated in order that there is no $x'y'$-term?

22. Introduction to Statistics

David P. Lewis/Shutterstock

▲ In Section 22.3, we see how statistical analysis was used in the design of the 12.9-km-long Confederation Bridge, which joins New Brunswick and Prince Edward Island.

LEARNING OUTCOMES

After completion of this chapter, the student should be able to:

- Understand the basic concepts of population, sample, parameter, statistic, and variable
- Construct frequency, relative frequency, and cumulative frequency tables for quantitative data
- Draw a histogram, a frequency polygon, and an ogive
- Use bar graphs and pie charts to display qualitative data
- Calculate measures of central tendency (mean, median, and mode) and measures of spread (range and standard deviation)
- Use Chebychev's theorem to draw conclusions about a data set
- Calculate relative frequencies using the normal distribution
- Construct large-sample confidence intervals for a population mean or for a population proportion
- Plot an $\bar{x}$ control chart, an R control chart, and a p control chart
- Find the equation of the least-squares line that best fits a given set of data
- Find the equation of a curve that best fits a given set of data by transforming the independent variable and using linear least squares

After the invention of the steam engine in the late 1700s by the Scottish engineer James Watt, the production of machine-made goods became widespread during the 1800s. However, it was not until the 1920s that much attention was paid to the quality control of the goods being produced. In 1924, Walter Shewhart of Bell Telephone Laboratories used a statistical chart for controlling product variables; in the 1940s, quality control was used in much of wartime production.

Quality control is one of the modern uses of *statistics*, the branch of mathematics in which data are collected, displayed, analysed, and interpreted. Today it is nearly impossible to read a newspaper or watch television news without seeing some type of study, in areas such as medicine and politics, that involves statistics. Other areas in which statistical methods are used include biology, physics, psychology, sociology, reliability engineering, actuarial science, economics, business, and education, to name but a few.

The first significant use of statistics was made in the 1660s by John Graunt, and in the 1690s by Edmund Halley (of Halley's Comet fame), when each published some conclusions about the population in England based on mortality tables. There was little development of statistics until the 1800s, when statistical measures became more widely used. For example, important contributions were made by the scientist Francis Galton, who used statistics in the study of human heredity, and by the nurse Florence Nightingale, who used statistical graphs to show that more soldiers died in the Crimean War (in the 1850s) from unsanitary conditions than from combat wounds.

In using statistics, we generally collect and summarize data (using methods from *descriptive statistics*) to make inferences based on those data (using methods from *inferential statistics*). The first two sections of this chapter are dedicated to descriptive statistics. After a discussion regarding the normal distribution, we dedicate the rest of the chapter to introducing some basic concepts of inferential statistics, including confidence intervals, statistical process control, and regression.

22.1 Tabular and Graphical Representation of Data

In statistics, a **population** is the complete collection of elements (people, DVDs, households, temperatures) that are of interest and about which information is desired. Typically, a researcher is interested in a numerical property of the population called a **parameter**. For example, if the population is all DVDs produced by a certain manufacturer over the course of a week, a parameter would be the proportion of defective DVDs in that lot. If the population is all Internet users, a parameter would be the average number of hours spent each week on the Internet.

Because of constraints on time, money, and other scarce resources, conclusions about the population are usually drawn after observing only a subset of the population, called a **sample**. Quantities computed from samples are called **statistics**. Statistics are used to estimate parameters in the population. For example, the average in a sample of Internet users can be used to estimate the average among all users. Similarly, the proportion of defectives in a sample can be used to estimate the proportion of defectives in the complete lot.

We are usually only interested in some of the characteristics that elements of the population have in common. A **variable** is any characteristic whose value changes from individual to individual in the population. A **quantitative variable** has a value that represents a numerical measurement. Examples of quantitative variables are weight, length, voltage, pressure, and number of children in a family. When the value of a variable is non-numerical, it is called a qualitative variable, or an **attribute**. Examples of attributes are colour, gender, and quality (measured as defective or nondefective).

Values of variables that have been recorded constitute data. *Data that have been collected but not yet organized are called* **raw data**. In order to obtain useful information from the data, it is necessary to organize it in some way. Often, a first step in organizing the data is to obtain its distribution. A **frequency distribution** summarizes all the values observed and their **frequencies**, that is, the number of times each value occurred in the data set.

SUMMARIZING AND GRAPHING QUANTITATIVE DATA

EXAMPLE 1 Distribution of exercise hours

In a sample of 50 university students, each student was asked to estimate carefully the number of hours they exercised per week, rounded to the nearest hour. Construct a frequency distribution of the data obtained:

3, 3, 8, 1, 4, 0, 4, 5, 0, 5, 10, 7, 0, 3, 2, 2, 4, 5, 5, 4, 12, 5, 5, 6, 16,
5, 4, 6, 0, 6, 14, 0, 10, 0, 4, 3, 3, 2, 5, 5, 2, 7, 2, 2, 8, 8, 7, 4, 1, 11

As we can see, no pattern is clear from the raw data. We arrange the possible observed values in numerical order and count their frequencies. The resulting distribution is shown in Table 22.1.

Table 22.1

Hours of Exercise per Week	Frequency
0	6
1	2
2	6
3	5
4	7
5	9
6	3
7	3
8	3
9	0
10	2
11	1
12	1
13	0
14	1
15	0
16	1

In Example 1, although a pattern is somewhat clearer from the table than from the raw data, a still clearer pattern can be found by *grouping the data*. Grouping is especially useful when working with large data sets and, although the raw data is lost, a much clearer understanding of the overall pattern of the data can usually be obtained.

The grouping of data is done by first dividing the number line into intervals of equal width (called **classes**), and then tabulating the frequency of each class. It is also possible to tabulate the **relative frequency** of each class, which is found by dividing the frequency by the total number of observations. This is illustrated in the following example.

EXAMPLE 2 Frequency distribution table

Obtain a frequency distribution and a relative frequency distribution with five classes for the exercise data in Example 1.

In order to obtain five classes of equal width, we form classes of 0–4 h, 4–7 h, and so on. We follow the convention that the *left endpoint, but not the right endpoint,* is included in the class. This gives us the following frequency distribution table of values, showing the number of students reporting the indicated number of hours of exercise in one week. Below the frequency we report the relative frequency, where the frequency has been divided by 50, the total number of students surveyed.

Hours of Exercise	0–4	4–8	8–12	12–16	16–20	Total
Frequency	19	22	6	2	1	50
Relative Frequency (%)	38	44	12	4	2	100

In this table, the values 0, 4, 8, 12, and 16 are the *lower class limits*, and 4, 8, 12, 16, and 20 are the *upper class limits*. The difference between two consecutive lower class limits is the *class width* (here the class width is four).

We can see from this table that most students in the sample exercise less than 8 hours per week, but there are a few students who exercise a lot more than that. The pattern of hours of exercise has become clearer after grouping.

Guidelines for Constructing Frequency Tables

- Make sure that classes are mutually exclusive, so that each observation belongs to one, and only one, class.
- Use between 5 and 20 classes. The principal consideration is that the relevant characteristics of the data should be clear.
- Ensure that all classes (except for open-ended classes) have the same width.
- Use class limits with convenient numbers. The first lower class limit is selected as the lowest value, or as a convenient number less than the lowest value.
- Be sure to include all classes, even if their frequency is zero.

Practice Exercise

1. Assuming the data in Example 1 are divided into classes of 0–3, 3–6 h, etc., for the 6–9 h class find:
   (a) the frequency
   (b) the relative frequency

■ A graphing utility may be used to display histograms and frequency polygons.

In order to represent *grouped* data graphically, where the raw data values are generally not all the same within a given class, we find it necessary to use a representative value for each class. For this, we use the **class mark**, *which is found by dividing the sum of the lower and upper class limits by* 2.

Using graphs is a very convenient method of visualizing distributions. There are several useful types of graphs for quantitative data. *Among the most important of these are the* **histogram** *and the* **frequency polygon**.

A histogram places classes on the horizontal axis. Each class is represented by a rectangle with height proportional to the frequency (or relative frequency) of the class. Each rectangle is usually labelled at the centre of its base by the *class mark*.

EXAMPLE 3 Histogram

Draw a histogram to represent the frequency distribution table on students' hours of exercise in Example 2.

From this table, the class marks are $(0 + 4)/2 = 2$, $(4 + 8)/2 = 6$, and so on. Using these values, a histogram representing these data is shown in Fig. 22.1.

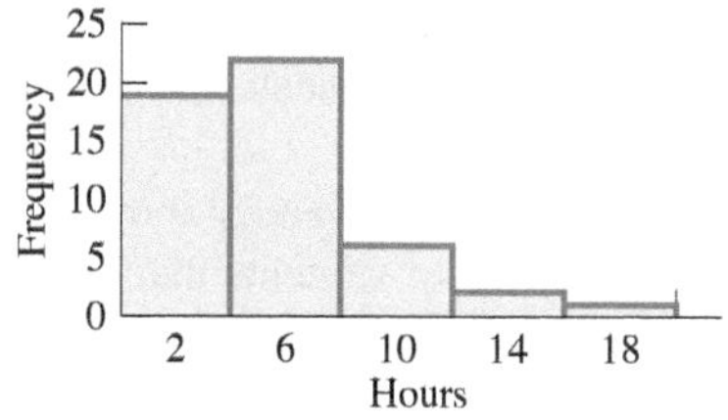

Fig. 22.1

■ Computer spreadsheets are very useful for this type of analysis.

A frequency polygon is used to represent a set of data by plotting the class marks as abscissas (x-values) and the frequencies as ordinates (y-values). The resulting points are joined by straight-line segments.

EXAMPLE 4 Frequency polygon

Draw a frequency polygon to represent the frequency distribution table on students' hours of exercise in Example 2.

Using the same class marks as Example 3, we obtain the frequency polygon in Fig. 22.2. Note how we have completed the polygon by connecting the first and last line segments to the adjacent (empty) class mark.

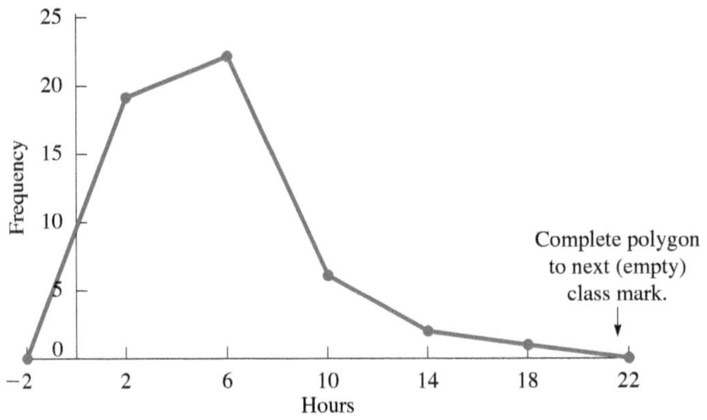

Fig. 22.2

Another way of analysing data is to use *cumulative* totals. The way this is generally done is to change the frequency into a "less than" **cumulative frequency**. To do this, we *add the class frequencies,* starting at the lowest class boundary. The graphical display that is generally used for cumulative frequency is called an **ogive** (pronounced oh-jive).

EXAMPLE 5 Cumulative frequency—ogive

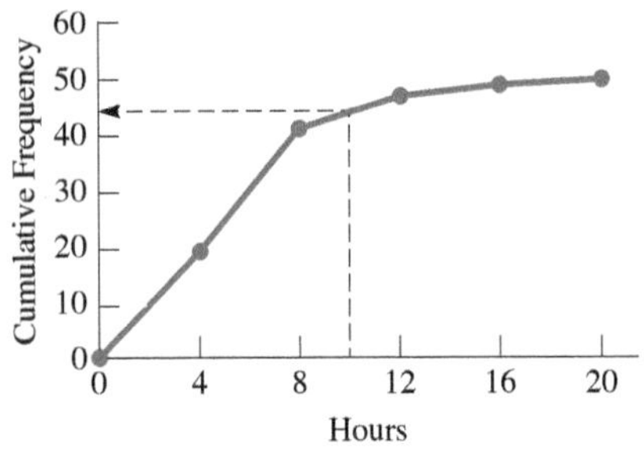

Fig. 22.3

Draw an ogive for the data on hours of exercise in Example 2.

The cumulative frequency is shown in the following table:

Hours of Exercise per Week	*Cumulative Frequency*
Less than 4	19
Less than 8	41
Less than 12	47
Less than 16	49
Less than 20	50

The ogive showing the cumulative frequency for the values in this table is shown in Fig. 22.3. The vertical scale shows the *frequency*, and the horizontal scale shows the *class boundaries.*

One important use of an ogive is to determine the number of values above or below a certain value. For example, to *approximate* the number of students that exercise *less than* 10 hours per week, we draw a line from the horizontal axis to the ogive and then to the vertical axis as shown in Fig. 22.3. From this, we see that *about* 43 students exercise *less than* 10 hours per week.

GRAPHING QUALITATIVE DATA

The two most common graphs used to display qualitative data are **bar graphs** and **pie charts**. Pie charts show the proportion of observations in each category, represented by a wedge in the pie, with area proportional to the relative frequency of the category. Because it is difficult to judge how much area is taken up by the slices of a pie chart, especially in the presence of many categories, bar graphs are usually preferred by statisticians for use in scientific settings. Moreover, only bar graphs can be used when observations fall in more than one category and relative frequencies add up to more than 100%.

The following two examples illustrate these graphs.

EXAMPLE 6 Bar graph—communication habits of high school students

In a survey of 1000 high school students, students were asked which type of communication they use daily to talk with friends and family. The table of frequencies and relative frequencies is shown below. Draw a relative frequency bar graph for these data.

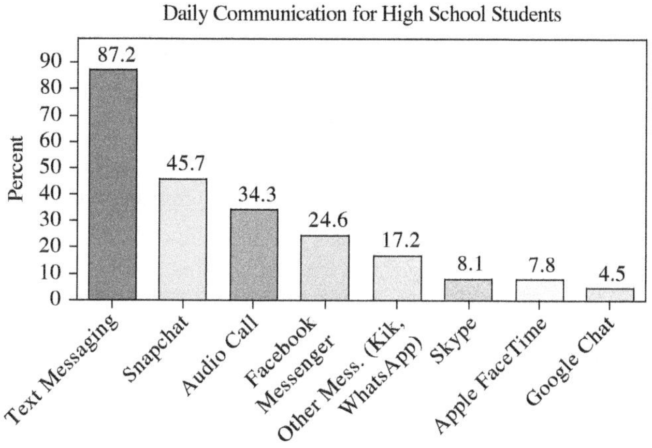

Fig. 22.4

Form of Communication	*Frequency*	*Relative Frequency (%)*
Text messaging	872	87.2
Snapchat	457	45.7
Audio call	343	34.3
Facebook Messenger	246	24.6
Other messaging (Kik, WhatsApp)	172	17.2
Skype	81	8.1
Apple FaceTime	78	7.8
Google Chat	45	4.5

The bar graph is shown in Fig 22.4. Note that, in contrast with a histogram, bars in this graph do not touch. Moreover, although we have chosen to order categories according to descending relative frequency, other choices for ordering categories are possible. Because a single student could use several forms of communication, relative frequencies add up to more than 100% and a pie chart *cannot* be used to represent these data.

EXAMPLE 7 Pie chart—market share among top video game consoles

For a certain month, the frequencies and relative frequencies of sales (in units sold) by the three leading video game consoles were tabulated as shown below. Draw a pie chart for these data.

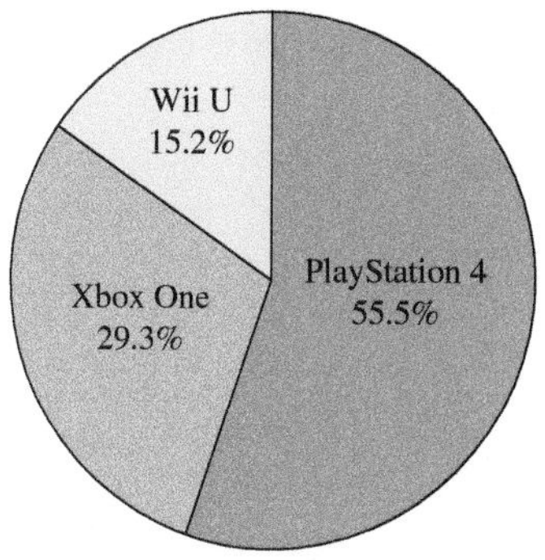

Fig. 22.5

Console	*Frequency*	*Relative Frequency (%)*
PlayStation 4	601 529	55.5
Xbox One	317 421	29.3
Wii U	167 702	15.2
Total	1 083 652	100

The pie chart is shown in Fig. 22.5. Note that use of the pie chart is appropriate here since the percentages add up to 100%.

EXERCISES 22.1

In Exercises 1–4, divide the data in Example 1 into six classes of hours (0–3, 3–6, etc.) of exercise per week and then do the following:

1. Form a frequency distribution table.
2. Find the relative frequencies.
3. Draw a histogram.
4. Draw an ogive.

In Exercises 5–8, indicate whether the variable is qualitative or quantitative.

5. The diameter of a bolt
6. A person's favourite genre of music
7. Whether or not a product passes inspection
8. The time it takes a worker to complete a task

In Exercises 9–16, use the following data. An automobile company tested a new engine, and found the following results in 20 tests of the number of litres of gasoline used by a certain model for each 100 km travelled.

5.3, 5.8, 5.6, 5.4, 5.9, 5.4, 6.0, 5.8, 5.8, 5.4,
6.3, 5.6, 5.7, 5.6, 5.7, 5.9, 5.5, 6.1, 5.9, 5.8

9. Form an ordered array and summarize it by finding the frequency of each number of litres used.
10. Find the relative frequency of each number of litres used.
11. Form a frequency distribution table with five classes.
12. Find the relative frequencies for the data in the frequency distribution table in Exercise 11.
13. Draw a histogram for the table of Exercise 11.
14. Draw a frequency polygon for the table of Exercise 11.
15. Form a cumulative frequency table for the table of Exercise 11.
16. Draw an ogive for the data of Exercise 11.

In Exercises 17–20, use the following data. In a random sample, 30 Android users were asked to record the number of apps that were installed on their phone. The resulting data are shown below:

112, 91, 101, 85, 76, 115, 93, 126, 78, 86, 105, 107, 58, 86, 109,
111, 103, 105, 97, 110, 92, 95, 107, 89, 101, 67, 103, 99, 93, 82

17. Form a frequency distribution table for these values using the class limits 50, 60, 70, . . . , 130.
18. For the table of Exercise 17, draw a histogram.
19. For the data of Exercise 17, draw a frequency polygon.
20. For the data of Exercise 17, form a relative frequency distribution table.

In Exercises 21–28, answer the given questions.

21. In testing a braking system, the distance required to stop a car from 110 km/h was measured in 120 trials. The results are shown in the following distribution table:

Stopping Distance (m)	47–49	50–52	53–55	56–58
Times Car Stopped	2	15	32	36

Stopping Distance (m)	59–61	62–64	65–67
Times Car Stopped	24	10	1

Form a relative frequency distribution table for these data.

22. For the data in Exercise 21, form a cumulative frequency distribution table.
23. For the data of Exercise 21, draw an ogive.
24. From the ogive in Exercise 23, estimate the number of cars that stopped in less than 57 m.
25. The dosage, in millisieverts (mSv), given by a particular X-ray machine, was measured 20 times, with the following readings:

0.425, 0.436, 0.396, 0.421, 0.444, 0.383, 0.437, 0.427, 0.433, 0.434, 0.415, 0.390, 0.441, 0.451, 0.418, 0.426, 0.429, 0.409, 0.436, 0.423

Form a histogram with six classes and the lowest class mark of 0.380 mSv.

26. For the data used for the histogram in Exercise 25, draw a frequency polygon.
27. The life of a certain type of battery was measured for a sample of batteries with the following results (in number of hours):

34, 30, 32, 35, 31, 28, 29, 30, 32, 25, 31, 30,
28, 36, 33, 34, 30, 31, 34, 29, 30, 32

Draw a frequency polygon using six classes.

28. For the data in Exercise 27, draw a cumulative frequency distribution table using six classes.

In Exercises 29–32, use the following data. In a random sample, 500 college students were asked which social networks they use on a daily basis. The results are summarized below:

Social Network	*Frequency*
Facebook	305
Instagram	255
Twitter	175
Google +	115
Pinterest	80
Vine	80

29. Make a bar graph of these data showing frequency on the vertical axis.
30. Find the relative frequencies for each social network.
31. Is it appropriate to use a pie chart for these data? Explain why or why not.
32. Make a bar graph of these data showing relative frequency on the vertical axis.

Answer to Practice Exercise

1. (a) 9 (b) 18%

22.2 Summarizing Data

Tables and graphical representations give a general description of data. However, it is also useful and convenient to find representative values for the location of the centre of the distribution, and other numbers to give a measure of the deviation from this central value. In this way, we can obtain a numerical summary of the data. We study some measures of centre and deviation (or spread) in this section.

> **LEARNING TIP**
> Note that all measures of central tendency and spread are usually rounded off to one more decimal place than was present in the original data.

MEASURES OF CENTRAL TENDENCY

The task of a *measure of central tendency* is to describe with a single value the location of the centre of the distribution. Since there are different ways of defining what centre is, there are several measures of central tendency.

The first of these measures of central tendency is the **median**. The median is the value that falls in the middle of an ordered data set, leaving as many observations above it as below it. If there is no middle observation, the median is the number halfway between the two numbers nearest to the middle of the ordered observations.

EXAMPLE 1 Median—odd or even number of values

Find the median of the numbers 5, 2, 6, 4, 7, 4, 7, 2, 8, 9, 4, 11, 9, 1, 3.

We first arrange the numbers in numerical order. This arrangement is

$$1, 2, 2, 3, 4, 4, 4, \underset{\text{middle number}}{5}, 6, 7, 7, 8, 9, 9, 11$$

Since there are 15 numbers, the middle number is the eighth. Since the eighth number is 5, the median is 5.

If the number 11 is not included in this set of numbers and there are only 14 numbers in all, the median is that number halfway between the seventh and eighth numbers. Since the seventh is 4 and the eighth is 5, the median is 4.5.

Another very widely applied measure of central tendency is the **arithmetic mean** (often referred to simply as the **mean**). *The mean is calculated by finding the sum of all the values and then dividing by the number of values.* (The mean is the number most people call the "average." However, in statistics the word *average* has the more general meaning of a measure of central tendency.)

EXAMPLE 2 Arithmetic mean

Find the arithmetic mean of the numbers given in Example 1.

We find the sum of all the numbers and then divide by 15. Therefore, by letting $\bar{x}$ (read as "x bar") represent the mean, we have

$$\bar{x} = \frac{5 + 2 + 6 + 4 + 7 + 4 + 7 + 2 + 8 + 9 + 4 + 11 + 9 + 1 + 3}{15}$$

$$= \frac{82}{15} = 5.5$$

(82: sum of values; 15: number of values)

Thus, the mean is 5.5.

If we wish to find the arithmetic mean of a large number of values, and if some of them appear more than once, the calculation can be simplified. The mean can be calculated by multiplying each value by its frequency, adding these results, and then dividing by the total number of values (the sum of the frequencies). Letting $\bar{x}$ represent the mean of the values $x_1, x_2, \ldots, x_n$, which occur with frequencies $f_1, f_2, \ldots, f_n$, respectively, we have

■ This is called a *weighted mean* since each value is given a weight based on the number of times it occurs.

$$\bar{x} = \frac{x_1 f_1 + x_2 f_2 + \cdots + x_n f_n}{f_1 + f_2 + \cdots + f_n} \qquad \textbf{(22.1)}$$

EXAMPLE 3 Arithmetic mean using frequencies

Use Eq. (22.1) to find the arithmetic mean of the numbers of Example 1.

We first set up a table of values and their respective frequencies, as follows:

Values	1	2	3	4	5	6	7	8	9	11
Frequency	1	2	1	3	1	1	2	1	2	1

We now calculate the arithmetic mean $\bar{x}$ by using Eq. (22.1):

multiply each value by its frequency and add results

$$\bar{x} = \frac{1(1) + 2(2) + 3(1) + 4(3) + 5(1) + 6(1) + 7(2) + 8(1) + 9(2) + 11(1)}{1 + 2 + 1 + 3 + 1 + 1 + 2 + 1 + 2 + 1}$$

sum of frequencies

$$= \frac{82}{15} = 5.5$$

We see that this agrees with the result of Example 2.

Summations such as those in Eq. (22.1) occur frequently in statistics and other branches of mathematics. In order to simplify writing these sums, the symbol Σ is used to indicate the process of summation. (Σ is the Greek capital letter sigma.) Σx means the sum of the x's.

EXAMPLE 4 Summation symbol Σ

We can show the sum of the numbers $x_1, x_2, x_3, \ldots, x_n$ as

$$\sum x = x_1 + x_2 + x_3 + \cdots + x_n$$

If these numbers are 3, 7, 2, 6, 8, 4, and 9, we have

$$\sum x = 3 + 7 + 2 + 6 + 8 + 4 + 9 = 39$$

Practice Exercises

For the following numbers, find the indicated value:
12, 17, 16, 12, 14, 18, 14, 12, 15, 18

1. The median **2.** The arithmetic mean

Using the summation symbol Σ, we can rewrite Eq. (22.1) for the arithmetic mean as

$$\bar{x} = \frac{x_1 f_1 + x_2 f_2 + x_3 f_3 + \cdots + x_n f_n}{f_1 + f_2 + f_3 + \cdots + f_n} = \frac{\sum xf}{\sum f} \qquad \textbf{(22.1)}$$

The summation notation Σx is an abbreviated form of the more general notation $\sum_{i=1}^{n} x_i$, which represents $x_1 + x_2 + x_3 + \cdots + x_n$.

EXAMPLE 5 Arithmetic mean using frequencies—smart devices

A sample of 50 people were asked how many smart devices they owned (including smartphones, tablets, and smartwatches). The grouped responses are shown below:

Number of Devices, x	0	1	2	3	4
Frequency, f	6	21	16	5	2

Find the mean using Eq. (22.1).

$$\bar{x} = \frac{\sum xf}{\sum f} = \frac{0(6) + 1(21) + 2(16) + 3(5) + 4(2)}{50}$$

$$= \frac{76}{50} = 1.5 \text{ smart devices} \quad \text{(rounded off to tenths)}$$

Another measure of central tendency is *the* **mode**, *which is the value that appears most frequently.* If two or more values appear with the same greatest frequency, each is a mode. If no value is repeated, there is no mode.

EXAMPLE 6 Mode

Find the mode of each data set.

(a) 1, 2, 2, 3, 4, 4, 4, 5, 6	4 appears more times than any other value.	Mode = 4
(b) 1, 2, 2, 2, 4, 4, 5, 5, 5	2 and 5 appear most, the same number of times.	Modes = 2, 5
(c) 1, 2, 4, 5, 6 , 7, 9	No value is repeated.	There is no mode.

LEARNING TIP

- The mean is useful for many statistical methods and is used extensively. Nevertheless, be aware that the mean is very sensitive to extreme observations so that a single extreme observation can change the value of the mean dramatically and give the wrong impression about the data.
- The median is not affected by extreme observations. Therefore, it is a good choice as a measure of centre in the presence of extreme values.
- The mean, the median, and the mode coincide when the distribution of data is symmetric. In those cases, the median or the mode (which are easy to calculate) can be used as estimates of the mean.

EXAMPLE 7 Measures of central tendency—force of friction

To find the frictional force between two specially designed surfaces, the force to move a block with one surface along an inclined plane with the other surface is measured 10 times. The results, with forces in newtons, are

2.2, 2.4, 2.1, 2.2, 2.5, 2.2, 2.4, 2.7, 2.1, 2.5

Find the mean, median, and mode of these forces.

To find the mean, we sum the values of the forces and divide this total by 10. This gives

$$\bar{F} = \frac{\sum F}{10} = \frac{2.2 + 2.4 + 2.1 + 2.2 + 2.5 + 2.2 + 2.4 + 2.7 + 2.1 + 2.5}{10}$$

$$= \frac{23.3}{10} = 2.33\ N$$

The median is found by arranging the values in order and finding the middle value. The values in order are

2.1, 2.1, 2.2, 2.2, 2.2, 2.4, 2.4, 2.5, 2.5, 2.7

Since there are 10 values, we see that the fifth value is 2.2 and the sixth is 2.4. The value midway between these is 2.3, which is the median. Therefore, the median force is 2.3 N.

The mode is 2.2 N, since this value appears three times, which is more than any other value.

MEASURES OF SPREAD

Measures of central tendency on their own are not very informative. They do not tell us whether values are grouped closely together or how spread out they are. Therefore, we also need some measure of the deviation, or spread, of the values from the centre. If the spread is small and the numbers are grouped closely together, the measure of central tendency is more reliable and descriptive of the data than in the case in which the spread is greater.

In statistics, there are several measures of spread that may be defined. The simplest one is the **range**, which is the difference between the highest value and the lowest value in the data set. For example, the range of the data in Example 7 is $2.7 - 2.1 = 0.6$. We will see how the range is applied to statistical process control in Section 22.5.

The most widely used measure of spread is the *standard deviation.* The **standard deviation** *of a set of* **sample** *values is defined by the equation*

$$s = \sqrt{\frac{\sum (x - \bar{x})^2}{n - 1}} \quad (22.2)$$

The definition of s shows that the following steps are used in computing its value.

LEARNING TIP

The standard deviation s is a positive number. It is a *deviation from the mean,* regardless of whether the individual numbers are greater than or less than the mean. Numbers close together will have a small standard deviation, whereas numbers further apart have a larger standard deviation. Therefore, *the standard deviation becomes larger as the spread of data increases.*

Steps for Calculating Standard Deviation

1. Find the arithmetic mean $\bar{x}$ of the numbers of the set.
2. Subtract the mean from each number of the set.
3. Square these differences.
4. Find the sum of these squares.
5. Divide this sum by $n - 1$.
6. Find the square root of this result.

Following the steps shown above, we use Eq. (22.2) for the calculation of standard deviation in the following examples.

LEARNING TIP

The *population standard deviation,* represented by the Greek letter σ (read "sigma"), is computed using n in the denominator of Eq. (22.2) instead of $n - 1$. The $n - 1$ in the denominator of Eq. (22.2) adjusts s so that it gives good estimates of the parameter σ when the standard deviation can only be measured from a sample.

Since we generally use data coming from samples, in this text we will always use Eq. (22.2), and we will refer to the sample standard deviation simply as the standard deviation.

EXAMPLE 8 Standard deviation—using Eq. (22.2)

Find the standard deviation of the following numbers: 1, 5, 4, 2, 6, 2, 1, 1, 5, 3.

A table of the necessary values is shown below, and steps 1–6 are indicated:

	x	$x - \bar{x}$ (step 2)	$(x - \bar{x})^2$ (step 3)
	1	−2	4
	5	2	4
	4	1	1
	2	−1	1
	6	3	9
	2	−1	1
	1	−2	4
	1	−2	4
	5	2	4
	3	0	0
Sum	30	0	32 (step 4)

$$\bar{x} = \frac{30}{10} = 3 \quad \text{step 1}$$

$$\frac{\sum (x - \bar{x})^2}{n - 1} = \frac{32}{10 - 1} = \frac{32}{9} \quad \text{step 5}$$

$$s = \sqrt{\frac{32}{9}} = 1.9 \quad \text{step 6 (rounded off to tenths)}$$

■ Note that the sum of differences is zero, which is why Eq. (22.2) uses the sum of squared differences.

If some of the values in the data are repeated, we can use the frequency of those values that occur more than once in calculating the standard deviation. This is illustrated in the following example.

It is possible to reduce the computational work required to find the standard deviation. Algebraically, it can be shown (although we will not do so here) that the following equation is another form of Eq. (22.2) and therefore gives the same results.

$$s = \sqrt{\frac{n\left(\sum x^2\right) - \left(\sum x\right)^2}{n(n - 1)}} \quad (22.3)$$

Although the form of this equation appears more involved, it does reduce the amount of calculation that is necessary. Consider the following examples.

EXAMPLE 9 Standard deviation—using Eq. (22.3)

Using Eq. (22.3), find s for the numbers in Example 8.

	x	x^2
	1	1
	5	25
	4	16
	2	4
	6	36
	2	4
	1	1
	1	1
	5	25
	3	9
Sum	30	122

$$n = 10$$
$$\Sigma x^2 = 122$$
$$(\Sigma x)^2 = 30^2 = 900$$
$$s = \sqrt{\frac{10(122) - 900}{10(9)}} = 1.9$$

Practice Exercise

3. Find the standard deviation of the first eight numbers in Example 8.

COMMON ERROR It is a common error to confuse Σx^2 and $(\Sigma x)^2$ in Eq. (22.3). Note that for Σx^2, we square the x values and then add the squares, whereas for $(\Sigma x)^2$, we first add the x values and then square the sum.

EXAMPLE 10 Standard deviation of resistance

An ammeter measures the electric current in a circuit. In an ammeter, two resistances are connected in parallel, with most of the current passing through a very low resistance called the *shunt*. The resistance of each shunt in a sample of 100 shunts was measured. The results were grouped, and the class mark and frequency for each class are shown in the following table. Calculate the arithmetic mean and the standard deviation of the resistances of the shunts.

■ Statistical measures such as $\bar{x}$, Σx, Σx^2, s_x, σ_x, and n can be obtained directly with technology, either with a graphing or scientific calculator, a spreadsheet, or statistical software.

R (ohms)	f	Rf	R^2f
0.200	1	0.200	0.0400
0.210	3	0.630	0.1323
0.220	5	1.100	0.2420
0.230	10	2.300	0.5290
0.240	17	4.080	0.9792
0.250	40	10.000	2.5000
0.260	13	3.380	0.8788
0.270	6	1.620	0.4374
0.280	3	0.840	0.2352
0.290	2	0.580	0.1682
	100	24.730	6.1421

$$\overline{R} = \frac{24.730}{100} = 0.2473\ \Omega$$
$$n = 100$$
$$\Sigma R^2 = 6.1421$$
$$(\Sigma R)^2 = 24.730^2$$
$$s = \sqrt{\frac{100(6.1421) - 24.730^2}{100(99)}} = 0.0163$$

The arithmetic mean of the resistances is 0.2473 Ω, with a standard deviation of 0.0163 Ω.

The mean and the standard deviation together can help us draw conclusions about the values in a data set. Thanks to a result known as **Chebychev's theorem**, we can state the percentage of data values that must be within a specific number of standard deviations from the mean.

Chebychev's Theorem
For **any** data set (population or sample), the proportion of observations that must be within k standard deviations of the mean is always at least $1 - \frac{1}{k^2}$ $(k > 1)$.

For the most common particular values of k, this is what Chebychev's theorem implies:

k	$1 - 1/k^2$	*Implication on the percentage of observations*
2	$1 - \frac{1}{2^2} = 0.75$	At least 75% are within two standard deviations of the mean
3	$1 - \frac{1}{3^2} = 0.89$	At least 89% are within three standard deviations of the mean
4	$1 - \frac{1}{4^2} = 0.94$	At least 94% are within four standard deviations of the mean

Note that since Chebychev's theorem is so general, it will underestimate the percentages for some distributions. In Section 22.3, we will obtain more precise percentages for the important case of the normal distribution.

EXAMPLE 11 Chebychev's theorem—computer malfunctions

A sample of computers of a certain brand had a mean time of 38 months without a hardware malfunction, with a standard deviation of 2.5 months. What percentage of the computers in the sample lasted between 33 and 43 months without a hardware malfunction?

We can write $33 = 38 - 2(2.5)$, and $43 = 38 + 2(2.5)$, so 33 and 43 are two standard deviations away from the mean, and we use Chebychev's theorem with $k = 2$. Therefore, *at least* 75% of the computers in the sample lasted between 33 and 43 months without a hardware malfunction.

In using the statistical measures we have discussed, we must be careful in using and interpreting such measures. Consider the following example.

EXAMPLE 12 Interpreting statistical measures

(a) The numbers 1, 2, 3, 4, 5 have a mean of 3, a median of 3, and a standard deviation of 1.6. These values fairly well describe the centre and distribution of the numbers in the set.

(b) The numbers 1, 2, 3, 4, 100 have a mean of 22, a median of 3, and a standard deviation of 44. The large difference between the median and the mean and the very large range of values within one standard deviation of the mean (−22 to 66) indicate that this set of measures does not describe this set of numbers well. In a case like this, the 100 should be checked to see if it is in error.

Example 12 illustrates how statistical measures can be misleading in the presence of extreme values. Misleading statistics can also come from the process of data collection. Consider the probable results of a survey to find the percent of persons in favour of raising income taxes for the wealthy if the survey is taken at the entrance to a welfare office or if it is taken at the entrance to a stock brokerage firm. There are many other considerations in the proper use and interpretation of statistical measures.

EXERCISES 22.2

In Exercises 1–4, delete the 5 from the data numbers given for Example 1 and then do the following with the resulting data.

1. Find the median.
2. Find the arithmetic mean using the definition, as in Example 2.
3. Find the arithmetic mean using Eq. (22.1), as in Example 3.
4. Find the mode, as in Example 6.

In Exercises 5 and 6, use the data in Example 8. Change the first 1 to 6 and the first 2 to 7 and then find the standard deviation of the resulting data as directed. In Exercise 7, answer the given question.

5. Find s from the definition, as in Example 8.
6. Find s using Eq. 22.3, as in Example 9.
7. In Example 11, change 33 to 28 and 43 to 48 and then find the percentage.

In Exercises 8–15, use the following sets of numbers.

A: 3, 6, 4, 2, 5, 4, 7, 6, 3, 4, 6, 4, 5, 7, 3

B: 25, 26, 23, 24, 25, 28, 26, 27, 23, 28, 25

C: 0.48, 0.53, 0.49, 0.45, 0.55, 0.49, 0.47, 0.55, 0.48, 0.57, 0.51, 0.46, 0.53, 0.50, 0.49, 0.53

D: 105, 108, 103, 108, 106, 104, 109, 104, 110, 108, 108, 104, 113, 106, 107, 106, 107, 109, 105, 111, 109, 108

In Exercises 8–11, determine (a) the mean, (b) the median, and (c) the mode of the numbers of the given set.

8. Set *A*
9. Set *B*
10. Set *C*
11. Set *D*

In Exercises 12–15, find the standard deviation s for the indicated set of numbers (a) using Eq. (22.2), and (b) using Eq. (22.3).

12. Set *A*
13. Set *B*
14. Set *C*
15. Set *D*

In Exercises 16–26, the required data are those in the specified exercises of Section 22.1.

16. Find the mean, the median, and the mode of L/100 km of fuel usage in Exercise 9.
17. Find the standard deviation of L/100 km of fuel usage in Exercise 9 using Eq. (22.2).
18. Find the standard deviation of L/100 km of fuel usage in Exercise 9 using Eq. (22.3).
19. Find the mean, the median, and the mode of number of apps in Exercise 17.
20. Find the range and standard deviation of number of apps in Exercise 17.
21. Find the mean and median of stopping distances in Exercise 21. (Use the class mark for each class.)
22. Find the standard deviation of stopping distances in Exercise 21. (Use the class mark for each class.)
23. Find the mean, the median, and the mode of X-ray dosages in Exercise 25.
24. Find the range and the standard deviation of X-ray dosages in Exercise 25.
25. Find the mean, the median, and the mode of battery lives in Exercise 27.
26. Find the standard deviation of battery lives in Exercise 27.

In Exercises 27–43, find the indicated measure of central tendency or of spread.

27. The weekly salaries (in dollars) for the workers in a small factory are as follows:

    600, 750, 625, 575, 525, 700, 550,
    750, 625, 800, 700, 575, 600, 700

    Find the median and the mode of the salaries.
28. Find the mean salary for the salaries in Exercise 27.
29. Find the range and the standard deviation of the salaries in Exercise 27.
30. In a particular month, the electrical usage, rounded to the nearest 400 MJ, of 1000 homes in a certain city, was summarized as follows:

Usage	2000	2400	2800	3200	3600	4000	4400	4800
No. Homes	22	80	106	185	380	122	90	15

    Find the mean of the electrical usage.
31. Find the median and mode of electrical usage in Exercise 30.
32. Find the standard deviation of electrical usage in Exercise 30.
33. A test of air pollution in a city gave the following readings of the concentration of sulfur dioxide (in parts per million) for 18 consecutive days:

    0.14, 0.18, 0.27, 0.19, 0.15, 0.22, 0.20, 0.18, 0.15,
    0.17, 0.24, 0.23, 0.22, 0.18, 0.32, 0.26, 0.17, 0.23

    Find the median and the mode of these readings.
34. Find the mean of the readings in Exercise 33.
35. Find the range and the standard deviation of air pollution data in Exercise 33.
36. The following data give the mean number of days of rain for Vancouver, British Columbia, for the 12 months of the year.

    20, 17, 17, 14, 12, 11, 7, 8, 9, 16, 19, 22

    Find the standard deviation.
37. The *midrange*, another measure of central tendency, is found by finding the sum of the lowest and the highest values and dividing this sum by 2. Find the midrange of the salaries in Exercise 27.
38. Find the midrange of the sulfur dioxide readings in Exercise 33. (See Exercise 37.)
39. Add \$100 to each of the salaries in Exercise 27. Then find the median, mean, and mode of the resulting salaries. State any conclusion that might be drawn from the results.
40. Multiply each of the salaries in Exercise 27 by 2. Then find the median, mean, and mode of the resulting salaries. State any conclusion that might be drawn from the results.
41. Change the final salary in Exercise 27 to \$4000, with all other salaries being the same. Then find the mean of these salaries. State any conclusion that might be drawn from the result. (The \$4000 here is called an *outlier,* which is an extreme value.)
42. Find the median and mode of the salaries indicated in Exercise 41. State any conclusion that might be drawn from the results.

43. The *k% trimmed mean* is a measure of central tendency that avoids the influence of extreme observations while still using most of the observations in the data set. It is computed by finding the mean of the data after the smallest $k\%$ and the largest $k\%$ of the data have been discarded. Find the 10% trimmed mean of X-ray dosages in Exercise 25 of Section 22.1.

In Exercises 44–46, solve the given problems.

44. Use Chebychev's theorem to find the percentage of values that are between 175 and 195 in a data set with mean 185 and standard deviation 5.

45. Use Chebychev's theorem to find the percentage of values that are between 55.7 and 68.3 in a data set with mean 62 and standard deviation 2.1.

46. The mean compressive strength of a sample of steel beams was 40 000 N/cm^2, with a standard deviation of 450 N/cm^2. What percent of the beams had compressive strength between 38 650 and 41 350 N/cm^2?

Answers to Practice Exercises

1. 14.5 2. 14.8 3. 2.0

22.3 Normal Distributions

In this section we discuss the **normal distribution**, the most important and most widely used distribution in statistics. The normal distribution is a continuous distribution, so we begin by discussing some generalities of continuous distributions.

In Section 22.1 we learned that for variables that take a limited number of values, the relative frequency of a value is obtained by dividing the frequency of that value by the total frequency of all values. Let us now consider variables that can be regarded as having an infinite number of possible values, such as weights, lengths, or durations for a very large population. (Such variables are said to be *continuous*.) When using the same procedure as before, the denominator becomes infinite, giving a relative frequency of zero for all values. How are we then to compute the relative frequency of intervals, if the relative frequency of all values is zero?

The answer lies in establishing a correspondence between relative frequency and *area*. To each continuous variable we associate a function, which we can graph as a curve on the plane. The relative frequency of a particular interval corresponds to the area under the curve in that interval. Because the relative frequency for the complete population must be 1, the total area under the curve must be 1 (100% of the data).

■ The first derivation of the normal distribution is due to Abraham de Moivre (1667–1754), who was interested in approximating quantities arising in gambling problems. It was also derived independently by Pierre-Simon Laplace (1749–1847), and by Carl Friedrich Gauss (1777–1855), both in the context of measurement errors.

The **normal distribution** is associated with the symmetric, bell-shaped curve shown in Fig. 22.6. Using advanced methods, its equation is found to be

$$y = \frac{e^{-(x-\mu)^2/2\sigma^2}}{\sigma\sqrt{2\pi}} \qquad (22.4)$$

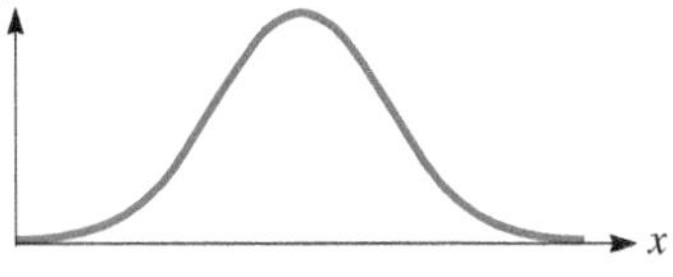

Fig. 22.6

Here, μ is the *population mean* and σ is the *population standard deviation,* and π and e are the familiar numbers first used in Chapters 2 and 12, respectively.

From Eq. (22.4), we can see that any particular normal distribution depends on the values of μ and σ. The horizontal location of the curve depends on μ, and the shape (how spread out the curve is) depends on σ, but the bell shape remains. This is illustrated in general in the following example.

EXAMPLE 1 Location and spread of a normal distribution

In Fig. 22.7, for the left curve, $\mu = 10$ and $\sigma = 5$, whereas for the right curve, $\mu = 20$ and $\sigma = 10$.

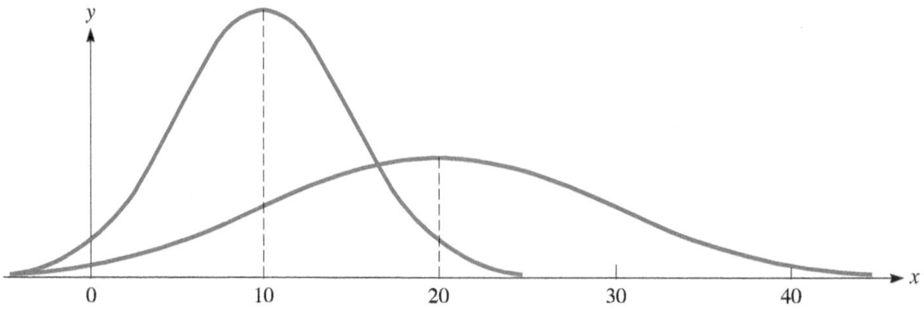

Fig. 22.7

EXAMPLE 2 Normal distribution—reliability in bridge design

■ See the chapter introduction.

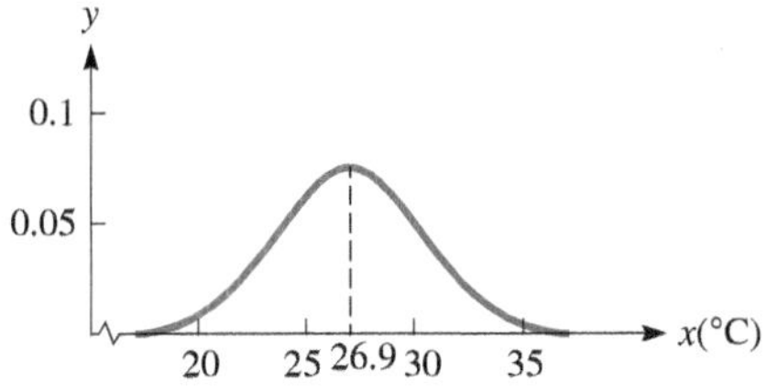

Fig. 22.8

The normal distribution has important applications in probabilistic reliability techniques for bridge design. For example, during the design of the Confederation Bridge joining New Brunswick and Prince Edward Island, it was concluded that the three-day temperature drop that was equalled or exceeded 100 times in 100 years is distributed normally, with mean 26.9°C and standard deviation 3.2°C. This distribution is shown in Fig. 22.8.

The normal distribution also had a role in the analysis of dead loads, live loads due to vehicles, wind loads, and ice loads for this bridge. (Source: J. G. MacGregor et al., "Design criteria and load and resistance factors for the Confederation Bridge," *Can. J. Civ..Eng.*, 24, 882–897 (1997).)

Properties of the Normal Curve

- The curve is symmetric about the mean.
- The curve is always above the x-axis. (y is always positive.)
- The x-axis is a horizontal asymptote. (As x increases numerically, y becomes very small.)
- The total area under the curve is 1.
- The curve is bell-shaped, as seen in Figs. 22.6–22.8.

■ A more complete and rigorous treatment of the material covered in this and the remaining sections of this chapter would require the study of probability theory, which is beyond the scope of this introductory chapter.

Fig. 22.9 shows areas under a normal curve with mean μ and standard deviation σ for particular regions. We can use the given information to find the percentage of data values that fall within one, two, and three standard deviations from the mean. This fact is called the **empirical rule** and is stated below.

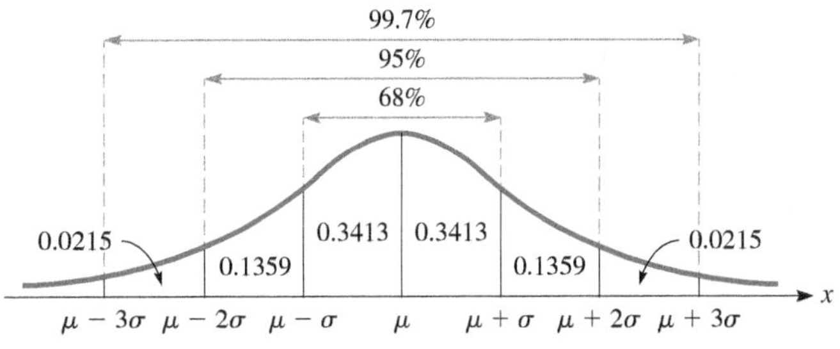

Fig. 22.9

Empirical Rule

Interval	*Percentage of Observations of the Normal Distribution*
$\mu - \sigma$ to $\mu + \sigma$	About 68% are within one standard deviation of the mean
$\mu - 2\sigma$ to $\mu + 2\sigma$	About 95% are within two standard deviations of the mean
$\mu - 3\sigma$ to $\mu + 3\sigma$	Almost all (99.7%) are within three standard deviations of the mean

We can compare these percentages with those given by Chebychev's theorem in Section 22.2—namely, that at least 75% of observations are within two standard deviations from the mean and that at least 89% of observations are within three standard deviations from the mean. We see that Chebychev's theorem heavily underestimates the true percentages in the case of the normal distribution.

We can use the areas from Fig. 22.9 to calculate relative frequencies for other intervals. This is illustrated in the following example.

EXAMPLE 3 Relative frequencies and areas under the curve—reliability

Consider the normal distribution of three-day temperature drops from Example 2, so that $\mu = 26.9°C$ and $\sigma = 3.2°C$. Find the percentage of values that lie between one standard deviation below the mean and two standard deviations above the mean. State the corresponding interval.

The area between $\mu - \sigma$ and $\mu + 2\sigma$ is $0.3413 + 0.3413 + 0.1359 = 0.8185$. Also,

$$\mu - \sigma = 26.9 - 3.2 = 23.7 \text{ and } \mu + 2\sigma = 26.9 + 2(3.2) = 33.3$$

Therefore, about 81.85% of values for this distribution are within 23.7°C and 33.3°C.

STANDARD NORMAL DISTRIBUTION

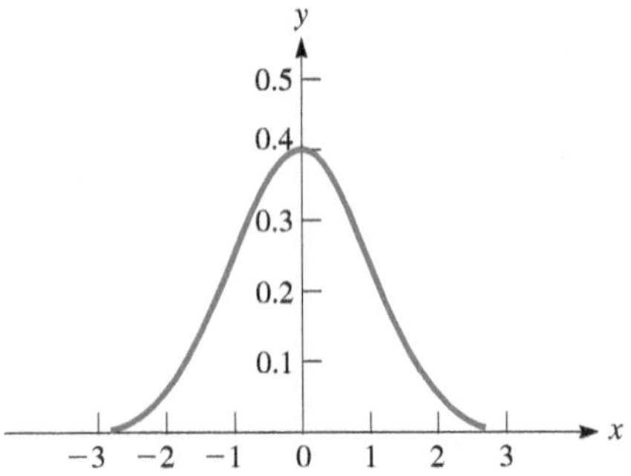

Fig. 22.10

As we have just seen, there are innumerable possible normal distributions. However, there is one of particular interest. *The* **standard normal distribution** *is the normal distribution for which the mean is* 0 *and the standard deviation is* 1. Making these substitutions in Eq. (22.4), we have

$$y = \frac{1}{\sqrt{2\pi}} e^{-x^2/2} \tag{22.5}$$

as the equation of the standard normal distribution curve. All the properties of a normal curve are satisfied by the standard normal distribution. In particular, since the mean is 0, the curve is symmetric with respect to the y-axis. The curve is also bell–shaped, as seen in Fig. 22.10. As we discuss below, areas under the standard normal curve are used to find relative frequencies for all other normal curves.

We can find the relative frequency of values for any normal distribution by use of the **standard score** z (or z-score), which is defined as

$$z = \frac{x - \mu}{\sigma} \tag{22.6}$$

For the standard normal distribution, where $\mu = 0$, if we let $x = \sigma$, then $z = 1$. If we let $x = 2\sigma$, $z = 2$. Therefore, we can see that *a value of z tells us the number of standard deviations the given value of x is above or below the mean.* From the discussion above, we can see that the value of z can tell us the area under the curve between the mean and the value of x corresponding to that value of z. In turn, *this tells us the relative frequency of all values between the mean and the value of x.*

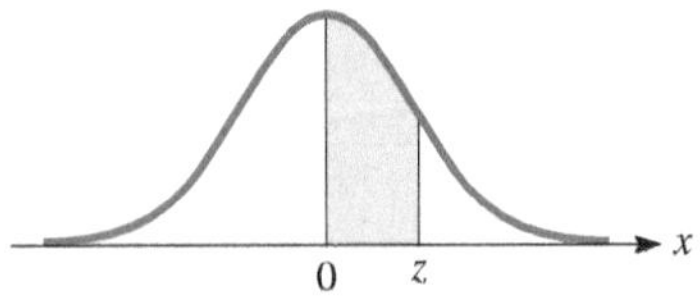

Fig. 22.11

Table 22.2 gives the area under the standard normal distribution curve between zero and the given values of z (see Fig. 22.11). The table includes values only to $z = 3$ since nearly all of the area is between $z = -3$ and $z = 3$. Since the curve is symmetric to the y-axis, the values shown are also valid for negative values of z.

Table 22.2 **Standard Normal (z) Distribution**

z	Area	z	Area	z	Area
0.0	0.0000	1.0	0.3413	2.0	0.4772
0.1	0.0398	1.1	0.3643	2.1	0.4821
0.2	0.0793	1.2	0.3849	2.2	0.4861
0.3	0.1179	1.3	0.4032	2.3	0.4893
0.4	0.1554	1.4	0.4192	2.4	0.4918
0.5	0.1915	1.5	0.4332	2.5	0.4938
0.6	0.2257	1.6	0.4452	2.6	0.4953
0.7	0.2580	1.7	0.4554	2.7	0.4965
0.8	0.2881	1.8	0.4641	2.8	0.4974
0.9	0.3159	1.9	0.4713	2.9	0.4981
1.0	0.3413	2.0	0.4772	3.0	0.4987

The procedure for finding relative frequencies using z-scores is summarized below. We will illustrate the procedure in the examples that follow the summary.

Finding Relative Frequencies Using z-Scores

1. Sketch the normal curve, labelling the mean and the given x-values. Identify the desired relative frequency as an area under the curve.
2. Use Eq. (22.6) to find the z-score for each x.
3. Look up the absolute value of each z-score in Table 22.2 to find its associated area.
4. Depending on the situation, proceed as follows:

Area	Procedure
Between two z-scores of the same sign	Subtract the smaller area from the larger one
Between two z-scores of different sign	Add both areas together
To the right of a positive z-score or to the left of a negative z-score	Subtract the area from 0.5
To the right of a negative z-score or to the left of a positive z-score	Add the area to 0.5

EXAMPLE 4 Normal score (z-score)

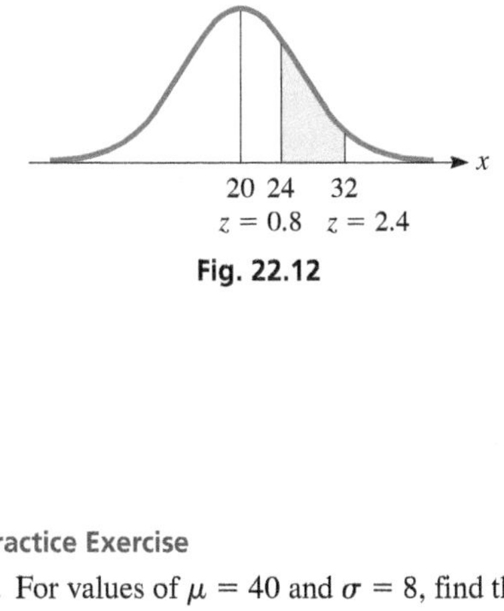

Fig. 22.12

For a normal distribution curve based on values of $\mu = 20$ and $\sigma = 5$, find the area between $x = 24$ and $x = 32$.

The area under the curve between $x = 24$ and $x = 32$ is shown in Fig. 22.12. We use Eq. (22.6) and find the z-scores

$$z = \frac{24 - 20}{5} = 0.8 \quad \text{and} \quad z = \frac{32 - 20}{5} = 2.4$$

For $z = 0.8$, the area in Table 22.2 is 0.2881, and for $z = 2.4$, the area is 0.4918. We have two z-scores of the same sign, so we subtract the smaller area (from 0 to 0.8) from the larger one (from 0 to 2.4) to obtain the area we want (from 0.8 to 2.4):

$$0.4918 - 0.2881 = 0.2037$$

The relative frequency of the values between $x = 24$ and $x = 32$ is thus 20.37%. If we had a large set of measured values with $\mu = 20$ and $\sigma = 5$, we would expect that about 20% of them would be between $x = 24$ and $x = 32$.

Practice Exercise

1. For values of $\mu = 40$ and $\sigma = 8$, find the area between $x = 36$ and $x = 48$.

EXAMPLE 5 z-score—battery lifetimes

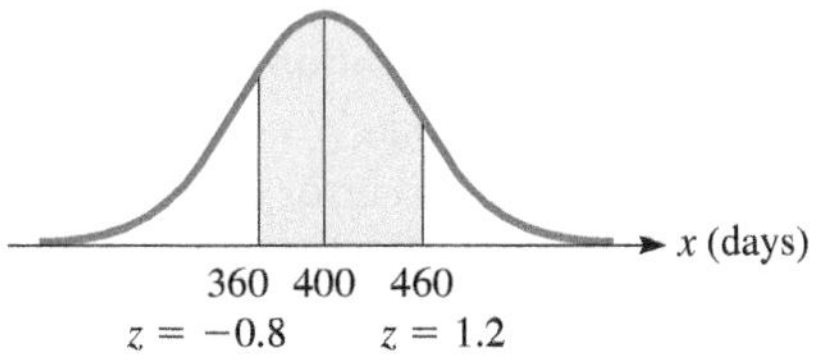

Fig. 22.13

The lifetimes of a certain type of watch battery are normally distributed. The mean lifetime is 400 days, and the standard deviation is 50 days (that is, $\mu = 400$ days and $\sigma = 50$ days). For a sample of 5000 new batteries, determine how many batteries are expected to last **(a)** between 360 days and 460 days, **(b)** more than 320 days, and **(c)** less than 280 days.

(a) The area under the curve between $x = 360$ and $x = 460$ is shown in Fig. 22.13. We use Eq. (22.6) and find the z-scores

$$z = \frac{360 - 400}{50} = -0.8 \quad \text{and} \quad z = \frac{460 - 400}{50} = 1.2$$

For $z = -0.8$, the area is to the left of 0. Since the curve is symmetric about 0, we can use the area from Table 22.2 for $z = 0.8$, that is, 0.2881. For $z = 1.2$, the area is 0.3849. We have two z-scores of different sign, so we add both areas together

$$0.2881 + 0.3849 = 0.6730$$

This means that 67.30% of the 5000 batteries, or 3365 of the batteries, are expected to last between 360 and 460 days. Because of variability within samples, not every sample will have exactly 3365 batteries that will last between 360 and 460 days. On average, however, a sample of 5000 batteries will have 3365 that will last that amount of time.

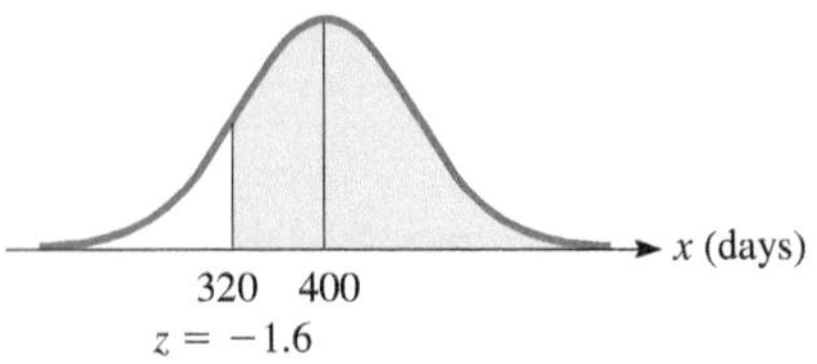

Fig. 22.14

(b) The area of interest is to the right of $x = 320$. See Fig. 22.14. The z-score for $x = 320$ is $z = (320 - 400)/50 = -1.6$. The area for -1.6 is the same as the area for $z = 1.6$, or 0.4452. The area to the right of -1.6 is the area between 0 and -1.6 plus 0.5 (half the area of the total area under the curve, which is 1). This is $0.4452 + 0.5 = 0.9452$. The relative frequency of 0.9452 means that $0.9452 \times 5000 = 4726$ batteries are expected to last more than 320 days.

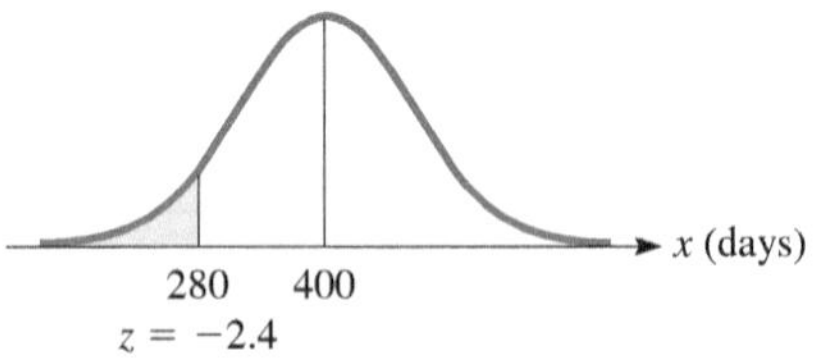

Fig. 22.15

(c) The area of interest is to the left of $x = 280$. See Fig. 22.15. The z-score for $x = 280$ is $z = (280 - 400)/50 = -2.4$. The area for -2.4 is the same as the area for $z = 2.4$, or 0.4918. The area to the left of -2.4 is the area between 0 and -2.4 subtracted from 0.5. This is $0.5 - 0.4918 = 0.0082$. The relative frequency of 0.0082 means that $0.0082 \times 5000 = 41$ are expected to last less than 280 days.

SAMPLING DISTRIBUTIONS

In Example 4, we assumed that the lifetimes of the batteries were normally distributed. Of course, for any set of 5000 batteries, or any number of batteries for that matter, the lifetimes that actually occur will not follow a normal distribution *exactly*. There will be some variation from the normal distribution, but for a large sample, this variation should be small. The mean and the standard deviation for any sample will vary somewhat from that of the population. When we consider the relative frequency distribution of the sample means obtained from all possible samples of the same size, we obtain what is called the **sampling distribution** of the sample means.

In the study of probability, it is shown that if we select all possible samples of size n from a population with a mean μ and standard deviation σ, the mean of the sample means is also μ. Also, *the standard deviation of the sample means,* denoted by $\sigma_{\bar{x}}$, and called **the standard error of the mean**, is

$$\sigma_{\bar{x}} = \frac{\sigma}{\sqrt{n}} \tag{22.7}$$

Moreover, when n is large (*large* in this situation is usually considered to be over 30), the sampling distribution of the sample means is approximately normal. In other words, the normal curve approximates the relative frequency distribution of the sample means, so that areas under the normal curve can be used to approximate relative frequencies of the sample means.

The normal distribution also approximates sampling distributions for qualitative data. For instance, suppose that we take samples of n items from a very large population containing a proportion p of defective items. If n is large (in this case, np and $n(1 - p)$ **must both be at least 5**), then the sampling distribution of the sample proportion $\hat{p}$ of defective items in each sample is also approximately normal. In this situation, the mean of the sample proportions is p, and the standard error of $\hat{p}$ is

$$\sigma_{\hat{p}} = \sqrt{\frac{p(1 - p)}{n}} \tag{22.8}$$

We can see from Eqs. (22.7) and (22.8) that as the sample size gets larger, the less variation there will be in the mean or the proportion obtained from a sample.

COMMON ERROR It is a common error to forget the $\sqrt{n}$ term in the denominator of the standard error formulas. It is very important to include it since it is what guarantees that variation between samples decreases as the sample size increases.

EXAMPLE 6 Sampling distribution of the sample mean

For the sample of 5000 watch batteries in Example 5 we know that $\sigma = 50$ days. Therefore, the standard error of the mean $\bar{x}$ is $50/\sqrt{5000} = 0.7$ day. This means that of all samples of 5000 batteries, about 68% should have a mean lifetime of 400 ± 0.7 day (between 399.3 days and 400.7 days). Also, about 95% of the samples should have a mean lifetime of $400 \pm 2(0.7)$ days (between 398.6 days and 401.4 days).

EXAMPLE 7 Sampling distribution of the sample proportion

Of the items in a very large population, 10% are defective. Samples of size 200 are taken from this population, and the proportion of defectives in each sample is recorded. The normal approximation applies since $np = 200(0.1) = 20$ and $n(1 - p) = 200(0.9) = 180$. Therefore, the sample proportion $\hat{p}$ is approximately normally distributed, with mean 0.1 and standard error $\sqrt{(0.10)(0.90)/200} = 0.021$, or 2.1%. This means that of all samples of size 200 taken from this population, about 68% would have a sample proportion of defectives of 0.1 ± 0.021 (i.e., between 7.9% and 12.1% of the sample would be defective in 68% of samples).

EXERCISES 22.3

In Exercises 1–4, make the given changes in the indicated examples of this section and then solve the indicated problem.

1. In Example 1, change the second σ from 10 to 5 and then describe the curve that would result in terms of either or both curves shown in Fig. 22.7.
2. In Example 3, change two to three and then find the resulting percent of values and the corresponding interval.
3. In Example 4, change $x = 32$ to $x = 33$ and then find the resulting area.
4. In Example 5(b), change 320 to 360 and then find the resulting number of batteries.

In Exercises 5–8, use the following information. If the weights of cement bags are normally distributed with a mean of 30 kg and a standard deviation of 2 kg, use the empirical rule to find the percent of the bags that weigh the following:

5. Between 26 kg and 34 kg
6. Between 28 kg and 32 kg
7. Less than 36 kg
8. More than 34 kg

In Exercises 9–12, use the following information. A standardized math test has a mean score of 200 and a standard deviation of 15. Find and interpret the z-scores of the following math test scores.

9. 218
10. 179
11. 164
12. 233

In Exercises 13–16, use the following data. Each AA battery in a sample of 500 batteries is checked for its voltage. It has been previously established for this type of battery (when newly produced) that the voltages are distributed normally with $\mu = 1.50$ V and $\sigma = 0.05$ V.

13. How many batteries are expected to have voltages between 1.45 V and 1.55 V?
14. How many batteries are expected to have voltages between 1.52 V and 1.58 V?
15. What percent of the batteries are expected to have voltages below 1.54 V?
16. What percent of the batteries are expected to have voltages above 1.64 V?

In Exercises 17–22, use the following data. The lifetimes of a certain type of automobile tire have been found to be distributed normally with a mean lifetime of 100 000 km *and a standard deviation of 10 000* km. *Answer the following questions for a sample of 5000 of these tires.*

17. How many tires are expected to last between 85 000 km and 100 000 km?

18. How many tires are expected to last between 95 000 km and 115 000 km?

19. How many tires are expected to last more than 118 000 km?

20. If the manufacturer guarantees to replace all tires that do not last 75 000 km, what percent of the tires may have to be replaced under this guarantee?

21. What is the standard error in the mean for all samples of 5000 of these tires? Explain the meaning of this result.

22. What percent of the samples of 5000 of these tires should have a mean lifetime of more than 100 282 km?

In Exercises 23–30, solve the given problems.

23. Find the standard error of the proportion of defective items in samples of size 500 taken from a very large population of which 12% of the items are defective. Explain the meaning of this result.

24. Of the 300 mL bottles filled by a certain filling machine, 1% contain less than 290 mL of juice. If samples of 600 bottles produced by this machine are selected, find the standard error of the sample proportion of bottles that contain less than 290 mL. Explain the meaning of this result.

25. With 75.8% of the area under the normal curve to the right of z, find the z-value.

26. With 21% of the area under the normal curve between z_1 and z_2, to the right of $z_1 = 0.8$, find z_2.

27. With 59% of the area under the normal curve between z_1 and z_2, to the left of $z_2 = 1.1$, find z_1.

28. With 5.8% of the area under the normal curve between z_1 and z_2, to the left of $z_2 = 2.0$, find z_1.

29. The residents of a city suburb live at a mean distance of 16.0 km from the centre of the city, with a standard deviation of 4.0 km. What percent of the residents live between 12.0 km and 18.0 km of the centre of the city?

30. For the data on the number of Android apps in Exercise 17 of Section 22.1, find the percent of the data that lie within one, two, and three standard deviations of the mean. Compare these percentages with the ones in the empirical rule.

Answer to Practice Exercise

1. $z = 0.5328$

22.4 Confidence Intervals

In this section we begin our study of inferential statistics, where information from a sample is used to make statements about a whole population. We focus on the problem of estimation of parameters, under the assumption that data are available for a random sample taken from a very large population.

When a parameter is being estimated, the estimate can be a single number (called a **point estimate**), or it can be a range of numbers (called a **confidence interval**). Consider the following example.

EXAMPLE 1 Two kinds of estimates

Consider the data on hours of exercise in Example 1 of Section 22.1. If the sample of 50 students constitutes a random sample of a large population of university students, then the information obtained from the sample can be used to estimate the population mean.

On the one hand, the sample mean $\bar{x} = 4.8$ is a point estimate of the population mean. Because it is a single number, this estimate does not convey information about the reliability of the estimation.

On the other hand, the interval $4.8 \pm 1.0 = (3.8\text{ h}, 5.8\text{ h})$ is a 95% confidence interval estimate of the population mean. This means that we are 95% confident that the true value of the population mean is between 3.8 h and 5.8 h. It is preferable to use a confidence interval because the length of the interval and the level of confidence attached to it give us an idea about the reliability of our estimation, in a sense that will be made precise below.

All confidence intervals are calculated by first selecting a **confidence level**, which measures the degree of certainty that the confidence interval will contain the population parameter. The most common values for the confidence level are 90%, 95%, and 99%, with the most common one being 95%. A confidence level of 95% means that, of all possible samples of size n taken from the same population, 95% of them will give an interval that will contain the population parameter, and 5% of them will not. For a particular sample, it is not possible to know whether it is one of the successful ones or not.

COMMON ERROR

It is a common error to interpret the level of confidence as measuring the likelihood that the parameter of interest will fall within a particular interval. There is nothing random about the parameter; its value is a constant (unfortunately unknown to us), and either our interval covers it or it does not. The randomness lies in the sample, so the level of confidence is the likelihood that a **sample of size *n*** will cover the parameter. At the 95% confidence level, 95% of the samples will, and 5% of the samples will not.

The method for constructing a confidence interval depends on the parameter(s) being estimated and on the characteristics of the sample. We will concentrate our attention on *large-sample confidence intervals for a single mean and a single proportion.*

LARGE-SAMPLE CONFIDENCE INTERVALS FOR THE MEAN

Suppose that we have a random sample of size n (n large) from a population with unknown mean μ. We begin by constructing a 95% confidence interval for the mean μ, under the assumption that the population standard deviation σ is known.

We know (from tables) that 95% of the observations from a normal distribution fall within 1.96 standard deviations of the mean. Because the sampling distribution of the sample mean is approximately normal, with mean μ and standard error $\sigma/\sqrt{n}$, it follows that 95% of all samples of size n will have a sample mean $\bar{x}$ that falls within 1.96 standard errors of the true mean. In other words, in 95% of samples, the sample mean $\bar{x}$ satisfies the inequality

■ According to the Empirical Rule, about 95% of observations from a normal distribution are within two standard deviations of the mean. The value of two results from rounding the more precise 1.96 value used here.

$$\mu - 1.96 \cdot \frac{\sigma}{\sqrt{n}} < \bar{x} < \mu + 1.96 \cdot \frac{\sigma}{\sqrt{n}}$$

We can manipulate this inequality in order to transform it into a statement about the unknown population mean μ. We have

$$\mu - 1.96 \cdot \frac{\sigma}{\sqrt{n}} < \bar{x} < \mu + 1.96 \cdot \frac{\sigma}{\sqrt{n}} \qquad \text{original inequality}$$

$$-1.96 \cdot \frac{\sigma}{\sqrt{n}} < \bar{x} - \mu < 1.96 \cdot \frac{\sigma}{\sqrt{n}} \qquad \text{subtract } \mu \text{ from each member}$$

$$-\bar{x} - 1.96 \cdot \frac{\sigma}{\sqrt{n}} < -\mu < -\bar{x} + 1.96 \cdot \frac{\sigma}{\sqrt{n}} \qquad \text{subtract } \bar{x} \text{ from each member}$$

$$\bar{x} + 1.96 \cdot \frac{\sigma}{\sqrt{n}} > \mu > \bar{x} - 1.96 \cdot \frac{\sigma}{\sqrt{n}} \qquad \text{multiply by } -1 \text{ (reverse the inequality)}$$

$$\bar{x} - 1.96 \cdot \frac{\sigma}{\sqrt{n}} < \mu < \bar{x} + 1.96 \cdot \frac{\sigma}{\sqrt{n}}$$

This last inequality is equivalent to the original one and is therefore satisfied for 95% of samples of size n. In other words,

$$\left(\bar{x} - 1.96 \cdot \frac{\sigma}{\sqrt{n}}, \bar{x} + 1.96 \cdot \frac{\sigma}{\sqrt{n}}\right) \qquad \textbf{(22.9)}$$

is a 95% confidence interval for μ.

The endpoints of the interval in Eq. (22.9) are often written in the form

$$\bar{x} \pm E, \text{ with } E = 1.96 \cdot \frac{\sigma}{\sqrt{n}} \qquad \textbf{(22.10)}$$

The quantity E is called the **margin of error**, and it represents the largest estimated difference between the estimate and the true value of the parameter.

EXAMPLE 2 A 95% confidence interval—σ known

A random sample of size $n = 100$ is taken from a population with $\sigma = 2.3$. Construct a 95% confidence interval for the population mean μ if the sample mean is $\bar{x} = 32.8$.

We substitute the given values of $\bar{x}$, σ, and n into Eq. (22.10). The resulting 95% confidence interval is

$$\bar{x} \pm 1.96 \cdot \frac{\sigma}{\sqrt{n}} = 32.8 \pm 1.96 \cdot \frac{2.3}{\sqrt{100}} = 32.8 \pm 0.451 = (32.3, 33.3)$$

Because the interval obtained is narrow, the estimation is quite precise.

When the standard deviation σ is unknown and the sample size is large, the sample standard deviation s can be used to estimate the population standard deviation σ. Therefore, a 95% confidence interval for μ when the standard deviation σ is unknown and the sample size is large is given by

■ Note that Eq. (22.11) is valid only when the sample size is large. A formula based on the *t*-distribution (beyond the scope of this text) is required if the sample size is small and the standard deviation of the normal distribution is unknown.

$$\bar{x} \pm E, \text{ with } E = 1.96 \cdot \frac{s}{\sqrt{n}} \quad \textbf{(22.11)}$$

EXAMPLE 3 A 95% confidence interval—σ estimated

Obtain a 95% confidence interval for the mean number of hours exercised per week from a random sample of 50 students (see Example 1 of Section 22.1). The sample mean and sample standard deviation for the 50 observations are $\bar{x} = 4.8$ h and $s = 3.6$ h, respectively.

Using Eq. (22.11), the resulting 95% confidence interval is

$$\bar{x} \pm 1.96\frac{s}{\sqrt{n}} = 4.8 \pm 1.96 \cdot \frac{3.6}{\sqrt{50}} = 4.8 \pm 1.0 = (3.8\text{ h}, 5.8\text{ h})$$

Practice Exercise

1. Find a 95% confidence interval for a mean μ if a sample with $n = 45$ gives $\bar{x} = 97.6$ and $s = 3.2$.

We now derive the formula for a large-sample confidence interval for a mean for an arbitrary confidence level $1 - \alpha$. Here we have written the confidence level as is standard for general formulas, expressing it as a decimal whose value is $1 - \alpha$ for some small α (for example, $\alpha = 0.05$ for a 95% confidence interval).

Let $z_{\alpha/2}$ denote the standard score such that the area under the standard normal curve between $-z_{\alpha/2}$ and $z_{\alpha/2}$ is $1 - \alpha$. See Fig. 22.16 and Table 22.3. By replacing 1.96 with $z_{\alpha/2}$ in the inequalities leading to Eq. (22.10), we obtain the following general formula.

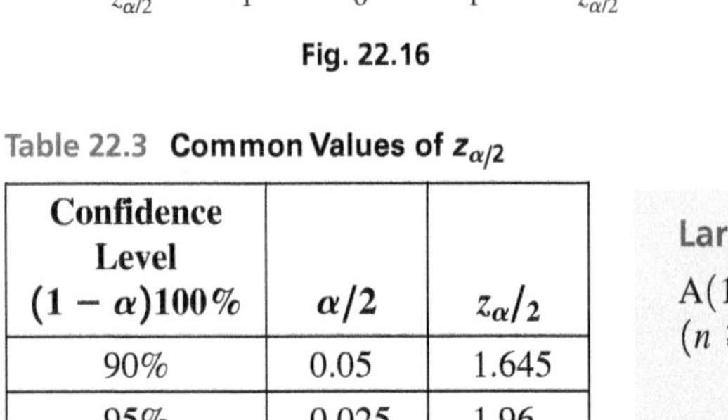

Fig. 22.16

Table 22.3 **Common Values of $z_{\alpha/2}$**

Confidence Level $(1-\alpha)100\%$	$\alpha/2$	$z_{\alpha/2}$
90%	0.05	1.645
95%	0.025	1.96
99%	0.005	2.575

Large-Sample Confidence Interval for a Mean

A$(1 - \alpha)$ 100% confidence interval for the mean μ when the sample size is large $(n \geq 30)$ is given by

$$\bar{x} \pm E, \text{ where } E = z_{\alpha/2} \cdot \frac{\sigma}{\sqrt{n}} \quad \text{if } \sigma \text{ is known}$$

$$E = z_{\alpha/2} \cdot \frac{s}{\sqrt{n}} \quad \text{if } \sigma \text{ is unknown} \quad \textbf{(22.12)}$$

EXAMPLE 4 A 99% confidence interval

Construct a 99% confidence interval for the hours of exercise data from Example 1 of Section 22.1. Recall that $\bar{x} = 4.8$ h, $s = 3.6$ h, and $n = 50$.

From Table 22.3 we get that $z_{\alpha/2} = 2.575$. We substitute the given values into Eq. (22.12). The desired 99% confidence interval is

$$\bar{x} \pm z_{\alpha/2}\frac{s}{\sqrt{n}} = 4.8 \pm 2.575 \cdot \frac{3.6}{\sqrt{50}} = 4.8 \pm 1.3 = (3.5\text{ h}, 6.1\text{ h})$$

Practice Exercise

2. Find a 90% confidence interval for a mean μ if a sample with $n = 45$ gives $\bar{x} = 97.6$ and $s = 3.2$.

COMMON ERROR Always state a confidence interval together with its confidence level. ***A confidence interval by itself is meaningless.***

By examining the formula for the margin of error in Eq. (22.12), we can find the general relationship between confidence level, margin of error, and sample size.

Confidence Level, Margin of Error, and Sample Size Relationships

- For a fixed sample size, a higher confidence level implies a larger margin of error. What we gain in confidence we lose in precision.
- For a fixed sample size, a smaller margin of error implies a lower confidence level. What we gain in precision we lose in confidence.
- The only way we can increase the confidence level while at the same time decreasing the margin of error is to increase the sample size.

The formula for E in Eq. (22.12) for known σ can be used to determine the sample size needed for a desired margin of error at a fixed confidence level. Suppose that a maximum margin of error E is desired with a confidence level $1 - \alpha$. We have

$$E = z_{\alpha/2} \cdot \frac{\sigma}{\sqrt{n}} \qquad \text{from Eq. (22.12)}$$

$$\sqrt{n} = \frac{z_{\alpha/2}\sigma}{E} \qquad \text{multiplying both sides by } \frac{\sqrt{n}}{E}$$

After squaring both sides, we get

$$n = \left[\frac{z_{\alpha/2}\sigma}{E}\right]^2 \qquad \textbf{(22.13)}$$

When the result of Eq. (22.13) is not an integer, it must always be **rounded up** to the next integer in order to guarantee the prescribed margin of error.

EXAMPLE 5 Determining the sample size

An estimate of the mean direct-current output voltage of a certain kind of AC adaptor is desired. If it can be assumed that $\sigma = 0.04$ V, find the sample size necessary to estimate that mean with a margin of error of 0.01 V with 95% confidence.

Substituting $\sigma = 0.04$, $E = 0.01$, and $z_{\alpha/2} = 1.96$, we get

$$n = \left[\frac{1.96(0.04)}{0.01}\right]^2 = 61.5$$

Therefore, a sample of 62 adaptors is necessary.

LARGE-SAMPLE CONFIDENCE INTERVALS FOR A PROPORTION

Consider a large population such that an unknown proportion p of its elements share a certain attribute. For example, p could be the proportion of defective items in a lot, or the proportion of incorrect entries in an account, or the proportion of components that will last a certain number of hours, or the proportion of voters who will vote for a certain candidate in the next election.

Suppose that a random sample of size n is taken. We calculate the sample proportion $\hat{p}$ by dividing the number of elements in the sample that share the attribute by the sample size n. We further assume that n is large, so that $np \geq 5$ and $n(1-p) \geq 5$. (Since p is unknown, we require that $n\hat{p} \geq 5$ and $n(1-\hat{p}) \geq 5$.)

As we saw in Section 22.3, under these conditions, the sampling distribution of $\hat{p}$ is approximately normal with mean p and standard error $\sqrt{p(1-p)/n}$. Therefore, if $z_{\alpha/2}$ is the value that leaves an area of $1-\alpha$ between $-z_{\alpha/2}$ and $z_{\alpha/2}$, then in $(1-\alpha)100\%$ of samples, $\hat{p}$ satisfies

$$p - z_{\alpha/2}\sqrt{\frac{p(1-p)}{n}} < \hat{p} < p + z_{\alpha/2}\sqrt{\frac{p(1-p)}{n}}$$

Manipulating this inequality and using the large sample-size condition gives the following general formula for a $(1-\alpha)100\%$ confidence interval for p.

Large-Sample Confidence Interval for a Proportion

Let $\hat{p}$ be the sample proportion obtained from a sample of size n such that $n\hat{p} \geq 5$ and $n(1-\hat{p}) \geq 5$. A $(1-\alpha)100\%$ confidence interval for the population proportion p is given by

$$\hat{p} \pm E, \text{ where } E = z_{\alpha/2}\sqrt{\frac{\hat{p}(1-\hat{p})}{n}} \tag{22.14}$$

EXAMPLE 6 A 90% confidence interval for *p*

A manufacturer wants to estimate the proportion of defective parts in a large lot produced by a particular machine. In a sample of 350 parts, 41 of them were found to be defective. Construct a 90% confidence interval for the proportion of defective parts in the lot.

We have

$$\hat{p} = \tfrac{41}{350},\ n\hat{p} = 41 \geq 5, \text{ and } n(1-\hat{p}) = 309 \geq 5$$

Therefore, we can apply Eq. (22.14) with $z_{\alpha/2} = 1.645$ (see Table 22.3). The desired 90% confidence interval is

$$\hat{p} \pm z_{\alpha/2}\sqrt{\frac{\hat{p}(1-\hat{p})}{n}} = \frac{41}{350} \pm 1.645\sqrt{\frac{\frac{41}{350}\cdot\frac{309}{350}}{350}} = 0.117 \pm 0.028 = (0.089, 0.145)$$

As we did with the mean, we can obtain the required sample size for a desired margin of error at a fixed confidence level by using the formula for margin of error from Eq. (22.14) and solving for n. For a confidence level $1-\alpha$, we get

$$n = \hat{p}(1-\hat{p})\left[\frac{z_{\alpha/2}}{E}\right]^2 \tag{22.15}$$

Note that Eq. (22.15) requires an estimate for $\hat{p}$. (It can be obtained from past data or from a pilot study.) When no such estimate is available, we can use the fact that $\hat{p}(1-\hat{p})$ is maximized when $\hat{p}=\frac{1}{2}$, so we use this worst-case scenario estimate in Eq. (22.15). The required sample size for a confidence level $1-\alpha$ becomes

$$n=\frac{1}{4}\left[\frac{z_{\alpha/2}}{E}\right]^2 \tag{22.16}$$

EXAMPLE 7 Determining sample size

Suppose that the manufacturer from Example 6 wishes to estimate the proportion of defectives with a maximum error of 0.025 with 90% confidence.

(a) How large a sample will he need if no information from the past is used?

For this case we use Eq. (22.16) with $E=0.025$ and $z_{\alpha/2}=1.645$. The required sample size is $n=\frac{1}{4}\left[\frac{z_{\alpha/2}}{E}\right]^2=\frac{1}{4}\left[\frac{1.645}{0.025}\right]^2=1082.41$, so that 1083 parts must be sampled.

(b) How large a sample will he need if the information from the sample of 350 parts is used? (Recall that $\hat{p}=\frac{41}{350}$.)

If $\hat{p}=\frac{41}{350}$ is known from the past, Eq. (22.15) gives

$$n=\hat{p}(1-\hat{p})\left[\frac{z_{\alpha/2}}{E}\right]^2=\frac{41}{350}\cdot\frac{309}{350}\left[\frac{1.645}{0.025}\right]^2=447.8$$

Therefore, 448 parts must be sampled. Note how information about the possible size of $\hat{p}$ substantially reduced the size of the required sample.

EXERCISES 22.4

In Exercises 1–5, make the given changes in the indicated examples of this section and then solve the resulting problems.

1. In Example 2, change $n=100$ to $n=140$ and then find the indicated confidence interval.
2. In Example 4, change the confidence level from 99% to 90% and then find the indicated confidence interval.
3. In Example 5, change $\sigma=0.04$ to $\sigma=0.03$ and find the required sample size.
4. In Example 6, change the 41 to a 62 and find the indicated confidence interval.
5. In Example 7(a), change 90% to 95% and find the required sample size.

In Exercises 6–8, use the following data. A random sample of size $n=300$ is taken from a large population with $\sigma=25.9$. The sample mean is $\bar{x}=247.1$.

6. Construct a 95% confidence interval for the population mean μ.
7. Construct a 99% confidence interval for the population mean μ.
8. How large a sample must be taken so that a 95% confidence interval for μ will have a maximum margin of error $E=2.4$?

In Exercises 9–12, use the following data. A random sample of size $n=215$ is taken from a large population, and 38 are found to be defective.

9. Construct a 95% confidence interval for the population proportion of defectives p.
10. Construct a 99% confidence interval for the population proportion of defectives p.
11. How large a sample must be taken so that a 95% confidence interval for p will have a maximum margin of error of 4.5%? Assume that the information from the sample is used.
12. How large a sample must be taken so that a 95% confidence interval for p will have a maximum margin of error of 4.5%? Assume that no prior information is used.

In Exercises 13–15, use the following information. A random sample of size n is taken from a large population with $\sigma = 3.14$. The sample mean is $\bar{x} = 83.7$.

13. Find a 90% confidence interval if the sample size n is 50.
14. Find a 90% confidence interval if the sample size n is 80. Compare the interval with the interval obtained in Exercise 13. How does increasing the sample size affect the margin of error?
15. Find a 95% confidence interval if the sample size is 80. Compare the interval with the interval obtained in Exercise 14. How does increasing the confidence level affect the margin of error?

In Exercises 16–22, solve the given problems.

16. A sample of 75 washing machines of a certain brand had a mean replacement time of 9.1 years, with a standard deviation of 2.7 years. Find a 95% confidence interval for the mean replacement time of all washing machines of this brand.
17. A test station measured the loudness of a random sample of 45 jets taking off from a certain airport. The mean was found to be 107.2 dB, with a standard deviation of 9.2 dB. Find a 90% confidence interval for the mean loudness of all jets taking off from this airport.
18. An airline wishes to estimate the mean time passengers have to wait for their luggage when arriving at a large airport. How many passengers must be sampled so that a 95% confidence interval for the true mean waiting time μ will have a maximum margin of error of 30 seconds? A similar study done in the past had a standard deviation of 2.16 minutes.
19. A toy manufacturer wishes to estimate the mean time it takes an adult to assemble a certain "easy to assemble" toy. How many adults must be sampled so that a 99% confidence interval for the true mean assembly time μ will have a maximum margin of error of 2.0 minutes? The standard deviation of assembly time for a similar model is known to be 5.9 minutes.
20. From a random sample of 60 bicycle helmets subjected to an impact test, 13 helmets showed some damage from the test. Find a 95% confidence interval for the true proportion of helmets that would show damage from this test.
21. Suppose that we want to estimate the proportion of drivers that exceed the 100 km/h speed limit by more than 10 km/h in a certain stretch of highway. How large a sample must be taken so that a 95% confidence interval for the true proportion p will have a maximum margin of error of 4%? Assume that no prior information is used.
22. Following are two confidence interval estimates of the true mean contents of certain 306 mL jars of sauce:

$$(306.2, 307.4)(306.3, 307.3)$$

The confidence level for one interval is 90%, and the confidence level for the other is 95%, with both intervals constructed from the same sample data. Which of the intervals is the 90% confidence interval? Explain.

Answers to Practice Exercises

1. (96.7, 98.5) with 95% confidence
2. (96.8, 98.4) with 90% confidence

22.5 Statistical Process Control

One of the most important uses of statistics in industry is statistical process control (SPC), which is used to maintain and improve product quality. Samples are tested during the production at specified intervals to determine whether the production process needs adjustment to meet quality requirements.

A particular industrial process is considered to be *in control* if it is stable and predictable, and sample measurements fall within upper and lower control limits. The process is *out of control* if it has an unpredictable amount of variation and there are sample measurements outside the control limits due to special causes.

■ This is intended only as a brief introduction to this topic. A complete development requires at least a chapter in a statistics book.

EXAMPLE 1 Process—in control/out of control

A manufacturer of 1.5-V batteries states that the voltage of its batteries is no less than 1.45 V or greater than 1.55 V and has designed the manufacturing process to meet these specifications.

If all samples of batteries that are tested have voltages in the proper range with only expected minor variations, the production process is *in control.*

■ Minor variations may be expected, for example, from very small fluctuations in voltage, temperature, or material composition. Special causes resulting in an out-of-control process could include line stoppage, material defect, or an incorrect applied pressure.

However, if some samples have batteries with voltages out of the proper range, the process is *out of control*. This would indicate some special cause for the problem, such as an improperly operating machine or an impurity getting into the process. The process would probably be halted until the cause is determined.

CONTROL CHARTS

An important device used in SPC is the *control chart*. It is used to show a trend of a production characteristic over time. In this section we study one type of **control chart for measurements** (when the variable involved is quantitative), and one type of **control chart for attributes** (when the variable involved is qualitative). In both cases, samples are observed at specified intervals of time to see if the sample values are within acceptable limits. The sample values are plotted on a chart to check for trends and abnormalities in the production process.

In making a control chart for measurements, we must determine what the mean should be. For a stable process for which previous data are known, it can be based on a production specification or on previous data. For a new or recently modified process, it may be necessary to use present data, although the value may have to be revised for future charts. On a control chart, *this value is used as the population mean,* μ.

It is also necessary to establish the upper and lower control limits. The standard generally used is that 99.7% of the sample measurements should fall within these control limits. This assumes a normal distribution, and we note that this is within three sample standard deviations of the population mean. We will establish these limits by use of a table or a formula that has been made using statistical measures developed in a more complete coverage of quality control. This does follow the normal practice of using a formula or a more complete table in setting up the control limits.

In Fig. 22.17, we show a sample control chart, and on the following pages, we illustrate how control charts are made.

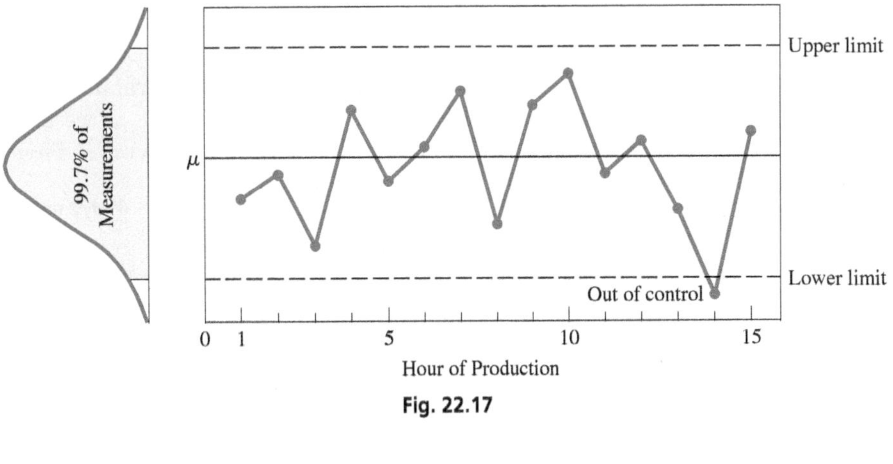

Fig. 22.17

EXAMPLE 2 Making $\bar{x}$ and R control charts

A pharmaceutical company makes a capsule of a prescription drug that contains 500 mg of the drug, according to the label. In a newly modified process of making the capsule, five capsules are tested every 15 min to check the amount of the drug in each capsule. Testing over a 5-h period gave the following results for the 20 subgroups of samples.

Subgroup	*Amount of Drug (in mg) of Five Capsules*					*Mean* $\bar{x}$	*Range R*
1	503	501	498	507	502	502.2	9
2	497	499	500	495	502	498.6	7
3	496	500	507	503	502	501.6	11
4	512	503	488	500	497	500.0	24
5	504	505	500	508	502	503.8	8
6	495	495	501	497	497	497.0	6
7	503	500	507	499	498	501.4	9
8	494	498	497	501	496	497.2	7
9	502	504	505	500	502	502.6	5
10	500	502	500	496	497	499.0	6
11	502	498	510	503	497	502.0	13
12	497	498	496	502	500	498.6	6
13	504	500	495	498	501	499.6	9
14	500	499	498	501	494	498.4	7
15	498	496	502	501	505	500.4	9
16	500	503	504	499	505	502.2	6
17	487	496	499	498	494	494.8	12
18	498	497	497	502	497	498.2	5
19	503	501	500	498	504	501.2	6
20	496	494	503	502	501	499.2	9
					Sum	9998.0	174
					Mean	499.9	8.7

From this table of values, make an $\bar{x}$ control chart and an R control chart. The $\bar{x}$ chart maintains a check on the average quality level, whereas the R chart maintains a check on the dispersion of the production process.

In order to define the **central line** of the $\bar{x}$ chart, which ideally is equivalent to the value of the population mean μ, we use the mean of the sample means $\bar{\bar{x}}$. For the central line of the R chart, we use $\overline{R}$. From the table, we see that

$$\bar{\bar{x}} = 499.9 \text{ mg} \quad \text{and} \quad \overline{R} = 8.7 \text{ mg}$$

The **upper control limit** (UCL) and the **lower control limit** (LCL) for each chart are defined in terms of the mean range $\overline{R}$ and an appropriate constant taken from a table of control chart factors. These factors, which are related to the sample size n, are determined by statistical considerations found in a more complete coverage of quality control. Below is a brief table of control chart factors (Table 22.4).

Table 22.4 **Control Chart Factors**

n	d_2	A	A_2	D_1	D_2	D_3	D_4
5	2.326	1.342	0.577	0.000	4.918	0.000	2.115
6	2.534	1.225	0.483	0.000	5.078	0.000	2.004
7	2.704	1.134	0.419	0.205	5.203	0.076	1.924

The UCL and LCL for the $\bar{x}$ chart are found as follows:

$$\text{UCL}(\bar{x}) = \bar{\bar{x}} + A_2\bar{R} = 499.9 + 0.577(8.7) = 504.9 \text{ mg} \qquad \text{(using Table 22.4 with } n = 5\text{)}$$
$$\text{LCL}(\bar{x}) = \bar{\bar{x}} - A_2\bar{R} = 499.9 - 0.577(8.7) = 494.9 \text{ mg}$$

The UCL and LCL for the R chart are found as follows:

$$\text{LCL}(R) = D_3\bar{R} = 0.000(8.7) = 0.0 \text{ mg}$$
$$\text{UCL}(R) = D_4\bar{R} = 2.115(8.7) = 18.4 \text{ mg}$$

Using these central lines and control limit lines, we now plot the $\bar{x}$ control chart in Fig. 22.18 and the R control chart in Fig. 22.19.

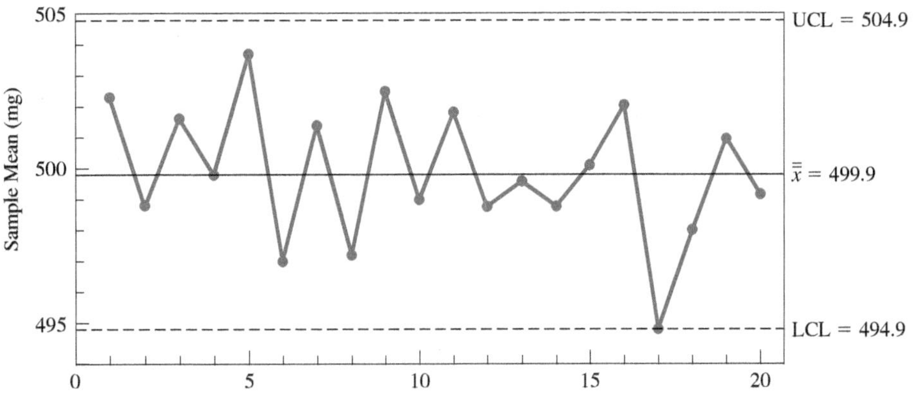

Fig. 22.18

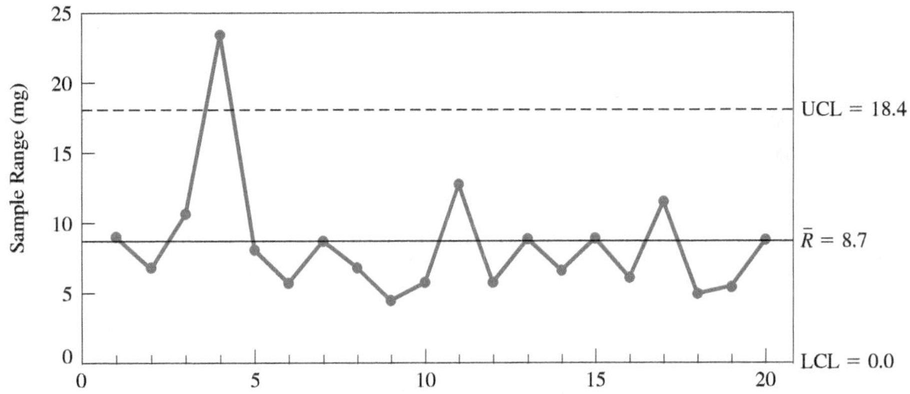

Fig. 22.19

This would be considered a *well-centred process* since $\bar{\bar{x}} = 499.9$ mg, which is very near the target value of 500.0 mg. We do note, however, that subgroup 17 was at a control limit and this might have been due to some special cause, such as the use of a substandard mixture of ingredients. We also note that the process was *out of control* due to some special cause since the range of subgroup 4 was above the upper control limit. We should keep in mind that there are numerous considerations, including human factors, that should be taken into account when making and interpreting control charts and that this is only a very brief introduction to this important industrial use of statistics.

We now discuss a control chart for attributes, for the case when each item tested is classified as being either acceptable or not acceptable. To monitor such an attribute in a production process, we obtain the *proportion* of defective parts by *dividing the number of defective parts in a sample by the total number of parts in the sample,* and then make a *p control chart.* This is illustrated in the following example.

EXAMPLE 3 Making a *p* control chart

A manufacturer of video discs has 1000 DVDs checked each day for defects (surface scratches, for example). The data for this procedure for 25 days are shown in the table at the left. Make a p control chart.

Day	*Defective DVDs*	*Proportion Defective*
1	22	0.022
2	16	0.016
3	14	0.014
4	18	0.018
5	12	0.012
6	25	0.025
7	36	0.036
8	16	0.016
9	14	0.014
10	22	0.022
11	20	0.020
12	17	0.017
13	26	0.026
14	20	0.020
15	22	0.022
16	28	0.028
17	17	0.017
18	15	0.015
19	25	0.025
20	12	0.012
21	16	0.016
22	22	0.022
23	19	0.019
24	16	0.016
25	20	0.020
Sum	490	

The central line for the p control chart is ideally equal to the true proportion of defectives in the population. This value is usually estimated from the data, using the average sample proportion $\bar{p}$, which in this case is

$$\bar{p} = \frac{490}{25\,000} = 0.0196$$

The control limits are each three standard deviations from $\bar{p}$. If n is the size of each sample, we obtain the standard error of $\bar{p}$ using Eq. (22.8) (with $\bar{p}$ in place of p). We get

$$\sigma_{\bar{p}} = \sqrt{\frac{\bar{p}(1-\bar{p})}{n}} = \sqrt{\frac{0.0196(1-0.0196)}{1000}} = 0.004\,38$$

Therefore, the control limits are

$$\text{UCL}(p) = 0.0196 + 3(0.004\,38) = 0.0327$$
$$\text{LCL}(p) = 0.0196 - 3(0.004\,38) = 0.0065$$

Using this central line and these control limit lines, we now plot the p control chart in Fig. 22.20.

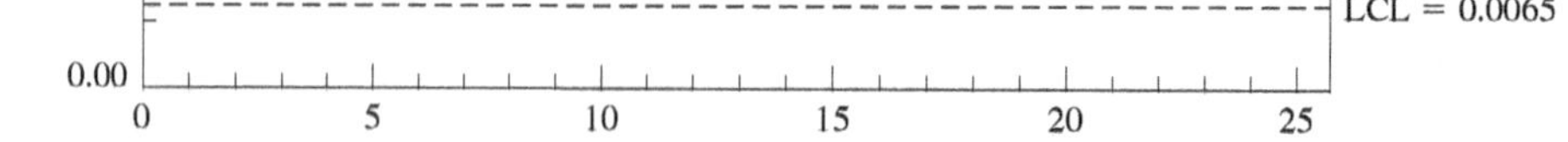

Fig. 22.20

According to the proportion mean of 0.0196, the process produces about 2% defective DVDs. We note that the process was out of control on Day 7. An adjustment to the production process was probably made to remove the special cause of the additional defective DVDs.

Practice Exercise

1. In Example 3, change Day 9 datum from 14 to 24 defective DVDs. Then find UCL(p) and LCL(p).

EXERCISES 22.5

In Exercises 1–4, in Example 2, change the first subgroup to 497, 499, 502, 493, and 498 and then proceed as directed.

1. Find UCL($\bar{x}$) and LCL($\bar{x}$).

2. Find LCL(R) and UCL(R).

3. How would the $\bar{x}$ control chart differ from Fig. 22.18?

4. How would the R control chart differ from Fig. 22.19?

In Exercises 5–8, use the following data.

Five automobile engines are taken from the production line each hour and tested for their torque (in N • m) when rotating at a constant frequency. The measurements of the sample torques for 20 h of testing are as follows:

Hour	*Torques (in N • m) of Five Engines*				
1	366	352	354	360	362
2	370	374	362	366	356
3	358	357	365	372	361
4	360	368	367	359	363
5	352	356	354	348	350
6	366	361	372	370	363
7	365	366	361	370	362
8	354	363	360	361	364
9	361	358	356	364	364
10	368	366	368	358	360
11	355	360	359	362	353
12	365	364	357	367	370
13	360	364	372	358	365
14	348	360	352	360	354
15	358	364	362	372	361
16	360	361	371	366	346
17	354	359	358	366	366
18	362	366	367	361	357
19	363	373	364	360	358
20	372	362	360	365	367

5. Find the central line, UCL, and LCL for the mean.

6. Find the central line, UCL, and LCL for the range.

7. Plot an $\bar{x}$ chart.

8. Plot an R chart.

In Exercise 9–12, use the following data.

Five AC adaptors that are used to charge batteries of a cellular phone are taken from the production line each 15 minutes and tested for their direct-current output voltage. The output voltages for 24 sample subgroups are as follows:

Subgroup	*Output Voltages of Five Adaptors*				
1	9.03	9.08	8.85	8.92	8.90
2	9.05	8.98	9.20	9.04	9.12
3	8.93	8.96	9.14	9.06	9.00
4	9.16	9.08	9.04	9.07	8.97
5	9.03	9.08	8.93	8.88	8.95
6	8.92	9.07	8.86	8.96	9.04
7	9.00	9.05	8.90	8.94	8.93
8	8.87	8.99	8.96	9.02	9.03
9	8.89	8.92	9.05	9.10	8.93
10	9.01	9.00	9.09	8.96	8.98
11	8.90	8.97	8.92	8.98	9.03
12	9.04	9.06	8.94	8.93	8.92
13	8.94	8.99	8.93	9.05	9.10
14	9.07	9.01	9.05	8.96	9.02
15	9.01	8.82	8.95	8.99	9.04
16	8.93	8.91	9.04	9.05	8.90
17	9.08	9.03	8.91	8.92	8.96
18	8.94	8.90	9.05	8.93	9.01
19	8.88	8.82	8.89	8.94	8.88
20	9.04	9.00	8.98	8.93	9.05
21	9.00	9.03	8.94	8.92	9.05
22	8.95	8.95	8.91	8.90	9.03
23	9.12	9.04	9.01	8.94	9.02
24	8.94	8.99	8.93	9.05	9.07

9. Find the central line, UCL, and LCL for the mean.

10. Find the central line, UCL, and LCL for the range.

11. Plot an $\bar{x}$ chart.

12. Plot an R chart.

In Exercises 13–16, use the following information.

For a production process for which there is a great deal of data since its last modification, the population mean μ and population standard deviation σ are assumed known. For such a process, we have the following values (using additional statistical analysis):

$\bar{x}$ chart: central line $= \mu$, UCL $= \mu + A\sigma$, LCL $= \mu - A\sigma$

R chart: central line $= d_2\sigma$, UCL $= D_2\sigma$, LCL $= D_1\sigma$

The values of A, d_2, D_2, and D_1 are found in the table of control chart factors in Example 2 (Table 22.4).

13. In the production of robot links and tests for their lengths, it has been found that $\mu = 2.725$ cm and $\sigma = 0.032$ cm. Find the central line, UCL, and LCL for the mean if the sample subgroup size is 5.

14. For the robot link samples of Exercise 13, find the central line, UCL, and LCL for the range.

15. After bottling, the volume of soft drink in six sample bottles is checked each 10 minutes. For this process $\mu = 750.0$ mL and $\sigma = 2.2$ mL. Find the central line, UCL, and LCL for the range.

16. For the bottling process of Exercise 15, find the central line, UCL, and LCL for the mean.

In Exercises 17 and 18, use the following data.

A telephone company rechecks the entries for 1000 of its new customers each week for name, address, and phone number. The data collected regarding the number of new accounts with errors, along with the proportion of these accounts with errors, is given in the following table for a 20-wk period:

Week	*Accounts with Errors*	*Proportion with Errors*
1	52	0.052
2	36	0.036
3	27	0.027
4	58	0.058
5	44	0.044
6	21	0.021
7	48	0.048
8	63	0.063
9	32	0.032
10	38	0.038
11	27	0.027
12	43	0.043
13	22	0.022
14	35	0.035
15	41	0.041
16	20	0.020
17	28	0.028
18	37	0.037
19	24	0.024
20	42	0.042
Total	738	

17. For a p chart, find the values for the central line, UCL, and LCL.

18. Plot a p chart.

In Exercises 19 and 20, use the following data.

A maker of electric fuses checks 500 fuses each day for defects. The number of defective fuses, along with the proportion of defective fuses for 24 days, is shown in the following table.

Day	*Number Defective*	*Proportion Defective*
1	26	0.052
2	32	0.064
3	37	0.074
4	16	0.032
5	28	0.056
6	31	0.062
7	42	0.084
8	22	0.044
9	31	0.062
10	28	0.056
11	24	0.048
12	35	0.070
13	30	0.060
14	34	0.068
15	39	0.078
16	26	0.052
17	23	0.046
18	33	0.066
19	25	0.050
20	25	0.050
21	32	0.064
22	23	0.046
23	34	0.068
24	20	0.040
Total	696	

19. For a p chart, find the values for the central line, UCL, and LCL.

20. Plot a p chart.

Answer to Practice Exercise

1. $\text{UCL}(p) = 0.0333$, $\text{LCL}(p) = 0.0067$

22.6 Linear Regression

In the previous sections of this chapter, we have discussed various statistical methods for graphically and numerically summarizing the values of a single variable. In this section, we will show how to describe the relationship between *two different* quantitative variables, which allows us to predict the value of one from the other. There are many situations where data points from paired data form a general straight-line pattern but don't line up exactly along a straight line. In these cases, we wish to find the equation of a line that ***best fits*** the data points. The process of finding such a line is called **linear regression**.

In this section, we will show a method of finding the equation of a straight line that best approximates passing through a set of points. We consider nonlinear regression in the next section.

EXAMPLE 1 Fitting a line to a set of points

All the students enrolled in a mathematics course took an entrance test. To study the reliability of this test as an indicator of future success, an instructor tabulated the test scores of 10 students (selected at random), along with their course averages at the end of the course, and made a table of the data, which is shown below. The instructor then plotted the data and noticed that in general, the higher the test score, the higher the course grade. She wondered if there might be a straight line that would *fit* the data points reasonably well so that a student's success in the course could be predicted based on his or her entrance test score. Fig. 22.21 shows two such possible lines.

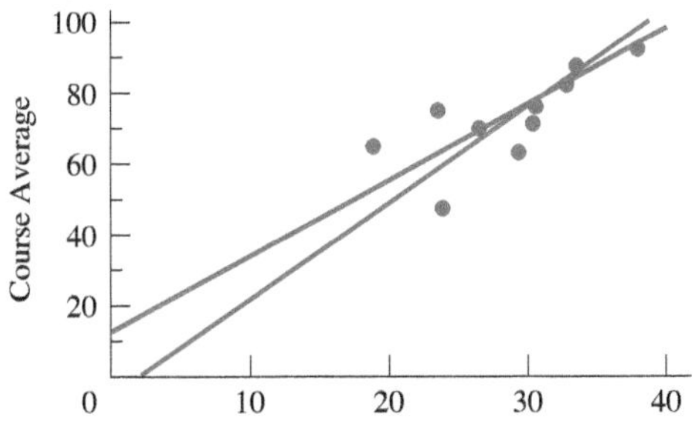

Fig. 22.21

Student	*Entrance Test Score, Based on 40*	*Course Average, Based on 100*
A	29	63
B	33	88
C	22	77
D	17	67
E	26	70
F	37	93
G	30	72
H	32	81
I	23	47
J	30	74

Since there are many ways of *visually* drawing a line through a set of points, we must establish *criteria* for determining which one fits the best.

There are a number of different methods of determining the straight line that best fits the given data points. We employ the method that is most widely used: the **method of least squares**. *The basic principle of this method is that the sum of the squares of the deviations of all data points from the best line (in accordance with this method) has the least value possible. By* **deviation**, *we mean the difference between the y-value of the line and the y-value for the point (of original data) for a particular value of x.*

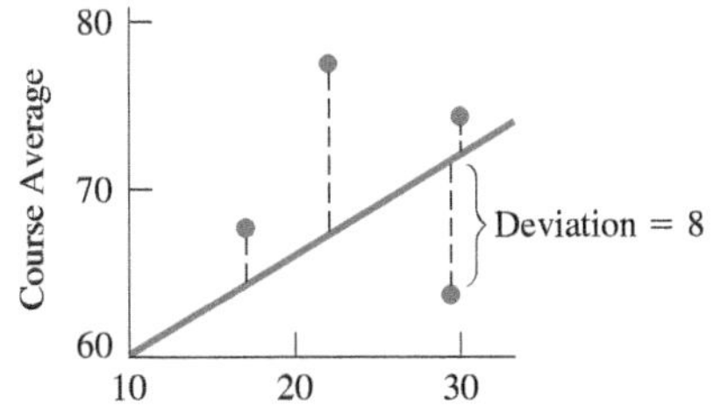

Fig. 22.22

EXAMPLE 2 Deviation

In Fig. 22.22, the deviations of some of the points of Example 1 are shown. The point (29, 63) (student A of Example 1) has a deviation of 8 from the indicated line in the figure. Thus, we square the value of this deviation to obtain 64. In order to find the equation of the straight line that best fits the given points, the method of least squares requires that the sum of all such squares be a minimum.

In applying this method of least squares, it is necessary to use the equation of a straight line and the coordinates of the points of the data. The deviations of all of these data points are determined, and these values are then squared. It is then necessary to determine the constants for the slope m and the y-intercept b in the equation of a straight line $y = mx + b$ for which the sum of the squared values is a minimum.

To do this requires certain methods of advanced mathematics. Using those methods, it is shown that *the equation of the* **least-squares line**

$$y = mx + b \tag{22.17}$$

can be found by calculating the values of the slope m and the y-intercept b by using the formulas

$$m = \frac{n\sum xy - \left(\sum x\right)\left(\sum y\right)}{n\sum x^2 - \left(\sum x\right)^2} \tag{22.18}$$

and

$$b = \frac{\left(\sum x^2\right)\left(\sum y\right) - \left(\sum xy\right)\left(\sum x\right)}{n\sum x^2 - \left(\sum x\right)^2} \tag{22.19}$$

In the above equations, ***the x's and y's are the values of the coordinates of the points in the given data,*** and n is the number of points of data.

The following examples illustrate finding the least-squares line by use of Eqs. (22.8) and (22.9). Note that much of the work involved is finding the required sums for x, y, xy, and x^2. All of the calculations can be done on a calculator, on a spreadsheet, or using statistical software.

COMMON ERROR

It is a common error to confuse $\sum x^2$ and $\left(\sum x\right)^2$ in the denominators of Eqs. (22.18) and (22.19). Note that for $\sum x^2$, we square the x values and then add the squares, whereas for $\left(\sum x\right)^2$, we first add the x values and then square the sum.

EXAMPLE 3 Finding the equation of the least-squares line

Find the equation of the least-squares line for the points indicated in the following table. Graph the line and data points on the same graph.

x	1	2	3	4	5
y	3	6	6	8	12

We see from Eqs. (22.18) and (22.19) that we need the sums of x, y, xy, and x^2 in order to find m and b. Thus, we set up a table for these values, along with the necessary calculations, as follows:

	x	y	xy	x^2
	1	3	3	1
	2	6	12	4
	3	6	18	9
	4	8	32	16
	5	12	60	25
sums →	15	35	125	55
	↑ $\sum x$	↑ $\sum y$	↑ $\sum xy$	↑ $\sum x^2$

$n = 5$ (5 points)

$$m = \frac{5(125) - (15)(35)}{5(55) - (15)^2} = \frac{100}{50} = 2$$

$$b = \frac{(55)(35) - (125)(15)}{50} = \frac{50}{50} = 1$$

This means that the equation of the least-squares line is $y = 2x + 1$. This line and the data points are shown in Fig. 22.23.

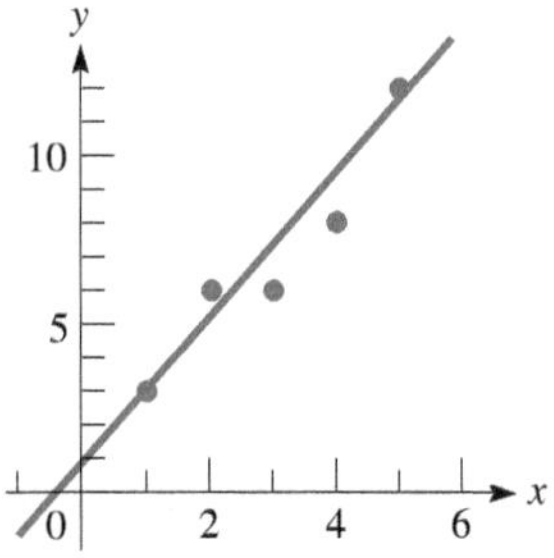

Fig. 22.23

EXAMPLE 4 Finding the equation of the least-squares line

Find the least-squares line for the data of Example 1.

Here, the x-values will be the entrance-test scores and the y-values are the course averages.

x	y	xy	x^2
29	63	1827	841
33	88	2904	1089
22	77	1694	484
17	67	1139	289
26	70	1820	676
37	93	3441	1369
30	72	2160	900
32	81	2592	1024
23	47	1081	529
30	74	2220	900
279	732	20 878	8101

$n = 10$

$$m = \frac{10(20\,878) - 279(732)}{10(8101) - 279^2} = 1.44$$

$$b = \frac{8101(732) - 20\,878(279)}{10(8101) - 279^2} = 33.1$$

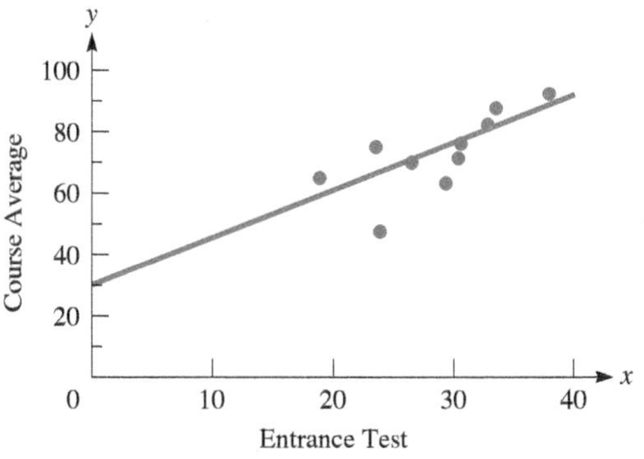

Fig. 22.24

Thus, the equation of the least-squares line is $y = 1.44x + 33.1$. The line and data points are shown in Fig. 22.24.

The regression line can be used to predict a student's expected course average based on their entrance test score. For example, to predict the course average for a student who scored 30 on the entrance test, we substitute 30 for x and evaluate: $y = 1.44(30) + 33.1 = 76$. The predicted course average is 76.

■ A graphing utility, a spreadsheet, or computer software can be used to find the slope and the intercept and to display the least-squares line.

EXAMPLE 5 Least-squares line—pharmacokinetics

In a research project to determine the amount of a drug that remains in the bloodstream after a given dosage, the amounts y (in mg of drug/dL of blood) were recorded after t hours, as shown in the following table. Find the least-squares line for these data, expressing y as a function of t. Sketch the graph of the line and data points.

The calculations are as follows:

t (h)	1.0	2.0	4.0	8.0	10.0	12.0
y (mg/dL)	7.6	7.2	6.1	3.8	2.9	2.0

t	y	ty	t^2
1.0	7.6	7.6	1.0
2.0	7.2	14.4	4.0
4.0	6.1	24.4	16.0
8.0	3.8	30.4	64.0
10.0	2.9	29.0	100
12.0	2.0	24.0	144
37.0	29.6	129.8	329

$n = 6$

$$m = \frac{6(129.8) - 37.0(29.6)}{6(329) - 37.0^2} = -0.523$$

$$b = \frac{(329)(29.6) - (129.8)(37.0)}{6(329) - 37.0^2} = 8.16$$

The equation of the least-squares line is $y = -0.523t + 8.16$. The line and the data points are shown in Fig. 22.25. This line is useful in determining the effectiveness of the drug. It can also be used to determine when additional medication may be administered.

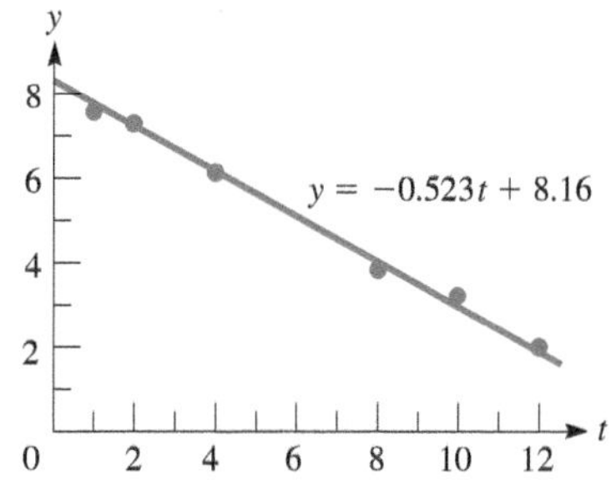

Fig. 22.25

EXERCISES 22.6

In Exercises 1–14, find the equation of the least-squares line for the given data. Graph the line and data points on the same graph.

1. In Example 3, replace the y-values with 3, 7, 9, 9, and 12. Then follow the instructions above.

2.

x	1	2	3	4	5	6	7
y	10	17	28	37	49	56	72

3.

x	20	26	30	38	48	60
y	160	145	135	120	100	90

4.

x	1	3	6	5	8	10	4	7	3	8
y	15	12	10	8	9	2	11	9	11	7

5. In Example 5, change the y (mg of drug/dL of blood) values to 8.7, 8.4, 7.7, 7.3, 5.7, and 5.2. Then proceed to find y as a function of t, as in Example 5.

6. The velocity v (in m/s) of a falling object was found each second by use of an electronic device, as shown in the following table. Find v as a function of t.

t (s)	1.00	2.00	3.00	4.00	5.00	6.00	7.00
v (m/s)	9.70	19.5	29.5	39.4	49.2	58.9	68.6

7. In an electrical experiment, the following data were found for the values of current and voltage for a particular element of the circuit. Find the voltage V as a function of the current i.

Current (mA)	15.0	10.8	9.30	3.55	4.60
Voltage (V)	3.00	4.10	5.60	8.00	10.50

8. The scale length x (mm) and body length y (mm) for a sample of pumpkinseed sunfish caught in the Otonabee River in Ontario in late September are shown in the table. Find the least-squares line of y as a function of x. (Source: Michael Fox, Trent University, unpublished data.)

x (mm)	0.4580	1.617	1.125	2.063	2.708	2.917	1.250
y (mm)	47	75	64	109	125	135	81

9. The fuel consumption x (in L/100 km) and the CO_2 emissions y (in g/km) for Canada's most fuel-efficient cars of 2017 are shown in the table (as reported by Natural Resources Canada). Find the least-squares line for y as a function of x.

x (L/100 km)	4.5	5.0	5.1	5.6	5.7
y (g/km)	105	114	121	131	132

10. In testing an air-conditioning system, the temperature T in a building was measured during the afternoon hours with the results shown in the table. Find the least-squares line for T as a function of the time t from noon.

t (h)	0.0	1.0	2.0	3.0	4.0	5.0
T(°C)	20.5	20.6	20.9	21.3	21.7	22.0

11. The pressure p was measured along an oil pipeline at different distances from a reference point, with results as shown. Find the least-squares line for p as a function of x.

x (m)	0	50	100	150	200
p (kPa)	4370	4240	4070	3970	3840

12. The heat loss L per hour through various thicknesses of a particular type of insulation was measured as shown in the table. Find the least-squares line for L as a function of t.

t (m)	3.0	4.0	5.0	6.0	7.0
L (MJ)	5.90	4.80	3.90	3.10	2.45

13. In an experiment on the photoelectric effect, the frequency of light being used was measured as a function of the stopping potential (the voltage just sufficient to stop the photoelectric effect) with the results given below. Find the least-squares line for V as a function of f. The frequency for $V = 0$ is known as the *threshold frequency*. From the graph determine the threshold frequency.

f (PHz)	0.550	0.605	0.660	0.735	0.805	0.880
V (V)	0.350	0.600	0.850	1.10	1.45	1.80

14. If gas is cooled under conditions of constant volume, it is noted that the pressure falls nearly proportionally as the temperature. If this were to happen until there was no pressure, the theoretical temperature for this case is referred to as *absolute zero*. In an elementary experiment, the following data were found for pressure and temperature under constant volume.

T (°C)	0.0	20	40	60	80	100
p (kPa)	133	143	153	162	172	183

Find the least-squares line for p as a function of T, and from the graph determine the value of absolute zero found in this experiment.

The linear coefficient of correlation, a measure of the strength of the linear relationship of two variables, is defined by $r = m(s_x/s_y)$, where s_x and s_y are the standard deviations of the x-values and y-values, respectively. Due to its definition, the values of r lie in the range $-1 \le r \le 1$. If r is near 1, the correlation is considered good. For the values of r between -0.5 and $+0.5$, the correlation is poor. If r is near -1, the variables are said to be negatively correlated; that is, one increases as the other decreases. In Exercises 15–18, compute r for the given data.

15. Exercise 1

16. Exercise 2

17. Exercise 4

18. Example 1

22.7 Nonlinear Regression

If the experimental points do not appear to be on a straight line, but we recognize them as being approximately on some other type of curve, then nonlinear regression must be used to fit a curve to the data points. For example, if the points are apparently on a parabola, we would want to fit a quadratic equation instead of a line.

Often, the method of linear least squares can be adapted to fit curves. The method is to create new variables from the available data. By using the appropriate transformation on the data, the curved function can be written as a linear function of the new variables. In particular, we consider nonlinear transformations $f(x)$ of the x variable, and extend the least-squares line to

$$y = m[f(x)] + b \quad \textbf{(22.20)}$$

Here, $f(x)$ must be calculated first, and then the problem can be treated as a least-squares line to find the values of m and b. Some of the functions $f(x)$ that may be considered for use are x^2, $1/x$, 10^x, and $\ln x$.

EXAMPLE 1 Fitting $y = mx^2 + b$ to a set of points

Find the least-squares curve $y = mx^2 + b$ for the following points:

x	0	1	2	3	4	5
y	1	5	12	24	53	76

In using Eq. (22.12), $f(x) = x^2$. Our first step is to calculate values of x^2, and then we use x^2 as we used x in finding the equation of the least-squares line.

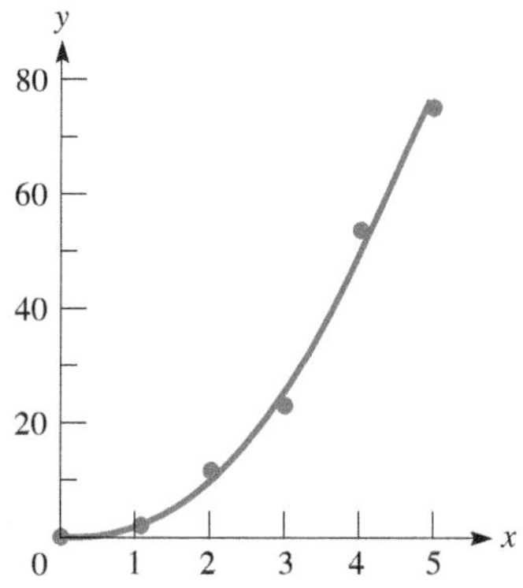

Fig. 22.26

x	$f(x) = x^2$	y	x^2y	$(x^2)^2$
0	0	1	0	0
1	1	5	5	1
2	4	12	48	16
3	9	24	216	81
4	16	53	848	256
5	25	76	1900	625
	55	171	3017	979

$$n = 6$$

$$m = \frac{6(3017) - 55(171)}{6(979) - 55^2} = 3.05$$

$$b = \frac{(979)(171) - (3017)(55)}{6(979) - 55^2} = 0.52$$

Therefore, the required equation is $y = 3.05x^2 + 0.52$. The graph of this equation and the data points are shown in Fig. 22.26.

EXAMPLE 2 Fitting $y = m(1/x) + b$ to a set of points

In a physics experiment, the pressure p and volume V of a gas were measured at constant temperature. When the points were plotted, they were seen to approximate the hyperbola $y = c/x$. Find the least-squares approximation to the hyperbola $y = m(1/x) + b$ for the given data. See Fig. 22.27.

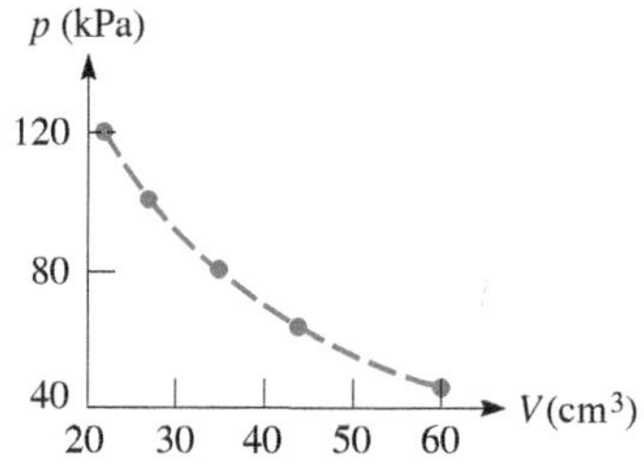

Fig. 22.27

p (kPa)	V (cm^3)	$x (= V)$	$f(x) = \frac{1}{x}$	$y (= p)$	$\left(\frac{1}{x}\right)y$	$\left(\frac{1}{x}\right)^2$
120.0	21.0	21.0	0.047 619 0	120.0	5.714 285 7	0.002 267 6
99.2	25.0	25.0	0.040 000 0	99.2	3.968 000 0	0.001 600 0
81.3	31.8	31.8	0.031 446 5	81.3	2.556 603 8	0.000 988 9
60.6	41.1	41.1	0.024 330 9	60.6	1.474 452 6	0.000 592 0
42.7	60.1	60.1	0.016 638 9	42.7	0.710 482 5	0.000 276 9
			0.160 035 3	403.8	14.423 824 6	0.005 725 4

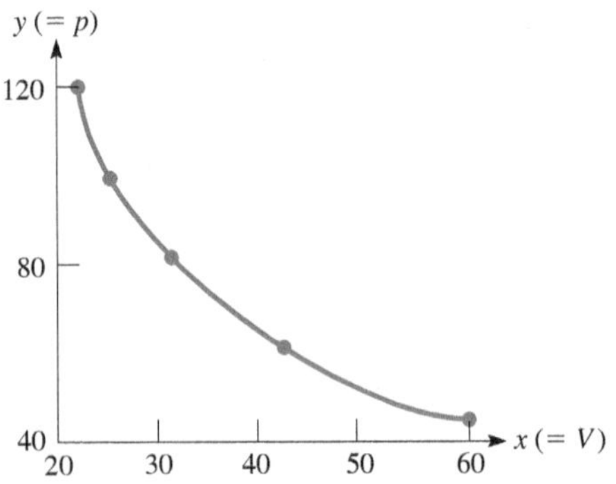

Fig. 22.28

(*Calculator note:* The final digits for the values shown may vary depending on the calculator and how the values are used. Here, all individual values are shown with eight digits (rounded off), although more digits were used. The value of $1/x$ was found from the value of x, with the eight digits shown. However, the values of $(1/x)y$ and $(1/x)^2$ were found from the value of $1/x$, using the extra digits. The sums were found using the rounded-off values shown. However, since the data contain only three digits, any variation in the final digits for $1/x$, $(1/x)y$, or $(1/x)^2$ will not matter.)

$$m = \frac{5(14.423\,824\,6) - 0.160\,035\,3(403.8)}{5(0.005\,725\,4) - 0.160\,035\,3^2} = 2490$$

$$b = \frac{(0.005\,725\,4)(403.8) - (14.423\,824\,6)(0.160\,035\,3)}{5(0.005\,725\,4) - 0.160\,035\,3^2} = 1.2$$

The equation of the hyperbola $y = m(1/x) + b$ is

$$y = \frac{2490}{x} + 1.2$$

This hyperbola and data points are shown in Fig. 22.28.

EXAMPLE 3 Fitting $y = m(10^x) + b$ to a set of points

It has been found experimentally that the tensile strength of brass (a copper–zinc alloy) increases (within certain limits) with the percent of zinc. The following table shows the values that have been found. See Fig. 22.29.

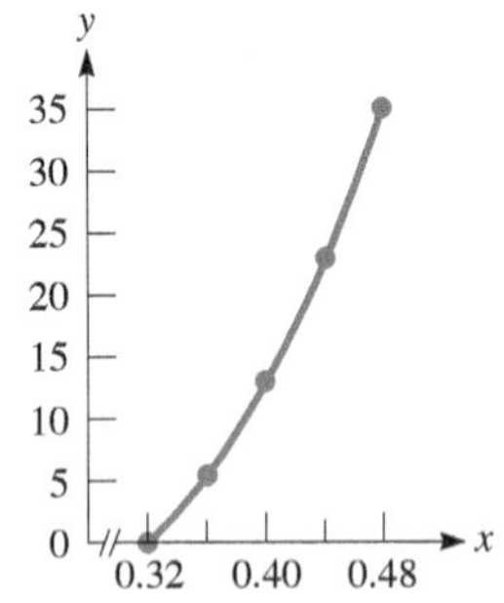

Fig. 22.29

Tensile Strength (GPa)	0.32	0.36	0.40	0.44	0.48
Percent of Zinc	0	5	13	22	34

Fit a curve of the form $y = m(10^x) + b$ to the data. Let $x =$ tensile strength and $y =$ percent of zinc.

x	$f(x) = 10^x$	y	$(10^x)y$	$(10^x)^2$
0.32	2.089 296 1	0	0.000 000	4.365 158 3
0.36	2.290 867 7	5	11.454 338	5.248 074 6
0.40	2.511 886 4	13	32.654 524	6.309 573 4
0.44	2.754 228 7	22	60.593 031	7.585 775 8
0.48	3.019 951 7	34	102.678 36	9.120 108 4
	12.666 230 6	74	207.380 25	32.628 690 5

(See the note on calculator use in Example 2.)

$$m = \frac{5(207.380\,25) - 12.666\,230\,6(74)}{5(32.628\,690\,5) - 12.666\,230\,6^2} = 36.8$$

$$b = \frac{32.628\,690\,5(74) - 207.380\,25(12.666\,230\,6)}{5(32.628\,690\,5) - 12.666\,230\,6^2} = -78.3$$

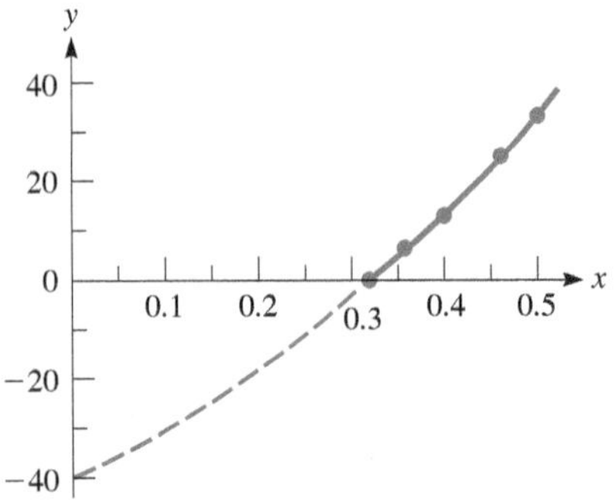

Fig. 22.30

The equation of the curve is $y = 36.8(10^x) - 78.3$. It must be remembered that for practical purposes, y must be positive. The graph of the equation is shown in Fig. 22.30, with the solid portion denoting the meaningful part of the curve. The points of the data are also shown.

As we noted in the previous section, a graphing utility, a spreadsheet, or computer software can be used to determine the equation of the regression curve, and to display its graph. The different models that can be considered include

Linear:	$y = ax + b$
Quadratic:	$y = ax^2 + bx + c$
Cubic:	$y = ax^3 + bx^2 + cx + d$
Quartic:	$y = ax^4 + bx^3 + cx^2 + dx + e$
Logarithmic:	$y = a + b \ln x$
Exponential:	$y = ab^x$
Power:	$y = ax^b$
Logistic:	$y = \dfrac{c}{1 - ae^{-bx}}$
Sinusoidal:	$y = a \sin(bx + c) + d$

EXERCISES 22.7

In the following exercises, find the equation of the indicated least-squares curve. Sketch the curve and plot the data points on the same graph.

1. In Example 1, replace the y-values with 2, 3, 10, 25, 44, and 65. Then follow the instructions above.

2. For the points in the following table, find the least-squares curve $y = m\sqrt{x} + b$.

x	0	4	8	12	16
y	1	9	11	14	15

3. In Example 2, change the V (volume of the gas) values to 19.9, 24.5, 29.4, 39.4, and 56.0. Then find $y\ (= p)$ as a function of $x\ (= V)$, as in Example 2.

4. In Example 3, change the y (percent of zinc) values to 2, 8, 15, 23, and 32. Then find y as a function of x (tensile strength), as in Example 3.

5. The following data were found for the distance y that an object rolled down an inclined plane in time t. Determine the least-squares curve $y = mt^2 + b$. Compare the equation with that using the *quadratic regression* feature on a graphing utility.

t (s)	1.0	2.0	3.0	4.0	5.0
y (cm)	6.0	23	55	98	148

6. The increase in length y of a certain metallic rod was measured in relation to particular increases x in temperature. Find the least-squares curve $y = mx^2 + b$.

x (°C)	50.0	100	150	200	250
y (cm)	1.00	4.40	9.40	16.4	24.0

7. The pressure p at which Freon, a refrigerant, vapourizes for temperature T is given in the following table. Find the least-squares curve $p = mT^2 + b$.

T(°C)	0	10	20	30	40
p (kPa)	480	600	830	1040	1400

8. A fraction f of annual hot-water loads at a certain facility are heated by solar energy. The fractions f for certain values of the collector area A are given in the following table. Find the least-squares curve $f = m\sqrt{A} + b$.

A(m^2)	0	12	27	56	90
f	0.0	0.2	0.4	0.6	0.8

9. The makers of a special blend of coffee found that the demand for the coffee depended on the price charged. The price P per pound and the monthly sales S are shown in the following table. Find the least-squares curve $P = m(1/S) + b$.

S (thousands)	240	305	420	480	560
P (dollars)	5.60	4.40	3.20	2.80	2.40

10. The resonant frequency f of an electric circuit containing a 4-μF capacitor was measured as a function of an inductance L in the circuit. The following data were found. Find the least-squares curve $f = m(1/\sqrt{L}) + b$.

L (H)	1.0	2.0	4.0	6.0	9.0
f (Hz)	490	360	250	200	170

11. The displacement y of an object at the end of a spring at given times t is shown in the following table. Find the least-squares curve $y = me^{-t} + b$.

t (s)	0.0	0.5	1.0	1.5	2.0	3.0
y (cm)	6.1	3.8	2.3	1.3	0.7	0.3

12. The average daily temperatures T (in °C) for each month in Montreal (Environment Canada Archives) are given in the following table:

t	J	F	M	A	M	J	J	A	S	O	N	D
T (°C)	−9	−7	−1	7	14	19	22	21	16	9	2	−6

Find the least-squares curve $T = m\cos\left[\frac{\pi}{6}(t - 0.5)\right] + b$. Assume the average temperature is for the 15th of each month. Then the values of t (in months) are 0.5, 1.5, . . . , 11.5.

CHAPTER 22 KEY FORMULAS AND EQUATIONS

Arithmetic mean	$\bar{x} = \dfrac{x_1 f_1 + x_2 f_2 + \cdots + x_n f_n}{f_1 + f_2 + \cdots + f_n}$	**(22.1)**
Standard deviation	$s = \sqrt{\dfrac{\sum (x - \bar{x})^2}{n - 1}}$	**(22.2)**
	$s = \sqrt{\dfrac{n\left(\sum x^2\right) - \left(\sum x\right)^2}{n(n - 1)}}$	**(22.3)**
Normal distribution	$y = \dfrac{e^{-(x-\mu)^2/2\sigma^2}}{\sigma\sqrt{2\pi}}$	**(22.4)**
Standard normal distribution	$y = \dfrac{1}{\sqrt{2\pi}} e^{-x^2/2}$	**(22.5)**
Standard (z) score	$z = \dfrac{x - \mu}{\sigma}$	**(22.6)**
Standard error of $\bar{x}$	$\sigma_{\bar{x}} = \dfrac{\sigma}{\sqrt{n}}$	**(22.7)**
Standard error of $\hat{p}$	$\sigma_{\hat{p}} = \sqrt{\dfrac{p(1 - p)}{n}}$	**(22.8)**
$(1 - \alpha)$ 100% confidence interval for μ	$\bar{x} \pm E$, where $E = z_{\alpha/2} \cdot \dfrac{\sigma}{\sqrt{n}}$ if σ is known $E = z_{\alpha/2} \cdot \dfrac{s}{\sqrt{n}}$ if σ is unknown	**(22.12)**
Sample size required (μ)	$n = \left[\dfrac{z_{\alpha/2}\sigma}{E}\right]^2$	**(22.13)**
$(1 - \alpha)$ 100% confidence interval for p	$\hat{p} \pm E$, where $E = z_{\alpha/2}\sqrt{\dfrac{\hat{p}(1 - \hat{p})}{n}}$	**(22.14)**
Sample size required (p)	$n = \hat{p}(1 - \hat{p})\left[\dfrac{z_{\alpha/2}}{E}\right]^2$ if an estimate $\hat{p}$ is available	**(22.15)**
	$n = \dfrac{1}{4}\left[\dfrac{z_{\alpha/2}}{E}\right]^2$ if no estimate $\hat{p}$ is available	**(22.16)**
Least-squares lines	$y = mx + b$	**(22.17)**
	$m = \dfrac{n\sum xy - \left(\sum x\right)\left(\sum y\right)}{n\sum x^2 - \left(\sum x\right)^2}$	**(22.18)**
	$b = \dfrac{\left(\sum x^2\right)\left(\sum y\right) - \left(\sum xy\right)\left(\sum x\right)}{n\sum x^2 - \left(\sum x\right)^2}$	**(22.19)**
Nonlinear curves	$y = m[f(x)] + b$	**(22.20)**

CHAPTER 22 REVIEW EXERCISES

In Exercises 1–10, use the following data. An airline's records showed that the percent of on-time flights each day for a 20-day period was as follows:

72, 75, 76, 70, 77, 73, 80, 75, 82, 85,
77, 78, 74, 86, 72, 77, 67, 78, 69, 80

1. Determine the median.

2. Determine the mode.

3. Determine the mean.

4. Determine the standard deviation.

5. Construct a frequency distribution table with five classes and a lowest class limit of 67.

6. Draw a frequency polygon for the data in Exercise 5.

7. Draw a histogram for the data in Exercise 5.

8. Construct a relative frequency table for the data in Exercise 5.

9. Construct a cumulative frequency table for the data of Exercise 5.

10. Draw an ogive for the data of Exercise 5.

In Exercises 11–16, use the following data. An important property of oil is its coefficient of viscosity, which gives a measure of how well it flows. In order to determine the viscosity of a certain motor oil, a refinery took samples from 12 different storage tanks and tested them at 50°C. The results (in pascal-seconds) were 0.24, 0.28, 0.29, 0.26, 0.27, 0.26, 0.25, 0.27, 0.28, 0.26, 0.26, 0.25.

11. Find the mean.

12. Find the median.

13. Find the standard deviation.

14. Draw a histogram.

15. Draw a frequency polygon.

16. Determine the range.

In Exercises 17–26, use the following data. A sample of wind generators was tested for power output when the wind speed was 30 km/h. The following table gives the class marks of the powers produced and the number of generators in each class.

Power (W)	650	660	670	680	690
No. Generators	3	2	7	12	27

Power (W)	700	710	720	730
No. Generators	34	15	16	5

17. Find the median.

18. Find the mean.

19. Find the mode.

20. Draw a histogram.

21. Find the standard deviation.

22. Draw a frequency polygon.

23. Make a cumulative frequency table.

24. Draw an ogive.

25. The mean and the standard deviation obtained from the raw data are $\bar{x} = 696$ and $s = 17.7$. Find a 95% confidence interval for the true mean power output produced by all generators of this kind when wind speed is 30 km/h.

26. How many more observations should be taken so that a 95% confidence interval for the true mean power output will have a maximum margin of error $E = 2.5$ W?

In Exercises 27–30, use the following data. A Geiger counter records the presence of high-energy nuclear particles. Even though no apparent radioactive source is present, some particles will be recorded. These are primarily cosmic rays, which are caused by very high-energy particles from outer space. In an experiment to measure the amount of cosmic radiation, the number of counts were recorded during 200 five-second intervals. The following table gives the number of counts and the number of five-second intervals having this number of counts. Draw a frequency curve for these data.

Counts	0	1	2	3	4	5	6	7	8	9	10
Intervals	3	10	25	45	29	39	26	11	7	2	3

27. Find the median.

28. Find the mean.

29. Draw a histogram.

30. Make a relative frequency table.

In Exercises 31–36, use the following data. Police radar on a city street recorded the speeds of 110 cars in a 65 km/h zone. The following table shows the class marks of the speeds recorded and the number of cars in each class.

Speed (km/h)	40	45	50	55	60	65	70	75	80	85
No. cars	3	4	4	5	8	22	48	10	4	2

31. Find the mean.

32. Find the median.

33. Find the standard deviation.

34. Draw an ogive.

35. The mean and the standard deviation obtained from the raw data are $\bar{x} = 66$ and $s = 9.1$. Find a 90% confidence interval for the true mean speed in this zone.

36. How many more observations should be taken so that a 90% confidence interval for the true mean speed will have a maximum margin of error $E = 1.0$ km/h?

In Exercises 37–38, solve the given problems.

37. Use Chebychev's theorem to find the percentage of values that are between 27.8 and 36.2 in a data set with mean 32 and standard deviation 2.1.

38. Use Chebychev's theorem to find the percentage of values that are between 174.2 and 189.8 in a data set with mean 182 and standard deviation 2.6.

In Exercises 39–42, use the following data. A random sample of size $n = 185$ is taken from a large population, and 36 are found to be defective.

39. Construct a 95% confidence interval for the population proportion of defectives p.

40. Construct a 90% confidence interval for the population proportion of defectives p.

41. How large a sample must be taken so that a 90% confidence interval for p will have a maximum margin of error of 3.2%? Assume that the information from the sample is used.

42. How large a sample must be taken so that a 90% confidence interval for p will have a maximum margin of error of 3.2%? Assume that no prior information is used.

In Exercises 43 and 44, use the following information. A company that makes electric light bulbs tests 500 bulbs each day for defects. The number of defective bulbs, along with the proportion of defective bulbs for 20 days, is shown in the following table.

Day	*Number Defective*	*Proportion Defective*
1	23	0.046
2	31	0.062
3	19	0.038
4	27	0.054
5	29	0.058
6	39	0.078
7	26	0.052
8	17	0.034
9	28	0.056
10	33	0.066
11	22	0.044
12	29	0.058
13	20	0.040
14	35	0.070
15	21	0.042
16	32	0.064
17	25	0.050
18	23	0.046
19	29	0.058
20	32	0.064
Total	540	

43. For a p chart, find the values of the central line, UCL, and LCL.

44. Plot a p chart.

In Exercises 45 and 46, use the following information. Five ball bearings are taken from the production line every 15 min and their diameters are measured. The diameters of the sample ball bearings for 16 successive subgroups are given in the following table.

Subgroup	*Diameters* (mm) *of Five Ball Bearings*				
1	4.98	4.92	5.02	4.91	4.93
2	5.03	5.01	4.94	5.06	5.07
3	5.05	5.03	5.00	5.02	4.96
4	5.01	4.92	4.91	4.99	5.03
5	4.92	4.97	5.02	4.95	4.94
6	5.02	4.95	5.01	5.07	5.15
7	4.93	5.03	5.02	4.96	4.99
8	4.85	4.91	4.88	4.92	4.90
9	5.02	4.95	5.06	5.04	5.06
10	4.98	4.98	4.93	5.01	5.00
11	4.90	4.97	4.93	5.05	5.02
12	5.03	5.05	4.92	5.03	4.98
13	4.90	4.96	5.00	5.02	4.97
14	5.09	5.04	5.05	5.02	4.97
15	4.88	5.00	5.02	4.97	4.94
16	5.02	5.09	5.03	4.99	5.03

45. Plot an $\bar{x}$ chart.

46. Plot an R chart.

In Exercises 47–50, use the following data. After analysing data for a long period of time, it was determined that samples of 500 readings of an organic pollutant for an area are distributed normally. For this pollutant, $\mu = 2.20\ \mu g/m^3$ and $\sigma = 0.50\ \mu g/m^3$.

47. In a sample, how many readings are expected to be between $1.50\ \mu g/m^3$ and $2.50\ \mu g/m^3$?

48. In a sample, how many readings are expected to be between $2.50\ \mu g/m^3$ and $3.50\ \mu g/m^3$?

49. In a sample, how many readings are expected to be above $1.00\ \mu g/m^3$?

50. In a sample, how many readings are expected to be below $2.00\ \mu g/m^3$?

In Exercises 51–60, find the indicated least-squares curve. Sketch the curve and data points on the same graph.

51. In a certain experiment, the resistance R of a certain resistor was measured as a function of the temperature T. The data found are shown in the following table. Find the least-squares line, expressing R as a function of T.

T (°C)	0.0	20.0	40.0	60.0	80.0	100
R (Ω)	25.0	26.8	28.9	31.2	32.8	34.7

52. An air-pollution monitoring station took samples of air each hour during the later morning hours and tested each sample for the number n of parts per million (ppm) of carbon monoxide. The results are shown in the table, where t is the number of hours after 6:00 A.M. Find the least-squares line for n as a function of t.

t (h)	0.0	1.0	2.0	3.0	4.0	5.0	6.0
n (ppm)	8.0	8.2	8.8	9.5	9.7	10.0	10.7

53. The *Mach number* of a moving object is the ratio of its speed to the speed of sound (1200 km/h). The following table shows the speed s of a jet aircraft, in terms of Mach numbers, and the time t after it starts to accelerate. Find the least-squares line of s as a function of t.

t (min)	0.00	0.60	1.20	1.80	2.40	3.00
s (Mach number)	0.88	0.97	1.03	1.11	1.19	1.25

54. In an experiment to determine the relation between the load x on a spring and the length y of the spring, the following data were found. Find the least-squares line that expresses y as a function of x.

Load (kg)	0.0	1.0	2.0	3.0	4.0	5.0
Length (cm)	10.0	11.2	12.3	13.4	14.6	15.9

55. The distance s of a missile above the ground at time t after being released from a plane is given by the following table. Find the least-squares curve of the form $s = mt^2 + b$ for these data.

t (s)	0.0	3.0	6.0	9.0	12.0	15.0	18.0
s (m)	3000	2960	2820	2600	2290	1900	1410

56. In an elementary experiment that measured the wavelength L of sound as a function of the frequency f, the following results were obtained.

Frequency (Hz)	240	320	400	480	560
Wavelength (cm)	140	107	81.0	70.0	60.0

Find the least-squares curve of the form $L = m(1/f) + b$ for these data.

57. Measurements were made on the current i (in A) in an electric circuit as a function of the time t (in s). The circuit contained a resistance of 5.00 Ω and an inductance of 10.0 H. The following data were found.

t (s)	0.00	2.00	4.00	6.00	8.00
i (A)	0.00	2.52	3.45	3.80	3.92

Find the least-squares curve $i = m(e^{-0.500t}) + b$ for these data. From the equation, determine the value of the current as t approaches infinity.

58. The power P (in W) generated by a wind turbine was measured for various wind velocities v (in km/h), as shown in the following table.

v (km/h)	10	15	20	25	30	40
P (W)	75	250	600	1200	2100	4800

Find the least-squares curve of the form $P = mv^3 + b$ for these data.

59. The vertical distance y of the cable of a suspension bridge above the surface of the bridge is measured at a horizontal distance x along the bridge from its centre. See Fig. 22.31. The results are as follows:

x (m)	0	100	200	300	400	500
y (m)	15	17	23	33	47	65

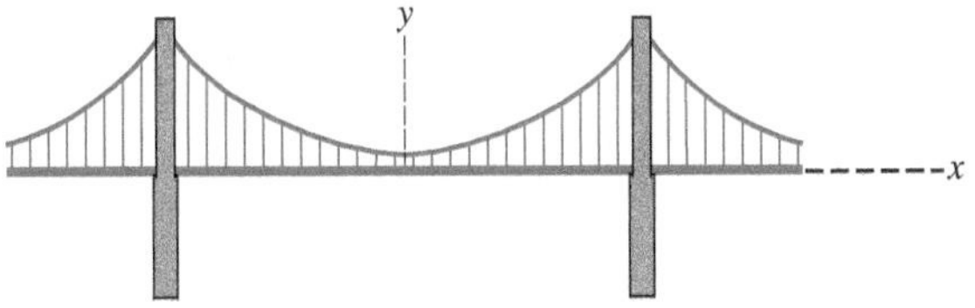

Fig. 22.31

Plot these points and choose an appropriate function $f(x)$ for $y = m[f(x)] + b$. Then find the equation of the least-squares curve.

60. After being heated, the temperature T of an insulated liquid is measured at times t as follows:

t (h)	0	2	4	6	8	10
T (°C)	100	85	72	63	54	48

Plot these points and choose an appropriate function $f(x)$ for $y = m[f(x)] + b$. Then find the equation of the least-squares curve.

In Exercises 61–64, use a graphing utility, a spreadsheet, or software to solve the given problems related to the following data. Using aerial photography, the area A (in km²) of an oil spill as a function of the time t (in h) after the spill was found to be as follows:

t (h)	1.0	2.0	4.0	6.0	8.0	10.0
A (km²)	1.4	2.5	4.7	6.8	8.8	10.2

61. Find the linear equation $y = ax + b$ to fit these data.

62. Find the quadratic equation $y = ax^2 + bx + c$ to fit these data.

63. Find the power equation $y = ax^b$ to fit these data.

64. Find the value of the linear coefficient of correlation r for these data.

In Exercises 65–71, solve the given problems.

65. With 30.5% of the area under the normal curve between $z_1 = 0.5$ and z_2, to the right of z_1, find z_2.

66. With 79.8% of the area under the normal curve between z_1 and $z_2 = 2.1$, find z_1.

67. The nth root of the product of n positive numbers is the *geometric mean* of the numbers. Find the geometric mean of the carbon monoxide readings in Exercise 52.

68. One use of the geometric mean (see Exercise 67) is to find an average ratio. By finding the geometric mean, find the average Mach number for the jet in Exercise 53.

69. Show that Eqs. (22.18) and (22.19) satisfy the equation $\bar{y} = m\bar{x} + b$.

70. Given that $\Sigma(x - \bar{x})^2 = \Sigma x^2 - n\bar{x}^2$, derive Eq. (22.3) from Eq. (22.2).

71. A research institute is planning a study of the effect of education on the income of workers. Write two or three paragraphs explaining what data should be collected and which of the measures discussed in this chapter would be useful in analysing the data.

CHAPTER 22 PRACTICE TEST

In Problems 1–3, use the following set of numbers.

5, 6, 1, 4, 9, 5, 7, 3, 8, 10, 5, 8, 4, 9, 6

1. Find the median.
2. Find the mode.
3. Draw a histogram with five classes and the lowest class limit at 1.

In Problems 4–8, use the following data. Two machine parts are considered satisfactorily assembled if their total thickness (to the nearest 0.01 cm) is between or equal to 0.92 cm and 0.94 cm. One hundred assemblies are tested, and the class mark of the thicknesses and the number of assemblies in each class are given in the following table.

Total Thickness (cm)	0.90	0.91	0.92	0.93	0.94	0.95	0.96
Number	3	9	31	38	12	5	2

4. Find the mean.
5. Find the standard deviation.
6. Draw a frequency polygon.
7. Make a relative frequency table.
8. Draw an ogive (less than).

In Problems 9–14,answer the given questions.

9. A random sample of size $n = 175$ is taken from a large population so that $\bar{x} = 143$ and $s = 5.2$. Construct a 95% confidence interval for the population mean μ.
10. A random sample of size $n = 120$ is taken from a large population and 33 are found defective. Construct a 99% confidence interval for the proportion p of defectives in the population.
11. For a set of values that are normally distributed, what percent of them is below the value (greater than the mean) for which the z-score is 0.2257?
12. The machine-part assemblies in Problems 4–8 were tested in groups of five each hour for 20 hours. Explain, in general, how to use the data from the test subgroups to plot an R chart.
13. Find the equation of the least-squares line for the points indicated in the following table. Graph the line and data points on the same graph.

x	1	3	5	7	9
y	5	11	17	20	27

14. The velocity (in m/s) of an object moving down an inclined plane was measured as a function of the distance (in m) it moved, with the following results:

Distance (m)	1.00	3.00	5.00	7.00	9.00
Velocity (m/s)	1.10	1.90	2.50	2.90	3.30

Find the equation of the least-squares curve of the form $y = m\sqrt{x} + b$, which expresses the velocity as a function of the distance.

23. The Derivative

Max Lindenthaler/Shutterstock

◀ Determining how fast an object is moving is important in physics and other areas of technology. In Section 23.4, we develop the method of finding the instantaneous velocity of a moving object.

LEARNING OUTCOMES

After completion of this chapter, the student should be able to:

- Determine the continuity of a function at a single point or over an interval
- Define the concept of a limit
- Evaluate a limit of a function
- Calculate average slope equations
- Find a derivative using the delta method/definition of derivative
- Interpret instantaneous slope
- Use derivative formulas to calculate derivatives of polynomials, and powers, products, and quotients of functions
- Apply implicit differentiation
- Calculate higher derivatives of functions

In the 1600s, the ability to understand and predict the motion of planets, projectiles, and other moving objects became the focus of many scientists and mathematicians. For example, the velocity of a falling object changes from one instant to the next, and particular interest was devoted to determining this *instantaneous velocity*. It was also noted that the geometric problem of finding the slope of a curve *at a specific point* was really equivalent to finding the instantaneous velocity of an object, since each involved an *instantaneous rate of change.*

It was well known at the time how to find an *average velocity* (distance travelled divided by time taken) and a slope (difference in y-values divided by difference in x-values). However, these methods do not work in finding an instantaneous velocity or slope since they would result in a division by zero. Therefore, a method for finding the slope of a tangent line to a curve at a given point was a major point of interest in mathematics in the 1600s.

This interest in motion and slope of a tangent line led to the development of *calculus*. In this chapter, we start developing *differential calculus*, which is concerned with determining the instantaneous rates of change of one quantity with respect to another. Other examples of an instantaneous rate of change are electric current (rate of change of electric charge with respect to time), and the rate of change of light intensity with respect to the distance from the source. In Chapter 25, we will study *integral calculus*, which involves finding the function for which the rate of change is known.

Isaac Newton, an English mathematician and physicist, and Gottfried Leibniz, a German mathematician and philosopher, are credited with the creation of the basic methods of calculus in the 1660s and 1670s. Others, including the French mathematician Pierre de Fermat, are known to have developed some of the topics related to calculus in the mid-1600s. In the 1700s and 1800s, many mathematicians further developed and refined the concepts of calculus.

The topic of this chapter, the *derivative*, is the basic concept of differential calculus that is used to measure an instantaneous rate of change. We will show some of the applications of the derivative in fields such as physical science and engineering in this chapter, and develop several important applications in the next chapter.

23.1 Limits

LEARNING TIP

Graphically, a function is continuous over an interval if you can draw its graph without lifting your pencil from the paper. If a "break" occurs in the graph because of an undefined point or instantaneous "jump" in the value of the function, or any domain gaps in the interval, then the function is discontinuous for some value or values in that interval.

CONTINUITY OF A FUNCTION

Before dealing with the rate of change of a function, we first take up the concept of a *limit*. We encountered a limit with infinite geometric series and with the asymptotes of a hyperbola. It is necessary to develop this concept further.

To develop the concept of a limit, we first consider the **continuity** of a function. *For a function to be* **continuous at a point**, *the function must exist at the point, and any small change in x must produce only a small change in* $f(x)$. In fact, if the function is continuous, the change in $f(x)$ can be made as small as we wish by restricting the change in x sufficiently. Also, *a function is said to be* **continuous over an interval** *if it is continuous at each point in the interval.*

If the domain of a function includes a point and an interval on only one side of this point, it is *continuous at the point* if the definition of continuity holds for that part of the domain.

EXAMPLE 1 Function continuous for all x

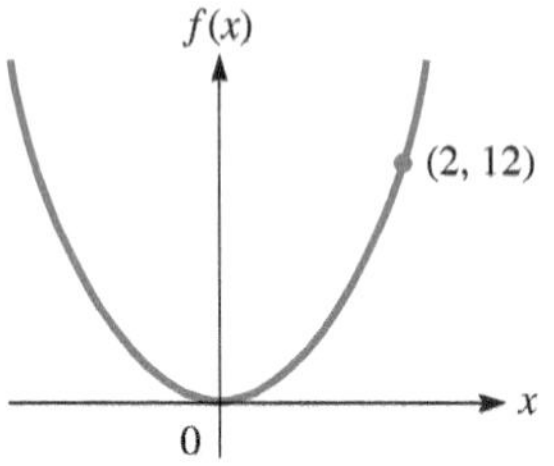

Fig. 23.1

Discuss the continuity of the function $f(x) = 3x^2$ at $x = 2$, and for all values of x.

The function is continuous at x if $f(x)$ is defined at x, and a small change in x produces only a small change in $f(x)$. If we choose $x = 2$, and then let x change by 0.1, 0.01, and so on, we obtain the values in the following table:

x	2	2.1	2.01	2.001
$f(x)$	12	13.23	12.1203	12.012 003
Change in x		0.1	0.01	0.001
Change in $f(x)$		1.23	0.1203	0.012 003

We see that the change in $f(x)$ is smaller for smaller changes in x. Also, $f(x)$ is defined at $x = 2$, so $f(x)$ is continuous at $x = 2$. Since the same analysis is valid for any other x we may choose, we see that $f(x)$ is continuous for all values of x.

The graph of the function $f(x) = 3x^2$ is shown in Fig. 23.1. Note that there are no breaks in the curve, confirming the continuity of this function over all the real numbers.

EXAMPLE 2 Function discontinuous at a point

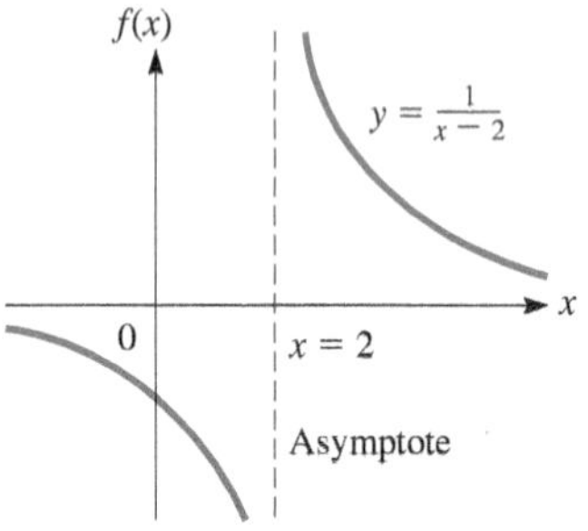

Fig. 23.2

Explain why the function $f(x) = \frac{1}{x-2}$ is not continuous at $x = 2$.

When we substitute 2 for x, we have division by zero. This means the function is not defined for the value $x = 2$. The condition of continuity—that the function must exist—is not satisfied.

The graph of the function is shown in Fig. 23.2. We see that there is a break in the curve at $x = 2$, confirming that the function is discontinuous at $x = 2$.

EXAMPLE 3 Continuity on an interval and discontinuity at a point

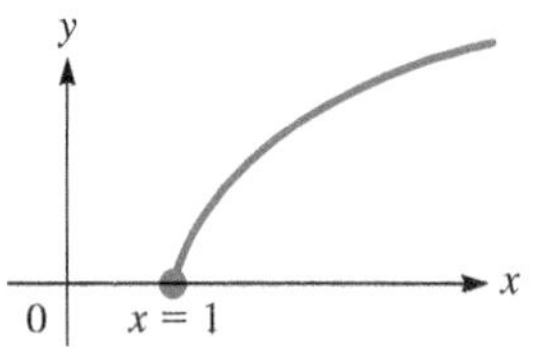

Fig. 23.3

Determine the continuity of each of the following functions,

(a) For the function represented in Fig. 23.3, the solid circle at $x = 1$ shows that the point is on the graph. Since it is continuous to the right of the point, it is also continuous at the point. Thus, the function is continuous for $x \geq 1$.

(b) The function represented by the graph in Fig. 23.4 is not continuous at $x = 1$. The function is defined (by the solid circle point) for $x = 1$. However, a small change from $x = 1$ may result in a change of at least 1.5 in $f(x)$, regardless of how small a change in x is made. The small change condition is not satisfied.

(c) The function represented by the graph in Fig. 23.5 is not continuous for $x = -2$. The open circle shows that the point is not part of the graph, and therefore $f(x)$ is not defined for $x = -2$.

Practice Exercise

1. Determine the continuity of the function $f(x) = \frac{3}{x(x+3)}$.

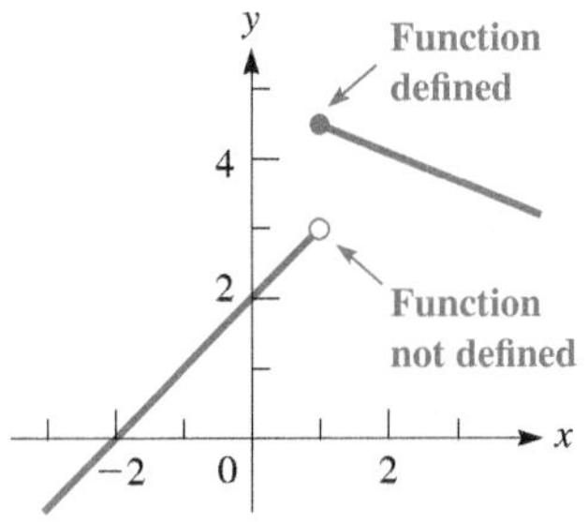

Fig. 23.4

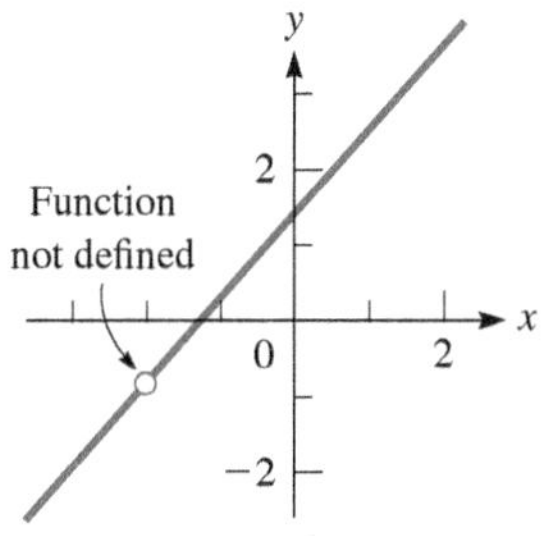

Fig. 23.5

EXAMPLE 4 Function defined differently over the domain

(a) We can define the function in Fig. 23.4 as

$$f(x) = \begin{cases} x+2 & \text{for } x < 1 \\ -\frac{1}{2}x + 5 & \text{for } x \geq 1 \end{cases}$$

where we note that the equation differs for different parts of the domain, and at $x = 1$, an instantaneous "jump" occurs in the graph, making it discontinuous at that point. This function is continuous on the intervals $-\infty < x < 1$ and $1 < x < \infty$ because no gaps or jumps exist in those intervals, but the function is discontinuous at the point $x = 1$.

(b) The graph of the function

$$g(x) = \begin{cases} 2x - 1 & \text{for } x \leq 2 \\ -x + 5 & \text{for } x > 2 \end{cases}$$

is shown in Fig. 23.6. We see that it is a continuous function even though the equation for $x \leq 2$ is different from that for $x > 2$.

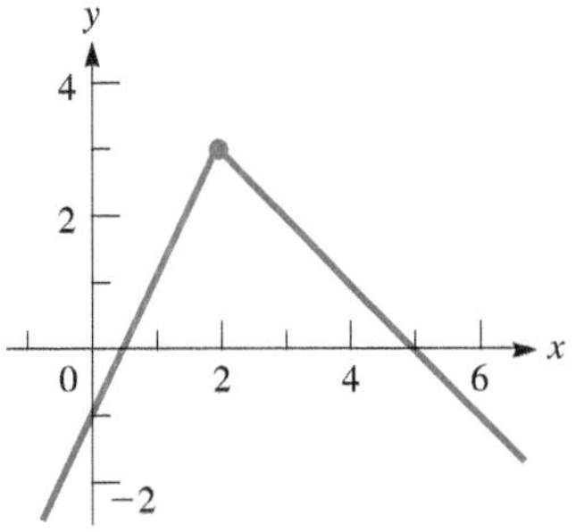

Fig. 23.6

LEARNING TIP

The notation used in limits can be interpreted as follows:

Symbol	Meaning
$x \to 2$	"as x approaches 2"; x can be any number as arbitrarily close to 2 as you want, but **cannot be exactly 2**
$x \to 2^+$	"x approaches 2 from above"; x can be any number as arbitrarily close to 2 as you want, but it **must be larger than 2**
$x \to 2^-$	"x approaches 2 from below"; x can be any number as arbitrarily close to 2 as you want, but it **must be less than 2**

EXAMPLE 5 Behaviour of a function as *x* approaches 2

Examine the behaviour of $f(x) = 2x + 1$ as $x \to 2$.

Since we are not to use $x = 2$, we use a calculator to set up tables in order to determine values of $f(x)$, as x gets close to 2:

x	1.000	1.500	1.900	1.990	1.999
$f(x)$	3.000	4.000	4.800	4.980	4.998

x	3.000	2.500	2.100	2.010	2.001
$f(x)$	7.000	6.000	5.200	5.020	5.002

values approach 5

We see that $f(x)$ approaches 5, as x approaches 2, from above 2 and from below 2.

LIMIT OF A FUNCTION

In Example 5, since $f(x) \to 5$ as $x \to 2$, the number 5 is called the limit of $f(x)$ as $x \to 2$. This leads to the meaning of the limit of a function. In general, *the* **limit of a function $f(x)$** *is that value which the function approaches as x approaches the given value a.* This is written as

$$\lim_{x \to a} f(x) = L \tag{23.1}$$

where L is the value of the limit of the function.

An important conclusion can be drawn from the limit in Example 5. The function $f(x)$ is a continuous function, and $f(2)$ equals the value of the limit as $x \to 2$. In general, it is true that

if $f(x)$ is continuous at $x = a$, then the limit as $x \to a$ equals $f(a)$.

In fact, looking back at our definition of continuity, we see that this is what the definition means. That is, a function $f(x)$ is continuous at $x = a$ if *all three* of the following conditions are satisfied:

1. $f(a)$ exists **2.** $\lim_{x \to a} f(x)$ exists **3.** $\lim_{x \to a} f(x) = f(a)$

Although we can evaluate the limit for a continuous function as $x \to a$ by evaluating $f(a)$, it is possible that a function is not continuous at $x = a$ and yet the limit exists and can be determined. Thus, we must be able to determine the value of a limit without finding $f(a)$. The following example illustrates the evaluation of such a limit.

EXAMPLE 6 Limit of a function using a table of values

Find $\lim_{x \to 2} \dfrac{2x^2 - 3x - 2}{x - 2}$.

We note immediately that the function is not continuous at $x = 2$, for division by zero is indicated. Thus, we cannot evaluate the limit by substituting $x = 2$ into the function. Using a calculator to set up tables we determine the value that $f(x)$ approaches, as x approaches 2 from either side:

x	1.000	1.500	1.900	1.990	1.999
$f(x)$	3.000	4.000	4.800	4.980	4.998

x	3.000	2.500	2.100	2.010	2.001
$f(x)$	7.000	6.000	5.200	5.020	5.002

values approach 5

We see that the values obtained are identical to those in Example 5. Since $f(x) \to 5$ as $x \to 2$, we have

$$\lim_{x \to 2} \frac{2x^2 - 3x - 2}{x - 2} = 5$$

Therefore, we see that the limit exists at $x \to 2$, although the function does not exist at $x = 2$.

The reason the functions in Examples 5 and 6 have the same limit as x approaches 2 is shown in the following example.

EXAMPLE 7 Factoring in limits

The function $y = \frac{2x^2 - 3x - 2}{x - 2}$ in Example 6 is the same as the function $y = 2x + 1$ in Example 5, except when $x = 2$. By factoring the numerator of the function of Example 6, and then cancelling, we have

$$\frac{2x^2 - 3x - 2}{x - 2} = \frac{(2x + 1)(x - 2)}{x - 2} = 2x + 1$$

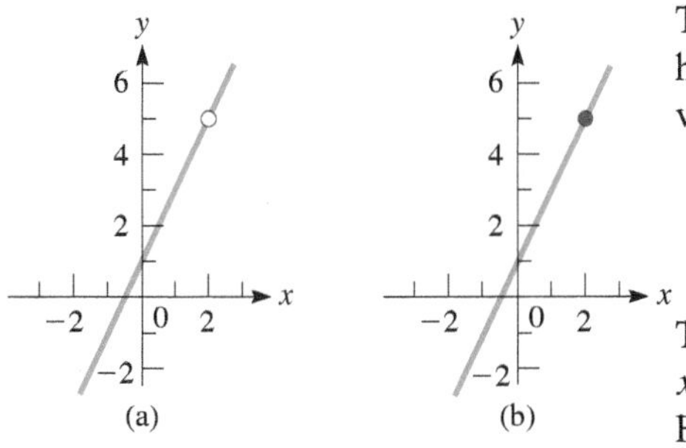

Fig. 23.7

The cancellation for this expression is valid, as long as x does not equal 2, for we have division by zero at $x = 2$. Also, in finding the limit as $x \to 2$, we do not use the value $x = 2$. Therefore,

$$\lim_{x\to 2} \frac{2x^2 - 3x - 2}{x - 2} = \lim_{x\to 2}(2x + 1) = 5$$

The limits of the two functions are equal, since, again, in finding the limit as x approaches 2, we do not let $x = 2$. The graphs of the two functions are shown in Fig. 23.7(a) and (b). We can see from the graphs that the limits are the same, although one of the functions is not continuous.

If $f(x) = 5$ for $x = 2$ is added to the definition of the function in Example 6, it is then the same as $f(x) = 2x + 1$, and its graph is that in Fig. 23.7(b).

The limit of the function in Example 6 was determined by calculating values near $x = 2$ and by means of an algebraic change in the function. This illustrates that limits may be found through the meaning and definition and through other procedures when the function is not continuous. The following is a step-by-step guide to evaluating limits. It is intended as an introduction to the most common limits that could be encountered, but it is by no means complete.

Procedure for Evaluating Limits

The following steps are listed in order of increasing difficulty. Try the simpler methods before moving on to the more difficult ones. Always remember that algebraic reduction in the initial stages of any problem can greatly simplify it.

1. **Direct substitution.** Substitute the limiting value of $x = a$ into the function. If you obtain a real result, and the function is continuous at that point, you have the limit L.
2. **Factoring.** If you obtain a $\frac{0}{0}$ indeterminate limit, try factoring the expression.
3. **Division by highest power.** If you obtain an $\frac{\infty}{\infty}$ indeterminate limit, try dividing both the numerator and the denominator by the highest power of x that appears in either the numerator or denominator.
4. **Multiplication by conjugate.** If you see square roots over individual terms in either the numerator or denominator, try conjugate multiplication of both the numerator and denominator.
5. **Rearrange inside the root sign.** If you see square roots over an entire numerator or denominator, try rearranging the function so that it is entirely under a root sign.
6. **Table.** Construct a table of values for values of x approaching the limiting value from above and below. If the limits from above and below agree, you have the limit L.

Remember, checking the valid domains for x in the cases where square roots are present is always important.

EXAMPLE 8 Limit for a continuous function

Find $\lim_{x\to 4}(x^2 - 7)$.

We start with direct substitution, substituting $x = 4$ in $f(x) = x^2 - 7$. We have $f(4) = 4^2 - 7 = 9$. Since the function is continuous at $x = 4$, we have

$$\lim_{x\to 4}(x^2 - 7) = 9$$

EXAMPLE 9 Limit for a discontinuous function, using factoring

Find $\lim_{t \to 2}\left(\frac{t^2 - 4}{t - 2}\right)$.

We start with direct substitution, substituting $t = 2$ into $f(t) = \frac{t^2 - 4}{t - 2}$. We have $f(2) = \frac{2^2 - 4}{2 - 2} = \frac{0}{0}$, which is indeterminate.

Next, we try factoring:

$$\frac{t^2 - 4}{t - 2} = \frac{(t - 2)(t + 2)}{t - 2} = t + 2$$

Since this expression is valid as long as $t \neq 2$, we find that

$$\lim_{t \to 2}\left(\frac{t^2 - 4}{t - 2}\right) = \lim_{t \to 2}(t + 2) = 4$$

Again, we do not have to be concerned with the fact that the cancellation is not valid for $t = 2$. In finding the limit, we do not consider the value of $f(t)$ at $t = 2$.

Practice Exercise

2. Find $\lim_{x \to 5} \frac{x^2 - 25}{x - 5}$.

LIMITS AS x APPROACHES INFINITY

Limits as x approaches infinity are also of importance. However, when dealing with these limits, we must remember the following.

COMMON ERROR

When we say the limit is infinite (∞), we mean the value of the function gets larger and larger, without bound. Infinity is not a specific value, a place, or a stop sign; it is a *concept*. ∞ does not represent a real number, and algebraic operations may not be performed on it.

Therefore, when we write $x \to \infty$, we know that we are to consider values of x that are becoming larger and larger without bound. We encountered this concept in Chapter 19 when discussing infinite geometric series, and in Chapter 21 when discussing the asymptotes of a hyperbola. The following examples illustrate the evaluation of this type of limit for algebraic expressions.

EXAMPLE 10 Limit as x approaches infinity

Find $\lim_{x \to \infty} \frac{x^2 + 1}{2x^2 + 3}$.

We note that as $x \to \infty$, both the numerator and the denominator become large without bound, giving an $\frac{\infty}{\infty}$ indeterminate limit. Hence we try dividing both the numerator and the denominator by x^2, which is the highest power of x that appears in either the numerator or the denominator. We have

$$\frac{x^2 + 1}{2x^2 + 3} = \frac{1 + \frac{1}{x^2}}{2 + \frac{3}{x^2}} \quad \text{terms} \to 0 \text{ as } x \to \infty$$

Here, we see that $1/x^2$ and $3/x^2$ both approach zero as $x \to \infty$. This means that the numerator approaches 1 and the denominator approaches 2. Therefore,

$$\lim_{x \to \infty} \frac{x^2 + 1}{2x^2 + 3} = \lim_{x \to \infty} \frac{1 + \frac{1}{x^2}}{2 + \frac{3}{x^2}} = \frac{1}{2}$$

Practice Exercise

3. Find $\lim_{x \to \infty} \frac{x + 2}{2x - 7}$.

EXAMPLE 11 Limits with conjugate multiplication

Find $\lim_{x \to 3} \frac{2 - \sqrt{1 + x}}{x - 3}$.

By direct substitution we obtain the indeterminate limit

$$\lim_{x \to 3} \frac{2 - \sqrt{1 + x}}{x - 3} = \frac{2 - \sqrt{4}}{3 - 3} = \frac{0}{0}$$

Since there is no way of factoring this expression, we instead try using conjugate multiplication. We switch the sign of the second term of the numerator to get $(2 + \sqrt{1 + x})$, and multiply both numerator and denominator by this conjugate. This will result in a **difference of squares** in the numerator, allowing for some cancellation of terms.

$$\begin{aligned}
\lim_{x \to 3} \frac{2 - \sqrt{1 + x}}{x - 3} &= \lim_{x \to 3} \frac{2 - \sqrt{1 + x}}{x - 3} \cdot \frac{2 + \sqrt{1 + x}}{2 + \sqrt{1 + x}} \\
&= \lim_{x \to 3} \frac{2^2 - \left(\sqrt{1 + x}\right)^2}{(x - 3)\left(2 + \sqrt{1 + x}\right)} \\
&= \lim_{x \to 3} \frac{4 - (1 + x)}{(x - 3)\left(2 + \sqrt{1 + x}\right)} \\
&= \lim_{x \to 3} \frac{3 - x}{-(3 - x)\left(2 + \sqrt{1 + x}\right)} \\
&= \lim_{x \to 3} -\frac{1}{\left(2 + \sqrt{1 + x}\right)}
\end{aligned}$$

$$\lim_{x \to 3} \frac{2 - \sqrt{1 + x}}{x - 3} = -\frac{1}{\left(2 + \sqrt{1 + 3}\right)} = -\frac{1}{4}$$

EXAMPLE 12 Limit with a square root over the entire function

Find $\lim_{x \to \infty} \frac{\sqrt{x^2 - 8x}}{2x + 1}$.

Because the square root covers the entire denominator, we will rearrange the function so that it is entirely under a root sign. Then the limit can move inside the square root sign, and the limit can be evaluated as $\lim_{x \to a} \sqrt{f(x)} = \sqrt{\lim_{x \to a} f(x)}$. We get

$$\begin{aligned}
\lim_{x \to \infty} \frac{\sqrt{x^2 - 8x}}{2x + 1} &= \lim_{x \to \infty} \frac{\sqrt{x^2 - 8x}}{\sqrt{(2x + 1)^2}} = \lim_{x \to \infty} \frac{\sqrt{x^2 - 8x}}{\sqrt{4x^2 + 4x + 1}} \\
&= \lim_{x \to \infty} \sqrt{\frac{x^2 - 8x}{4x^2 + 4x + 1}} = \sqrt{\lim_{x \to \infty} \frac{x^2 - 8x}{4x^2 + 4x + 1}}
\end{aligned}$$

We can now evaluate the limit inside, and take the square root of the answer at the end.

$$= \sqrt{\lim_{x \to \infty} \frac{\frac{x^2}{x^2} - \frac{8x}{x^2}}{\frac{4x^2}{x^2} + \frac{4x}{x^2} + \frac{1}{x^2}}} = \sqrt{\lim_{x \to \infty} \frac{1 - \frac{8}{x}}{4 + \frac{4}{x} + \frac{1}{x^2}}} = \sqrt{\frac{1 - 0}{4 + 0 + 0}} = \frac{1}{2}$$

EXAMPLE 13 Limit does not exist

Find $\lim_{x \to 2} \frac{1}{x - 2}$.

Note that $f(x)$ is not defined for $x = 2$, since we would have division by zero. Therefore, we set up the following table to see how $f(x)$ behaves as $x \to 2$:

x	3	2.5	2.1	2.01	2.001
$f(x)$	1	2	10	100	1000

$f(x) \to +\infty$

x	1	1.5	1.9	1.99	1.999
$f(x)$	-1	-2	-10	-100	-1000

$f(x) \to -\infty$

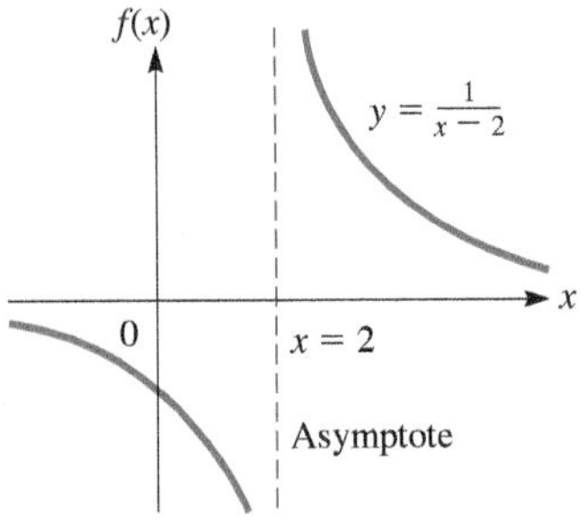

Fig. 23.2

We see that $f(x)$ gets larger as $x \to 2$ from above 2 and $f(x)$ gets smaller (large negative values) as $x \to 2$ from below 2. This may be written as $f(x) \to +\infty$ as $x \to 2^+$ and $f(x) \to -\infty$ as $x \to 2^-$, but we must remember that ∞ is not a real number. Therefore, the limit as $x \to 2$ does not exist. The graph of this function was displayed in Fig. 23.2, which is shown again here for reference.

EXAMPLE 14 Limit as a value approaches infinity

The efficiency of an engine is given by $\eta = 1 - Q_2/Q_1$, where Q_1 is the heat taken in and Q_2 is the heat ejected by the engine. ($Q_1 - Q_2$ is the work done by the engine.) If, in an engine cycle, $Q_2 = 500$ kJ, find η as Q_1 becomes large without bound.

We are to find

$$\lim_{Q_1 \to \infty} \left(1 - \frac{500}{Q_1}\right)$$

As Q_1 becomes larger and larger, $500/Q_1$ becomes smaller and smaller and approaches zero. This means $f(Q_1) \to 1$ as $Q_1 \to \infty$. Thus,

$$\lim_{Q_1 \to \infty} \left(1 - \frac{500}{Q_1}\right) = 1$$

We can verify our reasoning and the value of the limit by making a table of values for Q_1 and η as Q_1 becomes large:

Q_1	500	5000	50 000	500 000
η	0	0.9	0.99	0.999

values approach 1

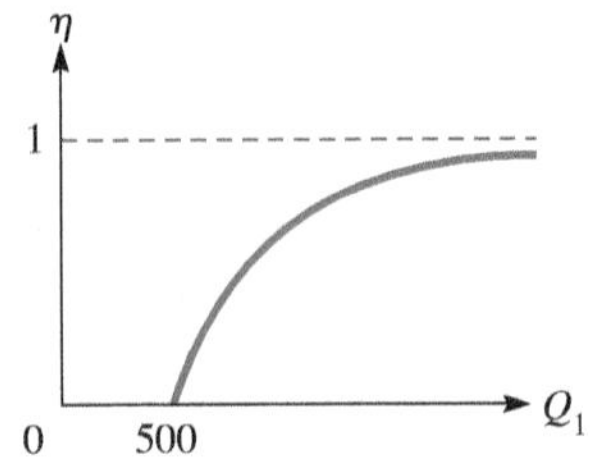

Fig. 23.8

Again, we see that $\eta \to 1$ as $Q_1 \to \infty$. See Fig. 23.8.

This is primarily a theoretical consideration, as there are obvious practical limitations as to how much heat can be supplied to an engine. An engine for which $\eta = 1$ would operate at 100% efficiency.

EXAMPLE 15 Value of the function exists—limit does not

Find $\lim_{x \to 0} \left(x\sqrt{x - 3}\right)$.

We see that $x\sqrt{x - 3} = 0$ if $x = 0$, but this function does not have real values for any values of x less than 3, other than $x = 0$. Therefore, *since x cannot approach zero, $f(x)$ does not* ***approach*** *zero, and the* ***limit does not exist***. The point of this example is that even if $f(a)$ exists, we cannot evaluate the limit simply by finding $f(a)$, unless $f(x)$ is continuous at $x = a$. In this case, $f(a)$ exists, but zero is the only value less than $x = 3$ for which $f(x)$ is defined, and therefore the limit does not exist.

■ Calculus did not have a sound mathematical basis until limits were properly developed by the French mathematician Augustin-Louis Cauchy (1789–1857) and others in the mid-1800s.

The definitions and development of continuity and of a limit presented in this section are not mathematically rigorous. However, the development is consistent with a more rigorous development, and the concept of a limit is the principal concern.

EXERCISES 23.1

In Exercises 1–4, make the given changes in the indicated examples of this section. Then solve the resulting problems.

1. In Example 2, change the denominator to $x + 2$ and then determine the continuity.
2. In Example 7, change the numerator to $3x^2 - 5x - 2$ and find the resulting limit. Disregard references to Examples 5 and 6.
3. In Example 9, change the denominator to $t + 2$ and then find the limit as $t \to -2$.
4. In Example 10, change the numerator to $4x^2 + 1$ and find the resulting limit.

In Exercises 5–10, determine the values of x for which the function is continuous. If the function is not continuous, determine the reason.

5. $f(x) = 6x - 15$ **6.** $f(x) = 9 - x^2$

7. $f(x) = \dfrac{2}{x^2 - 7x}$ **8.** $f(x) = \dfrac{1}{\sqrt{x + 6}}$

9. $f(x) = \sqrt{\dfrac{x}{x - 2}}$ **10.** $f(x) = \dfrac{\sqrt{x + 2}}{x - 7}$

In Exercises 11–16, determine the values of x for which the function, as represented by the graphs in Fig. 23.9, is continuous. If the function is not continuous, determine the reason.

11.

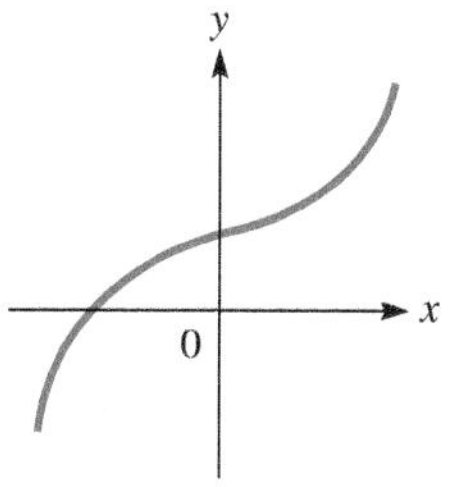

12.

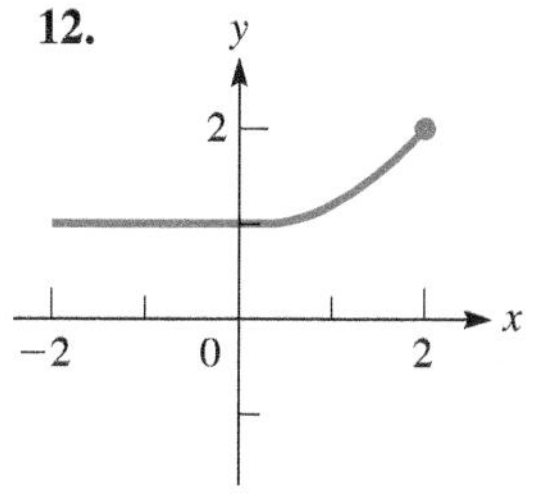

13.

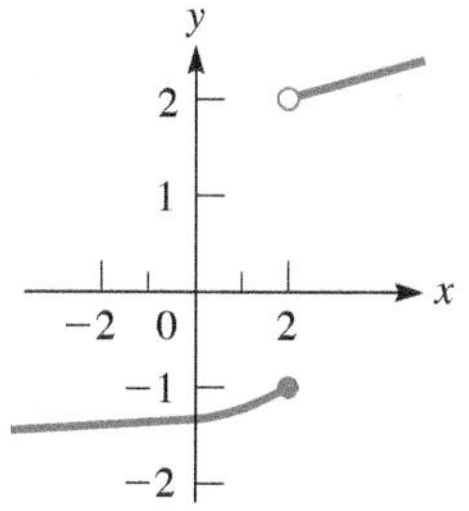

14.

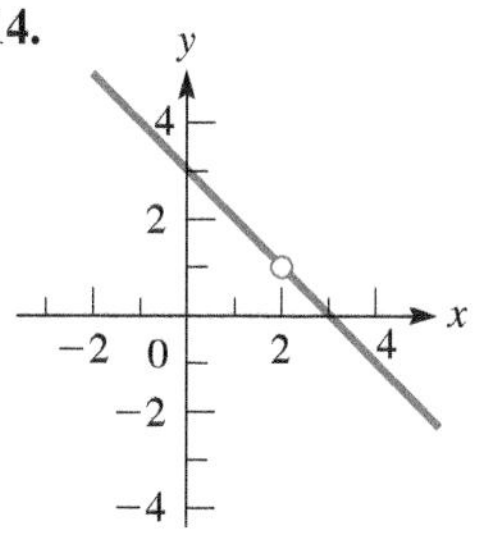

15.

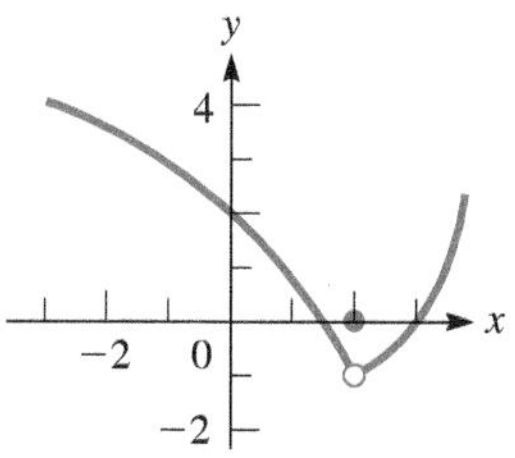

16.

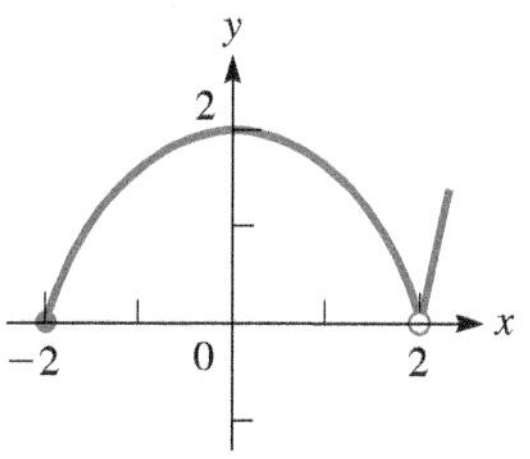

Fig. 23.9

In Exercises 17–20, for the function shown in the graph for the indicated exercise, find (a) f(2), and (b) $\lim_{x \to 2} f(x)$.

17. Exercise 13 **18.** Exercise 14

19. Exercise 15 **20.** Exercise 16

In Exercises 21–24, graph the function and determine the values of x for which the functions are continuous. Explain.

21. $f(x) = \begin{cases} x^2 & \text{for } x < 2 \\ 2 & \text{for } x \geq 2 \end{cases}$

22. $f(x) = \begin{cases} \dfrac{x^3 - x^2}{x - 1} & \text{for } x \neq 1 \\ 1 & \text{for } x = 1 \end{cases}$

23. $f(x) = \begin{cases} \dfrac{2x^2 - 18}{x - 3} & \text{for } x < 3 \text{ or } x > 3 \\ 12 & \text{for } x = 3 \end{cases}$

24. $f(x) = \begin{cases} \dfrac{x + 2}{x^2 - 4} & \text{for } x < -2 \\ \dfrac{x}{8} & \text{for } x > -2 \end{cases}$

In Exercises 25–30, evaluate the indicated limits by evaluating the function for values shown in the table and observing the values that are obtained. Do not change the form of the function.

25. Find $\lim_{x \to 1} \dfrac{x^3 - x}{x - 1}$.

x	0.900	0.990	0.999	1.001	1.010	1.100
$f(x)$						

26. Find $\lim_{x \to -3} \dfrac{x^3 + 2x^2 - 2x + 3}{x + 3}$.

x	−3.100	−3.010	−3.001	−2.999	−2.990	−2.900
$f(x)$						

27. Find $\lim_{x \to 2} \dfrac{2 - \sqrt{x + 2}}{x - 2}$.

x	1.900	1.990	1.999	2.001	2.010	2.100
$f(x)$						

28. Find $\lim_{x \to 0} \dfrac{e^x - 1}{x}$.

x	−0.1	−0.01	−0.001	0.001	0.01	0.1
$f(x)$						

29. Find $\lim_{x \to \infty} \dfrac{2x + 1}{5x - 3}$.

x	10	100	1000	10 000
$f(x)$				

30. Find $\lim_{x \to \infty} \dfrac{1 - x^2}{8x^2 + 5}$.

x	10	100	1000	10 000
$f(x)$				

In Exercises 31–48, evaluate the indicated limits by direct evaluation as in Examples 8–15. Change the form of the function where necessary.

31. $\lim_{x \to 3} (9x - 16)$

32. $\lim_{x \to 5} \sqrt{x^2 - 9}$

33. $\lim_{x \to 0} \dfrac{x^2 + x}{x}$

34. $\lim_{v \to 2} \dfrac{4v^2 - 8v}{v - 2}$

35. $\lim_{x \to -1} \dfrac{x^2 - 1}{3x + 3}$

36. $\lim_{x \to 3} \dfrac{x^2 - 2x - 3}{3 - x}$

37. $\lim_{h \to 3} \dfrac{h^3 - 27}{h - 3}$

38. $\lim_{x \to 1/3} \dfrac{3x - 1}{3x^2 + 5x - 2}$

39. $\lim_{x \to 1} \dfrac{(2x - 1)^2 - 1}{2x - 2}$

40. $\lim_{x \to 4} \dfrac{|x - 4|}{x - 4}$

41. $\lim_{p \to -1} \sqrt{p(p + 1.3)}$

42. $\lim_{x \to 1} (x - 1)\sqrt{x^2 - 4}$

43. $\lim_{x \to 1} \dfrac{\sqrt{x} - 1}{x - 1}$

44. $\lim_{x \to 8} \dfrac{x - 8}{\sqrt[3]{x} - 2}$

45. $\lim_{x \to \infty} \dfrac{3x^2 + 4.5}{x^2 - 1.5}$

46. $\lim_{x \to \infty} \dfrac{x - 1}{9x - 2}$

47. $\lim_{t \to \infty} \dfrac{\sqrt{16^2 + 16}}{t + 1}$

48. $\lim_{x \to \infty} \dfrac{1 - 2x^2}{(4x + 3)^2}$

In Exercises 49 and 50, evaluate the function at 0.1, 0.01, *and* 0.001 *from both sides of the value it approaches. In Exercises 51 and 52, evaluate the function for values of x of* 10, 100, *and* 1000. *From these values, determine the limit. Then, by using an appropriate change of algebraic form, evaluate the limit directly and compare values.*

49. $\lim_{x \to 0} \dfrac{x^2 - 3x}{x}$

50. $\lim_{x \to 3} \dfrac{2x^2 - 6x}{x - 3}$

51. $\lim_{x \to \infty} \dfrac{2x^2 + x}{x^2 - 3}$

52. $\lim_{x \to \infty} \dfrac{x^2 + 5}{\sqrt{64x^4 + 1}}$

In Exercises 53–56, solve the given problems involving limits.

53. A certain object, after being heated, cools at such a rate that its temperature T (in °C) decreases 10% each minute. If the object is originally heated to 100°C, find $\lim_{t \to 10} T$ and $\lim_{t \to \infty} T$, where t is the time (in min).

54. The area A (in mm^2) of the pupil of a certain person's eye is given by $A = \dfrac{36 + 24b^3}{1 + 4b^3}$, where b is the brightness (in lumens) of the light source. Between what values does A vary?

55. Velocity can be found by dividing the displacement s of an object by the elapsed time t in moving through the displacement. In a certain experiment, the following values were measured for the displacements and elapsed times for the motion of an object. Determine the limiting value of the velocity.

s (cm)	0.480 000	0.280 000	0.029 800	0.002 998 0	0.000 299 98
t (s)	0.200 000	0.100 000	0.010 000	0.001 000 0	0.000 100 00

56. A 5-Ω resistor and a variable resistor of resistance R are placed in parallel. The expression for the resulting resistance R_T is given by $R_T = \dfrac{5R}{5 + R}$. Determine the limiting value of R_T as $R \to \infty$.

In Exercises 57–60, construct a table of values to evaluate the indicated limits.

57. Approximate $\lim_{x \to 2} \dfrac{2^x - 4}{x - 2}$.

58. Approximate $\lim_{x \to 1} \dfrac{4^x - 4}{x - 1}$.

59. $\lim_{x \to 0} (1 + x)^{1/x}$ (Do you recognize the limiting value?)

60. $\lim_{x \to 0} \dfrac{\sin x}{x}$ (Use radian mode.)

In Exercises 61–66, solve the given problems involving one-sided limits.

61. For the function displayed in Exercise 13, find

(a) $\lim_{x \to 2^-} f(x)$ **(b)** $\lim_{x \to 2^+} f(x)$ **(c)** $\lim_{x \to 2} f(x)$

62. For the function displayed in Exercise 16, find

(a) $\lim_{x \to -2^-} f(x)$ **(b)** $\lim_{x \to -2^+} f(x)$ **(c)** $\lim_{x \to -2} f(x)$

63. Find $\lim_{x \to 4^-} x\sqrt{16 - x^2}$.

64. Explain why $\lim_{x \to 0^+} 2^{1/x} \neq \lim_{x \to 0^-} 2^{1/x}$.

65. For $f(x) = \dfrac{x}{|x|}$, find $\lim_{x \to 0^-} f(x)$ and $\lim_{x \to 0^+} f(x)$. Is $f(x)$ continuous at $x = 0$? Explain.

66. In Einstein's theory of relativity, the length L of an object moving at a velocity v is $L = L_0\sqrt{1 - \dfrac{v^2}{c^2}}$, where c is the speed of light and L_0 is the length of the object at rest. Find $\lim_{v \to c^-} L$ and explain why a limit from the left is used.

Answers to Practice Exercises

1. Discontinuous at $x = -3$ and $x = 0$

2. 10 **3.** 0.5

23.2 The Slope of a Tangent to a Curve

Having developed the basic operations with functions and the concept of a limit, we now turn our attention to a graphical interpretation of the rate of change of a function. This interpretation, basic to an understanding of calculus, deals with the slope of a line tangent to the curve of a function.

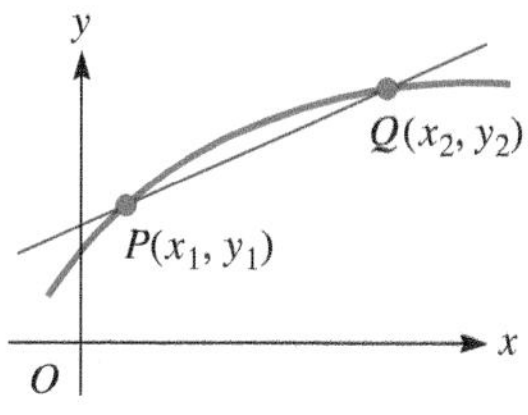

Fig. 23.10

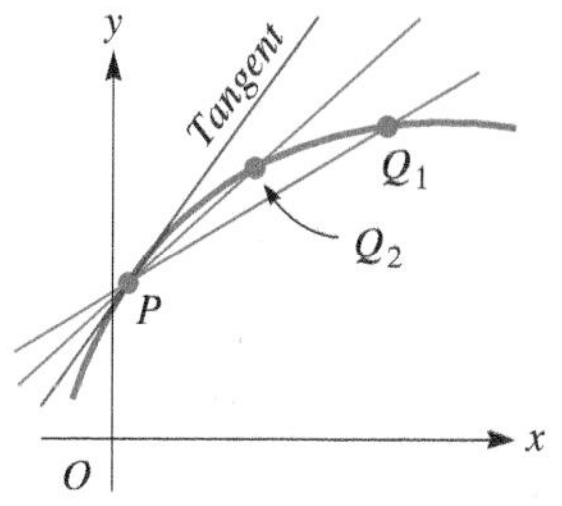

Fig. 23.11

Consider the points $P(x_1, y_1)$ and $Q(x_2, y_2)$ in Fig. 23.10. From Chapter 21, we know that the slope of the line through these points is given by

$$m = \frac{\text{rise}}{\text{run}} = \frac{y_2 - y_1}{x_2 - x_1}$$

This, however, represents the slope of the line through P and Q (a *secant line*) and no other line. We can also refer to this as the average rate of change over the interval, or the *average slope over the interval* from x_1 to x_2. If we now allow Q to be a point closer to P, the slope of PQ will more closely approximate the slope of a line drawn tangent to the curve at P (see Fig. 23.11). In fact, the closer Q is to P, the better this approximation becomes. It is not possible to allow Q to coincide with P, for then it would not be possible to define the slope of PQ in terms of two points. *The slope of the tangent line, often referred to as the slope of the curve, is the limiting value of the slope of the secant line PQ as Q approaches P.*

EXAMPLE 1 Limit of the slopes of secant lines

Find the slope of a line tangent to the curve $y = x^2 + 3x$ at the point $P(2, 10)$ by finding the limit of the slopes of the secant lines PQ as Q approaches P.

Let point Q have the x-values of 3.0, 2.5, 2.1, 2.01, and 2.001. Then, using a calculator, we tabulate the necessary values. Since P is the point $(2, 10)$, $x_1 = 2$ and $y_1 = 10$. Thus, using the values of x_2, we tabulate the values of y_2, $y_2 - 10$, $x_2 - 2$, and the slope m:

Point	Q_1	Q_2	Q_3	Q_4	Q_5	P
x_2	3.0	2.5	2.1	2.01	2.001	2
y_2	18.0	13.75	10.71	10.0701	10.007 001	10
$y_2 - 10$	8.0	3.75	0.71	0.0701	0.007 001	
$x_2 - 2$	1.0	0.5	0.1	0.01	0.001	
$m = \dfrac{y_2 - 10}{x_2 - 2}$	8.0	7.5	7.1	7.01	7.001	

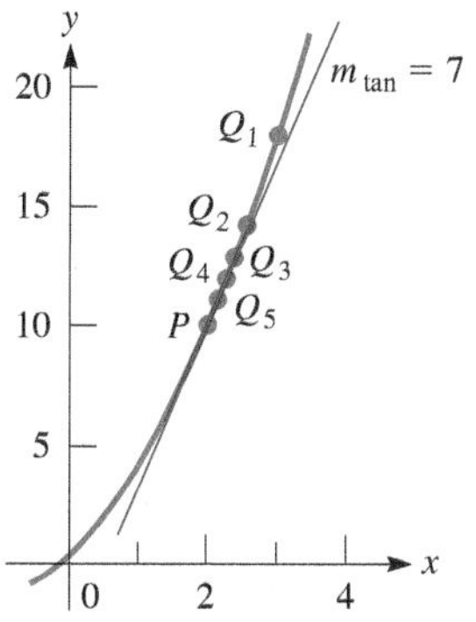

Fig. 23.12

We see that the slope of PQ approaches the value of 7 as Q approaches P. Therefore, the slope of the tangent line at $(2, 10)$ is 7. See Fig. 23.12.

Let a general point P be represented by $P(x, y)$. Using the appropriate notation, the coordinates of any other point $Q(x_2, y_2)$ on the curve of $f(x)$ can be expressed in terms of the coordinates of P. Define Δx and Δy as the x- and y-distances between the points P and Q. Then

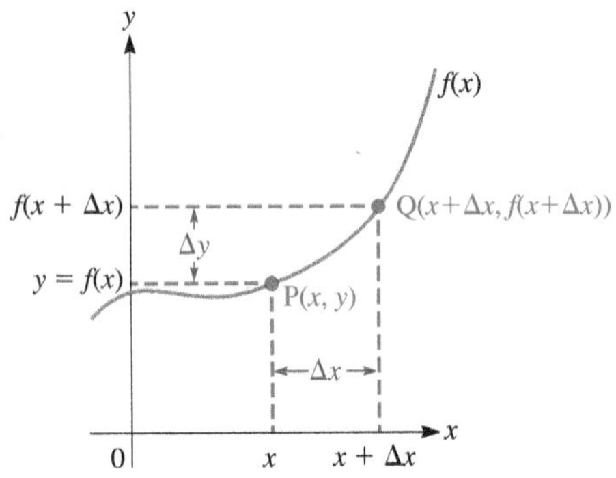

Fig. 23.13

$$\Delta x = x_2 - x \tag{23.2}$$

$$x_2 = x + \Delta x \tag{23.3}$$

For the function $y = f(x)$, the point $P(x, y)$ can be written as $P(x, f(x))$ and the point $Q(x_2, y_2)$ can be written as $Q(x + \Delta x, f(x + \Delta x))$. See Fig. 23.13.

You can see from the y-values in Fig. 23.13 that $y + \Delta y = f(x + \Delta x)$. Since this equation contains the quantities Δx and Δy, we can use it as a starting point for calculating the average slope, which we define as $\frac{\Delta y}{\Delta x}$.

To calculate the average slope of any function over an interval between any two points, we can use the delta method.

Delta Method for Finding an Average Slope

1. Write $y + \Delta y = f(x + \Delta x)$. Substitute $x + \Delta x$ into the function and simplify if possible.
2. Rearrange the equation to isolate Δy. Do this by subtracting y (the original function) from both sides of the equation. Make sure that y is expressed as a function of x during this subtraction. Simplify if possible.
3. Divide both sides of the equation by Δx to get $\frac{\Delta y}{\Delta x}$. This equation represents the average slope between any two points for the function. The first point is (x, y) and the interval width in question is Δx.

EXAMPLE 2 Average slope over an interval

For the function $y = x^2 + 1$, find the average slope from $x = 1$ to $x = 1.1$ using the delta method.

$$\begin{aligned} y + \Delta y &= f(x + \Delta x) && \text{Step 1} \\ &= (x + \Delta x)^2 + 1 && \text{substitute } x + \Delta x \text{ into } f \\ &= x^2 + 2x\Delta x + \Delta x^2 + 1 \\ \Delta y &= x^2 + 2x\Delta x + \Delta x^2 + 1 - y && \text{Step 2} \\ &= x^2 + 2x\Delta x + \Delta x^2 + 1 - (x^2 + 1) \\ &= 2x\Delta x + \Delta x^2 \\ \frac{\Delta y}{\Delta x} &= \frac{2x\Delta x + \Delta x^2}{\Delta x} = 2x + \Delta x && \text{Step 3} \end{aligned}$$

This is the general equation for the *average slope* between any two points for this function. If $x = 1$ and $\Delta x = 1.1 - 1 = 0.1$ then

$$\frac{\Delta y}{\Delta x} = 2(1) + 0.1 = 2.1$$

Using Eq. (23.3), along with the definition of slope, we can express the average slope over the interval from P to Q as

$$m_{PQ} = \frac{\Delta y}{\Delta x} = \frac{f(x + \Delta x) - f(x)}{\Delta x} \tag{23.4}$$

Notice that this summarizes the delta method in determining average slope. By our previous discussion, as Q approaches P, the slope of the tangent line will be approximated by Eq. (23.4).

EXAMPLE 3 Slope of the tangent line at a specific point

Find the slope of a line tangent to the curve of $y = x^2 + 3x$ at the point $(2, 10)$. (This is the same slope as in Example 1.)

As in Example 1, point P has the coordinates $(2, 10)$. Thus, the coordinates of any other point Q on the curve can be expressed as $(2 + \Delta x, f(2 + \Delta x))$. See Fig. 23.14. The slope of PQ then becomes

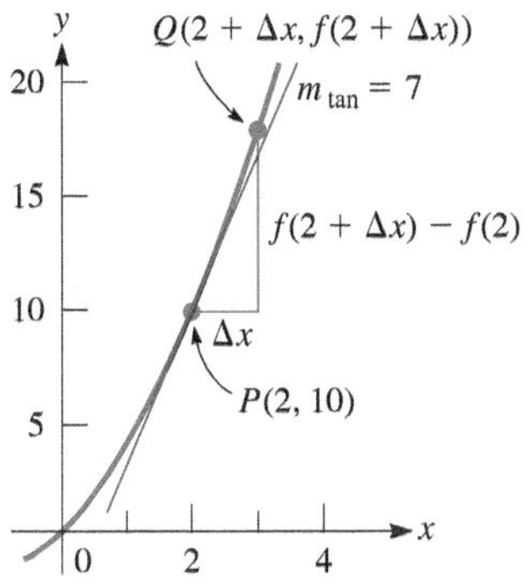

Fig. 23.14

$$\begin{aligned} m_{PQ} &= \frac{f(2 + \Delta x) - f(2)}{\Delta x} = \frac{[(2 + \Delta x)^2 + 3(2 + \Delta x)] - [2^2 + 3(2)]}{\Delta x} \\ &= \frac{(4 + 4\Delta x + \Delta x^2 + 6 + 3\Delta x) - (4 + 6)}{\Delta x} \\ &= \frac{7\Delta x + \Delta x^2}{\Delta x} = 7 + \Delta x \end{aligned}$$

From this expression, we can see that $m_{PQ} \to 7$ as $\Delta x \to 0$. Therefore, we can see that the slope of the tangent line is

$$m_{\tan} = \lim_{\Delta x \to 0} m_{PQ} = 7$$

We see that this result agrees with that found in Example 1.

We can define the slope of a tangent line at a single point P by letting the point Q approach P. As Q approaches P, we have $\Delta x \to 0$. Consequently, we add a final step to the delta method outlined above.

Delta Method for Finding a Tangent Line Slope

4. $m_{\tan} = \lim_{\Delta x \to 0} m_{PQ} = \lim_{\Delta x \to 0} \frac{\Delta y}{\Delta x}$

EXAMPLE 4 Slope of the tangent line at a general point

Find the slope of a line tangent to the curve of $y = 4x - x^2$ at a point (x, y) using the delta method.

$$\begin{aligned} m_{PQ} &= \frac{f(x + \Delta x) - f(x)}{\Delta x} \\ &= \frac{[4(x + \Delta x) - (x + \Delta x)^2] - (4x - x^2)}{\Delta x} \\ &= \frac{4x + 4\Delta x - x^2 - 2x \cdot \Delta x - \Delta x^2 - 4x + x^2}{\Delta x} \\ &= \frac{4\Delta x - 2x \cdot \Delta x - \Delta x^2}{\Delta x} = 4 - 2x - \Delta x \end{aligned}$$

Here, we see that $m_{PQ} \rightarrow 4 - 2x$ as $\Delta x \rightarrow 0$. Therefore,

$$m_{\tan} = 4 - 2x$$

This method has an advantage over that used in Example 3. We now have a general expression for the slope of a tangent line for any value x. If $x = -1$, $m_{\tan} = 6$ and if $x = 3$, $m_{\tan} = -2$. The tangent lines are shown in Fig. 23.15.

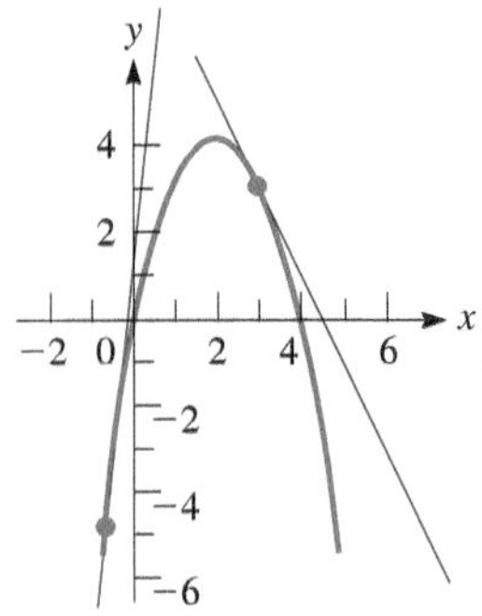

Fig. 23.15

■ Graphing utilities have a feature for drawing a line tangent to a curve.

EXAMPLE 5 Evaluating the slope of a tangent line

Find the expression for the slope of a line tangent to the curve of $y = x^3 + 2$ at the general point (x, y) and use this expression to find the slope when $x = 1/2$.

$$m_{PQ} = \frac{f(x + \Delta x) - f(x)}{\Delta x}$$

$$= \frac{[(x + \Delta x)^3 + 2] - (x^3 + 2)}{\Delta x}$$

$$= \frac{x^3 + 3x^2 \cdot \Delta x + 3x \cdot \Delta x^2 + \Delta x^3 + 2 - x^3 - 2}{\Delta x}$$

$$= \frac{3x^2 \cdot \Delta x + 3x \cdot \Delta x^2 + \Delta x^3}{\Delta x} = 3x^2 + 3x \cdot \Delta x + \Delta x^2$$

Taking the limit as $\Delta x \rightarrow 0$ yields the slope of the tangent line.

$$m_{\tan} = \lim_{\Delta x \to 0} (3x^2 + 3x \cdot \Delta x + \Delta x^2)$$

$$m_{\tan} = 3x^2$$

When $x = 1/2$, we find that the slope of the tangent is $3(1/4) = 3/4$. The curve and this tangent line are shown in Fig. 23.16.

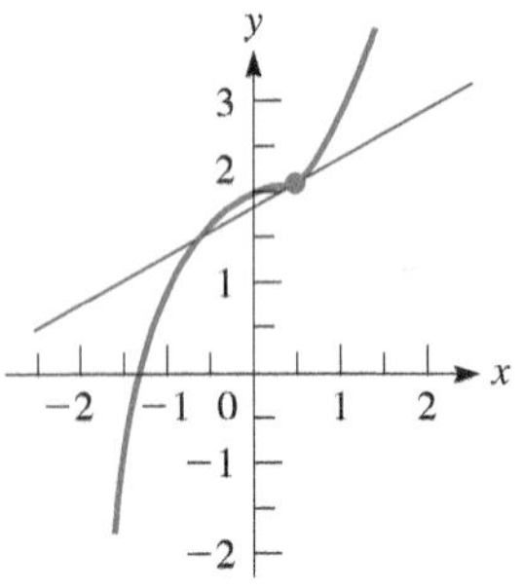

Fig. 23.16

Practice Exercise

1. Find the expression for the slope of a line tangent to the curve of $y = 4x^2$ at (x, y).

LEARNING TIP

The ratio of the change in f(x) to the change in x is the average rate of change of f(x) with respect to x.

As $\Delta x \rightarrow 0$, the limit of the ratio of the change in f(x) to the change in x is the instantaneous rate of change of f(x) with respect to x.

EXAMPLE 6 Slope as instantaneous rate of change

In Example 1, consider points $P(2, 10)$ and $Q(2.5, 13.75)$. From P to Q, x changes by 0.5 unit and $f(x)$ changes by 3.75 units. This means *the average change in $f(x)$ for a 1-unit change in x* is $3.75/0.5 = 7.5$ units. However, this is not the rate at which $f(x)$ is changing with respect to x ***at*** most points within this interval.

At point P, the slope of 7 of the tangent line tells us that $f(x)$ is changing 7 units for a 1-unit change in x. However, ***this is an instantaneous rate of change at point P*** and tells the rate at which $f(x)$ is changing with respect to x at P.

EXERCISES 23.2

In Exercises 1 and 2, make the given changes in the indicated examples of this section and then find the indicated slopes.

1. In Example 3, change the point $(2, 10)$ to $(3, 18)$.
2. In Example 4, change the $4x$ to $3x$.

In Exercises 3–6, use the method of Example 1 to calculate the slope of the line tangent to the curve of each of the given functions. Let Q_1, Q_2, Q_3, and Q_4 have the indicated x-values. Sketch the curve and tangent lines.

3. $y = x^2$; P is $(2, 4)$; let Q have x-values of 1.5, 1.9, 1.99, 1.999.
4. $y = 1 - \frac{1}{2}x^2$; P is $(2, -1)$; let Q have x-values of 1.5, 1.9, 1.99, 1.999.
5. $y = 2x^2 + 5x$; P is $(-2, -2)$; let Q have x-values of $-1.5, -1.9, -1.99, -1.999$.
6. $y = x^3 + 1$; P is $(-1, 0)$; let Q have x-values of -0.5, -0.9, -0.99, -0.999.

In Exercises 7–18, use the delta method to find a general expression for the slope of a tangent line to each of the indicated curves. Then find the slopes for the given values of x. Sketch the curves and tangent lines.

7. $y = x^2$; $x = 2$, $x = -1$
8. $y = 1 - \frac{1}{2}x^2$; $x = 2$, $x = -2$
9. $y = 2x^2 + 5x$; $x = -2$, $x = 0.5$

10. $y = 4.5 - 3x^2$; $x = 0, x = 2$

11. $y = x^2 + 4x + 2\pi$; $x = -3, x = 2$

12. $y = 2x^2 - 4x$; $x = 1, x = 1.5$

13. $y = 6x - x^2$; $x = -2, x = 3$

14. $y = x^3 - 2x$; $x = -1, x = 0, x = 1$

15. $y = 1.5x^4$; $x = 0, x = 0.5, x = 1$

16. $y = 4 - x^4$; $x = 0, x = 1, x = 2$

17. $y = x^5$; $x = 0, x = 0.5, x = 1$

18. $y = \frac{1}{x}$; $x = 0.5, x = 1, x = 2$

In Exercises 19–22, use a graphing utility to display the curve and the tangent line for the given values of x.

19. $y = 2x - 3x^2$; $x = 0, x = 0.5$

20. $y = 3x - x^3$; $x = -2, x = 0, x = 2$

21. $y = \frac{1}{3}x^6 + 2$; $x = -1, x = 0, x = 1$

22. $y = \dfrac{2}{x + 1}$; $x = -0.5, x = 0, x = 1$

In Exercises 23–26, find the average rate of change of y with respect to x from P to Q. Then compare this with the instantaneous rate of change of y with respect to x at P by finding m_{tan} *at P.*

23. $y = x^2 + 2$; $P(2, 6)$, $Q(2.1, 6.41)$

24. $y = 1 - 2x^2$; $P(1, -1)$, $Q(1.1, -1.42)$

25. $y = 9 - x^3$; $P(2, 1)$, $Q(2.1, -0.261)$

26. $y = x^3 - 6x$; $P(3, 9)$, $Q(3.1, 11.191)$

In Exercises 27 and 28, find the point(s) where the slope of a tangent line to the given curve has the given value. In Exercises 29–33, solve the given problems.

27. $y = 2x^2$, $m_{tan} = -4$

28. $y = x^3 + 3x^2$, $m_{tan} = 9$

29. Find the slope of a line perpendicular to the tangent of the curve of $y = 8 - 3x^2$ where $x = -1$.

30. Find the slopes of the tangent lines to the curve $y = \frac{1}{3}x^3 + x$ at points where $x = 1$ and $x = 2$. Then find the acute angle between these lines at the point where they cross.

31. In a computer game, at one point an airplane is diving along the curve of $y = -2x^2 + 10$. What is the angle of the dive (with the vertical) when $x = 1.5$?

32. At an amusement park, a waterslide follows the curve of $y = 8/(x + 1)$, for $x = 0$ m to $x = 7$ m. What is the angle (with the horizontal) of the slide when $x = 4.00$ m?

33. A structural support panel follows the curve $y = x^3 + 1$ (all dimensions are in metres) from $x = -2.00$ m to $x = 2.00$ m. Determine the slope of the structural panel at the point $x = -1.00$ m.

Answer to Practice Exercise

1. $m_{tan} = 8x$

23.3 The Derivative

In the preceding section, we found the slope of a line tangent to a curve at a point $(x, f(x))$ by calculating the limit (if it exists) of the difference $f(x + \Delta x) - f(x)$ divided by Δx as $\Delta x \to 0$. For brevity, we can define $h = \Delta x = x_2 - x$. We can then write the slope of the tangent line as

$$m_{tan} = \lim_{h \to 0} \frac{f(x + h) - f(x)}{h} \tag{23.5}$$

The limit on the right is defined as *the* **derivative** *of* $f(x)$ *at* x. This is one of the fundamental definitions of calculus.

Keep in mind that $\dfrac{dy}{dx}$ is simply a shorter way of writing $\lim_{\Delta x \to 0} \dfrac{\Delta y}{\Delta x}$. There are other ways of representing the derivative:

Derivative Slope of Tangent Line Slope at a Single Point Instantaneous Slope	$\dfrac{dy}{dx}$	$\dfrac{d}{dx}(y)$	$f'(x)$	$D_x y$	y'

Here $\dfrac{d}{dx}$ and D_x represent the *process* of finding the derivative with respect to a change in variable x.

The process of finding a derivative is called **differentiation**.

A four-step procedure for finding the derivative of a function by use of the definition is outlined here.

Summary of the Delta Process for Finding the Derivative of a Function

1. Find $f(x + h)$.
2. Subtract $f(x)$ from $f(x + h)$.
3. Divide the result of Step 2 by h.
4. For the result of Step 3, find the limit (if it exists) as $h \to 0$.

$$f'(x) = \lim_{h \to 0} \frac{f(x + h) - f(x)}{h} \qquad \textbf{(23.6)}$$

EXAMPLE 1 Using the definition to find a derivative

Find the derivative of $y = 2x^2 + 3x$ by using the definition.

With $y = f(x)$, using the above procedure to find the derivative $f'(x)$, we have the following:

$$f(x + h) = 2(x + h)^2 + 3(x + h) \qquad \text{Step 1}$$

$$f(x + h) - f(x) = 2(x + h)^2 + 3(x + h) - (2x^2 + 3x) \qquad \text{Step 2}$$

$$= 2x^2 + 4xh + 2h^2 + 3x + 3h - 2x^2 - 3x$$

$$= 4xh + 3h + 2h^2$$

$$\frac{f(x + h) - f(x)}{h} = \frac{4hx + 3h + 2h^2}{h} = 4x + 3 + 2h \qquad \text{Step 3}$$

$$\lim_{h \to 0} \frac{f(x + h) - f(x)}{h} = \lim_{h \to 0}(4x + 3 + 2h) = 4x + 3 \qquad \text{Step 4}$$

$$f'(x) = 4x + 3$$

We see that the derivative of the function $2x^2 + 3x$ is the function $4x + 3$. From the meanings of a slope of a tangent line and the derivative, this means we can find the slope of a tangent line for any point on the curve of $y = 2x^2 + 3x$ by substituting the x-coordinate into the expression $4x + 3$. For example, the slope of a tangent line is 5 if $x = 1/2$ (at the point $(1/2, 2)$). See Fig. 23.17.

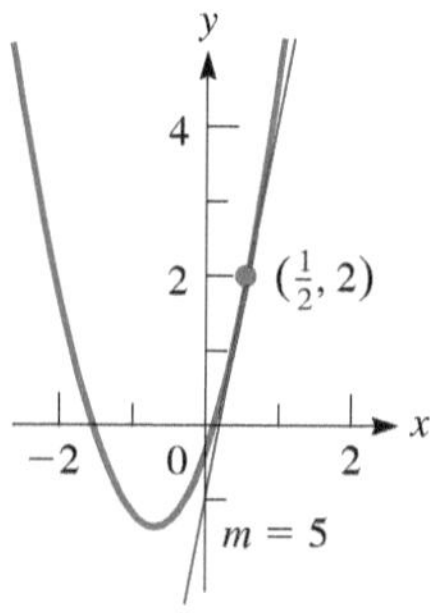

Fig. 23.17

Practice Exercise

1. Using the definition, find the derivative of $y = 5x - x^2$.

Since the derivative of a function is itself a function, it is possible that it may not be defined for all values of x. *If the value x_0 is in the domain of the derivative, then the function is said to be* **differentiable** *at x_0.* The examples that follow illustrate functions that are not differentiable for all values of x.

EXAMPLE 2 Using the definition—derivative of a fraction

Find the derivative of $y = \dfrac{3}{x+2}$ by using the definition.

> **LEARNING TIP**
> One might ask why, when finding a derivative, we take the limit as h approaches zero and do not just let h equal zero. If we did this, we would find that the ratio $[f(x+h) - f(x)]/h$ is exactly $0/0$, which requires division by zero. As we know, this is indeterminate, and therefore ***h* cannot equal zero.** However, it can equal any value as near zero as necessary. This idea is basic in the meaning of the word *limit.*

$$f(x+h) = \frac{3}{x+h+2} \qquad \text{Step 1}$$

$$f(x+h) - f(x) = \frac{3}{x+h+2} - \frac{3}{x+2} \qquad \text{Step 2}$$

$$f(x+h) - f(x) = \frac{3(x+2) - 3(x+h+2)}{(x+h+2)(x+2)} = \frac{-3h}{(x+h+2)(x+2)}$$

$$\frac{f(x+h) - f(x)}{h} = \frac{-3h}{h(x+h+2)(x+2)} = \frac{-3}{(x+h+2)(x+2)} \qquad \text{Step 3}$$

$$\lim_{h \to 0} \frac{f(x+h) - f(x)}{h} = \lim_{h \to 0} \frac{-3}{(x+h+2)(x+2)} = \frac{-3}{(x+2)^2} \qquad \text{Step 4}$$

$$f'(x) = \frac{-3}{(x+2)^2}$$

Note that neither the function nor the derivative is defined for $x = -2$. This means the function is not differentiable at $x = -2$.

> **COMMON ERROR** In Example 2, it was necessary to combine fractions in the process of finding the derivative. Such algebraic operations must be done with care. *One of the most common sources of errors is the improper handling of fractions.*

EXAMPLE 3 Derivative—proper handling of fractions

Find the derivative of $y = 4x^3 + \dfrac{5}{x}$ by using the definition.

$$f(x+h) = 4(x+h)^3 + \frac{5}{x+h}$$

$$f(x+h) - f(x) = 4(x+h)^3 + \frac{5}{x+h} - \left(4x^3 + \frac{5}{x}\right)$$

$$= 4(x^3 + 3x^2h + 3xh^2 + h^3) - 4x^3 + \frac{5}{x+h} - \frac{5}{x}$$

$$= 4x^3 + 12x^2h + 12xh^2 + 4h^3 - 4x^3 + \frac{5x - 5(x+h)}{x(x+h)}$$

$$= 12x^2h + 12xh^2 + 4h^3 - \frac{5h}{x(x+h)}$$

> **LEARNING TIP**
> The algebra is easier if the fractions are combined separately from the other terms.

$$\frac{f(x+h) - f(x)}{h} = \frac{12x^2h + 12xh^2 + 4h^3}{h} - \frac{5h}{hx(x+h)}$$

$$\lim_{h \to 0} \frac{f(x+h) - f(x)}{h} = \lim_{h \to 0} \left(12x^2 + 12xh + 4h^2 - \frac{5}{x(x+h)}\right) = 12x^2 - \frac{5}{x^2}$$

$$f'(x) = 12x^2 - \frac{5}{x^2}$$

Note that this function is not differentiable for $x = 0$.

Practice Exercise

2. Using the definition, find the derivative of $y = 3/x$.

EXAMPLE 4 Using the definition—derivative of the square root

Find dy/dx for the function $y = \sqrt{x}$ by using the definition.

When finding this derivative, we will employ the method of conjugate multiplication for evaluating limits. See Section 23.1 regarding this approach.

$$f(x+h) = \sqrt{x+h}$$

$$f(x+h) - f(x) = \sqrt{x+h} - \sqrt{x}$$

$$\frac{f(x+h) - f(x)}{h} = \frac{\sqrt{x+h} - \sqrt{x}}{h}$$

$$\lim_{h\to 0}\frac{f(x+h) - f(x)}{h} = \lim_{h\to 0}\frac{\sqrt{x+h} - \sqrt{x}}{h} = \frac{0}{0}\ \text{(indeterminate)}$$

$$\begin{aligned}\lim_{h\to 0}\frac{f(x+h) - f(x)}{h} &= \lim_{h\to 0}\frac{\sqrt{x+h} - \sqrt{x}}{h}\cdot\frac{\sqrt{x+h} + \sqrt{x}}{\sqrt{x+h} + \sqrt{x}}\\ &= \lim_{h\to 0}\frac{(\sqrt{x+h})^2 - (\sqrt{x})^2}{h(\sqrt{x+h} + \sqrt{x})}\\ &= \lim_{h\to 0}\frac{x+h-x}{h(\sqrt{x+h} + \sqrt{x})}\\ &= \lim_{h\to 0}\frac{h}{h(\sqrt{x+h} + \sqrt{x})}\\ &= \lim_{h\to 0}\frac{1}{\sqrt{x+h} + \sqrt{x}}\\ &= \frac{1}{\sqrt{x} + \sqrt{x}}\end{aligned}$$

$$\frac{dy}{dx} = \frac{1}{2\sqrt{x}}$$

The domain of $f(x)$ is $x \geq 0$. However, since x appears in the denominator of the derivative, the domain of the derivative is $x > 0$. Thus, the function is differentiable for $x > 0$.

■ Most graphing utilities have a feature for evaluating derivatives. For instance, the *numerical derivative* feature of some graphing calculators evaluates derivatives through numerical approximations.

Notation for Evaluating the Derivative	EXAMPLE 5
To denote the value of the derivative of $y = f(x)$ at the point $(a, f(a))$, we write $f'(a)$ or $\left.\frac{dy}{dx}\right\|_{x=a}$	Find the value of the derivative of $y = \frac{3}{x+2}$ at the point $(-1, 3)$. We saw in Example 2 that $f'(x) = \frac{-3}{(x+2)^2}$. Then
Note that only the x-coordinate of the point is needed to evaluate the derivative.	$f'(-1) = \left.\frac{dy}{dx}\right\|_{x=-1} = \frac{-3}{(-1+2)^2} = -3$

EXERCISES 23.3

In Exercises 1 and 2, make the given changes in the indicated examples of this section and then find the derivative by using the definition.

1. In Example 1, change $2x^2$ to $4x^2$.
2. In Example 2, change $x + 2$ in the denominator to $x - 2$.

In Exercises 3–26, find the derivative of each of the functions by using the definition.

3. $y = 5x - 2$
4. $y = 9x + 1$
5. $y = 1 - 2x$
6. $y = 2.3 - 5x$
7. $y = x^2 - 1$
8. $y = 4^2 - x^2$
9. $y = \pi x^3$
10. $y = -6x^2 + 9$
11. $y = x^2 - 7x$
12. $y = x^2 + 4ex$
13. $y = 8x - 2x^2$
14. $y = 3x - \frac{1}{2}x^2$
15. $y = x^3 + 4x - 3\pi$
16. $y = 5x^3 + 4$
17. $y = \dfrac{\sqrt{3}}{x + 2}$
18. $y = \dfrac{3}{5x + 3}$
19. $y = x + \dfrac{4}{3x}$
20. $y = \dfrac{x}{x - 3}$
21. $y = \dfrac{2}{x^2}$
22. $y = \dfrac{2e^2}{x^2 + 4}$
23. $y = x^4 + x^3 + x^2 + x$
24. $y = \frac{1}{3}x^3 + \frac{1}{2}x^2 + x$
25. $y = x^4 - \dfrac{2}{5x + 1}$
26. $y = \dfrac{1}{6x} + \dfrac{5}{x^2}$

In Exercises 27–30, find the derivative of each function by using the definition. Then evaluate the derivative at the given point.

27. $y = 3x^2 - 2x$; $(-1, 5)$
28. $y = 9x - x^3$; $(2, 10)$
29. $y = \dfrac{11}{3x + 2}$; $(3, 1)$
30. $y = x^2 - \dfrac{2}{x}$; $(-2, 5)$

In Exercises 31–34, find the derivative of each function by using the definition. Then determine the values for which the function is differentiable.

31. $y = 1 + \dfrac{2}{x}$
32. $y = \dfrac{5x}{2x - 5}$
33. $y = \dfrac{3}{x^2 - 1}$
34. $y = \dfrac{2}{x^2 + 1}$

In Exercises 35–44, solve the given problems.

35. Find the point(s) on the curve of $y = x^2 - 4x$ for which the slope of a tangent line is 6.
36. Find the point(s) on the curve of $y = 1/(x + 1)$ for which the slope of the tangent line is -1.
37. At what point on the curve of $y = 2x^2 - 16x$ is there a tangent line that is horizontal?
38. At what point on the curve of $y = 9 - 2x^2$ is there a tangent line that is parallel to the line $12x - 2y + 7 = 0$?
39. Find dy/dx for $y = \sqrt{x + 1}$ by the method of Example 4. For what values of x is the function differentiable? Explain.
40. Find dy/dx for $y = \sqrt{x^2 + 3}$ by the method of Example 4.
41. By noting the derivative for the function in Exercise 23, guess as to a formula for the derivative of $y = x^n$, where n is a positive integer.
42. By noting the derivative for the function in Exercise 26, guess as to a formula for the derivative of $y = x^n$, where n is a negative integer. Compare the result with that of Exercise 41.
43. A ski run follows the curve of $y = 0.01x^2 - 0.4x + 4$ from $x = 0$ m to $x = 20.0$ m. What is the angle between the ski run and the horizontal when $x = 10.0$ m? (Round to three significant digits.)
44. The cross-section of a hill can be approximated by the curve of $y = 0.3x - 0.00003x^3$ from $x = 0$ m to $x = 100$ m. The top of the hill is level. How high is the hill? (Round to three significant digits.)

Answers to Practice Exercises

1. $y' = 5 - 2x$ 2. $y' = -3/x^2$

23.4 The Derivative as an Instantaneous Rate of Change

In Section 23.2, we saw that the slope of a line tangent to a curve at point P was the limiting value of the slope of the line through points P and Q as Q approaches P. In Section 23.3, we defined the limit of the ratio $(f(x + h) - f(x))/h$ as $h \to 0$ as the derivative. Therefore, the first meaning we have given to *the derivative is the slope of a line tangent to a curve,* as we noted in Example 1 of Section 23.3. The following example further illustrates this meaning of the derivative.

EXAMPLE 1 Slope of a tangent line

Find the slope of the line tangent to the curve of $y = 4x - x^2$ at the point $(1, 3)$.

As we have noted, from Sections 23.2 and 23.3, we know that we must first find the derivative and then evaluate it at the given point.

$$f(x+h) = 4(x+h) - (x+h)^2$$

$$f(x+h) - f(x) = 4(x+h) - (x+h)^2 - (4x - x^2)$$

$$= 4x + 4h - x^2 - 2xh - h^2 - 4x + x^2 = 4h - 2xh - h^2$$

$$\frac{f(x+h) - f(x)}{h} = \frac{4h - 2xh - h^2}{h} = 4 - 2x - h$$

$$\lim_{h \to 0} \frac{f(x+h) - f(x)}{h} = \lim_{h \to 0}(4 - 2x - h) = 4 - 2x$$

$$\frac{dy}{dx} = 4 - 2x$$

$$\left.\frac{dy}{dx}\right|_{(1,3)} = 4 - 2(1) = 2$$

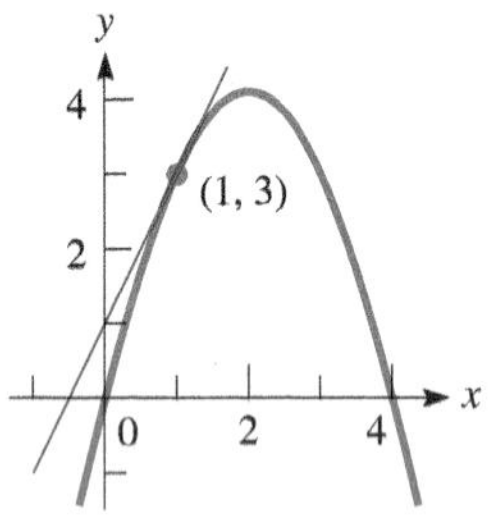

Fig. 23.18

The slope of the tangent line at $(1, 3)$ is 2. Note that only $x = 1$ was needed for the evaluation. The curve and tangent line are shown in Fig. 23.18.

We discussed earlier that the ratio $(f(x+h) - f(x))/h$ gives the rate of change of $f(x)$ with respect to x. In defining the derivative as the limit of this ratio as $h \to 0$, it is a measure of the rate of change of $f(x)$ with respect to x at point P. However, P may represent any point, which means the value of the derivative changes from one point on a curve to another point.

LEARNING TIP
The derivative gives the ***instantaneous rate of change*** of $f(x)$ with respect to x.

EXAMPLE 2 Rate of change of $f(x)$ for an exact value of x

In Examples 1 and 3 of Section 23.2, $f(x)$ is changing at the rate of 7 units for a 1-unit change in x, ***when x equals exactly* 2**. In Example 4 of Section 23.2, $f(x)$ is increasing 6 units for a 1-unit change in x, ***when x equals exactly* −1**, and $f(x)$ is decreasing 2 units for a 1-unit increase in x, ***when x equals exactly* 3**.

This gives us a more general meaning of the derivative. If a functional relationship exists between any two variables, then one can be taken to be varying with respect to the other, and the derivative gives us the instantaneous rate of change. There are many applications of this principle, one of which is the velocity of an object. We consider now the case of motion along a straight line, called *rectilinear motion.*

The **average velocity** of an object is found by dividing the change in displacement by the time interval required for this change. As the time interval approaches zero, the limiting value of the average velocity gives the value of the **instantaneous velocity**. Using symbols for the derivative, *the instantaneous velocity of an object moving in rectilinear motion at a specified time t is given by*

$$v = \lim_{h \to 0} \frac{s(t+h) - s(t)}{h} \qquad \textbf{(23.7)}$$

where $s(t)$ is the displacement as a function of the time t, and h is the time interval that approaches zero. In this case, the derivative has units of displacement divided by units of time, and we can denote it as ds/dt.

EXAMPLE 3 Instantaneous velocity

Find the instantaneous velocity, when $t = 4$ s (exactly), of a falling object for which the distance s (in m) fallen is the displacement given by $s = 4.90t^2$, by calculating average velocities between $t = 3.5$ s, 3.9 s, 3.99 s, 3.999 s, and $t = 4$ s, and then noting the apparent limiting value as $h \to 0$.

■ See the chapter introduction.

The values of h are found by subtracting the given times from 4 s. Also, since $s = 78.40$ m for $t = 4$ s, the differences in distance are found by subtracting the values of s for the given times from 78.40 m. Therefore, we have

t (s)	3.5	3.9	3.99	3.999
s (m)	60.025	74.529	78.00849	78.360805
$7840 - s$ (m)	18.375	3.871	0.3915	0.039195
$h = 4 - t$ (s)	0.5	0.1	0.01	0.001
$v = \dfrac{78.40 - s}{h}$ (m/s)	36.750	38.710	39.150	39.155

We can see that the value of v is approaching 39.20 m/s, which is therefore the instantaneous velocity when $t = 4$ s.

EXAMPLE 4 Instantaneous velocity from the derivative

■ See the chapter introduction.

Find the expression for the instantaneous velocity of the object of Example 3, for which $s = 4.90t^2$, where s is the displacement (in m) and t is the time (in s). Determine the instantaneous velocity for $t = 2$ s and $t = 4$ s.

The required expression is the derivative of s with respect to t.

$$f(t + h) = 4.90(t + h)^2$$

$$f(t + h) - f(t) = 4.90(t + h)^2 - 4.90t^2 = 9.80th + 4.90h^2$$

$$\frac{f(t + h) - f(t)}{h} = \frac{9.80th + 4.90h^2}{h} = 9.80t + 4.90h$$

$$v = \lim_{h \to 0} \frac{f(t + h) - f(t)}{h} = \lim_{h \to 0}(9.80t + 4.90h) = 9.80t$$

expression for instantaneous velocity

$$\left.\frac{ds}{dt}\right|_{t=2} = 9.80(2) = 19.6 \text{ m/s} \quad \text{and} \quad \left.\frac{ds}{dt}\right|_{t=4} = 9.80(4) = 39.2 \text{ m/s}$$

We see that the second result agrees with that found in Example 3.

Practice Exercise

1. Find the instantaneous velocity of an object moving such that $s = 6.00t^2$ for $t = 3.00$ s. Here, s is the displacement (in m) and t is the time (in s).

By finding $\lim_{h \to 0}(f(x + h) - f(x))/h$, we can find the instantaneous rate of change of $f(x)$ with respect to x. The expression $\lim_{h \to 0}(f(t + h) - f(t))/h$ gives the instantaneous velocity, or instantaneous rate of change of displacement *with respect to time.*

LEARNING TIP

The derivative can be interpreted as the instantaneous rate of change of the dependent variable with respect to the independent variable. This is true for a differentiable function, no matter what the variables represent.

EXAMPLE 5 Instantaneous rate of change of volume

A spherical balloon is being inflated. Find the expression for the instantaneous rate of change of the volume with respect to the radius. Evaluate this instantaneous rate of change for a radius of 2.00 m.

$$V = \tfrac{4}{3}\pi r^3 \qquad \text{volume of sphere}$$

$$f(r+h) = \tfrac{4}{3}\pi(r+h)^3 \qquad \text{find derivative}$$

$$f(r+h) - f(r) = \tfrac{4}{3}\pi(r+h)^3 - \tfrac{4}{3}\pi r^3$$

$$f(r+h) - f(r) = \tfrac{4}{3}\pi(r^3 + 3r^2h + 3rh^2 + h^3 - r^3) = \tfrac{4}{3}\pi(3r^2h + 3rh^2 + h^3)$$

$$\frac{f(r+h) - f(r)}{h} = \frac{4\pi}{3}\left(\frac{3r^2h + 3rh^2 + h^3}{h}\right) = \frac{4\pi}{3}(3r^2 + 3rh + h^2)$$

$$\frac{dV}{dr} = \lim_{h\to 0}\frac{f(r+h) - f(r)}{h} = \lim_{h\to 0}\left(\frac{4\pi}{3}(3r^2 + 3rh + h^2)\right) = 4\pi r^2$$

$$\left.\frac{dV}{dr}\right|_{r=2.00\text{ m}} = 4\pi(2.00)^2 = 16.0\pi = 50.3\text{ m}^2 \qquad \text{instantaneous rate of change when } r = 2.00\text{ m}$$

The instantaneous rate of change of the volume with respect to the radius (dV/dr) for $r = 2.00$ m is 50.3 m^3/m (this way of showing the units is more meaningful).

As r increases, dV/dr also increases. This should be expected as the volume of a sphere varies directly as the cube of the radius.

EXAMPLE 6 Instantaneous rate of change of power

The power P produced by an electric current i in a resistor varies directly as the square of the current. Given that 1.20 W of power are produced by a current of 0.500 A in a certain resistor, find an expression for the instantaneous rate of change of power with respect to current. Evaluate this rate of change for $i = 2.50$ A.

We must first find the functional relationship between power and current, by solving the indicated problem in variation:

$$P = ki^2 \quad 1.20 = k(0.500)^2 \quad k = 4.80\text{ W/A}^2 \qquad P = 4.8i^2$$

Now, knowing the function, we may determine the expression for the instantaneous rate of change of P with respect to i by finding the derivative.

$$f(i+h) = 4.8(i+h)^2$$

$$f(i+h) - f(i) = 4.8(i+h)^2 - 4.8i^2 = 4.8(2ih + h^2)$$

$$\frac{f(i+h) - f(i)}{h} = \frac{4.8(2ih + h^2)}{h} = 4.8(2i + h) \qquad \text{expression for instantaneous rate of change}$$

$$\frac{dP}{di} = \lim_{h\to 0}\frac{f(i+h) - f(i)}{h} = \lim_{h\to 0}[4.8(2i + h)] = 9.6i$$

$$\left.\frac{dP}{di}\right|_{i=2.50\text{ A}} = 9.60(2.50) = 24.0\text{ W/A} \qquad \text{instantaneous rate of change when } i = 2.50\text{ A}$$

This tells us that when $i = 2.50$ A, the rate of change of power with respect to current is 24.0 W/A. Also, we see that the larger the current is, the greater is the increase in power. This should be expected, since the power varies directly as the square of the current.

Practice Exercise

2. In Example 6, change 1.20 W of power to 1.60 W of power, and 0.500 A of current to 0.800 A of current, and then find the instantaneous rate of change of power with respect to current for $i = 3.00$ A.

EXERCISES 23.4

In Exercises 1 and 2, make the given changes in the indicated examples of this section and then solve the resulting problems.

1. In Example 4, change $4.90t^2$ to $14.00t - 4.90t^2$ and then evaluate the instantaneous velocity at the indicated times.

2. In Example 5, change "volume" in the second line to "surface area" and then evaluate the rate of change for $r = 2.00$ m.

In Exercises 3–6, find the slope of a line tangent to the curve of the given equation at the given point. Sketch the curve and the tangent line.

3. $y = x^2 - 1$; $(2, 3)$

4. $y = 2x - x^2$; $(-1, -3)$

5. $y = \frac{16}{3x + 1}$; $(-3, -2)$

6. $y = 3 - \frac{16}{x^2}$; $(2, -1)$

In Exercises 7–10, calculate the instantaneous velocity for the indicated value of the time (in s) of an object for which the displacement (in m) is given by the indicated function. Use the method of Example 3 and calculate values of the average velocity for the given values of t and note the apparent limit as the time interval approaches zero.

7. $s = 4t + 10$ when $t = 3$; use values of t of 2.0, 2.5, 2.9, 2.99, 2.999

8. $s = 6 - 3t$ when $t = 4$; use values of t of 3.0, 3.5, 3.9, 3.99, 3.999

9. $s = 3t^2 - 4t$ when $t = 2$; use values of t of 1.0, 1.5, 1.9, 1.99, 1.999

10. $s = 40t - 4.9t^2$ when $t = 0.5$; use values of t of 0.4, 0.45, 0.49, 0.499, 0.4999

In Exercises 11–14, use the definition to find an expression for the instantaneous velocity of an object moving with rectilinear motion according to the given functions (the same as those for Exercises 7–10) relating s (in m) and t (in s). Then calculate the instantaneous velocity for the given value of t, rounding answers to three significant digits.

11. $s = 4t + 10$; $t = 3$

12. $s = 6 - 3t$; $t = 4$

13. $s = 3t^2 - 4t$; $t = 2$

14. $s = 40t - 4.9t^2$; $t = 0.5$

In Exercises 15–20, use the definition to find an expression for the instantaneous velocity of an object moving with rectilinear motion according to the given functions relating s and t.

15. $s = 52t + 13$

16. $s = 3t^2 - 2t^3$

17. $s = 12t^2 - t^3$

18. $s = s_0 + v_0t + \frac{1}{2}at^2$ (s_0, v_0, and a are constants.)

19. $s = 3t - \frac{1}{3}\pi t^3$

20. $s = \frac{4t}{t + 6}$

In Exercises 21–24, use the definition to find an expression for the instantaneous acceleration of an object moving with rectilinear motion according to the given functions. The instantaneous acceleration of an object is defined as the instantaneous rate of change of the velocity with respect to time. Here, v is the velocity, s is the displacement, and t is the time.

21. $v = 6t^2 - 4t + 2$

22. $v = \sqrt{6t + 1}$

23. $s = t^3 + 15t$ (Find v, then find a.)

24. $s = s_0 + v_0t - \frac{1}{2}at^2$ (s_0, v_0, and a are constants.) (Find v, then find a.)

In Exercises 25–42, find the indicated instantaneous rates of change. Unless specified otherwise, round answers to three significant digits.

25. A circular metal ring is being cooled. Find the rate at which the circumference changes when the radius is 0.76 cm.

26. Liquid is poured into a vertical cylindrical tank of radius 12.8 cm. Find the instantaneous rate of change of volume with respect to the depth h.

27. The distance s (in m) above the ground for a projectile fired vertically upward with a velocity of 44.0 m/s as a function of time t (in s) is given by $s = 44.0t - 4.90t^2$. Find t for $v = 0$.

28. For the projectile in Exercise 27, find v for $t = 4.00$ s and for $t = 5.00$ s. What conclusion can be drawn?

29. The electric current i at a point in an electric circuit is the instantaneous rate of change of the electric charge q that passes the point, with respect to the time t. Find i in a circuit for which $q = 30 - 2t$.

30. A load L (in N) is distributed along a beam 10.0 m long such that $L = 5x - 0.5x^2$, where x is the distance from one end of the beam. Find the expression for the instantaneous rate of change of L with respect to x.

31. A rectangular metal plate contracts while cooling. Find the expression for the instantaneous rate of change of the area A of the plate with respect to its width w, if the length of the plate is constantly three times as long as the width.

32. A circular oil spill is increasing in size. Find the instantaneous rate of change of the area A of the spill with respect to its radius r for $r = 245$ m.

33. The total power P (in W) transmitted by an AM radio station is given by $P = 500 + 250m^2$, where m is the modulation index. Find the instantaneous rate of change of P with respect to m for $m = 0.920$.

34. The bottom of a soft-drink can is being designed as an inverted spherical segment, the volume of which is $V = \frac{1}{6}\pi h^3 + 2.00\pi h$, where h is the depth (in cm) of the segment. Find the instantaneous rate of change of V with respect to h for $h = 0.600$ cm.

35. The total solar radiation H (in W/m^2) on a particular surface during an average clear day is given by $H = \frac{5000}{t^2 + 10}$, where t ($-6 \le t \le 6$) is the number of hours from noon (6:00 A.M. is equivalent to $t = -6$ h). Find the instantaneous rate of change of H with respect to t at 3:00 P.M.

36. For the solar radiator in Exercise 35, find the average rate of change of H between 2:00 P.M. and 4:00 P.M. Compare with the instantaneous rate of change at 3:00 P.M.

37. The value (in thousands of dollars) of a certain car is given by the function $V = \frac{48}{t + 3}$, where t is measured in years. Find a general expression for the instantaneous rate of change of V with respect to t and evaluate this expression when $t = 3$ years.

38. For the car in Exercise 37, find the average rate of change of V between $t = 2$ years and $t = 4$ years. Compare with the instantaneous rate of change for $t = 3$ years.

39. Oil in a certain machine is stored in a conical reservoir, for which the radius and height are both 4.00 cm (see Fig. 23.19). Find the instantaneous rate of change of the volume V of oil in the reservoir with respect to the depth d of the oil.

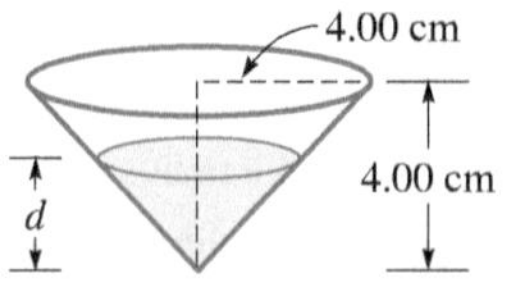

Fig. 23.19

40. The time t required to test a computer memory unit is directly proportional to the square of the number n of memory cells in the unit. For a particular type of unit, $n = 6400$ for $t = 25.0$ s. Find the instantaneous rate of change of t with respect to n for this type of unit for $n = 7600$.

41. A *holograph* (an image formed without using a lens) of concentric circles is formed. The radius r of each circle varies directly as the square root of the wavelength λ of the light used. If $r = 3.72$ cm for $\lambda = 592$ nm, find the expression for the instantaneous rate of change of r with respect to λ.

42. The force F between two electric charges varies inversely as the square of the distance r between them. For two charged particles, $F = 0.120$ N for $r = 0.0600$ m. Find the instantaneous rate of change of F with respect to r for $r = 0.120$ m.

Answers to Practice Exercises

1. 36.0 m/s **2.** 15.0 W/A

23.5 Derivatives of Polynomials

The task of finding a derivative can be considerably shortened by using the definition to arrive at certain basic formulas for derivatives. In this section, we derive the formulas that are used for finding the derivatives of polynomial functions of the form $f(x) = a_0x^n + a_1x^{n-1} + \cdots + a_n$.

First, we find the derivative of a constant. Letting $f(x) = c$ and then using the definition of a derivative, we find that $f(x + h) - f(x) = c - c = 0$. This means that

$$\lim_{h\to 0}\frac{f(x+h)-f(x)}{h} = \lim_{h\to 0}\left(\frac{0}{h}\right) = 0$$

From this we obtain the following *constant rule.*

Constant Rule	EXAMPLE 1
$\frac{dc}{dx} = 0$ **(23.8)** The derivative of a constant is zero.	If $y = -5$, $\frac{dy}{dx} = \frac{d(-5)}{dx} = 0$

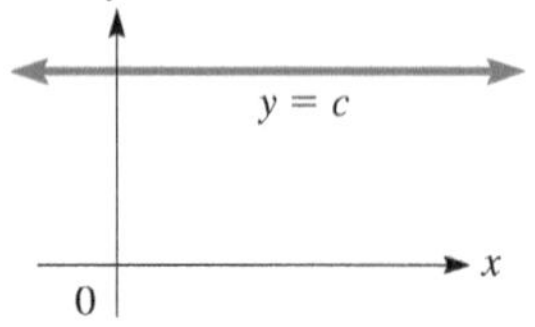

Fig. 23.20

The constant rule is consistent with the fact that the constant function $y = c$ is a horizontal line, which has a slope of zero. See Fig. 23.20.

Next, we find the derivative of a positive integral power of x. If $f(x) = x^n$, where n is a positive integer, by using the binomial theorem we have

$$f(x+h) - f(x) = (x+h)^n - x^n = x^n + nx^{n-1}h + \frac{n(n-1)}{2}x^{n-2}h^2 + \cdots + h^n - x^n$$

$$f(x+h) - f(x) = nx^{n-1}h + \frac{n(n-1)}{2}x^{n-2}h^2 + \cdots + h^n$$

$$\frac{f(x+h)-f(x)}{h} = nx^{n-1} + \frac{n(n-1)}{2}x^{n-2}h + \cdots + h^{n-1}$$

$$\lim_{h\to 0}\frac{f(x+h)-f(x)}{h} = nx^{n-1}$$

This formula for the derivative of the nth power of x is known as the *power rule.*

Power Rule	EXAMPLE 2
$$\frac{d(x^n)}{dx} = nx^{n-1} \quad \textbf{(23.9)}$$ The derivative of a power is found by multiplying the exponent by the numerical coefficient and then decreasing the exponent by 1.	If $y = x^{10}$, $$\frac{dy}{dx} = \frac{d(x^{10})}{dx} = 10x^{10-1} = 10x^9$$

Although our derivation of the power rule requires a positive integral power of x, we will see in Section 23.7 that it is in fact valid if the exponent is any rational number.

EXAMPLE 3 Derivative of a power of *r*

Find the derivative of the function $v = r$.

Here, the dependent variable is v, and the independent variable is r. Also, $n = 1$. Therefore,

$$\frac{dv}{dr} = \frac{d(r^1)}{dr} = (1)r^{1-1} = (1)r^0$$

Since $r^0 = 1$, we have

$$\frac{dv}{dr} = 1$$

This means that the slope of the line $v = r$ is always 1. This is consistent with our discussion of the slope of a straight line.

LEARNING TIP

From here on, we will be using functions that are combinations of simpler functions. Therefore, we must denote some functions by symbols other than $f(x)$. We will often be using u and v.

Next, we find the derivative of a constant times a function of x. We denote this function as u, or to show directly that it is a function of x, as $u(x)$. In finding the derivative of $c \cdot u$ with respect to x, we have

$$\frac{d}{dx}(c \cdot u) = \lim_{h \to 0} \frac{c \cdot u(x+h) - c \cdot u(x)}{h}$$

$$\frac{d(c \cdot u)}{dx} = c \cdot \lim_{h \to 0} \frac{u(x+h) - u(x)}{h} = c \cdot \frac{du}{dx}$$

We can thus state *the constant factor rule* for derivatives.

Constant Factor Rule	EXAMPLE 4
$$\frac{d(c \cdot u)}{dx} = c \cdot \frac{du}{dx} \quad \textbf{(23.10)}$$ The derivative of the product of a constant and a differentiable function of x is the product of the constant and the derivative of the function.	Find the derivative of $y = 3x^2$. In this case $c = 3$ and $u = x^2$. Thus, $du/dx = 2x$, and $$\frac{dy}{dx} = \frac{d(3x^2)}{dx} = 3 \cdot \frac{d(x^2)}{dx} = 3(2x)$$ $$\frac{dy}{dx} = 6x$$

Practice Exercise

1. Find the derivative of $y = 6x^3$.

COMMON ERROR

Occasionally, the derivative of a constant times a function of x is confused with the derivative of a constant that stands alone. It is necessary to distinguish clearly between a constant that multiplies a function and an isolated constant.

If the types of functions for which we have found derivatives are added, the result is a polynomial function with more than one term. The derivative is found by letting $y = u + v$, where u and v are functions of x. Using the definition, we have

$$y = u(x) + v(x)$$
$$\frac{dy}{dx} = \frac{d}{dx}[u(x) + v(x)] = \lim_{h\to 0}\frac{[u(x+h) + v(x+h)] - [u(x) + v(x)]}{h}$$
$$\frac{dy}{dx} = \lim_{h\to 0}\left[\frac{u(x+h) - u(x)}{h} + \frac{v(x+h) - v(x)}{h}\right]$$
$$\frac{dy}{dx} = \lim_{h\to 0}\left[\frac{u(x+h) - u(x)}{h}\right] + \lim_{h\to 0}\left[\frac{v(x+h) - v(x)}{h}\right] = \frac{du}{dx} + \frac{dv}{dx}$$

This gives us the *sum rule* for derivatives.

Sum Rule	EXAMPLE 5
$\frac{d(u+v)}{dx} = \frac{du}{dx} + \frac{dv}{dx}$ **(23.11)** The derivative of the sum of differentiable functions of x is the sum of the derivatives of the functions.	If $y = x^5 + 2x$, $\frac{dy}{dx} = \frac{d(x^5 + 2x)}{dx} = \frac{d(x^5)}{dx} + \frac{d(2x)}{dx}$ $= 5x^4 + 2$

The combination of the four rules obtained above allows us to take derivatives of any polynomial of the form $f(x) = a_0x^n + a_1x^{n-1} + \cdots + a_n$.

EXAMPLE 6 Evaluation of a derivative

Evaluate the derivative of $f(x) = 2x^4 - 6x^2 - 8x - 9$ at $(-2, 15)$.

First, finding the derivative, we have

$$f'(x) = \frac{d(2x^4)}{dx} - \frac{d(6x^2)}{dx} - \frac{d(8x)}{dx} - \frac{d(9)}{dx}$$
$$f'(x) = 8x^3 - 12x - 8$$

We now evaluate this derivative for $x = -2$.

$$f'(-2) = 8(-2)^3 - 12(-2) - 8 = -48$$

EXAMPLE 7 Slope of a tangent line

Find the slope of the line tangent to the curve of $y = 4x^7 - x^4$ at the point $(1, 3)$.

We must find and then evaluate the derivative for the value $x = 1$:

$$\frac{dy}{dx} = 28x^6 - 4x^3 \qquad \text{find derivative}$$
$$\left.\frac{dy}{dx}\right|_{x=1} = 28(1) - 4(1) = 24 \qquad \text{evaluate derivative}$$

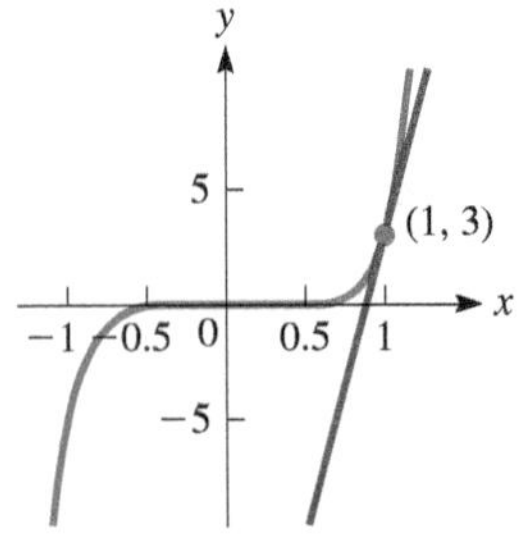

Fig. 23.21

Thus, the slope of the tangent line is 24. Again, we note that ***the substitution $x = 1$ must be made after the differentiation has been performed.*** The curve and the tangent line at $(1, 3)$ are shown in Fig. 23.21. Note that a slope of 24 results in a steep tangent line.

LEARNING TIP

Method for all general derivatives:

1. Identify the form.
2. Use the appropriate derivative formula to find the derivative.
3. Simplify algebraically and factor the solution.
4. Evaluate if required.

Practice Exercise

2. Find the expression for the instantaneous velocity if $s = 7t^4 - 4t + 5$, where s is the displacement and t is the time.

EXAMPLE 8 Evaluation of a derivative—instantaneous velocity

For each 4.00-s cycle, the displacement s (in cm) of a piston is given by $s = t^3 - 6t^2 + 8t$, where t is the time. Find the instantaneous velocity of the piston for $t = 2.60$ s.

$$s = t^3 - 6t^2 + 8t$$

$$\frac{ds}{dt} = 3t^2 - 12t + 8 \qquad \text{find derivative}$$

$$\left.\frac{ds}{dt}\right|_{t=2.6} = 3(2.60)^2 - 12(2.60) + 8 \qquad \text{evaluate derivative}$$

$$= -2.92 \text{ cm/s}$$

The piston is moving at -2.92 cm/s (it is moving in a negative direction) when $t = 2.60$ s.

EXERCISES 23.5

In Exercises 1–4, make the given changes in the indicated examples of this section, and then solve the resulting problem.

1. In Example 2, change the exponent 10 to 9.
2. In Example 4, change the coefficient 3 to 4.
3. In Example 6, change $2x^4$ to $2x^3$ and $-8x$ to $+8x$.
4. In Example 7, change $4x^7 - x^4$ to $4x^4 - x^7$ and evaluate the slope.

In Exercises 5–20, find the derivative of each of the given functions.

5. $y = x^8$
6. $y = x^{43}$
7. $f(x) = -4x^{11}$
8. $y = -7x^6$
9. $y = 5x^4 - 3\pi$
10. $s = 3t^5 + 4t - \frac{8}{t^2}$
11. $y = x^3 + 2x$
12. $y = x^5 - 1.5x^4$
13. $p = 5r^3 - 2r + 1$
14. $y = 6x^2 - 6x + 5$
15. $y = 25x^8 - 34x^5$
16. $u = 4v^4 - 12v + 9$
17. $f(x) = -6x^7 + 5x^3 + \pi^2$
18. $y = 13x^4 - 6x^3 - x - x^2$
19. $y = \frac{1}{3}x^3 + \frac{1}{2}x^2$
20. $f(z) = -\frac{1}{4}z^8 + \frac{1}{2}z^4 - 2^3$

In Exercises 21–24, evaluate the derivative of each of the given functions at the given point.

21. $y = 6x^2 - 8x + 1$; $(2, 9)$
22. $s = 2t^3 - 5t^2$; $(-1, -7)$
23. $y = 2x^3 + 9x - 7$; $(-2, -41)$
24. $y = -2x^4 + 9x^2 - 5x$; $(3, -96)$

In Exercises 25–28, find the slope of a line tangent to the curve of each of the given functions for the given values of x.

25. $y = 2x^6 - 4x^2 \quad (x = -1)$
26. $y = 3x^3 - \frac{9}{x} \quad (x = 1)$
27. $y = 35x - 2x^4 \quad (x = 2)$
28. $y = x^4 - \frac{1}{2}x^2 + 2 \quad (x = -2)$

In Exercises 29–32, determine an expression for the instantaneous velocity of objects moving with rectilinear motion according to the functions given, if s represents displacement in terms of time t.

29. $s = 6t^5 - 5t + 2$
30. $s = 20 + 60t - 4.9t^2$
31. $s = 2 - 6t - 2t^3$
32. $s = s_0 + v_0t + \frac{1}{2}at^2$

In Exercises 33–36, s represents the displacement and t represents the time for objects moving with rectilinear motion, according to the given functions. Find the instantaneous velocity for the given times, rounded to three significant digits.

33. $s = 2t^3 - 4t^2$; $t = 4$
34. $s = 8t^2 - 10t + 6$; $t = 5$
35. $s = 120 + 80t - 16t^2$; $t = 2.5$
36. $s = 0.5t^4 - 1.5t^2 + 2.5$; $t = 3$

In Exercises 37–57, solve the given problems by finding the appropriate derivative.

37. For what value(s) of x is the tangent to the curve of $y = 3x^2 - 6x$ parallel to the x-axis? (That is, where is the slope zero?)
38. Find the value of a if the tangent to the curve of $y = ax^2 + 2x$ has a slope of -4 for $x = 2$.
39. For what point(s) on the curve of $y = 3x^2 - 4x$ is the slope of a tangent line equal to 8?
40. Explain why the curve $y = 5x^3 + 4x - 3$ does not have a tangent line with a slope less than 4.
41. Find the point at which a tangent line to the parabola $y = 2x^2 - 7x$ is perpendicular to the line $x - 3y = 16$.
42. Display the graphs of $y = x^2$ and its derivative on a graphing utility. State any conclusions you can draw from the relationship of the two graphs.
43. For what value(s) of x is the slope of a line tangent to the curve of $y = 4x^2 + 3x$ equal to the slope of a line tangent to the curve of $y = 5 - 2x^2$?
44. For what value(s) of t is the instantaneous velocity of an object moving according to $s = 5t - 2t^2$ equal to the instantaneous velocity of an object moving according to $s = 3t^2 + 4$?
45. A metal cylinder is heated and then cools. If the radius always equals the height, find the expression for the instantaneous rate of change of the volume V with respect to the radius r.
46. A rectangular solid block of ice is melting such that the height is always twice the edge of the square base. Find the expression for the instantaneous rate of change of surface area A with respect to the edge of the base s.

47. The electric power P (in W) as a function of the current i (in A) in a certain circuit is given by $P = 16i^2 + 60i$. Find the instantaneous rate of change of P with respect to i for $i = 0.750$ A.

48. The torque T on the arm of a robotic control mechanism varies directly as the cube of the diameter d of the arm. If $T = 845$ N · m for $d = 0.925$ cm, find the expression for the instantaneous rate of change of T with respect to d.

49. The resistance R (in Ω) of a certain wire as a function of the temperature T (in °C) is given by $R = 16.0 + 0.450T + 0.0125T^2$. Find the instantaneous rate of change of R with respect to T when $T = 115$°C.

50. The deflection d of a diving board x m from the fixed end at the pool side is given by $d = kx^2(3L - x)$, where L is the length of the diving board and k is a positive constant. Find the expression for the instantaneous rate of change of d with respect to x.

51. The tensile strength S (in N) of a certain material as a function of the temperature T (in °C) is $S = 1600 - 0.000\,022T^2$. Find the instantaneous rate of change of S with respect to T for $T = 65.0$°C.

52. A tank containing 6000 L of water drains out in 30.0 min. The volume V of water in the tank after t min of draining is $V = 6000(1 - t/30)^2$. Find the instantaneous time rate of change of V after 15.0 min of draining.

53. The altitude h (in m) of a jet as a function of the horizontal distance x (in km) it has travelled is given by $h = 0.000\,104x^4 - 0.0417x^3 + 4.21x^2 - 8.33x$. Find the instantaneous rate of change of h with respect to x for $x = 125$ km.

54. The force F (in N) exerted by a cam on a lever is given by $F = x^4 - 12x^3 + 46x^2 - 60x + 25$, where x ($1 \le x \le 5$) is the distance (in cm) from the centre of rotation of the cam to the edge of the cam in contact with the lever (see Fig. 23.22). Find the instantaneous rate of change of F with respect to x when $x = 4.00$ cm.

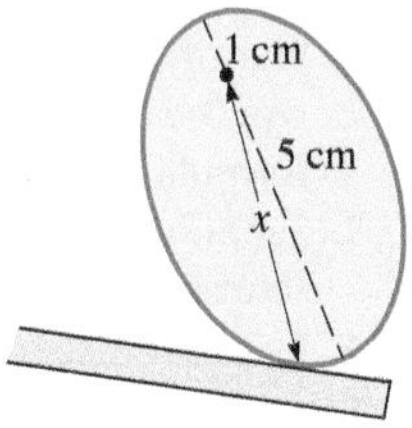

Fig. 23.22

55. Two ball bearings wear down such that the radius r of one is constantly 1.20 mm less than the radius of the other. Find the instantaneous rate of change of the total volume V_T of the two ball bearings with respect to r for $r = 3.30$ mm.

56. An open-top container is to be made from a rectangular piece of cardboard 6.00 cm by 8.00 cm. Equal squares of side x are to be cut from each corner, then the sides are to be bent up and taped together. Find the instantaneous rate of change of the volume V of the container with respect to x for $x = 1.75$ cm.

57. The head loss h_L (in m) for a fluid travelling in a pipe at average velocity v (in m/s) is given by $h_L = \dfrac{Kv^2}{2g}$, where K is a friction loss coefficient (constant), and $g = 9.81$ m/s². Determine the rate of change of head loss with respect to velocity when the velocity is 3.25 m/s and $K = 0.500$.

Answers to Practice Exercises

1. $dy/dx = 18x^2$ **2.** $v = 28t^3 - 4$

23.6 Derivatives of Products and Quotients of Functions

In the previous section, we considered polynomial functions. For functions that are not polynomials, some are products of simpler functions, some are quotients of simpler functions, and others are powers of simpler functions. In this section, we develop the formula for the derivative of a product of functions and the formula for the quotient of functions.

EXAMPLE 1 Product, quotient, and powers of functions

Combine the functions $f(x) = x^2 + 2$ and $g(x) = 3 - 2x$ as indicated.

(a) $p(x) = f(x) \cdot g(x) = (x^2 + 2)(3 - 2x)$ product of functions

(b) $q(x) = \dfrac{g(x)}{f(x)} = \dfrac{3 - 2x}{x^2 + 2}$ quotient of functions

(c) $F(x) = [g(x)]^2 = (3 - 2x)^2$ power of a function

If u and v are differentiable functions of x, the derivative of the function $u \cdot v$ is found by letting $y = u \cdot v$, and applying the definition as follows:

$$y = u(x) \cdot v(x)$$

$$\frac{dy}{dx} = \frac{d}{dx}[u(x) \cdot v(x)] = \lim_{h \to 0} \frac{u(x + h) \cdot v(x + h) - u(x) \cdot v(x)}{h}$$

By adding and subtracting $u(x + h) \cdot v(x)$ in the numerator of the last fraction, we can put it in a form that includes the derivatives of $u(x)$ and $v(x)$. Therefore,

$$\frac{d}{dx}[u(x) \cdot v(x)] = \lim_{h \to 0} \frac{u(x+h) \cdot v(x+h) - u(x+h) \cdot v(x) + u(x+h) \cdot v(x) - u(x) \cdot v(x)}{h}$$

$$\frac{d(u \cdot v)}{dx} = \lim_{h \to 0} \left[u(x+h) \frac{v(x+h) - v(x)}{h} + v(x) \frac{u(x+h) - u(x)}{h} \right] = u(x) \cdot \frac{dv(x)}{dx} + v(x) \cdot \frac{du(x)}{dx}$$

From this we conclude *the product rule for derivatives.*

Product Rule	**EXAMPLE 2**
$\frac{d(u \cdot v)}{dx} = u \cdot \frac{dv}{dx} + v \cdot \frac{du}{dx}$ **(23.12)** The derivative of the product of two differentiable functions is the first function times the derivative of the second function plus the second function times the derivative of the first function.	If $p(x) = (x^2 + 2) \cdot (3 - 2x)$, we have $u = x^2 + 2$ and $v = 3 - 2x$. Then $\frac{d(x^2+2) \cdot (3-2x)}{dx}$ $= (x^2 + 2) \cdot \frac{d(3-2x)}{dx}$ $+ (3 - 2x) \cdot \frac{d(x^2+2)}{dx}$ $= (x^2 + 2) \cdot (-2) + (3 - 2x) \cdot (2x)$ $= -2x^2 - 4 + 6x - 4x^2$ $p'(x) = -6x^2 + 6x - 4$

Practice Exercise

1. Find the derivative of $y = (3 - 2x^2)(x^4 - 1)$. Do not multiply factors together before finding the derivative.

EXAMPLE 3 Derivative of a product of functions with variable *r*

Find the derivative of $f(r) = 8r^3(2r^2 - 4r)$ using the product rule.

Recognize that the function $f(r) = 8r^3(2r^2 - 4r)$ is of the form $y = u \cdot v$ with $u = 8r^3$ and $v = 2r^2 - 4r$. Therefore,

$$\frac{df}{dr} = \frac{d(8r^3) \cdot (2r^2 - 4r)}{dr} = (8r^3) \cdot \frac{d(2r^2 - 4r)}{dr} + (2r^2 - 4r) \cdot \frac{d(8r^3)}{dr}$$

$$\frac{df}{dr} = (8r^3) \cdot (4r - 4) + (2r^2 - 4r) \cdot (24r^2)$$

$$\frac{df}{dr} = 32r^4 - 32r^3 + 48r^4 - 96r^3$$

$$\frac{df}{dr} = 80r^4 - 128r^3$$

$$\frac{df}{dr} = 16r^3(5r - 8)$$

LEARNING TIP

When the powers over the functions are larger than 1, it may be very tedious to multiply out and expand the functions first. The product rule will be very useful in combination with another rule, the chain rule, which discusses finding derivatives of powers of functions. This concept is covered in Section 23.7.

We will now find the derivative of the quotient of two differentiable functions by applying the definition to the function $y = u/v$, as shown below.

$$\frac{dy}{dx} = \frac{d}{dx}\left(\frac{u(x)}{v(x)}\right) = \lim_{h \to 0} \frac{\frac{u(x+h)}{v(x+h)} - \frac{u(x)}{v(x)}}{h} = \lim_{h \to 0} \frac{v(x) \cdot u(x+h) - u(x) \cdot v(x+h)}{h \cdot v(x+h) \cdot v(x)}$$

We can put the last fraction in a form that includes the derivatives of $u(x)$ and $v(x)$ by subtracting and adding $u(x) \cdot v(x)$ in the numerator. Therefore,

$$\frac{d}{dx}\left[\frac{u(x)}{v(x)}\right] = \lim_{h\to 0}\frac{v(x) \cdot u(x+h) - v(x) \cdot u(x) + v(x) \cdot u(x) - u(x) \cdot v(x+h)}{h \cdot v(x+h) \cdot v(x)}$$

$$\frac{d}{dx}(u/v) = \lim_{h\to 0}\frac{v(x) \cdot \dfrac{u(x+h)-u(x)}{h} - u(x) \cdot \dfrac{v(x+h)-v(x)}{h}}{v(x+h) \cdot v(x)}$$

$$\frac{d}{dx}(u/v) = \frac{v(x) \cdot \dfrac{du(x)}{dx} - u(x) \cdot \dfrac{dv(x)}{dx}}{[v(x)]^2}$$

Quotient Rule	EXAMPLE 4
$$\frac{d\left(\dfrac{u}{v}\right)}{dx} = \frac{v \cdot \dfrac{du}{dx} - u \cdot \dfrac{dv}{dx}}{v^2} \quad (23.13)$$ The derivative of the quotient of two differentiable functions equals the denominator times the derivative of the numerator minus the numerator times the derivative of the denominator, all divided by the square of the denominator.	If $q(x) = \dfrac{3-2x}{x^2+2}$, we have $u = 3 - 2x$ and $v = x^2 + 2$. Then $$q'(x) = \frac{(x^2+2) \cdot \dfrac{d(3-2x)}{dx} - (3-2x) \cdot \dfrac{d(x^2+2)}{dx}}{(x^2+2)^2}$$ $$= \frac{(x^2+2) \cdot (-2) - (3-2x) \cdot (2x)}{(x^2+2)^2}$$ $$= \frac{-2x^2 - 4 - 6x + 4x^2}{(x^2+2)^2}$$ $$q'(x) = \frac{2(x^2-3x-2)}{(x^2+2)^2}$$

COMMON ERROR Be careful when cancelling terms in derivatives of quotients. It is a common error to cancel terms that are not factors of both terms of the numerator. For instance, the factor $(x^2 + 2)$ in Example 4 is not a factor of both terms of the numerator and therefore cannot be cancelled.

Practice Exercise

2. Find the derivative of $y = \dfrac{2x^2}{x^4 - 1}$.

LEARNING TIP

Note that when taking derivatives of quotients, it is sometimes easier to rewrite the problem as a product, with the denominator changing to have a negative power when moved to the numerator. In other words, $y = \dfrac{u}{v}$ can be written in product form as $y = u \cdot v^{-1}$.

EXAMPLE 5 Derivative of a quotient—stress

The stress S on a hollow tube is given by

$$S = \frac{16DT}{\pi(D^4 - d^4)}$$

where T is the tension, D is the outer diameter, and d is the inner diameter of the tube. Find the expression for the instantaneous rate of change of S with respect to D, with the other values being constant.

We are to find the derivative of S with respect to D, and it is found as follows:

$$\frac{dS}{dD} = \frac{\pi(D^4 - d^4) \cdot (16T) - 16DT \cdot (\pi)(4D^3)}{\pi^2(D^4 - d^4)^2} = \frac{16\pi T(D^4 - d^4 - 4D^4)}{\pi^2(D^4 - d^4)^2}$$

$$\frac{dS}{dD} = \frac{-16T(3D^4 + d^4)}{\pi(D^4 - d^4)^2}$$

EXAMPLE 6 Evaluation of a derivative

Evaluate the derivative of $y = \dfrac{3x^2 + x}{1 - 4x}$ at $(2, -2)$.

$$\frac{dy}{dx} = \frac{(1 - 4x)(6x + 1) - (3x^2 + x)(-4)}{(1 - 4x)^2}$$

$$= \frac{6x + 1 - 24x^2 - 4x + 12x^2 + 4x}{(1 - 4x)^2}$$

$$= \frac{-12x^2 + 6x + 1}{(1 - 4x)^2}$$

$$\left.\frac{dy}{dx}\right|_{x=2} = \frac{-12(2^2) + 6(2) + 1}{[1 - 4(2)]^2} = \frac{-35}{49}$$

$$= -\frac{5}{7}$$

EXERCISES 23.6

In Exercises 1 and 2, make the given changes in the indicated examples of this section and then solve the resulting problems.

1. In Example 2, change u to $5 - 3x^2$.
2. In Example 6, change the numerator to $3x^2 - x$ and then find and evaluate the derivative at $x = 2$.

In Exercises 3–8, find the derivative of each function by using Eq. (23.12). Do not find the product before finding the derivative.

3. $y = 6x(3x^2 - 5x)$
4. $y = 2x^3(3x^4 + x)$
5. $s = (6t + 1)(3t - 4)$
6. $f(x) = (3x - 2)(4x^2 + 3)$
7. $y = (x^4 - 3x^2 + 3)(1 - 2x^3)$
8. $y = (x^3 - 6x)(2 - 4x^3)$

In Exercises 9–12, find the derivative of each function by using Eq. (23.12). Then multiply out each function and find the derivative by treating it as a polynomial. Compare the results.

9. $y = (2x - 7)(5 - 2x)$
10. $f(s) = (11s^2 + 3)(2s^2 - 1)$
11. $V = (h^3 - 1)(2h^2 - h - 1)$
12. $y = (3x^2 - 4x + 1)(5 - 6x^2)$

In Exercises 13–24, find the derivative of each function by using Eq. (23.13).

13. $y = \dfrac{x}{2x + 3}$
14. $u = \dfrac{4}{v^2}$
15. $y = \dfrac{\pi}{2x^2 + 1}$
16. $R = \dfrac{5i + 2}{2i + 3}$
17. $y = \dfrac{6x^2}{3 - 2x}$
18. $y = \dfrac{e^2}{3x^2 - 5x}$
19. $y = \dfrac{2x - 1}{3x^2 + 2}$
20. $P = \dfrac{2i^2}{4 - 3i}$
21. $f(x) = \dfrac{3x + 8}{x^2 + 4x + 2}$
22. $y = \dfrac{33x}{4x^5 - 3x - 4}$
23. $y = \dfrac{2x^2 - x - 1}{x^2(x + 2)}$
24. $y = \dfrac{3x^3 - 8x}{2x^2 - 5x + 4}$

In Exercises 25–32, evaluate the derivatives of the given functions for the given values of x.

25. $y = (3x - 1)(4 - 7x),\ x = 5$
26. $y = (3x^2 - 5)(2x^2 - 1),\ x = -1$
27. $y = (2x^2 - x + 1)(4 - 2x - x^2),\ x = -3$
28. $y = (4x^4 + 0.5x^2 + 1)(3x - 2x^2),\ x = 0.5$
29. $y = \dfrac{3x - 5}{2x + 3},\ x = -2$
30. $y = \dfrac{2x^2 - 5x}{3x + 2},\ x = 2$
31. $S = \dfrac{2n^3 - 3n + 8}{2n - 3n^4},\ n = -1$
32. $y = \dfrac{2x^3 - x^2 - 2}{4x + 3},\ x = 0.5$

In Exercises 33–57, solve the given problems by finding the appropriate derivatives, and round answers to three significant digits unless otherwise specified.

33. What text equation from Section 23.5 is equivalent to the product rule if one of the functions u and v is a constant?
34. By use of the quotient rule, derive a formula for the derivative of the function $1/v(x)$.
35. Using the product rule, find the point(s) on the curve of $y = (2x^2 - 1)(1 - 4x)$ for which the tangent line is $y = 4x - 1$.
36. Do the curves of $y = x^2$ and $y = 1/x^2$ cross at right angles? Explain.
37. If $f(x)$ is a differentiable function, find an expression for the derivative of $y = x^2 \cdot f(x)$.
38. If $f(x)$ is a differentiable function, find an expression for the derivative of $y = f(x)/x^2$.
39. Find the derivative of $y = \dfrac{x^2(1 - 2x)}{3x - 7}$ in each of the following two ways. (1) Do not multiply out the numerator before finding the derivative. (2) Multiply out the numerator before finding the derivative. Compare the results.
40. Find the derivative of $y = 4x^2 - \dfrac{1}{x - 1}$ in each of the following two ways. (1) Do not combine the terms over a common denominator before finding the derivative. (2) Combine the terms over a common denominator before finding the derivative. Compare the results.

41. Find the slope of a line tangent to the curve of the function $y = (4x + 1)(x^4 - 1)$ at the point $(-1, 0)$. Do not multiply the factors together before taking the derivative.

42. Find the slope of a line tangent to the curve of the function $y = (3x + 4)(1 - 4x)$ at the point $(2, -70)$. Do not multiply the factors together before taking the derivative.

43. For what value(s) of x is the slope of a tangent to the curve of $y = \dfrac{x}{x^2 + 1}$ equal to zero? View the graph on a graphing utility to verify the values found.

44. Determine the sign of the derivative of the function $y = \dfrac{2x - 1}{1 - x^2}$ for the following values of x: $-2, -1, 0, 1, 2$. Is the slope of a tangent line to this curve ever negative? View the graph on a graphing utility to verify your conclusion.

45. The power P (in W) in an electric circuit is the product of the voltage V and the current I (in A). If V and I vary with time t (in s) $(0 \le t \le 0.2\text{ s})$, according to $V = 0.0480 - 1.20t^2$ and $I = 2.00 - 0.800t$, find dP/dt when $t = 0.150$ s.

46. The thermodynamic temperature T (in K) varies jointly as the pressure p (in Pa) and volume V (in m^3). Find the expression for dT/dt if $p = 1200(1 - 0.0025t)$ and $V = 2.50(1 + 0.0048t^2)$, where t is the time (in s).

47. If a constant current of 2 A passes through the *current divider* parallel resistors shown in Fig. 23.23, the current i is given by $i = 8R/(7R + 12)$, where R is a variable resistor. Find di/dR.

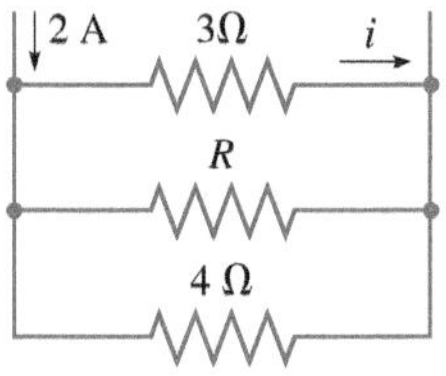

Fig. 23.23

48. The sales S of a product as a function of the time t (in weeks) are given by $S = \dfrac{2300(5t + 1)}{2t + 5}$. Find the instantaneous rate of change of S with respect to t for $t = 9$ weeks.

49. During each cycle, the vertical displacement s of the end of a robot arm is given by $s = (t^2 - 8t)(2t^2 + t + 1)$, where t is the time. Find the expression for the instantaneous velocity of the end of the robot arm. See Fig. 23.24.

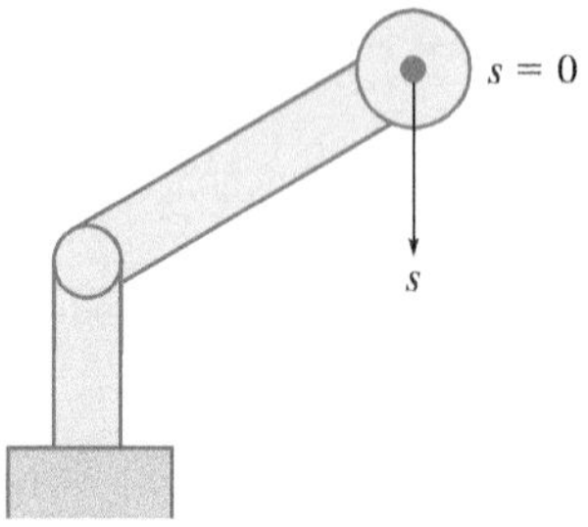

Fig. 23.24

50. The concentration c (in mg/L) of a certain drug in the bloodstream is found to be $c = 25t/(t^2 + 5)$, where t is the time (in h) after the drug is taken. Find dc/dt.

51. A computer, using data from a refrigeration plant, estimated that in the event of a power failure the temperature T (in °C) in the freezers would be given by $T = \dfrac{2t}{0.05t + 1} - 20$, where t is the number of hours after the power failure. Find the time rate of change of temperature after 6.00 h.

52. The voltage V across a resistor in an electric circuit is the product of the resistance and the current. If the current I (in A) varies with time t (in s) according to the relation $I = 5.00 + 0.01t^2$ and the resistance varies with time according to the relation $R = 15.00 - 0.10t$, find the time rate of change of the voltage when $t = 7.00$ s.

53. The frictional radius r_f of a disc clutch is given by the equation $r_f = \dfrac{2(R^2 + Rr + r^2)}{3(R + r)}$, where R and r are the outer radius and the inner radius of the clutch, respectively. Find the derivative of r_f with respect to R with r constant.

54. In thermodynamics, an equation relating the thermodynamic temperature T, the pressure p, and the volume V of a gas is $T = \left(p + \dfrac{a}{V^2}\right)\left(\dfrac{V - b}{R}\right)$, where a, b, and R are constants. Find the derivative of T with respect to V, assuming p is constant.

55. The electric power P produced by a certain source is given by $P = \dfrac{E^2 r}{R^2 + 2Rr + r^2}$, where E is the voltage of the source, R is the resistance of the source, and r is the resistance in the circuit. Find the derivative of P with respect to r, assuming that the other quantities remain constant.

56. In the theory of lasers, the power P radiated is given by the equation $P = \dfrac{kf^2}{\omega^2 - 2\omega f + f^2 + a^2}$, where f is the field frequency and a, k, and ω are constants. Find the derivative of P with respect to f.

57. The pressure p (in Pa) exerted at a depth h (in m) on the wall of a particular dam is given by $p = \dfrac{9800h^2}{h + 1}$. Determine the rate at which pressure is changing at the instant the depth is 48.0 m.

Answers to Practice Exercises

1. $dy/dx = -12x^5 + 12x^3 + 4x$ **2.** $\dfrac{dy}{dx} = \dfrac{-4x(x^4 + 1)}{(x^4 - 1)^2}$

23.7 The Derivative of a Power of a Function

In Example 1(c) of Section 23.6, we illustrated $y = (3 - 2x)^2$ as the power of a function of x, where $3 - 2x$ is the function. If we let $u = 3 - 2x$, we can write $y = u^2$, and in this way, y is a function of u, and u is a function of x. This means that *y is a function of a function of x, which is called a* **composite function.**

The formula for taking derivatives of composite functions is known as the **chain rule** *for derivatives*. It is one of the most important derivative formulas in calculus.

LEARNING TIP

The chain rule will be useful in many applications. It is critical for implicit differentiation (Section 23.8) and any differentiation involving functions expressed in terms of other functions.

Chain Rule

$$\frac{dy}{dx} = \frac{dy}{du} \cdot \frac{du}{dx} \qquad \textbf{(23.14)}$$

The derivative of a composite function y, where y is a function of u, and u is a function of x, is the product of the derivative of y with respect to u and the derivative of u with respect to x.

EXAMPLE 1

Find the derivative of $y = (3 - 2x)^2$.
Let $y = u^2$ and $u = 3 - 2x$:

$$y = u^2, \qquad \frac{dy}{du} = 2u$$

$$u = 3 - 2x, \qquad \frac{du}{dx} = -2$$

$$\frac{dy}{dx} = \frac{dy}{du} \cdot \frac{du}{dx}$$

$$= (2u)(-2)$$

$$= -4u = -4(3 - 2x)$$

EXAMPLE 2 Verifying the chain rule

Verify the derivative of $y = (3 - 2x)^2$ from Example 1 by finding the derivative with the product rule.

We write $y = (3 - 2x)(3 - 2x)$. Then

$$\frac{dy}{dx} = (3 - 2x) \cdot (-2) + (3 - 2x) \cdot (-2)$$

$$\frac{dy}{dx} = 2(-2)(3 - 2x) = -4(3 - 2x)$$

We see that this result is the same as the result using the chain rule in Example 1, verifying the chain rule. You can imagine that for powers higher than 2, using the chain rule is much simpler than using the product rule.

It is worth writing an explicit formula for the special case of power of functions. We use Eq. (23.14) for $y = u^n$, where u is a differentiable function of x, and obtain the following more general power rule.

General Power Rule

$$\frac{du^n}{dx} = nu^{n-1}\left(\frac{du}{dx}\right) \qquad \textbf{(23.15)}$$

EXAMPLE 3

Find the derivative of $y = (3 - 2x)^3$.
Here $n = 3$ and $u = 3 - 2x$. Therefore, $du/dx = -2$. This means

$$\frac{du^n}{dx} = n \quad u \quad {}^{n-1} \quad \left(\frac{du}{dx}\right)$$

$$\downarrow \quad \downarrow \quad \downarrow \quad \swarrow$$

$$\frac{dy}{dx} = 3(3 - 2x)^2(-2)$$

$$\frac{dy}{dx} = -6(3 - 2x)^2$$

Practice Exercise

1. Find the derivative of $y = (5x + 2)^4$.

COMMON ERROR When finding derivatives requiring the chain rule, the du/dx factor is often forgotten. The derivative is incomplete and therefore incorrect without this factor.

EXAMPLE 4 Do not forget du/dx

Find the derivative of $p(x) = (1 - 3x^2)^4$.

In this example, $n = 4$ and $u = 1 - 3x^2$, and $du/dx = -6x$.

$$\frac{du^n}{dx} = n \quad u \quad {}^{n-1} \quad \left(\frac{du}{dx}\right)$$

$$p'(x) = 4(1 - 3x^2)^3(-6x)$$
$$p'(x) = -24x(1 - 3x^2)^3$$

EXAMPLE 5 Product rule combined with power rule

Find the derivative of $y = 2x^3(3 - x^3)^4$.

Here, we must ***use the product rule in combination with the power rule.***

$$\frac{dy}{dx} = 2x^3 \cdot \left[4(3 - x^3)^3(-3x^2)\right] + (3 - x^3)^4 \cdot \left[2(3x^2)\right]$$

$$\frac{dy}{dx} = -24x^5(3 - x^3)^3 + 6x^2(3 - x^3)^4 = 6x^2(3 - x^3)^3\left[-4x^3 + (3 - x^3)\right]$$

$$\frac{dy}{dx} = 6x^2(3 - 5x^3)(3 - x^3)^3$$

Practice Exercise

2. Find the derivative of $y = 5x(2x + 7)^3$.

Until now, we have derived formulas for derivatives of differentiable functions of x raised to positive integral powers. We now show that these formulas are also valid for any rational number used as an exponent. If we raise each side of $y = u^{p/q}$ to the qth power, we have $y^q = u^p$. Applying the general power rule, we have

$$qy^{q-1}\left(\frac{dy}{dx}\right) = pu^{p-1}\left(\frac{du}{dx}\right)$$

$$\frac{dy}{dx} = \frac{pu^{p-1}(du/dx)}{qy^{q-1}} = \frac{p}{q}\frac{u^{p-1}}{(u^{p/q})^{q-1}}\frac{du}{dx} = \frac{p}{q}\frac{u^{p-1}}{u^{p-p/q}}\frac{du}{dx}$$

$$\frac{dy}{dx} = \frac{p}{q}u^{p-1-p+(p/q)}\frac{du}{dx}$$

General Power Rule (Rational Exponent)	EXAMPLE 6
$$\frac{du^{p/q}}{dx} = \frac{p}{q}u^{(p/q)-1}\frac{du}{dx} \quad \textbf{(23.16)}$$	Find the derivative of $y = \sqrt{x^2 + 1}$. First write $y = (x^2 + 1)^{1/2}$. Then $p/q = 1/2$ and $u = x^2 + 1$. Therefore, $du/dx = 2x$. This gives $\frac{dy}{dx} = \frac{1}{2}(x^2 + 1)^{-1/2}(2x)$ $\frac{dy}{dx} = \frac{x}{(x^2 + 1)^{1/2}}$

LEARNING TIP

We see that in finding the derivative, we multiply the function by the rational exponent and subtract 1 from it to find the exponent of the function in the derivative. *This is the same rule as derived for positive integral exponents in the general power rule.*

Practice Exercise

3. Find the derivative of $y = \sqrt[3]{4 - 9x}$.

In deriving Eqs. (23.15) and (23.16), we used the power rule, and we noted it was valid for positive exponents. We can show that the power rule is also valid for negative exponents by using the quotient rule on $1/x^n$, which is the same as x^{-n}. Therefore,

the general power rule for derivatives can be extended to include all rational exponents, positive or negative.

This, of course, includes all integral exponents, positive and negative. Also, we note that the power rule is a special case of the general power rule with $u = x$ (because $du/dx = 1$).

Having shown that we may use fractional exponents to find derivatives of roots of functions of x, we may also use them to find derivatives of roots of x itself.

EXAMPLE 7 Derivative using a fractional exponent

Find the derivative of $y = 6\sqrt[3]{x^2}$.

We can write this function as $y = 6x^{2/3}$. In finding the derivative, we may use the power rule with $n = \frac{2}{3}$. This gives us

$$y = 6x^{2/3} \qquad \frac{2}{3} - 1$$

$$\frac{dy}{dx} = 6\left(\frac{2}{3}\right)x^{-1/3} = \frac{4}{x^{1/3}}$$

LEARNING TIP

Note that we first rewrite the function in a different, more useful form. This is often an important step before taking the derivative.

We could also use the general power rule with $u = x$ and $n = \frac{2}{3}$. This gives us

$$\frac{du}{dx} = \frac{dx}{dx} = 1$$

$$\frac{dy}{dx} = 6\left(\frac{2}{3}\right)x^{-1/3}(1) = \frac{4}{x^{1/3}}$$

Note that the domain of the function is all real numbers, but the function is not differentiable for $x = 0$.

In the following examples, we illustrate the use of the power rule and the general power rule for the case in which n is a negative exponent. *Special care must be taken in the case of a negative exponent.*

EXAMPLE 8 Derivative using a negative exponent–electric resistance

The electric resistance R of a wire varies inversely as the square of its radius r. For a given wire, $R = 4.66\ \Omega$ for $r = 0.150$ mm. Find the derivative of R with respect to r for this wire.

Since R varies inversely as the square of r, we have $R = k/r^2$. Then, using the fact that $R = 4.66\ \Omega$ for $r = 0.150$ mm, we have

$$4.66 = \frac{k}{(0.150)^2}, \qquad k = 0.105\ \Omega \cdot \text{mm}^2$$

which means that

$$R = \frac{0.105}{r^2}$$

We could find the derivative by the quotient rule. However, when the numerator is constant, the derivative is easily found by using negative exponents and the power rule.

$$R = \frac{0.105}{r^2} = 0.105r^{-2}$$

$$\frac{dR}{dr} = 0.105(-2)r^{-3}$$

$$\frac{dR}{dr} = -\frac{0.210}{r^3}$$

Here, we used the power rule directly.

COMMON ERROR When subtracting 1 from a negative exponent, be sure to *decrease* the value (e.g., $-2 - 1 = -3$, or $-5 - 1 = -6$).

EXAMPLE 9 The chain rule using a negative exponent

Find the derivative of $y = \dfrac{1}{(1 - 4x)^5}$.

$$y = \frac{1}{(1-4x)^5} = (1-4x)^{-5} \quad \text{use negative exponent}$$

$$\frac{dy}{dx} = (-5)(1-4x)^{-6}(-4) \quad \text{use Eq. (23.15)}$$

$$\frac{dy}{dx} = \frac{20}{(1-4x)^6} \quad \text{express result with positive exponent}$$

Practice Exercise

4. Find the derivative of $y = \dfrac{3}{(6x + 5)^4}$.

We now see the value of fractional exponents in calculus. They are useful in many algebraic operations, but they are almost essential in calculus. Without fractional exponents, it would be necessary to develop additional formulas to find the derivatives of radical expressions. In order to find the derivative of an algebraic function, we need only those formulas we have already developed. Often, it is necessary to combine these formulas, as we saw in Example 5. Actually, most derivatives are combinations.

The key step in finding the derivative is ***recognizing the form of the function*** with which you are dealing. When you have recognized the form, completing the problem is only a matter of mechanics and algebra. You should now see the importance of being able to handle algebraic operations with ease.

EXAMPLE 10 Using the chain rule and the quotient rule

Using the quotient rule, find the derivative of $y = \dfrac{x}{(x^2 - 9)^3}$ and evaluate it at $x = 4$.

Notice that this function is a quotient because of the division of the two functions, but the denominator by itself is a power of a function and will require the chain rule to differentiate.

Recognize that the function $y = \dfrac{x}{(x^2 - 9)^3}$ is of the form $y = \dfrac{u}{v}$, with $u = x$ and $v = (x^2 - 9)^3$. In order to use the quotient rule, we need the derivative of each function by itself.

$u = x$ $\quad\quad v = (x^2 - 9)^3$

$\frac{du}{dx} = 1$ $\quad\quad \frac{dv}{dx} = 3(x^2 - 9)^2(2x)$ using chain rule

$$\frac{dv}{dx} = 6x(x^2 - 9)^2$$

$$\frac{dy}{dx} = \frac{v \cdot \frac{du}{dx} - u \cdot \frac{dv}{dx}}{v^2}$$ quotient rule

$$\frac{dy}{dx} = \frac{(x^2 - 9)^3 \cdot (1) - x \cdot 6x(x^2 - 9)^2}{(x^2 - 9)^6}$$

$$\frac{dy}{dx} = \frac{(x^2 - 9)^3 - 6x^2(x^2 - 9)^2}{(x^2 - 9)^6}$$

We can factor out a common factor of $(x^2 - 9)^2$ in the numerator, and it will divide out with the similar factor in the denominator:

$$\frac{dy}{dx} = \frac{(x^2 - 9)^2}{(x^2 - 9)^6}[(x^2 - 9) - 6x^2] = \frac{-1(5x^2 + 9)}{(x^2 - 9)^4}$$

Now evaluating at $x = 4$,

$$\left.\frac{dy}{dx}\right|_{x=4} = \frac{-1(5 \cdot 4^2 + 9)}{(4^2 - 9)^4} = \frac{-89}{2401} = -0.0371$$

EXAMPLE 11 Using the chain rule and the product rule

Using the product rule, find the derivative of $y = \frac{x}{(x^2 - 9)^3}$ and evaluate it at $x = 4$.

This is the same function as in Example 10.

Notice that this function is a quotient because of the division of the two functions, but we can change it into a product by bringing the denominator to the top of the fraction with a negative power before differentiating.

$$y = x(x^2 - 9)^{-3}$$

Recognize that the function $y = x(x^2 - 9)^{-3}$ is of the form $y = u \cdot v$, with $u = x$ and $v = (x^2 - 9)^{-3}$. In order to use the product rule, we need the derivative of each function by itself.

$u = x$ $\quad\quad v = (x^2 - 9)^{-3}$

$\frac{du}{dx} = 1$ $\quad\quad \frac{dv}{dx} = -3(x^2 - 9)^{-4}(2x)$ using chain rule

$$\frac{dv}{dx} = -6x(x^2 - 9)^{-4}$$

$$\frac{d(u \cdot v)}{dx} = u \cdot \frac{dv}{dx} + v \cdot \frac{du}{dx}$$ product rule

$$\frac{dy}{dx} = x \cdot (-6x)(x^2 - 9)^{-4} + (x^2 - 9)^{-3} \cdot (1)$$

$$\frac{dy}{dx} = -6x^2(x^2 - 9)^{-4} + (x^2 - 9)^{-3}$$

LEARNING TIP

In cases where the denominator of a quotient is a variable or a function with a power other than 1, it is sometimes easier to bring the denominator to the top of the fraction with a negative power. This creates a product rule, instead of a quotient rule, which is then combined with the chain rule. The simpler formula sometimes produces a solution for the derivative that requires less factoring to express it in its simplest form.

We can factor out a common factor between the two terms:

$$\frac{dy}{dx} = (x^2 - 9)^{-4}[-6x^2 + (x^2 - 9)]$$

$$\frac{dy}{dx} = (x^2 - 9)^{-4}[-5x^2 - 9] = \frac{-1(5x^2 + 9)}{(x^2 - 9)^4}$$

Now evaluating at $x = 4$,

$$\left.\frac{dy}{dx}\right|_{x=4} = \frac{-1(5 \cdot 4^2 + 9)}{(4^2 - 9)^4} = \frac{-89}{2401} = -0.0371$$

This is the same answer as in Example 10, though with somewhat simpler factoring involved.

EXAMPLE 12 The chain rule inside a product rule

Find the derivative of $y = (4x + 1)^2(2x^2 + 3)^3$.

Notice that y is a product because of the multiplication of the two functions, but each function by itself is a power of a function and will require the chain rule to differentiate.

Recognize that the function $y = (4x + 1)^2(2x^2 + 3)^3$ is of the form $y = u \cdot v$, with $u = (4x + 1)^2$ and $v = (2x^2 + 3)^3$. In order to use the product rule, we need the derivatives of each function by itself.

$$u = (4x + 1)^2 \qquad v = (2x^2 + 3)^3$$

$$\frac{du}{dx} = 2(4x + 1)^1(4) \qquad \frac{dv}{dx} = 3(2x^2 + 3)^2(4x)$$

$$\frac{du}{dx} = 8(4x + 1) \qquad \frac{dv}{dx} = 12x(2x^2 + 3)^2$$

$$\frac{d(u \cdot v)}{dx} = u \cdot \frac{dv}{dx} + v \cdot \frac{du}{dx} \qquad \text{product rule}$$

$$\frac{dy}{dx} = (4x + 1)^2 \cdot (12x)(2x^2 + 3)^2 + (2x^2 + 3)^3 \cdot 8(4x + 1)$$

There are some common algebraic factors between the two terms:

$$\frac{dy}{dx} = 4(4x + 1)(2x^2 + 3)^2[3x(4x + 1) + (2x^2 + 3) \cdot 2]$$

$$\frac{dy}{dx} = 4(4x + 1)(2x^2 + 3)^2[12x^2 + 3x + 4x^2 + 6]$$

$$\frac{dy}{dx} = 4(4x + 1)(2x^2 + 3)^2[16x^2 + 3x + 6]$$

EXAMPLE 13 Evaluation of a derivative

Evaluate the derivative of $y = \frac{x}{\sqrt{1 - 4x}}$ for $x = -2$.

Here we have a quotient, but the denominator function has a power not equal to 1. Therefore, we will take the denominator to the top as a negative power and use a product rule with a chain rule embedded in it.

$$y = x \cdot (1 - 4x)^{-1/2}$$

$$\frac{dy}{dx} = x \cdot \left(-\frac{1}{2}\right)(1 - 4x)^{-3/2} \cdot (-4) + (1 - 4x)^{-1/2} \cdot (1)$$

$$= \left(\frac{1}{2}\right)(1 - 4x)^{-3/2}[4x + 2(1 - 4x)^1]$$

$$= \left(\frac{1}{2}\right)(1 - 4x)^{-3/2}[4x + 2 - 8x]$$

$$= \left(\frac{1}{2}\right)(1 - 4x)^{-3/2}[2 - 4x] = (1 - 4x)^{-3/2}[1 - 2x]$$

$$\frac{dy}{dx} = \frac{1 - 2x}{(1 - 4x)^{3/2}}$$

Now, evaluating the derivative at $x = -2$, we have

$$\left.\frac{dy}{dx}\right|_{x=-2} = \frac{1 - 2(-2)}{[1 - 4(-2)]^{3/2}} = \frac{1 + 4}{(1 + 8)^{3/2}} = \frac{5}{9^{3/2}} = \frac{5}{27}$$

EXERCISES 23.7

In Exercises 1–4, make the given changes in the indicated examples of this section and then find the derivatives.

1. In Example 4, change $1 - 3x^2$ to $2 + 3x^3$.
2. In Example 5, change $3 - x^3$ to $2 + x^5$.
3. In Example 6, change $x^2 + 1$ to $2 - 3x^2$.
4. In Example 9, change the exponent 5 to 3.

In Exercises 5–32, find the derivative of each of the given functions.

5. $y = 6\sqrt{x}$
6. $y = \sqrt[4]{x^3}$
7. $v = \frac{3}{5t^3}$
8. $y = \frac{2}{x^{23}}$
9. $y = \frac{3}{\sqrt[3]{x}} + 4x^2$
10. $y = \frac{55}{\sqrt[5]{x^2}}$
11. $y = x\sqrt{x} - \frac{6}{x}$
12. $f(x) = 2x^{-3} - 3x^{-2}$
13. $y = (x^2 + 1)^5$
14. $y = (1 - 5x)^{12}$
15. $y = 2.25(7 - 4x^3)^8$
16. $s = 3(8t^2 - 7)^6$
17. $y = (2x^3 - 3)^{1/3}$
18. $y = 8(1 - 6x)^{1.5}$
19. $f(y) = \frac{3}{(4 - y^2)^4}$
20. $y = \frac{\pi^3}{\sqrt{1 - 3x}}$
21. $y = 4(2x^4 - 5)^{0.25}$
22. $r = 5(3\theta^6 - 4)^{2/3}$
23. $y = \sqrt[4]{1 - 8x^2}$
24. $y = 9\sqrt[3]{4x^6 + 2}$
25. $u = v\sqrt{8v + 5}$
26. $y = x^7(1 - 3x)^{15}$
27. $y = \frac{2\sqrt{1 - 6x}}{x^3}$
28. $R = \frac{2T^2}{\sqrt[3]{1 + 4T}}$
29. $y = \frac{2x\sqrt{x + 2}}{x + 4}$
30. $y = 8\sqrt{1 + \sqrt{x}}$
31. $f(R) = \sqrt{\frac{2R + 1}{4R + 1}}$
32. $y = \left(\frac{2x + 1}{3x - 2}\right)^2$

In Exercises 33–36, evaluate the derivatives of the given functions for the given values of x.

33. $y = \sqrt{3x + 4}, x = 7$
34. $y = (4 - x^2)^{-1}, x = -1$
35. $y = \frac{\sqrt{x}}{1 - x}, x = 4$
36. $y = x^2\sqrt[3]{3x + 2}, x = 2$

In Exercises 37–60, solve the given problems by finding the appropriate derivatives, and round answers to three significant digits unless otherwise specified.

37. Find the derivative of $x^{3/2}$ by writing $x^{3/2} = x(x^{1/2})$ and using the product rule.
38. Find the derivative of $x^{3/2}$ by writing $x^{3/2} = x^2/x^{1/2}$ and using the quotient rule.
39. Find the derivative of $y = 1/x^3$ as (a) a quotient and (b) a negative power of x and show that the results are the same.
40. Let $y = [u(x)]^2$ and find dy/dx, treating $[u(x)]^2$ as the product $u(x)u(x)$. (See Example 2.)
41. Find any values of x for which the derivative of $y = \frac{x^2}{\sqrt{x^2 + 1}}$ is zero. View the curve of the function on a graphing utility to verify the values found.

42. Find any values of x for which the derivative of $y = \frac{x}{\sqrt{4x-1}}$ is zero. View the curve of the function on a graphing calculator to verify the values found.

43. Is the line $x + 3y - 12 = 0$ ever perpendicular to a tangent to the graph of $y = \sqrt{2x+3}$?

44. Explain why the graph of $y_1 = \sqrt{x} + a$ cannot be tangent to the graph of $y_2 = \sqrt{4-2x}$, regardless of the value of a.

45. Find the slope of a line tangent to the parabola $y^2 = 4x$ at the point $(4, 4)$.

46. Find the slope of a line tangent to the circle $x^2 + y^2 = 25$ at the point $(4, 3)$.

47. The lowest flying speed v (in m/s) at which a certain airplane can fly varies directly as the square root of the wing load w (in Pa). If $v = 27.0$ m/s for $w = 775$ Pa, find the derivative of v with respect to w.

48. During and after a period of rain, the depth h (in m) of water behind a certain dam was given by $h = 75.0(x+2)/(x+3)$, where x is the number of days after the start of the rainy period. Find dh/dx for $x = 2.50$ days.

49. The displacement s (in cm) of a linkage joint of a robot is given by $s = (8t - t^2)^{2/3}$, where t is the time (in s). Find the velocity of the joint for $t = 6.25$ s.

50. Water is slowly rising in a horizontal drainage pipe. The width w of the water as a function of the depth h is $w = \sqrt{2rh - h^2}$, where r is the radius of the pipe. Find dw/dh for $h = 225$ mm and $r = 600$ mm.

51. When the volume of a gas changes very rapidly, an approximate relation is that the pressure p varies inversely as the 3/2 power of the volume. If p is 300 kPa when $V = 100\ \text{cm}^3$, find the derivative of p with respect to V. Evaluate this derivative for $V = 100\ \text{cm}^3$.

52. The power gain G of a certain antenna is inversely proportional to the square of the wavelength λ (in m) of the carrier wave. If $G = 5.00 \times 10^4$ for $\lambda = 0.110$ m, find the derivative of G with respect to λ for $\lambda = 0.110$ m.

53. In deep water, the velocity of a wave is $v = k\sqrt{\frac{l}{a} + \frac{a}{l}}$, where a and k are constants and l is the length of the wave. For what value of l is $dv/dl = 0$?

54. Due to air friction, the drag F on a plane is $F = c_1v^2 + c_2v^{-2}$, where v is the plane's velocity and c_1 and c_2 are positive constants. For what values of v is $dF/dv = 0$?

55. The total solar radiation H (in W/m^2) on a certain surface during an average clear day is given by

$$H = \frac{4000}{\sqrt{t^6 + 100}} \quad (-6 < t < 6)$$

where t is the number of hours from noon. Find the rate at which H is changing with time at 4:00 P.M.

56. In determining the time for a laser beam to go from S to P (see Fig. 23.25), which are in different mediums, it is necessary to find the derivative of the time

$$t = \frac{\sqrt{a^2 + x^2}}{v_1} + \frac{\sqrt{b^2 + (c-x)^2}}{v_2}$$

with respect to x, where a, b, c, v_1, and v_2 are constants. Here, v_1 and v_2 are the velocities of the laser beam in each medium. Find this derivative.

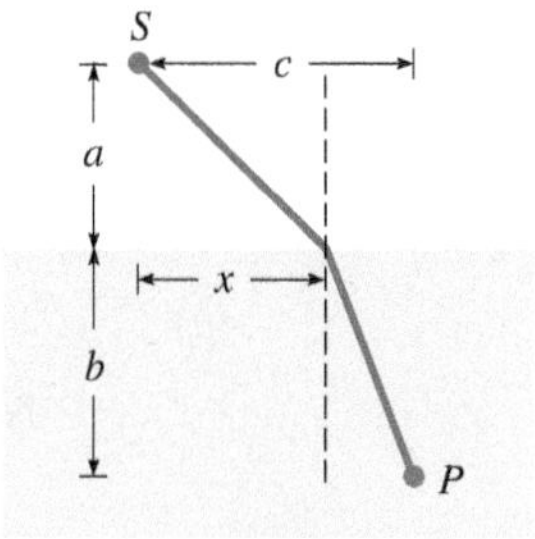

Fig. 23.25

57. The radio waveguide wavelength λ_r is related to its free-space wavelength λ by

$$\lambda_r = \frac{2a\lambda}{\sqrt{4a^2 - \lambda^2}}$$

where a is a constant. Find $d\lambda_r/d\lambda$.

58. The current I in a circuit containing a resistance R and an inductance L is found from the expression

$$I = \frac{V}{\sqrt{R^2 + (\omega L)^2}}$$

Find the expression for the instantaneous rate of change of current with respect to L, assuming that the other quantities remain constant.

59. The length l of a rectangular microprocessor chip is 2 mm longer than its width w. Find the derivative of the length of the diagonal D with respect to w.

60. The trapezoidal engineering support structure shown in Fig. 23.26 has an internal support of length l. Find the derivative of l with respect to x.

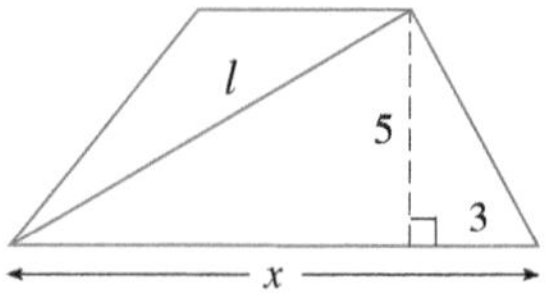

Fig. 23.26

Answers to Practice Exercises

1. $dy/dx = 20(5x+2)^3$ **2.** $dy/dx = 5(2x+7)^2(8x+7)$
3. $dy/dx = -3(4-9x)^{-2/3}$ **4.** $dy/dx = -72(6x+5)^{-5}$

23.8 Differentiation of Implicit Functions

To this point, all of the functions that have been differentiated have been of the form $y = f(x)$, where y is explicitly defined as a function of x. Sometimes, however, a relationship does not define the dependent variable explicitly in terms of the independent variable.

An equation in which y is not expressed explicitly in terms of x may determine one or more functions. *Any such function, where y is defined implicitly as a function of x, is called an* **implicit function**. Some equations defining implicit functions may be solved to determine the explicit functions, and for others it is not possible to solve for the explicit functions. Also, not all such equations define y as a function of x for real values of x. Even with this difficulty, the derivatives of implicitly defined relations can still be calculated using a process called **implicit differentiation**.

EXAMPLE 1 Illustrations of implicit functions

(a) The equation $3x + 4y = 5$ is an equation that defines a function, although it is not in explicit form. In solving for y as $y = -\frac{3}{4}x + \frac{5}{4}$, we have the explicit form of the function.

(b) The equation $y^2 + x = 3$ is an equation that defines two functions, although we do not have the explicit forms. When we solve for y, we obtain the explicit functions $y = \sqrt{3 - x}$ and $y = -\sqrt{3 - x}$.

(c) The equation $y^5 + xy^2 + 3x^2 = 5$ defines y as a function of x, although we cannot actually solve for the explicit algebraic form of the function.

(d) The equation $x^2 + y^2 + 4 = 0$ is not satisfied by any pair of real values of x and y.

Even when it is possible to determine the explicit form of a function given in implicit form, it is not always desirable to do so. In some cases, the implicit form is more convenient than the explicit form.

The derivative of an implicit function may be found directly without having to solve for the explicit function. Often, the chain rule must be used (since powers of y are powers of a function of x).

The following two steps can be used to determine the derivative at any point $\frac{dy}{dx}$ for implicit relations that are expressed in terms of x (the variable) and y (the relation).

Implicit Differentiation

1. **Differentiate with respect to x both sides of the implicit equation.** This means applying the process d/dx to every term of the equation. Remember that every time you differentiate a function y, you must introduce a dy/dx factor through using the chain rule. Moreover, remember that y depends on x, so you will need to use the product rule or the quotient rule if y (or y^n) is multiplied or divided by a function of x.
2. **Rearrange algebraically to isolate dy/dx.** It may appear in multiple terms, so algebraic factoring will be a common requirement. For implicitly defined relations, the derivative is itself an implicitly defined relation of x and y.

EXAMPLE 2 Implicit derivative

Find dy/dx if $y^2 + 2x^2 = 5$.

Here, we find the derivative of each term and then solve for dy/dx. Thus,

$$\frac{d(y^2)}{dx} + \frac{d(2x^2)}{dx} = \frac{d(5)}{dx}$$

$$2y^{2-1} \cdot \frac{dy}{dx} + 2\left(2x^{2-1} \cdot \frac{dx}{dx}\right) = 0$$

$$2y \cdot \frac{dy}{dx} + 4x = 0$$

$$\frac{dy}{dx} = -\frac{2x}{y}$$

EXAMPLE 3 Implicit derivative involving a product

Find dy/dx if $3y^4 + xy^2 + 2x^3 - 6 = 0$.

In finding the derivative, we note that the second term is a product of two functions of x, and we must use the product rule for derivatives on it. Thus, we have

$$\frac{d(3y^4)}{dx} + \frac{d(x \cdot y^2)}{dx} + \frac{d(2x^3)}{dx} - \frac{d(6)}{dx} = \frac{d(0)}{dx}$$

using product rule

$$12y^3 \cdot \frac{dy}{dx} + \left[x\left(2y \cdot \frac{dy}{dx}\right) + y^2(1)\right] + 6x^2 - 0 = 0$$

$$12y^3 \cdot \frac{dy}{dx} + 2xy \cdot \frac{dy}{dx} + y^2 + 6x^2 = 0 \quad \text{solve for } \frac{dy}{dx}$$

$$(12y^3 + 2xy)\frac{dy}{dx} = -y^2 - 6x^2$$

$$\frac{dy}{dx} = \frac{-y^2 - 6x^2}{12y^3 + 2xy}$$

EXAMPLE 4 Implicit derivative—product and power

Find dy/dx if $2x^3y + (y^2 + x)^3 = x^4$.

In this case, we use the product rule on the first term and the power rule on the second term:

$$\frac{d(2x^3 \cdot y)}{dx} + \frac{d(y^2 + x)^3}{dx} = \frac{d(x^4)}{dx}$$

product rule; power rule

$$2x^3\left(\frac{dy}{dx}\right) + y(6x^2) + 3(y^2 + x)^2\left(2y \cdot \frac{dy}{dx} + 1\right) = 4x^3$$

$$2x^3 \cdot \frac{dy}{dx} + 6x^2y + 3(y^2 + x)^2\left(2y \cdot \frac{dy}{dx}\right) + 3(y^2 + x)^2 = 4x^3$$

$$\left[2x^3 + 6y(y^2 + x)^2\right]\frac{dy}{dx} = 4x^3 - 6x^2y - 3(y^2 + x)^2$$

$$\frac{dy}{dx} = \frac{4x^3 - 6x^2y - 3(y^2 + x)^2}{2x^3 + 6y(y^2 + x)^2}$$

Practice Exercise

1. Find dy/dx if $2y^3 + xy + 1 = 0$.

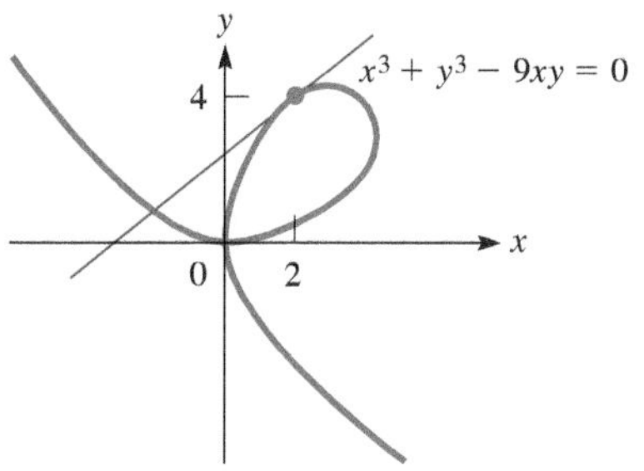

Fig. 23.27

EXAMPLE 5 Slope of a tangent line

Find the slope of a tangent line to the curve $x^3 + y^3 - 9xy = 0$ at the point (2, 4). This curve is known as a *folium*, which dates back to Descartes in the 1630s. See Fig. 23.27.

Here, we are to find dy/dx and evaluate it for $x = 2$ and $y = 4$.

$$\frac{d(x^3)}{dx} + \frac{d(y^3)}{dx} - \frac{d(9x \cdot y)}{dx} = \frac{d(0)}{dx}$$

$$3x^2 + 3y^2 \cdot \frac{dy}{dx} - 9\left(x \cdot \frac{dy}{dx} + y\right) = 0$$

$$(3y^2 - 9x)\frac{dy}{dx} = 9y - 3x^2$$

$$\frac{dy}{dx} = \frac{3y - x^2}{y^2 - 3x} \qquad \left.\frac{dy}{dx}\right|_{(2,4)} = \frac{3(4) - 2^2}{4^2 - 3(2)} = \frac{8}{10} = \frac{4}{5}$$

Therefore, the slope of the tangent line at (2, 4) is 4/5.

EXERCISES 23.8

In Exercises 1 and 2, make the given changes in the indicated examples of this section and then find dy/dx.

1. In Example 2, change y^2 to y^3.
2. In Example 3, change xy^2 to x^2y.

In Exercises 3–6, terms that might appear in an implicit function are shown. Differentiate each term with respect to x.

3. x^2y
4. $2xy^3$
5. $\frac{1}{xy}$
6. $\frac{y^2}{x+1}$

In Exercises 7–26, find dy/dx by differentiating implicitly. When applicable, express the result in terms of x and y.

7. $3x + 2y = 5$
8. $14x - 7y = 112$
9. $4y - 3x^2 = x$
10. $x^5 - 5y = 6 - x$
11. $x^2 - 4y^2 - 9 = 0$
12. $x^2 + 2y^2 - 11 = 0$
13. $y^7 = x^2 - 15$
14. $x^{2/3} + y^{2/3} = 5$
15. $y^2 + 6y = 3x^2 - 2$
16. $5y^3 - y = 11 - x^6$
17. $y + 3xy - 4 = 0$
18. $xy^3 + 3y + x^2 = 2\pi^2$
19. $x^2 = \frac{x - y}{x + y}$
20. $y^2x - \frac{5y}{x+1} + 3x = 4$
21. $\frac{3x^2}{y^2 + 1} + y = 3x + 1$
22. $\sqrt{xy} = \frac{x}{4} + \frac{1}{y^2}$
23. $(2y - x)^4 + x^2 = y + 3$
24. $(y^2 + 2)^9 = x^7y + e^2$
25. $2(x^2 + 1)^3 + (y^2 + 1)^2 = 17$
26. $(2x + 1)(1 - 3y) + y^2 = 13$

In Exercises 27–32, evaluate the derivatives of the given functions at the given points.

27. $3x^3y^2 - 2y^3 = -4$; $(1, 2)$
28. $2y + 5 - x^2 - y^3 = 0$; $(2, -1)$
29. $5y^4 + 7 = x^4 - 3y$; $(3, -2)$
30. $(xy - y^2)^3 = 5y^2 + 22$; $(4, 1)$
31. $xy^2 + 3x^2 - y^2 + 15 = 0$; $(-1, 3)$
32. $2(x + y)^3 - y^2/x = 15$; $(4, -2)$

In Exercises 33–48, solve the given problems by using implicit differentiation, and round answers to three significant digits if applicable.

33. At what point(s) does the graph of $x^2 + y^2 = 4x$ have a horizontal tangent?
34. Show that if $P(x, y)$ is any point on the circle $x^2 + y^2 = a^2$, then a tangent line at P is perpendicular to a line through P and the origin.
35. Show that two tangents to the curve $x^2 + xy + y^2 = 7$ at the points where it crosses the x-axis are parallel.
36. At what point(s) is the tangent to the curve $y^2 = 2x^3$ perpendicular to the line $4x - 3y + 1 = 0$?
37. Find the slope of a line tangent to the curve of the implicit function $xy + y^2 + 2 = 0$ at the point $(-3, 1)$. Use the derivative evaluation feature of a graphing utility to check your result.
38. Show that the graphs of $2x^2 + y^2 = 24$ and $y^2 = 8x$ are perpendicular at the point $(2, 4)$. Display the graphs on a graphing utility.
39. In an *RLC* circuit, the angular frequency ω at which the circuit resonates is given by $\omega^2 = 1/LC - R^2/L^2$. Find $d\omega/dL$.
40. A lens is described by the ellipse $x^2 - xy + y^2 = 7$. Find the slope of a light ray perpendicular to the lens at $(-1, 2)$.
41. The pressure p, volume V, and temperature T of a gas are related by $pV = n(RT + ap - bp/T)$, where a, b, n, and R are constants. For constant V, find dp/dT.
42. Oil moves through a pipeline such that the distance s it moves and the time t are related by $s^3 - t^2 = 7t$. Find the velocity of the oil for $s = 4.01$ m and $t = 5.25$ s.

43. The shelf support shown in Fig. 23.28 is 0.750 m long. Find the expression for dy/dx in terms of x and y.

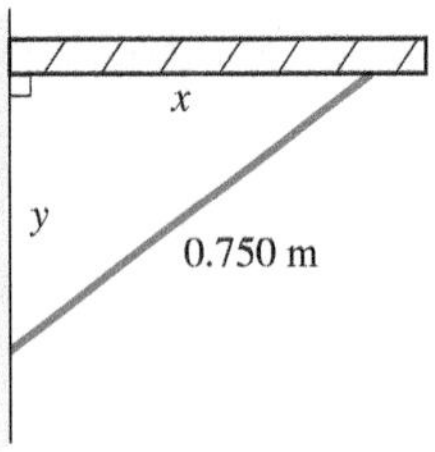

Fig. 23.28

44. An open (no top) right circular cylindrical container of radius r and height h has a total surface area of 936 cm^2. Find dr/dh in terms of r and h.

45. Two resistors, with resistances r and $r + 2$, are connected in parallel. Their combined resistance R is related to r by the equation $r^2 = 2rR + 2R - 2r$. Find dR/dr.

46. The polar moment of inertia I of a rectangular slab of concrete is given by $I = \frac{1}{12}(b^3h + bh^3)$, where b and h are the base and the height, respectively, of the slab. If I is constant, find the expression for db/dh.

47. A formula relating the length L and radius of gyration r of a steel column is $24C^3Sr^3 = 40C^3r^3 + 9LC^2r^2 - 3L^3$, where C and S are constants. Find dL/dr.

48. A computer is programmed to draw the graph of the implicit function $(x^2 + y^2)^3 = 64x^2y^2$ (see Fig. 23.29). Find the slope of a line tangent to this curve at (2.00, 0.56) and at (2.00, 3.07).

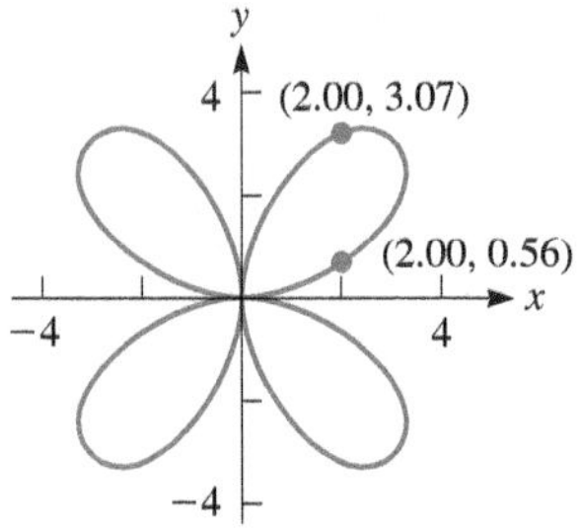

Fig. 23.29

Answer to Practice Exercise

1. $dy/dx = -y/(6y^2 + x)$

23.9 Higher Derivatives

Since the derivative of a function is itself a function, we may take its derivative. A *higher-order derivative* refers to repeating the differentiation process. This will become relevant to curve sketching, maximum and minimum problems, motion problems, and other applications.

The derivative of a function is called the **first derivative**. To find the **second derivative**, just differentiate the first derivative equation. To find the third derivative, just differentiate the second derivative, and so on (provided that each derivative is defined). The second derivative, third derivative, and so on are collectively known as **higher derivatives**, or **successive derivatives**.

Repeated use of the derivative process leads to some cumbersome notation. Consequently, many other more concise notations have been adopted to decrease the amount of writing required. Those used in this text are summarized in the chart below:

First Derivative	$\frac{dy}{dx}$	$\frac{dy}{dx}$	$f'(x)$	D_xy	y'
Second Derivative	$\frac{d}{dx}\left(\frac{dy}{dx}\right)$	$\frac{d^2y}{dx^2}$	$f''(x)$	D_x^2y	y''
Third Derivative	$\frac{d}{dx}\left(\frac{d}{dx}\left(\frac{dy}{dx}\right)\right)$	$\frac{d^3y}{dx^3}$	$f'''(x)$	D_x^3y	y'''
Fourth Derivative	$\frac{d}{dx}\left(\frac{d}{dx}\left(\frac{d}{dx}\left(\frac{dy}{dx}\right)\right)\right)$	$\frac{d^4y}{dx^4}$	$f^{(4)}(x)$	D_x^4y	$y^{(4)}$
Fifth Derivative	$\frac{d}{dx}\left(\frac{d}{dx}\left(\frac{d}{dx}\left(\frac{d}{dx}\left(\frac{dy}{dx}\right)\right)\right)\right)$	$\frac{d^5y}{dx^5}$	$f^{(5)}(x)$	D_x^5y	$y^{(5)}$

Note that the "primed" notation in the last column is a common method of denoting higher-order derivatives because it is the most concise. The disadvantage of "primed" notation is that it does not explicitly state the variable of differentiation.

EXAMPLE 1 Higher derivatives of a function

Find the higher derivatives of $y = 5x^3 - 2x$.

We find the first derivative as

$$\frac{dy}{dx} = 15x^2 - 2 \quad \text{or} \quad y' = 15x^2 - 2$$

Next, we obtain the second derivative by finding the derivative of the first derivative:

$$\frac{d^2y}{dx^2} = 30x \quad \text{or} \quad y'' = 30x$$

Continuing to find the successive derivatives, we have

$$\frac{d^3y}{dx^3} = 30 \quad \text{or} \quad y''' = 30$$

$$\frac{d^4y}{dx^4} = 0 \quad \text{or} \quad y^{(4)} = 0$$

Since the third derivative is a constant, the fourth derivative and all successive derivatives will be zero. This can be shown as $d^ny/dx^n = 0$ for $n \geq 4$.

EXAMPLE 2 Higher derivatives of a function

■ For reference, the product rule is $\frac{d(uv)}{dx} = u \cdot \frac{dv}{dx} + v \cdot \frac{du}{dx}$.

Find the higher derivatives of $f(x) = x(x^2 - 1)^2$.

Using the product rule to find the first derivative, we have

$$f'(x) = x(2)(x^2 - 1)(2x) + (x^2 - 1)^2(1)$$
$$f'(x) = (x^2 - 1)(4x^2 + x^2 - 1) = (x^2 - 1)(5x^2 - 1)$$
$$f'(x) = 5x^4 - 6x^2 + 1$$

Continuing to find the higher derivatives, we have

■ Note that when using the prime ($f'(x)$) notation the nth derivative may be shown as $f^{(n)}(x)$.

$$f''(x) = 20x^3 - 12x$$
$$f'''(x) = 60x^2 - 12$$
$$f^{(4)}(x) = 120x$$
$$f^{(5)}(x) = 120$$
$$f^{(n)}(x) = 0 \qquad \text{for } n \geq 6$$

All derivatives after the fifth derivative are equal to zero.

EXAMPLE 3 Evaluation of a second derivative

Evaluate the second derivative of $y = \frac{2}{1 - x}$ for $x = -2$.

We write the function as $y = 2(1 - x)^{-1}$ and then find the derivatives:

$$y = 2(1 - x)^{-1}$$
$$\frac{dy}{dx} = 2(-1)(1 - x)^{-2}(-1) = 2(1 - x)^{-2}$$
$$\frac{d^2y}{dx^2} = 2(-2)(1 - x)^{-3}(-1) = 4(1 - x)^{-3} = \frac{4}{(1 - x)^3}$$

Evaluating the second derivative for $x = -2$, we have

$$\left.\frac{d^2y}{dx^2}\right|_{x=-2} = \frac{4}{(1 + 2)^3} = \frac{4}{27}$$

The function is not differentiable for $x = 1$. Also, if we continue to find higher derivatives, the expressions will not become zero, as in Examples 1 and 2.

EXAMPLE 4 Second derivative of an implicit function

Find y'' for the implicit function defined by $2x^2 + 3y^2 = 6$.

Differentiating with respect to x, we have

$$2(2x) + 3(2yy') = 0$$

$$4x + 6yy' = 0 \quad \text{or} \quad 2x + 3yy' = 0 \qquad (1)$$

Before differentiating again, we see that $3yy'$ ***is a product***, and we note that the derivative of y' is y''. Thus, differentiating again, we have

differentiation of $3yy'$

$$2 + 3yy'' + 3y'(y') = 0$$

$$2 + 3yy'' + 3(y')^2 = 0 \qquad (2)$$

Now, solving Eq. (1) for y' and substituting this into Eq. (2), we have

$$y' = -\frac{2x}{3y}$$

$$2 + 3yy'' + 3\left(-\frac{2x}{3y}\right)^2 = 0$$

$$2 + 3yy'' + \frac{4x^2}{3y^2} = 0$$

$$6y^2 + 9y^3y'' + 4x^2 = 0$$

$$y'' = \frac{-4x^2 - 6y^2}{9y^3} = \frac{-2(2x^2 + 3y^2)}{9y^3}$$

Since $2x^2 + 3y^2 = 6$, we have

$$y'' = \frac{-2(6)}{9y^3} = -\frac{4}{3y^3}$$

Practice Exercise

1. Find the second derivative of $y = \frac{3}{x^2 + 4}$.

As mentioned earlier, higher derivatives are useful in certain applications. This is particularly true of the second derivative. The first and second derivatives are used in the next chapter for several types of applications, and higher derivatives are used when we discuss infinite series in Chapter 30. An important technical application of the second derivative is shown in the example that follows.

In Section 23.4, we briefly discussed the instantaneous velocity of an object, and in the exercises we mentioned acceleration. From that discussion, recall that the instantaneous velocity is the time rate of change of the displacement, and that *the* **instantaneous acceleration** *is the time rate of change of the instantaneous velocity*. Therefore, the *acceleration is found from the second derivative of the displacement with respect to time.*

EXAMPLE 5 Instantaneous acceleration

For the first 12 s after launch, the height s (in m) of a certain rocket is given by $s = 10\sqrt{t^2 + 25} - 50$. Find the vertical acceleration of the rocket when $t = 10.0$ s.

Since the velocity is found from the first derivative and the acceleration is found from the second derivative, we must find the second derivative and evaluate it for $t = 10.0$ s.

$$s = 10\sqrt{t^4 + 25} - 50$$

$$v = \frac{ds}{dt} = 10\left(\frac{1}{2}\right)(t^4 + 25)^{-1/2}(4t^3) = \frac{20t^3}{(t^4 + 25)^{1/2}}$$

$$a = \frac{dv}{dt} = \frac{d^2s}{dt^2} = \frac{(t^4 + 25)^{1/2}(60t^2) - 20t^3\left(\frac{1}{2}\right)(t^4 + 25)^{-1/2}(4t^3)}{t^4 + 25}$$

$$a = \frac{(t^4 + 25)(60t^2) - 40t^6}{(t^4 + 25)^{3/2}} = \frac{20t^6 + 1500t^2}{(t^4 + 25)^{3/2}}$$

multiply numerator and denominator by $(t^4 + 25)^{1/2}$

$$a = \frac{20t^2(t^4 + 75)}{(t^4 + 25)^{3/2}}$$

Finding the value of the acceleration when $t = 10.0$ s, we have

$$a\big|_{t=10.0} = \frac{20(10.0)^2(10.0^4 + 75)}{(10.0^4 + 25)^{3/2}} = 20.1 \text{ m/s}^2$$

In many applications, it is important to find at which values of x a specific derivative equals zero. If we factor derivatives whenever possible, the roots (zeros) of such derivatives can be easily obtained.

EXAMPLE 6 The zeros of a higher derivative

Determine all values of x for which $y'' = 0$ for the function

$$y = \frac{1}{12}x^4 - \frac{5}{6}x^3 + 3x^2 + 2x - 1.$$

$$y' = \frac{1}{12}(4x^3) - \frac{5}{6}(3x^2) + 3(2x) + 2$$

$$y' = \frac{1}{3}x^3 - \frac{5}{2}x^2 + 6x + 2$$

$$y'' = \frac{1}{3}(3x^2) - \frac{5}{2}(2x) + 6$$

$$y'' = x^2 - 5x + 6$$

Since $y'' = 0$,

$$0 = x^2 - 5x + 6$$

We can factor the quadratic in order to solve:

$$0 = (x - 2)(x - 3)$$

so $\quad x = 2 \quad \text{or} \quad x = 3$

EXERCISES 23.9

In Exercises 1 and 2, make the given changes in the indicated examples of this section and then solve the resulting problems.

1. In Example 1, change $2x$ to $2x^2$.
2. In Example 3, in the denominator change $1 - x$ to $1 + 2x$.

In Exercises 3–10, find all the higher derivatives of the given functions.

3. $y = x^3 + 7x^2$
4. $f(x) = 3x - x^4$
5. $f(x) = x^3 - 6x^4$
6. $s = 8t^5 + 5t^4$
7. $y = (1 - 2x)^4$
8. $f(x) = (3x + 2)^4$
9. $f(r) = r(5 - 2r)^3$
10. $y = x(x - 1)^3$

In Exercises 11–30, find the second derivative of each of the given functions.

11. $y = 2x^7 - x^6 - 3x$

12. $y = 6x - 2x^5$

13. $y = 2x + \sqrt{x}$

14. $r = 3\theta^2 - \frac{20}{\sqrt{\theta}}$

15. $f(x) = \sqrt[4]{8x - 3}$

16. $f(x) = \sqrt[3]{6x + 5}$

17. $f(p) = \frac{4.8\pi}{\sqrt{1 + 2p}}$

18. $f(x) = \frac{7.5}{\sqrt{3 - 4x}}$

19. $y = 2(2 - 5x)^4$

20. $y = (4x + 1)^{22}$

21. $y = (3x^2 - 1)^5$

22. $y = 3(2x^3 + 3)^4$

23. $f(x) = \frac{2\pi^2}{6 - x}$

24. $f(R) = \frac{1 - 3R}{1 + 3R}$

25. $u = \frac{v^2}{v + 15}$

26. $y = \frac{x}{\sqrt{1 - x^2}}$

27. $x^2 - y^2 = 9$

28. $2xy + y^2 = 16$

29. $x^2 - xy = 1 - y^2$

30. $4xy = y^2 + 2e^3$

In Exercises 31–36, evaluate the second derivative of the given function for the given value of x.

31. $f(x) = \sqrt{x^2 + 9}, x = 4$

32. $f(x) = x - \frac{2}{x^3}, x = -1$

33. $y = 3x^{2/3} - \frac{2}{x}, x = -8$

34. $y = 3(1 + 2x)^4, x = \frac{1}{2}$

35. $v = t(8 - t)^5, t = 2$

36. $y = \frac{x}{2 - 3x}, x = -\frac{1}{3}$

In Exercises 37–40, find the acceleration of an object for which the displacement s (in m) is given as a function of the time t (in s) for the given value of t.

37. $s = 26.0t - 4.90t^2, t = 3.00$ s

38. $s = 3(1 + 2t)^4, t = 0.500$ s

39. $s = \frac{16}{0.5t + 1}, t = 2.00$ s

40. $s = 250\sqrt{6t + 1}, t = 4.00$ s

In Exercises 41–53, solve the given problems by finding the appropriate derivatives; round answers to three significant digits.

41. Show that $\frac{d^2}{dx^2}(uv) = u\frac{d^2v}{dx^2} + 2\frac{du}{dx}\frac{dv}{dx} + \frac{d^2u}{dx^2}v$.

42. Show that $\frac{d^6(x^6)}{dx^6} = 6!$.

43. What is the instantaneous rate of change of the first derivative of y with respect to x for $y = (1 - 2x)^4$ for $x = 1$?

44. What is the instantaneous rate of change of the first derivative of y with respect to x for $2xy + y = 1$ for $x = 0.5$?

45. If the population of a city is $P(t) = 8000(1 + 0.02t + 0.005t^2)$ (t is in years from 2000), what is the acceleration in the size of the population?

46. The potential V (in V) of a certain electric charge is given by $V = 6/(t + 1)$, where t is the time (in s). Find d^2V/dt^2.

47. A bullet is fired vertically upward. Its distance s (in m) above the ground is given by $s = 655t - 4.90t^2$, where t is the time (in s). Find the acceleration of the bullet.

48. In testing the brakes on a new model car, it was found that the distance s (in m) it travelled after the brakes were applied was given by $s = 19.2 - 0.400t^3$, where t is the time (in s). What were the velocity and acceleration for $t = 4.00$ s?

49. The voltage V induced in an inductor in an electric circuit is given by $V = L(d^2q/dt^2)$, where L is the inductance (in H). Find the expression for the voltage induced in a 1.60-H inductor if $q = \sqrt{2t + 1} - 1$.

50. How fast is the rate of change of solar radiation changing on the surface in Exercise 35 of Section 23.4 at 3:00 P.M.?

51. The deflection y (in m) of a 5.00-m beam as a function of the distance x (in m) from one end is $y = 0.0001(x^5 - 25x^2)$. Find the value of d^2y/dx^2 (the rate of change at which the slope of the beam changes) where $x = 3.00$ m.

52. The force F (in N) on an object is $F = 12\,dv/dt + 2.0v + 5.0$, where v is the velocity (in m/s) and t is the time (in s). If the displacement is $s = 25t^{0.60}$, find F for $t = 3.50$ s.

53. A robotic arm moves according to the displacement s (in m) equation $s = \frac{1}{6}t^4 - \frac{7}{6}t^3 - 2t^2 + 3$, where t is time (in s), for $t > 0$. Determine the times at which the acceleration of the robotic arm will be zero.

Answer to Practice Exercise

1. $y'' = \frac{18x^2 - 24}{(x^2 + 4)^3}$

CHAPTER 23 KEY FORMULAS AND EQUATIONS

Limit of function	$\lim_{x \to a} f(x) = L$	**(23.1)**
Difference in x-coordinates	$\Delta x = x_2 - x$	**(23.2)**
	$x_2 = x + \Delta x$	**(23.3)**
Average slope	$m_{PQ} = \frac{\Delta y}{\Delta x} = \frac{f(x + \Delta x) - f(x)}{\Delta x}$	**(23.4)**

Slope of the tangent line	$m_{\tan} = \lim_{h \to 0} \frac{f(x+h) - f(x)}{h}$	**(23.5)**
Definition of derivative	$f'(x) = \lim_{h \to 0} \frac{f(x+h) - f(x)}{h}$	**(23.6)**
Instantaneous velocity	$v = \lim_{h \to 0} \frac{s(t+h) - s(t)}{h}$	**(23.7)**
Constant rule	$\frac{dc}{dx} = 0$	**(23.8)**
Power rule	$\frac{d(x^n)}{dx} = nx^{n-1}$	**(23.9)**
Constant factor rule	$\frac{d(c \cdot u)}{dx} = c \cdot \frac{du}{dx}$	**(23.10)**
Sum rule	$\frac{d(u+v)}{dx} = \frac{du}{dx} + \frac{dv}{dx}$	**(23.11)**
Product rule	$\frac{d(u \cdot v)}{dx} = u \cdot \frac{dv}{dx} + v \cdot \frac{du}{dx}$	**(23.12)**
Quotient rule	$\frac{d\left(\frac{u}{v}\right)}{dx} = \frac{v \cdot \frac{du}{dx} - u \cdot \frac{dv}{dx}}{v^2}$	**(23.13)**
Chain rule	$\frac{dy}{dx} = \frac{dy}{du} \cdot \frac{du}{dx}$	**(23.14)**
General power rule	$\frac{du^n}{dx} = nu^{n-1}\left(\frac{du}{dx}\right)$	**(23.15)**
	$\frac{du^{p/q}}{dx} = \frac{p}{q} u^{(p/q)-1} \frac{du}{dx}$	**(23.16)**

CHAPTER 23 REVIEW EXERCISES

In Exercises 1–12, evaluate the given limits.

1. $\lim_{x \to 4}(8 - 3x)$

2. $\lim_{x \to 3}(2x^2 - 10)$

3. $\lim_{x \to -2} \frac{|x+2|}{x+2}$

4. $\lim_{x \to 3} \sqrt{2x^2 - 18}$

5. $\lim_{x \to 3} \frac{4x - 12}{x^2 - 9}$

6. $\lim_{x \to 5} \frac{x^2 - 25}{3x - 15}$

7. $\lim_{x \to 3} \frac{x^2 - 5x + 6}{x^2 - 2x - 3}$

8. $\lim_{x \to 4} \frac{2 - \sqrt{x}}{x - 4}$

9. $\lim_{x \to \infty} \frac{2 + \frac{1}{x+4}}{3 - \frac{1}{x^2}}$

10. $\lim_{x \to \infty} \frac{3x^3 - 5x}{6x^2 + 3}$

11. $\lim_{x \to \infty} \frac{x - 2x^3}{(1+x)^3}$

12. $\lim_{x \to \infty} \frac{\sqrt{4x^2 + 3}}{x + 5}$

In Exercises 13–20, use the definition of the derivative/delta method to find the derivative of each of the given functions.

13. $y = 7 + 5x$

14. $y = 9 - 15x$

15. $y = 6 - 2x^2$

16. $y = 12x^2 - x^3$

17. $y = \frac{2}{(x-1)^2}$

18. $y = \frac{x}{1 - 4x}$

19. $y = \sqrt{x + 5}$

20. $y = \frac{1}{\sqrt{x}}$

In Exercises 21–36, find the derivative of each of the given functions.

21. $y = 2x^7 - 3x^2 + 5$

22. $y = 8x^{19} - 2^5 - 31x$

23. $y = 4\sqrt{x} - \frac{3}{x} + \sqrt{3}$

24. $R = \frac{3}{T^2} - 8\sqrt[4]{T}$

25. $f(y) = \frac{3y}{1 - 5y}$

26. $y = \frac{2x - 1}{x^2 + 1}$

27. $y = (2 - 7x)^4$

28. $y = (2x^2 - 3)^6$

29. $y = \frac{3\pi}{(5 - 2x^2)^{3/4}}$

30. $f(Q) = \frac{70}{(3Q + 1)^3}$

31. $v = \sqrt{1 + \sqrt{1 + \sqrt{1 + 8s}}}$

32. $y = (x - 1)^3(x^2 - 2)^2$

33. $y = \frac{\sqrt{4x + 3}}{2x}$

34. $R = \frac{\sqrt{t} + 4}{\sqrt{t} - 4}$

35. $(2x - 3y)^3 = x^2 - y$

36. $x^2y^2 = x^2 + y^2$

In Exercises 37–40, evaluate the derivatives of the given functions for the given values of x.

37. $y = \frac{4}{x} + 2\sqrt[3]{x}, x = 8$

38. $y = (3x - 5)^4, x = -2$

39. $y = 2x\sqrt{12x + 7}, x = 1.5$

40. $y = \frac{\sqrt{2x^2 + 1}}{3x}, x = 2$

In Exercises 41–44, find the second derivative of each of the given functions.

41. $y = 2x^6 - \frac{1}{x}$

42. $y = \sqrt{1 - 8x}$

43. $s = \frac{1 - 3t}{1 + 4t}$

44. $y = 2x(6x + 5)^4$

In Exercises 45–87, solve the given problems, and round answers to three significant digits unless otherwise specified.

45. As x approaches 0^+, which of the functions $1/x$, $1/x^2$, and $1/\sqrt{x}$ increases most rapidly (all become infinite)?

46. The parabola $y = ax^2 + bx + c$ passes through $(1, 2)$ and is tangent to the line $y = x$ at the origin. Find a, b, and c.

47. Find the acute angle between tangent lines to the parabolas $y = x^2$ and $y = (x - 2)^2$ at the point where they intersect.

48. Find the point(s) on the curve of $y = \frac{x}{x^2 + 1}$ where the tangent line is horizontal.

49. View the graph of $y = \frac{2(x^2 - 4)}{x - 2}$ with a graphing utility with display settings such that y can be evaluated exactly for $x = 2$. (Xmin = −1 (or 0), Xmax = 4, Ymin = 0, Ymax = 10 will probably work.) Using the *trace* feature, determine the value of y for $x = 2$. Comment on the accuracy of the view and the value found.

50. A continuous function $f(x)$ is positive at $x = 0$ and negative for $x = 1$. How many solutions does $f(x) = 0$ have between $x = 0$ and $x = 1$? Explain.

51. The velocity v (in m/s) of a weight falling in water is given by $v = \frac{6(t + 5)}{t + 1}$, where t is the time (in s). What are (a) the initial velocity and (b) the terminal velocity (as $t \rightarrow \infty$)?

52. Two lenses of focal lengths f_1 and f_2, separated by a distance d, are used in the study of lasers. The combined focal length f of this lens combination is $f = \frac{f_1f_2}{f_1 + f_2 - d}$. If f_2 and d remain constant, find the limiting value of f as f_1 continues to increase in value.

53. Find the slope of a line tangent to the curve of $y = 7x^4 - x^3$ at $(-1, 8)$.

54. Find the slope of a line tangent to the curve of $y = \sqrt[3]{3 - 8x}$ at $(-3, 3)$.

55. Find the point(s) at which a tangent line to the graph of $y = 1/\sqrt{3x^2 + 3}$ is parallel to the x-axis.

56. Find the point(s) on the graph of $y = 2(1 - 3x)^2$ at which a tangent line is parallel to the line $y = -2x + 5$.

57. If \$5000 is invested at interest rate i, compounded quarterly, in two years it will grow to an amount A given by $A = 5000(1 + 0.250i)^8$. Find dA/di.

58. The temperature T (in °C) of a rotating machine part that has been in operation for t hours is given by $T = \frac{100(t + 1)}{t + 5}$. Find dT/dt when $t = 4.00$ h.

59. Find the equations for (a) the velocity and (b) the acceleration if the displacement s (in m) of an object as a function of the time t (in s) is given by $s = \sqrt{1 + 8t}$.

60. Find the values of the velocity and acceleration for the object in Exercise 59 for $t = 3.00$ s.

61. The cable of a 200-m suspension bridge can be represented by $y = 0.0015x^2 + C$. At one point, the tension is directed along the line $y = 0.3x - 10$. Find the value of C.

62. The displacement s (in cm) of a piston during each 8.00-s cycle is given by $s = 8t - t^2$, where t is the time (in s). For what value(s) of t is the velocity of the piston 4.00 cm/s?

63. The reliability R of a computer system measures the probability that the system will be operating properly after t hours. For one system, $R = 1 - kt + \frac{k^2t^2}{2} - \frac{k^3t^3}{6}$, where k is a constant. Find the expression for the instantaneous rate of change of R with respect to t.

64. The distance s (in m) travelled by a subway train after the brakes are applied is given by $s = 20t - 2t^2$, where t is the time (in s). How far does it travel, after the brakes are applied, in coming to a stop?

65. The electric field E at a distance r from a point charge is $E = k/r^2$, where k is a constant. Find an expression for the instantaneous rate of change of the electric field with respect to r.

66. The velocity of an object moving with constant acceleration can be found from the equation $v = \sqrt{v_0^2 + 2as}$, where v_0 is the initial velocity, a is the acceleration, and s is the distance travelled. Find dv/ds.

67. The voltage induced in an inductor L is given by $E = L(dI/dt)$, where I is the current in the circuit and t is the time. Find the voltage induced in a 0.4-H inductor if the current I (in A) is related to the time (in s) by $I = t(0.01t + 1)^3$.

68. In studying the energy used by a mechanical robotic device, the equation $v = \frac{z}{\alpha(1 - z^2) - \beta}$ is used. If α and β are constants, find dv/dz.

69. The frictional radius r_f of a collar used in a braking system is given by $r_f = \frac{2(R^3 - r^3)}{3(R^2 - r^2)}$, where R is the outer radius and r is the inner radius. Find dr_f/dR if r is constant.

70. Water is being drained from a pond such that the volume V (in m^3) of water in the pond after t hours is given by $V = 5000(60 - t)^2$. Find the rate at which the pond is being drained after 4.00 h.

71. The energy output E of an electric heater is a function of the time t (in s) given by $E = t(1 + 2t)^2$ for $t < 10.0$ s. Find the power dE/dt (in W) generated by the heater for $t = 8.00$ s.

72. The amount n (in g) of a compound formed during a chemical change is $n = \frac{8t}{2t^2 + 3}$, where t is the time (in s). Find dn/dt for $t = 4.00$ s. What is the meaning of the result?

73. The deflection y of a 10-m beam is $y = kx(x^4 + 450x^2 - 950)$, where k is a constant and x is the horizontal distance from one end. Find the expression for the instantaneous rate of change of y with respect to x.

74. The kinetic energy K (in J) of a rotating flywheel varies directly as the square of its angular velocity ω (in rad/s). If $K = 125$ J for $\omega = 75.0$ rad/s, find $dK/d\omega$ for $\omega = 155$ rad/s.

75. The frequency f of a certain electronic oscillator is given by $f = \frac{1}{2\pi\sqrt{C(L+2)}}$, where C is a capacitance and L is an inductance. If C is constant, find df/dL.

76. The volume V of fluid produced in the retina of the eye in reaction to exposure to light of intensity I is given by $V = \frac{aI^2}{b - I}$, where a and b are constants. Find dV/dI.

77. The temperature T (in °C) in a freezer as a function of the time t (in h) is given by $T = \frac{10(1-t)}{0.5t + 1}$. Find dT/dt.

78. Under certain conditions, the efficiency η (in %) of an internal combustion engine is given by

$$\eta = 100\left(1 - \frac{1}{(V_1/V_2)^{0.4}}\right)$$

where V_1 and V_2 are the maximum and minimum volumes of air in a cylinder, respectively. Assuming that V_2 is kept constant, find the expression for the instantaneous rate of change of efficiency with respect to V_1.

79. The deflection y of a cantilever beam (clamped at one end and free at the other end) is $y = \frac{w}{24EI}(6L^2x^2 - 4Lx^3 + x^4)$. Here, L is the length of the beam, and w, E, and I are constants. Find the first four derivatives of y with respect to x. (Each of these derivatives is useful in analysing the properties of the beam.)

80. The number n of grams of a compound formed during a certain chemical reaction is given by $n = \frac{2t}{t+1}$, where t is the time (in min). Evaluate d^2n/dt^2 (the rate of increase of the amount of the compound being formed) when $t = 4.00$ min.

81. The area of a rectangular patio is to be 75.0 m^2. Express the perimeter p of the patio as a function of its width w and find dp/dw.

82. A water tank is being designed in the shape of a right circular cylinder with a volume of 100 m^3. Find the expression for the instantaneous rate of change of the total surface area A of the tank with respect to the radius r of the base.

83. An arch over a walkway can be described by the first-quadrant part of the parabola $y = 4 - x^2$. In order to determine the size and shape of rectangular objects that can pass under the arch, express the area A of a rectangle inscribed under the parabola in terms of x. Find dA/dx.

84. A computer analysis showed that a specialized piece of machinery has a value (in dollars) given by $V = 1\,500\,000/(2t + 10)$, where t is the number of years after the purchase. Calculate the value of dV/dt and d^2V/dt^2 for $t = 5.00$ years. What is the meaning of these values?

85. An airplane flies over an observer with a velocity of 400 km/h and at an altitude of 500 m. If the plane flies horizontally in a straight line, find the rate at which the distance r from the observer to the plane is changing 0.600 min after the plane passes over the observer. See Fig. 23.30.

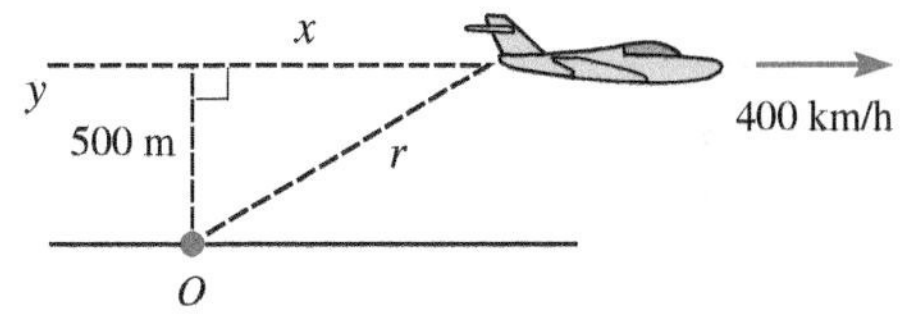

Fig. 23.30

86. The *radius of curvature* of $y = f(x)$ at the point (x, y) on the curve of $y = f(x)$ is given by $R = \frac{[1 + (y')^2]^{3/2}}{|y''|}$. A certain roadway follows the parabola $y = 1.2x - x^2$ for $0 < x < 1.2$, where x is measured in kilometres. Find R for $x = 0.200$ km and $x = 0.600$ km. See Fig. 23.31.

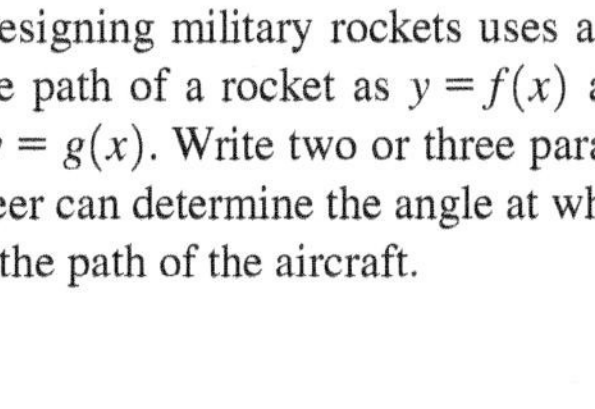

Fig. 23.31

87. An engineer designing military rockets uses a computer simulation to find the path of a rocket as $y = f(x)$ and the path of an aircraft to be $y = g(x)$. Write two or three paragraphs explaining how the engineer can determine the angle at which the path of the rocket crosses the path of the aircraft.

CHAPTER 23 PRACTICE TEST

1. Find $\lim_{x \to 1} \frac{x^2 - x}{x^2 - 1}$.

2. Find $\lim_{x \to \infty} \frac{1 - 4x^2}{x + 2x^2}$.

3. Find the slope of a line tangent to the curve of $y = 3x^2 - \frac{4}{x^2}$ at $(2, 11)$.

4. The displacement s (in cm) of a pumping machine piston in each cycle is given by $s = t\sqrt{10 - 2t}$, where t is the time (in s). Find the velocity of the piston for $t = 4.00$ s.

5. Find dy/dx. $y = 4x^6 - 2x^4 + \pi^3$

6. Find dy/dx. $y = 2x(5 - 3x)^4$

7. Find dy/dx. $(1 + y^2)^3 - x^2y = 7x$

8. Under certain conditions, due to the presence of a charge q, the electric potential V along a line is given by

$$V = \frac{kq}{\sqrt{x^2 + b^2}}$$

where k is a constant and b is the minimum distance from the charge to the line. Find the expression for the instantaneous rate of change of V with respect to x.

9. Find the second derivative of $y = \frac{2x}{3x + 2}$.

10. By using the definition, find the derivative of $y = 5x - 2x^2$ with respect to x.

24. Applications of the Derivative

▲ In Section 24.7, we see how to use the derivative in the design of cylindrical containers such as storage tanks.

LEARNING OUTCOMES

After completion of this chapter, the student should be able to:

- Find the equation of a line tangent or normal to a given curve
- Solve equations using Newton's method
- Find the velocity and acceleration of an object undergoing curvilinear motion
- Solve related rates problems
- Use derivatives to describe important features of the graph of a function such as maxima, minima, points of inflection, and concavity
- Sketch a curve using information about the function and its derivatives
- Solve applied maximum and minimum problems
- Use differentials to estimate errors in measurement
- Obtain the linear approximation of a function

Following the work of Newton and Leibniz, the development of the calculus proceeded rapidly but in a rather disorganized way. Much of the progress in the late 1600s and early 1700s was due to a desire to solve applied problems, particularly in some areas of physics. These included problems such as finding velocities in more complex types of motion, accurately measuring time by use of a pendulum, and finding the equation of a uniform cable hanging under its own weight.

A number of mathematicians, most of whom also studied in various areas of physics, contributed to these advances in calculus. Among them was the Swiss mathematician Leonhard Euler, the most prolific mathematician of all time. Throughout the mid- to late 1700s, he used the idea of a function to better organize the study of algebra, trigonometry, and calculus. In doing so, he fully developed the use of calculus on problems from physics in areas such as planetary motion, mechanics, and optics.

We have noted some of the problems in technology in which the derivative plays a key role in the solution. Another important type is finding the maximum values or minimum values of functions. Such values are useful, for example, in finding the maximum possible income from production or the least amount of material needed in making a product. In this chapter, we consider several of these kinds of applications of the derivative.

24.1 Tangents and Normals

The first application of the derivative we consider involves finding the equation of a line *tangent* to a given curve and the equation of a line *normal* (perpendicular) to a given curve.

Finding the Equation of a Tangent Line at a Given Point	EXAMPLE 1
	Find the tangent to $y = x^2 - 1$ at $(-2, 3)$.
1. Find the derivative of the function.	$\frac{dy}{dx} = 2x$
2. Evaluate at the point to obtain the slope.	$\left.\frac{dy}{dx}\right\rvert_{x=-2} = -4$
3. Find the point–slope form of the equation of the line, and simplify.	$y - 3 = -4(x + 2)$ $y = -4x - 5$ The parabola and the tangent line are shown in Fig. 24.1.

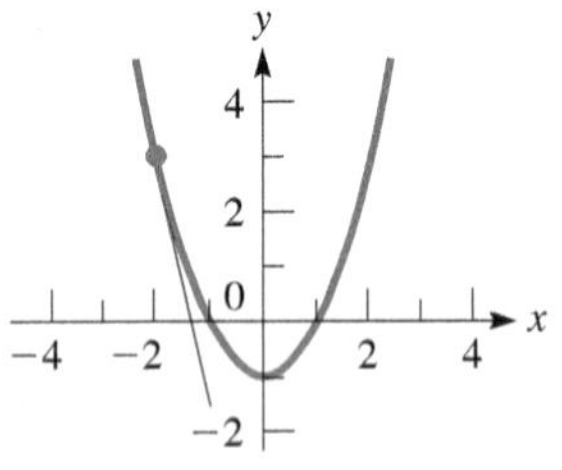

Fig. 24.1

EXAMPLE 2 Tangent line to an implicit function

Find the equation of the line tangent to the ellipse $4x^2 + 9y^2 = 40$ at the point $(1, 2)$.

Treating the equation as an implicit function, we have the following solution.

$$8x + 18yy' = 0 \quad \text{find derivative}$$

$$y' = -\frac{4x}{9y}$$

$$y'\rvert_{(1,2)} = -\frac{4}{18} = -\frac{2}{9} \quad \text{evaluate derivative to find slope of tangent line}$$

$$y - 2 = -\frac{2}{9}(x - 1) \quad \text{point–slope form of tangent line}$$

$$9y - 18 = -2x + 2$$

$$2x + 9y - 20 = 0 \quad \text{general form of tangent line}$$

The ellipse and the tangent line $2x + 9y - 20 = 0$ are shown in Fig. 24.2.

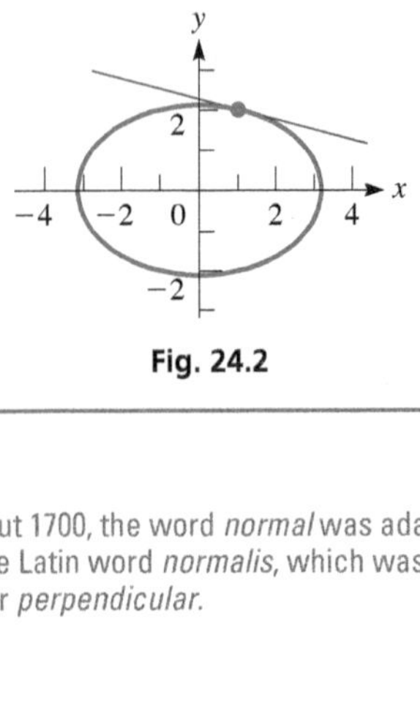

Fig. 24.2

NORMAL LINE

■ About 1700, the word *normal* was adapted from the Latin word *normalis*, which was being used for *perpendicular*.

A normal line to a curve at a given point is perpendicular to the tangent at that point. The method for finding a normal is as follows.

Finding the Equation of a Normal Line at a Given Point	EXAMPLE 3
	Find the normal to $y = 2/x$ at $(2, 1)$.
1. Find the derivative of the function.	$\frac{dy}{dx} = -\frac{2}{x^2}$
2. Evaluate at the point to obtain the slope of the tangent line.	$\left.\frac{dy}{dx}\right\rvert_{x=2} = -\frac{1}{2}$

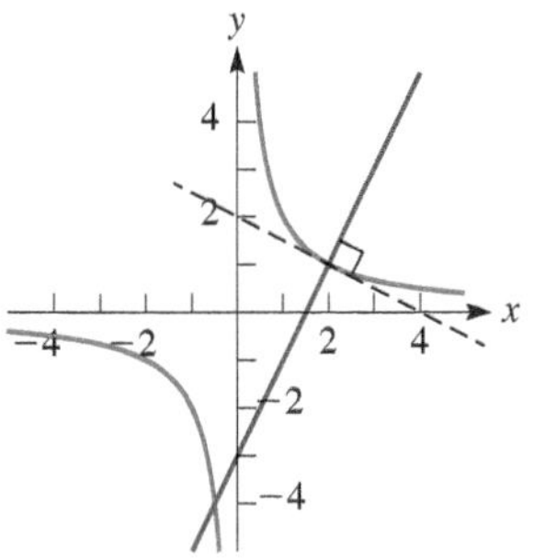

Fig. 24.3

3. Take the negative reciprocal of the number from Step 2 to find the slope of the normal line.	The negative reciprocal of $-\frac{1}{2}$ is 2.
4. Find the point–slope form of the equation of the line, and simplify.	$y - 1 = 2(x - 2)$ $y = 2x - 3$ The hyperbola and the normal line are shown in Fig. 24.3.

Many of the applications of tangents and normals are geometric. However, there are certain applications in technology, and one of these is shown in the following example. Others are shown in the exercises.

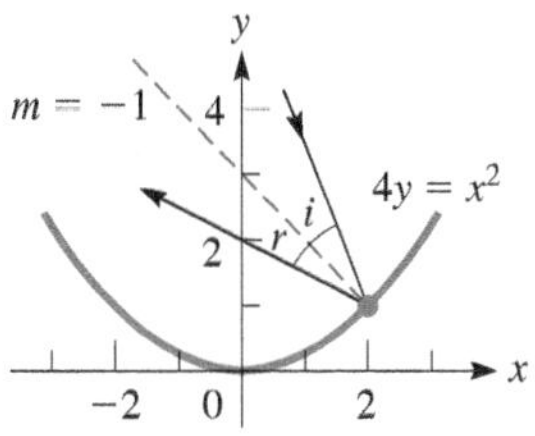

Fig. 24.4

EXAMPLE 4 Normal line—parabolic reflection

In Fig. 24.4, the cross-section of a parabolic solar reflector is shown, along with an incident ray of light and the reflected ray. The angle of incidence i is equal to the angle of reflection r where both angles are measured with respect to the normal to the surface. If the incident ray strikes at the point where the slope of the normal is -1 and the equation of the parabola is $4y = x^2$, what is the equation of the normal line?

Find the derivative: Rewrite as $y = \frac{1}{4}x^2$, so $\frac{dy}{dx} = \frac{1}{2}x$

The negative reciprocal of the slope of the normal will be the slope of the tangent: The negative reciprocal of -1 is 1.

Equate the slope of the tangent with the derivative to find x. Substitute in the function to find y: Equating dy/dx to 1 and solving for x,

$$\frac{dy}{dx} = 1 = \frac{1}{2}x$$

so $x = 2$ and $y = \frac{1}{4}(2)^2 = 1$.

Find the point–slope form of the normal line:

$$y - 1 = (-1)(x - 2)$$
$$y = -x + 3.$$

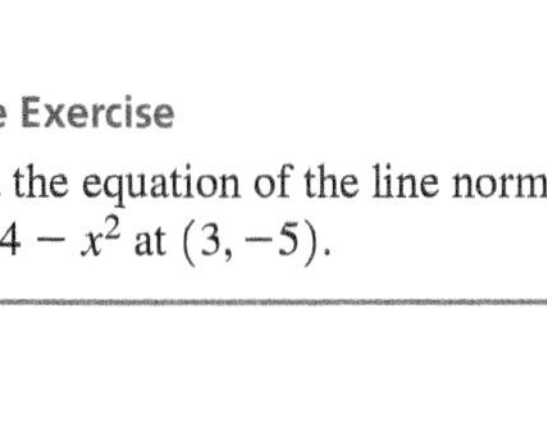

Fig. 24.5

If the incident ray is vertical, for which $i = 45°$ at the point $(2, 1)$, the reflected ray passes through $(0, 1)$, the focus of the parabola. See Fig. 24.5. This illustrates the important reflection property of a parabola that *any incident ray parallel to its axis passes through the focus.* We first noted this property in our discussion of the parabola in Example 4 of Section 21.4.

Practice Exercise

1. Find the equation of the line normal to $y = 4 - x^2$ at $(3, -5)$.

EXERCISES 24.1

In Exercises 1 and 2, make the given changes in the indicated examples of this section and then solve the resulting problems.

1. In Example 2, change $4x^2 + 9y^2$ to $x^2 + 4y^2$, change 40 to 17, and then find the equation of the tangent line.
2. In Example 3, change $2/x$ to $3/(x + 1)$ and then find the equation of the normal line.

In Exercises 3–6, find the equations of the lines tangent to the indicated curves at the given points. In Exercises 3 and 6, sketch the curve and the tangent line. In Exercises 4 and 5, use a graphing utility to view the curve and the tangent line.

3. $y = x^2 + 2$ at $(2, 6)$
4. $y = \frac{1}{3}x^3 - 5x$ at $(3, -6)$
5. $y = \frac{1}{x^2 + 1}$ at $\left(1, \frac{1}{2}\right)$
6. $x^2 + y^2 = 25$ at $(3, 4)$

In Exercises 7–10, find the equations of the lines normal to the indicated curves at the given points. In Exercises 7 and 10, sketch the curve and the normal line. In Exercises 8 and 9, use a graphing utility to view the curve and the normal line.

7. $y = 6x - 2x^2$ at $(2, 4)$

8. $y = 8 - x^3$ at $(-1, 9)$

9. $y = \dfrac{6}{(x^2 + 1)^2}$ at $\left(1, \frac{3}{2}\right)$

10. $4x^2 - y^2 = 20$ at $(-3, 4)$

In Exercises 11–14, find the equations of the lines tangent or normal to the given curves and with the given slopes. View the curves and lines on a graphing utility.

11. $y = x^2 - 2x$, tangent line with slope 2

12. $y = \sqrt{2x - 9}$, tangent line with slope 1

13. $y = (2x - 1)^3$, normal line with slope $-\frac{1}{24}$, $x > 0$

14. $y = \frac{1}{2}x^4 + 1$, normal line with slope 4

In Exercises 15–30, solve the given problems involving tangent and normal lines.

15. Find the equations of the tangent and normal lines to the parabola with vertex at $(0, 3)$ and focus at $(0, 0)$, where $x = -1$. Graph the curve and lines.

16. Find the equations of the tangent and normal lines to the ellipse with focus at $(4, 0)$, vertex at $(5, 0)$, and centre at the origin, where $x = 2$. Use a graphing utility to view the curve and lines.

17. Show that the line tangent to the graph of $y = x + 2x^2 - x^4$ at $(1, 2)$ is also tangent at $(-1, 0)$.

18. Show that the graphs of $y^2 = 4x + 4$ and $y^2 = 4 - 4x$ cross at right angles.

19. Without actually finding the points of intersection, explain why the parabola $y^2 = 4x$ and the ellipse $2x^2 + y^2 = 6$ intersect at right angles. (*Hint:* Call a point of intersection (a, b).)

20. Find the y-intercept of the line normal to the curve $y = x^{3/4}$, where $x = 16$.

21. Show that the equation of the tangent line to the circle $x^2 + y^2 = a^2$ at the point (x_1, y_1) is $x_1x + y_1y = a^2$.

22. At what point on the curve $y = x^4$ does the normal line have a slope of 16?

23. Heat flows normal to isotherms, curves along which the temperature is constant. Find the line along which heat flows through the point $(2, 1)$ if the isotherm is the graph of $2x^2 + y^2 = 9$.

24. The sparks from an emery wheel to sharpen blades fly off tangent to the wheel. Find the equation along which sparks fly from a wheel described by $x^2 + y^2 = 25$, at $(3, 4)$.

25. A certain suspension cable with supports on the same level is closely approximated as being parabolic in shape. If the supports are 80 m apart and the sag at the centre is 10 m, what is the equation of the line along which the tension acts (tangentially) at the right support? (Choose the origin of the coordinate system at the lowest point of the cable.)

26. In a video game, airplanes move from left to right along the path described by $y = 2 + 1/x$. They can shoot rockets tangent to the direction of flight at targets on the x-axis located at $x = 1, 2, 3$, and 4. Will a rocket fired from $(1, 3)$ hit a target?

27. In an electric field, the lines of force are perpendicular to the curves of equal electric potential. In a certain electric field, a curve of equal potential is $y = \sqrt{2x^2 + 8}$. If the line along which the force acts on an electron has an inclination of 135°, find its equation.

28. A radio wave reflects from a reflecting surface in the same way as a light wave (see Example 4). A certain horizontal radio wave reflects off a parabolic reflector such that the reflected wave is 43.60° below the horizontal, as shown in Fig. 24.6. If the equation of the parabola is $y^2 = 8x$, what is the equation of the normal line through the point of reflection?

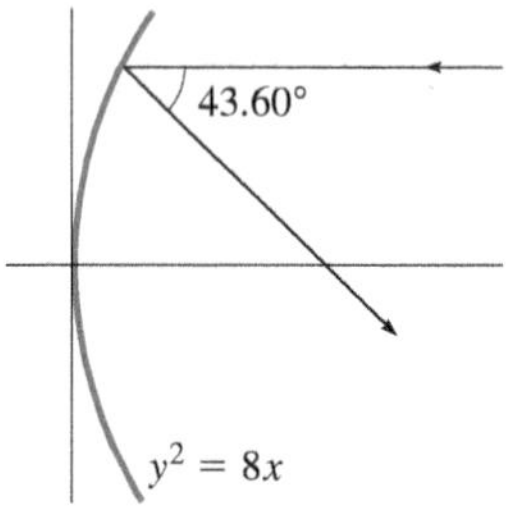

Fig. 24.6

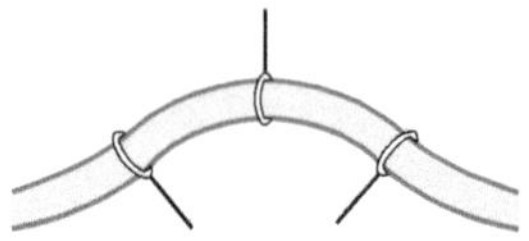

Fig. 24.7

29. In designing a flexible tubing system, the supports for the tubing must be perpendicular to the tubing. If a section of the tubing follows the curve $y = \dfrac{4}{x^2 + 1}$ $(-2 \text{ dm} < x < 2 \text{ dm})$, along which lines must the supports be directed if they are located at $x = -1$, $x = 0$, and $x = 1$? See Fig. 24.7.

30. On a particular drawing, a pulley wheel can be described by the equation $x^2 + y^2 = 100$ (units in cm). The pulley belt is directed along the lines $y = -10$ and $4y - 3x - 50 = 0$ when first and last making contact with the wheel. What are the first and last points on the wheel where the belt makes contact?

Answer to Practice Exercise

1. $x - 6y = 33$

24.2 Newton's Method for Solving Equations

Finding the roots of an equation $f(x) = 0$ is very important in mathematics and in many types of applications, and we have developed methods of solving many types of equations in the previous chapters. However, for a great many algebraic and nonalgebraic equations, there is no method for finding the roots exactly.

We have shown that the roots of an equation can be found with great accuracy on a graphing calculator. In this section, we develop **Newton's method**, which uses the derivative to find approximately, but also with great accuracy, the real roots of many kinds of algebraic and nonalgebraic equations.

■ Newton's method is another mathematical development by the English mathematician and physicist Isaac Newton (1642–1727).

Newton's method is an **iterative method**, which starts with a reasonable guess for the root, and then yields a new and better approximation. This, in turn, is used to obtain an even better approximation, and so on until an approximate answer with the required degree of accuracy is obtained. Iterative methods are easily programmable for use on a computer.

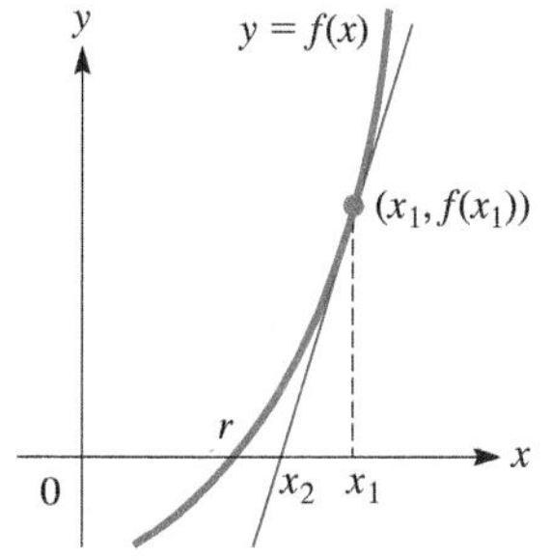

Fig. 24.8

Let us consider a section of the curve of $y = f(x)$ that (a) crosses the x-axis, (b) always has either a positive slope or a negative slope, and (c) has a slope that either becomes greater or becomes less as x increases. See Fig. 24.8. (When either (b) or (c) is not satisfied, Newton's method may fail to find the root. See Exercises 20 and 22.) The curve in the figure crosses the x-axis at $x = r$, which means that $x = r$ is a root of the equation $f(x) = 0$. If x_1 is sufficiently close to r, a line tangent to the curve at $[x_1, f(x_1)]$ will cross the x-axis at a point $(x_2, 0)$, with x_2 closer to r than x_1.

We know that the slope of the tangent line is the value of the derivative at x_1, or $m_{\tan} = f'(x_1)$. Therefore, the equation of the tangent line is

$$y - f(x_1) = f'(x_1)(x - x_1)$$

For the point $(x_2, 0)$ on this line, we have

$$-f(x_1) = f'(x_1)(x_2 - x_1)$$

Solving for x_2, we have

$$x_2 = x_1 - \frac{f(x_1)}{f'(x_1)}$$

Here, x_2 is a second approximation to the root. We can repeat the process starting with x_2 in place of x_1, obtaining an even better approximation x_3. The repetition of this process is what we call Newton's method, which we now summarize.

Newton's Method

Let $x = r$ be a root of the equation $f(x) = 0$, and let x_1 be a first approximation of r. We obtain approximations $x_2, x_3, \ldots$ by using

$$x_{n+1} = x_n - \frac{f(x_n)}{f'(x_n)} \qquad \textbf{(24.1)}$$

for $n = 1, 2, \ldots$. The number of iterations depends on the required accuracy. The initial approximation x_1 (which must be close to r) may be found by either of the following procedures:

1. Choose x_1 inside an interval (a, b) where $f(a)$ and $f(b)$ have opposite signs, with x_1 closer to the endpoint where the function is closer to zero.
2. Sketch the graph of the function and choose x_1 as an estimate of the x-intercept of the function. For some equations, it may be easier to choose x_1 as an estimate of the intersection of two functions.

EXAMPLE 1 Using Newton's method

Find the root of $x^2 - 3x + 1 = 0$ between $x = 0$ and $x = 1$.

Here, $f(x) = x^2 - 3x + 1, f(0) = 1$, and $f(1) = -1$. This indicates that the root may be near the middle of the interval. Therefore, we choose $x_1 = 0.5$.

The derivative is

$$f'(x) = 2x - 3$$

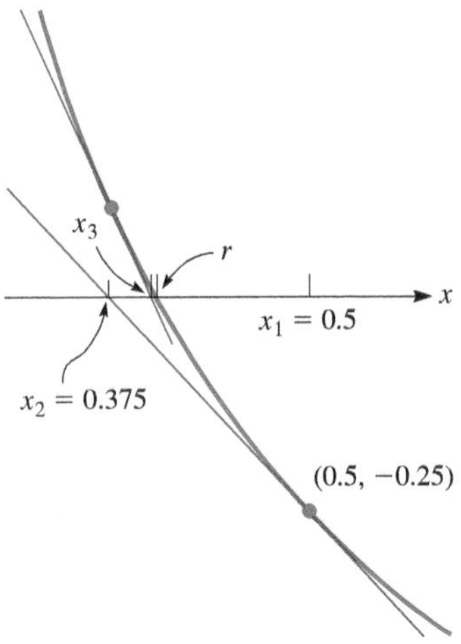

Fig. 24.9

Therefore, $f(0.5) = -0.25$ and $f'(0.5) = -2$, which gives us

$$x_2 = x_1 - \frac{f(x_1)}{f'(x_1)} = 0.5 - \frac{-0.25}{-2} = 0.375$$

This is a second approximation, which is closer to the actual value of the root. See Fig. 24.9. A second iteration using $x_2 = 0.375, f(0.375) = 0.015\,625$, and $f'(0.375) = -2.25$ gives us

$$x_3 = x_2 - \frac{f(x_2)}{f'(x_2)} = 0.375 - \frac{0.015\,625}{-2.25} = 0.381\,944\,4$$

We can check this particular result by using the quadratic formula. This tells us the root is $x = 0.381\,966\,0$. Our result using Newton's method is good to three decimal places, and additional accuracy may be obtained by using the method again as many times as needed.

Practice Exercise

1. In Example 1, let $x_1 = 0.3$, and find x_2.

EXAMPLE 2 Newton's method—spherical tank

A spherical water-storage tank holds 500.0 m^3. If the outside diameter is 10.0000 m, what is the thickness of the metal of which the tank is made?

Let $x =$ the thickness of the metal. We know that the outside radius of the tank is 5.0000 m. Therefore, using the formula for the volume of a sphere, we have

■ An explanation and an example of Newton's method due to Thomas Simpson (of Simpson's rule) and dating back to 1740 can be found in the text's companion website.

$$\frac{4\pi}{3}(5.0000 - x)^3 = 500.0$$

$$125.0 - 75.00x + 15.00x^2 - x^3 = 119.366$$

$$x^3 - 15.00x^2 + 75.00x - 5.634 = 0$$

$$f(x) = x^3 - 15.00x^2 + 75.00x - 5.634$$

$$f'(x) = 3x^2 - 30.00x + 75.00$$

Since $f(0) = -5.634$ and $f(0.1) = 1.717$, the root may be closer to 0.1 than to 0.0. Therefore, we let $x_1 = 0.07$. Setting up a table, we have these values:

n	x_n	$f(x_n)$	$f'(x_n)$	$x_n - \frac{f(x_n)}{f'(x_n)}$
1	0.07	−0.457 157	72.9147	0.076 269 750 8
2	0.076 269 750 8	−0.000 581 145	72.729 358 7	0.076 277 741 3

Since $x_2 = x_3 = 0.0763$ to four decimal places, the thickness is 0.0763 m. This means the inside radius of the tank is 4.9237 m, and this value gives an inside volume of 500.0 m^3. In using a calculator, the values in the table are more easily found if the values of $x_n, f(x_n)$, and $f'(x_n)$ are stored in memory for each step.

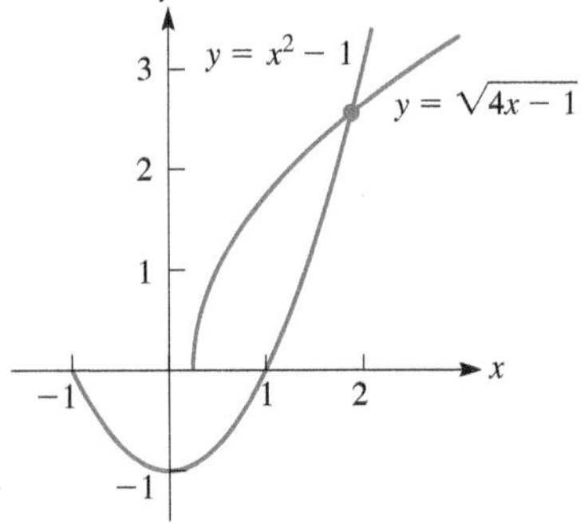

Fig. 24.10

EXAMPLE 3 Graphically locating x_1

Solve the equation $x^2 - 1 = \sqrt{4x - 1}$.

We can see approximately where the root is by sketching the graphs of $y_1 = x^2 - 1$ and $y_2 = \sqrt{4x - 1}$, as shown in Fig. 24.10. We see that they intersect between $x = 1$ and $x = 2$. Therefore, we choose $x_1 = 1.5$. With

$$f(x) = x^2 - 1 - \sqrt{4x - 1}$$

$$f'(x) = 2x - \frac{2}{\sqrt{4x - 1}}$$

we now find the values in the following table:

n	x_n	$f(x_n)$	$f'(x_n)$	$x_n - \dfrac{f(x_n)}{f'(x_n)}$
1	1.5	−0.986 067 98	2.105 572 8	1.968 313 4
2	1.968 313 4	0.252 568 59	3.173 759 8	1.888 733 2
3	1.888 733 2	0.007 052 69	2.996 295 7	1.886 379 4
4	1.886 379 4	0.000 006 20	2.991 026 5	1.886 377 3

Since $x_5 = x_4 = 1.886\,38$ to five decimal places, this is the required solution. (Here, rounded-off values of x_n are shown, although additional digits were carried and used.)

EXERCISES 24.2

In Exercises 1–4, find the indicated roots of the given quadratic equations by finding x_3 from Newton's method. Compare this root with that obtained by using the quadratic formula.

1. In Example 1, change the middle term from $-3x$ to $-5x$ and use the same x_1.
2. $2x^2 - x - 2 = 0$ (between 1 and 2)
3. $3x^2 - 5x - 1 = 0$ (between −1 and 0)
4. $x^2 + 4x + 2 = 0$ (between −4 and −3)

In Exercises 5–16, find the indicated roots of the given equations to at least four decimal places by using Newton's method.

5. $x^3 - 7x^2 + 11x - 3 = 0$ (between 0 and 1)
6. $x^3 - 4x^2 - 3x + 5 = 0$ (between 0 and 1)
7. $x^3 + 6x^2 + 9x + 2 = 0$ (the smallest root)
8. $2x^3 + 2x^2 - 11x + 3 = 0$ (the largest root)
9. $x^4 - x^3 - 3x^2 - x - 4 = 0$ (between 2 and 3)
10. $2x^4 - 2x^3 - 5x^2 - x - 3 = 0$ (between −2 and −1)
11. $x^4 - 2x^3 - 8x - 16 = 0$ (the negative root)
12. $3x^4 - 3x^3 - 11x^2 - x - 4 = 0$ (the negative root)
13. $2x^2 = \sqrt{2x + 1}$ (the positive real solution)
14. $x^3 = \sqrt{x + 1}$ (the real solution)
15. $x = \dfrac{1}{\sqrt{x + 2}}$ (the real solution)
16. $x^{3/2} = \dfrac{1}{2x + 1}$ (the real solution)

In Exercises 17–31, determine the required values to at least four decimal places by using Newton's method.

17. Find all the real roots of $x^3 - 2x^2 - 5x + 4 = 0$.
18. Find all the real roots of $x^4 - 2x^3 + 3x^2 + x - 7 = 0$.
19. Explain how to find $\sqrt[3]{4}$ by using Newton's method.
20. Explain why Newton's method does not work for finding the root of $2x^3 - 6x = 5$ if x_1 is chosen as 1.
21. Use Newton's method to find an expression for x_{n+1}, in terms of x_n and a, for the equation $x^2 - a = 0$. Such an equation can be used to find $\sqrt{a}$.
22. Use Newton's method on $f(x) = x^{1/3}$ with $x_1 = 1$. Calculate x_2, x_3, and x_4. What is happening as successive approximations are calculated?
23. To calculate reciprocals without dividing, a computer programmer applied Newton's method to the equation $1/x - a = 0$. Show that $x_2 = 2x_1 - ax_1^2$. From this, determine the expression for x_n.
24. The altitude h (in m) of a rocket is given by $h = -2t^3 + 84t^2 + 480t + 10$, where t is the time (in s) of flight. When does the rocket hit the ground?
25. A solid sphere of specific gravity s sinks in water to a depth h (in cm) given by $0.009\,26h^3 - 0.0833h^2 + s = 0$. Find h for $s = 0.786$ (when the diameter of the sphere is 6.00 cm).
26. A dome in the shape of a spherical segment is to be placed over the top of a sports stadium. If the radius r of the dome is to be 60.0 m and the volume V within the dome is 180 000 m^3, find the height h of the dome. See Fig. 24.11. $\left(V = \frac{1}{6}\pi h(h^2 + 3r^2).\right)$

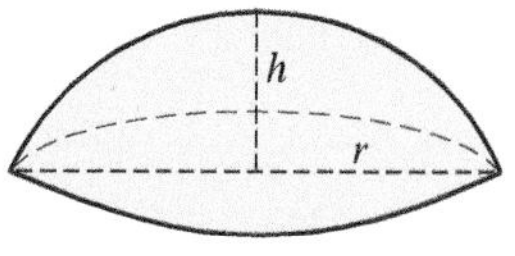

Fig. 24.11

27. The capacitances (in μF) of three capacitors in series are C, $C + 1.00$, and $C + 2.00$. If their combined capacitance is 1.00 μF, their individual values can be found by solving the equation
$$\frac{1}{C} + \frac{1}{C + 1.00} + \frac{1}{C + 2.00} = 1.00$$
Find these capacitances.
28. An oil-storage tank has the shape of a right circular cylinder with a hemisphere at each end. See Fig. 24.12. If the volume of the tank is 50.0 m^3 and the length l is 4.00 m, find the radius r.

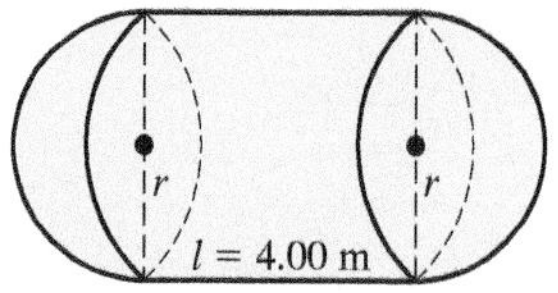

Fig. 24.12

29. A rectangular block of plastic with edges 2.00 cm, 2.00 cm, and 4.00 cm is heated until its volume doubles. By how much does each edge increase if each increases by the same amount?

30. Water flow in a wide channel approaches a bump. See Fig. 24.13. The water depth x (in m) over the bump is the largest root of the equation

$$x^3 - 1.015x^2 + 0.115 = 0.$$

Find x if it is between 0.750 m and 1.20 m.

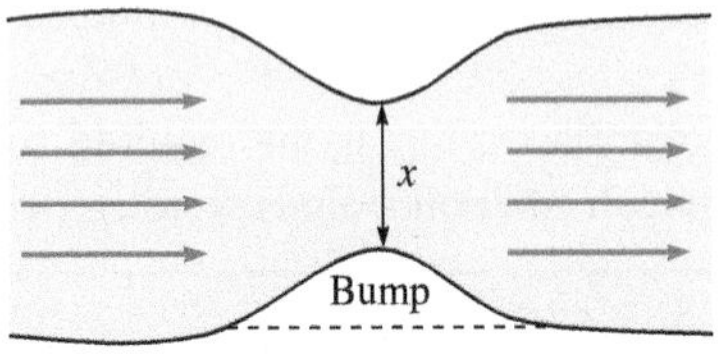

Fig. 24.13

31. The vibration analysis of a three-storey building requires that the following equation be solved:

$$\lambda^3 - 0.0012\lambda^2 + 9 \times 10^{-7}\lambda - 7.2 \times 10^{-11} = 0.$$

Find the root that lies between 0 and 4×10^{-4} to three significant digits.

Answer to Practice Exercise

1. $x_2 = 0.379167$

24.3 Curvilinear Motion

When velocity was introduced in Section 23.4, the discussion was limited to rectilinear motion, or motion along a straight line. A more general discussion of velocity is necessary when we discuss the motion of an object in a plane. There are many important applications of motion in a plane, a principal one being the motion of a projectile.

An important concept in developing this topic is that of a vector. The necessary fundamentals related to vectors are taken up in Chapter 9. Although vectors can be used to represent many physical quantities, we will restrict our attention to their use in describing the velocity and acceleration of an object moving in a plane along a specified path. Such motion is called **curvilinear motion.**

In describing an object undergoing curvilinear motion, it is common to express the x- and y-coordinates of its position separately as functions of time. Equations given in this form—that is, *x and y both given in terms of a third variable (in this case, t)—are said to be in* **parametric form**, which we encountered in Section 10.6. *The third variable, t, is called the* **parameter**.

To find the velocity of an object whose coordinates are given in parametric form, we can use the following steps.

Finding Resultant Velocity from Coordinates in Parametric Form	EXAMPLE 1
	Find the resultant velocity at $t = 2$ if $x = 3t^2$, $y = 1 - t^2$.
1. Find the x- and y-components of velocity by taking the derivatives with respect to t: $v_x = \frac{dx}{dt} \quad v_y = \frac{dy}{dt} \quad$ **(24.2)**	$v_x = \frac{dx}{dt} = 6t \quad v_y = \frac{dy}{dt} = -2t$
2. Evaluate the components at the given time.	$v_x\big\|_{t=2} = 6(2) = 12, \quad v_y\big\|_{t=2} = -2(2) = -4$
3. Find the magnitude of the resultant velocity: $v = \sqrt{v_x^2 + v_y^2} \quad$ **(24.3)**	$v = \sqrt{12^2 + (-4)^2} = 12.6$

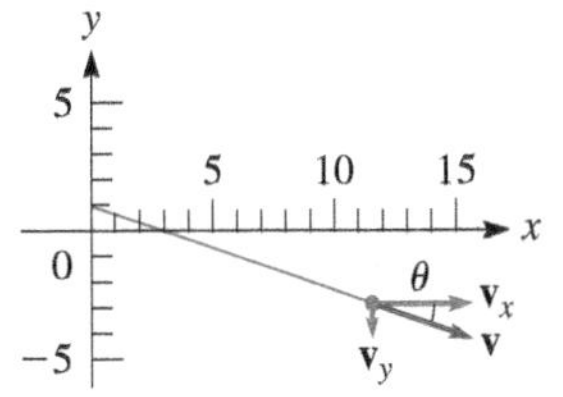

Fig. 24.14

4. Find the direction of the resultant velocity: $\tan\theta_v = \frac{v_y}{v_x}$ **(24.4)**	$\tan\theta_v = \frac{-4}{12}$ $\theta_v = -18.4°$
5. If required, sketch the path either by eliminating the parameter or by plotting a few points.	We eliminate the parameter t by substituting $t^2 = x/3$ (from the equation for x) into the equation for y. We get $y = 1 - x/3$, and the equation is a line. See Fig. 24.14.

EXAMPLE 2 Parametric form—resultant velocity

Find the velocity and direction of motion when $t = 2$ of an object moving such that its x- and y-coordinates of position are given by $x = 1 + 2t$ and $y = t^2 - 3t$.

Find the x- and y-components of velocity (Eqs. 24.2): $v_x = \frac{dx}{dt} = 2 \quad v_y = \frac{dy}{dt} = 2t - 3$

Evaluate the components at the given time: $v_x|_{t=2} = 2, \quad v_y|_{t=2} = 2(2) - 3 = 1$

Find the magnitude (Eq. 24.3): $v = \sqrt{2^2 + 1^2} = 2.24$

Find the direction (Eq. 24.4): $\tan\theta_v = \frac{1}{2} \quad \theta_v = 26.6°$

Sketch the path by plotting a few points: We plot the following points:

t	0	1	2	3
x	1	3	5	7
y	0	−2	−2	0

See Fig. 24.15.

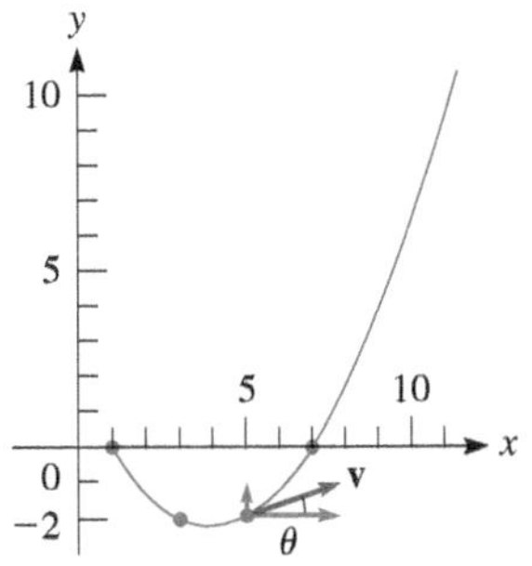

Fig. 24.15

COMMON ERROR Note that to find the resultant velocity, ***we first find the necessary derivatives and then evaluate them.*** This procedure should always be followed. When a derivative is to be found, it is incorrect to take the derivative of the evaluated expression (which is a constant).

Acceleration *is the time rate of change of velocity.* Therefore, from the displacement components, the acceleration is found as follows.

Finding the Acceleration from Coordinates in Parametric Form	EXAMPLE 3
	Find the acceleration at $t = 2$ if $x = t^3, y = 1 - t^2$.
1. Find the x- and y-components of the acceleration by taking the second derivative with respect to time (that is, the derivative of each of the velocity components): $a_x = \frac{dv_x}{dt} = \frac{d^2x}{dt} \quad a_y = \frac{dv_y}{dt} = \frac{d^2y}{dt^2}$ **(24.5)**	$v_x = \frac{dx}{dt} = 3t^2 \quad a_x = \frac{dv_x}{dt} = 6t$ $v_y = \frac{dy}{dt} = -2t \quad a_y = \frac{dv_y}{dt} = -2$
2. Evaluate the components at the given time.	$a_x\|_{t=2} = 6(2) = 12, \quad a_y\|_{t=2} = -2$

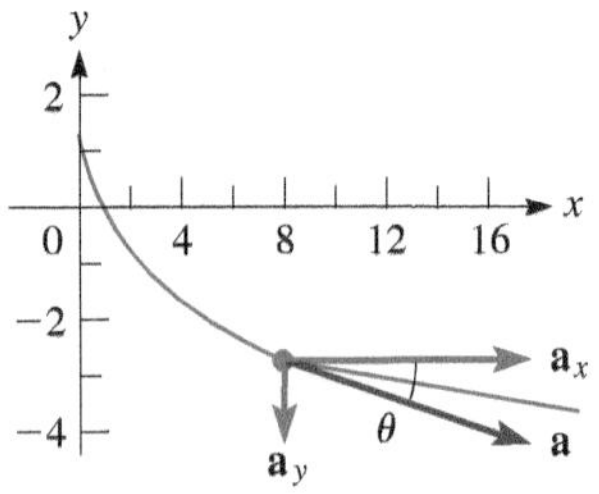

Fig. 24.16

3. Find the magnitude of the acceleration: $$a = \sqrt{a_x^2 + a_y^2} \qquad \textbf{(24.6)}$$	$a = \sqrt{12^2 + (-2)^2} = 12.2$
4. Find the direction of the acceleration: $$\tan\theta_a = \frac{a_y}{a_x} \qquad \textbf{(24.7)}$$	$\tan\theta_a = \frac{-2}{12} \qquad \theta_a = -9.5°$ At $t = 2$, $x = 8$ and $y = -3$. The path and the acceleration are shown in Fig. 24.16.

Practice Exercise

1. Solve for the acceleration when $t = 2$, if $x = 0.8\,t^{5/2}$, $y = 1 - t^2$.

In the following examples, we illustrate the use of Eqs. (24.2) to (24.7) in applied situations for which we know the equation of the path of motion given with y as a function of x. In those cases, both x and y are functions of time, even if it is not stated explicitly. When finding derivatives, we need to use the chain rule.

EXAMPLE 4 Velocity at a point along a path

■ For reference, the chain rule rewritten for y as a function of x, with x a function of t is
$$\frac{dy}{dt} = \frac{dy}{dx}\cdot\frac{dx}{dt}$$

In a physics experiment, a small sphere is constrained to move along a parabolic path described by $y = \frac{1}{3}x^2$. If the horizontal velocity v_x is constant at 6.00 cm/s, find the velocity at the point (2.00, 1.33). See Fig. 24.17.

Find the x- and y-components of velocity (Eqs. 24.2) using the chain rule:

$$v_x = \frac{dx}{dt} = 6.00$$
$$v_y = \frac{dy}{dt} = \frac{dy}{dx}\cdot\frac{dx}{dt} = \left(\frac{2}{3}x\right)\left(\frac{dx}{dt}\right)$$

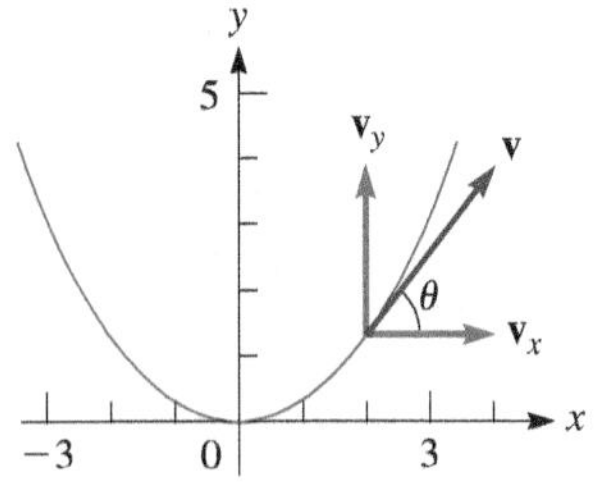

Fig. 24.17

Evaluate at the given time:

$$v_x\big|_{x=2.00} = 6.00 \text{ cm/s}$$
$$v_y\big|_{x=2} = \frac{2}{3}(2.00)(6.00) = 8.00 \text{ cm/s}$$

Find the magnitude (Eq. 24.3):

$$v = \sqrt{6.00^2 + 8.00^2} = 10.0 \text{ cm/s}$$

Find the direction (Eq. 24.4):

$$\tan\theta_v = \frac{8.00}{6.00} \qquad \theta_v = 53.1°$$

EXAMPLE 5 Velocity and acceleration—projectile motion

A helicopter is flying at 18.0 m/s and at an altitude of 120 m when a rescue marker is released from it. The marker maintains a horizontal velocity and follows a path given by $y = 120 - 0.0151x^2$, as shown in Fig. 24.18. Find the magnitude and direction of the velocity and of the acceleration of the marker 3.00 s after release.

From the given information, we know that $v_x = dx/dt = 18.0$ m/s. Taking derivatives with respect to time leads to this solution:

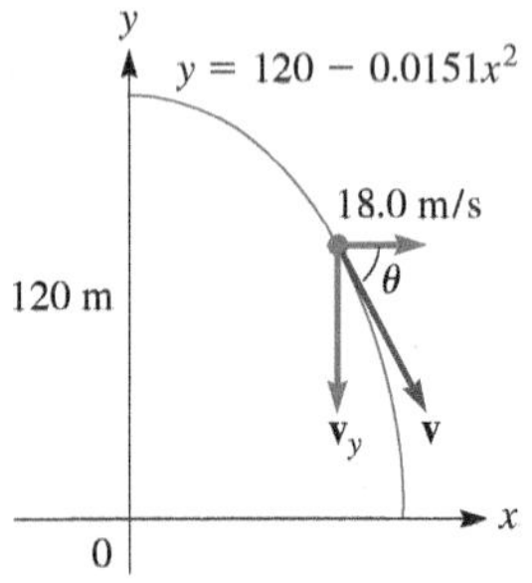

Fig. 24.18

$$y = 120 - 0.0151x^2$$
$$\frac{dy}{dt} = -0.0302x\frac{dx}{dt} \qquad \text{taking derivatives}$$
$$v_y = -0.0302xv_x \qquad \text{using Eq. (24.2)}$$
$$x = (3.00)(18.0) = 54.0 \text{ m} \qquad \text{evaluating at } t = 3.00 \text{ s}$$
$$v_y = -0.0302(54.0)(18.0) = -29.35 \text{ m/s}$$
$$v = \sqrt{18.0^2 + (-29.35)^2} = 34.4 \text{ m/s} \qquad \text{magnitude}$$
$$\tan\theta = \frac{-29.35}{18.0}, \qquad \theta = -58.5° \qquad \text{direction}$$

The velocity is 34.4 m/s and is directed at an angle of 58.5° below the horizontal.

To find the acceleration, we return to the equation $v_y = -0.0302xv_x$. Since v_x is constant, we can substitute 18.0 for v_x to get

$$v_y = -0.5436x$$

Again taking derivatives with respect to time, we have

$$\frac{dv_y}{dt} = -0.5436\frac{dx}{dt}$$

$$a_y = -0.5436v_x \qquad \text{using Eq. (24.5)}$$

$$a_y = -0.5436(18.0) = -9.78 \text{ m/s}^2 \qquad \text{evaluating}$$

We know that v_x is constant, which means that $a_x = 0$. Therefore, the acceleration is 9.78 m/s^2 and is directed vertically downward.

LEARNING TIP

When taking derivatives with respect to t, if a path is given with y as a function of x, **remember to use the chain rule,** so that the factor dx/dt is not neglected.

EXERCISES 24.3

In Exercises 1 and 2, make the given changes in the indicated examples of this section and then solve the resulting problems.

1. In Example 1, change $x = 3t^2$ to $x = 4t^2$.
2. In Example 4, change $y = x^2/3$ to $y = x^2/4$ and (2.00, 1.33) to (2.00, 1.00).

In Exercises 3–6, given that the x- and y-coordinates of a moving particle are defined by the indicated parametric equations, find the magnitude and direction of the velocity for the specific value of t to three significant digits. Sketch the curves and show the velocity and its components.

3. $x = 3t, y = 1 - t, t = 4$
4. $x = \dfrac{5t}{2t+1}, y = 0.1(t^2 + t), t = 2$
5. $x = t(2t+1)^2, y = \dfrac{6}{\sqrt{4t+3}}, t = 0.5$
6. $x = \sqrt{1+2t}, y = t - t^2, t = 4$

In Exercises 7–10, use the parametric equations and values of t of Exercises 3–6 to find the magnitude and direction of the acceleration in each case.

In Exercises 11–28, find the indicated velocities and accelerations.

11. The water from a fire hose follows a path described by $y = 2.0 + 0.80x - 0.20x^2$ (in metres). If v_x is constant at 5.0 m/s, find the resultant velocity at the point (5.0, 1.0).
12. A roller mechanism follows a path described by $y = \sqrt{5x+6}$, where units are in metres. If $v_x = 2x$, find the resultant velocity (in m/s) at the point (2.0, 4.0).
13. A float is used to test the flow pattern of a stream. It follows a path described by $x = 0.20t^2, y = -0.10t^3$ (x and y in m, t in min). Find the acceleration of the float after 2.0 min.
14. A radio-controlled model car is operated in a parking lot. The coordinates (in m) of the car are given by $x = 3.5 + 2.0t^2$ and $y = 8.5 + 0.25t^3$, where t is the time (in s). Find the acceleration of the car after 2.5 s.
15. An astronaut on Mars drives a golf ball that moves according to the equations $x = 25t$ and $y = 15t - 3.7t^2$ (x and y in metres, t in seconds). Find the resultant velocity and acceleration of the golf ball for $t = 6.0$ s.
16. A package of relief supplies is dropped and moves according to the parametric equations $x = 55t$ and $y = -4.9t^2$ (x and y in m, t in s). Find the velocity and acceleration when $t = 3.0$ s.
17. A spacecraft moves along a path described by the parametric equations $x = 10(\sqrt{1+t^4} - 1), y = 40t^{3/2}$ for the first 100 s after launch. Here, x and y are measured in metres, and t is measured in seconds. Find the magnitude and direction of the velocity of the spacecraft 10.0 s and 100 s after launch.
18. An electron moves in an electric field according to the equations $x = 8.0/\sqrt{1+t^2}$ and $y = 8.0t/\sqrt{1+t^2}$ (x and y in Mm and t in s). Find the velocity of the electron when $t = 0.50$ s.
19. In a computer game, an airplane starts at (1.00, 4.00) (in cm) on the curve $y = 3.00 + x^{-1.50}$ and moves with a constant horizontal velocity of 1.20 cm/s. What is the plane's velocity after 0.500 s?
20. A person on a hoverboard is riding up a ramp and follows a path described by $y = 0.15x^{1.2}$. If v_x is constant at 0.50 m/s, find v_y when $x = 8$.
21. Find the resultant acceleration of the spacecraft in Exercise 17 for the specified times.
22. A ski jump is designed to follow the path given by the equations $x = 3.50t^2$ and $y = 20.0 + 0.120t^4 - 3.00\sqrt{t^4+1}$ $(0 \le t \le 4.00 \text{ s})$ (x and y in m, t in s). Find the velocity and acceleration of a skier when $t = 4.00$ s. See Fig. 24.19.

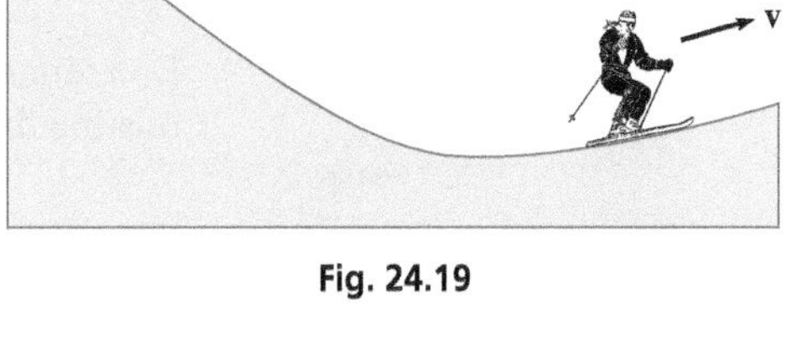

Fig. 24.19

23. A rocket follows a path given by $y = x - \frac{1}{90}x^3$ (distances in km). If the horizontal velocity is given by $v_x = x$, find the magnitude and direction of the velocity when the rocket hits the ground (assume level terrain) if time is in minutes.

24. A ship is moving around an island on a route described by $y = 3.0x^2 - 0.20x^3$. If $v_x = 1.2$ km/h, find the velocity of the ship where $x = 3.5$ km.

25. A computer's hard disk is 88.9 mm in diameter and rotates at 7200 r/min. With the centre of the disk at the origin, find the velocity components of a point on the rim for $x = 30.5$ mm, if $y > 0$ and $v_x > 0$.

26. A robot arm joint moves in an elliptical path (horizontal major axis 8.0 cm, minor axis 4.0 cm, centre at origin). For $y > 0$ and -2 cm $< x < 2$ cm, the joint moves such that $v_x = 2.5$ cm/s. Find its velocity for $x = -1.5$ cm.

27. An airplane ascends such that its gain h in altitude is proportional to the square root of the change x in horizontal distance travelled. If $h = 280$ m for $x = 400$ m and v_x is constant at 350 m/s, find the velocity at this point.

28. A meteor travelling toward the earth has a velocity inversely proportional to the square root of the distance from the earth's centre. State how its acceleration is related to its distance from the centre of the earth.

Answer to Practice Exercise

1. $a = 4.70$, $\theta = -25.2°$

24.4 Related Rates

Often, variables change with respect to time, and are therefore implicitly functions of time. If a relation is known to exist relating them, the time rate of change of one can be expressed in terms of the time rate of change of the other(s). This is done by taking the derivative with respect to time of the expression relating the variables, even if t does not appear in the expression, as in Examples 4 and 5 of Section 24.3. Since the time rates of change are related, this is referred to as a **related-rates** problem. The following steps for solving related rates problems are illustrated with several examples below.

Steps for Solving Related-Rates Problems

1. Identify the variables and rates in the problem.
2. Determine the equation relating the variables.
3. Differentiate with respect to time.
4. Solve for the required rate.
5. Evaluate the required rate.

COMMON ERROR Remember that all variables in related rates problems are functions of time, even if t does not appear explicitly in the expression. Once again, a common error is not to include the derivative with respect to t as a factor when applying the chain rule.

EXAMPLE 1 Related rates—volume and radius

A spherical balloon is being blown up such that its volume increases at the constant rate of 2.00 m³/min. Find the rate at which the radius is increasing when it is 3.00 m. See Fig. 24.20.

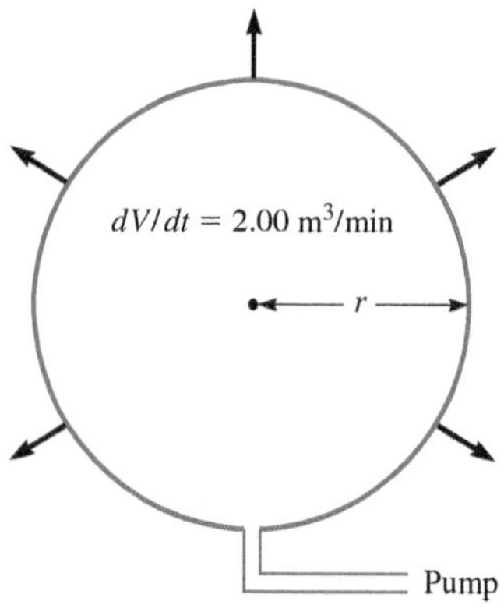

Fig. 24.20

Identify variables and rates: Given: $\frac{dV}{dt} = 2.00$ m³/min

Required: $\frac{dr}{dt}$ when $r = 3.00$ m

Determine the equation relating the variables: We can relate V to r using the expression for the volume of a sphere:

$$V = \frac{4}{3}\pi r^3$$

Differentiate with respect to time **using the chain rule**: $\frac{dV}{dt} = 4\pi r^2 \cdot \frac{dr}{dt}$

Solve for the required rate: $\frac{dr}{dt} = \frac{1}{4\pi r^2} \cdot \frac{dV}{dt}$

Evaluate the required rate: $\left.\frac{dr}{dt}\right|_{r=3.00} = \frac{1}{4\pi(3.00)^2} \cdot (2.00) = 0.0177 \text{ m/min}$

EXAMPLE 2 Related rates—voltage and temperature

The voltage E of a certain thermocouple as a function of the temperature T (in °C) is given by $E = 2.800T + 0.006T^2$. If the temperature is increasing at the rate of 1.00 °C/min, how fast is the voltage increasing when $T = 100$°C?

Identify variables and rates: Given: $\frac{dT}{dt} = 1.00\text{°C/min}$

Required: $\frac{dE}{dt}$ when $T = 100$°C

Determine the equation relating the variables: Given: $E = 2.800T + 0.006T^2$

Differentiate with respect to time **using the chain rule**: $\frac{dE}{dt} = 2.800\frac{dT}{dt} + 0.012T\frac{dT}{dt}$

Evaluate the required rate: $\left.\frac{dE}{dt}\right|_{T=100} = 2.800(1.00) + (0.012)(100)(1.00)$

$= 4.00 \text{ V/min}$

COMMON ERROR When working with related rates, the derivative must be taken **before** given values are substituted. A common error is to substitute before taking derivatives, so that the functions appear constant when they are not.

EXAMPLE 3 Related rates—distances

The distance q that an image is from a certain lens in terms of p, the distance of the object from the lens, is given by

$$q = \frac{10p}{p - 10}$$

If the object distance is increasing at the rate of 0.200 cm/s, how fast is the image distance changing when $p = 15.0$ cm? See Fig. 24.21.

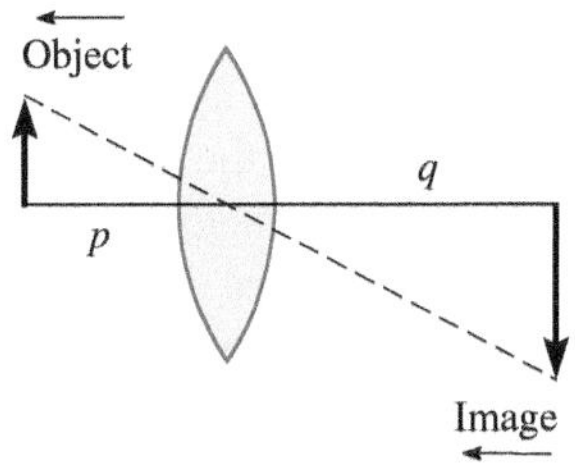

Fig. 24.21

Identify variables and rates: Given: $\frac{dp}{dt} = 0.200 \text{ cm/s}$

Required: $\frac{dq}{dt}$ when $p = 15.0$ cm

Determine the equation relating the variables: Given: $q = \frac{10p}{p - 10}$

Practice Exercise

1. In Example 3, change each 10 to 12 and then solve.

Differentiate with respect to time **using the chain rule:**

$$\frac{dq}{dt} = \frac{(p-10)\left(10 \cdot \frac{dp}{dt}\right) - 10p \cdot \frac{dp}{dt}}{(p-10)^2} = \frac{-100\frac{dp}{dt}}{(p-10)^2}$$

Evaluate the required rate:

$$\left.\frac{dq}{dt}\right|_{p=15.0} = \frac{-100(0.200)}{(15.0-10)^2} = -0.800 \text{ cm/s}$$

The minus sign implies that image distance is decreasing with time.

EXAMPLE 4 Related rates—force and distance

The force F of gravity of Earth on a spacecraft varies inversely as the square of the distance r of the spacecraft from the centre of Earth. A particular spacecraft weighs 4500 N on the launchpad ($F = 4500$ N for $r = 6370$ km). Find the rate at which F changes later as the spacecraft moves away from Earth at the rate of 12 km/s, where $r = 8500$ km.

We begin by setting up the equation relating F and r. Note that we substitute known values of F and r in order to obtain the constant of proportionality. As we said earlier, the values from the related rates problem are only substituted after taking derivatives.

$$F = \frac{k}{r^2} \quad \text{inverse variation}$$

$$4500 = \frac{k}{6370^2} \quad \text{substitute } F = 4500 \text{ N}, r = 6370 \text{ km}$$

$$k = 1.83 \times 10^{11} \text{ N} \cdot \text{km}^2 \quad \text{solve for } k$$

$$F = \frac{1.83 \times 10^{11}}{r^2} \quad \text{substitute for } k \text{ in equation}$$

$$\frac{dF}{dt} = (1.83 \times 10^{11})(-2)(r^{-3})\frac{dr}{dt} \quad \text{take derivatives with respect to time}$$

$$= \frac{-3.66 \times 10^{11}}{r^3}\frac{dr}{dt}$$

$$\left.\frac{dF}{dt}\right|_{r=8500 \text{ km}} = \frac{-3.66 \times 10^{11}}{8500^3}(12) \quad \text{evaluate derivative for } r = 8500 \text{ km}, dr/dt = 12 \text{ km/s}$$

$$= -7.2 \text{ N/s} \quad \text{gravitational force is decreasing}$$

EXAMPLE 5 Related rates—distances

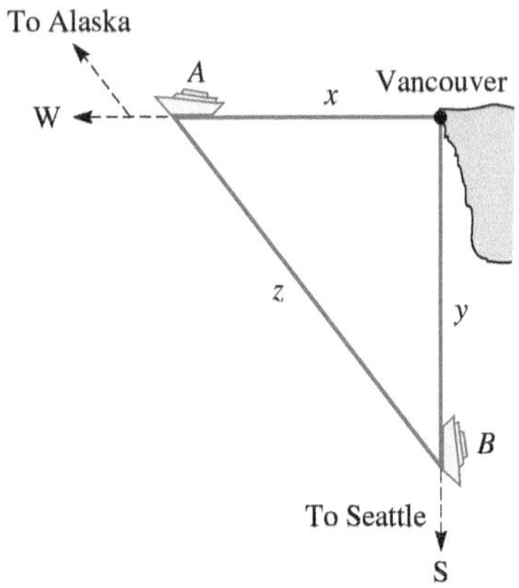

Fig. 24.22

Two cruise ships leave Vancouver, British Columbia, at noon. Ship A travels west at 12.0 km/h (before turning toward Alaska), and ship B travels south at 16.0 km/h (toward Seattle). How fast are they separating at 2:00 P.M.? See Fig. 24.22.

In Fig. 24.22, we let $x =$ the distance travelled by A and $y =$ the distance travelled by B. We can find the distance between them, z, from the Pythagorean theorem. Therefore, we are to find dz/dt for $t = 2.00$ h. Even though there are three variables, each is a function of time. This means we can find dz/dt by taking the derivative of each term with respect to time. This gives us

$$z^2 = x^2 + y^2 \quad \text{using the Pythagorean theorem}$$

$$2z\frac{dz}{dt} = 2x\frac{dx}{dt} + 2y\frac{dy}{dt} \quad \text{taking derivatives with respect to time}$$

$$\frac{dz}{dt} = \frac{x(dx/dt) + y(dy/dt)}{z} \quad \text{solve for } \frac{dz}{dt}$$

At 2:00 P.M., we have the values

$x = 24.0 \text{ km}, \quad y = 32.0 \text{ km}, \quad z = 40.0 \text{ km}$	$d = rt$ and Pythagorean theorem
$dx/dt = 12.0 \text{ km/h}, \quad dy/dt = 16.0 \text{ km/h}$	from statement of problem
$\left.\dfrac{dz}{dt}\right\|_{z=40} = \dfrac{(24.0)(12.0) + (32.0)(16.0)}{40.0} = 20.0 \text{ km/h}$	substitute values

EXERCISES 24.4

In Exercises 1 and 2, make the given changes in the indicated examples of this section and then solve the resulting problems.

1. In Example 1, change "volume" to "surface area" and change m^3/min to m^2/min.

2. In Example 2, change 0.006 to 0.012.

In Exercises 3–6, assume that all variables are implicit functions of time t. Find the indicated rates.

3. $y = 5x^2 - 4x$; $dx/dt = 0.5$ when $x = 8$; find dy/dt.

4. $x^2 + 2y^2 = 9$; $dx/dt = 6$ when $x = 1$ and $y = 2$; find dy/dt.

5. $x^2 + 3y^2 + 2y = 10$; $dx/dt = 4$ when $x = 3$ and $y = -1$; find dy/dt.

6. $z = 2x^2 - 3xy$; $dx/dt = -3$ and $dy/dt = 2$ when $x = 1$ and $y = 4$; find dz/dt.

In Exercises 7–44, solve the problems in related rates.

7. How fast is the slope of a tangent to the curve $y = 2/(x + 1)$ changing where $x = 3.0$ if $dx/dt = 0.50$ unit/s?

8. How fast is the slope of a tangent to the curve $y = 2(1 - 2x)^2$ changing where $x = 1.5$ if $dx/dt = 0.75$ unit/s?

9. The velocity v (in m/s) of a pulse travelling in a certain string is a function of the tension T (in N) in the string given by $v = 18\sqrt{T}$. Find dv/dt if $dT/dt = 0.20$ N/s when $T = 25$ N.

10. The force F (in N) on the blade of a certain wind generator as a function of the wind velocity v (in m/s) is given by $F = 0.024v^2$. Find dF/dt if $dv/dt = 0.75 \text{ m/s}^2$ when $v = 28$ m/s.

11. The electric resistance R (in Ω) of a certain resistor as a function of the temperature T (in °C) is $R = 4.000 + 0.003T^2$. If the temperature is increasing at the rate of 0.100°C/s, find how fast the resistance changes when $T = 150$°C.

12. The kinetic energy K (in J) of an object is given by $K = \frac{1}{2}mv^2$, where m is the mass (in kg) of the object and v is its velocity. If a 250-kg wrecking ball accelerates at 5.00 m/s^2, how fast is the kinetic energy changing when $v = 30.0$ m/s?

13. The length L (in cm) of a pendulum is slowly decreasing at the rate of 0.100 cm/s. What is the time rate of change of the period T (in s) of the pendulum when $L = 16.0$ cm, if the equation relating the period and length is $T = \pi\sqrt{L/245}$?

14. The voltage V that produces a current I (in A) in a wire of radius r (in mm) is $V = 0.030I/r^2$. If the current increases at 0.020 A/s in a wire of 0.040 mm radius, find the rate at which the voltage is increasing.

15. A plane flying at an altitude of 2.0 km is at a direct distance $D = \sqrt{4.0 + x^2}$ from an airport control tower, where x is the horizontal distance to the tower. If the plane's speed is 350 km/h, how fast is D changing when $x = 6.2$ km?

16. A variable resistor R and an 8-Ω resistor in parallel have a combined resistance R_T given by $R_T = \dfrac{8R}{8 + R}$. If R is changing at 0.30 Ω/min, find the rate at which R_T is changing when $R = 6.0$ Ω.

17. The radius r of a ring of a certain holograph (an image produced without using a lens) is given by $r = \sqrt{0.40\lambda}$, where λ is the wavelength of the light being used. If λ is changing at the rate of 0.10×10^{-7} m/s when $\lambda = 6.0 \times 10^{-7}$ m, find the rate at which r is changing.

18. An Earth satellite moves in a path that can be described by $\dfrac{x^2}{72.5} + \dfrac{y^2}{71.5} = 1$, where x and y are in thousands of kilometres. If $dx/dt = 12\,900$ km/h for $x = 3200$ km and $y > 0$, find dy/dt.

19. The magnetic field B due to a magnet of length l at a distance r is given by $B = \dfrac{k}{[r^2 + (l/2)^2]^{3/2}}$, where k is a constant for a given magnet. Find the expression for the time rate of change of B in terms of the time rate of change of r.

20. An approximate relationship between the pressure p and volume V of the vapour in a diesel engine cylinder is $pV^{1.4} = k$, where k is a constant. At a certain instant, $p = 4200$ kPa, $V = 75 \text{ cm}^3$, and the volume is increasing at the rate of $850 \text{ cm}^3/\text{s}$. What is the time rate of change of the pressure at this instant?

21. A swimming pool with a rectangular surface 18.0 m long and 12.0 m wide is being filled at the rate of $0.80 \text{ m}^3/\text{min}$. At one end it is 1.0 m deep, and at the other end it is 2.5 m deep, with a constant slope between ends. How fast is the height of water rising when the depth of water at the deep end is 1.0 m?

22. An engine cylinder 15.0 cm deep is being bored such that the radius increases by 0.100 mm/min. How fast is the volume V of the cylinder changing when the diameter is 9.50 cm?

23. Fatty deposits have decreased the circular cross-sectional opening of a person's artery. A test drug reduces these deposits such that the radius of the opening increases at the rate of 0.020 mm/month. Find the rate at which the area of the opening increases when $r = 1.2$ mm.

24. A computer program increases the side of a square image on the screen at the rate of 0.25 cm/s. Find the rate at which the area of the image increases when the edge is 6.50 cm.

25. A metal cube dissolves in acid such that an edge of the cube decreases by 0.500 mm/min. How fast is the volume of the cube changing when the edge is 8.20 mm?

26. A metal sphere is placed in seawater to study the corrosive effect of seawater. If the surface area decreases at 35 cm^2/year due to corrosion, how fast is the radius changing when it is 12 cm?

27. A uniform layer of ice covers a spherical water-storage tank. As the ice melts, the volume V of ice decreases at a rate that varies directly as the surface area A. Show that the outside radius decreases at a constant rate.

28. A light in a garage is 2.90 m above the floor and 3.65 m behind the door. If the garage door descends vertically at 0.45 m/s, how fast is the door's shadow moving toward the garage when the door is 0.60 m above the floor?

29. One statement of Boyle's law is that the pressure of a gas varies inversely as the volume for constant temperature. If a certain gas occupies 650 cm^3 when the pressure is 230 kPa and the volume is increasing at the rate of 20.0 cm^3/min, how fast is the pressure changing when the volume is 810 cm^3?

30. The tuning frequency f of an electronic tuner is inversely proportional to the square root of the capacitance C in the circuit. If $f = 920$ kHz for $C = 3.5$ pF, find how fast f is changing at this frequency if $dC/dt = 0.30$ pF/s.

31. The shadow of a 24-m high building is increasing at the rate of 18 cm/min when the shadow is 18 m long. How fast is the distance from the top of the building to the end of the shadow increasing?

32. The acceleration due to the gravity g on a spacecraft is inversely proportional to its distance from the centre of the earth. At the surface of the earth, $g = 9.80$ m/s^2. Given that the radius of the earth is 6370 km, how fast is g changing on a spacecraft approaching the earth at 1400 m/s at a distance of 41 000 km from the surface?

33. The intensity I of heat varies directly as the strength of the source and inversely as the square of the distance from the source. If an object approaches a heated object of strength 8.00 units at the rate of 50.0 cm/s, how fast is the intensity changing when it is 100 cm from the source?

34. The speed of sound v (in m/s) is $v = 331\sqrt{T/273}$, where T is the temperature (in K). If the temperature is 303 K (30°C) and is rising at 2.0°C/h, how fast is the speed of sound rising?

35. As a space shuttle moves into space, an astronaut's weight decreases. An astronaut weighing 650 N at sea level has a weight of $w = 650\left(\frac{6400}{6400 + h}\right)$ at h kilometres above sea level. If the shuttle is moving away from the earth at 6.0 km/s, at what rate is w changing when $h = 1200$ km?

36. The oil reservoir for the lubricating mechanism of a machine is in the shape of an inverted pyramid. It is being filled at the rate of 8.00 cm^3/s and the top surface is increasing at the rate of 6.00 cm^2/s. When the depth of oil is 6.50 cm and the top surface area is 22.5 cm^2, how fast is the level increasing?

37. A tank in the shape of an inverted cone has a height of 3.60 m and a radius at the top of 1.15 m. Water is flowing into the tank at the rate of 0.500 m^3/min. How fast is the level rising when it is 1.80 m deep?

38. The top of a ladder 4.00 m long is slipping down a vertical wall at the constant rate of 0.750 m/s. How fast is the bottom of the ladder moving along the ground away from the wall when it is 2.50 m from the wall?

39. A supersonic jet leaves an airfield travelling due east at 1600 km/h. A second jet leaves the same airfield at the same time and travels 1800 km/h along a line north of east such that it remains due north of the first jet. After a half-hour, how fast are the jets separating?

40. A car passes over a bridge at 15.0 m/s at the same time a boat passes under the bridge at a point 10.5 m directly below the car. If the boat is moving perpendicularly to the bridge at 4.00 m/s, how fast are the car and the boat separating 5.00 s later?

41. A rope attached to a boat is being pulled in at a rate of 2.50 m/s. If the water is 5.00 m below the level at which the rope is being drawn in, how fast is the boat approaching the wharf when 13.0 m of rope are yet to be pulled in? See Fig. 24.23.

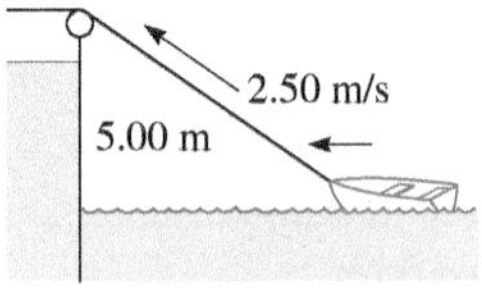

Fig. 24.23

42. A weather balloon leaves the ground 275 m from an observer and rises vertically at 12.0 m/s. How fast is the line of sight from the observer to the balloon increasing when the balloon is 450 m high? See Fig. 24.24.

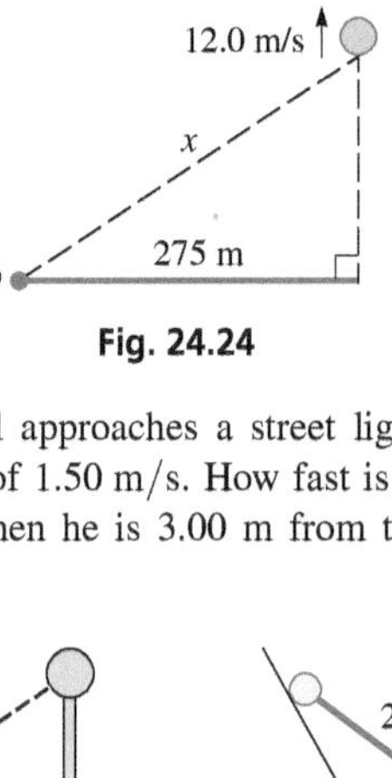

Fig. 24.24

43. A man 1.80 m tall approaches a street light 4.50 m above the ground at the rate of 1.50 m/s. How fast is the end of the man's shadow moving when he is 3.00 m from the base of the light? See Fig. 24.25.

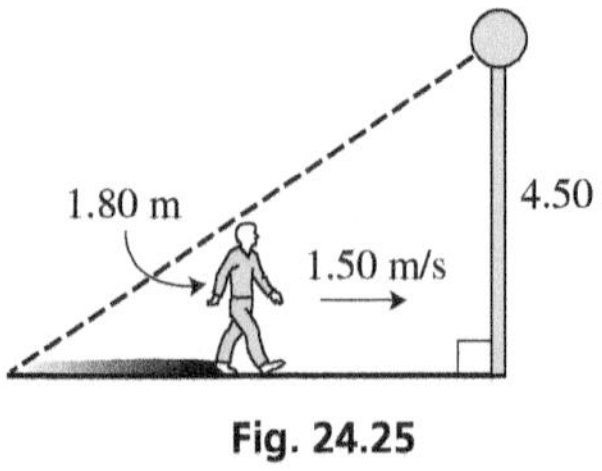

Fig. 24.25

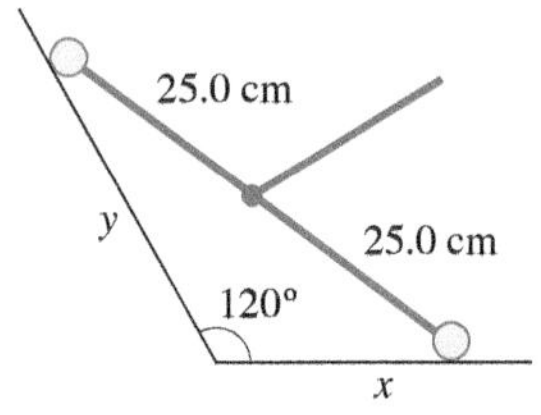

Fig. 24.26

44. A roller mechanism, as shown in Fig. 24.26, moves such that the right roller is always in contact with the bottom surface and the left roller is always in contact with the left surface. If the right roller is moving to the right at 1.50 cm/s when $x = 10.0$ cm, how fast is the left roller moving?

Answer to Practice Exercise

1. −3.20 cm/s

24.5 Using Derivatives in Curve Sketching

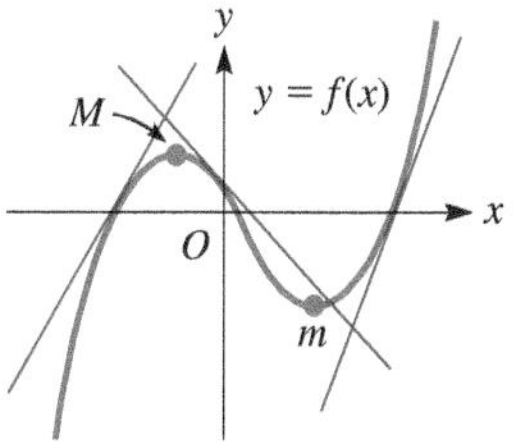

Fig. 24.27

Consider a continuous function $f(x)$ with continuous first and second derivatives over an interval of interest. In this section we will analyse what the signs of these derivatives tell us about the behaviour of the function ***as x increases from left to right.***

We begin by determining the intervals where a function rises (increases) or falls (decreases). For example, for the function in Fig. 24.27, we see that as x increases, y also increases until point M is reached. From M to m, y decreases. To the right of m, y again increases. Also, any tangent line left of M or right of m has a positive slope, and any tangent line between M and m has a negative slope. Since the derivative gives us the slope of a tangent line, we have the following relationship.

Sign of $f'(x)$ over Interval	Conclusion
$f'(x) > 0$	$f(x)$ increases
$f'(x) < 0$	$f(x)$ decreases

EXAMPLE 1 Function increasing and decreasing

Find those values of x for which the function $f(x) = x^3 - 3x^2$ is increasing and those values for which it is decreasing. See Fig. 24.28.

We find these values by taking the derivative and then finding the intervals where it is positive or negative, using the methods of Chapter 17. We have

$$f'(x) = 3x^2 - 6x = 3x(x - 2)$$

Setting each factor equal to zero and solving for x, we obtain the critical values $x = 0$ and $x = 2$. The analysis of signs of $f'(x)$ is as follows:

Interval	$3x$	$x - 2$	Sign of $f'(x)$	Conclusion
$x < 0$	$-$	$-$	$+$	$f'(x) > 0$ $f(x)$ increasing
$0 < x < 2$	$+$	$-$	$-$	$f'(x) < 0$ $f(x)$ decreasing
$x > 2$	$+$	$+$	$+$	$f'(x) > 0$ $f(x)$ increasing

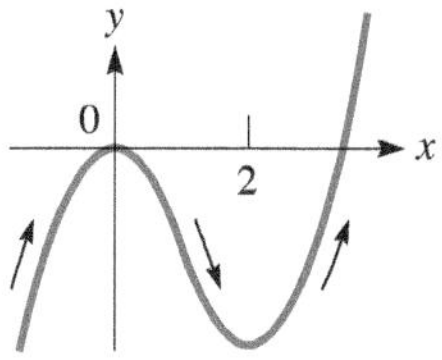

Fig. 24.28

Therefore, $f(x)$ is increasing if $x < 0$ or $x > 2$, and $f(x)$ is decreasing if $0 < x < 2$. See Fig. 24.28.

The point M in Fig. 24.27 is called a **local maximum point**. This means that no nearby points on the graph are higher than M. Although M does not necessarily have the greatest y-value of any point on the curve, it does have the greatest y-value for that part of the curve. Note that at M the function changes from increasing to decreasing, and hence the slopes of the tangent lines change from positive values to negative values, passing through the value of zero (at M, the tangent line is horizontal). Therefore we see that *a point is a local maximum if the derivative at that point is zero and the derivative changes sign from positive to negative when passing through that point.*

Similarly, the point m in Fig. 24.27 is called a **local minimum point**. This means that no nearby points on the graph are lower than m. Again, m does not necessarily have the smallest y-value of any point on the curve, but it does have the smallest y-value for that part of the curve. At m the function changes from decreasing to increasing, and hence the slopes of the tangent lines change from negative values to positive values, passing through the value of zero (at m, the tangent line is horizontal). Therefore we see that *a point is a local minimum if the derivative at that point is zero and the derivative changes sign from negative to positive when passing through that point.*

When taken together, these criteria are known as the **first-derivative test for maxima and minima**:

First-Derivative Test for Maxima and Minima

Local Maximum	Local Minimum
$f'(x) = 0$	$f'(x) = 0$
$f'(x)$ changes from + to −	$f'(x)$ changes from − to +

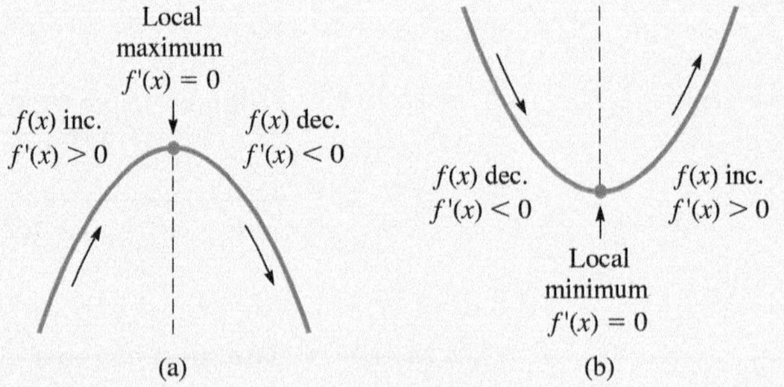

Fig. 24.29

Note that if $f'(x) = 0$ but the sign of the derivative does not change, we have neither a maximum nor a minimum point. Also, the sign changes of the first-derivative test remain valid even if $f'(x)$ is discontinuous at the extreme point.

EXAMPLE 2 Local maximum and minimum points

Find any local maximum points and any local minimum points of the graph of the function

$$y = 3x^5 - 5x^3$$

Finding the derivative and setting it equal to zero, we have

$$y' = 15x^4 - 15x^2 = 15x^2(x^2 - 1) = 15x^2(x - 1)(x + 1)$$

The derivative is zero for $x = 0$, $x = 1$ and $x = -1$. Therefore, as in Fig. 24.30,

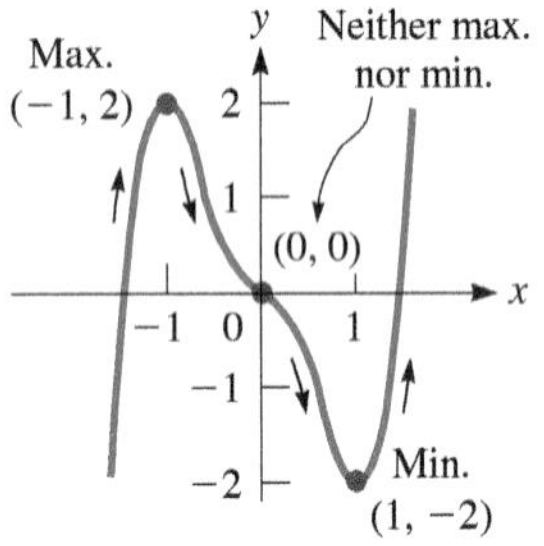

Fig. 24.30

Interval	$15x^2$	$x - 1$	$x + 1$	Sign of y'	Conclusion
$x < -1$	+	−	−	+	y' changes from + to − Maximum at $(-1, 2)$
$-1 < x < 0$	+	−	+	−	y' does not change Neither max. nor min. at $(0, 0)$
$0 < x < 1$	+	−	+	−	y' changes from − to + Minimum at $(1, -2)$
$x > 1$	+	+	+	+	

Practice Exercise

1. Find the local maximum and minimum points on the graph of $y = x^3 - 3x^2 - 9x$.

We now look again at the slope of a tangent drawn to a curve. In Fig. 24.31(a), consider the *change* in the values of the slope of a tangent at a point as the point moves from A to B. At A the slope is positive, and as the point moves toward M, the slope remains positive but becomes smaller until it becomes zero at M. To the right of M, the slope is negative and becomes more negative until it reaches I. Therefore, *from A to I, the slope continually decreases.* To the right of I, the slope remains negative but increases until it becomes zero again at m. To the right of m, the slope becomes positive and increases to point B. Therefore, *from I to B, the slope continually increases.* We say that *the curve is* **concave down** *from A to I and* **concave up** *from I to B.*

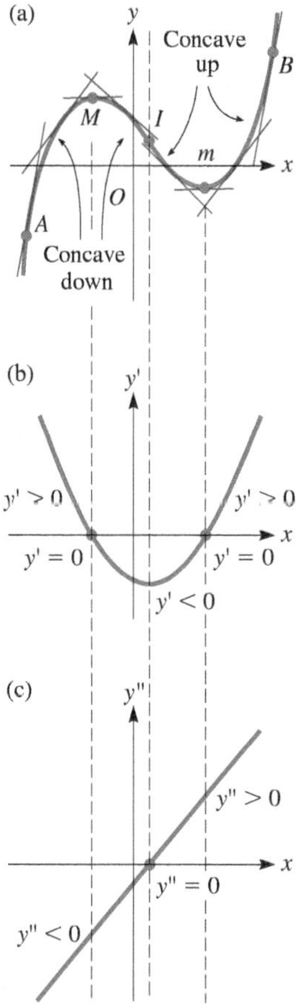

Fig. 24.31

The curve in Fig. 24.31(b) is that of the derivative, and it therefore indicates the values of the slope of $f(x)$. If the slope changes, we are dealing with the rate of change of slope or the rate of change of the derivative. This function is the second derivative. The curve in Fig. 24.31(c) is that of the second derivative. We see that *where the second derivative of a function is* ***negative****, the slope is decreasing, or the curve is* ***concave down*** *(opens down). Where the second derivative is* ***positive****, the slope is increasing, or the curve is* ***concave up*** *(opens up).* This may be summarized as follows:

Sign of $f''(x)$ over Interval	Conclusion
$f''(x) > 0$	$f(x)$ concave up
$f''(x) < 0$	$f(x)$ concave down

Combining this information with the fact that *a curve is concave down at a maximum point and concave up at a minimum point* (see Fig. 24.31a), we obtain what is known as the **second-derivative test for maxima and minima**:

Second-Derivative Test for Maxima and Minima

Local Maximum at $x = a$	Local Minimum at $x = a$
$f'(a) = 0$	$f'(a) = 0$
$f''(a) < 0$	$f''(a) > 0$

This test is often easier to use than the first-derivative test. However, if $f''(x) = 0$ at a maximum or minimum point, it is necessary to use the first-derivative test.

COMMON ERROR In using the second-derivative test, we should note that $f''(x)$ is ***negative*** **at a** ***maximum*** point and ***positive*** **at a** ***minimum*** point. This is contrary to a natural inclination to think of "maximum" and "positive" together or "minimum" and "negative" together.

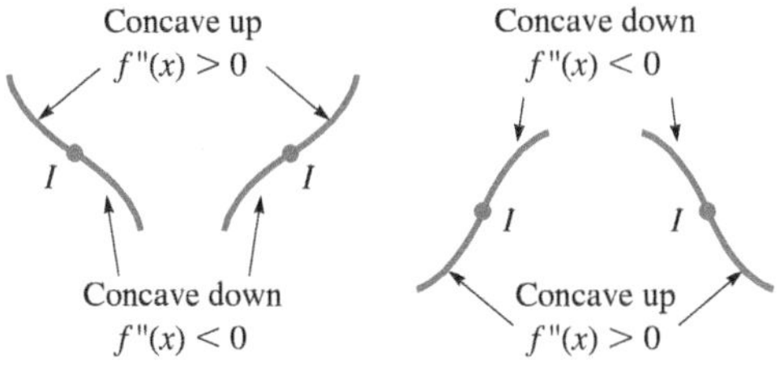

Points of inflection *I*

Fig. 24.32

The points at which the curve changes from concave up to concave down, or from concave down to concave up, are known as **points of inflection**. Thus, point *I* in Fig. 24.31(a) is a point of inflection. Inflection points are found by determining those values of x for which the second derivative changes sign. This is analogous to finding maximum and minimum points by the first-derivative test. In Fig. 24.32, various types of points of inflection are illustrated.

EXAMPLE 3 Concavity—points of inflection

Determine the concavity and find any points of inflection on the graph of the function $y = x^3 - 3x$.

This requires an inspection and analysis of the second derivative. Therefore, we find the first two derivatives:

$$y' = 3x^2 - 3$$
$$y'' = 6x$$

The second derivative is positive where the function is concave up, and this occurs if $x > 0$. The curve is concave down for $x < 0$, since y'' is negative. Thus, $(0, 0)$ is a point of inflection, since the concavity changes there. The graph of $y = x^3 - 3x$ is shown in Fig. 24.33.

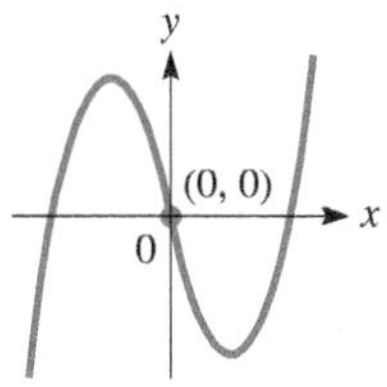

Fig. 24.33

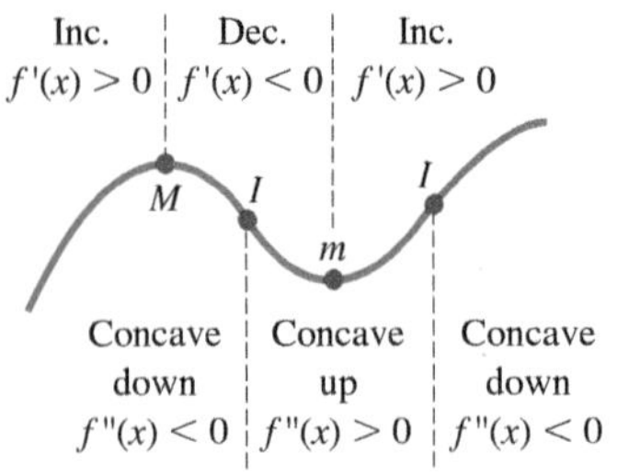

Fig. 24.34

Fig. 24.34 summarizes the information we have obtained from the first and second derivatives of a function $f(x)$. For continuous functions with continuous derivatives, this information is often sufficient to sketch a graph, and the following steps tell us how. Curve sketching for more general functions will be discussed in the next section.

Sketching the Graph of $y = f(x)$ Using Derivatives

1. Find y' and solve $y' = 0$.
2. Find y'' and evaluate at the x values found in Step 1. If possible, classify maxima and minima using the second-derivative test.
3. If necessary (or as check for Step 2), analyse the signs of y' to determine intervals of increase/decrease. Classify maxima and minima using the first-derivative test.
4. Solve $y'' = 0$. Analyse concavity from the signs of y''. Find any points of inflection.
5. Locate all the key points and connect them using the information obtained.

EXAMPLE 4 Sketching a curve using derivatives

Sketch the graph of $y = 6x - x^2$.

Find y' and solve $y' = 0$: $y' = 6 - 2x = 2(3 - x)$; $y' = 0$ if $x = 3$.

Find y'' and use second-derivative test: $y'' = -2$. Since $y'' < 0$ at $x = 3$, the point $(3, 9)$ is a maximum.

Analyse signs of y': For $x < 3$, $y' > 0$, so y is increasing. For $x > 3$, $y' < 0$, so y is decreasing. The point (3, 9) is indeed a maximum since y' changes from + to −.

Solve $y'' = 0$; analyse concavity and points of inflection: Since $y'' = -2$, y'' is never 0 and there are no points of inflection. The function is always concave down because $y'' < 0$.

Locate points and sketch: Draw a curve increasing for $x < 3$, decreasing for $x > 3$, and always concave down. The curve is the parabola shown in Fig. 24.35.

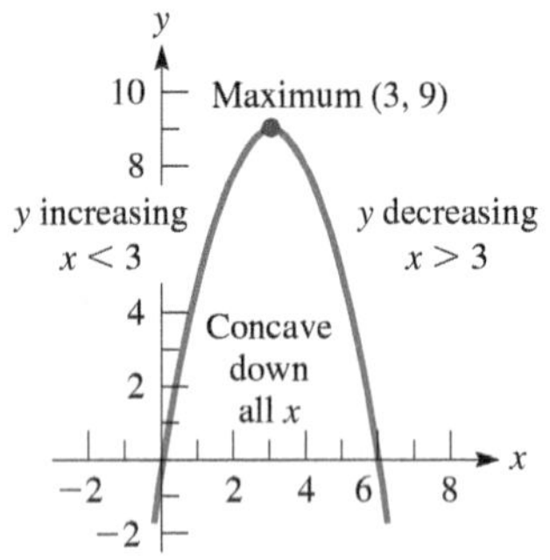

Fig. 24.35

Note that the same graph could have been obtained with the methods of Section 7.4 or Section 21.7.

Practice Exercise

2. Find the point(s) of inflection on the graph of $y = x^3 - 3x^2 - 9x$.

EXAMPLE 5 Sketching a curve using derivatives

Sketch the graph of $y = 2x^3 + 3x^2 - 12x$.

Find y' and solve $y' = 0$: $y' = 6x^2 + 6x - 12 = 6(x + 2)(x - 1)$; $y' = 0$ if $x = -2$ and $x = 1$.

Find y'' and use second-derivative test:

$y'' = 12x + 6 = 6(2x + 1)$

$y'' = -18 < 0$ at $x = -2$, so the point $(-2, 20)$ is a maximum.

$y'' = 18 > 0$ at $x = 1$, so the point $(1, -7)$ is a minimum.

Analyse signs of y':

For $x < -2$ or $x > 1$, $y' > 0$, so y is increasing.

For $-2 < x < 1$, $y' < 0$, so y is decreasing.

Solve $y'' = 0$; analyse concavity and points of inflection:

$y'' = 0$ if $x = -\frac{1}{2}$.

$y'' < 0$ if $x < -\frac{1}{2}$, so y is concave down;

$y'' > 0$ if $x > -\frac{1}{2}$, so y is concave up.

The point $\left(-\frac{1}{2}, \frac{13}{2}\right)$ is an inflection point.

Locate points and sketch:

Draw a concave down curve increasing to $(-2, 20)$, and then decreasing to $\left(-\frac{1}{2}, \frac{13}{2}\right)$. Continue decreasing but concave up to $(1, -7)$, while finally increasing for $x > 1$. The curve is shown in Fig. 24.36. For more precision, additional points may be used.

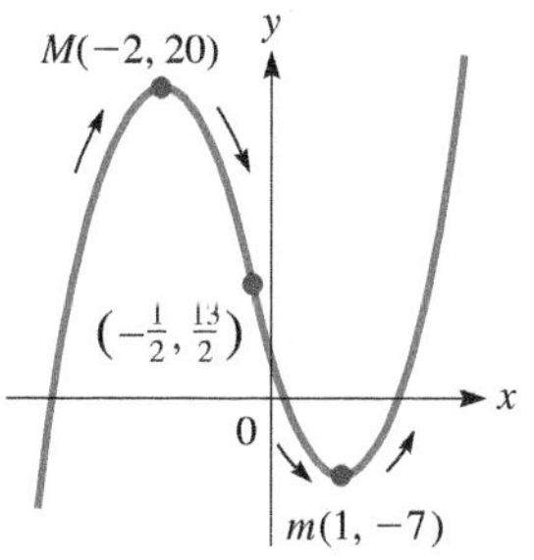

Fig. 24.36

EXAMPLE 6 Sketching a curve using derivatives

Sketch the graph of $y = x^5 - 5x^4$.

Find y' and solve $y' = 0$:

$y' = 5x^4 - 20x^3 = 5x^3(x - 4)$

$y' = 0$ if $x = 0$ and $x = 4$.

Find y'' and use second-derivative test:

$y'' = 20x^3 - 60x^2 = 20x^2(x - 3)$

$y'' = 320 > 0$ at $x = 4$, so the point $(4, -256)$ is a minimum.

$y'' = 0$ at $x = 0$, so the second derivative test does not apply.

Analyse signs of y':

For $x < 0$ or $x > 4$, $y' > 0$, so y is increasing.

For $0 < x < 4$, $y' < 0$, so y is decreasing.

The point $(0, 0)$ is a maximum because y' changes from $+$ to $-$.

Solve $y'' = 0$; analyse concavity and points of inflection:

$y'' = 0$ if $x = 0$ and $x = 3$.

$y'' < 0$ if $x < 0$ or $0 < x < 3$, so y is concave down; $y'' > 0$ if $x > 3$, so y is concave up. The point $(3, -162)$ is an inflection point because y'' changes sign.

Locate points and sketch:

Draw a concave down curve increasing to $(0, 0)$, and then decreasing to $(3, -162)$. Continue decreasing but concave up to $(4, -256)$, while finally increasing for $x > 4$. The curve is shown in Fig. 24.37. For more precision, additional points may be used.

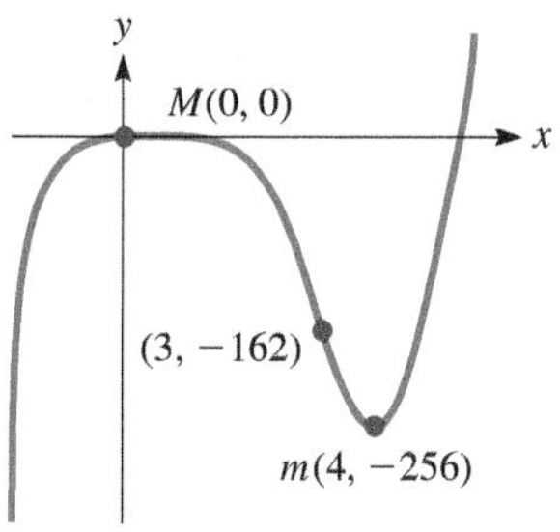

Fig. 24.37

Throughout this section it has been assumed that $f(x)$ and its derivatives are continuous functions. To show that this is necessary, see Exercise 54 of this section.

EXERCISES 24.5

In Exercises 1–4, make the given changes in the indicated examples of this section and then solve the resulting problems.

1. In Example 1, after the $-$ sign change $3x^2$ to $6x^2$ and then find the values of x for which $f(x)$ is increasing and those for which it is decreasing.
2. In Example 2, change the $5x^3$ to $15x$ and then find the local maximum and minimum points.
3. In Example 3, change the $-$ sign to $+$ and then determine the concavity and find any points of inflection.
4. In Example 4, change the $6x$ to $8x$ and then sketch the graph as in the example.

In Exercises 5–8, find those values of x for which the given functions are increasing and those values of x for which they are decreasing.

5. $y = x^2 + 2x$
6. $y = 2 + 6x - 3x^2$
7. $y = 2x^3 + 3x^2 - 36x - 10$
8. $y = x^4 - 6x^2$

In Exercises 9–12, find any local maximum or minimum points of the given functions. (These are the same functions as in Exercises 5–8.)

9. $y = x^2 + 2x$
10. $y = 2 + 6x - 3x^2$
11. $y = 2x^3 + 3x^2 - 36x - 10$
12. $y = x^4 - 6x^2$

In Exercises 13–16, find the values of x for which the given function is concave up, the values of x for which it is concave down, and any points of inflection. (These are the same functions as in Exercises 5–8.)

13. $y = x^2 + 2x$
14. $y = 2 + 6x - 3x^2$
15. $y = 2x^3 + 3x^2 - 36x - 10$
16. $y = x^4 - 6x^2$

In Exercises 17–20, sketch the graphs of the given functions by determining the appropriate information and points from the first and second derivatives (see Exercises 5–16).

17. $y = x^2 + 2x$
18. $y = 2 + 6x - 3x^2$
19. $y = 2x^3 + 3x^2 - 36x - 10$
20. $y = x^4 - 6x^2$

In Exercises 21–32, sketch the graphs of the given functions by determining the appropriate information and points from the first and second derivatives.

21. $y = 12x - 2x^2$
22. $y = 4x^2 - 16x - 20$
23. $y = 2x^3 + 6x^2 - 5$
24. $y = x^3 - 9x^2 + 15x + 1$
25. $y = x^3 + 3x^2 + 3x + 2$
26. $y = x^3 - 12x + 12$
27. $y = 4x^3 - 24x^2 + 36x$
28. $y = x(x - 4)^3$
29. $y = 4x^3 - 3x^4 + 6$
30. $y = x^5 - 20x^2$
31. $y = x^5 - 5x$
32. $y = x^4 + 32x + 2$

In Exercises 33 and 34, view the graphs of y, y′, and y″ together on a graphing utility. State how the graphs of y′ and y″ are related to the graph of y.

33. $y = x^3 - 12x$
34. $y = 24x - 9x^2 - 2x^3$

In Exercises 35–38, describe the indicated features of the given graphs.

35. Display the graph of $y = x^3 + cx$ for $c = -3, -1, 1$, and 3 on a graphing utility. Describe how the graph changes as c varies.
36. Follow the instructions of Exercise 35 for the function $y = x^4 + cx^2$.
37. Display the graph of $y = 2x^5 - 7x^3 + 8x$ on a graphing utility. Describe the relative locations of the left local maximum point and the local minimum points.
38. Describe the following features of the graph in Fig. 24.38 between or at the points A, B, C, D, E, F, and G. (a) Increasing and decreasing, (b) local maximum and minimum points, (c) concavity, and (d) points of inflection.

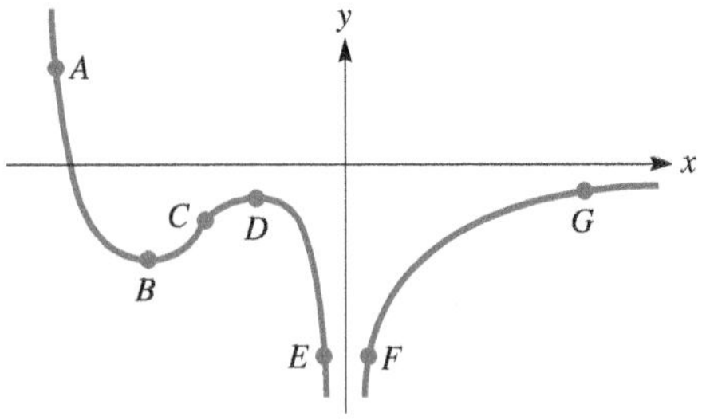

Fig. 24.38

In Exercises 39–50, sketch the indicated curves by the methods of this section. You may check the graphs by using a graphing calculator.

39. A batter hits a baseball that follows a path given by $y = x - 0.025x^2$, where distances are in metres. Sketch the graph of the path of the baseball.
40. The angle θ (in degrees) of a robot arm with the horizontal as a function of the time t (in s) is given by $\theta = 10 + 12t^2 - 2t^3$. Sketch the graph for $0 \le t \le 6$ s.
41. The power P (in W) in a certain electric circuit is given by $P = 4i - 0.5i^2$. Sketch the graph of P vs. i.
42. A computer analysis shows that the thrust T (in kN) of an experimental rocket motor is $T = 20 + 9t - 4t^3$, where t is the time (in min) after the motor is activated. Sketch the graph for the first 2 min.
43. The force F (in N) exerted by a cam on a lever is given by $F = x^4 - 12x^3 + 46x^2 - 60x + 25$, where x $(1 \le x \le 5)$ is the distance (in cm) from the centre of rotation of the cam to the edge of the cam in contact with the lever (see Fig. 23.22). For the force F exerted by the cam on the lever, sketch the graph of F vs. x. (*Hint:* Use methods of Chapter 15 to analyse the first derivative.)
44. The solar-energy power P (in W) produced by a certain solar system does not rise and fall uniformly during a cloudless day because of the system's location. An analysis of records shows that $P = -0.45(2t^5 - 45t^4 + 350t^3 - 1000t^2)$, where t is the time (in h) during which power is produced. Show that, during the solar-power production, the production flattens (inflection) in the middle and then peaks before shutting down. (*Hint:* The derivative is zero at integer values of t.)
45. An electric circuit is designed such that the resistance R (in Ω) is a function of the current i (in mA) according to $R = 75 - 18i^2 + 8i^3 - i^4$. Sketch the graph if $R \ge 0$ and i can be positive or negative.
46. A horizontal 12-m beam is deflected by a load such that it can be represented by the equation $y = 0.0004(x^3 - 12x^2)$. Sketch the curve followed by the beam.
47. The altitude h (in m) of a certain rocket is given by $h = -t^3 + 54t^2 + 480t + 20$, where t is the time (in s) of flight. Sketch the graph of $h = f(t)$.

48. An analysis of data showed that the mean density d (in mg/cm^3) of a calcium compound in the bones of women was given by $d = 0.00181x^3 - 0.289x^2 + 12.2x + 30.4$, where x represents the ages of women ($20 < x < 80$ years). (A woman probably has osteoporosis if $d < 115\ \text{mg/cm}^3$.) Sketch the graph.

49. A rectangular box is made from a piece of cardboard 8 cm by 12 cm by cutting equal squares from each corner and bending up the sides. See Fig. 24.39. Express the volume of the box as a function of the side of the square that is cut out and then sketch the curve of the resulting equation.

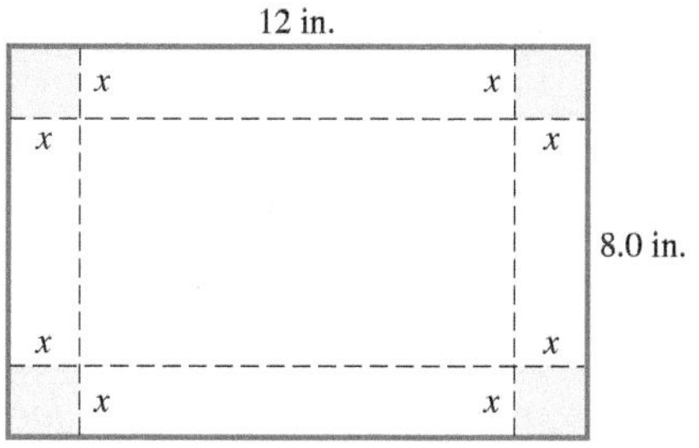

Fig. 24.39

50. A rectangular planter with a square end is to be made from 8.0 m^2 of redwood. Express the volume of soil the planter can hold as a function of the side of the square of the end. Sketch the curve of the resulting function.

In Exercises 51–53, sketch a continuous curve that has the given characteristics. In Exercise 54, answer the given question.

51. $f(1) = 0; f'(x) > 0$ for all x; $f''(x) < 0$ for all x

52. $f(0) = 1; f'(x) < 0$ for all x; $f''(x) < 0$ for $x < 0$; $f''(x) > 0$ for $x > 0$

53. $f(-1) = 0; f(2) = 2; f'(x) < 0$ for $x < -1$; $f'(x) > 0$ for $x > -1$; $f''(x) < 0$ for $0 < x < 2$; $f''(x) > 0$ for $x < 0$ or $x > 2$

54. Display the graph of $f(x) = x^{2/3}$ for $-2 < x < 2$ on a graphing utility. Determine the continuity of $f(x), f'(x)$, and $f''(x)$. Discuss the concavity of the curve in relation to the minimum point.

Answers to Practice Exercises

1. Max.$(-1, 5)$, Min.$(3, -27)$ **2.** Infl.$(1, -11)$

24.6 More on Curve Sketching

When graphing functions in earlier chapters, we often used information that was obtainable from the function itself. We will now use this type of information, along with that found from the derivatives, to graph functions. We will find that, in graphing any particular function, some types of information are of more value than others. The features we will consider in this section are as follows:

Features to Be Used in Graphing Functions

1. *Intercepts* Points for which the graph crosses (or is tangent to) each axis.
2. *Symmetry* For a review of symmetry, see Section 21.3.
3. *Behaviour as x Becomes Large* We will find what happens to the function as $x \to \infty$ and as $x \to -\infty$.
4. *Vertical Asymptotes* We can find vertical asymptotes by finding values that make factors in the denominator zero.
5. *Domain and Range* For a review of domain and range, see Section 3.3.
6. *Derivatives* We will find information from the first two derivatives as we did in Section 24.5.

EXAMPLE 1 Sketch a graph

Sketch the graph of $y = \dfrac{8}{x^2 + 4}$.

Intercepts If $x = 0$, $y = 2$, which means $(0, 2)$ is an intercept. If $y = 0$, there is no corresponding value of x, since $8/(x^2 + 4)$ is a fraction greater than zero for all x. This also indicates that all points on the curve are above the x-axis.

Symmetry The curve is symmetric to the y-axis since

$$y = \frac{8}{(-x)^2 + 4} \text{ is the same as } y = \frac{8}{x^2 + 4}.$$

The curve is not symmetric to the x-axis since

$$-y = \frac{8}{x^2+4} \text{ is not the same as } y = \frac{8}{x^2+4}.$$

The curve is not symmetric to the origin since

$$-y = \frac{8}{(-x)^2+4} \text{ is not the same as } y = \frac{8}{x^2+4}.$$

The value in knowing the symmetry is that we should find those portions of the curve on either side of the y-axis as reflections of the other. It is possible to use this fact directly or to use it as a check.

■ The curve for Example 1 is a special case (with $a = 1$) of the curve known as the *witch of Agnesi.* Its general form is

$$y = \frac{8a^3}{x^2 + 4a^2}$$

It is named for the Italian mathematician Maria Gaetana Agnesi (1718–1799). She wrote the first text that contained analytic geometry, differential and integral calculus, series (see Chapter 30), and differential equations (see Chapter 31). The word *witch* was used due to a mistranslation from Italian to English.

Behaviour as x Becomes Large We note that as $x \to \infty$, $y \to 0$ since $8/(x^2 + 4)$ is always a fraction that is greater than zero but which becomes smaller as x becomes larger. Therefore, we see that $y = 0$ is an asymptote. From either the symmetry or the function, we also see that $y \to 0$ as $x \to -\infty$.

Vertical Asymptotes An asymptote is a line that a curve approaches. *Vertical asymptotes, if any exist, are found by determining those values of x for which the denominator of any term is zero.* Since $x^2 + 4$ cannot be zero, this curve has no vertical asymptotes.

Domain and Range Since the denominator $x^2 + 4$ cannot be zero, x can take on any value. This means the domain of the function is all values of x. Also, we have noted that $8/(x^2 + 4)$ is a fraction greater than zero. Since $x^2 + 4$ is 4 or greater, y is 2 or less. This tells us that the range of the function is $0 < y \leq 2$.

Derivatives We now determine what the derivatives can also tell us about the curve. We start with the first derivative.

$$y = \frac{8}{(x^2+4)} = 8(x^2+4)^{-1}$$
$$y' = -8(x^2+4)^{-2}(2x)$$
$$= \frac{-16x}{(x^2+4)^2}$$

Since $(x^2 + 4)^2$ is positive for all values of x, the sign of y' is determined by the numerator. Thus, we note that $y' = 0$ for $x = 0$ and that $y' > 0$ for $x < 0$ and $y' < 0$ for $x > 0$. The curve, therefore, is increasing for $x < 0$, is decreasing for $x > 0$, and has a maximum point at $(0, 2)$.

Now, finding the second derivative, we have

$$y'' = \frac{(x^2+4)^2(-16) + 16x(2)(x^2+4)(2x)}{(x^2+4)^4} = \frac{-16(x^2+4) + 64x^2}{(x^2+4)^3}$$
$$= \frac{48x^2 - 64}{(x^2+4)^3} = \frac{16(3x^2-4)}{(x^2+4)^3}$$

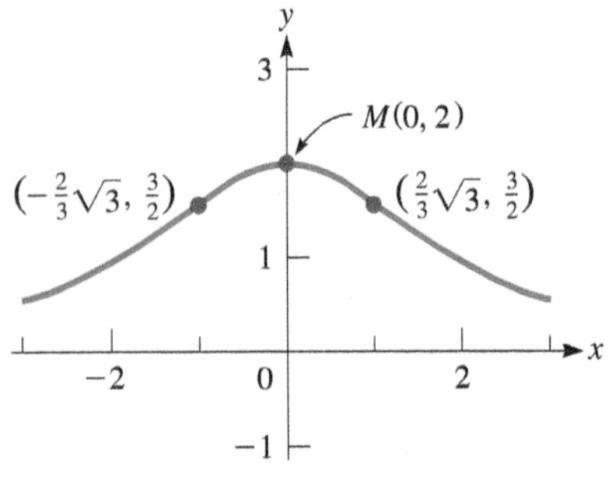

Fig. 24.40

We note that y'' is negative for $x = 0$, which confirms that $(0, 2)$ is a maximum point. Also, points of inflection are found for the values of x satisfying $3x^2 - 4 = 0$. Thus, $\left(-\frac{2}{3}\sqrt{3}, \frac{3}{2}\right)$ and $\left(\frac{2}{3}\sqrt{3}, \frac{3}{2}\right)$ are points of inflection. The curve is concave up if $x < -\frac{2}{3}\sqrt{3}$ or $x > \frac{2}{3}\sqrt{3}$, and the curve is concave down if $-\frac{2}{3}\sqrt{3} < x < \frac{2}{3}\sqrt{3}$.

Putting this information together, we sketch the curve shown in Fig. 24.40. Note that this curve could have been sketched primarily by use of the fact that $y \to 0$ as $x \to +\infty$ and as $x \to -\infty$ and the fact that a maximum point exists at $(0, 2)$. However, the other parts of the analysis, such as symmetry and concavity, serve as checks and make the curve more accurate.

EXAMPLE 2 Sketch a graph

Sketch the graph of $y = x + \frac{4}{x}$.

Intercepts If we set $x = 0$, y is undefined. This means that the curve is not *continuous* at $x = 0$ and there are no y-intercepts. If we set $y = 0$, $x + 4/x = (x^2 + 4)/x$ cannot be zero since $x^2 + 4$ cannot be zero. Therefore, there are no intercepts. This may seem to be of little value, but we must realize *this curve does not cross either axis*. This will be of value when we sketch the curve.

Symmetry In testing for symmetry, we find that the curve is not symmetric to either axis. However, this curve does possess symmetry to the origin. This is determined by the fact that when $-x$ replaces x and at the same time $-y$ replaces y, the equation does not change.

Behaviour as x Becomes Large As $x \to +\infty$ and as $x \to -\infty$, $y \to x$ since $4/x \to 0$. Thus, $y = x$ is an asymptote of the curve.

Vertical Asymptotes As we noted in Example 1, vertical asymptotes exist for values of x for which y is undefined. In this equation, $x = 0$ makes the second term on the right undefined, and therefore y is undefined. In fact, as $x \to 0$ from the positive side, $y \to +\infty$, and as $x \to 0$ from the negative side, $y \to -\infty$. This is derived from the sign of $4/x$ in each case.

Domain and Range Since x cannot be zero, the domain of the function is all x except zero. As for the range, the analysis from the derivatives will show it to be $y \leq -4$, $y \geq 4$.

Derivatives Finding the first derivative, we have

$$y' = 1 - \frac{4}{x^2} = \frac{x^2 - 4}{x^2}$$

The x^2 in the denominator indicates that the sign of the first derivative is the same as its numerator. The numerator is zero if $x = -2$ or $x = 2$. If $x < -2$ or $x > 2$, then $y' > 0$; and if $-2 < x < 2, x \neq 0, y' < 0$. Thus, y is increasing if $x < -2$ or $x > 2$, and also y is decreasing if $-2 < x < 2$, except at $x = 0$ (y is undefined). Also, $(-2, -4)$ is a local maximum point, and $(2, 4)$ is a local minimum point. The second derivative is $y'' = 8/x^3$. This cannot be zero, but it is negative if $x < 0$ and positive if $x > 0$. Thus, the curve is concave down if $x < 0$ and concave up if $x > 0$. Using this information, we have the curve shown in Fig. 24.41.

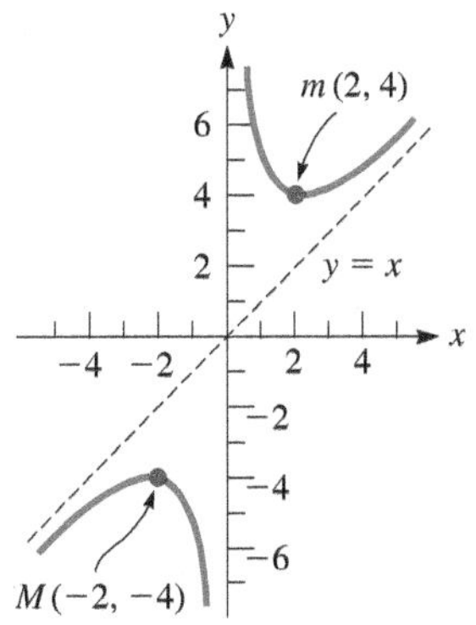

Fig. 24.41

EXAMPLE 3 Sketch a graph

Sketch the graph of $y = \frac{1}{\sqrt{1 - x^2}}$.

Intercepts If $x = 0$, $y = 1$. If $y = 0$, $1/\sqrt{1 - x^2}$ would have to be zero, but it cannot since it is a fraction with 1 as the numerator for all values of x. Thus, $(0, 1)$ is the only intercept.

Symmetry The curve is symmetric to the y-axis.

Behaviour as x Becomes Large The values of x cannot be considered beyond 1 or -1, for any value of $x < -1$ or $x > 1$ gives imaginary values for y. Thus, the curve does not exist for values of $x < -1$ or $x > 1$.

Vertical Asymptotes If $x = 1$ or $x = -1$, y is undefined. In each case, as $x \to 1$ and as $x \to -1$, $y \to +\infty$.

Domain and Range From the analysis of x becoming large and of the vertical asymptotes, we see that the domain is $-1 < x < 1$. Also, since $\sqrt{1 - x^2}$ is 1 or less, $1/\sqrt{1 - x^2}$ is 1 or more, which means the range is $y \geq 1$.

Derivatives

$$y' = -\frac{1}{2}(1 - x^2)^{-3/2}(-2x) = \frac{x}{(1 - x^2)^{3/2}}$$

We see that $y' = 0$ if $x = 0$. If $-1 < x < 0, y' < 0$, and also if $0 < x < 1$, $y' > 0$. Thus, the curve is decreasing if $-1 < x < 0$ and increasing if $0 < x < 1$. There is a minimum point at $(0, 1)$.

$$y'' = \frac{(1 - x^2)^{3/2} - x\left(\frac{3}{2}\right)(1 - x^2)^{1/2}(-2x)}{(1 - x^2)^3} = \frac{(1 - x^2) + 3x^2}{(1 - x^2)^{5/2}}$$

$$= \frac{2x^2 + 1}{(1 - x^2)^{5/2}}$$

The second derivative cannot be zero since $2x^2 + 1$ is positive for all values of x. The second derivative is also positive for all permissible values of x, which means the curve is concave up for these values.

Using this information, we sketch the graph in Fig. 24.42.

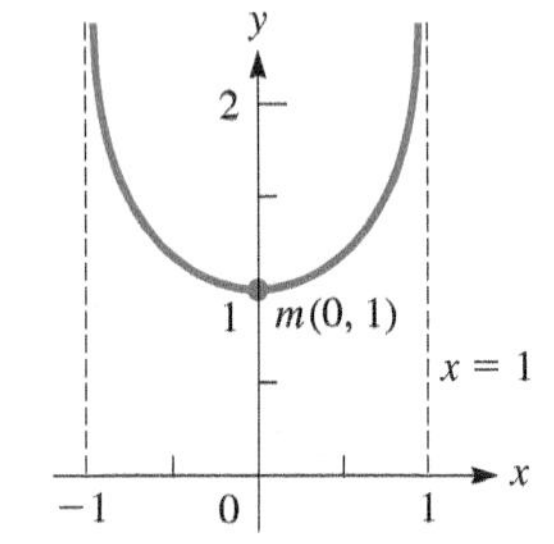

Fig. 24.42

EXAMPLE 4 Sketch a graph

Sketch the graph of $y = \frac{x}{x^2 - 4}$.

Intercepts If $x = 0, y = 0$, and if $y = 0, x = 0$. The only intercept is $(0, 0)$.

Symmetry The curve is not symmetric to either axis. However, since $-y = -x/[(-x)^2 - 4]$ is the same as $y = x/(x^2 - 4)$, it is symmetric to the origin.

Behaviour as x Becomes Large As $x \to +\infty$ and as $x \to -\infty$, $y \to 0$. This means that $y = 0$ is an asymptote.

Vertical Asymptotes If $x = -2$ or $x = 2$, y is undefined. As $x \to -2$, $y \to -\infty$ if $x < -2$ since $x^2 - 4$ is positive, and $y \to +\infty$ if $x > -2$ since $x^2 - 4$ is negative. As $x \to 2$, $y \to -\infty$ if $x < 2$, and $y \to +\infty$ if $x > 2$.

Domain and Range The domain is all real values of x except -2 and 2. As for the range, if $x < -2$, $y < 0$ (the numerator is negative and the denominator is positive). If $x > 2$, $y > 0$ (both numerator and denominator are positive). Since $(0, 0)$ is an intercept, we see that the range is all values of y.

Derivatives

$$y' = \frac{(x^2 - 4)(1) - x(2x)}{(x^2 - 4)^2} = -\frac{x^2 + 4}{(x^2 - 4)^2}$$

Since $y' < 0$ for all values of x except -2 and 2, the curve is decreasing for all values in the domain.

Now, finding the second derivative, we have

$$y'' = -\frac{(x^2 - 4)^2(2x) - (x^2 + 4)(2)(x^2 - 4)(2x)}{(x^2 - 4)^4}$$

$$= -\frac{2x(x^2 - 4) - 4x(x^2 + 4)}{(x^2 - 4)^3} = \frac{2x^3 + 24x}{(x^2 - 4)^3} = \frac{2x(x^2 + 12)}{(x^2 - 4)^3}$$

The sign of y'' depends on x and $(x^2 - 4)^3$. If $x < -2, y'' < 0$. If $-2 < x < 0$, $y'' > 0$. If $0 < x < 2, y'' < 0$. If $x > 2, y'' > 0$. This means the curve is concave down for $x < -2$ or $0 < x < 2$ and is concave up for $-2 < x < 0$ or $x > 2$.

The curve is sketched in Fig. 24.43.

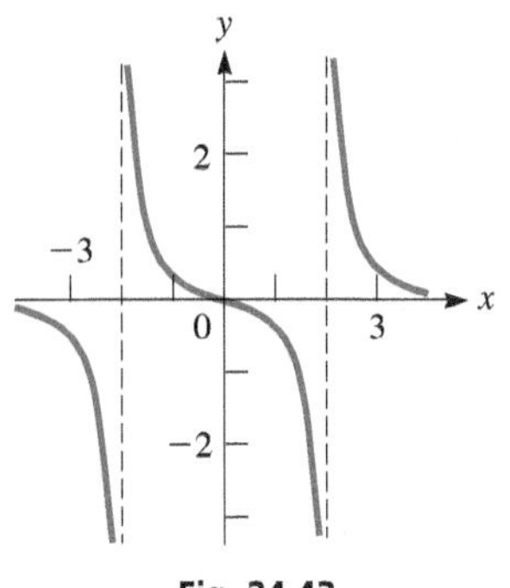

Fig. 24.43

EXERCISES 24.6

In Exercises 1–18, use the method of the examples of this section to sketch the indicated curves.

1. In Example 2, change the + to − before $4/x$ and then proceed.

2. $y = \dfrac{4}{x^2}$

3. $y = \dfrac{2}{x+1}$

4. $y = \dfrac{x}{x-2}$

5. $y = x^2 + \dfrac{2}{x}$

6. $y = x + \dfrac{4}{x^2}$

7. $y = x - \dfrac{1}{x}$

8. $y = 3x + \dfrac{1}{x^3}$

9. $y = \dfrac{x^2}{x+1}$

10. $y = \dfrac{9x^2}{x^2+9}$

11. $y = \dfrac{1}{x^2-1}$

12. $y = \dfrac{x^2-1}{x^3}$

13. $y = \dfrac{4}{x} - \dfrac{4}{x^2}$

14. $y = 4x + \dfrac{1}{\sqrt{x}}$

15. $y = x\sqrt{1-x^2}$

16. $y = \dfrac{x-1}{x^2-2x}$

17. $y = \dfrac{9x}{9-x^2}$

18. $y = \dfrac{x^2-4}{x^2+4}$

In Exercises 19–32, solve the given problems.

19. Display the graph of $y = \dfrac{cx}{1+c^2x^2}$ on a graphing utility for $c = -3, -1, 1,$ and 3. Describe how the graph changes as c varies.

20. Display the graph of $y = x + \dfrac{4}{x^n}$ on a graphing utility for $n = 1, 2, 3,$ and 4. Describe how the graph changes as n varies.

21. Sketch a continuous curve such that $f(0) = 2, f'(x) > 0$, and $f''(x) < 0$ for all x, and $f(x) \rightarrow 4$ as $x \rightarrow +\infty$.

22. For a continuous function $y = f(x)$, if for all x, $f(x) > 0$, $f'(x) < 0$, and $f''(x) > 0$, what do you conclude about the graph of $f(x)$?

23. In Exercise 10, first divide the numerator by the denominator. Does this simplify any of the graphing features?

24. The concentration C (in mg/cm^3) of a certain drug in a patient's bloodstream is $C = \frac{0.15t}{t^2+1}$, where t is the time (in h) after the drug is administered. Sketch the graph.

25. The combined capacitance C_T (in μF) of a 6-μF capacitance and a variable capacitance C in series is given by $C_T = \dfrac{6C}{6+C}$. Sketch the graph.

26. In thermodynamics, an equation relating the thermodynamic temperature T, the pressure p, and the volume V of a gas is $T = \left(p + \dfrac{a}{V^2}\right)\left(\dfrac{V-b}{R}\right)$, where a, b, and R are constants. It is known as van der Waal's equation. Sketch a graph of p vs. V from van der Waal's equation, assuming the following values: $R = T = a = 1$ and $b = 0$. For many gases the value of a is much greater than that of b. Even though the values of $R = T = 1$ are not realistic, the *shape* of the curve will be correct for the assumed value of $b = 0$.

27. The reliability R of a computer model is found to be $R = \dfrac{200}{\sqrt{t^2 + 40\,000}}$, where t is the time of operation in hours. ($R = 1$ is perfect reliability, and $R = 0.5$ means there is a 50% chance of a malfunction.) Sketch the graph.

28. The electric power P (in W) produced by a source is given by $P = \dfrac{36R}{R^2 + 2R + 1}$, where R is the resistance in the circuit. Sketch the graph.

29. Assuming that a raindrop is always spherical and as it falls its radius increases from 1 mm to r mm, its velocity v (in mm/s) is $v = k\left(r - \dfrac{1}{r^3}\right)$. With $k = 1$, sketch the graph.

30. If a positive electric charge of $+q$ is placed between two negative charges of $-q$ that are two units apart, Coulomb's law states that the force F on the positive charge is $F = -\dfrac{kq^2}{x^2} + \dfrac{kq^2}{(x-2)^2}$, where x is the distance from one of the negative charges. Let $kq^2 = 1$ and sketch the graph for $0 < x < 2$.

31. A cylindrical oil drum is to be made such that it will contain 20 kL. Sketch the area of sheet metal required for construction as a function of the radius of the drum.

32. A fence is to be constructed to enclose a rectangular area of 20 000 m^2. A previously constructed wall is to be used for one side. Sketch the length of fence to be built as a function of the side of the fence parallel to the wall. See Fig. 24.44.

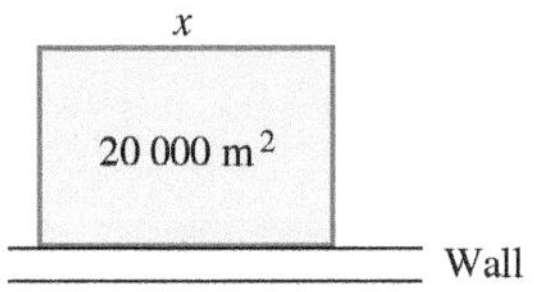

Fig. 24.44

24.7 Applied Maximum and Minimum Problems

Problems from various applied situations frequently occur that require finding a maximum or minimum value of some function. If the function is known, the methods we have already discussed can be used directly. This is discussed in the following example.

EXAMPLE 1 Finding the maximum efficiency

An automobile manufacturer, in testing a new engine on one of its new models, found that the efficiency η (in %) of the engine as a function of the speed s (in km/h) of the car was given by $\eta = 0.768s - 0.000\,04s^3$. What is the maximum efficiency of the engine?

In order to find a maximum value, we find the derivative of η with respect to s:

■ The first gasoline-engine automobile was built by the German engineer Karl Benz (1844–1929) in the 1880s.

$$\frac{d\eta}{ds} = 0.768 - 0.000\,12s^2$$

We then set the derivative equal to zero in order to find the value of s for which a maximum may occur:

$$0.768 - 0.000\,12s^2 = 0$$
$$0.000\,12s^2 = 0.768$$
$$s^2 = 6400$$
$$s = 80.0 \text{ km/h}$$

We know that s must be positive to have meaning in this problem. Therefore, the apparent solution of $s = -80$ is discarded. The second derivative is

$$\frac{d^2\eta}{ds^2} = -0.000\,24s$$

which is negative for any positive value of s. Therefore, we have a maximum for $s = 80.0$. Substituting $s = 80.0$ in the function for η, we obtain

$$\eta = 0.768(80.0) - 0.000\,04(80.0^3) = 61.44 - 20.48 = 40.96$$

The maximum efficiency is about 41.0%, which occurs for $s = 80.0$ km/h.

In many problems for which a maximum or minimum value is to be found, the function is not given. To solve such a problem, we use these steps:

Steps in Solving Applied Maximum and Minimum Problems

1. Introduce notation by identifying the quantity Q to be maximized or minimized, and all the variables in the problem.
2. If possible, draw a figure illustrating the problem.
3. Express Q in terms of the other variables. Write equations expressing known relationships between the other variables and use them to rewrite Q as a function of a single variable.
4. *Take the derivative of the function in Step 3.*
5. *Set the derivative equal to zero, and solve the resulting equation.*
6. *Check as to whether the value found in Step 5 makes Q a maximum or a minimum.* This might be clear from the statement of the problem, or it might require one of the derivative tests.
7. *Be sure the stated answer is the one the problem required.* Some problems require the maximum or minimum value, and others require values of other variables that give the maximum or minimum value.

The following examples illustrate several types of stated problems involving maximum and minimum values.

> **LEARNING TIP**
> ***The principal difficulty that arises in these problems is finding the proper function.*** We must carefully read the problem to find the information needed to set up the function.

EXAMPLE 2 Setting up an equation and finding the maximum

Find the number that exceeds its square by the greatest amount.

Introduce notation: Let x be any number, and D the difference between x and its square, x^2.

Write D as a function of x: $D = x - x^2$

Take the derivative: $\frac{dD}{dx} = 1 - 2x$

Set the derivative equal to zero and solve: $1 - 2x = 0, \quad x = \frac{1}{2}$

Check that it is a maximum: $\frac{d^2D}{dx^2} = -2$, so D is concave down and there is a maximum at $x = \frac{1}{2}$.

EXAMPLE 3 Finding maximum area—rectangular corral

A rectangular corral is to be enclosed with 1600 m of fencing. Find the maximum possible area of the corral.

Introduce notation: Let A be the area of a corral of sides x and y (Fig. 24.45).

Write A as a function of x and y. Use the relationship between x and y to write A as a function of one variable:

Area is $A = xy$.

Perimeter is known:

$2x + 2y = 1600$, so $y = 800 - x$

Substitute y to have area in terms of x:

$A = x(800 - x) = 800x - x^2$

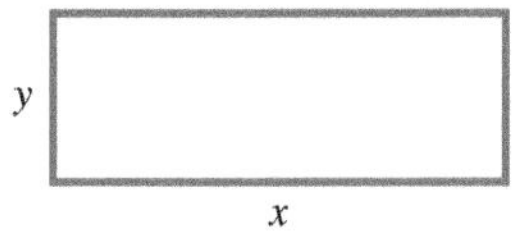

Fig. 24.45

Take the derivative: $\frac{dA}{dx} = 800 - 2x$

Set the derivative equal to zero and solve: $800 - 2x = 0, \quad x = 400$ m

Check that it is a maximum: $\frac{d^2A}{dx^2} = -2$, so A is concave down and there is a maximum at $x = 400$.

Sides of $x = 400$ m and $y = 400$ m give the maximum area of 160 000 m^2 for the corral.

EXAMPLE 4 Setting up an equation using variation—strength of a beam

The strength S of a beam with a rectangular cross-section is directly proportional to the product of the width w and the square of the depth d. Find the dimensions of the strongest beam that can be cut from a log with a circular cross-section that is 16.0 cm in diameter. See Fig. 24.46.

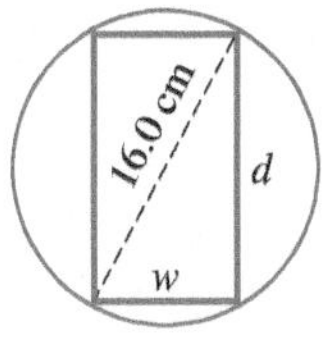

Fig. 24.46

Introduce notation: Let S be the strength of a beam with width w and depth d.

Write S as a function of w and d. Use the relationship between w and d to write S as a function of one variable: Direct variation:

$$S = kwd^2$$

The diagonal is known. By the Pythagorean theorem:

$$w^2 + d^2 = 16.0^2, \quad \text{so } d^2 = 256 - w^2$$

Substitute d^2 to have S in terms of w:

$$S = kw(256 - w^2) = k(256w - w^3)$$

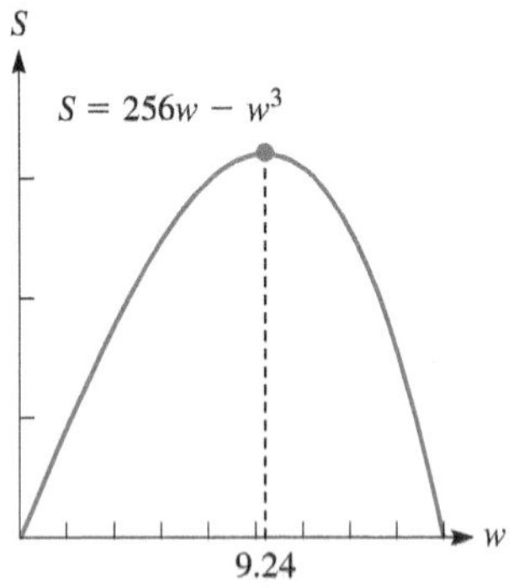

Fig. 24.47

Take the derivative:

$$\frac{dS}{dw} = k(256 - 3w^2)$$

Set the derivative equal to zero and solve:

$$k(256 - 3w^2) = 0, \; w = \frac{16}{\sqrt{3}} = 9.24 \text{ cm}$$

Check that it is a maximum: $\frac{d^2S}{dw^2} = -6kw < 0$ if $w > 0$, so S has a maximum at $w = 9.24$ cm (see Fig. 24.47, where $k = 1$ since *the value of k does not affect the solution*).

Substituting the maximizing w into the expression for d and solving, we get $d = \sqrt{256 - \frac{256}{3}} = 13.1$ cm. The strongest beam is about 9.24 cm wide and 13.1 cm deep.

Practice Exercise

1. In Example 4, change 16.0 cm to 12.0 cm and solve the resulting problem.

EXAMPLE 5 Setting up an equation using the distance formula

Find the point on the parabola $y = x^2$ that is nearest to the point $(6, 3)$.

Introduce notation: Let D be the distance between (x, y) and $(6, 3)$.

Write D as a function of x and y. Use D^2 instead of D (if distance is minimum, so is its square, and it is easier for derivatives). Use the relationship between x and y to write D^2 as a function of one variable:

Distance formula:

$$D = \sqrt{(x - 6)^2 + (y - 3)^2}$$

Square of distance:

$$D^2 = (x - 6)^2 + (y - 3)^2$$

Equation of the parabola:

$$y = x^2$$

Substitute y to have D^2 in terms of x:

$$\begin{aligned} D^2 &= (x - 6)^2 + (x^2 - 3)^2 \\ &= x^2 - 12x + 36 + x^4 - 6x^2 + 9 \\ &= x^4 - 5x^2 - 12x + 45 \end{aligned}$$

Take the derivative:

$$\frac{dD^2}{dx} = 4x^3 - 10x - 12$$

Set the derivative equal to zero and solve:

$$2x^3 - 5x - 6 = 0$$

There is exactly one positive solution (only one sign change). Testing the integer solutions (1, 2, 3, and 6), we find the solution $x = 2$.

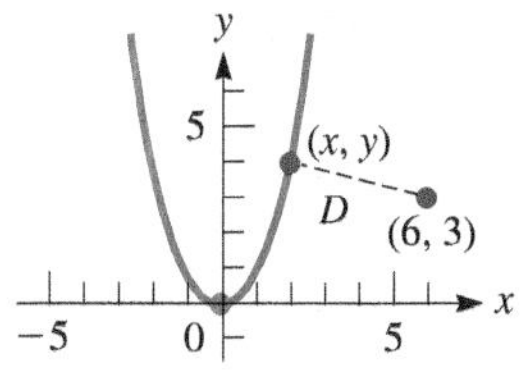

Fig. 24.48

Check that it is a minimum: $\frac{d^2D^2}{dx^2} = 6x^2 - 5 > 0$ at $x = 2$, so D^2 has a minimum at $x = 2$.

At $x = 2$, $y = 4$. The closest point on the parabola is $(2, 4)$. See Fig. 24.48.

EXAMPLE 6 Maximizing revenue

A company determines that it can sell 1000 units of a product per month if the price is \$5 for each unit. It also estimates that for each 1 cent reduction in the unit price, 10 more units can be sold. Under these conditions, what is the maximum possible income and what price per unit gives this income?

Introduce notation: Let I be the income for selling $1000 + x$ units at price p, where x is the number of units over 1000 sold.

Write I as a function of x and p. Use the relationship between x and p to write I as a function of one variable: Income: $I = (1000 + x)p$
Price reduced by \$0.01 from \$5 for each block of 10 units over 1000 units sold:

$$p = 5 - 0.01\left(\tfrac{x}{10}\right) = 5 - 0.001x$$

Substitute p to have I in terms of x:

$$I = (1000 + x)(5 - 0.001x)$$
$$= 5000 + 4x - 0.001x^2$$

Take the derivative:

$$\frac{dI}{dx} = 4 - 0.002x$$

Set the derivative equal to zero and solve: $4 - 0.002x = 0,\quad x = 2000$ units over 1000

Check that it is a maximum: The derivative is positive if $x < 2000$, and the derivative is negative if $x > 2000$, so there is a maximum at $x = 2000$.

State the answer required: Substituting $x = 2000$ into I and p, we find that maximum is \$9000 when 3000 units are sold at \$3 per unit.

Practice Exercise

2. In Example 6, change 1 cent to 2 cents and solve the resulting problem.

EXAMPLE 7 Minimizing surface area—oil-storage tank

■ See the chapter introduction.

Find the dimensions of a 700-kL cylindrical oil-storage tank that can be made with the least cost of sheet metal, assuming there is no wasted sheet metal.

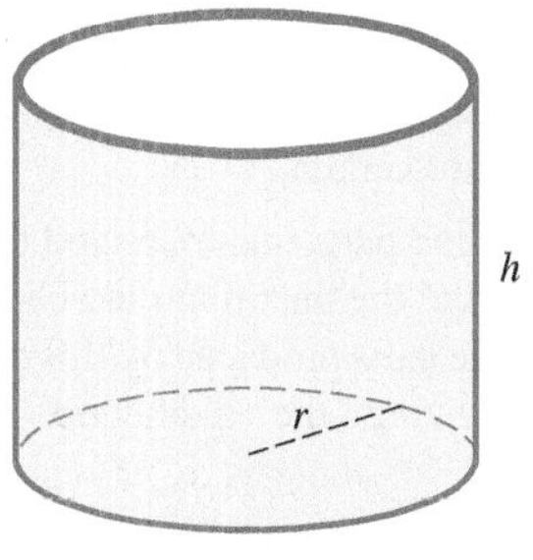

Fig. 24.49

Introduce notation: Let A be the surface area and V be the volume of a tank of radius r and height h (see Fig. 24.49).

Write A as a function of r and h. Use the relationship between r, h and the known volume V to write A as a function of one variable:

Area: $A = 2\pi r^2 + 2\pi rh$

Volume: $V = 700 = \pi r^2 h$, so $h = \frac{700}{\pi r^2}$

Substitute h to have A in terms of r:

$$A = 2\pi r^2 + 2\pi r\left(\frac{700}{\pi r^2}\right) = 2\pi r^2 + \frac{1400}{r}$$

Take the derivative: $\frac{dA}{dr} = 4\pi r - \frac{1400}{r^2}$

Set the derivative equal to zero and solve: $4\pi r - \frac{1400}{r^2} = 0, \quad 4\pi r^3 - 1400 = 0 \ (r \neq 0),$

$$r = \sqrt[3]{\frac{1400}{4\pi}} = 4.81 \text{ m}, \quad h = \frac{700}{\pi r^2} = 9.62 \text{ m}$$

Check that it is a minimum: The derivative is negative if $r < 4.81$, and the derivative is positive if $r > 4.81$, so there is a minimum at $r = 4.81$ m.

EXAMPLE 8 Minimizing a function—illuminance of light

The illuminance of a light source (in lux) at any point equals the strength of the source (in lumens) divided by the square of the distance (in metres) from the source. Two sources, of strengths 8 lumens and 1 lumen, are 100 m apart. Determine at what point between them the illuminance is the least, assuming that the illuminance at any point is the sum of the illuminances of the two sources.

Introduce notation: Let I be the sum of illuminances at x m from the source of strength 8 and y m from the source of strength 1.

Write I as a function of x and y. Use the relationship between x and y to write I as a function of one variable:

Illuminance: $I = \frac{8}{x^2} + \frac{1}{y^2}$

Distance: $x + y = 100$, so $y = 100 - x$

Substitute y to have I in terms of x:

$$I = \frac{8}{x^2} + \frac{1}{(100 - x)^2}$$

Take the derivative: $\frac{dI}{dx} = -\frac{16}{x^3} + \frac{2}{(100 - x)^3} = \frac{-16(100 - x)^3 + 2x^3}{x^3(100 - x)^3}$

Set the derivative equal to zero and solve: $2x^3 - 16(100 - x)^3 = 0, \quad x^3 = 8(100 - x)^3,$

$$x = 2(100 - x), \quad x = 66.7 \text{ m}$$

Check that it is a minimum: The derivative is negative if $x < 66.7$, and the derivative is positive if $x > 66.7$, so there is a minimum 66.7 m from the 8-lumen source.

EXERCISES 24.7

In Exercises 1–52, solve the given maximum and minimum problems.

1. In Example 3, change 1600 m to 2400 m and then proceed.
2. In Example 7, change 700 kL to 800 kL and then proceed.
3. The height (in m) of a flare shot upward from the ground is given by $s = 34.3t - 4.9t^2$, where t is the time (in s). What is the greatest height to which the flare goes?
4. A small oil refinery estimates that its daily profit P (in dollars) from refining x barrels of oil is $P = 8x - 0.02x^2$. How many barrels should be refined for maximum daily profit, and what is the maximum profit?
5. The power output P of a battery (in W) of a certain 12-V battery is given by $P = 12I - 5.0I^2$, where I is the current (in A). Find the current for which the power is a maximum.
6. In a random sample of 20 machine parts manufactured by a certain company, 3 are defective. Find the maximum likelihood estimate of the proportion of machine parts produced by this company that are defective by maximizing the likelihood function $L(p) = 1140p^3(1 - p)^{17}$.
7. A rectangular enclosure will be constructed using 100 m of fencing. Find the dimensions that will yield a maximum area.

8. If an airplane is moving at velocity v, the drag D on the plane is $D = av^2 + b/v^2$, where a and b are positive constants. Find the value(s) of v for which the drag is the least.

9. A company projects that its total savings S (in dollars) by converting to a solar-heating system with a solar-collector of area A (in m^2) will be $S = 360A - 0.10A^3$. Find the area that should give the maximum savings and find the amount of the maximum savings.

10. The altitude h (in m) of a jet that goes into a dive and then again turns upward is given by $h = 16t^3 - 240t^2 + 10\,000$, where t is the time (in s) after the start of the dive. What is the altitude of the jet when it turns up out of the dive?

11. If a resistance R and inductance L are in parallel with a capacitance C, the impedance is $Z = \sqrt{\dfrac{R^2 + \omega^2 L^2}{\omega^2 C^2 R^2 + (\omega^2 LC - 1)^2}}$, where ω is the angular frequency of the circuit impedance. For what value(s) of C is Z a maximum, if R and L are constant?

12. Test results show that, when coughing, the velocity v of the wind in a person's windpipe is $v = kr^2(a - r)$, where a is the radius of the windpipe when not coughing and k is a constant. Find r for the maximum value of v.

13. An alpha particle moves through a magnetic field along the parabolic path $y = x^2 - 4$. Determine the closest that the particle comes to the origin.

14. The electric potential V on the line $3x + 2y = 6$ is given by $V = 3x^2 + 2y^2$. At what point on this line is the potential a minimum?

15. In deep water, the velocity of a wave is $v = k\sqrt{\dfrac{l}{a} + \dfrac{a}{l}}$, where a and k are constants, and l is the length of the wave. What is the length of the wave that results in the minimum velocity?

16. The sum of the length l and width w of a rectangular table top is to be 280 cm. Determine l and w if the area of the table top is to be a maximum.

17. A rectangular hole is to be cut in a wall for a vent. If the perimeter of the hole is 48 cm and the length of the diagonal is a minimum, what are the dimensions of the hole?

18. When two electric resistors R_1 and R_2 are in series, their total resistance (the sum) is 32 Ω. If the same resistors are in parallel, their total resistance (the reciprocal of which equals the sum of the reciprocals of the individual resistances) is the maximum possible for two such resistors. What is the resistance of each?

19. A rectangular microprocessor chip is designed to have an area of 36 mm^2. What must its dimensions be if its perimeter is to be a minimum?

20. The rectangular animal display area in a zoo is enclosed by chain-link fencing and divided into two areas by internal fencing parallel to one of the sides. What dimensions will give the maximum area for the display if a total of 120 m of fencing are used?

21. What are the dimensions of the largest rectangular piece that can be cut from a semicircular metal sheet of diameter 14.0 cm?

22. A rectangular storage area is to be constructed along the side of a tall building. A security fence is required along the remaining three sides of the area. What is the maximum area that can be enclosed with 800 m of fencing?

23. Ship A is travelling due east at 18.0 km/h as it passes a point 40.0 km due south of ship B, which is travelling due south at 16.0 km/h. How much later are the ships nearest each other?

24. An architect is designing a rectangular building in which the front wall costs twice as much per linear metre as the other three walls. The building is to cover 1350 m^2. What dimensions must it have such that the cost of the walls is a minimum?

25. A computer is programmed to display a slowly changing right triangle with its hypotenuse always equal to 14.0 cm. What are the legs of the triangle when it has its maximum area?

26. Canada Post regulations require that the sum of the three dimensions of a rectangular package not exceed 3 m. What are the dimensions of the largest rectangular box with square ends that can be mailed?

27. Referring to Exercise 26, a cylinder-shaped package is considered to have width and height equal to its diameter. What are the dimensions of the largest cylindrical package that may be sent through the mail?

28. An architect designs a window in the shape of a rectangle surmounted by an equilateral triangle. If the perimeter of the window is to be 6.00 m, what dimensions of the rectangle give the window the largest area?

29. The printed area of a rectangular poster is 384 cm^2, with margins of 4.00 cm on each side and margins of 6.00 cm at the top and bottom. Find the dimensions of the poster with the smallest area.

30. A conical funnel, with a very small opening, is being designed such that the slant height of the cone is 4.00 cm. What is the maximum volume of liquid that the funnel will be able to hold?

31. A culvert designed with a semicircular cross-section of diameter 2.40 m is redesigned to have an isosceles trapezoidal cross-section by inscribing the trapezoid in the semicircle. See Fig. 24.50. What is the length of the bottom base b of the trapezoid if its area is to be maximum?

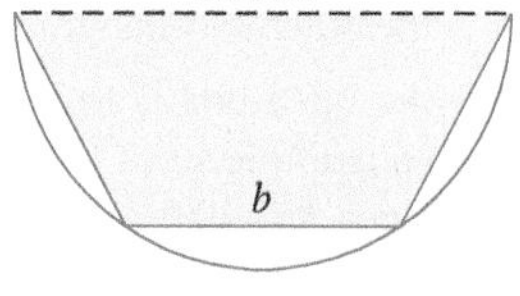

Fig. 24.50

32. A 36-cm-wide sheet of metal is bent into a rectangular trough as shown in Fig. 24.51. What dimensions give the maximum water flow?

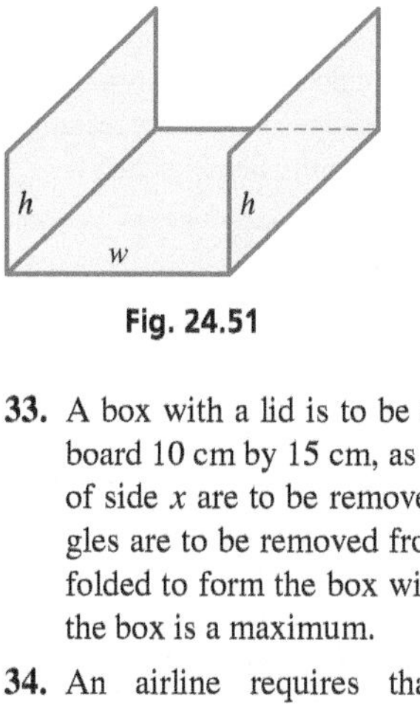

Fig. 24.51

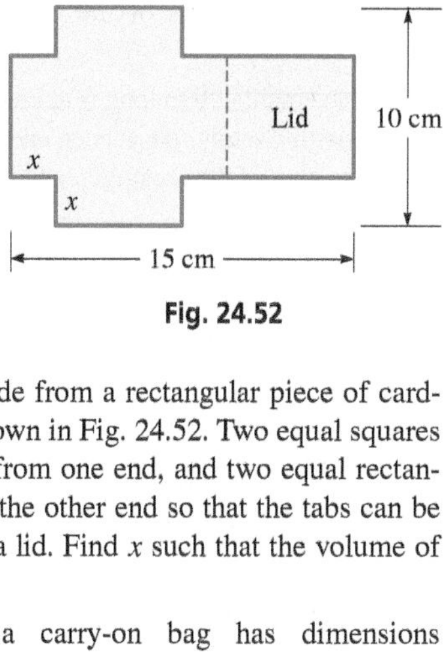

Fig. 24.52

33. A box with a lid is to be made from a rectangular piece of cardboard 10 cm by 15 cm, as shown in Fig. 24.52. Two equal squares of side x are to be removed from one end, and two equal rectangles are to be removed from the other end so that the tabs can be folded to form the box with a lid. Find x such that the volume of the box is a maximum.

34. An airline requires that a carry-on bag has dimensions (length + width + height) that do not exceed 114 cm. If a carry-on has a length 2.4 times the width, find the dimensions (to the nearest cm) of this type of carry-on that has the greatest volume.

35. What is the maximum slope of the curve $y = 6x^2 - x^3$?

36. What is the minimum slope of the curve $y = x^5 - 10x^2$?

37. The deflection y of a beam of length L at a horizontal distance x from one end is given by $y = k(2x^4 - 5Lx^3 + 3L^2x^2)$, where k is a constant. For what value of x does the maximum deflection occur?

38. The electric power P (in W) produced by a certain battery is given by $P = \dfrac{144r}{(r + 0.6)^2}$, where r is the resistance in the circuit. For what value of r is the power a maximum?

39. For raising a load, the efficiency E (in %) of a screw with square threads is $E = \dfrac{100T(1 - fT)}{T + f}$, where f is the coefficient of friction and T is the tangent of the pitch angle of the screw. If $f = 0.25$, what acute angle makes E the greatest?

40. Computer simulation shows that the drag F (in N) on a certain airplane is $F = 0.005\,00v^2 + 3.00 \times 10^8/v^2$, where v is the velocity (in km/h) of the plane. For what velocity is the drag the least?

41. Factories A and B are 8.0 km apart, with factory B emitting eight times the pollutants into the air as factory A. If the number n of particles of pollutants is inversely proportional to the square of the distance from a factory, at what point between A and B is the pollution the least?

42. The potential energy E of an electric charge q due to another charge q_1 at a distance of r_1 is proportional to q_1 and inversely proportional to r_1. If charge q is placed directly between two charges of 2.00 nC and 1.00 nC that are separated by 10.0 mm, find the point at which the total potential energy (the sum due to the other two charges) of q is a minimum.

43. An open box is to be made from a square piece of cardboard whose sides are 20.0 cm long, by cutting equal squares from the corners and bending up the sides. Determine the side of the square that is to be cut out so that the volume of the box may be a maximum. See Fig. 24.53.

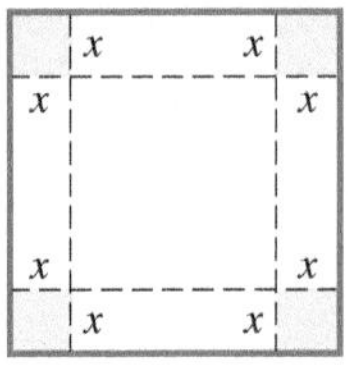

Fig. 24.53

44. A cone-shaped paper cup is to hold 100 cm^3 of water. Find the height and radius of the cup that can be made from the least amount of paper.

45. A race track 400 m long is to be built around an area that is a rectangle with a semicircle at each end. Find the open side of the rectangle if the area of the rectangle is to be a maximum. See Fig. 24.54.

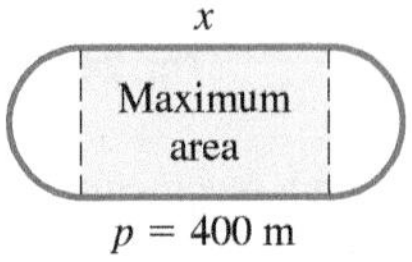

Fig. 24.54

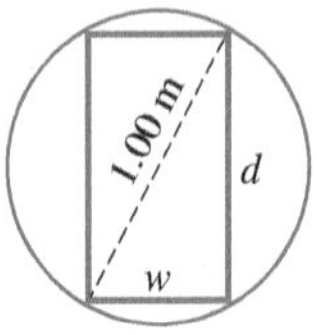

Fig. 24.55

46. A beam of rectangular cross-section is to be cut from a log 1.00 m in diameter. The stiffness of the beam varies directly as the width and the cube of the depth. What dimensions will give the beam maximum stiffness? See Fig. 24.55.

47. A company finds that there is a net profit of \$10 for each of the first 1000 units produced each week. For each unit over 1000 produced, there is 4 cents less profit per unit. How many units should be produced each week to net the greatest profit?

48. On a computer simulation, a target plane is projected to be at (1.20, 6.05) (distances in km), and a rocket is fired along the path $y = 8.00 - 2.00x^2$. How far from the target does the rocket pass from the target plane's projected position?

49. An oil pipeline is to be built from a refinery to a tanker loading area. The loading area is 10.0 km downstream from the refinery and on the opposite side of a river 2.5 km wide. The pipeline is to run along the river and then cross to the loading area. If the pipeline costs \$50 000 per kilometre alongside the river and \$80 000 per kilometre across the river, find the point P (see Fig. 24.56) at which the pipeline should be turned to cross the river if construction costs are to be a minimum.

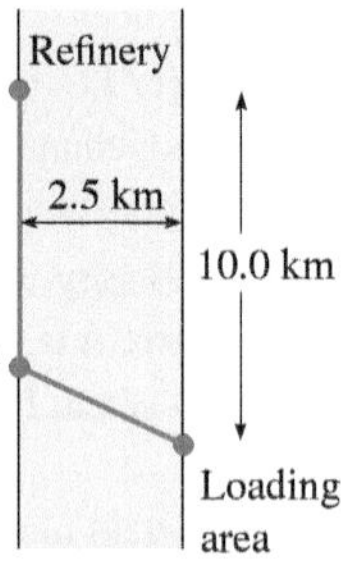

Fig. 24.56

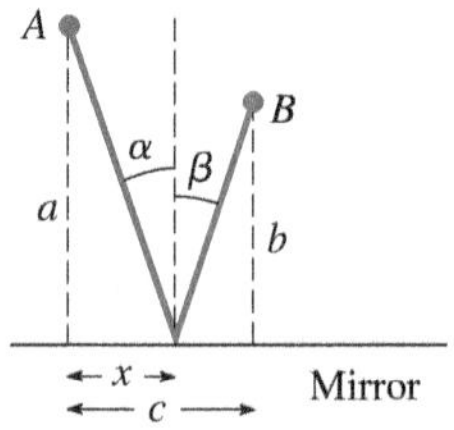

Fig. 24.57

50. A light ray follows a path of least time. If a ray starts at point A (see Fig. 24.57) and is reflected off a plane mirror to point B, show that the angle of incidence α equals the angle of reflection β. (*Hint*: Set up the expression in terms of x, which will lead to $\sin \alpha = \sin \beta$.)

51. A rectangular building covering 7000 m^2 is to be built on a rectangular lot as shown in Fig. 24.58. If the building is to be 10.0 m from the lot boundary on each side and 20.0 m from the boundary in front and back, find the dimensions of the building if the area of the lot is a minimum.

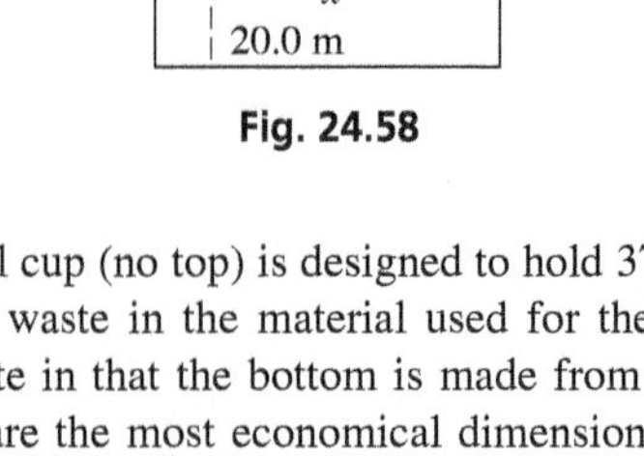

Fig. 24.58

52. A cylindrical cup (no top) is designed to hold 375 cm^3 (375 mL). There is no waste in the material used for the sides. However, there is waste in that the bottom is made from a square $2r$ on a side. What are the most economical dimensions for a cup made under these conditions?

Answers to Practice Exercises

1. 6.93 cm by 9.80 cm
2. \$6125 at \$3.50 per unit

24.8 Differentials and Linear Approximations

■ The symbol dy/dx for the derivative was first used by Leibniz.

To this point, we have used the dy/dx notation for the derivative of y with respect to x, but we have not considered it to be a ratio. In this section, we define the quantities dy and dx, called *differentials*, such that their ratio is equal to the derivative. Then we show that differentials have applications in approximating errors in measurement and in approximating values of functions. Also, in the next chapter, we use the differential notation in the development of integration, which is the inverse process of differentiation.

DIFFERENTIALS

We define the **differential** *of a function* $y = f(x)$ as

$$dy = f'(x)\,dx \qquad \textbf{(24.8)}$$

In Eq. (24.8), the quantity *dy is the differential of y, and dx is the differential of x.* In this way, we can interpret the derivative as the ratio of the differential of y to the differential of x.

EXAMPLE 1 Differential of a polynomial

Find the differential of $y = 3x^5 - x$.

Since $f(x) = 3x^5 - x$, we find $f'(x) = 15x^4 - 1$. This means that

$$dy = (15x^4 - 1)dx$$

($15x^4 - 1$ is $f'(x)$; dx is the differential of x)

EXAMPLE 2 Differential of a rational expression

Find the differential of $s = \dfrac{4t}{t^2 + 4}$.

$$ds = \frac{(t^2+4)(4) - (4t)(2t)}{(t^2+4)^2}dt \qquad \text{using derivative quotient rule}$$

$$= \frac{4t^2 + 16 - 8t^2}{(t^2+4)^2}dt = \frac{-4t^2+16}{(t^2+4)^2}dt$$

$$= \frac{-4(t^2-4)}{(t^2+4)^2}dt$$

don't forget the dt

Practice Exercise

1. Find the differential of $y = (2x - 1)^4$.

To understand more about differentials and their applications, we use the delta notation introduced in Chapter 23. Recall that we defined Δx as the difference between two values of x. (Here we call Δx the **increment** in x.) By choosing $\Delta x = dx$, if dx is small, then the differential in y, dy, closely approximates the increment in y, Δy.

This is the basis of the applications of the differential, and to understand this better, let us look at Fig. 24.59. We see that points $P(x, y)$ and $Q(x + \Delta x, y + \Delta y)$ lie on the graph of $f(x)$ and that the increments of x and y between the points are Δx and Δy. At point P, $f'(x) = dy/dx$, which means that the slope of a tangent line at P is indicated by dy/dx. With $dx = \Delta x$, we see that as dx becomes smaller, dy more nearly approximates Δy, which is the actual difference in the y-values. Therefore, *for small values of Δx, dy can be used to approximate Δy.*

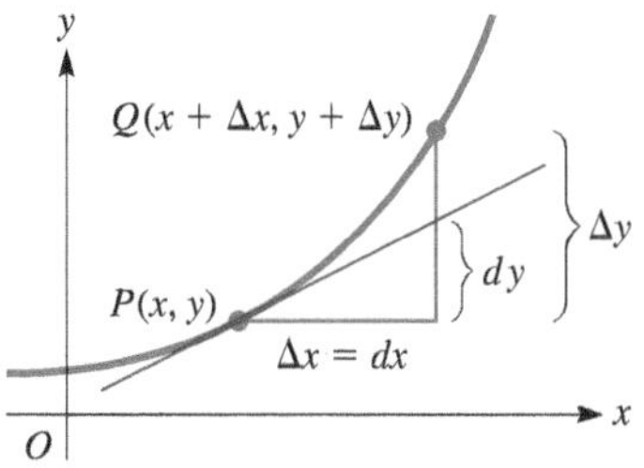

Fig. 24.59

EXAMPLE 3 Calculating the increment and the differential

Calculate Δy and dy for $y = x^3 - 2x$ for $x = 3$ and $\Delta x = 0.1$.

First, we find Δy by calculating $f(3.1) - f(3)$. Therefore,

$$\Delta y = f(3.1) - f(3) = [3.1^3 - 2(3.1)] - [3^3 - 2(3)] = 2.591$$

The differential of y is

$$dy = (3x^2 - 2)dx$$

Since $dx = \Delta x$, we have

$$dy = [3(9) - 2](0.1) = 2.5$$

Thus, $\Delta y = 2.591$ and $dy = 2.5$. In this case, dy is very nearly equal to Δy.

ESTIMATING ERRORS IN MEASUREMENT

The fact that dy can be used to approximate Δy is useful in finding the error in a result from a measurement, if the data are in error, or the equivalent problem of finding the change in the result if a change is made in the data. Even though such changes can be found by using a calculator, the differential can be used to set up a general expression for the change of a particular function.

EXAMPLE 4 Calculating the error in measurement

The edge of a cube of gold was measured to be 3.850 cm. From this value, the volume was found. Later, it was discovered that the value of the edge was 0.020 cm too small. By approximately how much was the volume in error?

The volume V of a cube, in terms of an edge s, is $V = s^3$. Since we wish to find the change in V for a given change in s, we want the value of dV for $s = 3.850$ cm and $ds = 0.020$ cm.

First, finding the general expression for dV, we have

$$dV = 3s^2ds$$

Now, evaluating this expression for the given values, we have

$$dV = 3(3.850)^2(0.020) = 0.89 \text{ cm}^3$$

In this case, the volume was in error by about 0.89 cm^3. As long as ds is small compared with s, we can calculate an error or change in the volume of a cube by calculating the value of $3s^2ds$.

Practice Exercise

2. In Example 4, approximate the error in the total surface area A of the cube.

Often, when considering the error of a given value or result, *the actual numerical value of the error, the* **absolute error**, is not as important as its size in relation to the size of the quantity itself. *The ratio of the absolute error to the size of the quantity itself is known as the* **relative error**, which is commonly expressed as a percent.

EXAMPLE 5 Absolute error and relative error

Referring to Example 4, we see that the absolute error in the edge was 0.020 cm. The relative error in the edge was

$$\frac{ds}{s} = \frac{0.020}{3.85} = 0.0052 = 0.52\%$$

The absolute error in the volume was 0.89 cm^3, and the original value of the volume was $3.850^3 = 57.07$ cm^3. This means the relative error in the volume was

$$\frac{dV}{V} = \frac{0.89}{57.07} = 0.016 = 1.6\%$$

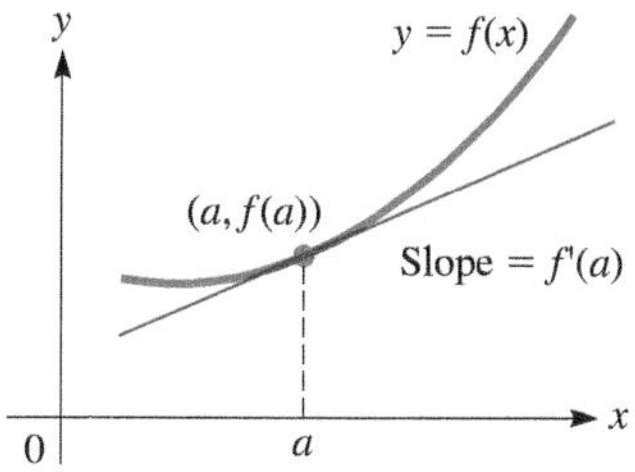

Fig. 24.60

LINEAR APPROXIMATIONS

Continuing our discussion related to Fig. 24.59, on the curve of the function $f(x)$ at the point $(a, f(a))$, the slope of a tangent line is $f'(a)$. See Fig. 24.60. For a point (x, y) on the tangent line, by using the point–slope form for the equation of a straight line, $y - y_1 = m(x - x_1)$, we have

$$f(x) = f(a) + f'(a)(x - a)$$

For the points on $y = f(x)$ near $x = a$, we can use the tangent line to approximate the function, as shown in Fig. 24.60. Therefore, the approximation $f(x) \approx L(x)$ *is the* **linear approximation** *of* $f(x)$ *near* $x = a$, *where the function*

$$L(x) = f(a) + f'(a)(x - a) \tag{24.9}$$

is called the **linearization** *of* $f(x)$ *at* $x = a$.

EXAMPLE 6 Linearizing a function

Find the linearization of the function $f(x) = \sqrt{2x + 1}$ at $x = 4$. Use it to approximate $\sqrt{9.06}$.

The solution is as follows:

$$f(x) = \sqrt{2x + 1}$$

$$f'(x) = \tfrac{1}{2}(2x + 1)^{-1/2}(2) = \frac{1}{\sqrt{2x + 1}}$$ find the derivative

$$f(4) = \sqrt{2(4) + 1} = 3 \qquad f'(4) = \frac{1}{\sqrt{2(4) + 1}} = \frac{1}{3}$$ evaluate $f(4)$ and $f'(4)$

$$L(x) = 3 + \tfrac{1}{3}(x - 4) = \frac{x + 5}{3}$$ use Eq. (24.9)

$$\sqrt{2x + 1} \approx \frac{x + 5}{3}$$ $2x + 1 = 9.06$, $x = 4.03$

$$\sqrt{9.06} \approx \frac{4.03 + 5}{3} = \frac{9.03}{3} = 3.01$$ by calculator, $\sqrt{9.06} = 3.009983389$

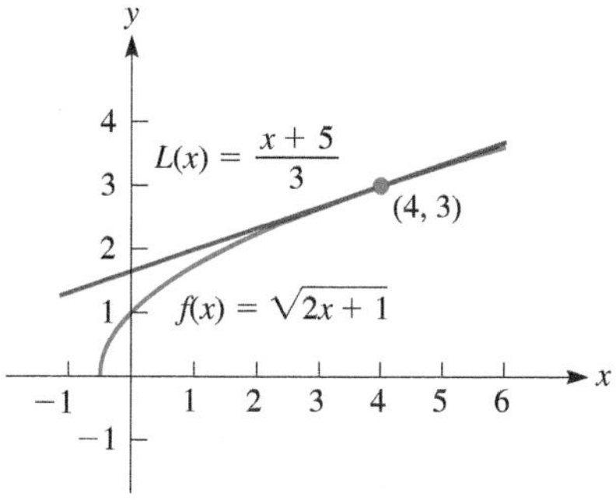

Fig. 24.61

In Fig. 24.61, the function $f(x) = \sqrt{2x + 1}$ and the tangent line are shown. We see that the tangent line gives a good approximation of the function when x is near 4.

EXERCISES 24.8

In Exercises 1–4, make the given changes in the indicated examples of this section and then solve the resulting problems.

1. In Example 2, change the t^2 to t^3.
2. In Example 3, change the $-$ before $2x$ to $+$.
3. In Example 5, change 0.020 to 0.025.
4. In Example 6, change $x = 4$ to $x = 12$; approximate $\sqrt{25.06}$.

In Exercises 5–16, find the differential of each of the given functions.

5. $y = x^5 + 7x$
6. $y = 3x^2 + 9x$
7. $V = \dfrac{2}{r^5} + 3\pi^2$
8. $y = 2\sqrt{x} - \dfrac{1}{8x}$
9. $s = 2(3t^2 - 5)^6$
10. $y = 5(4 + 3x)^{1/3}$
11. $y = \dfrac{12}{3x^2 + 1}$
12. $R = \sqrt{\dfrac{u}{1 + 2u}}$
13. $y = x^2(1 - x)^3$
14. $y = 6x\sqrt{1 - 4x}$
15. $y = \dfrac{x}{5x + 2}$
16. $y = \dfrac{3x + 1}{\sqrt{2x - 1}}$

In Exercises 17–20, find the values of Δy *and* dy *for the given values of* x *and* dx.

17. $y = 7x^2 + 4x$, $x = 4$, $\Delta x = 0.2$
18. $y = (x^2 + 2x)^3$, $x = 7$, $\Delta x = 0.01$
19. $y = x\sqrt{1 + 4x}$, $x = 12$, $\Delta x = 0.03$
20. $y = \dfrac{x}{\sqrt{6x - 1}}$, $x = 3.5$, $\Delta x = 0.025$

In Exercises 21–24, find the linearization $L(x)$ of the given functions for the given values of a.

21. $f(x) = x^2 + 2x, a = 0$

22. $f(x) = 2\sqrt[3]{x}, a = 8$

23. $f(x) = \dfrac{1}{2x + 1}, a = -1$

24. $g(x) = x\sqrt{2x + 8}, a = -2$

In Exercises 25–38, solve the given problems by finding the appropriate differential. Round answers to three significant digits.

25. If a spacecraft circles the earth at an altitude of 250 km, how much farther does it travel in one orbit than an airplane that circles the earth at a low altitude? The radius of the earth is 6370 km.

26. Approximate the amount of paint needed to apply one coat of paint 0.50 mm thick on a hemispherical dome 55 m in diameter.

27. The radius of a circular utility access hole cover is measured to be 40.6 ± 0.05 cm (this means the possible error in the radius is 0.05 cm). Estimate the possible relative error in the area of the top of the cover.

28. The side of a square microprocessor chip is measured as 0.950 cm, and later it is measured as 0.952 cm. What is the difference in the calculations of the area due to the difference in the measurements of the side? See Fig. 24.62.

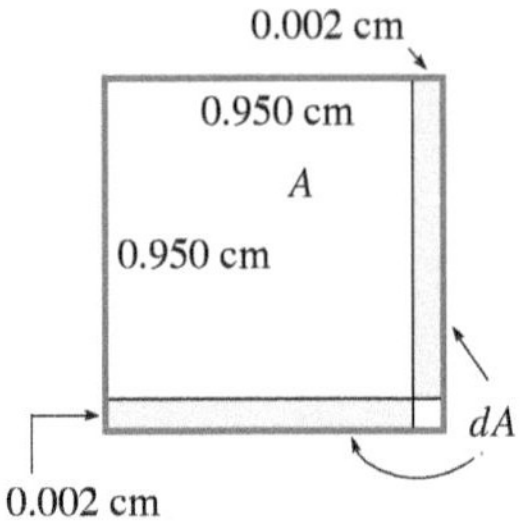

Fig. 24.62

29. The wavelength λ of light is inversely proportional to its frequency f. If $\lambda = 685$ nm for $f = 4.38 \times 10^{14}$ Hz, find the change in λ if f increases by 0.20×10^{14} Hz. (These values are for red light.)

30. The velocity of an object rolling down a certain inclined plane is given by $v = \sqrt{40 + 4.9h}$, where h is the distance (in m) travelled along the plane by the object. What is the increase in velocity (in m/s) of an object in moving from 20.0 m to 20.5 m along the plane? What is the relative change in the velocity?

31. If the diameter equals the height, what is the volume of the plastic that forms a closed cylindrical container for which the radius is 18.0 cm and the thickness is 2.00 mm?

32. The volume V of blood flowing through an artery is proportional to the fourth power of the radius r of the artery. Find how much a 5% increase in r affects V.

33. The radius r of a holograph is directly proportional to the square root of the wavelength λ of the light used. Show that $dr/r = \frac{1}{2}d\lambda/\lambda$.

34. The gravitational force F of the earth on an object is inversely proportional to the square of the distance r of the object from the centre of the earth. Show that $dF/F = -2dr/r$.

35. Show that an error of 2% in the measurement of the radius of a DVD results in an error of approximately 4% in the calculation of the area.

36. Show that an error of 1% in the measurement of the radius of a ball bearing results in an error of approximately 3% in the calculation of the volume.

37. Calculate $\sqrt{9.05}$, using differentials.

38. Explain how to evaluate 2.03^4.

In Exercises 39–44, solve the given linearization problems.

39. Show that the linearization of $f(x) = (1 + x)^k$ at $x = 0$ is $L(x) = 1 + kx$.

40. Use the result shown in Exercise 39 to approximate the value of $f(x) = \dfrac{1}{\sqrt{1 + x}}$ near zero.

41. Linearize $f(x) = \sqrt{2 - x}$ for $a = 1$ and use it to approximate the value of $\sqrt{1.9}$.

42. Explain how to evaluate $\sqrt[3]{8.03}$, using linearization.

43. The capacitance C (in μF) in an element of an electronic tuner is given by $C = \dfrac{3.6}{\sqrt{1 + 2V}}$, where V is the voltage. Linearize C for $V = 4.0$ V.

44. A 16-Ω resistor is put in parallel with a variable resistor of resistance R. The combined resistance of the two resistors is $R_T = \dfrac{16R}{16 + R}$. Linearize R_T for $R = 4.0\ \Omega$.

Answers to Practice Exercises

1. $dy = 8(2x - 1)^3dx$

2. $dA = 0.92 \text{ cm}^2$

CHAPTER 24 KEY FORMULAS AND EQUATIONS

Newton's method

$$x_{n+1} = x_n - \frac{f(x_n)}{f'(x_n)} \quad (24.1)$$

Curvilinear motion

$$v_x = \frac{dx}{dt} \qquad v_y = \frac{dy}{dt} \quad (24.2)$$

$$v = \sqrt{v_x^2 + v_y^2} \quad (24.3)$$

$$\tan \theta_v = \frac{v_y}{v_x} \quad (24.4)$$

$$a_x = \frac{dv_x}{dt} = \frac{d^2x}{dt} \qquad a_y = \frac{dv_y}{dt} = \frac{d^2y}{dt^2} \tag{24.5}$$

$$a = \sqrt{a_x^2 + a_y^2} \tag{24.6}$$

$$\tan \theta_a = \frac{a_y}{a_x} \tag{24.7}$$

Differential

$$dy = f'(x)dx \tag{24.8}$$

Linearization

$$L(x) = f(a) + f'(a)(x - a) \tag{24.9}$$

CHAPTER 24 REVIEW EXERCISES

In Exercises 1–6, find the equations of the tangent and normal lines.

1. Find the equation of the line tangent to the parabola $y = 3x - x^2$ at the point $(-1, -4)$. Graph the curve and the line.
2. Find the equation of the line tangent to the curve $y = x^2 - \frac{6}{x}$ at the point $(3, 7)$.
3. Find the equation of the line normal to $x^2 - 4y^2 = 9$ at the point $(5, 2)$. Graph the curve and the line.
4. Find the equation of the line normal to $y = \frac{4}{\sqrt{x-2}}$ at the point $(6, 2)$.
5. Find the equation of the line tangent to the curve $y = \sqrt{x^2 + 3}$ that has a slope of $\frac{1}{2}$.
6. Find the equation of the line normal to the curve $y = \frac{1}{2x + 1}$ that has a slope of $\frac{1}{2}$ if $x \geq 0$.

In Exercises 7–12, find the indicated velocities and accelerations. Round answers to three significant digits.

7. Given that the x- and y-coordinates of a moving particle are given as a function of time t by the parametric equations $x = \sqrt{t} + t$, $y = \frac{1}{12}t^3$, find the magnitude and direction of the velocity when $t = 4$.
8. If the x- and y-coordinates of a moving object as functions of time t are given by $x = 0.1t^2 + 1$, $y = \sqrt{4t + 1}$, find the magnitude and direction of the velocity when $t = 6$.
9. An object moves along the curve $y = 0.5x^2 + x$ such that $v_x = 0.5\sqrt{x}$. Find v_y at $(2, 4)$.
10. A particle moves along the curve of $y = \frac{1}{x + 2}$ with a constant velocity in the x-direction of 4 cm/s. Find v_y at $\left(2, \frac{1}{4}\right)$.
11. Find the magnitude and direction of the acceleration for the particle in Exercise 7.
12. Find the magnitude and direction of the acceleration for the particle in Exercise 10.

In Exercises 13–16, find the indicated roots of the given equations to at least four decimal places by use of Newton's method.

13. $x^3 - 3x^2 - x + 2 = 0$ (between 0 and 1)
14. $2x^3 - 4x^2 - 9 = 0$ (between 2 and 3)
15. $x^2 = \frac{3}{x}$ (the real solution)
16. $\sqrt{x - 2} = \frac{1}{(x - 6)^2}$ (the two real solutions)

In Exercises 17–24, sketch the graphs of the given functions by information obtained from the function as well as information obtained from the derivatives.

17. $y = 4x^2 + 16x$
18. $y = x^3 + 2x^2 + x + 1$
19. $y = 27x - x^3$
20. $y = x(6 - x)^3$
21. $y = x^4 - 32x$
22. $y = x^5 - 20x^2 + 10$
23. $y = \frac{x^2}{\sqrt{x^2 - 1}}$
24. $y = x^3 + \frac{3}{x}$

In Exercises 25–28, find the differential of each of the given functions.

25. $y = 4x^3 + \frac{1}{x}$
26. $y = \frac{1}{(2x - 1)^2}$
27. $y = x\sqrt[3]{1 - 3x}$
28. $s = \sqrt{\frac{2 + t}{2 - t}}$

In Exercises 29 and 30, evaluate $\Delta y - dy$ for the given functions and values.

29. $y = 4x^3 - 12$, $x = 2$, $\Delta x = 0.1$
30. $y = 6x^2 - x$, $x = 3$, $\Delta x = 0.2$

In Exercises 31 and 32, find the linearization of the given functions for the given values of a.

31. $f(x) = \sqrt{x^4 + 3x^2 + 8}$, $a = 2$
32. $f(x) = x^2(x + 1)^4$, $a = -2$

In Exercises 33–40, solve the given problems by finding the appropriate differentials.

33. A weather balloon 3.50 m in radius becomes covered with a uniform layer of ice 1.20 cm thick. What is the volume of the ice?
34. The total power P (in W) transmitted by an AM radio transmitter is $P = 460 + 230m^2$, where m is the modulation index. What is the change in power if m changes from 0.86 to 0.89?
35. A cylindrical silo with a flat top is 12.0 m in diameter and is 12.0 m high. By how much is the volume changed if the radius is increased by 0.100 m and the height is unchanged?

36. A ski slope follows a path that can be represented by $y = 0.010x^2 - 0.86x + 24$. What is the change in the slope of the path when x changes from 26 m to 28 m?

37. The impedance Z of an electric circuit as a function of the resistance R and the reactance X is given by $Z = \sqrt{R^2 + X^2}$. Derive an expression of the relative error in impedance for an error in R and a given value of X.

38. Evaluate $\sqrt{8.94}$.

39. Evaluate 3.02^5.

40. Show that the relative error in the calculation of the volume of a sphere is approximately three times the relative error in the measurement of the radius.

In Exercises 41–81, solve the given problems.

41. The parabolas $y = x^2 + 2$ and $y = 4x - x^2$ are tangent to each other. Find the equation of the line tangent to them at the point of tangency.

42. Find the equation of the line tangent to the curve of $y = x^4 - 8x$ and perpendicular to the line $4y - x + 5 = 0$.

43. The deflection y (in m) of a beam at a horizontal distance x (in m) from one end is given by $y = k(x^4 - 30x^3 + 1000x)$, where k is a constant. Observing the equation and using Newton's method, find the values of x where the deflection is zero, if the beam is 10.000 m long.

44. The edges of a rectangular water tank are 3.00 m, 5.00 m, and 8.00 m. By Newton's method, determine by how much each edge should be increased equally to double the volume of the tank.

45. A parachutist descends (after the parachute opens) in a path that can be described by $x = 8.0t$ and $y = -0.15t^2$, where distances are in metres and time is in seconds. Find the parachutist's velocity upon landing if the landing occurs when $t = 12$ s.

46. One of the curves on an automobile test track can be described by $y = 225/x$, where dimensions are in metres. For a car approaching the curve at a constant velocity of 120 km/h, find the x- and y-components of the velocity at $(10.0, 22.5)$.

47. A person walks 250 m north and turns west. Walking at the rate of 1.0 m/s, at what rate is the distance between the person and the starting point increasing 1.0 min after turning west?

48. A glass prism for refracting light has equilateral triangular ends and vertical cross-sections, and a volume of 45 cm^3. Find the edge of one of the ends such that the total surface area is a minimum.

49. In Fig. 24.63, the tension T supports the 40.0-N weight. The relation between the tension T and the deflection d is $d = \dfrac{1000}{\sqrt{T^2 - 400}}$.
If the tension is increasing at 2.00 N/s when $T = 28.0$ N, how fast is the deflection (in cm) changing?

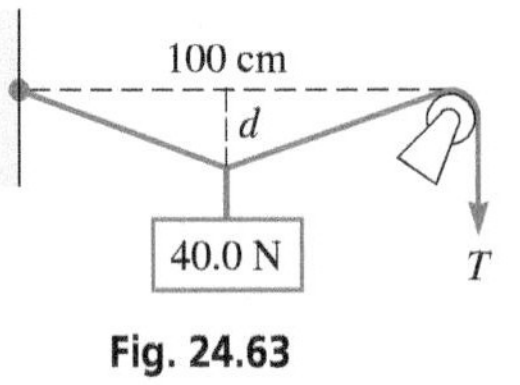

Fig. 24.63

50. The impedance Z (in Ω) in a particular electric circuit is given by $Z = \sqrt{48 + R^2}$, where R is the resistance. If R is increasing at a rate of 0.45 Ω/min for $R = 6.5$ Ω, find the rate at which Z is changing.

51. By using the methods of this chapter to graph $y = x^2 - \frac{2}{x}$, graph the solution of the inequality $x^2 > \frac{2}{x}$.

52. Display the graphs of $y_1 = x^2 - \frac{2}{x}$ and $y_2 = x^2$ on the same screen of a graphing utility, and explain why y_2 can be considered to be a nonlinear asymptote of y_1.

53. An analysis of the power output P (in kW/m^3) of a certain turbine showed that it depended on the flow rate r (in m^3/s) of water to the turbine according to the equation
$P = 0.030r^3 - 2.6r^2 + 71r - 200$ $(6.0 \le r \le 30\ \text{m}^3/\text{s})$.
Determine the rate for which P is a maximum.

54. The altitude h (in m) of a certain rocket as a function of the time t (in s) after launching is given by $h = 550t - 4.9t^2$. What is the maximum altitude the rocket attains?

55. Sketch the continuous curve having these characteristics:

$f(0) = 2 \quad f'(x) < 0$ for $x < 0 \quad f''(x) > 0$ for all x
$f'(x) > 0$ for $x > 0$

56. Sketch a continuous curve having these characteristics:

$f(0) = 1 \quad f'(0) = 0 \quad f''(x) < 0$ for $x < 0$
$f'(x) > 0$ for $|x| > 0 \quad f''(x) > 0$ for $x > 0$

57. A horizontal cylindrical oil tank (the length is parallel to the ground) of radius 2.00 m is being emptied. Find how fast the width w of the oil surface is changing when the depth h is 0.500 m and changing at the rate of 0.0500 m/min.

58. The current I (in A) in a circuit with a resistance R (in Ω) and a battery whose voltage is E and whose internal resistance is r (in Ω) is given by $I = E/(R + r)$. If R changes at the rate of 0.250 Ω/min, how fast is the current changing when $R = 6.25$ Ω, if $E = 3.10$ V and $r = 0.230$ Ω?

59. The radius of a circular oil spill is increasing at the rate of 15 m/min. How fast is the area of the spill changing when the radius is 400 m?

60. A special insulation strip is to be sealed completely around three edges of a rectangular solar panel. If 200 cm of the strip are used, what is the maximum area of the panel?

61. A baseball diamond is a square 90.0 ft on a side. See Fig. 24.64. As a player runs from first base toward second base at 18.0 ft/s, at what rate is the player's distance from home plate increasing when the player is 40.0 ft from first base? (1 ft = 0.3048 m)

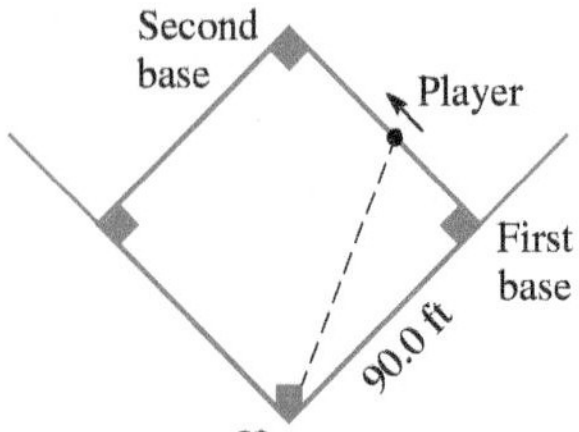

Fig. 24.64

62. A swimming pool with a rectangular surface of 130 m^2 is to have a cement border area that is 4.00 m wide at each end and 2.75 m wide at the sides. Find the surface dimensions of the pool if the total area covered is to be a minimum.

63. A study showed that the percent y of persons surviving burns to x percent of the body is given by $y = \dfrac{300}{0.0005x^2 + 2} - 50$. Linearize this function with $a = 50$ and sketch the graphs of y and $L(x)$.

64. A company estimates that the sales S (in dollars) of a new product will be $S = 5000t/(t+4)^2$, where t is the time (in months) after it is put into production. Sketch the graph of S vs. t.

65. An airplane flying horizontally at an altitude of 2400 m is moving toward a radar installation at 1110 km/h. If the plane is directly over a point on the ground 8.00 km from the radar installation, what is its actual speed? See Fig. 24.65.

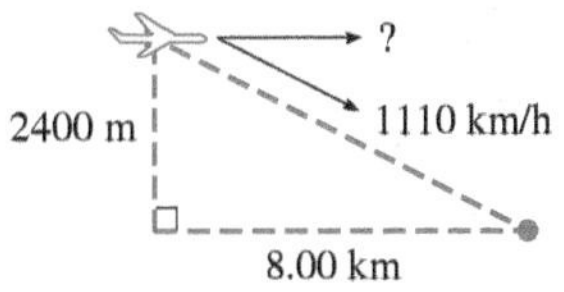

Fig. 24.65

66. The base of a conical machine part is being milled such that the height is decreasing at the rate of 0.050 cm/min. If the part originally had a radius of 1.0 cm and a height of 3.0 cm, how fast is the volume changing when the height is 2.8 cm?

67. The reciprocal of the total capacitance C_T of electrical capacitances in series equals the sum of the reciprocals of the individual capacitances. If the sum of two capacitances is 12 μF, find their values if their total capacitance in series is a maximum.

68. A cable is to be from point A to point B on a wall and then to point C. See Fig. 24.66. Where is B located if the total length of cable is a minimum?

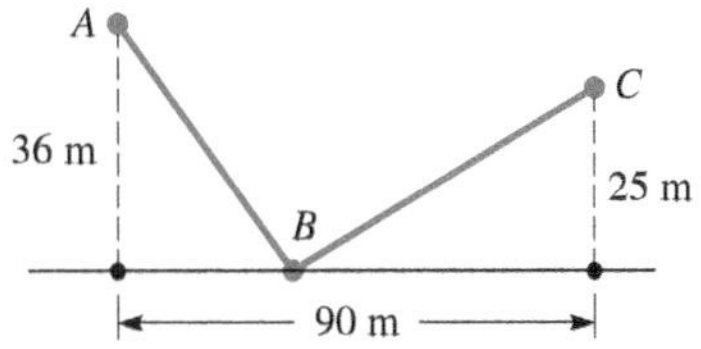

Fig. 24.66

69. A box with a square base and an open top is to be made of 27 dm^2 of cardboard. What is the maximum volume that can be contained within the box?

70. A car is travelling east at 90.0 km/h, and an airplane is travelling south at 450.0 km/h at an elevation of 2.00 km. At one instant the plane is directly above the car. At what rate are they separating 15.0 min later?

71. A person in a boat 4 km from the nearest point P on a straight shoreline wants to go to point A on the shoreline, 5 km from P. If the person can row at 3 km/h and walk at 5 km/h, at what point on the shoreline should the boat land in order that point A can be reached in the least time? See Fig. 24.67.

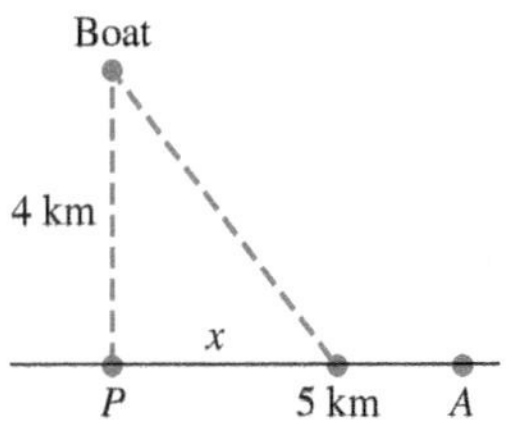

Fig. 24.67

72. A machine part is to be in the shape of a circular sector of radius r and central angle θ. Find r and θ if the area is one unit and the perimeter is a minimum. See Fig. 24.68.

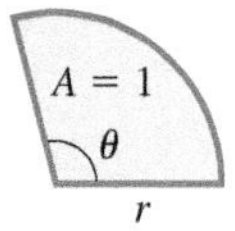

Fig. 24.68

73. An open drawer for small tools is to be made from a rectangular piece of heavy sheet metal 36.0 cm by 30.0 cm, by cutting out equal squares from two corners and bending up the three sides, as shown in Fig. 24.69. Find the side of the square that should be cut out so that the volume of the drawer is a maximum.

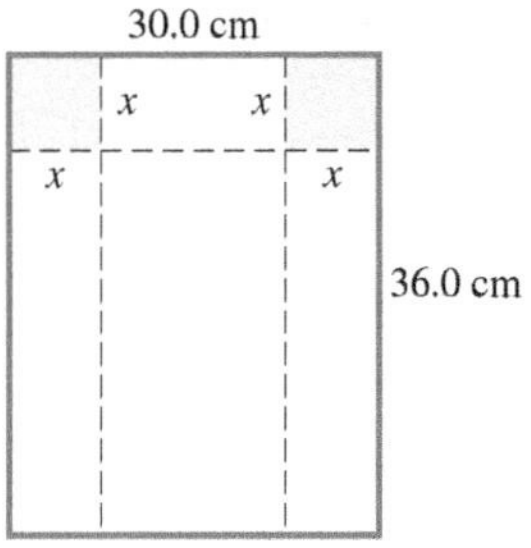

Fig. 24.69

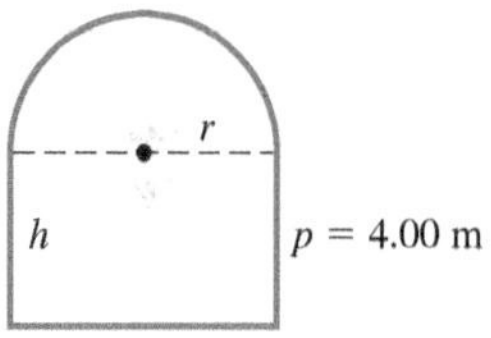

Fig. 24.70

74. A Norman window has the form of a rectangle surmounted by a semicircle. Find the dimensions (radius of circular part and height of rectangular part) of the window that will admit the most light if the perimeter of the window is 4.00 m. See Fig. 24.70.

75. A pile of sand in the shape of a cone has a radius that always equals the altitude. If 3.00 m^3 of sand are poured onto the pile each minute, how fast is the radius increasing when the pile is 2.50 m high?

76. A book is designed such that its (rectangular) pages have 2.5-cm margins at the top and bottom, 1.5-cm margins on the sides, and a total area of 320 cm^2. What are the page dimensions that give the maximum printed area?

77. A specially made cylindrical container is made of stainless steel sides and bottom and a silver top. If silver is 10 times as expensive as stainless steel, what are the most economical dimensions of the container if it is to hold 314 cm^3?

78. An object is moving in a horizontal circle of a radius 2.00 m at the rate of 2.00 rad/s. If the object is on the end of a string and the string breaks after 1.05 s, causing the object to travel along a line tangent to the circle, what is the equation of the path of the object after the string breaks? (Choose the origin of the coordinate system at the centre of the circle and assume the object started on the positive x-axis moving counterclockwise.)

79. A builder is designing a storage building with a total volume of 38.3 m^3, a rectangular base, and a flat roof. The width is to be 0.75 of the length. The cost per square foot is \$6.00 for the floor, \$9.00 for the sides, and \$4.50 for the roof. What dimensions will minimize the cost of the building?

80. City B is 16.0 km east and 12.0 km north of city A. City A is 8.00 km due south of a river that is 1.00 km wide. A road is to be built between A and B that goes straight across the river. See Fig. 24.71. Where should the bridge be located so that the road between A and B is as short as possible?

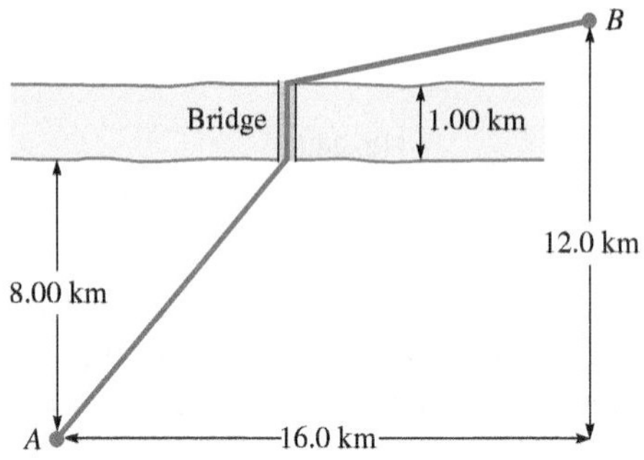

Fig. 24.71

81. A container manufacturer makes various sizes of closed cylindrical plastic containers for shipping liquid products. Write two or three paragraphs explaining how to determine the ratio of the height to radius of the container such that the least amount of plastic is used for each size. Include the reason why it is not necessary to specify the volume of the container in finding this ratio.

CHAPTER 24 PRACTICE TEST

1. Find the equation of the line tangent to the curve $y = x^4 - 3x^2$ at the point $(1, -2)$.

2. For $y = 3x^2 - x$, evaluate (a) Δy, (b) dy, and (c) $\Delta y - dy$ for $x = 3$ and $\Delta x = 0.1$.

3. If the x- and y-coordinates of a moving object as functions of time are given by the parametric equations $x = 3t^2$, $y = 2t^3 - t^2$, find the magnitude and direction of the acceleration when $t = 2$.

4. The electric power (in W) produced by a certain source is given by $P = \dfrac{144r}{(r + 0.6)^2}$, where r is the resistance (in Ω) in the circuit. For what value of r is the power a maximum?

5. Find the root of the equation $x^2 - \sqrt{4x + 1} = 0$ between 1 and 2 to four decimal places by use of Newton's method. Use $x_1 = 1.5$ and find x_3.

6. Linearize the function $y = \sqrt{2x + 4}$ for $a = 6$.

7. Sketch the graph of $y = x^3 + 6x^2$ by finding the values of x for which the function is increasing, decreasing, concave up, and concave down and by finding any maximum points, minimum points, and points of inflection.

8. Sketch the graph of $y = \dfrac{4}{x^2} - x$ by finding the same information as required in Problem 7, as well as intercepts, symmetry, behaviour as x becomes large, vertical asymptotes, and the domain and range.

9. Trash is being compacted into a cubical volume. The edge of the cube is decreasing at the rate of 0.10 m/s. When an edge of the cube is 1.25 m, how fast is the volume changing?

10. A rectangular field is to be fenced and then divided in half by a fence parallel to two opposite sides. If a total of 6000 m of fencing is used, what is the maximum area that can be fenced?

11. Calculate $(7.96)^{1/3}$ using differentials.

25. Integration

Comstock/Stockbyte/Getty Images

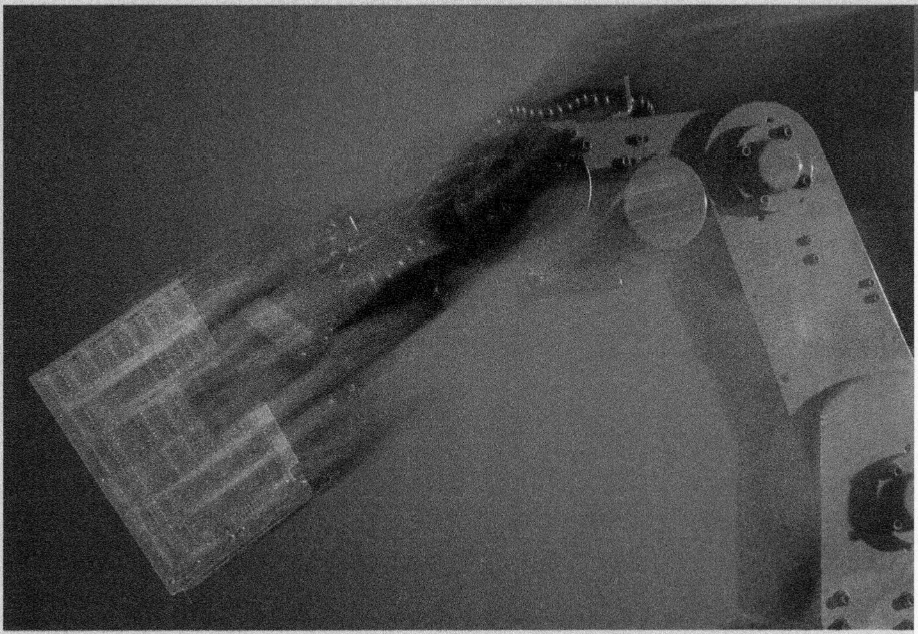

▲ In Section 25.2, we will find the displacement of a robot arm as a function of time by integrating its velocity.

LEARNING OUTCOMES

After completion of this chapter, the student should be able to:

- Understand the concept of the antiderivative of a function
- Integrate basic functions such as constants, polynomials, and powers
- Evaluate the constant of integration
- Approximate the area under a curve using inscribed rectangles
- Find the area under a curve using a definite integral
- Use the trapezoidal rule and Simpson's rule to approximate a definite integral

Finding areas of geometric figures had been studied by the ancient Greeks, and they had discovered how to find the area of any polygon. Also, they were able to calculate the area of a curved figure by inscribing polygons in the figure and then letting the number of sides of the polygon increase. There was little more progress in finding such areas until the 1600s when analytic geometry was developed.

Several mathematicians of the 1600s studied both the area problem and the tangent problem that was discussed in Chapter 23. These included the French mathematician Pierre de Fermat and the English mathematician Isaac Barrow, both of whom developed a few formulas by which tangents and areas could be found. However, Newton and Leibniz found that these two problems were related and determined general methods of finding them. For these reasons, Newton and Leibniz are credited with the creation of calculus.

Finding tangents and finding areas appear to be very different, but they are closely related. As we will show, finding an area uses the inverse process of finding the slope of a tangent line, which we have shown can be interpreted as an instantaneous rate of change.

In physical and technical applications, we often find information that gives us the instantaneous rate of change of a variable. With such information, we have to reverse the process of differentiation in order to find the function when we know its derivative. This procedure is known as *integration*, which is the inverse process of differentiation.

This means that areas are found by integration. There are also many applications of integration in science and technology. A few of these applications will be illustrated in this chapter, and several specific applications will be developed in the next chapter.

25.1 Antiderivatives

We now show how to reverse the process of finding a derivative or a differential. *This reverse process is known as* **antidifferentiation**. In the next section, we formalize the process, and it is only the basic idea that is the topic of this section.

EXAMPLE 1 Find a function, knowing its derivative

Find a function for which the derivative is $8x^3$. That is, find an antiderivative of $8x^3$.

When finding the derivative of a constant times a power of x, we multiply the constant coefficient by the power of x and reduce the power by 1. Therefore, in this case, the power of x must have been 4 before the differentiation was performed.

If we let the derivative function be $f(x) = 8x^3$ and then let its antiderivative function be $F(x) = ax^4$ (by increasing the power of x in $f(x)$ by 1), we can find the value of a by equating the derivative of $F(x)$ to $f(x)$. This gives us

$$F'(x) = 4ax^3 = 8x^3, \qquad 4a = 8, \qquad a = 2$$

This means that $F(x) = 2x^4$.

Practice Exercise

1. Find an antiderivative of $x^3 + 4x$.

EXAMPLE 2 Antiderivative of a polynomial

Find an antiderivative of $v^2 + 2v$.

For v^2, we know that the power of v required in an antiderivative is 3. Also, to make the coefficient correct, we must multiply by $\frac{1}{3}$. $2v$ should be recognized as the derivative of v^2. Therefore, we have as an antiderivative $\frac{1}{3}v^3 + v^2$.

In Examples 1 and 2, *we could add any constant to the antiderivative shown and still have a correct antiderivative.* This is true because the derivative of a constant is zero. This is discussed in the next section, and we will not show any such constants in this section.

When we find an antiderivative of a function, we obtain another function. Thus, *we can define an* **antiderivative** *of the function $f(x)$ to be a function $F(x)$ such that $F'(x) = f(x)$.*

EXAMPLE 3 Antiderivative of a square root and a power with a negative exponent

Find an antiderivative of the function $f(x) = \sqrt{x} - \dfrac{2}{x^3}$.

Since we wish to find an antiderivative of $f(x)$, we know that $f(x)$ is the derivative of the required function.

Considering the term $\sqrt{x}$, we first write it as $x^{1/2}$. To have x to the $\frac{1}{2}$ power in the derivative, we must have x to the $\frac{3}{2}$ power in the antiderivative. Knowing that the derivative of $x^{3/2}$ is $\frac{3}{2}x^{1/2}$, we write $x^{1/2}$ as $\frac{2}{3}\left(\frac{3}{2}x^{1/2}\right)$. Thus, the first term of the antiderivative is $\frac{2}{3}x^{3/2}$.

As for the term $-2/x^3$, we write it as $-2x^{-3}$. This we recognize as the derivative of x^{-2}, or $1/x^2$.

This means that an antiderivative of the function

function ↓ $\qquad f(x) = \sqrt{x} - \dfrac{2}{x^3}$ ← derivative

is the function

antiderivative → $F(x) = \dfrac{2}{3}x^{3/2} + \dfrac{1}{x^2}$ $\qquad$ function ↑

A great many functions of which we must find an antiderivative are not polynomials or simple powers of x. It is these functions that may cause more difficulty in the general process of antidifferentiation. Pay special attention to the following examples, for they illustrate a type of problem that you will find to be very important.

EXAMPLE 4 Antiderivative of a power of a function

Find an antiderivative of the function $f(x) = 3(x^3 - 1)^2(3x^2)$.

Noting that we have a power of $x^3 - 1$ in the derivative, it is reasonable that the antiderivative may include a power of $x^3 - 1$. Since, in the derivative, $x^3 - 1$ is raised to the power 2, the antiderivative would then have $x^3 - 1$ raised to the power 3. Noting that the derivative of $(x^3 - 1)^3$ is $3(x^3 - 1)^2(3x^2)$, the desired antiderivative is

$$F(x) = (x^3 - 1)^3$$

Practice Exercise

2. Find an antiderivative of $5(4x - 3)^4(4)$.

LEARNING TIP

Note that when finding the antiderivative of a power of a function, the factor representing the derivative of the function does not appear in the antiderivative. Nevertheless, **it must be present in the expression containing the power if we are to find a proper antiderivative.**

For instance, in Example 4, the factor $3x^2$ does not appear in the antiderivative. However, since the factor is part of the derivative of $(x^3 - 1)^3$, it must be present in the original function for $(x^3 - 1)^3$ to be a correct antiderivative.

EXAMPLE 5 Antiderivative of a power of a function

Find an antiderivative of the function $f(x) = (2x + 1)^{1/2}$.

Here, we note a power of $2x + 1$ in the derivative, which implies that the antiderivative has a power of $2x + 1$. Since, in finding a derivative, 1 is subtracted from the power of $2x + 1$, we should add 1 in finding the antiderivative. Thus, we should have $(2x + 1)^{3/2}$ as part of the antiderivative. Finding a derivative of $(2x + 1)^{3/2}$, we obtain $\frac{3}{2}(2x + 1)^{1/2}(2) = 3(2x + 1)^{1/2}$. This differs from the given derivative by the factor of 3. Thus, if we write $(2x + 1)^{1/2} = \frac{1}{3}[3(2x + 1)^{1/2}]$, we have the required antiderivative as

$$F(x) = \tfrac{1}{3}(2x + 1)^{3/2}$$

Checking, the derivative of $\frac{1}{3}(2x + 1)^{3/2}$ is $\frac{1}{3}\left(\frac{3}{2}\right)(2x + 1)^{1/2}(2) = (2x + 1)^{1/2}$.

EXERCISES 25.1

In Exercises 1–4, make the given changes in the indicated examples of this section and then solve the resulting problems.

1. In Example 1, change the coefficient 8 to 12.
2. In Example 2, change v^2 to v^3.
3. In Example 3, change $2/x^3$ to $3/x^4$.
4. In Example 5, change $2x$ to $4x$.

In Exercises 5–12, determine the value of a that makes $F(x)$ an antiderivative of $f(x)$.

5. $f(x) = 3x^2$, $F(x) = ax^3$
6. $f(x) = 5x^4$, $F(x) = ax^5$
7. $f(x) = 18x^5$, $F(x) = ax^6$
8. $f(x) = 40x^7$, $F(x) = ax^8$
9. $f(x) = 9\sqrt{x}$, $F(x) = ax^{3/2}$
10. $f(x) = 10x^{1/4}$, $F(x) = ax^{5/4}$
11. $f(x) = \dfrac{1}{x^2}$, $F(x) = \dfrac{a}{x}$
12. $f(x) = \dfrac{6}{x^4}$, $F(x) = \dfrac{a}{x^3}$

In Exercises 13–38, find antiderivatives of the given functions.

13. $f(x) = 3x^3$
14. $f(x) = \frac{4}{3}x^{1/3}$
15. $f(t) = 6t^3 + 15$
16. $f(x) = 32x^7 + 2x$
17. $f(x) = 2x^2 - x$
18. $f(x) = 6x^2 - 5$
19. $f(x) = 2\sqrt{x} + 3$
20. $f(s) = 9\sqrt[3]{s} - 3$
21. $f(x) = -\dfrac{7}{x^6} + \dfrac{1}{3^2}$
22. $f(x) = \dfrac{8}{x^9} - \pi$
23. $f(v) = 4v + 3\pi^2$
24. $f(x) = \dfrac{1}{2\sqrt{x}} + \sqrt{3}$
25. $f(x) = x^2 - 4 + x^{-2}$
26. $f(x) = x\sqrt{x} - x^{-3}$
27. $f(x) = 6(2x + 1)^5(2)$
28. $f(R) = 3(R^2 + 1)^2(2R)$
29. $f(p) = 4(p^2 - 1)^{11}(2p)$
30. $f(x) = 5(2x^4 + 1)^4(8x^3)$
31. $f(x) = x^3(2x^4 + 1)^9$
32. $f(x) = x(1 - x^2)^7$
33. $f(x) = \frac{3}{2}(6x + 1)^{1/2}(6)$
34. $f(y) = \frac{5}{4}(1 - y)^{1/4}(-1)$
35. $f(x) = (3x + 1)^{1/3}$
36. $f(x) = (4x + 3)^{1/5}$
37. $f(x) = \dfrac{-2}{(2x + 1)^2}$
38. $f(s) = \dfrac{4s}{(1 - s^2)^3}$

In Exercises 39 and 40, answer the given questions.

39. Why is $(x + 5)^3$ a correct antiderivative of $3(x + 5)^2$, whereas $(2x + 5)^3$ is not a correct antiderivative of $3(2x + 5)^2$?
40. Is $\dfrac{1}{(x + 5)^3}$ a correct antiderivative of $\dfrac{1}{3(x + 5)^2}$?

Answers to Practice Exercises

1. $\frac{1}{4}x^4 + 2x^2$
2. $F(x) = (4x - 3)^5$

25.2 The Indefinite Integral

In the previous section, in developing the basic technique of finding an antiderivative, we noted that the results given are not unique. That is, we could have added any constant to the answers and the result would still have been correct. Again, this is the case since the derivative of a constant is zero.

EXAMPLE 1 Antiderivatives are not unique

The derivatives of x^3, $x^3 + 4$, $x^3 - 7$, and $x^3 + 4\pi$ are all $3x^2$. This means that any of the functions listed, as well as innumerable others, would be a proper answer to the problem of finding an antiderivative of $3x^2$.

From Section 24.8, we know that the differential of a function $F(x)$ can be written as $d[F(x)] = F'(x)dx$. Therefore, since finding a differential of a function is closely related to finding the derivative, so is the antiderivative closely related to the process of finding the function for which the differential is known.

The notation used for finding the general form of the antiderivative, the **indefinite integral**, *is written in terms of the differential.* Thus, *the indefinite integral of a function $f(x)$, for which $dF(x)/dx = f(x)$, or $dF(x) = f(x)dx$, is defined as*

$$\int f(x)\,dx = F(x) + C \tag{25.1}$$

Here, $f(x)$ is called the **integrand**, *$F(x) + C$ is the indefinite integral, and C is an arbitrary constant, called the* **constant of integration**. It represents any of the constants that may be attached to an antiderivative to have a proper result. We must have additional information beyond a knowledge of the differential to assign a specific value to C. *The symbol $\int$ is the* **integral sign**, *and it indicates that the inverse of the differential is to be found. Determining the indefinite integral is called* **integration**, which we can see is essentially the same as finding an antiderivative.

EXAMPLE 2 Indefinite integral of a polynomial

In performing the integration

$$\int 5x^4 dx = x^5 + C$$

(integrand: $5x^4$; indefinite integral: $x^5 + C$; constant of integration: C)

we might think that the inclusion of this constant C would affect the derivative of the function x^5. However, the only effect of the C is to raise or lower the curve. The slope of $y = x^5 + 2$, $y = x^5 - 2$, or any function of the form $y = x^5 + C$ is the same for any given value of x. As Fig. 25.1 shows, tangents drawn to the curves are all parallel for the same value of x.

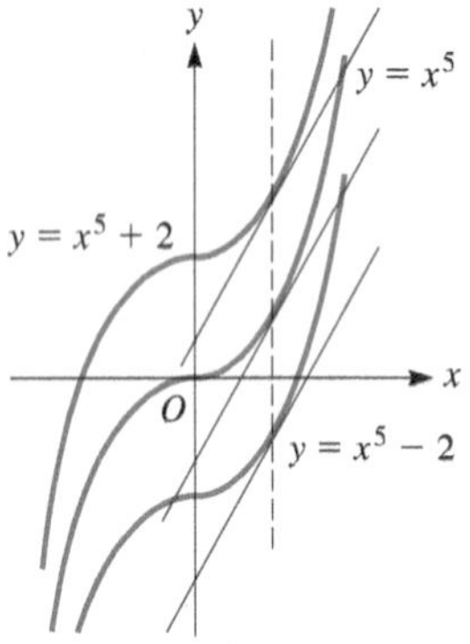

Fig. 25.1

COMMON ERROR When evaluating an indefinite integral, it is a common error to omit the constant of integration.

At this point, we shall derive some basic formulas for integration. We know that to find the differential of a power of a function, we multiply by the power, subtract 1 from it, and multiply by the differential of the function. *To find the integral, we reverse this procedure.*

LEARNING TIP

We must be able to recognize the proper form and the component parts to use these formulas. Unless you do this and have a good knowledge of differentiation, you will have trouble using Eq. (25.2). Most of the difficulty, if it exists, arises from an improper identification of *du*.

Power Rule for Integration	EXAMPLE 3
$$\int u^n du = \frac{u^{n+1}}{n+1} + C \quad (n \neq -1) \quad \textbf{(25.2)}$$	Integrate $\int x^2 dx$. We identify x as u, so $du = dx$, and $n = 2$. $\int u^n du$ $\downarrow\downarrow\downarrow$ u $n+1$ $\int x^2 dx = \frac{x^3}{3} + C$ ($3 \leftarrow n+1$)

Also, we know $d(cu)/dx = c(du/dx)$, where u is a function of x and c is a constant. Hence, multiplying constants can be moved across the integral sign.

Constant Factor Rule for Integration	EXAMPLE 4
$$\int c\,du = c\int du = cu + C \quad \textbf{(25.3)}$$	Integrate $\int 6x dx$. We note that 6 is a multiplying constant. Also, $du = x\,dx$, so we use the power rule with $n = 1$. Then $\int 6xdx = 6\int xdx = 6\left(\frac{x^2}{2}\right) + C = 3x^2 + C$

Since the derivative of a sum of functions equals the sum of their derivatives, the same is true for integrals.

Sum Rule for Integration	EXAMPLE 5
$$\int (du + dv) = u + v + C \quad \textbf{(25.4)}$$	Integrate $\int (x^2 + 6x)dx$. Using the results from Examples 3 and 4, we get $\int (x^2 + 6x)dx = \int x^2 dx + \int 6xdx = \frac{x^3}{3} + 3x^2 + C$

EXAMPLE 6 Integration of a polynomial

Integrate $\int (5x^3 - 6x^2 + 1)\,dx$.

Here, we must use a combination of Eqs. (25.2), (25.3), and (25.4). Therefore,

$$\int (5x^3 - 6x^2 + 1)\,dx = \int 5x^3\,dx + \int (-6x^2)\,dx + \int dx$$
$$= 5\int x^3\,dx - 6\int x^2\,dx + \int dx$$

In the first integral, $u = x$, $n = 3$, and $du = dx$. In the second, $u = x$, $n = 2$, and $du = dx$. The third uses Eq. (25.3) directly, with $c = 1$ and $du = dx$. This means

$$5\int x^3\,dx - 6\int x^2\,dx + \int dx = 5\left(\frac{x^4}{4}\right) - 6\left(\frac{x^3}{3}\right) + x + C$$
$$= \frac{5}{4}x^4 - 2x^3 + x + C$$

Practice Exercise

1. Integrate $\int (6x^2 - 5)\,dx$.

EXAMPLE 7 Integrating a square root and a power with a negative exponent

Integrate $\int\left(\sqrt{r} - \frac{1}{r^3}\right)dr$.

In order to use Eq. (25.2), we must first write $\sqrt{r} = r^{1/2}$ and $1/r^3 = r^{-3}$:

$$\int\left(\sqrt{r} - \frac{1}{r^3}\right)dr = \int r^{1/2}\,dr - \int r^{-3}\,dr = \frac{1}{\frac{3}{2}}r^{3/2} - \frac{1}{-2}r^{-2} + C$$

$$= \frac{2}{3}r^{3/2} + \frac{1}{2}r^{-2} + C = \frac{2}{3}r^{3/2} + \frac{1}{2r^2} + C$$

EXAMPLE 8 Integrating a power of a function

Integrate $\int (x^2 + 1)^3(2x\,dx)$.

We first note that $n = 3$, for this is the power involved in the function being integrated. If $n = 3$, then $x^2 + 1$ must be u. If $u = x^2 + 1$, then $du = 2x\,dx$. Thus, the integral is in proper form for integration *as it stands*. Using the power rule for integration,

$$\int (x^2 + 1)^3(2x\,dx) = \frac{(x^2 + 1)^4}{4} + C$$

Showing the use of u directly, we can write the integration as

$$\int (x^2 + 1)^3(2x\,dx) = \int u^3\,du = \frac{1}{4}u^4 + C = \frac{(x^2 + 1)^4}{4} + C$$

LEARNING TIP

It must be emphasized that the entire quantity **(2x dx)** ***must be equated to du.*** Normally, *u* and *n* are recognized first, and then *du* is derived from *u*.

EXAMPLE 9 Integrating a power of a function

Integrate $\int x^2\sqrt{x^3 + 2}\,dx$.

We first note that $n = \frac{1}{2}$ and u is then $x^3 + 2$. Since $u = x^3 + 2$, $du = 3x^2\,dx$. Now, we group $3x^2\,dx$ as du. Since there is no 3 under the integral sign, we introduce one. In order not to change the numerical value, we also introduce $\frac{1}{3}$, normally before the integral sign.

$$\int x^2\sqrt{x^3 + 2}\,dx = \frac{1}{3}\int 3x^2\sqrt{x^3 + 2}\,dx = \frac{1}{3}\int \sqrt{x^3 + 2}\,(3x^2\,dx)$$

Here, we indicate the proper grouping to have the proper form of Eq. (25.2):

$$\int x^2\sqrt{x^3 + 2}\,dx = \frac{1}{3}\int \sqrt{x^3 + 2}\,(3x^2\,dx) = \frac{1}{3}\left(\frac{2}{3}\right)(x^3 + 2)^{3/2} + C$$

$$= \frac{2}{9}(x^3 + 2)^{3/2} + C$$

The $1/\frac{3}{2}$ was written as $\frac{2}{3}$, since this form is more convenient with fractions.

With $u = x^3 + 2$ and using u directly in the integration, we can write

$$\int x^2\sqrt{x^3 + 2}\,dx = \int (x^3 + 2)^{1/2}(x^2\,dx)$$

$$= \int u^{1/2}\left(\frac{1}{3}du\right) = \frac{1}{3}\int u^{1/2}\,du \quad \text{integrating in terms of } u$$

$$= \frac{1}{3}\left(\frac{2}{3}\right)u^{3/2} + C = \frac{2}{9}u^{3/2} + C$$

$$= \frac{2}{9}(x^3 + 2)^{3/2} + C \quad \text{substituting } x^3 + 2 = u$$

■ We have integrated certain basic functions. Other methods are used to integrate other types of functions, and some of these are discussed in Chapter 28. Also, many functions cannot be integrated algebraically.

Practice Exercise

2. Integrate $\int 8x\sqrt{1 - 2x^2}\,dx$.

COMMON ERROR

Because a constant factor may be moved across the integral sign, we can always introduce a constant required to complete du, and also its reciprocal (so as not to change the numerical value of the integral), as in Example 9. We note, however, that **only constant factors may be moved across the integral sign.** If du is missing a variable factor, we **cannot introduce the variable factor and move its variable reciprocal across the integral sign.** For example, when integrating $\int (x^2+1)^2 dx$, we cannot integrate by setting $u = x^2 + 1$, $du = 2x\,dx$ because we cannot introduce the factor x and take $1/x$ out of the integral. Instead, we need to square the binomial and integrate $\int (x^4 + 2x^2 + 1)dx$.

EVALUATING THE CONSTANT OF INTEGRATION

To find the constant of integration, we need information such as a set of values that satisfy the function. A point through which the curve passes would provide the necessary information. This is illustrated in the following examples.

EXAMPLE 10 Evaluating the constant of integration

Find y in terms of x, given that $dy/dx = 3x - 1$ and the curve passes through $(1, 4)$.

The solution is as follows.

$$dy = (3x - 1)\,dx \quad \text{rewrite equation—solve for } dy \text{ in terms of } dx$$

$$\int dy = \int (3x - 1)\,dx \quad \text{set up integration}$$

$$y = \frac{3}{2}x^2 - x + C \quad \text{integrate}$$

$$4 = \frac{3}{2} - 1 + C \quad \text{or} \quad C = \frac{7}{2} \quad \text{evaluate } C\text{; point } (1, 4) \text{ satisfies equation}$$

$$y = \frac{3}{2}x^2 - x + \frac{7}{2} \quad \text{or} \quad 2y = 3x^2 - 2x + 7 \quad \text{see Fig. 25.2}$$

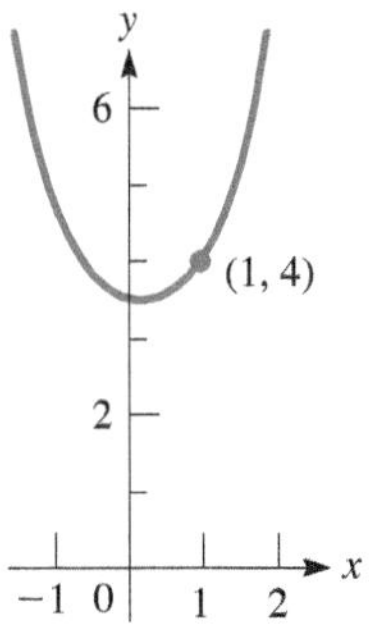

Fig. 25.2

EXAMPLE 11 Integral with a negative exponent

The time rate of change of the displacement (velocity) of a robot arm is $ds/dt = 8t/(t^2+4)^2$. Find the expression for the displacement as a function of time if $s = -1$ m when $t = 0$ s.

■ See the chapter introduction.

First, we write $ds = \frac{8t\,dt}{(t^2+4)^2}$ and then integrate. To integrate the expression on the right, we note that $n = -2$, $u = t^2 + 4$, and $du = 2t\,dt$. This means we need a 2 with $t\,dt$ to form the proper du. In turn, this means we place a $\frac{1}{2}$ before the integral sign. Also, we place the 8 in front of the integral sign. Therefore,

$$\int ds = \int \frac{8t\,dt}{(t^2+4)^2} = \left(\frac{1}{2}\right)(8)\int (t^2+4)^{-2}(2t\,dt) \quad \text{set up integration}$$

$$s = 4\left(\frac{1}{-1}\right)(t^2+4)^{-1} + C = \frac{-4}{t^2+4} + C \quad \text{integrate; } -2 + 1 = -1$$

$$-1 = \frac{-4}{0+4} + C \quad \text{or} \quad C = 0 \quad \text{evaluate } C \text{ given } s = -1 \text{ m when } t = 0 \text{ s}$$

$$s = \frac{-4}{t^2+4} \quad \text{expression for displacement}$$

EXERCISES 25.2

In Exercises 1–4, make the given changes in the indicated examples of this section, and then solve the resulting problems.

1. In Example 4, change the coefficient 6 to 8.

2. In Example 7, change $\sqrt{r}$ to $\sqrt[3]{r}$.

3. In Example 8, change the power 3 to 4.

4. In Example 10, change $(1, 4)$ to $(2, 3)$.

In Exercises 5–36, integrate each of the given expressions.

5. $\int 2x\,dx$ **6.** $\int 5x^4\,dx$ **7.** $\int x^{13}\,dx$

8. $\int 0.6y^5\,dy$ **9.** $\int \frac{x^{3/2}}{8}\,dx$ **10.** $\int 6\sqrt[3]{x}\,dx$

11. $\int 9R^{-4}\,dR$ **12.** $\int \frac{4}{\sqrt{x}}\,dx$ **13.** $\int (x^2 - x^5)\,dx$

14. $\int (1 - 3x)\,dx$ **15.** $\int (9x^2 + x + 3)\,dx$ **16.** $\int x(x - 2)^2\,dx$

17. $\int \left(\frac{t^2}{2} - \frac{2}{t^2}\right)dt$ **18.** $\int \frac{3x^2 - 4}{x^2}\,dx$

19. $\int \sqrt{x}(x^2 - x)\,dx$ **20.** $\int (3R\sqrt{R} - 5R^2)\,dR$

21. $\int (2x^{-2/3} + 3^{-2})\,dx$ **22.** $\int (x^{1/3} + x^{1/5} + x^{-1/7})\,dx$

23. $\int (1 + 12x^2)^2\,dx$ **24.** $\int (x^2 + 4x + 4)^{1/3}\,dx$

25. $\int (x^2 - 1)^9(2x\,dx)$ **26.** $\int (t^3 - 2)^6(3t^2\,dt)$

27. $\int (x^4 + 3)^4(4x^3\,dx)$ **28.** $\int (1 - 2x)^{1/3}(-2\,dx)$

29. $\int (2\theta^5 + 5)^7\theta^4\,d\theta$ **30.** $\int 6x^2(1 - x^3)^{4/3}\,dx$

31. $\int \sqrt{-6x + 1}\,dx$ **32.** $\int \frac{dV}{(0.3 + 7V)^3}$

33. $\int \frac{x\,dx}{\sqrt{6x^2 + 1}}$ **34.** $\int \frac{2x^2\,dx}{\sqrt{2x^3 + 1}}$

35. $\int \frac{4z - 4}{\sqrt{z^2 - 2z}}\,dz$

36. $\int (x^2 - x)\left(x^3 - \frac{3}{2}x^2\right)^8 dx$

In Exercises 37–40, find y in terms of x.

37. $\frac{dy}{dx} = 6x^2$, curve passes through $(0, 2)$

38. $\frac{dy}{dx} = 8x + 1$, curve passes through $(-1, 4)$

39. $\frac{dy}{dx} = x^2(1 - x^3)^5$, curve passes through $(1, 5)$

40. $\frac{dy}{dx} = 2x^3(x^4 - 6)^4$, curve passes through $(2, 10)$

In Exercises 41–62, solve the given problems.

41. Is $\int 3x^2\,dx = x^3$? Explain.

42. Can $\int (x^2 - 1)^2\,dx$ be integrated with $u = x^2 - 1$? Explain.

43. Can $\int (4x^3 + 3)^5 x^4\,dx$ be integrated with $u = 4x^3 + 3$ and $du = x^4\,dx$? Explain.

44. Is $\int \sqrt{2x + 1}\,dx = \frac{2}{3}(2x + 1)^{3/2}$? Explain.

45. Is $\int 3(2x + 1)^2\,dx = (2x + 1)^3 + C$? Explain.

46. Is $\int x^{-2}\,dx = -\frac{1}{3}x^{-3} + C$? Explain.

47. Find the equation of the curve whose slope is $-x\sqrt{1 - 4x^2}$ and that passes through $(0, 7)$.

48. Find the equation of the curve whose slope is $\sqrt{6x - 3}$ and that passes through $(2, -1)$.

49. Find $f(x)$ if $f'(x) = 4x - 5$ and $f(-1) = 10$.

50. Find $f(x)$ if $f'(x) = 2/\sqrt{x}$ and $f(9) = 8$.

51. Find the general form of the function whose second derivative is $\sqrt{x}$.

52. Find the general form of the expression for the displacement s of an object if its acceleration is $9.8\ \mathrm{m/s^2}$.

53. If the consumption of natural gas is $0.14 + 0.000\,28t$ billion $\mathrm{m^3}$/year, find the volume V that will be consumed in the next t years. (At the end of 2014, Canada had proven reserves of natural gas of about $1.9 \times 10^{12}\ \mathrm{m^3}$.)

54. The radius r (in m) of a circular oil spill is increasing at the rate given by $\frac{dr}{dt} = \frac{3}{\sqrt{4t + 1}}$, where t is in minutes. Find the radius as a function of t, if t is measured from the time of the spill.

55. The time rate of change of electric current in a circuit is given by $di/dt = 4t - 0.6t^2$. Find the expression for the current as a function of time if $i = 2\mathrm{A}$ when $t = 0$ s.

56. The rate of change of the frequency f of an electronic oscillator with respect to the inductance L is $df/dL = 80(4 + L)^{-3/2}$. Find f as a function of L if $f = 80$ Hz for $L = 0$ H.

57. The rate of change of the temperature T (in °C) from the centre of a blast furnace to a distance r (in m) from the centre is given by $dT/dr = -4500(r + 1)^{-3}$. Express T as a function of r if $T = 2500$°C for $r = 0$.

58. The rate of change of current i (in mA) in a circuit with a variable inductance is given by $di/dt = 300(5.0 - t)^{-2}$, where t (in ms) is the time the circuit is closed. Find i as a function of t if $i = 300$ mA for $t = 2.0$ ms.

59. At a given site, the rate of change of the annual fraction f of energy supplied by solar energy with respect to the solar-collector area A (in $\mathrm{m^2}$) is $\frac{df}{dA} = \frac{0.005}{\sqrt{0.01A + 1}}$. Find f as a function of A if $f = 0$ for $A = 0\ \mathrm{m^2}$.

60. An analysis of a company's records shows that in a day the rate of change of profit p (in dollars) in producing x generators is $\frac{dp}{dx} = \frac{600(30 - x)}{\sqrt{60x - x^2}}$. Find the profit in producing x generators if a loss of \$5000 is incurred if none are produced.

61. Find the equation of the curve for which the second derivative is 6. The curve passes through $(1, 2)$ with a slope of 8.

62. The second derivative of a function is $12x^2$. Explain how to find the function if its curve passes through the points $(1, 6)$ and $(2, 21)$. Find the function.

Answers to Practice Exercises

1. $2x^3 - 5x + C$ **2.** $-\frac{4}{3}(1 - 2x^2)^{3/2} + C$

25.3 The Area Under a Curve

In geometry, there are formulas and methods for finding the areas of regular figures. By means of integration, it is possible to find the area between curves for which we know the equations. The next example illustrates the basic idea behind the method.

EXAMPLE 1 Summing the areas of inscribed rectangles

Approximate the area in the first quadrant to the left of the line $x = 4$ and under the parabola $y = x^2 + 1$. Here, "under" means between the curve and the x-axis. First, make this approximation by inscribing two rectangles of equal width under the parabola and finding the sum of the areas of these rectangles. Then, improve the approximation by repeating the process with eight inscribed rectangles.

The area to be approximated is shown in Fig. 25.3(a). The area with two rectangles inscribed under the curve is shown in Fig. 25.3(b). The first approximation, admittedly small, of the area can be found by adding the areas of the two rectangles. Both rectangles have a width of 2. The left rectangle is 1 unit high, and the right rectangle is 5 units high. Thus, the area of the two rectangles is

$$A = 2(1 + 5) = 12$$

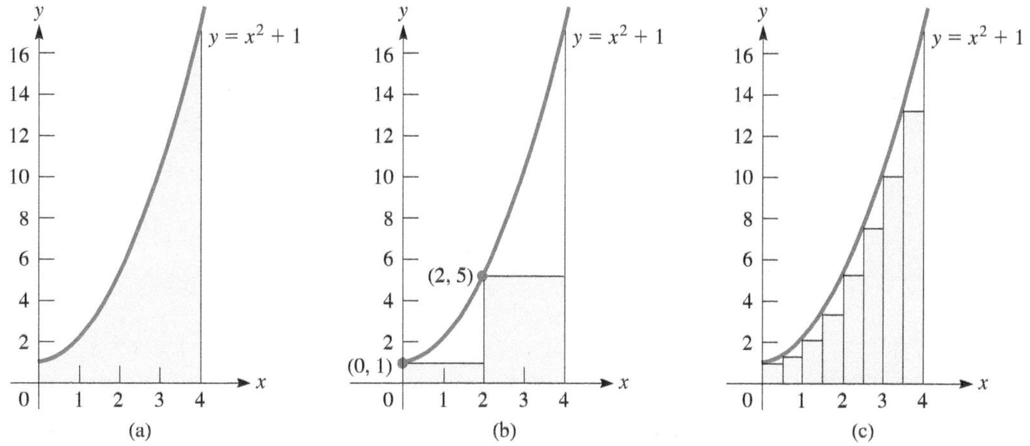

Fig. 25.3

Table 25.1

Number of Rectangles n	Total Area of Rectangles
8	21.5
100	25.0144
1000	25.301 344
10 000	25.330 134

A much better approximation is found by inscribing the eight rectangles as shown in Fig. 25.3(c). Each of these rectangles has a width of $\frac{1}{2}$. The leftmost rectangle has a height of 1. The next has a height of $\frac{5}{4}$, which is determined by finding y for $x = \frac{1}{2}$. The next rectangle has a height of 2, which is found by evaluating y for $x = 1$. Finding the heights of all rectangles and multiplying their sum by $\frac{1}{2}$ gives the area of the eight rectangles as

$$A = \frac{1}{2}\left(1 + \frac{5}{4} + 2 + \frac{13}{4} + 5 + \frac{29}{4} + 10 + \frac{53}{4}\right) = \frac{43}{2} = 21.5$$

An even better approximation could be obtained by inscribing more rectangles under the curve. The greater the number of rectangles, the closer the sum of their areas is to the area under the curve. See Table 25.1. By using integration later in this section, we determine the *exact* area to be $\frac{76}{3} = 25\frac{1}{3}$.

Practice Exercise

1. In Example 1, approximate the area by inscribing four rectangles.

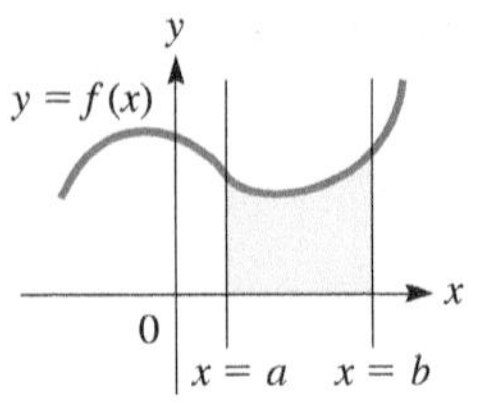

Fig. 25.4

We now develop the basic method used to find the area under a curve, which is the area bounded by the curve, the x-axis, and the lines $x = a$ and $x = b$. See Fig. 25.4. We assume here that $f(x)$ is never negative in the interval $a < x < b$. In Chapter 26, we will extend the method so that $f(x)$ may be negative.

In finding the area under a curve, we consider the sum of the areas of inscribed rectangles, as the number of rectangles is assumed to increase without bound. The reason for this last condition is that, as we saw in Example 1, as the number of rectangles increases, the approximation of the area is better.

EXAMPLE 2 Letting the number of rectangles approach infinity

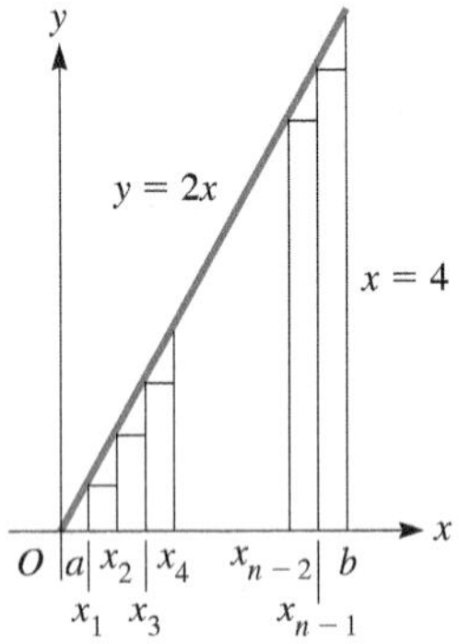

Fig. 25.5

Find the area under the straight line $y = 2x$, above the x-axis, and to the left of the line $x = 4$.

Since this figure is a right triangle, the area can easily be found. However, the *method* we use here is the important concept. We first subdivide the interval from $x = 0$ to $x = 4$ into n inscribed rectangles of width Δx. The endpoints of the intervals are labelled $a, x_1, x_2, \ldots, b\ (= x_n)$, as shown in Fig. 25.5, where $a = 0$, $b = 4$ and

$$x_1 = \Delta x \qquad x_2 = 2\Delta x, \ldots \qquad x_{n-1} = (n-1)\Delta x \qquad b = n\Delta x$$

The area of each of these n rectangles is as follows:

First: $f(a)\,\Delta x$, where $f(a) = f(0) = 2(0) = 0$ is the height.
Second: $f(x_1)\,\Delta x$, where $f(x_1) = 2(\Delta x) = 2\,\Delta x$ is the height.
Third: $f(x_2)\,\Delta x$, where $f(x_2) = 2(2\,\Delta x) = 4\,\Delta x$ is the height.
Fourth: $f(x_3)\,\Delta x$, where $f(x_3) = 2(3\,\Delta x) = 6\,\Delta x$ is the height.
⋮
Last: $f(x_{n-1})\,\Delta x$, where $f[(n-1)\,\Delta x] = 2(n-1)\,\Delta x$ is the height.

These areas are summed up as follows:

$$A_n = f(a)\,\Delta x + f(x_1)\,\Delta x + f(x_2)\,\Delta x + \cdots + f(x_{n-1})\,\Delta x \qquad \textbf{(25.5)}$$
$$= 0 + 2\,\Delta x(\Delta x) + 4\,\Delta x(\Delta x) + \cdots + 2[(n-1)\,\Delta x]\,\Delta x$$
$$= 2(\Delta x)^2[1 + 2 + 3 + \cdots + (n-1)]$$

Now, $b = n\,\Delta x$, or $4 = n\,\Delta x$, or $\Delta x = 4/n$. Thus,

$$A_n = 2\left(\frac{4}{n}\right)^2[1 + 2 + 3 + \cdots + (n-1)]$$

The sum of the arithmetic sequence $1 + 2 + 3 + \cdots + n - 1$ is

$$s = \frac{n-1}{2}(1 + n - 1) = \frac{n(n-1)}{2} = \frac{n^2 - n}{2}$$

Now, the expression for the sum of the areas can be written as

$$A_n = \frac{32}{n^2}\left(\frac{n^2 - n}{2}\right) = 16\left(1 - \frac{1}{n}\right)$$

LEARNING TIP
The area under the curve is the limit of the sum of the areas of the inscribed rectangles, as the number of rectangles approaches infinity.

This expression is an approximation of the actual area under consideration. The larger n becomes, the better the approximation. If we let $n \to \infty$ (which is equivalent to letting $\Delta x \to 0$), the limit of this sum will equal the area in question.

$$A = \lim_{n\to\infty} 16\left(1 - \frac{1}{n}\right) = 16 \qquad 1/n \to 0 \text{ as } n \to \infty$$

■ This checks with the geometric result.

The method indicated in Example 2 illustrates the interpretation of finding an area as a summation process, although it should not be considered as a proof. However, we will find that integration proves to be a much more useful method for finding an area. Let us now see how integration can be used directly.

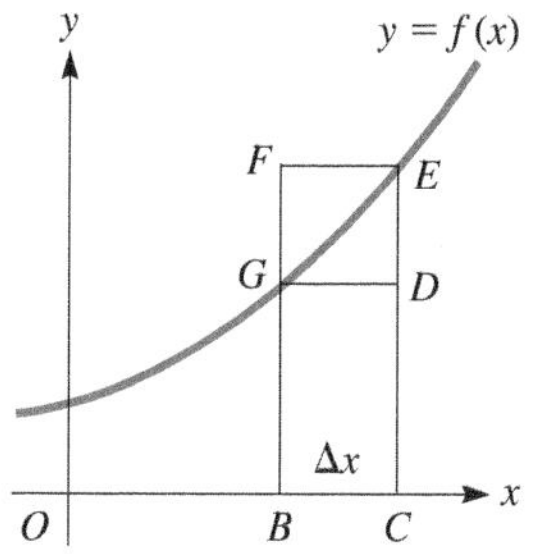

Fig. 25.6

Let ΔA represent the area $BCEG$ under the curve, as indicated in Fig. 25.6. We see that the following inequality is true for the indicated areas:

$$A_{BCDG} < \Delta A < A_{BCEF}$$

If the point G is now designated as (x, y) and E as $(x + \Delta x, y + \Delta y)$, we have $y\Delta x < \Delta A < (y + \Delta y)\Delta x$. Dividing through by Δx, we have

$$y < \frac{\Delta A}{\Delta x} < y + \Delta y$$

Now, we take the limit as $\Delta x \to 0$ (Δy then approaches zero). This results in

$$\frac{dA}{dx} = y \qquad \textbf{(25.6)}$$

This is true since the left member of the inequality is y and the right member approaches y. Also, in the definition of the derivative, Eq. (23.6), $f(x + h) - f(x)$ is equivalent to ΔA, and h is equivalent to Δx, which means

$$\lim_{\Delta x \to 0} \frac{\Delta A}{\Delta x} = \frac{dA}{dx}$$

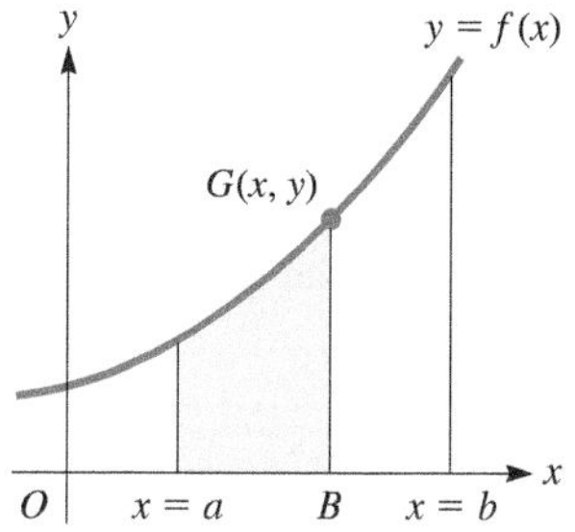

Fig. 25.7

We shall now use Eq. (25.6) to show the method of finding the complete area under a curve. We now let $x = a$ be the left boundary of the desired area and $x = b$ be the right boundary (Fig. 25.7). The area under the curve to the right of $x = a$ and bounded on the right by the line GB is now designated as A_{ax}. From Eq. (25.6), we have

$$dA_{ax} = [y\,dx]_a^x \quad \text{or} \quad A_{ax} = \left[\int y\,dx\right]_a^x = \left[\int f(x)\,dx\right]_a^x$$

where $[\quad]_a^x$ is the notation used to indicate the boundaries of the area. If the indefinite integral is given by $F(x) + C$, we have

$$A_{ax} = \left[\int f(x)\,dx\right]_a^x = [F(x) + C]_a^x \qquad \textbf{(25.7)}$$

But we know that if $x = a$, then $A_{aa} = 0$. Thus, $0 = F(a) + C$, or $C = -F(a)$. Therefore,

$$A_{ax} = \left[\int f(x)\,dx\right]_a^x = F(x) - F(a) \qquad \textbf{(25.8)}$$

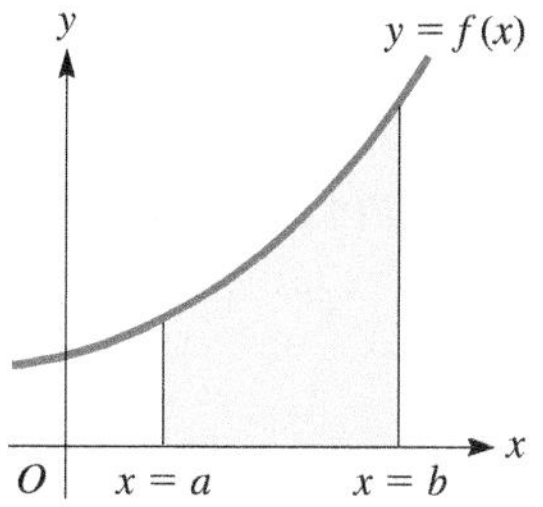

Fig. 25.8

Now, to find the area under the curve that reaches from a to b, we write

$$A_{ab} = F(b) - F(a) \qquad \textbf{(25.9)}$$

Thus, the area under the curve that reaches from a to b is given by

$$A_{ab} = \left[\int f(x)\,dx\right]_a^b = F(b) - F(a) \qquad \textbf{(25.10)}$$

LEARNING TIP

Eq. (25.10) tells us that the area under the curve may be found by integrating the function $f(x)$ to find the function $F(x)$, which is then evaluated at each boundary value. The area is the difference between these values of $F(x)$.
See Fig. 25.8.

Note that we do not have to include the constant of integration when using $F(x)$ in Eq. (25.10). Any constant added to $F(x)$ cancels out when $F(a)$ is subtracted from $F(b)$.

INTEGRATION AS SUMMATION

In Example 2, we found an area under a curve by finding the limit of the sum of the areas of the inscribed rectangles as the number of rectangles approaches infinity. Eq. (25.10) expresses the area under a curve in terms of integration. We can now see that we have obtained an area by summation and also expressed it in terms of integration. Therefore, we conclude that ***summations can be evaluated by integration.*** Also, we have seen the connection between the problem of finding the slope of a tangent to a curve (differentiation) and the problem of finding an area under a curve (integration).

We would not normally suspect that these two problems would have solutions that lead to reverse processes. We have also seen that the definition of integration has much more application than originally anticipated.

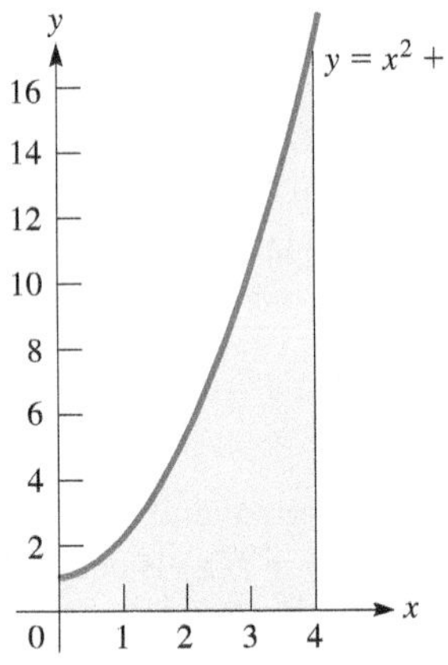

Fig. 25.3(a)

EXAMPLE 3 Area under a curve by integration

Find the area under the curve of $y = x^2 + 1$ between the y-axis and the line $x = 4$. This is the same area that we illustrated in Example 1 and showed in Fig. 25.3(a). This figure is shown again here for reference.

Identify $f(x)$ and obtain the indefinite integral $\int f(x)dx$: $f(x) = x^2 + 1$, so

$$\int (x^2 + 1)dx = \frac{x^3}{3} + x + C$$

Identify $F(x)$ and the values a and b: $F(x) = \frac{x^3}{3} + x,\ a = 0,\ b = 4$

Evaluate $F(x)$ at a and b and use Eq. (25.10):

$$A_{0,4} = F(4) - F(0)$$
$$= \left[\frac{4^3}{3} + 4\right] - \left[\frac{0^3}{3} + 0\right] = \frac{76}{3}$$

We note that $76/3$ is a little more than 25 square units and is therefore about 4 square units more than the value obtained using eight inscribed rectangles in Example 1. Therefore, from this result, we know that the *exact* area under the curve is $25\frac{1}{3}$, as stated at the end of Example 1.

Practice Exercise

2. In Example 3, change $x^2 + 1$ to $x^3 + 1$ and then find the area.

EXAMPLE 4 Area under a curve by integration

Find the area under the curve $y = x^3$ that is between the lines $x = 1$ and $x = 2$, as shown in Fig. 25.9.

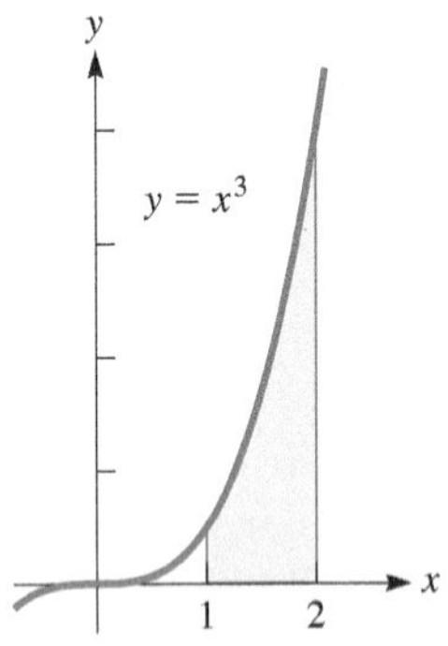

Fig. 25.9

Identify $f(x)$ and obtain the indefinite integral $\int f(x)dx$: $f(x) = x^3$, so

$$\int x^3dx = \frac{x^4}{4} + C$$

Identify $F(x)$ and the values a and b: $F(x) = \frac{x^4}{4},\ a = 1,\ b = 2$

Evaluate $F(x)$ at a and b and use Eq. (25.10):

$$A_{1,2} = F(2) - F(1)$$
$$= \frac{2^4}{4} - \frac{1^4}{4} = \frac{15}{4}$$

The calculated area of $15/4$ is the exact area, not an approximation.

EXERCISES 25.3

In Exercises 1–4, make the given changes in the indicated examples of this section and then solve the resulting problems.

1. In Example 1, change $x = 4$ to $x = 2$ and find the area of (a) two inscribed rectangles and (b) four inscribed rectangles.
2. In Example 3, change $x = 4$ to $x = 2$ and compare the results with those of Exercise 1.
3. In Example 4, change $x = 2$ to $x = 3$.
4. In Example 4, change $x = 1$ to $x = 2$ and $x = 2$ to $x = 3$. Note that the result added to the result of Example 4 is the same as the result for Exercise 3.

In Exercises 5–14, find the approximate area under the curves of the given equations by dividing the indicated intervals into n subintervals and then adding up the areas of the inscribed rectangles. There are two values of n for each exercise and therefore two approximations for each area. The height of each rectangle may be found by evaluating the function for the proper value of x. See Example 1. Round answers to three significant digits.

5. $y = 3x$, between $x = 0$ and $x = 3$, for (a) $n = 3$ ($\Delta x = 1$), (b) $n = 10$ ($\Delta x = 0.3$)
6. $y = 2x + 1$, between $x = 0$ and $x = 2$, for (a) $n = 4$ ($\Delta x = 0.5$), (b) $n = 10$ ($\Delta x = 0.2$)

7. $y = x^2$, between $x = 0$ and $x = 2$, for (a) $n = 5$ ($\Delta x = 0.4$), (b) $n = 10$ ($\Delta x = 0.2$)
8. $y = 9 - x^2$, between $x = 2$ and $x = 3$, for (a) $n = 5$ ($\Delta x = 0.2$), (b) $n = 10$ ($\Delta x = 0.1$)
9. $y = 4x - x^2$, between $x = 1$ and $x = 4$, for (a) $n = 6$, (b) $n = 10$
10. $y = 1 - x^2$, between $x = 0.5$ and $x = 1$, for (a) $n = 5$, (b) $n = 10$
11. $y = \frac{1}{x^2}$, between $x = 1$ and $x = 5$, for (a) $n = 4$, (b) $n = 8$
12. $y = \sqrt{x}$, between $x = 1$ and $x = 4$, for (a) $n = 3$, (b) $n = 12$
13. $y = \frac{1}{\sqrt{x+1}}$, between $x = 3$ and $x = 8$, for (a) $n = 5$, (b) $n = 10$
14. $y = 2x\sqrt{x^2+1}$, between $x = 0$ and $x = 6$, for (a) $n = 6$, (b) $n = 12$

In Exercises 15–24, find the exact area under the given curves between the indicated values of x. The functions are the same as those for which approximate areas were found in Exercises 5–14.

15. $y = 3x$, between $x = 0$ and $x = 3$
16. $y = 2x + 1$, between $x = 0$ and $x = 2$
17. $y = x^2$, between $x = 0$ and $x = 2$
18. $y = 9 - x^2$, between $x = 2$ and $x = 3$
19. $y = 4x - x^2$, between $x = 1$ and $x = 4$
20. $y = 1 - x^2$, between $x = 0.5$ and $x = 1$
21. $y = \frac{1}{x^2}$, between $x = 1$ and $x = 5$
22. $y = \sqrt{x}$, between $x = 1$ and $x = 4$
23. $y = \frac{1}{\sqrt{x+1}}$, between $x = 3$ and $x = 8$
24. $y = 2x\sqrt{x^2+1}$, between $x = 0$ and $x = 6$. Explain the reason for the difference between this result and the two values found in Exercise 14.

In developing the concept of the area under a curve, we first (in Examples 1 and 2) considered rectangles ***inscribed*** *under the curve. A more complete development also considers rectangles* ***circumscribed*** *above the curve and shows that the limiting area of the circumscribed rectangles equals the limiting area of the inscribed rectangles as the number of rectangles increases without bound. See Fig. 25.10 for an illustration of inscribed and circumscribed rectangles.*

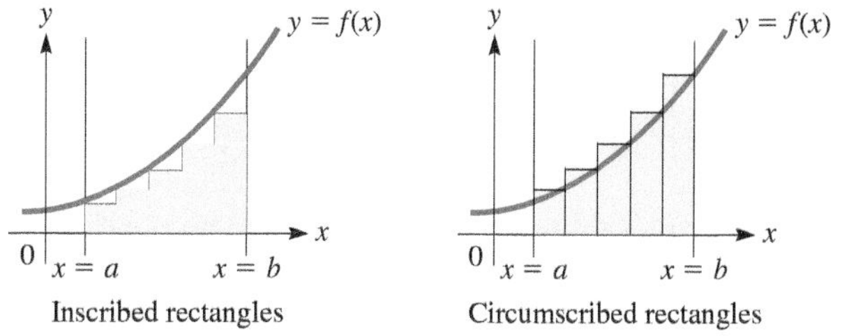

Fig. 25.10

In Exercises 25–28, find the sum of the areas of 10 circumscribed rectangles for each curve and show that the exact area (as shown in Exercises 15–18) is between the sum of the areas of the circumscribed rectangles and the inscribed rectangles [as found in Exercises 5(b)–8(b)]. Also, note that the mean of the two sums is close to the exact value.

25. $y = 3x$ between $x = 0$ and $x = 3$ (compare with Exercises 5(b) and 15). Why is the mean of the sums of the inscribed rectangles and circumscribed rectangles equal to the exact value?
26. $y = 2x + 1$ between $x = 0$ and $x = 2$ (compare with Exercises 6(b) and 16). Why is the mean of the sums of the inscribed rectangles and circumscribed rectangles equal to the exact value?
27. $y = x^2$ between $x = 0$ and $x = 2$ (compare with Exercises 7(b) and 17). Why is the mean of the sums of the inscribed rectangles and circumscribed rectangles greater than the exact value?
28. $y = 9 - x^2$ between $x = 2$ and $x = 3$ (compare with Exercises 8(b) and 18). Why is the mean of the sums of the inscribed rectangles and circumscribed rectangles less than the exact value?

Answers to Practice Exercises

1. $A = 18$ 2. $A = 68$

25.4 The Definite Integral

Using reasoning similar to that in the preceding section, *we define the* **definite integral** *of a function $f(x)$ as*

$$\int_a^b f(x)\,dx = F(b) - F(a) \qquad (25.11)$$

where $F'(x) = f(x)$. *We call this a* **definite integral** ***because the final result of integrating and evaluating is a*** **number**. (The *indefinite* integral had an arbitrary constant in the result.) *The numbers a and b are called the* **lower limit** *and the* **upper limit**, *respectively. We can see that the value of a definite integral is found by evaluating the function (found by integration) at the upper limit and subtracting the value of this function at the lower limit.*

■ That integration is equivalent to the limit of a sum is the reason Leibniz used an elongated *S* for the integral sign. It stands for the Latin word for *sum*.

From Section 25.3, we know that this definite integral can be interpreted as the area under the curve of $y = f(x)$ from $x = a$ to $x = b$, and in general as a summation. *This summation interpretation will be applied to many kinds of applied problems.*

EXAMPLE 1 Definite integral of a power of x

Evaluate the integral $\int_0^2 x^4\,dx$.

$$\int_0^2 x^4\,dx = \frac{x^5}{5}\bigg|_0^2 = \frac{2^5}{5} - 0 = \frac{32}{5}$$

upper limit; lower limit; $f(x)$; $F(x)$

Note that a vertical line—with the limits written at the top and the bottom—is the way the value is indicated after integration, but before evaluation.

The area that this definite integral represents is shown in Fig. 25.11.

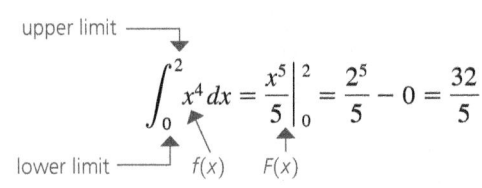

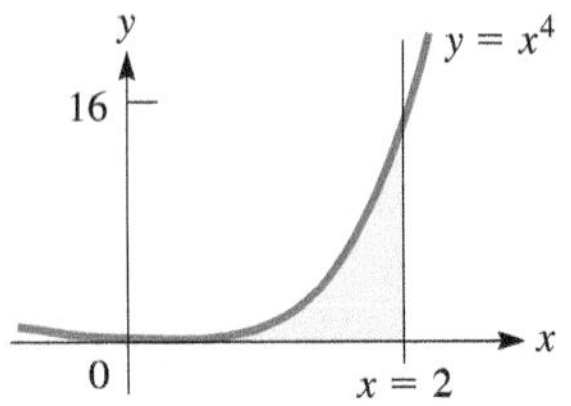

Fig. 25.11

EXAMPLE 2 Definite integral of a power with a negative exponent

Evaluate $\int_1^3 (x^{-2} - 1)\,dx$.

$$\int_1^3 (x^{-2} - 1)\,dx = -\frac{1}{x} - x\bigg|_1^3 = \left(-\frac{1}{3} - 3\right) - (-1 - 1)$$

$$= -\frac{10}{3} + 2 = -\frac{4}{3}$$

upper limit; subtract; lower limit

Practice Exercise

1. Evaluate $\int_1^2 (4x - x^{-3})\,dx$.

EXAMPLE 3 Definite integral of a power of a function

Evaluate $\int_0^1 5z(z^2 + 1)^5\,dz$.

For purposes of integration, $n = 5$, $u = z^2 + 1$, and $du = 2z\,dz$. Hence,

$$\int_0^1 5z(z^2 + 1)^5\,dz = \frac{5}{2}\int_0^1 (z^2 + 1)^5(2z\,dz)$$

$$= \frac{5}{2}\left(\frac{1}{6}\right)(z^2 + 1)^6\Big|_0^1$$

$$= \frac{5}{12}(2^6 - 1^6) = \frac{5(63)}{12} = \frac{105}{4}$$

EXAMPLE 4 Definite integral with a radical in the denominator

Evaluate $\int_{0.1}^{2.7} \frac{dx}{\sqrt{4x + 1}}$.

In order to integrate, we have $n = -\frac{1}{2}$, $u = 4x + 1$, and $du = 4\,dx$. Therefore,

$$\int_{0.1}^{2.7} \frac{dx}{\sqrt{4x + 1}} = \int_{0.1}^{2.7} (4x + 1)^{-1/2}\,dx = \frac{1}{4}\int_{0.1}^{2.7} (4x + 1)^{-1/2}(4\,dx)$$

$$= \frac{1}{4}\left(\frac{1}{\frac{1}{2}}\right)(4x + 1)^{1/2}\Big|_{0.1}^{2.7} = \frac{1}{2}(4x + 1)^{1/2}\Big|_{0.1}^{2.7} \quad \text{integrate}$$

$$= \frac{1}{2}\left(\sqrt{11.8} - \sqrt{1.4}\right) = 1.13 \quad \text{evaluate}$$

Practice Exercise

2. Evaluate $\int_0^6 \sqrt{4x + 1}\,dx$.

EXAMPLE 5 Integral involving a negative power of a function

Evaluate $\int_0^4 \frac{x+1}{(x^2+2x+2)^3}dx$.

For integrating, $n=-3$, $u=x^2+2x+2$, and $du=(2x+2)dx$.

$$\int_0^4 (x^2+2x+2)^{-3}(x+1)\,dx = \frac{1}{2}\int_0^4 (x^2+2x+2)^{-3}[2(x+1)\,dx]$$

$$= \frac{1}{2}\left(\frac{1}{-2}\right)(x^2+2x+2)^{-2}\Big|_0^4 \qquad \text{integrate}$$

$$= -\frac{1}{4}(16+8+2)^{-2} + \frac{1}{4}(0+0+2)^{-2} \qquad \text{evaluate}$$

$$= \frac{1}{4}\left(-\frac{1}{26^2}+\frac{1}{2^2}\right) = \frac{1}{4}\left(\frac{1}{4}-\frac{1}{676}\right)$$

$$= \frac{1}{4}\left(\frac{168}{676}\right) = \frac{21}{338}$$

The following example illustrates an application of the definite integral. In Chapter 26, we will see that the definite integral has many applications in science and technology.

EXAMPLE 6 Definite integral—stopping distance of a train

The driver of a light-rail train travelling at 20.0 m/s applies the brakes, and the velocity then decreases by 1.25 m/s each second. Find the distance travelled by the train while coming to a stop.

Note that the velocity (in m/s) after t seconds is given by $v(t) = 20.0 - 1.25t$. By setting this equal to zero and solving for t, we find that it takes 16.0 s for the train to come to a stop. The distance d travelled during this time can be found by evaluating the definite integral of the velocity from 0 s to 16 s.

$$d = \int_0^{16}(20.0-1.25t)dt = \left(20.0\,t - \frac{1.25t^2}{2}\right)\Big|_0^{16} \qquad \text{integrate}$$

$$= \left[20.0(16) - \frac{1.25(16)^2}{2}\right] - 0 = 160\text{ m} \qquad \text{evaluate}$$

Therefore, the train travels 160 m while coming to a stop.

Some graphing calculators are programmed to evaluate definite integrals and areas under curves. The manual should be consulted to determine how any particular model is used for integration.

EXERCISES 25.4

In Exercises 1 and 2, make the given changes in the indicated examples of this section and then solve the resulting problems.

1. In Example 2, change the upper limit from 3 to 4.

2. In Example 4, change $4x$ to $2x$.

In Exercises 3–34, evaluate the given definite integrals.

3. $\int_0^3 6x\,dx$ **4.** $\int_0^2 4x^3\,dx$ **5.** $\int_1^4 x^{5/2}\,dx$

6. $\int_4^9 (p^{3/2}-3)\,dp$ **7.** $\int_3^6 \left(\frac{1}{\sqrt{x}}-7\right)dx$ **8.** $\int_{1.2}^{1.6}\left(5+\frac{6}{x^4}\right)dx$

9. $\int_{-1.6}^{0.7} (1 - x)^{1/3}\,dx$

10. $\int_{1}^{5} \sqrt{3v + 1}\,dv$

11. $\int_{-2}^{2} (T - 2)(T + 2)\,dT$

12. $\int_{1}^{2} (3x^5 - 2x^3)\,dx$

13. $\int_{0.5}^{2.2} \left(\sqrt[3]{x} - 2\right) dx$

14. $\int_{2.7}^{5.3} \left(\frac{1}{x\sqrt{x}} + 4\right) dx$

15. $\int_{0}^{4} \left(1 - \sqrt{x}\right)^2 dx$

16. $\int_{1}^{4} \frac{y + 4}{\sqrt{y}}\,dy$

17. $\int_{-2}^{-1} 2x(4 - x^2)^4\,dx$

18. $\int_{0}^{1} x(3x^2 - 1)^3\,dx$

19. $\int_{0}^{4} \frac{x\,dx}{\sqrt{x^2 + 9}}$

20. $\int_{0.2}^{0.7} x^2(x^3 + 2)^{3/2}\,dx$

21. $\int_{2.75}^{3.25} \frac{dx}{\sqrt[3]{6x + 1}}$

22. $\int_{2}^{12} \frac{8\,du}{\sqrt{4u + 1}}$

23. $\int_{1}^{3} \frac{12x\,dx}{(2x^2 + 1)^3}$

24. $\int_{12.6}^{17.2} \frac{3\,dx}{(6x - 1)^2}$

25. $\int_{3}^{7} \sqrt{16t^2 + 8t + 1}\,dt$

26. $\int_{-5}^{1} \sqrt{6 - 2x}\,dx$

27. $\int_{0}^{2} 2x(9 - 2x^2)^2\,dx$

28. $\int_{-1}^{2} V(V^3 + 1)\,dV$

29. $\int_{-1}^{2} \frac{8x - 2}{(2x^2 - x + 1)^3}\,dx$

30. $\int_{2}^{3} \frac{x^2 + 1}{(x^3 + 3x)^2}\,dx$

31. $\int_{0}^{1} (x^2 + 3)(x^3 + 9x + 6)^2\,dx$

32. $\int_{-3}^{-2} (3x^2 - 2)\sqrt[3]{2x^3 - 4x + 1}\,dx$

33. $\int_{\sqrt{5}}^{3} 2z\sqrt[4]{z^4 + 8z^2 + 16}\,dz$

34. $\int_{-2}^{0} \left(\sqrt{2x + 4} - \sqrt[3]{3x + 8}\right) dx$

In Exercises 35–52, solve the given problems.

35. Show that $\int_{0}^{1} x^3\,dx + \int_{1}^{2} x^3\,dx = \int_{0}^{2} x^3\,dx$. In terms of area, explain the result.

36. Write $\int_{1}^{13} f(x)\,dx - \int_{9}^{13} f(x)\,dx$ as a single definite integral.

37. Evaluate $\int_{x=1}^{x=4} y\,dx$, when $y^2 = 4x\ (y > 0)$.

38. Show that $\int_{1}^{3} 4x\,dx = -\int_{3}^{1} 4x\,dx$.

39. Show that $\int_{0}^{1} x\,dx > \int_{0}^{1} x^2\,dx$ and $\int_{1}^{2} x\,dx < \int_{1}^{2} x^2\,dx$. In terms of area, explain the result.

40. Given that $\int_{0}^{9} \sqrt{x}\,dx = 18$, evaluate $\int_{0}^{9} 2\sqrt{t}\,dt$.

41. Evaluate $\int_{-1}^{1} t^{2k}\,dt$, where k is a positive integer.

42. Evaluate the following integral, which arises in the study of electricity: $\int_{0}^{L} \frac{1}{EI}\left(-\frac{1}{2}wx^2\right)(-x)\,dx$.

43. Evaluate the following integral, which arises in the study of hydrodynamics: $\int_{H}^{h} \frac{Ay^{-1/2}\,dy}{a\sqrt{2g}}$.

44. It is estimated that a newly discovered oil field will produce oil at the rate of $\frac{dR}{dt} = \frac{400t^2}{(t^3 + 20)^2} + 10$ thousand barrels per year. How much oil can be expected from the field in the next ten years?

45. Evaluate $\int_{-3}^{3} |x - 1|\,dx$ by geometrically finding the area represented.

46. Evaluate $\int_{-2}^{2} \sqrt{4 - x^2}\,dx$ by geometrically finding the area represented.

47. The work W (in N · m) in winding up an 80 m cable is $W = \int_{0}^{80} (1000 - 5x)\,dx$. Evaluate W.

48. The total volume V of liquid flowing through a certain pipe of radius R is $V = k\left(R^2 \int_{0}^{R} r\,dr - \int_{0}^{R} r^3\,dr\right)$, where k is a constant. Evaluate V and explain why R, but not r, can be to the left of the integral sign.

49. The surface area A (in m^2) of a certain parabolic radio-wave reflector is $A = 4\pi \int_{0}^{2} \sqrt{3x + 9}\,dx$. Evaluate A.

50. The total force (in N) on the circular end of a water tank is $F = 19\,600 \int_{0}^{5} y\sqrt{25 - y^2}\,dy$. Evaluate F.

51. In finding the average electron energy in a metal at very low temperatures, the integral $\frac{3N}{2E_F^{3/2}} \int_{0}^{E_F} E^{3/2}\,dE$ is used. Evaluate this integral.

52. In finding the electric field E caused by a surface electric charge on a disc, the equation $E = k\int_{0}^{R} \frac{r\,dr}{(x^2 + r^2)^{3/2}}$ is used. Evaluate the integral.

Answers to Practice Exercises

1. $45/8$ **2.** $62/3$

25.5 Numerical Integration: The Trapezoidal Rule

For data and functions that cannot be directly integrated by available methods, it is possible to develop numerical methods of integration. These numerical methods are of greater importance today since they are readily adaptable for use on a calculator or computer. There are a great many such numerical techniques for approximating the value of an integral. In this section, we develop one of these, the *trapezoidal rule.* In the following section, another numerical method is discussed.

We know from Sections 25.3 and 25.4 that we can interpret a definite integral as the area under a curve. We will therefore show how to approximate the value of the integral by approximating the appropriate area by a set of inscribed trapezoids. The basic idea here is very similar to that used when rectangles were inscribed under a curve. However, the use of trapezoids reduces the error and provides a better approximation.

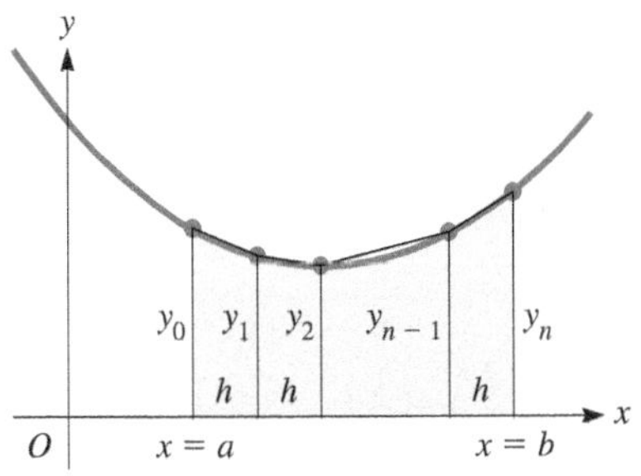

Fig. 25.12

The area to be found is subdivided into n intervals of equal width. Perpendicular lines are then dropped from the curve (or points, if only a given set of numbers is available). If the points on the curve are joined by straight-line segments, the area of successive parts under the curve is approximated by finding the areas of the trapezoids formed. However, if these points are not too far apart, the approximation will be very good (see Fig. 25.12). From geometry, recall that the area of a trapezoid equals one-half the product of the sum of the bases times the altitude. For these trapezoids, the bases are the y-coordinates, and the altitudes are h. Therefore, when we indicate the sum of these trapezoidal areas, we have

$$A_T = \frac{1}{2}(y_0 + y_1)h + \frac{1}{2}(y_1 + y_2)h + \frac{1}{2}(y_2 + y_3)h + \cdots$$
$$+ \frac{1}{2}(y_{n-2} + y_{n-1})h + \frac{1}{2}(y_{n-1} + y_n)h$$

We note, when this addition is performed, that the result is

$$A_T = h\left(\frac{1}{2}y_0 + y_1 + y_2 + \cdots + y_{n-1} + \frac{1}{2}y_n\right) \tag{25.12}$$

The y-values to be used either are derived from the function $y = f(x)$ or are the y-coordinates of a set of data.

Since A_T approximates the area under the curve, it also approximates the value of the definite integral. By factoring out $1/2$, we get what is known as the **trapezoidal rule.**

Trapezoidal Rule

$$\int_a^b f(x)\,dx \approx \frac{h}{2}(y_0 + 2y_1 + 2y_2 + \cdots + 2y_{n-1} + y_n) \tag{25.13}$$

The trapezoidal rule is essentially the same rule that we used in Chapter 2 to measure irregular geometric areas. Since the definite integral can be interpreted as the area under a curve, we can now use the trapezoidal rule to find the approximate value of a definite integral.

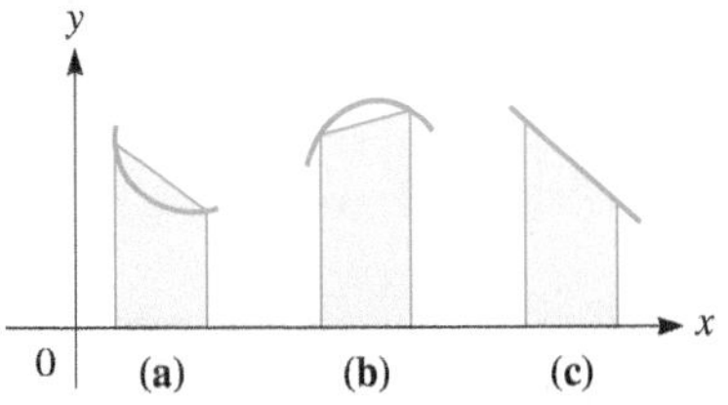

Fig. 25.13

Wherever the curve of the function being integrated is concave up, the approximating segments are above the curve and each trapezoid has slightly more area than the corresponding area under the curve (see Fig. 25.13a). If the curve is concave down, the approximating segments are below the curve, and each trapezoid has slightly less area than the corresponding area under the curve (see Fig. 25.13b). For a straight-line segment, the trapezoidal rule gives an exact value (see Fig. 25.13c).

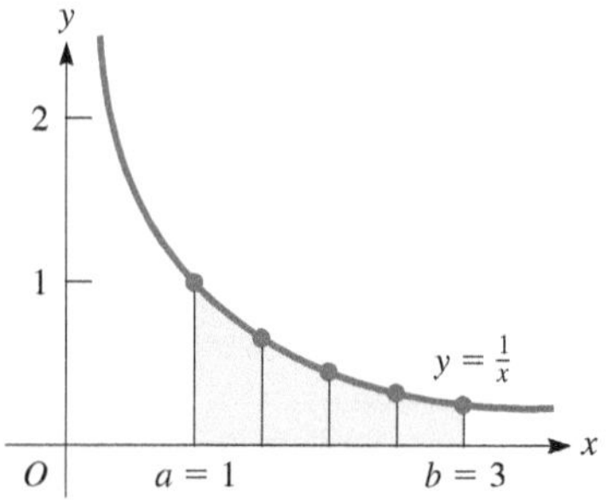

Fig. 25.14

Table 25.2

Number of Trapezoids n	Total Area of Trapezoids
4	1.116 666 7
100	1.098 641 9
1000	1.098 612 6
10 000	1.098 612 3

Practice Exercise

1. In Example 1, use the trapezoidal rule with $n = 2$.

EXAMPLE 1 Approximating an integral by the trapezoidal rule

Approximate the value of $\int_1^3 \frac{1}{x}dx$ by the trapezoidal rule. Let $n = 4$.

We are to approximate the area under $y = 1/x$ from $x = 1$ to $x = 3$ by dividing the area into four trapezoids. This area is found by applying Eq. (25.12), which is the approximate value of the integral, as shown in Eq. (25.13). Fig. 25.14 shows the graph. In this example, $f(x) = 1/x$, and

$$h = \frac{3-1}{4} = \frac{1}{2} \quad y_0 = f(a) = f(1) = 1$$

$$y_1 = f\left(\frac{3}{2}\right) = \frac{2}{3} \quad y_2 = f(2) = \frac{1}{2}$$

$$y_3 = f\left(\frac{5}{2}\right) = \frac{2}{5} \quad y_n = y_4 = f(b) = f(3) = \frac{1}{3}$$

$$A_T = \frac{1/2}{2}\left[1 + 2\left(\frac{2}{3}\right) + 2\left(\frac{1}{2}\right) + 2\left(\frac{2}{5}\right) + \frac{1}{3}\right]$$

$$= \frac{1}{4}\left(\frac{15 + 20 + 15 + 12 + 5}{15}\right) = \frac{1}{4}\left(\frac{67}{15}\right) = \frac{67}{60}$$

Therefore,

$$\int_1^3 \frac{1}{x}dx \approx \frac{67}{60} = 1.12$$

We cannot perform this integration directly by methods developed up to this point. As we increase the number of trapezoids, the value becomes more accurate. See Table 25.2. The actual value to seven decimal places is 1.098 612 3.

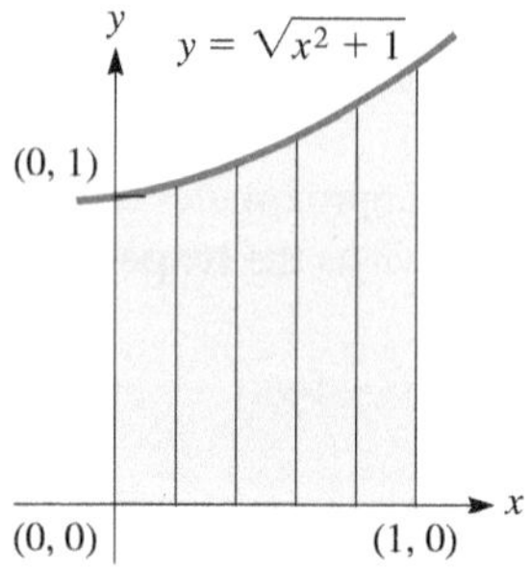

Fig. 25.15

EXAMPLE 2 Approximating an integral by the trapezoidal rule

Approximate the value of $\int_0^1 \sqrt{x^2 + 1}\,dx$ by the trapezoidal rule. Let $n = 5$.

Fig. 25.15 shows the graph. In this example,

$$h = \frac{1-0}{5} = 0.2$$

$$y_0 = f(0) = 1 \qquad y_1 = f(0.2) = \sqrt{1.04} = 1.019\,803\,9$$

$$y_2 = f(0.4) = \sqrt{1.16} = 1.077\,033\,0 \qquad y_3 = f(0.6) = \sqrt{1.36} = 1.166\,190\,4$$

$$y_4 = f(0.8) = \sqrt{1.64} = 1.280\,624\,8 \qquad y_5 = f(1) = \sqrt{2.00} = 1.414\,213\,6$$

Hence, we have

$$A_T = \frac{0.2}{2}[1 + 2(1.109\,803\,9) + 2(1.077\,033\,0) + 2(1.166\,190\,4)$$
$$+ 2(1.280\,624\,8) + 1.414\,213\,6]$$
$$= 1.15 \quad \text{(rounded off to three significant digits)}$$

This means that

$$\int_0^1 \sqrt{x^2 + 1}\,dx \approx 1.15 \quad \text{the actual value is 1.148 to three decimal places}$$

We note that the entire calculation can be done on a calculator without tabulating values by entering the formula directly in terms of square roots.

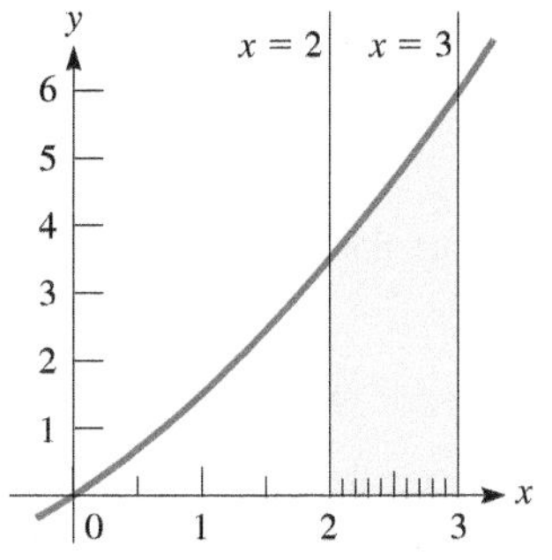

Fig. 25.16

EXAMPLE 3 Approximating an integral by the trapezoidal rule

Approximate the value of the integral $\int_2^3 x\sqrt{x+1}\,dx$ by using the trapezoidal rule. Use $n = 10$.

In Fig. 25.16, the graph of the function and the area used in the trapezoidal rule are shown. From the given values, we have $h = \frac{3-2}{10} = 0.1$. Therefore,

$$y_0 = f(2) = 2\sqrt{3} = 3.464\,101\,6 \qquad y_1 = f(2.1) = 2.1\sqrt{3.1} = 3.697\,431\,5$$
$$y_2 = f(2.2) = 2.2\sqrt{3.2} = 3.935\,479\,6 \qquad y_3 = f(2.3) = 2.3\sqrt{3.3} = 4.178\,157\,5$$
$$y_4 = f(2.4) = 2.4\sqrt{3.4} = 4.425\,381\,3 \qquad y_5 = f(2.5) = 2.5\sqrt{3.5} = 4.677\,071\,7$$
$$y_6 = f(2.6) = 2.6\sqrt{3.6} = 4.933\,153\,2 \qquad y_7 = f(2.7) = 2.7\sqrt{3.7} = 5.193\,553\,7$$
$$y_8 = f(2.8) = 2.8\sqrt{3.8} = 5.458\,204\,8 \qquad y_9 = f(2.9) = 2.9\sqrt{3.9} = 5.727\,041\,1$$
$$y_{10} = f(3) = 3\sqrt{4} = 6.000\,000\,0$$

$$A_T = \frac{0.1}{2}[3.464\,101\,6 + 2(3.697\,431\,5) + \cdots + 2(5.727\,041\,1) + 6.000\,000\,0]$$
$$= 4.6958$$

Therefore, $\int_2^3 x\sqrt{x+1}\,dx \approx 4.6958$. The actual value of the integral is 4.6954 to four decimal places.

EXAMPLE 4 Trapezoidal rule with empirical data—area of a park

In estimating the area of a proposed city park bounded by three straight streets, two parallel and perpendicular to the third, and a river (see Fig. 25.17), measurements were taken at 10.0 m intervals of the distance to the river, and the values found are in the following table. Find the area of the proposed park, using the trapezoidal rule.

x (m)	0	10	20	30	40	50
y (m)	56.8	67.5	73.2	73.5	68.8	62.4

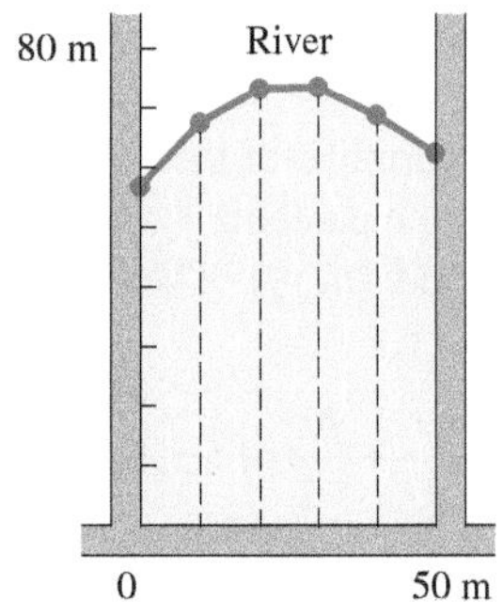

Fig. 25.17

To find the area, we use the values directly from the table. We note that $h = 10.0$ m.

$$A = \frac{10.0}{2}[56.8 + 2(67.5) + 2(73.2) + 2(73.5) + 2(68.8) + 62.4]$$
$$= 3426 \text{ m}^2$$

Rounding off to three significant digits, the accuracy of the data, the area of the proposed park is about 3430 m^2.

Although we do not know the mathematical form of the function, we can state that

$$\int_0^{50.0} f(x)dx \approx 3430$$

EXERCISES 25.5

In Exercises 1 and 2, make the given changes in the indicated examples of this section and then solve the resulting problems.

1. In Example 1, change n from 4 to 2.
2. In Example 3, change n from 10 to 5.

In Exercises 3–6, (a) approximate the value of each of the given integrals by use of the trapezoidal rule, using the given value of n, and (b) check by direct integration.

3. $\int_0^2 2x^2 dx,\ n = 4$
4. $\int_0^1 (1 - x^2)\,dx,\ n = 3$
5. $\int_1^4 (1 + \sqrt{x})\,dx,\ n = 6$
6. $\int_3^8 \sqrt{1+x}\,dx,\ n = 5$

In Exercises 7–14, approximate the value of each of the given integrals by use of the trapezoidal rule, using the given value of n. Round to three significant digits.

7. $\int_2^3 \frac{1}{2x}\,dx,\ n = 2$

8. $\int_2^6 \frac{dx}{x+3},\ n = 4$

9. $\int_0^5 \sqrt{25 - x^2}\,dx,\ n = 5$

10. $\int_0^2 \sqrt{x^3 + 1}\,dx,\ n = 4$

11. $\int_1^5 \frac{1}{x^2 + x}\,dx,\ n = 10$

12. $\int_2^4 \frac{1}{x^2 + 1}\,dx,\ n = 10$

13. $\int_0^4 2^x\,dx,\ n = 12$

14. $\int_0^{1.5} 10^x\,dx,\ n = 15$

In Exercises 15 and 16, approximate the values of the integrals defined by the given sets of points.

15. $\int_2^{14} y\,dx$

x	2	4	6	8	10	12	14
y	0.670	2.34	4.56	3.67	3.56	4.78	6.87

16. $\int_{1.4}^{3.2} y\,dx$

x	1.4	1.7	2.0	2.3	2.6	2.9	3.2
y	0.180	7.87	18.23	23.53	24.62	20.93	20.76

In Exercises 17–22, solve each given problem by using the trapezoidal rule. Round to three significant digits unless otherwise stated.

17. Explain why the approximate value of the integral in Exercise 5 is less than the exact value.

18. $\int_0^2 \sqrt{4 - x^2}\,dx = \pi$. Approximate the value of the integral with $n = 8$. Compare with π.

19. $\int_0^3 \frac{dx}{2x + 2} = \ln 2$. Approximate the value of the integral with $n = 6$. Compare with ln 2.

20. A force F that a distributed electric charge has on a point charge is $F = k\int_0^2 \frac{dx}{(4 + x^2)^{3/2}}$, where x is the distance along the distributed charge and k is a constant. With $n = 8$, evaluate F in terms of k.

21. The length L (in m) of telephone wire needed (considering the sag) between two poles exactly 100 m apart is $L = 2\int_0^{50}\sqrt{6.4 \times 10^{-7}x^2 + 1}\,dx$. With $n = 10$, evaluate L (to six significant digits).

22. The amount A (in standard pollution index) of a pollutant in the air in a city is measured to be $A = \frac{150}{1 + 0.25(t - 4.0)^2} + 25$, here t is the time (in h) after 6:00 A.M. With $n = 6$, find the total value of A between 6:00 A.M. and noon.

Answer to Practice Exercise

1. $7/6 = 1.1667$

25.6 Simpson's Rule

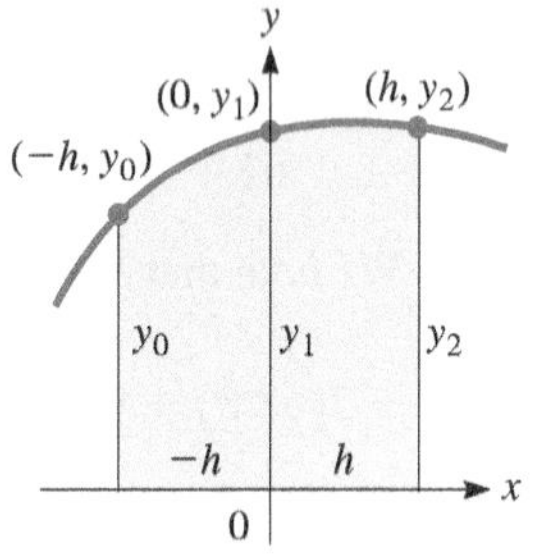

Fig. 25.18

The numerical method of integration developed in this section is also readily programmable for use on a computer or easily usable with the necessary calculations done on a calculator. It is obtained by interpreting the definite integral as the area under a curve, as we did in developing the trapezoidal rule, and by approximating the curve by a set of parabolic arcs. The use of parabolic arcs, rather than chords as with the trapezoidal rule, usually gives a better approximation.

Since we will be using parabolic arcs, we first derive a formula for the area that is under a parabolic arc. The curve shown in Fig. 25.18 represents the parabola $y = ax^2 + bx + c$. The points shown on this curve are $(-h, y_0)$, $(0, y_1)$, and (h, y_2). The area under the parabola is given by

$$A = \int_{-h}^{h} y\,dx = \int_{-h}^{h} (ax^2 + bx + c)\,dx = \frac{ax^3}{3} + \frac{bx^2}{2} + cx\Big|_{-h}^{h}$$
$$= \frac{2}{3}ah^3 + 2ch$$

$$A = \frac{h}{3}(2ah^2 + 6c) \qquad \textbf{(25.14)}$$

The coordinates of the three points also satisfy the equation $y = ax^2 + bx + c$. This means that

$$y_0 = ah^2 - bh + c$$
$$y_1 = c$$
$$y_2 = ah^2 + bh + c$$

By finding the sum of $y_0 + 4y_1 + y_2$, we have

$$y_0 + 4y_1 + y_2 = 2ah^2 + 6c \qquad \textbf{(25.15)}$$

LEARNING TIP

Note that the area under the parabolic arc depends only on the distance h and the three y-coordinates. The coefficients of the equation of the parabola are not required, so they are not calculated.

Substituting Eq. (25.15) into Eq. (25.14), we have

$$A = \frac{h}{3}(y_0 + 4y_1 + y_2) \qquad \textbf{(25.16)}$$

Now, let us consider the area under the curve in Fig. 25.19. If a parabolic arc is passed through the points (x_0, y_0), (x_1, y_1), and (x_2, y_2), we may use Eq. (25.16) to approximate the area under the curve between x_0 and x_2. We again note that the distance h is the difference in the x-coordinates. Therefore, the area under the curve between x_0 and x_2 is

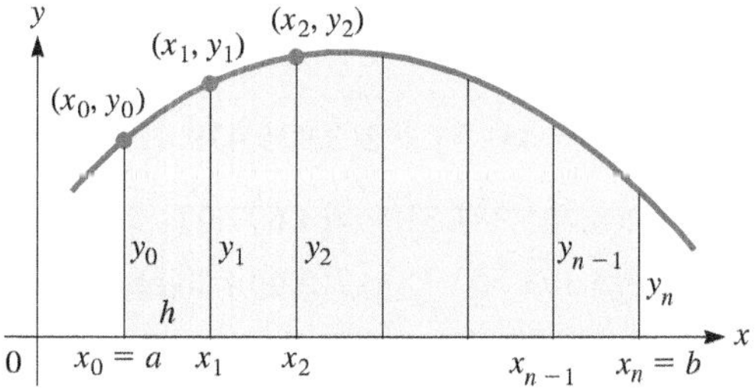

Fig. 25.19

$$A_1 = \frac{h}{3}(y_0 + 4y_1 + y_2)$$

Similarly, if a parabolic arc is passed through the three points starting with (x_2, y_2), the area between x_2 and x_4 is

$$A_2 = \frac{h}{3}(y_2 + 4y_3 + y_4)$$

The sum of these areas is

$$A_1 + A_2 = \frac{h}{3}(y_0 + 4y_1 + 2y_2 + 4y_3 + y_4) \qquad \textbf{(25.17)}$$

We can continue this procedure until the approximate value of the entire area has been found. We must note, however, that ***the number of intervals n of width h must be even.*** Therefore, generalizing Eq. (25.17) and recalling again that the value of the definite integral is the area under the curve, we have what is known as **Simpson's rule**.

Simpson's Rule

$$\int_a^b f(x)\,dx \approx \frac{h}{3}(y_0 + 4y_1 + 2y_2 + 4y_3 + 2y_4 + \cdots + 4y_{n-1} + y_n),\ n \text{ even} \qquad \textbf{(25.18)}$$

■ Although Simpson's rule is named for the English mathematician Thomas Simpson (1710–1761), he did not discover the rule. It was well known when he included it in some of his many books on mathematics.

As with the trapezoidal rule, we used Simpson's rule in Chapter 2 to measure irregular areas.

EXAMPLE 1 Approximating an integral by using Simpson's rule

Approximate the value of the integral $\int_0^1 \frac{dx}{x+1}$ by Simpson's rule. Let $n = 2$.

In Fig. 25.20, the graph of the function and the area used are shown. We are to approximate the integral by using Eq. (25.18). We therefore note that $f(x) = 1/(x+1)$. Also, $x_0 = a = 0$, $x_1 = 0.5$, and $x_2 = b = 1$. This is due to the fact that $n = 2$ and $h = 0.5$ since the total interval is 1 unit (from $x = 0$ to $x = 1$). Therefore,

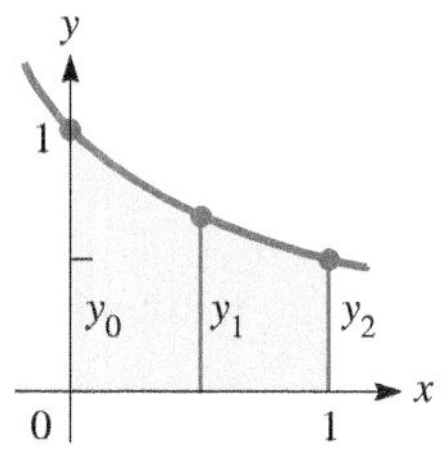

Fig. 25.20

$$y_0 = \frac{1}{0+1} = 1.0000 \qquad y_1 = \frac{1}{0.5+1} = 0.6667 \qquad y_2 = \frac{1}{1+1} = 0.5000$$

Substituting, we have

$$\int_0^1 \frac{dx}{x+1} = \frac{0.5}{3}[1.0000 + 4(0.6667) + 0.5000]$$
$$= 0.694$$

To three decimal places, the actual value of the integral is 0.693. We will consider the method of integrating this function in a later chapter.

Practice Exercise

1. In Example 1, use Simpson's rule with $n = 4$.

EXAMPLE 2 Approximating an integral by using Simpson's rule

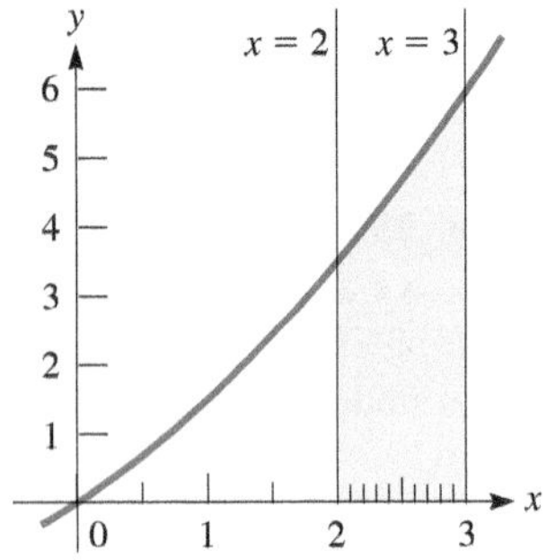

Fig. 25.21

Approximate the value of $\int_2^3 x\sqrt{x+1}\,dx$ by Simpson's rule. Use $n = 10$.

Since the necessary values for this function are shown in Example 3 of Section 25.5, we shall simply tabulate them here $(h = 0.1)$. See Fig. 25.21.

$$y_0 = 3.464\,101\,6 \quad y_1 = 3.697\,431\,5 \quad y_2 = 3.935\,479\,6 \quad y_3 = 4.178\,157\,5$$
$$y_4 = 4.425\,381\,3 \quad y_5 = 4.677\,071\,7 \quad y_6 = 4.933\,153\,2 \quad y_7 = 5.193\,553\,7$$
$$y_8 = 5.458\,204\,8 \quad y_9 = 5.727\,041\,1 \quad y_{10} = 6.000\,000\,0$$

Therefore, we evaluate the integral as follows:

$$\begin{aligned}\int_2^3 x\sqrt{x+1}\,dx &= \frac{0.1}{3}[3.464\,101\,6 + 4(3.697\,431\,5) + 2(3.935\,479\,6)\\ &\quad + 4(4.178\,157\,5) + 2(4.425\,381\,3) + 4(4.677\,071\,7)\\ &\quad + 2(4.933\,153\,2) + 4(5.193\,553\,7) + 2(5.458\,204\,8)\\ &\quad + 4(5.727\,041\,1) + 6.000\,000\,0]\\ &= \frac{0.1}{3}(140.861\,56) = 4.695\,385\,4\end{aligned}$$

This result agrees with the actual value to the eight significant digits shown. The value we obtained earlier with the trapezoidal rule was 4.6958.

EXAMPLE 3 Simpson's rule—area of an aircraft stabilizer

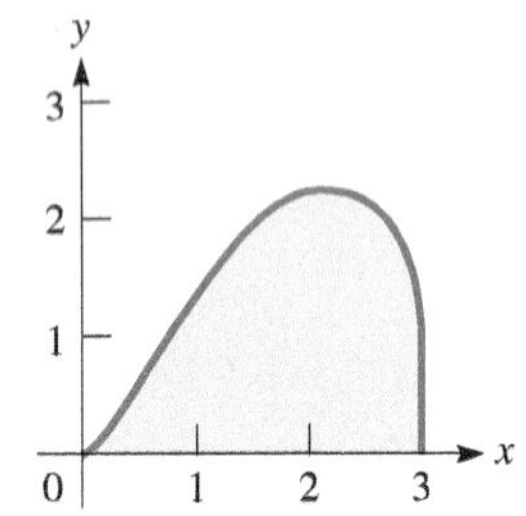

Fig. 25.22

The rear stabilizer of a certain aircraft is shown in Fig. 25.22. The area A (in m^2) of one side of the stabilizer is $A = \int_0^3 (3x^2 - x^3)^{0.6}dx$. Find this area, using Simpson's rule with $n = 6$.

Here, we note that $f(x) = (3x^2 - x^3)^{0.6}$, $a = 0$, $b = 3$, and $h = \frac{3-0}{6} = 0.5$. Therefore,

$$\begin{aligned}y_0 &= f(0) = [3(0)^2 - 0^3]^{0.6} = 0\\ y_1 &= f(0.5) = [3(0.5)^2 - 0.5^3]^{0.6} = 0.754\,272\,0\\ y_2 &= f(1) = [3(1)^2 - 1^3]^{0.6} = 1.515\,716\,6\\ y_3 &= f(1.5) = [3(1.5)^2 - 1.5^3]^{0.6} = 2.074\,742\,8\\ y_4 &= f(2) = [3(2)^2 - 2^3]^{0.6} = 2.297\,396\,7\\ y_5 &= f(2.5) = [3(2.5)^2 - 2.5^3]^{0.6} = 1.981\,116\,5\\ y_6 &= f(3) = [3(3)^2 - 3^3]^{0.6} = 0\end{aligned}$$

$$\begin{aligned}A &= \int_0^3 (3x^2 - x^3)^{0.6}dx = \frac{0.5}{3}[0 + 4(0.754\,272\,0) + 2(1.515\,716\,6)\\ &\quad + 4(2.074\,742\,8) + 2(2.297\,396\,7) + 4(1.981\,116\,5) + 0]\\ &= 4.477\,792\,0\ m^2\end{aligned}$$

Thus, the area of one side of the stabilizer is 4.8 m^2 (rounded off). As with the trapezoidal rule, the calculation can be done completely on a calculator, without recording the above values.

Just like the trapezoidal rule, Simpson's rule is especially useful when approximating integrals of functions for which only a given set of points is available. This is often the case for functions obtained from empirical data or from experimental measurements. Examples can be found in the exercises.

EXERCISES 25.6

In Exercises 1 and 2, make the given changes in the indicated examples of this section, and then solve the indicated problems.

1. In Example 1, change the denominator $x + 1$ to $x + 2$ and then find the approximate value of the integral.

2. In Example 2, change n such that $h = 0.2$ and explain why Simpson's rule cannot be used.

In Exercises 3–6, (a) approximate the value of each of the given integrals by use of Simpson's rule, using the given value of n, and (b) check by direct integration. Round to three significant digits.

3. $\int_0^2 (1 + x^3)\, dx,\ n = 2$ **4.** $\int_0^8 x^{1/3} dx,\ n = 2$

5. $\int_1^4 (2x + \sqrt{x})\, dx,\ n = 6$ **6.** $\int_0^2 x\sqrt{x^2 + 1}\, dx,\ n = 4$

In Exercises 7–12, approximate the value of each of the given integrals by use of Simpson's rule, using the given values of n. Exercises 8–10 are the same as Exercises 10–12 of Section 25.5. Round to three significant digits.

7. $\int_0^5 \sqrt{25 - x^2}\, dx,\ n = 4$ **8.** $\int_0^2 \sqrt{x^3 + 1}\, dx,\ n = 4$

9. $\int_1^5 \frac{1}{x^2 + x} dx,\ n = 10$ **10.** $\int_2^4 \frac{1}{x^2 + 1} dx,\ n = 10$

11. $\int_{-4}^5 (2x^4 + 1)^{0.1} dx,\ n = 6$ **12.** $\int_0^{2.4} \frac{dx}{(4 + \sqrt{x})^{3/2}},\ n = 8$

In Exercises 13 and 14, approximate the values of the integrals defined by the given sets of points by using Simpson's rule. These are the same as Exercises 15 and 16 of Section 25.5.

13. $\int_2^{14} y\, dx$

x	2	4	6	8	10	12	14
y	0.670	2.34	4.56	3.67	3.56	4.78	6.87

14. $\int_{1.4}^{3.2} y\, dx$

x	1.4	1.7	2.0	2.3	2.6	2.9	3.2
y	0.180	7.87	18.23	23.53	24.62	20.93	20.76

In Exercises 15–18, solve the given problems using Simpson's rule. Exercise 15 and 16 are the same as Exercises 18 and 19 of Section 25.5.

15. $\int_0^2 \sqrt{4 - x^2}\, dx = \pi$. Approximate the value of the integral with $n = 8$. Compare with π.

16. $\int_0^3 \frac{dx}{2x + 2} = \ln 2$. Approximate the value of the integral with $n = 6$. Compare with $\ln 2$.

17. The distance $\bar{x}$ (in cm) from one end of a barrel plug (with vertical cross-section) to its centre of mass, as shown in Fig. 25.23, is $\bar{x} = 0.9129\int_0^3 x\sqrt{0.3 - 0.1x}\,dx$. Find $\bar{x}$ with $n = 12$.

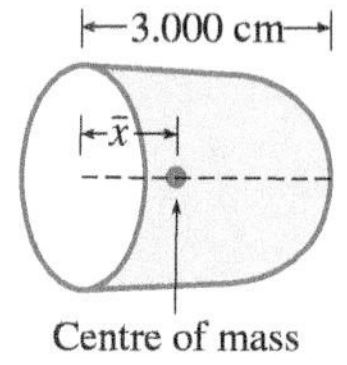

Fig. 25.23

18. The average value of the electric current i_{av} (in A) in a circuit for the first 4 s is $i_{av} = \frac{1}{4}\int_0^4 (4t - t^2)^{0.2} dt$. Find i_{av} with $n = 10$.

Answer to Practice Exercise

1. 0.693

CHAPTER 25 KEY FORMULAS AND EQUATIONS

Indefinite integral	$\int f(x)\, dx = F(x) + C$	**(25.1)**
Power rule	$\int u^n du = \frac{u^{n+1}}{n + 1} + C \quad (n \neq -1)$	**(25.2)**
Constant factor rule	$\int c\, du = c\int du = cu + C$	**(25.3)**
Sum rule	$\int (du + dv) = u + v + C$	**(25.4)**
Area under a curve	$A_{ab} = \left[\int f(x)\, dx\right]_a^b = F(b) - F(a)$	**(25.10)**
Definite integral	$\int_a^b f(x)\, dx = F(b) - F(a)$	**(25.11)**
Trapezoidal rule	$\int_a^b f(x)\, dx \approx \frac{h}{2}(y_0 + 2y_1 + 2y_2 + \cdots + 2y_{n-1} + y_n)$	**(25.13)**
Simpson's rule	$\int_a^b f(x)\, dx \approx \frac{h}{3}(y_0 + 4y_1 + 2y_2 + 4y_3 + 2y_4 + \cdots + 4y_{n-1} + y_n)$, n even	**(25.18)**

CHAPTER 25 REVIEW EXERCISES

In Exercises 1–24, evaluate the given integrals.

1. $\int (4x^3 - x)\,dx$

2. $\int (5 + 3t^2)\,dt$

3. $\int \sqrt{u}(u^2 + 8)\,du$

4. $\int x(x - 3x^4)\,dx$

5. $\int_1^4 \left(\frac{\sqrt{x}}{2} + \frac{2}{\sqrt{x}}\right)dx$

6. $\int_1^2 \left(x + \frac{4}{x^5}\right)dx$

7. $\int_0^2 x(4 - x)\,dx$

8. $\int_0^1 2t(2t + 1)^2 dt$

9. $\int \left(5 + \frac{6}{x^{13}}\right)dx$

10. $\int \left(3\sqrt{x} + \frac{1}{2\sqrt{x}} - \frac{1}{4}\right)dx$

11. $\int_{-2}^{5} \frac{dx}{\sqrt[3]{x^2 + 6x + 9}}$

12. $\int_{0.35}^{0.85} x\left(\sqrt{1 - x^2} + 1\right)dx$

13. $\int \frac{dn}{(9 - 5n)^3}$

14. $\int \frac{1}{x^2}\sqrt{6 + \frac{1}{x}}\,dx$

15. $\int 3(7 - 2x)^{3/4} dx$

16. $\int (y^3 + 3y^2 + 3y + 1)^{2/3} dy$

17. $\int_0^2 \frac{3x\,dx}{\sqrt[3]{1 + 2x^2}}$

18. $\int_1^6 \frac{12\,dx}{(3x - 2)^{3/4}}$

19. $\int x^2(1 - 2x^3)^9 dx$

20. $\int 3R^3(1 - 5R^4)\,dR$

21. $\int \frac{(2 - 3x^2)\,dx}{(2x - x^3)^2}$

22. $\int \frac{x^2 - 3}{\sqrt{6 + 9x - x^3}}dx$

23. $\int_1^3 (x^2 + x + 2)(2x^3 + 3x^2 + 12x)\,dx$

24. $\int_0^2 (4x + 18x^2)(x^2 + 3x^3)^2 dx$

In Exercises 25–38, solve the given problems.

25. Find the equation of the curve that passes through $(-1, 3)$ for which the slope is given by $3 - x^2$.

26. Find the equation of the curve that passes through $(1, -2)$ for which the slope is $x(x^2 + 1)^2$.

27. Perform the integration $\int (1 - 2x)\,dx$ (a) term by term, labelling the constant of integration as C_1, and then (b) by letting $u = 1 - 2x$, using the general power rule and labelling the constant of integration as C_2. Is $C_1 = C_2$? Explain.

28. Following the methods (a) and (b) in Exercise 27, perform the integration $\int (3x + 2)\,dx$. In (b) let $u = 3x + 2$. Is $C_1 = C_2$? Explain.

29. Write $\int_3^8 F(v)\,dv - \int_4^8 F(v)\,dv$ as a single definite integral.

30. Show that $\int_{-3}^{0} x^3 dx = -\int_0^3 x^3 dx$.

31. Find the general form of the function whose second derivative is $1/\sqrt{6x + 5}$.

32. Find the equation of the curve for which the second derivative is $-6x$ if the curve passes through $(-1, 3)$ with a slope of -3.

33. Show that $\int_0^1 x^3 dx = \int_1^2 (x - 1)^3 dx$. In terms of area, explain this result.

34. If $\int_0^1 [f(x) - g(x)]\,dx = 3$ and $\int_0^1 g(x)\,dx = -1$, find the value of $\int_0^1 2f(x)\,dx$.

35. Use Eq. (25.10) to find the area under $y = 8x - 5$ between $x = 1$ and $x = 3$.

36. Use Eq. (25.10) to find the first-quadrant area under $y = 8x - x^4$.

37. Given that $f(x)$ is continuous, $f(x) > 0$, and $f''(x) < 0$ for $a \le x \le b$, explain why the exact value of $\int_a^b f(x)\,dx$ is greater than the approximate value found by use of the trapezoidal rule.

38. It is shown in more advanced works that when evaluating $\int_a^b f(x)\,dx$, the maximum error in using Simpson's rule is $\frac{M(b - a)^5}{180n^4}$, where M is the greatest absolute value of the fourth derivative of $f(x)$ for $a \le x \le b$. Evaluate the maximum error for the integral $\int_2^3 \frac{dx}{2x}$ with $n = 4$.

In Exercises 39 and 40, solve the given problems by using the trapezoidal rule. In Exercises 41 and 42, solve the given problems using Simpson's rule.

39. Approximate $\int_1^3 \frac{dx}{2x - 1}$ with $n = 4$.

40. The water flow F (in m^3/s) passing through a cross-section of a stream is found by evaluating $F = \int_0^L g(x)\,dx$, where L is the width of the stream cross-section and $g(x)$ is the product of the depth and velocity of the stream x m from the bank. From the following table, estimate F with $L = 24$ m.

x (m)	0.0	3.0	6.0	9.0	12.0	15.0	18.0	21.0	24.0
$g(x)$	0.00	0.59	0.13	0.34	0.76	0.65	0.29	0.07	0.00

41. Approximate $\int_1^3 \frac{dx}{2x - 1}$ with $n = 4$ (see Exercise 39).

42. The velocity v (in km/h) of a car was recorded at 1-min intervals as shown. Estimate the distance travelled by the car.

t (min)	0	1	2	3	4	5	6	7	8	9	10
v (km/h)	60	62	65	69	72	74	76	77	77	75	76

In Exercises 43–48, use the function $y = x\sqrt[3]{2x^2 + 1}$ and approximate the area under the curve between $x = 1$ and $x = 4$ by the indicated method. See Fig. 25.24.

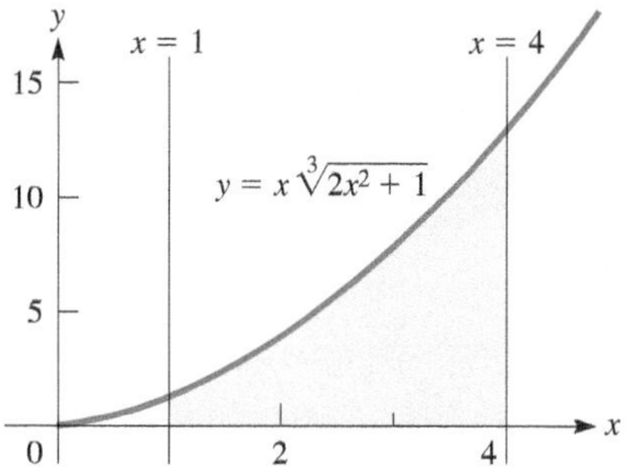

Fig. 25.24

43. Find the sum of the areas of three inscribed rectangles.

44. Find the sum of the areas of six inscribed rectangles.

45. Use the trapezoidal rule with $n = 3$.

46. Use the trapezoidal rule with $n = 6$.

47. Use Simpson's rule with $n = 6$.

48. Use integration (for the exact area).

In Exercises 49–51, find the area of the archway, as shown in Fig. 25.25, by the indicated method. The archway can be described as the area bounded by the elliptical arc $y = 4 + \sqrt{1 + 8x - 2x^2}$, $x = 0$, $x = 4$, and $y = 0$, where dimensions are in metres.

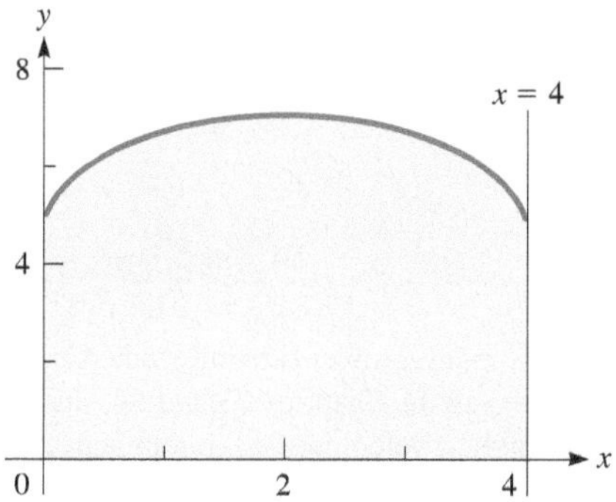

Fig. 25.25

49. Find the sum of the areas of eight inscribed rectangles.

50. Use the trapezoidal rule with $n = 8$.

51. Use Simpson's rule with $n = 8$.

In Exercises 52–58, solve the given problems by integration. In Exercise 59, provide the required information.

52. The charge Q (in C) transmitted in a certain electric circuit in 3 s is given by $Q = \int_0^3 6t^2 dt$. Evaluate Q.

53. The velocity ds/dt (in m/s) of a projectile is $ds/dt = -9.8t + 16$. Find the displacement s of the object after 4.0 s if the initial displacement is 48 m.

54. The deflection y of a certain beam at a distance x from one end is given by $dy/dx = k(2L^3 - 12Lx + 2x^4)$, where k is a constant and L is the length of the beam. Find y as a function of x if $y = 0$ for $x = 0$.

55. The total electric charge Q on a charged sphere is given by $Q = k\int\left(r^2 - \frac{r^3}{R}\right) dr$, where k is a constant, r is the distance from the centre of the sphere, and R is the radius of the sphere. Find Q as a function of r if $Q = Q_0$ for $r = R$.

56. Part of the deck of a boat is the parabolic area shown in Fig. 25.26. The area A (in m²) is $A = 2\int_0^{\sqrt{5}}\sqrt{5 - y}\, dy$. Evaluate A.

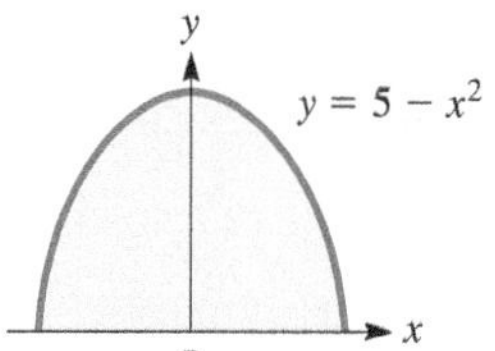

Fig. 25.26

57. The distance s (in cm) through which a cam follower moves in 4 s is $s = \int_0^4 t\sqrt{4 + 9t^2}\, dt$. Evaluate s.

58. The probability that the time of failure (in years) of a certain hydraulic component is between 3 and 6 years is given by $p = \int_3^6 \frac{32dx}{(x + 4)^3}$. Evaluate p.

59. A computer science student is writing a program to find a good approximation for the value of π by using the formula $A = \pi r^2$ for a circle. The value of π is to be found by approximating the area of a circle with a given radius. Write two or three paragraphs explaining how the value of π can be approximated in this way. Include any equations and values that may be used, but do not actually make the calculations.

CHAPTER 25 PRACTICE TEST

1. Find an antiderivative of $f(x) = 2x - (1 - x)^4$.

2. Integrate $\int x\sqrt{1 - 2x^2}\, dx$.

3. Find y in terms of x if $dy/dx = (6 - x)^4$ and the curve passes through $(5, 2)$.

4. Approximate the area under $y = \frac{1}{x + 2}$ between $x = 1$ and $x = 4$ (above the x-axis) by inscribing six rectangles and finding the sum of their areas.

5. Evaluate $\int_1^4 \frac{dx}{x + 2}$ by using the trapezoidal rule with $n = 6$.

6. Evaluate the definite integral of Problem 5 by using Simpson's rule with $n = 6$.

7. The total electric current i (in A) to pass a point in the circuit between $t = 1$ s and $t = 3$ s is $i = \int_1^3\left(t^2 + \frac{1}{t^2}\right)dt$. Evaluate i.

26. Applications of Integration

LEARNING OUTCOMES

After completion of this chapter, the student should be able to:

- Solve application problems involving indefinite integrals, including velocity and displacement and voltage across a capacitor
- Find the area between curves
- Obtain the volume of a solid of revolution
- Find the centre of mass, the moment of inertia, and the radius of gyration of a flat plate and of a solid of revolution
- Solve application problems involving definite integrals, including work, liquid pressure, and average value of a function

▲ In Section 26.2, we use integration to estimate the area of the glass walk portion of the Glacier Skywalk over Jasper National Park in Alberta.

With the development of the calculus, many problems being studied in the 1600s and later were much more easily solved. As we saw in Chapters 23 and 24, differential calculus led to the solution of problems such as finding velocities, maximum and minimum values, and various types of rates of change.

Integral calculus also led to the solution of many types of problems that were being studied. As Newton and Leibniz developed the methods of integral calculus, they were interested in finding areas and the problems that could be solved by finding these areas. They were also very interested in applications in which the rate of change was known and therefore led to the relation between the variables being studied.

One of the problems being studied in the 1600s by the French mathematician and physicist Blaise Pascal was that of the pressure within a liquid and the force on the walls of the container due to this pressure. Since pressure is a measure of the force on an area, finding the force became essentially solving an area problem. Also at the time, many mathematicians and physicists were studying various kinds of motion, such as motion along a curved path and the motion of a rotating object. When information about the velocity is known, the solution is found by integrating. Later, in the 1800s, since electric current is the time rate change of electric charge, the current could be found by integrating known expressions for the charge.

As it turns out, integration is useful in many areas of science, engineering, and technology. It has important applications in areas such as electricity, mechanics, architecture, machine design, statistics, and business, as well as other areas of physics and geometry.

In the first section of this chapter, we present some important applications of the indefinite integral, with emphasis on the motion of an object and the voltage across a capacitor. In the remaining sections, we show uses of the definite integral related to geometry, mechanics, work by a variable force, and force due to liquid pressure.

26.1 Applications of the Indefinite Integral

VELOCITY AND DISPLACEMENT

We first apply integration to the problem of finding the displacement and velocity as functions of time, when we know the relationship between acceleration and time, and certain values of displacement and velocity. As shown in Section 25.2, these values are needed for finding the constants of integration that are introduced.

■ Velocity as a first derivative and acceleration as a second derivative were introduced in Chapter 23.

Recalling that the acceleration a of an object is given by $a = dv/dt$, we can find the expression for the velocity v in terms of a, the time t, and the constant of integration. Therefore, we write $dv = a\,dt$, or

$$v = \int a\,dt \tag{26.1}$$

If the acceleration is constant, we have

$$v = at + C_1 \tag{26.2}$$

In general, Eq. (26.1) is used to find the velocity as a function of time when we know the acceleration as a function of time. Since the case of constant acceleration is often encountered, Eq. (26.2) can often be used.

EXAMPLE 1 Find the velocity, given the acceleration

Find the expression for the velocity if $a = 12t$, given that $v = 8$ when $t = 1$.

Using Eq. (26.1), we have

$$v = \int (12t)\,dt = 6t^2 + C_1$$

Substituting the known values, we have $8 = 6 + C_1$, or $C_1 = 2$. This means that

$$v = 6t^2 + 2$$

EXAMPLE 2 Find the velocity for a constant acceleration

For an object falling under the influence of gravity, the acceleration due to gravity is essentially constant. Its value is $-9.8\ \text{m/s}^2$. (The negative sign is chosen so that *all quantities directed up are positive and* ***all quantities directed down are negative***.) Find the expression for the velocity of an object under the influence of gravity if $v = v_0$ when $t = 0$.

We write

$$v = \int (-9.8)\,dt \quad \text{substitute } a = -9.8 \text{ into Eq. (26.1)}$$

$$= -9.8t + C_1 \quad \text{integrate}$$

$$v_0 = -9.8(0) + C_1 \quad \text{substitute given values}$$

$$C_1 = v_0 \quad \text{solve for } C_1$$

$$v = v_0 - 9.8t \quad \text{substitute}$$

The velocity v_0 is called the *initial velocity*. If the object is given an initial upward velocity of 40.0 m/s, $v_0 = 40.0$ m/s. If the object is dropped, $v_0 = 0$. If the object is given an initial downward velocity of 40.0 m/s, $v_0 = -40.0$ m/s.

Once we have the expression for velocity, we can then integrate to find the expression for displacement s in terms of time. Since $v = ds/dt$, we can write $ds = v\,dt$, or

$$s = \int v\,dt \qquad (26.3)$$

EXAMPLE 3 Find the initial velocity

A ball is thrown vertically from the top of a building 24.5 m high and hits the ground 5.0 s later. What initial velocity was the ball given?

Measuring vertical distances from the ground, we know that $s = 24.5$ m when $t = 0$ and that $v = v_0 - 9.8t$. Thus,

$$s = \int (v_0 - 9.8t)\,dt = v_0 t - 4.9t^2 + C \quad \text{integrate}$$

$$24.5 = v_0(0) - 4.9(0) + C, \qquad C = 24.5 \quad \text{evaluate } C$$

$$s = v_0 t - 4.9t^2 + 24.5$$

We also know that $s = 0$ when $t = 5.0$ s. Thus,

$$0 = v_0(5.0) - 4.9(5.0)^2 + 24.5 \quad \text{substitute given values}$$

$$5.0v_0 = 98.0$$

$$v_0 = 19.6 \text{ m/s}$$

This means that the initial velocity was 19.6 m/s upward. See Fig. 26.1.

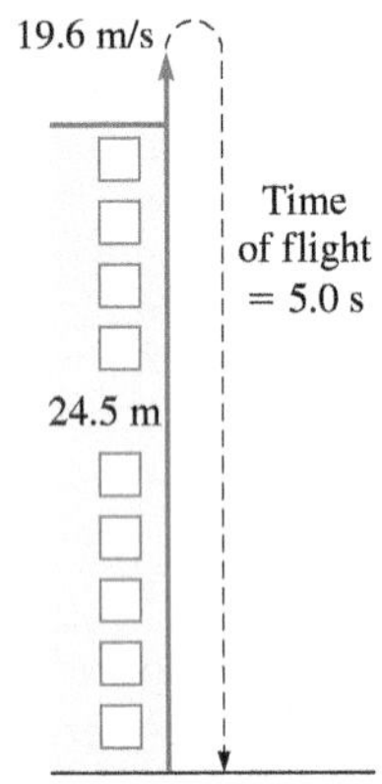

Fig. 26.1

EXAMPLE 4 Find the displacement, given the acceleration

During the initial stage of launching a spacecraft vertically, the acceleration a (in m/s^2) of the spacecraft is $a = 6t^2$. Find the height s of the spacecraft after 6.0 s if $s = 12$ m for $t = 0.0$ s and $v = 16$ m/s for $t = 2.0$ s.

First, we use Eq. (26.1) to get an expression for the velocity:

$$v = \int 6t^2 dt = 2t^3 + C_1 \quad \text{integrate}$$

$$16 = 2(2.0)^3 + C_1, \qquad C_1 = 0 \quad \text{evaluate } C_1$$

$$v = 2t^3$$

We now use Eq. (26.3) to get an expression for the displacement:

$$s = \int 2t^3\,dt = \tfrac{1}{2}t^4 + C_2 \quad \text{integrate}$$

$$12 = \tfrac{1}{2}(0.0)^4 + C_2, \qquad C_2 = 12 \quad \text{evaluate } C_2$$

$$s = \tfrac{1}{2}t^4 + 12$$

Now, finding s for $t = 6.0$ s, we have $s = \tfrac{1}{2}(6.0)^4 + 12 = 660$ m.

Practice Exercise

1. In Example 4, change the acceleration to $a = 4\ \text{m/s}^2$ and then find the height under the same given conditions.

VOLTAGE ACROSS A CAPACITOR

The second basic application of the indefinite integral we will discuss comes from the field of electricity. By definition, *the current i in an electric circuit equals the time rate of change of the charge q (in coulombs) that passes a given point in the circuit, or*

$$i = \frac{dq}{dt} \qquad (26.4)$$

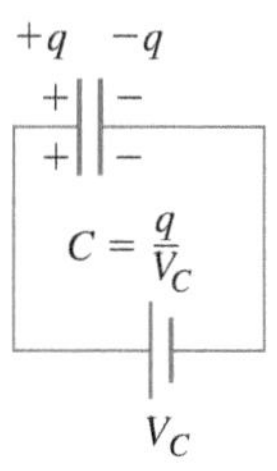

Fig. 26.2

Rewriting this expression $dq = i\,dt$ and integrating both sides, we have

$$q = \int i\,dt \tag{26.5}$$

Now, the voltage V_C across a capacitor with capacitance C (see Fig. 26.2) is given by $V_C = q/C$. By combining equations, the voltage V_C is given by

$$V_C = \frac{1}{C}\int i\,dt \tag{26.6}$$

Here, V_C is measured in volts, C in farads, i in amperes, and t in seconds.

LEARNING TIP

When using Eq. (26.6), all quantities must be in the units listed. If the quantities arising in a given problem involve different unit prefixes, **we must use the proper power of 10 that corresponds to each prefix.**

EXAMPLE 5 Find the electric charge, given the current

The current i (in A) in an electric circuit as a function of time t (in s) is given by $i = 6t^2 + 4$. Find an expression for the electric charge q (in C) that passes a point in the circuit as a function of t. If $q = 0$ C when $t = 0$ s, determine the total charge that passes the point in 2.0 s.

Since $q = \int i\,dt$, we have

$$q = \int (6t^2 + 4)\,dt \qquad \text{substitute into Eq. (26.5)}$$

$$= 2t^3 + 4t + C_1 \qquad \text{integrate}$$

■ Here, C represents coulombs and is not the *C* for capacitance or the constant of integration.

This is the required expression of charge as a function of time. When $t = 0$, $q = C_1$, which means the constant of integration represents the initial charge, or the charge that passed a given point before the timing started. Using q_0 to represent this charge, we have

$$q = 2t^3 + 4t + q_0$$

Returning to the second part of the problem, we note that $q_0 = 0$. Evaluating q for $t = 2.0$ s, we have $q = 2(2.0)^3 + 4(2.0) = 24$ C, which is the charge that passes any point in 2.0 s.

Practice Exercise

2. In Example 5, change the current to $i = 12t + 6$ and then find the charge under the same given conditions.

EXAMPLE 6 Find the voltage across a capacitor

The voltage across a 5.0-μF capacitor is zero. What is the voltage after 20 ms if a current of 75 mA charges the capacitor?

Since the current is 75 mA, we know that $i = 0.075\text{ A} = 7.5 \times 10^{-2}$ A. Since $5.0\ \mu\text{F} = 5.0 \times 10^{-6}$ F, we have

$$V_C = \frac{1}{5.0 \times 10^{-6}}\int 7.5 \times 10^{-2}dt \qquad \text{substituting into Eq. (26.6)}$$

$$= (1.5 \times 10^4)\int dt = (1.5 \times 10^4)t + C_1 \qquad \text{integrate}$$

From the given information, we know that $V_C = 0$ when $t = 0$. Thus,

$$0 = (1.5 \times 10^4)(0) + C_1 \quad \text{or} \quad C_1 = 0 \qquad \text{evaluate } C_1$$

This means that

$$V_C = (1.5 \times 10^4)t$$

Evaluating this expression for $t = 20 \times 10^{-3}$ s, we have

$$V_C = (1.5 \times 10^4)(20 \times 10^{-3})$$
$$= 30 \times 10 = 300 \text{ V}$$

EXAMPLE 7 Find the capacitance of a capacitor

A certain capacitor has 100 V across it. At this instant, a current $i = 0.06t^{1/2}$ is sent through the circuit. After 0.25 s, the voltage across the capacitor is 140 V. What is the capacitance?

Substituting $i = 0.06t^{1/2}$, we find that

$$V_C = \frac{1}{C}\int(0.06t^{1/2}dt) = \frac{0.06}{C}\int t^{1/2}dt \qquad \text{using Eq. (26.6)}$$

$$= \frac{0.04}{C}t^{3/2} + C_1 \qquad \text{integrate}$$

From the given information, we know that $V_C = 100$ V when $t = 0$. Thus,

$$100 = \frac{0.04}{C}(0) + C_1 \quad \text{or} \quad C_1 = 100 \text{ V} \qquad \text{evaluate}$$

$$V_C = \frac{0.04}{C}t^{3/2} + 100 \qquad \text{substituting 100 for } C_1$$

We also know that $V_C = 140$ V when $t = 0.25$ s. Therefore,

$$140 = \frac{0.04}{C}(0.25)^{3/2} + 100$$

$$40 = \frac{0.04}{C}(0.125)$$

$$C = 1.25 \times 10^{-4} \text{ F} = 125\ \mu\text{F}$$

EXERCISES 26.1

In Exercises 1 and 2, make the given changes in the indicated examples of this section and then solve the resulting problems.

1. In Example 3, change 5.0 s to 1.0 s and then solve the resulting problem.
2. In Example 7, change $0.06\,t^{1/2}$ to $0.06t$ and then solve the resulting problem.

In Exercises 3–32, solve the given problems.

3. What is the velocity (in m/s) of a rock that drops from the edge of Thor Peak on Baffin Island (the world's highest vertical drop) when it reaches the ground after 16.0 s?
4. A beach ball is rolled up a shallow slope with an initial velocity of 7.2 m/s. If the acceleration of the ball is 1.2 m/s² down the slope, find the velocity of the ball after 8.0 s.
5. A conveyor belt 8.00 m long moves at 0.25 m/s. If a package is placed at one end, find its displacement from the other end as a function of time.
6. During each cycle, the velocity v (in mm/s) of a piston is $v = 6t - 6t^2$, where t is the time (in s). Find the displacement s of the piston after 0.75 s if the initial displacement is zero.
7. While in the barrel of a tennis ball machine, the acceleration a (in m/s²) of a ball is $a = 30\sqrt{1 - 4t}$, where t is the time (in s). If $v = 0$ for $t = 0$, find the velocity of the ball as it leaves the barrel at $t = 0.25$ s.
8. A person skis down a slope with an acceleration (in m/s²) given by $a = \frac{600t}{(60 + 0.5t^2)^2}$, where t is the time (in s). Find the skier's velocity as a function of time if $v = 0$ when $t = 0$.
9. A motorcycle accelerates at a constant rate from 0 km/h to 105 km/h in 18.0 seconds. Assuming constant acceleration, how far (in m) does the motorcycle travel in this time?
10. The engine of a lunar lander is shut off when the lander is 5.0 m above the surface of the moon and descending at 2.0 m/s. If the acceleration due to gravity on the moon is 1.6 m/s², what is the speed of the lander just before it touches the surface?
11. If an aircraft is to attain a take-off velocity of 75 m/s after travelling 240 m along the flight deck of an aircraft carrier, find the aircraft's acceleration (assumed constant).
12. A stone is thrown straight up from the edge of a 45.0-m-high cliff. A loose stone at the edge of the cliff falls off 1.50 s later. What is the vertical velocity of the first stone, if the two stones reach the ground below at the same time?
13. What must be the nozzle velocity of the water from a fire hose if it is to reach a point 30 m directly above the nozzle?
14. An arrow is shot from the edge of a cliff with a vertical upward velocity of 35.0 m/s. If it strikes the plain below after 9.0 s, how high is the cliff?

15. In coming to a stop, the acceleration of a motorcycle is $-4.0t$. If it is travelling at 32 m/s when the brakes are applied, how far does it travel while stopping?

16. A hoist mechanism raises a crate with an acceleration (in m/s^2) $a = \sqrt{1 + 0.2t}$, where t is the time in seconds. Find the displacement of the crate as a function of time if $v = 0$ m/s and $s = 2$ m for $t = 0$ s.

17. The electric current in a microprocessor circuit is 0.230 μA. How many coulombs pass a given point in the circuit in 1.50 ms?

18. A capacitor in a microwave oven circuit has a charge of 0.10 μF. If it is further charged by a current of 0.25 μA, what is the charge on the capacitor 3.0 μs later?

19. In an amplifier circuit, the current i (in A) changes with time t (in s) according to $i = 0.06t\sqrt{1 + t^2}$. If 0.015 C of charge has passed a point in the circuit at $t = 0$, find the total charge to have passed the point at $t = 0.25$ s.

20. The current i (in μA) in a DVD player circuit is given by $i = 6.0 - 0.50t$, where t is the time (in μs) and $0 \leq t \leq 30$ μs. If $q_0 = 0$ C, for what value of t is $q = 0$ C? What interpretation can be given to this result?

21. The voltage across a 2.5-μF capacitor in a copying machine is zero. What is the voltage after 16 ms if a current of 25 mA charges the capacitor?

22. The voltage across an 8.50-nF capacitor in an FM receiver circuit is zero. Find the voltage after 2.00 μs if a current (in mA) $i = 0.042t$ charges the capacitor.

23. The voltage across a 3.75-μF capacitor in a television circuit is 4.50 mV. Find the voltage after 0.565 ms if a current (in μA) $i = \sqrt[3]{1 + 6t}$ further charges the capacitor.

24. A current $i = t/\sqrt{t^2 + 1}$ (in A) is sent through an electric dryer circuit containing a previously uncharged 2.0-μF capacitor. How long does it take for the capacitor voltage to reach 120 V?

25. The angular velocity ω is the time rate of change of the angular displacement θ of a rotating object. See Fig. 26.3. In testing the shaft of an engine, its angular velocity is $\omega = 16t + 0.50t^2$, where t is the time (in s) of rotation. Find the angular displacement through which the shaft goes in 10.0 s.

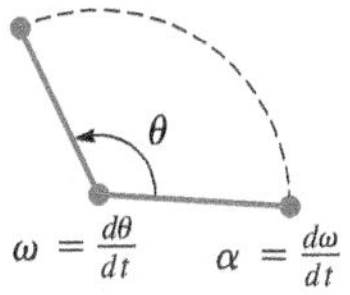

Fig. 26.3

26. The angular acceleration α is the time rate of change of angular velocity ω of a rotating object. See Fig. 26.3. When starting up, the angular acceleration of a helicopter blade is $\alpha = \sqrt{8t + 1}$. Find the expression for θ if $\omega = 0$ and $\theta = 0$ for $t = 0$.

27. An inductor in an electric circuit is essentially a coil of wire in which the voltage is affected by a changing current. By definition, the voltage caused by the changing current is given by $V_L = L(di/dt)$, where L is the inductance (in H). If $V_L = 12.0 - 0.2t$ for a 3.0-H inductor, find the current in the circuit after 20 s if the initial current was zero.

28. If the inner and outer walls of a container are at different temperatures, the rate of change of temperature with respect to the distance from one wall is a function of the distance from the wall. Symbolically, this is stated as $dT/dx = f(x)$, where T is the temperature. If x is measured from the outer wall, at 20°C, and $f(x) = 72x^2$, find the temperature at the inner wall if the container walls are 0.5 cm thick.

29. Surrounding an electrically charged particle is an electric field. The rate of change of electric potential with respect to the distance from the particle creating the field equals the negative of the value of the electric field. That is, $dV/dx = -E$, where E is the electric field. If $E = k/x^2$, where k is a constant, find the electric potential at a distance x_1 from the particle, if $V \to 0$ as $x \to \infty$.

30. The rate of change of the vertical deflection y with respect to the horizontal distance x from one end of a beam is a function of x. For a particular beam, the function is $k(x^5 + 1350x^3 - 7000x^2)$, where k is a constant. Find y as a function of x.

31. Freshwater is flowing into a brine solution, with an equal volume of mixed solution flowing out. The amount of salt in the solution decreases, but more slowly as time increases. Under certain conditions, the time rate of change of mass of salt (in g/min) is given by $-1/\sqrt{t + 1}$. Find the mass m of salt as a function of time if 1000 g were originally present. Under these conditions, how long would it take for all the salt to be removed?

32. A holograph of a circle is formed. The rate of change of the radius r of the circle with respect to the wavelength λ of the light used is inversely proportional to the square root of λ. If $dr/d\lambda = 3.55 \times 10^4$ and $r = 4.08$ cm for $\lambda = 574$ nm, find r as a function of λ.

Answers to Practice Exercises

1. 132 m **2.** 36 C

26.2 Areas by Integration

In Section 25.3, we introduced the method of finding the area under a curve by integration. We also showed that the area can be found by a summation process on the rectangles inscribed under the curve, which means that integration can be interpreted as a summation process. *The applications of the definite integral use this summation interpretation of the integral.* We now develop a general procedure for finding the area for which the bounding curves are known by summing the areas of inscribed rectangles and using integration for the summation.

Finding an Area Bounded by a Curve and an Axis	Using Vertical Elements	Using Horizontal Elements
1. Make a sketch of the area.	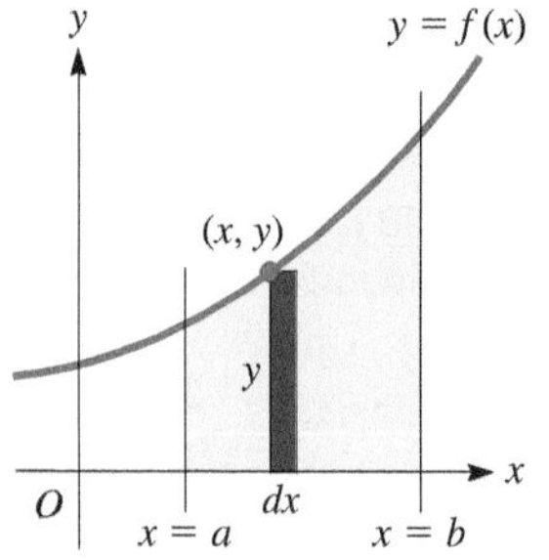 **Fig. 26.4**	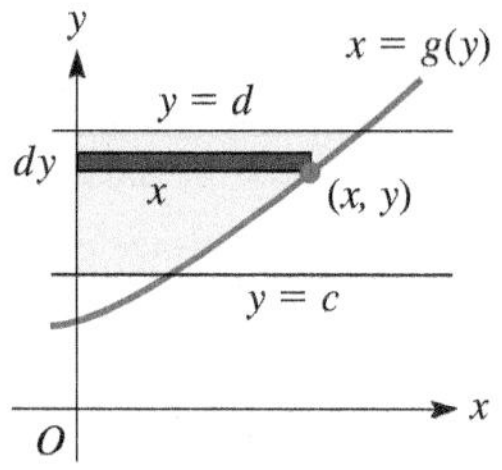 **Fig. 26.5**
2. Determine a representative **element of area** dA.	A vertical element has width dx. The length is y, the y-coordinate of the point on the curve. Hence $dA = y dx$	A horizontal element has width dy. The length is x, the x-coordinate of the point on the curve. Hence $dA = x dy$
3. Choose the limits of integration so that summation is done in the positive direction.	Left boundary: $x = a$ Right boundary: $x = b$ Integrate from left to right.	Lower boundary: $y = c$ Upper boundary: $y = d$ Integrate from bottom to top.
4. Write the sum of the elements as a definite integral. Integrate and evaluate to find the area.	$A = \int_a^b y dx = \int_a^b f(x) dx$ **(26.7)**	$A = \int_c^d x dy = \int_c^d g(y) dy$ **(26.8)**

The choice of vertical or horizontal elements is determined by (1) which one leads to the simplest solution, or (2) the form of the resulting integral. In some problems, it makes little difference which is chosen. Note that choosing horizontal elements requires rewriting the expression with x as a function of y.

EXAMPLE 1 Find an area using vertical elements

Find the area bounded by $y = x^2$, $x = 1$, and $x = 2$.

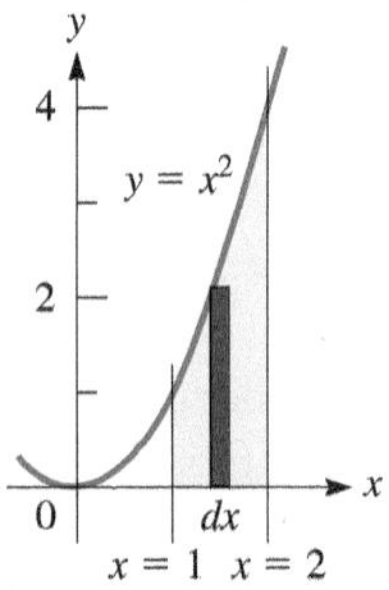

Fig. 26.6

Make a sketch of the area:	See Fig. 26.6.
For vertical elements, the representative element of area is $dA = y dx$:	Here $y = x^2$, so $dA = x^2 dx$
Choose the limits of integration:	Left boundary: $x = 1$ Right boundary: $x = 2$
Write as a definite integral using Eq. (26.7). Integrate and evaluate to find the area:	$A = \int_1^2 x^2 dx = \frac{1}{3}x^3 \Big\|_1^2$ $= \frac{1}{3}(8) - \frac{1}{3}(1) = \frac{7}{3}$

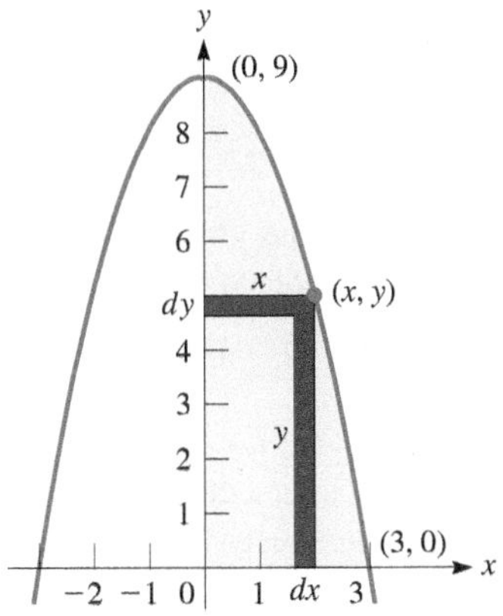

Fig. 26.7

Practice Exercise

1. Find the area in the first quadrant bounded by $y = 4 - x^2$.

EXAMPLE 2 Vertical and horizontal elements of an area

Find the area in the first quadrant bounded by $y = 9 - x^2$. See Fig. 26.7.

vertical element of length y and width dx		horizontal element of length x and width dy
$A = \int_0^3 y\,dx$	sum of areas of elements	$A = \int_0^9 x\,dy$
$= \int_0^3 (9 - x^2)\,dx$	substitute for y; substitute for x	$= \int_0^9 \sqrt{9 - y}\,dy = -\int_0^9 (9 - y)^{1/2}(-dy)$
$= \left(9x - \frac{x^3}{3}\right)\Big\|_0^3$	integrate	$= -\frac{2}{3}(9 - y)^{3/2}\Big\|_0^9$
$= (27 - 9) - 0 = 18$	evaluate	$= -\frac{2}{3}(9 - 9)^{3/2} + \frac{2}{3}(9 - 0)^{3/2} = 18$

Note that the limits are 0 to 3 for the vertical elements, and 0 to 9 for the horizontal elements. Note also that both solutions lead to the same area, 18 square units.

AREA BETWEEN TWO CURVES

It is also possible to find the area between two curves when one of the curves is not an axis. In such a case, the length of the element of area becomes the difference in the y- or x-coordinates, depending on whether a vertical element or a horizontal element is used.

Procedure for Finding an Area between Curves	Using Vertical Elements	Using Horizontal Elements
1. Make a sketch of the area.	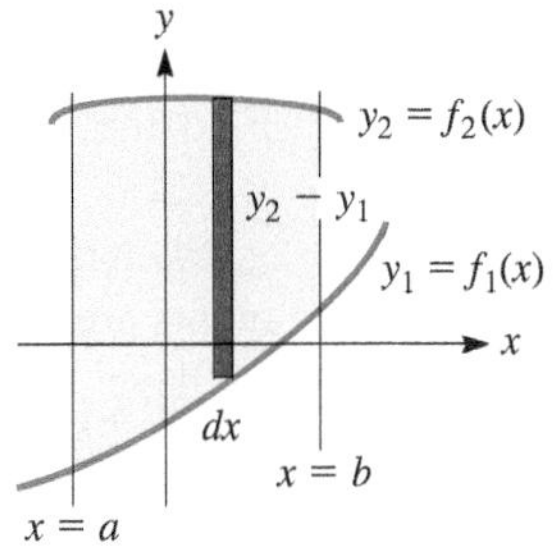 Fig. 26.8	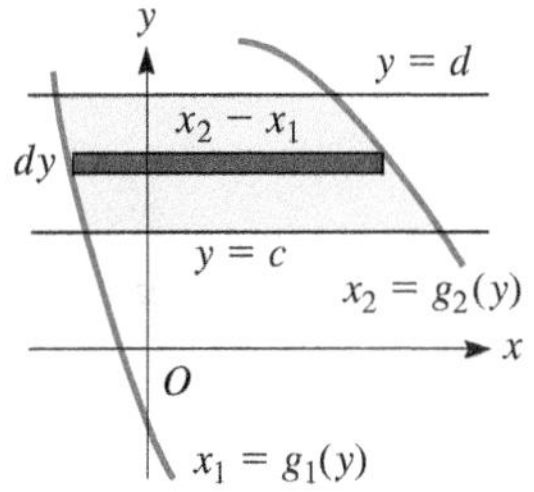 Fig. 26.9
2. Determine a representative element of area dA. *It is important that length be positive to ensure a positive area.*	A vertical element has width dx. The length is the difference between the ***upper curve*** y_2 ***minus the lower curve*** y_1. Hence $dA = (y_2 - y_1)\,dx$	A horizontal element has width dy. The length is the difference between the ***right curve*** x_2 ***minus the left curve*** x_1. Hence $dA = (x_2 - x_1)\,dy$
3. Choose the limits of integration. These are often the points of intersection of the curves.	Left boundary: $x = a$ Right boundary: $x = b$ Integrate from left to right.	Lower boundary: $y = c$ Upper boundary: $y = d$ Integrate from bottom to top.
4. Write the sum of the elements as a definite integral. Integrate and evaluate to find the area.	$A = \int_a^b (y_2 - y_1)\,dx$ **(26.9)**	$A = \int_c^d (x_2 - x_1)\,dy$ **(26.10)**

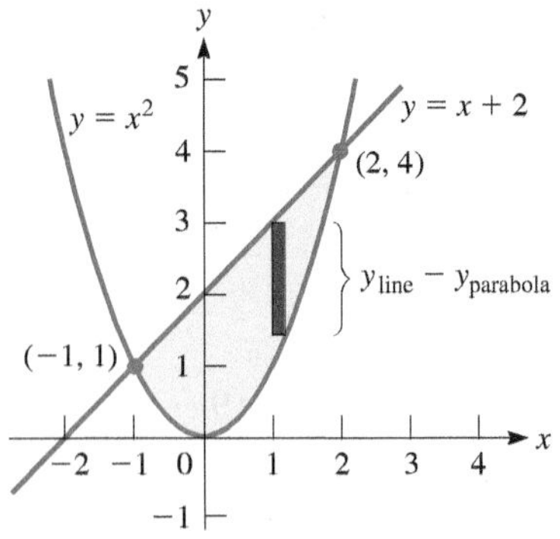

Fig. 26.10

EXAMPLE 3 Area between curves using vertical elements

Find the area bounded by $y = x^2$ and $y = x + 2$.

Make a sketch of the area: See Fig. 26.10.

For vertical elements, the length is the upper curve minus the lower curve, so

$$dA = (y_2 - y_1)\,dx$$

The upper curve is the line $y = x + 2$, and the lower curve is the parabola $y = x^2$. Hence

$$dA = (x + 2 - x^2)\,dx$$

Choose the limits of integration. Here they are the points of intersection of the two curves:

We find the points of intersection by solving the equations simultaneously. The process is shown at the left.

$x^2 = x + 2$
$x^2 - x - 2 = 0$
$(x + 1)(x - 2) = 0$
$x = -1, 2$

Left boundary: $x = -1$
Right boundary: $x = 2$

Write as a definite integral using Eq. (26.9). Integrate and evaluate to find the area:

$$A = \int_{-1}^{2} (x + 2 - x^2)\,dx = \left(\frac{x^2}{2} + 2x - \frac{x^3}{3}\right)\Bigg|_{-1}^{2}$$

$$= \left(2 + 4 - \frac{8}{3}\right) - \left(\frac{1}{2} - 2 + \frac{1}{3}\right)$$

$$= \frac{10}{3} + \frac{7}{6} = \frac{9}{2}$$

Note that we chose vertical elements since they require a single integral. Choosing horizontal elements would have required one integral from $y = 0$ to $y = 1$, and another integral from $y = 1$ to $y = 4$.

EXAMPLE 4 Area between curves using horizontal elements

Find the area bounded by the curve $y = x^3 - 3$ and the lines $x = 2$, $y = -1$, and $y = 3$.

Sketching the curve and lines, we show the area in Fig. 26.11. Horizontal elements are better, since they avoid having to evaluate the area in two parts. Therefore, we have

$$A = \int_{-1}^{3} (x_{\text{line}} - x_{\text{cubic}})\,dy = \int_{-1}^{3} \left(2 - \sqrt[3]{y + 3}\right)dy \quad \text{using Eq. (26.10)}$$

$$= 2y - \frac{3}{4}(y + 3)^{4/3}\Big|_{-1}^{3} = \left[6 - \frac{3}{4}(6^{4/3})\right] - \left[-2 - \frac{3}{4}(2^{4/3})\right]$$

$$= 8 - \frac{9}{2}\sqrt[3]{6} + \frac{3}{2}\sqrt[3]{2} = 1.71$$

As we see, the choice of horizontal elements leads to limits of -1 and 3. If we had chosen vertical elements, the limits would have been $\sqrt[3]{2}$ and $\sqrt[3]{6}$ for the area to the left of $(\sqrt[3]{6}, 3)$, and $\sqrt[3]{6}$ and 2 to the right of this point.

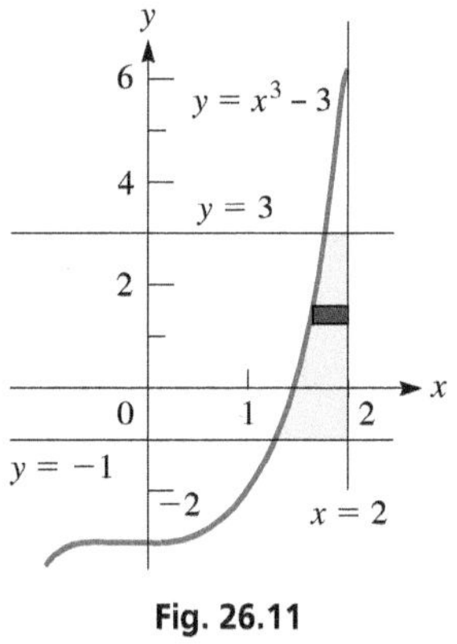

Fig. 26.11

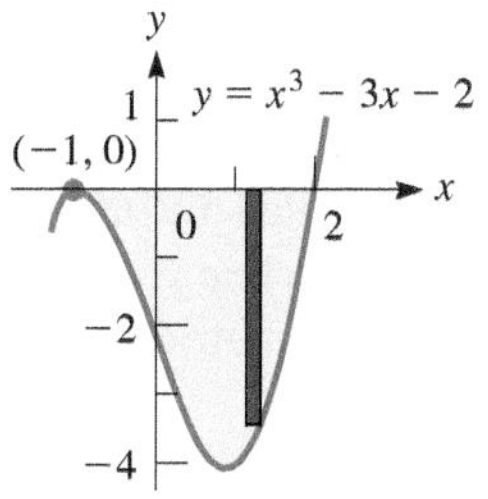

Fig. 26.12

EXAMPLE 5 Area below the *x*-axis

Find the area bounded by $x^3 - 3x - 2$ and the x-axis.

Sketching the graph, we find that $y = x^3 - 3x - 2$ has a maximum point at $(-1, 0)$, a minimum point at $(1, -4)$, and an intercept at $(2, 0)$. The curve is shown in Fig. 26.12, and we see that the area is *below* the x-axis. Using vertical elements, we see that the top is the x-axis ($y = 0$) and the bottom is the curve $y = x^3 - 3x - 2$. Therefore, we have

$$\begin{aligned} A &= \int_{-1}^{2} [0 - (x^3 - 3x - 2)]\,dx = \int_{-1}^{2} (-x^3 + 3x + 2)\,dx \\ &= -\frac{1}{4}x^4 + \frac{3}{2}x^2 + 2x\Big|_{-1}^{2} \\ &= \left[-\frac{1}{4}(2^4) + \frac{3}{2}(2^2) + 2(2)\right] - \left[-\frac{1}{4}(-1)^4 + \frac{3}{2}(-1)^2 + 2(-1)\right] \\ &= \frac{27}{4} = 6.75 \end{aligned}$$

If we had simply set up the area as $A = \int_{-1}^{2} (x^3 - 3x - 2)\,dx$, we would have found $A = -6.75$. The negative sign shows that the area is below the x-axis. Again, we avoid any complications with negative areas by making the length of the element positive.

Also, note that since

$$0 - (x^3 - 3x - 2) = -(x^3 - 3x - 2)$$

an area bounded on top by the x-axis can be found by setting up the area as being "under" the curve and using the negative of the function.

Practice Exercise

2. Find the area bounded by $y = x^2 - 4$ and the x-axis.

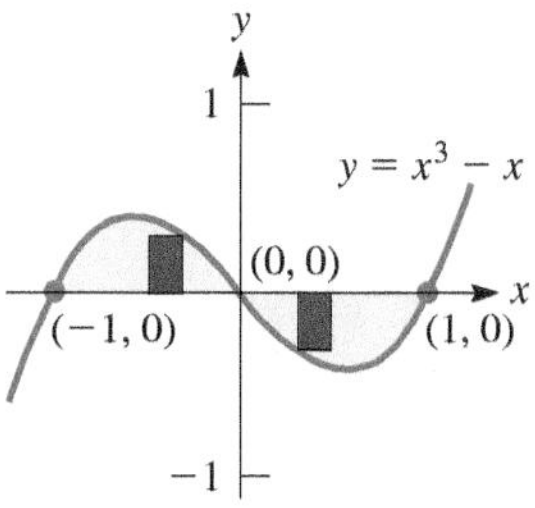

Fig. 26.13

EXAMPLE 6 Area above and below the *x*-axis

Find the area between $y = x^3 - x$ and the x-axis.

We note from Fig. 26.13 that the area to the left of the origin is above the axis and the area to the right is below. If we find the area from

$$\begin{aligned} A &= \int_{-1}^{1} (x^3 - x)\,dx = \frac{x^4}{4} - \frac{x^2}{2}\Big|_{-1}^{1} \\ &= \left(\frac{1}{4} - \frac{1}{2}\right) - \left(\frac{1}{4} - \frac{1}{2}\right) = 0 \end{aligned}$$

we see that the apparent area is zero. From the figure, we know this is not correct. Noting that the y-values (of the area) are negative to the right of the origin, we set up the integrals

$$\begin{aligned} A &= \int_{-1}^{0} (x^3 - x)\,dx + \int_{0}^{1} [0 - (x^3 - x)]\,dx \\ &= \left(\frac{x^4}{4} - \frac{x^2}{2}\right)\Big|_{-1}^{0} - \left(\frac{x^4}{4} - \frac{x^2}{2}\right)\Big|_{0}^{1} \\ &= 0 - \left(\frac{1}{4} - \frac{1}{2}\right) - \left(\frac{1}{4} - \frac{1}{2}\right) + 0 = \frac{1}{2} \end{aligned}$$

LEARNING TIP

We must *be very careful if the bounding curves of an area cross.* In such a case, for part of the area one curve is above the area, and for a different part of the area this same curve is below the area. When this happens, *two integrals must be used* to find the area. Example 6 illustrates the necessity of using this procedure.

EXAMPLE 7 Area between two curves—area of walkway

■ See the chapter introduction.

Find the area of the glass walkway of the Glacier Skywalk in Jasper National Park, Alberta. The steel frame of the walkway can be approximated by the parabolas $y = -0.11x^2 + 14$ and $y = -0.15x^2 + 11$, and is bounded by the line $y = 0.40x - 3.4$ on the right side (all distances in metres). The walkway is symmetrical on the other side. See Fig. 26.14.

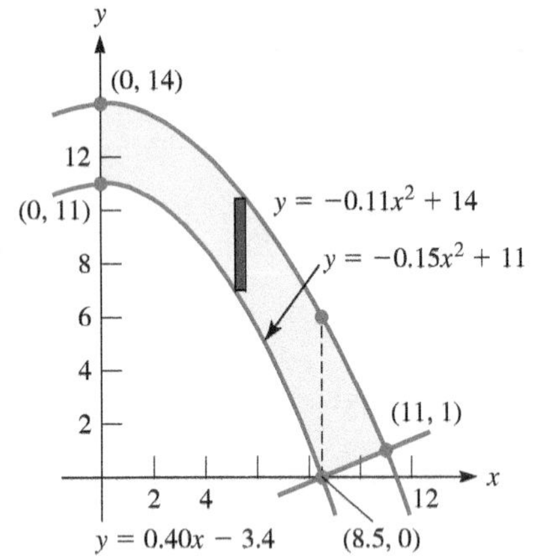

Fig. 26.14.

Because of symmetry, we can focus on the right side. We use vertical elements and obtain the area as the sum of two integrals. For the first integral, the area is between the two parabolas:

$$dA = (-0.11x^2 + 14 - (-0.15x^2 + 11))\,dx = (0.04x^2 + 3)\,dx$$

For the second integral, the upper curve is the parabola and the lower curve is the line:

$$\begin{aligned} dA &= (-0.11x^2 + 14 - (0.4x - 3.4))\,dx \\ &= (-0.11x^2 - 0.4x + 17.4)\,dx \end{aligned}$$

The area is thus

$$\begin{aligned} A &= \int_0^{8.5} (0.04x^2 + 3)\,dx + \int_{8.5}^{11} (-0.11x^2 - 0.4x + 17.4)\,dx \\ &= \left(\frac{0.04}{3}x^3 + 3x\right)\Bigg|_0^{8.5} + \left(-\frac{0.11}{3}x^3 - \frac{0.4}{2}x^2 + 17.4x\right)\Bigg|_{8.5}^{11} \\ &= 41\text{ m}^2 \end{aligned}$$

Taking into account the other side, the total area of the glass walkway is approximately $2 \times 41 = 82\text{ m}^2$.

EXERCISES 26.2

In Exercises 1 and 2, make the given changes in the indicated examples of this section and then find the resulting areas.

1. In Example 1, change $x = 2$ to $x = 3$.
2. In Example 3, change $y = x + 2$ to $y = 2x$.

In Exercises 3–30, find the areas bounded by the indicated curves.

3. $y = 4x, y = 0, x = 1$
4. $y = 3x^2, y = 0, x = 3$
5. $y = 8 - 2x^2, y = 0$
6. $y = \frac{1}{2}x^2 + 2, x = 0, y = 4 \quad (x > 0)$
7. $y = x^2 - 4, y = 0, x = -4 \quad (y > 0)$
8. $y = x^2 - 5x, y = 0$
9. $y = x^{-2}, y = 0, x = 2, x = 3$
10. $y = 16 - x^2, y = 0, x = -2, x = 3$
11. $y = \sqrt{x}, x = 0, y = 1, y = 3$
12. $y = 3\sqrt{x + 1}, x = 0, y = 6$
13. $y = 2/\sqrt{x}, x = 0, y = 1, y = 4$
14. $x = y^2 - y, x = 0$
15. $y = 6 - 3x, x = 0, y = 0, y = 3$
16. $y = x, y = 3 - x, x = 0$
17. $y = x - 2\sqrt{x}, y = 0$
18. $y = x^4 - 2x^3, y = 0$
19. $y = x^2, y = 2 - x, x = 0 \quad (x \geq 0)$
20. $y = x^2, y = 2 - x, y = 1$
21. $y = x^4 - 8x^2 + 16, y = 16 - x^4$
22. $y = \sqrt{x - 1}, y = 3 - x, y = 0$
23. $y = x^2 + 5x, y = 3 - x^2$
24. $y = x^3, y = x^2 + 4, x = -1$
25. $y = x^5, x = -1, x = 2, y = 0$
26. $y = x^2 + 2x - 8, y = x + 4$
27. $y = 4 - x^2, y = 4x - x^2, x = 0, x = 2$
28. $y = x^2, y = x^{1/3}$, between $x = -1$ and $x = 1$
29. $y = x^2\sqrt{1 - x^3}, y = 0$
30. $y = x - 1, y^2 = 2x + 6$

In Exercises 31–38, solve the given problems.

31. Describe a region for which the area is found by evaluating the integral $\int_1^2 (2x^2 - x^3)\,dx$.
32. Although the integral $\int_{-2}^{2} \sqrt{4 - x^2}\,dx$ cannot be integrated by methods we have developed to this point, by recognizing the region represented, it can be evaluated. Evaluate this integral.
33. Use integration to find the area of the triangle with vertices (0, 0) (4, 4), and (10, 0).

34. Show that the area bounded by the parabola $y = x^2$ and the line $y = b\,(b > 0)$ is two-thirds of the area of the rectangle that circumscribes it.

35. Show that the curve $y = x^n\,(n > 0)$ divides the unit square bounded by $x = 0$, $y = 0$, $x = 1$, and $y = 1$ into regions with areas in the ratio of $n/1$.

36. Why can the integral $\int_a^2 (2 + x - x^2)dx$ be used to find the area bounded by $x = a$, $y = 0$, and $y = 2 + x - x^2$ if $a = -1$, but not if $a = -2$?

37. Find the value of c such that the region bounded by $y = x^2$ and $y = 4$ is divided by $y = c$ into two regions of equal area.

38. Find the value(s) of c such that the region bounded by $y = x^2 - c^2$ and $y = -x^2 + c^2$ has an area of 576.

In Exercises 39–42, find the areas bounded by the indicated curves, using (a) vertical elements and (b) horizontal elements.

39. $y = 8x$, $x = 0$, $y = 4$

40. $y = x^3$, $x = 0$, $y = 3$

41. $y = x^4$, $y = 8x$

42. $y = 4x$, $y = x^3$

In Exercises 43–50, some applications of areas are shown.

43. Certain physical quantities are often represented as an area under a curve. By definition, power is the time rate of change of performing work. Thus, $P = dw/dt$, or $dw = P\,dt$. Therefore, if $P = 12t - 4t^2$, find the work (in J) performed in 3 s by finding the area under the curve of P vs. t. See Fig. 26.15.

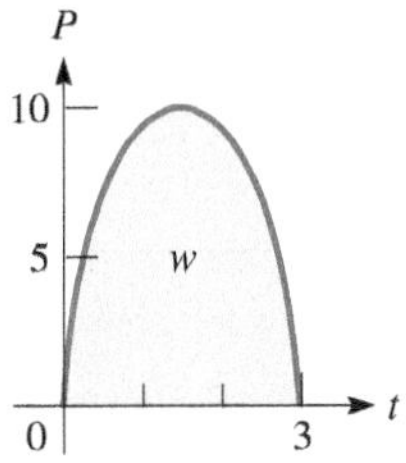

Fig. 26.15

44. The total electric charge Q (in C) to pass a point in the circuit from time t_1 to t_2 is $Q = \int_{t_1}^{t_2} i\,dt$, where i is the current (in A). Find Q if $t_1 = 1$ s, $t_2 = 4$ s, and $i = 0.0032t\sqrt{t^2 + 1}$.

45. Since the displacement s, velocity v, and time t of a moving object are related by $s = \int v\,dt$, it is possible to represent the change in displacement as an area. A rocket is launched such that its vertical velocity v (in km/s) as a function of time t (in s) is $v = 1 - 0.01\sqrt{2t + 1}$. Find the change in vertical displacement from $t = 10$ s to $t = 100$ s.

46. The total cost C (in dollars) of production can be interpreted as an area. If the cost per unit C' (in dollars per unit) of producing x units is given by $100/(0.01x + 1)^2$, find the total cost of producing 100 units by finding the area under the curve of C' vs. x.

47. A cam is designed such that one face of it is described as being the area between the curves $y = x^3 - 2x^2 - x + 2$ and $y = x^2 - 1$ (units in cm). Show that this description does not uniquely describe the face of the cam. Find the area of the face of the cam, if a complete description requires that $x \leq 1$.

48. Using CAD (computer-assisted design), an architect programs a computer to sketch the shape of a swimming pool designed between the curves $y = \dfrac{800x}{(x^2 + 10)^2}$, $y = 0.5x^2 - 4x$, and $x = 8$ (dimensions in m). Find the area of the surface of the pool.

49. A coffee-table top is designed to be the region between $y = 0.25x^4$ and $y = 12 - 0.25x^4$. What is the area (in dm^2) of the table top?

50. A window is designed to be the area between a parabolic section and a straight base, as shown in Fig. 26.16. What is the area of the window?

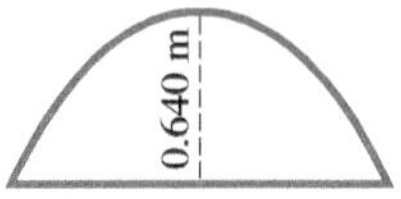

Fig. 26.16

Answers to Practice Exercises

1. 16/3 **2.** 32/3

26.3 Volumes by Integration

Consider a region in the xy-plane and its representative element of area, as shown in Figs. 26.17(a) and (b). When the region is revolved about one of the axes [the x-axis in Fig. 26.17(a), and the y-axis in Fig. 26.17(b)], it is said to generate a **solid of revolution**, which is also shown in the figures. We now show methods of finding the volume of solids that are generated in this way.

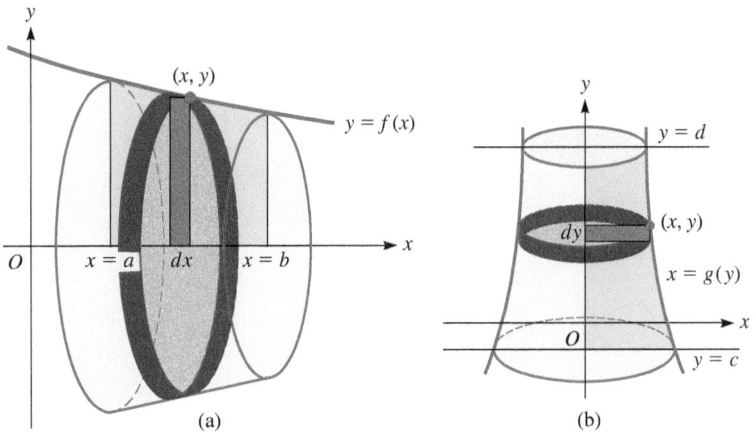

Fig. 26.17

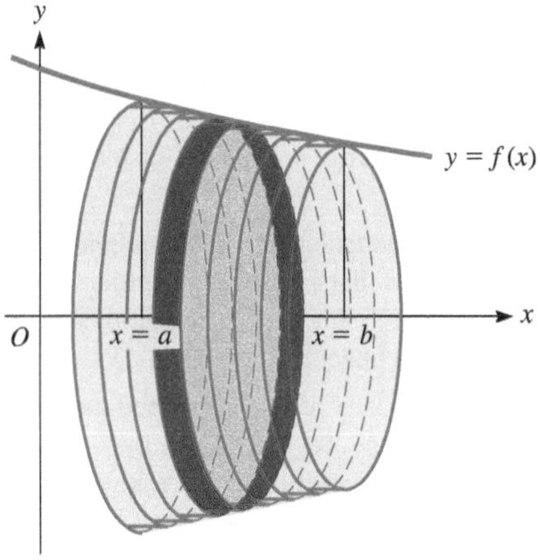

Fig. 26.18

■ Note that in Eq. (26.12), the function must be solved for x in terms of y: $x = g(y)$.

Finding the Volume of a Solid of Revolution Using Discs	Using Vertical Discs (revolving around x-axis)	Using Horizontal Discs (revolving around y-axis)
1. Make a sketch of the solid.	See Figs. 26.17(a) and 26.18.	See Fig. 26.17(b).
2. Determine a representative ***element of volume*** dV. Recall that the volume of a cylinder is π times its radius squared, times its thickness.	A vertical disc has thickness dx. The radius is y, the y-coordinate of the point on the curve. Hence $dV = \pi y^2 dx$	A horizontal disc has thickness dy. The length is x, the x-coordinate of the point on the curve. Hence $dV = \pi x^2 dy$
3. Choose the limits of integration so that summation is done in the positive direction.	Left boundary: $x = a$ Right boundary: $x = b$ Integrate from left to right.	Lower boundary: $y = c$ Upper boundary: $y = d$ Integrate from bottom to top.
4. Write the sum of the elements as a definite integral. Integrate and evaluate to find the volume.	$V = \pi \int_a^b y^2 dx = \pi \int_a^b [f(x)]^2 dx$ **(26.11)**	$V = \pi \int_c^d x^2 dy$ **(26.12)**

EXAMPLE 1 Find a volume using vertical discs

Find the volume of the solid generated by revolving the region bounded by $y = x^2$, $x = 2$, and $y = 0$ about the x-axis. See Fig. 26.19.

From the figure, we see that the radius of the disc is y and its thickness is dx. The elements are summed from left ($x = 0$) to right ($x = 2$):

$$V = \pi \int_0^2 y^2 dx \qquad \text{using Eq. (26.11)}$$

$$= \pi \int_0^2 (x^2)^2 dx = \pi \int_0^2 x^4 dx \qquad \text{substitute } x^2 \text{ for } y$$

$$= \frac{\pi}{5} x^5 \Big|_0^2 = \frac{32\pi}{5} \qquad \text{integrate and evaluate}$$

Since π is used in Eq. (26.11), it is common to leave results in terms of π. In applied problems, a decimal result would normally be given.

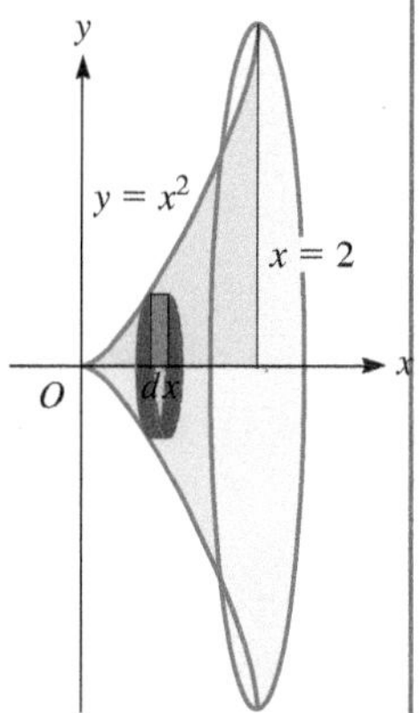

Fig. 26.19

EXAMPLE 2 Find a volume using horizontal discs

Find the volume of the solid generated by revolving the region bounded by $y = 2x$, $y = 6$, and $x = 0$ about the y-axis.

Fig. 26.20 shows the volume to be found. Note that the radius of the disc is x and its thickness is dy.

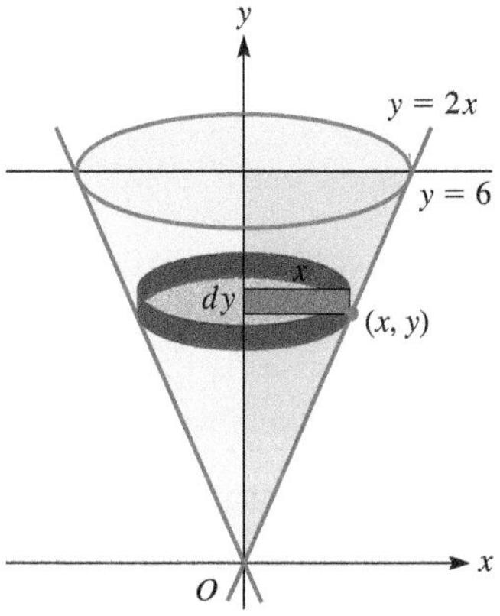

Fig. 26.20

$$V = \pi\int_0^6 x^2 dy \qquad \text{using Eq. (26.12)}$$

$$= \pi\int_0^6\left(\frac{y}{2}\right)^2 dy = \frac{\pi}{4}\int_0^6 y^2 dy \qquad \text{substituting } \frac{y}{2} \text{ for } x$$

$$= \frac{\pi}{12}y^3\Big|_0^6 = 18\pi \qquad \text{integrate and evaluate}$$

Since this volume is a right circular cone, it is possible to check the result:

$$V = \frac{1}{3}\pi r^2 h = \frac{1}{3}\pi(3^2)(6) = 18\pi$$

If the region in Fig. 26.21 is revolved about the y-axis, the element of area $y\,dx$ generates a different element of volume from that when it is revolved about the x-axis. In Fig. 26.22, this element of volume is a **cylindrical shell** (note the hole through its centre). *The total volume is made up of an infinite number of concentric shells.* When the volumes of these shells are summed, we have the total volume generated. Thus, we must now find the approximate volume dV of the representative shell.

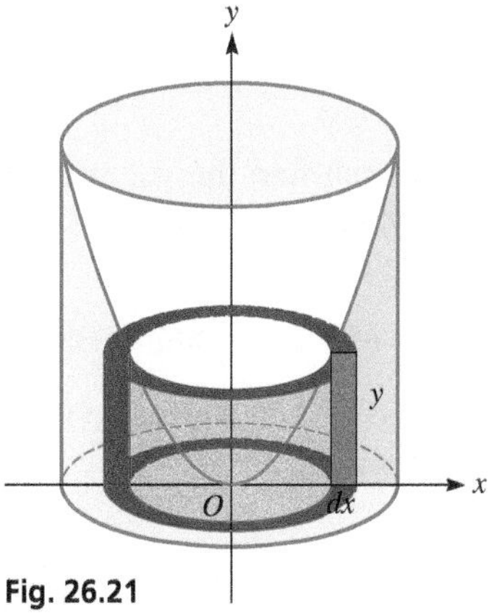

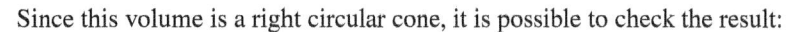

Fig. 26.21

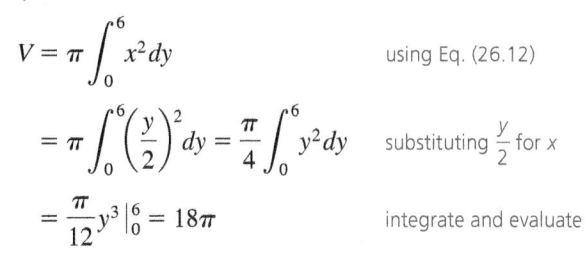

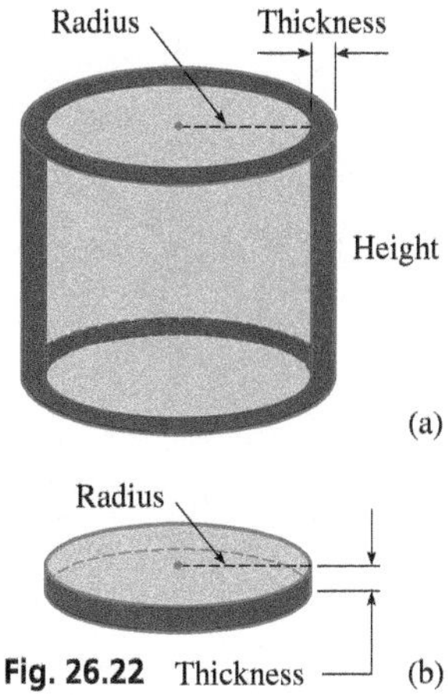

Fig. 26.22

By finding the circumference of the base and multiplying this by the height, we obtain an expression for the surface area of the shell. Then, by multiplying this by the thickness of the shell, we find its volume. The volume of the representative ***shell*** shown in Fig. 26.22(a) is

Shell

$$dV = 2\pi(\text{radius}) \times (\text{height}) \times (\text{thickness}) \qquad \textbf{(26.13)}$$

Similarly, when the element of volume is a disc (see Fig. 26.22b), its volume is given by

Disc

$$dV = \pi(\text{radius})^2 \times (\text{thickness}) \qquad \textbf{(26.14)}$$

LEARNING TIP

It is generally better to remember the formulas for the elements of volume in the general forms given in Eqs. (26.13) and (26.14), and not in the specific forms such as Eqs. (26.11) and (26.12) (both of these use *discs*). If we remember the formulas in this way, we can readily apply these methods to finding the volume of any solid of revolution.

EXAMPLE 3 Find a volume using cylindrical shells

Use the method of cylindrical shells to find the volume of the solid generated by revolving the first-quadrant region bounded by $y = 4 - x^2$, $x = 0$, and $y = 0$ about the y-axis.

From Fig. 26.23, we identify the radius, the height, and the thickness of the shell:

$$\text{radius} = x \qquad \text{height} = y \qquad \text{thickness} = dx$$

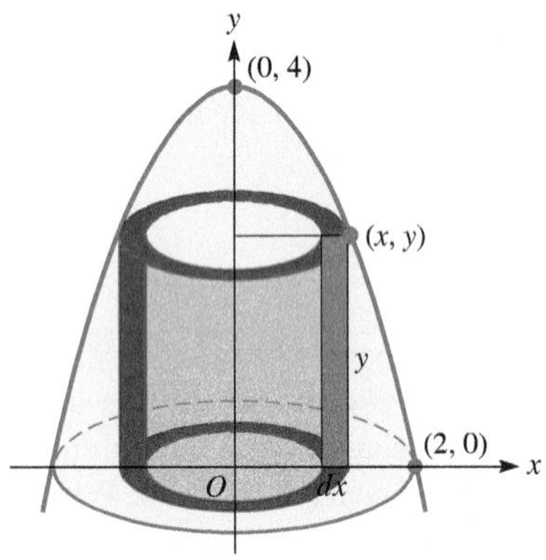

Fig. 26.23

The fact that the elements of area that generate the shells go from x = 0 to x = 2 determines the limits of integration as 0 and 2. Therefore,

$$V = 2\pi \int_0^2 x\, y\, dx \leftarrow \text{thickness} \qquad \text{using Eq. (26.13)}$$

(radius ↑ x, height ↑ y)

$$= 2\pi \int_0^2 x(4 - x^2)\, dx = 2\pi \int_0^2 (4x - x^3)\, dx \qquad \text{substitute } 4 - x^2 \text{ for } y$$

$$= 2\pi\left(2x^2 - \frac{1}{4}x^4\right)\Big|_0^2 \qquad \text{integrate}$$

$$= 8\pi \qquad \text{evaluate}$$

We can find the volume shown in Example 3 by using discs, as we show in the following example.

EXAMPLE 4 Same volume using horizontal discs

Use the method of discs to find the volume indicated in Example 3.

From Fig. 26.24, we identify the radius and the thickness of the disc:

$$\text{radius} = x \qquad \text{thickness} = dy$$

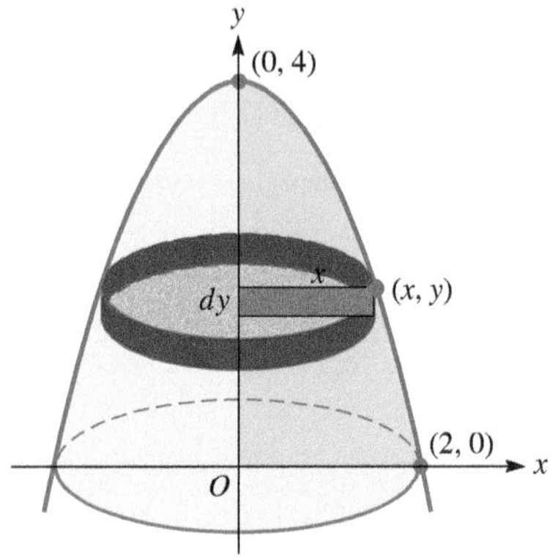

Fig. 26.24

Since the elements of area that generate the discs go from y = 0 to y = 4, the limits of integration are 0 and 4. Thus,

$$V = \pi \int_0^4 x^2 dy \qquad \text{using Eq. (26.14)}$$

(radius ↑ x^2, thickness ↑ dy)

$$= \pi \int_0^4 (4 - y)\, dy \qquad \text{substitute } \sqrt{4 - y} \text{ for } x$$

$$= \pi\left(4y - \frac{1}{2}y^2\right)\Big|_0^4 \qquad \text{integrate}$$

$$= 8\pi \qquad \text{evaluate}$$

We see that the volume of 8π using discs agrees with the result we obtained using shells in Example 3.

Practice Exercise

1. Find the volume of the solid generated by revolving the first-quadrant region bounded by $y = 4 - 2x$ about the x-axis. Use discs.

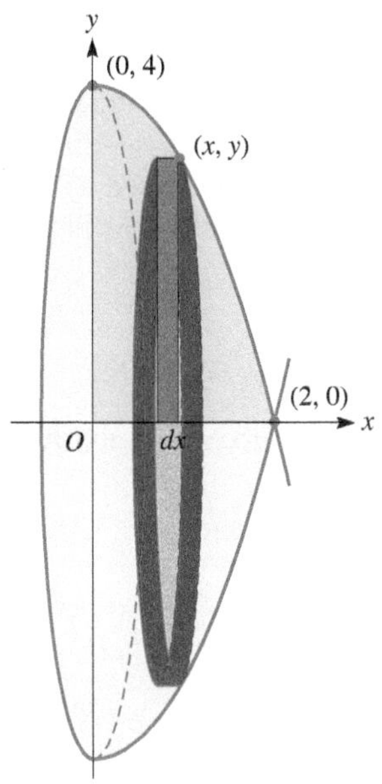

Fig. 26.25

EXAMPLE 5 Find a volume using vertical discs

Using discs, find the volume of the solid generated by revolving the first-quadrant region bounded by $y = 4 - x^2$, $x = 0$, and $y = 0$ about the x-axis. (This is the same region as used in Examples 3 and 4.)

For the disc in Fig. 26.25, we have

$$\text{radius} = y \qquad \text{thickness} = dx$$

and the limits of integration are $x = 0$ and $x = 2$. This gives us

$$V = \pi \int_0^2 y^2 dx \qquad \text{using Eq. (26.14)}$$

$$= \pi \int_0^2 (4 - x^2)^2 dx \qquad \text{substitute } 4 - x^2 \text{ for } y$$

$$= \pi \int_0^2 (16 - 8x^2 + x^4)\, dx$$

$$= \pi\left(16x - \frac{8}{3}x^3 + \frac{1}{5}x^5\right)\Bigg|_0^2 \qquad \text{integrate}$$

$$= \frac{256\pi}{15} \qquad \text{evaluate}$$

We now demonstrate how to set up the integral to find the volume of the solid shown in Example 5 by using cylindrical shells. As it turns out, we are not able at this point to integrate the expression that arises, but we are still able to set up the proper integral.

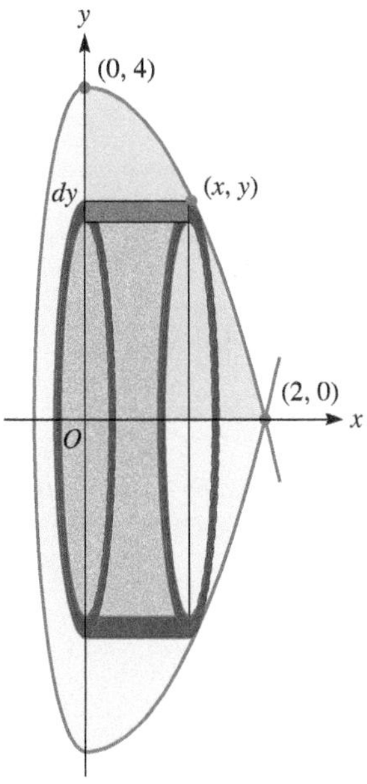

Fig. 26.26

EXAMPLE 6 Same volume using cylindrical shells

Use the method of cylindrical shells to find the volume indicated in Example 5.

From Fig. 26.26, we see that for the shell we have

$$\text{radius} = y \qquad \text{height} = x \qquad \text{thickness} = dy$$

Since the elements go from $y = 0$ to $y = 4$, the limits of integration are 0 and 4. Hence,

$$V = 2\pi \int_0^4 xy\, dy \qquad \text{using Eq. (26.13)}$$

$$= 2\pi \int_0^4 \sqrt{4 - y}(y\, dy) \qquad \text{substitute } \sqrt{4-y} \text{ for } x$$

$$= \frac{256\pi}{15}$$

The method of performing the integration $\int \sqrt{4 - y}(y\, dy)$ will be discussed in Section 28.7. We present the answer here for the reader's information to show that the volume found in this example is the same as that found in Example 5.

Practice Exercise

2. Find the volume of the solid generated by revolving the first-quadrant region bounded by $y = 4 - 2x$ about the x-axis. Use shells.

In the next example, we show how to find the volume of the solid generated if a region is revolved about a line other than one of the axes. We will see that a proper choice of the radius, height, and thickness for Eq. (26.13) leads to the result.

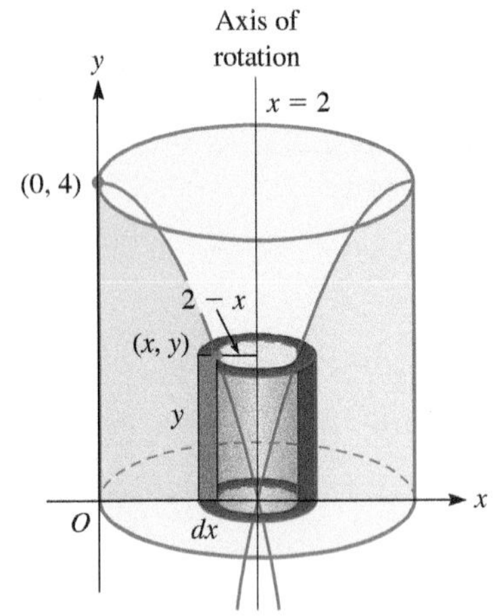

Fig. 26.27

EXAMPLE 7 Rotate a region about a line

Find the volume of the solid generated if the region in Examples 3–6 is revolved about the line $x = 2$.

Shells are convenient, since the volume of a shell can be expressed as a single integral. We can find the radius, height, and thickness of the shell from Fig. 26.27. Carefully note that ***the radius is not x but is 2 − x***, since the region is revolved about the line $x = 2$. This means

$$\text{radius} = 2 - x \qquad \text{height} = y \qquad \text{thickness} = dx$$

Since the elements that generate the shells go from $x = 0$ to $x = 2$, the limits of integration are 0 and 2. This means we have

$$V = 2\pi \int_0^2 (2 - x)y\, dx \quad \text{(radius } 2-x\text{, height } y\text{, thickness } dx\text{)} \qquad \text{using Eq. (25.13)}$$

$$= 2\pi \int_0^2 (2 - x)(4 - x^2)\, dx \qquad \text{substitute } 4 - x^2 \text{ for } y$$

$$= 2\pi \int_0^2 (8 - 2x^2 - 4x + x^3)\, dx$$

$$= 2\pi\left(8x - \frac{2}{3}x^3 - 2x^2 + \frac{1}{4}x^4\right)\Big|_0^2 \qquad \text{integrate}$$

$$= \frac{40\pi}{3} \qquad \text{evaluate}$$

(If the region had been revolved about the line $x = 3$, the only difference in the integral would have been that $r = 3 - x$. Everything else, including the limits, would have remained the same.)

EXERCISES 26.3

In Exercises 1 and 2, make the given changes in the indicated examples of this section and then find the indicated volumes.

1. In Example 1, change $y = x^2$ to $y = x^3$.
2. In Example 3, change $y = 4 - x^2$ to $y = 4 - x$.

In Exercises 3–6, find the volume generated by revolving the region bounded by $y = 2 - x$, $x = 0$, and $y = 0$ about the indicated axis, using the indicated element of volume.

3. x-axis (discs)
4. y-axis (discs)
5. y-axis (shells)
6. x-axis (shells)

In Exercises 7–16, find the volume generated by revolving the regions bounded by the given curves about the x-axis. Use the indicated method in each case.

7. $y = x$, $y = 0$, $x = 3$ (discs)
8. $y = \sqrt{x}$, $x = 0$, $y = 2$ (shells)
9. $y = 3\sqrt{x}$, $y = 0$, $x = 4$ (discs)
10. $y = 4x - x^2$, $y = 0$ (discs)
11. $y = x^3$, $y = 8$, $x = 0$ (shells)
12. $y = x^2$, $y = x$ (shells)
13. $y = x^2 + 1$, $x = 0$, $x = 3$, $y = 0$ (discs)
14. $y = 6 - x - x^2$, $x = 0$, $y = 0$ (quadrant I), (discs)
15. $x = 4y - y^2 - 3$, $x = 0$ (shells)
16. $y = x^4$, $x = 0$, $y = 1$, $y = 8$ (shells)

In Exercises 17–26, find the volume generated by revolving the regions bounded by the given curves about the y-axis. Use the indicated method in each case.

17. $y = x^{1/3}$, $x = 0$, $y = 2$ (discs)
18. $y = \sqrt{x^2 - 1}$, $y = 0$, $x = 3$ (shells)
19. $y = 2\sqrt{x}$, $x = 0$, $y = 3$ (discs)
20. $y^2 = x$, $y = 4$, $x = 0$ (discs)
21. $x^2 - 4y^2 = 4$, $x = 3$ (shells)
22. $y = 3x^2 - x^3$, $y = 0$ (shells)
23. $x = 6y - y^2$, $x = 0$ (discs)
24. $x^2 + 4y^2 = 4$ (quadrant I), (discs)
25. $y = \sqrt{4 - x^2}$ (quadrant I), (shells)
26. $y = 8 - x^3$, $x = 0$, $y = 0$ (shells)

In Exercises 27–40, find the indicated volumes by integration.

27. Describe a region that is revolved about the x-axis to generate a volume found by evaluating the integral $\pi \int_1^2 x^3\,dx$.

28. Describe a region that is revolved about the y-axis to generate a volume found by the integral in Exercise 27.

29. Find the volume generated if the region bounded by $y = 4 - x^2$, $y = 0$, and $x = 2$ is revolved about the line $x = 2$.

30. Find the volume generated if the region bounded by $y = \sqrt{x}$ and $y = x/2$ is revolved about the line $y = 4$.

31. Derive the formula for the volume of a right circular cone of radius r and height h by revolving the area bounded by $y = (r/h)x$, $y = 0$, and $x = h$ about the x-axis.

32. Explain how to derive the formula for the volume of a sphere by using the disc method.

33. The base of a solid is the region bounded by $y = x^2$, $y = 0$, and $x = 1$. All vertical cross-sections are *squares* with an edge in the xy-plane. Find the volume of the solid. (Note that this is not a solid generated by revolving a region about an axis.)

34. The oil in a spherical tank 20.0 m in diameter is 7.5 m deep. How much oil is in the tank?

35. A *drumlin* is an oval hill composed of relatively soft soil that was deposited beneath glacial ice. (The Halifax Citadel National Historic Site in Halifax, Nova Scotia, is built on a drumlin.) Computer analysis showed that the surface of a certain drumlin can be approximated by $y = 10(1 - 0.0001x^2)$ revolved 180° about the x-axis from $x = -100$ to $x = 100$ (see Fig. 26.28). Find the volume (in m³) of this drumlin.

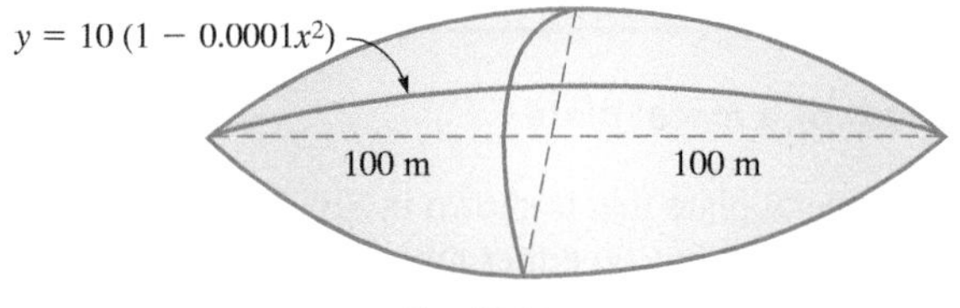

Fig. 26.28

36. If the area bounded by $y = 0$, $y = 2x$, and $x = 5$ is rotated about each axis, which volume is greater?

37. The ball used in Australian football is elliptical. Find its volume if it is 275 mm long and 170 mm wide.

38. A commercial dirigible used for outdoor advertising has a helium-filled balloon in the shape of an ellipse revolved about its major axis. If the balloon is 41.3 m long and 12.0 m in diameter, what volume of helium is required to fill it? See Fig. 26.29.

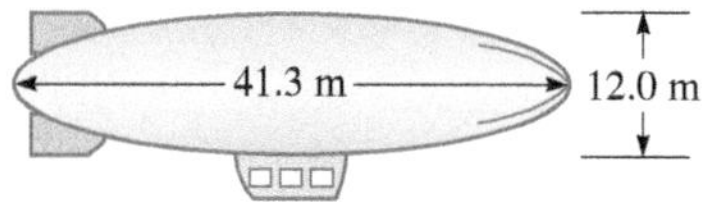

Fig. 26.29

39. A hole 2.00 cm in diameter is drilled through the centre of a spherical lead weight 6.00 cm in diameter. How much lead is removed? See Fig. 26.30.

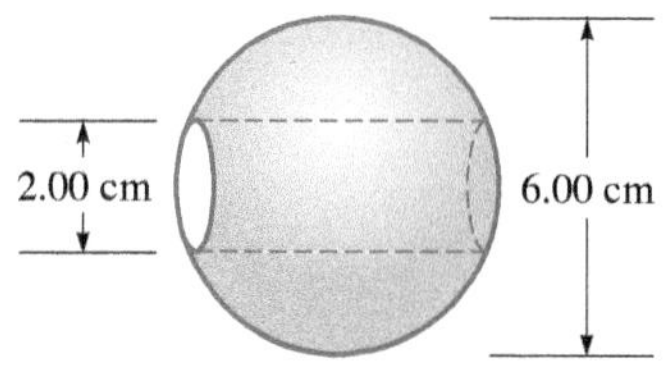

Fig. 26.30

40. All horizontal cross-sections of a keg 1.20 m tall are circular, and the sides of the keg are parabolic. The diameter at the top and bottom is 0.80 m, and the diameter in the middle is 1.0 m. Find the volume that the keg holds.

Answers to Practice Exercises

1. $32\pi/3$ **2.** $32\pi/3$

26.4 Centroids

In the study of mechanics, a very important property of an object is its centre of mass. In this section, we explain the meaning of centre of mass and then show how integration is used to determine the centre of mass for regions and solids of revolution.

If a mass m is at a distance d from a specified point O, the **moment** *of the mass about O is defined as md.* If several masses $m_1, m_2, \ldots, m_n$ are at distances $d_1, d_2, \ldots, d_n$, respectively, from point O, the total moment (as a group) about O is defined as $m_1d_1 + m_2d_2 + \cdots + m_nd_n$. *The* **centre of mass** *is that point $\overline{d}$ units from O at which all the masses could be concentrated to get the same total moment.* Therefore, $\overline{d}$ is defined by the equation

$$m_1d_1 + m_2d_2 + \cdots + m_nd_n = (m_1 + m_2 + \cdots + m_n)\overline{d} \qquad \textbf{(26.15)}$$

LEARNING TIP
When finding moments, we must always use **directed distances**.

The moment of a mass is a measure of its tendency to rotate about a point. A weight far from the point of balance of a long rod is more likely to make the rod turn than if the same weight were placed near the point of balance. It is easier to open a door if you push near the doorknob than if you push near the hinges. This is the type of physical property that the moment of mass measures.

EXAMPLE 1 Balancing masses

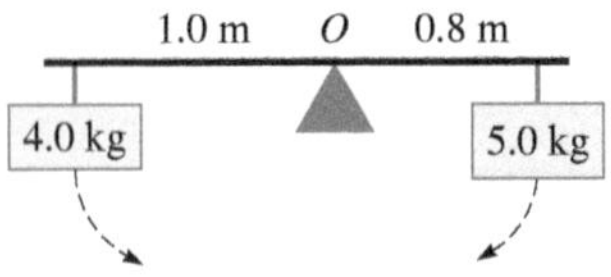

Fig. 26.31

One of the simplest and most basic illustrations of moments and centre of mass is seen in balancing a long rod (of negligible mass) with masses of different sizes, one on either side of the balance point.

In Fig. 26.31, a mass of 5.0 kg is hung from the rod 0.8 m to the right of point O. We see that this 5.0-kg mass tends to turn the rod clockwise. A mass placed on the opposite side of O will tend to turn the rod counterclockwise. Neglecting the mass of the rod, in order to balance the rod at O, the moments must be equal in magnitude but opposite in sign.

For instance, a 4-kg mass would have to be placed at a distance d_2, satisfying

$$5.0(0.8) + 4.0(d_2) = (5.0 + 4.0)(0) \quad \text{Eq. (26.15) with } \overline{d} = 0$$
$$d_2 = -1.0$$

In other words, the centre of mass of the combination of the 5.0-kg mass and the 4.0-kg mass is at O (with $\overline{d} = 0$) if the 4.0-kg mass is placed 1.0 m to the left of O.

EXAMPLE 2 Centre of mass of three masses

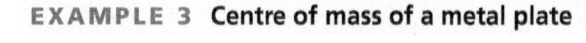

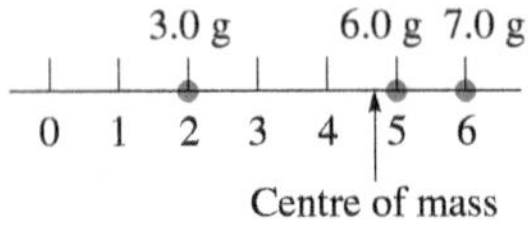

Fig. 26.32

A mass of 3.0 g is placed at (2.0, 0) on the x-axis (distances in cm). Another mass of 6.0 g is placed at (5.0, 0), and a third mass of 7.0 g is placed at (6.0, 0). See Fig. 26.32. Find the centre of mass of these three masses.

Taking the reference point as the origin, we find $d_1 = 2.0$ cm, $d_2 = 5.0$ cm, and $d_3 = 6.0$ cm. Thus, $m_1d_1 + m_2d_2 + m_3d_3 = (m_1 + m_2 + m_3)\overline{d}$ becomes

$$3.0(2.0) + 6.0(5.0) + 7.0(6.0) = (3.0 + 6.0 + 7.0)\overline{d} \quad \text{or} \quad \overline{d} = 4.9 \text{ cm}$$

This means that the centre of mass of the three masses is at (4.9, 0). Therefore, a mass of 16.0 g placed at this point has the same moment as the three masses as a unit.

EXAMPLE 3 Centre of mass of a metal plate

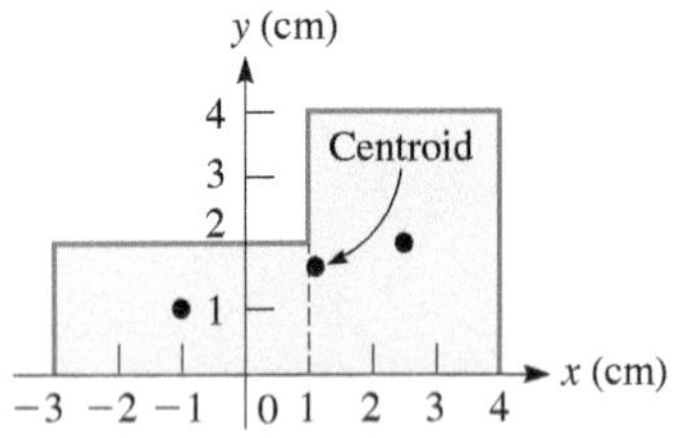

Fig. 26.33

Find the centre of mass of the flat metal plate that is shown in Fig. 26.33.

We first note that the centre of mass is not *on* either axis. This can be seen from the fact that the major portion of the area is in the first quadrant. *We will therefore measure the moments with respect to each axis to find the point that is the centre of mass. This point is also called the* **centroid** *of the plate.*

The easiest method of finding this centroid is to divide the plate into rectangles, as indicated by the dashed line in Fig. 26.33, and assume that we may consider the mass of each rectangle to be concentrated at its centre. In this way, the centre of the left rectangle is at (−1.0, 1.0) (distances in cm), and the centre of the right rectangle is at (2.5, 2.0). The mass of each rectangle area, assumed uniform, is proportional to its area. The area of the left rectangle is 8.0 cm^2, and that of the right rectangle is 12.0 cm^2. Thus, taking moments with respect to the y-axis, we have

$$8.0(-1.0) + 12.0(2.5) = (8.0 + 12.0)\overline{x}$$

where $\overline{x}$ is the x-coordinate of the centroid. Solving for $\overline{x}$, we have $\overline{x} = 1.1$ cm.

Now, taking moments with respect to the x-axis, we have

$$8.0(1.0) + 12.0(2.0) = (8.0 + 12.0)\overline{y}$$

where $\overline{y}$ is the y-coordinate of the centroid. Thus, $\overline{y} = 1.6$ cm. This means that the coordinates of the centroid, the centre of mass, are (1.1, 1.6). This may be interpreted as meaning that a plate of this shape would balance on a single support under this point. As an approximate check, we note from the figure that this point appears to be a reasonable balance point for the plate.

■ Since the centre of mass does not depend on the density of the metal, we have assumed the constant of proportionality to be 1.

CENTROID OF A THIN, FLAT PLATE BY INTEGRATION

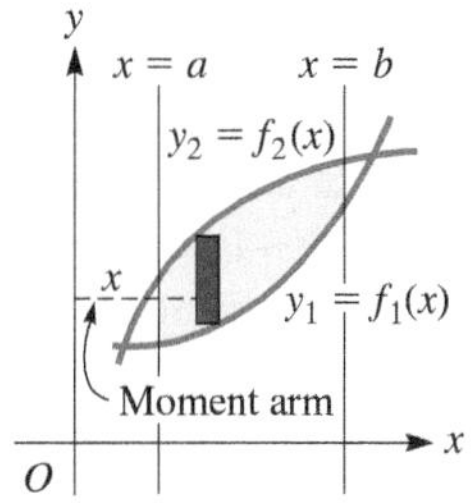

Fig. 26.34

If a thin, flat plate covers the region bounded by $y_1 = f_1(x)$, $y_2 = f_2(x)$, $x = a$, and $x = b$, as shown in Fig. 26.34, the moment of the mass of the element of area about the y-axis is given by $(\rho dA)x$, where ρ is the mass per unit area. In this expression, ρdA is the mass of the element, and x is its distance (moment arm) from the y-axis. The element dA may be written as $(y_2 - y_1)\,dx$, which means that the moment may be written as $\rho x(y_2 - y_1)\,dx$. If we then sum up the moments of all the elements and express this as an integral (which, of course, means sum), we have $\rho\int_a^b x(y_2 - y_1)\,dx$. If we consider all the mass of the plate to be concentrated at one point $\bar{x}$ units from the y-axis, the moment would be $(\rho A)\bar{x}$, where ρA is the mass of the entire plate and $\bar{x}$ is the distance the centre of mass is from the y-axis. By the previous discussion, these two expressions should be equal. This means $\rho\int_a^b x(y_2 - y_1)\,dx = \rho A\bar{x}$. Since ρ appears on each side of the equation, we divide it out (we are assuming that the mass per unit area is constant). The area A is found by the integral $\int_a^b (y_2 - y_1)\,dx$. Therefore, the x-coordinate of the centroid of the plate is given by

$$\bar{x} = \frac{\displaystyle\int_a^b x(y_2 - y_1)\,dx}{\displaystyle\int_a^b (y_2 - y_1)\,dx} \tag{26.16}$$

Eq. (26.16) gives us the x-coordinate of the centroid of the plate if vertical elements are used.

COMMON ERROR

The two integrals in Eq. (26.16) must be evaluated separately. We cannot cancel out the apparent common factor $y_2 - y_1$, and we cannot combine quantities and perform only one integration. The two integrals must be evaluated separately first. Then any possible cancellations of factors common to the numerator and the denominator may be made.

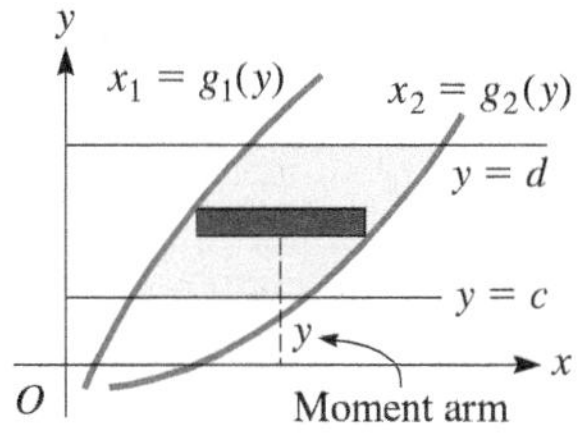

Fig. 26.35

Following the same reasoning that we used in developing Eq. (26.16), for a thin plate covering the region bounded by the functions $x_1 = g_1(y)$, $x_2 = g_2(y)$, $y = c$, and $y = d$, as shown in Fig. 26.35, the y-coordinate of the centroid of the plate is given by the equation

$$\bar{y} = \frac{\displaystyle\int_c^d y(x_2 - x_1)\,dy}{\displaystyle\int_c^d (x_2 - x_1)\,dy} \tag{26.17}$$

In this equation, horizontal elements are used.

LEARNING TIP

In applying Eqs. (26.16) and (26.17), we should keep in mind that each denominator on the right-hand side gives the area of the plate and that, once we have found this area, we may use it for both $\bar{x}$ and $\bar{y}$. In this way, we can avoid having to set up and perform one of the indicated integrations. Also, in finding the coordinates of the centroid, we should look for and utilize any symmetry the region may have.

EXAMPLE 4 Centroid of a thin plate by integration

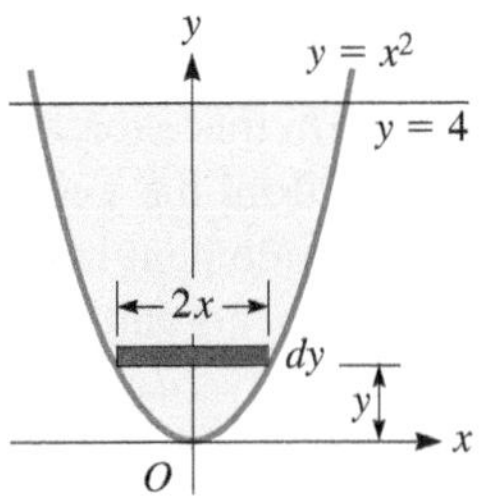

Fig. 26.36

Find the coordinates of the centroid of a thin plate covering the region bounded by the parabola $y = x^2$ and the line $y = 4$.

We sketch a graph indicating the region and an element of area (see Fig. 26.36). The curve is a parabola whose axis is the y-axis. Since the region is symmetric to the y-axis, the centroid must be on this axis. This means that the x-coordinate of the centroid is zero, or $\bar{x} = 0$. To find the y-coordinate of the centroid, we have

$$\bar{y} = \frac{\int_0^4 y(2x)\,dy}{\int_0^4 2x\,dy} \quad \text{using Eq. (26.17)}$$

(moment arm of element: y; area: $2x\,dy$)

$$= \frac{\int_0^4 y(2\sqrt{y})\,dy}{\int_0^4 2\sqrt{y}\,dy} = \frac{2\int_0^4 y^{3/2}dy}{2\int_0^4 y^{1/2}dy} = \frac{2\left(\frac{2}{5}\right)y^{5/2}\Big|_0^4}{2\left(\frac{2}{3}\right)y^{3/2}\Big|_0^4} \quad \text{integrate and evaluate numerator and denominator separately}$$

$$= \frac{\frac{4}{5}(32)}{\frac{4}{3}(8)} = \frac{128}{5} \times \frac{3}{32} = \frac{12}{5}$$

The coordinates of the centroid are $\left(0, \frac{12}{5}\right)$. This plate would balance if a single pointed support were to be put under this point.

EXAMPLE 5 Centroid of a triangular plate by integration

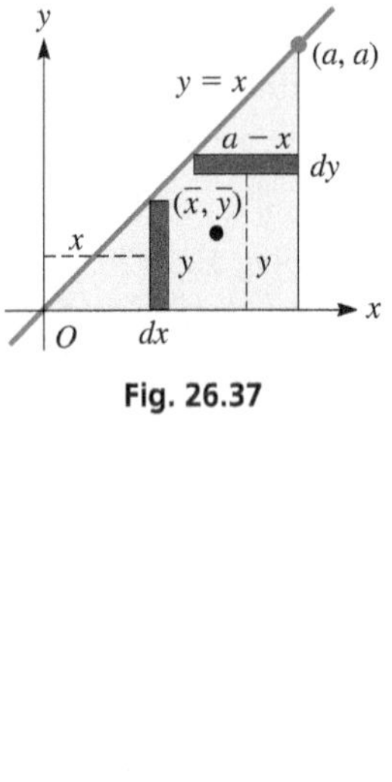

Fig. 26.37

Find the coordinates of the centroid of an isosceles right triangular plate with side a. See Fig. 26.37.

We must first set up the region in the xy-plane. The choice shown in Fig. 26.37 is to place the triangle with one vertex at the origin and the right angle on the x-axis. Since each side is a, the hypotenuse passes through the point (a, a). The equation of the hypotenuse is $y = x$. The x-coordinate of the centroid is found by using Eq. (26.16):

$$\bar{x} = \frac{\int_0^a xy\,dx}{\int_0^a y\,dx} = \frac{\int_0^a x(x)\,dx}{\int_0^a x\,dx} = \frac{\int_0^a x^2 dx}{\frac{1}{2}x^2\Big|_0^a} = \frac{\frac{1}{3}x^3\Big|_0^a}{\frac{a^2}{2}} = \frac{\frac{a^3}{3}}{\frac{a^2}{2}} = \frac{2a}{3}$$

The y-coordinate of the centroid is found by using Eq. (26.17):

$$\bar{y} = \frac{\int_0^a y(a-x)\,dy}{\frac{a^2}{2}} = \frac{\int_0^a y(a-y)\,dy}{\frac{a^2}{2}} = \frac{\int_0^a (ay - y^2)\,dy}{\frac{a^2}{2}}$$

$$= \frac{\left.\frac{ay^2}{2} - \frac{y^3}{3}\right|_0^a}{\frac{a^2}{2}} = \frac{\frac{a^3}{6}}{\frac{a^2}{2}} = \frac{a}{3}$$

Thus, the coordinates of the centroid are $\left(\frac{2}{3}a, \frac{1}{3}a\right)$. The results indicate that the centre of mass is $\frac{1}{3}a$ units from each of the equal sides.

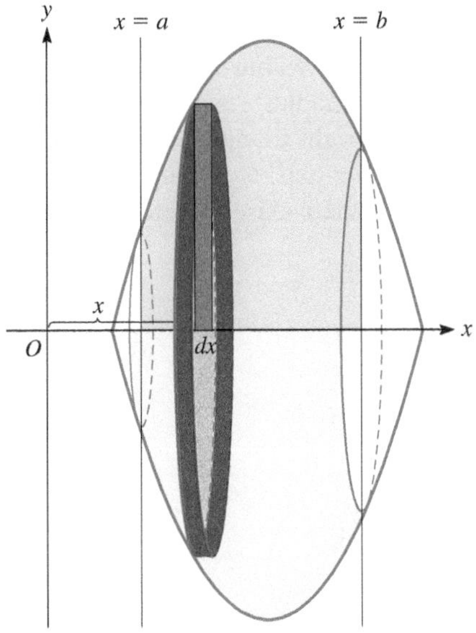

Fig. 26.38

CENTROID OF A SOLID OF REVOLUTION

Another figure for which we wish to find the centroid is a solid of revolution. If the density of the solid is constant, the centroid is on the axis of revolution. The problem that remains is to find just where on the axis the centroid is located.

If a region bounded by the x-axis, as shown in Fig. 26.38, is revolved about the x-axis, a vertical element of area generates a disc element of volume. The centre of mass of the disc is at its centre, and we may consider its mass concentrated there. The moment about the y-axis of a typical element is $x(\rho)(\pi y^2 dx)$, where x is the moment arm, ρ is the density, and $\pi y^2 dx$ is the volume. The sum of the moments of the elements can be expressed as an integral; it equals the volume times the density times the x-coordinate of the centroid of the volume. Since π and the density ρ would appear on each side of the equation, they cancel and need not be written. Therefore,

$$\bar{x} = \frac{\int_a^b xy^2\,dx}{\int_a^b y^2\,dx} \qquad \textbf{(26.18)}$$

is the equation for the x-coordinate of the centroid of a solid of revolution about the x-axis.

In the same manner, we may find that *the y-coordinate of the centroid of a solid of revolution about the y-axis is*

$$\bar{y} = \frac{\int_c^d yx^2\,dy}{\int_c^d x^2\,dy} \qquad \textbf{(26.19)}$$

EXAMPLE 6 Centroid of a solid by integration

Find the coordinates of the centroid of the volume generated by revolving the first-quadrant region under the curve $y = 4 - x^2$ about the y-axis as shown in Fig. 26.39.

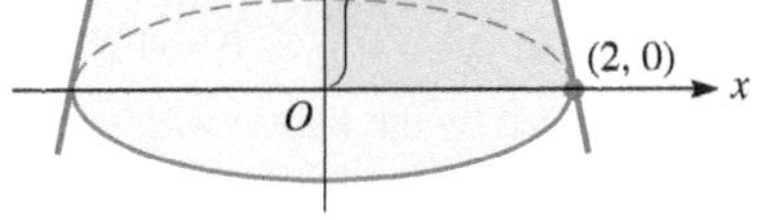

Fig. 26.39

Since the curve is rotated about the y-axis, $\bar{x} = 0$. The y-coordinate is

$$\bar{y} = \frac{\int_0^4 yx^2\,dy}{\int_0^4 x^2\,dy} \quad \text{(moment arm: } y\text{; using Eq. (26.19))}$$

$$= \frac{\int_0^4 y(4-y)\,dy}{\int_0^4 (4-y)\,dy} = \frac{\int_0^4 (4y-y^2)\,dy}{\int_0^4 (4-y)\,dy} = \frac{2y^2 - \frac{1}{3}y^3\Big|_0^4}{4y - \frac{1}{2}y^2\Big|_0^4}$$

$$= \frac{32 - \frac{64}{3}}{16 - 8} = \frac{4}{3}$$

The coordinates of the centroid are $\left(0, \frac{4}{3}\right)$.

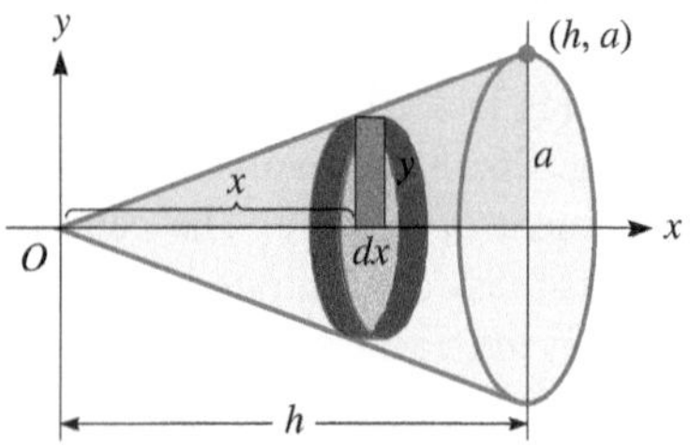

Fig. 26.40

EXAMPLE 7 Centroid of a right circular cone—machine parts

A company makes solid conical machine parts of various sizes. Show that the centroid of every part of any size is in the same relative position within the part. We can do this by finding the centroid of any right circular cone of radius r and height h.

To generate a right circular cone, we may revolve a right triangle about one of its legs (Fig. 26.40). Placing a leg of length h along the x-axis, we rotate the right triangle whose hypotenuse is given by $y = (a/h)x$ about the x-axis. Therefore,

$$\bar{x} = \frac{\int_0^h xy^2\,dx}{\int_0^h y^2\,dx} \quad \text{using Eq. (26.18)}$$

(moment arm: x)

$$= \frac{\int_0^h x\left[\left(\frac{a}{h}\right)x\right]^2 dx}{\int_0^h \left[\left(\frac{a}{h}\right)x\right]^2 dx} = \frac{\left(\frac{a^2}{h^2}\right)\left(\frac{1}{4}x^4\right)\Big|_0^h}{\left(\frac{a^2}{h^2}\right)\left(\frac{1}{3}x^3\right)\Big|_0^h} = \frac{3}{4}h$$

The centroid is located along the height $\frac{3}{4}$ of the way from the vertex to the base.

EXERCISES 26.4

In Exercises 1 and 2, make the given changes in the indicated examples of this section and then find the coordinates of the centroid.

1. In Example 4, change $y = x^2$ to $y = |x|$ ($y = x$ for $x \geq 0$, and $y = -x$ for $x < 0$).
2. In Example 6, change $y = 4 - x^2$ to $y = 4 - x$.

In Exercises 3–6, find the centre of mass (in cm) of the particles with the given masses located at the given points on the x-axis.

3. 5.0 g at (1.0, 0), 8.5 g at (4.2, 0), 3.6 g at (2.5, 0)
4. 2.3 g at (1.3, 0), 6.5 g at (5.8, 0), 1.2 g at (9.5, 0)
5. 42 g at (−3.5, 0), 24 g at (0, 0), 15 g at (2.6, 0), 84 g at (3.7, 0)
6. 550 g at (−42, 0), 230 g at (−27, 0), 470 g at (16, 0), 120 g at (22, 0)

In Exercises 7–10, find the coordinates (to 0.01 cm) of the centroids of the uniform flat-plate machine parts shown.

7.

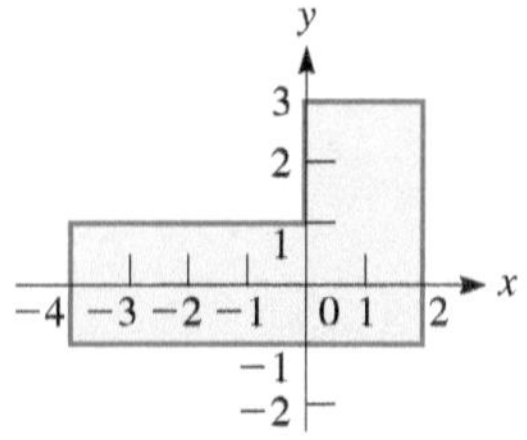

8.

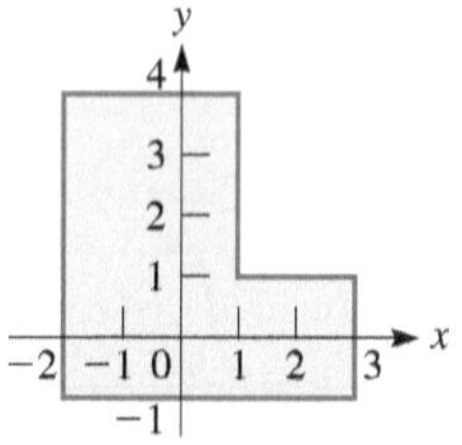

9.

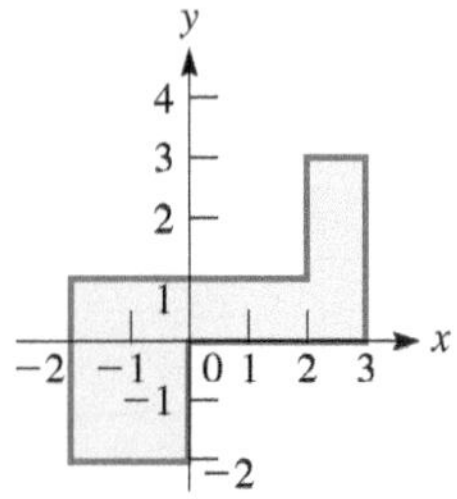

10.

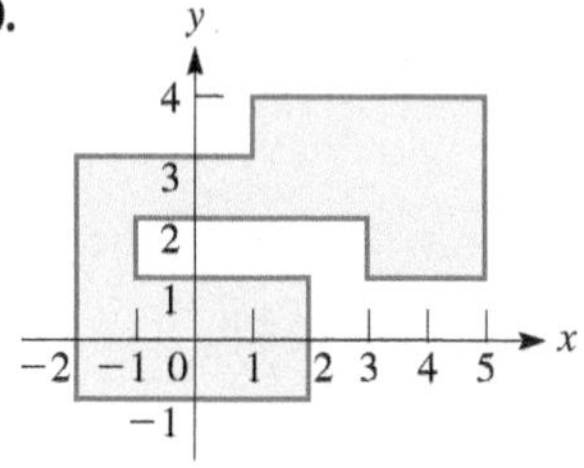

In Exercises 11–28, find the coordinates of the centroids of the given figures. In Exercises 11–22, each region is covered by a thin, flat plate.

11. The region bounded by $y = x^2$ and $y = 2$
12. The semicircular region in Fig. 26.41

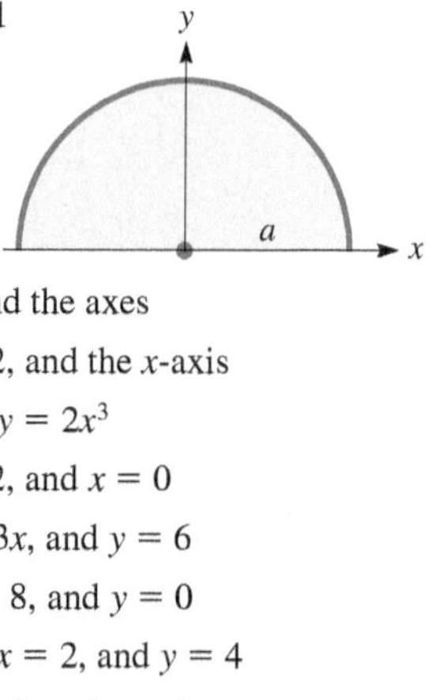

Fig. 26.41

13. The region bounded by $y = 4 - x$ and the axes
14. The region bounded by $y = x^3$, $x = 2$, and the x-axis
15. The region bounded by $y = 4x^2$ and $y = 2x^3$
16. The region bounded by $y^2 = x$, $y = 2$, and $x = 0$
17. The region bounded by $y = 2x$, $y = 3x$, and $y = 6$
18. The region bounded by $y = x^{2/3}$, $x = 8$, and $y = 0$
19. The region bounded by $y = 4 - 2x$, $x = 2$, and $y = 4$
20. The region bounded by $y = \sqrt{x}$, $y = 0$, and $x = 9$
21. The region bounded by $x^2 = 4py$ and $y = a$ if $p > 0$ and $a > 0$
22. The region above the x-axis, bounded by the ellipse with vertices $(a, 0)$ and $(-a, 0)$, and minor axis $2b$ (The area of an ellipse is πab.)
23. The solid generated by revolving the region bounded by $y = x^3$, $y = 0$, and $x = 1$ about the x-axis
24. The solid generated by revolving the region bounded by $y = 2 - 2x$, $x = 0$, and $y = 0$ about the y-axis
25. The solid generated by revolving the region in the first quadrant bounded by $y^2 = 4x$, $y = 0$, and $x = 1$ about the y-axis
26. The solid generated by revolving the region bounded by $y = x^2$, $x = 2$, and the x-axis about the x-axis

27. The solid generated by revolving the region bounded by $y^2 = 4x$ and $x = 1$ about the x-axis

28. The solid generated by revolving the region bounded by $x^2 - y^2 = 9$, $y = 4$, and the x-axis about the y-axis

In Exercises 29–34, answer the given question.

29. A sailboat has a right-triangular sail with a horizontal base 3.0 m long and a vertical side of 4.5 m high. Where is the centroid of the sail?

30. Find the location of the centroid of a hemisphere of radius a.

31. A lens with semielliptical vertical cross-sections and circular horizontal cross-sections is shown in Fig. 26.42. For proper installation in an optical device, its centroid must be known. Locate its centroid.

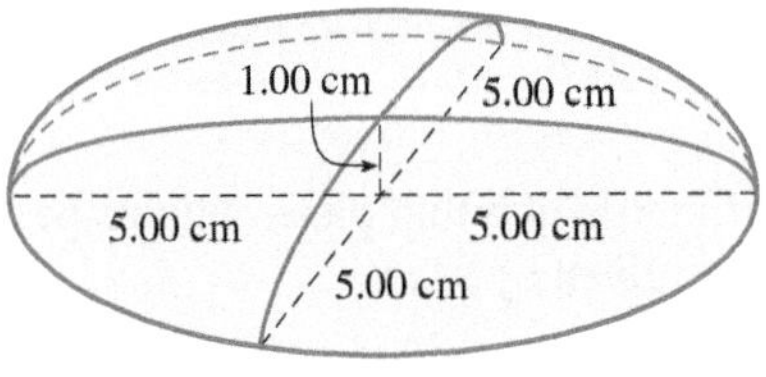

Fig. 26.42

32. A sanding machine disc can be described as the solid generated by rotating the region bounded by $y^2 = 4/x$, $y = 1$, $y = 2$, and the y-axis about the y-axis (measurements in cm). Locate the centroid of the disc.

33. A highway marking pylon has the shape of a frustum of a cone. Find its centroid if the radii of its bases are 5.00 cm and 20.0 cm and the height between bases is 60.0 cm.

34. A floodgate is in the shape of an isosceles trapezoid. Find the location of the centroid of the floodgate if the upper base is 20 m, the lower base is 12 m, and the height between bases is 6.0 m. See Fig. 26.43.

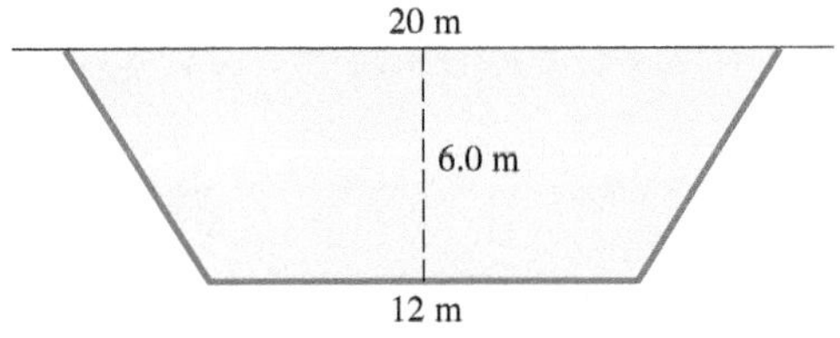

Fig. 26.43

26.5 Moments of Inertia

Important in the rotational motion of an object is its **moment of inertia**, which is analogous to the mass of a moving object. In each case, *the moment of inertia or mass is the measure of the tendency of the object to resist a change in motion.*

Suppose that a particle of mass m is rotating about some point: We define its moment of inertia as md^2, where d is the distance from the particle to the point. If a group of particles of masses $m_1, m_2, \ldots, m_n$ are rotating about an axis, as shown in Fig. 26.44, the moment of inertia I with respect to the axis of the group is

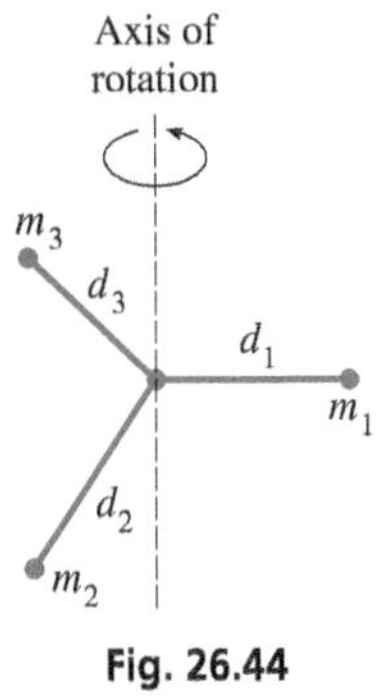

Fig. 26.44

$$I = m_1 d_1^2 + m_2 d_2^2 + \cdots + m_n d_n^2$$

where the d's are the respective distances of the particles from the axis. If all the masses were at the same distance R from the axis of rotation, so that the total moment of inertia were the same, we would have

$$m_1 d_1^2 + m_2 d_2^2 + \cdots + m_n d_n^2 = (m_1 + m_2 + \cdots + m_n)R^2 \qquad \textbf{(26.20)}$$

where R is called the **radius of gyration.**

EXAMPLE 1 Moment of inertia and radius of gyration

Find the moment of inertia and the radius of gyration of the array of three masses, one of 3.0 g at $(-2.0, 0)$, another of 5.0 g at $(1.0, 0)$, and the third of 4.0 g at $(4.0, 0)$, with respect to the origin (distances in cm). See Fig. 26.45.

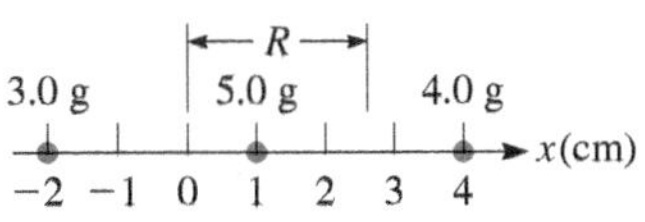

Fig. 26.45

The moment of inertia of the array is

$$I = 3.0(-2.0)^2 + 5.0(1.0)^2 + 4.0(4.0)^2 = 81 \text{ g} \cdot \text{cm}^2$$

The radius of gyration is found from $I = (m_1 + m_2 + m_3)R^2$. Thus,

$$81 = (3.0 + 5.0 + 4.0)R^2, \qquad R^2 = \frac{81}{12}, \qquad R = 2.6 \text{ cm}$$

Therefore, a mass of 12.0 g placed at $(2.6, 0)$ or at $(-2.6, 0)$ has the same rotational inertia about the origin as the array of masses as a unit.

Practice Exercise

1. In Example 1, interchange the positions of the 3.0-g and 4.0-g masses, and calculate R.

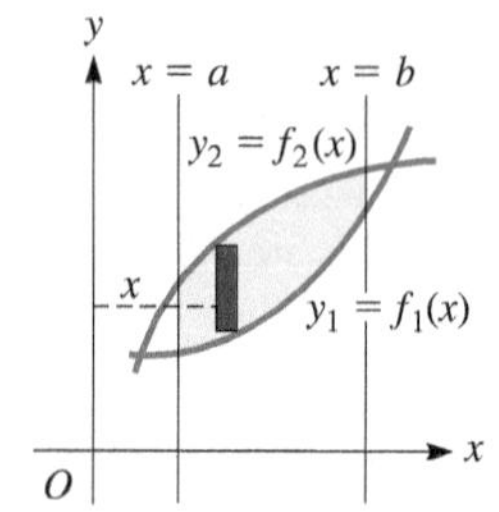

Fig. 26.46

MOMENT OF INERTIA OF A THIN, FLAT PLATE

If a thin, flat plate covering the region is bounded by the curves of the functions $y_1 = f_1(x)$, $y_2 = f_2(x)$, and the lines $x = a$ and $x = b$, as shown in Fig. 26.46, the moment of inertia of this plate with respect to the y-axis, I_y, is given by the sum of the moments of inertia of the individual elements. The mass of each element is $\rho(y_2 - y_1)dx$, where ρ is the mass per unit area and $(y_2 - y_1)dx$ is the area of the element. The distance of the element from the y-axis is x. Representing this sum as an integral, we have

$$I_y = \rho \int_a^b x^2(y_2 - y_1)\,dx \tag{26.21}$$

To find the radius of gyration of the plate with respect to the y-axis, R_y, first find the moment of inertia, divide this by the mass of the plate, and take the square root of this result.

In the same manner, the moment of inertia of a thin plate, with respect to the x-axis, bounded by $x_1 = g_1(y)$ and $x_2 = g_2(y)$ is given by

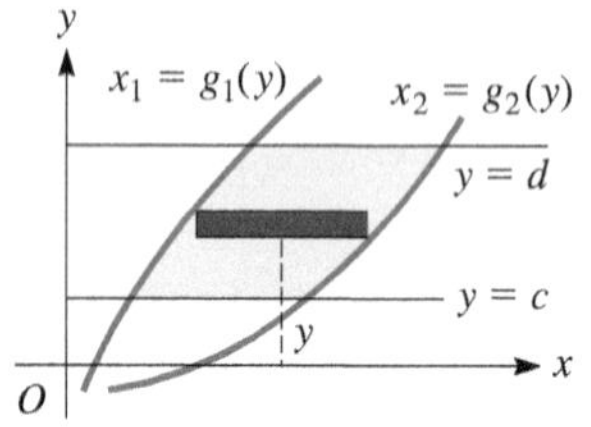

Fig. 26.47

$$I_x = \rho \int_c^d y^2(x_2 - x_1)\,dy \tag{26.22}$$

We find the radius of gyration of the plate with respect to the x-axis, R_x, in the same manner as we find it with respect to the y-axis (see Fig. 26.47).

EXAMPLE 2 Moment of inertia of a plate

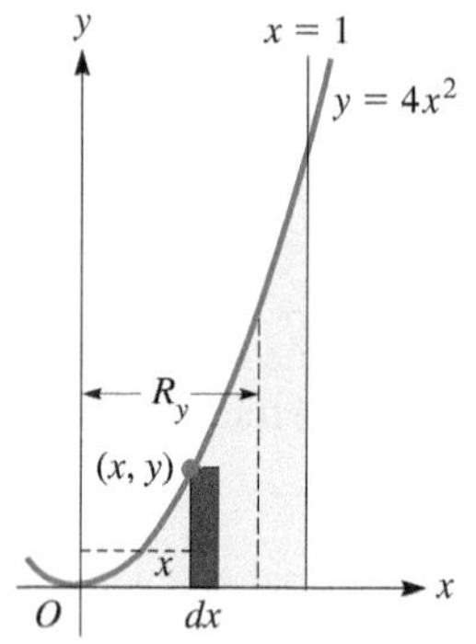

Fig. 26.48

Find the moment of inertia and radius of gyration of the plate covering the region bounded by $y = 4x^2$, $x = 1$, and the x-axis with respect to the y-axis.

We find the moment of inertia of this plate (see Fig. 26.48) as follows:

$$I_y = \rho \int_0^1 x^2 y\,dx \qquad \text{using Eq. (26.21)}$$

(distance from element to axis: x^2)

$$= \rho \int_0^1 x^2(4x^2)\,dx = 4\rho \int_0^1 x^4\,dx$$

$$= 4\rho\left(\frac{1}{5}x^5\right)\Big|_0^1 = \frac{4\rho}{5}$$

To find the radius of gyration, we first determine the mass of the plate:

$$m = \rho \int_0^1 y\,dx = \rho \int_0^1 (4x^2)\,dx \qquad m = \rho A$$

$$= 4\rho\left(\frac{1}{3}x^3\right)\Big|_0^1 = \frac{4\rho}{3}$$

$$R_y^2 = \frac{I_y}{m} = \frac{4\rho}{5} \times \frac{3}{4\rho} = \frac{3}{5} \qquad R_y^2 = I_y/m$$

$$R_y = \sqrt{\frac{3}{5}} = \frac{\sqrt{15}}{5} = 0.775$$

Therefore, if all of the mass of the plate were at a distance of 0.775 from the y-axis, the moment of inertia about the y-axis would be the same as the moment of inertia of the plate itself.

EXAMPLE 3 Moment of inertia of a triangular plate

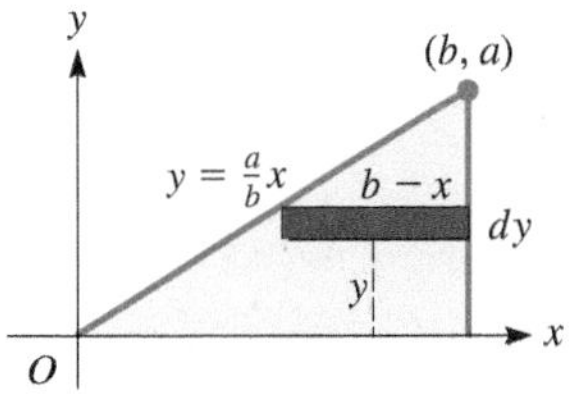

Fig. 26.49

Find the moment of inertia of a right triangular plate with sides a and b with respect to side b. Assume that $\rho = 1$.

Placing the triangle as shown in Fig. 26.49, we see that the equation of the hypotenuse is $y = (a/b)x$. The moment of inertia is

$$I_x = \int_0^a y^2(b - x)\,dy \quad \text{using Eq. (26.22)}$$

(distance from element to axis: y^2)

$$= \int_0^a y^2\left(b - \frac{b}{a}y\right)dy = b\int_0^a \left(y^2 - \frac{1}{a}y^3\right)dy$$

$$= b\left(\frac{1}{3}y^3 - \frac{1}{4a}y^4\right)\Bigg|_0^a = b\left(\frac{a^3}{3} - \frac{a^3}{4}\right) = \frac{ba^3}{12}$$

MOMENT OF INERTIA OF A SOLID

In applications, among the most important moments of inertia are those of solids of revolution. Since ***all parts of an element of mass should be at the same distance from the axis,*** the most convenient element of volume to use is the cylindrical shell. In Fig. 26.50, if the region bounded by the curves $y_1 = f_1(x)$, $y_2 = f_2(x)$, $x = a$, and $x = b$ is revolved about the y-axis, the moment of inertia of the element of volume is $\rho[2\pi x(y_2 - y_1)\,dx](x^2)$, where ρ is the density, $2\pi x(y_2 - y_1)\,dx$ is the volume of the element, and x^2 is the square of the distance from the y-axis. Expressing the sum of the elements as an integral, *the moment of inertia of the solid with respect to the y-axis, I_y, is*

■ Note carefully that Eq. (26.23) gives the moment of inertia with respect to the y-axis and that $(y_2 - y_1)$ is the height of the shell (see Fig. 26.21).

$$I_y = 2\pi\rho\int_a^b (y_2 - y_1)x^3\,dx \qquad \textbf{(26.23)}$$

The radius of gyration R_y is found by determining (1) the moment of inertia, (2) the mass of the solid, and (3) the square root of the quotient of the moment of inertia divided by the mass.

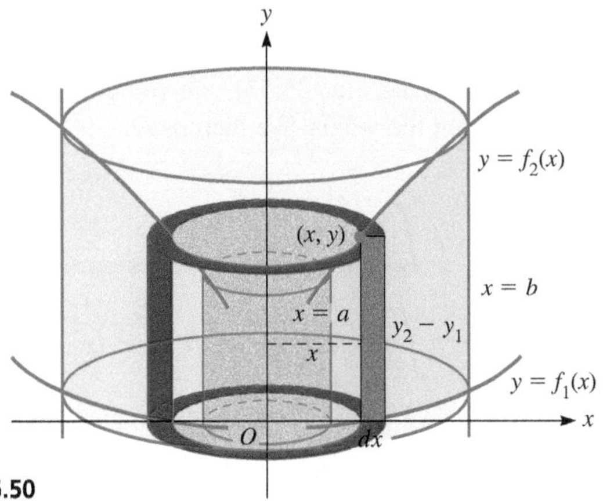

Fig. 26.50

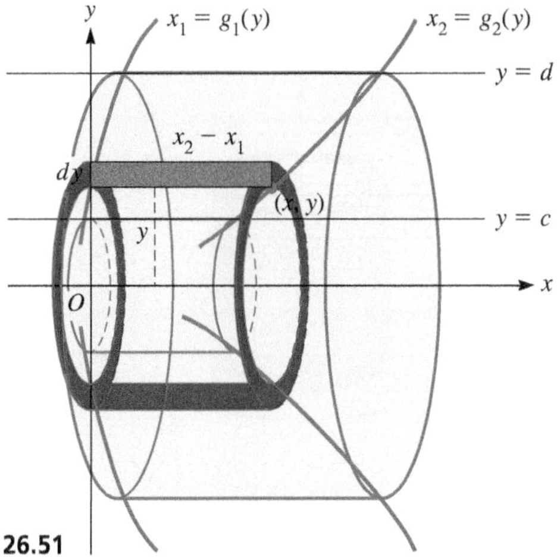

Fig. 26.51

The moment of inertia of the solid (see Fig. 26.51) generated by revolving the region bounded by $x_1 = g_1(y)$, $x_2 = g_2(y)$, $y = c$, and $y = d$ about the x-axis, I_x, is given by

■ Note carefully that Eq. (26.24) gives the moment of inertia with respect to the x-axis and that $(x_2 - x_1)$ is the height of the shell (see Fig. 26.21).

$$I_x = 2\pi\rho\int_c^d (x_2 - x_1)y^3\,dy \qquad \textbf{(26.24)}$$

The radius of gyration with respect to the x-axis, R_x, is found in the same manner as R_y.

EXAMPLE 4 Moment of inertia of a solid

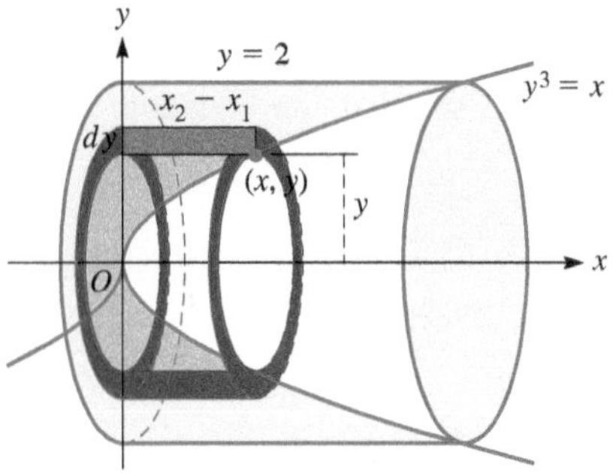

Fig. 26.52

Find the moment of inertia and radius of gyration with respect to the x-axis of the solid generated by revolving the region bounded by the curves of $y^3 = x$, $y = 2$, and the y-axis about the x-axis. See Fig. 26.52.

$$I_x = 2\pi\rho\int_0^2 (x_2 - x_1)y^3\,dy \quad \text{using Eq. (26.24); } y \text{ = distance from element to axis}$$

$$= 2\pi\rho\int_0^2 (y^3)y^3\,dy \quad x_2 - x_1 = y^3 - 0 = y^3$$

$$= 2\pi\rho\left(\frac{1}{7}y^7\right)\Bigg|_0^2 = \frac{256\pi\rho}{7}$$

$$m = 2\pi\rho\int_0^2 xy\,dy \quad \text{mass} = \rho \times \text{volume}$$

$$= 2\pi\rho\int_0^2 y^3 y\,dy = 2\pi\rho\left(\frac{1}{5}y^5\right)\Bigg|_0^2 = \frac{64\pi\rho}{5}$$

$$R_x^2 = \frac{256\pi\rho}{7} \times \frac{5}{64\pi\rho} = \frac{20}{7} \quad R_x^2 = I_x/m$$

$$R_x = \sqrt{\frac{20}{7}} = \frac{2}{7}\sqrt{35} = 1.69$$

EXAMPLE 5 Moment of inertia of a solid disc

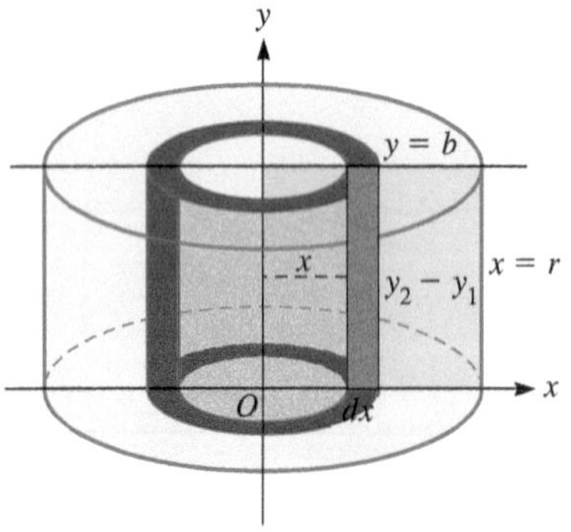

Fig. 26.53

As noted at the beginning of this section, the moment of inertia is important when studying the rotational motion of an object. For this reason, the moments of inertia of various objects are calculated, and the formulas tabulated. Such formulas are usually expressed in terms of the mass of the object.

Among the objects for which the moment of inertia is important is a solid disc. Find the moment of inertia of a disc with respect to its axis and rewrite it in terms of its mass.

To generate a disc (see Fig. 26.53), we rotate the region bounded by the axes, $x = r$ and $y = b$, about the y-axis. We then have

$$I_y = 2\pi\rho\int_0^r (y_2 - y_1)x^3\,dx \quad \text{using Eq. (26.23); } x \text{ = distance from element to axis}$$

$$= 2\pi\rho\int_0^r (b)x^3\,dx = 2\pi\rho b\int_0^r x^3\,dx \quad y_2 - y_1 = b - 0 = b$$

$$= 2\pi\rho b\left(\frac{1}{4}x^4\right)\Bigg|_0^r = \frac{\pi\rho b r^4}{2}$$

The mass of the disc is $\rho(\pi r^2)b$. Rewriting the expression for I_y, we have

$$I_y = \frac{(\pi\rho b r^2)r^2}{2} = \frac{mr^2}{2}$$

The rotation of a solid disc is important in the design and use of objects such as flywheels, pulley wheels, train wheels, engine rotors, various rollers such as are in printing presses, various types of machine tools, and numerous others.

Due to the limited methods of integration available at this point, we cannot integrate the expressions for the moments of inertia of circular areas or of a sphere. These will be introduced in Section 28.8 in the exercises, by which point the proper method of integration will have been developed.

EXERCISES 26.5

In Exercises 1 and 2, make the given changes in the indicated examples of this section, and then solve the resulting problems.

1. In Example 2, change $y = 4x^2$ to $y = 4x$.

2. In Example 4, change $y^3 = x$ to $y^2 = x$.

In Exercises 3–6, find the moment of inertia (in $g \cdot cm^2$) and the radius of gyration (in cm) with respect to the origin of each of the given arrays of masses located at the given points on the x-axis.

3. 5.0 g at (2.4, 0), 3.2 g at (3.5, 0)

4. 3.4 g at (−1.5, 0), 6.0 g at (2.1, 0), 2.6 g at (3.8, 0)

5. 45.0 g at (−3.80, 0), 90.0 g at (0.00, 0), 62.0 g at (5.50, 0)

6. 564 g at (−45.0, 0), 326 g at (−22.5, 0), 720 g at (15.4, 0), 205 g at (64.0, 0)

In Exercises 7–28, find the indicated moment of inertia or radius of gyration.

7. Find the moment of inertia of a plate covering the region bounded by $x = -1$, $x = 1$, $y = 0$, and $y = 1$ with respect to the x-axis.

8. Find the radius of gyration of a plate covering the region bounded by $x = 2$, $x = 4$, $y = 0$, and $y = 4$ with respect to the y-axis.

9. Find the moment of inertia of a plate covering the first-quadrant region bounded by $y^2 = x$, $x = 9$, and the x-axis with respect to the x-axis.

10. Find the moment of inertia of a plate covering the region bounded by $y = 2x$, $x = 1$, $x = 2$, and the x-axis with respect to the y-axis.

11. Find the radius of gyration of a plate covering the region bounded by $y = x^3$, $x = 3$, and the x-axis with respect to the y-axis.

12. Find the radius of gyration of a plate covering the first-quadrant region bounded by $y^2 = 1 - x$ with respect to the x-axis.

13. Find the moment of inertia of a right triangular plate with sides a and b with respect to side a in terms of the mass of the plate.

14. Find the moment of inertia of a rectangular sheet of metal of sides a and b with respect to side a. Express the result in terms of the mass of the metal sheet.

15. Find the radius of gyration of a plate covering the region bounded by $y = x^2$, $x = 3$, and the x-axis with respect to the x-axis.

16. Find the radius of gyration of a plate covering the region bounded by $y^2 = x^3$, $y = 8$, and the y-axis with respect to the y-axis.

17. Find the radius of gyration of the plate of Exercise 16 with respect to the x-axis.

18. Find the radius of gyration of a plate covering the first-quadrant region bounded by $x = 1$, $y = 2 - x$, and the y-axis with respect to the y-axis.

19. Find the moment of inertia with respect to its axis of the solid generated by revolving the region bounded by $y^2 = 4x$, $y = 2$, and the y-axis about the x-axis.

20. Find the radius of gyration with respect to its axis of the solid generated by revolving the first-quadrant region under the curve $y = 4 - x^2$ about the y-axis.

21. Find the radius of gyration with respect to its axis of the solid generated by revolving the region bounded by $y = 4x - x^2$ and the x-axis about the y-axis.

22. Find the radius of gyration with respect to its axis of the solid generated by revolving the region bounded by $y = 2x$ and $y = x^2$ about the y-axis.

23. Find the moment of inertia in terms of its mass of a right circular cone of radius r and height h with respect to its axis. See Fig. 26.54.

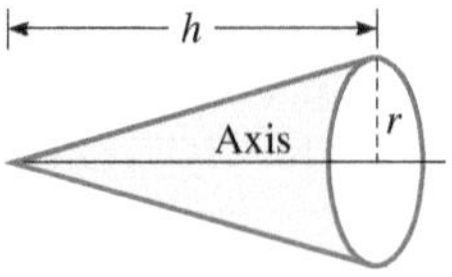

Fig. 26.54

24. Find the moment of inertia in terms of its mass of a circular hoop of radius r and of negligible thickness with respect to its centre.

25. A rotating drill head is in the shape of a right circular cone. Find the moment of inertia of the drill head with respect to its axis if its radius is 0.600 cm, its height is 0.800 cm, and its mass is 3.00 g. (See Exercise 23.)

26. Find the moment of inertia (in $kg \cdot m^2$) of a rectangular door 2 m high and 1 m wide with respect to its hinges if $\rho = 3\ kg/m^2$. (See Exercise 14.)

27. Find the moment of inertia of a flywheel with respect to its axis if its inner radius is 4.0 cm, its outer radius is 6.0 cm, and its mass is 1.2 kg. See Fig. 26.55.

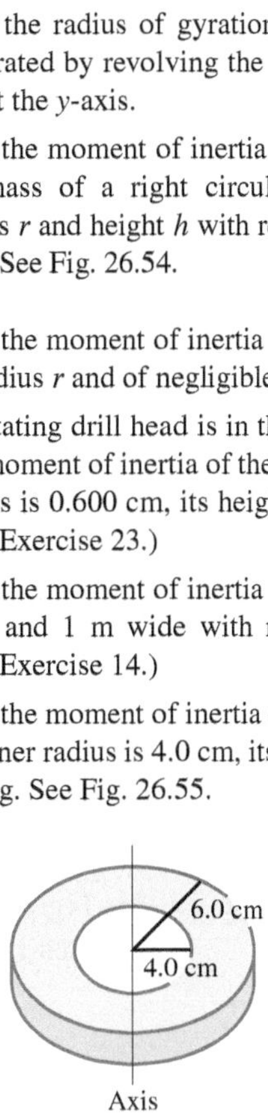

Fig. 26.55

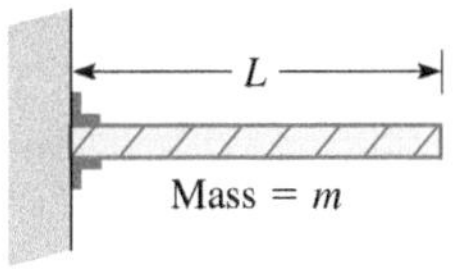

Fig. 26.56

28. A cantilever beam is supported only at its left end, as shown in Fig. 26.56. Explain how to find the formula for the moment of inertia of this beam with respect to a vertical axis through its left end if its length is L and its mass is m. (Consider the mass to be distributed evenly along the beam. This is not an area or volume type of problem.) Find the formula for the moment of inertia.

Answer to Practice Exercise

1. $R = 2.4$ cm

26.6 Other Applications

We have seen that the definite integral is used to find the exact measure of the sum of products in which one factor is an increment that approaches a limit of zero. This makes the definite integral a powerful mathematical tool in that a great many applications can be expressed in this form. The following examples show three more applications of the definite integral, and others are shown in the exercises.

WORK BY A VARIABLE FORCE

In physics, **work** *is defined as the product of a constant force times the distance through which it acts.* When we consider the work done in stretching a spring, the first thing we recognize is that the more the spring is stretched, the greater is the force necessary to stretch it. Thus, the force varies. However, if we are stretching the spring a distance Δx, where we are considering the limit as $\Delta x \to 0$, the force can be considered as constant over Δx. Adding the product of force$_1$ times Δx_1, force$_2$ times Δx_2, and so forth, we see that the total is the sum of these products. Thus, the work can be expressed as a definite integral in the form

$$W = \int_a^b f(x)\,dx \qquad \textbf{(26.25)}$$

where $f(x)$ is the force as a function of the distance the spring is stretched. The ***limits a and b refer to the initial and final distances the spring is stretched*** **from its normal length.**

■ Hooke's law is named for the English physicist Robert Hooke (1635–1703).

One problem remains: We must find the function $f(x)$. From physics, we learn that the force required to stretch a spring is proportional to the amount it is stretched (Hooke's law). If a spring is stretched x units from its normal length, then $f(x) = kx$. From conditions stated for a particular spring, the value of k may be determined. Thus, $W = \int_a^b kx\,dx$ is the formula for finding the total work done in stretching a spring. Here, a and b are the initial and final amounts the spring is stretched from its natural length.

EXAMPLE 1 Work done stretching a spring

A spring of natural length 12 cm requires a force of 6.0 N to stretch it 2.0 cm. See Fig. 26.57. Find the work done in stretching it 6.0 cm.

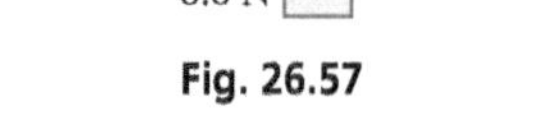

Fig. 26.57

From Hooke's law, we find the constant k for the spring, and therefore $f(x)$ as

$$f(x) = kx, \qquad 6.0 = k(2.0), \qquad k = 3.0\text{ N/cm}$$
$$= 3.0x$$

Since the spring is to be stretched 6.0 cm, $a = 0$ (it starts unstretched) and $b = 6.0$ (it is 6.0 cm longer than its normal length). Therefore, the work done in stretching it is

$$W = \int_0^{6.0} 3.0x\,dx = 1.5x^2 \Big|_0^{6.0} \qquad \text{using Eq. (26.25)}$$
$$= 54\text{ N} \cdot \text{cm}$$

Problems involving work by a variable force arise in many fields of technology. On the following page is an example from electricity that deals with the motion of an electric charge through an electric field created by another electric charge.

Electric charges are of two types, designated as positive and negative. A basic law is that charges of the same sign repel each other and charges of opposite signs attract each other. *The force between charges is proportional to the product of their charges, and inversely proportional to the square of the distance between them.*

The force $f(x)$ between electric charges is therefore given by

$$f(x) = \frac{kq_1q_2}{x^2} \qquad \textbf{(26.26)}$$

when q_1 and q_2 are the charges (in coulombs), x is the distance (in metres), the force is in newtons, and $k = 9.0 \times 10^9\ \text{N} \cdot \text{m}^2/\text{C}^2$. For other systems of units, the numerical value of k is different. We can find the work done when electric charges move toward each other or when they separate by use of Eq. (26.26) in Eq. (26.25).

EXAMPLE 2 Work done in moving α-particles

Find the work done when two α-particles, $q = 0.32$ aC each, move until they are 10 nm apart, if they were originally separated by 1.0 m.

From the given information, we have for each α-particle

$$q = 0.32\ \text{aC} = 0.32 \times 10^{-18}\text{C} = 3.2 \times 10^{-19}\ \text{C}$$

Since the particles start 1.0 m apart and are moved to 10 nm apart, $a = 1.0$ m and $b = 10 \times 10^{-9}\ \text{m} = 10^{-8}$ m. The work done is

$$W = \int_{1.0}^{10^{-8}} \frac{9.0 \times 10^9(3.2 \times 10^{-19})^2}{x^2}dx \qquad \text{using Eq. (26.25); } f(x) \text{ from Eq. (26.26)}$$

$$= 9.2 \times 10^{-28}\int_{1.0}^{10^{-8}} \frac{dx}{x^2} = 9.2 \times 10^{-28}\left(-\frac{1}{x}\right)\Bigg|_{1.0}^{10^{-8}}$$

$$= -9.2 \times 10^{-28}(10^8 - 1) = -9.2 \times 10^{-20}\ \text{J}$$

Since $10^8 \gg 1$, where $\gg$ means "much greater than," the 1 may be neglected in the calculation. The minus sign in the result means that work must be done *on* the system to move the particles toward each other. If free to move, they tend to separate.

The following is another type of problem involving work by a variable force.

EXAMPLE 3 Work done winding up a cable

Find the work done in winding up 60.0 m of a 100-m cable that weighs 4.00 N/m. See Fig. 26.58.

First, we let x denote the length of cable that has been wound up at any time. Then the force required to raise the remaining cable equals the weight of the cable that has not yet been wound up. This weight is the product of the unwound cable length, $100 - x$, and its weight per unit length, 4.00 N/m, or

$$f(x) = 4.00(100 - x)$$

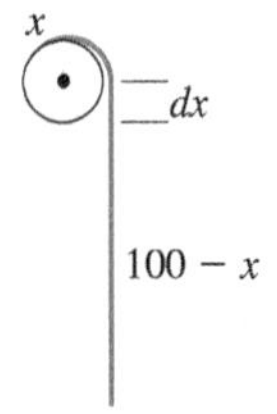

Fig. 26.58

Since 60.0 m of cable are to be wound up, $a = 0$ (none is initially wound up) and $b = 60.0$ m. The work done is

$$W = \int_0^{60.0} 4.00(100 - x)\,dx \qquad \text{using Eq. (26.25)}$$

$$= \int_0^{60.0} (400 - 4.00x)\,dx = 400x - 2.00x^2 \Big|_0^{60.0} = 16\ 800\ \text{N} \cdot \text{m}$$

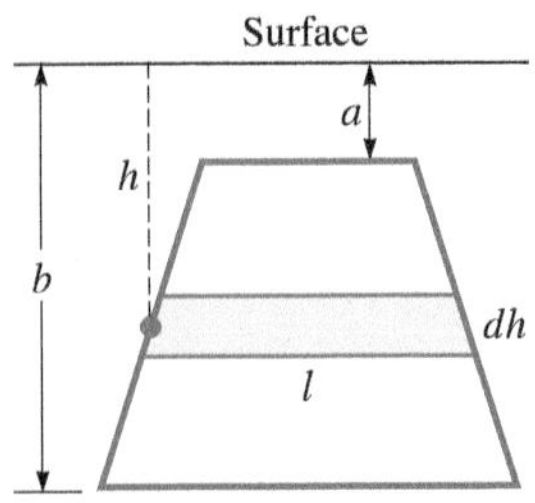

Fig. 26.59

FORCE DUE TO LIQUID PRESSURE

The second application of integration in this section deals with the force due to liquid pressure. The force F on an area A at the depth h in a liquid of density γ is $F = \gamma hA$. Let us assume that the plate shown in Fig. 26.59 is submerged vertically in the liquid. Using integration to sum the forces on the elements of area, *the total force on the plate is given by*

$$F = \gamma \int_a^b lh\,dh \tag{26.27}$$

LEARNING TIP

Note that integration in Eq. (26.27) is done from the top to the bottom of the submerged plate. If a problem is set up using coordinate axes, it is convenient to have the y-axis directed downward so that integration is done in the positive direction.

Here, l is the length of the element of area, h is the depth of the element of area, γ is the weight per unit volume of the liquid, a is the depth of the top, and b is the depth of the bottom of the area on which the force is exerted. When the liquid is water, we use $\gamma = 9800\ \text{N/m}^3$.

EXAMPLE 4 Force of water on the floodgate of a dam

A vertical floodgate of a dam is 3.00 m wide and 2.00 m high. Find the force on the floodgate if its upper edge is 1.00 m below the surface of the water. See Fig. 26.60.

Each element of area of the floodgate has a length of 3.00 m, which means that $l = 3.00$ m. Since the top of the gate is 1.00 m below the surface, $a = 1.00$ m, and since the gate is 2.00 m high, $b = 3.00$ m. The force on the gate is

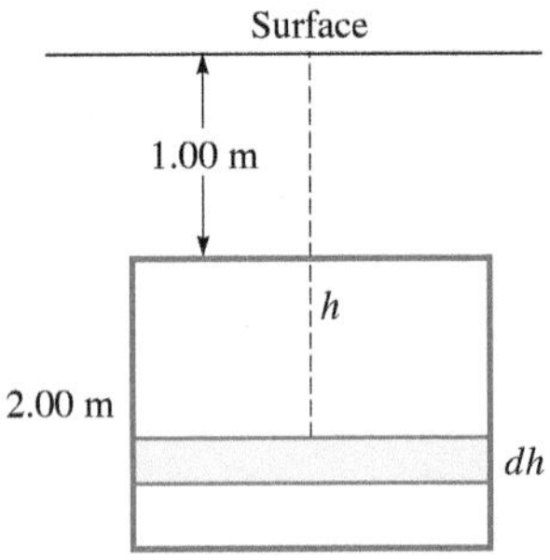

Fig. 26.60

$$F = 9800 \int_{1.00}^{3.00} 3.00h\,dh \qquad \text{using Eq. (26.27)}$$

$$= 29\,400 \int_{1.00}^{3.00} h\,dh$$

$$= 14\,700h^2 \Big|_{1.00}^{3.00} = 14\,700(9.00 - 1.00)$$

$$= 118\,000\ \text{N} = 118\ \text{kN}$$

EXAMPLE 5 Force on the end of a tank of water

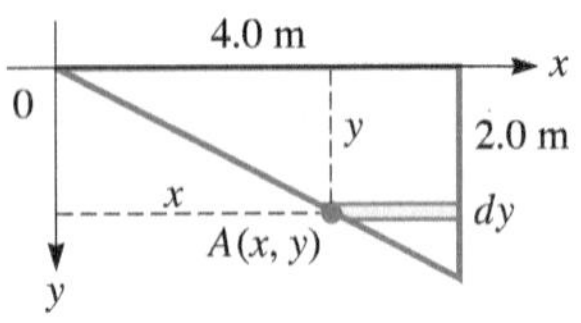

Fig. 26.61

The vertical end of a tank full of water is in the shape of a right triangle as shown in Fig. 26.61. (Note the y-axis directed downward.) What is the force on the end of the tank?

The equation of the line OA is $y = \frac{1}{2}x$. Thus, we see that the length of an element of area of the end of the tank is $4.0 - x$, the depth of the element of area is y, the top of the tank is $y = 0$, and the bottom is $y = 2.0$ m. Therefore, the force on the end of the tank is

$$F = 9800 \int_0^{2.0} \underbrace{(4.0 - x)}_{\text{length}}\underbrace{(y)}_{\text{depth}}(dy) \qquad \text{using Eq. (26.27)}$$

$$= 9800 \int_0^{2.0} (4.0 - 2y)(y\,dy)$$

$$= 19\,600 \int_0^{2.0} (2.0y - y^2)\,dy = 19\,600\left(1.0y^2 - \frac{1}{3}y^3\right)\Big|_0^{2.0}$$

$$= 26\,100\ \text{N}$$

AVERAGE VALUE OF A FUNCTION

The third application of integration shown in this section is that of the *average value* of a function. In general, *an average is found by summing up the quantities to be averaged and then dividing by the total number of them.* Generalizing on this and using integration for the summation, *the* **average value of a function** *y with respect to x from x = a to x = b is given by*

$$y_{av} = \frac{\int_a^b y\,dx}{b - a} \tag{26.28}$$

The following examples illustrate applications of the average value of a function.

EXAMPLE 6 Average value of velocity

The velocity v (in m/s) of an object falling under the influence of gravity as a function of time t (in s) is given by $v = 9.80t$. What is the average velocity of the object with respect to time for the first 3.0 s?

In this case, we want the average value of the function $v = 9.80t$ from $t = 0$ s to $t = 3.0$ s. This gives us

$$v_{av} = \frac{\int_0^{3.0} v\,dt}{3.0 - 0} \quad \text{using Eq. (26.28)}$$

$$= \frac{\int_0^{3.0} 9.80t\,dt}{3.0} = \frac{4.90t^2}{3.0}\Bigg|_0^{3.0}$$

$$= 14.7 \text{ m/s}$$

This result can be interpreted as meaning that an average velocity of 14.7 m/s for 3.0 s would result in the same distance, 44.1 m, being travelled by the object as that with the variable velocity. Since $s = \int v\,dt$, the numerator represents the distance travelled.

EXAMPLE 7 Average value of electric power

The power P (in W) developed in a certain resistor as a function of the current i (in A) is $P = 6.0i^2$. What is the average power with respect to the current as the current changes from 2.0 A to 5.0 A?

In this case, we are to find the average value of the function P from $i = 2.0$ A to $i = 5.0$ A. This average value of P is

$$P_{av} = \frac{\int_{2.0}^{5.0} P\,di}{5.0 - 2.0} \quad \text{using Eq. (26.28)}$$

$$= \frac{6.0\int_{2.0}^{5.0} i^2\,dt}{3.0} = \frac{2.0i^3\Big|_{2.0}^{5.0}}{3.0} = \frac{2.0(125 - 8.0)}{3.0} = 78 \text{ W}$$

In general, it might be noted that the average value of y with respect to x is that value of y which, when multiplied by the length of the interval for x, gives the same area as that under the curve of y as a function of x.

EXERCISES 26.6

In Exercises 1–4, make the given changes in the indicated examples of this section and then solve the resulting problems.

1. In Example 1, find the work done in stretching the spring from a length of 15.0 cm to a length of 18.0 cm.

2. In Example 3, find the work done in winding up all the cable.

3. In Example 4, find the force on the floodgate if the upper edge is 2.00 m below the surface.

4. In Example 6, find the average velocity of the object with respect to time between 3.00 s and 6.00 s.

In Exercises 5–38, solve the given problems.

5. The spring of a spring balance is 8.0 cm long when there is no weight on the balance, and it is 9.5 cm long with 6.0 N hung from the balance. How much work is done in stretching it from 8.0 cm to a length of 10.0 cm?

6. How much work is done in stretching the spring of Exercise 5 from a length of 10.0 cm to 12.0 cm?

7. A 640-N person compresses a bathroom scale 0.080 cm. If the scale obeys Hooke's law, how much work is done compressing the scale if a 720-N person stands on it?

8. A force F of 25 N on the spring in the lever–spring mechanism shown in Fig. 26.62 stretches the spring by 16 mm. How much work is done by the 25-N force in stretching the spring?

9. An electron has a 1.6×10^{-19} C negative charge. How much work is done in separating two electrons from 1.0 pm to 4.0 pm?

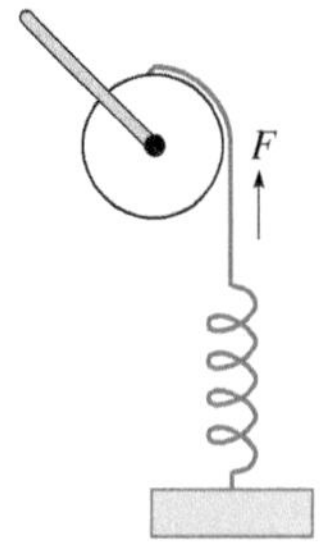

Fig. 26.62

10. How much work is done in separating an electron (see Exercise 9) and an oxygen nucleus, which has a positive charge of 1.3×10^{-18} C, from a distance of 2.0 μm to a distance of 1.0 m?

11. The gravitational force (in N) of attraction between two objects is given by $F = k/x^2$, where x is the distance between the objects. If the objects are 10 m apart, find the work required to separate them until they are 100 m apart. Express the result in terms of k.

12. Find the work done by winding up 20 m of a 25-m rope on which the force of gravity is 6.0 N/m.

13. A 6000-N elevator is suspended on cables that together weigh 48 N/m. How much work is done in raising the elevator from the basement to the second floor, a distance of 8.0 m?

14. A chain is being unwound from a winch. The force of gravity on it is 12.0 N/m. When 20 m have been unwound, how much work is done by gravity in unwinding another 30 m?

15. At liftoff, a rocket weighs 320 kN, including the weight of its fuel. During the first (vertical) stage of ascent, fuel is consumed at the rate of 12 kN per 1000 m of ascent. How much work is done in lifting the rocket to an altitude of 12 000 m? (Disregard the decrease in weight due to increasing elevation.)

16. While descending, a 550-N weather balloon enters a zone of freezing rain in which ice forms on the balloon at the rate of 7.50 N per 100 m of descent. Find the work done on the balloon during the first 1000 m of descent through the freezing rain.

17. A meteorite is 75 000 km from the centre of Earth and falls to the surface of Earth. From Newton's law of gravity, the force of gravity varies inversely as the square of the distance between the meteorite and the centre of Earth. Find the work done by gravity if the meteorite weighs 160 N at the surface, and the radius of Earth is 6400 km.

18. A rectangular swimming pool full of water is 5.50 m wide, 13.5 m long, and 1.75 m deep. Find the work done in pumping the water from the pool to a level 1.25 m above the top of the pool.

19. Find the work done in pumping the water out of the top of a cylindrical tank 2.0 m in radius and 4.0 m high, given that the tank is initially full and water weighs 9.8 kN/m^3. (*Hint:* If horizontal slices dx m thick are used, each element weighs $9800(\pi)(2.00^2)\,dx$ N, and each element must be raised $4.0 - x$ m, if x is the distance from the base to the element (see Fig. 26.63). In this way, the force, which is the weight of the slice, and the distance through which the force acts are determined. Thus, the products of force and distance are summed by integration.)

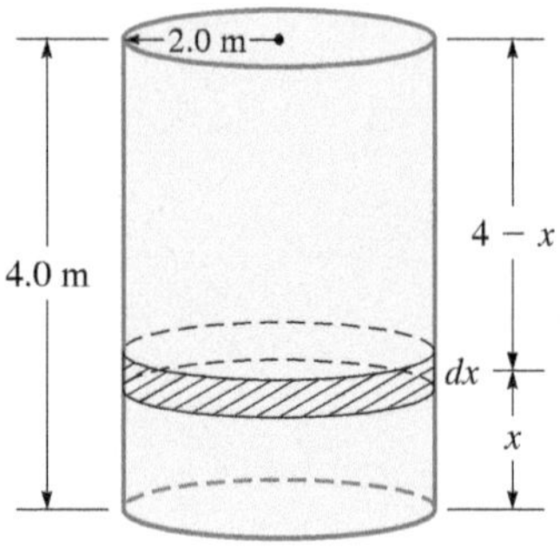

Fig. 26.63

20. A hemispherical tank of radius 3.0 m is full of water. Find the work done in pumping the water out of the top of the tank. (See Exercise 19. This problem is similar, except that the weight of each element is 9800π (radius)^2 (thickness), where the radius of each element is different. If we let x be the radius of an element and y be the distance the element must be raised, we have $9800\pi x^2 dy$ with $x^2 + y^2 = 9.0$.)

21. One end of a spa is a vertical rectangular wall 4.00 m wide. What is the force exerted on this wall by the water if it is 0.80 m deep?

22. Find the force on one side of a cubical container 6.0 cm on an edge if the container is filled with mercury. The density of mercury is 133 kN/m^3.

23. A rectangular sea aquarium observation window is 3.00 m wide and 2.00 m high. What is the force on this window if the upper edge is 1.50 m below the surface of the water? The density of seawater is 10.1 kN/m^3.

24. A horizontal tank has vertical circular ends, each with a radius of 2.00 m. It is filled to a depth of 2.00 m with oil of density 9400 N/m^3. Find the force on one end of the tank.

25. A swimming pool is 6.00 m wide and 15.0 m long. The bottom has a constant slope such that the water is 1.00 m deep at one end and 2.00 m deep at the other end. Find the force of the water on one of the sides of the pool.

26. Find the force on the lower half of the wall at the deep end of the swimming pool in Exercise 25.

27. A small dam is in the shape of the area bounded by $y = x^2$ and $y = 20$ (distances in m). Find the force on the area below $y = 4$ if the surface of the water is at the top of the dam.

28. The tank on a tanker truck has vertical elliptical ends with a vertical major axis. The major axis is 2.0 m and the minor axis 1.3 m. Find the force on one end of the tank when it is half-filled with fuel oil of density 7.8 kN/m^3.

29. A watertight cubical box with an edge of 2.00 m is suspended in water such that the top surface is 1.00 m below water level. Find the total force on the top of the box and the total force on the bottom of the box. What meaning can you give to the difference of these two forces?

30. Find the force on the region bounded by $x = 2y - y^2$ and the y-axis if the upper point of the area is at the surface of the water. All distances are in metres.

31. The electric current i (in A) as a function of the time t (in s) for a certain circuit is given by $i = 4t - t^2$. Find the average value of the current with respect to time for the first 4.0 s.

32. The temperature T (in °C) recorded in a city during a given day approximately followed the curve of $T = 0.001\,00t^4 - 0.280t^2 + 25.0$, where t is the number of hours from noon $(-12\text{ h} \le t \le 12\text{ h})$. What was the average temperature during the day?

33. The efficiency η (in %) of an automobile engine is given by $\eta = 0.768s - 0.000\,04s^3$, where s is the speed (in km/h) of the car. Find the average efficiency with respect to the speed for $s = 30.0$ km/h to $s = 90.0$ km/h. (See Example 1 of Section 24.7.)

34. Find the average value of the volume of a sphere with respect to the radius. Explain the meaning of the result.

35. The length of arc s of a curve from $x = a$ to $x = b$ is

$$s = \int_a^b \sqrt{1 + \left(\frac{dy}{dx}\right)^2}\,dx \qquad \textbf{(26.29)}$$

The cable of a bridge can be described by the equation $y = 0.04x^{3/2}$ from $x = 0$ to $x = 100$ m. Find the length of the cable. See Fig. 26.64.

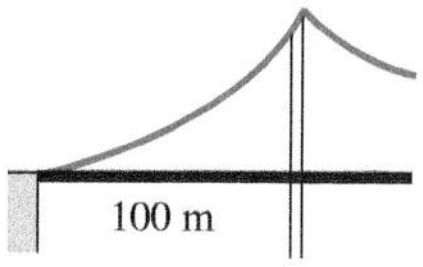

Fig. 26.64

36. A rocket takes off in a path described by the equation $y = \frac{2}{3}(x^2 - 1)^{3/2}$. Find the distance travelled by the rocket for $x = 1.0$ km to $x = 3.0$ km. (See Exercise 35.)

37. The area of a surface of revolution from $x = a$ to $x = b$ is

$$S = 2\pi \int_a^b y\sqrt{1 + \left(\frac{dy}{dx}\right)^2}\,dx$$

Find the formula for the lateral surface area of a right circular cone of radius r and height h.

38. The grinding surface of a grinding machine can be described as the surface generated by rotating the curve $y = 0.2x^3$ from $x = 0$ to $x = 2.0$ cm about the x axis. Find the grinding surface area. (See Exercise 37.)

CHAPTER 26 KEY FORMULAS AND EQUATIONS

Velocity

$$v = \int a\,dt \qquad \textbf{(26.1)}$$

$$v = at + C_1 \qquad \textbf{(26.2)}$$

Displacement

$$s = \int v\,dt \qquad \textbf{(26.3)}$$

Electric current

$$i = \frac{dq}{dt} \qquad \textbf{(26.4)}$$

Electric charge

$$q = \int i\,dt \qquad \textbf{(26.5)}$$

Voltage across a capacitor

$$V_C = \frac{1}{C}\int i\,dt \qquad \textbf{(26.6)}$$

Area

$$A = \int_a^b y\,dx = \int_a^b f(x)\,dx \qquad \textbf{(26.7)}$$

$$A = \int_c^d x\,dy = \int_c^d g(y)\,dy \qquad \textbf{(26.8)}$$

$$A = \int_a^b (y_2 - y_1)\,dx \qquad \textbf{(26.9)}$$

$$A = \int_c^d (x_2 - x_1)\,dy \qquad \textbf{(26.10)}$$

Volume	$V = \pi \int_a^b y^2\,dx = \pi \int_a^b [f(x)]^2\,dx$	**(26.11)**
	$V = \pi \int_c^d x^2\,dy$	**(26.12)**
Shell	$dV = 2\pi(\text{radius}) \times (\text{height}) \times (\text{thickness})$	**(26.13)**
Disc	$dV = \pi(\text{radius})^2 \times (\text{thickness})$	**(26.14)**
Centre of mass	$m_1d_1 + m_2d_2 + \cdots + m_nd_n = (m_1 + m_2 + \cdots + m_n)\overline{d}$	**(26.15)**
Centroid of a flat plate	$\overline{x} = \dfrac{\int_a^b x(y_2 - y_1)\,dx}{\int_a^b (y_2 - y_1)\,dx}$ **(26.16)** $\quad \overline{y} = \dfrac{\int_c^d y(x_2 - x_1)\,dy}{\int_c^d (x_2 - x_1)\,dy}$	**(26.17)**
Centroid of a solid of revolution	$\overline{x} = \dfrac{\int_a^b xy^2\,dx}{\int_a^b y^2\,dx}$ **(26.18)** $\quad \overline{y} = \dfrac{\int_c^d yx^2\,dy}{\int_c^d x^2\,dy}$	**(26.19)**
Radius of gyration	$m_1d_1^2 + m_2d_2^2 + \cdots + m_nd_n^2 = (m_1 + m_2 + \cdots + m_n)R^2$	**(26.20)**
Moment of inertia of a flat plate	$I_y = \rho \int_a^b x^2(y_2 - y_1)\,dx$	**(26.21)**
	$I_x = \rho \int_c^d y^2(x_2 - x_1)\,dy$	**(26.22)**
Moment of inertia of a solid of revolution	$I_y = 2\pi\rho \int_a^b (y_2 - y_1)x^3\,dx$	**(26.23)**
	$I_x = 2\pi\rho \int_c^d (x_2 - x_1)y^3\,dy$	**(26.24)**
Work	$W = \int_a^b f(x)\,dx$	**(26.25)**
Force between electric charges	$f(x) = \dfrac{kq_1q_2}{x^2}$	**(26.26)**
Force due to liquid pressure	$F = \gamma \int_a^b lh\,dh$	**(26.27)**
Average value	$y_{av} = \dfrac{\int_a^b y\,dx}{b - a}$	**(26.28)**
Length of arc	$s = \int_a^b \sqrt{1 + \left(\dfrac{dy}{dx}\right)^2}\,dx$	**(26.29)**

CHAPTER 26 REVIEW EXERCISES

1. A pitcher releases a baseball horizontally at 42.5 m/s. How far does it drop while travelling 17.1 m to home plate?
2. If the velocity v (m/s) of a subway train after the brakes are applied can be expressed as $v = \sqrt{400 - 20t}$, where t is the time in seconds, how far does it travel in coming to a stop?
3. A weather balloon is rising at the rate of 10.0 m/s when a small metal part drops off. If the balloon is 60.0 m high at this instant, when will the part hit the ground?
4. A float is dropped into a river at a point where it is flowing at 1.5 m/s. How far does the float travel in 30 s if it accelerates downstream at 0.010 m/s^2?
5. A golf ball is putted straight for the hole with an initial velocity of 2.50 m/s and acceleration of -0.750 m/s^2. Will the ball make it to the hole, which is 4.20 m away?
6. A block of ice breaks off from the top of a cliff 800 m high at the North Pole of Mars. If gravity on Mars is 3.71 m/s^2, find the velocity of the block when it hits the ground.
7. The electric current i (in A) in a circuit as a function of the time t (in s) is $i = 0.25(2\sqrt{t} - t)$. Find the total charge to pass a point in the circuit in 2.0 s.
8. The current i (in A) in a certain electric circuit is given by $i = \sqrt{1 + 4t}$, where t is the time (in s). Find the charge that passes a given point from $t = 1.0$ s to $t = 3.0$ s if $q_0 = 0$.
9. The voltage across a 5.5-nF capacitor in an FM radio receiver is zero. What is the voltage after 25 μs if a current of 12 mA charges the capacitor?
10. The initial voltage across a capacitor is zero, and $V_C = 2.50$ V after 8.00 ms. If a current $i = t/\sqrt{t^2 + 1}$, where i is the current (in A) and t is the time (in s), charges the capacitor, find the capacitance C of the capacitor.
11. The distribution of weight on a cable is not uniform. If the slope of the cable at any point is given by $dy/dx = 20 + 0.025x^2$ and if the origin of the coordinate system is at the lowest point, find the equation that gives the curve described by the cable.
12. The time rate of change of the reliability R (in %) of a computer system is $dR/dt = -2.5(0.05t + 1)^{-1.5}$, where t is the time (in h). If $R = 100$ for $t = 0$, find R for $t = 100$ h.
13. Find the area between $y = \sqrt{9 - x}$ and the coordinate axes.
14. Find the area bounded by $y = 3x^2 - x^3$ and the x-axis.
15. Find the area bounded by $y^2 = 2x$ and $y = x - 4$.
16. Find the area bounded by $y = 6/(x + 3)^2$, $y = 0$, $x = -1$, and $x = 3$.
17. Find the area between $y = x^2$ and $y = x^3 - 2x^2$.
18. Find the area between $y = x^2 + 8$ and $y = 3x^2$.
19. Show that the curve $y = x^{2n}$ $(n > 0)$ divides the square bounded by $x = 0$, $y = 0$, $x = 1$, and $y = 1$ into two regions, the areas of which are in the ratio $2n/1$.
20. Find the value of a such that the line $x = a$ bisects the area under the curve $y = 1/x^2$ between $x = 1$ and $x = 4$.
21. Find the volume generated by revolving the region bounded by $y = 3 + x^2$ and the line $y = 4$ about the x-axis.
22. Find the volume generated by revolving the region bounded by $y = 8x - x^4$ and the x-axis about the x-axis.
23. Find the volume generated by revolving the region bounded by $y = x^3 - 4x^2$ and the x-axis about the y-axis.
24. Find the volume generated by revolving the region bounded by $y = x$ and $y = 3x - x^2$ about the y-axis.
25. Find the volume generated by revolving an ellipse about its major axis.
26. A hole of radius 1.00 cm is bored along the diameter of a sphere of radius 4.00 cm. Find the volume of the material that is removed from the sphere.
27. Find the centre of mass of the following array of four masses in the xy-plane (distances in cm): 60 g at (4, 4), 160 g at $(-3, 6)$, 70 g at $(-5, -4)$, and 130 g at $(3, -5)$.
28. Find the centroid of the flat-plate machine part shown in Fig. 26.65. Each section is uniform, and the mass of the section to the right of the y-axis is twice that of the section to the left.

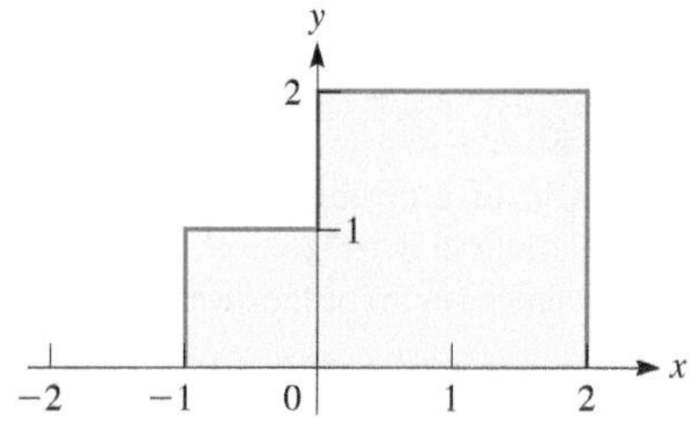

Fig. 26.65

29. Find the centroid of a flat plate covering the region bounded by $y^2 = x^3$ and $y = 3x$.
30. Find the centroid of a flat plate covering the region bounded by $y = 2x - 4$, $x = 1$, and $y = 0$.
31. Find the centroid of the volume generated by revolving the region bounded by $y = \sqrt{x}$, $x = 1$, $x = 4$, and $y = 0$ about the x-axis.
32. Find the centroid of the volume generated by revolving the region bounded by $yx^4 = 1$, $y = 1$, and $y = 4$ about the y-axis.
33. Find the moment of inertia of a flat plate covering the region bounded by $y = 3x - x^2$ and $y = x$ with respect to the y-axis.
34. Find the radius of gyration of a flat plate covering the region bounded by $y = 8 - x^3$, and the axes, with respect to the y-axis.
35. Find the moment of inertia with respect to its axis of a lead bullet that is defined by revolving the region bounded by $y = 3.00x^{0.10}$, $x = 0$, $x = 20.0$, and $y = 0$ about the x-axis (all measurements in mm). The density of lead is 0.0114 g/mm^3.
36. Find the radius of gyration with respect to its axis of a rotating machine part that can be defined by revolving the region bounded by $y = 1/x$, $x = 1.00$, $x = 4.00$, and $y = 0.25$ (all measurements in cm) about the x-axis.
37. A pail and its contents weigh 80 N. The pail is attached to the end of a 30-m rope that weighs 20 N and is hanging vertically. How much work is done in winding up the rope with the pail attached?
38. The gravitational force (in N) of the earth on a satellite (the weight of the satellite) is given by $f(x) = 10^{11}/x^2$, where x is the vertical distance (in km) from the centre of the earth to the satellite. How much work is done in moving the satellite from the earth's surface to an altitude of 3000 km? The radius of the earth is 6370 km.

39. A decorative glass table top is designed to be the region between the curves of $y = 0.0001x^4$ and $y = 110 - 0.0001x^4$. Find the area (in cm^2) of the table top.

40. A level putting green at a golf course can be approximated as the area bounded by $y = 0.003x^3 - 2x$ and $y = 1.5x - 0.001x^3$. Find the area (in m^2) of the green.

41. In a video game, a large rock is propelled up a slope at 12 m/s, but is accelerating down the slope at 6.0 m/s². What is the velocity of the rock after 5.0 s?

42. The vertical end of a trough is made of a right triangular section of concrete and a similar section of wood, as shown in Fig. 26.66. Find the force on each section if the trough is full of water.

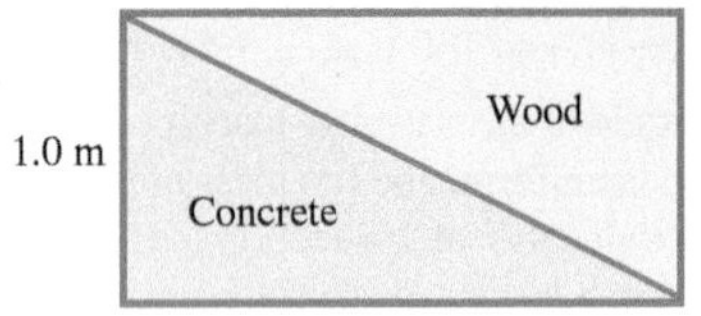

Fig. 26.66

43. The rear stabilizer of a certain aircraft can be described as the region under the curve $y = 3x^2 - x^3$, as shown in Fig. 26.67. Find the x-coordinate (in m) of the centroid of the stabilizer.

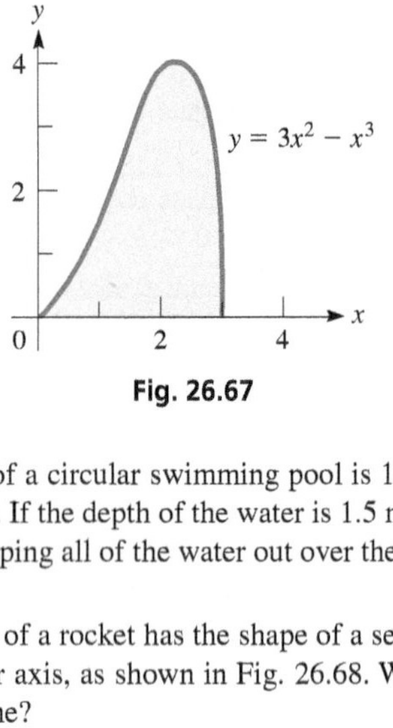

Fig. 26.67

44. The diameter of a circular swimming pool is 12 m, and the sides are 2.0 m high. If the depth of the water is 1.5 m, how much work is done in pumping all of the water out over the side?

45. The nose cone of a rocket has the shape of a semiellipse revolved about its major axis, as shown in Fig. 26.68. What is the volume of the nose cone?

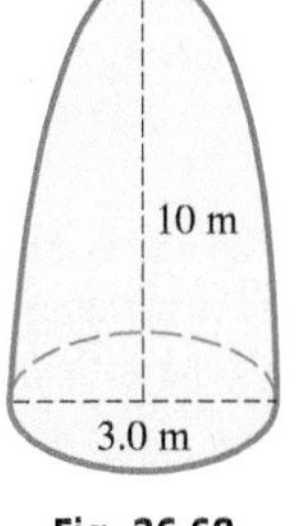

Fig. 26.68

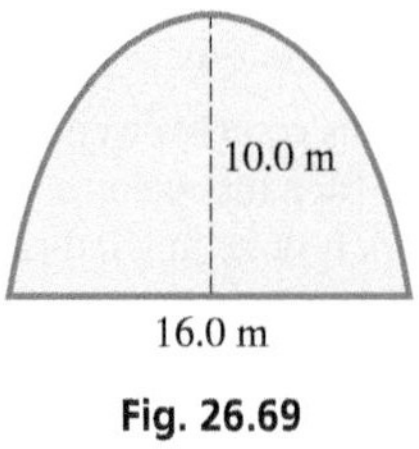

Fig. 26.69

46. The deck area of a boat is a parabolic section as shown in Fig. 26.69. What is the area of the deck?

47. The vertical ends of a fuel storage tank have a parabolic bottom section and a triangular top section, as shown in Fig. 26.70. What volume does the tank hold?

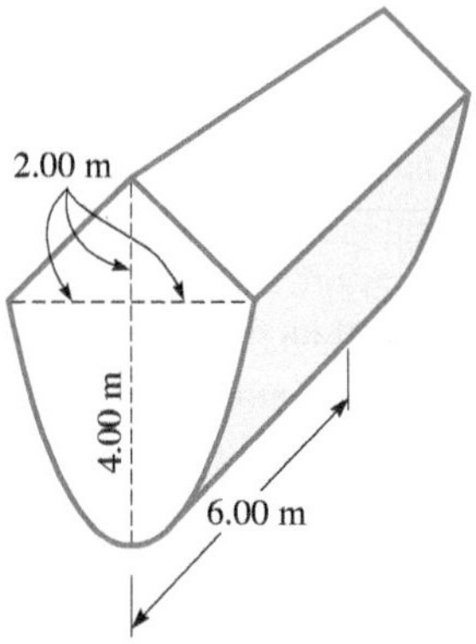

Fig. 26.70

48. The capillary tube shown in Fig. 26.71 has circular horizontal cross-sections of inner radius 1.1 mm. What is the volume of the liquid in the tube above the level of liquid outside the tube if the top of the liquid in the centre vertical cross-section is described by the equation $y = x^4 + 1.5$, as shown?

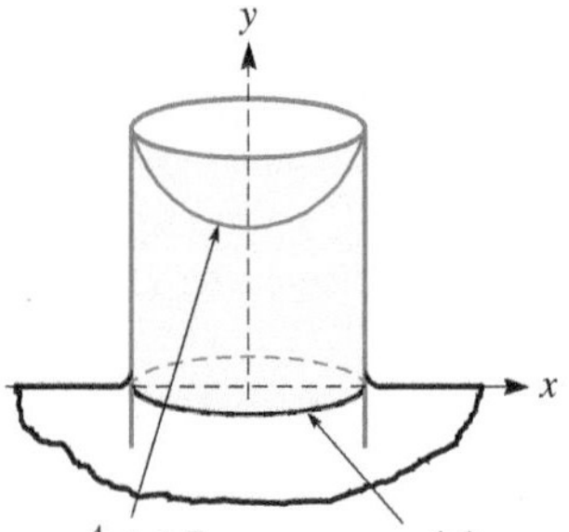

Fig. 26.71 $y = x^4 + 1.5$ $r = 1.1$ mm

49. A cylindrical chemical waste-holding tank 4.50 m in radius has a depth of 3.25 m. Find the total force on the circular side of the tank when it is filled with liquid with a density of 10.6 kN/m³.

50. A section of a dam is in the shape of a right triangle. The base of the triangle is 6.00 m and is in the surface of the water. If the triangular section goes to a depth of 4.00 m, find the force on it. See Fig. 26.72.

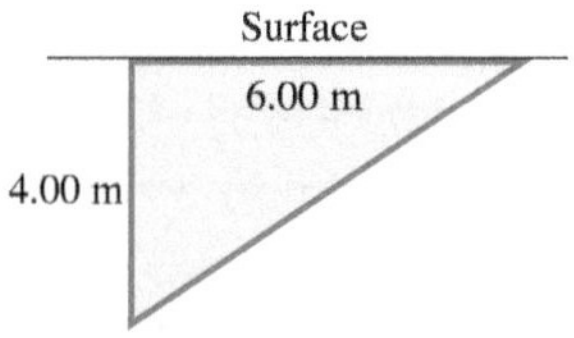

Fig. 26.72

51. The electric resistance of a wire is inversely proportional to the square of its radius. If a certain wire has a resistance of 0.30 Ω when its radius is 2.0 mm, find the average value of the resistance with respect to the radius as the radius changes from 2.0 mm to 2.1 mm.

52. The mass of Earth is 5.98×10^{24} kg, and the mass of the moon is 7.36×10^{22} kg. Assuming all of the mass of each is at its centre, find the centre of mass of the Earth–moon system, if their centres are 3.82×10^8 m apart. Compare this position with the radius of Earth, which is 6.37×10^6 m.

53. In the tube of an older television set, electrons are accelerated from rest with an acceleration of $5 \times 10^{14}\ \text{m/s}^2$. What is their velocity after travelling 2.5 cm?

54. A horizontal straight section of pipe is supported at its centre by a vertical wire as shown in Fig. 26.73. Find the formula for the moment of inertia of the pipe with respect to an axis along the wire if the pipe is of length L and mass m.

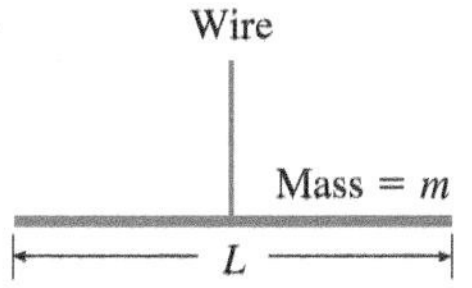

Fig. 26.73

55. The following table gives areas under the curve from 0 to z for the standard normal distribution, given by $y = \frac{1}{\sqrt{2\pi}} e^{-x^2/2}$.

Use the information in the table to find $\frac{1}{\sqrt{2\pi}} \int_1^2 e^{-x^2/2} dx$.

Standard Normal (z) Distribution

z	*Area*	z	*Area*	z	*Area*
0.0	0.0000	1.0	0.3413	2.0	0.4772
0.1	0.0398	1.1	0.3643	2.1	0.4821
0.2	0.0793	1.2	0.3849	2.2	0.4861
0.3	0.1179	1.3	0.4032	2.3	0.4893
0.4	0.1554	1.4	0.4192	2.4	0.4918
0.5	0.1915	1.5	0.4332	2.5	0.4938
0.6	0.2257	1.6	0.4452	2.6	0.4953
0.7	0.2580	1.7	0.4554	2.7	0.4965
0.8	0.2881	1.8	0.4641	2.8	0.4974
0.9	0.3159	1.9	0.4713	2.9	0.4981
1.0	0.3413	2.0	0.4772	3.0	0.4987

56. Use the information given in the table from Exercise 55 and the symmetry of the standard normal distribution to find $\frac{1}{\sqrt{2\pi}} \int_{-2}^{3} e^{-x^2/2} dx$ (see Exercise 55).

57. The float for a certain valve control has a circular top of radius a. All cross-sections of the float that are perpendicular to a fixed diameter of the top are squares. Write one or two paragraphs explaining how to derive the formula that gives the volume of the float. What is the formula?

CHAPTER 26 **PRACTICE TEST**

In Problems 1–3, use the region bounded by $y = \frac{1}{4}x^2$, $y = 0$, *and* $x = 2$.

1. Find the area.

2. Find the coordinates of the centroid of a flat plate that covers the region.

3. Find the volume if the given region is revolved about the x-axis.

In Problems 4 and 5, use the first-quadrant region bounded by $y = x^2$, $x = 0$, *and* $y = 9$.

4. Find the volume if the given region is revolved about the x-axis.

5. Find the moment of inertia of a flat plate that covers the region, with respect to the y-axis.

In Problems 6–10, answer the given questions.

6. The current i (in A) in an electric circuit is given by $i = 5.0 - 0.20t$, where t is time (in s). If 0 C of charge have passed a given point in the circuit when $t = 0$, how many coulombs pass the point after 2.0 s?

7. The velocity v of an object as a function of the time t is $v = 60 - 4t$. Find the expression for the displacement s if $s = 10$ for $t = 0$.

8. The natural length of a spring is 8.0 cm. A force of 12 N stretches it to a length of 10.0 cm. How much work is done in stretching it from a length of 10.0 cm to a length of 14.0 cm?

9. A vertical rectangular floodgate is 6.00 m wide and 2.00 m high. Find the force on the gate if its upper edge is 1.00 m below the surface of the water ($\gamma = 9.80\ \text{kN/m}^3$).

10. What is the average value of $f(x) = x^2 + 2x - 3$ for $0 \le x \le 6$?

27. Differentiation of Transcendental Functions

LEARNING OUTCOMES

After completion of this chapter, the student should be able to:

- Find the derivative of expressions involving trigonometric and inverse trigonometric functions
- Find the derivative of expressions involving logarithmic and exponential functions
- Find limits of indeterminate forms using L'Hospital's rule
- Solve application problems involving derivatives of transcendental functions

▲ In Sections 27.4 and 27.8, we use derivatives of transcendental functions in analysing the motion of a rocket.

While studying vibrations in a rod in the mid-1700s, the Swiss mathematician Euler noted that the trigonometric functions arose naturally as solutions to equations in which derivatives appeared. This was the first treatment of the trigonometric functions as functions of numbers essentially as we do today. Later, in 1755, Euler wrote a textbook on differential calculus in which he included differentiation of the trigonometric, inverse trigonometric, logarithmic, and exponential functions. Euler called these functions "transcendental," as "they transcend the power of algebraic methods." These functions are not algebraic in that they cannot be expressed using algebraic operations (addition, subtraction, multiplication, division, and taking roots) on polynomials.

Logarithms were developed as a tool for calculation. In establishing the calculus, Newton and Leibniz did some formulation of the logarithmic and exponential functions. However, the calculus of the transcendental functions was formulated mostly in the 1700s by Euler and a number of other mathematicians. This led to the rapid progress made later in many technical and scientific areas. For example, transcendental functions, along with their derivatives and integrals, have been of great importance in the development of the fields of electricity and electronics in the 1800s, 1900s, and 2000s, particularly with respect to alternating current.

In this chapter, we develop formulas for the derivatives of these transcendental functions, and in the next chapter, we will take up integration that involves these functions. Other areas in which we will show important applications include harmonic motion, rocket motion, monetary interest calculations, population growth, acoustics, and optics.

27.1 Derivatives of the Sine and Cosine Functions

We now find the derivative of the sine function. We will then be able to use it in finding the derivatives of the other trigonometric and inverse trigonometric functions.

Let $y = \sin x$, where x is expressed in radians. If x changes by an amount h, from the definition of the derivative, we have

$$\frac{dy}{dx} = \lim_{h \to 0} \frac{\sin(x + h) - \sin x}{h}$$

Referring now to Eq. (20.18), we have

■ For reference, Eq. (20.18) is $\sin x - \sin y = 2 \sin \frac{1}{2}(x - y) \cos \frac{1}{2}(x + y)$.

$$\frac{dy}{dx} = \lim_{h \to 0} \frac{2 \sin \frac{1}{2}(x + h - x) \cos \frac{1}{2}(x + h + x)}{h}$$

$$= \lim_{h \to 0} \frac{\sin(h/2) \cos(x + h/2)}{h/2}$$

Looking ahead to the next step of letting $h \to 0$, we see that the numerator and denominator both approach zero. This situation is precisely the same as that in which we were finding the derivatives of the algebraic functions. To find the limit, we must find

$$\lim_{h \to 0} \frac{\sin(h/2)}{h/2}$$

since these are the factors that cause the numerator and denominator to approach zero.

In finding this limit, we let $\theta = h/2$ for convenience of notation. This means that we are to determine $\lim_{\theta \to 0} \frac{\sin \theta}{\theta}$. Of course, it would be convenient to know before proceeding if this limit does actually exist. Therefore, by using a calculator, we can develop a table of values of $\frac{\sin \theta}{\theta}$ as θ becomes very small:

θ (radians)	0.5	0.1	0.05	0.01	0.001
$\frac{\sin \theta}{\theta}$	0.958 851 1	0.998 334 2	0.999 583 4	0.999 983 3	0.999 999 8

We see from this table that the limit of $\frac{\sin \theta}{\theta}$, as $\theta \to 0$, appears to be 1.

In order to prove that $\lim_{\theta \to 0} \frac{\sin \theta}{\theta} = 1$, we use a geometric approach. Considering Fig. 27.1, we see that the following inequality is true:

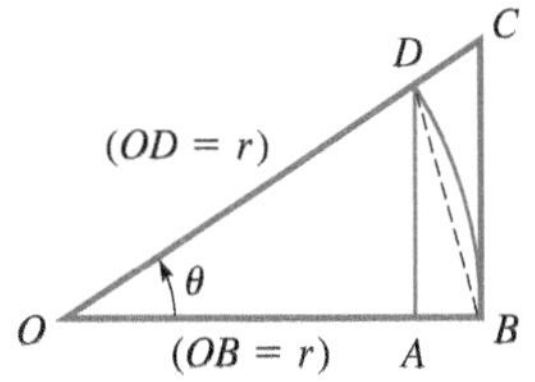

Fig. 27.1

Area triangle $OBD <$ area sector $OBD <$ area triangle OBC

$$\frac{1}{2}r(r \sin \theta) < \frac{1}{2}r^2\theta < \frac{1}{2}r(r \tan \theta) \quad \text{or} \quad \sin \theta < \theta < \tan \theta$$

Remembering that we want to find the limit of $(\sin \theta)/\theta$, we next divide through by $\sin \theta$ and then take reciprocals:

$$1 < \frac{\theta}{\sin \theta} < \frac{1}{\cos \theta} \qquad \text{or} \qquad 1 > \frac{\sin \theta}{\theta} > \cos \theta$$

When we consider the limit as $\theta \to 0$, we see that the left member remains 1 and the right member approaches 1. Thus, $(\sin \theta)/\theta$ must approach 1. This means

$$\lim_{\theta \to 0} \frac{\sin \theta}{\theta} = \lim_{h \to 0} \frac{\sin(h/2)}{h/2} = 1 \quad \textbf{(27.1)}$$

Using the result in Eq. (27.1) in the expression for dy/dx, we have

$$\lim_{h \to 0} \left[\cos(x + h/2) \frac{\sin(h/2)}{h/2} \right] = \cos x$$

$$\frac{dy}{dx} = \cos x \quad \textbf{(27.2)}$$

To find the derivative of $y = \sin u$, where u is a function of x, we use the chain rule, Eq. (23.14), which we repeat here for reference:

$$\frac{dy}{dx} = \frac{dy}{du} \cdot \frac{du}{dx} \quad \textbf{(27.3)}$$

Therefore, for $y = \sin u$, $dy/du = \cos u$, we have the following formula.

Derivative of the Sine of a Function	EXAMPLE 1
	Find the derivative of $y = \sin 2x$.
$\frac{d(\sin u)}{dx} = \cos u \frac{du}{dx}$ **(27.4)**	Here $u = 2x$, so $du/dx = 2$. Then $\frac{dy}{dx} = \frac{d(\sin 2x)}{dx} = \cos 2x \frac{d(2x)}{dx} = (\cos 2x)(2)$ ← $\frac{du}{dx}$ $= 2 \cos 2x$

EXAMPLE 2 Derivative of sin *u*

Find the derivative of $y = 2 \sin(x^2)$.

In this example, $u = x^2$, which means that $du/dx = 2x$. This means

$$\frac{dy}{dx} = 2[\cos(x^2)](2x) \quad \text{using Eq. (27.4)}$$
$$= 4x \cos(x^2)$$

Practice Exercise

1. Find the derivative of $y = 3 \sin(4x + 1)$.

LEARNING TIP

It is important here, just as it is in finding the derivatives of powers of all functions, to remember to include the factor du/dx.

EXAMPLE 3 Derivative of a power of sin *u*

Find the derivative of $r = \sin^2 \theta$.

This example is a combination of the use of the power rule, Eq. (23.15), and the derivative of the sine function, Eq. (27.4). Since $\sin^2 \theta$ means $(\sin \theta)^2$, in using the power rule we have $u = \sin \theta$. Thus,

$$\frac{dr}{d\theta} = 2(\sin \theta) \frac{d \sin \theta}{d\theta} \quad \text{using } \frac{du^n}{dx} = nu^{n-1}\left(\frac{du}{dx}\right)$$
$$= 2 \sin \theta \cos \theta \quad \text{using } \frac{d(\sin u)}{dx} = \cos u \frac{du}{dx}$$
$$= \sin 2\theta \quad \text{using identity } \sin 2\alpha = 2 \sin \alpha \cos \alpha$$

In order to find the derivative of the cosine function, we write it in the form $\cos u = \sin\left(\frac{\pi}{2} - u\right)$. Thus, if $y = \sin\left(\frac{\pi}{2} - u\right)$, we have

$$\frac{dy}{dx} = \cos\left(\frac{\pi}{2} - u\right)\frac{d\left(\frac{\pi}{2} - u\right)}{dx} = \cos\left(\frac{\pi}{2} - u\right)\left(-\frac{du}{dx}\right)$$

$$= -\cos\left(\frac{\pi}{2} - u\right)\frac{du}{dx}$$

Since $\cos\left(\frac{\pi}{2} - u\right) = \sin u$, we have the following result.

Derivative of the Cosine of a Function

$$\frac{d(\cos u)}{dx} = -\sin u\frac{du}{dx} \quad \textbf{(27.5)}$$

EXAMPLE 4

Find the derivative of $y = \cos\sqrt{x}$.

Here $u = \sqrt{x} = x^{1/2}$, so $du/dx = \frac{1}{2}x^{-1/2} = \frac{1}{2\sqrt{x}}$. Then

$$\frac{dy}{dx} = \frac{d\left(\cos\sqrt{x}\right)}{dx} = -\sin\sqrt{x}\frac{d\left(\sqrt{x}\right)}{dx}$$

$$= -\sin\sqrt{x}\left(\frac{1}{2\sqrt{x}}\right) = -\frac{\sin\sqrt{x}}{2\sqrt{x}}$$

$\frac{du}{dx}$ ↑

EXAMPLE 5 Derivative of cos *u*—power in an amplifier

The electric power P developed in a resistor of an amplifier circuit is $P = 25\cos^2 120\pi t$, where t is the time. Find the expression for the time rate of change of power.

From Chapter 23, we know that we are to find the derivative dP/dt. Therefore,

$$P = 25\cos^2 120\pi t$$

$$\frac{dP}{dt} = 25(2\cos 120\pi t)\frac{d\cos 120\pi t}{dt} \qquad \text{using } \frac{du^n}{dx} = nu^{n-1}\left(\frac{du}{dx}\right)$$

$$= 50\cos 120\pi t(-\sin 120\pi t)\frac{d(120\pi t)}{dt} \qquad \text{using } \frac{d(\cos u)}{dx} = -\sin u\frac{du}{dx}$$

$$= (-50\cos 120\pi t\sin 120\pi t)(120\pi)$$

$$= -6000\pi\cos 120\pi t\sin 120\pi t$$

$$= -3000\pi\sin 240\pi t \qquad \text{using } \sin 2\alpha = 2\sin\alpha\cos\alpha$$

EXAMPLE 6 Derivative of a root containing cos *u*

Find the derivative of $y = \sqrt{1 + \cos 2x}$.

$$y = (1 + \cos 2x)^{1/2}$$

$$\frac{dy}{dx} = \frac{1}{2}(1 + \cos 2x)^{-1/2}\frac{d(1 + \cos 2x)}{dx} \qquad \text{using } \frac{du^n}{dx} = nu^{n-1}\left(\frac{du}{dx}\right)$$

$$= \frac{1}{2}(1 + \cos 2x)^{-1/2}(-\sin 2x)(2) \qquad \text{using Eq. (27.5)}$$

$$= -\frac{\sin 2x}{\sqrt{1 + \cos 2x}}$$

Practice Exercise

2. Find the derivative of $y = 5(1 + \cos x^2)^2$.

EXAMPLE 7 Differential of the product sin *u* cos *v*

Find the differential of $y = \sin 2x \cos x^2$.

From Section 24.8, recall that the differential of a function $y = f(x)$ is $dy = f'(x)dx$. Thus, using the derivative product rule and the derivatives of the sine and cosine functions, we arrive at the following result:

$$y = \sin 2x \cos x^2 \qquad y = (\sin 2x) \cdot (\cos x^2)$$
$$dy = \left[\sin 2x(-\sin x^2)(2x) + \cos x^2(\cos 2x)(2)\right]dx$$
$$= (-2x \sin 2x \sin x^2 + 2 \cos 2x \cos x^2)dx$$

EXAMPLE 8 Slope of a tangent

Find the slope of a line tangent to the curve of $y = 5 \sin 3x$ at $x = 0.2$.

Here, we are to find the derivative of $y = 5 \sin 3x$ and then evaluate the derivative for $x = 0.2$. Therefore, we have the following:

$$y = 5 \sin 3x$$
$$\frac{dy}{dx} = 5(\cos 3x)(3) = 15 \cos 3x \qquad \text{find derivative}$$
$$\left.\frac{dy}{dx}\right|_{x=0.2} = 15 \cos 3(0.2) = 15 \cos 0.6 \qquad \text{evaluate}$$
$$= 12.38$$

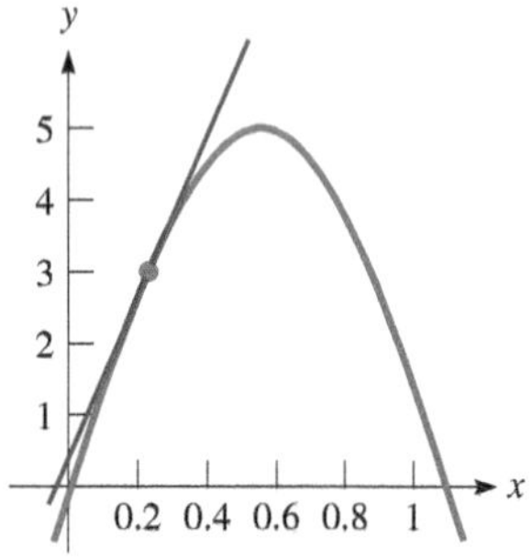

Fig. 27.2

In evaluating the slope, we must remember that $x = 0.2$ means the ***values are in radians.*** Therefore, the slope is 12.38. The curve and the tangent line at $x = 0.2$ are shown in Fig. 27.2.

EXERCISES 27.1

In Exercises 1 and 2, make the given changes in the indicated examples of this section and then find the derivatives.

1. In Example 3, in the given function, change θ to $2\theta^2$.
2. In Example 6, in the given function, change $2x$ to x^2.

In Exercises 3–34, find the derivatives of the given functions.

3. $y = \sin(x + 8)$
4. $y = 3 \sin 7x$
5. $y = 2 \sin(2x^3 - 1)$
6. $s = 5 \sin(7 - 3t)$
7. $y = 6 \cos \frac{1}{2}x$
8. $y = \cos(1 - x)$
9. $y = 2 \cos(3x - \pi)$
10. $y = 4 \cos(6x^2 + 5)$
11. $r = \sin^2 3\pi\theta$
12. $y = 3 \sin^3(2x^4 + 1)$
13. $y = 3 \cos^3(5x + 2)$
14. $y = 4 \cos^2 \sqrt{x}$
15. $y = x \sin 3x$
16. $v = 6t^2 \sin 3\pi t$
17. $y = 3x^5 \cos 5x$
18. $y = 0.5\theta \cos(2\theta + \pi/4)$
19. $u = 3 \sin v^2 \cos 5v$
20. $y = 6 \sin x \cos 4x$
21. $y = \sqrt{1 + \sin 4x}$
22. $y = (x - \cos^2 x)^4$
23. $r = \dfrac{\sin(3t - \pi/3)}{2t}$
24. $T = \dfrac{1 - 3z}{\sin \pi z}$
25. $y = \dfrac{2 \cos x^2}{3x - 1}$
26. $y = \dfrac{\cos^2 3x}{1 + 2 \sin^2 2x}$
27. $y = 2 \sin^2 3x \cos 2x$
28. $y = \cos^3 4x \sin^2 2x$
29. $s = \sin(\sin 2t)$
30. $z = 0.2 \cos(\sin 3\phi)$
31. $y = \sin^3 x - \cos 2x$
32. $y = 5x \sin 5x + \cos 5x$
33. $p = \dfrac{1}{\sin s} + \dfrac{1}{\cos s}$
34. $y = 2x \sin x + 2 \cos x - x^2 \cos x$

In Exercises 35–56, solve the given problems.

35. Using a graphing utility: (a) display the graph of $y = (\sin x)/x$ to verify that $(\sin \theta)/\theta \to 1$ as $\theta \to 0$, and (b) verify the values for $(\sin \theta)/\theta$ in the table at the beginning of Section 27.1.
36. Evaluate $\lim_{\theta \to 0}(\tan \theta)/\theta$. (Use the fact that $\lim_{\theta \to 0}(\sin \theta)/\theta = 1$.)

37. On a calculator, find the values of (a) $\cos 1.0000$ and (b) $(\sin 1.0001 - \sin 1.0000)/0.0001$. Compare the values and give the meaning of each in relation to the derivative of the sine function where $x = 1$.

38. On a calculator, find the values of (a) $-\sin 1.0000$ and (b) $(\cos 1.0001 - \cos 1.0000)/0.0001$. Compare the values and give the meaning of each in relation to the derivative of the cosine function where $x = 1$.

39. On the graph of $y = \sin x$ in Fig. 27.3, draw tangent lines at the indicated points and determine the slopes of these tangent lines. Then plot the values of these slopes for the same values of x and join the points with a smooth curve. Compare the resulting curve with $y = \cos x$. (Recall the meaning of the derivative as the slope of a tangent line.)

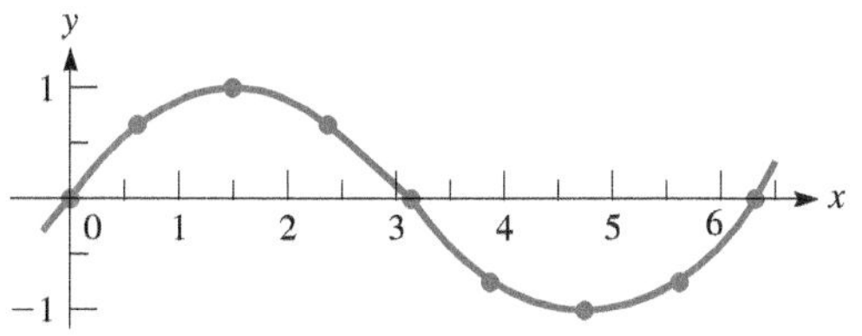

Fig. 27.3

40. Repeat the instructions given in Exercise 39 for the graph of $y = \cos x$ in Fig. 27.4. Compare the resulting curve with $y = \sin x$. (Be careful in this comparison and remember the difference between $y = \sin x$ and the derivative of $y = \cos x$. As in Exercise 39, note the meaning of the derivative as the slope of a tangent line.)

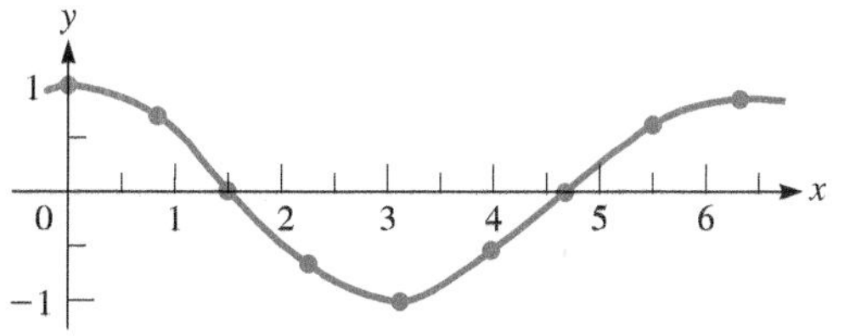

Fig. 27.4

41. Find the derivative of the implicit function $\sin(xy) + \cos 2y = x^2$.

42. Find the derivative of the implicit function $x \cos 2y + \sin x \cos y = 1$.

43. Show that $\dfrac{d^4 \sin x}{dx^4} = \sin x$.

44. Show that $y = A \sin kx + B \cos kx$ satisfies the equation $y'' + k^2 y = 0$.

45. Find the derivative of each member of the identity $\cos 2x = 2\cos^2 x - 1$ and thereby obtain another trigonometric identity.

46. Find values of x for which the following curves have horizontal tangents: (a) $y = x + \sin x$ and (b) $y = 4x + \cos \pi x$.

47. Use differentials to estimate the value of $\sin 31°$.

48. Find the linearization $L(x)$ of the function $f(x) = \sin(\cos x)$ for $a = \pi/2$.

49. Find the slope of a line tangent to the curve $y = \dfrac{2 \sin 3x}{x}$, where $x = 0.15$.

50. An object is oscillating vertically on the end of a spring such that its displacement d (in cm) is $d = 2.5 \cos 16t$, where t is the time (in s). What is the acceleration of the object after 1.5 s?

51. The displacement d (in mm) of a piano wire as a function of the time t (in s) is given by $d = 3.00 \sin 188t \cos 188t$. How fast is the displacement changing when $t = 2.00$ ms?

52. A water slide at an amusement park follows the curve (y in m) $y = 2.0 + 2.0\cos(0.53x + 0.40)$ for $0 \le x \le 5.0$ m. Find the angle with the horizontal of the slide for $x = 2.5$ m.

53. The blade of a sabre saw moves vertically up and down, and its displacement y (in cm) is given by $y = 1.85 \sin 36\pi t$, where t is the time (in s). Find the velocity of the blade for $t = 0.0250$ s.

54. The current i (in A) in an amplifier circuit as a function of the time t (in s) is given by $i = 0.10\cos(120\pi t + \pi/6)$. Find the expression for the voltage across a 2.0-mH inductor in the circuit. By definition, the voltage caused by the changing current is given by $V_L = L(di/dt)$, where L is the inductance (in H).

55. In testing a heat-seeking rocket, it is found to be always moving directly toward a remote-controlled aircraft. At a certain instant, the distance r (in km) from the rocket to the aircraft is $r = \dfrac{100}{1 - \cos\theta}$, where θ is the angle between their directions of flight. Find $dr/d\theta$ for $\theta = 120°$. See Fig. 27.5.

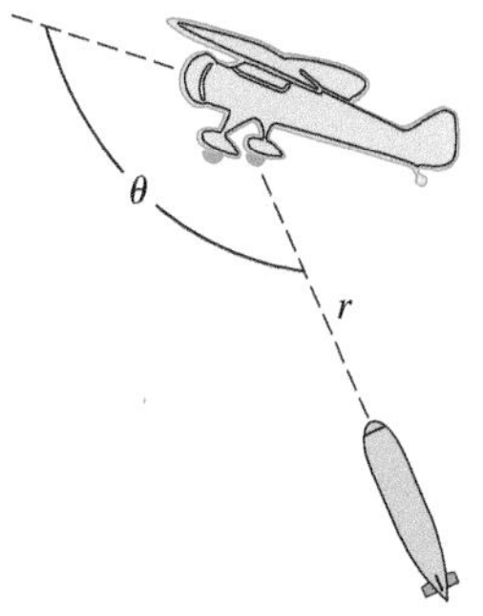

Fig. 27.5

56. The number N of reflections of a light ray passing through an optic fibre of length L and diameter d is $N = \dfrac{L \sin\theta}{d\sqrt{n^2 - \sin^2\theta}}$. Here, n is the index of refraction of the fibre, and θ is the angle between the light ray and the fibre's axis. Find $dN/d\theta$.

Answers to Practice Exercises

1. $y' = 12\cos(4x + 1)$ **2.** $y' = -20x \sin x^2(1 + \cos x^2)$

27.2 Derivatives of the Other Trigonometric Functions

We obtain the derivative of tan u by expressing tan u as $\sin u/\cos u$. Therefore, letting $y = \sin u/\cos u$, by employing the quotient rule, we have

$$\frac{dy}{dx} = \frac{\cos u[\cos u(du/dx)] - \sin u[-\sin u(du/dx)]}{\cos^2 u}$$

$$= \frac{\cos^2 u + \sin^2 u}{\cos^2 u}\frac{du}{dx} = \frac{1}{\cos^2 u}\frac{du}{dx} = \sec^2 u\frac{du}{dx}$$

$$\frac{d(\tan u)}{dx} = \sec^2 u\frac{du}{dx} \qquad \textbf{(24.6)}$$

We find the derivative of cot u by letting $y = \cos u/\sin u$, again using the quotient rule.

$$\frac{dy}{dx} = \frac{\sin u[-\sin u(du/dx)] - \cos u[\cos u(du/dx)]}{\sin^2 u}$$

$$= \frac{-\sin^2 u - \cos^2 u}{\sin^2 u}\frac{du}{dx}$$

$$\frac{d(\cot u)}{dx} = -\csc^2 u\frac{du}{dx} \qquad \textbf{(27.7)}$$

To obtain the derivative of sec u, we let $y = 1/\cos u$. Then,

$$\frac{dy}{dx} = -(\cos u)^{-2}\left[(-\sin u)\left(\frac{du}{dx}\right)\right] = \frac{1}{\cos u}\frac{\sin u}{\cos u}\frac{du}{dx}$$

$$\frac{d(\sec u)}{dx} = \sec u \tan u\frac{du}{dx} \qquad \textbf{(27.8)}$$

We obtain the derivative of csc u by letting $y = 1/\sin u$. And so,

$$\frac{dy}{dx} = -(\sin u)^{-2}\left(\cos u\frac{du}{dx}\right) = -\frac{1}{\sin u}\frac{\cos u}{\sin u}\frac{du}{dx}$$

$$\frac{d(\csc u)}{dx} = -\csc u \cot u\frac{du}{dx} \qquad \textbf{(27.9)}$$

EXAMPLE 1 Derivative of a power of sec *u*

Find the derivative of $y = 3\sec^2 4x$.

Using the power rule and Eq. (27.8), we have

$$\frac{dy}{dx} = 3(2)(\sec 4x)\frac{d(\sec 4x)}{dx} \qquad \text{using } \frac{du^n}{dx} = nu^{n-1}\frac{du}{dx}$$

$$= 6(\sec 4x)(\sec 4x \tan 4x)(4) \qquad \text{using } \frac{d(\sec u)}{dx} = \sec u \tan u\frac{du}{dx}$$

$$= 24\sec^2 4x \tan 4x$$

Practice Exercise

1. Find the derivative of $y = 3\tan 8x$.

EXAMPLE 2 Derivative of a product with csc *u*

Find the derivative of $y = t\csc^3 2t$.

Using the power rule, the product rule, and Eq. (27.9), we have

$$\begin{aligned}\frac{dy}{dt} &= t(3\csc^2 2t)(-\csc 2t\cot 2t)(2) + (\csc^3 2t)(1)\\ &= \csc^3 2t(-6t\cot 2t + 1)\end{aligned}$$

EXAMPLE 3 Derivative of a power with tan *u* and sec *u*

Find the derivative of $y = (\tan 2x + \sec 2x)^3$.

Using the power rule and Eqs. (27.6) and (27.8), we have

$$\begin{aligned}\frac{dy}{dx} &= 3(\tan 2x + \sec 2x)^2[\sec^2 2x(2) + \sec 2x\tan 2x(2)]\\ &= 3(\tan 2x + \sec 2x)^2(2\sec 2x)(\sec 2x + \tan 2x)\\ &= 6\sec 2x(\tan 2x + \sec 2x)^3\end{aligned}$$

Practice Exercise

2. Find the derivative of $y = 5(\cot 3x + \csc 3x)^2$.

EXAMPLE 4 Differential with sin *u* and tan *v*

Find the differential of $r = \sin 2\theta\tan\theta^2$.

Here, we are to find the derivative of the given function and multiply by $d\theta$. Therefore, using the product rule along with Eqs. (27.4) and (27.6), we have

$$\begin{aligned}dr &= [(\sin 2\theta)(\sec^2\theta^2)(2\theta) + (\tan\theta^2)(\cos 2\theta)(2)]\,d\theta\\ &= (2\theta\sin 2\theta\sec^2\theta^2 + 2\cos 2\theta\tan\theta^2)d\theta\end{aligned}$$

don't forget the $d\theta$

EXAMPLE 5 Derivative of an implicit function

Find dy/dx if $\cot 2x - 3\csc xy = y^2$.

In finding the derivative of this implicit function, we must be careful not to forget the factor dy/dx when it occurs. The derivative is found as follows:

$$\cot 2x - 3\csc xy = y^2$$

$$(-\csc^2 2x)(2) - 3(-\csc xy\cot xy)\left(x\frac{dy}{dx} + y\right) = 2y\frac{dy}{dx}$$

$$3x\csc xy\cot xy\frac{dy}{dx} - 2y\frac{dy}{dx} = 2\csc^2 2x - 3y\csc xy\cot xy$$

$$\frac{dy}{dx} = \frac{2\csc^2 2x - 3y\csc xy\cot xy}{3x\csc xy\cot xy - 2y}$$

EXAMPLE 6 Evaluation of a derivative at a point

Evaluate the derivative of $y = \dfrac{2x}{1 - \cot 3x}$, where $x = 0.25$.

Finding the derivative, we have

$$\begin{aligned}\frac{dy}{dx} &= \frac{(1 - \cot 3x)(2) - 2x(\csc^2 3x)(3)}{(1 - \cot 3x)^2}\\ &= \frac{2 - 2\cot 3x - 6x\csc^2 3x}{(1 - \cot 3x)^2}\end{aligned}$$

Now, substituting $x = 0.25$, we have

$$\left.\frac{dy}{dx}\right|_{x=0.25} = \frac{2 - 2\cot 0.75 - 6(0.25)\csc^2 0.75}{(1 - \cot 0.75)^2}$$
$$= -626.0$$

In the above calculation, we have used the calculator in *radian mode.* Moreover, we have calculated cot 0.75 and csc 0.75 as the reciprocals of tan 0.75 and sin 0.75, respectively.

EXERCISES 27.2

In Exercises 1 and 2, make the given changes in the indicated examples of this section and then find the derivatives.

1. In Example 1, in the given function, change $4x$ to x^2.

2. In Example 4, in the given function, change θ^2 to 3θ.

In Exercises 3–34, find the derivatives of the given functions.

3. $y = \tan 5x$

4. $y = 3\tan(3x + 2)$

5. $y = 5\cot(0.25\pi - \theta)$

6. $y = 22\cot 2x$

7. $u = 3\sec 9v$

8. $y = \sec\sqrt{1 - x}$

9. $y = -3\csc\sqrt{7x - 6}$

10. $h = 0.5\csc(1 - 2\pi t)$

11. $R = 5\tan^2 3\pi t$

12. $y = 2\tan^2(x^2)$

13. $y = 2\cot^4 \frac{1}{2}x$

14. $p = 3\cot^2(4 - 3r^2)$

15. $y = \sqrt{\sec 4x}$

16. $y = 0.8\sec^3 5u$

17. $y = 3\csc^4 7x$

18. $y = 7\csc^2(9x^{14})$

19. $r = t^2\tan 0.5t$

20. $y = 3x\sec 2\pi x$

21. $y = 4\cos x\csc x^2$

22. $y = \frac{1}{2}\sin 2x\sec x$

23. $y = \dfrac{\csc x}{x}$

24. $u = \dfrac{\cot 0.25z}{2z}$

25. $y = \dfrac{2\cos 4x}{1 + \cot 3x}$

26. $y = \dfrac{\tan^2 3x}{2 + \sin x^2}$

27. $y = \frac{1}{3}\tan^3 x - \tan x$

28. $y = 4\csc 4x - 2\cot 4x$

29. $r = \tan(\sin 2\pi\theta)$

30. $y = x\tan x + \sec^2 2x$

31. $y = \sqrt{2x + \tan 4x}$

32. $V = (1 - \csc^2 3r)^3$

33. $x\sec y - 2y = \sin 2x$

34. $3\cot(x + y) = \cos y^2$

In Exercises 35–38, find the differentials of the given functions.

35. $y = 4\tan^2 3x$

36. $y = 2.5\sec^3 6t$

37. $y = \tan 4x\sec 4x$

38. $y = 2x\cot \pi x$

In Exercises 39–50, solve the given problems.

39. On a calculator, find the values of (a) $\sec^2 1.0000$ and (b) $(\tan 1.0001 - \tan 1.0000)/0.0001$. Compare the values and give the meaning of each in relation to the derivative of $\tan x$ where $x = 1$.

40. On a calculator, find the values of (a) $\sec 1.0000 \tan 1.0000$ and (b) $(\sec 1.0001 - \sec 1.0000)/0.0001$. Compare the values and give the meaning of each in relation to the derivative of $\sec x$ where $x = 1$.

41. Find the derivative of each member of the identity $1 + \tan^2 x = \sec^2 x$ and show that the results are equal.

42. Find the points where a tangent to the curve of $y = \tan x$ is parallel to the line $y = 2x$ if $0 < x < 2\pi$.

43. Find the slope of a line tangent to the curve of $y = 2\cot 3x$ where $x = \pi/12$.

44. Find the slope of a line normal to the curve of $y = \csc\sqrt{2x + 1}$ where $x = 0.45$.

45. Show that $y = 2\tan x - \sec x$ satisfies $\dfrac{dy}{dx} = \dfrac{2 - \sin x}{\cos^2 x}$.

46. A helicopter takes off such that its height h (in m) above the ground is $h = 25\sec 0.16t$ for the first 8.0 s of flight. What is its vertical velocity after 6.0 s?

47. The vertical displacement y (in cm) of the end of an industrial robot arm for each cycle is $y = 2.0t^{1.5} - \tan 0.10t$, where t is the time (in s). Find its vertical velocity for $t = 15$ s.

48. The electric charge q (in C) passing a given point in a circuit is given by $q = t\sec\sqrt{0.20t^2 + 1.0}$, where t is the time (in s). Find the current i (in A) for $t = 0.80$ s. $(i = dq/dt)$

49. An observer to a rocket launch was 1000 m from the takeoff position. The observer found the angle of elevation of the rocket as a function of time to be $\theta = 3t/(2t + 10)$. Therefore, the height h (in m) of the rocket was $h = 1000\tan\dfrac{3t}{2t + 10}$. Find the time rate of change of height after 5.0 s. See Fig. 27.6.

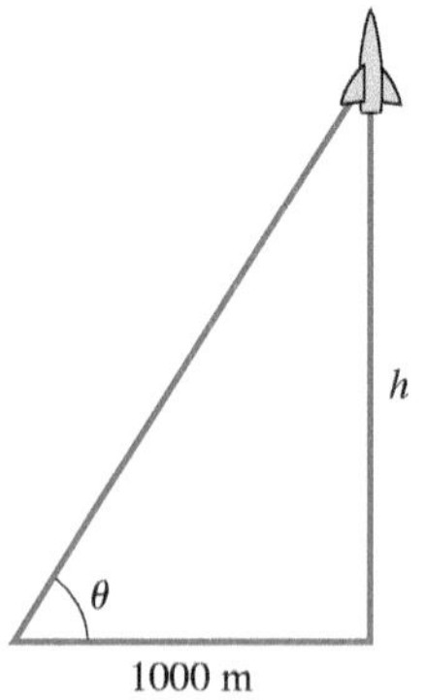

Fig. 27.6

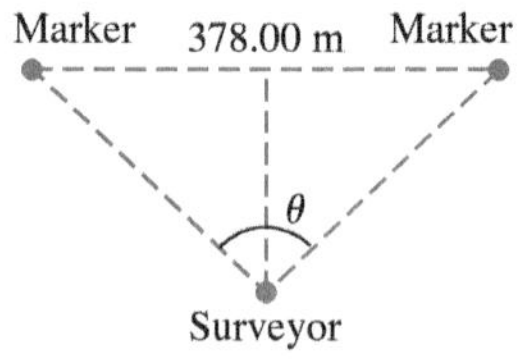

Fig. 27.7

50. A surveyor measures the distance between two markers to be 378.00 m. Then, moving along a line equidistant from the markers, the distance d from the surveyor to each marker is $d = 189.00\csc\frac{1}{2}\theta$, where θ is the angle between the lines of sight to the markers. See Fig. 27.7. By using differentials, find the change in d if θ changes from 98.20° to 98.45°.

Answers to Practice Exercises

1. $y' = 24\sec^2 8x$ **2.** $y' = -30\csc 3x(\cot 3x + \csc 3x)^2$

27.3 Derivatives of the Inverse Trigonometric Functions

To obtain the derivative of $y = \sin^{-1} u$, we first solve for u in the form $u = \sin y$, and then take the derivative with respect to x. This results in $\frac{du}{dx} = \cos y \frac{dy}{dx}$. Solving this for dy/dx, we have

$$\frac{dy}{dx} = \frac{1}{\cos y}\frac{du}{dx} = \frac{1}{\sqrt{1 - \sin^2 y}}\frac{du}{dx} = \frac{1}{\sqrt{1 - u^2}}\frac{du}{dx}$$

We choose the positive square root since $\cos y > 0$ for $-\frac{\pi}{2} < y < \frac{\pi}{2}$, which is the range of the defined values of $\sin^{-1} u$. Therefore, we obtain the following result:

Derivative of the Inverse Sine Function

$$\frac{d(\sin^{-1} u)}{dx} = \frac{1}{\sqrt{1 - u^2}}\frac{du}{dx} \qquad \textbf{(27.10)}$$

■ Note that the derivative of the inverse sine function is an algebraic function.

EXAMPLE 1

Find the derivative of $y = \sin^{-1} 4x$.

Here $u = 4x$, so $du/dx = 4$. Then

$$\frac{dy}{dx} = \frac{d(\sin^{-1} 4x)}{dx} = \frac{1}{\sqrt{1 - (4x)^2}}(4) \quad \leftarrow \frac{du}{dx}$$

$$= \frac{4}{\sqrt{1 - 16x^2}}$$

Practice Exercise

1. Find the derivative of $y = 5 \sin^{-1} x^2$.

We find the derivative of the inverse cosine function by letting $y = \cos^{-1} u$ and by following the same procedure as that used in finding the derivative of $\sin^{-1} u$:

$$u = \cos y, \qquad \frac{du}{dx} = -\sin y \frac{dy}{dx}$$

$$\frac{dy}{dx} = -\frac{1}{\sin y}\frac{du}{dx} = -\frac{1}{\sqrt{1 - \cos^2 y}}\frac{du}{dx}$$

$$\frac{d(\cos^{-1} u)}{dx} = -\frac{1}{\sqrt{1 - u^2}}\frac{du}{dx} \qquad \textbf{(27.11)}$$

The positive square root is chosen here since $\sin y > 0$ for $0 < y < \pi$, which is the range of the defined values of $\cos^{-1} u$. We note that the derivative of the inverse cosine is the negative of the derivative of the inverse sine.

By letting $y = \tan^{-1} u$, solving for u, and taking derivatives, we find the derivative of the inverse tangent function:

$$u = \tan y, \qquad \frac{du}{dx} = \sec^2 y \frac{dy}{dx}, \qquad \frac{dy}{dx} = \frac{1}{\sec^2 y}\frac{du}{dx} = \frac{1}{1 + \tan^2 y}\frac{du}{dx}$$

$$\frac{d(\tan^{-1} u)}{dx} = \frac{1}{1 + u^2}\frac{du}{dx} \qquad \textbf{(27.12)}$$

We can see that the derivative of the inverse tangent is also an algebraic function.

The inverse sine, inverse cosine, and inverse tangent prove to be the inverse trigonometric functions of greatest importance in applications and in further development of mathematics. Therefore, the formulas for the derivatives of the other inverse functions are not presented here, although they are included in the exercises.

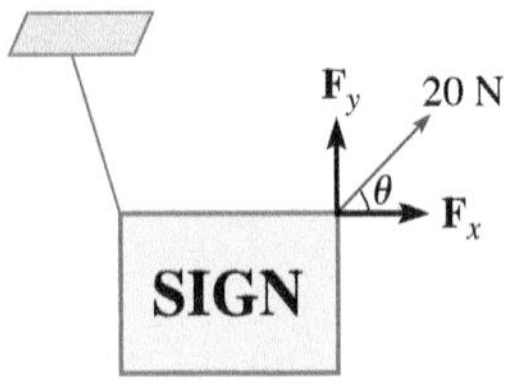

Fig. 27.8

EXAMPLE 2 Derivative of $\cos^{-1} u$—forces on a sign

A 20-N force acts on a sign as shown in Fig. 27.8. Express the angle θ as a function of the x-component, F_x, of the force, and then find the expression for the instantaneous rate of change of θ with respect to F_x.

From the figure, we see that $F_x = 20\cos\theta$. Solving for θ, we have $\theta = \cos^{-1}(F_x/20)$. To find the instantaneous rate of change of θ with respect to F_x, we are to take the derivative $d\theta/dF_x$:

$$\theta = \cos^{-1}\frac{F_x}{20} = \cos^{-1} 0.05F_x$$

$$\frac{d\theta}{dF_x} = -\frac{1}{\sqrt{1-(0.05F_x)^2}}(0.05) \quad \text{using Eq. (27.11)}$$

$$= \frac{-0.05}{\sqrt{1-0.0025F_x^2}}$$

EXAMPLE 3 Derivative of a product with $\tan^{-1} u$

Find the derivative of $y = (x^2+1)\tan^{-1} x - x$.

Using the product rule along with Eq. (27.12) on the first term, we have

$$\frac{dy}{dx} = (x^2+1)\left(\frac{1}{1+x^2}\right)(1) + (\tan^{-1} x)(2x) - 1 \quad \text{using Eq. (27.12)}$$

$$= 2x\tan^{-1} x$$

Practice Exercise

2. Find the derivative of $y = (\tan^{-1} 3x)^2$.

EXAMPLE 4 Differential with $\sin^{-1} u$

Find the derivative of $y = x\sin^{-1} 2x + \frac{1}{2}\sqrt{1-4x^2}$.

$$\frac{dy}{dx} = \underbrace{x\left(\frac{2}{\sqrt{1-4x^2}}\right)}_{\text{using Eq. (27.10)}} + \sin^{-1} 2x + \underbrace{\frac{1}{2}\left(\frac{1}{2}\right)(1-4x^2)^{-1/2}(-8x)}_{\text{using the general power rule}}$$

$$= \frac{2x}{\sqrt{1-4x^2}} + \sin^{-1} 2x - \frac{2x}{\sqrt{1-4x^2}}$$

$$= \sin^{-1} 2x$$

■ For reference, the general power rule is $\frac{du^n}{dx} = nu^{n-1}\left(\frac{du}{dx}\right)$.

EXAMPLE 5 Tangent line for a quotient with $\tan^{-1} u$

Find the slope of a tangent to the curve of $y = \frac{\tan^{-1} x}{x^2+1}$, where $x = 3.60$. In Fig. 27.9, the function and the tangent line are shown.

Here, we are to find the derivative and then evaluate it for $x = 3.60$.

$$\frac{dy}{dx} = \frac{(x^2+1)\left(\frac{1}{1+x^2}\right)(1) - (\tan^{-1} x)(2x)}{(x^2+1)^2} \quad \text{take derivative}$$

$$= \frac{1 - 2x\tan^{-1} x}{(x^2+1)^2}$$

$$\left.\frac{dy}{dx}\right|_{x=3.60} = \frac{1 - 2(3.60)(\tan^{-1} 3.60)}{(3.60^2+1)^2} = -0.0429 \quad \text{evaluate}$$

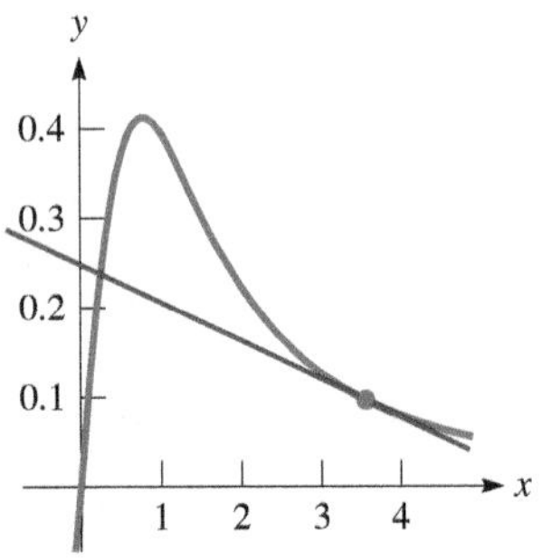

Fig. 27.9

EXERCISES 27.3

In Exercises 1 and 2, make the given changes in the indicated examples of this section and then find the derivatives.

1. In Example 1, in the given function, change $4x$ to x^2.
2. In Example 3, in the given function, change $(x^2 + 1)\tan^{-1}x$ to $(4x^2 + 1)\tan^{-1}2x$.

In Exercises 3–34, find the derivatives of the given functions.

3. $y = \sin^{-1} 7x$
4. $R = 3\sin^{-1}(4 - t^2)$
5. $y = 2\sin^{-1} 5x^5$
6. $y = \sin^{-1}\sqrt{1 - 2x}$
7. $y = 3.6\cos^{-1} 0.5s$
8. $\theta = 0.2\cos^{-1} 5t$
9. $y = 2\cos^{-1}\sqrt{2 - x}$
10. $y = 3\cos^{-1}(x^2 + 0.5)$
11. $V = 8\tan^{-1}\sqrt{s}$
12. $y = \tan^{-1}(1 - x)$
13. $y = 6x\tan^{-1}(1/x)$
14. $w = 4\tan^{-1}\pi u^4$
15. $y = 5x\sin^{-1} 2x + \sqrt{1 - 4x^2}$
16. $y = x^6\cos^{-1} x$
17. $v = 0.4u\tan^{-1} 2u$
18. $y = (x^2 + 1)\sin^{-1} 4x$
19. $T = \dfrac{3R - 1}{\sin^{-1} 2R}$
20. $\theta = \dfrac{\tan^{-1} 2r}{\pi r}$
21. $y = \dfrac{\sin^{-1} 2x}{\cos^{-1} 2x}$
22. $y = \dfrac{x^2 + 1}{\tan^{-1} x}$
23. $y = 2(\cos^{-1} 4x)^9$
24. $r = 0.5\,(\sin^{-1} 3t)^4$
25. $u = [\sin^{-1}(4t + 3)]^2$
26. $y = \sqrt{\sin^{-1}(x - 1)}$
27. $y = \tan^{-1}\left(\dfrac{1 - t}{1 + t}\right)$
28. $p = \dfrac{3}{\cos^{-1} 2w}$
29. $y = \dfrac{1}{1 + 4x^2} - \tan^{-1} 2x$
30. $y = \sin^{-1} x - \sqrt{1 - x^2}$
31. $y = 3(4 - \cos^{-1} 2x)^3$
32. $\sin^{-1}(x + y) + y = x^2$
33. $2\tan^{-1} xy + x = 3$
34. $y = \sqrt{2\pi - \sin^{-1} 4x}$

In Exercises 35–54, solve the given problems.

35. On a calculator, find the values of (a) $1/\sqrt{1 - 0.5^2}$ and (b) $(\sin^{-1} 0.5001 - \sin^{-1} 0.5000)/0.0001$. Compare the values and give the meaning of each in relation to the derivative of $\sin^{-1} x$ where $x = 0.5$.
36. On a calculator, find the values of (a) $1/(1 + 0.5^2)$ and (b) $(\tan^{-1} 0.5001 - \tan^{-1} 0.5000)/0.0001$. Compare the values and give the meaning of each in relation to the derivative of $\tan^{-1} x$ where $x = 0.5$.
37. Find the differential of the function $y = (\sin^{-1} x)^8$.
38. Find the linearization $L(x)$ of the function $f(x) = 2x\cos^{-1} x$ for $a = 0$.
39. Find the slope of a line tangent to the curve of $y = x/\tan^{-1} x$ at $x = 0.80$.
40. Explain what is wrong with a problem that requires finding the derivative of $y = \sin^{-1}(x^2 + 1)$.
41. Find the second derivative of $y = x\tan^{-1} x$.
42. Find the point(s) at which the line normal to $y = 2\sin^{-1} 0.5x$ is parallel to the line $y = 1 - x$.
43. Use a graphing utility to display the graphs of $y = \sin^{-1} x$ and $y = 1/\sqrt{1 - x^2}$. By roughly estimating slopes of tangent lines of $y = \sin^{-1} x$, note that $y = 1/\sqrt{1 - x^2}$ gives reasonable values for the derivative of $y = \sin^{-1} x$.
44. Use a graphing utility to display the graphs of $y = \tan^{-1} x$ and $y = 1/(1 + x^2)$. By roughly estimating slopes of tangent lines of $y = \tan^{-1} x$, note that $y = 1/(1 + x^2)$ gives reasonable values for the derivative of $y = \tan^{-1} x$.
45. Find the second derivative of the function $y = \tan^{-1} 2x$.
46. Show that $\dfrac{d(\cot^{-1} u)}{dx} = -\dfrac{1}{1 + u^2}\dfrac{du}{dx}$.
47. Show that $\dfrac{d(\sec^{-1} u)}{dx} = \dfrac{1}{\sqrt{u^2(u^2 - 1)}}\dfrac{du}{dx}$.
48. Show that $\dfrac{d(\csc^{-1} u)}{dx} = -\dfrac{1}{\sqrt{u^2(u^2 - 1)}}\dfrac{du}{dx}$.
49. In the analysis of the waveform of an AM radio wave, the equation $t = \dfrac{1}{\omega}\sin^{-1}\dfrac{A - E}{mE}$ arises. Find dt/dm, assuming that the other quantities are constant.
50. An equation that arises in the theory of solar collectors is $\alpha = \cos^{-1}\dfrac{2f - r}{r}$. Find the expression for $d\alpha/dr$ if f is constant.
51. When an alternating current passes through a series *RLC* circuit, the voltage and current are out of phase by angle θ (see Section 12.7). Here $\theta = \tan^{-1}[(X_L - X_C)/R]$, where X_L and X_C are the reactances of the inductor and capacitor, respectively, and R is the resistance. Find $d\theta/dX_C$ for constant X_L and R.
52. When passing through glass, a light ray is refracted (bent) such that the angle of refraction r is given by $r = \sin^{-1}[(\sin i)/\mu]$. Here, i is the angle of incidence, and μ is the index of refraction of the glass (see Fig. 27.10). For different types of glass, μ differs. Find the expression for dr for a constant value of i.

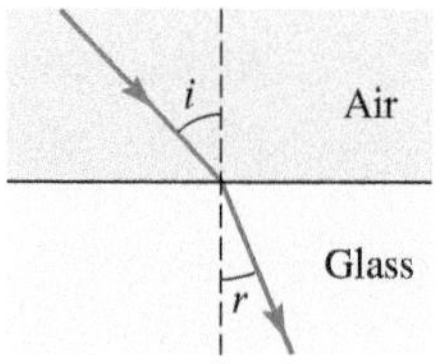

Fig. 27.10

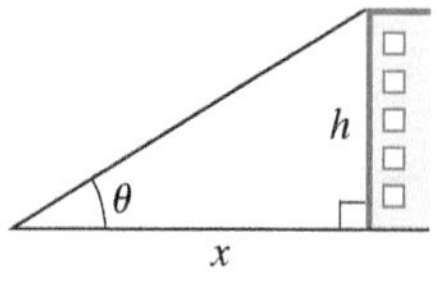

Fig. 27.11

53. As a person approaches a building of height h, the angle of elevation of the top of the building is a function of the person's distance from the building. Express the angle of elevation θ in terms of h and the distance x from the building and then find $d\theta/dx$. Assume the person's height is negligible to that of the building. See Fig. 27.11.
54. A triangular metal frame is designed as shown in Fig. 27.12. Express angle A as a function of x and evaluate dA/dx for $x = 6$ cm.

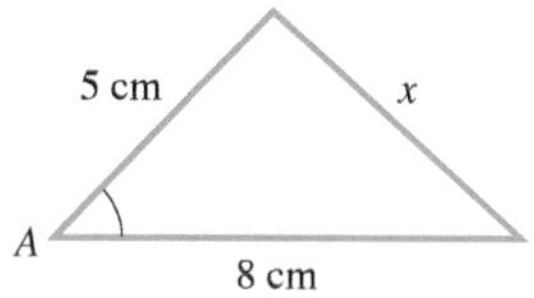

Fig. 27.12

Answers to Practice Exercises

1. $y' = 10x/\sqrt{1 - x^4}$ 2. $y' = (6\tan^{-1} 3x)/(1 + 9x^2)$

27.4 Applications

With our development of the formulas for the derivatives of the trigonometric and inverse trigonometric functions, it is now possible to use these derivatives in the same manner as we applied the derivatives of algebraic functions in Chapter 24.

EXAMPLE 1 Sketching a curve

Sketch the curve $y = \sin^2 x - \frac{x}{2}$ $(0 \leq x \leq 2\pi)$.

First, by setting $x = 0$, we see that the only easily obtainable intercept is $(0, 0)$. Replacing x by $-x$ and y by $-y$, we find that the curve is not symmetric to either axis or to the origin. Also, since x does not appear in a denominator, there are no vertical asymptotes. We are considering only the restricted domain $0 \leq x \leq 2\pi$. (Without this restriction, the domain is all x and the range is all y.)

We now find the information from the derivatives as in Section 24.5.

Find y' and solve $y' = 0$:

$y' = 2 \sin x \cos x - \frac{1}{2} = \sin 2x - \frac{1}{2}$

$y' = 0$ if $\sin 2x = \frac{1}{2}$, so

$$2x = \frac{\pi}{6}, \frac{5\pi}{6}, \frac{13\pi}{6}, \frac{17\pi}{6}, \quad \text{or} \quad x = \frac{\pi}{12}, \frac{5\pi}{12}, \frac{13\pi}{12}, \frac{17\pi}{12}$$

Find y''. Use second-derivative test:

$y'' = 2 \cos 2x$

$y'' > 0$ at $x = \frac{\pi}{12}$ and $x = \frac{13\pi}{12}$, so there are minimum points at $\left(\frac{\pi}{12}, -0.064\right)$ and $\left(\frac{13\pi}{12}, -1.63\right)$.

$y'' < 0$ at $x = \frac{5\pi}{12}$ and $x = \frac{17\pi}{12}$, so there are maximum points at $\left(\frac{5\pi}{12}, 0.279\right)$ and $\left(\frac{17\pi}{12}, -1.29\right)$.

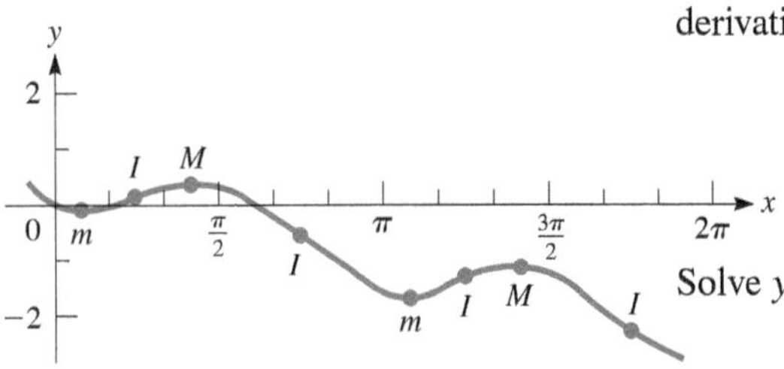

Fig. 27.13

Solve $y'' = 0$:

$y'' = 0$ if $\cos 2x = 0$, so

$$2x = \frac{\pi}{2}, \frac{3\pi}{2}, \frac{5\pi}{2}, \frac{7\pi}{2}, \quad \text{or} \quad x = \frac{\pi}{4}, \frac{3\pi}{4}, \frac{5\pi}{4}, \frac{7\pi}{4}$$

Analyse concavity and points of inflection:

Therefore, the points of inflection are $\left(\frac{\pi}{4}, 0.107\right)$, $\left(\frac{3\pi}{4}, -0.678\right)$, $\left(\frac{5\pi}{4}, -1.46\right)$, and $\left(\frac{7\pi}{4}, -2.25\right)$. At $x = 0$, $y'' > 0$ and the curve is concave up. It then changes concavity at every point of inflection.

Locate points and sketch:

The curve is shown in Fig. 27.13.

EXAMPLE 2 Solving an equation using Newton's method

By using Newton's method, solve the equation $2x - 1 = 3 \cos x$.

First, we locate the required root approximately by sketching $y_1 = 2x - 1$ and $y_2 = 3 \cos x$. As we can see in Fig. 27.14, they intersect between $x = 1$ and $x = 2$, near $x = 1.2$. Therefore, using $x_1 = 1.2$, with

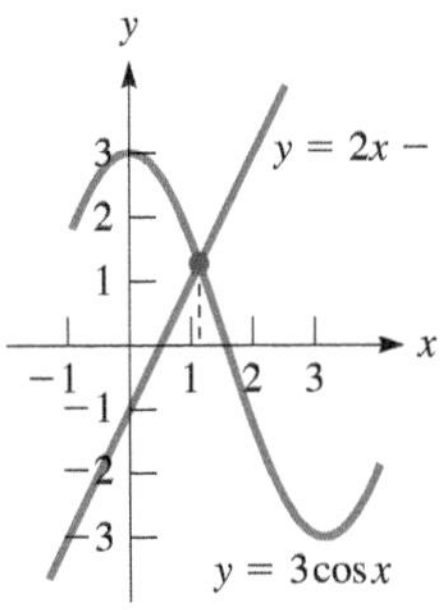

Fig. 27.14

$$f(x) = 2x - 1 - 3 \cos x$$
$$f'(x) = 2 + 3 \sin x$$

we use $x_{n+1} = x_n - \dfrac{f(x_n)}{f'(x_n)}$ with $n = 1$, which is

$$x_2 = x_1 - \frac{f(x_1)}{f'(x_1)}$$

To find x_2, we have

$$f(1.2) = 2(1.2) - 1 - 3\cos 1.2 = 0.312\,926\,7$$
$$f'(1.2) = 2 + 3\sin 1.2 = 4.796\,117\,3$$
$$x_2 = 1.2 - \frac{0.312\,926\,7}{4.796\,117\,3} = 1.134\,754\,2$$

Finding the next approximation, we find $x_3 = 1.134\,236\,6$, which is accurate to the value shown. Again, when using the calculator, it is not necessary to list the values of $f(x_1)$ and $f'(x_1)$, as the complete calculation can be done directly on the calculator.

EXAMPLE 3 Finding a maximum value—lumber area

Logs with a circular cross-section 1.20 m in diameter are cut in half lengthwise. Find the largest rectangular cross-sectional area that can then be cut from one of the halves.

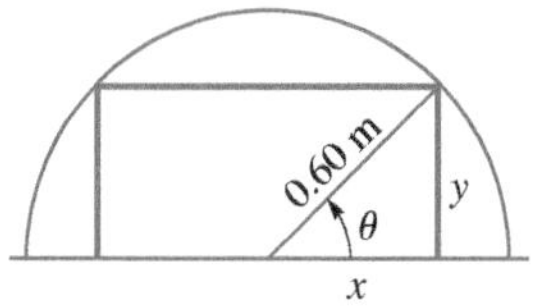

Fig. 27.15

Introduce notation: Let A be the area of a rectangular cross-section of sides $2x$ and y (see Fig. 27.15). Let θ be the angle between the horizontal and a radius to the vertex of the rectangle $(0 \le \theta \le \pi)$.

Write A as a function of x and y. Use the relationships between x, y, and θ to write A as a function of the variable θ:

Area:
$$A = (2x)y$$

Trigonometric functions of θ:
$$\sin\theta = \frac{y}{0.60} \text{ and } \cos\theta = \frac{x}{0.60}$$

Substitute x and y to have area in terms of θ:
$$A = 2(0.60\cos\theta)(0.60\sin\theta) = 0.72\cos\theta\sin\theta$$
$$= 0.36\sin 2\theta \quad \text{using identity } \sin 2\alpha = 2\sin\alpha\cos\alpha$$

Take the derivative:
$$\frac{dA}{d\theta} = (0.36\cos 2\theta)(2) = 0.72\cos 2\theta$$

Set the derivative equal to zero and solve:
$$0.72\cos 2\theta = 0, \quad 2\theta = \frac{\pi}{2}, \quad \theta = \frac{\pi}{4}$$

Check that it is a maximum:
$$\frac{d^2A}{d\theta^2} = -1.44\sin 2\theta, \quad \left.\frac{d^2A}{d\theta^2}\right|_{\theta=\frac{\pi}{4}} = -1.44 < 0,$$
so A has a maximum when $\theta = \pi/4$.

The largest rectangular cross-sectional area is $A = 0.36\sin 2\left(\frac{\pi}{4}\right)$
$$= 0.36\sin\frac{\pi}{2} = 0.36\text{ m}^2.$$

■ See the chapter introduction.

EXAMPLE 4 Related rates—rocket velocity

A rocket is taking off vertically at a distance of 6500 m from an observer, as shown in Fig. 27.16. If, when the angle of elevation is 38.4°, it is changing at 5.00°/s, how fast is the rocket ascending?

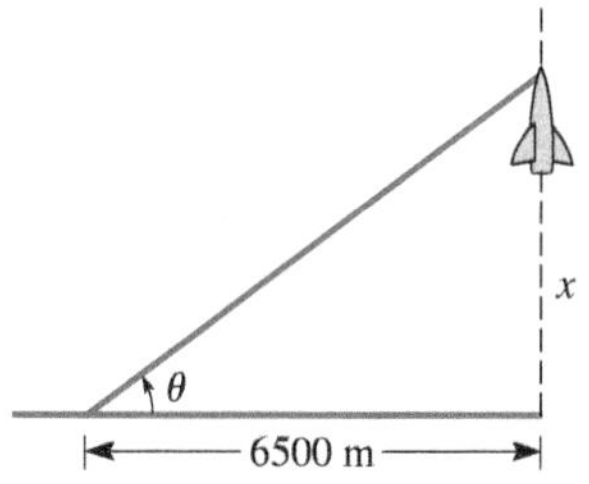

Fig. 27.16

Identify variables and rates:

Given: $\left.\frac{d\theta}{dt}\right|_{\theta=38.4°} = 5.00°/\text{s} = 0.0873\text{ rad/s}$

Required: $\frac{dx}{dt}$ when $\theta = 38.4°$

Determine the equation relating the variables: $\tan\theta = \dfrac{x}{6500}$, or $\theta = \tan^{-1}\left(\dfrac{x}{6500}\right)$

Differentiate with respect to time **using the chain rule**: $\dfrac{d\theta}{dt} = \dfrac{1}{1 + (x/6500)^2} \cdot \dfrac{dx/dt}{6500} = \dfrac{6500\, dx/dt}{6500^2 + x^2}$

Evaluate the required rate: When $\theta = 38.4°$, $x = 6500 \tan 38.4° = 5150$, and

$$0.0873 = \frac{6500\, dx/dt}{6500^2 + 5150^2}$$

$$\frac{dx}{dt} = 924 \text{ m/s}$$

EXAMPLE 5 Differential—error in calculated building height

On level ground, 180 m from the base of a building, the angle of elevation of the top of the building is 30.00°. What error in calculating the height h of the building would be caused by an error of 0.25° in the angle?

From Fig. 27.17, $h = 180 \tan\theta$. To find the error in h, we must find the differential dh.

$$dh = 180 \sec^2\theta\, d\theta$$

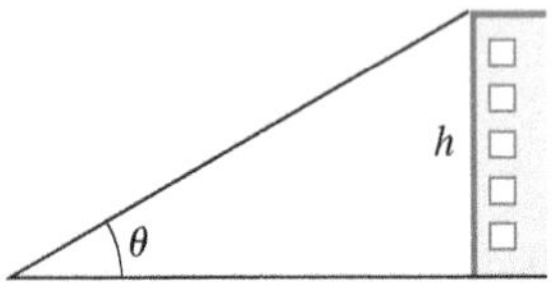

Fig. 27.17

The possible error in θ is 0.25°, which in *radian measure* is $0.25\pi/180$, which is the value we should use in the calculation of dh. Calculating dh, we have

$$dh = \frac{180(0.25\pi/180)}{\cos^2\dfrac{\pi}{6}} = 1.05 \text{ m}$$

An error of 0.25° in the angle results in an error of over 1 m in the calculated value of the height. In using the calculator, we divide by $\cos^2\dfrac{\pi}{6}$ since $\sec\theta = 1/\cos\theta$.

EXAMPLE 6 Curvilinear motion—parametric equations

A particle is rotating so that its x- and y-coordinates are given by $x = \cos 2t$ and $y = \sin 2t$. Find the magnitude and direction of its velocity when $t = \pi/8$.

Find the x- and y-components of velocity: $v_x = \dfrac{dx}{dt} = -2\sin 2t \qquad v_y = \dfrac{dy}{dt} = 2\cos 2t$

Evaluate the components at the given time:

$$v_x\big|_{t=\pi/8} = -2\sin 2\left(\frac{\pi}{8}\right) = -2\left(\frac{\sqrt{2}}{2}\right) = -\sqrt{2},$$

$$v_y\big|_{t=\pi/8} = 2\cos 2\left(\frac{\pi}{8}\right) = 2\left(\frac{\sqrt{2}}{2}\right) = \sqrt{2}$$

Find the magnitude: $v = \sqrt{v_x^2 + v_y^2} = \sqrt{2 + 2} = 2$

Find the direction:

$$\tan\theta_{\text{ref}} = \left|\frac{v_y}{v_x}\right| = \left|\frac{\sqrt{2}}{-\sqrt{2}}\right| = 1, \quad \theta_{\text{ref}} = \frac{\pi}{4},$$

θ_v is in QII: $\theta_v = \pi - \pi/4 = 3\pi/4$

Sketch by eliminating the parameter: $x^2 + y^2 = \cos^2 2t + \sin^2 2t = 1$, so the curve is a circle of radius 1.

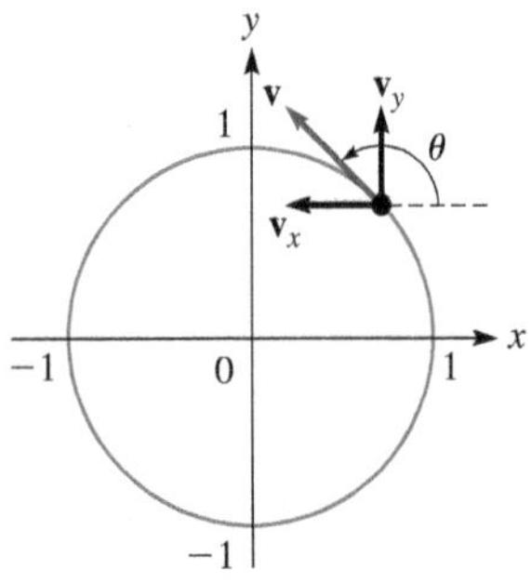

Fig. 27.18

As seen in Fig. 27.18, the particle is moving counterclockwise around the circle.

EXERCISES 27.4

In Exercises 1 and 2, make the given changes in the indicated examples of this section and then solve the resulting problems.

1. In Example 1, change $\sin^2 x$ to $\sin x$.

2. In Example 6, change $\cos 2t$ to $3 \cos 2t$, and $\sin 2t$ to $2 \sin 2t$.

In Exercises 3–40, solve the given problems. Where necessary, round answers to three significant digits.

3. Show that the slopes of the sine and cosine curves are negatives of each other at the points of intersection.

4. Show that the graph of the tangent function is always increasing (when the tangent is defined).

5. Show that the curve of $y = \tan^{-1} x$ is always increasing.

6. Sketch the graph of $y = \sin x + \cos x \ (0 \le x \le 2\pi)$.

7. Sketch the graph of $y = x - \tan x \ (-\frac{\pi}{2} < x < \frac{\pi}{2})$.

8. Sketch the graph of $y = 2 \sin x + \sin 2x \ (0 \le x \le 2\pi)$.

9. Find the equation of the line tangent to the curve of $y = x \sin^{-1} x$ at $x = 0.50$.

10. Find the equation of the line normal to the curve of $y = 3 \tan x^2$ at $x = 0.25$.

11. By Newton's method, find the positive root of the equation $x^2 - 4 \sin x = 0$ to at least four decimal places.

12. By Newton's method, find the smallest positive root of the equation $\tan x = 2x$ to at least four decimal places.

13. Find the minimum value of the function $y = 6 \cos x - 8 \sin x$.

14. Find the maximum value of the function $y = \tan^{-1}(1 + x) + \tan^{-1}(1 - x)$.

15. Power P is the time rate of change of work W. Find the equation for the power in a circuit for which $W = 8 \sin^2 2t$.

16. The phase shift ϕ in a certain electric circuit with a resistance R and variable capacitance C is $\phi = \tan^{-1} \omega RC$. Find the equation for the instantaneous rate of change of ϕ with respect to C.

17. In studying water waves, the vertical displacement y (in m) of a wave was determined to be $y = 0.50 \sin 2t + 0.30 \cos t$, where t is the time (in s). Find the velocity and the acceleration for $t = 0.40$ s.

18. At 45°N latitude, the number of hours h of daylight each day during the year is given approximately by the equation $h = 12.2 + 3.5 \sin\left[\frac{2\pi}{365}(t - 81)\right]$, where t is measured in days ($t = 37$ is February 6, etc.). Find the date of the longest day and the date of the shortest day. (Cities at 45°N are Ottawa, Ontario, and Venice, Italy.)

19. Find the time rate of change of the horizontal component T_x of the constant 46.6-N tension shown in Fig. 27.19 if $d\theta/dt = 0.36°/\text{s}$ for $\theta = 14.2°$.

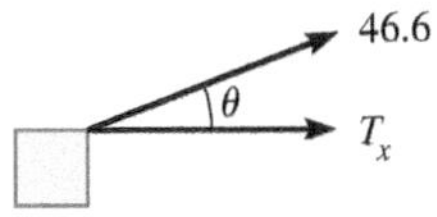

Fig. 27.19

20. The *apparent power* P_a (in W) in an electric circuit whose power is P and whose impedance phase angle is θ is given by $P_a = P \sec \theta$. Given that P is constant at 12 W, find the time rate of change of P_a if θ is changing at the rate of 0.050 rad/min, when $\theta = 40.0°$.

21. A point on the outer edge of a 38.0-cm wheel can be described by the equations $x = 19.0 \cos 6\pi t$ and $y = 19.0 \sin 6\pi t$. Find the velocity of the point for $t = 0.600$ s.

22. A machine is programmed to move an etching tool such that the position (in cm) of the tool is given by $x = 2 \cos 3t$ and $y = \cos 2t$, where t is the time (in s). Find the velocity of the tool for $t = 4.1$ s.

23. Find the acceleration of the tool of Exercise 22 for $t = 4.1$ s.

24. The volume V (in m^3) of water used each day by a community during the summer is found to be $V = 2500 + 480 \sin(\pi t/90)$, where t is the number of the summer day, and $t = 0$ is the first day of summer. On what summer day is the water usage the greatest?

25. A person observes an object dropped from the top of a building 40.0 m away. If the top of the building is 60.0 m above the person's eye level, how fast is the angle of elevation of the object changing after 1.0 s? (The distance the object drops is given by $s = 4.9t^2$.) See Fig. 27.20.

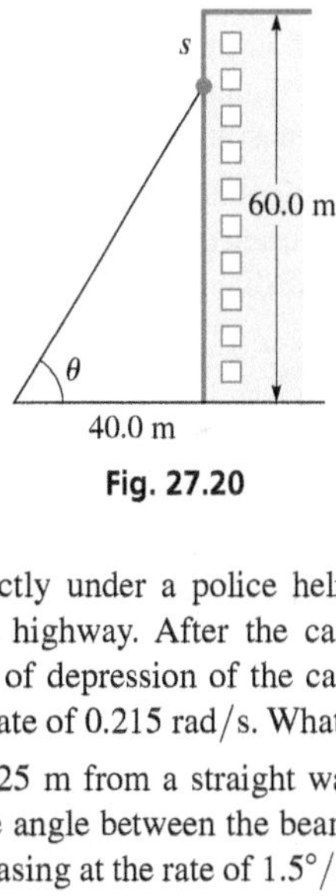

Fig. 27.20

26. A car passes directly under a police helicopter 150 m above a straight and level highway. After the car has travelled another 20.0 m, the angle of depression of the car from the helicopter is decreasing at the rate of 0.215 rad/s. What is the speed of the car?

27. A searchlight is 225 m from a straight wall. As the beam moves along the wall, the angle between the beam and the perpendicular to the wall is increasing at the rate of 1.5°/s. How fast is the length of the beam increasing when it is 315 m long? See Fig. 27.21.

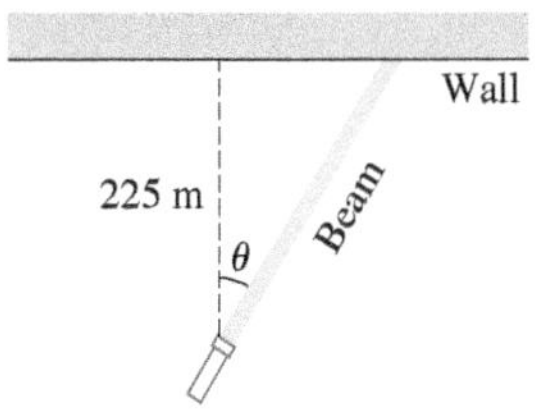

Fig. 27.21

28. A person standing 35.0 m from the base of the Skylon Tower in Niagara Falls observes one of the exterior elevators rising at a rate of 2.60 m/s. How fast is the angle of elevation of the line of sight to the elevator increasing when the elevator is 40.0 m above the ground?

29. A crate of weight w is being pulled along a level floor by a force F that is at an angle θ with the floor. The force is given by $F = \frac{0.25w}{0.25 \sin\theta + \cos\theta}$. Find θ for the minimum value of F.

30. The electric power P (in W) developed in a resistor in an FM receiver circuit is $P = 0.0307 \cos^2 120\pi t$, where t is the time (in s). Linearize P for $t = 0.0010$ s.

31. When an astronaut views the horizon of Earth from the International Space Station at an altitude of 400 km, the angle θ in Fig. 27.22 is found to be $70.2° \pm 0.5°$. Use differentials to approximate the possible error in the astronaut's calculation of Earth's radius.

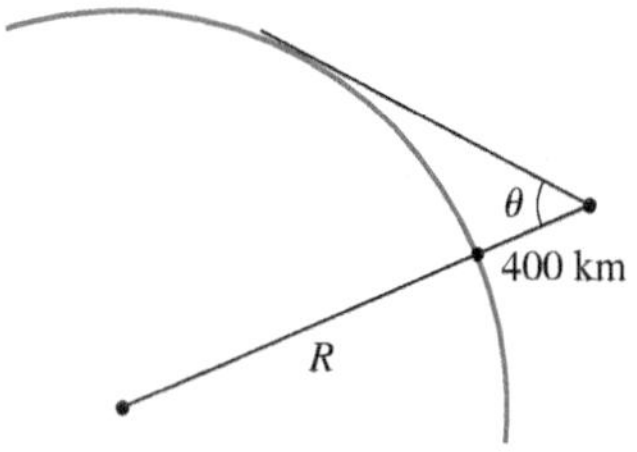

Fig. 27.22

32. A surveyor measures two sides and the included angle of a triangular parcel of land to be 82.04 m, 75.37 m, and 38.38°. What error is caused in the calculation of the third side by an error of 0.15° in the angle?

33. The volume V (in L) of air in a person's lungs during one normal cycle of inhaling and exhaling at any time t is $V = 0.48(1.2 - \cos 1.26t)$. What is the maximum flow rate (in L/s) of air?

34. To connect the four vertices of a square with the minimum amount of electric wire requires using the wiring pattern shown in Fig. 27.23. Find θ for the total length of wire ($L = 4x + y$) to be a minimum.

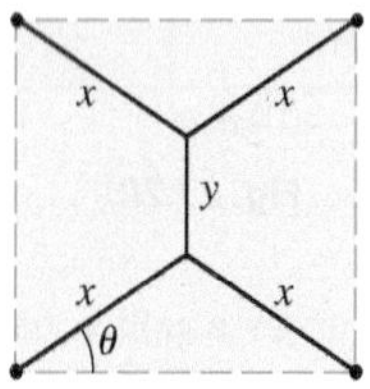

Fig. 27.23

35. The strength S of a rectangular beam is directly proportional to the product of its width w and the square of its depth d. Use trigonometric functions to find the dimensions of the strongest beam that can be cut from a circular log 16.0 cm in diameter.

36. An architect is designing a window in the shape of an isosceles triangle with a perimeter of 180 cm. What is the vertex angle of the window of greatest area?

37. A wall is 1.8 m high and 1.2 m from a building. What is the length of the shortest pole that can touch the building and the ground beyond the wall? (*Hint:* From Fig. 27.24, show that $y = 1.8 \csc\theta + 1.2 \sec\theta$.)

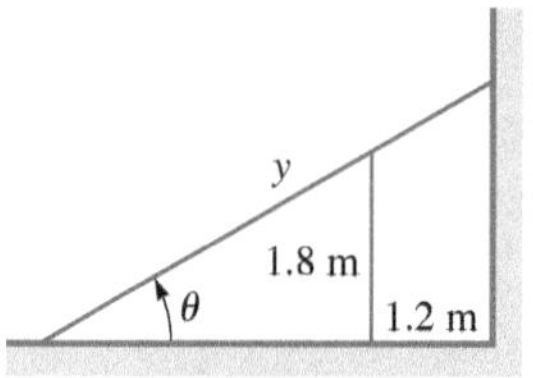

Fig. 27.24

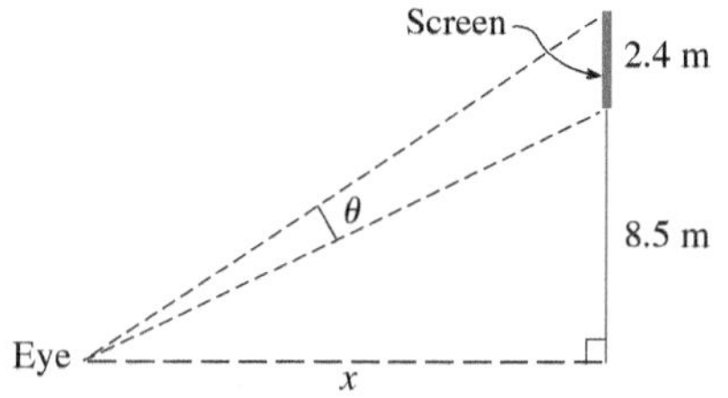

Fig. 27.25

38. The television screen at a sports arena is vertical and 2.4 m high. The lower edge is 8.5 m above an observer's eye level. If the best view of the screen is obtained when the angle subtended by the screen at eye level is a maximum, how far from directly below the screen must the observer's eye be? See Fig. 27.25.

39. A camera is on the starting line of a drag race 15.0 m from a racing car. After 1.5 s the car has travelled 30.0 m and the camera is rotating at 0.75 rad/s while filming the car. What is the speed of the car at this time?

40. What is the vertex angle at the bottom of an ice cream cone such that the cone holds a given amount of ice cream (within the cone itself) and the cone requires the least possible surface? (*Hint:* Set up equations using half the angle.)

27.5 Derivative of the Logarithmic Function

Using the definition of a derivative, we next find the derivative of the logarithmic function. Therefore, letting $y = \log_b x$, we have

$$\frac{dy}{dx} = \lim_{h\to 0}\frac{\log_b(x+h) - \log_b x}{h} = \lim_{h\to 0}\frac{\log_b \frac{x+h}{x}}{h}$$

$$= \lim_{h\to 0}\frac{1}{x}\frac{x}{h}\log_b\left(1 + \frac{h}{x}\right) \quad \text{multiply and divide by } x$$

$$= \frac{1}{x}\lim_{h\to 0}\log_b\left(1 + \frac{h}{x}\right)^{x/h} \quad \text{using the power rule for logarithms}$$

$$= \frac{1}{x}\log_b\left(\lim_{h\to 0}\left(1 + \frac{h}{x}\right)^{x/h}\right)$$

In the last line, we have interchanged the limit and the logarithm because the logarithm function is continuous. We can see that the exponent becomes unbounded, but the number being raised to this exponent approaches 1. Therefore, we will investigate this limiting value.

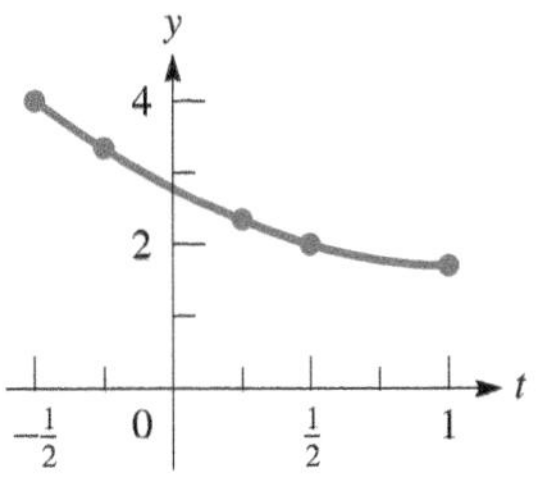

Fig. 27.26

To approximate the value, we graph the function $y = (1 + t)^{1/t}$ (for purposes of graphing, we let $h/x = t$). Constructing a table of values, we then graph this function in Fig. 27.26.

t	−0.5	−0.25	+0.25	+0.50	+1.00
y	4.00	3.16	2.44	2.25	2.00

Only these values are shown, since we are interested in the y-value corresponding to $t = 0$. We see from the graph that this value is approximately 2.7. Choosing very small values of t, we may obtain these values:

t	0.1	0.01	0.001	0.0001
y	2.5937	2.7048	2.7169	2.718 15

By methods developed in Chapter 29, it can be shown that this value is about 2.718 281 8. ***The limiting value is the irrational number e.*** This is the same number used in the exponential form of a complex number in Chapter 12 and as the base of natural logarithms in Chapter 13.

Returning to the derivative of the logarithmic function, we have

$$\frac{dy}{dx} = \frac{1}{x}\log_b\left(\lim_{h\to 0}\left(1 + \frac{h}{x}\right)^{x/h}\right) = \frac{1}{x}\log_b e$$

Now, for $y = \log_b u$, where u is a function of x, we use the chain rule. We obtain the following formula.

Derivative of the Logarithm Base *b* of a Function

$$\frac{d(\log_b u)}{dx} = \frac{1}{u}\log_b e\frac{du}{dx} \qquad \textbf{(27.13)}$$

EXAMPLE 1

Find the derivative of $y = \log 4x$.

Here $u = 4x$, so $du/dx = 4$. Then

$$\frac{dy}{dx} = \frac{d(\log 4x)}{dx} = \frac{1}{4x}(\log e)(4) \qquad \frac{du}{dx}$$

$$= \frac{1}{x}\log e \qquad \log e = 0.4343$$

If we choose e as the base of the logarithm, Eq. (27.13) is simplified.

LEARNING TIP

The choice of e as the base b makes $\log_e e = 1$; thus, this factor does not appear in Eq. (27.14). We now see why the number e is chosen as the base for the natural logarithms. The notation $\ln u$ is the same as that used in Chapter 13 for natural logarithms.

Derivative of the Natural Logarithm of a Function

$$\frac{d(\ln u)}{dx} = \frac{1}{u}\frac{du}{dx} \qquad \textbf{(27.14)}$$

EXAMPLE 2

Find the derivative of $s = \ln 3t^4$.

Here $u = 3t^4$, so $du/dx = 12t^3$. Then

$$\frac{ds}{dt} = \frac{d(\ln 3t^4)}{dt} = \frac{1}{3t^4}(12t^3) \qquad \frac{du}{dt}$$

$$= \frac{4}{t}$$

Practice Exercise

1. Find the derivative of $y = 2 \ln 5x^2$.

EXAMPLE 3 Derivative of ln tan u

Find the derivative of $y = \ln \tan 4x$.

Using Eq. (27.14), along with the derivative of the tangent, we have

$$\frac{dy}{dx} = \frac{1}{\tan 4x}\underbrace{(\sec^2 4x)(4)}_{\frac{d \tan 4x}{dx}}$$

$$= \frac{\cos 4x}{\sin 4x}\frac{4}{\cos^2 4x} \quad \text{using trigonometric relations}$$

$$= \frac{1}{\sin 4x}\frac{4}{\cos 4x} = 4 \csc 4x \sec 4x$$

LEARNING TIP

Often, finding the derivative of a logarithmic function is simplified by ***using the properties of logarithms to simplify the logarithmic expression before taking the derivative.***

EXAMPLE 4 Derivative using properties of logarithms

Find the derivative of $y = \ln\frac{x-1}{x+1}$.

In this example, it is easier to find the derivative if we write y in the form

$$y = \ln(x-1) - \ln(x+1)$$

by using the property $\log_b\left(\frac{x}{y}\right) = \log_b x - \log_b y$. Hence,

$$\frac{dy}{dx} = \frac{1}{x-1} - \frac{1}{x+1} = \frac{x+1-x+1}{(x-1)(x+1)}$$

$$= \frac{2}{x^2-1}$$

EXAMPLE 5 Derivative of ln u^3 and ln^3 u

(a) Find the derivative of $y = \ln(1-2x)^3$.

First, using the property $\log_b x^n = n \log_b x$, we rewrite the equation as $y = 3\ln(1-2x)$. Then we have

$$\frac{dy}{dx} = 3\left(\frac{1}{1-2x}\right)(-2) = \frac{-6}{1-2x}$$

(b) Find the derivative of $y = \ln^3(1-2x)$.

First, we note that

$$y = \ln^3(1-2x) = [\ln(1-2x)]^3$$

where $\ln^3(1-2x)$ is usually the preferred notation.

Next, we must be careful to distinguish this function from that in part (a). For $y = \ln^3(1-2x)$, it is the logarithm of $1-2x$ that is being cubed, whereas for $y = \ln(1-2x)^3$, it is $1-2x$ that is being cubed.

Now, finding the derivative of $y = \ln^3(1-2x)$, we have

$$\frac{dy}{dx} = 3[\ln^2(1-2x)]\underbrace{\left(\frac{1}{1-2x}\right)(-2)}_{\frac{d\ln(1-2x)}{dx}} \quad \text{using } \frac{du^n}{dx} = nu^{n-1}\left(\frac{du}{dx}\right)$$

$$= -\frac{6\ln^2(1-2x)}{1-2x}$$

Practice Exercise

2. Find the derivative of $y = \ln\frac{4x}{x+4}$.

EXAMPLE 6 Evaluation of the derivative of ln *u*

Evaluate the derivative of $y = \ln\left[(\sin 2x)\left(\sqrt{x^2+1}\right)\right]$ for $x = 0.375$.

First, using the product rule and the power rule for logarithms, we rewrite the function as

$$y = \ln \sin 2x + \frac{1}{2}\ln(x^2+1)$$

Now, we have

$$\frac{dy}{dx} = \frac{1}{\sin 2x}(\cos 2x)(2) + \frac{1}{2}\left(\frac{1}{x^2+1}\right)(2x) \quad \text{take the derivative}$$

$$= 2\cot 2x + \frac{x}{x^2+1}$$

$$\left.\frac{dy}{dx}\right|_{x=0.375} = 2\cot 0.750 + \frac{0.375}{0.375^2+1} = 2.48 \quad \text{evaluate}$$

EXERCISES 27.5

In Exercises 1 and 2, make the given changes in the indicated examples of this section and then find the derivatives.

1. In Example 3, in the given function, change tan to cos.
2. In Example 4, in the given function, change $x - 1$ to x^2.

In Exercises 3–34, find the derivatives of the given functions.

3. $y = \log x^4$
4. $y = \log_2 9x$
5. $y = 4\log_5(3 - x)$
6. $y = \log_7(x^2 + 1)$
7. $u = 2\ln(3 - x)^4$
8. $y = 2\ln(3x^2 - 1)$
9. $y = 2\ln\tan 2x$
10. $s = \ln\sin^2 t$
11. $R = \ln\sqrt{4T + 1}$
12. $y = \ln(4x - 3)^3$
13. $y = \ln(x - x^2)^5$
14. $s = 3\ln^2(7t^3 - 1)$
15. $v = 3(t + \ln t^2)^2$
16. $y = 6x^2\ln 5x$
17. $y = 3x\ln(6 - x)$
18. $y = \dfrac{8\ln x}{x}$
19. $y = \ln(\ln x)$
20. $y = \ln\dfrac{2x}{1 + x}$
21. $r = 0.5\ln\sin(\pi\theta^2)$
22. $y = \ln\left(x\sqrt{x + 1}\right)$
23. $y = \cos\ln x$
24. $y = \tan^{-1}\ln 2x$
25. $u = 3v\ln^2 2v$
26. $h = 0.1s\ln^4 s$
27. $y = \ln(x\tan x)$
28. $y = \ln\left(x + \sqrt{x^2 - 1}\right)$
29. $r = \ln\dfrac{v^2}{v + 2}$
30. $y = \sqrt{x + \ln 3x}$
31. $y = \sqrt{x^2 + 1} - \ln\dfrac{1 + \sqrt{x^2 + 1}}{x}$
32. $3\ln xy + \sin y = x^2$
33. $y = x - \ln^2(x + y)$
34. $y = \ln(x^2 + 2\ln x)$

In Exercises 35–56, solve the given problems.

35. On a calculator, find the value of $(\ln 2.0001 - \ln 2.0000)/0.0001$ and compare it with 0.5. Give the meanings of the value found and 0.5 in relation to the derivative of $\ln x$, where $x = 2$.
36. On a calculator, find the value of $(\ln 0.5001 - \ln 0.5000)/0.0001$ and compare it with 2. Give the meanings of the value found and 2 in relation to the derivative of $\ln x$, where $x = 0.5$.
37. Using a graphing utility, (a) display the graph of $y = (1 + x)^{1/x}$ to verify that $(1 + x)^{1/x} \to 2.718$ as $x \to 0$ and (b) verify the values for $(1 + x)^{1/x}$ in the tables at the beginning of Section 27.5.
38. (a) Display the graph of $y = \ln x$ on a graphing utility, and using the derivative feature, evaluate dy/dx for $x = 2$. (b) Display the graph of $y = 1/x$, and evaluate y for $x = 2$. (c) Compare the values in parts (a) and (b).
39. Given that $\ln\sin 45° = -0.3466$, use differentials to approximate $\ln\sin 44°$.
40. Find the second derivative of the function $y = x^2\ln x$.
41. Evaluate the derivative of $y = \sin^{-1}2x + \ln\sqrt{1 - 4x^2}$, where $x = 0.250$.
42. Evaluate the derivative of $y = \ln\sqrt{\dfrac{2x + 1}{3x + 1}}$, where $x = 2.75$.
43. Find the linearization $L(x)$ for the function $f(x) = 2\ln\tan x$ for $a = \pi/4$.
44. Find the differential of the function $y = 6\log_x 2$.
45. Find the slope of a line tangent to the curve of $y = \tan^{-1}2x + \ln(4x^2 + 1)$, where $x = 0.625$.
46. Find the slope of a line tangent to the curve of $y = x\ln 3x$ at $x = 4$.
47. Find the derivative of $y = x^x$ by first taking logarithms of each side of the equation. Explain why the general power rule equation $\dfrac{du^n}{dx} = nu^{n-1}\left(\dfrac{du}{dx}\right)$ cannot be used to find the derivative of this function.
48. Find the derivative of $y = (\sin x)^x$ by first taking logarithms of each side of the equation. Explain why the general power rule equation from Exercise 47 cannot be used to find the derivative of this function.
49. Find the derivatives of $y_1 = \ln(x^2)$ and $y_2 = 2\ln x$, and evaluate these derivatives for $x = -1$. Explain your results.

50. The inductance L (in μH) of a coaxial cable is given by $L = 0.032 + 0.15\log(a/x)$, where a and x are the radii of the outer and inner conductors, respectively. For constant a, find dL/dx.

51. If the loudness b (in decibels) of a sound of intensity I is given by $b = 10\log(I/I_0)$, where I_0 is a constant, find the expression for db/dt in terms of dI/dt.

52. The time t for a particular computer system to process N bits of data is directly proportional to $N \ln N$. Find the expression for dt/dN.

53. When a tractor-trailer turns a right-angle corner, the rear wheels follow a curve known as a *tractrix*, the equation for which is $y = \ln\left(\frac{1 + \sqrt{1 + x^2}}{x}\right) - \sqrt{1 - x^2}$. Find dy/dx.

54. When designing a computer to sort files on a hard disk, the equation $y = xA\log_x A$ arises. If A is constant, find dy/dx.

55. When air friction is considered, the time t (in s) it takes a certain falling object to attain a velocity v (in m/s) is given by $t = 5\ln\frac{5}{5 - 0.1v}$. Find dt/dv for $v = 10.0$ m/s.

56. The electric potential V at a point P at a distance x from an electric charge distributed along a wire of length $2a$ (see Fig. 27.27) is $V = k\ln\frac{\sqrt{a^2 + x^2} + a}{\sqrt{a^2 + x^2} - a}$, where k is a constant. Find the expression for the electric field E, where $E = -dV/dx$.

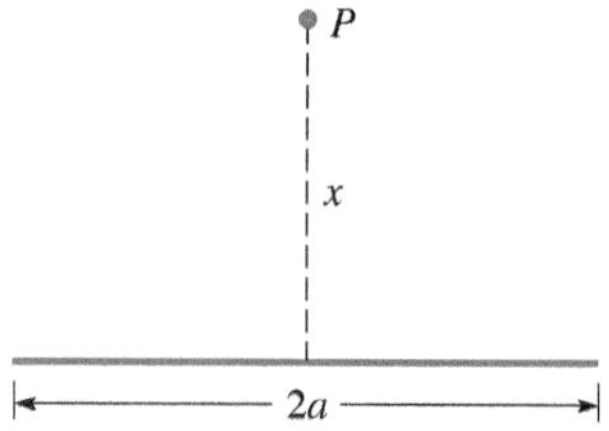

Fig. 27.27

Answers to Practice Exercises

1. $y' = 4/x$ **2.** $y' = 4/(x^2 + 4x)$

27.6 Derivative of the Exponential Function

To obtain the derivative of the exponential function, we let $y = b^u$, then take natural logarithms of both sides, and then take derivatives of both sides:

$$\ln y = \ln b^u = u \ln b$$

$$\frac{1}{y}\frac{dy}{dx} = \ln b \frac{du}{dx}$$

$$\frac{dy}{dx} = y \ln b \frac{du}{dx}$$

Substituting $y = b^u$, we can state the following.

Derivative of the Exponential Function Base b

$$\frac{d(b^u)}{dx} = b^u \ln b\left(\frac{du}{dx}\right) \quad \textbf{(27.15)}$$

EXAMPLE 1

Find the derivative of $y = 10^{2x}$.

Here $b = 10$, $u = 2x$, and $du/dx = 2$. Then

$$\frac{dy}{dx} = \frac{d(10^{2x})}{dx} = 10^{2x}(\ln 10)\frac{d(2x)}{dx}$$
$$= (10^{2x})(\ln 10)(2) = (2\ln 10)(10^{2x})$$

If we choose the base e, Eq. (27.15) is simplified.

Derivative of the Exponential Function Base e

$$\frac{d(e^u)}{dx} = e^u\left(\frac{du}{dx}\right) \quad \textbf{(27.16)}$$

EXAMPLE 2

Find the derivative of $y = e^x$.

Here $u = x$, so $du/dx = 1$. Then

$$\frac{dy}{dx} = \frac{d(e^x)}{dt} = e^x(1) = e^x$$

(annotation: (1) is $\frac{du}{dx}$)

We see from Example 2 that the derivative of the function e^x equals itself. This fact and the simplicity of Eq. (27.16) compared with Eq. (27.15) again show the advantage of choosing e as the base of natural logarithms. It is for this reason that e and the function e^x are widely used in applications of calculus.

LEARNING TIP

Note carefully that the power rule is used with a variable raised to a constant exponent, whereas the derivatives of exponential functions, Eqs. (27.15) and (27.16), are used with a constant raised to a variable exponent.

variable constant — constant variable

$$\frac{du^n}{dx} = nu^{n-1}\left(\frac{du}{dx}\right) \qquad \frac{db^u}{dx} = b^u \ln b\left(\frac{du}{dx}\right)$$

In the following example, we must note carefully this difference when choosing the rule to find the derivative.

EXAMPLE 3 Derivative of u^2 and 2^u

Find the derivatives of $y = (4x)^2$ and $y = 2^{4x}$.

Using the power rule, we have

$$y = (4x)^2 \qquad \frac{du}{dx}$$
$$\frac{dy}{dx} = 2(4x)^1(4)$$
$$= 32x$$

Using Eq. (27.15), we have

$$y = 2^{4x} \qquad \frac{du}{dx}$$
$$\frac{dy}{dx} = 2^{4x}(\ln 2)(4)$$
$$= (4 \ln 2)(2^{4x})$$

EXAMPLE 4 Derivative of ln cos e^u

Find the derivative of $y = \ln \cos e^{2x}$.

Here, we use Eq. (27.16) and the derivatives of the logarithmic and cosine functions.

$$\frac{dy}{dx} = \frac{1}{\cos e^{2x}}\frac{d \cos e^{2x}}{dx} \qquad \text{using } \frac{d \ln u}{dx} = \frac{1}{u}\frac{du}{dx}$$

$$= \frac{1}{\cos e^{2x}}(-\sin e^{2x})\frac{de^{2x}}{dx} \qquad \text{using } \frac{d \cos u}{dx} = -\sin u\frac{du}{dx}$$

$$= -\frac{\sin e^{2x}}{\cos e^{2x}}(e^{2x})(2) \qquad \text{using } \frac{de^u}{dx} = e^u\frac{du}{dx}$$

$$= -2e^{2x} \tan e^{2x} \qquad \text{using } \frac{\sin \theta}{\cos \theta} = \tan \theta$$

Practice Exercise

1. Find the derivative of $y = \ln(e^{3x} + 1)$.

EXAMPLE 5 Derivative of a product

Find the derivative of $r = \theta e^{\tan \theta}$.

Here, we use Eq. (27.16) with the derivatives of a product and the tangent:

$$\frac{dr}{d\theta} = \theta e^{\tan \theta}(\sec^2 \theta) + e^{\tan \theta}(1)$$
$$= e^{\tan \theta}(\theta \sec^2 \theta + 1)$$

Practice Exercise

2. Find the derivative of $y = 8e^{4x} \cos x$.

EXAMPLE 6 Using the laws of exponents

Find the derivative of $y = (e^{1/x})^2$.

One approach is to use the power rule directly on the function $y = u^2$, with $u = e^{1/x}$. We have $dy/du = 2u$, and Eq. (27.16) gives $du/dx = (e^{1/x})\left(-\frac{1}{x^2}\right)$. Hence

$$\frac{dy}{dx} = 2(e^{1/x})(e^{1/x})\left(-\frac{1}{x^2}\right) = \frac{-2(e^{1/x})^2}{x^2} = \frac{-2e^{2/x}}{x^2}$$

We can also find this derivative by first writing $y = e^{2/x}$ and then using Eq. (27.16) with $u = 2/x$, so $du/dx = -2/x^2$. This simplifies the steps significantly:

$$\frac{dy}{dx} = e^{2/x}\left(-\frac{2}{x^2}\right) = \frac{-2e^{2/x}}{x^2}$$

EXAMPLE 7 Evaluation of the slope of a tangent

Find the slope of a line tangent to the curve of $y = \frac{3e^{2x}}{x^2 + 1}$ at $x = 1.275$.

Here, we are to evaluate the derivative for $x = 1.275$. The solution is as follows:

$$\frac{dy}{dx} = \frac{(x^2 + 1)(3e^{2x})(2) - 3e^{2x}(2x)}{(x^2 + 1)^2} \quad \text{take the derivative}$$

$$= \frac{6e^{2x}(x^2 - x + 1)}{(x^2 + 1)^2}$$

$$\left.\frac{dy}{dx}\right|_{x=1.275} = \frac{6e^{2(1.275)}(1.275^2 - 1.275 + 1)}{(1.275^2 + 1)^2} = 15.05 \quad \text{evaluate}$$

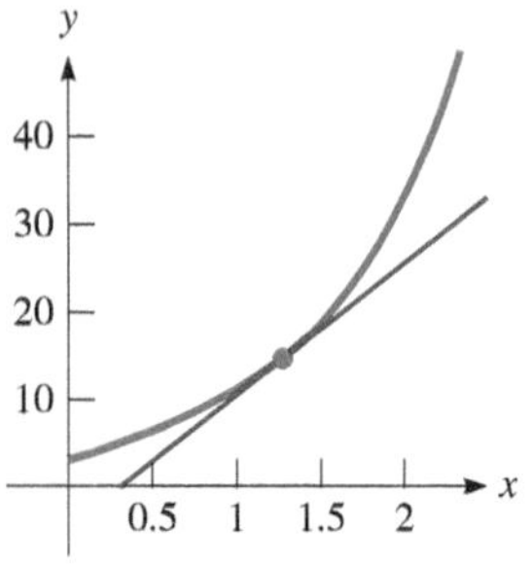

Fig. 27.28

The function and the tangent line are shown in Fig. 27.28.

EXAMPLE 8 Derivative of a power

Find the derivative of $y = (3e^{4x} + 4x^2 \ln x)^3$.

Using the general power rule, the derivative of the exponential function, the derivative of a product, and the derivative of a logarithm, we have

$$\frac{dy}{dx} = 3(3e^{4x} + 4x^2 \ln x)^2\left[12e^{4x} + 4x^2\left(\frac{1}{x}\right) + (\ln x)(8x)\right]$$

$$= 3(3e^{4x} + 4x^2 \ln x)^2(12e^{4x} + 4x + 8x \ln x)$$

EXERCISES 27.6

In Exercises 1 and 2, make the given changes in the indicated examples of this section and then solve the resulting problems.

1. In Example 4, in the given function, change cos to sin.
2. In Example 7, in the given function, change $x^2 + 1$ to $x + 1$.

In Exercises 3–32, find the derivatives of the given functions.

3. $y = 4^{6x}$
4. $y = 10^{x^6}$
5. $y = 6e^{\sqrt{x}}$
6. $r = 0.3e^{\theta^2}$
7. $y = 4e^t(e^{2t} - e^t)$
8. $y = 0.6 \ln(e^{5x} + 3)$
9. $R = Te^{-T}$
10. $y = 5x^2e^{2x}$
11. $y = xe^{\sin x}$
12. $y = 4e^x \sin \frac{1}{2}x$
13. $r = \frac{2(e^{2s} - e^{-2s})}{e^{2s}}$
14. $u = \frac{e^{0.5v}}{2v}$
15. $y = e^{-3x} \cos 4x$
16. $y = (\sin 2x)(e^{x^2-1})$
17. $y = \frac{2e^{3x}}{4x + 3}$
18. $y = \frac{7 \ln 3x}{e^{2x} + 8}$
19. $y = 0.5 \ln(e^{t^2} + 4)$
20. $p = (3e^{2n} + e^2)^8$
21. $y = (2e^{2x})^3 \sin x^2$
22. $y = (e^{3/x} \cos x)^2$
23. $u = 4\sqrt{\ln 2t + e^{2t}}$
24. $y = (2e^{x^2} + x^2)^{24}$
25. $y = e^{xy}$
26. $y = 4e^{-2/x} \ln y + 1$
27. $y = e^{2x} \ln x^3$
28. $r = 0.4e^{2\theta} \ln \cos \theta$
29. $I = \ln \sin 2e^{6t}$
30. $y = 6 \tan e^{4-x}$
31. $y = 2 \sin^{-1} e^{2x}$
32. $W = \tan^{-1} e^{3s}$

In Exercises 33–54, solve the given problems. Where necessary, round answers to three significant digits.

33. On a calculator, find the values of (a) e and (b) $(e^{1.0001} - e^{1.0000})/0.0001$. Compare the values and give the meaning of each in relation to the derivative of e^x, where $x = 1$.

34. On a calculator, find the values of (a) e^2 and (b) $(e^{2.0001} - e^{2.0000})/0.0001$. Compare the values and give the meaning of each in relation to the derivative of e^x, where $x = 2$.

35. Display the graph of $y = e^x$ on a graphing utility. Using the *derivative* feature, evaluate dy/dx for $x = 2$ and compare with the value of y for $x = 2$.

36. Find a formula for the nth derivative of $y = ae^{bx}$.

37. Find the slope of a line tangent to the curve of $y = e^{-x/2}\cos 4x$ for $x = 0.625$.

38. Find the slope of a line tangent to the curve of $y = \dfrac{e^{-x}}{1 + \ln 4x}$ for $x = 1.842$.

39. Find the differential of the function $y = \dfrac{12e^{4x}}{x + 6}$.

40. Find the linearization of the function $f(x) = \dfrac{6e^{4x}}{2x + 3}$ for $a = 0$.

41. Use a graphing utility to display the graph of $y = e^x$. By roughly estimating slopes of tangent lines, note that it is reasonable that these values are equal to the y-coordinates of the points at which these estimates are made. (*Remember:* For $y = e^x$, $dy/dx = e^x$ also.)

42. Use a graphing utility to display the graphs of $y = e^{-x}$ and $y = -e^{-x}$. By roughly estimating slopes of tangent lines of $y = e^{-x}$, note that $y = -e^{-x}$ gives reasonable values for the derivative of $y = e^{-x}$.

43. Show that $y = xe^{-x}$ satisfies the equation $(dy/dx) + y = e^{-x}$.

44. Show that $y = e^{-x}\sin x$ satisfies the equation
$$\frac{d^2y}{dx^2} + 2\frac{dy}{dx} + 2y = 0.$$

45. For $y = \dfrac{e^{2x} - 1}{e^{2x} + 1}$, show that $\dfrac{dy}{dx} = 1 - y^2$.

46. If $e^x + e^y = e^{x+y}$, show that $dy/dx = -e^{y-x}$.

47. For what values of m does the function $y = ae^{mx}$ satisfy the equation $y'' + y' - 6y = 0$?

48. For what values of m does the function $y = ae^{mx}$ satisfy the equation $y'' + 4y = 0$?

49. If $y = Ae^{kx} + Be^{-kx}$, show that $y'' = k^2y$.

50. The average energy consumption C (in MJ/year) of a certain model of refrigerator–freezer is approximately $C = 5350e^{-0.0748t} + 1800$, where t is measured in years, with $t = 0$ corresponding to 1990, and a newer model is produced each year. Assuming the function is continuous, use differentials to estimate the reduction in energy consumption of the 2012 model from that of the 2011 model.

51. The reliability R $(0 \leq R \leq 1)$ of a certain computer system is given by $R = e^{-0.002t}$, where t is the time of operation (in h). Find dR/dt for $t = 100$ h.

52. A thermometer is taken from a freezer at $-16°C$ and placed in a room at 24°C. The temperature T of the thermometer as a function of the time t (in min) after removal is given by $T = 8.0(3.0 - 5.0e^{-0.50t})$. How fast is the temperature changing when $t = 6.0$ min?

53. For the electric circuit shown in Fig. 27.29, the current i (in A) is given by $i = 4.42e^{-66.7t}\sin 226t$, where t is the time (in s). Find the expression for di/dt.

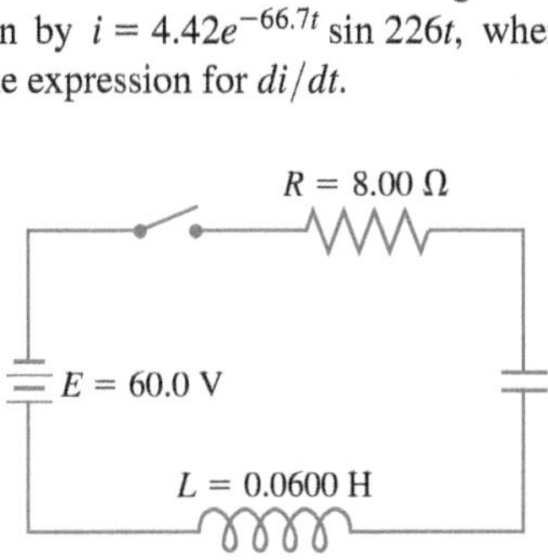

Fig. 27.29

54. Under certain assumptions of limitations to population growth, the population P (in billions) of the world is given by the *logistic equation* $P = \dfrac{10}{1 + 0.65e^{-0.036t}}$, where t is the number of years after the year 2010. Find the expression for dP/dt.

In Exercises 55–59, use the following information. The **hyperbolic sine** *of u is defined as* $\sinh u = \frac{1}{2}(e^u - e^{-u})$. *Fig. 27.30 shows the graph of* $y = \sinh x$. *The* **hyperbolic cosine** *of u is defined as* $\cosh u = \frac{1}{2}(e^u + e^{-u})$. *Fig. 27.31 shows the graph of* $y = \cosh x$. *These functions are called hyperbolic functions since, if* $x = \cosh u$ *and* $y = \sinh u$, *x and y satisfy the equation of the hyperbola* $x^2 - y^2 = 1$.

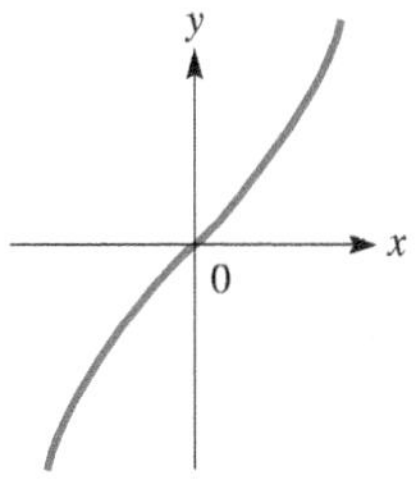

Fig. 27.30

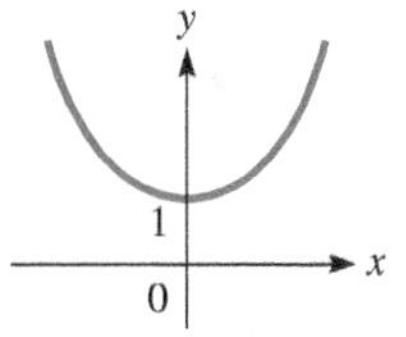

Fig. 27.31

55. Verify the fact that the exponential expressions for the hyperbolic sine and hyperbolic cosine given above satisfy the equation of the hyperbola.

56. Show that $\sinh u$ and $\cosh u$ satisfy the identity $\cosh^2 u - \sinh^2 u = 1$.

57. Show that $\dfrac{d}{dx}\sinh u = \cosh u\dfrac{du}{dx}$ and $\dfrac{d}{dx}\cosh u = \sinh u\dfrac{du}{dx}$ where u is a function of x.

58. Show that $\dfrac{d^2 \sinh x}{dx^2} = \sinh x$ and $\dfrac{d^2 \cosh x}{dx^2} = \cosh x$.

59. A telephone wire hangs so that its shape is described by $y = 50\cosh\dfrac{x}{50}$. Find the expression for dy/dx. (See Exercise 57.)

Answers to Practice Exercises

1. $y' = 3e^{3x}/(e^{3x} + 1)$
2. $y' = 8e^{4x}(4\cos x - \sin x)$

27.7 L'Hospital's Rule

We dedicate this section to finding limits of quotients where both numerator and denominator approach zero or where both numerator and denominator approach infinity. We have already encountered this type of expression in Section 23.1, where we evaluated the limit using algebraic techniques, and in Section 27.1, where we found that $\lim_{\theta \to 0} \frac{\sin \theta}{\theta} = 1$ using a geometric approach. Now we study a more general method to evaluate such limits by using derivatives.

For the limit of $u(x)/v(x)$ as $x \to a$, *if both* $u(x)$ *and* $v(x)$ *approach zero, the limit is called an* **indeterminate form of the type 0/0**. *If* $u(x)$ and $v(x)$ *approach infinity, the limit is called an* **indeterminate form of the type ∞/∞**.

To find these limits, we use ***L'Hospital's rule***. The proof of L'Hospital's rule can be found in texts covering advanced methods in calculus.

■ The method is named after the French mathematician the Marquis de l'Hospital (1661–1704), who incorporated it in the first textbook on differential calculus to appear in print (in 1696). However, the result was derived two years earlier by his calculus tutor, Johann Bernoulli (1667–1748).

L'Hospital's Rule

If $u(x)$ and $v(x)$ are differentiable such that $v'(x)$ is not zero, and if $\lim_{x \to a} u(x) = 0$ and $\lim_{x \to a} v(x) = 0$ or $\lim_{x \to a} u(x) = \pm\infty$ and $\lim_{x \to a} v(x) = \pm\infty$, then

$$\lim_{x \to a} \frac{u(x)}{v(x)} = \lim_{x \to a} \frac{u'(x)}{v'(x)}$$

Applying L'Hospital's Rule	EXAMPLE 1
	Find $\lim_{x \to 0} \frac{\sin x}{x}$.
1. Verify that the limit of $u(x)/v(x)$ is an indeterminate form $0/0$ or $\pm\infty/\pm\infty$.	$\lim_{x \to 0} \sin x \to 0$ and $\lim_{x \to 0} x = 0$, which means the limit is an indeterminate form $0/0$.
2. Take the derivative of $u(x)$ and $v(x)$ separately.	$\frac{d(\sin x)}{dx} = \cos x$ and $\frac{d(x)}{dx} = 1$
3. Find the limit of $u'(x)/v'(x)$. If this limit is finite, $+\infty$, or $-\infty$, then it is the limit of $u(x)/v(x)$. If the limit is an indeterminate form, then use L'Hospital's rule again.	$\lim_{x \to 0} \frac{\cos x}{1} = 1$ Since this limit is finite, $\lim_{x \to 0} \frac{\sin x}{x} = 1$ This agrees with the result in Section 27.1.

Practice Exercise

1. Find $\lim_{x \to 0} \frac{1 - e^{4x}}{2x}$.

EXAMPLE 2 L'Hospital's rule—indeterminate form ∞/∞

Evaluate $\lim_{x \to \infty} \frac{3e^{2x}}{5x}$.

Verify that the limit of $u(x)/v(x)$ is an indeterminate form $0/0$ or ∞/∞: $3e^{2x} \to \infty$ and $5x \to \infty$ as $x \to \infty$, which means the limit is an indeterminate form ∞/∞.

Take derivatives of $u(x)$ and $v(x)$: $\frac{d(3e^{2x})}{dx} = 6e^{2x}$ and $\frac{d(5x)}{dx} = 5$

Find the limit of $u'(x)/v'(x)$:

$$\lim_{x\to\infty} \frac{6e^{2x}}{5} = \infty$$

Therefore, $\lim_{x\to\infty} \frac{3e^{2x}}{5x} = \infty$. Note that this means the limit does not exist.

Practice Exercise

2. Find $\lim_{x\to\infty} \frac{\ln x}{x}$.

EXAMPLE 3 L'Hospital's rule—limit as $t \to \pi/2$

Evaluate $\lim_{t\to\pi/2} \frac{1 - \sin t}{\cos t}$.

Verify that the limit of $u(t)/v(t)$ is an indeterminate form $0/0$ or ∞/∞:

$\lim_{t\to\pi/2} (1 - \sin t) = 0$ and $\lim_{t\to\pi/2} \cos t \to 0$, so the limit is an indeterminate form $0/0$.

Take derivatives of $u(t)$ and $v(t)$:

$$\frac{d(1 - \sin t)}{dt} = -\cos t \text{ and } \frac{d(\cos t)}{dt} = -\sin t$$

Find the limit of $u'(t)/v'(t)$:

$$\lim_{t\to\pi/2} \frac{-\cos t}{-\sin t} = \frac{0}{-1} = 0$$

Since the limit is finite, $\lim_{t\to\pi/2} \frac{1 - \sin t}{\cos t} = 0$

EXAMPLE 4 Using L'Hospital's rule twice

Using L'Hospital's rule, find $\lim_{x\to\infty} \frac{x^2 + 1}{2x^2 + 3x}$.

Verify that the limit of $u(x)/v(x)$ is an indeterminate form $0/0$ or ∞/∞:

$x^2 + 1 \to \infty$ and $2x^2 + 3x \to \infty$ as $x \to \infty$, which means the limit is an indeterminate form ∞/∞.

Take derivatives of $u(x)$ and $v(x)$:

$$\frac{d(x^2 + 1)}{dx} = 2x \text{ and } \frac{d(2x^2 + 3x)}{dx} = 4x + 3$$

Find the limit of $u'(x)/v'(x)$. If the limit is an indeterminate form, then use L'Hospital's rule again.

$$\lim_{x\to\infty} \frac{2x}{4x + 3} = \frac{\infty}{\infty}$$

Since the limit is an indeterminate form ∞/∞, we apply L'Hospital's rule again. This time the derivatives are

$$\frac{d(2x)}{dx} = 2 \quad \text{and} \quad \frac{d(4x + 3)}{dx} = 4$$

This gives

$$\lim_{x\to\infty} \frac{2}{4} = \frac{1}{2}$$

Since the limit is finite, we get

$$\lim_{x\to\infty} \frac{x^2 + 1}{2x^2 + 3x} = \frac{1}{2}$$

COMMON ERROR

When using L'Hospital's rule, ***always verify that the limit fits the indeterminate form of 0/0 or ∞/∞ before you apply the rule.*** If you use the rule when it does not apply, it will lead to incorrect calculations.

EXAMPLE 5 Be careful to check the indeterminate form

Find $\lim_{x \to \pi^-} \frac{2 \sin x}{1 - \cos x}$.

Verify that the limit of $u(x)/v(x)$ is an indeterminate form $0/0$ or ∞/∞: As x approaches π from values below π,

$$\lim_{x \to \pi^-} 2 \sin x = 0 \text{ and } \lim_{x \to \pi^-} 1 - \cos x = 2.$$

These limits do not give an indeterminate form, so **we cannot apply L'Hospital's rule.**

Since this function is continuous as $x \to \pi^-$, we can use direct substitution to get

$$\lim_{x \to \pi^-} \frac{2 \sin x}{1 - \cos x} = \frac{2 \sin \pi}{1 - \cos \pi} = \frac{0}{1 - (-1)} = 0$$

Had we applied L'Hospital's rule, we would have obtained the **incorrect result**

$$\lim_{x \to \pi^-} \frac{2 \sin x}{1 - \cos x} = \lim_{x \to \pi^-} \frac{\frac{d}{dx} 2 \sin x}{\frac{d}{dx}(1 - \cos x)} = \lim_{x \to \pi^-} \frac{2 \cos x}{\sin x} = -\infty$$

We can verify the value of a limit found using L'Hospital's rule by successive evaluation. The table derived in Example 6 can be obtained using a spreadsheet or the *table* feature of some graphing calculators.

EXAMPLE 6 Numerical verification of a limit

Find $\lim_{x \to \pi} \frac{1 + \cos x}{(\pi - x)^2}$ and verify the value of the limit numerically using values of x that approach π.

Because $\cos \pi = -1$, we see that both the numerator and the denominator approach zero as x approaches π. Therefore, we have

$$\lim_{x \to \pi} \frac{1 + \cos x}{(\pi - x)^2} = \lim_{x \to \pi} \frac{\frac{d}{dx}(1 + \cos x)}{\frac{d}{dx}(\pi - x)^2} = \lim_{x \to \pi} \frac{-\sin x}{-2(\pi - x)}$$

indeterminate form $0/0$; use L'Hospital's rule again

$$= \lim_{x \to \pi} \frac{\frac{d}{dx}(-\sin x)}{\frac{d}{dx}[-2(\pi - x)]} = \lim_{x \to \pi} \frac{-\cos x}{2} = \frac{-(-1)}{2} = \frac{1}{2}$$

To verify the limit numerically, we form a table of values of $(1 + \cos x)/(\pi - x)^2$ using values of x that approach π. Such a table is shown in Fig. 27.32. We see that the limit $1/2$ is verified.

x	$\frac{1 + \cos x}{(\pi - x)^2}$
3.1376	0.499 999 336
3.1386	0.499 999 627
3.1396	0.499 999 835
3.1406	0.499 999 954
π	Error
3.1426	0.499 999 956
3.1436	0.499 999 832
3.1446	0.499 999 624
3.1456	0.499 999 331

Fig. 27.32

There are indeterminate forms other than $0/0$ and ∞/∞. One of these is illustrated in the following example, and others are covered in the exercises.

EXAMPLE 7 Indeterminate form $0 \cdot \infty$

Find $\lim_{x \to 0^+} x^2 \ln x$.

As $x \to 0^+$, we note that $x^2 \to 0$ and $\ln x \to -\infty$. This is the indeterminate form $0 \cdot \infty$. By writing x^2 as $1/(1/x^2)$ we can change this indeterminate form to ∞/∞. Making this change, and then applying L'Hospital's rule, we have

$$\lim_{x \to 0^+} x^2 \ln x = \lim_{x \to 0^+} \frac{\ln x}{\frac{1}{x^2}} = \lim_{x \to 0^+} \frac{\frac{d}{dx}\ln x}{\frac{d}{dx}x^{-2}} = \lim_{x \to 0^+} \frac{\frac{1}{x}}{-\frac{2}{x^3}} = \lim_{x \to 0^+}\left(-\frac{x^2}{2}\right) = 0$$

EXERCISES 27.7

In Exercises 1 and 2, make the given changes in the indicated examples of this section and then solve the resulting problems.

1. In Example 2, change the denominator $5x$ to $5 \ln x$.
2. In Example 5, change the denominator $1 - \cos x$ to $1 + \cos x$.

In Exercises 3–32, evaluate each limit (if it exists). Use L'Hospital's rule (if appropriate).

3. $\lim_{x \to 3} \frac{x - 3}{x^2 - 9}$
4. $\lim_{x \to 1} \frac{x^5 - 1}{x^3 - 1}$
5. $\lim_{\theta \to 0} \frac{\tan \theta}{\theta}$
6. $\lim_{x \to \infty} \frac{e^x}{x^2}$
7. $\lim_{x \to \infty} \frac{x \ln x}{x + \ln x}$
8. $\lim_{x \to 1} \frac{\ln x}{x - 1}$
9. $\lim_{t \to \pi/4} \frac{1 - \sin 2t}{\frac{\pi}{4} - t}$
10. $\lim_{x \to 0} \frac{\ln \cos x}{x}$
11. $\lim_{x \to 0} \frac{\ln 2x}{x^{-1}}$
12. $\lim_{\theta \to \pi/2} \frac{1 + \sec \theta}{\tan \theta}$
13. $\lim_{x \to 0} \frac{\sin x - x}{x^3}$
14. $\lim_{x \to \infty} \frac{x^2 + x}{e^x + 1}$
15. $\lim_{x \to 1} \frac{\sin \pi x}{x - 1}$
16. $\lim_{x \to 0} \frac{\tan^{-1} x}{x}$
17. $\lim_{x \to 0} x \cot x$
18. $\lim_{z \to +\infty} z e^{-z}$
19. $\lim_{x \to 0^+} x \ln \sin x$
20. $\lim_{x \to \pi/2} (1 - \sin x)\tan x$
21. $\lim_{x \to 0} \frac{2 \sin x}{5e^x}$
22. $\lim_{x \to \infty} \frac{e^{8x} - 20}{4x + 1}$
23. $\lim_{x \to +\infty} \frac{1 + e^{2x}}{2 + \ln x}$
24. $\lim_{x \to \pi/2} \frac{\frac{\pi}{2} - x}{1 + \sin x}$
25. $\lim_{x \to 0} \frac{\ln \sin x}{\ln \tan x}$
26. $\lim_{t \to +\infty} \frac{\ln \ln t}{\ln t}$
27. $\lim_{x \to 0^+} (\sin x)(\ln x)$
28. $\lim_{x \to 0^+} (\sin^{-1} x)(\ln x)$
29. $\lim_{x \to 2} \frac{2x^3 - 7x^2 + 11x - 10}{x^3 - 5x^2 + 7x - 2}$
30. $\lim_{x \to 1} \frac{x^4 - x^3 - 3x^2 + 5x - 2}{x^4 - 5x^3 + 9x^2 - 7x + 2}$
31. $\lim_{x \to 0} \frac{\sqrt{1 - x} - \sqrt{1 + x}}{x}$
32. $\lim_{x \to 0} \frac{e^x + e^{-x} - 2}{1 - \cos 2x}$

In Exercises 33–42, solve the given problems.

33. Another indeterminate form is $\infty - \infty$. Often, it is possible to make an algebraic or trigonometric change in the function so that it will take on the form of a $0/0$ or ∞/∞ indeterminate form. Find $\lim_{\theta \to \frac{\pi}{2}^-} (\sec \theta - \tan \theta)$.
34. Find $\lim_{x \to 0}\left(\frac{1}{\sin x} - \frac{1}{x}\right)$. (See Exercise 31.)
35. Three other indeterminate forms are 0^0, ∞^0, and 1^∞. For the function $y = [f(x)]^{g(x)}$ and the following limits, we have the indicated indeterminate form:

$\lim_{x \to a} f(x) = 0$ and $\lim_{x \to a} g(x) = 0$ (indet. form 0^0)

$\lim_{x \to a} f(x) = \infty$ and $\lim_{x \to a} g(x) = 0$ (indet. form ∞^0)

$\lim_{x \to a} f(x) = 1$ and $\lim_{x \to a} g(x) = \infty$ (indet. form 1^∞)

By taking the logarithm of $y = [f(x)]^{g(x)}$, we have

$\ln y = g(x) \ln f(x)$

and in each case the right-hand member is a type that can be solved by L'Hospital's rule. By knowing the limit of $\ln y$, we can find the limit of y.

Find $\lim_{x \to 0} x^x$.

36. Find $\lim_{x \to 0}\left(\frac{1}{x}\right)^{\sin x}$. (See Exercise 35.)
37. Find $\lim_{x \to +\infty} (1 + x^2)^{1/x}$. (See Exercise 35.)
38. Find $\lim_{\theta \to \pi/2} (\sin \theta)^{\tan \theta}$. (See Exercise 35.)
39. Explain why L'Hospital's rule cannot be applied to $\lim_{x \to \infty} x \sin x$.
40. By inspection, find $\lim_{x \to \frac{\pi}{2}^-} (\cos x)^{\tan x}$. What is the form of this limit? (Note that it is not one of the indeterminate forms noted in this section.)
41. If the force resisting the fall of an object of mass m through the atmosphere is directly proportional to the velocity v, then the velocity at time t is $v = \frac{mg}{k}(1 - e^{-kt/m})$, where g is the acceleration due to gravity and k is a positive constant. Find $\lim_{k \to 0^+} v$.
42. In Exercise 41, find $\lim_{m \to \infty} v$.

Answers to Practice Exercises

1. -2 2. 0

27.8 Applications

The following examples show applications of the logarithmic and exponential functions. These and other applications are included in the exercises.

EXAMPLE 1 Sketching a graph

Sketch the graph of the function $y = x \ln x$.

First, we note that x cannot be zero since $\ln x$ is not defined at $x = 0$. Since $\ln 1 = 0$, we have an intercept at $(1, 0)$. There is no symmetry to the axes or origin, and there are no vertical asymptotes. Also, because $\ln x$ is defined only for $x > 0$, the domain is $x > 0$.

We now find the information from the derivatives.

Find y' and solve $y' = 0$: $\quad y' = x\left(\frac{1}{x}\right) + (\ln x)(1) = 1 + \ln x$

$y' = 0$ if $\ln x = -1$, or $x = e^{-1}$.

Find y''. Use second-derivative test: $\quad y'' = \frac{1}{x}$

$y'' > 0$ at $x = e^{-1}$, so $(e^{-1}, -e^{-1})$ is a minimum point.

Solve $y'' = 0$. Analyse concavity and points of inflection: $y'' > 0$ for all x in the domain, so there are no points of inflection and the function is always concave up.

Locate points and sketch: The curve is shown in Fig. 27.33.

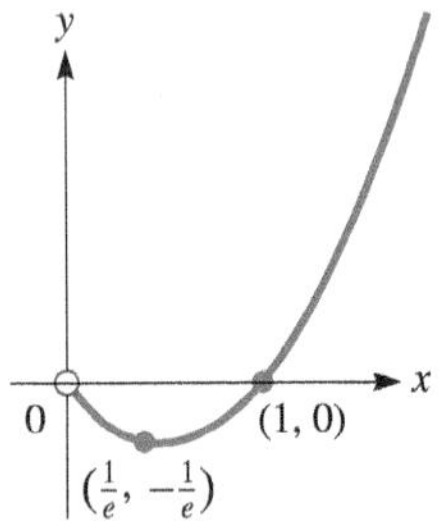

Fig. 27.33

EXAMPLE 2 Sketching a graph

Sketch the graph of the function $y = e^{-x} \cos x \ (0 \leq x \leq 2\pi)$.

This curve has intercepts for all values for which $\cos x$ is zero. Those values in the domain $0 \leq x \leq 2\pi$ for which $\cos x = 0$ are $x = \frac{\pi}{2}$ and $x = \frac{3\pi}{2}$. The factor e^{-x} is always positive, and $e^{-x} = 1$ for $x = 0$, which means $(0, 1)$ is also an intercept. There is no symmetry to the axes or the origin, and there are no vertical asymptotes.

The information from the first two derivatives is as follows.

Find y' and solve $y' = 0$: $\quad y' = -e^{-x} \sin x - e^{-x} \cos x = -e^{-x}(\sin x + \cos x)$

$y' = 0$ if $\sin x + \cos x = 0$, $\tan x = -1$, or $x = \frac{3\pi}{4}, \frac{7\pi}{4}$

(e^{-x} is never zero.)

Find y''. Use second-derivative test: $\quad y'' = -e^{-x}(\cos x - \sin x) - e^{-x}(-1)(\sin x + \cos x)$

$= 2e^{-x} \sin x$

$y'' > 0$ at $x = \frac{3\pi}{4}$ and $y'' < 0$ at $x = \frac{7\pi}{4}$

This means that $\left(\frac{3\pi}{4}, -0.067\right)$ is a minimum and $\left(\frac{7\pi}{4}, 0.003\right)$ is a maximum.

Solve $y'' = 0$. Analyse concavity and points of inflection: $y'' = 0$ for $x = 0$, π, and 2π. The function is concave up between 0 and π, and concave down between π and 2π.

Locate points and sketch: The curve is shown in Fig. 27.34.

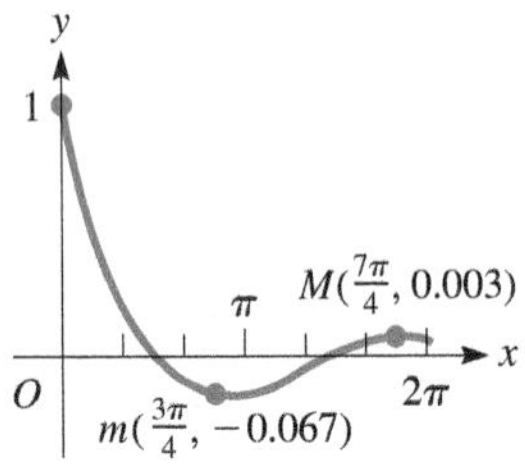

Fig. 27.34

EXAMPLE 3 Solving an equation using Newton's method

Find the root of the equation $e^{2x} - 4\cos x = 0$ that lies between 0 and 1, by using Newton's method.

Here,

$$f(x) = e^{2x} - 4\cos x$$
$$f'(x) = 2e^{2x} + 4\sin x$$

This means that $f(0) = -3$ and $f(1) = 5.2$. Therefore, we choose $x_1 = 0.5$. Using Eq. (24.1), which is

$$x_2 = x_1 - \frac{f(x_1)}{f'(x_1)}$$

we have these values:

$$f(x_1) = e^{2(0.5)} - 4\cos 0.5 = -0.792\,048\,4$$
$$f'(x_1) = 2e^{2(0.5)} + 4\sin 0.5 = 7.354\,265\,8$$
$$x_2 = 0.5 - \frac{-0.792\,048\,4}{7.354\,265\,8} = 0.607\,699\,2$$

Using the method again, we find $x_3 = 0.597\,975\,1$, which is correct to three decimal places.

Practice Exercise

1. In Example 3, choose $x_1 = 0.7$ and then find x_2.

EXAMPLE 4 Time rate of change—population growth

One model for population growth is that the population P at time t is given by $P = P_0e^{kt}$, where P_0 is the *initial* population ($t = 0$, when timing starts for the population being considered) and k is a constant. Show that the instantaneous time rate of change of population is directly proportional to the population present at time t.

To find the time rate of change, we find the derivative dP/dt:

$$\frac{dP}{dt} = (P_0e^{kt})(k) = kP_0e^{kt}$$
$$= kP \qquad \text{since } P = P_0e^{kt}$$

Thus, we see that population growth increases as population increases.

EXAMPLE 5 Acceleration of a rocket

A rocket is moving such that the only force acting on it is due to gravity and its mass is decreasing (because of the use of fuel) at a constant rate r. If it moves vertically, its velocity v as a function of time t is given by

■ See the chapter introduction.

$$v = v_0 - gt - k\ln\left(1 - \frac{rt}{m_0}\right)$$

where v_0 is the initial velocity, g is the acceleration due to gravity, t is the time, m_0 is the initial mass, and k is a constant. Determine the expression for the acceleration.

Since acceleration is the time rate of change of the velocity, we must find dv/dt. Therefore,

$$\frac{dv}{dt} = -g - k\frac{1}{1 - \frac{rt}{m_0}}\left(\frac{-r}{m_0}\right) = -g + \frac{km_0}{m_0 - rt}\left(\frac{r}{m_0}\right)$$
$$= \frac{kr}{m_0 - rt} - g$$

EXERCISES 27.8

In Exercises 1 and 2, make the given changes in the indicated examples of this section and then solve the resulting problems.

1. In Example 2, in the given function, change cos to sin.
2. In Example 3, in the given function, change e^{2x} to $2e^x$.

In Exercises 3–14, sketch the graphs of the given functions.

3. $y = \ln \cos x$
4. $y = \dfrac{2 \ln x}{x}$
5. $y = 3xe^{-x}$
6. $y = \dfrac{e^x}{x}$
7. $y = \ln \dfrac{1}{x^2 + 1}$
8. $y = 5 \ln \dfrac{e}{x}$
9. $y = 4e^{-x^2}$
10. $y = 4x - e^x$
11. $y = 8 \ln x - 2x$
12. $y = e^{-x} \sin x$
13. $y = \frac{1}{2}(e^x - e^{-x})$ (See Exercise 55 of Section 27.6.)
14. $y = \frac{1}{2}(e^x + e^{-x})$ (See Exercise 55 of Section 27.6.)

In Exercises 15–46, solve the given problems by finding the appropriate derivative.

15. Find the values of x for which the graphs of $y_1 = e^{2/x^2}$ and $y_2 = e^{-2/x^2}$ are increasing and decreasing. Explain why they differ as they do.
16. Find the values of x for which the graphs of $y_1 = \ln(x^2 + 1)$ and $y_2 = \ln(x^2 - 1)$ are concave up and concave down. Explain why they differ as they do.
17. Find the equation of the line tangent to the curve of $y = x^2 \ln x$ at the point $(1, 0)$.
18. Find the equation of the line tangent to the curve of $y = \tan^{-1} 2x$, where $x = 1$.
19. Find the equation of the line normal to the curve of $y = 2 \sin \frac{1}{2}x$, where $x = 3\pi/2$.
20. Find the equation of the line normal to the curve of $y = e^{2x}/x$, at $x = 1$.
21. By Newton's method, solve the equation $x^2 - 3 + \ln 4x = 0$ to at least four decimal places.
22. By Newton's method, find the value of x to at least four decimal places for which $y = e^{\cos x}$ is a minimum $(0 < x < 2\pi)$.
23. The electric current i (in A) through an inductor of 0.50 H as a function of time t (in s) is $i = e^{-5.0t} \sin 120\pi t$. The voltage across the inductor is given by $V_L = L\,(di/dt)$, where L is the inductance (in H). Find the voltage across the inductor for $t = 1.0$ ms.
24. A computer analysis showed that the population density D (in persons/km^2) at a distance r (in km) from the centre of a city is approximately $D = 200(1 + 5e^{-0.01r^2})$ if $r < 20$ km. At what distance from the city centre does the decrease in population density (dD/dr) itself start to decrease?
25. The power supply P (in W) in a satellite is $P = 100e^{-0.005t}$, where t is measured in days. Find the time rate of change of power after 100 days.
26. The number N of atoms of radium at any time t is given in terms of the number at $t = 0$, N_0, by $N = N_0 e^{-kt}$. Show that the time rate of change of N is proportional to N.
27. A metal bar is heated, and then allowed to cool. Its temperature T (in °C) is found to be $T = 15 + 75e^{-0.25t}$, where t (in min) is the time of cooling. Find the time rate of change of temperature after 5.0 min.
28. The insulation resistance R (in Ω/m) of a shielded cable is given by $R = k \ln(r_2/r_1)$. Here r_1 and r_2 are the inner and outer radii of the insulation. Find the expression for dR/dr_2 if k and r_1 are constant.
29. The vapour pressure p and thermodynamic temperature T of a gas are related by the equation $\ln p = \dfrac{a}{T} + b \ln T + c$, where a, b, and c are constants. Find the expression for dp/dT.
30. The charge q on a capacitor in a circuit containing a capacitor of capacitance C, a resistance R, and a source of voltage E is given by $q = CE(1 - e^{-t/RC})$. Show that this equation satisfies the equation $R\dfrac{dq}{dt} + \dfrac{q}{C} = E$.
31. Assuming that force is proportional to acceleration, show that a particle moving along the x-axis, so that its displacement is $x = Ae^{kt} + Be^{-kt}$, has a force acting on it which is proportional to its displacement.
32. The radius of curvature at a point on a curve is given by
$$R = \frac{\left[1 + (dy/dx)^2\right]^{3/2}}{d^2y/dx^2}.$$
A roller mechanism moves along the path defined by $y = \ln \sec x$ $(-1.5 \text{ dm} \le x \le 1.5 \text{ dm})$. Find the radius of curvature of this path for $x = 0.85$ dm.
33. Sketch the graph of $y = \ln \sec x$, marking that part which is the path of the roller mechanism of Exercise 32.
34. In an electronic device, the maximum current density i_m as a function of the temperature T is given by $i_m = AT^2e^{k/T}$, where A and k are constants. Find the expression for a small change in i_m for a small change in T.
35. The energy E (in J) dissipated by a certain resistor after t seconds is given by $E = \ln(t + 1) - 0.25t$. At what time is the energy dissipated the greatest?
36. In a study of traffic control, the number n of vehicles on a certain section of a highway from 2:00 P.M. to 8:00 P.M. was found to be $n = 200(1 + t^3e^{-t})$, where t is the number of hours after 2:00 P.M. At what time is the number of vehicles the greatest?
37. A meteorologist sketched the path of the jet stream on a map of the northern United States and southern Canada on which all latitudes were parallel and all longitudes were parallel and equally spaced. A computer analysis showed this path to be $y = 6.0e^{-0.020x} \sin 0.20x$ $(0 \le x \le 60)$, where the origin is 125.0°W, 45.0°N and $(60, 0)$ is 65.0°W, 45.0°N. Find the locations of the maximum and minimum latitudes of the jet stream between 65°W and 125°W for that day.
38. The reliability R $(0 \le R \le 1)$ of a certain computer system after t hours of operation is found from $R = 3e^{-0.004t} - 2e^{-0.006t}$. Use Newton's method to find how long the system operates to have a reliability of 0.8 (80% probability that there will be no system failure).

39. An object on the end of a spring is moving so that its displacement (in cm) from the equilibrium position is given by $y = e^{-0.5t}(0.4 \cos 6t - 0.2 \sin 6t)$. Find the expression for the velocity of the object. What is the velocity when $t = 0.26$ s? The motion described by this equation is called *damped harmonic motion*.

40. A package of weather instruments is propelled into the air to an altitude of about 7 km. A parachute then opens, and the package returns to the surface. The altitude y of the package as a function of the time t (in min) is given by $y = \frac{10t}{e^{0.4t} + 1}$. Find the vertical velocity of the package for $t = 8.0$ min.

41. The speed s of signaling by use of a certain communications cable is directly proportional to $x^2 \ln x^{-1}$, where x is the ratio of the radius of the core of the cable to the thickness of the surrounding insulation. For what value of x is s a maximum?

42. A computer is programmed to inscribe a series of rectangles in the first quadrant under the curve of $y = e^{-x}$. What is the area of the largest rectangle that can be inscribed?

43. The relative number N of gas molecules in a container that are moving at a velocity v can be shown to be $N = av^2e^{-bv^2}$, where a and b are constants. Find v for the maximum N.

44. A missile is launched and travels along a path that can be represented by $y = \sqrt{x}$. A radar tracking station is located 2.00 km directly behind the launch pad. Placing the launch pad at the origin and the radar station at $(-2.00, 0)$, find the largest angle of elevation required of the radar to track the missile.

45. The curve given by $y = \frac{1}{\sqrt{2\pi}}e^{-x^2/2}$, called the standard normal curve, is very important in statistics (see Section 22.3). Show that this curve has inflection points at $x = \pm 1$.

46. The St. Louis Gateway Arch (see Review Exercise 97) has a shape that is given approximately by (measurements in m) $y = -19.46(e^{x/38.92} + e^{-x/38.92}) + 230.9$. What is the maximum height of the arch?

Answer to Practice Exercise

1. $x_2 = 0.6068$

CHAPTER 27 KEY FORMULAS AND EQUATIONS

Limit of $\frac{\sin \theta}{\theta}$ as $\theta \to 0$

$$\lim_{\theta \to 0} \frac{\sin \theta}{\theta} = \lim_{h \to 0} \frac{\sin(h/2)}{h/2} = 1 \qquad (27.1)$$

Chain rule

$$\frac{dy}{dx} = \frac{dy}{du} \cdot \frac{du}{dx} \qquad (27.3)$$

Derivatives

$$\frac{d(\sin u)}{dx} = \cos u \frac{du}{dx} \qquad (27.4)$$

$$\frac{d(\cos u)}{dx} = -\sin u \frac{du}{dx} \qquad (27.5)$$

$$\frac{d(\tan u)}{dx} = \sec^2 u \frac{du}{dx} \qquad (27.6)$$

$$\frac{d(\cot u)}{dx} = -\csc^2 u \frac{du}{dx} \qquad (27.7)$$

$$\frac{d(\sec u)}{dx} = \sec u \tan u \frac{du}{dx} \qquad (27.8)$$

$$\frac{d(\csc u)}{dx} = -\csc u \cot u \frac{du}{dx} \qquad (27.9)$$

$$\frac{d(\sin^{-1} u)}{dx} = \frac{1}{\sqrt{1 - u^2}} \frac{du}{dx} \qquad (27.10)$$

$$\frac{d(\cos^{-1} u)}{dx} = -\frac{1}{\sqrt{1 - u^2}} \frac{du}{dx} \qquad (27.11)$$

$$\frac{d(\tan^{-1} u)}{dx} = \frac{1}{1+u^2}\frac{du}{dx} \tag{27.12}$$

$$\frac{d(\log_b u)}{dx} = \frac{1}{u}\log_b e\frac{du}{dx} \tag{27.13}$$

$$\frac{d(\ln u)}{dx} = \frac{1}{u}\frac{du}{dx} \tag{27.14}$$

$$\frac{d(b^u)}{dx} = b^u \ln b\frac{du}{dx} \tag{27.15}$$

$$\frac{d(e^u)}{dx} = e^u\frac{du}{dx} \tag{27.16}$$

CHAPTER 27 REVIEW EXERCISES

In Exercises 1–40, find the derivative of the given functions.

1. $y = 3\cos(4x - 1)$
2. $y = 4\sec(1 - x^3)$
3. $u = 0.2\tan\sqrt{3 - 2v}$
4. $y = 5\sin(1 - 6x)$
5. $y = \csc^2(3x + 2)$
6. $r = \cot^2 5\pi\theta$
7. $y = 3\cos^4 x^2$
8. $y = 2\sin^3\sqrt{x}$
9. $y = (e^{x-3})^2$
10. $y = 0.5e^{\sin 2x}$
11. $y = 3\ln(x^2 + 1)$
12. $R = \ln(3 + \sin T^2)$
13. $y = 10\tan^{-1}(x/5)$
14. $y = 0.4\cos^{-1}(2\pi t + 1)$
15. $\theta = \ln\sin^{-1} 0.1t$
16. $y = \sin(\tan^{-1} x)$
17. $y = \sqrt{\csc 4x + \cot 4x}$
18. $y = 3\cos^2(\tan 3x)$
19. $y = 7\ln(x - e^{-x})^2$
20. $h = \ln\sqrt[3]{\sin 6\theta}$
21. $y = \dfrac{\cos^2 x}{e^{3x} + \pi^2}$
22. $y = \sqrt{\dfrac{1 + \cos 2x}{2}}$
23. $v = \dfrac{u^2}{\tan^{-1} 2u}$
24. $y = \dfrac{\sin^{-1} x}{4x}$
25. $y = \ln(\csc x^2)$
26. $u = 0.5\ln\tan e^x$
27. $y = \ln^2(3 + \sin x)$
28. $y = \ln(3 + \sin\pi t)^2$
29. $L = 0.1e^{-2t}\sec\pi t$
30. $y = 5\,e^{3x}\ln x$
31. $y = \sqrt{\sin 2x + e^{4x}}$
32. $x + y\ln 2x = y^2$
33. $\tan^{-1}\dfrac{y}{x} = x^2e^y$
34. $3y + \ln xy = 2 + x^2$
35. $r = 0.5t(e^{2t} + 1)(e^{-2t} - 1)$
36. $y = (\ln 4x - \tan 4x)^3$
37. $\ln xy + ye^{-x} = 1$
38. $y = x(\sin^{-1} x)^2 + 2\sqrt{1 - x^2}\sin^{-1} x - 2x$
39. $y = x\cos^{-1} x - \sqrt{1 - x^2}$
40. $W = \ln(4s^2 + 1) + \tan^{-1} 2s$

In Exercises 41–44, sketch the graphs of the given functions.

41. $y = x - \cos 0.5x$
42. $y = 4\sin x + \cos 2x$
43. $y = x(\ln x)^2$
44. $y = 2\ln(3 + x)$

In Exercises 45–48, find the equations of the indicated tangent or normal lines.

45. Find the equation of the line tangent to the curve of $y = 4\cos^2(x^2)$ at $x = 1$.
46. Find the equation of the line tangent to the curve of $y = \ln\cos x$ at $x = \frac{\pi}{6}$.
47. Find the equation of the line normal to the curve of $y = e^{x^2}$ at $x = \frac{1}{2}$.
48. Find the equation of the line normal to the curve of $y = \tan^{-1} 4x$ at $x = 1/2$.

In Exercises 49–54, find the indicated limits by use of L'Hospital's rule.

49. $\lim\limits_{x\to 0}\dfrac{\sin 2x}{\sin 3x}$
50. $\lim\limits_{x\to 0}\dfrac{xe^x}{1 - e^x}$
51. $\lim\limits_{x\to\pi^+}\dfrac{\sin x}{x - \pi}$
52. $\lim\limits_{t\to+\infty}\dfrac{x^4 + 5x^2 + 1}{3x^4 + 4}$
53. $\lim\limits_{x\to\infty}\dfrac{\ln x}{\sqrt[3]{x}}$
54. $\lim\limits_{x\to\pi}(x - \pi)\cot x$

In Exercises 55–99, solve the given problems.

55. Find the derivative of each member of the identity $\sin^2 x + \cos^2 x = 1$ and show that the results are equal.
56. Find the derivative of each member of the identity $\sin(x + 1) = \sin x\cos 1 + \cos x\sin 1$ and show that the results are equal.
57. If $y = \sin 3x$, show that $\dfrac{d^2y}{dx^2} = -9y$.
58. If $y = e^{5x}(a + bx)$, show that $\dfrac{d^2y}{dx^2} - 10\dfrac{dy}{dx} + 25y = 0$.
59. By Newton's method, solve the equation $e^x - x^2 = 0$.
60. By Newton's method, find the nonzero solution to the equation $x^2 = \tan^{-1} x$.

61. Find the values of x for which the graph of $y = e^x - 2e^{-x}$ is concave up.

62. Find the values of x for which the graph of $y = 2e^{\sin(x/2)}$ has maximum or minimum points.

63. If a block is placed on a plane inclined with the horizontal at an angle θ such that the block stays at the same position, the coefficient of friction μ is given by $\mu = \tan\theta$. Use differentials to find the change in μ if θ changes from 18° to 20°.

64. The tensile strength S (in N) of a plastic is tested and found to change with the temperature T (in °C) according to the equation $S = 1800\ln(T + 25) - 40\text{T} + 8600$. For what temperature is the tensile strength the greatest?

65. If a 200-N crate is dragged along a horizontal floor by a force F (in N) acting along a rope at an angle θ with the floor, the magnitude of F is given by $F = \dfrac{200\,\mu}{\mu\sin\theta + \cos\theta}$, where μ is the coefficient of friction. Evaluate the instantaneous rate of change of F with respect to θ when $\theta = 15°$ if $\mu = 0.20$.

66. Find the date of the maximum number of hours of daylight in Toronto, Ontario. The hours h of daylight can be approximated by $h = 12.2 + 3.3\sin\left(\dfrac{2\pi}{365}t - \dfrac{160\pi}{365}\right)$, where t is measured in days ($t = 15$ is January 15, etc.).

67. Periodically, a robot moves a part vertically y cm in an automobile assembly line. If y as a function of the time t (in s) is $y = 0.75(\sec\sqrt{0.15t} - 1)$, find the velocity at which the part is moved after 5.0 s of each period.

68. The length L (in m) of the shadow of a tree 15 m tall is $L = 15\cot\theta$, where θ is the angle of elevation of the sun. Approximate the change in L if θ changes from 50° to 52°.

69. An analysis of temperature records for Sydney, Australia, indicates that the average daily temperature (in °C) during the year is given approximately by $T = 17.2 + 5.2\cos\left[\frac{\pi}{6}(x - 0.50)\right]$, where x is measured in months ($x = 0.5$ is January 15, etc.). What is the *daily* time rate of change of temperature on March 1? (*Hint*: 12 months/365 days = 0.033 month/day = dx/dt.)

70. An Earth-orbiting satellite is launched such that its altitude (in km) is given by $y = 240(1 - e^{-0.05t})$, where t is the time (in min). Find the vertical velocity of the satellite for $t = 10.0$ min.

71. Power P is the time rate of change of doing work W. If work is being done in an electric circuit according to $W = 25\sin^2 2t$, find P as a function of t.

72. The value V of a bank account in which \$1000 is deposited and then earns 6% annual interest, compounded continuously (daily compounding approximates this, and some banks actually use continuous compounding), is $V = 1000e^{0.06t}$ after t years. How fast is the account growing after exactly two years?

73. In determining how to divide files on the hard disk of a computer, we can use the equation $n = xN\log_x N$. Sketch the graph of n as a function of x for $1 < x \le 10$ if $N = 8$.

74. Under certain conditions, the potential V (in V) due to a magnet is given by $V = -k\ln\left(1 + \dfrac{L}{x}\right)$, where L is the length of the magnet and x is the distance from the point where the potential is measured. Find the expression for dV/dx.

75. In the theory of making images by holography, an expression used for the light-intensity distribution is $I = kE_0^2\cos^2\frac{1}{2}\theta$, where k and E_0 are constant and θ is the phase angle between two light waves. Find the expression for $dI/d\theta$.

76. Neglecting air resistance, the range R of a bullet fired at an angle θ with the horizontal is $R = \dfrac{v_0^2}{g}\sin 2\theta$, where v_0 is the initial velocity and g is the acceleration due to gravity. Find θ for the maximum range. See Fig. 27.35.

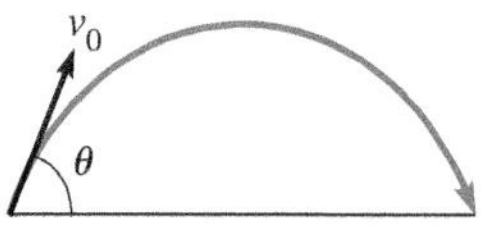

Fig. 27.35

77. In the design of a cone-type clutch, an equation that relates the cone angle θ and the applied force F is $\theta = \sin^{-1}(Ff/R)$, where R is the frictional resistance and f is the coefficient of friction. For constant R and f, find $d\theta/dF$.

78. If inflation makes the dollar worth 5% less each year, then the value of \$100 in t years will be $V = 100(0.95)^t$. What is the approximate change in the value during the fourth year?

79. An object attached to a cord of length l, as shown in Fig. 27.36, moves in a circular path. The angular velocity ω is given by $\omega = \sqrt{g/(l\cos\theta)}$. By use of differentials, find the approximate change in ω if θ changes from 32.50° to 32.75°, given that $g = 9.800\text{ m/s}^2$ and $l = 0.6375$ m.

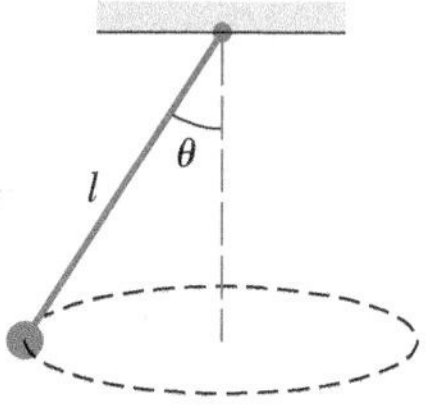

Fig. 27.36

80. An analysis of samples of air for a city showed that the number of parts per million p of sulfur dioxide on a certain day was $p = 0.05\ln(2 + 24t - t^2)$, where t is the hour of the day. Using differentials, find the approximate change in the amount of sulfur dioxide between 10:00 A.M. and noon.

81. According to Newton's law of cooling (Isaac Newton, again), the rate at which a body cools is proportional to the difference in temperature between it and the surrounding medium. By use of this law, the temperature T (in °C) of an engine coolant as a function of the time t (in min) is $T = 30 + 60(0.5)^{0.200t}$. The coolant was initially at 90°C and the air temperature was 30°C. Linearize this function for $t = 5.00$ min.

82. The charge q on a certain capacitor in an amplifier circuit as a function of time t is given by $q = e^{-0.1t}(0.2\sin 120\pi t + 0.8\cos 120\pi t)$. The current i in the circuit is the instantaneous time rate of change of the charge. Find the expression for i as a function of t.

83. A new 2017 car was purchased for \$32 000. Due to depreciation, its projected value V is given by $V = 32\,000e^{-0.14t}$, where t is the number of years after 2017. Find the instantaneous rate at which the value is changing after (a) one year and (b) five years. Round to two significant digits.

84. A football is thrown horizontally (very little arc) at 18 m/s parallel to the sideline. A TV camera is 31 m from the path of the football. Find $d\theta/dt$, the rate at which the camera must turn to follow the ball when $\theta = 15°$. See Fig. 27.37.

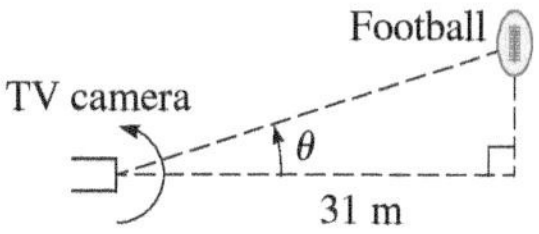

Fig. 27.37

85. An architect designs an arch of height y (in m) over a walkway by the curve of the equation $y = 3.00e^{-0.500x^2}$. What are the dimensions of the largest rectangular passage area under the arch?

86. A force P (in N) at an angle θ above the horizontal drags a 50-N box across a level floor. The coefficient of friction between the floor and the box is constant and equals 0.20. The magnitude of the force P is given by $P = \dfrac{(0.20)(50)}{0.20 \sin\theta + \cos\theta}$. Find θ such that P is a minimum.

87. A jet is flying at 220 m/s directly away from the control tower of an airport. If the jet is at a constant altitude of 1700 m, how fast is the angle of elevation of the jet from the control tower changing when it is 13.0°?

88. The current i in an electric circuit with a resistance R and an inductance L is $i = i_0 e^{-Rt/L}$, where i_0 is the initial current. Show that the time rate of change of the current is directly proportional to the current.

89. A silo constructed as shown in Fig. 27.38 is to hold 2880 m³ of silage when completely full. It can be shown (by you?) that the surface area S (in m²) (not including the base) is $S = 640 + 81\pi\left(\csc\theta - \frac{2}{3}\cot\theta\right)$. Find θ such that S is a minimum.

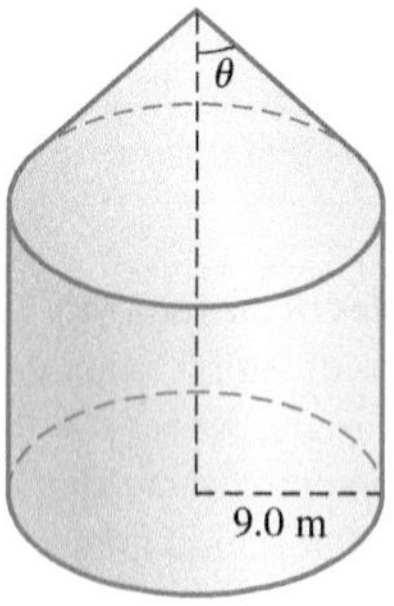

Fig. 27.38

90. Light passing through a narrow slit forms patterns of light and dark (see Fig. 27.39). The intensity I of the light at an angle θ is given by $I = I_0\left[\dfrac{\sin(k\sin\theta)}{k\sin\theta}\right]^2$, where k and I_0 are constants. Show that the maximum and minimum values of I occur for $k\sin\theta = \tan(k\sin\theta)$.

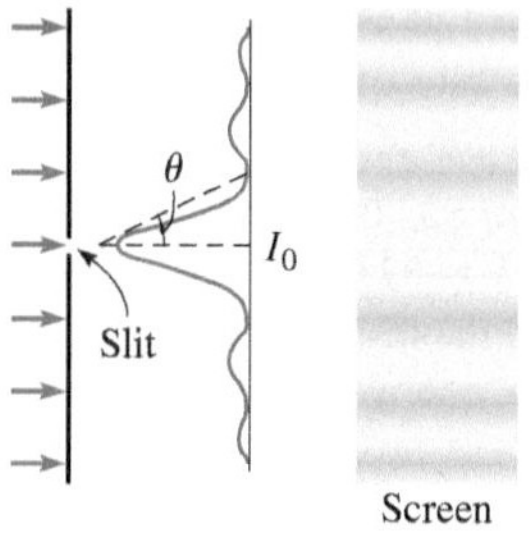

Fig. 27.39

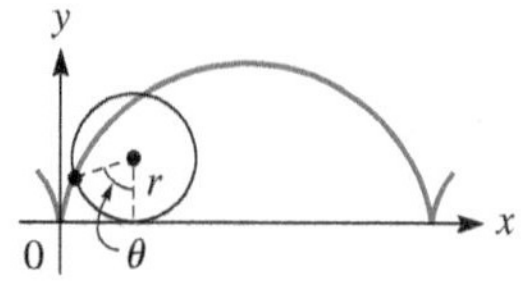

Fig. 27.40

91. When a wheel rolls along a straight line, a point P on the circumference traces a curve called a *cycloid.* See Fig. 27.40. The parametric equations of a cycloid are $x = r(\theta - \sin\theta)$ and $y = r(1 - \cos\theta)$. Find the velocity of the point on the rim of a wheel for which $r = 5.50$ cm and $d\theta/dt = 0.120$ rad/s for $\theta = 35.0°$. (An inverted cycloid is the path of least time of descent (the *brachistochrone*) of an object acted on only by gravity.)

92. In the study of atomic spectra, it is necessary to solve the equation $x = 5(1 - e^{-x})$ for x. Use Newton's method to find the solution.

93. The illuminance from a point source of light varies directly as the cosine of the angle of incidence (measured from the perpendicular) and inversely as the square of the distance r from the source. How high above the centre of a circle of radius 10.0 cm should a light be placed so that illuminance at the circumference will be a maximum? See Fig. 27.41.

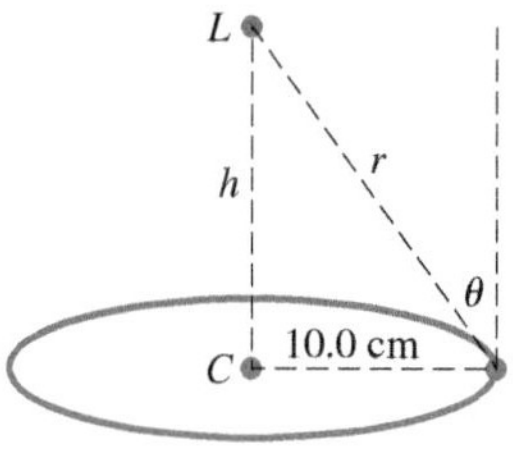

Fig. 27.41

94. A Y-shaped metal bracket is to be made such that its height is 10.0 cm and its width across the top is 6.00 cm. What shape will require the least amount of material? See Fig. 27.42.

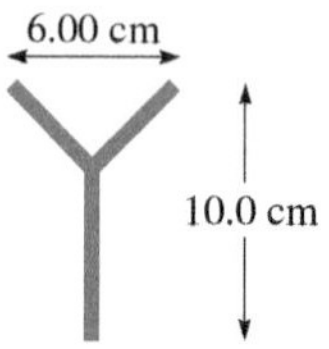

Fig. 27.42

95. A gutter is to be made from a sheet of metal 30.0 cm wide by turning up strips of width 10.0 cm along each side to make equal angles θ with the vertical. Sketch a graph of the cross-sectional area A as a function of θ. See Fig. 27.43.

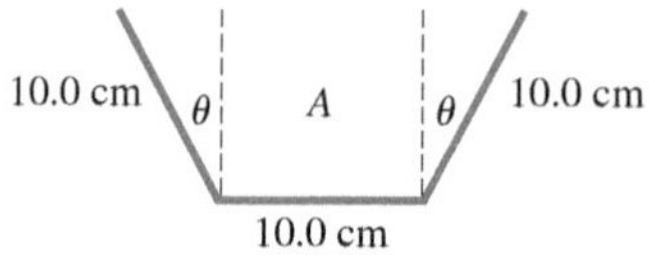

Fig. 27.43

96. The displacement y (in cm) of a weight on a spring in water is given by $y = 3.0te^{-0.20t}$, where t is the time (in s). What is the maximum displacement? (For this type of displacement, the motion is called *critically damped,* as the weight returns to its equilibrium position as quickly as possible without oscillating.)

97. Show that the equation of the hyperbolic cosine function $y = \dfrac{H}{w}\cosh\dfrac{wx}{H}$ (w and H are constants) satisfies the equation $\dfrac{d^2y}{dx^2} = \dfrac{w}{H}\sqrt{1 + \left(\dfrac{dy}{dx}\right)^2}$ (see Exercise 55 of Section 27.6).

A *catenary* is the curve of a uniform cable hanging under its own weight and is in the shape of a hyperbolic cosine curve. This shape (inverted) was chosen for the St. Louis Gateway Arch (shown in Fig. 27.44) and makes the arch self-supporting. Also, an igloo has the shape of a rotated inverted catenary (called a catenoid).

Fig. 27.44

98. A conical filter is made from a circular piece of wire mesh of radius 24.0 cm by cutting out a sector with central angle θ and then taping the cut edges of the remaining piece together (see Fig. 27.45). What is the maximum possible volume the resulting filter can hold?

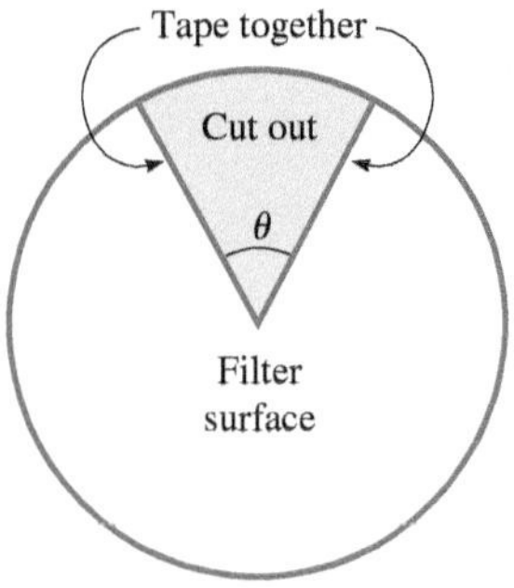

Fig. 27.45

99. To find the area of the largest rectangular microprocessor chip with a perimeter of 40 mm, it is possible to use either an algebraic function or a trigonometric function. Write two or three paragraphs to explain how each type of function can be used to find the required area.

CHAPTER 27 PRACTICE TEST

In Problems 1–3, find the derivative of each of the functions.

1. $y = \tan^3 2x + \tan^{-1} 2x$

2. $y = 2(3 + \cot 4x)^3$

3. $y \sec 2x = \sin^{-1} 3y$

4. Find the differential of the function $y = \dfrac{\cos^2(3x + 1)}{x}$.

5. Find the slope of a tangent to the curve of $y = \ln \dfrac{2x - 1}{1 + x^2}$ for $x = 2$.

6. Find the expression for the time rate of change of electric current that is given by the equation $i = 8e^{-t} \sin 10t$, where t is the time.

7. Sketch the graph of the function $y = xe^x$.

8. A balloon leaves the ground 250 m from an observer and rises at the rate of 5.0 m/s. How fast is the angle of elevation of the balloon increasing after 8.0 s?

9. Find $\lim\limits_{x \to 0} \dfrac{\tan^{-1} x}{2 \sin x}$ using L'Hospital's rule.

28. Methods of Integration

LEARNING OUTCOMES

After completion of this chapter, the student should be able to:

- Perform integrals by any of the following methods:
  - the general power rule
  - integration of basic logarithmic and exponential forms
  - integration of basic trigonometric and inverse trigonometric forms
  - integration by parts
  - trigonometric substitution
  - integration by partial fractions
  - using tables
- Solve application problems involving integration

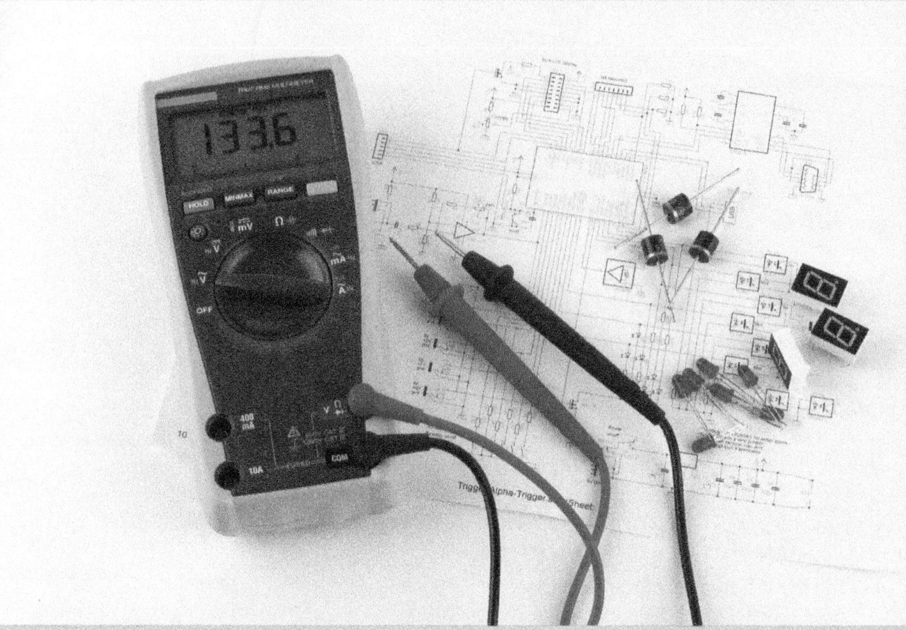

Ignard ten Have/123rf

▲ In Section 28.5, we use integration to find the root-mean-square (RMS) value of an alternating current. When using an AC voltmeter or ammeter, it is the RMS value of a voltage or current that is displayed.

In developing calculus, mathematicians saw that integration and differentiation were inverse processes and that many integrals could be formed by finding the antiderivative. However, many of the integrals that arose mathematically and from the study of mechanical systems did not fit a form from which the antiderivative could be found directly. This led to the creation of numerous methods to integrate various types of functions.

Some of these methods of integration had been developed by the early 1700s, and were used by many mathematicians, including Newton and Leibniz. By the mid-1700s, many of these methods were included in textbooks. One of these texts was written by the Italian mathematician Maria Agnesi. Her text included analytic geometry, differential calculus, integral calculus, and some advanced topics and was noted for clear and organized explanations with many examples. Another important set of textbooks was written by Euler, who wrote separate texts on precalculus topics, differential calculus, and integral calculus. These texts were also noted for clear and organized presentations and were widely used until the early 1800s. Euler's integral calculus text included most of the methods of integration presented in this chapter.

Being able to integrate functions by using special methods, as well as by directly using antiderivatives, made integral calculus much more useful in developing many areas of geometry, science, and technology in the 1800s and 1900s. In earlier chapters, we have noted a number of applications of integration, and additional examples are found in the examples and exercises of this chapter.

In this chapter, we expand the use of the general power formula for integration for use with integrands that include transcendental functions and several special methods of integration for integrands that do not directly fit standard forms. In using all forms of integration, *recognition of the integral form* is of great importance.

28.1 The Power Rule for Integration

The first method of integration we will discuss is the use of the power rule for integration (introduced in Chapter 25). We can now expand its use to cases where the integrand is a transcendental function.

Power Rule for Integration

$$\int u^n\,du = \frac{u^{n+1}}{n+1} + C \qquad (n \neq -1) \qquad \textbf{(28.1)}$$

EXAMPLE 1

Integrate $\int \sin^3 x \cos x\,dx$.

We identify $u = \sin x$, $du = \cos x\,dx$, and $n = 3$. Note that **$\cos x$ is a necessary part of the du** in order to have the proper form of integration. Therefore,

$$\int \sin^3 x \cos x\,dx = \int u^3\,du = \frac{u^4}{4} + C = \frac{1}{4}\sin^4 x + C$$

Practice Exercise

1. Integrate $\int \cos^2 x \sin x\,dx$.

LEARNING TIP

In applying Eq. (28.1) to transcendental integrands, as well as with algebraic integrands, ***we must properly recognize the quantities u, n, and du***. This requires familiarity with the differential forms of Chapters 23 and 27.

EXAMPLE 2 Trigonometric integrand

Integrate $\int 2\sqrt{1 + \tan\theta}\,\sec^2\theta\,d\theta$.

Here, we note that $d(\tan\theta) = \sec^2\theta\,d\theta$, which means that the integral fits the form of Eq. (28.1) with

$$u = 1 + \tan\theta \qquad du = \sec^2\theta\,d\theta \qquad n = \tfrac{1}{2}$$

The integral is of the form $\int u^{1/2}\,du$. Thus,

$$\int 2\sqrt{1+\tan\theta}\,(\sec^2\theta\,d\theta) = 2\int \underbrace{(1+\tan\theta)}_{u}{}^{1/2}\overbrace{(\sec^2\theta\,d\theta)}^{du}$$

$$= 2\left(\frac{2}{3}\right)(1+\tan\theta)^{3/2} + C$$

$$= \frac{4}{3}(1+\tan\theta)^{3/2} + C$$

EXAMPLE 3 Logarithmic integral

Integrate $\int \ln x\left(\frac{dx}{x}\right)$.

By noting that $d(\ln x) = \frac{dx}{x}$, we have: $u = \ln x \quad du = \frac{dx}{x} \quad n = 1$

This means that the integral is of the form $\int u\,du$. Thus,

$$\int \ln x\left(\frac{dx}{x}\right) = \frac{1}{2}(\ln x)^2 + C = \frac{1}{2}\ln^2 x + C$$

Practice Exercise

2. Integrate $\int \frac{2\ln 2x\,dx}{x}$.

EXAMPLE 4 Inverse trigonometric integrand

Find the value of $\int_0^{0.5} \frac{\sin^{-1} x}{\sqrt{1 - x^2}} dx$.

For purposes of integrating: $u = \sin^{-1} x \quad du = \frac{dx}{\sqrt{1 - x^2}} \quad n = 1$

$$\int_0^{0.5} \frac{\sin^{-1} x}{\sqrt{1 - x^2}} dx = \int_0^{0.5} \sin^{-1} x\left(\frac{dx}{\sqrt{1 - x^2}}\right) \qquad \int u\, du$$

$$= \frac{(\sin^{-1} x)^2}{2}\Bigg|_0^{0.5} \qquad \text{integrate}$$

$$= \frac{\left(\frac{\pi}{6}\right)^2}{2} - 0 = \frac{\pi^2}{72} \qquad \text{evaluate}$$

EXAMPLE 5 Inverse trigonometric integrand

Find the first-quadrant area bounded by $y = \frac{e^{2x}}{\sqrt{e^{2x} + 1}}$ and $x = 1.5$.

The area is shown in Fig. 28.1. Using a representative element of area $y\, dx$, the area is found by evaluating the integral

$$\int_0^{1.5} \frac{e^{2x}\, dx}{\sqrt{e^{2x} + 1}}$$

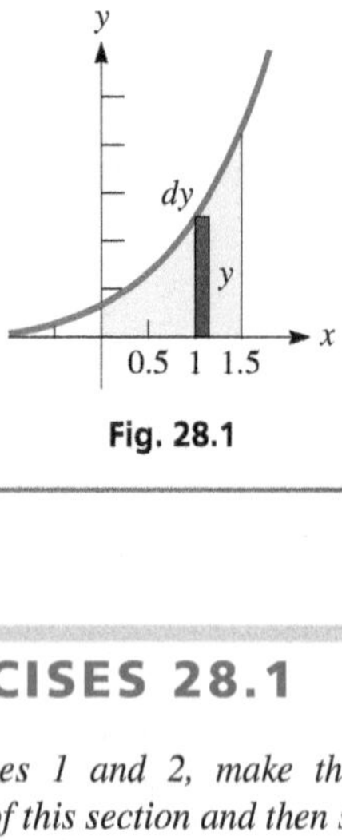

Fig. 28.1

For the purpose of integration, $n = -\frac{1}{2}$, $u = e^{2x} + 1$, and $du = 2e^{2x}\, dx$. Therefore,

$$\int_0^{1.5} (e^{2x} + 1)^{-1/2} e^{2x}\, dx = \frac{1}{2}\int_0^{1.5} (e^{2x} + 1)^{-1/2}(2e^{2x}\, dx)$$

$$= \frac{1}{2}(2)(e^{2x} + 1)^{1/2}\Bigg|_0^{1.5} = (e^{2x} + 1)^{1/2}\Bigg|_0^{1.5}$$

$$= \sqrt{e^3 + 1} - \sqrt{2} = 3.178$$

EXERCISES 28.1

In Exercises 1 and 2, make the given changes in the indicated examples of this section and then solve the given problems.

1. In Example 1, change $\sin^3 x \cos x$ to $\cos^3 x \sin x$ and then integrate.
2. In Example 3, change $\ln x$ to $\ln x^2$ and then integrate.

In Exercises 3–28, integrate each of the functions.

3. $\int (x^2 + 1)^3(2x\, dx)$
4. $\int (x^3 - 2)^6(3x^2\, dx)$
5. $\int \sin^4 x \cos x\, dx$
6. $\int \cos^{19} x(-\sin x\, dx)$
7. $\int 0.4\sqrt{\cos\theta}\sin\theta\, d\theta$
8. $\int 8\sin^{1/3} x \cos x\, dx$
9. $\int 4\tan^2 x \sec^2 x\, dx$
10. $\int \sec^7 x(\sec x \tan x)\, dx$
11. $\int_0^{\pi/8} \frac{\cos 4x}{\csc 4x} dx$
12. $\int_{\pi/6}^{\pi/4} 3\sqrt{\cot x}\csc^2 x\, dx$
13. $\int (\sin^{-1} x)^3\left(\frac{dx}{\sqrt{1 - x^2}}\right)$
14. $\int \frac{20(\cos^{-1} 2t)^4\, dt}{\sqrt{1 - 4t^2}}$
15. $\int \frac{5\tan^{-1} 5x}{1 + 25x^2} dx$
16. $\int \frac{\sin^{-1} 4x\, dx}{\sqrt{1 - 16x^2}}$
17. $\int [\ln(x + 1)]^2 \frac{dx}{x + 1}$
18. $\int 0.8(5 + 8\ln u)^3 \frac{du}{u}$
19. $\int_0^{1/2} \frac{\ln(2x + 3)}{2x + 3} dx$
20. $\int_1^e \frac{(1 - 2\ln x)\, dx}{x}$
21. $\int (4 + e^x)^{16} e^x\, dx$
22. $\int 2\sqrt{1 - e^{-x}}(e^{-x}\, dx)$
23. $\int \frac{e^{2t}\, dt}{(1 - e^{2t})^3}$
24. $\int \frac{(1 + 3e^{-2x})^4\, dx}{e^{2x}}$

25. $\int (1 + \sec^2 x)^4(\sec^2 x \tan x\, dx)$

26. $\int (e^x + e^{-x})^{1/4}(e^x - e^{-x})\, dx$

27. $\int_{\pi/6}^{\pi/4} (1 + \cot x)^2 \csc^2 x\, dx$ **28.** $\int_{\pi/3}^{\pi/2} \frac{\sin \theta\, d\theta}{\sqrt{1 + \cos \theta}}$

In Exercises 29–32, rewrite the given integrals so that they fit the form $\int u^n\, du$, *and identify u, n, and du.*

29. $\int \sec^5 x \sin x\, dx$ **30.** $\int \frac{\tan^3 x\, dx}{\cos^2 x}$

31. $\int \frac{dx}{x \ln^2 x}$ **32.** $\int \frac{\sqrt{(1 + e^{-r})(1 - e^{-r})}}{e^{2r}}\, dr$

In Exercises 33–42, solve the given problems by integration.

33. Find the first-quadrant area under the curve of $y = \ln^2 x/x$ from $x = 1$ to $x = 4$.

34. Find the first-quadrant area under the curve of $y = \sin^3 x \cos x$ from $x = 0$ to $x = \frac{\pi}{2}$.

35. Find the area under the curve $y = \frac{1 + \tan^{-1} 2x}{1 + 4x^2}$ from $x = 0$ to $x = 2$.

36. Find the first-quadrant area bounded by $y = \frac{\ln(4x + 1)}{4x + 1}$ and $x = 5$.

37. The general expression for the slope of a given curve is $(\ln x)^2/x$. If the curve passes through $(1, 2)$, find its equation.

38. Find the equation of the curve for which $dy/dx = (1 + \tan 2x)^2 \sec^2 2x$ if the curve passes through $(2, 1)$.

39. In the development of the expression for the total pressure p on a wall due to molecules with mass m and velocity v striking the wall, the equation $p = mnv^2 \int_0^{\pi/2} \sin \theta \cos^2 \theta\, d\theta$ is found. The symbol n represents the number of molecules per unit volume, and θ represents the angle between a perpendicular to the wall and the direction of the molecule. Find the expression for p.

40. The solar energy E passing through a hemispherical surface per unit time, per unit area, is $E = 2\pi I \int_0^{\pi/2} \cos \theta \sin \theta\, d\theta$, where I is the solar intensity and θ is the angle at which it is directed (from the perpendicular). Evaluate this integral.

41. After an electric power interruption, the current i in a circuit is given by $i = 3(1 - e^{-t})^2(e^{-t})$, where t is the time. Find the expression for the total electric charge q to pass a point in the circuit if $q = 0$ for $t = 0$.

42. A space vehicle is launched vertically from the ground such that its velocity v (in km/s) is given by $v = [\ln^2(t^3 + 1)]\frac{t^2}{t^3 + 1}$, where t is the time (in s). Find the altitude of the vehicle after 10.0 s.

Answers to Practice Exercises

1. $-\frac{1}{3}\cos^3 x + C$ **2.** $\ln^2 2x + C$

28.2 The Basic Logarithmic Form

The general power rule for integration, Eq. (28.1), is valid for all values of n except $n = -1$. If n were set equal to -1, this would cause the result to be undefined. When we obtained the derivative of the logarithmic function, we found

$$\frac{d(\ln u)}{dx} = \frac{1}{u}\frac{du}{dx}$$

This means the differential of the logarithmic form is $d(\ln u) = du/u$. Reversing the process, we then determine that $\int du/u = \ln u + C$. In other words, when the exponent of the expression being integrated is -1, the expression is a logarithmic form.

Logarithms are defined only for positive numbers. Thus, $\int du/u = \ln u + C$ is valid if $u > 0$. If $u < 0$, then $-u > 0$. In this case, $d(-u) = -du$, or $\int(-du)/(-u) = \ln(-u) + C$. However, $\int du/u = \int(-du)/(-u)$. These results can be combined into a single form using the absolute value of u.

Integration of a Logarithmic Form	EXAMPLE 1
	Integrate $\int \frac{dx}{x + 1}$.
$\int \frac{du}{u} = \ln\lvert u \rvert + C$ **(28.2)**	We identify $u = x + 1$, and $du = dx$. Then $\int \frac{dx}{x + 1} = \int \frac{du}{u} = \ln\lvert u \rvert + C = \ln\lvert x + 1 \rvert + C$

EXAMPLE 2 Algebraic integrand—cooling object

Newton's law of cooling states that the rate at which an object cools is directly proportional to the difference in its temperature T and the temperature of the surrounding medium. By use of this law, the time t (in min) a certain object takes to cool from 80°C to 50°C in air at 20°C is found to be

$$t = -9.8\int_{80}^{50}\frac{dT}{T-20}$$

Find the value of t.

We see that the integral fits Eq. (28.2) with $u = T - 20$ and $du = dT$. Thus,

$$\begin{aligned} t &= -9.8\int_{80}^{50}\frac{dT \leftarrow du}{T-20 \leftarrow u} \\ &= -9.8\ln|T-20|\Big|_{80}^{50} && \text{integrate} \\ &= -9.8(\ln 30 - \ln 60) && \text{evaluate} \\ &= -9.8\ln\frac{30}{60} = -9.8\ln(0.50) && \ln x - \ln y = \ln\frac{x}{y} \\ &= 6.8\text{ min} \end{aligned}$$

Practice Exercise

1. Integrate $\int\frac{3\,dx}{5-2x}$.

EXAMPLE 3 Trigonometric integrand

Integrate $\int\frac{\cos x}{\sin x}dx$.

We note that $d(\sin x) = \cos x\,dx$. This means that this integral fits the form of Eq. (28.2) with $u = \sin x$ and $du = \cos x\,dx$. Thus,

$$\begin{aligned} \int\frac{\cos x}{\sin x}dx &= \int\frac{\cos x\,dx \leftarrow du}{\sin x \leftarrow u} \\ &= \ln|\sin x| + C \end{aligned}$$

EXAMPLE 4 Comparing a logarithmic form with a power formula form

Integrate $\int\frac{x\,dx}{4-x^2}$.

This integral fits the form of Eq. (28.2) with $u = 4 - x^2$ and $du = -2x\,dx$. This means that we must introduce a factor of -2 into the numerator and a factor of $-\frac{1}{2}$ before the integral. Therefore,

$$\begin{aligned} \int\frac{x\,dx}{4-x^2} &= -\frac{1}{2}\int\frac{-2x\,dx \leftarrow du}{4-x^2 \leftarrow u} \\ &= -\frac{1}{2}\ln|4-x^2| + C \end{aligned}$$

We should note that if the quantity $4 - x^2$ were raised to any power other than -1, as in the example, we would have to employ the general power formula for integration. For example,

$$\begin{aligned} \int\frac{x\,dx}{(4-x^2)^2} &= -\frac{1}{2}\int\frac{-2x\,dx \leftarrow du}{(4-x^2)^2 \leftarrow u^2} \\ &= -\frac{1}{2}\frac{(4-x^2)^{-1}}{-1} + C = \frac{1}{2(4-x^2)} + C \end{aligned}$$

EXAMPLE 5 Exponential integrand

Integrate $\int \frac{e^{4x}\,dx}{1 + 3e^{4x}}$.

Since $d(1 + 3e^{4x})/dx = 12e^{4x}$, we see that we can use Eq. (28.2) with $u = 1 + 3e^{4x}$ and $du = 12e^{4x}\,dx$. Therefore, we write

$$\int \frac{e^{4x}\,dx}{1 + 3e^{4x}} = \frac{1}{12}\int \frac{12e^{4x}\,dx}{1 + 3e^{4x}} \quad \text{multiply and divide by 12}$$

$$= \frac{1}{12}\ln\left|1 + 3e^{4x}\right| + C \quad \text{integrate}$$

$$= \frac{1}{12}\ln(1 + 3e^{4x}) + C \quad 1 + 3e^{4x} > 0 \text{ for all } x$$

Practice Exercise

2. Integrate $\int \frac{\sin x\,dx}{1 - \cos x}$.

EXAMPLE 6 Definite integral with a trigonometric integrand

Evaluate $\int_0^{\pi/8} \frac{\sec^2 2\theta}{1 + \tan 2\theta}\,d\theta$.

Since $d(1 + \tan 2\theta) = 2\sec^2 2\theta\,d\theta$, we see that we can use Eq. (28.2) with $u = 1 + \tan 2\theta$ and $du = 2\sec^2 2\theta\,d\theta$. Therefore, we have

$$\int_0^{\pi/8} \frac{\sec^2 2\theta}{1 + \tan 2\theta}\,d\theta = \frac{1}{2}\int_0^{\pi/8} \frac{2\sec^2 2\theta\,d\theta}{1 + \tan 2\theta} \quad \text{multiply and divide by 2}$$

$$= \frac{1}{2}\ln\left|1 + \tan 2\theta\right|\Big|_0^{\pi/8} \quad \text{integrate}$$

$$= \frac{1}{2}(\ln|1 + 1| - \ln|1 + 0|) \quad \text{evaluate}$$

$$= \frac{1}{2}(\ln 2 - \ln 1) = \frac{1}{2}(\ln 2 - 0)$$

$$= \frac{1}{2}\ln 2$$

EXAMPLE 7 Find a volume using a logarithmic form

Find the volume within the piece of tapered tubing shown in Fig. 28.2, which can be described as the volume generated by revolving the region bounded by the curve of $y = \frac{3}{\sqrt{4x + 3}}$, $x = 2.50$ cm, and the axes about the x-axis.

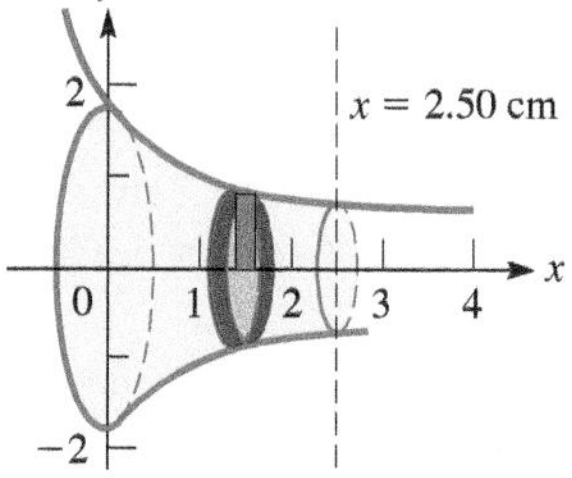

Fig. 28.2

The volume can be found by setting up only one integral by using a disc element of volume, as shown. The volume is found as follows:

$$V = \pi\int_0^{2.50} y^2\,dx = \pi\int_0^{2.50}\left(\frac{3}{\sqrt{4x + 3}}\right)^2 dx$$

$$= \pi\int_0^{2.50} \frac{9\,dx}{4x + 3} = \frac{9\pi}{4}\int_0^{2.50} \frac{4\,dx}{4x + 3}$$

$$= \frac{9\pi}{4}\ln(4x + 3)\Big|_0^{2.50} = \frac{9\pi}{4}(\ln 13.0 - \ln 3)$$

$$= \frac{9\pi}{4}\ln\frac{13.0}{3} = 10.4\text{ cm}^3$$

EXERCISES 28.2

In Exercises 1 and 2, make the given changes in the indicated examples of this section and then solve the given problems.

1. In Example 1, change $x + 1$ to $2x + 1$ and then integrate.
2. In Example 3, interchange $\cos x$ and $\sin x$ and then integrate.

In Exercises 3–30, integrate each of the given functions.

3. $\int \frac{dx}{1 + 4x}$
4. $\int \frac{dx}{4 - 9x}$
5. $\int \frac{2x\,dx}{4 - 3x^2}$
6. $\int \frac{4\sqrt{u}\,du}{1 + u\sqrt{u}}$
7. $\int_0^2 \frac{dx}{8 - 3x}$
8. $\int_{-1}^3 \frac{8x^3\,dx}{x^4 + 1}$
9. $\int \frac{0.4 \csc^2 2\theta\,d\theta}{\cot 2\theta}$
10. $\int \frac{\sin 3x}{\cos 3x}\,dx$
11. $\int_0^{\pi/2} \frac{\cos 4x\,dx}{1 + \sin 4x}$
12. $\int_0^{\pi/4} \frac{\sec^2 x\,dx}{15 + \tan x}$
13. $\int \frac{e^{-x}}{1 - e^{-x}}\,dx$
14. $\int \frac{5e^{3x}}{1 - e^{3x}}\,dx$
15. $\int \frac{1 + e^x}{x + e^x}\,dx$
16. $\int \frac{3\,e^t\,dt}{\sqrt{e^{2t} + 4e^t + 4}}$
17. $\int \frac{\sec x \tan x\,dx}{1 + 4\sec x}$
18. $\int \frac{\sin 2x}{1 - \sin^2 x}\,dx$
19. $\int_1^3 \frac{1 + x}{4x + 2x^2}\,dx$
20. $\int_1^2 \frac{4x + 6x^2}{x^2 + x^3}\,dx$
21. $\int \frac{0.5\,dr}{r \ln r}$
22. $\int \frac{dx}{x(1 + 2\ln x)}$
23. $\int \frac{2 + \sec^2 x}{2x + \tan x}\,dx$
24. $\int \frac{x + \cos 2x}{x^2 + \sin 2x}\,dx$
25. $\int \frac{6\,dx}{\sqrt{1 - 2x}}$
26. $\int \frac{4x\,dx}{(1 + x^2)^2}$
27. $\int \frac{x + 2}{x^2}\,dx$
28. $\int \frac{3v^2 - 2v}{v^2}\,dv$
29. $\int_0^{\pi/12} \frac{\sec^2 3x}{4 + \tan 3x}\,dx$
30. $\int_1^2 \frac{x^2 + 1}{x^3 + 3x}\,dx$

In Exercises 31–48, solve the given problems by integration.

31. Find the area bounded by $y(x + 1) = 1$, $x = 0$, $y = 0$, and $x = 2$. See Fig. 28.3.

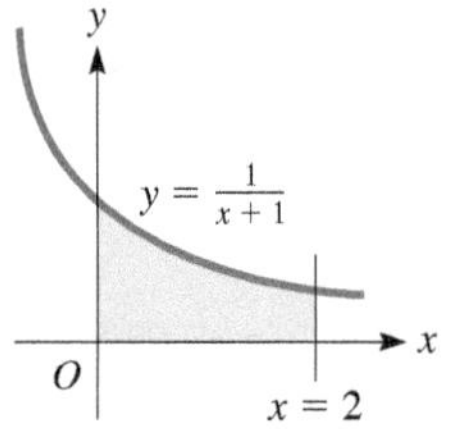

Fig. 28.3

32. Evaluate $\int_1^2 x^{-1}\,dx$ and $\int_2^4 x^{-1}\,dx$. Give a geometric interpretation of these two results.

33. Integrate $\int \frac{x - 6}{x + 6}\,dx$ by first using algebraic division to change the form of the integrand.

34. Integrate $\int \sec x\,dx$ by first multiplying the integrand by $\frac{\sec x + \tan x}{\sec x + \tan x}$ (a special form of 1).

35. Find the volume generated by revolving the region bounded by $y = 1/(x^2 + 1)$, $x = 0$, $x = 1$, and $y = 0$ about the y-axis. Use shells.

36. Find the volume of the solid generated by revolving the region bounded by $y = \frac{2}{\sqrt{3x + 1}}$, $x = 0$, $x = 3.5$, and $y = 0$ about the x-axis.

37. The general expression for the slope of a curve is $\frac{\sin x}{3 + \cos x}$. If the curve passes through the point $(\pi/3, 2)$, find its equation.

38. Show that $\int_0^3 \frac{dx}{2x + 2} = \ln 2$.

39. If $a > 0$ and $b > 0$, show that $\int_1^a \frac{du}{u} + \int_1^b \frac{du}{u} = \int_1^{ab} \frac{du}{u}$.

40. If $x > 0$, find $f(x)$ if $f''(x) = x^{-2}$, $f(1) = 0$, and $f(2) = 0$.

41. The acceleration a (in m/s^2) of an object is $a = (t + 4)^{-1}$. Find the velocity for $t = 4$ s, if the initial velocity is zero.

42. The pressure p (in kPa) and volume V (in cm^3) of a gas are related by $pV = 8600$. Find the average value of p from $V = 75$ cm^3 to $V = 95$ cm^3.

43. A hot metal rod with an initial temperature of 425°C is placed in a room with temperature 20°C. The time t (in min) required for the temperature T of the rod to cool to 175°C is given by $t = 8.72\int_{175}^{425} \frac{1}{T - 20.0}\,dT$. Find this time.

44. In determining the temperature that is absolute zero (0 K, or about −273°C), the equation $\ln T = -\int \frac{dr}{r - 1}$ is used. Here, T is the thermodynamic temperature and r is the ratio between certain specific vapour pressures. If $T = 273.16$ K for $r = 1.3361$, find T as a function of r (if $r > 1$ for all T).

45. The time t and electric current i for a certain circuit with a voltage E, a resistance R, and an inductance L is given by $t = L\int \frac{di}{E - iR}$. If $t = 0$ for $i = 0$, integrate and express i as a function of t.

46. Conditions are often such that a force proportional to the velocity tends to retard the motion of an object moving through a resisting medium. Under such conditions, the acceleration of a certain object moving down an inclined plane is given by $20 - v$. This leads to the equation $t = \int \frac{dv}{20 - v}$. If the object starts from rest, find the expression for the velocity as a function of time.

47. An architect designs a wall panel that can be described as the first-quadrant area bounded by $y = \dfrac{50}{x^2 + 20}$ and $x = 3.00$. If the area of the panel is 6.61 m^2, find the x-coordinate (in m) of the centroid of the panel.

48. The electric power P used by a robotic drilling machine is given by $P = 3\displaystyle\int \frac{\sin \pi t}{2 + \cos \pi t}\,dt$, where t is the time. Express P as a function of t.

Answers to Practice Exercises

1. $-\frac{3}{2}\ln|5 - 2x| + C$ **2.** $\ln|1 - \cos x| + C$

28.3 The Exponential Form

In working out the derivative for the exponential function, we obtained the result $de^u/dx = e^u(du/dx)$. This means that the differential of the exponential form is $d(e^u) = e^u\,du$. Reversing this form we find the proper form of the integral for the exponential function.

Integration of an Exponential Form

$$\int e^u\,du = e^u + C \qquad \textbf{(28.3)}$$

EXAMPLE 1

Integrate $\displaystyle\int xe^{x^2}\,dx$.

We identify $u = x^2$, and $du = 2x\,dx$. Then

$$\int xe^{x^2}\,dx = \frac{1}{2}\int e^{x^2}(2x\,dx)$$
$$= \frac{1}{2}\int e^u\,du$$
$$= \frac{1}{2}e^u + C = \frac{1}{2}e^{x^2} + C$$

Practice Exercise

1. Integrate $\displaystyle\int 6e^{-3x}\,dx$.

EXAMPLE 2 Exponential form—electric circuit

For an electric circuit containing a direct voltage source E, a resistance R, and an inductance L, the current i and time t are related by $ie^{Rt/L} = \dfrac{E}{L}\displaystyle\int e^{Rt/L}\,dt$. See Fig. 28.4. If $i = 0$ for $t = 0$, perform the integration and then solve for i as a function of t.

For this integral, we see that $u = \dfrac{Rt}{L}$, which means that $du = \dfrac{R\,dt}{L}$. The solution is then as follows:

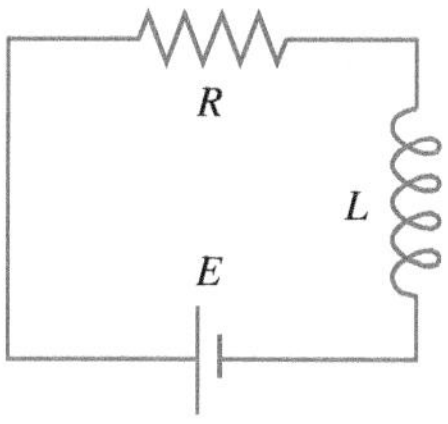

Fig. 28.4

$$ie^{Rt/L} = \frac{E}{L}\int e^{Rt/L}\,dt = \frac{E}{L}\left(\frac{L}{R}\right)\int e^{Rt/L}\left(\frac{R\,dt}{L}\right) \qquad \text{introduce factor } \frac{R}{L}$$

$$= \frac{E}{R}e^{Rt/L} + C \qquad \text{integrate}$$

$$0(e^0) = \frac{E}{R}e^0 + C, \qquad C = -\frac{E}{R} \qquad i = 0 \text{ for } t = 0\text{; evaluate } C$$

$$ie^{Rt/L} = \frac{E}{R}e^{Rt/L} - \frac{E}{R} \qquad \text{substitute for } C$$

$$i = \frac{E}{R} - \frac{E}{R}e^{-Rt/L} = \frac{E}{R}\left(1 - e^{-Rt/L}\right) \qquad \text{solve for } i$$

EXAMPLE 3 e^u in the denominator

Integrate $\int \frac{dx}{e^{3x}}$.

This integral can be put in proper form by writing it as $\int e^{-3x}\,dx$. In this form, $u = -3x$ and $du = -3\,dx$. Thus,

$$\int \frac{dx}{e^{3x}} = \int e^{-3x}\,dx = -\frac{1}{3}\int e^{-3x}(-3\,dx)$$
$$= -\frac{1}{3}e^{-3x} + C$$

EXAMPLE 4 Proper form using laws of exponents

Integrate $\int \frac{4e^{3x} - 3e^x}{e^{x+1}}\,dx$.

By dividing term by term and using the laws of exponents, this integral can be put in proper form for integration, and then integrated as follows:

$$\int \frac{4e^{3x} - 3e^x}{e^{x+1}}\,dx = \int \frac{4e^{3x}}{e^{x+1}}\,dx - \int \frac{3e^x}{e^{x+1}}\,dx$$
$$= 4\int e^{3x-(x+1)}\,dx - 3\int e^{x-(x+1)}\,dx \qquad \text{using } \frac{a^m}{a^n} = a^{m-n}$$
$$= 4\int e^{2x-1}\,dx - 3\int e^{-1}\,dx$$
$$= \frac{4}{2}\int e^{2x-1}(2\,dx) - \frac{3}{e}\int dx$$
$$= 2e^{2x-1} - \frac{3}{e}x + C$$

EXAMPLE 5 Definite integral with a trigonometric u

Evaluate: $\int_0^{\pi/2} (\sin 2\theta)(e^{\cos 2\theta})\,d\theta$.

With $u = \cos 2\theta$, $du = -2\sin 2\theta\,d\theta$, we have

$$\int_0^{\pi/2} (\sin 2\theta)(e^{\cos 2\theta})\,d\theta = -\frac{1}{2}\int_0^{\pi/2} (e^{\cos 2\theta})(-2\sin 2\theta\,d\theta)$$
$$= -\frac{1}{2}e^{\cos 2\theta}\Big|_0^{\pi/2} \qquad \text{integrate}$$
$$= -\frac{1}{2}\left(\frac{1}{e} - e\right) = 1.175 \qquad \text{evaluate}$$

EXAMPLE 6 Finding the equation of a curve

Find the equation of the curve for which $\frac{dy}{dx} = \frac{e^{\sqrt{x+1}}}{\sqrt{x+1}}$ if the curve passes through (0, 1).

The solution of this problem requires that we integrate the given function and then evaluate the constant of integration. Hence,

$$dy = \frac{e^{\sqrt{x+1}}}{\sqrt{x+1}}\,dx \qquad \int dy = \int \frac{e^{\sqrt{x+1}}}{\sqrt{x+1}}\,dx$$

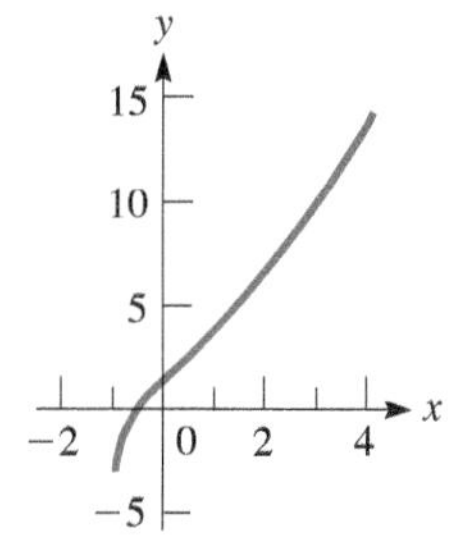

Fig. 28.5

For purposes of integrating the right-hand side,

$$u = \sqrt{x+1} \quad \text{and} \quad du = \frac{1}{2\sqrt{x+1}}\,dx$$

$$y = 2\int e^{\sqrt{x+1}}\left(\frac{1}{2\sqrt{x+1}}\,dx\right)$$

$$= 2e^{\sqrt{x+1}} + C$$

Letting $x = 0$ and $y = 1$, we have $1 = 2e + C$, or $C = 1 - 2e$. This means that the equation is

$$y = 2e^{\sqrt{x+1}} + 1 - 2e \qquad \text{see Fig. 28.5}$$

Practice Exercise

2. Integrate $\int \frac{e^{\sin x}}{\sec x}dx$.

EXERCISES 28.3

In Exercises 1 and 2, make the given changes in the indicated examples of this section and then solve the given problems.

1. In Example 1, change x to x^2 and x^2 to x^3 and then integrate.

2. In Example 4, change e^{x+1} to e^{x-1} and then integrate.

In Exercises 3–28, integrate each of the given functions.

3. $\int e^{7x}(7\,dx)$

4. $\int e^{x^4}(4x^3\,dx)$

5. $\int e^{9-5x}\,dx$

6. $\int 2e^{-11x}\,dx$

7. $\int_{-2}^{2} 6e^{s/2}\,ds$

8. $\int_{1}^{2} 3e^{4x}\,dx$

9. $\int 6x^2e^{x^3}\,dx$

10. $\int xe^{-x^2}\,dx$

11. $\int_{1}^{4} \frac{e^{\sqrt{x}}}{\sqrt{x}}\,dx$

12. $\int_{0}^{1} 4(\ln e^u)e^{-2u^2}\,du$

13. $\int 4(\sec\theta\tan\theta)e^{2\sec\theta}\,d\theta$

14. $\int(\csc^2 x)e^{\cot x}\,dx$

15. $\int\sqrt{e^{2y}+e^{3y}}dy$

16. $\int\frac{4\,dx}{x^2e^{1/x}}$

17. $\int_{1}^{3} 3e^{2x}(e^{-2x}-1)\,dx$

18. $\int_{0}^{0.5}\frac{3\,e^{3x+1}}{e^x}\,dx$

19. $\int\frac{2\,dx}{\sqrt{x}e^{\sqrt{x}}}$

20. $\int\frac{4\,dx}{e^{\sin x}\sec x}$

21. $\int\frac{e^{\tan^{-1}2x}}{4x^2+1}\,dx$

22. $\int\frac{e^{\sin^{-1}2x}\,dx}{\sqrt{1-4x^2}}$

23. $\int\frac{e^{\cos 3x}\,dx}{\csc 3x}$

24. $\int(e^x-e^{-x})^2\,dx$

25. $\int_{0}^{\pi}(\sin 2x)e^{\cos^2 x}\,dx$

26. $\int_{0}^{2}\frac{e^{2t}\,dt}{\sqrt{e^{2t}+4}}$

27. $\int\frac{6e^{4x}\,dx}{e^{4x}+1}$

28. $\int\frac{4e^x\,dx}{(1-e^x)^2}$

In Exercises 29–47, solve the given problems by integration.

29. Find the area bounded by $y = 3e^x$, $x = 0$, $y = 0$, and $x = 2$.

30. Find the area bounded by $x = a$, $x = b$, $y = 0$, and $y = e^x$. Explain the meaning of the result.

31. Integrate $\int 2e^{x^2+\ln x}\,dx$ by first showing that $e^{\ln x} = x$. (*Hint:* Let $y = e^{\ln x}$ and then take natural logarithms of both sides.)

32. Integrate $\int e^{x+\ln x}\frac{dx}{x}$. See Exercise 31.

33. Integrate $\int\frac{dx}{1+e^x}$ by first changing the form of the integrand by algebraic division.

34. Integrate $\int\frac{e^{2x}\,dx}{1+e^x}$ by first changing the form of the integrand by algebraic division.

35. Find the volume generated by revolving the region bounded by $y = e^{x^2}$, $x = 1$, $y = 0$, and $x = 2$ about the y-axis. See Fig. 28.6.

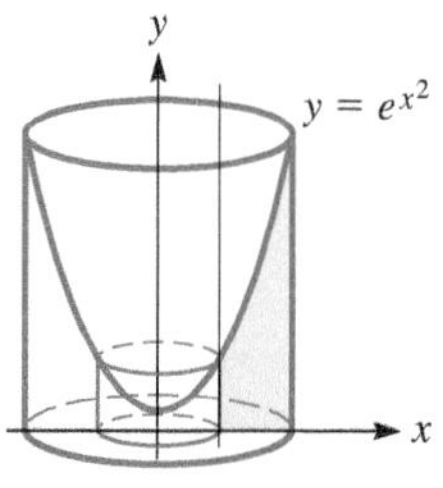

Fig. 28.6

36. Find the equation of the curve for which $dy/dx = \sqrt{e^{x+3}}$ if the curve passes through $(1, 0)$.

37. Find the average value of the function $y = 4e^{x/2}$ from $x = 0$ to $x = 4$.

38. Find the moment of inertia with respect to the y-axis of a flat plate that covers the first-quadrant region bounded by $y = e^{x^3}$, $x = 1$, and the axes.

39. Using $\frac{d(b^u)}{dx} = b^u \ln b\,\frac{du}{dx}$, show that

$$\int b^u\,du = \frac{b^u}{\ln b} + C\;(b > 0, b \neq 1).$$

40. Find the first-quadrant area bounded by $y = 2^x$ and $x = 3$. See Exercise 39.

41. For an electric circuit containing a voltage source E, a resistance R, and a capacitance C, an equation relating the charge q on the capacitor and the time t is $q e^{t/RC} = \frac{E}{R}\int e^{t/RC}\,dt$. See Fig. 28.7. If $q = 0$ for $t = 0$, perform the integration and then solve for q as a function of t.

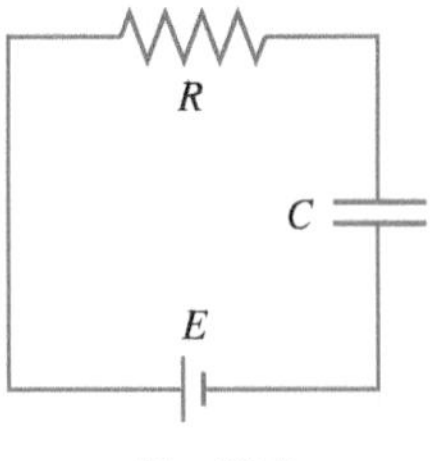

Fig. 28.7

42. In the theory dealing with energy propagation of lasers, the equation $E = a\int_0^{I_0} e^{-Tx}\,dx$ is used. Here, a, I_0, and T are constants. Evaluate this integral.

43. The St. Louis Gateway Arch (see Fig. 27.44) has a shape that is given approximately by (measurements in m) $y = -19.46(e^{x/38.92} + e^{-x/38.92}) + 230.9$. Find the total open area under the arch. (*Hint:* The limits of integration are the x-intercepts. To find them, set $y = 0$ and solve as an equation in quadratic form.)

44. The force F (in N) exerted by a robot programmed to staple carton sections together is given by $F = 6\int e^{\sin \pi t} \cos \pi t\, dt$, where t is the time (in s). Find F as a function of t if $F = 0$ for $t = 1.5$ s.

45. The average lifetime of a certain kind of emergency backup battery is 200 hours. The proportion of those batteries that are expected to last less than 100 hours is given by $p = \int_0^{100}(0.05)x^{-0.5}e^{-0.1x^{0.5}}dx$. Find p.

46. The expected number of hours per year that a certain wind turbine will not operate due to low wind speeds is given by $H = 8760\int_0^4 \frac{x}{64}e^{-\frac{x^2}{128}}\,dx$. Find H.

47. Two marathon runners are running in a race. The first runner's speed (in km/h) is $v = \frac{12.0}{0.200t + 1}$, where t is measured in hours. The second runner's speed (in km/h) is $v = 12.0e^{-t/6.00}$. Who is ahead (a) after 3.00 h, (b) after 4.00 h?

Answers to Practice Exercises

1. $-2e^{-3x} + C$ **2.** $e^{\sin x} + C$

28.4 Basic Trigonometric Forms

By noting the formulas for differentiating the six trigonometric functions, and the antiderivatives that are found using them, we have the following six integration formulas:

$$\int \sin u\, du = -\cos u + C \qquad (28.4)$$

$$\int \cos u\, du = \sin u + C \qquad (28.5)$$

$$\int \sec^2 u\, du = \tan u + C \qquad (28.6)$$

$$\int \csc^2 u\, du = -\cot u + C \qquad (28.7)$$

$$\int \sec u \tan u\, du = \sec u + C \qquad (28.8)$$

$$\int \csc u \cot u\, du = -\csc u + C \qquad (28.9)$$

EXAMPLE 1 Integration of $\sec^2 u\, du$

Integrate $\int x \sec^2 x^2\, dx$.

With $u = x^2$, $du = 2x\, dx$, we have

$$\int x \sec^2 x^2\, dx = \frac{1}{2}\int (\sec^2 x^2)(2x\, dx)$$

(du → $2x\,dx$; u → x^2)

$$= \frac{1}{2}\tan x^2 + C$$

using Eq. (28.6)

Practice Exercise

1. Integrate $\int \sin 5x\, dx$.

EXAMPLE 2 Integration of tan *u* sec *u* du

Integrate $\int \frac{\tan 2x}{\cos 2x} dx$.

By using the basic identity $\sec \theta = 1/\cos \theta$, we can transform this integral into the form $\int \sec 2x \tan 2x\, dx$. In this form, $u = 2x$ and $du = 2\, dx$. Therefore,

$$\int \frac{\tan 2x}{\cos 2x} dx = \int \sec 2x \tan 2x\, dx = \frac{1}{2}\int \sec \underset{u}{2x} \tan 2x\underset{du}{(2\, dx)}$$

$$= \frac{1}{2} \sec 2x + C \quad \text{using Eq. (28.8)}$$

Practice Exercise

2. Integrate $\int 6 \csc^2 3x\, dx$.

EXAMPLE 3 Integration of cos *u* du—velocity and displacement

The vertical velocity v (in cm/s) of the end of a vibrating rod is given by $v = 80 \cos 20\pi t$, where t is the time in seconds. Find the vertical displacement y (in cm) as a function of t if $y = 0$ for $t = 0$.

Since $v = dy/dt$, we have the following solution:

$$\frac{dy}{dt} = 80 \cos 20\pi t$$

$$\int dy = \int 80 \cos 20\pi t\, dt = \frac{80}{20\pi}\int (\cos 20\pi t)(20\pi\, dt) \quad \text{set up integration}$$

$$y = \frac{4}{\pi} \sin 20\pi t + C \quad \text{using Eq. (28.5)}$$

$$0 = 1.27 \sin 0 + C, \quad C = 0 \quad \text{evaluate } C$$

$$y = 1.27 \sin 20\pi t \quad \text{solution}$$

To find the integrals for the other trigonometric functions, we must change them to a form for which the integral can be determined by methods previously discussed. We can accomplish this by using the basic trigonometric relations.

The formula for $\int \tan u\, du$ is found by expressing the integral in the form $\int (\sin u/\cos u)\, du$. We recognize this as being a logarithmic form, where the u of the logarithmic form is $\cos u$ in this integral. The differential of $\cos u$ is $-\sin u\, du$. Therefore, we have

$$\int \tan u\, du = \int \frac{\sin u}{\cos u} du = -\int \frac{-\sin u\, du}{\cos u} = -\ln|\cos u| + C$$

The formula for $\int \cot u\, du$ is found by writing it in the form $\int (\cos u/\sin u)\, du$. In this manner, we obtain the result

$$\int \cot u\, du = \int \frac{\cos u}{\sin u} du = \int \frac{\cos u\, du}{\sin u} = \ln|\sin u| + C$$

The formula for $\int \sec u\, du$ is found by writing it in the form

$$\int \frac{\sec u(\sec u + \tan u)}{\sec u + \tan u} du$$

We see that this form is also a logarithmic form, since

$$d(\sec u + \tan u) = (\sec u \tan u + \sec^2 u)\, du$$

The right side of this equation is the expression appearing in the numerator of the integral. Thus,

$$\int \sec u\,du = \int \frac{\sec u(\sec u + \tan u)\,du}{\sec u + \tan u} = \int \frac{\sec u \tan u + \sec^2 u}{\sec u + \tan u}\,du$$

$$= \ln|\sec u + \tan u| + C$$

To obtain the formula for $\int \csc u\,du$, we write it in the form

$$\int \frac{\csc u(\csc u - \cot u)\,du}{\csc u - \cot u}$$

Thus, we have

$$\int \csc u\,du = \int \frac{\csc u(\csc u - \cot u)}{\csc u - \cot u}\,du = \int \frac{(-\csc u \cot u + \csc^2 u)\,du}{\csc u - \cot u}$$

$$= \ln|\csc u - \cot u| + C$$

Summarizing these results, we have the following integrals:

$$\int \tan u\,du = -\ln|\cos u| + C \quad \textbf{(28.10)}$$

$$\int \cot u\,du = \ln|\sin u| + C \quad \textbf{(28.11)}$$

$$\int \sec u\,du = \ln|\sec u + \tan u| + C \quad \textbf{(28.12)}$$

$$\int \csc u\,du = \ln|\csc u - \cot u| + C \quad \textbf{(28.13)}$$

EXAMPLE 4 Integration of tan *u* *du*

Integrate $\int \tan 4\theta\,d\theta$.

Noting that $u = 4\theta$, $du = 4\,d\theta$, we have

$$\int \tan 4\theta\,d\theta = \frac{1}{4}\int \tan 4\theta(4\,d\theta) \quad \text{multiply and divide by 4}$$

$$= -\frac{1}{4}\ln|\cos 4\theta| + C \quad \text{using Eq. (28.10)}$$

Practice Exercise

3. Integrate $\int 4x \cot x^2\,dx$.

EXAMPLE 5 Integration of sec *u* *du*

Integrate $\int \frac{\sec e^{-x}\,dx}{e^x}$.

In this integral, $u = e^{-x}$, $du = -e^{-x}\,dx$. Therefore,

$$\int \frac{\sec e^{-x}\,dx}{e^x} = -\int (\sec e^{-x})(-e^{-x}\,dx) \quad \text{introducing} - \text{sign}$$

$$= -\ln|\sec e^{-x} + \tan e^{-x}| + C \quad \text{using Eq. (28.12)}$$

EXAMPLE 6 Integration of csc *u du* and cot *u du*

Evaluate $\int_{\pi/6}^{\pi/4} \frac{1+\cos x}{\sin x}\,dx$.

$$\int_{\pi/6}^{\pi/4} \frac{1+\cos x}{\sin x}\,dx = \int_{\pi/6}^{\pi/4} \csc x\,dx + \int_{\pi/6}^{\pi/4} \cot x\,dx \qquad \frac{1}{\sin x} = \csc x, \quad \frac{\cos x}{\sin x} = \cot x$$

$$= \ln\left|\csc x - \cot x\right|\Big|_{\pi/6}^{\pi/4} + \ln\left|\sin x\right|\Big|_{\pi/6}^{\pi/4} \qquad \text{integrating}$$

$$= \ln\left|\sqrt{2}-1\right| - \ln\left|2-\sqrt{3}\right| + \ln\left|\frac{1}{2}\sqrt{2}\right| - \ln\left|\frac{1}{2}\right| \qquad \text{evaluating}$$

$$= \ln\frac{\left(\frac{1}{2}\sqrt{2}\right)\left(\sqrt{2}-1\right)}{\left(\frac{1}{2}\right)\left(2-\sqrt{3}\right)} = \ln\frac{2-\sqrt{2}}{2-\sqrt{3}} = 0.782$$

EXERCISES 28.4

In Exercises 1 and 2, make the given changes in the indicated examples of this section and then solve the given problems.

1. In Example 1, change x to x^2 and $\sec^2 x^2$ to $\sec^2 x^3$ and then integrate.
2. In Example 4, change tan to cot and then integrate.

In Exercises 3–26, integrate each of the given functions. You may need to use trigonometric identities to change some of the integrals into a basic trigonometric form.

3. $\int \cos 99x\,dx$
4. $\int 4\sin(2-x)\,dx$
5. $\int 0.3\sec^2 3\theta\,d\theta$
6. $\int \csc 8x \cot 8x\,dx$
7. $\int \sec \frac{1}{2}x \tan \frac{1}{2}x\,dx$
8. $\int e^{2x+1}\csc^2(e^{2x+1})\,dx$
9. $\int_{0.5}^{1} x^2 \cot x^3\,dx$
10. $\int_0^1 6\sin\frac{1}{2}t \sec\frac{1}{2}t\,dt$
11. $\int 3\phi \sec^2\phi^2 \cos\phi^2\,d\phi$
12. $\int (\cos^2 4x - \sin^2 4x)\,dx$
13. $\int \frac{\sin(1/x)}{x^2}\,dx$
14. $\int \frac{3\,dx}{\cos 4x}$
15. $\int_0^{\pi/6} \frac{dx}{\cos^2 2x}$
16. $\int_0^1 \frac{2e^s\,ds}{\sec e^s}$
17. $\int \sqrt{\frac{1-\cos x}{2}}\,dx\ \left(0 \le x < \frac{\pi}{2}\right)$
18. $\int \frac{\sin 2x}{\cos^2 x}\,dx$
19. $\int \sqrt{\tan^2 2x + 1}\,dx$
20. $\int 5(\tan u)(\ln\cos u)\,du$
21. $\int \frac{2\tan T}{1-\tan^2 T}\,dT$
22. $\int \frac{1-\cot^2 x}{\cos^2 x}\,dx$
23. $\int \frac{1-\sin x}{1+\cos x}\,dx$
24. $\int \frac{1+\sec^2 x}{x+\tan x}\,dx$
25. $\int_0^{\pi/9} \sin 3x(\csc 3x + \sec 3x)\,dx$
26. $\int_{\pi/4}^{\pi/3} (1+\sec x)^2\,dx$

In Exercises 27–40, solve the given problems by integration.

27. Find the area bounded by $y = 2\tan x$, $x = \frac{\pi}{4}$, and $y = 0$.
28. Find the area under the curve $y = \sin x$ from $x = 0$ to $x = \pi$.
29. Although $\int \frac{dx}{1+\sin x}$ does not appear to fit a form for integration, show that it can be integrated by multiplying the numerator and the denominator by $1 - \sin x$.
30. Change the integrand of $\int \sin^2 2x \cos x\,dx$ to a form that can be integrated by methods of this section.
31. Integrate $\int \sec^2 x \tan x\,dx$ (a) with $u = \tan x$, and (b) with $u = \sec x$. Explain why the answers appear to be different.
32. Evaluate $\int_a^{a+2\pi} \sin x\,dx$ for any real value of a. Show an interpretation of the result in terms of the area under a curve.
33. Find the volume generated by revolving the region bounded by $y = \sec x$, $x = 0$, $x = \frac{\pi}{3}$, and $y = 0$ about the x-axis.
34. Find the volume generated by revolving the region bounded by $y = \cos x^2$, $x = 0$, $y = 0$, and $x = 1$ about the y-axis.
35. The angular velocity ω (in rad/s) of a pendulum is $\omega = -0.25\sin 2.5t$. Find the angular displacement θ as a function of t if $\theta = 0.10$ for $t = 0$.
36. If the current i (in A) in a certain electric circuit is given by $i = 110\cos 377t$, find the expression for the voltage across a 500 μF capacitor as a function of time. The initial voltage is zero. Show that the voltage across the capacitor is 90° out of phase with the current.

37. A fin on a wind-direction indicator has a shape that can be described as the region bounded by $y = \tan x^2$, $y = 0$, and $x = 1$. Find the x-coordinate (in m) of the centroid of the fin if its area is $0.3984\ m^2$.

38. A force is given as a function of the distance from the origin as $F = \dfrac{2 + \tan x}{\cos x}$. Express the work done by this force as a function of x if $W = 0$ for $x = 0$.

39. The probability that the phase error in a tracking device is between 0 and $\pi/4$ is given by $p = \int_0^{\pi/4} \cos x\, dx$. Find p.

40. The number of hours h of daylight each day during the year in Toronto, Ontario, can be approximated by $h = 12.2 + 3.3 \sin\left(\frac{2\pi}{365}t - \frac{160\pi}{365}\right)$, where t is measured in days ($t = 15$ is January 15, etc.). Find the average number of daylight hours during the month of January.

Answers to Practice Exercises

1. $-\frac{1}{5}\cos 5x + C$ **2.** $-2\cot 3x + C$ **3.** $2\ln|\sin x^2| + C$

28.5 Other Trigonometric Forms

Integrals involving powers of trigonometric functions are integrated using the following trigonometric identities, which were developed in Chapter 20:

$$\cos^2 x + \sin^2 x = 1 \quad \textbf{(28.14)}$$
$$2\cos^2 x = 1 + \cos 2x \quad \textbf{(28.15)}$$
$$2\sin^2 x = 1 - \cos 2x \quad \textbf{(28.16)}$$
$$1 + \tan^2 x = \sec^2 x \quad \textbf{(28.17)}$$
$$1 + \cot^2 x = \csc^2 x \quad \textbf{(28.18)}$$

The identities in the first column are used to integrate products of integer powers of sines and cosines as summarized here.

Integrals of the Form $\int \sin^m x \cos^n x\, dx$

Case 1. m is odd	**1.** Write $\sin^m x$ as $\sin^{m-1} x \sin x$. **2.** Use Eq. (28.14) to write the even power of sines in terms of cosines. **3.** Use the substitution $u = \cos x$.
Case 2. n is odd	**1.** Write $\cos^n x$ as $\cos^{n-1} x \cos x$. **2.** Use Eq. (28.14) to write the even power of cosines in terms of sines. **3.** Use the substitution $u = \sin x$.
Case 3. Both m and n are even	Use Eqs. (28.15) and (28.16) to halve the even powers.

EXAMPLE 1 Integration with an odd power of sin *u*

Integrate $\int \sin^3 x \cos^2 x\, dx$.

Here $m = 3$ and $n = 2$. Because m is odd, we follow the steps for Case 1.

Write $\sin^3 x = \sin^2 x \sin x$: $\int \sin^3 x \cos^2 x\, dx = \int \sin^2 x \sin x \cos^2 x\, dx$

Use Eq. (28.14) to substitute $\sin^2 x = 1 - \cos^2 x$:

$$\begin{aligned}\int \sin^2 x \sin x \cos^2 x\, dx &= \int (1 - \cos^2 x) \sin x \cos^2 x\, dx \\ &= \int \cos^2 x \sin x\, dx - \int \cos^4 x \sin x\, dx\end{aligned}$$

Use the substitution $u = \cos x$, $du = -\sin x\, dx$ and integrate each term:

$$\int \sin^3 x \cos^2 x\, dx = -\int \cos^2 x(-\sin x\, dx) + \int \cos^4 x(-\sin x\, dx)$$
$$= -\frac{1}{3}\cos^3 x + \frac{1}{5}\cos^5 x + C$$

EXAMPLE 2 Integration with an odd power of cos *u*

Integrate $\int \cos^5 2x\, dx$.

Here $m = 0$ and $n = 5$. Because n is odd, we follow the steps for Case 2.

Write $\cos^5 2x = \cos^4 2x \cos 2x$: $\int \cos^5 2x\, dx = \int \cos^4 2x \cos 2x\, dx$

Use Eq. (28.14) to substitute $\cos^2 2x = 1 - \sin^2 2x$:

$$\int \cos^4 2x \cos 2x\, dx = \int (1 - \sin^2 2x)^2 \cos 2x\, dx$$
$$= \int (1 - 2\sin^2 2x + \sin^4 2x)\cos 2x\, dx$$
$$= \int \cos 2x\, dx - \int 2\sin^2 2x \cos 2x\, dx + \int \sin^4 2x \cos 2x\, dx$$

For the first term, use $u = 2x$, $du = 2\,dx$. For the others, use $u = \sin 2x$, $du = 2\cos 2x\, dx$.

$$\int \cos^5 2x\, dx = \frac{1}{2}\int \cos 2x(2dx) - \int \sin^2 2x(2\cos 2x\, dx) + \frac{1}{2}\int \sin^4 2x(2\cos 2x\, dx)$$
$$= \frac{1}{2}\sin 2x - \frac{1}{3}\sin^3 2x + \frac{1}{10}\sin^5 2x + C$$

EXAMPLE 3 Integration with an even power of sin *u*

Integrate $\int \sin^2 2x\, dx$.

Here $m = 2$ and $n = 0$. Because the only nonzero power is even, we follow the steps for Case 3. We use Eq. (28.16) with $2x$ instead of x, so that $\sin^2 2x = \frac{1}{2}(1 - \cos 4x)$. Then

$$\int \sin^2 2x\, dx = \int \left[\frac{1}{2}(1 - \cos 4x)\right] dx$$
$$= \frac{1}{2}\int dx - \frac{1}{8}\int \cos 4x(4\, dx)$$
$$= \frac{x}{2} - \frac{1}{8}\sin 4x + C$$

The following steps are used for integrating products of tangents and secants or of cotangents and cosecants.

Integrals of the Form $\int \tan^m x \sec^n x\, dx$ or $\int \cot^m x \csc^n x\, dx$

Case 1. m and n are both odd	**1.** Set aside a factor $\sec x \tan x$ or $\csc x \cot x$. **2.** Use Eq. (28.17) to write the even power of tangents in terms of secants, or Eq. (28.18) to write the even power of cotangents in terms of cosecants. **3.** Use the substitution $u = \sec x$ or $u = \csc u$.

Case 2. n is even	**1.** If possible, set aside a factor $\sec^2 x$ or $\csc^2 x$. If not possible, use Eq. (28.17) to change a factor $\tan^2 x$ into a term with $\sec^2 x$, or Eq. (28.18) to change a factor $\cot^2 x$ into a term with $\csc^2 x$. Repeat if necessary. **2.** Use Eq. (28.17) to write the even power of secants in terms of tangents, or Eq. (28.18) to write the even power of cosecants in terms of cotangents. **3.** Use the substitution $u = \tan x$ or $u = \cot u$.
Case 3. m is even, n is odd	**1.** Use Eq. (28.17) to write the even power of tangents in terms of secants, or Eq. (28.18) to write the even power of cotangents in terms of cosecants. **2.** Integrate the odd power of secants or cosecants by parts (see Section 28.7).

EXAMPLE 4 Integration with a power of sec *u*

Integrate $\int \tan t \sec^3 t\, dt$.

Here $m = 1$ and $n = 3$. Because both are odd, we follow the steps for Case 1.

Set aside a factor $\sec t \tan t$: $\int \tan t \sec^3 t\, dt = \int \sec^2 t(\sec t \tan t)\, dt$

No even power of secant is left, so no further substitution is needed.

Use the power rule with $u = \sec t$, $du = \sec t \tan t\, dt$:

$$\int \tan t \sec^3 t\, dt = \int \sec^2 t(\sec t \tan t)\, dt = \frac{1}{3}\sec^3 t\, dt + C$$

Practice Exercise

1. Integrate $\int \sec^4 x\, dx$.

EXAMPLE 5 Integration with a power of tan *u*

Integrate $\int \tan^5 x\, dx$.

Here $m = 5$ and $n = 0$. Because n is even, we follow the steps for Case 2.

Since we cannot set aside a factor $\sec^2 x$, we use Eq. (28.17) on $\tan^2 x$:

$$\begin{aligned}\int \tan^5 x\, dx &= \int \tan^3 x \tan^2 x\, dx\\ &= \int \tan^3 x(\sec^2 x - 1)\, dx\\ &= \int \tan^3 x \sec^2 x\, dx - \int \tan^3 x\, dx\end{aligned}$$

No further substitution is needed for the first term. Eq. (28.17) is used again for the second term:

$$\tan^3 x = \tan x \tan^2 x = \tan x(\sec^2 x - 1) = \tan x \sec^2 x - \tan x$$

Use $u = \tan x$, $du = \sec^2 x\, dx$ for the terms with $\sec^2 x$. Integrate $\tan x$ with Eq. (28.10).

$$\int \tan^5 x\, dx = \frac{1}{4}\tan^4 x - \frac{1}{2}\tan^2 x - \ln|\cos x| + C$$

EXAMPLE 6 Integration of another trigonometric form

Integrate $\int_0^{\pi/4} \frac{\tan^3 x}{\sec^3 x}\,dx$.

Of several possible ways in which the integrand can be transformed into an integrable form, among the easiest is the following:

$$\int_0^{\pi/4} \frac{\tan^3 x}{\sec^3 x}\,dx = \int_0^{\pi/4} \frac{\sin^3 x}{\cos^3 x}\frac{1}{\sec^3 x}\,dx \qquad \tan x = \sin x/\cos x$$

$$= \int_0^{\pi/4} \sin^3 x = \int_0^{\pi/4} (1 - \cos^2 x)\sin x\,dx \qquad \cos x \sec x = 1$$

$$= \int_0^{\pi/4} \sin x\,dx - \int_0^{\pi/4} \cos^2 x \sin x\,dx$$

$$= -\cos x + \frac{1}{3}\cos^3 x \Big|_0^{\pi/4} \qquad \text{integrate}$$

$$= -\frac{\sqrt{2}}{2} + \frac{1}{3}\left(\frac{\sqrt{2}}{2}\right)^3 - \left(-1 + \frac{1}{3}\right) \qquad \text{evaluate}$$

$$= \frac{8 - 5\sqrt{2}}{12} = 0.0774$$

ROOT-MEAN-SQUARE VALUE

The **root-mean-square (RMS) value** of a function $y = f(t)$ over a period T is defined by the following equation:

$$y_{\text{rms}} = \sqrt{\frac{1}{T}\int_0^T y^2\,dt} \qquad \textbf{(28.19)}$$

One of the principal applications of RMS values is to alternating currents, as in the next example.

EXAMPLE 7 Root-mean-square value of an AC current

■ See the chapter introduction.

Find the root-mean-square value of the electric current i (in A) used by a plasma TV, for which $i = 3.75 \cos 120\pi t$, for one period.

The period is $\frac{2\pi}{120\pi} = \frac{1}{60.0}$ s. Thus, we must find the square root of the integral

■ In most countries, the RMS voltage is 240 V. In Canada and the United States, it is 120 V.

$$\frac{1}{1/60.0}\int_0^{1/60.0} (3.75 \cos 120\pi t)^2\,dt = 844\int_0^{1/60.0} \cos^2 120\pi t\,dt$$

Evaluating this integral, we have

$$844\int_0^{1/60.0} \cos^2 120\pi t\,dt = 422\int_0^{1/60.0} (1 + \cos 240\pi t)\,dt$$

$$= 422t\Big|_0^{1/60.0} + \frac{422}{240\pi}\int_0^{1/60.0} \cos 240\pi t(240\pi\,dt)$$

$$= 7.033 + \frac{422}{240\pi}\sin 240\pi t\Big|_0^{1/60.0} = 7.033$$

This means the root-mean-square current is

$$i_{rms} = \sqrt{7.033} = 2.65 \text{ A}$$

This value of the current, often referred to as the *effective current*, is the value of direct current that would produce the same quantity of heat energy in the same time. It is important in the design of electronic equipment and electric appliances.

EXERCISES 28.5

In Exercises 1 and 2, answer the given questions related to the indicated examples of this section.

1. In Example 3, change $2x$ to $3x$ and then integrate.
2. In Example 5, change the exponent of 5 to a 3 and then integrate.

In Exercises 3–34, integrate each of the given functions.

3. $\int \sin^2 x \cos x\, dx$
4. $\int \sin x \cos^5 x\, dx$
5. $\int \sin^3 2x\, dx$
6. $\int 3 \cos^3 T\, dT$
7. $\int \tan^2 x \sec^2 x\, dx$
8. $\int \sin^3 x \cos^6 x\, dx$
9. $\int_0^{\pi/4} 5 \sin^5 x\, dx$
10. $\int_{\pi/3}^{\pi/2} 10 \sin t(1 - \cos 2t)^2\, dt$
11. $\int \sin^2 x\, dx$
12. $\int \cos^2 2x\, dx$
13. $\int 2(1 + \cos 3\phi)^2\, d\phi$
14. $\int_0^1 \sin^2 4x\, dx$
15. $\int \tan^3 x\, dx$
16. $\int \frac{6 \cot^2 y}{\tan y}\, dy$
17. $\int_0^{\pi/4} \tan x \sec^4 x\, dx$
18. $\int \cot 4x \csc^4 4x\, dx$
19. $\int \tan^4 2x\, dx$
20. $\int 4 \cot^4 x\, dx$
21. $\int 0.5 \sin s \sin 2s\, ds$
22. $\int \sqrt{\tan x} \sec^4 x\, dx$
23. $\int (\sin x + \cos x)^2\, dx$
24. $\int (\tan 2x + \cot 2x)^2\, dx$
25. $\int \frac{1 - \cot x}{\sin^4 x}\, dx$
26. $\int \frac{(\sin u + \sin^2 u)^2}{\sec u}\, du$
27. $\int_{\pi/6}^{\pi/4} \cot^5 p\, dp$
28. $\int_{\pi/6}^{\pi/3} \frac{2\, dx}{1 + \sin x}$
29. $\int \sec^6 x\, dx$
30. $\int \tan^7 x\, dx$
31. $\int \frac{\sin 2x}{\cos^3 x}\, dx$
32. $\int \frac{\sec^2 t \tan t}{4 + \sec^2 t}\, dt$
33. $\int \frac{\sec e^{-x}}{e^x}\, dx$
34. $\int_0^{\pi/4} \sqrt{1 + \cos 4x}\, dx$

In Exercises 35–52, solve the given problems by integration.

35. Using the identity $\sin \alpha \cos \beta = \frac{1}{2}[\sin(\alpha + \beta) + \sin(\alpha - \beta)]$, integrate $\int \sin 4x \cos 5x\, dx$.
36. Using the identity $\cos \alpha \cos \beta = \frac{1}{2}[\cos(\alpha + \beta) + \cos(\alpha - \beta)]$, integrate $\int \cos 3x \cos 4x\, dx$.
37. Find the volume generated by revolving the region bounded by $y = \sin x$ and $y = 0$, from $x = 0$ to $x = \pi$, about the x-axis.
38. Find the volume generated by revolving the region bounded by $y = \tan^3(x^2)$, $y = 0$, and $x = \frac{\pi}{4}$ about the y-axis.
39. Find the area bounded by $y = \sin x$, $y = \cos x$, and $x = 0$ in the first quadrant.
40. The length of an arc along a function is given by $s = \int_a^b \sqrt{1 + \left(\frac{dy}{dx}\right)^2}\, dx$. Find the length of the curve $y = \ln \cos x$ from $x = 0$ to $x = \frac{\pi}{3}$.
41. Show that $\int \sin x \cos x\, dx$ can be integrated in two ways. Explain the difference in the answers.
42. For $n > 0$, show that $\int \tan x \sec^n x\, dx = \frac{1}{n} \sec^n x + C$.
43. Show that $\int_0^{\pi} \sin^2 nx\, dx = \frac{1}{2}\pi$, where n is any positive integer.
44. The velocity v (in cm/s) of an object is $v = \cos^2 \pi t$. How far does the object move in 4.0 s?
45. The acceleration a (in m/s²) of an object is $a = \sin^2 t \cos t$. If the object starts at the origin with a velocity of 6 m/s, what is its position at time t?
46. Find the root-mean-square current in a circuit from $t = 0$ s to $t = 0.50$ s if $i = i_0 \sin t\sqrt{\cos t}$.
47. In the study of the rate of radiation by an accelerated charge, the following integral must be evaluated: $\int_0^{\pi} \sin^3 \theta\, d\theta$. Find the value of the integral.
48. In finding the volume of a special O-ring for a space vehicle, the integral $\int \frac{\sin^2 \theta}{\cos^2 \theta}\, d\theta$ must be evaluated. Perform this integration.
49. For a voltage $V = 340 \sin 120\pi t$, show that the root-mean-square voltage for one period is 240 V.
50. For a current $i = i_0 \sin \omega t$, show that the root-mean-square current for one period is $i_0/\sqrt{2}$.
51. In the analysis of the intensity of light from a certain source, the equation $I = A\int_{-a/2}^{a/2} \cos^2[b\pi(c - x)]\, dx$ is used. Here, A, a, b, and c are constants. Evaluate this integral. (The simplification is quite lengthy.)

52. In the study of the lifting force L due to a stream of fluid passing around a cylinder, the equation $L = k\int_0^{2\pi}(a\sin\theta + b\sin^2\theta - b\sin^3\theta)\,d\theta$ is used. Here, k, a, and b are constants and θ is the angle from the direction of flow. Evaluate the integral.

Answer to Practice Exercise

1. $\tan x + \frac{1}{3}\tan^3 x + C$

28.6 Inverse Trigonometric Forms

For reference, Eq. (27.10) is $\frac{d(\sin^{-1}u)}{dx} = \frac{1}{\sqrt{1-u^2}}\frac{du}{dx}$.

In this section we will integrate certain algebraic functions using inverse trigonometric functions. This is one of the principal uses of the inverse trigonometric functions in calculus.

We begin by finding the differential of $\sin^{-1}(u/a)$, where a is a constant. Using Eq. (27.10), we have

$$d\left(\sin^{-1}\frac{u}{a}\right) = \frac{1}{\sqrt{1-(u/a)^2}}\frac{du}{a} = \frac{a}{\sqrt{a^2-u^2}}\frac{du}{a} = \frac{du}{\sqrt{a^2-u^2}}$$

Noting this differentiation formula, and the antiderivative of the result, we have the following inverse sine integration formula:

Integral of an Inverse Sine Form

$$\int\frac{du}{\sqrt{a^2-u^2}} = \sin^{-1}\frac{u}{a} + C \qquad \textbf{(28.20)}$$

EXAMPLE 1

Integrate $\int\frac{dx}{\sqrt{9-x^2}}$.

We identify $u = x$, $du = dx$, and $a = 3$. Then

$$\int\frac{dx}{\sqrt{9-x^2}} = \int\frac{dx}{\sqrt{3^2-x^2}} = \sin^{-1}\frac{x}{3} + C$$

Practice Exercise

1. Integrate $\int\frac{dx}{\sqrt{9-4x^2}}$.

By finding the differential of $\tan^{-1}(u/a)$, we have

$$d\left(\tan^{-1}\frac{u}{a}\right) = \frac{1}{1+(u/a)^2}\frac{du}{a} = \frac{a^2}{a^2+u^2}\frac{du}{a} = \frac{a\,du}{a^2+u^2}$$

Again, noting the antiderivative, we have another important integration formula:

Integral of an Inverse Tangent Form

$$\int\frac{du}{a^2+u^2} = \frac{1}{a}\tan^{-1}\frac{u}{a} + C \qquad \textbf{(28.21)}$$

EXAMPLE 2

Integrate $\int\frac{dx}{25+4x^2}$.

We identify $u = 2x$, $du = 2dx$, and $a = 5$. Then

$$\int\frac{dx}{25+4x^2} = \frac{1}{2}\int\frac{2dx}{5^2+(2x)^2} = \frac{1}{2}\tan^{-1}\frac{2x}{5} + C$$

Practice Exercise

2. Integrate $\int\frac{dx}{4+9x^2}$.

EXAMPLE 3 Inverse tangent form—volume flow rate of a liquid

The volume flow rate Q (in m^3/s) of a constantly flowing liquid is given by $Q = 24\int_0^2 \frac{dx}{6 + x^2}$, where x is the distance from the centre of flow. Find the value of Q.

For the integral, we see that it fits Eq. (28.21) with $u = x$, $du = dx$, and $a = \sqrt{6}$.

$$Q = 24\int_0^2 \frac{dx}{6 + x^2} = 24\int_0^2 \frac{dx}{(\sqrt{6})^2 + x^2}$$

$$= \frac{24}{\sqrt{6}} \tan^{-1} \left. \frac{x}{\sqrt{6}} \right|_0^2$$

$$= \frac{24}{\sqrt{6}}\left(\tan^{-1}\frac{2}{\sqrt{6}} - \tan^{-1} 0\right)$$

$$= 6.71\ m^3/s$$

EXAMPLE 4 Completing the square to fit the $\tan^{-1} u$ form

Integrate $\int_{-1}^{3} \frac{dx}{x^2 + 6x + 13}$.

At first glance, it does not appear that this integral fits any of the forms presented up to this point. However, by writing the denominator in the form $(x^2 + 6x + 9) + 4 = (x + 3)^2 + 2^2$ (see Section 7.2), ***we recognize that*** $u = x + 3$, $du = dx$, and $a = 2$. Thus,

$$\int_{-1}^{3} \frac{dx}{x^2 + 6x + 13} = \int_{-1}^{3} \frac{dx}{(x + 3)^2 + 2^2}$$

(dx ← du; $x + 3$ ← u)

$$= \frac{1}{2}\tan^{-1} \left. \frac{x + 3}{2} \right|_{-1}^{3} \quad \text{integrate}$$

$$= \frac{1}{2}(\tan^{-1} 3 - \tan^{-1} 1) \quad \text{evaluate}$$

$$= 0.2318$$

Now, we can see the use of completing the square when we are transforming integrals into proper form.

EXAMPLE 5 Inverse tangent and logarithmic forms

Integrate $\int \frac{2r + 5}{r^2 + 9} dr$.

By writing this integral as the sum of two integrals, we may integrate each of these separately:

$$\int \frac{2r + 5}{r^2 + 9} dr = \int \frac{2r\,dr}{r^2 + 9} + \int \frac{5\,dr}{r^2 + 9}$$

The first integral is a logarithmic form, and the second is an inverse tangent form. For the first, $u = r^2 + 9$, $du = 2r\,dr$. For the second, $u = r$, $du = dr$, $a = 3$.

$$\int \frac{2r\,dr}{r^2 + 9} + 5\int \frac{dr}{r^2 + 9} = \ln|r^2 + 9| + \frac{5}{3}\tan^{-1}\frac{r}{3} + C$$

LEARNING TIP

The inverse trigonometric integral forms show very well the importance of ***proper recognition of the form of the integral.*** It is important that these forms are not confused with those of the general power rule or the logarithmic form.

EXAMPLE 6 Comparing the inverse sine, power, and logarithmic forms

The integral $\int \frac{dx}{\sqrt{1-x^2}}$ is of the inverse sine form with $u = x$, $du = dx$, and $a = 1$. Thus,

$$\int \frac{dx}{\sqrt{1-x^2}} = \sin^{-1} x + C$$

The integral $\int \frac{x\,dx}{\sqrt{1-x^2}}$ is not of the inverse sine form due to the factor of x in the numerator. It is integrated by use of the general power rule, with $u = 1 - x^2$, $du = -2x\,dx$, and $n = -\frac{1}{2}$. Thus,

$$\int \frac{x\,dx}{\sqrt{1-x^2}} = -\sqrt{1-x^2} + C$$

The integral $\int \frac{x\,dx}{1-x^2}$ is of the basic logarithmic form with $u = 1 - x^2$ and $du = -2x\,dx$. If $1 - x^2$ is raised to any power other than 1 in the denominator, we would use the general power rule. Thus,

$$\int \frac{x\,dx}{1-x^2} = -\frac{1}{2}\ln|1-x^2| + C$$

EXAMPLE 7 Recognizing different forms

Determine the form of each of the following integrals.

(a) $\int \frac{dx}{1+x^2}$	Inverse tangent form	$u = x$, $du = dx$
(b) $\int \frac{x\,dx}{1+x^2}$	Logarithmic form	$u = 1 + x^2$, $du = 2x\,dx$
(c) $\int \frac{x\,dx}{\sqrt{1+x^2}}$	General power form	$u = 1 + x^2$, $du = 2x\,dx$
(d) $\int \frac{dx}{1+x}$	Logarithmic form	$u = 1 + x$, $du = dx$
(e) $\int \frac{x\,dx}{\sqrt{1-x^4}}$	Inverse sine form	$u = x^2$, $du = 2x\,dx$
(f) $\int \frac{x\,dx}{1+x^4}$	Inverse tangent form	$u = x^2$, $du = 2x\,dx$

There are a number of integrals whose forms appear to be similar to those in Examples 6 and 7, but which do not fit the forms we have discussed. They include

$$\int \frac{dx}{\sqrt{x^2-1}} \qquad \int \frac{dx}{\sqrt{1+x^2}} \qquad \int \frac{dx}{1-x^2} \qquad \int \frac{dx}{x\sqrt{1+x^2}}$$

We will develop methods to integrate some of these forms, and all of them can be integrated by the tables discussed in Section 28.11.

EXERCISES 28.6

In Exercises 1 and 2, answer the given questions related to the indicated examples of this section.

1. In Example 1, change dx to $x\,dx$ and then integrate.
2. In Example 3, what change must be made in the integrand in order that the integration would lead to a result of $\ln(6 + x^2) + C$?

In Exercises 3–30, integrate each of the given functions.

3. $\displaystyle\int \frac{dx}{\sqrt{4 - x^2}}$
4. $\displaystyle\int \frac{dx}{\sqrt{49 - x^2}}$
5. $\displaystyle\int \frac{dx}{64 + x^2}$
6. $\displaystyle\int \frac{6p^2\,dp}{4 + p^6}$
7. $\displaystyle\int \frac{8x\,dx}{\sqrt{36 - 16x^4}}$
8. $\displaystyle\int_0^1 \frac{2\,dx}{\sqrt{9 - 4x^2}}$
9. $\displaystyle\int_0^2 \frac{3e^{-t}\,dt}{1 + 9e^{-2t}}$
10. $\displaystyle\int_1^3 \frac{4\,dx}{49 + 4x^2}$
11. $\displaystyle\int_0^{0.4} \frac{2\,dx}{\sqrt{4 - 5x^2}}$
12. $\displaystyle\int \frac{dx}{2\sqrt{x}\sqrt{1 - x}}$
13. $\displaystyle\int \frac{8x\,dx}{9x^2 + 16}$
14. $\displaystyle\int \frac{4y\,dy}{\sqrt{25 - 16y^2}}$
15. $\displaystyle\int_1^e \frac{3\,du}{u[1 + (\ln u)^2]}$
16. $\displaystyle\int_0^1 \frac{4x\,dx}{1 + x^4}$
17. $\displaystyle\int \frac{e^x\,dx}{\sqrt{81 - e^{2x}}}$
18. $\displaystyle\int \frac{\sec^2 x\,dx}{\sqrt{1 - \tan^2 x}}$
19. $\displaystyle\int \frac{dT}{T^2 + 2T + 2}$
20. $\displaystyle\int \frac{2\,dx}{x^2 + 8x + 17}$
21. $\displaystyle\int \frac{4\,dx}{\sqrt{-4x - x^2}}$
22. $\displaystyle\int \frac{0.3\,ds}{\sqrt{2s - s^2}}$
23. $\displaystyle\int_{\pi/6}^{\pi/2} \frac{2\cos 2\theta\,d\theta}{1 + \sin^2 2\theta}$
24. $\displaystyle\int_{-4}^{0} \frac{dx}{x^2 + 4x + 5}$
25. $\displaystyle\int \frac{2 - x}{\sqrt{4 - x^2}}\,dx$
26. $\displaystyle\int \frac{5 - 16x}{1 + 4x^2}\,dx$
27. $\displaystyle\int \frac{\sin^{-1} x}{\sqrt{1 - x^2}}\,dx$
28. $\displaystyle\int \frac{dx}{e^x + e^{-x}}$
29. $\displaystyle\int \frac{x^3 + 3x^5}{1 + x^6}\,dx$
30. $\displaystyle\int \frac{x\tan^{-1} x^2}{1 + x^4}\,dx$

In Exercises 31–34, identify the form of each integral as being inverse sine, inverse tangent, logarithmic, or general power, as in Examples 6 and 7. Do not integrate. In each part (a), explain how the choice was made.

31. (a) $\displaystyle\int \frac{2\,dx}{4 + 9x^2}$ (b) $\displaystyle\int \frac{2\,dx}{4 + 9x}$ (c) $\displaystyle\int \frac{2x\,dx}{\sqrt{4 + 9x^2}}$
32. (a) $\displaystyle\int \frac{2x\,dx}{4 - 9x^2}$ (b) $\displaystyle\int \frac{2\,dx}{\sqrt{4 - 9x}}$ (c) $\displaystyle\int \frac{2x\,dx}{4 + 9x^2}$
33. (a) $\displaystyle\int \frac{2x\,dx}{\sqrt{4 - 9x^2}}$ (b) $\displaystyle\int \frac{2\,dx}{\sqrt{4 - 9x^2}}$ (c) $\displaystyle\int \frac{2\,dx}{4 - 9x}$
34. (a) $\displaystyle\int \frac{2\,dx}{9x^2 + 4}$ (b) $\displaystyle\int \frac{2x\,dx}{\sqrt{9x^2 - 4}}$ (c) $\displaystyle\int \frac{2x\,dx}{9x^2 - 4}$

In Exercises 35–44, solve the given problems by integration.

35. Explain how to integrate $\displaystyle\int \frac{dx}{\sqrt{x}(1 + x)}$. What is the result?
36. Integrate $\displaystyle\int \sqrt{\frac{1 + x}{1 - x}}\,dx$ by first multiplying the numerator and denominator of the fraction under the radical by $1 + x$.
37. Find the area bounded by $y(1 + x^2) = 1$, $x = 0$, $y = 0$, and $x = 2$. See Fig. 28.8.

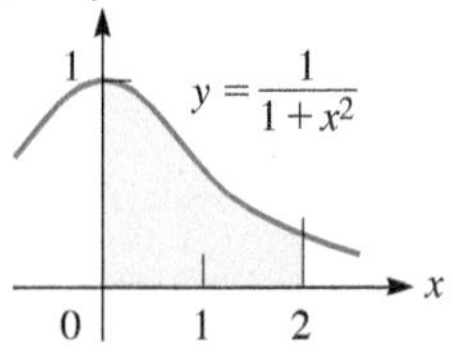

Fig. 28.8

38. A ball is rolling such that its velocity v (in cm/s) as a function of time t (in s) is $v = \dfrac{0.45}{0.25t^2 + 1}$. How far does it move in 10.0 s?
39. To find the electric field E from an electric charge distributed uniformly over the entire xy-plane at a distance d from the plane, it is necessary to evaluate the integral $kd\displaystyle\int \frac{dx}{d^2 + x^2}$. Here, x is the distance from the origin to the element of charge. Perform the indicated integration.
40. An oil-storage tank can be described as the volume generated by revolving the region bounded by $y = 24/\sqrt{16 + x^2}$, $x = 0$, $y = 0$, and $x = 3$ about the x-axis. Find the volume (in m^3) of the tank.
41. In dealing with the theory for simple harmonic motion, it is necessary to solve the equation $\dfrac{dx}{\sqrt{A^2 - x^2}} = \sqrt{\dfrac{k}{m}}\,dt$ (k, m, and A are constants). Determine the solution if $x = x_0$ when $t = 0$.
42. During each cycle, the velocity v (in m/s) of a robotic welding device is given by $v = 2t - \dfrac{12}{2 + t^2}$, where t is the time (in s). Find the expression for the displacement s (in m) as a function of t if $s = 0$ for $t = 0$.
43. Find the moment of inertia with respect to the y-axis for a flat plate covering the region bounded by $y = 1/(1 + x^6)$, the x-axis, $x = 1$, and $x = 2$.
44. The length of an arc along a function is given by $s = \displaystyle\int_a^b \sqrt{1 + \left(\frac{dy}{dx}\right)^2}\,dx$. Find the length of arc along the curve $y = \sqrt{1 - x^2}$ between $x = 0$ and $x = 1$.

Answers to Practice Exercises

1. $\frac{1}{2}\sin^{-1}\frac{2x}{3} + C$ **2.** $\frac{1}{6}\tan^{-1}\frac{3x}{2} + C$

28.7 Integration by Parts

In this section and the following one, we develop two general methods of integration. The method of *integration by parts* is discussed in this section.

Since the derivative of a product of functions is found by use of the formula

$$\frac{d(uv)}{dx} = u \cdot \frac{dv}{dx} + v \cdot \frac{du}{dx}$$

the differential of a product of functions is given by $d(uv) = u\,dv + v\,du$. Integrating both sides of this equation, we have $uv = \int u\,dv + \int v\,du$. Solving for $\int u\,dv$, we obtain what is known as the formula for **integration by parts**.

$$\int u\,dv = uv - \int v\,du \qquad \textbf{(28.22)}$$

EXAMPLE 1 Algebraic–trigonometric integrand

Integrate $\int x \sin x\,dx$.

This integral does not fit any of the previous forms we have discussed, since neither x nor $\sin x$ can be made a factor of a proper du. However, by choosing $u = x$ and $dv = \sin x\,dx$, integration by parts may be used. Thus,

$$u = x \qquad dv = \sin x\,dx$$

By finding the differential of u and integrating dv, we find du and v. This gives us

$$du = dx \qquad v = -\cos x + C_1$$

Now, substituting in Eq. (28.22), we have

$$\int \underset{\downarrow}{u}\ \underset{\downarrow}{dv} = \underset{\downarrow}{u}\ \underset{\downarrow}{v} - \int \underset{\downarrow}{v}\ \underset{\downarrow}{du}$$

$$\int (x)(\sin x\,dx) = (x)(-\cos x + C_1) - \int (-\cos x + C_1)(dx)$$

$$= -x\cos x + C_1x + \int \cos x\,dx - \int C_1\,dx$$

$$= -x\cos x + C_1x + \sin x - C_1x + C$$

$$= -x\cos x + \sin x + C$$

Other choices of u and dv may be made, but they are not useful. For example, if we choose $u = \sin x$ and $dv = x\,dx$, then $du = \cos x\,dx$ and $v = \frac{1}{2}x^2 + C_2$. This makes $\int v\,du = \int\left(\frac{1}{2}x^2 + C_2\right)(\cos x\,dx)$, which is more complex than the integrand of the original problem.

Practice Exercise

1. Integrate $\displaystyle\int x \sec^2 x\,dx$.

LEARNING TIP

We note that the constant C_1 that was introduced when we integrated dv does not appear in the final result. This constant will always cancel out, and therefore *we do not need a constant of integration when finding v.*

As in Example 1, there is often more than one choice as to the part of the integrand that is selected to be u and the part that is selected to be dv. There are no set rules that may be stated for the best choice of u and dv, but two guidelines may be stated.

Guidelines for Choosing *u* and *dv*

1. The quantity u is normally chosen such that du/dx is of simpler form than u.
2. The differential dv is normally chosen such that $\int dv$ is easily integrated.

Inverse trigonometric and logarithmic functions are difficult to integrate, so they usually are our first choice for u. Trigonometric and exponential functions are easy to integrate, so they are often chosen as part of dv.

Working examples, and thereby gaining experience in methods of integration, is the best way to determine when this method should be used and how to use it.

EXAMPLE 2 Algebraic integrand

Integrate $\int x\sqrt{1-x}\,dx$.

We see that this form does not fit the general power rule, for $x\,dx$ is not a factor of the differential of $1-x$. By choosing $u = x$ and $dv = \sqrt{1-x}\,dx$, we have $du/dx = 1$, and v can readily be determined. Thus,

$$u = x \qquad dv = \sqrt{1-x}\,dx = (1-x)^{1/2}\,dx$$

$$du = dx \qquad v = -\frac{2}{3}(1-x)^{3/2}$$

Substituting in Eq. (28.22), we have

$$\int \underbrace{x}_{u}\underbrace{[(1-x)^{1/2}\,dx]}_{dv} = \underbrace{x}_{u}\underbrace{\left[-\frac{2}{3}(1-x)^{3/2}\right]}_{v} - \int \underbrace{\left[-\frac{2}{3}(1-x)^{3/2}\right]}_{v}\underbrace{dx}_{du}$$

At this point, we see that we can complete the integration. Thus,

$$\begin{aligned}\int x(1-x)^{1/2}\,dx &= -\frac{2x}{3}(1-x)^{3/2} + \frac{2}{3}\int (1-x)^{3/2}\,dx \\ &= -\frac{2x}{3}(1-x)^{3/2} + \frac{2}{3}\left(-\frac{2}{5}\right)(1-x)^{5/2} + C \\ &= -\frac{2}{3}(1-x)^{3/2}\left[x + \frac{2}{5}(1-x)\right] + C \\ &= -\frac{2}{15}(1-x)^{3/2}(2+3x) + C\end{aligned}$$

EXAMPLE 3 Algebraic–logarithmic integrand

Integrate $\int \sqrt{x}\ln x\,dx$.

For this integral, we have

$$u = \ln x \qquad dv = x^{1/2}\,dx$$

$$du = \frac{1}{x}\,dx \qquad v = \frac{2}{3}x^{3/2}$$

$$\begin{aligned}\int \sqrt{x}\ln x\,dx &= \frac{2}{3}x^{3/2}\ln x - \frac{2}{3}\int x^{1/2}\,dx \\ &= \frac{2}{3}x^{3/2}\ln x - \frac{4}{9}x^{3/2} + C\end{aligned}$$

EXAMPLE 4 Inverse sine integrand

Integrate $\int \sin^{-1} x\, dx$.

We write

$$u = \sin^{-1} x \qquad dv = dx$$

$$du = \frac{dx}{\sqrt{1 - x^2}} \qquad v = x$$

$$\int \sin^{-1} x\, dx = x \sin^{-1} x - \int \frac{x\, dx}{\sqrt{1 - x^2}} = x \sin^{-1} x + \frac{1}{2}\int \frac{-2x\, dx}{\sqrt{1 - x^2}}$$

$$= x \sin^{-1} x + \sqrt{1 - x^2} + C$$

EXAMPLE 5 Algebraic-exponential integrand—electric current and charge

In a certain electric circuit, the current i (in A) is given by the equation $i = te^{-t}$, where t is the time (in s). Find the charge q to pass a point in the circuit between $t = 0$ s and $t = 1.0$ s.

Since $i = \dfrac{dq}{dt}$, we have $\dfrac{dq}{dt} = te^{-t}$. Thus, with $q = \int_0^{1.0} te^{-t}\, dt$, we are to solve for q.

For this integral,

$$u = t \qquad dv = e^{-t}\, dt$$

$$du = dt \qquad v = -e^{-t}$$

$$q = \int_0^{1.0} te^{-t}\, dt = -te^{-t}\Big|_0^{1.0} + \int_0^{1.0} e^{-t}\, dt = -te^{-t} - e^{-t}\Big|_0^{1.0}$$

$$= -e^{-1.0} - e^{-1.0} + 1.0 = 0.26 \text{ C}$$

Therefore, 0.26 C pass a given point in the circuit.

There are instances when integration by parts must be used more than once in order to complete an integral. Consider the following examples.

EXAMPLE 6 Integration by parts twice

Integrate $\int x^2 \cos x\, dx$.

Let $u = x^2$, $dv = \cos x\, dx$, $du = 2x\, dx$, $v = \sin x$. Then

$$\int x^2 \cos x\, dx = x^2 \sin x - 2\int x \sin x\, dx$$

We apply integration by parts a second time to the right-hand integral, with $u = x$, $dv = \sin x\, dx$, $du = dx$, $v = -\cos x$. Therefore,

$$\int x \sin x\, dx = -x \cos x + \int \cos x\, dx = -x \cos x + \sin x$$

Substituting, we get

$$\int x^2 \cos x\, dx = x^2 \sin x + 2x \cos x - 2 \sin x$$

LEARNING TIP

There are some integrals for which using integration by parts twice leads to an integral of the same form as the original integral but with a different coefficient. Although this may seem as if the integration took us back where we started, what we now have is an equation, which can be solved by combining the integrals of like form. See Example 7.

EXAMPLE 7 Integration leading to an equation

Integrate $\int e^x \sin x\, dx$.

Let $u = \sin x$, $dv = e^x\, dx$, $du = \cos x\, dx$, $v = e^x$.

$$\int e^x \sin x\, dx = e^x \sin x - \int e^x \cos x\, dx$$

At first glance, it appears that we have made no progress. Nevertheless, we carry on and apply integration by parts to the integral $\int e^x \cos x\, dx$:

$$u = \cos x \quad dv = e^x\, dx \quad du = -\sin x\, dx \quad v = e^x$$

And so $\int e^x \cos x\, dx = e^x \cos x + \int e^x \sin x\, dx$. Substituting this expression into the expression for $\int e^x \sin x\, dx$, we obtain

$$\int e^x \sin x\, dx = e^x \sin x - \left(e^x \cos x + \int e^x \sin x\, dx\right)$$

$$= e^x \sin x - e^x \cos x - \int e^x \sin x\, dx$$

Adding the right-hand integral to both sides gives

$$2\int e^x \sin x\, dx = e^x(\sin x - \cos x) + 2C$$

$$\int e^x \sin x\, dx = \frac{e^x}{2}(\sin x - \cos x) + C$$

Thus, by combining integrals of like form, we obtain the desired result.

EXERCISES 28.7

In Exercises 1 and 2, answer the given questions related to the indicated examples of this section.

1. In Example 2, do the choices $u = \sqrt{1-x}$ and $dv = x\, dx$ work for this integral? Explain.
2. In Example 7, do the choices $u = e^x$ and $dv = \sin x\, dx$ work for this integral? Explain.

In Exercises 3–22, integrate each of the given functions.

3. $\int \theta \cos\theta\, d\theta$
4. $\int x \sin 2x\, dx$
5. $\int 4xe^{2x}\, dx$
6. $\int 3xe^x\, dx$
7. $\int 3x \sec^2 x\, dx$
8. $\int_0^{\pi/4} x \sec x \tan x\, dx$
9. $\int 2 \tan^{-1} x\, dx$
10. $\int \ln s\, ds$
11. $\int_{-3}^{0} \frac{4t\, dt}{\sqrt{1-t}}$
12. $\int x\sqrt{x+1}\, dx$
13. $\int x \ln x\, dx$
14. $\int x^2 \ln 6x\, dx$
15. $\int \frac{\ln x}{x^3}\, dx$
16. $\int_0^1 r^2 e^{2r}\, dr$
17. $\int_0^{\pi/2} e^x \cos x\, dx$
18. $\int e^{-x} \sin 2x\, dx$
19. $\int (x+4)^{17}(2x+5)\, dx$
20. $\int \cos x \ln(\sin x)\, dx$
21. $\int \cos(\ln x)\, dx$
22. $\int \csc^3 x\, dx$

In Exercises 23–36, solve the given problems by integration.

23. To integrate $\int e^{-\sqrt{x}}\, dx$, the only choices possible of $u = e^{-\sqrt{x}}$ and $dv = dx$ do not work. However, if we first let $t = \sqrt{x}$, $dt = dx/2\sqrt{x}$, the integration can be done by parts. Perform this integration.
24. To integrate $\int x \ln(x+1)\, dx$, the substitution $t = x + 1$, $dt = dx$ leads to an integral that can be done readily by parts. Perform this integration in this way.
25. Find the area bounded by $y = xe^{-x}$, $y = 0$, and $x = 2$. See Fig. 28.9.

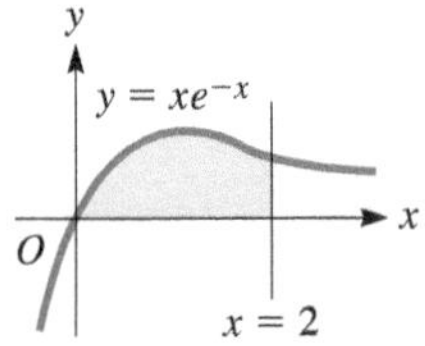

Fig. 28.9

26. Find the area bounded by $y = 2(\ln x)/x^2$, $y = 0$, and $x = 3$.
27. Find the volume generated by revolving the region bounded by $y = \tan^2 x$, $y = 0$, and $x = 0.5$ about the y-axis.
28. Find the volume generated by revolving the region bounded by $y = \sin x$ and $y = 0$ from $x = 0$ to $x = \pi$ about the y-axis.
29. Find the x-coordinate of the centroid of a flat plate covering the region bounded by $y = \cos x$ and $y = 0$ for $0 \leq x \leq \pi/2$.
30. Find the moment of inertia with respect to its axis of the solid generated by revolving the region bounded by $y = e^x$, $x = 1$, and the coordinate axes about the y-axis.
31. Find the root-mean-square value of the function $y = \sqrt{\sin^{-1} x}$ between $x = 0$ and $x = 1$. (See Section 28.5.)
32. The general expression for the slope of a curve is $dy/dx = x^3\sqrt{1 + x^2}$. Find the equation of the curve if it passes through the origin.
33. Computer simulation shows that the velocity v (in m/s) of a test car is $v = t^3/\sqrt{t^2 + 1}$ from $t = 0$ to $t = 8.0$ s. Find the expression for the distance travelled by the car in t seconds.
34. The nose cone of a rocket has the shape of the solid that is generated by revolving the region bounded by $y = \ln x$, $y = 0$, and $x = 9.5$ about the x-axis. Find the volume (in m³) of the nose cone. See Fig. 28.10.

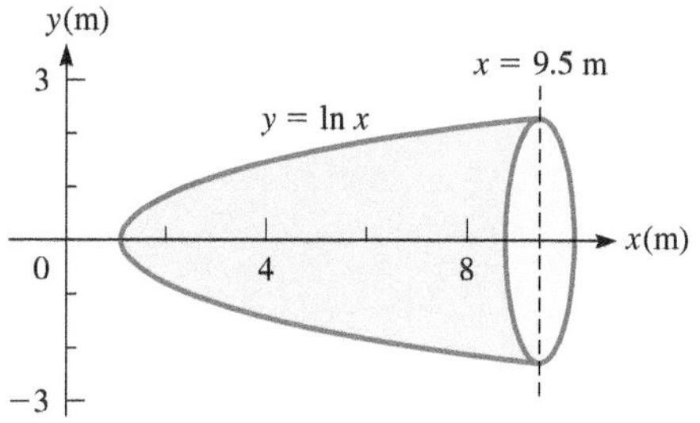

Fig. 28.10

35. The current in a given circuit is given by $i = e^{-2t}\cos t$. Find an expression for the amount of charge that passes a given point in the circuit as a function of the time, if $q_0 = 0$.
36. In finding the average length $\bar{x}$ (in nm) of a certain type of large molecule, we use the equation $\bar{x} = \lim_{b\to\infty}\left[0.1\int_0^b x^3 e^{-x^2/8}\,dx\right]$. Evaluate the integral and then use L'Hospital's rule (or a calculator) to show that $\bar{x} \to 3.2$ nm as $b \to \infty$.

Answer to Practice Exercise

1. $x\tan x + \ln|\cos x| + C$

28.8 Integration by Trigonometric Substitution

In this section, we show how trigonometric relations are useful in integrating algebraic integrals involving the expressions $\sqrt{a^2 - x^2}$, $\sqrt{a^2 + x^2}$, or $\sqrt{x^2 - a^2}$. We begin by summarizing the method, which is known as **trigonometric substitution**, and then consider some examples.

Trigonometric Substitution

1. Select the appropriate trigonometric substitution according to the following table and rewrite the radical as shown:

If the Integral Involves	Then Substitute	Use the Identity	To Rewrite the Radical as
$\sqrt{a^2 - x^2}$	$x = a\sin\theta$	$1 - \sin^2\theta = \cos^2\theta$	$a\cos\theta$
$\sqrt{a^2 + x^2}$	$x = a\tan\theta$	$1 + \tan^2\theta = \sec^2\theta$	$a\sec\theta$
$\sqrt{x^2 - a^2}$	$x = a\sec\theta$	$\sec^2\theta - 1 = \tan^2\theta$	$a\tan\theta$

2. Replace all factors of the integral with expressions in terms of θ, thus obtaining a trigonometric integral.
3. Integrate the trigonometric integral.
4. Express the answer in terms of the original variable x. The values of the different trigonometric functions may be read from a suitable reference triangle.

EXAMPLE 1 For $\sqrt{a^2 - x^2}$, let $x = a \sin \theta$

Integrate $\displaystyle\int \frac{dx}{x^2\sqrt{1-x^2}}$.

If we let $x = \sin\theta$, then $\sqrt{1-\sin^2\theta} = \cos\theta$, and the integral can be transformed into a trigonometric integral. Carefully ***replacing all factors of the integral with expressions in terms of θ***, we have $x = \sin\theta$, $\sqrt{1-x^2} = \cos\theta$, and $dx = \cos\theta\, d\theta$. Therefore,

$$\int \frac{dx}{x^2\sqrt{1-x^2}} = \int \frac{\cos\theta\, d\theta}{\sin^2\theta\sqrt{1-\sin^2\theta}} \quad \text{substituting}$$

$$= \int \frac{\cos\theta\, d\theta}{\sin^2\theta\cos\theta} = \int \csc^2\theta\, d\theta \quad \text{using trigonometric relations}$$

$$= -\cot\theta + C \quad \text{using Eq. (28.7)}$$

Making a triangle with an angle θ such that $\sin\theta = x/1$ (see Fig. 28.11), we may express any of the trigonometric functions in terms of x. (This is the method used with inverse trigonometric functions.) Thus,

$$\cot\theta = \frac{\sqrt{1-x^2}}{x}$$

Therefore, the result of the integration becomes

$$\int \frac{dx}{x^2\sqrt{1-x^2}} = -\cot\theta + C = -\frac{\sqrt{1-x^2}}{x} + C$$

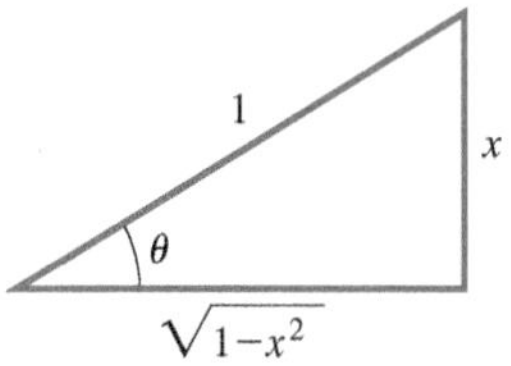

Fig. 28.11

EXAMPLE 2 For $\sqrt{a^2 + x^2}$, let $x = a \tan \theta$

Integrate $\displaystyle\int \frac{dx}{\sqrt{x^2+4}}$.

If we let $x = 2\tan\theta$, the radical in this integral becomes

$$\sqrt{x^2+4} = \sqrt{4\tan^2\theta + 4} = 2\sqrt{\tan^2\theta + 1} = 2\sqrt{\sec^2\theta} = 2\sec\theta$$

Therefore, with $x = 2\tan\theta$ and $dx = 2\sec^2\theta\, d\theta$, we have

$$\int \frac{dx}{\sqrt{x^2+4}} = \int \frac{2\sec^2\theta\, d\theta}{\sqrt{4\tan^2\theta+4}} = \int \frac{2\sec^2\theta\, d\theta}{2\sec\theta} \quad \text{substituting}$$

$$= \int \sec\theta\, d\theta = \ln|\sec\theta + \tan\theta| + C \quad \text{using Eq. (28.12)}$$

$$= \ln\left|\frac{\sqrt{x^2+4}}{2} + \frac{x}{2}\right| + C = \ln\left|\frac{\sqrt{x^2+4}+x}{2}\right| + C \quad \text{see Fig. (28.12)}$$

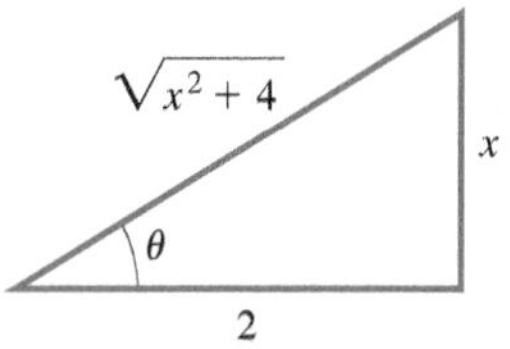

Fig. 28.12

This answer is acceptable, but by using the properties of logarithms, we have

$$\ln\left|\frac{\sqrt{x^2+4}+x}{2}\right| + C = \ln\left|\sqrt{x^2+4}+x\right| + (C - \ln 2)$$

$$= \ln\left|\sqrt{x^2+4}+x\right| + C'$$

■ Combining constants, as in $C' = C - \ln 2$, is a common practice in integration problems.

EXAMPLE 3 For $\sqrt{x^2 - a^2}$, let $x = a \sec\theta$

Integrate $\displaystyle\int \frac{2\,dx}{x\sqrt{x^2-9}}$.

If we let $x = 3\sec\theta$, the radical in this integral becomes

$$\sqrt{x^2-9} = \sqrt{9\sec^2\theta - 9} = 3\sqrt{\sec^2\theta - 1} = 3\sqrt{\tan^2\theta} = 3\tan\theta$$

Therefore, with $x = 3\sec\theta$ and $dx = 3\sec\theta\tan\theta\,d\theta$, we have

$$\int \frac{2\,dx}{x\sqrt{x^2-9}} = 2\int \frac{3\sec\theta\tan\theta\,d\theta}{3\sec\theta\sqrt{9\sec^2\theta-9}} = 2\int \frac{\tan\theta\,d\theta}{3\tan\theta}$$

$$= \frac{2}{3}\int d\theta = \frac{2}{3}\theta + C = \frac{2}{3}\sec^{-1}\frac{x}{3} + C$$

It is not necessary to refer to a triangle to express the result in terms of x. The solution is found by solving $x = 3\sec\theta$ for θ, as indicated.

EXAMPLE 4 Trigonometric substitution—arc length of a robot arm

The joint between two links of a robot arm moves back and forth along the curve defined by $y = 3.0\ln x$ from $x = 1.0$ cm to $x = 4.0$ cm. Find the distance the joint moves in one cycle.

The required distance is found by use of the equation for the length of arc, which is shown in the margin. Therefore, we find the derivative as $dy/dx = 3.0/x$, which means the total distance s moved in one cycle is

■ The arc length s along a curve is given by

$$s = \int_a^b \sqrt{1 + \left(\frac{dy}{dx}\right)^2}\,dx$$

$$s = 2\int_{1.0}^{4.0} \sqrt{\left[1 + \left(\frac{3.0}{x}\right)^2\right]}\,dx = 2\int_{1.0}^{4.0} \frac{\sqrt{x^2+9.0}}{x}\,dx$$

To integrate, we make the substitution $x = 3.0\tan\theta$ and $dx = 3.0\sec^2\theta\,d\theta$. Thus,

$$\int \frac{\sqrt{x^2+9.0}}{x}\,dx = \int \frac{\sqrt{9.0\tan^2\theta + 9.0}}{3.0\tan\theta}(3.0\sec^2\theta\,d\theta) = 3.0\int \frac{\sec\theta}{\tan\theta}(1+\tan^2\theta)d\theta$$

$$= 3.0\left(\int \frac{\sec\theta}{\tan\theta}\,d\theta + \int \tan\theta\sec\theta\,d\theta\right) = 3.0\left(\int \csc\theta\,d\theta + \int \tan\theta\sec\theta\,d\theta\right)$$ see Fig. 28.13

$$= 3.0(\ln|\csc\theta - \cot\theta| + \sec\theta) + C = 3.0\left[\ln\left|\frac{\sqrt{x^2+9.0}}{x} - \frac{3.0}{x}\right| + \frac{\sqrt{x^2+9.0}}{3.0}\right] + C$$

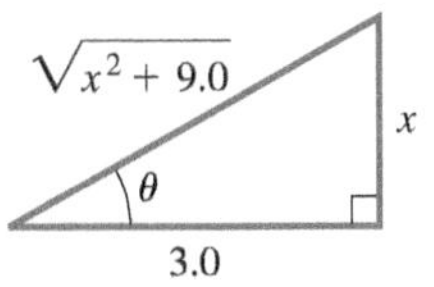

Fig. 28.13

Limits have not been included, due to the change in variables. Evaluating, we have

$$s = 2\int_{1.0}^{4.0} \frac{\sqrt{x^2+9.0}}{x}\,dx = 6.0\left[\ln\left|\frac{\sqrt{x^2+9.0} - 3.0}{x}\right| + \frac{\sqrt{x^2+9.0}}{3.0}\right]\Bigg|_{1.0}^{4.0}$$

$$= 6.0\left[\left(\ln 0.50 - \ln\frac{\sqrt{10.0}-3.0}{1.0}\right) + \frac{5.0}{3.0} - \frac{\sqrt{10.0}}{3.0}\right] = 10.4 \text{ cm}$$

For easy reference, we rewrite the radical forms and their appropriate trigonometric substitutions here:

$$\begin{aligned} &\text{For } \sqrt{a^2 - x^2}, && \text{use } x = a\sin\theta \\ &\text{For } \sqrt{a^2 + x^2}, && \text{use } x = a\tan\theta \\ &\text{For } \sqrt{x^2 - a^2}, && \text{use } x = a\sec\theta \end{aligned} \qquad \textbf{(28.23)}$$

EXERCISES 28.8

In Exercises 1 and 2, answer the given questions related to the indicated examples of this section.

1. In Example 1, how must the integrand be changed in order to have a result of $\sin^{-1} x + C$?

2. In Example 2, how must the integrand be changed in order to have a result of $\tan^{-1}(x/2) + C$?

In Exercises 3–8, give the proper trigonometric substitution and find the transformed integral, but do not integrate.

3. $\int \frac{\sqrt{9 - x^2}}{x^2} dx$

4. $\int \frac{dx}{\sqrt{x^2 - 16}}$

5. $\int \frac{dx}{x^2\sqrt{x^2 + 1}}$

6. $\int \frac{dx}{(1 - x^2)^{3/2}}$

7. $\int \frac{dx}{x\sqrt{x^2 - 1}}$

8. $\int \sqrt{121 + x^2}\, dx$

In Exercises 9–24, integrate each of the given functions.

9. $\int \frac{\sqrt{1 - x^2}}{x^2} dx$

10. $\int_0^4 \frac{dt}{(t^2 + 9)^{3/2}}$

11. $\int \frac{2\, dx}{\sqrt{x^2 - 16}}$

12. $\int \frac{\sqrt{x^2 - 49}}{x} dx$

13. $\int \frac{6\, dz}{z^2\sqrt{z^2 + 9}}$

14. $\int \frac{3\, dx}{x\sqrt{4 - x^2}}$

15. $\int \frac{4\, dx}{(4 - x^2)^{3/2}}$

16. $\int \frac{6p^3\, dp}{\sqrt{9 + p^2}}$

17. $\int_0^{0.5} \frac{x^3\, dx}{\sqrt{1 - x^2}}$

18. $\int_4^5 \frac{\sqrt{x^2 - 16}}{x^2} dx$

19. $\int \frac{5\, dx}{\sqrt{x^2 + 2x + 2}}$

20. $\int \frac{dx}{\sqrt{x^2 + 2x}}$

21. $\int_{2.5}^3 \frac{dy}{y\sqrt{4y^2 - 9}}$

22. $\int \sqrt{16 - x^2}\, dx$

23. $\int \frac{2\, dx}{\sqrt{e^{2x} - 1}}$

24. $\int \frac{12 \sec^2 u\, du}{(4 - \tan^2 u)^{3/2}}$

In Exercises 25–36, solve the given problems by integration.

25. Perform the integration $\int x\sqrt{1 - x^2}\, dx$ (a) by using the power rule, and (b) by trigonometric substitution. Compare results.

26. Perform the integration $\int \frac{x\, dx}{x^2 + 1}$ (a) by using the logarithmic formula, and (b) by trigonometric substitution. Compare results.

27. Find the area of a quarter circle of radius 2 by integrating $\int_0^2 \sqrt{4 - x^2}\, dx$.

28. Find the area bounded by $y = \frac{1}{x^2\sqrt{x^2 - 1}}$, $x = \sqrt{2}$, $x = \sqrt{5}$, and $y = 0$. See Fig. 28.14.

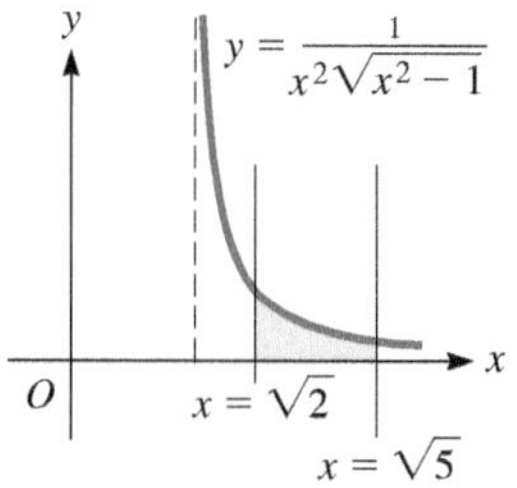

Fig. 28.14

29. Find the moment of inertia with respect to the y-axis of a flat plate covering the first-quadrant region under the circle $x^2 + y^2 = a^2$ in terms of its mass.

30. Find the moment of inertia of a sphere of radius a with respect to its axis in terms of its mass.

31. Find the volume generated by revolving the region bounded by $y = \frac{\sqrt{x^2 - 16}}{x^2}$, $y = 0$, and $x = 5$ about the y-axis.

32. The perimeter of the rudder of a boat can be described as the region bound by $y = -0.5x^2\sqrt{4 - x^2}$ and the x-axis. Find the area of one side of the rudder.

33. Find the x-coordinate of the centroid of the boat rudder in Exercise 32.

34. The vertical cross-section of a highway culvert is defined by the region within the ellipse $1.00x^2 + 9.00y^2 = 9.00$, where dimensions are in metres. Find the area of the cross-section of the culvert.

35. If an electric charge Q is distributed along a straight wire of length $2a$, the electric potential V at a point P, which is at a distance b from the centre of the wire, is $V = kQ\int_{-a}^{a} \frac{dx}{\sqrt{b^2 + x^2}}$. Here, k is a constant and x is the distance along the wire. Evaluate the integral.

36. An electric insulating ring for a machine part can be described as the volume generated by revolving the region bounded by $y = x^2\sqrt{x^2 - 4}$, $y = 0$, and $x = 2.5$ cm about the y-axis. Find the volume (in cm^3) of material in the ring.

Certain algebraic integrals can be transformed into integrable form with the appropriate ***algebraic substitution****. For an expression of the form $(ax + b)^{p/q}$, a substitution of the form $u = (ax + b)^{1/q}$ may put it into an integrable form. In Exercises 37–40, use this type of substitution for the given integrals.*

37. $\int x\sqrt{x + 1}\, dx$

38. $\int x\sqrt[3]{8 - x}\, dx$

39. $\int x(x - 4)^{2/3}\, dx$

40. $\int \frac{x^2\, dx}{(4x + 1)^{5/2}}$

28.9 Integration by Partial Fractions: Nonrepeated Linear Factors

We have seen how the *derivative of a product* and *trigonometric identities* are used to write integrals into a form that can be integrated. In this section and the next, we show an algebraic method by which integrands that are fractions can also be changed into an integrable form.

In algebra, we combine fractions into a single fraction by means of addition. However, if we wish to integrate an expression that contains rational fractions, in which both numerator and denominator are polynomials, it is often advantageous to reverse the process of addition and express the rational fraction as the sum of simpler fractions.

EXAMPLE 1 Illustrating partial fractions

In attempting to integrate $\int \frac{7-x}{x^2+x-2}\,dx$, we find that it does not fit any of the standard forms in this chapter. However, we can show that

$$\frac{7-x}{x^2+x-2} = \frac{2}{x-1} - \frac{3}{x+2}$$

This means that

$$\begin{aligned}\int \frac{7-x}{x^2+x-2}\,dx &= \int \frac{2\,dx}{x-1} - \int \frac{3\,dx}{x+2}\\ &= 2\ln|x-1| - 3\ln|x+2| + C\end{aligned}$$

We see that by writing the fraction in the original integrand as the sum of the simpler fractions, each resulting integrand can be integrated.

In Example 1, we saw that the integral is readily determined once the rational fraction $(7-x)/(x^2+x-2)$ is replaced by the simpler fractions. In this section and the next, we describe how certain rational fractions can be expressed in terms of simpler fractions and thereby be integrated. *This technique is called the* **method of partial fractions**.

There are four cases for the types of factors of the denominator. They are (1) *nonrepeated linear factors,* (2) *repeated linear factors,* (3) *nonrepeated quadratic factors,* and (4) *repeated quadratic factors.* We summarize the method of partial fractions for all four cases here. We will then illustrate each case with examples—nonrepeated linear factors in this section, and the other cases in the next section.

Method of Partial Fractions

To integrate a rational expression of the form $f(x)/g(x)$ by the method of partial fractions:

1. Make sure that the degree of the numerator $f(x)$ is less than the degree of the denominator $g(x)$. If it is not, divide the numerator by the denominator until the remainder is of the proper form.
2. Factor the denominator as completely as possible.
3. Select the term(s) that correspond to each factor in the denominator according to the following table. Keep in mind that if there is only one factor present, integration by substitution may be more convenient.

LEARNING TIP

In order to express the rational fraction $f(x)/g(x)$ in terms of simpler partial fractions, *the degree of the numerator $f(x)$ must be less than that of the denominator $g(x)$*. If this is not the case, we divide numerator by denominator until the remainder is of the proper form. Then the denominator $g(x)$ is factored into a product of linear and quadratic factors. The method of determining the partial fractions depends on the factors that are obtained.

Factor in Denominator	Term(s) in Partial Fractions Decomposition
$ax+b$	$\frac{A}{ax+b}$
$(ax+b)^n$	$\frac{A_1}{ax+b}+\frac{A_2}{(ax+b)^2}+\cdots+\frac{A_n}{(ax+b)^n}$
ax^2+bx+c	$\frac{Ax+B}{ax^2+bx+c}$
$(ax^2+bx+c)^n$	$\frac{A_1x+B_1}{ax^2+bx+c}+\frac{A_2x+B_2}{(ax^2+bx+c)^2}+\cdots+\frac{A_nx+B_n}{(ax^2+bx+c)^n}$

4. Determine the values of the coefficients either by substitution of convenient values of x, or by equating coefficients of like powers of x from each side.

5. Integrate.

EXAMPLE 2 Integration by partial fractions

Integrate $\int \frac{7-x}{x^2+x-2}\,dx$. (This is the same integral as in Example 1. Here, we see how the partial fractions are found.)

First, we note that the degree of the numerator is 1 (the highest-power term is x) and that of the denominator is 2 (the highest-power term is x^2). Since the degree of the denominator is higher, we may proceed to factoring it. Thus,

$$\frac{7-x}{x^2+x-2}=\frac{7-x}{(x-1)(x+2)}$$

There are two linear factors, $(x-1)$ and $(x+2)$, in the denominator, and they are different. This means that there are two partial fractions. Therefore, we write

$$\frac{7-x}{(x-1)(x+2)}=\frac{A}{x-1}+\frac{B}{x+2} \quad \textbf{(1)}$$

We are to determine constants A and B so that Eq. (1) is an identity. In finding A and B, we clear Eq. (1) of fractions by multiplying both sides by $(x-1)(x+2)$.

$$7-x=A(x+2)+B(x-1) \quad \textbf{(2)}$$

Eq. (2) is also an identity, which means that there are two ways of determining the values of A and B.

Solution by substitution: Since Eq. (2) is an identity, ***it is true for any value of x.*** Thus, in turn we pick $x=-2$ and $x=1$, for each of these values makes a factor on the right equal to zero, and the values of B and A are easily found. Therefore,

$$\text{For } x=-2:\quad 7-(-2)=A(-2+2)+B(-2-1)$$
$$9=-3B \qquad B=-3$$
$$\text{For } x=1:\quad 7-1=A(1+2)+B(1-1)$$
$$6=3A, \qquad A=2$$

Solution by equating coefficients: Since Eq. (2) is an identity, another way of finding the constants A and B is to ***equate coefficients of like powers of x from each side.*** Thus, writing Eq. (2) as

$$7-x=(2A-B)+(A+B)x$$

■ The method of partial fractions essentially reverses the process of combining fractions over a common denominator. Determining that

$$\frac{7-x}{x^2+x-2}=\frac{2}{x-1}-\frac{3}{x+2}$$

reverses the process by which

$$\frac{2}{x-1}-\frac{3}{x+2}=\frac{2(x+2)-3(x-1)}{(x-1)(x+2)}=\frac{7-x}{x^2+x-2}$$

Practice Exercise

1. Find the partial fractions for $\frac{3x+11}{x^2-2x-3}$.

we have

$$2A - B = 7 \quad \text{equating constants: } x^0 \text{ terms}$$
$$A + B = -1 \quad \text{equating coefficients of } x$$

■ $2A - B = 7$
$A + B = -1$
$3A = 6$
$A = 2$
$2(2) - B = 7$
$-B = 3$
$B = -3$

Now, using the values $A = 2$ and $B = -3$ (as found at the left), we have

$$\frac{7 - x}{(x-1)(x+2)} = \frac{2}{x-1} - \frac{3}{x+2}$$

We can now integrate term by term.

$$\int \frac{7-x}{x^2+x-2}\,dx = \int \frac{7-x}{(x-1)(x+2)}\,dx = \int \frac{2\,dx}{x-1} - \int \frac{3\,dx}{x+2}$$
$$= 2\ln|x-1| - 3\ln|x+2| + C$$

Using the properties of logarithms, we may write this as

$$\int \frac{7-x}{x^2+x-2}\,dx = \ln\left|\frac{(x-1)^2}{(x+2)^3}\right| + C$$

EXAMPLE 3 Integration by partial fractions

Integrate $\int \frac{6x^2 - 14x - 11}{(x+1)(x-2)(2x+1)}\,dx$.

The denominator is factored and is of degree 3 (when multiplied out, the highest-power term of x is x^3), which is higher than the degree 2 of the numerator. This means we have three nonrepeated linear factors.

■ The partial fractions corresponding to a quotient of polynomials can be found using the *expand* feature on some graphing calculators.

$$\frac{6x^2 - 14x - 11}{(x+1)(x-2)(2x+1)} = \frac{A}{x+1} + \frac{B}{x-2} + \frac{C}{2x+1}$$

Multiplying through by $(x+1)(x-2)(2x+1)$, we have

$$6x^2 - 14x - 11 = A(x-2)(2x+1) + B(x+1)(2x+1) + C(x+1)(x-2)$$

We determine the coefficients by substituting the convenient values of 2, $-\frac{1}{2}$ and -1. Therefore,

For $x = 2$: $\quad 6(4) - 14(2) - 11 = A(0)(5) + B(3)(5) + C(3)(0), \quad B = -1$

For $x = -\frac{1}{2}$: $\quad 6\left(\frac{1}{4}\right) - 14\left(-\frac{1}{2}\right) - 11 = A\left(-\frac{5}{2}\right)(0) + B\left(\frac{1}{2}\right)(0) + C\left(\frac{1}{2}\right)\left(-\frac{5}{2}\right), \quad C = 2$

For $x = -1$: $\quad 6(1) - 14(-1) - 11 = A(-3)(-1) + B(0)(-1) + C(0)(-3), \quad A = 3$

We can now integrate term by term.

$$\int \frac{6x^2 - 14x - 11}{(x+1)(x-2)(2x+1)}\,dx = \int \frac{3\,dx}{x+1} - \int \frac{dx}{x-2} + \int \frac{2\,dx}{2x+1}$$
$$= 3\ln|x+1| - \ln|x-2| + \ln|2x+1| + C_1 = \ln\left|\frac{(2x+1)(x+1)^3}{x-2}\right| + C_1$$

Here, we have let the constant of integration be C_1 since we used C as the numerator of the third partial fraction.

EXAMPLE 4 Integration by partial fractions

Integrate $\displaystyle\int \frac{2x^4 - x^3 - 9x^2 + x - 12}{x^3 - x^2 - 6x}\,dx.$

Since the numerator is of a higher degree than the denominator, we must first divide the numerator by the denominator. This gives

$$\frac{2x^4 - x^3 - 9x^2 + x - 12}{x^3 - x^2 - 6x} = 2x + 1 + \frac{4x^2 + 7x - 12}{x^3 - x^2 - 6x}$$

We must now express this rational fraction in terms of its partial fractions.

$$\frac{4x^2 + 7x - 12}{x^3 - x^2 - 6x} = \frac{4x^2 + 7x - 12}{x(x+2)(x-3)} = \frac{A}{x} + \frac{B}{x+2} + \frac{C}{x-3}$$

Clearing fractions, we have

$$4x^2 + 7x - 12 = A(x+2)(x-3) + Bx(x-3) + Cx(x+2)$$

Now, using values of x of -2, 3, and 0 for substitution, we obtain the values of $B = -1$, $C = 3$, and $A = 2$, respectively. Integrating term by term,

$$\begin{aligned}\int \frac{2x^4 - x^3 - 9x^2 + x - 12}{x^3 - x^2 - 6x}\,dx &= \int \left(2x + 1 + \frac{2}{x} - \frac{1}{x+2} + \frac{3}{x-3}\right)dx \\ &= x^2 + x + 2\ln|x| - \ln|x+2| + 3\ln|x-3| + C_1 \\ &= x^2 + x + \ln\left|\frac{x^2(x-3)^3}{x+2}\right| + C_1\end{aligned}$$

EXERCISES 28.9

In Exercises 1 and 2, make the given changes in the integrands of the indicated examples of this section, and then find the resulting fractions to be used in the integration. Do not integrate.

1. In Example 2, change the numerator to $10 - x$.
2. In Example 3, change the numerator to $x^2 - 12x - 10$.

In Exercises 3–6, write out the form of the partial fractions, similar to that shown in Eq. (1) of Example 2, that would be used to perform the indicated integrations. Do not evaluate the constants.

3. $\displaystyle\int \frac{3x+2}{x^2+x}\,dx$
4. $\displaystyle\int \frac{9-x}{x^2+2x-3}\,dx$
5. $\displaystyle\int \frac{x^2-6x-8}{x^3-4x}\,dx$
6. $\displaystyle\int \frac{2x^2-5x-7}{x^3+2x^2-x-2}\,dx$

In Exercises 7–24, integrate each of the given functions.

7. $\displaystyle\int \frac{x+3}{(x+1)(x+2)}\,dx$
8. $\displaystyle\int \frac{x+2}{x(x+1)}\,dx$
9. $\displaystyle\int \frac{dx}{x^2-4}$
10. $\displaystyle\int \frac{p-9}{2p^2-3p+1}\,dp$
11. $\displaystyle\int \frac{x^2+3}{x^2+3x}\,dx$
12. $\displaystyle\int \frac{x^3}{x^2+3x+2}\,dx$
13. $\displaystyle\int_0^1 \frac{2t+4}{3t^2+5t+2}\,dt$
14. $\displaystyle\int_1^3 \frac{x-1}{4x^2+x}\,dx$
15. $\displaystyle\int \frac{4x^2-10}{x(x+1)(x-5)}\,dx$
16. $\displaystyle\int \frac{4x^2+21x+6}{(x+2)(x-3)(x+4)}\,dx$
17. $\displaystyle\int \frac{6x^2-2x-1}{4x^3-x}\,dx$
18. $\displaystyle\int_2^3 \frac{dR}{R^3-R}$
19. $\displaystyle\int_1^2 \frac{x^3+7x^2+9x+2}{x(x^2+3x+2)}\,dx$
20. $\displaystyle\int \frac{2x^3+x-1}{x^3+x^2-4x-4}\,dx$
21. $\displaystyle\int \frac{dV}{(V^2-4)(V^2-9)}$
22. $\displaystyle\int \frac{5x^3-2x^2-15x+24}{x^4-2x^3-11x^2+12x}\,dx$
23. $\displaystyle\int \frac{2x\,dx}{x^4-3x^3+2x^2}$
24. $\displaystyle\int \frac{e^x\,dx}{e^{2x}+3e^x+2}$

In Exercises 25–34, solve the given problems by integration.

25. Derive the general formula $\displaystyle\int \frac{du}{u(a+bu)} = -\frac{1}{a}\ln\frac{a+bu}{u} + C.$
26. Derive the general formula $\displaystyle\int \frac{du}{u^2-a^2} = \frac{1}{2a}\ln\frac{u-a}{u+a} + C.$
27. Integrate $\displaystyle\int \frac{\cos\theta}{\sin^2\theta + 2\sin\theta - 3}\,d\theta$. (*Hint:* First, find the partial fractions for $1/(\sin^2\theta + 2\sin\theta - 3)$.)
28. Find the first-quadrant area bounded by $y = 1/(x^3 + 3x^2 + 2x)$, $x = 1$, and $x = 3$.

29. Find the volume generated if the region of Exercise 28 is revolved about the y-axis.

30. Find the x-coordinate of the centroid of a flat plate that covers the region bounded by $y(x^2 - 1) = 1$, $y = 0$, $x = 2$, and $x = 4$.

31. The general expression for the slope of a curve is $(3x + 5)/(x^2 + 5x)$. Find the equation of the curve if it passes through $(1, 0)$.

32. The current i (in A) as a function of the time t (in s) in a certain electric circuit is given by $i = (4t + 3)/(2t^2 + 3t + 1)$. Find the total charge that passes a given point in the circuit during the first second.

33. The force F (in N) applied by a stamping machine in making a certain computer part is $F = 4x/(x^2 + 3x + 2)$, where x is the distance (in cm) through which the force acts. Find the work done by the force from $x = 0$ to $x = 0.500$ cm.

34. Under specified conditions, the time t (in min) required to form x grams of a substance during a chemical reaction is given by $t = \int dx/[(4 - x)(2 - x)]$. Find the equation relating t and x if $x = 0$ g when $t = 0$ min.

Answer to Practice Exercise

1. $\frac{3x+11}{x^2 - 2x - 3} = \frac{5}{x - 3} - \frac{2}{x + 1}$

28.10 Integration by Partial Fractions: Other Cases

In the previous section, we introduced the method of partial fractions and considered the case of nonrepeated linear factors. In this section, we illustrate the use of partial fractions for the cases of repeated linear factors and nonrepeated quadratic factors. We also briefly discuss the case of repeated quadratic factors.

REPEATED LINEAR FACTORS

From the summary in Section 28.9, recall that for linear factors repeated n times, there will be n partial fractions

$$\frac{A_1}{ax + b} + \frac{A_2}{(ax + b)^2} + \cdots + \frac{A_n}{(ax + b)^n}$$

where $A_1, A_2, \ldots, A_n$ are constants to be determined.

EXAMPLE 1 Two factors—one repeated

Integrate $\int \frac{dx}{x(x + 3)^2}$.

Here, we see that the denominator has a factor of x and two factors of $x + 3$. For the factor of x, we use a partial fraction as in the previous section, for it is a nonrepeated factor. For the factor $x + 3$, we need two partial fractions, one with a denominator of $x + 3$ and the other with a denominator of $(x + 3)^2$. Thus, we write

$$\frac{1}{x(x + 3)^2} = \frac{A}{x} + \frac{B}{x + 3} + \frac{C}{(x + 3)^2}$$

Multiplying each side by $x(x + 3)^2$, we have

$$1 = A(x + 3)^2 + Bx(x + 3) + Cx \qquad (1)$$

Using the values of x of -3 and 0, we have

$$\text{For } x = -3: \quad 1 = A(0^2) + B(-3)(0) + (-3)C, \quad C = -\frac{1}{3}$$

$$\text{For } x = 0: \quad 1 = A(3^2) + B(0)(3) + C(0), \quad A = \frac{1}{9}$$

Since no other numbers make a factor in Eq. (1) equal to zero, we either choose some other value of x or equate coefficients of some power of x in Eq. (1). Since Eq. (1) is an identity, we may choose any value of x. With $x = 1$, we have

$$1 = A(4^2) + B(1)(4) + C(1)$$
$$1 = 16A + 4B + C$$

Practice Exercise

1. Find the partial fractions for $\frac{2}{x(x - 1)^2}$.

Using the known values of A and C, we have

$$1 = 16\left(\frac{1}{9}\right) + 4B - \frac{1}{3}, \qquad B = -\frac{1}{9}$$

This means that

$$\frac{1}{x(x+3)^2} = \frac{\frac{1}{9}}{x} + \frac{-\frac{1}{9}}{x+3} + \frac{-\frac{1}{3}}{(x+3)^2}$$

or

$$\begin{aligned}\int \frac{dx}{x(x+3)^2} &= \frac{1}{9}\int \frac{dx}{x} - \frac{1}{9}\int \frac{dx}{x+3} - \frac{1}{3}\int \frac{dx}{(x+3)^2} \\ &= \frac{1}{9}\ln|x| - \frac{1}{9}\ln|x+3| - \frac{1}{3}\left(\frac{1}{-1}\right)(x+3)^{-1} + C_1 \\ &= \frac{1}{9}\ln\left|\frac{x}{x+3}\right| + \frac{1}{3(x+3)} + C_1\end{aligned}$$

EXAMPLE 2 Two factors—one repeated

Integrate $\displaystyle\int \frac{3x^3 + 15x^2 + 21x + 15}{(x-1)(x+2)^3}\,dx.$

First, we set up the partial fractions as

$$\frac{3x^3 + 15x^2 + 21x + 15}{(x-1)(x+2)^3} = \frac{A}{x-1} + \frac{B}{x+2} + \frac{C}{(x+2)^2} + \frac{D}{(x+2)^3}$$

Next, we clear fractions:

$$\begin{aligned}3x^3 + 15x^2 + 21x + 15 = A(x+2)^3 &+ B(x-1)(x+2)^2 \\ &+ C(x-1)(x+2) + D(x-1)\end{aligned} \tag{1}$$

For $x = 1$: $3 + 15 + 21 + 15 = 27A, \quad 54 = 27A, \quad A = 2$
For $x = -2$: $3(-8) + 15(4) + 21(-2) + 15 = -3D, \quad 9 = -3D, \quad D = -3$

To find B and C, we equate coefficients of powers of x. Therefore, we write Eq. (1) as

$$\begin{aligned}3x^3 + 15x^2 + 21x + 15 = (A+B)x^3 &+ (6A + 3B + C)x^2 \\ &+ (12A + C + D)x + (8A - 4B - 2C - D)\end{aligned}$$

Coefficients of x^3: $3 = A + B, \quad 3 = 2 + B, \quad B = 1$

Coefficients of x^2: $15 = 6A + 3B + C, \quad 15 = 12 + 3 + C, \quad C = 0$

$$\frac{3x^3 + 15x^2 + 21x + 15}{(x-1)(x+2)^3} = \frac{2}{x-1} + \frac{1}{x+2} + \frac{0}{(x+2)^2} + \frac{-3}{(x+2)^3}$$

$$\begin{aligned}\int \frac{3x^3 + 15x^2 + 21x + 15}{(x-1)(x+2)^3}\,dx &= 2\int \frac{dx}{x-1} + \int \frac{dx}{x+2} - 3\int \frac{dx}{(x+2)^3} \\ &= 2\ln|x-1| + \ln|x+2| - 3\left(\frac{1}{-2}\right)(x+2)^{-2} + C_1 \\ &= \ln|(x-1)^2(x+2)| + \frac{3}{2(x+2)^2} + C_1\end{aligned}$$

If there is one repeated factor in the denominator and it is the only factor present in the denominator, a substitution is easier and more convenient than using partial fractions. This is illustrated in the following example.

EXAMPLE 3 Substitution instead of partial fractions

Integrate $\int \frac{x\,dx}{(x-2)^3}$.

This could be integrated by first setting up the appropriate partial fractions. However, the solution is more easily found by using the substitution $u = x - 2$. Using this, we have

$$u = x - 2 \qquad x = u + 2 \qquad dx = du$$

$$\int \frac{x\,dx}{(x-2)^3} = \int \frac{(u+2)(du)}{u^3} = \int \frac{du}{u^2} + 2\int \frac{du}{u^3}$$

$$= \int u^{-2}\,du + 2\int u^{-3}\,du = \frac{1}{-u} + \frac{2}{-2}u^{-2} + C$$

$$= -\frac{1}{u} - \frac{1}{u^2} + C = -\frac{u+1}{u^2} + C$$

$$= -\frac{x-2+1}{(x-2)^2} + C = \frac{1-x}{(x-2)^2} + C$$

NONREPEATED QUADRATIC FACTORS

For the case of nonrepeated quadratic factors, we have that corresponding to each irreducible quadratic factor $ax^2 + bx + c$ that occurs once in the denominator there is a partial fraction of the form

$$\frac{Ax + B}{ax^2 + bx + c}$$

where A and B are constants to be determined. (Here, an *irreducible quadratic factor* is one that cannot be further factored into linear factors involving only real numbers.)

EXAMPLE 4 Two factors—one quadratic

Integrate $\int \frac{4x+4}{x^3+4x}dx$.

In setting up the partial fractions, we note that the denominator factors as $x^3 + 4x = x(x^2 + 4)$. Here, the factor $x^2 + 4$ cannot be further factored. This means we have

$$\frac{4x+4}{x^3+4x} = \frac{4x+4}{x(x^2+4)} = \frac{A}{x} + \frac{Bx+C}{x^2+4}$$

Clearing fractions, we have

$$4x + 4 = A(x^2+4) + Bx^2 + Cx$$
$$= (A+B)x^2 + Cx + 4A$$

Equating coefficients of powers of x gives us

For x^2:	$0 = A + B$	
For x:	$4 = C$	
For constants:	$4 = 4A$,	$A = 1$

Therefore, we easily find that $B = -1$ from the first equation. This means that

$$\frac{4x+4}{x^3+4x} = \frac{1}{x} + \frac{-x+4}{x^2+4}$$

Practice Exercise

2. Find the partial fractions for $\frac{x^2-2}{x^3+2x}$.

and

$$\int \frac{4x+4}{x^3+4x}dx = \int \frac{1}{x}dx + \int \frac{-x+4}{x^2+4}dx$$

$$= \int \frac{1}{x}dx - \int \frac{x\,dx}{x^2+4} + \int \frac{4\,dx}{x^2+4}$$

$$= \ln|x| - \frac{1}{2}\ln|x^2+4| + 2\tan^{-1}\frac{x}{2} + C_1$$

We could use the properties of logarithms to combine the first two terms of the answer. Doing so, we have the following result:

$$\int \frac{4x+4}{x^3+4x}dx = \ln\frac{|x|}{\sqrt{x^2+4}} + 2\tan^{-1}\frac{x}{2} + C_1$$

EXAMPLE 5 One quadratic factor and a repeated linear factor

Integrate $\int \frac{x^3+3x^2+2x+4}{x^2(x^2+2x+2)}dx$.

In the denominator, we have a repeated linear factor, x^2, and a quadratic factor. Therefore,

$$\frac{x^3+3x^2+2x+4}{x^2(x^2+2x+2)} = \frac{A}{x} + \frac{B}{x^2} + \frac{Cx+D}{x^2+2x+2}$$

$$x^3+3x^2+2x+4 = Ax(x^2+2x+2) + B(x^2+2x+2) + Cx^3 + Dx^2$$

$$= (A+C)x^3 + (2A+B+D)x^2 + (2A+2B)x + 2B$$

Equating coefficients, we find that

For constants:	$2B = 4,$	$B = 2$	
For x:	$2A + 2B = 2,$	$A + B = 1,$	$A = -1$
For x^2:	$2A + B + D = 3,$	$-2 + 2 + D = 3,$	$D = 3$
For x^3:	$A + C = 1,$	$-1 + C = 1,$	$C = 2$

$$\frac{x^3+3x^2+2x+4}{x^2(x^2+2x+2)} = -\frac{1}{x} + \frac{2}{x^2} + \frac{2x+3}{x^2+2x+2}$$

$$\int \frac{x^3+3x^2+2x+4}{x^2(x^2+2x+2)}dx = -\int\frac{dx}{x} + 2\int\frac{dx}{x^2} + \int\frac{2x+3}{x^2+2x+2}dx$$

$$= -\ln|x| - 2\left(\frac{1}{x}\right) + \int\frac{2x+2+1}{x^2+2x+2}dx$$

$$= -\ln|x| - \frac{2}{x} + \int\frac{2x+2}{x^2+2x+2}dx + \int\frac{dx}{(x^2+2x+1)+1}$$

$$= -\ln|x| - \frac{2}{x} + \ln|x^2+2x+2| + \tan^{-1}(x+1) + C_1$$

Note the manner in which the integral with the quadratic denominator was handled for the purpose of integration. First, the numerator, $2x+3$, was written in the form $(2x+2)+1$ so that we could fit the logarithmic form with the $2x+2$. Then we completed the square in the denominator of the final integral so that it then fit an inverse tangent form.

REPEATED QUADRATIC FACTORS

Finally, considering the case of repeated quadratic factors, we have that corresponding to each irreducible quadratic factor $ax^2 + bx + c$ that occurs n times in the denominator there will be n partial fractions

$$\frac{A_1x + B_1}{ax^2 + bx + c} + \frac{A_2x + B_2}{(ax^2 + bx + c)^2} + \cdots + \frac{A_nx + B_n}{(ax^2 + bx + c)^n}$$

where $A_1, A_2, \ldots, A_n, B_1, B_2, \ldots, B_n$ are constants to be determined. The procedures that lead to the solution are the same as those for the other cases. Exercises 21 and 22 in this section are solved by using these partial fractions for repeated quadratic factors.

EXERCISES 28.10

In Exercises 1–4, make the given changes in the integrands of the indicated examples of this section. Then write out the equation that shows the partial fractions that would be used for the integration. Do not evaluate the constants.

1. In Example 1, change the numerator from 1 to 2.
2. In Example 2, change the denominator to $(x - 1)^2(x + 2)^2$.
3. In Example 4, change the denominator to $x^3 - 9x^2$.
4. In Example 5, change the denominator to $x^2(x^2 + 3x + 2)$.

In Exercises 5–22, integrate each of the given functions.

5. $\int \frac{1}{x^2(x + 1)}dx$
6. $\int \frac{x}{(x - 2)^2}dx$
7. $\int \frac{x - 8}{x^3 - 4x^2 + 4x}dx$
8. $\int \frac{dT}{T^3 - T^2}$
9. $\int \frac{2\,dx}{x^2(x^2 - 1)}$
10. $\int_1^3 \frac{3x^3 + 8x^2 + 10x + 2}{x(x + 1)^3}dx$
11. $\int_1^2 \frac{2s\,ds}{(s - 3)^3}$
12. $\int \frac{x\,dx}{(x + 2)^4}$
13. $\int \frac{x^3 - 2x^2 - 7x + 28}{(x + 1)^2(x - 3)^2}dx$
14. $\int \frac{4\,dx}{(x + 1)^2(x - 1)^2}$
15. $\int_0^2 \frac{x^2 + x + 5}{(x + 1)(x^2 + 4)}dx$
16. $\int \frac{v^2 + v - 1}{(v^2 + 1)(v - 2)}dv$
17. $\int \frac{5x^2 + 8x + 16}{x^2(x^2 + 4x + 8)}dx$
18. $\int \frac{2x^2 + x + 3}{(x^2 + 2)(x - 1)}dx$
19. $\int \frac{10x^3 + 40x^2 + 22x + 7}{(4x^2 + 1)(x^2 + 6x + 10)}dx$
20. $\int_3^4 \frac{5x^3 - 4x}{x^4 - 16}dx$
21. $\int \frac{-x^3 + x^2 + x + 3}{(x + 1)(x^2 + 1)^2}dx$
22. $\int \frac{2r^3}{(r^2 + 1)^2}dr$

In Exercises 23–32, solve the given problems by integration.

23. For the integral of Example 3, set up the integration by partial fractions, and then integrate. Compare results with Example 3.
24. Integrate $\int \frac{\cos x\,dx}{\sin x + \sin^3 x}$. (*Hint:* The numerator for the quadratic factor is $B \sin x + C$.)
25. Find the area bounded by $y = \frac{x - 3}{x^3 + x^2}$, $y = 0$, and $x = 1$.
26. Find the first-quadrant area bounded by $y = \frac{3x^2 + 2x + 9}{(x^2 + 9)(x + 1)}$ and $x = 2$.
27. Find the volume generated by revolving the first-quadrant region bounded by $y = 4/(x^4 + 6x^2 + 5)$ and $x = 2$ about the y-axis.
28. Find the volume generated by revolving the first-quadrant region bounded by $y = x/(x + 3)^2$ and $x = 3$ about the x-axis.
29. Under certain conditions, the velocity v (in m/s) of an object moving along a straight line as a function of the time t (in s) is given by $v = \frac{t^2 + 14t + 27}{(2t + 1)(t + 5)^2}$. Find the distance travelled by the object during the first 2.00 s.
30. By a computer analysis, the electric current i (in A) in a certain circuit is given by $i = \frac{0.0010(7t^2 + 16t + 48)}{(t + 4)(t^2 + 16)}$, where t is the time (in s). Find the total charge that passes a point in the circuit in the first 0.250 s.
31. Find the x-coordinate of the centroid of a flat plate covering the region bounded by $y = 4/(x^3 + x)$, $x = 1$, $x = 2$, and $y = 0$.
32. The slope of a curve is given by $\frac{dy}{dx} = \frac{29x^2 + 36}{4x^4 + 9x^2}$. Find the equation of the curve if it passes through $(1, 5)$.

Answers to Practice Exercises

1. $\frac{2}{x(x - 1)^2} = \frac{2}{x} - \frac{2}{x - 1} + \frac{2}{(x - 1)^2}$
2. $\frac{x^2 - 2}{x^3 + 2x} = -\frac{1}{x} + \frac{2x}{x^2 + 2}$

28.11 Integration by Use of Tables

In this chapter, we have introduced certain basic integrals and have also brought in some methods of reducing other integrals to these basic forms. Often, this transformation and integration requires a number of steps to be performed, and therefore integrals are tabulated for reference. The integrals found in tables have been derived by using the methods introduced thus far, as well as many other methods that can be used. Therefore, an understanding of the basic forms and some of the basic methods is very useful in finding integrals from tables. Such an understanding forms a basis for proper recognition of the forms that are used in the tables, as well as the types of results that may be expected. The following examples illustrate the use of the table of integrals found in Appendix B. More extensive tables are available in other sources.

LEARNING TIP

The use of the tables depends on proper recognition of the form and the variables and constants of the integral.

EXAMPLE 1 Integral fits the form of formula 6

Integrate $\int \frac{x\,dx}{\sqrt{2+3x}}$.

We first note that this integral fits the form of formula 6 of Appendix B, with $u = x$, $a = 2$, and $b = 3$. Therefore,

$$\int \frac{x\,dx}{\sqrt{2+3x}} = -\frac{2(4-3x)\sqrt{2+3x}}{27} + C$$

■ For reference, formula 6 is

$$\int \frac{u\,du}{\sqrt{a+bu}} = -\frac{2(2a-bu)\sqrt{a+bu}}{3b^2}$$

EXAMPLE 2 Integral fits the form of formula 18

Integrate $\int \frac{\sqrt{4-9x^2}}{x}dx$.

This fits the form of formula 18, with proper identification of constants; $u = 3x$, $du = 3\,dx$, $a = 2$. Hence,

$$\int \frac{\sqrt{4-9x^2}}{x}dx = \int \frac{\sqrt{4-9x^2}}{3x}3\,dx$$

$$= \sqrt{4-9x^2} - 2\ln\left(\frac{2+\sqrt{4-9x^2}}{3x}\right) + C$$

■ For reference, formula 18 is

$$\int \frac{\sqrt{a^2-u^2}}{u}du = \sqrt{a^2-u^2} - a\ln\left(\frac{a+\sqrt{a^2-u^2}}{u}\right)$$

EXAMPLE 3 Integral fits the form of formula 37

Integrate $\int 5\sec^3 2x\,dx$.

This fits the form of formula 37; $n = 3$, $u = 2x$, $du = 2\,dx$. And so,

$$\int 5\sec^3 2x\,dx = 5\left(\frac{1}{2}\right)\int \sec^3 2x(2\,dx)$$

$$= \frac{5}{2}\,\frac{\sec 2x\tan 2x}{2} + \frac{5}{2}\left(\frac{1}{2}\right)\int \sec 2x(2\,dx)$$

To complete this integral, we must use the basic form of Eq. (28.12). Thus, we complete it by

$$\int 5\sec^3 2x\,dx = \frac{5\sec 2x\tan 2x}{4} + \frac{5}{4}\ln|\sec 2x + \tan 2x| + C$$

■ For reference, formula 37 is

$$\int \sec^n u\,du = \frac{\sec^{n-2}u\tan u}{n-1} + \frac{n-2}{n-1}\int \sec^{n-2}u\,du$$

■ For reference, Eq. (28.12) is

$$\int \sec u\,du = \ln|\sec u + \tan u| + C$$

EXAMPLE 4 Area—integral fits the form of formula 46

Find the area bounded by $y = x^2 \ln 2x$, $y = 0$, and $x = e$.

From Fig. 28.15, we see that the area is

$$A = \int_{0.5}^{e} x^2 \ln 2x\, dx$$

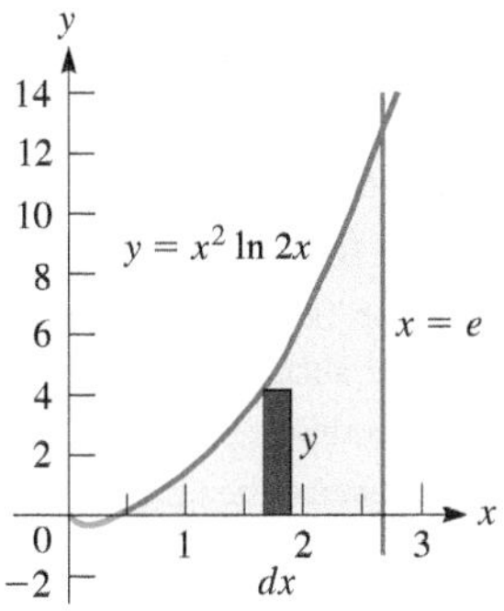

Fig. 28.15

This integral fits the form of formula 46 if $u = 2x$. Thus, we have

$$A = \frac{1}{8}\int_{0.5}^{e} (2x)^2 \ln 2x(2\,dx) = \frac{1}{8}(2x)^3\left[\frac{\ln 2x}{3} - \frac{1}{9}\right]_{0.5}^{e}$$

$$= e^3\left(\frac{\ln 2e}{3} - \frac{1}{9}\right) - \frac{1}{8}\left(\frac{\ln 1}{3} - \frac{1}{9}\right) = e^3\left(\frac{3\ln 2e - 1}{9}\right) + \frac{1}{72}$$

$$= 9.118$$

The proper identification of u and du is the key step in the use of tables. Therefore, for the integrals in the following example, the proper u and du, along with the appropriate formula from the table, are identified, but the integrations are not performed.

EXAMPLE 5 Identify the formula, *u*, and *du*

(a) $\int x\sqrt{1 - x^4}\, dx$ $\quad u = x^2, \quad du = 2x\, dx$ $\quad$ formula 15

(b) $\int \frac{(4x^6 - 9)^{3/2}}{x}\, dx$ $\quad u = 2x^3, \quad du = 6x^2\, dx$ $\quad$ formula 22

introduce a factor of x^2 into numerator and denominator

(c) $\int x^3 \sin x^2\, dx$ $\quad u = x^2, \quad du = 2x\, dx$ $\quad$ formula 47

EXERCISES 28.11

In Exercises 1 and 2, make the given changes in the indicated examples of this section, and then state which formula from Appendix B would be used to complete the integration.

1. In Example 1, change the denominator to $(2 + 3x)^2$.
2. In Example 2, in the numerator, change − to +.

In Exercises 3–8, identify u, du, and the formula from Appendix B that would be used to complete the integration. Do not integrate.

3. $\int \frac{4\, dy}{3y\sqrt{1 + 2y}}$
4. $\int \frac{x\, dx}{\sqrt{x^4 - 16}}$
5. $\int \frac{x\, dx}{(4 - x^4)^{3/2}}$
6. $\int x^5 \ln x^3\, dx$
7. $\int x \cos^2(x^2)\, dx$
8. $\int \frac{ds}{s(s^4 - 1)^{3/2}}$

In Exercises 9–52, integrate each function by using the table in Appendix B.

9. $\int \frac{3x\, dx}{2 + 5x}$
10. $\int \frac{4x\, dx}{(1 + x)^2}$
11. $\int_2^7 4x\sqrt{2 + x}\, dx$
12. $\int \frac{dx}{x^2 - 4}$
13. $\int \frac{dy}{(y^2 + 4)^{3/2}}$
14. $\int_0^{\pi/3} \sin^3 x\, dx$
15. $\int \sin 2x \sin 3x\, dx$
16. $\int 6 \sin^{-1} 3x\, dx$
17. $\int \frac{\sqrt{4x^2 - 9}}{x}\, dx$
18. $\int \frac{(9x^2 + 16)^{3/2}}{x}\, dx$
19. $\int \cos^5 4x\, dx$
20. $\int 0.2 \tan^2 2\phi\, d\phi$
21. $\int 6r \tan^{-1} r^2\, dr$
22. $\int 5xe^{4x}\, dx$
23. $\int_1^2 (4 - x^2)^{3/2}\, dx$
24. $\int \frac{3\, dx}{9 - 16x^2}$
25. $\int \frac{dx}{x\sqrt{4x^2 + 1}}$
26. $\int \frac{\sqrt{4 + x^2}}{x}\, dx$

27. $\displaystyle\int \frac{8\,dx}{x\sqrt{1-4x^2}}$

28. $\displaystyle\int \frac{dx}{x(1+4x)^2}$

29. $\displaystyle\int_0^{\pi/12} \sin\theta \cos 5\theta\, d\theta$

30. $\displaystyle\int_0^2 x^2 e^{3x}\, dx$

31. $\displaystyle\int x^5 \cos x^3\, dx$

32. $\displaystyle\int 5\sin^3 t \cos^2 t\, dt$

33. $\displaystyle\int \frac{2x\,dx}{(1-x^4)^{3/2}}$

34. $\displaystyle\int \frac{dx}{x(1-4x)}$

35. $\displaystyle\int_1^3 \frac{\sqrt{3+5x^2}\,dx}{x}$

36. $\displaystyle\int_{1/2}^1 \frac{\sqrt{9-4x^2}}{x}dx$

37. $\displaystyle\int x^3 \ln x^2\, dx$

38. $\displaystyle\int \frac{1.2u\,du}{u^2\sqrt{u^4-9}}$

39. $\displaystyle\int \frac{9x^2\,dx}{(x^6-1)^{3/2}}$

40. $\displaystyle\int x^7\sqrt{x^4+4}\,dx$

41. $\displaystyle\int t^2(t^6+1)^{3/2}\, dt$

42. $\displaystyle\int \frac{\sqrt{3+4x^2}\,dx}{x}$

43. $\displaystyle\int \sin^3 4x \cos^3 4x\, dx$

44. $\displaystyle\int 6\cot^4 2x\, dx$

45. Find the length of arc of the curve $y = x^2$ from $x = 0$ to $x = 1$. (Uses $\displaystyle\int_a^b \sqrt{1+\left(\frac{dy}{dx}\right)^2}\,dx$.)

46. Find the moment of inertia with respect to its axis of the solid generated by revolving the region bounded by $y = 3\ln x$, $x = e$, and the x-axis about the y-axis.

47. Find the area of an ellipse with a major axis $2a$ and a minor axis $2b$.

48. The voltage across a 5.0 μF capacitor in an electric circuit is zero. What is the voltage after 5.00 μs if a current i (in mA) as a function of the time t (in s) given by $i = \tan^{-1} 2t$ charges the capacitor?

49. Find the force (in N) on the region bounded by $x = 1/\sqrt{1+y}$, $y = 0$, $y = 3$, and the y-axis, if the surface of the water is at the upper edge of the area.

50. If 6.00 g of a chemical are placed in water, the time t (in min) it takes to dissolve half of the chemical is given by $t = 560\displaystyle\int_3^6 \frac{dx}{x(x+4)}$, where x is the amount of undissolved chemical at any time. Evaluate t.

51. The dome of a sports arena is the surface generated by revolving $y = 20.0\cos 0.0196x$ $(0 \le x \le 80.0\text{ m})$ about the y-axis. Find the volume within the dome.

52. If an electric charge Q is distributed along a wire of length $2a$, the force F exerted on an electric charge q placed at point P is $F = kqQ\displaystyle\int \frac{b\,dx}{(b^2+x^2)^{3/2}}$. Integrate to find F as a function of x.

CHAPTER 28 KEY FORMULAS AND EQUATIONS

Integrals

$$\int u^n\,du = \frac{u^{n+1}}{n+1} + C \quad (n \neq -1) \tag{28.1}$$

$$\int \frac{du}{u} = \ln|u| + C \tag{28.2}$$

$$\int e^u\,du = e^u + C \tag{28.3}$$

$$\int \sin u\,du = -\cos u + C \tag{28.4}$$

$$\int \cos u\,du = \sin u + C \tag{28.5}$$

$$\int \sec^2 u\,du = \tan u + C \tag{28.6}$$

$$\int \csc^2 u\,du = -\cot u + C \tag{28.7}$$

$$\int \sec u \tan u\,du = \sec u + C \tag{28.8}$$

$$\int \csc u \cot u\,du = -\csc u + C \tag{28.9}$$

$$\int \tan u\, du = -\ln|\cos u| + C \tag{28.10}$$

$$\int \cot u\, du = \ln|\sin u| + C \tag{28.11}$$

$$\int \sec u\, du = \ln|\sec u + \tan u| + C \tag{28.12}$$

$$\int \csc u\, du = \ln|\csc u - \cot u| + C \tag{28.13}$$

Trigonometric identities

$$\cos^2 x + \sin^2 x = 1 \tag{28.14}$$

$$2\cos^2 x = 1 + \cos 2x \tag{28.15}$$

$$2\sin^2 x = 1 - \cos 2x \tag{28.16}$$

$$1 + \tan^2 x = \sec^2 x \tag{28.17}$$

$$1 + \cot^2 x = \csc^2 x \tag{28.18}$$

Root-mean-square value

$$y_{\text{rms}} = \sqrt{\frac{1}{T}\int_0^T y^2\, dt} \tag{28.19}$$

Integrals

$$\int \frac{du}{\sqrt{a^2 - u^2}} = \sin^{-1}\frac{u}{a} + C \tag{28.20}$$

$$\int \frac{du}{a^2 + u^2} = \frac{1}{a}\tan^{-1}\frac{u}{a} + C \tag{28.21}$$

$$\int u\, dv = uv - \int v\, du \tag{28.22}$$

Trigonometric substitutions (28.23)

For $\sqrt{a^2 - x^2}$ use $x = a \sin\theta$

For $\sqrt{a^2 + x^2}$ use $x = a \tan\theta$

For $\sqrt{x^2 - a^2}$ use $x = a \sec\theta$

CHAPTER 28 REVIEW EXERCISES

In Exercises 1–42, integrate the given functions without using a table of integrals.

1. $\int e^{-8x}\, dx$

2. $\int e^{\cos 2x} \sin x \cos x\, dx$

3. $\int \frac{dx}{x(\ln 2x)^2}$

4. $\int_1^8 y^{1/3}\sqrt{y^{4/3} + 9}\, dy$

5. $\int_0^{\pi/2} \frac{4\cos\theta\, d\theta}{1 + \sin\theta}$

6. $\int \frac{\sec^2 x\, dx}{2 + \tan x}$

7. $\int \frac{2\, dx}{25 + 49x^2}$

8. $\int \frac{dx}{\sqrt{1 - 4x^2}}$

9. $\int_0^{\pi/2} \cos^3 2\theta\, d\theta$

10. $\int_0^{\pi/18} \sec^3 6x \tan 6x\, dx$

11. $\int_0^2 \frac{x\, dx}{4 + x^2}$

12. $\int_1^e \frac{\ln v^2\, dv}{\ln e^v}$

13. $\int (\sin t + \cos t)^2 \sin t\, dt$

14. $\int \frac{\sin^3 x\, dx}{\sqrt{\cos x}}$

15. $\int \frac{e^x\, dx}{1 + e^{2x}}$

16. $\int \frac{p + 25}{p^2 - 25}dp$

17. $\int \sec^4 3x\, dx$

18. $\int \frac{(1 - \cos^2\theta)\, d\theta}{1 + \cos 2\theta}$

19. $\int \frac{2x^2 + 6x + 1}{2x^3 - x^2 - x}dx$

20. $\int \frac{4 - e^{\sqrt{x}}}{\sqrt{x}\, e^{\sqrt{x}}}dx$

21. $\int \frac{3x\, dx}{4 + x^4}$

22. $\int_1^3 \frac{12\, dR}{\sqrt{R}(1 + R)}$

23. $\int \frac{4\, dx}{\sqrt{4x^2 - 9}}$

24. $\int \frac{x^2\, dx}{\sqrt{9 - x^2}}$

25. $\int \frac{e^{2x}\, dx}{\sqrt{e^{2x} + 1}}$

26. $\int \frac{x^2 - 2x + 3}{(x - 1)^3}dx$

27. $\int \frac{2x^2 + 3x + 18}{x^3 + 9x} dx$

28. $\int_{1/2}^{e/2} \frac{(4 + \ln 2u)^3 du}{u}$

29. $\int_0^{\pi/6} 3 \sin^2 3\phi \, d\phi$

30. $\int \sin^4 x \, dx$

31. $\int x \csc^2 2x \, dx$

32. $\int x \tan^{-1} x \, dx$

33. $\int \frac{3u^2 - 6u - 2}{3u^3 + u^2} du$

34. $\int \frac{R^2 + 3}{R^4 + 3R^2 + 2} dR$

35. $\int e^{2x} \cos e^{2x} dx$

36. $\int \frac{3\, dx}{x^2 + 6x + 10}$

37. $\int_1^e \frac{3 \cos(\ln x)\, dx}{x}$

38. $\int_1^3 \frac{2\, dx}{x^2 - 2x + 5}$

39. $\int \frac{u^2 - 8}{u + 3} du$

40. $\int \frac{\log_x 2\, dx}{x \ln x}$

41. $\int \frac{\sin x \cos^2 x}{5 + \cos^2 x} dx$

42. $\int \frac{e^{2x}\, dx}{16 + e^{4x}}$

In Exercises 43–84, solve the given problems by integration. In Exercise 85, answer the given question.

43. For the integral $\int \frac{dx}{x\sqrt{2 - x^2}}$, using a trigonometric substitution, find the transformed integral but do not integrate.

44. For the integral $\int \frac{dx}{\sqrt{x^2 + 4x + 3}}$, using a trigonometric substitution, find the transformed integral but do not integrate.

45. Perform the integrations $\int e^{\ln 4x} dx$ and $\int \ln e^{4x} dx$. Compare results.

46. Perform the integrations $\int \sin 6x \sin 5x \, dx$ using $\sin \alpha \sin \beta = \frac{1}{2}[\cos(\alpha - \beta) - \cos(\alpha + \beta)]$.

47. For the integral $\int \frac{\sqrt{16 + x^8}}{x} dx$, identify the formula found in Appendix B, u, and du that would be used for the integration. Do not integrate.

48. For the integral $\int x^9\sqrt{x^5 + 8}\, dx$, identify the formula found in Appendix B, u, and du, that would be used for the integration. Do not integrate.

49. Show that $\int e^x(e^x + 1)^2 dx$ can be integrated in two ways. Explain the difference in the answers.

50. Show that $\int \frac{1}{x}(1 + \ln x)\, dx$ can be integrated in two ways. Explain the difference in the answers.

51. Integrate $\int \frac{\sin x\, dx}{1 + \sin x}$ by first rewriting the integrand. (*Hint:* First divide and then note Exercise 29 in Section 28.4.)

52. Integrate $\int \frac{dx}{1 + e^x}$ by first rewriting the integrand. (*Hint:* It is possible to multiply the numerator and the denominator by an appropriate expression.)

53. The integral $\int \frac{x}{\sqrt{x^2 + 4}} dx$ can be integrated in more than one way. Explain what methods can be used and which is simpler.

54. Find the equation of the curve for which $dy/dx = e^x(2 - e^x)^2$, if the curve passes through $(0, 4)$.

55. Find the equation of the curve for which $dy/dx = \sec^4 x$, if the curve passes through the origin.

56. Find the equation of the curve for which $\frac{dy}{dx} = \frac{\sqrt{4 + x^2}}{x^4}$, if the curve passes through $(2, 1)$.

57. Find the area bounded by $y = 4e^{2x}$, $x = 1.5$, and the axes.

58. Find the area bounded by $y = x/(1 + x)^2$, the x-axis, and the line $x = 4$.

59. Find the area inside the circle $x^2 + y^2 = 25$ and to the right of the line $x = 3$.

60. Find the area bounded by $y = x\sqrt{x + 4}$, $y = 0$, and $x = 5$.

61. Find the area bounded by $y = \tan^{-1} 2x$, $x = 2$, and the x-axis.

62. In polar coordinates, the area A bounded by the curve $r = f(\theta)$, $\theta = \alpha$, and $\theta = \beta$ is found by evaluating the integral $A = \frac{1}{2}\int_\alpha^\beta r^2\, d\theta$. Find the area bounded by $\alpha = 0$, $\beta = \pi/2$, and $r = e^\theta$.

63. Find the volume generated by revolving the region bounded by $y = xe^x$, $y = 0$, and $x = 2$ about the y-axis.

64. Find the volume generated by revolving about the y-axis the region bounded by $y = x + \sqrt{x + 1}$, $x = 3$, and the axes.

65. Find the volume of the solid generated by revolving the region bounded by $y = e^x \sin x$ and the x-axis between $x = 0$ and $x = \pi$ about the x-axis.

66. Find the centroid of a flat plate that covers the region bounded by $y = \ln x$, $x = 2$, and the x-axis.

67. Find the length of arc along the curve of $y = \ln \sin x$ from $x = \pi/3$ to $x = 2\pi/3$. (Use $s = \int_a^b \sqrt{1 + \left(\frac{dy}{dx}\right)^2}\, dx$.)

68. Find the area of the surface generated by revolving the curve of $y = \sqrt{4 - x^2}$ from $x = -2$ to $x = 2$ about the x-axis. The area of a surface of revolution from $x = a$ to $x = b$ is

$$S = 2\pi \int_a^b y\sqrt{1 + \left(\frac{dy}{dx}\right)^2}\, dx$$

69. The force F (in N) on a nail by a hammer is $F = 5/(1 + 2t)$, where t is the time (in s). The impulse I of the force from $t = 0$ s to $t = 0.25$ s is $I = \int_0^{0.25} F\, dt$. Find the impulse.

70. The acceleration of a parachutist is given by $dv/dt = g - kv$, where k is a constant depending on the resisting force due to air friction. Find v as a function of the time t.

71. The change in the thermodynamic entity of entropy ΔS may be expressed as $\Delta S = \int (c_v/T)\, dT$, where c_v is the heat capacity at constant volume and T is the temperature. For increased accuracy, c_v is often given by the equation $c_v = a + bT + cT^2$, where a, b, and c are constants. Express ΔS as a function of temperature.

72. A certain type of chemical reaction leads to the equation $dt = \frac{dx}{k(a-x)(b-x)}$, where a, b, and k are constants. Solve for t as a function of x.

73. An electric transmission line between two towers has a shape given by $y = 16.0(e^{x/32} + e^{-x/32})$. Find the length of transmission line if the towers are 50.0 m apart (from $x = -25.0$ m to $x = 25.0$ m). (Use $s = \int_a^b \sqrt{1 + \left(\frac{dy}{dx}\right)^2}\, dx$.)

74. An object at the end of a spring is immersed in liquid. Its velocity (in cm/s) is then described by the equation $v = 2e^{-2t} + 3e^{-5t}$, where t is the time (in s). Such motion is called *overdamped.* Find the displacement s as a function of t if $s = -1.6$ cm for $t = 0$.

75. When we consider the resisting force of the air, the velocity v (in m/s) of a falling brick in terms of the time t (in s) is given by $dv/(9.8 - 0.1v) = dt$. If $v = 0$ when $t = 0$, find v as a function of t.

76. The power delivered to an electric circuit is given by $P = Ei$, where E and i are the instantaneous voltage and the instantaneous current in the circuit, respectively. The mean power, averaged over a period $2\pi/\omega$, is given by $P_{av} = \frac{\omega}{2\pi}\int_0^{2\pi/\omega} Ei\, dt$. If $E = 20 \cos 2t$ and $i = 3 \sin 2t$, find the average power over a period of $\pi/4$.

77. Find the root-mean-square value for one period of the electric current i if $i = 2 \sin t$.

78. In atomic theory, when finding the number n of atoms per unit volume of a substance, we use the equation $n = A\int_0^{\pi} e^{a\cos\theta} \sin\theta\, d\theta$. Perform the indicated integration.

79. In the study of the effects of an electric field on molecular orientation, the integral $\int_0^{\pi} (1 + k\cos\theta)\cos\theta \sin\theta\, d\theta$ is used. Evaluate this integral.

80. In finding the lift of the air flowing around an airplane wing, we use the integral $\int_{-\pi/2}^{\pi/2} \theta^2 \cos\theta\, d\theta$. Evaluate this integral.

81. Find the volume within the piece of tubing in an oil distribution line shown in Fig. 28.16. All cross-sections are circular.

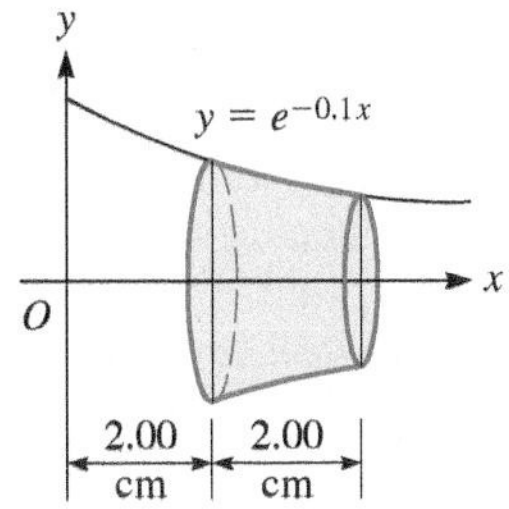

Fig. 28.16

82. A metal plate has a shape shown in Fig. 28.17. Find the x-coordinate of the centroid of the plate.

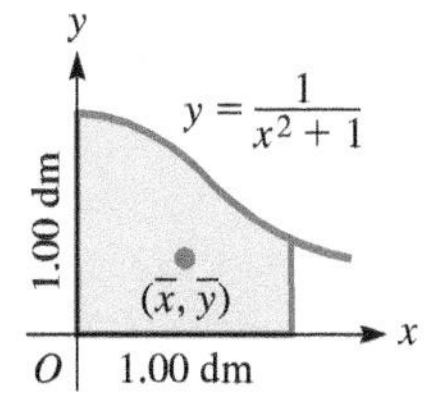

Fig. 28.17

83. The nose cone of a space vehicle is to be covered with a heat shield. The cone is designed such that a cross-section x metres from the tip and perpendicular to its axis is a circle of radius $1.5x^{2/3}$ metres. Find the surface area of the heat shield if the nose cone is 4.00 m long. See Fig. 28.18. (See Exercise 68.)

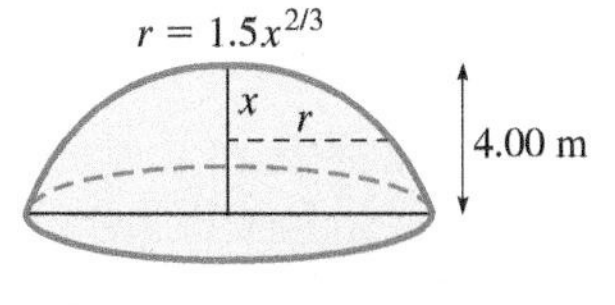

Fig. 28.18

84. A window has a shape of a semiellipse, as shown in Fig. 28.19. What is the area of the window?

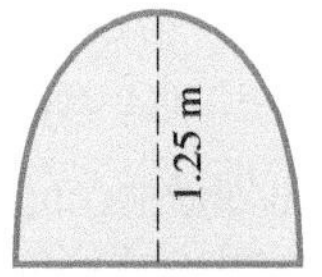

Fig. 28.19

85. The side of a blade designed to cut leather at a shoe factory can be described as the region bounded by $y = 4\cos^2 x$ and $y = 4$ from $x = 0$ to $x = 3.14$. Write two or three paragraphs explaining how the area of the blade may be found by integrating the appropriate integral by either of two methods of this chapter.

CHAPTER 28 PRACTICE TEST

In Problems 1–7, evaluate the given integrals.

1. $\int (\sec x - \sec^3 x \tan x)\, dx$

2. $\int \sin^3 x\, dx$

3. $\int \tan^3 2x\, dx$

4. $\int \cos^2 4\theta\, d\theta$

5. $\int \frac{dx}{x^2\sqrt{4 - x^2}}$

6. $\int xe^{-2x}\, dx$

7. $\int \frac{x^3 + 5x^2 + x + 2}{x^4 + x^2}\, dx$

8. The electric current in a certain circuit is given by $i = \int \frac{6t + 1}{4t^2 + 9}\, dt$, where t is the time. Integrate and find the resulting function if $i = 0$ for $t = 0$.

9. Find the first-quadrant area bounded by $y = \frac{1}{\sqrt{16 - x^2}}$ and $x = 3$.

29. Partial Derivatives and Double Integrals

LEARNING OUTCOMES

After completion of this chapter, the student should be able to:

- Set up and evaluate a function of two variables
- Use traces and sections to sketch the graph of a surface in the rectangular coordinate system in three dimensions
- Convert points and equations from rectangular to cylindrical coordinates and vice versa
- Find the first- and second-order partial derivatives of a function of several variables
- Evaluate double integrals
- Understand the geometric interpretation of the partial derivatives and the double integral of a function of two variables
- Solve application problems involving functions of two variables, their partial derivatives, and their integrals

▲ In Section 29.2, we see how the shape of the roof of the Saddledome arena in Calgary, Alberta, is determined by a function of two variables.

To this point, we have been dealing with functions that have a single independent variable. There are, however, numerous applications in which functions with more than one independent variable are used. Although a number of these functions involve three or more independent variables, many involve only two, and we shall be concerned primarily with these.

In the first two sections of this chapter, we establish the meaning of a function of two independent variables, and discuss how the graph of this type of function is shown. In the last two sections, we develop some of the basic concepts of the calculus of these functions, primarily *partial derivatives* and *double integrals.* In finding the partial derivative, we take the derivative of the function with respect to one independent variable, holding the other constant. Similarly, when integrating an expression with two differentials, we integrate with respect to one, holding the other constant, and then integrate with respect to the other.

While studying problems in motion and optics in the early 1700s, mathematicians including Leibniz developed the meaning and use of partial derivatives. By the middle of the eighteenth century, double integrals seem to have been known by the leading mathematicians of the time. However, their wide dissemination is due to Euler who, in 1769, published the first detailed explanation of them.

Partial derivatives and double integrals have applications in many areas of science and technology, including acoustics, electricity, electronics, mechanics, product design, and wave motion. Some of these applications are shown in the examples and exercises of this chapter.

29.1 Functions of Two Variables

Many familiar formulas express one variable in terms of two or more other variables. The following example illustrates one from geometry.

EXAMPLE 1 A function of two variables

The total surface area A of a right circular cylinder is a function of the radius r and the height h of the cylinder. That is, the area will change if either or both of these change. The formula for the total surface area is

$$A = 2\pi r^2 + 2\pi rh$$

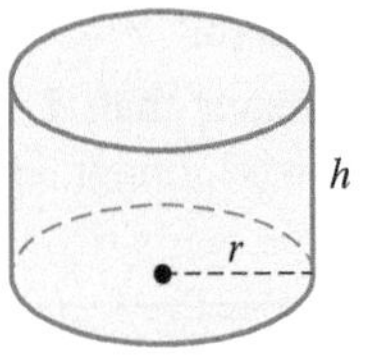

Fig. 29.1

We say that A is a function of r and h. See Fig. 29.1.

We define a function of two variables as follows: *If z is uniquely determined for given values of x and y, then z is a function of x and y.* The notation used is similar to that used for one independent variable. It is $z = f(x, y)$, where both x and y are independent variables. Therefore, it follows that $f(a, b)$ *means "the value of the function when $x = a$ and $y = b$."*

EXAMPLE 2 Illustrating the notation $z = f(x, y)$

If $f(x, y) = 3x^2 + 2xy - y^3$, find $f(-1, 2)$.

Substituting -1 for x and 2 for y, we have

$$\begin{aligned} f(-1, 2) &= 3(-1)^2 + 2(-1)(2) - (2)^3 \\ &= 3 - 4 - 8 = -9 \end{aligned}$$

EXAMPLE 3 Current as a function of two variables

For a certain electric circuit, the current i (in A), in terms of the voltage E and resistance R (in Ω) is given by

$$i = \frac{E}{R + 0.25}$$

Find the current for $E = 1.50$ V and $R = 1.20\ \Omega$ and for $E = 1.60$ V and $R = 1.05\ \Omega$.

Substituting the first values, we have

$$i = \frac{1.50}{1.20 + 0.25} = 1.03 \text{ A}$$

For the second pair of values, we have

$$i = \frac{1.60}{1.05 + 0.25} = 1.23 \text{ A}$$

For this circuit, the current generally changes if either or both of E and R change.

EXAMPLE 4 Using $f(x, y)$ notation

If $f(x, y) = 2xy^2 - y$, find $f(x, 2x) - f(x, x^2)$.

We note that in each evaluation, the x factor remains as x, but that we are to substitute $2x$ for y and subtract the function for which x^2 is substituted for y.

$$\begin{aligned} f(x, 2x) - f(x, x^2) &= \left[2x(2x)^2 - (2x)\right] - \left[2x(x^2)^2 - x^2\right] \\ &= \left[8x^3 - 2x\right] - \left[2x^5 - x^2\right] \\ &= 8x^3 - 2x - 2x^5 + x^2 \\ &= -2x^5 + 8x^3 + x^2 - 2x \end{aligned}$$

This type of difference of functions is important in Section 29.4.

Practice Exercise

1. If $f(x, y) = 4xy^2 - 3x^2y$, find $f(1, 4) - f(x, 2x)$.

Restricting values of the function to real numbers means that *values of either x or y or both that lead to division by zero or to imaginary values for z are not permissible.*

EXAMPLE 5 Restrictions on independent variables

If $f(x, y) = \frac{3xy}{(x - y)(x + 3)}$, all values of x and y are permissible except those for which $x = y$ and $x = -3$. Each of these would indicate division by zero.

If $f(x, y) = \sqrt{4 - x^2 - y^2}$, neither x nor y may be greater than 2 in absolute value, and the sum of their squares may not exceed 4. Otherwise, imaginary values of the function would result.

Following is an example of setting up a function of two variables from stated conditions. As with word problems with one unknown, although no general rules can be given for this procedure, a careful analysis of the statement should lead to the required function.

EXAMPLE 6 Function of two variables—fish tank surface area

An open rectangular fish tank is to hold 0.25 m^3 of water when completely full. Express the total surface area of glass required to make the tank as a function of the length and width of the tank.

An "open" tank is one that has no top. Therefore, the surface area S of the tank is

$$S = lw + 2lh + 2hw$$

where l is the length, w the width, and h the height of the tank (see Fig. 29.2). However, this equation contains three independent variables. Using the condition that the volume of water is 0.25 m^3, we have $lwh = 0.25$. Since we wish to have only l and w, we solve this equation for h and find that $h = 0.25/lw$. Substituting for h in the equation for the surface area, we have

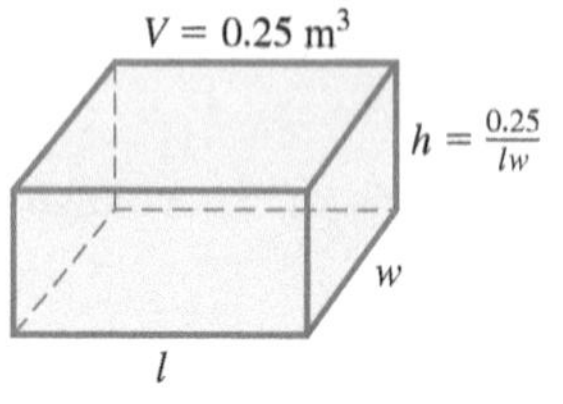

Fig. 29.2

$$S = lw + 2l\left(\frac{0.25}{lw}\right) + 2\left(\frac{0.25}{lw}\right)w$$

$$= lw + \frac{0.50}{w} + \frac{0.50}{l} \quad \text{the required function}$$

EXERCISES 29.1

In Exercises 1 and 2, make the given changes in the indicated examples of this section and then solve the resulting problems.

1. In Example 2, change $f(-1, 2)$ to $f(-2, 1)$.
2. In Example 4, change $f(x, 2x) - f(x, x^2)$ to $f(2x, x) - f(x^2, x)$.

In Exercises 3–8, determine the indicated function.

3. Express the volume V of a right circular cylinder as a function of the radius r and height h.
4. Express the length of a diagonal of a rectangle as a function of the length and the width.
5. A cylindrical can is to be made to contain a volume V. Express the total surface area (including the top) of the can as a function of V and the radius of the can.
6. The angle between two forces $\mathbf{F}_1$ and $\mathbf{F}_2$ is 30.0°. Express the magnitude of the resultant $\mathbf{R}$ in terms of F_1 and F_2. See Fig. 29.3.

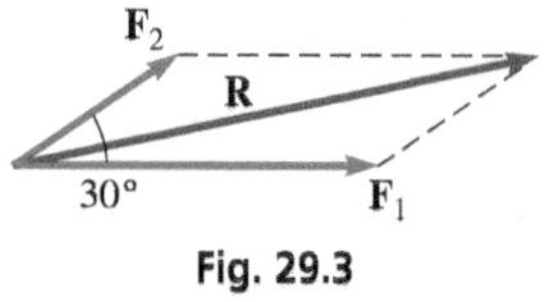

Fig. 29.3

7. A right circular cylinder is to be inscribed in a sphere of radius r. Express the volume of the cylinder as a function of the height h of the cylinder and r. See Fig. 29.4.

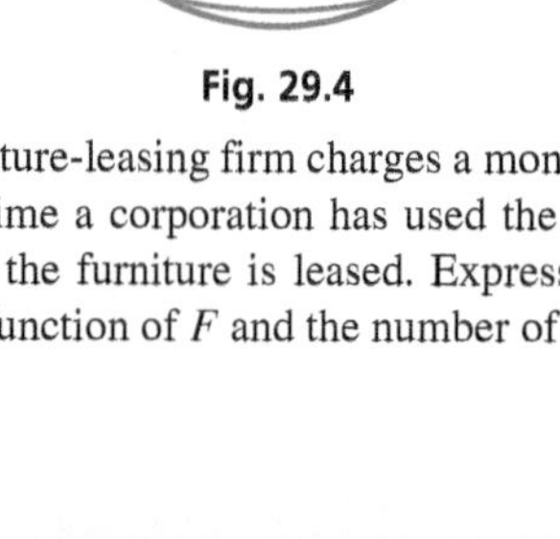

Fig. 29.4

8. An office-furniture-leasing firm charges a monthly fee F based on the length of time a corporation has used the service, plus \$100 for each week the furniture is leased. Express the total monthly charge T as a function of F and the number of weeks w the furniture is leased.

In Exercises 9–24, evaluate the given functions.

9. $f(x, y) = 5x - 4y$; find $f(0, -3)$.

10. $F(x, y) = x^2 - 5y + y^2$; find $F(2, -2)$.

11. $g(r, s) = r - 2rs - r^2s$; find $g(-2, 1)$.

12. $f(r, \theta) = 2r(r \tan\theta - \sin 2\theta)$; find $f(3, \pi/4)$.

13. $Y(y, t) = \frac{12 + 5y}{t - 1} + 2y^2t$; find $Y(y, 2)$.

14. $f(r, t) = r^3 - 3r^2t + 3rt^2 - t^3$; find $f(3, t)$.

15. $X(x, t) = -6xt + xt^2 - t^3$; find $X(x, -t)$.

16. $g(y, z) = 2yz^2 - 6y^2z - y^2z^2$; find $g(y, 2y)$.

17. $H(p, q) = p - \frac{p - 2q^2 - 5q}{p + q}$; find $H(p, q + k)$.

18. $g(x, z) = z\tan^{-1}(x^2 + xz)$; find $g(-x, z)$.

19. $f(x, y) = x^2 - 2xy - 4x$; find $f(x + h, y + k) - f(x, y)$.

20. $g(y, z) = 4yz - z^3 + 4y$; find $g(y + 1, z + 2) - g(y, z)$.

21. $f(x, y) = xy + x^2 - y^2$; find $f(x, x) - f(x, 0)$.

22. $f(x, y) = 4x^2 - xy - 2y$; find $f(x, x^2) - f(x, 1)$.

23. $g(y, z) = 3y^3 - y^2z + 5z^2$; find $g(3z^2, z) - g(z, z)$.

24. $X(x, t) = 2x - \frac{t^2 - 2x^2}{x}$; find $X(2t, t) - X(2t^2, t)$.

In Exercises 25–28, determine which values of x and y, if any, are not permissible. In Exercise 27, explain your answer.

25. $f(x, y) = \frac{\sqrt{y}}{2x}$

26. $f(x, y) = \frac{x^2 - 4y^2}{x^2 - 16}$

27. $f(x, y) = \sqrt{x^2 + y^2 - x^2y - y^3}$

28. $f(x, y) = \frac{1}{xy - y}$

In Exercises 29–40, solve the given problems. When necessary, round answers to three significant digits.

29. The voltage V across a resistor R in an electric circuit is given by $V = iR$, where i is the current. What is the voltage if the current is 3 A and the resistance is 6 Ω?

30. The centripetal acceleration a of an object moving in a circular path is $a = v^2/R$, where v is the velocity of the object and R is the radius of the circle. What is the centripetal acceleration of an object moving at 6 m/s in a circular path of radius 4 m?

31. The pressure p (in Pa) of a gas as a function of its volume V and temperature T is $p = nRT/V$. If $n = 3.00$ mol and $R = 8.31$ J/mol · K, find p for $T = 300$ K and $V = 50.0\text{ m}^3$.

32. The power P (in W) supplied to a resistance R in an electric circuit by a current i is $P = i^2R$. Find the power supplied by a current of 4.0 A to a resistance of 2.4 Ω.

33. The atmospheric temperature T near ground level in a certain region is $T = ax^2 + by^2$, where a and b are constants. What type of curve is each isotherm (along which the temperature is constant) in this region?

34. The pressure p exerted by a force F on an area A is $p = F/A$. If a given force is doubled on an area that is 2/3 of a given area, what is the ratio of the initial pressure to the final pressure?

35. The current i (in A) in a certain electric circuit is a function of the time t (in s) and a variable resistor R (in Ω), given by $i = \frac{6.0 \sin 0.01t}{R + 0.12}$. Find i for $t = 0.75$ s and $R = 1.50$ Ω.

36. The reciprocal of the image distance q from a lens as a function of the object distance p and the focal length f of the lens is

$$\frac{1}{q} = \frac{1}{f} - \frac{1}{p}.$$

Find the image distance of an object 30 cm from a lens whose focal length is 10 cm.

37. A rectangular solar cell panel has a perimeter p and a width w. Express the area A of the panel in terms of p and w and evaluate the area for $p = 250$ cm and $w = 55$ cm. See Fig. 29.5.

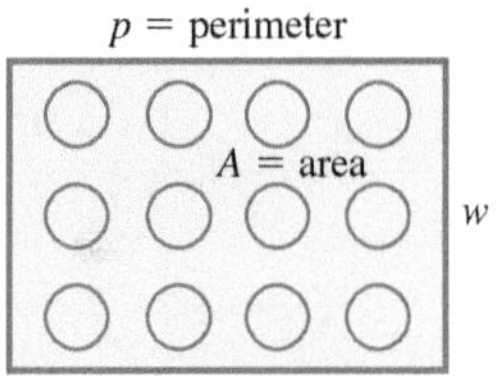

Fig. 29.5

38. A gasoline storage tank is in the shape of a right circular cylinder with a hemisphere at each end, as shown in Fig. 29.6. Express the volume V of the tank in terms of r and h and then evaluate the volume for $r = 1.25$ m and $h = 4.17$ m.

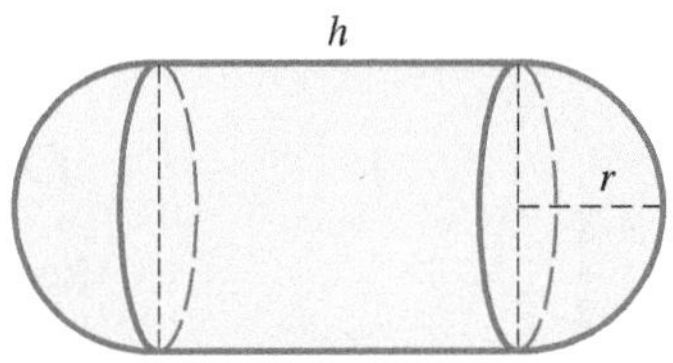

Fig. 29.6

39. The crushing load L of a pillar varies as the fourth power of its radius r and inversely as the square of its length l. Express L as a function of r and l for a pillar 6.0 m tall and 1.0 m in diameter that is crushed by a load of 20 Mg.

40. The resonant frequency f (in Hz) of an electric circuit containing an inductance L and capacitance C is inversely proportional to the square root of the product of the inductance and the capacitance. If the resonant frequency of a circuit containing a 4-H inductor and a 64-μF capacitor is 10 Hz, express f as a function of L and C.

Answer to Practice Exercise

1. $f(1, 4) - f(x, 2x) = 52 - 10x^3$

29.2 Curves and Surfaces in Three Dimensions

We will now undertake a brief description of the graphical representation of a function of two variables. We shall show first a method of representation in the rectangular coordinate system in two dimensions. The following example illustrates the method.

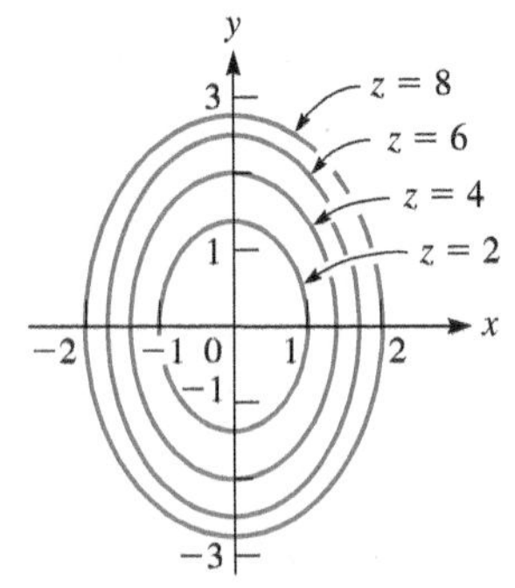

Fig. 29.7

EXAMPLE 1 Representing $f(x, y)$ in two dimensions

In order to represent $z = 2x^2 + y^2$, we will assume various values of z and sketch the curve of the resulting equation in the xy-plane. For example, if $z = 2$ we have

$$2x^2 + y^2 = 2$$

We recognize this as an ellipse with its major axis along the y-axis and vertices at $(0, \sqrt{2})$ and $(0, -\sqrt{2})$. The ends of the minor axis are at $(1, 0)$ and $(-1, 0)$. However, the ellipse $2x^2 + y^2 = 2$ represents the function $z = 2x^2 + y^2$ only for the value of $z = 2$. If $z = 4$, we have $2x^2 + y^2 = 4$, which is another ellipse. In fact, for all positive values of z, an ellipse is the resulting curve. Negative values of z are not possible since neither x^2 nor y^2 may be negative. Fig. 29.7 shows the ellipses that are obtained by using the indicated values of z.

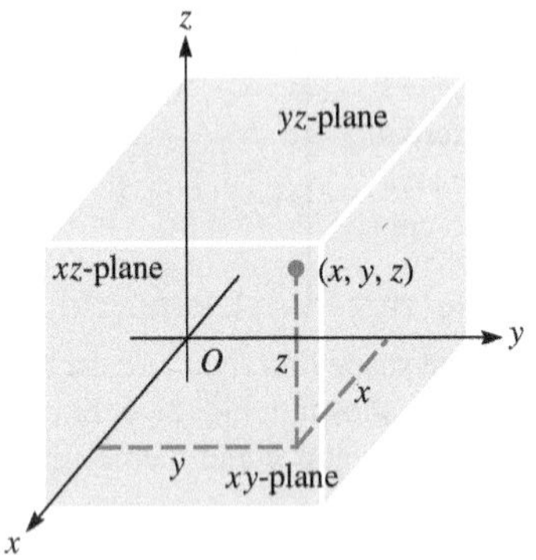

Fig. 29.8

The method of representation illustrated in Example 1 is useful if only a few specific values of z are to be used, or at least if the various curves do not intersect in such a way that they cannot be distinguished. If a general representation of z as a function of x and y is desired, it is necessary to use three coordinate axes, one each for x, y, and z. The most widely applicable system of this kind is to place a third coordinate axis at right angles to each of the x- and y-axes. In this way, we employ three dimensions for the representation.

The three mutually perpendicular coordinate axes—the x-axis, the y-axis, and the z-axis—are the basis of the **rectangular coordinate system in three dimensions**. Together they form three mutually perpendicular planes in space: the xy-plane, the yz-plane, and the xz-plane. To every point in space of the coordinate system is associated a set of numbers (x, y, z). The point at which the axes meet is the *origin*. The positive directions of the axes are indicated in Fig. 29.8. *That part of space in which all values of the coordinates are positive is called the first* **octant**. Numbers are not assigned to the other octants.

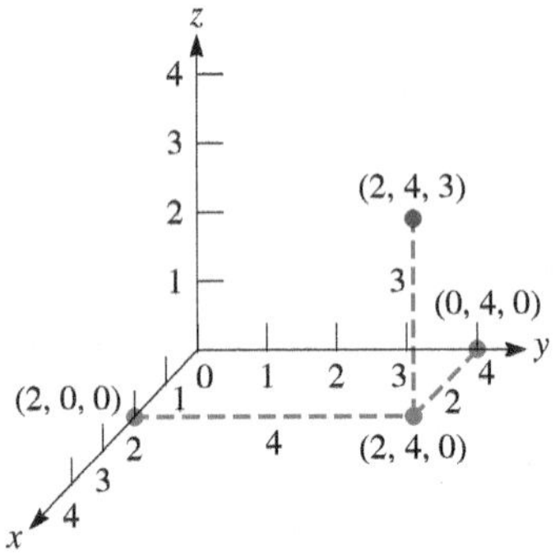

Fig. 29.9

EXAMPLE 2 Representing (x, y, z) in rectangular coordinates

Represent the point $(2, 4, 3)$ in rectangular coordinates.

We first note that when representing a point in three dimensions, a certain distortion is necessary to represent values of x reasonably, since the x-axis "comes out" of the plane of the page. Units that are $\sqrt{2}/2$ (≈ 0.7) as long as those used on the other axes give a good representation. With this in mind, we draw a line 4 units long from the point $(2, 0, 0)$ on the x-axis in the xy-plane, parallel to the y-axis. This locates the point $(2, 4, 0)$. From this point, a line 3 units long is drawn vertically upward, and this locates the point $(2, 4, 3)$. See the points and dashed lines in Fig. 29.9.

This point $(2, 4, 3)$ may also be located by starting from $(0, 4, 0)$ on the y-axis, then proceeding 2 units *parallel* to the x-axis to $(2, 4, 0)$, and then proceeding vertically 3 units upward to $(2, 4, 3)$.

It may also be located by starting from the point $(0, 0, 3)$ on the z-axis, although it is generally preferred to start on the x-axis, or possibly the y-axis.

We now show certain basic techniques by which three-dimensional figures may be drawn. We start by showing the general equation of a plane.

In Chapter 21, we showed that the graph of the equation $Ax + By + C = 0$ in two dimensions is a straight line. By the following example, we will verify that *the graph of the equation*

$$Ax + By + Cz + D = 0 \tag{29.1}$$

is a **plane** *in three dimensions.*

EXAMPLE 3 Illustrating the graph of a plane

Show that the graph of the linear equation $2x + 3y + z - 6 = 0$ in the rectangular coordinate system of three dimensions is a plane.

If we let any of the three variables take on a specific value, we obtain a linear equation in the other two variables. For example, the point $\left(\frac{1}{2}, 1, 2\right)$ satisfies the equation and therefore lies on the graph of the equation. For $x = \frac{1}{2}$, we have

$$3y + z - 5 = 0$$

which is the equation of a straight line. This means that all pairs of values of y and z that satisfy this equation, along with $x = \frac{1}{2}$, satisfy the given equation. Thus, for $x = \frac{1}{2}$, the straight line $3y + z - 5 = 0$ lies on the graph of the equation.

For $z = 2$, we have

$$2x + 3y - 4 = 0$$

which is also a straight line. By similar reasoning, this line lies on the graph of the equation. Since two lines through a point define a plane, these lines through $\left(\frac{1}{2}, 1, 2\right)$ define a plane. This plane is the graph of the equation (see Fig. 29.10).

For any point on the graph of this equation, there is a straight line on the graph that is parallel to one of the coordinate planes. Therefore, there are intersecting straight lines through the point. Thus, the graph is a plane. A similar analysis can be made for any equation of the same linear form.

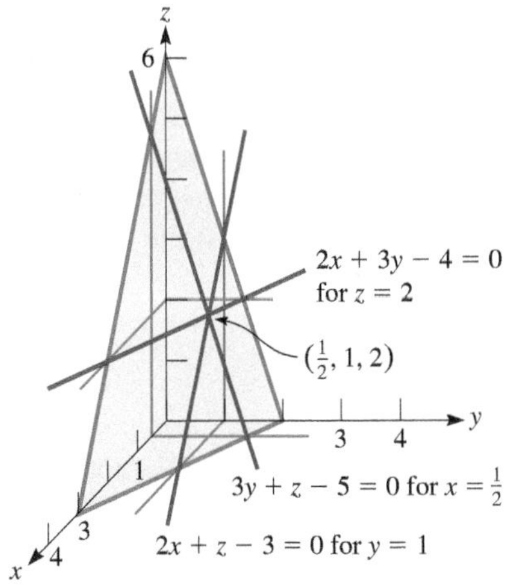

Fig. 29.10

LEARNING TIP

Since we know that the graph of an equation of the form of Eq. (29.1) is a plane, its graph can be found by determining its three intercepts, and the plane can then be represented by drawing in the lines between these intercepts. If the plane passes through the origin, by letting two of the variables in turn be zero, two straight lines that define the plane are found.

EXAMPLE 4 Sketching the graph of a plane

Sketch the graph of $3x - y + 2z - 4 = 0$.

As in the case of two-dimensional graphs, the intercepts of a graph in three dimensions are those points where it crosses the respective axes. Therefore, by letting two of the variables at a time equal zero, we obtain the intercepts. For the graph of the given equation, the intercepts are $\left(\frac{4}{3}, 0, 0\right)$, $(0, -4, 0)$, and $(0, 0, 2)$. These points are located (see Fig. 29.11), and lines drawn between them to represent the plane.

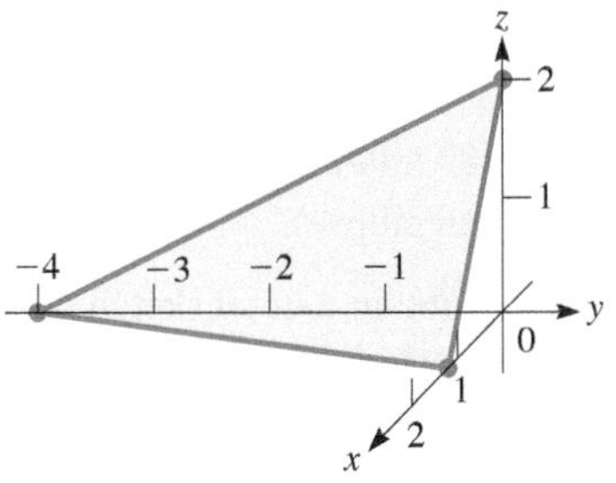

Fig. 29.11

The intersection of two surfaces is a **curve** *in space.* This has been seen in Examples 3 and 4, since the intersections of the given planes and the coordinate planes are lines (which in the general sense are curves). *We define the* **traces** *of a surface to be the curves of intersection of the surface and the coordinate planes.* The traces of a plane are those lines drawn between the intercepts to represent the plane. Many surfaces may be sketched by finding their traces and intercepts.

LEARNING TIP

The graphs of planes and of functions with two independent variables are examples of **surfaces** in space. More generally, the graph of an equation relating three variables is also a surface.

EXAMPLE 5 Using intercepts and traces to sketch a graph

Find the intercepts and traces for the graph of the equation $z = 4 - x^2 - y^2$ and then sketch the graph.

The intercepts of the graph of the equation are $(2, 0, 0)$, $(-2, 0, 0)$, $(0, 2, 0)$, $(0, -2, 0)$, and $(0, 0, 4)$.

Since the traces of a surface lie within the coordinate planes, for each trace one of the variables is zero. Thus, by ***letting each variable in turn be zero, we find the trace of the surface in the plane of the other two variables.*** Therefore, the traces of this surface are

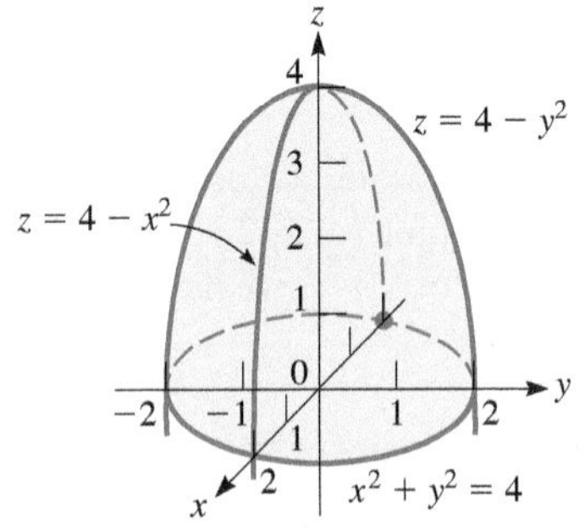

Fig. 29.12

in the *yz*-plane:	$z = 4 - y^2$	(a parabola)
in the *xz*-plane:	$z = 4 - x^2$	(a parabola)
in the *xy*-plane:	$x^2 + y^2 = 4$	(a circle)

Using the intercepts and sketching the traces, we obtain the surface represented by the equation as shown in Fig. 29.12. This figure is called a **circular paraboloid**.

There are numerous techniques for analysing the equation of a surface in order to obtain its graph. Another that we shall discuss here, which is closely associated with a trace, is that of a **section**. *By assuming a specific value of one of the variables, we obtain an equation in two variables, the graph of which lies in a plane parallel to the coordinate plane of the two variables.* The following examples illustrate sketching a surface by use of intercepts, traces, and sections.

EXAMPLE 6 Using traces and sections to sketch a graph

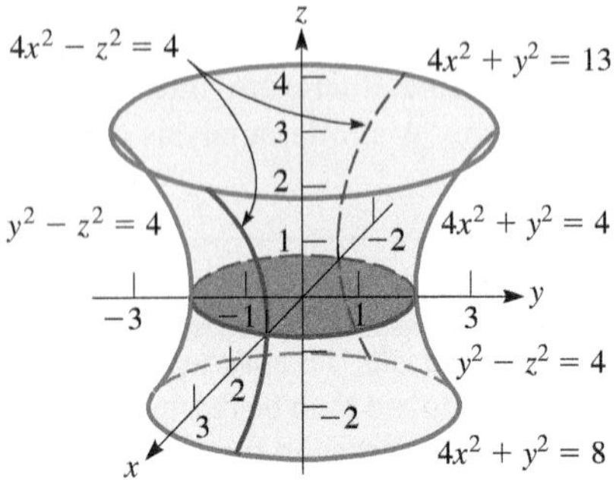

Fig. 29.13

Sketch the graph of $4x^2 + y^2 - z^2 = 4$.

The intercepts are $(1, 0, 0)$, $(-1, 0, 0)$, $(0, 2, 0)$, and $(0, -2, 0)$. We note that there are no intercepts on the *z*-axis, for this would necessitate $z^2 = -4$.

The traces are

in the *yz*-plane:	$y^2 - z^2 = 4$	(a hyperbola)
in the *xz*-plane:	$4x^2 - z^2 = 4$	(a hyperbola)
in the *xy*-plane:	$4x^2 + y^2 = 4$	(an ellipse)

The surface is reasonably defined by these curves, but by assuming suitable values of *z*, we may indicate its shape better. For example, if $z = 3$, we have $4x^2 + y^2 = 13$, which is an ellipse. In using it, we must remember that it is valid for $z = 3$ and therefore should be drawn 3 units above the *xy*-plane. If $z = -2$, we have $4x^2 + y^2 = 8$, which is also an ellipse. Thus, we have the following sections:

for $z = 3$:	$4x^2 + y^2 = 13$	(an ellipse)
for $z = -2$:	$4x^2 + y^2 = 8$	(an ellipse)

Other sections could be found, but these are sufficient to obtain a good sketch of the graph (see Fig. 29.13). The figure is called an **elliptic hyperboloid**.

Practice Exercise

1. What is the trace of $x^2 + 4y^2 - 4z^2 = 4$ in the (a) *xy*-plane? (b) *xz*-plane?

EXAMPLE 7 Sketching a graph—saddle-shaped cable roofs

■ See the chapter introduction.

Sketch the graph of $z = y^2 - x^2$ (a hyperbolic paraboloid).

There is only one intercept at the origin. The traces are

in the *yz*-plane:	$z = y^2$	(a parabola opening upward)
in the *xz*-plane:	$z = -x^2$	(a parabola opening downward)
in the *xy*-plane:	$y = \pm x$	(two intersecting lines)

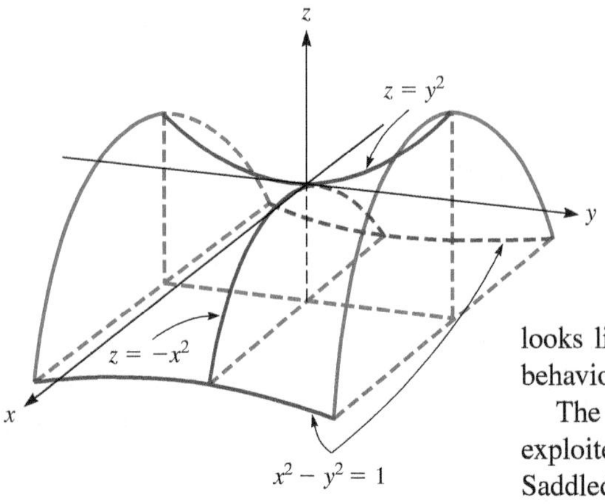

Fig. 29.14

Sections in the *yz*-plane are all parabolas opening upward, and sections in the *xz*-plane are all parabolas opening downward. Finally, sections in the horizontal *xy*-plane are hyperbolas of the form $y^2 - x^2 = k$. For example, if $z = -1$, we have the hyperbola $x^2 - y^2 = 1$.

This surface is not easy to draw, but using the information above, we obtain a sketch of the surface, as shown in Fig. 29.14. Note that the origin looks like a minimum from one direction but like a maximum from another. This behaviour at the origin is what gives this surface its characteristic saddle shape.

The shape of a hyperbolic paraboloid has intrinsic structural strength that is often exploited in the building of large roofs. For example, the cable net roof of the Saddledome arena in Calgary, Alberta, is a hyperbolic paraboloid (we will examine the equation that approximates its shape in the next section).

Other structures with saddle-shaped cable net roofs include the Stadium of Peace and Friendship in Athens, Greece, and the velodrome in London, England (host of the London 2012 Summer Olympics track cycling events).

Having developed the rectangular coordinate system in three dimensions, we can compare the graphs of a function using two dimensions and three dimensions. The next example shows the surface for the function of Example 1.

EXAMPLE 8 Comparing a 2D graph and a 3D graph

Sketch the graph of $z = 2x^2 + y^2$.

The only intercept is $(0, 0, 0)$. The traces are

in the yz-plane: $z = y^2$ (a parabola)

in the xz-plane: $z = 2x^2$ (a parabola)

in the xy-plane: the origin

The trace in the xy-plane is only the point of origin, since $2x^2 + y^2 = 0$ may be written as $y^2 = -2x^2$, which is true only for $x = 0$ and $y = 0$.

To get a better graph, we should use some positive values for z. As we noted in Example 1, negative values of z cannot be used. Since we used $z = 2$, $z = 4$, $z = 6$, and $z = 8$ in Example 1, we shall use these values here. Therefore,

for $z = 2$: $2x^2 + y^2 = 2$ for $z = 4$: $2x^2 + y^2 = 4$

for $z = 6$: $2x^2 + y^2 = 6$ for $z = 8$: $2x^2 + y^2 = 8$

Each of these sections is an ellipse. The surface, called an **elliptic paraboloid**, is shown in Fig. 29.15. Compare with Fig. 29.7.

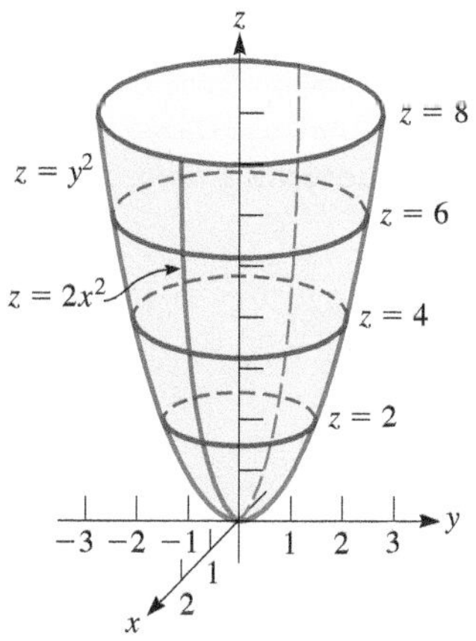

Fig. 29.15

Example 8 illustrates how *topographic maps* may be drawn. These maps represent three-dimensional terrain in two dimensions. For example, if Fig. 29.15 represents an excavation in the surface of the earth, then Fig. 29.7 represents the curves of constant elevation, or *contours*, with equally spaced elevations measured from the bottom of the excavation.

An equation with only two variables may represent a surface in space. Since only two variables are included in the equation, the surface is independent of the other variable. Another interpretation is that all sections, for all values of the variable not included, are the same. That is, ***all sections parallel to the coordinate plane of the included variables are the same as the trace in that plane.***

EXAMPLE 9 Comparing a linear equation in 2D and 3D

Sketch the graph of $x + y = 2$ in the rectangular coordinate system in three dimensions and in two dimensions.

Since z does not appear in the equation, we can consider the equation to be $x + y + 0z = 2$. Therefore, we see that *for any value of z*, the section is the straight line $x + y = 2$. Thus, the graph is a vertical plane as shown in Fig. 29.16(a). The graph as a straight line in two dimensions is shown in Fig. 29.16(b).

■ Whether an equation in two variables, like $x + y = 2$, represents a two- or three-dimensional graph must be determined from the context of the problem.

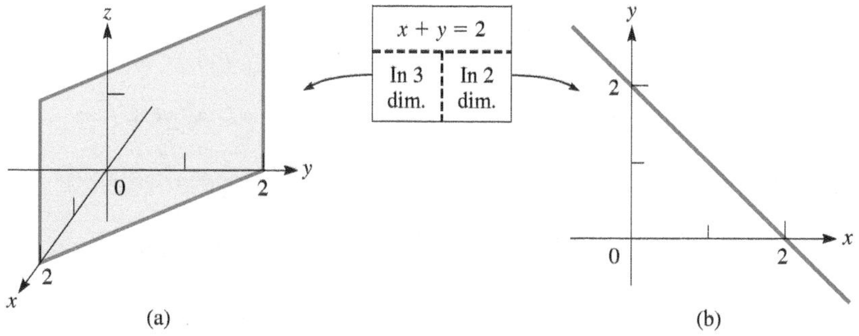

Fig. 29.16

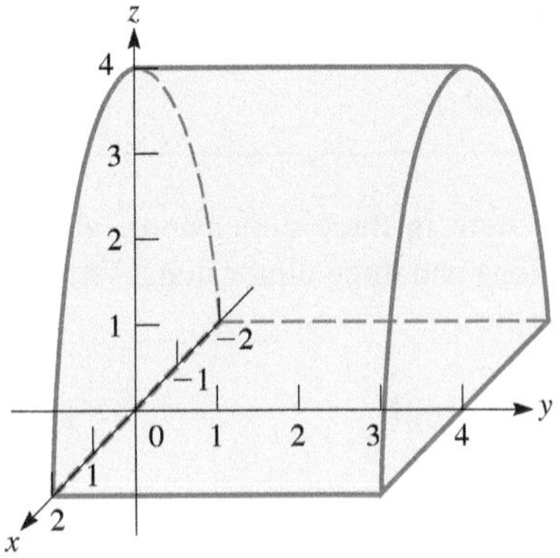

Fig. 29.17

EXAMPLE 10 Sketching a cylindrical surface

The graph of the equation $z = 4 - x^2$ in three dimensions is a surface whose sections, for all values of y, are given by the parabola $z = 4 - x^2$. The surface is shown in Fig. 29.17.

The surface in Example 10 is known as a **cylindrical surface**. In general, a cylindrical surface is one that can be generated by a line moving parallel to a fixed line while passing through a plane curve.

It must be realized that most of the figures shown extend beyond the ranges indicated by the traces and sections. However, these traces and sections are convenient for representing and visualizing these surfaces.

There are various computer programs, and some graphing calculators, that can be used to display three-dimensional surfaces. They generally use sections in planes that are perpendicular to both the x- and y-axes. Such computer-drawn surfaces are of great value for visualizing all types of surfaces, especially those of a complex nature. Fig. 29.18 shows the graph of

$$z = \frac{\sin(2x^2 + y^2)}{x^2 + 1}$$

drawn by a computer program.

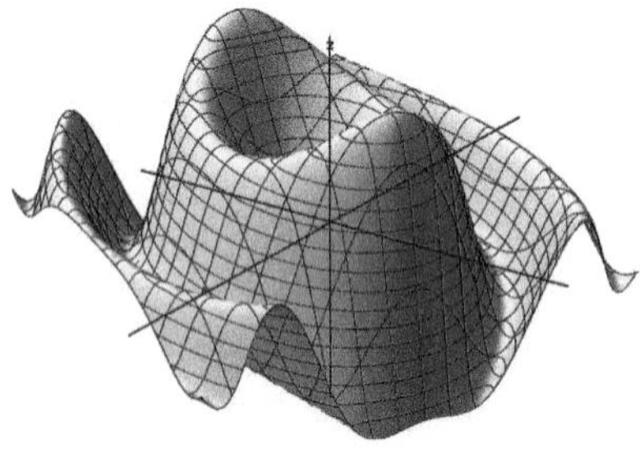

Fig. 29.18

CYLINDRICAL COORDINATES

Another set of coordinates that can be used to display a three-dimensional figure are ***cylindrical coordinates**, which are polar coordinates combined with the z-axis.* In using cylindrical coordinates, every point in space is designated by the coordinates (r, θ, z), as shown in Fig. 29.19. The equations relating the rectangular coordinates (x, y, z) and the cylindrical coordinates (r, θ, z) of a point are

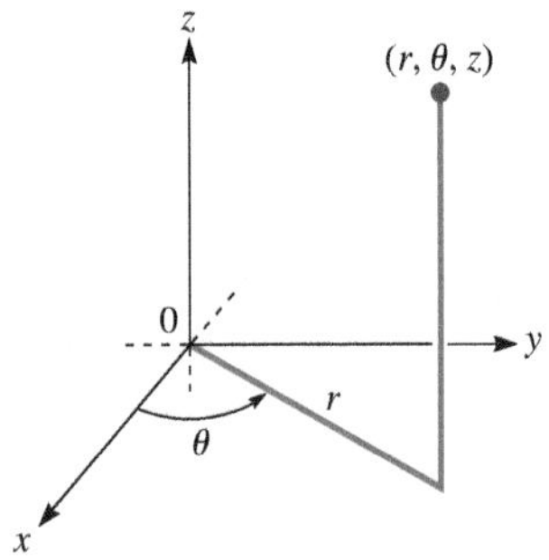

Fig. 29.19

$$x = r\cos\theta \qquad y = r\sin\theta \qquad z = z$$
$$r^2 = x^2 + y^2 \qquad \tan\theta = \frac{y}{x} \tag{29.2}$$

The following examples illustrate the use of cylindrical coordinates.

EXAMPLE 11 Plotting points in cylindrical coordinates

(a) Plot the point with cylindrical coordinates $(4, 2\pi/3, 2)$ and find the corresponding rectangular coordinates.

This point is shown in Fig. 29.20. From Eq. (29.2), we have

$$x = 4\cos\frac{2\pi}{3} = 4\left(-\frac{1}{2}\right) = -2 \qquad y = 4\sin\frac{2\pi}{3} = 4\left(\frac{\sqrt{3}}{2}\right) = 2\sqrt{3} \qquad z = 2$$

Therefore, the rectangular coordinates of the point are $(-2, 2\sqrt{3}, 2)$.

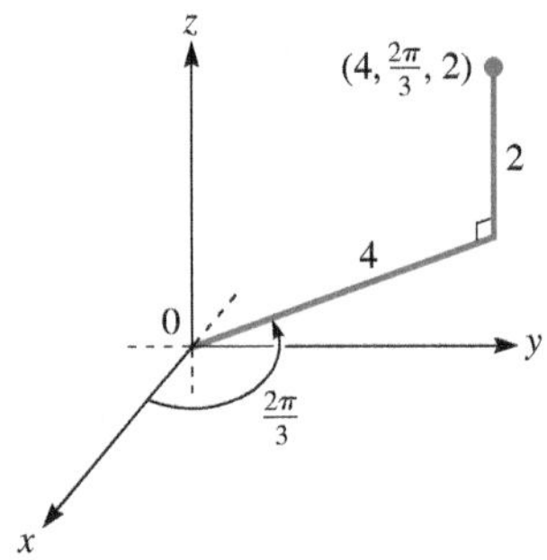

Fig. 29.20

(b) Find the cylindrical coordinates of the point with rectangular coordinates $(2, -2, 6)$.

Using Eq. (29.2), we have

$$r = \sqrt{2^2 + (-2)^2} = 2\sqrt{2} \qquad z = 6$$

$$\tan\theta = \frac{-2}{2} = -1 \quad \text{(quadrant IV)} \qquad \theta = 2\pi - \frac{\pi}{4} = \frac{7\pi}{4}$$

The cylindrical coordinates are $\left(2\sqrt{2}, 7\pi/4, 6\right)$. As with polar coordinates, the value of θ can be $7\pi/4 + 2n\pi$, where n is an integer.

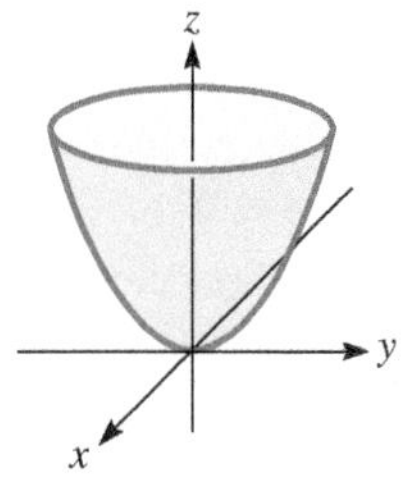

Fig. 29.21

EXAMPLE 12 Sketching a surface in cylindrical coordinates

Find the equation for the surface $z = 4x^2 + 4y^2$ in cylindrical coordinates and sketch the surface.

Factoring the 4 from the terms on the right, we have $z = 4(x^2 + y^2)$. We can now use the fact that $x^2 + y^2 = r^2$ to write this equation in cylindrical coordinates (r, θ, z) as $z = 4r^2$. From this equation, we see that $z \geq 0$ and that as r increases, we have circular sections (parallel to the xy-plane) of increasing radius. Therefore, we see that it is a **circular paraboloid**. See Fig. 29.21.

LEARNING TIP

If the centre of the trace of a circular cylinder in the *xy*-plane is at the origin, the equation of the cylinder in cylindrical coordinates is simply $r = a$. This is why these coordinates are called "cylindrical" coordinates.

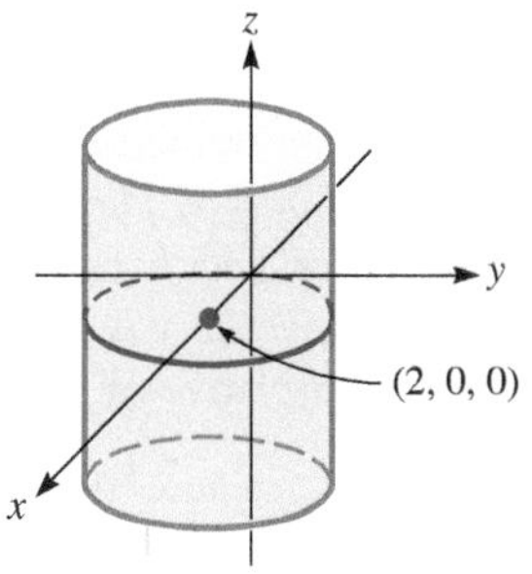

Fig. 29.22

EXAMPLE 13 Sketching a surface in cylindrical coordinates

Find the equation for the surface $r = 4\cos\theta$ in rectangular coordinates and sketch the surface.

If we multiply each side of the equation by r, we obtain $r^2 = 4r\cos\theta$. Now, from Eq. (29.2), we have $r^2 = x^2 + y^2$ and $r\cos\theta = x$, which means that $x^2 + y^2 = 4x$, or

$$x^2 - 4x + 4 + y^2 = 4$$
$$(x - 2)^2 + y^2 = 4$$

We recognize this as a cylinder that has as its trace in the xy-plane a circle with centre at $(2, 0, 0)$, as shown in Fig. 29.22.

EXERCISES 29.2

In Exercises 1 and 2, make the given changes in the indicated examples of this section and then solve the resulting problems.

1. In Example 4, change the -4 to $+6$.
2. In Example 6, change the $-$ sign before z^2 to $+$.

In Exercises 3 and 4, use the method of Example 1 and show the graphs of the given equations for the given values of z.

3. $z = x^2 + y^2$, $z = 1$, $z = 4$, $z = 9$
4. $z = y - x^2$, $z = 0$, $z = 2$, $z = 4$

In Exercises 5–24, sketch the graphs of the given equations in the rectangular coordinate system in three dimensions.

5. $x + y + 2z - 4 = 0$
6. $2x - y - z + 6 = 0$
7. $4x - 2y + z - 8 = 0$
8. $3x + 3y - 2z - 6 = 0$
9. $z = y - 2x - 2$
10. $z = x - 4y$
11. $x + 2y = 4$
12. $2x - 3z = 6$
13. $x^2 + y^2 + z^2 = 4$
14. $2x^2 + 2y^2 + z^2 = 8$
15. $z = 4 - 4x^2 - y^2$
16. $z = x^2 + y^2$
17. $z = 2x^2 + y^2 + 2$
18. $x^2 + y^2 - 4z^2 = 4$
19. $x^2 - y^2 - z^2 = 9$
20. $z^2 = 9x^2 + 4y^2$
21. $x^2 + y^2 = 16$
22. $4z = x^2$
23. $y^2 + 9z^2 = 9$
24. $xy = 2$

In Exercises 25–30, perform the indicated operations involving cylindrical coordinates.

25. Find the rectangular coordinates of the points whose cylindrical coordinates are (a) $(3, \pi/4, 5)$, (b) $(2, \pi/2, 3)$, (c) $(4, \pi/3, 2)$.
26. Find the cylindrical coordinates of the points whose rectangular coordinates are (a) $(3, 4, 5)$, (b) $(8, 15, -6)$, (c) $(\sqrt{3}, -2, 1)$.
27. Describe the surface for which the cylindrical coordinate equation is (a) $r = 2$, (b) $\theta = 2$, (c) $z = 2$.
28. Write the equation $x^2 + y^2 + 4z^2 = 4$ in cylindrical coordinates and sketch the surface.
29. Write the equation $r^2 = 4z$ in rectangular coordinates and sketch the surface.
30. Write the equation $r = 2\sin\theta$ in rectangular coordinates and sketch the surface.

In Exercises 31–38, sketch the indicated curves and surfaces.

31. Curves that represent a constant temperature are called *isotherms*. The temperature at a point (x, y) of a flat plate is t (°C), where $t = 4x - y^2$. In two dimensions, draw the isotherms for $t = -4, 0, 8$.
32. At a point (x, y) in the xy-plane, the electric potential V (in volts) is given by $V = y^2 - x^2$. Draw the lines of equal potential for $V = -9, 0, 9$.

33. An electric charge is so distributed that the electric potential at all points on an imaginary surface is the same. Such a surface is called an *equipotential surface*. Sketch the graph of the equipotential surface whose equation is $2x^2 + 2y^2 + 3z^2 = 6$.

34. The surface of a small hill can be roughly approximated by the equation $z(2x^2 + y^2 + 100) = 1500$, where the units are metres. Draw the surface of the hill and the contours for $z = 3$ m, $z = 6$ m, $z = 9$ m, $z = 12$ m, and $z = 15$ m.

35. The pressure p (in kPa), volume V (in m^3), and temperature T (in K) for a certain gas are related by the equation $p = T/2V$. Sketch the p-V-T surface by using the z-axis for p, the x-axis for V, and the y-axis for T. Use units of 100 K for T and 10 m for V. Sections must be used for this surface, *a thermodynamic surface*, since none of the variables may equal zero.

36. Sketch the line in space defined by the intersection of the planes $x + 2y + 3z - 6 = 0$ and $2x + y + z - 4 = 0$.

37. Sketch the graph of $x^2 + y^2 - 2y = 0$ in three dimensions and in two dimensions.

38. Sketch the curve in space defined by the intersection of the surfaces $x^2 + (z - 1)^2 = 1$ and $z = 4 - x^2 - y^2$.

Answer to Practice Exercise

1. (a) $x^2 + 4y^2 = 4$ (ellipse) (b) $x^2 - 4z^2 = 4$ (hyperbola)

29.3 Partial Derivatives

In Chapter 23, when showing the derivative to be the instantaneous rate of change of one variable with respect to another, only one independent variable was present. To extend the derivative to functions of two (or more) variables, we find the derivative of the function with respect to one variable, while holding the other variable(s) constant.

If $z = f(x, y)$ and y is held constant, z becomes a function of only x. The derivative of $f(x, y)$ with respect to x is termed the **partial derivative** *of z with respect to x. Similarly, if x is held constant, the derivative of $f(x, y)$ with respect to y is the* **partial derivative** *of z with respect to y.* The notations for the partial derivative of $z = f(x, y)$ with respect to x include

$$\frac{\partial z}{\partial x} \qquad \frac{\partial f}{\partial x} \qquad f_x \qquad \frac{\partial}{\partial x}f(x, y) \qquad f_x(x, y)$$

■ The symbol ∂ was introduced by the German mathematician Carl Jacobi (1804–1851).

Similarly, $\partial z/\partial y$ denotes the partial derivative of z with respect to y. In speaking, this is often shortened to "the partial of z with respect to y."

EXAMPLE 1 Finding partial derivatives

If $z = 4x^2 + xy - y^2$, find $\partial z/\partial x$ and $\partial z/\partial y$.

Finding the partial derivatives of z, we have

$$z = 4x^2 + xy - y^2$$

$$\frac{\partial z}{\partial x} = 8x + y \qquad \text{treat } y \text{ as a constant}$$

$$\frac{\partial z}{\partial y} = x - 2y \qquad \text{treat } x \text{ as a constant}$$

EXAMPLE 2 Finding partial derivatives

If $z = \frac{x \ln y}{x^2 + 1}$, find $\partial z/\partial x$ and $\partial z/\partial y$.

$$\frac{\partial z}{\partial x} = \frac{(x^2 + 1)(\ln y) - (x \ln y)(2x)}{(x^2 + 1)^2} = \frac{(1 - x^2)\ln y}{(1 + x^2)^2}$$

$$\frac{\partial z}{\partial y} = \left(\frac{x}{x^2 + 1}\right)\left(\frac{1}{y}\right) = \frac{x}{y(x^2 + 1)}$$

We note that in finding $\partial z/\partial x$, it is necessary to use the quotient rule, since x appears in both numerator and denominator. However, when finding $\partial z/\partial y$, the only derivative needed is that of $\ln y$.

Practice Exercise

1. If $z = 4x^2 + x \sin y$, find $\partial z/\partial x$ and $\partial z/\partial y$.

EXAMPLE 3 Evaluating a partial derivative

For the function $f(x, y) = x^2y\sqrt{2 + xy^2}$, find $f_y(2, 1)$.

The notation $f_y(2, 1)$ means the partial derivative of f with respect to y, evaluated for $x = 2$ and $y = 1$. Thus, first finding $f_y(x, y)$, we have

$$f(x, y) = x^2y(2 + xy^2)^{1/2}$$

$$f_y(x, y) = x^2y\left(\frac{1}{2}\right)(2 + xy^2)^{-1/2}(2xy) + (2 + xy^2)^{1/2}(x^2)$$

$$= \frac{x^3y^2}{(2 + xy^2)^{1/2}} + x^2(2 + xy^2)^{1/2}$$

$$= \frac{x^3y^2 + x^2(2 + xy^2)}{(2 + xy^2)^{1/2}} = \frac{2x^2 + 2x^3y^2}{(2 + xy^2)^{1/2}}$$

$$f_y(2, 1) = \frac{2(4) + 2(8)(1)}{(2 + 2)^{1/2}} = 12$$

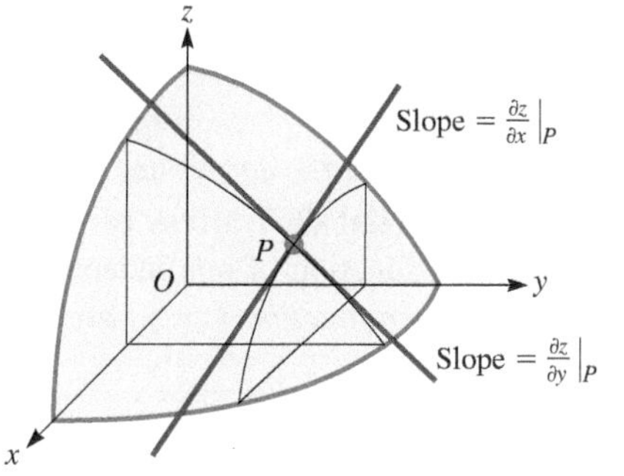

Fig. 29.23

To determine the geometric interpretation of a partial derivative, assume that $z = f(x, y)$ is the surface shown in Fig. 29.23. Choosing a point P on the surface, we then draw a plane through P parallel to the xz-plane. On this plane through P, the value of y is constant. The intersection of this plane and the surface is the curve as indicated. *The partial derivative of z with respect to x represents the slope of a line tangent to this curve.* When the values of the coordinates of point P are substituted into the expression for this partial derivative, it gives the slope of the tangent line at that point. In the same way, the partial derivative of z with respect to y, evaluated at P, gives the slope of the line tangent to the curve that is found from the intersection of the surface and the plane parallel to the yz-plane through P.

EXAMPLE 4 Slopes of lines tangent to a surface—Saddledome roof

■ See the chapter introduction.

The saddle-shaped roof surface of the Saddledome arena in Calgary, Alberta, is given approximately by the equation $z = \frac{1}{327}y^2 - \frac{1}{763}x^2$, with the origin at the centre of the roof and all measurements in metres. Find the slopes of the tangent lines at the point (50.0, 40.0, 1.62) that are parallel to the xz- and yz-planes (see Fig. 29.24).

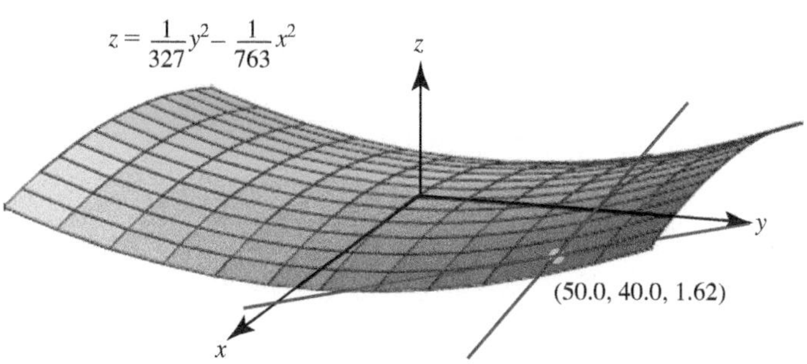

Fig. 29.24

Finding the partial derivatives with respect to x and y, we have

$$\frac{\partial z}{\partial x} = -\frac{2}{763}x \quad \text{and} \quad \frac{\partial z}{\partial y} = \frac{2}{327}y$$

These partial derivatives, when evaluated at the point (50.0, 40.0, 1.62), give us the slopes of the tangent lines parallel to the xz- and yz-planes, respectively:

$$\left.\frac{\partial z}{\partial x}\right|_{(50.0,\,40.0,\,1.62)} = -\frac{2}{763}(50.0) = -0.131 \quad \text{slope of tangent parallel to } xz\text{-plane}$$

$$\left.\frac{\partial z}{\partial y}\right|_{(50.0,\,40.0,\,1.62)} = \frac{2}{327}(40.0) = 0.245 \quad \text{slope of tangent parallel to } yz\text{-plane}$$

Therefore, in the x-direction, the roof is descending by 0.131 m per horizontal metre and in the y-direction, it is rising by 0.245 m per horizontal metre.

LEARNING TIP

The general interpretation of the partial derivative follows that of a derivative of a function with one independent variable. *The partial derivative $f_x(x_0, y_0)$ is the instantaneous rate of change of the function $f(x, y)$ with respect to x, with y held constant at the value of y_0.* This holds regardless of what the variables represent.

EXAMPLE 5 Partial derivatives—rates of change of wind turbine power

The power P (in W) in the wind hitting a certain style of wind turbine is given by

$$P = 0.750r^2v^3,$$

where r is the length of the blades (in m) and v is the wind velocity (in m/s). Find and interpret $\frac{\partial P}{\partial r}$ and $\frac{\partial P}{\partial v}$ when $r = 30.0$ m and $v = 5.00$ m/s.

In finding and evaluating the partial derivatives, we have

$$\frac{\partial P}{\partial r} = 1.50rv^3 \qquad \left.\frac{\partial P}{\partial r}\right|_{r=30.0,\, v=5.00} = 1.50(30.0)(5.00)^3 = 5625 \text{ W/m}$$

$$\frac{\partial P}{\partial v} = 2.25r^2v^2 \qquad \left.\frac{\partial P}{\partial v}\right|_{r=30.0,\, v=5.00} = 2.25(30.0)^2(5.00)^2 = 50\,625 \text{ W/(m/s)}$$

Therefore, when the blades are 30.0 m long and the wind velocity is 5.00 m/s, the power is increasing at a rate of 5625 W per metre of increase in the blade length. Also, the power is increasing at a rate of 50 625 W for each m/s increase in wind velocity.

■ The values of these partial derivatives show us that an increase of 1 m/s in the wind velocity has a much larger impact on the power than an increase of 1 m in the length of the blades.

Since the partial derivatives $\partial f/\partial x$ and $\partial f/\partial y$ are functions of x and y, we can take partial derivatives of each of them. This gives rise to **partial derivatives of higher order**, in a manner similar to the higher derivatives of a function of one independent variable. *The possible* **second-order partial derivatives** *of a function $f(x, y)$ are*

$$\frac{\partial^2 f}{\partial x^2} = \frac{\partial}{\partial x}\left(\frac{\partial f}{\partial x}\right) \qquad \frac{\partial^2 f}{\partial y^2} = \frac{\partial}{\partial y}\left(\frac{\partial f}{\partial y}\right)$$

$$\frac{\partial^2 f}{\partial x \partial y} = \frac{\partial}{\partial x}\left(\frac{\partial f}{\partial y}\right) \qquad \frac{\partial^2 f}{\partial y \partial x} = \frac{\partial}{\partial y}\left(\frac{\partial f}{\partial x}\right)$$

EXAMPLE 6 Second-order partial derivatives

Find the second-order partial derivatives of $z = x^3y^2 - 3xy^3$.

First, we find $\partial z/\partial x$ and $\partial z/\partial y$:

$$\frac{\partial z}{\partial x} = 3x^2y^2 - 3y^3 \quad \text{or} \quad \frac{\partial z}{\partial y} = 2x^3y - 9xy^2$$

Therefore, we have the following second-order partial derivatives:

$$\frac{\partial^2 z}{\partial x^2} = \frac{\partial}{\partial x}\left(\frac{\partial z}{\partial x}\right) = 6xy^2 \qquad \frac{\partial^2 z}{\partial y^2} = \frac{\partial}{\partial y}\left(\frac{\partial z}{\partial y}\right) = 2x^3 - 18xy$$

$$\frac{\partial^2 z}{\partial x \partial y} = \frac{\partial}{\partial x}\left(\frac{\partial z}{\partial y}\right) = 6x^2y - 9y^2 \qquad \frac{\partial^2 z}{\partial y \partial x} = \frac{\partial}{\partial y}\left(\frac{\partial z}{\partial x}\right) = 6x^2y - 9y^2$$

Practice Exercise

2. If $z = 2x^2y - 5x^2y^4$, find $\frac{\partial^2 z}{\partial x \partial y}$.

In Example 6, we note that

$$\frac{\partial^2 z}{\partial x \partial y} = \frac{\partial^2 z}{\partial y \partial x} \qquad \textbf{(29.3)}$$

In general, this is true if the function and partial derivatives are continuous.

EXAMPLE 7 Equality of second-order partial derivatives

For $f(x, y) = \tan^{-1}\frac{y}{x^2}$, show that $\frac{\partial^2 f}{\partial x\,\partial y} = \frac{\partial^2 f}{\partial y\,\partial x}$.

Finding $\partial f/\partial x$ and $\partial f/\partial y$, we have

$$\frac{\partial f}{\partial x} = \frac{1}{1 + \left(\frac{y}{x^2}\right)^2}\left(\frac{-2y}{x^3}\right) = \frac{x^4}{x^4 + y^2}\left(-\frac{2y}{x^3}\right) = \frac{-2xy}{x^4 + y^2}$$

$$\frac{\partial f}{\partial y} = \frac{1}{1 + \left(\frac{y}{x^2}\right)^2}\left(\frac{1}{x^2}\right) = \frac{x^4}{x^4 + y^2}\left(\frac{1}{x^2}\right) = \frac{x^2}{x^4 + y^2}$$

Now, finding $\partial^2 f/\partial x\partial y$ and $\partial^2 f/\partial y\,\partial x$, we have

$$\frac{\partial^2 f}{\partial x\,\partial y} = \frac{(x^4 + y^2)(2x) - x^2(4x^3)}{(x^4 + y^2)^2} = \frac{-2x^5 + 2xy^2}{(x^4 + y^2)^2}$$

$$\frac{\partial^2 f}{\partial y\,\partial x} = \frac{(x^4 + y^2)(-2x) - (-2xy)(2y)}{(x^4 + y^2)^2} = \frac{-2x^5 + 2xy^2}{(x^4 + y^2)^2}$$

We see that they are equal.

EXERCISES 29.3

In Exercises 1 and 2, make the given changes in the indicated examples of this section and then solve the resulting problems.

1. In Example 2, change x^2 to y^2.
2. In Example 4, change the equation to $2z = 2x^2 + y^2$ and the point $(2, 1, 3)$ to $(1, 2, 3)$.

In Exercises 3–24, find the partial derivative of the dependent variable or function with respect to each of the independent variables.

3. $z = 9x^6 - 3y^5$
4. $z = 3x^2y^3 - 3x + 4y$
5. $f(x, y) = x^2e^{5xy}$
6. $z = 3y\cos 2x$
7. $f(x, y) = \frac{2 + \cos x}{1 - \sec 3y}$
8. $f(x, y) = \frac{\tan^{-1} y}{2 + x^2}$
9. $\phi = r\sqrt{1 + 2rs}$
10. $w = uv^2\sqrt{1 - u}$
11. $z = (x^2 + xy^3)^4$
12. $f(x, y) = (2xy - x^2)^5$
13. $z = \cos xy$
14. $y = \tan^{-1}\left(\frac{x}{t}\right)$
15. $y = \ln(r^9 + s)$
16. $u = e^{x+2y}$
17. $f(x, y) = \frac{2\sin^3 2x}{1 - 3y}$
18. $f(x, y) = \frac{3x + \ln y}{x^2 + y^2}$
19. $z = \frac{\sin^{-1} xy}{3 + x^2}$
20. $z = \frac{\sqrt{1 - \tan xy}}{xy + y^2}$
21. $z = \sin x + \cos xy - \cos y$
22. $t = 2re^{rs^2} - \tan(2r + s)$
23. $f(x, y) = e^x\cos xy + e^{-2x}\tan y$
24. $u = \ln\frac{y^2}{x - y} + e^{-x}(\sin y - \cos 2y)$

In Exercises 25–28, evaluate the indicated partial derivatives at the given points.

25. $z = 3xy - x^2$, $\left.\frac{\partial z}{\partial x}\right|_{(1,-2,-7)}$
26. $z = x^2\cos 4y$, $\left.\frac{\partial z}{\partial y}\right|_{(2,\frac{\pi}{2},4)}$
27. $z = x\sqrt{x^2 - y^2}$, $\left.\frac{\partial z}{\partial x}\right|_{(5,3,20)}$
28. $z = e^y\ln xy$, $\left.\frac{\partial z}{\partial y}\right|_{(e,1,e)}$

In Exercises 29–32, find all second partial derivatives.

29. $z = 2xy^3 - 3x^2y$
30. $F(x, y) = y\ln(x + 2y)$
31. $z = \frac{x}{y} + e^x\sin y$
32. $f(x, y) = \frac{2 + \cos y}{1 + x^2}$

In Exercises 33–46, solve the given problems.

33. Find the slope of a line tangent to the surface $z = 9 - x^2 - y^2$ and parallel to the yz-plane that passes through $(1, 2, 4)$. Repeat the instructions for the line through $(2, 2, 1)$. Draw an appropriate figure.
34. A metal plate in the shape of a circular segment of radius r expands by being heated. Express the width w (straight dimension) as a function of r and the height h. Then find both $\partial w/\partial r$ and $\partial w/\partial h$.
35. Two resistors R_1 and R_2, placed in parallel, have a combined resistance R_T given by $\frac{1}{R_T} = \frac{1}{R_1} + \frac{1}{R_2}$. Find $\frac{\partial R_T}{\partial R_1}$.
36. Find $\partial z/\partial y$ for the function $z = 4x^2 - 8$. Explain your result. Draw an appropriate figure.

37. A metallic machine part contracts while cooling. It is in the shape of a hemisphere attached to a cylinder, as shown in Fig. 29.25. Find the rate of change of volume with respect to r when $r = 2.65$ cm and $h = 4.20$ cm.

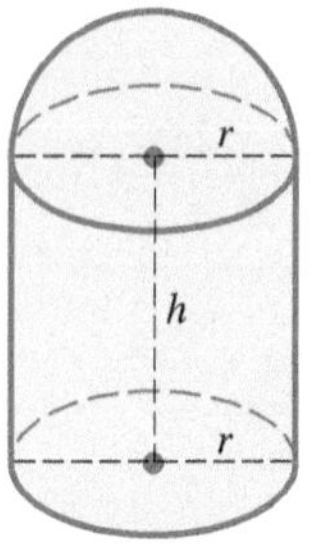

Fig. 29.25

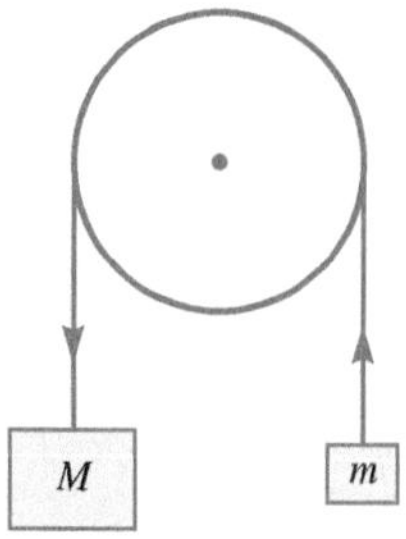

Fig. 29.26

38. Two masses M and m are attached as shown in Fig. 29.26. If $M > m$, the downward acceleration a of mass M is given by $a = \dfrac{M - m}{M + m}g$, where g is the acceleration due to gravity. Show that $M\dfrac{\partial a}{\partial M} + m\dfrac{\partial a}{\partial m} = 0$.

39. In quality testing, a rectangular sheet of vinyl is stretched. Set up the length of the diagonal d of the sheet as a function of the sides x and y. Find the rate of change of d with respect to x for $x = 2.20$ m if y remains constant at 1.55 m.

40. If an observer and a source of sound are moving toward or away from each other, the observed frequency of sound is different from that emitted. This is known as the *Doppler effect*. The equation relating the frequency f_o the observer hears and the frequency f_s emitted by the source (a constant) is $f_o = f_s\left(\dfrac{v + v_o}{v - v_s}\right)$, where v is the velocity of sound in air (a constant), v_o is the velocity of the observer, and v_s is the velocity of the source. Show that $f_s\dfrac{\partial f_o}{\partial v_s} = f_o\dfrac{\partial f_o}{\partial v_o}$. Explain the meaning of $\partial f_o/\partial v_s$.

41. The *mutual conductance* (in $1/\Omega$) of a certain electronic device is defined as $g_m = \partial i_b/\partial V_c$. Under certain circumstances, the current i_b (in μA) is given by $i_b = 50(V_b + 5V_c)^{1.5}$. Find g_m when $V_b = 200$ V and $V_c = -20$ V.

42. The *amplification factor* of the electronic device of Exercise 41 is defined as $\mu = -\partial V_b/\partial V_c$. For the device of Exercise 41, under the given conditions, find the amplification factor.

43. The temperature u in a metal bar depends on the distance x from one end and the time t. Show that $u(x, t) = 5e^{-t}\sin 4x$ satisfies the *one-dimensional heat-conduction equation* $\dfrac{\partial u}{\partial t} = k\dfrac{\partial^2 u}{\partial x^2}$, where k is called the *diffusivity*. Here $k = 1/16$.

44. The displacement y at any point in a taut, flexible string depends on the distance x from one end of the string and the time t. Show that $y(x, t) = 2\sin 2x\cos 4t$ satisfies the *wave equation* $\dfrac{\partial^2 y}{\partial t^2} = a^2\dfrac{\partial^2 y}{\partial x^2}$ with $a = 2$.

45. The steady-state temperature u of a thin, flat plate satisfies *Laplace's equation*, $\dfrac{\partial^2 u}{\partial x^2} + \dfrac{\partial^2 u}{\partial y^2} = 0$. Show that the function $u(x, y) = e^{-x}\sin y$ satisfies Laplace's equation.

46. Find $\partial L/\partial I$ for an electric circuit involving the equation

$$I = \frac{V}{\sqrt{R^2 + L^2C^2}}$$

Answers to Practice Exercises

1. $\partial z/\partial x = 8x + \sin y,\ \partial z/\partial y = x\cos y$

2. $\dfrac{\partial^2 z}{\partial x\,\partial y} = 4x - 40xy^3$

29.4 Double Integrals

We now turn our attention to integration in the case of a function of two variables. The analysis has similarities to that of partial differentiation, in that an operation is performed while holding one of the independent variables constant.

If $z = f(x, y)$ and we wish to integrate with respect to x and y, we first consider either x or y constant and integrate with respect to the other. After this integral is evaluated, we then integrate with respect to the variable first held constant. We shall now define this type of integral and then give an appropriate geometric interpretation.

If $z = f(x, y)$ the **double integral** *of the function over x and y is defined as*

$$\int_a^b\left[\int_{g(x)}^{G(x)} f(x, y)\,dy\right]dx$$

Since it is customary not to include the brackets in writing a double integral, we write

$$\int_a^b\left[\int_{g(x)}^{G(x)} f(x, y)\,dy\right]dx = \int_a^b\int_{g(x)}^{G(x)} f(x, y)\,dy\,dx \qquad (29.4)$$

LEARNING TIP

- When the outer differential of a double integral is *dx*, the limits on the inner integral are functions of *x*, and those on the outer integral are explicit values of *x*. The inner integral is performed first, integrating with respect to *y* while keeping *x* constant. This results in an integral with *x* only, which can then be evaluated.
- When the outer differential of a double integral is *dy*, the limits on the inner integral are functions of *y*, and those on the outer integral are explicit values of *y*. The inner integral is performed first, integrating with respect to *x* while keeping *y* constant. This results in an integral with *y* only, which can then be evaluated.

EXAMPLE 1 Evaluating a double integral

Evaluate $\int_0^1 \int_{x^2}^x xy\,dy\,dx$.

First, we integrate the inner integral with *y* as the variable and *x* as a constant.

treat as constant

$$\int_{x^2}^x xy\,dy = \left(x\frac{y^2}{2}\right)\Bigg|_{x^2}^x = x\left(\frac{x^2}{2} - \frac{x^4}{2}\right) = \frac{1}{2}(x^3 - x^5)$$

This means

$$\int_0^1 \int_{x^2}^x xy\,dy\,dx = \int_0^1 \frac{1}{2}(x^3 - x^5)dx = \frac{1}{2}\left(\frac{x^4}{4} - \frac{x^6}{6}\right)\Bigg|_0^1$$
$$= \frac{1}{2}\left(\frac{1}{4} - \frac{1}{6}\right) - \frac{1}{2}(0) = \frac{1}{24}$$

EXAMPLE 2 Evaluating a double integral

Evaluate $\int_0^{\pi/2} \int_0^{\sin y} e^{2x}\cos y\,dx\,dy$.

Since the inner differential is *dx*, we first integrate with *x* as the variable and *y* as a constant. The second integration is with *y* as the variable.

$$\int_0^{\pi/2} \int_0^{\sin y} e^{2x}\cos y\,dx\,dy = \int_0^{\pi/2}\left[\frac{1}{2}e^{2x}\cos y\right]_0^{\sin y} dy$$
$$= \frac{1}{2}\int_0^{\pi/2}(e^{2\sin y}\cos y - \cos y)\,dy$$
$$= \frac{1}{2}\left[\frac{1}{2}e^{2\sin y} - \sin y\right]_0^{\pi/2} = \frac{1}{2}\left(\frac{1}{2}e^2 - 1\right) - \frac{1}{2}\left(\frac{1}{2} - 0\right)$$
$$= \frac{1}{4}e^2 - \frac{1}{2} - \frac{1}{4} = \frac{1}{4}(e^2 - 3) = 1.097$$

Practice Exercise

1. Evaluate $\int_0^1 \int_0^x 2xy\,dy\,dx$.

LEARNING TIP

We interpret a double integral as the **volume under a surface,** in the same way as the integral was interpreted as the area of a plane figure.

For the geometric interpretation of a double integral, consider the surface shown in Fig. 29.27(a). An **element of volume** (dimensions of *dx*, *dy*, and *z*) extends from the *xy*-plane to the surface. With *x* a constant, sum (integrate) these elements of volume from the left boundary, $y = g(x)$, to the right boundary, $y = G(x)$. Now, the volume of the vertical slice is a function of *x*, as shown in Fig. 29.27(b). By summing (integrating) the volumes of these slices from $x = a$ ($x = 0$ in the figure) to $x = b$, we have the complete volume, as shown in Fig. 29.27(c).

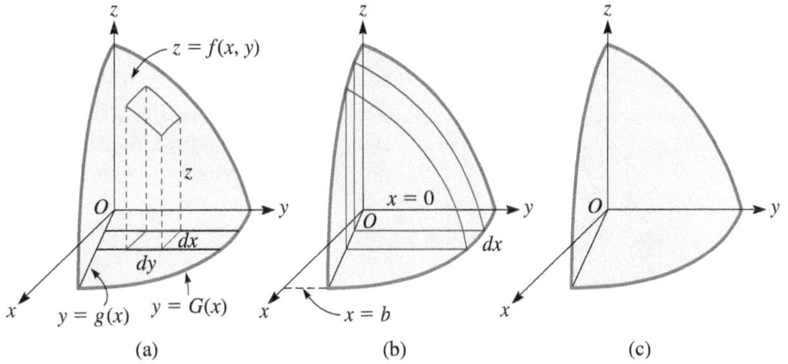

Fig. 29.27

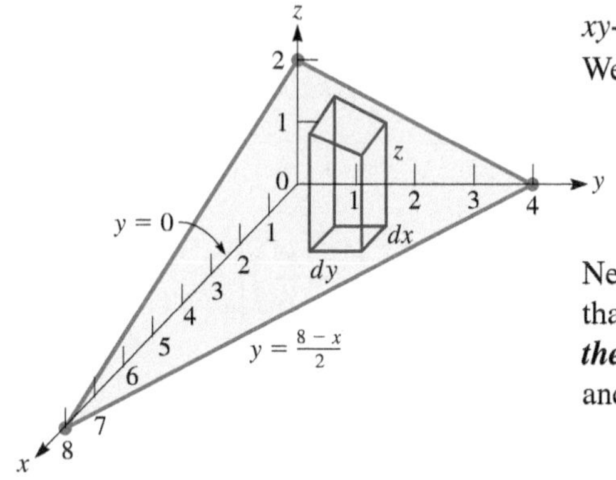

Fig. 29.28

EXAMPLE 3 Volume under a plane

Find the volume that is in the first octant and under the plane $x + 2y + 4z - 8 = 0$. See Fig. 29.28.

This figure is a tetrahedron, for which $V = \frac{1}{3}Bh$. Assuming the base is in the xy-plane, $B = \frac{1}{2}(4)(8) = 16$, and $h = 2$. Therefore, $V = \frac{1}{3}(16)(2) = \frac{32}{3}$ cubic units. We shall use this value to check the one we find by double integration.

To find $z = f(x, y)$, we solve the given equation for z. Thus,

$$z = \frac{8 - x - 2y}{4}$$

Next, we must find the limits on y and x. Choosing to integrate over y first, we see that y goes from $y = 0$ to $y = (8 - x)/2$. ***This last limit is the trace of the surface in the xy-plane.*** Next, we note that x goes from $x = 0$ to $x = 8$. Therefore, we set up and evaluate the integral:

$$V = \int_0^8 \int_0^{(8-x)/2} \left(\frac{8 - x - 2y}{4}\right) dy\, dx$$

$$= \int_0^8 \left[\frac{1}{4}(8y - xy - y^2)\Big|_0^{(8-x)/2}\right] dx$$

$$= \frac{1}{4}\int_0^8 \left[8\left(\frac{8-x}{2}\right) - x\left(\frac{8-x}{2}\right) - \left(\frac{8-x}{2}\right)^2\right] dx$$

$$= \frac{1}{4}\int_0^8 \left(32 - 4x - 4x + \frac{x^2}{2} - 16 + 4x - \frac{x^2}{4}\right) dx$$

$$= \frac{1}{4}\int_0^8 \left(16 - 4x + \frac{x^2}{4}\right) dx$$

$$= \frac{1}{4}\left(16x - 2x^2 + \frac{x^3}{12}\right)\Bigg|_0^8 = \frac{1}{4}\left(128 - 128 + \frac{512}{12}\right)$$

$$= \frac{1}{4}\left(\frac{128}{3}\right) = \frac{32}{3} \text{ cubic units}$$

We see that the values obtained by the two different methods agree.

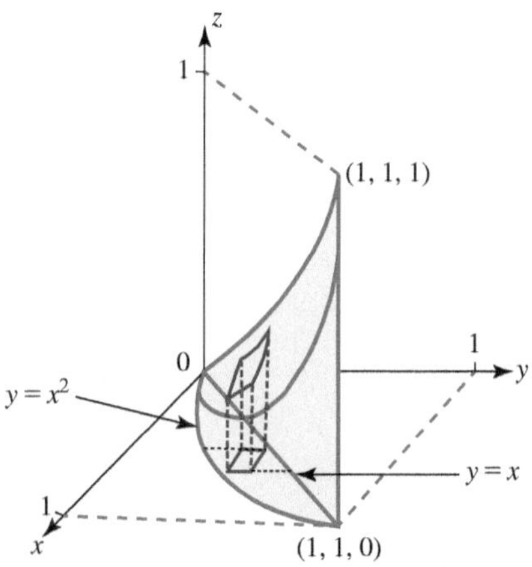

Fig. 29.29

EXAMPLE 4 Volume under a surface

Find the volume above the xy-plane, below the surface $z = xy$, and enclosed by the cylinder $y = x^2$ and the plane $y = x$.

Because the cylinders are perpendicular to the xy-plane, the volume of interest lies under the function $z = xy$ and above the region on the xy-plane bounded by the parabola $y = x^2$ and the line $y = x$ (see Fig. 29.29). The limits of integration can be found directly from the graph of this region without having to plot the surface. Integrating over y first, the limits on y are $y = x^2$ to $y = x$. The corresponding limits on x are $x = 0$ to $x = 1$. Therefore, the double integral to be evaluated is

$$V = \int_0^1 \int_{x^2}^{x} xy\, dy\, dx$$

This integral has already been evaluated in Example 1 of this section, and we can now see the geometric interpretation of that integral. Using the result from Example 1, we see that the required volume is $1/24$ cubic unit.

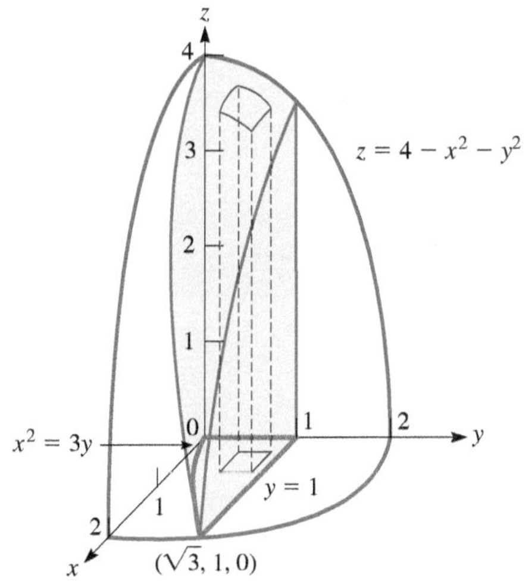

Fig. 29.30

EXAMPLE 5 Volume under a surface

Find the volume in the first octant that is under the surface $z = 4 - x^2 - y^2$, and is between the cylinder $x^2 = 3y$ and the plane $y = 1$. See Fig. 29.30.

We find the limits of integration from the shaded region on the xy-plane. Integrating over x first, we have

$$\begin{aligned} V &= \int_0^1 \int_0^{\sqrt{3y}} (4 - x^2 - y^2)\,dx\,dy \\ &= \int_0^1 \left[4x - \frac{x^3}{3} - y^2x\right]_0^{\sqrt{3y}} dy \\ &= \int_0^1 (4\sqrt{3y} - \sqrt{3}y^{3/2} - \sqrt{3}y^{5/2})\,dy \\ &= \sqrt{3}\left[4\left(\frac{2}{3}\right)y^{3/2} - \frac{2}{5}y^{5/2} - \frac{2}{7}y^{7/2}\right]\Bigg|_0^1 \\ &= \sqrt{3}\left(\frac{8}{3} - \frac{2}{5} - \frac{2}{7}\right) = \frac{208\sqrt{3}}{105} = 3.43 \text{ cubic units} \end{aligned}$$

If we integrate over y first, we arrive at the same result evaluating the double integral

$$V = \int_0^{\sqrt{3}} \int_{x^2/3}^{1} (4 - x^2 - y^2)\,dy\,dx$$

EXERCISES 29.4

In Exercises 1 and 2, make the given changes in the indicated examples of this section and then solve the resulting problems.

1. In Example 1, change the integrand xy to $(x + y)$.

2. In Example 2, delete the e^{2x}.

In Exercises 3–16, evaluate the given double integrals.

3. $\displaystyle\int_2^4 \int_0^1 xy^2\,dx\,dy$

4. $\displaystyle\int_0^2 \int_0^1 \frac{y}{(xy + 1)^2}\,dx\,dy$

5. $\displaystyle\int_1^2 \int_0^{y^2} xy^2 dx\,dy$

6. $\displaystyle\int_0^4 \int_1^{\sqrt{y}} (x - y)dx\,dy$

7. $\displaystyle\int_0^1 \int_0^{\sqrt{1-x^2}} y\,dy\,dx$

8. $\displaystyle\int_4^9 \int_0^x \sqrt{x - y}\,dy\,dx$

9. $\displaystyle\int_0^{\pi/6} \int_{\pi/3}^{y} \sin x\,dx\,dy$

10. $\displaystyle\int_0^{\sqrt{3}} \int_{x^2/3}^{1} (4 - x^2)dy\,dx$

11. $\displaystyle\int_1^e \int_1^y \frac{1}{x}dx\,dy$

12. $\displaystyle\int_{-1}^1 \int_1^{e^x} \frac{1}{xy}\,dy\,dx$

13. $\displaystyle\int_1^2 \int_0^x yx^3 e^{xy^2}\,dy\,dx$

14. $\displaystyle\int_0^{\pi/6} \int_0^1 y \sin x\,dy\,dx$

15. $\displaystyle\int_0^{\ln 3} \int_0^x e^{2x+3y}\,dy\,dx$

16. $\displaystyle\int_0^{1/2} \int_y^{y^2} \frac{dx\,dy}{\sqrt{y^2 - x^2}}$

In Exercises 17–26, find the indicated volumes by double integration.

17. The first-octant volume under the plane $x + y + z - 4 = 0$

18. The first-octant volume under the surface $z = y^2$ and bounded by the planes $x = 2$ and $y = 3$

19. The volume above the xy-plane and under the surface $z = 4 - x^2 - y^2$

20. The volume above the xy-plane, below the surface $z = x^2 + y^2$, and inside the cylinder $x^2 + y^2 = 4$

21. The first-octant volume bounded by the xy-plane, the planes $x = y$, $y = 2$, and $z = 2 + x^2 + y^2$

22. The volume bounded by the planes $x + 3y + 2z - 6 = 0$, $2x = y$, $x = 0$, and $z = 0$

23. The first-octant volume under the plane $z = x + y$ and inside the cylinder $x^2 + y^2 = 9$

24. The volume above the xy-plane and bounded by the cylinders $x = y^2$, $y = 8x^2$, and $z = x^2 + 1$. Integrate over y first and then check by integrating over x first.

25. A wedge is to be made in the shape shown in Fig. 29.31. (All vertical cross-sections are equal right triangles.) By double integration, find the volume of the wedge.

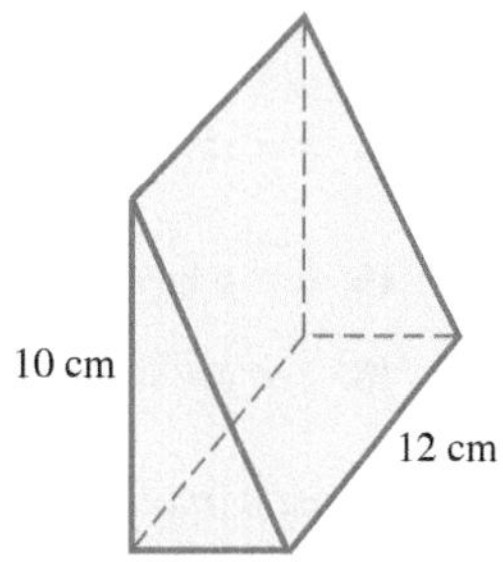

Fig. 29.31

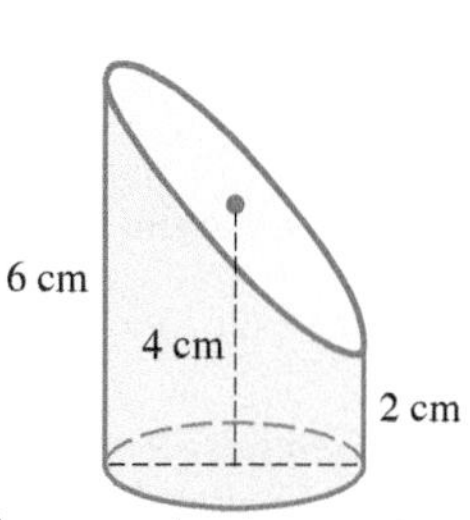

Fig. 29.32

26. A circular piece of pipe is cut as shown in Fig. 29.32. Find the volume within the pipe. Describe how to set up the coordinate system in order to determine the required volume.

In Exercises 27 and 28, draw the appropriate figure.

27. Draw the appropriate figure indicating a volume that is found from the integral $\int_0^1 \int_{x^2}^1 (4 - x - 2y)\,dy\,dx$.

28. Repeat Exercise 27 for the integral $\int_1^2 \int_0^{2-y} \sqrt{1 + x^2 + y^2}\,dx\,dy$.

Answer to Practice Exercise

1. 1/4

CHAPTER 29 KEY FORMULAS AND EQUATIONS

Equation of a plane $\quad Ax + By + Cz + D = 0 \quad$ **(29.1)**

Cylindrical coordinates $\quad x = r\cos\theta \quad y = r\sin\theta \quad z = z \quad$ **(29.2)**

$r^2 = x^2 + y^2 \quad \tan\theta = \frac{y}{x}$

Partial derivatives $\quad \frac{\partial^2 z}{\partial x\,\partial y} = \frac{\partial^2 z}{\partial y\,\partial x} \quad$ **(29.3)**

Double integral $\quad \int_a^b \left[\int_{g(x)}^{G(x)} f(x, y)\,dy\right] dx = \int_a^b \int_{g(x)}^{G(x)} f(x, y)\,dy\,dx \quad$ **(29.4)**

CHAPTER 29 REVIEW EXERCISES

In Exercises 1–4, evaluate the given functions.

1. $f(x, y) = 3x^2y - y^3$, find $f(-1, 4)$

2. $f(r, \theta) = r^2\cos 2\theta - r\sin\theta$, find $f(3, \pi)$

3. $f(s, t) = \frac{2st^2 - t}{s}$, find $f(4, s^2)$

4. $f(x, y) = \frac{4x}{xy - 2}$, find $f(x^2, 2x)$

In Exercise 5–8, sketch the graphs of the given equations in the rectangular coordinate system in three dimensions.

5. $x - y + 2z - 4 = 0$

6. $2y + 3z = 6$

7. $z = x^2 + 4y^2$

8. $x^2 + y^2 - 4z^2 - 4 = 0$

In Exercises 9–18, find the partial derivatives of the given functions with respect to each of the independent variables.

9. $z = 5x^3y^2 - 2xy^4$

10. $z = 2x\sqrt{y} - x^2y$

11. $z = \sqrt{x^2 - 3y^2}$

12. $u = \frac{r}{(r - 3s)^2}$

13. $z = \frac{2x - 3y}{x^2y + 1}$

14. $z = x(y^2 + xy + 2)^4$

15. $u = y\ln\sin(x^2 + 2y)$

16. $q = p\ln(r + 1) - \frac{rp}{r + 1}$

17. $z = \sin^{-1}\sqrt{x + y}$

18. $z = ye^{xy}\sin(2x - y)$

In Exercises 19 and 20, find all of the second partial derivatives of the given functions.

19. $z = 3x^2y - y^3 + 2xy$

20. $z = x\sqrt{2y + 1} + y^2(x - 2)^3$

In Exercises 21–28, evaluate each of the given double integrals.

21. $\int_0^2 \int_1^2 (3y + 2xy)\,dx\,dy$

22. $\int_2^7 \int_0^1 x\sqrt{2 + x^2y}\,dx\,dy$

23. $\int_0^3 \int_1^x (x + 2y)\,dy\,dx$

24. $\int_1^2 \int_0^{\pi/4} r\sec^2\theta\,d\theta\,dr$

25. $\int_0^1 \int_0^{2x} x^2e^{xy}\,dy\,dx$

26. $\int_{\pi/4}^{\pi/2} \int_1^{\sqrt{\cos\theta}} r\sin\theta\,dr\,d\theta$

27. $\int_1^e \int_1^x \frac{\ln y}{xy}\,dy\,dx$

28. $\int_1^3 \int_0^x \frac{2}{x^2 + y^2}\,dy\,dx$

In Exercises 29–50, solve the given problems.

29. Sketch the surface representing $z = \sqrt{x^2 + 4y^2}$.

30. For the function of Exercise 29, find the equation of a line tangent to the surface at $(2, 1, 2\sqrt{2})$ that is parallel to the yz-plane.

31. Sketch the surface representing $z = e^{x+y}$.

32. For the function of Exercise 31, find the volume in the first octant under the surface and inside the planes $x = 1$ and $y = x$.

33. Describe the surface for which the cylindrical coordinate equation is (a) $\theta = 3$, (b) $z = r^2$.

34. Write the cylindrical coordinate equation $r = 2(\sin\theta + \cos\theta)$ in rectangular coordinates and sketch the surface.

35. In a simple series electric circuit, with two resistors r and R connected across a voltage source E, the voltage V across r is $V = rE/(r + R)$. Assuming E to be constant, find $\partial V/\partial r$ and $\partial v/\partial R$.

36. For a gas, the volume expansivity is defined as $\beta = \frac{1}{V}\frac{\partial V}{\partial T}$, where V is the volume and T is the temperature of the gas. If V as a function of T and the pressure p is given by $V = a + bT/p - c/T^2$, where a, b, and c are constants, find β.

37. In the theory dealing with transistors, the current gain α of a transistor is defined as $\alpha = \partial i_c/\partial i_e$, where i_c is the collector current and i_e is the emitter current. If i_c is a function of i_e and the collector voltage V_c given by $i_c = i_e(1 - e^{-2V_c})$, find α if V_c is 2 V.

38. Young's modulus, which measures the ratio of the stress to strain in a stretched wire, is defined as $Y = \frac{L}{A}\frac{\partial F}{\partial L}$, where L is the length of the wire, A is its cross-sectional area, and F is the tension in the wire. Find Y (in Pa) if $F = -0.0100\,T/L^2$ for $L = 1.10$ m, $T = 300$ K, and $A = 1.00 \times 10^{-6}$ m^2.

39. The period T of the pendulum as a function of its length l and the acceleration due to gravity g is given by $T = 2\pi\sqrt{l/g}$. Show that $\partial T/\partial l = T/2l$.

40. The volume of a right circular cone of radius r and height h is given by $V = \frac{1}{3}\pi r^2 h$. Show that $\partial V/\partial r = 2V/r$.

41. The *coefficient of linear expansion* α of a wire whose length L is a function of the tension and temperature T is given by $\alpha = \frac{1}{L}\left(\frac{\partial L}{\partial T}\right)$. If L is a function of T and the tension F given by $L = L_0 + k_1F + k_2T + k_3FT^2$, where k_1, k_2, and k_3 are constants, find the expression for α.

42. An *isothermal process* is one during which the temperature does not change. If the volume V, pressure p, and temperature T of an ideal gas are related by the equation $pV = nRT$, where n and R are constants, find the expression for $\partial p/\partial V$, which is the rate of change of pressure with respect to volume for an isothermal process.

43. For the ideal gas of Exercise 42, show that $\left(\frac{\partial V}{\partial T}\right)\left(\frac{\partial T}{\partial p}\right)\left(\frac{\partial p}{\partial V}\right) = -1$.

44. Find the volume in the first octant below the plane $x + y + z - 6 = 0$ and inside the cylinder $y = 4 - x^2$.

45. Find the first-octant volume bounded by $x^2 + y^2 = 16$ and $x + z = 8$. Describe each of the bounding surfaces.

46. Find the first-octant volume below the surface $z = 4 - y^2$ and inside the plane $x + y = 2$.

47. An architectural design student determined the area of a patio could be described by the double integral $\int_0^4\int_0^{\sqrt{y}} f(x, y)\,dx\,dy$. Write the integration with the order of integration interchanged. Write one or two paragraphs to explain your method on interchanging the order of integration.

48. A spherical cap (the smaller part of a sphere cut through by a plane) has a volume $V = \frac{1}{3}\pi h^2(3r - h)$, where r is the radius of the sphere and h is the height of the cap. What is the volume of Earth above 30° north latitude ($r = 6380$ km)? (*Hint:* 30° north latitude is 30° above the plane passing through the equator measured from the centre of Earth.)

49. The area bounded by $y = 0$, $y = 2x$, and $x = 2$ is shaded on a calculator screen. Does $\int_0^4\int_{y/2}^2 dx\,dy$ represent the area?

50. In analysing the frictional resistance of a bearing, the integral $M = \frac{k}{\pi(b^2 - a^2)}\int_0^{2\pi}\int_a^b r^2\,dr\,d\theta$ arises. Evaluate this integral.

CHAPTER 29 **PRACTICE TEST**

In Problems 1–8, answer the given questions.

1. Given $f(x, y) = \frac{2y}{x^2 - y^2}$, find $f(-1, 3)$.

2. Sketch the surface representing the function $z = 4 - x^2 - 4y^2$.

3. Given $z = xe^{2xy}$, find $\frac{\partial z}{\partial x}$ and $\frac{\partial z}{\partial y}$.

4. Given $z = 3x^3y + 2x^2y^4$, find $\frac{\partial^2 z}{\partial x \partial y}$.

5. Evaluate $\int_0^2\int_{x^2}^{2x}(x^3 + 4y)\,dy\,dx$.

6. Evaluate $\int_1^{\ln 8}\int_0^{\ln y} e^{x+y}\,dx\,dy$.

7. Find the volume in the first octant bounded by the coordinate planes and the cylinders $x^2 + y^2 = 9$ and $y^2 + z^2 = 9$.

8. The fundamental frequency of vibration f of a string varies directly as the square root of the tension T and inversely as the length L. If a string 60 cm long is under a tension of 65 N and has a fundamental frequency of 30 Hz, find the partial derivative of f with respect to T and evaluate it for the given values.

30. Expansion of Functions in Series

LEARNING OUTCOMES

After completion of this chapter, the student should be able to:

- Find the terms of sequences and series
- Decide whether a geometric series is convergent or divergent and, if convergent, find its sum
- Find the Maclaurin series expansion of a function
- Use algebraic, trigonometric, or calculus procedures on known series to obtain other series expansions
- Find the Taylor series expansion of a function
- Approximate the value of functions by using series
- Find the Fourier series expansion of a periodic function
- Find the half-range Fourier series of a function
- Solve application problems using series

▲ Fourier series are used in electronic synthesizers to generate waveforms that mimic various musical instruments. In Section 30.7, we see how a square wave, which produces an organ-like sound, can be created by adding different sine waves together.

In the mid-1600s, mathematicians found that transcendental functions can be represented by polynomials and that by using these polynomials it was possible to calculate the values of these functions more easily. In this chapter, we show how a given function may be expressed in terms of a polynomial and how this polynomial is used to evaluate the function.

The polynomials that we will develop are known as *power series,* and they can be expressed with an unlimited number of terms. Although power series were first noted for their usefulness in calculating values of transcendental functions, many mathematicians, including Newton, used them extensively in their contributions to various areas of mathematics. In fact, the French mathematician Joseph-Louis Lagrange (1736–1813) attempted to make power series the basis for the development of all methods in calculus.

Another type of series was developed by the French physicist and mathematician Jean Baptiste Joseph Fourier (1768–1830) in the study of heat conduction. In 1822, he showed that a function can be expressed in a series of sine and cosine terms. Today, these series are very important in the study of electricity and electronics. They are also useful in the study of mechanical vibrations and other applications that are periodic in nature. We study these series in the last two sections of this chapter.

We see again that a concept first used to ease calculation became very important in the later development of mathematics. Also, a concept developed for the study of heat, long before the advent of electronics, has become important in electronics.

In Chapter 19, we discussed arithmetic and geometric sequences and the concept of an infinite geometric sequence. In the first section of this chapter, we develop these topics further for use in the sections that follow.

30.1 Infinite Series

In addition to arithmetic sequences and geometric sequences, there are many other ways of generating sequences of numbers. The squares of the integers 1, 4, 9, 16, 25 . . . form a sequence. Also, the successive approximations $x_1, x_2, x_3, \ldots$ found by using Newton's method in solving a particular equation form a sequence.

In general, *a* **sequence** *(or* **infinite sequence***) is an infinite succession of numbers. Each of the numbers is a* **term** *of the sequence.* Each term of the sequence is associated with a positive integer, although at times it is convenient to associate the first term with zero (or some specified positive integer). We shall use a_n to designate the term of the sequence corresponding to the integer n.

EXAMPLE 1 Find the terms of a sequence, given the general term

Find the first three terms of the sequence for which the general term is $a_n = 2n + 1$, $n = 1, 2, 3, \ldots$.

Substituting the values of n, we obtain the values

$$a_1 = 2(1) + 1 = 3 \qquad a_2 = 2(2) + 1 = 5 \qquad a_3 = 2(3) + 1 = 7, \ldots$$

Therefore, we have the sequence 3, 5, 7, ...

Given $a_n = 2n + 1$ for $n = 0, 1, 2, \ldots$, the sequence is 1, 3, 5, ...

As we stated in Chapter 19, *the sum of the terms of a sequence is called an* **infinite series**. Thus, for the sequence

$$a_1, a_2, a_3, \ldots, a_n, \ldots$$

the associated infinite series is

$$a_1 + a_2 + a_3 + \cdots + a_n + \cdots$$

■ The *sequence* feature on some graphing utilities can be used to find the terms of a sequence.

Using the summation sign Σ to indicate the sum, we have

Infinite Series

$$\sum_{n=1}^{\infty} a_n = a_1 + a_2 + a_3 + \cdots + a_n + \cdots \qquad \textbf{(30.1)}$$

Practice Exercise

1. Find the first three terms of the sequence for which $a_n = \dfrac{n+3}{n^2+1}$, $n = 1, 2, 3, \ldots$.

We define the sum for an infinite series in terms of a limit. For the infinite series of Eq. (30.1), we let S_n represent the sum of the first n terms. Therefore,

$$S_1 = a_1$$
$$S_2 = a_1 + a_2$$
$$S_3 = a_1 + a_2 + a_3$$
$$S_n = a_1 + a_2 + a_3 + \cdots + a_n$$

The numbers $S_1, S_2, S_3, \ldots, S_n, \ldots$ form a sequence. *Each term of this sequence is called a* **partial sum**. *We say that the infinite series, Eq. (30.1), is* **convergent** *and has the sum S given by*

$$S = \lim_{n \to \infty} S_n = \lim_{n \to \infty} \sum_{i=1}^{n} a_i \qquad \textbf{(30.2)}$$

if this limit exists. If the limit does not exist, the series is **divergent**.

EXAMPLE 2 Convergence of partial sums

For the infinite series

$$\sum_{n=0}^{\infty} \frac{1}{5^n} = \frac{1}{5^0} + \frac{1}{5^1} + \frac{1}{5^2} + \cdots + \frac{1}{5^n} + \cdots$$

the first six partial sums are

$$S_0 = 1 \qquad \text{first term}$$

$$S_1 = 1 + \frac{1}{5} = 1.2 \qquad \text{sum of first two terms}$$

$$S_2 = 1 + \frac{1}{5} + \frac{1}{25} = 1.24 \qquad \text{sum of first three terms}$$

$$S_3 = 1 + \frac{1}{5} + \frac{1}{25} + \frac{1}{125} = 1.248$$

$$S_4 = 1 + \frac{1}{5} + \frac{1}{25} + \frac{1}{125} + \frac{1}{625} = 1.2496$$

$$S_5 = 1 + \frac{1}{5} + \frac{1}{25} + \frac{1}{125} + \frac{1}{625} + \frac{1}{3125} = 1.24992$$

These values can be found using the standard calculational features of a calculator, or by the *cumulative sum* (cumSum) feature. Here, it appears that the sequence of partial sums approaches the value 1.25. We therefore conclude that this infinite series converges and that its sum is approximately 1.25. (In Example 4 of this section, we show that this infinite series does in fact converge and that its sum is 1.25.)

EXAMPLE 3 Divergent series

(a) The infinite series

$$\sum_{n=1}^{\infty} 5^n = 5 + 5^2 + 5^3 + \cdots + 5^n + \cdots$$

is a divergent series. The first four partial sums are

$$S_1 = 5 \qquad S_2 = 30 \qquad S_3 = 155 \qquad S_4 = 780$$

Obviously, they are increasing without bound.

(b) The infinite series

$$\sum_{n=0}^{\infty} (-1)^n = 1 + (-1) + 1 + (-1) + \cdots + (-1)^n + \cdots$$

has as its first five partial sums

$$S_0 = 1 \qquad S_1 = 0 \qquad S_2 = 1 \qquad S_3 = 0 \qquad S_4 = 1$$

The values of these partial sums do not approach a limiting value, and therefore, the series diverges.

Since convergent series are those that have a value associated with them, they are the ones that are of primary use to us. However, generally, it is not easy to determine whether a given series is convergent, and many types of tests have been developed for this purpose. These tests for convergence may be found in most textbooks that include the more advanced topics in calculus.

One important series for which we are able to determine the convergence, and its sum if convergent, is the geometric series. For this series, the nth partial sum is

$$S_n = a_1 + a_1 r + a_1 r^2 + \cdots + a_1 r^{n-1}$$

LEARNING TIP

If $r = 1$, we see that the series is $a_1 + a_1 + a_1 + \cdots + a_1 + \cdots$ and is, therefore, divergent. If $r = -1$, the series is $a_1 - a_1 + a_1 - a_1 + \cdots$ and is also divergent. If $|r| > 1$, $\lim_{n\to\infty} r^n$ is unbounded. Therefore, ***the geometric series is convergent only if*** $|r| < 1$ *and has the value given by Eq. (30.3).*

where r is the fixed number by which we multiply a given term to get the next term. In Chapter 19, we determined that if $|r| < 1$, the sum S of the infinite geometric series is

$$S = \lim_{n\to\infty} S_n = \frac{a_1}{1-r} \tag{30.3}$$

EXAMPLE 4 Geometric series

Show that the infinite series

$$\sum_{n=0}^{\infty} \frac{1}{5^n} = \frac{1}{5^0} + \frac{1}{5^1} + \frac{1}{5^2} + \cdots + \frac{1}{5^n} + \cdots$$

is convergent and find its sum. This is the same series as in Example 2.

This is a geometric series with $r = \frac{1}{5}$. Since $|r| < 1$, the series converges. The sum is

$$S = \frac{1}{1-\frac{1}{5}} = \frac{1}{\frac{4}{5}} = \frac{5}{4} = 1.25 \quad \text{using Eq. (30.3)}$$

Practice Exercise

2. Show that the infinite series $\sum_{n=0}^{\infty} \frac{2}{3^n}$ is convergent and find its sum.

EXERCISES 30.1

In Exercises 1 and 2, make the given changes in the indicated examples of this section and then solve the given problems.

1. In Example 3(a), change 5^n to 0.5^n. What other changes occur?
2. In Example 4, change $n = 0$ to $n = 1$. What is the value of S?

In Exercises 3–6, give the first four terms of the sequences for which a_n is given.

3. $a_n = n^2, n = 1, 2, 3, \ldots$
4. $a_n = \dfrac{2^{n+1}}{n!}, n = 1, 2, 3, \ldots$
5. $a_n = \dfrac{1}{n+2}, n = 0, 1, 2, \ldots$
6. $a_n = \dfrac{n^2+1}{2n+1}, n = 0, 1, 2, \ldots$

In Exercises 7–10, give (a) the first four terms of the sequence for which a_n is given and (b) the first four terms of the infinite series associated with the sequence.

7. $a_n = \left(-\dfrac{2}{5}\right)^n, n = 1, 2, 3, \ldots$
8. $a_n = \dfrac{1}{n} + \dfrac{1}{n+1}, n = 1, 2, 3, \ldots$
9. $a_n = \cos\dfrac{n\pi}{2}, n = 0, 1, 2, \ldots$
10. $a_n = \dfrac{n^n}{n!}, n = 2, 3, 4 \ldots$

In Exercises 11–14, find the nth term of the given infinite series for which $n = 1, 2, 3, \ldots$.

11. $\dfrac{1}{2} + \dfrac{1}{3} + \dfrac{1}{4} + \dfrac{1}{5} + \cdots$
12. $\dfrac{1}{2} + \dfrac{1}{4} + \dfrac{1}{8} + \dfrac{1}{16} + \cdots$
13. $\dfrac{1}{2\times 3} - \dfrac{1}{3\times 4} + \dfrac{1}{4\times 5} - \dfrac{1}{5\times 6} + \cdots$
14. $\dfrac{1}{\sqrt{2}} - \dfrac{1}{2} + \dfrac{\sqrt{2}}{4} - \dfrac{1}{4} + \cdots$

In Exercises 15–24, find the first five partial sums of the given series and determine whether the series appears to be convergent or divergent. If it is convergent, find its approximate sum.

15. $1 + \dfrac{1}{8} + \dfrac{1}{27} + \dfrac{1}{64} + \dfrac{1}{125} + \cdots$
16. $1 + 2 + 5 + 10 + 17 + \cdots$
17. $1 + \dfrac{1}{2} + \dfrac{2}{3} + \dfrac{3}{4} + \dfrac{4}{5} + \cdots$
18. $\dfrac{1}{3} - \dfrac{1}{9} + \dfrac{1}{27} - \dfrac{1}{81} + \dfrac{1}{243} - \cdots$
19. $\sum_{n=0}^{\infty} \sqrt{n}$
20. $\sum_{n=1}^{\infty} \dfrac{2}{n(n+1)}$
21. $\sum_{n=1}^{\infty} \dfrac{2n+1}{n^2(n+1)^2}$
22. $\sum_{n=1}^{\infty} \dfrac{n}{2n+1}$
23. $\sum_{n=1}^{\infty} \dfrac{\sin n}{4^n}$
24. $\sum_{n=3}^{\infty} \dfrac{\ln n}{e^n}$

In Exercises 25–32, test each of the given geometric series for convergence or divergence. Find the sum of each series that is convergent.

25. $1 + 2 + 4 + \cdots + 2^n + \cdots$
26. $1 + \dfrac{1}{2} + \dfrac{1}{4} + \cdots + \dfrac{1}{2^n} + \cdots$
27. $1 - \dfrac{1}{3} + \dfrac{1}{9} - \cdots + \left(-\dfrac{1}{3}\right)^n + \cdots$
28. $1 - \dfrac{5}{2} + \dfrac{25}{4} - \cdots + \left(-\dfrac{5}{2}\right)^n + \cdots$
29. $10 + 9 + 8.1 + 7.29 + 6.561 + \cdots$

30. $3 + 1 + \frac{1}{3} + \frac{1}{9} + \frac{1}{27} + \cdots$

31. $512 - 64 + 8 - 1 + \frac{1}{8} - \cdots$

32. $16 + 12 + 9 + \frac{27}{4} + \frac{81}{16} + \cdots$

In Exercises 33 and 34, find the values of x for which the given series converge.

33. $\sum_{n=0}^{\infty} (x-4)^n$

34. $\sum_{n=2}^{\infty} \frac{x^n}{5^n}$

In Exercises 35–46, solve the given problems as indicated.

35. Using a calculator, take successive square roots of 2 and find at least 20 approximate values for the terms of the sequence $2^{1/2}, 2^{1/4}, 2^{1/8}, 2^{1/16}, \ldots$. From the values that are obtained, (a) what do you observe about the value of $\lim_{n\to\infty} 2^{1/2^n}$? (b) Determine whether the infinite series for this sequence converges or diverges.

36. Using a calculator, (a) take successive square roots of 0.01 and then (b) take successive square roots of 100. From these sequences of square roots, state any general conclusions that might be drawn.

37. Referring to Chapter 19, we see that the sum of the first n terms of a geometric sequence is

$$S_n = \frac{a_1(1-r^n)}{1-r} \quad (r \neq 1) \qquad \text{Eq. (19.6)}$$

where a_1 is the first term and r is the common ratio. We can visualize the corresponding infinite series by graphing the function $f(x) = a_1(1-r^x)/(1-r)$ $(r \neq 1)$ (or using a graphing utility that can graph a sequence). The graph represents the sequence of partial sums for values where $x = n$, since $f(n) = S_n$.

Use a graphing utility to visualize the first five partial sums of the series

$$\frac{1}{2} + \frac{1}{4} + \frac{1}{8} + \cdots$$

What value does the infinite series approach? (*Remember:* Only points for which x is an integer have real meaning.)

38. Following Exercise 37, use a graphing utility to show that the sum of the infinite series of Example 4 is 1.25. (*Be careful:* Because of the definition of the series, $x = 1$ corresponds to $n = 0$.)

39. The value V (in dollars) of a certain investment after n years can be expressed as

$$V = 100(1.05 + 1.05^2 + 1.05^3 + \cdots + 1.05^n)$$

(a) By finding partial sums, determine whether this series converges or diverges. (b) Following Exercise 37, use a graphing utility to visualize the first 10 partial sums. (See Example 7 of Section 19.2.)

40. If an electric discharge is passed through hydrogen gas, a spectrum of isolated parallel lines, called the Balmer series, is formed. See Fig. 30.1. The wavelengths λ (in nm) of the light for these lines is given by the formula

$$\frac{1}{\lambda} = 1.097 \times 10^{-2}\left(\frac{1}{2^2} - \frac{1}{n^2}\right) \qquad (n = 3, 4, 5, \ldots)$$

Find the wavelengths of the first three lines and the shortest wavelength of all the lines of the series.

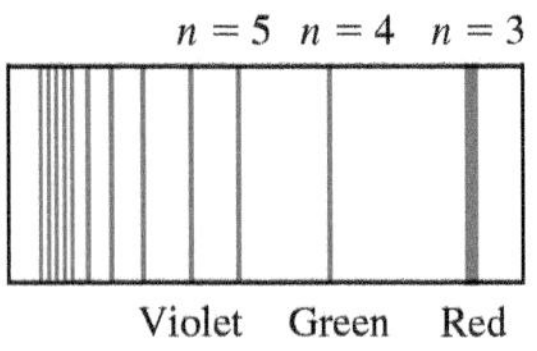

Fig. 30.1

41. Use geometric series to show that $\sum_{n=0}^{\infty} x^n = \frac{1}{1-x}$ for $|x| < 1$.

42. Use geometric series to show that $\sum_{n=0}^{\infty} (-1)^n x^n = \frac{1}{1+x}$ for $|x| < 1$.

43. If term a_1 is given along with a rule to find term a_{n+1} from term a_n, the sequence is said to be defined *recursively*. If $a_1 = 2$ and $a_{n+1} = (n+1)a_n$, find the first five terms of the sequence.

44. A sequence is defined recursively (see Exercise 43) by $x_1 = \frac{N}{2}$, $x_{n+1} = \frac{1}{2}\left(x_n + \frac{N}{x_n}\right)$. With $N = 10$, find x_6 and compare the value with $\sqrt{10}$. It can be seen that $\sqrt{N}$ can be approximated using this recursion sequence.

45. Write out the first four terms of the series $\sum_{n=1}^{\infty}\left(\frac{1}{n} - \frac{1}{n+1}\right)$. Then find the first four partial sums. Does this series appear to converge? Why or why not?

46. When two dice are rolled repeatedly, the probability of rolling a sum of 4 before rolling a sum of 7 is given by

$\frac{3}{36}\left[1 + \frac{27}{36} + \left(\frac{27}{36}\right)^2 + \left(\frac{27}{36}\right)^3 + \cdots\right]$. Find this probability.

Answers to Practice Exercises

1. $2, 1, 3/5, \ldots$ **2.** $r = 1/3 < 1, S = 3$

30.2 Maclaurin Series

In this section, we study a basic representation of a function in what is called its **power-series expansion**. More precisely, given a function $f(x)$, we will try to represent it in the form

$$f(x) = a_0 + a_1x + a_2x^2 + \cdots + a_nx^n + \cdots \qquad \textbf{(30.4)}$$

The representation is valid for a given value of x if, when we substitute that value into the series and also into the function, the series will converge to the value of the function. The values of x for which the series converges form what is called the **interval of convergence** of the power series.

As you can see, a power-series expansion is in the form of a polynomial, except that it has an infinite number of terms. Having $f(x)$ represented in this form can give us a simpler way to perform algebraic operations involving $f(x)$. It can also make it easier to evaluate (at least approximately), differentiate, and integrate $f(x)$. A further study of calculus shows other uses of power series.

We begin by obtaining the power-series expansion of a simple algebraic function by using long division. We will then use calculus to obtain representations for more general functions.

EXAMPLE 1 An algebraic function represented by a series

Find a power-series expansion of the function $\frac{2}{2-x}$ by using long division. Determine the interval of convergence of the power series.

$$\begin{array}{r} 1+\frac{x}{2} \\ 2-x\overline{)2\phantom{-\frac{x^2}{2}}} \\ \underline{2-x}\phantom{-\frac{x^2}{2}} \\ x\phantom{-\frac{x^2}{2}} \\ \underline{x-\frac{x^2}{2}} \\ \frac{x^2}{2} \end{array}$$

The first steps of the long division can be seen at the left. At every step, the residual is divided by the first term of the denominator, which is 2. We start by dividing 2 by 2, so the first term in the series is 1 and the residual is x. Next, we divide x by 2, so the next term of the series is $x/2$, and we have residual $x^2/2$. For the next term, we would divide $x^2/2$ by 2. As we continue, we obtain the series

$$\frac{2}{2-x} = 1 + \frac{1}{2}x + \frac{1}{4}x^2 + \cdots + \left(\frac{1}{2}x\right)^{n-1} + \cdots \tag{1}$$

where n is the number of the term of the expression on the right. Since x represents a number, the right-hand side of Eq. (1) becomes a geometric series.

From Eq. (30.3), we know that the sum of a geometric series with first term a_1 and common ratio r is

$$S = \frac{a_1}{1-r}$$

when $|r| < 1$ and the series converges.

If $x = 1$, the right-hand side of Eq. (1) is

$$1 + \frac{1}{2} + \frac{1}{4} + \cdots + \left(\frac{1}{2}\right)^{n-1} + \cdots$$

For this series, $r = \frac{1}{2}$ and $a_1 = 1$, which means that the series converges and $S = 2$. Moreover, the left side of Eq. (1) is also 2 when $x = 1$, so the function may be represented by the series for this value of x.

If $x = 3$, the right-hand side of Eq. (1) is

$$1 + \frac{3}{2} + \frac{9}{4} + \cdots + \left(\frac{3}{2}\right)^{n-1} + \cdots$$

which diverges since $r > 1$. Therefore, the left-hand side of Eq. (1) cannot be represented by the series when $x = 3$.

As we can see, the function and the series agree when the series converges. Since the series converges as long as $|x| < 2$, the interval of convergence is $|x| < 2$.

At this point, we will assume that unless otherwise noted, the functions with which we will be dealing may be properly represented by a power-series expansion (it takes more advanced methods to prove that this is generally possible), for appropriate intervals of convergence. We will find that the methods of calculus are very useful in developing the method of general representation. Thus, assuming that $f(x)$ has a power-series expansion $f(x) = a_0 + a_1x + a_2x^2 + \cdots$, we need to find the coefficients a_0, a_1, a_2, and so on.

Finding a_0: Since $f(x) = a_0 + a_1x + a_2x^2 + a_3x^3 + a_4x^4 + \cdots$, we have

$$f(0) = a_0 + a_1(0) + a_2(0) + a_3(0) + a_4(0) + \cdots = a_0, \quad \text{and } a_0 = f(0)$$

Finding a_1: Taking the derivative of $f(x)$ gives us

$$f'(x) = a_1 + 2a_2x + 3a_3x^2 + 4a_4x^3 + \cdots$$

Therefore,

$$f'(0) = a_1 + 2a_2(0) + 3a_3(0) + 4a_4(0) + \cdots = a_1, \quad \text{and } a_1 = f'(0)$$

Finding a_2: Taking the second derivative of $f(x)$,

$$f''(x) = 2a_2 + (2)3a_3x + (3)4a_4x^2 + \cdots$$

Therefore,

$$f''(0) = 2a_2 + (2)3a_3(0) + (3)4a_4(0) + \cdots = 2a_2, \quad \text{and } a_2 = \frac{f''(0)}{2!}$$

Finding a_3: Similarly, we find $f'''(x)$ and then evaluate $f'''(0)$,

$$f'''(x) = (2)3a_3 + (2)(3)4a_4x + \cdots, \quad f'''(0) = (2)3a_3, \quad \text{and } a_3 = \frac{f'''(0)}{3!}$$

Continuing this way and substituting into the expression for $f(x)$, we get the following expression, which is known as the **Maclaurin series expansion** of a function.

Maclaurin Series

$$f(x) = f(0) + f'(0)x + \frac{f''(0)x^2}{2!} + \frac{f'''(0)x^3}{3!} + \cdots + \frac{f^n(0)x^n}{n!} + \cdots \qquad \textbf{(30.5)}$$

■ The Maclaurin series is named for the Scottish mathematician Colin Maclaurin (1698–1746).

For a function to be represented by a Maclaurin expansion, the function and all of its derivatives must exist at $x = 0$. Also, we note that the factorial notation introduced in Section 19.4 is used in writing the Maclaurin series expansion.

As we mentioned earlier, one of the uses we will make of series expansions is that of determining the values of functions for particular values of x. If x is sufficiently small, successive terms become smaller and smaller and the series will converge rapidly. This is considered in the sections that follow.

The following examples illustrate Maclaurin expansions for algebraic, exponential, and trigonometric functions.

EXAMPLE 2 Maclaurin series for an algebraic function

Find the first four terms of the Maclaurin series expansion of $f(x) = \dfrac{2}{2-x}$.

■ Compare with Example 1.

$$f(x) = \frac{2}{2-x} \quad f(0) = 1 \quad f''(x) = \frac{4}{(2-x)^3} \quad f''(0) = \frac{1}{2}$$

$$f'(x) = \frac{2}{(2-x)^2} \quad f'(0) = \frac{1}{2} \quad f'''(x) = \frac{12}{(2-x)^4} \quad f'''(0) = \frac{3}{4}$$

find derivatives and evaluate each at $x = 0$

$$f(x) = 1 + \frac{1}{2}x + \frac{1}{2}\left(\frac{x^2}{2!}\right) + \frac{3}{4}\left(\frac{x^3}{3!}\right) + \cdots$$

using Eq. (30.5)

$$\frac{2}{2-x} = 1 + \frac{1}{2}x + \frac{1}{4}x^2 + \frac{1}{8}x^3 + \cdots$$

Practice Exercise

1. Find the first four terms of the Maclaurin series expansion for $f(x) = \dfrac{1}{1+x}$.

EXAMPLE 3 Maclaurin series for an exponential function

Find the first four terms of the Maclaurin series expansion of $f(x) = e^{-x}$.

$$f(x) = e^{-x} \quad f(0) = 1 \quad f''(x) = e^{-x} \quad f''(0) = 1$$
$$f'(x) = -e^{-x} \quad f'(0) = -1 \quad f'''(x) = -e^{-x} \quad f'''(0) = -1$$

find derivatives and evaluate each at $x = 0$

$$f(x) = 1 + (-1)x + 1\left(\frac{x^2}{2!}\right) + (-1)\left(\frac{x^3}{3!}\right) + \cdots$$

using Eq. (30.5)

$$e^{-x} = 1 - x + \frac{x^2}{2!} - \frac{x^3}{3!} + \cdots$$

EXAMPLE 4 Maclaurin series for a trigonometric function

Find the first three nonzero terms of the Maclaurin series expansion of $f(x) = \sin 2x$.

$$f(x) = \sin 2x \quad f(0) = 0 \quad f'''(x) = -8\cos 2x \quad f'''(0) = -8$$
$$f'(x) = 2\cos 2x \quad f'(0) = 2 \quad f^{\text{iv}}(x) = 16\sin 2x \quad f^{\text{iv}}(0) = 0$$
$$f''(x) = -4\sin 2x \quad f''(0) = 0 \quad f^{\text{v}}(x) = 32\cos 2x \quad f^{\text{v}}(0) = 32$$

$$f(x) = 0 + 2x + 0 + (-8)\frac{x^3}{3!} + 0 + 32\frac{x^5}{5!} + \cdots$$

$$\sin 2x = 2x - \frac{4}{3}x^3 + \frac{4}{15}x^5 - \cdots$$

■ The series in Example 4 is called an **alternating series** since every other term is negative.

EXAMPLE 5 Maclaurin series—critically damped motion

Frictional forces in the spring shown in Fig. 30.2 are just sufficient so that the lever does not oscillate after being depressed. Such motion is called *critically damped.* The displacement y as a function of the time t for one case is $y = (1 + t)e^{-t}$. To study the motion for small values of t, a Maclaurin expansion of $y = f(t)$ is to be used. Find the first four terms of the expansion.

$$f(t) = (1 + t)e^{-t} \qquad f(0) = 1$$
$$f'(t) = (1 + t)e^{-t}(-1) + e^{-t} = -te^{-t} \qquad f'(0) = 0$$
$$f''(t) = te^{-t} - e^{-t} \qquad f''(0) = -1$$
$$f'''(t) = -te^{-t} + e^{-t} + e^{-t} = 2e^{-t} - te^{-t} \qquad f'''(0) = 2$$
$$f^{\text{iv}}(t) = -2e^{-t} + te^{-t} - e^{-t} = te^{-t} - 3e^{-t} \qquad f^{\text{iv}}(0) = -3$$

$$f(t) = 1 + 0 + (-1)\frac{t^2}{2!} + 2\frac{t^3}{3!} + (-3)\frac{t^4}{4!} + \cdots$$

$$(1 + t)e^{-t} = 1 - \frac{t^2}{2} + \frac{t^3}{3} - \frac{t^4}{8} + \cdots$$

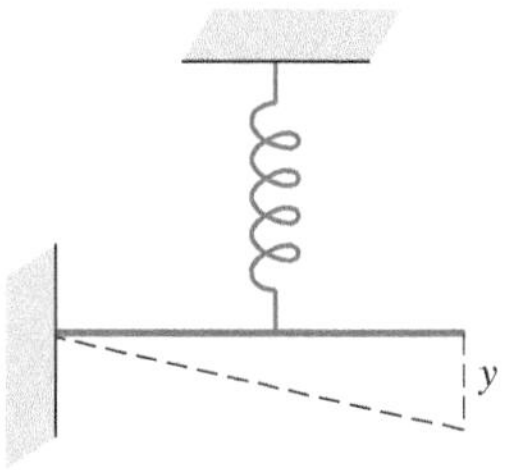

Fig. 30.2

EXERCISES 30.2

In Exercises 1 and 2, make the given changes in the indicated examples of this section and then find the resulting series.

1. In Example 2, in $f(x)$, change the denominator to $2 + x$.
2. In Example 4, in $f(x)$, change $2x$ to $(-2x)$.

In Exercises 3–20, find the first three nonzero terms of the Maclaurin expansion of the given functions.

3. $f(x) = e^x$
4. $f(x) = \sin x$
5. $f(x) = \cos x$
6. $f(x) = \ln(1 + x)$
7. $f(x) = \sqrt{1 + x}$
8. $f(x) = \sqrt[3]{1 + x}$
9. $f(x) = e^{-2x}$
10. $f(x) = \dfrac{1}{\sqrt{1 + x}}$
11. $f(x) = \cos 4\pi x$
12. $f(x) = e^x \sin x$
13. $f(x) = \dfrac{1}{1 - x}$
14. $f(x) = \dfrac{1}{(1 + x)^2}$
15. $f(x) = \ln(1 - 2x)$
16. $f(x) = (1 + x)^{3/2}$
17. $f(x) = \cos^2 x$
18. $f(x) = \ln(1 + 4x)$
19. $f(x) = \sin(x + \frac{\pi}{4})$
20. $f(x) = (2x - 1)^2$

In Exercises 21–28, find the first two nonzero terms of the Maclaurin expansion of the given functions.

21. $f(x) = \tan^{-1} x$ **22.** $f(x) = \cos x^2$

23. $f(x) = \tan x$ **24.** $f(x) = \sec x$

25. $f(x) = \ln \cos x$ **26.** $f(x) = xe^{\sin x}$

27. $f(x) = \sqrt{1 + \sin x}$ **28.** $f(x) = xe^{-x^2}$

In Exercises 29–42, solve the given problems.

29. Is it possible to find a Maclaurin expansion for (a) $f(x) = \csc x$ or (b) $f(x) = \ln x$? Explain.

30. Use long division to find a series expansion for $f(x) = \dfrac{1}{(1+x)^2}$. Compare the results with Exercise 14.

31. Find the first three nonzero terms of the Maclaurin expansion for (a) $f(x) = e^x$ and (b) $f(x) = e^{x^2}$. Compare these expansions.

32. By finding the Maclaurin expansion of $f(x) = (1 + x)^n$, derive the first four terms of the binomial series, which is Eq. (19.10). Its interval of convergence is $|x| < 1$ for all values of n.

33. If $f(x) = e^{3x}$, compare the Maclaurin expansion with the linearization for $a = 0$.

34. The hyperbolic sine function is defined as $\sinh x = \frac{1}{2}(e^x - e^{-x})$. Find the Maclaurin series for $y = \sinh x$.

35. The hyperbolic cosine function is defined as $\cosh x = \frac{1}{2}(e^x + e^{-x})$. Find the Maclaurin series for $y = \cosh x$.

36. Find the Maclaurin series for $f(x) = \cos^2 x$, by using the identity $\cos^2 x = \frac{1}{2}(1 + \cos 2x)$. Compare the result with that of Exercise 17.

37. If $f(x) = x^2$, show that this function is obtained when a Maclaurin expansion is found.

38. If $f(x) = x^4 + 3x^2 + 5x$, show that this function is obtained when a Maclaurin expansion is found.

39. The displacement y (in cm) of an object hung vertically from a spring and allowed to oscillate is given by the equation $y = 4e^{-0.2t} \cos t$, where t is the time (in s). Find the first three terms of the Maclaurin expansion of this function.

40. For the circuit shown in Fig. 30.3, after the switch is closed, the transient current i (in A) is given by $i = 2.5(1 + e^{-0.1t})$. Find the first three terms of the Maclaurin expansion of this function.

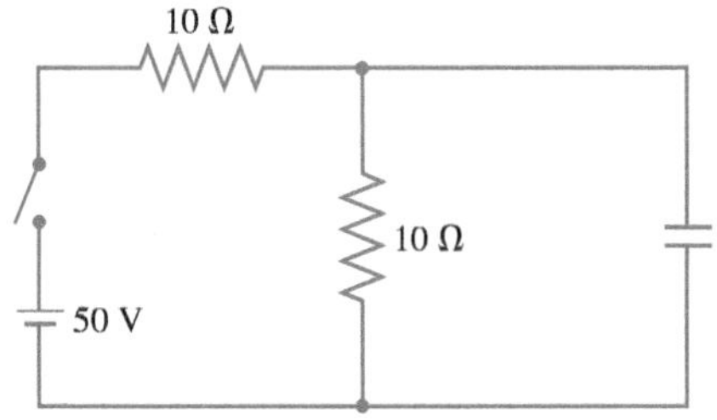

Fig. 30.3

41. The reliability R $(0 \le R \le 1)$ of a certain computer system is $R = e^{-0.001t}$, where t is the time of operation (in min). Express $R = f(t)$ in polynomial form by using the first three terms of the Maclaurin expansion.

42. In the analysis of the optical paths of light from a narrow slit S to a point P, as shown in Fig. 30.4, the law of cosines is used to obtain the equation

$$c^2 = a^2 + (a + b)^2 - 2a(a + b)\cos\frac{s}{a}$$

where s is part of the circular arc $\widehat{AB}$. By using two nonzero terms of the Maclaurin expansion of $\cos\frac{s}{a}$, simplify the right side of the equation. (In finding the expansion, let $x = \frac{s}{a}$ and then substitute back into the expansion.)

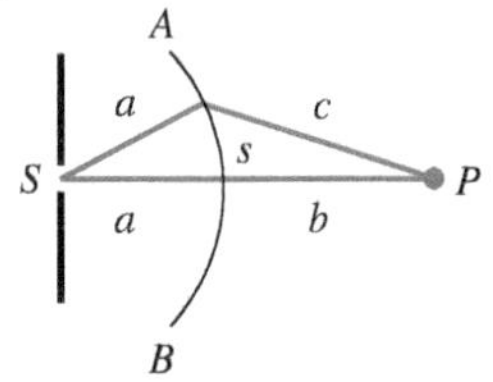

Fig. 30.4

Answer to Practice Exercise

1. $\dfrac{1}{1+x} = 1 - x + x^2 - x^3 + \cdots$

30.3 Operations with Series

In this section we will learn to derive the power-series expansion of a function using another function for which we already have a series expansion. The basic series we will use are given here, together with their intervals of convergence. These series were derived as exercises in Section 30.2.

$$e^x = 1 + x + \frac{x^2}{2!} + \frac{x^3}{3!} + \cdots \quad \text{(all } x\text{)} \tag{30.6}$$

$$\sin x = x - \frac{x^3}{3!} + \frac{x^5}{5!} - \cdots \quad \text{(all } x\text{)} \tag{30.7}$$

$$\cos x = 1 - \frac{x^2}{2!} + \frac{x^4}{4!} - \cdots \quad \text{(all } x\text{)} \tag{30.8}$$

$$\ln(1 + x) = x - \frac{x^2}{2} + \frac{x^3}{3} - \frac{x^4}{4} + \cdots \quad (|x| < 1) \tag{30.9}$$

$$(1 + x)^n = 1 + nx + \frac{n(n-1)}{2!}x^2 + \cdots \quad (|x| < 1) \tag{30.10}$$

LEARNING TIP

When we discussed functions in Chapter 3, we mentioned functions such as $f(2x)$ and $f(-x)$. *By using functional notation and the series in Eqs. (30.6)–(30.10), we can find the series expansions of many other series without using direct expansion.* This can often save time in finding a desired series.

EXAMPLE 1 Series formed using functional notation

Find the Maclaurin expansion of e^{2x}.

From Eq. (30.6), we know the expansion of e^x. Hence,

$$f(x) = 1 + x + \frac{x^2}{2!} + \frac{x^3}{3!} + \cdots$$

Since $e^{2x} = f(2x)$, we replace x by $2x$ in the expansion to get

$$f(2x) = 1 + (2x) + \frac{(2x)^2}{2!} + \frac{(2x)^3}{3!} + \cdots$$

$$e^{2x} = 1 + 2x + 2x^2 + \frac{4x^3}{3} + \cdots$$

EXAMPLE 2 Series formed using functional notation

Find the Maclaurin expansion of $\sin x^2$.

From Eq. (30.7), we know the expansion of $\sin x$. Therefore,

$$f(x) = x - \frac{x^3}{3!} + \frac{x^5}{5!} - \cdots$$

$$f(x^2) = (x^2) - \frac{(x^2)^3}{3!} + \frac{(x^2)^5}{5!} - \cdots \quad \text{in } f(x)\text{, replace } x \text{ by } x^2$$

$$\sin x^2 = x^2 - \frac{x^6}{3!} + \frac{x^{10}}{5!} - \cdots$$

Direct expansion of this series is quite lengthy.

Practice Exercise

1. Using the Maclaurin series for $\ln(1 + x)$, find the first four terms of the Maclaurin expansion of $\ln(1 - 2x)$.

The basic algebraic operations may be applied to series in the same manner they are applied to polynomials. That is, we may add, subtract, multiply, or divide series in order to obtain other series. The interval of convergence for the resulting series is that which is common to those of the series being used. The multiplication of series is illustrated in the following example.

EXAMPLE 3 Series formed by multiplication

Multiply the series expansion for e^x by the series expansion for $\cos x$ to obtain the series expansion for $e^x \cos x$.

Using the series expansion for e^x and $\cos x$ as shown in Eqs. (30.6) and (30.8), we have the following indicated multiplication:

$$e^x \cos x = \left(1 + x + \frac{x^2}{2!} + \frac{x^3}{3!} + \frac{x^4}{4!} + \cdots\right)\left(1 - \frac{x^2}{2!} + \frac{x^4}{4!} - \cdots\right)$$

By multiplying the series on the right, we have the following result, considering through the x^4 terms in the product.

$$1\left(1 - \frac{x^2}{2!} + \frac{x^4}{4!}\right) \quad x\left(1 - \frac{x^2}{2!}\right) \quad \frac{x^2}{2!}\left(1 - \frac{x^2}{2!}\right) \quad \left(\frac{x^3}{3!} + \frac{x^4}{4!}\right)(1)$$

$$e^x \cos x = 1 - \frac{x^2}{2} + \frac{x^4}{24} + x - \frac{x^3}{2} + \frac{x^2}{2} - \frac{x^4}{4} + \frac{x^3}{6} + \frac{x^4}{24} + \cdots$$

$$= 1 + x - \frac{1}{3}x^3 - \frac{1}{6}x^4 + \cdots$$

It is also possible to use the operations of differentiation and integration to obtain series expansions, although the proof of this is found in more advanced texts. Consider the following example.

EXAMPLE 4 Series formed by differentiation

Show that by differentiating the series for $\ln(1+x)$ term by term, the result is the same as the series for $\frac{1}{1+x}$.

The series for $\ln(1+x)$ is shown in Eq. (30.9) as

$$\ln(1+x) = x - \frac{x^2}{2} + \frac{x^3}{3} - \frac{x^4}{4} + \cdots$$

Differentiating, we have

$$\begin{aligned}\frac{1}{1+x} &= 1 - \frac{2x}{2} + \frac{3x^2}{3} - \frac{4x^3}{4} + \cdots \\ &= 1 - x + x^2 - x^3 + \cdots\end{aligned}$$

Using the binomial expansion for $\frac{1}{1+x} = (1+x)^{-1}$, we have

$$\begin{aligned}(1+x)^{-1} &= 1 + (-1)x + \frac{(-1)(-2)}{2!}x^2 + \frac{(-1)(-2)(-3)}{3!}x^3 + \cdots \\ &= 1 - x + x^2 - x^3 + \cdots\end{aligned}$$

using Eq. (30.10) with $n = -1$

We see that the results are the same.

We can use algebraic operations on series to verify that the definition of the exponential form of a complex number, $re^{j\theta} = r(\cos\theta + j\sin\theta)$, is consistent with other definitions. The only assumption required here is that the Maclaurin expansions for e^x, $\sin x$, and $\cos x$ are also valid for complex numbers. This is shown in advanced calculus. Thus,

$$e^{j\theta} = 1 + j\theta + \frac{(j\theta)^2}{2!} + \frac{(j\theta)^3}{3!} + \cdots = 1 + j\theta - \frac{\theta^2}{2!} - j\frac{\theta^3}{3!} + \cdots \quad \textbf{(30.11)}$$

$$j\sin\theta = j\theta - j\frac{\theta^3}{3!} + \cdots \quad \textbf{(30.12)}$$

$$\cos\theta = 1 - \frac{\theta^2}{2!} + \cdots \quad \textbf{(30.13)}$$

When we add the terms of Eq. (30.12) to those of Eq. (30.13), the result is the series given in Eq. (30.11). Thus,

$$e^{j\theta} = \cos\theta + j\sin\theta \quad \textbf{(30.14)}$$

■ Eq. (30.14) is known as Euler's Formula. If $\theta = \pi$, we have $e^{j\pi} = -1$, which can be written as

$$e^{j\pi} + 1 = 0$$

This equation connects the five fundamental numbers e, j, π, 1, and 0, and it has been called a "beautiful" equation by mathematicians.

By multiplying both sides of Eq. (30.14) by r, we get the exponential form of a complex number: $re^{j\theta} = r(\cos\theta + j\sin\theta)$.

An additional use of power series is now shown. Many integrals that occur in practice cannot be integrated by methods given in the preceding chapters. However, power series can be very useful in giving excellent approximations to some definite integrals.

EXAMPLE 5 Using series for integration—area of a cutting blade

The shape of a special cutting blade can be described by the region bounded by the axes, the line $x = 0.500$ cm, and the curve $y = \sqrt{1 + x^3}$. Find the area of the blade (in cm^2).

From Fig. 30.5, we see that the area is

$$A = \int_0^{0.5} \sqrt{1 + x^3}\,dx$$

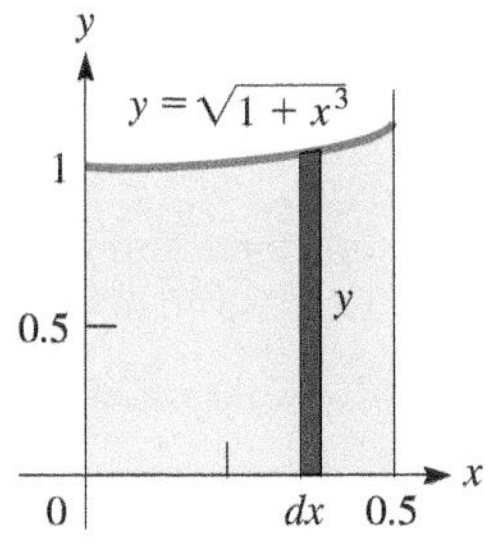

Fig. 30.5

This integral does not fit any form we have used. However, its value can be closely approximated by using the binomial expansion for $\sqrt{1 + x^3}$ and then integrating.

Using the binomial expansion to find the first three terms of the expansion for $\sqrt{1 + x^3}$, we have

$$\sqrt{1 + x^3} = (1 + x^3)^{0.5} = 1 + 0.5x^3 + \frac{0.5(-0.5)}{2}(x^3)^2 + \cdots$$
$$= 1 + 0.5x^3 - 0.125x^6 + \cdots$$

Substituting in the integral, we have

$$A = \int_0^{0.5} (1 + 0.5x^3 - 0.125x^6 + \cdots)\,dx$$
$$= x + \frac{0.5}{4}x^4 - \frac{0.125}{7}x^7 + \cdots \Big|_0^{0.5}$$
$$= 0.5 + 0.007\,812\,5 - 0.000\,139\,5 + \cdots = 0.507\,673 + \cdots$$

We can see that each of the terms omitted was very small. Because the data given ($x = 0.500$ cm) is accurate to three significant digits, we conclude that the area of the blade is $A = 0.508\ cm^2$.

EXAMPLE 6 Using series for integration

Evaluate $\int_0^{0.1} e^{-x^2}\,dx$.

$$e^{-x^2} = 1 + (-x^2) + \frac{(-x^2)^2}{2!} + \cdots \quad \text{using Eq. (30.6)}$$

$$\int_0^{0.1} e^{-x^2}\,dx = \int_0^{0.1}\left(1 - x^2 + \frac{x^4}{2} - \cdots\right)dx \quad \text{substitute}$$

$$= \left(x - \frac{x^3}{3} + \frac{x^5}{10} - \cdots\right)\Big|_0^{0.1} \quad \text{integrate}$$

$$= 0.1 - \frac{0.001}{3} + \frac{0.000\,01}{10} = 0.099\,667\,7 \quad \text{evaluate}$$

This answer is correct to the indicated accuracy.

LEARNING TIP

For small values of x, a Maclaurin series gives good accuracy with a very few terms. In this case, the series *converges* rapidly. For this reason, a Maclaurin series is of particular use for small values of x. For larger values of x, a function is usually expanded in a Taylor series (see Section 30.5). Of course, if we omit any term in a series, there is some error in the calculation.

The question of accuracy now arises. The integrals just evaluated indicate that the more terms used, the greater the accuracy of the result. To show the accuracy involved graphically, Fig. 30.6 depicts the graphs of $y = \sin x$ and the graphs of

$$y = x \qquad y = x - \frac{x^3}{3!} \qquad y = x - \frac{x^3}{3!} + \frac{x^5}{5!}$$

which are the first three approximations of $y = \sin x$. We can see that each term added gives a better fit to the curve of $y = \sin x$. Also, this gives a graphical representation of the meaning of a series expansion.

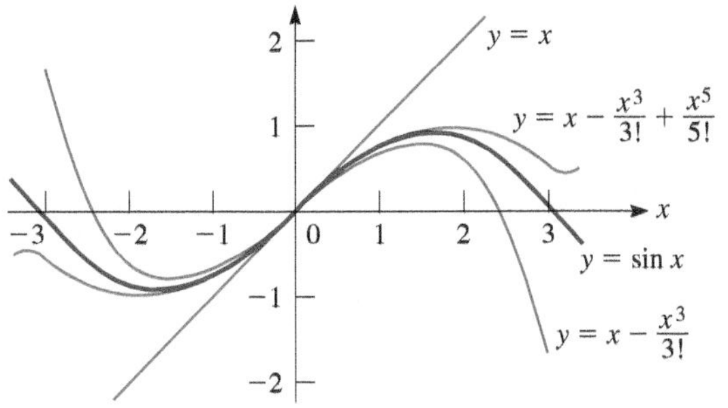

Fig. 30.6

EXERCISES 30.3

In Exercises 1 and 2, make the given changes in the indicated examples of this section, and then find the resulting series.

1. In Example 1, change e^{2x} to e^{2x^2}.

2. In Example 3, change e^x to e^{-x}.

In Exercises 3–10, find the first four nonzero terms of the Maclaurin expansions of the given functions by using Eqs. (30.6) to (30.10).

3. $f(x) = e^{3x}$ **4.** $f(x) = e^{-4x}$

5. $f(x) = \sin \frac{1}{2}x$ **6.** $f(x) = \sin x^4$

7. $f(x) = x \cos 8x$ **8.** $f(x) = \sqrt{1 - x^4}$

9. $f(x) = \ln(1 + x^2)$ **10.** $f(x) = x^2 \ln(1 - x)$

In Exercises 11–14, evaluate the given integrals by using three terms of the appropriate series. Round answers to three significant digits.

11. $\int_0^1 \sin x^2\, dx$ **12.** $\int_0^{0.4} \sqrt[4]{1 - 2x^2}\, dx$

13. $\int_0^{0.2} \cos\sqrt{x}\, dx$ **14.** $\int_{0.1}^{0.2} \frac{\cos x - 1}{x}\, dx$

In Exercises 15–28, find the indicated series by the given operation.

15. Find the first four terms of the Maclaurin expansion of the function $f(x) = \frac{2}{1 - x^2}$ by adding the terms of the series for the functions $\frac{1}{1 - x}$ and $\frac{1}{1 + x}$.

16. Find the first four nonzero terms of the expansion of the function $f(x) = \frac{1}{2}(e^x - e^{-x})$ by subtracting the terms of the appropriate series. The result is the series for $\sinh x$.

17. Find the first three terms of the expansion for $e^x \sin x$ by multiplying the proper expansions together, term by term.

18. Find the first three nonzero terms of the expansion for $f(x) = \tan x$ by dividing the series for $\sin x$ by that for $\cos x$.

19. By using the properties of logarithms and the series for $\ln(1 + x)$, find the series for $x^2 \ln(1 - x)^2$.

20. By using the properties of logarithms and the series for $\ln(1 + x)$, find the series for $\ln \frac{1 + x}{1 - x}$.

21. Find the first three terms of the expansion for $\ln(1 + \sin x)$ by using the expansions for $\ln(1 + x)$ and $\sin x$.

22. Show that by differentiating term by term the expansion for $\cos x$, the result is the expansion for $-\sin x$.

23. Show that by differentiating term by term the expansion for e^x, the result is also the expansion for e^x.

24. Find the expansion for $\sin x + x \cos x$ by differentiating term by term the expansion for $x \sin x$.

25. Show that by integrating term by term the expansion for $\cos x$, the result is the expansion for $\sin x$.

26. Show that by integrating term by term the expansion for $-1/(1 - x)$ (see Exercise 13 of Section 30.2), the result is the expansion for $\ln(1 - x)$.

27. By multiplication of series, find the first three terms of the expansion for the displacement of the oscillating object of Exercise 39 of Section 30.2.

28. By using the series for e^x, find the first three terms of the expansion of the electric current given in Exercise 40 of Section 30.2.

In Exercises 29–40, solve the given problems. Where necessary, round answers to three significant digits.

29. Evaluate $\int_0^1 e^x\, dx$ directly and compare the result obtained by using four terms of the series for e^x and then integrating.

30. Evaluate $\lim_{x \to 0} \frac{\sin x}{x}$ by using the series expansion for $\sin x$. Compare the result with Eq. (27.1).

31. Evaluate $\lim_{x \to 0} \frac{\sin x - x}{x^3}$ by using the expansion for $\sin x$.

32. Find the approximate area bounded by $y = \sin x$, $y = 0$, and $x = \pi/6$ by using two terms of the expansion for $\sin x$. Compare the result with that found by direct integration.

33. Find the approximate value of the area bounded by $y = x^2 e^x$, $x = 0.2$, and the x-axis by using three terms of the appropriate Maclaurin series.

34. Find the approximate area under the graph of $y = \frac{1}{\sqrt{2\pi}} e^{-x^2/2}$ from $x = -1$ to $x = 1$ by using three terms of the appropriate series. See Fig. 30.7. Compare with the value obtained from Table 22.1.

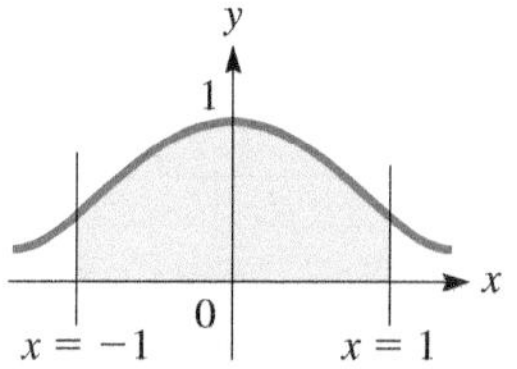

Fig. 30.7

35. The *Fresnel integral* $\int_0^x \cos t^2\, dt$ is used in the analysis of beam displacements (and in optics). Evaluate this integral for $x = 0.2$ by using two terms of the appropriate series.

36. The dome of a sports arena is designed as the surface generated by revolving the curve of $y = 20.0 \cos 0.0196x$ $(0 \le x \le 80.0 \text{ m})$ about the y-axis. Find the volume within the dome by using three terms of the appropriate series.

37. In the theory of relativity, when studying the kinetic (moving) energy of an object, the equation $K = \left[\left(1 - \frac{v^2}{c^2}\right)^{-1/2} - 1\right] mc^2$ is used. Here, for a given object, K is the kinetic energy, v is its velocity, and c is the velocity of light. If v is much smaller than c, show that $K = \frac{1}{2}mv^2$, which is the classical expression for K.

38. The charge q on a capacitor in a certain electric circuit is given by $q = ce^{-at} \sin 6at$, where t is the time. By multiplication of series, find the first four nonzero terms of the expansion for q.

39. By differentiating the expansion for $\frac{1}{1 - x}$, show that $\sum_{n=1}^{\infty} nx^{n-1} = \frac{1}{(1 - x)^2}$ for $|x| < 1$.

40. In a torture test, a machine drops a mobile phone repeatedly until it breaks. If the probability that the phone will break any time it is dropped is 0.04, the average number of times it can be dropped before it breaks is given by $\mu = 0.04\sum_{n=1}^{\infty} n(0.96)^{n-1}$. Find μ. (*Hint*: Use Exercise 39.)

In Exercises 41–44, use a graphing utility to display (a) the given function and (b) the first three series approximations of the function in the same display. Each display will be similar to that in Fig. 30.6 for the function $y = \sin x$ and its first three approximations. Be careful in choosing the appropriate display settings.

41. $y = e^x$ **42.** $y = \cos x$

43. $y = \ln(1 + x) \quad (|x| < 1)$

44. $y = \sqrt{1 + x} \quad (|x| < 1)$

Answer to Practice Exercise

1. $\ln(1 - 2x) = -2x - 2x^2 - \frac{8}{3}x^3 - 4x^4 - \cdots$

30.4 Computations by Use of Series Expansions

Power-series expansions can be used to compute numerical values of exponential functions, trigonometric functions, logarithms, powers, and roots. By including a sufficient number of terms in the expansion, we can calculate these values to any degree of accuracy that may be required.

It is through such calculations that tables of values can be made, and decimal approximations of numbers such as e and π can be found. Also, many of the values found on a calculator or a computer are calculated by using series expansions that have been programmed into the chip which is in the calculator or computer.

EXAMPLE 1 Exponential value

Calculate the value of $e^{0.1}$.

In order to evaluate $e^{0.1}$, we substitute 0.1 for x in the expansion for e^x. The more terms that are used, the more accurate a value we can obtain. The limit of the partial sums would be the actual value. However, since $e^{0.1}$ is irrational, we cannot express the exact value in decimal form.

Therefore, the value is found as follows:

$$e^x = 1 + x + \frac{x^2}{2!} + \cdots \qquad \text{Eq. (30.6)}$$

$$e^{0.1} = 1 + 0.1 + \frac{(0.1)^2}{2} + \cdots \qquad \text{substitute 0.1 for } x$$

$$= 1.105 \qquad \text{using 3 terms}$$

Using a calculator, we find that $e^{0.1} = 1.105\ 170\ 918$, which shows that our answer is valid to the accuracy shown.

EXAMPLE 2 Trigonometric value

Calculate the value of sin 2°.

In finding trigonometric values, we must be careful to ***express the angle in radians***. Thus, the value of sin 2° is found as follows:

$$\sin x = x - \frac{x^3}{3!} + \cdots \qquad \text{Eq. (30.7)}$$

$$\sin 2^\circ = \left(\frac{\pi}{90}\right) - \frac{(\pi/90)^3}{6} + \cdots \qquad 2^\circ = \frac{\pi}{90}\text{ rad}$$

$$= 0.034\ 899\ 496\ 3 \qquad \text{using 2 terms}$$

A calculator gives the value 0.034 899 496 7. Here, we note that the second term is much smaller than the first. In fact, a good approximation of 0.0349 can be found by using just one term. We now see that $\sin\theta \approx \theta$ for small values of θ, as we noted in Section 8.4.

EXAMPLE 3 Trigonometric value

Calculate the value of cos 0.5429.

Since the angle is expressed in radians, we have

$$\cos 0.5429 = 1 - \frac{0.5429^2}{2} + \frac{0.5429^4}{4!} - \cdots \quad \text{using Eq. (30.8)}$$

$$= 0.856\ 249\ 5 \quad \text{using 3 terms}$$

A calculator shows that cos 0.5429 = 0.856 214 082 4. Since the angle is not small, additional terms are needed to obtain this accuracy. With one more term, the value 0.856 213 9 is obtained.

Practice Exercise

1. Using two terms of the appropriate series, calculate the value of cos 2°.

EXAMPLE 4 Logarithmic value

Calculate the value of ln 1.2.

$$\ln(1 + x) = x - \frac{x^2}{2} + \frac{x^3}{3} - \cdots \quad \text{Eq. (30.9)}$$

$$\ln 1.2 = \ln(1 + 0.2)$$

$$= 0.2 - \frac{(0.2)^2}{2} + \frac{(0.2)^3}{3} - \cdots = 0.1827$$

To four significant digits, ln (1.2) = 0.1823. One more term is required to obtain this accuracy.

We now illustrate the use of series in error calculations and measurement approximations. We also discussed these as applications of differentials. A series solution allows as close a value of the calculated error or approximation as needed, whereas with differentials, the calculation is limited to one term.

EXAMPLE 5 Approximation of error—velocity of a falling object

The velocity v of an object that has fallen h m is $v = 4.43\sqrt{h}$. Find the approximate error in calculating the velocity of an object that has fallen 100.0 m with a possible error of 2.0 m.

If we ***let*** $\boldsymbol{v = 4.43\sqrt{100.0 + x}}$***, where x is the error in h***, we may express v as a Maclaurin expansion in x:

$$f(x) = 4.43(100.0 + x)^{1/2} \qquad f(0) = 44.3$$

$$f'(x) = 2.22(100.0 + x)^{-1/2} \qquad f'(0) = 0.222$$

$$f''(x) = -1.11(100.0 + x)^{-3/2} \qquad f''(0) = -0.001\ 11$$

Therefore,

$$v = 4.43\sqrt{100.0 + x} = 4.43 + 0.222x - 0.000\ 56x^2 + \cdots$$

Since the calculated value of v for $x = 0$ is 44.3, the error E in the value of v is

$$E = 0.222x - 0.000\ 56x^2 + \cdots$$

Calculating, the error for $x = 2.0$ is

$$E = 0.222(2.0) - 0.000\ 56(4.0) = 0.444 - 0.002 = 0.442 \text{ m/s}$$

The value 0.444 is that which is found using differentials. The additional terms are corrections to this term. The additional term in this case shows that the first term is a good approximation to the error. Although this problem can be done numerically, a series solution allows us to find the error for any value of x.

EXAMPLE 6 Approximation of a tangent to earth's surface

From a point on the surface of Earth, a laser beam is aimed tangentially toward a vertical rod 2 km distant. How far up on the rod does the beam touch? (Assume Earth is a perfect sphere of radius 6400 km.)

From Fig. 30.8, we see that

$$x = 6400 \sec \theta - 6400$$

Finding the series for $\sec \theta$, we have

$$f(\theta) = \sec \theta \qquad f(0) = 1$$
$$f'(\theta) = \sec \theta \tan \theta \qquad f'(0) = 0$$
$$f''(\theta) = \sec^3 \theta + \sec \theta \tan^2 \theta \qquad f''(0) = 1$$

Thus, the first two nonzero terms are $\sec \theta = 1 + (\theta^2/2)$. Therefore,

$$x = 6400(\sec \theta - 1)$$
$$= 6400\left(1 + \frac{\theta^2}{2} - 1\right) = 3200\,\theta^2$$

The first two terms of the expansion for $\tan \theta$ are $\theta + \theta^3/3$, which means that $\tan \theta \approx \theta$, since θ is small (see Section 8.4). From Fig. 30.8, $\tan \theta = 2/6400$, and therefore $\theta = 1/3200$. Therefore, we have

$$x = 3200\left(\frac{1}{3200}\right)^2 = \frac{1}{3200} = 0.0003 \text{ km}$$

This means the 2-km-long beam touches the rod only 30 cm above the surface.

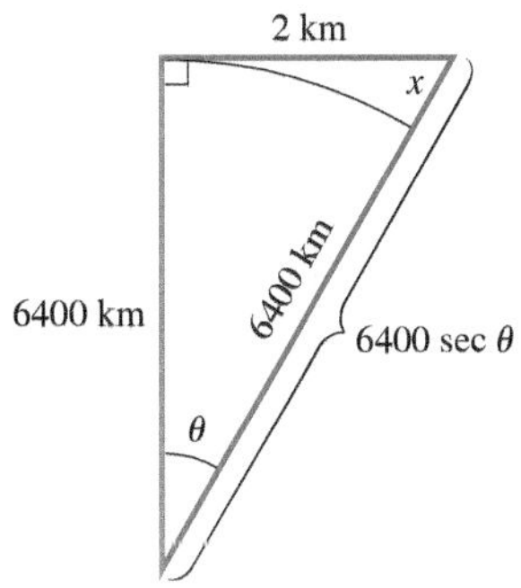

Fig. 30.8

EXERCISES 30.4

In Exercises 1 and 2, make the given changes in the indicated examples of this section, and then solve the resulting problems.

1. In Example 1, change $e^{0.1}$ to $e^{-0.1}$.

2. In Example 4, change ln 1.2 to ln 0.8.

In Exercises 3–20, calculate the value of each of the given functions. Use the indicated number of terms of the appropriate series. Compare with the value found directly on a calculator.

3. $e^{0.2}$ (3) **4.** 1.01^{-1} (4)

5. sin 0.125 (2) **6.** cos 0.05 (2)

7. e (7) **8.** $e^{-0.5}$ (5)

9. $\cos \pi^\circ$ (2) **10.** sin 7.6° (3)

11. ln 1.4 (4) **12.** ln 0.986 (4)

13. sin 0.3625 (3) **14.** cos 1 (4)

15. ln 0.9372 (5) **16.** ln 1.0534 (3)

17. 1.015^6 (3) **18.** 0.9982^8 (3)

19. $1.1^{-0.2}$ (3) **20.** 0.931^{-1} (3)

In Exercises 21–24, calculate the value of each of the given functions. In Exercises 21 and 22, use the expansion for $\sqrt{1+x}$, and in Exercises 23 and 24, use the expansion for $\sqrt[3]{1+x}$. Use three terms of the appropriate series.

21. $\sqrt{1.1076}$ **22.** $\sqrt{0.7915}$

23. $\sqrt[3]{0.9628}$ **24.** $\sqrt[3]{1.1392}$

In Exercises 25–28, calculate the maximum error of the values calculated in the indicated exercises. If a series is alternating (every other term is negative), the maximum possible error in the calculated value is the value of the first term omitted.

25. Exercise 5 **26.** Exercise 4

27. Exercise 9 **28.** Exercise 11

In Exercises 29–40, solve the given problems by using series expansions.

29. Evaluate $\sqrt{3.92}$ by noting that $\sqrt{3.92} = \sqrt{4 - 0.08} = 2\sqrt{1 - 0.02}$.

30. Evaluate sin 32° by first finding the expansion for $\sin(x + \pi/6)$.

31. We can evaluate π by use of $\frac{1}{4}\pi = \tan^{-1}\frac{1}{2} + \tan^{-1}\frac{1}{3}$, along with the series for $\tan^{-1} x$. The first three terms are $\tan^{-1} x = x - \frac{1}{3}x^3 + \frac{1}{5}x^5$. Using these terms, expand $\tan^{-1}\frac{1}{2}$ and $\tan^{-1}\frac{1}{3}$ and approximate the value of π.

32. Use the fact that $\frac{1}{4}\pi = \tan^{-1}\frac{1}{7} + 2\tan^{-1}\frac{1}{3}$ to approximate the value of π. (See Exercise 31.)

33. Explain why $e^x > 1 + x + \frac{1}{2}x^2$ for $x > 0$.

34. Using a calculator, determine how many terms of the expansion for $\ln(1 + x)$ are needed to give the value of ln 1.3 accurate to five decimal places.

35. The time t (in years) for an investment to increase by 10% when the interest rate is 6% is given by $t = \dfrac{\ln 1.1}{0.06}$. Evaluate this expression by using the first four terms of the appropriate series.

36. The period T of a pendulum of length L is given by

$$T = 2\pi\sqrt{\frac{L}{g}}\left(1 + \frac{1}{4}\sin^2\frac{\theta}{2} + \frac{9}{64}\sin^4\frac{\theta}{2} + \cdots\right)$$

where g is the acceleration due to gravity and θ is the maximum angular displacement. If $L = 1.000$ m and $g = 9.800\ \text{m/s}^2$, calculate T for $\theta = 10.0°$ (a) if only one term (the 1) of the series is used and (b) if two terms of the indicated series are used. In the second term, substitute one term of the series for $\sin^2(\theta/2)$.

37. The current in a circuit containing a resistance R, an inductance L, and a battery whose voltage is E is given by the equation $i = \frac{E}{R}(1 - e^{-Rt/L})$, where t is the time. Approximate this expression by using the first three terms of the appropriate exponential series. Under what conditions will this approximation be valid?

38. The image distance q from a certain lens as a function of the object distance p is given by $q = 20p/(p - 20)$. Find the first three nonzero terms of the expansion of the right side. From this expression, calculate q for $p = 2.00$ cm and compare it with the value found by substituting 2.00 in the original expression.

39. At what height above the shoreline of Lake Ontario must an observer be in order to see a point 15 km distant on the surface of the lake? (The radius of the earth is 6400 km.)

40. The efficiency η (in %) of an internal combustion engine in terms of its compression ratio c is given by $\eta = 100(1 - c^{-0.40})$. Determine the possible approximate error in the efficiency for a compression ratio measured to be 6.00 with a possible error of 0.50. (*Hint:* Set up a series for $(6 + x)^{-0.40}$.)

Answer to Practice Exercise

1. $\cos 2° = 0.999\ 390\ 8$

30.5 Taylor Series

To obtain accurate values of a function for values of x that are not close to zero, it is usually necessary to use many terms of a Maclaurin expansion. However, we can use another type of series, called a **Taylor series,** *which is a more general expansion than a Maclaurin expansion.* Also, functions for which a Maclaurin series may not be found may have a Taylor series.

The basic assumption in formulating a Taylor expansion is that a function may be expanded in a polynomial of the form

$$f(x) = c_0 + c_1(x - a) + c_2(x - a)^2 + \cdots \qquad (30.15)$$

Following the same line of reasoning as in deriving the Maclaurin expansion, we may find the constants $c_0, c_1, c_2, \ldots$. That is, derivatives of Eq. (30.15) are taken, and the function and its derivatives are evaluated at $x = a$. This leads to the following expression, which is called the **Taylor series expansion** *of a function.*

Taylor Series

$$f(x) = f(a) + f'(a)(x - a) + \frac{f''(a)(x - a)^2}{2!} + \cdots \qquad (30.16)$$

■ The Taylor series is named for the English mathematician Brook Taylor (1685–1731).

A Taylor series converges rapidly for values of x that are close to a, and this is illustrated in Examples 3 and 4.

EXAMPLE 1 Taylor series for e^x

Expand $f(x) = e^x$ in a Taylor series with $a = 1$.

$$f(x) = e^x \qquad f(1) = e \qquad \text{find derivatives and evaluate each at } x = 1$$
$$f'(x) = e^x \qquad f'(1) = e$$
$$f''(x) = e^x \qquad f''(1) = e$$
$$f'''(x) = e^x \qquad f'''(1) = e$$

$$f(x) = e + e(x - 1) + e\frac{(x - 1)^2}{2!} + e\frac{(x - 1)^3}{3!} + \cdots \qquad \text{using Eq. (30.16)}$$

$$e^x = e\left[1 + (x - 1) + \frac{(x - 1)^2}{2} + \frac{(x - 1)^3}{6} + \cdots\right]$$

This series can be used in evaluating e^x for values of x near 1.

Practice Exercise

1. Expand $f(x) = e^x$ in a Taylor series with $a = 3$.

EXAMPLE 2 Taylor series for $\sqrt{x}$

Expand $f(x) = \sqrt{x}$ in powers of $(x - 4)$.

Another way of stating this is to find the Taylor series for $f(x) = \sqrt{x}$, with $a = 4$. Thus,

$$f(x) = x^{1/2} \qquad f(4) = 2 \qquad \text{find derivatives and evaluate each at } x = 4$$

$$f'(x) = \frac{1}{2x^{1/2}} \qquad f'(4) = \frac{1}{4}$$

$$f''(x) = -\frac{1}{4x^{3/2}} \qquad f''(4) = -\frac{1}{32}$$

$$f'''(x) = \frac{3}{8x^{5/2}} \qquad f'''(4) = \frac{3}{256}$$

$$f(x) = 2 + \frac{1}{4}(x - 4) - \frac{1}{32}\frac{(x - 4)^2}{2!} + \frac{3}{256}\frac{(x - 4)^3}{3!} - \cdots \qquad \text{using Eq. (30.16)}$$

$$\sqrt{x} = 2 + \frac{(x - 4)}{4} - \frac{(x - 4)^2}{64} + \frac{(x - 4)^3}{512} - \cdots$$

This series would be used to evaluate square roots of numbers near 4.

Fig. 30.9 shows the graphs of $y = \sqrt{x}$ and $y = 1 + x/4$ (the first two terms of the series, and the linearization of the function at $x = 4$). We see that each curve passes through $(4, 2)$, and they have nearly equal values of y for values of x near 4.

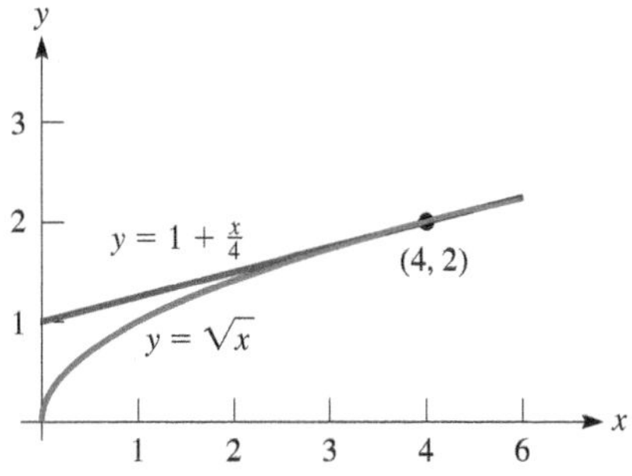

Fig. 30.9

In the last section, we evaluated functions by using Maclaurin series. In the following examples, we use Taylor series to evaluate functions.

EXAMPLE 3 Evaluating a square root using Taylor series

By using Taylor series, evaluate $\sqrt{4.5}$.

Using the four terms of the series found in Example 2, we have

$$\sqrt{4.5} = 2 + \frac{(4.5 - 4)}{4} - \frac{(4.5 - 4)^2}{64} + \frac{(4.5 - 4)^3}{512} \qquad \text{substitute 4.5 for } x$$

$$= 2 + \frac{(0.5)}{4} - \frac{(0.5)^2}{64} + \frac{(0.5)^3}{512}$$

$$= 2.121\,337\,891$$

The value found directly on a calculator is 2.121 320 344. Therefore, the value found by these terms of the series expansion is correct to four decimal places.

LEARNING TIP

In Example 3, we see that successive terms become small rapidly. If a value of x is chosen such that $x - a$ is larger, the successive terms may not become small rapidly, and many terms may be required. Therefore, ***we should choose the value of a as conveniently close as possible to the x-values that will be used.*** Also, we should note that a Maclaurin expansion for $\sqrt{x}$ cannot be used since the derivatives of $\sqrt{x}$ are not defined for $x = 0$.

EXAMPLE 4 Evaluating a sine value using Taylor series

Calculate the approximate value of sin 29° by using three terms of the appropriate Taylor expansion.

The *taylor* feature on some graphing utilities can be used to find the Taylor expansion of a function.

Since the value of sin 30° is known to be $\frac{1}{2}$, we let $a = \frac{\pi}{6}$ (remember, we must use values expressed in radians) when we evaluate the expansion for $x = 29°$. When expressed in radians, the quantity $(x - a)$ is $-\frac{\pi}{180}$ (equivalent to $-1°$). This means that its numerical values are small and become smaller when it is raised to higher powers. Therefore,

$$f(x) = \sin x \qquad f\left(\frac{\pi}{6}\right) = \frac{1}{2} \qquad \text{find derivatives and evaluate each at } x = \frac{\pi}{6}$$

$$f'(x) = \cos x \qquad f'\left(\frac{\pi}{6}\right) = \frac{\sqrt{3}}{2}$$

$$f''(x) = -\sin x \qquad f''\left(\frac{\pi}{6}\right) = -\frac{1}{2}$$

$$f(x) = \frac{1}{2} + \frac{\sqrt{3}}{2}\left(x - \frac{\pi}{6}\right) - \frac{1}{4}\left(x - \frac{\pi}{6}\right)^2 - \cdots \qquad \text{using Eq. (30.16)}$$

$$\sin x = \frac{1}{2} + \frac{\sqrt{3}}{2}\left(x - \frac{\pi}{6}\right) - \frac{1}{4}\left(x - \frac{\pi}{6}\right)^2 - \cdots \qquad f(x) = \sin x$$

$$\sin 29° = \sin\left(\frac{\pi}{6} - \frac{\pi}{180}\right) \qquad 29° = 30° - 1° = \frac{\pi}{6} - \frac{\pi}{180}$$

$$= \frac{1}{2} + \frac{\sqrt{3}}{2}\left(\frac{\pi}{6} - \frac{\pi}{180} - \frac{\pi}{6}\right) - \frac{1}{4}\left(\frac{\pi}{6} - \frac{\pi}{180} - \frac{\pi}{6}\right)^2 - \cdots \qquad \text{substitute } \frac{\pi}{6} - \frac{\pi}{180} \text{ for } x$$

$$= \frac{1}{2} + \frac{\sqrt{3}}{2}\left(-\frac{\pi}{180}\right) - \frac{1}{4}\left(-\frac{\pi}{180}\right)^2 - \cdots$$

$$= 0.484\,808\,850\,9$$

The value found directly on a calculator is 0.484 809 620 2.

EXERCISES 30.5

In Exercises 1 and 2, make the given changes in the indicated examples of this section, and then solve the resulting problems.

1. In Example 2, change $(x - 4)$ to $(x - 1)$.
2. In Example 4, change sin 29° to sin 31°.

In Exercises 3–10, evaluate the given functions by using the series developed in the examples of this section.

3. $e^{1.2}$ 4. $e^{0.7}$ 5. $\sqrt{4.2}$ 6. $\sqrt{3.5}$
7. sin 32° 8. sin 28° 9. sin 29.53° 10. $\sqrt{3.8527}$

In Exercises 11–22, find the first three nonzero terms of the Taylor expansion for the given function and given value of a.

11. e^{-x} $(a = 2)$
12. $\cos x$ $\left(a = \frac{\pi}{4}\right)$
13. $\sin x$ $\left(a = \frac{\pi}{3}\right)$
14. $\ln x$ $(a = 3)$
15. $\sqrt[3]{x}$ $(a = 8)$
16. $\frac{1}{x}$ $(a = 2)$
17. $\tan x$ $\left(a = \frac{\pi}{4}\right)$
18. $\ln \sin x$ $\left(a = \frac{\pi}{2}\right)$
19. $e^x \sin x$ $\left(a = \frac{\pi}{2}\right)$
20. xe^{-x} $(a = -1)$
21. $\frac{1}{x+2}$ $(a = 3)$
22. $\frac{1}{(1+x)^2}$ $(a = -2)$

In Exercises 23–30, evaluate the given functions by using three terms of the appropriate Taylor series. Compare with the value found directly on a calculator.

23. e^{π} 24. ln 3.089 25. $\sqrt{9.28}$ 26. 2.034^{-1}
27. $\sqrt[3]{8.3}$ 28. tan 46° 29. sin 61° 30. cos 42°

In Exercises 31–38, solve the given problems.

31. By completing the steps indicated before Eq. (30.16) in the text, complete the derivation of Eq. (30.16).
32. Find the first three terms of the Taylor expansion of $f(x) = \ln x$ with $a = 1$. Compare this Taylor expansion with the linearization $L(x)$ of $f(x)$ with $a = 1$. Compare the graphs of $f(x)$, $L(x)$, and the Taylor expansion on a graphing utility.
33. Show that the polynomial $2x^3 + x^2 - 3x + 5$ can be written as $2(x - 1)^3 + 7(x - 1)^2 + 5(x - 1) + 5$.
34. Calculate $\sqrt{3}$ using the series in Example 2 and compare with the value using the series in Exercise 1. Which is the better approximation?

35. Calculate sin 31° by using three terms of the Maclaurin expansion for $\sin x$. Also, calculate sin 31° by using three terms of the Taylor expansion in Example 4 (see Exercise 2). Compare the accuracy of the values obtained with that found directly on a calculator.

36. Referring to Eq. (30.16), show that a Taylor series can be expressed in the form

$$f(a+h) = f(a) + f'(a)h + \frac{f''(a)}{2!}h^2 + \cdots$$

37. The current i in a certain electric circuit is $i = 6 \sin \pi t$. Write the first three terms of the Taylor series of this function about $t = \pi/2$.

38. In the analysis of the electric potential of an electric charge distributed along a straight wire of length L, the expression $\ln \frac{x+L}{x}$ is used. Find three terms of the Taylor expansion of this expression in powers of $(x - L)$.

In Exercises 39–42, use a graphing utility to display (a) the function in the indicated exercise of this set and (b) the first two terms of the Taylor series found for that exercise in the same display. Describe how closely the graph in part (b) fits the graph in part (a). Use the given values of x as limits for the display settings.

39. Exercise 13: $(\sin x)$, $x = 0$ to $x = 2$

40. Exercise 15: $(\sqrt[3]{x})$, $x = 0$ to $x = 16$

41. Exercise 16: $(1/x)$, $x = 0$ to $x = 4$

42. Exercise 17: $(\tan x)$, $x = 0$ to $x = 1.5$

Answer to Practice Exercise

1. $f(x) = e^3\left[1 + (x-3) + \frac{(x-3)^2}{2} + \frac{(x-3)^3}{6} + \cdots\right]$

30.6 Introduction to Fourier Series

■ The Fourier series is named for the French mathematician and physicist Jean Baptiste Joseph Fourier (1768–1830).

Many problems encountered in the various fields of science and technology involve functions that are periodic. *A periodic function is one for which $F(x + P) = F(x)$, where P is the period.* We noted that the trigonometric functions are periodic when we discussed their graphs in Chapter 10. There are numerous applied problems that involve periodic functions, including alternating-current voltages, mechanical oscillations, and sound waves. The main focus of this section is to show how periodic functions like these, which often have complicated waveforms, can be expressed as an ***infinite sum of sine and cosine waves***, called a **Fourier series**.

We will begin by discussing how to find a Fourier series for a function with a period of 2π. (In Section 30.7, we will show how to find a Fourier series for a function with a period different from 2π.) Since both $\sin(nx)$ and $\cos(nx)$ repeat every 2π units for integer values of n, a function with a period of 2π can be represented by a series of the following form, where a_n and b_n are constant coefficients.

Fourier Series with Period 2π

$$f(x) = a_0 + a_1\cos x + a_2 \cos 2x + \cdots + a_n \cos nx + \cdots$$
$$+ b_1 \sin x + b_2 \sin 2x + \cdots + b_n \sin nx + \cdots \quad \textbf{(30.17)}$$

Although we will not show the derivation here (details are provided online), the coefficients are given by the following integrals.

■ For a complete derivation of the Fourier series coefficients, please visit MyMathLab.

$$a_0 = \frac{1}{2\pi}\int_{-\pi}^{\pi} f(x)\,dx \quad \textbf{(30.18)}$$

$$a_n = \frac{1}{\pi}\int_{-\pi}^{\pi} f(x)\cos nx\,dx \quad \textbf{(30.19)}$$

$$b_n = \frac{1}{\pi}\int_{-\pi}^{\pi} f(x)\sin nx\,dx \quad \textbf{(30.20)}$$

By finding the values of these coefficients and inserting them into Eq. (30.17), we obtain the Fourier series for a given function.

If we use some, but not all, of the terms of the series, we will get an approximation of the function. The more terms we use, the better the approximation will be. Fourier series provide us a good approximation over a greater interval than Maclaurin and Taylor series, which are only accurate near specific values. With a reasonable number of terms, a Fourier series can accurately represent a function throughout its entire domain.

The following examples illustrate the method of finding Fourier series for various functions.

EXAMPLE 1 Fourier series for a square wave function

LEARNING TIP

When $f(x)$ is defined differently for the intervals of x indicated, ***it requires two integrals for each coefficient.***

Find the Fourier series for the square wave function

$$f(x) = \begin{cases} -1 & -\pi \le x < 0 \\ 1 & 0 \le x < \pi \end{cases}$$

(Many of the functions we shall expand in Fourier series are discontinuous (not continuous) like this one. See Section 23.1 for a discussion of continuity.)

using Eq. (30.18)

$$a_0 = \frac{1}{2\pi}\int_{-\pi}^{0}(-1)\,dx + \frac{1}{2\pi}\int_{0}^{\pi}(1)\,dx = -\frac{x}{2\pi}\bigg|_{-\pi}^{0} + \frac{x}{2\pi}\bigg|_{0}^{\pi} = -\frac{1}{2} + \frac{1}{2} = 0$$

using Eq. (30.19)

$$a_n = \frac{1}{\pi}\int_{-\pi}^{0}(-1)\cos nx\,dx + \frac{1}{\pi}\int_{0}^{\pi}(1)\cos nx\,dx = -\frac{1}{n\pi}\sin nx\,\Big|_{-\pi}^{0} + \frac{1}{n\pi}\sin nx\Big|_{0}^{\pi} = 0 + 0 = 0$$

for all values of n, since $\sin n\pi = 0$;

using Eq. (30.20) with $n = 1$

$$b_1 = \frac{1}{\pi}\int_{-\pi}^{0}(-1)\sin x\,dx + \frac{1}{\pi}\int_{0}^{\pi}(1)\sin x\,dx = \frac{1}{\pi}\cos x\,\Big|_{-\pi}^{0} - \frac{1}{\pi}\cos x\Big|_{0}^{\pi}$$

$$= \frac{1}{\pi}(1+1) - \frac{1}{\pi}(-1-1) = \frac{4}{\pi}$$

using Eq. (30.20) with $n = 2$

$$b_2 = \frac{1}{\pi}\int_{-\pi}^{0}(-1)\sin 2x\,dx + \frac{1}{\pi}\int_{0}^{\pi}(1)\sin 2x\,dx = \frac{1}{2\pi}\cos 2x\,\Big|_{-\pi}^{0} - \frac{1}{2\pi}\cos 2x\Big|_{0}^{\pi}$$

$$= \frac{1}{2\pi}(1-1) - \frac{1}{2\pi}(1-1) = 0$$

using Eq. (30.20) with $n = 3$

$$b_3 = \frac{1}{\pi}\int_{-\pi}^{0}(-1)\sin 3x\,dx + \frac{1}{\pi}\int_{0}^{\pi}(1)\sin 3x\,dx = \frac{1}{3\pi}\cos 3x\,\Big|_{-\pi}^{0} - \frac{1}{3\pi}\cos 3x\Big|_{0}^{\pi}$$

$$= \frac{1}{3\pi}(1+1) - \frac{1}{3\pi}(-1-1) = \frac{4}{3\pi}$$

In general, if n is even, $b_n = 0$, and if n is odd, then $b_n = 4/n\pi$. Therefore,

$$f(x) = \frac{4}{\pi}\sin x + \frac{4}{3\pi}\sin 3x + \frac{4}{5\pi}\sin 5x + \cdots = \frac{4}{\pi}\left(\sin x + \frac{1}{3}\sin 3x + \frac{1}{5}\sin 5x + \cdots\right)$$

A graph of the function as defined, and the curve found by using the first three terms of the Fourier series, are shown in Fig. 30.10.

Since functions found by Fourier series have a period of 2π, they can represent functions with this period. If the function $f(x)$ were defined to be periodic with period 2π, with the same definitions as originally indicated, we would graph the function as shown in Fig. 30.11. The Fourier series representation would follow it as in Fig. 30.10. If more terms were used, the fit would be closer.

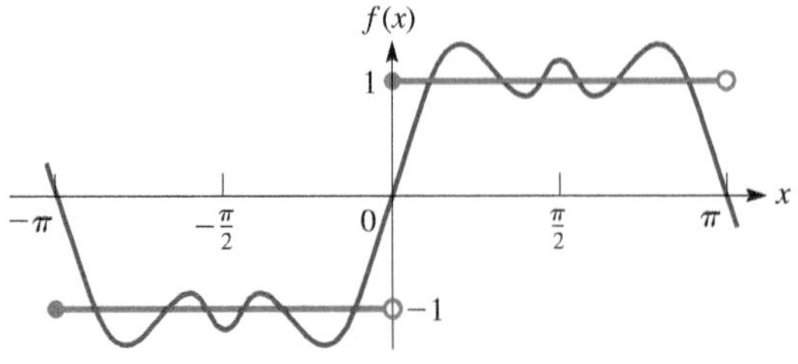

Fig. 30.10

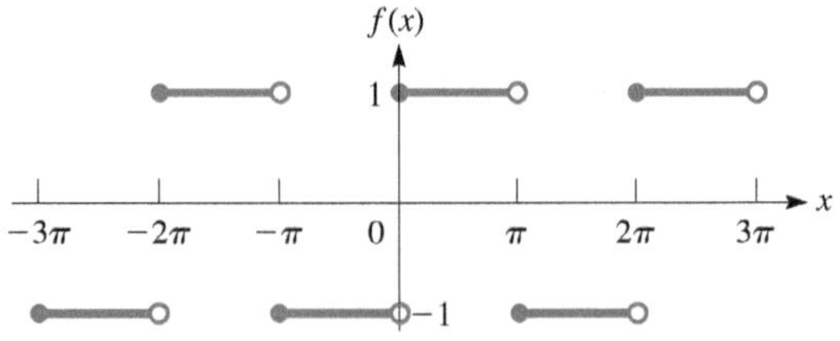

Fig. 30.11

EXAMPLE 2 Finding a Fourier series

Find the Fourier series for the function

$$f(x) = \begin{cases} 1 & -\pi \le x < 0 \\ x & 0 \le x < \pi \end{cases}$$

For the periodic function, let $f(x + 2\pi) = f(x)$ for all x.

A graph of three periods of this function is shown in Fig. 30.12.

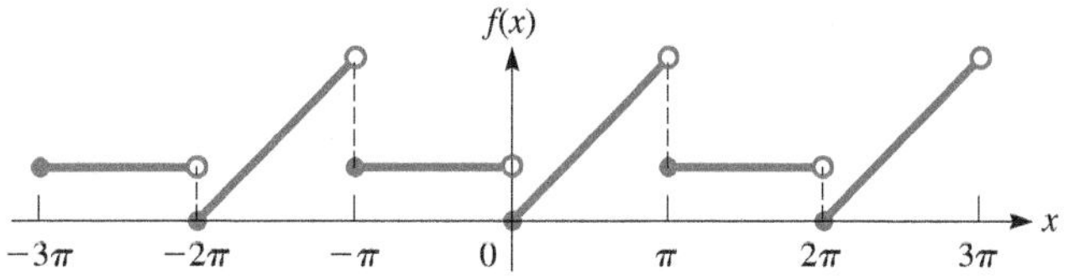

Fig. 30.12

Now, finding the coefficients, we have

$$a_0 = \frac{1}{2\pi}\int_{-\pi}^{0} dx + \frac{1}{2\pi}\int_{0}^{\pi} x\,dx = \frac{x}{2\pi}\Big|_{-\pi}^{0} + \frac{x^2}{4\pi}\Big|_{0}^{\pi}$$ using Eq. (30.18)

$$= \frac{1}{2} + \frac{\pi}{4} = \frac{2+\pi}{4}$$

$$a_1 = \frac{1}{\pi}\int_{-\pi}^{0} \cos x\,dx + \frac{1}{\pi}\int_{0}^{\pi} x\cos x\,dx$$ using Eq. (30.19) with $n = 1$

$$= \frac{1}{\pi}\sin x\,\Big|_{-\pi}^{0} + \frac{1}{\pi}(\cos x + x\sin x)\,\Big|_{0}^{\pi} = -\frac{2}{\pi}$$

$$a_2 = \frac{1}{\pi}\int_{-\pi}^{0} \cos 2x\,dx + \frac{1}{\pi}\int_{0}^{\pi} x\cos 2x\,dx$$ using Eq. (30.19) with $n = 2$

$$= \frac{1}{2\pi}\sin 2x\,\Big|_{-\pi}^{0} + \frac{1}{4\pi}(\cos 2x + 2x\sin 2x)\Big|_{0}^{\pi} = 0$$

$$a_3 = \frac{1}{\pi}\int_{-\pi}^{0} \cos 3x\,dx + \frac{1}{\pi}\int_{0}^{\pi} x\cos 3x\,dx$$ using Eq. (30.19) with $n = 3$

$$= \frac{1}{3\pi}\sin 3x\,\Big|_{-\pi}^{0} + \frac{1}{9\pi}(\cos 3x + 3x\sin 3x)\,\Big|_{0}^{\pi} = -\frac{2}{9\pi}$$

$$b_1 = \frac{1}{\pi}\int_{-\pi}^{0} \sin x\,dx + \frac{1}{\pi}\int_{0}^{\pi} x\sin x\,dx$$ using Eq. (30.20) with $n = 1$

$$= -\frac{1}{\pi}\cos x\,\Big|_{-\pi}^{0} + \frac{1}{\pi}(\sin x - x\cos x)\,\Big|_{0}^{\pi} = \frac{\pi - 2}{\pi}$$

$$b_2 = \frac{1}{\pi}\int_{-\pi}^{0} \sin 2x\,dx + \frac{1}{\pi}\int_{0}^{\pi} x\sin 2x\,dx$$ using Eq. (30.20) with $n = 2$

$$= -\frac{\cos 2x}{2\pi}\Big|_{-\pi}^{0} + \frac{\sin 2x - 2x\cos 2x}{4\pi}\Big|_{0}^{\pi} = -\frac{1}{2}$$

Therefore, the first few terms of the Fourier series are

$$f(x) = \frac{2+\pi}{4} - \frac{2}{\pi}\cos x - \frac{2}{9\pi}\cos 3x - \cdots + \left(\frac{\pi - 2}{\pi}\right)\sin x - \frac{1}{2}\sin 2x + \cdots$$

EXAMPLE 3 Fourier series for a half-wave rectifier

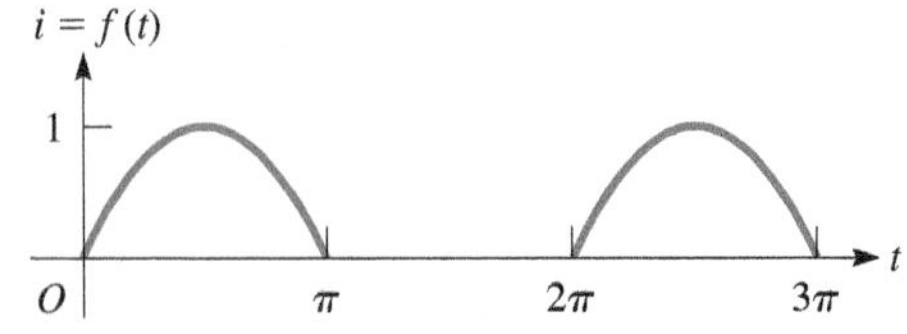

Fig. 30.13

Certain electronic devices allow an electric current to pass through in only one direction. When an alternating current is applied to the circuit, the current exists for only half the cycle. Figure 30.13 is a representation of such a current as a function of time. This type of electronic device is called a *half-wave rectifier.* Derive the Fourier series for a rectified wave for which half is defined by $f(t) = \sin t\ (0 \leq t \leq \pi)$ and for which the other half is defined by $f(t) = 0$.

In finding the Fourier coefficients, we first find a_0 as

$$a_0 = \frac{1}{2\pi}\int_0^{\pi} \sin t\, dt = \frac{1}{2\pi}(-\cos t)\Big|_0^{\pi} = \frac{1}{2\pi}(1 + 1) = \frac{1}{\pi}$$

In the previous example, we evaluated each of the coefficients individually. Here, we show how to set up a general expression for a_n and another for b_n. Once we have determined these, we can substitute values of n in the formula to obtain the individual coefficients:

$$a_n = \frac{1}{\pi}\int_0^{\pi} \sin t \cos nt\, dt = -\frac{1}{2\pi}\left[\frac{\cos(1-n)t}{1-n} + \frac{\cos(1+n)t}{1+n}\right]_0^{\pi}$$

$$= -\frac{1}{2\pi}\left[\frac{\cos(1-N)\pi}{1-n} + \frac{\cos(1+N)\pi}{1+n} - \frac{1}{1-n} - \frac{1}{1+n}\right]$$

Use this formula: $\int \sin au \cos bu\, du = -\frac{\cos(a-b)u}{2(a-b)} - \frac{\cos(a+b)u}{2(a+b)}$. It is valid for all values of n except $n = 1$. Now, we write

$$a_1 = \frac{1}{\pi}\int_0^{\pi} \sin t \cos t\, dt = \frac{1}{2\pi}\sin^2 t\Big|_0^{\pi} = 0$$

$$a_2 = -\frac{1}{2\pi}\left(\frac{-1}{-1} + \frac{-1}{3} - \frac{1}{-1} - \frac{1}{3}\right) = -\frac{2}{3\pi}$$

$$a_3 = -\frac{1}{2\pi}\left(\frac{1}{-2} + \frac{1}{4} - \frac{1}{-2} - \frac{1}{4}\right) = 0$$

$$a_4 = -\frac{1}{2\pi}\left(\frac{-1}{-3} + \frac{-1}{5} - \frac{1}{-3} - \frac{1}{5}\right) = -\frac{2}{15\pi}$$

$$b_n = \frac{1}{\pi}\int_0^{\pi} \sin t \sin nt\, dt = \frac{1}{2\pi}\left[\frac{\sin(1-n)t}{1-n} - \frac{\sin(1+n)t}{1+n}\right]_0^{\pi}$$

$$= -\frac{1}{2\pi}\left[\frac{\sin(1-N)\pi}{1-n} - \frac{\sin(1+N)\pi}{1+n}\right]$$

Use this formula: $\int \sin au \sin bu\, du = -\frac{\sin(a-b)u}{2(a-b)} - \frac{\sin(a+b)u}{2(a+b)}$. It is valid for all values of n except $n = 1$.

Therefore, we have

$$b_1 = \frac{1}{\pi}\int_0^{\pi} \sin t \sin t\, dt = \frac{1}{\pi}\int_0^{\pi} \sin^2 t\, dt = \frac{1}{2\pi}(t - \sin t \cos t)\Big|_0^{\pi} = \frac{1}{2}$$

We see that $b_n = 0$ if $n > 1$, since each is evaluated in terms of the sine of a multiple of π.

Therefore, the Fourier series for the rectified wave is

$$f(t) = \frac{1}{\pi} + \frac{1}{2}\sin t - \frac{2}{\pi}\left(\frac{1}{3}\cos 2t + \frac{1}{15}\cos 4t + \cdots\right)$$

The graph of these terms of the Fourier series and the original function are shown in Fig. 30.14.

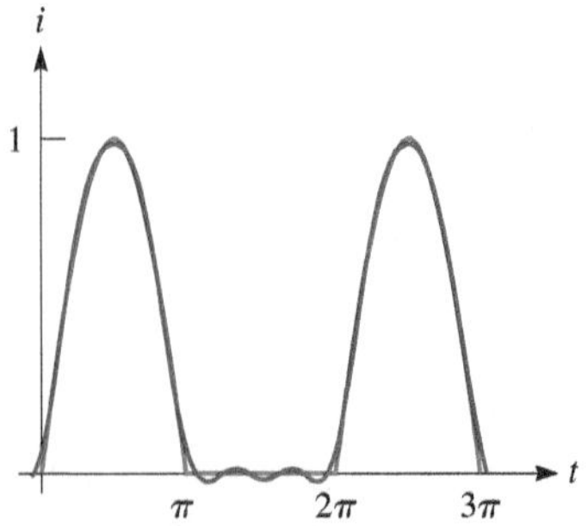

Fig. 30.14

All the types of periodic functions included in this section (as well as many others) may actually be seen on an oscilloscope when the proper signal is sent into it. In this way, the oscilloscope may be used to analyse the periodic nature of such phenomena as sound waves and electric currents.

EXERCISES 30.6

In Exercises 1 and 2, make the given changes in Example 1 of this section and then find the resulting Fourier series.

1. Change the -1 to -2, and the 1 to 2.

2. Change the -1 to 0.

In Exercises 3–14, find at least three nonzero terms (including a_0 and at least two cosine terms and two sine terms if they are not all zero) of the Fourier series for the given periodic functions and sketch at least three periods of the function.

3. $f(x) = \begin{cases} 1 & -\pi \le x < 0 \\ 0 & 0 \le x < \pi \end{cases}$

4. $f(x) = \begin{cases} 0 & -\pi \le x < -\frac{\pi}{2}, \frac{\pi}{2} \le x < \pi \\ 2 & -\frac{\pi}{2} \le x < \frac{\pi}{2} \end{cases}$

5. $f(x) = \begin{cases} 1 & -\pi \le x < 0 \\ 2 & 0 \le x < \pi \end{cases}$

6. $f(x) = \begin{cases} 0 & -\pi \le x < 0, \dfrac{\pi}{2} < x < \pi \\ 1 & 0 \le x \le \dfrac{\pi}{2} \end{cases}$

7. $f(x) = \begin{cases} 0 & -\pi \le x < 0 \\ x & 0 \le x < \pi \end{cases}$

8. $f(x) = x \quad -\pi \le x < \pi$

9. $f(x) = \begin{cases} -1 & -\pi \le x < 0 \\ 0 & 0 \le x < \dfrac{\pi}{2} \\ 1 & \dfrac{\pi}{2} \le x < \pi \end{cases}$

10. $f(x) = x^2 \quad -\pi \le x < \pi$

11. $f(x) = \begin{cases} -x & -\pi \le x < 0 \\ x & 0 \le x < \pi \end{cases}$

12. $f(x) = \begin{cases} 0 & -\pi \le x < 0 \\ x^2 & 0 \le x < \pi \end{cases}$

13. $f(x) = e^x \quad -\pi \le x < \pi$

14. $f(x) = \begin{cases} \pi + x & -\pi \le x < 0 \\ \pi - x & 0 < x < \pi \end{cases}$

In Exercises 15–20, use a graphing utility to display the terms of the Fourier series given in the indicated example or answer for the indicated exercise. Compare with the sketch of the function. Use Xmin $= -8$ and Xmax $= 8$ for the display settings.

15. Example 1 **16.** Example 2 **17.** Exercise 5

18. Exercise 7 **19.** Exercise 11 **20.** Exercise 10

In Exercises 21–24, solve the given problems.

21. The periodic force F (in N) applied in testing a spring system can be represented by $F = 0$ for $-\pi \le t < 0$ and $F = t^2 + t$ for $0 < t < \pi$, where the time t is in seconds. Find the Fourier series that represents this force.

22. Another representation for a half-wave rectifier (see Example 3) is $f(t) = \cos t\ (-\pi/2 \le t < \pi/2)$, $f(t) = 0$ $(-\pi < t < -\pi/2, \pi/2 < t < \pi)$. Find the Fourier series for this half-wave rectifier.

23. Find the Fourier expansion of the electronic device known as a *full-wave rectifier.* This is found by using as the function for the current $f(t) = -\sin t$ for $-\pi \le t \le 0$ and $f(t) = \sin t$ for $0 < t \le \pi$. The graph of this function is shown in Fig. 30.15. The portion of the curve to the left of the origin is dashed because from a physical point of view we can give no significance to this part of the wave, although mathematically we can derive the proper form of the Fourier expansion by using it.

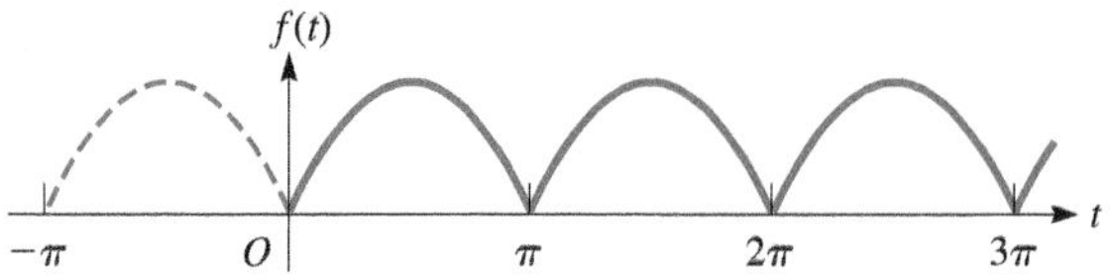

Fig. 30.15

24. The loudness L (in decibels) of a certain siren as a function of time t (in s) can be described by the function

$$\begin{aligned} L &= 0 & -\pi \le t < 0 \\ L &= 100t & 0 \le t < \pi/2 \\ L &= 100(\pi - t) & \pi/2 \le t < \pi \end{aligned}$$

with a period of 2π seconds (where only positive values of t have physical significance). Find a_0, the first nonzero cosine term, and the first two nonzero sine terms of the Fourier expansion for the loudness of the siren. See Fig. 30.16.

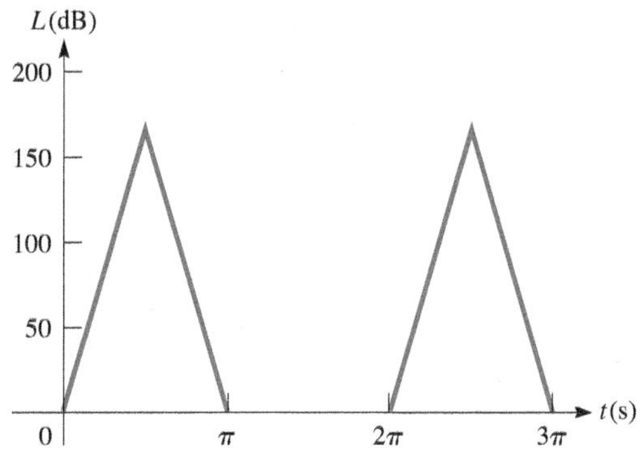

Fig. 30.16

30.7 More About Fourier Series

When finding the Fourier expansion of some functions, it may turn out that all the sine terms evaluate to be zero or that all the cosine terms evaluate to be zero. In fact, in Example 1 of Section 30.6, we see that all of the cosine terms were zero and that the expansion contained only sine terms. We now show how to quickly determine if an expansion will contain only sine terms, or only cosine terms.

EVEN FUNCTIONS AND ODD FUNCTIONS

In Chapter 21 (Section 21.3), we showed that when $-x$ replaces x in a function $f(x)$, and the function does not change, the curve of the function is symmetric to the y-axis. *Such a function is called an* **even function**.

EXAMPLE 1 cos x is an even function

We can show that the function $y = \cos x$ is an even function by using the Maclaurin expansions for $\cos x$ and $\cos(-x)$. These are

$$\cos x = 1 - \frac{x^2}{2} + \frac{x^4}{24} - \cdots$$

$$\cos(-x) = 1 - \frac{(-x)^2}{2} + \frac{(-x)^4}{24} \cdots = 1 - \frac{x^2}{2} + \frac{x^4}{24} - \cdots$$

Because the expansions are the same, $\cos x$ is an even function (and so is $\cos nx$ for all n).

Again referring to Chapter 21, we recall that if $-x$ replaces x and $-y$ replaces y at the same time, and the function does not change, then the function is symmetric to the origin. *Such a function is called an* **odd function**.

Practice Exercises

Determine whether the following functions are even or odd or neither:

1. $f(x) = x^{1/3}$ **2.** $f(x) = \sin^2 x$

EXAMPLE 2 sin x is an odd function

We can show that the function $y = \sin x$ is an odd function by using the Maclaurin expansions for $\sin x$ and $-\sin(-x)$ (the $-$ sign before $\sin(-x)$ is equivalent to making y negative). These are

$$\sin x = x - \frac{x^3}{6} + \frac{x^5}{120} - \cdots$$

$$-\sin(-x) = -\left[(-x) - \frac{(-x)^3}{6} + \frac{(-x)^5}{120} - \cdots\right] = x - \frac{x^3}{6} + \frac{x^5}{120} - \cdots$$

Since $\sin x = -\sin(-x)$, $\sin x$ is an odd function (and so is $\sin nx$ for all n).

LEARNING TIP

- The Fourier series expansion of an even function contains only cosine terms (and possibly a constant term). When finding the Fourier series for such a function, we do not have to find any sine terms.
- The Fourier series expansion of an odd function contains only sine terms (and no constant term). When finding the Fourier series for such a function, we do not have to find any cosine terms.

It follows from the definitions that the product of two odd functions is even and that the product of an even function and an odd function is also odd. Moreover, if $f(x)$ is odd, $\int_{-\pi}^{\pi} f(x) = 0$ (the area from $-\pi$ to 0 cancels out with the area from 0 to π).

These properties combined with the results of Examples 1 and 2 imply that when $f(x)$ is even, the integrand in Eq. (30.20) is odd, and all b_n terms are 0. Similarly, when $f(x)$ is odd, the integrands in Eqs. (30.18) and (30.19) are odd, and all a_n terms are 0. We summarize this in the Learning Tip to the left.

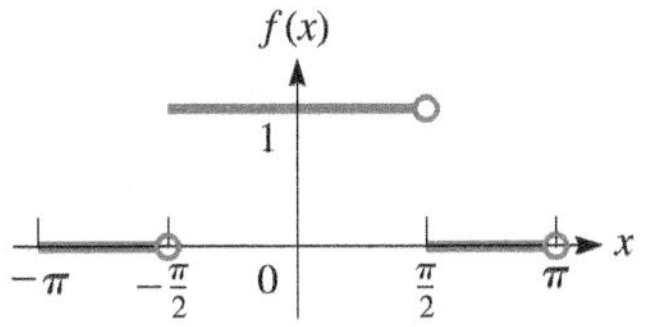

Fig. 30.17

EXAMPLE 3 Fourier series of an even function

The function

$$f(x) = \begin{cases} 0 & -\pi \le x < -\pi/2,\ \pi/2 \le x < \pi \\ 1 & -\pi/2 \le x < \pi/2 \end{cases}$$

is even (the symmetry to the y-axis can be seen in Fig. 30.17). Therefore, its Fourier series expansion contains only cosine terms (and a constant). We find the series to be

$$f(x) = \frac{1}{2} + \frac{2}{\pi}\left(\cos x - \frac{1}{3}\cos 3x + \frac{1}{5}\cos 5x - \cdots\right).$$

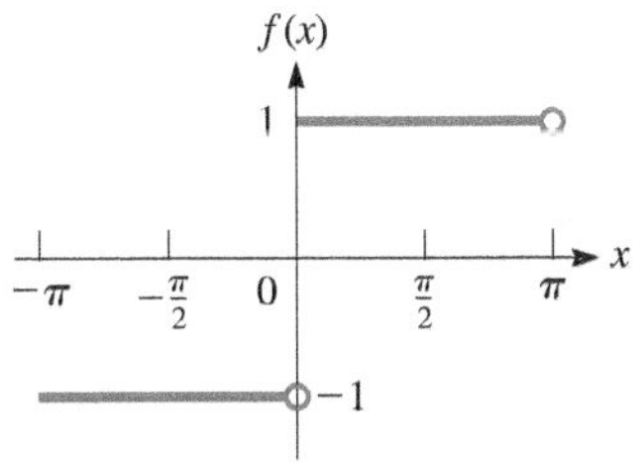

Fig. 30.18

EXAMPLE 4 Fourier series of an odd function

The function

$$f(x) = \begin{cases} -1 & -\pi < x < 0 \\ 1 & 0 \le x < \pi \end{cases}$$

is odd (the symmetry to the origin can be seen in Fig. 30.18). Therefore, its Fourier series expansion contains only sine terms and no constant. As we showed in Example 1, the series is $f(x) = \frac{4}{\pi}\left(\sin x + \frac{1}{3}\sin 3x + \frac{1}{5}\sin 5x + \cdots\right)$.

LEARNING TIP

If a constant k is added to a function $f_1(x)$, the resulting function $f(x)$ is

$$f(x) = k + f_1(x)$$

Therefore, if we know the Fourier series expansion for $f_1(x)$, *the Fourier series expansion of $f(x)$ is found by adding k to the Fourier series expansion of $f_1(x)$.*

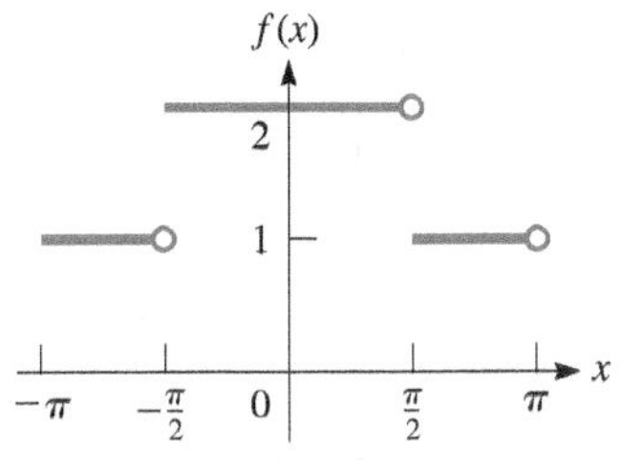

Fig. 30.19

EXAMPLE 5 Constant added to a Fourier series

The values of the function

$$f(x) = \begin{cases} 1 & -\pi \le x < -\pi/2,\ \pi/2 \le x < \pi \\ 2 & -\pi/2 \le x < \pi/2 \end{cases}$$

are all 1 greater than those of the function of Example 3. Therefore, denoting the function of Example 3 as $f_1(x)$, we have $f(x) = 1 + f_1(x)$. This means that the Fourier series for $f(x)$ is

$$\begin{aligned} f(x) &= 1 + \left[\frac{1}{2} + \frac{2}{\pi}\left(\cos x - \frac{1}{3}\cos 3x + \frac{1}{5}\cos 5x - \cdots\right)\right] \\ &= \frac{3}{2} + \frac{2}{\pi}\left(\cos x - \frac{1}{3}\cos 3x + \frac{1}{5}\cos 5x - \cdots\right) \end{aligned}$$

In Fig. 30.19, we see that the graph of $f(x)$ is shifted up vertically by 1 unit from the graph of $f_1(x)$ in Fig. 30.17. This is equivalent to a vertical translation of axes. We also note that $f(x)$ is an even function.

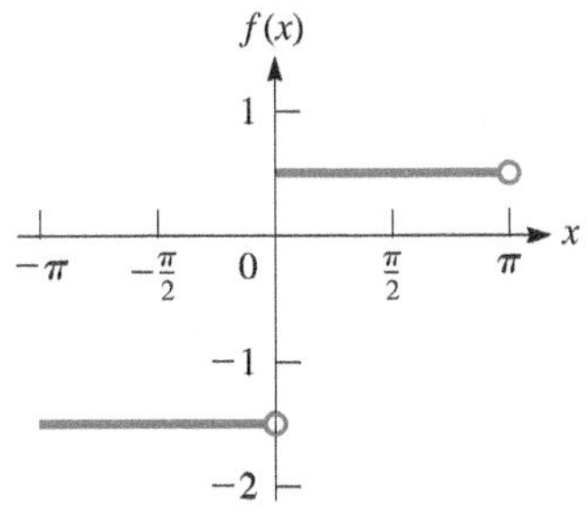

Fig. 30.20

LEARNING TIP

A function that is obtained as a vertical shift of an odd function will no longer be odd. However, its series expansion will still contain only sine terms, as seen in Example 6.

EXAMPLE 6 Constant subtracted from a Fourier series

The values of the function

$$f(x) = \begin{cases} -\frac{3}{2} & -\pi \le x < 0 \\ \frac{1}{2} & 0 \le x < \pi \end{cases}$$

are all $\frac{1}{2}$ less than those of the function of Example 4. Therefore, denoting the function of Example 4 as $f_1(x)$, we have $f(x) = -\frac{1}{2} + f_1(x)$. This means that the Fourier series for $f(x)$ is

$$f(x) = -\frac{1}{2} + \frac{4}{\pi}\left(\sin x - \frac{1}{3}\sin 3x + \frac{1}{5}\sin 5x - \cdots\right)$$

In Fig. 30.20, we see that the graph of $f(x)$ is shifted vertically down by $\frac{1}{2}$ unit from the graph of $f_1(x)$ in Fig. 30.18. Because of the presence of the constant term, $f(x)$ is not an odd function. However, since $f(x)$ is obtained as a shift of an odd function, its series expansion contains only sine terms.

FOURIER SERIES WITH PERIOD 2L

To this point, we have discussed how to find a Fourier series for a function with a period of 2π, defined over the interval $x = -\pi$ to $x = \pi$. We now show how to find such a series for a function with a period of $2L$, defined from $x = -L$ to $x = L$. Since functions of the form $\sin(n\pi x/L)$ and $\cos(n\pi x/L)$ repeat every $2L$ units, the Fourier series will consist of these types of terms as shown below:

Fourier Series with Period 2L

$$f(x) = a_0 + a_1\cos(\pi x/L) + a_2\cos(2\pi x/L) + \cdots + a_n\cos(n\pi x/L) + \cdots$$
$$+ b_1\sin(\pi x/L) + b_2\sin(2\pi x/L) + \cdots + b_n\sin(n\pi x/L) + \cdots \quad \textbf{(30.21)}$$

$$a_0 = \frac{1}{2L}\int_{-L}^{L} f(x)\,dx \quad \textbf{(30.22)}$$

$$a_n = \frac{1}{L}\int_{-L}^{L} f(x)\cos\frac{n\pi x}{L}\,dx \quad \textbf{(30.23)}$$

$$b_n = \frac{1}{L}\int_{-L}^{L} f(x)\sin\frac{n\pi x}{L}\,dx \quad \textbf{(30.24)}$$

Note that the conclusions regarding the Fourier series of odd and even functions also apply when the period is $2L$.

EXAMPLE 7 Fourier series with period of 8—synthesizer square wave

■ See the chapter introduction.

A musical note (B, octave 2) played on a synthesizer has a square wave that is given approximately by the function

$$f(t) = \begin{cases} 0 & -4 \le t < 0 \\ 2 & 0 \le t < 4 \end{cases}$$

where t is in ms and the period is 8 ms. See Fig. 30.21. Find the Fourier series expansion of this function.

Since the period is 8 ms, $L = 4$ ms. Next, we note that $f(t) = 1 + f_1(t)$, where $f_1(t)$ is an odd function (from the definition of $f(t)$, and from Fig. 30.21 we can see the symmetry to the point $(0, 1)$). Therefore, *the constant is* 1 *and there are no cosine terms* in the Fourier series for $f(t)$. Now, finding the sine terms, we have

$$b_n = \frac{1}{4}\int_{-4}^{0} 0\sin\frac{n\pi t}{4}\,dt + \frac{1}{4}\int_{0}^{4} 2\sin\frac{n\pi t}{4}\,dt \qquad \text{using Eq. (30.24)}$$

$$= \frac{1}{2}\left(\frac{4}{n\pi}\right)\int_0^4 \sin\frac{n\pi t}{4}\left(\frac{n\pi\,dt}{4}\right) = -\frac{2}{n\pi}\cos\frac{n\pi t}{4}\bigg|_0^4$$

$$= -\frac{2}{n\pi}(\cos n\pi - \cos 0) = \frac{2}{n\pi}(1 - \cos n\pi)$$

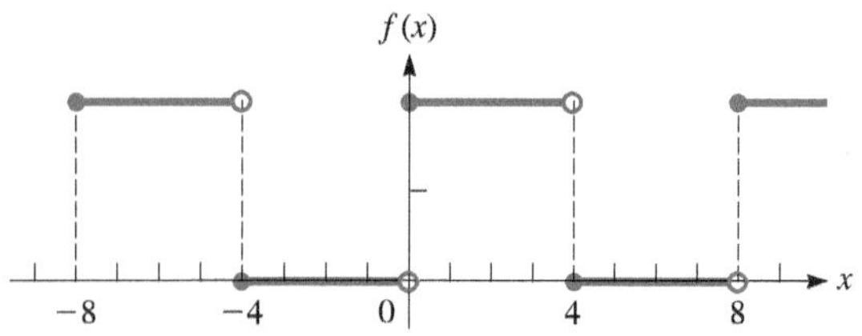

Fig. 30.21

$$b_1 = \frac{2}{\pi}[1 - (-1)] = \frac{4}{\pi} \qquad b_2 = \frac{2}{2\pi}(1 - 1) = 0$$

$$b_3 = \frac{2}{3\pi}[1 - (-1)] = \frac{4}{3\pi} \qquad b_4 = \frac{2}{4\pi}(1 - 1) = 0$$

Therefore, the Fourier series is

$$f(t) = 1 + \frac{4}{\pi}\sin\frac{\pi t}{4} + \frac{4}{3\pi}\sin\frac{3\pi t}{4} + \cdots$$

EXAMPLE 8 Fourier series with period of 2

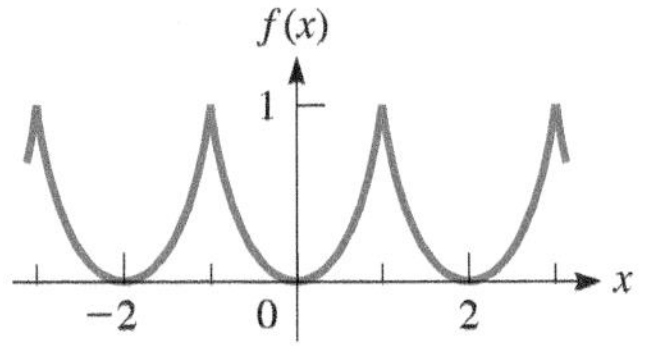

Fig. 30.22

Find the Fourier series for the function

$$f(x) = x^2 \quad -1 \le x < 1$$

for which the period is 2. See Fig. 30.22.

Since the period is 2, $L = 1$. Next, we note that $f(x) = f(-x)$, which means it is an even function. Therefore, there are no sine terms in the Fourier series. Finding the constant and the cosine terms, we have

$$a_0 = \frac{1}{2(1)}\int_{-1}^{1} x^2 dx = \frac{1}{6}x^3\Big|_{-1}^{1} = \frac{1}{6}(1+1) = \frac{1}{3}$$

$$a_n = \frac{1}{1}\int_{-1}^{1} x^2 \cos\frac{n\pi x}{1}\,dx = \int_{-1}^{1} x^2 \cos n\pi x\,dx$$ integrating by parts: $u = x^2, du = 2xdx$, $dv = \cos n\pi x\,dx, v = (1/n\pi)\sin n\pi x$

$$= x^2\left(\frac{1}{n\pi}\sin n\pi x\right)\Big|_{-1}^{1} - \frac{2}{n\pi}\int_{-1}^{1} x \sin n\pi x\,dx$$ integrating by parts: $u = x, du = dx$, $dv = \sin n\pi x\,dx, v = (-1/n\pi)\cos n\pi x$

$\sin n\pi = 0$ $$= \frac{1}{n\pi}\sin n\pi - \frac{1}{n\pi}\sin(-n\pi) - \frac{2}{n\pi}\left[x\left(-\frac{1}{n\pi}\cos n\pi x\right)\Big|_{-1}^{1} - \left(-\frac{1}{n\pi}\int_{-1}^{1}\cos n\pi x\,dx\right)\right]$$

$\sin n\pi = 0$ $$= \frac{2}{n^2\pi^2}\left[\cos n\pi + \cos(-n\pi)\right] + \frac{1}{n^2\pi^2}\sin n\pi x\big|_{-1}^{1} = \frac{2}{n^2\pi^2}(2\cos n\pi) = \frac{4}{n^2\pi^2}\cos n\pi$$

$$a_1 = \frac{4}{\pi^2}\cos\pi = -\frac{4}{\pi^2} \qquad a_2 = \frac{4}{4\pi^2}\cos 2\pi = \frac{4}{4\pi^2} \qquad a_3 = \frac{4}{9\pi^2}\cos 3\pi = -\frac{4}{9\pi^2}$$

Therefore, the Fourier series is

$$f(x) = \frac{1}{3} - \frac{4}{\pi^2}\left(\cos\pi x - \frac{1}{4}\cos 2\pi x + \frac{1}{9}\cos 3\pi x - \cdots\right)$$

Practice Exercise

3. Find the Fourier series for the function $f(x) = x^2 + 2 \quad -1 \le x < 1$.

HALF-RANGE EXPANSIONS

We have seen that the Fourier series expansion for an even function contains only cosine terms (and possibly a constant), and the expansion of an odd function contains only sine terms. It is also possible to force a function to be even or odd, so that the expansion will contain only cosine terms or only sine terms.

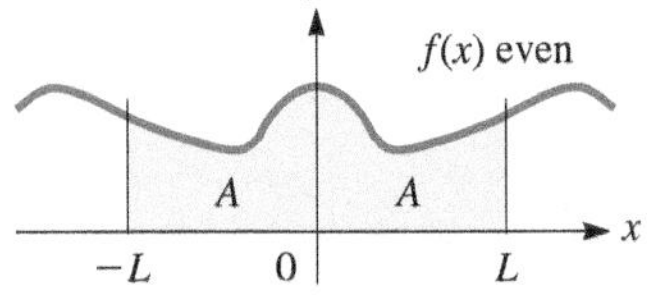

Fig. 30.23

Considering the symmetry of an even function, the area under the curve from $-L$ to 0 is the same as the area under the curve from 0 to L (see Fig. 30.23). This means the value of the integral from $-L$ to 0 equals the value of the integral from 0 to L. Therefore, the value of the integral from $-L$ to L equals twice the value of the integral from 0 to L, or

$$\int_{-L}^{L} f(x)\,dx = 2\int_{0}^{L} f(x)\,dx \qquad f(x) \text{ even}$$

Therefore, to obtain the Fourier coefficients for an expression from $-L$ to L for an even function, we can multiply the coefficients obtained using Eqs. (30.22) and (30.23) from 0 to L by 2. Similar reasoning shows that the Fourier coefficients for an expansion from $-L$ to L for an odd function may be found by multiplying the coefficients obtained using Eq. (30.24) from 0 to L by 2.

A **half-range Fourier cosine series** *is a series that contains only cosine terms,* and *a* **half-range Fourier sine series** *is a series that contains only sine terms.* To find the half-range expansion for a function $f(x)$, it is defined for interval 0 to L (*half* of the interval from $-L$ to L) and then specified as odd or even, thereby clearly defining the function in the interval from $-L$ to 0. This means that *the Fourier coefficients for a half-range cosine series are given by*

$$a_0 = \frac{1}{L}\int_0^L f(x)\,dx \quad \text{and} \quad a_n = \frac{2}{L}\int_0^L f(x)\cos\frac{n\pi x}{L}dx \quad (n = 1, 2, \ldots) \tag{30.25}$$

Similarly, *the Fourier coefficients for a half-range sine series are given by*

$$b_n = \frac{2}{L}\int_0^L f(x)\sin\frac{n\pi x}{L}\,dx \qquad (n = 1, 2, \ldots) \tag{30.26}$$

EXAMPLE 9 Half-range cosine series

Find $f(x) = x$ in a half-range cosine series for $0 \le x < 2$.

Since we are to have a cosine series, we extend the function to be an even function with its graph as shown in Fig. 30.24. The red portion between $x = 0$ and $x = L$ shows the given function as defined, and blue portions show the extension that makes it an even function. Now, by use of Eqs. (30.25), we find the Fourier expansion coefficients, with $L = 2$.

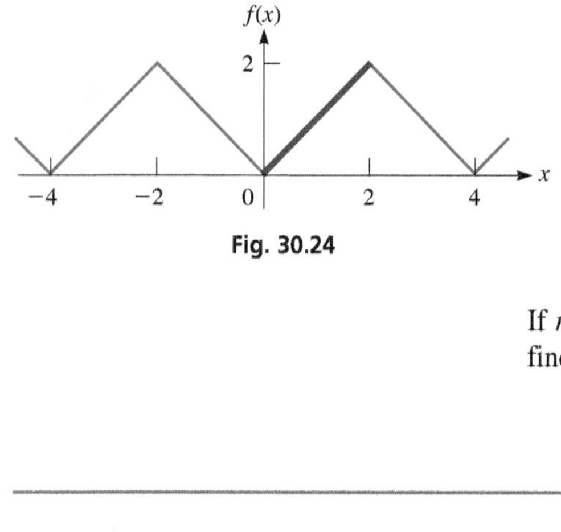

Fig. 30.24

$$a_0 = \frac{1}{2}\int_0^2 x\,dx = \frac{1}{4}x^2\Big|_0^2 = 1$$

$$a_n = \frac{2}{2}\int_0^2 x\cos\frac{n\pi x}{2}\,dx = x\left[\left(\frac{2}{n\pi}\sin\frac{n\pi x}{2}\right) - \left(\frac{-4}{n^2\pi^2}\cos\frac{n\pi x}{2}\right)\right]\Big|_0^2$$

$$= \frac{4}{n^2\pi^2}(\cos n\pi - 1) \qquad (n \ne 0)$$

If n is even, $\cos n\pi - 1 = 0$. Therefore, we evaluate a_n for the odd values of n, and find the expansion is

$$f(x) = 1 - \frac{8}{\pi^2}\left(\cos\frac{\pi x}{2} + \frac{1}{9}\cos\frac{3\pi x}{2} + \frac{1}{25}\cos\frac{5\pi x}{2} + \cdots\right)$$

EXAMPLE 10 Half-range sine series

Expand $f(x) = x$ in a half-range sine series for $0 \le x < 2$.

Since we are to have a sine series, we extend the function to be an odd function with its graph as shown in Fig. 30.25. Again, the red portion shows the given function as defined, and blue portions show the extension that makes it an odd function. By using Eq. (30.26), we find the Fourier expansion coefficients, with $L = 2$.

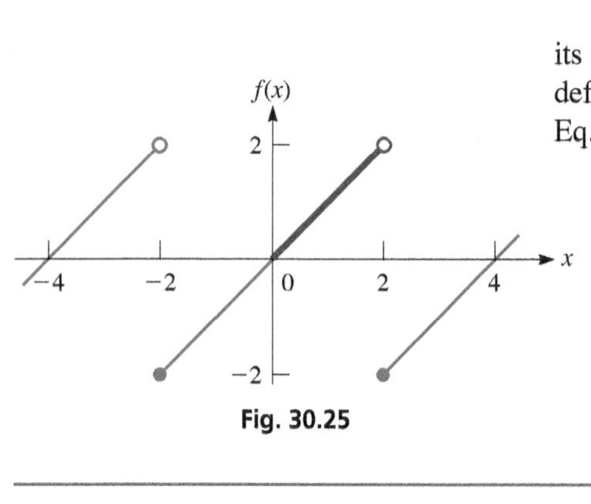

Fig. 30.25

$$b_n = \frac{2}{2}\left(\int_0^2 x\sin\frac{n\pi x}{2}\right)dx$$

$$= x\left[\left(\frac{-2}{n\pi}\cos\frac{n\pi x}{2}\right) - \left(\frac{-4}{n^2\pi^2}\sin\frac{n\pi x}{2}\right)\right]\Big|_0^2 = -\frac{4}{n\pi}\cos n\pi$$

$$f(x) = \frac{4}{\pi}\left(\sin\frac{\pi x}{2} - \frac{1}{2}\sin\pi x + \frac{1}{3}\sin\frac{3\pi x}{2} - \cdots\right)$$

EXERCISES 30.7

In Exercises 1–4, write the Fourier series for each function by comparing it to an appropriate function given in an example of this section. Do not use any of the formulas for a_0, a_n, or b_n.

1. $f(x) = \begin{cases} 2 & -\pi \le x < -\pi/2,\ \pi/2 \le x < \pi \\ 3 & -\pi/2 \le x < \pi/2 \end{cases}$

2. $f(x) = \begin{cases} -\frac{1}{2} & -\pi \le x < 0 \\ \frac{3}{2} & 0 \le x < \pi \end{cases}$

3. $f(x) = \begin{cases} -2 & -4 \le x < 0 \\ 0 & 0 \le x < 4 \end{cases}$

4. $f(x) = \begin{cases} -\frac{1}{3} & -\pi \le x < -\pi/2,\ \pi/2 \le x < \pi \\ \frac{2}{3} & -\pi/2 \le x < \pi/2 \end{cases}$

In Exercises 5–12, determine whether the given function is even, or odd, or neither. One period is defined for each function.

5. $f(x) = \begin{cases} 5 & -3 \le x < 0 \\ 0 & 0 \le x < 3 \end{cases}$

6. $f(x) = \begin{cases} -1 & -2 \le x < 0 \\ 1 & 0 \le x < 2 \end{cases}$

7. $f(x) = \begin{cases} 2 & -1 \le x < 1 \\ 0 & -2 \le x < -1,\ 1 \le x < 2 \end{cases}$

8. $f(x) = \begin{cases} 0 & -2 \le x < 0,\ 1 \le x < 2 \\ 1 & 0 \le x < 1 \end{cases}$

9. $f(x) = |x| \quad -4 \le x < 4$

10. $f(x) = \begin{cases} 0 & -1 \le x < 0 \\ e^x & 0 \le x < 1 \end{cases}$

11. $f(x) = -x \cos 3x \quad -3 \le x < 3$

12. $f(x) = x \cos 2x \quad -4 \le x < 4$

In Exercises 13–16, determine whether the Fourier series of the given functions will include only sine terms, only cosine terms, or both sine terms and cosine terms.

13. $f(x) = 2 - x \quad -4 \le x < 4$

14. $f(x) = \cos(\sin x) \quad -\pi \le x < \pi$

15. $f(x) = \begin{cases} 0 & -\pi \le x < 0 \\ \cos x & 0 \le x < \pi \end{cases}$

16. $f(x) = \begin{cases} -3 & -3 \le x < 0 \\ 0 & 0 \le x < 3 \end{cases}$

In Exercises 17–22, find at least three nonzero terms (including a_0 and at least two cosine terms and two sine terms if they are not all zero) of the Fourier series for the function from the indicated exercise of this section.

17. Exercise 5
18. Exercise 6
19. Exercise 7
20. Exercise 8
21. Exercise 9
22. Exercise 10

In Exercises 23–28, solve the given problems.

23. Expand $f(x) = 1$ in a half-range sine series for $0 \le x < 4$.

24. Expand $f(x) = 1$ $(0 \le x < 2)$, $f(x) = 0$ $(2 \le x < 4)$ in a half-range cosine series for $0 \le x < 4$.

25. Expand $f(x) = x^2$ in a half-range cosine series for $0 \le x < 2$.

26. Expand $f(x) = x^2$ in a half-range sine series for $0 \le x < 2$.

27. Each pulse of a pulsating force F of a pressing machine is 8 N. The force lasts for 1 s, followed by a 3-s pause. Thus, it can be represented by $F = 0$ for $-2 \le t < 0$ and $1 \le t < 2$, and $F = 8$ for $0 \le t < 1$, with a period of 4 s (only positive values of t have physical significance). Find the Fourier series for the force.

28. A pulsating electric current i (in mA) with a period of 2 s can be described by $i = e^{-t}$ for $-1 \le t < 1$ s for one period (only positive values of t have physical significance). Find the Fourier series that represents this current.

Answers to Practice Exercises

1. odd 2. even

3. $f(x) = \frac{7}{3} - \frac{4}{\pi^2}\left(\cos \pi x - \frac{1}{4}\cos 2\pi x + \cdots\right)$

CHAPTER 30 KEY FORMULAS AND EQUATIONS

Infinite series
$$\sum_{n=1}^{\infty} a_n = a_1 + a_2 + a_3 + \cdots + a_n + \cdots \tag{30.1}$$

Sum of series
$$S = \lim_{n\to\infty} S_n = \lim_{n\to\infty} \sum_{i=1}^{n} a_i \tag{30.2}$$

Sum of geometric series
$$S = \lim_{n\to\infty} S_n = \frac{a_1}{1-r} \quad \text{for } |r| < 1 \tag{30.3}$$

Power series
$$f(x) = a_0 + a_1x + a_2x^2 + \cdots + a_nx^n + \cdots \tag{30.4}$$

Maclaurin series
$$f(x) = f(0) + f'(0)x + \frac{f''(0)x^2}{2!} + \frac{f'''(0)x^3}{3!} + \cdots + \frac{f^n(0)x^n}{n!} + \cdots \tag{30.5}$$

Special series

$$e^x = 1 + x + \frac{x^2}{2!} + \frac{x^3}{3!} + \cdots \quad (\text{all } x) \tag{30.6}$$

$$\sin x = x - \frac{x^3}{3!} + \frac{x^5}{5!} - \cdots \quad (\text{all } x) \tag{30.7}$$

$$\cos x = 1 - \frac{x^2}{2!} + \frac{x^4}{4!} - \cdots \quad (\text{all } x) \tag{30.8}$$

$$\ln(1 + x) = x - \frac{x^2}{2} + \frac{x^3}{3} - \frac{x^4}{4} + \cdots \quad (|x| < 1) \tag{30.9}$$

$$(1 + x)^n = 1 + nx + \frac{n(n-1)}{2!}x^2 + \cdots \quad (|x| < 1) \tag{30.10}$$

Taylor series

$$f(x) = f(a) + f'(a)(x - a) + \frac{f''(a)(x-a)^2}{2!} + \cdots \tag{30.16}$$

Fourier series with period 2π

$$f(x) = a_0 + a_1 \cos x + a_2 \cos 2x + \cdots + a_n \cos nx + \cdots$$
$$+ b_1 \sin x + b_2 \sin 2x + \cdots + b_n \sin nx + \cdots \tag{30.17}$$

$$a_0 = \frac{1}{2\pi}\int_{-\pi}^{\pi} f(x)\,dx \tag{30.18}$$

$$a_n = \frac{1}{\pi}\int_{-\pi}^{\pi} f(x)\cos nx\,dx \tag{30.19}$$

$$b_n = \frac{1}{\pi}\int_{-\pi}^{\pi} f(x)\sin nx\,dx \tag{30.20}$$

Fourier series with period $2L$

$$f(x) = a_0 + a_1 \cos(\pi x/L) + a_2 \cos(2\pi x/L) + \cdots + a_n \cos(n\pi x/L) + \cdots$$
$$+ b_1 \sin(\pi x/L) + b_2 \sin(2\pi x/L) + \cdots + b_n \sin(n\pi x/L) + \cdots \tag{30.21}$$

$$a_0 = \frac{1}{2L}\int_{-L}^{L} f(x)\,dx \tag{30.22}$$

$$a_n = \frac{1}{L}\int_{-L}^{L} f(x)\cos\frac{n\pi x}{L}\,dx \tag{30.23}$$

$$b_n = \frac{1}{L}\int_{-L}^{L} f(x)\sin\frac{n\pi x}{L}\,dx \tag{30.24}$$

Half-range expansions

$$a_0 = \frac{1}{L}\int_{0}^{L} f(x)\,dx \quad \text{and} \quad a_n = \frac{2}{L}\int_{0}^{L} f(x)\cos\frac{n\pi x}{L}\,dx \; (n = 1, 2, \ldots) \tag{30.25}$$

$$b_n = \frac{2}{L}\int_{0}^{L} f(x)\sin\frac{n\pi x}{L}\,dx \quad (n = 1, 2, \cdots) \tag{30.26}$$

CHAPTER 30 REVIEW EXERCISES

In Exercises 1–10, find the first three nonzero terms of the Maclaurin expansion of the given functions.

1. $f(x) = \dfrac{1}{1 + e^x}$

2. $f(x) = e^{\cos x}$

3. $f(x) = \sin 2x^2$

4. $f(x) = \dfrac{1}{(1 - x)^2}$

5. $f(x) = (x + 1)^{1/3}$

6. $f(x) = \dfrac{x^2}{1 + x^2}$

7. $f(x) = \sin^{-1} x$

8. $f(x) = \dfrac{1}{1 - \sin x}$

9. $f(x) = \cos(a + x)$

10. $f(x) = \ln(a + x)$

In Exercises 11–22, calculate the value of each of the given functions. Use three terms of the appropriate series. Round answers to three significant digits.

11. $e^{-0.2}$ **12.** $\ln(1.10)$ **13.** $\sqrt[3]{1.3}$

14. $\sin 3.5°$ **15.** 1.086^{-1} **16.** 0.9839^{10}

17. $\ln 0.8172$ **18.** $\cos 0.1376$ **19.** $\tan 43.62°$

20. $\sqrt[4]{260}$ **21.** $\sqrt{148}$ **22.** $\cos 47°$

In Exercises 23 and 24, evaluate the given integrals by using three terms of the appropriate series.

23. $\displaystyle\int_{0.1}^{0.2} \frac{\cos x}{\sqrt{x}}\, dx$ **24.** $\displaystyle\int_{0}^{0.1} \sqrt[3]{1 + x^2}\, dx$

In Exercises 25 and 26, find the first three terms of the Taylor expansion for the given function and value of a.

25. $\cos x \quad (a = \pi/3)$ **26.** $\ln \cos x \quad (a = \pi/4)$

In Exercises 27–30, write the Fourier series for each function by comparing it to an appropriate function in an example of either Section 30.6 or 30.7. One period is given for each function. Do not use any formulas for a_0, a_n, or b_n.

27. $f(x) = \begin{cases} 0 & -\pi \le x < 0 \\ x - 1 & 0 \le x < \pi \end{cases}$

28. $f(x) = x^2 - 1 \quad -1 \le x < 1$

29. $f(x) = \begin{cases} \pi - 1 & -4 \le x < 0 \\ \pi + 1 & 0 \le x < 4 \end{cases}$

30. $f(x) = \begin{cases} 1 & -\pi \le x < 0 \\ 1 + \sin x & 0 \le x < \pi \end{cases}$

In Exercises 31–34, find at least three nonzero terms (including a_0 and at least two cosine terms and two sine terms if they are not all zero) of the Fourier series for the given function. One period is given for each function.

31. $f(x) = \begin{cases} 0 & -\pi \le x < -\pi/2,\ \pi/2 \le x < \pi \\ 1 & -\pi/2 \le x < \pi/2 \end{cases}$

32. $f(x) = \begin{cases} -x & -\pi \le x < 0 \\ 0 & 0 \le x < \pi \end{cases}$

33. $f(x) = x \quad -2 \le x < 2$

34. $f(x) = \begin{cases} -2 & -3 \le x < 0 \\ 2 & 0 \le x < 3 \end{cases}$

In Exercises 35–75, solve the given problems.

35. Test the series $1000 + 800 + 640 + 512 + \cdots$ for convergence or divergence. If convergent, find its sum.

36. Test the series $1 + 1.1 + 1.21 + 1.331 + \cdots$ for convergence or divergence. If convergent, find its sum.

37. Find the sum of the series $64 + 48 + 36 + 27 + \cdots$.

38. Find the first five partial sums of the series $\displaystyle\sum_{n=1}^{\infty} \frac{n}{3n + 1}$ and determine whether it appears to be convergent or divergent.

39. Express the integration of the indefinite integral $\int \sin(x^2)\, dx$ as an infinite series.

40. Express the integration of the indefinite integral $\int (1 + x^4)^{-1} dx$ as an infinite series.

41. Find the first three terms of the Taylor series for $f(x) = \tan x$ with $a = \pi/4$.

42. Integrate the series found in Exercise 41 to find the Taylor expansion for $\ln \cos x$ with $a = \pi/4$.

43. Differentiate the series found in Exercise 41 to find the Taylor expansion for $\sec^2 x$ with $a = \pi/4$.

44. Use the series for $(1 - x^2)^{-1/2}$ to find the Maclaurin series for $\sin^{-1} x$.

45. If h is small, show that $\sin(x + h) - \sin(x - h) = 2h \cos x$.

46. Find the first three nonzero terms of the Maclaurin expansion of the function $\sin x + x \cos x$ by differentiating the expansion term by term for $x \sin x$.

47. Using the properties of logarithms and Eq. (30.9), find four terms of the Maclaurin expansion of $\ln(1 + x)^4$.

48. By multiplication of series, show that the first two terms of the Maclaurin series for $2 \sin x \cos x$ are the same as those of the series for $\sin 2x$.

49. Find the first four nonzero terms of the expansion for $\sin^2 x$ by using the identity $\sin^2 x = \frac{1}{2}(1 - \cos 2x)$ and the series for $\cos x$.

50. Evaluate the integral $\int_0^1 x \sin x\, dx$ (a) using integration by parts and (b) using three terms of the series for $\sin x$. Compare results.

51. Find the first three terms of the Maclaurin expansion for $\sec x$ by finding the reciprocal of the series for $\cos x$.

52. By simplifying the sum of the squares for the Maclaurin series for $\sin x$ and $\cos x$, verify (to this extent) that $\sin^2 x + \cos^2 x = 1$.

53. From the Maclaurin series for $f(x) = 1/(1 - x)$ (see Exercise 13 of Section 30.2), find the series for $1/(1 + x)$.

54. Show that the Maclaurin expansions for $\cos x$ and $\cos(-x)$ are the same.

55. Expand $f(x) = x^2$ in a half-range cosine series for $0 < x \le 1$.

56. Expand $f(x) = 2 - x$ in a half-range sine series for $0 < x \le 2$.

57. Calculate $e^{0.9}$ by using four terms of the Maclaurin expansion for e^x. Also, calculate $e^{0.9}$ by using the first three terms of the Taylor expansion in Example 1 of Section 30.5. Compare the accuracy of the values obtained with that found directly on a calculator.

58. Find the volume generated by revolving the region bounded by $y = e^{-x}$, $y = 0$, $x = 0$, and $x = 0.1$ about the y-axis by using three terms of the appropriate series.

59. Find the approximate area between the curve of $y = \dfrac{x - \sin x}{x^2}$ and the x-axis between $x = 0.1$ and $x = 0.2$.

60. Find the approximate value of the moment of inertia with respect to its axis of the solid generated by revolving the smaller region bounded by $y = \sin x$, $x = 0.3$, and the x-axis about the y-axis. Use two terms of the appropriate series.

61. Find three terms of the Maclaurin series for $\tan^{-1} x$ by integrating the series for $1/(1 + x^2)$, term by term.

62. The current i in an electric circuit containing a resistance R and an inductance L is given by $i = Ie^{-Rt/L}$, where I is the current at $t = 0$. Express i as an infinite series.

63. A piano wire vibrates with a displacement y (in mm) given by $y = 3.2 \cos 880\pi t$, where t is the time (in s). Express y as an infinite series.

64. An empty underground cubical tank was later filled with water. The amount of water needed to fill the tank was 30.0 m^3 with a possible error of 1.00 m^3. Use a series to estimate the error in calculating the length of one side, s, of the tank ($s = V^{1/3}$). [*Hint:* Find a series for $(30 + x)^{1/3}$.]

65. The number N of radioactive nuclei in a radioactive sample is $N = N_0 e^{-\lambda t}$. Here, t is the time, N_0 is the number at $t = 0$, and λ is the *decay constant.* By using four terms of the appropriate series, express the right side of this equation as a polynomial.

66. The length of Lake Erie is a great circle arc of 390 km. If the lake is assumed to be flat, use series to find the error in calculating the distance from the centre of the lake to the centre of the earth. The radius of the earth is 6400 km.

67. From what height can a person see a point 10 km distant on Earth's surface?

68. The vertical displacement y of a mass at the end of a spring is given by $y = \sin 3t - \cos 2t$, where t is the time. By subtraction of series, find the first four nonzero terms of the series for y.

69. The electric potential V at a distance x along a certain surface is given by $V = \ln \dfrac{1 + x}{1 - x}$. Find the first four terms of the Maclaurin series for V.

70. If a mass M is hung from a spring of mass m, the ratio of the masses is $m/M = k\omega \tan k\omega$, where k is a constant and ω is a measure of the frequency of vibration. By using two terms of the appropriate series, express m/M as a polynomial in terms of ω.

71. In the study of electromagnetic radiation, the expression $\dfrac{N_0}{1 - e^{-k/T}}$ is used. Here, T is the thermodynamic temperature, and N_0 and k are constants. Show that this expression can be written as $N_0(1 + e^{-k/T} + e^{-2k/T} + \cdots)$. (*Hint:* Let $x = e^{-k/T}$.)

72. In the analysis of reflection from a spherical mirror, it is necessary to express the x-coordinate on the surface shown in Fig. 30.26 in terms of the y-coordinate and the radius R. Using the equation of the semicircle shown, solve for x (note that $x \leq R$). Then express the result as a series. (Note that the first approximation gives a parabolic surface.)

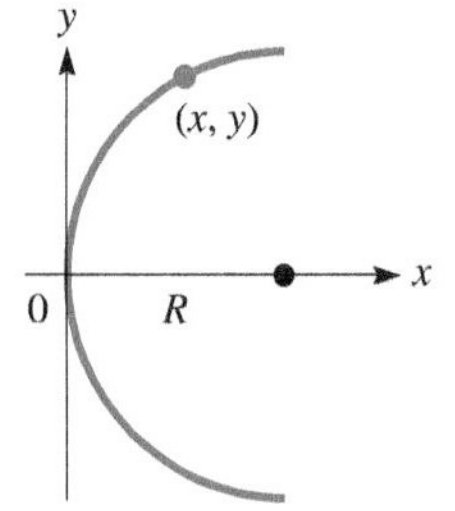

Fig. 30.26

73. A certain electric current is pulsating so that the current as a function of time is given by $f(t) = 0$ if $-\pi \leq t < 0$ and $\pi/2 < t < \pi$. If $0 < t < \pi/2$, $f(t) = \sin t$. Find the Fourier expansion for this pulsating current and sketch three periods.

74. The force F applied to a spring system as a function of the time t is given by $F = t/\pi$ if $0 \leq t \leq \pi$ and $F = 0$ if $\pi < t < 2\pi$. If the period of the force is 2π, find the first few terms of the Fourier series that represents the force.

75. A computer science class is assigned to write a program to find the Maclaurin series for $\sin^2 x$, using only the series for $\sin x$ and/or $\cos x$, and any algebraic, trigonometric, or calculus procedures. Write a paragraph or two explaining how this can be done in at least four different ways.

CHAPTER 30 PRACTICE TEST

In Problems 1–7, answer the given questions.

1. By direct expansion, find the first four nonzero terms of the Maclaurin expansion for $f(x) = (1 + e^x)^2$.

2. Find the first three nonzero terms of the Taylor expansion for $f(x) = \cos x$, with $a = \pi/3$.

3. Evaluate ln 0.96 by using four terms of the expansion for $\ln(1 + x)$.

4. Find the first three nonzero terms of the expansion for $f(x) = \dfrac{1}{\sqrt{1 - 2x}}$ by using the binomial series.

5. Evaluate $\int_0^1 x \cos x \, dx$ by using three terms of the appropriate series.

6. An electric current is pulsating such that it is a function of the time with a period of 2π. If $f(t) = 2$ for $0 \leq t < \pi$ and $f(t) = 0$ for the other half-cycle, find the first three nonzero terms of the Fourier series for this current.

7. $f(x) = x^2 + 2$ for $-2 \leq x < 2$ (period = 4). Is $f(x)$ an even function, an odd function, or neither? Expressing the Fourier series as $F(x)$, what is the Fourier series for the function $g(x) = x^2 - 1$ for $-2 \leq x < 2$ (period = 4)? (Do *not* use integration to derive specific terms for either series.)

31. Differential Equations

Jason Cheever/Shutterstock

▲ In Section 31.6, we show how differential equations are used in the dating of geological samples (such as those from the Dome Glacier in Alberta) by using carbon dating.

LEARNING OUTCOMES

After completion of this chapter, the student should be able to:

- Show that a function is a solution of a differential equation
- Solve first-order differential equations by separation of variables or by recognizing integrating combinations
- Solve first-order linear differential equations using an integrating factor
- Solve first-order differential equations numerically by Euler's method or by the Runge–Kutta method
- Solve homogeneous linear differential equations of higher order
- Solve higher-order nonhomogeneous linear differential equations by the method of undetermined coefficients
- Find the Laplace transform and the inverse Laplace transform of a function
- Solve differential equations using Laplace transforms
- Solve application problems involving differential equations

Many of the physical problems being studied in the 1700s, such as velocity and light, led to equations that involved derivatives or differentials. These equations are called *differential equations.* Solving these differential equations became a very important topic of mathematical development during the eighteenth century.

In this chapter, we study some of the basic methods of solving differential equations. Many of these methods were first developed by the famous Swiss mathematician Leonhard Euler (1707–1783). He is undoubtedly the most prolific mathematician of all time, in that his work in mathematics and other fields filled over 70 large volumes.

The final topic covered in this chapter is the solution of differential equations by *Laplace transforms.* They are named for the French mathematician Pierre Laplace (1749–1827). Actually, Laplace had devised a mathematical method in the late 1700s that the English electrical engineer Oliver Heaviside (1850–1925) refined and developed into its present useful form in the late 1800s. The Laplace transform is particularly useful for solving problems involving electrical circuits and mechanical systems.

Here, we see again that a field of mathematics was developed in response to the need for solving real-life physical problems. Also, we see that a method developed for purely mathematical reasons was then very usefully applied 100 years later in electricity, an area of study that did not exist when the method was first devised.

We actually solved a few simple differential equations in earlier chapters when we started a solution with the expression for the slope of a tangent line or the velocity of an object. Also, we have noted the applications of differential equations in electrical circuits, mechanical systems, and the study of light. Other areas of application include chemical reactions, interest calculations, changes in pressure and temperature, population growth, forces on beams and structures, and nuclear energy.

31.1 Solutions of Differential Equations

A **differential equation** *is an equation that contains derivatives or differentials.* Most differential equations we shall consider contain first and/or second derivatives, although some will have higher derivatives. *An equation that contains only first derivatives is called a* **first-order** *differential equation. An equation that contains second derivatives, and possibly first derivatives, is called a* **second-order** *differential equation.* In general, *the* **order** *of the differential equation is that of the highest derivative in the equation, and the* **degree** *is the highest power of that derivative.*

EXAMPLE 1 Illustrations of differential equations

(a) The equation $dy/dx + x = y$ is a first-order, first-degree differential equation since it contains only a first derivative raised to the first power.

(b) The equations $\frac{d^2y}{dx^2} + y = 3x^2$ and $\frac{d^2y}{dx^2} + 2\frac{dy}{dx} = x$ are second-order differential equations since each contains a second derivative and no higher derivatives. The dy/dx in the second equation does not affect the order. Both equations have degree one.

(c) The equation $\frac{d^2y}{dx^2} + \left(\frac{dy}{dx}\right)^4 - y = 6$ is a differential equation of the second order and first degree. That is, the highest derivative that appears is the second, and it is raised to the first power. Since the second derivative appears, the fourth power of the first derivative does not affect the degree.

In our discussion of differential equations, we will restrict our attention to equations of the first degree.

A **solution** *of a differential equation is a relation between the variables that satisfies the differential equation.* That is, when this relation is substituted into the differential equation, an algebraic identity results. *A solution containing a number of independent arbitrary constants equal to the order of the differential equation is called the* **general solution** *of the equation. When specific values are given to at least one of these constants, the solution is called a* **particular solution**.

EXAMPLE 2 Independent arbitrary constants

Obtain the general solution to the differential equation $d^2y/dx^2 = 0$ from the solution $y = c_1x + c_2 + c_3x$.

This solution is not the general solution since it contains three arbitrary constants, whereas the general solution should contain only two (the order of the differential equation). We combine like terms by letting $c_4 = c_1 + c_3$, so the general solution can be written as $y = c_2 + c_4x$.

EXAMPLE 3 Illustration of the general solution

$y = c_1e^{-x} + c_2e^{2x}$ is the general solution of the differential equation

$$\frac{d^2y}{dx^2} - \frac{dy}{dx} = 2y$$

The order of this differential equation is 2, and there are two independent arbitrary constants in the solution. The equation $y = 4e^{-x}$ is a particular solution. It can be derived from the general solution by letting $c_1 = 4$ and $c_2 = 0$. Each of these solutions can be shown to satisfy the differential equation by taking two derivatives and substituting.

To solve a differential equation, we have to find some method of transforming the equation so that each term may be integrated. Some of these methods will be considered after this section.

LEARNING TIP

To show that a given equation is a solution of the differential equation, *we take the required derivatives and show that an identity results after substitution.*

EXAMPLE 4 Showing that an equation is the general solution

Show that $y = c_1 \sin x + c_2 \cos x$ is the general solution of the differential equation $y'' + y = 0$.

The function and its first two derivatives are

$$y = c_1 \sin x + c_2 \cos x$$
$$y' = c_1 \cos x - c_2 \sin x$$
$$y'' = -c_1 \sin x - c_2 \cos x$$

Substituting these into the differential equation, we have

$$\underbrace{(-c_1 \sin x - c_2 \cos x)}_{y''} + \underbrace{(c_1 \sin x + c_2 \cos x)}_{y} = 0 \quad \text{or} \quad 0 = 0$$

We know that this must be the general solution, since there are two independent arbitrary constants and the order of the differential equation is 2.

Practice Exercise

1. Show that $y = 2 + ce^{2x}$ is a solution of $y'' - 2y' = 0$. Is it the general solution?

EXAMPLE 5 Showing that an equation is a particular solution

Show that $y = 3x + x^2$ is a solution of the differential equation $xy' - y = x^2$.

Taking one derivative of the function and substituting in the differential equation, we have

$$y = 3x + x^2 \quad \text{(particular solution—no arbitrary constants)}$$
$$xy' - y = x^2$$
$$y' = 3 + 2x$$
$$x(3 + 2x) - (3x + x^2) = x^2 \quad \text{or} \quad x^2 = x^2$$

EXERCISES 31.1

In Exercises 1 and 2, show that the indicated solutions are, in fact, solutions of the differential equations in the indicated examples.

1. In Example 3, two solutions are shown for the given differential equation. Show that each is a solution.
2. In Example 4, show that $y_1 = c \sin x + 5 \cos x$ and $y_2 = 2 \sin x - 3 \cos x$ are solutions of the given differential equation.

In Exercises 3–6, determine whether the given equation is the general solution or a particular solution of the given differential equation.

3. $\dfrac{dy}{dx} + 2xy = 0, \quad y = e^{-x^2}$
4. $y' \ln x - \dfrac{y}{x} = 0, \quad y = c \ln x$
5. $y'' + 3y' - 4y = 3e^x, \quad y = c_1e^x + c_2e^{-4x} + \frac{3}{5}xe^x$
6. $\dfrac{d^2y}{dx^2} + 4y = 8, \quad y = c \sin 2x + 3 \cos 2x + 2$

In Exercises 7–10, show that each function $y = f(x)$ is a solution of the given differential equation.

7. $\dfrac{dy}{dx} - y = 1; \quad y = e^x - 1, \quad y = 5e^x - 1$
8. $\dfrac{dy}{dx} = 2xy^2; \quad y = -\dfrac{1}{x^2}, \quad y = -\dfrac{1}{x^2 + c}$
9. $y'' + 4y = 0; \quad y = 3 \cos 2x, \quad y = c_1 \sin 2x + c_2 \cos 2x$
10. $y'' = 2y'; \quad y = 3e^{2x}, \quad y = 2e^{2x} - 5$

In Exercises 11–32, show that the given equation is a solution of the given differential equation.

11. $\dfrac{dy}{dx} = 2x, \quad y = x^2 + 25$
12. $xy' = 2y, \quad y = cx^2$
13. $\dfrac{dy}{dx} = 8 - 12x^2, \quad y = 5 + 8x - 4x^3$
14. $\dfrac{dy}{dx} = 3y + 2x, \quad y = ce^{3x} - \dfrac{2}{3}x - \dfrac{2}{9}$

15. $y' + 2y = 2x, \quad y = ce^{-2x} + x - \frac{1}{2}$

16. $y'' = 18x + 10, \quad y = 3x^3 + 5x^2 + c$

17. $y'' + 9y = 4\cos x, \quad 2y = \cos x$

18. $y'' - 4y' + 4y = e^{2x}, \quad y = e^{2x}\left(c_1 + c_2x + \frac{x^2}{2}\right)$

19. $x^2y' + y^2 = 0, \quad xy = cx + cy$

20. $xy' - 3y = x^2, \quad y = cx^3 - x^2$

21. $x\frac{d^2y}{dx^2} + \frac{dy}{dx} = 0, \quad y = c_1 \ln x + c_2$

22. $y'' + 4y = 10e^x, \quad y = c_1 \sin 2x + c_2 \cos 2x + 2e^x$

23. $y' + y = 2\cos x, \quad y = \sin x + \cos x - e^{-x}$

24. $(x + y) - xy' = 0, \quad y = x\ln x - cx$

25. $y'' - 3y' + 2y = 3, \quad y = c_1e^x + c_2e^{2x} + 3/2$

26. $xy'' + y' = 16x^3, \quad y = x^4 + c_1 + c_2 \ln x$

27. $\cos x\frac{dy}{dx} + \sin x = 1 - y, \quad y = \frac{x + c}{\sec x + \tan x}$

28. $2xyy' + x^2 = y^2, \quad x^2 + y^2 = cx$

29. $(y')^2 + xy' = y, \quad y = cx + c^2$

30. $x^4(y')^2 - xy' = y, \quad y = c^2 + \frac{c}{x}$

31. $\frac{d^3y}{dx^3} = \frac{d^2y}{dx^2}, \quad y = c_1 + c_2x + c_3e^x$

32. $\frac{d^3y}{dx^3} + 4\frac{d^2y}{dx^2} + 4\frac{dy}{dx} = 0, \quad y = c_1 + c_2e^{-2x} + xe^{-2x}$

In Exercises 33–35, solve the given problems.

33. The general solution of the differential equation $y'' - y'/x = 3x$ is $y = x^3 + c_1x^2 + c_2$. Find the particular solution if the graph of the solution passes through the point $(0, -4)$.

34. For a given electric circuit, the charge q in the circuit as a function of the time t is $q = 0.01(1 - \cos 316t)$. Show that this satisfies the differential equation $0.1\frac{d^2q}{dt^2} + 10^4q = 100$.

35. A differential equation that arises in the study of radioactivity is $dN/dt = kN$. Show that $N = N_0e^{kt}$ is the general solution.

Answer to Practice Exercise

1. $(4ce^{2x}) - 2(2ce^{2x}) = 0$, No (two constants required).

31.2 Separation of Variables

We will now solve differential equations of the first order and first degree. Of the many methods for solving such equations, a few are presented in this and the next two sections. The first of these is *the method of* **separation of variables.**

A differential equation of the first order and first degree contains the first derivative to the first power. That is, it may be written as $dy/dx = f(x, y)$. This type of equation is more commonly expressed in its differential form,

$$M(x, y)\,dx + N(x, y)\,dy = 0 \qquad \textbf{(31.1)}$$

where $M(x, y)$ and $N(x, y)$ may represent constants, functions of either x or y, or functions of x and y.

If it is possible to rewrite Eq. (31.1) as

$$A(x)\,dx + B(y)\,dy = 0 \qquad \textbf{(31.2)}$$

where $A(x)$ does not contain y and $B(y)$ does not contain x, then *we may find the solution by integrating each term and adding the constant of integration.* Many differential equations can be solved in this way.

Solving a First Order, First Degree Equation by Separation of Variables	EXAMPLE 1
	Solve $dx - 4xy^3dy = 0$.
1. Write the equation in the form of Eq. (31.1). Verify that $M(x, y)$ and $N(x, y)$ are both products of a function involving only x and a function involving only y. If this is not the case, separation of variables cannot be used.	We write the equation as $(1)\,dx + (-4xy^3)\,dy = 0$ so that $M(x, y) = 1$ and $N(x, y) = -4xy^3$. Both are products of a function involving only x and a function involving only y.

LEARNING TIP

It is important to note that, if division is used when separating variables, the solution is not valid for values that make the divisor zero.

2. Separate the x and y terms by multiplying or dividing the equation by an appropriate expression.	We remove x from the coefficient of dy by dividing each term by x: $\frac{dx}{x} - 4y^3dy = 0$
3. Integrate each term and add the constant of integration, which becomes the arbitrary constant of the solution. Whenever possible, use the properties of logarithms to simplify.	Performing the integration, we obtain the general solution $\ln \lvert x \rvert - y^4 = c$

LEARNING TIP

In Example 1, we showed the integration of dx/x as $\ln|x|$, which follows our discussion in Section 28.2. We know $\ln|x| = \ln x$ if $x > 0$ and $\ln|x| = \ln(-x)$ if $x < 0$. Since we know the values being used when we find a particular solution, *we generally will not use the absolute value notation when integrating logarithmic forms.* We would show the integration of dx/x as $\ln x$, with the understanding that we know $x > 0$. When using negative values of x, we would express it as $\ln(-x)$.

■ For reference, the power rule for logarithms is $\log_b x^n = n \log_b x$, and the product rule for logarithms is $\log_b xy = \log_b x + \log_b y$.

Practice Exercise

1. Find the general solution of the differential equation $dx + 2y \sec x\, dy = 0$.

EXAMPLE 2 Use properties of logarithms to rewrite solutions

Solve the differential equation $xy\,dx + (x^2 + 1)\,dy = 0$.

Verify that $M(x, y)$ and $N(x, y)$ are products of a function of x and a function of y:	We identify $M(x, y) = xy$ and $N(x, y) = x^2 + 1$, both of which are products of a function of x and a function of y.
Separate the x and y terms:	Divide each term by $y(x^2 + 1)$: $\frac{x\,dx}{x^2+1} + \frac{dy}{y} = 0$
Integrate each term and add the constant of integration. Use the properties of logarithms to simplify:	Integrating, $\frac{1}{2}\ln(x^2 + 1) + \ln y = c$ Using the power rule for logarithms, we have $\ln(x^2 + 1)^{1/2} + \ln y = c.$ Now using the product rule for logarithms, $\ln\left[(x^2 + 1)^{1/2} y\right] = c.$ This means $(x^2 + 1)^{1/2} y = c_1$, where the new constant is $c_1 = e^c$.

EXAMPLE 3 Solving a differential equation containing exponential and trigonometric functions

Solve the differential equation $2e^{3x} \sin y\, dx + e^x \csc y\, dy = 0$.

We identify $M(x, y) = 2e^x \sin y$ and $N(x, y) = e^x \csc y$, both of which are products of a function of x and a function of y. To separate variables, we divide by $e^x \sin y$. Using properties of trigonometric and exponential functions, we have

$$\frac{2e^{3x}\sin y\,dx}{e^x \sin y} + \frac{e^x \csc y\,dy}{e^x \sin y} = 0 \qquad \text{divide by } e^x \sin y$$

$$2e^{2x}dx + \csc^2 y\,dy = 0 \qquad \text{variables separated}$$

$$e^{2x}(2\,dx) + \csc^2 y\,dy = 0 \qquad \text{form for integrating}$$

$$e^{2x} - \cot y = c \qquad \text{integrate}$$

LEARNING TIP

Any expression that represents a constant may be chosen as the constant of integration and leads to a correct solution. In checking answers, we must remember that a different choice of constant will lead to a different form of the solution. Thus, two different-appearing answers may both be correct. *Often there is more than one reasonable choice of a constant, and different forms of the solution may be expected.*

EXAMPLE 4 Many choices for the arbitrary constant

Solve the differential equation $\frac{d\theta}{dt} = \frac{\theta}{t^2 + 4}$.

The solution proceeds as follows:

$$\frac{d\theta}{\theta} = \frac{dt}{t^2 + 4}$$ separate variables by multiplying by dt and dividing by θ

$$\ln \theta = \frac{1}{2}\tan^{-1}\frac{t}{2} + \frac{c}{2}$$ integrate

$$2 \ln \theta = \tan^{-1}\frac{t}{2} + c$$

$$\ln \theta^2 = \tan^{-1}\frac{t}{2} + c$$

Note the different forms of the result using $c/2$ as the constant of integration. These forms would differ somewhat had we chosen c as the constant.

The choice of $\ln c$ as the constant of integration is also reasonable. On the left, it would lead to the result $2 \ln c\theta = \tan^{-1}(t/2)$. On the right, after solving explicitly for θ using the exponential function, it would lead to the expression $\theta = ce^{\frac{1}{2}\tan^{-1}(\frac{t}{2})}$.

FINDING PARTICULAR SOLUTIONS

To find a particular solution, we need information that allows us to evaluate the constant of integration. We now show how to find particular solutions, and graphically show the difference between the general solution and a particular solution.

EXAMPLE 5 Finding a particular solution

Solve the equation $(x^2 + 1)^2\,dy + 4x\,dx = 0$, subject to the condition that $x = 1$ when $y = 3$.

$$dy + \frac{4x\,dx}{(x^2 + 1)^2} = 0$$ dividing by $(x^2 + 1)^2$

$$y - \frac{2}{x^2 + 1} = c$$ integrating

$$y = \frac{2}{x^2 + 1} + c$$ general solution

$$3 = \frac{2}{1 + 1} + c, \quad c = 2$$ use $x = 1$, $y = 3$ to evaluate c

$$y = \frac{2}{x^2 + 1} + 2$$ particular solution

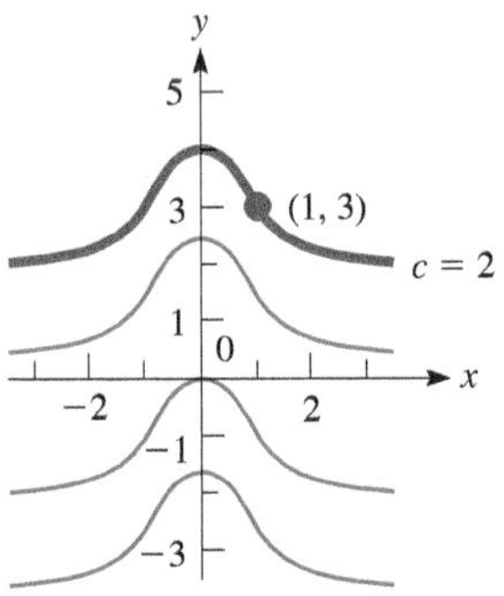

Fig. 31.1

The general solution defines a *family* of curves, one member of the family for each value of c. A few of these curves are shown in Fig. 31.1. When c is specified as in the particular solution, we have the specific curve shown.

EXAMPLE 6 The form of the constant does not affect the result

Find the particular solution in Example 2, given that $x = 0$ when $y = e$.

Using the solution $\frac{1}{2}\ln(x^2 + 1) + \ln y = c$, we have

$$\tfrac{1}{2}\ln(0 + 1) + \ln e = c, \quad \tfrac{1}{2}\ln 1 + 1 = c, \quad c = 1$$

$$\tfrac{1}{2}\ln(x^2 + 1) + \ln y = 1$$ substitute $c = 1$

$$\ln(x^2 + 1)^{1/2}y = 1$$ using properties of logarithms

$$(x^2 + 1)^{1/2}y = e$$ exponential form of particular solution

Using the general solution $(x^2 + 1)^{1/2} y = c_1$, we have

$$(0 + 1)^{1/2} e = c_1, \quad c_1 = e$$
$$(x^2 + 1)^{1/2} y = e$$

which is precisely the same solution as above. This shows that *the choice of the form of the constant does not affect the final result, and the constant is truly arbitrary.*

Practice Exercise

2. In Example 4, find the particular solution if $x = 0$ when $y = \pi/4$.

EXERCISES 31.2

In Exercises 1 and 2, make the given changes in the indicated examples of this section and then solve the resulting differential equations.

1. In Example 2, change the first term to $2xy\,dx$.
2. In Example 3, change the second term to $e^{2x} \csc y\,dy$.

In Exercises 3–36, solve the given differential equations. Explain your method of solution for Exercise 19.

3. $\dfrac{dy}{dx} = \dfrac{x}{y^2}$
4. $dy = k(20 + y)dt$
5. $\dfrac{ds}{dt} = \dfrac{\cos t}{s - 1}$
6. $\dfrac{dp}{dx} = \sqrt{\dfrac{p}{x}}$
7. $2x\,dx + dy = 0$
8. $y^2 dy + x^3 dx = 0$
9. $y^2 dx + dy = 0$
10. $y\,dt + t\,dy = 3ty\,dt$
11. $\dfrac{dV}{dP} = -\dfrac{V}{P^2}$
12. $\dfrac{2\,dy}{dx} = \dfrac{y(2x + 7)}{x}$
13. $x^2 + (x^3 + 5)y' = 0$
14. $xyy' + \sqrt{1 + y^2} = 0$
15. $dy + \ln xy\,dx = (4x + \ln y)\,dx$
16. $r\sqrt{1 - \theta^2}\dfrac{dr}{d\theta} = \theta + 4$
17. $e^{x^2} dy = x\sqrt{1 - y}\,dx$
18. $\sqrt{1 + 4x^2}\,dy = y^3 x\,dx$
19. $e^{x+y} dx + dy = 0$
20. $e^{2x} dy + e^x dx = 4\,dx$
21. $y' - y = 4$
22. $ds - s^2 dt = 9\,dt$
23. $x\dfrac{dy}{dx} = y^2 + y^2 \ln x$
24. $(yx^2 + y)\dfrac{dy}{dx} = \tan^{-1} x$
25. $y \tan x\,dx + \cos^2 x\,dy = 0$
26. $\sin x \sec y\,dx = dy$
27. $yx^2\,dx = y\,dx - x^2\,dy$
28. $y\sqrt{1 - x^2}\,dy + 2\,dx = 0$
29. $e^{\cos\theta} \tan\theta\,d\theta + \sec\theta\,dy = 0$
30. $(x^3 + x^2)\,dx + (xy + x + y + 1)\,dy = 0$
31. $xe^{x^2-y}\,dx = dy$
32. $\cos^2\phi + y \csc\phi\dfrac{dy}{d\phi} = 0$
33. $2 \ln t\,dt + t\,di = 0$
34. $y^2 e^t + (e^t + 4)y' = 0$
35. $2y(x^3 + 1)\,dy + 3x^2(y^2 - 1)\,dx = 0$
36. $V + 1 + (1 + \sin T) \sec T\dfrac{dV}{dT} = 0$

In Exercises 37–42, find the particular solution of the given differential equation for the indicated values.

37. $\dfrac{dy}{dx} + yx^2 = 0$; $x = 0$ when $y = e$
38. $\dfrac{ds}{dt} = \sec s$; $t = 0$ when $s = 0$
39. $y' = (1 - y) \cos x$; $x = \pi/6$ when $y = 0$
40. $x\,dy = y \ln y\,dx$; $x = 2$ when $y = e$
41. $y^2 e^x\,dx + e^{-x}\,dy = y^2\,dx$; $x = 0$ when $y = 2$
42. $2y \cos y\,dy - \sin y\,dy = y \sin y\,dx$; $x = 0$ when $y = \pi/2$

In Exercise 43–48, solve the given problems.

43. In Example 6, show that the particular solution is the same if the constant of integration is $\ln c$, rather than c.
44. Find $f(x)$ if $f'(x) = x^2 f(x)$ and $f(0) = 1$.
45. The temperature reading T (in °C) at time t (in s) of a thermometer initially reading 40°C and then placed in water at 10°C is found by solving the equation $dT + 0.15(T - 10)dt = 0$. Solve for T as a function of t.
46. The current i (in A) in an electric circuit changes with time t (in s) according to the equation $di + 10i\,dt = 6\,dt$. Find i as a function of t if the initial current is zero.
47. In the study of fluid mechanics, streamlines are often drawn to visualize the flow field. (A streamline is tangent to the velocity vector.) Find an expression for the streamlines of a fluid flow if these satisfy the equation $\dfrac{dy}{dx} = -\dfrac{x}{y}$. What shape do these lines have?
48. Find an expression for the streamlines of a fluid flow if these satisfy the equation $\dfrac{dy}{dx} = -\dfrac{y}{x}$. What shape do these lines have? (See Exercise 47.)

Answers to Practice Exercises

1. $\sin x + y^2 = c$ 2. $\cot y = e^{2x}$

31.3 Integrating Combinations

For different equations that cannot be solved by separation of variables, other methods have been developed. One is based on the fact that *certain combinations of differentials can be integrated as a unit.* The following differentials show some of the possible combinations:

$$d(xy) = x\,dy + y\,dx \tag{31.3}$$

$$d(x^2 + y^2) = 2(x\,dx + y\,dy) \tag{31.4}$$

$$d\left(\frac{y}{x}\right) = \frac{x\,dy - y\,dx}{x^2} \tag{31.5}$$

$$d\left(\frac{x}{y}\right) = \frac{y\,dx - x\,dy}{y^2} \tag{31.6}$$

Eq. (31.3) suggests that if the combination $x\,dy + y\,dx$ occurs in a differential equation, we should look for a function of xy as a solution. Eq. (31.4) suggests that if the combination $x\,dx + y\,dy$ occurs, we should look for a function of $x^2 + y^2$. Eqs. (31.5) and (31.6) suggest that if either of the combinations $x\,dy - y\,dx$ or $y\,dx - x\,dy$ occurs, we should look for a function of y/x or x/y.

EXAMPLE 1 Recognizing the differential of *xy*

Solve the differential equation $x\,dy + y\,dx + xy\,dy = 0$.

By dividing through by xy, we have

$$\frac{x\,dy + y\,dx}{xy} + dy = 0$$

The left term is the differential of xy divided by xy. Thus, it integrates to $\ln xy$.

$$\frac{d(xy)}{xy} + dy = 0$$

for which the solution is

$$\ln xy + y = c$$

EXAMPLE 2 Recognizing the differential of *y*/*x*

Solve the differential equation $y\,dx - x\,dy + x\,dx = 0$.

The combination of $y\,dx - x\,dy$ suggests that this equation might make use of either Eq. (31.5) or (31.6). This would require dividing through by x^2 or y^2. If we divide by y^2, the last term cannot be integrated, but ***division by x^2 still allows integration of the last term***. Performing this division, we obtain

$$\frac{y\,dx - x\,dy}{x^2} + \frac{dx}{x} = 0$$

This left combination is the negative of Eq. (31.5). Thus, we have

$$-d\left(\frac{y}{x}\right) + \frac{dx}{x} = 0$$

for which the solution is $-\dfrac{y}{x} + \ln x = c$.

Practice Exercise

1. Find the general solution of the differential equation $(4x - y)dx = x\,dy$.

EXAMPLE 3 Two combinations in the same equation

Find the general solution of the differential equation

$$(x^3 + xy^2 + 2y)\,dx + (y^3 + x^2y + 2x)\,dy = 0$$

which satisfies the condition that $x = 1$ when $y = 0$.

Regrouping the terms of the equation, we have

$$x(x^2 + y^2)\,dx + y(x^2 + y^2)\,dy + 2(y\,dx + x\,dy) = 0$$

Factoring $x^2 + y^2$ from each of the first two terms gives

$$(x^2 + y^2)(x\,dx + y\,dy) + 2(y\,dx + x\,dy) = 0 \quad d(x^2+y^2)\ d(xy)$$

$$\frac{1}{2}(x^2 + y^2)(2x\,dx + 2y\,dy) + 2(y\,dx + x\,dy) = 0$$

$$\frac{1}{2}\left(\frac{1}{2}\right)(x^2 + y^2)^2 + 2xy + \frac{c}{4} = 0 \quad \text{integrating}$$

$$(x^2 + y^2)^2 + 8xy + c = 0$$

LEARNING TIP

The use of integrating combinations depends on proper recognition of the forms. It may take two or three arrangements to find the combination that leads to the solution. Of course, many equations cannot be arranged so as to give integrable combinations in all terms.

EXAMPLE 4 Particular solution—differential of $x^2 + y^2$

Find the particular solution of the differential equation $(x^2 + y^2 + x)\,dx + y\,dy = 0$ that satisfies the condition $x = 1$ when $y = 0$.

Regrouping the terms of this equation, we have

$$(x^2 + y^2)\,dx + (x\,dx + y\,dy) = 0 \quad \text{divide each term by } x^2 + y^2$$

$$dx + \frac{x\,dx + y\,dy}{x^2 + y^2} = 0$$

The right term now can be put in the form of du/u (with $u = x^2 + y^2$) by multiplying each of the terms of the numerator by 2. This leads to

$$dx + \left(\frac{1}{2}\right)\frac{2x\,dx + 2y\,dy}{x^2 + y^2} = 0 \quad d(x^2+y^2) = 2x\,dx + 2y\,dy$$

$$x + \frac{1}{2}\ln(x^2 + y^2) = \frac{c}{2} \quad \text{or} \quad 2x + \ln(x^2 + y^2) = c$$

Using the condition that $x = 1$ when $y = 0$, $2(1) + \ln(1^2 + 0^2) = c$, or $c = 2$. The particular solution is then $2x + \ln(x^2 + y^2) = 2$.

EXERCISES 31.3

In Exercises 1 and 2, make the given changes in the indicated examples of this section and then solve the resulting differential equations.

1. In Example 1, change the third term to $2xy^2dy$.
2. In Example 2, change the third term to $2\,dx$.

In Exercises 3–18, solve the given differential equations.

3. $x\,dy + y\,dx + x\,dx = 0$
4. $(2y + t)\,dy + y\,dt = 0$
5. $y\,dx - x\,dy + x^3\,dx = 2\,dx$
6. $x\,dy - y\,dx + y^2\,dx = 0$
7. $A^3\,dr + A^2r\,dA + r\,dA = A\,dr$
8. $\sec(xy)\,dx + (x\,dy + y\,dx) = 0$
9. $\sin x\,dy = (1 - y\cos x)dx$
10. $x(x + e^{xy})dy + y(2x + e^{xy})dx = 0$
11. $\sqrt{x^2 + y^2}\,dx - 2y\,dy = 2x\,dx$
12. $R\,dR + (R^2 + T^2 + T)\,dT = 0$
13. $\tan(x^2 + y^2)dy + x\,dx + y\,dy = 0$
14. $(x^2 + y^3)^2\,dy + 2x\,dx + 3y^2\,dy = 0$
15. $y\,dy + (y^2 - x^2)dx = x\,dx$ (Explain your solution.)
16. $e^{x+y}\,(dx + dy) + 4x\,dx = 0$
17. $10x\,dy + 5y\,dx + 3y\,dy = 0$
18. $2(u\,dv + v\,du)\ln uv + 3u^3v\,du = 0$

In Exercises 19–24, find the particular solutions to the given differential equations that satisfy the given conditions.

19. $2(x\,dy + y\,dx) + 3x^2dx = 0$; $x = 1$ when $y = 2$
20. $t\,dt + s\,ds = 2(t^2 + s^2)dt$; $t = 1$ when $s = 0$
21. $y\,dx - x\,dy = y^3dx + y^2x\,dy$; $x = 2$ when $y = 4$
22. $e^{x/y}(x\,dy - y\,dx) = y^4dy$; $x = 0$ when $y = 2$
23. $2\csc(xy)dx + x\,dy + y\,dx = 0$; $x = 0$ when $y = \pi/2$
24. $\sqrt[3]{x^2 + y^2}\,dy = 3(x\,dx + y\,dy)$; $x = 0$ when $y = 8$

In Exercises 25 and 26, rewrite each equation such that each resulting term or combination is in an integrable form, then solve the equation.

25. $e^{-x}\,dy - 2y\,dy = ye^{-x}\,dx$
26. $\cos(x + 2y)dx + 2\cos(x + 2y)dy - x\,dy = y\,dx$

Answer to Practice Exercise

1. $2x^2 - xy = c$

31.4 The Linear Differential Equation of the First Order

There is one type of differential equation of the first order and first degree for which an integrable combination can always be found. *It is the* **linear differential equation** *of the first order and is of the form*

$$dy + Py\,dx = Q\,dx \tag{31.7}$$

where P and Q are functions of x only. This type of equation occurs widely in applications.

If each side of Eq. (31.7) is multiplied by $e^{\int P\,dx}$, it becomes integrable, since the left side becomes of the form du with $u = ye^{\int P\,dx}$ and the right side is a function of x only. This is shown by finding the differential of $ye^{\int P\,dx}$. Thus,

$$d(ye^{\int P\,dx}) = e^{\int P\,dx}(dy + Py\,dx)$$

In finding the differential of $\int P\,dx$, we use the fact that, by definition, the integral and the differential are reverse processes. Thus, $d(\int P\,dx) = P\,dx$. Therefore, if each side is multiplied by $e^{\int P\,dx}$, the left side may be immediately integrated to $ye^{\int P\,dx}$, and the right-side integration may be indicated. The solution becomes

$$ye^{\int P\,dx} = \int Qe^{\int P\,dx}dx + c \tag{31.8}$$

LEARNING TIP

In finding the factor $e^{\int P dx}$, we often obtain an expression of the form $e^{\ln u}$. Using the properties of logarithms, we recall how $e^{\ln u} = u$:

$$\text{Let } y = e^{\ln u}.$$
$$\ln y = \ln e^{\ln u} = \ln u(\ln e) = \ln u$$
$$y = u \quad \text{or} \quad e^{\ln u} = u$$

EXAMPLE 1 Quadratic factor

Solve the differential equation $dy + \left(\frac{2}{x}\right)y\,dx = 4x\,dx$.

This equation fits the form of Eq. (31.7) with $P = 2/x$ and $Q = 4x$. The first expression to find is $e^{\int P\,dx}$. In this case, this is

$$e^{\int (2/x)\,dx} = e^{2\ln x} = e^{\ln x^2} = x^2 \qquad \text{see the Learning Tip to the left}$$

The left side integrates to yx^2, while the right side becomes $\int 4x(x^2)dx$. Thus,

$$ye^{\int Pdx} = \int Qe^{\int Pdx}dx + c$$

$$y(x^2) = \int (4x)(x^2)dx + c \qquad \text{using Eq. (31.8)}$$

$$yx^2 = \int 4x^3\,dx + c = x^4 + c \qquad \text{integrating}$$

$$y = x^2 + cx^{-2}$$

LEARNING TIP

In finding $e^{\int Pdx}$, the constant of integration in the exponent $\int Pdx$ can always be taken as zero, as we did in Example 1. To show why this is so, let $P = 2/x$ as in Example 1:

$$e^{\int (2/x)dx} = e^{\ln x^2 + c} = (e^{\ln x^2})(e^c) = x^2e^c$$

The solution to the differential equation, as given in Eq. (31.8), is then

$$y(x^2)(e^c) = \int 4x(x^2)(e^c)dx + c_1e^c$$

Regardless of the value of c, the factor e^c can be divided out. Therefore, it is convenient to let $c = 0$ and have $e^c = 1$.

EXAMPLE 2 Power function factor

Solve the differential equation $x\,dy - 3y\,dx = x^3\,dx$.

Putting this equation in the form of Eq. (31.7) by dividing through by x gives $dy - (3/x)y\,dx = x^2\,dx$. Here, $P = -3/x$, $Q = x^2$, and the factor $e^{\int P\,dx}$ becomes

$$e^{\int (-3/x)dx} = e^{-3\ln x} = e^{\ln x^{-3}} = x^{-3}$$

Therefore,

$$ye^{\int P dx} = \int Qe^{\int P dx} dx + c$$

$$yx^{-3} = \int x^2(x^{-3})dx + c \qquad \text{using Eq. (31.8)}$$

$$= \int x^{-1}\,dx + c = \ln x + c$$

$$y = x^3(\ln x + c)$$

Practice Exercise

1. Find the general solution of the differential equation $x\,dy - 2y\,dx = 2x^4\,dx$.

EXAMPLE 3 Exponential factor

Solve the differential equation $dy + y\,dx = x\,dx$.

Here, $P = 1$, $Q = x$, and $e^{\int P dx} = e^{\int (1) dx} = e^x$. Therefore,

$$ye^x = \int xe^x\,dx + c = e^x(x - 1) + c$$

$$y = x - 1 + ce^{-x} \qquad \text{using Eq. (31.8) and integrating by parts or using tables}$$

EXAMPLE 4 Trigonometric factor

Solve the differential equation $\cos x\dfrac{dy}{dx} = 1 - y\sin x$.

Writing this in the form of Eq. (31.7), we have

$$dy + y\tan x\,dx = \sec x\,dx \qquad \text{dividing by } \cos x$$

Thus, with $P = \tan x$, we have

$$e^{\int P\,dx} = e^{\int \tan x\,dx} = e^{-\ln \cos x} = \sec x \qquad \text{using } e^{\ln u} = u$$

$$y\sec x = \int \sec^2 x\,dx = \tan x + c \qquad \text{using Eq. (31.8)}$$

$$y = \sin x + c\cos x$$

EXAMPLE 5 Finding a particular solution

For the differential equation $dy = (1 - 2y)x\,dx$, find the particular solution such that $x = 0$ when $y = 2$.

The solution proceeds as follows:

$$dy + 2xy\,dx = x\,dx \qquad \text{form of Eq. (31.7)}$$

$$e^{\int P\,dx} = e^{\int 2x\,dx} = e^{x^2} \qquad \text{find } e^{\int P\,dx}$$

$$ye^{x^2} = \int xe^{x^{2}d}\,dx \qquad \text{using Eq. (31.8)}$$

$$= \frac{1}{2}e^{x^2} + c \qquad \text{general solution}$$

$$(2)(e^0) = \frac{1}{2}(e^0) + c, \quad 2 = \frac{1}{2} + c, \quad c = \frac{3}{2} \qquad x = 0, y = 2;\ \text{evaluate } c$$

$$ye^{x^2} = \frac{1}{2}e^{x^2} + \frac{3}{2} \qquad \text{substitute } c = \frac{3}{2}$$

$$y = \frac{1}{2}(1 + 3e^{-x^2}) \qquad \text{particular solution}$$

EXERCISES 31.4

In Exercises 1 and 2, make the given changes in the indicated examples of this section and then solve the resulting differential equations.

1. In Example 1, change the right side to $3\,dx$.
2. In Example 3, change the right side to $2\,dx$.

In Exercises 3–28, solve the given differential equations.

3. $dy + y\,dx = e^{-x}\,dx$
4. $dy + 3y\,dx = e^{-3x}\,dx$
5. $dy + 2y\,dx = e^{-4x}\,dx$
6. $di + i\,dt = e^{-t}\cos t\,dt$
7. $\dfrac{dy}{dx} - 5y = 10$
8. $2\dfrac{dy}{dx} = 5 - 6y$
9. $dy = 3x^2(2 - y)\,dx$
10. $x\,dy + 3y\,dx = dx$
11. $2x\,dy + y\,dx = 8x^3\,dx$
12. $3x\,dy - y\,dx = 9x\,dx$
13. $dr + r\cot\theta\,d\theta = d\theta$
14. $y' = x^2y + 3x^2$
15. $\sin x\dfrac{dy}{dx} = 1 - y\cos x$
16. $\dfrac{dv}{dt} - \dfrac{v}{t} = 4\ln t$
17. $y' + y = x + e^x$
18. $y' + 2y = \sin x$
19. $ds = (te^{4t} + 4s)dt$
20. $y' - 2y = 8e^{6x}$
21. $y' = x^3(1 - 4y)$
22. $y' + y\tan x = -\sin x$
23. $x\dfrac{dy}{dx} = y + (x^2 - 1)^2$
24. $dy = dt - \dfrac{y\,dt}{(1 + t^2)\tan^{-1} t}$
25. $\sqrt{1 + x^2}\,dy + x(1 + y)dx = 0$
26. $(1 + x^2)dy + xy\,dx = x\,dx$
27. $\tan\theta\dfrac{dr}{d\theta} - r = \tan^2\theta$
28. $y' + y = y^2$ (Solve by letting $y = 1/u$ and solving the resulting linear equation for u.)

In Exercises 29 and 30, solve the given differential equations. Explain how each can be solved using either of two different methods.

29. $y' = 2(1 - y)$
30. $x\,dy = (2x - y)\,dx$

In Exercises 31–36, find the indicated particular solutions of the given differential equations.

31. $\dfrac{dy}{dx} + 2y = e^{-x}$; $x = 0$ when $y = 4$
32. $dq - 4q\,du = 2\,du$; $q = 2$ when $u = 0$
33. $y' + 2y\cot x = 4\cos x$; $x = \pi/2$ when $y = 1/3$
34. $y'\sqrt{x} + \frac{1}{2}y = e^{\sqrt{x}}$; $x = 1$ when $y = 3$
35. $(\sin x)y' + y = \tan x$; $x = \pi/4$ when $y = 0$
36. $f(x)dy + 2yf'(x)dx = f(x)f'(x)dx$; $f(x) = -1$ when $y = 3$

In Exercises 37–42, solve the given problems.

37. The differential equation $y' + P(x)y = Q(x)y^2$ is not linear. Show that the substitution $u = y^{-1}$ will transform it into a linear equation.
38. An equation used in the analysis of rocket motion is $m\,dv + kv\,dt = 0$, where m and k are positive constants. Solve this equation for v as a function of t in two ways.
39. The electric current i in a circuit with a voltage V, resistance R, and inductance L, is a function of the time t given by $\dfrac{di}{dt} + \dfrac{R}{L}i = \dfrac{V}{L}$. Solve for i as a function of t if the initial current is zero.
40. If A dollars are placed in an account that pays 5% interest, compounded *continuously*, and A dollars are added to the account each year, the number of dollars n in the account after t years is given by $dn/dt = A + 0.05n$. Solve for n as a function of t.
41. In a certain national forest, dead leaves fall and accumulate on the ground at the rate of 20 kg per square metre per year. Once on the ground, the leaves decompose at the rate of 80% per year. This leads to the differential equation $\dfrac{dL}{dt} = 20 - 0.8L$, where L is the amount of leaves on the ground (in kg/m^2) and t is the time in years. Express L as a function of t.
42. A drug is given to a patient intravenously at a rate of 0.5 mg per hour. The person's body continuously removes 2.0% of the drug from the blood per hour through absorption. If y is the amount of the drug (in mg) in the blood at time t (in h), then $\dfrac{dy}{dt} = 0.5 - 0.02y$. (a) Express y as a function of t if $y = 0$ when $t = 0$. (b) What is the limiting value of y as $t \to \infty$?

Answer to Practice Exercise

1. $y = x^4 + cx^2$

31.5 Numerical Solutions of First-Order Equations

Many differential equations do not have exact solutions. Therefore, in this section, we show one basic method and one more advanced method of solving such equations numerically.

EULER'S METHOD

To find an approximate solution to a differential equation of the form $dy/dx = f(x, y)$, that passes through a known point (x_0, y_0), we write the equation as $dy = f(x, y)dx$ and then approximate dy as $y_1 - y_0$, and replace dx with Δx. From Section 24.8, we recall that Δy closely approximates dy for a small dx and that $dx = \Delta x$. This gives us

■ Euler's method is named for the Swiss mathematician Leonhard Euler (1707–1783).

$$y_1 = y_0 + f(x_0, y_0)\Delta x \qquad \text{and} \qquad x_1 = x_0 + \Delta x$$

Therefore, we now know another point (x_1, y_1) that is on (or very nearly on) the curve of the solution. We can now repeat this process using (x_1, y_1) as a known point to obtain a next point (x_2, y_2). Continuing this process, we can get a series of points that are approximately on the solution curve. The method is called ***Euler's method***.

EXAMPLE 1 Euler's method

For the differential equation $dy/dx = x + y$, use Euler's method to find the y-values of the solution for $x = 0$ to $x = 0.5$ with $\Delta x = 0.1$, if the curve of the solution passes through $(0, 1)$.

Using the method outlined above, we have $x_0 = 0$, $y_0 = 1$, and

$$y_1 = 1 + (0 + 1)(0.1) = 1.1 \quad \text{and} \quad x_1 = 0 + 0.1 = 0.1$$

This tells us that the curve passes (or nearly passes) through the point $(0.1, 1.1)$. Assuming this point is correct, we use it to find the next point on the curve.

$$y_2 = 1.1 + (0.1 + 1.1)(0.1) = 1.22 \quad \text{and} \quad x_2 = 0.1 + 0.1 = 0.2$$

Therefore, the next approximate point is $(0.2, 1.22)$. Continuing this process, we find a set of points that would approximately satisfy the function that is the solution of the differential equation. Tabulating results, we have the following table:

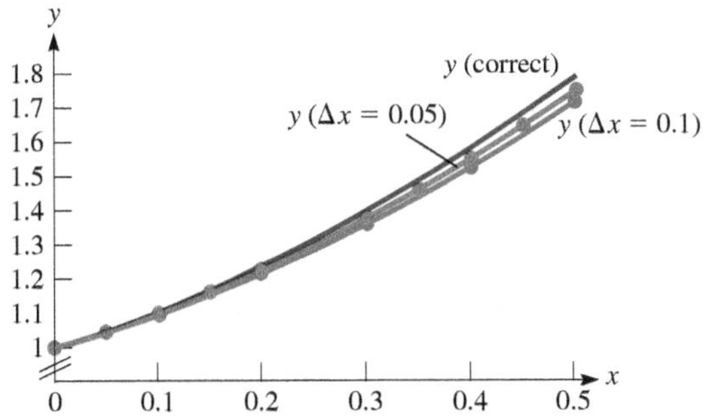

Fig. 31.2

x	y	Correct Value of y
0.0	1.0000	1.0000
0.1	1.1000	1.1103
0.2	1.2200	1.2428
0.3	1.3620	1.3997
0.4	1.5282	1.5836
0.5	1.7210	1.7974

the values shown have been rounded off, although more digits were carried in the calculations

In this case, we are able to find the correct values since the equation can be written as $dy/dx - y = x$, and the solution is $y = 2e^x - x - 1$. Although numerical methods are generally used with equations that cannot be solved exactly, we chose this equation so that we could compare values obtained with known values.

We can see that as x increases, the error in y increases. More accurate values can be found by using smaller values of Δx. In Fig. 31.2 the solution curves using $\Delta x = 0.1$ and $\Delta x = 0.05$ are shown along with the correct values of y.

Euler's method is easy to use and understand, but it is less accurate than other methods. We will now discuss one of the more accurate methods.

RUNGE–KUTTA METHOD

■ The Runge–Kutta method is named for the German mathematicians Carl Runge (1856–1927) and Martin Kutta (1867–1944).

For more accurate numerical solutions of a differential equation, the ***Runge–Kutta method*** is often used. Starting at a first point (x_0, y_0), the coordinates of the second point (x_1, y_1) are found by using a weighted average of the slopes calculated at the points where $x = x_0$, $x = x_0 + \frac{1}{2}\Delta x$, and $x = x_0 + \Delta x$. The formulas for y_1 and x_1 are

$$y_1 = y_0 + \frac{1}{6}H(J + 2K + 2L + M) \qquad \text{and} \qquad x_1 = x_0 + H \qquad (\text{for convenience, } H = \Delta x)$$

where

$$J = f(x_0, y_0)$$
$$K = f(x_0 + 0.5H, y_0 + 0.5HJ)$$
$$L = f(x_0 + 0.5H, y_0 + 0.5HK)$$
$$M = f(x_0 + H, y_0 + HL)$$

We have used uppercase letters to correspond to calculator use. Traditional sources normally use *h* for *H* and a lowercase letter (such as *k*) with subscripts for *J*, *K*, *L*, and *M*, and express 0.5 as 1/2.

As with Euler's method, once (x_1, y_1) is determined, we use the formulas again to find (x_2, y_2) by replacing (x_0, y_0) with (x_1, y_1). The following example illustrates the use of the Runge–Kutta method.

EXAMPLE 2 Runge–Kutta method

For the differential equation $dy/dx = x + \sin xy$, use the Runge–Kutta method to find *y*-values of the solution for $x = 0$ to $x = 0.5$ with $\Delta x = 0.1$, if the curve of the solution passes through $(0, 0)$.

Using the formulas and method outlined above, we have the following solution, with calculator notes to the right of the equations. Also, calculator symbols are used on the right sides of the equations to indicate the way in which they should be entered.

$x_0 = 0$ store as *X*

$y_0 = 0$ store as *Y*

$H = 0.1$ store as *H*

$J = X + \sin XY = 0$ store as *J*

$K = X + 0.5H + \sin[(X + 0.5H)(Y + 0.5HJ)] = 0.05$ store as *K*

$L = X + 0.5H + \sin[(X + 0.5H)(Y + 0.5HK)] = 0.050\,125$ store as *L*

$M = X + H + \sin[(X + H)(Y + HL)] = 0.100\,501\,25$ store as *M*

$y_1 = Y + (H/6)(J + 2K + 2L + M) = 0.005\,012\,520\,8$ store as *Y*

$x_1 = X + H = 0.1$ store as *X*

We now use (x_1, y_1) as we just used (x_0, y_0) to get the next point (x_2, y_2), which is then used to find (x_3, y_3), and so on. A table showing calculator values and a graph of these values is given in Fig. 31.3.

x	*y*
0.0	0.0
0.1	0.0050125208
0.2	0.0202013395
0.3	0.0460278455
0.4	0.0832868181
0.5	0.1331460062

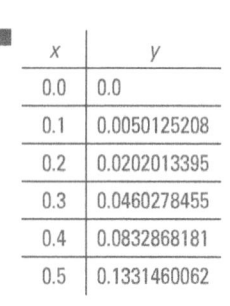

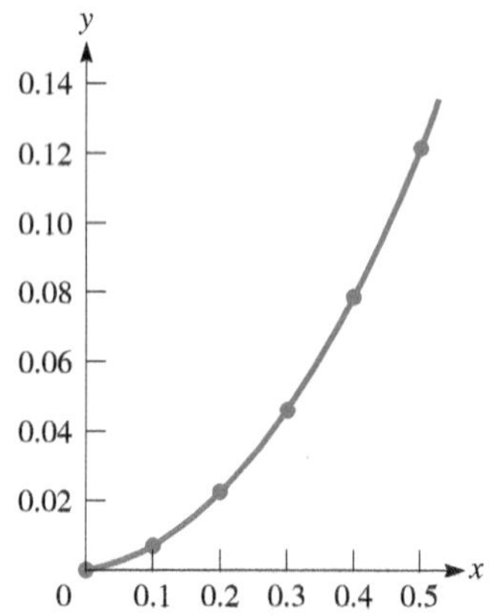

Fig. 31.3

EXERCISES 31.5

In Exercises 1–8, use Euler's method to find y-values of the solution for the given values of x and Δx, if the curve of the solution passes through the given point. Check the results against known values by solving the differential equations exactly.

1. $\frac{dy}{dx} = x + 1$; $x = 0$ to $x = 1$; $\Delta x = 0.2$; $(0, 1)$
2. $\frac{dy}{dx} = \sqrt{2x + 1}$; $x = 0$ to $x = 1.2$; $\Delta x = 0.3$; $(0, 2)$
3. $\frac{dy}{dx} = y(0.4x + 1)$; $x = -0.2$ to $x = 0.3$; $\Delta x = 0.1$; $(-0.2, 2)$
4. $\frac{dy}{dx} = y + e^x$; $x = 0$ to $x = 0.5$; $\Delta x = 0.1$; $(0, 0)$
5. The differential equation of Exercise 1 with $\Delta x = 0.1$
6. The differential equation of Exercise 2 with $\Delta x = 0.1$
7. The differential equation of Exercise 3 with $\Delta x = 0.05$
8. The differential equation of Exercise 4 with $\Delta x = 0.05$

In Exercises 9–14, use the Runge–Kutta method to find y-values of the solution for the given values of x and Δx, if the curve of the solution passes through the given point.

9. $\frac{dy}{dx} = xy + 1$; $x = 0$ to $x = 0.4$; $\Delta x = 0.1$; $(0, 0)$
10. $\frac{dy}{dx} = x^2 + y^2$; $x = 0$ to $x = 0.4$; $\Delta x = 0.1$; $(0, 1)$

11. $\frac{dy}{dx} = e^{xy}$; $x = 0$ to $x = 1$; $\Delta x = 0.2$; $(0, 0)$

12. $\frac{dy}{dx} = \sqrt{1 + xy}$; $x = 0$ to $x = 0.2$; $\Delta x = 0.05$; $(0, 1)$

13. $\frac{dy}{dx} = \cos(x + y)$; $x = 0$ to $x = 0.6$; $\Delta x = 0.1$; $(0, \pi/2)$

14. $\frac{dy}{dx} = y + \sin x$; $x = 0.5$ to $x = 1.0$; $\Delta x = 0.1$; $(0.5, 0)$

In Exercises 15–18, solve the given problems.

15. In Example 1, use Euler's method to find the y-values from $x = 0$ to $x = 3$ with $\Delta x = 1$. Compare with the value found using the exact solution. Comment on the use of Euler's method in finding the value of y for $x = 3$.

16. For the differential equation $dy/dx = x + 1$, if the curve of the solution passes through $(0, 0)$, calculate the y-value for $x = 0.04$ with $\Delta x = 0.01$. Find the exact solution, and compare the result using three terms of the Maclaurin series that represents the solution.

17. An electric circuit contains a 1-H inductor, a 2-Ω resistor, and a voltage source of $\sin t$. The resulting differential equation relating the current i and the time t is $di/dt + 2i = \sin t$. Find i after 0.5 s by Euler's method with $\Delta t = 0.1$ s if the initial current is zero. Solve the equation exactly and compare the values.

18. An object is being heated such that the rate of change of the temperature T (in °C) with respect to time t (in min) is $dT/dt = \sqrt[3]{1 + t^3}$. Find T for $t = 5$ min by using the Runge–Kutta method with $\Delta t = 1$ min, if the initial temperature is 0°C.

31.6 Elementary Applications

The differential equations of the first order and first degree we have discussed thus far have numerous applications in geometry and the various fields of technology. In this section, we illustrate some of these applications.

EXAMPLE 1 Find a curve given its slope

The slope of a curve is given by the expression $6xy$. Find the equation of the curve if it passes through the point $(2, 1)$.

Since the slope is $6xy$, the differential equation for the curve is

$$\frac{dy}{dx} = 6xy$$

We now want to find the particular solution of this equation for which $x = 2$ when $y = 1$. The solution follows:

$$\frac{dy}{y} = 6x\,dx \qquad \text{separate variables}$$

$$\ln y = 3x^2 + c \qquad \text{general solution}$$

$$\ln 1 = 3(2^2) + c \qquad \text{evaluate } c$$

$$0 = 12 + c, \qquad c = -12$$

$$\ln y = 3x^2 - 12 \qquad \text{particular solution}$$

The graph of this solution is shown in Fig. 31.4.

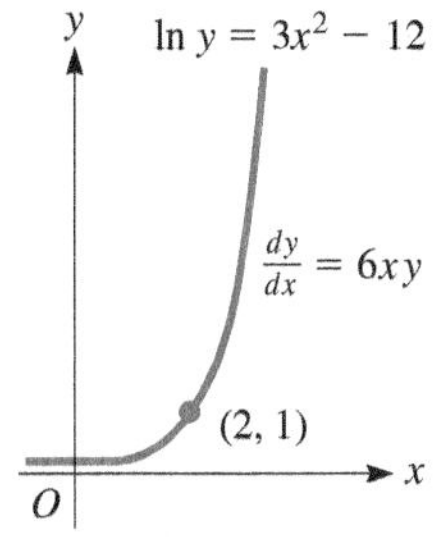

Fig. 31.4

EXAMPLE 2 Orthogonal trajectories

A curve that intersects all members of a family of curves at right angles is called an **orthogonal trajectory** *of the family.* Find the equations of the orthogonal trajectories of the parabolas $x^2 = ky$. As before, each value of k gives us a particular member of the family.

The derivative of the given equation is $dy/dx = 2x/k$. This equation contains the constant k, which depends on the point (x, y) on the parabola. ***Eliminating this constant between the equations of the parabolas and the derivative***, we have

$$k = \frac{x^2}{y} \qquad \frac{dy}{dx} = \frac{2x}{k} = \frac{2x}{x^2/y} \quad \text{or} \quad \frac{dy}{dx} = \frac{2y}{x}$$

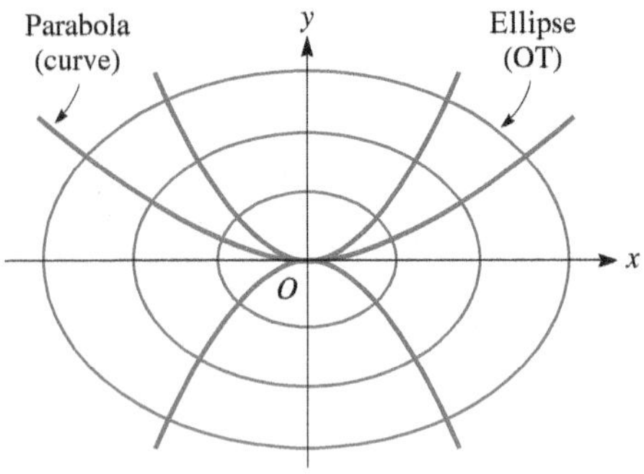

Fig. 31.5

This equation gives a general expression for the slope of any of the members of the family. For a curve to be perpendicular, its slope must equal the negative reciprocal of this expression, or the slope of the orthogonal trajectories must be

$$\left.\frac{dy}{dx}\right|_{OT} = -\frac{x}{2y} \qquad \text{this equation must not contain the constant } k$$

Solving this differential equation gives the family of orthogonal trajectories.

$$2y\,dy = -x\,dx$$
$$y^2 = -\frac{x^2}{2} + \frac{c}{2}$$
$$2y^2 + x^2 = c \qquad \text{orthogonal trajectories}$$

Thus, the orthogonal trajectories are ellipses. Note in Fig. 31.5 that each parabola intersects each ellipse at right angles.

EXAMPLE 3 Radioactivity

■ See the chapter introduction.

■ Radioactivity was discovered in 1898 by the French physicist Henri Becquerel (1852–1908). Carbon dating was developed in 1947 by the U.S. chemist Willard Libby (1908–1980).

Radioactive elements decay at rates that are proportional to the amount of the element present. Carbon-14 decays such that one-half of an original amount decays into other forms in about 5730 years. By measuring the proportion of carbon-14 in remains at a given site, the approximate age of the remains can be determined. This method is called *carbon dating*.

The analysis of some detrital wood washed from the Dome Glacier in Alberta showed that the concentration of carbon-14 was 47.7% of the concentration that new wood would have. Determine the equation relating the amount of carbon-14 present with the time and then determine the age of the wood from the Dome Glacier. (Scientists use paleoecological samples of this kind to study past climate change.)

Let N_0 be the original amount and N be the amount present at any time t (in years). The rate of decay can be expressed as a derivative. Therefore, since the rate of change is proportional to N, we have the equation

$$\frac{dN}{dt} = kN$$

Solving this differential equation, we have

$$\frac{dN}{N} = k\,dt \qquad \text{separate variables}$$
$$\ln N = kt + \ln c \qquad \text{general solution}$$
$$\ln N_0 = k(0) + \ln c \qquad N = N_0 \text{ for } t = 0$$
$$c = N_0 \qquad \text{solve for } c$$
$$\ln N = kt + \ln N_0 \qquad \text{substitute } N_0 \text{ for } c$$
$$\ln N - \ln N_0 = kt \qquad \text{use properties of logarithms}$$
$$\ln\frac{N}{N_0} = kt$$
$$N = N_0 e^{kt} \qquad \text{exponential form}$$

Now, using the condition that one-half of carbon-14 decays in 5730 years, we have $N = N_0/2$ when $t = 5730$ years. This gives

$$\frac{N_0}{2} = N_0 e^{5730k} \quad \text{or} \quad \frac{1}{2} = (e^k)^{5730} \text{ and}$$

$$e^k = 0.5^{1/5730} \qquad \text{both sides to the power } 1/5730$$

Therefore, the equation relating N and t is

$$N = N_0(0.5)^{t/5730} \quad \text{substitute in the exponential form}$$

Since the present concentration is $N = 0.477N_0$, we can solve for t:

$$0.477N_0 = N_0(0.5)^{t/5730} \quad \text{or} \quad 0.477 = 0.5^{t/5730}$$

$$\ln 0.477 = \ln 0.5^{t/5730}, \qquad \ln 0.477 = \frac{t}{5730}\ln 0.5 \quad \text{solving an exponential equation}$$

$$t = 5730\frac{\ln 0.477}{\ln 0.5} = 6120 \text{ years}$$

Therefore, the wood from the Dome Glacier is about 6120 years old.

EXAMPLE 4 Electric circuits

The general equation relating the current i, voltage E, inductance L, capacitance C, and resistance R in a simple electric circuit (see Fig. 31.6) is

$$L\frac{di}{dt} + Ri + \frac{q}{C} = E \tag{31.9}$$

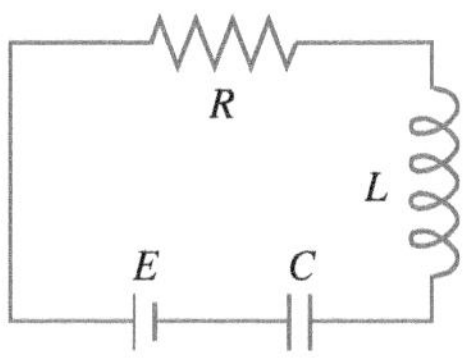

Fig. 31.6

where q is the charge on the capacitor. Find the general expression for the current in a circuit containing an inductance, a resistance, and a voltage source if $i = 0$ when $t = 0$.

The differential equation for this circuit is

$$L\frac{di}{dt} + Ri = E$$

Using the method of the linear differential equation of the first order, we have the equation

$$di + \frac{R}{L}i\,dt = \frac{E}{L}dt$$

The factor $e^{\int P\,dt}$ is $e^{\int (R/L)\,dt} = e^{(R/L)t}$. This gives

$$ie^{(R/L)t} = \frac{E}{L}\int e^{(R/L)t}\,dt = \frac{E}{R}e^{(R/L)t} + c$$

Letting the current be zero for $t = 0$, we have $c = -E/R$. The result is

$$ie^{(R/L)t} = \frac{E}{R}e^{(R/L)t} - \frac{E}{R}$$

$$i = \frac{E}{R}(1 - e^{-(R/L)t})$$

(We can see that $i \to E/R$ as $t \to \infty$. In practice, the exponential term becomes negligible very quickly.)

EXAMPLE 5 Mixing problem—salt solution

Fifty litres of brine originally containing 3.00 kg of salt are in a tank into which 2.00 L of water run each minute with the same amount of mixture running out each minute. How much salt is in the tank after 10.0 min?

Let x = the number of kilograms of salt in the tank after t minutes. Each litre of brine contains $x/50$ kg of salt, and in time dt, $2\,dt$ L *of mixture leave the tank with* $(x/50)(2\,dt)$ kg *of salt.* The amount of salt that is leaving may also be written as $-dx$ (the minus sign is included to show that x is decreasing). Thus,

$$-dx = \frac{2x\,dt}{50} \quad \text{or} \quad \frac{dx}{x} = -\frac{dt}{25}$$

This leads to $\ln x = -(t/25) + \ln c$. Using the fact that $x = 3.00$ kg when $t = 0$, we find that $\ln 3.00 = \ln c$, or $c = 3.00$. Therefore,

$$x = 3.00e^{-t/25}$$

is the general expression for the amount of salt in the tank at time t. Therefore, when $t = 10.0$ min, we have

$$x = 3.00e^{-10/25} = 3.00e^{-0.4} = 3.00(0.670) = 2.01 \text{ kg}$$

There are 2.01 kg of salt in the tank after 10.0 min. (Although the data were given with three significant digits, we did not use all significant digits in writing the equations that were used.)

EXAMPLE 6 Motion in a resisting medium

An object moving through (or across) a resisting medium often experiences a retarding force approximately proportional to the velocity as well as the force that causes the motion. An example is a ball falling due to the force of gravity, with air resistance producing a retarding force. Applying Newton's laws of motion (from physics) to the ball leads to the equation

■ There is at least some resistance to the motion of any object moving through a medium (not a vacuum). The exact nature of the resistance is not usually known.

$$m\frac{dv}{dt} = F - kv \qquad \textbf{(31.10)}$$

where m is the mass of the object, v is the velocity of the object, t is the time, F is the force causing the motion, and k $(k > 0)$ is a constant. The quantity kv is the retarding force.

■ Slug is the unit of mass when force is expressed in pounds.

We assume that these conditions hold for a falling object whose mass is 5.00 kg and experiences a force (its own weight) of 49.0 N. The object starts from rest, and the air causes a retarding force numerically equal to 0.200 times the velocity.

Substituting in Eq. (31.10) and then solving the differential equation, we have

$$5\frac{dv}{dt} = 49 - 0.2\,v$$

$$\frac{5\,dv}{49 - 0.2v} = dt \qquad \text{separate variables}$$

$$-25\ln(49 - 0.2v) = t - 25\ln c \qquad \text{integrate}$$

$$\ln(49 - 0.2v) = -\frac{t}{25} + \ln c \qquad \text{solve for } v$$

$$\ln\frac{49 - 0.2v}{c} = -\frac{t}{25}$$

$$49 - 0.2v = ce^{-t/25}$$

$$0.2v = 49 - ce^{-t/25}$$

$$v = 5(49 - ce^{-t/25}) \qquad \text{general solution}$$

Since the object started from rest, $v = 0$ when $t = 0$. Thus,

■ The data were given to three significant digits, but not all significant digits were used in writing the equations.

$$0 = 5(49 - c) \quad \text{or} \quad c = 49 \qquad \text{evaluate } c$$

$$v = 245(1 - e^{-t/5}) \qquad \text{particular solution}$$

$$= 245(1 - e^{-0.200}) = 245(1 - 0.819) \qquad \text{evaluating } v \text{ for } t = 5.00 \text{ s}$$

$$= 44.4 \text{ m/s}$$

After 5.00 s, the velocity is 44.4 m/s. Without the air resistance, the velocity would be about 49.0 m/s.

EXERCISES 31.6

In Exercises 1–4, make the given changes in the indicated examples of this section, and then solve the resulting problems.

1. In Example 2, change $x^2 = ky$ to $y^2 = kx$.
2. In Example 4, if the term $L\frac{di}{dt}$ is deleted (no inductance) in Eq. (31.9), find the expression for the charge in the circuit. (There is a capacitance C, and $i = dq/dt$.) The initial charge is zero.
3. In Example 5, change 2.00 L to 1.00 L.
4. In Example 6, change 0.200 to 0.100 (for the retarding force).

In Exercises 5–8, find the equation of the curve for the given slope and point through which it passes. Use a graphing calculator to display the curve.

5. Slope given by $2x/y$; passes through $(2, 3)$
6. Slope given by $-y/(x + y)$; passes through $(-1, 3)$
7. Slope given by $y + x$; passes through $(0, 1)$
8. Slope given by $-2y + e^{-x}$; passes through $(0, 2)$

In Exercises 9–12, find the equation of the orthogonal trajectories of the curves for the given equations. Use a graphing calculator to display at least two members of the family of curves and at least two of the orthogonal trajectories.

9. The exponential curves $y = ke^x$
10. The hyperbolas $x^2 - y^2 = a^2$
11. The curves $y = k(\sec x + \tan x)$
12. The family of circles, all with centres at the origin

In Exercises 13–50, solve the given problems by solving the appropriate differential equation.

13. The isotope cobalt-60, with half-life of 5.27 years, is used in treating cancerous tumours. What percent of an initial amount remains after 2.00 years?
14. Radium-226 decays such that 10% of the original amount disintegrates in 246 years. Find the half-life (the time for one-half of the original amount to disintegrate) of radium-226.
15. A possible health hazard in the home is radon gas. It is radioactive and about 90.0% of an original amount disintegrates in 12.7 days. Find the half-life of radon gas. (The problem with radon is that it is a gas and is being continually produced by the radioactive decay of minute amounts of radioactive radium found in the soil and rocks of an area.)
16. Noting Example 3, another element used to date more recent events is helium-3, which has a half-life of 12.3 years. If a building has 5.0% of helium-3 that a new building would have, about how old is the building?
17. A radioactive element leaks from a nuclear power plant at a constant rate r, and it decays at a rate that is proportional to the amount present. Find the relation between the amount N present in the environment in which it leaks and the time t, if $N = 0$ when $t = 0$.
18. Most use of the pesticide DDT was banned in Canada in 1970. It has been found that 19% of an initial amount of DDT is degraded into harmless products in 10 years. If no DDT has been used in an area since 1970, in what year will the concentration become only 20% of the 1970 amount?
19. The growth of the population P of a nation with a constant immigration rate I may be expressed as $\frac{dP}{dt} = kP + I$, where t is in years. If the population of Canada in 2017 was 36.7 million and about 0.250 million immigrants enter Canada each year, what will the population of Canada be in 2030, given that the growth rate k is about 1.0% (0.010) annually?
20. Assuming that the natural environment of the earth is limited and that the maximum population it can sustain is M, the rate of growth of the population P is given by the *logistic* differential equation $\frac{dP}{dt} = kP(M - P)$. Using this equation for the earth, if $P = 7.4$ billion in 2016, $k = 0.00040$, $M = 25$ billion, what will be the population of the earth in 2026?
21. The rate of change of the radial stress S on the walls of a pipe with respect to the distance r from the axis of the pipe is given by $r\frac{dS}{dr} = 2(a - S)$, where a is a constant. Solve for S as a function of r.
22. The velocity v of a meteor approaching the earth is given by $v\frac{dv}{dr} = -\frac{GM}{r^2}$, where r is the distance from the centre of the earth, M is the mass of the earth, and G is a universal gravitational constant. If $v = 0$ for $r = r_0$, solve for v as a function of r.
23. Assume that the rate at which highway construction increases is directly proportional to the total length M of all highways already completed at time t (in years). Solve for M as a function of t if $M = 5250$ km for a certain region when $t = 0$ and $M = 5460$ km for $t = 2.00$ years.
24. The marginal profit function gives the change in the total profit P of a business due to a change in the business, such as adding new machinery or reducing the size of the sales staff. A company determines that the marginal profit dP/dx is $e^{-x^2} - 2Px$, where x is the amount invested in new machinery. Determine the total profit (in thousands of dollars) as a function of x, if $P = 0$ for $x = 0$.
25. According to Newton's law of cooling, the rate at which a body cools is proportional to the difference in temperature between it and the surrounding medium. Assuming Newton's law holds, how long will it take a cup of hot water, initially at 90°C, to cool to 40°C if the room temperature is 25°C, if it cools to 60°C in 5.0 min?
26. One model of animal growth is that the rate of growth is proportional to remaining growth expected to develop. If a certain animal has a length x (in m) at age t (in years), find the equation relating the animal's length as a function of t, if the expected length is L.
27. If interest in a bank account is compounded continuously, the amount grows at a rate that is proportional to the amount present in the account. Interest that is compounded daily very closely approximates this situation. Determine the amount in an account after one year if \$1000 is placed in the account and it pays 4% interest per year, compounded continuously.
28. The rate of change in the intensity I of light below the surface of the ocean with respect to the depth y is proportional to I. If the intensity at 5.0 m is 50% of the intensity I_0 at the surface, at what depth is the intensity 15% of I_0?

29. For a DNA sample in a liquid containing a solute of constant concentration c_0, the rate at which the concentration $c(t)$ of solute in the sample changes is proportional to $c_0 - c(t)$. Find $c(t)$ if $c(0) = 0$.

30. In a town of N persons, during a flu epidemic, it was determined that the rate dS/dt at which persons were being infected was proportional to the product of the number S of infected persons and the number $N - S$ of healthy persons. Find S as a function of t.

31. If the current in an RL circuit with a voltage source E is zero when $t = 0$ (see Example 4), show that $\lim_{t \to \infty} i = E/R$. See Fig. 31.7.

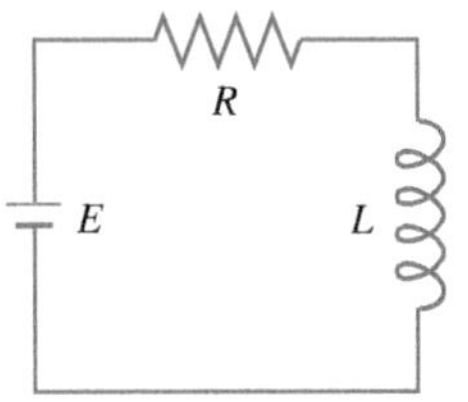

Fig. 31.7

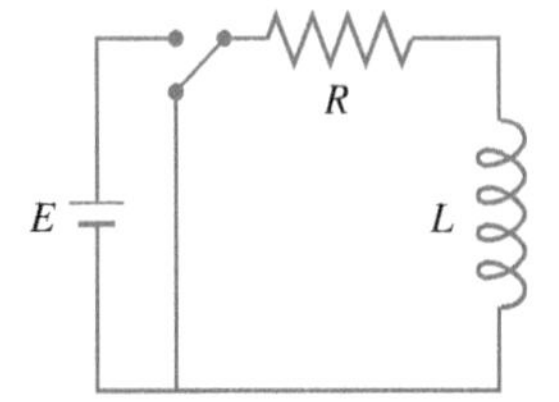

Fig. 31.8

32. If a circuit contains only an inductance and a resistance, with $L = 2.0$ H and $R = 30\ \Omega$, find the current i as a function of time t if $i = 0.020$ A when $t = 0$. See Fig. 31.8.

33. An amplifier circuit contains a resistance R, an inductance L, and a voltage source $E \sin \omega t$. Express the current in the circuit as a function of the time t if the initial current is zero.

34. A radio transmitter circuit contains a resistance of 2.0 Ω, a variable inductor of $100 - t$ henrys, and a voltage source of 4.0 V. Find the current i in the circuit as a function of the time t for $0 \le t \le 100$ s if the initial current is zero.

35. If a circuit contains only a resistance and a capacitance C, find the equation relating the charge q on the capacitor in terms of the time t if $i = dq/dt$ and $q = q_0$ when $t = 0$. See Fig. 31.9.

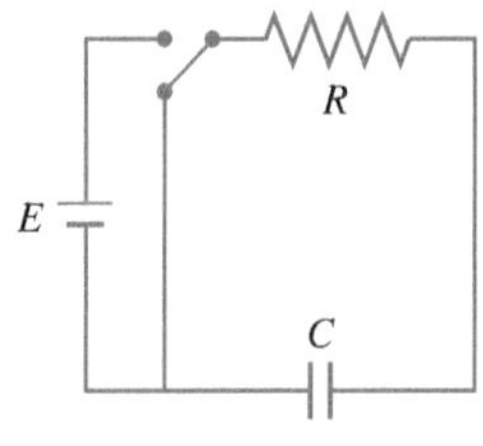

Fig. 31.9

36. One hundred litres of brine originally containing 4.0 kg of salt are in a tank into which 5.0 L of water run each minute. The same amount of mixture from the tank leaves each minute. How much salt is in the tank after 20 min?

37. An object falling under the influence of gravity has a variable acceleration given by $9.8 - v$, where v represents the velocity. If the object starts from rest, find an expression for the velocity in terms of the time. Also, find the limiting value of the velocity (find $\lim_{t \to \infty} v$).

38. In a ballistics test, a bullet is fired into a sandbag. The acceleration of the bullet within the sandbag is $-15\sqrt{v}$, where v is the velocity (in m/s). When will the bullet stop if it enters the sandbag at 300 m/s?

39. A boat with a mass of 150 kg is being towed at 8.0 km/h. The tow rope is then cut, and a motor that exerts a force of 80 N on the boat is started. If the water exerts a retarding force that numerically equals twice the velocity, what is the velocity of the boat 3.0 min later?

40. A parachutist is falling at a rate of 60.0 m/s when her parachute opens. If the air resists the fall with a force equal to $5v^2$, find the velocity as a function of time. The person and equipment have a combined mass of 100 kg (weight is 980 N).

41. For each cycle, a roller mechanism follows a path described by $y = 2x - x^2$, $y \ge 0$, such that $dx/dt = 6t - 3t^2$. Find x and y (in cm) in terms of the time t (in s) if x and y are zero for $t = 0$.

42. In studying the flow of water in a stream, it is found that an object follows the hyperbolic path $y(x + 1) = 10$ such that $(t + 1)\,dx = (x - 2)\,dt$. Find x and y (in m) in terms of time t (in s) if $x = 4$ m and $y = 2$ m for $t = 0$.

43. The rate of change of air pressure p (in kPa) with respect to height h (in m) is approximately proportional to the pressure. If the pressure is 100 kPa when $h = 0$ and $p = 80$ kPa when $h = 2000$ m, find the expression relating pressure and height.

44. Water flows from a vertical cylindrical storage tank through a hole of area A at the bottom of the tank. The rate of flow is $2.6\,A\sqrt{h}$, where h is the distance (in m) from the surface of the water to the hole. If h changes from 9.0 m to 8.0 m in 16 min, how long will it take the tank to empty? See Fig. 31.10.

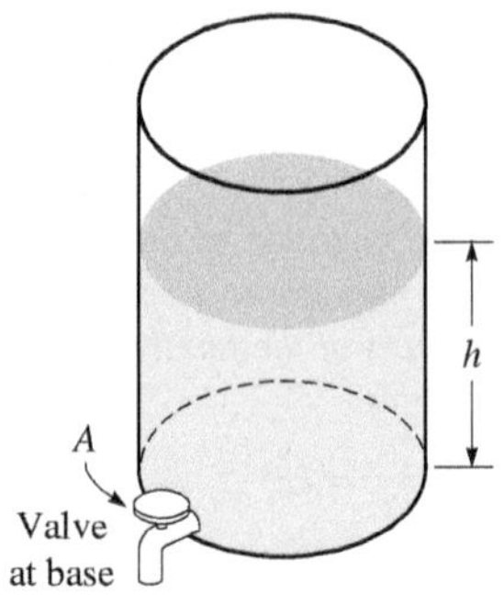

Fig. 31.10

45. Assume that the rate of depreciation of an object is proportional to its value at any time t. If a car costs \$33 000 new and its value 3 years later is \$19 700, what is its value 11 years after it was purchased?

46. Assume that sugar dissolves at a rate proportional to the undissolved amount. If there are initially 525 g of sugar and 225 g remain after 4.00 min, how long does it take to dissolve 375 g?

47. Fresh air is being circulated into a room whose volume is 120 m³. Under specified conditions the number of cubic metres x of carbon dioxide present at any time t (in min) is found by solving the differential equation $dx/dt = 1 - 5.0\,x$. Find x as a function of t if $x = 0.35$ m³ when $t = 0$.

48. Moisture evaporates from a surface at a rate proportional to the amount of moisture present at any time. If 75% of the moisture evaporates from a certain surface in 1.00 h, how long did it take for 50% to evaporate?

49. On a certain weather map, the *isobars* (curves of equal barometric pressure) are given by $y = e^{x/2} + k$. Find the equation of the orthogonal trajectories (curves that show the wind direction), and display a few of each on a calculator.

50. The lines of equal potential in a field of force are all at right angles to the lines of force. In an electric field of force caused by charged particles, the lines of force are given by $x^2 + y^2 = kx$. Find the equation of the lines of equal potential. Use a graphing calculator to view a few members of the lines of force and those of equal potential.

31.7 Higher-Order Homogeneous Equations

Another important type of differential equation is the linear differential equation of higher order with constant coefficients. First, we shall briefly describe the general higher-order equation and the notation we shall use with this type of equation.

The general **linear differential equation of the** ***n*****th order** *is of the form*

$$a_0\frac{d^n y}{dx^n} + a_1\frac{d^{n-1}y}{dx^{n-1}} + \cdots + a_{n-1}\frac{dy}{dx} + a_n y = b \qquad \textbf{(31.11)}$$

where the a's and b are either functions of x or constants.

For convenience of notation, the nth derivative with respect to the independent variable will be denoted by D^n. Here, D is called the **operator**, since it denotes the *operation* of differentiation. Using this notation with x as the independent variable, Eq. (31.11) becomes

$$a_0 D^n y + a_1 D^{n-1} y + \cdots + a_{n-1} Dy + a_n y = b \qquad \textbf{(31.12)}$$

If $b = 0$, *the general linear equation is called* **homogeneous**, *and if* $b \neq 0$, *it is called* **nonhomogeneous**. Both types of equations have important applications.

EXAMPLE 1 Illustrating the operator *D*

Write the following differential equation using the operator form of Eq. (31.12):

$$\frac{d^3y}{dx^3} - 3\frac{d^2y}{dx^2} + 4\frac{dy}{dx} - 2y = e^x \sec x$$

Letting D^n denote the nth derivative, we get

$$D^3y - 3D^2y + 4Dy - 2y = e^x \sec x$$

This equation is nonhomogeneous since $b = e^x \sec x$.

Although the a's may be functions of x, we shall restrict our attention to the cases in which they are constants. We shall, however, consider both homogeneous equations and nonhomogeneous equations. Also, since second-order linear equations are the most commonly found in elementary applications, we shall devote most of our attention to them. The methods used to solve second-order equations may be applied to equations of higher order, and we shall consider certain of these higher-order equations.

SECOND-ORDER HOMOGENEOUS EQUATIONS WITH CONSTANT COEFFICIENTS

Using the operator notation, *a second-order, linear, homogeneous differential equation with constant coefficients is one of the form*

$$a_0 D^2 y + a_1 Dy + a_2 y = 0 \qquad \textbf{(31.13)}$$

where the a's are constants. The following example indicates the kind of solution we should expect for this type of equation.

EXAMPLE 2 Solution for a second-order equation

Solve the differential equation $D^2y - Dy - 2y = 0$.

First, we put this equation in the form $(D^2 - D - 2)y = 0$. This is another way of saying that we are to take the second derivative of y, subtract the first derivative, and finally subtract twice the function. This expression may now be factored as $(D - 2)(D + 1)y = 0$. (We will not develop the algebra of the operator D. However, most such algebraic operations can be shown to be valid.) This formula tells us to find the first derivative of the function and add this to the function. Then twice this result is to be subtracted from the derivative of this result. If we let $z = (D + 1)y$, which is valid since $(D + 1)y$ is a function of x, we have $(D - 2)z = 0$. This equation is easily solved by separation of variables. Thus,

$$\frac{dz}{dx} - 2z = 0 \qquad \frac{dz}{z} - 2\,dx = 0 \qquad \ln z - 2x = \ln c_1$$

$$\ln\frac{z}{c_1} = 2x \quad \text{or} \quad z = c_1e^{2x}$$

Replacing z by $(D + 1)y$, we have

$$(D + 1)y = c_1e^{2x}$$

This is a linear equation of the first order. Then,

$$dy + y\,dx = c_1e^{2x}\,dx$$

The factor $e^{\int P\,dx}$ is $e^{\int dx} = e^x$. And so,

$$ye^x = \int c_1e^{3x}\,dx = \frac{c_1}{3}e^{3x} + c_2 \qquad \text{using Eq. (31.8)}$$

$$y = c_1'e^{2x} + c_2e^{-x}$$

where $c_1' = \frac{1}{3}c_1$. This example indicates that solutions of the form e^{mx} result for this equation.

Based on the result of Example 2, assume that Eq. (31.13) has a particular solution ce^{mx}. Substituting into Eq. (31.13) gives

$$a_0cm^2e^{mx} + a_1cme^{mx} + a_2ce^{mx} = 0$$

Since $e^{mx} > 0$ for all real x, this equation will be satisfied when m is a root of the quadratic equation $a_0m^2 + a_1m + a_2 = 0$. This leads us to the following method of solution.

■ The auxiliary equation is formed directly from Eq. (31.13) by replacing the second derivative with m^2, the first derivative with m, and the original function with 1.

Solving a Second Order Linear Homogeneous Equation (Distinct Roots)	EXAMPLE 3
	Solve $y'' - 5y' + 6y = 0$.
1. Use operator notation and write the equation in the form $a_0D^2y + a_1Dy + a_2y = 0$ **(31.13)**	We write the equation as $D^2y - 5Dy + 6y = 0$
2. Use the coefficients from Step 1 to write the **auxiliary equation** $a_0m^2 + a_1m + a_2 = 0$ **(31.14)**	The auxiliary equation is $m^2 - 5m + 6 = 0$

3. Solve the quadratic equation from Step 2. Let m_1 and m_2 be the two **distinct real roots**. (The solution for repeated or complex roots will be given in the next section).	We factor as $(m-3)(m-2)=0$, which gives us the roots $$m_1 = 3, m_2 = 2$$
4. Write the general solution as $$y = c_1 e^{m_1 x} + c_2 e^{m_2 x} \qquad \textbf{(31.15)}$$ It makes no difference which constant is written with each exponential term.	We obtain the general solution $$y = c_1 e^{3x} + c_2 e^{2x}$$

EXAMPLE 4 The auxiliary equation has zero as a root

Solve the differential equation $y'' = 6y'$.

Rewrite using operator notation:	$D^2y - 6Dy = 0$
Write the auxiliary equation Eq. (31.14):	The auxiliary equation is $$m^2 - 6m = 0$$
Solve to find the two distinct real roots:	We factor as $m(m-6)=0$, which gives us the roots $m_1 = 0$, $m_2 = 6$.
Write the general solution as Eq. (31.15) and simplify:	We obtain the general solution $$y = c_1 e^{0x} + c_2 e^{6x} = c_1 + c_2 e^{6x}$$

EXAMPLE 5 A third-order equation

Solve the differential equation $2\dfrac{d^3y}{dx^3} + \dfrac{d^2y}{dx^2} - 7\dfrac{dy}{dx} = 0$.

Although this is a third-order equation, the *method* of solution is the same as in the previous examples.

Rewrite using operator notation:	$2D^3y + D^2y + -7Dy = 0$
Write the auxiliary equation:	The auxiliary equation is $$2m^3 + m^2 - 7m = 0$$
Solve to find the *three* distinct real roots:	We factor as $m(2m^2 + m - 7) = 0$, which gives us the roots $m_1 = 0$, $m_2 = \dfrac{-1+\sqrt{57}}{4}$, and $m_3 = \dfrac{-1-\sqrt{57}}{4}$.
Write the general solution as indicated in Eq. (31.15), with three arbitrary constants:	We obtain the general solution $$y = c_1 e^{0x} + c_2 e^{(-1+\sqrt{57})x/4} + c_3 e^{(-1-\sqrt{57})x/4}$$ $$= c_1 + e^{-x/4}\left(c_2 e^{x\sqrt{57}/4} + c_3 e^{-x\sqrt{57}/4}\right)$$

Practice Exercise

1. Solve the differential equation $2D^2y - Dy - 3y = 0$.

In all examples and exercises of this section, all the roots of the auxiliary equation are different, and they do not include complex numbers. For such roots, the solutions have a different form. They are the topic of the following section.

EXAMPLE 6 A particular solution

Solve the differential equation $D^2y - 2\,Dy - 15y = 0$ and find the particular solution that satisfies the conditions $Dy = 2$ and $y = -1$, when $x = 0$. (It is necessary to give two conditions since there are two constants to evaluate.)

We have

$$m^2 - 2m - 15 = 0, \qquad (m-5)(m+3) = 0$$
$$m_1 = 5 \qquad m_2 = -3$$
$$y = c_1e^{5x} + c_2e^{-3x}$$

This equation is the general solution. In order to evaluate the constants c_1 and c_2, *we use the given conditions to find two simultaneous equations in* c_1 *and* c_2. These are then solved to determine the particular solution. Thus,

$$y' = 5c_1e^{5x} - 3c_2e^{-3x}$$

Using the given conditions in the general solution and its derivative, we have

$$c_1 + c_2 = -1 \qquad y = -1 \text{ when } x = 0$$
$$5c_1 - 3c_2 = 2 \qquad Dy = 2 \text{ when } x = 0$$

The solution to this system of equations is $c_1 = -\frac{1}{8}$ and $c_2 = -\frac{7}{8}$. The particular solution becomes

$$y = -\frac{1}{8}e^{5x} - \frac{7}{8}e^{-3x} \quad \text{or} \quad 8y + e^{5x} + 7e^{-3x} = 0$$

EXERCISES 31.7

In Exercises 1 and 2, make the given changes in the indicated examples of this section and then solve the resulting differential equations.

1. In Example 3, delete the $+6y$ term.
2. In Example 4, add $16y$ to the right side.

In Exercises 3–26, solve the given differential equations.

3. $\frac{d^2y}{dx^2} - \frac{dy}{dx} - 6y = 0$
4. $\frac{d^2y}{dx^2} + \frac{dy}{dx} = 0$
5. $3\frac{d^2y}{dx^2} + 4\frac{dy}{dx} + y = 0$
6. $\frac{d^2y}{dx^2} - 2\frac{dy}{dx} - 8y = 0$
7. $D^2y - 3\,Dy = 0$
8. $D^2y = 25y$
9. $2\,D^2y - 3y = Dy$
10. $D^2y + 7\,Dy + 6y = 0$
11. $3\,D^2y + 12y = 20\,Dy$
12. $4\,D^2y + 12\,Dy = 7y$
13. $3y'' + 8y' - 3y = 0$
14. $8y'' + 6y' = 9y$
15. $3y'' + 2y' - y = 0$
16. $2y'' - 7y' + 6y = 0$
17. $2\frac{d^2y}{dx^2} - 4\frac{dy}{dx} + y = 0$
18. $\frac{d^2y}{dx^2} + \frac{dy}{dx} = 5y$
19. $4\,D^2y - 3\,Dy = 2y$
20. $2\,D^2y - 3\,Dy - y = 0$
21. $y'' = 3y' + y$
22. $5y'' - y' = 3y$
23. $y'' + y' = 8y$
24. $8y'' = y' + y$
25. $2D^2y + 5aDy - 12a^2y = 0 \quad (a > 0)$
26. $3k^4D^2y + 14k^2Dy - 5y = 0$

In Exercises 27–30, find the particular solutions of the given differential equations that satisfy the given conditions.

27. $D^2y - 4\,Dy - 21y = 0;\quad Dy = 0$ and $y = 2$ when $x = 0$
28. $4\,D^2y - Dy = 0;\quad Dy = 2$ and $y = 4$ when $x = 0$
29. $D^2y - Dy = 12y;\quad y = 0$ when $x = 0$, and $y = 1$ when $x = 1$
30. $2\,D^2y + 5\,Dy = 0;\quad y = 0$ when $x = 0$, and $y = 2$ when $x = 1$

In Exercises 31–34, solve the given third- and fourth-order differential equations.

31. $y''' - 2y'' - 3y' = 0$
32. $D^3y - 6\,D^2y + 11\,Dy - 6y = 0$
33. $D^4y - 5\,D^2y + 4y = 0$
34. $D^4y - D^3y - 9\,D^2y + 9\,Dy = 0$

In Exercises 35–37, solve the given problems.

35. The voltage v at a distance s along a transmission line is given by $d^2v/ds^2 = a^2v$, where a is called the *attenuation constant.* Solve for v as a function of s.
36. Following the method of Example 2, solve the differential equation $D^2y - 3Dy + 2y = 0$. Do not use Eqs. (31.14) and (31.15).
37. The displacement y (in cm) of an object at the end of a spring is described by the equation $d^2y/dt^2 + 4\,dy/dt + 3y = 0$, where t is the time (in s). Find $y = f(t)$ if $f(0) = 0$ and $f(1) = 2.00$ cm.

Answer to Practice Exercise

1. $y = c_1e^{-x} + c_2e^{3x/2}$

31.8 Auxiliary Equation with Repeated or Complex Roots

In solving higher-order homogeneous differential equations in the previous section, we purposely avoided repeated or complex roots of the auxiliary equation. In this section, we develop the solutions for such equations. The following example indicates the type of solution that results from the case of repeated roots.

EXAMPLE 1 Solution for a double root

Solve the differential equation $D^2y - 4Dy + 4y = 0$.

Using the method of Example 2 of the previous section, we have the following steps:

$$(D^2 - 4D + 4)y = 0, \qquad (D-2)(D-2)y = 0, \qquad (D-2)z = 0$$

where $z = (D - 2)y$. The solution to $(D - 2)z = 0$ is found by separation of variables. And so,

$$\frac{dz}{dx} - 2z = 0 \qquad \frac{dz}{z} - 2dx = 0$$

$$\ln z - 2x = \ln c_1 \quad \text{or} \quad z = c_1e^{2x}$$

Substituting back, we have $(D - 2)y = c_1e^{2x}$, which is a linear equation of the first order. Then

$$dy - 2y\,dx = c_1e^{2x}\,dx \qquad e^{\int -2\,dx} = e^{-2x}$$

This leads to

$$ye^{-2x} = c_1\int dx = c_1x + c_2 \quad \text{or} \quad y = c_1xe^{2x} + c_2e^{2x}$$

This example indicates the type of solution that results when the auxiliary equation has repeated roots. If the method of the previous section were to be used, the solution of the above example would be $y = c_1e^{2x} + c_2e^{2x}$. This would not be the general solution, since both terms are similar, which means that there is only one independent constant. The constants can be combined to give a solution of the form $y = ce^{2x}$, where $c = c_1 + c_2$. This solution would contain only one constant for a second-order equation.

Based on the result of Example 1, we have the following general solution for repeated roots.

Solving a Second Order Linear Homogeneous Equation (Repeated Roots)	EXAMPLE 2
	Solve $y'' + 4y' + 4y = 0$.
1. Write the homogeneous equation with operator notation.	We write the equation as $D^2y + 4Dy + 4y = 0$
2. Write the auxiliary equation $a_0m^2 + a_1m + a_2 = 0$ **(31.14)**	The auxiliary equation is $m^2 + 4m + 4 = 0$
3. Solve the quadratic equation from Step 2. Let m be the **repeated real root.**	We can factor $(m + 2)^2 = 0$, which gives us the repeated root $m = -2$.
4. Write the general solution as $y = e^{mx}(c_1 + c_2x)$ **(31.16)**	We obtain the general solution $y = e^{-2x}(c_1 + c_2x)$

EXAMPLE 3 Solving an equation with a double root

Solve the differential equation $\frac{d^2y}{dx^2} + 25y = 10\frac{dy}{dx}$.

Rewrite with operator notation: $D^2y - 10Dy + 25y = 0$

Write the auxiliary equation: The auxiliary equation is

$$m^2 - 10m + 25 = 0$$

Solve to find the repeated real root: We can factor $(m-5)^2 = 0$, which gives us the repeated root $m_1 = 5$.

Write the general solution as Eq. (31.16). We obtain the general solution

$$y = e^{5x}(c_1 + c_2x)$$

Practice Exercise

1. Solve the differential equation $D^2y + 8Dy + 16y = 0$.

When the auxiliary equation has complex roots, it can be solved by the method of the previous section and the solution can be put in a more useful form. For complex roots of the auxiliary equation $m = \alpha \pm j\beta$, the solution is of the form

$$y = c_1e^{(\alpha + j\beta)x} + c_2e^{(\alpha - j\beta)x} = e^{\alpha x}(c_1e^{j\beta x} + c_2e^{-j\beta x})$$

Using the exponential form of a complex number, Eq. (12.11), we have

$$\begin{aligned} y &= e^{\alpha x}[c_1 \cos \beta x + jc_1 \sin \beta x + c_2 \cos(-\beta x) + jc_2 \sin(-\beta x)] \\ &= e^{\alpha x}(c_1 \cos \beta x + c_2 \cos \beta x + jc_1 \sin \beta x - jc_2 \sin \beta x) \\ &= e^{\alpha x}(c_3 \cos \beta x + c_4 \sin \beta x) \end{aligned}$$

where $c_3 = c_1 + c_2$ and $c_4 = jc_1 - jc_2$.

The resulting procedure for solution is as follows.

Solving a Second Order Linear Homogeneous Equation (Complex Roots)	EXAMPLE 4
	Solve $D^2y - Dy + y = 0$.
1. Write the homogeneous equation with operator notation.	The equation is already in the correct form.
2. Write the auxiliary equation $a_0m^2 + a_1m + a_2 = 0$ **(31.14)**	The auxiliary equation is $m^2 - m + 1 = 0$
3. Solve the quadratic equation from Step 2. Let the **two complex conjugate solutions** be of the form $\alpha \pm j\beta$.	The solutions are $m = \frac{1 \pm j\sqrt{3}}{2}$. We identify $\alpha = 1/2$, and $\beta = \sqrt{3}/2$.
4. Write the general solution as $y = e^{\alpha x}(c_1 \sin \beta x + c_2 \cos \beta x)$ **(31.17)**	We obtain the general solution $y = e^{x/2}\left(c_1 \sin \frac{\sqrt{3}}{2}x + c_2 \cos \frac{\sqrt{3}}{2}x\right)$

EXAMPLE 5 Solving a third-order equation

Solve the differential equation $D^3y + 4Dy = 0$.

Write the auxiliary equation: The auxiliary equation is

$$m^3 + 4m = 0$$

Solve to find the *three* roots (real or complex):	We can factor $m(m^2 + 4) = 0$, which gives us the roots $m_1 = 0$, $m_2 = 2j$, and $m_3 = -2j$, so $\alpha = 0$ and $\beta = 2$.
Write the general solution with *three* arbitrary constants and simplify:	We obtain the general solution $y = c_1e^{0x} + e^{0x}(c_2 \sin 2x + c_3 \cos 2x)$ $= c_1 + c_2 \sin 2x + c_3 \cos 2x$

Practice Exercise

2. Solve the differential equation $D^2y + 2Dy + 5y = 0$.

EXAMPLE 6 Complex roots—a particular solution

Solve the differential equation $y'' - 2y' + 12y = 0$, if $y' = 2$ and $y = 1$ when $x = 0$.

$$D^2y - 2Dy + 12y = 0 \quad \text{using operator } D \text{ notation}$$

$$m^2 - 2m + 12 = 0 \quad \text{auxiliary equation}$$

$$m = \frac{2 \pm \sqrt{4 - 48}}{2} = 1 \pm j\sqrt{11} \quad \text{complex roots: } \alpha = 1, \beta = \sqrt{11}$$

$$y = e^x\left(c_1 \cos\sqrt{11}x + c_2 \sin\sqrt{11}x\right) \quad \text{general solution}$$

Using the condition that $y = 1$ when $x = 0$, we have

$$1 = e^0(c_1 \cos 0 + c_2 \sin 0) \quad \text{or} \quad c_1 = 1$$

Since $y' = 2$ when $x = 0$, we find the derivative and then evaluate c_2.

$$y' = e^x\left(c_1 \cos\sqrt{11}x + c_2 \sin\sqrt{11}x - \sqrt{11}c_1 \sin\sqrt{11}x + \sqrt{11}c_2 \cos\sqrt{11}x\right)$$

$$2 = e^0\left(\cos 0 + c_2 \sin 0 - \sqrt{11} \sin 0 + \sqrt{11}c_2 \cos 0\right) \quad y' = 2 \text{ when } x = 0$$

$$2 = 1 + \sqrt{11}c_2, \quad c_2 = \frac{1}{11}\sqrt{11} \quad \text{solve for } c_2$$

$$y = e^x\left(\cos\sqrt{11}x + \frac{1}{11}\sqrt{11} \sin\sqrt{11}x\right) \quad \text{particular solution}$$

If the root of the auxiliary equation is repeated more than once—for example, a *triple root*—an additional term with another arbitrary constant and the next higher power of x is added to the solution for each additional root. Also, if a pair of complex roots is repeated, an additional term with a factor of x and another arbitrary constant is added for each root of the pair. These are illustrated in the following example.

EXAMPLE 7 Root repeated more than once

(a) Solve the differential equation $D^3y + 3D^2y + 3Dy + y = 0$.
The auxiliary equation is

$$m^3 + 3m^2 + 3m + 1 = 0, \qquad (m + 1)^3 = 0$$

Each of the three roots is $m = -1$. The equation is a third-order equation, which means there are three arbitrary constants. Therefore, the general solution is

$$y = e^{-x}(c_1 + c_2x + c_3x^2)$$

(b) Solve the differential equation $D^4y + 8D^2y + 16y = 0$.
The auxiliary equation is

$$m^4 + 8m^2 + 16 = 0, \quad (m^2 + 4)^2 = 0$$

With two factors of $m^2 + 4$, the roots are $2j, 2j, -2j$, and $-2j$. The fourth-order equation, and four roots, indicate four arbitrary constants. Since $e^{0x} = 1$, the general solution is

$$y = (c_1 + c_2x)\sin 2x + (c_3 + c_4x)\cos 2x$$

Knowing the various types of possible solutions, it is possible to determine the differential equation if the solution is known. Consider the following example.

EXAMPLE 8 Determine the equation from the solution

(a) A solution of $y = c_1e^x + c_2e^{2x}$ indicates an auxiliary equation with roots of $m_1 = 1$ and $m_2 = 2$. Thus, the auxiliary equation is $(m - 1)(m - 2) = 0$, and the simplest form of the differential equation is $D^2y - 3Dy + 2y = 0$.

(b) A solution of $y = e^{2x}(c_1 + c_2x)$ indicates repeated roots $m_1 = m_2 = 2$ of the auxiliary equation $(m - 2)^2 = 0$, and a differential equation is $D^2y - 4Dy + 4y = 0$.

Let us summarize the type of solution that results from each of the cases discussed in the last two sections.

The Homogeneous Linear Differential Equation

The terms of the general solution to a homogeneous linear differential equation of constant coefficients are related to the roots of the auxiliary equation as follows.

Root of the Auxiliary Equation	Terms in the General Solution
m a single real root	c_1e^{mx}
m a double real root	$e^{mx}(c_1 + c_2x)$
m an n-fold real root	$e^{mx}(c_1 + c_2x + \cdots + c_nx^{n-1})$
$m = \alpha \pm \beta j$ complex conjugate roots	$e^{\alpha x}(c_1 \sin \beta x + c_2 \cos \beta x)$

EXERCISES 31.8

In Exercises 1–4, make the given changes in the indicated examples of this section, and then solve the resulting differential equations.

1. In Example 3, change the sign of the term $10\frac{dy}{dx}$.
2. In Example 3, delete the term $10\frac{dy}{dx}$.
3. In Example 3, delete the term $25y$.
4. In Example 4, change the $-$ sign to $+$.

In Exercises 5–32, solve the given differential equations.

5. $\frac{d^2y}{dx^2} - 2\frac{dy}{dx} + y = 0$
6. $\frac{d^2y}{dx^2} - 6\frac{dy}{dx} + 9y = 0$
7. $D^2y + 12Dy + 36y = 0$
8. $16D^2y + 8Dy + y = 0$
9. $\frac{d^2y}{dx^2} + 9y = 0$
10. $\frac{d^2y}{dx^2} + y = 0$
11. $D^2y + Dy + 2y = 0$
12. $D^2y + 4y = 2Dy$
13. $D^4y - y = 0$
14. $4D^2y = 12Dy - 9y$
15. $4D^2y + y = 0$
16. $9D^2y + 4y = 0$
17. $16y'' - 24y' + 9y = 0$
18. $9y'' - 24y' + 16y = 0$
19. $25y'' + 4y = 0$
20. $y'' + 5y = 4y'$
21. $2D^2y + 5y = 4Dy$
22. $D^2y + 4Dy + 6y = 0$
23. $25y'' + 16y = 40y'$
24. $9y''' + 0.6y'' + 0.01y' = 0$
25. $2D^2y - 3Dy - y = 0$
26. $D^2y - 5Dy = 14y$
27. $3D^2y + 12Dy = 2y$
28. $36D^2y = 25y$
29. $D^3y - 6D^2y + 12Dy - 8y = 0$
30. $D^4y - 2D^3y + 2D^2y - 2Dy + y = 0$
31. $D^4y + 2D^2y + y = 0$
32. $16D^4y - y = 0$

In Exercises 33–36, find the particular solutions of the given differential equations that satisfy the given conditions.

33. $y'' + 2y' + 10y = 0$; $y = 0$ when $x = 0$ and $y = e^{-\pi/6}$ when $x = \pi/6$
34. $9D^2y + 16y = 0$; $Dy = 0$ and $y = 2$ when $x = \pi/2$

35. $D^2y + 16y = 8Dy$; $Dy = 2$ and $y = 4$ when $x = 0$

36. $D^4y + 3D^3y + 2D^2y = 0$; $y = 0$ and $Dy = 4$ and $D^2y = -8$ and $D^3y = 16$ when $x = 0$

In Exercises 37–40, find the simplest form of the second-order homogeneous linear differential equation that has the given solution. In Exercises 38 and 39, explain how the equation is found.

37. $y = c_1e^{3x} + c_2e^{-3x}$

38. $y = c_1e^{3x} + c_2xe^{3x}$

39. $y = c_1 \cos 3x + c_2 \sin 3x$

40. $y = c_1e^{2x} \cos x + c_2e^{2x} \sin x$

In Exercises 41–43, solve the given problems.

41. Find the solution of the equation $D^2y + ay = 0$, if $y = 0$ for $x = 0$ and $x = 1$, (a) if $a = 0$, and (b) $a < 0$.

42. What is the solution of the equation in Exercise 41, with the same conditions, if $0 < a < \pi^2$?

43. The displacement y (in cm) of an object at the end of a spring is described by the equation $d^2y/dt^2 + 4\,dy/dt + 4y = 0$, where t is the time (in s). Find $y = f(t)$ if $f(0) = 0$ and $f(1) = 0.50$ cm.

Answers to Practice Exercises

1. $y = e^{-4x}(c_1 + c_2x)$ 2. $y = e^{-x}(c_1 \sin 2x + c_2 \cos 2x)$

31.9 Solutions of Nonhomogeneous Equations

We now consider the solution of a nonhomogeneous linear equation of the form

$$a_0D^2y + a_1Dy + a_2y = b \tag{31.18}$$

where b is a function of x or is a constant. When the solution is substituted into the left side, we must obtain b. Solutions found from the methods of Sections 31.7 and 31.8 give zero when substituted into the left side, but they do contain the arbitrary constants necessary in the solution. If we could find a particular solution that when substituted into the left side produced b, it could be added to the solution containing the arbitrary constants. Therefore, *the solution is of the form*

$$y = y_c + y_p \tag{31.19}$$

where y_c, called the **complementary solution**, *is obtained by solving the corresponding homogeneous equation and where y_p is the* **particular solution** *necessary to produce the expression b of Eq. (31.18).* It should be noted that y_p satisfies the differential equation, but it has no arbitrary constants and therefore cannot be the general solution. The arbitrary constants are part of y_c.

EXAMPLE 1 Complementary and particular solutions

The differential equation $D^2y - Dy - 6y = e^x$ has the solution

$$y = c_1e^{3x} + c_2e^{-2x} - \tfrac{1}{6}e^x$$

where the complementary solution y_c and particular solution y_p are

$$y_c = c_1e^{3x} + c_2e^{-2x} \qquad y_p = -\tfrac{1}{6}e^x$$

The complementary solution y_c is obtained by solving the corresponding homogeneous equation $D^2y - Dy - 6y = 0$, and we shall discuss below the method of finding y_p. Again, we note that y_c contains the arbitrary constants, y_p contains the expression needed to produce the e^x on the right, and therefore both are needed to have the complete general solution.

The method that is used to find a particular solution y_p is called the **method of undetermined coefficients.**

Solving Nonhomogeneous Equations by Undetermined Coefficients

1. Obtain the complementary solution y_c by solving the corresponding homogeneous equation.
2. **Make a guess as to the form** of a particular solution y_p. Since a combination of the particular solution and its derivatives must form the function b on the right side of the equation, y_p **should contain all possible forms of b and its derivatives**. Leave the coefficients of y_p undetermined.
3. Substitute the chosen form of y_p into the differential equation and equate the coefficients of like terms. Find the values of the coefficients by solving the resulting equation(s).
4. Complete the solution by adding y_p to y_c. Note that some adjustments will be necessary in Step 3 if the proposed y_p and y_c have similar terms (see Example 8).

EXAMPLE 2 Forms of particular solutions

Find the form of a particular solution y_p for each of the following functions b by considering the form of b and of its derivative(s).

Function b	Form of the Derivative(s)	Particular Solution y_p
(a) $4x$	A	$y_p = A + Bx$
(b) e^{2x}	Ce^{2x}	$y_p = Ce^{2x}$ (note that only one e^{2x} term is needed)
(c) $4x + e^{2x}$	$A,\ Ce^{2x}$	$y_p = A + Bx + Ce^{2x}$
(d) $x^2 + e^{-x}$	$A,\ Bx,\ Ee^{-x}$	$y_p = A + Bx + Cx^2 + Ee^{-x}$ (again, only one e^{-x} term is needed)
(e) $xe^{-2x} - 5$	$Ae^{-2x},\ Bxe^{-2x}$	$y_p = Ae^{-2x} + Bxe^{-2x} + C$
(f) $x \sin x$	$A \sin x,\ B \cos x,\ Cx \sin x,\ Ex \cos x$	$y_p = A \sin x + B \cos x + Cx \sin x + Ex \cos x$
(g) $e^x + xe^x$	$Ae^x,\ Bxe^x$	$y_p = Ae^x + Bxe^x$
(h) $xe^{-2x} - 5$	$Ae^{-2x},\ Bxe^{-2x}$	$y_p = Ae^{-2x} + Bxe^{-2x} + C$

Practice Exercise

1. Find the form of y_p if b is of the form $x + \cos x$.

EXAMPLE 3 Solving a nonhomogeneous equation

Solve the differential equation $D^2y - Dy - 6y = e^x$.

In this case, the solution of the auxiliary equation $m^2 - m - 6 = 0$ gives us the roots $m_1 = 3$ and $m_2 = -2$. Thus,

$$y_c = c_1e^{3x} + c_2e^{-2x}$$

The proper form of the particular solution is $y_p = Ae^x$. This means that $Dy_p = Ae^x$ and $D^2y_p = Ae^x$. Substituting y_p and its derivatives into the differential equation, we have

$$Ae^x - Ae^x - 6Ae^x = e^x$$

To produce equality, the coefficients of e^x must be the same on each side of the equation. Thus,

$$-6A = 1 \quad \text{or} \quad A = -1/6$$

Therefore, $y_p = -\frac{1}{6}e^x$. This gives the complete solution $y = y_c + y_p$,

$$y = c_1e^{3x} + c_2e^{-2x} - \frac{1}{6}e^x \qquad \text{see Example 1}$$

This solution checks when substituted into the original differential equation.

EXAMPLE 4 Solving a nonhomogeneous equation

Solve the differential equation $D^2y + 4y = x - 4e^{-x}$.

In this case, we have $m^2 + 4 = 0$, which give us $m_1 = 2j$ and $m_2 = -2j$. Therefore, $y_c = c_1 \sin 2x + c_2 \cos 2x$.

The proper form of the particular solution is $y_p = A + Bx + Ce^{-x}$. Finding two derivatives and then substituting into the differential equation gives

$$y_p - A + Bx + Ce^{-x} \qquad Dy_p - B - Ce^{-x} \qquad D^2y_p = Ce^{-x}$$

$$D^2y + 4y = x - 4e^{-x} \qquad \text{differential equation}$$

$$(Ce^{-x}) + 4(A + Bx + Ce^{-x}) = x - 4e^{-x} \qquad \text{substituting}$$

$$Ce^{-x} + 4A + 4Bx + 4Ce^{-x} = x - 4e^{-x}$$

$$(4A) + (4B)x + (5C)e^{-x} = 0 + (1)x + (-4)e^{-x} \qquad \text{note coefficients}$$

Equating the constants and the coefficients of x and e^{-x} on either side gives

$$4A = 0 \qquad 4B = 1 \qquad 5C = -4$$

$$A = 0 \qquad B = 1/4 \qquad C = -4/5$$

This means that the particular solution is

$$y_p = \frac{1}{4}x - \frac{4}{5}e^{-x}$$

In turn, this tells us that the complete solution is

$$y = c_1 \sin 2x + c_2 \cos 2x + \frac{1}{4}x - \frac{4}{5}e^{-x}$$

Substitution into the original differential equation verifies this solution.

Practice Exercise

2. Find y_p for the differential equation $D^2y + 4y = 8xe^{2x}$.

EXAMPLE 5 Solving a nonhomogeneous equation

Solve the differential equation $D^3y - 3\,D^2y + 2\,Dy = 10 \sin x$.

$$m^3 - 3m^2 + 2m = 0 \qquad m(m-1)(m-2) = 0 \qquad m_1 = 0 \qquad m_2 = 1 \qquad m_3 = 2 \qquad \text{auxiliary equation}$$

$$y_c = c_1 + c_2e^x + c_3e^{2x} \qquad \text{complementary solution}$$

We now find the particular solution:

$$y_p = A \sin x + B \cos x \qquad \text{particular solution form}$$

$$Dy_p = A \cos x - B \sin x \qquad D^2y_p = -A \sin x - B \cos x \qquad D^3y_p = -A \cos x + B \sin x \qquad \text{find three derivatives}$$

$$(-A \cos x + B \sin x) - 3(-A \sin x - B \cos x) + 2(A \cos x - B \sin x) = 10 \sin x \qquad \text{substitute into differential equation}$$

$$(3A - B)\sin x + (A + 3B)\cos x = 10 \sin x$$

$$3A - B = 10 \qquad A + 3B = 0 \qquad \text{equate coefficients}$$

The solution of this system is $A = 3$, $B = -1$.

$$y_p = 3 \sin x - \cos x \qquad \text{particular solution}$$

$$y = c_1 + c_2e^x + c_3e^{2x} + 3 \sin x - \cos x \qquad \text{complete general solution}$$

This solution checks when substituted into the differential equation.

EXAMPLE 6 Particular solution

Find the particular solution of $y'' + 16y = 2e^{-x}$ if $Dy = -2$ and $y = 1$ when $x = 0$.

In this case, we must not only find y_c and y_p, but we must also evaluate the constants of y_c from the given conditions. The solution is as follows:

$$D^2y + 16y = 2e^{-x} \quad \text{operator } D \text{ form}$$
$$m^2 + 16 = 0, \quad m = \pm 4j \quad \text{auxiliary equation}$$
$$y_c = c_1 \sin 4x + c_2 \cos 4x \quad \text{complementary solution}$$
$$y_p = Ae^{-x} \quad \text{particular solution form}$$
$$Dy_p = -Ae^{-x} \qquad D^2y_p = Ae^{-x}$$
$$Ae^{-x} + 16Ae^{-x} = 2e^{-x} \quad \text{substituting}$$
$$17Ae^{-x} = 2e^{-x}, \qquad A = \frac{2}{17} \quad \text{equate coefficients}$$
$$y_p = \frac{2}{17}e^{-x}$$
$$y = c_1 \sin 4x + c_2 \cos 4x + \frac{2}{17}e^{-x} \quad \text{complete general solution}$$

We now evaluate c_1 and c_2 from the given conditions:

$$Dy = 4c_1 \cos 4x - 4c_2 \sin 4x - \frac{2}{17}e^{-x}$$
$$1 = c_1(0) + c_2(1) + \frac{2}{17}(1), \qquad c_2 = \frac{15}{17} \quad y = 1 \text{ when } x = 0$$
$$-2 = 4c_1(1) - 4c_2(0) - \frac{2}{17}(1), \qquad c_1 = -\frac{8}{17} \quad Dy = -2 \text{ when } x = 0$$
$$y = -\frac{8}{17}\sin 4x + \frac{15}{17}\cos 4x + \frac{2}{17}e^{-x} \quad \text{required particular solution}$$

This solution checks when substituted into the differential equation.

FOURIER SERIES SOLUTIONS

The particular solution of a non-homogeneous differential equation can also be obtained when the function b in Eq. (31.18) is periodic. By expanding b in a Fourier series, the particular solution can be found by undetermined coefficients, as we see in the next example.

EXAMPLE 7 A solution using Fourier series

Solve the differential equation $D^2y + 4y = f(x)$ where

$$f(x) = \begin{cases} x & 0 \le x < 2 \\ 4 - x & 2 \le x < 4 \end{cases}, \; f(x+4) = f(x), \text{ and } f(x) \text{ is extended to be even.}$$

See Fig. 31.11.

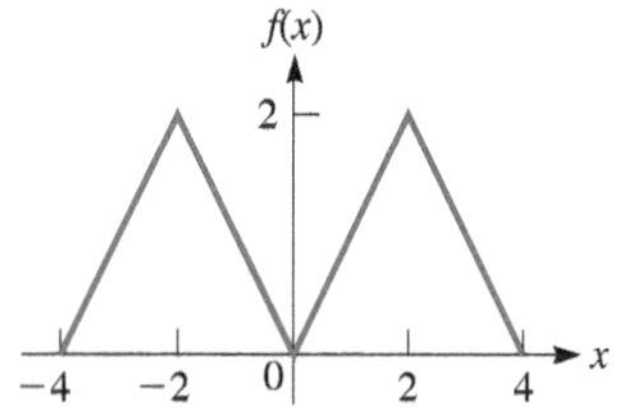

Fig. 31.11

We have seen in Example 4 that the complementary solution is of the form $y_c = c_1 \sin 2x + c_2 \cos 2x$. We now find a particular solution.

We begin by writing $f(x)$ in a half-range cosine series ($f(x)$ is even). We have already done that in Example 9 of Section 30.7. We have

$$f(x) = 1 - \frac{8}{\pi^2}\left(\cos\frac{\pi x}{2} + \frac{1}{9}\cos\frac{3\pi x}{2} + \frac{1}{25}\cos\frac{5\pi x}{2} + \cdots\right)$$

Since the differential equation contains only y and its second derivatives, a particular solution involving only cosines is appropriate. Therefore we choose a particular solution of the form

$$y_p = A_0 + \sum_{n=1}^{\infty} A_n \cos\frac{n\pi x}{2}$$

Therefore,

$$D^2y_p = \sum_{n=1}^{\infty}\left(-\frac{n^2\pi^2}{4}\right)A_n \cos\frac{n\pi x}{2}$$

We substitute y_p and D^2y_p into the differential equation and equate the coefficients of like terms. If $n = 0$ we have $4A_0 = 1$, so $A_0 = 1/4$. If n is even, $A_n = 0$, and if n is odd,

$$\left(-\frac{n^2\pi^2}{4} + 4\right)A_n = -\frac{8}{\pi^2 n^2} \quad \text{or} \quad A_n = \frac{32}{\pi^2(\pi^2 n^4 - 16n^2)}$$

Therefore, the particular solution is

$$y_p = \frac{1}{4} + \frac{32}{\pi^2}\left(\frac{1}{\pi^2 - 16}\cos\frac{\pi x}{2} + \frac{1}{81\pi^2 - 144}\cos\frac{3\pi x}{2} + \cdots\right)$$

The complete solution is

$$y = c_1 \sin 2x + c_2 \cos 2x + \frac{1}{4} + \frac{32}{\pi^2}\left(\frac{1}{\pi^2 - 16}\cos\frac{\pi x}{2} + \frac{1}{81\pi^2 - 144}\cos\frac{3\pi x}{2} + \cdots\right)$$

LEARNING TIP

If a term of the proposed y_p is similar to a term of y_c, any term of the proposed y_p included to account for the similar term of the function b must be multiplied by the smallest possible integral power of x such that any resulting term y_p is not similar to the term of y_c.

A SPECIAL CASE

It may happen that a term of the proposed particular solution y_p is similar to a term of the complementary solution y_c. Since any term of y_c gives zero when substituted in the differential equation, so will that term of the proposed y_p. This means the proposed y_p must be modified as indicated in the Learning Tip to the left. The following example shows that it is not as involved as it sounds.

EXAMPLE 8 y_p and y_c have similar terms

Solve the differential equation $D^2y - 2Dy + y = x + e^x$.

We find that the auxiliary equation and complementary solution are

$$m^2 - 2m + 1 = 0, \quad (m - 1)^2 = 0 \quad m_1 = 1 \quad m_2 = 1$$
$$y_c = e^x(c_1 + c_2 x)$$

Based on the function b on the right side, the *proposed* form of y_p is

$$y_p = A + Bx + Ce^x \quad \text{proposed form}$$

We now note that the term Ce^x is similar to the term c_1e^x of y_c. Therefore, we must multiply the term Ce^x by the smallest power of x such that it is not similar to any term of y_c. If we multiply by x, the term becomes similar to c_2xe^x. Therefore, we must multiply Ce^x by x^2 to get

■ Note that the $A + Bx$ are not multiplied by x^2 since they are not included in y_p to account for the e^x in the function b to the right.

$$y_p = A + Bx + Cx^2e^x \quad \text{correct modified form}$$

Using this form of y_p, we now complete the solution.

$$Dy_p = B + Cx^2e^x + 2Cxe^x \quad D^2y_p = Cx^2e^x + 2Cxe^x + 2Ce^x + 2Cxe^x = 2Ce^x + 4Cxe^x + Cx^2e^x$$
$$(2Ce^x + 4Cxe^x + Cx^2e^x) - 2(B + Cx^2e^x + 2Cxe^x) + (A + Bx + Cx^2e^x) = x + e^x$$
$$(A - 2B) + Bx + 2Ce^x = x + e^x$$
$$A - 2B = 0 \quad B = 1 \quad 2C = 1 \quad A = 2 \quad B = 1 \quad C = 1/2$$
$$y_p = 2 + x + \tfrac{1}{2}x^2e^x$$
$$y = e^x(c_1 + c_2x) + 2 + x + \tfrac{1}{2}x^2e^x$$

EXERCISES 31.9

In Exercises 1–4, make the given changes in the indicated examples of this section and then solve the given problems.

1. In Example 2(g), add $2x$ to the form b and then determine the form of y_p.
2. In Example 3, change e^x to e^{2x} and then find the solution.
3. In Example 4, on the right side, change x to $\sin x$, and then find the solution.
4. In Example 6, change the right side to $x^2 + xe^x$ and then find the proper form for y_p.

In Exercises 5–16, solve the given differential equations. The form of y_p is given.

5. $D^2y - Dy - 2y = 4$ (Let $y_p = A$.)
6. $D^2y - Dy - 6y = 4x$ (Let $y_p = A + Bx$.)
7. $D^2y - y = 2 + x^2$ (Let $y_p = A + Bx + Cx^2$.)
8. $D^2y + 4Dy + 3y = 2 + e^x$ (Let $y_p = A + Be^x$.)
9. $y'' - 3y' = 2e^x + xe^x$ (Let $y_p = Ae^x + Bxe^x$.)
10. $y'' + y' - 2y = 8 + 4x + 2xe^{2x}$ (Let $y_p = A + Bx + Ce^{2x} + Exe^{2x}$.)
11. $9D^2y - y = \sin x$ (Let $y_p = A\sin x + B\cos x$.)
12. $D^2y + 4y = \sin x + 4$ (Let $y_p = A + B\sin x + C\cos x$.)
13. $\dfrac{d^2y}{dx^2} - 2\dfrac{dy}{dx} + y = 2x + x^2 + \sin 3x$ (Let $y_p = A + Bx + Cx^2 + E\sin 3x + F\cos 3x$.)
14. $D^2y - y = e^{-x}$ (Let $y_p = Axe^{-x}$.)
15. $D^2y + 4y = -12\sin 2x$ (Let $y_p = Ax\sin 2x + Bx\cos 2x$.)
16. $y'' - 2y' + y = 3 + e^x$ (Let $y_p = A + Bx^2e^x$.)

In Exercises 17–32, solve the given differential equations.

17. $\dfrac{d^2y}{dx^2} - \dfrac{dy}{dx} - 30y = 10$
18. $2\dfrac{d^2y}{dx^2} + 11\dfrac{dy}{dx} - 6y = 8x$
19. $3\dfrac{d^2y}{dx^2} - \dfrac{dy}{dx} - 4y = 5e^{3x}$
20. $\dfrac{d^2y}{dx^2} + 4y = 2\sin 3x$
21. $D^2y - 4y = \sin x + 2\cos x$
22. $6D^2y + Dy - y = e^x - e^{-x}$
23. $D^2y + y = 4 + \sin 2x$
24. $D^2y - Dy + y = x + \sin x$
25. $D^2y + 5Dy + 4y = xe^x + 4$
26. $3D^2y + Dy - 2y = 4 + 2x + e^x$
27. $y''' - y' = \sin 2x$
28. $D^4y - y = x$
29. $D^2y + y = \cos x$
30. $4y'' - 4y' + y = 4e^{x/2}$
31. $D^2y + 2Dy = 8x + e^{-2x}$
32. $D^3y - Dy = 4e^{-x} + 3e^{2x}$

In Exercises 33–36, find the particular solution of each differential equation for the given conditions.

33. $D^2y - Dy - 6y = 5 - e^x$; $Dy = 4$ and $y = 2$ when $x = 0$
34. $3y'' - 10y' + 3y = xe^{-2x}$; $y' = -\frac{9}{35}$ and $y = -\frac{13}{35}$ when $x = 0$
35. $y'' + y = x + \sin 2x$; $y' = 1$ and $y = 0$ when $x = \pi$
36. $D^2y - 2Dy + y = xe^{2x} - e^{2x}$; $Dy = 4$ and $y = -2$ when $x = 0$

In Exercises 37 and 38, solve the given problems.

37. Solve the first-order equation $Dy - y = x^2$ by the method of undetermined coefficients. For this equation, why is this method easier than the method of Section 31.4?
38. The displacement y (in cm) of an object at the end of a spring is described by the equation $d^2y/dt^2 + 4\,dy/dt + 4y = 0$, where t is the time (in s). Find $y = f(t)$ if $f(0) = 0$ and $f(1) = 0.50$ cm.

In Exercises 39–40, find a particular solution of the equation $D^2y + 4y = f(x)$ for the given function. (One period is defined for each function.) See Example 7.

39. $f(x) = \begin{cases} 1 & 0 \le x < 2 \\ -1 & 2 \le x < 4 \end{cases}$ and $f(x)$ is extended to be odd.
40. $f(x) = \begin{cases} 1 & 0 \le x < 4 \\ 8 - x & 4 \le x < 8 \end{cases}$ and $f(x)$ is extended to be even.

Answers to Practice Exercises

1. $y_p = A + Bx + C\sin x + D\cos x$ 2. $y_p = -\dfrac{1}{2}e^{2x} + xe^{2x}$

31.10 Applications of Higher-Order Equations

We now show important applications of second-order differential equations to simple harmonic motion and simple electric circuits. Also, we will show an application of a fourth-order differential equation to the deflection of a beam.

EXAMPLE 1 Simple harmonic motion

Simple harmonic motion may be defined as motion in a straight line for which the acceleration is proportional to the displacement and in the opposite direction. Examples of this type of motion are a weight on a spring, a simple pendulum, and an object bobbing in water. If x represents the displacement, d^2x/dt^2 is the acceleration.

Using the definition of simple harmonic motion, we have

$$\frac{d^2x}{dt^2} = -k^2x$$

(We chose k^2 for convenience of notation in the solution.) We write this equation in the form

$$D^2x + k^2x = 0 \qquad \text{here, } D = d/dt$$

The roots of the auxiliary equation are kj and $-kj$, and the solution is

$$x = c_1 \sin kt + c_2 \cos kt$$

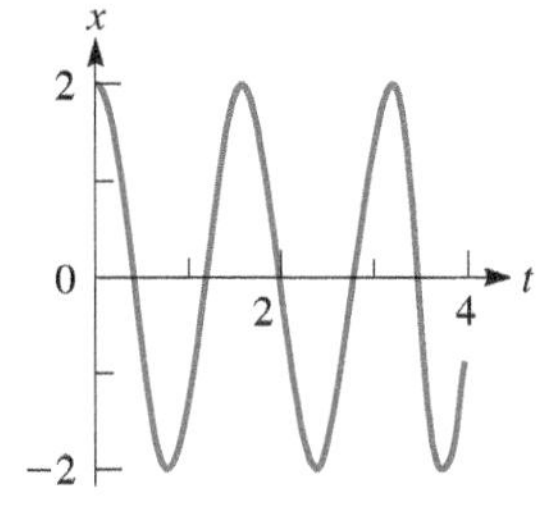

Fig. 31.12

This solution indicates an oscillating motion, which is known to be the case. If, for example, $k = 4$ and we know that $x = 2$ and $Dx = 0$ (which means the velocity is zero) for $t = 0$, we have

$$Dx = 4c_1 \cos 4t - 4c_2 \sin 4t$$

$$2 = c_1(0) + c_2(1) \qquad x = 2 \text{ for } t = 0$$

$$0 = 4c_1(1) - 4c_2(0) \qquad Dx = 0 \text{ for } t = 0$$

which gives $c_1 = 0$ and $c_2 = 2$. Therefore,

$$x = 2 \cos 4t$$

is the equation relating the displacement and time; Dx is the velocity and D^2x is the acceleration. See Fig. 31.12.

Practice Exercise

1. In Example 1, find the solution if $x = 0$ and $Dx = 2$ when $t = 0$.

EXAMPLE 2 Damped simple harmonic motion

■ Newton's second law states that the net force acting on an object is equal to its mass times its acceleration. (This is one of Newton's best-known contributions to physics.)

In practice, an object moving with simple harmonic motion will in time cease to move due to unavoidable frictional forces. A "freely" oscillating object has a retarding force that is approximately proportional to the velocity. The differential equation for this case is $D^2x = -k^2x - bDx$. This results from applying (from physics) Newton's second law of motion (see the margin note at the left). Again, using the operator $D^2x = d^2x/dt^2$, the term D^2x represents the acceleration of the object, the term $-k^2x$ is a measure of the restoring force (of the spring, for example), and the term $-bDx$ represents the retarding (damping) force. This equation can be written as

$$D^2x + bDx + k^2x = 0$$

The auxiliary equation is $m^2 + bm + k^2 = 0$, for which the roots are

$$m = \frac{-b \pm \sqrt{b^2 - 4k^2}}{2}$$

If $k = 3$ and $b = 4$, $m = -2 \pm j\sqrt{5}$, which means the solution is

$$x = e^{-2t}\left(c_1 \sin\sqrt{5}t + c_2 \cos\sqrt{5}t\right) \qquad \textbf{(1)}$$

Here, $4k^2 > b^2$, and this case is called **underdamped**. In this case, the object oscillates as the amplitude becomes smaller.

If $k = 2$ and $b = 5$, $m = -1, -4$, which means the solution is

$$x = c_1e^{-t} + c_2e^{-4t} \qquad \textbf{(2)}$$

Here, $4k^2 < b^2$, and the motion is called **overdamped**. Note that the motion is not oscillatory, since no sine or cosine terms appear. In this case, the object returns slowly to equilibrium without oscillating.

If $k = 2$ and $b = 4$, $m = -2, -2$, which means the solution is

$$x = e^{-2t}(c_1 + c_2 t) \qquad \textbf{(3)}$$

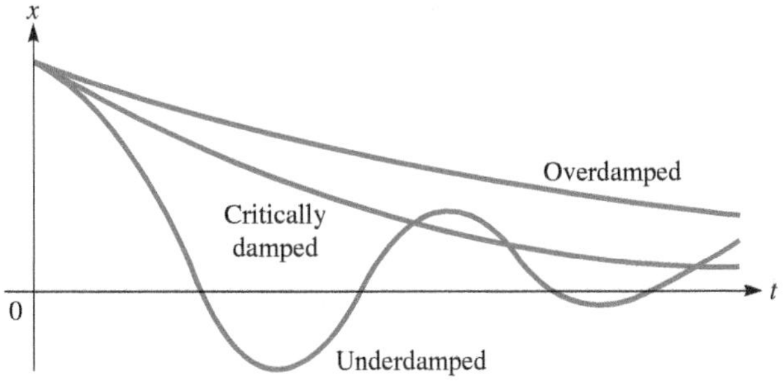

Fig. 31.13

Here, $4k^2 = b^2$, and the motion is called **critically damped**. Again the motion is not oscillatory. In this case, there is just enough damping to prevent any oscillations. The object returns to equilibrium in the minimum time.

See Fig. 31.13, in which Eqs. (1), (2), and (3) are represented in general. The actual values depend on c_1 and c_2, which in turn depend on the conditions imposed on the motion.

EXAMPLE 3 Underdamped harmonic motion

In testing the characteristics of a particular type of spring, it is found that a weight of 4.90 N stretches the spring 0.490 m when the weight and spring are placed in a fluid that resists the motion with a force equal to twice the velocity. If the weight is brought to rest and then given a velocity of 12.0 m/s, find the equation of motion. See Fig. 31.14.

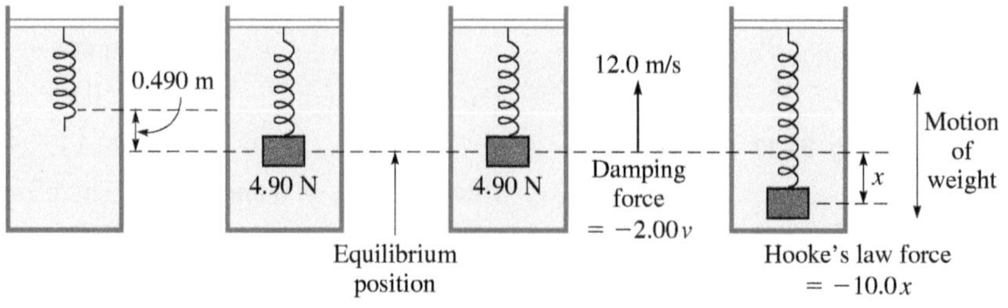

Fig. 31.14

In order to find the equation of motion, we use Newton's second law of motion (see Example 2). The weight (one force) at the end of the spring is offset by the equilibrium position force exerted by the spring, in accordance with Hooke's law (see Section 26.6). Therefore, the net force acting on the weight is the sum of the Hooke's law force due to the displacement from the equilibrium position and the resisting force. Using Newton's second law, we have

mass × acceleration = resisting force + Hooke's law force

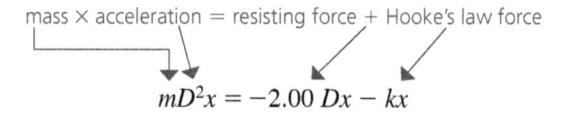

$$mD^2x = -2.00\,Dx - kx$$

The mass of an object is its weight divided by the acceleration due to gravity. The weight is 4.90 N, and the acceleration due to gravity is 9.80 m/s^2. Thus the mass m is

$$m = \frac{4.90 \text{ N}}{9.80 \text{ m/s}^2} = 0.500 \text{ kg}$$

where the kilogram is the unit of mass.

The constant k for the Hooke's law force is found from the fact that the spring stretches 0.490 m for a force of 4.90 N. Thus, using Hooke's law,

$$4.90 = k(0.490), \qquad k = 10.0 \text{ N/m}$$

This means that the differential equation to be solved is

$$0.500D^2x + 2.00Dx + 10.0x = 0$$

or

$$1.00D^2x + 4.00Dx + 20.0x = 0$$

Solving this equation, we have

$$1.00m^2 + 4.00m + 20.0 = 0 \qquad \text{auxiliary equation}$$

$$m = \frac{-4.00 \pm \sqrt{16.0 - 4(20.0)(1.00)}}{2.00}$$

$$= -2.00 \pm 4.00j \qquad \text{complex roots}$$

$$x = e^{-2.00t}(c_1 \cos 4.00t + c_2 \sin 4.00t) \qquad \text{general solution}$$

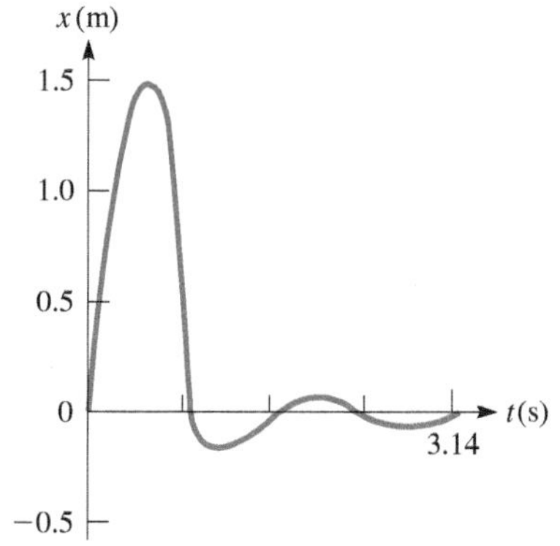

Fig. 31.15

Since the weight started from the equilibrium position with a velocity of 12.0 m/s, we know that $x = 0$ and $Dx = 12.0$ for $t = 0$. Thus,

$$0 = e^0(c_1 + 0c_2) \quad \text{or} \quad c_1 = 0 \qquad x = 0 \text{ for } t = 0$$

Thus, since $c_1 = 0$, we have

$$\begin{aligned} x &= c_2e^{-2.00t} \sin 4.00t \\ Dx &= c_2e^{-2.00t}(\cos 4.00t)(4.00) + c_2 \sin 4.00t(e^{-2.00t})(-2.00) \\ 12.0 &= c_2e^0(1)(4.00) + c_2(0)(e^0)(-2.00) \qquad Dx = 12.0 \text{ for } t = 0 \\ c_2 &= 3.00 \end{aligned}$$

This means that the equation of motion is

$$x = 3.00e^{-2.00t} \sin 4.00t$$

The motion is underdamped; the graph is shown in Fig. 31.15.

It is possible to have an additional force acting on a weight such as the one in Example 3. For example, a vibratory force may be applied to the support of the spring. In such a case, called ***forced vibrations***, this additional external force is added to the other net force. This means that the added force $F(t)$ becomes a nonzero function on the right side of the differential equation, and we must then solve a nonhomogeneous equation.

EXAMPLE 4 Electric circuits

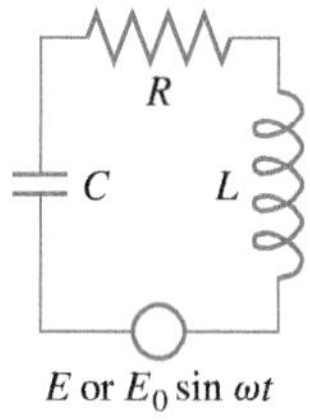

Fig. 31.16

The impressed voltage in the electric circuit shown in Fig. 31.16 equals the sum of the voltages across the components of the circuit. For this circuit with a resistance R, an inductance L, a capacitance C, and a voltage source E, we have

$$L\frac{d^2q}{dt^2} + R\frac{dq}{dt} + \frac{q}{C} = E \tag{31.20}$$

By definition, q represents the electric charge, $dq/dt = i$ is the current, and d^2q/dt^2 is the time rate of change of current. This equation may be written as

$$LD^2q + RDq + q/C = E$$

The auxiliary equation is $Lm^2 + Rm + 1/C = 0$. The roots are

$$\begin{aligned} m &= \frac{-R \pm \sqrt{R^2 - 4L/C}}{2L} \\ &= -\frac{R}{2L} \pm \sqrt{\frac{R^2}{4L^2} - \frac{1}{LC}} \end{aligned}$$

If we let $a = R/2L$ and $\omega = \sqrt{1/LC - R^2/4L^2}$, we have (assuming complex roots, which corresponds to realistic values of R, L, and C)

$$q_c = e^{-at}(c_1 \sin \omega t + c_2 \cos \omega t)$$

This indicates an oscillating charge, or an alternating current. However, the exponential term usually is such that the current dies out rapidly unless there is a source of voltage in the circuit.

LEARNING TIP

To find the steady-state solution for Eq. (31.20), we need find only the particular solution.

If there is no source of voltage in the circuit of Example 4, we have a homogeneous differential equation to solve. If we have a constant voltage source, the particular solution is of the form $q_p = A$. If there is an alternating voltage source, the particular solution is of the form $q_p = A \sin \omega_1 t + B \cos \omega_1 t$, where ω_1 is the angular velocity of the source. After a very short time, the exponential factor in the complementary solution makes it negligible. For this reason, it is referred to as the **transient** term, and *the particular solution is the* **steady-state** *solution.*

It should be noted that the complementary solutions of the mechanical and electric cases are of identical form. There is also an equivalent mechanical case to that of an impressed sinusoidal voltage source in the electric case. This arises in the case of forced vibrations, when an external force affecting the vibrations is applied to the system. Thus, we may have transient and steady-state solutions to mechanical and other nonelectric situations.

EXAMPLE 5 Electric circuit

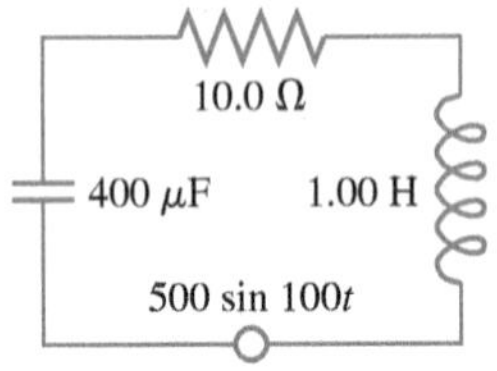

Fig. 31.17

Find the steady-state solution for the current in a circuit containing the following elements: $C = 400\ \mu\text{F}$, $L = 1.00$ H, $R = 10.0\ \Omega$, and a voltage source of $500 \sin 100t$. See Fig. 31.17.

This means the differential equation to be solved is

$$\frac{d^2q}{dt^2} + 10\frac{dq}{dt} + \frac{10^4}{4}q = 500 \sin 100t$$

Since we wish to find the steady-state solution, we must find q_p, from which we may find i_p by finding a derivative. The solution now follows:

$$q_p = A \sin 100t + B \cos 100t \qquad \text{particular solution form}$$

$$\frac{dq_p}{dt} = 100A \cos 100t - 100B \sin 100t$$

$$\frac{d^2q_p}{dt^2} = -10^4A \sin 100t - 10^4B \cos 100t$$

$$-10^4A \sin 100t - 10^4B \cos 100t + 10^3A \cos 100t - 10^3B \sin 100t \qquad \text{substitute into differential equation}$$

$$+ \frac{10^4}{4}A \sin 100t + \frac{10^4}{4}B \cos 100t = 500 \sin 100t$$

$$(-0.75 \times 10^4A - 10^3B)\sin 100t + (-0.75 \times 10^4B + 10^3A)\cos 100t = 500 \sin 100t$$

$$-7.5 \times 10^3A - 10^3B = 500 \qquad \text{equate coefficients of } \sin 100t$$

$$10^3A - 7.5 \times 10^3B = 0 \qquad \text{equate coefficients of } \cos 100t$$

Solving these equations, we obtain

$$B = -8.73 \times 10^{-3} \quad \text{and} \quad A = -65.5 \times 10^{-3}$$

Therefore,

$$q_p = -65.5 \times 10^{-3} \sin 100t - 8.73 \times 10^{-3} \cos 100t$$

$$i_p = \frac{dq_p}{dt} = -6.55 \cos 100t + 0.87 \sin 100t$$

which is the required solution. (We assumed three significant digits for the data but did not use all of them in most equations of the solution.)

The solutions to the second-order differential equations for the applications of simple harmonic motion and electric circuits generally include sines and cosines, because of the oscillatory nature of these applications. We now consider problems involving the deflections of beams, which involve fourth-order differential equations and algebraic functions.

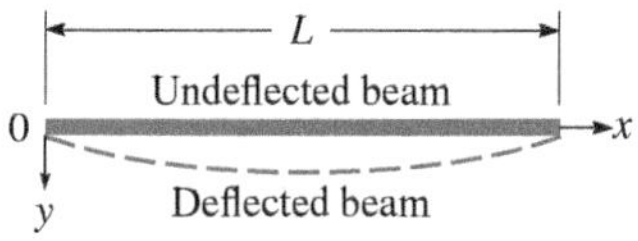

Fig. 31.18

In the study of the strength of materials and elasticity, it is shown that the deflection y of a beam of length L satisfies the differential equation $EI\, d^4y/dx^4 = w(x)$, where EI is a measure of the stiffness of the beam and $w(x)$ is the weight distribution along the beam. See Fig. 31.18. Since this is a fourth-order equation, it is necessary to specify four conditions to obtain a solution. These conditions are determined by the way in which the ends, where $x = 0$ and where $x = L$, are held. For an end held in the specified manner, these conditions are *clamped:* $y = 0$ and $y' = 0$; *hinged:* $y = 0$ and $y'' = 0$; *free:* $y'' = 0$ and $y''' = 0$ ($y' = 0$ indicates no change in alignment; $y'' = 0$ indicates no curvature; $y''' = 0$ indicates no shearing force). Since the conditions are given for specific positions, this kind of problem is called a *boundary value problem.* Consider the following example.

EXAMPLE 6 Deflection of a beam

A uniform beam of length L is hinged at both ends and has a constant load distribution of w due to its own weight. Find the deflection y of the beam in terms of the distance x from one end of the beam.

Using the differential equation given above, we have $EI\, d^4y/dx^4 = w$. For convenience in the solution, let $k = w/EI$. Thus, the solution is as follows:

$$D^4y = k$$

$$m^4 = 0 \qquad m_1 = m_2 = m_3 = m_4 = 0$$

Since the four roots of the auxiliary equation are equal,

$$y_c = c_1 + c_2x + c_3x^2 + c_4x^3$$

The form of y_p indicates that we must multiply the *proposed* $y_p = A$ by x^4 so that it is not similar to any of the terms of y_c. This gives us $y_p = x^4$. Therefore, we now find the general solution as follows:

$$y_p = Ax^4 \qquad Dy_p = 4Ax^3 \qquad D^2y_p = 12Ax^2 \qquad D^3y_p = 24Ax \qquad D^4y_p = 24A$$

$$24A = k, \qquad A = k/24$$

$$y = c_1 + c_2x + c_3x^2 + c_4x^3 + \frac{k}{24}x^4 \qquad \text{general solution}$$

From the discussion about beams before this example, we now find four boundary conditions in order to evaluate the four constants in y_c. For a beam hinged at both ends, we know that $y = 0$ and $D^2y = 0$ for both $x = 0$ and $x = L$. Therefore, we now find two derivatives, use these conditions, and thereby find y_c.

$$Dy = c_2 + 2c_3x + 3c_4x^2 + \frac{k}{6}x^3 \qquad \text{find derivatives}$$

$$D^2y = 2c_3 + 6c_4x + \frac{k}{2}x^2$$

$$\text{At } x = 0, y = 0: \quad c_1 = 0; \qquad \text{At } x = 0, D^2y = 0: \quad c_3 = 0 \qquad \text{use conditions to evaluate constants}$$

$$\text{At } x = L, D^2y = 0: \quad 0 = 6c_4L + \frac{kL^2}{2}; \qquad c_4 = -\frac{kL}{12}$$

$$\text{At } x = L, y = 0: \quad 0 = c_2L + c_4L^3 + \frac{kL^4}{24}$$

$$0 = c_2L + \left(-\frac{kL}{12}\right)L^3 + \frac{kL^4}{24}; \qquad c_2 = \frac{kL^3}{24}$$

Therefore, substituting the values of the four constants in the general solution above, the particular solution that satisfies these conditions is

$$y = \frac{kL^3}{24}x - \frac{kL}{12}x^3 + \frac{kx^4}{24} = \frac{k}{24}(L^3x - 2Lx^3 + x^4)$$

$$= \frac{w}{24EI}(L^3x - 2Lx^3 + x^4) \qquad k = w/EI$$

EXERCISES 31.10

In Exercises 1 and 2, make the given changes in the indicated examples of this section and then solve the resulting problems.

1. In Example 1, change the conditions that $x = 2$ and $Dx = 0$ for $t = 0$ to $x = 0$ and $Dx = 2$ for $t = 0$.
2. In Example 5, change the voltage source to $500 \cos 100t$.

In Exercises 3–28, solve the given problems.

3. An object moves with simple harmonic motion according to $D^2x + 0.2Dx + 100x = 0$, $D = d/dt$. Find the displacement as a function of time, subject to the conditions $x = 4$ and $Dx = 0$ when $t = 0$.
4. What must be the value of b so that the motion of an object given by the equation $D^2x + bDx + 100x = 0$ is critically damped?
5. When the angular displacement θ of a pendulum is small (less than about 6°), the pendulum moves with simple harmonic motion closely approximated by $D^2\theta + \frac{g}{l}\theta = 0$. Here, $D = d/dt$, g is the acceleration due to gravity, and l is the length of the pendulum. Find θ as a function of time (in s) if $g = 9.8 \text{ m/s}^2$, $l = 1.0$ m, $\theta = 0.1$, and $D\theta = 0$ when $t = 0$. Sketch the curve.
6. A block of wood floating in oil is depressed from its equilibrium position such that its equation of motion is $D^2y + 8Dy + 3y = 0$, where y is the displacement (in cm) and $D = d/dt$. Find its displacement after 12 s if $y = 6.0$ cm and $Dy = 0$ when $t = 0$.
7. A car suspension is depressed from its equilibrium position such that its equation of motion is $D^2y + b\,Dy + 25y = 0$, where y is the displacement and $D = d/dt$. What must be the value of b if the motion is critically damped?
8. In an electric circuit, if a capacitor discharges through a negligible resistance, the current i is related to the time t by the equation $d^2i/dt^2 = -a^2i$, where a is a constant. Find the frequency of the current if $a = 1000$.
9. For an elastic band that is stretched vertically, with one end fixed and a mass m at the other end, the displacement s of the mass is given by $m\frac{d^2s}{dt^2} = -\frac{mg}{e}(s - L)$, where L is the natural length of the band and e is the elongation due to the weight mg. Find s if $s = s_0$ and $ds/dt = 0$ when $t = 0$.
10. A mass of 0.820 kg stretches a given spring by 0.250 m. The mass is pulled down 0.150 m below the equilibrium position and released. Find the equation of motion of the mass if there is no damping.
11. A 4.00-N weight stretches a certain spring 5.00 cm. With this weight attached, the spring is pulled 10.0 cm longer than its equilibrium length and released. Find the equation of the resulting motion, assuming no damping.
12. Find the solution for the spring of Exercise 11 if a damping force numerically equal to the velocity is present.
13. Find the solution for the spring of Exercise 11 if no damping is present but an external force of $4 \sin 2t$ is acting on the spring.
14. Find the solution for the spring of Exercise 11 if the damping force of Exercise 12 and the impressed force of Exercise 13 are both acting.
15. Find the equation relating the charge and the time in an electric circuit with the following elements: $L = 0.200$ H, $R = 8.00\ \Omega$, $C = 1.00\ \mu$F, and $E = 0$. In this circuit, $q = 0$ and $i = 0.500$ A when $t = 0$.
16. For a given electric circuit, $L = 2$ mH, $R = 0$, $C = 50$ nF, and $E = 0$. Find the equation relating the charge and the time if $q = 10^5$ C and $i = 0$ when $t = 0$.
17. For a given circuit, $L = 0.100$ H, $R = 0$, $C = 100\ \mu$F, and $E = 100$ V. Find the equation relating the charge and the time if $q = 0$ and $i = 0$ when $t = 0$.
18. Find the relation between the current and the time for the circuit of Exercise 17.
19. For a radio tuning circuit, $L = 0.500$ H, $R = 10.0\ \Omega$, $C = 200\ \mu$F, and $E = 120 \sin 120\pi t$. Find the equation relating the charge and time.
20. Find the steady-state current for the circuit of Exercise 19.
21. In a given electric circuit $L = 8.00$ mH, $R = 0$, $C = 0.500\ \mu$F, and $E = 20.0e^{-200t}$ mV. Find the relation between the current and the time if $q = 0$ and $i = 0$ for $t = 0$.
22. Find the current as a function of time for a circuit in which $L = 0.400$ H, $R = 60.0\ \Omega$, $C = 0.200\ \mu$F, and $E = 0.800e^{-100t}$ V, if $q = 0$ and $i = 5.00$ mA for $t = 0$.
23. Find the steady-state current for a circuit with $L = 1.00$ H, $R = 5.00\ \Omega$, $C = 150\ \mu$F, and $E = 120 \sin 100t$ V.
24. Find the steady-state solution for the current in an electric circuit containing the following elements: $C = 20.0\ \mu$F, $L = 2.00$ H, $R = 20.0\ \Omega$, and $E = 200 \sin 10t$ V.
25. A *cantilever* beam is clamped at the end $x = 0$ and is free at the end $x = L$. Find the equation for the deflection y of the beam in terms of the distance x from one end if it has a constant load distribution of w due to its own weight. See Fig. 31.19.

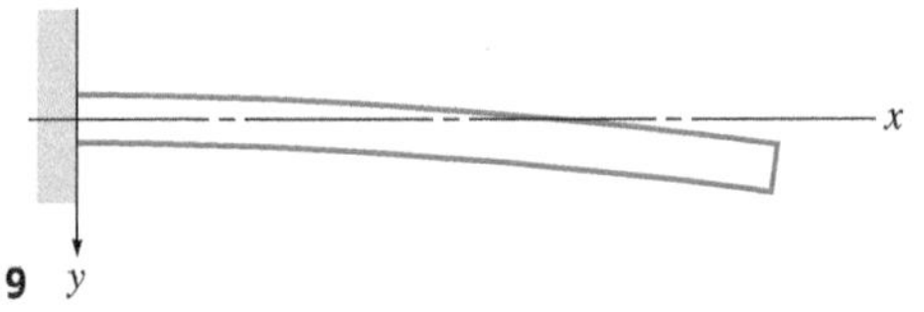

Fig. 31.19

26. A beam 10 m in length is hinged at both ends and has a variable load distribution of $w = kEIx$, where $k = 7.2 \times 10^{-4}$/m and x is the distance from one end. Find the equation of the deflection y in terms of x.
27. A mass of $\frac{1}{16}$ kg stretches a spring for which $k = 4$ N/m. A periodic external force $f(x)$ is applied to the spring, with one period defined by $f(t) = \begin{cases} \pi t & 0 \le t < 1 \\ \pi t - 2\pi & 1 \le t < 2 \end{cases}$, and $f(t)$ is extended to be odd. Express the displacement y of the object as a function of time. (See Example 7 in Section 31.9.)
28. For the pendulum in Exercise 5, what values, if any, of the length l give a critically or overdamped solution? Explain.

Answer to Practice Exercise

1. $x = \frac{1}{2}\sin 4t$

31.11 Laplace Transforms

Laplace transforms *provide an algebraic method of obtaining a* ***particular*** *solution of a differential equation from stated initial conditions.* Since this is frequently what is wanted, Laplace transforms are often used in engineering and electronics. The treatment in this text is intended only as an introduction to Laplace transforms.

■ The Laplace transform is named for the French mathematician and astronomer Pierre Laplace (1749–1827).

The Laplace transform of a function $f(t)$ is defined as the function $F(s)$ such that

$$F(s) = \int_0^\infty e^{-st} f(t)\,dt \tag{31.21}$$

By writing the transform as $F(s)$, we show that the result of integrating and evaluating is a function of s. To denote that we are dealing with "the Laplace transform of the function $f(t)$," the notation $\mathcal{L}(f)$ is used. Thus,

$$F(s) = \mathcal{L}(f) = \int_0^\infty e^{-st} f(t)\,dt \tag{31.22}$$

We shall see that both notations are quite useful.

In Eqs. (31.21) and (31.22), we note that the upper limit is ∞, which means it is unbounded. This integral is one type of what is known as an **improper integral**. In evaluating at the upper limit, it is necessary to find the limit of the resulting function as the upper limit approaches infinity. This may be shown as

$$\lim_{c \to \infty} \int_0^c e^{-st} f(t)\,dt$$

where we substitute c for t in the resulting function and determine the limit as $c \to \infty$ to determine the result for the upper limit. This also means that the Laplace transform, $F(s)$, is defined only for those values of s for which the limit is defined.

EXAMPLE 1 Finding a transform from the definition

Find the Laplace transform of the function $f(t) = t$, $t > 0$.

By the definition of the Laplace transform,

$$\mathcal{L}(f) = \mathcal{L}(t) = \int_0^\infty e^{-st} t\,dt$$

This may be integrated by parts or by the integration formula $\int u e^{au}\,du = \dfrac{e^{au}(au - 1)}{a^2}$. Using the formula, we have

$$\mathcal{L}(t) = \int_0^\infty te^{-st}\,dt = \lim_{c \to \infty} \int_0^c te^{-st}\,dt = \lim_{c \to \infty} \frac{e^{-st}(-st - 1)}{s^2}\Bigg|_0^c$$

$$= \lim_{c \to \infty}\left[\frac{e^{-sc}(-sc - 1)}{s^2}\right] + \frac{1}{s^2}$$

■ For reference, if $\lim\limits_{c \to \infty} u(c) \lim\limits_{c \to \infty} v(c) = \infty$, then by L'Hospital's rule $\lim\limits_{c \to \infty} \dfrac{u(c)}{v(c)} = \lim\limits_{c \to \infty} \dfrac{u'(c)}{v'(c)}$.

For $s > 0$, we can rewrite the quotient so as to have an ∞/∞ indeterminate form and use L'Hospital's rule (see Section 27.7) to evaluate the limit. We have

$$\mathcal{L}(t) = -\lim_{c \to \infty}\left[\frac{sc + 1}{s^2 e^{sc}}\right] + \frac{1}{s^2} \quad \text{rewriting to have } \infty/\infty$$

$$= -\lim_{c \to \infty}\left[\frac{s}{s^3 e^{sc}}\right] + \frac{1}{s^2} \quad \text{differentiate numerator and denominator with respect to } c$$

$$= 0 + \frac{1}{s^2} = \frac{1}{s^2} \quad \text{defined for } s > 0$$

EXAMPLE 2 Finding a transform from the definition

Find the Laplace transform of the function $f(t) = \cos at$.

By definition,

$$\mathcal{L}(f) = \mathcal{L}(\cos at) = \int_0^\infty e^{-st} \cos at \, dt$$

Using the integration formula $\int e^{au} \cos bu \, du = \dfrac{e^{au}(a \cos bu + b \sin bu)}{a^2 + b^2}$, we have

$$\begin{aligned}\mathcal{L}(\cos at) &= \int_0^\infty e^{-st}\cos at \, dt = \lim_{c\to\infty} \int_0^c e^{-st} \cos at \, dt \\ &= \lim_{c\to\infty} \frac{e^{-st}(-s \cos at + a \sin at)}{s^2 + a^2}\Bigg|_0^c \\ &= \lim_{c\to\infty} \frac{e^{-sc}(-s \cos ac + a \sin ac)}{s^2 + a^2} - \left(-\frac{s}{s^2 + a^2}\right) \\ &= 0 + \frac{s}{s^2 + a^2} = \frac{s}{s^2 + a^2} \qquad (s > 0)\end{aligned}$$

Therefore, the Laplace transform of the function $\cos at$ is

$$\mathcal{L}(\cos at) = \frac{s}{s^2 + a^2}$$

In both examples, the resulting transform was an algebraic function of s.

Table 31.1 is a short table of Laplace transforms. They are sufficient for our work in this chapter. More complete tables are available in many references.

Table 31.1 Laplace Transforms

	$f(t) = \mathcal{L}^{-1}(F)$	$\mathcal{L}(f) = F(s)$		$f(t) = \mathcal{L}^{-1}(F)$	$\mathcal{L}(f) = F(s)$
1.	1	$\dfrac{1}{s}$	11.	te^{-at}	$\dfrac{1}{(s+a)^2}$
2.	$\dfrac{t^{n-1}}{(n-1)!}$	$\dfrac{1}{s^n}(n = 1, 2, 3, \ldots)$	12.	$t^{n-1}e^{-at}$	$\dfrac{(n-1)!}{(s+a)^n}$
3.	e^{-at}	$\dfrac{1}{s+a}$	13.	$e^{-at}(1 - at)$	$\dfrac{s}{(s+a)^2}$
4.	$1 - e^{-at}$	$\dfrac{a}{s(s+a)}$	14.	$[(b-a)t + 1]e^{-at}$	$\dfrac{s+b}{(s+a)^2}$
5.	$\cos at$	$\dfrac{s}{s^2+a^2}$	15.	$\sin at - at \cos at$	$\dfrac{2a^3}{(s^2+a^2)^2}$
6.	$\sin at$	$\dfrac{a}{s^2+a^2}$	16.	$t \sin at$	$\dfrac{2as}{(s^2+a^2)^2}$
7.	$1 - \cos at$	$\dfrac{a^2}{s(s^2+a^2)}$	17.	$\sin at + at \cos at$	$\dfrac{2as^2}{(s^2+a^2)^2}$
8.	$at - \sin at$	$\dfrac{a^3}{s^2(s^2+a^2)}$	18.	$t \cos at$	$\dfrac{s^2-a^2}{(s^2+a^2)^2}$
9.	$e^{-at} - e^{-bt}$	$\dfrac{b-a}{(s+a)(s+b)}$	19.	$e^{-at} \sin bt$	$\dfrac{b}{(s+a)^2+b^2}$
10.	$ae^{-at} - be^{-bt}$	$\dfrac{s(a-b)}{(s+a)(s+b)}$	20.	$e^{-at} \cos bt$	$\dfrac{s+a}{(s+a)^2+b^2}$

An important property of transforms is the **linearity property**,

$$\mathcal{L}[af + bg] = a\mathcal{L}(f) + b\mathcal{L}(g) \tag{31.23}$$

We state this property here since it determines that the transform of a sum of functions is the sum of the transforms. This is of definite importance when dealing with a sum of functions. This property is a direct result of the definition of the Laplace transform.

Another Laplace transform important to the solution of a differential equation is the transform of the derivative of a function. Let us first find the Laplace transform of the first derivative of a function.

By definition,

$$\mathcal{L}(f') = \int_0^{\infty} e^{-st} f'(t)\,dt$$

To integrate by parts, let $u = e^{-st}$ and $dv = f'(t)\,dt$, so $du = -se^{-st}dt$ and $v = f(t)$ (the integral of the derivative of a function is the function). Therefore,

$$\begin{aligned}\mathcal{L}(f') &= \lim_{c\to\infty} e^{-st}f(t)\Big|_0^c + s\int_0^{\infty} e^{-st}f(t)\,dt \\ &= 0 - f(0) + s\mathcal{L}(f)\end{aligned}$$

It is noted that the integral in the second term on the right is the Laplace transform of $f(t)$ by definition. Therefore, *the Laplace transform of the first derivative of a function* is

$$\mathcal{L}(f') = s\mathcal{L}(f) - f(0) \tag{31.24}$$

Applying the same analysis, we may find *the Laplace transform of the second derivative of a function.* It is

$$\mathcal{L}(f'') = s^2\mathcal{L}(f) - sf(0) - f'(0) \tag{31.25}$$

Here, it is necessary to integrate by parts twice to derive the result. The transforms of higher derivatives are found in a similar manner.

Eqs. (31.24) and (31.25) allow us to express the transform of each derivative in terms of s and the transform itself. This is illustrated in the following example.

Practice Exercise

1. Given that $f(0) = 1$ and $f'(0) = 0$, express the transform of $f''(t) - f'(t)$ in terms of s and the transform of $f(t)$.

EXAMPLE 3 Linearity property—transforms of derivatives

Given that $f(0) = 0$ and $f'(0) = 1$, express the transform of $f''(t) - 2f'(t)$ in terms of s and the transform of $f(t)$.

By using the linearity property and the transforms of the derivatives, we have

$$\begin{aligned}\mathcal{L}[f'' - 2f'] &= \mathcal{L}(f'') - 2\mathcal{L}(f') && \text{using Eq. (31.23)}\\ &= [s^2\mathcal{L}(f) - sf(0) - f'(0)] - 2[s\mathcal{L}(f) - f(0)] && \text{using Eqs. (31.25) and (31.24)}\\ &= [s^2\mathcal{L}(f) - s\cdot 0 - 1] - 2[s\mathcal{L}(f) - 0] && \text{substitute given values}\\ &= (s^2 - 2s)\mathcal{L}(f) - 1\end{aligned}$$

INVERSE TRANSFORMS

If the Laplace transform of a function is known, it is then possible to find the function by finding the **inverse transform**,

$$\mathcal{L}^{-1}(F) = f(t) \tag{31.26}$$

where $\mathcal{L}^{-1}$ denotes the inverse transform.

EXAMPLE 4 Inverse transform from Table 31.1

If $F(s) = \frac{s}{s^2 + a^2}$, from Transform 5 of the table, we see that

$$\mathcal{L}^{-1}(F) = \mathcal{L}^{-1}\left(\frac{s}{s^2 + a^2}\right) = \cos at$$

$$f(t) = \cos at$$

EXAMPLE 5 Inverse transform from Table 31.1

If $(s^2 - 2s)\mathcal{L}(f) - 1 = 0$, then

$$\mathcal{L}(f) = \frac{1}{s^2 - 2s} \quad \text{or} \quad F(s) = \frac{1}{s(s-2)}$$

Therefore, we have

$$f(t) = \mathcal{L}^{-1}(F) = \mathcal{L}^{-1}\left[\frac{1}{s(s-2)}\right] \quad \text{inverse transform}$$

$$= -\frac{1}{2}\mathcal{L}^{-1}\left[\frac{-2}{s(s-2)}\right] \quad \text{fit form of Transform 4}$$

$$= -\frac{1}{2}(1 - e^{2t}) \quad \text{use Transform 4}$$

Practice Exercise

2. Find $f(t)$ if $F(s) = \frac{6}{s^2 + 9}$.

The introduction of the factor -2 in Example 5 illustrates that it often takes some algebra steps to get $F(s)$ to match the proper form in the table. Another algebraic step that may be useful is *completing the square.* For a review of this algebraic method, see Section 7.2 (in particular, see Example 3 of that section). The following example illustrates its use in finding an inverse transform.

EXAMPLE 6 Inverse transform by completing the square

If $F(s) = \frac{s+5}{s^2 + 6s + 10}$, then

$$\mathcal{L}^{-1}(F) = \mathcal{L}^{-1}\left[\frac{s+5}{s^2 + 6s + 10}\right]$$

It appears that this function does not fit any of the forms given. However,

$$s^2 + 6s + 10 = (s^2 + 6s + 9) + 1 = (s+3)^2 + 1$$

By writing $F(s)$ as

$$F(s) = \frac{(s+3) + 2}{(s+3)^2 + 1} = \frac{s+3}{(s+3)^2 + 1} + \frac{2}{(s+3)^2 + 1}$$

we can find the inverse of each term. Therefore,

$$\mathcal{L}^{-1}(F) = e^{-3t}\cos t + 2e^{-3t}\sin t \quad \text{using Transforms 20 and 19}$$

$$f(t) = e^{-3t}(\cos t + 2\sin t)$$

The following example shows how *partial fractions* can be used to find the inverse transform of $F(s)$. For a review of the method of expressing a given algebraic fraction in terms of partial fractions, refer to Sections 28.9 and 28.10.

EXAMPLE 7 Inverse transform by partial fractions

If $F(s) = \dfrac{5s^2 - 17s + 32}{s^3 - 8s^2 + 16s}$, then

$$\mathcal{L}^{-1}(F) = \mathcal{L}^{-1}\left[\frac{5s^2 - 17s + 32}{s^3 - 8s^2 + 16s}\right]$$

To fit forms in the table, we will now use partial fractions.

$$\frac{5s^2 - 17s + 32}{s^3 - 8s^2 + 16s} = \frac{5s^2 - 17s + 32}{s(s-4)^2} = \frac{A}{s} + \frac{B}{s-4} + \frac{C}{(s-4)^2}$$ factor of s, repeated factor $s - 4$

$$5s^2 - 17s + 32 = A(s-4)^2 + Bs(s-4) + Cs$$ multiply each side by $s(s-4)^2$

$s = 0$: $32 = 16A$, $A = 2$

$s = 4$: $5(4^2) - 17(4) + 32 = 4C$, $C = 11$

s^2 terms: $5 = A + B$, $5 = 2 + B$, $B = 3$

$$\mathcal{L}^{-1}(F) = \mathcal{L}^{-1}\left[\frac{2}{s} + \frac{3}{s-4} + \frac{11}{(s-4)^2}\right]$$ substitute in $F(s)$

$$\mathcal{L}^{-1}(F) = f(t) = 2 + 3e^{4t} + 11te^{4t}$$ using Transforms 1, 3, and 11

EXERCISES 31.11

In Exercises 1–4, make the given changes in the indicated examples of this section, and then solve the resulting problems.

1. In Example 1, change the function $f(t)$. Let $f(t) = 1$.

2. In Example 2, change the function $f(t)$. Let $f(t) = \sin at$.

3. In Example 3, interchange the values of $f(0)$ and $f'(0)$.

4. In Example 4, in the function $F(s)$, change the numerator to a.

In Exercises 5–12, find the transforms of the given functions by use of the table.

5. $f(t) = e^{3t}$

6. $f(t) = 1 - \cos 2t$

7. $f(t) = 5t^3e^{-2t}$

8. $f(t) = 8e^{-3t}\sin 4t$

9. $f(t) = \cos 2t - \sin 2t$

10. $f(t) = 2t\sin 3t + e^{-3t}\cos t$

11. $f(t) = 3 + 2t\cos 3t$

12. $f(t) = t^3 - 3te^{-t}$

In Exercises 13–16, express the transforms of the given expressions in terms of s and $\mathcal{L}(f)$.

13. $f'' + f'$, $f(0) = 0$, $f'(0) = 0$

14. $f'' - 3f'$, $f(0) = 2$, $f'(0) = -1$

15. $2f'' - f' + f$, $f(0) = 1$, $f'(0) = 0$

16. $f'' - 3f' + 2f$, $f(0) = -1$, $f'(0) = 2$

In Exercises 17–28, find the inverse transforms of the given functions of s.

17. $F(s) = \dfrac{2}{s^3}$

18. $F(s) = \dfrac{6}{s^2 + 4}$

19. $F(s) = \dfrac{15}{2s + 6}$

20. $F(s) = \dfrac{3}{s^4 + 4s^2}$

21. $F(s) = \dfrac{1}{s^3 + 3s^2 + 3s + 1}$

22. $F(s) = \dfrac{s^2 - 1}{s^4 + 2s^2 + 1}$

23. $F(s) = \dfrac{s + 2}{(s^2 + 9)^2}$

24. $F(s) = \dfrac{s + 3}{s^2 + 4s + 13}$

25. $F(s) = \dfrac{4s^2 - 8}{(s+1)(s-2)(s-3)}$

26. $F(s) = \dfrac{3s + 1}{(s-1)(s^2+1)}$

27. $F(s) = \dfrac{2s + 3}{s^2 - 2s + 5}$

28. $F(s) = \dfrac{3s^4 + 3s^3 + 6s^2 + s + 1}{s^5 + s^3}$ (Explain your method of solution.)

In Exercises 29 and 30, find the indicated Laplace transforms.

29. For the Laplace transform $F(s)$ of the function $f(t)$, it can be shown that $\mathcal{L}\{tf(t)\} = -\frac{d}{ds}F(s)$. Verify this relationship by deriving Transform 11 from Transform 3.

30. Using the equation given in Exercise 29, derive the transform for $t^2 \cos at$ from Transform 18.

Answers to Practice Exercises

1. $(s^2 - s)\mathcal{L}(f) - s + 1$ **2.** $2\sin 3t$

31.12 Solving Differential Equations by Laplace Transforms

We will now show how certain differential equations can be solved by using Laplace transforms. *It must be remembered that these solutions are the* ***particular*** *solutions of the equations subject to the given conditions.* The necessary operations were developed in the preceding section. The following examples illustrate the method.

EXAMPLE 1 Solution of a first-order equation

Using Laplace transforms, solve the differential equation $2y' - y = 0$, if $y(0) = 1$. (Note that we are using y to denote the function.)

Taking transforms of each term in the equation, we have

$$\mathcal{L}(2y') - \mathcal{L}(y) = \mathcal{L}(0)$$
$$2\mathcal{L}(y') - \mathcal{L}(y) = 0$$

$\mathcal{L}(0) = 0$ by direct use of the definition of the transform. Now, using Eq. (31.24), $\mathcal{L}(y') = s\mathcal{L}(y) - y(0)$, we have

$$2[s\mathcal{L}(y) - 1] - \mathcal{L}(y) = 0 \qquad y(0) = 1$$

Solving for $\mathcal{L}(y)$, we obtain

$$2s\mathcal{L}(y) - \mathcal{L}(y) = 2$$
$$\mathcal{L}(y) = \frac{2}{2s-1} = \frac{1}{s-\frac{1}{2}}$$

Finding the inverse transform, we have

$$y = e^{t/2} \qquad \text{using Transform 3}$$

You should check this solution with that obtained by methods developed earlier.

LEARNING TIP

It should be noted that the solution in Example 1 was essentially an algebraic one. This points out the power and usefulness of Laplace transforms. ***We are able to change a differential equation into an algebraic form,*** which can in turn be translated into the solution of the differential equation. Thus, we can solve a differential equation by using algebra and specific algebraic forms.

EXAMPLE 2 Solution of a second-order equation

Using Laplace transforms, solve the differential equation $y'' + 2y' + 2y = 0$, if $y(0) = 0$ and $y'(0) = 1$.

Using the same steps as outlined in Example 1, we have

$$\mathcal{L}(y'') + 2\mathcal{L}(y') + 2\mathcal{L}(y) = 0 \qquad \text{take transforms}$$
$$[s^2\mathcal{L}(y) - sy(0) - y'(0)] + 2[s\mathcal{L}(y) - y(0)] + 2\mathcal{L}(y) = 0 \qquad \text{using Eqs. (31.25) and (31.24)}$$
$$[s^2\mathcal{L}(y) - s(0) - 1] + 2[s\mathcal{L}(y) - 0] + 2\mathcal{L}(y) = 0 \qquad \text{substitute given values}$$
$$s^2\mathcal{L}(y) - 1 + 2s\mathcal{L}(y) + 2\mathcal{L}(y) = 0$$
$$(s^2 + 2s + 2)\mathcal{L}(y) = 1 \qquad \text{solve for } \mathcal{L}(y)$$
$$\mathcal{L}(y) = \frac{1}{s^2 + 2s + 2} = \frac{1}{(s+1)^2 + 1} \qquad \text{fit transform form}$$
$$y = e^{-t}\sin t \qquad \text{take inverse transform using Transform 19}$$

Practice Exercise

1. In Example 2, find the solution if $y(0) = 1$ and $y'(0) = 0$.

EXAMPLE 3 Solution of a second-order equation

Solve the differential equation $y'' + y = \cos t$, if $y(0) = 1$ and $y'(0) = 2$.

$$\mathcal{L}(y'') + \mathcal{L}(y) = \mathcal{L}(\cos t) \qquad \text{take transforms}$$

$$\left[s^2\mathcal{L}(y) - s(1) - 2\right] + \mathcal{L}(y) = \frac{s}{s^2+1} \qquad \text{using Eq. (31.25) and Transform 5}$$

$$(s^2+1)\mathcal{L}(y) = \frac{s}{s^2+1} + s + 2$$

$$\mathcal{L}(y) = \frac{s}{(s^2+1)^2} + \frac{s}{s^2+1} + \frac{2}{s^2+1} \qquad \text{do not combine the fractions}$$

$$y = \frac{t}{2}\sin t + \cos t + 2\sin t \qquad \text{using Transforms 16, 5, and 6}$$

EXAMPLE 4 Application—simple harmonic motion

A spring is stretched 0.31 m by a weight of 4.9 N (mass of 0.5 kg). The medium resists the motion with a force of $4v$, where v is the velocity of the motion. The differential equation describing the displacement y of the weight is

$$\frac{1}{2}\frac{d^2y}{dt^2} + 4\frac{dy}{dt} + 16y = 0 \qquad \text{see Example 3 of Section 31.10}$$

Find y as a function of time t, if $y(0) = 1$ and $dy/dx = 0$ for $t = 0$.

Clearing fractions and denoting derivatives by y'' and y', we have the following differential equation and solution.

$$y'' + 8y' + 32y = 0$$

$$\mathcal{L}(y'') + 8\mathcal{L}(y') + 32\mathcal{L}(y) = 0 \qquad \text{take transforms}$$

$$\left[s^2\mathcal{L}(y) - s \cdot 1 - 0\right] + 8\left[s\mathcal{L}(y) - 1\right] + 32\mathcal{L}(y) = 0 \qquad \text{substitute given values}$$

$$(s^2 + 8s + 32)\mathcal{L}(y) = s + 8 \qquad \text{solve for } \mathcal{L}(y)$$

$$\mathcal{L}(y) = \frac{s+8}{(s+4)^2+4^2} = \frac{s+4}{(s+4)^2+4^2} + \frac{4}{(s+4)^2+4^2} \qquad \text{fit transform forms}$$

$$y = e^{-4t}\cos 4t + e^{-4t}\sin 4t = e^{-4t}(\cos 4t + \sin 4t) \qquad \text{take inverse transforms}$$

The graph of this solution is shown in Fig. 31.20.

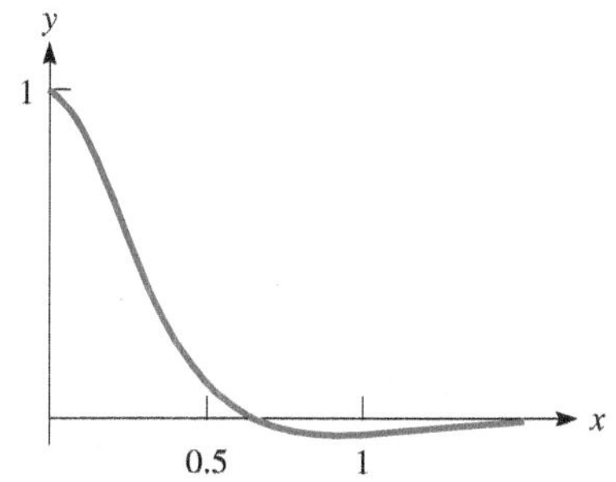

Fig. 31.20

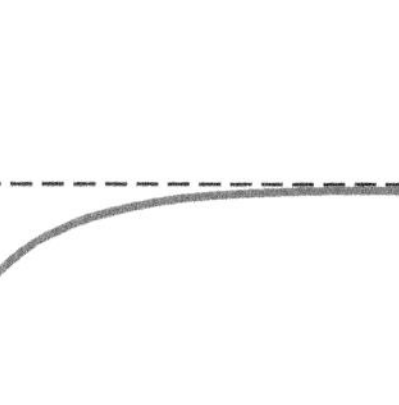

Fig. 31.21

EXAMPLE 5 Application—electric current

The initial current in the circuit shown in Fig. 31.21 is zero. Find the current as a function of the time t.

Setting up the differential equation, and then solving it, we have

$$\frac{di}{dt} + 10i = 6 \qquad \text{using Eq. (31.20)}$$

$$\mathcal{L}\left(\frac{di}{dt}\right) + 10\mathcal{L}(i) = \mathcal{L}(6) \qquad \text{take transforms}$$

$$\left[s\mathcal{L}(i) - 0\right] + 10\mathcal{L}(i) = \frac{6}{s} \qquad \text{substitute given values and find transform on right}$$

$$\mathcal{L}(i) = \frac{6}{s(s+10)} \qquad \text{solve for } \mathcal{L}(i)$$

$$i = 0.6(1 - e^{-10t}) \qquad \text{take inverse transform}$$

The graph of this solution is shown in Fig. 31.22.

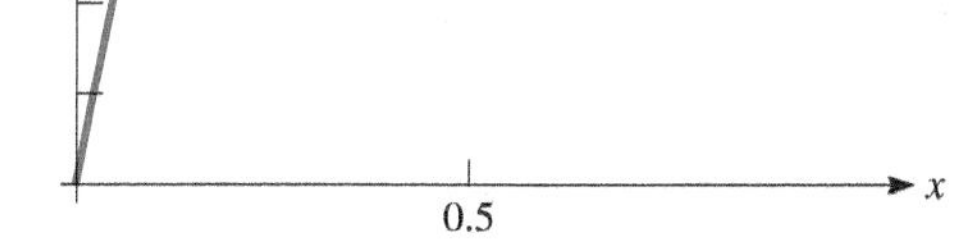

Fig. 31.22

EXAMPLE 6 Application—electric current

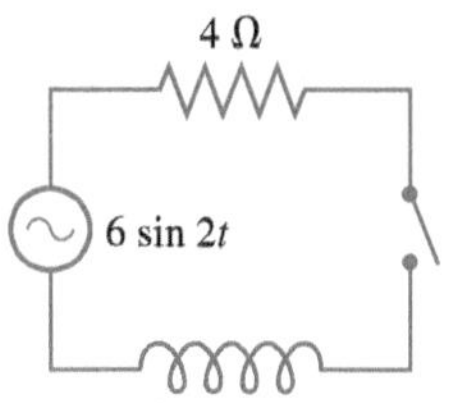

Fig. 31.23

An electric circuit in an FM radio transmitter contains a 1-H inductor and a 4-Ω resistor. It is being tested using a voltage source of 6 sin $2t$. If the initial current is zero, find the current i as a function of time t. See Fig. 31.23.

The solution is as follows:

$$(1)Di + 4i = 6 \sin 2t \qquad \text{differential equation, } D = d/dt$$

$$\mathcal{L}(Di) + 4\mathcal{L}(i) = 6\mathcal{L}(\sin 2t) \qquad \text{take transforms}$$

$$[s\mathcal{L}(i) - 0] + 4\mathcal{L}(i) = \frac{6(2)}{s^2 + 4} \qquad i(0) = 0$$

$$\mathcal{L}(i) = \frac{12}{(s + 4)(s^2 + 4)} = \frac{A}{s + 4} + \frac{Bs + C}{s^2 + 4} \qquad \text{use partial fractions}$$

$$12 = A(s^2 + 4) + B(s^2 + 4s) + C(s + 4)$$

■ $s = 0$: $12 = 4A + 4C$, $3 = A + C$
s terms: $0 = 4B + C$
s^2 terms: $0 = A + B$

The equations that give us the following values are shown in the margin at the left.

$$A = 0.6 \qquad B = -0.6 \qquad C = 2.4$$

$$\mathcal{L}(i) = 0.6\left(\frac{1}{s + 4}\right) - 0.6\left(\frac{s}{s^2 + 4}\right) + 1.2\left(\frac{2}{s^2 + 4}\right) \qquad \text{fit transform forms}$$

$$\mathcal{L} = 0.6e^{-4t} - 0.6 \cos 2t + 1.2 \sin 2t \qquad \text{take inverse transforms}$$

This is checked by showing that $i(0) = 0$ and that it satisfies the original equation.

EXAMPLE 7 Application—electric current

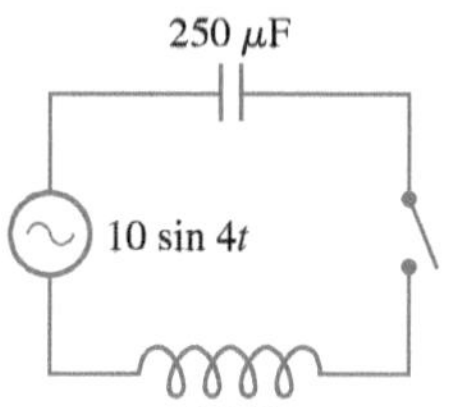

Fig. 31.24

An electric circuit contains a 0.1-H inductor, a 250-μF capacitor, a voltage source of 10 sin $4t$, and negligible resistance (assume $R = 0$). See Fig. 31.24. If the initial charge on the capacitor is zero, and the initial current is also zero, find the current in the circuit as a function of the time t.

The solution is as follows:

$$0.1D^2q + \frac{1}{250 \times 10^{-6}}q = 10 \sin 4t \qquad \text{differential equation, } D = d/dt$$

$$D^2q + 40\,000q = 100 \sin 4t$$

$$\mathcal{L}(D^2q) + 40\,000\mathcal{L}(q) = 100\mathcal{L}(\sin 4t) \qquad \text{take transforms}$$

$$[s^2\mathcal{L}(q) - sq(0) - Dq(0)] + 40\,000\mathcal{L}(q) = \frac{400}{s^2 + 16} \qquad q(0) = 0,\ D(q) = 0$$

$$\mathcal{L}(q) = \frac{400}{(s^2 + 200^2)(s^2 + 16)} = \frac{As + B}{s^2 + 200^2} + \frac{Cs + E}{s^2 + 16} \qquad \text{use partial fractions}$$

$$400 = (As + B)(s^2 + 16) + (Cs + E)(s^2 + 200^2)$$

■ $s = 0$: $400 = 16B + 200^2E$
s terms: $0 = 16A + 200^2C$
s^2 terms: $0 = B + E$
s^3 terms: $0 = A + C$

The equations that give us the following values are shown in the margin at the left.

$$A = 0 \qquad B = -0.010 \qquad C = 0 \qquad E = 0.010$$

$$\mathcal{L}(q) = \frac{0.010}{s^2 + 16} - \frac{0.010}{s^2 + 200^2} = \frac{0.010}{4}\left(\frac{4}{s^2 + 16}\right) - \frac{0.010}{200}\left(\frac{200}{s^2 + 200^2}\right)$$

$$q = 0.0025 \sin 4t - 5.0 \times 10^{-5} \sin 200t \qquad \text{take inverse transforms}$$

$$i = dq/dt = 0.010 \cos 4t - 0.010 \cos 200t \qquad \text{take derivative}$$

EXERCISES 31.12

In Exercises 1–4, make the given changes in the indicated examples of this section, and then solve the resulting problems.

1. In Example 1, change the function $y(0)$ from 1 to 2.
2. In Example 2, interchange the values of $y(0)$ and $y'(0)$.
3. In Example 3, interchange the values of $y(0)$ and $y'(0)$.
4. In Example 5, change the initial current to 1 A.

In Exercises 5–37, solve the given differential equations by Laplace transforms. The function is subject to the given conditions.

5. $y' + y = 0$, $y(0) = 1$
6. $y' - 2y = 0$, $y(0) = 2$
7. $2y' - 3y = 0$, $y(0) = -1$
8. $y' + 2y = 1$, $y(0) = 0$
9. $y' + 3y = e^{-3t}$, $y(0) = 1$
10. $y' + 2y = te^{-2t}$, $y(0) = 0$
11. $y'' + 4y = 0$, $y(0) = 0$, $y'(0) = 1$
12. $9y'' - 4y = 0$, $y(0) = 1$, $y'(0) = 0$
13. $4y'' + 4y' + 5y = 0$, $y(0) = 1$, $y'(0) = -1/2$
14. $y'' + 2y' + y = 0$, $y(0) = 0$, $y'(0) = -2$
15. $y'' - 4y' + 5y = 0$, $y(0) = 1$, $y'(0) = 2$
16. $4y'' + 4y' + y = 0$, $y(0) = 1$, $y'(0) = 0$
17. $y'' + y = 1$, $y(0) = 1$, $y'(0) = 1$
18. $9y'' + 4y = 2t$, $y(0) = 0$, $y'(0) = 0$
19. $y'' + 2y' + y = e^{-t}$, $y(0) = 1$, $y'(0) = 2$
20. $2y'' + 8y = 3\sin 2t$, $y(0) = 0$, $y'(0) = 0$
21. $y'' - 4y = 10e^{3t}$, $y(0) = 5$, $y'(0) = 0$
22. $y'' - 2y' + y = e^{2t}$, $y(0) = 1$, $y'(0) = 3$
23. $y'' - y = 5\sin 2t$, $y(0) = 0$, $y'(0) = 1$
24. $2y'' + y' - y = \sin 3t$, $y(0) = 0$, $y'(0) = 0$
25. A constant force of 6 N moves a 2-kg mass through a medium that resists the motion with a force equal to the velocity v. The equation relating the velocity and the time is $2\dfrac{dv}{dt} = 6 - v$. Find v as a function of t if the object starts from rest.
26. A pendulum moves with simple harmonic motion according to the differential equation $D^2\theta + 20\theta = 0$, where θ is the angular displacement and $D = d/dt$. Find θ as a function of t if $\theta = 0$ and $D\theta = 0.40$ rad/s when $t = 0$.
27. The end of a certain vibrating metal rod oscillates according to $D^2y + 6400y = 0$ (assuming no damping), where $D^2 = d^2/dt^2$. If $y = 4$ mm and $Dy = 0$ when $t = 0$, find the equation of motion.
28. If there is a retarding force of $0.2Dy$ to the motion of the rod in Exercise 27, find the equation of motion.
29. A 50-Ω resistor, a 4.0-μF capacitor, and a 40-V battery are connected in series. Find the charge on the capacitor as a function of time t if the initial charge is zero.
30. A 2-H inductor, an 80-Ω resistor, and an 8-V battery are connected in series. Find the current in the circuit as a function of time if the initial current is zero.
31. A 10-H inductor, a 40-μF capacitor, and a voltage supply whose voltage is given by $100\sin 50t$ are connected in series in an electric circuit. Find the current as a function of the time if the initial charge on the capacitor is zero and the initial current is zero.
32. A 20-mH inductor, a 40-Ω resistor, a 50-μF capacitor, and a voltage source of $100e^{-1000t}$ are connected in series in an electric circuit. Find the charge on the capacitor as a function of time t, if $q = 0$ and $i = 0$ when $t = 0$.
33. The weight on a spring undergoes forced vibrations according to the equation $D^2y + 9y = 18\sin 3t$. Find its displacement y as a function of the time t, if $y = 0$ and $Dy = 0$ when $t = 0$.
34. A spring is stretched 1 m by a 20-N weight. The spring is stretched 0.5 m below the equilibrium position with the weight attached and then released. If it is in a medium that resists the motion with a force equal to $12v$, where v is the velocity, find the displacement y of the weight as a function of the time.
35. For the electric circuit shown in Fig. 31.25, find the current as a function of the time t if the initial current is zero.

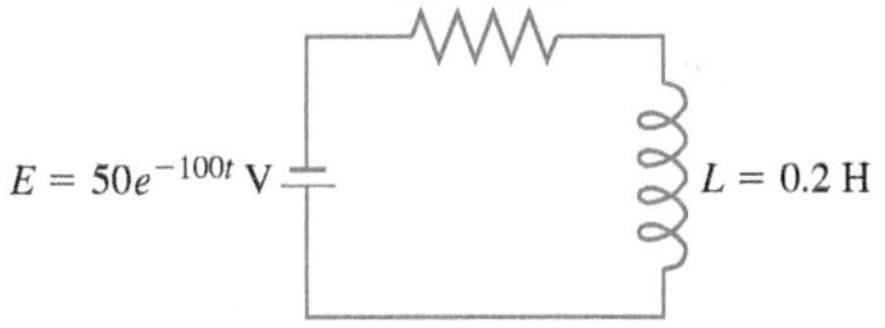

Fig. 31.25

36. For the electric circuit shown in Fig. 31.26, find the current as a function of time t if the initial charge on the capacitor is zero and the initial current is zero.

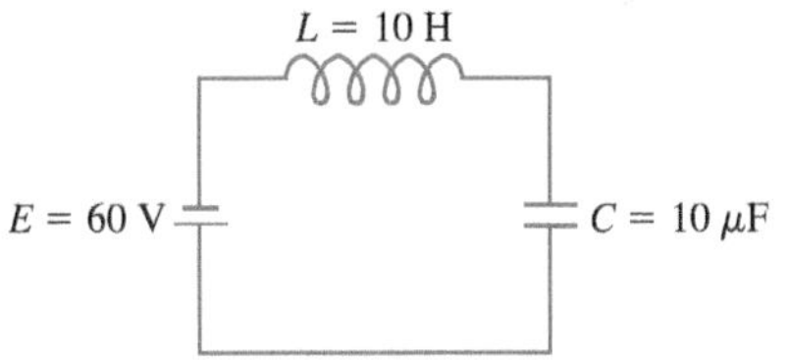

Fig. 31.26

37. For the beam in Example 6 of Section 31.10, find the deflection y as a function of x using Laplace transforms. The Laplace transform of the fourth derivative y^{iv} is given by

$$\mathcal{L}(y^{iv}) = s^4\mathcal{L}(y) - s^3y(0) - s^2y'(0) - sy''(0) - y'''(0)$$

Also, since $y'(0)$ and $y'''(0)$ are not given, but are constants, assume $y'(0) = a$, and $y'''(0) = b$. It is then possible to evaluate a and b to obtain the solution.

Answer to Practice Exercise

1. $y = e^{-t}(\sin t + \cos t)$

CHAPTER 31 KEY FORMULAS AND EQUATIONS

Separation of variables	$M(x, y)dx + N(x, y)dy = 0$	**(31.1)**
	$A(x)dx + B(y)dy = 0$	**(31.2)**
Integrating combinations	$d(xy) = x\,dy + y\,dx$	**(31.3)**
	$d(x^2 + y^2) = 2(x\,dx + y\,dy)$	**(31.4)**
	$d\left(\frac{y}{x}\right) = \frac{x\,dy - y\,dx}{x^2}$	**(31.5)**
	$d\left(\frac{x}{y}\right) = \frac{y\,dx - x\,dy}{y^2}$	**(31.6)**
Linear differential equation of first order	$dy + Py\,dx = Q\,dx$	**(31.7)**
	$ye^{\int P\,dx} = \int Qe^{\int P\,dx}dx + c$	**(31.8)**
Electric circuit	$L\frac{di}{dt} + Ri + \frac{q}{C} = E$	**(31.9)**
Motion in resisting medium	$m\frac{dv}{dt} = F - kv$	**(31.10)**
General linear differential equation	$a_0\frac{d^ny}{dx^n} + a_1\frac{d^{n-1}y}{dx^{n-1}} + \cdots + a_{n-1}\frac{dy}{dx} + a_ny = b$	**(31.11)**
	$a_0D^ny + a_1D^{n-1}y + \cdots + a_{n-1}Dy + a_ny = b$	**(31.12)**
Homogeneous linear differential equation	$a_0D^2y + a_1Dy + a_2y = 0$	**(31.13)**
Auxiliary equation	$a_0m^2 + a_1m + a_2 = 0$	**(31.14)**
Distinct roots	$y = c_1e^{m_1x} + c_2e^{m_2x}$	**(31.15)**
Repeated roots	$y = e^{mx}(c_1 + c_2x)$	**(31.16)**
Complex roots	$y = e^{\alpha x}(c_1 \sin \beta x + c_2 \cos \beta x)$	**(31.17)**
Nonhomogeneous linear differential equation	$a_0D^2y + a_1Dy + a_2y = b$	**(31.18)**
	$y = y_c + y_p$	**(31.19)**
Electric circuit	$L\frac{d^2q}{dt^2} + R\frac{dq}{dt} + \frac{q}{C} = E$	**(31.20)**
Laplace transforms	$F(s) = \int_0^\infty e^{-st}f(t)dt$	**(31.21)**
	$F(s) = \mathcal{L}(f) = \int_0^\infty e^{-st}f(t)dt$	**(31.22)**
	$\mathcal{L}[af + bg] = a\mathcal{L}(f) + b\mathcal{L}(g)$	**(31.23)**
	$\mathcal{L}(f') = s\mathcal{L}(f) - f(0)$	**(31.24)**
	$\mathcal{L}(f'') = s^2\mathcal{L}(f) - sf(0) - f'(0)$	**(31.25)**
Inverse transform	$\mathcal{L}^{-1}(F) = f(t)$	**(31.26)**

CHAPTER 31 REVIEW EXERCISES

In Exercises 1–32, find the general solution to the given differential equations.

1. $4xy^3\,dx + (x^2 + 1)\,dy = 0$

2. $\dfrac{dy}{dx} = e^{x-y}$

3. $\sin 2x\,dx + y\sin x\,dy = \sin x\,dx$

4. $x\,dy + y\,dx = y\,dy$

5. $2D^2y + Dy = 0$

6. $2D^2y - 5Dy + 2y = 0$

7. $16y'' - 8y' + y = 0$

8. $y'' + 2y' + 2y = 0$

9. $(x + y)dx + (x + y^3)dy = 0$

10. $R\ln L\,dL = L\,dR$

11. $V\dfrac{dP}{dV} - 5P = V^2$

12. $dy - 2y\,dx = (x - 2)e^x\,dx$

13. $dy = 2y\,dx + y^2\,dx$

14. $x^2y\,dy = (1 + x)\csc y\,dx$

15. $D^2y + 2Dy + 6y = 0$

16. $4D^2y - 4Dy + y = 0$

17. $y' + 4y = 2e^{-2x}$

18. $2uv\,du = (2v - \ln v)\,dv$

19. $\sin x\dfrac{dy}{dx} + y\cos x + x = 0$

20. $y\,dy = (x^2 + y^2 - x)\,dx$

21. $2\dfrac{d^2s}{dt^2} + \dfrac{ds}{dt} - 3s = 6$

22. $\dfrac{d^2y}{dx^2} + 6\dfrac{dy}{dx} + 9y = 3x$

23. $y'' + y' - y = 2e^x$

24. $4D^3y + 9Dy = xe^x$

25. $9D^2y - 18Dy + 8y = 16 + 4x$

26. $9y'' + 4y = 4\cos 2x$

27. $D^3y - D^2y + 9Dy - 9y = \sin x$

28. $y'' + y' = e^x + \cos 2x$

29. $y'' - 7y' - 8y = 2e^{-x}$

30. $3y'' - 6y' = 4 + xe^x$

31. $D^2y + 25y = 50\cos 5x$

32. $D^2y + 4y = 8x\sin 2x$

In Exercises 33–40, find the indicated particular solution of the given differential equations.

33. $3y' = 2y\cot x$; $x = \dfrac{\pi}{2}$ when $y = 2$

34. $T\,dV - V\,dT = V^3\,dV$; $T = 1$ when $V = 3$

35. $y' = 4x - 2y$; $x = 0$ when $y = -2$

36. $xy^2\,dx + e^x\,dy = 0$; $x = 0$ when $y = 2$

37. $\dfrac{d^2v}{dt^2} + \dfrac{dv}{dt} + 4v = 0$; $\dfrac{dv}{dt} = \sqrt{15}$, $v = 0$ when $t = 0$

38. $5y'' + 7y' - 6y = 0$; $y' = 10$, $y = 2$ when $x = 0$

39. $D^2y + 4Dy + 4y = 4\cos x$; $Dy = 1$, $y = 0$ when $x = 0$

40. $y'' - 2y' + y = e^x + x$; $y = 0$, $y' = 0$ when $x = 0$

In Exercises 41–48, solve the given differential equations by using Laplace transforms, where the function is subject to the given conditions.

41. $4y' - y = 0$, $y(0) = 1$

42. $2y' - y = 4$, $y(0) = 1$

43. $y' - 3y = e^t$, $y(0) = 0$

44. $y' + 2y = e^{-2t}$, $y(0) = 2$

45. $y'' + y = 0$, $y(0) = 0$, $y'(0) = -4$

46. $y'' + 4y' + 5y = 0$, $y(0) = 1$, $y'(0) = 1$

47. $16y'' + 9y = 3e^x$, $y(0) = 0$, $y'(0) = 0$

48. $y'' - 2y' + y = e^x + x$, $y(0) = 0$, $y'(0) = 1$

In Exercises 49–92, solve the given problems.

49. Use Euler's method to find the y-values of the solution of the equation $dy/dx = 1 + y^2$ from $x = 0$ to $x = 0.4$, with $\Delta x = 0.1$, if the curve passes through $(0, 0)$.

50. Solve the equation of Exercise 49, subject to the same conditions, using the Runge–Kutta method.

51. Find the particular solution of the equation $dy/dx - 2y = e^{3x}$, if $y = 1$ when $x = 0$, (a) as a first-order linear equation, and (b) using Laplace transforms.

52. The current i (in A) in an electric circuit changes with time t (in s) according to the equation $di + 10i\,dt = 6\,dt$. Find i as a function of t if the initial current is zero.

53. An object moves along a hyperbolic path described by $xy = 1$, such that $dx/dt = 2t$. Express x and y in terms of t if $x = 1$, $y = 1$ when $t = 0$.

54. An object moves along a parabolic path described by $y = x^2 + x$, such that $dx/dt = 4t + 1$. Express x and y in terms of t, if both x and y are zero when $t = 0$.

55. The time rate of change of volume of an evaporating substance is proportional to the surface area. Express the radius of an evaporating sphere of ice as a function of time. Let $r = r_0$ when $t = 0$. (*Hint:* Express both V and A in terms of the radius r.)

56. An insulated tank is filled with a solution containing radioactive cobalt. Due to the radioactivity, energy is released and the temperature T (in °C) of the solution rises with the time t (in h). The following equation expresses the relation between temperature and time for a specific case:

$$56\,600 = 262(T - 70) + 20\,200\frac{dT}{dt}$$

If the initial temperature is 70°C, what is the temperature 24 h later?

57. In a certain chemical reaction, the velocity of the reaction is proportional to the mass m of the chemical that remains unchanged. If m_0 is the initial mass and dm/dt is the velocity of the reaction, find m as a function of the time t.

58. Under proper conditions, bacteria grow at a rate proportional to the number present. In a certain culture, there were 10^4 bacteria present at a given time, and there were 3.0×10^5 bacteria present after 10 h. How many were present after 5.0 h?

59. An object with a mass of 1.00 kg slides down a long inclined plane. The effective force of gravity is 4.00 N, and the motion is retarded by a force numerically equal to the velocity. If the object starts from rest, what is the velocity (in m/s) 4.00 s later?

60. A 760-N object falls from rest under the influence of gravity. Find the equation for the velocity at any time t (in s) if the air resists the motion with a force numerically equal to twice the velocity.

61. A particle is moving along a path $y = f(x)$ such that the slope of the path is $y/(y - x)$, and the path passes through the point $(-1, 2)$. Find the equation of the path $(y > 0)$.

62. On a certain weather map, the lines indicating equal temperature (*isotherms*) are given by $y = x^3 + k$. Find the equation of the orthogonal trajectories of the isotherms, the curves that show the direction of heat flow.

63. After 10.0 s, it is noted that 15.9% of the radioactive isotope neon-23 has decayed. Find the half-life of neon-23.

64. The isotope iodine-131, with a half-life of 8.04 days, is used in nuclear medicine to study the thyroid gland. Of an original amount, what percent of iodine-131 remains after 21 days?

65. Radioactive potassium-40 with a half-life of 1.28×10^9 years is used for dating rock samples. If a given rock sample has 75% of its original amount of potassium-40, how old is the rock?

66. When a gas undergoes an adiabatic change (no gain or loss of heat), the rate of change of pressure with respect to volume is directly proportional to the pressure and inversely proportional to the volume. Express the pressure in terms of the volume.

67. Under ideal conditions, the natural law of population change is that the population increases at a rate proportional to the population at any time. Under these conditions, project the population of the world in 2025 if it reached 6.9 billion in 2010 and 7.3 billion in 2015.

68. A spherical balloon is being blown up such that its volume V increases at a rate proportional to its surface area. Show that this leads to the differential equation $dV/dt = kV^{2/3}$ and solve for V as a function of t.

69. Find the orthogonal trajectories of the family of curves $y = kx^5$.

70. Find the equation of the curves for which their normals at all points are in the direction of the lines connecting the points and the origin.

71. Find the temperature after 1.0 h of an object originally 100°C, if it cools to 90° in 5.0 min in air that is at 20°C. (See Exercise 25 in Section 31.6.)

72. If a circuit contains a resistance R, a capacitance C, and a source of voltage E, express the charge q on the capacitor as a function of time.

73. A 2-H inductor, a 40-Ω resistor, and a 20-V battery are connected in series. Find the current in the circuit as a function of time if the initial current is zero.

74. A hollow cylinder moves vertically up and down in water according to the equation $D^2y + 6.5y = 0$, where $D = d/dt$. Find the displacement y as a function of the time t, if $y = 8.0$ cm and $y' = 0$ cm/s when $t = 0$ s.

75. A certain spring stretches 0.5 m by a 40-N weight. With this weight suspended on it, the spring is stretched 0.5 m beyond the equilibrium position and released. Find the equation of the resulting motion if the medium in which the weight is suspended retards the motion with a force equal to 16 times the velocity. Classify the motion as underdamped, critically damped, or overdamped. Explain your choice.

76. The end of a vibrating rod moves according to the equation $D^2y + 0.2Dy + 4000y = 0$, where y is the displacement and $D = d/dt$. Find y as a function of t if $y = 3.00$ cm and $Dy = -0.300$ cm/s when $t = 0$.

77. A 0.5-H inductor, a 6-Ω resistor, and a 20-mF capacitor are connected in series with a generator for which $E = 24 \sin 10t$. Find the charge on the capacitor as a function of time if the initial charge and initial current are zero.

78. A 5.00-mH inductor and a 10.0-μF capacitor are connected in series with a voltage source of $0.200e^{-200t}$ V. Find the charge on the capacitor as a function of time if $q = 0$ and $i = 4.00$ mA when $t = 0$.

79. Find the equation for the current as a function of time if a resistor of 20 Ω, an inductor of 4 H, a capacitor of 100 μF, and a battery of 100 V are in series. The initial charge on the capacitor is 10 mC, and the initial current is zero.

80. If an electric circuit contains an inductance L, a capacitor with a capacitance C, and a sinusoidal source of voltage $E_0 \sin \omega t$, express the charge q on the capacitor as a function of the time. Assume $q = 0$, $i = 0$ when $t = 0$.

81. The differential equation relating the current and time for a certain electric circuit is $2\,di/dt + i = 12$. Solve this equation by use of Laplace transforms, given that the initial current is zero. Evaluate the current for $t = 0.300$ s.

82. A 6-H inductor and a 30-Ω resistor are connected in series with a voltage source of $10 \sin 20t$. Find the current as a function of time if the initial current is zero. Use Laplace transforms.

83. A 0.25-H inductor, a 4.0-Ω resistor, and a 100-μF capacitor are connected in series. If the initial charge on the capacitor is 400 μC and the initial current is zero, find the charge on the capacitor as a function of time. Use Laplace transforms.

84. An inductor of 0.5 H, a resistor of 6 Ω, and a capacitor of 200 μF are connected in series. If the initial charge on the capacitor is 10 mC and the initial current is zero, find the charge on the capacitor as a function of time after the switch is closed. Use Laplace transforms.

85. A mass of 0.25 kg stretches a spring for which $k = 16$ N/m. An external force of $\cos 8t$ is applied to the spring. Express the displacement y of the object as a function of time if the initial displacement and velocity are zero. Use Laplace transforms.

86. A spring is stretched 1.00 m by a mass of 5.00 kg (assume the weight to be 50.0 N). Find the displacement y of the object as a function of time if $y(0) = 1$ m and $dy/dt = 0$ when $t = 0$. Use Laplace transforms.

87. Air containing 20% oxygen passes into a 5.00-L container initially filled with 100% oxygen. A uniform mixture of the air and oxygen then passes from the container at the same rate. What volume of oxygen is in the container after 5.00 L of air have passed into it?

88. When a circular disc of mass m and radius r is suspended by a wire at the centre of one of its flat faces and the disc is twisted through an angle θ, torsion in the wire tends to turn the disc back in the opposite direction. The differential equation for this case is $\frac{1}{2}mr^2\frac{d^2\theta}{dt^2} = -k\theta$, where k is a constant. Determine the equation of motion if $\theta = \theta_0$ and $d\theta/dt = \omega_0$ when $t = 0$. See Fig. 31.27.

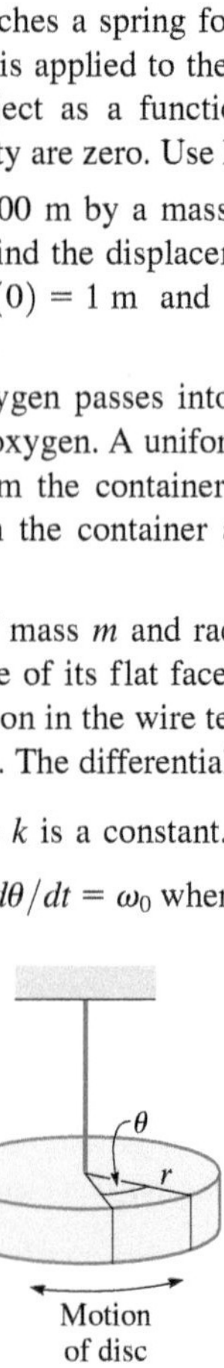

Fig. 31.27

89. The approximate differential equation relating the displacement y of a beam at a horizontal distance x from one end is $EI\frac{d^2y}{dx^2} = M$, where E is the modulus of elasticity, I is the moment of inertia of the cross-section of the beam perpendicular to its axis, and M is the bending moment at the cross-section. If $M = 2000x - 40x^2$ for a particular beam of length L for which $y = 0$ when $x = 0$ and when $x = L$, express y in terms of x. Consider E and I as constants.

90. The gravitational acceleration of an object is inversely proportional to the square of its distance r from the centre of the earth. Use the chain rule to show that the acceleration is $\frac{dv}{dt} = v\frac{dv}{dr}$, where $v = \frac{dr}{dt}$ is the velocity of the object. Then solve for v as a function of r if $dv/dt = -g$ and $v = v_0$ for $r = R$, where R is the radius of the earth. Finally, show that a spacecraft must have a velocity of at least $v_0 = \sqrt{2gR}$ (11.2 km/s) in order to escape from the earth's gravitation. (Note the expression for v^2 as $r \to \infty$.)

91. For the electric circuit shown in Fig. 31.28, find the current as a function of the time t, if the initial charge on the capacitor is zero and the initial current is zero.

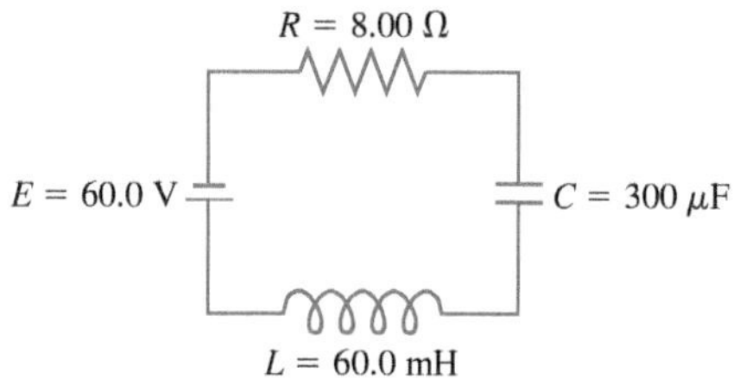

Fig. 31.28

92. An electric circuit contains an inductor L, a resistor R, and a battery of voltage E. The initial current in the circuit is zero. Write three or four paragraphs explaining how the differential equation for the current in the circuit is solved using (a) separation of variables, (b) the linear differential equation of the first order, and (c) Laplace transforms.

CHAPTER 31 PRACTICE TEST

In Problems 1–6, find the general solution of each of the given differential equations.

1. $x\frac{dy}{dx} + 2y = 4$

2. $y'' + 2y' + 5y = 0$

3. $x\,dx + y\,dy = x^2\,dx + y^2\,dx$

4. $2D^2y - Dy = 2\cos x$

5. $\frac{d^2y}{dx^2} - 4\frac{dy}{dx} + 4y = 3x$

6. $D^2y - 2Dy - 8y = 4e^{-2x}$

In Problems 7–12, answer the given questions.

7. Find the particular solution of the differential equation $(xy + y)\frac{dy}{dx} = 2$, if $y = 2$ when $x = 0$.

8. If interest in a bank account is compounded continuously, the amount grows at a rate that is proportional to the amount present. Derive the equation for the amount A in an account with continuous compounding in which the initial amount is A_0 and the interest rate is r as a function of the time t after A_0 is deposited.

9. Using Laplace transforms, solve the differential equation $y'' + 9y = 9$, if $y(0) = 0$ and $y'(0) = 1$.

10. Using Laplace transforms, solve the differential equation $D^2y - Dy - 2y = 12$, if $y(0) = 0$ and $y'(0) = 0$.

11. Find the equation for the current as a function of the time (in s) in a circuit containing a 2-H inductance, an 8-Ω resistor, and a 6-V battery in series, if $i = 0$ when $t = 0$.

12. A mass of 0.5 kg stretches a spring for which $k = 32$ N/m. With this weight attached, the spring is pulled 0.3 m longer than its equilibrium length and released. Find the equation of the resulting motion, assuming no damping. (The acceleration due to gravity is 9.8 m/s^2.)

APPENDIX A Solving Word Problems

Drill-type problems require a working knowledge of the methods presented, and some algebraic steps to change the algebraic form may be required to complete the solution. *Word problems,* however, require a proper interpretation of the statement of the problem before they can be put in a form for solution.

We have to put word problems in symbolic form in order to solve them, and it is this procedure that most students find difficult. Because such problems require more than going through a certain routine, they demand more analysis and appear to be more difficult. Among the reasons for the student's difficulty at solving word problems are (1) unsuccessful previous attempts at solving word problems, leading the student to believe that all word problems are "impossible," (2) a poorly organized approach to the solution, and (3) failure to read the problem carefully, thereby having an improper and incomplete interpretation of the statement given. These can be overcome with proper attitude and care.

A specific procedure for solving word problems is shown in Section 1.12, when word problems are first covered in our study of algebra. There are over 120 completely worked examples of word problems (as well as numerous other examples that show a similar analysis) throughout this text, illustrating proper interpretations and approaches to these problems.

RISERS

The procedure shown in Section 1.12 is similar to that used by most instructors and texts. One of the variations that a number of instructors use is called *RISERS.* This is a word formed from the first letters (an *acronym*) of the words that outline the procedure. These are *Read, Imagine, Sketch, Equate, Relate,* and *Solve.* We now briefly outline this procedure here.

Read the statement of the problem carefully.

Imagine. Take time to get a mental image of the situation described.

Sketch a figure.

Equate, *on the sketch,* the known and unknown quantities.

Relate the known and unknown quantities with an equation.

Solve the equation.

There are problems where a *sketch* may simply be words and numbers placed so that we may properly *equate* the known and unknown quantities. For example, in Example 2 in Section 1.12, the *sketch, equate,* and *relate* steps might look like this:

sketch		34 lights	1000 W
	25 W lights	40 W lights	
equate	x	$34 - x$	
wattage	$25x$	$40(34 - x)$	1000
relate	$25x$	$+ 40(34 - x) =$	1000

If you follow the method in Section 1.12, or this RISERS variation, or any appropriate step-by-step method, and *write out the solution neatly,* you will find that word problems lend themselves to solution more readily than you have previously found.

APPENDIX

B

A Table of Integrals

The basic forms of Chapter 28 are not included. The constant of integration is omitted.

Forms containing $a + bu$ and $\sqrt{a + bu}$

1. $$\int \frac{u\,du}{a + bu} = \frac{1}{b^2}\left[(a + bu) - a \ln(a + bu)\right]$$

2. $$\int \frac{du}{u(a + bu)} = -\frac{1}{a}\ln\frac{a + bu}{u}$$

3. $$\int \frac{u\,du}{(a + bu)^2} = \frac{1}{b^2}\left(\frac{a}{a + bu} + \ln(a + bu)\right)$$

4. $$\int \frac{du}{u(a + bu)^2} = \frac{1}{a(a + bu)} - \frac{1}{a^2}\ln\frac{a + bu}{u}$$

5. $$\int u\sqrt{a + bu}\,du = -\frac{2(2a - 3bu)(a + bu)^{3/2}}{15b^2}$$

6. $$\int \frac{u\,du}{\sqrt{a + bu}} = -\frac{2(2a - bu)\sqrt{a + bu}}{3b^2}$$

7. $$\int \frac{du}{u\sqrt{a + bu}} = \frac{1}{\sqrt{a}}\ln\left(\frac{\sqrt{a + bu} - \sqrt{a}}{\sqrt{a + bu} + \sqrt{a}}\right), \quad a > 0$$

8. $$\int \frac{\sqrt{a + bu}}{u}\,du = 2\sqrt{a + bu} + a\int \frac{du}{u\sqrt{a + bu}}$$

Forms containing $\sqrt{u^2 \pm a^2}$ and $\sqrt{a^2 - u^2}$

9. $$\int \frac{du}{u^2 - a^2} = \frac{1}{2a}\ln\frac{u - a}{u + a}$$

10. $$\int \frac{du}{\sqrt{u^2 \pm a^2}} = \ln\left(u + \sqrt{u^2 \pm a^2}\right)$$

11. $$\int \frac{du}{u\sqrt{u^2 + a^2}} = -\frac{1}{a}\ln\left(\frac{a + \sqrt{u^2 + a^2}}{u}\right)$$

12. $$\int \frac{du}{u\sqrt{u^2 - a^2}} = \frac{1}{a}\sec^{-1}\frac{u}{a}$$

13. $$\int \frac{du}{u\sqrt{a^2 - u^2}} = -\frac{1}{a}\ln\left(\frac{a + \sqrt{a^2 - u^2}}{u}\right)$$

14. $$\int \sqrt{u^2 \pm a^2}\,du = \frac{u}{2}\sqrt{u^2 \pm a^2} \pm \frac{a^2}{2}\ln\left(u + \sqrt{u^2 \pm a^2}\right)$$

15. $$\int \sqrt{a^2 - u^2}\,du = \frac{u}{2}\sqrt{a^2 - u^2} + \frac{a^2}{2}\sin^{-1}\frac{u}{a}$$

16. $\int \frac{\sqrt{u^2 + a^2}}{u}\,du = \sqrt{u^2 + a^2} - a \ln\left(\frac{a + \sqrt{u^2 + a^2}}{u}\right)$

17. $\int \frac{\sqrt{u^2 - a^2}}{u}\,du = \sqrt{u^2 - a^2} - a \sec^{-1}\frac{u}{a}$

18. $\int \frac{\sqrt{a^2 - u^2}}{u}\,du = \sqrt{a^2 - u^2} - a \ln\left(\frac{a + \sqrt{a^2 - u^2}}{u}\right)$

19. $\int (u^2 \pm a^2)^{3/2}\,du = \frac{u}{4}(u^2 \pm a^2)^{3/2} \pm \frac{3a^2u}{8}\sqrt{u^2 \pm a^2} + \frac{3a^4}{8}\ln\left(u + \sqrt{u^2 \pm a^2}\right)$

20. $\int (a^2 - u^2)^{3/2}\,du = \frac{u}{4}(a^2 - u^2)^{3/2} + \frac{3a^2u}{8}\sqrt{a^2 - u^2} + \frac{3a^4}{8}\sin^{-1}\frac{u}{a}$

21. $\int \frac{(u^2 + a^2)^{3/2}}{u}\,du = \frac{1}{3}(u^2 + a^2)^{3/2} + a^2\sqrt{u^2 + a^2} - a^3 \ln\left(\frac{a + \sqrt{u^2 + a^2}}{u}\right)$

22. $\int \frac{(u^2 - a^2)^{3/2}}{u}\,du = \frac{1}{3}(u^2 - a^2)^{3/2} - a^2\sqrt{u^2 - a^2} + a^3 \sec^{-1}\frac{u}{a}$

23. $\int \frac{(a^2 - u^2)^{3/2}}{u}\,du = \frac{1}{3}(a^2 - u^2)^{3/2} - a^2\sqrt{a^2 - u^2} + a^3 \ln\left(\frac{a + \sqrt{a^2 - u^2}}{u}\right)$

24. $\int \frac{du}{(u^2 \pm a^2)^{3/2}} = \pm\frac{u}{a^2\sqrt{u^2 \pm a^2}}$

25. $\int \frac{du}{(a^2 - u^2)^{3/2}} = \frac{u}{a^2\sqrt{a^2 - u^2}}$

26. $\int \frac{du}{u(u^2 + a^2)^{3/2}} = \frac{1}{a^2\sqrt{u^2 + a^2}} - \frac{1}{a^3}\ln\left(\frac{a + \sqrt{u^2 + a^2}}{u}\right)$

27. $\int \frac{du}{u(u^2 - a^2)^{3/2}} = -\frac{1}{a^2\sqrt{u^2 - a^2}} - \frac{1}{a^3}\sec^{-1}\frac{u}{a}$

28. $\int \frac{du}{u(a^2 - u^2)^{3/2}} = \frac{1}{a^2\sqrt{a^2 - u^2}} - \frac{1}{a^3}\ln\left(\frac{a + \sqrt{a^2 - u^2}}{u}\right)$

Trigonometric forms

29. $\int \sin^2 u\,du = \frac{u}{2} - \frac{1}{2}\sin u \cos u$

30. $\int \sin^3 u\,du = -\cos u + \frac{1}{3}\cos^3 u$

31. $\int \sin^n u\,du = -\frac{1}{n}\sin^{n-1} u \cos u + \frac{n-1}{n}\int \sin^{n-2} u\,du$

32. $\int \cos^2 u\,du = \frac{u}{2} + \frac{1}{2}\sin u \cos u$

33. $\int \cos^3 u\,du = \sin u - \frac{1}{3}\sin^3 u$

34. $\int \cos^n u\,du = \frac{1}{n}\cos^{n-1} u \sin u + \frac{n-1}{n}\int \cos^{n-2} u\,du$

35. $\int \tan^n u\,du = \frac{\tan^{n-1} u}{n-1} - \int \tan^{n-2} u\,du$

36. $\int \cot^n u\, du = -\frac{\cot^{n-1} u}{n-1} - \int \cot^{n-2} u\, du$

37. $\int \sec^n u\, du = \frac{\sec^{n-2} u \tan u}{n-1} + \frac{n-2}{n-1}\int \sec^{n-2} u\, du$

38. $\int \csc^n u\, du = \frac{\csc^{n-2} u \cot u}{n-1} + \frac{n-2}{n-1}\int \csc^{n-2} u\, du$

39. $\int \sin au \sin bu\, du = \frac{\sin (a-b)u}{2(a-b)} - \frac{\sin (a+b)u}{2(a+b)}$

40. $\int \sin au \cos bu\, du = -\frac{\cos (a-b)u}{2(a-b)} - \frac{\cos (a+b)u}{2(a+b)}$

41. $\int \cos au \cos bu\, du = \frac{\sin (a-b)u}{2(a-b)} + \frac{\sin (a+b)u}{2(a+b)}$

42. $\int \sin^m u \cos^n u\, du = \frac{\sin^{m+1} u \cos^{n-1} u}{m+n} + \frac{n-1}{m+n}\int \sin^m u \cos^{n-2} u\, du$

43. $\int \sin^m u \cos^n u\, du = -\frac{\sin^{m-1} u \cos^{n+1} u}{m+n} + \frac{m-1}{m+n}\int \sin^{m-2} u \cos^n u\, du$

Other forms

44. $\int u e^{au}\, du = \frac{e^{au}(au-1)}{a^2}$

45. $\int u^2 e^{au}\, du = \frac{e^{au}}{a^3}(a^2u^2 - 2au + 2)$

46. $\int u^n \ln u\, du = u^{n+1}\left(\frac{\ln u}{n+1} - \frac{1}{(n+1)^2}\right)$

47. $\int u \sin u\, du = \sin u - u \cos u$

48. $\int u \cos u\, du = \cos u + u \sin u$

49. $\int e^{au} \sin bu\, du = \frac{e^{au}(a \sin bu - b \cos bu)}{a^2 + b^2}$

50. $\int e^{au} \cos bu\, du = \frac{e^{au}(a \cos bu + b \sin bu)}{a^2 + b^2}$

51. $\int \sin^{-1} u\, du = u \sin^{-1} u + \sqrt{1 - u^2}$

52. $\int \tan^{-1} u\, du = u \tan^{-1} u - \frac{1}{2}\ln(1 + u^2)$

Answers to Odd-Numbered Exercises

Since statements will vary for writing exercises ✎, answers here are in abbreviated form. Answers are not included for end-of-chapter writing exercises.

Chapter 1: Basic Algebraic Operations

Exercises 1.1, page 6

1. Change $\frac{5}{1}$ to $\frac{-3}{1}$ and $\frac{-19}{1}$ to $\frac{14}{1}$. **3.** $-6 < -4$ should be read "-6 is less than -4" (-6 is to the left of -4 on the number line). **5.** 3: integer, rational, real. $\sqrt{-4}$: imag. **7.** $-\frac{\pi}{6}$: irrational, real. $\frac{1}{8}$: rational, real. **9.** 3, 4, $\frac{\pi}{2}$ **11.** $6 < 8$ **13.** $\pi > -3.2$

15. $-4 < -|-3|$ **17.** $-\frac{1}{3} > -\frac{1}{2}$ **19.** $\frac{1}{3}, -\frac{\sqrt{3}}{4}, \frac{b}{y}$

21.

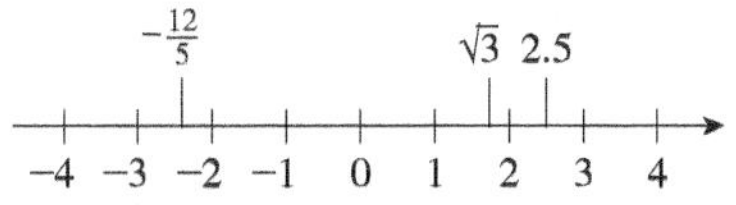

23. No, $|0| = 0$ **25.** The number itself

27. $\frac{3}{23}$ **29.** $-3.5, -|-3|, -1, \sqrt{5}, \pi, |-8|, 9$

31. (a) Positive integer (b) Negative integer (c) Positive rational number less than 1

33. (a) Yes (b) Yes

35. (a) To the right of zero (b) To the left of -4

37. Positive number less than 1 **39.** All real values of a; $b = 0$.

41. 0.000 80 F **43.** $N = 1000an$ bits

45. Yes; $-30°C$ is to the left of $-20°C$. **47.** $m_P > m_T$ or $m_T < m_P$

Exercises 1.2, page 12

1. 22 **3.** -4 **5.** 4 **7.** 6 **9.** -3 **11.** -24

13. 35 **15.** 20 **17.** 40 **19.** -1 **21.** -32

23. -17 **25.** Undefined **27.** 20 **29.** 16

31. -6 **33.** 24 **35.** -6 **37.** 3

39. Commutative law of multiplication

41. Distributive law **43.** Associative law of addition

45. Associative law of multiplication **47.** d **49.** b **51.** =

53. (a) Positive (b) Negative

55. Correct. (For $x > 0$, x is positive; for $x < 0$, $-x$ is positive.)

57. (a) Negative reciprocals of each other (b) All values of x and y with $x \neq y$ **59.** -0.29

61. -2.4 kW · h **63.** $-2.0°C$ **65.** 10 V

67. 100 m + 200 m = 200 m + 100 m; commutative law of addition

69. $7(25 + 15)$ OR $7(25) + 7(15)$; distributive law

Exercises 1.3, page 21

1. 0.3900 has four significant digits. **3.** 75.7 **5.** 1 000 000 Hz

7. 0.001 m **9.** 1000 volts **11.** 0.001 amperes

13. 100 000 cm **15.** 0.000 02 Ms **17.** 0.000 25 m^2

19. 80 000 L (with three sig. digits)

21. 4500 cm/s (with three sig. digits) **23.** 3 530 000 cm/min^2

25. 90 000 000 ms (with two significant digits)

27. 0.0150 mC/V **29.** 16.78 mi **31.** 40.97 lb/ft^2

33. 110 000 km/h **35.** 56 000 cm^3 **37.** 24 000 km/h

39. 0.0112 m^2 **41.** 1000 g/L

43. 1200 km/h (with 3 significant digits)

45. 0.135 J/(s · cm^2) **47.** 120 000 mA/cm^2

49. 8 cylinders: exact; 55 km/h: approximate **51.** Both are exact.

53. 107 has three sig. digits; 3004 has four sig. digits; 1040 has three sig. digits. **55.** 6.80 has three sig. digits; 6.08 has three sig. digits; 0.068 has two sig. digits. **57.** 3000 has one sig. digit; 3000.1 has five sig. digits; 3000.10 has six sig. digits.

59. (a) 0.01 (b) 30.8 **61.** (a) Same (b) 78.0

63. (a) 0.004 (b) Same **65.** (a) 4.94. (b) 4.9

67. (a) -50.9 (b) -51 **69.** (a) 9550 (b) 9500

71. (a) 0.945 (b) 0.94 **73.** (a) 12 (b) 12.20

75. (a) 13 (b) 13 **77.** (a) 6 (b) 5.57 to three sig. digits

79. (a) 3.0 (b) 2.91 to three sig. digits **81.** 15.8788 **83.** -204.2

85. Measurements can be more than 2.745 MHz and less than 2.755.

87. Too many sig. digits; time has only 2 sig. digits.

89. (a) 19.3 (b) 27 **91.** (a) 2 (b) 2 (c) -2 (d) 0 (e) Undefined

93. Example: $(231\,465 - 164\,352) \div 9 = 7457$; integer

95. (a) $\pi < 3.1416$ (b) $\pi < (22 \div 7)$

97. (a) $1 \div 3 = 0.333\,333\ldots$ (b) $5 \div 11 = 0.454\,545\ldots$ (c) $2 \div 5 = 0.400\,000\ldots$ (0 repeats)

99. 95.3 MJ **101.** 262 144 bytes **103.** 59.14%

Exercises 1.4, page 27

1. $9k^2$ **3.** $\frac{a^6x^2}{b^4t^2}$ **5.** x^7 **7.** $2b^6$ **9.** m^2 **11.** $\frac{-1}{7n^4}$

13. P^8 **15.** $8\pi^3$ **17.** $a^{30}T^{60}$ **19.** $\frac{8}{b^3}$ **21.** $\frac{x^8}{16}$

23. 1 **25.** -3 **27.** $\frac{1}{6}$ **29.** R^2 **31.** $-t^{14}$

33. $-L^2$ **35.** $\frac{1}{8}$ **37.** 1 **39.** $\frac{a}{x^2}$ **41.** $\frac{64s^6}{g^2}$ **43.** $\frac{x^3}{64a^3}$

45. $\frac{5n^3}{T}$ **47.** $\frac{y^4}{2^{16}x^4}$ **49.** -53 **51.** 253 **53.** -0.421

55. 114 **57.** Yes **59.** 625 **61.** 1, provided that $x \neq 0$

63. $\frac{G^2k^5T^5}{h}$ **65.** $\frac{r}{6}$ **67.** \$3212.27

Exercises 1.5, page 30

1. 8060 **3.** 45 000 **5.** 0.002 01 **7.** 3.23 **9.** 18.6

11. 4×10^3 **13.** 8.7×10^{-3} **15.** 6.09×10^8

17. 6.3×10^{-2} **19.** 1×10^0 **21.** 5.6×10^{13}

23. 2.2×10^8 **25.** 3.2×10^{-34} **27.** 1.728×10^{87}

29. 4.85×10^{10} **31.** 1.59×10^7 **33.** 9.965×10^{-3}

35. 3.38×10^{16} **37.** 2×10^6 kW **39.** 3×10^{-6} W

41. 1.2×10^9 Hz **43.** 1.2×10^{10} m^2

45. 0.000 000 000 001 6 W **47.** 2 GW **49.** 3 μW
51. 1.2 GHz **53.** (a) 8.09×10^6 (b) 809×10^3 (c) 80.9×10^{-3}
55. 10^{21} **57.** $2^{30} \approx 1 \times 10^9$ **59.** 3.29×10^{-5} °C/d
61. (a) 8.64×10^4 s (b) $3.155\,760\,0 \times 10^9$ s
63. 4.8×10^2 W **65.** 2.998×10^5 km/s

Exercises 1.6, page 33

1. −4 **3.** 12 **5.** 7 **7.** $-3\sqrt{5}$ **9.** −8 **11.** 0.3
13. 5 **15.** −6 **17.** 5 **19.** 47 **21.** 53 **23.** $3\sqrt{2}$
25. $4\sqrt{21}$ **27.** $2\sqrt{5}$ **29.** 4 **31.** $\frac{50\sqrt{2}}{81}$
33. 10 **35.** $3\sqrt{10}$ **37.** 9.24 **39.** 0.9029
41. (a) 60.00 (b) 84.00 **43.** (a) 0.0388 (b) 0.0246
45. 98 km/h **47.** 1450 m/s **49.** 107 cm **51.** 190 m/s
53. No, not true if $a < 0$ **55.** (a) 12.9 (b) −0.598
57. (a) Imaginary (b) Real **59.** 50.2 Hz

Exercises 1.7, page 37

1. $3x - 3y$ **3.** $4ax + 5s$ **5.** $8x$ **7.** $y + 4x$
9. $5F - 3T - 2$ **11.** $-a^2b - a^2b^2$ **13.** $3s - 4$
15. $-v + 5x - 4$ **17.** $5a - 5$ **19.** $-5a + 2$
21. $-2t + 5u$ **23.** $7r + 8s$ **25.** $19j - 50$
27. $7n - 7$ **29.** $-2t^2 + 18$ **31.** $6a$ **33.** $2aZ + 1$
35. $4c - 6$ **37.** $8p - 5q$ **39.** $-4x^2 + 22$ **41.** $7V^2 - 3$
43. $-6t + 13$ **45.** $-24R + 4Z$ **47.** $2D + d$
49. $3B - 2\alpha$ **51.** $(12x + 200)$ terabytes
53. (a) $x^2 + 2y + 2a - b$ (b) $3x^2 - 4y + 2a + b$
55. $|(b - a)|$

Exercises 1.8, page 40

1. $-8s^8t^{13}$ **3.** $x^2 - 5x + 6$ **5.** a^3x **7.** $-a^4c^3x^3$
9. $-8a^3x^5$ **11.** $i^2R + 2i^2r$ **13.** $-15s^4 + 10st$
15. $5m^3n + 15m^2n$ **17.** $-3M^2 - 3MN + 6M$
19. $acx^4 + acx^3y^3$ **21.** $x^2 + 2x - 15$ **23.** $2x^2 + 9x - 5$
25. $y^2 - 64$ **27.** $6a^2 - 7ab + 2b^2$ **29.** $6s^2 + 11st - 35t^2$
31. $2x^3 + 5x^2 - 2x - 5$ **33.** $x^2 + 4y^2 - 4xy - 16$
35. $2a^2 - 16a - 18$ **37.** $18T^2 - 15T - 18$
39. $-2L^3 + 6L^2 + 8L$ **41.** $4x^2 - 20x + 25$
43. $x_1^2 + 6x_1x_2 + 9x_2^2$ **45.** $x^2y^2z^2 - 4xyz + 4$
47. $2x^2 + 32x + 128$ **49.** $-x^3 + 2x^2 + 5x - 6$
51. $6T^3 + 9T^2 - 6T$ **53.** (a) $49 \neq 9 + 16$ (b) $1 \neq 9 - 16$
55. If $1 < x < 9$, then $x^2 - 1$ can be factored to $(x - 1)(x + 1)$.
57. $(x + y)^3 = x^3 + 3x^2y + 3y^2x + y^3$
59. $0.0001r^2P + 0.02rP + P$ **61.** $3R^2 - 4RX$
63. $n^2 + 200n + 10\,000$ **65.** $R_1^2 - R_2^2$

Exercises 1.9, page 43

1. $\frac{3}{y^3}$ **3.** $3x - 2$ **5.** $-4x^2y$ **7.** $\frac{4t^4}{r^2}$ **9.** $-\frac{y^4}{2x^5}$
11. $\frac{20n^{15}}{m^7}$ **13.** $4x^2$ **15.** $-6a$ **17.** $a^2 + 2y$
19. $-2rt^2 + t$ **21.** $-4q^3 + 2p + q$ **23.** $\frac{2L}{R} - R$
25. $-\frac{1}{2b} + \frac{1}{a} - \frac{3a}{2b^2}$ **27.** $3y^n - 2ay$ **29.** $x + 5$
31. $2x + 1$ **33.** $x - 1$ **35.** $4x^2 - x - 1 - \frac{3}{2x - 3}$
37. $Z - 2 - \frac{1}{4Z + 3}$ **39.** $x^2 + x - 6$ **41.** $2a^2 + 8$
43. $x^2 - 2x + 4$ **45.** $x - y$ **47.** $t - 2$ **49.** −5
51. $\frac{x^4 + 1}{x + 1} = x^3 - x^2 + x - 1 + \frac{2}{x + 1} \neq x^3$
53. $A + \frac{\mu^2E^2}{2A} - \frac{\mu^4E^4}{8A^3}$ **55.** $\frac{GMm}{R}$
57. $s^2 + 2s + 6 + \frac{16s + 16}{s^2 - 2s - 2}$

Exercises 1.10, page 47

1. (a) −9 (b) −15 (c) −36 (d) −4 **3.** $\frac{1}{8}$ **5.** 9
7. −1 **9.** −10 **11.** 20 **13.** −5 **15.** −3 **17.** 4
19. $-\frac{7}{2}$ **21.** −8 **23.** $\frac{-2}{3}$ **25.** 4 **27.** $-\frac{5}{2}$ **29.** 0
31. 8 **33.** 11 or −11 **35.** 9.5 **37.** −2.1 **39.** 5.7
41. 0.85 **43.** 16.5 **45.** (a) Identity (b) Conditional
47. 3.75 **49.** 55 km/h **51.** 120 °C **53.** 750 L
55. 820 km **57.** 32 min

Exercises 1.11, page 49

1. $\frac{v - v_0}{t}$ **3.** $\frac{V_0 + bTV_0 - V}{bV_0}$ **5.** $\frac{E}{I}$ **7.** $g_2 - rL$
9. $\frac{12B}{TWL}$ **11.** $\frac{p - p_a}{dg}$ **13.** $\frac{mv^2}{F_c}$ **15.** $5T(S_T - 0.05d)$
17. $\frac{-ct^2 + 0.3t}{c}$ **19.** $Tv - c$ **21.** $\frac{K_1m_1 - K_2m_1}{K_2}$
23. $\frac{2gm - 2am}{a}$ **25.** $\frac{C_0^2 - C_1^2}{2C_1^2}$ **27.** $\frac{Ar - N}{r}$
29. $100T_1 - 100T_2$ **31.** $\frac{Q_1 + PQ_1}{P}$ **33.** $\frac{N + N_2 - N_2T}{T}$
35. $\frac{L - \pi r_2 - 2x_1 - 2x_2}{\pi}$ **37.** $\frac{V_1^2 + gJP}{V_1}$ **39.** $\frac{Cd(k_1 + k_2)}{2Ak_1k_2}$
41. $\frac{CN - NV}{C}$ **43.** 1070 K **45.** 32.3°C **47.** 3.22 Ω
49. $\frac{d - v_2(4\text{ h}) - v_1(2\text{ h})}{v_1}$

Exercises 1.12, page 54

1. 16 25-W lights and 15 40-W lights **3.** 3.000 h
5. \$32 500 6 years ago, \$37 500 today
7. 1.9×10^5 the first year, 2.6×10^5 the second year
9. 50 hectares at \$200 per hectare and 90 hectares at \$300 per hectare
11. 60 ppm/h **13.** 20 girders
15. −2.3 μA, −4.6 μA, 6.9 μA **17.** 6.9 km, 9.5 km
19. 32 flat panels and 22 custom panels **21.** 900 m
23. 84.2 km/h, 92.2 km/h **25.** 395 s, first car
27. 146 km from A **29.** 4.0 L **31.** 79 km/h **33.** 11%

Review Exercises for Chapter 1, page 55

1. -10 **3.** -20 **5.** -22 **7.** -25 **9.** -4

11. 5 **13.** $4r^2t^4$ **15.** $-\frac{24t}{m^2n}$ **17.** $\frac{8T^3}{N}$ **19.** $3\sqrt{5}$

21. (a) 3 (b) 8800 **23.** (a) 4 (b) 9.0 **25.** 18.0

27. 1.3×10^{-4} **29.** $-2ab - a$ **31.** $7LC - 3$

33. $2x^2 + 9x - 5$ **35.** $x^2 + 16x + 64$ **37.** $-3h^2k^4 + hk$

39. $7R - 6r$ **41.** $13xy - 10z$ **43.** $2x^3 - x^2 - 7x - 3$

45. $-3x^2y + 24xy^2 - 48y^3$ **47.** $18p^2q - 9p^2 + 3pq$

49. $\frac{3q^4}{p^3} + \frac{6q}{p} - 2$ **51.** $2x - 5$ **53.** $x^2 - 2x + 3$

55. $4x^3 - 2x^2 + 6x, R = -1$ **57.** $15r - 3s - 3t$

59. $y^2 + 5y - 1, R = 4$ **61.** $-\frac{9}{2}$ **63.** $\frac{21}{10}$

65. $-\frac{7}{3}$ **67.** 3 **69.** $-\frac{19}{5}$ **71.** 1.0

73. (a) 6×10^{10} bytes (b) 60 gigabytes

75. (a) 1.54×10^{10} km (b) 15.4 Tm

77. (a) 40 500 000 000 000 km (b) 40.5 Pm

79. (a) 0.000 000 000 001 W/m^2 (b) 1 pW/m^2

81. (a) 0.15 Bq/L (b) 150×10^{-3} mBq/L

83. $\frac{V}{\pi r^2}$ **85.** $\frac{L^2P}{\pi^2 I}$ **87.** $\frac{Rr - Pp}{Q}$ **89.** $\frac{d + A}{A}$

91. $\frac{N_1 - N_3 + N_3T}{T}$ **93.** $\frac{HR + AT_1}{A}$ **95.** $\frac{d - 3bkx^2 + kx^3}{3kx^2}$

97. 8.2×10^5 **99.** 1.25 **101.** 0.0188 Ω

103. $101x + 198a$ cm **105.** $-2t^2 - 2h^2 - 4ht + 4t + 4h$

107. Yes (18 to 0) **109.** Identity **111.** $-(y - x)^3$

113. 4×10^{-7} **115.** \$59, \$131 **117.** 160 cm^3, 80 cm^3, 320 cm^3

119. 1900 Ω, 3100 Ω **121.** 6.7 h **123.** 15.3 h

125. 400 L of 0.50% grade oil and 600 L of 0.75% grade oil

127. 27 m^2 **129.** 5.500%

Chapter 2: Geometry

Exercises 2.1, page 63

1. 90° **3.** four pairs **5.** $\angle EBD$ and $\angle DBC$ **7.** $\angle ABC$

9. 25° **11.** BD and BC **13.** 140° **15.** 40° **17.** 35°

19. 62° **21.** 28° **23.** 118° **25.** 134° **27.** 44°

29. 46° **31.** 4.53 m **33.** 3.40 m **35.** $\angle BHC, \angle DGC$

37. $\angle BCH, \angle GHC, \angle HGC, \angle GCD$ **39.** $\angle AHF, \angle EGI$

41. 133° **43.** 882 m **45.** 180° **47.** Sum of angles is 180°

Exercises 2.2, page 70

1. 65° **3.** 7.02 m **5.** 56° **7.** 48° **9.** 8.4 m^2

11. 32 300 cm^2 **13.** 4.41 cm^2 **15.** 0.390 m^2 **17.** 942 cm

19. 64.5 cm **21.** 26.6 mm **23.** 522 cm **25.** 67°

27. 227.2 cm **29.** 45° **31.** An equilateral triangle

33. The smaller triangles are similar.

35. $\angle LMK = \angle OMN$; $\angle KLM = \angle MON$; $\Delta MKL \sim \Delta MNO$

37. 8 **39.** 65° **41.** 1150 cm^2 **43.** 9.6 m^2 **45.** 5.7 m

47. 23.1 m **49.** 7.5 m, 9.0 m **51.** 3.4 m **53.** 20.0 km

Exercises 2.3, page 73

1. 8900 m^2 **3.** 260 m **5.** 3.324 mm **7.** 12.8 m

9. 214.4 dm **11.** 7.3 mm^2 **13.** 140 ft^2 **15.** 0.683 km^2

17. 9.2 m^2 **19.** 2.00×10^3 dm^2 **21.** $p = 2b + 4a$

23. $A = bh + a^2$ **25.** Rectangle **27.** 288 cm^2

29. The diagonal always divides the rhombus into two congruent triangles. All outer sides are always equal. **31.** 348 m

33. 2000 mm, 8000 mm **35.** 2.4 L **37.** 3.04 km^2

39. 360°. A diagonal divides a quadrilateral into two triangles, and the sum of the interior angles of each triangle is 180°.

41. 59 m^2

Exercises 2.4, page 78

1. 18° **3.** $p = 11.6$ cm, $A = 8.30$ cm^2

5. (a) AD (b) AF **7.** (a) $AF \perp OE$ (b) ΔOCE

9. 1730 cm **11.** 72.6 mm **13.** 0.0285 km^2

15. 4.26 m^2 **17.** 128 cm^2 **19.** 25° **21.** 25°

23. 120° **25.** 40° **27.** 0.393 rad **29.** 2.185 rad

31. $\frac{\pi r}{2} + 2r$ **33.** $\frac{1}{4}\pi r^2 - \frac{1}{2}r^2$ **35.** 0.314 m^2

37. All are on the same diameter. **39.** 10.3 cm^2

41. 20.9 cm^2 **43.** 40 060 km **45.** 35.7 cm

47. 105 N/cm^2 **49.** 9.7×10^7 mm^2

51. Horizontally and opposite to original direction **53.** 630 km

Exercises 2.5, page 82

1. Simpson's rule. The use of smaller intervals improves the approximation.

3. Simpson's rule. It accounts better for the arcs between points on the curve.

5. 84 m^2 **7.** 0.45 m^2 **9.** 9.8 km^2

11. 12 000 km^2 **13.** 380 km^2 **15.** 8.0×10^3 m^2

17. 2.73 cm^2. All of the trapezoids are inscribed.

19. 2.98 cm^2. The ends of the areas are curved so they can get closer to the boundary.

Exercises 2.6, page 86

1. The volume increases by a factor of 4. **3.** 771 cm^3

5. 366 cm^3 **7.** 3.99×10^6 m^2 **9.** 3.358 m^2

11. 20 500 cm^2 **13.** 72.3 cm^2 **15.** 2.5×10^5 cm^3

17. 23.3 dm^2 **19.** 2.83 m^3 **21.** 153 000 mm^3

23. 604 cm^2 **25.** 2.4×10^5 m^3 **27.** 114 cm^2

29. 145 cm^2 **31.** 2.6×10^6 m^3 **33.** 66 600 m^3

35. $3\pi r^3$ **37.** 2.94×10^8 N **39.** $\frac{6}{1}$

41. 1.10 cm^3 **43.** 130 cm^3 **45.** 7330 cm^3

Review Exercises for Chapter 2, page 89

1. 32° **3.** 32° **5.** 41 **7.** 700 **9.** 7.36 **11.** 21.1

13. 25.5 mm **15.** 3.06 m^2 **17.** 309 mm **19.** 3320 cm^2

21. 6190 cm^3 **23.** 1.60×10^5 m^3 **25.** 1.62 m^2

27. 66.6 mm^2 **29.** 25° **31.** 65° **33.** 53° **35.** 2.4

37. $p = b + \sqrt{b^2 + 4a^2} + \pi a$ **39.** $A = ab + \frac{1}{2}\pi a^2$

41. Yes **43.** n^2; $A = \pi(nr)^2 = n^2(\pi r^2)$ **45.** $\frac{a}{d} = \frac{b}{c}$

47. 71° **49.** 7.89 m **51.** 12 cm **53.** 30 m **55.** 6.0 m

57. 5.91 km **59.** 42 100 km **61.** 2.7×10^6 mm^2

63. 1.1×10^6 m^2 **65.** 193 m^3 **67.** 10.000 m **69.** 873 L

71. $h = 52.5$ cm, $w = 93.3$ cm **73.** 136 000 m^2

Chapter 3: Functions and Graphs

Exercises 3.1, page 95

1. $(-1, 1)$ **3.** $A(2, 1), B(-1, 2), C(-2, -3)$

5.

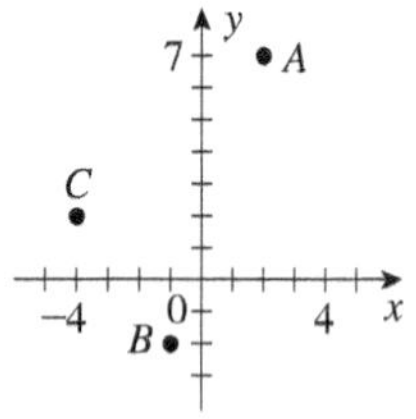

7. Isosceles triangle

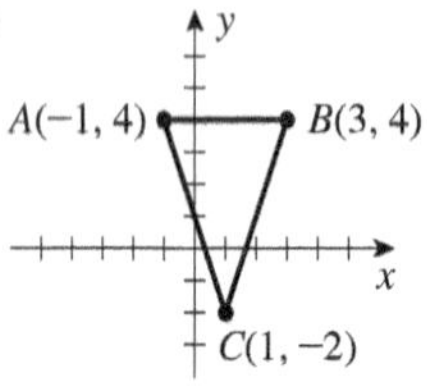

9. Rectangle

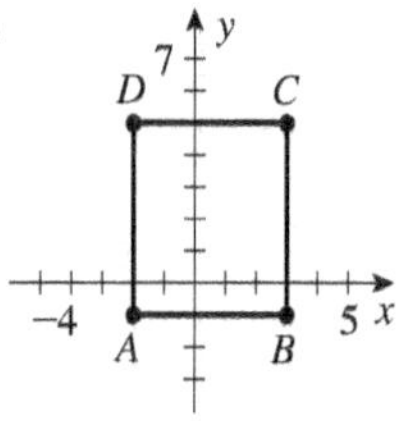

11. $(5, 4)$ **13.** $(3, -2)$ **15.** Quadrant II **17.** Quadrant III

19. On a vertical line one unit to the right of the y-axis

21. On a horizontal line three units above the x-axis

23. On a 45° line through the origin

25. 0 **27.** To the right of the y-axis

29. To the left of a line parallel to the y-axis, one unit to its left

31. Quadrant I or Quadrant III **33.** On the x- or y-axis

35. In Quadrant III **37.** $\sqrt{32}$

Exercises 3.2, page 99

1. -13 **3.** $f(T - 10) = 9.1 + 0.08T + 0.001T^2$

5. (a) $A(r) = \pi r^2$ (b) $A(d) = \frac{\pi d^2}{4}$

7. $d(V) = \sqrt[3]{\frac{6V}{\pi}}$ **9.** $A(s) = s^2, s(A) = \sqrt{A}$

11. $A(r) = 4r^2 - \pi r^2$ **13.** $3, -1$ **15.** $5, 5$

17. $-2.66, -\frac{1}{2}$ **19.** $\frac{a}{4} + \frac{a^2}{2}, 0$

21. $3s^2 + s + 6, 12s^2 - 2s + 6$ **23.** -8

25. $62.9, 2.60 \times 10^2$ **27.** -299.67

29. Square the value of the independent variable and add 2 to the result.

31. Multiply the value of the independent variable by 6, cube the value of the independent variable, and subtract the second result from the first.

33. Multiply the value of the independent variable by 2 and then add 5. Multiply this result by 3, then subtract 1.

35. Multiply the value of the independent variable by 2 and then subtract 3. Divide this result by the sum of the independent variable and 2.

37. $A = 5s^2$
$f(s) = 5s^2$

39. $A = 6.00 - 0.25t$
$f(t) = 6.00 - 0.25t$

41. 10.4 m **43.** 13.2 m, $0.4v + 0.32v^2$, 40.8 m, 40.8 m

45. $d(t) = 55t$

47. (a) $f[f(x)]$ means "function of the function of x." (b) $8x^4$

Exercises 3.3, page 104

1. $f(x) = -x^2 + 2$ is defined for all real values of x.
Since x^2 cannot be negative, the maximum value of $f(x)$ is 2.
Range: all real numbers $f(x) \leq 2$

3. 4 mA, 0 mA

5. Domain: all real numbers
Range: all real numbers

7. Domain: all real numbers $R \neq 0$, or $(-\infty, 0)$ and $(0, \infty)$
Range: all real numbers $G(R) \neq 0$, or $(-\infty, 0)$ and $(0, \infty)$

9. Domain: all real numbers $s \neq 0$, or $(-\infty, 0)$ and $(0, \infty)$
Range: all real numbers $f(s) > 0$, or $(0, \infty)$

11. Domain: all real numbers $h \geq 0$, or $[0, \infty)$
Range: all real numbers $H(h) \geq 1$, or $[1, \infty)$

13. Domain: all real numbers
Range: all real numbers $y \geq 0$, or $[0, \infty)$

15. Domain: all real numbers $y > 2$, or $(2, \infty)$

17. Domain: all real numbers $D \neq 2, -4$, or $(-\infty, -4), (-4, 2)$, and $(2, \infty)$

19. 2, does not exist **21.** 2, 0.75 **23.** $d(t) = 120 + 80t$

25. $w(t) = 5500 - 2t$ **27.** $m(h) = 0.5h - 390$ for $h > 1000$ m

29. $C(l) = 5l + 250$ **31.** (a) $y(x) = \frac{1200 - 0.1x}{0.4}$ (b) 2900 L

33. $y(x) = \frac{2750 - 15x}{25}$ **35.** $A(p) = \frac{p^2}{16} + \frac{(60 - p)^2}{4\pi}$

37. $d(h) = \sqrt{h^2 + 14\,400}$
Domain: all real values $h \geq 0$, or $[0, \infty)$
Range: all real values $d \geq 120$ m, or $[120, \infty)$

39. $v(t) = \frac{300}{t}$
Domain: all real values $t > 0$, or $(0, \infty)$
Range: all real numbers $0 < v < c$, or $(0, c)$

41. Domain: all real values $C > 0$, or $(0, \infty)$

43. $m(h) = \begin{cases} 110 & \text{for } 0 \leq h \leq 1000 \text{ m} \\ 0.5h - 390 & \text{for } h > 1000 \text{ m} \end{cases}$

45. (a) $V(w) = 10w^2 - 150w + 500$
(b) Domain: $w > 10$ cm, or $(10 \text{ cm}, \infty)$

47. 1 **49.** Range: all real values $f(x) \geq 2$, or $[2, \infty)$

Exercises 3.4, page 110

1.
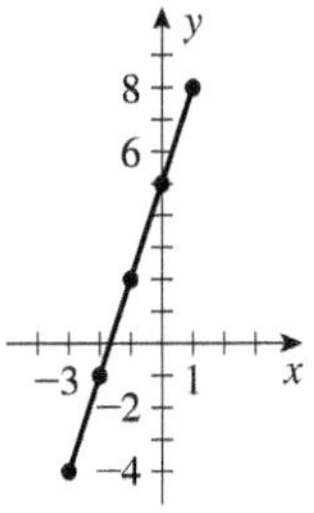

3.
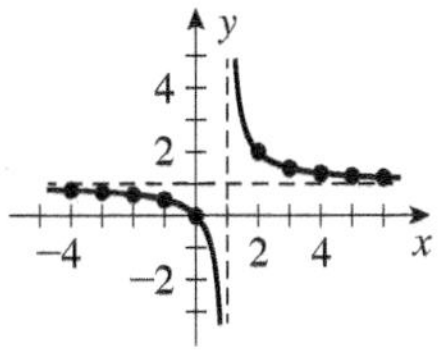

5.
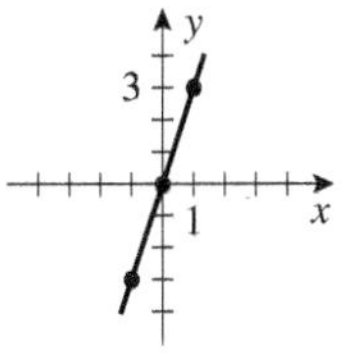

7.
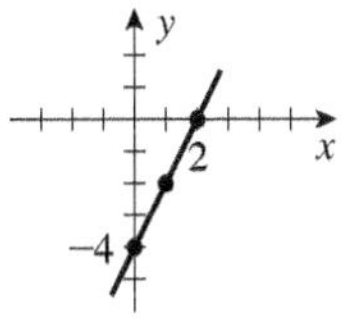

9.
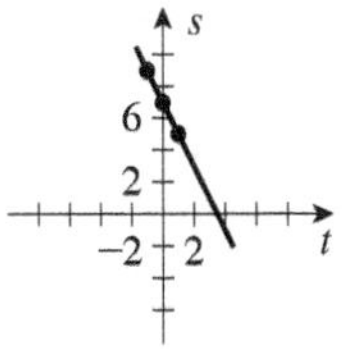

11.
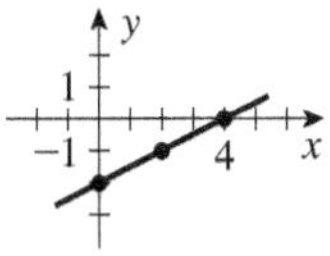

13.
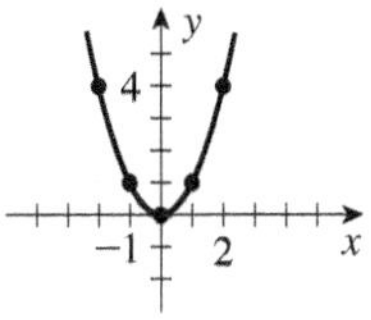

15.
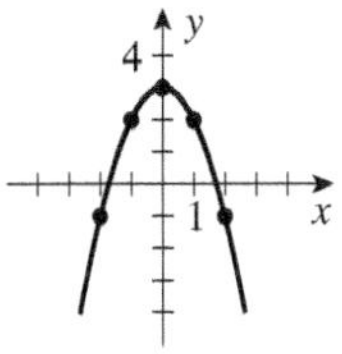

17.
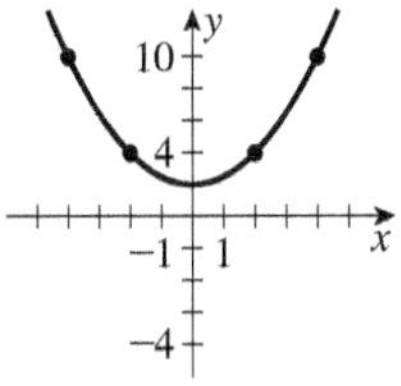

19.
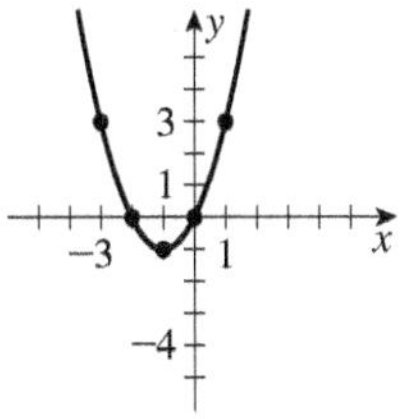

21.
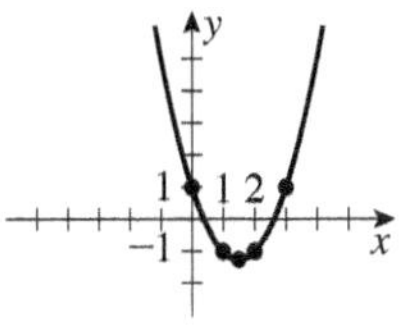

23.
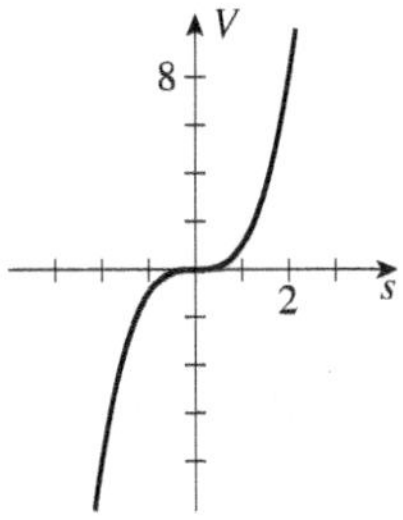

25.
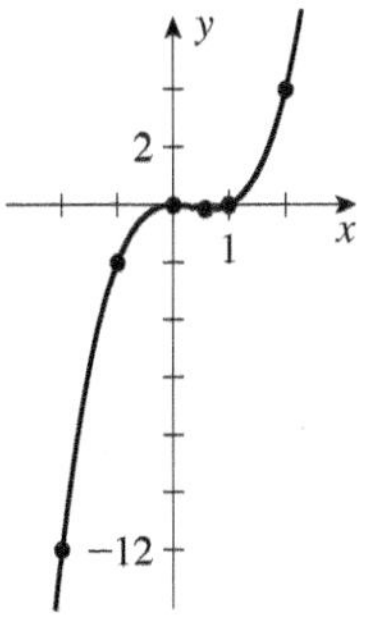

27.
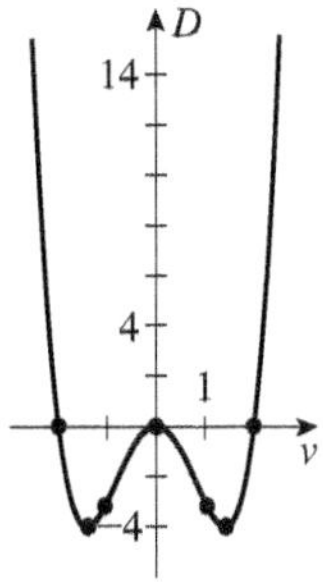

29.
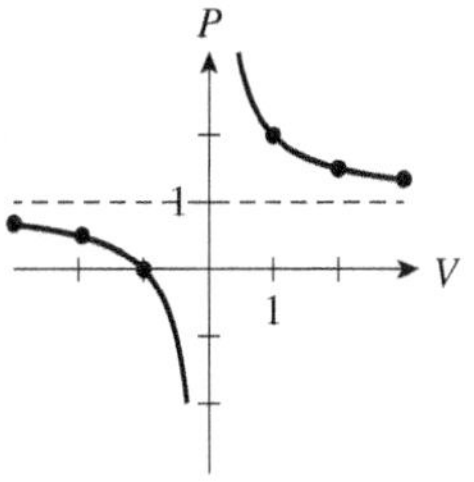

31.
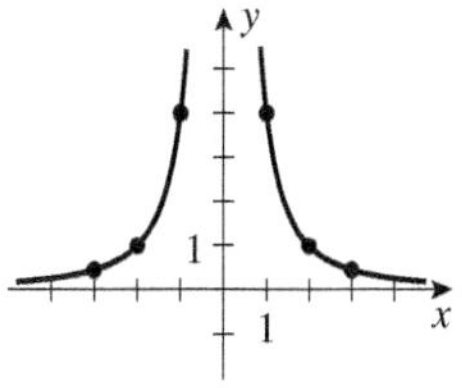

33.
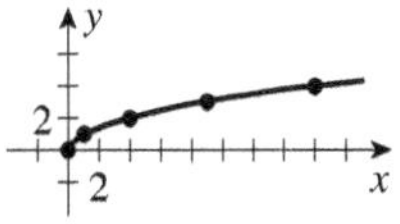

35.
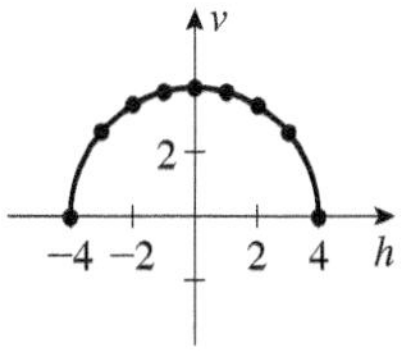

37.
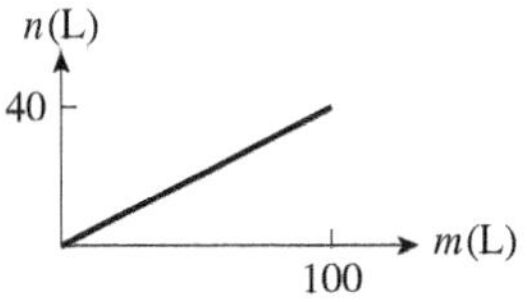

39.
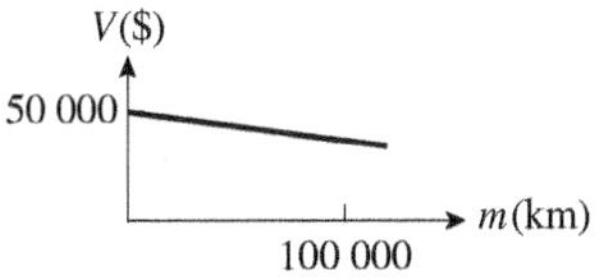

41.
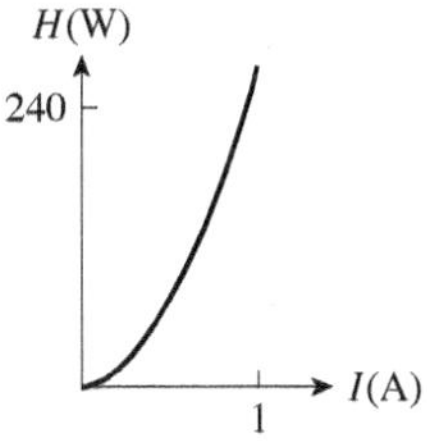

43.
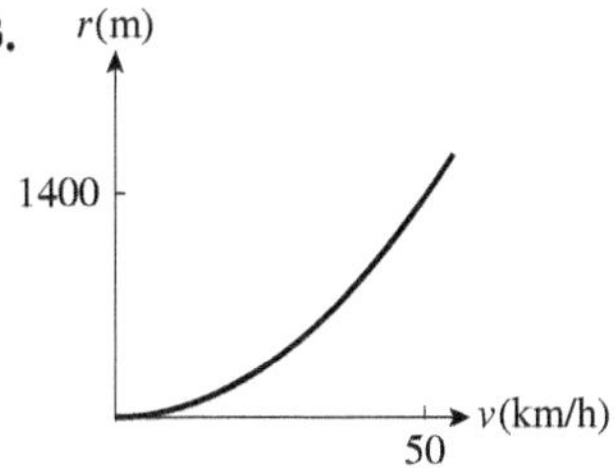

45.
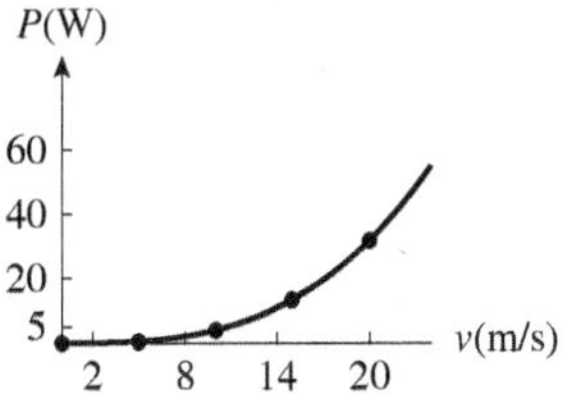

47.
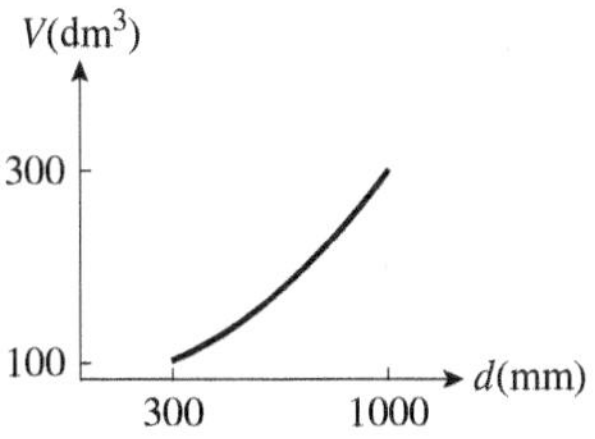

49.
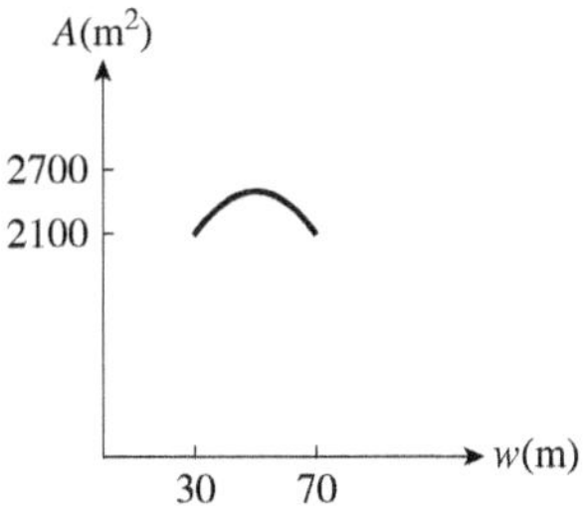

51.

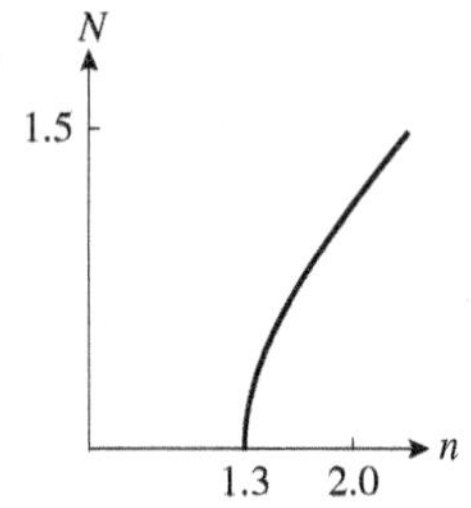

53.

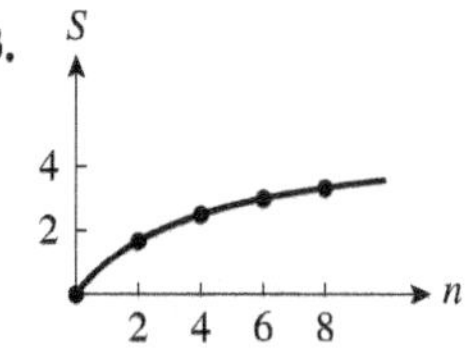

55. No. $f(1) = 2$ says nothing about $f(2)$.

57.

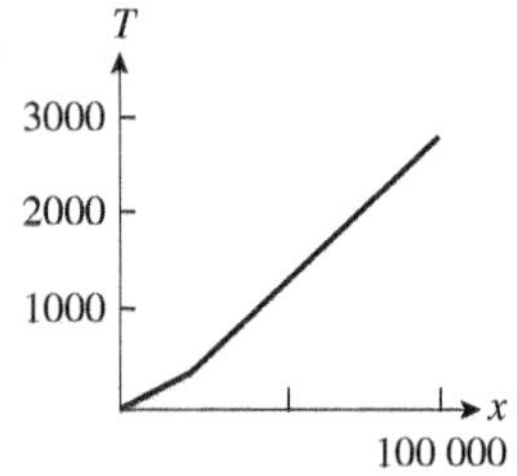

59. $y = x$ is the same as $y = |x|$ for $x \geq 0$.
$y = |x|$ is the same as $y = -x$ for $x < 0$.

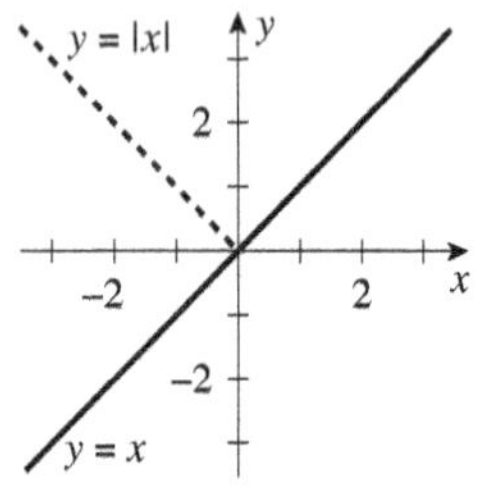

61.

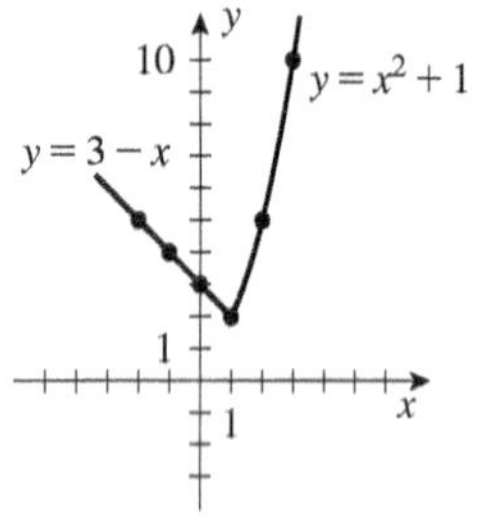

63. (a)

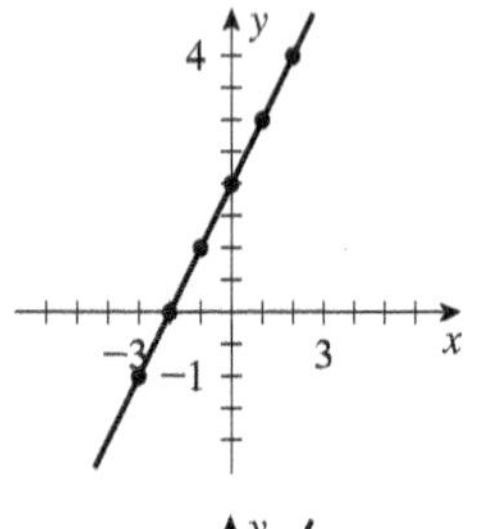

(b)

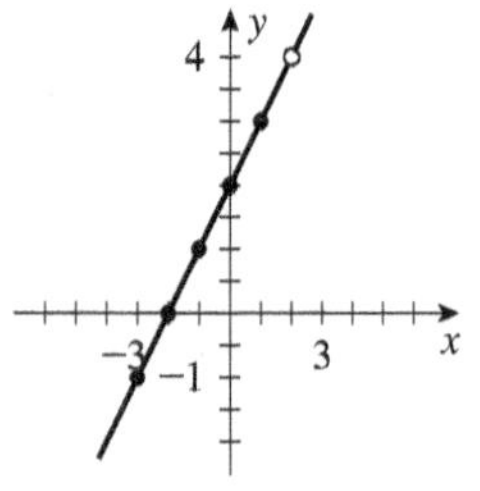

The graphs are identical except the second curve is undefined at a single point $x = 2$.

65. Yes **67.** No

Exercises 3.5, page 115

1. $-2.4, 0.4$

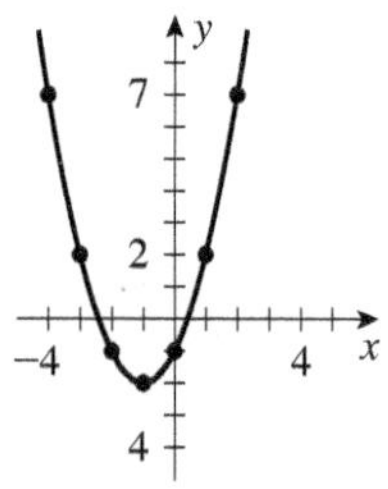

3. 4.49 cm, 4.49 cm, 1.49 cm

5.

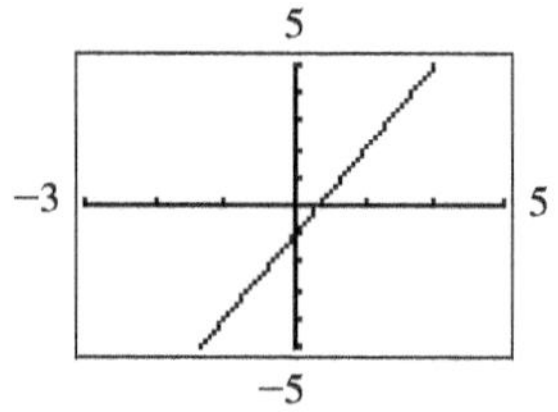

7.

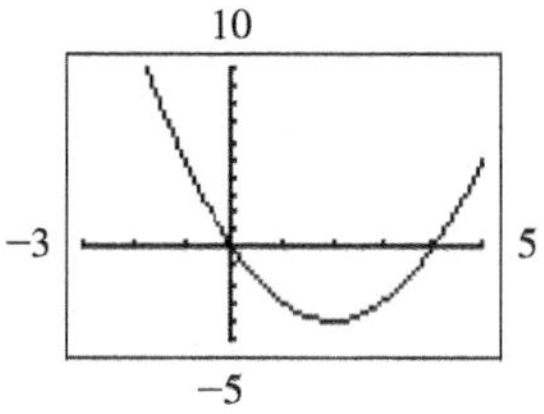

9.

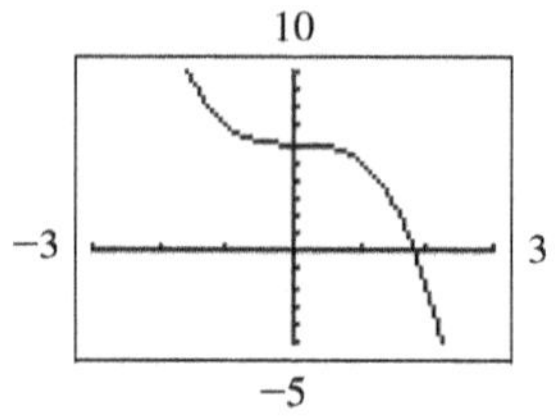

11.

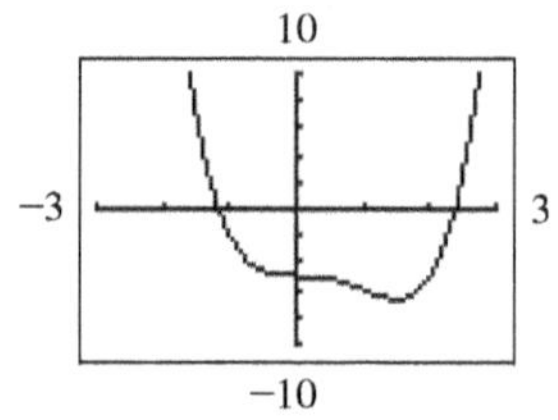

13.

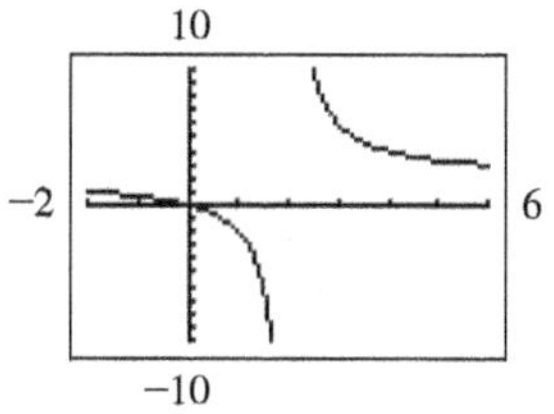

15.

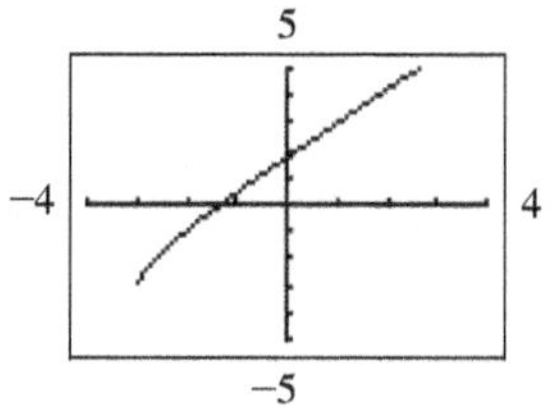

17.

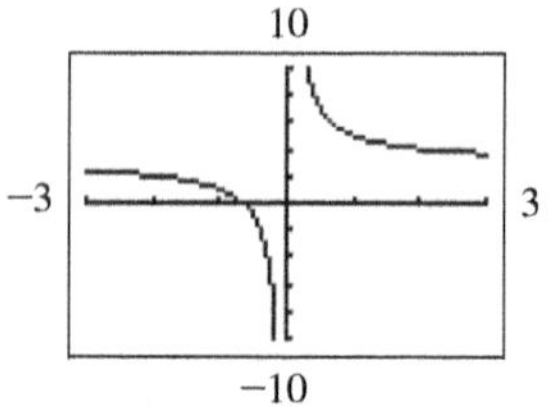

19. $-2.791, 1.791$ **21.** 2.104 **23.** 4.321

25. 1.400 **27.** No solution **29.** Range: all values $y \leq 7$

31. Range: all real values $y > 0$ or $y \leq -1$

33. Range: all real values $y \leq -4$ or $y \geq 0$

35. Range: all real values $Y(y) \geq 3.464$ (approx.)

37. Range: all real numbers

39. $y = 3x + 1$

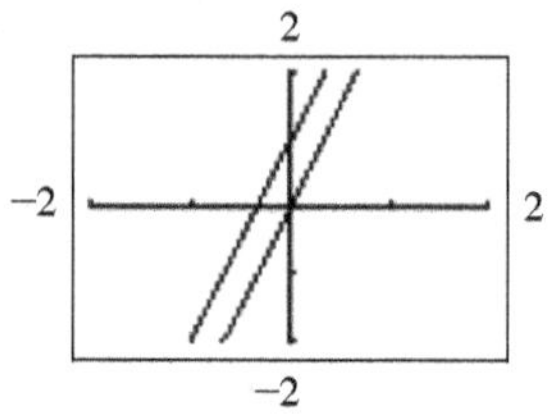

41. $y = \sqrt{x - 3}$

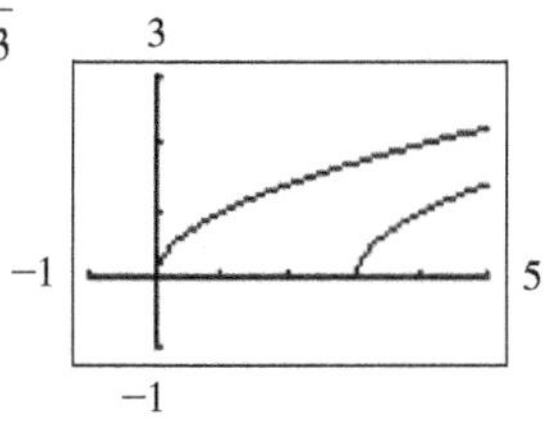

43. $y = -2(x + 2)^2 - 3$

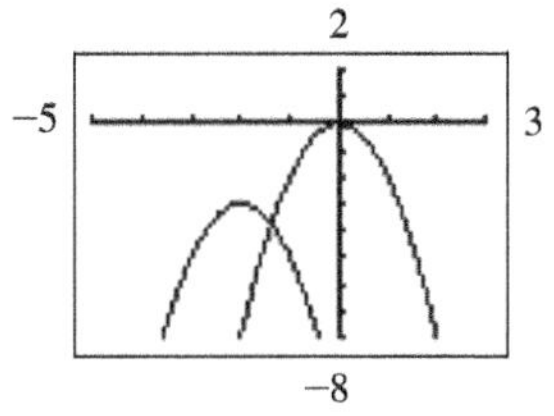

45. $y = \sqrt{2x + 3} + 1$

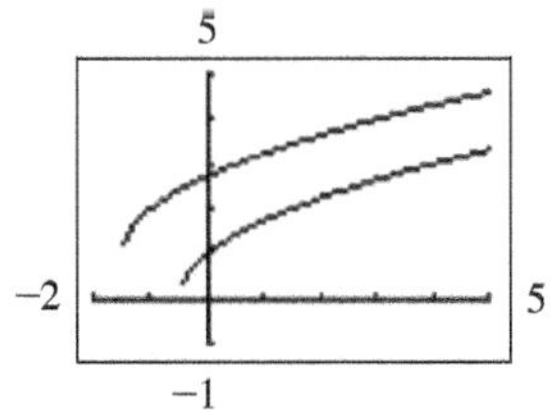

47.

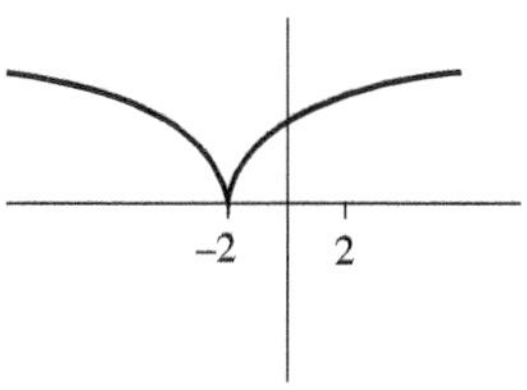

49.

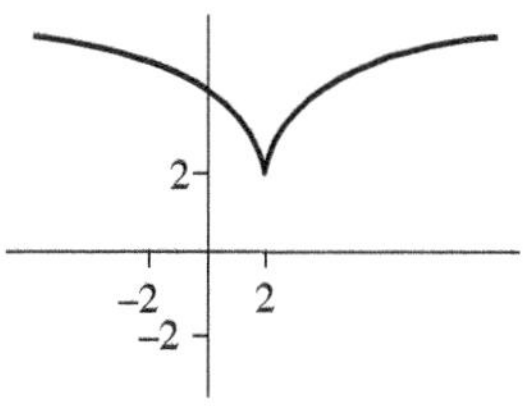

51. 6.00 V **53.** 24.4 cm, 29.4 cm

55. $w = 17.6$ cm, $l = 29.6$ cm **57.** 66.7 m

59. 3.9362 cm, 3.9362 cm, and 1.9362 cm **61.** 0.250 cm/min

63. (a) The graph will look similar to the graph in (b).

(b)

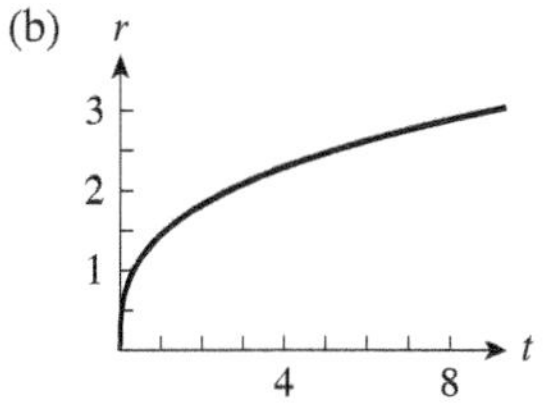

65. Curves are reflected about the x-axis.

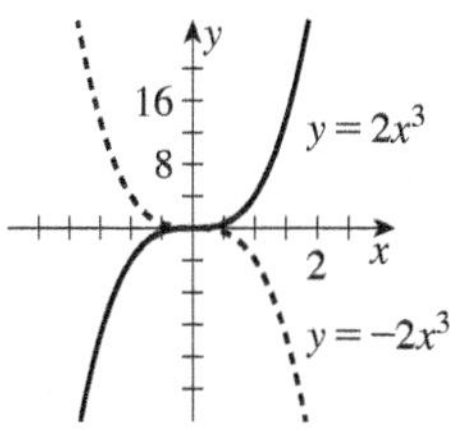

Exercises 3.6, page 118

1. *Production* (1000s of litres)

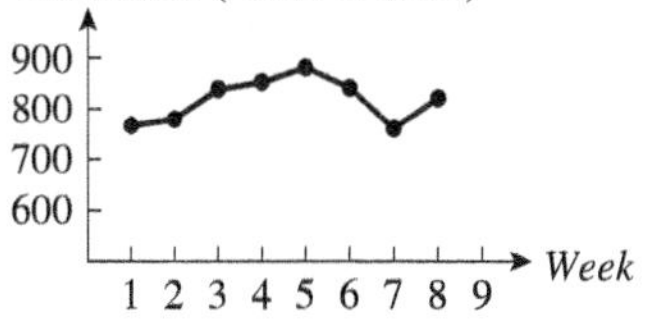

3. *Material* (cm^2)

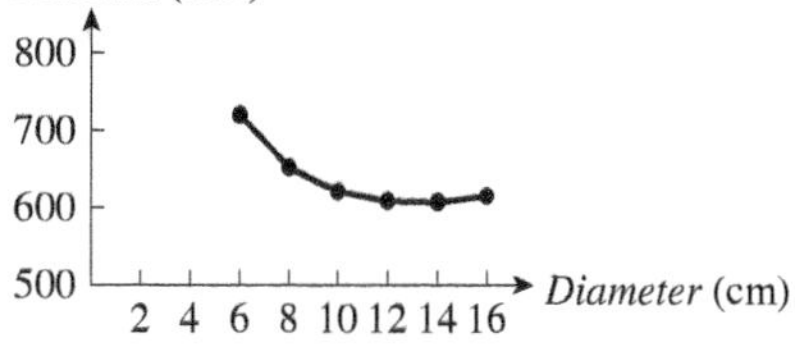

5. *M. ind.* (H)

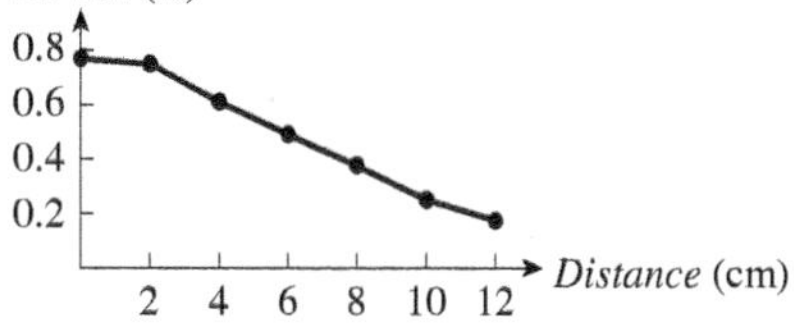

7. *Time* (yrs)

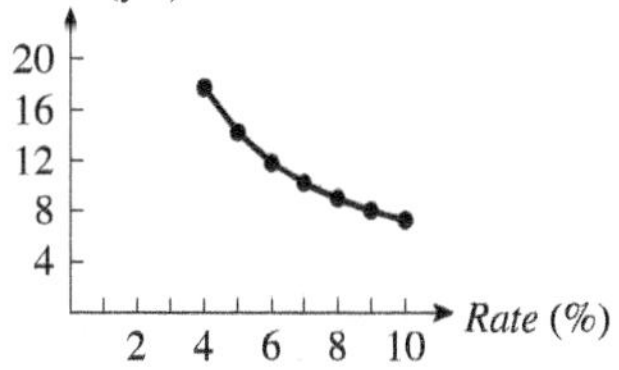

9. 132°C **11.** 0.7 min **13.** (a) 1.5 mm (b) 3.2 mm

15. 0.30 H **17.** 7.2%

19. (a) 170 cm

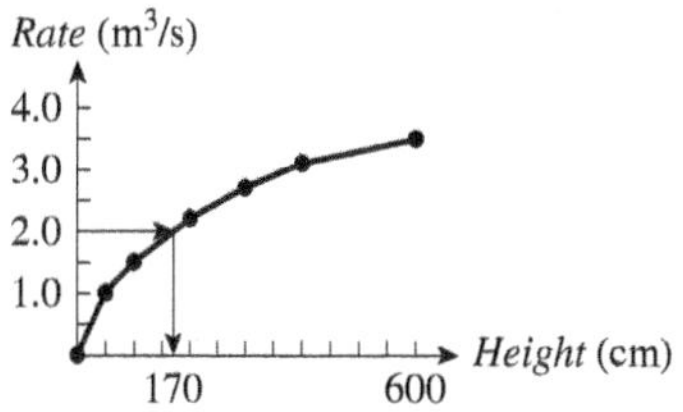

(b) 2.4 m^3/s

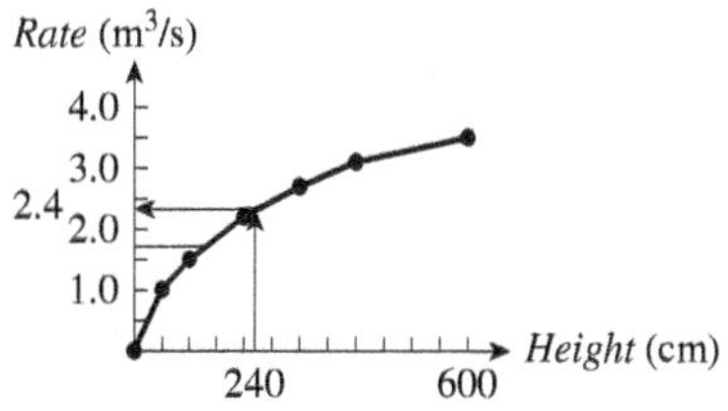

21. 1.3 m^3/s **23.** 0.34 **25.** 76 m^2

27. 130.3°C **29.** 3.7 m^3/s

Review Exercises for Chapter 3, page 119

1. $A(t) = 4\pi t^2$ **3.** $y(x) = \dfrac{2000 - 0.05x}{0.04}$ **5.** 16, −47

7. 3, $\sqrt{1 - 4h}$ **9.** $h^3 + 11h^2 + 36h$ **11.** −3

13. −3.67, 16.6 **15.** 0.165 03, −0.214 76

17. Domain: all real numbers, or $(-\infty, \infty)$
Range: all real numbers $f(x) \geq 1$, or $[1, \infty)$

19. Domain: all real numbers $t > -4$, or $(-4, \infty)$
Range: all real numbers $g(t) > 0$, or $(0, \infty)$

21. Domain: all real numbers $n \neq 5$, or $(-\infty, 5)$ and $(5, \infty)$
Range: all real numbers $f(n) > 1$, or $(1, \infty)$

23.

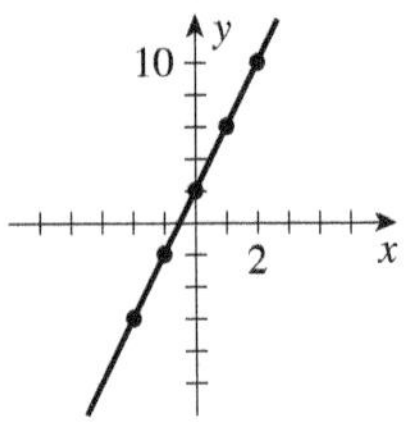

25.

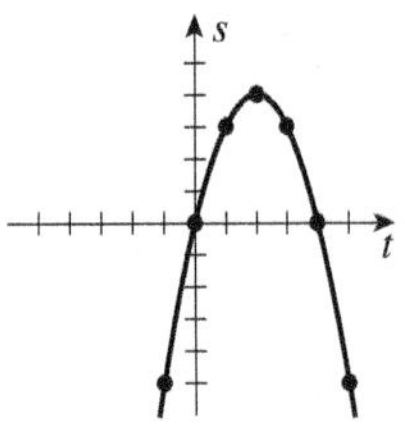

27.

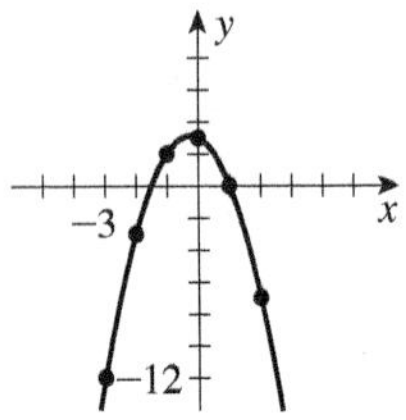

29.

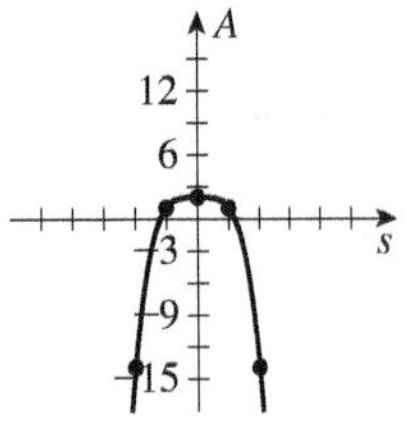

31.

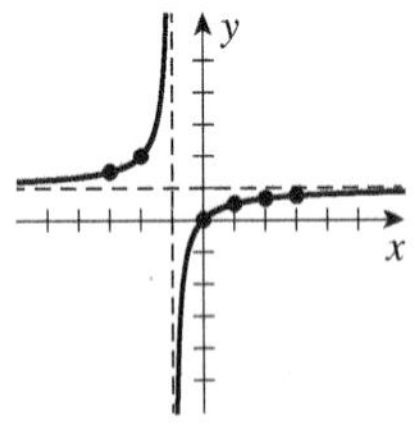

33. 0.43 **35.** 0.17, 5.83 **37.** 1.35 **39.** −0.71, 0.71

41. Range: all real values $y \geq -6.25$ or $[-6.25, \infty)$

43. Range: all real values $A \leq -2.83$ or $A \geq 2.83$, or $(-\infty, -2.83]$ or $[2.83, \infty)$

45. a and b have opposite signs. **47.** $(1, \pm\sqrt{3})$

49. All values of (x, y) that are not on the x-axis or y-axis.

51. There are many possibilities. Two are shown.

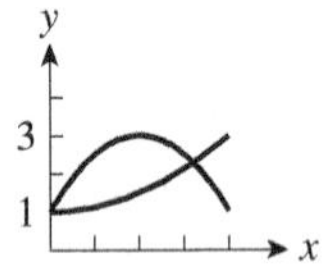

53. $y = \sqrt{x + 1} + 1$

55. The graphs are reflections of each other across the y-axis.

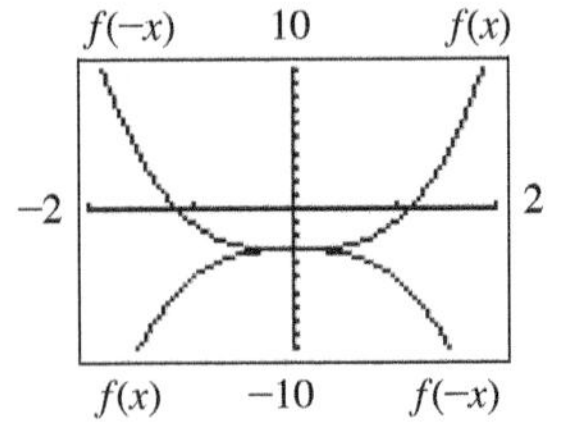

57. Yes; this passes the vertical line test.

59. 13.4 **61.** 72°

63.

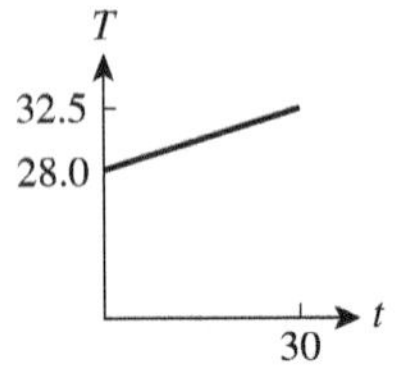

65.

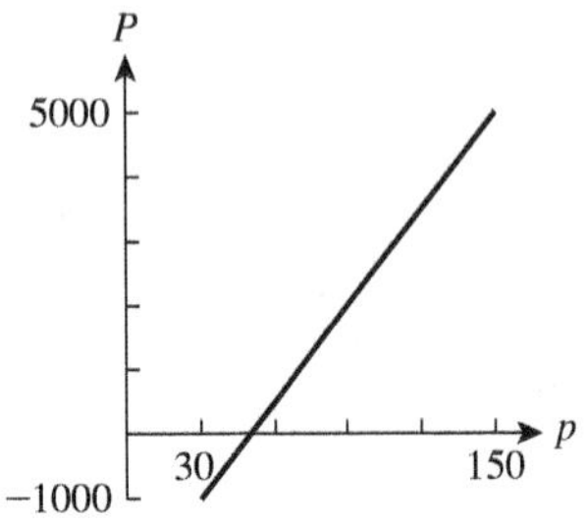

67. $L = f(r) = 2\pi r + 12$

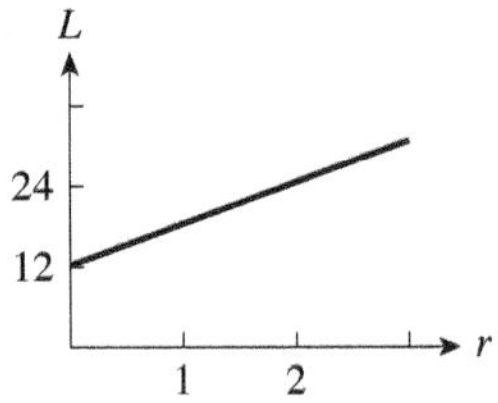

69. 2.63%

71.

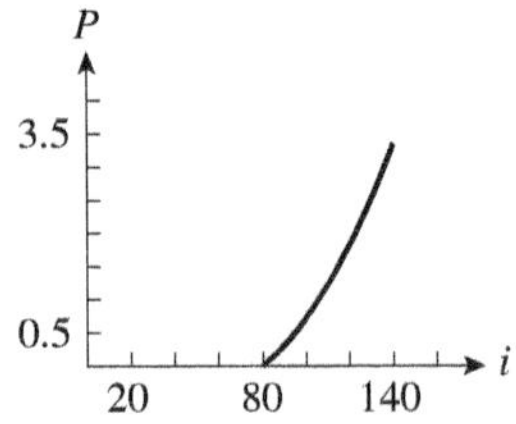

73.

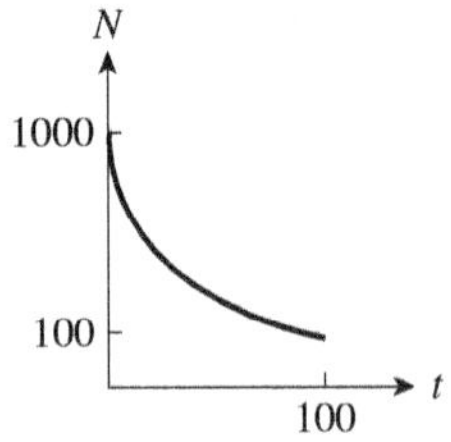

75.

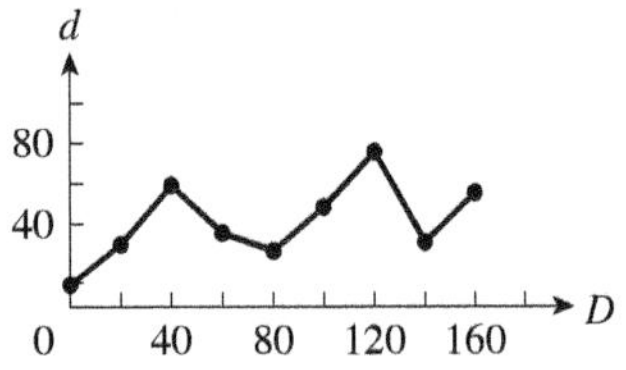

77. 3.46 m

79. 3.5 h

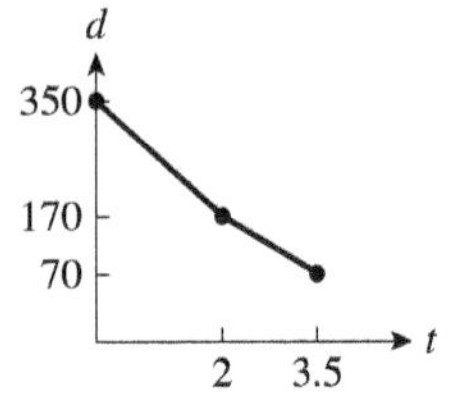

81. 32.8 °C **83.** 0.389 m **85.** 6.53 hours

Chapter 4: The Trigonometric Functions

Exercises 4.1, page 125

1. 1225.6°, or −934.4° **3.** −135°

5.

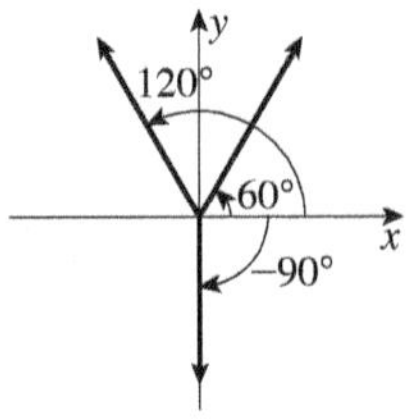

7.

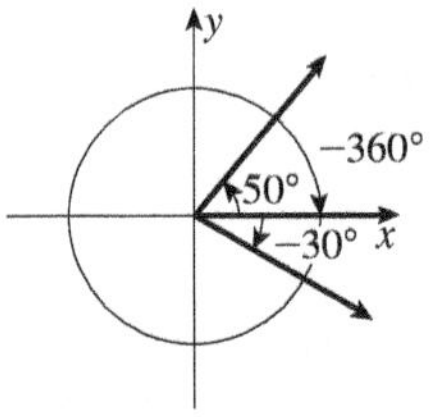

9. 495°, −225° **11.** 210°, −510° **13.** 390°7′, −329°53′

15. 638.1°, −81.9° **17.** 37.36° **19.** 82.91° **21.** 13.69°

23. −235.49° **25.** 1.25 rad **27.** 6.72 rad **29.** 54°42′

31. −5°37′ **33.** 15.20° **35.** 301.27°

37.

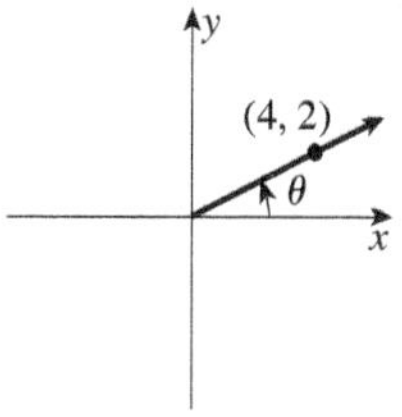

39.

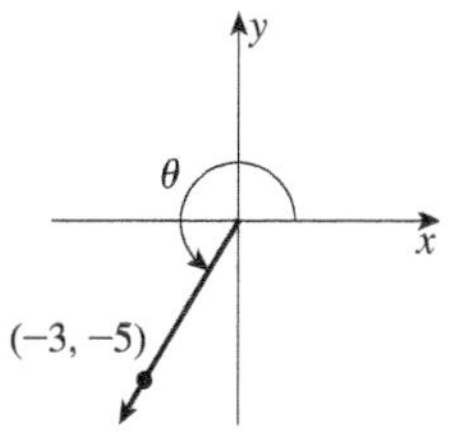

41.

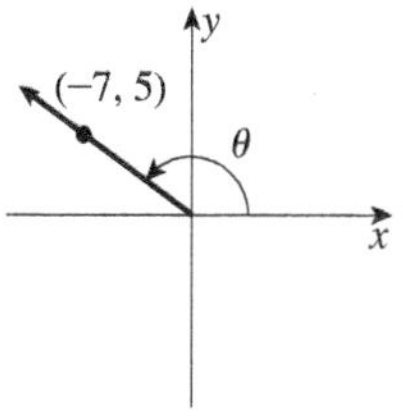

43.

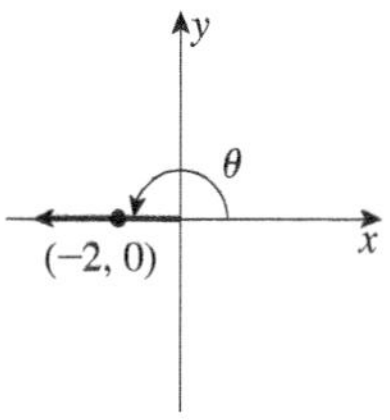

45. First-quadrant angle, fourth-quadrant angle

47. First-quadrant angle, quadrantal angle

49. First-quadrant angle, second-quadrant angle

51. Third-quadrant angle, first-quadrant angle

53. 21.710° **55.** 86°16′26″ **57.** −1260°

Exercises 4.2, page 129

1. $\sin\theta = \frac{3}{5}$ $\csc\theta = \frac{5}{3}$
$\cos\theta = \frac{4}{5}$ $\sec\theta = \frac{5}{4}$
$\tan\theta = \frac{3}{4}$ $\cot\theta = \frac{4}{3}$

3. $\sin\theta = \frac{7\sqrt{65}}{65}$ $\csc\theta = \frac{\sqrt{65}}{7}$
$\cos\theta = \frac{4\sqrt{65}}{65}$ $\sec\theta = \frac{\sqrt{65}}{4}$
$\tan\theta = \frac{7}{4}$ $\cot\theta = \frac{4}{7}$

5. $\sin\theta = \frac{8}{17}$ $\csc\theta = \frac{17}{8}$
$\cos\theta = \frac{15}{17}$ $\sec\theta = \frac{17}{15}$
$\tan\theta = \frac{8}{15}$ $\cot\theta = \frac{15}{8}$

7. $\sin\theta = \frac{40}{41}$ $\csc\theta = \frac{41}{40}$
$\cos\theta = \frac{9}{41}$ $\sec\theta = \frac{41}{9}$
$\tan\theta = \frac{40}{9}$ $\cot\theta = \frac{9}{40}$

9. $\sin\theta = \frac{\sqrt{15}}{4}$ $\csc\theta = \frac{4}{\sqrt{15}}$
$\cos\theta = \frac{1}{4}$ $\sec\theta = 4$
$\tan\theta = \sqrt{15}$ $\cot\theta = \frac{1}{\sqrt{15}}$

11. $\sin\theta = \frac{1}{\sqrt{2}}$ $\csc\theta = \sqrt{2}$
$\cos\theta = \frac{1}{\sqrt{2}}$ $\sec\theta = \sqrt{2}$
$\tan\theta = 1$ $\cot\theta = 1$

13. $\sin\theta = \frac{2}{\sqrt{29}}$ $\csc\theta = \frac{\sqrt{29}}{2}$
$\cos\theta = \frac{5}{\sqrt{29}}$ $\sec\theta = \frac{\sqrt{29}}{5}$
$\tan\theta = \frac{2}{5}$ $\cot\theta = \frac{5}{2}$

15. $\sin\theta = 0.808$ $\csc\theta = 1.24$
$\cos\theta = 0.589$ $\sec\theta = 1.70$
$\tan\theta = 1.37$ $\cot\theta = 0.729$

17. $\frac{2\sqrt{14}}{9}, \frac{5\sqrt{14}}{28}$ **19.** $\frac{2}{\sqrt{5}}, \sqrt{5}$ **21.** 0.882, 1.33

23. 0.246, 3.94 **25.** $\sin\theta = \frac{4}{5}, \tan\theta = \frac{4}{3}$

27. $\tan\theta = \frac{1}{3}, \sec\theta = \frac{\sqrt{10}}{3}$ **29.** 1 **31.** 0.96, 1.04

33. 1 **35.** $\cos\theta = \sqrt{1 - y^2}$ **37.** $-\frac{7}{3}$ **39.** $\sec\theta$

Exercises 4.3, page 134

1. 20.65° **3.** 71.0°

5. sin 40° = 0.65 csc 40° = 1.5
cos 40° = 0.76 sec 40° = 1.3
tan 40° = 0.86 cot 40° = 1.2

7. sin 15° = 0.26 csc 15° = 3.8
cos 15° = 0.97 sec 15° = 1.0
tan 15° = 0.27 cot 15° = 3.7

9. 0.410 **11.** 1.58 **13.** 0.9626 **15.** 1.00 **17.** 0.4085

19. 1.32 **21.** 116.9 **23.** 0.07063 **25.** 70.97°

27. 65.70° **29.** 11.7° **31.** 49.453° **33.** 53.44°

35. 86.53° **37.** No solution **39.** 6.95° **41.** 0.956 = 0.956

43. 2.7 = 2.7

45. y is always less than or equal to r, and the minimum value of y is 0.

47. As θ increases from 0° to 90°, the radius vector rotates counterclockwise and x decreases.

49. 0.8885 **51.** 0.93614 **53.** 32.1° **55.** 87.5 dB

57. 48.6°

Exercises 4.4, page 138

1. $\sin A = 0.868$, $\cos A = 0.496$, $\tan A = 1.75$, $\sin B = 0.496$
$\cos B = 0.868$, $\tan B = 0.571$

3.

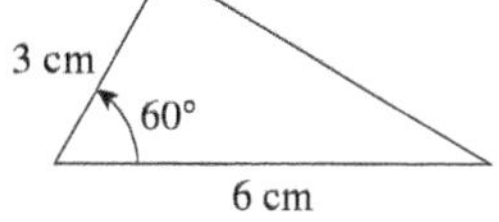

5.

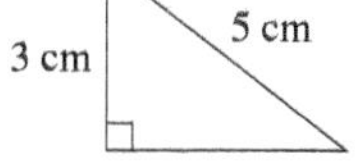

7. $c = 7750$
$b = 1520$
$B = 11.3°$

9. $b = 417$
$A = 16.4°$
$B = 73.6°$

11. $b = 12.6$
$a = 20.2$
$A = 57.9°$

13. $a = 32$
$A = 21°$
$B = 69°$

15. $a = 30.21$
$b = 48.16$
$B = 57.90°$

17. $c = 71.85$
$A = 52.15°$
$B = 37.85°$

19. $c = 1.09$
$b = 0.661$
$A = 52.5°$

21. $c = 1.922$
$a = 0.5239$
$A = 15.82°$

23. $c = 648.46$
$A = 65.883°$
$B = 24.117°$

25. $c = 14.61$
$a = 0.7584$
$B = 87.025°$

27. $b = 0.0162$
$a = 0.0959$
$A = 80.44°$

29. The given information does not determine a unique triangle.

31. 4.45 **33.** 40.24° **35.** 43.1° **37.** 788

39. 108π **41.** 551.8 m

Exercises 4.5, page 142

1. 639 m **3.** 98.7 m **5.** 4.40 m **7.** 0.390° **9.** 850.1 cm **11.** 7610 mm **13.** 0.37 km **15.** 26.6°, 63.4° **17.** 3.4° **19.** 23.5° **21.** 8.12° **23.** 3.07 cm **25.** 651 m **27.** 30.2° **29.** 6.28 cm **31.** 47.3 m **33.** (a) 270 m (b) 282 m **35.** $d = 2x\tan(\frac{\theta}{2})$ **37.** $A = a\sin\theta(b + a\cos\theta)$ **39.** 35.3° **41.** 51.3°, 9.43 m

Review Exercises for Chapter 4, page 146

1. 377.0°, −343.0° **3.** 142.5°, −577.5° **5.** 31.90° **7.** −38.10° **9.** 17°30′ **11.** 749°42′

13. $\sin\theta = \frac{7}{25}$ $\csc\theta = \frac{25}{7}$
$\cos\theta = \frac{24}{25}$ $\sec\theta = \frac{25}{24}$
$\tan\theta = \frac{7}{24}$ $\cot\theta = \frac{24}{7}$

15. $\sin\theta = \frac{1}{\sqrt{2}}$ $\csc\theta = \sqrt{2}$
$\cos\theta = \frac{1}{\sqrt{2}}$ $\sec\theta = \sqrt{2}$
$\tan\theta = 1$ $\cot\theta = 1$

17. 0.923, 2.40 **19.** 0.447, 1.12 **21.** 0.952 **23.** 42.12 **25.** 1.05 **27.** 0.00 **29.** 18.2° **31.** 57.57° **33.** 12.25° **35.** 87.7° **37.** 7.998° **39.** 88.85°

41. $B = 73.0°$, $a = 1.83$, $c = 6.27$ **43.** $c = 104$, $A = 51.5°$, $B = 38.5°$ **45.** $B = 52.5°$, $b = 15.6$, $c = 19.7$

47. $a = 4.006$, $A = 31.61°$, $B = 58.39°$ **49.** $B = 40.33°$, $a = 0.6292$, $b = 0.5341$ **51.** $b = 10.196$, $A = 48.813°$, $B = 41.187°$

53. 4.6 **55.** 61.2 **57.** 10.5

59.
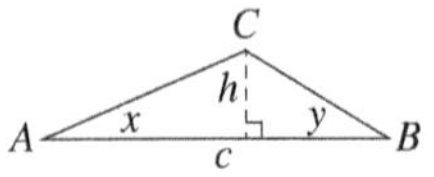

$\cot A = \frac{x}{h}$, $\cot B = \frac{y}{h}$, $c = x + y$, $c = h\cot A + h\cot B$

61. $\frac{\sqrt{x^2+1}}{x}$ **63.** 16.7° **65.** 80.3° **67.** 12.0°

69. (a) $A = \frac{1}{2}bh = \frac{1}{2}b(a\sin C) = \frac{1}{2}ab\sin C$
(b) 679.2 m^2

71. 4.92 km **73.** 0.977 m^2 **75.** 15.3 m^2 **77.** 4.43 m **79.** 34 m **81.** 56% **83.** 10.2 cm

85. End: 6.0 km
Middle: 6.0 km

87. 1.83 km

89. (a) 147 000 mm^2
(b) 155 000 mm^2
(c) The flat surface approximation does not account for the curved surface of the soccer ball.

91. 464 m **93.** 73.3 cm **95.** 35 m^2

Chapter 5: Systems of Linear Equations; Determinants

Exercises 5.1, page 156

1. $x = 4, y = 4$ **3.** Slope: $\frac{2}{3}$; y-intercept: $(0, -\frac{4}{3})$ **5.** Linear **7.** Not linear **9.** Yes, no **11.** Yes, yes **13.** $-3, -\frac{21}{2}$ **15.** $\frac{1}{4}, -0.6$ **17.** 4 **19.** −5 **21.** $\frac{2}{7}$ **23.** 0.5 **25.** 1.298

27.
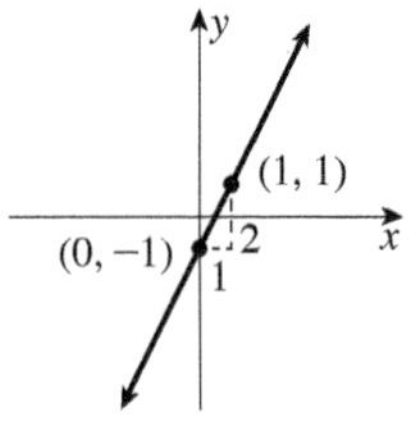

29.
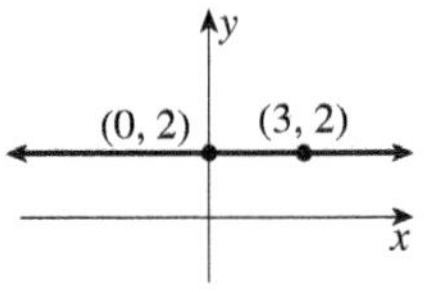

31.
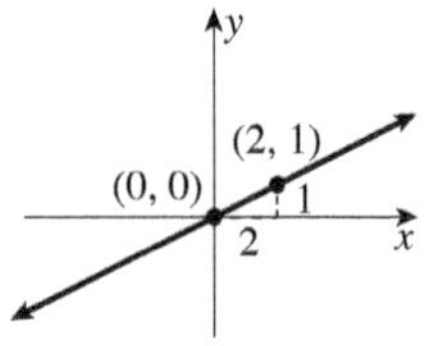

33.
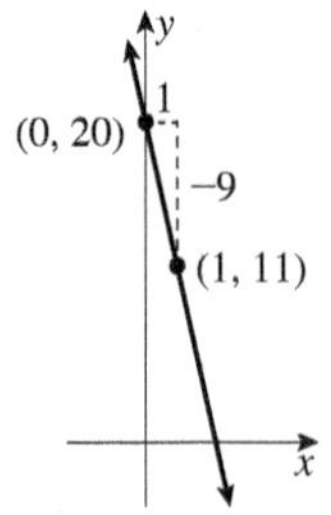

35.
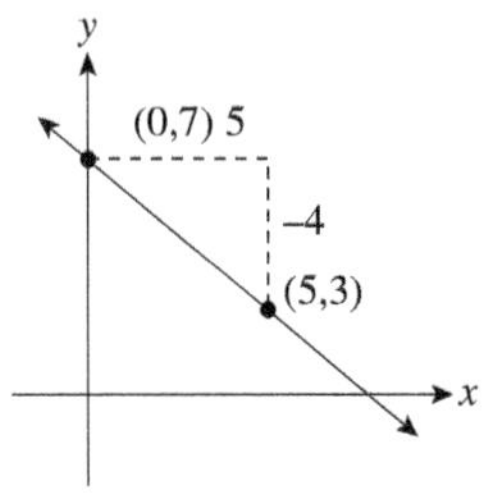

37. $m = -2, b = 1$
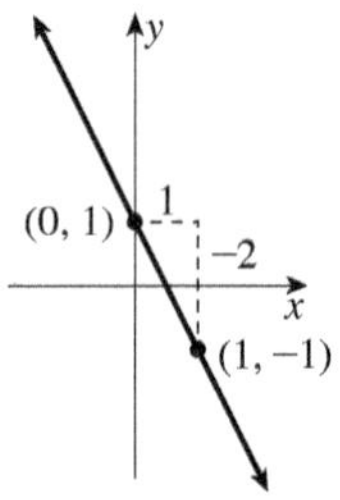

39. $m = 1, b = -4$
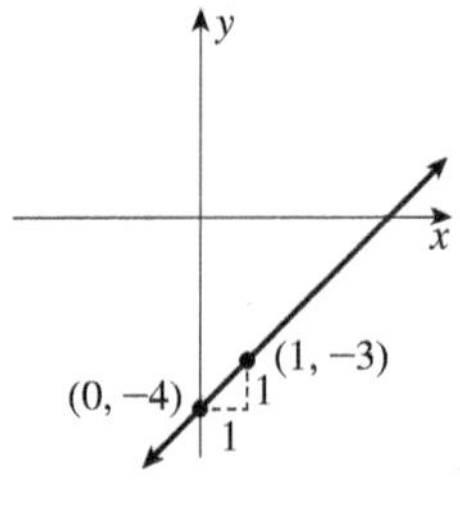

41. $m = \frac{5}{2}, b = -20$
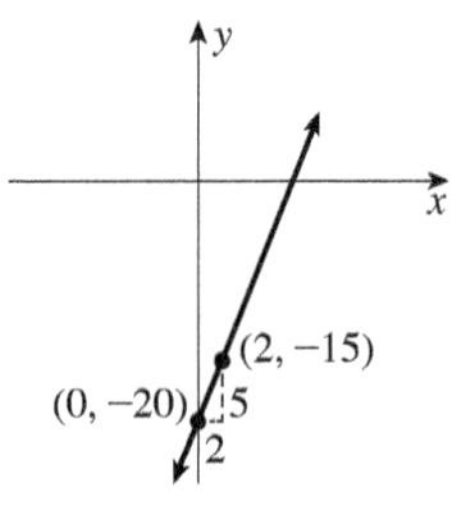

43. $m = -\frac{3}{5}, b = \frac{3}{8}$
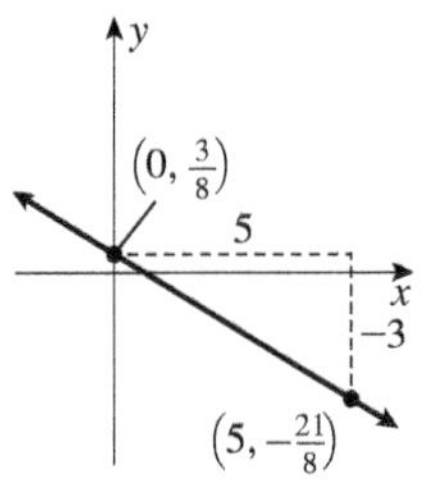

45.
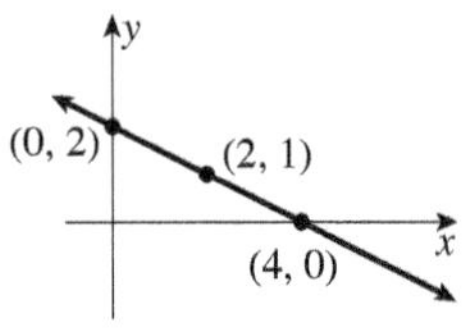

47.
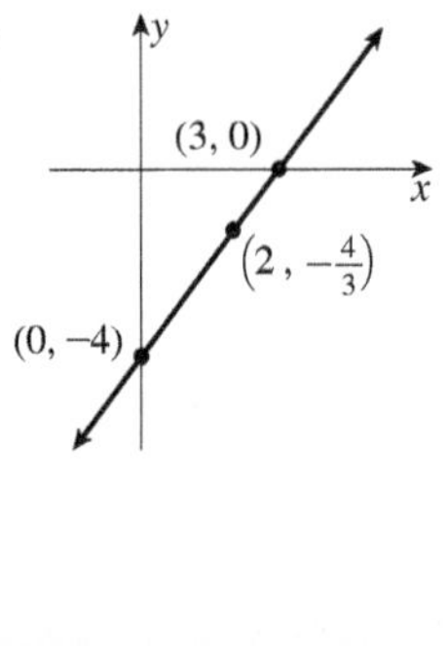

49.

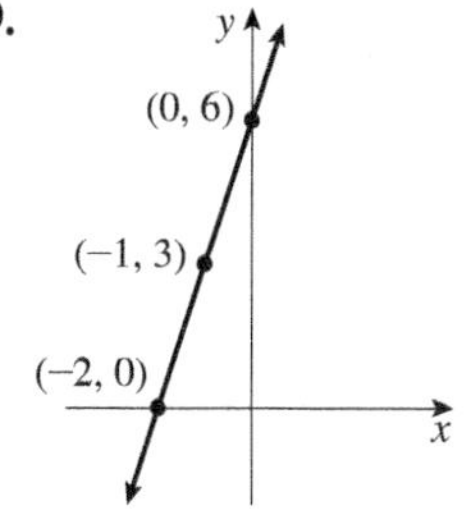

51.

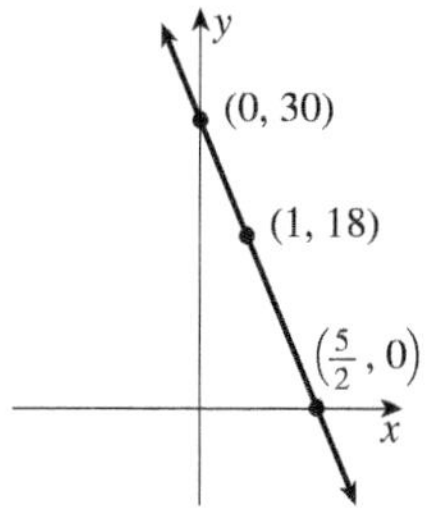

53. $(a, 0)$, $(0, b)$ **55.** No

57.

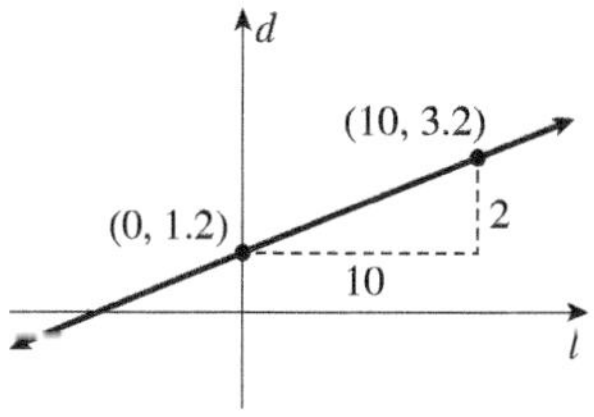

59.

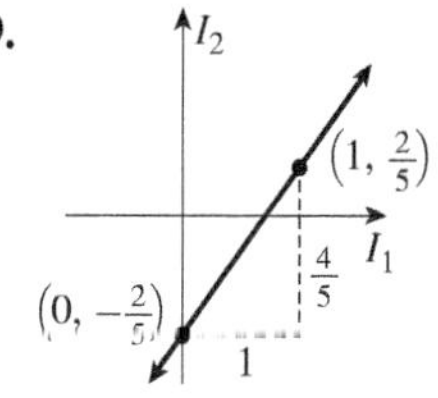

61. $-\dfrac{E_a}{R}$, $\ln A$

Exercises 5.2, page 160

1. Yes **3.** Dependent **5.** Yes **7.** No **9.** No
11. Yes **13.** $x = 3.0, y = 1.0$ **15.** $x = 3.0, y = 0.0$
17. $x = 2.2, y = -0.3$ **19.** $x = -0.9, y = -2.3$
21. $t = -1.7, s = 1.1$ **23.** $x = 0, y = 3$
25. $x = -14.0, y = -5.0$ **27.** $r_1 = 4.0, r_2 = 7.5$
29. $x = 2.1, y = -0.4$ **31.** $x = -3.600, y = -1.400$
33. $x = 1.111, y = 0.841$ **35.** Dependent
37. $t = -1.887, v = -3.179$ **39.** $x = 1.500, y = 4.500$
41. Inconsistent **43.** Yes **45.** No
47. $T_1 = 50\,\text{N}, T_2 = 47\,\text{N}$ **49.** 2.6 m, 3.4 m **51.** \$18.00, \$4.00

Exercises 5.3, page 166

1. $x = -3, y = -3$ **3.** $x = \dfrac{16}{13}, y = -\dfrac{2}{13}$
5. $x = 1, y = -2$ **7.** $p = 3, V = 7$ **9.** $x = -1, y = -4$
11. $x = \frac{1}{2}, y = 2$ **13.** $x = 1, y = \frac{1}{2}$ **15.** $x = 3, y = 1$
17. $x = -1, y = -2$ **19.** $t = -\frac{1}{3}, y = 2$
21. Inconsistent **23.** $x = -\frac{14}{5}, y = -\frac{16}{5}$ **25.** $x = \frac{1}{2}, y = -4$
27. $x = -\frac{2}{3}, y = 0$ **29.** $x = \frac{3}{5}, y = \frac{1}{5}$ **31.** $V = -2, C = -1$
33. $A = -1, B = 3$ **35.** $a = \frac{1}{3}, b = -7$ **37.** $x = \frac{69}{29}, y = \frac{13}{29}$
39. $s = \frac{39}{10}, t = \frac{16}{5}$ **41.** $x = \frac{17}{3}, y = \frac{1}{6}$
43. $V_1 = 9\,\text{V}, V_2 = 6\,\text{V}$ **45.** $x = 6250$ L, $y = 3750$ L **47.** 54
49. $W_f = 9580$ N, $W_r = 8120$ N **51.** $t_1 = 32\,\text{s}, t_2 = 20\,\text{s}$
53. 34 at \$900 per month, 20 at \$1250 per month
55. 7.00 kW, 9.39 kW
57. This is an inconsistent system. There could be an error in the sales figures, and/or the conclusion is in error.
59. $a \neq b$ **61.** $(-3, -4)$, $(9, 0)$

Exercises 5.4, page 172

1. 86 **3.** $x = -13, y = -27$ **5.** −10 **7.** 29
9. 32 **11.** 9300 **13.** 1.083 **15.** 96 **17.** $x = 3, y = 1$
19. $x = -1, y = -2$ **21.** $t = -\dfrac{1}{3}, y = 2$ **23.** Inconsistent
25. $x = -\dfrac{14}{5}, y = -\dfrac{16}{5}$ **27.** $x = \dfrac{69}{29}, y = \dfrac{13}{29}$
29. $s = \dfrac{39}{10}, t = \dfrac{16}{5}$ **31.** $x = -11.2, y = -9.26$
33. $x = -1.0, y = -2.0$ **35.** 0 **37.** 0
39. $F_1 = 15$ N, $F_2 = 6.0$ N **41.** $x = 73.6$ L, $y = 70.4$ L
43. 160 3-bedroom homes, 80 4-bedroom homes
45. 210 phones, 110 detectors **47.** \$2000, 6.0%
49. 2.5 h, 2.1 h **51.** $V(i) = 4.5i - 3.2$

Exercises 5.5, page 177

1. $x = 1, y = -4, z = \frac{1}{3}$ **3.** $x = 2, y = -1, z = 1$
5. $x = 4, y = -3, z = 3$ **7.** $l = \frac{1}{2}, w = \frac{2}{3}, h = \frac{1}{6}$
9. $x = \frac{2}{3}, y = -\frac{1}{3}, z = 1$ **11.** $x = \frac{4}{15}, y = -\frac{3}{5}, z = \frac{1}{3}$
13. $x = \frac{3}{4}, y = 1, z = -\frac{1}{2}$ **15.** $x = 2.1, y = 4.3, z = -1.7$
17. $r = 0, s = 0, t = 0, u = -1$ **19.** $f(x) = x^2 - 3x + 5$
21. $P = 800$ h, $M = 125$ h, $I = 225$ h
23. $F_1 = 9.43$ N, $F_2 = 8.33$ N, $F_3 = 1.67$ N
25. $A = 22.5°, B = 45.0°, C = 112.5°$
27. $A = 19\,400, B = 20\,000, C = 19\,900$
29. $a = 0.295, b = -5.53, c = 24.2$
31. $x = 70.0$ kg, $y = 100.0$ kg, $z = 30.0$ kg
33. Infinite number of solutions. A possible solution is $x = -10, y = -6, z = 0$ **35.** Inconsistent

Exercises 5.6, page 182

1. −38 **3.** 122 **5.** 651 **7.** −439
9. 202 **11.** 29 440 **13.** 0.128 **15.** $-\dfrac{17\pi}{48}$
17. $x = -1, y = 2, z = 0$ **19.** $x = 2, y = -1, z = 1$
21. $x = 4, y = -3, z = 3$ **23.** $l = \frac{1}{2}, w = \frac{2}{3}, h = \frac{1}{6}$
25. $x = \frac{2}{3}, y = -\frac{1}{3}, z = 1$ **27.** $x = \frac{4}{15}, y = -\frac{3}{5}, z = \frac{1}{3}$
29. $p = -2, q = \frac{2}{3}, r = \frac{1}{3}$ **31.** $x = \frac{3}{4}, y = 1, z = -\frac{1}{2}$
33. Collinear **35.** Value changes from 19 to −19.
37. 19 (no change) **39.** $A = 125$ N, $B = 60$ N, $C = 75$ N
41. $s_0 = 2.00$ m, $v_0 = 5.00$ m/s, $a = 4.00$ m/s^2
43. $V = f(T) = 5.00 + 0.500T + 0.100T^2$
45. 79% nickel, 16% iron, 5% molybdenum
47. $v_c = 45.9$ km/h, $v_j = 551$ km/h, $v_t = 30.9$ km/h
49. $F_1 = 27.9$ N, $F_2 = 44.5$ N, $F_3 = 52.6$ N

Review Exercises for Chapter 5, page 185

1. −17 **3.** −1485 **5.** −4 **7.** $\dfrac{2}{7}$

9. $m = -2, b = 4$

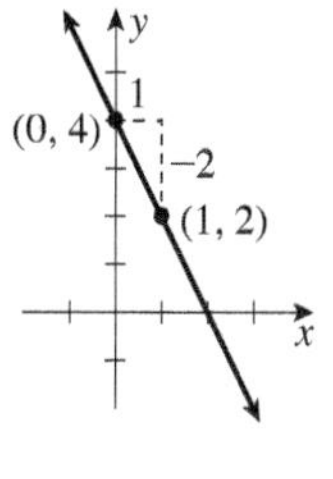

11. $m = 4, b = -\dfrac{5}{2}$

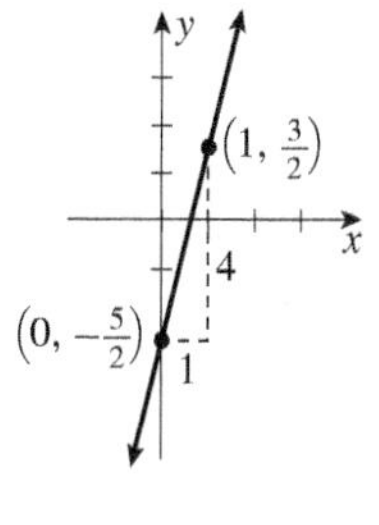

13. $x = 2, y = 0$ **15.** $A = 2.2, B = 2.7$
17. $x = 1.5, y = -1.9$ **19.** $M = 1.5, N = 0.4$
21. $x = 1, y = 2$ **23.** $x = \frac{1}{2}, y = -2$ **25.** $i = 2, v = -\frac{1}{3}$
27. $x = -\frac{6}{19}, y = \frac{36}{19}$ **29.** $x = \frac{43}{39}, y = \frac{7}{13}$ **31.** $x = 1, y = 2$
33. $x = \frac{1}{2}, y = -2$ **35.** $i = 2, v = -\frac{1}{3}$
37. $x = -\frac{6}{19}, y = \frac{36}{19}$ **39.** $x = \frac{43}{39}, y = \frac{7}{13}$
41. 33 (the second equation is already solved for y)
43. 40 (the coefficients don't indicate another method)
45. -115 **47.** 230.08 **49.** $x = 2, y = -1, z = 1$
51. $r = 3, s = -1, t = \frac{3}{2}$ **53.** $x = -0.168, y = 0.156, z = 2.41$
55. $x = 2, y = -1, z = 1$ **57.** $r = 3, s = -1, t = \frac{3}{2}$
59. $x = -0.168, y = 0.156, z = 2.41$ **61.** 4 **63.** $-\frac{15}{7}$
65. $x = \frac{8}{3}, y = -8$ **67.** $x = 1, y = 3$ **69.** -6 **71.** $-\frac{4}{3}$
73. $F_1 = 21\,000$ N, $F_2 = 2400$ N, $F_3 = 18\,000$ N
75. $p_1 = 42\%, p_2 = 58\%$ **77.** \$40 000, \$4000
79. 27.6 tonnes of 6.0% copper, 14.4 tonnes of 2.4% copper
81. $a = 444$ m · °C, $b = 9.56$ °C **83.** 22 800 km/h, 1400 km/h
85. $R_1 = 0.500\ \Omega, R_2 = 1.50\ \Omega$ **87.** $L = 10.0$ N, $w = 40.0$ N
89. $A = 75.0°, B = 65.0°, C = 40.0°$
91. $A = 68.0$ MB, $B = 24.0$ MB, $C = 48.0$ MB
93. 4.26 N, 1.74 N **95.** 8.84 m³/h, 4.65 m³/h

Chapter 6: Factoring and Fractions

Exercises 6.1, page 193

1. $9r^2 - 4s^2$ **3.** $x^2 + 2xy + y^2 - 6x - 6y + 9$
5. $30x - 30y$ **7.** $7x^4 - 35x^3$ **9.** $T^2 - 16$
11. $16v^2 - 9$ **13.** $25x^2 - 16y^2$ **15.** $144 - 25a^2b^2$
17. $16f^2 + 40f + 25$ **19.** $4x^2 + 52x + 169$
21. $L^4 - 2L^2 + 1$ **23.** $16a^2 + 56axy + 49x^2y^2$
25. $0.25s^2 - 0.10st + 0.01t^2$ **27.** $x^2 + 6x + 5$
29. $18K^2 + 39KC^2 + 6C^4$ **31.** $20x^2 - 21x - 5$
33. $40v^2 + 138v - 45$ **35.** $600x^2 - 130xy - 63y^2$
37. $2x^2 - 8$ **39.** $8a^3 - 2a$ **41.** $6ax^2 + 24abx + 24ab^2$
43. $20n^4 + 100n^3 + 125n^2$ **45.** $16R^4 - 72R^2r^2 + 81r^4$
47. $x^2 + 2xy + y^2 + 2x + 2y + 1$
49. $9 - 6x - 6y + x^2 + 2xy + y^2$
51. $-t^3 + 15t^2 - 75t + 125$
53. $27L^3 + 189L^2R + 441LR^2 + 343R^3$
55. $w^2 + 2wh + h^2 - 1$ **57.** $x^3 + 8$ **59.** $64 - 27x^3$
61. $x^4 - 2x^2y^2 + y^4$ **63.** $P_1P_0c + P_1G$
65. $4p^2 + 8pDA + 4D^2A^2$ **67.** $\frac{1}{2}\pi R^2 - \frac{1}{2}\pi r^2$
69. $\frac{L}{6}x^3 - \frac{L}{2}ax^2 + \frac{L}{2}a^2x - \frac{L}{6}a^3$
71. $L_0 + aL_0T - aL_0T_0$
73. 2499 **75.** $4x^2 - 9$
77. (a) $(x + y)^2$ (b) $x^2 + 2xy + y^2$ (Eq. 6.2)
79. $4x^2 - y^2 - 4y - 4$ **81.** $0.01wl - 0.121$

Exercises 6.2, page 197

1. $2ax(2x - 1)$ **3.** $5(x + 3)(x - 3)$ **5.** $7(x - y)$
7. $6(b - 4)$ **9.** $3x(5x - 1)$ **11.** $7w(wh - 4)$
13. $24n(12n + 1)$ **15.** $2(x + 2y - 4z)$
17. $3ab(b - 2 + 4b^2)$ **19.** $pq(6q - 5 - 14q^2)$
21. $2(a^2 - b^2 + 2c^2 - 3d^2)$ **23.** $(x + 3)(x - 3)$
25. $(10 + 3A)(10 - 3A)$ **27.** $36a^4 + 1$
29. $2(9s + 5t)(9s - 5t)$ **31.** $(12n + 13p^2)(12n - 13p^2)$
33. $(x + y + 3)(x + y - 3)$ **35.** $2(4 + 3x)(4 - 3x)$
37. $300(x + 3z)(x - 3z)$ **39.** $2(I - 1)(I - 5)$
41. $(x^2 + 4)(x + 2)(x - 2)$
43. $(x^4 + 1)(x^2 + 1)(x + 1)(x - 1)$
45. $(v(i_1 - 4) + 5i_2)(v(i_1 - 4) - 5i_2)$
47. $\frac{b + 3}{2 - b}$ **49.** $\frac{3}{2(t - 1)}$ **51.** $\frac{5}{2}$ **53.** $(x - y)(b + 3)$
55. $(a + x)(a - b)$ **57.** $(x + 3)(x + 2)(x - 2)$
59. $(x - y)(x + y + 1)$ **61.** 16 777 216
63. $n^2 + n = n(n + 1)$; the product is even. **65.** $2(Q^2 + 1)$
67. $Rv(1 + v + v^2)$ **69.** $r(R + r)(R - r)$
71. $r^2(\pi - 2)$ **73.** $4\pi(r_2 - r_1)(r_2 + r_1)$
75. $\frac{i_2R_2}{R_1 + R_2}$ **77.** $\frac{5Y}{3(3S - Y)}$ **79.** $\frac{ER}{A(T_0 - T_1)}$

Exercises 6.3, page 203

1. $(x + 3)(x + 1)$ **3.** $(2x - 1)(x - 5)$
5. $(x + 1)(x + 5)$ **7.** $(s - 8)(s + 7)$ **9.** $(t + 7)(t - 4)$
11. $(x + 1)^2$ **13.** $(L - 2K)^2$ **15.** $(3x + 1)(x - 2)$
17. $4(3y + 1)(y - 3)$ **19.** $(2s + 11)(s + 1)$
21. $(3f^2 - 1)(f^2 - 5)$ **23.** $(2t - 3)(t + 5)$
25. $(3t - 4u)(t - u)$ **27.** $(4x - 7)(x + 1)$
29. $(x + y)(9x - 2y)$ **31.** $(2m + 5)^2$ **33.** $2(2x - 3)^2$
35. $(3t - 4)(3t - 1)$ **37.** $(2b - 1)(4b^2 + 2b + 1)(b^3 + 4)$
39. $(4p - q)(p - 6q)$ **41.** $(12x - y)(x + 4y)$
43. $2(x - 1)(x - 6)$ **45.** $2x(2x^2 - 1)(x^2 + 4)$
47. $ax(x + 6a)(x - 2a)$ **49.** $(a + b + 2)(a + b - 2)$
51. $(5a + 5x + y)(5a - 5x - y)$ **53.** $(4x^n - 3)(x^n + 4)$
55. $16(t - 4)(t - 1)$ **57.** $(d^2 - 8)(d^2 - 2)$
59. $100(2n + 3)(n - 12)$ **61.** $(V - nB)^2$
63. $wx^2(x - 3L)(x - 2L)$ **65.** $Ad(3u - v)(u - v)$
67. $k = -1, (2x + 1)^2$ **69.** $(x^2 + 2x + 2)(x^2 - 2x + 2)$
71. 4 m, 2 m

Exercises 6.4, page 206

1. $(x - 2)(x^2 + 2x + 4)$ **3.** $(x + 1)(x^2 - x + 1)$
5. $(2 - t)(4 + 2t + t^2)$ **7.** $(3x - 2y)(9x^2 + 6yx + 4y^2)$
9. $4(x + 2)(x^2 - 2x + 4)$ **11.** $7n^2(n - 1)(n^2 + n + 1)$
13. $6x^3y(3 - y)(9 + 3y + y^2)$ **15.** $x^3y^3(x + y)(x^2 - xy + y^2)$
17. $3a^2(a^2 + 1)(a - 1)(a + 1)$
19. $0.001(R - 4r)(R^2 + 4Rr + 16r^2)$

21. $27L^3(L+2)(L^2-2L+4)$

23. $(a+b+4)(a^2+2ab+b^2-4a-4b+16)$

25. $(2+x)(2-x)(16+4x^2+x^4)$

27. $-h^2w^2(2w-5)(4w^2+10w+25)$

29. $4x(2-x)(4+2x+x^2)$ **31.** $D(D-d)(D^2+Dd+d^2)$

33. $QH(H+Q)(H^2-HQ+Q^2)$

35. $x^5-y^5=(x-y)(x^4+x^3y+x^2y^2+xy^3+y^4)$,
$x^7-y^7=(x-y)(x^6+x^5y+x^4y^2+x^3y^3+x^2y^4+xy^5+y^6)$

37. $(x+y)(x-y)(x^4+x^2y^2+y^4)$

39. $n^3+1=(n+1)(n^2-n+1)$; $(n+1)$ is a factor of n^3+1 so n^3+1 is not prime.

Exercises 6.5, page 210

1. $\frac{x+2}{x-2}, x\neq -2$ **3.** $\frac{14}{21}$ **5.** $\frac{2ax^2}{2xy}$

7. $\frac{2x-4}{x^2+x-6}$ **9.** $\frac{ax^2-ay^2}{x^2-xy-2y^2}$ **11.** $\frac{7}{11}$

13. $\frac{2xy}{4y^2}, x\neq 0$ **15.** $\frac{2}{R+1}, R\neq 1$ **17.** $\frac{s-5}{2s-1}, s\neq -2$

19. $9xy$ **21.** a^2-25 **23.** $2x$ **25.** $1, x\neq b$

27. $\frac{1}{4}, a\neq 0$ **29.** $\frac{3x}{4}, x, y\neq 0$ **31.** $\frac{1}{5a}, a+b\neq 0$

33. $\frac{2(a-b)}{2a-b}$ **35.** $\frac{4x^2+1}{(2x+1)(2x-1)}$ **37.** $3x, x\neq 2$

39. $\frac{1}{2y^2}, y\neq -\frac{3}{2}$ **41.** $\frac{x-5}{x+5}, x\neq 5$ **43.** $\frac{2w^2-1}{w^2+8}$

45. $\frac{5x+4}{x(x+3)}, x\neq 2$ **47.** $\frac{(N^2+4)(N+2)}{8}, N\neq 2$

49. $\frac{1}{2t+1}, t\neq -4$ **51.** $\frac{x+3}{x-3}, x\neq 1$ **53.** $-\frac{y+x}{2}, x\neq y$

55. $\frac{x+1}{x-1}, x\neq \pm 1$ **57.** $\frac{(x+5)(x-3)}{(5-x)(x+3)}, x\neq \pm 2$

59. $\frac{x^2-xy+y^2}{2}, x\neq -y$ **61.** $\frac{2x}{9x^2-3x+1}, x\neq -\frac{1}{3}$

63. (a) $\frac{x^2(x+2)}{x^2+4}$ (x^2+4 does not factor)

(b) $\frac{x^2}{(x+2)(x-2)}$ (there are no more common factors)

65. (a) $\frac{(x-2)(x+1)}{x(x-1)}$ (no common factor)

(b) $\frac{x-2}{x}, x\neq -1$ (no common factor)

67. $u+v, u\neq v$ **69.** $\frac{E^2(R-r)}{(R+r)^3}$ **71.** $\frac{v+v_0}{t}, v\neq v_0, t\neq 0$

Exercises 6.6, page 214

1. $\frac{(x-2)(x+3)}{3x-1}, x\neq 5/2$ **3.** $\frac{3}{28}$ **5.** $6xy, y\neq 0$

7. $\frac{7}{18}$ **9.** $\frac{xy^2}{bz^2}, a\neq 0$ **11.** $4t, x\neq -3$

13. $3(u+v)(u-v), u\neq -2v$ **15.** $\frac{50}{3(a+4)}, a\neq -4$

17. $\frac{x^2-3}{x^2(x^2+3)}$ **19.** $\frac{3x}{5a}, x\neq 0, 3, -\frac{1}{2}$ and $a\neq 0$

21. $\frac{(x+1)(x-1)(x-4)}{4(x+2)}, x\neq 0$

23. $\frac{x^2}{a+x}, x\neq -a, b\neq \frac{cx}{2}$ **25.** $\frac{15}{4}, a\neq -\frac{5}{7}, -\frac{11}{4}$

27. $\frac{3}{4x+3}, x\neq 5, 1, -\frac{7}{2}$

29. $\frac{2(3T+N)(5V-6)}{(V-7)(4T+N)}, T\neq \frac{N}{2}, V\neq -\frac{5}{2}$ **31.** $\frac{7x^4}{3a^4}, x\neq 0$

33. $\frac{4t(2t-1)(t+5)}{(2t+1)^2}, t\neq 5, -\frac{1}{2}$

35. $\frac{x+y}{2} x\neq \pm y, x^2+xy+y^2\neq 0$

37. $(x+y)(3p+7q), a\neq -b, p\neq q$

39. $\frac{x-2}{6x}, x\neq -2, 0$ **41.** $-\frac{2(2x-1)}{5(x-2)}, x\neq -\frac{1}{2}, -2$

43. $\frac{c(\lambda+\lambda_0)(\lambda-\lambda_0)}{\lambda^2+\lambda_0^2}, \lambda_0\neq 0$ **45.** $\frac{\pi}{2}, a, b, \lambda\neq 0, a\neq -b$

47. $v=\sqrt{\frac{GM}{r}}, m,r\neq 0$

Exercises 6.7, page 220

1. $12a^2b^3$ **3.** $\frac{12}{s(s+4)}$ **5.** 2 **7.** $\frac{8}{x}$ **9.** $\frac{5}{4}$

11. $\frac{3+7ax+8x}{4x}$ **13.** $\frac{ax-b}{x^2}$ **15.** $\frac{30+ax^2}{25x^3}$

17. $\frac{14-a^2}{10a}$ **19.** $\frac{xy+2y+7x-8}{4xy}$ **21.** $\frac{3x+8}{6(x-1)}$

23. $\frac{5-3x}{2x(x+1)}$ **25.** $\frac{-3}{4(s-3)}$ **27.** $\frac{7R+6}{3(R+3)(R-3)}$

29. $\frac{2x-5}{(x-4)^2}$ **31.** $\frac{-4}{(v+1)(v-3)}$ **33.** $\frac{9x^2+x-2}{(3x-1)(x-4)}$

35. $\frac{-t(2t^2-3t-15)}{(t+3)^2(t-3)(t+2)}$ **37.** $\frac{-w(2w^2-w+1)}{(w+1)(w^2-w+1)}$

39. $\frac{1}{x-1}, x\neq 0$ **41.** $\frac{x-y}{y}, x\neq 0, -y$

43. $-\frac{(3x+4)(x-1)}{2x}, x\neq 0, 1, -1$ **45.** $\frac{h}{(x+1)(x+h+1)}$

47. $\frac{-h(2x+h)}{x^2(x+h)^2}$ **49.** $\frac{r^2+y^2-rx}{r^2}$ **51.** $\frac{2a-1}{a^2}$

53. $\frac{a^2+2a-1}{a+1}$ **55.** $ab, a\neq -b$ **57.** $\frac{2mn}{m^2+n^2}$

59. $\frac{3(H-H_0)}{4\pi H}$ **61.** $\frac{n(2n+1)}{2(n-1)(n+2)}$

63. $\frac{P^2(9x^2+L^2)}{4L^4}$ **65.** $\frac{Ls+R}{CLs^2+CRs+1}$

67. $\frac{\omega^2L^2+\omega^4L^2c^2R^2-2\omega^2LcR^2+R^2}{R^2\omega^2L^2}$

69. $\frac{10v}{v^2-w^2}$ h

Exercises 6.8, page 225

1. $\frac{2}{b-1}$ **3.** $\frac{arv^2}{M(2a-r)}$ **5.** 4 **7.** −3 **9.** $\frac{13}{2}$

11. $\frac{16}{21}$ **13.** −9 **15.** $-\frac{2}{13}$ **17.** $\frac{5}{3}$ **19.** −2

21. $-\frac{1}{2}$ **23.** $\frac{37}{6}$ **25.** 7 **27.** $\frac{63}{8}$ **29.** No solution

31. $\frac{2}{3}$ **33.** $\frac{3b}{1-2b}$ **35.** $\frac{(2b-1)(b+6)}{2(b-1)}$

37. $\frac{2(s-s_0)-tv_0}{t}$ **39.** $\frac{8.0(V-6.0)}{15.6-V}$ **41.** $\frac{jX}{1-g_mZ}$

43. $\frac{pV^3-pV^2b+aV-ab}{RV^2}$ **45.** $\frac{CC_3+CC_2-C_2C_3}{C_2-C}$

47. $\frac{f(1-n)(R_2)}{nf-f-R_2}$ **49.** 3.08 h **51.** 3.24 min

53. 220 m **55.** 80.1 km/h **57.** 2.23 V

59. $A=3, B=-2$

Review Exercises for Chapter 6, page 227

1. $12ax+15a^2$ **3.** $4a^2-49b^2$ **5.** $4a^2+4a+1$

7. $b^2+3b-28$ **9.** $2x^2-13x-45$ **11.** $4c^{2a}-d^2$

13. $3(s+3t)$ **15.** $a^2(x^2+1)$

17. $b^x(Wb+12)(Wb-12)$

19. $(4x+8+t^2)(4x+8-t^2)$ **21.** $4(3t-1)^2$

23. $(5t+1)^2$ **25.** $(x+8)(x-7)$

27. $(t+3)(t-3)(t^2+4)$ **29.** $(2k-9)(k+4)$

31. $2(2x+5)(2x-7)$ **33.** $(5b-1)(2b+5)$

35. $4(x+4y)(x-4y)$ **37.** $2(5-2y^2)(25+10y^2+4y^4)$

39. $(2x+3)(4x^2-6x+9)$ **41.** $(a-3)(b^2+1)$

43. $(x+5)(n-x+5)$ **45.** $\frac{16x^2}{3a^2}, a, x, y \neq 0$

47. $\frac{3x+1}{2x-1}, x \neq \frac{3}{2}$ **49.** $\frac{16}{5x(x-y)}, x \neq 0, -y$

51. $\frac{-6}{L-5}, L \neq \pm 3$ **53.** $\frac{x+2}{2x(7x-1)}, x \neq 0, -2$

55. $\frac{1}{x-1}, x \neq 0$ **57.** $\frac{16x-15}{36x^2}$ **59.** $\frac{5y+6}{2xy}$

61. $\frac{-2(2a+3)}{a(a+2)}$ **63.** $\frac{4x^2-x+1}{2x(x+3)(x-1)}$

65. $\frac{12x^2-7x-4}{2(4x-1)(x+1)(x-1)}$

67. $\frac{x^3+6x^2-2x+2}{x(x+3)(x-1)}$

69. $(x+\sqrt{5})(x-\sqrt{5})$ **71.** $(\sqrt{x}+1)(\sqrt{x}-1)$

73. $x\left(1+\frac{y}{x}\right)$ **75.** $2\left(\frac{x}{2}-y\right)$ **77.** 2

79. $-\frac{(a-1)^2}{2a}$ **81.** $\frac{10}{3}$ **83.** −6

85. (a) Changes the sign of the fraction.
(b) Leaves the sign of the fraction unchanged.

87. $\frac{1}{4}[(x+y)^2-(x-y)^2]$
$=\frac{1}{4}[x^2+2xy+y^2-(x^2-2xy+y^2)]$
$=\frac{1}{4}[x^2+2xy+y^2-x^2+2xy-y^2]$
$=\frac{1}{4}[4xy]$
$=xy$

89. $2zS^2+2zS$ **91.** $4b^2+4bn\lambda-4b\lambda+n^2\lambda^2-2n\lambda^2+\lambda^2$

93. $(T_2-T_1)(c+R)$ **95.** $R(3R-4r)$

97. $8n^6+36n^5+66n^4+63n^3+33n^2+9n+1$

99. $10aT-10at+aT^2-2aTt+at^2$

101. $4(3x^2+12x+16)$ **103.** $\frac{12wv^2\pi^2D}{gn^2t}$

105. $\frac{R^2+r^2}{2}, R \neq r, -r$ **107.** $\frac{120-60d^2+5d^4-d^6}{120}$

109. $\frac{4k^2+k-2}{4k(k-1)}$ **111.** $\frac{4r^3-3ar^2-a^3}{4r^3}$

113. $\frac{c^2(u^2-2gx)}{1-u^2c^2+2gc^2x}$ **115.** $\frac{W}{g(h_2-h_1)}$ **117.** $\frac{RHw}{w-RH}$

119. $\frac{p^2}{2E-2V_0-(m+M)V^2}$ **121.** $-\frac{(s^2b^2m+kL^2)}{sb^2}$

123. $\frac{i}{sV-V_0}$ **125.** 3.43 h **127.** 600 Hz

129. 11.3 **131.** 18.0 Ω

Chapter 7: Quadratic Equations

Exercises 7.1, page 234

1. $-\frac{1}{2}, 4$ **3.** $a=1, b=-2, c=-4$

5. No x^2 term so it is not quadratic. **7.** $a=1, b=-1, c=0$

9. −5, 5 **11.** $-\frac{3}{2}, \frac{3}{2}$ **13.** 7, −2 **15.** 4, 3 **17.** $0, \frac{5}{2}$

19. $\frac{1}{3}, -\frac{1}{3}$ **21.** $4, \frac{1}{3}$ **23.** $-1, \frac{4}{7}$ **25.** $\frac{2}{3}, \frac{3}{2}$ **27.** $\frac{1}{2}, -\frac{3}{2}$

29. $\frac{b}{2}, \frac{-2b}{3}$ **31.** $\frac{5}{2}, -\frac{9}{2}$ **33.** 0, −2 **35.** $b-a, -b-a$

37. $-a-b, b-a$ **39.** $\frac{1}{2}+3=\frac{7}{2}=-\frac{-7}{2}=-\frac{b}{a}$

41. −6.00 A or 2.00 A **43.** 2:00 a.m., 10:00 a.m.

45. −1, 0, 1 **47.** $4, \frac{3}{2}$ **49.** −9

51. 3 N/cm, 6 N/cm **53.** 30 km/h going, 40 km/h returning

Exercises 7.2, page 237

1. $-3 \pm \sqrt{17}$ **3.** ±5 **5.** $\pm\sqrt{7}$ **7.** −3, 7

9. $-3 \pm \sqrt{7}$ **11.** 3, −5 **13.** −2, −1

15. $2 \pm \sqrt{2}$ **17.** −5, 3 **19.** $-3, \frac{1}{2}$ **21.** $\frac{1}{2} \pm \frac{\sqrt{33}}{6}$

23. $\frac{1}{4} \pm \frac{\sqrt{17}}{4}$ **25.** $1 \pm \frac{\sqrt{5}}{5}$ **27.** $-\frac{1}{3}$ (double root)

29. $-b \pm \sqrt{b^2-c}$ **31.** $(x+3)^2+2^2$

33. 5.00°C, 15.0°C **35.** 5.61 m

Exercises 7.3, page 241

1. $-2, -3$ **3.** $3, -5$ **5.** $-2, -1$ **7.** $2 \pm \sqrt{2}$

9. $-5, 3$ **11.** $-3, \frac{1}{2}$ **13.** $\frac{1}{6}(3 \pm \sqrt{33})$

15. $\frac{1 \pm \sqrt{17}}{4}$ **17.** $-\frac{8}{5}, \frac{5}{6}$

19. $\frac{7 \pm \sqrt{-11}}{10}$ (complex roots; no real roots)

21. $\frac{1}{6}\left(1 \pm \sqrt{109}\right)$ **23.** $\frac{11}{5}, -\frac{11}{5}$ **25.** $\frac{3}{4}, -\frac{5}{8}$

27. $-0.540, 0.740$ **29.** $-0.256, 2.43$ **31.** $-c \pm \sqrt{c^2 + 1}$

33. $\frac{b + 1 \pm \sqrt{4ab^2 - 3b^2 + 2b + 1}}{2b^2}$

35. Unequal, no real roots **37.** Real, irrational, and unequal

39. 4 **41.** $\pm 2, \pm 1$ **43.** 4.38 cm

45. $\frac{15 \pm \sqrt{33}}{16}L$ **47.** 1.618 **49.** $\frac{-R \pm \sqrt{R^2 - 4L/C}}{2L}$

51. 1.78 cm **53.** $l = 23.8$ m, $w = 11.0$ m **55.** 1.11 m

57. 2.25 cm

Exercises 7.4, page 245

1.

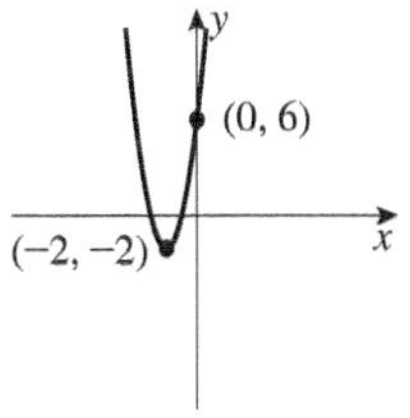

3.

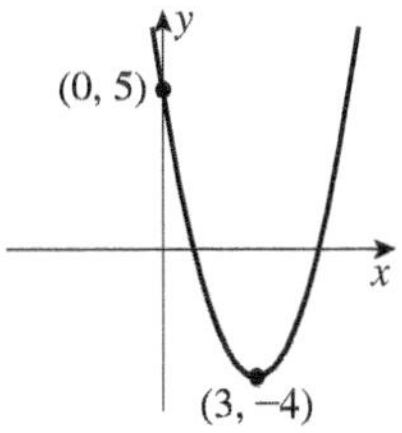

5.

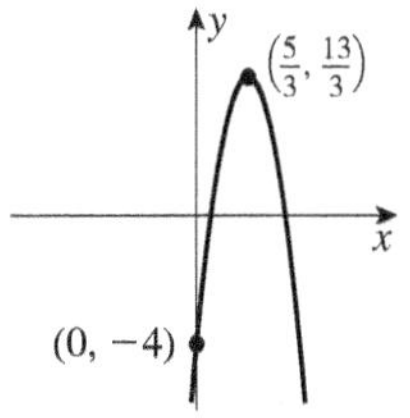

7.

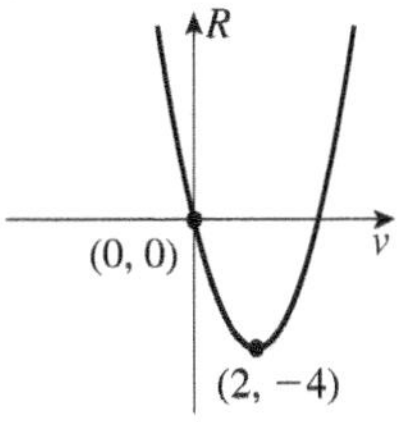

9.

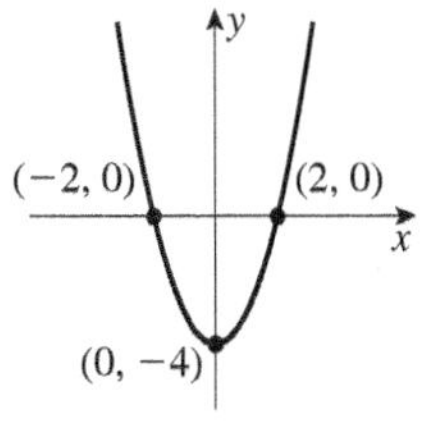

11.

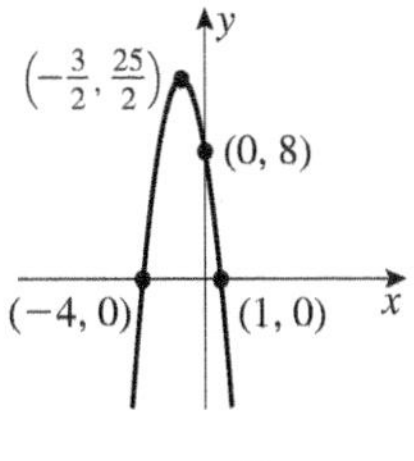

13.

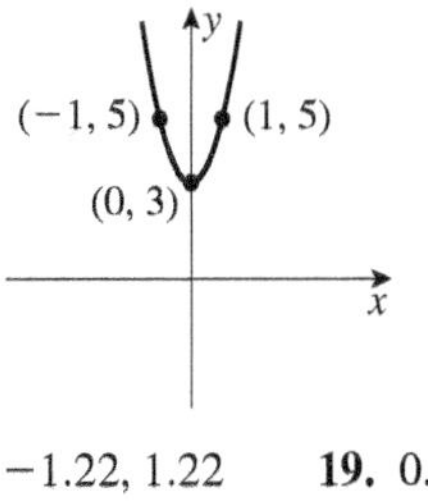

15.

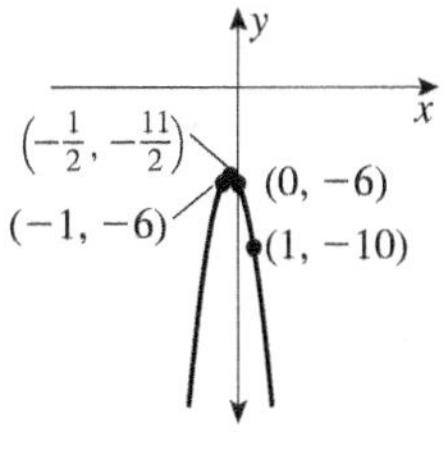

17. $-1.22, 1.22$ **19.** $0.532, 3.14$

21. No real solutions **23.** $-2.41, 1.24$

25. The parabola $y = x^2 + 3$ is shifted up $+3$ units (minimum point $(0, 3)$).
The parabola $y = x^2 - 3$ is shifted down -3 units (minimum point $(0, -3)$).

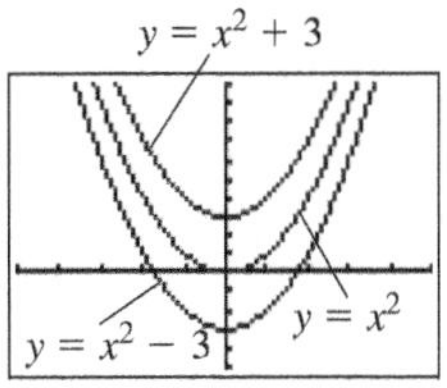

27. $y = (x - 2)^2 + 3$ is $y = x^2$ shifted right 2 and up 3.
$y = (x + 2)^2 - 3$ is $y = x^2$ shifted left 2 and down 3.

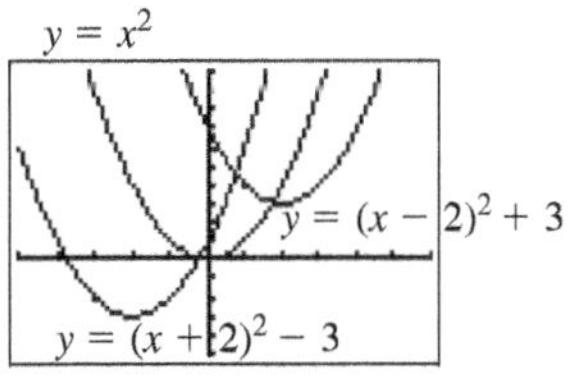

29.

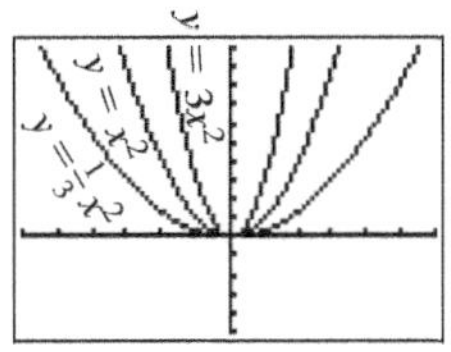

The graph of $y = 3x^2$ is the graph of $y = x^2$ narrowed. The graph of $y = \frac{1}{3}x^2$ is the graph of $y = x^2$ broadened.

31. 7 **33.** -2

35.

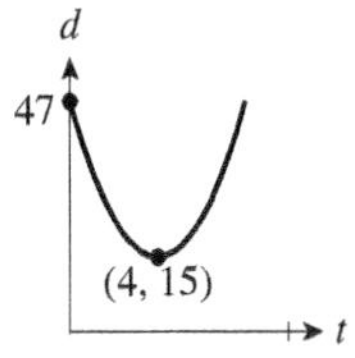

37.

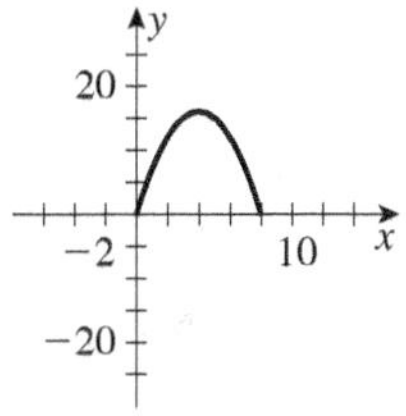

39.

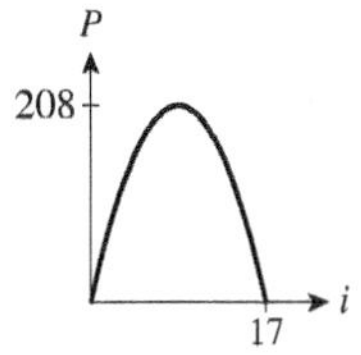

41. 6.9 s **43.** 4.53 cm

45. 21.6 m by 92.6 m, or 32.4 m by 61.7 m

Review Exercises for Chapter 7, page 246

1. $-4, 1$ **3.** $3, 7$ **5.** $-4, \frac{1}{3}$ **7.** $\frac{1}{2}, \frac{5}{3}$

9. $0, \frac{9}{2}$ **11.** $-\frac{3}{2}, \frac{7}{2}$ **13.** $-10, 11$

15. $-1 \pm \sqrt{7}$ **17.** $\frac{9}{2}, -4$ **19.** $\frac{3 \pm \sqrt{41}}{8}$

21. $\frac{-2.3 \pm \sqrt{-40.91}}{4.2}$ (no real roots)

23. $\frac{-2 \pm \sqrt{58}}{6}$ **25.** $-2 \pm 2\sqrt{2}$ **27.** $\frac{-4 \pm \sqrt{10}}{3}$

29. $-1, \frac{5}{4}$ **31.** $\frac{-3 \pm \sqrt{-47}}{4}$ (complex roots)

33. $\dfrac{-1 \pm \sqrt{-1}}{a}$ ($a \neq 0$, complex roots) **35.** $\dfrac{-3 \pm \sqrt{9 + 4a^2}}{2a}$

37. $-5, 6$ **39.** $\dfrac{1 \pm \sqrt{33}}{4}$ **41.** $3 \pm \sqrt{7}$ **43.** 0

45.

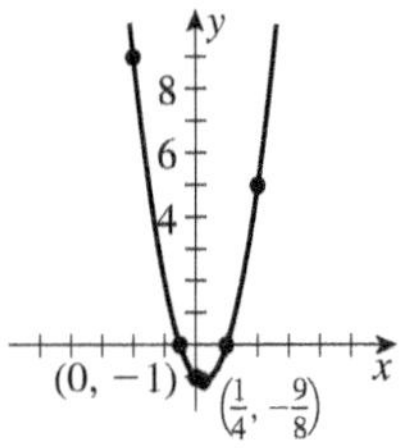

47.

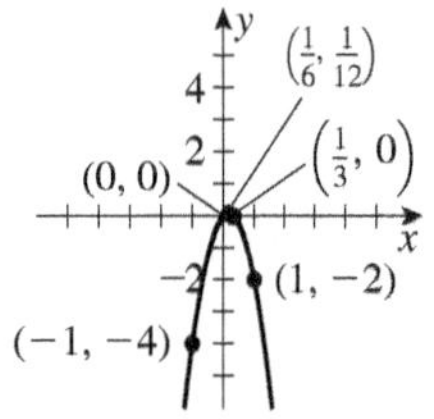

49. $-1.69, 1.19$ **51.** No real roots. **53.** -4

55. $0, L$ **57.** 17 **59.** 95.8 °C **61.** 0.55 s, 2.22 s

63. 8000 **65.** $\dfrac{-\pi h + \sqrt{\pi^2 h^2 + 2\pi A}}{2\pi}$

67. $\dfrac{r + 1 \pm \sqrt{(r + 1)^2 - 4rp_2}}{2r}$

69.

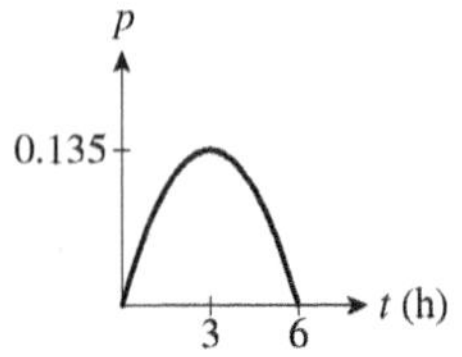

71. 81.4 m and 61.4 m **73.** 9.88 cm

75. 4.00 cm thick, 15.3 cm long, 11.3 cm wide

77. 40.7 cm × 55.2 cm **79.** 25

81.

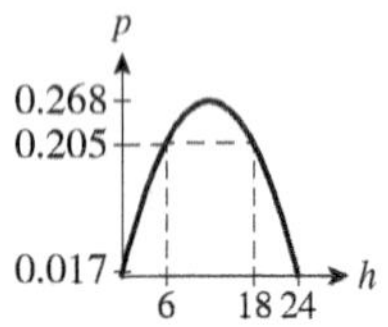

83. 3.56

Chapter 8: Trigonometric Functions of Any Angle

Exercises 8.1, page 252

1. $-, +, +, -, +, -$ **3.** $-, +$ **5.** $-, -$ **7.** $+, +$

9. $-, +$ **11.** $+, -$ **13.** $-, -$ **15.** $+, +$

17. $\sin\theta = \dfrac{1}{\sqrt{5}}$, $\cos\theta = \dfrac{2}{\sqrt{5}}$, $\tan\theta = \dfrac{1}{2}$,

$\csc\theta = \sqrt{5}$, $\sec\theta = \dfrac{\sqrt{5}}{2}$, $\cot\theta = 2$

19. $\sin\theta = \dfrac{-3}{\sqrt{13}}$, $\cos\theta = \dfrac{-2}{\sqrt{13}}$, $\tan\theta = \dfrac{3}{2}$,

$\csc\theta = -\dfrac{\sqrt{13}}{3}$, $\sec\theta = -\dfrac{\sqrt{13}}{2}$, $\cot\theta = \dfrac{2}{3}$

21. $\sin\theta = \dfrac{12}{13}$, $\cos\theta = -\dfrac{5}{13}$, $\tan\theta = -\dfrac{12}{5}$,

$\csc\theta = \dfrac{13}{12}$, $\sec\theta = -\dfrac{13}{5}$, $\cot\theta = -\dfrac{5}{12}$

23. $\sin\theta = \dfrac{-2}{\sqrt{29}}$, $\cos\theta = \dfrac{5}{\sqrt{29}}$, $\tan\theta = -\dfrac{2}{5}$,

$\csc\theta = -\dfrac{\sqrt{29}}{2}$, $\sec\theta = \dfrac{\sqrt{29}}{5}$, $\cot\theta = -\dfrac{5}{2}$

25. I, II **27.** II, IV **29.** II, III **31.** II **33.** II

35. IV **37.** I **39.** II **41.** $-$ **43.** $-$

Exercises 8.2, page 258

1. (a) $-\sin 20° = -0.342$
(b) $-\tan 30° = -0.577$
(c) $-\cos 85° = -0.0872$
(d) $-\cot 60° = -0.577$
(e) $\sec 16° = 1.04$
(f) $\sin 37° = 0.602$

3. $\sin 25°, -\cos 40°$ **5.** $-\tan 75°, -\csc 32°$

7. $\sec 65°, -\sin 20°$ **9.** $-\sin 15° = -0.259$

11. $-\cos 73.7° = -0.2807$ **13.** $\sec 31.67° = 1.1750$

15. $\tan 70.9° = 2.8878$ **17.** 0.4586 **19.** -0.76199

21. -3.9096 **23.** -0.6188 **25.** 237.99°, 302.01°

27. 66.40°, 293.60° **29.** 90.7°, 270.7° **31.** 119.5°

33. 263° **35.** 312.95° **37.** 299.24° **39.** -0.7002

41. -0.777 **43.** $<$ **45.** $=$ **47.** -0.9659

49. $\cot\theta$ **51.** 0.0183 A **53.** 12.6 cm

Exercises 8.3, page 262

1. 160° **3.** $\dfrac{\pi}{12}, \dfrac{2\pi}{3}$ **5.** $\dfrac{5\pi}{12}, \dfrac{11\pi}{6}$ **7.** $\dfrac{7\pi}{6}, \dfrac{11\pi}{20}$

9. $4\pi, -\dfrac{\pi}{20}$ **11.** 108°, 270° **13.** 100°, 315°

15. 70°, 150° **17.** $-12°$, 27° **19.** 1.47 rad

21. 4.40 rad **23.** -5.821 rad **25.** 8.351 rad

27. 43.0° **29.** 195.2° **31.** 710° **33.** $-940.8°$

35. 0.7071 **37.** 3.732 **39.** -0.8660

41. -8.3265 **43.** 0.9056 **45.** -0.890

47. -2.09 **49.** 0.149 **51.** 0.3141, 2.827

53. 2.932, 6.074 **55.** 0.8309, 5.452

57. 2.442, 3.841 **59.** 1.25 **61.** -0.3827

63. $-\cot\theta$ **65.** 612 mil **67.** 10.1 rad

69. 0.030 J **71.** 2900 m

Exercises 8.4, page 268

1. 2.36 cm **3.** 14.4 m/s **5.** 5.97 cm **7.** 385 mm

9. 0.1849 km^2 **11.** 48 cm^2 **13.** 0.0647 m **15.** 8.21 m

17. 43.1 m **19.** 1.07 rad, 8.56 cm^2 **21.** 12:32:44

23. 627 m^2 **25.** 0.52 rad/s **27.** 34.73 m^2 **29.** 0.704 m

31. 22.6 m^2 **33.** 369 m^3 **35.** 0.4 km **37.** 5.42 m/s

39. 188 rad/s^2 **41.** 3.87 km/s **43.** 75.8 r/min

45. 9.41 m **47.** 1260 mm/s **49.** 649 r/min

51. 1070 km/h **53.** 11.0×10^3 m^2 **55.** 14.9 m^3

57. The sequences both converge to 1. **59.** 1.15×10^8 km

Review Exercises for Chapter 8, page 271

1. $\sin\theta = \dfrac{4}{5}$, $\cos\theta = \dfrac{3}{5}$, $\tan\theta = \dfrac{4}{3}$,

$\csc\theta = \dfrac{5}{4}$, $\sec\theta = \dfrac{5}{3}$, $\cot\theta = \dfrac{3}{4}$

3. $\sin\theta = -\dfrac{2}{\sqrt{53}}$, $\cos\theta = \dfrac{7}{\sqrt{53}}$, $\tan\theta = -\dfrac{2}{7}$,
$\csc\theta = -\dfrac{\sqrt{53}}{2}$, $\sec\theta = \dfrac{\sqrt{53}}{7}$, $\cot\theta = -\dfrac{7}{2}$

5. $-\cos 48°$, $\tan 14°$ **7.** $-\sin 71°$, $\sec 15°$ **9.** $\dfrac{2\pi}{9}, \dfrac{17\pi}{20}$

11. $\dfrac{34\pi}{15}, \dfrac{9\pi}{8}$ **13.** 252°, 130° **15.** 12°, 330° **17.** 32.1°

19. −2067° **21.** 1.78 rad **23.** 0.3534 rad **25.** 4.5736 rad

27. −11.10 rad **29.** −0.5388 **31.** −0.466 **33.** −1.0799

35. −0.04885 **37.** −1.638 **39.** 4.1398 **41.** −0.5878

43. −0.8660 **45.** 0.5569 **47.** 1.197 **49.** 10.30°, 190.30°

51. 118.23°, 241.77° **53.** 0.5759, 5.707 **55.** 4.187, 5.238

57. 227.78° **59.** 246.78° **61.** 10.8 cm **63.** 3.23 rad

65. 14.2 m **67.** 2100 cm^2

69. $A = \frac{1}{2}\theta r^2 - \frac{1}{2}r(r\sin\theta) = \frac{1}{2}r^2(\theta - \sin\theta)$, 3.66 cm^2

71. 1040 m^2 **73.** 0.0562 W **75.** 45.8°

77. 21.8 r/min **79.** 3600 km/h

81. (a) 6670 km
(b) 9170 km; The distance over the north pole is shorter.

83. 47.1 m/s **85.** 2.70 m^2 **87.** 302 m/s

89. 138 m^2 **91.** 3.58×10^5 km **93.** 5970 km/h

Chapter 9: Vectors and Oblique Triangles

Exercises 9.1, page 278

1.
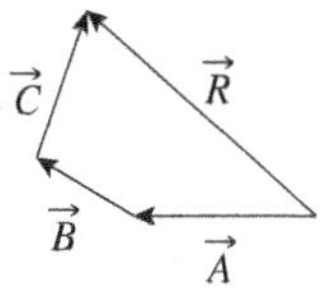

3.
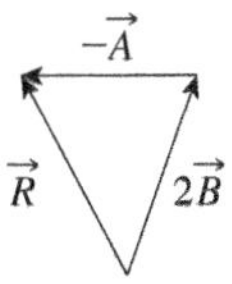

5. (a) Scalar, no direction given (b) Vector, magnitude and direction both given

7. (a) Vector; it has magnitude and direction
(b) Scalar; it has magnitude but not direction

9.
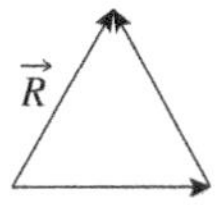

11.
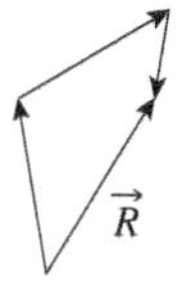

13.
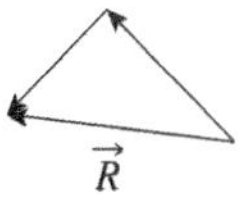

15. 5.6 cm, 50°
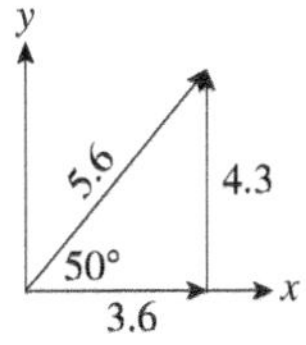

17. 4.3 cm, 156°
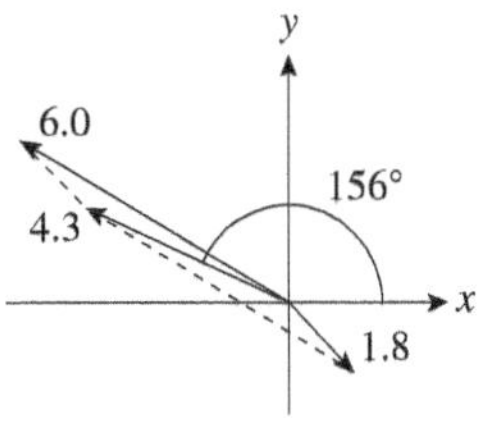

19.
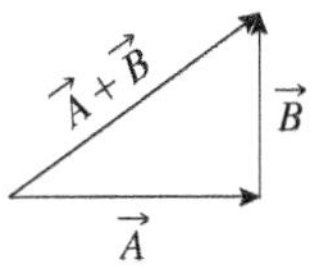

21.
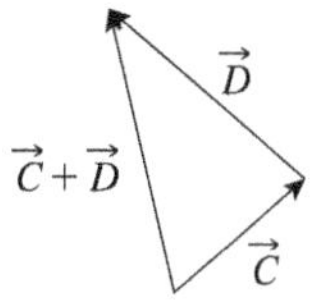

23.
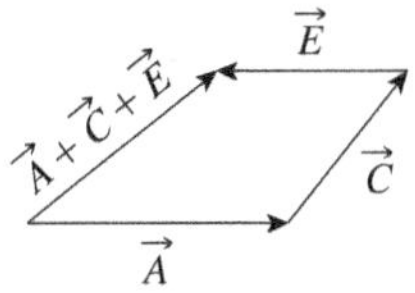

25.
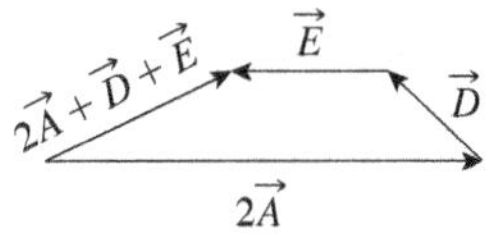

27.
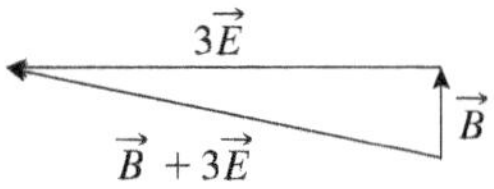

29.
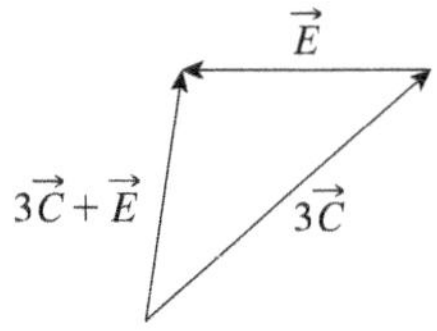

31.
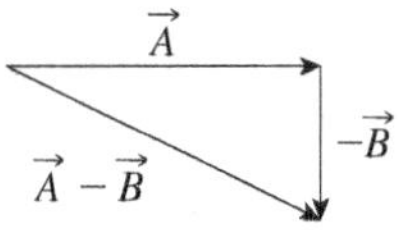

33.
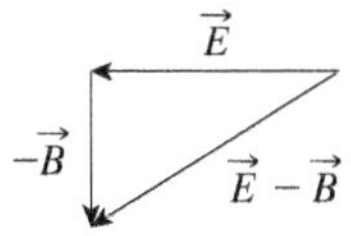

35.
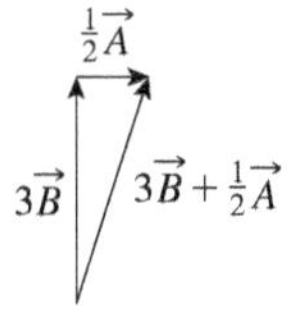

37.
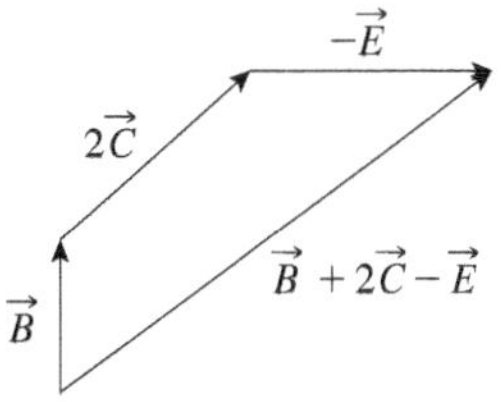

39.
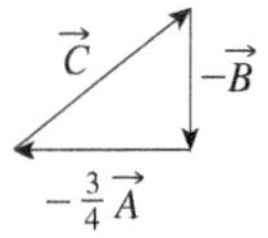

41.
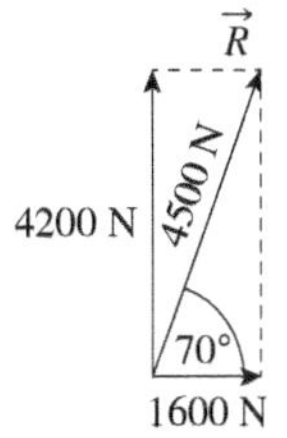

43.
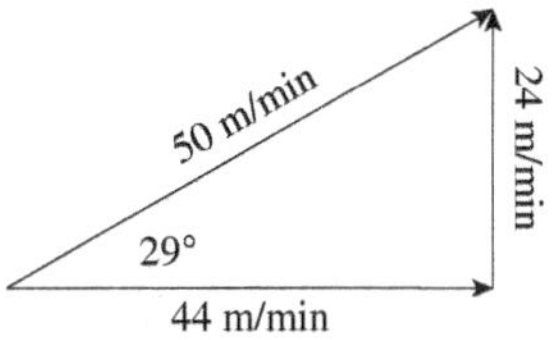

45.
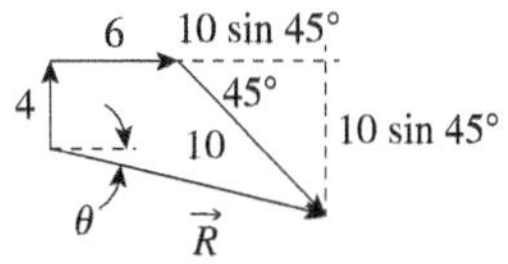

47.
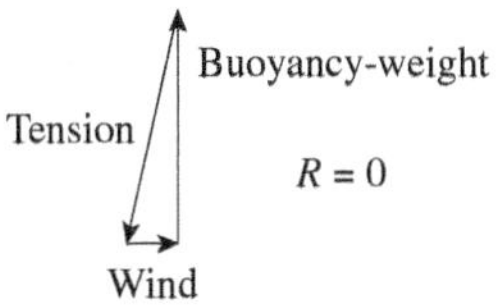

Exercises 9.2, page 281

1. $V_x = -11.6$, $V_y = -8.46$ **3.** $A_x = 0$, $A_y = -375.4$

5. 662, 352 **7.** −349, −664 **9.** −750, 0

11. 3.22 N, 7.97 N **13.** −62.9 m/s, 44.1 m/s

15. 0 mm/s^2, −9040 mm/s^2 **17.** −2.53 mN, −0.788 mN

19. −0.8088 dm, 0.3296 dm **21.** 23.9 km/h, 7.43 km/h

23. 12 400 N, 757 N **25.** 116 km/h **27.** 0.237 m, 1.18 m

29. 89 N, −190 N **31.** 290 N

33. 0.57 (km/h)/m, 0.48 (km/h)/m

Exercises 9.3, page 286

1. $R = 1660$, $\theta = 73.1°$ **3.** $R = 24.2$, $\theta = 52.6°$

5. $R = 7.781$, $\theta = 66.63°$ (with $\vec{A}$) **7.** $R = 10.0$, $\theta = 58.8°$

9. $R = 2.74, \theta = 111.0°$ **11.** $R = 2130, \theta = 107.7°$

13. $R = 7052, \theta = 349.82°$ **15.** $R = 29.2, \theta = 10.8°$

17. $R = 5920, \theta = 88.4°$ **19.** $R = 27.27, \theta = 33.14°$

21. $R = 50.2, \theta = 50.3°$ **23.** $R = 0.242, \theta = 285.9°$

25. $R = 532, \theta = 95.7°$ **27.** $R = 235$ N, $\theta = 121.7°$

29. $R = 68\,000$ N, $\theta = 345°$ **31.** 40 N

33. 24.3 kg · m/s **35.** $R = 230$ N, $\theta = 7.3°$

Exercises 9.4, page 291

1. 47.18 km, 10.31° N of E **3.** 39.6 N, 29.5° from 34.5-N force

5. 11 000 N, 44° above horizontal **7.** 3070 m, 17.8° S of W

9. 229.5 m, 72.81° N of E **11.** 3750 N, 56.3° from 2080-N force

13. 25.3 km/h, 29.6° S of E **15.** 9.6 m/s^2, 7.0° from vertical

17. 781 N, 9.3° above horizontal

19. 540 km/h, 6.2° from direction of plane

21. 29 180 km/h, 0.03° from direction of launcher

23. 27.0 km at a bearing of 192.4° **25.** 9.3 km/h

27. 910 N **29.** 584 N **31.** 138 km, 65.0° N of E

33. 79.0 m/s, 75.6° from vertical

35. 4.06 A/m, 11.6° with magnet

Exercises 9.5, page 298

1. $b = 76.01, C = 40.77°, c = 50.01$

3. $b = 38.1, C = 66.0°, c = 46.1$

5. $b = 2620, C = 108.0°, c = 2800$

7. $B = 12.20°, C = 149.57°, c = 7.448$

9. $A = 149.7°, a = 11\,050, C = 9.574°$

11. $A = 125.6°, a = 0.0776, c = .005\,66$

13. $A = 99.4°, b = 55.1, c = 24.4$

15. $A = 68.01°, a = 5520, c = 5376$

17. $A_1 = 61.36°, c_1 = 5.628, C_1 = 70.51°$; $A_2 = 118.64°, c_2 = 1.366, C_2 = 13.23°$

19. $A_1 = 107.3°, a_1 = 5280, C_1 = 41.3°$, $A_2 = 9.9°, a_2 = 950, C_2 = 138.7°$

21. No solution **23.** 373 m **25.** 9.4° **27.** 6.0 m

29. 880 N **31.** 20.4 m **33.** 13.94 cm **35.** 27 300 km

37. 77.4° with bank downstream

39. 62 m downhill at 62° with the hillside slope **41.** 105 m

Exercises 9.6, page 303

1. $A = 14.0°, B = 21.0°, c = 107$

3. $A = 50.3°, B = 75.7°, c = 6.31$

5. $A = 70.9°, B = 11.1°, c = 4750$

7. $A = 34.73°, B = 40.66°, C = 104.61°$

9. $A = 18.21°, B = 22.28°, C = 139.51°$

11. $A = 5.96°, B = 16.0°, c = 1150$

13. $A = 82.3°, b = 2160, C = 11.4°$,

15. $A = 36.24°, a = 97.22, B = 39.09°$

17. $A = 46.94°, B = 61.82°, C = 71.24°$

19. $A = 138°, B = 33.7°, C = 8.3°$

21. $b = 3700, C = 25°, c = 2400$ **23.** 1.8 m **25.** $c^2 = a^2 + b^2$

27. 69.4 km **29.** 47° **31.** 57.3°, 141.7° **33.** 16.5 mm

35. 5.09 km/h, 91.6° with respect to the land

37. 17.8 km **39.** 13.6 m **41.** 16.4 cm

Review Exercises for Chapter 9, page 305

1. $A_x = 57.4, A_y = 30.5$ **3.** $A_x = -0.7485, A_y = -0.5357$

5. $R = 602, \theta = 32.9°$ with B **7.** $R = 5960, \theta = 33.60°$ with A

9. $R = 963, \theta = 8.53°$ **11.** $R = 26.12, \theta = 146.03°$

13. $R = 71.93, \theta = 336.5°$ **15.** $R = 0.994, \theta = 359.6°$

17. $b = 181, C = 64.0°, c = 175$

19. $A = 21.2°, b = 34.8, c = 51.5$

21. $a = 17\,340, b = 24\,660, C = 7.99°$

23. $A = 39.88°, a = 5193, C = 30.03°$

25. $A_1 = 54.8°, a_1 = 12.7, A_2 = 12.0°, a_2 = 3.24$, $B_1 = 68.6°, B_2 = 111.4°$

27. $A = 32.3°, b = 267, C = 17.7°$

29. $A = 176.4°, B = 1.1°, c = 5.41$

31. $a = 1782, b = 1920, C = 16.00°$

33. $A = 37°, B = 25°, C = 118°$

35. $A = 20.6°, B = 35.4°, C = 124°$

37. Add the 3 forms of the law of cosines together; simplify; divide by $2abc$.

39. 1300 m **41.** $A_t = \frac{1}{2}ab;\ b = \dfrac{a \sin B}{\sin A}$; substitute

43. −155.7 N, 81.14 N **45.** 630 m/s **47.** 9.8 km

49. $R = 2700$ N, $\theta = 107°$ **51.** 6.1 m/s, 35° with horizontal

53. 0.11 N **55.** 1.5 pm **57.** 2.30 m, 2.49 m

59. 0.0390 km **61.** 52 900 km **63.** 2.65 km

65. 293 km **67.** 1270 m or 1680 m (ambiguous)

69. 810 N, 36° N of E **71.** 46 m

Chapter 10: Graphs of the Trigonometric Functions

Exercises 10.1, page 311

1. $y = 3 \cos x$

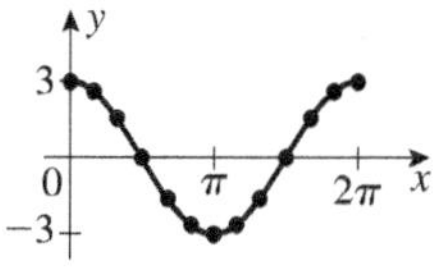

3. 0, −0.7,−1,−0.7, 0, 0.7, 1, 0.7, 0, −0.7,−1,−0.7, 0, 0.7, 1, 0.7, 0

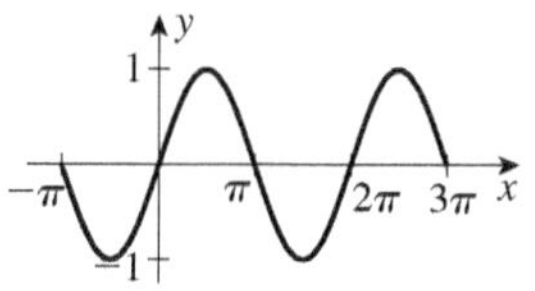

5. −3, −2.1, 0, 2.1, 3, 2.1, 0, −2.1, −3, −2.1, 0, 2.1, 3, 2.1, 0, −2.1, −3

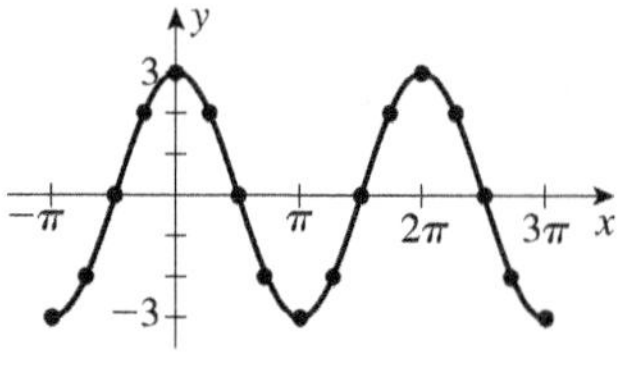

7. 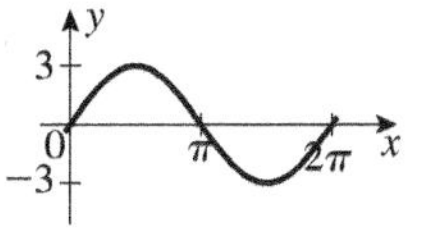

9. 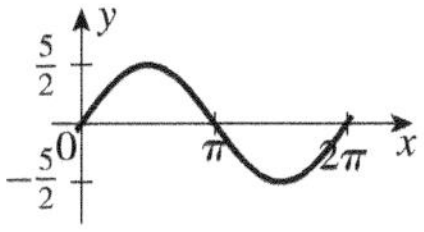

11. 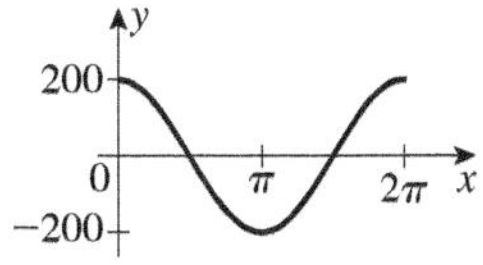

13. 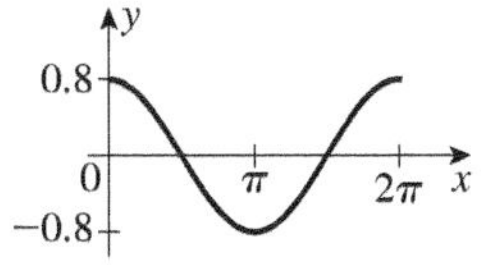

15. 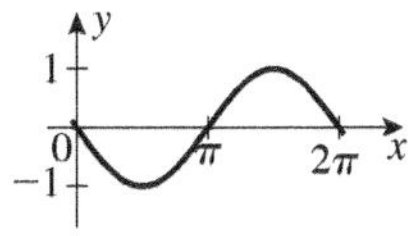

17. 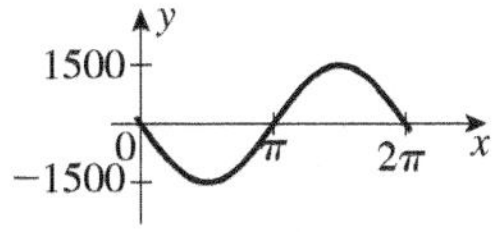

19. 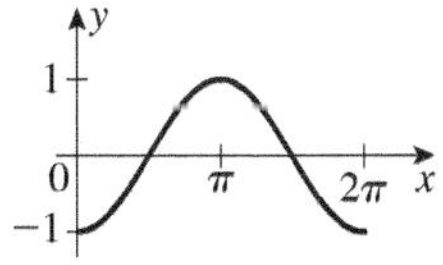

21. 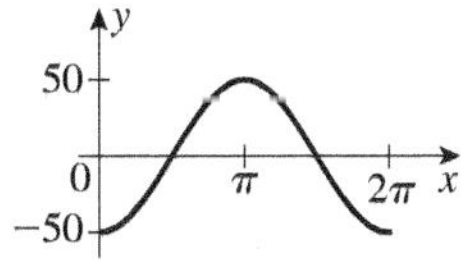

23. 0, 0.84, 0.91, 0.14, −0.76, −0.96, −0.28, 0.66

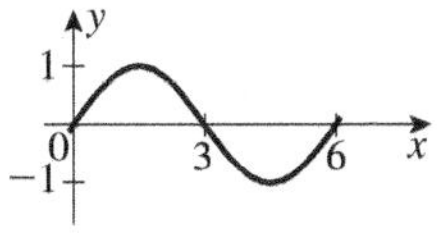

25. 12, 6.48, −4.99, −11.9, −7.84, 3.40, 11.5, 9.05

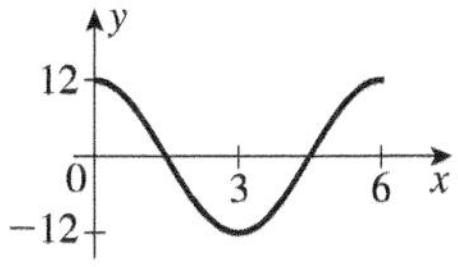

27. 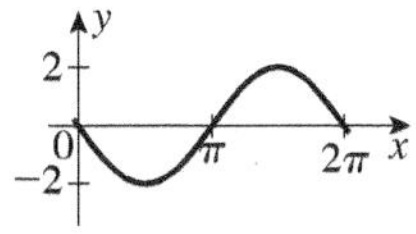

29. $y = -2\cos x$

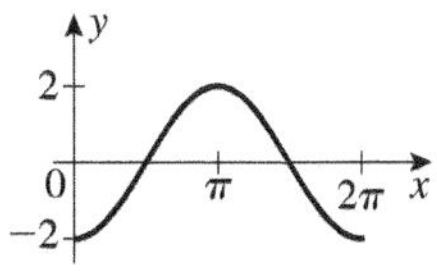

31. 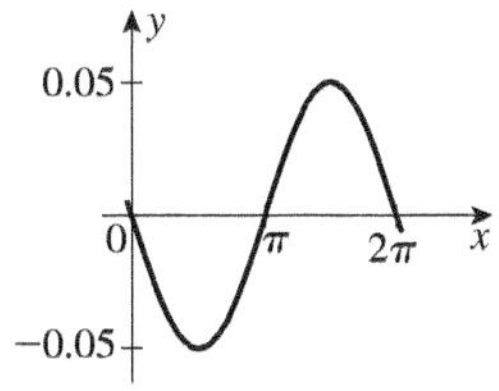

33. $y = 4\sin x$ **35.** $y = -1.5\cos x$

37. $y = -2.50\sin x$ **39.** $y = -2.50\cos x$

Exercises 10.2, page 315

1. $y = 3\sin 6x$

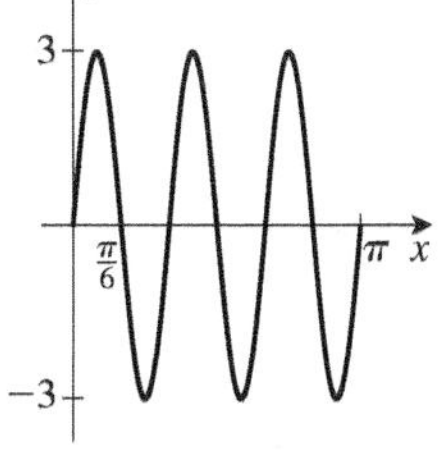

3. 2, $\pi/3$

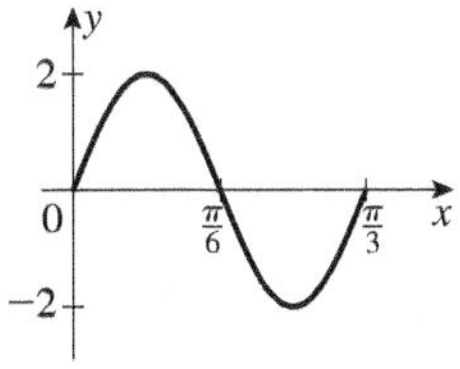

5. 3, $\pi/4$

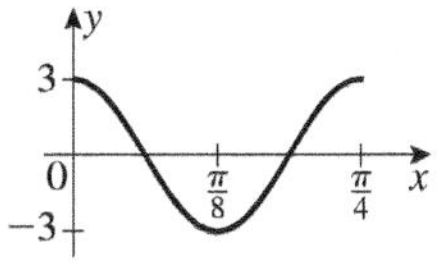

7. 2, $\pi/6$

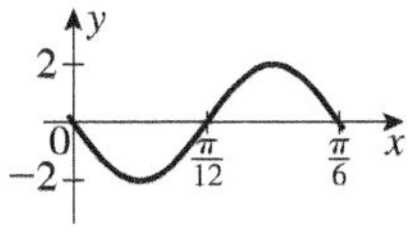

9. 1, $\pi/8$

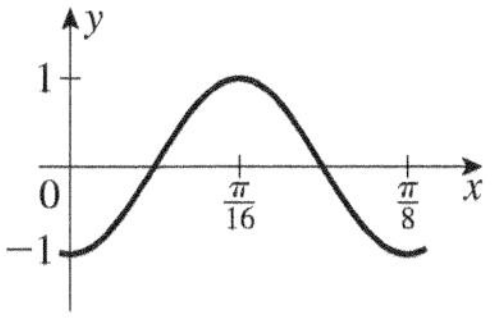

11. 520, 1

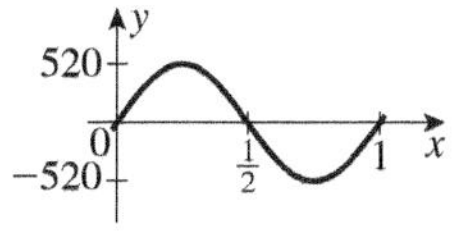

13. 3, $1/2$

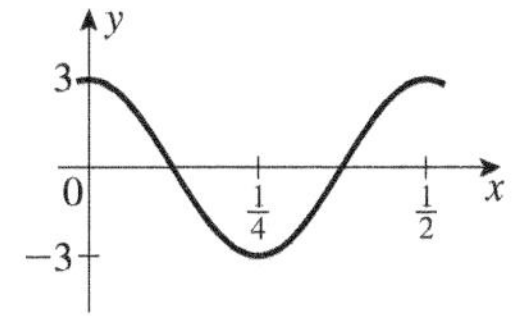

15. 15, 6π

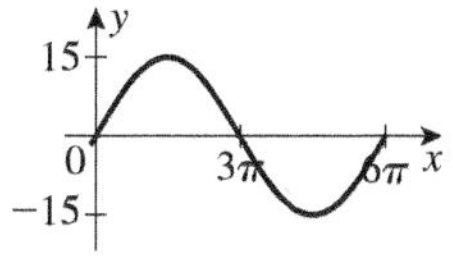

17. $1/2$, 3π

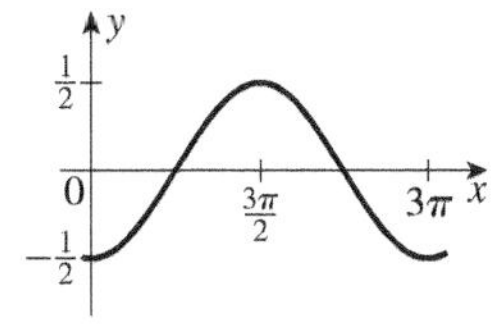

19. 0.4, 3

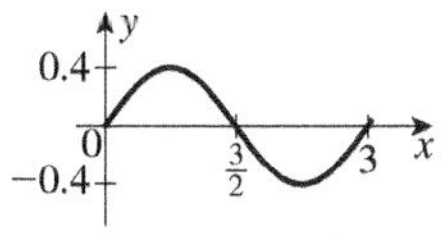

21. 3.3, $2/\pi$

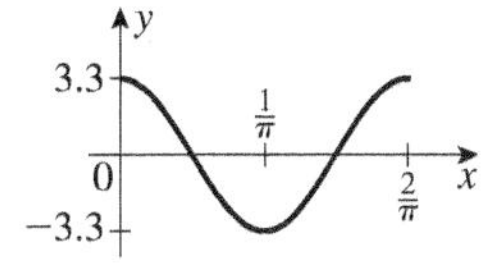

23. $y = \sin 6x$ **25.** $y = \sin 6\pi x$

27. 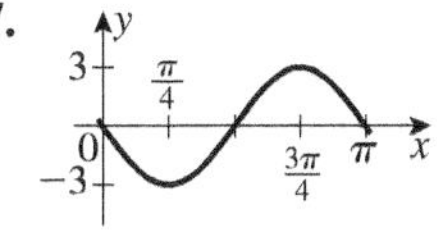

29. 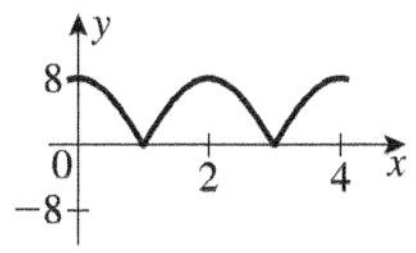

31. 2π

33. $y = -2\sin 2x$

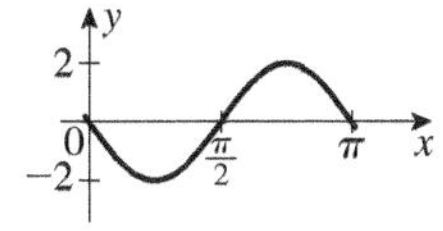

35. $y = 2\cos\frac{1}{2}x$

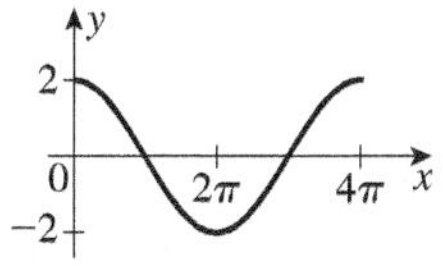

37. 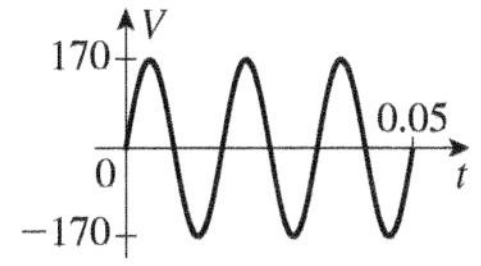

39. 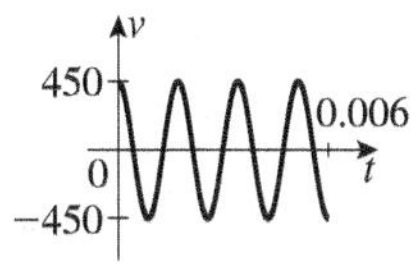

41. 105 MHz **43.** $y = \frac{1}{2}\cos 2x$ **45.** $y = -4\sin \pi x$

Exercises 10.3, page 320

1. $y = -\cos\left(2x - \frac{\pi}{6}\right)$

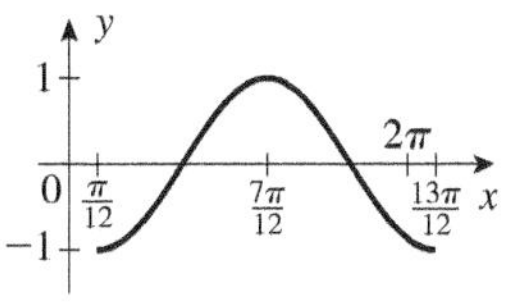

3. 1, 2π, $\frac{\pi}{6}$

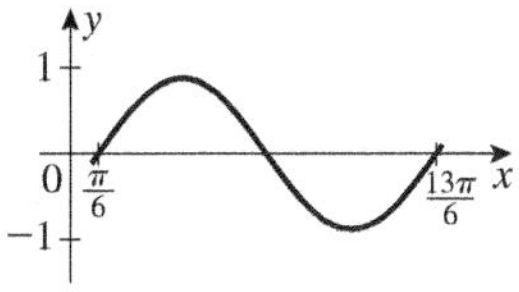

5. $1, 2\pi, -\frac{\pi}{6}$

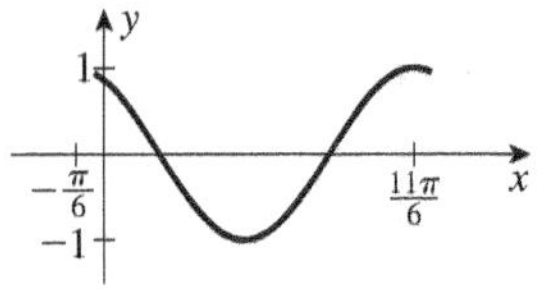

7. $0.2, \pi, -\frac{\pi}{4}$

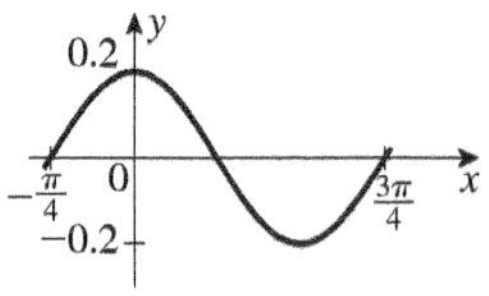

9. $1, \pi, \frac{\pi}{2}$

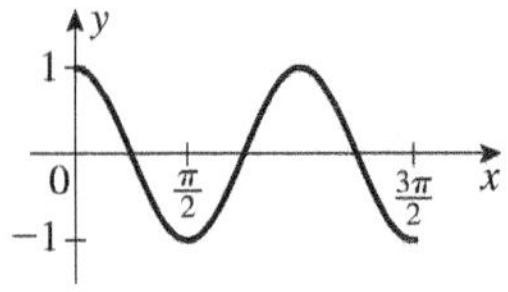

11. $\frac{1}{2}, 4\pi, \frac{\pi}{2}$

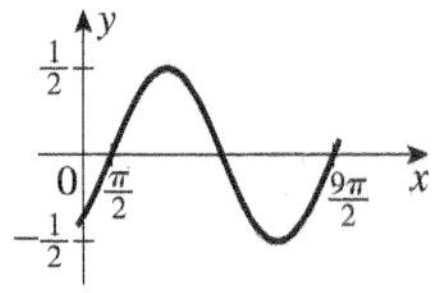

13. $30, 6\pi, -\pi$

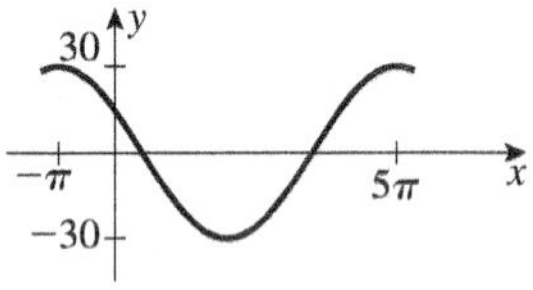

15. $1, 2, -\frac{1}{8}$

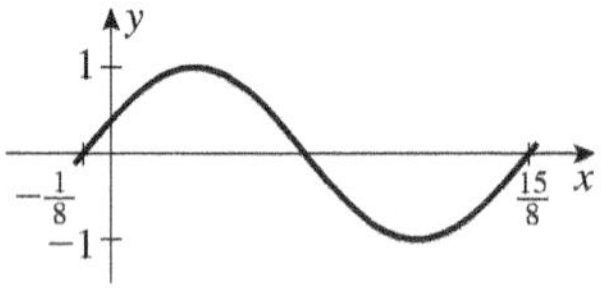

17. $0.08, \frac{1}{2}, \frac{1}{20}$

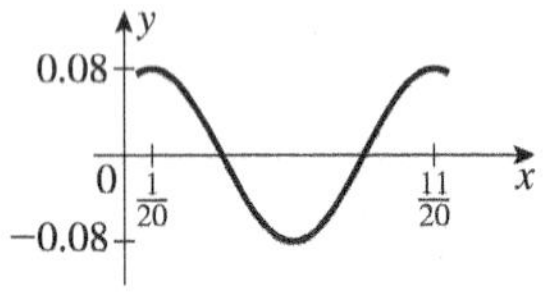

19. $0.6, 1, \frac{1}{2\pi}$

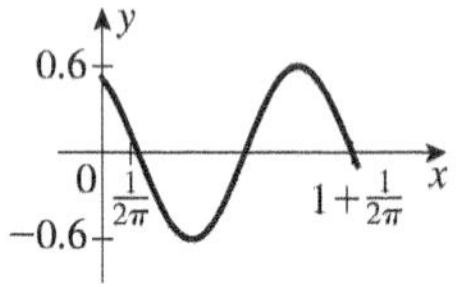

21. $40, \frac{2}{3}, -\frac{2}{3\pi}$

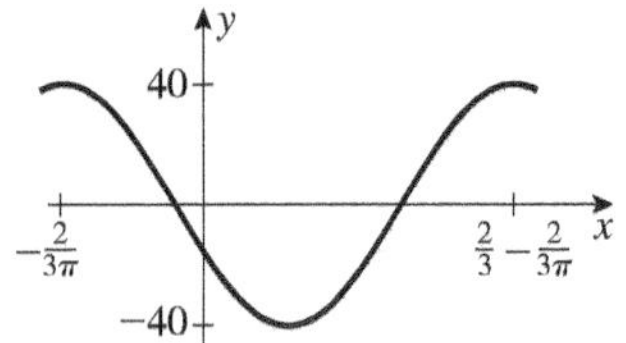

23. $1, \frac{2}{\pi}, \frac{1}{\pi}$

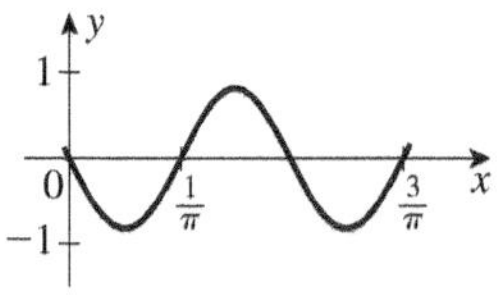

25. $\frac{3}{2}, 2, -\frac{\pi}{6}$

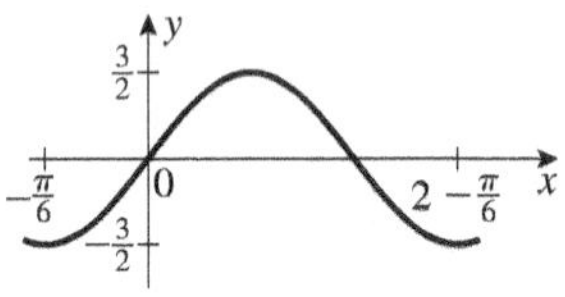

27. $y = 4 \sin\left(\frac{2}{3}x + \frac{\pi}{6}\right)$ **29.** $y = 12 \cos\left(4\pi x - \frac{\pi}{2}\right)$

31.

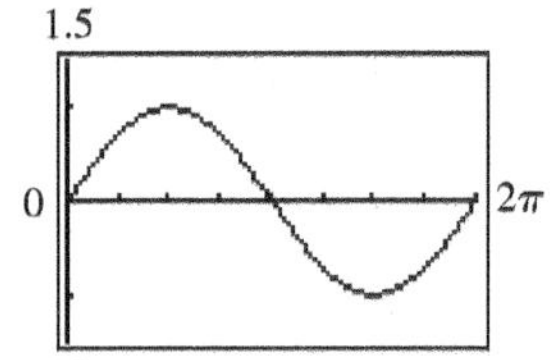

33. $v = 18 \sin(120\pi t - 60°)$, $\frac{1}{360}$ s

35.

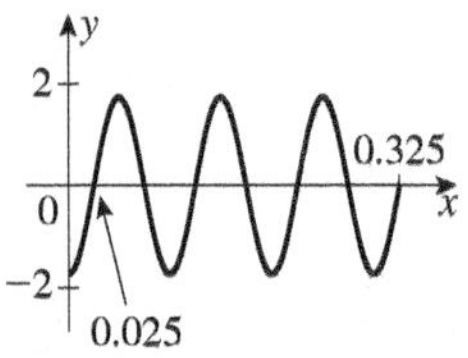

37.

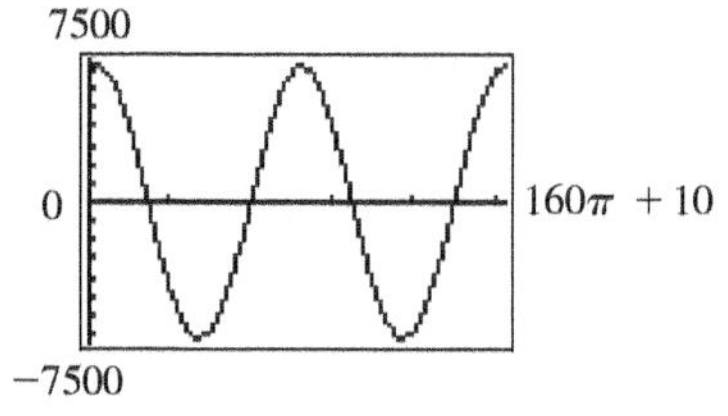

39. $h = 2.25 \sin\left(\frac{2\pi}{365}t - 4.54\right) + 12.2$

41. $y = 5 \sin\left(\frac{\pi}{8}x + \frac{\pi}{8}\right)$. As shown: amplitude = 5; period $= \frac{2\pi}{b} = 16$, $b = \frac{\pi}{8}$; displacement $= -\frac{c}{b} = -1$, $c = \frac{\pi}{8}$

43. $y = -0.8 \cos 2x$. As shown: amplitude $= |-0.8| = 0.8$; period $= \frac{2\pi}{b} = \pi$, $b = 2$; displacement $= -\frac{c}{b} = 0$, $c = 0$

Exercises 10.4, page 323

1. $y = 5 \cot 2x$

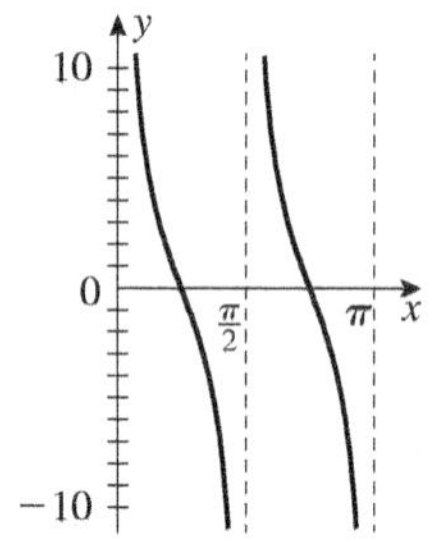

3. Undef., −1.7, −1, −0.58, 0, 0.58, 1, 1.7, undef., −1.7, −1, −0.58, 0

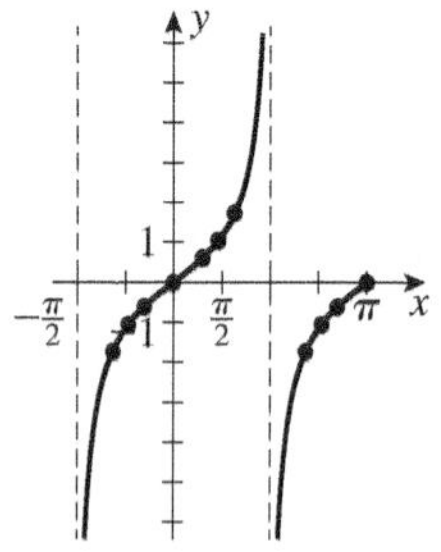

5. Undef., 2, 1.4, 1.2, 1, 1.2, 1.4, 2, undef., −2, −1.4, −1.2, −1

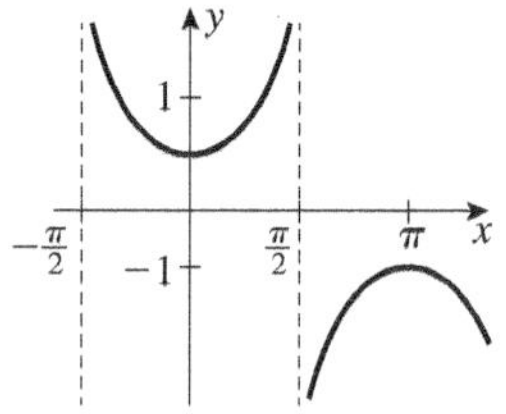

7.

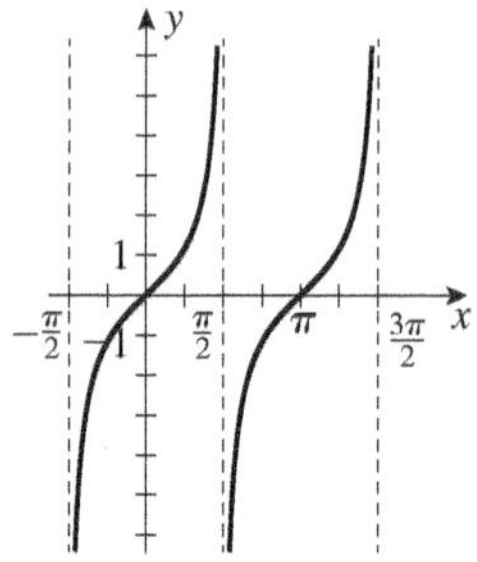

9.

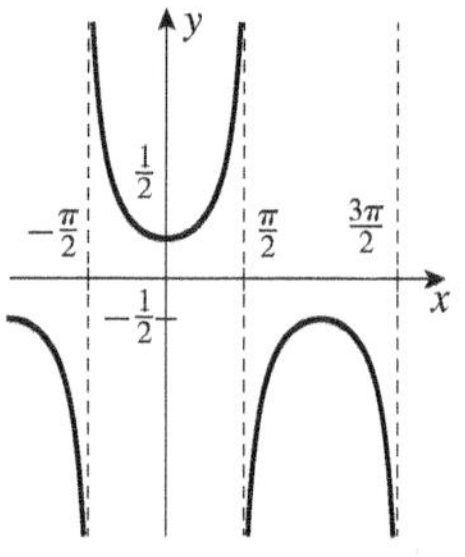

11.

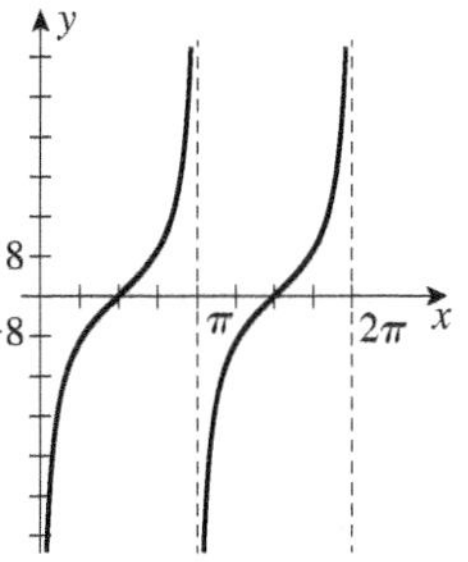

13.

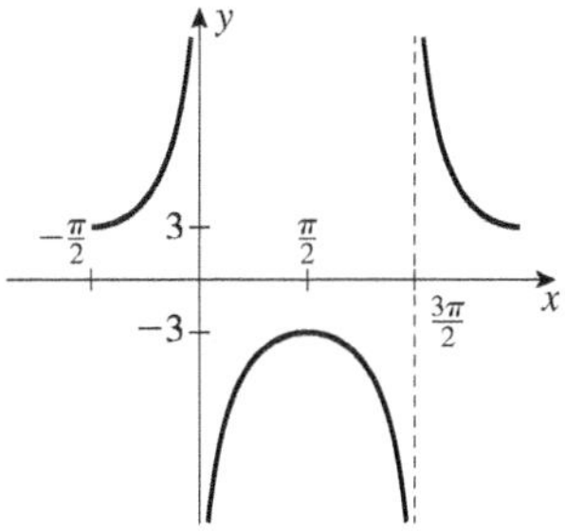

15.

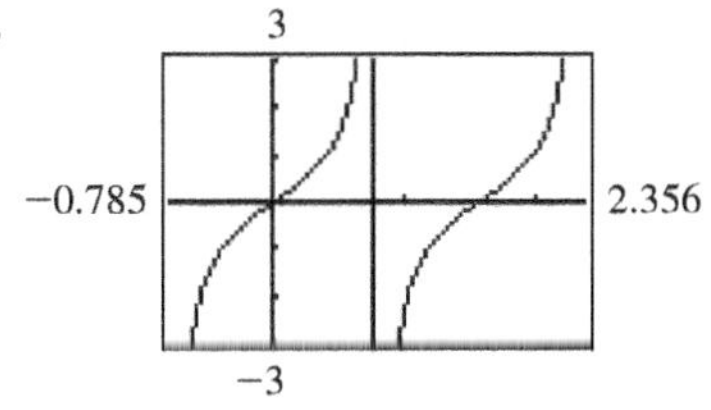

17.

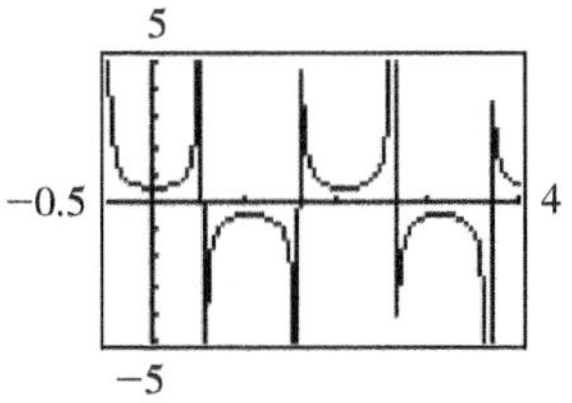

19.

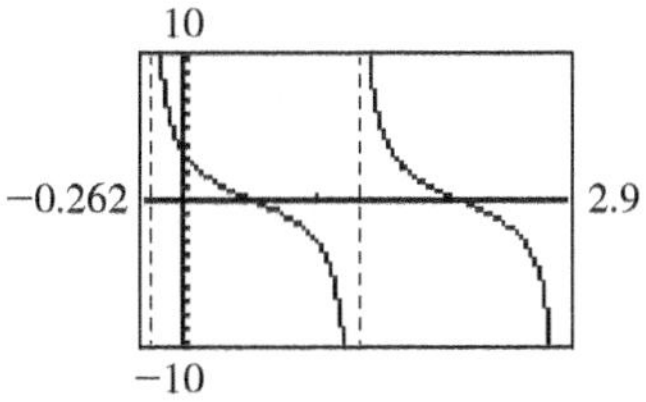

21.

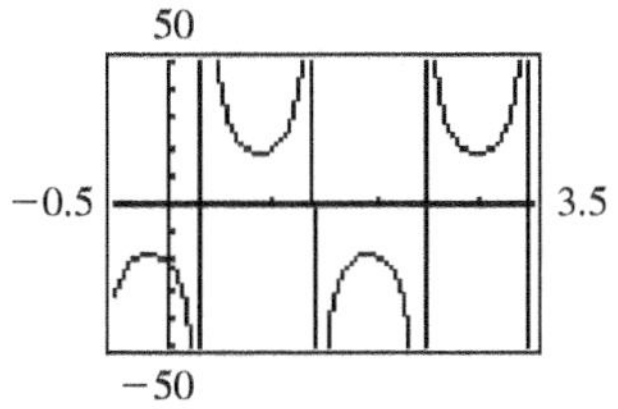

23.

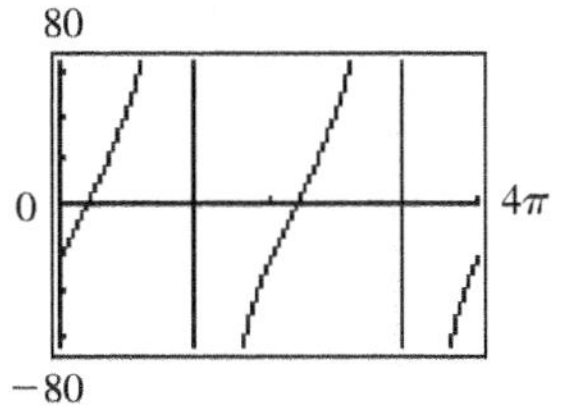

25. $y = -3\sec(x/2)$

27.

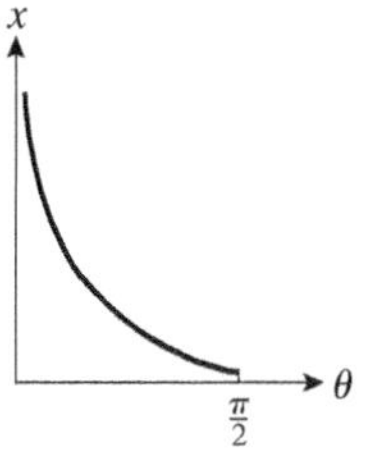

29.

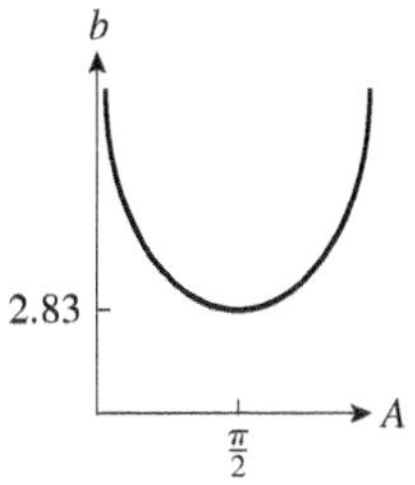

Exercises 10.5, page 326

1. $d = R\cos\omega t$

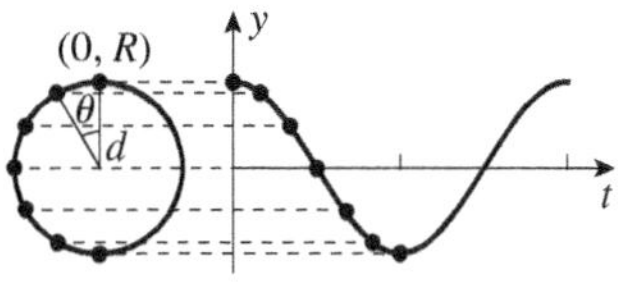

3.

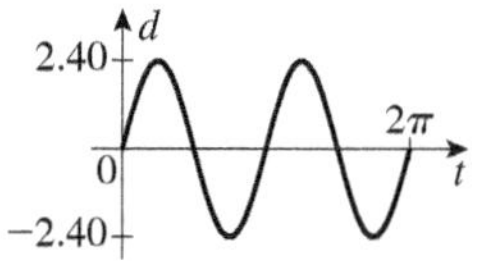

5.

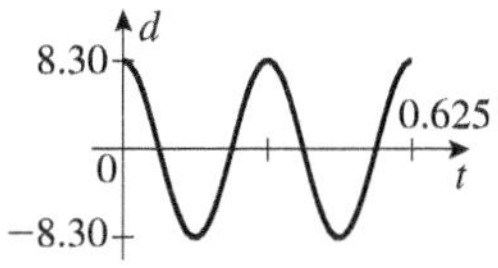

7.

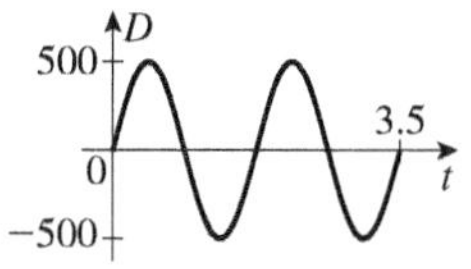

9.

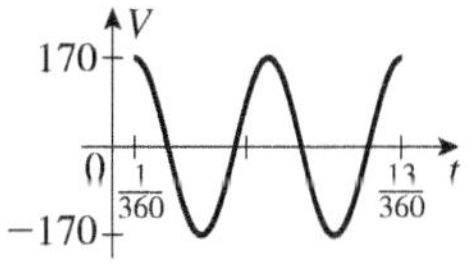

11.

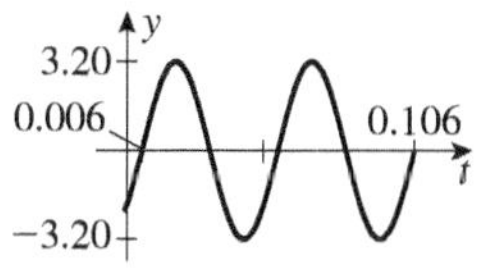

13.

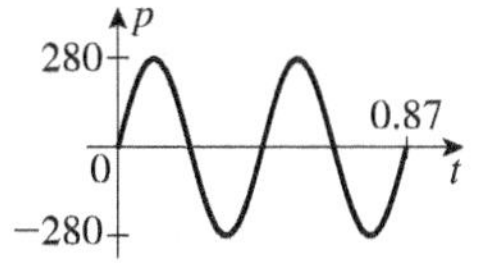

15.

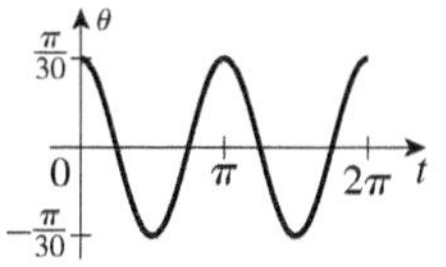

17.

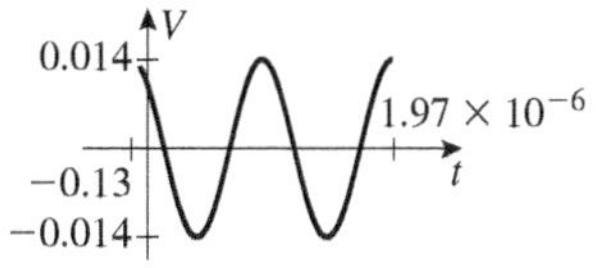

19. (a)

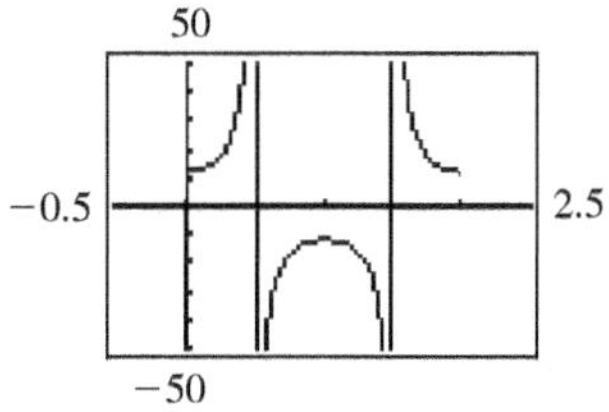

(b) $L > 0$, above x-axis

21. $d = 1.98\sin(\pi t)$

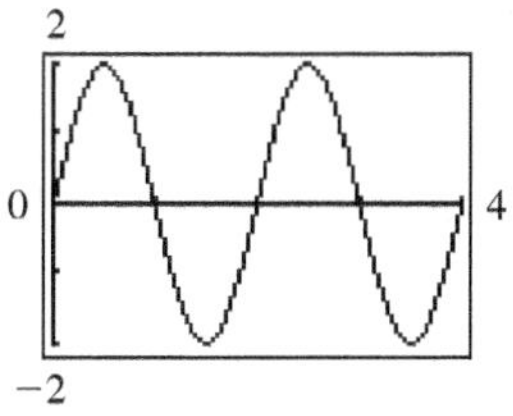

Exercises 10.6, page 329

1.

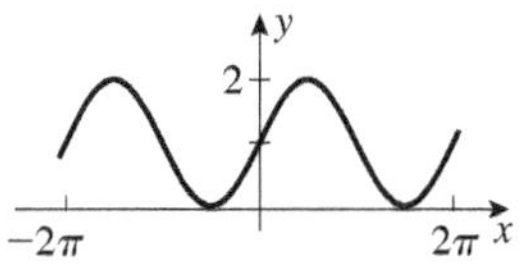

3.

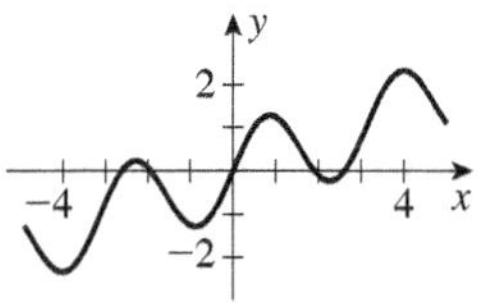

5.

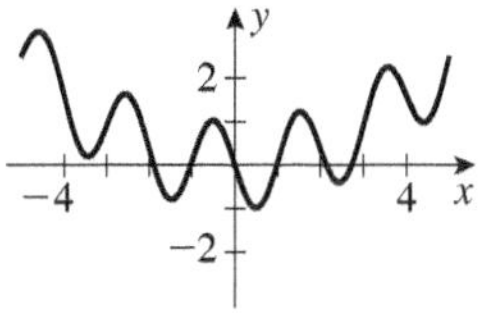

7.

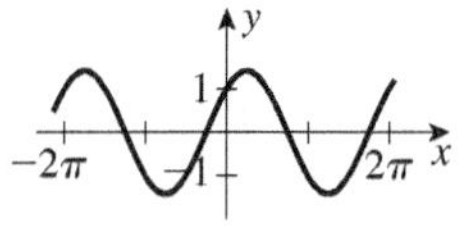

9.
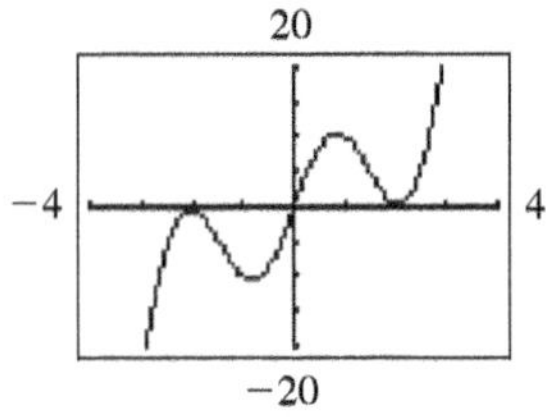

11.
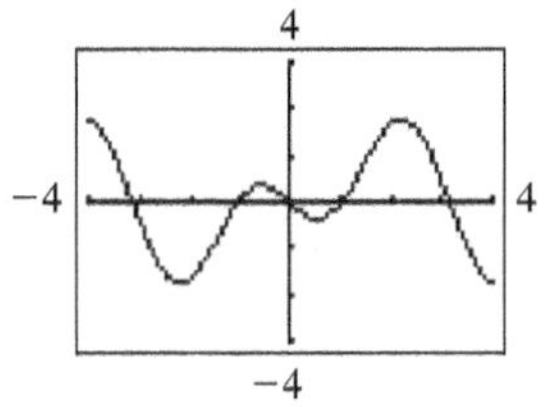

13.
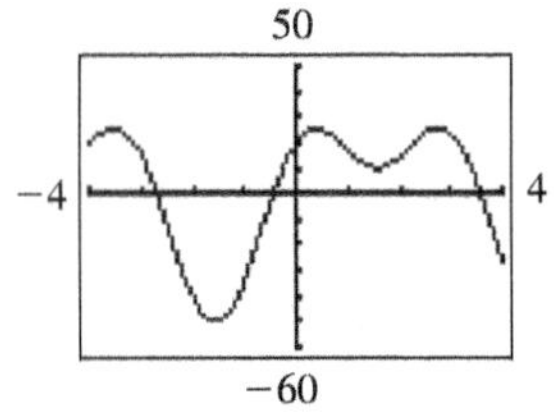

15.
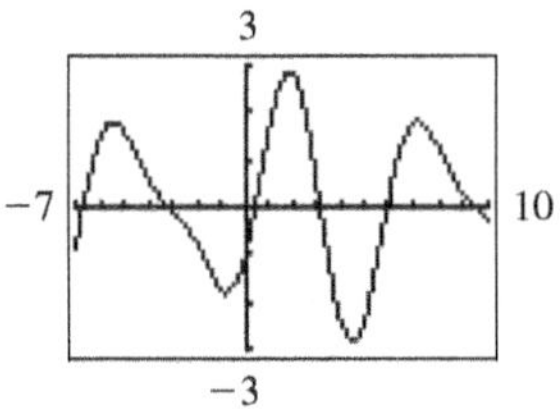

17.
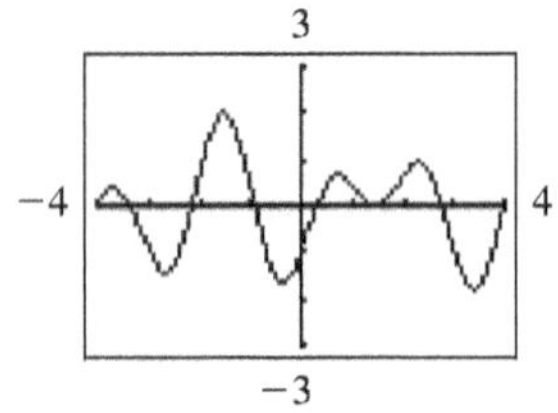

19.
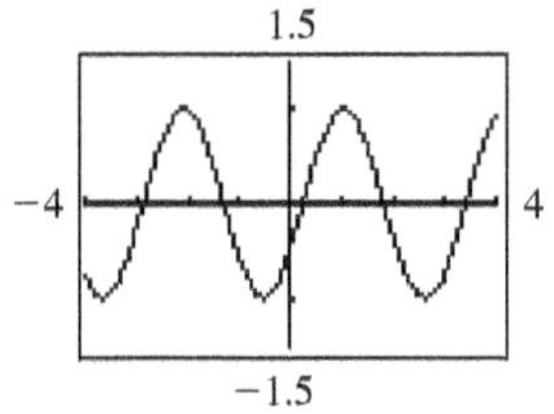

21.
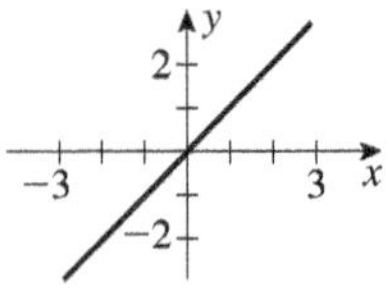

23.
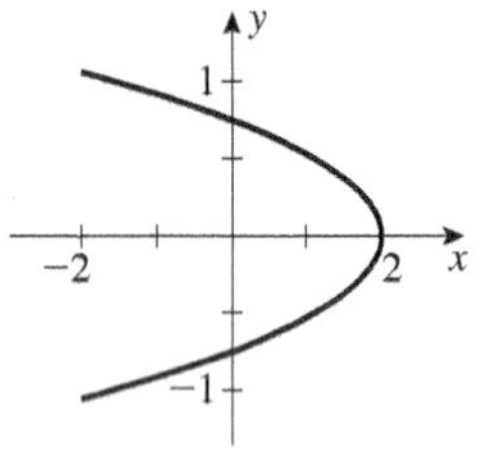

25.
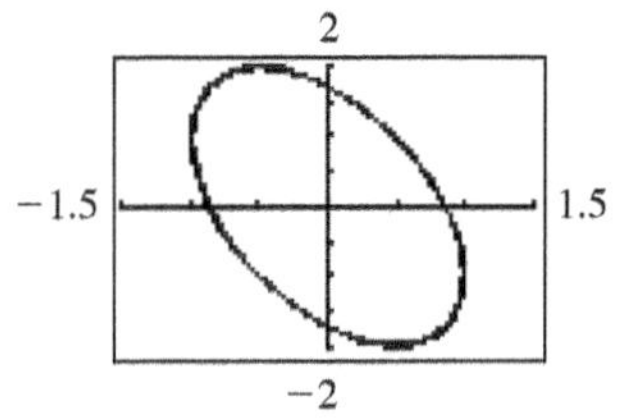

27.
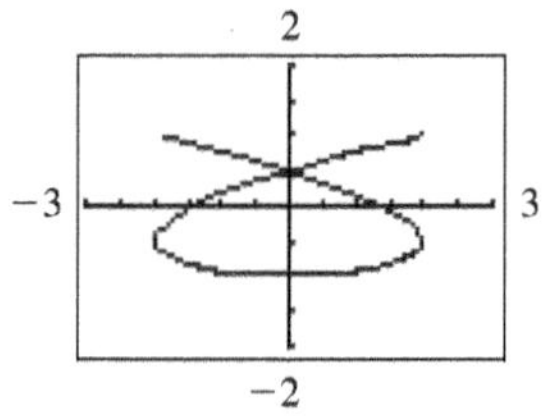

29.
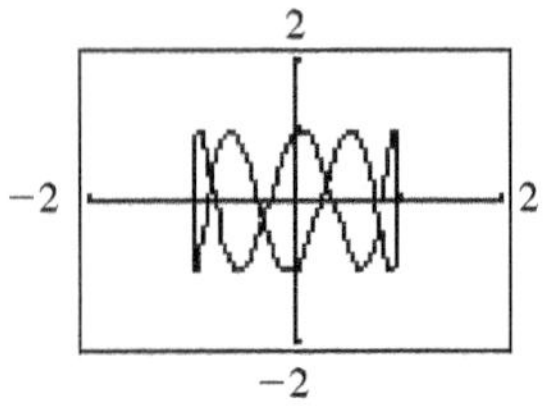

31.
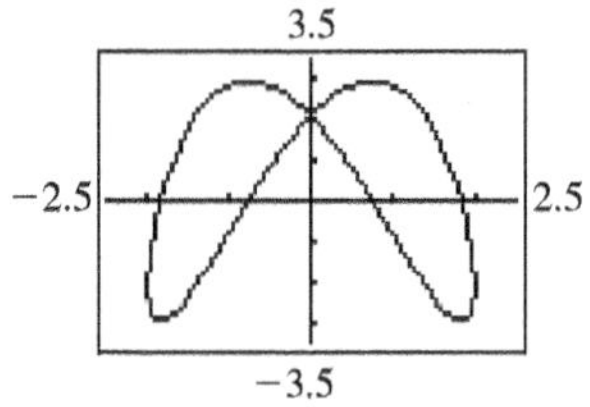

33.
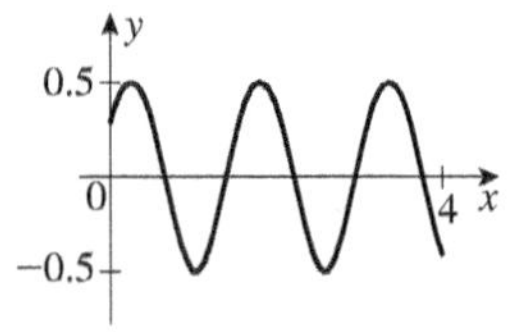

35.
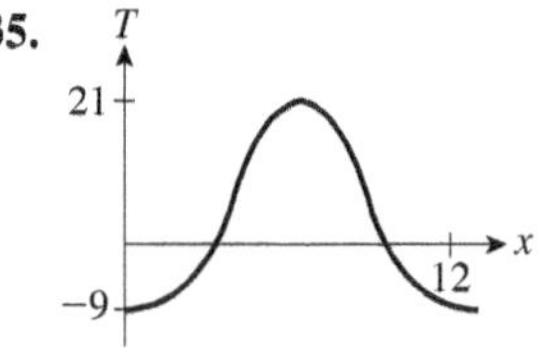

37. $p = 100 + 20 \cos 2\pi t$
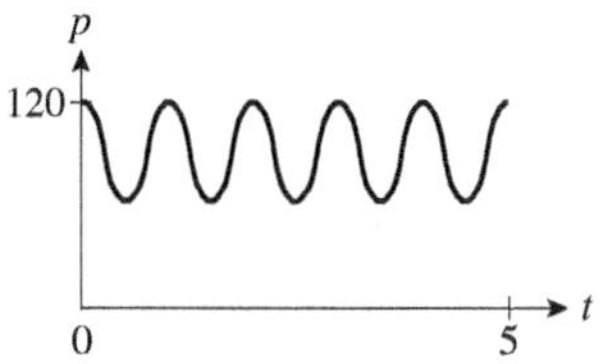

39.
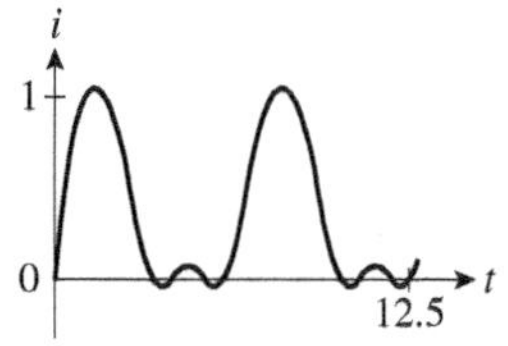

41.
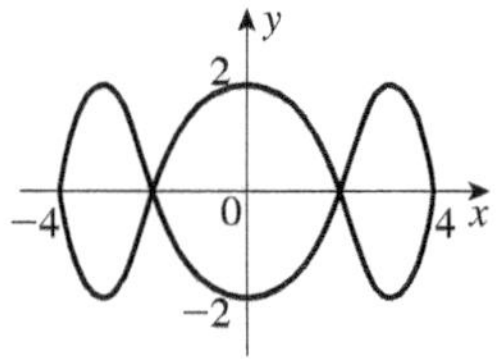

43.
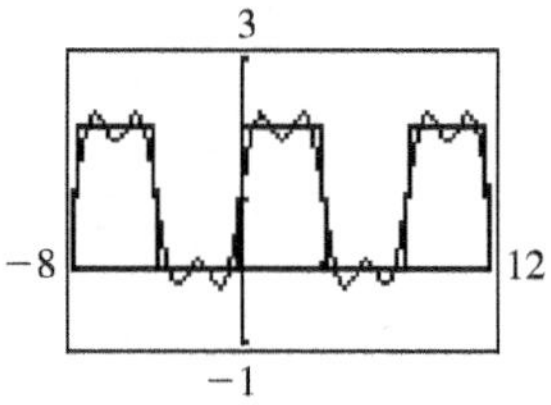

Review Exercises for Chapter 10, page 331

1.
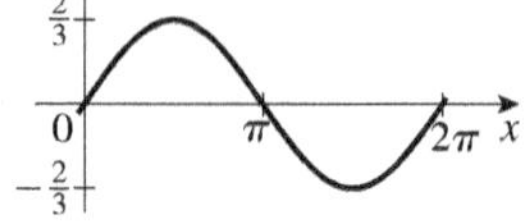

3.
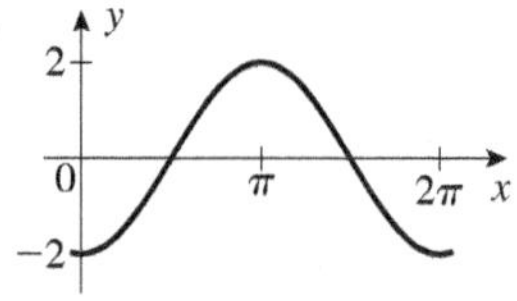

5.
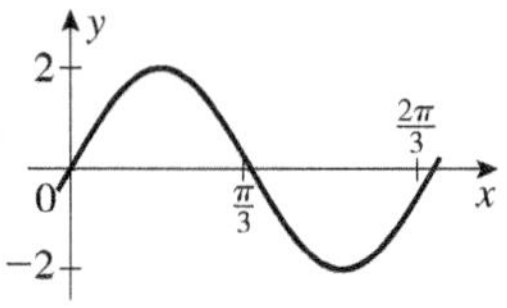

7.
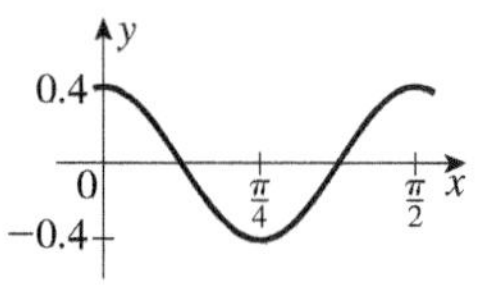

9.
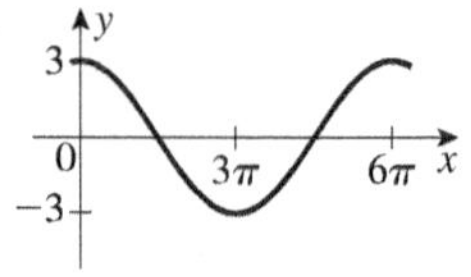

11.
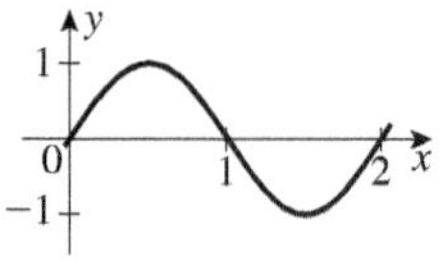

13.
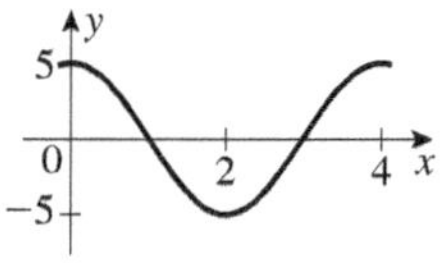

15.
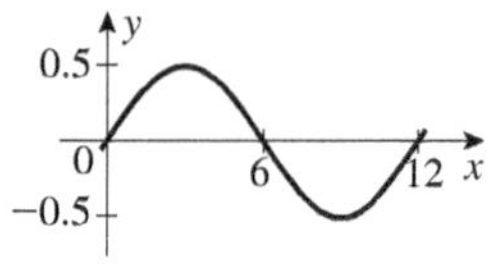

17.
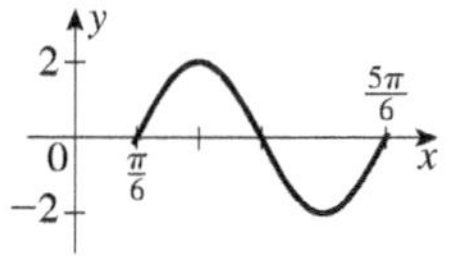

19.
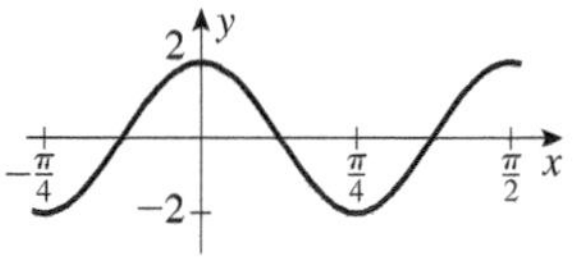

21.
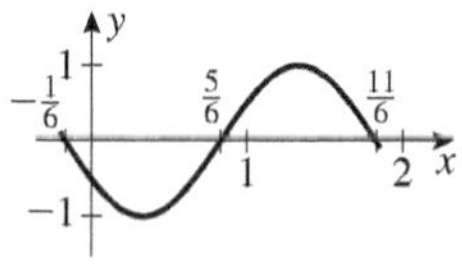

23.
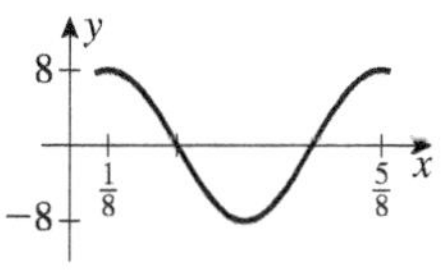

25.
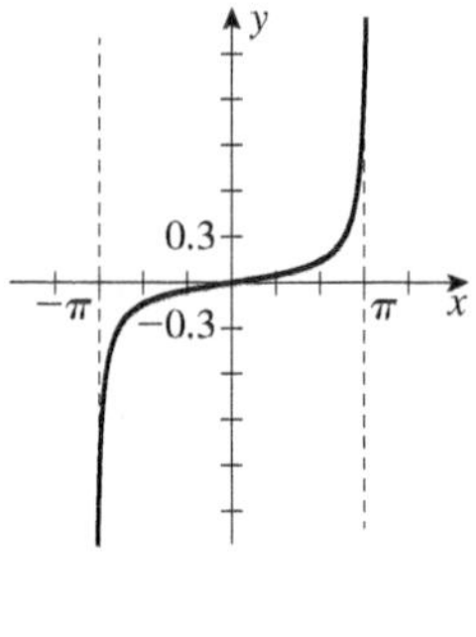

27.
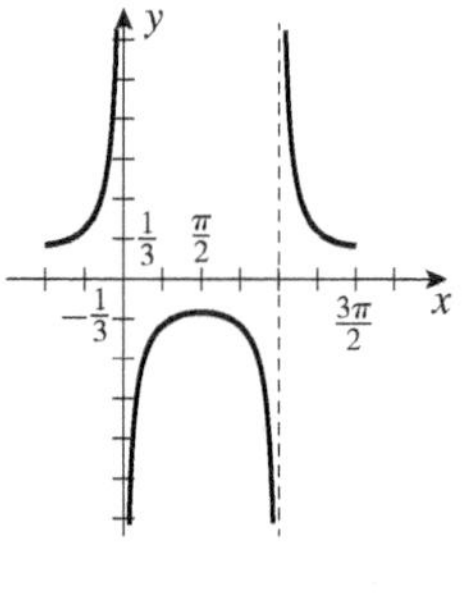

29.
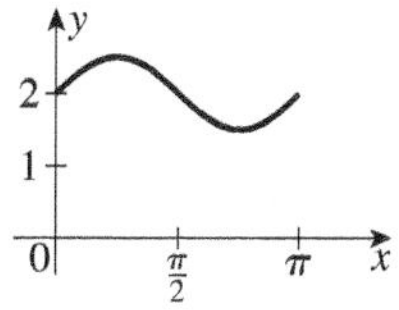

31.
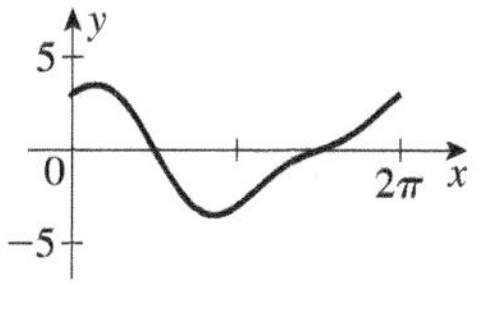

33.
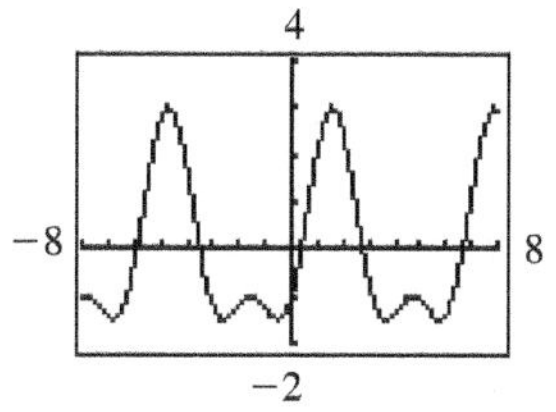

35.
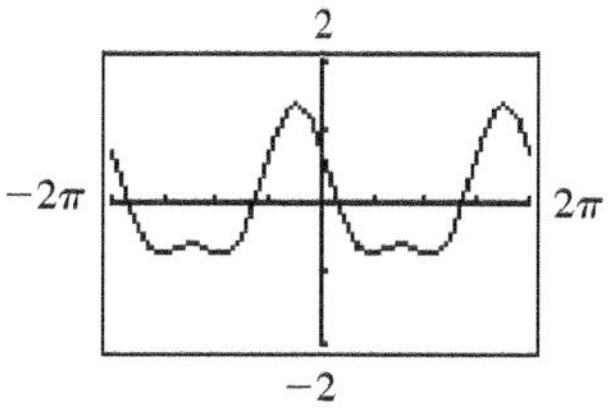

37.
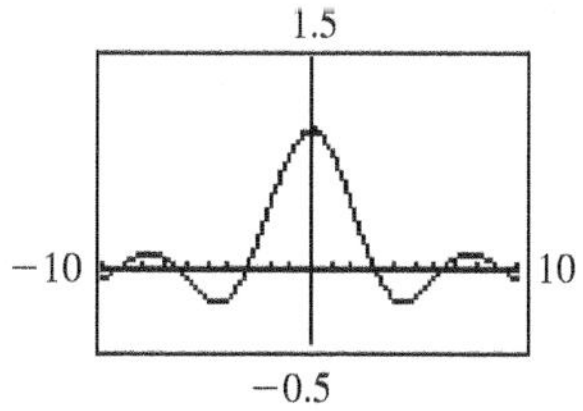

39.
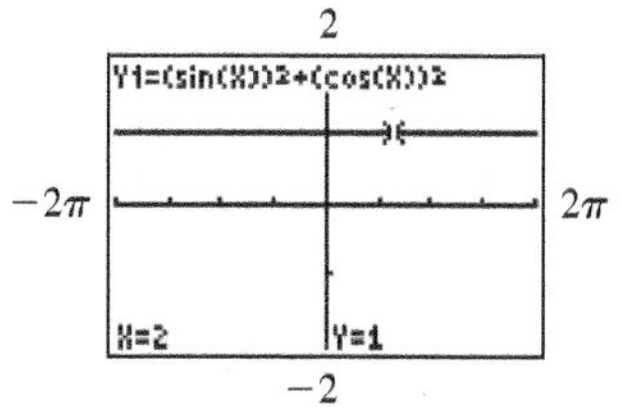

41. $y = 2\sin\left(2x + \frac{\pi}{2}\right)$ **43.** $y = \cos\left(\frac{\pi}{4}x - \frac{3\pi}{4}\right)$

45.
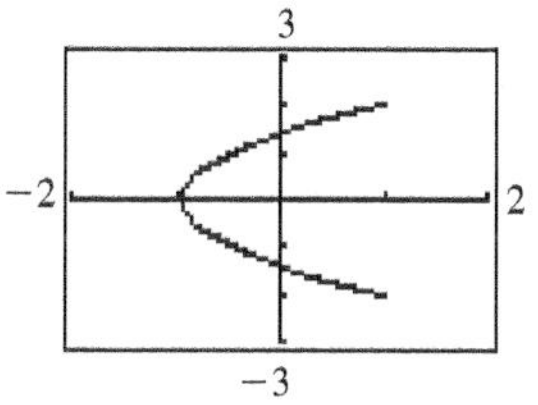

47.
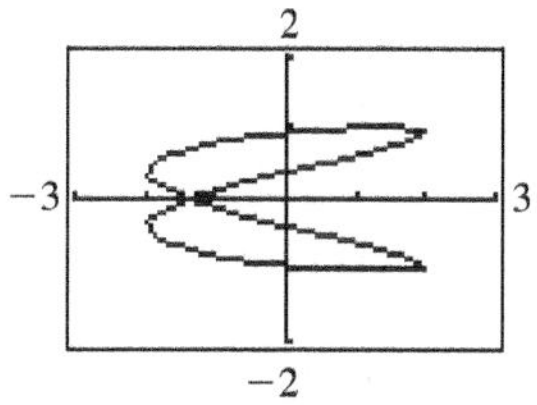

49.
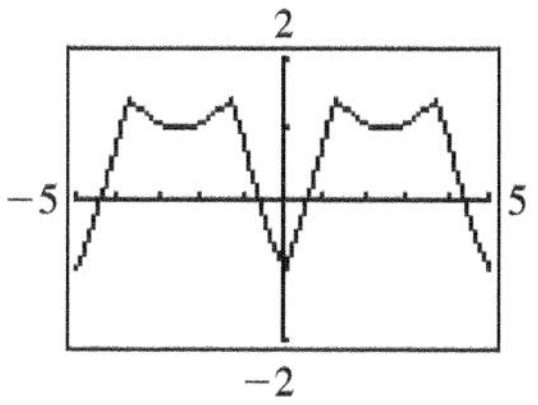

51.
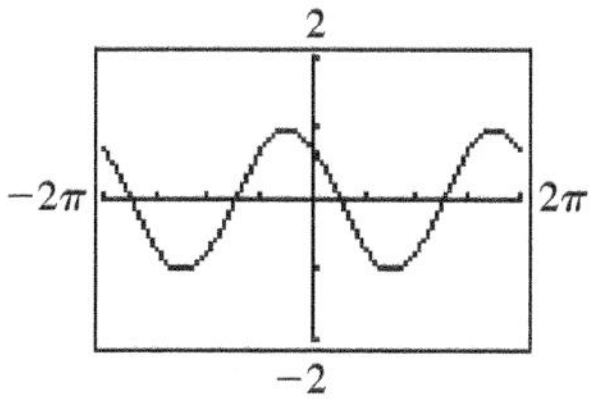

53. 4π

55. $y = 3\sin x$
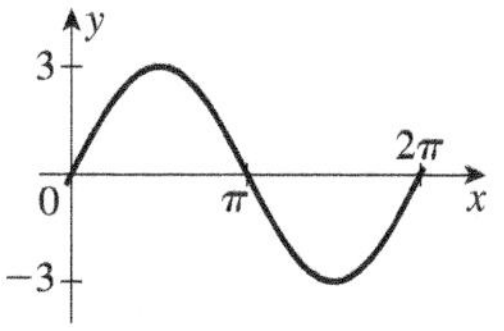

57. $y = 3\cos 3x$
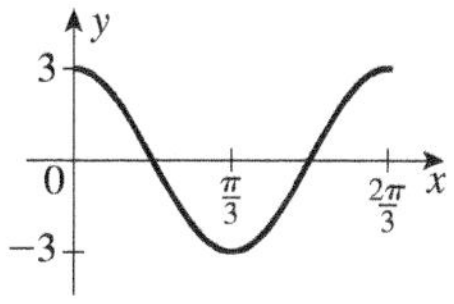

59. $y = 3\sin(\pi x + 0.25\pi)$
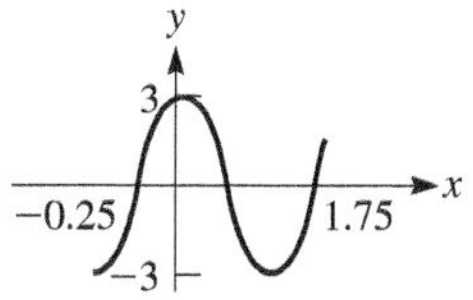

61.
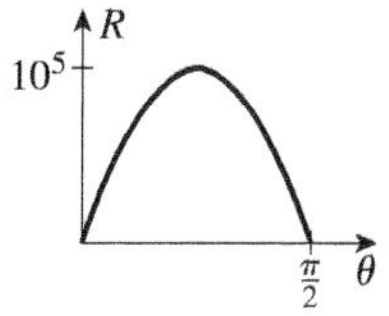

63.
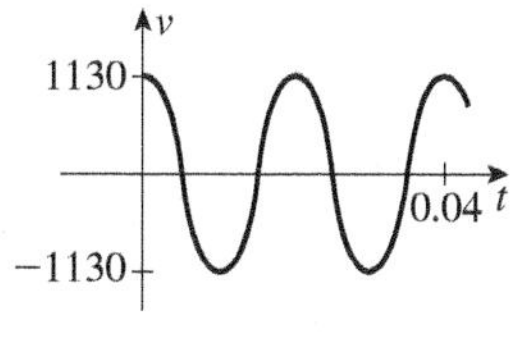

65.
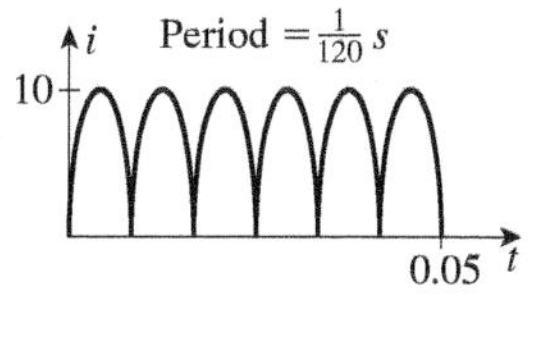

67.
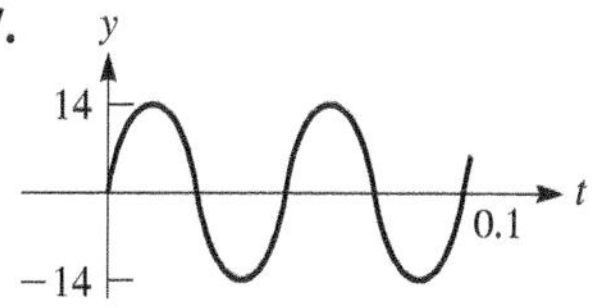

69.
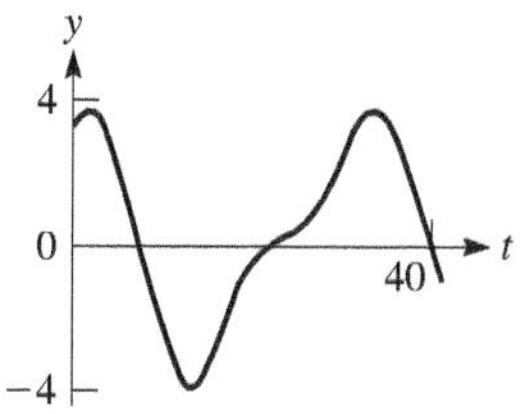

71.
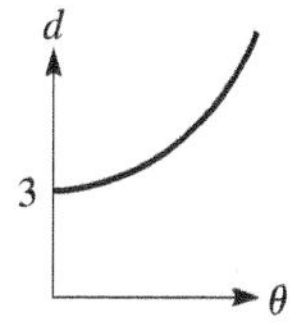

73.
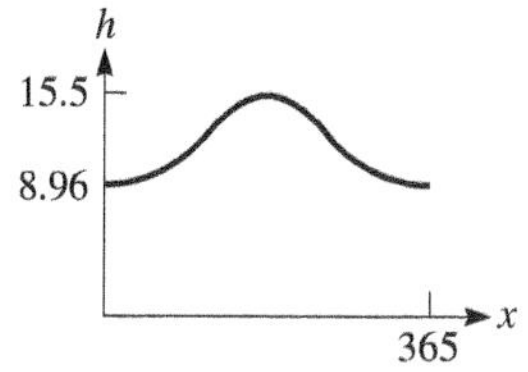

75.
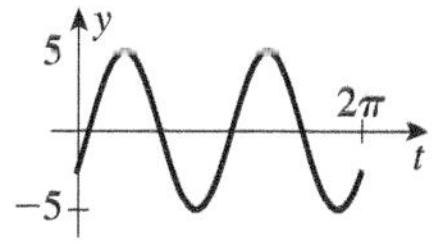

77.
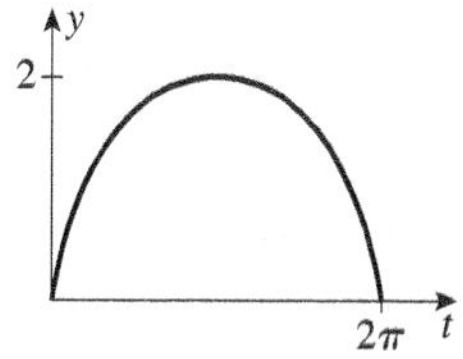

79.
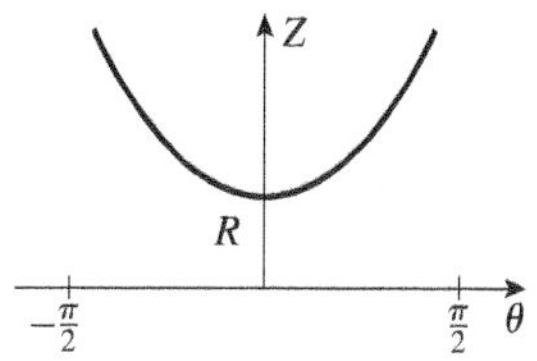

81.
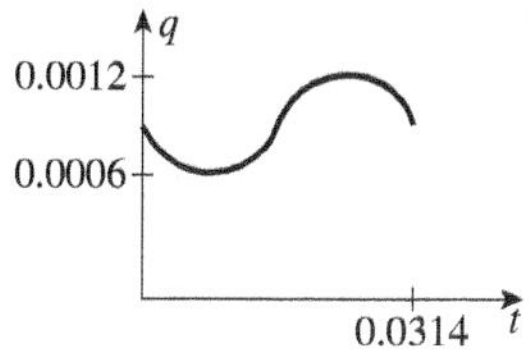

Chapter 11: Exponents and Radicals

Exercises 11.1, page 339

1. $\frac{y^2}{4x^2}$ **3.** $\frac{3}{416}$ **5.** x^4 **7.** $\frac{2}{a^4}$ **9.** c^9 **11.** x^5y^2

13. $\frac{1}{125}$ **15.** $\frac{9\pi^2}{x^6}$ **17.** $\frac{2n^2}{5a}$ **19.** 1 **21.** −7

23. $\frac{6}{x^4}$ **25.** $\frac{a^3}{343x^3}$ **27.** $\frac{1}{16a^4b^{12}}$ **29.** $\frac{n^9}{8}$ **31.** $\frac{3}{a^3b^6}$

33. $\frac{1}{a+b}$ **35.** $\frac{2x^2+3y^2}{x^2y^2}$ **37.** $\frac{8}{3a^n}$ **39.** $\frac{b^3}{432a}, a \neq 0$

41. $\frac{4}{t^4V^4}, t, V \neq 0$ **43.** $\frac{2a^6+16}{a^8}$ **45.** $\frac{10}{9}$ **47.** $\frac{R_1R_2}{R_1+R_2}$

49. $\frac{4n^2-4n+1}{n^4}$ **51.** $\frac{8}{99}$ **53.** $\frac{x+y}{xy}, x, y \neq 0, x \neq y$

55. $\frac{t^2+t+2}{t^2}$ **57.** $\frac{2D}{D^2-1}$ **59.** No

61. (a) 4^5 (b) 2^{10}

63. (a) $\left(\frac{a}{b}\right)^{-n} = \frac{1}{\left(\frac{a}{b}\right)^n} = \frac{1}{\frac{a^n}{b^n}} = \frac{b^n}{a^n} = \left(\frac{b}{a}\right)^n$

(b) 303.551 82 = 303.551 82

65. True **67.** $16(4^a)$ **69.** 7 **71.** N • m **73.** $\mathrm{J/s^3}$

75. 1 **77.** $\frac{p[(1+i)^n - 1]}{i(1+i)^n}$ **79.** −5

Exercises 11.2, page 343

1. 16 **3.**

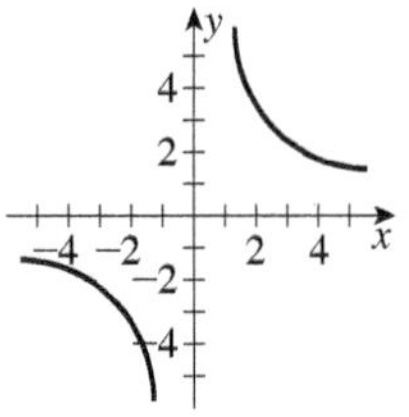

5. 6 **7.** 3 **9.** 10^{25}

11. $\frac{1}{2}$ **13.** $\frac{1}{16}$ **15.** $5\sqrt{5}$ **17.** 81 **19.** −200

21. $\frac{\sqrt[5]{(21)^4}}{9}$ **23.** −2 **25.** $\frac{39}{1000}$ **27.** 0.6914

29. 2.059 **31.** 0.538 91 **33.** $B^{22/15}$ **35.** $\frac{1}{-y^{9/10}}$

37. $\frac{1}{x^{3/2}}$ **39.** $2ab^2$ **41.** $\frac{1}{8a^3b^{9/4}}$ **43.** $a^{1/12}$

45. $\frac{4x}{(4x^2+1)^{1/2}}$ **47.** $-y,\ y \neq 1$ **49.** $\frac{T}{(T+2)^{1/2}}$

51. $\frac{a^2+1}{a^4}$ **53.** $\frac{a+1}{a^{1/2}}$ **55.** $\frac{x(5x-2)}{(2x-1)^{1/2}}$

57.

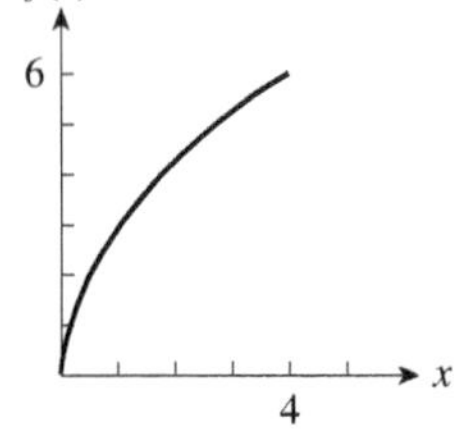

59.

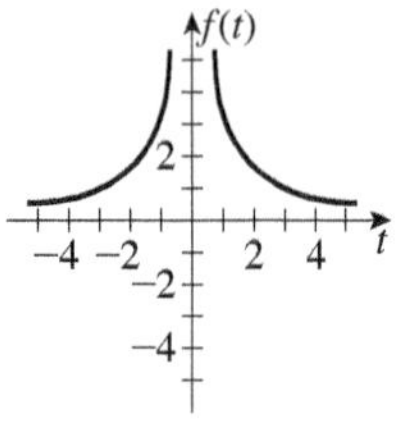

61. $\sqrt[3]{x^2}$

63. If $(A/S)^{-1/4} = 0.5 = 1/2$, then $(A/S)^{1/4} = 2$. Raise each to the fourth power and get $A/S = 16$.

65. $R = \frac{T^{2/3}}{k^{1/3}} - d$ **67.** 1.91 mA **69.** 37.1 mg

Exercises 11.3, page 347

1. $ab^2\sqrt{a}$ **3.** $\frac{\sqrt[4]{54}}{3}$ **5.** $2\sqrt{7}$ **7.** $6\sqrt{2}$ **9.** $x^2y^3\sqrt{y}$

11. $x^2yz^2\sqrt{z}$ **13.** $3RT^3V\sqrt{6RV}$ **15.** $2\sqrt[3]{2}$ **17.** $2\sqrt[5]{3}$

19. $2\sqrt[3]{a^2}$ **21.** $2st\sqrt[4]{4r^3t}$ **23.** −2 **25.** $P\sqrt[3]{V}$

27. $\frac{2\sqrt{3}}{3}$ **29.** $\frac{1}{2}\sqrt{6}$ **31.** $\frac{1}{2}\sqrt[3]{6}$ **33.** $\frac{1}{3}\sqrt[5]{27}$

35. $2\sqrt{5}$ **37.** 2 **39.** 200 **41.** 2000 **43.** $\sqrt{2a}$

45. $\frac{1}{2}\sqrt{2}$ **47.** $\sqrt[3]{2}$ **49.** $\sqrt[8]{n}$ **51.** $\frac{2u\sqrt{7uv}}{v^3}$

53. $4\sqrt{13}$ **55.** $\frac{\sqrt{6x}}{3c^2}$ **57.** $\frac{3}{4}\sqrt{2}$ **59.** $\frac{\sqrt{xy(x^2+y^2)}}{xy}$

61. $\frac{\sqrt{(C-2)(C+2)}}{(C+2)}$ **63.** $\sqrt{a^2+b^2}$ **65.** $3x - 1$

67. $\sqrt[6]{a^3};\ \sqrt[6]{b^2};\ \sqrt[6]{c}$ **69.**

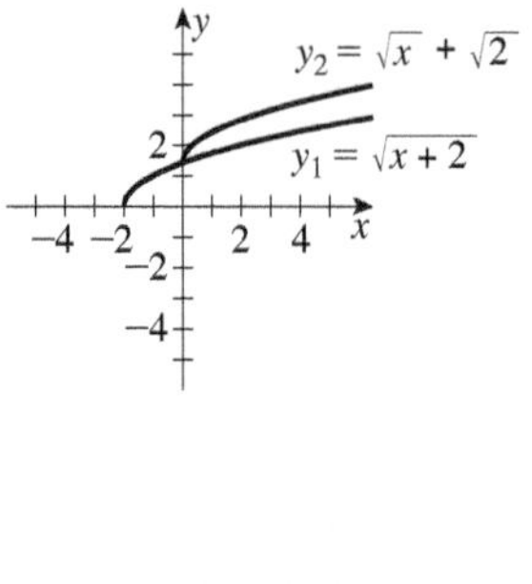

71. $\frac{\sqrt{T\mu}}{2L\mu}$ **73.** $\sqrt{2ag}$ **75.** $\frac{8Af\sqrt{f^2+f_0^2}}{\pi^2(f^2+f_0^2)}$

Exercises 11.4, page 349

1. $16\sqrt{5}$ **3.** $7\sqrt{11}$ **5.** $\sqrt{5} - 2\sqrt{2}$ **7.** $3\sqrt{5}$

9. $-4t\sqrt{3}$ **11.** $-2\sqrt{2a}$ **13.** $19\sqrt{7}$ **15.** $-20\sqrt{2}$

17. $\sqrt{R}(23\sqrt{3} - 6\sqrt{2})$ **19.** $\frac{7}{3}\sqrt{15}$ **21.** $-9\sqrt{2}$

23. $13\sqrt[3]{3}$ **25.** $\sqrt[4]{2}$ **27.** $(5a - 2b^2)\sqrt{ab}$

29. $(3 - 2a)\sqrt{10}$ **31.** $(2b - a)\sqrt[3]{3a^2b}$

33. $\sqrt{RH}\left(\frac{1}{H^3} - \frac{1}{R^2}\right)$ **35.** $\frac{(a-2b)\sqrt[3]{ab^2}}{ab}$

37. $\frac{-2V\sqrt{T^2-V^2}}{T^2-V^2}$ **39.** $\frac{\sqrt{6}T^{1/4}}{6R^{5/12}}$ **41.** $8\sqrt{x+2}$

43. $15\sqrt{3} - 11\sqrt{5} = 1.384\ 014\ 4$ **45.** $\frac{1}{6}\sqrt{6} = 0.408\ 248\ 3$

47. $\frac{7}{2}\sqrt[3]{2} = 4.409\ 723\ 7$ **49.** $3\sqrt{3}$

51. Positive, $\sqrt{1000} = 10\sqrt{10}$, $\sqrt{11} > \sqrt{10}$

53. $6\sqrt{2} + 2\sqrt{6}$ units **55.** $5400 + 900\sqrt{2} = 6670$ mm

Exercises 11.5, page 353

1. $3\sqrt{10} - 16$ **3.** $\sqrt{3} - \sqrt{2}$ **5.** $\sqrt{21}$ **7.** $2\sqrt{5}$

9. 2 **11.** 50 **13.** $2\sqrt{5}$ **15.** $\sqrt{6} - \sqrt{15}$ **17.** −1

19. $498 + 27\sqrt{10}$ **21.** $66 + 13\sqrt{11x} - 5x$

23. $20 - 9x$ **25.** $a\sqrt{b} + c\sqrt{ac}$ **27.** $\frac{2-\sqrt{6}}{2}$

29. $2a - 3b + 2\sqrt{2ab}$ **31.** $\sqrt[6]{72}$ **33.** $\frac{1}{4}(\sqrt{7} - \sqrt{3})$

35. $\frac{1}{11}(\sqrt{7} + 3\sqrt{2} - 6 - \sqrt{14})$ **37.** $\frac{1}{17}(-56 + 9\sqrt{15})$

39. $\frac{2(x+\sqrt{5x})}{x-5}$ **41.** 1 **43.** $-\frac{(R+2)(R-1)}{R}$

45. $-\frac{\sqrt{x^2-y^2} + \sqrt{x^2+xy}}{y}$ **47.** $a - 1 + \sqrt{a(a-2)}$

49. $-1 - \sqrt{66} = -9.124\ 038\ 4$

51. $-\frac{16+5\sqrt{30}}{26} = -1.668\ 697\ 2$ **53.** $\frac{2x+1}{\sqrt{x}}$

55. $\frac{x(5x+2)}{\sqrt{2x+1}}$ **57.** $\frac{17}{15\sqrt{10} - 12\sqrt{5}}$ **59.** $\frac{1}{\sqrt{x+h} + \sqrt{x}}$

61. $(1 - \sqrt{2})^2 - 2(1 - \sqrt{2}) - 1$
$= 1 - 2\sqrt{2} + 2 - 2 + 2\sqrt{2} - 1 = 0$

63. $a = c$ **65.** $\frac{\sqrt[3]{x} - 1}{x - 1}$

67. $\left[\frac{1}{2}\left(\sqrt{b^2-4k^2} - b\right)\right]^2 + b\left[\frac{1}{2}\left(\sqrt{b^2-4k^2} - b\right)\right] + k^2 = 0$

69. $\frac{2500 - 50\sqrt{V}}{2500 - V}$ **71.** $2Q\sqrt{\sqrt{2} + 1}$ **73.** $\frac{\sqrt{C(L - R^2C)}}{LC}$

Review Exercises for Chapter 11, page 354

1. $\frac{2}{a^2}$ **3.** $\frac{9d^3}{c}$ **5.** 375 **7.** $\frac{1}{8000}$ **9.** $\frac{t^4}{9}$ **11.** −28

13. $64a^2b^5$ **15.** $-8m^9n^6$ **17.** $\frac{2(C - 2L^2)}{CL^2}$

19. $\dfrac{2y}{x+2y}, x \neq 0$ **21.** $\dfrac{b}{ab-3}$ **23.** $\dfrac{(x^3y^3-1)^{1/3}}{y}$

25. $\dfrac{1}{W+H}$ **27.** $\dfrac{-2(x+1)}{(x-1)^3}$ **29.** $2\sqrt{17}$ **31.** $b^2c\sqrt{ab}$

33. $3ab^2\sqrt{a}$ **35.** $\dfrac{2t\sqrt{21st}}{u}$ **37.** $\dfrac{5\sqrt{2s}}{2s}$ **39.** $\dfrac{1}{9}\sqrt{33}$

41. $mn^2\sqrt{m}\sqrt[4]{8n}$ **43.** $\sqrt{2}$ **45.** 0 **47.** $-7\sqrt{7}$

49. $3ax\sqrt{2x}$ **51.** $(2a+b)\sqrt[3]{a}$ **53.** $25(6-\sqrt{7})$

55. $4(\sqrt{3}-\sqrt{5})$ **57.** $6-51B-7\sqrt{17B}$

59. $42-7\sqrt{7a}-3a$ **61.** $\dfrac{6x+\sqrt{3xy}}{12x-y}$ **63.** $-\dfrac{8+\sqrt{6}}{29}$

65. $\dfrac{13-2\sqrt{35}}{29}$ **67.** $\dfrac{6x-13a\sqrt{x}+5a^2}{9x-25a^2}$ **69.** $\sqrt{4b^2+1}$

71. $x(9-4x)$ **73.** $1+\sqrt{6}$ **75.** $\dfrac{15-2\sqrt{15}}{4}$

77. $\dfrac{3+n+\sqrt{n(n+3)}}{3}$ **79.** $51-7\sqrt{105}$

81. $\sqrt{\sqrt{2}-1}\,(\sqrt{2}+1) = \sqrt{(\sqrt{2}-1)(\sqrt{2}+1)^2}$
$= \sqrt{(\sqrt{2}-1)(3+2\sqrt{2})} = \sqrt{\sqrt{2}+1} = 1.553\,774\,0$

83. $12\sqrt{2}$ units **85.** $\dfrac{33}{4}-2\sqrt{3}$ **87.** 2.528%

89. (a) $v = k(P/w)^{1/3}$ (b) $v = \dfrac{k\sqrt[3]{Pw^2}}{w}$ **91.** $\dfrac{100(1+2\sqrt{x})}{x}$

93. $\dfrac{c}{(c^2-v^2)^{1/2}}$ **95.** $\dfrac{1}{\sqrt{A+h}+\sqrt{A}}$ **97.** $6\sqrt{2}$ cm

99. $\dfrac{\sqrt{LC_1C_2(C_1+C_2)}}{2\pi LC_1C_2}$

Chapter 12: Complex Numbers

Exercises 12.1, page 360

1. $-\sqrt{5}$ **3.** -1 **5.** $9j$ **7.** $-2j$ **9.** $0.6j$ **11.** $8j\sqrt{2}$

13. $\dfrac{1}{2}j\sqrt{7}$ **15.** $-2ej$ **17.** (a) -7 (b) 7 **19.** (a) 4 (b) -4

21. $-\dfrac{3}{5}$ **23.** $10j$ **25.** (a) 1 (b) -1 **27.** 0 **29.** $-2j$

31. $-j$ **33.** $2+3j$ **35.** $-7j$ **37.** $2+2j$ **39.** $-2+3j$

41. $3\sqrt{2}-2j\sqrt{2}$ **43.** -1 **45.** (a) $6+7j$ (b) $8-j$

47. (a) $-2j$ (b) -4 **49.** $x=2, y=-2$ **51.** $x=10, y=-6$

53. $x=-2, y=3$ **55.** $\pm 4j\sqrt{2}$ **57.** $-1 \pm j\sqrt{6}$

59. (a) No change (b) Sign changes **61.** Yes **63.** 0

65. Yes; imag. part is zero. **67.** Each is x.

Exercises 12.2, page 363

1. $1-5j$ **3.** $\dfrac{1}{25}(29+22j)$ **5.** $5-8j$ **7.** $-25+10j$

9. $-0.23+0.86j$ **11.** $-36+21j$ **13.** $7+49j$

15. $22+3j$ **17.** $-18j\sqrt{2}$ **19.** $-28j$ **21.** $3\sqrt{7}+3j$

23. $-40-42j$ **25.** $-67+36j$ **27.** $\dfrac{1}{29}(-30+12j)$

29. $-\dfrac{1}{3}(1+j)$ **31.** $\dfrac{1}{11}(-13+8j\sqrt{2})$ **33.** $\dfrac{-1+3j}{5}$

35. $\dfrac{-45}{13}+\dfrac{48j}{13}$ **37.** $-8j$ **39.** $80j$ **41.** $-\dfrac{7}{8}+4j$

43. $(-1-j)^2+2(-1-j)+2 = 1+2j-1-2-2j+2 = 0$

45. 10 **47.** $\dfrac{3}{10}+\dfrac{1}{10}j$ **49.** $-1+j$ **51.** -4

53. $\dfrac{1}{10}(11+27j)$ **55.** $281+35.2j$ volts

57. $0.016+0.037j$ amperes **59.** (a) Real (b) Pure imaginary

61. Product is the sum of squares of two real numbers
$(a+bj)(a-bj) = a^2+b^2$

Exercises 12.3, page 365

1. $3-j$

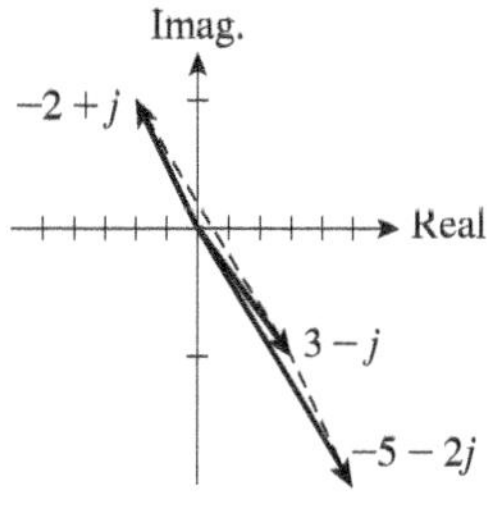

3.

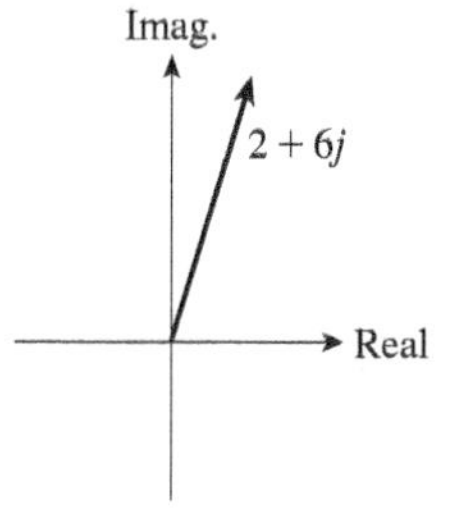

5.

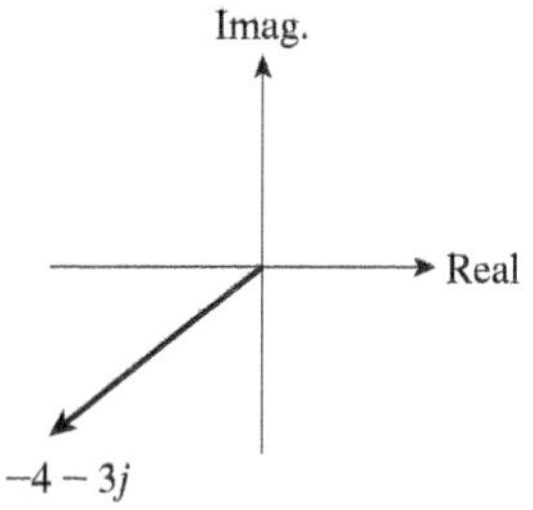

7. $-3j$

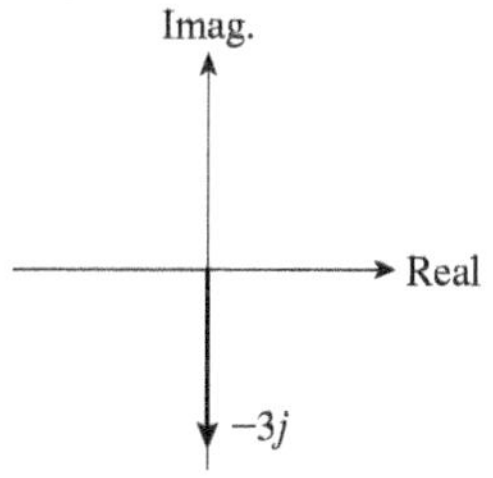

9. $5+4j$

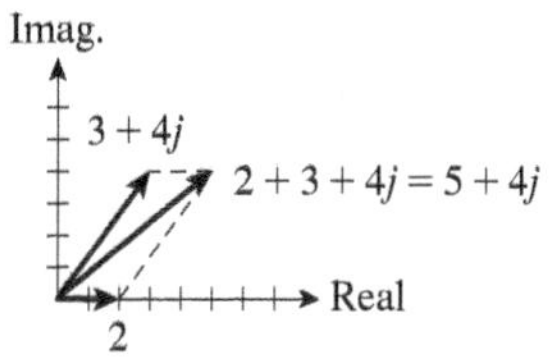

11. $8+j$

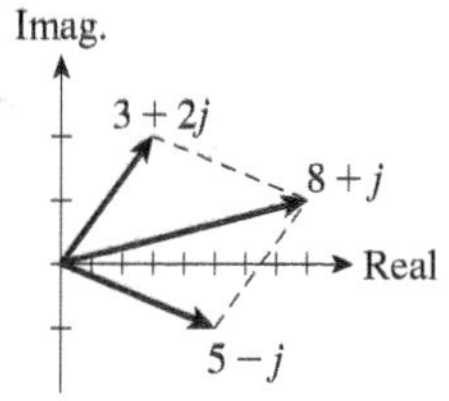

13. $-1+9j$

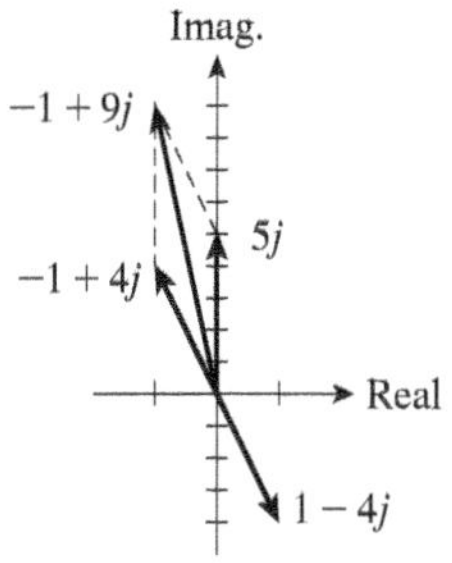

15. $-3j$

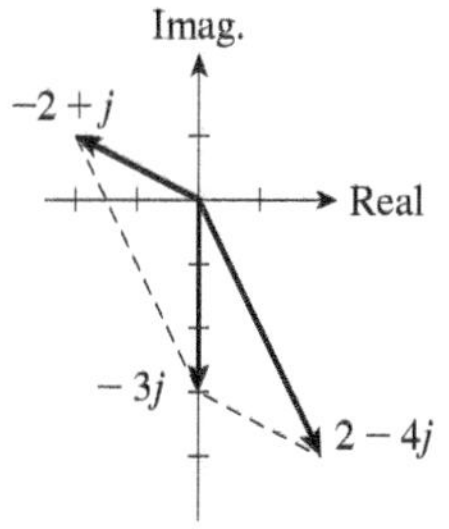

17. $-1+4j$

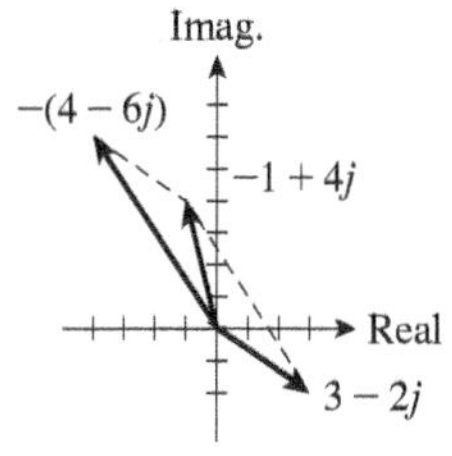

19. $-180+150j$

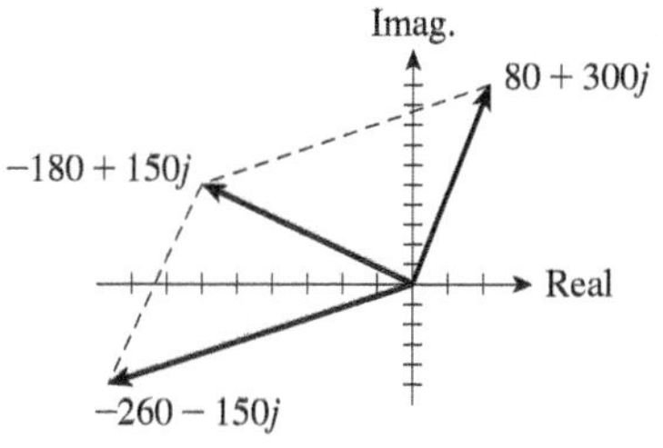

21. $4.5 + 2.0j$

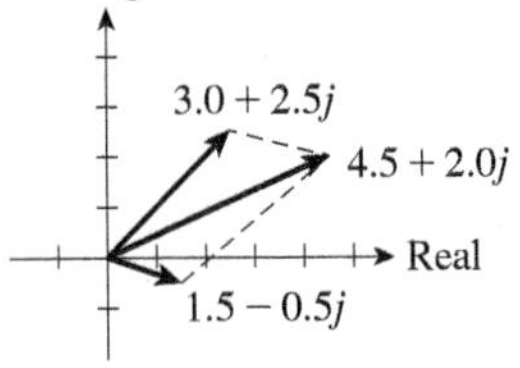

23. $4 + 2j$

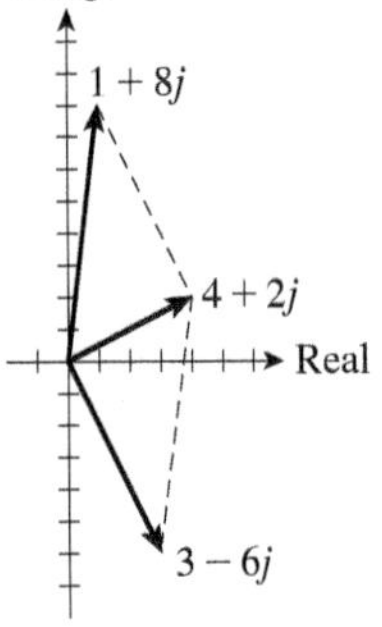

25. $-2j$

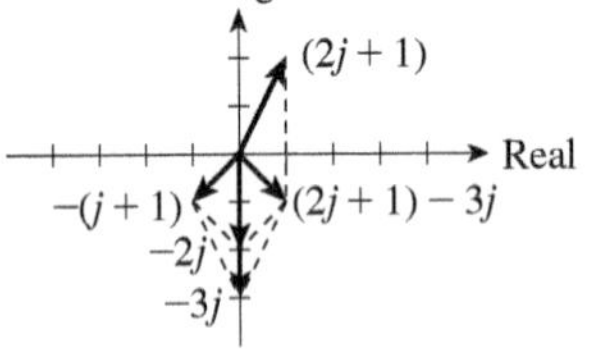

27. $-13 + j$

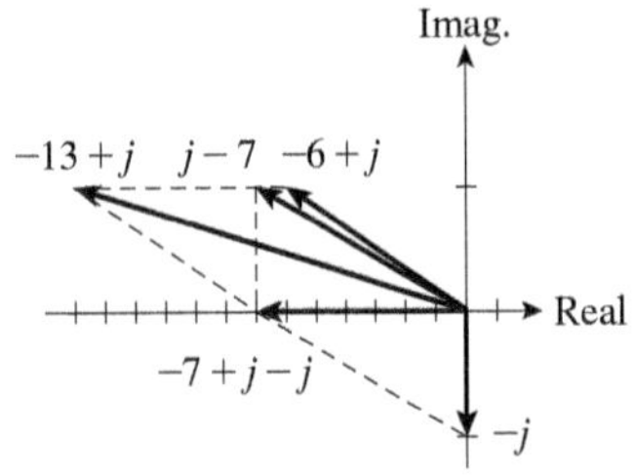

29.

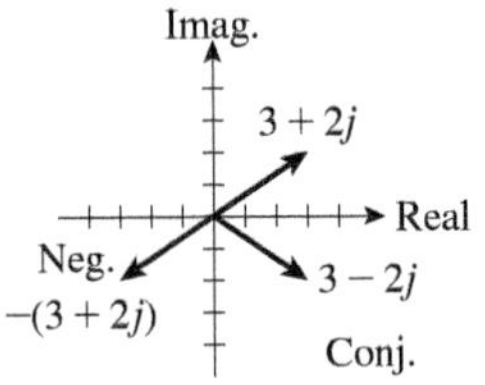

31.

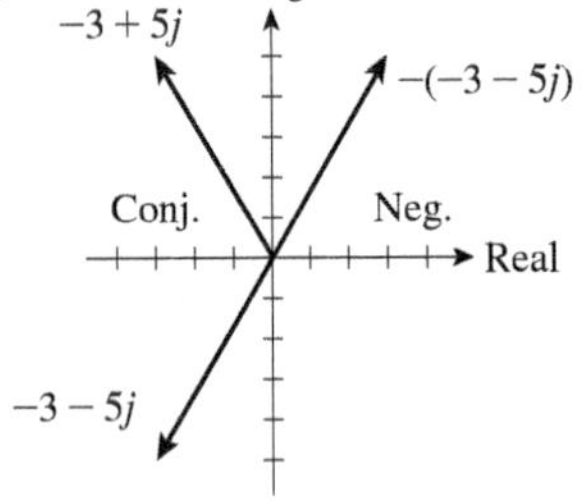

33.

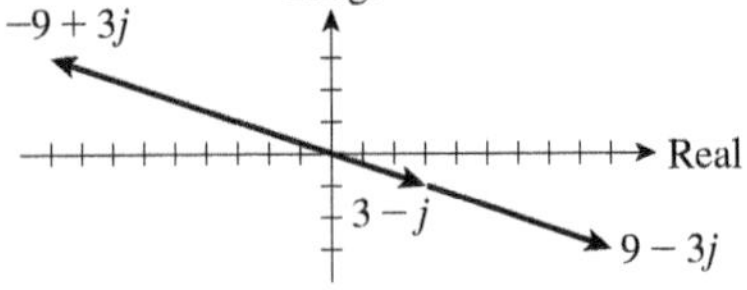

35. Reflections of each other about the real axis.

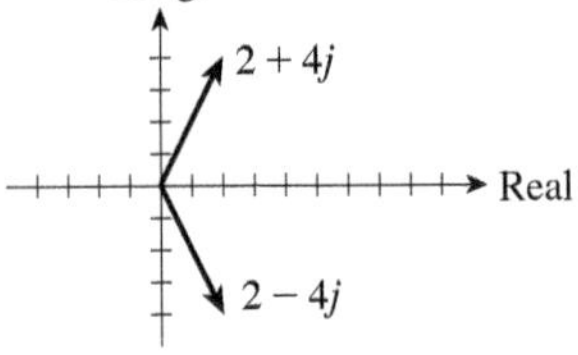

37. Subtracting the conjugate from the complex number results in an imaginary number.

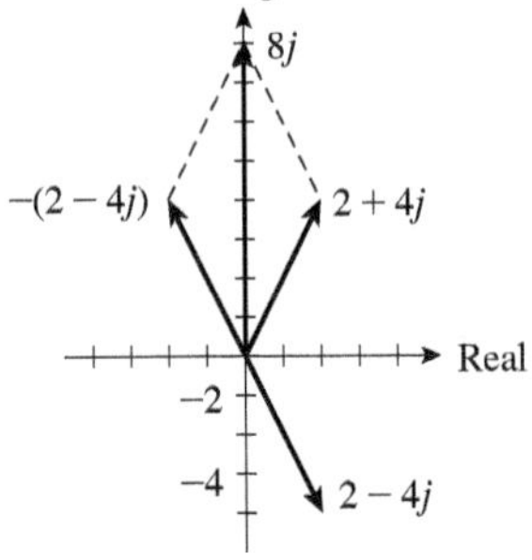

39. $90 - 15j$ N

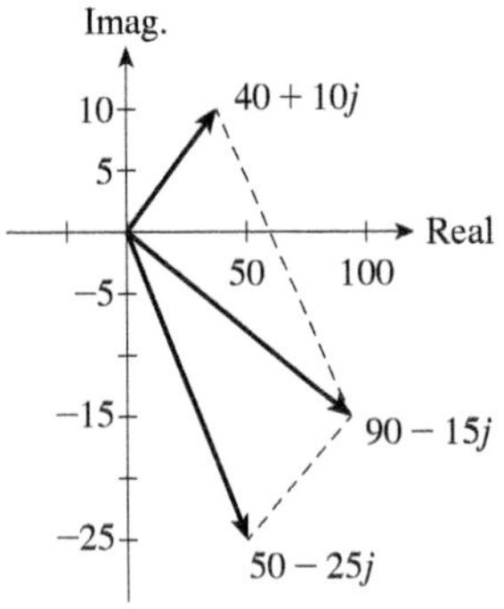

Exercises 12.4, page 368

1. $5(\cos 126.9° + j\sin 126.9°)$

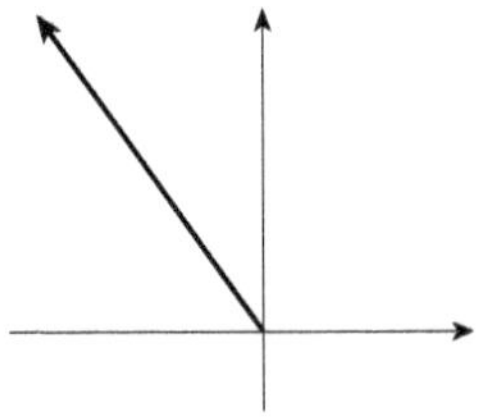

3. $10(\cos 36.9° + j\sin 36.9°)$

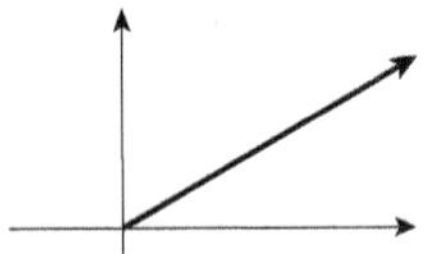

5. $50(\cos 306.9° + j\sin 306.9°)$

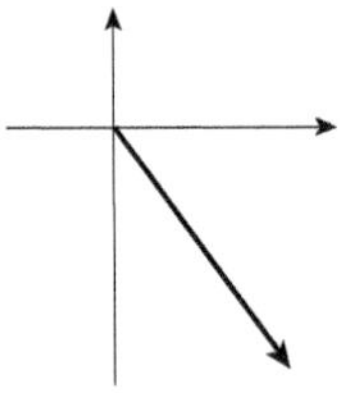

7. $3.61(\cos 123.7° + j\sin 123.7°)$

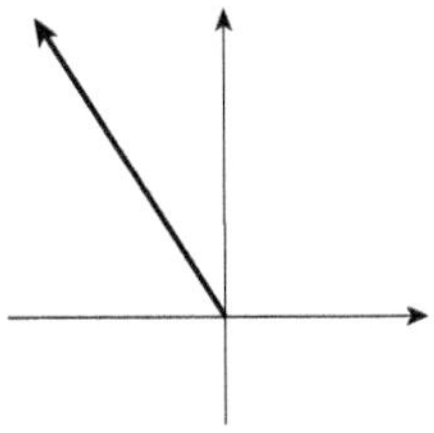

9. $0.60(\cos 204° + j\sin 204°)$

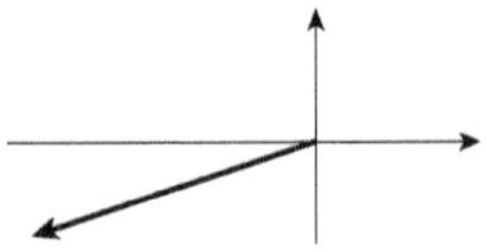

11. $2(\cos 60° + j\sin 60°)$

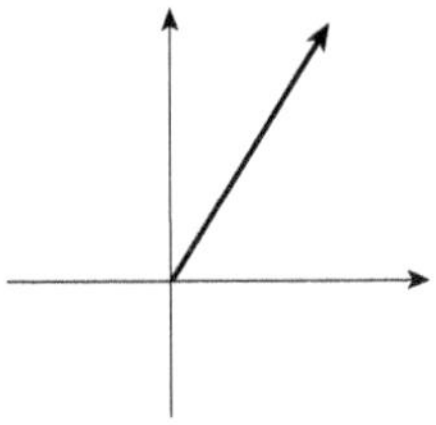

13. $8.062(\cos 295.84° + j\sin 295.84°)$

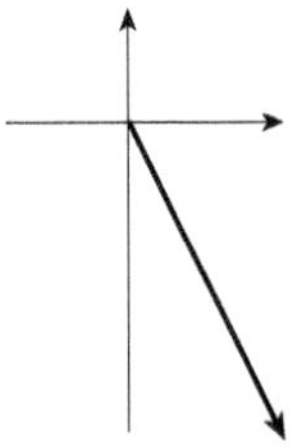

15. $3(\cos 180° + j\sin 180°)$

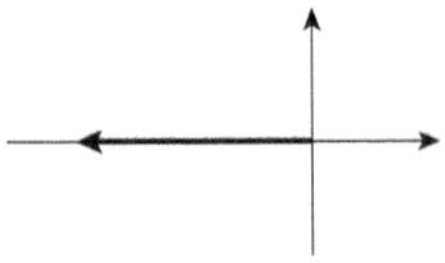

17. $9(\cos 90° + j\sin 90°)$

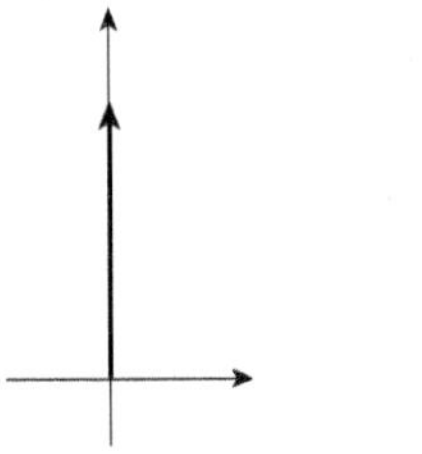

19. $2.94 + 4.05j$

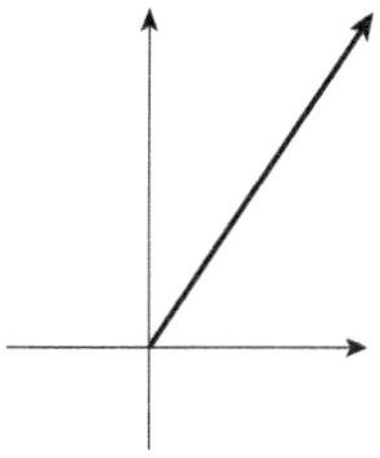

21. $-140 + 80j$

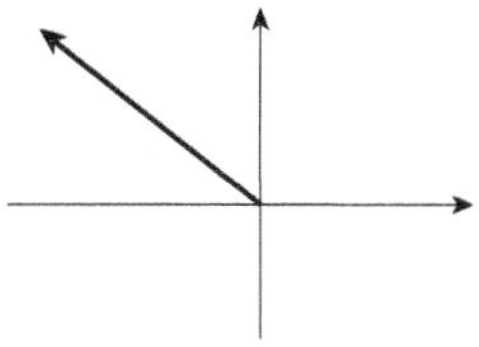

23. $-1.85 - 2.36j$

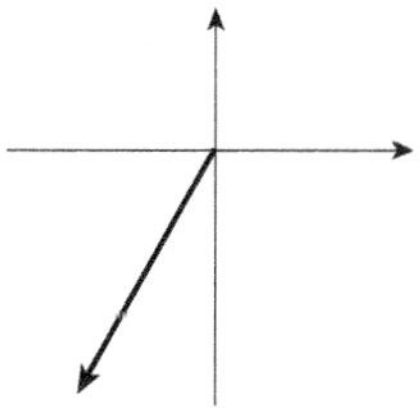

25. 0.08

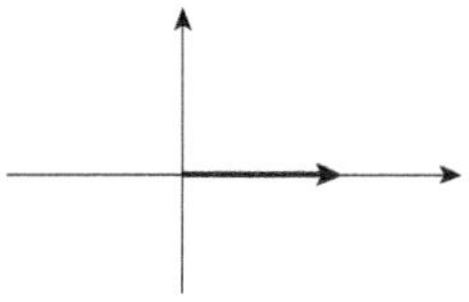

27. $-120j$

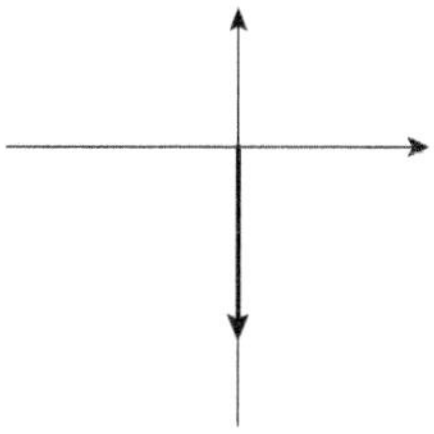

29. $-4.71 + 0.595j$

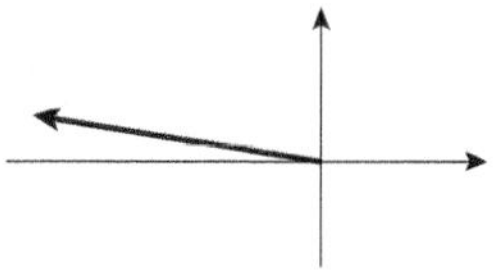

31. $-0.6052 - 0.7096j$

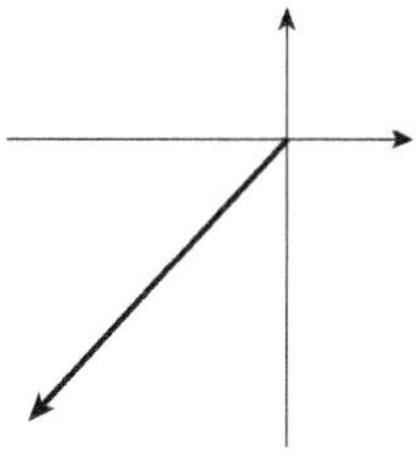

33. $7.32j$

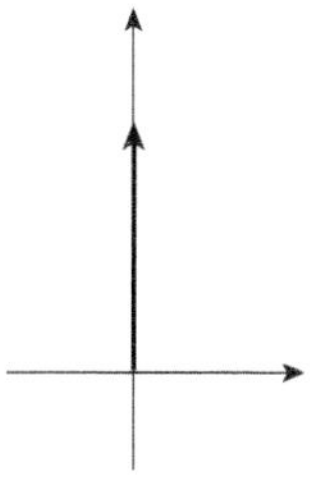

35. $-6.961 + 86.14j$

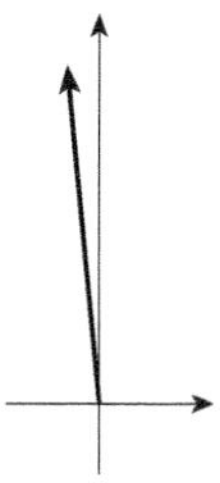

37. 180°

39. conj. of $r(\cos\theta + j\sin\theta) = r(\cos\theta - j\sin\theta)$
$= r(\cos(-\theta) + j\sin(-\theta))$

41. $3.03\angle 339.5^\circ$ kV **43.** $2.51 + 12.1j$ V/m

Exercises 12.5, page 370

1. $8.50e^{3.95j}$ **3.** $3.00e^{1.05j}$ **5.** $0.450e^{4.93j}$ **7.** $375.5e^{4.617j}$
9. $0.5150e^{3.461j}$ **11.** $4.06e^{-1.07j} = 4.06e^{5.21j}$ **13.** $9245e^{5.171j}$
15. $5e^{5.36j}$ **17.** $36.1e^{2.55j}$ **19.** $6.37e^{0.386j}$ **21.** $825.7e^{3.836j}$
23. $3.00\angle 28.6^\circ$, $2.63 + 1.44j$ **25.** $464\angle 106.0^\circ$, $-128 + 446j$
27. $3.20\angle 50.0^\circ$, $2.06 + 2.45j$
29. $0.1724\angle 137.0^\circ$, $-0.1261 + 0.1176j$
31. $-18.2 + 9.95j$, $20.7\angle 151.3^\circ$
33. $-0.891 - 27.5j$, $27.5\angle 268^\circ$ **35.** $391e^{0.288j}$ ohms; $391\,\Omega$
37. $3.17 \times 10^{-4}e^{0.478j}\ 1/\Omega$

Exercises 12.6, page 375

1. $5.09(\cos 101.3^\circ + j\sin 101.3^\circ)$
3. $613(\cos 281.5^\circ + j\sin 281.5^\circ)$ **5.** $8(\cos 80^\circ + j\sin 80^\circ)$
7. $3(\cos 250^\circ + j\sin 250^\circ)$ **9.** $2(\cos 35^\circ + j\sin 35^\circ)$
11. $2.4\angle 170^\circ$ **13.** $0.008(\cos 105^\circ + j\sin 105^\circ)$
15. $256(\cos 0^\circ + j\sin 0^\circ)$ **17.** $0.8\angle 273^\circ$ **19.** $5\angle 87^\circ$
21. $1.73\angle 79.8^\circ$ **23.** $11\,750\angle 115.91^\circ$
25. $65.0(\cos 345.7^\circ + j\sin 345.7^\circ) = 63 - 16j$
27. $61.4(\cos 343.9^\circ + j\sin 343.9^\circ); 59 - 17j$
29. $2.21(\cos 71.6^\circ + j\sin 71.6^\circ); \frac{7}{10} + \frac{21}{10}j$
31. $3.85(\cos 120.5^\circ + j\sin 120.5^\circ) = -1.95 + 3.31j$
33. $625(\cos 212.5^\circ + j\sin 212.5^\circ) = -527 - 336j$
35. $609.7(\cos 281.5^\circ + j\sin 281.5^\circ); 122 - 597j$
37. $2(\cos 30^\circ + j\sin 30^\circ); 2(\cos 210^\circ + j\sin 210^\circ)$
39. $-0.364 + 1.67j, -1.26 - 1.15j, 1.63 - 0.520j$
41. $1.10 + 0.455j, -1.10 - 0.455j$
43. $1, -1, j, -j$ **45.** $3j, -\frac{3}{2}(\sqrt{3} + j), \frac{3}{2}(\sqrt{3} - j)$
47. $1.62 + 1.18j, -0.618 + 1.90j, -2, -0.618 - 1.90j, 1.62 - 1.18j$
49. $-5, \frac{5}{2} + j\frac{5\sqrt{3}}{2}, \frac{5}{2} - j\frac{5\sqrt{3}}{2}$
51.
$$\left[\frac{1}{2}(1 - j\sqrt{3})\right]^3 = \frac{1}{8}[1 - 3(j\sqrt{3}) + 3(j\sqrt{3})^2$$
$$-(j\sqrt{3})^3] = \frac{1}{8}[1 - 3j\sqrt{3} - 9 + 3j\sqrt{3}]$$
$$= \frac{1}{8}(-8) = -1$$
53. $-1, \frac{1}{2} + j\frac{\sqrt{3}}{2}, \frac{1}{2} - j\frac{\sqrt{3}}{2}$ **55.** $p = 0.479\angle 40.5^\circ$ watts
57. $43.3\angle 165^\circ$, 43.3 V

Exercises 12.7, page 380

1. $V_R = 24.0$ V, $V_L = 32.0$ V, $V_{RL} = 40.0$ V, $\theta = 53.1^\circ$ (voltage leads current)
3. 12.9 V **5.** (a) $2850\,\Omega$ (b) 37.9° (c) 16.4 V
7. (a) $14.6\,\Omega$ (b) -90.0° **9.** (a) $47.8\,\Omega$ (b) 19.8°
11. 38.0 V **13.** $54.4\,\Omega$, -62.3° **15.** 0.682 H
17. 208 kHz **19.** 1.30×10^{-11} F = 13 pF
21. 1.02 mW **23.** $21.4 - 33.9j\,\Omega$

Review Exercises for Chapter 12, page 382

1. $10 - j$ **3.** $6 + 2j$ **5.** $9 + 2j$ **7.** $-12 + 66j$
9. $\frac{1}{85}(21 + 18j)$ **11.** $-2 - 3j$ **13.** $\frac{1}{10}(-12 + 9j)$
15. $\frac{1}{5}(13 + 11j)$ **17.** $x = -3, y = -2$
19. $x = \frac{10}{13}, y = \frac{11}{13}$

21. $3 + 11j$

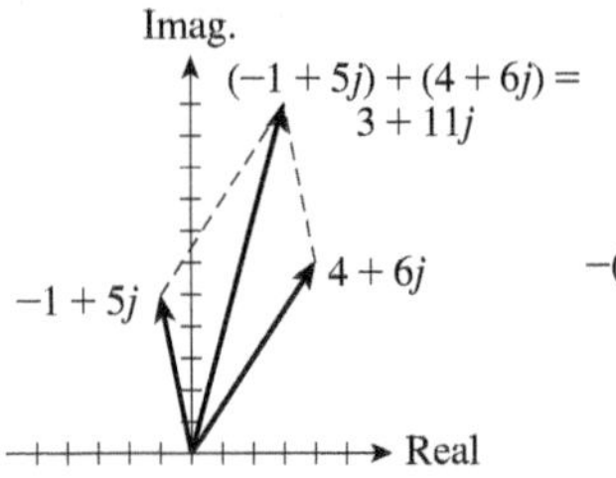

23. $4 + 8j$

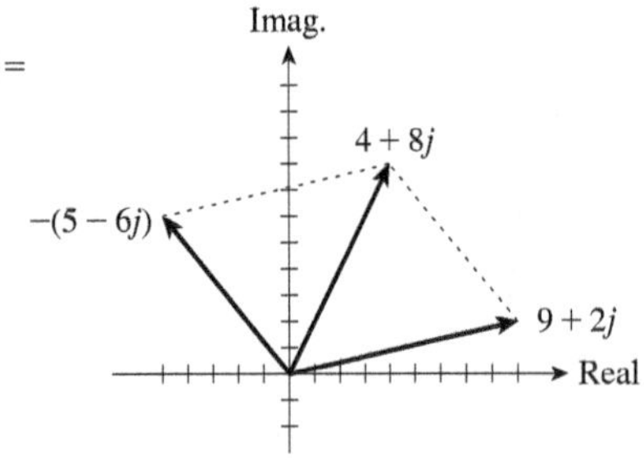

25. $\sqrt{2}(\cos 315° + j\sin 315°) = \sqrt{2}e^{5.50j}$

27. $80.1(\cos 254.1° + j\sin 254.1°) = 80.1e^{4.43j}$

29. $4.67(\cos 76.8° + j\sin 76.8°); 4.67e^{1.34j}$

31. $5000\underline{/0°}, 5000e^{0j}$ **33.** $-\sqrt{2} - j\sqrt{2}$ **35.** $-2.789 + 4.163j$

37. $0.19 - 0.59j$ **39.** $26.31 - 6.427j$ **41.** $1.94 + 0.495j$

43. $-728.1 + 1017j$ **45.** $15(\cos 84° + j\sin 84°)$

47. $20\underline{/263°}$ **49.** $8(\cos 59° + j\sin 59°)$ **51.** $14.29\underline{/133.61°}$

53. $1.26\underline{/59.7°}$ **55.** $9682\underline{/249.5°}$

57. $1024(\cos 160° + j\sin 160°)$ **59.** $343\underline{/331.5°}$

61. $32(\cos 270° + j\sin 270°) = -32j$

63. $\frac{625}{2}(\cos 270° + j\sin 270°) = -\frac{625}{2}j$

65. $1 + j\sqrt{3}, -2, 1 - j\sqrt{3}$

67. $0.383 + 0.924j, -0.924 + 0.383j, -0.383 - 0.924j, 0.924 - 0.383j$

69. $40 + 9j, 41(\cos 12.7° + j\sin 12.7°)$

71. $-15.0 - 10.9j, 18.5(\cos 216.0° + j\sin 216.0°)$

73. $15 - 16j$ **75.** $-j, -2j$ **77.** $x^2 - 4x + 5 = 0$

79. $1 - j$, Yes; $-1 - j$, No **81.** $\frac{1}{2}$ **83.** $2 + \frac{9}{2}j$ **85.** 60 V

87. −21.6° **89.** 22.9 Hz **91.** $5500(\cos 53° + j\sin 53°)$ N

93. $\dfrac{\mu - j\omega n}{\mu^2 + \omega^2 n^2}$ **95.** $e^{j\pi} = \cos\pi + j\sin\pi = -1$

Chapter 13: Exponential and Logarithmic Functions

Exercises 13.1, page 388

1. $-\frac{1}{4}$ **3.** 11.7 **5.** 0.006 77 **7.** (a) Yes (b) Yes

9. (a) No (base cannot be negative) (b) Yes

11. $\sqrt{7}$ **13.** $\frac{1}{49}$ **15.** $\frac{1}{7^{3/2}}$

17.

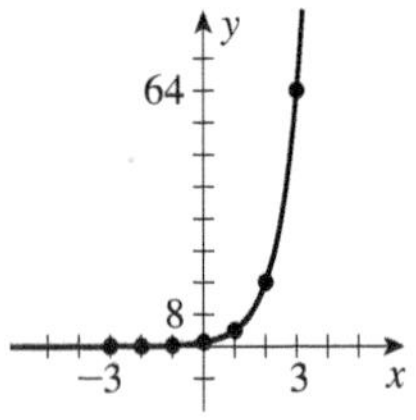

19.

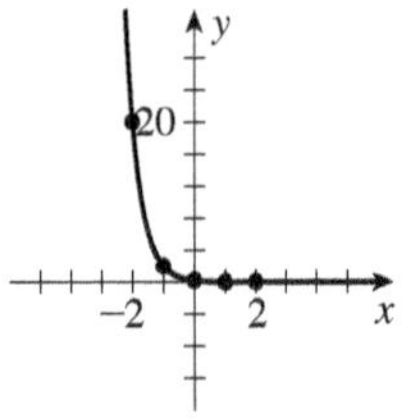

21.

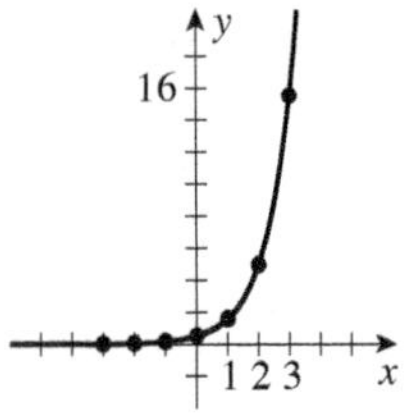

23.

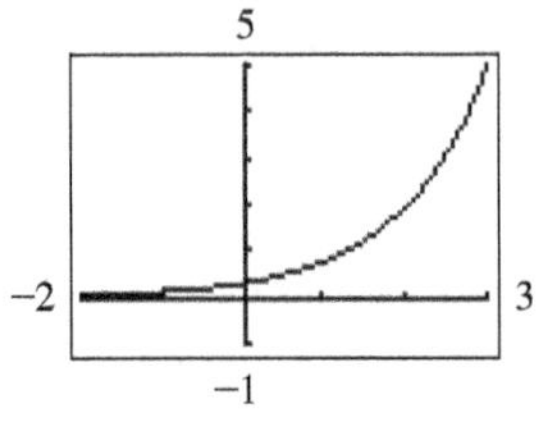

25.

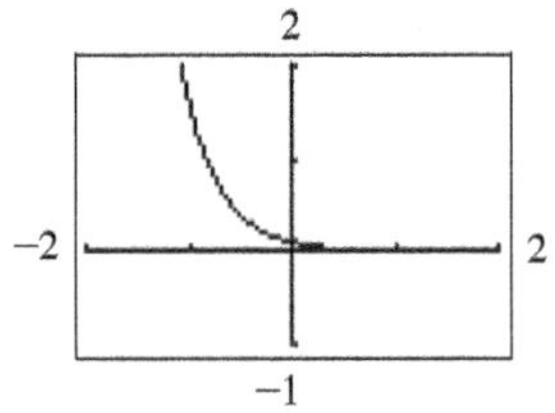

27.

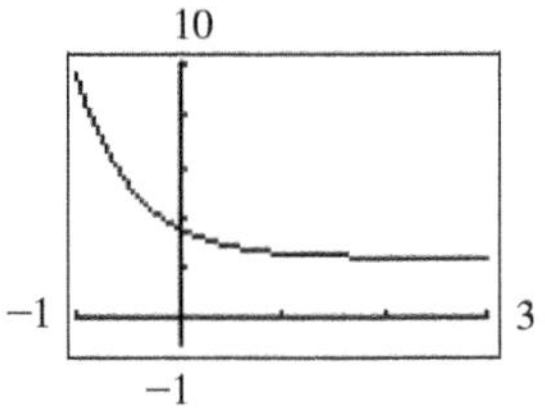

29. 4 **31.** $f(c + d) = b^{c+d} = b^c b^d = f(c) \cdot f(d)$

33.

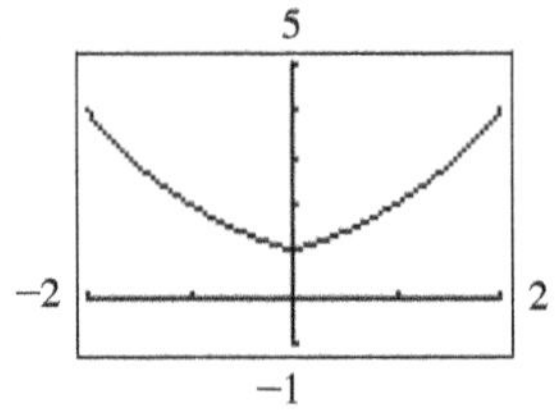

35. −0.767, 2, 4 **37.** \$304

39. 0.001 25 mA **41.**

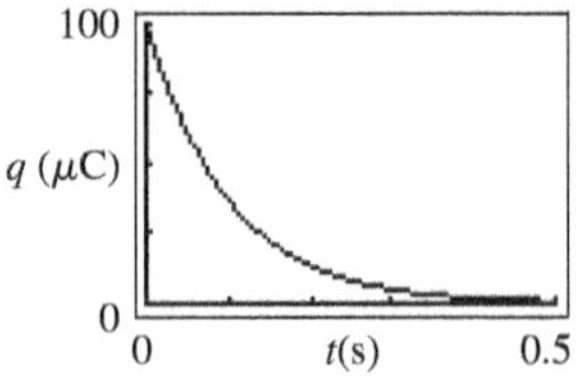

Exercises 13.2, page 392

1. $32^{4/5} = 16$ in logarithmic form is $\frac{4}{5} = \log_{32} 16$.

3. If $x = 16$, $y = \log_4 16$ means $y = 2$, since $4^2 = 16$. If $x = \frac{1}{16}$, $y = \log_4\left(\frac{1}{16}\right)$ means $y = -2$, since $4^{-2} = \frac{1}{16}$.

5. $\log_3 81 = 4$ **7.** $\log_4 1024 = 5$ **9.** $\log_7\left(\frac{1}{49}\right) = -2$

11. $\log_2\left(\frac{1}{64}\right) = -6$ **13.** $\log_8 2 = \frac{1}{3}$ **15.** $\log_{1/4}\left(\frac{1}{16}\right) = 2$

17. $81 = 3^4$ **19.** $9 = 9^1$ **21.** $5 = 25^{1/2}$ **23.** $3 = 243^{0.2}$

25. $0.1 = 10^{-1}$ **27.** $16 = (0.5)^{-4}$ **29.** 2 **31.** −2

33. 343 **35.** $\frac{9}{4}$ **37.** $\sqrt{5}$ **39.** $\frac{1}{64}$ **41.** 0.2 **43.** −4

45.

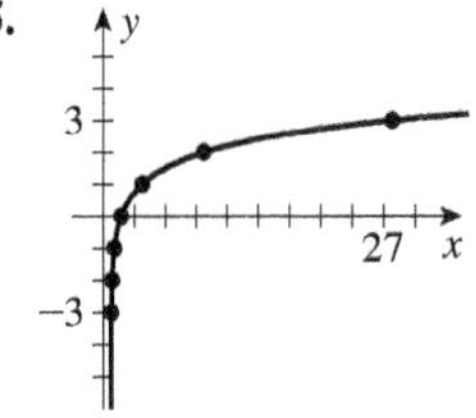

47.

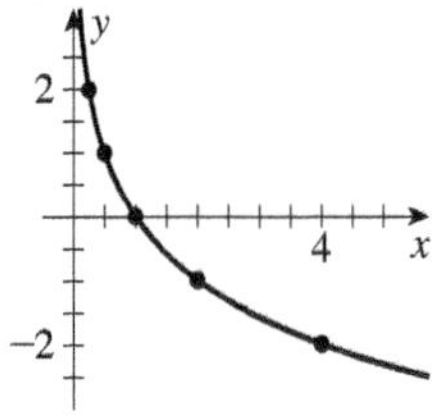

49.

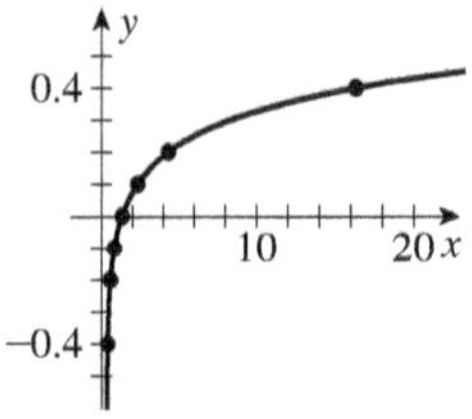

51.

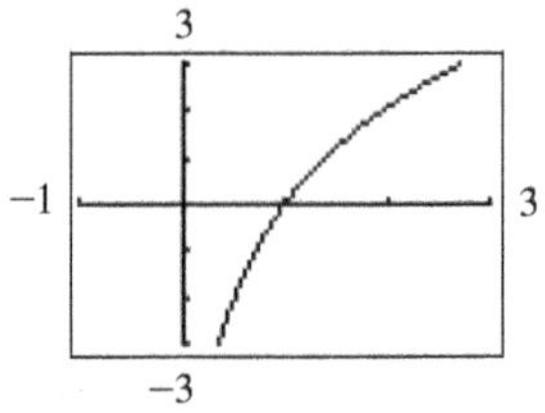

53.

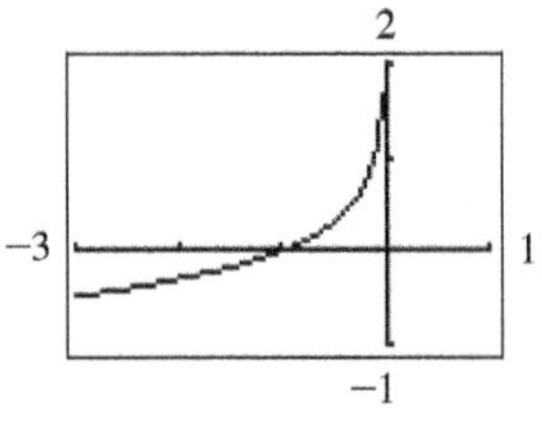

55. (a) -3 (b) No value (x cannot be negative)

57. (a) $\frac{1}{2}$ (b) Not defined

59.

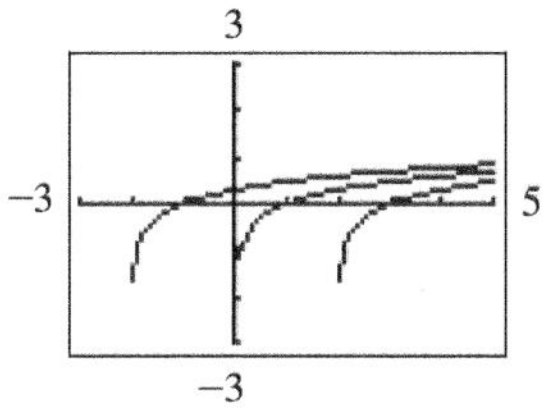

61. 0.0102, 2.38

63. $b_2 = b_1 10^{0.4(m_1 - m_2)}$ **65.** $N = N_0 e^{-kt}$ **67.** $V_1 = V_2 e^{-W/k}$

69.

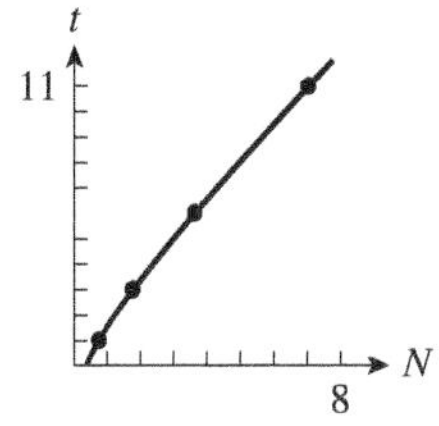

71. They are the same.

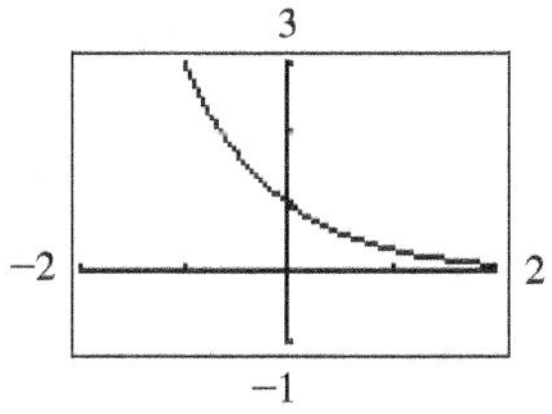

73.

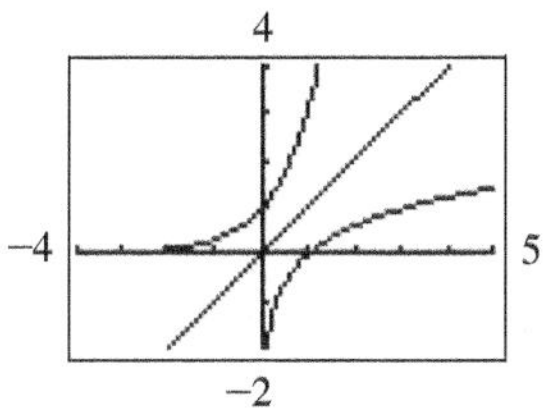

75.

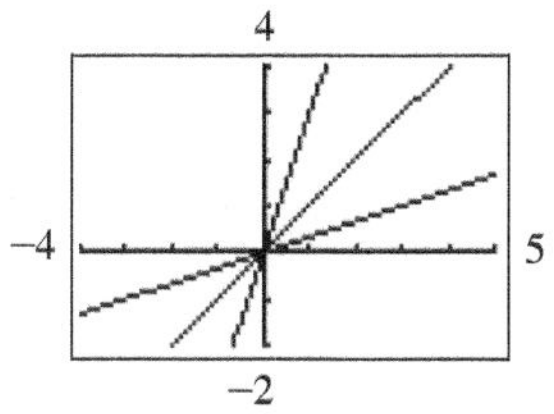

Exercises 13.3, page 397

1. $\log_4 3 + \log_4 7$ **3.** $\log_4\left(\frac{5}{x}\right)$ **5.** 3 **7.** $\frac{1}{2}$

9. $\log_5 2 + \log_5 17$ **11.** $\log_7 11 - 2\log_7 3$

13. $4\log_7 a$ **15.** $2\log_5 a + \log_5 b + \log_5 c$ **17.** $6\log_7 y$

19. $\frac{1}{2}\log_3 x - 5\log_3 a$ **21.** $\log_b HM$ **23.** $\log_3 2$

25. $\log_b x^{3/2}$ **27.** $\log_e\left[\frac{4\pi^3}{3}\right]$ **29.** -5 **31.** 2.5

33. 3 **35.** 8 **37.** $2 + \log_3 2$ **39.** $-1 - \log_2 3$

41. $\frac{1}{2}(1 + \log_3 2)$ **43.** $3 + \log_{10} 3$ **45.** $y = 3x$

47. $y = \frac{2x}{7}$ **49.** $y = \frac{49}{x^3}$ **51.** $y = 2(2ax)^{1/5}$

53. $y = \frac{2}{x}$ **55.** $y = \left(\frac{x^2}{9}\right)^{1/\log_5 3}$

57. $\log_{10} x + \log_{10} 3 = \log_{10} 3x$

59. $y = \log_e(e^2 x) = 2\log_e e + \log_e x = 2 + \log_e x$

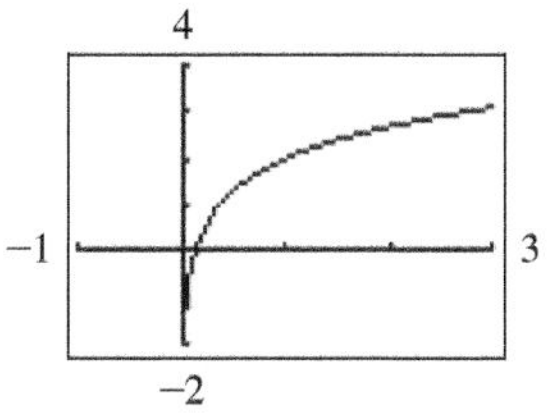

61. 8

63. $y_1 = y_2$

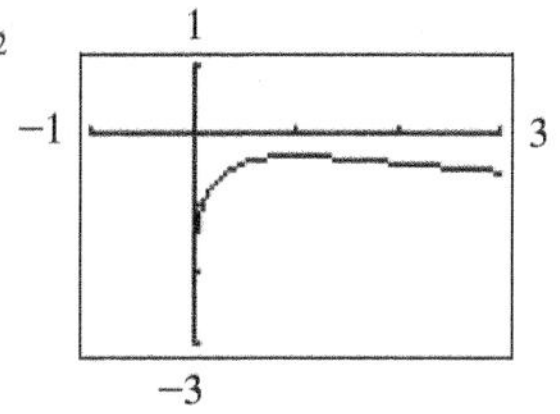

65. $D = ae^{cr^2 - br}$

67. $T = 90.0e^{-0.23t}$

Exercises 13.4, page 400

1. -0.4372 **3.** 2.829 **5.** 5.632 **7.** -0.7224

9. 1.325 **11.** 5.52×10^7 **13.** 0.035 46 **15.** 2000.4

17. 0.005 788 2 **19.** 85.5 **21.** 7.37×10^{101}

23. $1.1461 - 0.3010 = 0.8451$ **25.** $1.9085 = 1.9085$

27. 8.954 **29.** -13.89 **31.** -30.040 **33.** 2.65×10^7 m/s

35. 3.2×10^{-19} J **37.** 2 **39.** 15.2 dB **41.** 7.9

43. $2^{400} = 2.58 \times 10^{120}$

Exercises 13.5, page 403

1. 5.298 **3.** 3.258 **5.** 0.4460 **7.** $-4.916\,23$

9. 1.92 **11.** 4.806 **13.** 1.795 **15.** 3.809

17. 1.3646 **19.** -0.2589 **21.** $-11.847\,15$

23. 1.6549 **25.** $-0.164\,13$ **27.** 8.94

29. 1.0085 **31.** 0.4757 **33.** 6.20×10^{-11}

35.

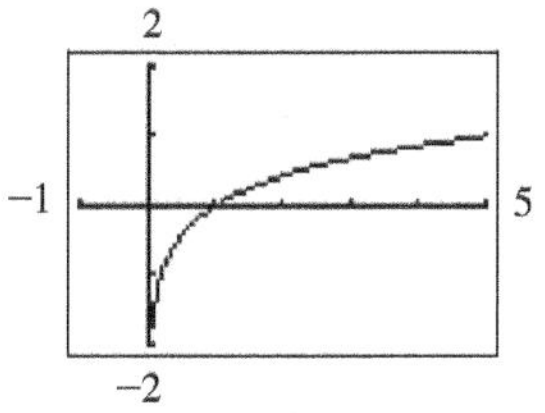

37.

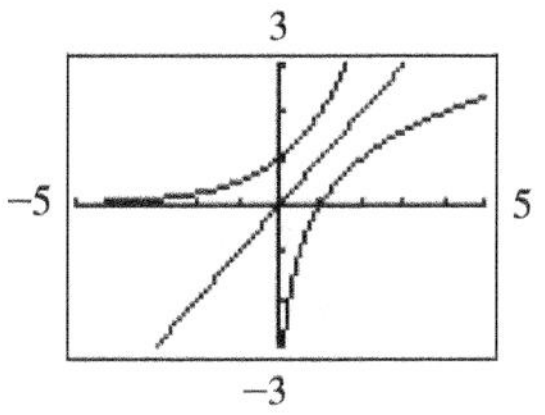

39. $1.6094 + 2.0794 = 3.6889$ **41.** $4.3944 = 4.3944$

43. 3 **45.** 10 **47.** $2x + y$ **49.** 2.45×10^9 Hz

51. 8.2% **53.** 0.38 s **55.** 21.7 s

Exercises 13.6, page 407

1. -0.535 **3.** 4 **5.** -0.7 **7.** 0.133 **9.** 6.71

11. 1.74 **13.** $0, \frac{\ln 0.3}{\ln 4} = -0.9$ **15.** 0.330 **17.** 1, 100

19. 3 **21.** -0.162 **23.** 250 **25.** 4 **27.** 2

29. 20.1 **31.** -0.104 **33.** 0.203 **35.** 0.974

37. 10.9 **39.** ± 0.9624 **41.** 0.0025 **43.** $(-1.21, 0)$

45. 28.0 **47.** Noon of previous day (36 h earlier)

49. 1.72×10^{-5} mol/L **51.** 3.55×10^{22}

53. $c = 15e^{-0.20t}$ **55.** $P = P_0(0.999)^t$

57.

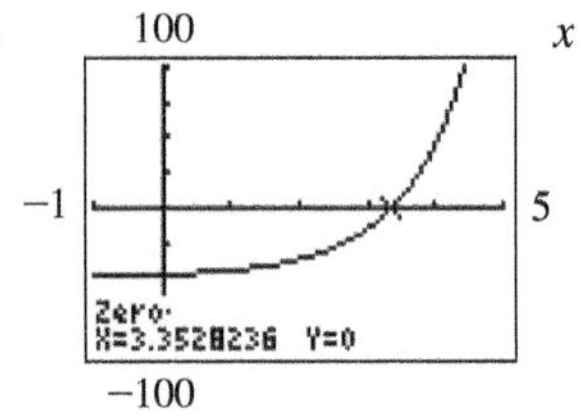

$x = 3.35$

59. ± 3.66 m

Exercises 13.7, page 411

1.

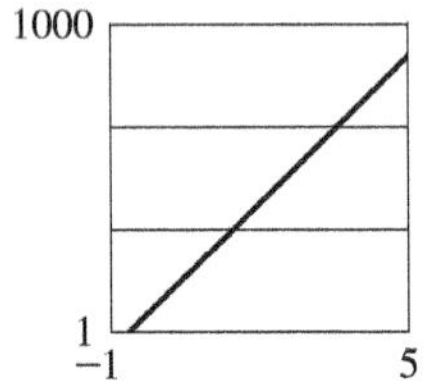

3.

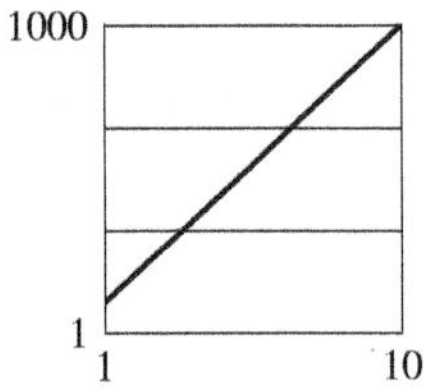

5.

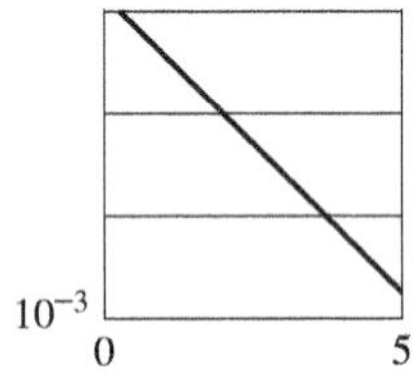

7.

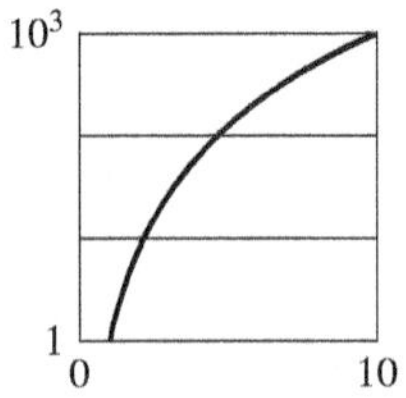

9.

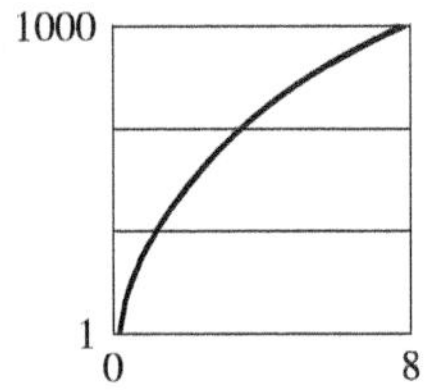

11.

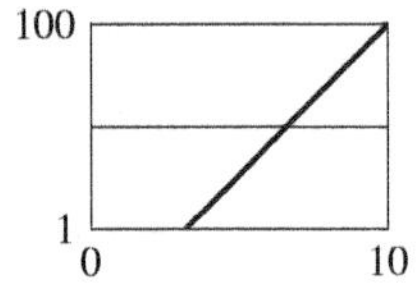

13.

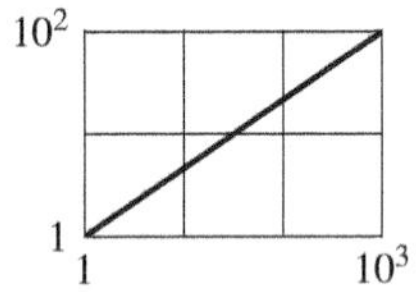

15.

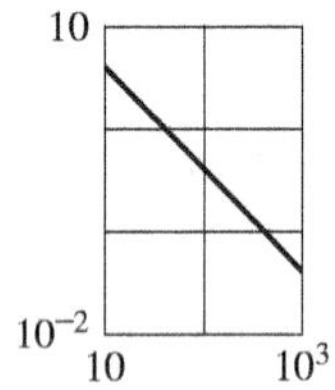

17.

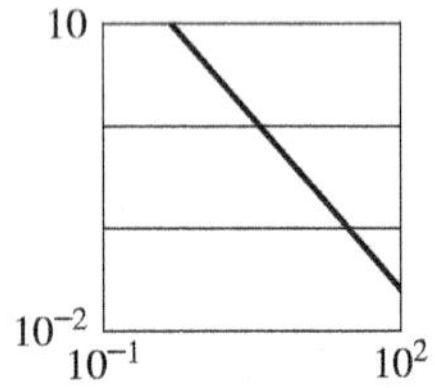

19. Semilog scale

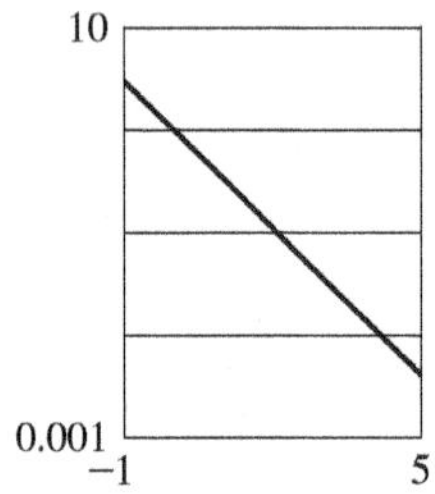

21. Log-log scale

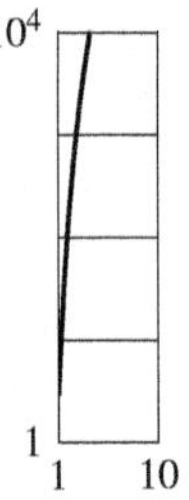

23. Semilog scale

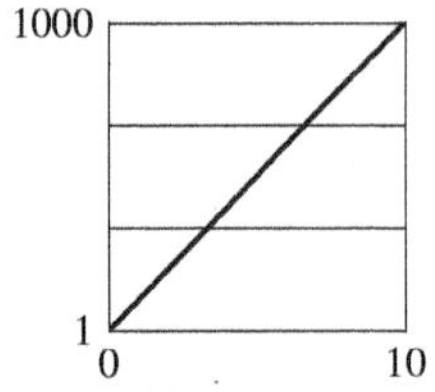

25. Log-log scale

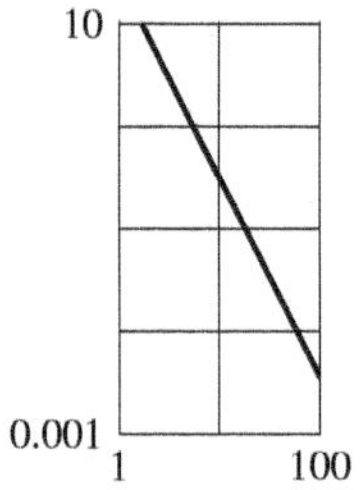

27. (a)

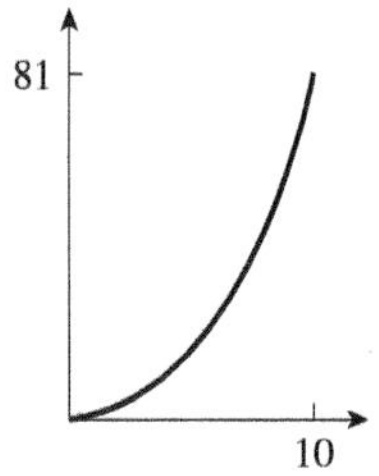

27. (b)

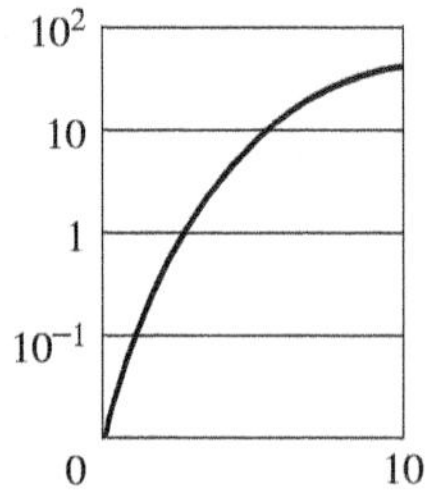

29.

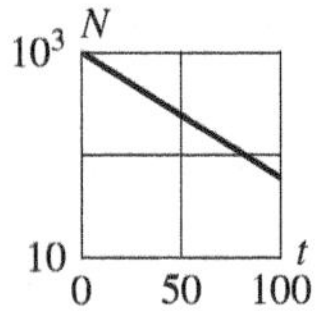

31.

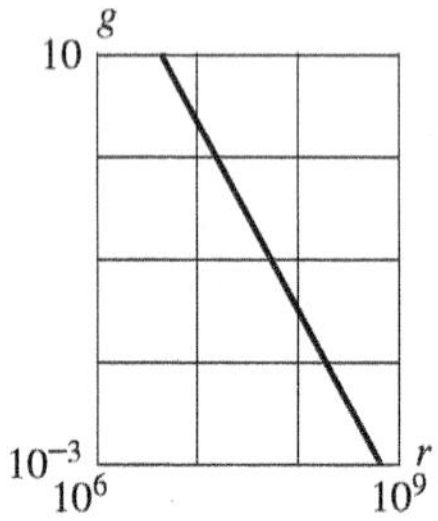

33.

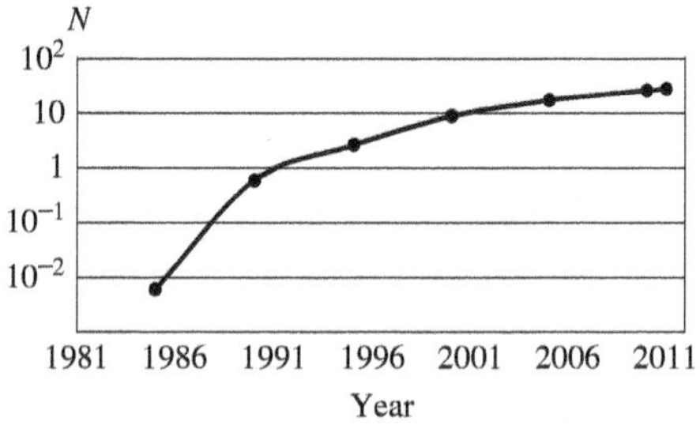

35.

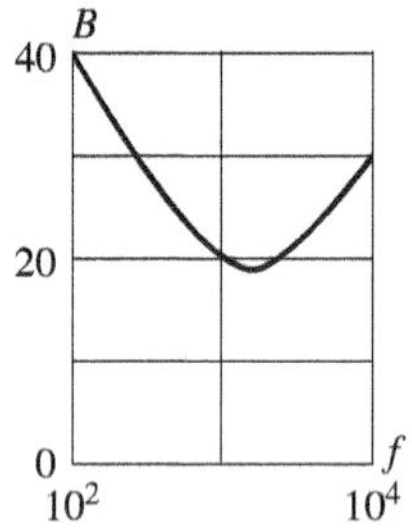

37.

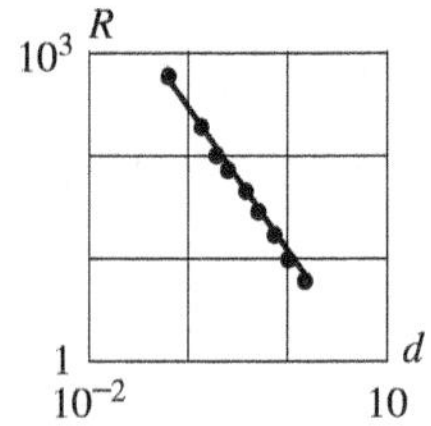

39.

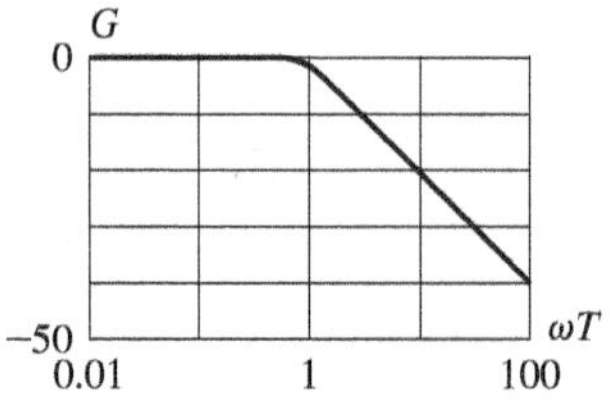

Review Exercises for Chapter 13, page 413

1. 10 000 **3.** $\frac{1}{5}$ **5.** 6 **7.** $\frac{5}{3}$ **9.** 6 **11.** 100

13. $\log_3 2 + \log_3 x$ **15.** $2 \log_3 t$ **17.** $2 + \log_2 7$

19. $1 + \frac{1}{2}\log_4 3$ **21.** $2 - \log_3 x$ **23.** $3 + 4 \log_{10} x$

25. $y = \frac{4}{x}$ **27.** $y = e^{2/3}x$ **29.** $y = \sqrt{7x}$

31. $y = \frac{15}{x}$ **33.** $y = 8x^3$ **35.** $y = \frac{x}{\ln 2}$

37.

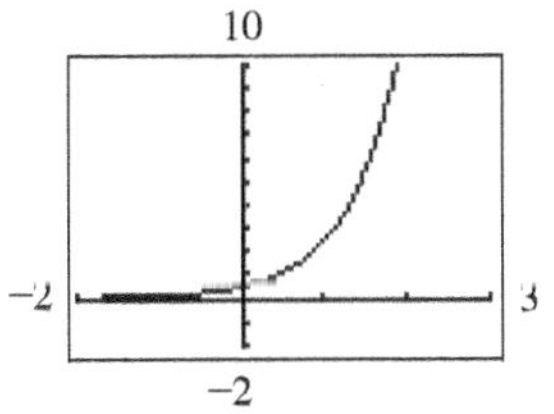

39.

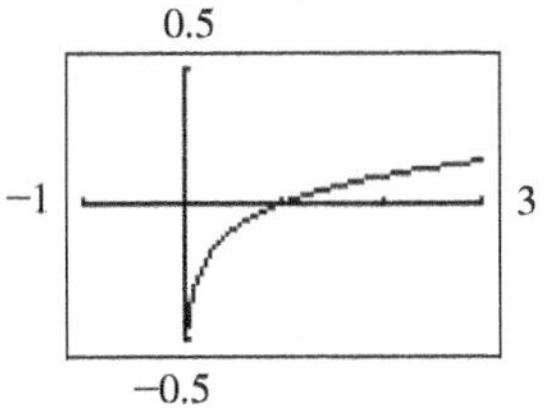

41.

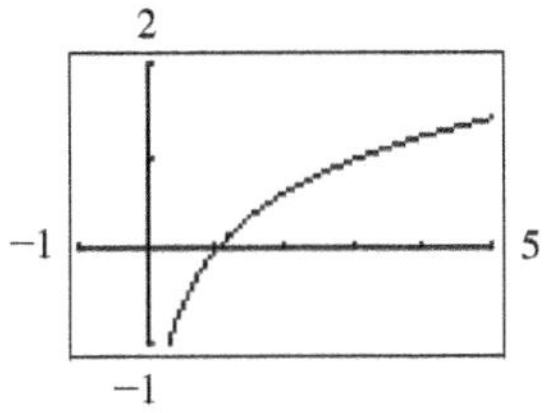

43.

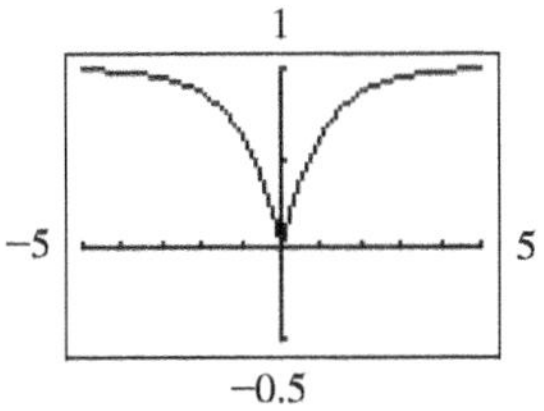

45. 2.182 **47.** −3.32 **49.** 1.8188 **51.** −2.7657

53. 0.805 **55.** 4.30 **57.** 2 **59.** −0.2

61.

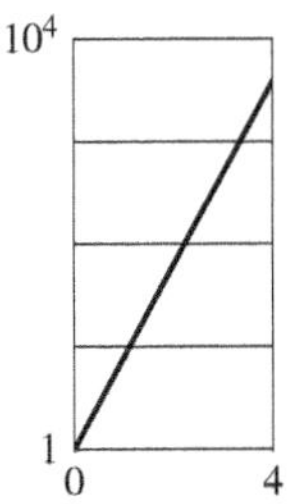

63.

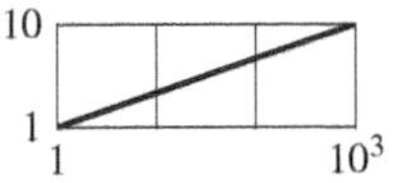

65. 4 **67.** 12 **69.** $0.9542 - 0.7782 = 0.1761$ **71.** 0

73. 1 **75.** 0.706 **77.** 1.28×10^{-5} or 3.925

79. $V = 1000(1.03)^{2t}$, $t = \frac{\log(V/1000)}{2 \log 1.03}$ **81.** $I = I_0 e^{-\beta h}$

83. 4.60×10^6 N · m **85.** 5.42 hours

87.

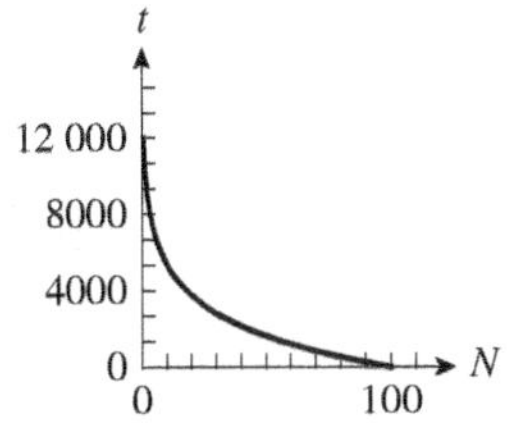

89. $\sin\theta = \frac{l\omega^2}{3g}$ **91.** $R = 2^{C/B} - 1$ **93.** 910 times brighter

95. 1.17 min **97.** 0.813 cm **99.** $\dfrac{\ln \frac{p}{p_0}}{\ln \frac{pT_0}{p_0 T}}$

101. $n = 20e^{-0.04t}$ **103.**

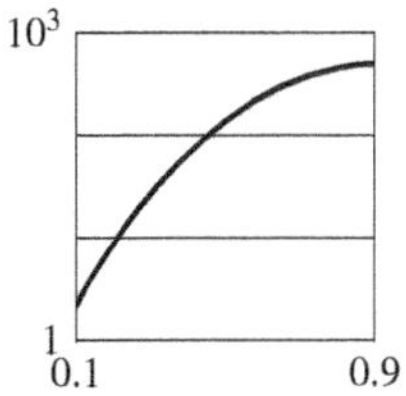

Chapter 14: Additional Types of Equations and Systems of Equations

Exercises 14.1, page 419

1. $x^2 - y^2 = 25$

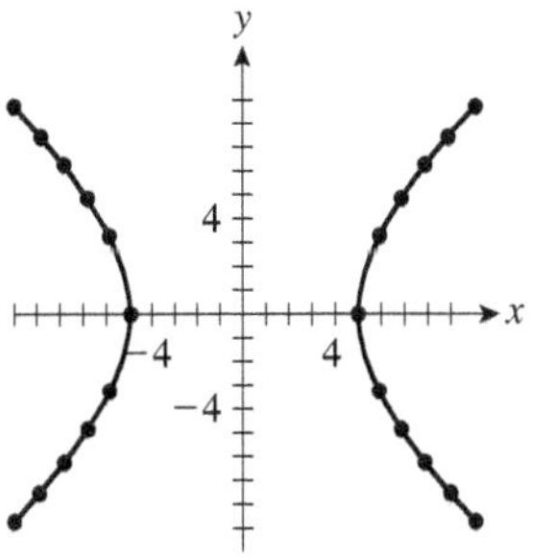

3. $x = -1.2, y = 0.27; x = 0.76, y = 2.3$

5. $x = 1.79, y = 3.58; x = -1.79, y = -3.58$

7. $x = 0.00, y = -2.00; x = 2.67, y = -0.67$

9. $x = 1.50, y = 0.25$ **11.** $x = 2.79, y = 10.74$

13. $x = 1.10, y = 2.79; x = -1.10, y = 2.79;$ $x = 2.41, y = -1.80; x = -2.41, y = -1.80$

15. No real solutions

17. $x = -2.83, y = -1.00; x = 2.83, y = 1.00;$ $x = 2.83, y = -1.00; x = -2.83, y = 1.00$

19. $x = -0.81, y = 2.52; x = 2.56, y = 0.66$

21. $x = 0.00, y = 0.00; x = 0.88, y = 0.77$

23. $x = 0.48, y = 0.62$ **25.** $x = 16.34, y = 16.12$

27. $x = 3.64, y = 0.97$ **29.** $x = -2.06, y = 4.23; x = 1.06, y = 1.12$

31.

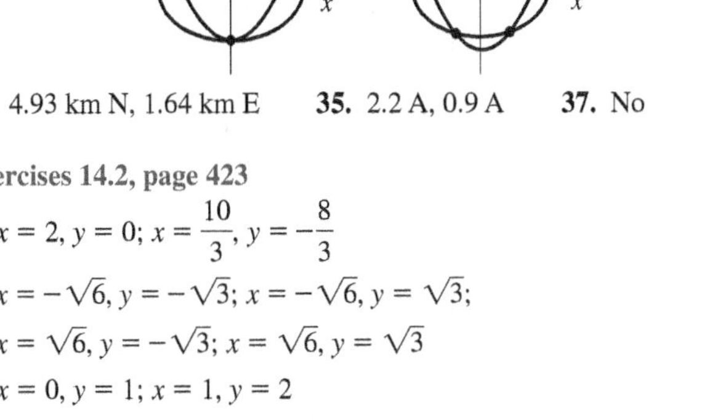

33. 4.93 km N, 1.64 km E **35.** 2.2 A, 0.9 A **37.** No

Exercises 14.2, page 423

1. $x = 2, y = 0; x = \frac{10}{3}, y = -\frac{8}{3}$

3. $x = -\sqrt{6}, y = -\sqrt{3}; x = -\sqrt{6}, y = \sqrt{3};$ $x = \sqrt{6}, y = -\sqrt{3}; x = \sqrt{6}, y = \sqrt{3}$

5. $x = 0, y = 1; x = 1, y = 2$

7. $x = -\frac{19}{5}, y = \frac{17}{5}; x = 5, y = -1$ **9.** $x = 1, y = 0$

11. $x = \frac{2}{7}(3 + \sqrt{2}), y = \frac{2}{7}(-1 + 2\sqrt{2});$
$x = \frac{2}{7}(3 - \sqrt{2}), y = \frac{2}{7}(-1 - 2\sqrt{2})$

13. $w = 1, h = 1$ **15.** $x = -\frac{3}{2}, y = -\frac{8}{3}; x = 2, y = 2$

17. $x = -5, y = 25; x = 5, y = 25$

19. $x = 1, y = 2; x = -1, y = 2;$
$x = 2j, y = -3; x = -2j, y = -3$

21. $D = 1, R = 0; D = -1, R = 0; D = \frac{1}{2}\sqrt{6}, R = \frac{1}{2};$
$D = -\frac{1}{2}\sqrt{6}, R = \frac{1}{2}$

23. $x = \sqrt{19}, y = \sqrt{6}; x = \sqrt{19}, y = -\sqrt{6};$
$x = -\sqrt{19}, y = \sqrt{6}; x = -\sqrt{19}, y = -\sqrt{6}$

25. $x = -5, y = -2; x = -5, y = 2; x = 5, y = -2; x = 5, y = 2$

27. $x = -3, y = 2; x = -1, y = -2$ **29.** $x = a, y = b$

31. $x = 50$ km, $h = 25$ km **33.** 2.00 cm, 4.00 cm

35. 13.0 cm, 16.6 cm **37.** 2.19 m, 0.21 m **39.** 12, 10

41. 12.0 cm, 18.0 cm **43.** 130 km/h

Exercises 14.3, page 426

1. $\pm\frac{1}{2}j\sqrt{2}, \pm 2$ **3.** $-3, -1, 1, 3$ **5.** $-\frac{3}{2}, \frac{1}{3}$

7. $-\frac{1}{2}, \frac{1}{2}, \frac{1}{6}j\sqrt{6}, -\frac{1}{6}j\sqrt{6}$ **9.** $1, \frac{25}{4}$ **11.** $\frac{64}{729}, 1$

13. $-27, 125$ **15.** 81 **17.** 26 **19.** $-2, -1, 3, 4$

21. 18 **23.** $1, -1, j\sqrt{2}, -j\sqrt{2}$ **25.** 0 **27.** $\pm 2, \pm 4$

29. 1, 4 **31.** 10, 100 **33.** $-2, -1, 1, 2$

35. $R_1 = 2.62\ \Omega, R_2 = 1.62\ \Omega$ **37.** 0.610

39. 40.4 cm, 55.4 cm

Exercises 14.4, page 430

1. $\frac{13}{12}$ **3.** 10 **5.** 12 **7.** 3 **9.** $\frac{2}{3}$ **11.** -1

13. 15 **15.** 4, 9 **17.** ± 6

19. 12 (Extraneous root introduced in squaring both sides of $\sqrt{x + 4} = x - 8$.)

21. 16 **23.** $\frac{1}{2}$ **25.** $7, -1$ **27.** 0 **29.** 5

31. 25 **33.** 258 **35.** 4 **37.** 4 **39.** 16

41. $x = 2$; For $x = 5$, squaring $3 - x$ is squaring a negative number. A different extraneous root is introduced.

43. $-r \pm \sqrt{r^2 + R^2}$ **45.** $r_1^2 = \left(kC + A - \sqrt{R_1^2 - R_2^2}\right)^2 + r_2^2$

47. \$228 000 **49.** 9.2 km **51.** 1.4 km

Review Exercises for Chapter 14, page 431

1. $x = -0.93, y = 3.44; x = 0.81, y = 2.60$

3. $x = 2.00, y = 0.00; x = 1.60, y = 0.60$

5. $x = -0.56, y = 1.32; x = 0.56, y = 1.32$

7. $x = -2, y = 7; x = 2, y = 7$

9. $x = 0.00, y = 0.00; x = 2.38, y = 0.91$

11. $x = 0, y = 0; x = 2, y = 16$

13. $L = \sqrt{2}, R = 1; L = -\sqrt{2}, R = 1;$
$L = j\sqrt{6}, R = -3; L = -j\sqrt{6}, R = -3$

15. $u = \frac{1}{12}(1 + \sqrt{97}), v = \frac{1}{18}(5 - \sqrt{97});$
$u = \frac{1}{12}(1 - \sqrt{97}), v = \frac{1}{18}(5 + \sqrt{97})$

17. $x = 7, y = 5; x = 7, y = -5; x = -7, y = 5;$
$x = -7, y = -5$

19. $x = -2, y = 2; x = \frac{2}{3}, y = \frac{10}{9}$ **21.** $-4, -2, 2, 4$ **23.** 1, 16

25. $\frac{1}{3}, -\frac{1}{7}$ **27.** 1, 1.65 **29.** $\sqrt{3}, -\sqrt{3}, \frac{1}{2}j\sqrt{3}, -\frac{1}{2}j\sqrt{3}$

31. 6 **33.** 8 **35.** $\frac{9}{16}$ **37.** $\frac{1}{3}(11 - 4\sqrt{15})$ **39.** 2

41. 4 **43.** 5 **45.** $-1, 2$ **47.** $x = a + 1, y = a$

49. 25 **51.** $x = -2, y = -3; x = 3, y = 2$

53. $l = \frac{1}{2}\left(-1 + \sqrt{1 + 16\pi^2L^2/h^2}\right)$

55. $m = \frac{1}{2}\left(-y \pm \sqrt{2s^2 - y^2}\right)$ **57.** 0.40 s, 0.80 s **59.** 70 m

61. 20 dm, 26 dm, 26 dm **63.** 3.0 mm, 1.0 mm

65. 34 mm, 52 mm **67.** 0.353 cm, 6.29 cm

69. 27.0 km/h, 23.8 km/h

Chapter 15: Equations of Higher Degree

Exercises 15.1, page 438

1. -25 **3.** Coefficients 1 0 0 −4 11, $R = -26$

5. 0 **7.** -40 **9.** -43 **11.** 51 **13.** -28

15. 14 **17.** Yes **19.** No **21.** No

23. $x^2 + 3x + 2, R = 0$ **25.** $x^2 + x - 4, R = 8$

27. $p^5 + 2p^4 + 4p^3 + 2p^2 + 2p + 4, R = 2$

29. $x^6 + 2x^5 + 4x^4 + 8x^3 + 16x^2 + 32x + 64, R = 0$

31. $x^3 + x^2 + 2x + 1, R = 0$ **33.** No **35.** No **37.** Yes

39. No **41.** Yes **43.** Yes **45.** $2x^2 - 5x + 1 - \frac{8}{x + 4}$

47. $(4x^3 + 8x^2 - x - 2) \div (2x - 1) = 2x^2 + 5x + 2$; no, because the coefficient of x in $2x - 1$ is 2, not 1.

49. -7 **51.** $x^2 - (3 + j)x + 3j, R = 0$

53. Yes: If r is a zero of $f(x)$, $f(r) = 0 = -g(r)$, so r is a zero of $-g(x)$.

55. (a) No (b) Yes **57.** Yes **59.** $2w - 1$

Exercises 15.2, page 443

(Note: Unknown roots listed)

1. $1, 1, 1, -1, -1$ **3.** $-3, 2, -2$ **5.** $-3, -3, 2j, -2j$

7. $-1, 4$ **9.** $-2, -2$ **11.** $-j, -\frac{2}{3}$ **13.** $2j, -2j$

15. $-1, \frac{1}{2}$ **17.** $-2, 1$ **19.** $1 - j, \frac{1}{2}, -2$ **21.** $j, -j$

23. $-j, -1, 1$ **25.** $-2j, 3, -3$

27. Complex roots occur in conjugate pairs. Two real and one complex is not a possibility for a third-degree polynomial.

29. $x^3 - 2x^2 + 2x$ **31.** (2, 4)

Exercises 15.3, page 449

1. No more than two positive roots and three negative roots

3. $1, -2, -4$ **5.** $2, -1, -3$

7. $1, \frac{1}{6}(-3 + j\sqrt{15}), \frac{1}{6}(-3 - j\sqrt{15})$ **9.** $\frac{1}{2}, -1, 2$

11. $-2, -2, 2 \pm \sqrt{3}$ **13.** $-2, \frac{2}{5}, 1 + j\sqrt{3}, 1 - j\sqrt{3}$

15. $-3, -\frac{2}{3}, -\frac{1}{2}, \frac{1}{2}$ **17.** $2, 2, -1, -1, -3$

19. $2, 2, 2, \frac{1}{2}j, -\frac{1}{2}j$ **21.** $-2.17, 0.39, 1.78$

23. $-3.01, -1.49, -0.33, 0.33$ **25.** 0.59 **27.** -0.77

29. $x = -2, y = -28; x = 2 - \sqrt{3}, y = 20 - 12\sqrt{3};$
$x = 2 + \sqrt{3}, y = 20 + 12\sqrt{3}$ **31.** $-\frac{3}{4}, \sqrt{5}, -\sqrt{5}$

33. 1.76 s, 4.51 s **35.** $0, L$ **37.** 76 %

39. 1.23 cm or 2.14 cm **41.** 3.0 mm, 4.0 mm; 5.0 mm, 6.0 mm

43. 2800 Ω, 4200 Ω, 8400 Ω

45. Two positive, one negative; zero positive, one negative, two complex

Review Exercises for Chapter 15, page 450

1. 1 **3.** -107 **5.** Yes **7.** No

9. $x^2 + 4x + 10, R = 11$ **11.** $2x^2 - 7x + 10, R = -17$

13. $x^3 - 3x^2 - 2x + 10, R = -4$

15. $2m^4 + 10m^3 + 4m^2 + 21m + 105, R = 516$

17. No **19.** Yes **21.** (unlisted roots) $-2, 1$

23. (unlisted roots) 3, 4 **25.** (unlisted roots) $-1 + j\sqrt{2}, -1 - j\sqrt{2}$

27. (unlisted roots) $-j, \frac{1}{2}, -\frac{3}{2}$ **29.** (unlisted roots) $2, -2$

31. (unlisted roots) $-2 - j, \frac{1}{2}(-1 \pm j\sqrt{3})$ **33.** $1, 2, -4$

35. $-\frac{1}{2}, 1, 1$ **37.** $-1, -\frac{1}{2}, \frac{5}{3}$

39. $\frac{1}{2}, \frac{3}{2}, -1 + \sqrt{2}, -1 - \sqrt{2}$ **41.** 1, 3, 5 **43.** 2 or 4

45. Use k as second coefficient in synthetic division, equate remainder to 0; $k = 4$.

47. $x^3 - 5x^2 + x - 5 = 0$ **49.** $x = -1, y = -2; x = 2, y = 1$

51. 1.91 **53.** 11.22, 1.44 **55.** 0.75 cm **57.** 2 cm

59. 5.1 m, 8.3 m **61.** $h = 2.25$ m, $w = 1.20$ m **63.** 16.34 cm

Chapter 16: Matrices; Systems of Linear Equations

Exercises 16.1, page 455

1. $\begin{bmatrix} 5 & 7 & -1 & 9 \\ 6 & 4 & 1 & 12 \end{bmatrix}$ **3.** $a = 1, b = -3, c = 4, d = 7$

5. $x = -2, y = 5, z = -9, r = 48, s = 4, t = -1$

7. $C = 3, D = 2, E = -2$

9. Elements cannot be equated; different number of rows.

11. $\begin{bmatrix} 1 & 10 \\ 0 & 2 \end{bmatrix}$ **13.** $\begin{bmatrix} -5 & 0 \\ 11 & 71 \\ 11 & -5 \end{bmatrix}$ **15.** $\begin{bmatrix} 9 & 9 \\ -5 & -11 \end{bmatrix}$

17. Cannot be added **19.** $\begin{bmatrix} -8 & -5 & -1 \\ -2 & -5 & 19 \end{bmatrix}$

21. $\begin{bmatrix} 9 & 1 & 8 \\ 0 & 11 & -30 \end{bmatrix}$ **23.** $\begin{bmatrix} 11 & -7 & 22 \\ -4 & 23 & -52 \end{bmatrix}$

25. $\begin{bmatrix} -15 & 21 \\ -21 & 6 \end{bmatrix}$ **27.** $\begin{bmatrix} 18 & -9 \\ 12 & -15 \end{bmatrix}$

29. $\begin{bmatrix} 20 & 31 & -25 \\ 14 & -9 & -13 \end{bmatrix}$ **31.** $\begin{bmatrix} 18 & -63 \\ 56 & -1 \end{bmatrix}$ **33.** $\begin{bmatrix} -42 & -60 \\ 38 & 56 \end{bmatrix}$

35. $A + B = B + A = \begin{bmatrix} 3 & 1 & 0 & 7 \\ 5 & -3 & -2 & 5 \\ 10 & 10 & 8 & 0 \end{bmatrix}$

37. $-(A - B) = B - A = \begin{bmatrix} 5 & -3 & -6 & -7 \\ 5 & 3 & 0 & -3 \\ -8 & 12 & 8 & 4 \end{bmatrix}$

39. $v_w = 30.9$ km/h, $v_p = 249$ km/h

41. $\begin{bmatrix} 288 & 225 & 0 & 0 \\ 186 & 132 & 72 & 0 \\ 0 & 105 & 204 & 234 \end{bmatrix}$

43. $B = \begin{bmatrix} 15 & 25 & 10 & 25 \\ 10 & 10 & 10 & 45 \end{bmatrix}, J = \begin{bmatrix} 15 & 30 & 3 & 3 \\ 0 & 100 & 2 & 2 \end{bmatrix}$

Exercises 16.2, page 461

1. $\begin{bmatrix} 5 & 6 & 7 & -10 \\ -9 & 0 & -3 & 12 \\ 1 & 12 & 11 & -8 \end{bmatrix}$ **3.** $[-8 \quad -12]$ **5.** $\begin{bmatrix} 305 \\ -325 \end{bmatrix}$

7. $\begin{bmatrix} -\frac{1}{2} & \frac{111}{4} \\ -\frac{121}{8} & -\frac{101}{2} \\ \frac{1}{10} & 22 \end{bmatrix}$ **9.** $\begin{bmatrix} 33 & -22 \\ 31 & -12 \\ 15 & 13 \\ 50 & -41 \end{bmatrix}$ **11.** $\begin{bmatrix} -15 & 15 & -26 \\ 8 & 5 & -13 \end{bmatrix}$

13. $\begin{bmatrix} -49.43 & 55.20 \\ -53.02 & 79.16 \end{bmatrix}$ **15.** $AB = [40], BA = \begin{bmatrix} -1 & 3 & -8 \\ 5 & -15 & 40 \\ 7 & -21 & 56 \end{bmatrix}$

17. $AB = \begin{bmatrix} 45 \\ 327 \end{bmatrix}$, BA not defined **19.** $AI = IA = A$

21. $AI = IA = A$ **23.** $B = A^{-1}$ **25.** $B = A^{-1}$

27. Yes **29.** No **31.** $A^2 = A$ **33.** $B^3 = B$

35. $\begin{bmatrix} ac + bd & ad + bc \\ bc + ad & bd + ac \end{bmatrix}$

37. $\begin{bmatrix} -1 & 0 \\ 0 & -1 \end{bmatrix}\begin{bmatrix} -1 & 0 \\ 0 & -1 \end{bmatrix} = \begin{bmatrix} 1 & 0 \\ 0 & 1 \end{bmatrix}$

39. $A^2 - I = (A + I)(A - I) = \begin{bmatrix} 15 & 28 \\ 21 & 36 \end{bmatrix}$

41. $\begin{bmatrix} 0 & -j \\ j & 0 \end{bmatrix}\begin{bmatrix} 0 & -j \\ j & 0 \end{bmatrix} = \begin{bmatrix} 1 & 0 \\ 0 & 1 \end{bmatrix}$

43. $V_2 = V_1, i_2 = -V_1/R + i_1$

45. Ellipse **47.** $\begin{bmatrix} 2 & 0 & 1 & 0 & 1 \\ 0 & 2 & 1 & 1 & 0 \\ 1 & 1 & 2 & 0 & 0 \\ 0 & 1 & 0 & 2 & 1 \\ 1 & 0 & 0 & 1 & 2 \end{bmatrix}$

Exercises 16.3, page 466

1. $\begin{bmatrix} -\frac{5}{2} & \frac{3}{2} \\ -2 & 1 \end{bmatrix}$ **3.** $\begin{bmatrix} -2 & -\frac{5}{2} \\ -1 & -1 \end{bmatrix}$ **5.** $\begin{bmatrix} -\frac{1}{3} & \frac{1}{6} \\ \frac{2}{15} & \frac{1}{30} \end{bmatrix}$

7. $\begin{bmatrix} \frac{3}{4} & \frac{1}{2} \\ -\frac{1}{4} & 0 \end{bmatrix}$ **9.** $\begin{bmatrix} -\frac{8}{283} & -\frac{9}{566} \\ \frac{13}{1415} & \frac{5}{283} \end{bmatrix}$ **11.** $\begin{bmatrix} -3 & 2 \\ 2 & -1 \end{bmatrix}$

13. $\begin{bmatrix} -\frac{1}{2} & -2 \\ \frac{1}{2} & 1 \end{bmatrix}$ **15.** $\begin{bmatrix} \frac{2}{9} & -\frac{5}{9} \\ \frac{1}{9} & \frac{2}{9} \end{bmatrix}$ **17.** $\begin{bmatrix} \frac{7}{25} & \frac{3}{5} \\ -\frac{1}{5} & -\frac{1}{2} \end{bmatrix}$

19. $\begin{bmatrix} -18 & -7 & 5 \\ -3 & -1 & 1 \\ -5 & -2 & 1 \end{bmatrix}$ **21.** $\begin{bmatrix} 2 & 4 & \frac{7}{2} \\ -1 & -2 & -\frac{3}{2} \\ 1 & 1 & \frac{1}{2} \end{bmatrix}$

23. $\begin{bmatrix} \frac{5}{2} & -2 & -2 \\ -1 & 1 & 1 \\ \frac{7}{4} & -\frac{3}{2} & -1 \end{bmatrix}$ **25.** $\begin{bmatrix} 0.3 & -0.4 \\ 0.05 & 0.1 \end{bmatrix}$

27. $\begin{bmatrix} 2 & 4 & 3.5 \\ -1 & -2 & -1.5 \\ 1 & 1 & 0.5 \end{bmatrix}$ **29.** $\begin{bmatrix} 2.5 & -2 & -2 \\ -1 & 1 & 1 \\ 1.75 & -1.5 & -1 \end{bmatrix}$

31. $\begin{bmatrix} 1 & 2 & 3 & 1 \\ 1 & 3 & 3 & 2 \\ 2 & 4 & 3 & 3 \\ 1 & 1 & 1 & 1 \end{bmatrix}$ **33.** $\begin{bmatrix} 2.537 & -0.950 & 0.159 & 0.470 \\ -0.213 & 0.687 & 0.290 & -0.272 \\ -1.006 & 0.870 & 0.496 & -0.532 \\ 0.113 & -0.123 & -0.033 & 0.385 \end{bmatrix}$

35. $\begin{vmatrix} 1 & 1 \\ 1 & 1 \end{vmatrix} = 0$ means the inverse does not exist.

37. $\frac{1}{ad-bc}\begin{bmatrix} ad-bc & -ba+ab \\ cd-dc & -bc+ad \end{bmatrix} = \begin{bmatrix} 1 & 0 \\ 0 & 1 \end{bmatrix}$

39. $\begin{bmatrix} \frac{1}{2} & \frac{1}{2} & \frac{1}{2} \\ \frac{1}{2} & -\frac{3}{2} & -\frac{5}{2} \\ \frac{1}{2} & -\frac{5}{2} & -\frac{7}{2} \end{bmatrix}$ which is symmetric **41.** $\begin{bmatrix} 0.8 & 0.0 & 0.6 \\ 0.0 & 1.0 & 0.0 \\ -0.6 & 0.0 & 0.8 \end{bmatrix}$

Exercises 16.4, page 470

1. $x = 2, y = -3$ **3.** $x = \frac{1}{2}, y = 3$ **5.** $x = 1, y = 3$

7. $x = 2, y = -2$ **9.** $x = -1, y = 0, z = 3$

11. $x = -\frac{3}{2}, y = -2$ **13.** $x = 1.6, y = -2.5$

15. $x = 2, y = -4, z = 1$ **17.** $x = 2, y = -\frac{1}{2}, z = 3$

19. $x = 1, y = -2, z = -3$ **21.** $u = 2, v = -5, w = 4$

23. $x = 2, y = 3, z = -2, t = 1$

25. $v = 2, w = -1, x = \frac{1}{2}, y = \frac{3}{2}, z = -3$ **27.** $x = \pm 2, y = -2$

29. $x = 1, y = -2$; The three lines meet in the point $(1, -2)$.

31. 1180 N, 1860 N **33.** 10 V, 8 V **35.** 40 mL, 8 mL

37. 6.4 L, 1.6 L, 2.0 L

39. $x = 2.5$ ha, $y = 3.2$ ha, $z = 1.5$ ha, $t = 4$ ha

Exercises 16.5, page 475

1. $x = -\frac{11}{7}, y = \frac{6}{7}$ **3.** $x = 2, y = 1$ **5.** $x = \frac{17}{14}, y = \frac{19}{14}$

7. Inconsistent **9.** $x = -2, y = -\frac{2}{3}, z = \frac{1}{3}$

11. $w = 1, x = 0, y = -2, z = 3$

13. Unlimited: $x = -3, y = -1, z = 1$; $x = 12, y = 0, z = -10$

15. Inconsistent **17.** $x = 4, y = \frac{2}{3}, z = -2$

19. $x = \frac{1}{2}, y = -\frac{3}{2}$ **21.** $x = \frac{1}{6}, y = -\frac{1}{2}$

23. Inconsistent **25.** Inconsistent

27. $r = 0, s = 0, t = 0, u = -1$

29. $x = \dfrac{c_1b_2 - c_2b_1}{a_1b_2 - a_2b_1}, y = \dfrac{a_1c_2 - a_2c_1}{a_1b_2 - a_2b_1}$ **31.** 1080 km, 1290 km

33. 250 parts/h, 220 parts/h, 180 parts/h

Exercises 16.6, page 480

1. $\begin{vmatrix} 3 & 0 & 0 \\ 1 & 1 & 0 \\ 2 & 1 & 3 \end{vmatrix} = 9, \begin{vmatrix} 0 & 0 & 3 \\ 0 & 1 & 1 \\ 3 & 1 & 2 \end{vmatrix} = -9$ **3.** -60 **5.** 0

7. -40 **9.** -40 **11.** 39 **13.** 57 **15.** -13

17. -72 **19.** 0 **21.** 39 **23.** 57 **25.** -13

27. -72 **29.** 0 **31.** $x = -1, y = 0, z = 2, t = 1$

33. $x = 1, y = 2, z = -1, t = 3$

35. $x = 2, y = -1, z = -1, t = 3$

37. $D = 1, E = 2, F = -1, G = -2$ **39.** 0 **41.** 8

43. $\frac{33}{16}$A, $\frac{11}{8}$A, $-\frac{5}{8}$A, $-\frac{15}{8}$A, $-\frac{15}{16}$A **45.** 20.8 km^2

47. ppm SO_2: 0.5, NO: 0.3, NO_2: 0.2, CO: 5.0

Review Exercises for Chapter 16, page 481

1. $a = 4, b = -1$

3. $x = 2, y = -3, z = \frac{5}{2}, a = -1, b = -\frac{7}{2}, c = \frac{1}{2}$

5. $x = -1, y = \frac{1}{2}, a = -\frac{1}{2}, b = -\frac{3}{2}$

7. $\begin{bmatrix} 1 & -3 \\ 8 & -5 \\ -8 & -2 \\ 3 & -10 \end{bmatrix}$ **9.** $\begin{bmatrix} -3 & 3 \\ 0 & -7 \\ 2 & -2 \\ -1 & -4 \end{bmatrix}$ **11.** $\begin{bmatrix} 7 & -6 \\ -4 & 20 \\ -1 & 6 \\ 1 & 15 \end{bmatrix}$

13. $\begin{bmatrix} 0 & 0 \\ 0 & 0 \end{bmatrix}$ **15.** $\begin{bmatrix} 0.3 & 0.1 & 0.0 \\ 0.0 & -0.1 & 0.1 \\ 0.0 & -0.2 & 0.2 \end{bmatrix}$ **17.** $\begin{bmatrix} -2 & \frac{5}{2} \\ -1 & 1 \end{bmatrix}$

19. $\begin{bmatrix} \frac{40}{3} & \frac{5}{3} \\ -\frac{20}{3} & \frac{35}{3} \end{bmatrix}$ **21.** $\begin{bmatrix} 11 & 10 & 3 \\ -4 & -4 & -1 \\ 3 & 3 & 1 \end{bmatrix}$ **23.** $\begin{bmatrix} \frac{1}{2} & -\frac{1}{2} & -1 \\ -3 & 2 & 1 \\ -4 & 3 & 2 \end{bmatrix}$

25. $x = -3, y = 1$ **27.** $x = 10, y = -15$

29. $u = -1, v = -3, w = 0$ **31.** $x = 1, y = \frac{1}{2}, z = -\frac{1}{3}$

33. $x = -3, y = 1$ **35.** $u = -1, v = -3, w = 0$

37. $x = 1, y = \frac{1}{2}, z = -\frac{1}{3}$ **39.** Unlimited number of solutions

41. $u = -1, v = -3, w = 0$ **43.** $x = 1, y = \frac{1}{2}, z = -\frac{1}{3}$

45. $x = 3, y = 1, z = -1$ **47.** $x = 1, y = 2, z = -3, t = 1$

49. $x = -\frac{1}{3}, y = 3, z = \frac{2}{3}, t = -4$

51. $r = \frac{11}{10}, s = \frac{-92}{5}, t = \frac{3}{20}, u = \frac{64}{5}, v = \frac{3}{10}$

53. $\begin{bmatrix} 1 & 0 \\ 15 & 16 \end{bmatrix}, \begin{bmatrix} 1 & 0 \\ 63 & 64 \end{bmatrix}, \begin{bmatrix} 1 & 0 \\ 255 & 256 \end{bmatrix}$

55. $B^3 = \begin{bmatrix} 1 & 0 & 0 \\ 0 & 1 & 0 \\ 0 & 0 & 1 \end{bmatrix}$ **57.** 77 **59.** 323

61. 77 **63.** 323 **65.** $N^{-1} = -N = \begin{bmatrix} 0 & 1 \\ -1 & 0 \end{bmatrix}$

67. $\begin{bmatrix} n & 1+n \\ 1-n & -n \end{bmatrix}\begin{bmatrix} n & 1+n \\ 1-n & -n \end{bmatrix} = \begin{bmatrix} n^2 + (1-n^2) & (n+n^2) - (n+n^2) \\ (n-n^2) - (n-n^2) & (1-n^2) + n^2 \end{bmatrix}$

69. $(A+B)(A-B) = \begin{bmatrix} -6 & 2 \\ 4 & 2 \end{bmatrix}, A^2 - B^2 = \begin{bmatrix} -10 & -4 \\ 8 & 6 \end{bmatrix}$

71. $\begin{bmatrix} \frac{1}{2} & \frac{1}{3} \\ 0 & \frac{1}{6} \end{bmatrix} = \frac{1}{2}\begin{bmatrix} 1 & \frac{2}{3} \\ 0 & \frac{1}{3} \end{bmatrix}$ **73.** $R_1 = 4\ \Omega, R_2 = 6\ \Omega$

75. $F = 303$ N, $T = 175$ N **77.** $R_1 = 4\ \Omega, R_2 = 6\ \Omega$

79. $F = 303$ N, $T = 175$ N

81. 0.20 h after police pass intersection **83.** 30 g, 50 g, 20 g

85. $x = 0.58, y = 0.22, z = 0.20$

87. 830 g, 880 g, 290 g **89.** 1575 kJ, 581 kJ, 741 kJ

Chapter 17: Inequalities

Exercises 17.1, page 489

1. $x > 1$, or $(1, \infty)$ **3.** $2 < 5$ **5.** $9 < 14$ **7.** $16 < 36$

9. $-4 > -9$ **11.** $16 < 81$ **13.** $x > -2$ **15.** $x \le 45$

17. $1 < x < 7$ **19.** $x < -9$ or $x \ge -4$

21. $x < 1$ or $3 < x \le 5$ **23.** $-2 < x < 2$ or $3 \le x < 4$

25. x is greater than 0 and less than or equal to 2.

27. x is less than -10, or greater than or equal to 10 and less than 20.

29. $(-\infty, 3)$

31. $(-\infty, -1]$ or $(0.5, \infty)$

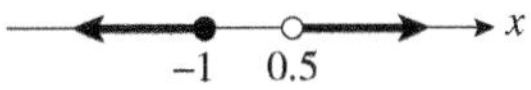

33. $[0, 5)$

35. $[-3, 5)$

37. $(-\infty, -1)$ or $[1, 4)$

39. $(-3, -1)$ or $(1, 3]$

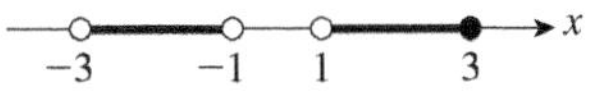

41. $t = -5$

43. $(3, 5]$ or $[8, 10)$

45. Absolute inequality **47.** No, $a - b < 0$

49. $|x + y| < |x| + |y|$

51. From step (4) to step (5), both sides divided by log 0.5, which is negative. Sign in step (5) should be $>$.

53. $2000 \le M \le 1\,000\,000$

55. $29\,000 < v < 40\,000$ km/h

57. $0 < n \le 2565$ steps **59.** $4.5\ h \le t \le 5$ h

Exercises 17.2, page 493

1. $x \le 3, (-\infty, 3]$

3. $1 < x < \frac{9}{2}, \left(1, \frac{9}{2}\right)$

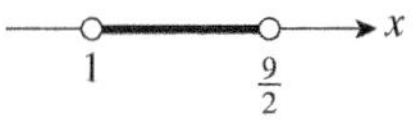

5. $x > -1, (-1, \infty)$

7. $x < 64, (-\infty, 64)$

9. $x \le -2, (-\infty, -2]$

11. $y < -2, (-\infty, -2)$

13. $x \le \frac{5}{2}, \left(-\infty, \frac{5}{2}\right]$

15. $T < -\frac{177}{20}, \left(-\infty, -\frac{177}{20}\right)$

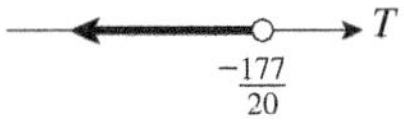

17. $x > -1.80, (-1.80, \infty)$

19. $L > -\frac{7}{9}, \left(-\frac{7}{9}, \infty\right)$

21. $-1 \le x \le 1, [-1, 1]$

23. $2 < x \le 5, (2, 5]$

25. $-3 \le x < -1, [-3, -1)$

27. No values

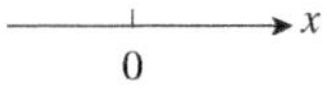

29. $x \ge 5, [5, \infty)$

31. $x \ge -5, [-5, \infty)$

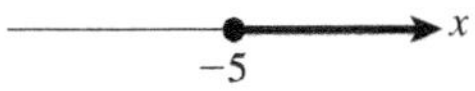

33. $-6 < k < 6, (-6, 6)$

35. $n > 35$ h, $(35, \infty)$

37. $50° < F < 68°, (50, 68)$

39. 0.542 m $< w < 0.813$ m, $(0.542, 0.813)$

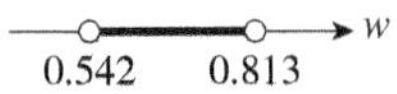

41. 0.400 h $< t < 2.60$ h, $(0.400, 2.60)$

43. $0 \le x \le 500, 200 \le y \le 700, [200, 700]$

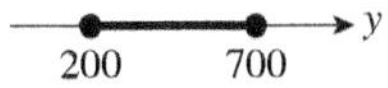

45. 2 min $\le$ stop times $\le$ 4 min, $[2, 4]$

Exercises 17.3, page 499

1. $x < 1$ or $x > 3$
$(-\infty, 1)$ or $(3, \infty)$,

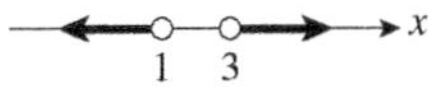

3. $-3 < x < 2$ or $x > 4$,
$(-3, 2)$ or $(4, \infty)$

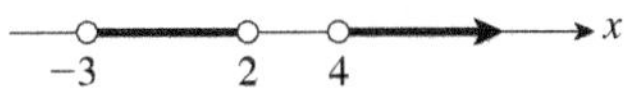

5. $-4 < x < 4$, $(-4, 4)$

7. $0 \le x \le 2$, $[0, 2]$

9. $-4 \le x \le \frac{3}{2}$, $\left[-4, \frac{3}{2}\right]$

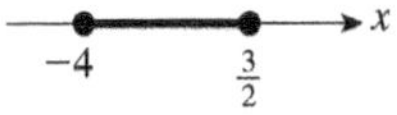

11. $x = -2$

13. All R, $(-\infty, \infty)$

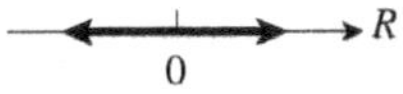

15. $-\infty < x < -2$, $0 < x < 1$, $(-\infty, -2)$ or $(0, 1)$

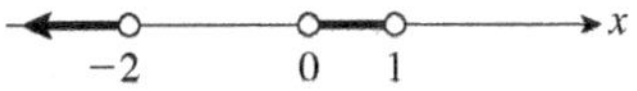

17. $-2 \le s \le -1$, $s \ge 1$, $[-2, -1]$ or $[1, \infty)$

19. $x \le -2$, $x \ge \frac{1}{3}$, $(-\infty, -2]$ or $\left[\frac{1}{3}, \infty\right)$

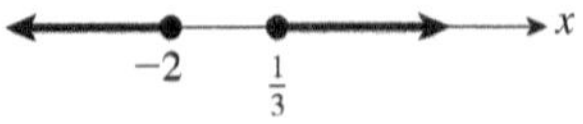

21. $-6 < x \le \frac{3}{2}$, $\left(-6, \frac{3}{2}\right]$

23. $-5 < x < -1$, $x > 7$, $(-5, -1)$ or $(7, \infty)$

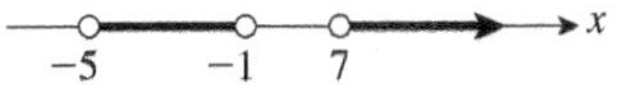

25. $x < -1$, $(-\infty, -1)$

27. $T < 3$, $T \ge 8$, $(-\infty, 3)$ or $[8, \infty)$

29. $-1 < x < \frac{3}{4}$, $x \ge 6$, $\left(-1, \frac{3}{4}\right)$ or $[6, \infty)$

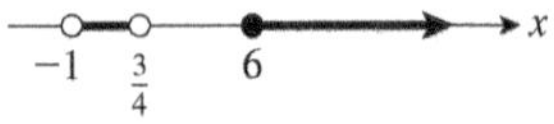

31. $2 < x < 4$, $5 < x < 9$, $(2, 4)$ or $(5, 9)$

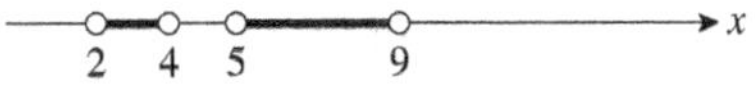

33. $x \le -2, x \ge 1$, $(-\infty, -2]$ or $[1, \infty)$ **35.** $-1 \le x \le 0$, $[-1, 0]$

37. $x > 1.52$, $(1.52, \infty)$

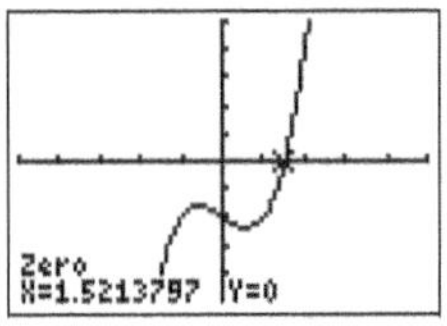

39. $-1.39 < x < -0.43$, $(-1.39, -0.43)$

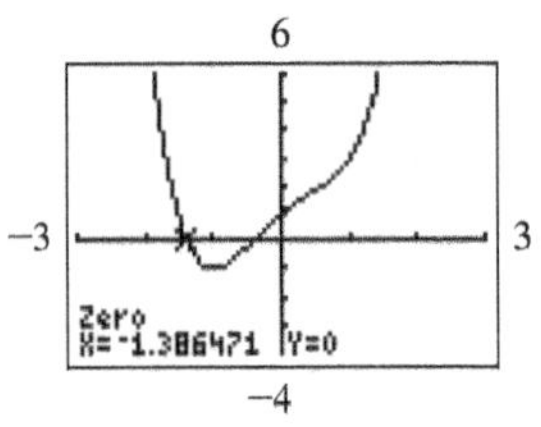

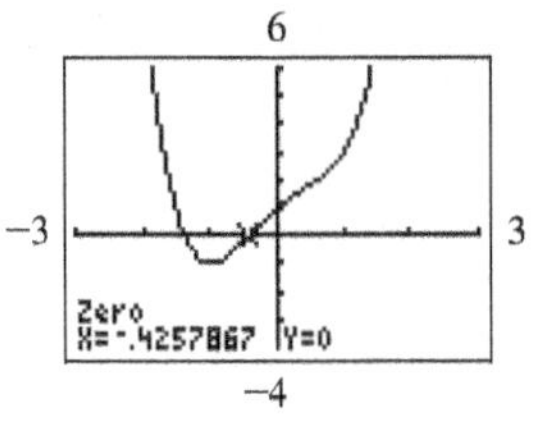

41. $x < -1.69$, $x > 2.00$, $(-\infty, -1.69)$ or $(2.00, \infty)$

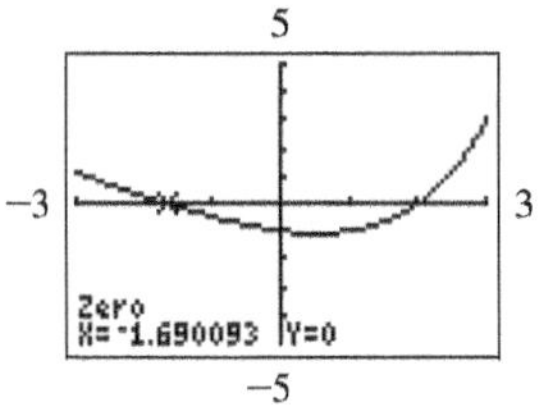

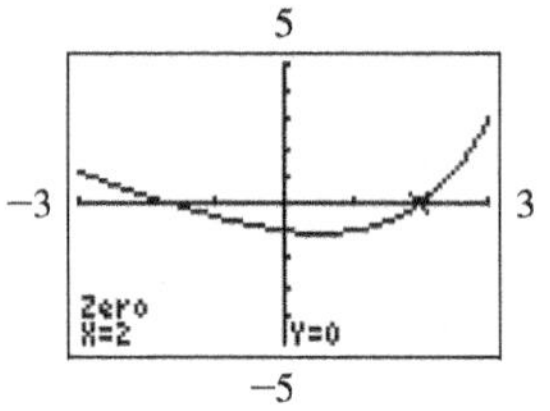

43. $x < -4.43$, $-3.11 < x < -1.08$, $x > 3.15$, $(-\infty, -4.43)$ or $(-3.11, -1.08)$ or $(3.15, \infty)$

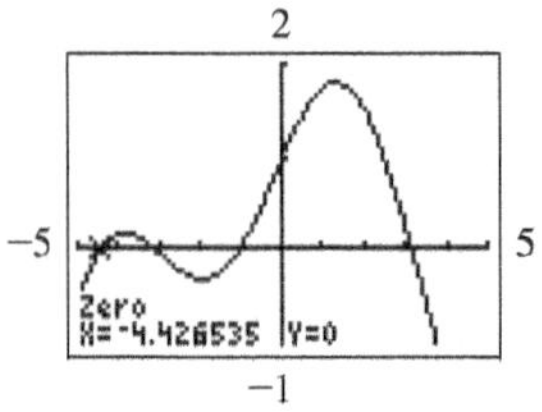

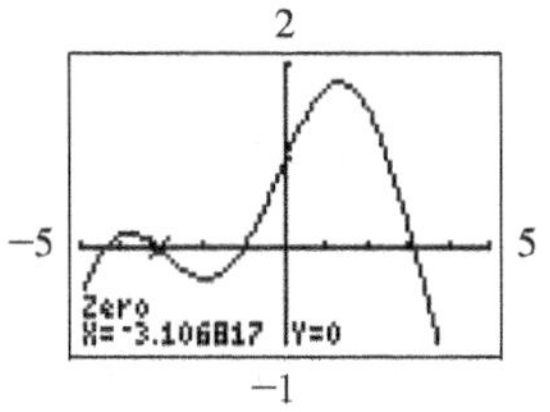

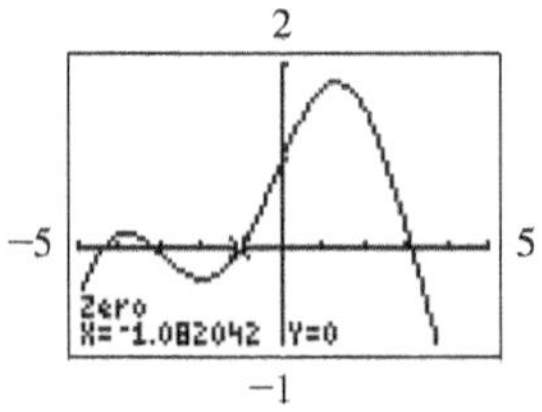

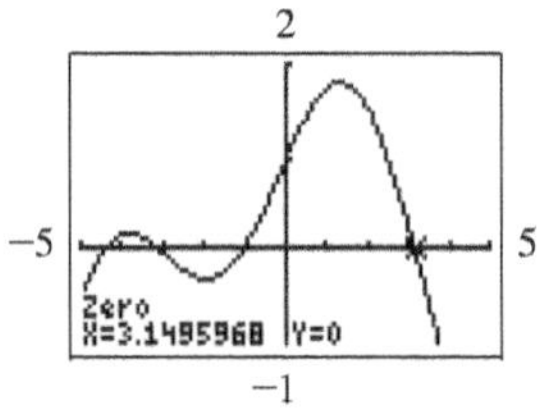

45. No; not true if $0 \le x \le 1$ **47.** $x^2 - 3x - 4 < 0$

49. $x < 3.11$, $(-\infty, 3.11)$ **51.** $a \ne b$

53. $0.5 \text{ A} < i < 1 \text{ A}$, $(0.5 \text{ A}, 1 \text{ A})$ **55.** $4 \le t \le 16$, $[4, 16]$

57. $C_1 > \frac{4}{3}\mu\text{F}$, $(4/3\ \mu\text{F}, \infty)$

59. $h > 2640$ km, $(2640 \text{ km}, \infty)$

61. $3.0 \le w < 5.0$ mm, $[3.0 \text{ mm}, 5.0 \text{ mm})$

63. $0 \le t < 0.92$ h, $[0, 0.92 \text{ h})$

Exercises 17.4, page 502

1. $-2 < x < 3$, $(-2, 3)$

3. $3 < x < 5$, $(3, 5)$

5. $x < -2$, $x > \frac{2}{5}$, $(-\infty, 2,)$ or $\left(\frac{2}{5}, \infty\right)$

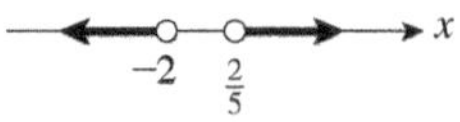

7. $\frac{1}{6} \le x \le \frac{3}{2}$, $\left[\frac{1}{6}, \frac{3}{2}\right]$

9. $x < 0$, $x > \frac{3}{2}$, $(-\infty, 0)$ or $\left(\frac{3}{2}, \infty\right)$

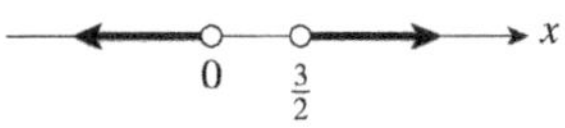

11. $-26 < t < 24$, $(-26, 24)$

13. $-6.4 \le x \le -2.1$, $[-6.4, -2.1]$

15. $x < -18$, $x > 66$, $(-\infty, -18)$ or $(66, \infty)$

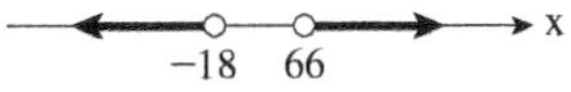

17. $1 < x < 2$, $(1, 2)$

19. $x \le \dfrac{1}{10}, x \ge \dfrac{7}{10}$, $(-\infty, \frac{1}{10}]$ or $[\frac{7}{10}, \infty)$

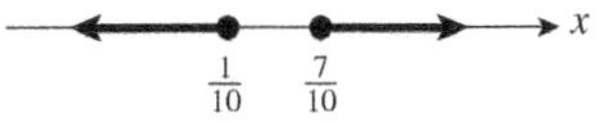

21. $-15 < R < \dfrac{35}{3}, \left(-15, \dfrac{35}{3}\right)$

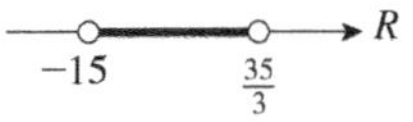

23. $x \le 8.4$, $x \ge 17.6$, $(-\infty, 8.4]$ or $[17.6, \infty)$

25. $x < -3, -2 < x < 1, x > 2$ $(-\infty, -3)$ or $(-2, 1)$ or $(2, \infty)$

27. $-3 < x < -2, 1 < x < 2$, $(-3, -2)$ or $(1, 2)$

29. No solutions; $|x|$ is never less than 0.

31. $x > \dfrac{a - c}{|b|}$ **33.** 4 km, 50 km

35. $|p - 2\,000\,000| \le 200\,000$; production is at least 1 800 000 barrels, but not greater than 2 200 000 barrels.

37. $|T - 70| \le 20$ **39.** $|d - 3.765| \le 0.002$

41. $|m - 1607.3| \le 65.3$ **43.** (8.87 s, 29.9 s) or (44.7 s, 50.8 s]

Exercises 17.5, page 505

1. $y < 3 - x$

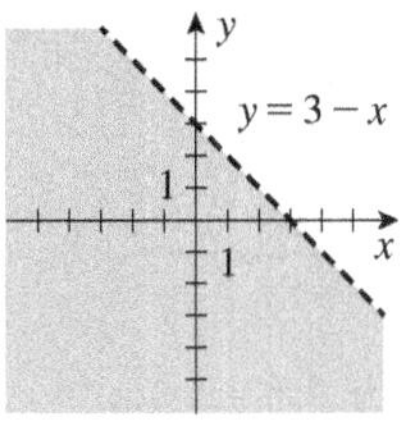

3.

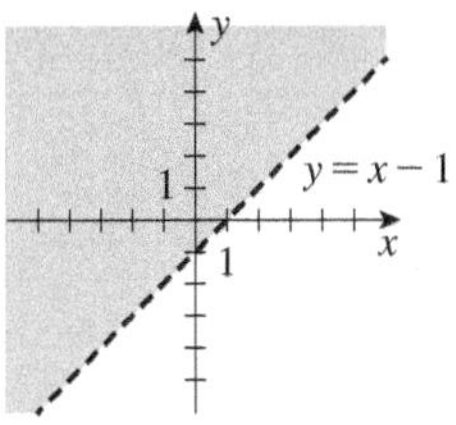

5.

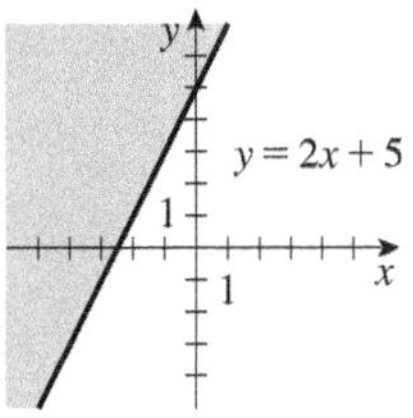

7.

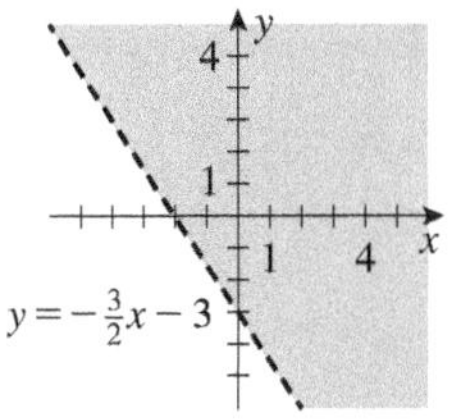

9.

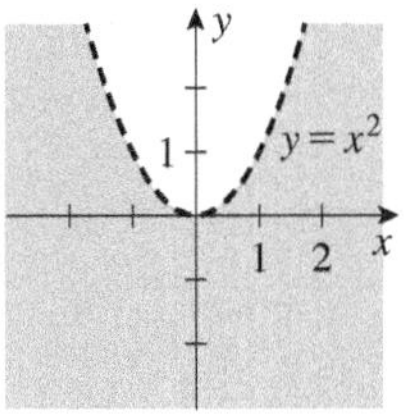

11.

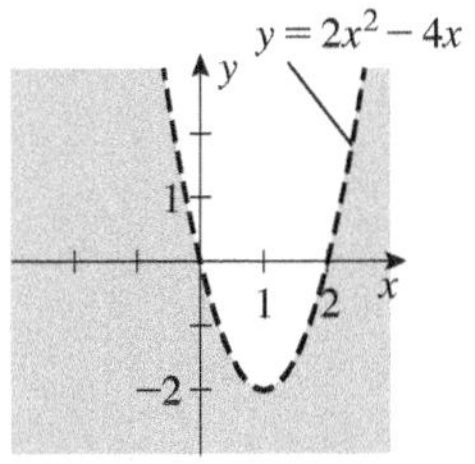

13.

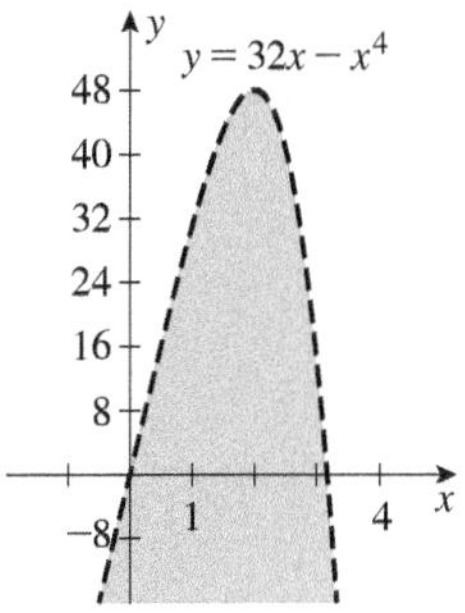

15.

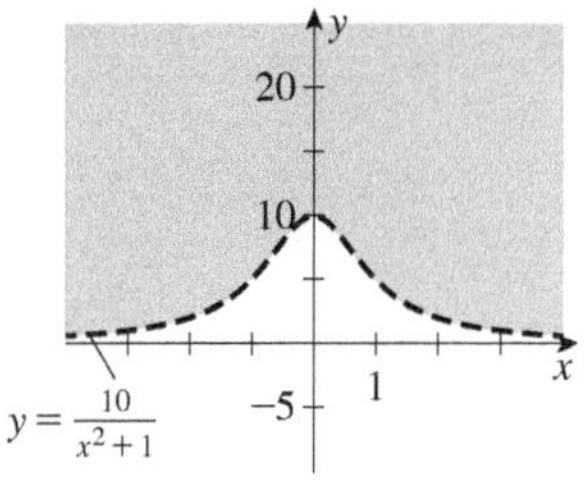

17.

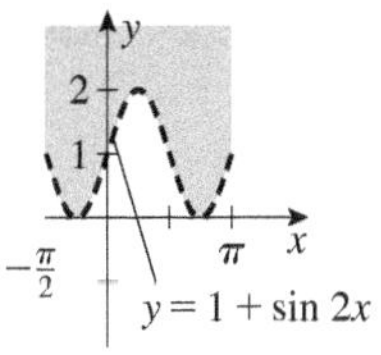

19.

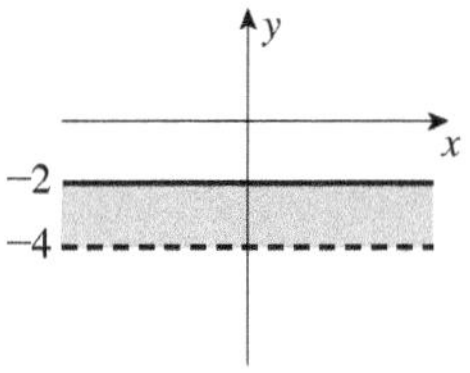

21.

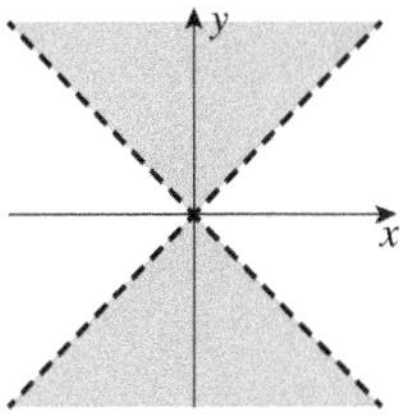

23.

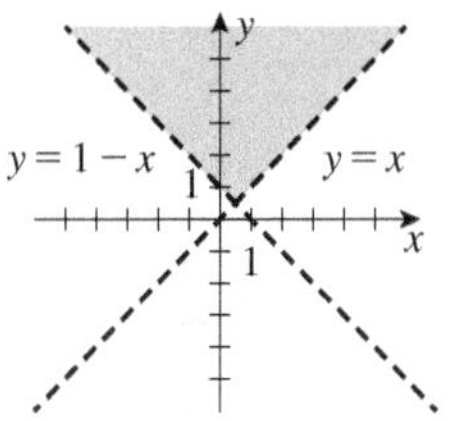

25.

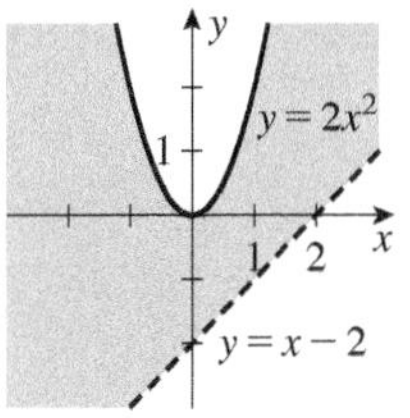

27.

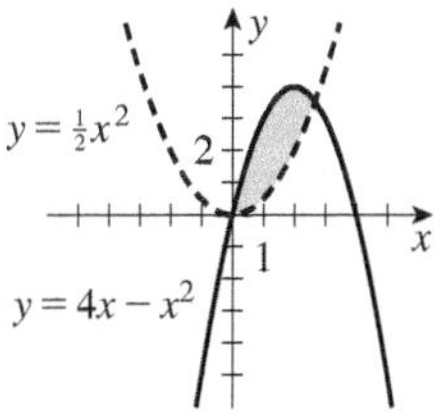

29.

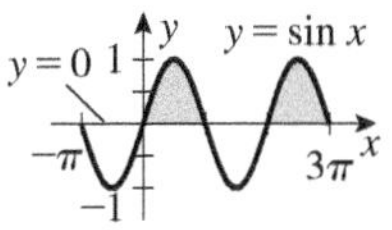

31.

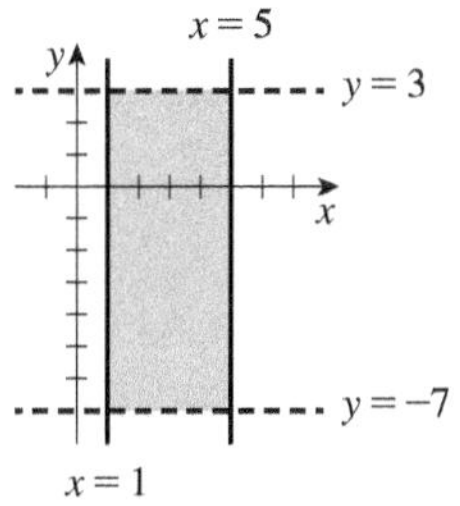

33.

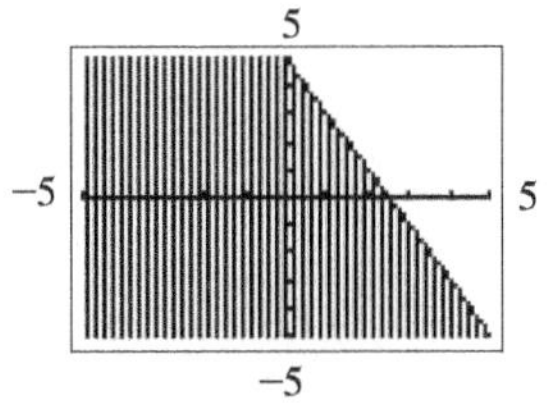

35.

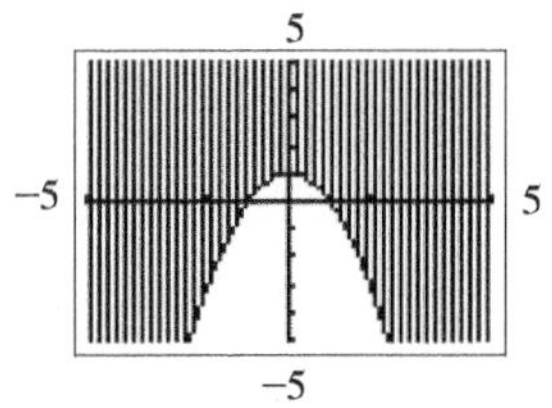

37.

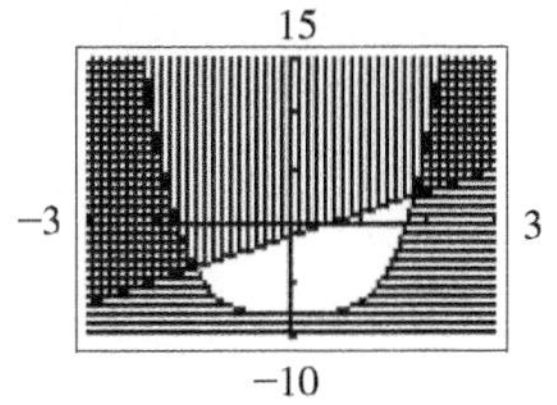

39.

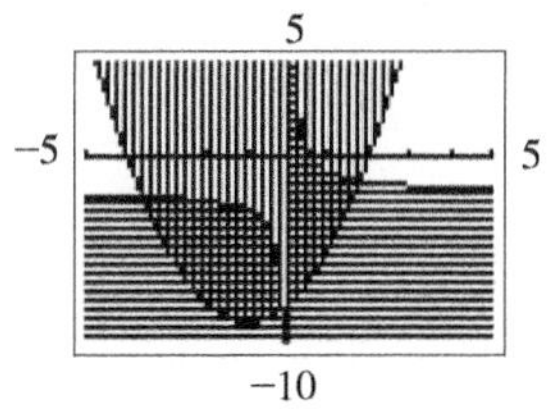

41.

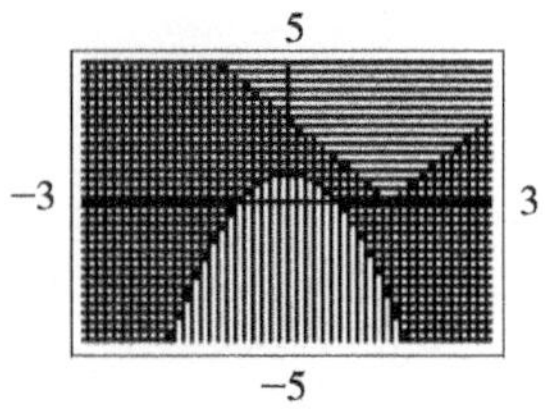

43. $y < 2x + \frac{5}{2}$

45. Below

47.

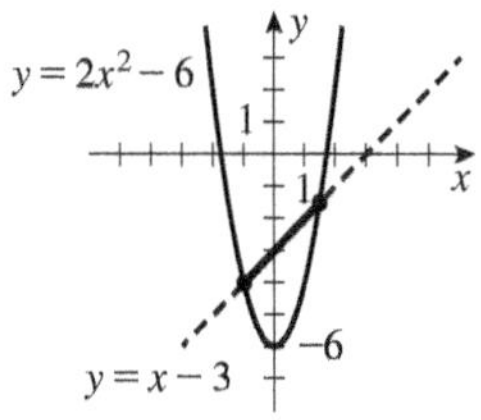

49.

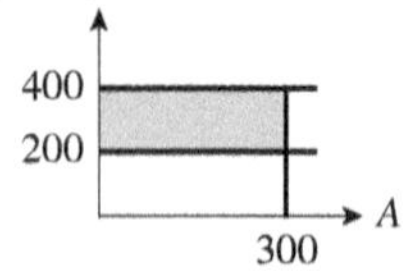

51.

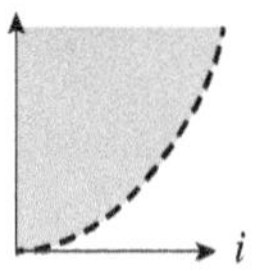

53.

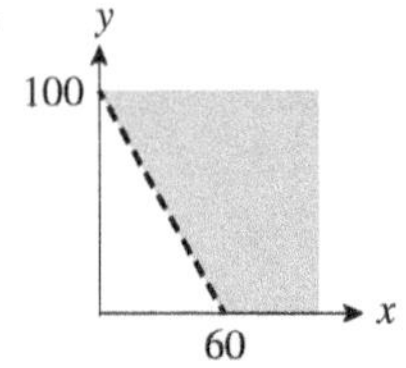

Exercises 17.6, page 509

1. Max $F = 18$ at $(0, 6)$

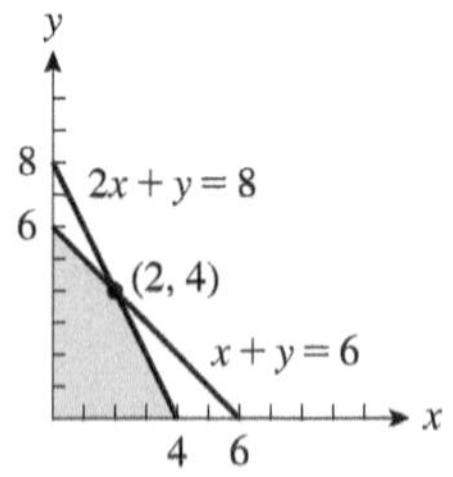

3. Max $F = 58$ at $(10, 7)$
Min $F = 0$ at $(0, 0)$

5. Max $P = 30$ at $(0, 6)$

7. Max $P = 30$ at $(6, 0)$

9. Min $C = 24$ at $(3, 2)$

11. Min $F = 7$ at $(1, 2)$
Max $F = \frac{47}{3}$ at $\left(\frac{5}{3}, \frac{14}{3}\right)$

13. Max $P = 27$ at $(3, 0)$

15. Min $C = 40$ at $(4, 4)$

17. \$6000 at 6%, \$3000 at 5%

19. 40 business models
60 graphing models

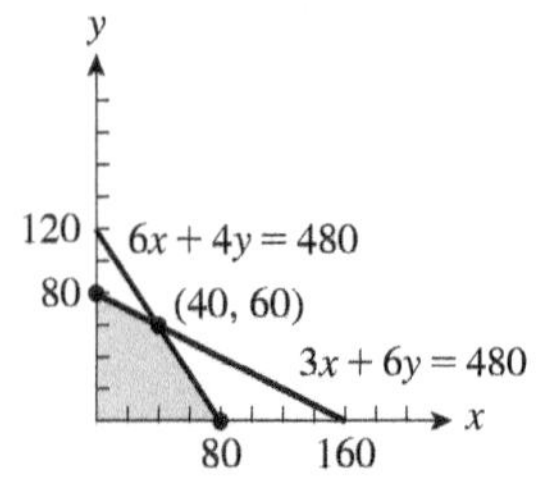

21. $4\frac{2}{7}$ servings of $A = \frac{900}{7}$ g of A
$2\frac{6}{7}$ servings of $B = \frac{600}{7}$ g of B

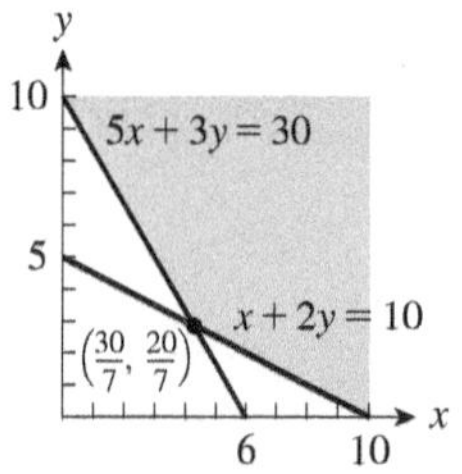

Review Exercises for Chapter 17, page 510

1. $x > 6$, $(6, \infty)$

3. $x > \frac{5}{3}$, $\left(\frac{5}{3}, \infty\right)$

5. $\frac{5}{2} < x < 6$, $\left(\frac{5}{2}, 6\right)$

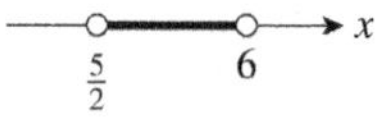

7. $2 \le n < \frac{11}{4}$, $\left[2, \frac{11}{4}\right)$

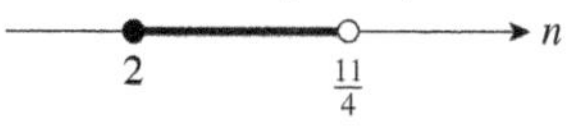

9. $-2 < x < \frac{1}{5}$, $\left(-2, \frac{1}{5}\right)$

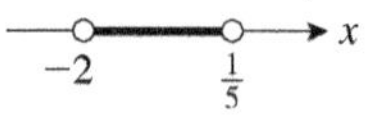

11. $n < -\frac{7}{3}$ or $n > \frac{5}{2}$, $\left(-\infty, -\frac{7}{3}\right)$ or $\left(\frac{5}{2}, \infty\right)$

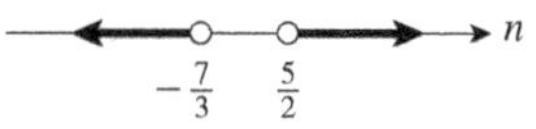

13. $x < -4, \frac{1}{2} < x < 3$
$(-\infty, -4)$ or $(1/2, 3)$

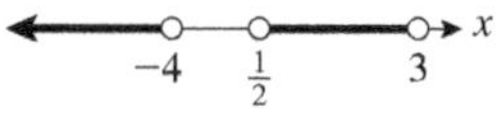

15. $x < 0, x > 4$, $(-\infty, 0)$ or $(4, \infty)$

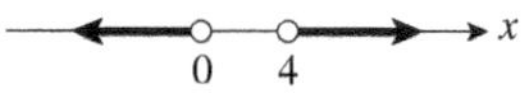

17. $R < -\frac{1}{2}, R \ge 8$, $\left(-\infty, -\frac{1}{2}\right)$ or $[8, \infty)$

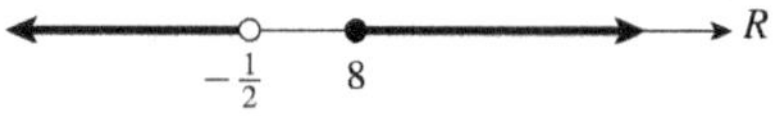

19. $-2 \le x \le \frac{2}{3}$, $\left[-2, \frac{2}{3}\right]$

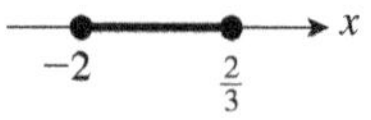

21. $x < -\frac{4}{5}, x > 2$, $\left(-\infty, -\frac{4}{5}\right)$ or $(2, \infty)$

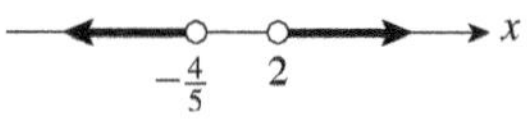

23. $x < -18, x > 78$, $(-\infty, 18)$ or $(78, \infty)$

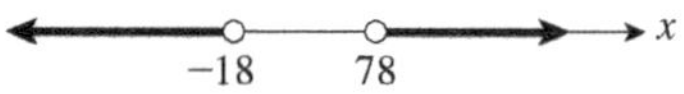

25. $x < -0.68$, $(-\infty, -0.68)$

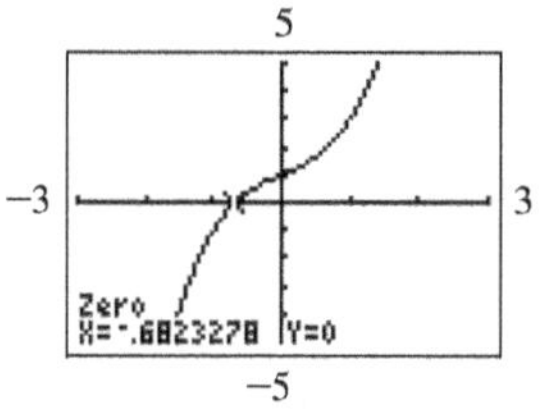

27. $x < 0.69$, $(-\infty, 0.69)$

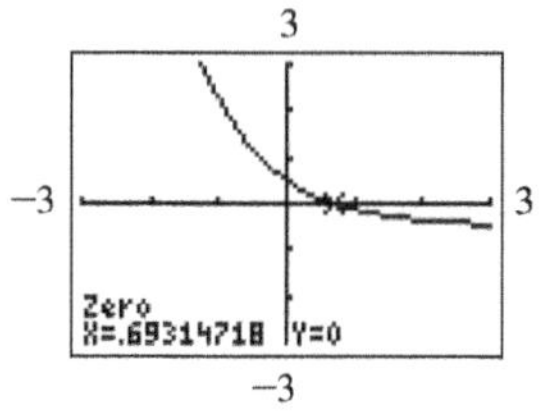

29.

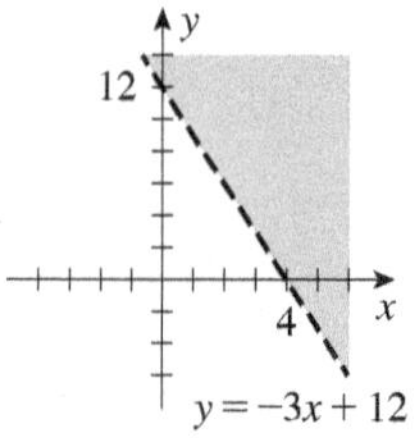

31.

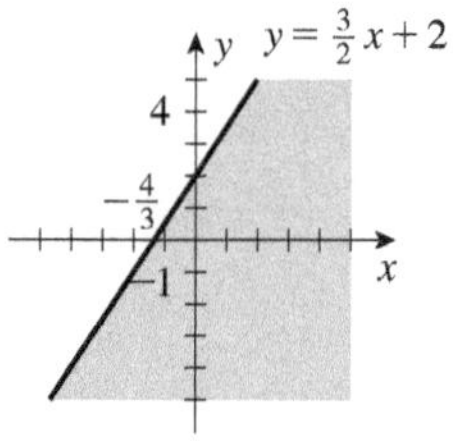

33.

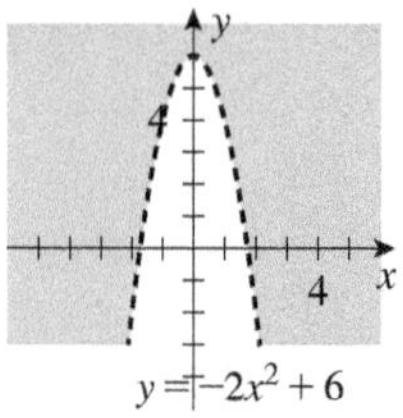

35.

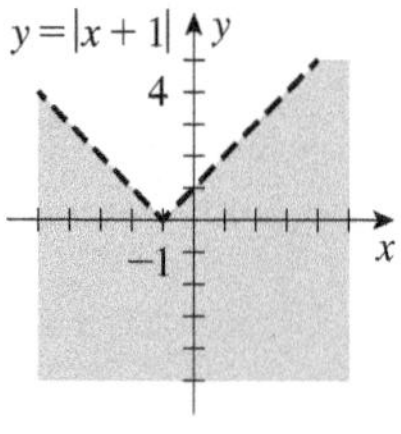

37.

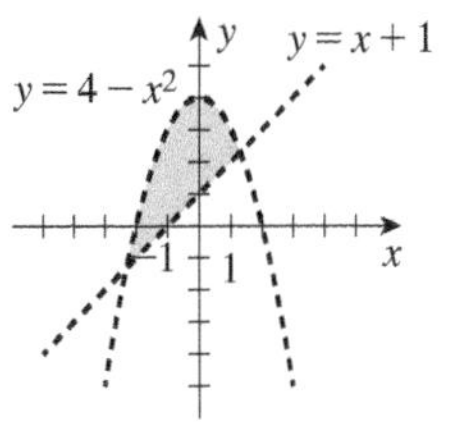

39.

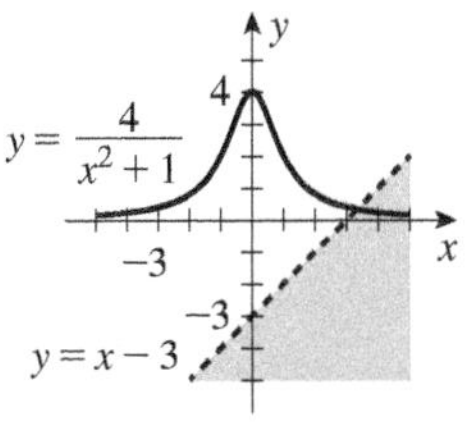

41.

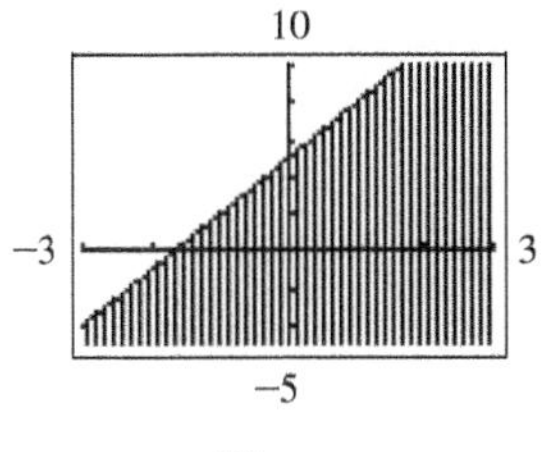

43.

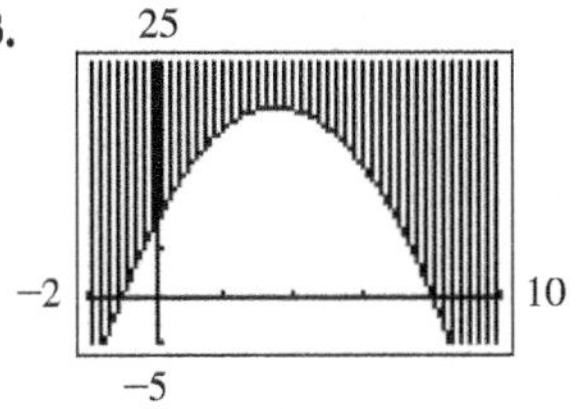

45.

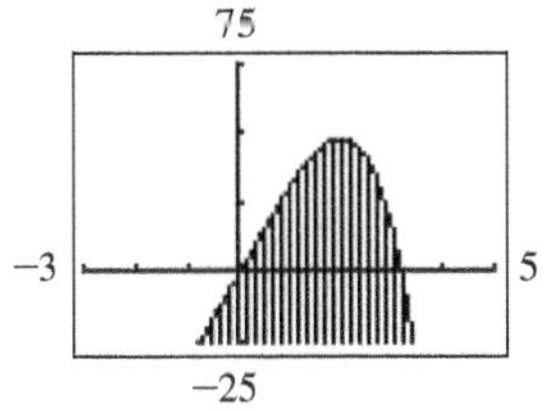

47.

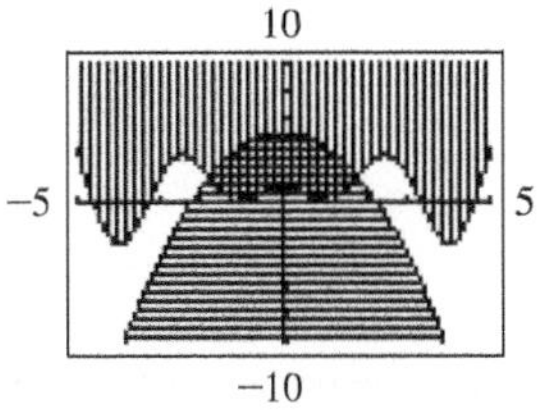

49. $x \le 3, (-\infty, 3]$ **51.** $x \le -4, x \ge 0, (-\infty, 4]$ or $[0, \infty)$

53. Max $P = 29$ at $(1, 3)$ **55.** Min $C = \frac{65}{8}$ at $\left(\frac{3}{8}, \frac{7}{4}\right)$

57. a and b must have different signs.

59. $a + \frac{1}{a} - 2 > 0, a^2 + 1 - 2a > 0, (a-1)^2 > 0$

61. $x^2 - 3x - 10 < 0$ **63.**

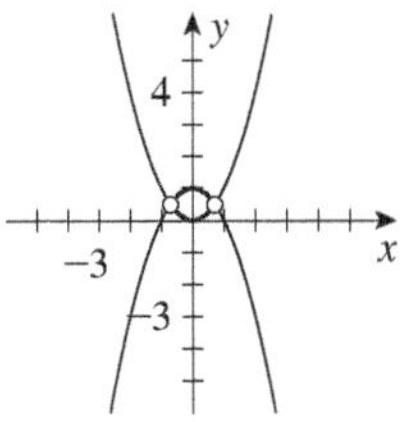

65. $f(x) > 0$ for $x < 2$ or $x > 3$
$f(x) < 0$ for $2 < x < 3$
$f(x) = 0$ for $x = 2, x = 3$
$f(0) = 6, f(5) = 6$

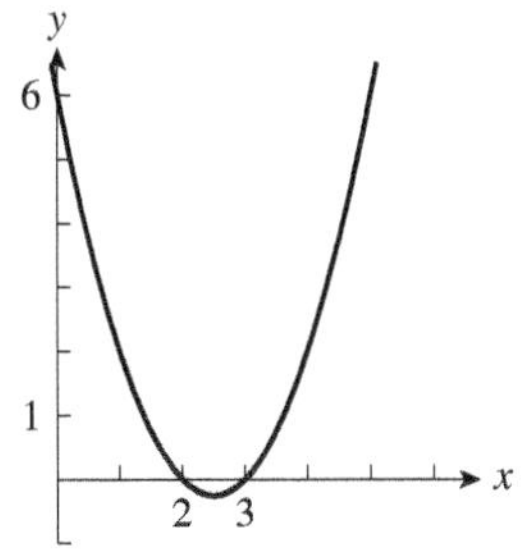

67. $0 < x \le 8.0$ cm **69.** Between \$5 and \$12.50

71. $d > 39.5$ m **73.** $168 \le B \le 400$ MJ

75. $0.456 < i < 0.816$ A **77.** $r > 5.7$

79.

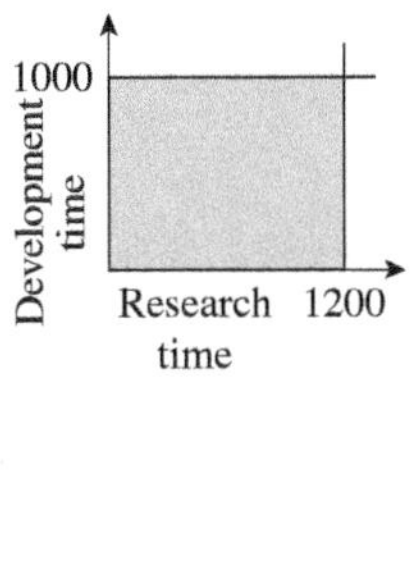

81. 300 regular 150 deluxe

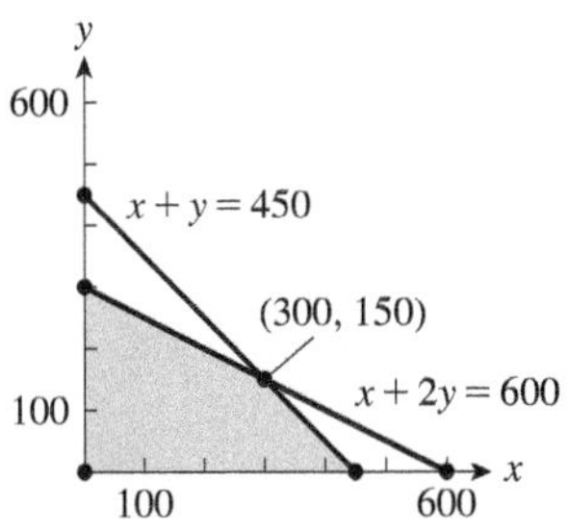

Chapter 18: Variation

Exercises 18.1, page 516

1. 56 km **3.** 6 **5.** $\frac{4}{3}$ **7.** 25 **9.** 40 **11.** 0.27 = 27%

13. 0.41 **15.** 0.025 μF **17.** 5.56 **19.** 0.931 **21.** 3.05%

23. 863 kg **25.** 0.103 m^3 **27.** 1.44 Ω **29.** 0.45 ha

31. 237 600 s **33.** 2.50×10^3 cm^2 **35.** 0.0324 kL/h

37. 23 400 cm^3 **39.** 8.8 m **41.** 3.0 m, 4.5 m

43. 19.67 kg **45.** 17 500 chips **47.** 112 cm **49.** 1.20%

Exercises 18.2, page 521

1. $c = kd, k = \pi$ **3.** 0.254 N/cm^2

5. $v = kr$ **7.** $R = \frac{k}{d^2}$ **9.** $p = \frac{k}{\sqrt{A}}$ **11.** $S = kwd^3$

13. A varies directly as the square of r.

15. f varies directly as L and inversely as the square root of m.

17. $V = \frac{H^2}{2048}$ **19.** $p = \frac{16q}{r^3}$ **21.** 125 **23.** 0.033

25. 180 **27.** 2.56×10^5

29. $A = k_1x, B = k_2x; A + B = (k_1 + k_2)x$ **31.** 10.2 kB/s

33. 209 kJ **35.** 2.40 cm **37.** 560 kJ **39.** 1.4 h

41. (a) Inverse (b) $a = 60/m$

43. 7.65 MW **45.** $F = 2.32Av^2$ **47.** 0.536 N

49. 480 m/s **51.** $R = \frac{2.60 \times 10^{-5}l}{A}$ **53.** 80.0 W

55. $G = \frac{5.9d^2}{\lambda^2}$ **57.** −6.58 cm/s^2 **59.** 0.288 W/m^2

61. (a) $p = \frac{21.0T}{V}$ (b) 265 kPa

Review Exercises for Chapter 18, page 524

1. 200 **3.** 1.5 **5.** 3.1417 **7.** 5.6 **9.** 7.39 N/cm^2

11. 2260 J/g **13.** 5.5% **15.** $\frac{a}{b} + \frac{b}{b} = \frac{c}{d} + \frac{d}{d} = \frac{a}{b} + \frac{c}{d}$

17. 630 km **19.** 2.66×10^{-3} kJ **21.** 380 pages

23. 140 mL **25.** 90.6 m **27.** 4500, 7500 **29.** 30.2 kg

31. 115 bolts, 207 bolts **33.** $y = 3x^2$ **35.** $v = \frac{128x}{y^3}$

37. 11.7 cm **39.** 10 cm **41.** $R = 13$ A

43. 1500 bacteria/h **45.** $F = \frac{5500}{L}$ **47.** 2.22 s

49. 4.3 μC **51.** 18.0 kW **53.** 533 m **55.** 1.4

57. 48.7 Hz **59.** 2.99×10^8 m/s **61.** 759 N **63.** 3.25 cm

65. 78 m **67.** 47 m **69.** 150% **71.** \$125.00, \$600.13

73. 4.80 MJ **75.** 0.023 W/m^2 **77.** 290 km

Chapter 19: Sequences and the Binomial Theorem

Exercises 19.1, page 532

1. −3 **3.** 4, 6, 8, 10, 12 **5.** 2.5, 1.5, 0.5, −0.5, −1.5

7. 28 **9.** $-\frac{9\pi}{4}$ **11.** 30.9 **13.** $49b$ **15.** 544

17. $-\frac{85}{2}$ **19.** $n = 6, S_6 = 150$ **21.** $d = -\frac{2}{19}, a_{20} = -\frac{1}{3}$

23. $a_1 = 19, a_{40} = 136$ **25.** $n = 62, S_{62} = -486.7$

27. $n = 13, a_{13} = 11k$ **29.** $n = 8, d = \frac{1}{14}(b + 2c)$

31. $a_1 = 360, d = 40, S_{10} = 5400$ **33.** Yes, $d = \ln 2$; $a_5 = \ln 48$

35. $a_n = 2n + 1$ **37.** $d = \frac{b-a}{2}$

39. $2b - c, b, c, 2c - b, 3c - 2b$ **41.** 5050 **43.** −8

45. 2700 m^2 **47.** 195 logs **49.** 10 rows

51. 13 years, \$11 700 **53.** 185 m

55. Marketing company, \$3000 more

57. $S_n = \frac{1}{2}n\left[2a_1 + (n-1)d\right]$ **59.** $a_1 = 1, a_n = n$

Exercises 19.2, page 536

1. 243 **3.** 6400, 1600, 400, 100, 25 **5.** $\frac{1}{6}, \frac{1}{2}, \frac{3}{2}, \frac{9}{2}, \frac{27}{2}$

7. 128 **9.** $\frac{1}{125}$ **11.** $\frac{100}{729}$ **13.** 1 **15.** $\frac{341}{8}$

17. 762 **19.** $\frac{3(2^{10} - k^{10})}{16(2+k)}$ **21.** $a_6 = 64, S_6 = \frac{1365}{16}$

23. $a_1 = 16, a_5 = 81$ **25.** $a_1 = 1, r = 3$

27. $n = 7, S_n = \frac{58\,593}{625}$ **29.** Yes; $r = 3^x$; $a_{20} = 3^{19x+1}$ **31.** 2

33. (a) a_1, (b) 0 **35.** 3.3% **37.** 1.09 mA **39.** \$275.18

41. 5800 °C **43.** 43% **45.** 21.1 °C **47.** \$671 088.64

49. The second option pays \$1 600 000 more.

51. $S_n = \frac{a_1 - ra_n}{1 - r}$ **53.** Yes, the ratio is squared.

55. 8, 1, −6; 8, 16, 24; 1, −6, 36; 16, 24, 36

Exercises 19.3, page 540

1. $\frac{32}{7}$ **3.** 8 **5.** 0.2 **7.** $\frac{400}{21}$ **9.** $\frac{8}{3}$ **11.** $\frac{10\,000}{9999}$

13. $\frac{1}{2}(5 + 3\sqrt{3})$ **15.** $\frac{1}{3}$ **17.** 0.5 **19.** $\frac{60}{99}$ **21.** $\frac{2}{11}$

23. $\frac{91}{3330}$ **25.** $\frac{11}{30}$ **27.** $\frac{100\,741}{999\,000}$ **29.** $\frac{125}{2}, \frac{125}{3}$

31. 350 L **33.** 346 g **35.** 100 m **37.** $-\frac{1}{4}$ **39.** 0.368

Exercises 19.4, page 544

1. $32x^5 + 240x^4 + 720x^3 + 1080x^2 + 810x + 243$

3. $t^3 + 12t^2 + 48t + 64$ **5.** $81x^4 - 108x^3 + 36x^2 - 12x + 1$

7. 8445.963 01

9. $n^5 + 10\pi n^4 + 40\pi^2 n^3 + 80\pi^3 n^2 + 80\pi^4 n + 32\pi^5$

11. $64a^6 - 192a^5b^2 + 240a^4b^4 - 160a^3b^6 + 60a^2b^8 - 12ab^{10} + b^{12}$

13. $625x^4 - 1500x^3 + 1350x^2 - 540x + 81$

15. $64a^6 + 192a^5 + 240a^4 + 160a^3 + 60a^2 + 12a + 1$

17. $x^{10} + 20x^9 + 180x^8 + 960x^7 + \cdots$

19. $128a^7 - 448a^6 + 672a^5 - 560a^4 + \cdots$

21. $x^6 - 48x^{11/2}y + 1056x^5y^2 - 14\,080x^{9/2}y^3 + \cdots$

23. $b^{40} + 10b^{37} + \frac{95}{2}b^{34} + \frac{285}{2}b^{31} + \cdots$

25. 1.264 **27.** 1.015

29. $1 + 8x + 28x^2 + 56x^3 + \cdots$

31. $1 + 6x + 27x^2 + 108x^3 + \cdots$

33. $1 + \frac{1}{2}x - \frac{1}{8}x^2 + \frac{1}{16}x^3 - \cdots$

35. $\frac{1}{3}\left[1 + \frac{1}{2}x + \frac{3}{8}x^2 + \frac{5}{16}x^3 + \cdots\right]$

37. (a) 3.557×10^{14} (b) 5.109×10^{19} (c) 8.536×10^{15} (d) 2.480×10^{96}

39. $n! = n(n-1)(n-2)(\cdots)(2)(1) = n \times (n-1)!$; for $n = 1$, $1! = 1 \times 0!$ Since $1! = 1$, $0!$ must $= 1$.

41. $56a^3b^5$ **43.** $10\,264\,320x^8b^4$

45. $n!$ contains factors 2 and 5.

47. $n = 1, 1 - 1 = 0$; $n = 2, 1 - 2 + 1 = 0$; $n = 3$, $1 - 3 + 3 - 1 = 0$, etc.

49. 0.171 **51.** 2.45

53. $V = A(1 - 5r + 10r^2 - 10r^3 + 5r^4 - r^5)$

55. $1 - \frac{x}{a} + \frac{x^3}{2a^3} - \cdots$

57. $(1-t)^4P_0 + 4(1-t)^3tP_1 + 6(1-t)^2t^2P_2 + 4(1-t)t^3P_3 + t^4P_4$

Review Exercises for Chapter 19, page 545

1. 81 **3.** 1.28×10^{-3} **5.** 107 **7.** $\frac{16}{243}$ **9.** $\frac{195}{2}$

11. $\frac{5461}{512}$ **13.** 81 **15.** 32 **17.** −15 **19.** 1 **21.** 186

23. $\frac{455}{2}$ (as), 127 (gs), or 43 (gs) **25.** 2.7 **27.** 51

29. $\frac{1}{33}$ **31.** $\frac{3}{110}$ **33.** $x^4 - 8x^3 + 24x^2 - 32x + 16$

35. $x^{10} + 20x^8 + 160x^6 + 640x^4 + 1280x^2 + 1024$

37. $a^{10} + 20a^9e + 180a^8e^2 + 960a^7e^3 + \cdots$

39. $p^{18} - \frac{3}{2}p^{16}q + p^{14}q^2 - \frac{7}{18}p^{12}q^3 + \cdots$

41. $1 + 12x + 66x^2 + 220x^3 + \cdots$

43. $1 + \frac{1}{2}x^2 - \frac{1}{8}x^4 + \frac{1}{16}x^6 - \cdots$

45. $1 - \frac{1}{2}a^2 - \frac{1}{8}a^4 - \frac{1}{16}a^6 - \cdots$

47. $\frac{1}{8} + \frac{3}{4}x + 3x^2 + 10x^3 + \cdots$

49. 1 001 000 **51.** $4b - 3a$ **53.** No

55. $a_n = -3n - 2$ **57.** 0.716 **59.** 5.475

61. 11th **63.** 12.6 mm **65.** 7690 mm

67. \$4700 **69.** 1.65×10^{10} cm = 165 000 km

71. 191 m **73.** 100 cm **75.** \$47 340.80

77. \$6.40 **79.** 4.0 °C **81.** $1 + \frac{1}{2}am^2 + \frac{1}{8}am^4$

83. 10^{-20} atm **85.** Five applications **87.** 22 years

89. $a_{mid} = a_{(n+1)/2} = a + \frac{n-1}{2}d$

$= \frac{1}{n}\left[\frac{n}{2}(a + (a + (n-1)d))\right] = \frac{S_n}{n}$

91. Yes; term to term ratios are equal.

Chapter 20: Additional Topics in Trigonometry

Exercises 20.1, page 553

(*Note:* "Answers" to trigonometric identities are intermediate steps of suggested reductions of the left member.)

1. $\sin x = \frac{\tan x}{\sec x} = \frac{\frac{\sin x}{\cos x}}{\frac{1}{\cos x}} = \frac{\sin x}{\cos x} \cdot \frac{\cos x}{1} = \sin x$

3. $\frac{0.883}{0.469} = 1.88$ **5.** $\left(-\frac{1}{2}\sqrt{3}\right)^2 + \left(-\frac{1}{2}\right)^2 = \frac{3}{4} + \frac{1}{4} = 1$

7. $\sin x - 1$ **9.** $\cot\theta - 2\cos\theta$ **11.** $\sec^2 u$

13. $\tan x \sec x$ **15.** $\sin t \cos t$ **17.** $\cot^2 y(\csc^2 y + 1)$

19. $\frac{\sin x}{\frac{\sin x}{\cos x}} = \frac{\sin x}{1}\left(\frac{\cos x}{\sin x}\right)$ **21.** $\sin x\left(\frac{1}{\cos x}\right)$

23. $\csc^2 x(\sin^2 x)$ **25.** $\sin x(\csc^2 x) = \sin x\left(\frac{1}{\sin^2 x}\right)$

27. $\cos\theta\left(\frac{\cos\theta}{\sin\theta}\right) + \sin\theta = \frac{\cos^2\theta + \sin^2\theta}{\sin\theta} = \frac{1}{\sin\theta}$

29. $\cot\theta(\sec^2\theta - 1) = \cot\theta\tan^2\theta = (\cot\theta\tan\theta)\tan\theta$

31. $\frac{\sin x}{\cos x} + \frac{\cos x}{\sin x} = \frac{\sin^2 x + \cos^2 x}{\cos x \sin x} = \frac{1}{\cos x \sin x}$

33. $(1 - \sin^2 x) - \sin^2 x$ **35.** $\frac{\sin\theta}{\frac{1}{\sin\theta}} + \frac{\cos\theta}{\frac{1}{\cos\theta}} = \sin^2\theta + \cos^2\theta$

37. $(2\sin^2 x - 1)(\sin^2 x - 1)$

39. $\frac{\sin\pi t}{2}\left(\frac{\sin^2\pi t + (1 - \cos\pi t)^2}{(1 - \cos\pi t)\sin\pi t}\right)$

$= \frac{\sin^2\pi t + 1 - 2\cos\pi t + \cos^2\pi t}{2(1 - \cos\pi t)} = \frac{2(1 - \cos\pi t)}{2(1 - \cos\pi t)}$

41. Geometric series with $a_1 = 1$, $r = \sin^2 x$

43. $\cot x$ **45.** $\sin x$ **47.** $\sec x$ **49.** $\cos x$

51.

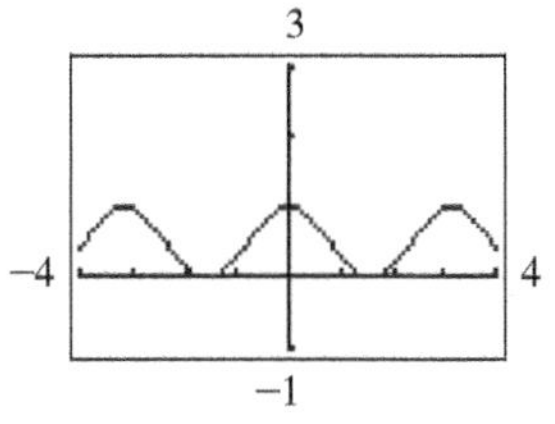

53.

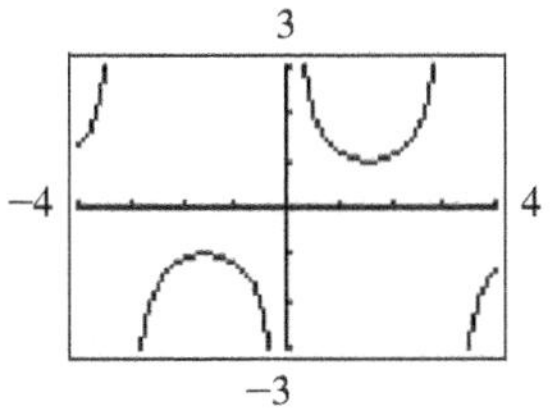

55. Yes

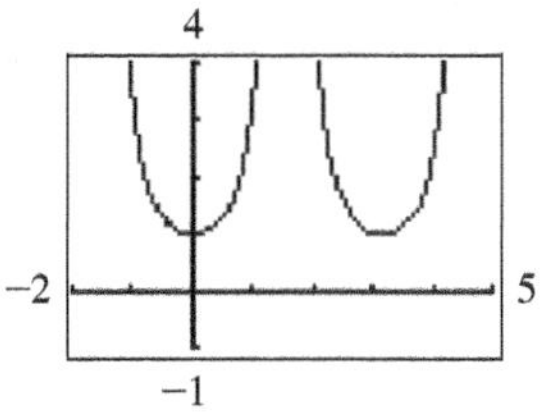

57. No

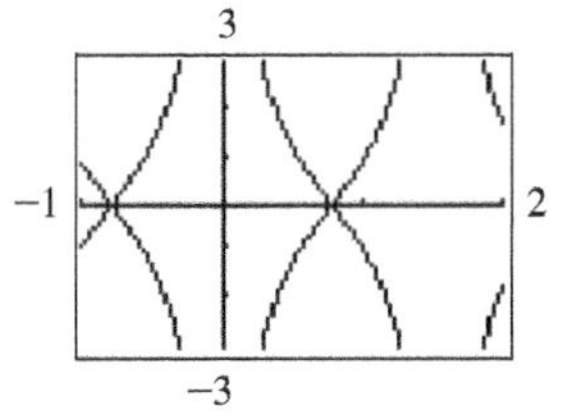

59. $0 = \cos A\cos B\cos C + \sin A\sin B$,

$\cos C = -\frac{\sin A\sin B}{\cos A\cos B}$

61. $l = a\csc\theta + a\sec\theta = a\left(\frac{1}{\sin\theta} + \frac{\tan\theta}{\sin\theta}\right)$

63. Write $\tan\theta$ in terms of $\sin\theta$ and $\cos\theta$. Use $\cos\theta$ as LCD and then Eq. (20.6) to simplify.

65. $\sin^2 x - \sin^2 x\sec^2 x + \cos^2 x + \cos^2 x\sec^4 x$
$= \sin^2 x - \tan^2 x + \cos^2 x + \sec^2 x = 2$

67. $\left(\frac{r}{x}\right)^2 + \left(\frac{r}{y}\right)^2 = \frac{r^2(x^2 + y^2)}{x^2y^2}$

69. $\sqrt{1 - \cos^2\theta} = \sqrt{\sin^2\theta}$

71. $\sqrt{64 + 64\tan^2\theta} = 8\sqrt{1 + \tan^2\theta}$

Exercises 20.2, page 557

1. $\frac{16}{65}$

3. $\sin 105° = \sin 60°\cos 45° + \cos 60°\sin 45°$

$= \frac{\sqrt{3}}{2}\frac{\sqrt{2}}{2} + \frac{1}{2}\frac{\sqrt{2}}{2} = 0.966$

5. $\cos 15° = \cos(60° - 45°) = \cos 60°\cos 45° + \sin 60°\sin 45°$

$= \left(\frac{1}{2}\right)\left(\frac{1}{2}\sqrt{2}\right) + \left(\frac{1}{2}\sqrt{3}\right)\left(\frac{1}{2}\sqrt{2}\right)$

$= \frac{1}{4}\sqrt{2} + \frac{1}{4}\sqrt{6} = \frac{1}{4}(\sqrt{2} + \sqrt{6}) = 0.966$

7. $-\frac{33}{65}$ **9.** $-\frac{56}{65}$ **11.** $\sin 3x$ **13.** $-\cos x$

15. $-\sin x$ **17.** $\tan x$ **19.** 0 **21.** 1 **23.** 0

25. $(\sin x\cos y + \cos x\sin y)(\sin x\cos y - \cos x\sin y)$
$= \sin^2 x\cos^2 y - \cos^2 x\sin^2 y$
$= \sin^2 x(1 - \sin^2 y) - (1 - \sin^2 x)\sin^2 y$

27. $(\cos\alpha\cos\beta - \sin\alpha\sin\beta) + (\cos\alpha\cos\beta + \sin\alpha\sin\beta)$

29.

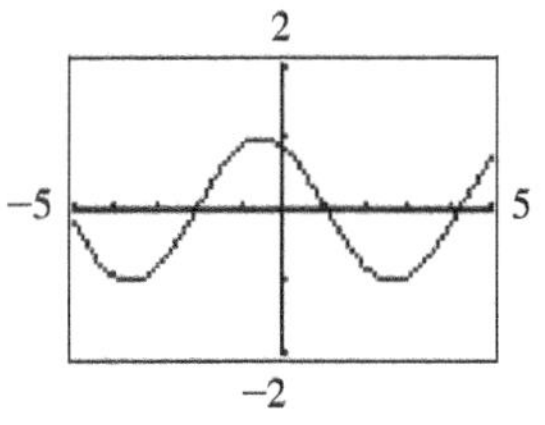

31.

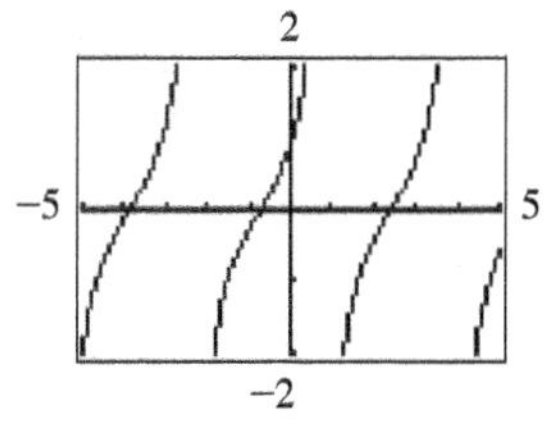

33, 35, 37, 39. Use the indicated method.

41. $\frac{\sin(x + x)}{\sin x} = \frac{\sin x\cos x + \cos x\sin x}{\sin x}$

$= \frac{2\sin x\cos x}{\sin x}$

43. $\cos A\cos B\cos C - \sin A\sin B\cos C - \sin A\cos B\sin C$
$- \cos A\sin B\sin C$

45. $20\sqrt{2}(\sin 120\pi t\cos\pi/4 + \sin\pi/4\cos 120\pi t)$

$= 20\sqrt{2}\left(\frac{1}{\sqrt{2}}\right)(\sin 120\pi t + \cos 120\pi t)$

47. $i_0\sin(\omega t + \alpha) = i_0(\sin\omega t\cos\alpha + \cos\omega t\sin\alpha)$

49. $\tan\alpha(R + \cos\beta) = \sin\beta$, $R = \frac{\sin\beta - \tan\alpha\cos\beta}{\tan\alpha}$

$= \frac{\sin\beta\cos\alpha - \cos\beta\sin\alpha}{\cos\alpha\tan\alpha}$

Exercises 20.3, page 561

1. $-\sqrt{3}$ **3.** $-\dfrac{24}{25}$

5. $\sin 60° = \sin 2(30°) = 2 \sin 30° \cos 30°$

$= 2\left(\dfrac{1}{2}\right)\left(\dfrac{1}{2}\sqrt{3}\right) = \dfrac{1}{2}\sqrt{3}$

7. $\tan 120° = \dfrac{2 \tan 60°}{1 - \tan^2 60°} = \dfrac{2\sqrt{3}}{1 - (\sqrt{3})^2} = -\sqrt{3}$

9. $\sin 258° = 2 \sin 129° \cos 129° = -0.978$

11. $\cos 96° = \cos^2 48° - \sin^2 48° = -0.105$

13. 3.08 **15.** $\dfrac{24}{25}$ **17.** $-\sqrt{3}$ **19.** $3 \sin 10x$ **21.** $\cos 8x$

23. $\cos x$ **25.** $-2 \cos 4x$ **27.** $2 \cos 2\theta$ **29.** 2

31. $\cos^2 \alpha - (1 - \cos^2 \alpha)$

33. $\dfrac{\cos x - (\sin x/\cos x) \sin x}{1/\cos x} = \cos^2 x - \sin^2 x$

35. $\dfrac{2 \sin \theta \cos \theta}{1 + 2 \cos^2 \theta - 1} = \dfrac{\sin \theta}{\cos \theta}$

37. $1 - (1 - 2 \sin^2 2\theta) = \dfrac{2}{\csc^2 \theta}$

39. $\ln\dfrac{1 - \cos 2x}{1 + \cos 2x} = \ln\dfrac{2 \sin^2 x}{2 \cos^2 x} = \ln \tan^2 x$

41.

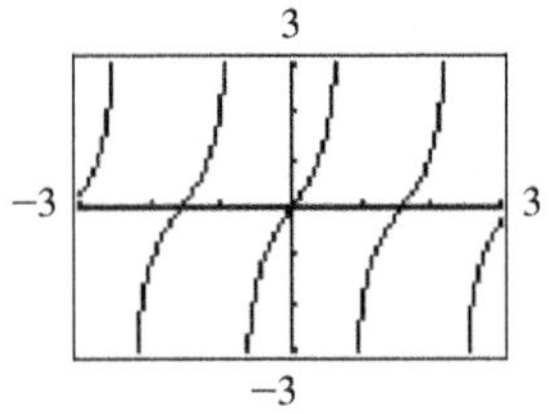

43.

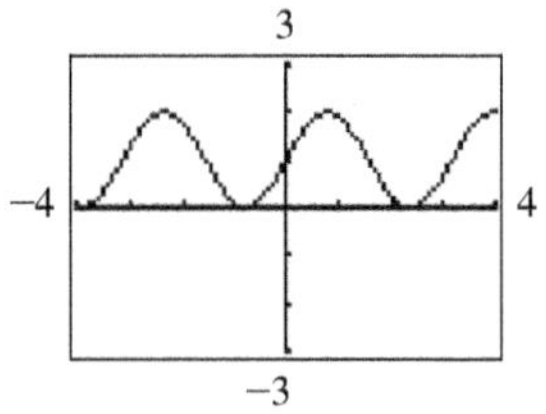

45. $3 \sin x - 4 \sin^3 x$ **47.** $8 \cos^4 x - 8 \cos^2 x + 1$

49. 0 **51.** amp. = 2, per. = π; write equation as $y = 2(2 \sin x \cos x) = 2 \sin 2x$.

53. $\sqrt{\sin^2 x + 2 \sin x \cos x + \cos^2 x} = \sqrt{1 + \sin 2x}$

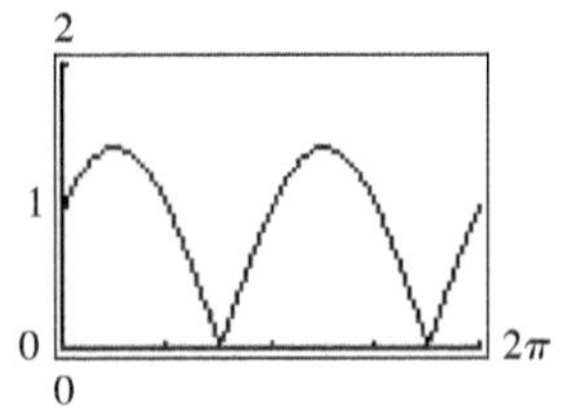

55. 474 m **57.** $R = v\left(\dfrac{2v \sin \alpha}{g}\right) \cos \alpha = \dfrac{v^2(2 \sin \alpha \cos \alpha)}{g}$

59. $vi \sin \omega t \sin\left(\omega t - \dfrac{\pi}{2}\right)$

$= vi \sin \omega t\left(\sin \omega t \cos \dfrac{\pi}{2} - \cos \omega t \sin \dfrac{\pi}{2}\right)$

$= vi \sin \omega t[-(\cos \omega t)(1)] = -\dfrac{1}{2}vi(2 \sin \omega t \cos \omega t)$

Exercises 20.4, page 564

1. $\cos 57°$

3. $\cos 15° = \cos\dfrac{1}{2}(30°) = \sqrt{\dfrac{1 + \cos 30°}{2}} = \sqrt{\dfrac{1.8660}{2}}$

$= 0.966$

5. $\sin 105° = \sin \dfrac{1}{2}(210°) = \sqrt{\dfrac{1 - \cos 210°}{2}}$

$= \sqrt{\dfrac{1.8660}{2}} = 0.966$

7. -0.924 **9.** $\sin 96° = 0.994\ 521\ 9$

11. $\sqrt{2\left(\dfrac{1 + \cos 164°}{2}\right)} = \sqrt{2} \cos 82° = 0.196\ 820\ 5$

13. $\sin 3x$ **15.** $4 \cos 2x$ **17.** $2\sqrt{2} \sin 5\theta$ **19.** 1

21. $\dfrac{1}{4}$ **23.** 0.142 **25.** $\pm\sqrt{\dfrac{2 \sec \alpha}{\sec \alpha - 1}}$

27. $\tan \dfrac{1}{2}\alpha = \dfrac{1 - \cos \alpha}{\sin \alpha}$

29. $\dfrac{1 - \cos \alpha}{2 \sin \frac{1}{2}\alpha} = \dfrac{1 - \cos \alpha}{2\sqrt{\frac{1}{2}(1 - \cos \alpha)}} = \sqrt{\dfrac{1 - \cos \alpha}{2}}$

31. $2\sqrt{2\dfrac{(1 + \cos x)(1 + \cos x)}{2(2)(1 + \cos x)}} = \dfrac{2(1 + \cos x)}{2} \cdot \sqrt{\dfrac{2}{1 + \cos x}}$

33.

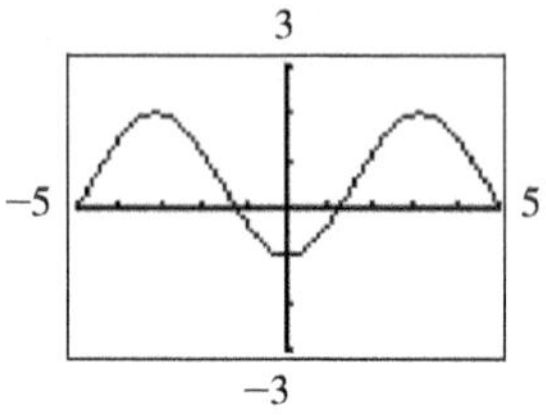

35.

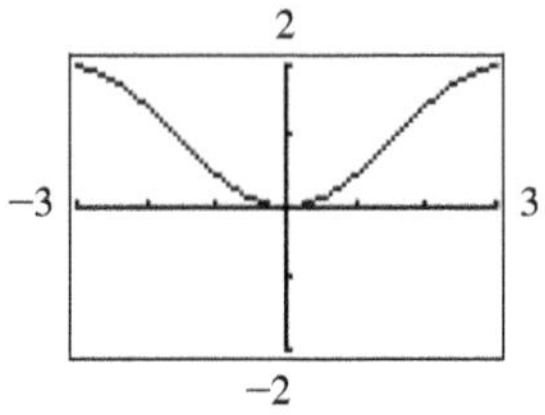

37. $\pm\dfrac{840}{41}$ **39.** $4 \sin \dfrac{\theta}{2}$

41. $\sin^2 \omega t = \left(\sqrt{\dfrac{1 - \cos 2\omega t}{2}}\right)^2$

$= \dfrac{1 - \cos 2\omega t}{2}$

43. $\dfrac{\sqrt{\dfrac{1 - \cos(A + \phi)}{2}}}{\sqrt{\dfrac{1 - \cos A}{2}}} = \sqrt{\dfrac{1 - \cos(A + \phi)}{1 - \cos A}}$

Exercises 20.5, page 568

1. $\dfrac{\pi}{4}, \dfrac{5\pi}{4}$ **3.** $0, \dfrac{\pi}{3}, \dfrac{5\pi}{3}$ **5.** $\dfrac{\pi}{2}$ **7.** $\dfrac{7\pi}{6}, \dfrac{11\pi}{6}$

9. $\dfrac{\pi}{3}, \dfrac{2\pi}{3}, \dfrac{4\pi}{3}, \dfrac{5\pi}{3}$ **11.** $0, \dfrac{\pi}{6}, \dfrac{5\pi}{6}, \pi$ **13.** $\dfrac{\pi}{2}, \dfrac{3\pi}{2}$

15. $\dfrac{\pi}{4}, \dfrac{3\pi}{4}, \dfrac{5\pi}{4}, \dfrac{7\pi}{4}$ **17.** 0.262, 1.31, 3.40, 4.45

19. $0, \pi$ **21.** $\dfrac{3\pi}{4}, \dfrac{7\pi}{4}$ **23.** 1.98, 4.30

25. $\dfrac{\pi}{3}, \dfrac{2\pi}{3}, \dfrac{4\pi}{3}, \dfrac{5\pi}{3}$ **27.** $\dfrac{\pi}{12}, \dfrac{\pi}{4}, \dfrac{5\pi}{12}, \dfrac{3\pi}{4}, \dfrac{13\pi}{12}, \dfrac{5\pi}{4}, \dfrac{17\pi}{12}, \dfrac{7\pi}{4}$

29. $0, \dfrac{\pi}{3}, \pi, \dfrac{5\pi}{3}$ **31.** 3.569, 5.856 **33.** $\dfrac{\pi}{4}, \dfrac{5\pi}{4}, 1.25, \pi + 1.25$

35. $\dfrac{3\pi}{8}, \dfrac{7\pi}{8}, \dfrac{11\pi}{8}, \dfrac{15\pi}{8}$ **37.** $0, \dfrac{\pi}{2}, \pi, \dfrac{3\pi}{2}$ **39.** $0, \dfrac{\pi}{2}, \pi, \dfrac{3\pi}{2}$

41. No; $\cot \theta > 1$, $\sec \theta > 1$, $\csc \theta > 1$ for $0 < \theta < \dfrac{\pi}{2}$

43. $\theta = 0, r = 0; \theta = \frac{\pi}{3}, r = \frac{\sqrt{3}}{2}; \theta = \pi, r = 0; \theta = \frac{5\pi}{3}, r = -\frac{\sqrt{3}}{2}$

45. 37.8° **47.** 10.2 s, 15.7 s, 21.2 s, 47.1 s

49. 6.6×10^{-4} rad **51.** 300.0 N, 400.0 N

53. −2.28, 0.00, 2.28 **55.** 0.29, 0.95

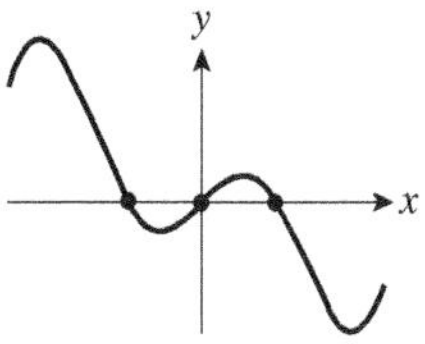

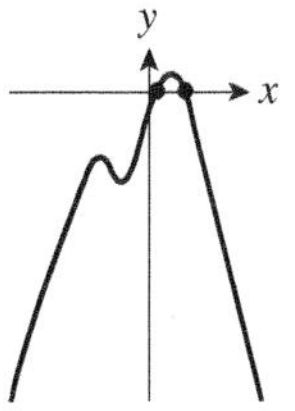

57. 2.10 **59.** 1.08

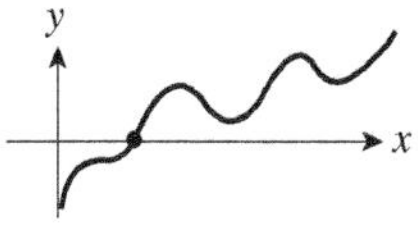

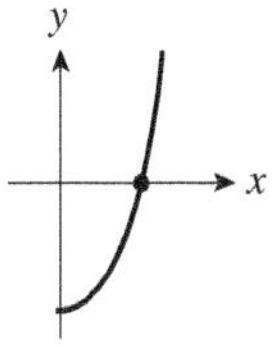

Exercises 20.6, page 573

1. y is the angle whose tangent is $3A$. **3.** 0

5. y is the angle whose cotangent is $3x$.

7. y is twice the angle whose sine is x.

9. y is five times the angle whose cosine is $2x - 1$.

11. $\frac{\pi}{3}$ **13.** $\frac{\pi}{4}$ **15.** $-\frac{\pi}{3}$ **17.** No value **19.** $-\frac{\pi}{4}$

21. $\frac{1}{2}$ **23.** $\frac{\pi}{4}$ **25.** $\frac{1}{\sqrt{26}}$ **27.** −1 **29.** $2/\sqrt{21}$

31. $2/\sqrt{3}$ **33.** −1.2640 **35.** 0.0219 **37.** −1.239

39. −0.2239 **41.** $x = \frac{1}{5}\sin^{-1} y$ **43.** $x = 4 \tan y$

45. $x = 1 - \cos(1 - y)$ **47.** $\frac{x}{\sqrt{1 - x^2}}$

49. $xy + \sqrt{1 - x^2 - y^2 + x^2y^2}$ **51.** $\frac{3x}{\sqrt{9x^2 - 1}}$

53. $2x\sqrt{1 - x^2}$ **55.** No; $\sin^{-1}(\sin x) = x$ for $-\frac{\pi}{2} \le x \le \frac{\pi}{2}$.

57. $t = \frac{1}{2\omega}\cos^{-1}\frac{y}{A} - \frac{\phi}{\omega}$ **59.** $t = \frac{1}{\omega}\left(\sin^{-1}\frac{i}{I_m} - \alpha - \phi\right)$

61. $\sin\left(\sin^{-1}\frac{3}{5} + \sin^{-1}\frac{5}{13}\right) = \frac{3}{5}\cdot\frac{12}{13} + \frac{4}{5}\cdot\frac{5}{13} = \frac{56}{65}$

63. $\frac{\pi}{2}$ **65.** 0 **67.** $\sin^{-1}\left(\frac{a}{c}\right)$ **69.** $\tan^{-1}\left(\frac{b \tan B}{a}\right)$

71. Let y = height to top of base; $\tan \alpha = \frac{2.7 + y}{d}$, $\tan \beta = \frac{y}{d}$; $\tan \alpha = \frac{2.7 + d \tan \beta}{d}$

73. $\theta = \tan^{-1}\left(\frac{y + 50}{x}\right) - \tan^{-1}\left(\frac{y}{x}\right)$

Review Exercises for Chapter 20, page 575

1. $\sin(90° + 30°) = \sin 90° \cos 30° + \cos 90° \sin 30°$
$= (1)\left(\frac{1}{2}\sqrt{3}\right) + (0)\left(\frac{1}{2}\right) = \frac{1}{2}\sqrt{3}$

3. $\sin(360° - 45°) = \sin 360° \cos 45° - \cos 360° \sin 45°$
$= 0\left(\frac{1}{2}\sqrt{2}\right) - 1\left(\frac{1}{2}\sqrt{2}\right) = -\frac{1}{2}\sqrt{2}$

5. $\cos\left(2\frac{\pi}{2}\right) = \cos^2\frac{\pi}{2} - \sin^2\frac{\pi}{2} = 0 - 1^2 = -1$

7. $\tan 2(30°) = \frac{2 \tan 30°}{1 - \tan^2 30°} = \frac{2(\sqrt{3}/3)}{1 - (\sqrt{3}/3)^2} = \sqrt{3}$

9. $\sin 52° = 0.788$ **11.** $\frac{1}{2}$ **13.** $\cos 215° = -0.819$

15. $2 \tan 24° = 0.890$ **17.** $\sin 5x$ **19.** $4 \sin 12x$

21. $2 \cos 12x$ **23.** $2 \cos x$ **25.** $-\frac{\pi}{2}$ **27.** 0.2619

29. $-\frac{1}{3}\sqrt{3}$ **31.** $-\frac{\pi}{6}$ **33.** $\sec^2 y - \tan^2 y = 1$

35. $\sin x \csc x - \sin^2 x = 1 - \sin^2 x = \cos^2 x$

37. $\frac{(\sec^2 x - 1)(\sec^2 x + 1)}{\tan^2 x} = \sec^2 x + 1$

39. $2\left(\frac{1}{\sin 2x}\right)\left(\frac{\cos x}{\sin x}\right) = 2\left(\frac{1}{2 \sin x \cos x}\right)\left(\frac{\cos x}{\sin x}\right) = \frac{1}{\sin^2 x}$

41. $\frac{\cos^2 \theta}{\sin^2 \theta}$ **43.** $\frac{1}{2}\left(2 \sin\frac{\theta}{2}\cos\frac{\theta}{2}\right)$ **45.** $\cot x$

47. $\sec x$ **49.** $\sin x$ **51.** $\sin x$

53.

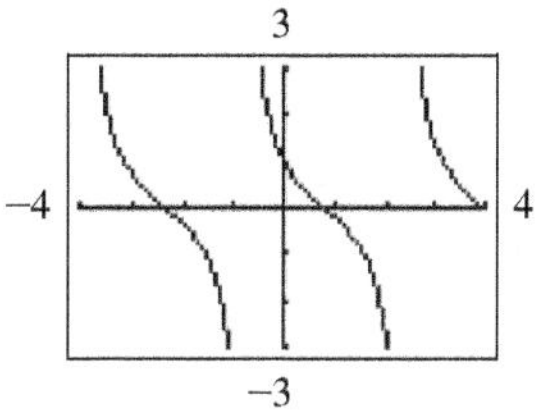

55.

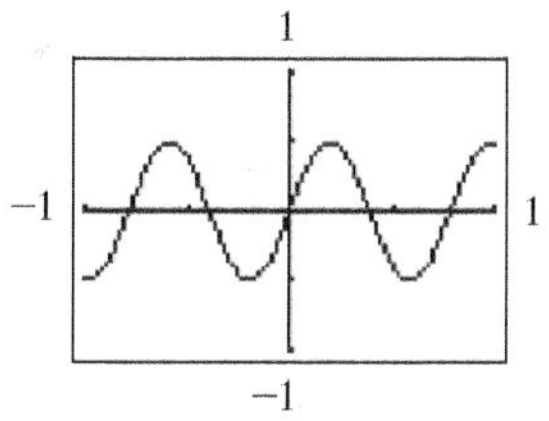

57.

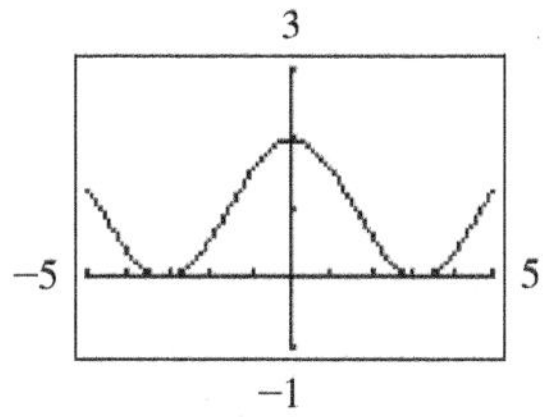

59.

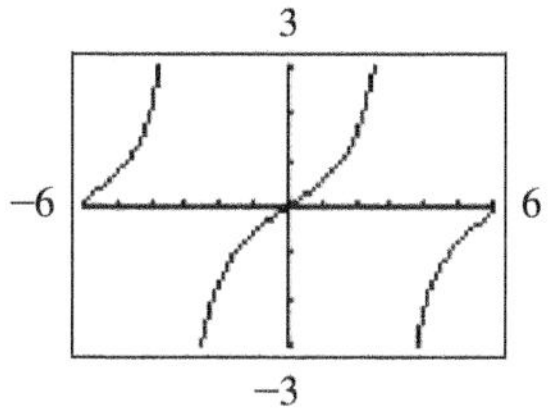

61. $x = \frac{1}{2}\cos^{-1}\frac{1}{2}y$ **63.** $x = \frac{1}{5}\sin\frac{1}{3}\left(\frac{1}{4}\pi - y\right)$

65. 1.29, 4.43 **67.** $\frac{\pi}{6}, \frac{5\pi}{6}, \frac{7\pi}{6}, \frac{11\pi}{6}$ **69.** $0, \frac{\pi}{3}, \frac{5\pi}{3}$

71. $0, \frac{2\pi}{3}$ **73.** $\frac{\pi}{10}, \frac{\pi}{2}, \frac{9\pi}{10}, \frac{13\pi}{10}, \frac{3\pi}{2}, \frac{17\pi}{10}$ **75.** 0

77. Identity: $\tan x + \frac{1}{\tan x} = \frac{\tan^2 x + 1}{\tan x} = \sec^2 x(\cot x) = \frac{\sec^2 x \cos x}{\sin x}$

79. $\frac{\pi}{2}, \pi$

81. 1.56, 2.16, 3.46

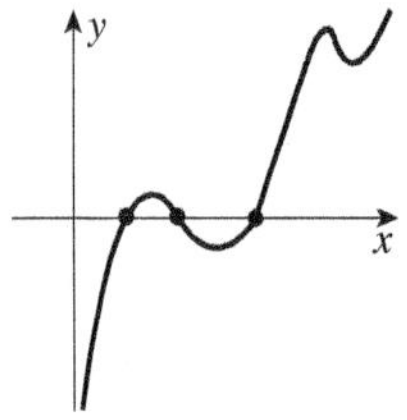

83. −2.31, 1.14

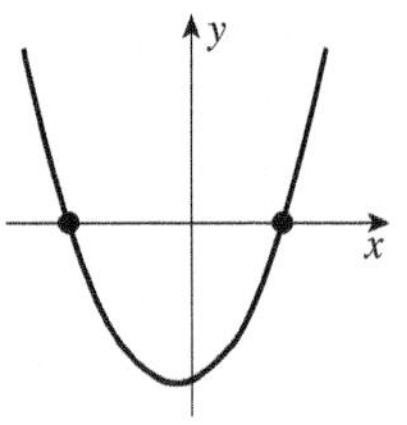

85. $\dfrac{1}{x}$ **87.** $2x\sqrt{1-x^2}$ **89.** $\dfrac{\sqrt{1-x^2}-xy}{\sqrt{1+y^2}}$

91. $2\sqrt{1-\cos^2\theta}$ **93.** $\dfrac{\tan\theta}{\sqrt{1+\tan^2\theta}}=\dfrac{\tan\theta}{\sec\theta}$

95. $(\cos\theta + j\sin\theta)^2 = (\cos^2\theta - \sin^2\theta) + j(2\sin\theta\cos\theta)$

97. $\dfrac{\pi}{2} < x < \dfrac{3\pi}{2}$

99. $C\left(\dfrac{A}{C}\sin 2t + \dfrac{B}{C}\cos 2t\right)$
$= C(\cos\alpha\sin 2t + \sin\alpha\cos 2t)$

101. $R = \sqrt{(A\cos\theta - B\sin\theta)^2 + (A\sin\theta + B\cos\theta)^2}$
$= \sqrt{A^2(\cos^2\theta + \sin^2\theta) + B^2(\sin^2\theta + \cos^2\theta)}$

103. $\dfrac{k}{2}\cdot\dfrac{1}{\sin^2\frac{\theta}{2}} = \dfrac{k}{2}\cdot\dfrac{1}{\frac{1-\cos\theta}{2}}$ **105.** $\theta = a + R\sin\omega t$

107. $\dfrac{\cos 2\alpha}{2\cos^2\alpha}$ **109.** $P = VI\cos\omega t[\cos(\phi + \omega t)]$

111. 6.8 m **113.** 54.7°

Chapter 21: Plane Analytic Geometry

Exercises 21.1, page 582

1. $\sqrt{61}$ **3.** 150° **5.** $2\sqrt{29}$ **7.** 3 **9.** 55

11. $2\sqrt{53}$ **13.** 2.86 **15.** $\dfrac{5}{2}$ **17.** Undefined

19. $-\dfrac{3}{4}$ **21.** $-\dfrac{5}{9}$ **23.** 0.747 **25.** $\dfrac{1}{3}\sqrt{3}$ **27.** −0.306

29. 20.0° **31.** 98.5° **33.** Parallel **35.** Perpendicular

37. 8, −2 **39.** −3 **41.** Two sides equal $2\sqrt{10}$.

43. $m_1 = \dfrac{5}{12}, m_2 = \dfrac{4}{3}$ **45.** 10 **47.** $4\sqrt{10} + 4\sqrt{2} = 18.3$

49. (1, 5) **51.** (−2.8, 4.2) **53.** $x^2 + y^2 = 9$

55. $m_1 = 1, m_2 = -1$ **57.** $\left(-\dfrac{11}{3}, 0\right)$ **59.** $\left(0, -\dfrac{7}{9}\right)$

61. $0, \sqrt{3}, -\sqrt{3}$

Exercises 21.2, page 587

1. $2y - x - 2 = 0$ **3.** $2, \left(0, \dfrac{5}{2}\right)$

5. $4x - y + 20 = 0$

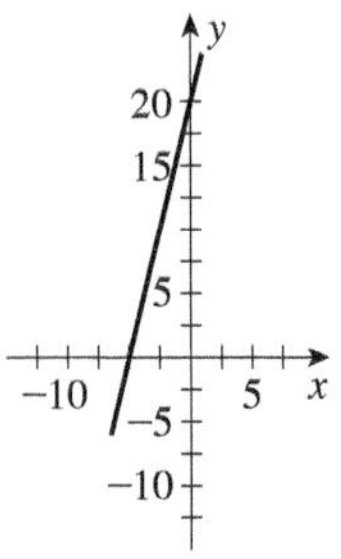

7. $7x - 2y - 24 = 0$

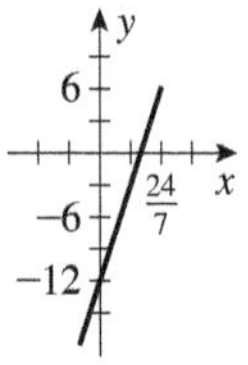

9. $x - y + 19 = 0$

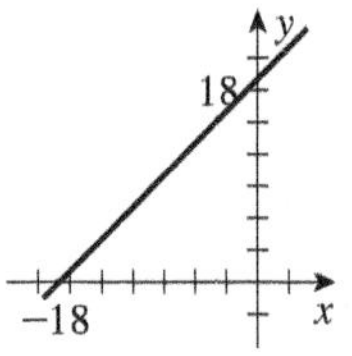

11. $y = -2.7$

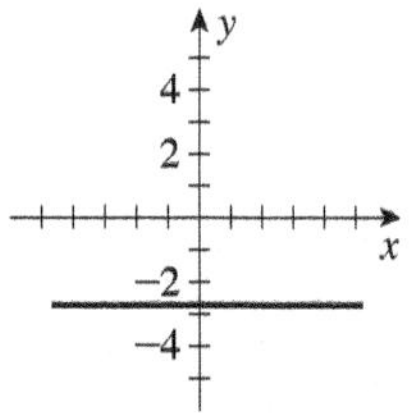

13. $x = -3$

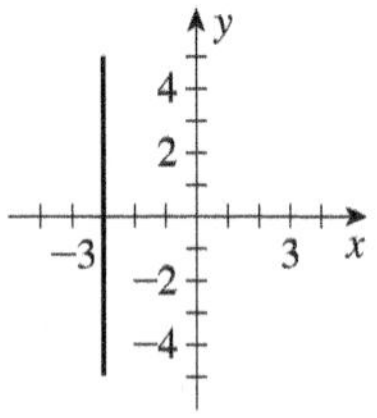

15. $x - 3y - 7 = 0$

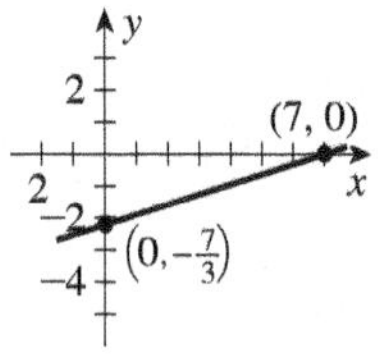

17. $x + y - 7 = 0$

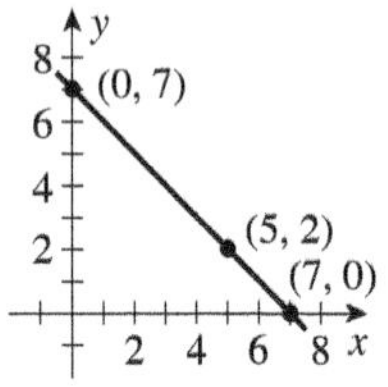

19. $3x + y - 18 = 0$

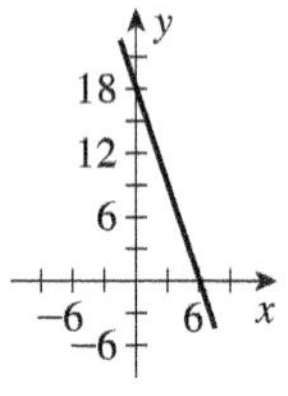

21. $y = 4x - 8; m = 4, (0, -8)$

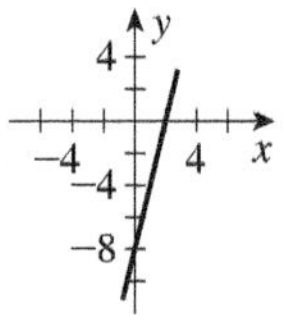

23. $y = -\dfrac{3}{5}x + 2; m = -\dfrac{3}{5}, (0, 2)$

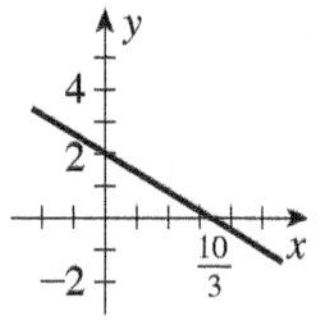

25. $y = \dfrac{3}{2}x - \dfrac{1}{2}; m = \dfrac{3}{2}, \left(0, -\dfrac{1}{2}\right)$

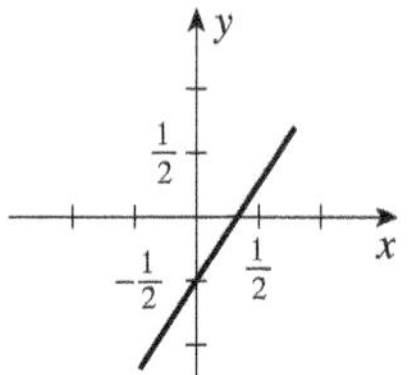

27. $y = 3.5x + 0.5; m = 3.5, (0, 0.5)$

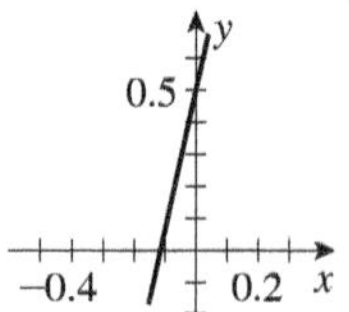

29. Parallel **31.** Perpendicular **33.** Neither

35. Perpendicular **37.** −2

39. The slope of the first line is 3. A line perpendicular to it has a slope of $-\frac{1}{3}$. The slope of the second line is $-\frac{k}{3}$, so $k = 1$.

41. $b - a$ **43.** 5

45. $m_1 = -\dfrac{4}{5}, m_2 = \dfrac{2}{3}, m_3 = \dfrac{2}{3}; m_4 = -\dfrac{4}{5}; m_1 = m_4, m_2 = m_3$

47. $m_1 = -\frac{a}{b}$, $m_2 = -\frac{a}{b}$, $m_1 = m_2$; $m_3 = -\frac{a}{b}$, $m_4 = \frac{b}{a}$, $m_3 = -1/m_4$

49. $C = \frac{5}{4}R$ **51.** $v = 0.607T + 331$ **53.** $T = \frac{4}{3}x + 3$

55. $600t$ for $0 \le t \le 5$; $1500 + 300t$ for $5 \le t \le 20$; 7500 for $t \ge 20$

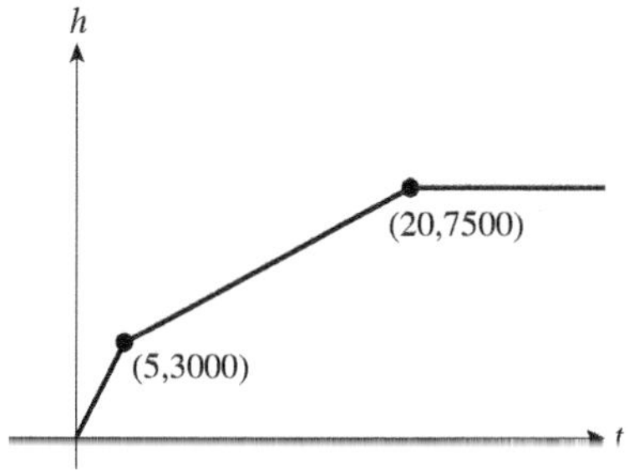

57. $y = 10^{-5}(2.4 - 5.6x)$

59. $n = \frac{7}{6}t + 10$; at 6:30, $n = 10$; at 8.30, $n = 150$

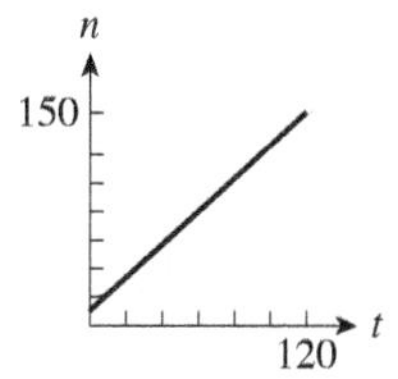

61.

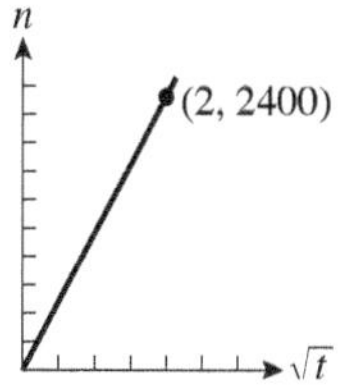

63.

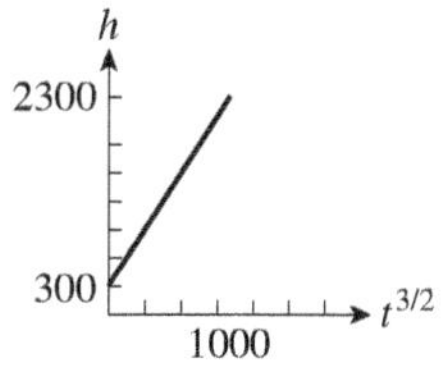

65.

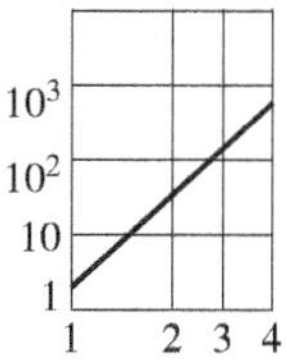

67.

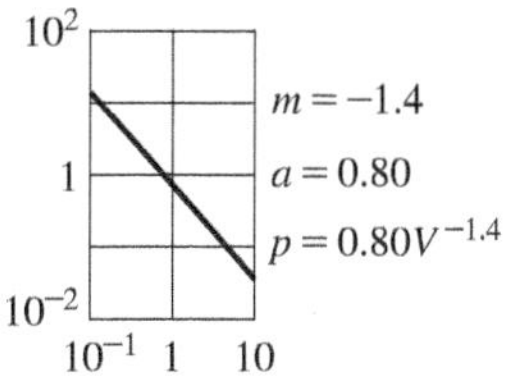

Exercises 21.3, page 592

1. $C(1, -1)$, $r = 4$

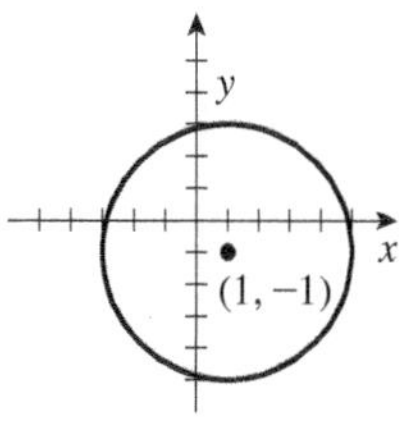

3. $C(3, 4)$, $r = 7$

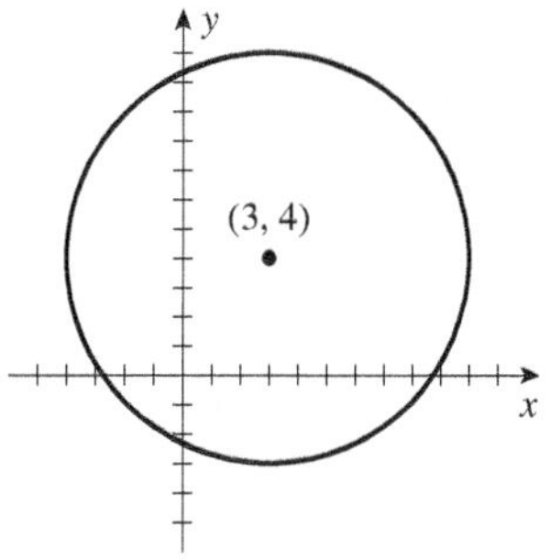

5. $(2, 1)$, $r = 5$ **7.** $C(-1, 0)$, $r = \frac{11}{2}$ **9.** $x^2 + y^2 = 9$

11. $(x - 2)^2 + (y - 3)^2 = 16$, or $x^2 + y^2 - 4x - 6y - 3 = 0$

13. $(x - 12)^2 + (y + 15)^2 = 324$, or $x^2 + y^2 - 24x + 30y + 45 = 0$

15. $(x + 3)^2 + (y - 4)^2 = 25$ **17.** $(x - 2)^2 + (y - 1)^2 = 8$

19. $(x + 3)^2 + (y - 5)^2 = 25$, or $x^2 + y^2 + 6x - 10y + 9 = 0$

21. $(x + 2)^2 + (y - 2)^2 = 4$ or $x^2 + y^2 + 4x - 4y + 4 = 0$ **23.** $x^2 + y^2 = 2$

25. $(0, 3)$, $r = 2$

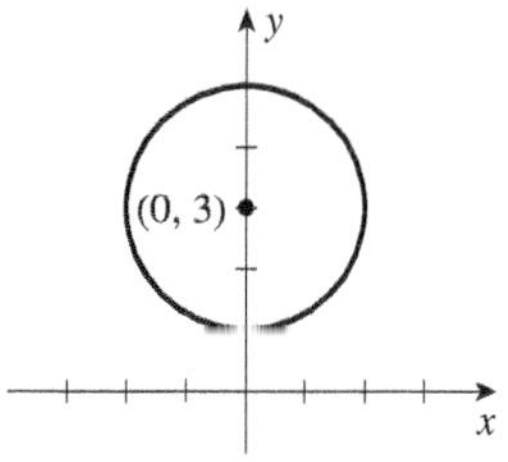

27. $(-1, 5)$, $r = \frac{9}{2}$

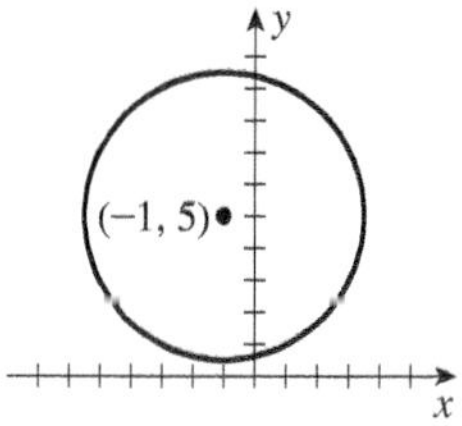

29. $(1, 0)$, $r = 3$

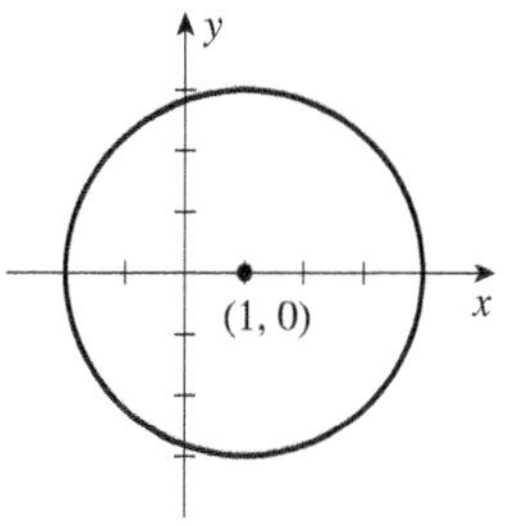

31. $(-2.1, 1.3)$, $r = 3.1$

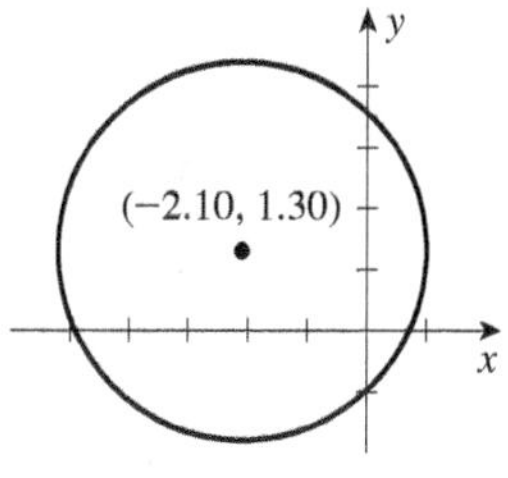

33. $(0, 2)$, $r = \frac{5}{2}$

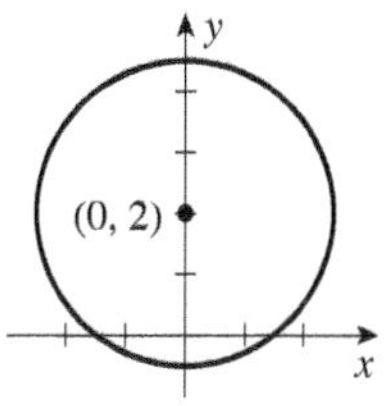

35. $(1, 2)$, $r = \frac{1}{2}\sqrt{22}$

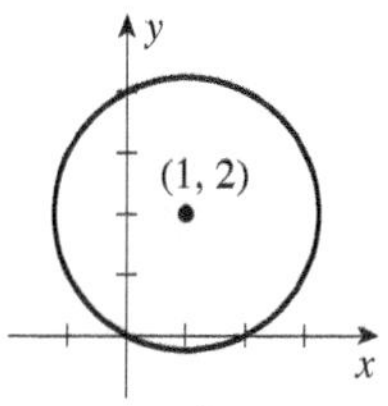

37. Symmetric to both axes and origin

39. Symmetric to y-axis **41.** 64 **43.** $(7, 0)$, $(-1, 0)$

45. $3x^2 + 3y^2 + 4x + 8y - 20 = 0$, circle

47. Graph $y = -2.5 \pm \sqrt{10.25 - x^2}$

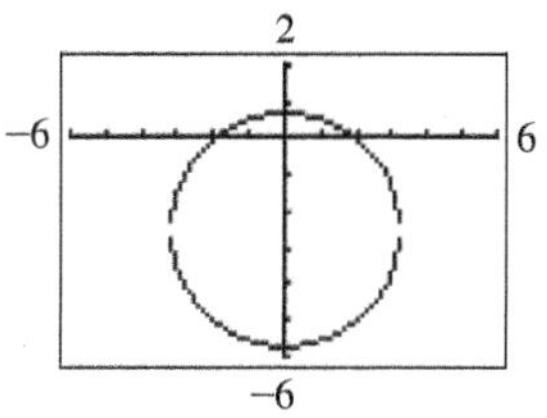

49. (a) Semicircle (b) Semicircle

(c) Yes, there is only one value of y for each x in the domain.

51. Outside

53. $p > 0$, circle; $p = 0$, point; $p < 0$, does not exist

55. 0.0912 cm **57.** 2.82 cm **59.** $x^2 + y^2 = 0.0100$

61. $x^2 + (y + 2.4)^2 = 0.16$

63. $(x - 500 \times 10^{-6})^2 + y^2 = 0.16 \times 10^{-6}$ **65.** 27 m

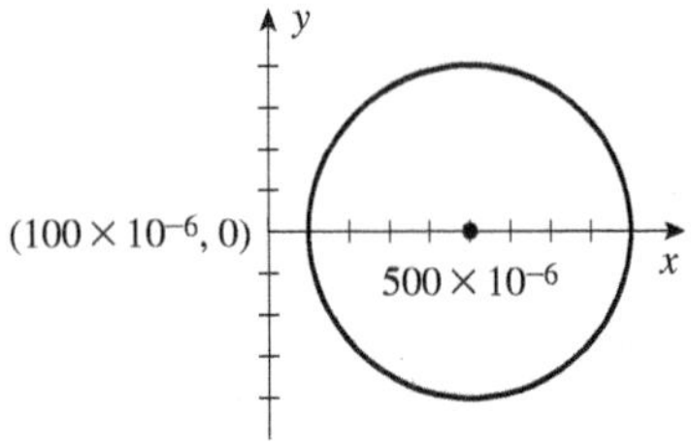

Exercises 21.4, page 597

1. $F(5, 0), x = -5$

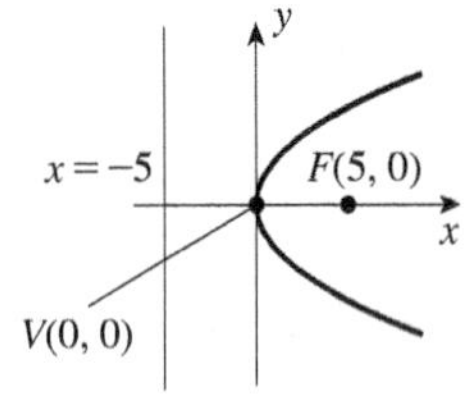

3. $F\left(0, -\frac{3}{2}\right), y = \frac{3}{2}$

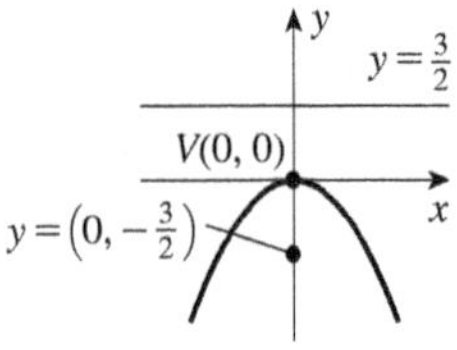

5. $F(1, 0), x = -1$

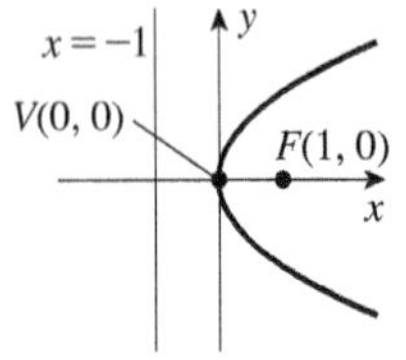

7. $F(-1, 0), x = 1$

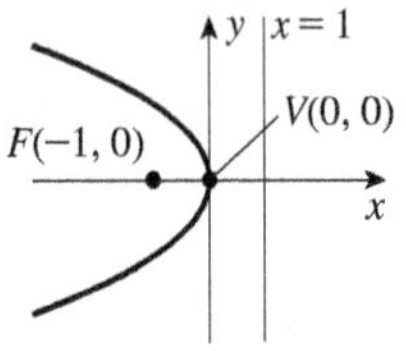

9. $F(0, 18), y = -18$

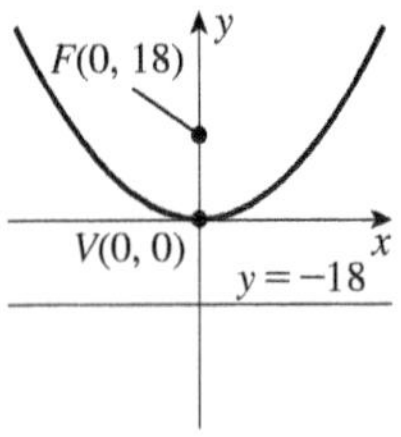

11. $F(0, -1), y = 1$

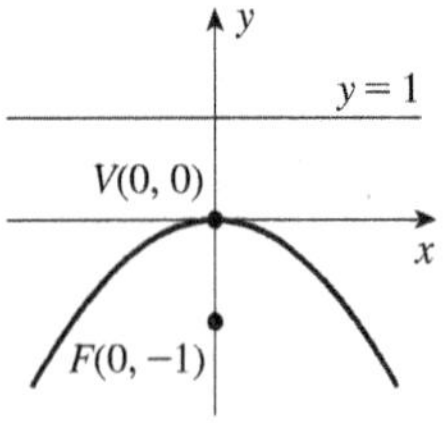

13. $F\left(\frac{5}{8}, 0\right), x = -\frac{5}{8}$

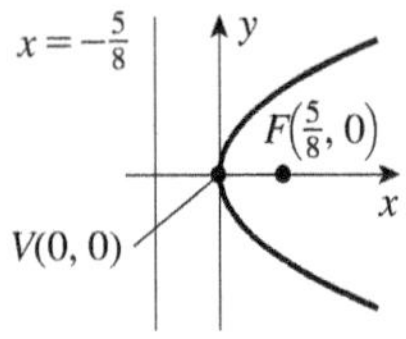

15. $F\left(0, \frac{25}{48}\right), y = -\frac{25}{48}$

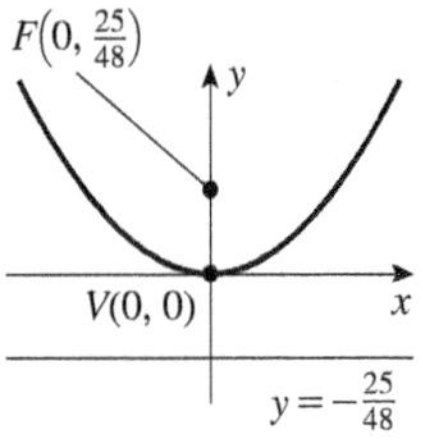

17. $y^2 = 12x$ **19.** $x^2 = -2y$ **21.** $x^2 = 0.64y$ **23.** $x^2 = 336y$

25. $x^2 = \frac{1}{8}y$ **27.** $y^2 = \frac{25}{3}x$ **29.** $y^2 = 3x$

31.

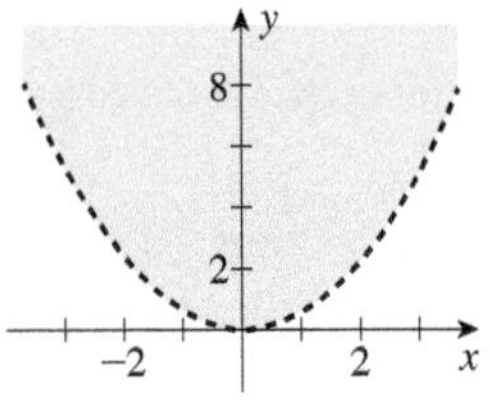

33. $(0, 0), (8, -4)$

35. $y^2 - 2y - 12x + 37 = 0$

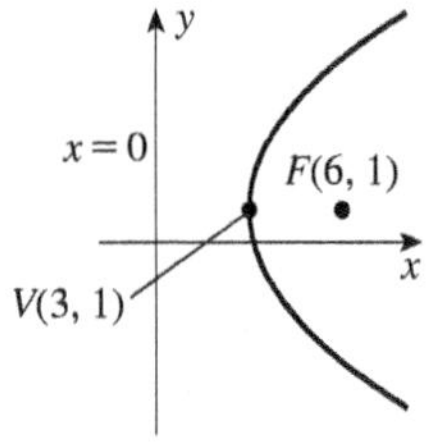

37. Graph $y = -4 \pm \sqrt{3 - 2x}$

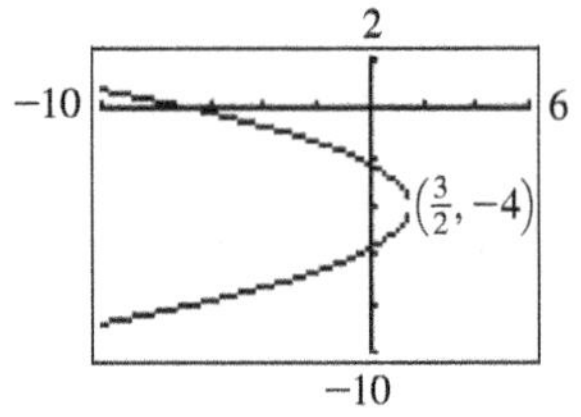

39.

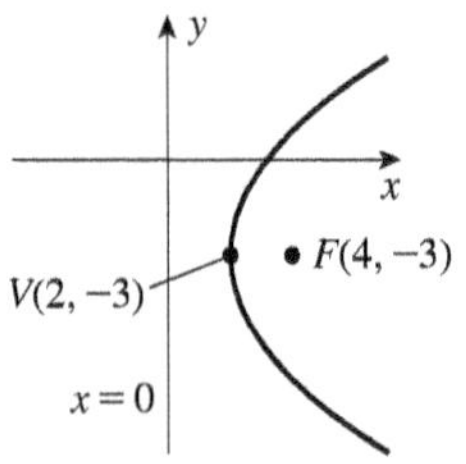

41. $4|p|$

43. The parabola becomes broader as $|p|$ increases.

45. $x^2 = 1020y$ **47.** 32 cm **49.**

51. 57.6 m

53. No (a course correction is necessary) **55.** 2.16 cm from vertex

57.

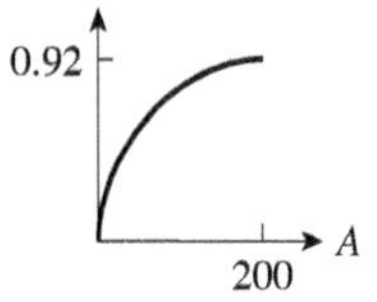

59. $x^2 = 8y$ with vertex midway between island and shore

Exercises 21.5, page 603

1. $V(0, 6), V(0, -6)$,
ends minor axis $(5, 0), (-5, 0)$
foci $(0, \sqrt{11}), (0, -\sqrt{11})$

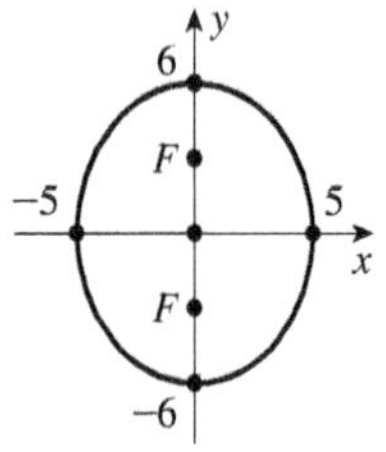

3. $V(2, 0), V(-2, 0)$,
$F(\sqrt{3}, 0), F(-\sqrt{3}, 0)$

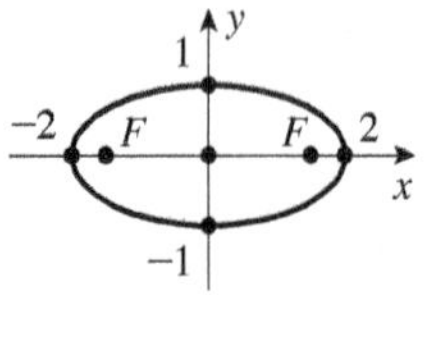

5. $V(0, 12), V(0, -12)$,
$F(0, \sqrt{119}), F(0, -\sqrt{119})$

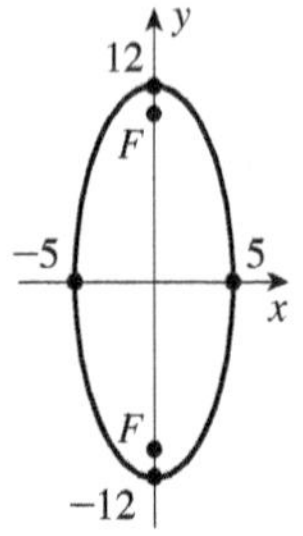

7. $V\left(\frac{5}{2}, 0\right), V\left(-\frac{5}{2}, 0\right)$,
$F\left(\frac{3}{2}, 0\right), F\left(-\frac{3}{2}, 0\right)$

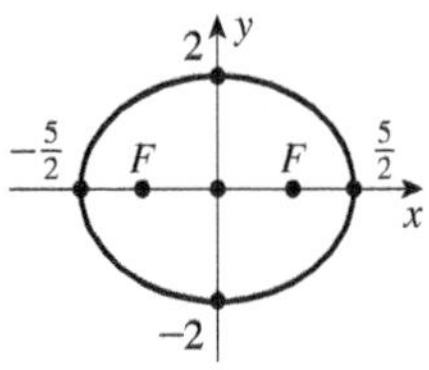

9. $V(9, 0), V(-9, 0),$
$F(3\sqrt{5}, 0), F(-3\sqrt{5}, 0)$

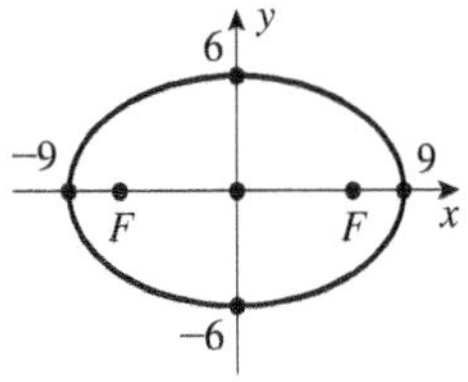

11. $V(0, 7), V(0, -7),$
$F(0, \sqrt{45}), F(0, -\sqrt{45})$

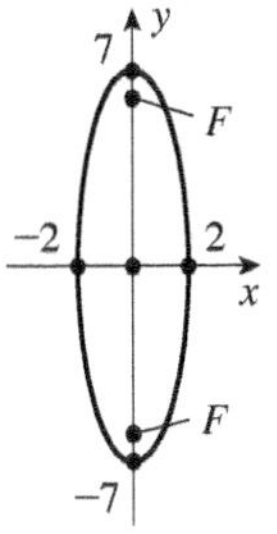

13. $V(0, 4), V(0, -4),$
$F(0, \sqrt{14}), F(0, -\sqrt{14})$

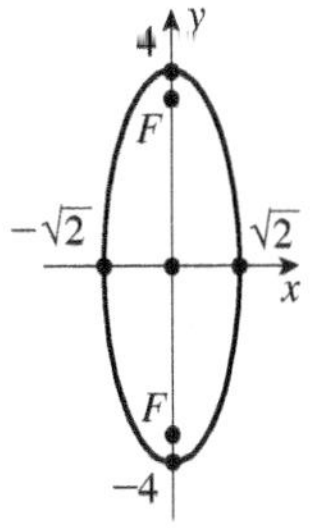

15. $V(0.25, 0), V(-0.25, 0),$
$F(0.23, 0), F(-0.23, 0)$

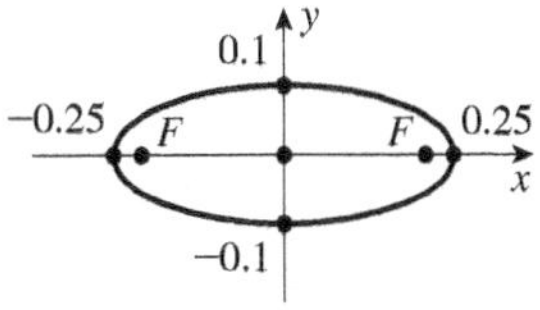

17. $\dfrac{x^2}{225} + \dfrac{y^2}{144} = 1$, or $144x^2 + 225y^2 = 32\,400$

19. $\dfrac{x^2}{225} + \dfrac{y^2}{289} = 1$, or $289x^2 + 225y^2 = 65\,025$

21. $\dfrac{x^2}{208} + \dfrac{y^2}{144} = 1$, or $144x^2 + 208y^2 = 29\,952$

23. $\dfrac{x^2}{64} + \dfrac{15y^2}{144} = 1$, or $3x^2 + 20y^2 = 192$

25. $\dfrac{x^2}{5} + \dfrac{y^2}{20} = 1$, or $4x^2 + y^2 = 20$

27. $\dfrac{x^2}{100} + \dfrac{y^2}{64} = 1$, or $16x^2 + 25y^2 = 1600$

29. $(1, -2), (1, 2)$

31. $16x^2 + 25y^2 - 32x - 50y - 359 = 0$

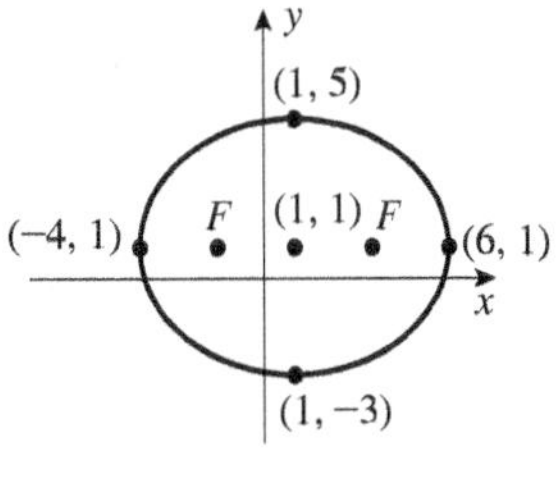

33. Graph $y = 3 \pm \dfrac{\sqrt{-12x^2 - 48x - 12}}{3}$

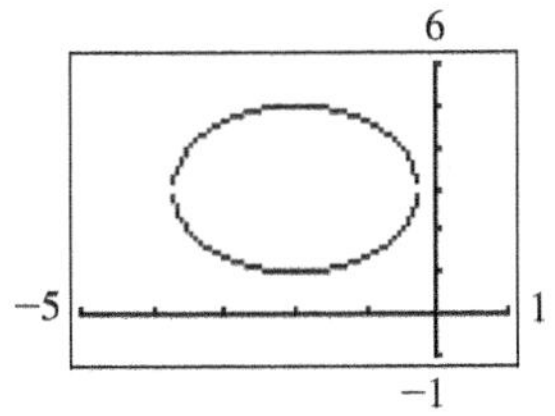

35.

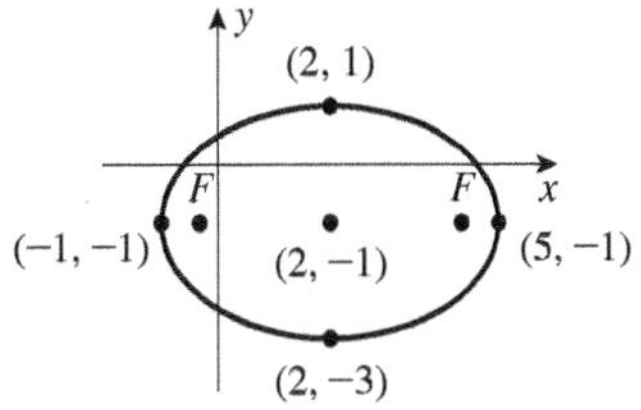

37. Write equation as $\dfrac{x^2}{1} + \dfrac{y^2}{1/k^2} = 1$. Thus, $\dfrac{1}{k^2} > 1$, or $|k| < 1$.

39. $2x^2 + 3y^2 - 8x - 4 = 2x^2 + 3(-y)^2 - 8x - 4$

41.

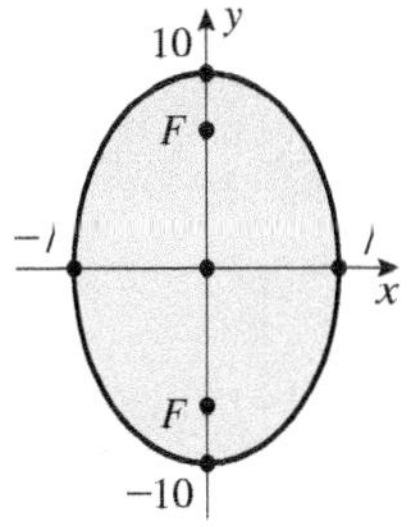

43. $\sin^2 t + \cos^2 t = \dfrac{x^2}{4} + \dfrac{y^2}{9} = 1$

45. $i_1^2 + 4i_2^2 = 32$

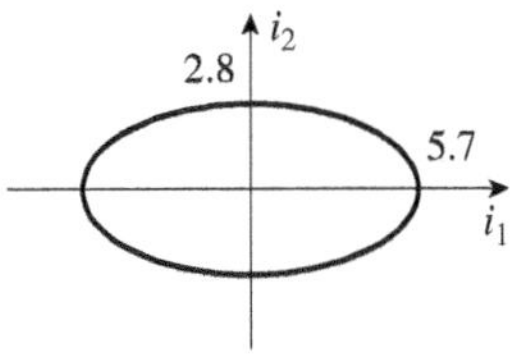

47. 0.44

49. $7x^2 + 16y^2 = 112$

51. 27.5 m

53. 4.0 m

55. 843 m³

Exercises 21.6, page 608

1. $V(0, -4), V(0, 4),$
conj. axis $(-2, 0), (2, 0)$
$F(0, -2\sqrt{5}), F(0, 2\sqrt{5})$

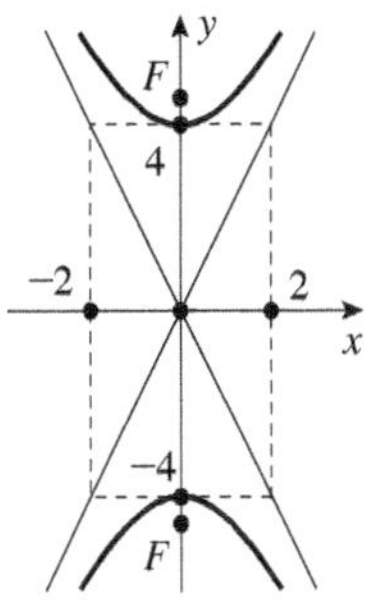

3. $V(5, 0), V(-5, 0),$
$F(13, 0), F(-13, 0)$

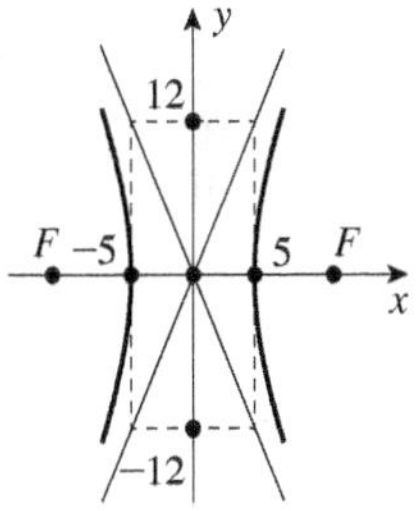

5. $V(0, 3), V(0, -3),$
$F(0, \sqrt{10}), F(0, -\sqrt{10})$

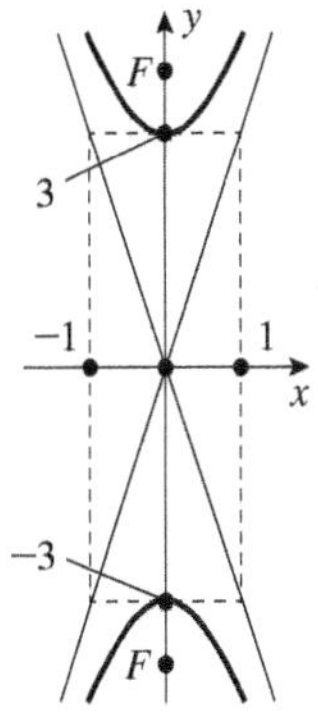

7. $V\left(-\dfrac{5}{2}, 0\right), V\left(\dfrac{5}{2}, 0\right),$
$F\left(-\dfrac{1}{2}\sqrt{41}, 0\right), F\left(\dfrac{1}{2}\sqrt{41}, 0\right)$

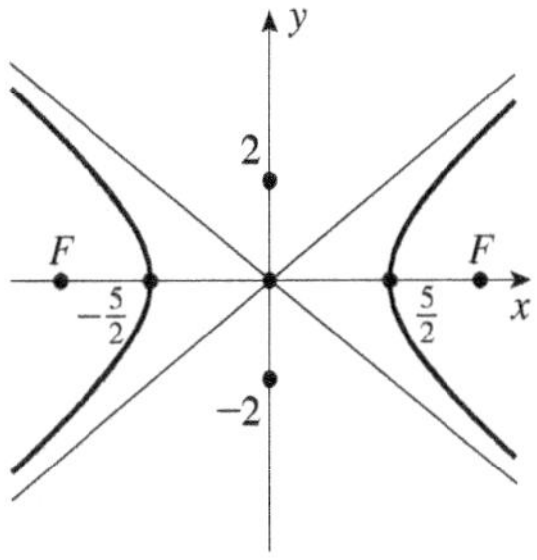

9. $V(1, 0)$, $V(-1, 0)$, $F(\sqrt{5}, 0)$, $F(-\sqrt{5}, 0)$

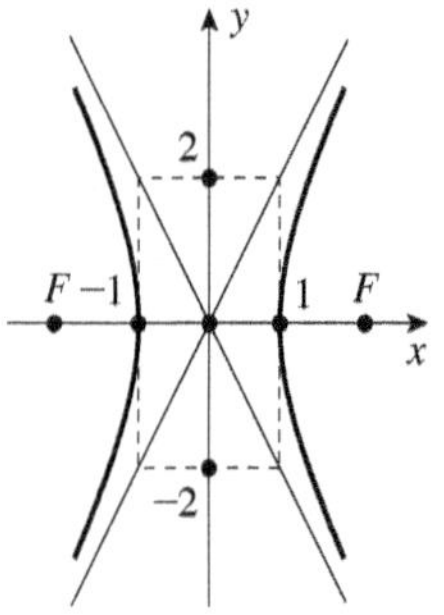

11. $V(0, \sqrt{5})$, $V(0, -\sqrt{5})$, $F(0, \sqrt{7})$, $F(0, -\sqrt{7})$

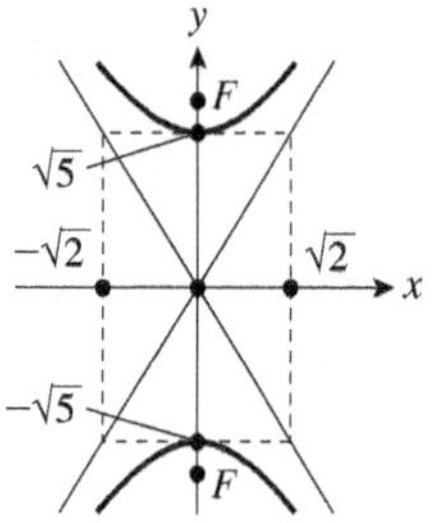

13. $V(0, 2)$, $V(0, -2)$, $F(0, \sqrt{5})$, $F(0, -\sqrt{5})$

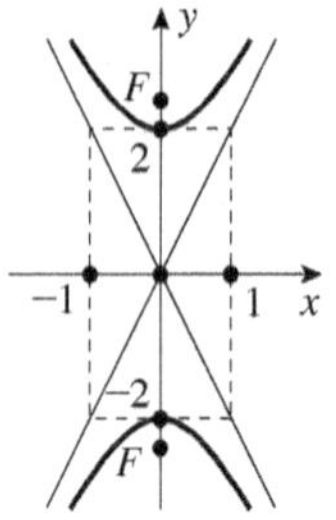

15. $V(0.4, 0)$, $V(-0.4, 0)$, $F(0.9, 0)$, $F(-0.9, 0)$

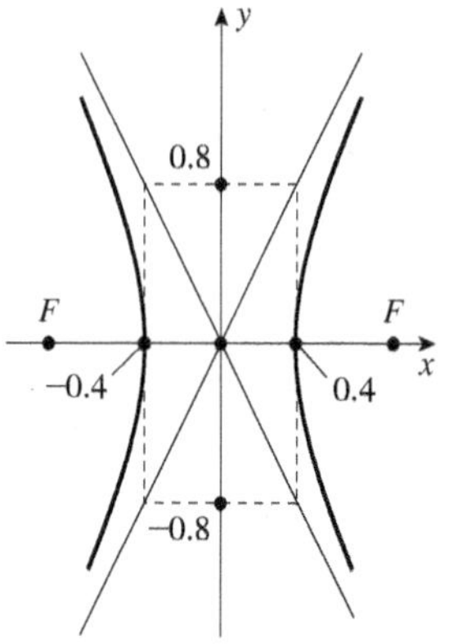

17. $\frac{x^2}{9} - \frac{y^2}{16} = 1$, or $16x^2 - 9y^2 = 144$

19. $\frac{y^2}{100} - \frac{x^2}{576} = 1$, or $144y^2 - 25x^2 = 14\,400$

21. $\frac{x^2}{1} - \frac{y^2}{3} = 1$, or $3x^2 - y^2 = 3$

23. $\frac{x^2}{5} - \frac{y^2}{4} = 1$, or $4x^2 - 5y^2 = 20$

25. $x^2 - \frac{y^2}{4} = 1$, or $4x^2 - y^2 = 4$

27. $\frac{x^2}{36} - \frac{y^2}{64} = 1$, or $16x^2 - 9y^2 = 576$

29.

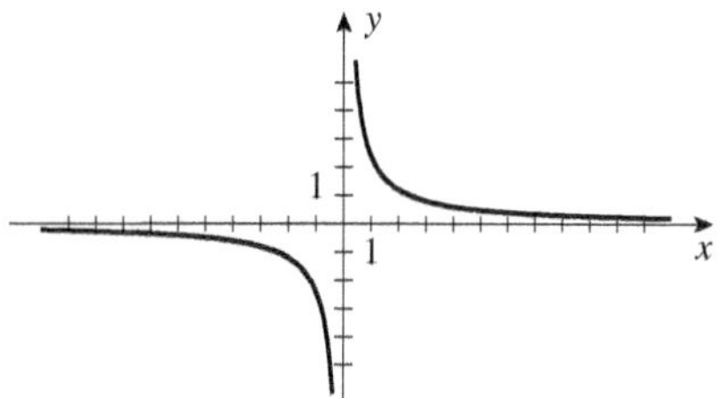

31. $\sec^2 t - \tan^2 t = x^2 - y^2 = 1$

33. $(-2, -3)$, $(-2, 3)$, $(2, -3)$, $(2, 3)$,

35. $9x^2 - 16y^2 - 108x + 64y + 116 = 0$

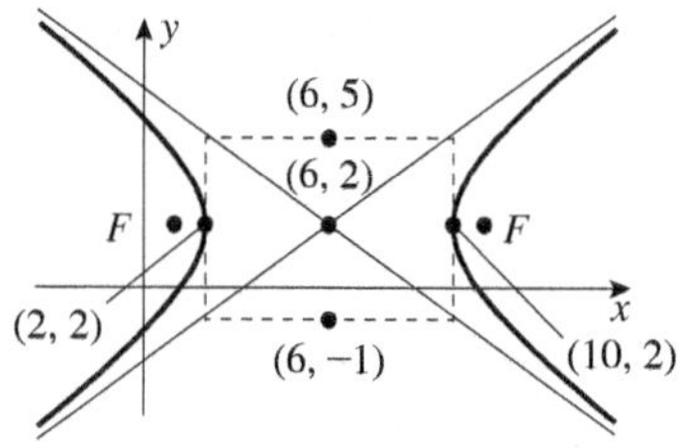

37. Graph $y = 4 \pm 0.5\sqrt{x^2 + 4x}$

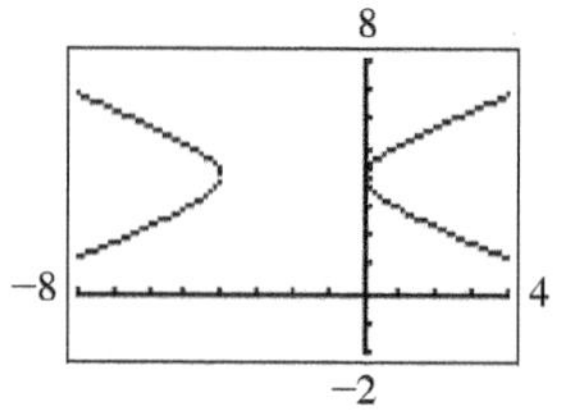

39.

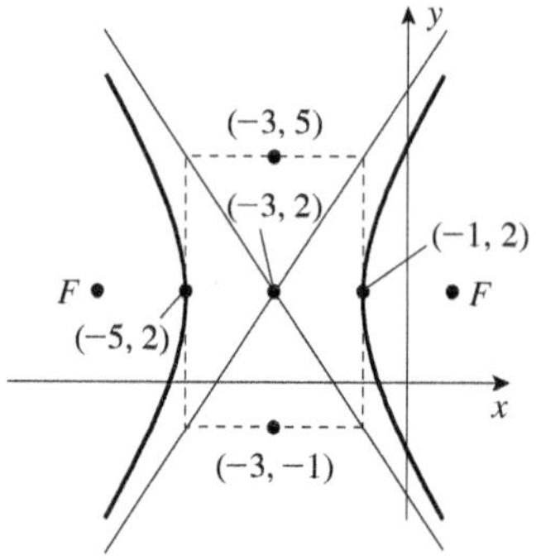

41. $x^2 - 2y^2 = 2$

43.

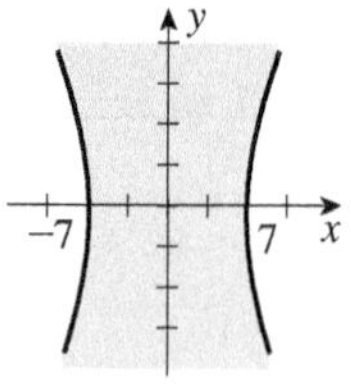

45. $9x^2 - 16y^2 = 144$

47. $3y^2 - x^2 = 27$

49.

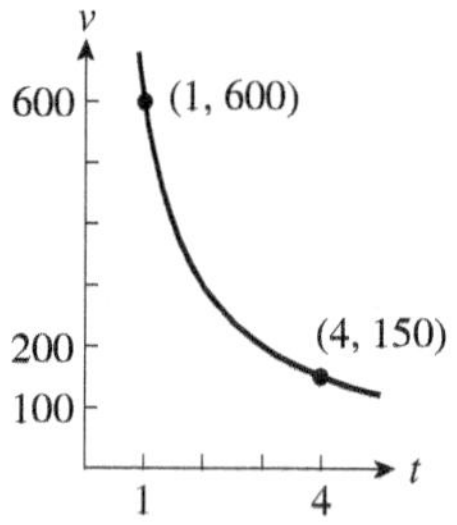

51. $i = 6.00/R$

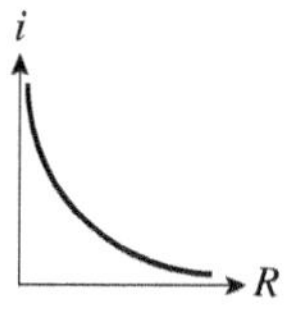

53.

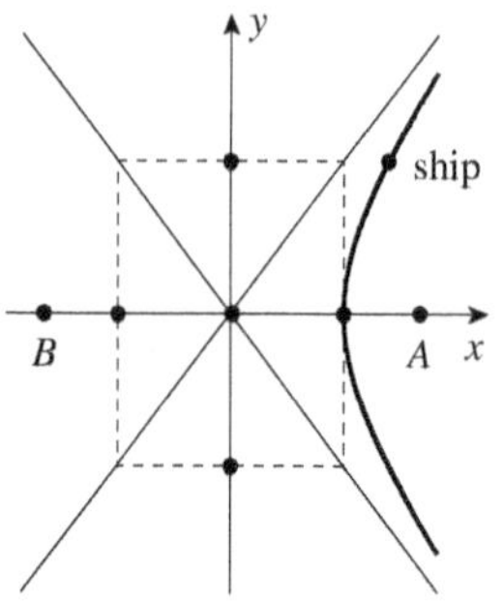

Exercises 21.7, page 612

1. Hyperbola, $C(3, 2)$, transverse axis parallel to x-axis $a = 5$, $b = 3$

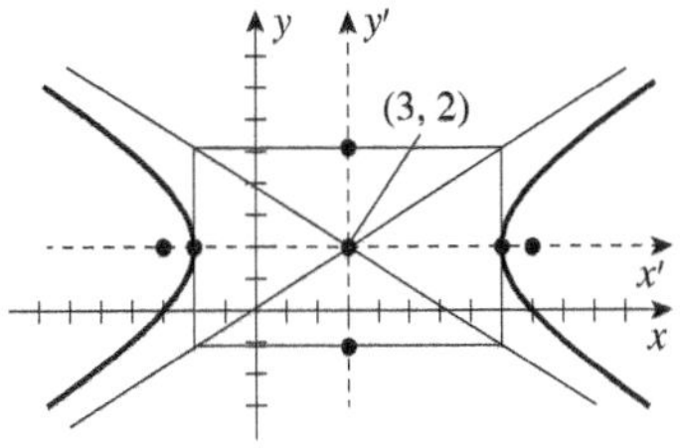

3. Parabola, $(-1, 2)$

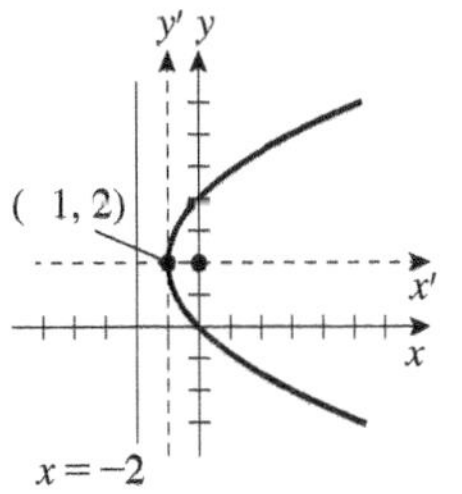

5. Hyperbola, $(1, 2)$

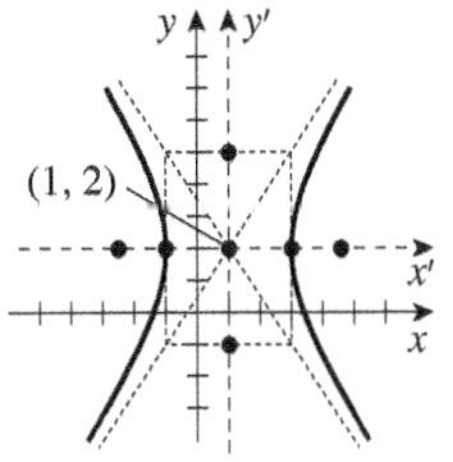

7. Ellipse, $(-1, 0)$

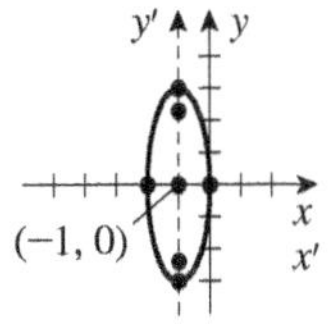

9. Parabola, $(-3, 1)$

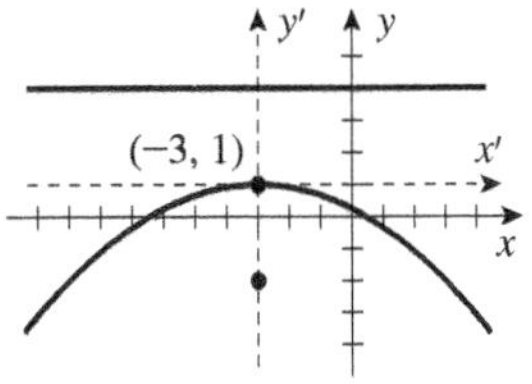

11. $(y - 3)^2 = 16(x + 1)$, or $y^2 - 6y - 16x - 7 = 0$

13. $y^2 = 24(x - 6)$

15. $\dfrac{(x+2)^2}{25} + \dfrac{(y-2)^2}{16} = 1$, or $16x^2 + 25y^2 + 64x - 100y - 236 = 0$

17. $\dfrac{(y-1)^2}{16} + \dfrac{(x+2)^2}{4} = 1$, or $4x^2 + y^2 + 16x - 2y + 1 = 0$

19. $\dfrac{(y-2)^2}{1} - \dfrac{(x+1)^2}{3} = 1$, or $x^2 - 3y^2 + 2x + 12y - 8 = 0$

21. $\dfrac{(x+1)^2}{9} - \dfrac{(y-1)^2}{16} = 1$, or $16x^2 - 9y^2 + 32x + 18y - 137 = 0$

23. Parabola, $(-1, -1)$

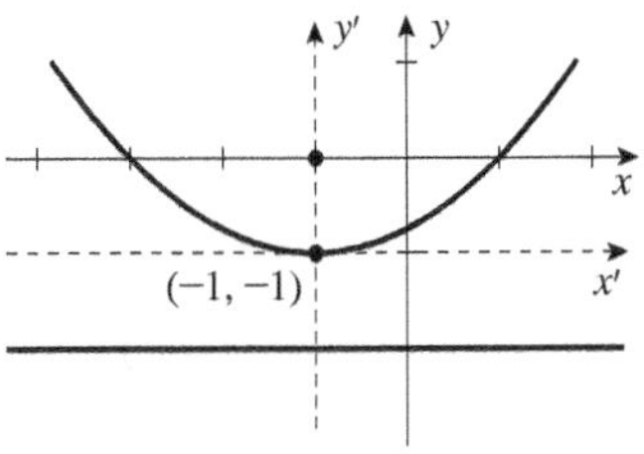

25. Parabola, $(0, 6)$

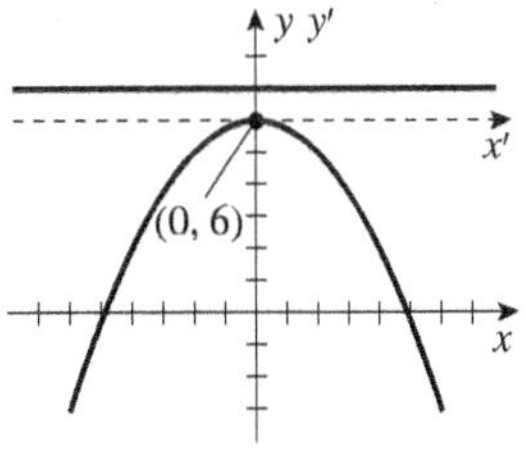

27. Ellipse, $(-3, 0)$

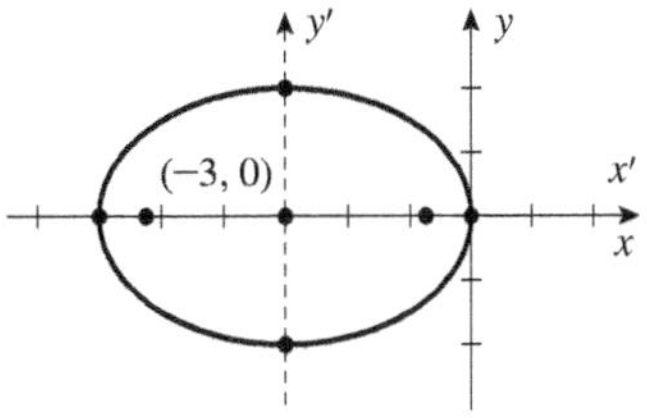

29. Hyperbola, $(0, 4)$

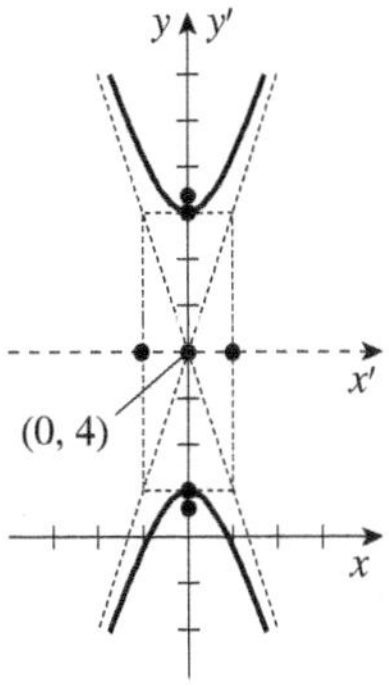

31. Hyperbola, $(-2, 1)$

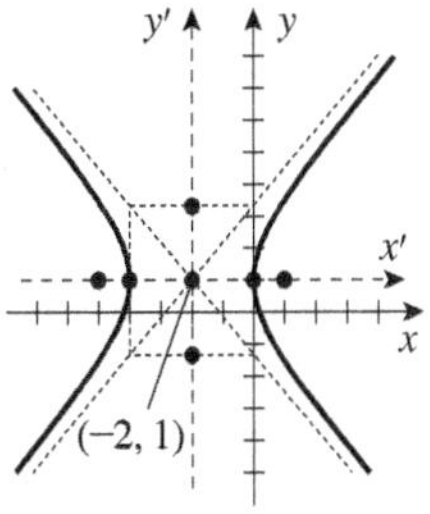

33. Hyperbola, $(-4, 5)$

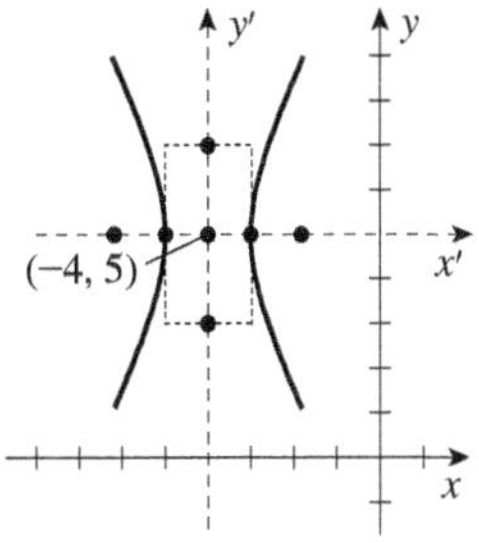

35. Ellipse, $(1, -2)$

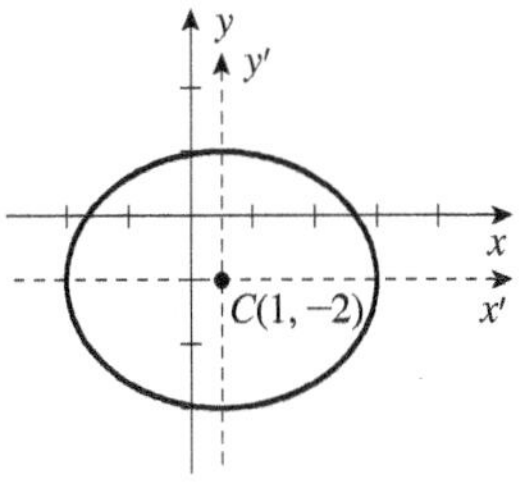

37. Hyperbola, $(4, 0)$

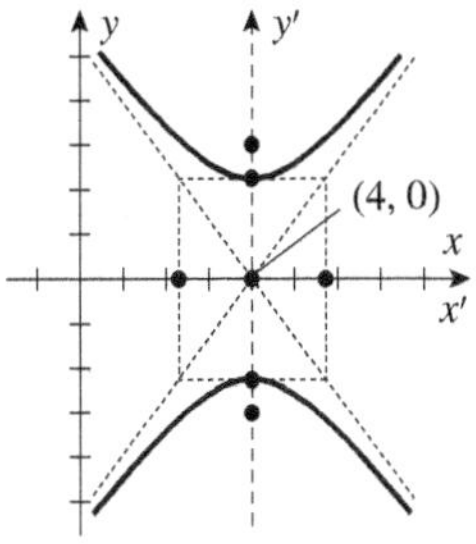

39. Circle, $\left(\dfrac{1}{3}, \dfrac{4}{3}\right)$

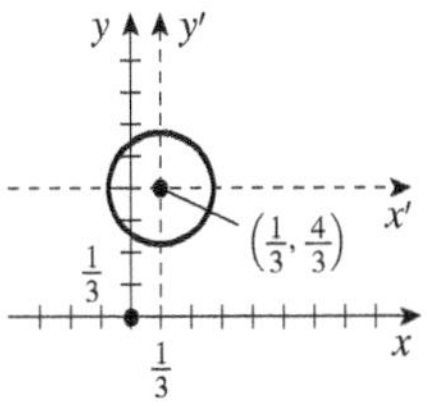

41. $x^2 - y^2 + 4x - 2y - 22 = 0$ **43.** $y^2 + 4x - 4 = 0$

45. $y^2 = 4p(x - h)$ **47.** $i' = \sin 2\pi t'$

49. $(x - 28)^2 = -\dfrac{28^2}{18}(y - 18)$

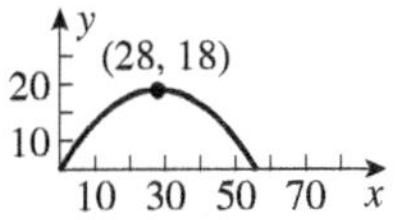

51. $\dfrac{x^2}{9.0} + \dfrac{y^2}{16} = 1$, $\dfrac{(x - 7.0)^2}{16} + \dfrac{y^2}{9.0} = 1$

Exercises 21.8, page 615

1. Hyperbola **3.** Ellipse **5.** Hyperbola **7.** Circle
9. Parabola **11.** Hyperbola **13.** Circle
15. None (straight line) **17.** Hyperbola **19.** Ellipse
21. None (point at origin) **23.** Ellipse
25. Parabola: $V(-4, 0)$; $F(-4, 2)$

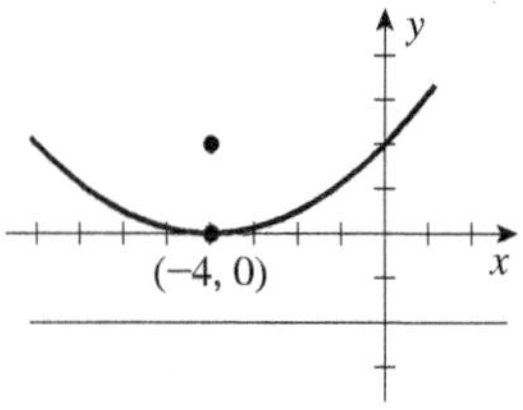

27. Hyperbola; $C(1, -2)$; $V(1, -2 \pm \sqrt{2})$

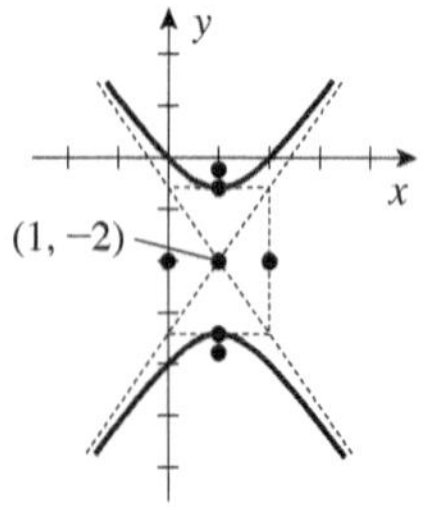

29. Ellipse; $C(5, 0)$; $V(5, \pm 2\sqrt{2})$

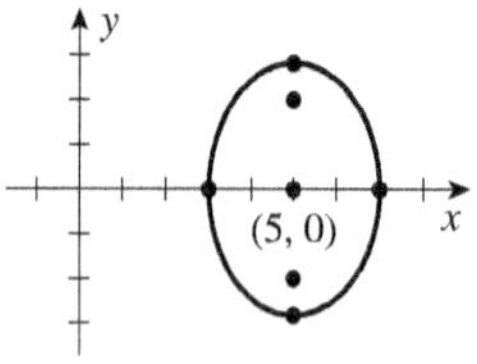

31. Ellipse; graph $y = -3 \pm 0.5\sqrt{-2x^2 + 8x + 8}$

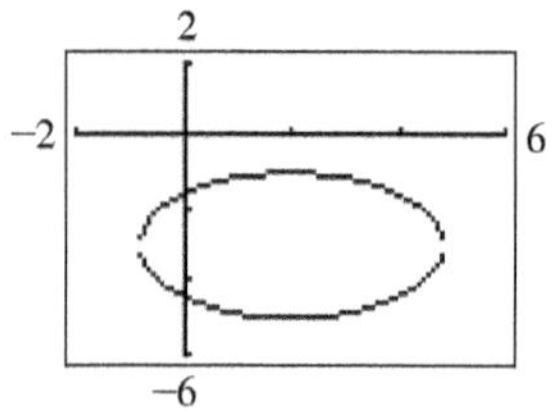

33. Parabola; graph $y = \dfrac{5 \pm \sqrt{16x + 8}}{2}$

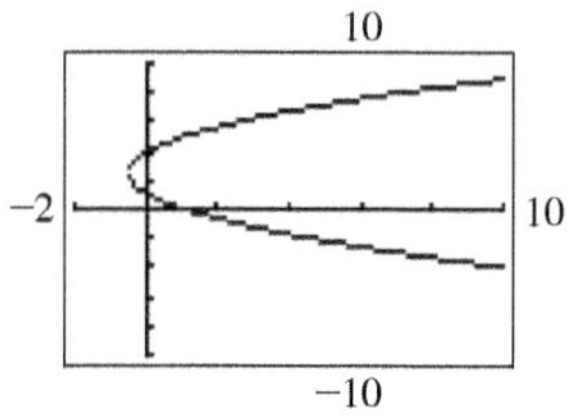

35. Hyperbola; graph $y = -3 \pm \dfrac{\sqrt{x^2 - 4x + 8}}{2}$

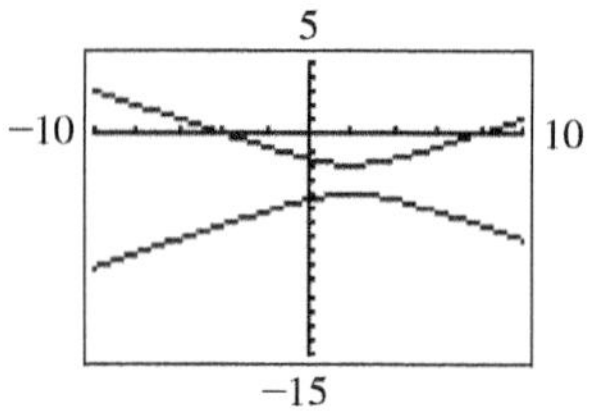

37. (a) Circle (b) Hyperbola (c) Ellipse **39.** Point at the origin
41. Straight line **43.** Parabola
45. Circle **47.** One branch of a hyperbola

Exercises 21.9, page 619

1. Hyperbola; $2x'y' + 25 = 0$

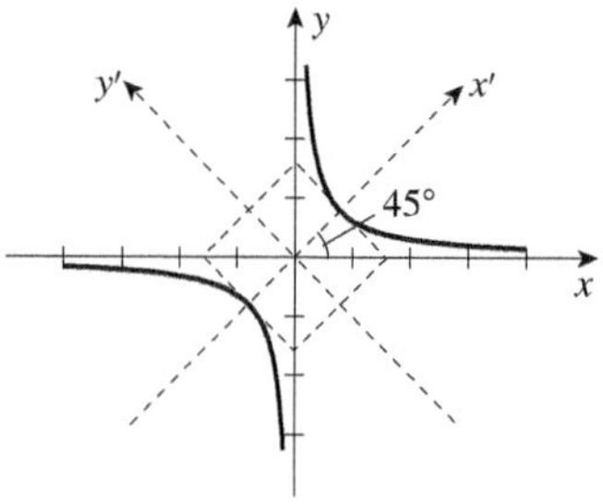

3. Ellipse; $4x'^2 + 9y'^2 = 36$

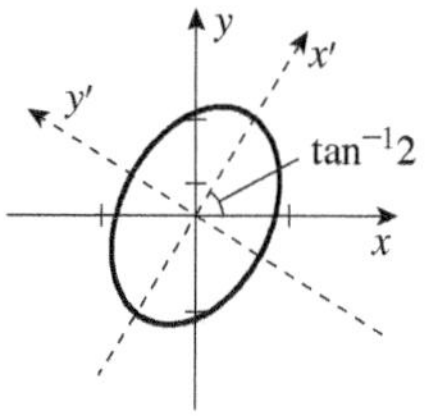

5. Hyperbola **7.** Parabola **9.** Ellipse
11. Parabola; $x'^2 + \sqrt{2}y' = 0$

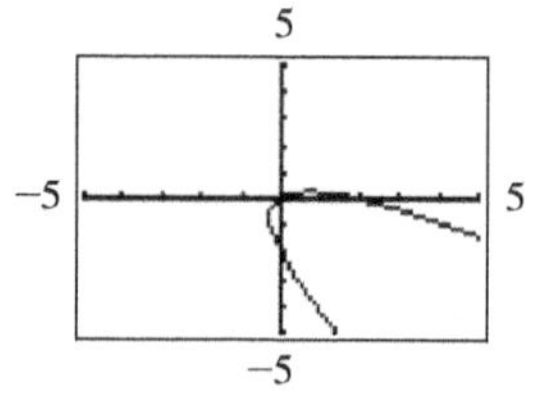

13. Hyperbola; $4x'^2 - y'^2 = 4$

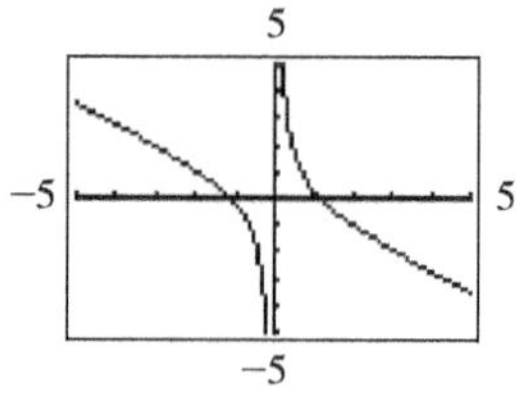

15. Ellipse; $x'^2 + 2y'^2 = 2$

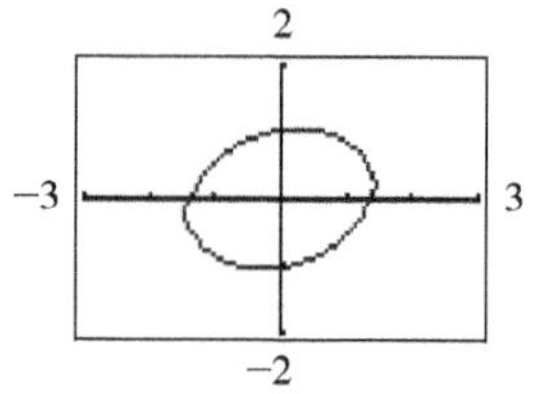

17. Parabola $y'^2 - 4x' + 16 = 0$
$y''^2 = 4x''$

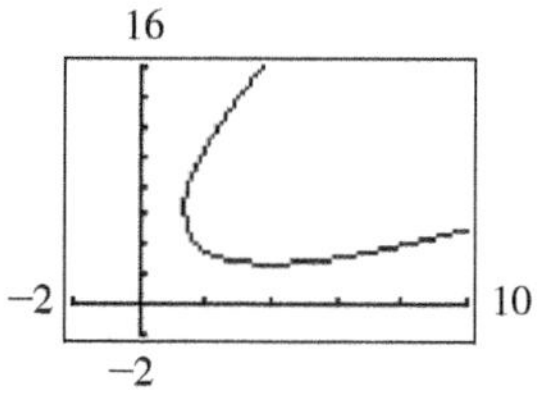

19. (a) Hyperbola (b) Hyperbola

Exercises 21.10, page 622

1. $\left(3, \dfrac{-5\pi}{3}\right), \left(3, -\dfrac{2\pi}{3}\right)$ **3.** (5.83, 2.11)

5.

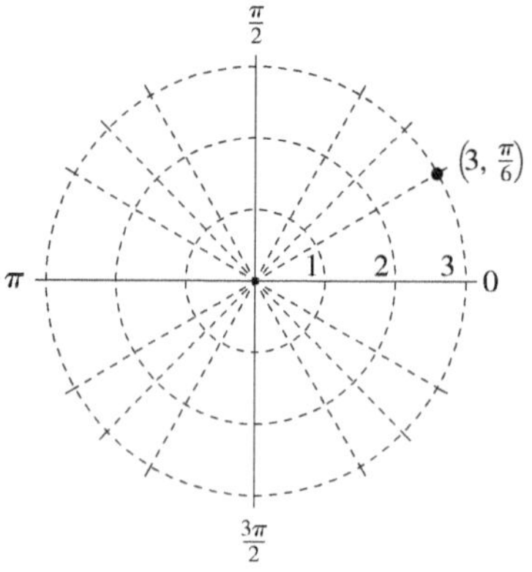

7.

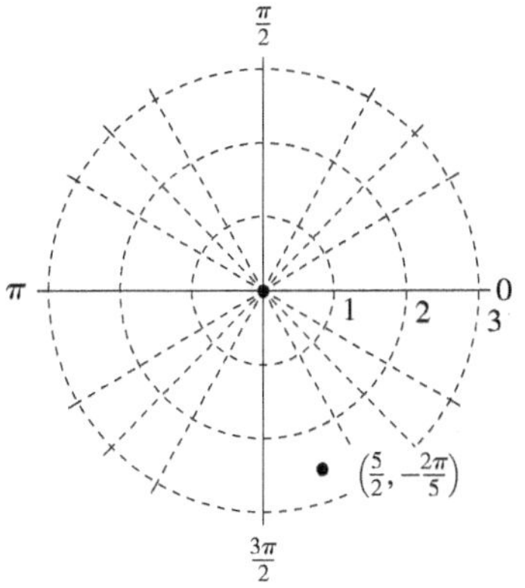

9.

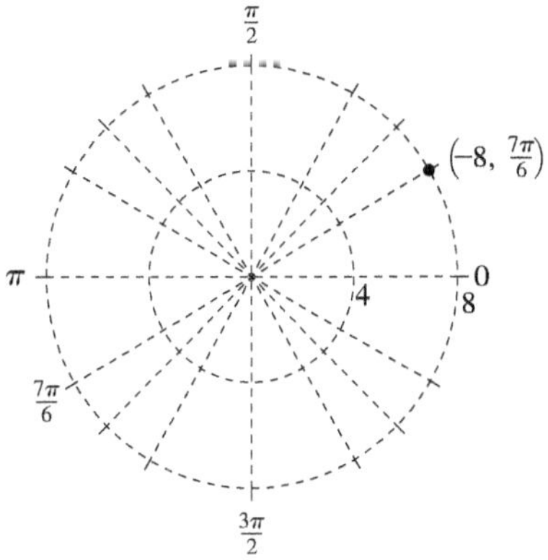

11.

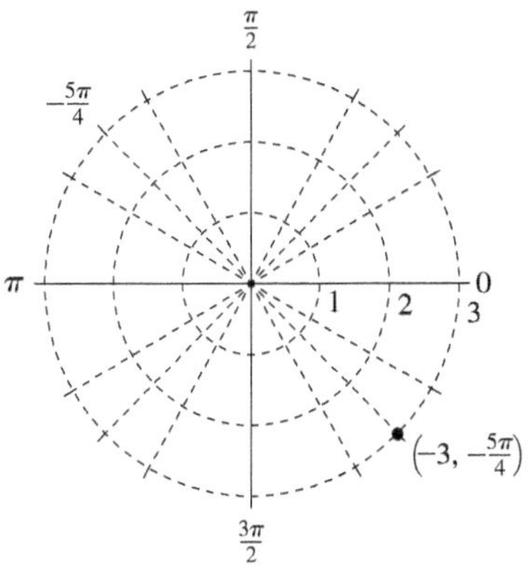

13.

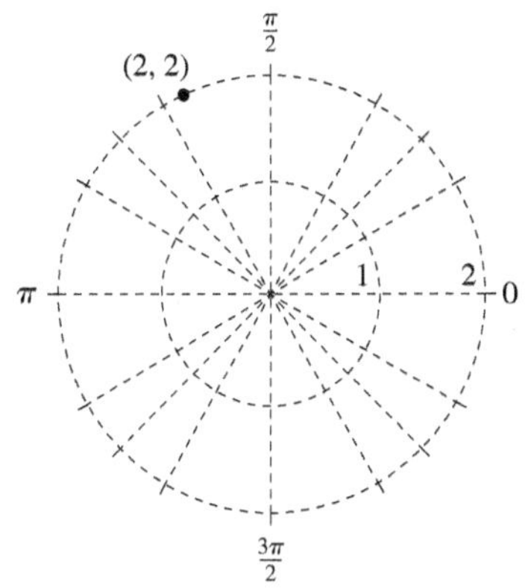

15.

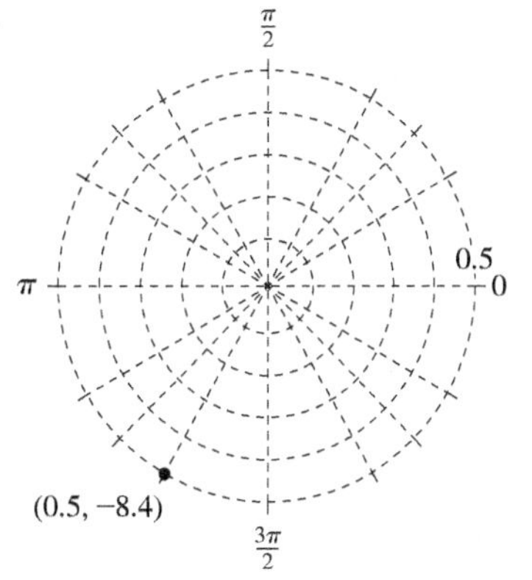

17. $\left(2, \frac{\pi}{6}\right)$ **19.** $\left(1, \frac{7\pi}{6}\right)$ **21.** $(4, \pi/2)$ **23.** $(-4, -4\sqrt{3})$

25. $(2.76, -1.17)$ **27.** $(-1.16, -7.91)$ **29.** $r = 3 \sec \theta$

31. $r = \dfrac{-3}{\cos \theta + 2 \sin \theta}$ **33.** $r = 4 \sin \theta$

35. $r^2 = \dfrac{4}{1 + 3 \sin^2 \theta}$ **37.** $r = 6 \sin \theta$

39. $r = \dfrac{2 \sin 2\theta}{\cos^3 \theta + \sin^3 \theta}$ **41.** $x^2 + y^2 - y = 0$, circle

43. $x = 4$, straight line **45.** $x - 3y - 2 = 0$, straight line

47. $x^2 + y^2 - 4x - 2y = 0$, circle

49. $x^4 + y^4 - 4x^3 + 2x^2y^2 - 4xy^2 - 4y^2 = 0$

51. $(x^2 + y^2)^2 = 2xy$ **53.** Yes $\left[\text{as}\left(-2, \frac{7\pi}{4}\right)\right]$

55. $x^2 + y^2 - bx - ay = 0$

57. $(2, 0), \left(2, \frac{\pi}{3}\right), \left(2, \frac{2\pi}{3}\right), \left(2, \frac{4\pi}{3}\right), \left(2, \frac{5\pi}{3}\right)$

59. $B_x = -\dfrac{k \sin \theta}{r}$, $B_y = \dfrac{k \cos \theta}{r}$

61. $x^4 + y^4 + 2x^2y^2 + 2x^2y + 2y^3 - 9x^2 - 8y^2 = 0$

63. 13.8 km

Exercises 21.11, page 625

1. $\theta = \dfrac{5\pi}{6}$

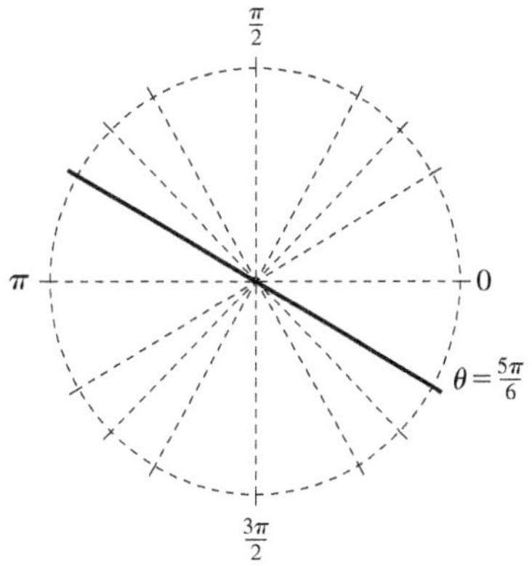

3. $r = 2 \sin 2\theta$

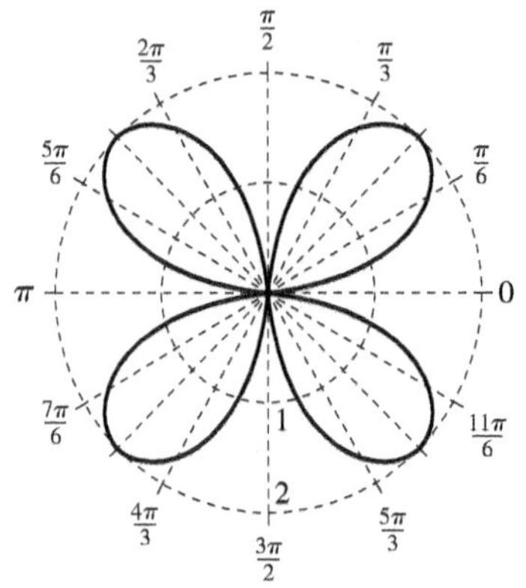

5.

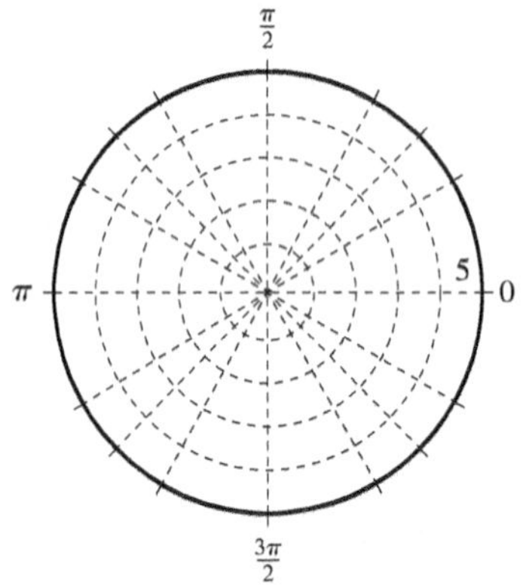

7.

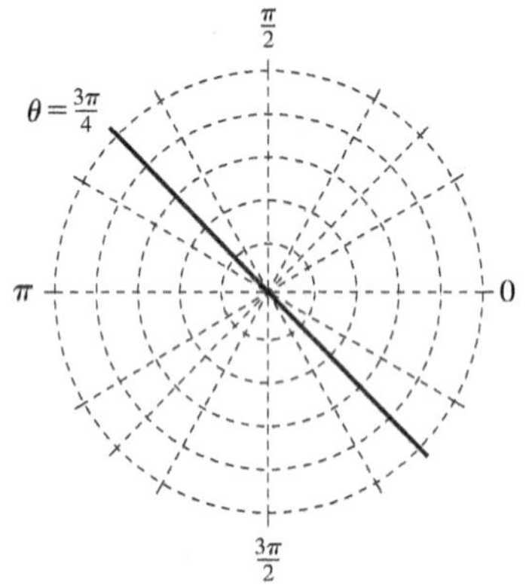

9.

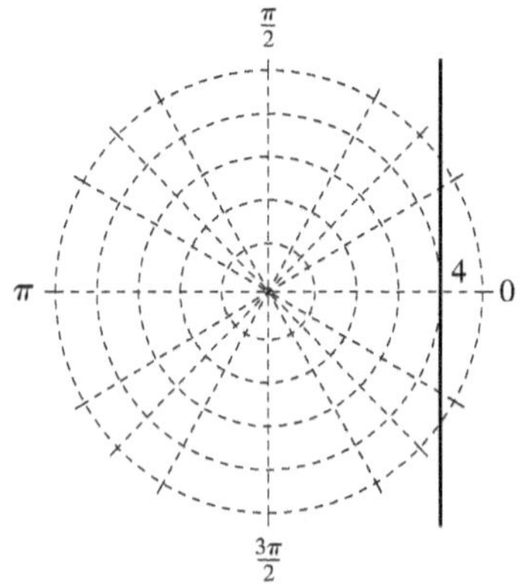

11.

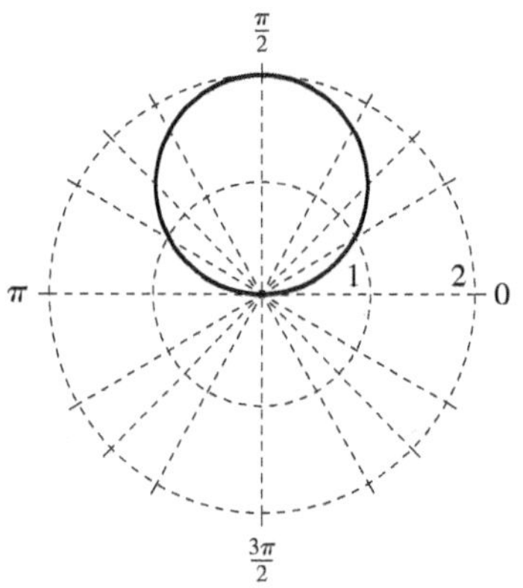

13.

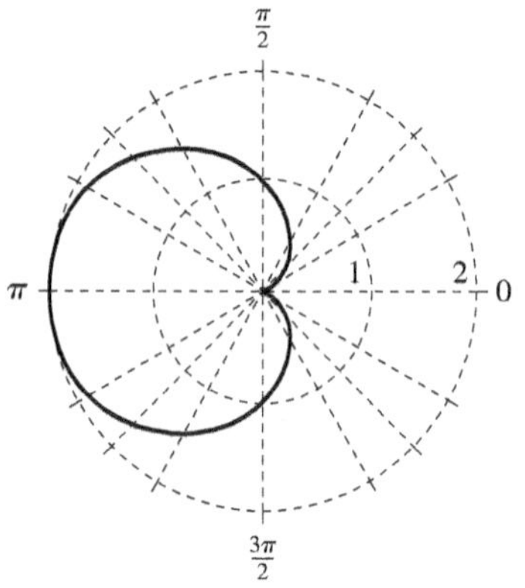

15.

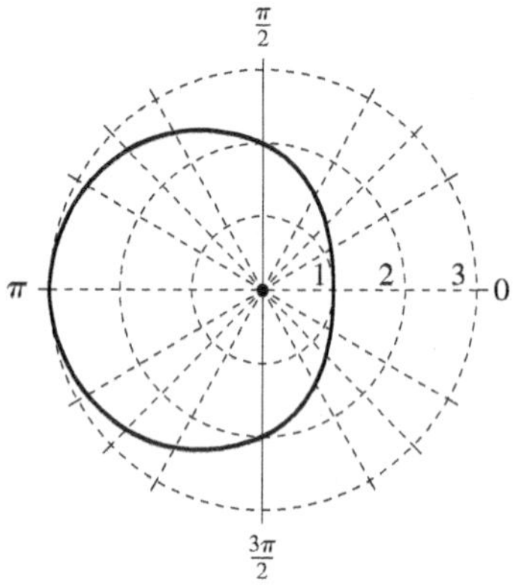

17.

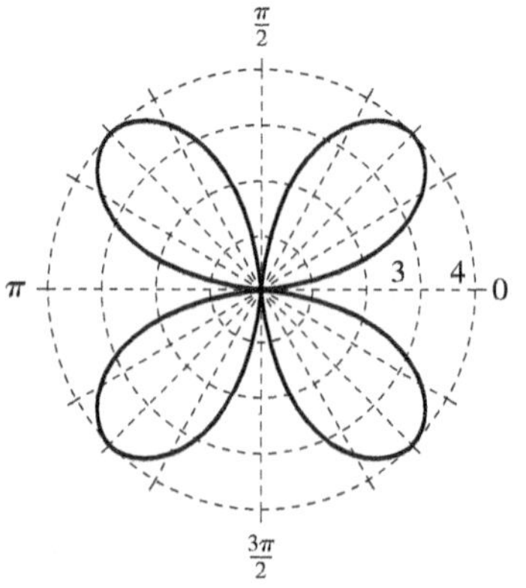

19.

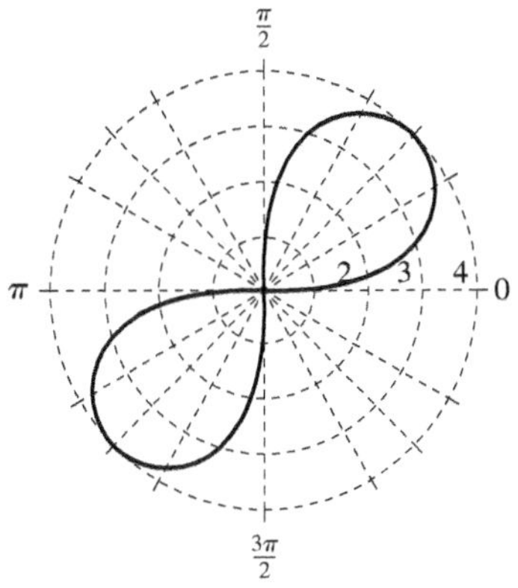

21.

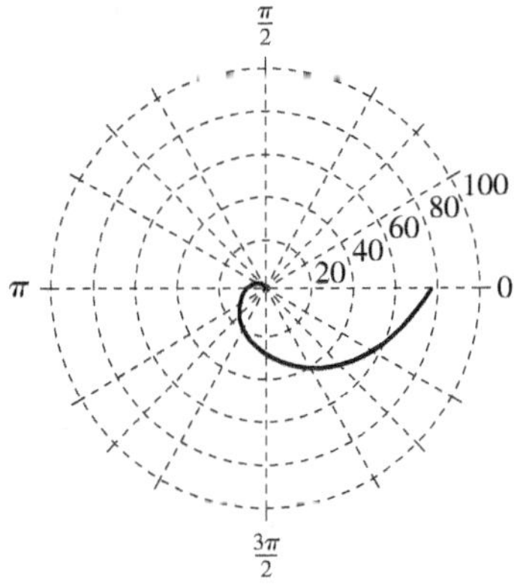

23.

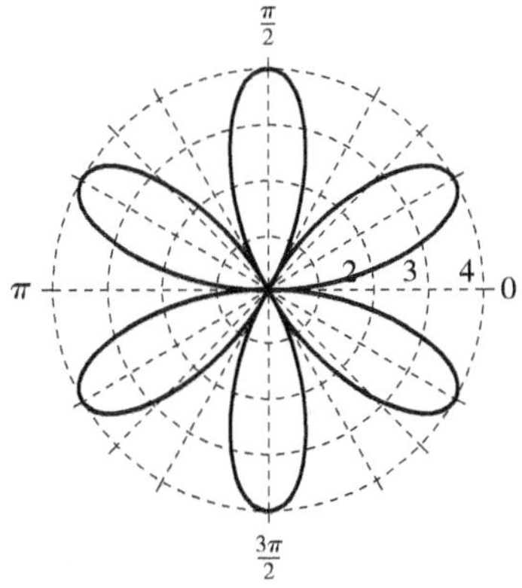

25.

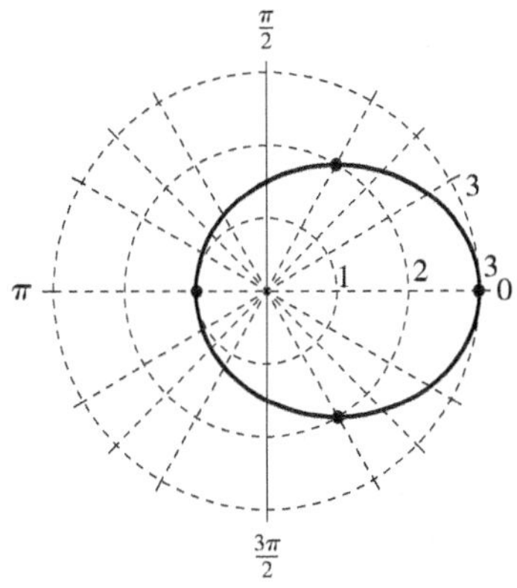

27.

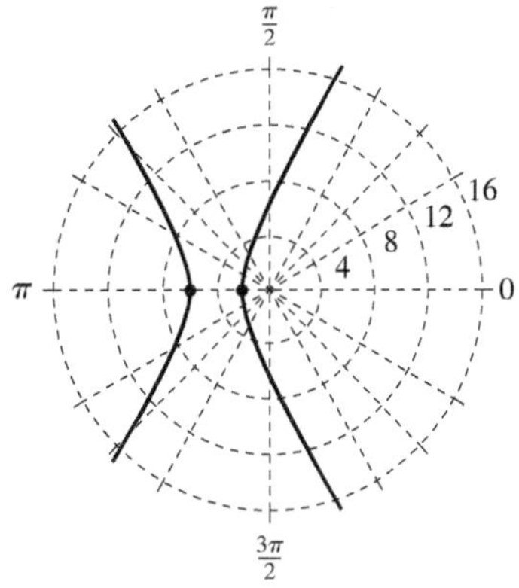

29.

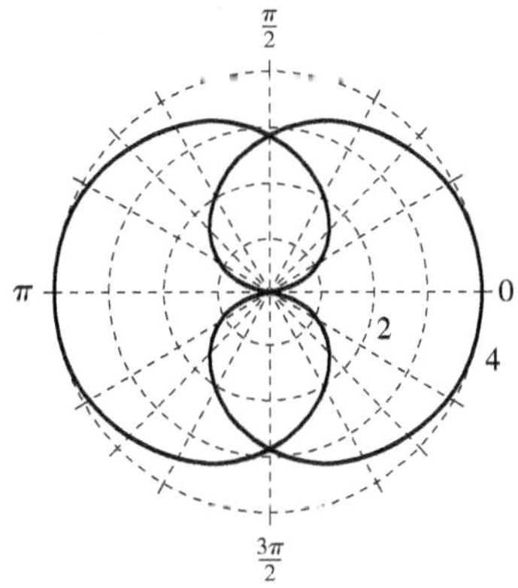

31.

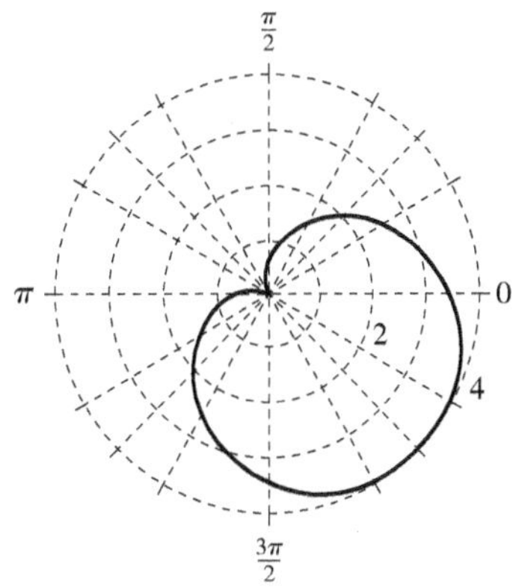

33.

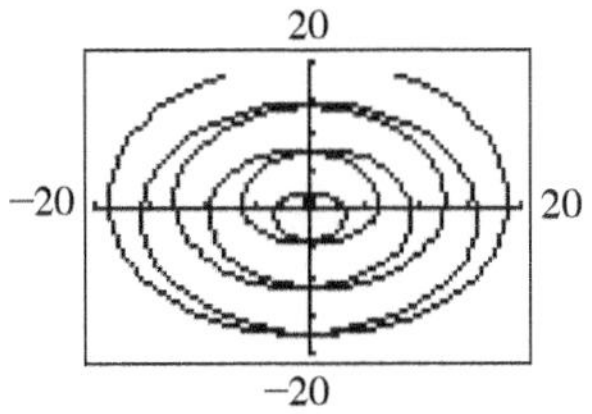

35.

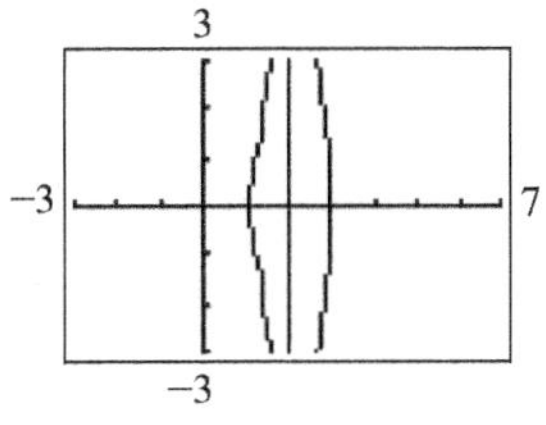

37.

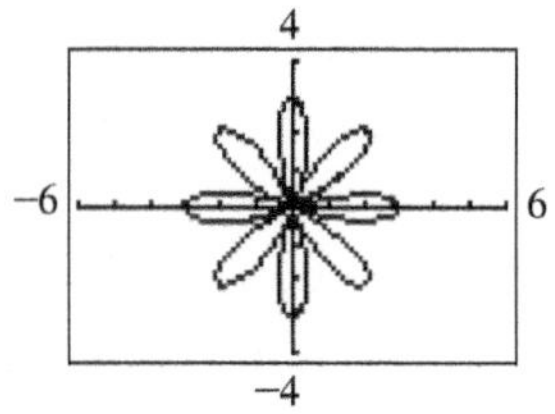

39.

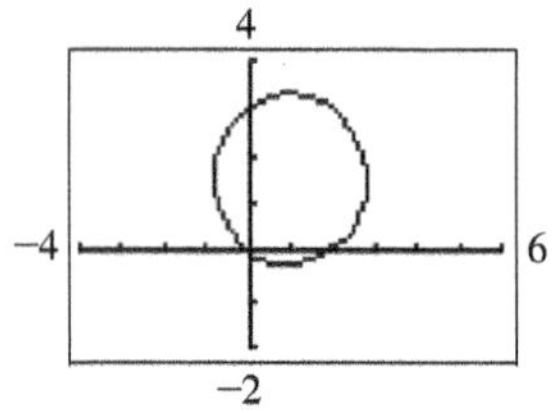

41.

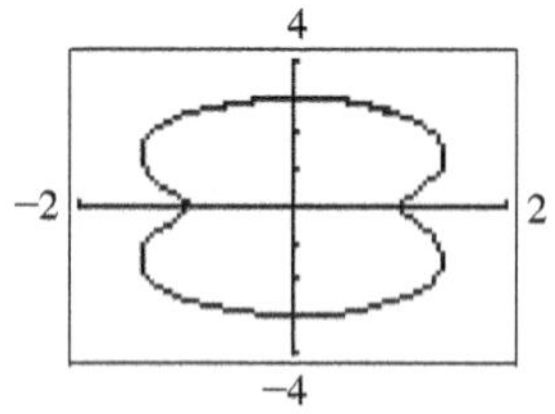

43. Straight line

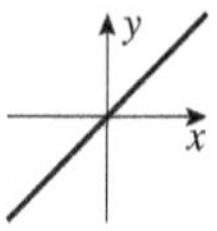

45.

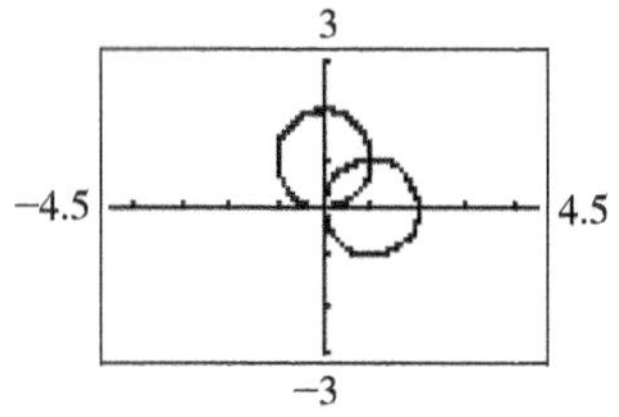

47.

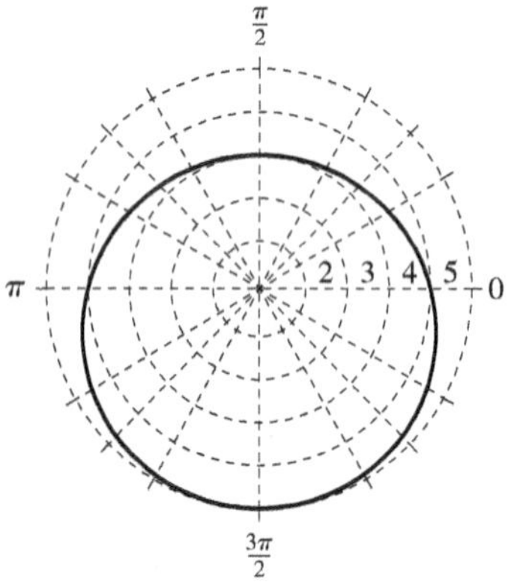

49.

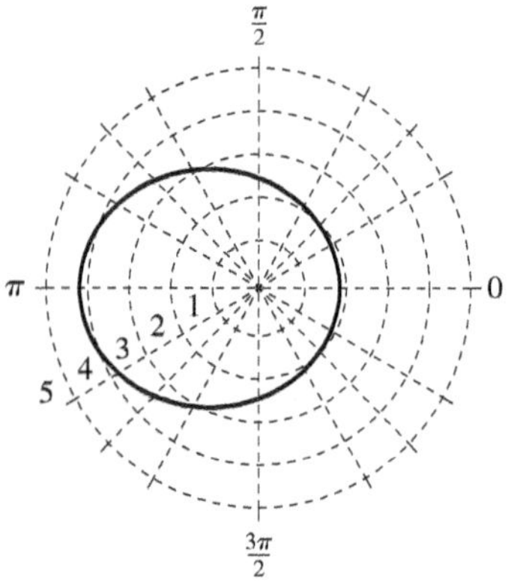

51.

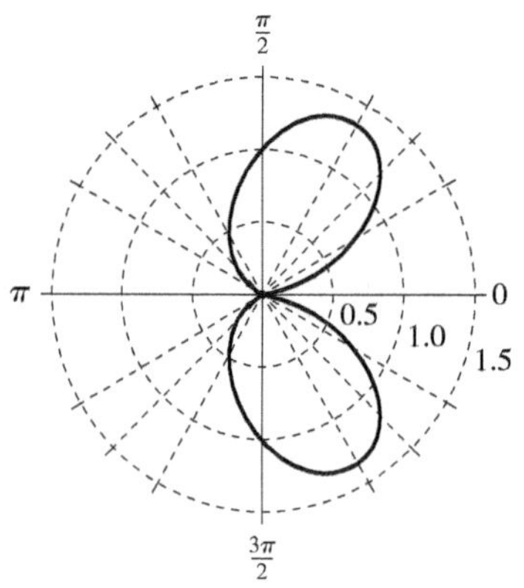

53. If n is odd, there are n loops. If n is even, there are $2n$ loops.

Review Exercises for Chapter 21, page 627

1. $4x - y - 11 = 0$

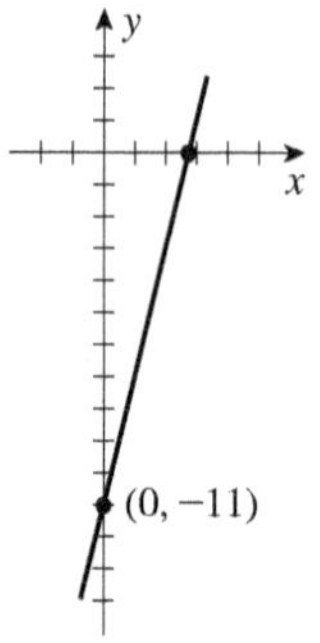

3. $2x + 3y + 3 = 0$

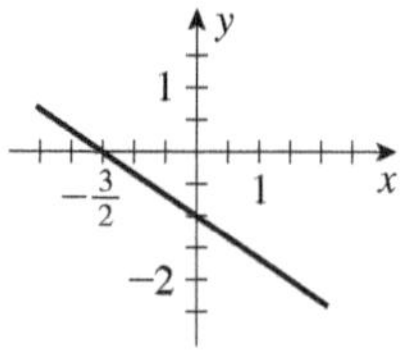

5. $x^2 - 6x + y^2 - 1 = 0$

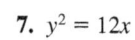

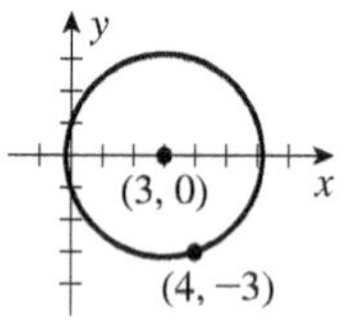

7. $y^2 = 12x$

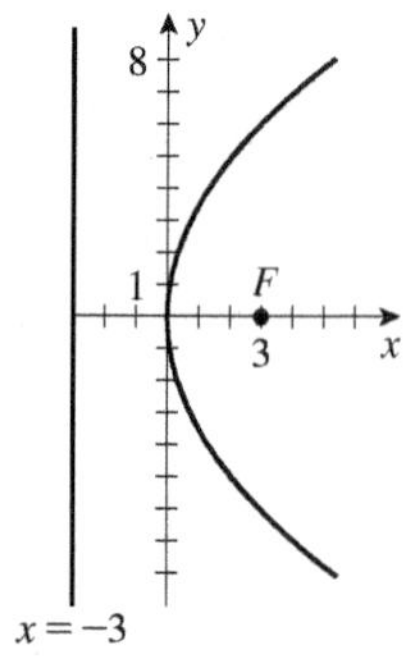

9. $9x^2 + 25y^2 = 900$

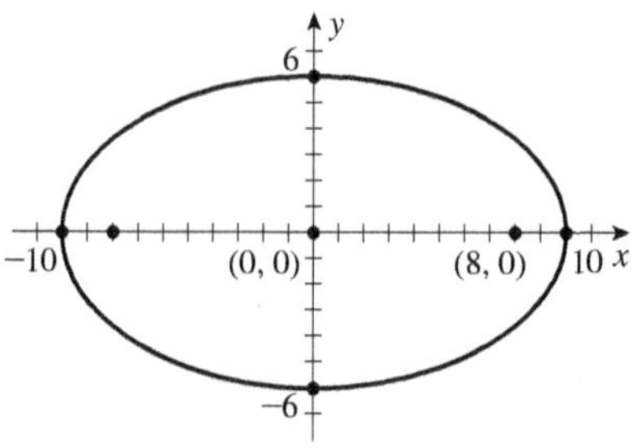

11. $144y^2 - 169x^2 = 24\,336$

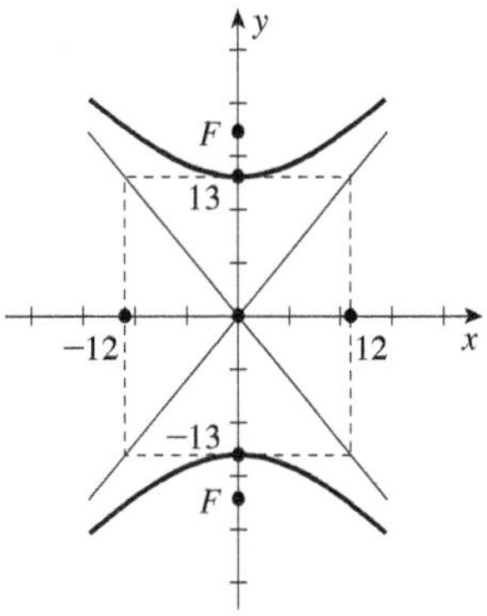

13. $(-3, 0),\ r = 4$

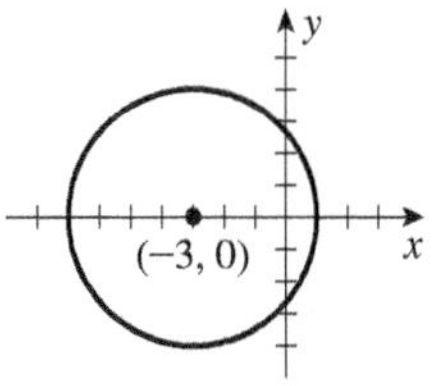

15. $(0, -5),\ y = 5$

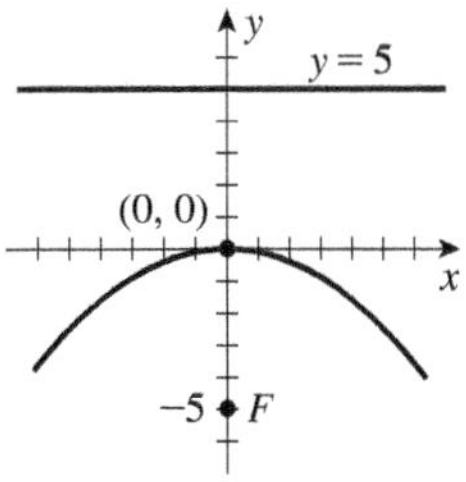

17. $V(0, 1),\ V(0, -1),\ F\left(0, \frac{1}{2}\sqrt{3}\right),$
$F\left(0, -\frac{1}{2}\sqrt{3}\right)$

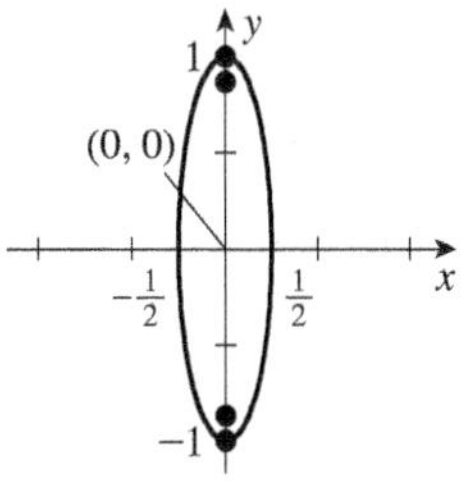

19. $V\left(\frac{1}{2\sqrt{2}}, 0\right),\ V\left(-\frac{1}{2\sqrt{2}}, 0\right),\ F\left(\frac{\sqrt{70}}{20}, 0\right),\ F\left(-\frac{\sqrt{70}}{20}, 0\right)$

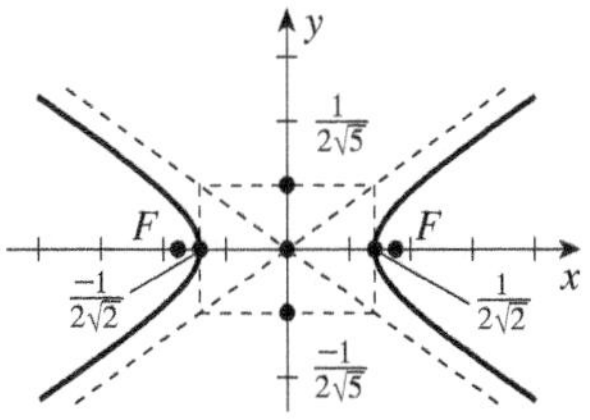

21. $V(4, -8),\ F(4, -7)$

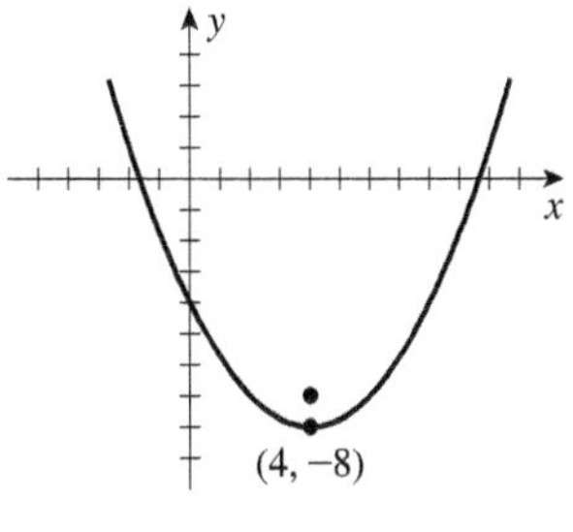

23. $(2, -1)$

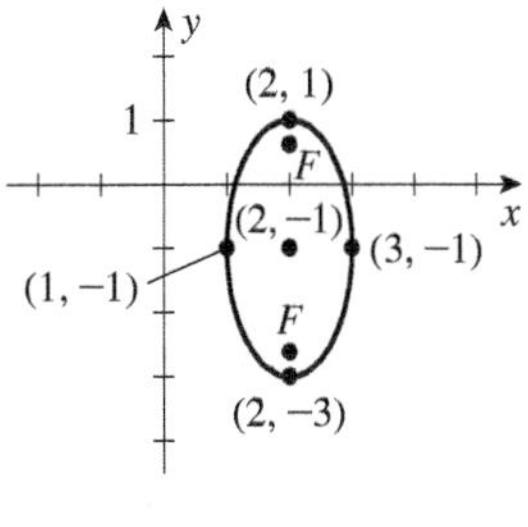

25. $(0, 0)$

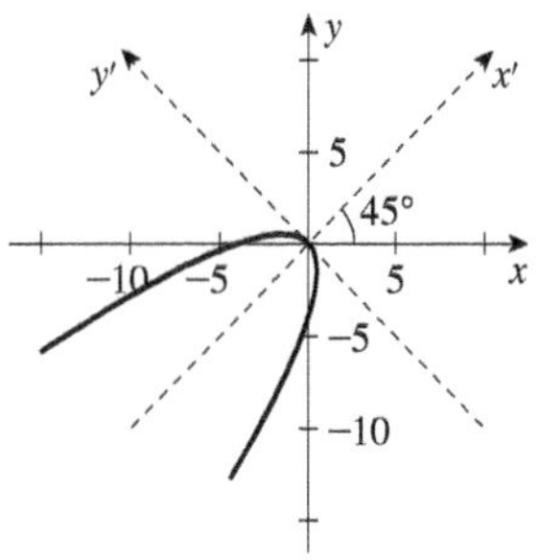

27.

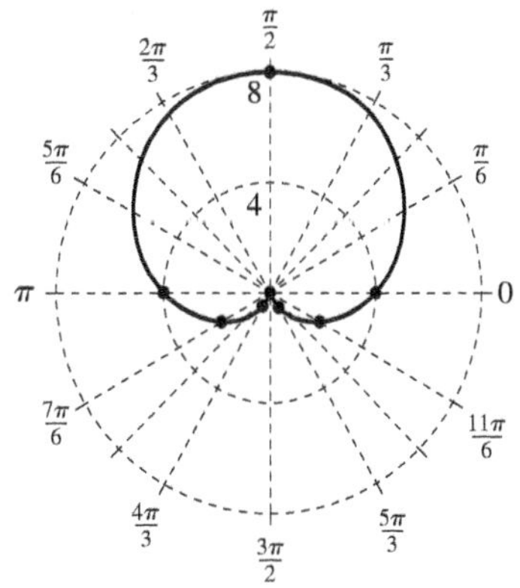

29.

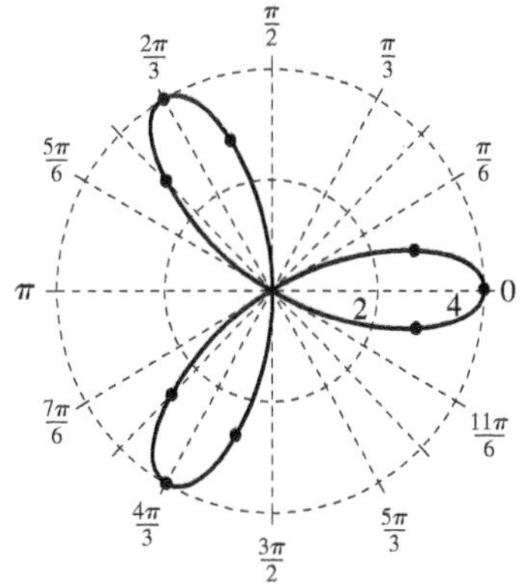

31.

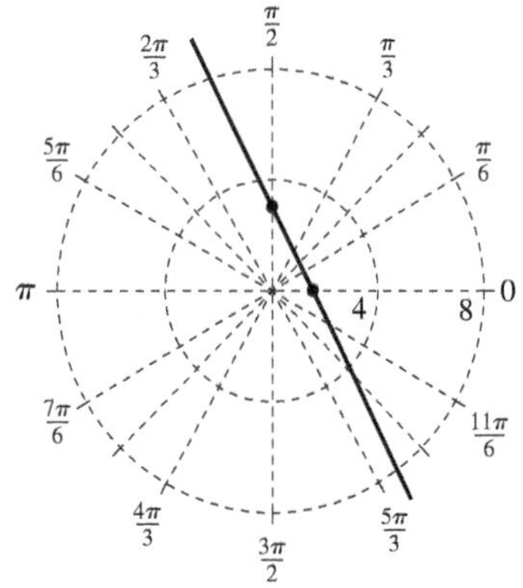

33.

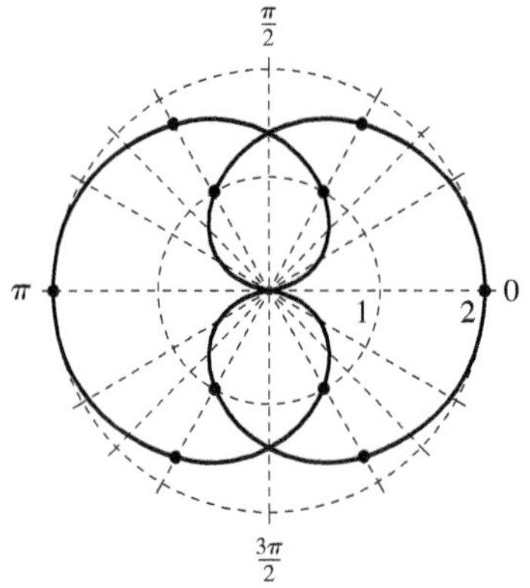

35. $\theta = \tan^{-1} 2 = 1.11$

37. $r^2 = \dfrac{2}{1 + \sin\theta\cos\theta}$ **39.** $(x^2 + y^2)^3 = 16x^2y^2$

41. $3x^2 + 4y^2 - 8x - 16 = 0$ **43.** 4 **45.** 2 **47.** 2

49. Graph $y = \dfrac{2}{3}x - \dfrac{1}{3}$

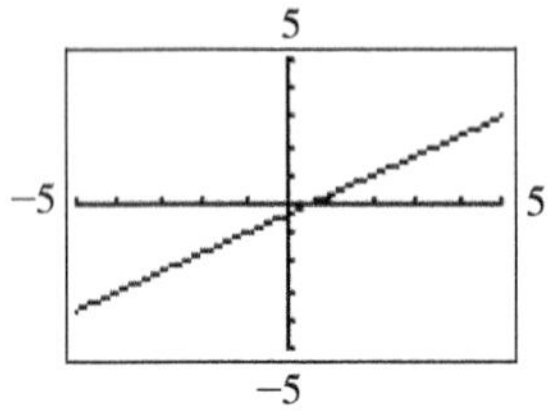

51. Graph $y = -1 \pm \dfrac{1}{2}\sqrt{10 - 4x^2}$

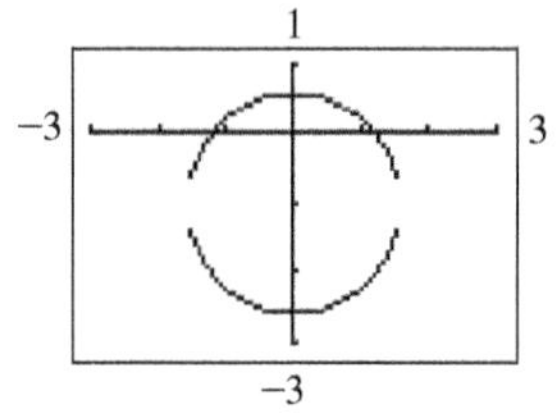

53. Graph $y = \pm\sqrt{0.25x^2 + x - 3} + 3$

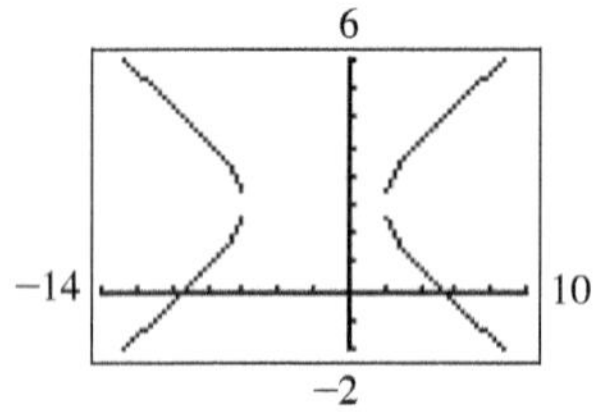

55. Graph $r = 2 - \dfrac{3}{\sin\theta}$

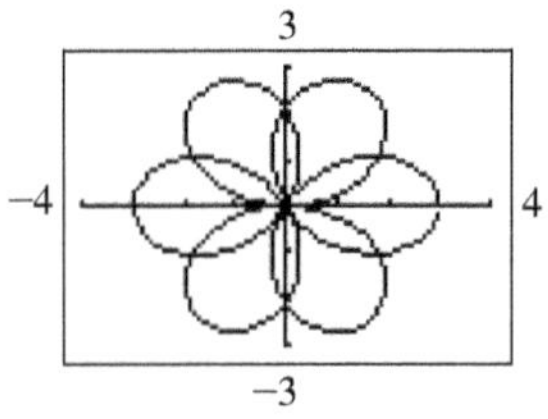

57.

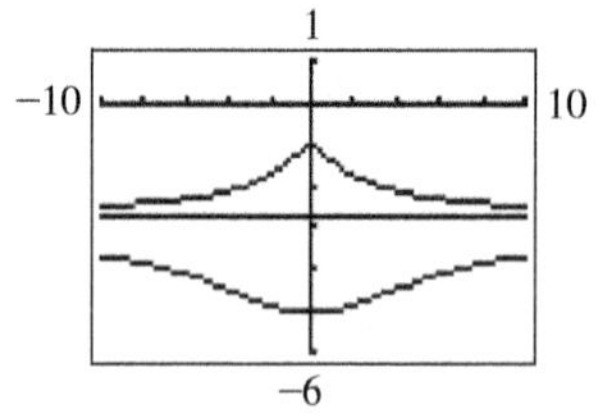

59. $x^2 + y^2 - 6x + 8y + 9 = 0$

61. $\dfrac{(x-4)^2}{16} + \dfrac{(y+3)^2}{7} = 1$ or

$7x^2 + 16y^2 - 56x + 96y + 144 = 0$

63. Circle **65.** 1 **67.** $\sqrt{45}$

69. $m_1 = -\dfrac{12}{5}, m_2 = \dfrac{5}{12}$; $13^2 + 13^2 = 338$

71.

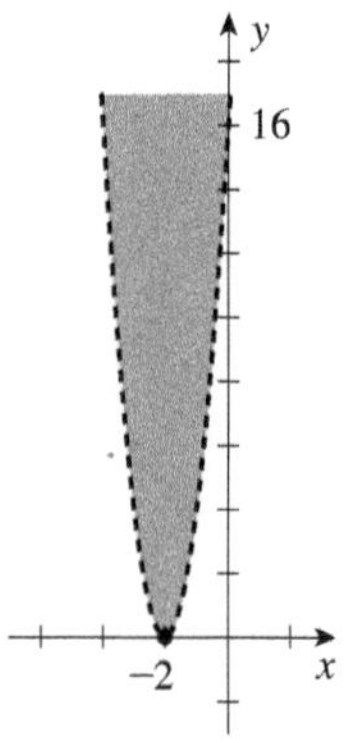

73. Hyperbola **75.** 8

77. $x^2 - 6x - 8y + 1 = 0$

79. $R_T = R + 2.5$

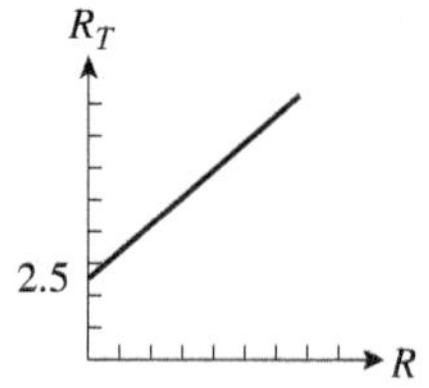

81. $v = 1.92 + 0.778t$

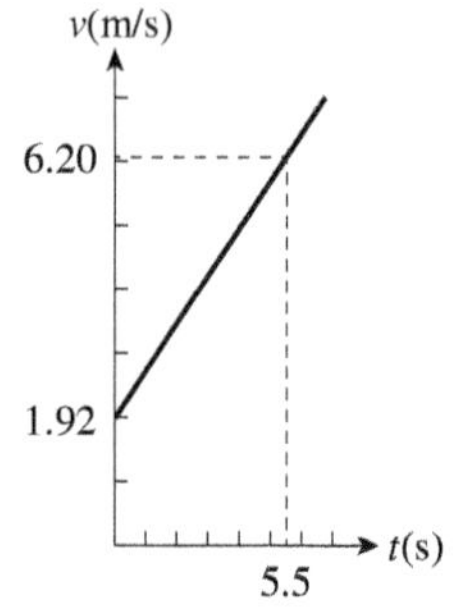

83. $y = 100.5T - 10\,050$

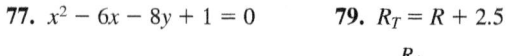

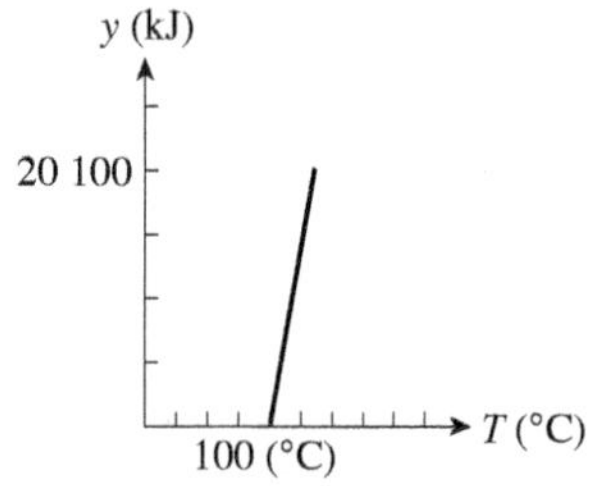

85. 1400 m^2 **87.** $y = -\dfrac{1}{80}x^2$ **89.** $y^2 = 32x$

91.

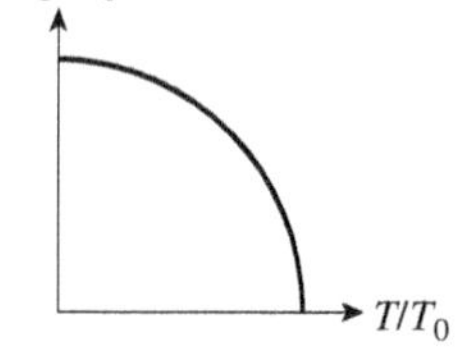

93.

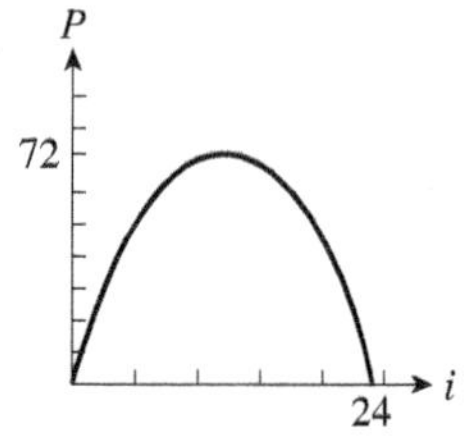

95. 700 m^2 **97.** 18 cm, 8 cm **99.** 11.3 m

101. Dist. from rifle to P − dist. from target to P = constant (related to dist. from rifle to target)

103.

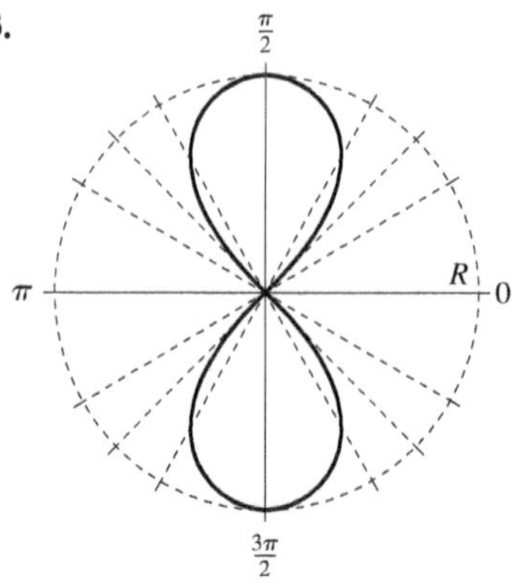

105.

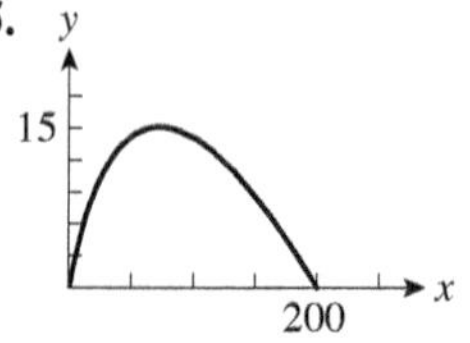

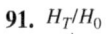

Chapter 22: Introduction to Statistics

Exercises 22.1, page 636

1.

Exercise (h)	0–3	3–6	6–9	9–12	12–15	15–18
f	14	21	9	3	2	1

3.

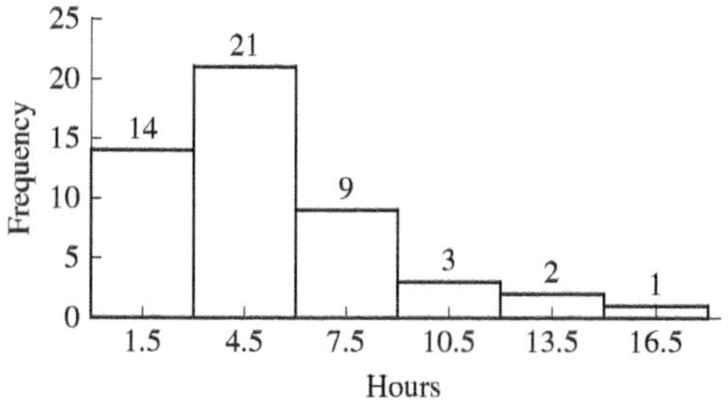

5. Quantitative **7.** Qualitative

9.

No.	5.3	5.4	5.5	5.6	5.7	5.8	5.9	6.0	6.1	6.2	6.3
f	1	3	1	3	2	4	3	1	1	0	1

11.

No.	5.1–5.4	5.4–5.7	5.7–6.0	6.0–6.3	6.3–6.6
f	1	7	9	2	1

13.

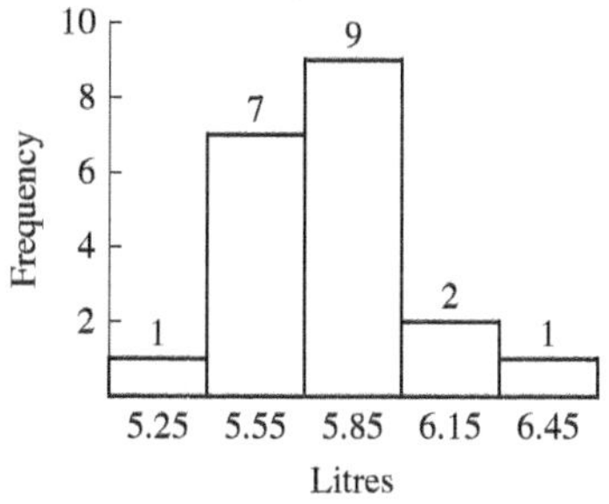

15.

No.	<5.4	<5.7	<6.0	<6.3	<6.6
Cum. f	1	8	17	19	20

17.

No. Apps	*Freq.*
50–60	1
60–70	1
70–80	2
80–90	5
90–100	7
100–110	9
110–120	4
120–130	1

19.

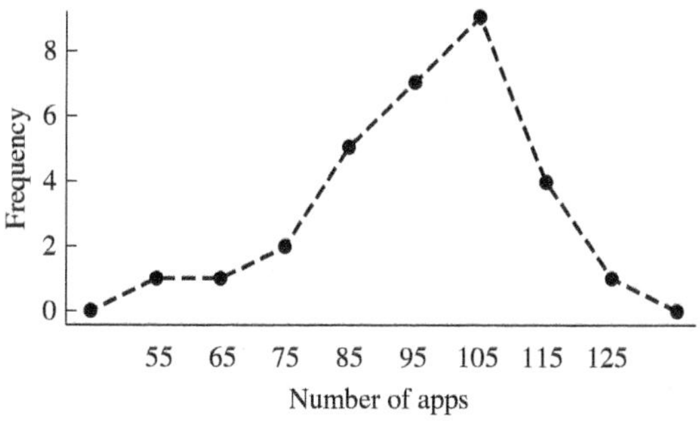

21.

Dist. (m)	47–49	50–52	53–55	56–58	59–61	62–64	65–67
Rel. Freq. (%)	1.7	12.5	26.7	30.0	20.0	8.3	0.8

23.

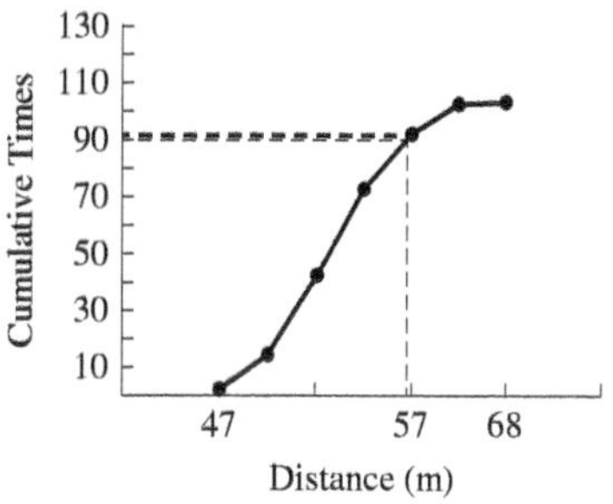

25.

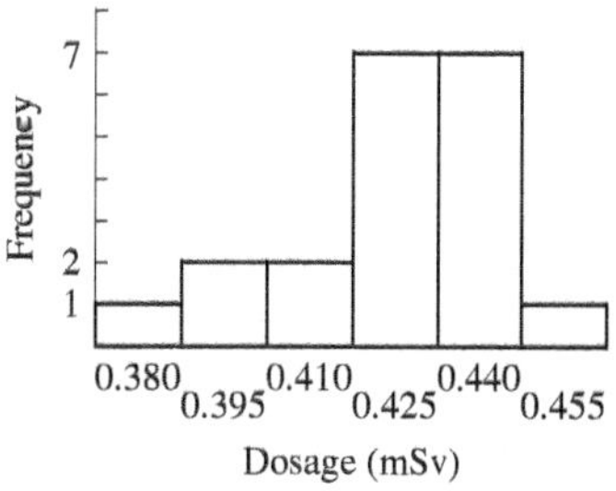

27.

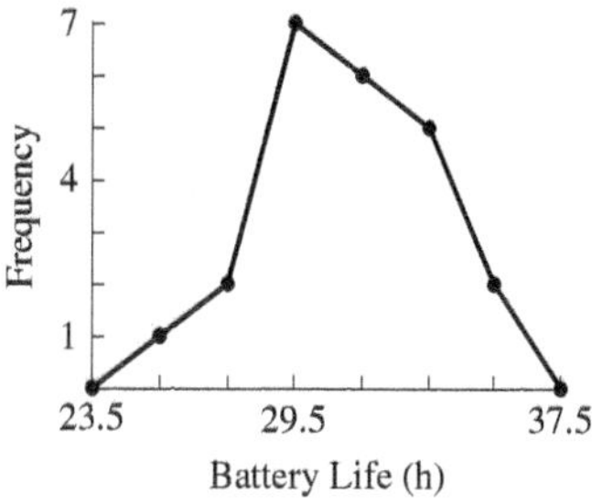

29.

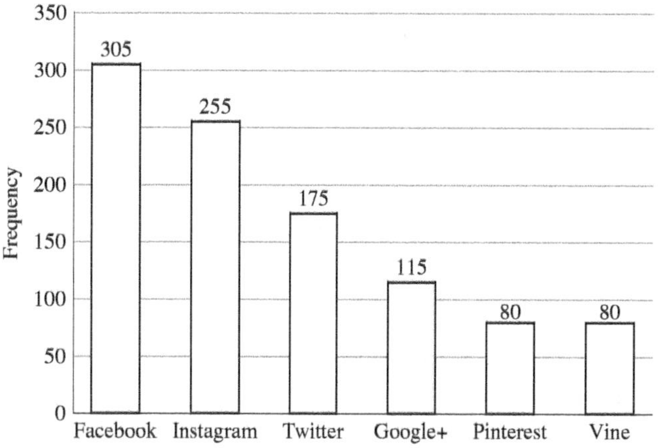

31. No; categories not mutually exclusive

Exercises 22.2, page 643

1. 5 **3.** 5.5 **5.** 2.2 **7.** 94% **9.** (a) 25.5 (b) 25 (c) 25

11. (a) 107.2 (b) 107.5 (c) 108 **13.** (a) 1.8 (b) 1.8

15. (a) 2.500 (b) 2.500 **17.** 0.257 L/100 km

19. Mean: 96.1 apps, median: 98 apps, modes: 86, 93, 101, 103, 105, 107 apps

21. Mean: 56.5 m, median: 57 m

23. Mean: 0.4237 mSv, median: 0.4265 mSv, mode: 0.436 mSv

25. Mean: 31.2 h, median: 31 h, mode: 30 h

27. Median: \$625, mode: \$700

29. Range \$275, $s = \$84.62$

31. Median: 3600 MJ, mode: 3600 MJ

33. Median: 0.195 ppm, mode: 0.18 ppm

35. Range: 0.18 ppm, $s = 0.0474$ ppm **37.** \$662.50

39. Median: \$725, mean: \$748.2, mode: \$800. When adding a constant, the mean, the median, and the mode are all increased by the same constant.

41. \$883.90; The mean no longer represents the centre of the data.

43. 0.4254 mSv **45.** At least 89%

Exercises 22.3, page 649

1. $\mu = 10, \alpha = 5$ and $\mu = 20, \alpha = 5$ result in the same curve, with the first centred at $x = 10$ and the second centred at $x = 20$.

3. 0.2072 **5.** About 95% **7.** 99.85%

9. 1.2; 1.2 st. dev. above mean **11.** −2.4; 2.4 st. dev. below mean

13. 341 **15.** 78.81% **17.** 2166 **19.** 179

21. 141; about 68% of all samples of size 5000 have a mean lifetime from 99 859 km to 100 141 km.

23. 0.0145; 68% of all samples of size 500 will have a proportion of defectives that is within (0.1055, 0.1345).

25. −0.7 **27.** −0.6 **29.** 35%

Exercises 22.4, page 655

1. (32.4, 33.2) **3.** 35 **5.** 1537 **7.** (243.2, 251.0)

9. (0.126, 0.228) **11.** 277 **13.** (83.0, 84.4)

15. (83.0, 84.4). Margin of error has increased as well.

17. (104.9, 109.5) **19.** 58 **21.** 601

Exercises 22.5, page 661

1. $UCL(\bar{x}) = 504.7$ mg, $LCL(\bar{x}) = 494.7$ mg

3. The first point would be below the $\bar{\bar{x}}$ line. The UCL and LCL lines would be 0.2 unit lower.

5. $\bar{\bar{x}} = 361.8$ N · m, UCL = 368.9 N · m, LCL = 354.6 N · m

7.

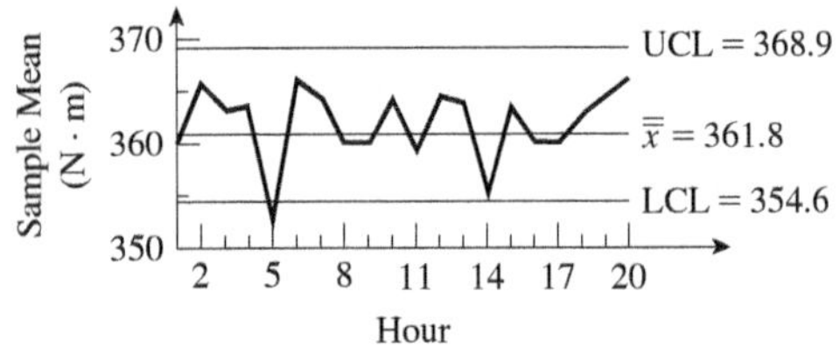

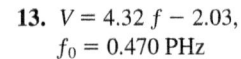

9. $\bar{\bar{x}} = 8.986$ V, UCL = 9.081 V, LCL = 8.891 V

11.

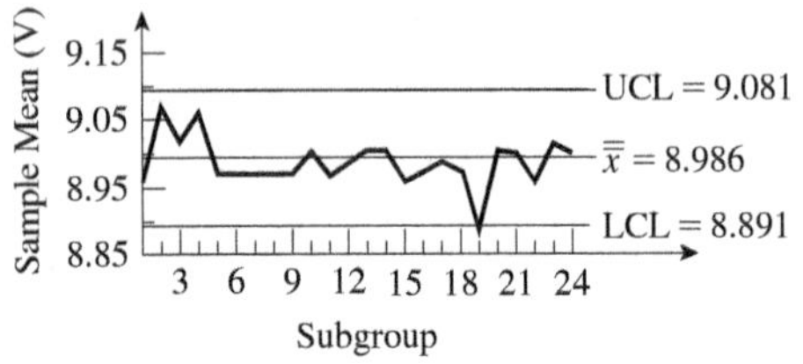

13. $\mu = 2.725$ cm, UCL = 2.729 cm, LCL = 2.721 cm

15. $\mu = 5.57$ mL, UCL = 11.17 mL, LCL = 0.00 mL

17. $\bar{p} = 0.0369$, UCL = 0.0548, LCL = 0.0190

19. $\bar{p} = 0.0580$, UCL = 0.0894, LCL = 0.0266

Exercises 22.6, page 666

1. $y = 2x + 2$

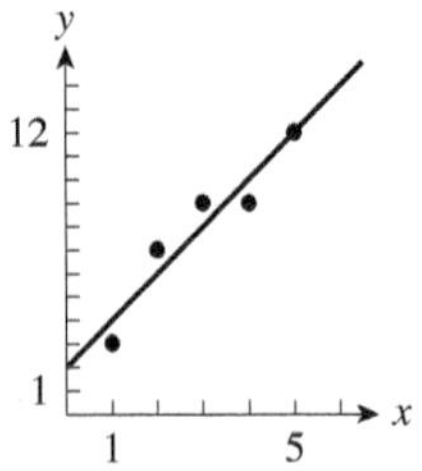

3. $y = -1.77x + 191$

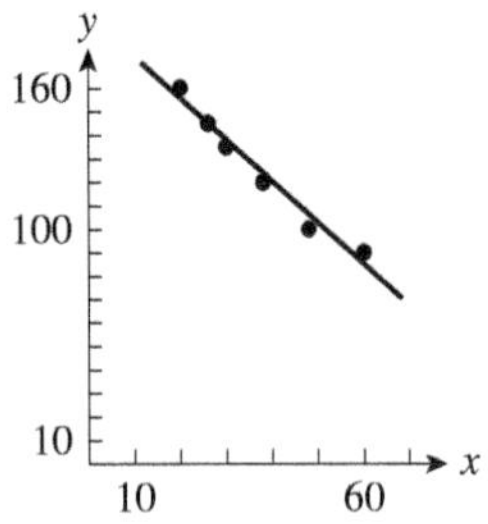

5. $y = -0.308t + 9.07$

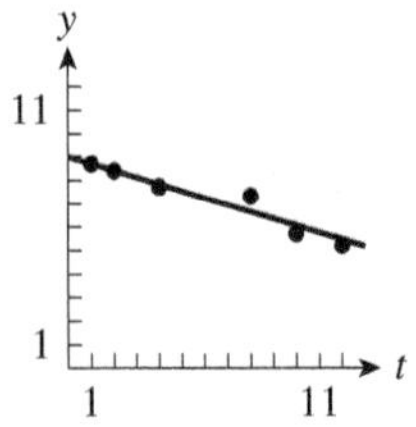

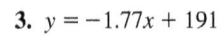

7. $V = -0.590i + 11.3$

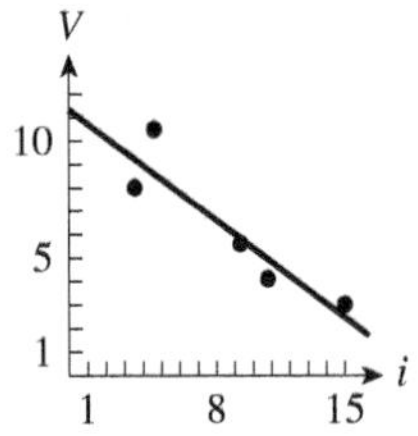

9. $y = 23.3x + 0.61$

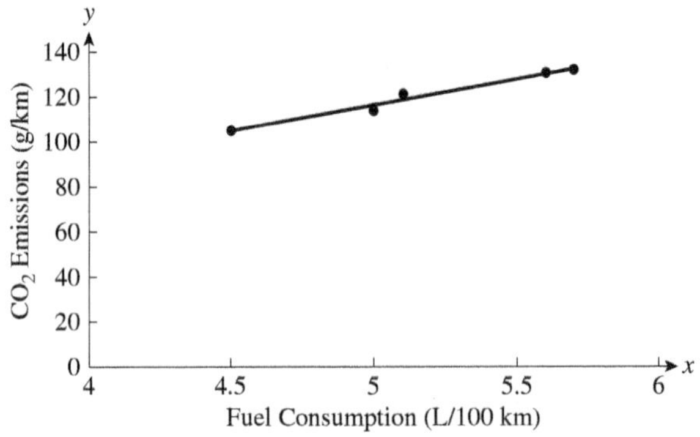

11. $p = -2.66x + 4364$

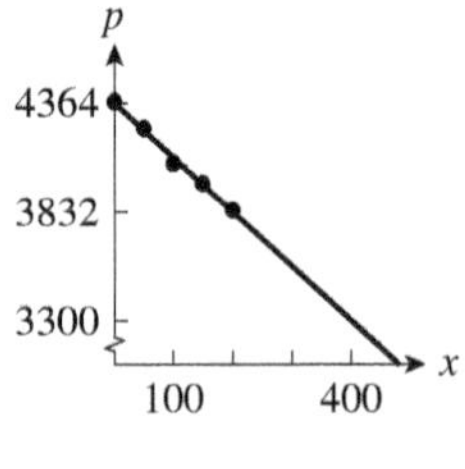

13. $V = 4.32 f - 2.03$, $f_0 = 0.470$ PHz

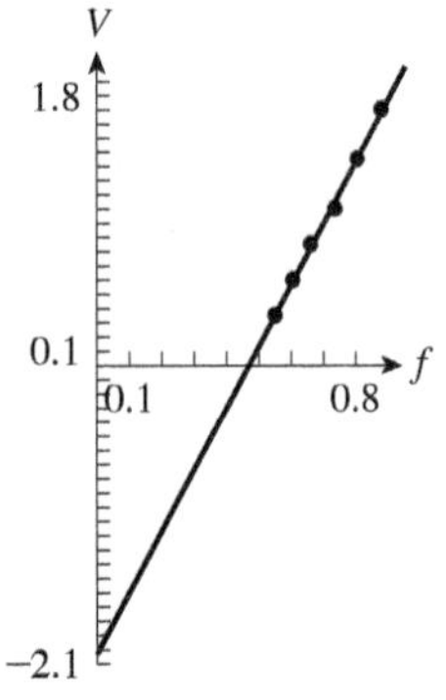

15. 0.953 **17.** −0.901

Exercises 22.7, page 669

1. $y = 2.59x^2 + 1.07$

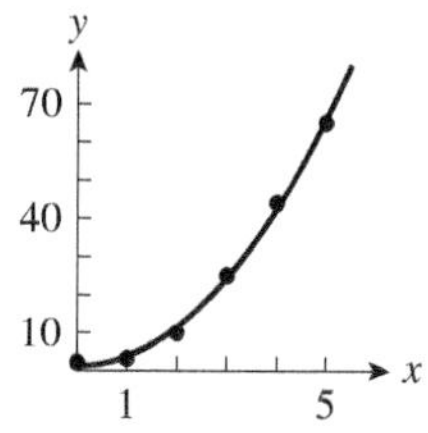

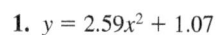

3. $y = \dfrac{2410}{x} - 0.234$

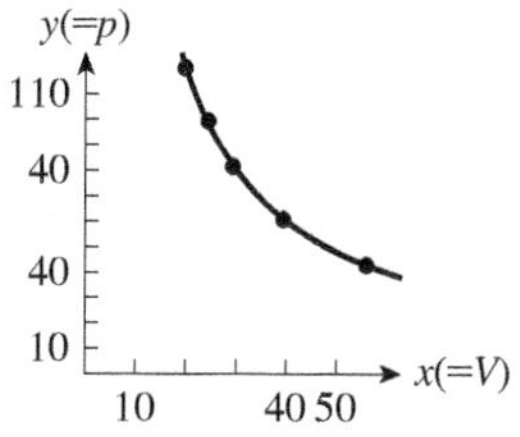

5. $y = 5.97t^2 + 0.38$

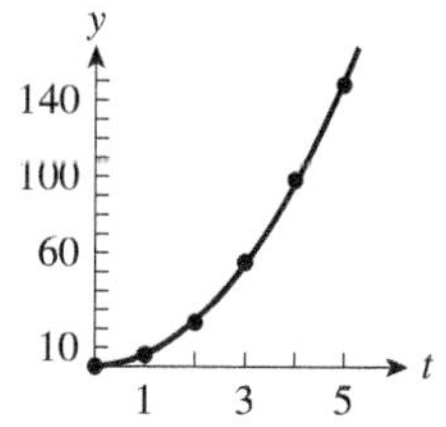

7. $p = 0.551T^2 + 540$

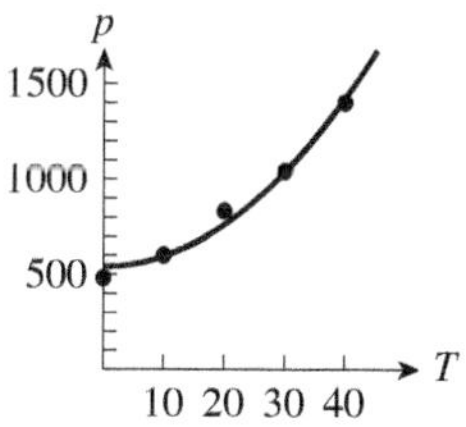

9. $P = \dfrac{1343}{S}$

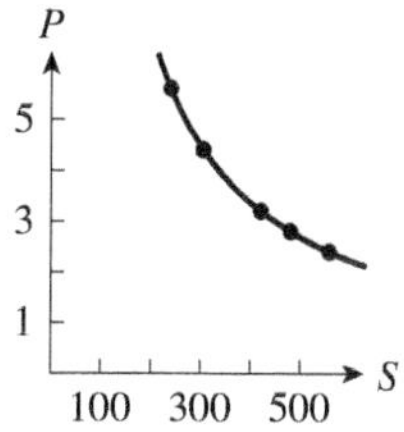

11. $y = 6.20e^{-t} - 0.05$

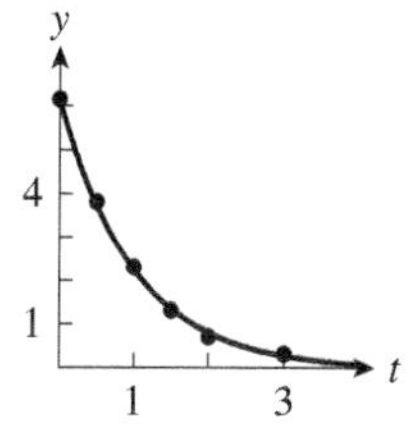

Review Exercises for Chapter 22, page 671

1. 76.5 **3.** 76.2

5.

No.	67–70	71–74	75–78	79–82	83–86
f	3	4	8	3	2

7.

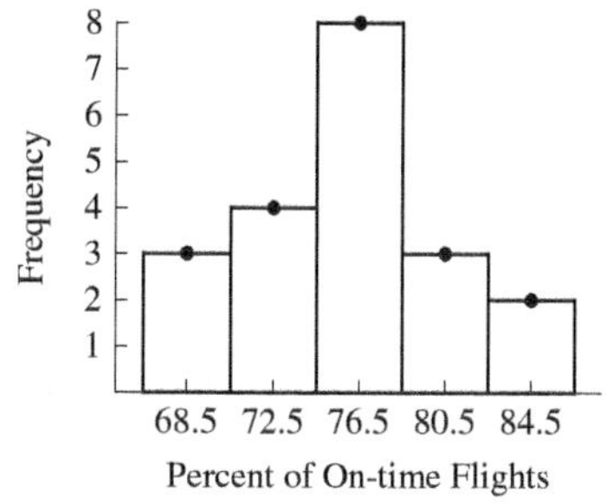

9.

No.	<71	<75	<79	<83	<87
f	3	7	15	18	20

11. 0.264 Pa · s **13.** 0.014 Pa · s

15.

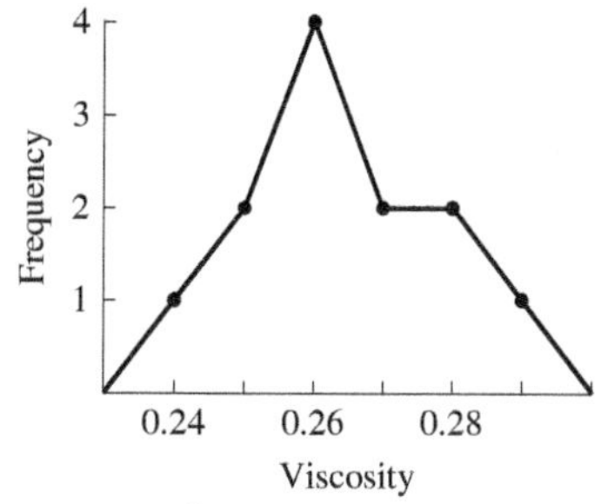

17. 700 W **19.** 700 W **21.** 17.3 W

23.

Power (W)	<660	<670	<680	<690
Cum. f	3	5	12	24

Power (W)	<700	<710	<720	<730	<740
Cum. f	51	85	100	116	121

25. (692.8, 699.2) **27.** 4

29.

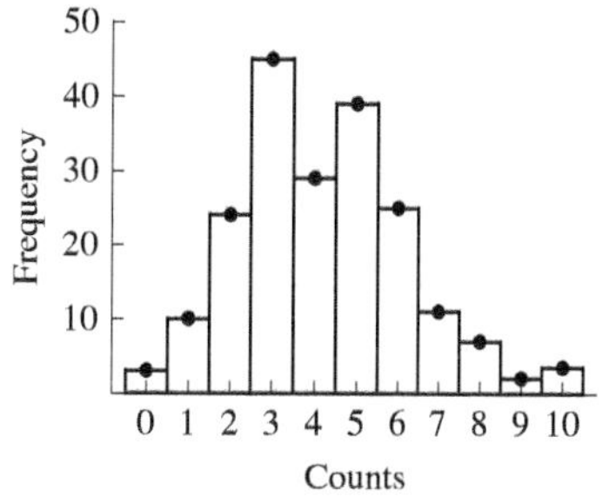

31. 66.2 km/h **33.** 9.0 km/h **35.** (64.6, 67.4)

37. At least 75% **39.** (0.138, 0.252) **41.** 415

43. $\bar{p} = 0.0540$, UCL = 0.0843, LCL = 0.0237

45.

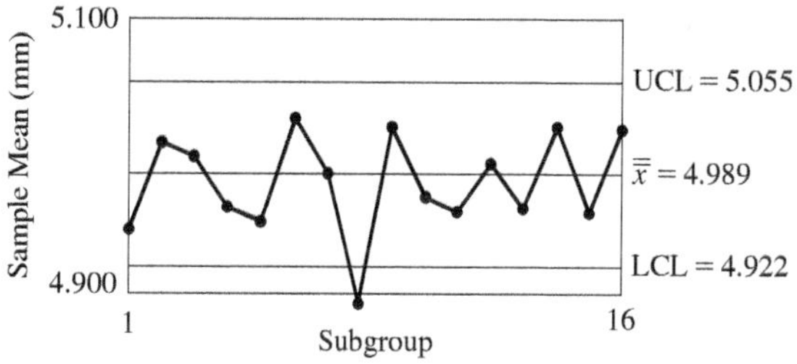

47. 322 **49.** 495

51. $R = 0.0983T + 25.0$

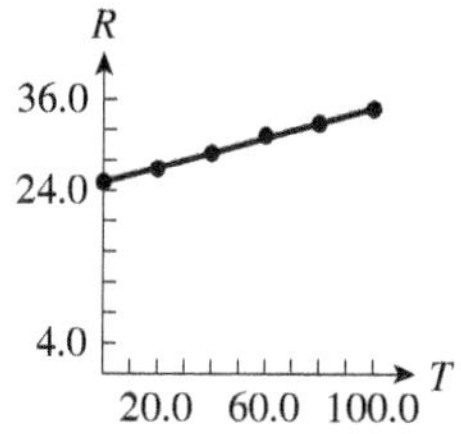

53. $s = 0.123t + 0.887$

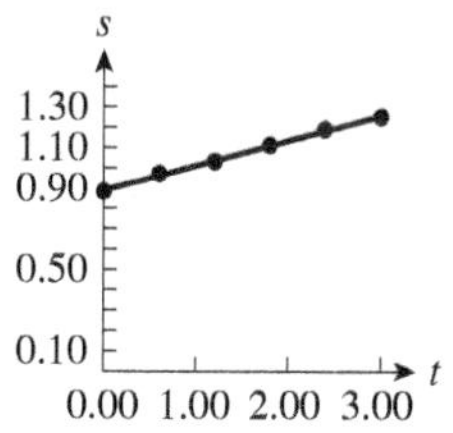

55. $s = -4.90t^2 + 3000$

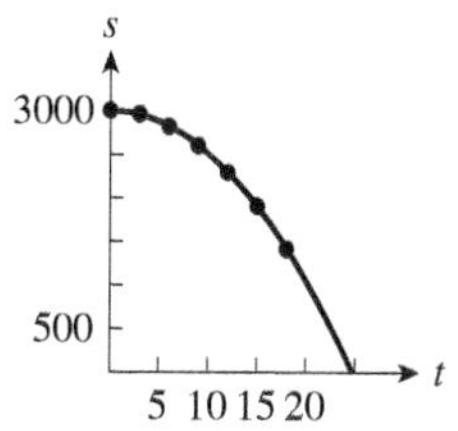

57. $i = 3.99(1 - e^{-0.500t})$, 3.99 A

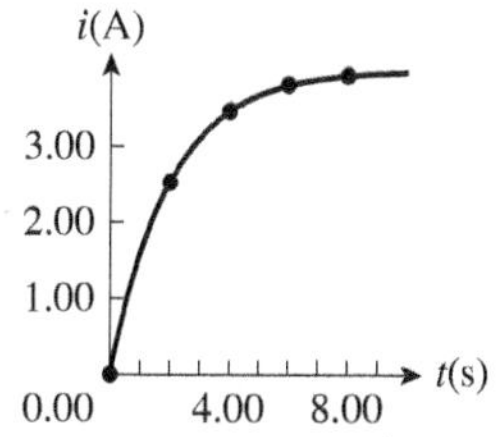

59. $y = 0.0002x^2 + 15$

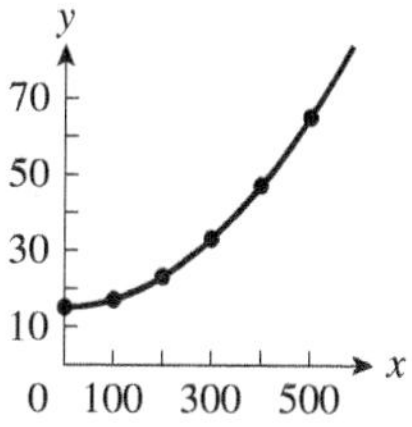

61. $y = 0.997x + 0.581$ **63.** $y = 1.39x^{0.878}$

65. 2.7 **67.** 9.2 ppm

69. Substitute expressions for m, b, and $\bar{x}$. Simplify to $\bar{y} = \bar{y}$.

Chapter 23: The Derivative

Exercises 23.1, page 683

1. Not cont. at $x = -2$ **3.** -4 **5.** Cont. all x

7. Not cont. $x = 0$ and $x = 7$, div. by zero

9. Cont. $x \leq 0$, $x > 2$; function not defined **11.** Cont. all x

13. Not cont. $x = 2$, jump in the graph

15. Not cont. $x = 2$, $f(2) \neq \lim_{x\to 2} f(x)$

17. (a) -1 (b) $\lim_{x\to 2} f(x)$ does not exist

19. (a) 0 (b) $\lim_{x\to 2} f(x) = -1$

21. Not cont. $x = 2$, jump in the graph **23.** Cont. all x

25.

x	0.900	0.990	0.999	1.001
$f(x)$	1.7100	1.9701	1.9970	2.0030

x	1.010	1.100
$f(x)$	2.0301	2.3100

$\lim_{x\to 1} f(x) = 2$

27.

x	1.900	1.990	1.999	2.001
$f(x)$	-0.2516	-0.2502	$-0.250\,02$	$-0.249\,98$

x	2.010	2.100
$f(x)$	-0.2498	-0.2485

$\lim_{x\to 2} f(x) = -0.25$

29.

x	10	100	1000
$f(x)$	0.4468	0.4044	0.4004

$\lim_{x\to\infty} f(x) = 0.4$

31. 11 **33.** 1 **35.** $-\frac{2}{3}$ **37.** 27 **39.** 2

41. Does not exist **43.** $\frac{1}{2}$ **45.** 3 **47.** 4

49.

x	-0.1	-0.01	-0.001	0.001
$f(x)$	-3.1	-3.01	-3.001	-2.999

x	0.01	0.1
$f(x)$	-2.99	-2.9

$\lim_{x\to 0} f(x) = -3$: use factoring

51.

x	10	100	1000
$f(x)$	2.1649	2.0106	2.0010

$\lim_{x\to\infty} f(x) = 2$; divide numerator and denominator by x^2

53. 34.9 °C, 0 °C **55.** 3 cm/s **57.** 2.77 **59.** e

61. (a) -1 (b) 2, (c) Does not exist **63.** 0

65. -1; $+1$; no, $\lim_{x\to 0^+} f(x) \neq \lim_{x\to 0^-} f(x)$

Exercises 23.2, page 688

1. 9 **3.** (Slopes) 3.5, 3.9, 3.99, 3.999; $m = 4$ **5.** (Slopes) -2, -2.8, -2.98, -2.998; $m = -3$

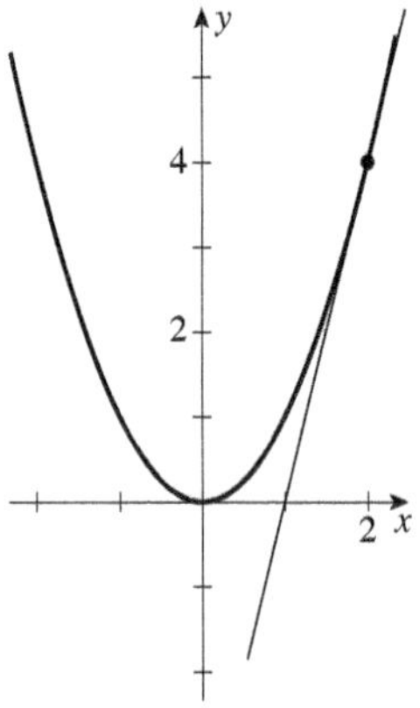

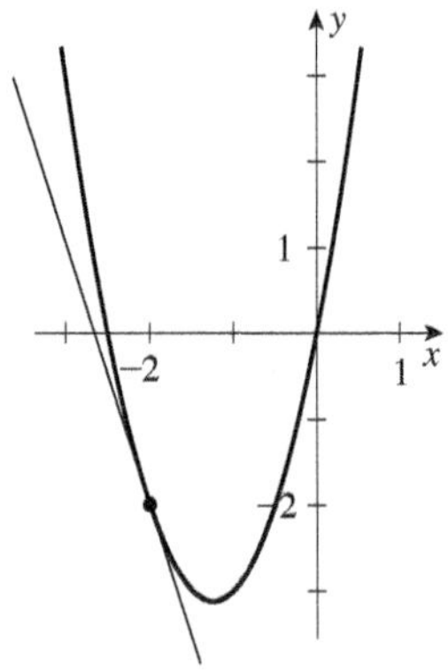

7. $m_{\tan} = 2x$; 4, -2

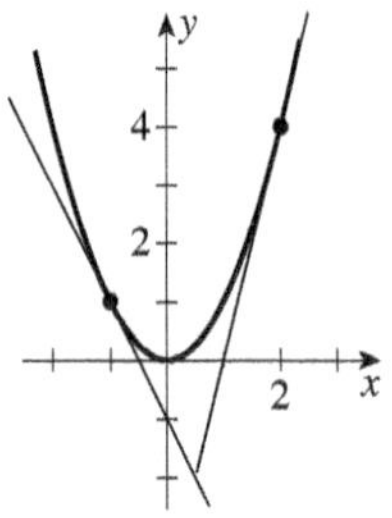

9. $m_{\tan} = 4x + 5$; -3, 7

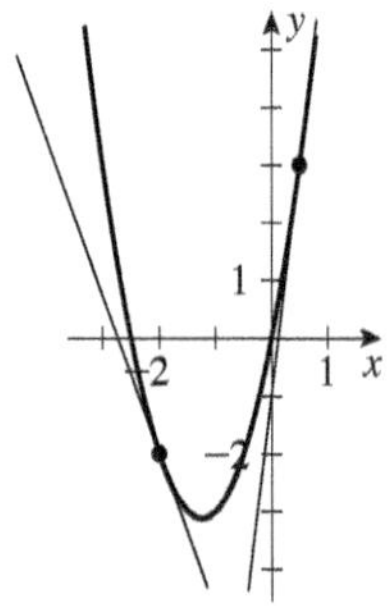

11. $m_{\tan} = 2x + 4$; -2, 8

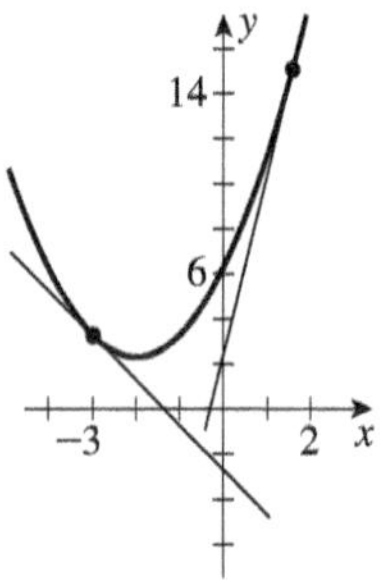

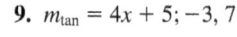

13. $m_{\tan} = 6 - 2x$; 10, 0

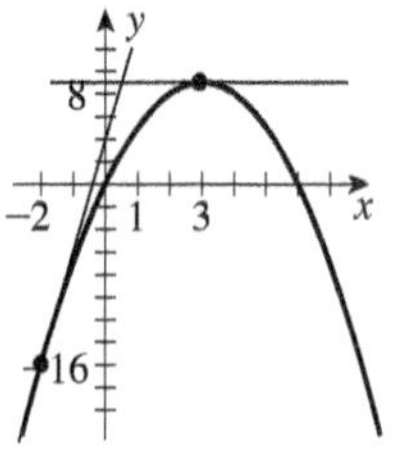

15. $m_{\tan} = 6x^3$; 0, 0.75, 6

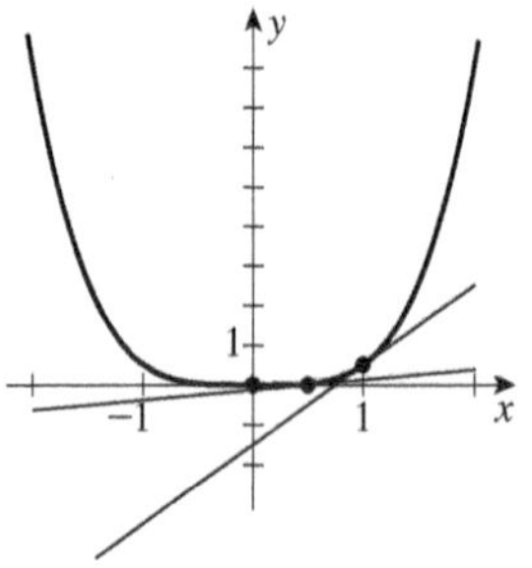

17. $m_{\tan} = 5x^4$; 0, 0.31, 5

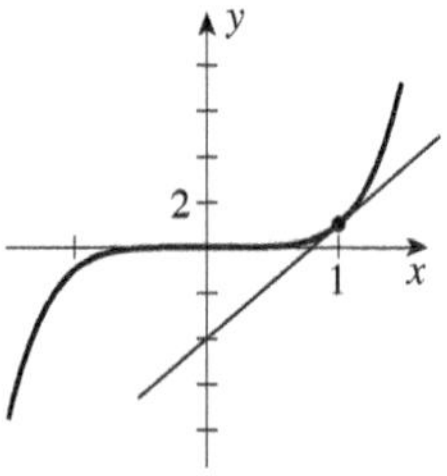

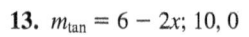

19. 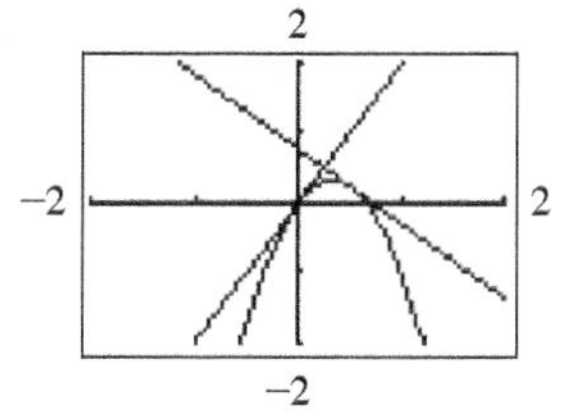

21. 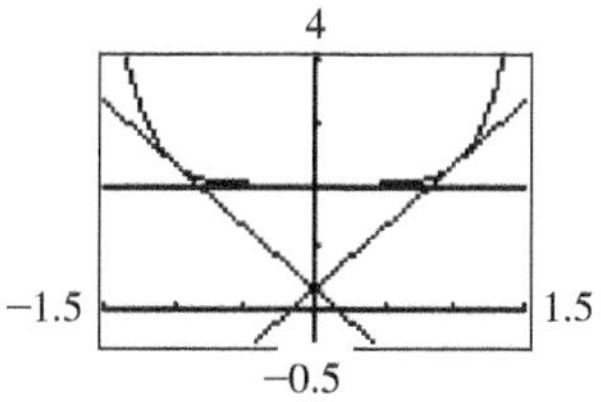

23. av. ch. = 4.1, $m_{tan} = 4$ **25.** av. ch. = −12.6, $m_{tan} = -12.0$

27. $(-1, 2)$ **29.** $-\dfrac{1}{6}$ **31.** 9.46° **33.** 3

Exercises 23.3, page 693

1. $8x + 3$ **3.** 5 **5.** −2 **7.** $2x$ **9.** $3\pi x^2$

11. $2x - 7$ **13.** $4(2 - x)$ **15.** $3x^2 + 4$

17. $-\dfrac{\sqrt{3}}{(x+2)^2}$ **19.** $1 - \dfrac{4}{3x^2}$ **21.** $-\dfrac{4}{x^3}$

23. $4x^3 + 3x^2 + 2x + 1$ **25.** $4x^3 + \dfrac{10}{(5x+1)^2}$

27. $2(3x - 1); -8$ **29.** $-\dfrac{33}{(3x+2)^2}; -\dfrac{3}{11}$

31. $-\dfrac{2}{x^2}$; all real numbers except 0

33. $\dfrac{-6x}{(x^2-1)^2}$; all real numbers except −1 and 1

35. $(5, 5)$ **37.** $(4, -32)$

39. $\dfrac{1}{2\sqrt{x+1}}$; differentiable for $x > -1$ since derivative and function are both defined for these values.

41. $\dfrac{dy}{dx} = nx^{n-1}$ **43.** −11.3°

Exercises 23.4, page 697

1. $\dfrac{ds}{dt} = 14.0 - 9.80t$; −5.60 m/s, −25.2 m/s

3. $\left.\dfrac{dy}{dx}\right|_{x=2} = 4$ **5.** $\left.\dfrac{dy}{dx}\right|_{x=-3} = -\dfrac{3}{4}$

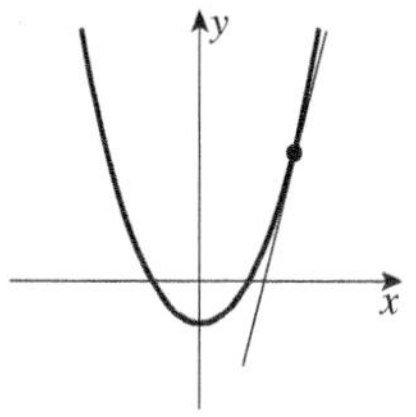

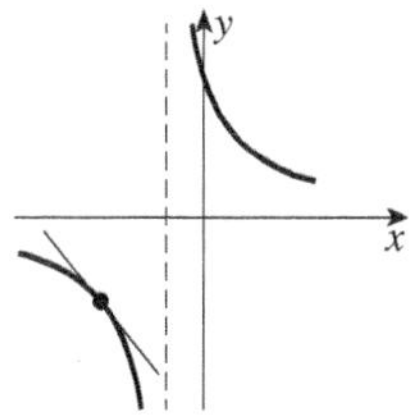

7. 4.00, 4.00, 4.00, 4.00, 4.00; $\lim_{t\to 3} v = 4.00$ m/s

9. 5, 6.5, 7.7, 7.97, 7.997; $\lim_{t\to 2} v = 8.00$ m/s

11. 4; 4.00 m/s **13.** $6t - 4$; 8.00 m/s **15.** 52

17. $3t(8 - t)$ **19.** $3 + \dfrac{6}{5t^2}$ **21.** $12t - 4$

23. $6t$ **25.** 6.28 cm/cm **27.** 4.49 s **29.** −2

31. $6w$ **33.** 0.460 kW **35.** −83.1 W/(m² · h)

37. $-\dfrac{48}{(t+3)^2}$; −\$1330/year **39.** πd^2 **41.** $24.2/\sqrt{\lambda}$ m/m

Exercises 23.5, page 701

1. $9x^8$ **3.** $6x^2 - 12x + 8$; 56 **5.** $8x^7$ **7.** $44x^{10}$

9. $20x^3$ **11.** $3x^2 + 2$ **13.** $15r^2 - 2$ **15.** $200x^7 - 170x^4$

17. $-42x^6 + 15x^2$ **19.** $x^2 + x$ **21.** 16 **23.** 33

25. −4 **27.** −29 **29.** $5(6t^4 - 1)$ **31.** $-6(1 + t^2)$

33. 64.0 m/s **35.** 0.00 m/s **37.** 1 **39.** $(2, 4)$

41. $(1, -5)$ **43.** $-\frac{1}{4}$ **45.** $3\pi r^2$ **47.** 84.0 W/A

49. 3.32 Ω/°C **51.** −0.002 86 N/°C **53.** −98.0 m/km

55. 391 mm³/mm **57.** 0.166 m/(m/s)

Exercises 23.6, page 705

1. $2(9x^2 - 9x - 5)$ **3.** $6x(9x - 10)$ **5.** $36t - 21$

7. $2x(-7x^5 + 15x^3 + 2x^2 - 9x - 3)$

9. $-8(x - 3)$ **11.** $10h^4 - 4h^3 - 3h^2 - 4h + 1$

13. $\dfrac{3}{(2x+3)^2}$ **15.** $-\dfrac{4\pi x}{(2x^2+1)^2}$ **17.** $\dfrac{12x(3-x)}{(3-2x)^2}$

19. $\dfrac{-6x^2 + 6x + 4}{(3x^2+2)^2}$ **21.** $\dfrac{-3x^2 - 16x - 26}{(x^2+4x+2)^2}$

23. $\dfrac{-2x^3 + 2x^2 + 5x + 4}{x^3(x+2)^2}$ **25.** −191 **27.** 75.0

29. 19 **31.** −5.64 **33.** Eq. (23.10)

35. $(0, -1)$ **37.** $x^2f'(x) + 2xf(x)$

39. (1) $\dfrac{-12x^3 + 45x^2 - 14x}{(3x-7)^2}$ (2) $\dfrac{-12x^3 + 45x^2 - 14x}{(3x-7)^2}$

41. 12.0 **43.** 1, −1 **45.** −0.694 W/s

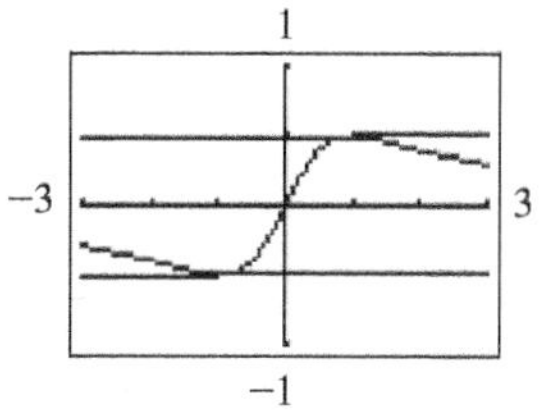

47. $\dfrac{96}{(7R+12)^2}$ **49.** $8t^3 - 45t^2 - 14t - 8$ **51.** 1.18 °C/h

53. $\dfrac{2R(R+2r)}{3(R+r)^2}$ **55.** $\dfrac{E^2(R-r)}{(R+r)^3}$ **57.** 9.80×10^3 Pa/m

Exercises 23.7, page 713

1. $36x^2(2 + 3x^3)^3$ **3.** $-\dfrac{3x}{(2-3x^2)^{1/2}}$ **5.** $\dfrac{3}{x^{1/2}}$ **7.** $-\dfrac{9}{5t^4}$

9. $-\dfrac{1}{x^{4/3}} + 8x$ **11.** $\dfrac{3}{2}x^{1/2} + \dfrac{6}{x^2}$ **13.** $10x(x^2 + 1)^4$

15. $-216x^2(7 - 4x^3)^7$ **17.** $\dfrac{2x^2}{(2x^3-3)^{2/3}}$ **19.** $\dfrac{24y}{(4-y^2)^5}$

21. $\dfrac{8x^3}{(2x^4-5)^{0.75}}$ **23.** $\dfrac{-4x}{(1-8x^2)^{3/4}}$ **25.** $\dfrac{12v+5}{(8v+5)^{1/2}}$

27. $\dfrac{6(5x-1)}{x^4(1-6x)^{1/2}}$ **29.** $\dfrac{x^2 + 12x + 16}{(x+4)^2(x+2)^{1/2}}$

31. $\dfrac{-1}{\sqrt{2R+1}(4R+1)^{3/2}}$ **33.** $\dfrac{3}{10}$ **35.** $\dfrac{5}{36}$

37. $\frac{d}{dx}x(x^{1/2}) = x\left(\frac{1}{2}x^{-1/2}\right) + x^{1/2}(1) = \frac{3}{2}x^{1/2}$

39. $\frac{x^3(0) - 1(3x^2)}{x^6} = -3x^{-4}$ **41.** $x = 0$

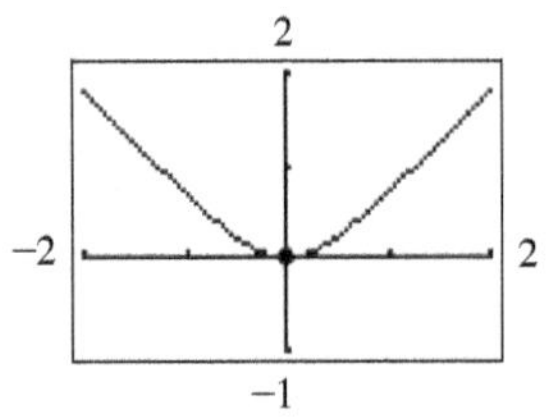

43. Yes, at $\left(-\frac{13}{9}, \frac{1}{3}\right)$ **45.** $\frac{1}{2}$ **47.** $\frac{0.485}{\sqrt{w}}$

49. -1.35 cm/s **51.** $\frac{-450\,000}{V^{5/2}}$, -4.50 kPa/cm^3

53. $l = a$ **55.** -45.2 W/(m^2 · h)

57. $\frac{8a^3}{(4a^2 - \lambda^2)^{3/2}}$ **59.** $\frac{\sqrt{2}(w + 1)}{\sqrt{w^2 + 2w + 2}}$

Exercises 23.8, page 717

1. $-\frac{4x}{3y^2}$ **3.** $x^2\frac{dy}{dx} + 2xy$ **5.** $-\frac{x\frac{dy}{dx} + y}{(xy)^2}$ **7.** $-\frac{3}{2}$

9. $\frac{6x + 1}{4}$ **11.** $\frac{x}{4y}$ **13.** $\frac{2x}{7y^6}$ **15.** $\frac{3x}{y + 3}$ **17.** $\frac{-3y}{3x + 1}$

19. $\frac{y - x^3 - 2x^2y - xy^2}{x}$ **21.** $\frac{3(y^2 + 1)(y^2 - 2x + 1)}{(y^2 + 1)^2 - 6x^2y}$

23. $\frac{4(2y - x)^3 - 2x}{8(2y - x)^3 - 1}$ **25.** $\frac{-3x(x^2 + 1)^2}{y(y^2 + 1)}$ **27.** 3

29. $-\frac{108}{157}$ **31.** $\frac{1}{4}$ **33.** $(2, 2), (2, -2)$

35. At $(\sqrt{7}, 0)$ and $(-\sqrt{7}, 0)$, $m_{\tan} = -2$

37. 1 **39.** $\frac{1}{2\omega L^3}\left(-\frac{L}{C} + 2R^2\right)$ **41.** $\frac{nRT^2 + bnP}{VT^2 - anT^2 + bnT}$

43. $-\frac{x}{y}$ **45.** $\frac{r - R + 1}{r + 1}$ **47.** $\frac{2C^2r(12CSr - 20Cr - 3L)}{3(Cr + L)(Cr - L)}$

Exercises 23.9, page 721

1. $y' = 15x^2 - 4x$, $y'' = 30x - 4$, $y''' = 30$, $y^{(n)} = 0\ (n \geq 4)$

3. $y' = 3x^2 + 14x$, $y'' = 6x + 14$, $y''' = 6$, $y^{(n)} = 0\ (n \geq 4)$

5. $f'(x) = 3x^2 - 24x^3$, $f''(x) = 6x - 72x^2$,
$f'''(x) = 6 - 144x$, $f^{(4)}(x) = -144$, $f^{(n)}(x) = 0\ (n \geq 5)$

7. $y' = -8(1 - 2x)^3$, $y'' = 48(1 - 2x)^2$,
$y''' = -192(1 - 2x)$, $y^{(4)} = 384$, $y^{(n)} = 0\ (n \geq 5)$

9. $f'(r) = (5 - 2r)^2(-8r + 5)$
$f''(r) = 4(5 - 2r)(12r - 15)$
$f'''(r) = 24(-8r + 15)$, $f^{(4)}(r) = -192$
$f^{(n)}(r) = 0 (n \geq 5)$

11. $84x^5 - 30x^4$ **13.** $-\frac{1}{4x^{3/2}}$ **15.** $-\frac{12}{(8x - 3)^{7/4}}$

17. $\frac{14.4\pi}{(1 + 2p)^{5/2}}$ **19.** $600(2 - 5x)^2$

21. $30(27x^2 - 1)(3x^2 - 1)^3$ **23.** $\frac{4\pi^2}{(6 - x)^3}$

25. $\frac{450}{(v + 15)^3}$ **27.** $-\frac{9}{y^3}$ **29.** $-\frac{6(x^2 - xy + y^2)}{(2y - x)^3}$

31. $\frac{9}{125}$ **33.** $-\frac{13}{384}$ **35.** 800 **37.** -9.80 m/s^2

39. 1.00 m/s^2 **41.** $\frac{d}{dx}\left(u\frac{dv}{dx} + v\frac{du}{dx}\right) = u\frac{d^2v}{dx^2} + \frac{2dv}{dx}\frac{du}{dx} + v\frac{d^2u}{dx^2}$

43. 48.0 **45.** $\frac{d^2P}{dt^2} = 80.0$ **47.** -9.80 m/s^2

49. $-\frac{1.60}{(2t + 1)^{3/2}}$ **51.** 0.0490 **53.** 4.00 s

Review Exercises for Chapter 23, page 723

1. -4 **3.** Does not exist **5.** $\frac{2}{3}$ **7.** $\frac{1}{4}$ **9.** $\frac{2}{3}$

11. -2 **13.** 5 **15.** $-4x$ **17.** $-\frac{4}{(x - 1)^3}$

19. $\frac{1}{2\sqrt{x + 5}}$ **21.** $2x(7x^5 - 3)$ **23.** $\frac{1}{x^2}(2x^{3/2} + 3)$

25. $\frac{3}{(1 - 5y)^2}$ **27.** $-28(2 - 7x)^3$ **29.** $\frac{9\pi x}{(5 - 2x^2)^{7/4}}$

31. $\frac{1}{\sqrt{1 + \sqrt{1 + \sqrt{1 + 8s}}}\sqrt{1 + \sqrt{1 + 8s}}\sqrt{1 + 8s}}$

33. $\frac{-2x - 3}{2x^2(4x + 3)^{1/2}}$ **35.** $\frac{2x - 6(2x - 3y)^2}{1 - 9(2x - 3y)^2}$ **37.** $\frac{5}{48}$

39. 13.6 **41.** $\frac{2(30x^7 - 1)}{x^3}$ **43.** $\frac{56}{(1 + 4t)^3}$

45. $\frac{1}{\sqrt{x}}$ **47.** 53.1°

49. It appears to be 8, but using *trace*, there is no value shown for $x = 2$.

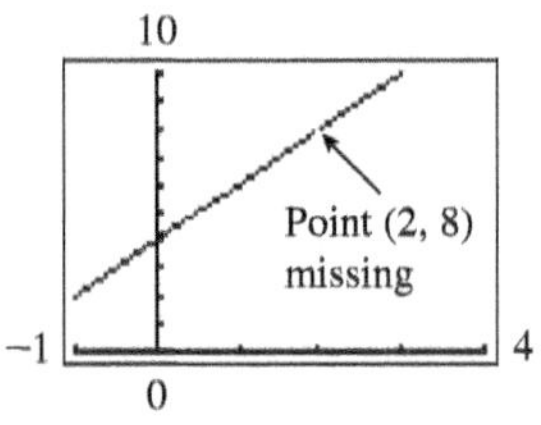

51. (a) 30.0 m/s (b) 6.00 m/s **53.** -31.0

55. $\left(0, \frac{1}{3}\sqrt{3}\right)$ **57.** $10\,000(1 + 0.250i)^7$

59. (a) $v = \frac{4}{(1 + 8t)^{1/2}}$ (b) $a = \frac{-16}{(1 + 8t)^{3/2}}$

61. 5 **63.** $-k + k^2t - \frac{1}{2}k^3t^2$ **65.** $-\frac{2k}{r^3}$

67. $0.4(0.01t + 1)^2(0.04t + 1)$ **69.** $\frac{2R(R + 2r)}{3(R + r)^2}$

71. 833 W **73.** $5k(x^4 + 270x^2 - 190)$

75. $-\frac{1}{4\pi\sqrt{C}(L + 2)^{3/2}}$ **77.** $\frac{-15}{(0.5t + 1)^2}$

79. $y' = \frac{w}{6EI}(3L^2x - 3Lx^2 + x^3)$

$y'' = \frac{w}{2EI}(L - x)^2$

$y''' = \frac{w}{EI}(x - L)$

$y^{iv} = \frac{w}{EI}$

81. $p = 2w + \frac{150}{w}, \frac{dp}{dw} = 2 - \frac{150}{w^2}$

83. $A = 4x - x^3, \frac{dA}{dx} = 4 - 3x^2$

85. 397 km/h

Chapter 24: Applications of the Derivative

Exercises 24.1, page 729

1. $x + 8y - 17 = 0$

3. $4x - y - 2 = 0$

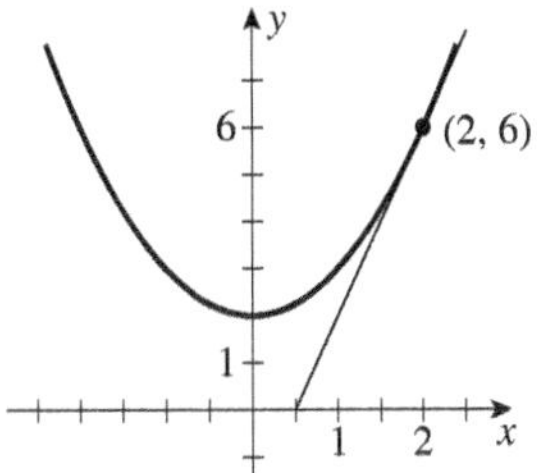

5. $2y + x - 2 = 0$

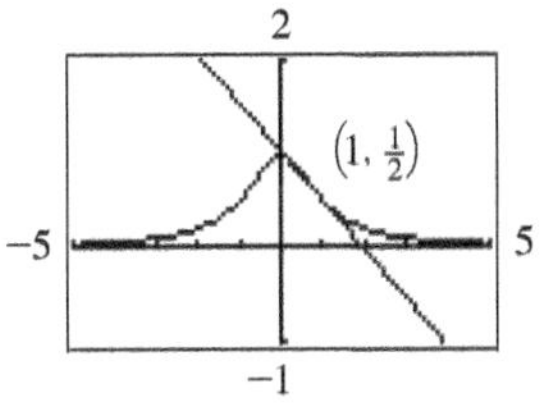

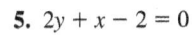

7. $x - 2y + 6 = 0$

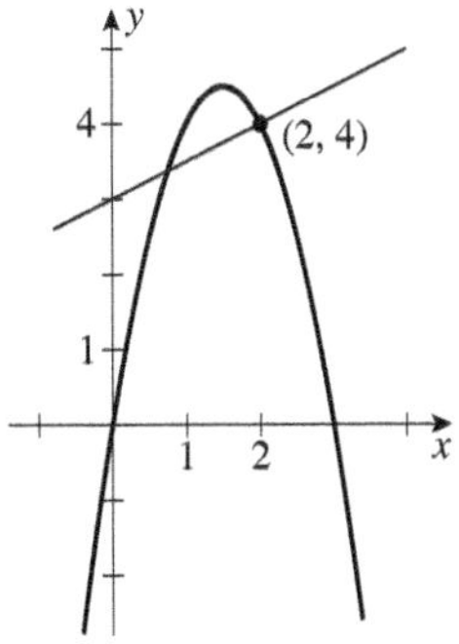

9. $2x - 6y + 7 = 0$

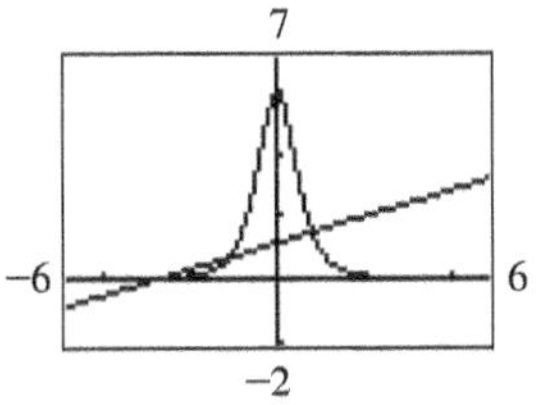

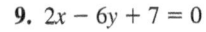

11. $y = 2x - 4$

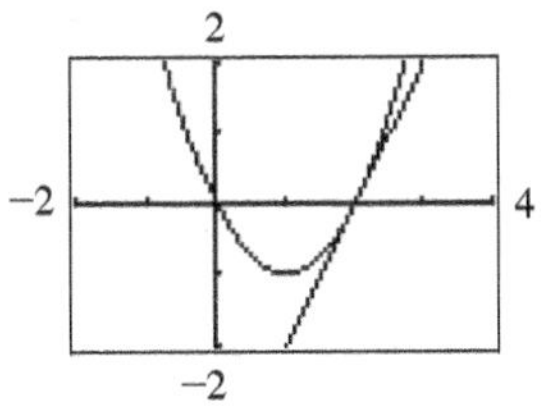

13. $y - 8 = -\frac{1}{24}\left(x - \frac{3}{2}\right)$, or $2x + 48y - 387 = 0$

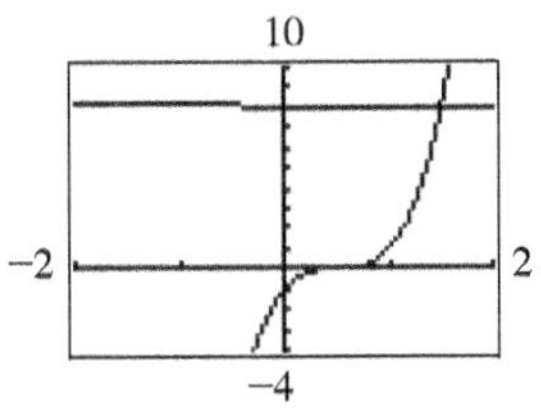

15. $2x - 12y + 37 = 0$

$72x + 12y + 37 = 0$

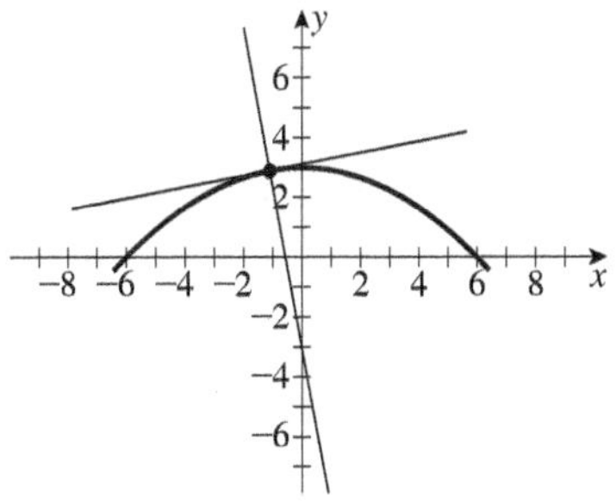

17. The line is $y - x - 1 = 0$.

19. Take derivatives; evaluate at (a, b); show that product $m_1m_2 = -4a/b^2 = -1$.

21. $y - y_1 = -\frac{x_1}{y_1}(x - x_1)$; $y_1y - y_1^2 = -x_1x + x_1^2$;

$x_1x + y_1y = x_1^2 + y_1^2 = a^2$

23. $x - 4y + 2 = 0$ **25.** $x - 2y - 20 = 0$

27. $x + y - 6 = 0$

29. $x + 2y - 3 = 0$, $x = 0$, $x - 2y + 3 = 0$

Exercises 24.2, page 733

1. $x_3 = 0.2086$; calculator: 0.2087 (to three decimal places)

3. −0.180 460 4 **5.** 0.344 56 **7.** −3.7321

9. 2.561 6 **11.** −1.236 068 **13.** 0.917 543 3

15. 0.618 034 0 **17.** −1.855 772 5, 0.678 362 8, 3.177 409 7

19. Find the real root of $x^3 - 4 = 0$; 1.587

21. $x_{n+1} = \frac{1}{2}x_n + \frac{a}{2x_n}$ **23.** $x_n = 2x_{n-1} - ax_{n-1}^2, n \geq 2$

25. 4.21 cm **27.** 2.214 μF, 3.214 μF, 4.214 μF

29. 0.629 cm **31.** 1.00×10^{-4}

Exercises 24.3, page 737

1. 16.5, −14.0°

3. 3.16, 341.6°

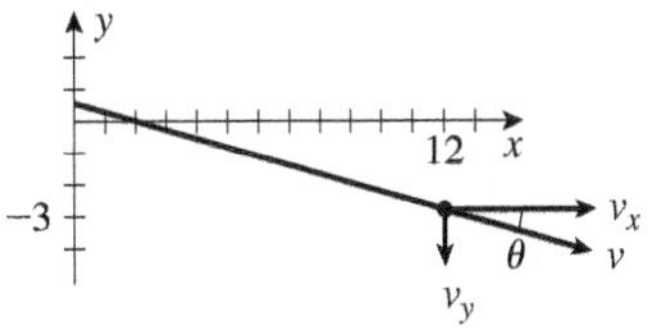

5. 8.07, 352.4°

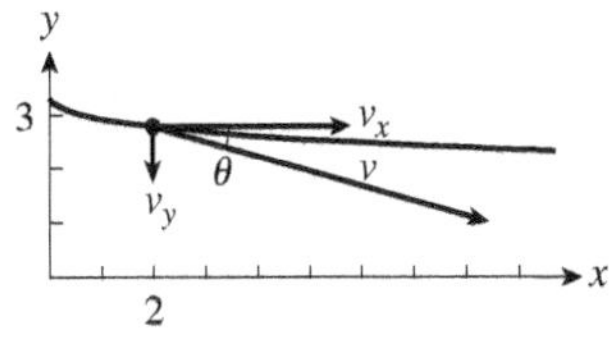

7. $a = 0$ **9.** 20.0, 3.68° **11.** 7.81 m/s, 310°

13. 1.3 m/min², 288° **15.** 39 m/s, 310° and 7.4 m/s², 270°

17. 276 m/s, 43.5°; 2090 m/s, 16.7° **19.** 1.32 cm/s, −24.9°

21. 22.1 m/s², 25.4°; 20.2 m/s², 8.5° **23.** 21.2 km/min, 296.6°

25. $v_x = 1460$ m/min, $v_y = -1380$ m/min **27.** 370 m/s, 19°

Exercises 24.4, page 741

1. 0.0265 m/min **3.** 38 **5.** 6 **7.** 0.031 units/s
9. 0.36 m/s^2 **11.** 0.0900 Ω/s **13.** −0.002 51 s/s
15. 330 km/h **17.** 4.1×10^{-6} m/s **19.** $\dfrac{dB}{dt} = \dfrac{-3kr(dr/dt)}{[r^2 + (l/2)^2]^{5/2}}$
21. 0.0056 m/min **23.** 0.15 mm^2/month **25.** −101 mm^3/min
27. $\dfrac{dV}{dt} = -kA; \dfrac{dr}{dt} = -k$ **29.** −4.6 kPa/min
31. 10.8 cm/min **33.** $I = \dfrac{8k}{x^2}; \dfrac{dI}{dt} = 0.000\,800k$ units/s
35. −0.43 N/s **37.** 0.481 m/min **39.** 820 km/h
41. 2.71 m/s **43.** 2.50 m/s

Exercises 24.5, page 748

1. Inc. $x < 0, x > 4$, dec. $0 < x < 4$
3. Conc. down $x < 0$, conc. up $x > 0$, infl. $(0, 0)$
5. Inc. $x > -1$, dec. $x < -1$
7. Inc. $x < -3, x > 2$; dec. $-3 < x < 2$
9. Min. $(-1, -1)$
11. Max. $(-3, 71)$; min. $(2, -54)$
13. Conc. up all x
15. Conc. up, $x > -\frac{1}{2}$; conc. down, $x < -\frac{1}{2}$; infl. $\left(-\frac{1}{2}, \frac{17}{2}\right)$

17.

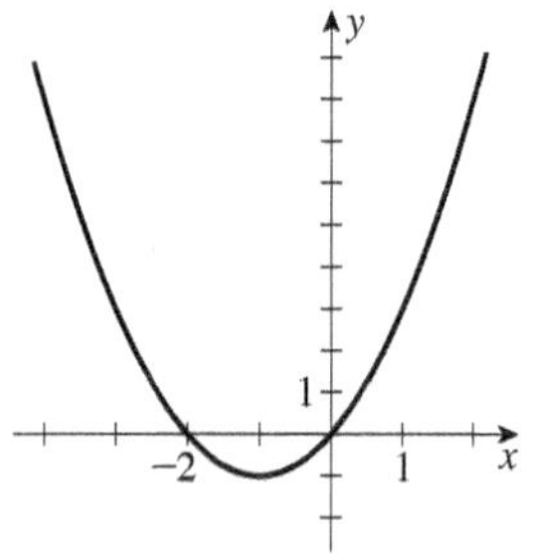

19.

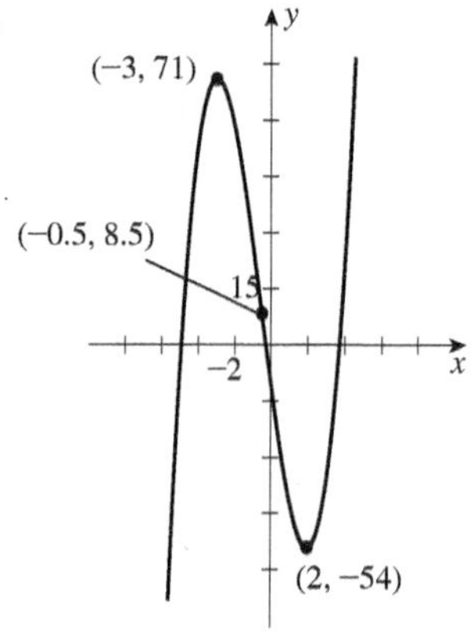

21. Max. $(3, 18)$, conc. down all x

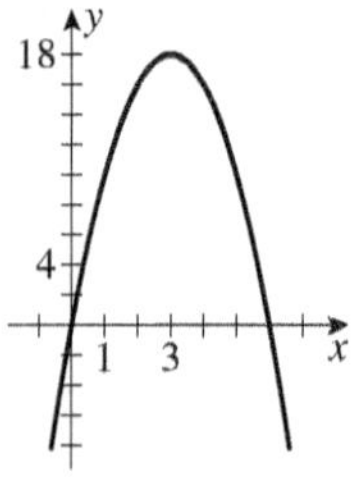

23. Max. $(-2, 3)$, min. $(0, -5)$, infl. $(-1, -1)$

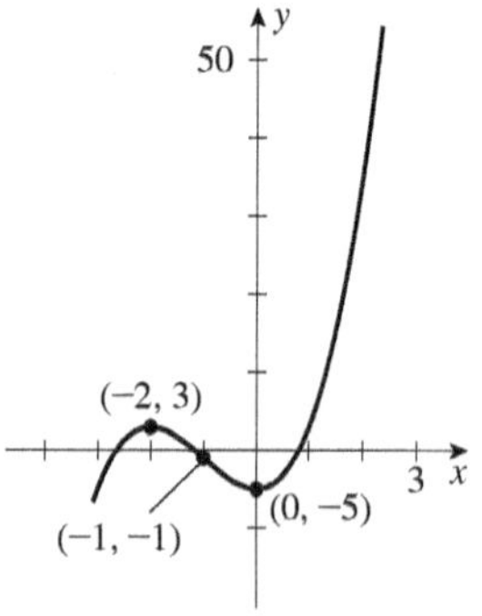

25. No max. or min., infl. $(-1, 1)$

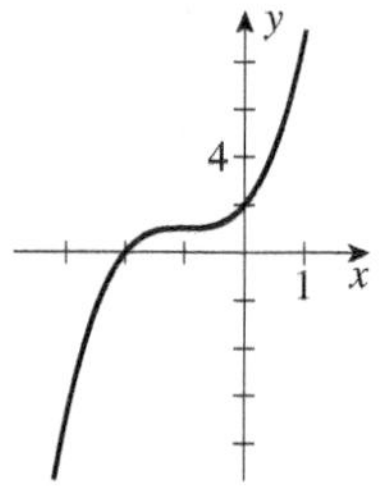

27. Max. $(1, 16)$, min. $(3, 0)$, infl. $(2, 8)$

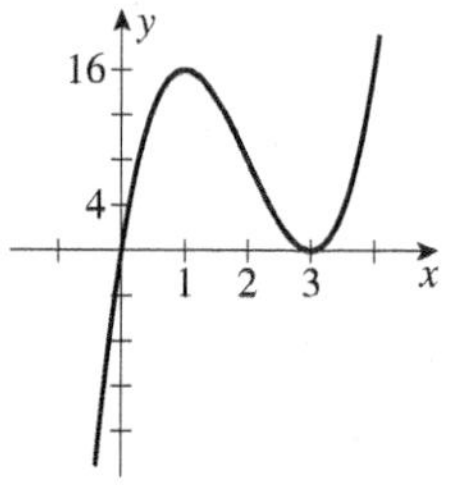

29. Max. $(1, 7)$, infl. $(0, 6)$, $\left(\frac{2}{3}, \frac{178}{27}\right)$

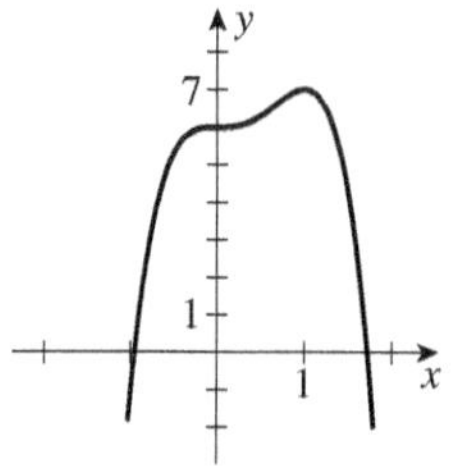

31. Max. $(-1, 4)$, min. $(1, -4)$, infl. $(0, 0)$

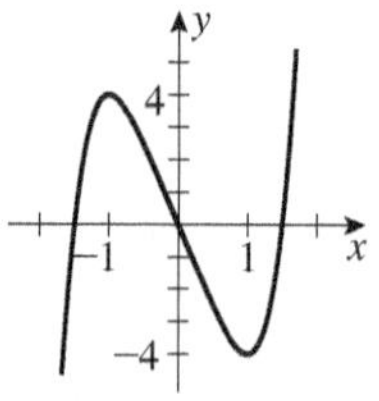

33. Where $y' > 0$, y inc.
$y' = 0$, y has a max. or min.
$y' < 0$, y dec.
$y'' > 0$, y conc. up
$y'' = 0$, y has infl.
$y'' < 0$, y conc. down

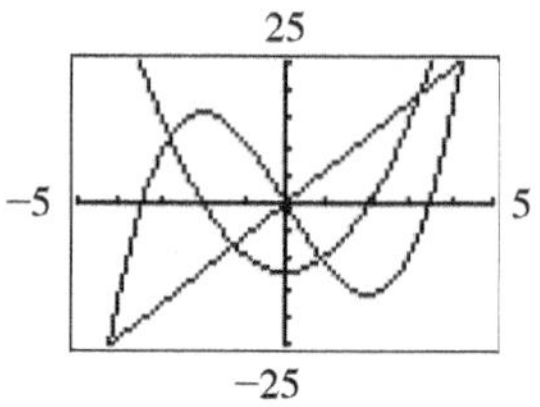

35. Curve has a maximum and a minimum for $c < 0$ but is always increasing for $c > 0$.

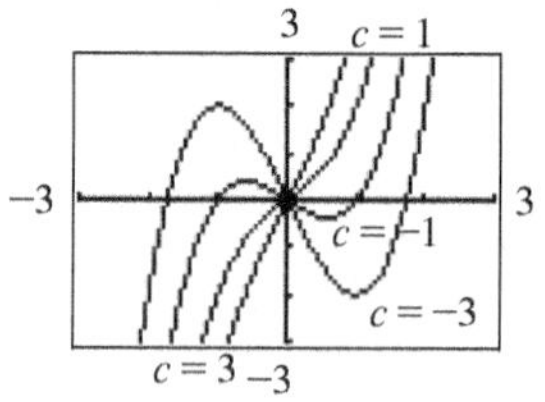

37. The left relative maximum point is above the left relative minimum point but below the right relative minimum point.

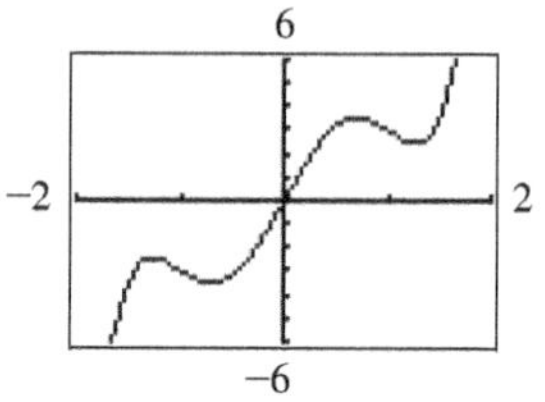

39. Max. $(20, 10)$

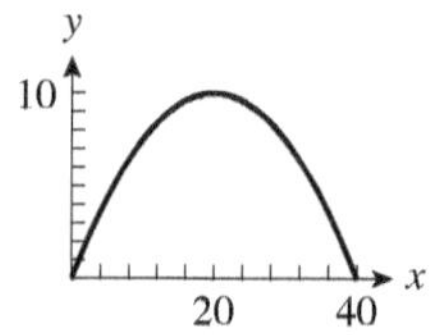

41. Max. $(4, 8)$

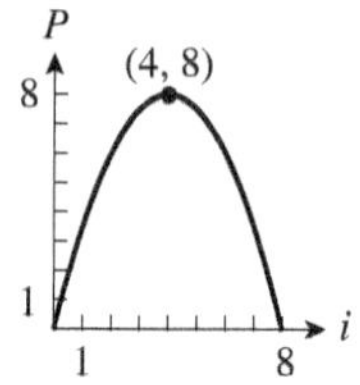

43. Max. $(3, 16)$

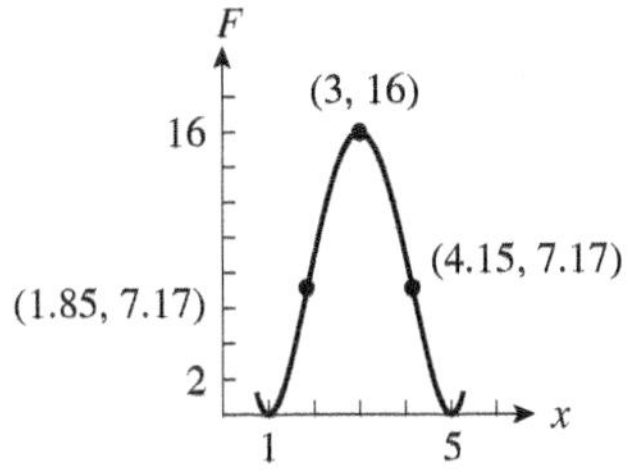

45. Max. $(0, 75)$, infl. $(1, 64)$, $(3, 48)$

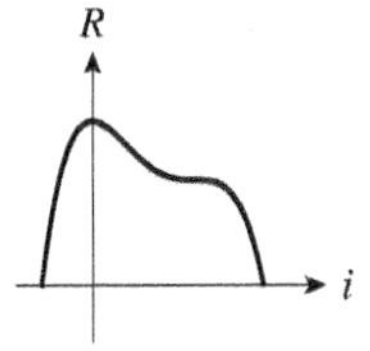

47. Max. $(40, 41\,620)$

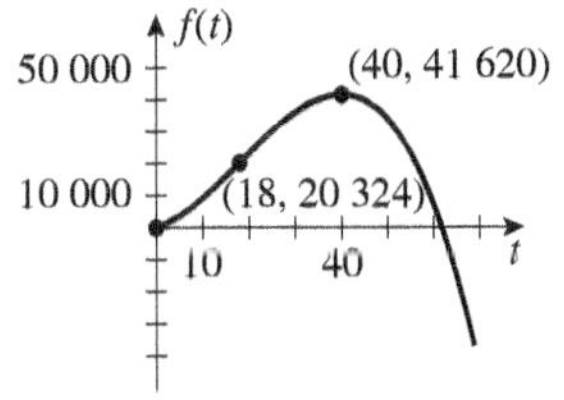

49. $V = 4x^3 - 40x^2 + 96x$, max. $(1.57, 67.6)$, infl. $\left(\frac{10}{3}, 23.7\right)$

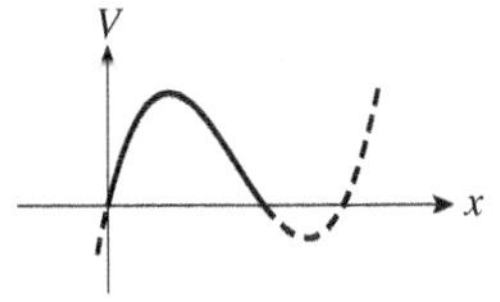

51.

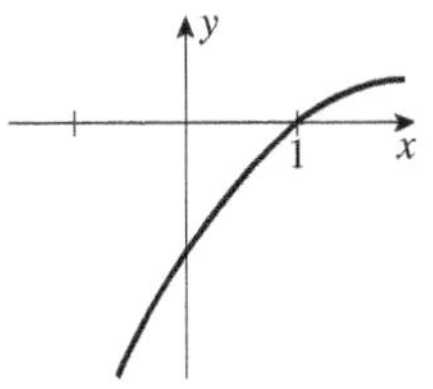

53.

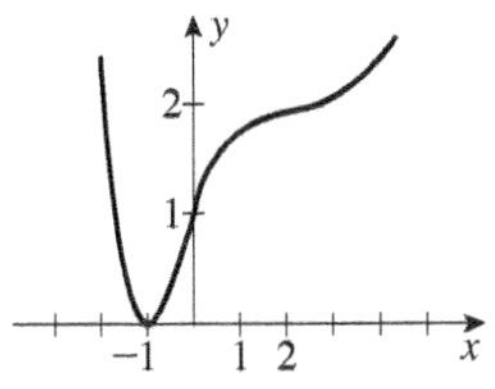

Exercises 24.6, page 753

1. $y = x - \dfrac{4}{x}$, int. $(2, 0)$, $(-2, 0)$; inc. all x, except $x = 0$; conc. up $x < 0$, conc. down $x > 0$

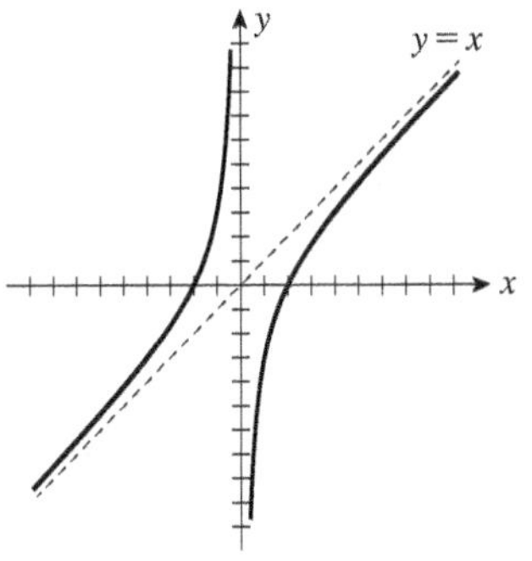

3. Dec. $x < -1$, $x > -1$, conc. up $x > -1$, conc. down $x < -1$, int. $(0, 2)$, asym. $x = -1$, $y = 0$

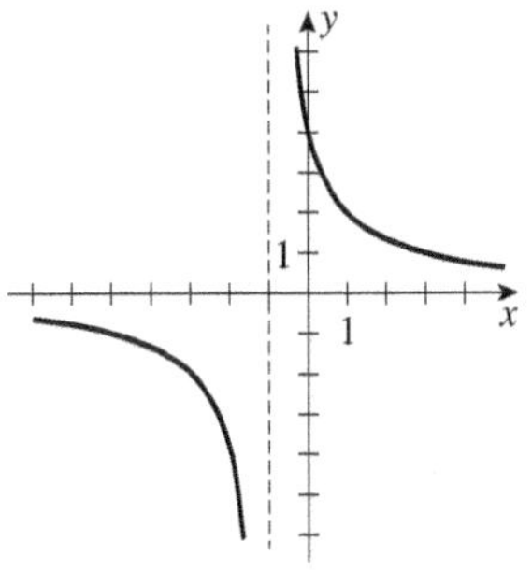

5. Int. $(-\sqrt[3]{2}, 0)$, min. $(1, 3)$, infl. $(-\sqrt[3]{2}, 0)$, asym. $x = 0$

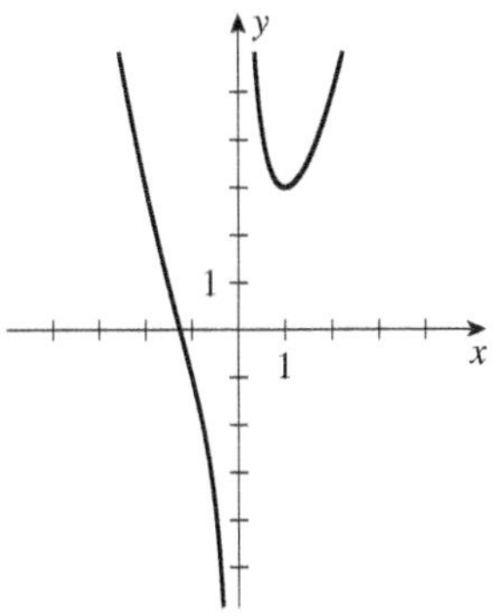

7. Int. $(1, 0)$, $(-1, 0)$, asym. $x = 0$, $y = x$, conc. up $x < 0$, conc. down $x > 0$

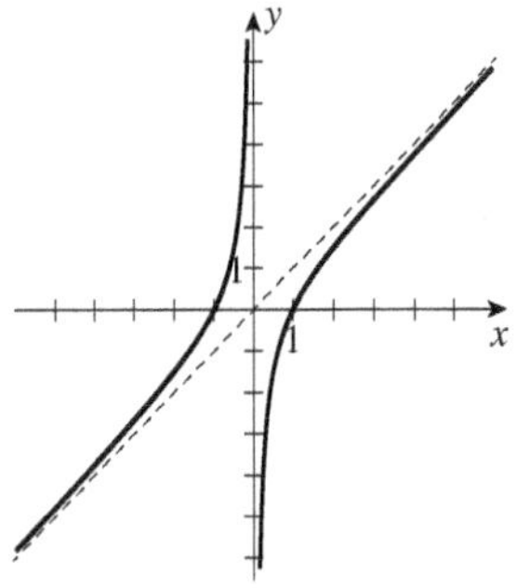

9. Int. $(0, 0)$, max. $(-2, -4)$, min. $(0, 0)$, asym. $x = -1$

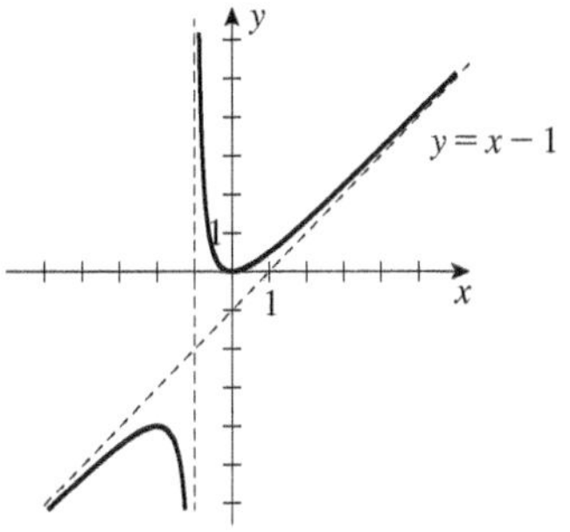

11. Int. $(0, -1)$, max. $(0, -1)$, asym. $x = 1$, $x = -1$, $y = 0$

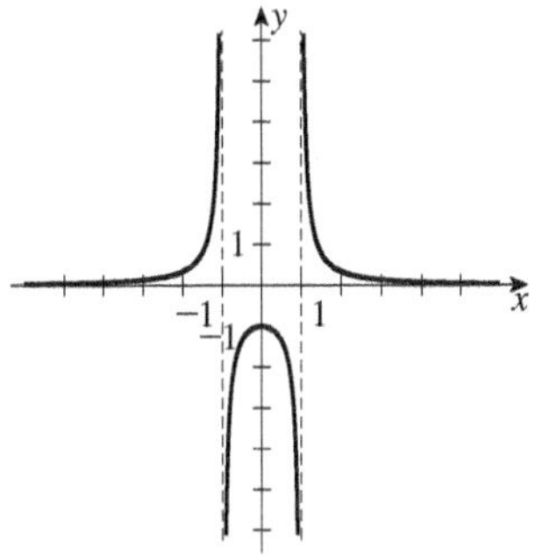

13. Int. $(1, 0)$, max. $(2, 1)$, infl. $\left(3, \frac{8}{9}\right)$, asym. $x = 0$, $y = 0$

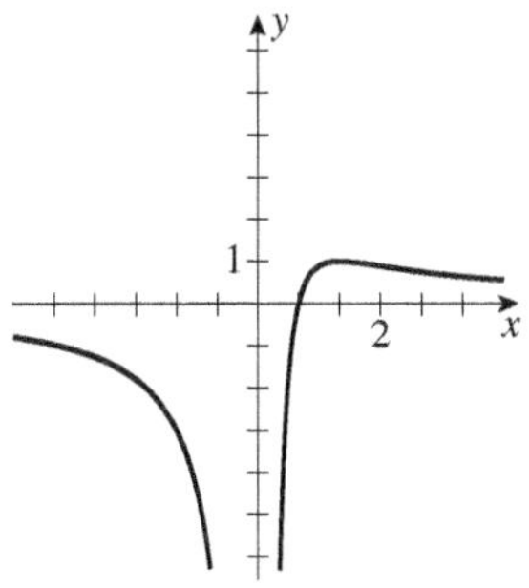

15. Int. $(0, 0)$, $(1, 0)$, $(-1, 0)$, max. $\left(\frac{1}{2}\sqrt{2}, \frac{1}{2}\right)$, min. $\left(-\frac{1}{2}\sqrt{2}, -\frac{1}{2}\right)$

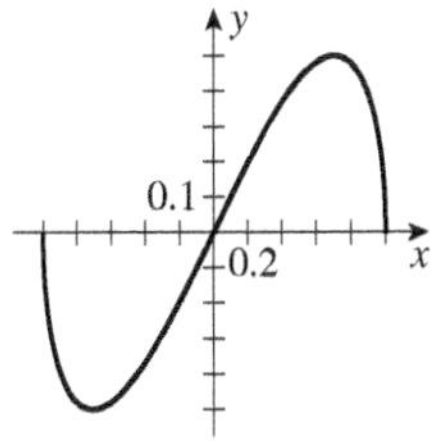

17. Int. $(0, 0)$, infl. $(0, 0)$, asym. $x = -3$, $x = 3$, $y = 0$

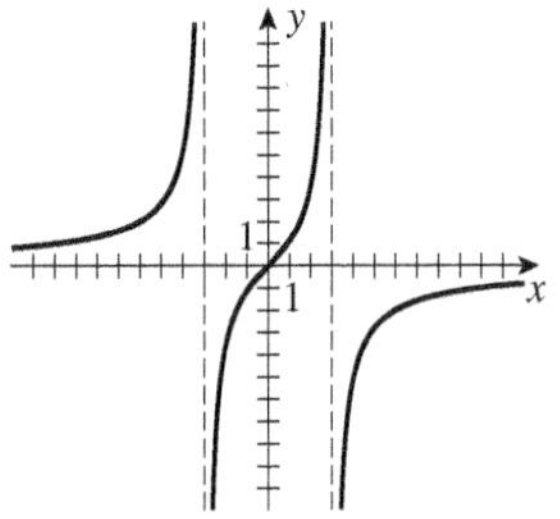

19. As c goes from -3 to $+3$, the graph goes from second and fourth quadrants to the third and first quadrants.

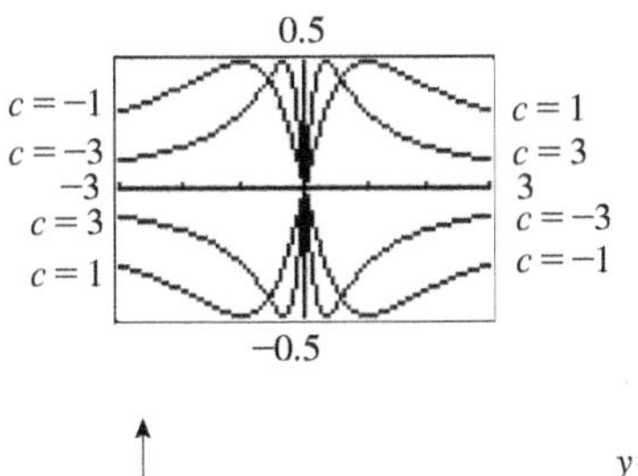

21.

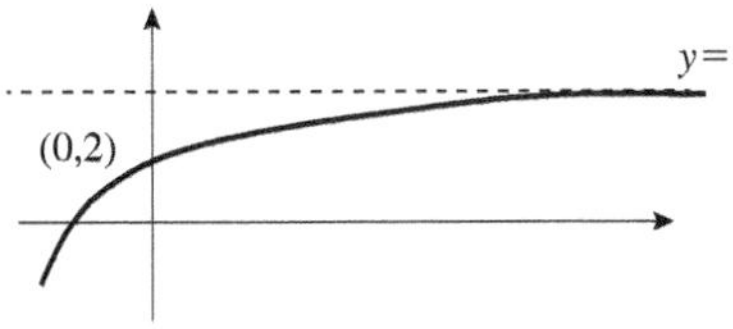

23. Dividing simplifies analysing behaviour as x becomes large and finding the range.

25. Int. $(0, 0)$, asym. $C_T = 6$, inc. $C \geq 0$, conc. down $C \geq 0$

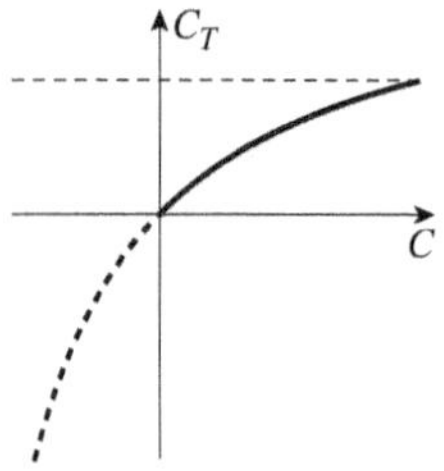

27. Int. $(0, 1)$, max. $(0, 1)$, infl. $(141, 0.82)$, asym. $R = 0$

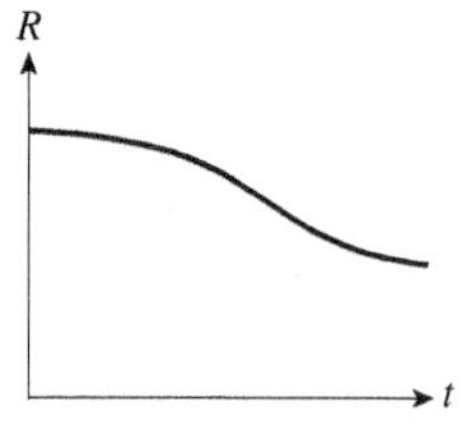

29. Int. $(1, 0)$, $(-1, 0)$; inc. all x, except $x = 0$; conc. up $x < 0$, conc. down $x > 0$

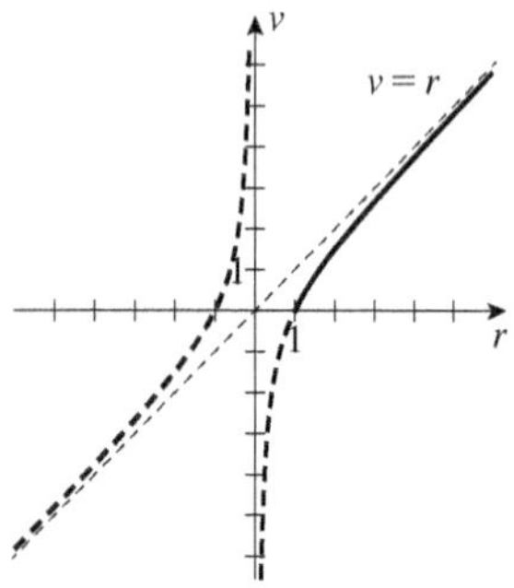

31. $A = 2\pi r^2 + \dfrac{40}{r}$, min. $(1.47, 40.8)$, asym. $r = 0$

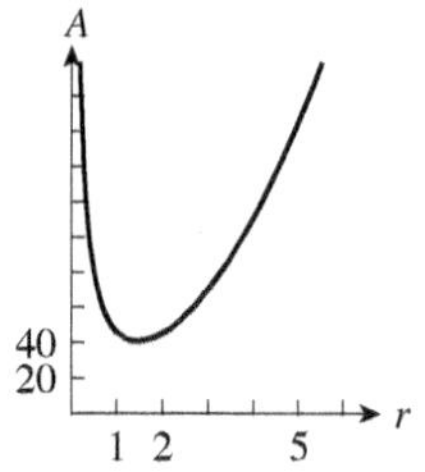

Exercises 24.7, page 758

1. 360 000 m² **3.** 60 m **5.** 1.2 A **7.** 25 m by 25 m **9.** 35 m², \$8300 **11.** $\dfrac{L}{R^2 + \omega^2 L^2}$ **13.** 1.94 units **15.** a **17.** 12 cm by 12 cm **19.** 6 mm, 6 mm **21.** $xy = A$, $p = 2x + 2y$; $p = 2x + \frac{2A}{x}$; $\frac{dp}{dx} = 2 - \frac{2A}{x^2}$; $x = \sqrt{A}$, $y = \sqrt{A}$; $x = y$ (square) **23.** 1.10 h **25.** 9.90 cm, 9.90 cm **27.** $r = 0.5$ cm, $l = 1$ cm, $V = 0.8$ m³ **29.** 24.0 cm, 36.0 cm **31.** 1.20 m **33.** 2 cm **35.** 12 **37.** $0.58L$ **39.** 38.0° **41.** 8/3 km from A **43.** 3.3 cm **45.** 100 m **47.** 1125 units per week **49.** 8.0 km from refinery **51.** 59.2 m, 118 m

Exercises 24.8, page 763

1. $\dfrac{-8(t^3 - 2)}{(t^3 + 4)^2}dt$ **3.** $\dfrac{dx}{x} = 0.0065 = 0.65\%$; $\dfrac{dV}{V} = 0.019 = 1.9\%$ **5.** $(5x^4 + 7)dx$ **7.** $\dfrac{-10dr}{r^6}$ **9.** $72t(3t^2 - 5)^5 dt$ **11.** $\dfrac{-72xdx}{(3x^2 + 1)^2}$ **13.** $x(1 - x)^2(-5x + 2)dx$ **15.** $\dfrac{2dx}{(5x + 2)^2}$ **17.** 12.28, 12 **19.** 0.3129, 0.3131 **21.** $L(x) = 2x$ **23.** $L(x) = -2x - 3$ **25.** 1570 km **27.** 0.246% **29.** −31.3 nm **31.** 1235 cm³ **33.** $\dfrac{dr}{r} = \dfrac{1}{2} \cdot \dfrac{d\lambda}{\lambda}$ **35.** $\dfrac{dA}{A} = \dfrac{2dr}{r}$ **37.** 3.008 33

39. $L(x) = 1^k + k(1^{k-1})x = 1 + kx$ **41.** $L(x) = -\frac{1}{2}x + \frac{3}{2}$; 1.45 **43.** $L(V) = 1.73 - 0.133V$

Review Exercises for Chapter 24, page 765

1. $5x - y + 1 = 0$

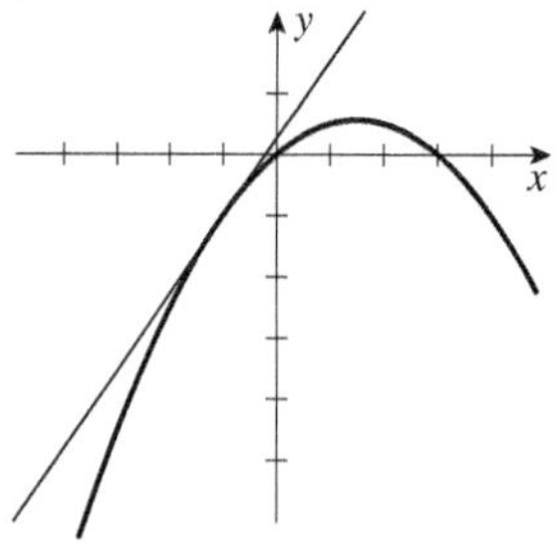

3. $8x + 5y - 50 = 0$

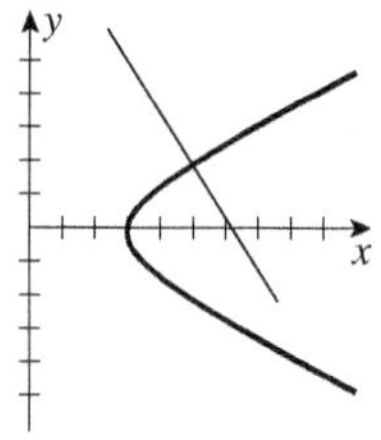

5. $x - 2y + 3 = 0$ **7.** 4.19, 72.6° **9.** 2.12 **11.** 2.00, 90.9° **13.** 0.745 9 **15.** 1.442 249 6

17. Min. $(-2, -16)$, conc. up all x

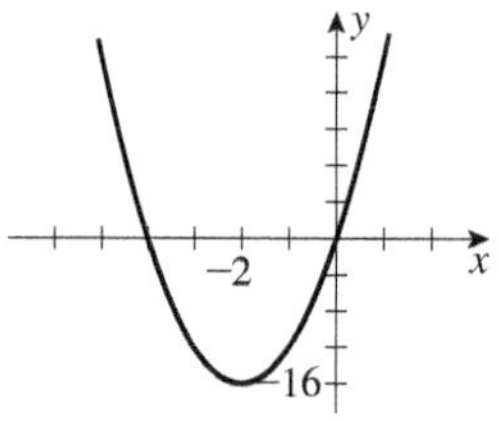

19. Int. $(0, 0)$, $(\pm 3\sqrt{3}, 0)$; max. $(3, 54)$, min. $(-3, -54)$; infl. $(0, 0)$

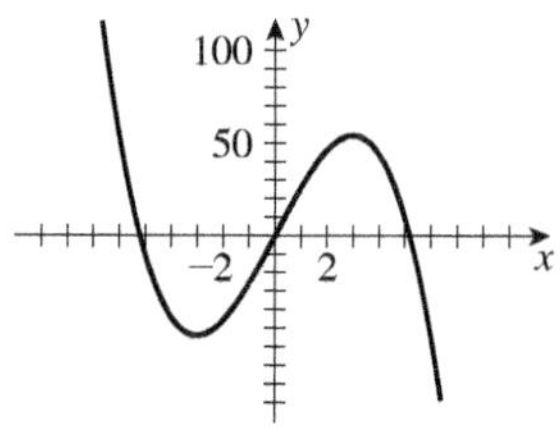

21. Min. $(2, -48)$; conc. up $x < 0$, $x > 0$, int. $(0, 0)$, $(2\sqrt[3]{4}, 0)$

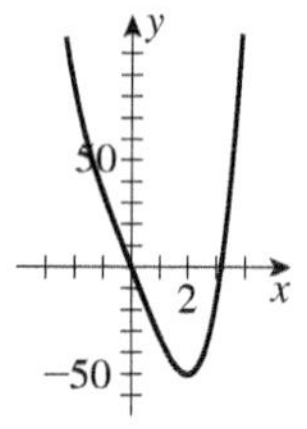

23. Min. $(\pm\sqrt{2}, 2)$; asym. $x = \pm 1$; conc. up $x < 1$, $x > 1$

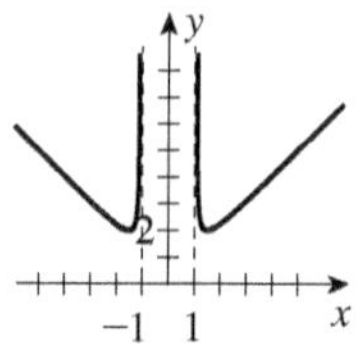

25. $\left(12x^2 - \dfrac{1}{x^2}\right)dx$ **27.** $\dfrac{(1 - 4x)dx}{(1 - 3x)^{2/3}}$ **29.** 0.244 **31.** $L(x) = \dfrac{1}{3}(11x - 4)$ **33.** 1.85 m³ **35.** 45.2 m³ **37.** $\dfrac{RdR}{R^2 + X^2}$ **39.** 251.1 **41.** $2x - y + 1 = 0$ **43.** 0.0 m, 6.527 m **45.** 8.8 m/s, 336° **47.** 0.23 m/s **49.** −7.44 cm/s

51.

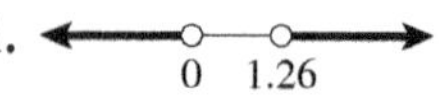

53. 22 m³/s

55.

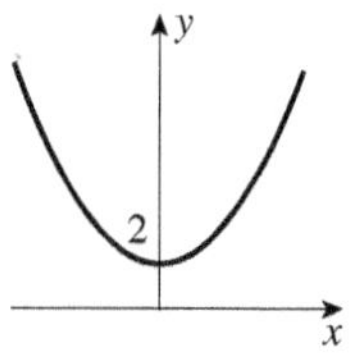

57. −0.113 m/min **59.** 38 000 m²/min **61.** 7.30 ft/s

63. Max. (0, 100); infl. (37, 63); int. (0, 100), (89, 0), $L(x) = 113 - 1.42x$

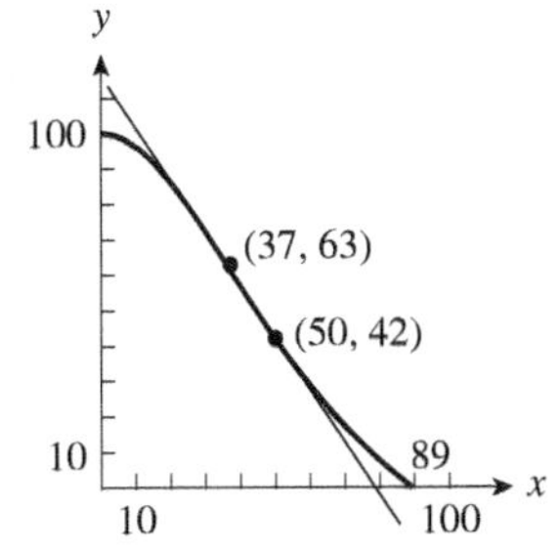

65. 1160 km/h **67.** 6 μF, 6 μF **69.** 13.5 dm^3

71. 3 km **73.** 6.56 cm **75.** 0.153 m/min

77. $r = 2.09$ cm; $h = 23.0$ cm

79. $l = 4.67$ m, $w = 3.50$ m, $h = 2.34$ m

Chapter 25: Integration

Exercises 25.1, page 771

1. $F(x) = 3x^4$ **3.** $F(x) = \frac{2}{3}x^{3/2} + \frac{1}{x^3}$ **5.** 1 **7.** 3

9. 6 **11.** -1 **13.** $\frac{3}{4}x^4$ **15.** $\frac{3}{2}t^4 + 15t$ **17.** $\frac{2}{3}x^3 - \frac{1}{2}x^2$

19. $\frac{4}{3}x^{3/2} + 3x$ **21.** $\frac{7}{5x^5} + \frac{1}{9}x$ **23.** $2v^2 + 3\pi^2 v$

25. $\frac{1}{3}x^3 - 4x - \frac{1}{x}$ **27.** $(2x + 1)^6$ **29.** $\frac{1}{3}(p^2 - 1)^{12}$

31. $\frac{1}{80}(2x^4 + 1)^{10}$ **33.** $(6x + 1)^{3/2}$ **35.** $\frac{1}{4}(3x + 1)^{4/3}$

37. $\frac{1}{2x + 1}$

39. $(2x + 5)^3$ is an antiderivative of $6(2x + 5)^2$. Factor of 2 from $2x + 5$ is needed.

Exercises 25.2, page 776

1. $4x^2 + C$ **3.** $\frac{(x^2 + 1)^5}{5} + C$ **5.** $x^2 + C$ **7.** $\frac{1}{14}x^{14} + C$

9. $\frac{1}{20}x^{5/2} + C$ **11.** $-\frac{3}{R^3} + C$ **13.** $\frac{1}{3}x^3 - \frac{1}{6}x^6 + C$

15. $3x^3 + \frac{1}{2}x^2 + 3x + C$ **17.** $\frac{1}{6}t^3 + \frac{2}{t} + C$

19. $\frac{2}{7}x^{7/2} - \frac{2}{5}x^{5/2} + C$ **21.** $6x^{1/3} + \frac{1}{9}x + C$

23. $x + 8x^3 + \frac{144}{5}x^5 + C$ **25.** $\frac{1}{10}(x^2 - 1)^{10} + C$

27. $\frac{1}{5}(x^4 + 3)^5 + C$ **29.** $\frac{1}{80}(2\theta^5 + 5)^8 + C$

31. $-\frac{1}{9}(-6x + 1)^{3/2} + C$ **33.** $\frac{1}{6}\sqrt{6x^2 + 1} + C$

35. $4\sqrt{z^2 - 2z} + C$ **37.** $y = 2x^3 + 2$

39. $y = 5 - \frac{1}{18}(1 - x^3)^6$

41. No. The result should include the constant of integration.

43. No. With $u = 4x^3 + 3$, $du = 12x^2dx$, and the x^4 would have to be x^2.

45. No. There should be a factor of 1/2 in the result.

47. $12y = 83 + (1 - 4x^2)^{3/2}$ **49.** $2x^2 - 5x + 3$

51. $\frac{4}{15}x^{5/2} + C_1x + C_2$ **53.** $V = 0.14t + 0.000\,14t^2$

55. $i = 2t^2 - 0.2t^3 + 2$ **57.** $T = 2250(r + 1)^{-2} + 250$

59. $f = \sqrt{0.01A + 1} - 1$ **61.** $y = 3x^2 + 2x - 3$

Exercises 25.3, page 780

1. (a) 3 (b) $\frac{15}{4}$ **3.** 20 **5.** 9, 12.2 **7.** 1.92, 2.28

9. 7.62, 8.21 **11.** 0.464, 0.599 **13.** 1.92, 1.96

15. 13.5 **17.** $\frac{8}{3}$ **19.** 9 **21.** 0.8 **23.** 2

25. 14.85; the extra area above $y = 3x$ using circumscribed rectangles is the same as the omitted area under $y = 3x$ using inscribed rectangles.

27. 3.08; the extra area above $y = x^2$ using circumscribed rectangles is greater than the omitted area under $y = x^2$ using inscribed rectangles.

Exercises 25.4, page 783

1. $-\frac{9}{4}$ **3.** 27 **5.** $\frac{254}{7}$ **7.** $2\sqrt{6} - 2\sqrt{3} - 21 = -19.57$

9. 2.53 **11.** $-\frac{32}{3}$ **13.** $\frac{33}{20}\sqrt[3]{\frac{11}{5}} - \frac{3}{8}\sqrt[3]{\frac{1}{2}} - \frac{17}{5} = -1.55$ **15.** $\frac{4}{3}$

17. $-\frac{243}{5}$ **19.** 2 **21.** $\frac{1}{4}(20.5^{2/3} - 17.5^{2/3}) = 0.188$

23. $\frac{176}{1083} = 0.162$ **25.** 84 **27.** $\frac{364}{3}$ **29.** $\frac{33}{784} = 0.0421$

31. $\frac{3880}{9}$ **33.** $\frac{2}{3}(13\sqrt{13} - 27) = 13.2$

35. $\frac{1}{4} + \frac{15}{4} = 4$; under $y = x^3$, the area from $x = 0$ to $x = 1$ plus the area from $x = 1$ to $x = 2$ equals the area from $x = 0$ to $x = 2$.

37. $\frac{28}{3}$

39. $\frac{1}{2} > \frac{1}{3}$ and $\frac{3}{2} < \frac{7}{3}$; from $x = 0$ to $x = 1$ the area under $y = x$ is greater than the area under $y = x^2$. Between $x = 1$ and $x = 2$, the area under $y = x$ is less than the area under $y = x^2$.

41. $\frac{2}{2k + 1}$ **43.** $\frac{A\sqrt{2}}{a\sqrt{g}}(\sqrt{h} - \sqrt{H})$ **45.** 10

47. 64 000 N · m **49.** 86.8 m^2 **51.** $\frac{3NE_F}{5}$

Exercises 25.5, page 787

1. $\frac{7}{6}$ **3.** $\frac{11}{2} = 5.50$, $\frac{16}{3} = 5.33$ **5.** 7.66, $\frac{23}{3} = 7.67$ **7.** 0.204

9. 19.0 **11.** 0.520 **13.** 21.7 **15.** 45.4

17. The tops of all trapezoids are below $y = 1 + \sqrt{x}$.

19. 0.702, ln 2 = 0.693 **21.** 100.027 m

Exercises 25.6, page 791

1. 0.406 **3.** (a) 6 (b) 6 **5.** (a) 19.7 (b) 19.7 **7.** 19.3

9. 0.511 **11.** 13.1 **13.** 44.6 **15.** 3.12 **17.** 1.19 cm

Review Exercises for Chapter 25, page 792

1. $x^4 - \frac{1}{2}x^2 + C$ **3.** $\frac{2}{7}u^{7/2} + \frac{16}{3}u^{3/2} + C$ **5.** $\frac{19}{3}$ **7.** $\frac{16}{3}$

9. $5x - \frac{1}{2x^{12}} + C$ **11.** 3 **13.** $\frac{1}{6(8 - 3n)^2} + C$

15. $-\frac{6}{7}(7 - 2x)^{7/4} + C$ **17.** $\frac{9}{8}(3\sqrt[3]{3} - 1)$

19. $-\frac{1}{60}(1 - 2x^3)^{10} + C$ **21.** $-\frac{1}{2x - x^3} + C$ **23.** $\frac{3350}{3}$

25. $y = 3x - \frac{1}{3}x^3 + \frac{17}{3}$

27. (a) $x - x^2 + C_1$

(b) $-\frac{1}{4}(1 - 2x)^2 + C_2 = x - x^2 + C_2 - \frac{1}{4}$; $C_1 = C_2 - \frac{1}{4}$

29. $\int_3^4 F(v)dv$ **31.** $\frac{1}{27}(6x + 5)^{3/2} + C_1x + C_2$

33. $0.25 = 0.25$; $y = x^3$ shifted 1 unit to the right is $y = (x - 1)^3$. Therefore, areas are the same.

35. 22 **37.** The graph of $f(x)$ is concave down. This means the tops of the trapezoids are below $y = f(x)$.

39. 0.842 **41.** 0.811 **43.** 13.6 **45.** 19.30 **47.** 19.04

49. 25.7 m^2 **51.** 25.8 m^2 **53.** − 14.4 m

55. $Q = \dfrac{k}{12R}(4r^3R - 3r^4 - R^4) + Q_0$ **57.** 66.4 m

Chapter 26: Applications of Integration

Exercises 26.1, page 798

1. -19.6 m/s **3.** -157 m/s **5.** $s = 8.00 - 0.25t$

7. 5.0 m/s **9.** 263 m **11.** 12 m/s^2 **13.** 24 m/s

15. 85 m **17.** 0.345 nC **19.** 0.015 C **21.** 160 V

23. 4.68 mV **25.** 970 rad **27.** 67 A **29.** $\dfrac{k}{x_1}$

31. $m = 1002 - 2\sqrt{t+1}$, 2.51×10^5 min

Exercises 26.2, page 804

1. $\frac{26}{3}$ **3.** 2 **5.** $\frac{64}{3}$ **7.** $\frac{32}{3}$ **9.** $\frac{1}{6}$ **11.** $\frac{26}{3}$

13. 3 **15.** $\frac{9}{2}$ **17.** $\frac{8}{3}$ **19.** $\frac{7}{6}$ **21.** $\frac{256}{15}$ **23.** $\frac{343}{24}$

25. $\frac{65}{6}$ **27.** 4 **29.** $\frac{2}{9}$

31. The area bounded by $x = 1$, $y = 2x^2$, and $y = x^3$ or the area bounded by $x = 1$, $y = 0$, and $y = 2x^2 - x^3$.

33. 20 **35.** $\dfrac{A_1}{A_2} = \dfrac{1 - \frac{1}{n+1}}{\frac{1}{n+1}} = \dfrac{n}{n+1}\dfrac{n+1}{1} = \dfrac{n}{1}$

37. $4^{2/3}$ **39.** 1 **41.** $\frac{48}{5}$ **43.** 18 J **45.** 80.8 km

47. 4 cm^2 **49.** 42.5 dm^2

Exercises 26.3, page 810

1. $\frac{128\pi}{7}$ **3.** $\frac{8}{3}\pi$ **5.** $\frac{8}{3}\pi$ **7.** 9π **9.** 72π

11. $\frac{768}{7}\pi$ **13.** $\frac{348}{5}\pi$ **15.** $\frac{16}{3}\pi$ **17.** $\frac{128}{7}\pi$ **19.** $\frac{243}{80}\pi$

21. $\dfrac{10\pi}{3}\sqrt{5}$ **23.** $\frac{1296}{5}\pi$ **25.** $\frac{16}{3}\pi$

27. The region bounded by $y = x^{3/2}$, $y = 0$, $x = 1$, and $x = 2$

29. $\frac{128}{3}\pi$ **31.** $\frac{1}{3}\pi r^2 h$ **33.** $\frac{1}{5}$ **35.** 16 800 m^3

37. 4.2×10^6 mm^3 **39.** 18.3 cm^3

Exercises 26.4, page 816

1. $\left(0, \frac{8}{3}\right)$ **3.** 2.9 cm **5.** 1.2 cm **7.** (-0.5 cm, 0.5 cm)

9. (0.32 cm, 0.23 cm) **11.** $\left(0, \frac{6}{5}\right)$ **13.** $\left(\frac{4}{3}, \frac{4}{3}\right)$

15. (1.20, 5.49) **17.** $\left(\frac{5}{3}, 4\right)$ **19.** $\left(\frac{4}{3}, \frac{8}{3}\right)$ **21.** $\left(0, \frac{3}{5}a\right)$

23. $\left(\frac{7}{8}, 0\right)$ **25.** $\left(0, \frac{5}{6}\right)$ **27.** $\left(\frac{2}{3}, 0\right)$ **29.** $\left(2, \dfrac{3}{2}\right)$

31. 0.375 cm above centre of base **33.** 19.3 cm from larger base

Exercises 26.5, page 821

1. $I_y = \rho$, $R_y = \dfrac{1}{2}\sqrt{2}$ **3.** 68 g · cm^2, 2.9 cm

5. 2530 g · cm^2, 3.58 cm **7.** $\frac{2}{3}\rho$ **9.** $\frac{162}{5}\rho$ **11.** $\sqrt{6}$

13. $\frac{1}{6}mb^2$ **15.** $\frac{9}{7}\sqrt{7}$ **17.** $\frac{8}{11}\sqrt{55}$ **19.** $\frac{16}{3}\pi\rho$

21. $\frac{4}{5}\sqrt{10}$ **23.** $\frac{3}{10}mr^2$ **25.** 0.324 g · cm^2

27. 31.2 kg · cm^2

Exercises 26.6, page 826

1. 41 N · cm **3.** 176 kN **5.** 8.0 N · cm

7. 32.4 N · cm **9.** 1.7×10^{-16} J **11.** $0.09k$ N · m

13. 50 000 N · m **15.** 3.00×10^6 kN · m

17. 9.37×10^5 N · km **19.** 9.8×10^5 N · m

21. 12.5 kN **23.** 152 kN **25.** 1.72×10^5 N

27. 1.84 MN **29.** 3.92×10^4 N, 1.18×10^5 N; buoyant force

31. 2.7 A **33.** 35.3% **35.** 109 m **37.** $S = \pi r\sqrt{r^2 + h^2}$

Review Exercises for Chapter 26, page 829

1. 0.784 m **3.** 4.7 s **5.** No **7.** 0.44 C **9.** 55 V

11. $y = 20x + \frac{1}{120}x^3$ **13.** 18 **15.** 18 **17.** $\frac{27}{4}$

19. $A_{top}/A_{bot} = \left(\dfrac{2n}{2n+1}\right)\Big/\left(\dfrac{1}{2n+1}\right) = \dfrac{2n}{1}$

21. $\frac{48}{5}\pi$ **23.** $\frac{512}{5}\pi$ **25.** $\frac{4}{3}\pi ab^2$ **27.** (-0.5 cm, 0.6 cm)

29. $\left(\frac{30}{7}, \frac{45}{4}\right)$ **31.** $\left(\frac{14}{5}, 0\right)$ **33.** $\frac{8}{5}\rho$ **35.** 68.7 g · mm^2

37. 2700 N · m **39.** 4790 cm^2 **41.** -18 m/s

43. 1.8 m **45.** 47 m^3 **47.** 88 m^3 **49.** 1580 kN

51. 0.29 Ω **53.** 5×10^6 m/s **55.** 0.1359

Chapter 27: Differentiation of Transcendental Functions

Exercises 27.1, page 836

1. $8\theta \sin 2\theta^2 \cos 2\theta^2 = 4\theta \sin 4\theta^2$ **3.** $\cos(x + 8)$

5. $12x^2\cos(2x^3 - 1)$ **7.** $-3\sin\frac{1}{2}x$ **9.** $-6\sin(3x - \pi)$

11. $6\pi\sin 3\pi\theta\cos 3\pi\theta = 3\pi\sin 6\pi\theta$

13. $-45\cos^2(5x + 2)\sin(5x + 2)$

15. $\sin 3x + 3x\cos 3x$ **17.** $15x^4\cos 5x - 15x^5\sin 5x$

19. $6v\cos(5v)\cos v^2 - 15\sin(5v)\sin v^2$ **21.** $\dfrac{2\cos 4x}{\sqrt{1 + \sin 4x}}$

23. $\dfrac{3t\cos(3t - \pi/3) - \sin(3t - \pi/3)}{2t^2}$

25. $\dfrac{4x(1 - 3x)\sin x^2 - 6\cos x^2}{(3x - 1)^2}$

27. $4\sin 3x(3\cos 3x\cos 2x - \sin 3x\sin 2x)$

29. $2\cos 2t\cos(\sin 2t)$ **31.** $3\sin^2 x\cos x + 2\sin 2x$

33. $-\dfrac{\cos s}{\sin^2 s} + \dfrac{\sin s}{\cos^2 s}$

35. (a)

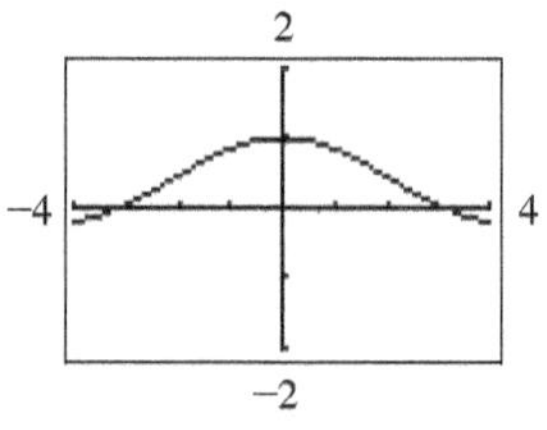

(b) See the table.

37. (a) 0.540 302 3, value of derivative
(b) 0.540 260 2, slope of secant line

39. Resulting curve is $y = \cos x$. **41.** $\dfrac{2x - y\cos xy}{x\cos xy - 2\sin 2y}$

43. $\dfrac{d\sin x}{dx} = \cos x$, $\dfrac{d^2\sin x}{dx^2} = -\sin x$, $\dfrac{d^3\sin x}{dx^3} = -\cos x$, $\dfrac{d^4\sin x}{dx^4} = \sin x$

45. $\sin 2x = 2 \sin x \cos x$ **47.** 0.515 **49.** -2.65
51. 412 mm/s **53.** -199 cm/s **55.** -38.5 km

Exercises 27.2, page 840

1. $12x \sec^2 x^2 \tan x^2$ **3.** $5 \sec^2 5x$ **5.** $5 \csc^2(0.25\pi - \theta)$

7. $27 \sec 9v \tan 9v$ **9.** $\dfrac{21}{2\sqrt{7x-6}} \csc\sqrt{7x-6} \cot \sqrt{7x-6}$

11. $30\pi \tan 3\pi t \sec^2 3\pi t$ **13.** $-4 \cot^3 \frac{1}{2}x \csc^2 \frac{1}{2}x$

15. $2 \tan 4x\sqrt{\sec 4x}$ **17.** $-84 \csc^4 7x \cot 7x$

19. $0.5t^2 \sec^2 0.5t + 2t \tan 0.5t$

21. $-4 \csc x^2(2x \cos x \cot x^2 + \sin x)$ **23.** $-\dfrac{\csc x(x \cot x + 1)}{x^2}$

25. $\dfrac{2(-4 \sin 4x - 4 \sin 4x \cot 3x + 3 \cos 4x \csc^2 3x)}{(1 + \cot 3x)^2}$

27. $\sec^2 x(\tan^2 x - 1)$ **29.** $2\pi \cos 2\pi\theta \sec^2 (\sin 2\pi\theta)$

31. $\dfrac{1 + 2 \sec^2 4x}{\sqrt{2x + \tan 4x}}$ **33.** $\dfrac{2 \cos 2x - \sec y}{x \sec y \tan y - 2}$

35. $24 \tan 3x \sec^2 3x\,dx$ **37.** $4 \sec 4x(\tan^2 4x + \sec^2 4x)\,dx$

39. (a) 3.425 518 8, value of derivative
(b) 3.426 052 4, slope of secant line

41. $2 \tan x \sec^2 x = 2 \sec x(\sec x \tan x)$ **43.** -12

45. $2 \sec^2 x - \sec x \tan x = \dfrac{2}{\cos^2 x} - \dfrac{\sin x}{\cos^2 x}$

47. -8.4 cm/s **49.** 140 m/s

Exercises 27.3, page 843

1. $\dfrac{2x}{\sqrt{1-x^4}}$ **3.** $\dfrac{7}{\sqrt{1-49x^2}}$ **5.** $\dfrac{50x^4}{\sqrt{1-25x^{10}}}$

7. $\dfrac{-1.8}{\sqrt{1-0.25s^2}}$ **9.** $\dfrac{1}{\sqrt{(x-1)(2-x)}}$ **11.** $\dfrac{4}{\sqrt{s}(1+s)}$

13. $6 \tan^{-1}\left(\dfrac{1}{x}\right) - \dfrac{6x}{x^2+1}$ **15.** $\dfrac{6x}{\sqrt{1-4x^2}} + 5 \sin^{-1} x$

17. $\dfrac{0.8u}{1+4u^2} + 0.4\tan^{-1} 2u$ **19.** $\dfrac{3\sqrt{1-4R^2} \sin^{-1} 2R - 6R + 2}{\sqrt{1-4R^2}(\sin^{-1} 2R)^2}$

21. $\dfrac{2(\cos^{-1} 2x + \sin^{-1} 2x)}{\sqrt{1-4x^2}(\cos^{-1} 2x)^2}$ **23.** $\dfrac{-72(\cos^{-1} 4x)^8}{\sqrt{1-16x^2}}$

25. $\dfrac{4 \sin^{-1}(4t+3)}{\sqrt{-(4t^2+6t+2)}}$ **27.** $\dfrac{-1}{t^2+1}$ **29.** $\dfrac{-2(2x+1)^2}{(1+4x^2)^2}$

31. $\dfrac{18(4 - \cos^{-1} 2x)^2}{\sqrt{1-4x^2}}$ **33.** $-\dfrac{x^2y^2 + 2y + 1}{2x}$

35. (a) 1.154 700 5, value of derivative
(b) 1.154 739 0, slope of secant line

37. $\dfrac{8(\sin^{-1} x)^7\,dx}{\sqrt{1-x^2}}$ **39.** 0.41 **41.** $\dfrac{2}{(1+x^2)^2}$

43.

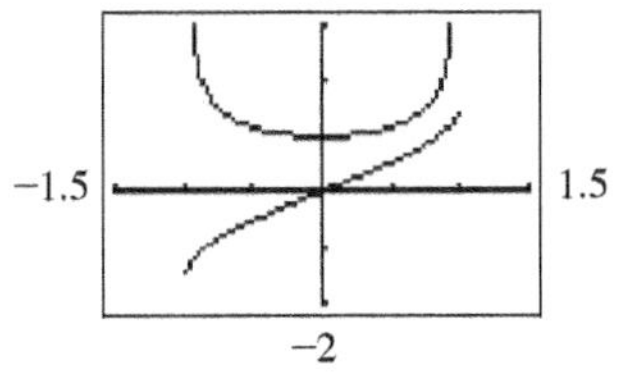

45. $\dfrac{-16x}{(1+4x^2)^2}$

47. Let $y = \sec^{-1} u$; solve for u; take derivatives; substitute.

49. $\dfrac{E-A}{\omega m\sqrt{m^2E^2 - (A-E)^2}}$ **51.** $-\dfrac{R}{R^2 + (X_L - X_C)^2}$

53. $\theta = \tan^{-1} \dfrac{h}{x}; \dfrac{d\theta}{dx} = \dfrac{-h}{x^2+h^2}$

Exercises 27.4, page 847

1. Max. $\left(\frac{\pi}{3}, 0.34\right)$;
min. $\left(\frac{5\pi}{3}, -3.48\right)$;
infl. $(0, 0)$, $\left(\pi, -\frac{\pi}{2}\right)$

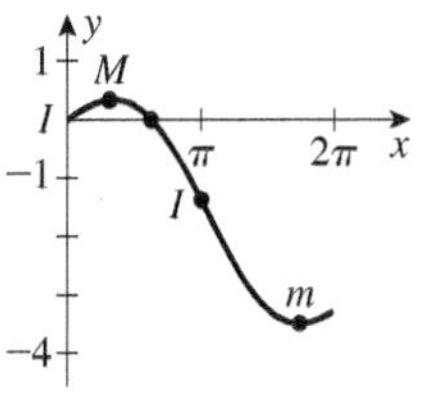

3. $d \sin x/dx = \cos x$ and $d \cos x/dx = -\sin x$, and $\sin x = \cos x$ at points of intersection.

5. $\dfrac{1}{x^2+1}$ is always positive.

7.

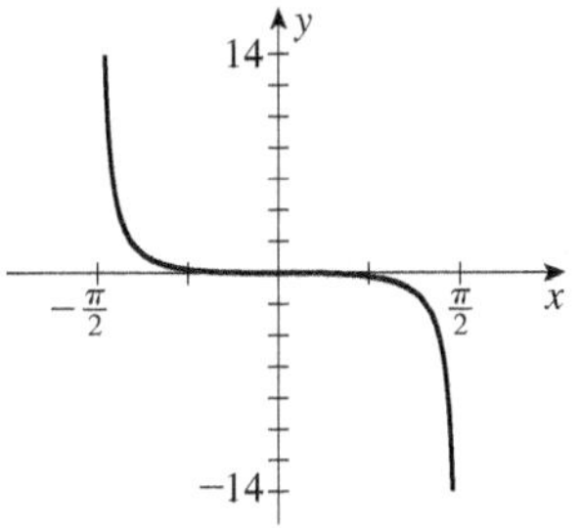

Dec. $0 < x < \frac{\pi}{2}, -\frac{\pi}{2} < x < 0$;
infl. $(0, 0)$;
asym. $x = \frac{\pi}{2}, x = -\frac{\pi}{2}$

9. $y = 1.10x - 0.29$ **11.** 1.933 753 8 **13.** -10
15. $32 \sin 2t \cos 2t$ **17.** 0.58 m/s, -1.7 m/s^2
19. -0.0718 N/s **21.** 358 cm/s, 18.0°
23. 17.4 cm/s^2, 176° **25.** -0.085 rad/s
27. 8.08 m/s **29.** 14.0° **31.** 338 km **33.** 1.06 L/s
35. $w = 9.24$ cm, $d = 13.1$ cm **37.** 4.21 m **39.** 56 m/s

Exercises 27.5, page 851

1. $-4 \tan 4x$ **3.** $\dfrac{4 \log e}{x}$ **5.** $\dfrac{4 \log_5 e}{x-3}$ **7.** $\dfrac{8}{x-3}$

9. $\dfrac{4 \sec^2 2x}{\tan 2x} = 4 \sec 2x \csc 2x$ **11.** $\dfrac{2}{4T+1}$ **13.** $\dfrac{5(1-2x)}{x-x^2}$

15. $\dfrac{6(t+2)(t + \ln t^2)}{t}$ **17.** $3 \ln(6-x) - \dfrac{3x}{6-x}$

19. $\dfrac{1}{x \ln x}$ **21.** $\pi\theta \cot(\pi\theta^2)$ **23.** $\dfrac{-\sin \ln x}{x}$

25. $6 \ln 2v + 3 \ln^2 2v$ **27.** $\dfrac{x \sec^2 x + \tan x}{x \tan x}$ **29.** $\dfrac{v+4}{v(v+2)}$

31. $\dfrac{\sqrt{x^2+1}}{x}$ **33.** $\dfrac{x + y - 2 \ln(x+y)}{x + y + 2 \ln(x+y)}$

35. 0.5 is value of derivative; 0.499 987 5 is slope of secant line.

37. (a)

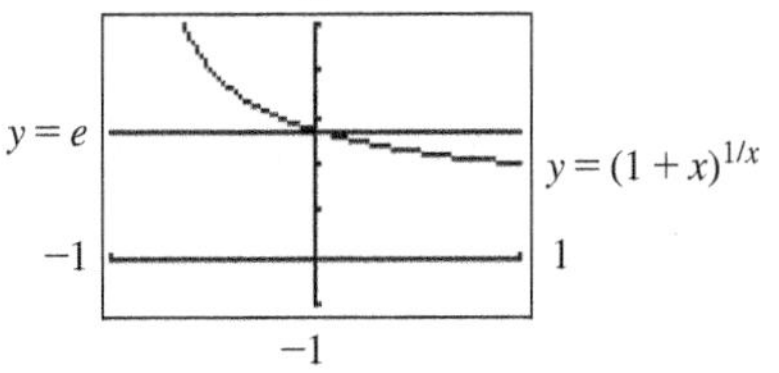

(b) See the table.

39. -0.3640 **41.** 0.976 **43.** $L(x) = 4x - \pi$ **45.** 2.73

47. $x^x(\ln x + 1)$. In the general power rule the exponent is constant. For x^x, both the base and the exponent are variables.

49. $\ln(x^2)$ is defined for all x, $x \neq 0$.

Therefore $\left.\dfrac{dy_1}{dx}\right|_{x=-1} = \left.\dfrac{2}{x}\right|_{x=-1} = -2.$

$\ln x$ is not defined for $x \leq 0$. Therefore $\left.\dfrac{dy_2}{dx}\right|_{x=-1} = \left.\dfrac{2}{x}\right|_{x=-1}$ does not exist.

51. $\dfrac{10 \log e}{I}\dfrac{dI}{dt}$ **53.** $\dfrac{x}{x^2 + 1 + \sqrt{1 + x^2}} - \dfrac{1}{x} + \dfrac{x}{\sqrt{1 - x^2}}$

55. $0.125\ \mathrm{s^2/m}$

57. $\dfrac{d}{dx}\left[\dfrac{1}{2}(e^u - e^{-u})\right] = \dfrac{1}{2}(e^u + e^{-u})\dfrac{du}{dx}$;

$\dfrac{d}{dx}\left[\dfrac{1}{2}(e^u + e^{-u})\right] = \dfrac{1}{2}(e^u - e^{-u})\dfrac{du}{dx}$

59. $\sinh\dfrac{x}{50}$

Exercises 27.6, page 854

1. $2e^{2x} \cot e^{2x}$ **3.** $(6 \ln 4)4^{6x}$ **5.** $\dfrac{3e^{\sqrt{x}}}{\sqrt{x}}$ **7.** $4e^{2t}(3e^t - 2)$

9. $e^{-T}(1 - T)$ **11.** $e^{\sin x}(x \cos x + 1)$ **13.** $8e^{-4s}$

15. $-e^{-3x}(4 \sin 4x + 3 \cos 4x)$ **17.** $\dfrac{2e^{3x}(12x + 5)}{(4x + 3)^2}$

19. $\dfrac{te^{t^2}}{e^{t^2} + 4}$ **21.** $16e^{6x}(x \cos x^2 + 3 \sin x^2)$

23. $\dfrac{2(1 + 2te^{2t})}{t\sqrt{\ln 2t + e^{2t}}}$ **25.** $\dfrac{y^2}{1 - xy}$ **27.** $\dfrac{3e^{2x}}{x} + 6e^{2x} \ln x$

29. $12e^{6t} \cot 2e^{6t}$ **31.** $\dfrac{4e^{2x}}{\sqrt{1 - e^{4x}}}$

33. (a) 2.718 281 8, value of derivative
(b) 2.718 417 7, slope of secant line

35.

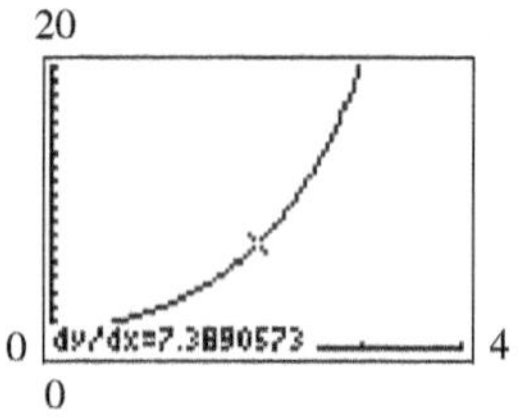

37. -1.46

39. $\dfrac{12e^{4x}(4x + 23)\,dx}{(x + 6)^2}$

41.

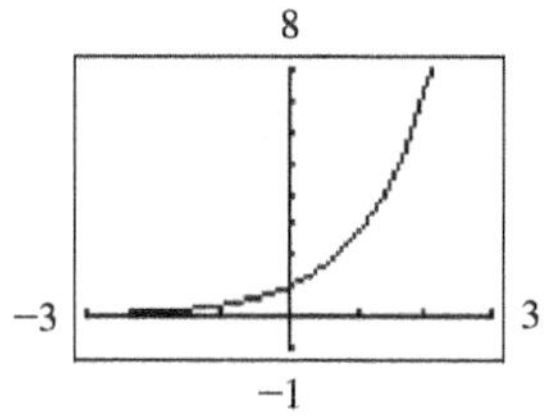

43. $(-xe^{-x} + e^{-x}) + (xe^{-x}) = e^{-x}$

45. $\dfrac{2e^{2x}(e^{2x} + 1) - 2e^{2x}(e^{2x} - 1)}{(e^{2x} + 1)^2} = \dfrac{4e^{2x}}{(e^{2x} + 1)^2}$

$= \dfrac{(e^{2x} + 1)^2 - (e^{2x} - 1)^2}{(e^{2x} + 1)^2}$

47. $-3, 2$ **49.** $Ak^2e^{kx} + Bk^2e^{-kx} = k^2(Ae^{kx} + Be^{-kx})$

51. $-0.001\ 64/\mathrm{h}$ **53.** $e^{-66.7t}(999 \cos 226t - 295 \sin 226t)$

55. Substitute and simplify.

Exercises 27.7, page 859

1. ∞ **3.** $\frac{1}{6}$ **5.** 1 **7.** ∞ **9.** 0 **11.** 0 **13.** $-\frac{1}{6}$

15. $-\pi$ **17.** 1 **19.** 0 **21.** 0 **23.** ∞ **25.** 1

27. 0 **29.** -7 **31.** -1 **33.** 0 **35.** 1 **37.** 1

39. $\sin x$ varies between 1 and -1 as $x \to \infty$. **41.** gt

Exercises 27.8, page 862

1. Int. $(0, 0)$, $(\pi, 0)$, $(2\pi, 0)$;
max. $\left(\frac{\pi}{4}, 0.322\right)$;
min. $\left(\frac{5\pi}{4}, -0.014\right)$;
infl. $\left(\frac{\pi}{2}, 0.208\right)$, $\left(\frac{3\pi}{2}, -0.009\right)$

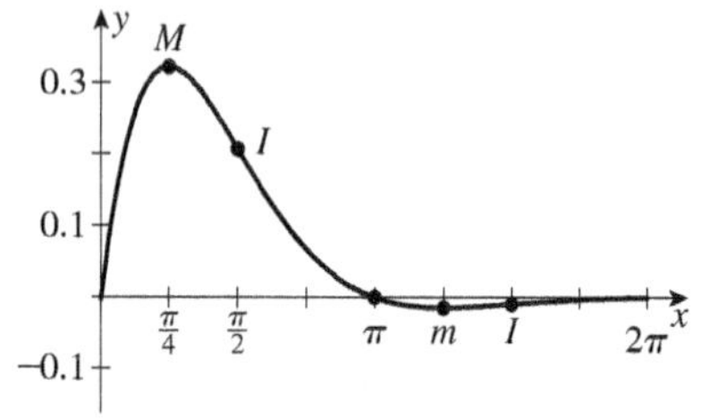

3. Int. $(0, 0)$, max. $(0, 0)$,
not defined for $\cos x < 0$,
asym. $x = -\frac{1}{2}\pi, \frac{1}{2}\pi, \ldots$

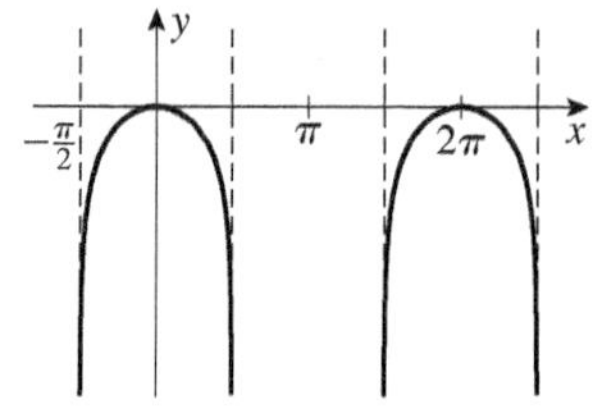

5. Int. $(0, 0)$, max. $\left(1, \dfrac{1}{e}\right)$,
infl. $\left(2, \dfrac{6}{e^2}\right)$, asym. $y = 0$

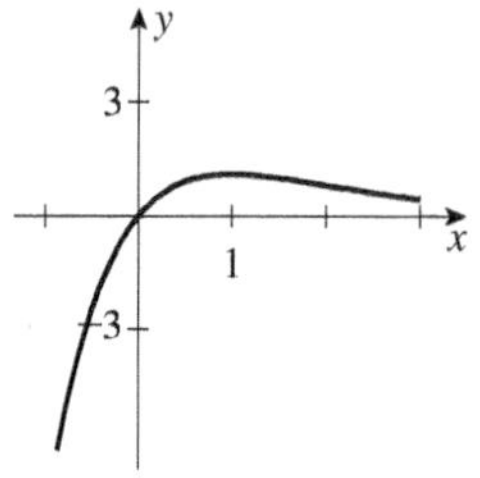

7. Int. $(0, 0)$, max. $(0, 0)$,
infl. $(-1, -\ln 2)$, $(1, -\ln 2)$

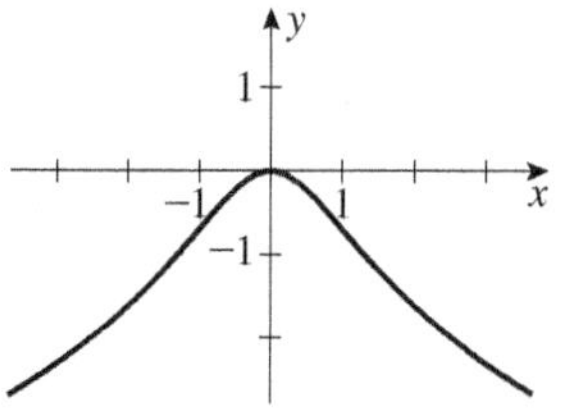

9. Int. $(0, 4)$; max. $(0, 4)$;
infl. $\left(\dfrac{1}{2}\sqrt{2}, \dfrac{4}{e}\sqrt{e}\right), \left(-\dfrac{1}{2}\sqrt{2}, \dfrac{4}{e}\sqrt{e}\right)$;
asym. $y = 0$

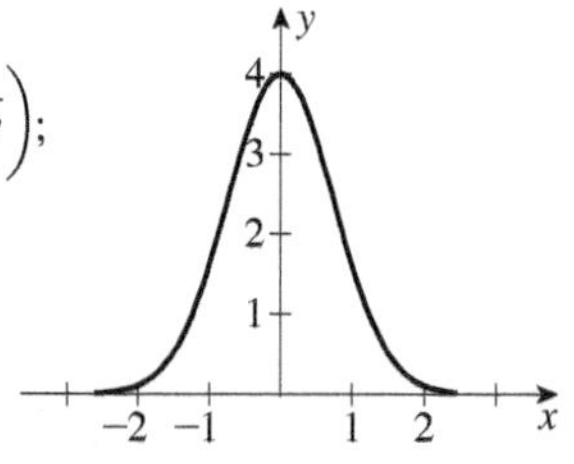

11. Max. $(4, 3.09)$, asym. $x = 0$

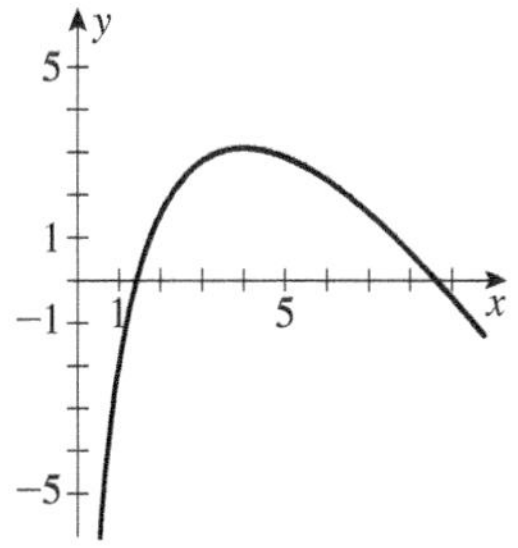

13. Int. $(0, 0)$, infl. $(0, 0)$, inc. all x

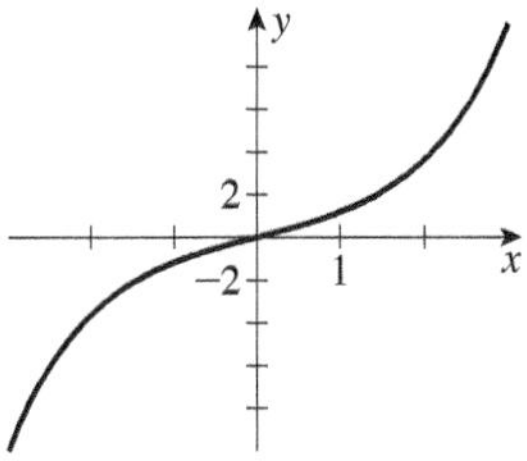

15. y_1 inc. $x < 0$, y_1 dec. $x > 0$; y_2 inc. $x > 0$, y_2 dec. $x < 0$; y_1 and y_2 are reciprocals.

17. $y = x - 1$ **19.** $2\sqrt{2}x - 2y + 2\sqrt{2} - 3\pi\sqrt{2} = 0$

21. 1.197 333 8 **23.** 174 V **25.** -0.303 W/day

27. -5.4 °C/min **29.** $\dfrac{p(-a + bT)}{T^2}$ **31.** $a = k^2x$

33. Int. $(0, 0)$; min. $(0, 0)$, $(2\pi, 0)$, $\cdots$; asym. $x = -\dfrac{\pi}{2}, \dfrac{\pi}{2}, \dfrac{3\pi}{2}, \cdots$
The path of the roller mechanism is the dark curve $(-1.5 < x < 1.5)$

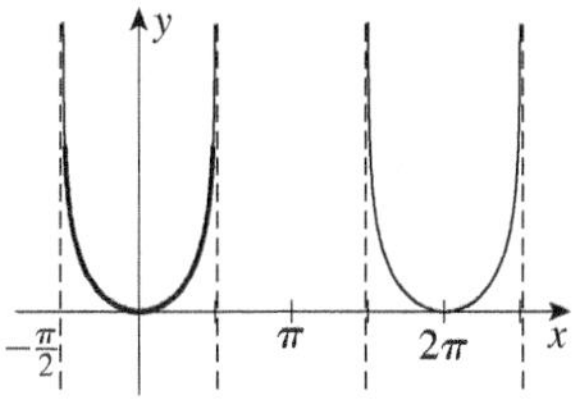

35. 3.0 s

37. Max. (117.6° W, 50.2° N), (86.2° W, 47.7° N); min. (101.9° W, 41.2° N), (70.5° W, 43.0° N)

39. $v = -e^{-0.5t}(1.4\cos 6t + 2.3\sin 6t), -2.03$ cm/s

41. $1/\sqrt{e} = 0.607$ **43.** $\dfrac{1}{\sqrt{b}}$

45. $\dfrac{d^2y}{dx^2} = -\dfrac{e^{-x^2/2}}{\sqrt{2\pi}}(1 - x^2)$; $\dfrac{d^2y}{dx^2}$ changes sign at $x = \pm 1$.

Review Exercises for Chapter 27, page 864

1. $-12\sin(4x - 1)$ **3.** $\dfrac{-0.2\sec^2\sqrt{3 - 2v}}{\sqrt{3 - 2v}}$

5. $-6\csc^2(3x + 2)\cot(3x + 2)$ **7.** $-24x\cos^3 x^2 \sin x^2$

9. $2e^{2(x-3)}$ **11.** $\dfrac{6x}{x^2 + 1}$ **13.** $\dfrac{50}{25 + x^2}$

15. $\dfrac{0.1}{\sin^{-1} 0.1t\sqrt{1 - 0.01t^2}}$ **17.** $(-2\csc 4x)\sqrt{\csc 4x + \cot 4x}$

19. $\dfrac{14(1 + e^{-x})}{x - e^{-x}}$ **21.** $-\dfrac{2\sin x\cos x}{e^{3x} + \pi^2} - \dfrac{3e^{3x}\cos^2 x}{(e^{3x} + \pi^2)^2}$

23. $\dfrac{2u(1 + 4u^2)(\tan^{-1} 2u) - 2u^2}{(1 + 4u^2)(\tan^{-1} 2u)^2}$ **25.** $-2x\cot x^2$

27. $\dfrac{2\cos x\ln(3 + \sin x)}{3 + \sin x}$ **29.** $0.1e^{-2t}\sec \pi t(-2 + \pi\tan \pi t)$

31. $\dfrac{\cos 2x + 2e^{4x}}{\sqrt{\sin 2x + e^{4x}}}$ **33.** $\dfrac{2x^3e^y + 2xy^2e^y + y}{x - x^4e^y - x^2y^2e^y}$

35. $0.5[e^{-2t}(1 - 2t) - e^{2t}(1 + 2t)]$ **37.** $\dfrac{y(xye^{-x} - 1)}{x(ye^{-x} + 1)}$

39. $\cos^{-1} x$

41. Inc. all x, infl. $x = (2n + 1)\pi$

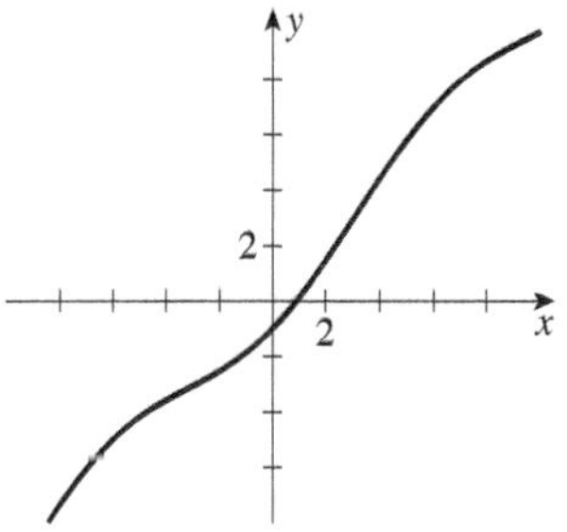

43. Max. $(e^{-2}, 4e^{-2})$, min. $(1, 0)$, infl. (e^{-1}, e^{-1})

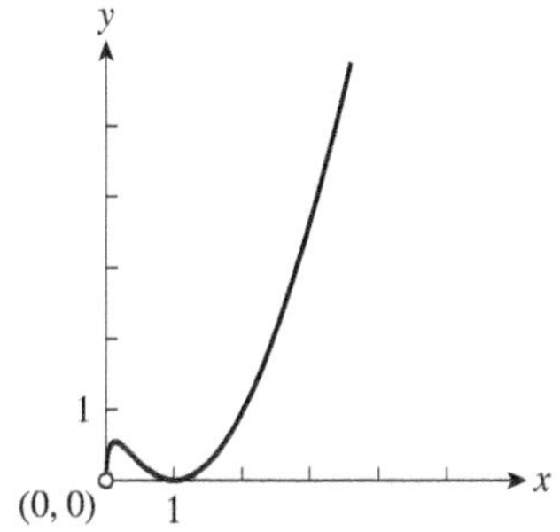

45. $7.27x + y - 8.44 = 0$

47. $2x + 2.57y - 4.30 = 0$ **49.** $\frac{2}{3}$ **51.** -1 **53.** 0

55. $2\sin x\cos x - 2\cos x\sin x = 0$

57. $-9\sin 3x = -9(\sin 3x)$ **59.** $-0.703\,467\,4$

61. $x > \frac{1}{2}\ln 2\ (= 0.3466)$ **63.** 0.039 **65.** 2.5 N

67. 0.12 cm/s **69.** -0.064 °C/day **71.** $50\sin 4t$

73.

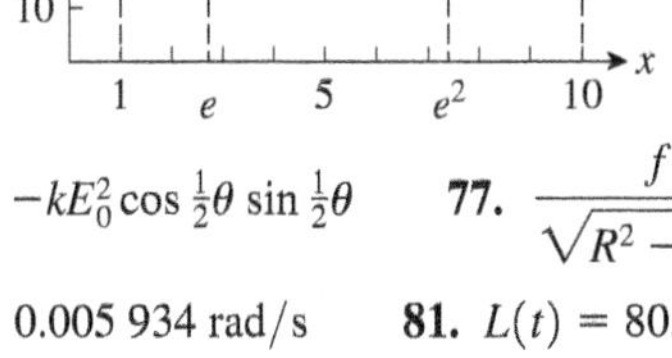

$n = \dfrac{(8\ln 8)x}{\ln x}$, min. $(e, 8e\ln 8)$, asym. $x = 1$

75. $-kE_0^2\cos\frac{1}{2}\theta\sin\frac{1}{2}\theta$ **77.** $\dfrac{f}{\sqrt{R^2 - F^2f^2}}$

79. 0.005 934 rad/s **81.** $L(t) = 80.8 - 4.16t$

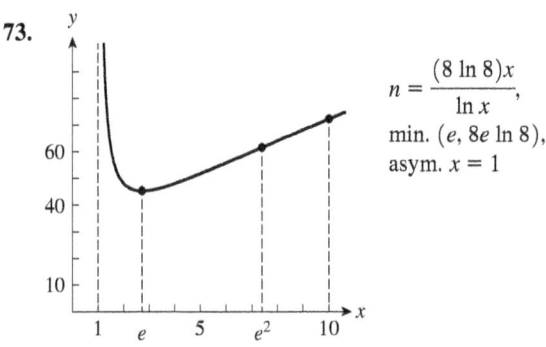

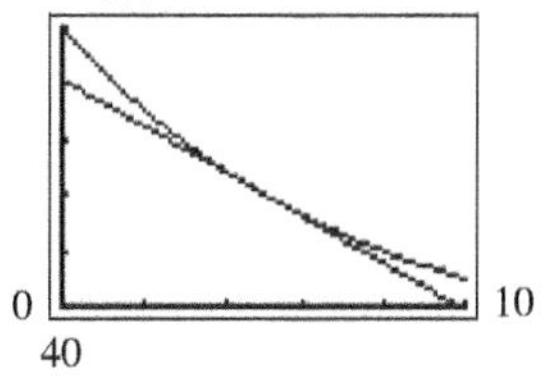

83. (a) -3900 \$/yr (b) -2200 \$/yr **85.** 2.00 m, 1.82 m
87. -0.0065 rad/s **89.** 48.2° **91.** 0.397 cm/s, 72.6°
93. 7.07 cm **95.** $A = 50\cos\theta(1 + 2\sin\theta)$, max. (0.635, 88.0)

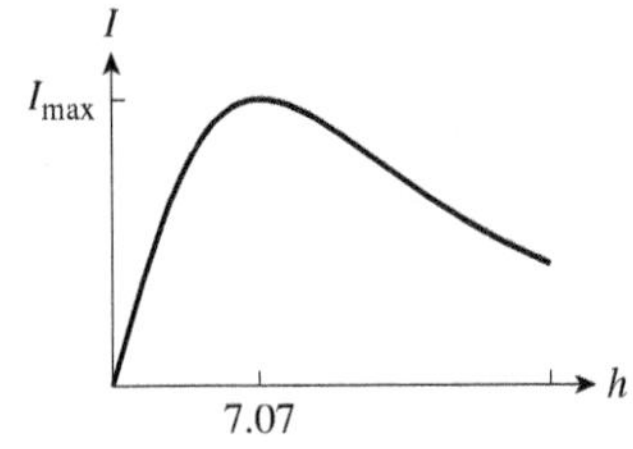

97. $\frac{w}{H}\cosh\frac{wx}{H} = \frac{w}{H}\sqrt{1 + \sinh^2\frac{wx}{H}}$

Chapter 28: Methods of Integration

Exercises 28.1, page 870

1. $-\frac{1}{4}\cos^4 x + C$ **3.** $\frac{1}{4}(x^2 + 1)^4 + C$ **5.** $\frac{1}{5}\sin^5 x + C$
7. $-\frac{4}{15}(\cos\theta)^{3/2} + C$ **9.** $\frac{4}{3}\tan^3 x + C$ **11.** $\frac{1}{8}$
13. $\frac{1}{4}(\sin^{-1} x)^4 + C$ **15.** $\frac{1}{2}(\tan^{-1} 5x)^2 + C$
17. $\frac{1}{3}[\ln(x + 1)]^3 + C$ **19.** 0.179 **21.** $\frac{1}{17}(4 + e^x)^{17} + C$
23. $\frac{1}{4(1 - e^{2t})^2} + C$ **25.** $\frac{1}{10}(1 + \sec^2 x)^5 + C$ **27.** 4.13
29. $-\int\frac{-\sin x\,dx}{\cos^5 x}$; $u = \cos x$, $du = -\sin x\,dx$, $n = -5$
31. $\int\frac{1}{\ln^2 x}\frac{dx}{x}$, $u = \ln x$, $du = \frac{dx}{x}$, $n = -2$
33. 0.888 **35.** 1.10 **37.** $y = \frac{1}{3}(\ln x)^3 + 2$ **39.** $\frac{1}{3}mnv^2$
41. $q = (1 - e^{-t})^3$

Exercises 28.2, page 874

1. $\frac{1}{2}\ln|2x + 1| + C$ **3.** $\frac{1}{4}\ln|1 + 4x| + C$
5. $-\frac{1}{3}\ln|4 - 3x^2| + C$ **7.** $\frac{1}{3}\ln 4 = 0.462$
9. $-0.2\ln|\cot 2\theta| + C$ **11.** $\frac{1}{5}\ln 2 = 0.139$
13. $\ln|1 - e^{-x}| + C$ **15.** $\ln|x + e^x| + C$
17. $\frac{1}{4}\ln|1 + 4\sec x| + C$ **19.** $\frac{1}{4}\ln 5 = 0.402$
21. $0.5\ln|\ln r| + C$ **23.** $\ln|2x + \tan x| + C$
25. $-6\sqrt{1 - 2x} + C$ **27.** $\ln|x| - \frac{2}{x} + C$
29. $\frac{1}{3}\ln\left(\frac{5}{4}\right) = 0.0744$ **31.** 1.10
33. $\int\frac{x - 6}{x + 6}dx = \int\left(1 - \frac{12}{x + 6}\right)dx = x - 12\ln|x + 6| + C$
35. $\pi\ln 2 = 2.18$ **37.** $y = \ln\left(\frac{3.5}{3 + \cos x}\right) + 2$
39. $\ln a + \ln b = \ln ab$ **41.** ln 2 m/s **43.** 8.38 min
45. $i = \frac{E}{R}(1 - e^{-Rt/L})$ **47.** 1.41 m

Exercises 28.3, page 877

1. $\frac{1}{3}e^{x^3} + C$ **3.** $e^{7x} + C$ **5.** $-\frac{1}{5}e^{9-5x} + C$ **7.** 28.2
9. $2e^{x^3} + C$ **11.** $2(e^2 - e) = 9.34$ **13.** $2e^{2\sec\theta} + C$
15. $\frac{2}{3}(1 + e^y)^{3/2} + C$ **17.** $6 - \frac{3(e^6 - e^2)}{2} = -588$
19. $-\frac{4}{e^{\sqrt{x}}} + C$ **21.** $\frac{1}{2}e^{\tan^{-1}2x} + C$ **23.** $-\frac{1}{3}e^{\cos 3x} + C$
25. 0 **27.** $\frac{3}{2}\ln(e^{4x} + 1) + C$ **29.** $3e^2 - 3 = 19.2$
31. $y = e^{\ln x}$; $\ln y = \ln e^{\ln x}$; $y = x$; $e^{x^2} + C$
33. $\int\left(1 - \frac{e^x}{e^x + 1}\right)dx = x - \ln(e^x + 1) + C$
35. $\pi(e^4 - e) = 163$ **37.** $2(e^2 - 1) = 12.8$
39. $\ln b\int b^u\,du = b^u + C_1$ **41.** $q = EC(1 - e^{-t/RC})$
43. 26 610 m^2 **45.** 0.632
47. (a) second runner (b) first runner

Exercises 28.4, page 881

1. $\frac{1}{3}\tan x^3 + C$ **3.** $\frac{1}{99}\sin 99x + C$ **5.** $0.1\tan 3\theta + C$
7. $2\sec\frac{1}{2}x + C$ **9.** 0.6365 **11.** $\frac{3}{2}\ln|\sec\phi^2 + \tan\phi^2| + C$
13. $\cos\left(\frac{1}{x}\right) + C$ **15.** $\frac{1}{2}\sqrt{3}$ **17.** $-2\cos\frac{x}{2} + C$
19. $\frac{1}{2}\ln|\sec 2x + \tan 2x| + C$ **21.** $-\frac{1}{2}\ln\cos 2T + C$
23. $\csc x - \cot x - \ln|\csc x - \cot x| + \ln|\sin x| + C$
25. $\frac{1}{9}\pi + \frac{1}{3}\ln 2 = 0.580$ **27.** 0.693
29. Integral changes to $\int(\sec^2 x - \sec x\tan x)\,dx$
31. (a) $\frac{1}{2}\tan^2 x + C_1$, (b) $\frac{1}{2}\sec^2 x + C_2$; $C_1 = C_2 + \frac{1}{2}$
33. $\pi\sqrt{3} = 5.44$ **35.** $\theta = 0.10\cos 2.5t$
37. 0.7726 m **39.** 0.707

Exercises 28.5, page 886

1. $\frac{x}{2} - \frac{1}{12}\sin 6x + C$ **3.** $\frac{1}{3}\sin^3 x + C$
5. $-\frac{1}{2}\cos 2x + \frac{1}{6}\cos^3 2x + C$ **7.** $\frac{1}{3}\tan^3 x + C$
9. $\frac{1}{24}(64 - 43\sqrt{2}) = 0.133$ **11.** $\frac{1}{2}x - \frac{1}{4}\sin 2x + C$
13. $\frac{1}{3}(9\phi + 4\sin 3\phi + \sin 3\phi\cos 3\phi) + C$
15. $\frac{1}{2}\tan^2 x + \ln|\cos x| + C$ **17.** $\frac{3}{4}$
19. $\frac{1}{6}\tan^3 2x - \frac{1}{2}\tan 2x + x + C$ **21.** $\frac{1}{3}\sin^3 s + C$
23. $x - \frac{1}{2}\cos 2x + C$ **25.** $\frac{1}{4}\cot^4 x - \frac{1}{3}\cot^3 x + \frac{1}{2}\cot^2 x - \cot x + C$
27. $1 + \frac{1}{2}\ln 2 = 1.35$ **29.** $\frac{1}{5}\tan^5 x + \frac{2}{3}\tan^3 x + \tan x + C$
31. $2\sec x + C$ **33.** $-\ln|\sec e^{-x} + \tan e^{-x}| + C$
35. $\frac{1}{2}\cos x - \frac{1}{18}\cos 9x + C$ **37.** $\frac{1}{2}\pi^2 = 4.93$
39. $\sqrt{2} - 1 = 0.414$
41. $\int\sin x\cos x\,dx = \frac{1}{2}\sin^2 x + C_1 = -\frac{1}{2}\cos^2 x + C_2$; $C_2 = C_1 + \frac{1}{2}$
43. $\int_0^{\pi}\sin^2 nx\,dx = \frac{1}{2}\int_0^{\pi}(1 - \cos 2nx)\,dx$
$= \frac{1}{2}x - \sin 2nx\Big|_0^{\pi} = \frac{\pi}{2}$
45. $s = -\frac{1}{3}\cos t + \frac{1}{9}\cos^3 t + 6t + \frac{2}{9}$
47. $\frac{4}{3}$ **49.** $V = \sqrt{\frac{1}{1/60.0}\int_0^{1/60.0}(340\ 120\pi t)^2\,dt} = 240$ V
51. $\frac{aA}{2} + \frac{A}{2b\pi}\sin ab\pi\cos 2bc\pi$

Exercises 28.6, page 890

1. $-\sqrt{9-x^2}+C$ **3.** $\sin^{-1}\frac{1}{2}x+C$ **5.** $\frac{1}{8}\tan^{-1}\frac{1}{8}x+C$

7. $\sin^{-1}\frac{4x^2}{6}+C$ **9.** 0.863 **11.** $\frac{2}{5}\sqrt{5}\sin^{-1}\frac{1}{5}\sqrt{5}=0.415$

13. $\frac{4}{9}\ln|9x^2+16|+C$ **15.** 2.36 **17.** $\sin^{-1}\frac{e^x}{9}+C$

19. $\tan^{-1}(T+1)+C$ **21.** $4\sin^{-1}\frac{1}{2}(x+2)+C$

23. -0.714 **25.** $2\sin^{-1}\left(\frac{1}{2}x\right)+\sqrt{4-x^2}+C$

27. $\frac{1}{2}(\sin^{-1}x)^2+C$ **29.** $\frac{1}{3}\tan^{-1}x^3+\frac{1}{2}\ln(x^6+1)+C$

31. (a) Inverse tangent, $\int\frac{du}{a^2+u^2}$ where $u=3x$, $du=3\,dx$, $a=2$; numerator cannot fit du of denominator. Positive $9x^2$ leads to inverse tangent form.

(b) Logarithmic, $\int\frac{du}{u}$ where $u=4+9x$, $du=9\,dx$

(c) General power, $\int u^{-1/2}du$ where $u=4+9x^2$, $du=18x\,dx$

33. (a) General power, $\int u^{-1/2}du$ where $u=4-9x^2$, $du=-18x\,dx$; numerator can fit du of denominator. Square root becomes $-1/2$ power. Does not fit inverse sine form.

(b) Inverse sine, $\int\frac{du}{\sqrt{a^2-u^2}}$ where $u=3x$, $du=3\,dx$, $a=2$

(c) Logarithmic, $\int\frac{du}{u}$ where $u=4-9x$, $du=-9\,dx$

35. Form fits inverse tangent integral with $u=\sqrt{x}$, $du=\frac{dx}{2\sqrt{x}}$. Result is $2\tan^{-1}\sqrt{x}+C$.

37. $\tan^{-1}2=1.11$ **39.** $k\tan^{-1}\frac{x}{d}+C$

41. $\sin^{-1}\frac{x}{A}=\sqrt{\frac{k}{m}}t+\sin^{-1}\frac{x_0}{A}$ **43.** 0.220ρ

Exercises 28.7, page 894

1. No. Integral $\int v\,du$ is more complex than the given integral.

3. $\cos\theta+\theta\sin\theta+C$ **5.** $2xe^{2x}-e^{2x}+C$

7. $3\ln|\cos x|+3x\tan x+C$ **9.** $2x\tan^{-1}x-\ln(1+x^2)+C$

11. $-\frac{32}{3}$ **13.** $\frac{1}{2}x^2\ln x-\frac{1}{4}x^2+C$

15. $\frac{1}{2}\phi\sin 2\phi-\frac{1}{4}(2\phi^2-1)\cos 2\phi+C$ **17.** $\frac{1}{2}(e^{\pi/2}-1)=1.91$

19. $\frac{1}{18}(2x+5)(x+4)^{18}-\frac{1}{171}(x+4)^{19}+C$

21. $\frac{1}{2}x\left[\cos(\ln x)+\sin(\ln x)\right]+C$ **23.** $-2e^{-\sqrt{x}}(\sqrt{x}+1)+C$

25. $1-\frac{3}{e^2}=0.594$ **27.** 0.110 **29.** $\frac{1}{2}\pi-1=0.571$

31. 0.756 **33.** $s=\frac{1}{3}\left[(t^2-2)\sqrt{t^2+1}+2\right]$

35. $q=\frac{1}{5}\left[e^{-2t}(\sin t-2\cos t)+2\right]$

Exercises 28.8, page 898

1. Delete the x^2 before the radical in the denominator.

3. $x=3\sin\theta$, $\int\cot^2\theta\,d\theta$ **5.** $x=\tan\theta$, $\int\csc\theta\cot\theta\,d\theta$

7. $x=\sec\theta$, $\int d\theta$ **9.** $-\frac{\sqrt{1-x^2}}{x}-\sin^{-1}x+C$

11. $2\ln|x+\sqrt{x^2-16}|+C$ **13.** $-\frac{2\sqrt{z^2+9}}{3z}+C$

15. $\frac{x}{\sqrt{4-x^2}}+C$ **17.** $\frac{16-9\sqrt{3}}{24}=0.017$

19. $5\ln|\sqrt{x^2+2x+2}+x+1|+C$ **21.** 0.0400

23. $2\sec^{-1}e^x+C$

25. (a) $-\frac{1}{3}(1-x^2)^{3/2}+C$, (b) $-\frac{1}{3}(1-x^2)^{3/2}+C$

27. π **29.** $\frac{1}{4}ma^2$ **31.** 2.68 **33.** 1.36

35. $kQ\ln\frac{\sqrt{a^2+b^2}+a}{\sqrt{a^2+b^2}-a}$ **37.** $\frac{2(x+1)^{5/2}}{5}-\frac{2(x+1)^{3/2}}{3}+C$

39. $\frac{3}{8}(x-4)^{8/3}+\frac{12}{5}(x-4)^{5/3}+C$

Exercises 28.9, page 902

1. $\frac{3}{x-1}-\frac{4}{x+2}$ **3.** $\frac{A}{x}+\frac{B}{x+1}$ **5.** $\frac{A}{x}+\frac{B}{x+2}+\frac{C}{x-2}$

7. $\ln\left|\frac{(x+1)^2}{x+2}\right|+C$ **9.** $\frac{1}{4}\ln\left|\frac{x-2}{x+2}\right|+C$

11. $x+\ln\left|\frac{x}{(x+3)^4}\right|+C$ **13.** 1.06

15. $\ln\left|\frac{x^2(x-5)^3}{x+1}\right|+C$ **17.** $\frac{1}{4}\ln\left|\frac{x^4(2x+1)^3}{2x-1}\right|+C$

19. $1+\ln\frac{16}{3}=2.674$ **21.** $\frac{1}{60}\ln\left|\frac{(V+2)^3(V-3)^2}{(V-2)^3(V+3)^2}\right|+C$

23. $\ln\left|\frac{x(x-2)}{(x-1)^2}\right|+C$

25. $\frac{1}{u(a+bu)}=\frac{A}{u}+\frac{B}{a+bu}$; $1=A(a+bu)+Bu$; $A=\frac{1}{a}$, $B=-\frac{b}{a}$

27. $-\frac{1}{4}\ln\left|\frac{\sin\theta+3}{\sin\theta-1}\right|+C$ **29.** $2\pi\ln\frac{6}{5}=1.15$

31. $y=\ln\left|\frac{x(x+5)^2}{36}\right|$ **33.** 0.163 N · cm

Exercises 28.10, page 907

1. $\frac{2}{x(x+3)^2}=\frac{A}{x}+\frac{B}{x+3}+\frac{C}{(x+3)^2}$

3. $\frac{4x+4}{x^3-9x^2}=\frac{A}{x}+\frac{B}{x^2}+\frac{C}{x-9}$ **5.** $\ln\left|\frac{x+1}{x}\right|-\frac{1}{x}+C$

7. $\frac{3}{x-2}+2\ln\left|\frac{x-2}{x}\right|+C$ **9.** $\frac{2}{x}+\ln\left|\frac{x-1}{x+1}\right|+C$

11. $-\frac{5}{4}$ **13.** $-\frac{2}{x+1}-\frac{1}{x-3}+\ln|x+1|+C$

15. $\frac{1}{8}\pi+\ln 3=1.49$ **17.** $-\frac{2}{x}+\frac{3}{2}\tan^{-1}\frac{x+2}{2}+C$

19. $\frac{1}{4}\ln(4x^2+1)+\ln|x^2+6x+10|+\tan^{-1}(x+3)+C$

21. $\tan^{-1}x+\frac{x}{x^2+1}+\ln|x+1|-\frac{1}{2}\ln|x^2+1|+C$

23. $\dfrac{1-x}{(x-2)^2}+C$ **25.** $2+4\ln\dfrac{2}{3}=0.378$

27. $\pi\ln\dfrac{25}{9}=3.21$ **29.** 0.919 m **31.** 1.37

Exercises 28.11, page 909

1. Formula 3 **3.** $u=y$, $du=dy$, Formula 7

5. Formula 25; $u=x^2$, $du=2x\,dx$

7. $u=x^2$, $du=2x\,dx$, Formula 32

9. $\frac{3}{25}\left[2+5x-2\ln|2+5x|\right]+C$ **11.** $\frac{3544}{15}=236$

13. $\dfrac{y}{4\sqrt{y^2+4}}+C$ **15.** $\frac{1}{2}\sin x-\frac{1}{10}\sin 5x+C$

17. $\sqrt{4x^2-9}-3\sec^{-1}\left(\dfrac{2x}{3}\right)+C$

19. $\frac{1}{20}\cos^4 4x\sin 4x+\frac{1}{5}\sin 4x-\frac{1}{15}\sin^3 4x+C$

21. $3r^2\tan^{-1}r^2-\frac{3}{2}\ln(1+r^4)+C$

23. $\frac{1}{4}(8\pi-9\sqrt{3})=2.39$

25. $-\ln\left(\dfrac{1+\sqrt{4x^2+1}}{2x}\right)+C$ **27.** $-8\ln\left(\dfrac{1+\sqrt{1-4x^2}}{2x}\right)+C$

29. 0.0208 **31.** $\frac{1}{3}(\cos x^3+x^3\sin x^3)+C$

33. $\dfrac{x^2}{\sqrt{1-x^4}}+C$ **35.** 4.89

37. $\frac{1}{4}x^4\left(\ln x^2-\frac{1}{2}\right)+C$ **39.** $-\dfrac{3x^3}{\sqrt{x^6-1}}+C$

41. $\dfrac{t^3}{12}(t^6+1)^{3/2}+\dfrac{t^3}{8}\sqrt{t^6+1}+\dfrac{1}{8}\ln\left(t^3+\sqrt{t^6+1}\right)+C$

43. $\dfrac{1}{24}\sin^4 4x\cos^2 4x+\dfrac{1}{48}\sin^4 4x+C$

45. $\frac{1}{4}\left[2\sqrt{5}+\ln(2+\sqrt{5})\right]=1.479$

47. πab **49.** 32.7 kN **51.** 187 000 m^3

Review Exercises for Chapter 28, page 911

1. $-\frac{1}{8}e^{-8x}+C$ **3.** $-\dfrac{1}{\ln 2x}+C$ **5.** $4\ln 2=2.77$

7. $\frac{2}{35}\tan^{-1}\frac{7}{5}x+C$ **9.** 0 **11.** $\frac{1}{2}\ln 2=0.347$

13. $\frac{2}{3}\sin^3 t-\cos t+C$ **15.** $\tan^{-1}e^x+C$

17. $\frac{1}{9}\tan^3 3x+\frac{1}{3}\tan 3x+C$ **19.** $\ln\left|\dfrac{(x-1)^3}{x(2x+1)}\right|+C$

21. $\frac{3}{4}\tan^{-1}\dfrac{x^2}{2}+C$ **23.** $2\ln\left|2x+\sqrt{4x^2-9}\right|+C$

25. $\sqrt{e^{2x}+1}+C$ **27.** $2\ln|x|+\tan^{-1}\dfrac{x}{3}+C$ **29.** $\dfrac{\pi}{4}$

31. $-\frac{1}{2}x\cot 2x+\frac{1}{4}\ln|\sin 2x|+C$ **33.** $\dfrac{2}{u}+\ln|3u+1|+C$

35. $\frac{1}{2}\sin e^{2x}+C$ **37.** $3\sin 1=2.52$

39. $\frac{1}{2}u^2-3u+\ln|u+3|+C$

41. $\sqrt{5}\tan^{-1}\left(\dfrac{\sqrt{5}\cos x}{5}\right)-\cos x+C$

43. $x=\sqrt{2}\sin\theta;\ \dfrac{1}{\sqrt{2}}\displaystyle\int\csc\theta\,d\theta$ **45.** $2x^2+C$, $2x^2+C$

47. Formula 16, $u=x^4$, $du=4x^3dx$

49. $\frac{1}{3}(e^x+1)^3+C_1=\frac{1}{3}e^{3x}+e^{2x}+e^x+C_2$; $C_2=C_1+\frac{1}{3}$

51. $x-\tan x+\sec x+C$

53. Power rule with $u=x^2+4$, $du=2x\,dx$, and $n=-1/2$ is easier to use than a trigonometric substitution with $x=2\tan\theta$.

55. $y=\frac{1}{3}\tan^3 x+\tan x$ **57.** $2(e^3-1)=38.2$ **59.** 11.2

61. $\frac{1}{4}(8\tan^{-1}4-\ln 17)=1.94$ **63.** $4\pi(e^2-1)=80.3$

65. $\frac{1}{8}\pi(e^{2\pi}-1)=210$ **67.** $\ln 3=1.10$ **69.** 1.01 N · s

71. $\Delta S=a\ln T+bT+\frac{1}{2}cT^2+C$ **73.** 55.2 m

75. $v=98(1-e^{-0.1t})$ **77.** $\sqrt{2}$ **79.** $\frac{2}{3}k$

81. 3.47 cm^3 **83.** 73.0 m^2

Chapter 29: Partial Derivatives and Double Integrals

Exercises 29.1, page 916

1. 7 **3.** $V=\pi r^2h$ **5.** $A=\dfrac{2V}{r}+2\pi r^2$

7. $V=\dfrac{1}{4}\pi h(4r^2-h^2)$ **9.** 12 **11.** -2

13. $12+5y+4y^2$ **15.** $6xt+xt^2+t^3$

17. $\dfrac{p^2+pq+kp-p+2q^2+4kq+2k^2+5q+5k}{p+q+k}$

19. $2hx-2kx-2hy+h^2-2hk-4h$ **21.** 0

23. $81z^6-9z^5-2z^3$ **25.** $x=0$, $y<0$

27. $y>1$ **29.** 18 V **31.** 150 Pa

33. For a, b, and T with same sign: circle if $a=b$, ellipse if $a\neq b$; for a and b of different signs, hyperbola

35. 0.0278 A **37.** $A=\dfrac{pw-2w^2}{2}$, 3850 cm^2

39. $L=\dfrac{1.2\times 10^4}{l^2}r^4$

Exercises 29.2, page 923

1.

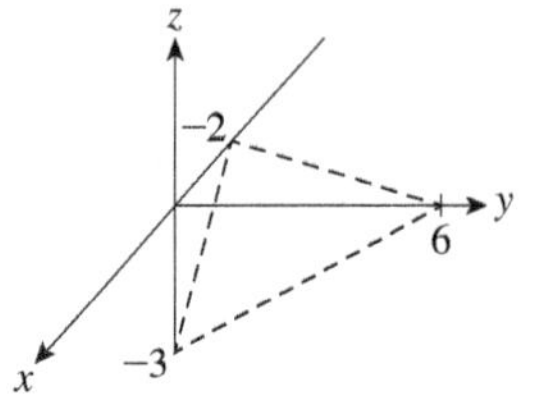

3.

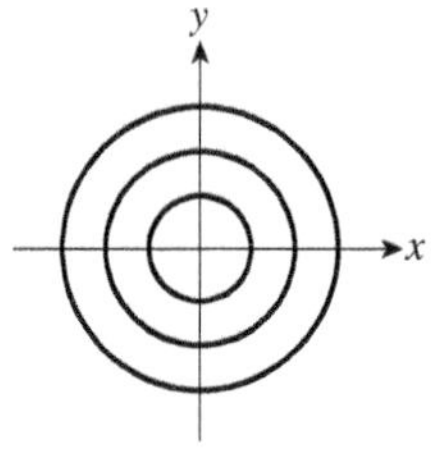

5.

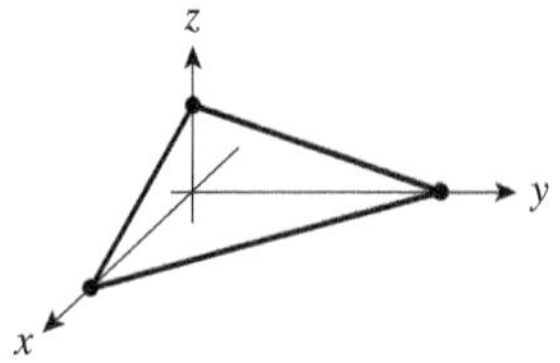

7.

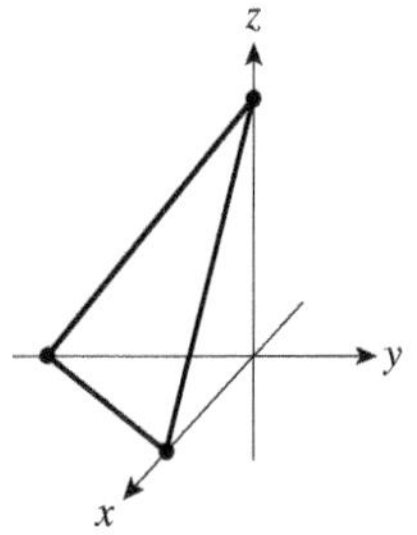

9.

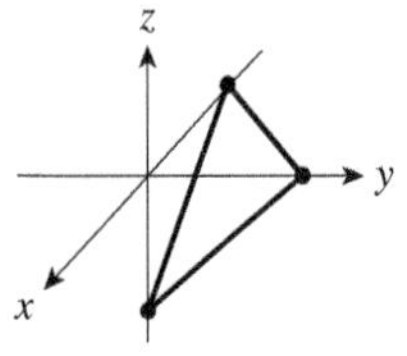

11.

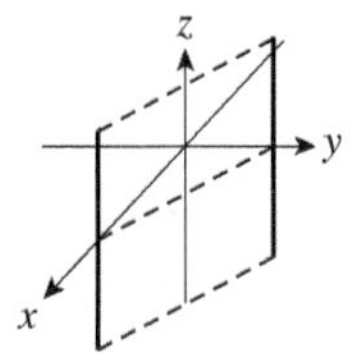

13.

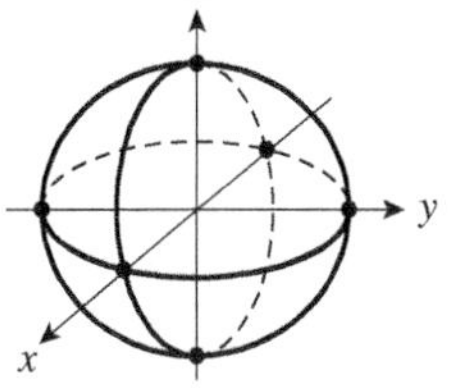

15.

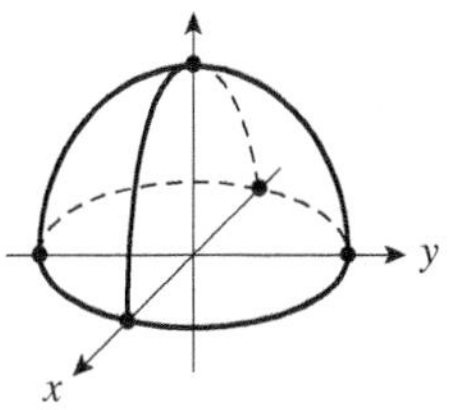

17.

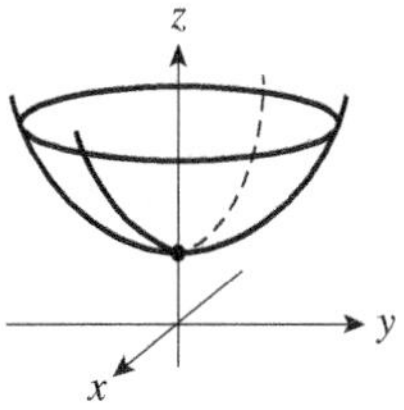

19.

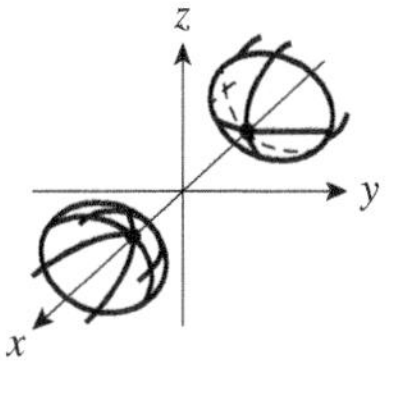

21.

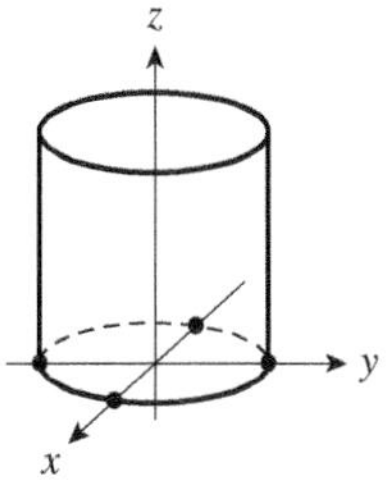

23.

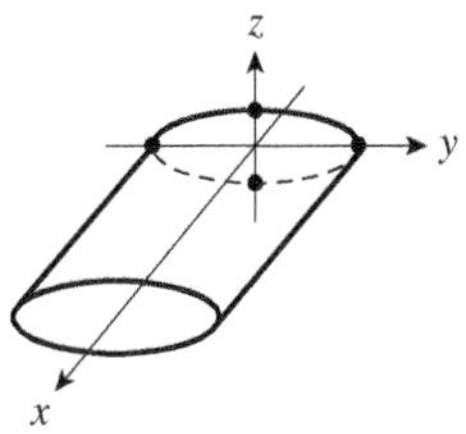

25. (a) $\left(\frac{3}{2}\sqrt{2}, \frac{3}{2}\sqrt{2}, 5\right)$ (b) $(0, 2, 3)$ (c) $(2, 2\sqrt{3}, 2)$

27. (a) cylinder, axis is z-axis, $r = 2$ (b) plane $\theta = 2$ for all r and z (c) plane $z = 2$ for all r and θ

29. $x^2 + y^2 = 4z$

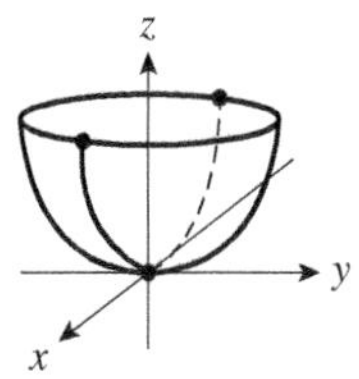

31.

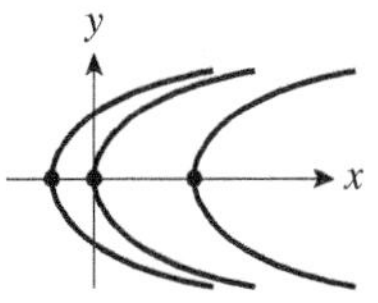

33.

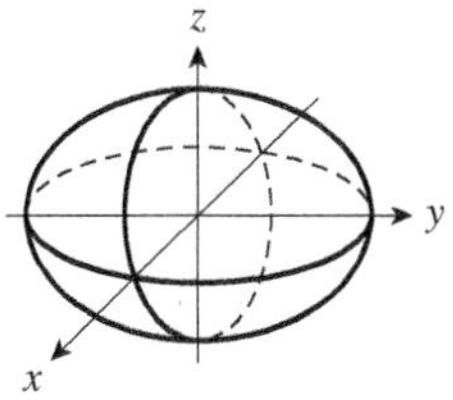

35.

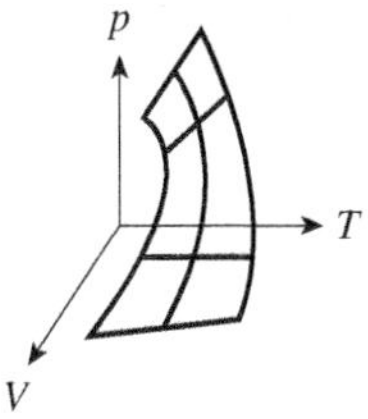

37.

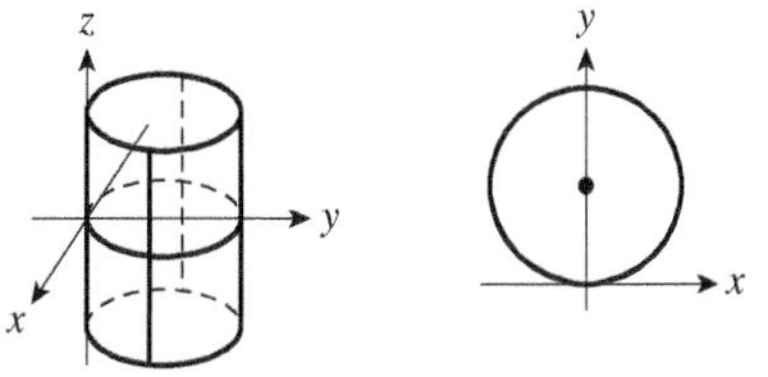

Exercises 29.3, page 927

1. $\dfrac{\partial z}{\partial x} = \dfrac{\ln y}{y^2 + 1}, \dfrac{\partial z}{\partial y} = \dfrac{x(y^2 + 1) - 2xy^2 \ln y}{y(y^2 + 1)^2}$

3. $\dfrac{\partial z}{\partial x} = 54x^5, \dfrac{\partial z}{\partial y} = -15y^4$

5. $\dfrac{\partial f}{\partial x} = 5x^3ye^{5xy} + 3x^2e^{5xy}, \dfrac{\partial f}{\partial y} = 5x^4e^{5xy}$

7. $\dfrac{\partial f}{\partial x} = -\dfrac{\sin x}{1 - \sec 3y}, \dfrac{\partial f}{\partial y} = \dfrac{3(2 + \cos x) \sec 3y \tan 3y}{(1 - \sec 3y)^2}$

9. $\dfrac{\partial \phi}{\partial r} = \dfrac{1 + 3rs}{\sqrt{1 + 2rs}}, \dfrac{\partial \phi}{\partial s} = \dfrac{r^2}{\sqrt{1 + 2rs}}$

11. $\dfrac{\partial z}{\partial x} = 4(2x + y^3)(x^2 + xy^3)^3, \dfrac{\partial z}{\partial y} = 12xy^2(x^2 + xy^3)^3$

13. $\dfrac{\partial z}{\partial x} = -y \sin xy, \dfrac{\partial z}{\partial y} = -x \sin xy$

15. $\dfrac{\partial y}{\partial r} = \dfrac{5r^4}{r^5 + s}, \dfrac{\partial y}{\partial s} = \dfrac{1}{r^5 + s}$

17. $\dfrac{\partial f}{\partial x} = \dfrac{12 \sin^2 2x \cos 2x}{1 - 3y}, \dfrac{\partial f}{\partial y} = \dfrac{6 \sin^3 2x}{(1 - 3y)^2}$

19. $\dfrac{\partial z}{\partial x} = \dfrac{3y + x^2y - 2x\sqrt{1 - x^2y^2} \sin^{-1}xy}{(3 + x^2)^2\sqrt{1 - x^2y^2}},$

$\dfrac{\partial z}{\partial y} = \dfrac{x}{(3 + x^2)\sqrt{1 - x^2y^2}}$

21. $\dfrac{\partial z}{\partial x} = \cos x - y \sin xy, \dfrac{\partial z}{\partial y} = -x \sin xy + \sin y$

23. $\dfrac{\partial f}{\partial x} = e^x(\cos xy - y \sin xy) - 2e^{-2x} \tan y,$

$\dfrac{\partial f}{\partial y} = -xe^x \sin xy + e^{-2x} \sec^2 y$

25. -8 **27.** $\frac{41}{4}$ **29.** $\dfrac{\partial^2 z}{\partial x^2} = -6y, \dfrac{\partial^2 z}{\partial y^2} = 12xy, \dfrac{\partial^2 z}{\partial x\,\partial y} = 6y^2 - 6x$

31. $\dfrac{\partial^2 z}{\partial x^2} = e^x \sin y, \dfrac{\partial^2 z}{\partial y^2} = \dfrac{2x}{y^3} - e^x \sin y,$

$\dfrac{\partial^2 z}{\partial x\,\partial y} = \dfrac{\partial^2 z}{\partial y\,\partial x} = -\dfrac{1}{y^2} + e^x \cos y$

33. $-4, -4$

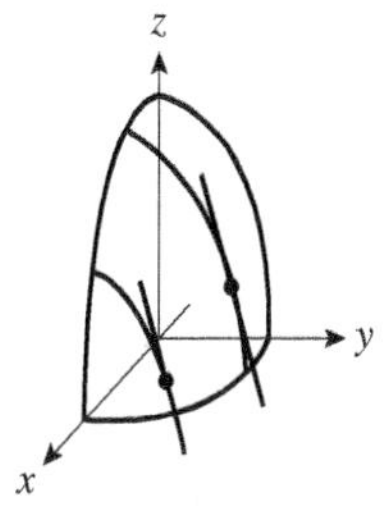

35. $\left(\dfrac{R_2}{R_1 + R_2}\right)^2$

37. 114 cm^2

39. 0.817

41. $3.75 \times 10^{-3}\ 1/\Omega$

43. $-5e^{-t} \sin 4x = \frac{1}{16}(-80e^{-t} \sin 4x)$

45. $\dfrac{\partial^2 u}{\partial x^2} = e^{-x} \sin y, \dfrac{\partial^2 u}{\partial y^2} = -e^{-x} \sin y$

Exercises 29.4, page 931

1. $\dfrac{3}{20}$ **3.** $\dfrac{28}{3}$ **5.** $\dfrac{127}{14}$ **7.** $\dfrac{1}{3}$ **9.** $\dfrac{\pi - 6}{12}$ **11.** 1

13. 495 **15.** $\dfrac{74}{5}$ **17.** $\dfrac{32}{3}$ **19.** 8π **21.** $\dfrac{28}{3}$ **23.** 18

25. 300 cm^3 **27.**

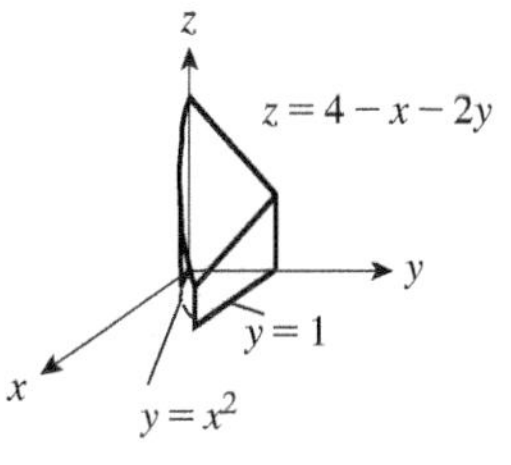

Review Exercises for Chapter 29, page 932

1. -52 **3.** $\dfrac{8s^4 - s^2}{4}$

5.

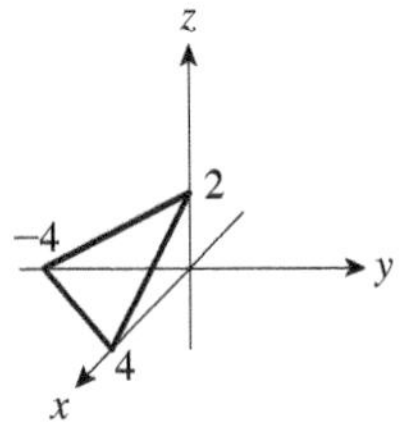

7.

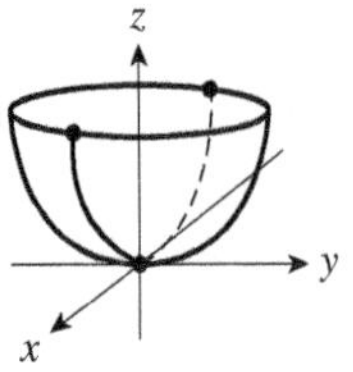

9. $\dfrac{\partial z}{\partial x} = 15x^2y^2 - 2y^4, \dfrac{\partial z}{\partial y} = 10x^3y - 8xy^3$

11. $\dfrac{\partial z}{\partial x} = \dfrac{x}{\sqrt{x^2 - 3y^2}}, \dfrac{\partial z}{\partial y} = \dfrac{-3y}{\sqrt{x^2 - 3y^2}}$

13. $\dfrac{\partial z}{\partial x} = \dfrac{2 - 2x^2y + 6xy^2}{(x^2y + 1)^2}, \dfrac{\partial z}{\partial y} = -\dfrac{3 + 2x^3}{(x^2y + 1)^2}$

15. $\dfrac{\partial u}{\partial x} = 2xy\cot(x^2 + 2y),$

$\dfrac{\partial u}{\partial y} = 2y\cot(x^2 + 2y) + \ln\sin(x^2 + 2y)$

17. $\dfrac{\partial z}{\partial x} = \dfrac{\partial z}{\partial y} = \dfrac{1}{2\sqrt{(x + y)(1 - x - y)}}$

19. $\dfrac{\partial^2 z}{\partial x^2} = 6y, \dfrac{\partial^2 z}{\partial y^2} = -6y, \dfrac{\partial^2 z}{\partial x \partial y} = 6x + 2$

21. 12 **23.** $\dfrac{21}{2}$ **25.** $\dfrac{e^2 - 3}{4} = 1.10$ **27.** $\dfrac{1}{6}$

29.

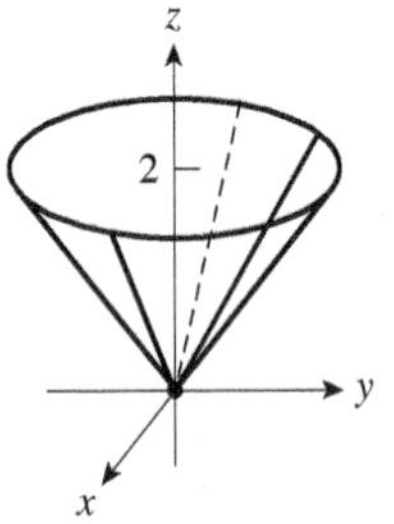

31.

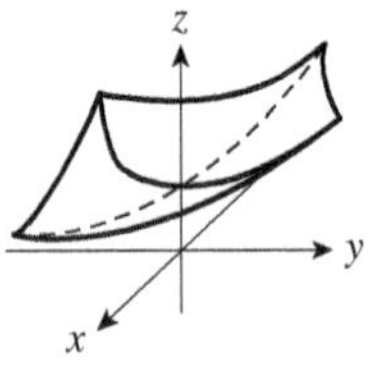

33. (a) $\theta = 3$ represents a vertical plane.
(b) $z = r^2$ represents a circular paraboloid.

35. $\dfrac{\partial v}{\partial r} = \dfrac{ER}{(r + R)^2}, \dfrac{\partial v}{\partial R} = \dfrac{-rE}{(r + R)^2}$ **37.** 0.982

39. $\dfrac{\pi}{\sqrt{gl}} = \dfrac{2\pi\sqrt{l/g}}{2l}$ **41.** $\dfrac{k_2 + 2k_3FT}{L_0 + k_1F + k_2T + k_3FT^2}$

43. $\left(\dfrac{\partial V}{\partial T}\right)\left(\dfrac{\partial T}{\partial p}\right)\left(\dfrac{\partial p}{\partial V}\right) = -\left(\dfrac{nR}{p}\right)\left(\dfrac{V}{nR}\right)\left(\dfrac{pV}{V^2}\right) = -1$

45. $32\left(\pi - \dfrac{2}{3}\right) = 79.20$. The bounding surfaces are a cylinder with axis along the z-axis, and a plane parallel to the y-axis

47. 1.70×10^{11} km^3 **49.** $\dfrac{2k(b^2 + ab + a^2)}{3(b + a)}$

Chapter 30: Expansion of Functions in Series

Exercises 30.1, page 937

1. Converges; $S_1 = 0.5, S_2 = 0.75, S_3 = 0.875, S_4 = 0.9375$

3. 1, 4, 9, 16 **5.** $\frac{1}{2}, \frac{1}{3}, \frac{1}{4}, \frac{1}{5}$

7. (a) $-\frac{2}{5}, \frac{4}{25}, -\frac{8}{125}, \frac{16}{625}$ (b) $-\frac{2}{5} + \frac{4}{25} - \frac{8}{125} + \frac{16}{625} - \cdots$

9. (a) $1, 0, -1, 0$ (b) $1 + 0 - 1 + 0 + \cdots$

11. $a_n = \dfrac{1}{n + 1}$ **13.** $a_n = \dfrac{(-1)^{n+1}}{(n + 1)(n + 2)}$

15. 1, 1.125, 1.162 037 0, 1.177 662 0, 1.185 662 0; convergent; 1.2

17. 1, 1.5, 2.166 666 7, 2.916 666 7, 3.716 666 7; divergent

19. 0, 1, 2.4142, 4.1463, 6.1463; divergent

21. 0.75, 0.888 888 9, 0.937 500 0, 0.960 000 0, 0.972 222 2; convergent; 1

23. 0.210 367 7, 0.267 198 8, 0.269 403 8, 0.266 447 6, 0.265 511 1; convergent; 0.26

25. Divergent **27.** Convergent, $S = \frac{3}{4}$

29. Convergent, $S = 100$ **31.** Convergent, $S = \frac{4096}{9}$

33. $3 < x < 5$ **35.** (a) 1 (b) Diverges

37. 1

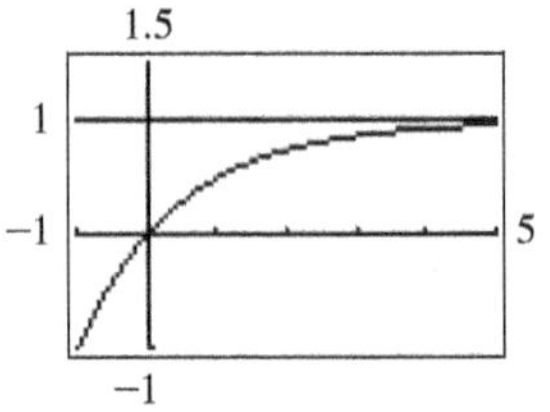

39. (a) Diverges
(b) $y = 2100(1.05^x - 1)$

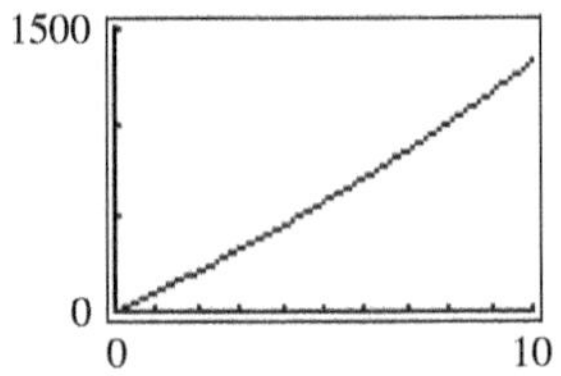

41. $r = x; S = \dfrac{1}{1 - x}$ **43.** 2, 4, 12, 48, 240

45. Converges: partial sums converge to 1.

Exercises 30.2, page 941

1. $\dfrac{2}{2 + x} = 1 - \dfrac{1}{2}x + \dfrac{1}{4}x^2 - \dfrac{1}{8}x^3 + \cdots$

3. $1 + x + \frac{1}{2}x^2 + \cdots$ **5.** $1 - \frac{1}{2}x^2 + \frac{1}{24}x^4 - \cdots$

7. $1 + \frac{1}{2}x - \frac{1}{8}x^2 + \cdots$ **9.** $1 - 2x + 2x^2 - \cdots$

11. $1 - 8\pi^2x^2 + \frac{32}{3}\pi^4x^4 - \cdots$ **13.** $1 + x + x^2 + \cdots$

15. $-2x - 2x^2 - \frac{8}{3}x^3 - \cdots$ **17.** $1 - x^2 + \frac{1}{3}x^4 - \cdots$

19. $\dfrac{\sqrt{2}}{2}\left(1 + x - \dfrac{1}{2}x^2 - \cdots\right)$ **21.** $x - \frac{1}{3}x^3 + \cdots$

23. $x + \frac{1}{3}x^3 + \cdots$ **25.** $-\frac{1}{2}x^2 - \frac{1}{12}x^4 - \cdots$

27. $1 + \frac{1}{2}x - \cdots$

29. No. Functions are not defined at $x = 0$.

31. $e^x = 1 + x + \dfrac{x^2}{2} + \cdots, e^{x^2} = 1 + x^2 + \dfrac{x^4}{2} + \cdots$

33. $f(x) = 1 + 3x + \frac{9}{2}x^2 + \cdots; L(x) = 1 + 3x$

35. $1 + \dfrac{x^2}{2!} + \dfrac{x^4}{4!} + \cdots$ **37.** $f(x) = x^2$

39. $4 - 0.8t - 1.92t^2 + \cdots$

41. $R = e^{-0.001t} = 1 - 0.001t + (5 \times 10^{-7})t^2 - \cdots$

Exercises 30.3, page 946

1. $e^{2x^2} = 1 + 2x^2 + 2x^4 + \frac{4}{3}x^6 + \cdots$

3. $1 + 3x + \frac{9}{2}x^2 + \frac{9}{2}x^3 + \cdots$

5. $\dfrac{x}{2} - \dfrac{x^3}{2^3 3!} + \dfrac{x^5}{2^5 5!} - \dfrac{x^7}{2^7 7!} + \cdots$

7. $x - 32x^3 + \frac{512}{3}x^5 - \frac{16\,384}{45}x^7 + \cdots$

9. $x^2 - \frac{1}{2}x^4 + \frac{1}{3}x^6 - \frac{1}{4}x^8 + \cdots$ **11.** 0.310 **13.** 0.190

15. $2(1 + x^2 + x^4 + x^6 + \cdots)$ **17.** $x + x^2 + \frac{1}{3}x^3 + \cdots$

19. $-2x^3 - x^4 - \dfrac{2x^5}{3} - \dfrac{x^6}{2} - \cdots$

21. $x - \frac{1}{2}x^2 + \frac{1}{6}x^3 + \frac{1}{6}x^4 - \cdots$

23. $\dfrac{d}{dx}\left(1 + x + \dfrac{1}{2}x^2 + \dfrac{1}{6}x^3 + \cdots\right) = 1 + x + \dfrac{1}{2}x^2 + \cdots$

25. $\displaystyle\int \cos x\,dx = x - \frac{x^3}{3!} + \cdots$

27. $4 - 0.8t - 1.92t^2 + 0.4t^3 + \cdots$

29. $\displaystyle\int_0^1 e^x\,dx = 1.718\,281\,8,$

$\displaystyle\int_0^1 \left(1 + x + \tfrac{1}{2}x^2 + \tfrac{1}{6}x^3\right)dx = 1.708\,333\,3$

31. $-\frac{1}{6}$ **33.** 0.003 10 **35.** 0.200

37. $\left[\left(1 - \dfrac{v^2}{c^2}\right)^{-1/2} - 1\right]mc^2$

$= \left[1 + \left(-\dfrac{1}{2}\right)\left(-\dfrac{v^2}{c^2}\right) + \dfrac{\left(-\frac{1}{2}\right)\left(-\frac{3}{2}\right)}{2}\left(-\dfrac{v^2}{c^2}\right)^2 + \cdots - 1\right]mc^2$

$= 1 + \dfrac{1}{2}mv^2 + \dfrac{3}{8}m\dfrac{v^4}{c^2} + \cdots - 1 = \dfrac{1}{2}mv^2$, if v is much smaller than c.

39. $\dfrac{1}{1-x} = \displaystyle\sum_{n=0}^{\infty} x^n$; differentiate on the left by the power rule and on the right term by term.

41.

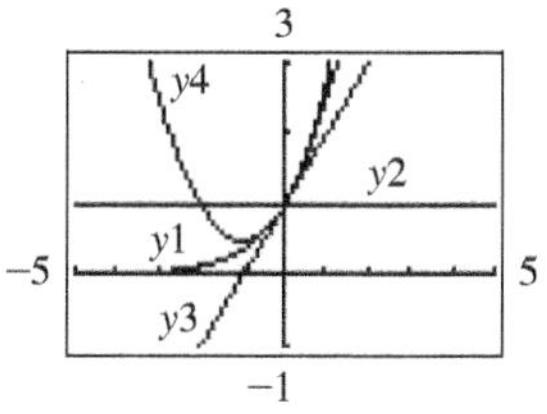

43.

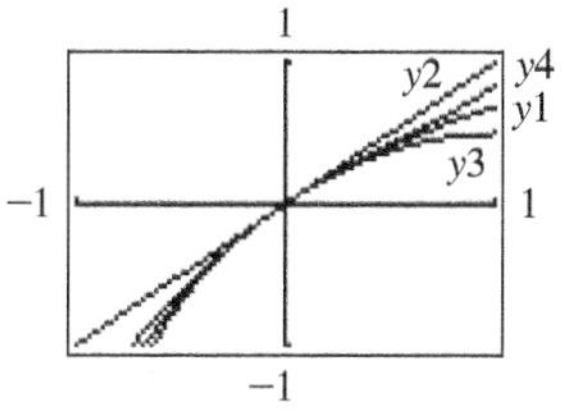

Exercises 30.4, page 949

1. 0.905 **3.** 1.22, 1.221 402 8 **5.** 0.124 674 5, 0.124 674 733

7. 2.718 055 6, 2.718 281 8 **9.** 0.998 496 77, 0.998 497 15

11. 0.334 933 3, 0.336 472 2 **13.** 0.354 613 0, 0.354 612 9

15. −0.064 858 561 5, −0.064 858 572 4

17. 1.093 375, 1.093 443 3 **19.** 0.981 16, 0.981 118 50

21. 1.052 352 8 **23.** 0.987 446 2 **25.** 2.54×10^{-7}

27. 3.77×10^{-7} **29.** 1.9799 **31.** 3.146

33. The terms of the expansion for e^x after those on the right side of the inequality have a positive value.

35. 1.59 years **37.** $i = \dfrac{E}{L}\left(t - \dfrac{Rt^2}{2L}\right)$; small values of t

39. 18 m

Exercises 30.5, page 952

1. $\sqrt{x} = 1 + \dfrac{1}{2}(x-1) - \dfrac{1}{8}(x-1)^2 + \dfrac{1}{16}(x-1)^3 - \cdots$

3. 3.32 **5.** 2.049 **7.** 0.5299 **9.** 0.492 88

11. $e^{-2}\left[1 - (x-2) + \dfrac{(x-2)^2}{2!} - \cdots\right]$

13. $\dfrac{1}{2}\left[\sqrt{3} + \left(x - \dfrac{1}{3}\pi\right) - \dfrac{\sqrt{3}}{2!}\left(x - \dfrac{1}{3}\pi\right)^2 - \cdots\right]$

15. $2 + \frac{1}{12}(x-8) - \frac{1}{288}(x-8)^2 + \cdots$

17. $1 + 2\left(x - \frac{1}{4}\pi\right) + 2\left(x - \frac{1}{4}\pi\right)^2 + \cdots$

19. $e^{\pi/2}\left[1 + \left(x - \dfrac{\pi}{2}\right) - \dfrac{1}{3}\left(x - \dfrac{\pi}{2}\right)^3 + \cdots\right]$

21. $\dfrac{1}{5} - \dfrac{x-3}{25} + \dfrac{(x-3)^2}{125} - \cdots$

23. 23.1308 **25.** 3.0496 **27.** 2.0247 **29.** 0.874 62

31. Use the indicated method.

33. $2x^3 + x^2 - 3x + 5 = 5 + 5(x-1) + 14\dfrac{(x-1)^2}{2} + 12\dfrac{(x-1)^3}{6}$

35. 0.515 040 8, 0.515 038 8, 0.515 038 1

37. $6\sin\dfrac{\pi^2}{2} + 6\pi\cos\dfrac{\pi^2}{2}\left(t - \dfrac{\pi}{2}\right) - 3\pi^2\sin\dfrac{\pi^2}{3}\left(t - \dfrac{\pi}{2}\right)^2 + \cdots$

39. Graph of part (b) fits well near $x = \pi/3$.

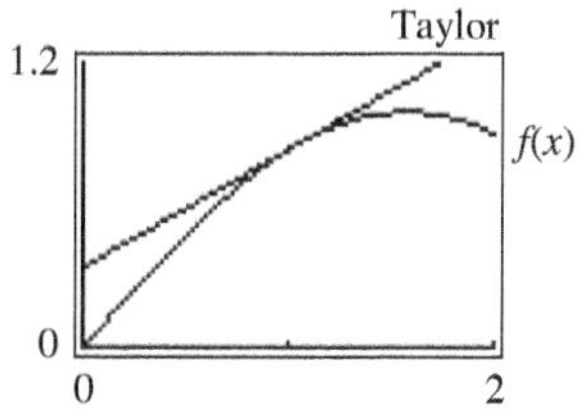

41. Graph of part (b) fits well near $x = 2$.

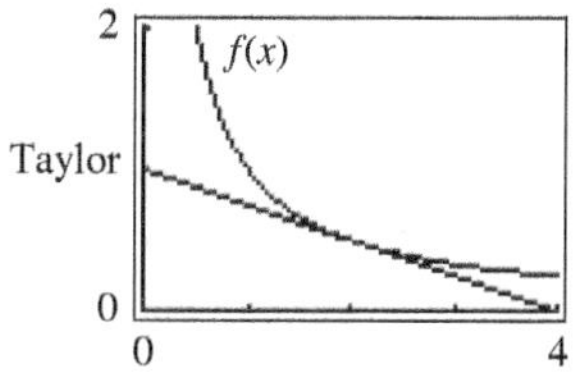

Exercises 30.6, page 957

1. $f(x) = \dfrac{8}{\pi}\left(\sin x + \dfrac{1}{3}\sin 3x + \dfrac{1}{5}\sin 5x + \cdots\right)$

3. $f(x) = \dfrac{1}{2} - \dfrac{2}{\pi}\sin x - \dfrac{2}{3\pi}\sin 3x - \cdots$

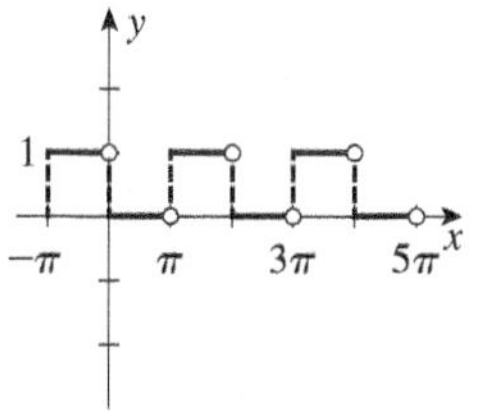

5. $f(x) = \dfrac{3}{2} + \dfrac{2}{\pi}\sin x + \dfrac{2}{3\pi}\sin 3x + \cdots$

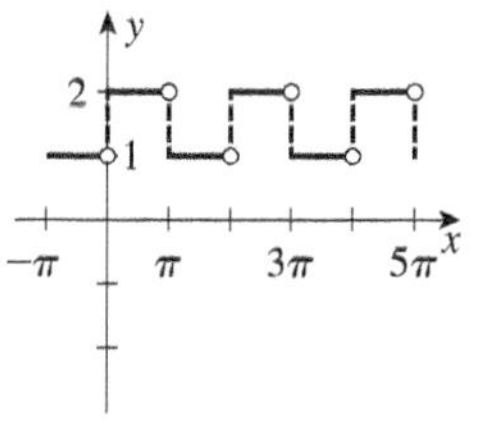

7. $f(x) = \frac{\pi}{4} - \frac{2}{\pi}\left(\cos x + \frac{1}{9}\cos 3x + \cdots\right)$
$+ \left(\sin x - \frac{1}{2}\sin 2x + \cdots\right)$

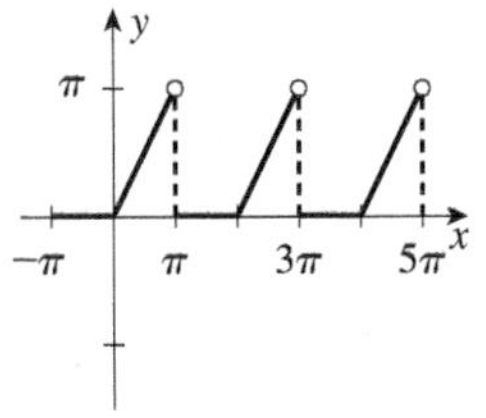

9. $f(x) = -\frac{1}{4} - \frac{1}{\pi}\cos x + \frac{1}{3\pi}\cos 3x - \cdots$
$+ \frac{3}{\pi}\sin x - \frac{1}{\pi}\sin 2x + \frac{1}{\pi}\sin 3x - \cdots$

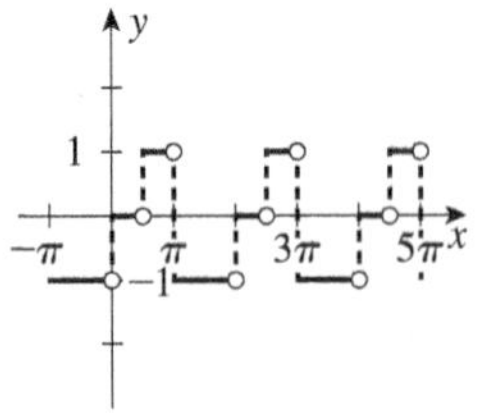

11. $f(x) = \frac{\pi}{2} - \frac{4}{\pi}\cos x - \frac{4}{9\pi}\cos 3x - \cdots$

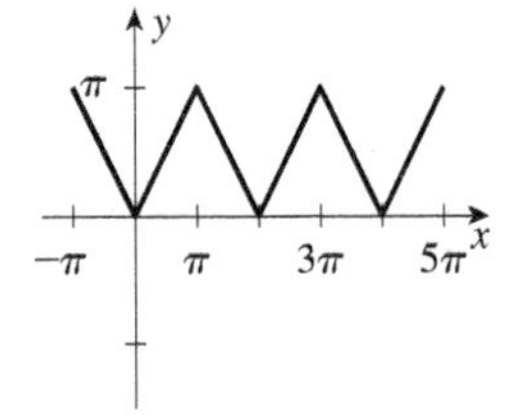

13. $\frac{e^{\pi} - e^{-\pi}}{2\pi}\left(1 - \cos x + \frac{2}{5}\cos 2x - \cdots + \sin x - \frac{4}{5}\sin 2x \cdots\right)$

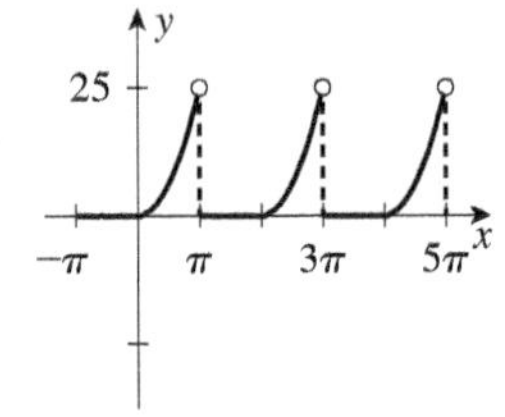

15.

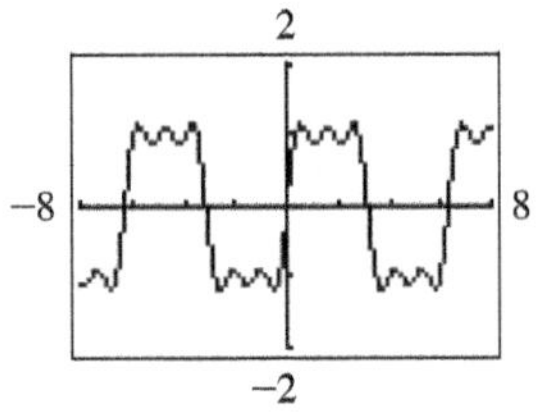

17.

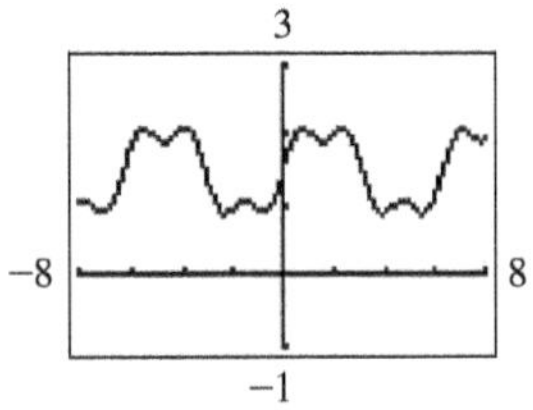

19.

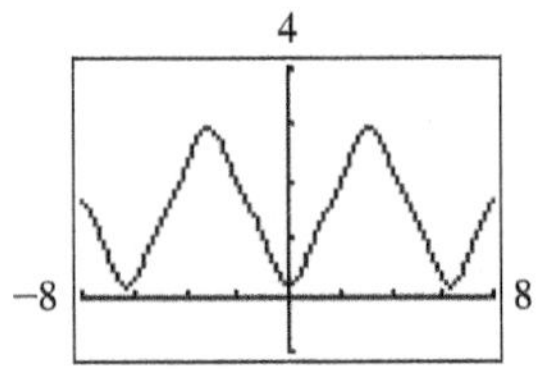

21. $\frac{\pi(2\pi + 3)}{12} - \frac{2\pi + 2}{\pi}\cos t + \frac{1}{2}\cos 2t - \cdots$
$+ \frac{\pi^2 + \pi - 4}{\pi}\sin t - \frac{\pi + 1}{2}\sin 2t + \cdots$

23. $f(t) = \frac{2}{\pi} - \frac{4}{3\pi}\cos 2t - \frac{4}{15\pi}\cos 4t - \cdots$

Exercises 30.7, page 963

1. $\frac{5}{2} + \frac{2}{\pi}(\cos x - \frac{1}{3}\cos 3x + \frac{1}{5}\cos 5x - \cdots)$

3. $-1 + \frac{4}{\pi}\sin\frac{\pi x}{4} + \frac{4}{3\pi}\sin\frac{3\pi x}{4} + \cdots$

5. Neither **7.** Even **9.** Even **11.** Odd

13. Sine terms **15.** Sine terms and cosine terms

17. $f(x) = \frac{5}{2} - \frac{10}{\pi}\left(\sin\frac{\pi x}{3} + \frac{1}{3}\sin \pi x - \cdots\right)$

19. $f(x) = 1 + \frac{4}{\pi}\cos\frac{\pi x}{2} - \frac{4}{3\pi}\cos\frac{3\pi x}{2} + \cdots$

21. $f(x) = 2 - \frac{16}{\pi^2}\left(\cos\frac{\pi x}{4} + \frac{1}{9}\cos\frac{3\pi x}{4} + \cdots\right)$

23. $f(x) = \frac{4}{\pi}\left(\sin\frac{\pi x}{4} + \frac{1}{3}\sin\frac{3\pi x}{4} + \frac{1}{5}\sin\frac{5\pi x}{4} + \cdots\right)$

25. $f(x) = \frac{4}{3} - \frac{16}{\pi^2}\left(\cos\frac{\pi x}{2} - \frac{1}{4}\cos \pi x + \frac{1}{9}\cos\frac{3\pi x}{2} - \cdots\right)$

27. $f(t) = 2 + \frac{8}{\pi}\left(\cos\frac{\pi}{2}t - \frac{1}{3}\cos\frac{3\pi}{2}t + \cdots + \sin\frac{\pi}{2}t\right.$
$\left. + \sin \pi t + \frac{1}{3}\sin\frac{3\pi}{2}t + \cdots\right)$

Review Exercises for Chapter 30, page 964

1. $\frac{1}{2} - \frac{1}{4}x + \frac{1}{48}x^3 - \cdots$ **3.** $2x^2 - \frac{4}{3}x^6 + \frac{4}{15}x^{10} - \cdots$

5. $1 + \frac{1}{3}x - \frac{1}{9}x^2 + \cdots$ **7.** $x + \frac{1}{6}x^3 + \frac{3}{40}x^5 + \cdots$

9. $\cos a - (\sin a)x - (\cos a)\frac{x^2}{2} + \cdots$

11. 0.82 **13.** 1.09 **15.** 0.921 **17.** −0.202

19. 0.953 **21.** 12.1655 **23.** 0.259

25. $\frac{1}{2} - \frac{1}{2}\sqrt{3}\left(x - \frac{1}{3}\pi\right) - \frac{1}{4}\left(x - \frac{1}{3}\pi\right)^2 + \cdots$

27. $f(x) = \frac{\pi - 2}{4} - \frac{2}{\pi}\left(\cos x + \frac{1}{9}\cos 3x + \cdots\right)$
$+ \left(\frac{\pi - 2}{\pi}\right)\sin x - \frac{1}{2}\sin 2x + \cdots$

29. $f(x) = \pi + \frac{4}{\pi}\left(\sin\frac{\pi x}{4} + \frac{1}{3}\sin\frac{3\pi x}{4} + \cdots\right)$

31. $f(x) = \frac{1}{2} + \frac{2}{\pi}\left(\cos x - \frac{1}{3}\cos 3x + \cdots\right)$

33. $f(x) = \frac{4}{\pi}\left(\sin\frac{\pi x}{2} - \frac{1}{2}\sin \pi x + \frac{1}{3}\sin\frac{3\pi x}{2} - \cdots\right)$

35. Convergent, $S = 5000$ **37.** $S = 256$

39. $\frac{1}{3}x^3 - \frac{1}{42}x^7 + \cdots$ **41.** $1 + 2\left(x - \frac{\pi}{4}\right) + 2\left(x - \frac{\pi}{2}\right)^2 + \cdots$

43. $2 + 4\left(x - \frac{\pi}{4}\right) + \cdots$

45. $(x+h) - \frac{(x+h)^3}{3!} + \cdots - (x-h) + \frac{(x-h)^3}{3!} - \frac{1}{315}x^8 + \cdots$

$= 2h - \frac{2hx^2}{2!} + \cdots = 2h\left(1 - \frac{x^2}{2!} + \cdots\right)$

47. $4x - 2x^2 + \frac{4}{3}x^3 - x^4 + \cdots$ **49.** $x^2 - \frac{1}{3}x^4 + \frac{2}{45}x^6 - \cdots$

51. $1 + \frac{x^2}{2} + \frac{5x^4}{24} + \cdots$ **53.** $1 - x + x^2 - \cdots$

55. $\frac{1}{3} + \frac{4}{\pi^2}\left(-\cos \pi x + \frac{1}{4}\cos 2\pi x - \frac{1}{9}\cos 3\pi x + \cdots\right)$

57. 2.4265, 2.4600, 2.459 603 1 **59.** 0.002 496 88

61. $x - \frac{x^3}{3} + \frac{x^5}{5} - \cdots$ **63.** $3.2\left[1 - \frac{(880\pi t)^2}{2} + \frac{(880\pi t)^4}{24} - \cdots\right]$

65. $N_0\left(1 - \lambda t + \frac{\lambda^2 t^2}{2} - \frac{\lambda^3 t^3}{6} + \cdots\right)$

67. 7.8 m **69.** $2x + \frac{2}{3}x^3 + \frac{2}{5}x^5 + \frac{2}{7}x^7 + \cdots$

71. $N_0[1 + e^{-k/T} + (e^{-k/T})^2 + \cdots] = N_0(1 + e^{-k/T} + e^{-2k/T} + \cdots)$

73. $f(t) = \frac{1}{2\pi} + \frac{1}{\pi}\left(\frac{1}{2}\cos t - \frac{1}{3}\cos 2t + \cdots\right) + \frac{1}{4}\sin t + \frac{2}{3\pi}\sin 2t + \cdots$

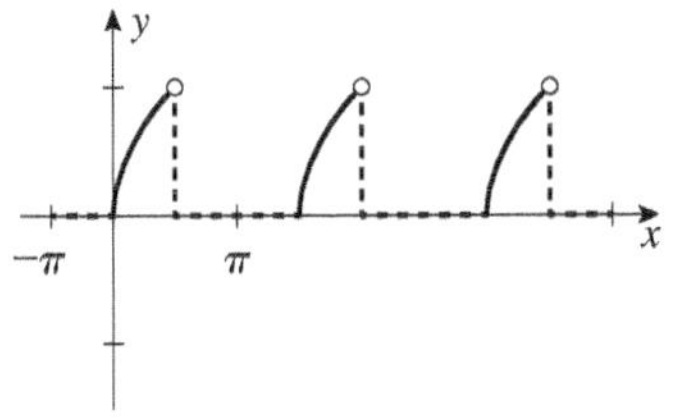

Chapter 31: Differential Equations

Exercises 31.1, page 969

1. $(c_1e^{-x} + 4c_2e^{2x}) - (-c_1e^{-x} + 2c_2e^{2x}) = 2(c_1e^{-x} + c_2e^{2x})$; $(4e^{-x}) - (-4e^{-x}) = 2(4e^{-x})$

3. Particular solution

5. General solution
(The following "answers" are the unsimplified expressions obtained by substituting functions and derivatives.)

7. $e^x - (e^x - 1) = 1$; $5e^x - (5e^x - 1) = 1$

9. $-12\cos 2x + 4(3\cos 2x) = 0$; $(-4c_1 \sin 2x - 4c_2 \cos 2x) + 4(c_1 \sin 2x + c_2 \cos 2x) = 0$

11. $2x = 2x$ **13.** $8 - 12x^2 = 8 - 12x^2$

15. $(-2ce^{-2x} + 1) + 2\left(ce^{-2x} + x - \frac{1}{2}\right) = 2x$

17. $-\frac{1}{2}\cos x + \frac{9}{2}\cos x = 4\cos x$

19. $x^2\left[-\frac{c^2}{(x-c)^2}\right] + \left[\frac{cx}{(x-c)}\right]^2 = 0$ **21.** $x\left(-\frac{c_1}{x^2}\right) + \frac{c_1}{x} = 0$

23. $(\cos x - \sin x + e^{-x}) + (\sin x + \cos x - e^{-x}) = 2\cos x$

25. $(c_1e^x + 4c_2e^{2x}) - 3(c_1e^x + 2c_2e^{2x}) + 2\left(c_1e^x + c_2e^{2x} + \frac{3}{2}\right) = 3$

27. $\cos x\left[\frac{(\sec x + \tan x) - (x + c)(\sec x \tan x + \sec^2 x)}{(\sec x + \tan x)^2}\right] + \sin x = 1 - \frac{x + c}{\sec x + \tan x}$

29. $c^2 + cx = cx + c^2$ **31.** $c_3e^x = c_3e^x$

33. $y = x^3 + c_1x^2 - 4$ **35.** $kN_0e^{kt} = kN_0e^{kt}$

Exercises 31.2, page 973

1. $y(x^2 + 1) = c$ **3.** $y^3 = \frac{3}{2}x^2 + c$

5. $\frac{1}{2}s^2 - s = \sin t + c$ **7.** $y = c - x^2$ **9.** $x - \frac{1}{y} = c$

11. $\ln V = \frac{1}{P} + c$ **13.** $\ln(x^3 + 5) + 3y = c$

15. $y = 2x^2 + x - x\ln x + c$ **17.** $4\sqrt{1 - y} = e^{-x^2} + c$

19. $e^x - e^{-y} = c$; e^{x+y} is the same as e^xe^y. Divide each term by e^y and integrate.

21. $\ln(y + 4) = x + c$ **23.** $y(1 + \ln x)^2 + cy + 2 = 0$

25. $\tan^2 x + 2\ln y = c$ **27.** $x^2 + 1 + x\ln y + cx = 0$

29. $y - e^{\cos\theta} = c$ **31.** $e^{x^2} = 2e^y + c$

33. $i = c - (\ln t)^2$ **35.** $(y^2 - 1)(x^3 + 1) = c$

37. $3\ln y + x^3 = 3$ **39.** $2\ln(1 - y) = 1 - 2\sin x$

41. $e^{2x} - \frac{2}{y} = 2(e^x - 1)$

43. $\frac{1}{2}\ln(x^2 + 1) + \ln y = \ln c$; $\frac{1}{2}\ln 1 + \ln e = \ln c$; $c = e$

45. $T = 10 + 30e^{-0.15t}$ **47.** $y^2 + x^2 = c$; circles of radius $\sqrt{c}$

Exercises 31.3, page 975

1. $\ln xy + y^2 = c$ **3.** $2xy + x^2 = c$ **5.** $x^3 - 2y = cx - 4$

7. $A^2r - r = cA$ **9.** $y\sin x = x + c$ **11.** $2\sqrt{x^2 + y^2} = x + c$

13. $y = c - \frac{1}{2}\ln\sin(x^2 + y^2)$

15. $\ln(y^2 - x^2) + 2x = c$; subtract $x\,dx$ from each side and divide through by $y^2 - x^2$.

17. $5xy^2 + y^3 = c$ **19.** $2xy + x^3 = 5$ **21.** $2x = 2xy^2 - 15y$

23. $2x + 1 = \cos xy$ **25.** $ye^{-x} = y^2 + c$

Exercises 31.4, page 978

1. $y = x + cx^{-2}$ **3.** $y = e^{-x}(x + c)$ **5.** $y = -\frac{1}{2}e^{-4x} + ce^{-2x}$

7. $y = -2 + ce^{5x}$ **9.** $y = ce^{-x^3} + 2$

11. $y = \frac{8}{7}x^3 + \frac{c}{\sqrt{x}}$ **13.** $r = -\cot\theta + c\csc\theta$

15. $y = (x + c)\csc x$ **17.** $y = x + \frac{1}{2}e^x - 1 + ce^{-x}$

19. $2s = e^{4t}(t^2 + c)$ **21.** $y = \frac{1}{4} + ce^{-x^4}$

23. $3y = x^4 - 6x^2 - 3 + cx$ **25.** $y = ce^{-\sqrt{x^2+1}} - 1$

27. $r = \sin\theta[\ln(\tan\theta + \sec\theta)] + c\sin\theta$

29. Can solve by separation of variables: $\frac{dy}{1 - y} = 2\,dx$. Can also solve as linear differential equation of first order: $dy + 2y\,dx = 2\,dx$; $y = 1 + ce^{-2x}$

31. $y = e^{-x} + 3e^{-2x}$ **33.** $y = \frac{4}{3}\sin x - \csc^2 x$

35. $y(\csc x - \cot x) = \ln\frac{(\sqrt{2} - 1)(\csc 2x - \cot 2x)}{\csc x - \cot x}$

37. $u' - P(x)u = -Q(x)$ **39.** $i = \frac{V}{R}(1 - e^{-Rt/L})$

41. $L = 25 + ce^{-0.8t}$

Exercises 31.5, page 980

1.

x	0.0	0.2	0.4	0.6	0.8	1.0
y	1.00	1.20	1.44	1.72	2.04	2.40
y (correct)	1.00	1.22	1.48	1.78	2.12	2.50

$y = \frac{1}{2}x^2 + x + 1$

3.

x	−0.2	−0.1	0.0	0.1	0.2	1.3
y	2.0000	2.1840	2.3937	2.6330	2.9069	3.2208
y (correct)	2.00	2.20	2.42	2.68	2.98	3.33

$y = 2.4233e^{0.2x^2+x}$

5.

x	0.0	0.1	0.2	0.3	0.4	0.5	0.6	0.7
y	1.00	1.10	1.21	1.33	1.46	1.60	1.75	1.91

x	0.8	0.9	1.0
y	2.08	2.26	2.45

7. (Not all values shown)

x	−0.2	−0.1	0.0	0.1	0.2	0.3
y	2.000	2.1903	2.4079	2.6573	2.9436	3.2732

9.

x	0.0	0.1	0.2	0.3	0.4
y	0.0000	0.1003	0.2027	0.3092	0.4220

11.

x	0.0	0.2	0.4	0.6	0.8	1.0
y	0.0000	0.2027	0.4232	0.6884	1.0588	1.7722

13.

x	0.0	0.1	0.2	0.3	0.4	0.5	0.6
y	1.5708	1.5660	1.5521	1.5302	1.5011	1.4656	1.4244

15. $y_3 = 12$; $y_{\text{actual}} = 36.2$. Euler method is too inaccurate for larger values of x or Δx.

17. $i_{\text{approx}} = 0.0804\text{A}$, $i_{\text{exact}} = 0.0898\text{A}$

Exercises 31.6, page 985

1. $y^2 + 2x^2 = c$ **3.** 2.46 kg

5. $y^2 = 2x^2 + 1$

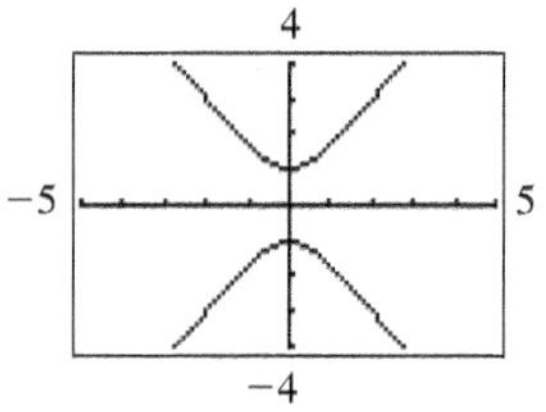

7. $y = 2e^x - x - 1$

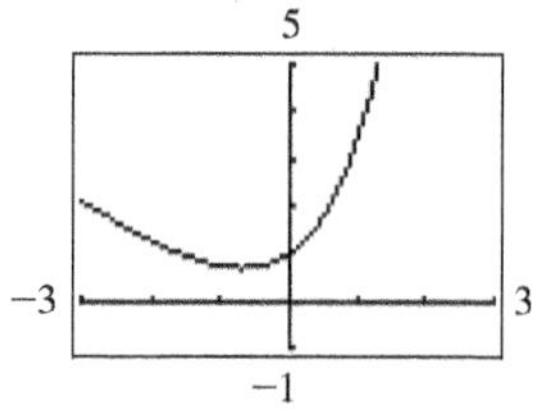

9. $y^2 = c - 2x$

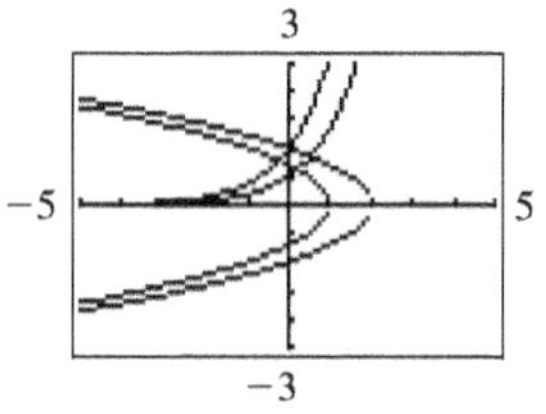

11. $y^2 = c - 2\sin x$

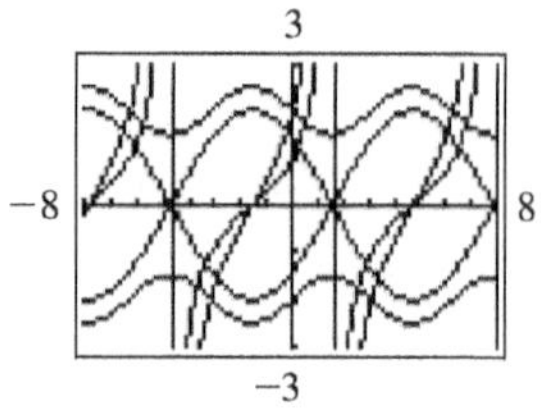

13. 76.9% **15.** 3.82 days **17.** $N = \frac{r}{k}(1 - e^{-kt})$

19. 45.3 million **21.** $S = a + \frac{c}{r^2}$ **23.** $5250e^{0.0196t}$

25. 12 min **27.** \$1040.81 **29.** $c = c_0(1 - e^{-kt})$

31. $\lim_{t\to\infty}\frac{E}{R}(1 - e^{-Rt/L}) = \frac{E}{R}$

33. $i = \frac{E}{R^2 + \omega^2L^2}(R\sin\omega t - \omega L\cos\omega t + \omega Le^{-Rt/L})$

35. $q = q_0e^{-t/RC}$ **37.** $v = 9.8(1 - e^{-t})$, 9.8 **39.** 37 m/s

41. $x = 3t^2 - t^3$, $y = 6t^2 - 2t^3 - 9t^4 + 6t^5 - t^6$

43. $p = 100(0.8)^{h/2000}$ **45.** \$4980 **47.** $x = 0.20 + 0.15^{-5.0t}$

49. $y = 4e^{-x/2} + c_1$

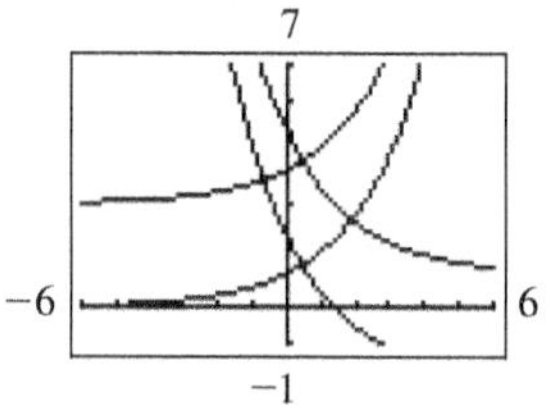

Exercises 31.7, page 990

1. $y = c_1 + c_2e^{5x}$ **3.** $y = c_1e^{3x} + c_2e^{-2x}$

5. $y = c_1e^{-x} + c_2e^{-x/3}$ **7.** $y = c_1 + c_2e^{3x}$

9. $y = c_1e^{3x/2} + c_2e^{-x}$ **11.** $y = c_1e^{6x} + c_2e^{2x/3}$

13. $y = c_1e^{x/3} + c_2e^{-3x}$ **15.** $y = c_1e^{x/3} + c_2e^{-x}$

17. $y = e^x(c_1e^{x\sqrt{2}/2} + c_2e^{-x\sqrt{2}/2})$

19. $y = e^{3x/8}(c_1e^{x\sqrt{41}/8} + c_2e^{-x\sqrt{41}/8})$

21. $y = e^{3x/2}(c_1e^{x\sqrt{13}/2} + c_2e^{-x\sqrt{13}/2})$

23. $y = e^{-x/2}(c_1e^{x\sqrt{33}/2} + c_2e^{-x\sqrt{33}/2})$

25. $y = c_1e^{3ax/2} + c_2e^{-4ax}$ **27.** $y = \frac{1}{5}(3e^{7x} + 7e^{-3x})$

29. $y = \frac{e^3}{e^7 - 1}(e^{4x} - e^{-3x})$ **31.** $y = c_1 + c_2e^{-x} + c_3e^{3x}$

33. $y = c_1e^x + c_2e^{-x} + c_3e^{2x} + c_4e^{-2x}$

35. $v = c_1e^{as} + c_2e^{-as}$ **37.** $y = \frac{2}{e^{-1} - e^{-3}}(e^{-3x} - e^{-x})$

Exercises 31.8, page 994

1. $y = e^{-5x}(c_1 + c_2x)$ **3.** $y = c_1 + c_2e^{10x}$

5. $y = (c_1 + c_2x)e^x$ **7.** $y = (c_1 + c_2x)e^{-6x}$

9. $y = c_1\sin 3x + c_2\cos 3x$

11. $y = e^{-x/2}(c_1\sin\frac{1}{2}\sqrt{7}x + c_2\cos\frac{1}{2}\sqrt{7}x)$

13. $y = c_1e^x + c_2e^{-x} + c_3\sin x + c_4\cos x$

15. $y = c_1\sin\frac{1}{2}x + c_2\cos\frac{1}{2}x$ **17.** $y = (c_1 + c_2x)e^{3x/4}$

19. $y = c_1\sin\frac{2}{5}x + c_2\cos\frac{2}{5}x$

21. $y = e^x(c_1\cos\frac{1}{2}\sqrt{6}x + c_2\sin\frac{1}{2}\sqrt{6}x)$

23. $y = (c_1 + c_2x)e^{4x/5}$ **25.** $y = e^{3x/4}(c_1e^{x\sqrt{17}/4} + c_2e^{-x\sqrt{17}/4})$

27. $y = c_1e^{x(-6+\sqrt{42})/3} + c_2e^{x(-6-\sqrt{42})/3}$

29. $y = e^{2x}(c_1 + c_2x + c_3x^2)$

31. $y = (c_1 + c_2x)\sin x + (c_3 + c_4x)\cos x$

33. $y = e^{-x}\sin 3x$ **35.** $y = e^{4x}(4 - 14x)$ **37.** $D^2y - 9y = 0$

39. $D^2y + 9y = 0$. The sum of cos $3x$ and sin $3x$ with no exponential factor indicates imaginary roots with $\alpha = 0$ and $\beta = 3$.

41. (a) $y = 0$ (b) $y = 0$ **43.** $y = 0.5e^2te^{-2t}$

Exercises 31.9, page 1000

1. $y_p = A + Bx + Cx^2 + Ee^{-x}$

3. $y = c_1\sin 2x + c_2\cos 2x + \frac{1}{3}\sin x - \frac{4}{5}e^{-x}$

5. $y = c_1e^{2x} + c_2e^{-x} - 2$ **7.** $y = c_1e^{-x} + c_2e^{x} - 4 - x^2$

9. $y = c_1 + c_2e^{3x} - \frac{3}{4}e^{x} - \frac{1}{2}xe^{x}$ **11.** $y = c_1e^{x/3} + c_2e^{-x/3} - \frac{1}{10}\sin x$

13. $y = (c_1 + c_2x)e^{x} + 10 + 6x + x^2 - \frac{2}{25}\sin 3x + \frac{3}{50}\cos 3x$

15. $y = c_1 \sin 2x + c_2 \cos 2x + 3x \cos 2x$

17. $y = c_1e^{-5x} + c_2e^{6x} - \frac{1}{3}$ **19.** $y = c_1e^{4x/3} + c_2e^{-x} + \frac{1}{4}e^{3x}$

21. $y = c_1e^{2x} + c_2e^{-2x} - \frac{1}{5}\sin x - \frac{2}{5}\cos x$

23. $y = c_1 \sin x + c_2 \cos x - \frac{1}{3}\sin 2x + 4$

25. $y = c_1e^{-x} + c_2e^{-4x} - \frac{7}{100}e^{x} + \frac{1}{10}xe^{x} + 1$

27. $y = c_1 + c_2e^{x} + c_3e^{-x} + \frac{1}{10}\cos 2x$

29. $y = c_1 \sin x + c_2 \cos x + \frac{1}{2}x \sin x$

31. $y = c_1 + c_2e^{-2x} + 2x^2 - 2x - \frac{1}{2}xe^{-2x}$

33. $y = \frac{1}{6}(11e^{3x} + 5e^{-2x} + e^{x} - 5)$

35. $y = -\frac{2}{3}\sin x + \pi \cos x + x - \frac{1}{3}\sin 2x$

37. $y = ce^{x} - x^2 - 2x - 2$; If solved as a first-order linear equation, integration is more complex.

39. $\frac{16}{\pi}\left(\frac{1}{16 - \pi^2}\sin\frac{\pi x}{2} + \frac{1}{32 - 8\pi^2}\sin\frac{3\pi x}{2} + \cdots\right)$

Exercises 31.10, page 1006

1. $x = \frac{1}{2}\sin 4t$ **3.** $x = e^{-0.1t}(0.04 \sin 10t + 4 \cos 10t)$

5. $\theta = 0.1 \cos 3.1t$ **7.** 10

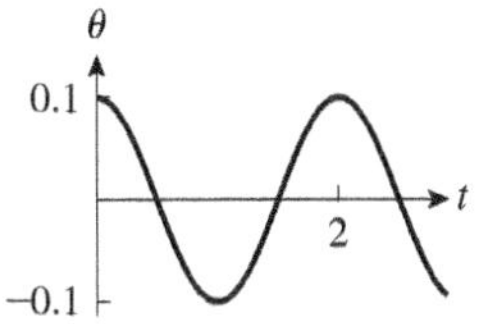

9. $s = (s_0 - L)\cos\sqrt{\frac{g}{e}}\,t + L$ **11.** $y = 0.100 \cos 14.0t$

13. $y = -\frac{7}{960}\sin 14t + 0.100 \cos 14t + \frac{49}{960}\sin 2t$

15. $q = 2.24 \times 10^{-4}e^{-20t}\sin 2240t$

17. $q = 0.01(1 - \cos 316t)$

19. $q = e^{-10t}(c_1 \sin 99.5t + c_2 \cos 99.5t)$
$- 1.81 \times 10^{-3}\sin 120\pi t$
$- 1.03 \times 10^{-4}\cos 120\pi t$

21. $i = 10^{-6}(2.00 \cos(1.58 \times 10^{4}t)$
$+ 158 \sin(1.58 \times 10^{4}t) - 2.00e^{-200t})$

23. $i_p = 0.528 \sin 100t - 3.52 \cos 100t$

25. $y = \frac{w}{24EI}(6L^2x^2 - 4Lx^3 + x^4)$

27. $y = c_1 \sin 8t + c_2 \cos 8t + \sum_{n=1}^{\infty}\frac{32(-1)^{n+1}}{n(64 - \pi^2n^2)}\sin n\pi t$

Exercises 31.11, page 1011

1. $F(s) = \int_0^{\infty} e^{-st}dt = -\frac{1}{s}e^{-st}\Big|_0^{\infty} = \frac{1}{s}$

3. $(s^2 - 2s)\mathcal{L}(f) - s + 2$ **5.** $\frac{1}{s - 3}$ **7.** $\frac{30}{(s + 2)^4}$

9. $\frac{s - 2}{s^2 + 4}$ **11.** $\frac{3}{s} + \frac{2(s^2 - 9)}{(s^2 + 9)^2}$ **13.** $s^2\mathcal{L}(f) + s\mathcal{L}(f)$

15. $(2s^2 - s + 1)\mathcal{L}(f) - 2s + 1$ **17.** t^2 **19.** $\frac{15}{2}e^{-3t}$

21. $\frac{1}{2}t^2e^{-t}$ **23.** $\frac{1}{54}(9t \sin 3t + 2 \sin 3t - 6t \cos 3t)$

25. $-\frac{1}{3}e^{-t} - \frac{8}{3}e^{2t} + 7e^{3t}$ **27.** $\frac{1}{2}e^{t}(4 \cos 2t + 5 \sin 2t)$

29. $-\frac{d}{ds}\left(\frac{1}{s + a}\right) = \frac{1}{(s + a)^2}$

Exercises 31.12, page 1015

1. $y = 2e^{t/2}$ **3.** $y = \frac{t}{2}\sin t + \sin t + 2 \cos t$ **5.** $y = e^{-t}$

7. $y = -e^{3t/2}$ **9.** $y = (1 + t)e^{-3t}$ **11.** $y = \frac{1}{2}\sin 2t$

13. $y = e^{-t/2}\cos t$ **15.** $y = e^{2t}\cos t$ **17.** $y = 1 + \sin t$

19. $y = e^{-t}\left(\frac{1}{2}t^2 + 3t + 1\right)$ **21.** $y = 2e^{3t} + 3e^{-2t}$

23. $y = \frac{3}{2}e^{t} - \frac{3}{2}e^{-t} - \sin 2t$ **25.** $v = 6(1 - e^{-t/2})$

27. $y = 4 \cos 80t$ **29.** $q = 1.6 \times 10^{-4}(1 - e^{-5000t})$

31. $i = 5t \sin 50t$ **33.** $y = \sin 3t - 3t \cos 3t$

35. $i = 5.0e^{-50t} - 5.0e^{-100t}$ **37.** $y = \frac{w}{24EI}(L^3x - 2Lx^3 + x^4)$

Review Exercises for Chapter 31, page 1017

1. $2 \ln(x^2 + 1) - \frac{1}{2y^2} = c$ **3.** $y^2 = 2x - 4 \sin x + c$

5. $y = c_1 + c_2e^{-x/2}$ **7.** $y = (c_1 + c_2x)e^{x/4}$

9. $2x^2 + 4xy + y^4 = c$ **11.** $P = cV^5 - \frac{1}{3}V^2$

13. $y = \frac{2c}{1 - ce^{2x}}$ **15.** $y = e^{-x}(c_1 \sin \sqrt{5}x + c_2 \cos \sqrt{5}x)$

17. $y = e^{-2x} + ce^{-4x}$ **19.** $y = \frac{1}{2}(c - x^2)\csc x$

21. $s = c_1e^{t} + c_2e^{-3t/2} - 2$

23. $y = e^{-x/2}(c_1e^{x\sqrt{5}/2} + c_2e^{-x\sqrt{5}/2}) + 2e^{x}$

25. $y = c_1e^{2x/3} + c_2e^{4x/3} + \frac{1}{2}x + \frac{25}{8}$

27. $y = c_1e^{x} + c_2 \sin 3x + c_3 \cos 3x - \frac{1}{16}(\sin x + \cos x)$

29. $y = c_1e^{-x} + c_2e^{8x} - \frac{2}{9}xe^{-x}$

31. $y = c_1 \sin 5x + c_2 \cos 5x + 5x \sin 5x$ **33.** $y^3 = 8 \sin^2 x$

35. $y = 2x - 1 - e^{-2x}$ **37.** $v = 2e^{-t/2}\sin\left(\frac{1}{2}\sqrt{15}\,t\right)$

39. $y = \frac{1}{25}[16 \sin x + 12 \cos x - 3e^{-2x}(4 + 5x)]$

41. $y = e^{t/4}$ **43.** $y = \frac{1}{2}(e^{3t} - e^{t})$ **45.** $y = -4 \sin t$

47. $y = \frac{1}{25}\left(3e^{x} - 3 \cos \frac{3x}{4} - 4 \sin \frac{3x}{4}\right)$

49.

x	0	0.1	0.2	0.3	0.4
y	0	0.100	0.201	0.305	0.414

51. (a) and (b) $y = e^{3x}$ **53.** $x = t^2 + 1$; $y = \frac{1}{t^2 + 1}$

55. $r = r_0 + kt$ **57.** $m = m_0e^{kt}$ **59.** 3.93 m/s

61. $y^2 - 2xy - 8 = 0$ **63.** 40.0 s **65.** 5.31×10^8 years

67. 8.2 billion **69.** $5y^2 + x^2 = c$ **71.** 36°C

73. $i = 0.5(1 - e^{-20t})$

75. $y = 0.25e^{-2t}(2 \cos 4t + \sin 4t)$, underdamped

77. $q = e^{-6t}(0.4 \cos 8t + 0.3 \sin 8t) - 0.4 \cos 10t$

79. $i = 0$ **81.** $i = 12(1 - e^{-t/2})$; $i(0.3) = 1.67$ A

83. $q = 10^{-4}e^{-8t}(4.0 \cos 200t + 0.16 \sin 200t)$ **85.** $y = 0.25t \sin 8t$

87. 2.47 L **89.** $y = \frac{10}{3EI}[100x^3 - x^4 + xL^2(L - 100)]$

91. $i = 4.42e^{-66.7t}\sin(226t)$

Solutions to Practice Test Problems

Chapter 1

1. $\sqrt{9+36} = \sqrt{45} = \sqrt{3^2 \cdot 5} = 3\sqrt{5}$

2. $\dfrac{(5)(-31)(6)}{(-4)(0)}$ is undefined (division by zero).

3. $\dfrac{5.279 \times 10^7}{4.393 \times 10^{-6}} = 1.202 \times 10^{13}$

4. $\dfrac{(-4)(3) - (-4)(-2)}{|-3-1|} = \dfrac{-12-8}{|-4|} = \dfrac{-20}{4} = -5$

5. $\dfrac{392.4 - 57.9}{486.2} - \dfrac{0.675^3}{(2.75)(0.113)} = \dfrac{334.5}{486.2} - \dfrac{0.307547}{0.31075} = -0.3017$

6. $(5a^4b^0c^{-5})^{-2} = 5^{-2}a^{4(-2)}b^{0(-2)}c^{-5(-2)} = \dfrac{c^{10}}{5^2a^8}$

7. $(5x-2)^2 = (5x-2)(5x-2) = 25x^2 - 20x + 4$

8. $4k^3(bk - 5k^2) = 4bk^4 - 20k^5$

9. $\dfrac{8a^3x^2 - 4a^2x^4}{-2ax^2} = \dfrac{8a^3x^2}{-2ax^2} - \dfrac{4a^2x^4}{-2ax^2} = -4a^2 - (-2ax^2)$
$= -4a^2 + 2ax^2$

10.
$$\begin{array}{r} 3x - 5 \quad \text{(quotient)} \\ 2x-1\overline{)6x^2 - 13x + 7} \\ \underline{6x^2 - 3x} \\ -10x + 7 \\ \underline{-10x + 5} \\ 2 \quad \text{(remainder)} \end{array}$$

11. $(2x-3)(x+7)$
$= 2x(x) + 2x(7) + (-3)(x) + (-3)(7)$
$= 2x^2 + 14x - 3x - 21 = 2x^2 + 11x - 21$

12. $[3x - (4x-3)] - 2x = [3x - 4x + 3] - 2x$
$= [-x+3] - 2x = -3x + 3$

13. $3y - 2(3y-4) = 10$
$3y - 6y + 8 = 10$
$-3y + 8 = 10$
$-3y = 2$
$y = -\dfrac{2}{3}$

14. $10(2x-3) = 2x - (5-3d)$
$20x - 30 = 2x - 5 + 3d$
$18x = 25 + 3d$
$x = \dfrac{25+3d}{18}$
$x = \dfrac{25}{18} + \dfrac{d}{6}$

15. $0.000\,41 = 4.1 \times 10^{-4}$ (four places to right)

16.

$-\pi$	-3	0.3	$\sqrt{2}$	$\lvert -4 \rvert$	(order)
-3.14	-3	0.3	1.41	4	(value)

17. $3(5+8) = 3(5) + 3(8)$ illustrates distributive law.

18. (a) 5 (b) 3.0 (zero is significant)

19. Evaluation:
$2000(1 + 0.02/12)^{12(2)} = \$2081.55$

20. $8(100-x)^2 + x^2 = 8(100-x)(100-x) + x^2$
$= 8(10\,000 - 200x + x^2) + x^2$
$= 80\,000 - 1600x + 8x^2 + x^2$
$= 80\,000 - 1600x + 9x^2$

21. $L = L_0[1 + \alpha(t_2 - t_1)]$
$= L_0[1 + \alpha t_2 - \alpha t_1]$
$= L_0 + \alpha L_0 t_2 - \alpha L_0 t_1$
$L - L_0 + \alpha L_0 t_1 = \alpha L_0 t_2$
$t_2 = \dfrac{L - L_0 + \alpha L_0 t_1}{\alpha L_0}$

22. Let n = number of newtons of second alloy
$0.3(20) + 0.8n = 0.6(n+20)$
$6 + 0.8n = 0.6n + 12$
$0.2n = 6$
$n = 30$ N

Chapter 2

1. $\angle 1 + \angle 3 + 90° = 180°$ (sum of angles of a triangle)
$\angle 3 = 52°$ (vertical angles)
$\angle 1 = 180° - 90° - 52° = 38°$

2. $\angle 2 + \angle 4 = 180°$ (straight angle)
$\angle 4 = 52°$ (corresponding angles)
$\angle 2 + 52° = 180°$
$\angle 2 = 128°$

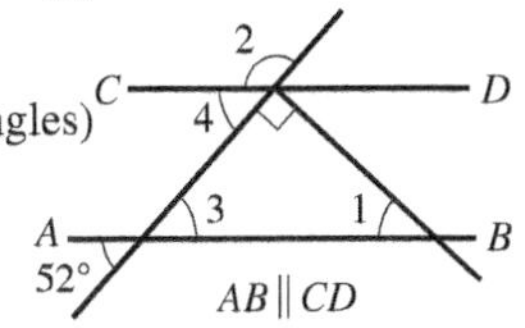

3.

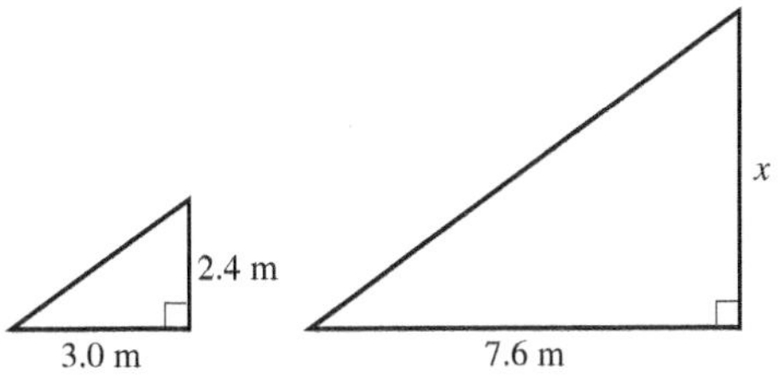

$\dfrac{x}{2.4} = \dfrac{7.6}{3.0} \qquad x = \dfrac{2.4(7.6)}{3.0} = 6.1$ m

4. Use Hero's formula:
$s = \frac{1}{2}(2.46 + 3.65 + 4.07) = 5.09$ cm
$A = \sqrt{5.09(5.09 - 2.46)(5.09 - 3.65)(5.09 - 4.07)}$
$= 4.43$ cm^2

5. $d^2 = 3810^2 + 5180^2$
$d = \sqrt{3810^2 + 5180^2}$
$= 6430$ mm

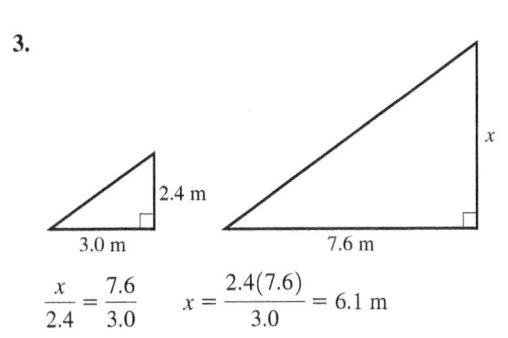

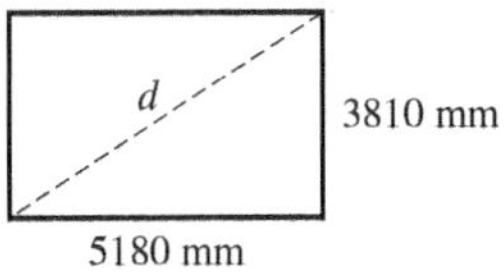

6. $A = \frac{1}{2}h(b_1 + b_2)$
$= \frac{1}{2}(1.3)(3.0 + 5.0) = 5.2$ m^2

7. (a) $A = 6s^2 = 6(4.50)^2 = 122$ cm^2
(b) $V = s^3 = 4.50^3 = 91.1$ cm^3

8. $c = 2\pi r \qquad A = 4\pi r^2$
$21.0 = 2\pi r \qquad = 4\pi\left(\dfrac{10.5}{\pi}\right)^2 = \dfrac{4(10.5^2)}{\pi}$
$r = \dfrac{10.5}{\pi} \qquad = 140$ cm^2

9. $V = \frac{1}{3}\pi r^2 h = \frac{1}{3}\pi(2.08^2)(1.78)$
$= 8.06$ m^3

10. $\angle ACO + 64° = 90°$ (tangent perpendicular to radius)
$\angle ACO = 26°$
$\angle A = \angle ACO = 26°$ (isosceles triangle)
$\frac{1}{2}\widehat{CD} = 26°$ (intercepted arc)
$\widehat{CD} = 52°$
$\angle 1 = 52°$ (central angle)

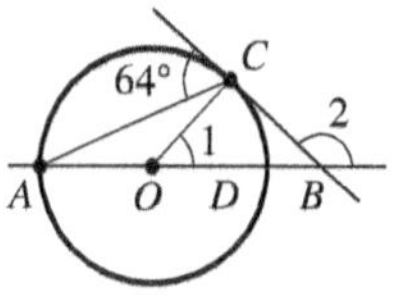

11. $\angle CBO + \angle 1 + 90° = 180°$ (sum of angles of triangle)
$\angle CBO + 52° + 90° = 180°$
$\angle CBO = 180° - 52° - 90° = 38°$
$\angle 2 + \angle CBO = 180°$ (straight angle)
$\angle 2 + 38° = 180°$
$\angle 2 = 142°$

12. $r = \frac{1}{2}(2.25)$ cm
$p = 3(2.25) + \frac{1}{2}(2\pi)\left[\frac{1}{2}(2.25)\right]$
$= 10.3$ cm

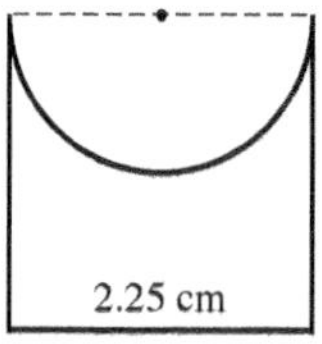

13. $A = 2.25^2 - \frac{1}{2}\pi\left[\frac{1}{2}(2.25)\right]^2 = 3.07\ \text{cm}^2$

14. $A = \frac{1}{3}(50)[0 + 2(90) + 2(145) + 2(260)$
$+2(205) + 2(110) + 20]$
$= 41\,000\ \text{m}^2$

Chapter 3

1. $f(x) = 2x - x^2 + \dfrac{8}{x}$

$f(-4) = 2(-4) - (-4)^2 + \dfrac{8}{-4}$
$= -8 - 16 - 2$
$= -26$

$f(x-4) = 2(x-4) - (x-4)^2 + \dfrac{8}{x-4}$
$= 2x - 8 - x^2 + 8x - 16 + \dfrac{8}{x-4}$
$= -x^2 + 10x - 24 + \dfrac{8}{x-4}$

2. $m = 2000 - 10t$

3. $f(x) = 4 - 2x$
$y = 4 - 2x$
$y = 4 - 2(-1) = 6$
$y = 4 - 2(0) = 4$
$y = 4 - 2(1) = 2$
$y = 4 - 2(2) = 0$
$y = 4 - 2(3) = -2$
$y = 4 - 2(4) = -4$

x	y
−1	6
0	4
1	2
2	0
3	−2
4	−4

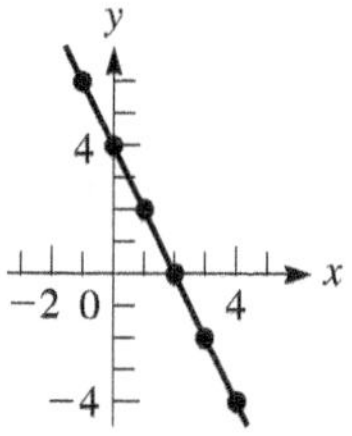

4. $y = 2x^2 - 3x - 3$

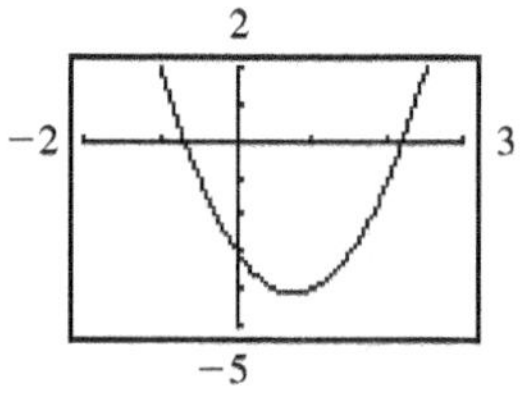

$x = -0.7$ and $x = 2.2$

5. $y = \sqrt{4 + 2x}$
$y = \sqrt{4 + 2(-2)} = 0$
$y = \sqrt{4 + 2(-1)} = 1.4$
$y = \sqrt{4 + 2(0)} = 2$
$y = \sqrt{4 + 2(1)} = 2.4$
$y = \sqrt{4 + 2(2)} = 2.8$
$y = \sqrt{4 + 2(4)} = 3.5$

x	y
−2	0
−1	1.4
0	2
1	2.4
2	2.8
4	3.5

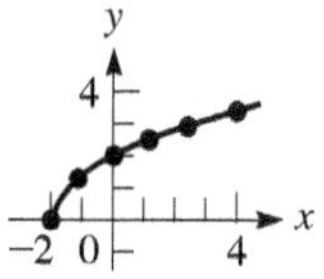

6. On negative x-axis

7. $f(x) = \sqrt{6 - x}$
Domain: $x \leq 6$, or $(-\infty, 6]$; x cannot be greater than 6 to have real values of $f(x)$.
Range: $f(x) \geq 0$, or $[0, \infty)$; $\sqrt{6 - x}$ is the principal square root of $6 - x$ and cannot be negative.

8. Shifting $y = 2x^2 - 3$ to the right 1 and up 3 gives
$y = 2(x-1)^2 - 3 + 3 = 2(x-1)^2 = 2x^2 - 4x + 2.$

9. Range: $y \leq -8.9$ or $y \geq 0.9$, or $(-\infty, -8.9]$ or $[0.9, \infty)$

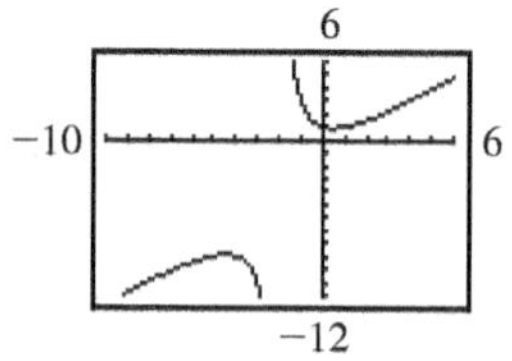

10. Let r = radius of circular part
area = square + semicircle
$A = (2r)(2r) + \frac{1}{2}(\pi r^2)$
$= 4r^2 + \frac{1}{2}\pi r^2$

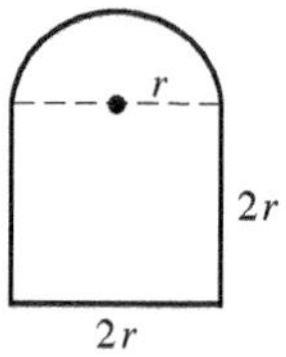

11.

Voltage V	10.0	20.0	30.0	40.0	50.0	60.0
Current i	145	188	220	255	285	315

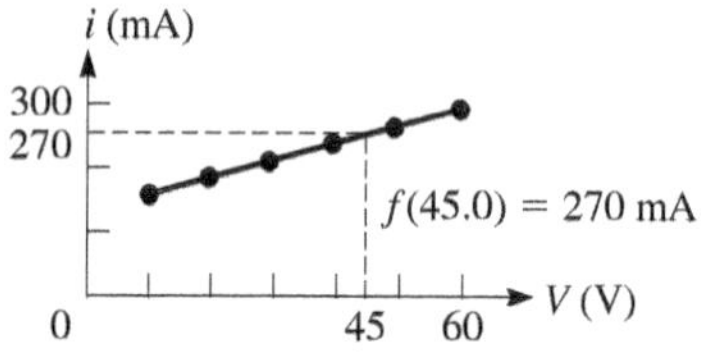

12. $V = 20.0$ V for $i = 188$ mA
$V = 30.0$ V for $i = 220$ mA
$\dfrac{x}{10} = \dfrac{12}{32}$, $x = 3.8$ (rounded off)
$V = 20.0 + 3.8 = 23.8$ V
for $i = 200$ mA

V i
20.0 188
x[ ]12
200
10 30.0 220 32

Chapter 4

1. $137°29'33'' = 137 + \left(\dfrac{29}{60}\right)^\circ + \left(\dfrac{33}{3600}\right)^\circ = 137.4925°$

2. $\theta = \cot^{-1}(6.2) = \tan^{-1}\left(\dfrac{1}{6.2}\right) = 9.16°$

3. $\sin\theta = 0.3726$; $\theta = 21.88°$

4. Let x = distance east of the course

$$\frac{x}{22.62} = \sin 4.05°$$

$$x = 22.62 \sin 4.05° = 1.598 \text{ km}$$

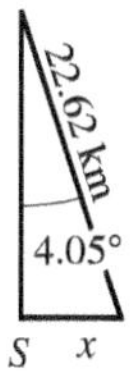

5. $\sin\theta = \dfrac{2}{3}$

$$x = \sqrt{3^2 - 2^2} = \sqrt{5}$$

$$\tan\theta = \frac{2}{\sqrt{5}}$$

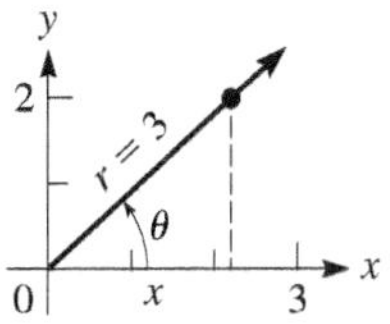

6. $\tan\theta = 1.294;\ \sec(\tan^{-1}(1.294)) = \dfrac{1}{\cos(\tan^{-1}(1.294))} = 1.635$

7. $A = 90° - 37.4° = 52.6°$ $\quad \dfrac{52.8}{c} = \cos 37.4°$

$$\frac{b}{52.8} = \tan 37.4° \qquad c = \frac{52.8}{\cos 37.4°}$$

$$b = 52.8 \tan 37.4° = 40.4 \qquad = 66.5$$

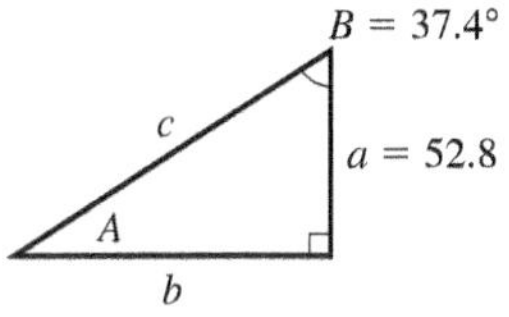

8. $2.49^2 + b^2 = 3.88^2$

$$b = \sqrt{3.88^2 - 2.49^2} = 2.98$$

$$\sin A = \frac{2.49}{3.88}$$

$$A = 39.9° \qquad B = 90° - 39.9° = 50.1°$$

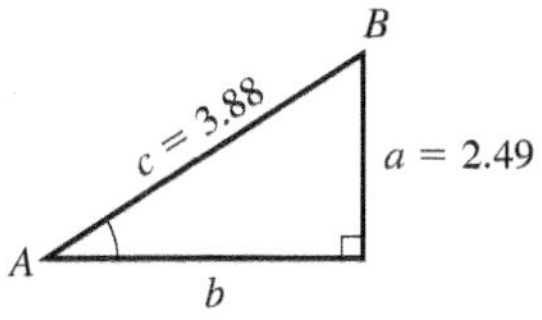

9.

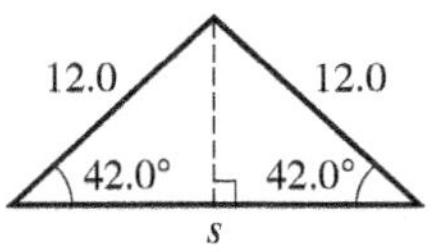

$$\frac{s/2}{12.0} = \cos 42.0°$$

$$s = 24.0 \cos 42.0° = 17.8$$

10.

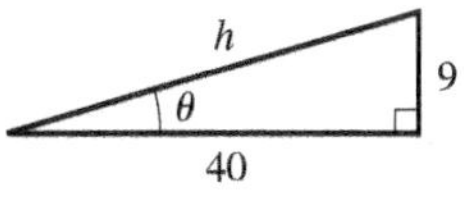

$$h = \sqrt{9^2 + 40^2} = 41$$

$$\sin\theta = \frac{9}{41},\ \cos\theta = \frac{40}{41}$$

$$\frac{\sin\theta}{\cos\theta} = \frac{9/41}{40/41} = \frac{9}{40}$$

11. $\lambda = d \sin\theta$

$$= 30.05 \sin 1.167° = 0.6120\ \mu\text{m}$$

12. $r = \sqrt{5^2 + 2^2} = \sqrt{29}$

$$\sin\theta = \frac{2}{\sqrt{29}} = 0.3714 \qquad \csc\theta = \frac{\sqrt{29}}{2} = 2.693$$

$$\cos\theta = \frac{5}{\sqrt{29}} = 0.9285 \qquad \sec\theta = \frac{\sqrt{29}}{5} = 1.077$$

$$\tan\theta = \frac{2}{5} = 0.4000 \qquad \cot\theta = \frac{5}{2} = 2.500$$

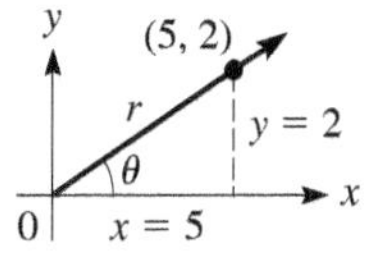

13. $\sin\theta = 1.0/3.2$

$$\theta = 18°$$

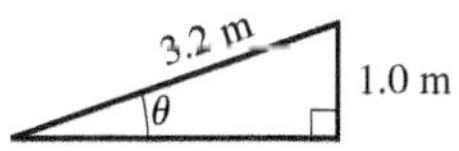

14. Distance between points is $x - y$.

$$\frac{18.525}{x} = \tan 13.500° \qquad \frac{18.525}{y} = \tan 21.375°$$

$$x - y = \frac{18.525}{\tan 13.500°} - \frac{18.525}{\tan 21.375°} = 29.831 \text{ m}$$

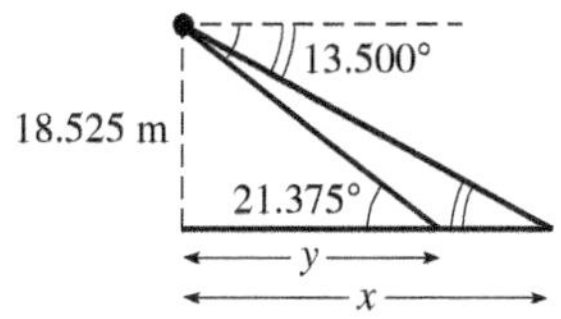

Chapter 5

1. Points $(2, -5)$ and $(-1, 4)$

$$m = \frac{4 - (-5)}{-1 - 2} = \frac{9}{-3} = -3$$

2.

$$\begin{aligned} 2x + 5y &= 11 \\ y - 5x &= 12 \\ x = -2,\ y &= 3 \end{aligned} \qquad \begin{aligned} &2(-2) + 5(3) = 11 \quad \text{OK} \\ &3 - 5(-2) = 13 \quad \text{No} \end{aligned}$$

Not a solution. Values do not satisfy second equation.

3.

$$\begin{vmatrix} -1 & 3 & -2 \\ 4 & -3 & 0 \\ 5 & -4 & 2 \end{vmatrix} \begin{matrix} -1 & 3 \\ 4 & -3 \\ 5 & -4 \end{matrix} = 6 + 0 + 32 - 30 - 0 - 24 = -16$$

4.

$$\begin{aligned} x + 2y &= 5 \\ 4y &= 3 - 2x \\ x &= 5 - 2y \end{aligned} \qquad \begin{aligned} 4y &= 3 - 2(5 - 2y) \\ 4y &= 3 - 10 + 4y \\ 0 &= -7 \end{aligned}$$

Inconsistent

5. $3x - 2y = 4$

$2x + 5y = -1$

$$x = \frac{\begin{vmatrix} 4 & -2 \\ -1 & 5 \end{vmatrix}}{\begin{vmatrix} 3 & -2 \\ 2 & 5 \end{vmatrix}} = \frac{20 - 2}{15 - (-4)} = \frac{18}{19}$$

$$y = \frac{\begin{vmatrix} 3 & 4 \\ 2 & -1 \end{vmatrix}}{19} = \frac{-3 - 8}{19} = -\frac{11}{19}$$

6. $2x + y = 4$

$y = -2x + 4$

$m = -2 \qquad b = 4$

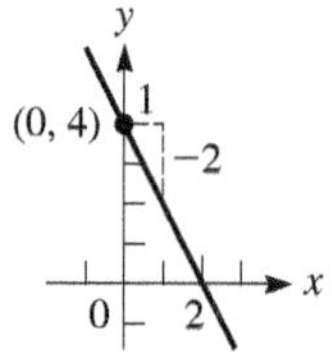

7. $2l + 2w = 24$ (perimeter)

$$\begin{aligned} l &= w + 6.0 \\ l + w &= 12 \\ l - w &= 6.0 \\ 2l &= 18 \\ l &= 9 \text{ km} \\ 9 &= w + 6 \\ w &= 3 \text{ km} \end{aligned}$$

8.

$$\begin{aligned} 6N - 2P &= 13 \\ 4N + 3P &= -13 \\ 18N - 6P &= 39 \\ 8N + 6P &= -26 \\ 26N &= 13 \\ N &= \frac{1}{2} \\ 6\left(\frac{1}{2}\right) - 2P &= 13 \\ -2P &= 10 \\ P &= -5 \end{aligned}$$

9. $C = 310 - 2d$

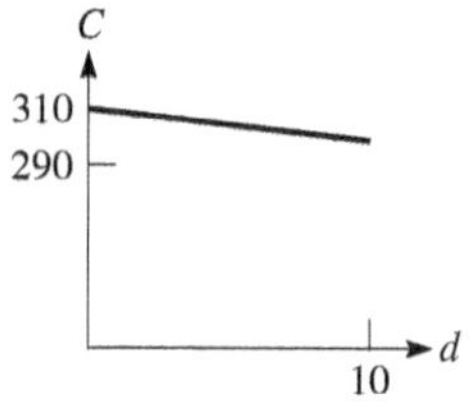

10. $2x - 3y = 6$ $\qquad 4x + y = 4$

$x = 0{:}\ y = -2$ $\qquad x = 0{:}\ y = 4$

$y = 0{:}\ x = 3$ $\qquad y = 0{:}\ x = 1$

Int: $(0, -2), (3, 0)$ $\qquad$ Int: $(0, 4), (1, 0)$

$x = 1.3 \qquad y = -1.1$

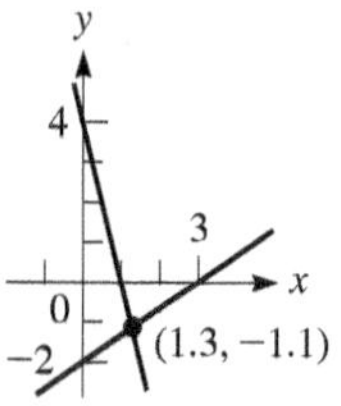

11.

$$\begin{aligned} x + y + z &= 4 \\ x + y - z &= 6 \\ 2x + y + 2z &= 5 \\ 2x + 2y &= 10 \\ 4x + 3y &= 17 \\ -y &= -3 \\ y &= 3 \\ x &= 2 \\ z &= -1 \end{aligned}$$

12. Let x = vol. of first alloy

y = vol. of second alloy

z = vol. of third alloy

$$\begin{aligned} x + y + z &= 100 && \text{total vol.} \\ 0.6x + 0.5y + 0.3z &= 40 && \text{copper} \\ 0.3x + 0.3y &= 15 && \text{zinc} \\ x + y + z &= 100 \\ 6x + 5y + 3z &= 400 \\ 3x + 3y &= 150 \\ 3x + 2y &= 100 \\ 3x + 3y &= 150 \\ y &= 50 \text{ cm}^3 \\ x &= 0 \text{ cm}^3 \\ z &= 50 \text{ cm}^3 \end{aligned}$$

13.

$$\begin{aligned} 3x + 2y - z &= 4 \\ 2x - y + 3z &= -2 \\ x + 4z &= 5 \end{aligned}$$

$$y = \frac{\begin{vmatrix} 3 & 4 & -1 \\ 2 & -2 & 3 \\ 1 & 5 & 4 \end{vmatrix}}{\begin{vmatrix} 3 & 2 & -1 \\ 2 & -1 & 3 \\ 1 & 0 & 4 \end{vmatrix}} = \frac{-24 + 12 - 10 - 2 - 45 - 32}{-12 + 6 + 0 - 1 - 0 - 16} = \frac{-101}{-23} = \frac{101}{23}$$

Chapter 6

1. $2x(2x - 3)^2 = 2x\left[(2x)^2 - 2(2x)(3) + 3^2\right] = 2x(4x^2 - 12x + 9) = 8x^3 - 24x^2 + 18x$

2.

$$\begin{aligned} \frac{1}{R} &= \frac{1}{R_1 + r} + \frac{1}{R_2} \\ \frac{RR_2(R_1 + r)}{R} &= \frac{RR_2(R_1 + r)}{R_1 + r} + \frac{RR_2(R_1 + r)}{R_2} \\ R_2(R_1 + r) &= RR_2 + R(R_1 + r) \\ R_1R_2 + rR_2 &= RR_2 + RR_1 + rR \\ R_1R_2 - RR_1 &= RR_2 + rR - rR_2 \\ R_1(R_2 - R) &= RR_2 + rR - rR_2 \\ R_1 &= \frac{RR_2 + rR - rR_2}{R_2 - R} \end{aligned}$$

3. $\dfrac{2x^2 + 5x - 3}{2x^2 + 12x + 18} = \dfrac{(2x - 1)(x + 3)}{2(x + 3)^2} = \dfrac{2x - 1}{2(x + 3)}, x \neq -3$

4. $4x^2 - 16y^2 = 4(x^2 - 4y^2) = 4(x + 2y)(x - 2y)$

5. $pb^3 + 8a^3p = p(b^3 + 8a^3) = p(b + 2a)(b^2 - 2ab + 4a^2)$

6. $2a - 4T - ba + 2bT = 2(a - 2T) - b(a - 2T) = (2 - b)(a - 2T)$

7. $36x^2 + 14x - 16 = 2(18x^2 + 7x - 8) = 2(9x + 8)(2x - 1)$

8.

$$\begin{aligned} \frac{3}{4x^2} - \frac{2}{x^2 - x} - \frac{x}{2x - 2} &= \frac{3}{4x^2} - \frac{2}{x(x - 1)} - \frac{x}{2(x - 1)} \\ &= \frac{3(x - 1) - 2(4x) - x(2x^2)}{4x^2(x - 1)} \\ &= \frac{3x - 3 - 8x - 2x^3}{4x^2(x - 1)} \\ &= \frac{-2x^3 - 5x - 3}{4x^2(x - 1)} \end{aligned}$$

9.

$$\begin{aligned} \frac{x^2 + x}{2 - x} \div \frac{x^2}{x^2 - 4x + 4} &= \frac{x^2 + x}{2 - x} \times \frac{x^2 - 4x + 4}{x^2} \\ &= \frac{x(x + 1)}{2 - x} \times \frac{(x - 2)^2}{x^2} \\ &= -\frac{x(x + 1)(x - 2)^2}{(x - 2)(x^2)} \\ &= -\frac{(x + 1)(x - 2)}{x}, x \neq 2 \end{aligned}$$

10. $\dfrac{1-\dfrac{3}{2x+2}}{\dfrac{x}{5}-\dfrac{1}{2}} = \dfrac{\dfrac{2(x+1)-3}{2(x+1)}}{\dfrac{2x-5}{10}}$

$= \dfrac{2x+2-3}{2(x+1)} \times \dfrac{10}{2x-5}$

$= \dfrac{5(2x-1)}{(x+1)(2x-5)}$

11. Let t = time working together

$\dfrac{t}{12} + \dfrac{t}{16} = 1$

LCD of 12 and 16 is 48.

$\dfrac{48t}{12} + \dfrac{48t}{16} = 48$

$4t + 3t = 48$

$t = \dfrac{48}{7} = 6.9$ days

12. $\dfrac{3}{2x^2-3x} + \dfrac{1}{x} = \dfrac{3}{2x-3}$ LCD $= x(2x-3), x \neq 0, 3/2$

$\dfrac{3x(2x-3)}{x(2x-3)} + \dfrac{x(2x-3)}{x} = \dfrac{3x(2x-3)}{2x-3}$

$3 + (2x-3) = 3x$

$x = 0$

No solution.

Chapter 7

1. $2x^2 + 5x = 12$

$2x^2 + 5x - 12 = 0$

$(2x-3)(x+4) = 0$

$2x - 3 = 0 \quad x + 4 = 0$

$x = \dfrac{3}{2}$ or $x = -4$

2. $x^2 = 3x + 5$

$x^2 - 3x - 5 = 0$

$a = 1, \quad b = -3, \quad c = -5$

$x = \dfrac{-(-3) \pm \sqrt{(-3)^2 - 4(1)(-5)}}{2(1)}$

$= \dfrac{3 \pm \sqrt{29}}{2}$

3. $4x^2 - 5x - 3 = 0$

On calculator

$Y_1 = 4X^2 - 5X - 3$

From zero feature

$x = -0.44$ or $x = 1.69$

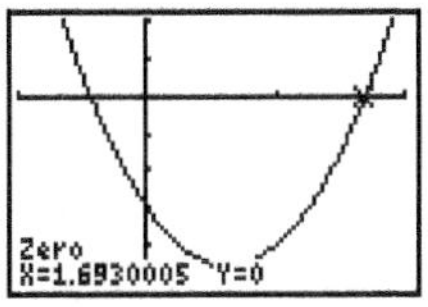

4. $2x^2 - x = 6 - 2x(3 - x)$

$2x^2 - x = 6 - 6x + 2x^2$

$5x = 6$ (not quadratic)

$x = \dfrac{6}{5}$

5. $\dfrac{3}{x} - \dfrac{2}{x+2} = 1, x \neq 0, -2$

$\dfrac{3x(x+2)}{x} - \dfrac{2x(x+2)}{x+2} = x(x+2)$

$3(x+2) - 2x = x^2 + 2x$

$3x + 6 - 2x = x^2 + 2x$

$0 = x^2 + x - 6$

$(x+3)(x-2) = 0$

$x = -3, 2$

6. $y = 2x^2 + 8x + 5$

$\dfrac{-b}{2a} = \dfrac{-8}{2(2)} = -2$

$y = 2(-2)^2 + 8(-2) + 5 = -3$

Min. pt. ($a > 0$) is $(-2, -3)$,

$c = 5$, y-intercept is $(0, 5)$.

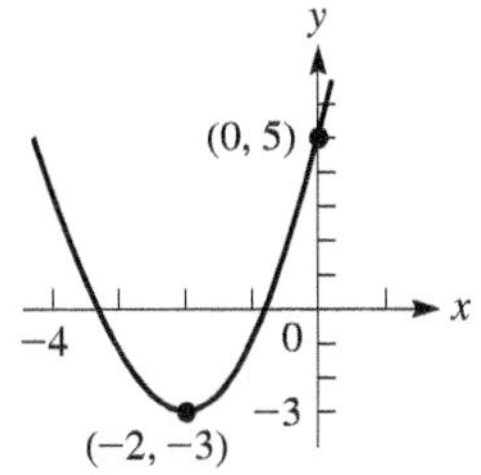

7. $P = EI - RI^2$

$RI^2 - EI + P = 0$

$I = \dfrac{-(-E) \pm \sqrt{(-E)^2 - 4RP}}{2R}$

$= \dfrac{E \pm \sqrt{E^2 - 4RP}}{2R}$

8. $x^2 - 6x - 9 = 0$

$x^2 - 6x \quad = 9$

$x^2 - 6x + 9 = 9 + 9$

$(x-3)^2 = 18$

$x - 3 = \pm\sqrt{18}$

$x = 3 \pm 3\sqrt{2}$

9. Let w = width of window

h = height of window

$2w + 2h = 8.4$ (perimeter)

$w + h = 4.2 \quad h = 4.2 - w$

$w(4.2 - w) = 3.8$ (area)

$-w^2 + 4.2w = 3.8$

$w^2 - 4.2w + 3.8 = 0$

$w = \dfrac{-(-4.2) \pm \sqrt{(-4.2)^2 - 4(3.8)}}{2}$

$= 2.88, 1.32$

$w = 1.3$ m, $h = 2.9$ m or

$w = 2.9$ m, $h = 1.3$ m

10. $y = x^2 - 8x + 8$

$a = 1, b = -8$

$\dfrac{-b}{2a} = \dfrac{-(-8)}{2(1)} = 4$

$y = 4^2 - 8(4) + 8 = -8$

$V(4, -8)$, $c = 8$, y-int $(0, 8)$

For $x = 8$, $y = 8$

$y = 0$ for $x = 1.2, 6.8$

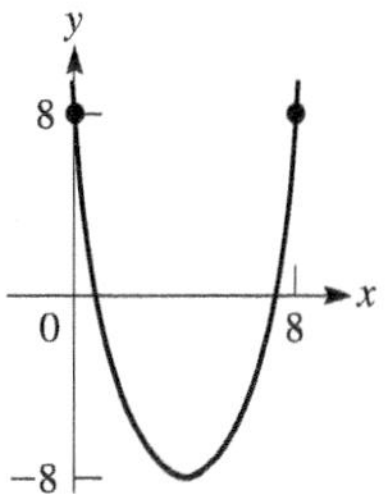

Chapter 8

1. $150° = \left(\dfrac{\pi}{180}\right)(150) = \dfrac{5\pi}{6}$

2. (a) + (b) − (c) +

3. $\sin 205° = -\sin(205° - 180°) = -\sin 25°$

4. $x = -9, \quad y = 12$

$r = \sqrt{(-9)^2 + 12^2} = 15$

$\sin\theta = \dfrac{12}{15} = \dfrac{4}{5}$

$\sec\theta = \dfrac{15}{-9} = -\dfrac{5}{3}$

5. $r = 1.40$ m

$\omega = 2200$ r/min $= (2200$ r/min$)(2\pi$ rad/r$)$

$= 4400\pi$ rad/min

$v = \omega r = (4400\pi)(1.40) = 19\,000$ m/min

6. $3.572 = 3.572\left(\dfrac{180°}{\pi}\right) = 204.7°$

7. $\tan\theta = 0.2396$

$\theta_{ref} = 13.47°$

$\theta = 13.47°$ or

$\theta = 180° + 13.47° = 193.47°$

8. $\cos\theta = -0.8244$, $\csc\theta < 0$;
$\cos\theta$ negative, $\csc\theta$ negative,
θ in third quadrant
$\theta_{ref} = 0.6017, \quad \theta = 3.7432$

9. $s = r\theta$, $16.0 = 4.25\theta$

$$\theta = \frac{16.0}{4.25} = 3.76 \text{ rad}$$

$$A = \frac{1}{2}\theta r^2 = \frac{1}{2}(3.76)(4.25)^2$$
$$= 34.0 \text{ m}^2$$

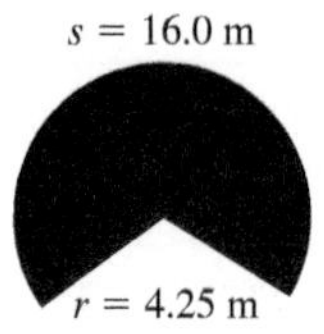

10. $A = \frac{1}{2}\theta r^2$

$A = 38.5 \text{ cm}^2$, $d = 12.2$ cm, $r = 6.10$ cm

$$38.5 = \frac{1}{2}\theta(6.10)^2, \ \theta = \frac{2(38.5)}{6.10^2} = 2.07$$

$s = r\theta$, $s = 6.10(2.07) = 12.6$ cm

Chapter 9

1.

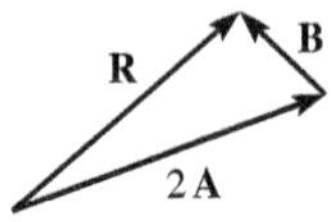

2. $b^2 = a^2 + c^2 - 2ac\cos B$

$$b = \sqrt{22.5^2 + 30.9^2 - 2(22.5)(30.9)\cos 78.6^\circ}$$
$$= 34.4$$

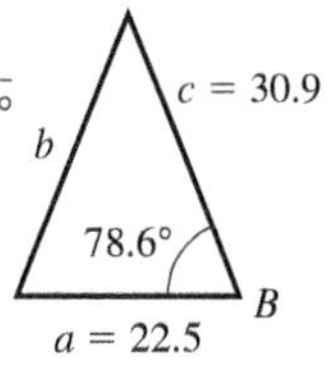

3. $x^2 = 36.50^2 + 21.38^2 - 2(36.50)(21.38)\cos 45.00^\circ$

$$x = 26.19 \text{ m} \qquad \frac{21.38}{\sin\alpha} = \frac{26.19}{\sin 45.00^\circ}$$

$$\sin\alpha = \frac{21.38 \sin 45.00^\circ}{26.19}, \quad \alpha = 35.26^\circ$$

$\theta = 45.00^\circ - 35.26^\circ = 9.74^\circ$
Displacement is 26.19 m,
9.74° N of E.

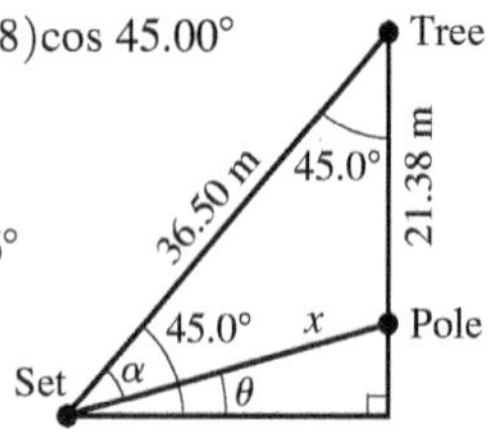

4. $C = 180^\circ - (18.9^\circ + 104.2^\circ) = 56.9^\circ$

$$\frac{c}{\sin C} = \frac{a}{\sin A}$$

$$c = \frac{426 \sin 56.9^\circ}{\sin 18.9^\circ} = 1100$$

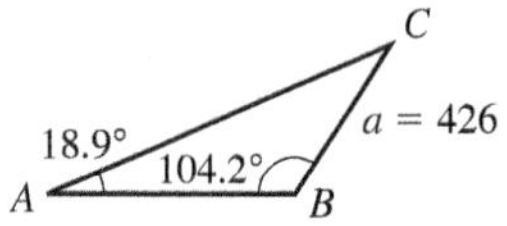

5. Since a is longest side, find A first.
$a^2 = b^2 + c^2 - 2bc\cos A$

$$\cos A = \frac{b^2 + c^2 - a^2}{2bc} = \frac{3.29^2 + 8.44^2 - 9.84^2}{2(3.29)(8.44)}$$

$A = 105.4^\circ$

$$\frac{b}{\sin B} = \frac{a}{\sin A}$$

$$\sin B = \frac{b \sin A}{a} = \frac{3.29 \sin 105.4^\circ}{9.84}$$

$B = 18.8^\circ$
$C = 180^\circ - (105.4^\circ + 18.8^\circ) = 55.8^\circ$

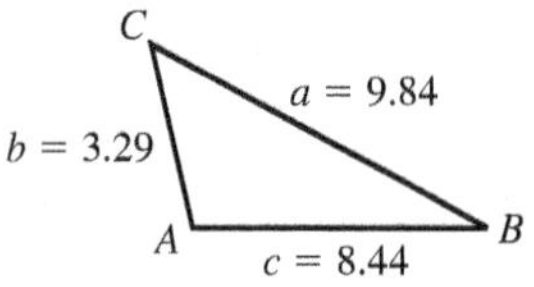

6. $R_x = -235$, $R_y = 152$

$$R = \sqrt{(-235)^2 + 152^2} = 280$$

$$\tan\theta_{ref} = \left|\frac{152}{-235}\right| = \frac{152}{235}, \qquad \theta_{ref} = 32.9^\circ$$

Quad. II: $\theta = 180^\circ - 32.9^\circ = 147.1^\circ$

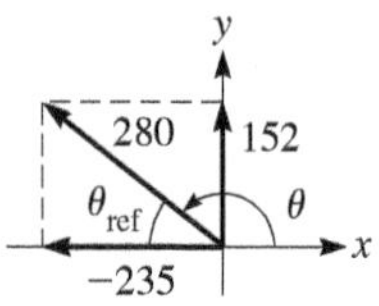

7. $A_x = 871 \cos 284.3^\circ = 215$
$A_y = 871 \sin 284.3^\circ = -844$

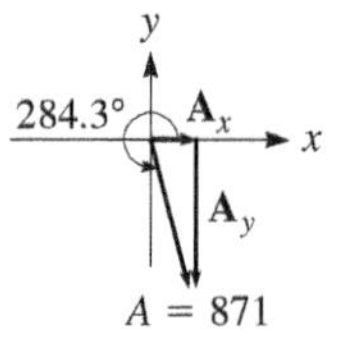

8.
$$\frac{63.0}{\sin 148.5^\circ} = \frac{42.0}{\sin A}$$

$$\sin A = \frac{42.0 \sin 148.5^\circ}{63.0}$$

$A = 20.4^\circ$
$C = 180^\circ - (148.5^\circ + 20.4^\circ) = 11.1^\circ$

$$\frac{c}{\sin 11.1^\circ} = \frac{63.0}{\sin 148.5^\circ}$$

$$c = \frac{63.0 \sin 11.1^\circ}{\sin 148.5^\circ}$$
$$= 23.2 \text{ km}$$

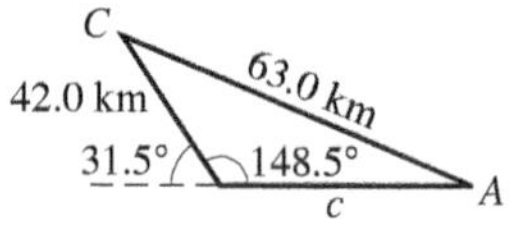

9. $A_x = 449 \cos 74.2^\circ \qquad B_x = 285 \cos 208.9^\circ$
$A_y = 449 \sin 74.2^\circ \qquad B_y = 285 \sin 208.9^\circ$
$R_x = A_x + B_x = 449 \cos 74.2^\circ + 285 \cos 208.9^\circ$
$= -127.3$
$R_y = A_y + B_y = 449 \sin 74.2^\circ + 285 \sin 208.9^\circ$
$= 294.3$

$$R = \sqrt{(-127.3)^2 + 294.3^2} = 321$$

$$\tan\theta_{ref} = \frac{294.3}{127.3}, \qquad \theta_{ref} = 66.6^\circ, \qquad \theta = 113.4^\circ$$

θ is in second quadrant, since R_x is negative and R_y is positive.

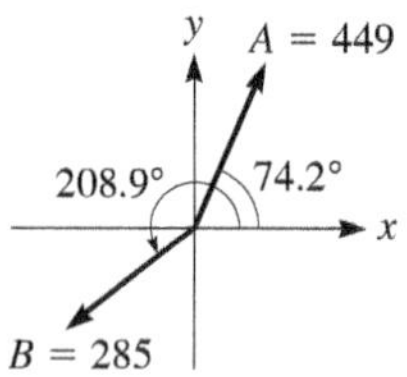

10. $29.6 \sin 36.5^\circ = 17.6$
$17.6 < 22.3 < 29.6$ means two solutions.

$$\frac{29.6}{\sin B} = \frac{22.3}{\sin 36.5^\circ}, \qquad \sin B = \frac{29.6 \sin 36.5^\circ}{22.3}$$

$B_1 = 52.1^\circ \qquad C_1 = 180^\circ - 36.5^\circ - 52.1^\circ = 91.4^\circ$
$B_2 = 180^\circ - 52.1^\circ = 127.9^\circ$,
$C_2 = 180^\circ - 36.5^\circ - 127.9^\circ = 15.6^\circ$

$$\frac{c_1}{\sin 91.4^\circ} = \frac{22.3}{\sin 36.5^\circ}, \qquad c_1 = \frac{22.3 \sin 91.4^\circ}{\sin 36.5^\circ} = 37.5$$

$$\frac{c_2}{\sin 15.6^\circ} = \frac{22.3}{\sin 36.5^\circ}, \qquad c_2 = \frac{22.3 \sin 15.6^\circ}{\sin 36.5^\circ} = 10.1$$

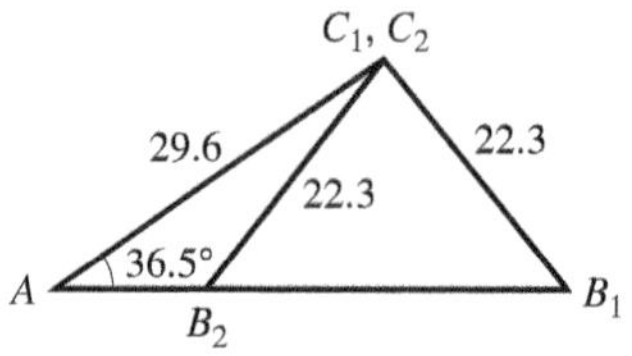

11. Sum of x-components = 0:
$185 \cos 305.6^\circ + T_2 \cos 90^\circ + T_1 \cos 221.7^\circ = 0$

$$T_1 = \frac{-185 \cos 305.6^\circ}{\cos 221.7^\circ} = 144 \text{ N}$$

Sum of y-components = 0:
$185 \sin 305.6^\circ + T_2 \sin 90^\circ + T_1 \sin 221.7^\circ = 0$
$185 \sin 305.6^\circ + T_2 + (144.24) \sin 221.7^\circ = 0$
$-T_2 = -246.38$
$T_2 = 246$ N

Chapter 10

1. $y = -3\sin(4\pi x - \pi/3)$
$a = -3,\ b = 4\pi,\ c = -\pi/3$ amplitude $= |-3| = 3$
period $= \dfrac{2\pi}{4\pi} = \dfrac{1}{2}$ displacement $= -\dfrac{-\frac{\pi}{3}}{4\pi} = \dfrac{1}{12}$

2. $y = 0.5\cos\dfrac{\pi}{2}x$
Amp. $= 0.5$, disp. $= 0$,
per. $= \dfrac{2\pi}{\pi/2} = 4$

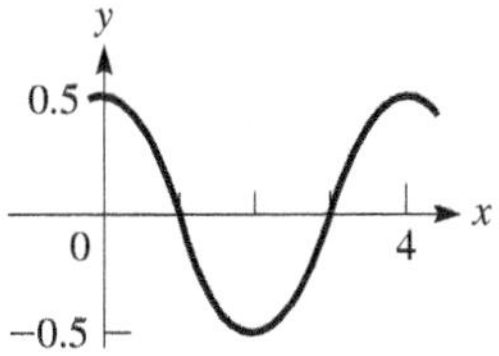

x	0	1	2	3	4
y	0.5	0	−0.5	0	0.5

3. $y = 2 + 3\sin x$
For $y_1 = 3\sin x$,
amp. $= 3$, per. $= 2\pi$,
disp. $= 0$

x	0	$\frac{\pi}{2}$	π	$\frac{3\pi}{2}$	2π
$y_1 = 3\sin x$	0	3	0	−3	0
$y = 2 + 3\sin x$	2	5	2	−1	2

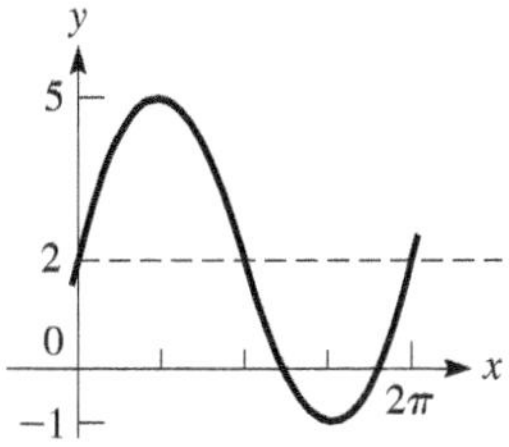

4. $y = 3\sec x$
$\sec x = \dfrac{1}{\cos x}$

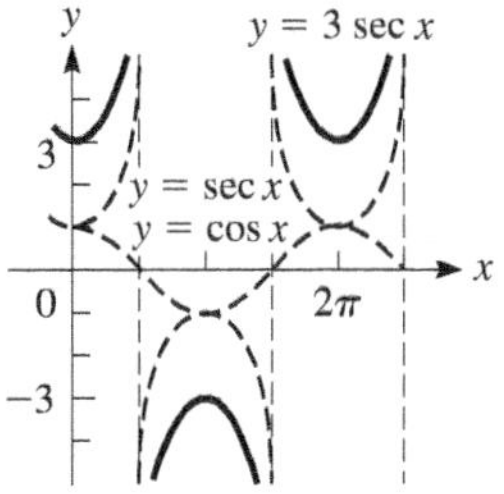

5. $y = 2\sin\left(2x - \dfrac{\pi}{3}\right)$
Amp. $= 2$, per. $= \dfrac{2\pi}{2} = \pi$,
disp. $= -\dfrac{-\pi/3}{2} = \dfrac{\pi}{6}$

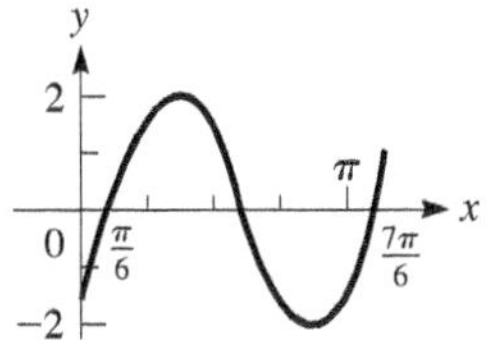

x	$\frac{\pi}{6}$	$\frac{5\pi}{12}$	$\frac{2\pi}{3}$	$\frac{11\pi}{12}$	$\frac{7\pi}{6}$
y	0	2	0	−2	0

$\frac{\pi}{6} + \frac{\pi}{4} = \frac{5\pi}{12}, \frac{\pi}{6} + \frac{\pi}{2} = \frac{2\pi}{3}, \frac{\pi}{6} + \frac{3\pi}{4} = \frac{11\pi}{12}$

6. $y = A\cos\frac{2\pi}{T}t$
$A = 0.200$ cm, $T = 0.100$ s,
$y = 0.200\cos 20\pi t$,
amp. $= 0.200$ cm, per. $= 0.100$ s

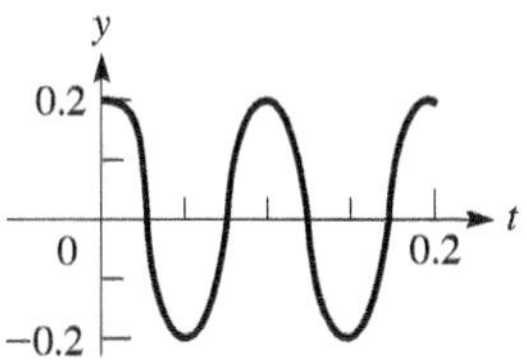

7. $y = 2\sin x + \cos 2x$
For $y_1 = 2\sin x$,
amp. $= 2$, per. $= 2\pi$, disp. $= 0$
For $y_2 = \cos 2x$,
amp. $= 1$, per. $= \frac{2\pi}{2} = \pi$, disp. $= 0$

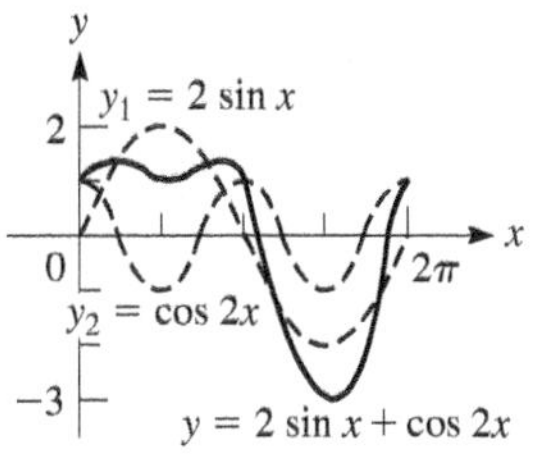

8. $x = \sin\pi t,\ y = 2\cos 2\pi t$

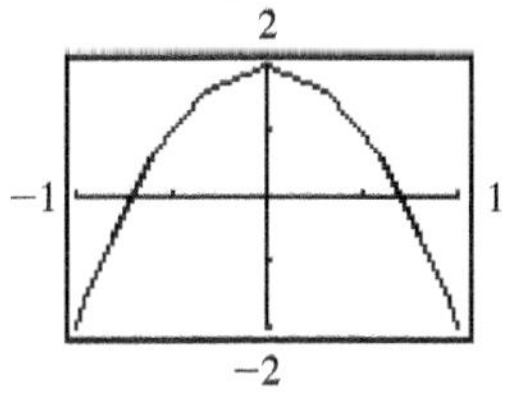

9. $d = R\sin\left(\omega t + \frac{\pi}{6}\right)$
$\omega = 2.00$ rad/s
$d = R\sin\left(2.00t + \frac{\pi}{6}\right)$
Amp. $= R$, per. $= \dfrac{2\pi}{2.00} = \pi$ s $= 3.14$ s,
disp. $= \dfrac{-\pi/6}{2.00} = -\dfrac{\pi}{12}$ s $= -0.26$ s

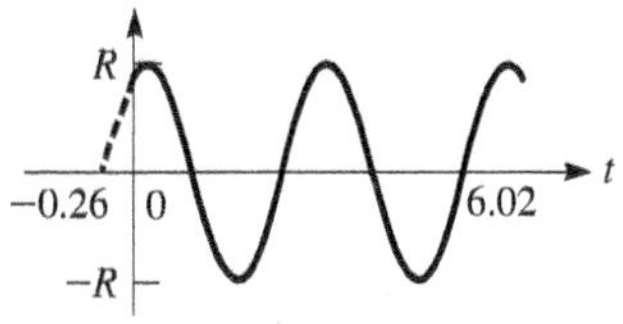

10. $y = 2\sin bx$
$\left(\frac{\pi}{3}, 2\right)$ is first max. with $x > 0$
period $= \frac{2\pi}{b}, \frac{1}{4}\left(\frac{2\pi}{b}\right) = \frac{\pi}{3}, \frac{\pi}{2b} = \frac{\pi}{3}, b = \frac{3}{2}$
$y = 2\sin\frac{3x}{2}$

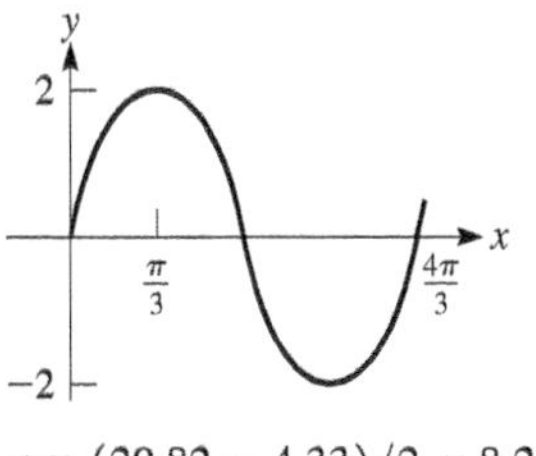

11. $a = (20.82 - 4.33)/2 = 8.25$. The period is 365 days, so $2\pi/b = 365$ and $b = 2\pi/365$. Maximum is at $365/4 = 91.25$. For these data, maximum is at $t = 172$ (June 21). Hence we shift the sine function to the right $172 - 91.25 = 80.75$ days. This gives $-c/b = 80.75$, and $c = -80.75(2\pi/365) = -1.39$. Finally, the range is $[4.33, 20.82]$. We shift the sine function $20.82 - 8.25 = 12.57$ hours vertically by setting $d = 12.57$. The function is $h = 8.25\sin\left(\frac{2\pi}{365}t - 1.39\right) + 12.57$.

Chapter 11

1. $-5y^0 = -5(1) = -5$

2. $(3\pi x^{-4})^{-2} = (3\pi)^{-2}(x^{-4})^{-2} = \dfrac{x^8}{(3\pi)^2} = \dfrac{x^8}{9\pi^2}$

3. $\dfrac{100^{3/2}}{8^{-2/3}} = (100^{3/2})(8^{2/3}) = \left[(100^{1/2})^3\right]\left[(8^{1/3})^2\right]$

$= (10^3)(2^2) = 4000$

4. $(as^{-1/3}t^{3/4})^{12} = a^{12}s^{(-1/3)(12)}t^{(3/4)(12)} = a^{12}s^{-4}t^9 = \dfrac{a^{12}t^9}{s^4}$

5. $2\sqrt{20} - \sqrt{125} = 2\sqrt{4 \times 5} - \sqrt{25 \times 5}$

$= 2(2\sqrt{5}) - 5\sqrt{5}$

$= 4\sqrt{5} - 5\sqrt{5} = -\sqrt{5}$

6. $(2x^{-1} + y^{-2})^{-1} = \dfrac{1}{2x^{-1} + y^{-2}} = \dfrac{1}{\dfrac{2}{x} + \dfrac{1}{y^2}} = \dfrac{1}{\dfrac{2y^2 + x}{xy^2}}$

$= \dfrac{xy^2}{2y^2 + x}$

7. $\left(\sqrt{2x} - 3\sqrt{y}\right)^2 = \left(\sqrt{2x}\right)^2 - 2\sqrt{2x}\left(3\sqrt{y}\right) + \left(3\sqrt{y}\right)^2$

$= 2x - 6\sqrt{2xy} + 9y$

8. $\sqrt[3]{\sqrt[4]{4}} = \sqrt[12]{4} = \sqrt[12]{2^2} = 2^{2/12} = 2^{1/6} = \sqrt[6]{2}$

9. $\dfrac{3 - 2\sqrt{2}}{2\sqrt{x}} = \dfrac{3 - 2\sqrt{2}}{2\sqrt{x}} \times \dfrac{\sqrt{x}}{\sqrt{x}} = \dfrac{3\sqrt{x} - 2\sqrt{2x}}{2x}$

10. $\sqrt{27a^4b^3} = \sqrt{9 \times 3 \times (a^2)^2(b^2)(b)} = 3a^2b\sqrt{3b}$

11. $2\sqrt{2}(3\sqrt{10} - \sqrt{6}) = 2\sqrt{2}(3\sqrt{10}) - 2\sqrt{2}(\sqrt{6})$

$= 6\sqrt{20} - 2\sqrt{12}$

$= 6(2\sqrt{5}) - 2(2\sqrt{3})$

$= 12\sqrt{5} - 4\sqrt{3}$

12. $(2x + 3)^{1/2} + (x + 1)(2x + 3)^{-1/2}$

$= (2x + 3)^{1/2} + \dfrac{x + 1}{(2x + 3)^{1/2}}$

$= \dfrac{(2x + 3)^{1/2}(2x + 3)^{1/2} + x + 1}{(2x + 3)^{1/2}} = \dfrac{2x + 3 + x + 1}{(2x + 3)^{1/2}}$

$= \dfrac{3x + 4}{(2x + 3)^{1/2}}$

13. $\left(\dfrac{4a^{-1/2}b^{3/4}}{b^{-2}}\right)\left(\dfrac{b^{-1}}{2a}\right) = \dfrac{4a^{-1/2}b^{3/4-1}}{2ab^{-2}} = \dfrac{2b^{2-1/4}}{a^{1+1/2}} = \dfrac{2b^{7/4}}{a^{3/2}}$

14. $\dfrac{2\sqrt{15} + \sqrt{3}}{\sqrt{15} - 2\sqrt{3}} = \dfrac{2\sqrt{15} + \sqrt{3}}{\sqrt{15} - 2\sqrt{3}} \times \dfrac{\sqrt{15} + 2\sqrt{3}}{\sqrt{15} + 2\sqrt{3}}$

$= \dfrac{2(15) + 4\sqrt{45} + \sqrt{45} + 2(3)}{15 - 4(3)}$

$= \dfrac{36 + 15\sqrt{5}}{3}$

$= 12 + 5\sqrt{5}$

15. $\dfrac{3^{-1/2}}{2} = \dfrac{1}{2 \times 3^{1/2}} = \dfrac{1}{2\sqrt{3}}$

$= \dfrac{\sqrt{3}}{2\sqrt{3}\sqrt{3}} = \dfrac{\sqrt{3}}{6}$

16. $0.220N^{-1/6} \quad N = 64 \times 10^6$

$0.220(64 \times 10^6)^{-1/6} = \dfrac{0.220}{(64 \times 10^6)^{1/6}} = \dfrac{0.220}{2 \times 10} = 0.011$

Chapter 12

1. $\left(3 - \sqrt{-4}\right) + \left(5\sqrt{-9} - 1\right) = (3 - 2j) + \left[5(3j) - 1\right]$

$= 3 - 2j + 15j - 1$

$= 2 + 13j$

2. $(2\underline{/130^\circ})(3\underline{/45^\circ}) = (2)(3)\underline{/130^\circ + 45^\circ} = 6\underline{/175^\circ}$

3. $2 - 7j$: $\quad r = \sqrt{2^2 + (-7)^2} = \sqrt{53} = 7.28$

$\tan\theta = \dfrac{-7}{2} = -3.500, \quad \theta_{ref} = 74.1^\circ, \quad \theta = 285.9^\circ$

$2 - 7j = 7.28(\cos 285.9^\circ + j\sin 285.9^\circ)$

$= 7.28\underline{/285.9^\circ}$

4. (a) $-\sqrt{-64} = -(8j) = -8j$

(b) $-j^{15} = -j^{12}j^3 = (-1)(-j) = j$

5. $(4 - 3j) + (-1 + 4j) = 3 + j$

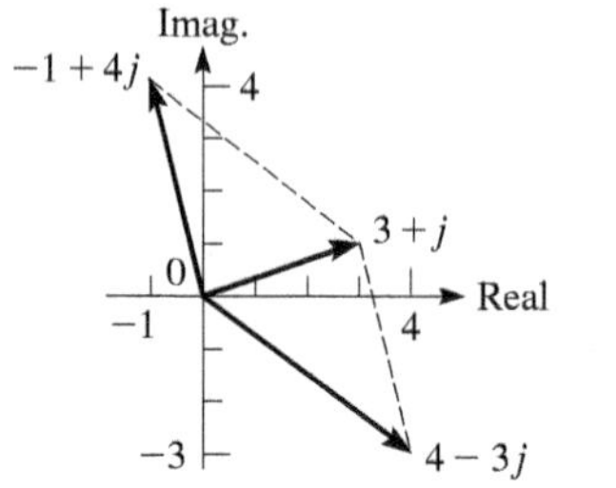

6. $\dfrac{2 - 4j}{5 + 3j} = \dfrac{(2 - 4j)(5 - 3j)}{(5 + 3j)(5 - 3j)} = \dfrac{10 - 26j + 12j^2}{25 - 9j^2}$

$= \dfrac{10 - 26j - 12}{25 - 9(-1)} = \dfrac{-2 - 26j}{34} = -\dfrac{1 + 13j}{17}$

7. $2.56(\cos 125.2^\circ + j\sin 125.2^\circ) = 2.56e^{2.185j}$

$125.2^\circ = \dfrac{125.2\pi}{180} = 2.185\text{ rad}$

8. $R = 3.50\ \Omega,\ X_L = 6.20\ \Omega,\ X_C = 7.35\ \Omega$

$|Z| = \sqrt{R^2 + (X_L - X_C)^2}$

$= \sqrt{3.50^2 + (6.20 - 7.35)^2} = 3.68\Omega$

$\tan\theta = \dfrac{X_L - X_C}{R} = \dfrac{6.20 - 7.35}{3.50} = \dfrac{-1.15}{3.50}$

$\theta = -18.2^\circ$

9. $3.47 - 2.81j = 4.47e^{5.60j}$

$R = \sqrt{3.47^2 + (-2.81)^2} = 4.47$

$\tan\theta = \dfrac{-2.81}{3.47}, \quad \theta_{ref} = 39.0^\circ$

$\theta = 321.0^\circ = 5.60\text{ rad}$

10. $x + 2j - y = yj - 3xj$

$x - y + 3xj - yj = -2j$

$(x - y) + (3x - y)j = 0 - 2j$

$x - y = 0$

$3x - y = -2$

$2x = -2$

$x = -1, \quad y = -1$

11. $L = 8.75\text{ mH} = 8.75 \times 10^{-3}\text{ H}$

$f = 600\text{ kHz} = 6.00 \times 10^5\text{ Hz}$

$2\pi fL = \dfrac{1}{2\pi fC}$

$C = \dfrac{1}{(2\pi f)^2 L} = \dfrac{1}{(2\pi)^2(6.00 \times 10^5)^2(8.75 \times 10^{-3})}$

$= 8.04 \times 10^{-12} = 8.04\text{ pF}$

12. $j = 1(\cos 90° + j\sin 90°)$

$$j^{1/3} = 1^{1/3}\left(\cos\frac{90°}{3} + j\sin\frac{90°}{3}\right)$$
$$= \cos 30° + j\sin 30° = 0.8660 + 0.5000j$$
$$= 1^{1/3}\left(\cos\frac{90° + 360°}{3} + j\sin\frac{90° + 360°}{3}\right)$$
$$= \cos 150° + j\sin 150° = -0.8660 + 0.5000j$$
$$= 1^{1/3}\left(\cos\frac{90° + 720°}{3} + j\sin\frac{90° + 720°}{3}\right)$$
$$= \cos 270° + j\sin 270° = -j$$

Cube roots of j: $0.8660 + 0.5000j$
$-0.8660 + 0.5000j$
$-j$

Chapter 13

1. $\log_9 x = -\dfrac{1}{2}$

$$x = 9^{-1/2}$$
$$= \frac{1}{9^{1/2}} = \frac{1}{3}$$

2. $\log_3 x - \log_3 2 = 2$

$$\log_3\frac{x}{2} = 2$$
$$\frac{x}{2} = 3^2$$
$$x = 2(3^2) = 18$$

3. $\log_x 64 = 3$

$$64 = x^3$$
$$4^3 = x^3$$
$$x = 4$$

4. $3^{3x+1} = 8$

$$(3x + 1)\log 3 = \log 8$$
$$3x + 1 = \frac{\log 8}{\log 3}$$
$$x = \frac{1}{3}\left(\frac{\log 8}{\log 3} - 1\right) = 0.298$$

5. $y = 2\log_4 x$

x	$\frac{1}{4}$	1	4	16
y	-2	0	2	4

$\log_4 \frac{1}{4} = -1$, $\log_4 16 = 2$

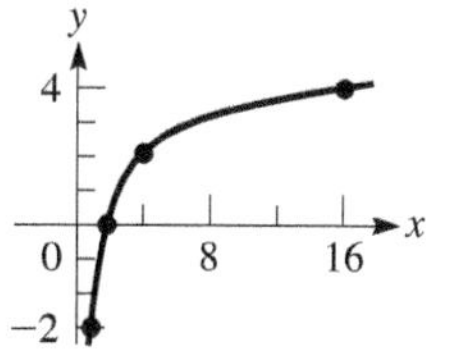

6. $y = 2(3^x)$

x	-1	0	1	2
y	0.7	2	6	18

x	3	4	5
y	54	162	486

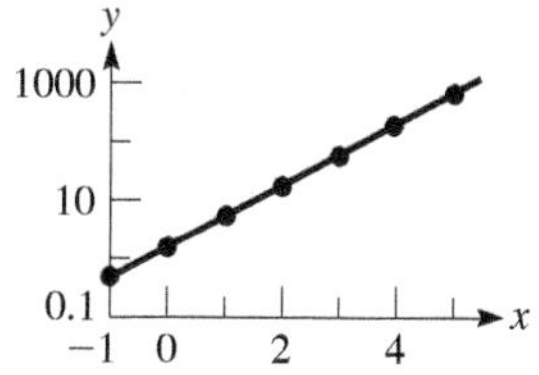

7. $\log_5\left(\dfrac{4a^3}{7}\right) = \log_5 4a^3 - \log_5 7$

$$= \log_5 4 + \log_5 a^3 - \log_5 7$$
$$= \log_5 4 + 3\log_5 a - \log_5 7$$
$$= 2\log_5 2 + 3\log_5 a - \log_5 7$$

8. $3\log_7 x - \log_7 y = 2$

$$\log_7 x^3 - \log_7 y = 2$$
$$\log_7\frac{x^3}{y} = 2 \qquad \frac{x^3}{y} = 7^2$$
$$x^3 = 49y \qquad y = \tfrac{1}{49}x^3$$

9. $\ln i - \ln I = -t/RC$

$$\ln\frac{i}{I} = -t/RC$$
$$\frac{i}{I} = e^{-t/RC} \qquad i = Ie^{-t/RC}$$

10. $\dfrac{2\ln 0.9523}{\log 6066} = -0.025\,84$

11. $\log_b x = \dfrac{\log_a x}{\log_a b}$

$$\log_5 7.32 = \frac{\log 7.32}{\log 5} = 1.237$$

12. $A = A_0e^{0.08t} \qquad A = 2A_0$

$$2A_0 = A_0e^{0.08t} \qquad 2 = e^{0.08t}$$
$$\ln 2 = \ln e^{0.08t} = 0.08t \qquad t = \frac{\ln 2}{0.08} = 8.66 \text{ years}$$

Chapter 14

1. $x^{1/2} - 2x^{1/4} = 3$

Let $y = x^{1/4}$

$$y^2 - 2y - 3 = 0$$
$$(y - 3)(y + 1) = 0$$
$$y = 3, -1$$
$$x^{1/4} \neq -1$$
$$x^{1/4} = 3, \qquad x = 81$$

Check: $81^{1/2} - 2(81^{1/4}) = 3$
$9 - 6 = 3$

Solution: $x = 81$

2. $3\sqrt{x - 2} - \sqrt{x + 1} = 1$

$$3\sqrt{x - 2} = 1 + \sqrt{x + 1}$$
$$9(x - 2) = 1 + 2\sqrt{x + 1} + (x + 1)$$
$$8x - 20 = 2\sqrt{x + 1}$$
$$4x - 10 = \sqrt{x + 1}$$
$$16x^2 - 80x + 100 = x + 1$$
$$16x^2 - 81x + 99 = 0$$
$$x = \frac{81 \pm \sqrt{81^2 - 4(16)(99)}}{32}$$
$$= \frac{81 \pm 15}{32} = 3, \frac{33}{16}$$

Check: $x = 3$: $3\sqrt{3 - 2} - \sqrt{3 + 1} \stackrel{?}{=} 1$
$3 - 2 = 1$
$x = \frac{33}{16}$: $3\sqrt{\frac{33}{16} - 2} - \sqrt{\frac{33}{16} + 1} \stackrel{?}{=} 1$
$\frac{3}{4} - \frac{7}{4} \neq 1$

Solution: $x = 3$

3. $x^4 - 17x^2 + 16 = 0$

Let $y = x^2$

$$y^2 - 17y + 16 = 0$$
$$(y - 1)(y - 16) = 0$$
$$y = 1, 16$$
$$x^2 = 1, 16$$
$$x = -1, 1, -4, 4$$

All values check.

4. $x^2 - 2y = 5$

$\underline{2x + 6y = 1}$

$y = \frac{1}{2}(x^2 - 5)$

$2x + 6\left(\frac{1}{2}\right)(x^2 - 5) = 1$

$3x^2 + 2x - 16 = 0$

$(3x + 8)(x - 2) = 0$

$x = -\frac{8}{3}, 2$

$x = -\frac{8}{3}$: $y = \frac{1}{2}\left(\frac{64}{9} - 5\right) = \frac{19}{18}$

$x = 2$: $y = \frac{1}{2}(4 - 5) = -\frac{1}{2}$

$x = -\frac{8}{3}, y = \frac{19}{18}$

or $x = 2, y = -\frac{1}{2}$

5. $\sqrt[3]{2x + 5} = 5$

$2x + 5 = 125$

$2x = 120$

$x = 60$

6. $v = \sqrt{v_0^2 + 2gh}$

$v^2 = v_0^2 + 2gh$

$2gh = v^2 - v_0^2$

$h = \dfrac{v^2 - v_0^2}{2g}$

7. $x^2 - y^2 = 4 \quad xy = 2$

$y = \pm\sqrt{x^2 - 4} \quad y = \dfrac{2}{x}$

x	y
± 2	0
± 3	± 2.2
± 4	± 3.5

x	y
-4	$-\frac{1}{2}$
-2	-1
-1	-2
0	—
1	2
2	1
4	$\frac{1}{2}$

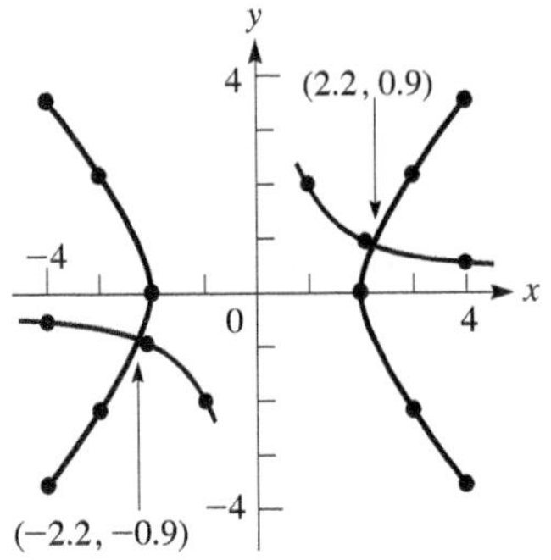

$x = 2.2, \quad y = 0.9; \quad x = -2.2, \quad y = -0.9$

8. Let l = length, w = width

$2l + 2w = 14$

$\underline{lw = 10}$

$l + w = 7$

$l = 7 - w$

$(7 - w)w = 10$

$7w - w^2 = 10$

$w^2 - 7w + 10 = 0$

$(w - 5)(w - 2) = 0$

$w = 5, 2$

$l = 5.00$ m, $w = 2.00$ m

(or $l = 2.00$ m, $w = 5.00$ m)

Chapter 15

1. $f(x) = 2x^3 + 3x^2 + 7x - 6$

$f(-3) = 2(-3)^3 + 3(-3)^2 + 7(-3) - 6$

$= -54 + 27 - 21 - 6 = -54$

$f(-3) \neq 0, \quad -3$ is not a zero

2. $x^4 - 2x^3 - 7x^2 + 20x - 12 = 0$

$$\begin{array}{rrrrr|l} 1 & -2 & -7 & 20 & -12 & \underline{2} \\ & 2 & 0 & -14 & 12 & \\ \hline 1 & 0 & -7 & 6 & & \underline{2} \\ & 2 & 4 & -6 & & \\ \hline 1 & 2 & -3 & & & \end{array}$$

$x^2 + 2x - 3 = (x + 3)(x - 1)$

Other roots: $x = -3, 1$

3. $(x^3 - 5x^2 + 4x - 9) \div (x - 3)$

$$\begin{array}{rrrr|l} 1 & -5 & 4 & -9 & \underline{3} \\ & 3 & -6 & -6 & \\ \hline 1 & -2 & -2 & -15 & \end{array}$$

Quotient: $x^2 - 2x - 2$

Remainder $= -15$

4. $f(x) = 2x^4 + 15x^3 + 23x^2 - 16$

$2x + 1 = 2\left(x + \frac{1}{2}\right)$

$$\begin{array}{rrrrr|l} 2 & 15 & 23 & 0 & -16 & \underline{-\frac{1}{2}} \\ & -1 & -7 & -8 & 4 & \\ \hline 2 & 14 & 16 & -8 & -12 & \end{array}$$

Remainder is not zero;

$2x + 1$ is not a factor.

5. $(x^3 + 4x^2 + 7x - 9) \div (x + 4)$

$f(x) = x^3 + 4x^2 + 7x - 9$

$f(-4) = (-4)^3 + 4(-4)^2 + 7(-4) - 9$

$= -64 + 64 - 28 - 9 = -37$

Remainder $= -37$

6. $2x^4 - x^3 + 5x^2 - 4x - 12 = 0$

$f(x) = 2x^4 - x^3 + 5x^2 - 4x - 12$

$f(-x) = 2x^4 + x^3 + 5x^2 + 4x - 12$

$n = 4$; 4 roots

No more than 3 positive roots

One negative root

Rational roots: factors of 12 divided by factors of 2

Possible rational roots: $\pm 1, \pm 2, \pm 3, \pm 4, \pm 6, \pm 12, \pm\frac{1}{2}, \pm\frac{3}{2}$

$$\begin{array}{rrrrr|l} 2 & -1 & 5 & -4 & -12 & \underline{2} \\ & 4 & 6 & 22 & 36 & \\ \hline 2 & 3 & 11 & 18 & 24 & \end{array}$$

2 is too large.

$$\begin{array}{rrrrr|l} 2 & -1 & 5 & -4 & -12 & \underline{\frac{3}{2}} \\ & 3 & 3 & 12 & 12 & \\ \hline 2 & 2 & 8 & 8 & 0 & \underline{-1} \\ & -2 & 0 & -8 & & \\ \hline 2 & 0 & 8 & 0 & & \end{array}$$

$2x^2 + 8 = 0, \quad x^2 + 4 = 0, \quad x = \pm 2j$

roots: $\frac{3}{2}, -1, 2j, -2j$

7. $y = kx^2(x^3 + 436x - 4000)$

$y = 0$, $kx^2(x^3 + 436x - 4000) = 0$

$x = 0$, $x^3 + 436x - 4000 = 0$

$$\begin{array}{rrrr|l} 1 & 0 & 436 & -4000 & \underline{8} \\ & 8 & 64 & 4000 & \\ \hline 1 & 8 & 500 & 0 & \end{array}$$

$y = 0$ for $x = 0$ m, $x = 8$ m

8. Let x = length of edge.

$V = x^3, \quad 2V = (x+1)^3$

$2x^3 = x^3 + 3x^2 + 3x + 1$

$x^3 - 3x^2 - 3x - 1 = 0$

$y = x^3 - 3x^2 - 3x - 1$

$x = 3.8$ mm

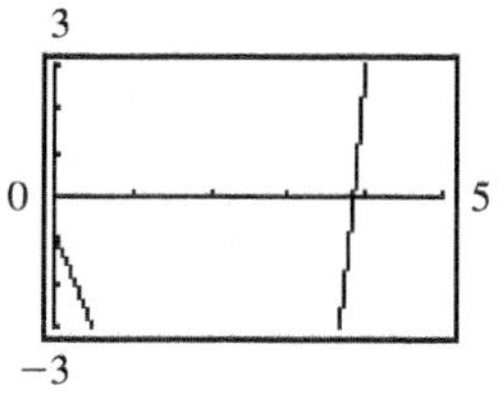

Chapter 16

1. $A = \begin{bmatrix} 3 & -1 & 4 \\ 2 & 0 & -2 \end{bmatrix} \quad B = \begin{bmatrix} 1 & 4 & 5 \\ -1 & -2 & 3 \end{bmatrix}$

$2B = \begin{bmatrix} 2 & 8 & 10 \\ -2 & -4 & 6 \end{bmatrix}$

$A - 2B = \begin{bmatrix} 3-2 & -1-8 & 4-10 \\ 2+2 & 0+4 & -2-6 \end{bmatrix}$

$= \begin{bmatrix} 1 & -9 & -6 \\ 4 & 4 & -8 \end{bmatrix}$

2. $\begin{bmatrix} 2x & x-y & z \\ x+z & 2y & y+z \end{bmatrix} = \begin{bmatrix} 6 & -2 & 4 \\ a & b & c \end{bmatrix}$

$2x = 6 \qquad x - y = -2 \qquad z = 4$

$x = 3 \qquad 3 - y = -2$

$y = 5$

$x + z = a \qquad 2y = b \qquad y + z = c$

$3 + 4 = a \qquad 2(5) = b \qquad 5 + 4 = c$

$a = 7 \qquad b = 10 \qquad c = 9$

3. $CD = \begin{bmatrix} 1 & 0 & 4 \\ 2 & -2 & 1 \\ -1 & 3 & 2 \end{bmatrix}\begin{bmatrix} 2 & -2 \\ 4 & -5 \\ 6 & 1 \end{bmatrix}$

$= \begin{bmatrix} 2+0+24 & -2+0+4 \\ 4-8+6 & -4+10+1 \\ -2+12+12 & 2-15+2 \end{bmatrix}$

$= \begin{bmatrix} 26 & 2 \\ 2 & 7 \\ 22 & -11 \end{bmatrix}$

$DC = \begin{bmatrix} 2 & -2 \\ 4 & -5 \\ 6 & 1 \end{bmatrix}\begin{bmatrix} 1 & 0 & 4 \\ 2 & -2 & 1 \\ -1 & 3 & 2 \end{bmatrix}$ not defined, since D has 2 columns and C has 3 rows,

4. $A = \begin{bmatrix} 2 & -5 \\ 1 & -2 \end{bmatrix} \quad B = \begin{bmatrix} -2 & 5 \\ -1 & 2 \end{bmatrix}$

$AB = \begin{bmatrix} 2 & -5 \\ 1 & -2 \end{bmatrix}\begin{bmatrix} -2 & 5 \\ -1 & 2 \end{bmatrix} = \begin{bmatrix} -4+5 & 10-10 \\ -2+2 & 5-4 \end{bmatrix}$

$= \begin{bmatrix} 1 & 0 \\ 0 & 1 \end{bmatrix} = I$

$BA = \begin{bmatrix} -2 & 5 \\ -1 & 2 \end{bmatrix}\begin{bmatrix} 2 & -5 \\ 1 & -2 \end{bmatrix} = \begin{bmatrix} -4+5 & 10-10 \\ -2+2 & 5-4 \end{bmatrix}$

$= \begin{bmatrix} 1 & 0 \\ 0 & 1 \end{bmatrix} = I$

$AB = BA = I, B = A^{-1}$

5. $\left[\begin{array}{ccc|ccc} 1 & 0 & 4 & 1 & 0 & 0 \\ 2 & -2 & 1 & 0 & 1 & 0 \\ -1 & 3 & 2 & 0 & 0 & 1 \end{array}\right] \rightarrow \left[\begin{array}{ccc|ccc} 1 & 0 & 4 & 1 & 0 & 0 \\ 0 & -2 & -7 & -2 & 1 & 0 \\ 0 & 3 & 6 & 1 & 0 & 1 \end{array}\right] \rightarrow$

$\left[\begin{array}{ccc|ccc} 1 & 0 & 4 & 1 & 0 & 0 \\ 0 & 1 & \frac{7}{2} & 1 & -\frac{1}{2} & 0 \\ 0 & 3 & 6 & 1 & 0 & 1 \end{array}\right] \rightarrow \left[\begin{array}{ccc|ccc} 1 & 0 & 4 & 1 & 0 & 0 \\ 0 & 1 & \frac{7}{2} & 1 & -\frac{1}{2} & 0 \\ 0 & 0 & -\frac{9}{2} & -2 & \frac{3}{2} & 1 \end{array}\right] \rightarrow$

$\left[\begin{array}{ccc|ccc} 1 & 0 & 4 & 1 & 0 & 0 \\ 0 & 1 & \frac{7}{2} & 1 & -\frac{1}{2} & 0 \\ 0 & 0 & 1 & \frac{4}{9} & -\frac{1}{3} & -\frac{2}{9} \end{array}\right] \rightarrow \left[\begin{array}{ccc|ccc} 1 & 0 & 0 & -\frac{7}{9} & \frac{4}{3} & \frac{8}{9} \\ 0 & 1 & 0 & -\frac{5}{9} & \frac{2}{3} & \frac{7}{9} \\ 0 & 0 & 1 & \frac{4}{9} & -\frac{1}{3} & -\frac{2}{9} \end{array}\right]$

$C^{-1} = \begin{bmatrix} -\frac{7}{9} & \frac{4}{3} & \frac{8}{9} \\ -\frac{5}{9} & \frac{2}{3} & \frac{7}{9} \\ \frac{4}{9} & -\frac{1}{3} & -\frac{2}{9} \end{bmatrix}$

6. $2x - 3y = 11$

$x + 2y = 2$

$A = \begin{bmatrix} 2 & -3 \\ 1 & 2 \end{bmatrix} \quad A^{-1} = \begin{bmatrix} \frac{2}{7} & \frac{3}{7} \\ -\frac{1}{7} & \frac{2}{7} \end{bmatrix} \quad C = \begin{bmatrix} 11 \\ 2 \end{bmatrix}$

$A^{-1}C = \begin{bmatrix} \frac{2}{7} & \frac{3}{7} \\ -\frac{1}{7} & \frac{2}{7} \end{bmatrix}\begin{bmatrix} 11 \\ 2 \end{bmatrix} = \begin{bmatrix} \frac{22}{7} + \frac{6}{7} \\ -\frac{11}{7} + \frac{4}{7} \end{bmatrix} = \begin{bmatrix} 4 \\ -1 \end{bmatrix}$

$x = 4 \qquad y = -1$

7. $\left[\begin{array}{cccc|c} 1 & 1 & -1 & 2 & 6 \\ 2 & -2 & 3 & -1 & 0 \\ -1 & 3 & -5 & 1 & 1 \\ 0 & 5 & -1 & 4 & 3 \end{array}\right] \rightarrow \left[\begin{array}{cccc|c} 1 & 1 & -1 & 2 & 6 \\ 0 & -4 & 5 & -5 & -12 \\ 0 & 4 & -6 & 3 & 7 \\ 0 & 5 & -1 & 4 & 3 \end{array}\right] \rightarrow$

$\left[\begin{array}{cccc|c} 1 & 1 & -1 & 2 & 6 \\ 0 & 1 & -\frac{5}{4} & \frac{5}{4} & 3 \\ 0 & 0 & -1 & -2 & -5 \\ 0 & 0 & \frac{21}{4} & -\frac{9}{4} & -12 \end{array}\right] \rightarrow \left[\begin{array}{cccc|c} 1 & 1 & -1 & 2 & 6 \\ 0 & 1 & -\frac{5}{4} & \frac{5}{4} & 3 \\ 0 & 0 & 1 & 2 & 5 \\ 0 & 0 & 0 & -\frac{51}{4} & -\frac{153}{4} \end{array}\right] \rightarrow$

$\left[\begin{array}{cccc|c} 1 & 1 & -1 & 2 & 6 \\ 0 & 1 & -\frac{5}{4} & \frac{5}{4} & 3 \\ 0 & 0 & 1 & 2 & 5 \\ 0 & 0 & 0 & 1 & 3 \end{array}\right]$

$t = 3$, and using back-substitution, $z = -1$, $y = -2$, $x = 1$

8. $\begin{vmatrix} 2 & -4 & -3 \\ -3 & 6 & 2 \\ 5 & -1 & 5 \end{vmatrix} = 2\begin{vmatrix} 6 & 2 \\ -1 & 5 \end{vmatrix} - (-4)\begin{vmatrix} -3 & 2 \\ 5 & 5 \end{vmatrix} + (-3)\begin{vmatrix} -3 & 6 \\ 5 & -1 \end{vmatrix}$

$= 2[30 - (-2)] + 4[-15 - 10] - 3[3 - 30]$

$= 2(32) + 4(-25) - 3(-27) = 64 - 100 + 81 = 45$

9. $\begin{vmatrix} 2 & -4 & -3 \\ -3 & 6 & 2 \\ 5 & -1 & 5 \end{vmatrix} = \begin{vmatrix} 2 & 0 & -3 \\ -3 & 0 & 2 \\ 5 & 9 & 5 \end{vmatrix} = -\begin{vmatrix} 5 & 9 & 5 \\ -3 & 0 & 2 \\ 2 & 0 & -3 \end{vmatrix} = \begin{vmatrix} 9 & 5 & 5 \\ 0 & -3 & 2 \\ 0 & 2 & -3 \end{vmatrix}$

$= 9\begin{vmatrix} -3 & 2 \\ 2 & -3 \end{vmatrix} = 9(9 - 4) = 45$

(1) 2 × col. 1 added to col. 2; (2) rows 1 and 3 interchanged; (3) cols. 1 and 2 interchanged; (4) expanded by col. 1

10. $\left[\begin{array}{rrr|rrr} 7 & -2 & 1 & 1 & 0 & 0 \\ 2 & 3 & -4 & 0 & 1 & 0 \\ 4 & -5 & 2 & 0 & 0 & 1 \end{array}\right] \rightarrow \left[\begin{array}{rrr|rrr} 1 & -\frac{2}{7} & \frac{1}{7} & \frac{1}{7} & 0 & 0 \\ 2 & 3 & -4 & 0 & 1 & 0 \\ 4 & -5 & 2 & 0 & 0 & 1 \end{array}\right] \rightarrow$

$\left[\begin{array}{rrr|rrr} 1 & -\frac{2}{7} & \frac{1}{7} & \frac{1}{7} & 0 & 0 \\ 0 & \frac{25}{7} & -\frac{30}{7} & -\frac{2}{7} & 1 & 0 \\ 0 & -\frac{27}{7} & \frac{10}{7} & -\frac{4}{7} & 0 & 1 \end{array}\right] \rightarrow \left[\begin{array}{rrr|rrr} 1 & -\frac{2}{7} & \frac{1}{7} & \frac{1}{7} & 0 & 0 \\ 0 & 1 & -\frac{6}{5} & -\frac{2}{25} & \frac{7}{25} & 0 \\ 0 & -\frac{27}{7} & \frac{10}{7} & -\frac{4}{7} & 0 & 1 \end{array}\right] \rightarrow$

$\left[\begin{array}{rrr|rrr} 1 & 0 & -\frac{1}{5} & \frac{3}{25} & \frac{2}{25} & 0 \\ 0 & 1 & -\frac{6}{5} & -\frac{2}{25} & \frac{7}{25} & 0 \\ 0 & 0 & -\frac{16}{5} & -\frac{22}{25} & \frac{27}{25} & 1 \end{array}\right] \rightarrow \left[\begin{array}{rrr|rrr} 1 & 0 & 0 & \frac{7}{40} & \frac{1}{80} & -\frac{1}{16} \\ 0 & 1 & 0 & \frac{1}{4} & -\frac{1}{8} & -\frac{3}{8} \\ 0 & 0 & 1 & \frac{11}{40} & -\frac{27}{80} & -\frac{5}{16} \end{array}\right]$

$$A^{-1}C = \begin{bmatrix} \frac{7}{40} & \frac{1}{80} & -\frac{1}{16} \\ \frac{1}{4} & -\frac{1}{8} & -\frac{3}{8} \\ \frac{11}{40} & -\frac{27}{80} & -\frac{5}{16} \end{bmatrix} \begin{bmatrix} 6 \\ 6 \\ 10 \end{bmatrix}$$

$$= \begin{bmatrix} \frac{1}{2} \\ -3 \\ -\frac{7}{2} \end{bmatrix}$$

The solution is $x = \frac{1}{2}$, $y = -3$, $z = -\frac{7}{2}$.

11. Let A = price per share of stock A
B = price per share of stock B

$50A + 30B = 2600$
$30A + 40B = 2000$

$5A + 3B = 260$
$3A + 4B = 200$

Let C = coefficient matrix

$$C = \begin{bmatrix} 5 & 3 \\ 3 & 4 \end{bmatrix}, \quad \begin{vmatrix} 5 & 3 \\ 3 & 4 \end{vmatrix} = 20 - 9 = 11$$

$$C^{-1} = \frac{1}{11}\begin{bmatrix} 4 & -3 \\ -3 & 5 \end{bmatrix} = \begin{bmatrix} \frac{4}{11} & -\frac{3}{11} \\ -\frac{3}{11} & \frac{5}{11} \end{bmatrix}$$

$$\begin{bmatrix} A \\ B \end{bmatrix} = \begin{bmatrix} \frac{4}{11} & -\frac{3}{11} \\ -\frac{3}{11} & \frac{5}{11} \end{bmatrix} \begin{bmatrix} 260 \\ 200 \end{bmatrix} = \begin{bmatrix} \frac{4}{11}(260) - \frac{3}{11}(200) \\ -\frac{3}{11}(260) + \frac{5}{11}(200) \end{bmatrix}$$

$$= \begin{bmatrix} 40 \\ 20 \end{bmatrix}$$

$A = \$40 \qquad B = \$20$

12.
$p + 2r + 2s = 6$
$-p + 4s + 3t = 1$
$r + s - 3t = 5$
$4p + 6t = 2$

$$\begin{vmatrix} 1 & 2 & 2 & 0 \\ -1 & 0 & 4 & 3 \\ 0 & 1 & 1 & -3 \\ 4 & 0 & 0 & 6 \end{vmatrix} = 72$$

$$p = \frac{\begin{vmatrix} 6 & 2 & 2 & 0 \\ 1 & 0 & 4 & 3 \\ 5 & 1 & 1 & -3 \\ 2 & 0 & 0 & 6 \end{vmatrix}}{72} = \frac{144}{72} = 2 \qquad r = \frac{\begin{vmatrix} 1 & 6 & 2 & 0 \\ -1 & 1 & 4 & 3 \\ 0 & 5 & 1 & -3 \\ 4 & 2 & 0 & 6 \end{vmatrix}}{72} = \frac{36}{72} = \frac{1}{2}$$

$$s = \frac{\begin{vmatrix} 1 & 2 & 6 & 0 \\ -1 & 0 & 1 & 3 \\ 0 & 1 & 5 & -3 \\ 4 & 0 & 2 & 6 \end{vmatrix}}{72} = \frac{108}{72} = \frac{3}{2} \qquad w = \frac{\begin{vmatrix} 1 & 2 & 2 & 6 \\ -1 & 0 & 4 & 1 \\ 0 & 1 & 1 & 5 \\ 4 & 0 & 0 & 2 \end{vmatrix}}{72} = \frac{-72}{72} = -1$$

The solution is $p = 2$, $r = 1/2$, $s = 3/2$, $t = -1$.

13.
$0.7x + 0.4y + 0.3z = 50$
$0.3x + 0.3y + 0.6z + 0.2w = 32$
$0.3y + 0.4w = 13$
$0.1z + 0.4w = 5$

$$A = \begin{bmatrix} 0.7 & 0.4 & 0.3 & 0 \\ 0.3 & 0.3 & 0.6 & 0.2 \\ 0 & 0.3 & 0 & 0.4 \\ 0 & 0 & 0.1 & 0.4 \end{bmatrix}, C = \begin{bmatrix} 50 \\ 32 \\ 13 \\ 5 \end{bmatrix}, A^{-1}C = \begin{bmatrix} 50 \\ 30 \\ 10 \\ 10 \end{bmatrix}$$

The solution is $x = 50$ g, $y = 30$ g, $z = 10$ g, $w = 10$ g.

Chapter 17

1. $x < 0, y > 0$

2. $\dfrac{-x}{2} \geq 3$
$-x \geq 6$
$x \leq -6$
$(-\infty, -6]$

3. $3x + 1 < -5$
$3x < -6$
$x < -2$
$(-\infty, -2)$

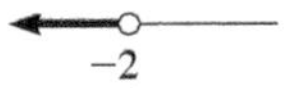

4. $-1 < 1 - 2x < 5$
$-2 < -2x < 4$
$1 > x > -2$
$-2 < x < 1$
$(-2, 1)$

5. $\dfrac{x^2 + x}{x - 2} \leq 0$, $\dfrac{x(x+1)}{x-2} \leq 0$

Interval	$\frac{x(x+1)}{x-2}$	Sign
$x < -1$	$\frac{- \ -}{-}$	$-$
$-1 < x < 0$	$\frac{- \ +}{-}$	$+$
$0 < x < 2$	$\frac{+ \ +}{-}$	$-$
$x > 2$	$\frac{+ \ +}{+}$	$+$

Solution: $x \leq -1$ or $0 \leq x < 2$
$(-\infty, -1]$ or $[0, 2)$

(x cannot equal 2)

6. $|2x + 1| \geq 3$
$2x + 1 \geq 3$, $\quad 2x + 1 \leq -3$
$2x \geq 2 \quad\quad 2x \leq -4$
$x \geq 1 \quad$ or $\quad x \leq -2$
$[1, \infty) \quad\quad (-\infty, -2]$

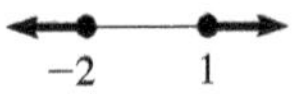

7. $|2 - 3x| < 8$
$-8 < 2 - 3x < 8$
$-10 < -3x < 6$
$\dfrac{10}{3} > x > -2$
$-2 < x < \dfrac{10}{3}$
$(-2, 10/3)$

8.

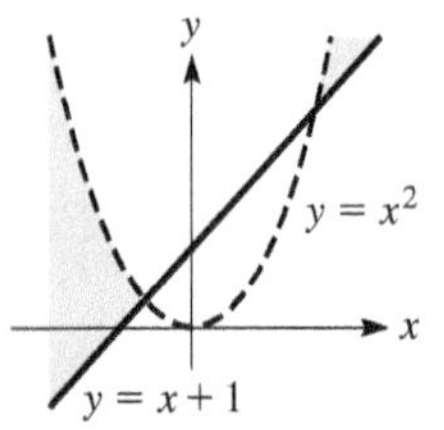

9. If $\sqrt{x^2 - x - 6}$ is real, then $x^2 - x - 6 \geq 0$.
$(x - 3)(x + 2) \geq 0$
$x \leq -2$ or $x \geq 3$
$(-\infty, -2]$ or $[3, \infty)$

Interval	$(x-3)(x+2)$	Sign
$x < -2$	$(-)(-)$	$+$
$-2 < x < 3$	$(-)(+)$	$-$
$x > 3$	$(+)(+)$	$+$

10. Let $w =$ width, $l =$ length

$$l = w + 20 \qquad w^2 + 20w - 4800 \geq 0$$
$$wl \geq 4800 \qquad (w + 80)(w - 60) \geq 0$$
$$w(w + 20) \geq 4800 \qquad w \geq 60 \text{ m}$$
$$[60, \infty)$$

11. Let $A =$ length of type A wire
$B =$ length of type B wire
$0.10A + 0.20B < 5.00$
$A + 2B < 50$

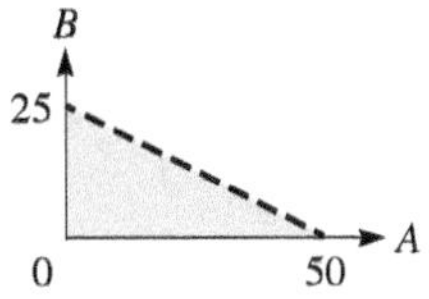

12. $|\lambda - 550 \text{ nm}| < 150 \text{ nm}$ (within 150 nm of $\lambda = 550$ nm)

13. Graph $y = x^2 + x - 12$. We see that $x^2 + x - 12 > 0$ for $x < -4$ or $x > 3$, or $(-\infty, -4)$ or $(3, \infty)$

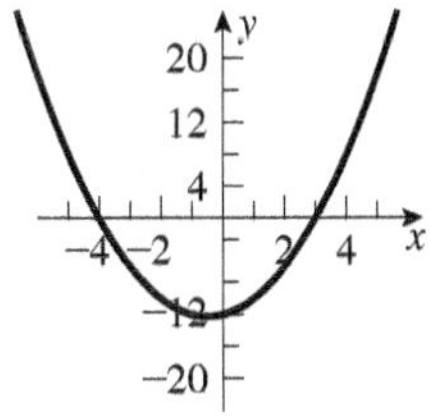

14. $P = 5x + 3y$
$x \geq 0, y \geq 0$
$2x + 3y \leq 12$
$4x + y \leq 8$

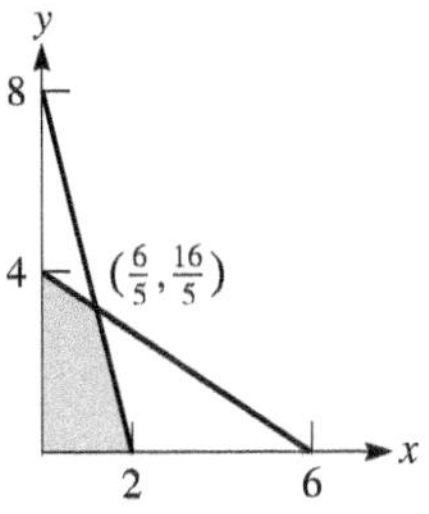

Vertices:	(0, 0)	(2, 0)	$\left(\frac{6}{5}, \frac{16}{5}\right)$	(0, 4)
$P = 5x + 3y$	0	10	15.6	12

Max. value of $P = 15.6$ at $\left(\frac{6}{5}, \frac{16}{5}\right)$

Chapter 18

1. $\dfrac{180 \text{ s}}{4 \text{ min}} = \dfrac{180 \text{ s}}{240 \text{ s}} = \dfrac{3}{4}$

2. $\dfrac{1.8 \text{ m}}{20.0 \text{ mm}} = \dfrac{x}{14.5 \text{ mm}}$

$$20.0x = 1.8(14.5)$$
$$x = 1.3 \text{ m}$$

3. $L = kT$

$L = 2.7$ cm for $T = 150$ °C

$$2.7 = k(150)$$
$$k = 0.018 \text{ cm/°C}$$
$$L = 0.018T$$

4. Let $x =$ length in inches

$$\frac{1.00 \text{ in.}}{2.54 \text{ cm}} = \frac{x}{7.24 \text{ cm}}$$
$$x = \frac{7.24}{2.54} = 2.85 \text{ in.}$$

5.
$$2l + 2w = 210.0$$
$$l = 105.0 - w$$
$$\frac{l}{w} = \frac{7}{3}$$
$$\frac{105.0 - w}{w} = \frac{7}{3}$$
$$315.0 - 3w = 7w$$
$$10w = 315.0$$
$$w = 31.5 \text{ cm}$$
$$l = 105.0 - 31.5$$
$$= 73.5 \text{ cm}$$

6. $p = kdh$

Using values of water,

$$1.96 = k(1000)(0.200)$$
$$k = 0.009\,80 \text{ kPa} \cdot \text{m}^2/\text{kg}$$

For alcohol,

$$p = 0.009\,80(800)(0.300)$$
$$= 2.35 \text{ kPa}$$

7. Let $L_1 =$ crushing load of first pillar
$L_2 =$ crushing load of second pillar

$$L_2 = \frac{kr_2^4}{l_2^2} \qquad L_1 = \frac{k(2r_2)^4}{(3l_2)^2}$$

$$\frac{L_1}{L_2} = \frac{\dfrac{k(2r_2)^4}{(3l_2)^2}}{\dfrac{kr_2^4}{l_2^2}} = \frac{16kr_2^4}{9l_2^2} \times \frac{l_2^2}{kr_2^4} = \frac{16}{9}$$

Chapter 19

1. (a) $a_1 = 8, d = -1/2$; arith. seq. $8, 15/2, 7, 13/2, 6, \ldots$

(b) $a_1 = 8, r = -1/2$; geom. seq. $8, -4, 2, -1, 1/2, \ldots$

2. $6, -2, \frac{2}{3}, \ldots;$

geometric sequence

$$a_1 = 6 \qquad r = \frac{-2}{6} = -\frac{1}{3}$$

$$S_7 = \frac{6\left[1 - \left(-\frac{1}{3}\right)^7\right]}{1 - \left(-\frac{1}{3}\right)}$$

$$= \frac{6\left(1 + \dfrac{1}{3^7}\right)}{\frac{4}{3}} = \frac{1094}{243}$$

3. $a_1 = 6, d = 4, s_n = 126;$

arithmetic sequence

$$126 = \tfrac{n}{2}(6 + a_n)$$
$$a_n = 6 + (n - 1)4 = 2 + 4n$$
$$126 = \tfrac{n}{2}[6 + (2 + 4n)]$$
$$252 = n(8 + 4n) = 4n^2 + 8n$$
$$n^2 + 2n - 63 = 0$$
$$(n + 9)(n - 7) = 0$$
$$n = 7$$

4. $0.454\,545\ldots = 0.45 + 0.0045 + 0.000\,045 + \cdots$

$$a = 0.45 \qquad r = 0.01$$
$$S = \frac{0.45}{1 - 0.01} = \frac{0.45}{0.99} = \frac{5}{11}$$

5. $\sqrt{1 - 4x} = (1 - 4x)^{1/2}$

$$= 1 + \left(\tfrac{1}{2}\right)(-4x) + \frac{\frac{1}{2}\left(\frac{1}{2} - 1\right)}{2}(-4x)^2 + \cdots$$
$$= 1 - 2x - 2x^2 + \cdots$$

6. $(2x - y)^5 = (2x)^5 + 5(2x)^4(-y) + \dfrac{5(4)}{2}(2x)^3(-y)^2$

$$+ \frac{5(4)(3)}{2(3)}(2x)^2(-y)^3 + \frac{5(4)(3)(2)}{2(3)(4)}(2x)(-y)^4 + (-y)^5$$
$$= 32x^5 - 80x^4y + 80x^3y^2 - 40x^2y^3 + 10xy^4 - y^5$$

7. $5\% = 0.05$
Value after 1 year is
$2500 + 2500(0.05) = 2500(1.05)$
$V_{20} = 2500(1.05)^{20}$
$= \$6633.24$

8. $2 + 4 + \cdots + 200$
$a_1 = 2, \quad a_{100} = 200, \quad n = 100$
$S_{100} = \frac{100}{2}(2 + 200) = 10\,100$

9. Ball falls 8.00 m, rises 4.00 m, falls 4.00 m, etc.
Distance $= 8.00 + (4.00 + 4.00) + (2.00 + 2.00) + \cdots$
$= 8.00 + 8.00 + 4.00 + 2.00 + \cdots$
$= 8.00 + \dfrac{8.00}{1 - 0.5} = 8.00 + 16.0 = 24.0 \text{ m}$

Chapter 20

1. $\sec\theta - \dfrac{\tan\theta}{\csc\theta} = \dfrac{1}{\cos\theta} - \dfrac{\frac{\sin\theta}{\cos\theta}}{\frac{1}{\sin\theta}} = \dfrac{1}{\cos\theta} - \dfrac{\sin^2\theta}{\cos\theta}$
$= \dfrac{1 - \sin^2\theta}{\cos\theta} = \dfrac{\cos^2\theta}{\cos\theta} = \cos\theta$
$\cos\theta = \cos\theta$

2. $\sin 2x + \sin x = 0$
$2\sin x\cos x + \sin x = 0$
$\sin x(2\cos x + 1) = 0$
$\sin x = 0 \qquad \cos x = -\dfrac{1}{2}$
$x = 0, \pi, \frac{2\pi}{3}, \frac{4\pi}{3}$

3. $\theta = \sin^{-1} x$
$\cos\theta = \cos(\sin^{-1} x)$
$= \sqrt{1 - x^2}$

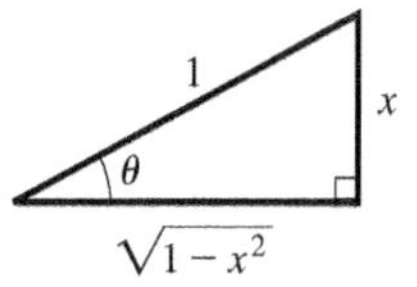

4. $i = 8.00e^{-20t}(1.73\cos 10.0t - \sin 10.0t)$
Since $8.00e^{-20t}$ will not equal zero, $i = 0$ if
$1.73\cos 10.0t - \sin 10.0t = 0$
$1.73 = \tan 10.0t$
$10.0t = \tan^{-1} 1.73$
$t = 0.105 \text{ s}$

5. $\dfrac{\tan\alpha + \tan\beta}{\tan\alpha - \tan\beta} = \dfrac{\frac{\sin\alpha}{\cos\alpha} + \frac{\sin\beta}{\cos\beta}}{\frac{\sin\alpha}{\cos\alpha} - \frac{\sin\beta}{\cos\beta}} = \dfrac{\sin\alpha\cos\beta + \sin\beta\cos\alpha}{\sin\alpha\cos\beta - \sin\beta\cos\alpha}$
$= \dfrac{\sin(\alpha + \beta)}{\sin(\alpha - \beta)}$

6. $\cot^2 x - \cos^2 x = \dfrac{\cos^2 x}{\sin^2 x} - \cos^2 x =$
$\cos^2 x(\csc^2 x - 1) = \cos^2 x\cot^2 x;$
$\cos^2 x\cot^2 x = \cos^2 x\cot^2 x$

7. $\sin x = -\frac{3}{5}$
$\cos x = \frac{4}{5}$

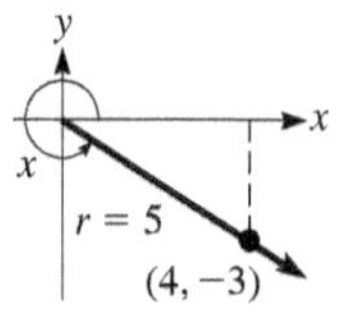

$\cos\dfrac{x}{2} = -\sqrt{\dfrac{1 + \left(\frac{4}{5}\right)}{2}} = -\sqrt{\dfrac{9}{10}} = -0.9487$
since $270° < x < 360°$, $135° < \frac{x}{2} < 180°$;
$\frac{x}{2}$ is in second quadrant, where $\cos\frac{x}{2}$ is negative.

8. $I = I_0 \sin 2\theta\cos 2\theta$
$\dfrac{2I}{I_0} = 2\sin 2\theta\cos 2\theta = \sin 4\theta$
$4\theta = \sin^{-1}\dfrac{2I}{I_0}$
$\theta = \dfrac{1}{4}\sin^{-1}\dfrac{2I}{I_0}$

9. $y = x - 2\cos x - 5$
$x = 3.02, 4.42, 6.77$

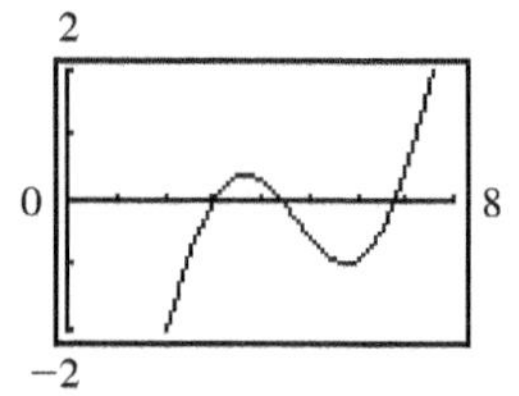

10. $\cos^{-1} x = -\tan^{-1}(-1)$
$\tan^{-1}(-1) = -\frac{\pi}{4}$
$\cos^{-1} x = -\left(-\frac{\pi}{4}\right) = \frac{\pi}{4}$
$x = \cos\frac{\pi}{4} = \frac{1}{2}\sqrt{2}$

Chapter 21

1. Points: $(4, -1), (6, 3)$
(a) $d = \sqrt{(6 - 4)^2 + [3 - (-1)]^2} = \sqrt{4 + 16} = 2\sqrt{5}$
(b) Between points $m = \dfrac{3 - (-1)}{6 - 4} = 2$; perp. line, $m = -\dfrac{1}{2}$

2. $2(x^2 + x) = 1 - y^2$
$2x^2 + y^2 + 2x - 1 = 0$
$A \neq C$ (same sign)
$B = 0$: ellipse

3. $4x - 2y + 5 = 0$
$y = 2x + \frac{5}{2}$
$m = 2 \quad b = \frac{5}{2}$

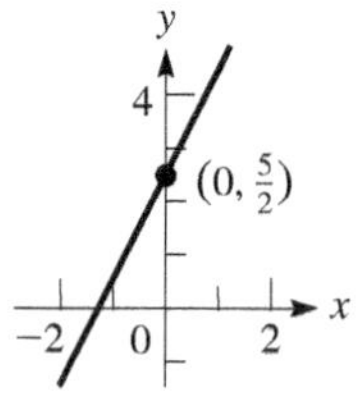

4. $x^2 = 2x - y^2$
$x^2 + y^2 = 2x$
$r^2 = 2r\cos\theta$
$r = 2\cos\theta$

5. $x^2 = -12y$
$4p = -12, \qquad p = -3$
$V(0, 0) \qquad F(0, -3)$

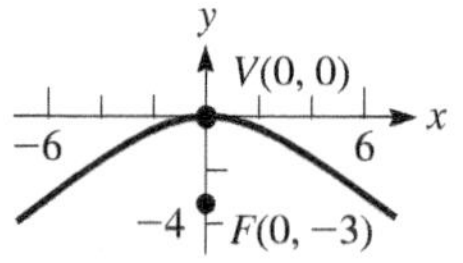

6. Centre $(-1, 2)$; $h = -1, k = 2$
$r = \sqrt{(2 + 1)^2 + (3 - 2)^2}$
$= \sqrt{10}$
$(x + 1)^2 + (y - 2)^2 = 10$
or
$x^2 + y^2 + 2x - 4y - 5 = 0$

7. $m = \dfrac{-2 - 1}{2 + 4} = -\dfrac{1}{2}$
$y - 1 = -\frac{1}{2}(x + 4)$
$2y - 2 = -x - 4$
$x + 2y + 2 = 0$

8. $x^2 = 4py$
$(6.00)^2 = 4p(4.00)$
$p = 2.25$
Focus is 2.25 cm from vertex.

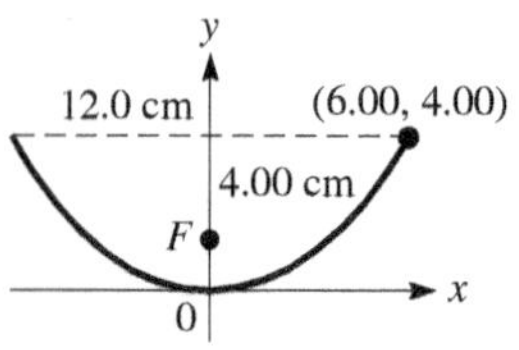

9. $a = 3 \qquad b = 1$

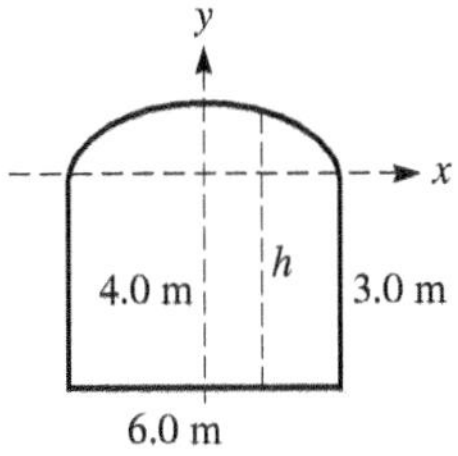

$$\frac{x^2}{9} + \frac{y^2}{1} = 1$$

Find y for $x = 2$ m:

$$\frac{4}{9} + \frac{y^2}{1} = 1$$

$y^2 = 5/9 \qquad y = 0.7$ m

$h = 3.0 + 0.7 = 3.7$ m

10. $r = 3 + \cos\theta$

r	0	$\frac{\pi}{4}$	$\frac{\pi}{2}$	$\frac{3\pi}{4}$	π	$\frac{5\pi}{4}$	$\frac{3\pi}{2}$	$\frac{7\pi}{4}$	2π
θ	4.0	3.7	3.0	2.3	2.0	2.3	3.0	3.7	4.0

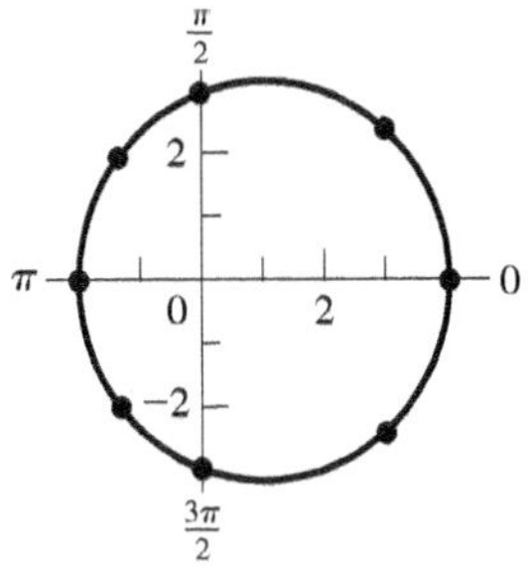

11. $4y^2 - x^2 - 4x - 8y - 4 = 0$

$4(y^2 - 2y \qquad) - (x^2 + 4x \qquad) = 4$

$4(y^2 - 2y + 1) - (x^2 + 4x + 4) = 4 + 4 - 4$

$$\frac{(y-1)^2}{1^2} - \frac{(x+2)^2}{2^2} = 1$$

$C(-2, 1) \qquad V(-2, 0) \qquad V(-2, 2)$

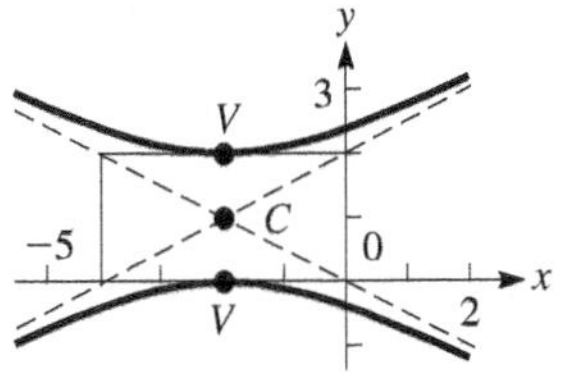

12. Equation of curve: $8x^2 - 4xy + 5y^2 = 36$; $A = 8$, $B = -4$, $C = 5$

(a) $B^2 - 4AC = (-4)^2 - 4(8)(5) = 16 - 160 < 0$; ellipse

(b) $\tan 2\theta = \frac{B}{A - C} = \frac{-4}{8 - 5} = -\frac{4}{3}$; $\theta = \frac{1}{2}\tan^{-1}\left(-\frac{4}{3}\right) = 63.4°$

Chapter 22

1.

Number	1	2	3	4	5	6	7	8	9	10
Frequency	1	0	1	2	3	2	1	2	2	1

$\Sigma f = 15$

Median is eighth number; median = 6.

2. 5 appears three times, and no other number appears more than twice; mode = 5.

3. (left endpoint included)

Number	1–3	3–5	5–7	7–9	9–11
Frequency	1	3	5	3	3

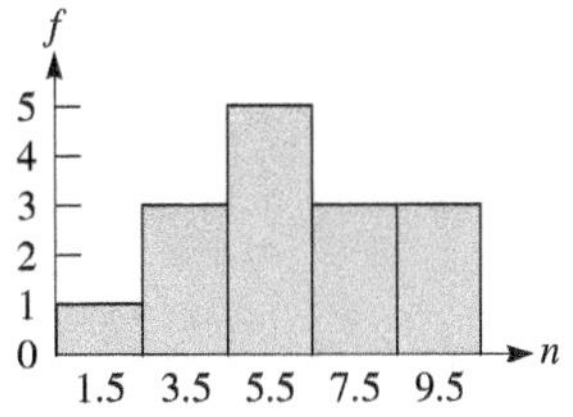

4. Let t = thickness

$$\bar{t} = \frac{3(0.90) + 9(0.91) + 31(0.92) + 38(0.93) + 12(0.94) + 5(0.95) + 2(0.96)}{3 + 9 + 31 + 38 + 12 + 5 + 2}$$

$$= \frac{92.7}{100} = 0.927 \text{ cm}$$

5. $\Sigma x^2 = 3(0.90)^2 + 9(0.91)^2 + 31(0.92)^2 + 38(0.93)^2 + 12(0.94)^2 + 5(0.95)^2 + 2(0.96)^2 = 85.9464$

$(\Sigma x)^2 = 92.7^2$

$$s = \sqrt{\frac{100(85.9464) - 92.7^2}{100(99)}} = 0.117 \text{ cm}$$

6.

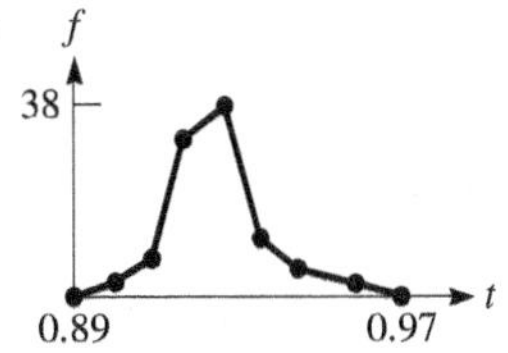

7.

t (cm)	0.90	0.91	0.92	0.93	0.94	0.95	0.96
%	3	9	31	38	12	5	2

8.

t (cm)	<0.905	<0.915	<0.925	<0.935
cum f	3	12	43	81

t (cm)	<0.945	<0.955	<0.965
cum f	93	98	100

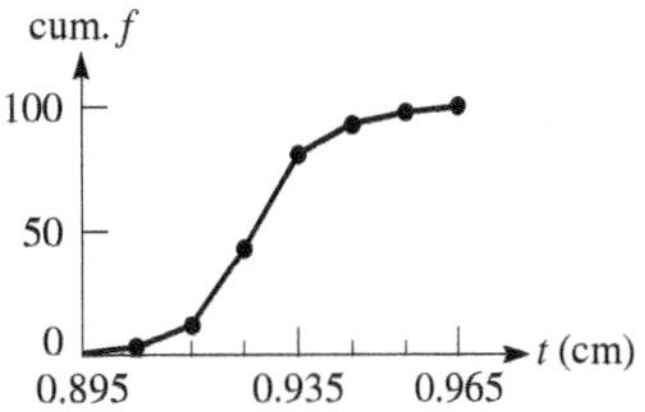

9. A 95% confidence interval is given by

$$\bar{x} \pm z_{\alpha/2} \cdot \frac{s}{\sqrt{n}} = 143 \pm 1.96 \cdot \frac{5.2}{\sqrt{175}}$$

$$= 143 \pm 0.77$$

$$= (142.2, 143.8)$$

10. We have $\hat{p} = \frac{33}{120}$, $n\hat{p} = 33 \geq 5$, and $n(1 - \hat{p}) = 87 \geq 5$. Therefore, we can apply Eq. (22.14) with $z_{\alpha/2} = 2.575$. The desired 99% confidence interval is

$$\hat{p} \pm z_{\alpha/2}\sqrt{\frac{\hat{p}(1-\hat{p})}{n}} = \frac{33}{120} \pm 2.575\sqrt{\frac{\frac{33}{120} \cdot \frac{87}{120}}{120}}$$

$$= 0.275 \pm 0.105$$

$$= (0.170, 0.380)$$

11. $0.5000 + 0.2257 = 0.7257$ (total area to the left) $= 72.57\%$

12. Find the range of each subgroup (subtract the lowest value from the largest value). Find the mean R of these ranges (divide their sum by 20). This is the value of the central line. Multiply R by the appropriate control chart factors to obtain the upper control limit (UCL) and the lower control limit (LCL). Draw the central line, the UCL line, and the LCL line on a graph. Plot the points for each R corresponding to its subgroup and join successive points by straight-line segments.

13.

x	y	xy	x^2
1	5	5	1
3	11	33	9
5	17	85	25
7	20	140	49
9	27	243	81
25	80	506	165

$n = 5$

$$m = \frac{5(506) - (25)(80)}{5(165) - 25^2} = 2.65$$

$$b = \frac{165(80) - (506)(25)}{5(165) - 25^2} = 2.75$$

$y = 2.65x + 2.75$

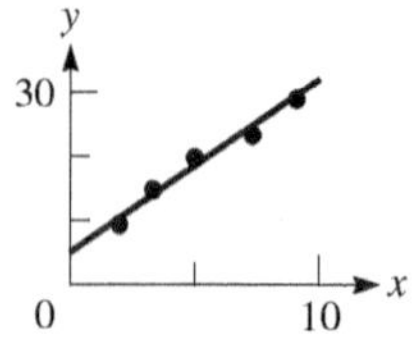

14.

x	$\sqrt{x}$	y	$\sqrt{x}y$	$(\sqrt{x})^2 = x$
1.00	1.000	1.10	1.1000	1.000
3.00	1.732	1.90	3.2908	3.000
5.00	2.236	2.50	5.5900	5.000
7.00	2.646	2.90	7.6734	7.000
9.00	3.000	3.30	9.9000	9.000
	10.614	11.70	27.5542	25.000

$n = 5$

$$m = \frac{5(27.5542) - (10.614)(11.70)}{5(25.000) - (10.614)^2} = 1.10$$

$$b = \frac{(25.000)(11.70) - (27.5542)(10.614)}{5(25.000) - (10.614)^2} = 0.00$$

(rounded off)

$y = 1.10\sqrt{x}$

Chapter 23

1. $\displaystyle \lim_{x\to 1}\frac{x^2 - x}{x^2 - 1} = \lim_{x\to 1}\frac{x(x-1)}{(x+1)(x-1)} = \lim_{x\to 1}\frac{x}{x+1} = \frac{1}{2}$

2. $\displaystyle \lim_{x\to\infty}\frac{1 - 4x^2}{x + 2x^2} = \lim_{x\to\infty}\frac{\frac{1}{x^2} - 4}{\frac{1}{x} + 2} = -2$

3. $y = 3x^2 - \dfrac{4}{x^2}$

$$\frac{dy}{dx} = 6x + \frac{8}{x^3}$$

$$\left.\frac{dy}{dx}\right|_{x=2} = 6(2) + \frac{8}{2^3} = 13$$

$m_{\tan} = 13$

4. $s = t\sqrt{10 - 2t}$

$$v = \frac{ds}{dt} = t\left(\tfrac{1}{2}\right)(10 - 2t)^{-1/2}(-2) + (10 - 2t)^{1/2}(1)$$

$$= \frac{-t}{(10 - 2t)^{1/2}} + (10 - 2t)^{1/2} = \frac{10 - 3t}{(10 - 2t)^{1/2}}$$

$$v\big|_{t=4.00} = \frac{10 - 3(4.00)}{[10 - 2(4.00)]^{1/2}} = \frac{10 - 12.00}{2.00^{1/2}} = -1.41 \text{ cm/s}$$

5. $y = 4x^6 - 2x^4 + \pi^3$

$$\frac{dy}{dx} = 4(6x^5) - 2(4x^3) \qquad \pi^3 \text{ is constant}$$

$$= 24x^5 - 8x^3$$

6. $y = 2x(5 - 3x)^4$

$$\frac{dy}{dx} = 2x[4(5 - 3x)^3(-3)] + (5 - 3x)^4(2)$$

$$= -24x(5 - 3x)^3 + 2(5 - 3x)^4$$

$$= 2(5 - 3x)^3(-12x + 5 - 3x)$$

$$= 2(5 - 15x)(5 - 3x)^3$$

$$= 10(1 - 3x)(5 - 3x)^3$$

7. $(1 + y^2)^3 - x^2y = 7x$

$$3(1 + y^2)^2(2yy') - x^2y' - y(2x) = 7$$

$$6y(1 + y^2)^2y' - x^2y' = 7 + 2xy$$

$$y' = \frac{7 + 2xy}{6y(1 + y^2)^2 - x^2}$$

8. $V = \dfrac{kq}{\sqrt{x^2 + b^2}} = kq(x^2 + b^2)^{-1/2}$

$$\frac{dV}{dx} = kq\left(-\frac{1}{2}\right)(x^2 + b^2)^{-3/2}(2x) = \frac{-kqx}{(x^2 + b^2)^{3/2}}$$

9. $y = \dfrac{2x}{3x + 2}$

$$\frac{dy}{dx} = \frac{(3x + 2)(2) - 2x(3)}{(3x + 2)^2} = \frac{4}{(3x + 2)^2} = 4(3x + 2)^{-2}$$

$$\frac{d^2y}{dx^2} = -2(4)(3x + 2)^{-3}(3) = \frac{-24}{(3x + 2)^3}$$

10. $y = 5x - 2x^2$

$$f(x + h) = 5(x + h) - 2(x + h)^2$$

$$f(x + h) - f(x) = 5(x + h) - 2(x + h)^2 - (5x - 2x^2)$$

$$= 5h - 4xh - 2h^2$$

$$\frac{f(x + h) - f(x)}{h} = \frac{5h - 4xh - 2h^2}{h} = 5 - 4x - 2h$$

$$\lim_{h\to 0}\frac{f(x + h) - f(x)}{h} = 5 - 4x$$

Chapter 24

1. $y = x^4 - 3x^2$

$$\frac{dy}{dx} = 4x^3 - 6x$$

$$\left.\frac{dy}{dx}\right|_{x=1} = 4(1^3) - 6(1)$$

$$= -2$$

$$y - (-2) = -2(x - 1)$$

$$y = -2x$$

2. $y = 3x^2 - x$

$$\Delta y = f(3.1) - f(3)$$

$$[3(3.1)^2 - 3.1] - [3(3^2) - 3] = 1.73$$

$$dy = (6x - 1)dx$$
$$= [6(3) - 1](0.1) = 1.7$$

$$\Delta y - dy = 0.03$$

3. $x = 3t^2 \qquad y = 2t^3 - t^2$

$$v_x = \frac{dx}{dt} = 6t \qquad v_y = \frac{dy}{dt} = 6t^2 - 2t$$

$$a_x = \frac{dv_x}{dt} = \frac{d^2x}{dt^2} = 6 \qquad a_y = \frac{dv_y}{dt} = \frac{d^2y}{dt^2} = 12t - 2$$

$$a_x\big|_{t=2} = 6 \qquad a_y\big|_{t=2} = 12(2) - 2 = 22$$

$$a\big|_{t=2} = \sqrt{6^2 + 22^2} = 22.8 \qquad \tan\theta = \tfrac{22}{6}, \quad \theta = 74.7°$$

4. $P = \dfrac{144r}{(r + 0.6)^2}$

$$\frac{dP}{dr} = \frac{144[(r + 0.6)^2(1) - r(2)(r + 0.6)(1)]}{(r + 0.6)^4}$$

$$= \frac{144[(r + 0.6) - 2r]}{(r + 0.6)^3} = \frac{144(0.6 - r)}{(r + 0.6)^3}$$

$$\frac{dP}{dr} = 0; \quad 0.6 - r = 0, \quad r = 0.6\,\Omega$$

$$\left(r < 0.6, \frac{dP}{dr} > 0;\ r > 0.6, \frac{dP}{dr} < 0\right)$$

5. $x^2 - \sqrt{4x + 1} = 0; \quad f(x) = x^2 - \sqrt{4x + 1}$

$$f'(x) = 2x - \tfrac{1}{2}(4x + 1)^{-1/2}(4) = 2x - \frac{2}{(4x + 1)^{1/2}}$$

n	x_n	$f(x_n)$	$f'(x_n)$	$x_n - \dfrac{f(x_n)}{f'(x_n)}$
1	1.5	−0.395 751 3	2.244 071 1	1.676 354 2
2	1.676 354 2	0.034 300 1	2.632 211 8	1.663 323 3

$x_3 = 1.6633$

6. $y = \sqrt{2x + 4},\ a = 6$

$$\frac{dy}{dx} = \frac{1}{2}(2x + 4)^{-1/2}(2) = \frac{1}{\sqrt{2x + 4}}$$

$$\frac{dy}{dx}\bigg|_{x=6} = \frac{1}{\sqrt{2(6) + 4}} = \frac{1}{4}$$

$$f(6) = \sqrt{2(6) + 4} = 4$$

$$L(x) = 4 + \tfrac{1}{4}(x - 6) = \tfrac{1}{4}x + \tfrac{5}{2}$$

7. $y = x^3 + 6x^2$

$y' = 3x^2 + 12x = 3x(x + 4)$

$y'' = 6x + 12 = 6(x + 2)$

$x < -4$	y inc.	Max. $(-4, 32)$
$-4 < x < 0$	y dec.	Min. $(0, 0)$
$x > 0$	y inc.	Infl. $(-2, 16)$
$x < -2$	y conc. down	
$x > -2$	y conc. up	

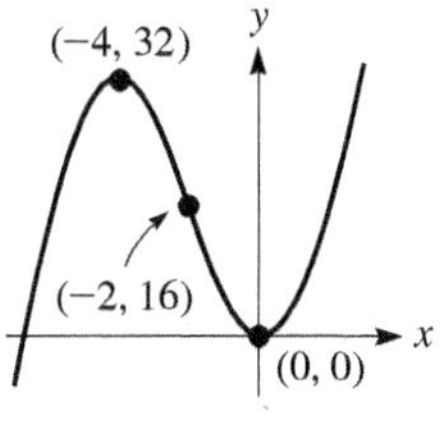

8. $y = \dfrac{4}{x^2} - x$

$$y' = -\frac{8}{x^3} - 1 = -\frac{8 + x^3}{x^3}$$

$$y'' = \frac{24}{x^4}$$

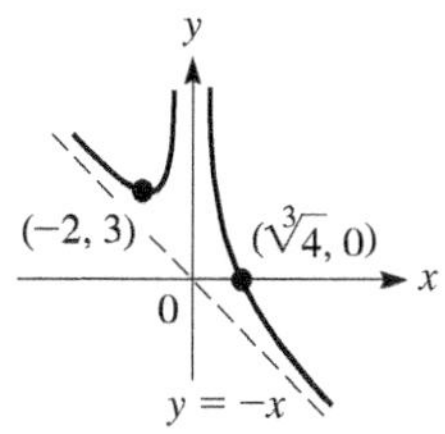

$x < -2$	y dec.
$-2 < x < 0$	y inc.
$x > 0$	y dec., conc. up
$x < 0$	y conc. up

Min. $(-2, 3)$, no infl., int. $(\sqrt[3]{4}, 0)$, sym. none; as $x \to \pm\infty$, $y \to -x$, asym. $y = -x$, $x = 0$. Domain: all real x except 0; range: all real y.

9. Let V = volume of cube
s = edge of cube

$$V = s^3 \qquad \frac{dV}{dt} = 3s^2\frac{ds}{dt}$$

$$\frac{dV}{dt}\bigg|_{s=1.25} = 3(1.25)^2(-0.10)$$
$$= -0.47\ \text{m}^3/\text{s}$$

10.

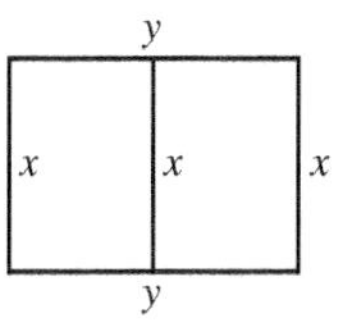

$$3x + 2y = 6000$$

$$y = \frac{6000 - 3x}{2}$$

$$A = xy = x\left(\frac{6000 - 3x}{2}\right)$$
$$= 3000x - \tfrac{3}{2}x^2$$

$$\frac{dA}{dx} = 3000 - 3x$$

$3000 - 3x = 0,\ x = 1000$ m

$y = 1500$ m

$A_{max} = (1000)(1500) = 1.5 \times 10^6\ \text{m}^2$

$$\left(x < 1000, \frac{dA}{dx} > 0;\ x > 1000, \frac{dA}{dx} < 0\right)$$

11. To estimate $(7.96)^{1/3}$, let $y = x^{1/3}$. We approximate the increment Δy by the differential dy when $x = 8.00$ and $\Delta x = -0.04$:

$$dy = \tfrac{1}{3}x^{-2/3}dx$$

$$y\big|_{x=8.00} = 2.00$$

$$dy\bigg|_{\substack{x=8.00 \\ \Delta x=-0.04}} = \tfrac{1}{3}(8.00)^{-2/3}(-0.04) = -\tfrac{1}{300}$$

$$y + dy = 2.00 - \tfrac{1}{300} = 1.9967$$

Chapter 25

1. Power of x required for $2x$ is 2. Therefore, multiply by $1/2$. Antiderivative of $2x = \frac{1}{2}(2x^2) = x^2$. Power of $(1 - x)^4$ required is 5. Derivative of $(1 - x)^5$ is $5(1 - x)^4(-1)$. Writing $-(1 - x)^4$ as $\frac{1}{5}[5(1 - x)^4(-1)]$, the antiderivative of $-(1 - x)^4$ is $\frac{1}{5}(1 - x)^5$. Therefore, the antiderivative of $2x - (1 - x)^4$ is $x^2 + \frac{1}{5}(1 - x)^5$.

2. $\int x\sqrt{1 - 2x^2}dx = \int x(1 - 2x^2)^{1/2}dx$

$$u = 1 - 2x^2 \quad du = -4xdx \quad n = \tfrac{1}{2} \quad n + 1 = \tfrac{3}{2}$$

$$\int x(1 - 2x^2)^{1/2}dx = -\tfrac{1}{4}\int(1 - 2x^2)^{1/2}(-4xdx)$$
$$= -\tfrac{1}{4}\left(\tfrac{2}{3}\right)(1 - 2x^2)^{3/2} + C$$
$$= -\tfrac{1}{6}(1 - 2x^2)^{3/2} + C$$

3. $\frac{dy}{dx} = (6-x)^4,\ dy = (6-x)^4\,dx$

$\int dy = \int (6-x)^4 dx$

$y = -\int (6-x)^4(-dx) = -\frac{1}{5}(6-x)^5 + C$

$2 = -\frac{1}{5}(6-5) + C,\ C = \frac{11}{5}$

$y = -\frac{1}{5}(6-x)^5 + \frac{11}{5}$

4. $y = \frac{1}{x+2},\quad n = 6,\quad \Delta x = \frac{4-1}{6} = \frac{1}{2}$

$A = \frac{1}{2}\left(\frac{2}{7} + \frac{1}{4} + \frac{2}{9} + \frac{1}{5} + \frac{2}{11} + \frac{1}{6}\right) = 0.6532$

x	1	$\frac{3}{2}$	2	$\frac{5}{2}$	3	$\frac{7}{2}$	4
y	$\frac{1}{3}$	$\frac{2}{7}$	$\frac{1}{4}$	$\frac{2}{9}$	$\frac{1}{5}$	$\frac{2}{11}$	$\frac{1}{6}$

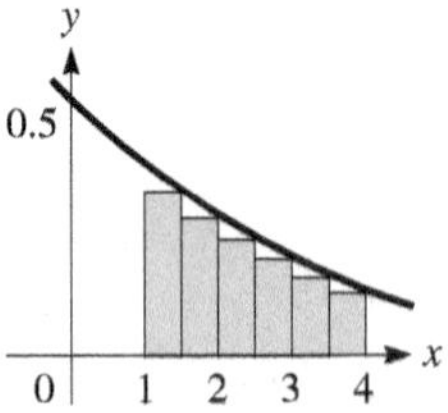

5. (See values for Problem 4.)

$$\int_1^4 \frac{dx}{x+2} = \frac{1}{4}\left[\frac{1}{3} + 2\left(\frac{2}{7}\right) + 2\left(\frac{1}{4}\right) + 2\left(\frac{2}{9}\right) + 2\left(\frac{1}{5}\right) + 2\left(\frac{2}{11}\right) + \frac{1}{6}\right]$$

$= 0.6949$

6. (See values for Problem 4.)

$$\int_1^4 \frac{dx}{x+2} = \frac{1}{6}\left[\frac{1}{3} + 4\left(\frac{2}{7}\right) + 2\left(\frac{1}{4}\right) + 4\left(\frac{2}{9}\right) + 2\left(\frac{1}{5}\right) + 4\left(\frac{2}{11}\right) + \frac{1}{6}\right] = 0.6932$$

7. $i = \int_1^3 \left(t^2 + \frac{1}{t^2}\right)dt = \frac{1}{3}t^3 - \frac{1}{t}\Big|_1^3$

$= \frac{1}{3}(27) - \frac{1}{3} - \left(\frac{1}{3} - 1\right) = 9.3\ \text{A}$

Chapter 26

1. $A = \int_0^2 \frac{1}{4}x^2 dx$

$= \frac{1}{12}x^3\Big|_0^2 = \frac{2}{3}$

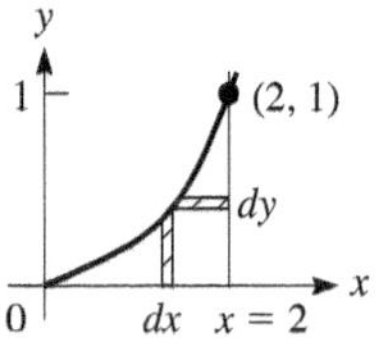

2. $\bar{x} = \frac{\int_0^2 x\left(\frac{1}{4}x^2\right)dx}{\frac{2}{3}} = \frac{\frac{1}{4}\int_0^2 x^3 dx}{\frac{2}{3}} = \frac{\frac{1}{16}x^4\big|_0^2}{\frac{2}{3}} = \frac{1}{\frac{2}{3}} = \frac{3}{2}$

$\bar{y} = \frac{\int_0^1 y\left(2 - 2\sqrt{y}\right)dy}{\frac{2}{3}} = \frac{2\int_0^1 (y - y^{3/2})dy}{\frac{2}{3}}$

$= \frac{2\left(\frac{1}{2}y^2 - \frac{2}{5}y^{5/2}\right)\big|_0^1}{\frac{2}{3}}$

$= \frac{1 - \frac{4}{5}}{\frac{2}{3}} = \frac{1}{5} \times \frac{3}{2} = \frac{3}{10}$

3. $V = \pi \int_0^2 \left(\frac{1}{4}x^2\right)^2 dx = \frac{\pi}{16}\int_0^2 x^4 dx = \frac{\pi}{80}x^5\Big|_0^2$

$= \frac{32\pi}{80} = \frac{2\pi}{5}$

4. $V = \pi \int_0^3 9^2\,dx - \pi \int_0^3 (x^2)^2\,dx = 81\pi x\Big|_0^3 - \frac{\pi}{5}x^5\Big|_0^3$

$= 243\pi - \frac{243\pi}{5} = \frac{972\pi}{5}$

or $V = 2\pi \int_0^9 xy\,dy = 2\pi \int_0^9 y^{1/2}y\,dy = \frac{4\pi}{5}y^{5/2}\Big|_0^9$

$= \frac{4\pi}{5}(3^5 - 0) = \frac{972\pi}{5}$

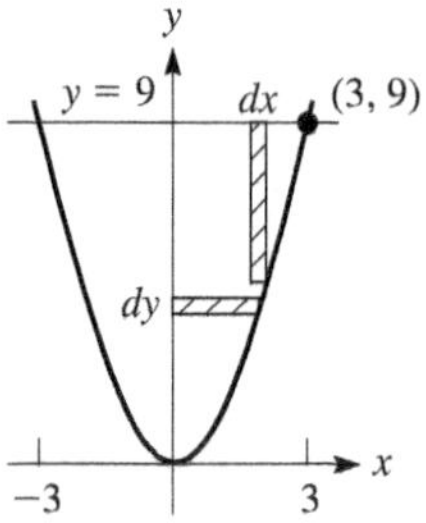

5. $I_y = \rho \int_0^3 x^2(9 - x^2)dx = \rho \int_0^3 (9x^2 - x^4)dx$

$= \rho\left(3x^3 - \frac{1}{5}x^5\right)\Big|_0^3 = \rho\left(81 - \frac{243}{5}\right) = \frac{162\rho}{5}$

6. $q = \int (5.0 - 0.20t)dt$

$q = 5.0t - 0.10t^2 + C$

$0 = C$

$q = 5.0t - 0.10t^2$

At $t = 2.0$, $q = 5.0(2.0) - 0.10(2.0)^2 = 9.6\ \text{C}$

7. $s = \int (60 - 4t)dt = 60t - 2t^2 + C$

$s = 10$ for $t = 0$, $\quad 10 = 60(0) - 2(0^2) + C, \quad C = 10$

$s = 60t - 2t^2 + 10$

8. $F = kx,\ 12 = k(2.0),\ k = 6.0\ \text{N/cm}$

$W = \int_{2.0}^{6.0} 6.0x\,dx = 3.0x^2\Big|_{2.0}^{6.0} = 3.0(36.0 - 4.0)$

$= 96\ \text{N} \cdot \text{cm}$

9. $F = 9.80\int_{1.00}^{3.00} 6.00h\,dh = (9.80)(6.00)\frac{1}{2}h^2\Big|_{1.00}^{3.00}$

$= (9.80)(3.00)(9.00 - 1.00) = 235\ \text{kN}$

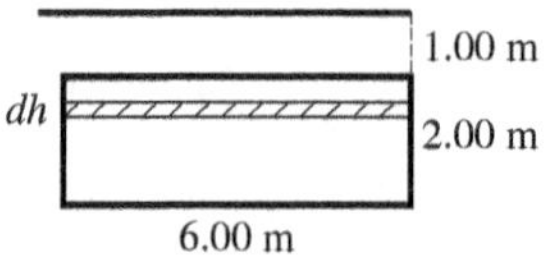

10. $\frac{1}{6}\int_0^6 (x^2 + 2x - 3)dx = \frac{1}{6}\left(\frac{1}{3}x^3 + x^2 - 3x\right)\Big|_0^6 = 15$

Chapter 27

1. $y = \tan^3 2x + \tan^{-1} 2x$

$\frac{dy}{dx} = 3(\tan^2 2x)(\sec^2 2x)(2) + \frac{2}{1 + (2x)^2}$

$= 6\tan^2 2x \sec^2 2x + \frac{2}{1 + 4x^2}$

2. $y = 2(3 + \cot 4x)^3$

$$\frac{dy}{dx} = 2(3)(3 + \cot 4x)^2(-\csc^2 4x)(4)$$
$$= -24(3 + \cot 4x)^2 \csc^2 4x$$

3. $y \sec 2x = \sin^{-1} 3y$

$$y(\sec 2x \tan 2x)(2) + (\sec 2x)(y') = \frac{3y'}{\sqrt{1 - (3y)^2}}$$
$$\sqrt{1 - 9y^2}\, \sec 2x(2y \tan 2x + y') = 3y'$$
$$y' = \frac{2y\sqrt{1 - 9y^2}\, \sec 2x \tan 2x}{3 - \sqrt{1 - 9y^2}\, \sec 2x}$$

4. $y = \dfrac{\cos^2(3x + 1)}{x}$

$$dy = \frac{x\{2 \cos (3x + 1)[-\sin(3x + 1)(3)]\} - \cos^2(3x + 1)(1)}{x^2}\,dx$$
$$= \frac{-6x \cos(3x + 1) \sin(3x + 1) - \cos^2(3x + 1))}{x^2}\,dx$$

5. $y = \ln \dfrac{2x - 1}{1 + x^2}$

$$= \ln (2x - 1) - \ln (1 + x^2)$$
$$\frac{dy}{dx} = \frac{2}{2x - 1} - \frac{2x}{1 + x^2}$$
$$m_{\tan} = \left.\frac{dy}{dx}\right|_{x=2} = \frac{2}{4 - 1} - \frac{4}{1 + 4} = \frac{2}{3} - \frac{4}{5} = -\frac{2}{15}$$

6. $i = 8e^{-t} \sin 10t$

$$\frac{di}{dt} = 8[(e^{-t} \cos 10t)(10) + (\sin 10t)(-e^{-t})]$$
$$= 8e^{-t}(10 \cos 10t - \sin 10t)$$

7. $y = xe^x$

$y' = xe^x + e^x = e^x(x + 1)$
$y'' = xe^x + e^x + e^x = e^x(x + 2)$
$x < -1$ y dec.
$x > -1$ y inc.
$x < -2$ y conc. down
$x > -2$ y conc. up
Int. $(0, 0)$, min. $(-1, -e^{-1})$,
infl. $(-2, -2e^{-2})$, asym. $y = 0$

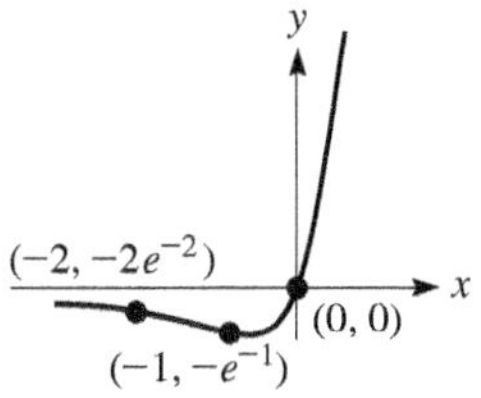

8. For $t = 8.0$ s, $x = 40$ m

$$\theta = \tan^{-1} \frac{x}{250}$$
$$\frac{d\theta}{dt} = \frac{1}{1 + \dfrac{x^2}{250^2}}\,\frac{dx/dt}{250} = \frac{250^2}{250^2 + x^2}\,\frac{dx/dt}{250}$$
$$\left.\frac{d\theta}{dt}\right|_{t=8.0} = \frac{250}{250^2 + 40^2}(5.0) = 0.020 \text{ rad/s}$$

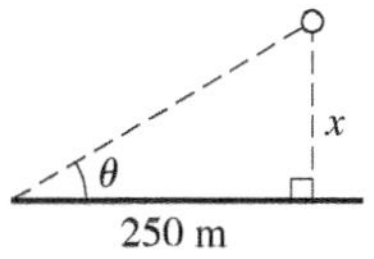

9. Function is $0/0$ indeterminate form.

$$\lim_{x\to 0} \frac{\tan^{-1} x}{2 \sin x} = \lim_{x\to 0} \frac{\frac{d}{dx}\tan^{-1} x}{\frac{d}{dx}(2 \sin x)} = \lim_{x\to 0} \frac{\frac{1}{1 + x^2}}{2 \cos x} = \frac{1}{2}$$

Chapter 28

1. $\displaystyle\int (\sec x - \sec^3 x \tan x)dx$

$$= \int \sec x\, dx - \int \sec^2 x(\sec x \tan x)dx$$
$$= \ln|\sec x + \tan x| - \tfrac{1}{3} \sec^3 x + C$$

2. $\displaystyle\int \sin^3 x\, dx = \int \sin^2 x \sin x\, dx$

$$= \int (1 - \cos^2 x)\sin x\, dx$$
$$= \int \sin x\, dx - \int \cos^2 x \sin x\, dx$$
$$= -\cos x + \tfrac{1}{3} \cos^3 x + C$$

3. $\displaystyle\int \tan^3 2x\, dx = \int \tan 2x(\tan^2 2x)dx$

$$= \int \tan 2x(\sec^2 2x - 1)dx$$
$$= \tfrac{1}{2} \int \tan 2x \sec^2 2x(2\, dx) \quad - \tfrac{1}{2} \int \tan 2x(2\, dx)$$
$$= \tfrac{1}{4} \tan^2 2x + \tfrac{1}{2} \ln|\cos 2x| + C$$

4. $\displaystyle\int \cos^2 4\theta\, d\theta = \tfrac{1}{2} \int (1 + \cos 8\theta)d\theta$

$$= \tfrac{1}{2} \int d\theta + \tfrac{1}{16} \int \cos 8\theta(8\, d\theta)$$
$$= \tfrac{1}{2}\theta + \tfrac{1}{16} \sin 8\theta + C$$

5. Let $x = 2 \sin \theta$,
$dx = 2 \cos \theta\, d\theta$.

$$\int \frac{dx}{x^2\sqrt{4 - x^2}} = \int \frac{2 \cos \theta\, d\theta}{4 \sin^2 \theta\sqrt{4 - 4 \sin^2 \theta}}$$
$$= \frac{1}{4}\int \frac{\cos \theta\, d\theta}{\sin^2 \theta\sqrt{\cos^2 \theta}} = \tfrac{1}{4} \int \csc^2 \theta\, d\theta$$
$$= -\tfrac{1}{4} \cot \theta + C$$
$$= -\frac{\sqrt{4 - x^2}}{4x} + C$$

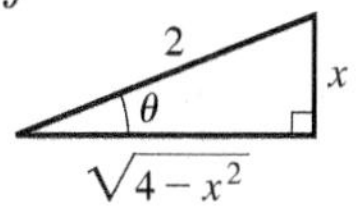

6. $\displaystyle\int xe^{-2x}\, dx$; $u = x$, $du = dx$, $dv = e^{-2x}\, dx$, $v = -\tfrac{1}{2}e^{-2x}$

$$\int xe^{-2x}\, dx = x\left(-\tfrac{1}{2}e^{-2x}\right) - \int \left(-\tfrac{1}{2}e^{-2x}\right)dx$$
$$= -\tfrac{1}{2}xe^{-2x} - \tfrac{1}{4}e^{-2x} + C$$

7. $\displaystyle\int \frac{x^3 + 5x^2 + x + 2}{x^4 + x^2}\,dx$

$$\frac{x^3 + 5x^2 + x + 2}{x^4 + x^2} = \frac{x^3 + 5x^2 + x + 2}{x^2(x^2 + 1)}$$
$$= \frac{A}{x} + \frac{B}{x^2} + \frac{Cx + D}{x^2 + 1}$$

$x^3 + 5x^2 + x + 2 = Ax(x^2 + 1) + B(x^2 + 1) + Cx^3 + Dx^2$

x^3-terms: $1 = A + C$ $\quad$ x^2-terms: $5 = B + D$

x-terms: $1 = A$ $\quad$ $x = 0$: $2 = B$

$A = 1, B = 2, C = 0, D = 3$

$$\int \frac{x^3 + 5x^2 + x + 2}{x^4 + x^2}\,dx = \int \frac{dx}{x} + \int \frac{2\, dx}{x^2} + \int \frac{3\, dx}{x^2 + 1}$$
$$= \ln|x| - \frac{2}{x} + 3 \tan^{-1}x + C$$

8. $i = \int \frac{6t+1}{4t^2+9}\,dt = \int \frac{6t\,dt}{4t^2+9} + \int \frac{dt}{4t^2+9}$

$= \frac{6}{8}\int \frac{8t\,dt}{4t^2+9} + \frac{1}{2}\int \frac{2\,dt}{9+(2t)^2}$

$= \frac{3}{4}\ln(4t^2+9) + \frac{1}{2}\left(\frac{1}{3}\right)\tan^{-1}\frac{2t}{3} + C$

$i = 0$ for $t = 0$: $0 = \frac{3}{4}\ln 9 + \frac{1}{6}\tan^{-1}0 + C,\ C = -\frac{3}{4}\ln 9$

$i = \frac{3}{4}\ln(4t^2+9) + \frac{1}{6}\tan^{-1}\frac{2t}{3} - \frac{3}{4}\ln 9$

$= \frac{3}{4}\ln\frac{4t^2+9}{9} + \frac{1}{6}\tan^{-1}\frac{2t}{3}$

9. $y = \frac{1}{\sqrt{16-x^2}}$; $A = \int_0^3 y\,dx$

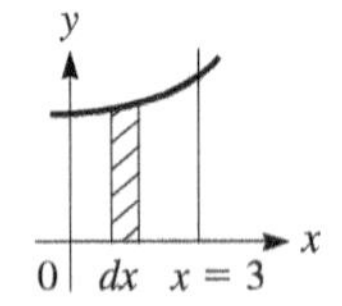

$= \int_0^3 \frac{dx}{\sqrt{16-x^2}}$

$= \sin^{-1}\frac{x}{4}\Big|_0^3$

$= \sin^{-1}\frac{3}{4} = 0.8481$

Chapter 29

1. $f(x, y) = \frac{2y}{x^2 - y^2}$; $f(-1, 3) = \frac{2(3)}{(-1)^2 - 3^2} = \frac{6}{-8} = -\frac{3}{4}$

2. $z = 4 - x^2 - 4y^2$

Intercepts: $(2, 0, 0)$, $(-2, 0, 0)$, $(0, 1, 0)$, $(0, -1, 0)$, $(0, 0, 4)$

Traces:

in yz-plane: $z = 4 - 4y^2$ (parabola)

in xz-plane: $z = 4 - x^2$ (parabola)

in xy-plane: $x^2 + 4y^2 = 4$ (ellipse)

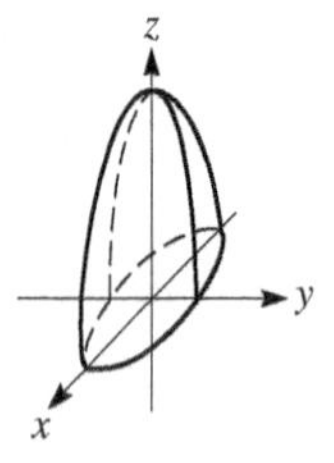

3. $z = xe^{2xy}$; $\frac{\partial z}{\partial x} = x(e^{2xy})(2y) + e^{2xy}(1) = e^{2xy}(2xy + 1)$

$\frac{\partial z}{\partial y} = xe^{2xy}(2x) = 2x^2e^{2xy}$

4. $z = 3x^3y + 2x^2y^4$; $\frac{\partial z}{\partial y} = 3x^3(1) + 2x^2(4y^3) = 3x^3 + 8x^2y^3$

$\frac{\partial^2 z}{\partial x\,\partial y} = 9x^2 + (16x)y^3 = 9x^2 + 16xy^3$

5. $\int_0^2 \int_{x^2}^{2x} (x^3 + 4y)\,dy\,dx = \int_0^2 (x^3y + 2y^2)\Big|_{x^2}^{2x} dx$

$= \int_0^2 (2x^4 + 8x^2 - x^5 - 2x^4)\,dx = \int_0^2 (8x^2 - x^5)\,dx$

$= \left(\frac{8}{3}x^3 - \frac{1}{6}x^6\right)\Big|_0^2 = \frac{8}{3}(8) - \frac{1}{6}(64) = \frac{64}{3} - \frac{32}{3} = \frac{32}{3}$

6. $\int_1^{\ln 8} \int_0^{\ln y} e^{x+y}\,dx\,dy = \int_1^{\ln 8} e^{x+y}\Big|_0^{\ln y} dy$

$= \int_1^{\ln 8} (e^{y+\ln y} - e^y)\,dy$

$= \int_1^{\ln 8} (e^y e^{\ln y} - e^y)\,dy = \int_1^{\ln 8} (ye^y - e^y)\,dy$

$= e^y(y-1) - e^y\Big|_1^{\ln 8} = e^{\ln 8}(\ln 8 - 1) - e^{\ln 8} + e$

$= 8(\ln 8 - 1) - 8 + e = 8\ln 2^3 - 8 - 8 + e$

$= e - 16 + 24\ln 2 = 3.354$

7. $V = \int_0^3 \int_0^{\sqrt{9-y^2}} \sqrt{9-y^2}\,dx\,dy$

$= \int_0^3 x\sqrt{9-y^2}\Big|_0^{\sqrt{9-y^2}} dy$

$= \int_0^3 (9 - y^2)\,dy$

$= 9y - \frac{1}{3}y^3\Big|_0^3 = 18$

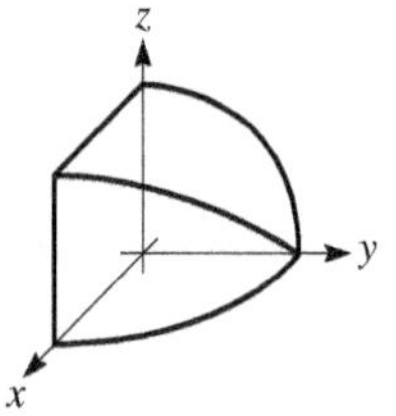

8. $f = \frac{k\sqrt{T}}{L}$; $30\text{ Hz} = \frac{k\sqrt{65\text{ N}}}{60\text{ cm}}$; $k = \frac{1800\text{ Hz}\cdot\text{cm}}{\sqrt{65}\ \ \text{N}^{1/2}}$

$\frac{\partial f}{\partial T} = \frac{k}{L}\frac{1}{2\sqrt{T}} = \frac{1800}{\sqrt{65}L}\frac{1}{2\sqrt{T}} = \frac{900}{L\sqrt{65T}}$

$\frac{\partial f}{\partial T}\Big|_{L=60,\,T=65} = \frac{900}{60\sqrt{65\times 65}} = \frac{15}{65} = 0.23\text{ Hz/N}$

Chapter 30

1.

$f(x) = (1 + e^x)^2$ $\qquad f(0) = (1+1)^2 = 4$

$f'(x) = 2(1 + e^x)(e^x)$ $\qquad f'(0) = 2(1+1)(1) = 4$

$= 2e^x + 2e^{2x}$ $\qquad f''(0) = 2(1) + 4(1) = 6$

$f''(x) = 2e^x + 4e^{2x}$ $\qquad f'''(0) = 2(1) + 8(1) = 10$

$f'''(x) = 2e^x + 8e^{2x}$

$(1 + e^x)^2 = 4 + 4x + \frac{6}{2}x^2 + \frac{10}{6}x^3 + \cdots$

$= 4 + 4x + 3x^2 + \frac{5}{3}x^3 + \cdots$

2.

$f(x) = \cos x$ $\qquad f\left(\frac{\pi}{3}\right) = \frac{1}{2}$

$f'(x) = -\sin x$ $\qquad f'\left(\frac{\pi}{3}\right) = -\frac{\sqrt{3}}{2}$

$f''(x) = -\cos x$ $\qquad f''\left(\frac{\pi}{3}\right) = -\frac{1}{2}$

$\cos x = \frac{1}{2} - \frac{\sqrt{3}}{2}\left(x - \frac{\pi}{3}\right) - \frac{\frac{1}{2}\left(x - \frac{\pi}{3}\right)^2}{2} + \cdots$

$= \frac{1}{2}\left[1 - \sqrt{3}\left(x - \frac{\pi}{3}\right) - \frac{1}{2}\left(x - \frac{\pi}{3}\right)^2 + \cdots\right]$

3. $\ln(1 + x) = x - \frac{x^2}{2} + \frac{x^3}{3} - \frac{x^4}{4} + \cdots$

$\ln 0.96 = \ln(1 - 0.04)$

$= -0.04 - \frac{(-0.04)^2}{2} + \frac{(-0.04)^3}{3} - \frac{(-0.04)^4}{4}$

$= -0.040\,822\,0$

4. $f(x) = \dfrac{1}{\sqrt{1-2x}} = (1-2x)^{-1/2}$

$$(1+x)^n = 1 + nx + \frac{n(n-1)}{2}x^2 + \cdots$$

$$(1-2x)^{-1/2} = 1 + \left(-\frac{1}{2}\right)(-2x) + \frac{-\frac{1}{2}\left(-\frac{3}{2}\right)}{2}(-2x)^2 + \cdots$$

$$= 1 + x + \tfrac{3}{2}x^2 + \cdots$$

5. $$\int_0^1 x\cos x\,dx = \int_0^1 x\left(1 - \frac{x^2}{2} + \frac{x^4}{24}\right)dx$$

$$= \int_0^1 \left(x - \frac{x^3}{2} + \frac{x^5}{24}\right)dx$$

$$= \frac{1}{2}x^2 - \frac{x^4}{8} + \frac{x^6}{144}\Big|_0^1$$

$$= \tfrac{1}{2} - \tfrac{1}{8} + \tfrac{1}{144} - 0 = 0.3819$$

6. $f(t) = 0 \quad -\pi \le t < 0$

$f(t) = 2 \quad 0 \le t < \pi$

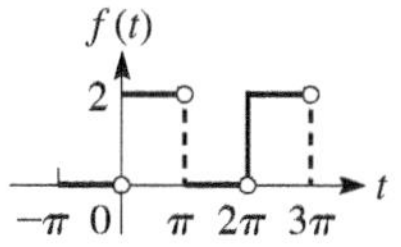

$$a_0 = \frac{1}{2\pi}\int_0^{\pi} 2\,dt = \frac{1}{\pi}t\Big|_0^{\pi} = 1$$

$$a_n = \frac{1}{\pi}\int_0^{\pi} 2\cos nt\,dt = \frac{2}{n\pi}\sin nt\Big|_0^{\pi}$$

$= 0$ for all n

$$b_n = \frac{1}{\pi}\int_0^{\pi} 2\sin nt\,dt = \frac{2}{n\pi}(-\cos nt)\Big|_0^{\pi}$$

$$= \frac{2}{n\pi}(1 - \cos n\pi)$$

$$b_1 = \frac{2}{\pi}(1+1) = \frac{4}{\pi} \quad b_2 = \frac{2}{2\pi}(1-1) = 0$$

$$b_3 = \frac{2}{3\pi}(1+1) = \frac{4}{3\pi}$$

$$f(t) = 1 + \frac{4}{\pi}\sin t + \frac{4}{3\pi}\sin 3t + \cdots$$

7. $f(x) = x^2 + 2, f(-x) = (-x)^2 + 2 = x^2 + 2, -f(-x) \ne f(x)$

Since $f(x) = f(-x)$, $f(x)$ is an even function.

Since $f(x) \ne -f(-x)$, $f(x)$ is not an odd function.

$g(x) = x^2 - 1$, $g(x) = f(x) - 3$

Fourier series for $g(x)$ is $F(x) - 3$.

Chapter 31

1. $x\dfrac{dy}{dx} + 2y = 4$

$$dy + \frac{2}{x}y\,dx = \frac{4}{x}dx$$

$$e^{\int \frac{2}{x}dx} = e^{2\ln x} = x^2$$

$$yx^2 = \int \frac{4}{x}x^2\,dx = \int 4x\,dx = 2x^2 + c$$

$$y = 2 + \frac{c}{x^2}$$

2. $y'' + 2y' + 5y = 0$

$m^2 + 2m + 5 = 0$

$$m = \frac{-2 \pm \sqrt{4-20}}{2} = -1 \pm 2j$$

$y = e^{-x}(c_1 \sin 2x + c_2 \cos 2x)$

3. $x\,dx + y\,dy = x^2\,dx + y^2\,dx = (x^2+y^2)dx$

$$\frac{x\,dx + y\,dy}{x^2+y^2} = dx$$

$\tfrac{1}{2}\ln(x^2+y^2) = x + \tfrac{1}{2}c \qquad \ln(x^2+y^2) = 2x + c$

4. $2D^2y - Dy = 2\cos x$

$2m^2 - m = 0 \qquad m = 0, \tfrac{1}{2}$

$y_c = c_1 + c_2e^{x/2}$

$y_p = A\sin x + B\cos x$

$Dy_p = A\cos x - B\sin x$

$D^2y_p = -A\sin x - B\cos x$

$2(-A\sin x - B\cos x) - (A\cos x - B\sin x) = 2\cos x$

$-2A + B = 0 \quad -2B - A = 2 \quad A = -\tfrac{2}{5} \quad B = -\tfrac{4}{5}$

$y = c_1 + c_2e^{x/2} - \tfrac{2}{5}\sin x - \tfrac{4}{5}\cos x$

5. $\dfrac{d^2y}{dx^2} - 4\dfrac{dy}{dx} + 4y - 3x$

$m^2 - 4m + 4 = 0 \qquad m = 2, 2$

$y_c = (c_1 + c_2x)e^{2x}$

$y_p = A + Bx \qquad Dy_p = B \qquad D^2y_p = 0$

$0 - 4B + 4(A + Bx) = 3x$

$4A - 4B = 0 \qquad 4B = 3$

$B = \tfrac{3}{4} \qquad A = \tfrac{3}{4}$

$y = (c_1 + c_2x)e^{2x} + \tfrac{3}{4} + \tfrac{3}{4}x$

6. $D^2y - 2Dy - 8y = 4e^{-2x}$

$m^2 - 2m - 8 = 0 \qquad (m+2)(m-4) = 0 \qquad m = -2, 4$

$y_c = c_1e^{-2x} + c_2e^{4x}$

$y_p = Axe^{-2x}$ (factor of x necessary due to first term of y_c)

$Dy_p = Ae^{-2x} - 2Axe^{-2x}$

$D^2y_p = -2Ae^{-2x} - 2Ae^{-2x} + 4Axe^{-2x} = -4Ae^{-2x} + 4Axe^{-2x}$

$(-4Ae^{-2x} + 4Axe^{-2x}) - 2(Ae^{-2x} - 2Axe^{-2x})$

$-8Axe^{-2x} = 4e^{-2x}$

$-6Ae^{-2x} = 4e^{-2x} \qquad A = -\tfrac{2}{3}$

$y = c_1e^{-2x} + c_2e^{4x} - \tfrac{2}{3}xe^{-2x}$

7. $(xy + y)\dfrac{dy}{dx} = 2$

$y(x+1)dy = 2\,dx$

$$y\,dy = \frac{2\,dx}{x+1}$$

$\tfrac{1}{2}y^2 = 2\ln(x+1) + c$

$y = 2$ when $x = 0$

$\tfrac{1}{2}(4) = 2\ln(1) + c, \quad c = 2$

$\tfrac{1}{2}y^2 = 2\ln(x+1) + 2$

$y^2 = 4\ln(x+1) + 4$

8. $\dfrac{dA}{dt} = rA$

$$\frac{dA}{A} = rdt$$

$\ln A = rt + \ln c$

$$\ln\frac{A}{c} = rt$$

$A = ce^{rt}$

$A_0 = ce^0$

$c = A_0$

$A = A_0e^{rt}$

9. $y'' + 9y = 9$

$y(0) = 0 \quad y'(0) = 1$

$\mathcal{L}(y'') + 9\mathcal{L}(y) = \mathcal{L}(9)$

$s^2\mathcal{L}(y) - s(0) - 1 + 9\mathcal{L}(y) = \dfrac{9}{s}$

$(s^2 + 9)\mathcal{L}(y) = \dfrac{9}{s} + 1$

$\mathcal{L}(y) = \dfrac{9}{s(s^2 + 9)} + \dfrac{1}{s^2 + 9}$

$y = 1 - \cos 3t + \frac{1}{3}\sin 3t$

10. $D^2y - Dy - 2y = 12 \quad y(0) = 0,\ y'(0) = 0$

$\mathcal{L}(y'') - \mathcal{L}(y') - 2\mathcal{L}(y) = \mathcal{L}(12)$

$s^2\mathcal{L}(y) - s(0) - 0 - [s\mathcal{L}(y) - 0] - 2\mathcal{L}(y) = \dfrac{12}{s}$

$\mathcal{L}(y)(s^2 - s - 2) = \dfrac{12}{s}$

$\mathcal{L}(y) = \dfrac{12}{s(s^2 - s - 2)} = \dfrac{12}{s(s + 1)(s - 2)}$

$= \dfrac{A}{s} + \dfrac{B}{s + 1} + \dfrac{C}{s - 2}$

$12 = A(s + 1)(s - 2) + Bs(s - 2) + Cs(s + 1)$

$s = 0$: $\quad 12 = -2A,\ A = -6$

$s = -1$: $\quad 12 = 3B,\ B = 4$

$s = 2$: $\quad 12 = 6C,\ C = 2$

$\mathcal{L}(y) = -\dfrac{6}{5} + \dfrac{4}{s + 1} + \dfrac{2}{s - 2}$

$y = -6 + 4e^{-t} + 2e^{2t}$

11. $L\dfrac{d^2q}{dt^2} + R\dfrac{dq}{dt} = E \qquad L\dfrac{di}{dt} + Ri = E$

$L = 2\text{ H} \qquad R = 8\ \Omega \qquad E = 6\text{ V}$

$2\dfrac{di}{dt} + 8i = 6 \qquad di + 4i\,dt = 3\,dt$

$e^{\int 4\,dt} = e^{4t} \qquad ie^{4t} = \displaystyle\int 3e^{4t}\,dt = \frac{3}{4}e^{4t} + c$

$i = 0$ for $t = 0 \qquad 0 = \frac{3}{4} + c \qquad c = -\frac{3}{4}$

$ie^{4t} = \frac{3}{4}e^{4t} - \frac{3}{4}$

$i = \frac{3}{4}(1 - e^{-4t}) = 0.75(1 - e^{-4t})$

12. $m = 0.5\text{ kg},\ k = 32\text{ N/m}$

$mD^2x = -kx, \quad 0.5D^2x + 32x = 0$

$D^2x + 64x = 0$

$m^2 + 64 = 0; \qquad m = \pm 8j$

$x = c_1 \sin 8t + c_2 \cos 8t$

$Dx = 8c_1 \cos 8t - 8c_2 \sin 8t$

$x = 0.3\text{ m} \qquad Dx = 0 \quad \text{for} \quad t = 0$

$0.3 = c_1 \sin 0 + c_2 \cos 0, \qquad c_2 = 0.3$

$0 = 8c_1 \cos 0 - 8c_2 \sin 0, \qquad c_1 = 0$

$x = 0.3 \cos 8t$

Index of Applications

Computer

Construction

Design

Electricity and Electronics

Energy Technology

Environmental Science

Fire Science

Geodesy, Geology, Seismology

Hydrodynamics and Hydrostatics

Machine Technology

Materials

Measurement

Meteorology

Optics

Petroleum

Photography

Physics

Police Science

Refrigeration and Air Conditioning

Space Technology

Statics

Strength of Materials

Surveying

Thermal

Thermodynamics

Wastewater Technology

Index

D

M

N

O

S

CALCULUS

Derivatives

$$\frac{dc}{dx} = 0$$

$$\frac{dx^n}{dx} = nx^{n-1}$$

$$\frac{du^n}{dx} = nu^{n-1}\left(\frac{du}{dx}\right)$$

$$\frac{d(u+v)}{dx} = \frac{du}{dx} + \frac{dv}{dx}$$

$$\frac{d(u \cdot v)}{dx} = u \cdot \frac{dv}{dx} + v \cdot \frac{du}{dx}$$

$$\frac{d\left(\frac{u}{v}\right)}{dx} = \frac{v \cdot \frac{du}{dx} - u \cdot \frac{dv}{dx}}{v^2}$$

$$\frac{dy}{dx} = \frac{dy}{du} \cdot \frac{du}{dx}$$

$$\frac{d(\sin u)}{dx} = \cos u \frac{du}{dx}$$

$$\frac{d(\cos u)}{dx} = -\sin u \frac{du}{dx}$$

$$\frac{d(\tan u)}{dx} = \sec^2 u \frac{du}{dx}$$

$$\frac{d(\cot u)}{dx} = -\csc^2 u \frac{du}{dx}$$

$$\frac{d(\sec u)}{dx} = \sec u \tan u \frac{du}{dx}$$

$$\frac{d(\csc u)}{dx} = -\csc u \cot u \frac{du}{dx}$$

$$\frac{d(\sin^{-1} u)}{dx} = \frac{1}{\sqrt{1-u^2}} \frac{du}{dx}$$

$$\frac{d(\tan^{-1} u)}{dx} = \frac{1}{1+u^2} \frac{du}{dx}$$

$$\frac{d(\ln u)}{dx} = \frac{1}{u} \frac{du}{dx}$$

$$\frac{d(e^u)}{dx} = e^u \frac{du}{dx}$$

Integrals

$$\int u^n \, du = \frac{u^{n+1}}{n+1} + C \qquad n \neq -1$$

$$\int \frac{du}{u} = \ln|u| + C$$

$$\int e^u \, du = e^u + C$$

$$\int \sin u \, du = -\cos u + C$$

$$\int \cos u \, du = \sin u + C$$

$$\int \tan u \, du = -\ln|\cos u| + C$$

$$\int \cot u \, du = \ln|\sin u| + C$$

$$\int \frac{du}{\sqrt{a^2 - u^2}} = \sin^{-1}\frac{u}{a} + C$$

$$\int \frac{du}{a^2 + u^2} = \frac{1}{a}\tan^{-1}\frac{u}{a} + C$$

$$\int u \, dv = uv - \int v \, du$$

$$\int \sec^2 u \, du = \tan u + C$$

$$\int \csc^2 u \, du = -\cot u + C$$

$$\int \sec u \tan u \, du = \sec u + C$$

$$\int \csc u \cot u \, du = -\csc u + C$$

$$\int \sec u \, du = \ln|\sec u + \tan u| + C$$

$$\int \csc u \, du = \ln|\csc u - \cot u| + C$$